| Element | Symbol | Atomic Number | Average Atomic Mass[a] |
|---|---|---|---|
| Actinium | Ac | 89 | [227] |
| Aluminum | Al | 13 | 26.982 |
| Americium | Am | 95 | [243] |
| Antimony | Sb | 51 | 121.76 |
| Argon | Ar | 18 | 39.948 |
| Arsenic | As | 33 | 74.922 |
| Astatine | At | 85 | [210] |
| Barium | Ba | 56 | 137.33 |
| Berkelium | Bk | 97 | [247] |
| Beryllium | Be | 4 | 9.0122 |
| Bismuth | Bi | 83 | 208.98 |
| Bohrium | Bh | 107 | [270] |
| Boron | B | 5 | 10.811 |
| Bromine | Br | 35 | 79.904 |
| Cadmium | Cd | 48 | 112.41 |
| Calcium | Ca | 20 | 40.078 |
| Californium | Cf | 98 | [251] |
| Carbon | C | 6 | 12.011 |
| Cerium | Ce | 58 | 140.12 |
| Cesium | Cs | 55 | 132.91 |
| Chlorine | Cl | 17 | 35.453 |
| Chromium | Cr | 24 | 51.996 |
| Cobalt | Co | 27 | 58.933 |
| Copernicium | Cn | 112 | [285] |
| Copper | Cu | 29 | 63.546 |
| Curium | Cm | 96 | [247] |
| Darmstadtium | Ds | 110 | [281] |
| Dubnium | Db | 105 | [268] |
| Dysprosium | Dy | 66 | 162.50 |
| Einsteinium | Es | 99 | [252] |
| Erbium | Er | 68 | 167.26 |
| Europium | Eu | 63 | 151.96 |
| Fermium | Fm | 100 | [257] |
| Flerovium | Fl | 114 | [289] |
| Fluorine | F | 9 | 18.998 |
| Francium | Fr | 87 | [223] |
| Gadolinium | Gd | 64 | 157.25 |
| Gallium | Ga | 31 | 69.723 |
| Germanium | Ge | 32 | 72.63 |
| Gold | Au | 79 | 196.97 |

| Element | Symbol | Atomic Number | Average Atomic Mass[a] |
|---|---|---|---|
| Hafnium | Hf | 72 | 178.49 |
| Hassium | Hs | 108 | [269] |
| Helium | He | 2 | 4.0026 |
| Holmium | Ho | 67 | 164.93 |
| Hydrogen | H | 1 | 1.0079 |
| Indium | In | 49 | 114.82 |
| Iodine | I | 53 | 126.90 |
| Iridium | Ir | 77 | 192.22 |
| Iron | Fe | 26 | 55.845 |
| Krypton | Kr | 36 | 83.798 |
| Lanthanum | La | 57 | 138.91 |
| Lawrencium | Lr | 103 | [262] |
| Lead | Pb | 82 | 207.2 |
| Lithium | Li | 3 | 6.941 |
| Livermorium | Lv | 116 | [293] |
| Lutetium | Lu | 71 | 174.97 |
| Magnesium | Mg | 12 | 24.305 |
| Manganese | Mn | 25 | 54.938 |
| Meitnerium | Mt | 109 | [276] |
| Mendelevium | Md | 101 | [258] |
| Mercury | Hg | 80 | 200.59 |
| Molybdenum | Mo | 42 | 95.96 |
| Moscovium | Mc | 115 | [290] |
| Neodymium | Nd | 60 | 144.24 |
| Neon | Ne | 10 | 20.180 |
| Neptunium | Np | 93 | [237] |
| Nickel | Ni | 28 | 58.693 |
| Nihonium | Nh | 113 | [284] |
| Niobium | Nb | 41 | 92.906 |
| Nitrogen | N | 7 | 14.007 |
| Nobelium | No | 102 | [259] |
| Oganesson | Og | 118 | [294] |
| Osmium | Os | 76 | 190.23 |
| Oxygen | O | 8 | 15.999 |
| Palladium | Pd | 46 | 106.42 |
| Phosphorus | P | 15 | 30.974 |
| Platinum | Pt | 78 | 195.08 |
| Plutonium | Pu | 94 | [244] |
| Polonium | Po | 84 | [209] |

| Element | Symbol | Atomic Number | Average Atomic Mass[a] |
|---|---|---|---|
| Potassium | K | 19 | 39.098 |
| Praseodymium | Pr | 59 | 140.91 |
| Promethium | Pm | 61 | [145] |
| Protactinium | Pa | 91 | 231.04 |
| Radium | Ra | 88 | [226] |
| Radon | Rn | 86 | [222] |
| Rhenium | Re | 75 | 186.21 |
| Rhodium | Rh | 45 | 102.91 |
| Roentgenium | Rg | 111 | [280] |
| Rubidium | Rb | 37 | 85.468 |
| Ruthenium | Ru | 44 | 101.07 |
| Rutherfordium | Rf | 104 | [265] |
| Samarium | Sm | 62 | 150.36 |
| Scandium | Sc | 21 | 44.956 |
| Seaborgium | Sg | 106 | [271] |
| Selenium | Se | 34 | 78.96 |
| Silicon | Si | 14 | 28.086 |
| Silver | Ag | 47 | 107.87 |
| Sodium | Na | 11 | 22.990 |
| Strontium | Sr | 38 | 87.62 |
| Sulfur | S | 16 | 32.065 |
| Tantalum | Ta | 73 | 180.95 |
| Technetium | Tc | 43 | [98] |
| Tellurium | Te | 52 | 127.60 |
| Tennessine | Ts | 117 | [294] |
| Terbium | Tb | 65 | 158.93 |
| Thallium | Tl | 81 | 204.38 |
| Thorium | Th | 90 | 232.04 |
| Thulium | Tm | 69 | 168.93 |
| Tin | Sn | 50 | 118.71 |
| Titanium | Ti | 22 | 47.867 |
| Tungsten | W | 74 | 183.84 |
| Uranium | U | 92 | 238.03 |
| Vanadium | V | 23 | 50.942 |
| Xenon | Xe | 54 | 131.29 |
| Ytterbium | Yb | 70 | 173.05 |
| Yttrium | Y | 39 | 88.906 |
| Zinc | Zn | 30 | 65.38 |
| Zirconium | Zr | 40 | 91.224 |

[a]Average atomic mass values for most elements are from *Pure Appl. Chem.* (2011) 83, 359. Those for B, C, Cl, H, Li, N, O, Si, S, and Tl are from *Pure Appl. Chem.* (2009) 81, 2131, and are within the ranges cited in the first reference. Atomic masses in brackets are the mass numbers of the longest-lived isotopes of elements with no stable isotopes.

THIRD EDITION

# Chemistry

## An Atoms-Focused Approach

**Thomas R. Gilbert**
NORTHEASTERN UNIVERSITY

**Rein V. Kirss**
NORTHEASTERN UNIVERSITY

**Stacey Lowery Bretz**
MIAMI UNIVERSITY

**Natalie Foster**
LEHIGH UNIVERSITY (EMERITA)

W. W. NORTON & COMPANY
*Independent Publishers Since 1923*

W. W. Norton & Company has been independent since its founding in 1923, when William Warder Norton and Mary D. Herter Norton first published lectures delivered at the People's Institute, the adult education division of New York City's Cooper Union. The firm soon expanded its program beyond the Institute, publishing books by celebrated academics from America and abroad. By midcentury, the two major pillars of Norton's publishing program—trade books and college texts—were firmly established. In the 1950s, the Norton family transferred control of the company to its employees, and today—with a staff of five hundred and hundreds of trade, college, and professional titles published each year—W. W. Norton & Company stands as the largest and oldest publishing house owned wholly by its employees.


Printed in Canada

Editor: Erik Fahlgren
Developmental Editor: Andrew Sobel
Project Editor: David Bradley
Associate Director of Production: Benjamin Reynolds
Media Editor: Cailin Barrett-Bressack
Associate Media Editor: Grace Tuttle
Media Project Editor: Jesse Newkirk
Media Editorial Assistant: Sarah McGinnis
Managing Editor, College: Marian Johnson
Managing Editor, College Digital Media: Kim Yi
Ebook Production Manager: Danielle Lehmann
Marketing Manager, Chemistry: Stacy Loyal
Design Director: Jillian Burr
Designer: Lisa Buckley
Director of College Permissions: Megan Schindel
College Permissions Assistant: Patricia Wong
Photo Editor: Mike Cullen
Composition: GraphicWorld/Gary Clark, Project Manager
Illustration Studio: Imagineeringart.com
Manufacturing: Transcontinental—Beauceville, QC

Permission to use copyrighted material is included at the back of the book beginning on page C-1.

ISBN: **978-0-393-67402-6**

W. W. Norton & Company, Inc., 500 Fifth Avenue, New York, NY 10110

www.wwnorton.com

W. W. Norton & Company Ltd., 15 Carlisle Street, London W1D 3BS

1 2 3 4 5 6 7 8 9 0

# Brief Contents

# Contents

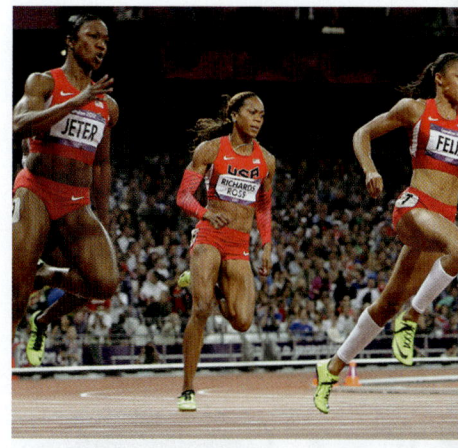

BANQUE PO
BANQUE POPULAIRE

## 13 Chemical Kinetics: Clearing the Air 590

## 14 Chemical Equilibrium: Equal but Opposite Reaction Rates 656

## 15 Acid–Base Equilibria: Proton Transfer in Biological Systems 712

7108368

# 18 The Solid State: A Particulate View 868

# 19 Organic Chemistry: Fuels, Pharmaceuticals, and Modern Materials 916

# 20 Biochemistry: The Compounds of Life 976

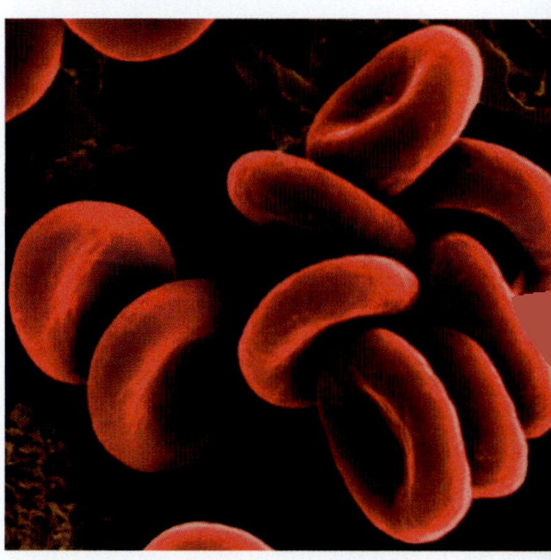



# 21 Nuclear Chemistry: The Risks and Benefits 1018

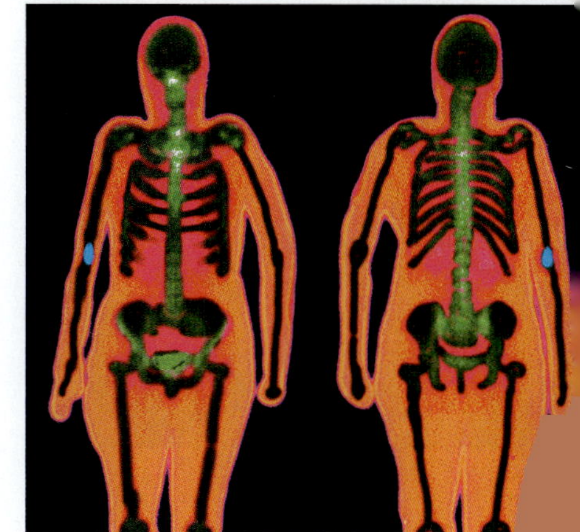

SPRING VALLEY
VITAMIN #1 BRAND
Easy to Swallow
Natural
Calcium
600 mg
with Vitamin D
BONE HEALTH**
Dietary Supplement
100 TABLETS



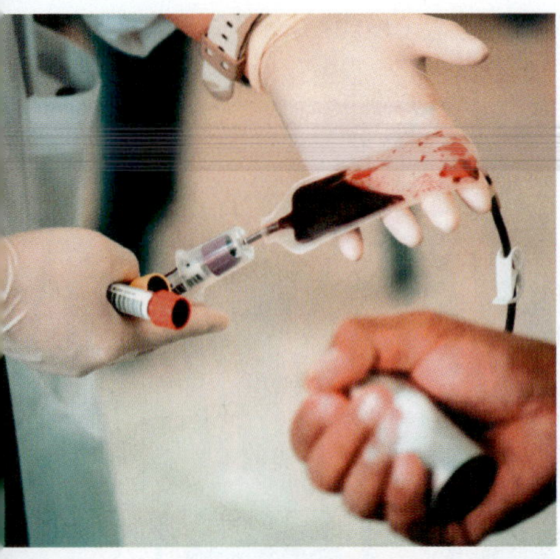

# Applications

# Animations

# About the **Authors**

**Thomas R. Gilbert** has a BS in chemistry from Clarkson and a PhD in analytical chemistry from MIT. After 10 years with the Research Department of the New England Aquarium in Boston, he joined the faculty of Northeastern University, where he is associate professor of chemistry and chemical biology. His research interests are in chemical and science education. He teaches general chemistry and science education courses and conducts professional development workshops for K–12 teachers. He has won Northeastern's Excellence in Teaching Award and Outstanding Teacher of First-Year Engineering Students Award. He is a Fellow of the American Chemical Society and in 2012 was elected to the American Chemical Society's board of directors.

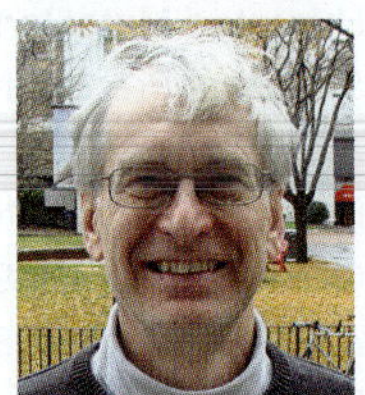

**Rein V. Kirss** received both a BS in chemistry and a BA in history as well as an MA in chemistry from SUNY Buffalo. He received his PhD in inorganic chemistry from the University of Wisconsin, Madison, where the seeds for this textbook were undoubtedly planted. After two years of postdoctoral study at the University of Rochester, he spent a year at Advanced Technology Materials Inc. before returning to academics at Northeastern University in 1989. He has won the Northeastern University College of Science Excellence in Teaching Award and was the recipient of the John A. Timm Award from the New England Association of Chemistry Teachers in 2019. He is an associate professor of chemistry with an active research interest in organometallic chemistry.

**Stacey Lowery Bretz** is a University Distinguished Professor in the Department of Chemistry and Biochemistry at Miami University in Oxford, Ohio. She earned her BA in chemistry from Cornell University, an MS from Pennsylvania State University, and a PhD in chemistry education research from Cornell University. She spent one year at the University of California, Berkeley, as a postdoc in the Department of Chemistry. Her research expertise includes the development of assessments to measure chemistry students' thinking with multiple representations (particulate, symbolic, and macroscopic) and to promote meaningful and inquiry learning in the chemistry laboratory. She is a Fellow of the American Chemical Society and a Fellow of the American Association for the Advancement of Science. She has been honored with both of Miami University's highest teaching awards: the E. Phillips Knox Award for Undergraduate Teaching and the Distinguished Teaching Award for Excellence in Graduate Instruction and Mentoring. Stacey won the prestigious, international award from the American Chemical Society for Achievement in Research on Teaching and Learning of Chemistry in 2020.

**Natalie Foster** is emerita professor of chemistry at Lehigh University in Bethlehem, Pennsylvania. She received a BS in chemistry from Muhlenberg College and MS, DA, and PhD degrees from Lehigh University. Her research interests included studying poly(vinyl alcohol) gels by nuclear magnetic resonance as part of a larger interest in porphyrins and phthalocyanines as candidate contrast enhancement agents for magnetic resonance imaging. She taught both semesters of the introductory chemistry class to engineering, biology, and other nonchemistry majors and a spectral analysis course at the graduate level. She is the recipient of the Christian R. and Mary F. Lindback Foundation Award for distinguished teaching and a Fellow of the American Chemical Society.

# Preface

Dear Student,

At first glance you may have thought that the cover of your book showed three spoonfuls of caviar. However, the cover has a clue that would help you realize that those red beads are actually Sriracha pearls made through a cooking technique called cold oil spherification, which uses some of the chemistry you will learn in this course. Each of those tiny spheres will burst in your mouth and release an explosion of flavor—in this case, the spicy hot sauce extracted from red chili peppers. The primary compound responsible for the burning sensation is capsaicin, which is represented by the molecular structure on the cover.

Our cover illustrates a central message of this book: the properties of substances are directly linked to their atomic and molecular structures. We start with the smallest particles of matter and assemble them into more elaborate structures: from subatomic particles to single atoms to monatomic ions and polyatomic ions, and from atoms to small molecules to bigger ones to truly gigantic polymers. By constructing this layered particulate view of matter, we hope our book helps you visualize the structure and properties of substances and the changes they undergo during chemical reactions. We believe, because education research shows, that being able to visualize chemistry at the macro, particulate, and symbolic levels will help you better understand the material and apply that understanding when you're solving a problem.

With that in mind, we begin each chapter with a **Particulate Review** and **Particulate Preview**. The goal of these tools is to prepare you for the material in the chapter. The Particulate Review assesses important prior knowledge that you need to interpret particulate images in the chapter. The Particulate Preview asks you to expand your prior knowledge and to speculate about the new concepts you will see in the chapter. It is also designed to focus your reading by asking you to look out for key terms and concepts.

As you develop your ability to visualize atoms and molecules, you will find that you don't have to resort to memorizing formulas and reactions as a strategy for surviving general chemistry. Instead,

PARTICULATE **REVIEW**

***Polar Bonds versus Polar Molecules***

In Chapter 6, we explore the connections between the structure of molecules and how they interact with each other. Here are representations of four molecules: carbon dioxide, oxygen, water, and ozone.

- Which molecule is polar and contains polar bonds?
- Which molecule is nonpolar despite containing polar bonds?
- Which molecule is polar despite containing nonpolar bonds?

(Review Section 5.3 if you need help.)

*(Answers to Particulate Review questions are in the back of the book.)*

PARTICULATE **PREVIEW**

***Bonds and Functional Groups***

Each molecule shown here contains one or two carbon–oxygen bonds. As you read Chapter 6, look for ideas that will help you answer these questions:

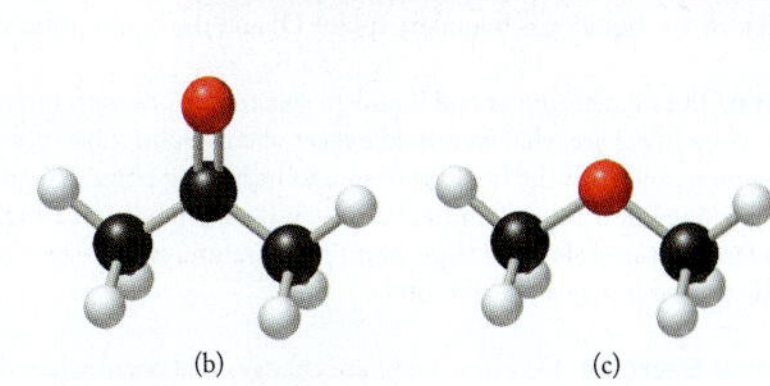


- What is the functional group in each molecule?
- What intermolecular forces exist between molecules of (a)? Between molecules of (b)? Between molecules of (c)?

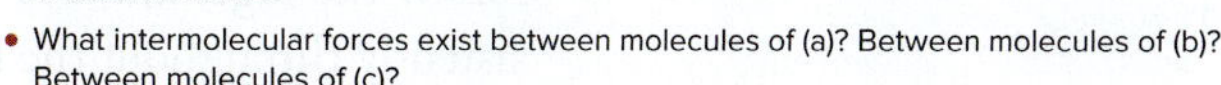

- Rank these three compounds from lowest to highest boiling point.

you will be able to understand why elements combine to form compounds with particular formulas and why substances react with each other the way they do.

## Context

Although our primary goal is for you to be able to interpret, explain, and even predict the physical and chemical properties of substances on the basis of their atomic and molecular structures, we would also like you to understand how chemistry is linked to other disciplines as well as your life. We illustrate these connections by using contexts drawn from biology, medicine, environmental science, materials science, and engineering. We hope that this approach helps you better understand how, for example, chefs use chemistry in their kitchens, and scientists apply the principles of chemistry to treat and cure diseases, to make more efficient use of natural resources, and to minimize the negative impact of human activity on our planet and its people. We also hope you will see how chemistry is all around you, in the air you breathe and the food you eat, and how understanding chemistry will help you solve problems in this course and beyond.

## Problem-Solving Strategies

Building on that theme, another primary goal of this book is to help you improve your problem-solving skills. To do this, you first need to recognize the connections between the information provided in a problem and the answer you are asked to find. Sometimes the hardest part of solving a problem is distinguishing between information that is relevant and information that is not. Once you are clear on where you are starting and where you are going, planning for and carrying out a solution become much easier.

To help you hone your problem-solving skills, we have developed a framework that we introduce in Chapter 1. It is a four-step approach we call COAST, which is our acronym for (1) **C**ollect and **O**rganize, (2) **A**nalyze, (3) **S**olve, and (4) **T**hink About It. We use these four steps in *every* Sample Exercise and in the solutions to *odd-numbered* problems in the *Student Solutions Manual.* They are also used in the hints and feedback embedded in the Smartwork5 online homework program. To summarize the four steps:

**Collect and Organize** helps you understand where to begin to solve the problem. In this step, we often rephrase the problem and what we are trying to find, and we identify the relevant information given in the problem statement or available elsewhere in the book.

**Analyze** is where we map out a strategy for solving the problem. As part of that strategy, we often estimate what a reasonable answer might be.

**Solve** applies our strategy from the second step to the information and relationships identified in the first step to actually solve the problem. We walk you through each step in the solution so that you can follow the logic and the math, and we use dimensional analysis consistently throughout the book.

**SAMPLE EXERCISE 6.5** Interpreting Phase Diagrams **LO4**

Describe the phase changes that take place when the pressure on a sample of water is increased from 0.0001 to 100 atm at a constant temperature of −25°C, and the sample is then warmed from −25°C to 350°C at a constant pressure of 100 atm.

**Collect and Organize** We are asked to describe the phase changes that a sample of water undergoes as its pressure increases at constant temperature and its temperature increases at constant pressure. Figure 6.25 shows which phases of water are stable at various combinations of temperature and pressure.

**Analyze** The change in pressure at constant temperature defines two points on the phase diagram of water with the coordinates (−25°C, 0.0001 atm) and (−25°C, 100 atm). Connecting those points will give us a vertical line. If the line crosses a phase boundary, a change in physical state will occur. The change in temperature at constant pressure defines a third point (350°C, 100 atm) on the phase diagram, which will connect to the second point with a horizontal line.

**Solve** **Figure 6.29** shows a plot of our sample's changes in pressure and temperature. At the bottom end of vertical line 1, water is a gas (vapor). As pressure increases along line 1, it crosses the boundary between gas and solid (point A), which means that vapor turns directly into solid ice, which is stable beyond 100 atm at −25°C (point B). As the temperature of the ice increases along horizontal line 2, it intersects the solid–liquid boundary (point C) and the ice melts. At even higher temperatures, just below 350°C, the line intersects the liquid–gas boundary (point D) and the liquid water vaporizes.

**Think About It** The solid-to-liquid and liquid-to-gas transitions with increasing temperature along line 2 are what we would expect when a solid substance is warmed to its melting point and then the liquid is heated to its boiling point at a given pressure. The transition along line 1 is less familiar because it is caused by increasing the pressure on a gas at a temperature below the triple point temperature, so the vapor never condenses. Instead, it is deposited as a solid.

**Practice Exercise** Describe the phase changes that occur when the temperature of $CO_2$ is increased from −100°C to 50°C at a pressure of 25 atm and the pressure is then increased to 100 atm.

**Think About It** reminds us that an answer is not the last step in solving a problem. We should check the accuracy of the solution and think about the value of a quantitative answer. Is it realistic? Are the units correct? Is the number of significant figures appropriate? How does it compare with our estimate from the Analyze step?

Sample Exercises that require a single-step solution are often streamlined by combining Collect, Organize, and Analyze steps, but the essential COAST features are always maintained.

Many students use the **Sample Exercises** more than any other part of the book. Sample Exercises start with the concepts being discussed and model how to apply them to solve problems. We think that consistently using the COAST framework will help you refine your problem-solving skills, and we hope that the approach will become habit-forming for you. After you finish each Sample Exercise, you'll find a **Practice Exercise** to try on your own. The next few pages describe how to use the tools built into each chapter to gain a conceptual understanding of chemistry and to connect the microscopic structure of substances to their observable physical and chemical properties.

## Chapter Structure

As mentioned earlier, each chapter begins with the Particulate Review and Particulate Preview to help you prepare for the material ahead.

If you are trying to decide what is most important in a chapter, check the **Learning Outcomes** listed on the first page. Whether you are reading the chapter from first page to last or reviewing it for an exam, the Learning Outcomes will help you focus on the key information you need and the skills you should develop. Please notice how Learning Outcomes are linked to Sample Exercises in the chapter.

As you study each chapter, you will find **key terms** in boldface in the text and in a running glossary in the margin. All key terms are also defined in the glossary in the back of the book. We have deliberately duplicated these definitions so that you can continue reading without interruption but quickly refer to them when doing homework or studying.

Many concepts build on others described earlier in the book. We point out these relationships with **Connection** icons in the margins. We hope they enable you to draw your own connections between major themes covered in the book.

**CONNECTION** Isomers are introduced and defined in Section 5.6.

To help you develop your own microscale view of matter, we use **molecular art** to enhance photos and figures and to illustrate what is happening at the atomic and molecular levels.

If you're looking for additional help visualizing a concept, we have approximately 140 animations and simulations denoted by the ChemTour and Stepwise Animation icons in the book, and available online at https://digital.wwnorton.com/atoms3. Both the ChemTours and Stepwise Animations demonstrate dynamic processes and chemical concepts and help you visualize events at the molecular level. Many of the ChemTours allow you to manipulate variables and observe the resulting changes.

CONCEPT **TEST**

Ethane and hydrazine have similar molecular structures (**Figure 6.11**), yet the boiling point of hydrazine (387 K) is more than 200 K greater than that of ethane (184 K). What intermolecular interactions account for this huge difference in boiling points?

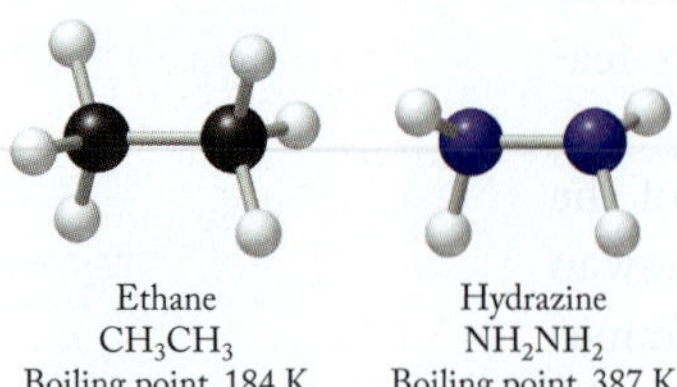


**FIGURE 6.11** Ethane and hydrazine have similar structures but different boiling points.

**Concept Tests** are short, conceptual questions that serve as self-checks by asking you to stop and answer questions related to what you just read. We designed them to help you see for yourself whether you have grasped a key concept and can apply it. We have included an average of one Concept Test per section, and many have visual components. The answers to all the Concept Tests are in the back of the book.

At the end of each chapter is a special Sample Exercise that draws on several key concepts from the chapter and occasionally others from preceding chapters to solve a problem that is framed in the context of a real-world scenario or incident. We call these **Integrated Sample Exercises**. You may find them more challenging than most of the exercises that precede them in each chapter, but please invest your time in working through them because they represent authentic, real-world scenarios that will enhance your problem-solving skills.

Also at the end of each chapter are a thematic **Summary** and a **Problem-Solving Summary**. The first is a synopsis of the chapter, organized by learning outcomes and connected to sections of the chapter. Key figures provide visual cues as you review. The Problem-Solving Summary is unique to this general chemistry book—it outlines the types of problems you should be able to solve, indicates where to find examples of them in the Sample Exercises, and reminds you of key concepts and equations.

## PROBLEM-SOLVING SUMMARY

| Type of Problem | Concepts and Equations | Sample Exercises |
|---|---|---|
| **Identifying intermolecular forces** | All atoms and molecules experience London dispersion forces. Ions in aqueous solution interact with water molecules through ion–dipole interactions. Molecules with permanent dipoles interact through a combination of London dispersion forces and dipole–dipole interactions. The strongest dipole–dipole interactions are hydrogen bonds, which form between H atoms bonded to N, O, and F atoms and other N, O, and F atoms. | **6.1** |
| **Explaining differences and trends in boiling points of liquids** | Because of London dispersion forces, substances made of large molecules usually have higher boiling points than those with smaller molecules. Because of dipole–dipole interactions, polar compounds have higher boiling points than nonpolar compounds of similar molar mass. Compounds whose molecules form hydrogen bonds have even higher boiling points because H bonds are especially strong dipole–dipole interactions. | **6.2** |
| **Predicting solubility in water** | Polar molecules are more soluble in water than nonpolar molecules. Molecules that form hydrogen bonds are more soluble in water than molecules that cannot form these bonds. Like dissolves like. | **6.3** |
| **Explaining solubility trends for ionic compounds** | The solubility of any compound in a solvent depends on the relative strengths of the solute–solute, solvent–solvent, and solute–solvent interactions. | **6.4** |
| **Interpreting phase diagrams** | Locate the combination of temperature and pressure of interest on the phase diagram, and determine which physical state exists at that point. Changes in pressure at constant temperature are represented by vertical paths, and changes in temperature at constant pressure are represented by horizontal paths. If a path crosses a phase boundary line, a change in phase occurs. | **6.5** |

After the summaries are groups of questions and problems. The first group consists of **Visual Problems**. In many of them, you are asked to interpret a molecular view of a sample or a graph of experimental data. The last Visual Problem in each chapter contains a **Visual Problem Matrix**. This grid consists of nine images followed by questions that will test your ability to identify the similarities and differences among the macroscopic, particulate, and symbolic images.

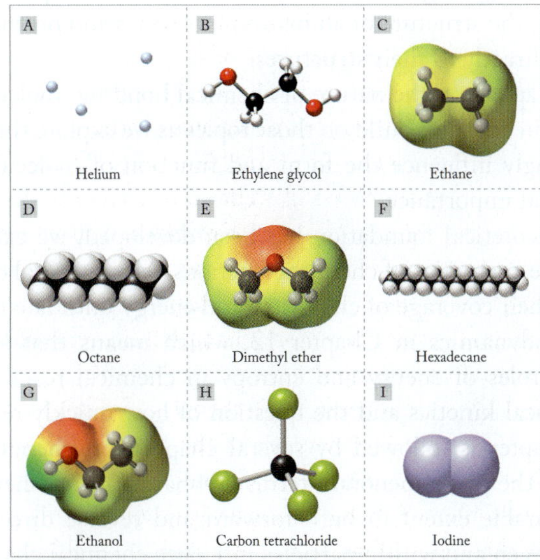


**FIGURE P6.16**

6.16. Use representations [A] through [I] in Figure P6.16 to answer questions a–f.
  a. What are the intermolecular forces between the molecules in representations C, E, and G?
  b. Which of the nine substances has the lowest boiling point?
  c. Which representations are isomers of one another?
  d. Which is more soluble in water: ethylene glycol or iodine? Which is more soluble in carbon tetrachloride? Explain your answers in terms of the intermolecular forces involved.
  e. Which hydrocarbon has the higher boiling point?
  f. Which substance does not form hydrogen bonds between its molecules but can form hydrogen bonds to another molecule?

**Concept Review Questions and Problems** come next, arranged by topic in the same order as they appear in the chapter. Concept Reviews are qualitative and often ask you to explain why or how something happens. Problems are paired and can be quantitative, conceptual, or both. **Contextual Problems** have a title that describes the context in which the problem is placed. Finally, **Additional Problems** can come from any section or combination of sections in the chapter. Some problems incorporate concepts from previous chapters. Problems marked with an asterisk (*) are more challenging and often take multiple steps to solve.

We want you to have confidence in using the answers in the back of the book as well as the *Student Solutions Manual*, so we used a rigorous triple-check accuracy program for this book. Each end-of-chapter question or problem was solved independently by the *Solutions Manual* author, Karen Brewer, and by two additional PhD chemistry instructors. Karen compared her solutions with those from the two reviewers and resolved any discrepancies. This process is designed to ensure clearly written problems and accurate answers in the appendices and *Solutions Manual.*

***6.27.** The compound with the molecular structure in Figure P6.27 (b) melts at a higher temperature than the compound in (a). Explain why, using the different types and strengths of the intermolecular forces experienced by the molecules.

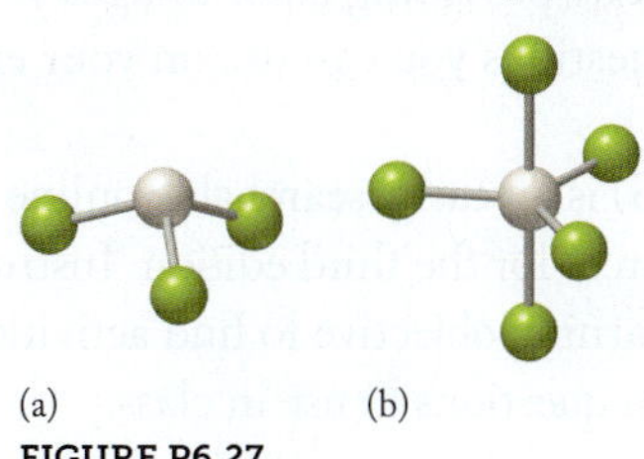


**FIGURE P6.27**

***6.57.** **Improving the Solubility of Drugs** Many common over-the-counter medications such as pseudoephedrine and acetylsalicylic acid are formulated with either sodium cations or chloride anions to improve their water solubility. Why would the structures in Figure P6.57 have improved solubility in water?

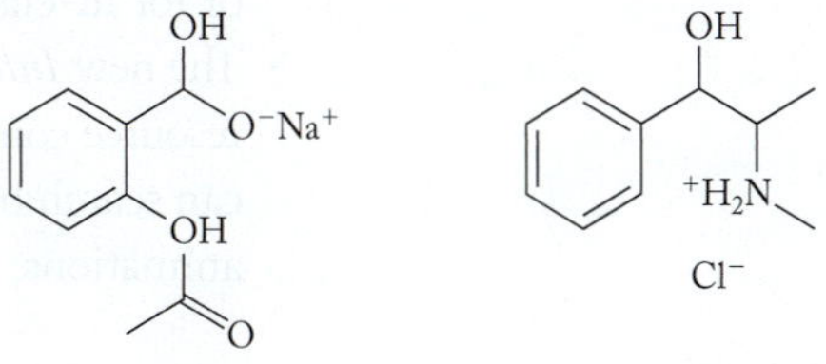

OH $O^-Na^+$ OH O — Sodium acetylsalicyclate

OH $^+H_2N$ $Cl^-$ — Pseudoephedrine hydrochloride

**FIGURE P6.57**

Dear Instructor,

This book adopts what has traditionally been called an atoms-first approach to teaching chemistry. Consequently, the sequence of chapters in the book and the sequence of topics in many of the chapters differ, on the basis of our experience in the classroom, from some other atoms-first general chemistry textbooks, thus our subtitle, *An Atoms-Focused Approach*. For example, we devote the early chapters to giving students an in-depth view of the particulate nature of matter, including the structure of atoms and molecules and how the properties of substances link directly to their structures.

After two chapters on the nature of chemical bonding, molecular shape, and theories to explain both, we build on those topics as we explore the intermolecular forces that strongly influence the form and function of molecules, particularly those of biological importance.

Once this theoretical foundation has been developed, we examine chemical reactivity and the energetics of chemical reactions. Most general chemistry books don't complete their coverage of chemistry and energy until late in the book. We examine thermodynamics in Chapter 12, which means that students already understand the roles of energy and entropy in chemical reactions before they encounter chemical kinetics and the question of how quickly reactions happen. The kinetics chapter is followed by several chapters on chemical equilibrium, which introduce the phenomenon in terms of what happens when reactions proceed to a measurable extent in both forward and reverse directions, and how interactions between and within particles influence chemical changes.

## Changes in the Third Edition

As authors of a textbook, we are often asked: "Why is a new edition necessary? Has the chemistry changed that much since the previous edition?" Although chemistry is a vigorous and dynamic field, most basic concepts presented in an introductory course have not changed dramatically. However, two areas that are tightly intertwined in this text—pedagogy and context—have changed significantly, and those areas drive this new edition. Here are some of the most noteworthy changes we made throughout this edition:

- Adopters enthusiastically embraced the visualization pedagogy introduced in the second edition and then asked for additional tools to help them use this research-based pedagogy to assess students. The teaching tools accompanying the third edition include:
  - In-class activities, often including premade handouts, help you facilitate active learning with a focus on visualization, in any size classroom.
  - Projected visualization questions help you assess students on their understanding and support the Particulate Review, Particulate Preview, and Visual Matrix problems in the book by providing color images you can project in your class, accompanied by questions you can use on your exam or for in-class quizzes.
  - The new *Interactive Instructor's Guide* (IIG) is an easily searchable online resource containing all the teaching resources for the third edition. Instructors can search by chapter, phrase, topic, or learning objective to find activities, animations, simulations, and visualization questions to use in class.

- On the basis of her chemistry education research, Stacey Lowery Bretz evaluated and revised the art in the book and the media to be more pedagogically effective and address student misconceptions.
- In Chapter 5, we've removed any reference to *d* orbital hybridization when discussing molecular geometry, updating our explanations for molecules with more than an octet of electrons around a central atom to be grounded in molecular orbital theory.
- The gases chapter, now Chapter 9, is introduced before thermochemistry. We believe this will allow instructors to develop the concept of pressure in Chapter 9 and then build on that understanding to discuss pressure–volume work in Chapter 10.
- Several new Sample Exercises were added to the acid–base and equilibria chapters on the basis of reviewer feedback and to provide more detailed discussions of titrations and buffers.
- End-of-chapter problems have been refined to better align with the text's Learning Outcomes and provide a more balanced selection in terms of level and concept coverage. We estimate that 10% to 15% of the end-of-chapter problems have been replaced or revised.
- More than 600 problems have been added to the Smartwork5 course, including additional end-of-chapter, algorithmic, and pooled problems to provide more than 5000 problems in the third edition course.
- Powered by Knewton, adaptive functionality in Smartwork5 has been continually improved, resulting in more intuitive experiences for students as they work through personalized learning paths and the availability of better data for instructors to review their students' mastery of chosen learning objectives.
- Revised ChemTour Animations, also found in the ebook and Smartwork5, are now easier to navigate and assign individual sections and have been revised to reflect Stacey Lowery Bretz's visualization pedagogy. ChemTours are available streaming in the ebook, in Smartwork5 questions, as remediation in adaptive assignments, and in the *Interactive Instructor's Guide*, which includes discussion questions and activity ideas for each.
- Our new partnership with the Squarecap classroom response system gives you book-specific clicker questions in an easy-to-use system for a low cost to students.

# Teaching and Learning Resources

## Smarkwork5 Online Homework For General Chemistry

**digital.wwnorton.com/atoms3**

Smartwork5 is the most intuitive online tutorial and homework management system available for general chemistry. The many question types, including graded molecule drawing, math and chemical equations, ranking tasks, and interactive figures, help students develop and apply their understanding of fundamental concepts in chemistry. Adaptive functionality, powered by Knewton, gives students personalized coaching, allowing them to master assigned concepts at their own pace.

Every problem in Smartwork5 includes response-specific feedback and general hints often organized by the steps in COAST. Links to the ebook version of *Chemistry: An Atoms-Focused Approach,* Third Edition, take students to the specific place in the text where the concept is explained. All problems in Smartwork5 use the same language and notation as the textbook.

Smartwork5 also features Tutorial Problems. If students ask for help in a Tutorial Problem, the system breaks the problem down into smaller steps, coaching them with hints, answer-specific feedback, and probing questions within each part. At any point in a Tutorial, students can return to and answer the original problem.

Assigning, editing, and administering homework within Smartwork5 is easy. Smartwork5 allows the instructor to search for problems by text section, learning objectives, question type, difficulty, and Bloom's taxonomy. Instructors can use premade assignment sets provided by Norton authors, modify those assignments, or create their own. Instructors can also make changes in the problems at the question level. All instructors have access to our WYSIWYG (what you see is what you get) authoring tools—the same ones Norton authors use. Those intuitive tools make it easy to modify existing problems or to develop content that meets the specific needs of your course.

Wherever possible, Smartwork5 uses algorithmic variables so that students see slightly different versions of the same problem. Assignments are graded automatically, and Smartwork5 includes sophisticated yet flexible tools for managing class data. Instructors can use the class activity report to assess students' performance on specific problems within an assignment. Instructors can also review individual students' work on problems.

Smartwork5 for *Chemistry*: *An Atoms-Focused Approach,* Third Edition, features the following problem types:

- End-of-Chapter Problems. These problems, which use algorithmic variables when appropriate, all have hints and answer-specific feedback to coach students through mastering single- and multiconcept problems based on chapter content. They use all of Smartwork5's answer-entry tools.
- ChemTour Problems. Many of our ChemTour animations have been redesigned to make them easier than ever to navigate, making it simple to assign Smartwork5 questions on specific interactives.
- Visual and Graphing Problems. These problems challenge students to identify chemical phenomena and to interpret graphs by using Smartwork5's Drag-and-Drop functionality.
- Ranking Task Problems. These problems ask students to make comparative judgments between items in a set.
- Nomenclature Problems. New matching and multiple-choice problems help students master course vocabulary.
- Multistep Tutorials. These problems offer students who demonstrate a need for help a series of linked, step-by-step subproblems to work. They are based on the Concept Review problems at the end of each chapter.
- Math Review Problems. These problems can be used by students for practice or by instructors to diagnose the mathematical ability of their students.

## Ebook

**digital.wwnorton.com/atoms3**

An affordable and convenient alternative to the print text, the Norton Ebook lets students access the entire book and much more: they can search, highlight, and take notes with ease. The Norton Ebook allows instructors to share their notes with students. And the ebook can be viewed on most devices—laptop, tablet, even a public computer—and will stay synced between devices.

The online version of *Chemistry: An Atoms-Focused Approach*, Third Edition, also offers students one-click access to nearly 100 ChemTour animations.

The online ebook is available bundled with the print text and Smartwork5 at no extra cost, or it may be purchased bundled with Smartwork5 access.

Norton also offers a downloadable PDF version of the ebook.

## *Student Solutions Manual*

**by Karen Brewer, Hamilton University**

The *Student Solutions Manual* gives students fully worked solutions to select end-of-chapter problems by using the COAST four-step method (**C**ollect and **O**rganize, **A**nalyze, **S**olve, and **T**hink About It). The *Student Solutions Manual* contains several pieces of art for each chapter, designed to help students visualize ways to approach problems. This artwork is also used in the hints and feedback within Smartwork.

## Test Bank

**by Rebecca Gibbons, Malcolm X College, and Kathie Snyder, Winthrop University**

Norton uses an innovative, evidence-based model to deliver high-quality and pedagogically effective quizzing and testing materials. Each chapter of the Test Bank is structured around an expanded list of student learning objectives and evaluates student knowledge on six distinct levels based on Bloom's taxonomy: remembering, understanding, applying, analyzing, evaluating, and creating.

Questions are further classified by text section and difficulty, making it easy to construct tests and quizzes that are meaningful and diagnostic, according to each instructor's needs. New questions are marked and sortable, making it easy to find brand-new problems to add to your exams. More than 2500 questions are divided into multiple-choice and short-answer categories.

The Test Bank is available with ExamView Test Generator software, allowing instructors to effortlessly create, administer, and manage assessments. The convenient and intuitive test-making wizard makes it easy to create customized exams with no software learning curve. Other key features include the ability to create paper exams with algorithmically generated variables and export files directly to Blackboard, Canvas, Desire2Learn, and Moodle.

## Projected Visualization Questions

**by Rebecca Gibbons, Malcolm X College, and Kathie Snyder, Winthrop University**

Stacey Lowery Bretz believes that you must include visualization questions on exams if you want students to take learning this skill seriously. To overcome the challenge of not being able to distribute four-color exams to her class, she includes several problems on her exams that require students to look at a particulate, macro, or symbolic image that is projected at the front of the room. Our new projected visualization problems allow instructors to use this approach to ask questions on full-color images on exams or as in-class activities.

## *Instructor's Solutions Manual*

**by Karen Brewer, Hamilton University**

The *Instructor's Solutions Manual* gives instructors fully worked solutions to every end-of-chapter Concept Review and Problem. Each solution uses the COAST four-step method (**C**ollect and **O**rganize, **A**nalyze, **S**olve, and **T**hink About It).

## *Interactive Instructor's Guide*

**iig.wwnorton.com/atoms3/full**
**by Spencer Berger, Western Washington University, Thomas Pentecost, Grand Valley State University, and Elizabeth Raymond, Western Washington University**

The *Interactive Instructor's Guide* will help instructors use our unique visualization pedagogy in class and for assessment by compiling the many valuable teaching resources available with *Chemistry: An Atoms-Focused Approach*, Third Edition, in an easily searchable online format. In-class activities, backed by chemical education research and written by active learning experts who teach atoms first, emphasize visualization and are designed to promote collaboration. New projected visualization problems, developed based on how Stacey Lowery Bretz gives her own exams, allow instructors to ask exam questions that use full-color images. The ChemTours are also accompanied by new activity ideas, discussion questions, and clicker questions, making them easier than ever to use. All resources are searchable by chapter, keyword, or learning objective.

## *Clickers in Action: Increasing Student Participation in General Chemistry*

**by Margaret Asirvatham, University of Colorado, Boulder**

This instructor-oriented resource describes implementing clickers in general chemistry courses. *Clickers in Action* contains more than 250 class-tested, lecture-ready questions, with histograms showing student responses, as well as insights and suggestions for implementation. Question types include macroscopic observation, symbolic representation, and atomic/molecular views of processes.

## Downloadable Instructor's Resources

**digital.wwnorton.com/atoms3**

This password-protected site for instructors includes the following:

- Stepwise animations and classroom response questions are included. Developed by Jeffrey Macedone of Brigham Young University and his team, these animations, which use native PowerPoint functionality and textbook art, help instructors walk students through nearly 100 chemical concepts and processes. Where appropriate, the slides contain two types of questions for students to answer in class: questions that ask them to predict what will happen next and why, and questions that ask them to apply knowledge gained from watching the animation. Self-contained notes help instructors adapt these materials to their own classrooms.
- Lecture PowerPoints are available.
- All ChemTours are included.
- Test bank is available in PDF, Word, and ExamView Assessment Suite formats.
- *Solutions Manual* is offered in PDF and Word so that instructors may edit solutions.
- All end-of-chapter Questions and Problems are available in Word along with the key equations.
- Labeled and unlabeled photographs, drawn figures, and tables from the text are available in PowerPoint and JPEG formats.
- Clicker questions, including those from *Clickers in Action,* are included.
- Course cartridges: Available for the most common learning management systems, course cartridges include access to the ChemTours and Stepwise animations as well as links to the ebook and Smartwork5.

## Acknowledgments

Our thanks begin with our publisher, W. W. Norton, for supporting us in writing a book that is written the way we prefer to teach general chemistry. We especially acknowledge the vision, hard work, and dedication of our editor/motivator/taskmaster, Erik Fahlgren. Erik has been an indefatigable source of guidance, perspective, persuasion, and inspiration to all of us. His patience was almost limitless as we worked to make deadlines and debate the intricacies of chemistry. Erik's leadership is the single greatest reason for this book's completion, and our greatest thanks are far too humble an offering for his unwavering vision and commitment. He is the consummate professional and a valued friend.

We are grateful beyond measure for the contributions of our outstanding developmental editor, Andrew Sobel. His questions kept us focused, and his analyses improved our craft; we are all better writers for having benefited from his mentorship. Andrew's guidance, patience, negotiation skills, creativity, and detailed knowledge of chemistry and the editing process combine to create the human element essential for the success of this book.

David Bradley is our project editor, who kept the trains running on time and smoothly. We also thank production manager Ben Reynolds for his work behind the scenes; Julie Tesser and Mike Cullen for finding just the right photos; Grace Tuttle for recruiting a group of first-rate authors to write and update the teaching and learning materials that accompany the third edition; Cailin Barrett-Bressack for her dedication to producing high-quality content in the media program and especially in Smartwork5; and Stacy Loyal for her creative marketing ideas and honest, positive outlook. The entire Norton team is staffed by skilled, dedicated professionals who are delightful colleagues to work with and, as a bonus, to relax with, as the occasion allows.

Many reviewers, listed here, contributed to the development and production of this book. We owe an extra-special thanks to Karen Brewer for her dedicated work on the solutions manuals and for her invaluable suggestions on how to improve the inventory and organization of problems and concept questions at the end of each chapter. She, along with Jon Moldenhauer and Steve Trail, comprised the triple-check accuracy team who helped ensure the quality of the back-of-book answers and solutions manuals. Finally, we acknowledge the care and thoroughness of George Barnes, Suzanne Bart, Mary Ellen Biggin, Chris Culbertson, Narayan Hosmane, Monica Ilies, Chris Lawrence, Dan Lawson, Laurie Lazinski, Mitchell Millan, Matt Mongelli, Dan Moriarty, Chris Nelson, Chad Reznyak, Margaret Scheuermann, Kim Shih, Kim Simons, Brandon Tenn, and Andrea Van Duzor for checking the accuracy of the myriad facts that frame the contexts and the science in the pages that follow.

Thomas R. Gilbert
Rein V. Kirss
Stacey Lowery Bretz

## Third Edition Reviewers

George Barnes, Siena College
Suzanne Bart, Purdue University
Bart Bartlett, University of Michigan
Mary Ellen Biggin, Augustana College
David Boatright, University of West Georgia
Ian Butler, Concordia University
Vanessa Castleberry, Baylor University
Siman Chahal, Saint Paul College
Timothy Chapp, Allegheny College
Daniel Collins, Texas A&M University
Chris Culbertson, Kansas State University
Milagros Delgado, Florida International University
Soren Eustis, Bowdoin College
Anna George, Augustana College
Peter Golden, Sandhills Community College
Elizabeth Griffith, University of Maryland
Shubo Han, Fayetteville State University
Michelle Hatley, Sandhills Community College
Narayan Hosmane, Northern Illinois University
Monica Ilies, Drexel University
Gerald Korenowski, Rensselaer Polytechnic Institute
Jakub Kostal, George Washington University
Heather-Rose Lacy, San Francisco State University
Dan Lawson, University of Michigan, Dearborn
Chris Lawrence, Grand Valley State University
Laurie Lazinski, Fulton-Montgomery Community College
Min Li, California University, Pennsylvania
Oksana Love, University of North Carolina, Asheville
Hrvoje Lusic, William Peace University
Will McWhorter, Clemson University
Mitchel Millan, Casper College
Jon Moldenhauer, Tennessee Tech University
Matt Mongelli, Kean University
Daniel Moriarty, Siena College
Chris Nelson, Chemeketa Community College
James Patterson, Brigham Young University
Chad Rezsnyak, Tennessee Tech University
Jon Rienstra-Kiracofe, Purdue University
Margaret Scheuermann, Western Washington University
Tom Schmedake, University of North Carolina
Dee Sigismond, Merced College
Kim Simons, Emporia State University
Penny Starkey, Saint Paul College
Chad Thibodeaux, Northwestern State University
Steve Trail, Elgin Community College
Andrea Van Duzor, Chicago State University

## Previous Edition Reviewers

Kevin Alliston, Wichita State University
Ioan Andricioaei, University of California, Irvine
Merritt Andrus, Brigham Young University
David Arnett, Northwestern College
Daniel Autrey, Fayetteville State University
Christopher Babayco, Columbia College
Carey Bagdassarian, University of Wisconsin, Madison
Nathan Barrows, Grand Valley State University
Craig Bayse, Old Dominion University
Chris Bender, The University of South Carolina Upstate
Vladimir Benin, University of Dayton
Philip Bevilacqua, Pennsylvania State University
Robert Blake, Glendale Community College
Petia Bobadova-Parvanova, Rockhurst University
Randy A. Booth, Colorado State University
Simon Bott, University of Houston
Stephanie Boussert, DePaul University
John C. Branca, Wichita State University
Jonathan Breitzer, Fayetteville State University
Drew Brodeur, Worcester Polytechnic Institute
Jasmine Bryant, University of Washington
Michael Bukowski, Pennsylvania State University
Charles Burns, Wake Technical Community College
Jerry Burns, Pellissippi State Community College
Jon Camden, University of Tennessee at Knoxville
Tara Carpenter, University of Maryland, Baltimore County
Andrea Carroll, University of Washington
David Carter, Angelo State University
Allison Caster, Colorado School of Mines
Christina Chant, Saint Michael's College
Ramesh Chinnasamy, New Mexico State University
Travis Clark, Wright State University
David Cleary, Gonzaga University
Colleen Craig, University of Washington
Gary Crosson, University of Dayton
Guy Dadson, Fullerton College
David Dearden, Brigham Young University
Danilo DeLaCruz, Southeast Missouri State University
Anthony Diaz, Central Washington University
Keying Ding, Middle Tennessee State University
John DiVincenzo, Middle Tennessee State University
Greg Domski, Augustana College
Jacqueline Drak, Bellevue Community College
Stephen Drucker, University of Wisconsin, Eau Claire
Michael Ducey, Missouri Western State University
Sheryl Ann Dykstra, Pennsylvania State University
Lisa Dysleski, Colorado State University
Mark Eberhart, Colorado School of Mines
Jack Eichler, University of California, Riverside
Amina El-Ashmawy, Collin College
Doug English, Wichita State University
Michael Evans, Georgia Institute of Technology
Renee Falconer, Colorado School of Mines
Hua-Jun Fan, Prairie View A&M University
Jim Farrar, University of Rochester
MD Abul Fazal, College of Saint Benedict & Saint John's University
Anthony Fernandez, Merrimack College
Max Fontus, Prairie View A&M University
Carol Fortney, University of Pittsburgh
Lee Friedman, University of Maryland
Matthew Gerner, University of Arkansas
Arthur Glasfeld, Reed College
Maojun Gong, Wichita State University
Daniel Groh, Grand Valley State University
Megan Grunert, Western Michigan University
Margaret Haak, Oregon State University
Benjamin Hafensteiner, University of Rochester
Tracy Hamilton, University of Alabama at Birmingham
David Hanson, Stony Brook University
Roger Harrison, Brigham Young University
Julie Henderleiter, Grand Valley State University
David Henderson, Trinity College
Carl Hoeger, University of California, San Diego
Amanda Holton, University of California, Irvine
Adam Jacoby, Southeast Missouri State University
James Jeitler, Marietta College
Amy Johnson, Eastern Michigan University
Christina Johnson, University of California, San Diego
Crisjoe Joseph, University of California, Santa Barbara
Marc Knecht, University of Miami
Colleen Knight, College of Coastal Georgia
Maria Kolber, University of Colorado
Regis Komperda, Wright State University
Jeffrey Kovac, University of Tennessee at Knoxville
Ava Kreider-Mueller, Clemson University
John Krenos, Rutgers University
Maria Krisch, Trinity College
Jeremy Kua, University of California, San Diego
Robin Lammi, Winthrop University
Annie Lee, Rockhurst University
Willem Leenstra, University of Vermont
Joseph Lodmell, College of Coastal Georgia
Ted Lorance, Vanguard University
Richard Lord, Grand Valley State University
Charity Lovitt, Bellevue Community College
Suzanne Lunsford, Wright State University
Jeffrey Macedone, Brigham Young University
Sudha Madhugiri, Collin College, Preston Ridge
Douglas Magde, University of California, San Diego
Rita Maher, Richland College
Anna Victoria Martinez-Saltzberg, San Francisco State University
Jason Matthews, Florida State College at Jacksonville
Thomas McGrath, Baylor University
Heather McKechney, Monroe Community College

Anna McKenna, College of Saint Benedict & Saint John's University
Claude Mertzenich, Luther College
Gellert Mezei, Western Michigan University
Alice Mignerey, University of Maryland
Todd Miller, Augustana College
Katie Mitchell-Koch, Emporia State University
Stephanie Morris, Pellissippi State Community College
Nancy Mullins, Florida State College at Jacksonville
Stephanie Myers, Augusta University
Joseph Nguyen, Mount Mercy University
Anne-Marie Nickel, Milwaukee School of Engineering
Sherine Obare, Western Michigan University
Edith Osborne, Angelo State University
Ruben Parra, DePaul University
Robert Parson, University of Colorado
Brad Parsons, Creighton University
Garry Pennycuff, Pellissippi State Community College
Thomas Pentecost, Grand Valley State University
Sandra Peszek, DePaul University
John Pollard, University of Arizona
Gretchen Potts, University of Tennessee at Chattanooga
William Quintana, New Mexico State University
Cathrine Reck, Indiana University, Bloomington
Chad Richardson, Collin College, Preston Ridge
Alan Richardson, Oregon State University
Hope Rindal, Western Washington University
Jason Ritchie, The University of Mississippi
James Roach, Emporia State University
Jill Robinson, Indiana University
Lary Sanders, Wright State University
Perminder Sandhu, Bellevue Community College
Allan Scruggs, Gonzaga University
James Silliman, Texas A&M University, Corpus Christi
Joseph Simard, University of New England
Sergei Smirnov, New Mexico State University
Thomas Sorensen, University of Wisconsin, Milwaukee
Justin Stace, Belmont University
John Stubbs, The University of New England
Alyssa Thomas, Utica College
Lucas Tucker, Siena College
Gabriele Varani, University of Washington
Jess Vickery, SUNY Adirondack
Rebecca Weber, University of North Texas
Karen Wesenberg-Ward, Montana Tech of the University of Montana
Wayne Wesolowski, University of Arizona
Amanda Wilmsmeyer, Augustana College
Thao Yang, University of Wisconsin, Eau Claire
Eric Zuckerman, Augusta University

# Chemistry

## An Atoms-Focused Approach

# 1

# Matter and Energy

## An Atomic Perspective

**TENNIS RACKET TECHNOLOGY** This tennis racket is both lightweight and exceptionally stiff and strong. Its frame is reinforced by graphene, a form of carbon.

PARTICULATE **REVIEW**

### *Solids, Liquids, and Gases*

In Chapter 1, we explore the particulate nature of matter. Throughout this book, colored spheres represent the fundamental particles of matter known as atoms. (*Note*: This book's inside back cover shows the colors that represent the atoms of different elements.) The spheres in these three images represent atoms of the element mercury, which is a liquid at room temperature but turns into a solid at low temperatures and into a vapor (or gas) at high temperatures.

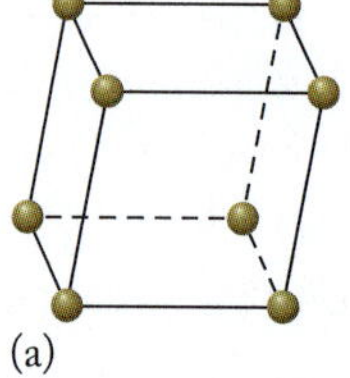
(a)

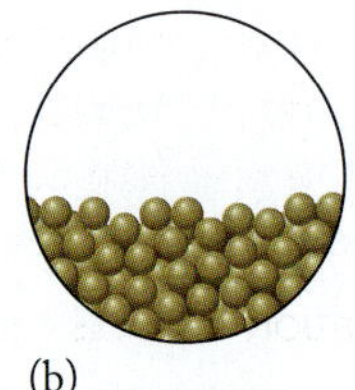
(b)

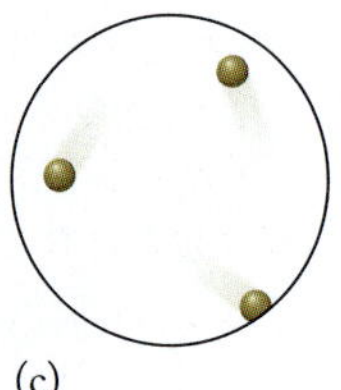
(c)

- Which representation depicts liquid mercury?
- Which representation depicts solid mercury?
- Which representation depicts mercury vapor?

*(Answers to Particulate Review questions are in the back of the book.)*

PARTICULATE **PREVIEW**

### *Elements versus Compounds*

The tennis player on this page is using a racket whose frame is strengthened by graphene, which is composed of two-dimensional arrays of atoms of the element carbon. As you read Chapter 1, look for ideas that will help you answer these questions:

- Which representation depicts molecules of a compound?
- Which representation depicts a mixture of molecular elements?
- Which representation depicts a compound consisting of an array of ions?
- Which representation depicts an element consisting of an array of atoms?

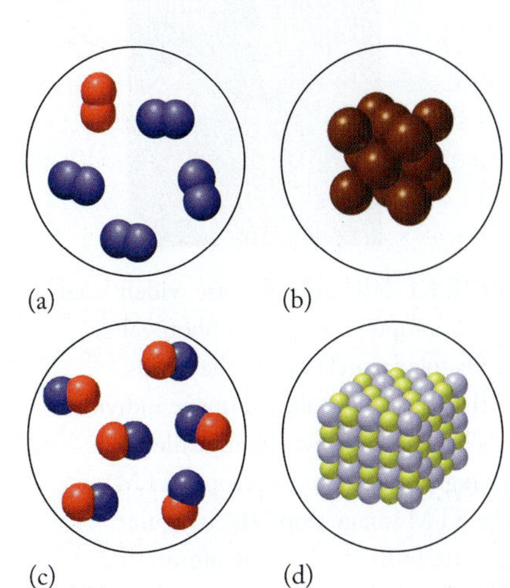

## Learning Outcomes

**LO1** Describe scientific methods

**LO2** Apply the COAST approach to solving problems
**Sample Exercises 1.1–1.12**

**LO3** Distinguish between the classes of matter and between the physical and chemical properties of pure substances
**Sample Exercises 1.1, 1.2**

**LO4** Describe the states of matter and how their physical properties can be explained by the particulate nature of matter
**Sample Exercise 1.3**

**LO5** Distinguish between heat, work, potential energy, and kinetic energy, and describe the law of conservation of energy

**LO6** Use molecular formulas and molecular models to describe the elemental composition and three-dimensional arrangement of the atoms in compounds

**LO7** Distinguish between exact and uncertain values and express uncertain values with the appropriate number of significant figures
**Sample Exercises 1.4–1.6**

**LO8** Accurately convert values from one set of units to another
**Sample Exercises 1.7–1.10**

**LO9** Analyze and express experimental results to convey their certainty, that is, how precisely and with what accuracy they are known
**Sample Exercises 1.11, 1.12**

## 1.1 Exploring the Particulate Nature of Matter

The tennis racket in this chapter's opening photograph is both lightweight and exceptionally stiff and strong. Those properties are linked to the chemical composition of its frame, which is reinforced by a substance called graphene. Viewed at the particle level, graphene consists of sheets of carbon atoms bonded together. The strength of the carbon–carbon bonds and the rigidity of the resulting network contribute to the racket's notable strength and stiffness. Graphene was discovered only recently, but carbon networks like those in its structure have existed in nature for billions of years. Modern scientists can characterize the properties of graphene because they understand the carbon–carbon bonds that form its structure.

In this chapter we begin an exploration of how the properties of materials are linked to their particulate structure. As we do, we need to go back in time and acknowledge the philosophers of ancient Greece who espoused *atomism*, a belief that all forms of matter are composed of extremely tiny, indestructible building blocks called **atoms**. Atomism is an example of a natural philosophy; it is not a **scientific theory**. The difference between the two is that although both seek to explain natural phenomena, scientific theories do so through concise, testable explanations based on observation and experimentation. An important quality of a valid scientific theory is that it accurately predicts the results of experiments and can even serve as a guide to designing those experiments. The ancient Greeks did not have the technology to test whether matter really is made of atoms—but now we do.

Consider the images in **Figure 1.1**. On the bottom is a photograph of silicon wafers, the material used to make computer chips and photovoltaic cells. Above the photograph is a magnified view of a silicon wafer produced by an instrument called a scanning tunneling microscope.[1] The fuzzy spheres are individual atoms of silicon.

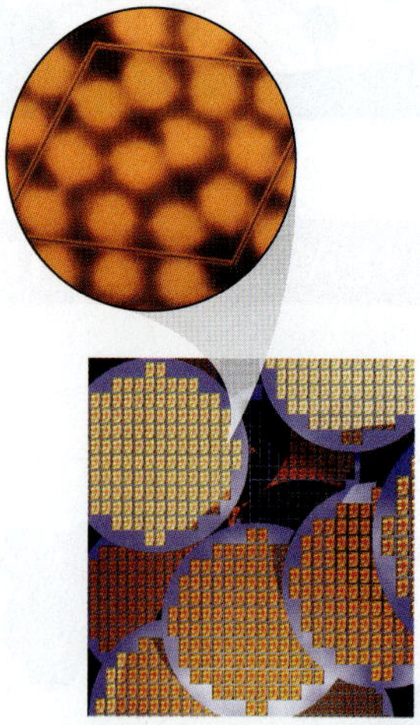

**FIGURE 1.1** Silicon wafers are widely used to make computer chips and photovoltaic cells for solar panels. Since the 1980s, scientists have been able to image individual atoms by using an instrument called a scanning tunneling microscope (STM). In the STM image (top), the irregular shapes are individual silicon atoms. The radius of each atom is 117 picometers (pm), or 117 trillionths of a meter.

[1]German physicist Gerd Binnig (b. 1947) and Swiss physicist Heinrich Rohrer (1933–2013) shared the 1986 Nobel Prize in Physics for developing scanning tunneling microscopy.

## Atomic Theory: Scientific Methods in Action

Scanning tunneling microscopes have been used to image atoms since the early 1980s, but the scientific theory that matter was composed of atoms evolved two centuries earlier, when chemists in France and England made enormous advances in our understanding of the composition of matter. Among those researchers was French chemist Antoine Lavoisier (1743–1794), who published the first modern chemistry textbook in 1789. That book contained a list of substances that he believed could not be separated into simpler substances. Today we call such "simple" substances **elements** (**Figure 1.2**). The silicon (Si) in Figure 1.1 is an element, as is carbon (C). Every element has a one- or two-letter symbol, shown in the periodic table of the elements inside this book's front cover.

Lavoisier and other scientists conducted experiments that examined patterns in how elements combined with other elements to form **compounds**. Those experiments followed systematic approaches known as **scientific methods** to investigate and understand natural phenomena (**Figure 1.3**). When such investigations reveal consistent patterns and relationships, they may be used to formulate concise descriptions of fundamental scientific truths. Those descriptions are known as **scientific laws**.

When French chemist Joseph Louis Proust (1754–1826) studied the composition of compounds containing different metals and oxygen, he concluded that those compounds always contained the same proportions of their component elements. His **law of definite proportions** applies to all compounds. An equivalent law, known as the **law of constant composition**, states that a compound always has the same *elemental composition* by mass no matter what its source. Thus, the composition of pure water is always the same: 11.2% by mass hydrogen and 88.8% by mass oxygen.

When Proust published his law of definite proportions, some leading chemists of the time refused to believe it. Their own experiments seemed to show, for example, that in samples of the compound that tin formed with oxygen, the content of tin varied. Those scientists did not realize that their samples were actually

**atom** the smallest particle of an element that cannot be chemically or mechanically divided into smaller particles.

**scientific theory** a concise, extensively tested explanation of widely observed natural phenomena.

**element** a pure substance that cannot be separated into simpler substances.

**compound** a pure substance composed of two or more elements that are chemically bonded in fixed proportion.

**scientific methods** approaches to acquiring knowledge by observing phenomena, developing a testable hypothesis, and carrying out additional experiments that test the validity of the hypothesis.

**scientific law** a concise and generally applicable statement of a fundamental scientific principle.

**law of definite proportions** the principle that a compound always contains the same proportion of its component elements.

**law of constant composition** the principle that all samples of a given compound have the same elemental composition.

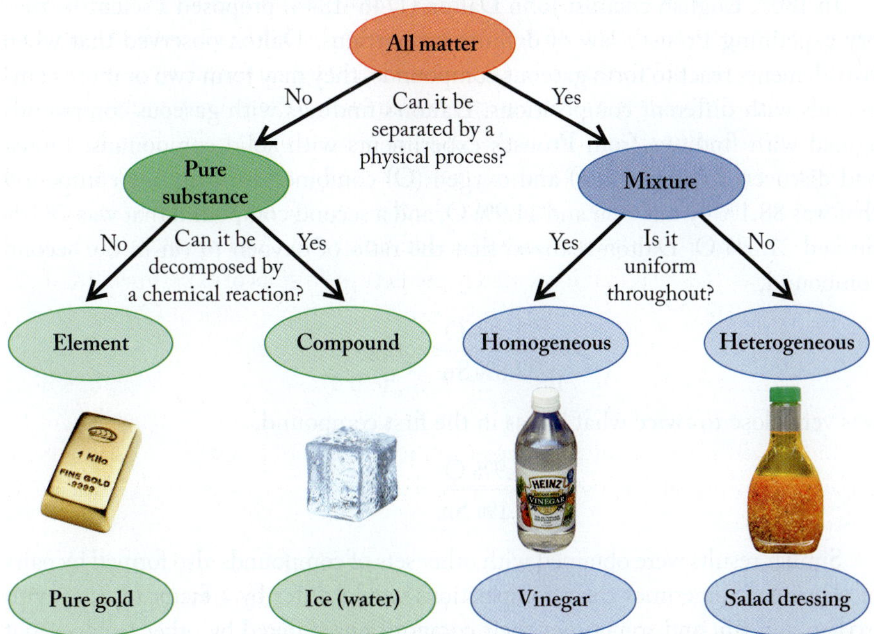


**FIGURE 1.2** Matter is classified as shown in this diagram. The two principal categories are pure substances and mixtures. A pure substance may be a compound (such as water) or an element (such as gold). When the substances making up a mixture are distributed uniformly, as in vinegar (a mixture of acetic acid and water), the mixture is homogeneous. When the substances making up a mixture are not distributed uniformly, as in salad dressing, the mixture is heterogeneous. We explore those categories in Section 1.3.

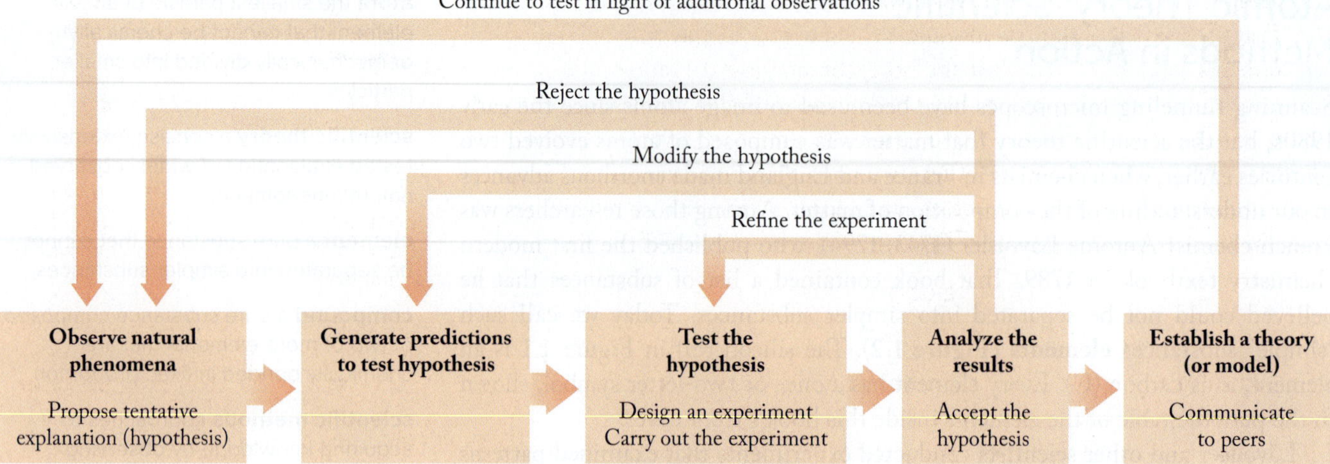


**FIGURE 1.3** Using scientific methods, observations lead to a tentative explanation, or hypothesis, which leads to more observations and testing, which in turn may lead to the formulation of a succinct, comprehensive explanation called a theory. That process is rarely linear; it often involves looping back because the results of one test lead to additional tests and a revised hypothesis. Science, when done right, is a dynamic and self-correcting process.

mixtures of two compounds with different compositions, which Proust was able to demonstrate. Still, accepting Proust's law required more than corroborating results from other scientists; the law also needed to be explained by a scientific theory. That is, scientists needed a convincing argument that explained *why* the composition of a compound is always the same, no matter its source.

Scientific laws and theories complement each other in that scientific laws describe natural phenomena and relationships, and scientific theories explain *why* those phenomena and relationships are always observed. Scientific theories usually start out as tentative explanations of why a set of experimental results was obtained or why a particular phenomenon is consistently observed. Such a tentative explanation is called a **hypothesis**. An important feature of a hypothesis is that it can be tested through additional observations and experiments. A hypothesis also enables scientists to accurately predict the likely outcomes of observations and experiments. Further testing and observation might support a hypothesis, disprove it, or require that it be modified. A hypothesis that withstands the tests of many experiments, accurately explaining further observations and accurately predicting the results of additional experimentation, may be elevated to the rank of scientific theory.

In 1803, English chemist John Dalton (1746–1844) proposed a scientific theory explaining Proust's law of definite proportions. Dalton observed that when two elements react to form gaseous compounds, they may form two or more compounds with different compositions. Dalton's findings with gaseous compounds agreed with findings from Proust's experiments with solid compounds. Proust had discovered that tin (Sn) and oxygen (O) combined to form one compound that was 88.1% by mass Sn and 11.9% O, and a second compound that was 78.8% Sn and 21.2% O. Dalton realized that the ratio of oxygen to tin in the second compound,

$$\frac{21.2\%\ \mathrm{O}}{78.8\%\ \mathrm{Sn}} = 0.269$$

was very close to twice what it was in the first compound,

$$\frac{11.9\%\ \mathrm{O}}{88.1\%\ \mathrm{Sn}} = 0.135$$

Similar results were obtained with other sets of compounds also formed by pairs of elements. Sometimes their compositions would differ by a factor of 2, as with oxygen and tin, and sometimes their compositions differed by other factors—but

Dalton showed that the compositions always differed *by ratios of small whole numbers*. That pattern led Dalton to formulate the **law of multiple proportions**: when two elements combine to make two (or more) compounds, the ratio of the masses of one of the elements that combine with a given mass of the second element is always a ratio of small whole numbers. For example, 15 grams of oxygen combines with 10 grams of sulfur under one set of reaction conditions, whereas only 10 grams of oxygen combines with 10 grams of sulfur to form a different compound under a different set of reaction conditions. The ratio of the two masses of oxygen

$$\frac{15\ \cancel{\text{g oxygen}}}{10\ \cancel{\text{g oxygen}}} = \frac{3}{2}$$

is indeed a ratio of two small whole numbers and is consistent with Dalton's law of multiple proportions.

To explain the laws of definite proportions and multiple proportions, Dalton proposed the scientific theory that *elements are composed of atoms*. Thus, Proust's compound with the O:Sn ratio of 0.135 contains one atom of oxygen for each atom of tin, whereas his compound with twice that O:Sn ratio (0.269) contains *two* atoms of O per atom of Sn. Those atomic ratios are reflected in the **chemical formulas** of the two compounds: SnO and $SnO_2$, in which the subscripts after the symbols represent the relative number of atoms of each element in the substance. The absence of a subscript means the formula contains one atom of the preceding element. Similarly, the two compounds that sulfur and oxygen form have an oxygen ratio of 3:2 because their chemical formulas are $SO_3$ and $SO_2$, respectively.

Since the early 1800s, scientists have learned much more about the atomic, and even subatomic, structure of the matter that makes up our world and the universe that surrounds us. Although the laws developed two centuries ago are still useful, Dalton's atomic theory, like many theories, has undergone revisions as new discoveries have been made. Dalton assumed, for example, that all the atoms of a particular element were the same. However, we will learn in Chapter 2 that atoms have internal components and structures, only some of which are the same for all the atoms of a given element. Atoms can differ in other ways, too, that the scientists of centuries ago could not have observed or even imagined.

## 1.2 COAST: A Framework for Solving Problems

Throughout this chapter and book, you will find Sample Exercises designed to help you better understand chemical concepts and develop your problem-solving skills. Each Sample Exercise follows a systematic approach to problem solving and is followed by a Practice Exercise that can be solved using a similar approach. We also encourage you to apply that approach to the end-of-chapter problems. We use the acronym COAST (**C**ollect and **O**rganize, **A**nalyze, **S**olve, and **T**hink about the answer) to represent the four steps in our problem-solving approach. As you read about it here and use it later, keep in mind that COAST is merely a *framework* for solving problems, not a recipe. Use it as a guide to develop your own approach to solving problems.

**Collect and Organize** The first step in solving a problem is to decide how to use the given information. Identify the key concepts of the problem and define the key terms used to express those concepts. Restating the problem in your own words might be useful. Sort through the information given in the problem to

**hypothesis** a tentative and testable explanation for an observation or a series of observations.

**law of multiple proportions** the principle that, when two masses of one element react with a given mass of another element to form two compounds, the two masses of the first element have a ratio of two small whole numbers.

**chemical formula** a notation that uses the symbols of the elements to represent the elemental composition of a pure substance; subscripts indicate the relative number of atoms of each element in the substance.

separate what is relevant from what is not. Then assemble the relevant information along with any supplemental information that may be needed, such as equations, definitions, and the values of constants.

**Analyze** The next step is to analyze the information you have collected to determine how to connect it to the answer you seek. Sometimes working backward to create the relationships may be easier: consider the nature of the answer first and think about how you might get to it from the information given in the problem and other sources. That approach might mean finding some intermediate quantity to use in a later step. If the problem is quantitative and requires a numerical answer, the units of the initial values and the final answer may help you identify how they are connected and which equation(s) may be useful. Consider rearranging equations to solve for an unknown.

For some problems, drawing a sketch based on molecular models or on an experimental setup may help you visualize how the starting points and final answer can be connected. You also should look at the numbers involved and estimate an answer. Having an order-of-magnitude ("ballpark") estimate of your final answer before entering numbers into your calculator can serve as a check on the accuracy of your calculated answer.

Some Sample Exercises test your understanding of a single concept or require only a single-step calculation. In those exercises we may combine the Collect and Organize with the Analyze steps.

**Solve** For most conceptual questions, the solution flows directly from your analysis of the problem. To solve quantitative problems, you need to insert the starting values and the appropriate constants into the relevant equations and then calculate the answer. Make sure that units are consistent and cancel as needed and that the certainty of the answer is reflected by an appropriate number of *significant figures* (we discuss significant figures in Section 1.7 and conversion factors in Section 1.8).

**Think About It** Finally, you need to think about your result. Does your answer make sense in the context of your own experience and what you have just learned? Is the value for a quantitative answer reasonable—is it close to your estimate from the Analyze step? Are the units correct and the number of significant figures appropriate? Then ask yourself how confident you are that you could solve another problem, perhaps drawn from another context but based on the same chemical concept. You may also think about how the problem relates to other observations you may have made about matter in your daily life.

The COAST approach should help you solve problems logically and avoid certain pitfalls, such as grabbing an equation that seems to have the right variables and simply plugging numbers into it or resorting to trial and error. As you study the steps in each Sample Exercise, try to answer these questions: **What** is done in this step? **How** is it done? **Why** is it done? After answering those questions, you will be ready to solve the Practice Exercises and end-of-chapter problems systematically.

## 1.3 Classes and Properties of Matter

Everything that we can see, touch, smell, or feel—from the air we breathe to the ground we walk on—is a form of matter. Scientists define **matter** as everything in the universe that has **mass (*m*)** and occupies space. **Chemistry** is the study of the composition, structure, and properties of matter and the changes it undergoes.

**matter** anything that has mass and occupies space.

**mass (*m*)** the property that defines the quantity of matter in an object.

**chemistry** the study of the composition, structure, and properties of matter and of the energy consumed or given off when matter undergoes a change.

**pure substance** matter that has a constant composition and cannot be broken down to simpler matter by any physical process.

**physical process** a transformation of a sample of matter, such as a change in its physical state, that does not alter the chemical identity of any substance in the sample; also called *physical change*.

**intensive property** a property that is independent of the amount of substance.

**extensive property** a property that varies with the amount of substance.

**physical property** a property of a substance that can be observed without changing the substance into another substance.

Matter is classified according to its composition, as we saw in Figure 1.2. The simplest forms of matter—elements and compounds—are **pure substances**, such as gold or water, that cannot be separated into simpler substances by any physical process. A **physical process** (or *physical change*) in a sample of matter refers to a transformation that does not alter the chemical identities of any substance in the sample, such as melting ice.

Pure substances have distinctive properties. Pure gold, for example (**Figure 1.4a**), has a characteristic color, is a soft metal, is malleable (it can be hammered into very thin sheets called gold leaf), is ductile (it can be drawn into thin wires), and melts at 1064°C. Those properties, which characterize a pure substance but are independent of the amount of the substance in a sample, are called **intensive properties**. Other properties, such as the particular length, width, mass, and volume of an ingot of gold, are called **extensive properties** because they depend on how much of the substance is present in a particular sample.

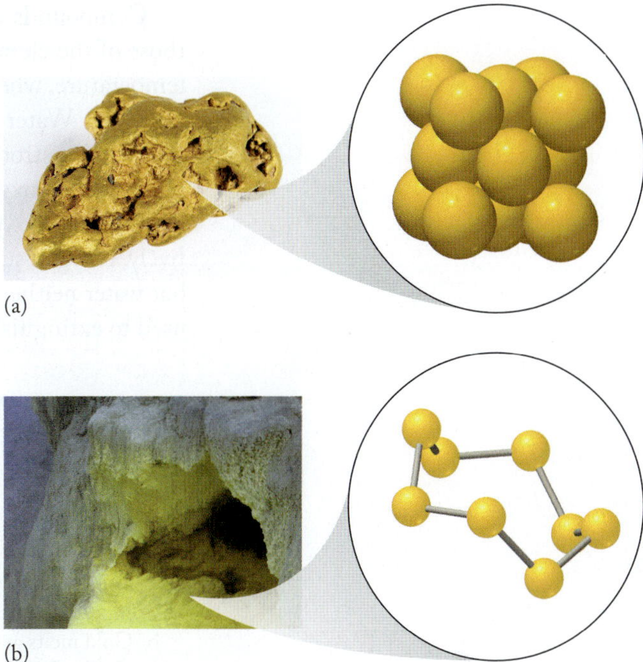


**FIGURE 1.4** (a) Gold and (b) sulfur are among the few elements that occur in nature uncombined with other elements.

The properties of substances are either *physical* or *chemical*. **Physical properties**, such as those just described for gold, can be observed or measured without changing the substance into another substance. Another physical property is **density (*d*)**, which is the ratio of the mass (*m*) of a substance or object to its volume (*V*):

$$d = \frac{m}{V} \tag{1.1}$$

## CONCEPT TEST

Which properties of a sample of pure iron are intensive? (a) mass; (b) density; (c) volume; (d) hardness

*(Answers to Concept Tests are in the back of the book.)*

Gold is one of the few elements that occurs in nature uncombined with other elements, as is sulfur (**Figure 1.4b**). The tendencies of most elements to react with other substances—that is, to be transformed in **chemical reactions** into compounds with different chemical identities and properties—represent the **chemical properties** of the elements. Chemical properties include whether or not a particular element reacts with another element or with a particular compound. They also include how rapidly the reactions take place and what products are formed.

Throughout this book we will see many examples of how the properties of substances are linked to the behavior of the particles that form them, including how those particles interact with the particles that form other substances. Those particles may be atoms, but they may also be groups of atoms held together in a characteristic pattern by forces called **chemical bonds**. Many of those groups of atoms are *neutral* **molecules**, meaning that they have no net electrical charge. However, some atoms or molecules acquire a net positive or negative electrical charge, and therefore we refer to them as monatomic (single-atom) or polyatomic (many-atom) **ions**. We will discuss how and why ions form in Chapter 2, but for now the key point is that the particles that make up elements and compounds may be atoms, molecules, or ions.

**density (*d*)** the ratio of the mass (*m*) of an object to its volume (*V*).

**chemical reaction** the conversion of one or more substances into one or more different substances; also called *chemical change*.

**chemical property** a property of a substance that can be observed only by reacting the substance with something else to form another substance.

**chemical bond** a force that holds two atoms or ions in a molecule or a compound together.

**molecule** a collection of chemically bonded atoms.

**ion** a particle consisting of one or more atoms that has a net positive or negative charge.

Compounds have physical and chemical properties that are different from those of the elements that make them up. For example, water is a liquid at room temperature, whereas hydrogen and oxygen, the elements that react to form water, are gases. Water is a liquid at room temperature because molecules of water are much more strongly attracted to each other than are molecules of hydrogen or molecules of oxygen. Liquid water expands when it freezes at 0°C, but liquid hydrogen and oxygen contract when they freeze at −259°C and −219°C, respectively. Oxygen supports combustion reactions, and hydrogen is highly flammable, but water neither supports combustion nor is flammable. Indeed, water is widely used to extinguish fires.

**SAMPLE EXERCISE 1.1** Distinguishing Physical and Chemical Properties **LO3**

Which properties of gold are chemical and which are physical?

a. Gold metal, which is insoluble in water, can be made soluble by reacting it with a mixture of nitric and hydrochloric acids known as aqua regia.
b. Gold melts at 1064°C.
c. Gold can be hammered into sheets so thin that light passes through them.
d. Gold metal can be recovered from gold ore by treating the ore with a solution containing cyanide, which reacts with and dissolves gold.

**Collect, Organize, and Analyze** Chemical properties describe how a substance reacts with other substances, whereas physical properties can be observed or measured without changing one substance into another. Properties (a) and (d) describe reactions that chemically change gold metal into compounds of gold that dissolve in water. Properties (b) and (c) describe processes in which elemental gold remains elemental gold. When it melts, gold changes its physical state from solid to liquid, but not its chemical identity. When gold is hammered flat, it is still solid, elemental gold.

**Solve** Properties (a) and (d) are chemical properties, whereas (b) and (c) are physical properties.

**Think About It** When possible, rely on your experiences and observations. Gold jewelry does not dissolve in water, so dissolving gold metal requires a change in its chemical identity: it can no longer be elemental gold. However, physical processes such as melting do not alter the chemical identity of the gold. Gold can be melted and then cooled to produce solid gold again.

**Practice Exercise** Which of the following properties of water are chemical and which are physical?

a. Water normally freezes at 0.0°C.
b. Water and carbon dioxide combine during photosynthesis, forming sugar and oxygen.
c. A cork floats on water, but a piece of copper sinks.
d. During digestion, starch reacts with water to form sugar.

*(Answers to Practice Exercises are in the back of the book.)*

Most of the matter in nature exists as mixtures of elements and compounds. **Mixtures** are composed of two or more pure substances and are classified as either *homogeneous* or *heterogeneous* (see Figure 1.2). The substances in a **homogeneous mixture** are distributed uniformly, and the composition and appearance of the mixture are uniform. Homogeneous mixtures are also called **solutions**, a term that scientists apply to homogeneous mixtures of gases and solids as well as liquids.

**mixture** a combination of pure substances in various proportions in which the individual substances retain their chemical identities and can be separated from one another by a physical process.

**homogeneous mixture** a mixture in which the components are distributed uniformly and the composition and appearance are uniform.

**solution** another name for a *homogeneous mixture*. Solutions are often liquids, but they may also be solids or gases.

**heterogeneous mixture** a mixture in which the components are not distributed uniformly, so that the mixture contains regions of different compositions.

**immiscible liquids** combinations of liquids that do not mix with, or dissolve in, each other.

**distillation** a process using evaporation and condensation to separate a mixture of substances with different volatilities.

**volatility** a measure of how readily a substance vaporizes.

**filtration** a process for separating solid particles from a liquid or gaseous sample by passing the sample through a porous material that retains the solid particles.

In contrast, the substances in a **heterogeneous mixture** are not distributed uniformly and contain distinct regions of different composition. Heterogeneous mixtures include mixtures of **immiscible liquids**, which, like the oil and water that separate into distinct layers in salad dressing, do not dissolve in each other. Nearly all the forms of matter we encounter, including the air we breathe and the food and drink we consume, are mixtures.

## Separating Mixtures

The pure substances in mixtures can be separated from one another on the basis of differences in their physical properties. The differences are closely linked to how strongly the particles in each substance interact with one another and with the particles of the other substances in the mixture. For example, the water in seawater can be made drinkable by separating it from the salts (mostly sodium chloride) dissolved in it. One process for doing so is **distillation** (**Figure 1.5**), in which a component of the mixture (water) is separated by evaporation, and the resulting vapor is recovered by condensation. Distillation works as a separation technique whenever the individual components of a mixture have a different **volatility**—that is, one or more of them vaporizes more readily than the others. The volatility of a substance is inversely proportional to the strength of the interactions between its particles: as the strength of the interactions increases, the probability decreases that particles of the substance will have enough energy to break away from adjacent particles in the liquid phase and become particles of vapor.

In addition to dissolved salts, seawater may also contain single-celled plants called phytoplankton (**Figure 1.6**). Scientists who study those microscopic algae separate them from seawater by **filtration**. That process works in that case because the phytoplankton cells, which are many micrometers in diameter, are larger than the pores in the filter (**Figure 1.7**). The cells are trapped on the filter, whereas the water and dissolved salts readily pass through. Filtration is a useful technique for removing particles suspended in gases and in liquids. For example, the air in

**FIGURE 1.5** (a) In a solar-powered distillation apparatus used in survival gear to yield freshwater from seawater, sunlight passes through the transparent dome and heats a pool of seawater. Water vapor rises from the pool, contacts the inside of the cooler transparent dome, and condenses. The distilled water collects in the depression around the rim and then passes into the attached tube, from which one may drink. (b) A solar-powered distillation apparatus in use.

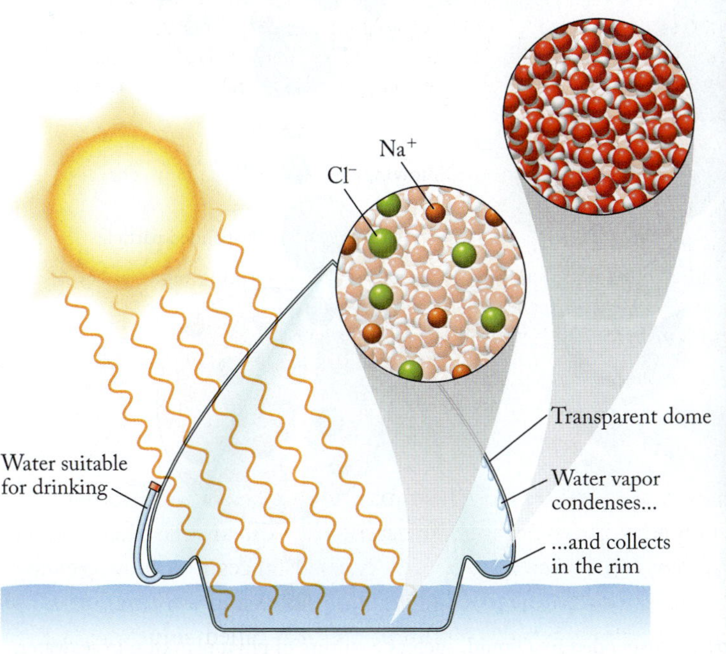

**FIGURE 1.6** (a) Very high concentrations (called "blooms") of phytoplankton known as *coccolithophores* in the Black Sea were photographed by a NASA satellite. During intense blooms, coccolithophores turn the sea color a milky aquamarine. The color comes from chlorophyll and other pigments in the phytoplankton; the milkiness comes from sunlight scattering off the organisms' textured exterior shells. (b) A single coccolithophore cell.

hospitals and in the clean rooms of laboratories is typically purified using HEPA (high-efficiency particulate air) filters to remove dust, bacteria, and even viruses.

The components of the phytoplankton caught on the filter in Figure 1.7 can be further separated by soaking the wet filter in acetone (**Figure 1.8a**), which dissolves some of the compounds inside the phytoplankton cells. Those compounds include chlorophyll and other pigments that have distinctive colors. Once dissolved in acetone, the pigments can be separated from one another by using **chromatography** (**Figure 1.8b**). In a chromatographic separation, the components of a liquid or gaseous mixture are distributed between two phases: a stationary solid (or liquid-coated solid) and a moving liquid or gas. The more strongly the particles in the

**chromatography** a process involving a stationary and a mobile phase for separating a mixture of substances according to their different affinities for the two phases.

**solid** a phase of matter that has a definite shape and volume.

**liquid** a phase of matter that occupies a definite volume but flows to assume the shape of its container.

**gas** a phase of matter that has neither definite volume nor definite shape and that expands to fill its container; also called *vapor*.

**FIGURE 1.7** Particles suspended in a liquid, such as a culture of phytoplankton in seawater, can be separated from the liquid by filtration. The suspended particles and filter pores are many thousands of times larger than water molecules.

mobile phase interact with the stationary solid, the more slowly they move. Those particles that interact weakly with the stationary solid move more rapidly.

**CONCEPT TEST**

Which physical process—distillation, filtration, or chromatography—would you use to perform each of these separations?

a. Removing particles of rust from drinking water.

b. Separating the different coloring agents in a sample of ink.

c. Separating volatile compounds normally found in natural gas that have dissolved in a sample of crude oil.

*(Answers to Concept Tests are in the back of the book.)*

**SAMPLE EXERCISE 1.2** Distinguishing between Classes of Matter **LO3**

Classify each material as an element, a compound, a homogeneous mixture, or a heterogeneous mixture: (a) fruit salad; (b) filtered air in a scuba tank; (c) helium gas inside a party balloon; (d) dry ice (solid carbon dioxide).

**Collect, Organize, and Analyze** Fruit salad is a mixture of several fruits. Because of their large size, the pieces of fruit are not mixed uniformly. Filtered air is a mixture of gases that are uniformly distributed throughout the inner volume of the tank. Helium is an element. Dry ice is solid carbon dioxide, which, given its name, is a compound composed of molecules of chemically bonded carbon and oxygen.

**Solve**

a. Fruit salad is a heterogeneous mixture.
b. Filtered air is a homogeneous mixture.
c. Helium is an element.
d. Dry ice (solid carbon dioxide) is a compound.

**Think About It** Distinguishing between heterogeneous and homogeneous mixtures can be challenging. Some mixtures, such as murky river water, may look homogeneous; however, if the water is allowed to stand in a glass, solid particles may settle to the bottom, showing that the mixture is actually heterogeneous.

**Practice Exercise** Which of the following are pure substances? (a) oxygen gas; (b) human bones; (c) a brick wall; (d) a wooden baseball bat

*(Answers to Practice Exercises are in the back of the book.)*

(a)

(b)

**FIGURE 1.8** (a) Chlorophyll and other pigments can be extracted from a phytoplankton sample trapped on a filter by using the solvent acetone. (b) Dissolved pigments can be separated from one another by using chromatography, producing a characteristic pattern of colored bands.

## 1.4 States of Matter

The forms of matter described in Section 1.3 can exist in one of three phases, or physical states: solid, liquid, or gas. Their characteristic properties are as follows:

- A **solid** has a definite volume and shape.
- A **liquid** has a definite volume but not a definite shape. Instead, the liquid takes the shape of its container.
- A **gas** (or *vapor*) has neither a definite volume nor a definite shape. Rather, gas expands to occupy the entire volume and shape of its container. Unlike solids and liquids, gases are highly compressible, which means they can be squeezed into smaller volumes.

**sublimation** the transformation of a solid directly into a gas (vapor).

**deposition** the transformation of a gas (vapor) directly into a solid.

Consider the three states of water shown with photographs and particle-view magnifications in **Figure 1.9**. All three phases contain the same particles: molecules made up of two atoms of hydrogen chemically bonded to a central atom of oxygen. In solid ice (Figure 1.9a), each water molecule is surrounded by four others and locked in place in a hexagonal array of molecules that extends in all three spatial dimensions. Molecules in the array may vibrate a little, depending on their temperature, but they do *not* move past the molecules that surround them. Thus, ice is rigid at both the molecular and macroscopic levels. The molecules in liquid water (Figure 1.9b), by contrast, are more randomly ordered and flow past one another. They are still close to each other, but their nearest neighbors change over time. Although little space exists between the molecules of water in ice and liquid water, the molecules of water vapor (Figure 1.9c) are widely separated. The volume of the particles themselves is negligible in comparison with the volume occupied by the vapor. Most of the volume is made up of nothing—just empty space between the particles of gas. That empty space between particles explains why gases are so compressible.

The state of a substance may change if it is heated or cooled. Consider the changes that water undergoes in each process in **Figure 1.10**. The arrows that point up on the left side of the figure mean that energy is absorbed when (1) an icicle *melts*, (2) ice cubes in a frost-free freezer slowly become water vapor in a process called **sublimation**, and (3) bubbles of water vapor form in boiling water as the liquid *vaporizes*. The arrows that point down on the right side mean that heat is released when (1) water vapor *condenses* into drops of liquid water on a glass holding a cold drink on a hot day, (2) liquid water *freezes* on the surface of a pond in early winter, and (3) water vapor forms solid ice directly in a process called **deposition**. In Section 1.5, we examine how energy transfer affects the behavior of the particles that make up solids, liquids, and gases.

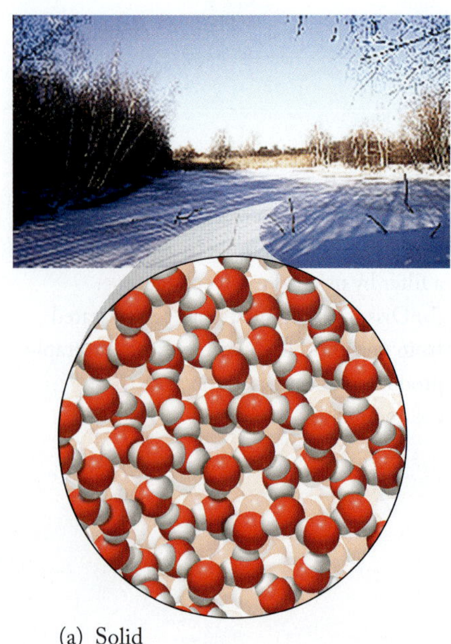

(a) Solid

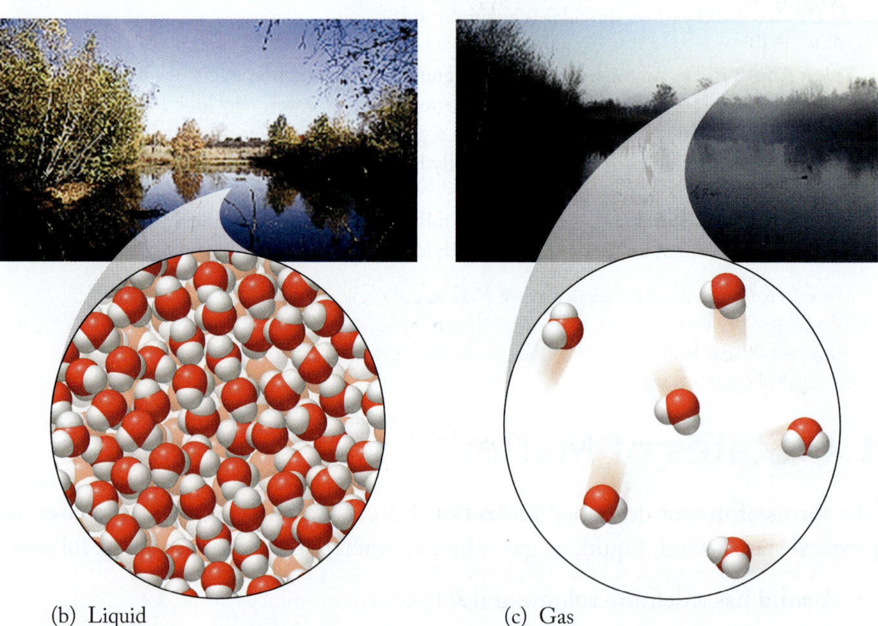

(b) Liquid

(c) Gas

**FIGURE 1.9** The three states of water. (a) In the solid state, each water molecule in ice is held in place in a rigid three-dimensional array. (b) In the liquid state, the molecules are close together but tumble over one another. (c) In the gas state, the molecules are far apart and move freely. Water vapor is invisible, but we can see the clouds and fog that form when atmospheric water vapor condenses into liquid droplets.

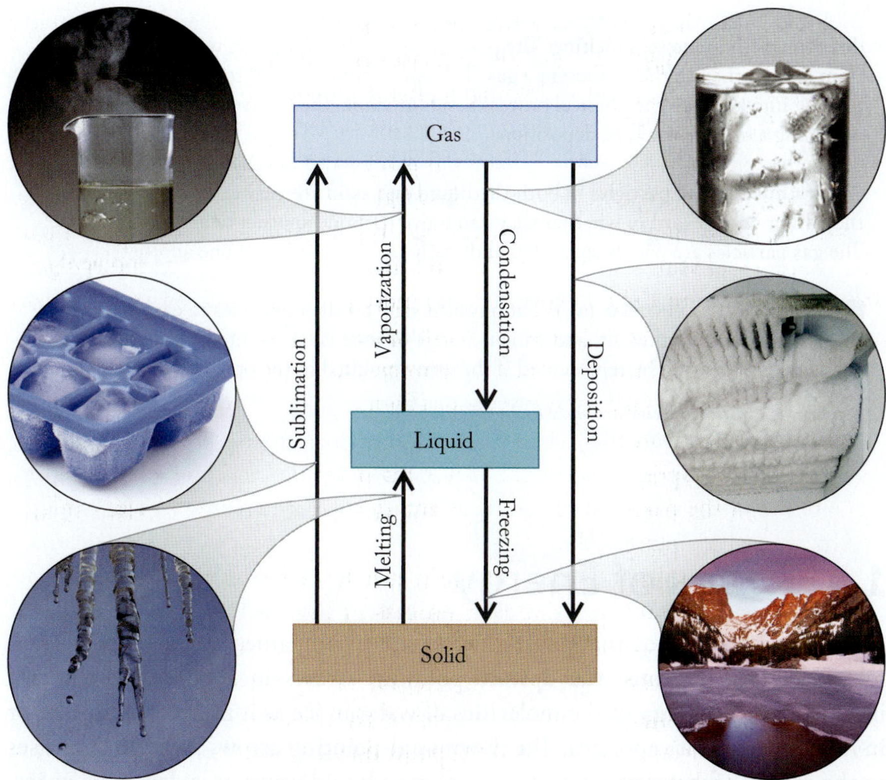


**FIGURE 1.10** Matter may change from one state to another when energy is added or removed. Arrows pointing upward represent transformations that require adding energy, whereas arrows pointing downward represent transformations that release energy.

## CONCEPT TEST

Ice cream vendors often use dry ice (solid carbon dioxide, $CO_2$) to keep their ice cream frozen. Over time, the dry ice disappears as solid $CO_2$ turns into $CO_2$ gas.

a. What is the name of that change in physical state?

b. What is the name of the reverse process, in which dry ice is produced from $CO_2$ gas?

*(Answers to Concept Tests are in the back of the book.)*

## SAMPLE EXERCISE 1.3 Distinguishing between Particulate Views of the States of Matter LO4

Which physical state does each part of **Figure 1.11** represent? What change of state does each arrow indicate? What would the changes of state be if both arrows pointed in the opposite direction?

**Collect, Organize, and Analyze** We need to examine the spacing between the particles:

- An ordered arrangement of particles with little space between the particles represents a solid.
- A less-ordered arrangement with little space between the particles represents a liquid.
- Particles widely separated from one another represent a gas.

**Solve** The particles in Figure 1.11(a) are close together but not ordered, so they represent a liquid. The particles in both Figure 1.11(b) and Figure 1.11(c) are ordered and have little space between them, so they both represent solids. The arrow from (a) to (b) represents a liquid turning into a solid, so the physical process is freezing. An arrow in the opposite

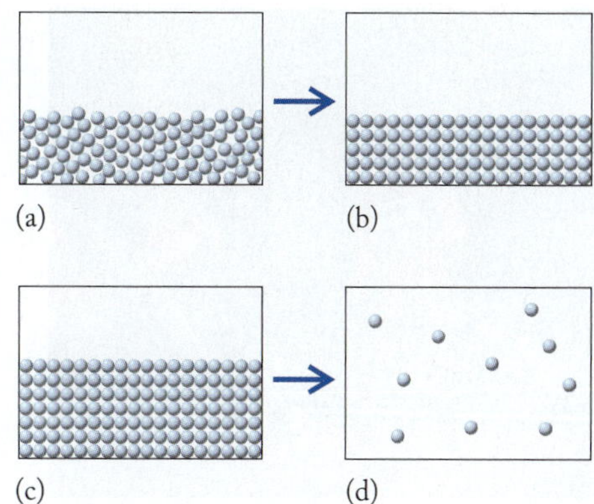


**FIGURE 1.11** Changes in the physical state of matter.

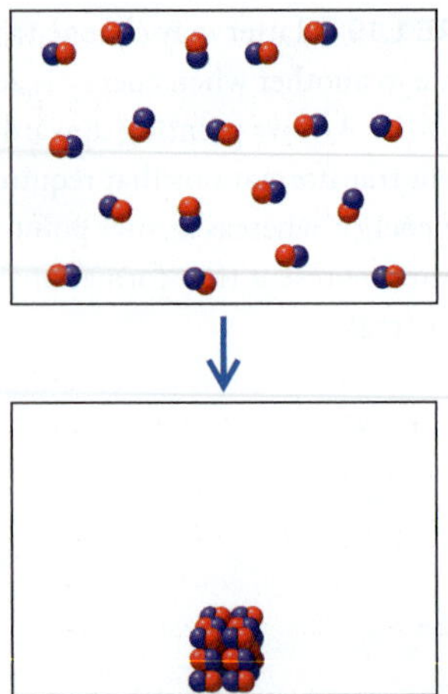

FIGURE 1.12 A change in the physical state of matter.

direction would represent melting. The particles in Figure 1.11(d) are widely separated from one another, so they represent a gas. The arrow from (c) to (d) represents a solid turning into a gas, so the physical process is sublimation. The reverse process—a gas becoming a solid—would be deposition.

**Think About It** The particles in both the liquid and solid are very close together, but they differ with regard to whether they can move past one another or are fixed in place. The gas particles are widely separated and easily move around past one another.

**Practice Exercise** (a) What physical state do the molecules in each part of **Figure 1.12** represent, and which change of state is represented? (b) Which change of state would be represented if the arrow pointed in the opposite direction?

*(Answers to Practice Exercises are in the back of the book.)*

## 1.5 Forms of Energy

In Section 1.4 we noted that phase changes are accompanied by the absorption or release of energy. Thus, the upward-pointing arrows in Figure 1.10 indicate increases in the energy of the molecules of water in ice as it melts or sublimes, or in liquid water as it vaporizes. The downward-pointing arrows indicate decreases in the energy of water vapor as it condenses to liquid water or is deposited as ice, or as liquid water freezes. Here we examine several forms of energy and how and why they change during physical and chemical changes.

For example, consider the sprinters in **Figure 1.13**, who rapidly expend energy as they run. In the physical sciences, **energy** is defined as the capacity to do work, and **work** ($w$) is the exertion of a force ($F$) through a distance ($d$):

$$w = F \times d \qquad (1.2)$$

As the sprinters use their muscles to move their bodies, they do work as they run 100 meters in about 11 seconds.

The energy inside a sprinter's body (and yours) is derived from chemical reactions fueled by glucose (blood sugar). The chemical energy stored in glucose is an example of **potential energy (PE)**, which is the energy stored in an object because of its position or composition. When the combustion of glucose releases energy during vigorous exercise, such as running the 100-meter dash, some of that energy is transformed into **kinetic energy (KE)**: the energy of motion. The amount of kinetic energy in a moving object is the product of the mass ($m$) of the object and the square of its speed ($u$):

$$\text{KE} = \tfrac{1}{2}mu^2 \qquad (1.3)$$

Equation 1.3 confirms that a heavy object has more kinetic energy than a lighter one moving at the same speed. Similarly, two objects having the same mass but traveling at different speeds have different kinetic energies. If one object is moving twice as fast as the other, its kinetic energy is $2^2$, or four times the kinetic energy of the other.

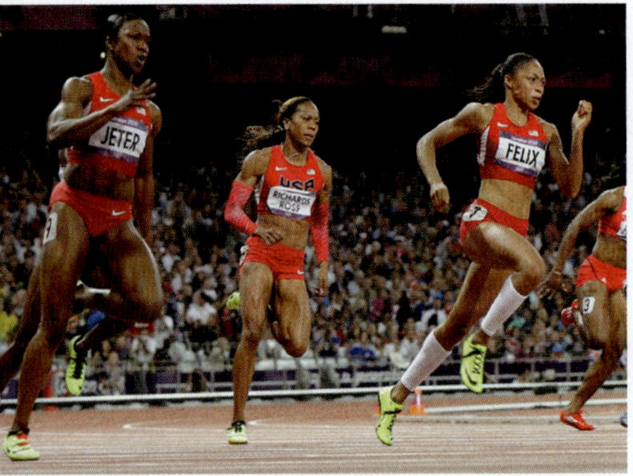


FIGURE 1.13 Olympic sprinters converting chemical energy into kinetic energy.

Other portions of the chemical energy released in a sprinter's body do other types of work, such as pumping blood through the circulation system. Some of the energy is also given off as heat. **Heat** is the transfer of energy that takes place because of a difference in temperatures. Heat spontaneously transfers from a warm object, such as a sprinter's skin, to a cooler one, such as the sprinter's surroundings. The dispersion of the energy released by biochemical processes in the body—some from doing work of various kinds and some as heat transfer—happens in accordance with the **law of conservation of energy**. That law states that energy cannot be created or destroyed, but it can be converted from one form to another.

The particles that make up all forms of matter also have kinetic energy that depends on their temperature and physical state. For example, the tiny molecules of oxygen in the air you breathe are moving at supersonic speeds. The microscopic views of the three states of water in Figure 1.9 show that the water molecules in ice are locked in place and have limited kinetic energy. They have more when they can tumble over one another in liquid water, and they have a lot more when they are free to move as individual gas molecules in water vapor.

**energy** the capacity to do work ($w$).

**work ($w$)** the exertion of a force ($F$) through a distance ($d$): $w = F \times d$.

**potential energy (PE)** the energy stored in an object because of its position or composition.

**kinetic energy (KE)** the energy of an object in motion because of its mass ($m$) and its speed ($u$).

**heat** the transfer of energy between objects that occurs because of differences in their temperatures.

**law of conservation of energy** the principle that energy cannot be created or destroyed but can change from one form to another.

**molecular formula** a chemical formula that indicates how many atoms of each element are in one molecule of a pure substance.

**structural formula** a representation of a molecule that uses short lines between the symbols of elements to show chemical bonds between atoms.

CONCEPT **TEST**

If the speed of a vehicle increases by 22%, by what factor does its kinetic energy increase?

*(Answers to Concept Tests are in the back of the book.)*

## 1.6 Formulas and Models

As we noted in Section 1.4, a molecule is a collection of atoms that chemical bonds hold together in a characteristic pattern and proportion. Figure 1.9 showed several molecular views of water. **Figure 1.14(a)** depicts molecules of the compound carbon dioxide. Each molecule is composed of a central carbon atom, represented by a black sphere, which is bonded to two atoms of oxygen, the red spheres. (The atomic color palette inside the back cover of this book shows the standard colors used to represent atoms of the elements we study most often.) Some pure elements also exist as molecules. The molecular views of hydrogen, nitrogen, and oxygen gas in **Figure 1.14(b)** show that those elements exist as *diatomic* (two-atom) molecules: $H_2$, $N_2$, and $O_2$, respectively. Fluorine, chlorine, bromine, and iodine also form diatomic molecules: $F_2$, $Cl_2$, $Br_2$, and $I_2$, respectively.

(a)

(b)

**FIGURE 1.14** Molecular views of (a) carbon dioxide gas and (b) a mixture of three gases: hydrogen (pairs of white spheres), oxygen (pairs of red spheres), and nitrogen (pairs of blue spheres).

The chemical formulas of molecular compounds are also called *molecular* formulas. The element symbols and subscripts in a **molecular formula** indicate how many atoms of each element are present in one molecule of the compound. The molecular formulas of acetone, a common solvent, and acetic acid, the ingredient that gives vinegar its distinctive aroma and taste, are shown in **Figure 1.15(a)**. Those formulas indicate the number of atoms of carbon, hydrogen, and oxygen in a molecule of each compound, but they do not tell us how the atoms are bonded, nor do they tell us anything about the shape of the molecules. One way to show how the atoms are connected is to draw a **structural formula** (**Figure 1.15b**), which represents the chemical bonds between atoms. *Condensed* structural formulas (**Figure 1.15c**) omit common structural components, such as C—H bonds or C atoms bonded to other C atoms. For example, the structural formula of acetic acid shows that three of the four hydrogen atoms are bonded to the same carbon atom. The same information is conveyed in the condensed structural formula by grouping three H atoms before the first C atom in the formula ($H_3C$–).

*Ball-and-stick* models (**Figure 1.15d**) are three-dimensional representations of molecules. The models use balls to represent atoms and sticks to represent chemical bonds. Although ball-and-stick models accurately show the angles between the bonds in a molecule, the sticks make the atoms seem far apart when they

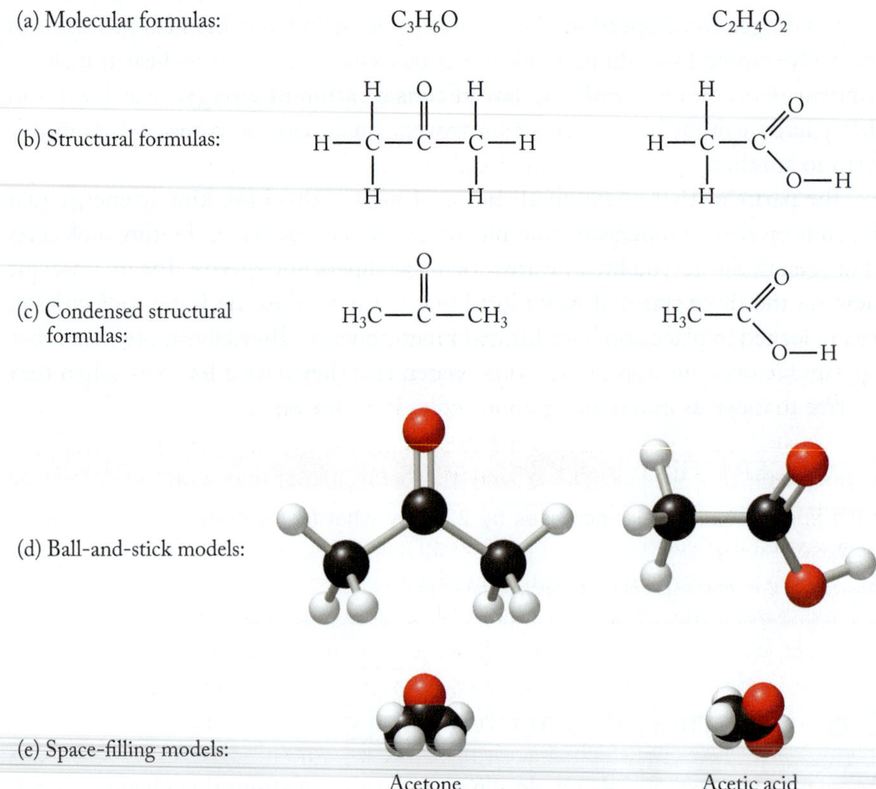


**FIGURE 1.15** Five ways to represent the arrangement of atoms in molecules of acetone and acetic acid: (a) molecular formulas; (b) structural formulas; (c) condensed structural formulas; (d) ball-and-stick models, in which white spheres represent hydrogen atoms, black spheres represent carbon atoms, and red spheres represent oxygen atoms; (e) space-filling models.

actually overlap. *Space-filling* models (**Figure 1.15e**) more accurately show how the atoms are arranged in a molecule and its overall three-dimensional shape, but seeing all the atoms and the angles between the bonds is sometimes hard, especially in molecules with many atoms.

**FIGURE 1.16** Space-filling model of methanol.

### CONCEPT TEST

**Figure 1.16** shows a space-filling model of methanol (also called methyl alcohol and wood alcohol). What is that compound's molecular formula?

*(Answers to Concept Tests are in the back of the book.)*

Not all compounds are molecular. Some consist of positively and negatively charged ions that are attracted to each other because they have opposite electrical charges. (Particles with the same charge repel each other.) One of the most common **ionic compounds**, and the principal component of table salt, is sodium chloride (NaCl). It consists of ordered three-dimensional arrays (**Figure 1.17**), or crystals, of sodium ($Na^+$) ions and chloride ($Cl^-$) ions, where the superscripts in their symbols indicate the electrical charges on the ions. The lack of subscripts in the chemical formula of sodium chloride tells us that the ratio of sodium ions to chloride ions in a crystal of NaCl is 1:1. Because no "molecules" of NaCl exist, we never refer to NaCl as a molecular formula. Rather, NaCl represents the simplest whole-number ratio of the ions in its three-dimensional structure. That ratio is referred to as the **empirical formula** of a compound. Scientists use the term *empirical* to describe information obtained through experimentation. Here, if we were to analyze the chemical composition of sodium chloride, we would discover that the ratio of sodium ions to chloride ions is 1:1, which would lead to the conclusion that the empirical formula of the compound is NaCl.

**FIGURE 1.17** Crystals of sodium chloride consist of ordered three-dimensional arrays of $Na^+$ ions and $Cl^-$ ions.

## 1.7 Expressing Experimental Results

Advances in scientific inquiry in the late 18th century, including those that led to the atomic theory of matter, created a heightened awareness of the need for accurate measurements and an international system of units to express the results of those measurements. In 1791, French scientists proposed a standard unit of length, which they called the **meter (m)** after the Greek *metron*, which means "measure." They based the length of the meter on 1/10,000,000 of the distance along an imaginary line running from the North Pole to the equator. By 1794, hard work by teams of surveyors had determined the length of the meter still used today.

The French scientists also settled on a decimal-based system for designating lengths that are multiples or fractions of a meter (**Table 1.1**). They chose Greek prefixes such as *kilo-* for lengths much greater than 1 meter (1 kilometer = 1000 m) and Latin prefixes such as *centi-* for lengths much smaller than a meter (1 centimeter = 0.01 m).

Since 1960 scientists have by international agreement used a modern version of the French *metric* system of units: the *Système International d'Unités*, commonly referred to as **SI units. Table 1.2** lists six SI base units; many others are derived from them. For example, a common SI unit for volume, the cubic meter ($m^3$), is derived from the base unit for length, the meter. A common SI unit for speed, meters per second (m/s), is derived from the base units for length and time. The SI unit for energy is the **joule (J)**, which is equivalent to 1 kg · $(m/s)^2$. That unit is consistent with the relationship introduced in Equation 1.3 (Section 1.5) between kinetic energy and mass and speed: KE $= \frac{1}{2} mu^2$. When mass ($m$) is expressed in kilograms and speed ($u$) is in meters per second, the units of KE are kg · $(m/s)^2$, or joules.

**ionic compound** a compound that consists of a characteristic ratio of positive ions and negative ions.

**empirical formula** a chemical formula in which the subscripts represent the simplest whole-number ratio of the atoms or ions in a compound.

**meter (m)** the standard unit of length, equivalent to 39.37 inches.

**SI units** a set of base and derived units used worldwide to express distances and quantities of matter and energy.

**joule (J)** the SI unit of energy, equivalent to 1 kg · $(m/s)^2$.

**CHEMTOUR**

Scientific Notation

**TABLE 1.1** Commonly Used Prefixes for SI Units

| PREFIX | | VALUE | | |
|---|---|---|---|---|
| Name | Symbol | Numerical | Exponential | Example |
| zetta | Z | 1,000,000,000,000,000,000,000 | $10^{21}$ | 1 Zm = $10^{21}$ m |
| exa | E | 1,000,000,000,000,000,000 | $10^{18}$ | 1 Em = $10^{18}$ m |
| peta | P | 1,000,000,000,000,000 | $10^{15}$ | 1 Pm = $10^{15}$ m |
| tera | T | 1,000,000,000,000 | $10^{12}$ | 1 Tm = $10^{12}$ m |
| giga | G | 1,000,000,000 | $10^{9}$ | 1 Gm = $10^{9}$ m |
| mega | M | 1,000,000 | $10^{6}$ | 1 Mm = $10^{6}$ m |
| kilo | k | 1000 | $10^{3}$ | 1 km = $10^{3}$ m |
| hecto | h | 100 | $10^{2}$ | 1 hm = $10^{2}$ m |
| deka | da | 10 | $10^{1}$ | 1 dam = 10 m |
| deci | d | 0.1 | $10^{-1}$ | 1 dm = $10^{-1}$ m |
| centi | c | 0.01 | $10^{-2}$ | 1 cm = $10^{-2}$ m |
| milli | m | 0.001 | $10^{-3}$ | 1 mm = $10^{-3}$ m |
| micro | μ | 0.000001 | $10^{-6}$ | 1 μm = $10^{-6}$ m |
| nano | n | 0.000000001 | $10^{-9}$ | 1 nm = $10^{-9}$ m |
| pico | p | 0.000000000001 | $10^{-12}$ | 1 pm = $10^{-12}$ m |
| femto | f | 0.000000000000001 | $10^{-15}$ | 1 fm = $10^{-15}$ m |
| atto | a | 0.000000000000000001 | $10^{-18}$ | 1 am = $10^{-18}$ m |
| zepto | z | 0.000000000000000000001 | $10^{-21}$ | 1 zm = $10^{-21}$ m |

**TABLE 1.2** Six SI Base Units

| Quantity or Dimension | Unit Name | Unit Abbreviation |
|---|---|---|
| Mass | kilogram | kg |
| Length | meter | m |
| Temperature | kelvin | K |
| Time | second | s |
| Electric current | ampere | A |
| Quantity of a substance | mole | mol |

**TABLE 1.3** Equivalencies for SI and Customary U.S. Units

| Quantity or Dimension | Equivalent Units |
|---|---|
| Mass (*m*) | 1 kg = 2.205 pounds (lb) = 35.27 ounces (oz) |
| Length (distance) | 1 m = 39.37 inches (in) = 3.281 feet (ft) = 1.094 yards (yd)<br>1 km = 0.6214 mile (mi)<br>1 in = 2.54 cm (exactly) |
| Volume (*V*) | 1 $m^3$ = 35.31 $ft^3$ = 1000 liters (L) (exactly)<br>1 L = 0.2642 gallon (gal) = 1.057 quarts (qt) |
| Temperature (*T*) | $T$(K) = $T$(°C) + 273.15<br>$T$(°C) = (5/9)[$T$(°F) − 32] |

**CHEMTOUR**
Temperature Conversion

**Table 1.3** lists some SI units and their equivalents in the customary U.S. system of units, such as the important exact unit equivalence 2.54 cm = 1 inch. The table also lists the liter, a unit of volume common throughout the world. One liter is exactly 1 cubic decimeter (a cube 1/10 meter on a side). Because exactly 10 decimeters are in a meter, exactly $10^3$ $dm^3$, or $10^3$ L, are in 1 cubic meter (1 $m^3$).

The temperature ($T$) equivalencies in Table 1.3 also deserve our attention because temperature has been the most often measured quantity since the first thermometers were developed nearly 500 years ago. Modern thermometers based on the thermal expansion of liquids such as mercury and alcohol were introduced in the early 18th century by German scientist (and glassblower) Daniel Gabriel Fahrenheit (1686–1736). He later developed a temperature scale with a zero point corresponding to the freezing point of a concentrated salt solution and an upper value (100°) corresponding to the average internal temperature of the human body. The Fahrenheit scale, still widely used in the United States, is a slightly modified version of the one based on those two reference temperatures. Later in the 18th century, Swedish astronomer Anders Celsius (1701–1744) proposed an alternative temperature scale based on pure water's freezing point (0°C) and typical boiling point at sea level (100°C). In 1848 British scientist William Thomson (later called Lord Kelvin, 1824–1907) proposed another temperature scale not based on the physical properties of any substance but rather on the notion that temperature has a lower limit, called **absolute zero (0 K)**. That temperature is the zero point of the Kelvin scale named in his honor. **Figure 1.18** compares those three temperature scales.

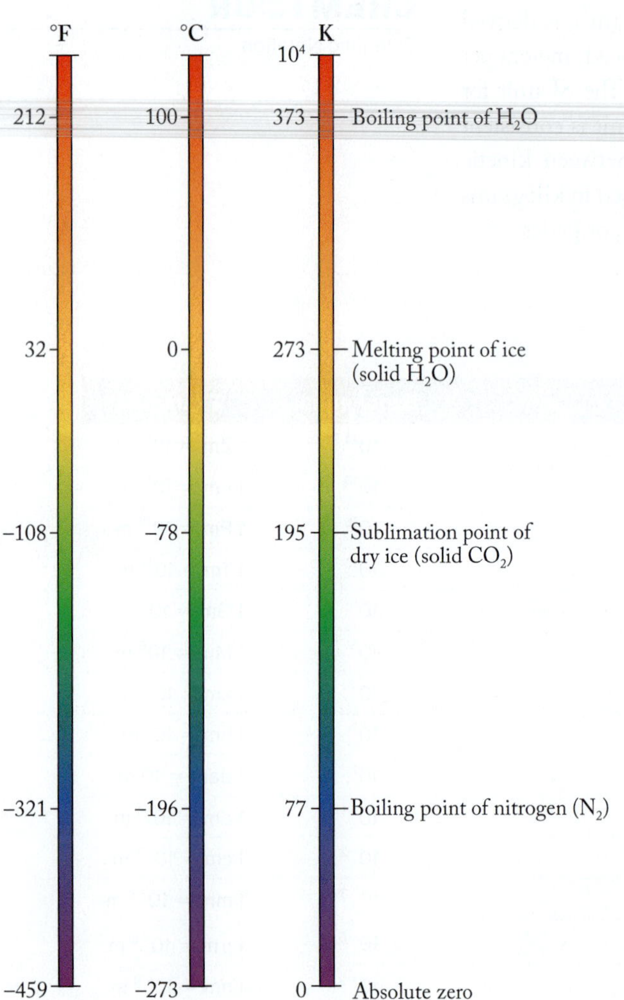


**FIGURE 1.18** Three temperature scales are commonly used today, although the Fahrenheit scale is rarely used in scientific work.

To conduct science today, the length of the meter and the dimensions of the other SI units must be known or defined by quantities that are much more constant and precisely known than, for example, the distance from the North Pole to the equator. Two such quantities are the speed of light ($c$) in a vacuum and time. In 1983, 1 meter was redefined as the distance traveled in 1/299,792,458 of a second in a vacuum by the light emitted from a helium–neon laser. That modern definition of the meter is consistent with the one adopted in France in 1794.

All scientific measurements are limited in how well we can know the results. Nobody is perfect, and neither is any analytical method. Therefore, measured quantities and the values derived from them always have some degree of uncertainty. However, some quantities are known exactly because they can be determined by counting, such as the number of steps in a ladder or the number of eggs in a carton. No uncertainty is present in such a value.

**absolute zero (0 K)** the zero point on the Kelvin temperature scale; theoretically the lowest temperature possible.

**precision** the extent to which repeated measurements of the same variable agree.

**SAMPLE EXERCISE 1.4** Distinguishing Exact from Uncertain Values **LO7**

Which quantities associated with the Washington Monument in Washington, DC (**Figure 1.19**), are exact values, and which are inexact?

a. The monument is made of 36,941 white marble blocks.
b. The monument is 169 m tall.
c. The monument has 893 stair steps to the top.
d. The mass of the aluminum apex is 2.8 kg.
e. The area of the foundation is 1487 $m^2$.

**Collect and Organize** We must distinguish between values based on an exact number, such as 12 eggs in a dozen, from those that are not exact numbers, such as the mass of an egg.

**Analyze** One way to distinguish exact from inexact values is to answer the question, "Which values represent quantities that can be counted?" Exact values can be counted.

**Solve** The number of marble blocks (a) and the number of stairs (c) are quantities we can count, so they are exact numbers. The other three quantities are based on measurements of length (b), mass (d), and area (e), so they are inexact.

**Think About It** The "you can count them" property of exact numbers was used in this exercise, which assumes that no uncertainty exists in a counted value. Can you think of a type of counting whose certainty is sometimes challenged?

**Practice Exercise** Which statistics associated with the Golden Gate Bridge in San Francisco, CA (**Figure 1.20**), are exact numbers, and which have some inherent uncertainty?

a. The roadway is six lanes wide.
b. The bridge is 27.4 m wide.
c. The bridge has a mass of 381 million kg.
d. The bridge is 2740 m long.
e. The "FasTrack" toll for a car traveling south is $7.00 (as of July 1, 2018).

*(Answers to Practice Exercises are in the back of the book.)*

**FIGURE 1.19** The Washington Monument.

**FIGURE 1.20** The Golden Gate Bridge.

## Precision and Accuracy

Two terms—*precision* and *accuracy*—describe the certainty with which we know a measured quantity or a value calculated from a measured quantity. **Precision** indicates how repeatable a measurement is. Suppose we used the balance on the top in **Figure 1.21** to determine the mass of a penny over and over again, and suppose the reading on the balance was always 2.53 g. Those results tell us that the mass of the penny is precisely 2.53 g. We can also say that we are certain of its mass to the nearest 0.01 g.

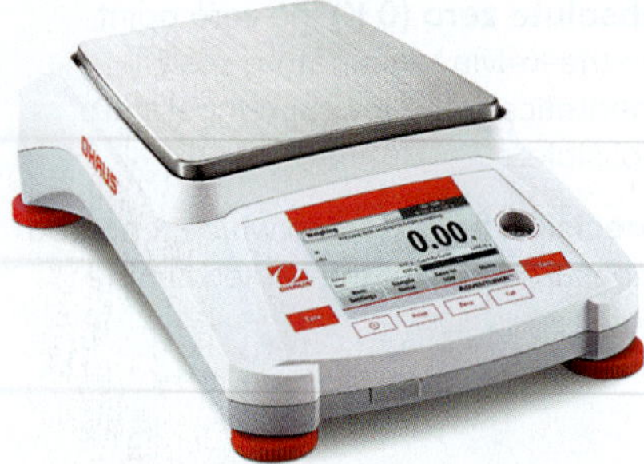


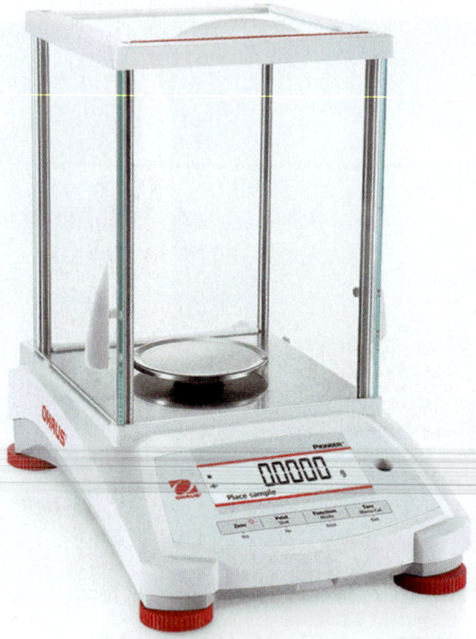

**FIGURE 1.21** The mass of a penny can be measured to the nearest 0.01 g with the balance on the top and to the nearest 0.0001 g with the balance on the bottom.

Now suppose we used the balance on the bottom in Figure 1.21 to determine the mass of the same penny five more times and obtained the following values:

| Measurement | Mass (g) |
|---|---|
| 1 | 2.5270 |
| 2 | 2.5271 |
| 3 | 2.5272 |
| 4 | 2.5271 |
| 5 | 2.5271 |

The small variability in the last decimal place is not unusual when using a balance that can report masses to the nearest 0.0001 g. Those results are consistent with one another, so we can say that the balance is precise. Many other factors—particles of dust landing on the balance, vibrations of the laboratory bench, or the transfer of moisture from our fingers to the penny as we handled it—could have produced a change in mass of 0.0001 g or more.

One way to express the precision of those results is to cite the range between the highest and lowest values—here, 2.5270 to 2.5272. Range also can be expressed using the average value (2.5271 g) and the range above (0.0001) and below (0.0001) the average that includes all the observed results. A convenient way to express the observed range here is $2.5271 \pm 0.0001$, where the symbol ± means "plus or minus" the value that follows it.

Whereas precision relates to the agreement among repeated measurements, **accuracy** reflects how close the measured value is to the true value. Suppose the true mass of our penny is 2.5267 g. That means the average result obtained with the bottom balance in Figure 1.21, 2.5271 g, is 0.0004 g too high. Thus, the measurements made on this balance may be *precise* to within 0.0001 g, but they are not *accurate* to within 0.0001 g of the true value. One way to visualize the difference between accuracy and precision is presented in **Figure 1.22**.

How can we be sure that the results of measurements are accurate? We can check a balance's accuracy by weighing objects of known mass. A thermometer can be calibrated by measuring the temperature at which a substance changes state. For example, an ice-water bath should have a temperature of 0.0°C. At sea level, liquid water boils at 100.0°C, so an accurate thermometer dipped into boiling water reads that temperature. A measurement validated by calibration with an accepted standard material is considered accurate.

**STEPWISE ANIMATION**
Precision and Accuracy

## Significant Figures

Let's now learn a way to express how well we know the results of measurements or calculated values derived from measurements. Our approach uses **significant figures**. The number of significant figures used to report a value indicates how certain we are of the value. For example, if the two balances shown in Figure 1.21 are working properly, we can use the top balance to weigh a penny to the nearest 0.01 g, obtaining a mass value of 2.53 g. The bottom balance can measure the mass of the same penny to the nearest 0.0001 g, or 2.5271 g. The mass obtained with the top balance has three significant figures: the 2, 5, and 3 are considered *significant* because we are confident in their values. The mass obtained using the bottom balance has five significant figures (2, 5, 2, 7, and 1). The mass of the penny can be determined with greater certainty by using the bottom balance.

**accuracy** agreement between an experimentally measured value and the true value.

**significant figures** all the certain digits in a measured value plus one estimated digit. The greater the number of significant figures, the greater the certainty with which the value is known.

Now suppose an aspirin tablet is placed on the bottom balance in Figure 1.21, and the display reads 0.0810 g. How many significant figures are in that value? You might be tempted to say five because five digits are displayed. However, the first two zeros are *not* considered significant because they serve only to determine the location of the decimal point. Those two zeros function the way exponents do when we express values in scientific notation. For example, expressing 0.0810 g in scientific notation gives us $8.10 \times 10^{-2}$ g. Only the three digits in "8.10" indicate how precisely we know the value; the "−2" in the exponent conveys how many places the decimal was moved (two) and that it moved to the right (the negative sign). (See Appendix 1 to review of how to express values in scientific notation.)

Why is the rightmost zero in 0.0810 g significant? The balance, if operating correctly, can measure masses to the nearest 0.0001 g, so we may assume that the last digit is significant. If we dropped the zero and recorded a value of only 0.081 g, we would be implying that we knew the value only to the nearest 0.001 g, which is not the case. The following guidelines will help you determine which zeros (highlighted in green) are significant and which are not:

1. Zeros at the beginning of a value, as in **0.0**592, are never significant: they just set the decimal place.
2. Zeros after a decimal point and after a nonzero digit, as in 3.**00** $\times 10^8$, are always significant.
3. Zeros at the end of a value that contains no decimal point, as in 96,5**00**, may or may not be significant.[2] We can use scientific notation to indicate whether those terminal zeros are significant. If those zeros are not significant and serve only to set the decimal point, the value in scientific notation is $9.65 \times 10^4$, where the positive 4 exponent means the decimal point was moved four places to the left from its original location.
4. Zeros between significant digits, as in 1**0**1.3, are always significant.

(a)

(b)

(c)

**FIGURE 1.22** (a) Three dart throws meant to hit the center of the target are both accurate and precise. (b) Three throws meant to hit the center of the target are precise but not accurate. (c) This set of three throws is neither precise nor accurate.

CONCEPT **TEST**

How many significant figures are in the values used as examples in guidelines 1, 2, and 4?

*(Answers to Concept Tests are in the back of the book.)*

## Significant Figures in Calculations

Now let's consider how significant figures are used to express the results of calculations involving measured quantities. The rules for significant figures should be used *only at the end of a calculation, never on intermediate results.* That policy helps avoid errors that may happen when we round off intermediate values.

Rounding off means that we drop the *insignificant digits* (all digits to the right of the last significant digit) and then either increase the value of the last significant digit by 1 or leave it unchanged, depending on the value of the first insignificant digit. For example, if we need to round 76.4523 to three significant figures, 5 is the first insignificant digit that must be dropped. If that first insignificant digit had been greater than 5, we would have rounded up (thus, 76.46 would be rounded to 76.5). If it had been less than 5, we would have rounded down (thus, 76.44 would be rounded to 76.4). Because it is 5, we check whether any nonzero

[2]Some books add decimal points after terminal zeros (for example, "1000.") to indicate that the zeros are significant. We do not follow that practice, in part because it does not work for values in which only some of the terminal zeros are significant.

digits are to the right of it. If so (as here), we round up. Therefore, 76.4521 is rounded to 76.5.

Had no nonzero digits been to the right of the 5 (for example, in 76.45 or 76.450), we would have needed a tie-breaking rule for rounding down or up. A good rule to follow is to round to the nearest even number. Thus, rounding 76.45 (or 76.450) to three significant figures makes both values 76.4 because the 4 in the tenths place is the nearest even number. However, 76.55 (or 76.550) would be rounded to 76.6 because the 6 in the tenths place is the nearest even number.

Significant Figures

To see how significant figures affect calculations involving measured values, let's calculate the time sunlight takes to travel from the sun to Earth in early July, when the distance between them is 152 million km, or $1.52 \times 10^{11}$ m. We divide that distance by $c$, the speed of light in a vacuum (a reasonable description of the space between the sun and Earth), because speed is a ratio of distance to time. Expressing the distance in scientific notation and using the value for $c$ from the inside back cover of this book ($2.998 \times 10^8$ m/s), we calculate a travel time of:

$$t = \frac{1.52 \times 10^{11} \cancel{\text{m}}}{2.998 \times 10^{8} \cancel{\text{m}}/\text{s}} = 507 \text{ s}$$

We have expressed the results of the calculation to three significant figures. Is that an appropriate number? The answer is yes because we know the distance to three significant figures and the speed to four. Therefore, we may express time to only three significant figures. Why? Because *when measured numbers (or numbers derived from measurements) are multiplied or divided, the result has the same number of significant figures as the value used to calculate it with the smallest number of significant figures.* That statement is derived from this more general rule: *We can know the result of a calculation only as well as we know the least well-known value used in the calculation.* That rule is called the weak-link principle because a chain is only as strong as its weakest link. We will shortly apply that rule to calculations involving addition and subtraction.

Now let's consider the results of measurements conducted by a class of chemistry students as part of the undergraduate laboratory safety program at their school. Suppose their task was to check the rate of airflow at the face of a fume hood they used for handling hazardous substances. Six teams of students measured the airflow rate at different times during a 3-hour lab period with the same handheld anemometer. They obtained these results:

| Team number | 1 | 2 | 3 | 4 | 5 | 6 |
|---|---|---|---|---|---|---|
| Airflow rate (m/s) | 0.51 | 0.50 | 0.51 | 0.49 | 0.50 | 0.52 |

The students then calculated an average flow rate by adding the results of the six measurements and dividing the sum by 6:

$$\frac{0.51 + 0.50 + 0.51 + 0.49 + 0.50 + 0.52}{6} \text{ m/s} = \frac{3.03}{6} \text{ m/s} = 0.505 \text{ m/s}$$

Note that the results of each measurement contain two significant digits, but their sum (3.03 m/s) has three. Is that OK? The answer is yes because the weak-link principle is that *when measured numbers (or numbers derived from measurements) are added or subtracted, the result has the same number of digits to the right of the decimal as the measured number with the fewest digits to the right of the decimal.* Because each measurement of flow rate was reported to the nearest 0.01 m/s, the sum of the flow rates also should be expressed to the nearest 0.01 m/s.

Next the average value for flow rate was calculated by dividing the sum by 6. That value (0.505 m/s) also is expressed with three significant figures, consistent with the weak-link principle for division because the sum of the measurements has three significant figures and the divisor (6) is an exact number, not a measured value.

**SAMPLE EXERCISE 1.5** Using Significant Figures in Calculations I **LO7**

Suppose we add a penny with a mass of 2.5271 g to 49 other pennies with a combined mass of 124.01 g. What is the combined mass of the 50 pennies?

**Collect and Organize** We are asked to calculate the combined mass of 49 pennies that were weighed to the nearest 0.01 g and a single penny that was weighed to the nearest 0.0001 g. To express the results of that calculation, we must follow the weak-link principle: we can know a combination of measured values only as well as we know the least well-known measured value.

**Analyze** We sum two values here, so the weak-link value is the one with the fewer digits to the right of its decimal point, 124.01 g, the combined mass of the 49 pennies.

**Solve** Adding the mass of the 50th penny to the mass of the other 49, we have

$$\begin{array}{r} 124.01\text{ g} \\ +\ \ 2.5271\text{ g} \\ \hline 126.5371\text{ g} = 126.54\text{ g} \end{array}$$

**Think About It** We can know the value of the sum only to the nearest 0.01 g, so we round 126.5371 to 126.54. We round the "3" up to 4 because the first digit to be dropped is the "7" highlighted in red.

**Practice Exercise** According to the rules of golf, a golf ball cannot weigh more than 45.97 g, and it must be at least 4.267 cm in diameter. The volume of a sphere is $(4/3)\pi r^3$, where $r$ is the radius (half the diameter).

a. What is the maximum density of a golf ball, expressed in grams per cubic centimeter to the appropriate number of significant figures?
b. Is such a golf ball more dense or less dense than water? (You may need to look up the density of water—unless you are a golfer, in which case you probably already know the answer.)

*(Answers to Practice Exercises are in the back of the book.)*

**SAMPLE EXERCISE 1.6** Using Significant Figures in Calculations II **LO7**

Suppose you are asked to determine whether a nugget is made of gold on the basis of its density. You weigh the nugget and find that it has a mass of 163 grams (g). Then you carefully add it to a graduated cylinder that contains 50.0 milliliters (mL) of water (**Figure 1.23**) and observe that the water level in the cylinder rises to 58.5 mL. What is the density of the nugget in grams per cubic centimeter? Is it made of gold?

**Collect and Organize** In Section 1.3 we learned that density is the ratio of the mass of an object to its volume (Equation 1.1). Here, we are given the mass of the nugget and the volume of water it displaces. Appendix Table A3.2 lists the density of pure gold as 19.3 g/cm$^3$. We know from the table of units inside the back cover of this book that 1 mL = 1 cm$^3$ (cubic centimeter).

**FIGURE 1.23** When a nugget believed to be pure gold is placed in a graduated cylinder containing 50.0 mL of water, the volume rises to 58.5 mL.

**Analyze** We know the mass of the nugget to three significant figures. The volume of the nugget is equal to the volume of water it displaces, which is the difference between 58.5 and 50.0 mL.

**Solve** The volume of the nugget is:

$$\begin{array}{r} 58.5\ \text{mL} \\ -\ 50.0\ \text{mL} \\ \hline 8.5\ \text{mL} \end{array}$$

and the density of the nugget, expressed in grams per cubic centimeter, is

$$d = \frac{m}{V} = \frac{163\ \text{g}}{8.5\ \text{mL}} = 19.2\ \frac{\text{g}}{\text{mL}} = 19.2\ \frac{\text{g}}{\text{cm}^3}$$

That density value contains three significant figures, but the volume used to calculate it is known only to two because the two starting values were known to the nearest tenth of a milliliter. Therefore, the difference between them is known only to the nearest 0.1 mL. The two significant figures in the volume of the nugget are the weak link in that calculation, which means we should express the results of the calculation to only two significant figures, or 19 g/cm$^3$. That value matches the density of pure gold if expressed only to two significant figures.

**Think About It** The nugget *could* be gold, and it is very likely to be gold, given its appearance, but we cannot claim for certain that it is gold. If our calculated density had been 18 or 21 g/cm$^3$, we could have said with some certainty that the nugget was not gold. Additional tests would need to be conducted on the nugget to determine whether it is indeed gold.

**Practice Exercise** A silver-colored metallic cube is 2.54 cm on a side and has a mass of 146 g. What is its density?

*(Answers to Practice Exercises are in the back of the book.)*

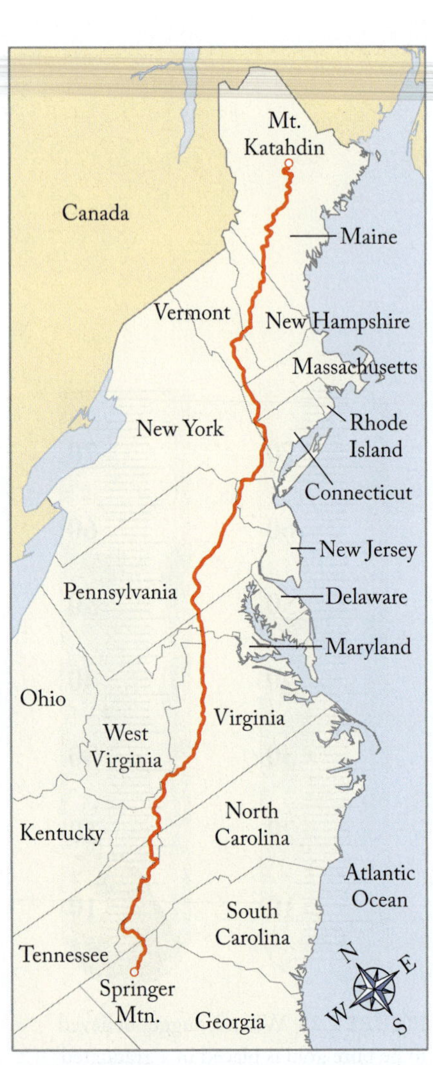


**FIGURE 1.24** Map of the Appalachian Trail.

## CONCEPT TEST

In the summer of 2018, Belgian dentist Karl Sabbe set a record by hiking the Appalachian Trail (**Figure 1.24**) from its southern terminus at Springer Mountain, Georgia, to its northern terminus on the peak of Mt. Katahdin, Maine, in 45 days, 7 hours, and 39 minutes. His hike was 3523 km (2189 miles) long.

a. What was Sabbe's average speed in kilometers per day and miles per day?

b. Which do you think is the weak link in calculating his average speed: the actual distance hiked or the time that it took?

*(Answers to Concept Tests are in the back of the book.)*

## 1.8 Unit Conversions and Dimensional Analysis

On July 23, 1983, Air Canada Flight 143, a Boeing 767 that had been in service only a few months, ran out of fuel near the halfway point on a flight from Montreal, Quebec, to Edmonton, Alberta. Without power, the plane became a large, heavy glider. Fortunately, the pilot of Flight 143 was also an experienced glider pilot, and he made a successful emergency landing at an abandoned airbase in Gimli, Manitoba. No serious injuries occurred, even though the nose gear collapsed during the landing (**Figure 1.25**). The plane was repaired and flown out of Gimli and back into service, where it remained until its retirement in 2008. During its 25 years of operation, the plane became widely known in aviation circles as the *Gimli Glider*.

**FIGURE 1.25** Air Canada Flight 143 after landing safely in Gimli, Manitoba.

Flight 143 ran out of fuel for several reasons, including faulty communications between ground and flight crews about a malfunctioning fuel gauge. However, the most immediate reason was an error in calculating how much fuel the plane needed to fly from Montreal to Edmonton. The error occurred when Canada and its national airline were converting from a unit system called the *Imperial System*, which expresses masses in ounces and pounds (as in the United States today), to the SI system, whose base unit of mass is the kilogram. The Boeing 767 that landed in Gimli was among the first in the Air Canada fleet to use the SI system, so its fuel capacity and consumption were expressed in kilograms. The plane was supposed to take off from Montreal with 22,300 kg of fuel in its tanks, but it left with less than half that. Let's explore why.

To make sure the plane had enough fuel, but not too much (to avoid transporting excess weight), the refueling crew in Montreal measured the *volume* of fuel already in its tanks and found that they contained 7700 L. Next, they converted that volume of jet fuel into an equivalent *mass* of fuel because they knew the plane needed 22,300 kg of fuel to fly to Edmonton. To convert volume in liters to mass in kilograms, they should have multiplied 7700 L by the density of jet fuel, which is 0.80 kg/L:

Dimensional Analysis

$$7700\ \cancel{\text{L}} \times \frac{0.80\ \text{kg}}{1\ \cancel{\text{L}}} = 6200\ \text{kg}$$

Note how the volume units cancel out and we obtain 6200 kg for the mass of 7700 L of fuel. In that calculation density is used as a **conversion factor**, a ratio of equivalent quantities expressed using the initial and desired final units of measure. That equivalency means that multiplying a quantity by a conversion factor is like multiplying by 1: the quantity does not change, but its units do. Equation 1.4 illustrates how any conversion factor (highlighted in blue) can be used to convert the initial units on a quantity to the desired units:

$$\cancel{\text{Initial units}} \times \frac{\text{desired units}}{\cancel{\text{initial units}}} = \text{desired units} \tag{1.4}$$

The Montreal refueling crew next had to calculate the mass of jet fuel to be added by subtracting the mass of fuel onboard from that needed for the flight:

$$\begin{array}{r} 22{,}300\ \text{kg needed} \\ -6200\ \text{kg onboard} \\ \hline 16{,}100\ \text{kg to be added} \end{array}$$

Then they had to calculate the *volume* of jet fuel to be added because jet fuel is pumped into airplanes the way gasoline is pumped into automobiles: with a metering system based on volume. The refueling crew calculated the volume of jet fuel in liters, corresponding to the mass of fuel needed. In that calculation the reciprocal of the density serves as the conversion factor so that mass units cancel out:

$$16{,}100\ \cancel{\text{kg}} \times \frac{1\ \text{L}}{0.80\ \cancel{\text{kg}}} = 20{,}100\ \text{L} \approx 20{,}000\ \text{L}$$

Had that volume of fuel been added, Flight 143 would have made it to Edmonton. Unfortunately, the refueling crew added only 4900 L of fuel! Why? Because they used the wrong density value to convert volume to mass and again

**conversion factor** fraction in which the numerator is equivalent to the denominator but is expressed in different units, making the fraction equivalent to 1.

when they converted mass to volume. The value they used was 1.77 instead of 0.80. They had used 1.77 for many years because that is the density of jet fuel expressed in pounds per liter, not kilograms per liter. Using the wrong conversion factor meant that they overestimated the mass of fuel already onboard the airplane (they multiplied by 1.77 instead of by 0.80), and then they underestimated how much fuel to add by using the reciprocal of 1.77 instead of 0.80 to calculate the volume of fuel to pump into the plane's tanks. As a result, Flight 143 left Montreal with less than half as much fuel as it should have carried.

The flight of the *Gimli Glider* is not the only disaster (or near disaster) caused by mixing up units. In 1999 the *Mars Climate Orbiter* space probe crashed onto the surface of Mars instead of orbiting it because the programmers who wrote the software that controlled the spacecraft used one set of units to express the thrust of its rocket engines, whereas the engineers who designed the engines used another. Those examples show that we must take great care to use the correct units when expressing the quantities of substances. Our goal here in Section 1.8 is to help you become more skilled at converting measured and calculated values from one set of units to another. You will have an opportunity to develop that skill in the following sample and practice exercises.

**SAMPLE EXERCISE 1.7** Interconverting Customary U.S. Units and SI Units I **LO8**

A pediatrician prescribes the antibiotic amoxicillin to treat a 2-year-old child suffering from an ear infection (**Figure 1.26**). The therapeutic dose is 75 mg of the drug per kilogram of body mass of the patient per day. The drug is administered twice each day in the form of a flavored liquid that contains 125 mg of amoxicillin per milliliter. If the child weighs 28 lb, how many milliliters of the liquid should be administered in each dose?

**FIGURE 1.26** A child pushes on her ear as a result of the pain of an ear infection.

**Collect and Organize** We need to calculate the volume of liquid amoxicillin in milliliters to be administered twice each day to provide a daily dose of 75 mg/kg of patient mass. We know the patient's mass (28 lb) and the concentration of liquid amoxicillin (125 mg/mL). We know from Table 1.3 that 1 kg = 2.205 lb.

**Analyze** Let's start with the mass of the patient: 28 lb. Dosage is expressed in milligrams of amoxicillin per kilogram of the patient's mass, so we need to convert 28 lb to kilograms and then multiply that mass by the daily dosage in milligrams of amoxicillin per kilogram of patient mass to calculate the milligrams of amoxicillin to be given each day. To convert that mass to an equivalent volume of the amoxicillin solution, we invert the concentration of the liquid from "mg of amoxicillin/mL" to "mL/mg of amoxicillin" so that the mass units cancel out. Finally, we need to divide the daily volume in two because the drug is administered twice a day. The resulting conversion factors are

$$\frac{1\ \text{kg}}{2.205\ \text{lb}} \quad \frac{75\ \text{mg}}{\text{kg} \cdot \text{day}} \quad \frac{1\ \text{mL}}{125\ \text{mg}} \quad \frac{1\ \text{day}}{2\ \text{doses}}$$

To estimate the result of our calculation, we note that a kilogram is a little more than two pounds, so a 28-lb child weighs a little less than 14 kg—say, 12 kg. The product of 12 kg × 75 mg of amoxicillin/kg of patient mass is about 1000 mg of amoxicillin. Dividing that value by the concentration of the liquid (125 mg/mL) gives a volume of about 8 mL per day, or 4 mL twice a day.

**Solve**

$$28\ \cancel{\text{lb}} \times \frac{1\ \cancel{\text{kg}}}{2.205\ \cancel{\text{lb}}} \times \frac{75\ \cancel{\text{mg}}}{\cancel{\text{kg}} \cdot \cancel{\text{day}}} \times \frac{1\ \text{mL}}{125\ \cancel{\text{mg}}} \times \frac{1\ \cancel{\text{day}}}{2\ \text{doses}} = 3.8\ \frac{\text{mL}}{\text{dose}}$$

**Think About It** The answer nearly matches our estimate. We rounded the result to two significant figures because the child's weight and the therapeutic daily dose both have only two significant figures. The number of doses per day (two) is an exact number and thus has no uncertainty.

**Practice Exercise** A student planning a party has \$20 to spend on her favorite soft drink. It is on sale at store A for \$1.29 for a 2-L bottle (plus 10-cent deposit); at store B the price of a 12-pack of 12 fl oz cans is \$2.99 (plus a 5-cent deposit per can). At which store can she buy the most of her favorite soft drink for no more than \$20? (A volume of 29.57 mL is 1 U.S. fl oz, which is a unit of volume, not mass.)

*(Answers to Practice Exercises are in the back of the book.)*

## SAMPLE EXERCISE 1.8 Interconverting Customary U.S. Units and SI Units II LO8

Natural gas is the most widely used fuel in the United States for generating electricity, for cooking (**Figure 1.27**), and for heating homes and water. Natural gas also is used in industry as a starting material (also called a *feedstock*) for making fertilizer, fabrics, and plastics. The United States consumed 24.77 trillion cubic feet of natural gas for electric power generation and for industrial, commercial, and residential uses in one calendar year (2017). What is that rate of consumption expressed in cubic meters of natural gas per second?

**FIGURE 1.27** Cooking with natural gas.

**Collect and Organize** We need to convert a rate expressed in trillions of cubic feet ($ft^3$) per year into cubic meters ($m^3$) per second. One trillion units = $10^{12}$ units. Also, 1 foot = 12 inches (exactly), and 1 m = $10^2$ cm (exactly).

**Analyze** To use our distance equivalencies, we will need to cube both sides of each equation to obtain equivalent volumes. Doing so means cubing everything on both sides of each equation—that is, the coefficients as well as the units:

$$(1\ \text{ft})^3 = (12\ \text{in})^3 = 1728\ \text{in}^3$$
$$\text{and}\quad (1\ \text{in})^3 = (2.54\ \text{cm})^3 = 16.387064\ \text{cm}^3$$
$$\text{and}\quad (1\ \text{m})^3 = (10^2\ \text{cm})^3 = 10^6\ \text{cm}^3$$

One meter is a little longer than 1 yard or 3 feet; 1 cubic meter should be a bit more than $3^3$, or about 30 cubic feet. One day has 24 h × 60 min/h × 60 s/min, or about $10^5$ seconds. Therefore, the calculated rate value should be about $(25 \times 10^{12})/(30 \times 10^5)$, or about $8 \times 10^6$.

**Solve** Let's handle the volume and time conversions separately and then combine them:

$$24.77 \times 10^{12}\ \cancel{\text{ft}^3} \times \frac{1728\ \cancel{\text{in}^3}}{1\ \cancel{\text{ft}^3}} \times \frac{16.387064\ \cancel{\text{cm}^3}}{1\ \cancel{\text{in}^3}} \times \frac{1\ \text{m}^3}{10^6\ \cancel{\text{cm}^3}} = 7.0141 \times 10^{11}\ \text{m}^3$$

$$1\ \cancel{\text{d}} \times \frac{24\ \cancel{\text{h}}}{1\ \cancel{\text{d}}} \times \frac{60\ \cancel{\text{min}}}{1\ \cancel{\text{h}}} \times \frac{60\ \text{s}}{1\ \cancel{\text{min}}} = 86{,}400\ \text{s}$$

Taking the quotient of the calculated volume and time values:

$$\text{rate} = \frac{7.0141 \times 10^{11}\ \text{m}^3}{86{,}400\ \text{s}} = 8.12 \times 10^6\ \text{m}^3/\text{s}$$

**Think About It** The calculated value is reassuringly close to our estimate. It is expressed with three significant figures because the initial rate value has that many. All the conversion factors used are based on exact values and do not affect how well we know the calculated rate. Many of the volume conversion factors were cubed, for example, $(2.54\ \text{cm/in})^3 = 16.387064\ \text{cm}^3/\text{in}^3$. Those volume conversion factors also are exact values. However, if 2.54 cm/in were not an exact value, cubing it would result in a volume conversion factor that had only three significant figures—the same number as in 2.54 before it was cubed.

**Practice Exercise** Suppose the heart of an adult male at rest takes exactly 1 minute to recirculate all the blood (5.6 L) in his body. How rapidly does his heart pump his blood? Express your answer in cubic centimeters per second and cubic inches per second.

*(Answers to Practice Exercises are in the back of the book.)*

**STEPWISE ANIMATION**
Temperature Scales

## SAMPLE EXERCISE 1.9 Converting between Temperature Scales — LO8

The temperature of interstellar space is 2.73 K. What is that temperature on the Celsius and Fahrenheit scales?

**Collect and Organize** We are asked to convert a temperature from the Kelvin scale to degrees Celsius and degrees Fahrenheit. Table 1.3 lists equations that relate Kelvin and Celsius temperatures, and Celsius and Fahrenheit temperatures.

**Analyze** We will first convert 2.73 K into an equivalent Celsius temperature and then calculate an equivalent Fahrenheit temperature. We can estimate that the value in degrees Celsius should be close to absolute zero (about −273°C). Because a Fahrenheit degree is about half the size of a Celsius degree, the temperature on the Fahrenheit scale should be a little less than twice the absolute value of the temperature on the Celsius scale (or around −500°F).

**Solve** Converting 2.73 K to degrees Celsius:

$$\begin{aligned} T(\text{K}) &= T(^\circ\text{C}) + 273.15 \\ T(^\circ\text{C}) &= T(\text{K}) - 273.15 \\ &= 2.73 - 273.15 = -270.42^\circ\text{C} \end{aligned}$$

Converting that Celsius temperature to Fahrenheit:

$$\begin{aligned} T(^\circ\text{C}) &= (5/9)[T(^\circ\text{F}) - 32] \\ T(^\circ\text{F}) &= (9/5)T(^\circ\text{C}) + 32 \\ &= 9/5(-270.42) + 32 = -454.76^\circ\text{F} \end{aligned}$$

**Think About It** Converting Kelvin to Celsius temperatures is a matter of subtracting their zero points, which differ by 273.15. The sizes of the degrees on both scales are the same. However, the Celsius and Fahrenheit scales differ in both their zero points and the sizes of their degrees, so the math involved requires two steps. The calculated Celsius value of −270.42°C is only a few degrees above absolute zero, as we estimated. The Fahrenheit value is within 10% of our estimate, so it is reasonable, too.

**Practice Exercise** The temperature of the moon's surface varies from −233°C at night to 123°C during the day. What are those temperatures on the Kelvin and Fahrenheit scales?

*(Answers to Practice Exercises are in the back of the book.)*

## SAMPLE EXERCISE 1.10 Calculating the Kinetic Energy of a Moving Object — LO8

One of tennis star Serena Williams's serves was clocked at 207.0 km/h (128.6 mph) during the 2013 Australian Open. If the mass of the tennis ball she served was 58.0 g, what was its kinetic energy in joules?

**Collect and Organize** We know the mass ($m$) and speed ($u$) of a served tennis ball and are asked to calculate its kinetic energy (KE). Equation 1.3 relates those variables: $KE = \frac{1}{2}mu^2$. One joule of energy is equivalent to 1 kg · (m/s)$^2$.

**Analyze** Before using Equation 1.3, we need to convert the speed given from 207.0 km/h to meters per second and to convert the mass of the tennis ball given from 58.0 g to kilograms.

**Solve** The mass of the tennis ball:

$$58.0\ \cancel{\text{g}} \times \frac{1\ \text{kg}}{1000\ \cancel{\text{g}}} = 0.0580\ \text{kg}$$

The speed of the ball:

$$\frac{207.0\ \cancel{\text{km}}}{1\ \cancel{\text{h}}} \times \frac{1000\ \text{m}}{1\ \cancel{\text{km}}} \times \frac{1\ \cancel{\text{h}}}{60\ \cancel{\text{min}}} \times \frac{1\ \cancel{\text{min}}}{60\ \text{s}} = 57.50\ \text{m/s}$$

The kinetic energy of the ball:

$$KE = \tfrac{1}{2}mu^2 = \tfrac{1}{2}(0.0580\ \text{kg})(57.50\ \text{m/s})^2 = 95.9\ \text{kg} \cdot (\text{m/s})^2 = 95.9\ \text{J}$$

**Think About It** The calculated value is expressed with three significant figures because three significant figures are in the mass of the ball and four in its speed. Therefore, the mass value is the weak link in that calculation. The time conversion factors have only two digits, but they are exact numbers and do not affect how well we know the final calculated value.

**Practice Exercise** On May 20, 2018, Jordan Hicks of the St. Louis Cardinals tied a Major League Baseball record by throwing a pitch clocked at 169.1 km/h (105.1 mph). If the mass of the baseball he threw was 146 g, what was its kinetic energy in joules?

*(Answers to Practice Exercises are in the back of the book.)*

## 1.9 Analyzing Experimental Results

Recall from Section 1.7 that experimental results are considered accurate when they have been validated by checking them with an accepted standard material of known composition. In many laboratories those accuracy checks are made by periodically analyzing *control samples*, which contain known quantities of the substance(s) of interest. Control samples are used in clinical chemistry labs to ensure the accuracy of routine analyses of samples of blood serum and urine. One substance included in comprehensive blood tests is creatinine, a product of protein metabolism that doctors use to monitor patients' kidney function (**Figure 1.28**). Suppose that a control sample known to contain 0.681 mg of creatinine per deciliter of sample is analyzed after every 10 patient samples and that the results of five control sample analyses are 0.685, 0.676, 0.669, 0.688, and 0.692 mg/dL. How precise (repeatable) are those results, and do they mean that the analyses of the control sample (and those later conducted on the real samples) are accurate?

A widely used method to express the precision of such data involves calculating the average of all the values and then calculating how much each value deviates from the average. The average of the five control samples is:

$$\frac{0.685 + 0.676 + 0.669 + 0.688 + 0.692}{5}\ \text{mg/dL} = \frac{3.410}{5}\ \text{mg/dL} = 0.6820\ \text{mg/dL}$$

We can write a general equation to represent that calculation for any number of measurements $n$ of any parameter $x$. Individual results are identified by the generic

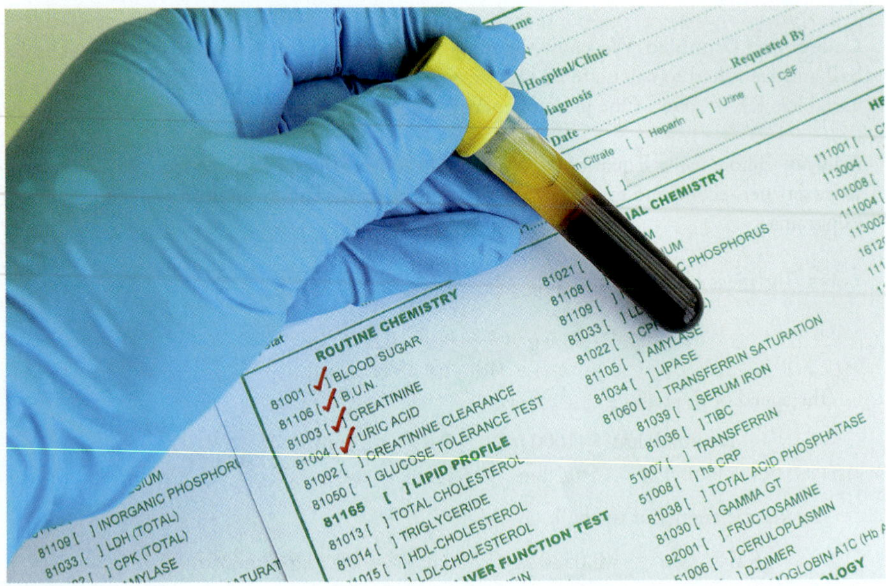


**FIGURE 1.28** The levels of creatinine in the blood indicate how well kidneys function.

symbol $x_i$, where $i$ can be any integer from 1 to $n$, the number of measurements made. The capital Greek sigma ($\Sigma$) with an $i$ subscript, $\Sigma_i(x_i)$, represents the *sum* of all the individual $x_i$ values. That kind of average is called the **mean** or **arithmetic mean**, represented by the symbol $\bar{x}$ in Equation 1.5:

$$\bar{x} = \frac{\Sigma_i(x_i)}{n} \tag{1.5}$$

To evaluate the precision of the five analyses of the control sample, we first calculate how much each of the values, $x_i$, deviates from the mean value: $(x_i - \bar{x})$. To average those deviations, we first square each deviation: $(x_i - \bar{x})^2$. Then we sum all those squared values, divide the sum by $(n - 1)$, and finally take the square root of the quotient. The results of those calculations are shown in **Table 1.4**, where the final result is called the **standard deviation (*s*)** of the control sample values. Equation 1.6 puts the steps for calculating $s$ in equation form:

$$s = \sqrt{\frac{\Sigma_i(x_i - \bar{x})^2}{n - 1}} \tag{1.6}$$

The standard deviation is sometimes reported with the mean value, using a ± sign to indicate the uncertainty in the mean value. For the five control samples, that expression is 0.6820 ± 0.0094 mg/dL, where the calculated $s$ value is rounded off to match the last decimal place of the mean. The smaller the value of $s$, the

**TABLE 1.4 Calculating the Standard Deviation of the Control Sample Creatinine Values (mg/dL)**

| $x_i$ | $x_i - \bar{x}$ | $(x_i - \bar{x})^2$ | $s$ |
|---|---|---|---|
| 0.685 | 0.003 | 0.000009 | $\Sigma_i(x_i - \bar{x})^2 = 0.000350$ |
| 0.676 | −0.006 | 0.000036 | |
| 0.669 | −0.013 | 0.000169 | $s = \sqrt{\frac{\Sigma_i(x_i - \bar{x})^2}{n - 1}}$ |
| 0.688 | 0.006 | 0.000036 | |
| 0.692 | 0.010 | 0.000100 | $= \sqrt{\frac{0.000350}{5 - 1}}$ |
| | | | $= 0.00935$ |

more tightly clustered the measurements are around the mean and the more precise the data are.

Now that we have evaluated the precision of the control sample data, we should evaluate how accurate they are—that is, how well they agree with the actual creatinine value of the control sample, 0.681 mg/dL. The mean of the five values is within 0.001 mg/dL of the actual value, so the results certainly seem accurate, but a way is available to quantitatively express the certainty that they are accurate. Doing so involves calculating the **confidence interval**, a range of values around a calculated mean (0.6820 mg/dL in our control sample analyses) that probably contains the true mean value, $\mu$. Once we have calculated the size of that range, we can determine whether the actual creatinine value lies within it. If so, the analyses are considered accurate.

**mean (arithmetic mean)** an average calculated by summing all the values in a series and then dividing the sum by the number of values.

**standard deviation ($s$)** a measure of the amount of variation, or dispersion, in a set of related values.

**confidence interval** a range of values that has a specified probability of containing the true value of a measurement.

To calculate the confidence interval, we use a statistical tool called the $t$-distribution and Equation 1.7:

$$\mu = \bar{x} \pm \frac{t\,s}{\sqrt{n}} \tag{1.7}$$

A table of $t$ values is located in Appendix 1. **Table 1.5** contains a portion of it. The values are arranged based on two parameters: the number of values in a set of data (actually, $n - 1$) and the confidence level we wish to use in our decision making. A commonly used confidence level in chemical analysis is 95%. Using it means that the range we calculate, using Equation 1.7, has a 95% chance of containing the true mean value (here, the amount of the creatinine in our control sample). The 95% $t$ value for $n - 1 = 5 - 1 = 4$ in Table 1.5 is 2.776, so using the mean ($\bar{x}$) and standard deviation ($s$) values calculated previously gives us the following value for the true mean value ($\mu$):

$$\mu = \bar{x} \pm \frac{t\,s}{\sqrt{n}} = \left(0.6820 \pm \frac{2.776 \times 0.00935}{\sqrt{5}}\right) \text{mg/dL} = (0.6820 \pm 0.0116)\ \text{mg/dL}$$

Thus, we can say with 95% certainty that the true mean of our control sample data is between 0.6704 and 0.6936 mg/dL. Because that range includes the actual creatinine concentration (0.681 mg/dL) of the control sample, we can report with 95% confidence that those five control analyses (and the analyses of the patients' samples) are accurate.

**TABLE 1.5** Values of $t$

| | CONFIDENCE LEVEL (%) | | |
|---|---|---|---|
| ($n - 1$) | 90 | 95 | 99 |
| 3 | 2.353 | 3.182 | 5.841 |
| 4 | 2.132 | 2.776 | 4.604 |
| 5 | 2.015 | 2.571 | 4.032 |
| 10 | 1.812 | 2.228 | 3.169 |
| 20 | 1.725 | 2.086 | 2.845 |
| ∞ | 1.645 | 1.960 | 2.576 |

**SAMPLE EXERCISE 1.11** Calculating the Means, Standard Deviation, and Confidence Interval of a Set of Data **LO9**

A group of students collected a sample of water from a river near their campus and divided the sample among five other groups of students. All six groups independently determined the concentration of dissolved oxygen in their samples. Their results were 9.2, 8.6, 9.0, 9.3, 9.1, and 8.9 mg of $O_2$/L. Calculate the mean, standard deviation, and 95% confidence interval of those results.

**Collect, Organize, and Analyze** Equations 1.5, 1.6, and 1.7 can be used to calculate the mean, standard deviation, and 95% confidence interval, respectively, of the results of the six analyses. The $t$ value (Table 1.5) to use in Equation 1.7 for $n - 1 = 5$ is 2.571.

**Solve**

a. Using Equation 1.5 to calculate the mean:

$$\bar{x} = \frac{\Sigma_i(x_i)}{n} = \left(\frac{9.2 + 8.6 + 9.0 + 9.3 + 9.1 + 8.9}{6}\right) \text{mg/L} = \frac{54.1}{6} = 9.02\ \text{mg/L}$$

b. We use a data table like Table 1.4 with Equation 1.6 to calculate the standard deviation ($s$) value:

| Dissolved $O_2$ (mg/L) | $x_i - \bar{x}$ | $(x_i - \bar{x})^2$ | $s$ |
|---|---|---|---|
| 9.2 | 0.18 | 0.0324 | $\Sigma_i(x_i - \bar{x})^2 = 0.3084$ |
| 8.6 | −0.42 | 0.1764 | |
| 9.0 | −0.02 | 0.0004 | $s = \sqrt{\dfrac{\Sigma_i(x_i - \bar{x})^2}{n-1}}$ |
| 9.3 | 0.28 | 0.0784 | |
| 9.1 | 0.08 | 0.0064 | $= \sqrt{\dfrac{0.3084}{6-1}}$ |
| 8.9 | −0.12 | 0.0144 | $= 0.248$ |

c. Using Equation 1.7 to calculate the 95% confidence interval:

$$\mu = \bar{x} \pm \frac{t\,s}{\sqrt{n}} = \left(9.02 \pm \frac{2.571 \times 0.248}{\sqrt{6}}\right) \text{mg/L} = (9.02 \pm 0.26) \text{ mg/L}$$

**Think About It** The arithmetic mean of 9.02 mg/L is consistent with the fact that three measurements are less than the mean and three measurements are greater than the mean. Note that two measurements (8.6 and 9.3 mg/L) fall outside the 95% confidence interval.

**Practice Exercise** Analyses of a sample of water from the Dead Sea produced the following results for the concentration of sodium ions: 35.8, 36.6, 36.3, 36.8, and 36.4 mg/L. What are the mean, standard deviation, and 95% confidence interval of those results?

*(Answers to Practice Exercises are in the back of the book.)*

An assumption built into calculating means, standard deviations, and confidence intervals is that the variability in the data is random. *Random* means that data points are as likely to be above the mean value as below it and that a greater probability exists that values will lie close to the mean than far away from it. That kind of distribution is called a *normal* distribution. Large numbers of such data produce a distribution profile called a bell curve, such as the one shown in **Figure 1.29**, which is based on a study conducted by the U.S. Centers for Disease Control and Prevention. The data in Figure 1.29 depict the weights of 19-year-old American males and have a mean of 175 lb and a standard deviation of 41 lb (79 ± 19 kg). In randomly distributed data, 68% of the values—represented by the area under the curve highlighted in pale red—are within one standard deviation of the mean.

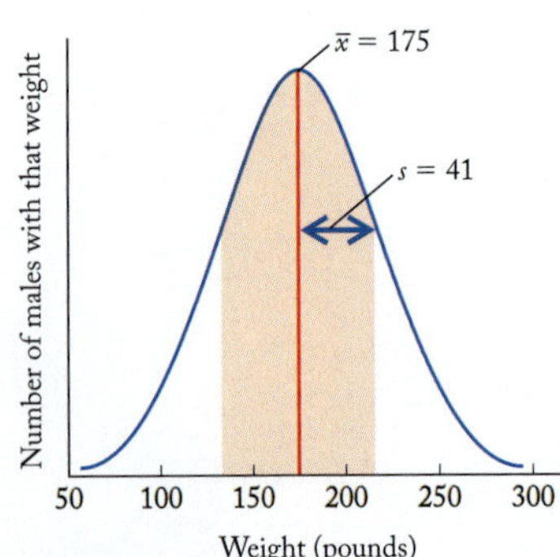


**FIGURE 1.29** The distribution of body weight values among 19-year-old U.S. males is an example of a normal distribution.

Sometimes a set of data contains an **outlier**—an individual value that is much farther away from the mean than any other values. You may be tempted to simply ignore such a value, but without a valid reason for doing so, such as accidentally leaving out a step in an analytical procedure, disregarding a value just because it is unexpected or not similar to the others is unethical.

Suppose a college freshman discovers a long-forgotten piggy bank full of pennies and, being short of money, decides to pack them into rolls of 50 to cash them in at a bank. To avoid having to count hundreds of pennies, the student decides to use a balance in a general chemistry lab that can weigh up to 300 g to the nearest 0.001 g. The student individually weighs 10 pennies from the piggy bank to determine the average mass of a penny, intending to multiply the average by 50 to calculate how many to weigh out for each roll. **Table 1.6** lists the results of the 10 measurements from lowest to highest values.

The results reveal that nine of the ten masses are close to 2.5 g, but the tenth is considerably heavier. Did the student make an error in the measurement, or is something unusual about that tenth penny? To answer such questions—and the broader one of whether an unusually high or low value is statistically different enough from the others in a set of data to be labeled an outlier—we analyze the data by using *Grubbs' test of a single outlier*, or **Grubbs' test**. In that test, the absolute difference between the suspected outlier and the mean of a set of data is divided by the standard deviation of the data set. The result is a statistical parameter that has the symbol $Z$:

$$Z = \frac{|x_i - \bar{x}|}{s} \tag{1.8}$$

If that calculated $Z$ value is greater than the reference $Z$ value (**Table 1.7**) for a given number of data points and a particular confidence level—usually 95%—the suspect data point is determined to be an outlier and can be discarded.

To apply Grubbs' test to the mass of the tenth and heaviest penny, we first calculate the mean and standard deviation of the masses of all 10 pennies. Those values are 2.562 ± 0.191 g. We use them in Equation 1.8 to calculate $Z$:

$$Z = \frac{|x_i - \bar{x}|}{s} = \frac{|3.107 - 2.562|}{0.191} = 2.85$$

Next, we check the reference $Z$ values in Table 1.7 for $n = 10$ data points, and we find that our calculated $Z$ value is greater than both 2.290 and 2.482—the $Z$ values above which we can conclude with 95% and 99% confidence, respectively, that the data point of 3.107 g is an outlier. Stated another way, the probability that this data point *is not* an outlier is less than 1%. Note that Grubbs' test can be used *only once* to identify *only one* outlier in a set of data.

The tenth penny probably weighed much more than the others because U.S. pennies minted since 1983 weigh 2.50 g new, but those minted before 1983 weighed 3.11 g new. The older pennies are 95% copper and 5% zinc, whereas the newer pennies are 97.5% zinc and are coated with a thin layer of copper. (Copper is about 25% more dense than zinc.) Thus, a piggy bank containing hundreds of pennies will probably have a few that are heavier than most of the others.

**TABLE 1.6** Masses of 10 Circulated Pennies

| Mass |
|---|
| 2.486 g |
| 2.495 g |
| 2.500 g |
| 2.502 g |
| 2.502 g |
| 2.505 g |
| 2.506 g |
| 2.507 g |
| 2.515 g |
| 3.107 g |

**TABLE 1.7** Reference $Z$ Values for Grubbs' Test

| | CONFIDENCE LEVEL (%) | |
|---|---|---|
| $n$ | 95 | 99 |
| 3 | 1.155 | 1.155 |
| 4 | 1.481 | 1.496 |
| 5 | 1.715 | 1.764 |
| 6 | 1.887 | 1.973 |
| 7 | 2.020 | 2.139 |
| 8 | 2.126 | 2.274 |
| 9 | 2.215 | 2.387 |
| 10 | 2.290 | 2.482 |
| 11 | 2.355 | 2.564 |
| 12 | 2.412 | 2.636 |

**SAMPLE EXERCISE 1.12** Testing Whether a Data Point Should Be Considered an Outlier **LO9**

Use Grubbs' test to determine whether the lowest value in the set of sodium ion concentration data from Practice Exercise 1.11 should be considered an outlier at the 95% confidence level.

**Collect, Organize, and Analyze** We want to determine whether the lowest value in the following set of five results is an outlier: 35.8, 36.6, 36.3, 36.8, and 36.4 mg/L. Grubbs' test (Equation 1.8) is used to determine whether a data point should be deemed an outlier. If the resulting $Z$ value is equal to or greater than the reference $Z$ values for $n = 5$ from Table 1.7, the suspect value is an outlier.

**outlier** a data point that is distant from the other observations.

**Grubbs' test** a statistical test used to detect an outlier in a set of data.

**Solve** Using the statistics functions of a programmable calculator, we find that the mean and standard deviation of the data set are 36.38 ± 0.38 mg/L. We calculate the value of $Z$ by using Equation 1.8:

$$Z = \frac{|x_i - \bar{x}|}{s} = \frac{|35.8 - 36.38|}{0.38} = 1.5$$

That calculated $Z$ value is less than 1.715, the reference $Z$ value for $n = 5$ at the 95% confidence level (Table 1.7). Therefore, the lowest value is *not* an outlier, and it should be included with the other four values in any analysis of the data.

**Think About It** Although the lowest value may have appeared to be considerably lower than the others in the data set, Grubbs' test tells us that it is not *significantly* lower at the 95% confidence level.

**Practice Exercise** Duplicate determinations of the cholesterol concentration in a sample of blood serum produce the following results: 181 and 215 mg/dL. The patient's doctor is concerned about the difference between the two results and the fact that values above 200 mg/dL are considered "borderline high," so she orders a third cholesterol test. The result is 185 mg/dL. Should the doctor consider the 215 mg/dL value an outlier at either the 95% or 99% confidence interval and discard it?

*(Answers to Practice Exercises are in the back of the book.)*

## SAMPLE EXERCISE 1.13 Integrating Concepts: Shopping for Gasoline

On a weekend in April 2018, one Canadian dollar (CAD) was worth 0.7947 U.S. dollars (USD), and the average prices of regular-grade gasoline in the bordering cities of Detroit, Michigan, and Windsor, Ontario, were 2.55 USD per gallon and 1.226 CAD per liter, respectively. How much money (on average) would a Windsor resident save by driving 5.0 km to a gasoline station in Detroit to fill the gas tank in her car with 12.5 gallons of gasoline? Assume that the car consumes gasoline at the rate of 7.50 liters per 100 km and that the cost of the round trip requires paying Detroit–Windsor Tunnel tolls. (The toll is 4.75 CAD to drive from Windsor to Detroit, and the toll is 6.25 CAD to drive from Detroit to Windsor.)

**Collect and Organize** The problem gives gasoline prices in Detroit and Windsor that are expressed in different currencies for different quantities of gasoline, and it asks how much money a Windsor resident would save by driving to Detroit to fill the tank in her car. We also are given the distance to a Detroit station in kilometers, the fuel efficiency of the car in liters/100 km, and the Detroit–Windsor Tunnel toll in CAD. According to Table 1.3, 1 liter contains 0.2642 gallons.

**Analyze** We have to assume that gas is cheaper at the U.S. station; otherwise, why make the trip? That assumption is supported by these unit conversion estimates: the currency exchange rate is approximately 4 USD = 5 CAD, so the price of gas at a Detroit station is about 3.0 CAD/gal, or 0.8 CAD/L. That price is about 2/3 of that in Windsor. A logical strategy for solving the problem is to calculate the cost (in CAD) of filling up at stations in Windsor and Detroit and then to subtract the costs of the round-trip drive to a station in Detroit.

**Solve**

1. Cost in CAD of buying gasoline at a station in Detroit:

$$12.5\ \cancel{\text{gal}} \times \frac{2.55\ \cancel{\text{USD}}}{\cancel{\text{gal}}} \times \frac{1\ \text{CAD}}{0.7947\ \cancel{\text{USD}}} = 40.11\ \text{CAD}$$

2. Cost of buying the same volume of gasoline at a station in Windsor:

$$12.5\ \cancel{\text{gal}} \times \frac{1\ \cancel{\text{L}}}{0.2642\ \cancel{\text{gal}}} \times \frac{1.226\ \text{CAD}}{\cancel{\text{L}}} = 58.01\ \text{CAD}$$

3. Difference in fill-up costs: 58.01 − 40.11 = 17.90 CAD
4. Cost of gasoline to drive to and from Detroit:

$$2 \times 5.0\ \cancel{\text{km}} \times \frac{7.50\ \cancel{\text{L}}}{100\ \cancel{\text{km}}} \times \frac{1.226\ \text{CAD}}{\cancel{\text{L}}} = 0.92\ \text{CAD}$$

5. Net savings factoring in the tunnel tolls:

$$17.90 - (0.92 + 6.25 + 4.75)\ \text{CAD} = 5.98\ \text{CAD}$$

**Think About It** Would you be willing to drive to another country to save about $6 on a tank of gasoline? How about driving the same distance within your own country to save half as much? What additional considerations besides those in this Sample Exercise would influence your decision?

## SUMMARY

**LO1** **Scientific methods** start with observations of natural phenomena and/or the results of laboratory experiments; next, a tentative explanation, or **hypothesis**, is developed that explains the observations and results; then the hypothesis is tested through further experimentation before a **scientific theory** is formulated that explains all the results and observations available. A **scientific law** is a comprehensive, succinct description of a phenomenon or process. Dalton's atomic theory explains Proust's **law of definite proportions** and Dalton's **law of multiple proportions**. (Section 1.1)

**LO2** This book's COAST framework for solving problems has four components: **C**ollect and **O**rganize information and ideas, **A**nalyze the information to determine how it can be used to obtain the answer, **S**olve the problem (often the math-intensive step), and **T**hink about the answer. (Section 1.2)

**LO3** The principal classes of **matter** are **mixtures** and **pure substances**. Pure substances may be either **elements** or **compounds** (elements chemically bonded). A **chemical formula** indicates the proportion of elements in a substance. The properties of a substance are either **intensive properties**, which are independent of quantity, or **extensive properties**, which are related to the quantity of the substance. The **physical properties** of a substance can be observed without changing the substance into another one, whereas the **chemical properties** of a substance (such as flammability) can be observed only through **chemical reactions** involving the substance. The **density (*d*)** of an object or substance is the ratio of its **mass (*m*)** to its volume. Mixtures may be **homogeneous** (such mixtures are also called **solutions**) or **heterogeneous**, and they can be separated by **physical processes** such as **distillation**, **filtration**, and **chromatography**. Distillation separates substances of different **volatility**. (Sections 1.1, 1.3)

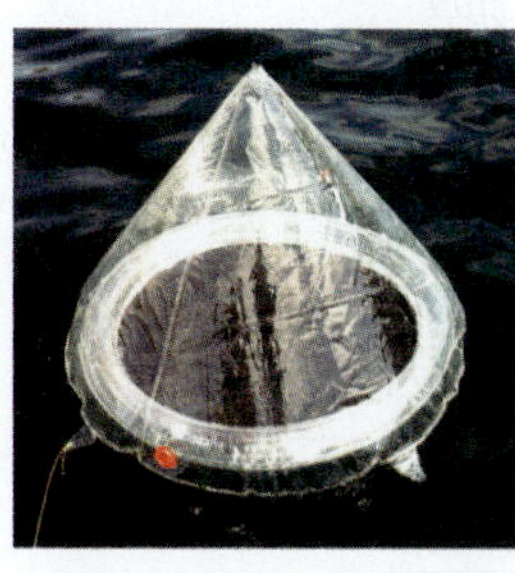

**LO4** The states (or phases) of matter include **solid**, in which the particles have an ordered structure; **liquid**, in which the particles are free to move past one another; and **gas** (or *vapor*), in which the particles have the most freedom and fill their container. Familiar phase changes include melting, freezing, vaporization, and condensation. The transformation of a solid directly into a gas is **sublimation**; the reverse process is **deposition**. (Section 1.4)

**LO5** **Energy** can be defined as the ability to do **work**. **Heat** is an energy transfer due to a difference in temperature. **Potential energy (PE)** is the energy in an object as a result of its position or composition. **Kinetic energy (KE)** is the energy of motion. According to the **law of conservation of energy**, energy cannot be created or destroyed. (Section 1.5)

**LO6** The composition of molecular compounds is described by their **molecular formulas**. The three-dimensional arrangements of their atoms may be represented by **structural formulas**, ball-and-stick models, and space-filling models. **Ionic compounds** do not consist of molecules but rather are formed from bonds between positive and negative ions. Ionic compounds have an **empirical formula** that indicates the ratio of positive ions to negative ions. (Section 1.6)

**LO7** Measured quantities and the values derived from them are inherently uncertain. Exact values include those derived from counting objects or values that are defined, such as 60 seconds in a minute. The appropriate number of **significant figures** is used to express the certainty in the result of a measurement or calculation. The **precision** of any set of measurements indicates how repeatable the measurement is, whereas the **accuracy** of a measurement indicates how close to the true value the measured value is. (Section 1.7)

**LO8** The International System of Units (SI), in which the **meter (m)** is the standard unit of length, evolved from the metric system and is widely used in science to express the results of measurements. Prefixes naming powers of 10 are used with SI base units to express quantities much larger or much smaller than the base units. Dimensional analysis uses **conversion factors** (fractions in which the numerators and denominators have different units but represent the same quantity) to convert a value from one unit into another unit. (Sections 1.7 and 1.8)

**LO9** The average value and variability in repeated measurements or analyses are determined by calculating the **arithmetic mean** ($\bar{x}$), **standard deviation** ($s$), and **confidence interval**. An **outlier** in a data set may be identified on the basis of the results of **Grubbs' test**. (Section 1.9)

## PARTICULATE PREVIEW WRAP-UP

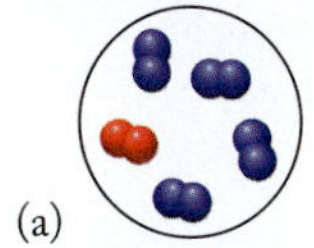

(a) Mixture of molecular elements

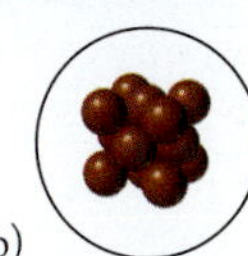

(b) Atoms of an element

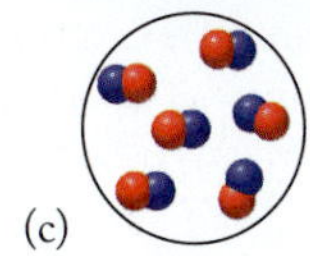

(c) Molecules of a compound

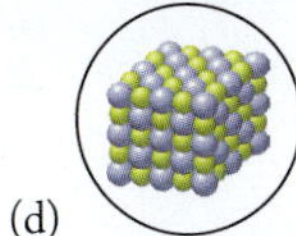

(d) Compound consisting of an array of ions

## PROBLEM-SOLVING SUMMARY

| Type of Problem | Concepts and Equations | Sample Exercises |
|---|---|---|
| **Distinguishing exact from uncertain values** | Quantities that can be counted are exact. Measured quantities or conversion factors that are not exact values are inherently uncertain. | **1.4** |
| **Using significant figures in calculations** | Apply the weak-link principle: The number of significant figures in a calculated quantity involving multiplication or division can be no greater than the number of significant figures in the least certain value used to calculate it. For adding or subtracting values, the number of digits after the decimal point in the sum or difference can be no greater than in the value that has the fewest digits after its decimal point. | **1.5, 1.6** |
| **Converting units by using dimensional analysis** | Converting units involves multiplication by one or more conversion factors, which are set up so that the original units cancel out. | **1.7, 1.8** |
| **Converting between temperature scales** | $T(\text{K}) = T(°\text{C}) + 273.15$ <br> $T(°\text{C}) = (5/9)[T(°\text{F}) - 32]$ | **1.9** |
| **Calculating kinetic energy** | $\text{KE} = \frac{1}{2}mu^2$ (1.3) | **1.10** |
| **Calculating the mean ($\bar{x}$), standard deviation ($s$), and confidence interval of a set of data** | $\bar{x} = \frac{\Sigma_i(x_i)}{n}$ (1.5) <br> $s = \sqrt{\frac{\Sigma_i(x_i - \bar{x})^2}{n-1}}$ (1.6) <br> $\mu = \bar{x} \pm \frac{ts}{\sqrt{n}}$ (1.7) | **1.11** |
| **Testing whether a suspect data point ($x_i$) is an outlier** | Use Grubbs' test: calculate the value of $Z$: <br> $Z = \frac{\lvert x_i - \bar{x}\rvert}{s}$ (1.8) <br> and compare it with the appropriate reference value in Table 1.7. If the calculated $Z$ is greater, the suspect data point is an outlier. | **1.12** |

## VISUAL PROBLEMS

*(Answers to boldface end-of-chapter questions and problems are in the back of the book.)*

*Note:* This book's inside back cover shows the colors used to represent the atoms of different elements.

**1.1.** For each image in Figure P1.1, identify what class of matter is depicted (an element, a compound, a homogeneous mixture, or a heterogeneous mixture) and identify the physical state(s).

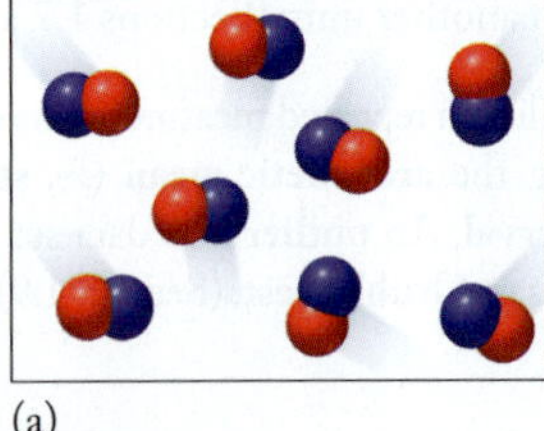

(a)

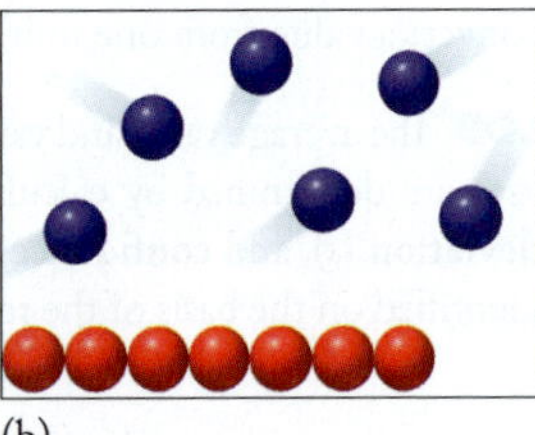

(b)

**FIGURE P1.1**

1.2. For each image in Figure P1.2, identify what class of matter is depicted (an element, a compound, a homogeneous mixture, or a heterogeneous mixture) and identify the physical state(s).

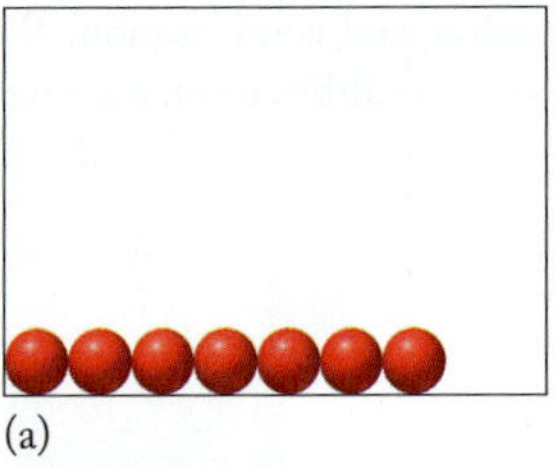

(a)

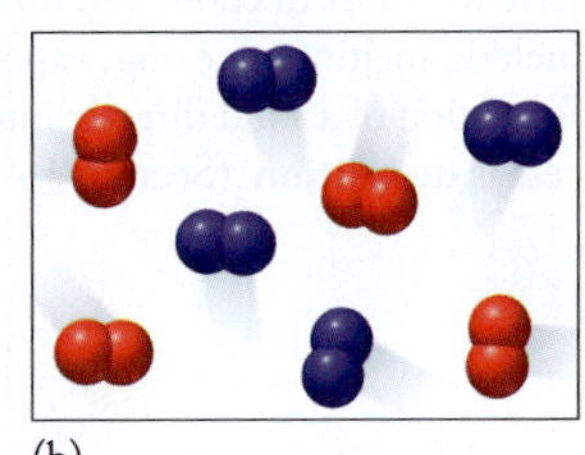

(b)

**FIGURE P1.2**

**1.3.** Which of the following statements best describes the change depicted in Figure P1.3?

a. A mixture of two gaseous elements undergoes a chemical reaction, forming a gaseous compound.
b. A mixture of two gaseous elements undergoes a chemical reaction, forming a solid compound.
c. A mixture of two gaseous elements undergoes deposition.
d. A mixture of two gaseous elements condenses.

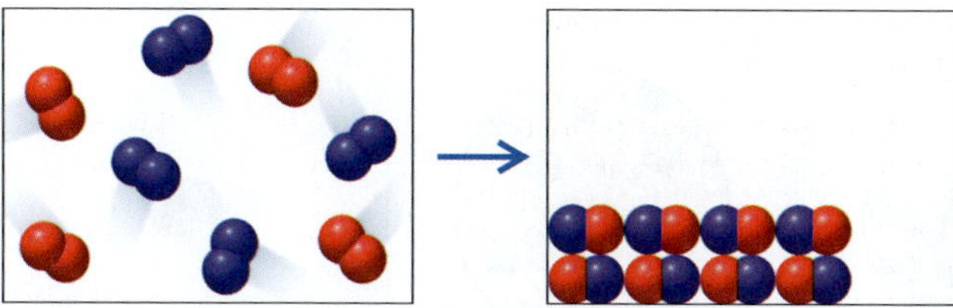

**FIGURE P1.3**

1.4. Which of the following statements best describes the change depicted in Figure P1.4?

a. A mixture of two gaseous elements is cooled to a temperature at which one of them condenses.
b. A mixture of two gaseous compounds is heated to a temperature at which one of them decomposes.
c. A mixture of two gaseous elements undergoes deposition.
d. A mixture of two gaseous elements reacts to form two compounds, one of which is a liquid.

**FIGURE P1.4**

**1.5.** The $CO_2$ gas represented by the molecules in the box on the left in Figure P1.5 is cooled from room temperature to 180 K. Describe the particulate image of $CO_2$ at that temperature in the box on the right.

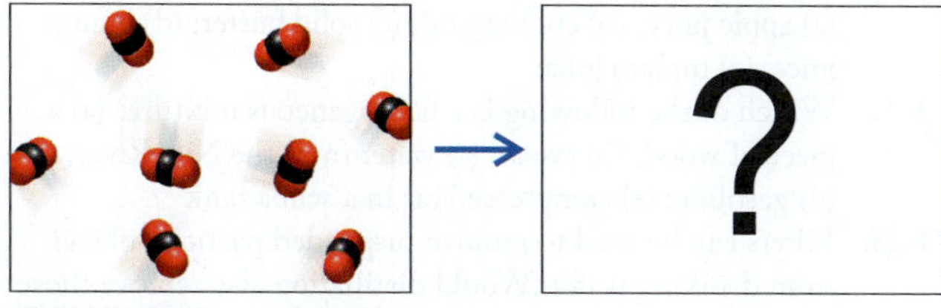


**FIGURE P1.5**

1.6. The temperature of the mixture of substances represented by the molecules in the box on the left in Figure P1.6 is reduced from 298 to 70 K. Describe the particulate image of the mixture present at 70 K in the box on the right.

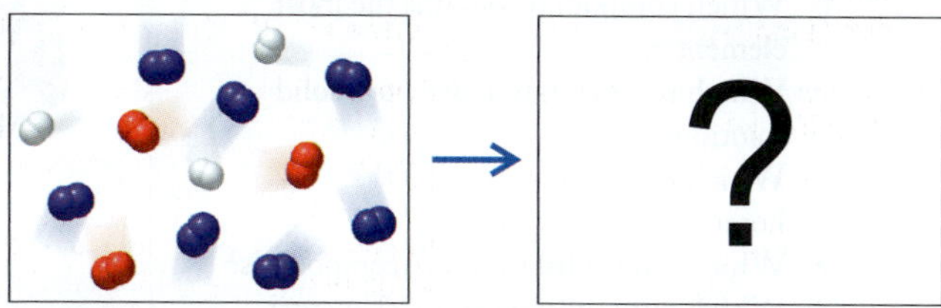


**FIGURE P1.6**

**1.7.** What are the chemical formulas of the three compounds with the molecular structures shown in Figure P1.7?

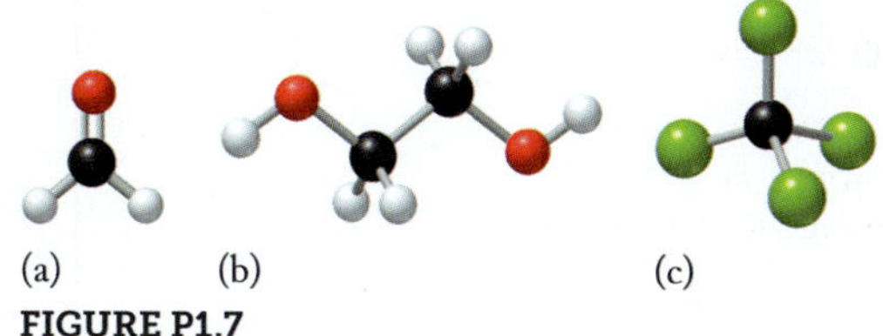

(a) (b) (c)

**FIGURE P1.7**

1.8. What are the chemical formulas of the three compounds with the molecular structures shown in Figure P1.8?

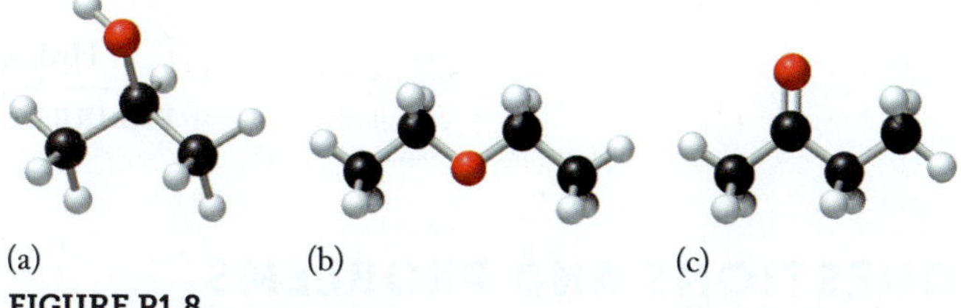

(a) (b) (c)

**FIGURE P1.8**

**1.9.** A pharmaceutical company checks the quality control process involved in manufacturing pills of one of its medicines by taking the mass of two samples of four pills each. Figure P1.9 shows graphs of the masses of the pills in each sample. The pill is supposed to weigh 3.25 mg. Label each sample as *both precise and accurate*, *precise but not accurate*, *accurate but not precise*, or *neither precise nor accurate*.

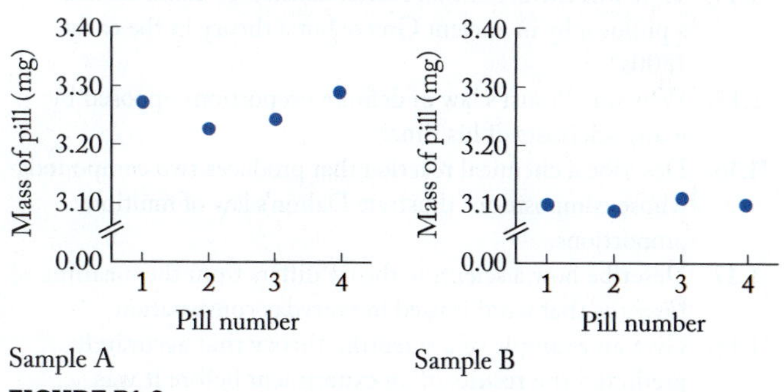


**FIGURE P1.9**

1.10. Use representations [A] through [I] in Figure P1.10 to answer questions a–f.
  a. Which molecule contains the most atoms?
  b. Which compound contains the most elements?
  c. Which representation depicts a solid solution?
  d. Which representation depicts a homogeneous mixture?
  e. Which pure substances are compounds?
  f. Which pure substances are elements?

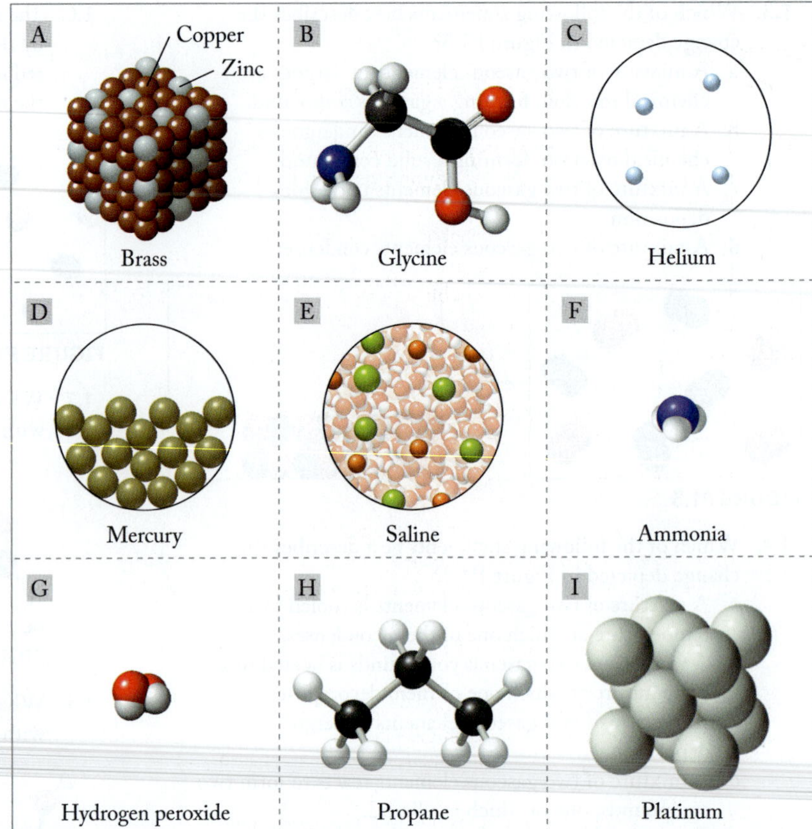


**FIGURE P1.10**

# QUESTIONS AND PROBLEMS

## *Atomic Theory: The Scientific Method in Action*

### Concept Review

**1.11.** How does a hypothesis differ from a scientific theory?

1.12. How does a hypothesis become a theory?

**1.13.** Describe how Dalton's atomic theory supported his law of multiple proportions.

1.14. Why was the belief that matter consists of atoms considered a philosophy in ancient Greece but a theory in the early 1800s?

**1.15.** Why was Proust's law of definite proportions opposed by many scientists of his time?

*1.16. Describe a chemical reaction that produces two compounds whose compositions illustrate Dalton's law of multiple proportions.

**1.17.** Describe how a scientific theory differs from the meaning of *theory* as that word is used in everyday conversation.

*1.18. Give an example of a scientific theory that accurately predicted the results of an experiment before it was carried out.

**1.19.** Can a scientific hypothesis be disproven?

1.20. Can a theory be proven?

## *Classes of Matter*

### Concept Review

**1.21.** Which of the following foods is a heterogeneous mixture? (a) bottled water; (b) a Snickers bar; (c) grape juice; (d) an uncooked hamburger

1.22. Which of the following foods is a homogeneous mixture? (a) freshly brewed coffee; (b) vinegar; (c) a slice of white bread; (d) a slice of ham

**1.23.** Which of the following foods is a heterogeneous mixture? (a) apple juice; (b) cooking oil; (c) solid butter; (d) orange juice; (e) tomato juice

1.24. Which of the following is a homogeneous mixture? (a) a piece of wood; (b) sweat; (c) water from the Nile River; (d) gasoline; (e) compressed air in a scuba tank

***1.25.** Filters can be used to remove suspended particles of soil from drinking water. Would distillation also remove those particles? If so, suggest a reason why it is not widely used.

1.26. Which of the colored compounds in the photograph in Figure 1.8(b) interact more strongly with the stationary phase of the separation: those at the very top or those below them? Assume that the liquid phase migrates upward.

## Properties of Matter

### Concept Review

**1.27.** List one chemical property and four physical properties of gold.

1.28. Describe three physical properties that gold and silver have in common and three physical properties that distinguish them.

**1.29.** Give three properties that enable a person to distinguish between table sugar, water, and oxygen.

1.30. Give three properties that enable a person to distinguish between table salt, sand, and copper.

**1.31.** Indicate whether each of the following is a physical or chemical property of sodium (Na):
   a. Its density is greater than that of kerosene and less than that of water.
   b. It has a lower melting point than that of most other metals.
   c. It is an excellent conductor of heat and electricity.
   d. It is soft and can be easily cut with a knife.
   e. Freshly cut sodium is shiny, but it rapidly tarnishes when it comes in contact with air.
   f. It reacts vigorously with water, releasing hydrogen gas ($H_2$).

1.32. Indicate whether each of the following is a physical or chemical property of hydrogen gas ($H_2$):
   a. At room temperature, its density is less than that of any other gas.
   b. It reacts vigorously with oxygen ($O_2$) to form water.
   c. Liquefied $H_2$ boils at −253°C.
   d. $H_2$ gas does not conduct electricity.

**1.33.** Can an extensive property be used to identify a substance? Explain why or why not.

1.34. Which of the following are intensive properties of a sample of a substance? (a) freezing point; (b) heat content; (c) temperature

***1.35.** Is carbon dioxide's ability to extinguish fires linked to its chemical properties, its physical properties, or both? Explain your answer.

1.36. The stainless steel used to make kitchen knives and many other tools gets its name from the metal's ability to resist corrosion and, therefore, *stain less*. Is that a chemical or physical property of stainless steel?

## States of Matter

### Concept Review

**1.37.** In what ways are the arrangements of water molecules in ice and liquid water similar, and in what ways are they different?

1.38. What occupies the space between the particles that make up a gas?

**1.39.** Substances have characteristic *triple points*, unique combinations of temperature and pressure at which substances can simultaneously exist as solids, liquids, and gases. In which of those three states do the particles of a substance at its triple point have the greatest motion, and in which state do they have the least motion?

1.40. A pot of water on a stove is heated to a rapid boil. Identify the gas inside the bubbles that forms in the boiling water.

**1.41.** A brief winter storm leaves a dusting of snow on the ground. During the sunny but very cold day after the storm, the snow disappears even though the air temperature never gets above freezing. If the snow didn't melt, where did it go?

1.42. Equal masses of water undergo condensation, deposition, evaporation, and sublimation.
   a. Which of the processes is accompanied by the *release* of the most energy?
   b. In which of the processes is the most energy *absorbed*?

## Forms of Energy

### Concept Review

**1.43.** How are energy and work related?

1.44. Explain the difference between potential energy and kinetic energy.

**1.45.** Which of the following statements about heat are true?
   a. Heat is the transfer of energy from a warmer place to a cooler one.
   b. Heat flows faster from a full container of hot coffee than from a half-full container of coffee at the same temperature.
   c. A cup of hot coffee loses heat faster than the same cup full of warm coffee.

1.46. Describe three examples of energy transfer that happen when you speak on a cell phone to a friend.

### Problems

**1.47.** A subcompact car with a mass of 1400 kg and a loaded dump truck with a mass of 18,000 kg are traveling at the same speed. How many times more kinetic energy does the dump truck have than the car?

1.48. **Speed of Baseball Pitches** Major League Baseball pitchers throw a pitch called a *changeup*, which looks like a fastball leaving the pitcher's hand but has less speed than a typical fastball. How much more kinetic energy does a 148 km/h (92 mph) fastball have than a 125 km/h (78 mph) changeup? Express your answer as a percentage of the kinetic energy of the changeup.

## Unit Conversions, Dimensional Analysis, and Significant Figures

### Concept Review

**1.49.** Describe in general terms how the SI and customary U.S. unit systems differ.

1.50. Suggest two reasons why SI units are so widely used in science.

**1.51.** Both the Fahrenheit and Celsius scales are based on reference temperatures that are 100 degrees apart. Suggest a reason why scientists prefer the Celsius scale.

1.52. How are the Celsius and Kelvin scales similar, and how are they different?

**1.53.** What is meant by an *absolute* temperature scale?

1.54. Can a temperature in °C ever have the same value in °F?

## Problems

*Note:* Some physical properties of the elements are listed in Appendix 3. Be sure to express the results of all calculations with the appropriate number of significant figures.

**1.55.** Write conversion factor(s) for each of the following unit conversions: (a) liters to milliliters; (b) kilometers to meters; (c) centigrams to kilograms; (d) cubic meters to cubic centimeters.

1.56. Write conversion factor(s) for each of the following unit conversions: (a) picoseconds to seconds; (b) meters to centimeters; (c) nanoseconds to milliseconds; (d) square meters to square kilometers.

---

**1.57.** A single strand of natural silk may be as long as $4.0 \times 10^3$ m. (a) How many significant figures are in that value? (b) Express that length in kilometers, centimeters, inches, and feet.

1.58. **Olympic Mile** An Olympic "mile" is actually 1500.0 m. (a) How many significant figures are in that value? (b) Express that distance in kilometers, centimeters, inches, and feet.

---

**1.59. Kentucky Derby Record** In 1973 a horse named Secretariat ran the fastest Kentucky Derby in history, taking 1 minute and 59.4 seconds to run 2.01 km (1.25 miles). (a) How many significant figures are in that time when expressed in seconds? (b) How many significant figures are in that distance value? (c) What was Secretariat's average speed in meters per second? (d) Express that speed in miles per hour.

1.60. Suppose a wheelchair-marathon racer moving at 8.28 m/s expends energy at a rate of 665 Calories per hour. (a) How many significant figures are in those speed and energy values? (b) How much energy in Calories would be required to complete a marathon (42.195 km) at that pace?

---

***1.61.** The level of water in an Olympic-size swimming pool (50.0 m long, 25.0 m wide, and 2.30 m deep) is to be lowered by 25 cm. If water is pumped out at a rate of 5.2 L per second, how long will lowering the water level by 25 cm take?

*1.62. The price of a popular soft drink is 2.17 CAD for 2.0 L in Canada and 1.00 USD for 24 fluid ounces (fl oz) in the United States. If 1 CAD = 0.7947 USD, which beverage is a better buy? (1 qt = 32 fl oz.)

---

**1.63.** A chemist needs 35.0 g of concentrated sulfuric acid for an experiment. The density of concentrated sulfuric acid at room temperature is 1.84 g/mL. What volume of the acid is required?

1.64. What is the mass of 65.0 mL of ethanol? (Its density at room temperature is 0.789 g/mL.)

---

**1.65.** What volume of gold would be equal in mass to a piece of copper with a volume of 125 $cm^3$?

*1.66. A small hot-air balloon is filled with $1.00 \times 10^6$ L of air at a temperature at which the density of air is 1.18 g/L. As the air in the balloon is heated further, it expands and $9 \times 10^4$ L escapes out the open bottom of the balloon. What is the density of the heated air remaining inside the balloon?

---

***1.67. Density of Venus** The average density of Earth is 5.5 g/$cm^3$. The mass of Venus is 81.5% of Earth's mass, and the volume of Venus is 88% of Earth's volume. What is the density of Venus?

1.68. **Volume of Earth** Earth has a mass of $6.0 \times 10^{27}$ g and an average density of 5.5 g/$cm^3$.

a. What is the volume of Earth in cubic kilometers?

*b. Geologists sometimes express the "natural" density of Earth after doing a calculation that corrects for gravitational squeezing (compression of the core because of high pressure). Is Earth's natural density more or less than 5.5 g/$cm^3$?

---

***1.69. Utility Boats for the Navy** A plastic material called high-density polyethylene (HDPE) was once evaluated for use in impact-resistant hulls of small utility boats for the U.S. Navy. A cube of that material measures $1.20 \times 10^{-2}$ m on a side and has a mass of $1.70 \times 10^{-3}$ kg. Seawater at the surface of the ocean has a density of 1.03 g/$cm^3$. Will that cube float on water?

1.70. **Dimensions of the Sun** The sun is a sphere with an estimated mass of $2 \times 10^{30}$ kg. If the radius ($r$) of the sun is $7.0 \times 10^5$ km, what is the average density of the sun in units of grams per cubic centimeter? The volume of a sphere is $\frac{4}{3}\pi r^3$.

---

**1.71.** The Golden Jubilee diamond (Figure P1.71) has a mass of 545.67 carats (1 carat = 0.200 g). What is the mass of the diamond in (a) grams and (b) ounces? (1 pound = 16 ounces.)

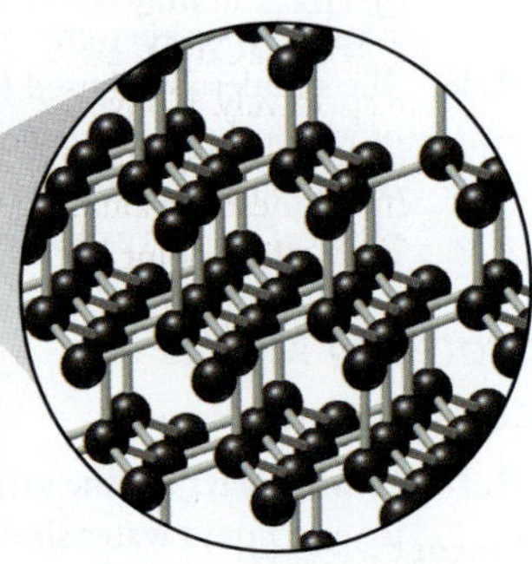

**FIGURE P1.71**

1.72. The density of diamond is 3.51 g/$cm^3$. What is the volume of the Golden Jubilee diamond in Figure P1.71?

**1.73.** Perform each of the following calculations, and express the answer with the correct number of significant figures:
a. $3.15 \times 2255 / 7.7 =$
b. $(6.7399 \times 10^{-18}) \times (1.0135 \times 10^{3}) / (52.67 + 0.144) =$
c. $(4.7 + 58.69)/(6.022 \times 10^{23} \times 6.864) =$
d. $(76.2 - 60.0)/[43.53 \times (9.988 \times 10^{4})] =$

1.74. Perform each of the following calculations, and express the answer with the correct number of significant figures:
a. $[(12 \times 60.0) + 55.3]/(5.000 \times 10^{3}) =$
b. $3.1416 \times (2.031)^2 \times 3.75 \times 8.00 =$
c. The number of cubic centimeters in 389 cubic inches
d. The average (mean) of 8.7, 8.5, 8.5, 8.9, and 8.8

**1.75.** **Topical Anesthetic** Ethyl chloride acts as a mild topical anesthetic because it chills the skin when sprayed on it. It dulls the pain of injury and is sometimes used to make removing splinters easier. The boiling point of ethyl chloride is 12.3°C. What is its boiling point on the Fahrenheit and Kelvin scales?

1.76. **Dry Ice** The temperature of the dry ice (solid carbon dioxide) in ice cream vending carts is −78°C. What is that temperature on the Fahrenheit and Kelvin scales?

**1.77.** **Record Low** The lowest temperature ever measured on Earth is −128.6°F, recorded at Vostok, Antarctica, in July 1983. What is that temperature on the Celsius and Kelvin scales?

1.78. **Record High** The highest temperature ever recorded in the United States is 134°F at Greenland Ranch, Death Valley, California, on July 13, 1913. What is that temperature on the Celsius and Kelvin scales?

**1.79.** **Critical Temperature** The discovery of "high temperature" superconducting materials in the mid-1980s spurred a race to prepare the material with the highest superconducting temperature. The critical temperatures ($T_c$)—the temperatures at which the material becomes superconducting—of $YBa_2Cu_3O_7$, $Nb_3Ge$, and $HgBa_2CaCu_2O_6$ are 93.0 K, −250.0°C, and −231.1°F, respectively. Convert those temperatures into a single temperature scale, and determine which superconductor has the highest $T_c$ value.

1.80. The boiling point of $O_2$ is −183°C, whereas the boiling point of $N_2$ is 77 K. As air is cooled, which gas condenses first?

## Analyzing Experimental Results

### Concept Review

**1.81.** How many suspect data points can be identified in a data set by using Grubbs' test?

1.82. Which confidence interval is the largest for a given value of $n$: 50%, 90%, or 95%?

**1.83.** The concentration of ammonia in an aquarium tank is determined each day for a week. Which of these measures of the variability in the results of those analyses is greater: (a) mean ± standard deviation or (b) 95% confidence interval? Explain your selection.

1.84. If the results of Grubbs' test indicate that a suspect data point is not an outlier at the 95% confidence level, could it be one at the 99% confidence level?

### Problems

***1.85.** The widths of copper lines in printed circuit boards must be close to a design value. Three manufacturers were asked to prepare circuit boards with copper lines that are 0.500 μm (micrometers) wide (1 μm = $1 \times 10^{-6}$ m). Each manufacturer's quality control department reported the following line widths on five sample circuit boards (given in micrometers):

| Manufacturer 1 | Manufacturer 2 | Manufacturer 3 |
|---|---|---|
| 0.512 | 0.514 | 0.500 |
| 0.508 | 0.513 | 0.501 |
| 0.516 | 0.514 | 0.502 |
| 0.504 | 0.514 | 0.502 |
| 0.513 | 0.512 | 0.501 |

a. What are the mean and the standard deviation of the data provided by each manufacturer?
b. For which of the three sets of data does the 95% confidence interval include 0.500 μm?
c. Which of the data sets fit the description "precise and accurate," and which is "precise but not accurate"?

1.86. **Diabetes Test** Glucose concentrations in the blood above 110 mg/dL can be an early indication of several medical conditions, including diabetes. Suppose analyses of several blood samples from a patient at risk of diabetes produce the following results: 106, 99, 109, 108, and 105 mg/dL.
a. What are the mean and the standard deviation of the data?
b. Patients with blood glucose levels above 120 mg/dL are considered diabetic. Is that value within the 95% confidence interval of those data?

**1.87.** Use Grubbs' test to decide whether the value 3.41 should be considered an outlier in the following data set from the analyses of portions of the same sample conducted by six groups of students: 3.15, 3.03, 3.09, 3.11, 3.12, and 3.41.

1.88. Use Grubbs' test to decide whether any one of the values in the following set of replicate measurements should be considered an outlier: 61, 75, 64, 65, 64, and 66.

## Additional Problems

***1.89.** **Agricultural Runoff** A farmer applies 1.50 metric tons of a fertilizer that contains 10% nitrogen to his fields each year (1 metric ton = 1000 kg). Fifteen percent of the fertilizer washes into a stream that runs through the farm. If the stream flows at an average rate of 1.4 cubic meters per minute, what is the additional concentration of nitrogen (expressed in milligrams of nitrogen per liter) in the stream water that results from the farmer's yearly application of fertilizer?

1.90. Your laboratory instructor has given you two shiny, light-gray metal cylinders (A and B). Your assignment is to determine which one is made of aluminum ($d = 2.699$ g/mL) and which one is made of titanium ($d = 4.54$ g/mL). The mass of each cylinder was determined on a balance to five significant figures. The volume of each was determined by immersing it in a partially filled graduated cylinder as shown in Figure P1.90.

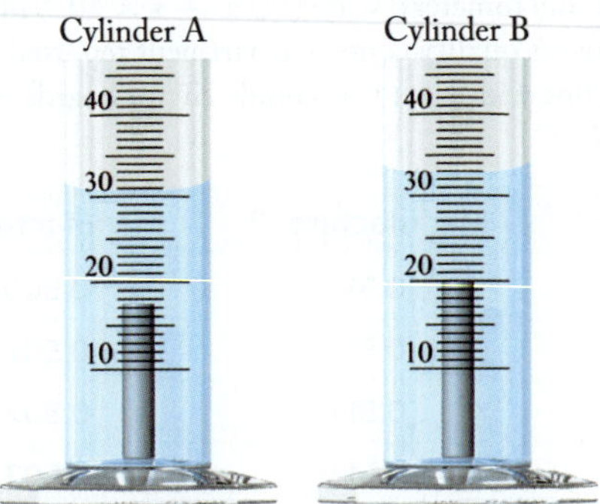


**FIGURE P1.90**

The initial volume of water was 25.0 mL in each graduated cylinder. The following data were collected:

| Cylinder | Mass (g) | Height (cm) | Inside Diameter (cm) |
|---|---|---|---|
| A | 15.560 | 5.1 | 1.2 |
| B | 35.536 | 5.9 | 1.3 |

a. Calculate the volume of each cylinder by using the dimensions of the cylinder only. The volume of a cylinder is $\pi r^2 h$, where $r$ is the radius and $h$ is the height.
b. Calculate the volume from the water displacement method.
c. Which volume measurement allows for more significant figures in the calculated densities?
d. Express the density of each cylinder to the appropriate number of significant figures.

***1.91.** Table salt contains 1.54 g of chlorine (as chloride ions) for every 1.00 g of sodium ions. Which of the following mixtures would react to produce NaCl with no sodium or chlorine left over?
a. 11.0 g of sodium and 17.0 g of chlorine
b. 6.5 g of sodium and 10.0 g of chlorine
c. 6.5 g of sodium and 12.0 g of chlorine
d. 6.5 g of sodium and 8.0 g of chlorine

*1.92. The wood of the black ironwood tree (*Krugiodendron ferreum*, Figure P1.92), which grows in the West Indies and coastal areas of South Florida, is so dense that it sinks in seawater. Does it sink in freshwater, too? Explain your answer.

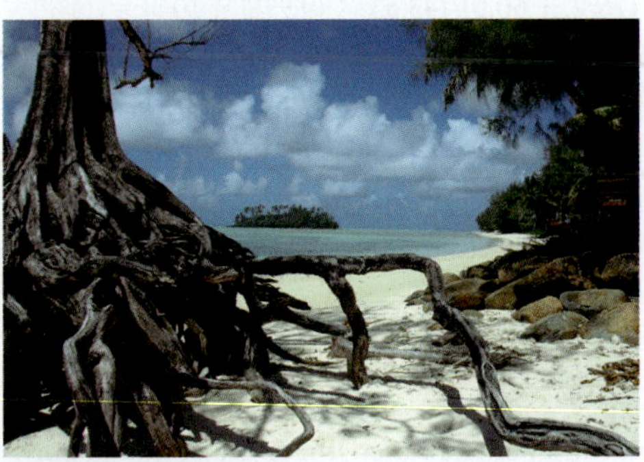

**FIGURE P1.92**

**1.93.** Manufacturers of trail mix have to control the distribution of ingredients in their products. Deviations of more than 2% from specifications may cause production delays and supply problems. A favorite trail mix is supposed to contain 67% peanuts and 33% raisins. Bags of trail mix were sampled from the production line on different days with the following results:

| Day | Number of Peanuts | Number of Raisins |
|---|---|---|
| 1 | 50 | 32 |
| 11 | 56 | 26 |
| 21 | 48 | 34 |
| 31 | 52 | 30 |

On which day(s) did the product meet the specification of 65% to 69% peanuts?

*1.94. Gasoline and water are immiscible. Regular-grade (87 octane) gasoline has a lower density (0.73 g/mL) than that of water (1.00 g/mL). A 100-mL graduated cylinder with an inside diameter of 3.2 cm contains 34.0 g of gasoline and 34.0 g of water. What is the combined height of the two liquid layers in the cylinder? The volume of a cylinder is $\pi r^2 h$, where $r$ is the radius and $h$ is the height.

**1.95.** **Drug Overdose** In 1999, an incident of drug overdose occurred when a prescription that called for a patient to receive 0.5 *grain* of the powerful sedative phenobarbital each day was misread, and the patient was given 0.5 gram of the drug. Actually, four intravenous injections of 130 mg each were administered each day for 3 days. How many times as much phenobarbital was administered with respect to the prescribed amount? (1 grain = 64.79891 mg.)

*1.96. **Mercury in Dental Fillings** The controversy over human exposure to mercury from dental fillings (Figure P1.96) is linked to concerns that mercury may volatilize from fillings made of a combination of silver and mercury and may then be breathed into the lungs and absorbed into the blood. In 1995 the U.S. Environmental Protection Agency set a safe exposure level for mercury vapor in air of 0.3 μg/$m^3$. Typically, an adult breathes in 0.5 L of air 15 times per minute.

a. What rate of volatilization of mercury (in micrograms/minute) from dental fillings would create an exposure level of 0.3 μg/$m^3$ in the air entering the lungs of an adult?
b. The safe exposure level to inhaled mercury vapor adopted by Health Canada is only 0.06 μg/$m^3$. What rate of volatilization of mercury (in micrograms/minute) from dental fillings would create that exposure level in air entering the lungs of a child who breathes in 0.35 L of air 18 times per minute?

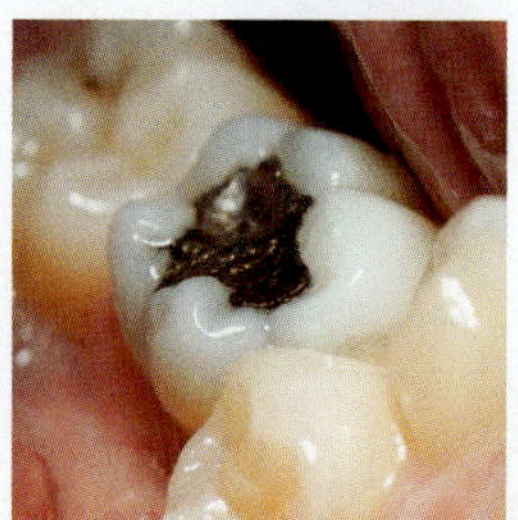

**FIGURE P1.96**

***1.97.** The accuracy of digital thermometers used in a hospital is evaluated by immersing them in an ice-water bath at 0.0°C and then in boiling water at 100.0°C. The following results were obtained for three thermometers (A, B, and C):

| | Measured Temperature (°C) | |
|---|---|---|
| Thermometer | Ice Water | Boiling Water |
| A | −0.8 | 99.4 |
| B | 0.2 | 99.8 |
| C | 0.4 | 101.0 |

a. Which of the three thermometers, if any, would detect an increase of 0.1°C in the temperature of a patient?
b. Which of the three thermometers, if any, would accurately give a reading of 36.8°C (under-the-tongue temperature) for a patient without a fever?

1.98. **Deepwater Horizon Oil Spill** According to the U.S. government, 4.9 billion barrels of crude oil flowed into the Gulf of Mexico after the explosion that destroyed the Deepwater Horizon drilling rig in April 2010. Express that volume of crude oil in liters and in cubic kilometers. (1 barrel of oil = 42 gallons.)

**1.99.** ***New Horizons*** As the *New Horizons* spacecraft approached Pluto (Figure P1.99) in June 2015, the craft was traveling away from the sun at 14.51 km/s. What was that speed in miles per hour?

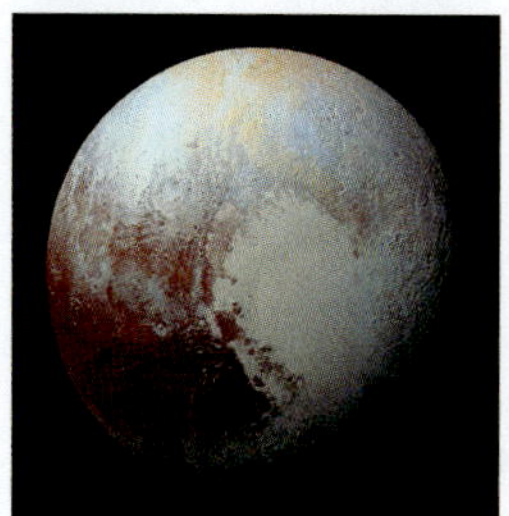

**FIGURE P1.99**

1.100. **Sodium in Candy Bars** Three analytical techniques were used to determine the quantity of sodium in a Mars Milky Way candy bar. Each technique was used to analyze five portions of the same candy bar, with the following results (expressed in milligrams of sodium per candy bar):

| Technique 1 | Technique 2 | Technique 3 |
|---|---|---|
| 109 | 110 | 114 |
| 111 | 115 | 115 |
| 110 | 120 | 116 |
| 109 | 116 | 115 |
| 110 | 113 | 115 |

The actual quantity of sodium in the candy bar was 115 mg. Which techniques would you describe as precise, which as accurate, and which as both? What is the range of the values (the difference between the highest and lowest measurements) for each technique?

# 2

# Atoms, Ions, and Molecules

## The Building Blocks of Matter

**MAGNETIC RESONANCE IMAGING** Magnetic resonance imaging (MRI) is used extensively to diagnose injuries and diseases in soft tissue. MRI signals are produced by the protons in the nuclei of the hydrogen atoms in molecules of $H_2O$.

PARTICULATE **REVIEW**

### *Atoms and Molecules*

In Chapter 2, we explore the structure of atoms and begin our discussion of why they form molecules and ions. The decomposition of hydrogen peroxide produces water and oxygen, as shown in the two images here.

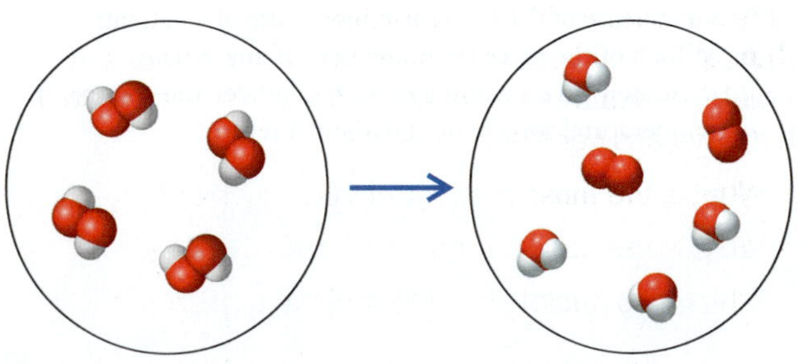

- How many atoms are in one molecule of hydrogen peroxide?
- How many oxygen molecules are depicted in the two images?
- What is the ratio of hydrogen atoms to oxygen atoms in hydrogen peroxide?
- What is the ratio of hydrogen atoms to oxygen atoms in water?

(Review Sections 1.1, 1.3, and 1.4 if you need help. This book's inside back cover shows the colors that represent the atoms of different elements.)

*(Answers to Particulate Review questions are in the back of the book.)*

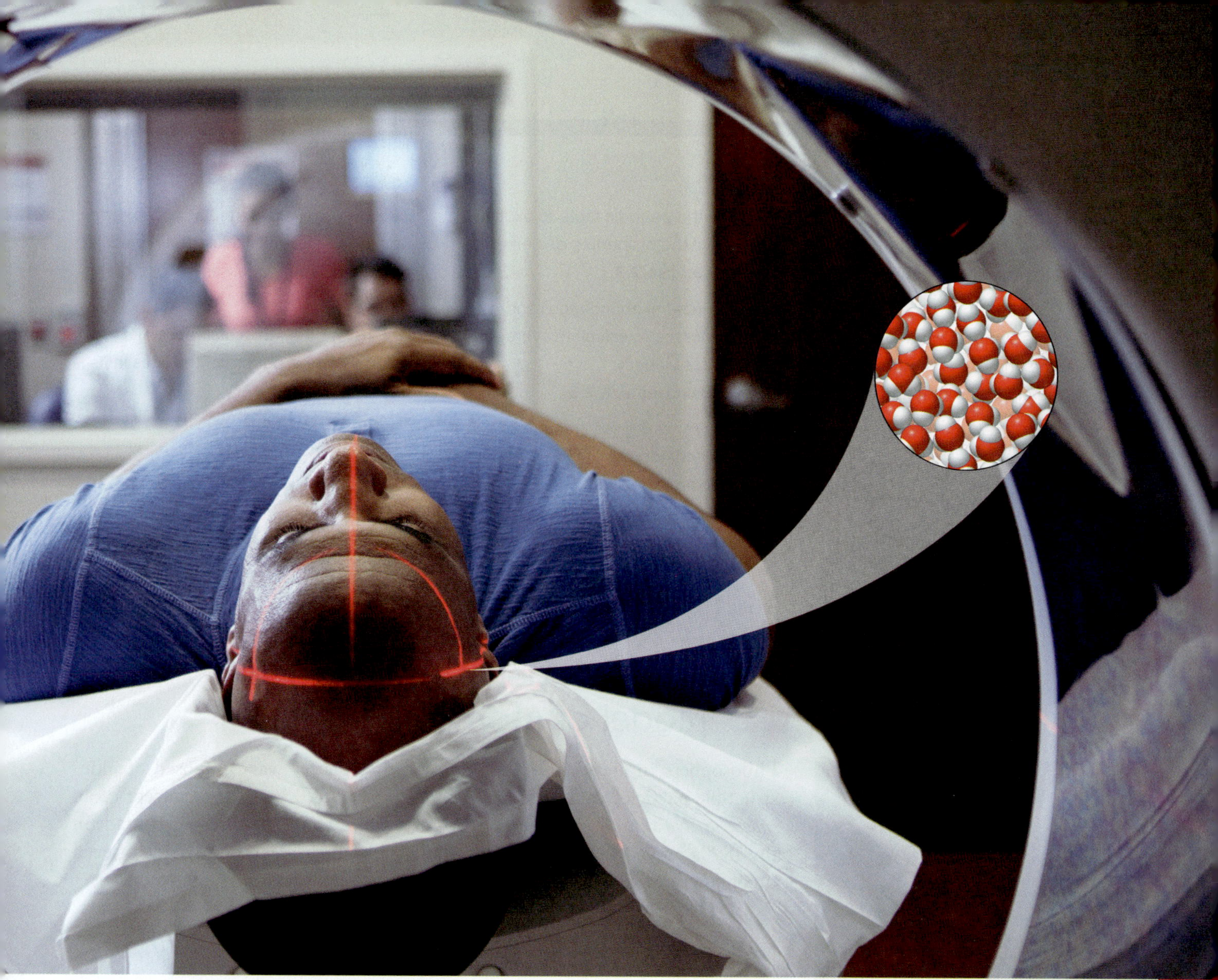

PARTICULATE **PREVIEW**

### *Nuclei and Nucleons*

Three atomic nuclei are depicted here. Assume that all the particles that make up each nucleus are visible. As you read Chapter 2, look for ideas that will help you answer these questions:

- What is the mass number of each nucleus?
- What is the atomic number of each nucleus?
- Which two nuclei are isotopes of each other?

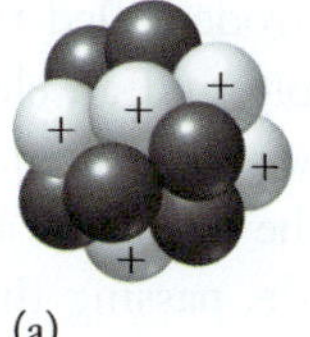
(a)

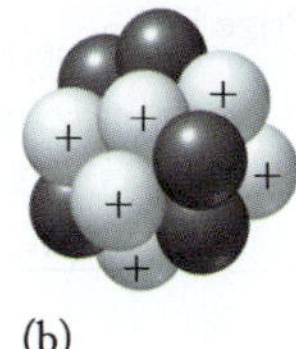
(b)

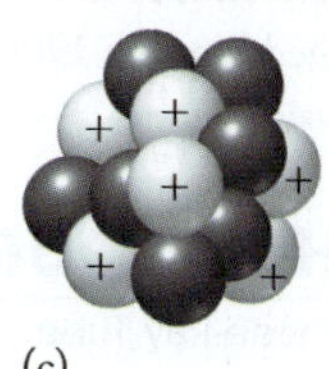
(c)

## Learning Outcomes

**LO1** Explain how the experiments of Thomson, Millikan, and Rutherford contributed to our understanding of atomic structure

**LO2** Relate the symbols used to describe nuclides to the number of protons, neutrons, and electrons in atoms and monatomic ions
**Sample Exercises 2.1, 2.2**

**LO3** Use the periodic table to explain and predict the properties of elements
**Sample Exercise 2.3**

**LO4** Use natural abundances and isotopic masses to calculate average atomic mass values
**Sample Exercises 2.4, 2.5**

**LO5** Interconvert mass, number of particles, and number of moles
**Sample Exercises 2.6–2.9**

**LO6** Use chemical formulas and average atomic masses to calculate the molecular masses and molar masses of compounds
**Sample Exercise 2.10**

**LO7** Relate the molecular mass of a molecule to the pattern of the lines in its mass spectrum
**Sample Exercise 2.11**

# 2.1 When Projectiles Bounced Off Tissue Paper: The Rutherford Model of Atomic Structure

Philosophers in ancient Greece proposed that matter was composed of indestructible particles called atoms. For more than 2000 years, the concept of an *indestructible* atom also meant an *indivisible* atom. However, discoveries made in the 1890s proved that atoms are not indivisible. Instead, they are made of even tinier **subatomic particles**: neutrons, protons, and electrons.

**FIGURE 2.1** In 1897, J. J. Thomson discovered electrons when he studied how gases conduct electricity. The research earned him the 1906 Nobel Prize in Physics.

**CHEMTOUR**
Cathode-Ray Tube

**subatomic particles** the neutrons, protons, and electrons in an atom.

**cathode rays** streams of electrons emitted by the cathode in a partially evacuated tube.

**electron** a subatomic particle that has a negative charge and negligible mass.

## Electrons

Discovery of the first subatomic particle occurred in 1897 in the laboratory of English physicist Joseph John (J. J.) Thomson (1856–1940; **Figure 2.1**). During the 1890s, Thomson's research included investigations of how gases conduct electricity. In that work he used a device called a cathode-ray tube, or CRT (**Figure 2.2**), which consists of a glass tube from which most of the air has been removed. Electrodes within the tube are attached to the poles of a high-voltage power supply. An electrode called the cathode is connected to the negative terminal of the power supply, and an electrode called the anode is connected to the positive terminal. When those connections are made, electricity travels the length of the glass tube in the form of a beam of **cathode rays** that flows from the cathode toward the anode, passing through a hole cut in the center of the anode. Cathode rays are invisible to the naked eye, but when the end of the tube is coated with a phosphorescent material, a glowing spot appears where the beam hits the coating.

Thomson discovered that cathode-ray beams could be deflected by magnetic (Figure 2.2a) and electric (Figure 2.2b) fields. That behavior revealed that cathode rays were not rays of energy but rather charged particles of matter. The directions of the deflections meant that their charges were negative. By adjusting the strengths of the electric and magnetic fields, Thomson balanced out the deflections (Figure 2.2c) to force the particles to pass straight through the tube. From the strengths of the opposing electric and magnetic fields, he calculated the mass-to-charge ratio of the particles. He also observed that the deflection pattern and the calculated mass-to-charge ratio were always the same no matter what cathode material he used to generate the beam of particles. That observation convinced Thomson that those particles, which he called *corpuscles* but which we now know as **electrons**, were fundamental particles that occur in all forms of matter.

In 1909, American physicist Robert Millikan (1868–1953) advanced Thomson's work by determining the charge of an electron and, indirectly, its mass. **Figure 2.3** illustrates Millikan's experimental apparatus. It consisted of two chambers filled with air (that is, mostly $N_2$ and $O_2$). A fine spray of oil drops produced in the top chamber fell through a hole into the bottom chamber. Highly energetic X-rays also passed through the bottom chamber, colliding with the $N_2$ and $O_2$ molecules.

The X-rays removed electrons ($e^-$) from the $N_2$ and $O_2$ molecules, generating *molecular ions* with a positive charge:

$$N_2 \rightarrow N_2^+ + e^- \qquad (2.1)$$

$$O_2 \rightarrow O_2^+ + e^- \qquad (2.2)$$

The superscripts in Equations 2.1 and 2.2 indicate the electrical charges on the nitrogen molecular ion (1+), the oxygen molecular ion (1+), and the electron (1−).

Particles of matter not only can lose electrons; they also can gain electrons. As Millikan's oil drops fell through the bottom chamber, they collided with and absorbed free electrons, thereby acquiring a negative charge. He observed the rate of fall of the negatively charged oil drops with a microscope through the side of the bottom chamber. Because like charges repel one another and opposite charges attract, Millikan could adjust the drops' rate of fall by using charged metal plates located both above and below the bottom chamber, thereby creating a vertical electric field within it. By measuring the strength of the electric field and the rate

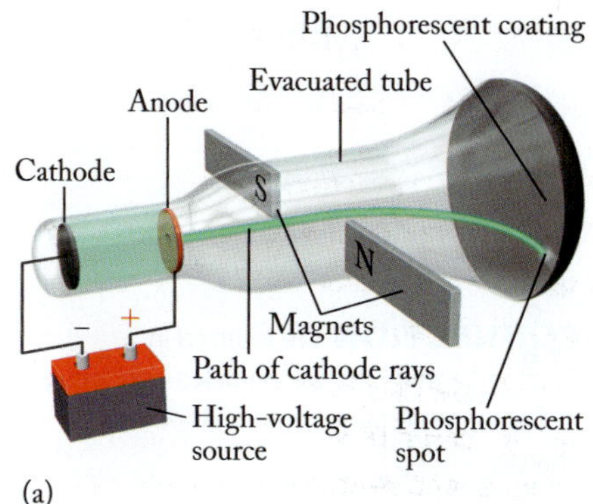


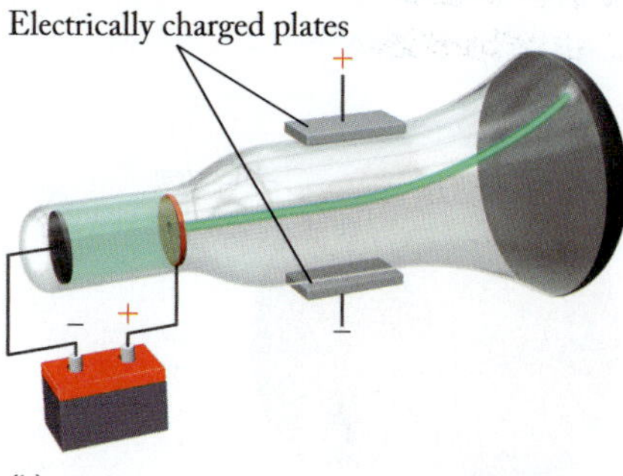


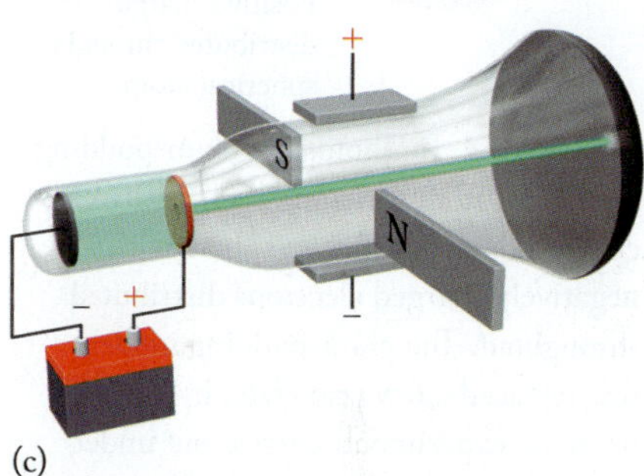


**FIGURE 2.2** A cathode "ray"—actually a beam of electrons—carries electricity through this partially evacuated tube. Though invisible, the path of the beam can be inferred by the bright spot it makes on a phosphorescent coating on the end of the tube. (a) The beam is deflected in one direction by a magnetic field, (b) deflected in the opposite direction by an electric field, and (c) not deflected at all if the electric and magnetic fields are tuned to balance out the deflections.

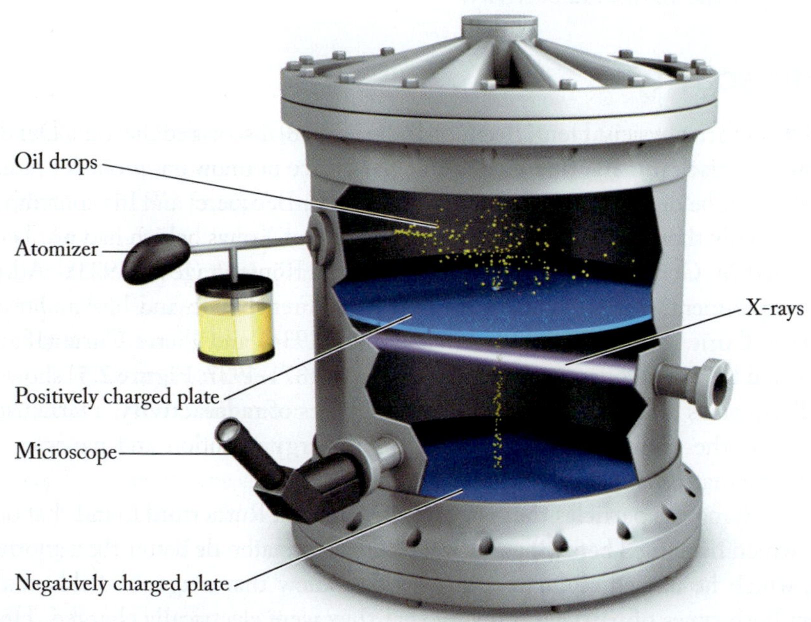


**FIGURE 2.3** In Millikan's oil-drop experiment, X-rays ionized the air in the lower chamber, producing electrons absorbed by tiny drops of oil falling through the chamber. The descent of the electrically charged drops could be slowed, stopped, or even reversed by applying an electric field with charged plates above and below the chamber. A microscope allowed Millikan to follow the descent of the oil drops.

**CHEMTOUR**
Millikan Oil-Drop Experiment

of fall of the drops, Millikan could calculate the charges on the drops. He discovered that the charge on each drop was always a whole-number multiple of a minimum value, and he concluded that this minimum value must be the charge on one electron. Millikan's experiments and calculations yielded a value for the charge of an electron ($e$) that is within 1% of today's experimentally determined value. By using Thomson's value of the electron's mass-to-charge ratio, Millikan calculated the electron's mass ($m_e$). We can reproduce Millikan's calculation by using today's experimentally determined value for an electron's charge ($-1.602 \times 10^{-19}$ C, where C is the abbreviation for *coulomb*, the SI unit of charge) and the electron's mass-to-charge ratio, which is $-5.686 \times 10^{-12}$ kg/C:

$$m_e = e \times (m_e/e) = (-1.602 \times 10^{-19} \cancel{\text{C}}) \times (-5.686 \times 10^{-12} \text{ kg}/\cancel{\text{C}})$$
$$= 9.109 \times 10^{-31} \text{ kg}$$

**CONNECTION** We learned in Section 1.3 that ions are particles consisting of one or more atoms with an electrical charge.

CONCEPT **TEST**

By adjusting the electrical charges on the top and bottom plates of the lower chamber of his apparatus, Millikan could slow, stop, or even reverse the fall of oil drops. Which of the two plates must have been positively charged?

*(Answers to Concept Tests are in the back of the book.)*

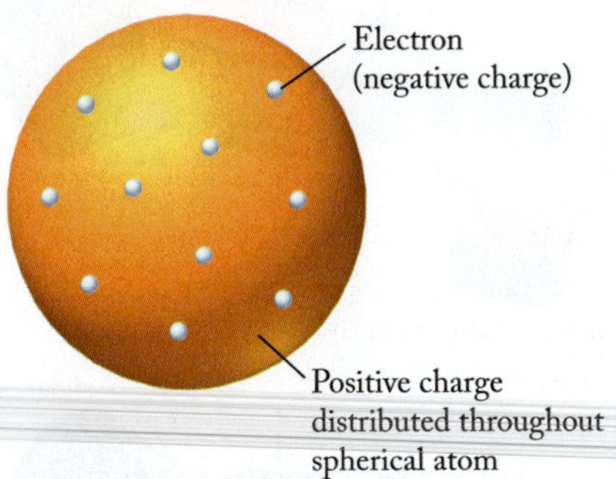


**FIGURE 2.4** In Thomson's plum-pudding model, an atom was a massive, positively charged, but very diffuse sphere with negatively charged electrons distributed throughout. The plum-pudding model was replaced a few years later by a model based on experiments carried out under the direction of Thomson's former student, Ernest Rutherford.

Scientists knew that matter was electrically neutral, so the discovery of the electron raised the possibility that positively charged subatomic particles must exist, too. At the time, however, scientists didn't know where the electrons and positive charges were located inside atoms. Thomson proposed a model of the atom (**Figure 2.4**) in which electrons were distributed throughout the atom like raisins in an English plum pudding (or blueberries in a muffin). Thomson's "plum pudding" model was replaced after only a few years as a result of another scientific discovery of the 1890s: radioactivity.

## Radioactivity

In 1896, French physicist Henri Becquerel (1852–1908) discovered that pitchblende, a brownish-black mineral that is the principal source of uranium, produces radiation that can be detected using photographic plates. Becquerel and his contemporaries initially thought that this radiation consisted of X-rays, which had just been discovered by German scientist Wilhelm Conrad Röntgen (1845–1923).[1] Additional experiments by Becquerel, by the Polish and French wife-and-husband team of Marie Curie (born Marie Skłodowska, 1867–1934) and Pierre Curie (1859–1906), and by British scientist Ernest Rutherford (1871–1937; **Figure 2.5**) showed that Becquerel's radiation was actually several types of **radioactivity**, a term used to describe the spontaneous emission of high-energy radiation and particles by radioactive materials such as pitchblende.

**FIGURE 2.5** Ernest Rutherford was born in New Zealand. He was awarded a scholarship in 1894 to attend Trinity College in Cambridge, England, where he was a research assistant in the laboratory of J. J. Thomson. Rutherford's contributions include characterizing the properties of α and β particles. By 1907 he was a professor at the University of Manchester, where his gold-foil experiments led to our modern view of atomic structure. He was awarded the Nobel Prize in Chemistry in 1908.

In studying the particles that pitchblende emitted, Rutherford found that one type, which he named **beta (β) particles**, penetrated materials better than another type, which he named **alpha (α) particles**. He knew that magnetic fields could deflect both types of particles, proving that they were electrically charged. How

[1]Röntgen discovered X-rays in experiments with a cathode-ray tube much like the apparatus used by J. J. Thomson. After encasing the tube in a black carton, Röntgen discovered that invisible rays had escaped the carton and could be detected by a photographic plate. Because he knew so little about the rays, he called them X-rays.

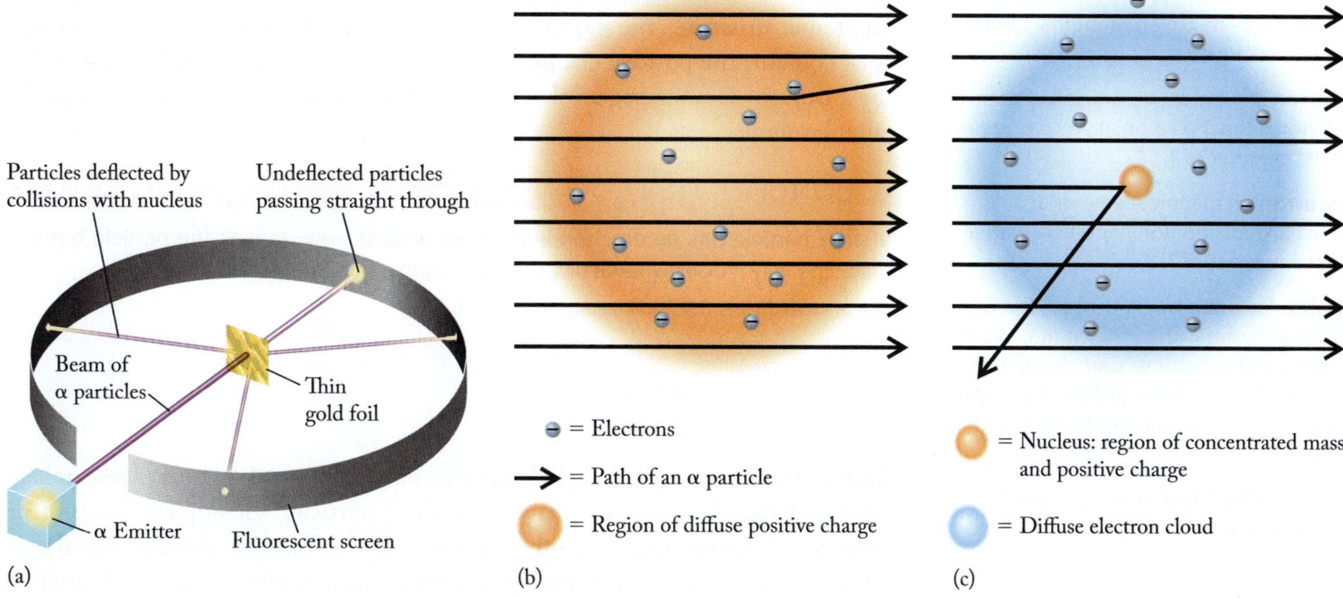


**FIGURE 2.6** (a) The design of Rutherford's gold-foil experiments. (b) If Thomson's plum-pudding model had been correct, most of the α particles would have passed straight through the thin gold foil, though a few might have been deflected slightly. (c) In fact, most did pass straight through, but a few were scattered widely as shown in (a). That unexpected result led to the theory that an atom has a small, positively charged nucleus that contains most of the atom's mass.

much they were deflected by fields of different strengths allowed him to calculate the particles' mass-to-charge ratios. He found that the ratio for β particles matched the mass-to-charge ratio of the electron that Thomson had determined, which suggested that β particles were simply high-energy electrons.

The direction in which α particles are deflected in a magnetic field is opposite that of β particles, which means α particles are positively charged. If the β particle (an electron) is assigned a relative charge of 1−, the corresponding charge of an α particle is 2+. Rutherford also discovered that α particles are about 10,000 times more massive than β particles.

In 1909, Rutherford directed two of his students at Manchester University—Hans Geiger (1882–1945, for whom the Geiger counter was named) and Ernest Marsden (1889–1970)—to test Thompson's plum-pudding model by bombarding a thin foil of gold with a beam of α particles (**Figure 2.6a**). If the plum-pudding model was correct, most of the particles would pass straight through the diffuse spheres of positive charge that made up the gold atoms, though a few might interact with the electrons (the "raisins" embedded in the pudding) enough to be deflected slightly (**Figure 2.6b**).

Geiger and Marsden observed that most of the α particles passed straight through the gold, as Rutherford expected. However, about 1 in every 8000 particles was deflected by an average angle of 90° (**Figure 2.6c**), and a very few bounced almost straight back at their source. Rutherford later described his amazement at the result: "It was almost as incredible as if you had fired a 15-inch shell[2] at a piece of tissue paper and it came back and hit you."

Thomson's plum-pudding model could not explain such large angles of deflection. Rutherford concluded that the deflections occurred because a tiny fraction of the α particles encountered small regions of high positive charge and large mass. On the basis of the relative numbers of α particles that were deflected in that way, Rutherford determined that the diameter of a gold atom was more than 10,000 times greater than the diameter of the region of positive charge at its

[2]The most widely used "heavy" guns on British battleships in the early 20th century fired shells with a diameter of 15 inches (38 cm).

**CHEMTOUR**
Rutherford Experiment

**radioactivity** the spontaneous emission of high-energy radiation and particles by materials.

**beta (β) particle** a particle that is emitted during radioactive decay and is equivalent in mass and charge to a high-energy electron.

**alpha (α) particle** a particle that is emitted during radioactive decay and is equivalent in mass and charge to a $^4$He nucleus.

**nucleus** (of an atom) the positively charged center of an atom that contains nearly all the atom's mass.

**proton** a subatomic particle in the nuclei of atoms that has a positive charge and a mass number of 1.

**neutron** an electrically neutral (uncharged) subatomic particle with a mass number of 1.

**unified atomic mass unit (u)** the unit used to express the relative masses of atoms and subatomic particles that is exactly 1/12 the mass of 1 atom of carbon with 6 protons and 6 neutrons in its nucleus.

**dalton (Da)** a unit of mass equal to 1 unified atomic mass unit.

**CONNECTION** In Section 1.1, we learned about John Dalton's experiments that led to the law of multiple proportions.

center. Rutherford's model of the atom is the basis for our current understanding of atomic structure. The model assumes that an atom consists of a tiny **nucleus** that contains the positive charge and most of the atom's mass and is surrounded by a diffuse cloud of negatively charged electrons that accounts for most of the atom's volume.

**CONCEPT TEST**

If an α particle hits an electron in an atom of gold, why doesn't the particle bounce back the way it does when it hits the nucleus of a gold atom?

*(Answers to Concept Tests are in the back of the book.)*

## The Nuclear Atom

In the decade after the gold-foil experiments, Rutherford and other researchers observed that bombarding elements with α particles sometimes changed, or *transmuted*, the elements into other elements. The researchers also discovered that hydrogen nuclei were often produced during transmutation reactions. By 1920, a consensus was growing that hydrogen nuclei, which Rutherford called **protons** (from the Greek *protos,* meaning "first"), were part of all nuclei. For example, to account for the mass and charge of an α particle, Rutherford assumed that it was made of four protons, two of which had combined with two electrons to form two electrically neutral particles, which he called **neutrons**. Repeated attempts to produce neutrons by neutralizing protons with electrons were unsuccessful. However, in 1932 one of Rutherford's former students, James Chadwick (1891–1974), was the first to successfully detect and characterize free neutrons. With the discovery of neutrons, the current model of atomic structure was complete, as illustrated by the gold atom in **Figure 2.7**.

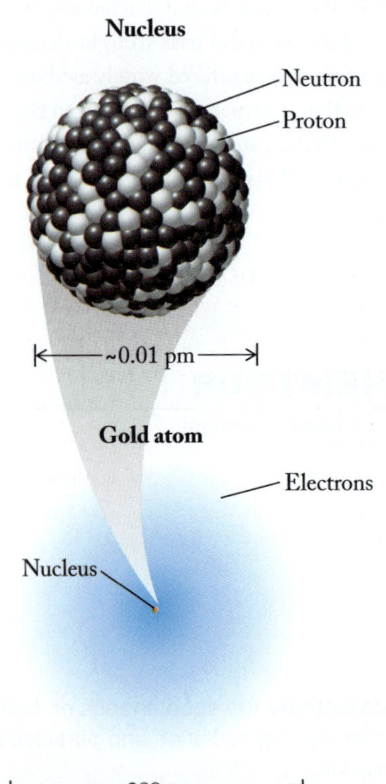


**FIGURE 2.7** The modern view of Rutherford's model of the gold atom includes a nucleus that is about 1/10,000 the overall size of the atom. The nucleus would be too small to see if drawn to scale in the lower drawing.

**Table 2.1** summarizes the properties of neutrons, protons, and electrons. Their masses are expressed in kilograms (the SI standard unit of mass) and in **unified atomic mass units (u)**. One u is exactly 1/12 the mass of a carbon atom that has 6 protons and 6 neutrons in its nucleus. Unified atomic mass units are also called **daltons (Da)** to honor John Dalton, who published the first table of atomic masses in 1803. Dividing the mass of any of the particles in kilograms by its mass in u yields the result that 1 u is equivalent to $1.66054 \times 10^{-27}$ kg. We use that equality in several calculations later in this chapter.

According to the data in Table 2.1, the masses of neutrons and protons are similar and are much greater than the mass of an electron. The relative masses of those particles are reflected in their *mass numbers*, which are not real mass values but instead reflect the number of subatomic particles in an atom. Neutrons and protons, which make up nearly all of an atom's mass, each have a mass number of 1. Because the mass of an electron is negligible in comparison with the mass of a proton or neutron, the electron is assigned a mass number of 0, reflecting its minimal contribution to the mass of an atom.

Rutherford's pioneering work marked the beginning of decades of research on the composition and properties of atomic nuclei. For example, in the 1930s it was discovered that atoms with nuclei that have odd numbers of protons and/or neutrons can absorb and reemit tiny quantities of energy when placed inside magnetic fields. That discovery led to the development of a technique for the medical imaging of soft tissue known as magnetic resonance imaging, or MRI (see the chapter-opening figure).

**TABLE 2.1** Properties of Subatomic Particles

| Particle | Symbol | Mass (u) | Mass Number | Mass (kg) | Charge (relative value) | Charge (C) |
|---|---|---|---|---|---|---|
| Neutron | ${}^{1}_{0}\text{n}$ | 1.00866 | 1 | $1.67493 \times 10^{-27}$ | 0 | 0 |
| Proton | ${}^{1}_{1}\text{p}$ | 1.00728 | 1 | $1.67262 \times 10^{-27}$ | 1+ | $+1.60218 \times 10^{-19}$ |
| Electron | ${}^{0}_{-1}\text{e}$ | $5.48580 \times 10^{-4}$ | 0 | $9.10938 \times 10^{-31}$ | 1− | $-1.60218 \times 10^{-19}$ |

## 2.2 Nuclides and Their Symbols

As Thomson continued to experiment with cathode-ray tubes, he modified them so that he could study the beams of positively charged particles that flowed from the anode toward the cathode as cathode rays (electrons) flowed in the opposite direction. In 1912, Thomson and his research assistant, Francis W. Aston (1877–1945), observed that when they passed an electric current through a tube that contained small quantities of neon gas, two bright patches formed on a photographic plate, as shown in **Figure 2.8**. The combination of electric and magnetic fields surrounding the tubes of those positive-ray analyzers had deflected positively charged ions according to their charges and masses. For example, if all the ions in a particular beam had a charge of 1+ but different masses, the ions with the greatest mass would be deflected the least, and those with the smallest mass would be deflected the most.

Thomson and Aston assumed that the patches were produced by neon atoms that had lost electrons in the tube, forming positively charged $Ne^+$ ions. The presence of two patches meant that the $Ne^+$ ions—and, by extension, their parent Ne atoms—had different masses. The ions with lower mass produced the brighter patch, indicating that more of them hit the plate and that they were more abundant than the ions with higher mass.

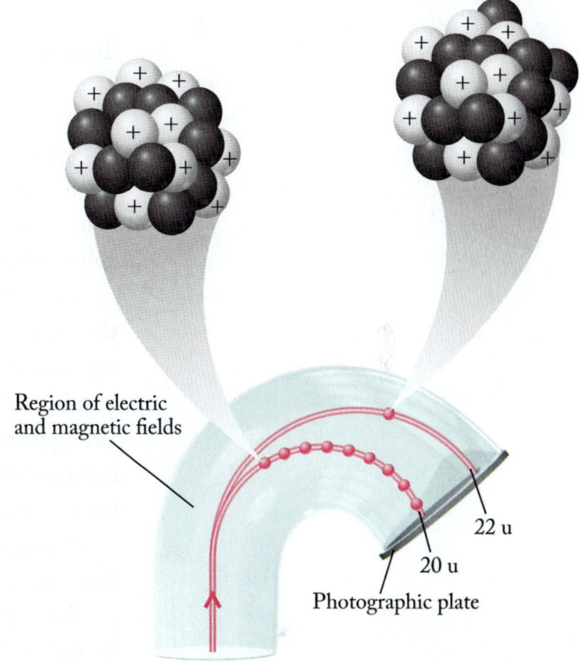


**FIGURE 2.8** Aston's positive-ray analyzer. A beam of $Ne^+$ ions passing through electric and magnetic fields separates into two beams. Ions with a mass of about 20 u—90% of the total—are deflected more than the other 10%, which have a mass of about 22 u. Aston's positive-ray analyzer was the forerunner of the modern mass spectrometer.

Today we know that the $Ne^+$ ions detected by Thomson and Aston came from two **isotopes** of neon. Isotopes are atoms of the same element that have the same number of protons (10 for neon) in their nuclei but different numbers of neutrons. The lighter Ne isotope has 10 neutrons per nucleus, giving it a total mass of about 20 u. Atoms of the less abundant isotope have 12 neutrons in their nuclei, giving them a mass of about 22 u. The term **nuclide** refers to any atom of any element that has a particular number of neutrons in its nucleus.

Remember that since the time of John Dalton, scientists had defined an element as *matter composed of identical atoms, all of which have the same mass.* The work of Thomson, Aston, Chadwick, and others required that definition to be modified: an element was defined as *matter composed of atoms all having the same number of protons in their nuclei.* That number of protons is called the **atomic number (*Z*)** of the element. The **mass number (*A*)** indicates the total number of **nucleons** (neutrons and protons) in the nucleus of an atom. The isotopes of a given element thus all have the same atomic number, *Z*, but different mass numbers, *A*.

The symbol we use to represent a particular nuclide has the generic form

$$ {}^{A}_{Z}\text{X} $$

**isotopes** atoms of an element containing the same number of protons but different numbers of neutrons.

**nuclide** a specific isotope of an element.

**atomic number (*Z*)** the number of protons in the nucleus of an atom.

**mass number (*A*)** the number of nucleons in an atom.

**nucleon** a proton or neutron in a nucleus.

**STEPWISE ANIMATION**
Parts of the Atom

where X represents the one- or two-letter symbol for the element, $Z$ is the element's atomic number, and $A$ is the mass number of the particular nuclide. For example, two of the isotopes of oxygen (O) and two of the isotopes of lead (Pb) are written

$$^{16}_{8}\text{O} \qquad ^{18}_{8}\text{O} \qquad ^{206}_{82}\text{Pb} \qquad ^{208}_{82}\text{Pb}$$

Because $Z$ and X provide the same information—each by itself identifies the element—the subscript $Z$ may be omitted, so that the nuclide symbol may be simplified to

$$^{A}\text{X}$$

Therefore, the nuclides of oxygen and lead shown above also can be written as

$$^{16}\text{O} \qquad ^{18}\text{O} \qquad ^{206}\text{Pb} \qquad ^{208}\text{Pb}$$

The name of a nuclide may also be spelled out as the name of the element followed by the mass number of the nuclide. For example, the names of the two isotopes of neon detected by Thomson and Aston may be written *neon-20* and *neon-22* or abbreviated as *Ne-20* and *Ne-22*, respectively.

Note that the nuclide symbol does not have a feature that explicitly represents the number of electrons. However, in a neutral atom, the number of electrons equals the number of protons, so the number of electrons is equal to the atomic number. If an atom becomes an ion, we can write its nuclide by using the symbol $^{A}\text{X}^{Q}$, where $Q$ represents the positive or negative charge on the ion. We can then compare the charge of the ion with its atomic number, $Z$ (which equals the number of protons), to determine how many electrons are in the ion. For example, the most common ion of oxygen has a charge of 2− and the nuclide symbol $^{16}\text{O}^{2-}$. Because oxygen has an atomic number of 8, that ion has 8 + 2 = 10 electrons.

The $^{A}_{Z}\text{X}$ symbol can also be used to represent subatomic particles, as shown in Table 2.1. As with nuclides, the superscripts of the symbols are the particles' mass numbers. The subscripts, however, represent the relative charges of the particles: 0 for a neutron, 1 for a proton, and 1− for an electron. That notation for the subscripts makes sense because an element's atomic number, $Z$, is simply the number of protons in the nucleus of an atom of that element, which is the positive charge of that nucleus.

**SAMPLE EXERCISE 2.1** Relating the Symbols of Nuclides to the Composition of Their Nuclei **LO2**

Several nuclides of gaseous elements are useful for the MRI of pulmonary (lung) function. Write symbols in the form $^{A}_{Z}\text{X}$ for the nuclides that have (a) 2 protons and 1 neutron, (b) 36 protons and 47 neutrons, and (c) 54 protons and 75 neutrons.

**Collect, Organize, and Analyze** We are given the number of protons in the nucleus of an atom of element X, which defines its atomic number ($Z$) and defines its one- or two-letter symbol. The sum of the number of nucleons (protons plus neutrons) is the mass number ($A$) of the nuclide. We are asked to write symbols in the form $^{A}_{Z}\text{X}$.

**Solve**

a. The nuclide has 2 protons, so $Z = 2$, the atomic number of helium. Two protons plus 1 neutron gives a mass number of 3. Therefore, the nuclide is helium-3, or $^{3}_{2}\text{He}$.

b. The nuclide has 36 protons, so $Z = 36$, the atomic number of krypton. Thirty-six protons plus 47 neutrons gives a mass number of 83. The isotope is krypton-83, or $^{83}_{36}Kr$.

c. The nuclide has 54 protons, so $Z = 54$, the atomic number of xenon. The mass number is $54 + 75 = 129$, so the isotope is xenon-129, or $^{129}_{54}Xe$.

**Think About It** Remember that in a neutral atom, the number of electrons equals the number of protons. $A$ will always be equal to or greater than $Z$ because of the presence of neutrons, except for the $^{1}_{1}H$ nuclide (which has no neutrons in its nucleus).

**Practice Exercise** Use the $^{A}_{Z}X$ format to write the symbols of the nuclides whose atoms each have (a) 26 protons and 30 neutrons, (b) 7 protons and 8 neutrons, (c) 17 protons and 20 neutrons, and (d) 19 protons and 20 neutrons.

*(Answers to Practice Exercises are in the back of the book.)*

## SAMPLE EXERCISE 2.2 Identifying Ionic Nuclides and Writing Their Isotopic Symbols — LO2

In March 2011, an earthquake and tsunami crippled nuclear reactors at a power station in Fukushima, Japan. The resulting explosions and fires released radiation into the atmosphere and ocean. That radiation included two ionic nuclides, $^{134}Cs^{+}$ and $^{131}I^{-}$, and two other single-atom ions, labeled (c) and (d) in the table. Complete the table by determining the numbers of protons, neutrons, and electrons in each $^{134}Cs^{+}$ ion and $^{131}I^{-}$ ion and by writing the $^{A}X^{Q}$ symbols of ionic nuclides (c) and (d), where $Q$ is the charge of each ion.

| | Symbol | Protons | Neutrons | Electrons |
|---|---|---|---|---|
| **(a)** | $^{134}Cs^{+}$ | | | |
| **(b)** | $^{131}I^{-}$ | | | |
| **(c)** | | 55 | 82 | 54 |
| **(d)** | | 94 | 145 | 90 |

**Collect, Organize, and Analyze** We need to connect the numbers of neutrons, protons, and electrons to the $^{A}X^{Q}$ symbols for four single-atom ions. The number of protons in the nucleus of an atom or monatomic ion defines its atomic number, which in turn determines the identity of the element (X). The sum of the number of protons plus neutrons is the mass number ($A$), and the charge ($Q$) is the difference between the number of protons and the number of electrons.

**Solve**

a. Cs is the symbol of cesium, which has an atomic number of 55. Therefore, the nucleus of a $^{134}Cs^{+}$ ion has 55 protons and $134 - 55 = 79$ neutrons. The ion has a 1+ charge, which means it has one fewer electron than protons, or 54 electrons.

b. I is the symbol of iodine, whose atomic number is 53. Therefore, the nucleus of a $^{131}I^{-}$ ion has 53 protons and $131 - 53 = 78$ neutrons. The charge on the ion is 1−, which means it has one more electron than protons, or $53 + 1 = 54$ electrons.

c. The ion has 55 protons, which makes it cesium (Cs). Its mass number ($A$) is $55 + 82 = 137$, and its charge is $55 - 54 = 1+$. The charge symbol of an ion with a single positive charge is simply +, so the symbol of the ion is $^{137}Cs^{+}$.

d. The ion has 94 protons, which makes it plutonium (Pu). Its mass number ($A$) is $94 + 145 = 239$, and its charge is $94 - 90 = 4+$. The symbol of the ion is $^{239}Pu^{4+}$.

Use those results to fill in the table:

| | Symbol | Protons | Neutrons | Electrons |
|---|---|---|---|---|
| (a) | $^{134}Cs^{+}$ | 55 | 79 | 54 |
| (b) | $^{131}I^{-}$ | 53 | 78 | 54 |
| (c) | $^{137}Cs^{+}$ | 55 | 82 | 54 |
| (d) | $^{239}Pu^{4+}$ | 94 | 145 | 90 |

**Think About It** The value of $Q$ is the sum of the positive charges in the nucleus and the negative charges of the electrons surrounding the nucleus. If an atom or ion has more electrons than protons, $Q$ has a negative value; if fewer electrons than protons are present, $Q$ has a positive value; and when numbers of electrons and protons are equal, $Q$ is zero and the particle is a neutral atom.

**Practice Exercise** Complete the table below by determining the number of protons, neutrons, and electrons or by writing the $^{A}X^{Q}$ symbol for each of these four ions present in seawater.

| | Symbol | Protons | Neutrons | Electrons |
|---|---|---|---|---|
| (a) | $^{40}Ca^{2+}$ | | | |
| (b) | $^{79}Br^{-}$ | | | |
| (c) | | 17 | 18 | 18 |
| (d) | | 11 | 12 | 10 |

*(Answers to Practice Exercises are in the back of the book.)*

## 2.3 Navigating the Periodic Table

**periodic table of the elements** a chart of the elements in order of their atomic numbers and in a pattern based on their physical and chemical properties.

**period** (of elements) all the elements in a row of the periodic table.

**group** or **family** (of elements) all the elements in a column of the periodic table.

**metals** elements that are typically shiny, malleable, ductile solids that conduct heat and electricity well and tend to form positive ions.

**nonmetals** elements with properties opposite those of metals, including poor conductivity of heat and electricity.

**metalloids** elements that tend to have some properties of metals and some properties of nonmetals.

Long before they knew about subatomic particles and the concept of atomic numbers, chemists knew that groups of elements such as Li, Na, and K or F, Cl, and Br had similar properties and that, when the elements were arranged by increasing atomic mass, repeating patterns of similar properties emerged. That *periodicity* in the properties of the elements inspired several 19th-century scientists to create tables of the elements in which the elements were arranged in patterns based on similarities in their chemical properties.

By far the most successful of those scientists was Russian chemist Dmitri Mendeleev (1834–1907). In 1872, he published a table (**Figure 2.9a**) that is widely considered the forerunner of the modern **periodic table of the elements** (inside the front cover and **Figure 2.9b**). In addition to organizing all the elements known at the time, Mendeleev chose to leave empty cells in his table for elements unknown at the time so that he could align the known elements with similar chemical properties in the same columns. On the basis of the locations of those empty cells, Mendeleev predicted the chemical properties of the unknown elements. Those predictions greatly facilitated the later discovery of those elements by other scientists. Note that Mendeleev arranged the elements in his periodic table in order of increasing atomic mass, but in the modern periodic table the elements appear in order of their atomic numbers.

## CONCEPT TEST

Suggest a reason why the elements in Mendeleev's version of the periodic table are in order of atomic mass and not atomic number.

*(Answers to Concept Tests are in the back of the book.)*

The elements in modern periodic tables are arranged in seven horizontal rows (**periods**) and 18 columns (**groups** or **families**). The periods are numbered at the far left of each row, whereas the group numbers appear at the top of each column. The periodic table inside the front cover has a second set of column headings consisting of a number followed by the letter A or B. Those secondary headings, though not used in this text, were widely used in earlier versions of the periodic table.

The elements in the periodic table can be categorized broadly as metals, nonmetals, and metalloids (or semimetals), as shown in Figure 2.9(b). **Metals** (tan cells) tend to conduct heat and electricity well; they are malleable (can be shaped by hammering) and ductile (can be drawn out into wires), and all but mercury (Hg) are shiny solids at room temperature. **Nonmetals** (blue cells) are poor conductors of heat and electricity. Most are gases at room temperature; the solids among them, such as iodine ($I_2$), tend to be brittle, and bromine ($Br_2$) is a liquid at room temperature. **Metalloids** (green cells) are so named because they have some properties of metals and some properties of nonmetals.

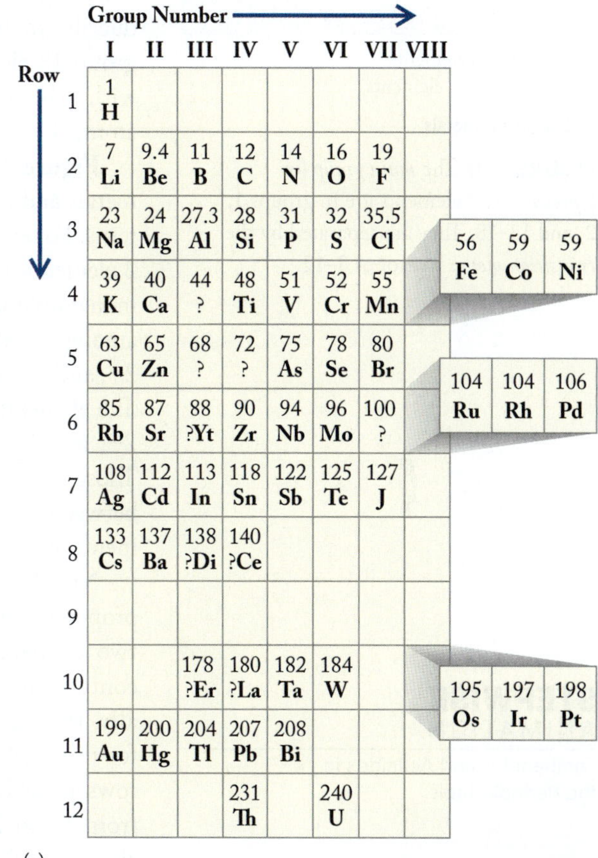

Group Number →

| Row | I | II | III | IV | V | VI | VII | VIII |
|---|---|---|---|---|---|---|---|---|
| 1 | 1 H | | | | | | | |
| 2 | 7 Li | 9.4 Be | 11 B | 12 C | 14 N | 16 O | 19 F | |
| 3 | 23 Na | 24 Mg | 27.3 Al | 28 Si | 31 P | 32 S | 35.5 Cl | |
| 4 | 39 K | 40 Ca | 44 ? | 48 Ti | 51 V | 52 Cr | 55 Mn | 56 Fe, 59 Co, 59 Ni |
| 5 | 63 Cu | 65 Zn | 68 ? | 72 ? | 75 As | 78 Se | 80 Br | |
| 6 | 85 Rb | 87 Sr | 88 ?Yt | 90 Zr | 94 Nb | 96 Mo | 100 ? | 104 Ru, 104 Rh, 106 Pd |
| 7 | 108 Ag | 112 Cd | 113 In | 118 Sn | 122 Sb | 125 Te | 127 J | |
| 8 | 133 Cs | 137 Ba | 138 ?Di | 140 ?Ce | | | | |
| 9 | | | | | | | | |
| 10 | | | 178 ?Er | 180 ?La | 182 Ta | 184 W | | 195 Os, 197 Ir, 198 Pt |
| 11 | 199 Au | 200 Hg | 204 Tl | 207 Pb | 208 Bi | | | |
| 12 | | | | 231 Th | | 240 U | | |

(a)

Atomic number / Symbol for element

| Period | 1 | 2 | 3 | 4 | 5 | 6 | 7 | 8 | 9 | 10 | 11 | 12 | 13 | 14 | 15 | 16 | 17 | 18 |
|---|---|---|---|---|---|---|---|---|---|---|---|---|---|---|---|---|---|---|
| 1 | 1 H | | | | | | | | | | | | | | | | | 2 He |
| 2 | 3 Li | 4 Be | | | | | | | | | | | 5 B | 6 C | 7 N | 8 O | 9 F | 10 Ne |
| 3 | 11 Na | 12 Mg | | | | | | | | | | | 13 Al | 14 Si | 15 P | 16 S | 17 Cl | 18 Ar |
| 4 | 19 K | 20 Ca | 21 Sc | 22 Ti | 23 V | 24 Cr | 25 Mn | 26 Fe | 27 Co | 28 Ni | 29 Cu | 30 Zn | 31 Ga | 32 Ge | 33 As | 34 Se | 35 Br | 36 Kr |
| 5 | 37 Rb | 38 Sr | 39 Y | 40 Zr | 41 Nb | 42 Mo | 43 Tc | 44 Ru | 45 Rh | 46 Pd | 47 Ag | 48 Cd | 49 In | 50 Sn | 51 Sb | 52 Te | 53 I | 54 Xe |
| 6 | 55 Cs | 56 Ba | 57 La | 72 Hf | 73 Ta | 74 W | 75 Re | 76 Os | 77 Ir | 78 Pt | 79 Au | 80 Hg | 81 Tl | 82 Pb | 83 Bi | 84 Po | 85 At | 86 Rn |
| 7 | 87 Fr | 88 Ra | 89 Ac | 104 Rf | 105 Db | 106 Sg | 107 Bh | 108 Hs | 109 Mt | 110 Ds | 111 Rg | 112 Cn | 113 Nh | 114 Fl | 115 Mc | 116 Lv | 117 Ts | 118 Og |

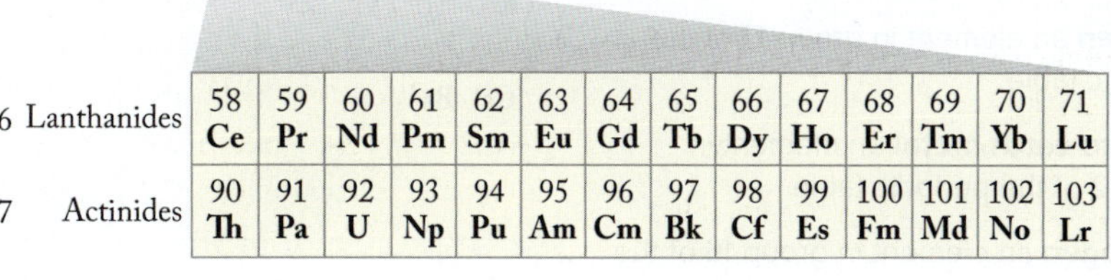

| | | | | | | | | | | | | | | |
|---|---|---|---|---|---|---|---|---|---|---|---|---|---|---|
| 6 Lanthanides | 58 Ce | 59 Pr | 60 Nd | 61 Pm | 62 Sm | 63 Eu | 64 Gd | 65 Tb | 66 Dy | 67 Ho | 68 Er | 69 Tm | 70 Yb | 71 Lu |
| 7 Actinides | 90 Th | 91 Pa | 92 U | 93 Np | 94 Pu | 95 Am | 96 Cm | 97 Bk | 98 Cf | 99 Es | 100 Fm | 101 Md | 102 No | 103 Lr |

(b)

**FIGURE 2.9** Two views of the periodic table. (a) Mendeleev organized his periodic table according to similar properties and atomic masses. He assigned three elements with similar properties to group VIII in rows 4, 6, and 10. As a result, his rows 4 and 5 together contain spaces for 18 elements, corresponding to the 18 groups in the modern periodic table. (b) In the modern table, elements are arranged in order of atomic number ($Z$). The elements in tan are *metals*, those in blue are *nonmetals*, and those in green are *metalloids* (also called *semimetals*).

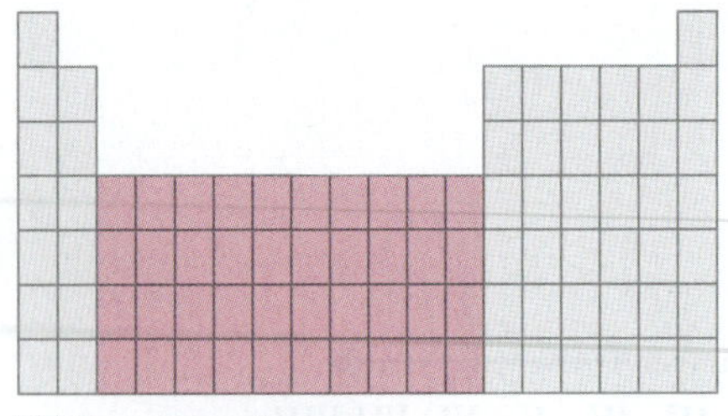

□ Main group elements (representative elements)

■ Transition metals

**FIGURE 2.10** The *main group* (or *representative*) elements are in groups 1, 2, and 13–18. They are separated by the *transition metals* in groups 3–12.

Groups 1, 2, and 13–18 are referred to collectively as **main group elements** or **representative elements** (**Figure 2.10**). They include the most abundant elements in the solar system and many of the most abundant on Earth. They are the "A" elements in the former group numbering system. The elements in groups 3–12 are called **transition metals**; they are the former "B" elements. All except mercury exhibit the classic physical properties of metals: they are hard, shiny, ductile, malleable solids and excellent conductors of heat and electricity. The group 18 elements are called **noble gases** because they mostly do not interact with other elements. Becquerel's α particle is actually the nucleus of a helium atom, the first element in the group.

**Figure 2.11** summarizes the charges of common monatomic ions for many metals and nonmetals. It also shows that those charges vary across groups in patterns. For example, atoms of group 1 elements always form 1+ ions; atoms of group 2 elements always form 2+ ions, and elements in groups 3 and 4 *commonly* form monatomic ions with 3+ and 4+ charges, respectively. A similar pattern exists among the elements in groups 11, 12, and 13, whose atoms commonly form 1+, 2+, and 3+ ions, respectively. We refer to those positively charged ions as **cations**. The elements in groups 15, 16, and 17 form monatomic ions with charges of 3−, 2−, and 1−, respectively. We refer to those negatively charged ions as **anions**. Those trends in charges are summarized in **Table 2.2**. We explore the connection between the charges of all those ions and the structure of the atoms of those elements, particularly the number of electrons surrounding their nuclei, in Chapter 3.

Now let's consider in more detail the structure of the periodic table and the properties of the elements in it. The first row of the periodic table contains only two elements—hydrogen and helium—and the second and third rows each contain only eight. All 18 columns are full starting with the fourth row. Actually, the sixth and seventh rows contain more elements than space is available for in an 18-column array. The additional elements appear in the two separate rows at the bottom of the main table. Elements in the row with atomic numbers from 58 to 71 are called the lanthanides (after element 57, lanthanum), and elements with atomic numbers between 90 and 103 are called actinides (after element 89, actinium). All the isotopes of the elements with atomic numbers above 83, including all the actinide elements, are radioactive. Those **radionuclides** spontaneously emit high-energy radiation and particles and are transformed into

**STEPWISE ANIMATION**

Lanthanides and Actinides in the Periodic Table

**main group elements** or **representative elements** the elements in groups 1, 2, and 13–18 of the periodic table.

**transition metals** the elements in groups 3–12 of the periodic table.

**noble gases** the elements in group 18 of the periodic table.

**cation** a positively charged ion.

**anion** a negatively charged ion.

**radionuclide** a radioactive (unstable) nuclide.

**alkali metal** an element in group 1 of the periodic table.

**halogen** an element in group 17 of the periodic table.

**alkaline earth metal** an element in group 2 of the periodic table.

**chalcogen** an element in group 16 of the periodic table.

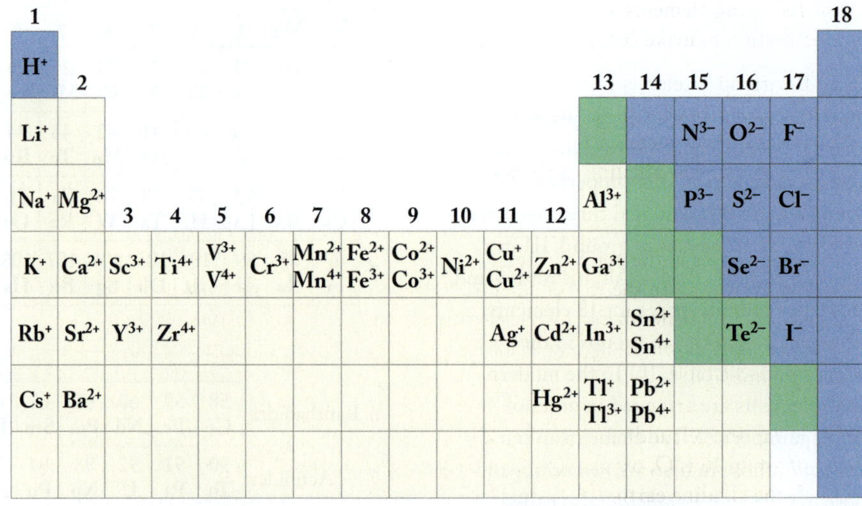


**FIGURE 2.11** Periodic trends in the charges of common monatomic ions.

**TABLE 2.2** Common[a] Charges of Monatomic Ions

| Group Number | 1 | 2 | 3 | 4 | 11 | 12 | 13 | 15 | 16 | 17 |
|---|---|---|---|---|---|---|---|---|---|---|
| Charge of Monatomic Ions | 1+ | 2+ | 3+ | 4+ | 1+ | 2+ | 3+ | 3− | 2− | 1− |

[a]The charges of the ions in groups 1 and 2 are *always* 1+ and 2+, respectively.

other nuclides. Elements with $Z > 94$ are not found in nature, and elements 93 and 94 occur in only extremely small (trace) amounts in minerals that also contain uranium ($Z = 92$).

Several groups of elements have names in addition to numbers—names based on chemical properties common to the elements in that group (**Figure 2.12**). For example, the group 1 elements (except for hydrogen) are called **alkali metals**, and the elements of group 17 are called **halogens**. The word *halogen* is derived from the Greek word for "salt former." Chlorine is a typical halogen. It forms 1:1 binary (two-element) ionic compounds with the group 1 elements; a familiar example is sodium chloride, NaCl (the principal ingredient in table salt). The ratio is 1:1 because all the group 1 elements form cations that have a 1+ charge, and all the group 17 elements form 1− anions (as shown in Figure 2.11). An equal number of those anions and cations means that those ionic compounds are neutral substances even though they consist of charged particles.

**CONNECTION** The structure of NaCl as an ordered three-dimensional array of sodium ($Na^+$) ions and chloride ($Cl^-$) ions is described in Figure 1.17.

Group 2 elements are called **alkaline earth metals**. They form ionic compounds with halogens in which the ratio of cations to anions is 1:2 because all the group 2 cations have 2+ charges, so a neutral compound must have twice as many anions with 1− charges as alkaline earth cations. A common example of such a compound is calcium chloride, $CaCl_2$, which is widely used in regions with cold winters to melt ice and snow from exterior steps and sidewalks. Alkaline earth metals form ionic compounds with the group 16 elements in a 1:1 cation-to-anion ratio because the group 16 elements form 2− anions. The group 16 elements are sometimes referred to as the **chalcogens**, a name derived from the association of O, S, Se, and Te with copper (*chalco-*) in minerals.

**SAMPLE EXERCISE 2.3** Navigating the Periodic Table **LO3**

The following elements are the major components of Portland cement, used in construction to make concrete and mortar. Which elements are they?

a. The group 13 element in the third row (period)
b. The group 16 element with the smallest atomic number
c. The third-row metalloid
d. The fourth-row alkaline earth element

**Collect, Organize, and Analyze** We are asked to identify elements from the locations of their symbols in the periodic table. In parts a, c, and d, we are given the row number; in b we are not given the row number directly, but we know that the element has the lowest atomic number of all the elements in its group. Information about group locations comes from the group number in a and b, the type of element in c, and the group name in d.

**Solve** (a) The cell address is group 13, row 3: Al, aluminum; (b) the element at the top of group 16 is O, oxygen; (c) the only metalloid in row 3 is in group 14: Si, silicon; (d) all the alkaline earths, including the one in row 4, are in group 2: Ca, calcium.

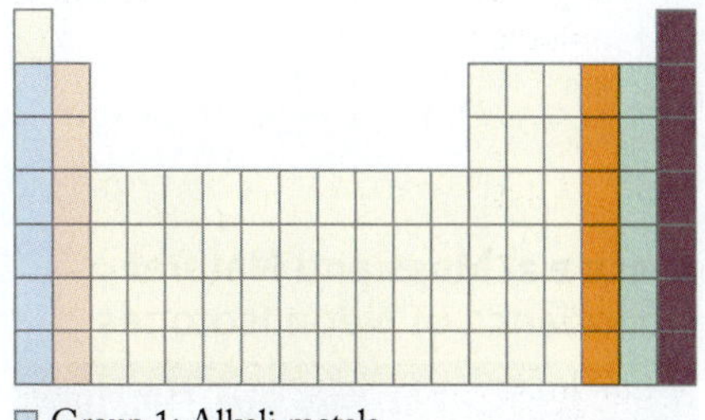


**FIGURE 2.12** The commonly used names of groups 1, 2, and 16–18 of the periodic table.

**Think About It** Each element has a unique location in the periodic table that is linked to both its atomic number and its chemical properties. Chemists usually describe element locations in terms of their group number or name (the column they're in) and their row number.

 **Practice Exercise** What are the symbol and name of each of the following elements?

a. The metalloid in group 15 closest in mass to the noble gas krypton
b. A representative element in the third row that is an alkaline earth metal
c. A transition metal in the sixth row and in the same group as zinc (Zn)

*(Answers to Practice Exercises are in the back of the book.)*

## 2.4 The Masses of Atoms, Ions, and Molecules

The number above each symbol in the periodic table is the element's atomic number, $Z$; the value below the symbol is the element's **average atomic mass**. More precisely, the value below the symbol is the *weighted* average of the masses of all the element's isotopes.

To understand the meaning of a weighted average, let's consider the masses and **natural abundances** of the three isotopes of neon in **Table 2.3**. Natural abundances are usually expressed as percentages. Thus, 90.4838% of all neon atoms are neon-20, 9.2465% are neon-22, and only 0.2696% are neon-21. To calculate the average atomic mass of neon, we multiply the mass of each of its isotopes by its natural abundance (in the language of mathematics, we *weight* the isotope's mass by using natural abundance as the *weighting factor*) and then sum the three weighted masses. To simplify the calculation, we first convert the percent abundance values into their decimal equivalents:

$$\begin{aligned} \text{Average atomic mass of neon} = \quad & 19.9924 \text{ u} \times 0.904838 = 18.08988 \text{ u} \\ & + 20.9940 \text{ u} \times 0.002696 = 0.056600 \text{ u} \\ & + 21.9914 \text{ u} \times 0.092465 = 2.03344 \text{ u} \\ & \overline{\qquad\qquad\qquad\qquad\qquad 20.17992 \text{ u}} \end{aligned}$$

To avoid rounding errors in that calculation, we carry one more digit (highlighted in red) than the significant-figure rules for multiplication allow. Rounding the sum to the appropriate number of significant figures gives us an average atomic mass of 20.1799 u. No single atom of neon has that average atomic mass; instead, every atom of neon in the universe has the mass of one of the three neon isotopes.

That method for calculating average atomic mass works for all elements. The general equation for this calculation is

$$m_X = a_1m_1 + a_2m_2 + a_3m_3 + \cdots \tag{2.3}$$

where $m_X$ is the average atomic mass of element X, which has isotopes with masses $m_1, m_2, m_3, \ldots$, for which the natural abundances expressed in decimal form are $a_1, a_2, a_3, \ldots$.

Note that the natural abundance of neon-21 is so small that Thomson and Aston could not detect it with their positive-ray analyzer. However, modern

**TABLE 2.3** Mass and Natural Abundance of Neon Isotopes

| Isotope | Mass (u) | Natural Abundance (%) |
|---|---|---|
| Neon-20 | 19.9924 | 90.4838 |
| Neon-21 | 20.9940 | 0.2696 |
| Neon-22 | 21.9914 | 9.2465 |

versions of their instrument, called *mass spectrometers*, can ionize atoms and molecules and can separate and detect their ions with much more sensitivity. Those instruments can yield precise natural abundance values for even low-abundance isotopes, as illustrated by the data for neon. We should also consider that the term *natural abundance* does not mean that the relative abundances of the isotopes of all the elements are the same no matter where they occur in nature. For some elements, natural abundances vary depending on where on the planet the sample was gathered. For example, a sample of seawater near the coast of Japan contained chloride ions that were 75.644% $^{35}Cl^-$ and 24.356% $^{37}Cl^-$. If we look up the masses of chlorine-35 and chlorine-37 in Appendix 3 (Table A3.3), we can use the decimal equivalents of those abundances to calculate the average atomic mass of chlorine in the sample:

$$m_{Cl} = a_{^{35}Cl^-} \cdot m_{^{35}Cl^-} + a_{^{37}Cl^-} \cdot m_{^{37}Cl^-}$$
$$= 0.75644 \times 34.9689 \text{ u} + 0.24356 \times 36.9659 \text{ u} = 35.455 \text{ u}$$

That average atomic mass of chlorine is slightly higher than the 35.453 u in the periodic table inside the front cover of this book, which indicates that the seawater sample was slightly enriched in $^{37}Cl$ isotope in comparison with $^{35}Cl$. Note that we use the average atomic masses of chlorine for the chloride ions in that calculation. We do so because gaining an electron has very little impact on the mass of an atom; nearly all its mass is concentrated in its nucleus.

Sample Exercise 2.4 provides more practice with those ideas and calculations with strontium, another element with various isotopic abundances. The data in Sample Exercise 2.4 come from analysis of the isotopes in a sample known to contain strontium that was carefully purified, blended, and analyzed by the U.S. National Institute of Standards and Technology (NIST). NIST sells portions of it to scientists who wish to check the accuracy of isotope analyses in their own laboratories using mass spectrometry.

**average atomic mass** the weighted average of the masses of all isotopes of an element, calculated by multiplying the natural abundance of each isotope by its mass in unified atomic mass units and then summing the products.

**natural abundance** the proportion of a particular isotope, usually expressed as a percentage, relative to all the isotopes of that element in a natural sample.

**CONNECTION** The use of reference materials with known composition to check the accuracy of chemical analyses was described in Section 1.9.

**SAMPLE EXERCISE 2.4** Calculating Average Atomic Mass **LO4**

The table below contains the atomic masses and experimentally determined relative abundances of the isotopes of strontium in a sample of NIST Standard Reference Material Number 987:

| Isotope | Mass (u) | Natural Abundance (%) |
|---|---|---|
| $^{84}Sr$ | 83.9134 | 0.557 |
| $^{86}Sr$ | 85.9092 | 9.857 |
| $^{87}Sr$ | 86.9089 | 7.002 |
| $^{88}Sr$ | 87.9056 | 82.584 |

What is the average atomic mass of strontium in the sample? Compare the calculated average with the atomic mass of Sr in the periodic table (87.62 u) in the front of this book.

**Collect, Organize, and Analyze** We know the masses and the relative abundances of the four stable isotopes of strontium in a sample, and we are asked to calculate the average atomic mass of Sr in the sample. Using Equation 2.3, we multiply the mass of each isotope by its abundance (expressed as a decimal) and then add the products together.

**Solve** The weighted average atomic mass of Sr in the sample is:

$$
\begin{aligned}
{}^{84}\text{Sr: } 83.9134\text{ u} \times 0.00557 &= 0.4674\text{ u}\\
{}^{86}\text{Sr: } 85.9092\text{ u} \times 0.09857 &= 8.4681\text{ u}\\
{}^{87}\text{Sr: } 86.9089\text{ u} \times 0.07002 &= 6.0854\text{ u}\\
{}^{88}\text{Sr: } 87.9056\text{ u} \times 0.82584 &= \underline{72.5960\text{ u}}\\
&\phantom{=} 87.6169\text{ u}
\end{aligned}
$$

Note that to avoid rounding errors, we carried one more digit (highlighted in red) than required by the significant-figure rules in Chapter 1. Rounding the sum to the allowed number of significant figures gives us 87.617 u. Rounding that value to the nearest 0.01 u yields the same average atomic mass of Sr (87.62 u) as in the periodic table.

**Think About It** The abundance values of the isotopes contain fewer significant digits (3, 4, 4, and 5 for $^{84}$Sr, $^{86}$Sr, $^{87}$Sr, and $^{88}$Sr, respectively) than do their masses (6 in every value). Therefore, isotopic abundance is the weak link in calculating that average atomic mass. The four abundance values should add up to 1.000, and they do. Sometimes that is not the case (check the neon abundances in the calculation earlier in this section). Uncertainties in the last decimal place may be due to uncertainties in measured values or in rounding them off. To avoid rounding errors, consider using your calculator's memory function to store and sum intermediate values.

**Practice Exercise** We briefly discussed the application of $^{83}$Kr to MRI in Sample Exercise 2.1. Krypton's six stable isotopes with natural abundances are listed below. Use these data to calculate the average atomic mass of krypton.

| Isotope | Mass (u) | Natural Abundance (%) |
|---|---|---|
| $^{78}$Kr | 77.920 | 0.35 |
| $^{80}$Kr | 79.916 | 2.25 |
| $^{82}$Kr | 81.913 | 11.60 |
| $^{83}$Kr | 82.914 | 11.50 |
| $^{84}$Kr | 83.912 | 57.00 |
| $^{86}$Kr | 85.911 | 17.30 |

*(Answers to Practice Exercises are in the back of the book.)*

**SAMPLE EXERCISE 2.5** Calculating Natural Isotope Abundances **LO4**

Silver (Ag) has two stable isotopes: $^{107}$Ag, 106.90 u and $^{109}$Ag, 108.90 u. The average atomic mass of silver is 107.87 u. What is the natural abundance of each isotope?

**Collect and Organize** We know the masses of both isotopes of silver and its average atomic mass. We are asked to calculate the natural abundance of each.

**Analyze** Unlike with Sample Exercise 2.4, we can't insert values directly into Equation 2.3 because we do not know the coefficients $a_1$ and $a_2$. However, if we let $x = a_1$ (the natural abundance of one of the isotopes expressed as a decimal), the natural abundance of the other isotope ($a_2$) = 1 − $x$. The average mass (107.87 u) is approximately halfway between the exact masses of the two isotopes. That suggests that the two isotopes are present in nearly equal proportions, so the values of $a_1$ and $a_2$ should be nearly the same.

**Solve** Substituting $x$ and $1 - x$ into Equation 2.3 gives us an equation that we can solve for $x$:

$$m_{Ag} = xm_{Ag\text{-}107} + (1 - x)m_{Ag\text{-}109}$$
$$107.89 \text{ u} = x(106.90 \text{ u}) + (1 - x)(108.90 \text{ u})$$
$$x = 0.5050 \qquad (1 - x) = 0.4950$$

Therefore, the percent natural abundances are the following:

$$^{107}\text{Ag: } 0.5050 \times 100 = 50.5\% \text{ and } ^{109}\text{Ag: } 0.4950 \times 100 = 49.5\%$$

**Think About It** We predicted that the natural abundances of $^{107}Ag$ and $^{109}Ag$ would be nearly equal, and the calculated abundances are consistent with our prediction. Note that this approach works only for elements that have two isotopes.

**Practice Exercise** Iridium has two stable isotopes: iridium-191 (190.96 u) and iridium-193 (192.96 u). The average atomic mass of iridium is 192.22 u. What is the natural abundance of each isotope?

*(Answers to Practice Exercises are in the back of the book.)*

**molecular mass** the mass in u of one molecule of a molecular compound.

**formula unit** the smallest electrically neutral unit of an ionic compound.

**formula mass** the mass in u of one formula unit of an ionic compound.

Let's extend the concept of particle mass from single atoms to the mass of molecular compounds. Just as the atoms of an element have an average atomic mass, each molecule of a molecular compound has a **molecular mass** that is the sum of the average atomic masses of the atoms that are chemically bonded together in it. Like average atomic masses, molecular masses are expressed in u. To calculate the molecular mass of a compound, we simply add the average atomic masses of the atoms in each of its molecules. For example, the molecular mass of sulfur dioxide ($SO_2$), a gas released during volcanic eruptions (**Figure 2.13**), is the sum of the average atomic mass of one atom of sulfur and two atoms of oxygen:

$$\left(\frac{1 \text{ atom S}}{1 \text{ molecule SO}_2} \times \frac{32.065 \text{ u}}{1 \text{ atom S}}\right) + \left(\frac{2 \text{ atoms O}}{1 \text{ molecule SO}_2} \times \frac{15.999 \text{ u}}{1 \text{ atom O}}\right)$$
$$= \frac{64.063 \text{ u}}{\text{molecule SO}_2}$$

How does the concept of a molecular mass apply to ionic compounds, such as sodium chloride (NaCl), that consist of three-dimensional arrays of positive and negative ions, not molecules? The lack of molecules means that ionic compounds do not have molecular masses. However, the formula of an ionic compound defines a quantity called the **formula unit**, the smallest electrically neutral unit in an ionic compound. Thus, the formula unit of NaCl consists of one $Na^+$ ion and one $Cl^-$ ion. The formula of calcium chloride is $CaCl_2$, so its formula unit consists of one calcium ion, $Ca^{2+}$, and *two* chloride ions, $Cl^-$. A formula unit has a particular **formula mass**: the sum of the average atomic masses of the cations and anions that make up a neutral formula unit. The formula mass for $CaCl_2$ is the combined mass of 1 $Ca^{2+}$ ion and 2 $Cl^-$ ions:

$$\left(\frac{1 \text{ Ca}^{2+} \text{ ion}}{1 \text{ formula unit}} \times \frac{40.078 \text{ u}}{1 \text{ Ca}^{2+} \text{ ion}}\right) + \left(\frac{2 \text{ Cl}^- \text{ ions}}{1 \text{ formula unit}} \times \frac{35.453 \text{ u}}{1 \text{ Cl}^- \text{ ion}}\right)$$
$$= \frac{110.984 \text{ u}}{\text{formula unit}}$$

Note that we use the average atomic masses of calcium and chlorine from the periodic table for the masses of the $Ca^{2+}$ and $Cl^-$ ions in that calculation

**FIGURE 2.13** Volcanic eruptions, such as this one of Kilauea in Hawaii in May 2018, can introduce thousands of metric tons (1 metric ton = 1000 kg) of $SO_2$ into the atmosphere each day.

because losing or gaining an electron or two has very little impact on the mass of an atom. Moreover, the masses of the electrons lost when atoms form cations are balanced by the masses gained when atoms form anions, so the overall mass of a formula unit is the same as the sum of the average atomic masses of the elements in it.

## 2.5 Moles and Molar Masses

Individual atoms are tiny particles, which is why we need powerful instruments such as scanning tunneling microscopes to see them (see Figure 1.1). Atoms also have small masses. The average mass of a gold atom—196.97 u—is heavy for an atom, but it wouldn't even register on a conventional scale. What, then, is 196.97 u in grams? We know that 1 u = $1.66054 \times 10^{-27}$ kg, so the mass of an average gold atom is too small to weigh:

$$196.97 \text{ \sout{u} Au} \times \frac{1.66054 \times 10^{-27} \text{ \sout{kg}}}{1 \text{ \sout{u}}} \times \frac{1000 \text{ g}}{1 \text{ \sout{kg}}} = 3.2708 \times 10^{-22} \text{ g Au}$$

In our macroscopic (visible) world, chemists usually work with quantities of substances that they can see, transfer from one container to another, and weigh on balances. Inevitably those quantities of substances contain enormous numbers of atoms, ions, or molecules.

To deal with such large numbers, chemists need a unit that relates macroscopic quantities of substances, such as their masses expressed in grams, to the number of particles they contain. That unit is the **mole (mol)**, the SI base unit for expressing quantities of substances (see Table 1.2).

One mole of a substance is defined as the quantity of it that contains exactly $6.02214076 \times 10^{23}$ elementary entities—that is, atoms, ions, or molecules. That enormous number is the fixed numerical value of the **Avogadro constant**, $N_A$, named in honor of the Italian scientist Amedeo Avogadro (1776–1856), whose research enabled other scientists to accurately determine the atomic masses of the elements. The Avogadro constant is usually expressed "per mole," in units of $mol^{-1}$. In calculations that use the Avogadro constant, we will usually round it off to four significant figures: $6.022 \times 10^{23}$.

Avogadro Constant

To put a number of that magnitude in perspective: it would take more than 2 trillion computer flash drives, each capable of storing 256 gigabytes ($2.56 \times 10^{11}$ bytes) of data, to store a mole of bytes. In a different analogy, some estimates put the number of cells in the human body at 37 trillion ($3.7 \times 10^{13}$). The Avogadro constant corresponds to the number of cells in 16 billion people, or about twice Earth's current population.

We can use the Avogadro constant to convert between number of particles and moles, and vice versa. Dividing the number of particles in a sample by the Avogadro constant yields the number of moles of those particles. For example, a U.S. penny minted before 1962 contained about $2.4 \times 10^{22}$ atoms of Cu. That number is equivalent to $4.0 \times 10^{-2}$ moles of copper:

$$2.4 \times 10^{22} \text{ \sout{atoms} Cu} \times \frac{1 \text{ mol}}{6.022 \times 10^{23} \text{ \sout{atoms}}} = 4.0 \times 10^{-2} \text{ mol Cu}$$

Conversely, multiplying a number of moles by the Avogadro constant gives us the number of particles in that many moles. For example, silicon wafers used in solar panels are typically 10–20 cm on a side and 200–300 μm thick. A

**mole (mol)** an amount of a substance that contains a number of particles (atoms, ions, molecules, or formula units) equal to the Avogadro constant.

**Avogadro constant ($N_A$)** the number of elementary entities—atoms, ions, or molecules—in one mole of a substance. The exact value of the constant is $6.02214076 \times 10^{23}$ $mol^{-1}$.

silicon wafer with dimensions 10.0 cm by 10.0 cm by $2.00 \times 10^{-2}$ cm contains 0.167 moles of silicon. The number of silicon atoms in such a wafer is

$$0.167\ \cancel{\text{mol}}\ \text{Si} \times \frac{6.022 \times 10^{23}\ \text{atoms}}{1\ \cancel{\text{mol}}} = 1.00 \times 10^{23}\ \text{atoms Si}$$

Atom-to-mole conversions are illustrated in **Figure 2.14**, and the quantities of some common elements equivalent to one mole are shown in **Figure 2.15**.

## CONCEPT **TEST**

How does a unit of measure such as 500 facial tissues in a box relate to the concept of the mole?

*(Answers to Concept Tests are in the back of the book.)*

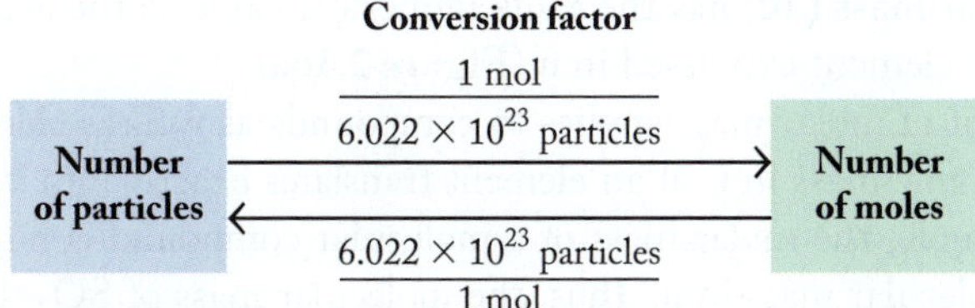


**FIGURE 2.14** Converting between a number of particles and an equivalent number of moles (or vice versa) is a matter of dividing (or multiplying) by the Avogadro constant.

### SAMPLE EXERCISE 2.6 Converting Number of Moles into Number of Particles **LO5**

The silicon used to make computer chips must be extremely pure. For example, it must contain less than $3 \times 10^{-10}$ moles of phosphorus (a common impurity in silicon) per mole of silicon. What is that level of impurity expressed in atoms of phosphorus per mole of silicon?

**Collect and Organize** The problem states the maximum number of moles of phosphorus allowed per mole of silicon and asks us to calculate the equivalent number of atoms of phosphorus per mole of silicon. The Avogadro constant defines the number of atoms in one mole of an element: $6.022 \times 10^{23}$.

**Analyze** We can convert a number of moles into an equivalent number of atoms by multiplying by the Avogadro constant because its denominator contains our initial unit (mol) and its numerator contains the units we seek (atoms). We can estimate the correct answer by rounding each starting value to the nearest power of 10 and combining exponents: $10^{24} \times 10^{-10} = 10^{14}$.

**Solve**

$$\frac{3 \times 10^{-10}\ \cancel{\text{mol P}}}{1\ \text{mol Si}} \times \frac{6.022 \times 10^{23}\ \text{atoms P}}{1\ \cancel{\text{mol P}}} = \frac{2 \times 10^{14}\ \text{atoms P}}{\text{mol Si}}$$

**Think About It** The result reveals that even a very small number of moles of an impurity translates to a very large number of atoms. The exponent of our calculated value matches our estimate, so our answer is reasonable.

**Practice Exercise** If 1.0 mL of seawater contains about $2.5 \times 10^{-14}$ moles of dissolved gold, how many atoms of gold are in the seawater?

*(Answers to Practice Exercises are in the back of the book.)*

**FIGURE 2.15** The quantities shown represent one mole of each element: helium gas in the balloon and (from left to right) copper wire, iron powder, solid sulfur, and liquid mercury.

**molar mass ($\mathcal{M}$)** the mass of one mole of a substance.

## Molar Mass

The mole serves as an important link between the atomic mass values in the periodic table and measurable masses of elements and compounds. To see how that link works, let's convert the average mass of an atom of gold, 196.97 u, into an equivalent mass expressed in grams per mole of gold:

$$\frac{196.97\ \cancel{u}\ Au}{1\ \cancel{atom}\ Au} \times \frac{6.022 \times 10^{23}\ \cancel{atoms}}{1\ mol} \times \frac{1.66054 \times 10^{-27}\ \cancel{kg}}{1\ \cancel{u}} \times \frac{1000\ g}{1\ \cancel{kg}}$$

$$= \frac{196.97\ g\ Au}{mol\ Au}$$

| Atomic mass of Au | 196.97 u/atom |
|---|---|
| Mass of 1 mol of Au | 196.97 g |
| Molar mass of Au | 196.97 g/mol |

(a)

Note that the numerical value of the atomic mass of gold expressed in u/atom *is the same* as the mass in grams of one mole of gold. That equality holds for all elements: the mass in grams of one mole of an element, a quantity called the element's **molar mass ($\mathcal{M}$)**, has the same numerical value as the average mass of an atom of the element expressed in u (**Figure 2.16a**).

| Molecular mass of $SO_3$ | 80.062 u/molecule |
|---|---|
| Mass of 1 mol of $SO_3$ | 80.062 g |
| Molar mass of $SO_3$ | 80.062 g/mol |

(b)

**FIGURE 2.16** (a) The atomic mass (in u/atom) and the molar mass (in grams per mole) of gold have the same numerical value. (b) The molecular mass (in u/molecule) and the molar mass (in grams per mole) of $SO_3$ have the same numerical value.

The concept of molar mass applies to compounds as well as elements. Just as the average atomic mass in u of an element translates exactly into its molar mass in grams per mole, the molar mass of a molecular compound is numerically the same as its molecular mass in u. Thus, the molecular mass of $SO_2$ that we calculated in the previous section, 64.063 u, is numerically the same as its molar mass of 64.063 g/mol. Molecules of sulfur trioxide, $SO_3$, have one more atom of oxygen (mass = 15.999 u) per molecule, giving sulfur trioxide a molecular mass of (64.063 + 15.999) = 80.062 u and a molar mass of 80.062 g/mol (**Figure 2.16b**). Similarly, the formula masses of ionic compounds, expressed in u, have the same values as the molar masses of the compounds expressed in grams per mole. As we saw in the previous section, the formula mass of $CaCl_2$ is 110.984 u, which means its molar mass is 110.984 g/mol. If we didn't already know the molecular mass or formula mass of a compound, we could calculate its molar mass directly from the molar masses of the elements in it. For example, the molar mass of carbon dioxide, $CO_2$, is the sum of the masses of one mole of carbon atoms and 2 moles of oxygen atoms:

$$\left(\frac{1\ \cancel{mol\ C}}{1\ mol\ CO_2} \times \frac{12.011\ g}{1\ \cancel{mol\ C}}\right) + \left(\frac{2\ \cancel{mol\ O}}{1\ mol\ CO_2} \times \frac{15.999\ g}{1\ \cancel{mol\ O}}\right) = \frac{44.009\ g}{mol\ CO_2}$$

The mole enables us to know the number of particles in any sample of a given substance simply by knowing the mass of the sample. We can do that because the mole represents both a fixed number of particles (the Avogadro constant) and a specific mass (the molar mass) of the substance. You may find it useful to think about moles and mass by comparing 1 dozen grapes to 1 dozen apples. Twelve grapes and twelve apples have very different masses, but each grouping contains 12 pieces of fruit (**Figure 2.17**). Indeed, the mole is sometimes referred to as the "chemist's dozen." **Figure 2.18** summarizes how to use the Avogadro constant, chemical formulas, and molar masses to convert between the mass, the number of moles, and the number of particles in a given quantity of a substance.

**FIGURE 2.17** A dozen grapes weighs less than a dozen apples, but both quantities contain the same number of pieces of fruit.

**SAMPLE EXERCISE 2.7** Calculating the Molar Mass of a Compound **LO5**

Volcanic eruptions are often accompanied by the release of $SO_2$ and water vapor (see Figure 2.13). When those two gases combine, they form sulfurous acid, $H_2SO_3$, which is a molecular compound. What is the molar mass of $H_2SO_3$?

**Collect and Organize** We are asked to calculate the molar mass of $H_2SO_3$. The molar mass of a molecular compound is the sum of the molar masses of the elements in its formula, each multiplied by the number of atoms of that element in a molecule of the compound. The molar masses of the relevant elements are 1.0079 g/mol of H, 32.065 g/mol of S, and 15.999 g/mol of O.

**Analyze** One mole of $H_2SO_3$ contains two moles of H atoms, one mole of S atoms, and three moles of O atoms. To estimate the molar mass of $H_2SO_3$, we can multiply those numbers of moles by the molar masses of the three elements rounded to whole numbers: (2 mol H × 1 g/mol) + (1 mol S × 32 g/mol) + (3 mol O × 16 g/mol) = 82 g/mol $H_2SO_3$.

**Solve**

$$\left(\frac{2\ \cancel{\text{mol H}}}{1\ \text{mol H}_2\text{SO}_3} \times \frac{1.0079\ \text{g}}{1\ \cancel{\text{mol H}}}\right) + \left(\frac{1\ \cancel{\text{mol S}}}{1\ \text{mol H}_2\text{SO}_3} \times \frac{32.065\ \text{g}}{1\ \cancel{\text{mol S}}}\right) + \left(\frac{3\ \cancel{\text{mol O}}}{1\ \text{mol H}_2\text{SO}_3} \times \frac{15.999\ \text{g}}{1\ \cancel{\text{mol O}}}\right)$$
$$= 82.078\ \text{g/mol H}_2\text{SO}_3$$

**Think About It** The calculated molar mass agrees well with our whole-number estimate because the molar masses of H, S, and O are all close to whole-number values, which is not the case for many other elements.

**Practice Exercise** During photosynthesis, green plants convert water and carbon dioxide into glucose ($C_6H_{12}O_6$) and oxygen. What is the molar mass of glucose?

*(Answers to Practice Exercises are in the back of the book.)*

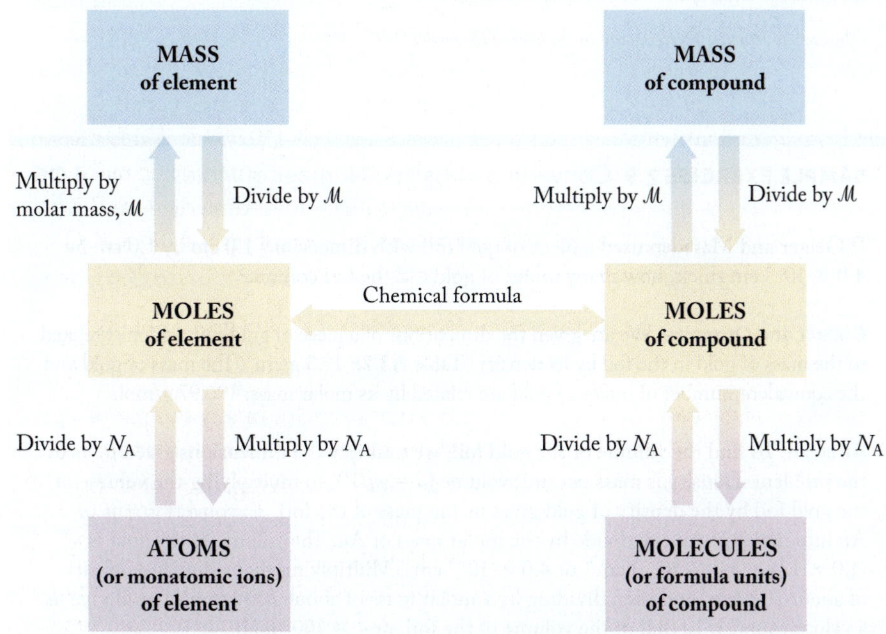


**FIGURE 2.18** The mass of a pure substance can be converted into the equivalent number of moles or number of particles (atoms, ions, or molecules) and vice versa.

## SAMPLE EXERCISE 2.8 Converting Number of Moles into Mass LO5

In 2014, the World Anti-Doping Agency announced that xenon gas would be added to the list of banned substances for athletes. Apparently since 2004, Russian athletes had been inhaling xenon while training. The physiological effect of Xe is to boost the capacity of blood to carry oxygen, an important factor in aerobically demanding sports. If an athlete inhaled an average of $1.68 \times 10^{-4}$ moles of Xe per breath while sleeping, how many grams of Xe were inhaled during 7.0 hours of sleep at an average respiration rate of 10.0 breaths per minute?

**Collect and Organize** We are asked to use the given breathing rate, time, and average number of moles of Xe per breath to find the mass of Xe inhaled. The molar mass of Xe is 131.29 g/mol.

**Analyze** To calculate the total quantity of Xe inhaled, we need to multiply the total number of breaths taken by the quantity of Xe in each breath and then convert that quantity (in moles) into an equivalent mass by multiplying by the molar mass (in grams per mole). To estimate the final answer, we note that 10 breaths per minute means 600 breaths per hour, or 4200 breaths over 7 hours. Each breath has nearly $2 \times 10^{-4}$ mol of Xe, or about 0.8 mol of Xe in 4200 breaths. The molar mass of Xe is about 132 g/mol, so the mass of inhaled Xe should be about 100 g.

**Solve**

$$7.0\ \cancel{\text{h}} \times \frac{60\ \cancel{\text{min}}}{1\ \cancel{\text{h}}} \times \frac{10.0\ \cancel{\text{breaths}}}{1\ \cancel{\text{min}}} \times \frac{1.68 \times 10^{-4}\ \cancel{\text{mol Xe}}}{1\ \cancel{\text{breath}}} \times \frac{131.29\ \text{g Xe}}{1\ \cancel{\text{mol Xe}}} = 93\ \text{g Xe}$$

**Think About It** Our calculated answer is close to our estimate. In general, we multiply by molar mass to convert a number of moles into an equivalent mass in grams. The small number of moles of Xe per breath suggests that the athletes did not need to breathe a high concentration of xenon. The mixture in question contained about 6 mL of Xe per liter of air.

**Practice Exercise** Some of the cylindrical silicon wafers used in the semiconductor industry are 10.00 cm in diameter and contain $3.42 \times 10^{-3}$ mol of Si. If the density of silicon is 2.33 g/cm$^3$, how thick is one of those wafers? Hint: The formula for the volume of a cylinder is $\pi r^2 h$.

*(Answers to Practice Exercises are in the back of the book.)*

## SAMPLE EXERCISE 2.9 Converting Mass into Number of Moles LO5

If Geiger and Marsden used a piece of gold foil with dimensions 1.0 cm by 1.0 cm by $4.0 \times 10^{-5}$ cm thick, how many moles of gold did the foil contain?

**Collect and Organize** We are given the dimensions of a piece of gold foil, which is related to the mass of gold in the foil by its density (Table A3.2): 19.3 g/cm$^3$. The mass of gold and the equivalent number of moles of gold are related by its molar mass: 196.97 g/mol.

**Analyze** To find the volume of the gold foil, we multiply the dimensions given to us in the problem. Density is mass per unit volume ($d = m/V$), so multiplying the volume of the gold foil by the density of gold gives us the mass of the foil. To convert grams of Au into moles of Au, we divide by the molar mass of Au. The volume of the gold is $(1.0 \times 1.0 \times 4.0 \times 10^{-5})$ cm$^3$, or $4.0 \times 10^{-5}$ cm$^3$. Multiplying that value by a density of about 20 g/cm$^3$ and then dividing by a molar mass of about 200 g/mol should give us a value around 1/10 that of the volume of the foil, or $4 \times 10^{-6}$ mol.

**Solve** Volume of gold foil:

$$1.0\ \text{cm} \times 1.0\ \text{cm} \times 4.0 \times 10^{-5}\ \text{cm} = 4.0 \times 10^{-5}\ \text{cm}^3$$

Moles of gold:

$$4.0 \times 10^{-5}\ \cancel{\text{cm}^3\ \text{Au}} \times \frac{19.3\ \cancel{\text{g Au}}}{1\ \cancel{\text{cm}^3\ \text{Au}}} \times \frac{1\ \text{mol Au}}{196.97\ \cancel{\text{g Au}}} = 3.9 \times 10^{-6}\ \text{mol Au}$$

**Think About It** The foil sample is very thin, so we would expect it to contain only a small fraction of a mole of Au. The answer reflects that fact and agrees with our predicted value.

**Practice Exercise** The mass of the diamond in **Figure 2.19** is 3.25 carats (1 carat = 0.200 g). If we assume that diamonds are nearly pure carbon, how many moles and how many atoms of carbon are in the diamond?

*(Answers to Practice Exercises are in the back of the book.)*

**FIGURE 2.19** A 3.25-carat diamond.

## CONCEPT TEST

Which contains more atoms: 1 gram of gold (Au) or 1 gram of silver (Ag)?

*(Answers to Concept Tests are in the back of the book.)*

### SAMPLE EXERCISE 2.10 Interconverting Grams, Moles, Molecules, and Formula Units LO6

The ionic compound calcium carbonate, $CaCO_3$, is the active ingredient in a popular antacid tablet.

a. How many moles of $CaCO_3$ are in a tablet with a mass of 502 mg?
b. How many formula units of $CaCO_3$ are in that tablet?

**Collect and Organize** We are asked to convert a given mass of $CaCO_3$ into the number of moles and formula units in that mass. The molar masses of the elements in $CaCO_3$ are 40.078 g/mol of Ca, 12.011 g/mol of C, and 15.999 g/mol of O. The number of formula units in a mole of an ionic compound is equal to the Avogadro constant ($6.022 \times 10^{23}$).

**Analyze** To convert a mass of a compound in milligrams into an equivalent number of moles, we need to convert the mass from milligrams into grams and then divide by the molar mass. The molar mass of $CaCO_3$ is the sum of the molar masses of Ca and C plus three times the molar mass of O. Multiplying the number of moles by the Avogadro constant will give us the number of formula units. To estimate our answers, let's round the mass of the tablet to 500 mg, or 0.5 g, and approximate the molar mass of $CaCO_3$ by using whole-number values for the molar masses of Ca, C, and O:

$$(1 \text{ mol Ca} \times 40 \text{ g/mol}) + (1 \text{ mol C} \times 12 \text{ g/mol}) + (3 \text{ mol O} \times 16 \text{ g/mol}) = 100 \text{ g/mol } CaCO_3$$

Therefore, the answer to part a should be about 0.5 g/100 g/mol = 0.005 mol, and the answer to part b should be about 0.005 mol × ($6 \times 10^{23}$ formula units/mol), or about $3 \times 10^{21}$ formula units.

**Solve**

a. The molar mass of $CaCO_3$ is

$$\left(\frac{1 \text{ mol Ca}}{1 \text{ mol } CaCO_3} \times \frac{40.078 \text{ g}}{1 \text{ mol Ca}}\right) + \left(\frac{1 \text{ mol C}}{1 \text{ mol } CaCO_3} \times \frac{12.011 \text{ g}}{1 \text{ mol C}}\right) + \left(\frac{3 \text{ mol O}}{1 \text{ mol } CaCO_3} \times \frac{15.999 \text{ g}}{1 \text{ mol O}}\right)$$
$$= 100.086 \text{ g/mol } CaCO_3$$

Converting from milligrams of $CaCO_3$ to moles of $CaCO_3$:

$$502 \text{ mg } CaCO_3 \times \frac{1 \text{ g}}{1000 \text{ mg}} \times \frac{1 \text{ mol } CaCO_3}{100.086 \text{ g } CaCO_3} = 0.00502 \text{ mol } CaCO_3$$

b. Multiplying by the Avogadro constant:

$$0.00502 \text{ mol } CaCO_3 \times \frac{6.022 \times 10^{23} \text{ formula units}}{1 \text{ mol}} = 3.02 \times 10^{21} \text{ formula units } CaCO_3$$

**Think About It** The calculated values agree well with our estimates. As expected, the number of formula units in a relatively small mass is so enormous that we have no practical way to "count them." However, combining the concept of molar mass with the Avogadro constant allows us to calculate how many formula units are in a sample of known mass.

**Practice Exercise** A standard aspirin tablet contains 325 mg of aspirin, which has the molecular formula $C_9H_8O_4$. How many moles and how many molecules of aspirin are in one tablet?

*(Answers to Practice Exercises are in the back of the book.)*

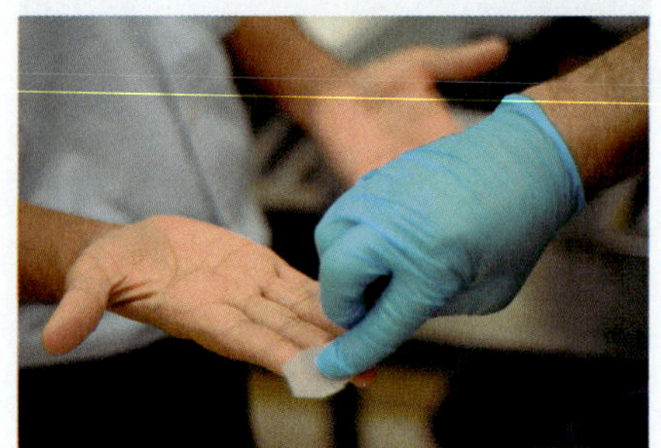

(a)

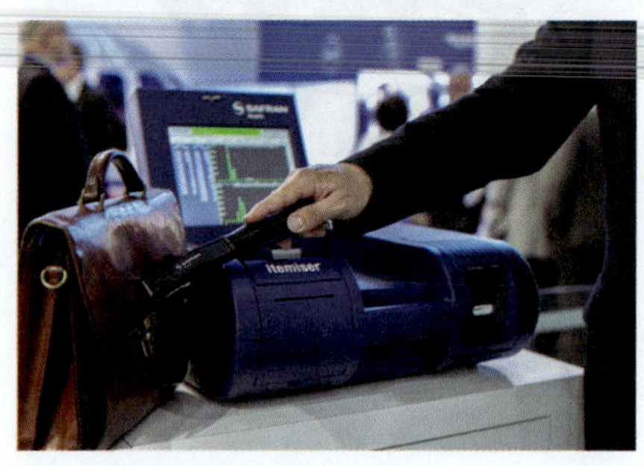

(b)

**FIGURE 2.20** Airport security officials screen passengers for explosives by (a) swabbing a passenger's hands with a piece of cloth and (b) using a mass spectrometer to analyze the cloth.

## 2.6 Mass Spectrometry: Determining Molecular Masses

When you check in for a flight at many airports, a security worker may wipe the handle on your luggage or your hands with a small piece of cloth to check for the presence of explosives (**Figure 2.20a**). The instrument used to rapidly analyze the swab (**Figure 2.20b**) is a **mass spectrometer**, a technologically advanced version of Aston's positive-ray analyzer. In this section we explore how mass spectrometers can be used to determine the molar mass and even the identities of compounds, enabling the detection of explosives at airports or banned substances in athletes.

Inside mass spectrometers, atoms and molecules are converted into ions that are then separated according to the ratio of their masses ($m$) to their electric charges ($z$). A common way to produce ions in a mass spectrometer involves vaporizing the sample and then bombarding the vapor with a beam of high-energy electrons. When one of those electrons collides with a molecule, it may knock out one of the molecule's own electrons, forming a **molecular ion ($M^+$)**. That is what happens to the molecule of benzene depicted in **Figure 2.21**.

Other collisions in the spectrometer may cause molecules to break apart into fragment ions. When those ions and the molecular ion are separated according to their $m/z$ values and then reach a detector, the resulting signals are used to create a graphical display called a **mass spectrum**, in which the $m/z$ values of the ions are plotted on the horizontal axis and the intensity (the number of ions with a particular $m/z$ value) on the vertical axis. Often the charge on every ion is 1+, so the $m/z$ ratio is simply $m$. Therefore, the mass of a molecular ion or fragment ion can be read directly from the position of its peak on the horizontal axis.

**Figure 2.22** shows mass spectra for acetylene ($C_2H_2$) and benzene ($C_6H_6$). The information in mass spectra such as those allows scientists to know the molecular mass of compounds with high precision and accuracy. For now, we concentrate on the molecular-ion peak, which is often the prominent peak in a mass spectrum with the largest mass.

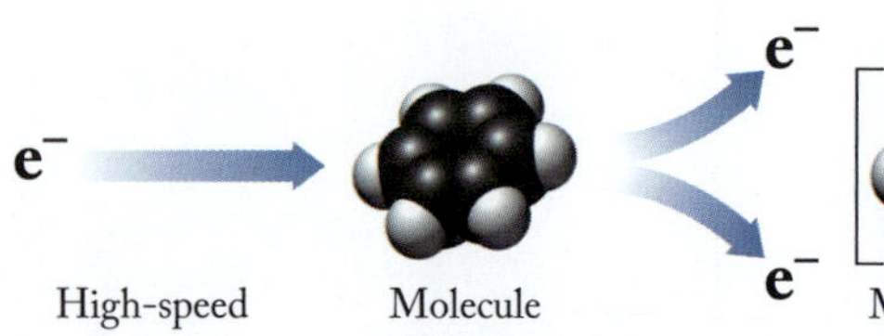


**FIGURE 2.21** In many mass spectrometers, atoms or molecules—such as the benzene molecule shown here—are bombarded with a beam of high-energy electrons to make atomic or molecular ions.

For acetylene (Figure 2.22a), the molecular-ion peak is at 26 u, which corresponds to the molecular mass of $C_2H_2$:

$$\left(\frac{2\ \cancel{\text{atoms C}}}{\text{molecule } C_2H_2} \times \frac{12.001\ \text{u}}{1\ \cancel{\text{atom C}}}\right) + \left(\frac{2\ \cancel{\text{atoms H}}}{\text{molecule } C_2H_2} \times \frac{1.0079\ \text{u}}{1\ \cancel{\text{atom H}}}\right) = \frac{26.038\ \text{u}}{\text{molecule } C_2H_2}$$

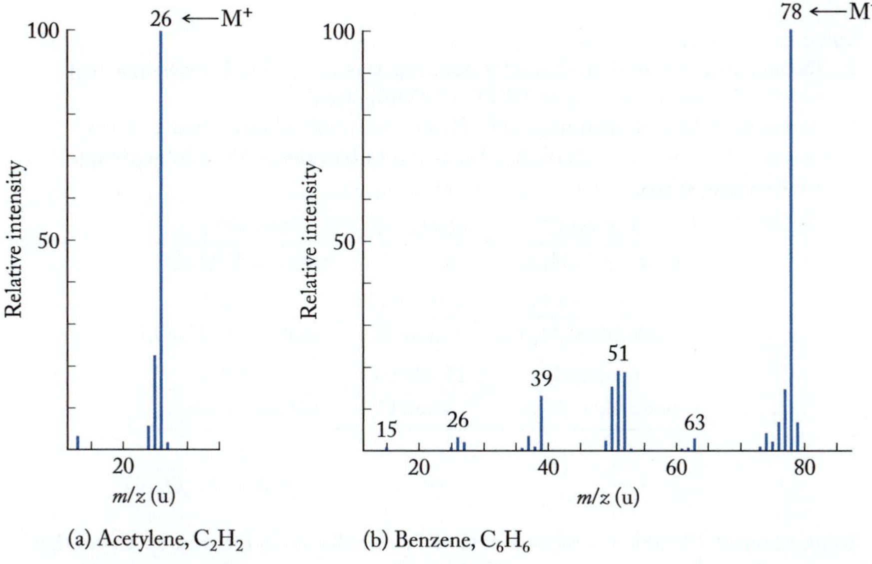


**FIGURE 2.22** Mass spectra of (a) acetylene and (b) benzene.

**mass spectrometer** a device that separates and counts ions according to their ratio of mass (*m*) to charge (*z*), *m*/*z*.

**molecular ion (M⁺)** an ion formed in a mass spectrometer when a molecule loses an electron after being bombarded with high-energy electrons. The molecular ion has a charge of 1+ and has essentially the same molecular mass as the molecule from which it came.

**mass spectrum** a graph of data from a mass spectrometer featuring peaks whose heights are measures of the numbers of ions of different masses produced from a sample in the spectrometer's ion source.

**STEPWISE ANIMATION**

A Simplified Mass Spectrometer

For benzene (Figure 2.22b), the molecular-ion peak has a mass of 78 u, consistent with the molecular mass of $C_6H_6$:

$$\left(\frac{6 \text{ atoms C}}{\text{molecule } C_6H_6} \times \frac{12.001 \text{ u}}{1 \text{ atom C}}\right) + \left(\frac{6 \text{ atoms H}}{\text{molecule } C_6H_6} \times \frac{1.0079 \text{ u}}{1 \text{ atom H}}\right) = \frac{78.113 \text{ u}}{\text{molecule } C_6H_6}$$

The other peaks in mass spectra such as those in Figure 2.22 are also useful in confirming the identity of a compound. Those fragment ion peaks at lower *m*/*z* values represent fragments of the molecule that survived electron bombardment intact, except for the loss of an electron. Distinctive fragmentation patterns help scientists confirm molecular structures.

**SAMPLE EXERCISE 2.11** Determining Molecular Mass by Mass Spectrometry **LO7**

The explosive compound TATP (triacetone triperoxide) is a major concern for law enforcement officials because it can be synthesized from readily available ingredients. Fortunately for airport security, TATP can be detected by its mass spectrum, shown in **Figure 2.23**.

a. What is the mass of the molecular-ion peak in Figure 2.23?
b. Show that this mass is consistent with the formula of TATP: $C_9H_{18}O_6$.

**Collect, Organize, and Analyze** The peak with the highest mass in a mass spectrum is often the molecular-ion peak. Its mass is the molecular mass of the compound. Once we determine the molecular mass, we compare it with the molecular mass derived from the chemical formula. To obtain a compound's molecular mass, we need the average atomic masses of its elements: 1.0079 u/atom of H, 12.011 u/atom of C, and 15.999 u/atom of O.

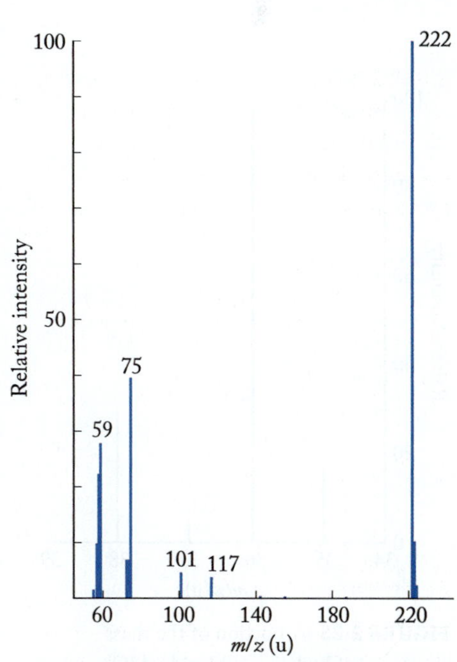


**FIGURE 2.23** A portion of the mass spectrum of the explosive compound TATP.

**nanoparticle** approximately spherical sample of matter with dimensions less than 100 nm ($1 \times 10^{-7}$ m).

**Solve**

a. The molecular-ion peak for TATP is observed at $m/z = 222$ u. If we assume that $z = +1$, the molecular mass of TATP is 222 u/molecule.

b. To calculate the molecular mass of $C_9H_{18}O_6$, we simply add the average atomic masses of the atoms in each of its molecules: 9 carbon atoms, 18 hydrogen atoms, and 6 oxygen atoms:

$$\frac{9 \text{ atoms C}}{\text{molecule } C_9H_{18}O_6} \times \frac{12.011 \text{ u C}}{\text{atom C}} = \frac{108.099 \text{ u}}{\text{molecule } C_9H_{18}O_6}$$

$$+ \frac{18 \text{ atoms H}}{\text{molecule } C_9H_{18}O_6} \times \frac{1.0079 \text{ u H}}{\text{atom H}} = \frac{18.1422 \text{ u}}{\text{molecule } C_9H_{18}O_6}$$

$$+ \frac{6 \text{ atoms O}}{\text{molecule } C_9H_{18}O_6} \times \frac{15.999 \text{ u O}}{\text{atom O}} = \frac{95.994 \text{ u}}{\text{molecule } C_9H_{18}O_6}$$

$$\frac{222.235 \text{ u}}{\text{molecule } C_9H_{18}O_6}$$

**Think About It** Detection of the $m/z$ ratio of a molecular ion in the mass spectrum that corresponds to the molecular mass of TATP is strong evidence of the presence of the compound. The fragmentation pattern of the sample should also match that of TATP when analyzed in a comparable mass spectrometer, confirming the presence of TATP in the sample.

**Practice Exercise** Mass spectrometry also is used to detect banned substances in athletes. The mass spectrum of a substance suspected to be testosterone is shown in **Figure 2.24**. Find the molecular-ion peak in the mass spectrum. Does its mass match the molecular mass of testosterone, whose formula is $C_{19}H_{28}O_2$?

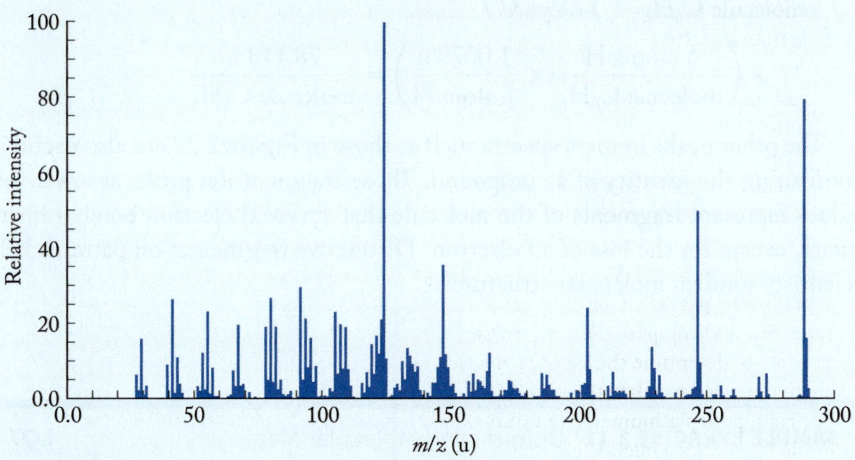


**FIGURE 2.24** A portion of the mass spectrum of testosterone.

*(Answers to Practice Exercises are in the back of the book.)*

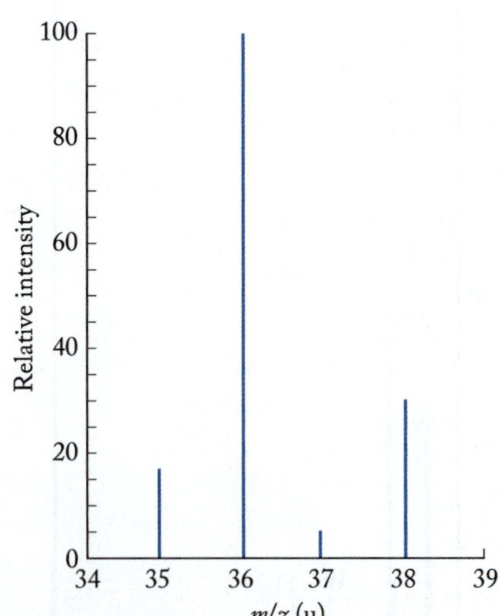


**FIGURE 2.25** A portion of the mass spectrum of hydrogen chloride, HCl.

## CONCEPT TEST

The mass spectrum of HCl is shown in **Figure 2.25**, and the peak at $m/z = 36$ is the molecular ion. Use the fact that chlorine has two isotopes ($^{35}Cl$, with 76% abundance, and $^{37}Cl$, with 24% abundance) to explain (a) the peak at $m/z = 38$ and (b) the relative intensity of the peak at $m/z = 38$ to that of the molecular ion.

*(Answers to Concept Tests are in the back of the book.)*

## SAMPLE EXERCISE 2.12 Integrating Concepts: Gold/Platinum Nanoparticles

For centuries, people believed that metallic gold had medicinal properties. In recent years, tiny gold **nanoparticles** with diameters less than $10^{-7}$ m have been used in medicine. The therapeutic properties of gold nanoparticles by themselves are limited, but mixing gold with platinum yields materials with antibiotic properties.

a. Do gold (Au) and platinum (Pt) belong to the same group in the periodic table?
b. Are gold and platinum best described as metals, nonmetals, or transition metals?
c. Platinum has six isotopes with the following natural abundances:

| Isotope | Mass (u) | Natural Abundance (%) |
|---|---|---|
| $^{190}Pt$ | 189.96 | 0.014 |
| $^{192}Pt$ | 191.96 | 0.782 |
| $^{194}Pt$ | 193.96 | 32.967 |
| $^{195}Pt$ | 194.97 | 33.832 |
| $^{196}Pt$ | 195.97 | 25.242 |
| $^{198}Pt$ | 197.97 | 7.163 |

Use these data to calculate the average atomic mass of platinum.
d. Which has more neutrons, the most abundant isotope of platinum or $^{197}Au$?
e. Will more atoms of gold or platinum be present in a nanoparticle containing 50.0% gold and 50.0% platinum by mass?
f. Suppose we have two cubes that are both 1.00 mm on a side. One is pure gold and the other, pure platinum. How many atoms are in each cube?

**Collect and Organize** We need to locate two elements, Au and Pt, in the periodic table, classify them according to their properties, and compare the composition of their nuclei. We are asked to determine the average atomic mass of platinum. Finally, we need to relate the number of atoms in samples of pure gold, pure platinum, and a mixture of the two metals. The atomic number of a nuclide is equal to the number of protons in its nucleus; its mass number is equal to the number of nucleons (protons plus neutrons). Equation 2.3 may be used to calculate the average atomic mass of an element, given the exact masses and natural abundances of its stable isotopes. The densities of the two metals (Table A3.2) are 19.3 g/cm$^3$ of Au and 21.45 g/cm$^3$ of Pt.

**Analyze** In part e we have a nanoparticle that contains equal masses of gold and platinum. More atoms will be present of the element with the smaller atomic mass because it will take more of them to have the same mass as the other metal. In part f we first need to calculate the masses of two cubes that each have a volume of 1.00 mm$^3$. Doing so involves converting each volume into cubic centimeters and then multiplying by density values expressed in grams per cubic centimeter. The number of atoms in each cube is calculated by dividing its mass by the appropriate molar mass to get the number of moles and then multiplying that value by the Avogadro constant.

**Solve**

a. Platinum and gold are in different columns of the periodic table, which means they are in different groups: group 10 for platinum and group 11 for gold.
b. Both Pt and Au are classified as metals. More specifically, both are transition metals.
c. Using Equation 2.3 to calculate the average atomic mass of platinum (carrying one more digit than allowed under the rules concerning significant figures):

Average atomic mass =

$$\begin{aligned}
&189.96\text{ u} \times 0.00014 = 0.0266\text{ u}\\
&+191.96\text{ u} \times 0.00782 = 1.5011\text{ u}\\
&+193.96\text{ u} \times 0.32967 = 63.9428\text{ u}\\
&+194.97\text{ u} \times 0.33832 = 65.9622\text{ u}\\
&+195.97\text{ u} \times 0.25242 = 49.4667\text{ u}\\
&+197.97\text{ u} \times 0.07163 = 14.1806\text{ u}\\
&\phantom{+197.97\text{ u} \times 0.07163 =}\ 195.0800\text{ u}
\end{aligned}$$

Rounding the sum to the appropriate number of significant figures gives 195.080 u.

d. The nuclei of atoms of Pt and Au contain 78 and 79 protons, respectively. The most abundant platinum isotope is $^{195}Pt$. The numbers of neutrons in the nuclei of platinum-195 and gold-197 atoms are

$$^{195}\text{Pt} = 195 - 78 = 117 \text{ neutrons}$$
$$^{197}\text{Au} = 197 - 79 = 118 \text{ neutrons}$$

Gold-197 has one more neutron than platinum-195.

e. Equal masses of Au and Pt are present in each nanoparticle, but each atom of Au has, on average, a larger mass than an atom of Pt. Therefore, each nanoparticle has more Pt atoms than Au atoms.
f. Each cube is 1.0 mm, or 0.10 cm, on a side. Therefore, each has a volume of

$$0.10 \text{ cm} \times 0.10 \text{ cm} \times 0.10 \text{ cm} = 1.0 \times 10^{-3} \text{ cm}^3$$

Calculating the number of atoms in the gold and platinum cubes:

$$1.0 \times 10^{-3} \text{ cm}^3 \text{ Au} \times \frac{19.3 \text{ g Au}}{1 \text{ cm}^3 \text{ Au}} \times \frac{1 \text{ mol Au}}{196.97 \text{ g Au}} \times \frac{6.022 \times 10^{23} \text{ atoms Au}}{1 \text{ mol Au}} = 5.9 \times 10^{19} \text{ atoms Au}$$

$$1.0 \times 10^{-3} \text{ cm}^3 \text{ Pt} \times \frac{21.45 \text{ g Pt}}{1 \text{ cm}^3 \text{ Pt}} \times \frac{1 \text{ mol Pt}}{195.08 \text{ g Pt}} \times \frac{6.022 \times 10^{23} \text{ atoms Pt}}{1 \text{ mol Pt}} = 6.6 \times 10^{19} \text{ atoms Pt}$$

The cube of platinum contains more atoms than the cube of gold.

**Think About It** A nanoparticle containing equal masses of Au and Pt has fewer gold atoms than platinum atoms because it takes more Pt atoms to have the same mass as a given mass of Au.

## SUMMARY

**LO1** The values of the charge and mass of the electron were determined from J. J. Thomson's studies using cathode-ray tubes and from Robert Millikan's oil-drop experiments. Ernest Rutherford's group bombarded thin gold foil with **alpha (α) particles** and discovered that the positive charge and nearly all the mass of an atom are contained in its nucleus. (Section 2.1)

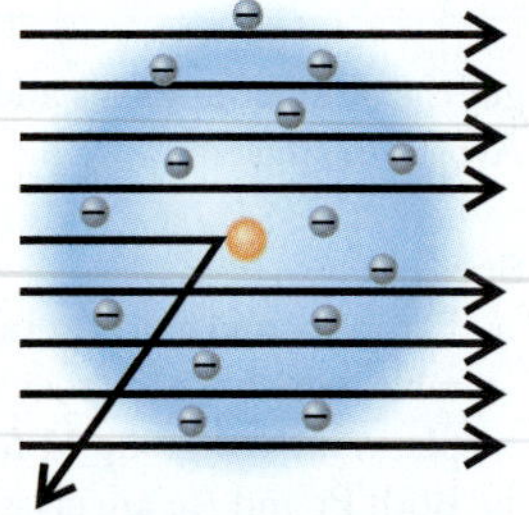

**LO2** Atoms are composed of negatively charged **electrons** surrounding a **nucleus**, which contains positively charged **protons** and electrically neutral **neutrons**. The number of protons in the nucleus of an element defines its **atomic number (*Z*)**; the number of **nucleons** (protons + neutrons) in the nucleus defines the element's **mass number (*A*)**. The **isotopes** of an element consist of atoms with the same number of protons per nucleus but different numbers of neutrons. Symbols for subatomic particles and atoms list the symbol for the particle (X) with the value of *A* as a superscript and the value of *Z* as a subscript: $^{A}_{Z}X$. (Sections 2.1 and 2.2)

**LO3** Elements are arranged in the **periodic table of the elements** in order of increasing atomic number and in a pattern based on their chemical properties, including the charges of the monatomic ions they form. Elements in the same column are in the same **group** and have similar properties. An atom or group of atoms having a net charge is called an ion. If the charge is positive, it is a **cation**; if the charge is negative, it is an **anion**. Elements in groups 1, 2, and 13–18 are **main group** (or **representative**) **elements**. The **transition metals** are in groups 3–12. **Metals** are mostly malleable, ductile solids; they form cations and are good conductors of heat and electricity. **Nonmetals** include elements in all three physical states; they form anions and are poor conductors of heat and electricity. **Metalloids** have the physical properties of metals and chemical properties of nonmetals. (Section 2.3)

**LO4** To calculate the **average atomic mass** of an element, multiply the mass of each of its stable isotopes by the **natural abundance** of that isotope as a percentage and then sum the products. (Section 2.4)

**LO5** The **mole (mol)** is the SI base unit for quantity of substances. One mole of a substance consists of $6.022 \times 10^{23}$ (the **Avogadro constant, $N_A$**) particles of the substance. The mass of one mole of a substance is its **molar mass (ℳ)**. The Avogadro constant and a substance's molar mass can be used to interconvert that substance's mass in grams, its quantity in moles, and its number of particles. (Section 2.5)

**LO6** The **molecular mass** of a compound is the sum of the average atomic mass of each atom in one of its molecules. The formula of an ionic compound defines the simplest combination of its ions that gives a neutral **formula unit** of the compound, which has a corresponding **formula mass.** (Sections 2.4 and 2.5)

**LO7** The *m*/*z* values for the **molecular ions, M⁺**, in the **mass spectra** of atoms and molecules allow us to determine their molar masses. (Section 2.6)

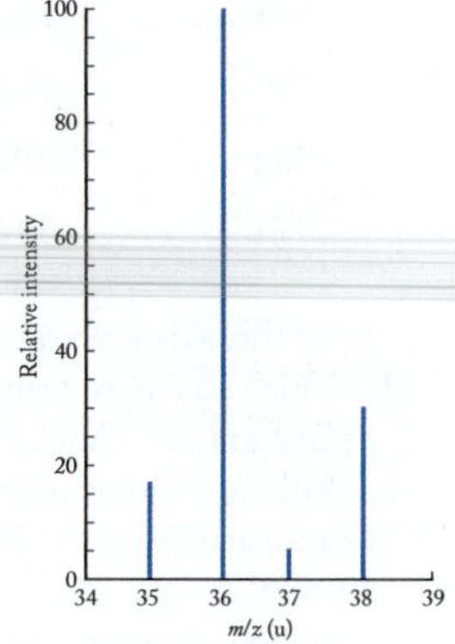


## PARTICULATE PREVIEW WRAP-UP

Counting the protons and neutrons in the three nuclei in the figure yields the values in the following table, in which each mass number (*A*) is the sum of the numbers of protons and neutrons, and each atomic number (*Z*) is equal to the number of protons in that nucleus. Nuclei (b) and (c) have the same atomic number (6), which makes them isotopes of the same element (carbon).

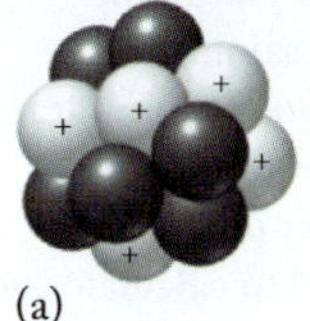
(a)

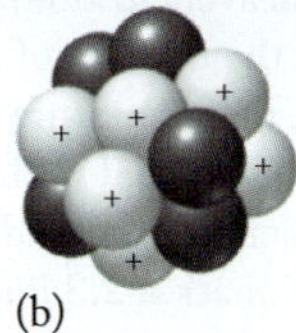
(b)

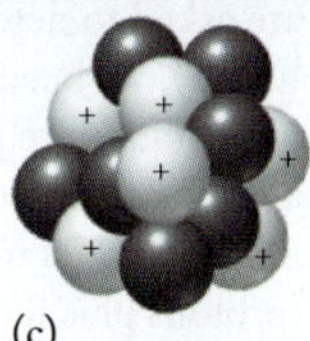
(c)

| Nucleus | Number of protons | Number of neutrons | *A* | *Z* |
|---|---|---|---|---|
| (a) | 5 | 6 | 11 | 5 |
| (b) | 6 | 5 | 11 | 6 |
| (c) | 6 | 7 | 13 | 6 |

## PROBLEM-SOLVING SUMMARY

| Type of Problem | Concepts and Equations | Sample Exercises |
|---|---|---|
| **Writing symbols of nuclides and ions** | Place a superscript for the mass number (*A*) and a subscript for the atomic number (*Z*) to the left of the element symbol. If the particle is a monatomic ion, add its charge as a superscript after the symbol. | **2.1, 2.2** |
| **Navigating the periodic table** | Use row numbers to identify periods in the periodic table, and use column numbers to identify groups. Groups with special names include the alkali metals (group 1), alkaline earth metals (group 2), chalcogens (group 16), halogens (group 17), and noble gases (group 18). | **2.3** |
| **Calculating the average atomic mass of an element** | Multiply the mass (*m*) of each stable isotope of the element by the natural abundance (*a*) of that isotope; then sum the products: $m_X = a_1m_1 + a_2m_2 + a_3m_3 + \cdots$ (2.3) | **2.4, 2.5** |
| **Converting number of particles into number of moles (or vice versa)** | Convert number of particles into number of moles by dividing by the Avogadro constant. Convert number of moles into number of particles by multiplying by the Avogadro constant. | **2.6, 2.10** |
| **Converting mass of a substance into number of moles (or vice versa)** | Convert mass of the substance to number of moles by dividing by the molar mass ($\mathcal{M}$) of the substance. Convert number of moles of the substance to mass by multiplying by the molar mass ($\mathcal{M}$) of the substance. | **2.8–2.10** |
| **Calculating the molar mass of a compound** | Sum the molar masses of the elements in the compound's formula, with each element multiplied by the number of atoms of that element in one molecule or formula unit of the compound. | **2.7, 2.10** |
| **Determining molecular mass by mass spectrometry** | Identify the mass spectrum's molecular-ion peak. Its mass is the molecular mass of the compound. | **2.11** |

## VISUAL PROBLEMS

*(Answers to boldface end-of-chapter questions and problems are in the back of the book.)*

**2.1.** Atoms of which highlighted element in Figure P2.1 have the fewest protons per nucleus? Which element is that?

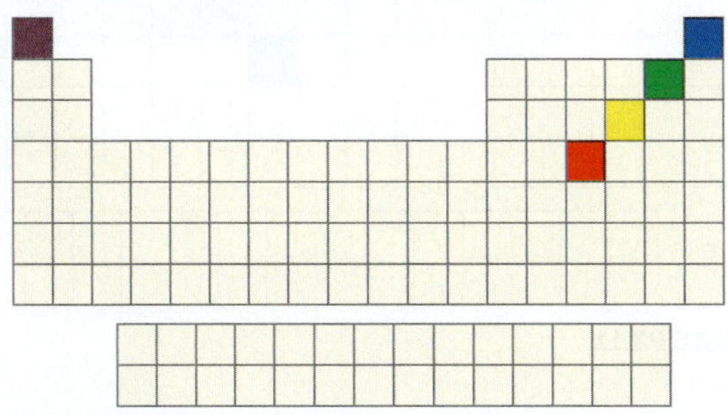

FIGURE P2.1

2.2. Atoms of which highlighted element in Figure P2.1 have, on average, the most neutrons?

**2.3.** Which highlighted element in Figure P2.1 has a stable isotope with no neutrons in its nucleus?

2.4. Which highlighted element in Figure P2.4 has no stable isotopes?

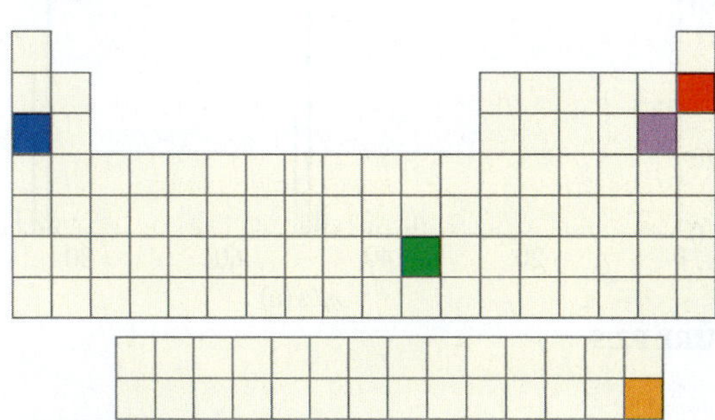

FIGURE P2.4

**2.5.** Which highlighted element in Figure P2.4 is (a) a transition metal; (b) an alkali metal; (c) a halogen; (d) a chemically inert gas?

2.6. Write nuclide symbols of the form $^{A}X$ for the nuclei represented in Figure P2.6. Assume that all nucleons are visible.

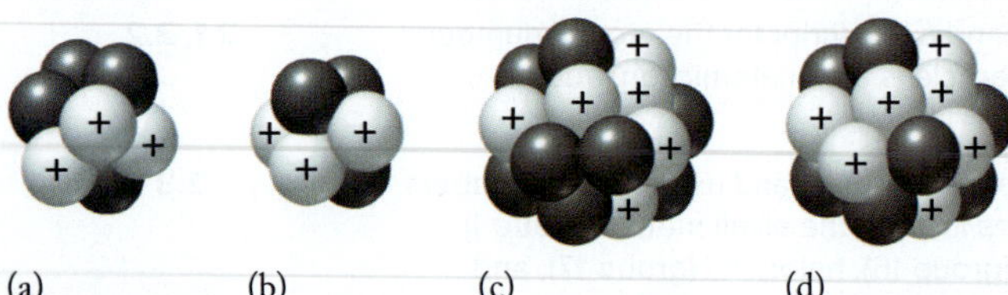


**FIGURE P2.6**

**2.7.** Alpha and beta particles emitted by a sample of pitchblende escape through a narrow channel in the shielding surrounding the sample and into an electric field, as shown in Figure P2.7. Identify which colored arrow corresponds to each of those forms of radiation.

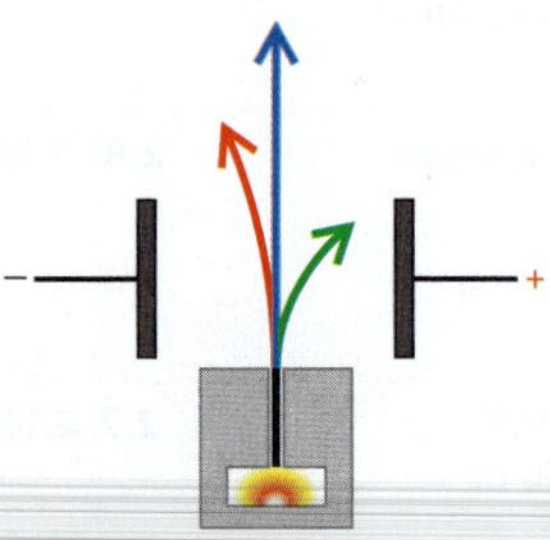


**FIGURE P2.7**

2.8. Which subatomic particle would curve in the same direction as the red arrow in Figure P2.7?

***2.9.** Dichloromethane ($CH_2Cl_2$, 84.93 g/mol) and cyclohexane ($C_6H_{12}$, 84.15 g/mol) have nearly the same molar masses. Which compound produced the mass spectrum shown in Figure P2.9? Explain your selection.

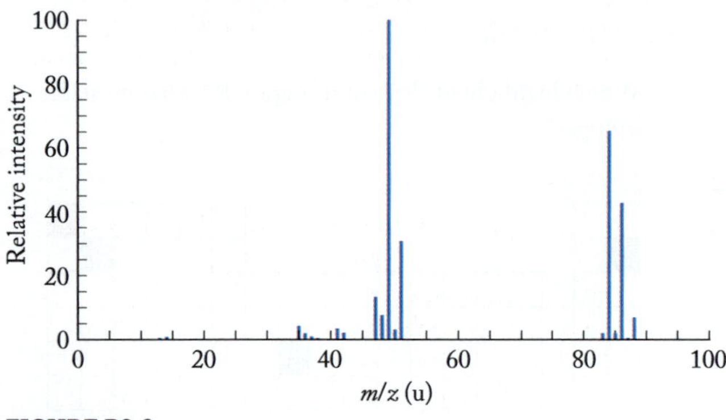


**FIGURE P2.9**

2.10. Krypton has six stable isotopes. How many neutrons are in the most abundant isotope of krypton according to the mass spectrum in Figure P2.10? Write the $^{A}X$ form of the symbol of that isotope.

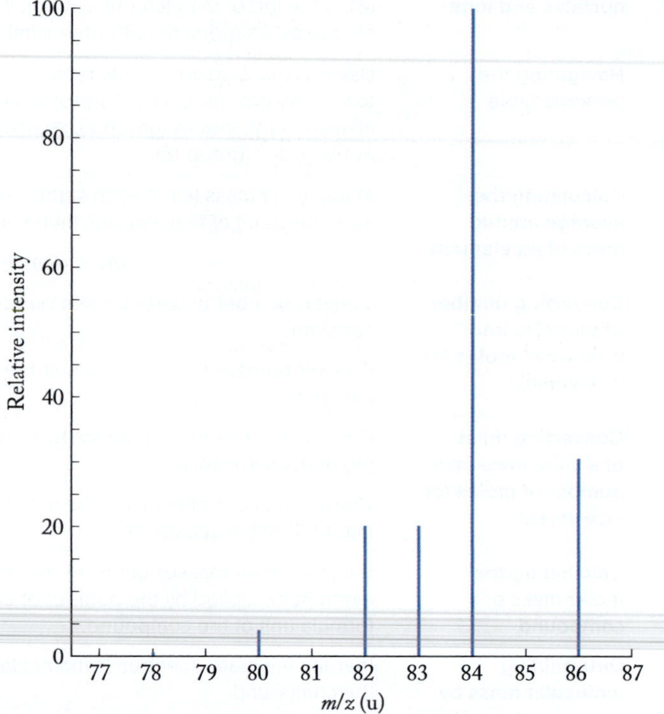


**FIGURE P2.10**

**2.11.** Which highlighted element in Figure P2.11 forms monatomic ions with a charge of (a) 1+; (b) 2+; (c) 3+; (d) 1−; (e) 2−?

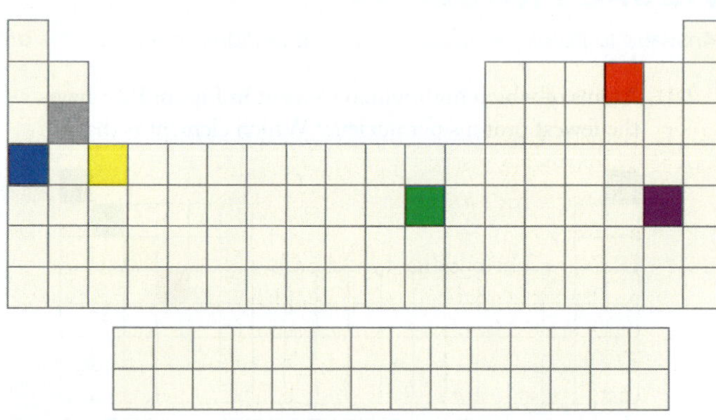

**FIGURE P2.11**

2.12. Use representations [A] through [I] in Figure P2.12 to answer questions a–f. (This book's inside back cover shows the colors used to represent the atoms of different elements.)

a. According to the ratio of cations to anions in representations [A] and [I], which compound is potassium iodide and which is potassium oxide?
b. Order the nuclei from the fewest neutrons to the most neutrons.
c. Which representations depict isotopes?
d. What is the mass of the molecule in representation [E]?
e. Which would contain more molecules, 100 g of [C] or 100 g of [G]?
f. Which would contain more sulfur atoms, 100 g of [C] or 100 g of [G]?

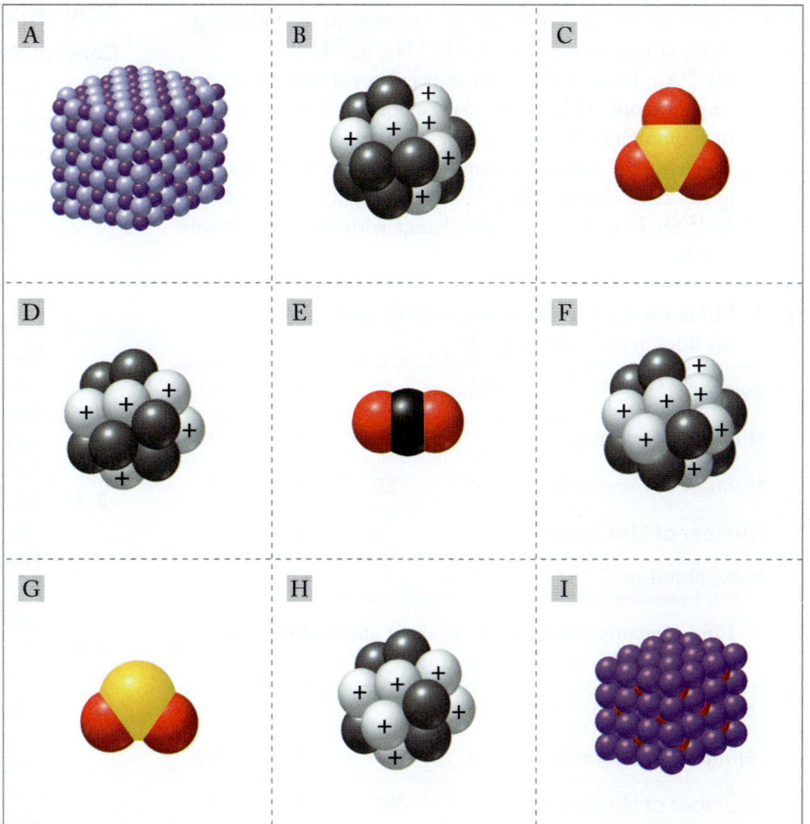

**FIGURE P2.12**

## QUESTIONS AND PROBLEMS

### *The Rutherford Model of Atomic Structure*

**Concept Review**

**2.13.** Explain how the results of the gold-foil experiment led Rutherford to dismiss the plum-pudding model of the atom and create his own model based on a nucleus surrounded by electrons.

2.14. Had the plum-pudding model been valid, how would the results of the gold-foil experiment have differed from what Geiger and Marsden actually observed?

**2.15.** What properties of cathode rays led Thomson to conclude that they were not rays of energy but rather particles with an electric charge?

2.16. Describe two ways in which α particles and β particles differ.

***2.17.** **Helium in Pitchblende** The element helium was first discovered on Earth in a sample of pitchblende, an ore of radioactive uranium oxide. How did helium get in the ore?

2.18. How might using a thicker piece of gold foil have affected the scattering pattern of α particles observed by Rutherford's students?

***2.19.** What would have happened to the gold atoms in Rutherford's experiment if their nuclei had absorbed β particles?

*2.20. In addition to gold foil, Geiger and Marsden tried silver and aluminum foils in their experiment. Why might foils of those metals have deflected fewer α particles than gold foil?

### *Nuclides and Their Symbols*

**Concept Review**

**2.21.** If the mass number of a nuclide is more than twice the atomic number, is the neutron-to-proton ratio less than, greater than, or equal to 1?

2.22. Oxygen has three stable isotopes: $^{16}O$, $^{17}O$, and $^{18}O$. Compare those isotopes with regard to the number of (a) neutrons; (b) protons; (c) nucleons; (d) electrons.

**2.23.** Nearly all stable nuclides have at least as many neutrons as protons in their nuclei. Which common nuclide is an exception?

2.24. Why does the mass number of an atom of $^{19}F$ not change when it acquires an electron and becomes $^{19}F^-$?

**Problems**

**2.25.** How many protons, neutrons, and electrons are in the following atoms? (a) $^{14}C$; (b) $^{59}Fe$; (c) $^{90}Sr$; (d) $^{210}Pb$

2.26. How many protons, neutrons, and electrons are in the following atoms? (a) $^{11}B$; (b) $^{19}F$; (c) $^{131}I$; (d) $^{222}Rn$

**2.27.** Calculate the *ratio* of neutrons to protons in the following stable atomic nuclei: (a) $^{4}$He; (b) $^{23}$Na; (c) $^{59}$Co; and (d) $^{197}$Au. Each of those elements exists naturally as a single isotope. What trend do you observe for the neutron-to-proton ratio as $Z$ increases?

2.28. Calculate the *ratio* of neutrons to protons in the following group 15 nuclei: (a) $^{14}$N; (b) $^{31}$P; (c) $^{75}$As; (d) $^{121}$Sb; and (e) $^{123}$Sb. How does the ratio change with increasing atomic number?

**2.29.** Fill in the missing information about atoms of the four nuclides in the following table.

| Symbol | $^{23}$Na | ? | ? | ? |
|---|---|---|---|---|
| Number of Protons | ? | 39 | ? | 79 |
| Number of Neutrons | ? | 50 | ? | ? |
| Number of Electrons | ? | ? | 50 | ? |
| Mass Number | ? | ? | 118 | 197 |

2.30. Fill in the missing information about atoms of the four nuclides in the following table.

| Symbol | $^{27}$Al | ? | ? | ? |
|---|---|---|---|---|
| Number of Protons | ? | 42 | ? | 92 |
| Number of Neutrons | ? | 56 | ? | ? |
| Number of Electrons | ? | ? | 60 | ? |
| Mass Number | ? | ? | 143 | 238 |

**2.31.** Fill in the missing information about the monatomic ions in the following table.

| Symbol | $^{37}Cl^{-}$ | ? | ? | ? |
|---|---|---|---|---|
| Number of Protons | ? | 11 | ? | 88 |
| Number of Neutrons | ? | 12 | 46 | ? |
| Number of Electrons | ? | 10 | 36 | 86 |
| Mass Number | ? | ? | 81 | 226 |

2.32. Fill in the missing information about the monatomic ions in the following table.

| Symbol | $^{137}Ba^{2+}$ | ? | ? | ? |
|---|---|---|---|---|
| Number of Protons | ? | 30 | ? | 40 |
| Number of Neutrons | ? | 34 | 16 | ? |
| Number of Electrons | ? | 28 | 18 | 36 |
| Mass Number | ? | ? | 32 | 90 |

## Navigating the Periodic Table

### Concept Review

**2.33.** Mendeleev arranged the elements on the left side of his periodic table on the basis of the formulas of the binary compounds they formed with oxygen, and he used the formulas as column labels. For example, group 1 in a modern periodic table was labeled "$R_2O$" in Mendeleev's table, where "R" represented one of the elements in the group. What labels did Mendeleev use for groups 2–4 from the modern periodic table?

2.34. Mendeleev arranged the elements on the right side of his periodic table according to the formulas of the binary compounds they formed with hydrogen and used those formulas as column labels. Which groups in the modern periodic table were labeled "HR," "$H_2R$," and "$H_3R$," where "R" represented one of the elements in the group?

**2.35.** Mendeleev left empty spaces in his periodic table for elements he suspected existed but had yet to be discovered. However, he left no spaces for the noble gases (group 18 in the modern periodic table). Suggest a reason why he left no spaces for them.

2.36. Describe how the appearance of the modern periodic table inside the front cover of this book would change if the elements in the columns labeled "A" and "B" were arranged in separate rows and stacked so that, for example, the $n$A element of the fourth row was directly above the $n$B element of the fourth row: K (1A) above Cu (1B), Ca (2A) above Zn (2B), Ga (3A) above Sc (3B), and so on. Compare such an arrangement of elements with Mendeleev's table in Figure 2.9(a).

### Problems

**2.37.** **TNT** Molecules of the explosive TNT contain atoms of hydrogen and second-row elements in groups 14–16. Which three elements are they?

2.38. **Chemical Weapons** Phosgene was used as a chemical weapon during World War I. Despite the name, phosgene molecules contain no atoms of phosphorus. Instead, they contain atoms of carbon and the group 16 element in the second row of the periodic table and the group 17 element in the third row. What are the identities and atomic numbers of the two elements?

**2.39.** **Catalytic Converters** The catalytic converters used to remove pollutants from automobile exhaust contain compounds of several fairly expensive elements, including those described in the following list. Which elements are they?

a. The group 10 transition metal in the fifth row of the periodic table
b. The transition metal with one fewer proton in each of its atoms than your answer to part a
c. The transition metal with 32 more protons in each of its atoms than your answer to part a

2.40. **Swimming Pool Chemistry** Compounds containing chlorine have long been used to disinfect the water in swimming pools, but in recent years a compound of a less corrosive halogen has become a popular alternative disinfectant. What is the name of that fourth-row element?

**2.41.** How many metallic elements are in the third row of the periodic table?

2.42. Which third-row element in the periodic table has chemical properties of a nonmetal but physical properties of a metal?

## The Masses of Atoms, Ions, and Molecules

### Concept Review

**2.43.** What is meant by a *weighted average*?

2.44. A hypothetical element consists of two isotopes (X and Y) with masses $m_X$ and $m_Y$. If the natural abundance of the X isotope is exactly 50%, what is the average atomic mass of the element?

**2.45.** In calculating the formula masses of binary ionic compounds, we use the average masses of neutral atoms, not ions. Why?

2.46. Boron, lithium, and nitrogen each have two stable isotopes. From the average atomic masses of those elements, predict which isotope in each of the following pairs of stable isotopes is the more abundant. (a) $^{10}B$ or $^{11}B$; (b) $^{6}Li$ or $^{7}Li$; (c) $^{14}N$ or $^{15}N$

**2.47.** Argon atoms exist as one of three isotopes: $^{36}Ar$, $^{38}Ar$, and $^{40}Ar$. Which one is the most abundant?

2.48. The two most common isotopes of hydrogen have either no neutrons or 1 neutron. The two most common isotopes of oxygen have either 8 or 10 neutrons. What are the possible values for the molecular weight (in u) of a single water molecule?

***2.49.** The average mass of platinum is 195.08 u, yet the natural abundance of $^{195}Pt$ is only 33.8%. Propose an explanation for that observation.

*2.50. The average atomic mass of europium (Eu, 151.96 u), is only 0.04 u less than a whole number. Can we conclude that europium has only one stable isotope, $^{152}Eu$? Why or why not?

### Problems

**2.51.** A sample of hydrogen gas ($H_2$) from a natural gas well in Kansas contains 99.99745% $^{1}H$ (1.007825 u) and 0.00255% $^{2}H$ (2.014102 u). What is the average atomic mass of hydrogen in the sample? Compare your result with the average atomic mass of hydrogen in the periodic table inside the front cover of this book.

2.52. A water sample collected from a volcanic crater lake in southeastern Australia contains 81.07% $^{11}B$ (11.0093 u) and 18.93% $^{10}B$ (10.0129 u). What is the average atomic mass of the boron in the sample? Compare your result with the average atomic mass of boron in the periodic table inside the front cover of this book.

**2.53.** **Chemistry of Mars** Chemical analyses conducted by the first Mars Rover robotic vehicle in its 1997 mission produced the magnesium isotope data shown in the following table. Is the average atomic mass of magnesium in that Martian sample the same as on Earth (24.31 u)?

| Isotope | Mass (u) | Natural Abundance (%) |
|---|---|---|
| $^{24}Mg$ | 23.9850 | 78.70 |
| $^{25}Mg$ | 24.9858 | 10.13 |
| $^{26}Mg$ | 25.9826 | 11.17 |

2.54. A sample of groundwater from coastal South Carolina contains 7.73% $^{6}Li$ (6.015123 u) and 92.27% $^{7}Li$ (mass = 7.016003 u). What is the average atomic mass of lithium in the sample? Compare your result with the average atomic mass of lithium in the periodic table inside the front cover of this book.

**2.55.** Use the data in the following table of abundances and masses of the five stable titanium isotopes to calculate the atomic mass of $^{48}Ti$.

| Isotope | Mass (u) | Natural Abundance (%) |
|---|---|---|
| $^{46}Ti$ | 45.9526 | 8.25 |
| $^{47}Ti$ | 46.9518 | 7.44 |
| $^{48}Ti$ | ? | 73.72 |
| $^{49}Ti$ | 48.94787 | 5.41 |
| $^{50}Ti$ | 49.94479 | 5.18 |
| Average | 47.867 | |

2.56. Use the following table of abundances and masses of the stable isotopes of zirconium to calculate the atomic mass of $^{92}Zr$.

| Isotope | Mass (u) | Natural Abundance (%) |
|---|---|---|
| $^{90}Zr$ | 89.905 | 51.45 |
| $^{91}Zr$ | 90.906 | 11.22 |
| $^{92}Zr$ | ? | 17.15 |
| $^{94}Zr$ | 93.906 | 17.38 |
| $^{96}Zr$ | 95.908 | 2.80 |
| Average | 91.224 | |

**2.57.** What are the masses of the formula units of each of the following ionic compounds? (a) NaI; (b) CaS; (c) $Al_2O_3$; (d) $NH_4Cl$

2.58. What are the masses of the formula units of each of the following ionic compounds? (a) LiCl; (b) $TiO_2$; (c) $MgCO_3$; (d) $CaHPO_4$

**2.59.** How many carbon atoms are in one molecule of each of the following compounds? (a) $CO_2$; (b) $HOCH_2CH_2OH$; (c) $CH_3COOH$; (d) $C_2H_5OH$

2.60. How many oxygen atoms are in each molecule in Problem 2.59?

---

**2.61.** Rank the following compounds in order of increasing molecular mass. (a) $CS_2$; (b) HI; (c) $NO_2$; (d) HClO; (e) $C_4H_{10}$

2.62. Rank the following compounds in order of decreasing molecular mass. (a) $H_2O$; (b) $CO_2$; (c) $N_2O$; (d) $CH_2Cl_2$; (e) $BF_3$

## Moles and Molar Masses

### Concept Review

**2.63.** In principle, we could use the more familiar unit *dozen* in place of mole when expressing the quantities of particles (atoms, ions, or molecules). What would be the disadvantage in doing so?

2.64. Do equal masses of two isotopes of an element contain the same number of atoms?

### Problems

**2.65.** Earth's atmosphere contains trace concentrations of many volatile substances. The following quantities of trace gases were found in a 1.0 mL sample of air. Calculate the number of moles of each gas in the sample.

a. $4.4 \times 10^{14}$ atoms of Ne
b. $4.2 \times 10^{13}$ molecules of $CH_4$
c. $2.5 \times 10^{12}$ molecules of $O_3$
d. $4.9 \times 10^{9}$ molecules of $NO_2$

2.66. The following quantities of trace gases were found in a 1.0 mL sample of air. Calculate the number of moles of each compound in the sample.

a. $1.4 \times 10^{13}$ molecules of $H_2$
b. $1.5 \times 10^{14}$ atoms of He
c. $7.7 \times 10^{12}$ molecules of $N_2O$
d. $3.0 \times 10^{12}$ molecules of CO

---

**2.67.** How many moles of nitrogen are in one mole of the following compounds? (a) NO; (b) $NO_2$; (c) $NH_4NO_3$; (d) $N_2O_5$

2.68. How many moles of $K^+$ ions are in one mole of the following compounds? (a) KBr; (b) $K_2SO_4$; (c) $KH_2PO_4$; (d) $K_2O$

---

**2.69.** Which substance in each of the following pairs of quantities contains more moles of oxygen?

a. 1 mol of $MnO_2$ or 1 mol of $Fe_2O_3$
b. 1.5 mol of $CO_2$ or 1 mol of $N_2O_5$
c. 3 mol of NO or 2 mol of $SO_2$

2.70. Which substance in each of the following pairs of quantities contains more moles of oxygen?

a. 1 mol of $N_2O$ or 1 mol of $NO_2$
b. 1 mol of $H_2SO_4$ or 1.5 mol of $H_2SO_3$
c. 1 mol of $Ca_3(PO_4)_2$ or 1.5 mol of $Ca(HCO_3)_2$

---

**2.71.** Calculate the molar masses of the following gases. (a) $SO_2$; (b) $O_3$; (c) $CO_2$; (d) $N_2O_5$

2.72. Calculate the molar masses of the following minerals.

a. rhodonite, $MnSiO_3$
b. scheelite, $CaWO_4$
c. ilmenite, $FeTiO_3$
d. magnesite, $MgCO_3$

---

**2.73.** **Flavors** Calculate the molar masses of the following common flavors in food.

a. vanillin, $C_8H_8O_3$
b. oil of cloves, $C_{10}H_{12}O_2$
c. anise oil, $C_{10}H_{12}O$
d. oil of cinnamon, $C_9H_8O$

2.74. **Sweeteners** Calculate the molar masses of the following common sweeteners.

a. sucrose, $C_{12}H_{22}O_{11}$
b. saccharin, $C_7H_5O_3NS$
c. aspartame, $C_{14}H_{18}N_2O_5$
d. fructose, $C_6H_{12}O_6$

---

**2.75.** **Quartz Crystal** Quartz is a crystalline form of silicon dioxide, $SiO_2$. How many moles of $SiO_2$ are in the quartz crystal in Figure P2.75, which has a mass of 7.474 grams?

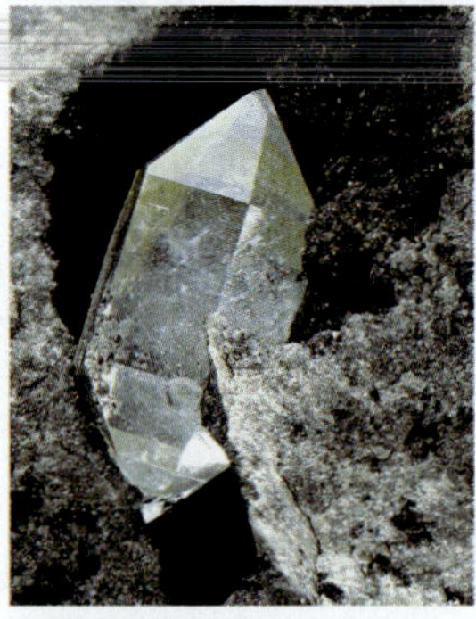

**FIGURE P2.75**

2.76. **Washington Monument** An aluminum apex weighing 2.8 kg was installed at the top of the Washington Monument (see Figure 1.19) when it was built in 1884. How many moles of Al were in the apex?

---

**2.77.** How many moles of $Ca^{2+}$ ions are in 0.25 mol of $CaTiO_3$? What is the mass in grams of the $Ca^{2+}$ ions?

2.78. How many moles of $O^{2-}$ ions are in 0.55 mol of $Al_2O_3$? What is the mass in grams of the $O^{2-}$ ions?

---

**2.79.** Suppose pairs of balloons are filled with 10.0 g of the following pairs of gases. Which balloon in each pair has more particles? (a) $CO_2$ or NO; (b) $CO_2$ or $SO_2$; (c) $O_2$ or Ar

2.80. If you had equal masses of the substances in the following pairs of compounds, which of the two would contain more ions? (a) NaBr or KCl; (b) NaCl or $MgCl_2$; (c) $CrCl_3$ or $Na_2S$

---

**2.81.** Rank the following quantities in order of increasing amounts of aluminum. (a) 1.0 mol of $Al_2O_3$; (b) 76 g of $Al_2O_3$; (c) 27 g of Al; (d) $1.1 \times 10^{24}$ atoms of Al; (e) 1.2 mol of $AlCl_3$

*2.82. Aluminum ($d = 2.70$ g/cm$^3$) and strontium ($d = 2.64$ g/cm$^3$) have nearly the same density. Given two cubes, one consisting of 1 mol of Al and the other consisting of 1 mol of Sr, which cube will be larger? What are the dimensions of the cube?

## Mass Spectrometry

### Concept Review

**2.83.** How does mass spectrometry provide information on the molecular mass of a compound?

2.84. The mass spectrum for $CH_4$ has a molecular ion peak at $m/z = 16$ u. Offer a possible explanation for a smaller peak at $m/z = 15$ u.

**2.85.** Would you expect the mass spectra of $CO_2$ and $C_3H_8$ to have molecular ions with the same mass (to the nearest u)?

*2.86. Would you expect the mass spectra of $CO_2$ and $C_3H_8$ to be the same? Why or why not?

### Problems

**2.87.** **Screening for Explosives** Many of the explosive materials of concern to airport security contain nitrogen and oxygen. Calculate the masses of the molecular ions formed by the following compounds. (a) $C_3H_6N_6O_6$; (b) $C_4H_8N_8O_8$; (c) $C_5H_8N_4O_{12}$; (d) $C_{14}H_6N_6O_{12}$

*2.88. **Landfill Gas** Mass spectrometry has been used to analyze the gases emitted from landfills. The principal component is methane ($CH_4$), but small amounts of dimethylsulfide ($C_2H_6S$) and dichloroethene ($C_2H_2Cl_2$) are often present, too. Calculate the masses of the molecular ions formed by those three compounds in a mass spectrometer.

**2.89.** **Sewer Gas** Hydrogen sulfide, $H_2S$, is a foul-smelling and toxic gas that may be present in wastewater sewers. Although the human nose detects $H_2S$ in low concentrations, prolonged exposure to $H_2S$ deadens our sense of smell, making it particularly dangerous to sewer workers who work in poorly ventilated areas. Explain how the peaks in the mass spectrum of $H_2S$ (Figure P2.89) reflect the sequential loss of H atoms from molecules of $H_2S$.

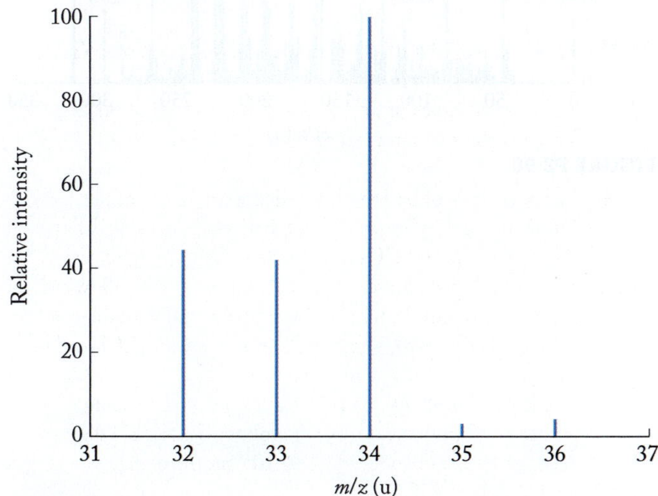


**FIGURE P2.89**

2.90. **Bar Code Readers** Arsine, $AsH_3$, is a hazardous gas used in the manufacture of electronic devices, including the bar code readers used at the checkout counters of many stores. The mass spectrum of arsine is shown in Figure P2.90. Arsenic has only one stable isotope. What are the formulas of the ions responsible for the four peaks in the mass spectrum?

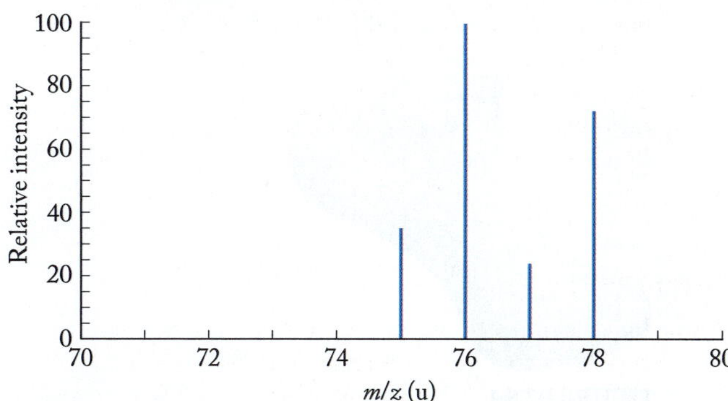


**FIGURE P2.90**

## Additional Problems

**2.91.** In April 1897, J. J. Thomson presented the results of his experiment with cathode-ray tubes, in which he proposed that the rays were actually beams of negatively charged particles, which he called "corpuscles."

a. What is the name we use for the particles today?
b. Why did the beam deflect when passed between electrically charged plates?
c. If the polarity of the plates was switched, how would the position of the light spot on the phosphorescent screen change?

2.92. Magnesium has three stable isotopes. Their masses are 23.9850, 24.9858, and 25.9826 u. If the average atomic mass of magnesium is 24.3050 u and the natural abundance of the lightest isotope is 78.99%, what are the natural abundances of the other two isotopes?

**2.93.** Without consulting a periodic table, give the atomic number ($Z$) for each highlighted element in Figure P2.93.

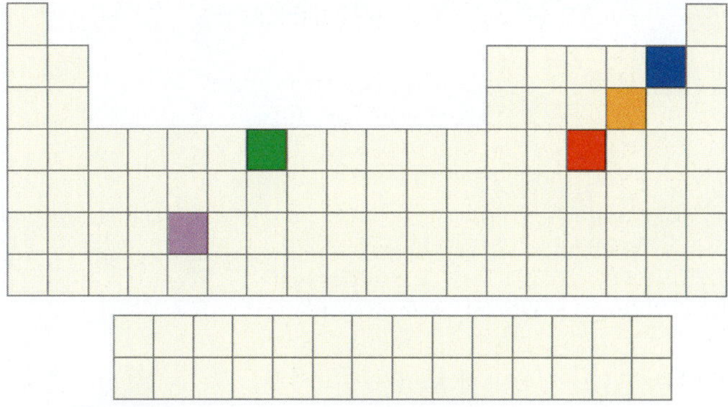

**FIGURE P2.93**

2.94. **Silver Nanoparticles in Clothing** The antimicrobial properties of silver metal have led to the use of silver nanoparticles in clothing (Figure P2.94) to reduce odors. If a spherical silver nanoparticle with a diameter of $1.00 \times 10^{-7}$ m contains $4.8 \times 10^{7}$ atoms of Ag, how many nanoparticles are in 1.00 g?

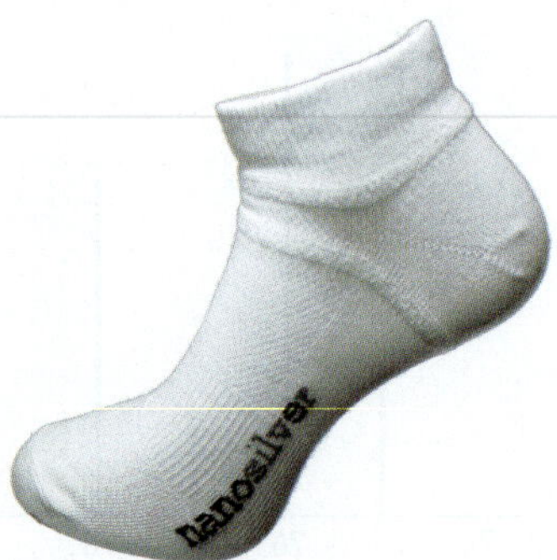


**FIGURE P2.94**

**2.95. HD Television** Some newer television sets use nanoparticles of cadmium sulfide (CdS) and cadmium selenide (CdSe), called "quantum dots," to produce the colors on the screen. Different-sized quantum dots lead to different colors.

a. Calculate the formula masses of CdS and CdSe.
b. If a nanoparticle of CdSe contains $2.7 \times 10^{7}$ atoms of Cd, how many atoms of Se are in the particle?
c. If a nanoparticle of CdS weighs $4.3 \times 10^{-15}$ g, how many grams of Cd and how many grams of S does it contain?

*2.96. **Greenhouse Gases** Samples of air are analyzed weekly for $CO_2$ content at the Mauna Loa Observatory in Hawaii. In May 2018, the average of those $CO_2$ determinations was 411.31 μmol ($10^{-6}$ mol) of $CO_2$ per mole of air—the highest monthly average since those analyses began in 1958. If the average molar mass of the gases in air is 28.8 g/mol, how many micrograms of $CO_2$ per gram of air were in those samples?

**2.97. Detecting Illegal Drugs** In July 2015, researchers in Britain reported a new method for using mass spectrometry to detect cocaine on people's fingertips. The mass spectrum of cocaine is shown in Figure P2.97. What is the molar mass of cocaine?

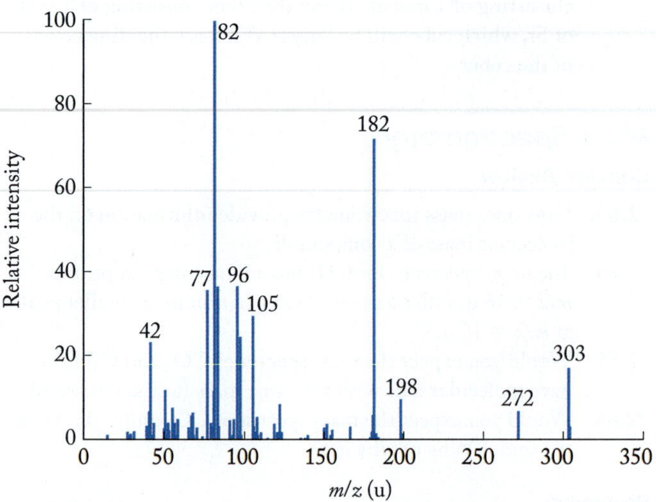


**FIGURE P2.97**

2.98. **Performance-Enhancing Drugs** Mass spectrometry is used to detect performance-enhancing drugs in body fluids. Included on the list of banned substances for Olympic athletes is tetrahydrogestrinone, a compound that mimics the effects of the steroid testosterone and can be used to build muscle. The mass spectrum of tetrahydrogestrinone is shown in Figure P2.98. Identify the molecular ion and show that it has a mass consistent with the formula $C_{21}H_{28}O_2$.

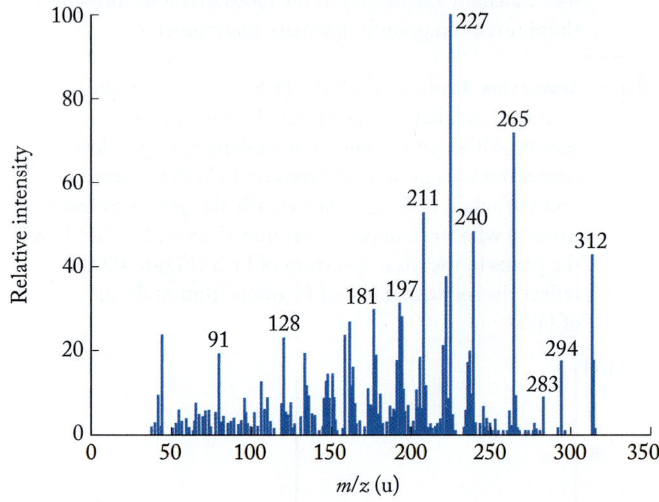


**FIGURE P2.98**

**2.99. Hope Diamond** The Hope Diamond (Figure P2.99) at the Smithsonian National Museum of Natural History has a mass of 45.52 carats.

a. How many moles of carbon are in the Hope Diamond (1 carat = 200 mg)?
b. How many carbon atoms are in the diamond?

FIGURE P2.99

2.100. **Earth's Crust** Minerals called feldspars make up more than 40% of Earth's continental crust. Those minerals are mixtures of the compounds listed below. Rank the following compounds in order of increasing aluminum content (%Al by mass). (a) $NaAlSi_3O_8$; (b) $KAlSi_3O_8$; (c) $CaAl_2Si_2O_8$

# 3

# Atomic Structure

## Explaining the Properties of Elements

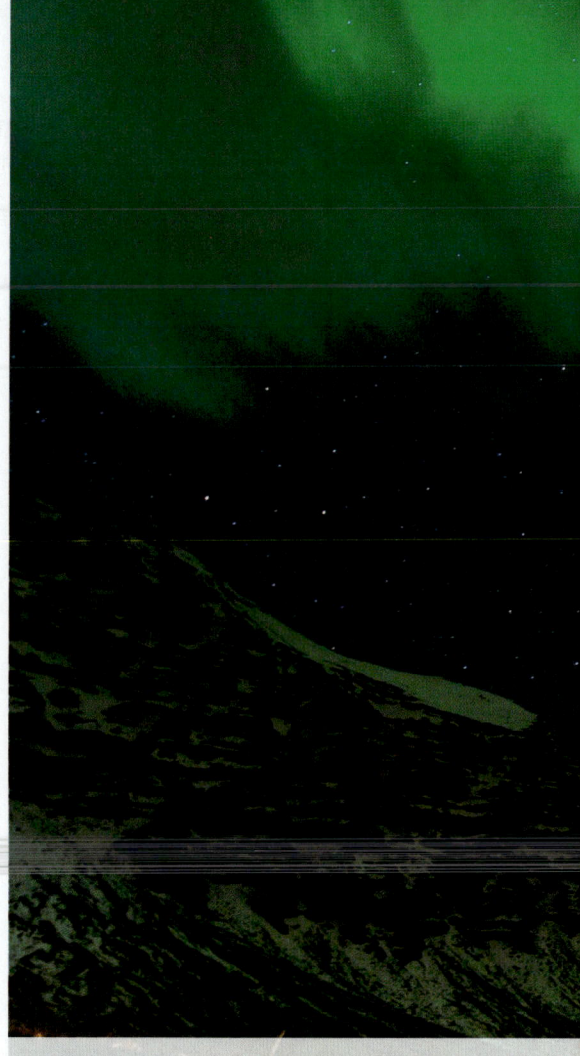

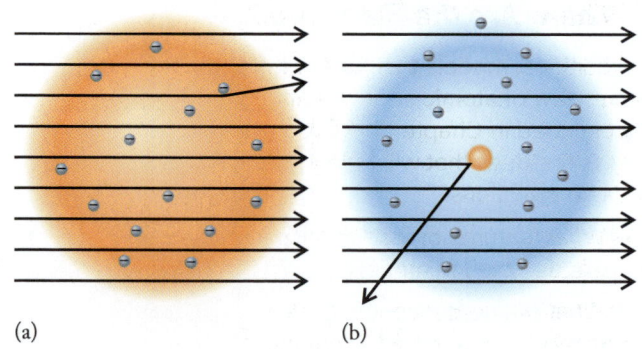

**COLORS OF THE AURORA** Some of the red and green colors of an aurora display are produced when atoms of oxygen in the upper atmosphere collide with high-speed charged particles emitted by the sun.

PARTICULATE **REVIEW**

### *Rutherford's Atom*

In Chapter 3, we explore the structure of the atom in detail. We will see how some of the most brilliant scientists of the 20th century built on the understanding of atomic structure first established through Ernest Rutherford's gold-foil experiment.

- How many protons, neutrons, and electrons are in one atom of $^{197}Au$?
- Which image depicts the model of the atom at the time when Rutherford began his experiments?
- What feature of the other image was inconsistent with the prevailing model and led Rutherford to propose a new one?

(a) (b)

(Review Figure 2.6 if you need help.)

*(Answers to Particulate Review questions are in the back of the book.)*

Oxygen
atomic
emission

Sunlight

400 500 600 700

Wavelength (nm)

PARTICULATE **PREVIEW**

### *Where Are the Electrons?*

Additional models that further refined our view of atomic structure are described in Chapter 3. As you read, look for ideas that will help you answer these questions:

- What features of Rutherford's model are depicted in model (a)?
- What do the concentric circles in model (a) suggest about the arrangement of electrons in atoms?
- What key limitation of model (a) does model (b) address?

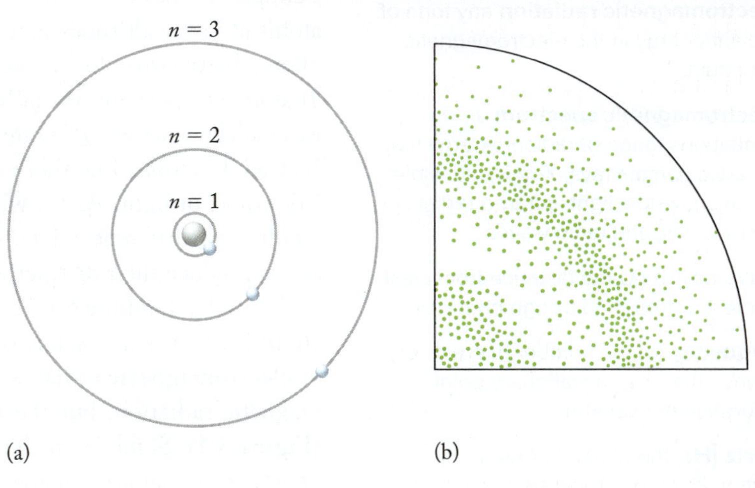


(a) (b)

## Learning Outcomes

**LO1** Interconvert the energies, wavelengths, and frequencies of electromagnetic radiation and link their values to the appropriate regions of the electromagnetic spectrum
**Sample Exercises 3.1, 3.2**

**LO2** Describe quantum theory and use it to explain the photoelectric effect
**Sample Exercise 3.3**

**LO3** Calculate the energies and wavelengths of photons absorbed and emitted by atoms for electron transitions between atomic energy levels
**Sample Exercises 3.4, 3.5**

**LO4** Apply the Heisenberg uncertainty principle to particles in motion and calculate their de Broglie wavelengths
**Sample Exercises 3.6, 3.7**

**LO5** Assign quantum numbers to orbitals and use their values to describe the sizes, energies, and orientations of orbitals
**Sample Exercises 3.8, 3.9**

**LO6** Use the aufbau principle and Hund's rule to write electron configurations and draw orbital diagrams of atoms and monatomic ions
**Sample Exercises 3.10–3.12**

**LO7** Use atomic orbitals and the concept of effective nuclear charge to explain ionization energies, to predict the relative sizes of atoms and monatomic ions, and to interpret photoelectron spectra
**Sample Exercises 3.13–3.15**

**LO8** Relate the electron affinities of the elements to their positions in the periodic table

## 3.1 Nature's Fireworks and the Electromagnetic Spectrum

**CHEMTOUR**
Electromagnetic Radiation

The opening photo of this chapter shows the aurora borealis, one of nature's most colorful light displays. Auroras are produced by electrons and positively charged particles flowing out from the sun at speeds approaching 1000 km/s. Earth's magnetic poles capture some of them and draw them into the upper atmosphere, where they collide with atoms and molecules of oxygen, nitrogen, and other atmospheric gases.

Some of the collisions between high-speed electrons and molecules produce molecular ions, in much the way that beams of high-energy electrons ionize molecules in mass spectrometers. Sometimes these collisions are not energetic enough to ionize atoms or molecules, but the energy may still be absorbed only to be reemitted as visible light. Different particles emit different colors of light. For example, atoms of oxygen in the upper atmosphere may emit red light, whereas O atoms at lower altitudes may emit green light, as shown in the chapter-opening photo. These two colors make up the visible emission spectrum of atomic oxygen. That *atomic* spectrum is not like the visible-light spectrum emitted by the sun or by artificial sources of "white" light, which contain all the colors of the rainbow. Instead, O atoms, like those of other elements, produce relatively few characteristic colors of light. As we will see in later sections of this chapter, different elements emit different colors. More important, we will also learn how and why atoms produce their distinctive colors of light.

To understand the origins of the colors of auroras, we need to understand how atoms interact with the forms of energy that are collectively called *radiant* energy or **electromagnetic radiation**. Visible light is the most familiar form of electromagnetic radiation, but the **electromagnetic spectrum** includes several forms (**Figure 3.1**). Some forms have more energy than visible light does, such as the gamma rays that accompany nuclear reactions, the X-rays used in medical imaging, and the ultraviolet rays that can give you a sunburn. Other forms have less

**electromagnetic radiation** any form of radiant energy in the electromagnetic spectrum.

**electromagnetic spectrum** a continuous range of radiant energy that includes gamma rays, X-rays, ultraviolet radiation, visible light, infrared radiation, microwaves, and radio waves.

**wavelength ($\lambda$)** the distance from crest to crest or trough to trough on a wave.

**frequency ($\nu$)** the number of crests of a wave that pass a stationary point of reference per second.

**hertz (Hz)** the SI unit of frequency with units of reciprocal seconds: 1 Hz = 1 $s^{-1}$ = 1 cycle per second (cps).

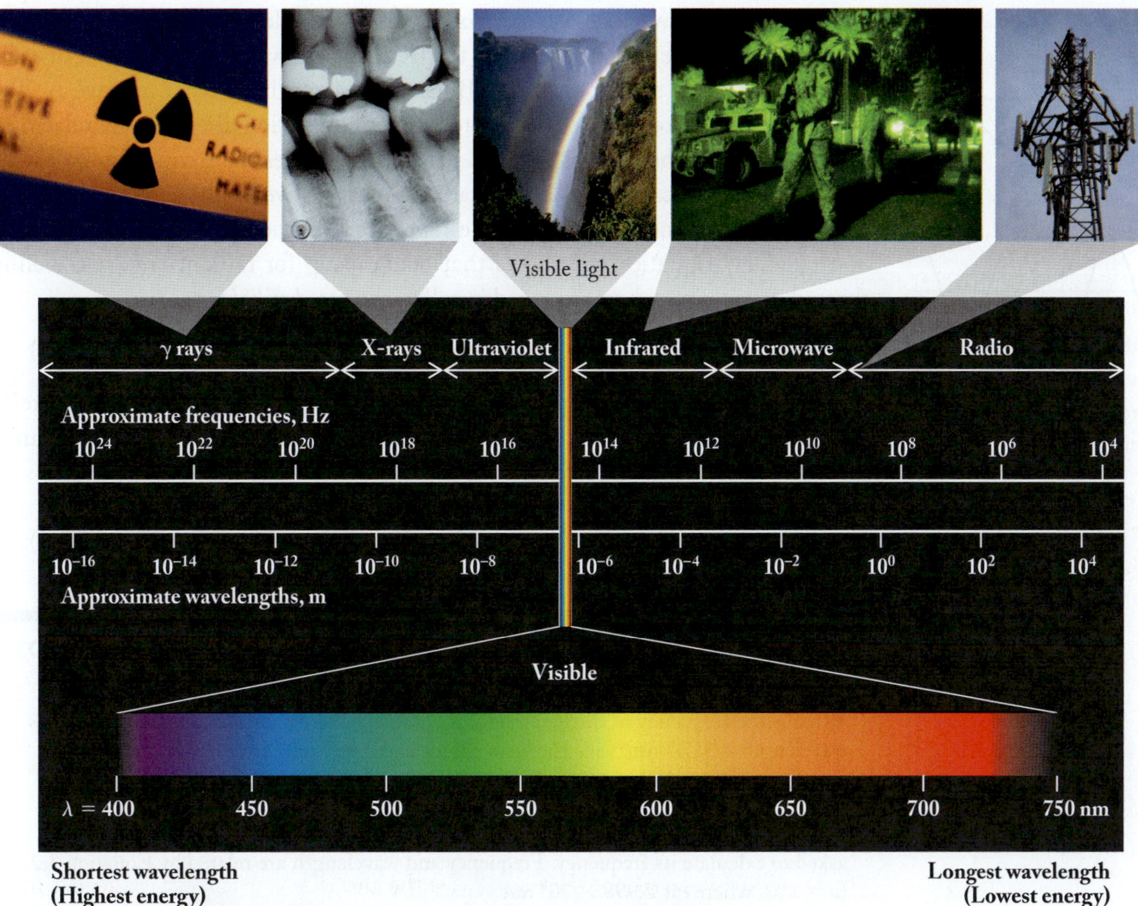


**FIGURE 3.1** Visible light occupies a tiny fraction of the electromagnetic spectrum, which ranges from ultrashort-wavelength, high-frequency gamma (γ) rays to long-wavelength, low-frequency radio waves. Note that frequencies increase from right to left, whereas wavelengths increase from left to right.

energy, including the infrared radiation we feel as heat flowing from warm objects, the microwaves used in ovens, and the radio waves that connect devices in wireless networks.

All the forms of radiation in Figure 3.1 are considered *electromagnetic* because of a theory about their properties developed by Scottish scientist James Clerk Maxwell (1831–1879). According to Maxwell's theory, electromagnetic radiation moves through space (or any transparent medium) as waves with two perpendicular components: an oscillating *electric* field and an oscillating *magnetic* field (**Figure 3.2**). Maxwell derived a set of equations based on his oscillating-wave model that accurately describes nearly all the observed properties of light and the other forms of radiant energy.

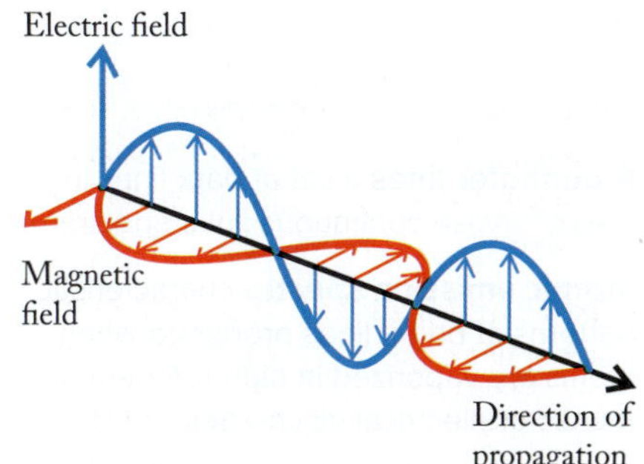


**FIGURE 3.2** Electromagnetic waves consist of electric fields and magnetic fields that oscillate in planes oriented at right angles to each other.

A wave of electromagnetic radiation, like any wave traveling through any medium, has a characteristic **wavelength (λ)**, which is the distance from one wave crest to the next, as shown in **Figure 3.3**. Note that wave A in Figure 3.3 has twice the wavelength of wave B. Each wave also has a characteristic **frequency (ν)**, which is the number of crests that pass a stationary point in space per second. Frequency has units of waves per second, or simply *cycles per second* ($s^{-1}$). Scientists also use the frequency unit called **hertz (Hz)**, which is equal to 1 cycle per second ($1\ Hz = 1\ s^{-1}$).

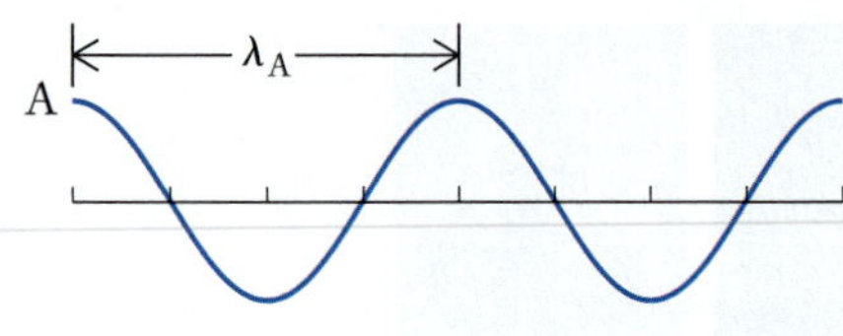


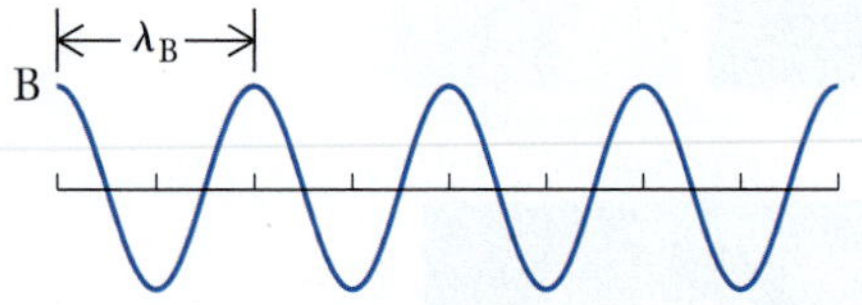


**FIGURE 3.3** Every wave has a characteristic wavelength ($\lambda$) and frequency ($\nu$). Wave A has a longer wavelength (and lower frequency) than wave B.

Figure 3.3 shows that wave A completes two cycles over the same distance that wave B completes four cycles, so the frequency of wave B is *twice* that of wave A. Figure 3.3 also shows that wave B has a wavelength *half* that of wave A. The wavelengths and frequencies of electromagnetic radiation are inversely related in exactly that way: the shorter the wavelength, the higher the frequency. That relationship occurs because all forms of radiant energy travel at the same speed in the same transparent medium, and a wave's speed is the product of its wavelength and its frequency. Equation 3.1 shows that relationship for radiant energy traveling through a vacuum, where its speed is called the *speed of light*, *c*:

$$\lambda\nu = c \tag{3.1}$$

The value of $c$ to four significant figures is $2.998 \times 10^8$ m/s. The reciprocal relationship between wavelength and frequency may be more evident if we rearrange the terms in Equation 3.1 as follows:

$$\nu = \frac{c}{\lambda} \tag{3.2}$$

**SAMPLE EXERCISE 3.1** Calculating Frequency from Wavelength **LO1**

What is the frequency of the green light that oxygen atoms omit in auroras? The light's wavelength is 557.7 nm. (See the visible lines in oxygen's atomic emission spectrum in this chapter's opening photo.)

**Collect and Organize** We are given the wavelength of a color of visible light and asked to calculate its frequency. Frequency and wavelength are related by Equation 3.2 ($\nu = c/\lambda$), where $c$ is $2.998 \times 10^8$ m/s.

**Analyze** The value $c$ is in meters per second, but the wavelength is given in nanometers. Therefore, we need to convert nanometers to meters by using the equality 1 nm = $10^{-9}$ m as a conversion factor. The unit labels in Figure 3.1 indicate that frequencies of visible light are around $10^{14}$ Hz, or $10^{14}$ s$^{-1}$, so we expect our answer to have an exponent in that range.

**Solve** Substituting the given values and appropriate unit conversion factor into Equation 3.2:

$$\nu = \frac{c}{\lambda} = \frac{2.998 \times 10^8 \text{ m s}^{-1}}{557.7 \text{ nm} \times \dfrac{10^{-9} \text{ m}}{\text{nm}}} = 5.376 \times 10^{14} \text{ s}^{-1}$$

**Think About It** The large value of $\nu$ is in the range that we expect for visible light. Remember that $\lambda$ and $\nu$ are inversely proportional: as one increases, the other decreases.

**Practice Exercise** If the waves transmitted by a radio station have a frequency of 90.9 MHz, what is their wavelength (in meters)?

*(Answers to Practice Exercises are in the back of the book.)*

**Fraunhofer lines** a set of dark lines in the otherwise continuous solar spectrum.

**atomic emission spectra** characteristic patterns of bright lines produced when atoms are vaporized in high-temperature flames or electrical discharges.

**atomic absorption spectra** characteristic patterns of dark lines produced when an external source of radiation passes through free gaseous atoms.

CONCEPT **TEST**

The ultraviolet (UV) region of the electromagnetic spectrum contains radiation with wavelengths from about $10^{-7}$ to $10^{-9}$ m; the infrared (IR) region contains radiation with wavelengths from about $10^{-4}$ to $10^{-6}$ m. Is UV radiation higher or lower in frequency than IR radiation?

*(Answers to Concept Tests are in the back of the book.)*

## 3.2 Atomic Spectra

In this section, we explore how atoms and electromagnetic radiation interact. We start with observations made more than 200 years ago by English scientist William Hyde Wollaston (1766–1828). In 1802, Wollaston was studying the spectrum of sunlight by using carefully ground glass prisms when he discovered that the spectrum was not completely continuous. Instead, it contained a series of very narrow dark lines. Using even better prisms, German physicist Joseph von Fraunhofer (1787–1826) resolved and mapped the wavelengths of more than 500 of those lines, now called **Fraunhofer lines** (**Figure 3.4**).

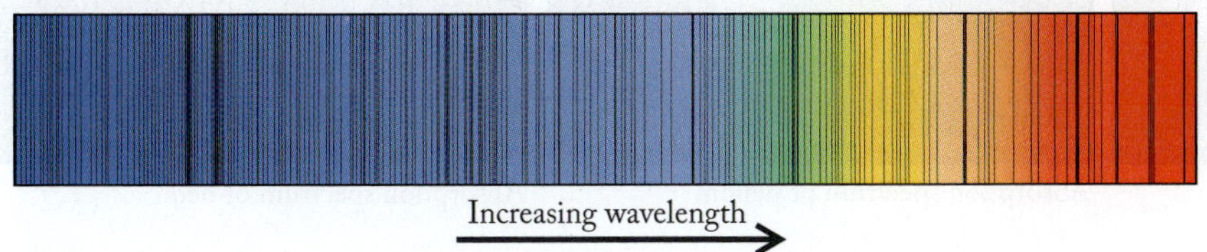


**FIGURE 3.4** The spectrum of sunlight is not continuous but contains many narrow gaps, which appear as dark lines called Fraunhofer lines.

Fraunhofer did not know why those narrow lines were present in the sun's spectrum. That understanding came nearly half a century later from discoveries by two other Germans, Robert Wilhelm Bunsen (1811–1899) and Gustav Robert Kirchhoff (1824–1887), using an instrument called a spectroscope (**Figure 3.5a**). They discovered that the light emitted by elements vaporized in the transparent flame of a gas burner developed by Bunsen produced spectra that were the opposite of Fraunhofer's. Whereas the sun's spectrum was nearly continuous except for a set of narrow dark lines, the spectra produced by the elements consisted of only a few narrow bright lines on a dark background. Bunsen and Kirchhoff discovered that the lines in the **atomic emission spectra** of certain elements matched the wavelengths of some of the Fraunhofer lines in the spectrum of sunlight. For example, the Fraunhofer D line matched the bright yellow-orange line in the emission spectrum produced by hot sodium vapor (**Figure 3.5b**).

Those experiments and others employing light sources called gas-discharge tubes showed that at very high temperatures the atoms of each element emit a characteristic spectrum (**Figure 3.6**). Conversely, atoms of elements in the gaseous state absorb electromagnetic radiation when illuminated by an external source of radiation. When that radiation is passed through a narrow slit and then a prism, as shown schematically in **Figure 3.7(a)**, **atomic absorption spectra** of narrow dark lines in otherwise continuous spectra are produced. For any given element (**Figure 3.7b**), the dark lines in its absorption spectrum are at the same wavelengths as the bright lines in its emission spectrum. The phenomenon of atomic absorption explains the Fraunhofer lines: gaseous atoms in the outer regions of the sun absorb characteristic wavelengths of the sunlight passing through them on its way to Earth.

(a)

(b)

**FIGURE 3.5** (a) Kirchhoff and Bunsen used spectroscopes such as this one to observe the atomic emission spectra of elements introduced to the flame of the burner on the right. (b) Emission from sodium atoms when viewed through a spectroscope.

**STEPWISE ANIMATION**

Absorption of Light

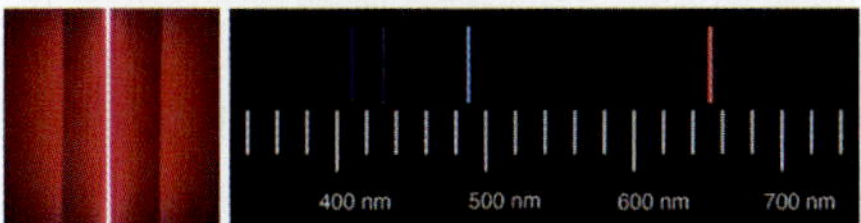


(a) Emission spectrum of hydrogen

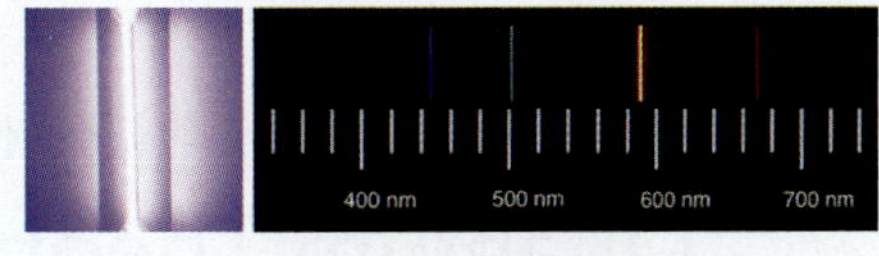


(b) Emission spectrum of helium

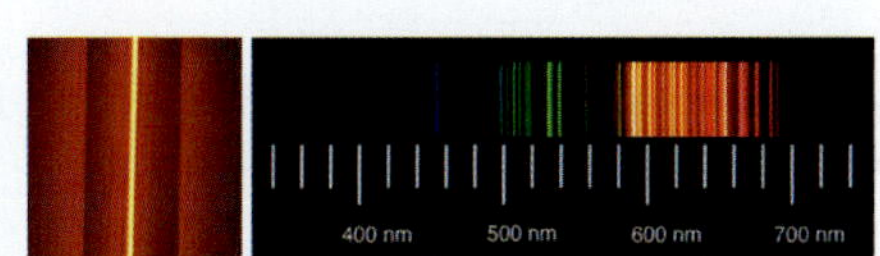


(c) Emission spectrum of neon

**FIGURE 3.6** The colored light emitted from gas-discharge tubes filled with (a) hydrogen, (b) helium, and (c) neon produces atomic emission spectra that are characteristic of the element.

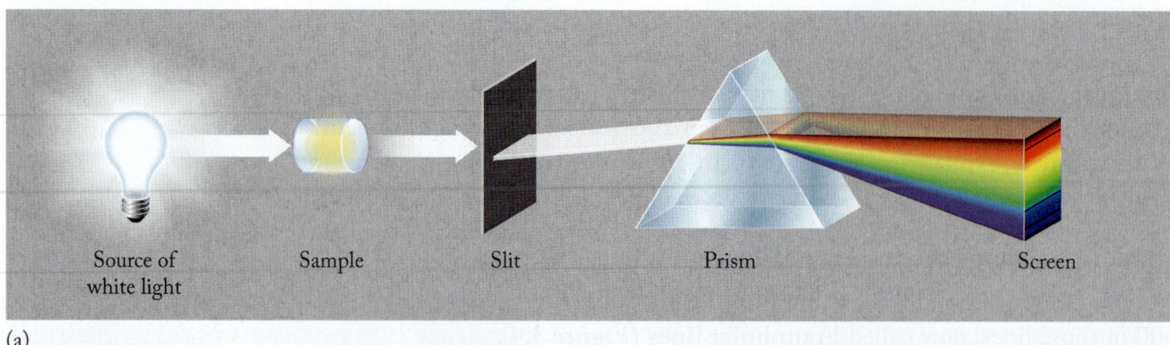


(a)

400 nm 500 nm 600 nm 700 nm
Absorption spectrum of hydrogen

400 nm 500 nm 600 nm 700 nm
Absorption spectrum of helium

400 nm 500 nm 600 nm 700 nm
Absorption spectrum of neon

(b)

**FIGURE 3.7** (a) When gaseous atoms of hydrogen, helium, and neon are illuminated by an external source of white light (containing all colors of the visible spectrum), the resultant atomic absorption spectra contain dark lines that are characteristic of the elements. (b) The dark lines have the same wavelengths as the bright lines in the elements' atomic emission spectra, shown in Figure 3.6.

### CONCEPT TEST

Mercury lamps are used to illuminate large spaces such as sports arenas. The top spectrum in **Figure 3.8** is the emission spectrum of mercury vapor. Select the absorption spectrum of mercury vapor from the four spectra labeled (a)–(d).

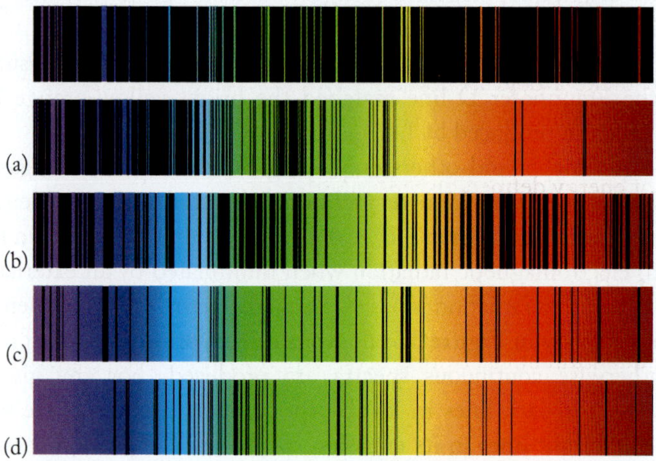


**FIGURE 3.8** Atomic emission spectrum of mercury vapor (top) and four atomic absorption spectra (a–d).

*(Answers to Concept Tests are in the back of the book.)*

**quantum** (plural, *quanta*) the smallest discrete quantity of a particular form of energy.

**Planck constant (*h*)** the proportionality constant between the energy and frequency of electromagnetic radiation expressed in $E = h\nu$; $h = 6.626 \times 10^{-34}$ J · s.

## 3.3 Particles of Light: Quantum Theory

In addition to his studies of the narrow-line emission spectra of atoms, Kirchhoff also studied the continuous emission spectra produced by hot objects such as metals heated to incandescence (**Figure 3.9**).

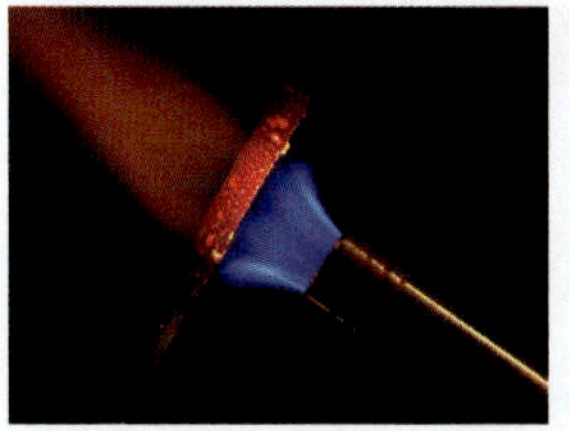
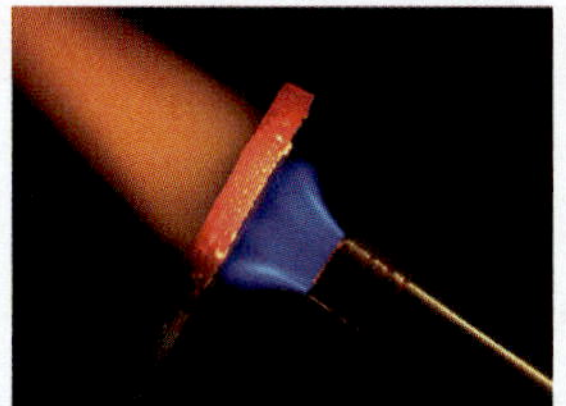

**FIGURE 3.9** When a metal rod is heated, it glows red at first, then orange, and finally becomes white-hot as its temperature increases and light of shorter wavelengths is emitted.

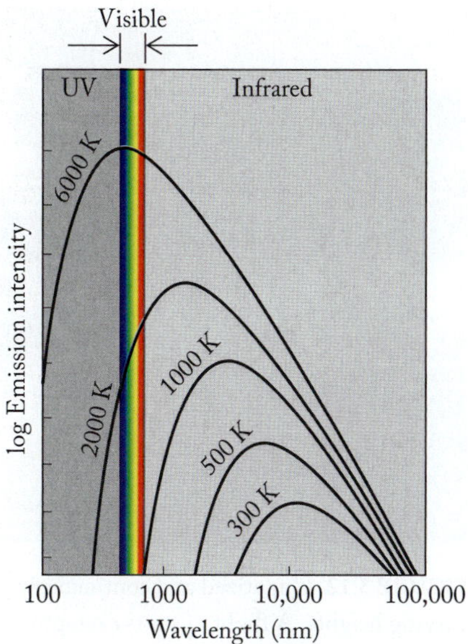


**FIGURE 3.10** The solid curves show how the intensities and wavelengths of radiation emitted by blackbody radiators change with changing temperature. Logarithmic scales are used on both axes to include a wide range of intensities and wavelengths.

Kirchhoff and others recorded the spectra of sources of radiant energy that are called *blackbody* radiators because, when cold, their surfaces are jet black and they absorb all the light that strikes them. Those perfect absorbers of light when cold become perfect radiators of light when hot. The spectra they emit depend only on temperature: as they become hotter, they emit more radiation at shorter and shorter wavelengths, as shown in the emission intensity profiles in **Figure 3.10**. Those profiles presented a challenge for physicists in the late 19th century because none of the laws describing the behavior of electromagnetic radiation could explain them. A new model describing how radiant energy and matter interact was needed, and yet another German physicist developed it. His name was Max Planck (1858–1947; **Figure 3.11**).

## Photons of Energy

While Planck was a professor at the University of Berlin in 1900, electric companies commissioned him to maximize the output of the incandescent lightbulbs that Thomas Edison had invented. Planck's attempts at developing a theoretical model that accurately described their incandescence led him to discard classical physics and make a bold assumption about radiant energy: no matter what its source, it can never be truly continuous. Instead, Planck proposed that objects emit electromagnetic radiation only in integral multiples of an elementary unit, or **quantum**, of energy defined by the equation

$$E = h\nu \tag{3.3}$$

where $\nu$ is the frequency of the radiation and $h$ is the **Planck constant**, $6.626 \times 10^{-34}$ J · s. Combining Equations 3.2 and 3.3 yields an equation that relates the energy of a quantum of electromagnetic radiation to its wavelength:

$$E = \frac{hc}{\lambda} \tag{3.4}$$

Because Planck's model is characterized by those *quanta*, it has become known as **quantum theory**.

To visualize the meaning of Planck's quantum of energy, consider the difference between taking the steps or the ramp to get from the sidewalk to the entrance of a building (**Figure 3.12**). If you walk up the steps, you can stand at only discrete heights above the sidewalk. You cannot stand at a height between adjacent steps because nothing is there to stand on. If you walk up the ramp, however, you can stop at any height between the sidewalk and the entrance because height is continuous on the ramp. On the steps, height is **quantized**: the discrete changes in height serve as a model of Planck's hypothesis that energy is released (analogous to walking down the steps) or absorbed (walking up the steps) in discrete packets, or quanta, of energy.

**FIGURE 3.11** German scientist Max Karl Ernst Ludwig Planck is considered the father of quantum physics. He won the 1918 Nobel Prize in Physics for his pioneering work on the quantized nature of electromagnetic radiation.

**quantum theory** a model of matter and energy based on the principle that energy is absorbed or emitted in discrete packets, or quanta.

**quantized** having values restricted to whole-number multiples of a specific base value.

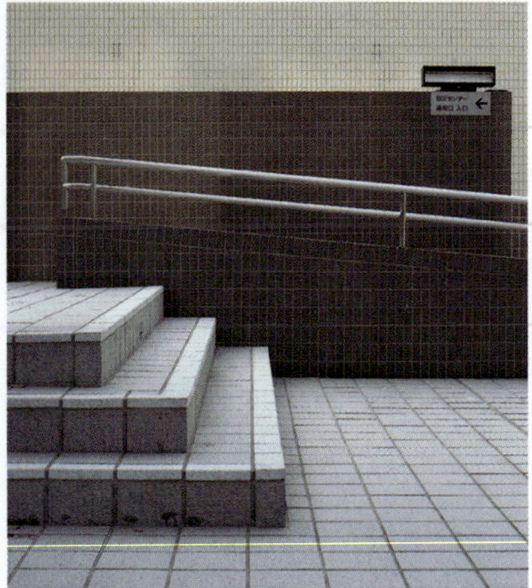

**FIGURE 3.12** Quantized and continuously varying heights. A flight of stairs exemplifies quantization: each step rises by a discrete height to the next step. In contrast, the heights on a ramp are not quantized—they are continuous.

## CONCEPT TEST

Which of the following are quantized?

a. The volume of water in the Atlantic Ocean
b. The number of eggs remaining in a carton
c. The time you typically take to get ready for class in the morning
d. The number of red lights encountered when driving the length of Fifth Avenue in New York City

*(Answers to Concept Tests are in the back of the book.)*

Given the extremely tiny value of the Planck constant ($h$), the energy values we obtain using Equation 3.4 tend to be very small because $E$ represents the number of joules in a single quantum of energy. Today we call the tiny packets of radiant energy **photons**. They represent elementary building blocks of electromagnetic radiation in much the same way that atoms represent the building blocks of matter. The observed brightness of a source of radiant energy is the sum of the energies of the enormous number of photons it produces per unit of time.

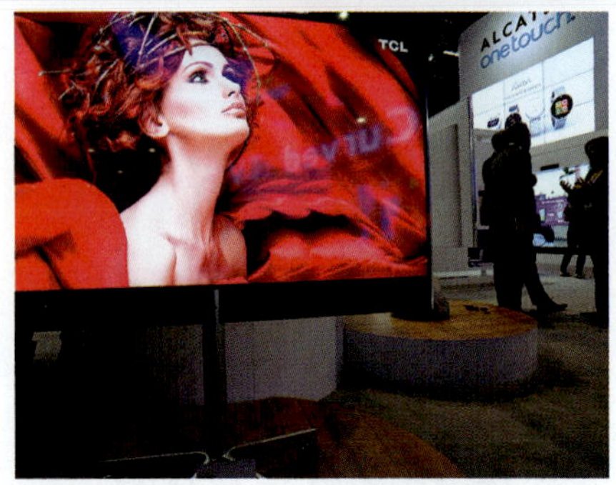

**FIGURE 3.13** An ultra-high-definition television incorporating quantum dot technology.

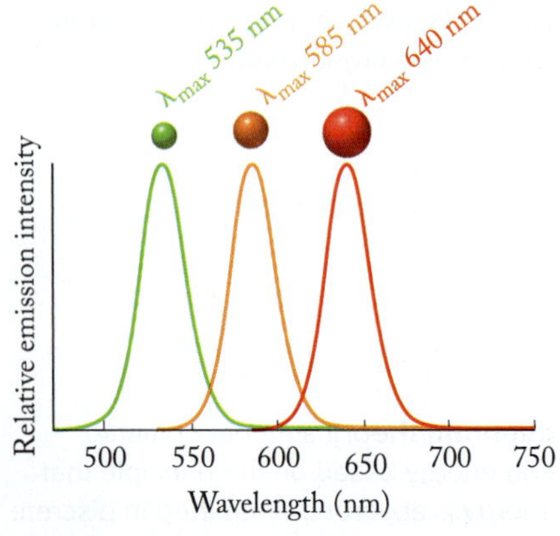


**FIGURE 3.14** Emission intensity versus wavelength for three quantum dots of different size.

### SAMPLE EXERCISE 3.2 Calculating the Energy of a Photon — LO1

Some manufacturers of ultra-high-definition televisions (**Figure 3.13**) use tiny particles called *quantum dots* to more accurately reproduce the colors of transmitted images. The wavelengths of the light produced by quantum dots depend on their size. **Figure 3.14** shows a plot of emission intensity as a function of wavelength for three quantum dots whose relative sizes are represented by the sizes of the spheres.

a. What is the energy of one photon of light from the largest quantum dot, the one with a wavelength of 640 nm ($6.40 \times 10^2$ nm)? What is the energy of one mole of those photons?
b. Is that energy greater than or less than the energies of the photons produced by smaller quantum dots?

**Collect and Organize** We need to calculate the energy of a photon of electromagnetic radiation starting with its wavelength. Equation 3.4 ($E = hc/\lambda$) relates the energy of a photon to its wavelength. The value of the Planck constant ($h$) is $6.626 \times 10^{-34}$ J · s, and the speed of light ($c$) is $2.998 \times 10^8$ m/s.

**Analyze** The wavelength is given in nanometers, but the value of the speed of light has units of meters per second, so we need to convert nanometers to meters. The value of $h$ is extremely small, so the energy of one photon, even if we factor in the speed of light, should be very small, too. The energy for one mole of photons, given the large value of the Avogadro constant, should be much larger.

**Solve**

a. $$E = \frac{hc}{\lambda} = \frac{(6.626 \times 10^{-34}\ \text{J} \cdot \text{s})(2.998 \times 10^{8}\ \cancel{\text{m}}\ \text{s}^{-1})}{6.40 \times 10^{2}\ \cancel{\text{nm}} \times \dfrac{10^{-9}\ \cancel{\text{m}}}{\cancel{\text{nm}}}} = 3.10 \times 10^{-19}\ \text{J}$$

That is the energy of one photon. To calculate the energy of one mole of photons, we multiply that result by $N_A$:

$$E = \frac{3.10 \times 10^{-19}\ \text{J}}{\cancel{\text{photon}}} \times \frac{6.022 \times 10^{23}\ \cancel{\text{photons}}}{1\ \text{mol photons}} = 1.87 \times 10^{5}\ \text{J/mol photons}$$

b. Figure 3.14 shows that smaller quantum dots produce photons with shorter wavelengths. The energy of a photon is inversely proportional to its wavelength, so those shorter-wavelength photons have more energy than the photon at 640 nm.

**Think About It** The calculated energy for one photon is indeed small, as it should be, because a photon is an atomic-level particle of radiant energy.

**Practice Exercise** Some instruments differentiate individual quanta of electromagnetic radiation on the basis of their energies. Assume that such an instrument has been adjusted to detect photons that have $5.00 \times 10^{-19}$ J of energy. What is the wavelength of the detected radiation? Give your answer in nanometers and in meters.

*(Answers to Practice Exercises are in the back of the book.)*

**photon** a quantum of electromagnetic radiation.

**photoelectric effect** the release of electrons from a material as a result of electromagnetic radiation striking it.

**threshold frequency ($\nu_0$)** the minimum frequency of light required to produce the photoelectric effect.

## The Photoelectric Effect

Although Planck's quantum model explained the emission spectra of hot objects, no experimental evidence existed in 1900 to support the existence of quanta of energy. In 1905, Albert Einstein (1879–1955) supplied that evidence. It came from his studies of a phenomenon called the **photoelectric effect**, in which electrons are emitted from metals and semiconductor materials when they are illuminated by and absorb electromagnetic radiation. Because light releases those electrons, they are called *photoelectrons*, derived from the Greek *photo*-, meaning "light."

**STEPWISE ANIMATION**
The Photoelectric Effect

Photoelectrons are emitted when the frequency of incident radiation exceeds some minimum **threshold frequency ($\nu_0$) (Figure 3.15)**. Radiation at frequencies less than the threshold value produces no photoelectrons, no matter how intense the radiation is. However, even a dim source of radiant energy produces at least a few photoelectrons when the frequencies it emits are equal to or greater

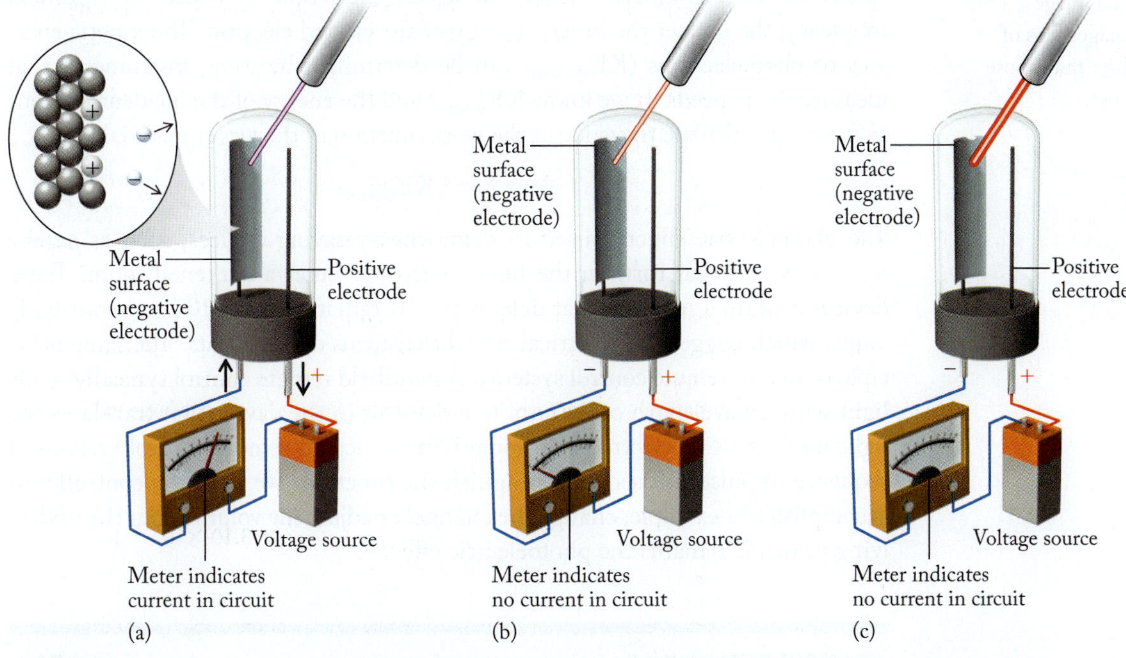


**FIGURE 3.15** A phototube includes a positive electrode and a negative metal electrode. (a) If radiation of high enough frequency and energy (represented by the violet beam of light) illuminates the negative electrode, electrons are dislodged from the surface and flow toward the positive electrode. That flow of electrons completes the circuit and produces an electric current. The size of the current is proportional to the intensity of the radiation—to the number of photons per unit time striking the negative electrode. (b) Photons of lower frequency (red beam) and hence lower energy do not have enough energy to dislodge electrons and do not produce the photoelectric effect, no matter how many of them (c) bombard the surface of the metal. The circuit is not complete, and no current exists.

**FIGURE 3.16** Night vision goggles incorporate photoelectric sensors and electronic amplifiers called image intensifiers to produce images tens of thousands of times brighter than those perceived by the human eye.

than the threshold frequency. For example, some night vision goggles used by military and law enforcement personnel (**Figure 3.16**) can detect extremely low levels of light because they incorporate photoelectric sensors and amplifier circuits that produce more than 10,000 electrons for each photoelectron initially emitted. When such large numbers of electrons strike a phosphorescent screen (such as the ones used in cathode-ray tubes, as in Figure 2.2), they produce images tens of thousands of times brighter than the original.

Einstein used Planck's quantum model to explain the ability of photoelectric materials to emit photoelectrons. He proposed that every photoelectric material has a characteristic threshold frequency, $\nu_0$, associated with the minimum quantum of absorbed energy needed to remove a single electron from the material's surface. The threshold frequency of a sensor used in night vision goggles, for example, is in the infrared region of the electromagnetic spectrum. When photons of radiation with at least that frequency are focused on one of the sensors, photoelectrons are emitted.

The minimum quantity of energy needed to emit photoelectrons from a photoelectric material is called the material's **work function (Φ)**:

$$\Phi = h\nu_0 \tag{3.5}$$

The value of Φ is related to the strength of the attraction between the nuclei of the photoelectric material's atoms and the electrons surrounding those nuclei. In Einstein's model, a photoelectron is emitted when a quantum of radiant energy (a photon) gives an electron enough energy to break free of the surface of a photoelectric material. If the incoming beam includes photons with frequencies above the threshold frequency ($\nu > \nu_0$), each photon has more than enough energy to dislodge an electron. That extra energy in excess of the work function is imparted to each ejected electron as kinetic energy: the higher the frequency above the threshold frequency, the greater the kinetic energy of the ejected electron. The kinetic energies of photoelectrons ($KE_{electron}$) can be determined by using instruments that measure their speeds. If we know $KE_{electron}$ and the energy of the incident photons ($h\nu$), we can calculate the value of the work function of the target material:

$$\Phi = h\nu - KE_{electron} \tag{3.6}$$

The photoelectric effect is used in many energy-saving devices, such as *passive infrared devices* that turn on the lights when you enter a darkened room. These devices contain a material that detects the IR radiation ($\lambda \approx 10^{-5}$ m) your body emits, which triggers an electrical signal that turns on the lights. The same principle is used in remote control systems. A handheld remote control typically sends light with a wavelength of 940 nm to a detector in the device that translates the light into a flow of electrons. In today's sophisticated remote control systems, a sequence of pulses is used to distinguish the functions we wish the controller to accomplish (for example, change the channel or adjust the volume), but the underlying principle remains the photoelectric effect.

**work function (Φ)** the amount of energy needed to dislodge an electron from the surface of a material.

### SAMPLE EXERCISE 3.3 Relating Work Functions and Threshold Frequencies — LO2

The work function of mercury is $7.22 \times 10^{-19}$ J.

a. What is the minimum frequency of radiation required to eject photoelectrons from a mercury surface?
b. Could visible light produce the photoelectric effect in mercury?

**Collect, Organize, and Analyze** We are asked to use the value of the work function ($\Phi$) of a metal to find its corresponding threshold (minimum) frequency ($\nu_0$). Equation 3.5 relates those parameters. Figure 3.1 shows the frequencies and wavelengths of the regions of the electromagnetic spectrum; the frequencies of visible light are between $10^{14}$ and $10^{15}$ Hz.

**Solve**

a. Rearrange Equation 3.5 to solve for the threshold frequency:

$$\nu_0 = \frac{\Phi}{h} = \frac{7.22 \times 10^{-19}\ \text{J}}{6.626 \times 10^{-34}\ \text{J} \cdot \text{s}} = 1.09 \times 10^{15}\ \text{s}^{-1}$$

b. The threshold frequency is greater than $10^{15}$ Hz and, according to Figure 3.1, is in the ultraviolet region of electromagnetic radiation. The frequencies of visible light are below the threshold frequency, so visible light cannot generate photoelectrons from mercury.

**Think About It** The calculated threshold frequency is close to, but still greater than, that of the highest-frequency (violet) visible light. Photons of visible light simply lack the energy needed to dislodge electrons from atoms on the surface of a drop of mercury.

**Practice Exercise** The work function of silver is $7.59 \times 10^{-19}$ J. What is the longest wavelength (in nanometers) of electromagnetic radiation that can eject a photoelectron from the surface of a piece of silver?

*(Answers to Practice Exercises are in the back of the book.)*

## 3.4 The Hydrogen Spectrum and the Bohr Model

In formulating his quantum theory, Planck was influenced by the results of investigations of the emission spectra produced by free, gas-phase atoms (as described in Section 3.2). Those line spectra led him to question whether any spectrum, even that of an incandescent lightbulb, was truly continuous. Among the early investigators of atomic emission spectra was a Swiss mathematician and schoolteacher named Johann Balmer (1825–1898). His work focused on the pattern of emission lines produced by high-temperature hydrogen atoms. In 1885, Balmer formulated an empirical equation that accurately predicted the wavelengths of the four brightest atomic emission lines in the visible spectrum of hydrogen. To express the wavelengths in nanometers, the units most often used for visible light, we can use the following form of the Balmer equation:

$$\lambda = 364.56\ \text{nm}\left(\frac{m^2}{m^2 - n^2}\right) \tag{3.7}$$

where $m$ is an integer greater than 2 (the values 3, 4, 5, and 6 predict the wavelengths of the four visible hydrogen lines) and $n = 2$. If, for example, we let $m = 3$ and solve for $\lambda$, we get

$$\lambda = 364.56\ \text{nm}\left(\frac{3^2}{3^2 - 2^2}\right) = 364.56 \times \frac{9}{5} = 656.21\ \text{nm}$$

which is the wavelength of the red line in the atomic emission spectrum of hydrogen (Figure 3.6a). By using $m$ values of 4, 5, and 6 in Equation 3.7, we get the wavelengths of the blue-green (486.08 nm), blue (434.00 nm), and violet (410.13 nm) lines, respectively.

The Balmer equation is called an *empirical* equation because he derived it strictly from experimental data—namely, the wavelengths of light in the visible atomic emission spectrum of hydrogen. The equation had no theoretical foundation. No one in 1885 could explain why his equation fit the hydrogen spectrum, though other scientists of the time used it successfully to search for hydrogen lines corresponding to $m$ values greater than 6, which no one could see because they are in the ultraviolet region of the electromagnetic spectrum.

In 1888, Swedish physicist Johannes Robert Rydberg (1854–1919) published a more general empirical equation for predicting the wavelengths of hydrogen's spectral lines:

$$\frac{1}{\lambda} = R_H\left(\frac{1}{n_1^2} - \frac{1}{n_2^2}\right) \tag{3.8}$$

where $n_1$ and $n_2$ are any positive integers (provided that $n_2 > n_1$) and $R_H$ is the Rydberg constant. The term on the left in Rydberg's equation is the reciprocal of wavelength ($1/\lambda$), which is called the *wavenumber* of a spectral line. To understand why Rydberg found wavenumbers useful in formulating his equation, remember that the energy of a photon is inversely proportional to its wavelength. Therefore, its energy is directly proportional to its wavenumber. Though Rydberg did not know it in 1888, the values of $n_1$ and $n_2$ in his equation correspond to energy levels inside hydrogen atoms, and the energy of a photon that a hydrogen atom absorbs or emits is the same as the difference in energy between a pair of energy levels.

The value of Rydberg's constant, $R_H$, is $1.0974 \times 10^7\ m^{-1}$, which we can convert to more suitable units as needed. When hydrogen lines are in the ultraviolet or visible region of the electromagnetic spectrum (as happens when $n_1 = 1$ or 2), their wavelengths are usually expressed in nanometers, and the most convenient form of the Rydberg equation has $R_H$ in units of nanometers$^{-1}$:

$$\frac{1}{\lambda} = 1.0974 \times 10^{-2}\ nm^{-1}\left(\frac{1}{n_1^2} - \frac{1}{n_2^2}\right) \tag{3.9}$$

When we let $n_1 = 2$ and $n_2 = 3$, 4, 5, or 6 in Equation 3.9 and then solve for $\lambda$, we get four wavelength values that match those we obtained with the Balmer equation when the value of his $m$ parameter was 3, 4, 5, or 6. They are the wavelengths of the four lines in the visible emission spectrum of hydrogen. Actually, the Balmer equation represents a special case of the Rydberg equation for the hydrogen lines that correspond to $n_1 = 2$. An advantage of Rydberg's equation was that it allowed scientists to predict the wavelengths of other series of hydrogen emission lines for which $n_1 \neq 2$. None of those additional lines were in the visible region. In 1908, German physicist Friedrich Paschen (1865–1947) discovered a series of hydrogen emission lines in the infrared region with wavelengths corresponding to Rydberg equation values of $n_1 = 3$ and $n_2 = 4$, 5, 6, and so on. A few years later, Theodore Lyman (1874–1954) at Harvard University discovered another series of hydrogen emission lines in the ultraviolet region corresponding to $n_1 = 1$. By the 1920s the $n_1 = 4$ and $n_1 = 5$ series had been discovered. Like the $n_1 = 3$ series, they also are in the infrared region.

**SAMPLE EXERCISE 3.4** Calculating the Wavelength of a Line in the Hydrogen Spectrum **LO3**

What is the wavelength in nanometers of the line in the hydrogen spectrum that corresponds to $m = 7$ in the Balmer equation (Equation 3.7)? Check your answer by also calculating that wavelength with the Rydberg equation.

**Collect and Organize** We need to calculate the wavelength of a line in hydrogen's atomic emission spectrum from its $m$ value in the Balmer equation. The $n$ value in the Balmer equation is always 2. We will then use the Rydberg equation to check the accuracy of our calculation.

**Analyze** To calculate a wavelength by using the Balmer equation, we insert the appropriate values of $m$ and $n$ (7 and 2 here) and solve for $\lambda$. In the corresponding calculation based on the Rydberg equation, we let $n_2 = 7$ because it is the greater integer and $n_1 = 2$. Our answer should be near 400 nm because the wavelength corresponding to $m = 6$ in the Balmer equation is 410 nm, and the wavelengths generated by the Balmer equation become both smaller and closer to each other as $m$ increases.

**Solve** Using the Balmer equation,

$$\lambda = 364.56 \text{ nm}\left(\frac{m^2}{m^2 - n^2}\right) = 364.56 \text{ nm}\left(\frac{7^2}{7^2 - 2^2}\right) = 396.97 \text{ nm}$$

Using the Rydberg equation (Equation 3.9),

$$\frac{1}{\lambda} = (1.0974 \times 10^{-2} \text{ nm}^{-1})\left(\frac{1}{{n_1}^2} - \frac{1}{{n_2}^2}\right)$$

$$= (1.0974 \times 10^{-2} \text{ nm}^{-1})\left(\frac{1}{2^2} - \frac{1}{7^2}\right)$$

$$= (1.0974 \times 10^{-2} \text{ nm}^{-1})(0.22959) = 2.5195 \times 10^{-3} \text{ (nm}^{-1})$$

$$\lambda = 396.90 \text{ nm}$$

**Think About It** The results of the two calculations are nearly, but not quite, the same. The difference is due to a slight error in the constant in Balmer's equation that he developed in 1885. The modern value of Balmer's constant is 364.51 nm. Had we used that value, the two calculated wavelengths would have differed by only 0.01 nm.

**Practice Exercise** What is the wavelength of the photon emitted by a hydrogen atom that corresponds to $m = 12$ in the Balmer equation? Could Balmer have seen that line?

*(Answers to Practice Exercises are in the back of the book.)*

**CONNECTION** We discussed Rutherford's gold-foil experiments and the model of the atom that evolved from them in Chapter 2.

## The Bohr Model

**CHEMTOUR**
Emission Spectra and the Bohr Model of the Atom

Rutherford suggested that electrons might orbit the nucleus in the way that planets orbit the sun. However, classical physics predicted that negative electrons orbiting a positive nucleus should emit energy in the form of electromagnetic radiation and eventually spiral into the nucleus. If that happened, no atom would be stable. Later, Planck proposed that the discrete wavelengths of Balmer's and Rydberg's emission lines meant that hydrogen atoms lost and gained only discrete quanta of energy, which indicated to him that discrete energy levels were present inside the atoms. But classical physics could not explain the existence of quantized energy levels. A new model of the atom was needed to explain what classical physics could not.

In 1913, Danish scientist Niels Bohr (1885–1962) proposed a model for the hydrogen atom in which the electron revolves around the nucleus in one of an array of concentric orbits. Each orbit represents a discrete energy level inside the atom. Bohr assigned each orbit a number, $n$, starting with $n = 1$ for the orbit

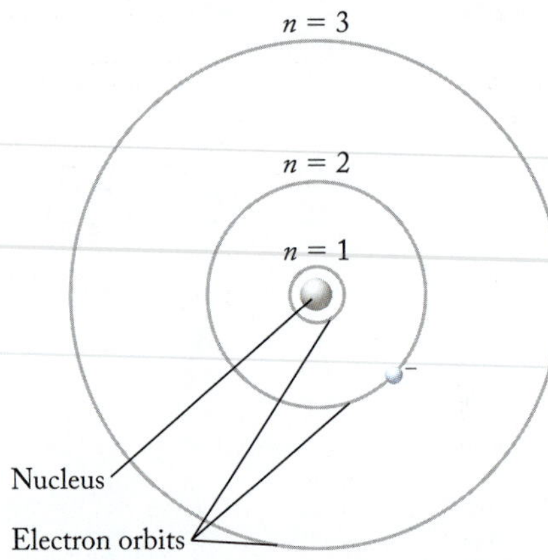


**FIGURE 3.17** In the Bohr model of the hydrogen atom, its electron revolves around the nucleus in one of several concentric orbits. Each orbit represents an allowed energy level. An electron in the orbit closest to the nucleus ($n = 1$) has the lowest energy.

closest to the nucleus (**Figure 3.17**). In his model, orbits farther from the nucleus have larger values of $n$, and the electrons in them have higher (less negative) energies according to the following equation:

$$E = -2.178 \times 10^{-18}\,\text{J}\left(\frac{1}{n^2}\right) \tag{3.10}$$

where $n = 1, 2, 3, \ldots, \infty$. According to Equation 3.10, an electron in the orbit closest to the nucleus ($n = 1$) has the lowest (most negative) energy:

$$E = -2.178 \times 10^{-18}\,\text{J}\left(\frac{1}{1^2}\right) = -2.178 \times 10^{-18}\,\text{J}$$

An electron in the next-closest ($n = 2$) orbit has an energy of

$$E = -2.178 \times 10^{-18}\,\text{J}\left(\frac{1}{2^2}\right) = -5.445 \times 10^{-19}\,\text{J}$$

Note that this value is less negative than the value for the electron in the $n = 1$ orbit. The values of $n$ for orbits farther and farther from the nucleus approach $\infty$, and as they do, $E$ approaches zero:

$$E = -2.178 \times 10^{-18}\,\text{J}\left(\frac{1}{\infty^2}\right) = 0$$

Zero energy means that the electron at $n = \infty$ is no longer part of the hydrogen atom and that the atom no longer exists as a single entity. Rather, it has become two separate particles: a $H^+$ ion and a free electron.

Why does Equation 3.10 have a negative sign? What does it mean for an electron to have negative energy? To understand that, let's assume that the addition of energy is required to pry a negatively charged electron away from a positively charged nucleus. That idea is logical because oppositely charged particles are attracted to each other, and that attraction must be overcome to separate them. An electron that has been separated from an atom no longer interacts with the nucleus, and no energy of attraction exists between the particles ($E = 0$). So, if we had to add energy to get to zero energy, the initial energy of the electron in the atom must have been less than zero, as we would calculate using Equation 3.10.

An important feature of the Bohr model is that it offers a theoretical framework for explaining the experimental observations of Balmer, Rydberg, and others we have discussed. To see the connection, consider what happens when an electron moves between two allowed energy levels in Bohr's model. If we label the energy level where the electron starts as $n_{\text{initial}}$ and the level where the electron ends up as $n_{\text{final}}$, the change in energy of the electron is

$$\Delta E = -2.178 \times 10^{-18}\,\text{J}\left(\frac{1}{n_{\text{final}}^{2}} - \frac{1}{n_{\text{initial}}^{2}}\right) \tag{3.11}$$

Here we use the capital Greek letter delta ($\Delta$) to represent change (and we will do so throughout this book).

If the electron moves to an orbit farther from the nucleus, then $n_{\text{final}} > n_{\text{initial}}$, and the overall value of the terms inside the parentheses in Equation 3.11 is negative because $(1/n_{\text{final}}^2) < (1/n_{\text{initial}}^2)$. That negative value multiplied by the negative coefficient gives us a positive $\Delta E$ and represents an increase in electron energy. By contrast, if an electron moves from an outer orbit to one closer to the nucleus, then $n_{\text{final}} < n_{\text{initial}}$, and the sign of $\Delta E$ is negative. That means the electron loses energy.

**ground state** the most stable, lowest-energy state of a particle.

**excited state** any energy state above the ground state.

**electron transition** movement of an electron between energy levels.

When the electron in a hydrogen atom is in the lowest ($n = 1$) energy level, the atom is said to be in its **ground state**. According to the Bohr model, the electron cannot have any less energy than it has in the ground state, which means that it can't lose more energy and spiral into the nucleus.

If the electron in a hydrogen atom is in an energy level above $n = 1$, the atom is said to be in an **excited state**. An electron can move from the $n = 1$ (ground state) orbit to a higher level (for example, $n = 3$) by absorbing a quantum of energy that matches the energy difference ($\Delta E$) between the two states. Similarly, an electron in an excited state can move to an even higher energy level by absorbing a quantum of energy that matches the energy difference between those two states. However, an electron in an excited state can move to a lower-energy excited state, or all the way down to the ground state, by emitting a quantum of energy that matches the energy difference between those two states. The movement of an electron between any two energy levels is called an **electron transition**.

The energy-level diagram in **Figure 3.18** shows some of the transitions that an electron in a hydrogen atom can make. The black arrow pointing upward represents the absorption of enough energy to remove the electron from a hydrogen atom (ionization). The downward-pointing colored arrows represent decreases in the internal energy of the hydrogen atom that occur when photons are emitted as the electron moves from a higher energy level to a lower energy level. If the colored arrows pointed up, they would represent the absorption of photons and would lead to increases in the internal energy of the atom. The energy of the photon absorbed or emitted always matches the absolute value of $\Delta E$.

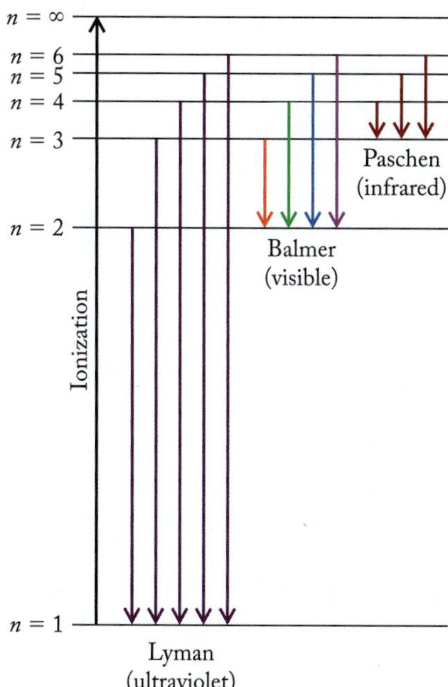


**FIGURE 3.18** An energy-level diagram showing some possible electron transitions in a hydrogen atom. The arrow pointing up represents ionization. Arrows pointing down represent electron transitions from a higher to a lower energy level, in which the atom loses a particular quantity of energy that may be emitted as a photon of electromagnetic radiation. The series of lines ending at $n = 2$ corresponds to the visible emission lines studied by Balmer as described in Equation 3.7, whereas the series ending at $n = 1$ and $n = 3$ were studied by Lyman and Paschen, respectively.

## CONCEPT TEST

On the basis of the lengths of the arrows in Figure 3.18, rank the following transitions in order from the greatest change in electron energy to the smallest change:

a. $n = 4 \rightarrow n = 2$

b. $n = 3 \rightarrow n = 2$

c. $n = 2 \rightarrow n = 1$

d. $n = 4 \rightarrow n = 3$

*(Answers to Concept Tests are in the back of the book.)*

## SAMPLE EXERCISE 3.5 Calculating the Energy Change of an Electron Transition in the Hydrogen Atom LO3

a. What is the change in energy associated with the electron transition from $n = 3$ to $n = 2$ in Figure 3.18?

b. How much energy is required to ionize a ground-state hydrogen atom? That is, what is the *ionization energy* of hydrogen?

**Collect and Organize** We are asked to determine the energy associated with two electron transitions in the hydrogen atom. Equation 3.11 enables us to calculate the energy change associated with any such electron transition.

**Analyze** To use Equation 3.11, we need to identify the initial ($n_{\text{initial}}$) and final ($n_{\text{final}}$) energy levels of each electron. For the first electron, $n_{\text{initial}} = 3$ and $n_{\text{final}} = 2$. For the electron that begins in the ground state of a H atom, $n_{\text{initial}} = 1$, and because the atom is ionized, $n_{\text{final}} = \infty$; the electron is no longer associated with the nucleus.

**Solve**

a.
$$\Delta E = -2.178 \times 10^{-18}\,\text{J}\left(\frac{1}{n_{\text{final}}^{\ 2}} - \frac{1}{n_{\text{initial}}^{\ 2}}\right)$$
$$= -2.178 \times 10^{-18}\,\text{J}\left(\frac{1}{2^2} - \frac{1}{3^2}\right)$$
$$= -3.025 \times 10^{-19}\,\text{J}$$

b.
$$\Delta E = -2.178 \times 10^{-18}\,\text{J}\left(\frac{1}{\infty^2} - \frac{1}{1^2}\right)$$

Dividing by $\infty^2$ yields zero, so the value inside the parentheses simplifies to −1, which gives us $\Delta E = 2.178 \times 10^{-18}$ J.

**Think About It** Our $\Delta E$ values in parts a and b have opposite signs. The value in part a corresponds to hydrogen's visible red line, and it is negative because the associated electron loses energy when it emits a photon. The sign of ionization energy is positive because energy must be added to overcome the attraction between the negatively charged electron and the positively charged nucleus.

**Practice Exercise** Calculate the energy, in joules, required to ionize a hydrogen atom when its electron is initially in the $n = 3$ energy level. Before doing the calculation, predict whether that energy is greater than or less than the $2.178 \times 10^{-18}$ J needed to ionize a ground-state hydrogen atom.

*(Answers to Practice Exercises are in the back of the book.)*

Before we end this section, let's compare Equation 3.11, which describes the differences in energy between pairs of energy levels in hydrogen atoms, and the Rydberg equation (Equation 3.9), which describes the wavenumbers for the lines in hydrogen's atomic emission and absorption spectra. Notice how similar they are. The coefficients in them are different because of the different units used to express wavenumber and energy, but the empirical Rydberg equation has the same form as the theoretical equation Bohr developed to explain the internal structure of the hydrogen atom. Thus, atomic emission and absorption spectra reveal the energies of electrons inside hydrogen atoms and, as we will see, the atoms of other elements.

When we extend our exploration of atomic spectra and internal energies of atoms to other elements, we discover that the Bohr model of electrons revolving around nuclei in stable circular orbits works only for atoms or ions with only one electron. In multielectron atoms, the electrons interact with one another in ways that the Bohr model does not take into account. Thus, the picture of the atom in Bohr's model is limited. However, it was an important stepping-stone to developing more complete models of atomic structure based on quantum theory.

## 3.5 Electrons as Waves

**CHEMTOUR**
De Broglie Wavelength

A decade after Bohr published his model of the hydrogen atom, a French graduate student named Louis de Broglie (1892–1987) developed another explanation for the stability of electrons orbiting the nuclei of hydrogen atoms. He based his hypothesis on the assumption that electrons could behave like *waves of matter* as well as particles of matter. That dual nature of electrons was modeled after the behavior of electromagnetic radiation, which could be explained not only by assigning it wavelike properties, as James Maxwell and many others had done, but also by treating it as though it was made up of particles, or quanta, as Einstein had done to explain the photoelectric effect.

**matter wave** the wave associated with any moving particle.

## De Broglie Wavelengths

An electron traveling through space or around an atom's nucleus has a kinetic energy that, as we saw in Chapter 1, is related to its mass and the square of its speed. If the electron can also behave as a wave, its wavelength is inversely proportional to its energy. Those two perspectives on the energy of a moving electron are contained in the de Broglie equation for calculating the wavelength of an electron (or any particle) in motion:

$$\lambda = \frac{h}{mu} \tag{3.12}$$

In that equation, $m$ is the mass of the particle, $u$ is its speed, and $h$ is Planck's constant. Note the inverse relationship between the particle's wavelength and the product of its mass times its speed. That product is the particle's *momentum*. The more momentum a particle has, the shorter its *de Broglie wavelength*.

Equation 3.12 tells us that any moving particle has wavelike properties; that is, the particle behaves as a **matter wave**. De Broglie predicted that moving particles much bigger than electrons, such as atomic nuclei, molecules, and even tennis balls and airplanes, have characteristic wavelengths that can be calculated using Equation 3.12. The wavelengths of large objects are extremely small, given the tiny size of the Planck constant and their considerable mass, so we never notice the wave nature of large objects in motion.

### SAMPLE EXERCISE 3.6 Calculating the Wavelength of a Particle in Motion — LO4

Calculate the de Broglie wavelength of a 142-g baseball thrown at 44 m/s (98 mph), and compare that wavelength with the diameter of the ball, 7.35 cm.

**Collect and Organize** We know the mass and speed of a baseball and need to calculate its de Broglie wavelength so that we can compare its value with the size of the baseball. Equation 3.12 may be used to calculate that wavelength.

**Analyze** Given how small $h$ is, the wavelength of a pitched baseball should be only a tiny fraction of the size of the baseball. The right side of Equation 3.12 has units of joule-seconds in the numerator and of mass and speed in the denominator. To combine the units in a way that gives us a unit of length, we need to use the following conversion factor:

$$1\ \text{J} = 1\ \text{kg} \cdot \text{m}^2/\text{s}^2$$

To use that equality, we must convert the mass of the baseball from grams to kilograms:

$$142\ \text{g} = 0.142\ \text{kg}$$

**Solve** The de Broglie wavelength of the baseball is

$$\lambda = \frac{h}{mu} = \frac{6.626 \times 10^{-34}\ \cancel{\text{J}} \cdot \cancel{\text{s}}}{(0.142\ \cancel{\text{kg}})(44\ \text{m}/\cancel{\text{s}})} \times \frac{1\ \cancel{\text{kg}} \cdot \text{m}^2/\cancel{\text{s}^2}}{1\ \cancel{\text{J}}} = 1.06 \times 10^{-34}\ \text{m}$$

This wavelength is

$$\frac{1.06 \times 10^{-34}\ \text{m}}{7.35\ \cancel{\text{cm}} \times (1\ \text{m}/100\ \cancel{\text{cm}})} \times 100\% = 1.4 \times 10^{-33}\%$$

of the ball's diameter.

**Think About It** The wavelength of the matter wave of the baseball is much too small to be observed, so its character contributes nothing to the behavior of the baseball. That is what we expected: large moving objects behave like large moving objects, not like waves.

**Practice Exercise** The speed of the electron in the ground state of the hydrogen atom is $2.2 \times 10^6$ m/s. What is the wavelength of the electron in meters?

*(Answers to Practice Exercises are in the back of the book.)*

**FIGURE 3.19** Linear and circular standing waves. (a) The wavelength of a standing wave in a violin string fixed at both ends is related to the distance $L$ between the ends of the string by the equation $L = n(\lambda/2)$. In the standing waves shown, $n = 1, 2$, and 3. An $n$ value of $2\frac{1}{2}$ does not produce a standing wave because a standing wave can have no string motion at either end. (b) The circular waves proposed by de Broglie account for the stability of the energy levels in Bohr's model of the hydrogen atom. Each stable wave must have a circumference equal to $n\lambda$, with $n$ restricted to being an integer, such as the $n = 3$ wave shown here. (c) If the circumference is not an exact multiple of $\lambda$, as shown in the $n = 2\frac{1}{4}$ image, no standing wave occurs.

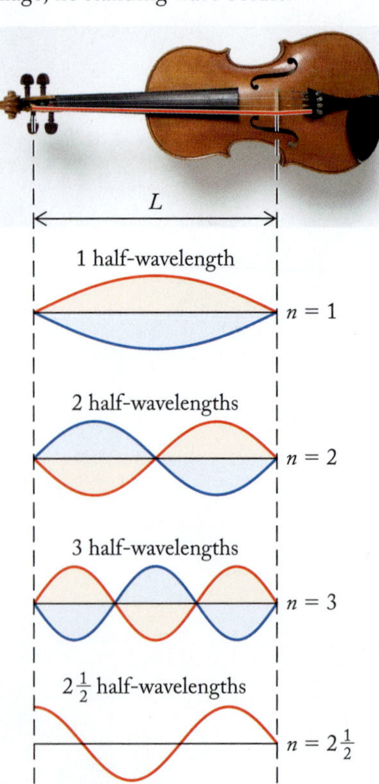


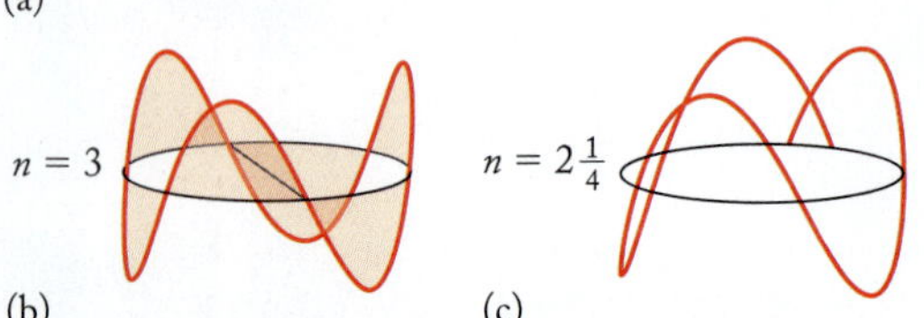


De Broglie used matter waves to explain the stability of the electron levels in Bohr's model of the hydrogen atom, proposing that the orbiting electron behaves like a circular wave oscillating around the nucleus. To understand the implications of de Broglie's proposal, we need to examine the features of a stable, circular wave. Consider a vibrating violin string of length $L$ (**Figure 3.19a**). Because the string is fixed at both ends, no vibration occurs at the ends and the vibration is at maximum in the middle. The wave created by that combination of fixed ends and maximum vibration in the middle is a **standing wave**: a wave that oscillates back and forth within a fixed space, contrary to the way waves of light travel through space. Points on a standing wave that have zero displacement, such as the two ends of the string, are called **nodes**. The sound wave produced in that way is called the *fundamental* of the violin string, and the wavelength of the fundamental is $2L$. The wave has the lowest frequency and longest wavelength possible for that string. The length of the string is one-half the wavelength of the fundamental ($L = \lambda/2$).

If the string is held down in the middle (creating a third node) and plucked halfway between the middle and one end, a new, higher-frequency wave called the *first harmonic* is produced. The wavelength of that harmonic is equal to $L$ because $L = 2(\lambda/2) = \lambda$. We can continue generating higher frequencies with wavelengths related to the length of the string by the equation $L = n(\lambda/2)$, where $n$ is a whole number and $L = 3(\lambda/2), 4(\lambda/2), 5(\lambda/2)$, and so on.

The standing-wave pattern for a circular wave generated by an electron differs slightly from that of a vibrating violin string in that the circular wave has no defined stationary ends. Instead, the electron vibrates in an endless series of waves. However, standing waves are produced only if, as shown in **Figure 3.19(b)**, the circumference of the circle equals a whole-number multiple of the electron's wavelength:

$$\text{Circumference} = n\lambda \tag{3.13}$$

If the circumference is not a whole-number multiple, the oscillating wave is discontinuous (**Figure 3.19c**) and unstable.

Equation 3.13 gives a new meaning to Bohr's orbit label $n$: it represents the number of matter-wave wavelengths in that orbit's circumference. De Broglie's proposal also solved the problem of a negatively charged electron spiraling into the positively charged nucleus. Because $n = 1$ represents the minimum circumference of the circular wave of a moving electron, the electron must always remain a minimum distance from the nucleus. Furthermore, the lowest energy possible for an electron in a hydrogen atom occurs for a standing wave with circumference $\lambda$, so the electron cannot get any closer to the nucleus than that.

De Broglie's matter-wave hypothesis created a quandary for the graduate faculty at the University of Paris, where he studied. Bohr's model of electrons moving between allowed energy levels had been widely criticized as an arbitrary suspension of well-tested physical laws. De Broglie's rationalization of Bohr's model seemed even more outrageous to many scientists. Before the faculty would accept his doctoral thesis based on matter waves, they wanted another opinion, so they sent it to Albert Einstein for review. Einstein wrote back that he found de Broglie's work "quite interesting." That endorsement was good enough for the faculty: de Broglie's thesis was accepted in 1924 and immediately submitted for publication.

**standing wave** a wave confined to a given space, with a wavelength $\lambda$ related to the length $L$ of the space by $L = n(\lambda/2)$, where $n$ is a whole number.

**node** a location in a standing wave that experiences no displacement. In the context of orbitals, nodes are locations at which electron density goes to zero.

**Heisenberg uncertainty principle** the principle that one cannot simultaneously know the exact position and the exact momentum of an electron.

## The Heisenberg Uncertainty Principle

After de Broglie proposed that electrons exhibited both particle-like and wave-like behavior, questions arose about how wave behavior affected our ability to locate the electron, given that a wave by its very nature is spread out in space. The question, "Where is the electron?" was addressed by German physicist Werner Heisenberg (1901–1976), who proposed the following thought experiment: What if we tried to watch an electron orbiting the nucleus of an atom? We would need an extremely powerful microscope to see a particle as tiny as an electron. No such microscope exists, but if it did, it would need to use gamma rays for illumination because they are the only part of the electromagnetic spectrum with wavelengths short enough to match the diminutive size of electrons. Longer-wavelength radiation would pass right by an electron without being reflected by it. Unfortunately, the ultrashort wavelengths and high frequencies of gamma rays mean that they have enormous energies—so large that any gamma ray striking an electron would knock the electron off course. The only way not to affect the electron's motion would be to use a much lower-energy, longer-wavelength source of radiation to illuminate it, but then we would not be able to see the tiny electron clearly.

That situation presents an experimental dilemma. The only means for clearly observing an electron make it impossible to know the electron's motion or, more precisely, its momentum, defined as an object's velocity multiplied by its mass. Therefore, we can never simultaneously know exactly both the position and the momentum of the electron. That conclusion, the **Heisenberg uncertainty principle**, is mathematically expressed as

$$\Delta x \cdot m\Delta u \geq \frac{h}{4\pi} \tag{3.14}$$

where $\Delta x$ is the uncertainty in the position of the electron, $m$ is its mass, $\Delta u$ is the uncertainty in its velocity, and $h$ is the Planck constant. To Heisenberg, that uncertainty was the essence of quantum theory. Its message for us is that limits exist on what we can observe, measure, and therefore know about particles the size of electrons.

**SAMPLE EXERCISE 3.7** Calculating Heisenberg Uncertainties **LO4**

Use the data in Sample Exercise 3.6 to compare the uncertainties in the speeds of a thrown baseball and an electron. Assume that the position of the baseball is known to within one wavelength of red light ($\Delta x_{\text{baseball}} = 6.80 \times 10^{-7}$ m = 680 nm) and that the position of the electron is known to within the radius of the hydrogen atom ($\Delta x_{\text{electron}} = 5.3 \times 10^{-11}$ m).

**Collect and Organize** We are asked to calculate the uncertainty in the speeds of two particles and are given the uncertainties in their positions. Equation 3.14 provides a mathematical connection between the variables. From Sample Exercise 3.6 we know that the mass of a baseball is 0.142 kg. The mass of an electron is $9.109 \times 10^{-31}$ kg.

**Analyze** According to Equation 3.14, the uncertainty in the speed and position of a particle is inversely proportional to its mass. Therefore, we can expect little uncertainty in the speed of the baseball but much greater uncertainty in the speed of the electron. We need to rearrange the terms in the equation to solve for the uncertainty in speed ($\Delta u$):

$$\Delta u \geq \frac{h}{4\pi \Delta x \cdot m}$$

**Solve** For the baseball:

$$\Delta u \geq \frac{6.626 \times 10^{-34}\ \text{J} \cdot \text{s}}{4\pi(6.80 \times 10^{-7}\ \text{m})(0.142\ \text{kg})} \times \frac{1\ \text{kg} \cdot \text{m}^2/\text{s}^2}{1\ \text{J}} \geq 5.46 \times 10^{-28}\ \text{m/s}$$

For the electron:

$$\Delta u \geq \frac{6.626 \times 10^{-34}\ \text{J} \cdot \text{s}}{4\pi(5.3 \times 10^{-11}\ \text{m})(9.109 \times 10^{-31}\ \text{kg})} \times \frac{1\ \text{kg} \cdot \text{m}^2/\text{s}^2}{1\ \text{J}} \geq 1.1 \times 10^{6}\ \text{m/s}$$

**Think About It** The uncertainty in the speed of the baseball is insignificant compared with, for example, the speed of a Major League fastball (> 40 m/s). That result is expected for objects in the macroscopic world. The uncertainty in the speed of the electron is many orders of magnitude larger because its mass (which is in the denominator of the expression we derived from Equation 3.14) is many orders of magnitude smaller.

**Practice Exercise** What is the uncertainty, in meters, in the position of an electron moving near a nucleus at a speed of $8 \times 10^{7}$ m/s? Assume that the relative uncertainty in the speed of the electron is 1%—that is, $8 \times 10^{5}$ m/s.

*(Answers to Practice Exercises are in the back of the book.)*

When he proposed his uncertain principle, Heisenberg was working with Bohr at the University of Copenhagen. The two scientists had widely different views about the significance of the uncertainty principle and the idea that particles could behave like waves. To Heisenberg, uncertainty was a fundamental characteristic of nature. To Bohr, it was merely a mathematical consequence of the wave–particle duality of electrons; an electron's position and path had no physical meaning. The debate between the two gifted scientists was heated at times. Heisenberg later wrote about one particularly emotional debate:

> At the end of the discussion I went alone for a walk in the neighboring park [and] repeated to myself again and again the question: "Can nature possibly be as absurd as it seems . . . ?"[1]

## 3.6 Quantum Numbers

Many leading scientists of the 1920s would not accept de Broglie's model of electrons as waves until it had a stronger theoretical foundation. Those scientists wanted a mathematical model that accurately described the behavior of matter

[1]Werner Heisenberg, *Physics and Philosophy: The Revolution in Modern Science* (Harper & Row, 1958), p. 42.

waves and accounted for the atomic spectra of hydrogen. During a Christmas vacation in the Swiss Alps in 1925, Austrian physicist Erwin Schrödinger (1887–1961) created that mathematical foundation by developing what came to be called **wave mechanics** or **quantum mechanics**.

Schrödinger's mathematical description of electron waves is known as the **Schrödinger wave equation**. We do not examine it in detail in this book, but you should know that solutions to it are called **wave functions** ($\psi$): mathematical expressions that describe how the matter wave of an electron in an atom varies in both time and location inside the atom. Wave functions define the energy levels in the hydrogen atom. They can be simple trigonometric functions, such as sine or cosine waves, or they can be complex.

What is the physical significance of a wave function? Schrödinger believed that a wave function depicted the "smearing" of an electron through three-dimensional space. However, that notion of subdividing a discrete particle was later rejected in favor of the model developed by German physicist Max Born (1882–1970), who proposed that the square of a wave function, $\psi^2$, indicates the probability of finding an electron within an atom. The region of space within an atom in which a high probability exists of finding an electron is called an **orbital**. Born later showed that his interpretation could be used to calculate the probability of electron transitions between orbitals, as happens when an atom absorbs or emits a quantum of energy.

To help visualize the probabilistic meaning of $\psi^2$, consider what happens when we spray ink onto a flat surface as in **Figure 3.20**. If we then draw a circle encompassing most of the ink spots, we are identifying the region of maximum probability for finding the spots.

Quantum mechanical orbitals are *not* two-dimensional concentric orbits, as in Bohr's model of the hydrogen atom, or even two-dimensional circles, as in the pattern of ink drops in Figure 3.20. Instead, they are three-dimensional regions of space with distinctive shapes, orientations, and average distances from the nucleus. Each orbital is a solution to Schrödinger's wave equation and is identified by a unique combination of three integers called **quantum numbers**, whose values flow directly from the mathematical solutions to the wave equation. The quantum numbers are as follows:

- The **principal quantum number, *n***, is like Bohr's $n$ value for the hydrogen atom in that it is a positive integer that indicates the relative size and energy of an orbital or of a group of orbitals in an atom. Orbitals (and the electrons in them) with the same value of $n$ are in the same *shell*. Orbitals with larger values of $n$ are in shells farther from the nucleus and have higher energies than those with lower values of $n$. In a hydrogen atom, an orbital's $n$ value defines the energy of an electron in the orbital, consistent with Bohr's model. In multielectron atoms, the relationship between energy levels and orbitals is more complex, but increasing values of $n$ generally represent higher energy levels.
- The **angular momentum quantum number**, $\ell$, is an integer with a value ranging from zero to $(n - 1)$ that defines the shape of an orbital. Orbitals with the same values of $n$ and $\ell$ (and the electrons that reside in them) are in the same *subshell* and have the same energy. Orbitals with a given value of $\ell$ are identified with a letter according to the following scheme:

| Value of $\ell$ | 0 | 1 | 2 | 3 | 4 |
|---|---|---|---|---|---|
| Letter Identifier | *s* | *p* | *d* | *f* | *g* |

**wave mechanics** or **quantum mechanics** a mathematical description of the wavelike behavior of particles on the atomic level.

**Schrödinger wave equation** a description of how the electron matter wave varies with location and time around the nucleus of a hydrogen atom.

**wave function ($\psi$)** a solution to the Schrödinger wave equation.

**orbital** a region around the nucleus of an atom where the probability of finding an electron is high; each orbital is defined by the square of the wave function ($\psi^2$).

**quantum number** one of four related numbers that specify the energy, shape, and orientation of orbitals in an atom and the spin orientation of electrons in the orbitals.

**principal quantum number (*n*)** a positive integer describing the relative size and energy of an atomic orbital or group of orbitals in an atom.

**angular momentum quantum number ($\ell$)** an integer having any value from 0 to $(n - 1)$ that defines the shape of an orbital.

**CHEMTOUR**
Quantum Numbers

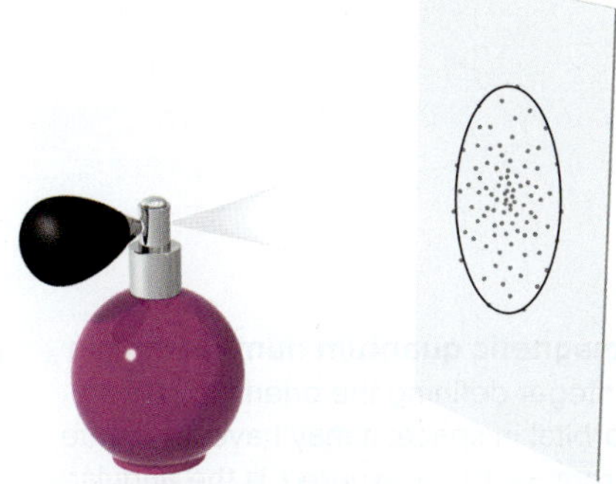

**FIGURE 3.20** The probability of finding an ink spot in the pattern produced by a source of ink spray decreases with increasing distance from the center of the pattern.

- The **magnetic quantum number,** $m_\ell$, is an integer with a value from $-\ell$ to $+\ell$. It defines the orientation of an orbital in the space around the nucleus of an atom.

Each subshell has a two-part label that contains the appropriate value of $n$ and a letter designation for $\ell$. For example, orbitals with $n = 3$ and $\ell = 1$ are called 3*p* orbitals, and electrons in 3*p* orbitals are called 3*p* electrons. How many 3*p* orbitals are there? We can answer that question by finding all possible values of $m_\ell$. Because *p* orbitals are those for which $\ell = 1$, they have $m_\ell$ values of −1, 0, and +1. The three values mean that three 3*p* orbitals exist, each with a unique combination of $n$, $\ell$, and $m_\ell$ values. All possible combinations of those three quantum numbers for the orbitals of the first four shells are listed in **Table 3.1.**

**TABLE 3.1** Quantum Numbers of the Orbitals in the First Four Shells

| Value of $n$ | Allowed Values of $\ell$ | Subshell Letter | Allowed Values of $m_\ell$ | Number of Orbitals in: Subshell | Shell |
|---|---|---|---|---|---|
| 1 | 0 | *s* | 0 | 1 | 1 |
| 2 | 0 | *s* | 0 | 1 | |
| | 1 | *p* | −1, 0, +1 | 3 | 4 |
| 3 | 0 | *s* | 0 | 1 | |
| | 1 | *p* | −1, 0, +1 | 3 | |
| | 2 | *d* | −2, −1, 0, +1, +2 | 5 | 9 |
| 4 | 0 | *s* | 0 | 1 | |
| | 1 | *p* | −1, 0, +1 | 3 | |
| | 2 | *d* | −2, −1, 0, +1, +2 | 5 | |
| | 3 | *f* | −3, −2, −1, 0, +1, +2, +3 | 7 | 16 |

**SAMPLE EXERCISE 3.8** Identifying the Subshells and Orbitals in an Energy Level **LO5**

a. What are the names of all the subshells in the $n = 4$ shell?
b. How many orbitals are in all the subshells of the $n = 4$ shell?

**Collect, Organize, and Analyze** We are asked to name the subshells in the fourth shell and to determine how many orbitals are in all the subshells. Subshell designations (Table 3.1) are based on the possible values of the quantum numbers $n$ and $\ell$. The allowed values of $\ell$ depend on the value of $n$ because $\ell$ is an integer between 0 and $(n - 1)$. The number of orbitals in a subshell depends on the number of possible values of $m_\ell$, which range from $-\ell$ to $+\ell$.

**Solve**

a. The allowed values of $\ell$ for $n = 4$ range from 0 to $(n - 1)$—that is, from 0 to 3—so they are 0, 1, 2, and 3. The $\ell$ values correspond to the subshell designations *s, p, d,* and *f*, respectively. The appropriate subshell names are thus 4*s*, 4*p*, 4*d*, and 4*f*.
b. The possible values of $m_\ell$ from $-\ell$ to $+\ell$ are
   - $\ell = 0$; $m_\ell = 0$. That combination of $\ell$ and $m_\ell$ values for the $n = 4$ shell represents a single 4*s* orbital.
   - $\ell = 1$; $m_\ell = -1, 0,$ or $+1$. Those three combinations of $\ell$ and $m_\ell$ values for the $n = 4$ shell represent the three 4*p* orbitals.

**magnetic quantum number ($m_\ell$)** an integer defining the orientation of an orbital in space; it may have any value from $-\ell$ to $+\ell$, where $\ell$ is the angular momentum quantum number.

**spin quantum number ($m_s$)** either $+\frac{1}{2}$ or $-\frac{1}{2}$, indicating the spin orientation of an electron.

- $\ell = 2$; $m_\ell = -2, -1, 0, +1$, or $+2$. Those five combinations of $\ell$ and $m_\ell$ values represent the five $4d$ orbitals.
- $\ell = 3$; $m_\ell = -3, -2, -1, 0, +1, +2$, or $+3$. Those seven combinations of $\ell$ and $m_\ell$ values represent the seven $4f$ orbitals.

Thus, $1 + 3 + 5 + 7 = 16$ orbitals are in the $n = 4$ shell.

**Think About It** We determined that the fourth shell has 16 orbitals. The number of orbitals in each shell is equal to $n^2$, the square of the principal quantum number of the shell.

 **Practice Exercise** How many orbitals are in the $n = 5$ shell? What are the names of all the subshells in the $n = 5$ shell?

*(Answers to Practice Exercises are in the back of the book.)*

The following relationships are worth noting in the quantum numbering system:

- The $n$th shell has $n$ subshells: one subshell ($1s$) in the $n = 1$ shell, two subshells ($2s$ and $2p$) in the $n = 2$ shell, and so on.
- The $n$th shell has $n^2$ orbitals: $1^2 = 1$ in the $n = 1$ shell, $2^2 = 4$ in the $n = 2$ shell, and so on.
- Each subshell has $(2\ell + 1)$ orbitals: one $s$ orbital $(2 \times 0 + 1 = 1)$ in each $s$ subshell, three $p$ orbitals $(2 \times 1 + 1 = 3)$ in each $p$ subshell, five $d$ orbitals $(2 \times 2 + 1 = 5)$ in each $d$ subshell, and so on.

The Schrödinger equation accounts for most, but not all, aspects of atomic spectra. The emission spectrum of hydrogen, for example, when viewed through a high-resolution spectrometer, contains a pair of red lines at 656 nm, where Balmer saw only one (Figure 3.7b). Pairs of lines also exist in the spectra of multielectron atoms that have a single electron in their outermost shells.

In 1925, two students at the University of Leiden in the Netherlands, Samuel Goudsmit (1902–1978) and George Uhlenbeck (1900–1988), proposed that the pairs of lines, called *doublets*, were caused by a property they called *electron spin*. In their model, electrons spin in one of two directions, designated "spin up" and "spin down." A moving electron (or any charged particle) creates a magnetic field by moving through space. The spinning motion produces a second magnetic field oriented up or down. To account for the two spin orientations, Goudsmit and Uhlenbeck proposed a fourth quantum number, the **spin quantum number, $m_s$**. Two values are possible for $m_s$: $+\frac{1}{2}$ for spin up and $-\frac{1}{2}$ for spin down.

Even before Goudsmit and Uhlenbeck proposed the electron-spin hypothesis, two other scientists, Otto Stern (1888–1969) and Walther Gerlach (1889–1979), observed the effect of electron spin when they shot a beam of silver atoms through a magnetic field (**Figure 3.21**). Those atoms in which the net electron spin was "up" were deflected in one direction by the field, whereas those in which the net electron spin was "down" were deflected in the opposite direction.

In 1925, Austrian physicist Wolfgang Pauli (1900–1958) proposed that no two electrons in a multielectron atom can have the same set of values for the four

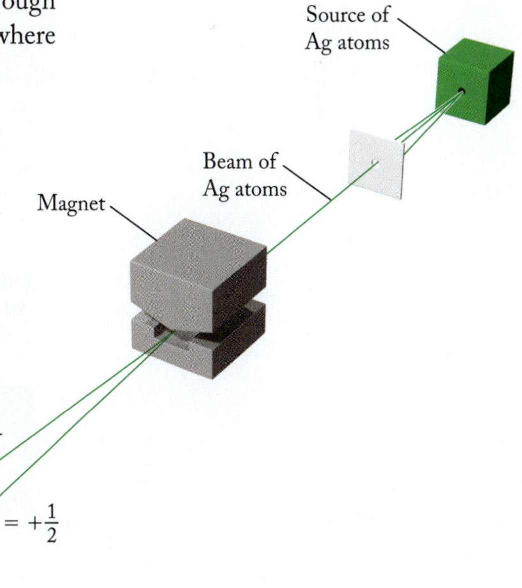


**FIGURE 3.21** A narrow beam of silver atoms passed through a magnetic field is split into two beams because of the interactions between the field and the spinning electrons in the atoms. That observation led to the proposal of a fourth quantum number, $m_s$.

quantum numbers $n$, $\ell$, $m_\ell$, and $m_s$. That idea is known as the **Pauli exclusion principle**. We have seen how unique combinations of the three quantum numbers from Schrödinger's wave equation define each orbital in an atom. The fourth (spin) quantum number serves as a unique address for each electron in each orbital.

**SAMPLE EXERCISE 3.9** Identifying Valid Sets of Quantum Numbers **LO5**

Which of these five combinations of quantum numbers are valid?

| | $n$ | $\ell$ | $m_\ell$ | $m_s$ |
|---|---|---|---|---|
| (a) | 1 | 0 | −1 | $+\frac{1}{2}$ |
| (b) | 3 | 2 | −2 | $+\frac{1}{2}$ |
| (c) | 2 | 2 | 0 | 0 |
| (d) | 2 | 0 | 0 | $-\frac{1}{2}$ |
| (e) | −3 | −2 | −1 | $-\frac{1}{2}$ |

**Collect, Organize, and Analyze** Table 3.1 contains valid combinations of the quantum numbers $n$, $\ell$, and $m_\ell$ for the first four shells. The principal quantum number ($n$) can be any positive integer; the valid values of $\ell$ in a given shell are integers from 0 to $(n - 1)$, and the values of $m_\ell$ in a given subshell include all integers from $-\ell$ to $+\ell$, including 0. The only two options for $m_s$ are $+\frac{1}{2}$ and $-\frac{1}{2}$.

**Solve**

a. Because $n = 1$, the maximum (and only) value of $\ell$ is $(n - 1) = 1 - 1 = 0$. Therefore, the values of $n$ and $\ell$ are valid. However, if $\ell = 0$, $m_\ell$ must be 0; it cannot be −1. Therefore, that set is not valid. The spin quantum number, $+\frac{1}{2}$, is valid, however.
b. Because $n = 3$, $\ell$ can be 2 and $m_\ell$ can be −2. Also, $m_s = +\frac{1}{2}$ is a valid choice for the spin quantum number. That set is valid.
c. Because $n = 2$, $\ell$ cannot be 2, making that set invalid. Also, $m_s = 0$ is an invalid value.
d. Because $n = 2$, $\ell$ can be 0, and for that value of $\ell$, $m_\ell$ must be 0, too. The value of $m_s$ also is valid, and so is the set.
e. That set contains two impossible values, $n = -3$ and $\ell = -2$, so it is invalid.

**Think About It** The values of $n$, $\ell$, and $m_\ell$ are related mathematically, and $m_s$ can be either $+\frac{1}{2}$ or $-\frac{1}{2}$. Each electron in an atom has a unique combination of these four values.

**Practice Exercise** Write all the possible sets of quantum numbers for an electron in the $n = 3$ shell that has an angular momentum quantum number $\ell = 1$ and a spin quantum number $m_s = +\frac{1}{2}$.

*(Answers to Practice Exercises are in the back of the book.)*

## 3.7 The Sizes and Shapes of Atomic Orbitals

We have learned that atomic orbitals have three-dimensional shapes that are graphical representations of $\psi^2$. In this section, we examine how the shapes of orbitals affect the energies of the electrons in them.

**Pauli exclusion principle** the principle that no two electrons in an atom can have the same set of four quantum numbers.

## s Orbitals

**Figure 3.22** shows four representations of the 1*s* orbital of hydrogen. In Figure 3.22(a), electron density is plotted against distance from the nucleus, and the graph seems to show that density decreases with increasing distance. However, Figure 3.22(b) shows a more useful profile of electron distribution. To understand why, think of the hydrogen atom as a tiny onion, made of many concentric spherical layers all of the same thickness. A cross section of that image of the atom is shown in Figure 3.22(c). What is the probability of finding the electron in one of the spherical layers? A layer very close to the nucleus has a very small radius, so it accounts for only a small fraction of the total volume of the atom. A layer with a larger radius makes up a much larger fraction of the atom's volume because the volume of the layers increases as a function of $r^2$. Even though electron densities are higher closer to the nucleus (as Figure 3.22a shows), the volumes of the spherical shells closest to the nucleus are so small that the chances that the electron will be near the center of an atom are extremely low. That low probability is shown in Figure 3.22(b), where the curve starts off at essentially zero for electron distribution values at distances very close to the nucleus. Farther from the nucleus, electron densities are lower but the volumes of the layers are much larger, so the probability of the electron's being in one of the layers is relatively high, represented by the peak in the curve of Figure 3.22(b). At greater distances, volumes of the layers are very large but $\psi^2$ drops to nearly zero (see Figure 3.22a); therefore, the chances of finding the electron in layers far from the nucleus are very small.

Thus, Figure 3.22(b) represents a combination of two competing factors—increasing layer volume and decreasing probability of the electron's being far from the nucleus. That combination produces a *radial distribution profile* for the electron. Figure 3.22(b) is a plot not of $\psi^2$ versus distance from the nucleus, as in Figure 3.22(a), but rather of $4\pi r^2\psi^2$ versus distance from the nucleus. In geometry, $4\pi r^2$ is the formula for the surface area of a sphere, but here it represents the volume of one of the thin spherical layers in Figure 3.22(c). The maximum value of the curve in Figure 3.22(b), at 53 pm, corresponds to the most likely radial distance of a 1*s* electron from the nucleus.

Figure 3.22(d) shows the spherical shape of this (or any other) *s* orbital. The surface of the sphere encloses the volume within which the probability of finding a 1*s* electron is 90%. That type of depiction, called a *boundary–surface representation*, is one of the most useful ways to view the relative sizes, shapes, and orientations of orbitals. All *s* orbitals are spheres, so they have no angular dependence on orientation (that is, no matter what direction you "travel" out from the center, it's the same).

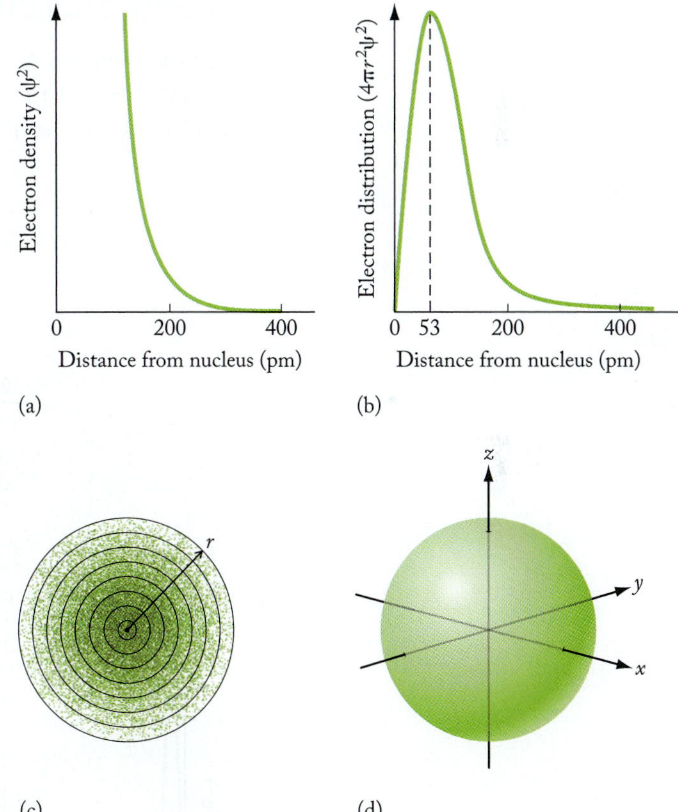


**FIGURE 3.22** (a) Probable electron density in the 1*s* orbital of the hydrogen atom represented by a plot of electron density ($\psi^2$) versus distance from the nucleus. (b) Electron distribution in the 1*s* orbital versus distance from the nucleus. The distribution is essentially zero at both very short and very long distances from the nucleus. (c) Cross section through the hydrogen atom, with the space surrounding the nucleus divided into an arbitrary number of thin, concentric hollow layers. Each layer has a unique value for radius *r*. The probability of finding an electron in a particular layer of radius *r* depends on the volume of the layer and the density of electrons in the layer. (d) Boundary–surface representation of a sphere within which the probability of finding a 1*s* electron is 90%.

Radial electron distribution profiles of hydrogen's 1*s*, 2*s*, and 3*s* orbitals are shown in **Figure 3.23**. Note that orbital size increases with increasing values of the principal quantum number, $n$. Note also that the quadrants above the profile curves have bands in which the density of dots is high. The dots represent the probability of finding an electron at those locations, and each dark band represents a local maximum in electron distribution. All three profiles have a local maximum close to the nucleus. That means that electrons in *s* orbitals, even *s* orbitals with high values of $n$, have some probability of being close to the nucleus.

The local maxima in any *s* orbital are separated from other local maxima by nodes. The number of nodes in any *s* orbital is equal to $n - 1$. Nodes have the same meaning here as they do in one-dimensional standing waves: they are places where the wave has zero amplitude. In the context of electrons as three-dimensional matter waves, nodes are locations at which electron density goes to zero.

## *p* and *d* Orbitals

All shells with $n \geq 2$ have a subshell containing three *p* orbitals ($\ell = 1$; $m_\ell = -1, 0, +1$). Each of the three *p* orbitals has two lobes oriented on both sides of the nucleus along one of the three perpendicular Cartesian axes *x*, *y*, and *z*. The true

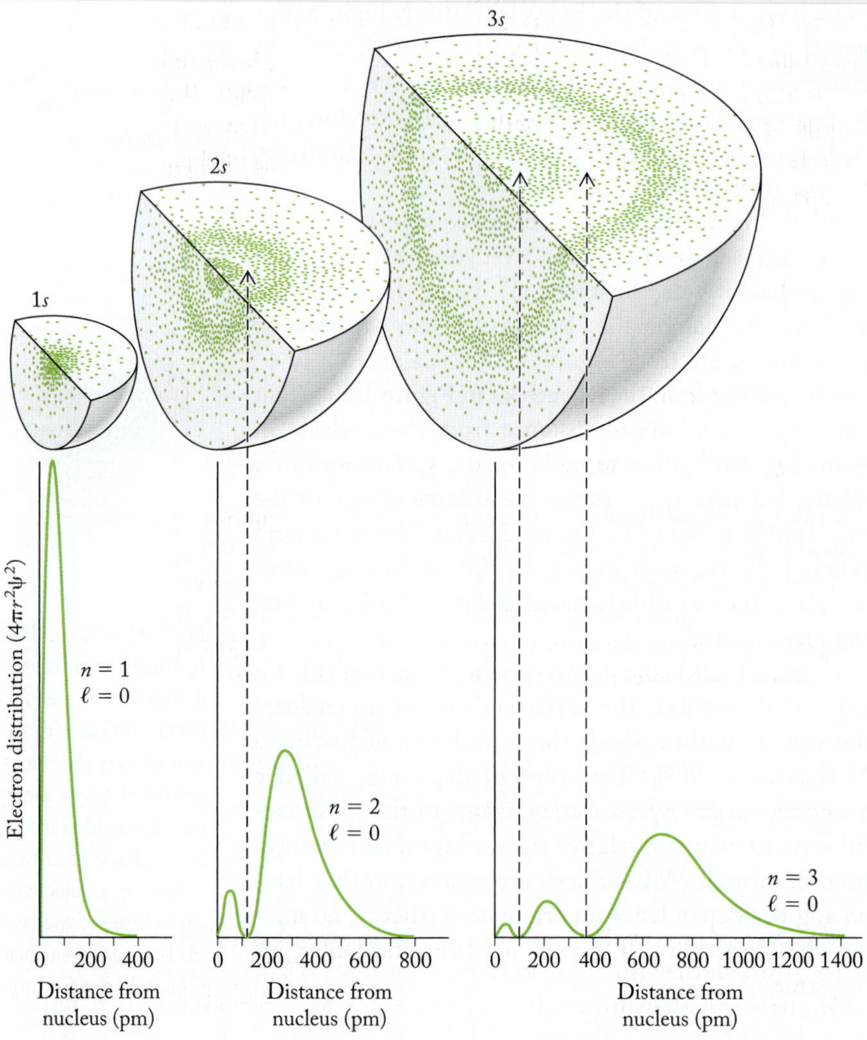


**FIGURE 3.23** These radial distribution profiles of 1*s*, 2*s*, and 3*s* orbitals have 0, 1, and 2 nodes, respectively, identifying (with dashed arrows) locations of zero electron density. Electrons in all these *s* orbitals have some probability of being close to the nucleus, but 3*s* electrons are more likely to be farther from the nucleus than 2*s* electrons, which are more likely to be farther away than 1*s* electrons.

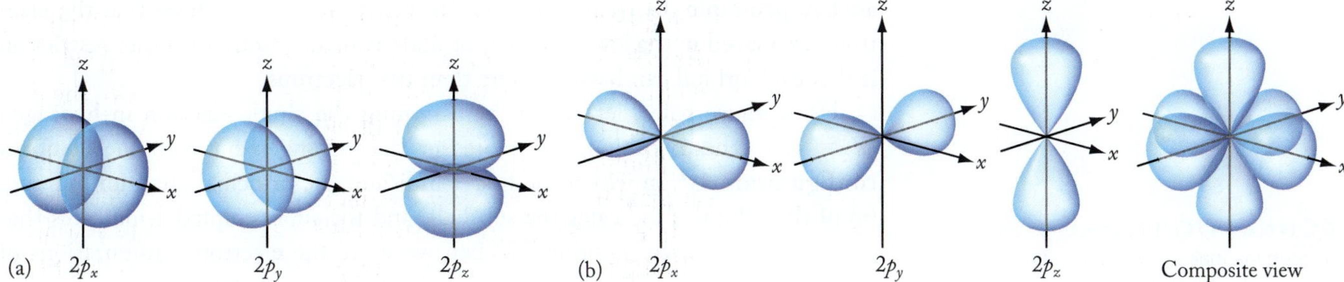


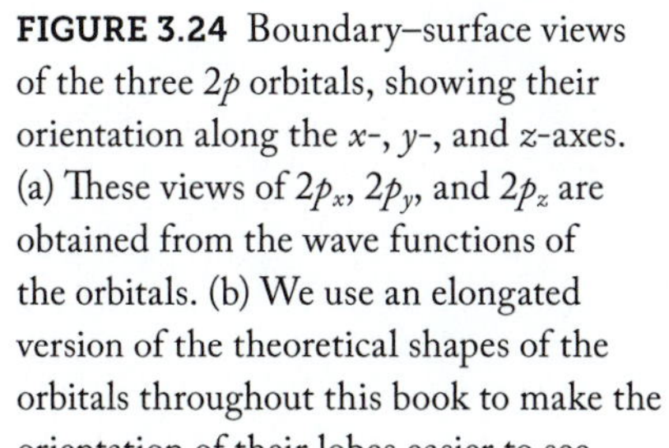
**FIGURE 3.24** Boundary–surface views of the three $2p$ orbitals, showing their orientation along the $x$-, $y$-, and $z$-axes. (a) These views of $2p_x$, $2p_y$, and $2p_z$ are obtained from the wave functions of the orbitals. (b) We use an elongated version of the theoretical shapes of the orbitals throughout this book to make the orientation of their lobes easier to see.

shape of the lobes is squashed and roundish, like mushroom caps (**Figure 3.24a**), but in **Figure 3.24(b)** we draw them in an elongated teardrop shape to more clearly show their orientation. The orbitals are designated $p_x$, $p_y$, and $p_z$, depending on the axis along which the lobes are aligned. The two lobes of a $p$ orbital are sometimes labeled with plus or minus signs to indicate the sign of the wave function that defines each lobe (but don't confuse the signs with electrical charges—all electrons have negative charges).

Shells with principal quantum numbers of 3 or higher have five $d$ orbitals ($\ell = 2$; $m_\ell = -2, -1, 0, +1, +2$). Their shapes are shown in **Figure 3.25**. Four of them have teardrop-shaped lobes oriented like the leaves in a four-leaf clover. The lobes of three of the four, designated $d_{xy}$, $d_{xz}$, and $d_{yz}$, lie *between* the Cartesian axes. The lobes of the fourth orbital, $d_{x^2-y^2}$, are centered *on* the $x$- and $y$-axes. The fifth $d$ orbital, $d_{z^2}$, is mathematically equivalent to the other four but has a much different shape, with two teardrop-shaped lobes oriented along the $z$-axis and a doughnut shape called a *torus* in the $x$–$y$ plane that surrounds the middle of the two lobes. We will not address the shapes and geometries of $f$ orbitals here because they are not part of our discussions of chemical bonding in the chapters to come.

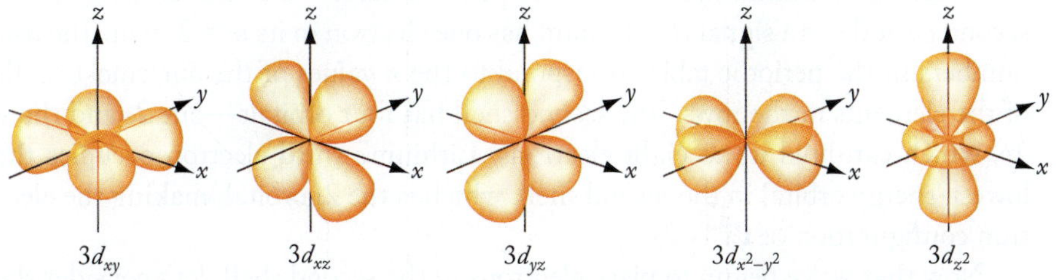


**FIGURE 3.25** Boundary–surface views of the five $3d$ orbitals, showing their orientation in relation to the $x$-, $y$-, and $z$-axes. As with the $2p$ orbitals in Figure 3.24(b), the theoretical boundary surfaces of the $3d$ orbitals are elongated to make the orientation of their lobes easier to see. The $d_{xy}$, $d_{xz}$, and $d_{yz}$ orbitals are not aligned along any axis; the $d_{x^2-y^2}$ orbital lies along the $x$- and $y$-axes; the $d_{z^2}$ orbital consists of two teardrop-shaped lobes along the $z$-axis with a donut-shaped torus ringing the point where the two lobes meet.

## 3.8 The Periodic Table and Filling Orbitals

In this section, we explore how the sizes and shapes of orbitals determine the order in which they fill with electrons as we work our way through the periodic table, starting with hydrogen. In assigning electrons to orbitals, we will follow the

Electron Configuration

**aufbau principle** (German *aufbauen*, "to build up"), which states that the electrons are placed in the lowest-energy orbitals available. Our only other restriction is that each orbital can have no more than two electrons.

Using those rules, let's begin by assigning the single electron in hydrogen ($Z = 1$) to the 1*s* orbital. We represent that arrangement with the **electron configuration** $1s^1$, in which the first 1 indicates the principal quantum number ($n$) of the orbital, *s* indicates the subshell, and the superscripted 1 indicates that *one* electron is in the 1*s* orbital. When we write the electron configurations of elements, we assume that their atoms are in their ground states.

CONNECTION In Chapter 2, we defined the main group elements as those in groups 1, 2, and 13–18 in the periodic table.

The atomic number of helium is 2, which tells us that two protons are in the nucleus surrounded by two electrons in the neutral atom. Using the aufbau principle, we simply add another electron to the 1*s* orbital. The 1*s* orbital already contains one electron, so it has space for one more. The Pauli exclusion principle dictates that the spin quantum number for the second electron cannot be the same as that of the first electron: one spin number must be $+\frac{1}{2}$ and the other must be $-\frac{1}{2}$. The two electrons with opposite spin quantum numbers are said to be *spin-paired*. Their presence gives helium a ground-state electron configuration of $1s^2$. With the two electrons, the 1*s* orbital is filled, and so is the $n = 1$ shell.

The concept of filled shells and subshells is crucial to understanding the chemical properties of the elements. Helium and the other group 18 elements are composed of atoms that have filled *s*, or *s* and *p*, orbitals in their outermost shells. Those elements, the noble gases, are chemically stable and generally unreactive. Other main group elements have partially filled sets of *s* and *p* orbitals in their outermost shell. They engage in chemical reactions in which their atoms lose, gain, or share electrons in ways that produce filled outermost *s* and *p* subshells. In that way they acquire the stable electron configurations of group 18 elements.

**aufbau principle** the concept of building up ground-state atoms so that their electrons occupy the lowest-energy orbitals available.

**electron configuration** the distribution of electrons among the orbitals of an atom or ion.

**effective nuclear charge ($Z_{eff}$)** the attraction toward the nucleus experienced by an electron in an atom; the positive charge on the nucleus is reduced by the extent to which other electrons in the atom shield the electron from the nucleus.

**core electrons** the electrons in the filled, inner shells in an atom or ion; they are not involved in chemical reactions.

**valence electrons** the electrons in the outermost occupied shell of an atom; they have the most influence on the atom's chemical behavior.

**valence shell** the outermost occupied shell of an atom.

**Hund's rule** the principle that the lowest-energy electron configuration of an atom has the maximum number of unpaired electrons, all of which have the same spin, in degenerate orbitals.

**orbital diagram** a depiction of the arrangement of electrons in an atom or ion, using boxes to represent orbitals.

## Effective Nuclear Charge

The location of lithium ($Z = 3$) in the periodic table—the first element in the second period—is a signal that lithium has one electron in its $n = 2$ shell. The row numbers in the periodic table correspond to the *n* values of the outermost shells of the elements in the rows. The second shell has four orbitals—one 2*s* and three 2*p*, which can hold up to eight electrons. Lithium's third electron occupies the lowest-energy orbital in the second shell, which is the 2*s* orbital, making the electron configuration of Li $1s^22s^1$.

Now that we've begun to place electrons in the second shell, let's consider the forces that an electron in the second shell experiences. An electron in a 2*s* or 2*p* orbital has an electrostatic attraction to the positively charged nucleus, but that electron is also partially shielded from the nucleus by the negative charges of the two 1*s* electrons (**Figure 3.26**). That shielding effect reduces the electrostatic attraction between the nucleus and the 2*s* or 2*p* electron, and we say that the second-shell electron experiences an **effective nuclear charge** ($Z_{eff}$).

The third electron in a ground-state lithium atom occupies the 2*s* orbital rather than a 2*p* orbital because the 2*s* orbital is lower in energy than any of the 2*p* orbitals. When we compare their radial distribution profiles in **Figure 3.27**, the small peak on the 2*s* curve near the nucleus means that an electron in the 2*s* orbital is closer to the nucleus more often than an electron in a 2*p* orbital, which has no such secondary peak. That proximity means that the 2*s* electron can get closer to the nucleus and experience a larger effective nuclear charge than a 2*p* electron would. Therefore, a 2*s* electron is lower in energy than a 2*p* electron, which is why the 2*s* orbital fills first.

## Condensed Electron Configurations

We can simplify the electron configurations of Li and all the elements that follow it in the periodic table by writing their *condensed* electron configurations. In that format, the symbols representing all the electrons in orbitals that were filled in the rows above the element of interest are replaced by the symbol of the group 18 element at the end of the row above the element. For example, the condensed electron configuration of Li is [He]$2s^1$. Condensed electron configurations are useful because they eliminate the symbols of the **core electrons** in filled shells and subshells. Core electrons are not involved in the chemistries of the elements, so they are of less interest to us than electrons that reside in the outermost shells and subshells. The outermost electrons are the ones involved in bond formation and are called **valence electrons**. The outermost shell is called the **valence shell**. Notice that lithium has a single electron in its valence-shell *s* orbital, as does hydrogen, the element directly above it in the periodic table. Therefore, both elements have the same general valence-shell configuration, $ns^1$, where $n$ represents both the number of the row in which the element is located and the principal quantum number of the valence shell.

Beryllium ($Z = 4$) is the fourth element in the periodic table and the first in group 2. Its electron configuration is $1s^22s^2$ or, in condensed form, [He]$2s^2$. The other elements in group 2 also have two spin-paired electrons in the *s* orbital of their valence shell. The second shell is not full at this point because it also has three *p* orbitals, which are all empty and fill as we move to the next elements in the periodic table.

Boron ($Z = 5$) is the first element in group 13. Its fifth electron is in one of its three $2p$ orbitals, resulting in the condensed electron configuration [He]$2s^22p^1$. Which of the three $2p$ orbitals contains the fifth electron does not matter because all three orbitals have the same energy. Chemists call orbitals with the same energy *degenerate* orbitals.

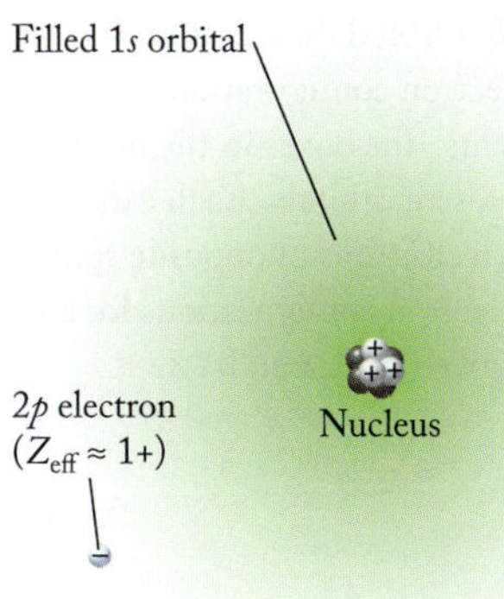


**FIGURE 3.26** The effective nuclear charge ($Z_{eff}$) experienced by a $2p$ electron in an excited-state Li atom is approximately the sum of the actual nuclear charge (3+) and the shielding effect of the negative charges of the two $1s$ electrons (2−). That shielding produces a net $Z_{eff}$ of about 1+.

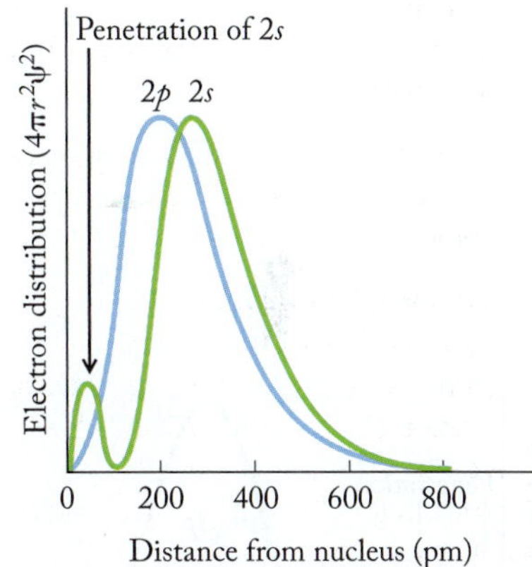


**FIGURE 3.27** Radial distribution profiles of electrons in $2s$ and $2p$ orbitals. The $2s$ orbital is lower in energy because electrons in it penetrate more closely to the nucleus, as indicated by the local maximum in electron distribution about 50 pm from the nucleus. As a result, $2s$ electrons experience a greater effective nuclear charge than $2p$ electrons.

## Hund's Rule and Orbital Diagrams

The next element is carbon ($Z = 6$). It has another electron in one of its $2p$ orbitals, giving it the condensed electron configuration [He]$2s^22p^2$. The second $2p$ electron resides in a different $2p$ orbital from that of the first electron. Why? Because electrons repel each other, and they will repel each other less if they are in separate orbitals rather than in the same orbital. Furthermore, experiments indicate that both $2p$ electrons have the same spin. That separation of the $2p$ electrons into different orbitals is an application of **Hund's rule**, named after German physicist Friedrich Hund (1896–1997), which states that the lowest-energy electron configuration for degenerate orbitals, like the three in the $2p$ subshell, is the one with the maximum number of unpaired valence electrons, all of which have the same spin.

We use **orbital diagrams** to show in detail how electrons, represented by single-headed arrows, are distributed among orbitals, which are represented by boxes. A single-headed arrow pointing upward represents an electron with spin up ($m_s = +\frac{1}{2}$), and a downward-pointing single-headed arrow represents an electron with spin down ($m_s = -\frac{1}{2}$). To obey Hund's rule, the orbital diagram for carbon must be

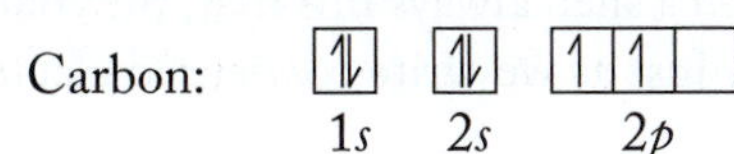

**FIGURE 3.28** Orbital diagrams and condensed electron configurations for the first 10 elements. The boxes in the orbital diagrams represent orbitals. Each can hold up to two electrons of opposite spin. Condensed electron configurations for all elements are given in Appendix 3.

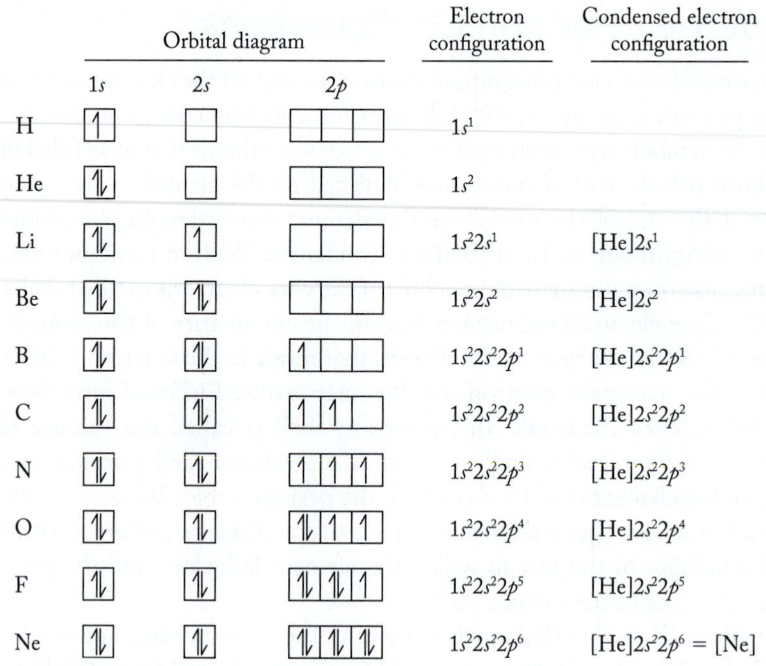

| | Orbital diagram | | | Electron configuration | Condensed electron configuration |
|---|---|---|---|---|---|
| | 1$s$ | 2$s$ | 2$p$ | | |
| H | [↑] | [ ] | [ \| \| ] | $1s^1$ | |
| He | [↑↓] | [ ] | [ \| \| ] | $1s^2$ | |
| Li | [↑↓] | [↑] | [ \| \| ] | $1s^22s^1$ | $[He]2s^1$ |
| Be | [↑↓] | [↑↓] | [ \| \| ] | $1s^22s^2$ | $[He]2s^2$ |
| B | [↑↓] | [↑↓] | [↑\| \| ] | $1s^22s^22p^1$ | $[He]2s^22p^1$ |
| C | [↑↓] | [↑↓] | [↑\|↑\| ] | $1s^22s^22p^2$ | $[He]2s^22p^2$ |
| N | [↑↓] | [↑↓] | [↑\|↑\|↑] | $1s^22s^22p^3$ | $[He]2s^22p^3$ |
| O | [↑↓] | [↑↓] | [↑↓\|↑\|↑] | $1s^22s^22p^4$ | $[He]2s^22p^4$ |
| F | [↑↓] | [↑↓] | [↑↓\|↑↓\|↑] | $1s^22s^22p^5$ | $[He]2s^22p^5$ |
| Ne | [↑↓] | [↑↓] | [↑↓\|↑↓\|↑↓] | $1s^22s^22p^6$ | $[He]2s^22p^6$ = [Ne] |

The two 2$p$ electrons are unpaired (in separate orbitals). By convention, the spin arrows of the single electrons in those orbitals are drawn pointed up.

The condensed electron configuration of the next element, nitrogen ($Z = 7$), is $[He]2s^22p^3$. According to Hund's rule, the third 2$p$ electron resides alone in the third 2$p$ orbital, so that the electron distribution is

Nitrogen: [↑↓] 1$s$ [↑↓] 2$s$ [↑|↑|↑] 2$p$

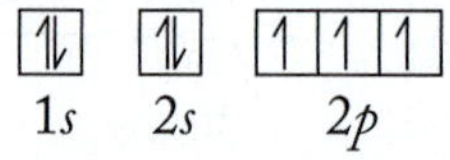

with all three spin arrows pointed up. As we proceed across the second row to neon ($Z = 10$), we fill the 2$p$ orbitals as shown in **Figure 3.28**. The last three 2$p$ electrons added (in oxygen, fluorine, and neon) pair up with the first three so that the spin orientations of each pair have opposite directions. All three 2$p$ orbitals are filled in an atom of neon, and so is the $n = 2$ shell.

Sodium ($Z = 11$) follows neon and is the third element in group 1 and the first in the third row. Its position means that its outermost electron is in the third shell. The third shell has three types of orbitals: 3$s$, 3$p$, and 3$d$. Which type gets sodium's 11th electron? According to the radial distribution profiles of electrons in those orbitals, shown in **Figure 3.29**, the 3$s$ orbital gets the electron. Note the two peaks in the 3$s$ profile and one of the peaks in the 3$p$ profile that are all close to the nucleus. Those peaks mean that electrons in 3$s$ and 3$p$ orbitals penetrate through the filled orbitals of the first two shells and experience greater effective nuclear charge than do 3$d$ electrons. Therefore, their relative energies are lower, with 3$s$ the lowest of all. So, in the third shell, the 3$s$ orbital fills first and the 3$d$ orbitals fill last. The same sequence applies to all the shells with $n > 3$. Shells for which $n \geq 4$ also contain $f$ orbitals, which have higher energies than the $d$ orbitals in the same shell, so the $f$ orbitals fill last.

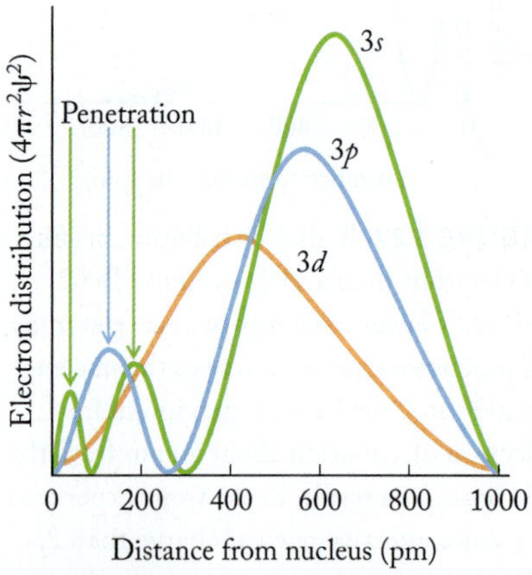

**FIGURE 3.29** Radial distribution profiles of electrons in 3$s$, 3$p$, and 3$d$ orbitals. The highlighted maxima in the 3$s$ and 3$p$ profiles mean that electrons in those orbitals penetrate closer to the nucleus, are less shielded by the electrons in filled inner shells, and have lower energy than 3$d$ electrons.

Because the $s$ orbital in a shell always fills first, the condensed electron configuration of Na is $[Ne]3s^1$. Just as we write condensed electron configurations that focus on the valence shell, we can draw condensed orbital diagrams that do the same thing. The one for sodium is

Sodium: [Ne] [↿] $3s$

That diagram reinforces the message that a sodium atom has a neon core of 10 electrons plus one more in its $3s$ orbital. Sodium has the same $ns^1$ valence-shell configuration as lithium and hydrogen, where $n$ is the principal quantum number of the outermost shell. That pattern holds throughout the periodic table: all the elements in a given group have the same generic valence-shell electron configuration. For instance, the electron configuration of magnesium ($Z = 12$) is [Ne]$3s^2$, and the condensed electron configuration of every other element in group 2 consists of the immediately preceding noble gas core followed by $ns^2$.

### CONCEPT TEST

What is the generic valence-shell configuration of the halogen (group 17) elements?

*(Answers to Concept Tests are in the back of the book.)*

The next six elements in the periodic table—aluminum, [Ne]$3s^23p^1$, through argon, [Ne]$3s^23p^6$—contain increasing numbers of $3p$ electrons until they achieve a filled $3p$ subshell in argon. As noted when we discussed helium, the filled $3s$ and $3p$ subshells of argon impart a chemical stability in keeping with that of the other noble gas elements.

Before leaving the third row, let's revisit the condensed electron configuration of sodium, [Ne]$3s^1$. That configuration represents a ground-state sodium atom because all the electrons, and most importantly its valence electron, occupy the lowest-energy orbitals available. Now think back to the discussion about atomic emission spectra in Section 3.2 and the distinctive yellow-orange glow that sodium makes in the flames of Bunsen burners, as shown in Figure 3.5(b) and here again in **Figure 3.30(a)**. Each sodium atom absorbs from the Bunsen burner a quantum of energy that raises the valence electron from the ground state to an excited state (a transition represented by the yellow-orange arrow pointing to the right in **Figure 3.31**). The easiest excited state to populate is the one with the smallest energy above the ground state. The $3p$ orbitals fill after the $3s$ orbital because they have the next lowest energy, so the lowest-energy (or *first*) excited state of sodium is the one in which its $3s$ electron has moved up to a $3p$ orbital. That excited state has the electron configuration [Ne]$3p^1$ (see Figure 3.31) and also has a very short lifetime. The electron typically takes less than a nanosecond to fall back to the ground state in a transition represented by the left-pointing yellow-orange arrow in Figure 3.31. That transition emits a quantum of energy ($h\nu$) equal to the difference in energy between the $3p$ and $3s$ orbitals in a Na atom—the energy of a photon of yellow-orange light.

### CONCEPT TEST

Which of the following could describe the radiation emitted by sodium atoms initially in an excited state with the electron configuration [Ne]$4p^1$ as they transition to the ground state? (a) ultraviolet radiation; (b) the same yellow-orange light in Figure 3.5(b); (c) red light; (d) infrared radiation

*(Answers to Concept Tests are in the back of the book.)*

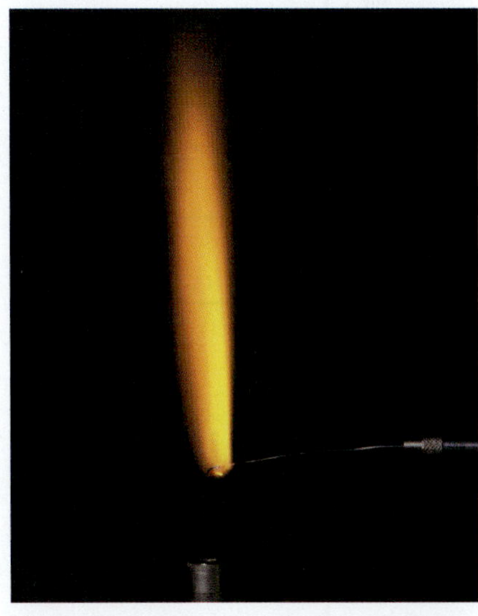

(a)

(b)

**FIGURE 3.30** The alkali metals produce characteristic colors in Bunsen burner flames because the high flame temperatures produce excited-state atoms of those elements. (a) The yellow-orange glow of Na atoms. (b) The lavender color of K atoms.

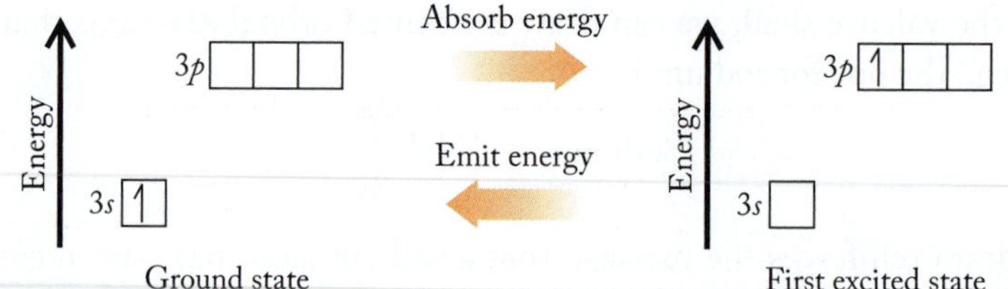


**FIGURE 3.31** A ground-state Na atom absorbs a quantum of energy as its valence electron moves from the 3*s* orbital to a 3*p* orbital. This 3*p* electron in the excited-state atom spontaneously falls back to the empty 3*s* orbital, emitting a photon of yellow-orange light. The energy of the photon matches the difference in energy between the 3*p* and 3*s* orbitals of Na atoms.

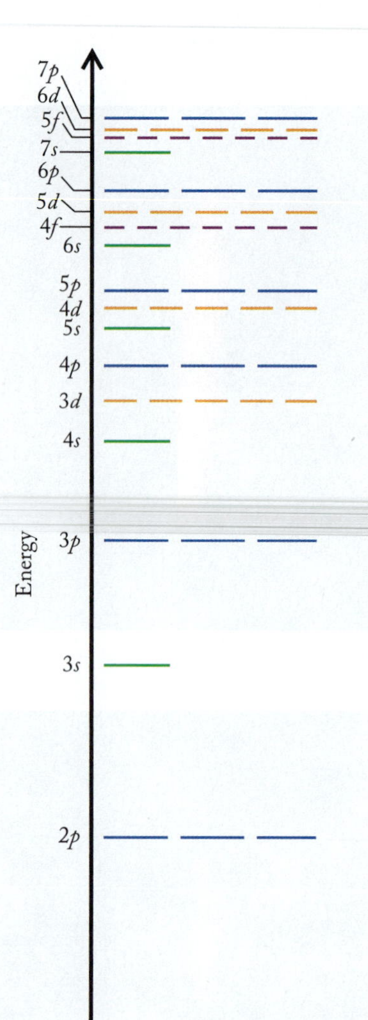


Potassium ($Z = 19$) is the first element of the fourth row. Like all group 1 elements, its generic valence-shell electron configuration is $ns^1$. For potassium, that translates into the condensed electron configuration $[Ar]4s^1$. Similarly, the condensed electron configuration of the next element, calcium ($Z = 20$), is $[Ar]4s^2$. At this point, the 4*s* orbital is filled. However, the 3*d* subshell *is still empty*. Why does the 4*s* orbital fill before the 3*d* subshell? The answer to that question is linked to the relative energies of electrons in the orbitals. Although energy levels do increase with increasing *n* values, the increases get smaller as *n* values get larger, as shown in **Figure 3.32**. The differences in energy between orbitals in the third and fourth shells are so small that the 3*d* subshell is slightly higher in energy than the 4*s* orbital. Therefore, the 4*s* orbital fills first.

Only after the 4*s* orbital is full do the 3*d* orbitals begin to fill, starting with scandium ($Z = 21$), the first transition metal in the fourth row. Scandium has the condensed electron configuration $[Ar]3d^14s^2$. Note how the valence-shell orbitals are arranged in order of increasing principal quantum number in that electron configuration, not in the order in which they were filled. The reason we use such sequences will become clear in the next section, where we discuss which valence electrons are lost when transition metals form cations.

The 3*d* orbitals are filled in the fourth-row transition metals from scandium (group 3) to zinc (group 12). That pattern of filling the *d* orbitals of the shell whose principal quantum number is one less than the row number, abbreviated $(n - 1)d$, is followed throughout the periodic table. Thus, the 4*d* orbitals are filled in the transition metals of the fifth row, the 5*d* orbitals in the sixth row, and so on, as shown in **Figure 3.33**.

Titanium ($Z = 22$) has one more 3*d* electron than scandium, so titanium has the condensed electron configuration $[Ar]3d^24s^2$. At this point, you may feel that you can accurately predict the electron configurations of the remaining transition metals in the fourth period. However, because the energies of the 3*d* and 4*s* orbitals are so close together (see Figure 3.32), the sequence of *d*-orbital filling sometimes deviates from the pattern you might expect. The first deviation appears in the electron configuration of chromium ($Z = 24$). You might

**FIGURE 3.32** The energy levels in multielectron atoms increase with increasing values of *n* and with increasing values of $\ell$ within a shell. The difference in energy between adjacent shells decreases with increasing values of *n*, which may cause the energies of subshells in two adjacent shells to overlap. For example, electrons in 3*d* orbitals have slightly higher energy than those in the 4*s* orbital, resulting in the order of subshell filling $4s \rightarrow 3d \rightarrow 4p$.

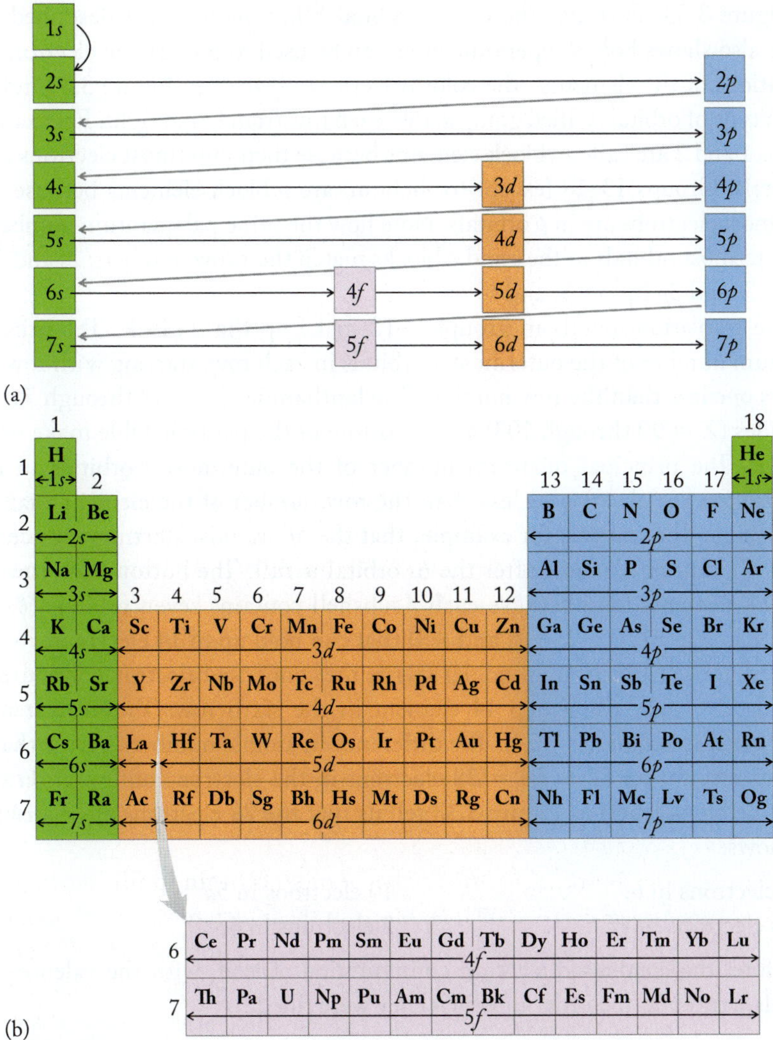


**FIGURE 3.33** (a) The sequence in which atomic orbitals fill. (b) The same color coding is used in this version of the periodic table to highlight the four "blocks" of elements in which valence-shell *s* (green), *p* (blue), *d* (orange), and *f* (purple) orbitals are filled with increasing atomic number across a row of the table.

expect it to be [Ar]$3d^4 4s^2$, but it is [Ar]$3d^5 4s^1$ with one electron in each of the five *d* orbitals:

Chromium: [Ar] ↿ ↿ ↿ ↿ ↿ (3$d^5$) ↿ (4$s^1$)

This half-filled set of *d* orbitals represents a lower-energy electron configuration than [Ar]$3d^4 4s^2$. Its stability compensates for the energy needed to raise a 4*s* electron to a 3*d* orbital.

Another deviation from the expected filling pattern happens near the end of each row of transition metals. For example, the electron configuration of copper ($Z = 29$) is [Ar]$3d^{10} 4s^1$ instead of [Ar]$3d^9 4s^2$ because the electron configuration with a filled *d* subshell represents a lower-energy configuration. By zinc ($Z = 30$), the 3*d* subshell is full and the next six electrons are added to 4*p* orbitals to reach the end of the fourth row, giving krypton ($Z = 36$) the condensed electron configuration [Ar]$3d^{10} 4s^2 4p^6$. The pattern we have just described for the fourth row is repeated, though with additional deviations from the expected pattern, in the fifth row. In the fifth row, the deviations are due to the similar energies of 5*s* and 4*d* orbitals.

Figure 3.33 illustrates the overall orbital filling pattern just described. The figure also shows how the periodic table can be used to predict the electron configurations of the elements. The color patterns and labels in Figure 3.33 indicate which type of orbital is filled going across each row from left to right. For example, groups 1 and 2 are called *s* block elements because their outermost electrons are in *s* orbitals. Groups 13–18 (except for helium) are *p* block elements because their outermost electrons are in *p* orbitals. Note how the principal quantum numbers of the outermost orbitals in the *s* and *p* blocks match their row numbers: 2*s* and 2*p* in row 2, 3*s* and 3*p* in row 3, and so on.

The transition metals in groups 3–12 make up the *d* block. The principal quantum number of the outermost *d* orbitals in each row, starting with row 4, is always one less than the row number. The lanthanides ($Z = 58$ through 71) and actinides ($Z = 90$ through 103) at the bottom of the periodic table make up the *f* block. The principal quantum number of the outermost *f* orbitals in each of those rows is always *two* less than the row number of the element that precedes them. That means, for example, that the 4*f* orbitals, starting with cerium ($Z = 58$), do not fill until after the 6*s* orbital is full. The bottom two rows are each 14 elements long because each *f* subshell contains seven orbitals ($\ell = 3$, $m_\ell = -3, -2, -1, 0, +1, +2,$ and $+3$) that can hold up to 14 electrons.

With those patterns in mind, let's write the condensed electron configuration of lead ($Z = 82$), the group 14 element in the sixth row. The nearest noble gas above it is xenon ($Z = 54$). The difference in atomic numbers means that we need to account for $82 - 54 = 28$ electrons in the electron configuration symbols. The block labels in Figure 3.33 tell us that the 28 electrons are distributed as follows:

| | |
|---|---|
| 2 electrons in 6*s* | 10 electrons in 5*d* |
| 14 electrons in 4*f* | 2 electrons in 6*p* |

Therefore, the condensed electron configuration of lead, with the valence-shell orbitals written in the order in which they were filled, is

$$\text{Pb:} \quad [\text{Xe}]4f^{14}5d^{10}6s^{2}6p^{2}$$

**SAMPLE EXERCISE 3.10** Writing Electron Configurations **LO6**

Write the condensed electron configuration of silver ($Z = 47$).

**Collect and Organize** In a condensed electron configuration, the filled sets of inner-shell orbitals are represented by the atomic symbol of the noble gas immediately preceding the element of interest in the periodic table.

**Analyze** Silver ($Z = 47$) is the group 11 element in the fifth row of the periodic table. The noble gas at the end of the preceding row is krypton ($Z = 36$). The difference between the atomic numbers ($47 - 36$) means that we need to assign 11 electrons to orbitals.

**Solve** At first, we would predict that the first two of the 11 electrons would be in the 5*s* orbital and that the next nine would be in 4*d* orbitals, resulting in a condensed electron configuration of $[\text{Kr}]4d^{9}5s^{2}$. However, a filled set of *d* orbitals is more stable than a partially filled set, as we saw for copper, the element just above Ag in the periodic table. Therefore, silver has 10 electrons in its 4*d* orbitals and only one in its 5*s* orbital, making its condensed electron configuration $[\text{Kr}]4d^{10}5s^{1}$.

**Think About It** We can generate a tentative electron configuration by simply moving across a row in Figure 3.33 until we come to the element of interest. However, in the transition metals, we have to remember the special stability of half-filled and filled *d* orbitals and then appropriately adjust our configuration.

**Practice Exercise** Write the condensed electron configuration of cobalt ($Z = 27$).

*(Answers to Practice Exercises are in the back of the book.)*

# 3.9 Electron Configurations of Ions

To write the electron configuration of monatomic ions, we begin with the electron configuration of the parent element. If the ion has a positive charge, we remove the appropriate number of electrons from the orbital(s) with the highest principal quantum number. If the ion has a negative charge, we add the appropriate number of electrons to one or more partially filled outer-shell orbitals.

## Ions of the Main Group Elements

The *s* block elements (see Figure 3.33) form monatomic cations by losing all their outer-shell electrons, leaving their ions with the electron configuration of the noble gas immediately preceding them in the periodic table. For example, an atom of sodium forms a $Na^+$ ion by losing its single 3*s* electron:

$$Na \rightarrow Na^+ + e^-$$

$$[Ne]3s^1 \rightarrow [Ne] + e^-$$

An element of the *p* block that forms a monatomic anion does so by gaining enough electrons to fill its valence-shell *p* orbitals. The ion it forms has the electron configuration of the noble gas at the end of its row in the periodic table. For example, an atom of fluorine ($[He]2s^22p^5$) forms a fluoride ($F^-$) ion by gaining one electron, which fills its set of three 2*p* orbitals and gives it the electron configuration of neon:

$$F + e^- \rightarrow F^-$$

$$[He]2s^22p^5 + e^- \rightarrow [He]2s^22p^6 = [Ne]$$

Thus, a $Na^+$ ion and a $F^-$ ion have the same electron configuration as an atom of Ne. We say that the three species, $Na^+$, $F^-$, and Ne, are **isoelectronic**, meaning that they have the same electron configuration.

**isoelectronic** describes atoms or ions that have identical electron configurations.

**SAMPLE EXERCISE 3.11** Determining Isoelectronic Species in Main Group Ions **LO6**

a. Write the condensed electron configuration of each of the following ions: $Mg^{2+}$, $Cl^-$, $Ca^{2+}$, and $O^{2-}$.

b. Which ions in part a are isoelectronic with neon?

**Collect and Organize** In part a, we are asked to determine the electron configurations of four ions. In part b, we are asked to identify which of the part a ions have the same electron configuration as neon, which has 10 electrons filling its $n = 1$ and $n = 2$ shells.

**Analyze** The ions include

- two from group 2, Mg and Ca, which form 2+ cations;
- one from group 16, O, which forms a 2− anion; and
- one from group 17, Cl, which forms a 1− anion.

Let's arrange the atoms and ions in a table in order of increasing atomic number, remembering that atoms form cations by losing electrons and form anions by gaining them:

| Element | Electron Configuration | Atomic Number (*Z*) | Charge on Ion | Electrons/Ion |
|---|---|---|---|---|
| O | $[He]2s^22p^4$ | 8 | 2− | 10 |
| Mg | $[Ne]3s^2$ | 12 | 2+ | 10 |
| Cl | $[Ne]3s^23p^5$ | 17 | 1− | 18 |
| Ca | $[Ar]4s^2$ | 20 | 2+ | 18 |

**Solve**

a. The electron configurations for the ions are

$O^{2-}$ $[He]2s^22p^6$ (same as Ne)
$Mg^{2+}$ [Ne]
$Cl^-$ $[Ne]3s^22p^6$ (same as Ar)
$Ca^{2+}$ [Ar]

b. Two of the ions formed, $O^{2-}$ and $Mg^{2+}$, are isoelectronic with Ne and with each other. The other two, $Cl^-$ and $Ca^{2+}$, are isoelectronic with Ar and with each other.

**Think About It** Each ion has the stability we associate with the electron configuration of a noble gas.

 **Practice Exercise** Write the condensed electron configurations of $K^+$, $I^-$, $Ba^{2+}$, $S^{2-}$, and $Al^{3+}$. Which of those ions are isoelectronic with Ar?

*(Answers to Practice Exercises are in the back of the book.)*

## Transition Metal Cations

As with the main group elements, writing the electron configurations of transition metal cations begins with the atoms from which the cations form. Nickel atoms, like those of many transition metals, form ions with 2+ charges by losing both electrons from their valence-shell *s* orbital:

$$\mathrm{Ni} \rightarrow \mathrm{Ni}^{2+} + 2e^-$$

$$[\mathrm{Ar}]3d^84s^2 \rightarrow [\mathrm{Ar}]3d^8 + 2e^-$$

We might have expected Ni atoms to form cations by losing 3*d* electrons, reasoning that the last orbitals to be filled are the highest in energy and so should be the first to be emptied. But that does not happen for Ni and the other transition metals. Among the reasons are the following:

- The differences in energy between the valence-shell *s* orbitals (*ns*) and the $(n - 1)d$ orbitals of transition metals are very small.

- As the $(n - 1)d$ orbitals fill with increasing atomic number, the effective nuclear charge ($Z_{eff}$) felt by their electrons increases more than the $Z_{eff}$ felt by the *ns* electrons. That happens because the $(n - 1)d$ electrons in a transition metal atom are shielded less than the *ns* electrons in the next higher shell. As a result, the *ns* electrons have higher energy and ionize first.

The rule that the electrons in orbitals with the highest *n* value ionize first applies to transition metals, too. Preferential loss of outer-shell *s* electrons explains why the charge most often encountered on transition metal ions is 2+.

Many transition metals form ions with charges greater than 2+ by losing *d* electrons in addition to their valence-shell *s* electrons. Atoms of scandium, $[Ar]3d^14s^2$, for example, lose both 4*s* electrons *and* their single 3*d* electron as they form $Sc^{3+}$ ions. The chemistry of titanium, $[Ar]3d^24s^2$, is dominated by its tendency to lose two 4*s* and two 3*d* electrons, forming $Ti^{4+}$ ions. In general, transition metals form 1+ and 2+ cations by losing all their valence-shell *s* electrons (some, such as Ag, have only one to lose). They form cations with charges greater than 2+ by also losing $(n - 1)d$ electrons.

**SAMPLE EXERCISE 3.12** Writing Electron Configurations of Transition Metal Ions **LO6**

What are the condensed electron configurations of $Fe^{2+}$ and $Fe^{3+}$?

**Collect and Organize** We are asked to write the condensed electron configuration of two ions formed by iron ($Z = 26$). Iron is the group 8 element of the fourth row of the periodic table. Transition metals preferentially lose their outermost *s* electrons when they form ions.

**Analyze** Its location on the periodic table means that iron has two 4*s* and six 3*d* electrons built on an argon core:

$$Fe: [Ar]3d^64s^2$$

**Solve** We remove the two 4*s* electrons to form $Fe^{2+}$. We remove the two 4*s* electrons and one of the 3*d* electrons to form $Fe^{3+}$:

$$Fe^{2+}: [Ar]3d^6 \qquad Fe^{3+}: [Ar]3d^5$$

**Think About It** The electron configuration of $Fe^{3+}$ means that each of its 3*d* orbitals contains one electron. That half-filled set of 3*d* orbitals should have enhanced stability. Actually, both $Fe^{2+}$ and $Fe^{3+}$ are common in nature. Most of the iron in your blood and tissues is $Fe^{2+}$.

**Practice Exercise** Write the condensed electron configurations for $Mn^{3+}$ and $Mn^{4+}$.

*(Answers to Practice Exercises are in the back of the book.)*

**CONCEPT TEST**

The electron configuration of Eu ($Z = 63$) is $[Xe]4f^76s^2$, but the electron configuration of Gd ($Z = 64$) is $[Xe]4f^75d^16s^2$. Suggest a reason why the additional electron in Gd is not in a 4*f* orbital, which would result in the electron configuration $[Xe]4f^86s^2$.

*(Answers to Concept Tests are in the back of the book.)*

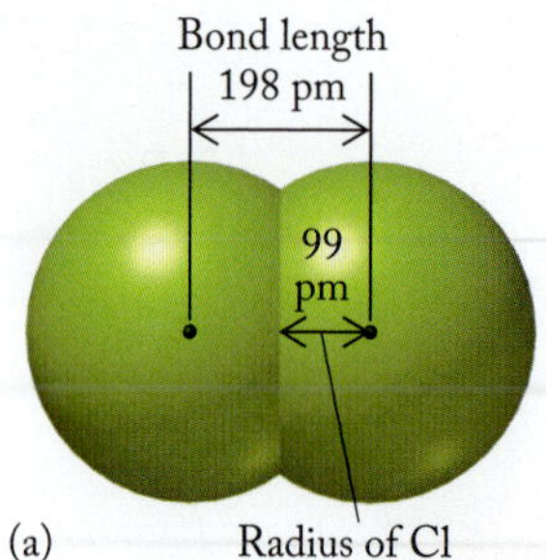


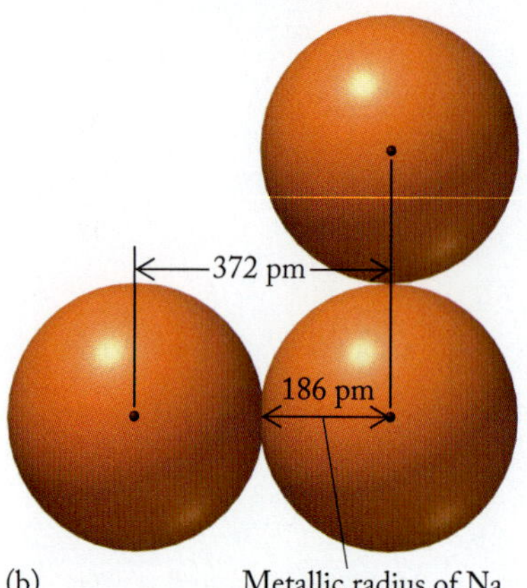


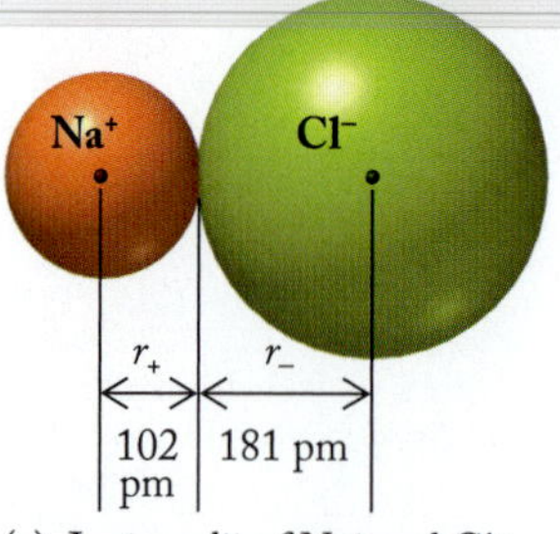


**FIGURE 3.34** A comparison of atomic, metallic, and ionic radii. (a) An atomic radius is half the distance between identical nuclei in a molecule, such as the distance between chlorine nuclei in $Cl_2$. (b) A metallic radius is based on the distance of closest approach of adjacent atoms in a solid metal. (c) An ionic radius is determined by comparisons among ionic compounds containing the ion of interest.

## 3.10 The Sizes of Atoms and Ions

We usually describe the sizes of atoms in terms of their radii. Given the wavelike behavior of electrons, the "edge" of a single atom cannot be exactly defined, so we must look at interactions between atoms to determine their radii. The atomic radius of an element that exists as a diatomic molecule, such as $N_2$, $O_2$, or $Cl_2$, is simply half the distance between the nuclear centers in the molecule (**Figure 3.34a**). The atomic radius of a metal, also called its *metallic* radius, is half the distance between the nuclear centers in the solid metal (**Figure 3.34b**). The values of ionic radii are derived from the distances between nuclear centers in solid ionic compounds (**Figure 3.34c**).

### Trends in Atomic Size

The radial distribution profiles in Figure 3.23 showed us how the distance from the nucleus to the *s* orbitals increases as the principal quantum number *n* increases. That same trend is observed for *p, d,* and *f* orbitals as well; therefore, we expect the atomic radii of elements in the same group of the periodic table to increase with increasing atomic number. For example, the atomic radii of the halogens increase as we move from the top to the bottom of group 17 of the periodic table:

| Element | Atomic Number | *n* Value of Valence Shell | Atomic Radius (pm) |
|---|---|---|---|
| F | 9 | 2 | 71 |
| Cl | 17 | 3 | 99 |
| Br | 35 | 4 | 114 |
| I | 53 | 5 | 133 |
| At | 85 | 6 | 140 |

As we move across a row (period) in the periodic table, each successive element has one additional electron. We might expect that the addition of electrons across a row would mean a corresponding increase in the size of the atom. Surprisingly, however, experimental data do not support that prediction. Atomic radii tend to *decrease* as the atomic number increases across a row of the periodic table. That pattern is particularly evident as we move from left to right across the main group elements (**Figure 3.35**). To explain those data, we need to carefully consider two competing interactions:

1. **Increasing effective nuclear charge.** Each time the atomic number increases, so does the positive charge of the nucleus. For example, an atom of sodium-23 has 11 protons, 12 neutrons, and 11 electrons, whereas an atom of magnesium-24 has 12 protons, 12 neutrons, and 12 electrons. The valence electrons in Na and Mg atoms are all in the 3*s* orbital, which means they are all about the same distance from the nuclei of these atoms. Each valence electron in Mg therefore experiences a larger effective nuclear charge than the valence electron in Na does because the electrostatic attraction of a 3*s* electron to 12 protons is stronger than its attraction to 11 protons. As $Z_{\text{eff}}$ increases, the size of atoms decreases.
2. **Increasing repulsion between valence electrons.** As atomic number increases, so does the number of valence electrons. More electrons generate more electron–electron repulsions (because two negatively charged particles repel each other), which tend to increase the size of the atoms.

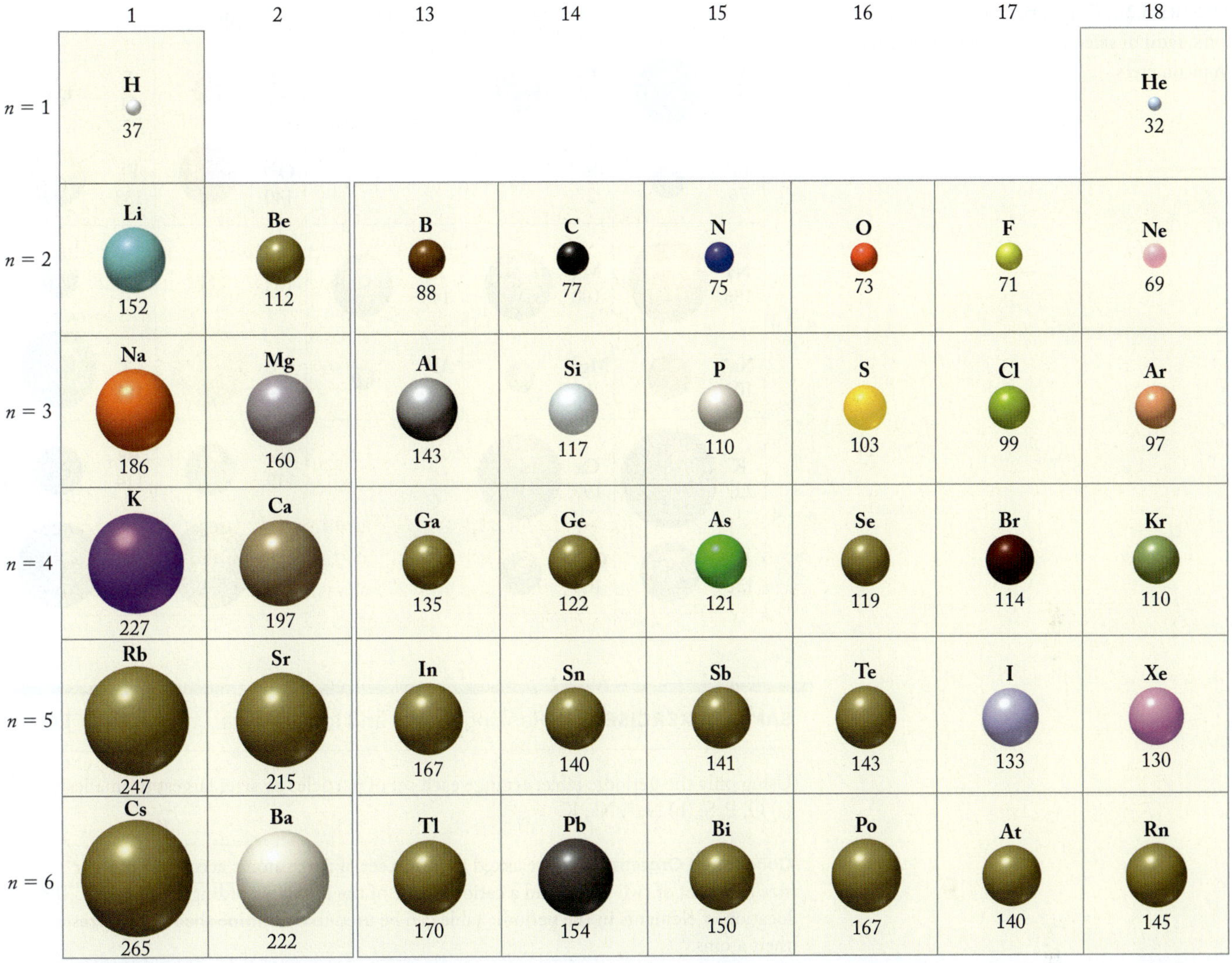

| | 1 | 2 | 13 | 14 | 15 | 16 | 17 | 18 |
|---|---|---|---|---|---|---|---|---|
| $n = 1$ | H 37 | | | | | | | He 32 |
| $n = 2$ | Li 152 | Be 112 | B 88 | C 77 | N 75 | O 73 | F 71 | Ne 69 |
| $n = 3$ | Na 186 | Mg 160 | Al 143 | Si 117 | P 110 | S 103 | Cl 99 | Ar 97 |
| $n = 4$ | K 227 | Ca 197 | Ga 135 | Ge 122 | As 121 | Se 119 | Br 114 | Kr 110 |
| $n = 5$ | Rb 247 | Sr 215 | In 167 | Sn 140 | Sb 141 | Te 143 | I 133 | Xe 130 |
| $n = 6$ | Cs 265 | Ba 222 | Tl 170 | Pb 154 | Bi 150 | Po 167 | At 140 | Rn 145 |

**FIGURE 3.35** Atomic radii in picometers of the main group elements. Size generally increases from top to bottom in any group and generally decreases from left to right across any period.

The experimental data tabulated in Figure 3.35 show that atomic size decreases with increasing $Z$ across each row of main group elements. That means that interaction 1, the effective nuclear charge, more than offsets interaction 2, the repulsions between the increasing number of valence electrons.

## Trends in Ionic Size

Periodic Trends

The cations of the main group elements are much smaller than their parent atoms, but the anions of the main group elements are much larger (**Figure 3.36**). To understand the trends, let's revisit what happens when a Na atom forms a $Na^+$ ion. The atom loses its only valence-shell (3*s*) electron, forming an ion with a much smaller neon core of 1*s*, 2*s*, and 2*p* electrons. However, when a Cl atom acquires an electron and forms a $Cl^-$ ion, it contains more electrons than protons; hence, the attractive force per electron decreases as electron–electron repulsion increases. As a result, $Cl^-$ ions and all monatomic anions are larger than the atoms from which they form.

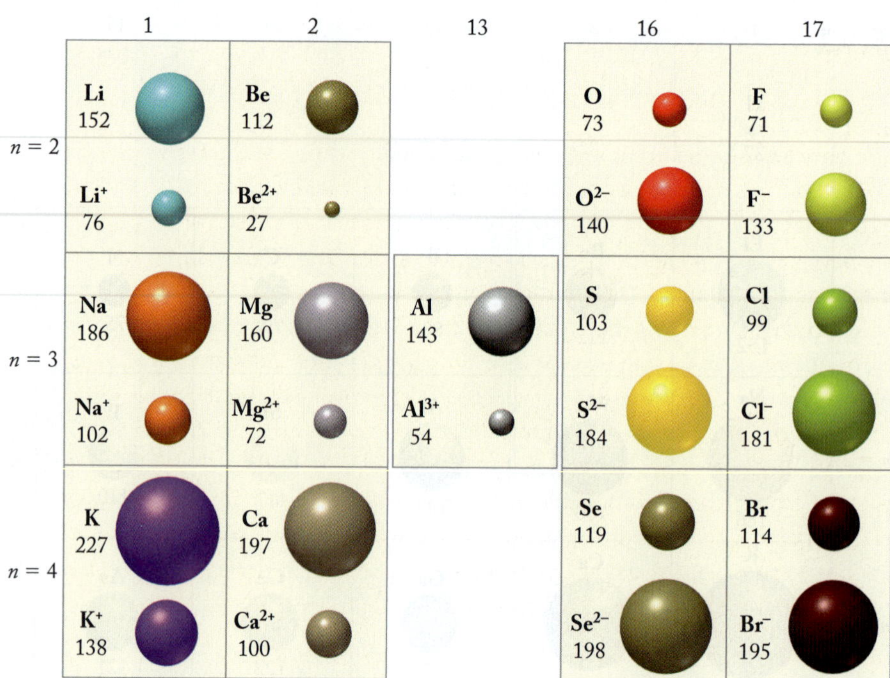


**FIGURE 3.36** Comparison of atomic and ionic radii of selected main group elements, in picometers.

## SAMPLE EXERCISE 3.13 Ranking Atoms and Ions by Size — LO7

Using only the periodic table, arrange each set of particles by size, largest to smallest: (a) O, P, S; (b) $Na^+$, Na, K.

**Collect and Organize** We are asked to rank a set of three atoms according to their size and a set of two atoms and a cation of one of the atoms according to their size. The location of elements in the periodic table can be used to determine the relative sizes of their atoms.

**Analyze** All the atoms and ions are of main group elements. Atomic radii of the elements decrease with increasing atomic number across rows because their valence electrons experience greater $Z_{eff}$ as $Z$ increases. Atomic radii increase with increasing atomic number down groups of elements because their valence electrons are in orbitals farther from the nucleus. Main group cations are always smaller than their parent atoms because they have lost their valence-shell electrons.

**Solve**

a. Sulfur is below oxygen in group 16, so S atoms are larger than O atoms. Phosphorus is to the left of sulfur in row 3, so P atoms are larger than S atoms. Therefore, the size order is P > S > O.

b. Cations are smaller than their parent atoms, so Na > $Na^+$. Size increases down a group of elements and K is below Na, so K > Na. Therefore, K > Na > $Na^+$.

**Think About It** The trend in the size of the atoms in set (a) reflects decreasing atomic size and increasing $Z_{eff}$ across a row of elements in the periodic table, as well as increasing atomic size with increasing $n$ values of their valence shells down a group. The relative sizes of the particles in set (b) are linked to increasing atomic size down a group of elements in the periodic table and to a cation being smaller than its parent atom.

**Practice Exercise** Arrange each set in order of increasing size (smallest to largest): (a) $Cl^-$, $F^-$, $Na^+$; (b) $P^{3-}$, $Al^{3+}$, $Mg^{2+}$.

*(Answers to Practice Exercises are in the back of the book.)*

# 3.11 Ionization Energies and Photoelectron Spectroscopy

**ionization energy (IE)** the amount of energy needed to remove one mole of electrons from one mole of ground-state atoms or ions in the gas phase.

We have seen how atomic emission spectra led to theories about the structure of atoms and the presence of quantized energy levels inside them. Another type of experimental evidence for energy levels in atoms and the electron configurations we have been exploring is obtained from the different energies required to remove electrons from atoms—that is, their ionization energies.

**Ionization energy (IE)** is the energy needed to remove one mole of electrons from one mole of gas-phase atoms or ions in their ground states. Removing the electrons always consumes energy because a negatively charged electron is always attracted to a positively charged nucleus, and overcoming that attraction requires energy. The amount of energy needed to remove one mole of electrons from one mole of atoms to make one mole of 1+ cations is called the *first ionization energy* ($IE_1$); the energy needed to remove one mole of electrons from one mole of 1+ cations to make one mole of 2+ cations is the *second ionization energy* ($IE_2$), and so forth. For example, for Mg atoms in the gas phase,

$$Mg \rightarrow Mg^+ + 1\ e^- \qquad IE_1 = 738\ \text{kJ/mol}$$
$$Mg^+ \rightarrow Mg^{2+} + 1\ e^- \qquad IE_2 = 1451\ \text{kJ/mol}$$

The total energy necessary to make one mole of $Mg^{2+}$ cations from one mole of Mg atoms in the gas phase is the sum of the two ionization energies:

$$Mg \rightarrow Mg^{2+} + 2\ e^-$$
$$\text{Total IE} = (738 + 1451)\ \text{kJ/mol} = 2189\ \text{kJ/mol}$$

**Figure 3.37** shows how the first ionization energies vary in the main group elements. The $IE_1$ of hydrogen is 1312 kJ/mol, and the $IE_1$ of helium is nearly twice as big: 2372 kJ/mol. That difference seems reasonable because He atoms have two protons per nucleus, whereas H atoms have only one. In general, first ionization energies increase from left to right across a period; the easiest element to remove an electron from is in group 1, and the hardest is in group 18. That pattern arises because $Z_{eff}$ increases with increasing atomic number across a row, and so do the strengths of the attractive forces between nuclei and valence electrons.

Two anomalies occur in the general trend of increasing $IE_1$ across each row. The first anomaly shows up between the group 2 and 13 elements that are next to each other in the second and third rows: a decrease in $IE_1$ occurs between Be and B and between Mg and Al. The decrease occurs because B and Al lose a *p* electron when they ionize, whereas Be and Mg must lose an *s* electron. Recall from Figures 3.27 and 3.29 that electrons in an *s* orbital penetrate closer to the nucleus than electrons in the *p* orbitals in the same shell ($n > 2$). Therefore, a *p* electron experiences proportionately less $Z_{eff}$ than an *s* electron in its shell and requires less energy to be removed from the atom. The second anomaly occurs in the values of $IE_1$ between the group 15 and 16 elements. Recall that an

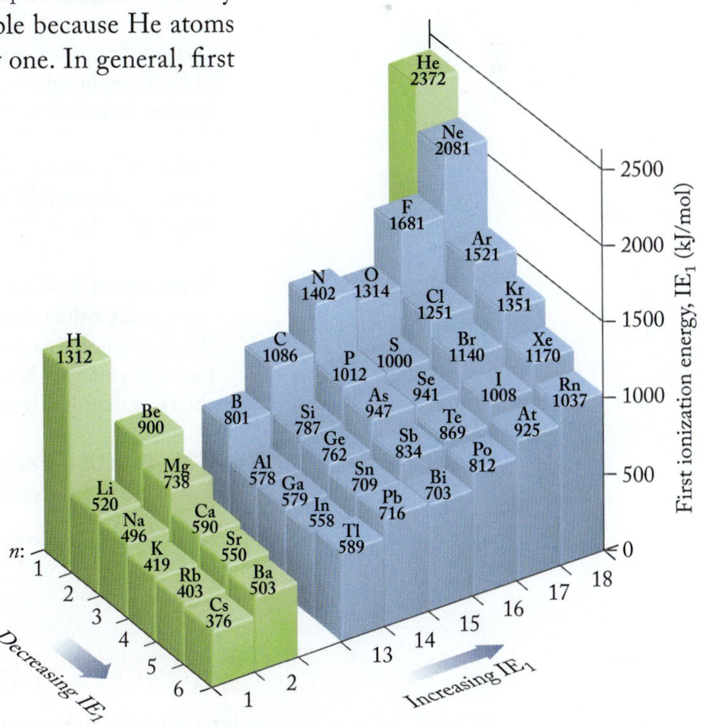


**FIGURE 3.37** The first ionization energies of the main group elements generally increase from left to right in a period and decrease from top to bottom in a group.

electron lost by a group 15 element, such as nitrogen, was originally alone in a half-filled orbital:

$$\text{N: } [\uparrow\downarrow]_{1s}\ [\uparrow\downarrow]_{2s}\ [\uparrow|\uparrow|\uparrow]_{2p} \rightarrow \text{N}^+\text{: } [\uparrow\downarrow]_{1s}\ [\uparrow\downarrow]_{2s}\ [\uparrow|\uparrow|\ \ ]_{2p} + e^-$$

However, the electron lost by a group 16 element, such as oxygen, was paired with another in a filled orbital:

$$\text{O: } [\uparrow\downarrow]_{1s}\ [\uparrow\downarrow]_{2s}\ [\uparrow\downarrow|\uparrow|\uparrow]_{2p} \rightarrow \text{O}^+\text{: } [\uparrow\downarrow]_{1s}\ [\uparrow\downarrow]_{2s}\ [\uparrow|\uparrow|\uparrow]_{2p} + e^-$$

Repulsion between the paired electrons raises their energy, which means less energy must be added to ionize one of them.

To see how ionization energy changes as we go down a group of elements in the periodic table, let's begin with group 1 and the $IE_1$ values of hydrogen and lithium. A lithium atom has three times the nuclear charge of a hydrogen atom, so we might expect its ionization energy to be about three times larger. However, its $IE_1$ value is only 520 kJ/mol, or less than half that of hydrogen (1312 kJ/mol), because the 2*s* electron in a lithium atom is shielded from the nucleus by the two electrons in the 1*s* orbital, so the $Z_{eff}$ felt by the lithium valence electron is much less than 3+. In general, the combination of larger atomic size and more shielding by inner-shell electrons leads to decreasing $IE_1$ values from top to bottom in every group of the periodic table.

### SAMPLE EXERCISE 3.14 Recognizing Trends in Ionization Energies LO7

Without referring to Figure 3.37, arrange argon, magnesium, and phosphorus in order of increasing first ionization energy.

**Collect, Organize, and Analyze** We are asked to order three elements in the third row of the periodic table on the basis of their $IE_1$ values. First ionization energies generally increase from left to right across a row because the values of $Z_{eff}$ increase.

**Solve** If increasing ionization energy with increasing atomic number is the dominant factor for those elements, their order of increasing first ionization energies should be Mg < P < Ar.

**Think About It** Magnesium forms stable 2+ cations, so we expect its ionization energy to be smaller than the values for phosphorus and argon, which do not form cations. Argon is a noble gas with a stable valence-shell electron configuration, so its first ionization energy should be the highest in the set. We can check our prediction against the $IE_1$ values in Figure 3.37: Mg, 738 kJ/mol; P, 1012 kJ/mol; Ar, 1521 kJ/mol.

**Practice Exercise** Without referring to Figure 3.37, arrange cesium, calcium, and neon in order from largest first ionization energy to smallest.

*(Answers to Practice Exercises are in the back of the book.)*

### CONCEPT **TEST**

Why does ionization energy decrease with increasing atomic number down the group 18 elements?

*(Answers to Concept Tests are in the back of the book.)*

Another perspective on energy levels in atoms comes from looking at the energies required to take more than one electron away from an atom. Consider the successive ionization energies for the first 10 elements, as shown in **Table 3.2**. For multielectron atoms, the energy required to remove an electron from an ion that has already lost one electron—that is, the second ionization energy, $IE_2$—is always greater than $IE_1$ because the second electron is being removed from an ion that already has a positive charge. Because electrons carry a negative charge, they are held more strongly in a cation than in the atom from which the cation is formed. The energy needed to remove a third electron is greater still because it is being removed from a 2+ ion.

Superimposed on that trend are much more dramatic increases in ionization energy (defined by the red line in Table 3.2) when all the valence electrons in ionizing atoms have been removed and the next electrons must come from inner shells. A core electron experiences much less shielding and a much greater $Z_{eff}$, so it requires much more energy to be removed from an atom than does an outer-shell electron. The first and second ionization energies of Na and Mg show the effect of that energy difference between valence-shell and inner-shell electrons:

| | Na | Mg |
|---|---|---|
| **$IE_1$ (kJ/mol) (orbital of lost electron)** | 496 (3s) | 738 (3s) |
| **$IE_2$ (kJ/mol) (orbital of lost electron)** | 4562 (2*p*) | 1451 (3s) |

The $IE_2$ of Na is nearly 10 times its $IE_1$ because Na has only one 3*s* electron to lose. The second electron must come from an inner-shell 2*p* orbital. The $IE_2$ of Mg is only twice its $IE_1$ because the two electrons lost in forming a $Mg^{2+}$ ion come from the same valence-shell 3*s* orbital.

While the data and red line in Table 3.2 provide evidence for the existence of core electrons and valence-shell electrons, they offer no insight into energy differences between electrons within the same shell, such as between 2*s* electrons and 2*p* electrons. We now turn to a technique called photoelectron spectroscopy that provides just the data needed to explore those differences.

**TABLE 3.2 Successive Ionization Energies[a] of the First 10 Elements**

| Element | *Z* | $IE_1$ | $IE_2$ | $IE_3$ | $IE_4$ | $IE_5$ | $IE_6$ | $IE_7$ | $IE_8$ | $IE_9$ | $IE_{10}$ |
|---|---|---|---|---|---|---|---|---|---|---|---|
| H | 1 | 1312 | | | | | | | | | |
| He | 2 | 2372 | 5249 | | | | | | | | |
| Li | 3 | 520 | 7296 | 12,040 | | | | | | | |
| Be | 4 | 900 | 1758 | 15,050 | 21,070 | | | | | | |
| B | 5 | 801 | 2426 | 3660 | 24,682 | 32,508 | | | | | |
| C | 6 | 1086 | 2348 | 4617 | 6201 | 37,926 | 46,956 | | | | |
| N | 7 | 1402 | 2860 | 4581 | 7465 | 9391 | 52,976 | 64,414 | | | |
| O | 8 | 1314 | 3383 | 5298 | 7465 | 10,956 | 13,304 | 71,036 | 84,280 | | |
| F | 9 | 1681 | 3371 | 6020 | 8428 | 11,017 | 15,170 | 17,879 | 92,106 | 106,554 | |
| Ne | 10 | 2081 | 3949 | 6140 | 9391 | 12,160 | 15,231 | 19,986 | 23,057 | 115,584 | 131,236 |

[a]In kJ/mol.

**photoelectron spectroscopy (PES)** a technique for determining the binding energies of electrons with an atom.

**photoelectron spectrum** a graph plotting the number of electrons ejected by a metal as a function of binding energy.

## Photoelectron Spectroscopy

In Section 3.3, we discussed how Einstein's studies of the photoelectric effect supported Planck's quantum model and the concept that electromagnetic radiation consists of discrete packets, or quanta, of energy. Measuring the kinetic energy of ejected electrons when light with enough energy strikes the surface of a metal serves as a measure of the strength of the attraction between the atomic nuclei of that metal and the electrons surrounding them. That principle also can be used to explore the distribution of electrons within the shells and subshells of atoms.

Suppose that the surface of a piece of lithium metal is irradiated with high-energy X-rays. Each X-ray photon that collides with a Li atom has more than enough energy to dislodge an electron not only from the atom's outermost 2*s* orbital but also from its core 1*s* orbital. As you might imagine, ejecting a 1*s* electron requires significantly more energy because it experiences much greater effective nuclear charge than the 2*s* electron—that is, the 1*s* electron experiences a much larger *binding energy* to the nucleus.

As with the photoelectric effect, any energy of the X-ray in excess of that needed to remove an electron from a lithium (or any other) atom is imparted to the ejected electron as kinetic energy: the smaller the binding energy that had to be overcome, the greater the kinetic energy of the emitted photoelectron. Therefore, electrons closer to the nucleus have larger binding energies. **Photoelectron spectroscopy (PES)** is a technique to experimentally determine kinetic energies and, on the basis of the energy of the X-rays used to produce the photoelectrons, can be used to calculate the binding energy of the emitted photoelectrons. Those data are transformed into graphs called **photoelectron spectra** in which the number of electrons ejected by a sample is plotted as a function of binding energy. Note that the *x*-axis is drawn such that binding energies *decrease* from left to right. Electrons closer to the nucleus have a larger binding energy; therefore, as you move from left to right on the *x*-axis, the electrons are farther from the nucleus.

The photoelectron spectrum for lithium is shown in **Figure 3.38(a)**. The height of each peak is proportional to the number of electrons with that binding energy.

**FIGURE 3.38** Photoelectron spectra of (a) lithium and (b) aluminum.

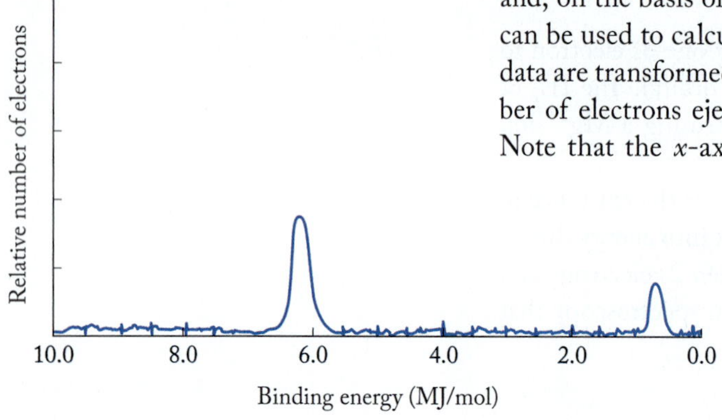


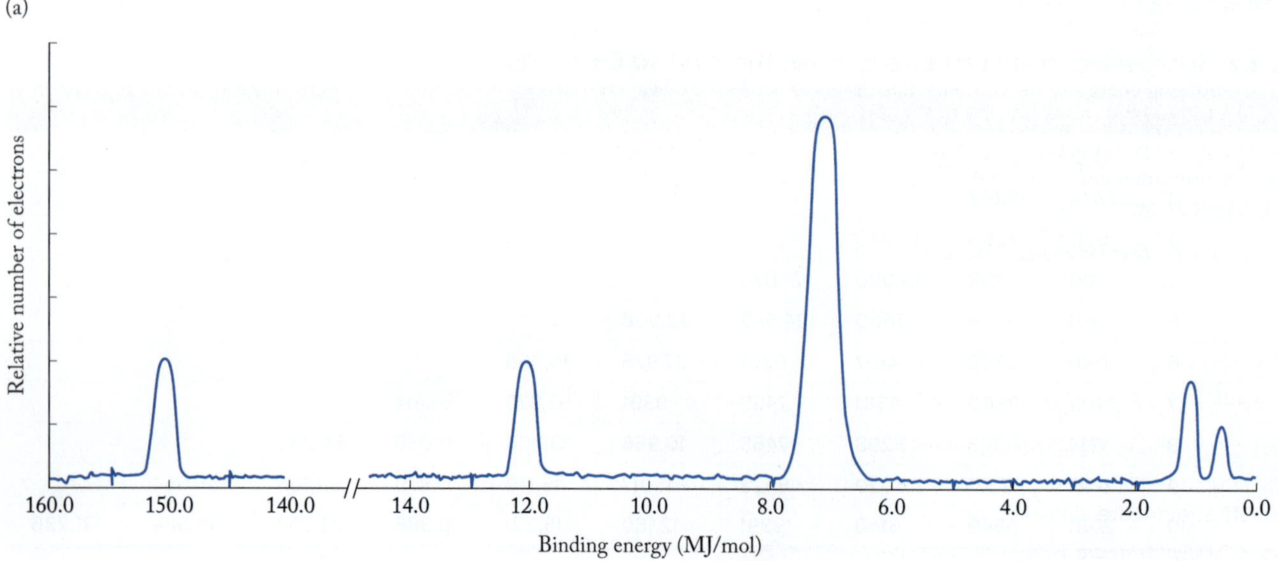

Two peaks are present in the Li spectrum at binding energies of 6.26 million joules (MJ) per mole and 0.52 MJ/mol. The relative heights of the peaks are 2:1, indicating that twice as many electrons have the larger binding energy. Those data are consistent with the predicted ground-state electron configuration for lithium: $1s^2 2s^1$.

As you might expect, larger atoms with electrons in more subshells (and therefore more types of orbitals) have more complex photoelectron spectra. For example, the photoelectron spectrum of aluminum (**Figure 3.38b**) has five peaks, which we will examine in Sample Exercise 3.15.

## SAMPLE EXERCISE 3.15 Interpreting Photoelectron Spectroscopy Data to Determine Electron Configurations LO7

Assign each peak in the photoelectron spectrum of aluminum metal in Figure 3.38(b) to the orbital(s) whose electrons have that binding energy.

**Collect, Organize, and Analyze** We know that electrons in orbitals closer to aluminum atoms' nuclei experience a greater $Z_{eff}$ and are bound more strongly to the nuclei; that is, those electrons have greater binding energies. Therefore, the leftmost peaks on the spectrum are due to inner-shell electrons, whereas those farthest to the right are produced by valence-shell electrons. According to the orbital filling sequence from Figure 3.33, the predicted electron configuration of an Al atom ($Z = 13$) is $1s^2 2s^2 2p^6 3s^2 3p^1$.

**Solve** The orbital assignments for the binding energies (BE) of the five peaks are

| BE (MJ/mol) | 151 | 12.1 | 7.19 | 1.09 | 0.58 |
|---|---|---|---|---|---|
| Orbital(s) | 1s | 2s | 2p | 3s | 3p |

**Think About It** The sequence of orbitals in the above table agrees with the principle that binding energies decrease with increasing distance from the nucleus and increasing shielding of outer-shell electrons by those in inner shells. Those assignments also correspond to aluminum's electron configuration: the peak assigned to the three filled 2*p* orbitals is three times the height of those assigned to the filled 1*s*, 2*s*, and 3*s* orbitals, which are each twice the height of the peak assigned to the single electron in the 3*p* orbital.

**Practice Exercise** Photoelectron spectroscopy also can be used to analyze the binding energies in monatomic ions, such as the $Al^{3+}$ and $O^{2-}$ ions in $Al_2O_3$. (a) In the PES spectrum of $Al_2O_3$, which peaks from Figure 3.38b are not present? Explain your choice. (b) The peaks for $Al^{3+}$ are shifted in comparison with those of aluminum metal. Do the $Al^{3+}$ peaks shift to smaller or larger energies? Explain your answer.

*(Answers to Practice Exercises are in the back of the book.)*

## CONCEPT TEST

In Figure 3.38, why is the largest binding energy in the Al spectrum so much greater than the largest binding energy in the Li spectrum, whereas the smallest binding energies in both spectra are similar?

*(Answers to Concept Tests are in the back of the book.)*

**electron affinity (EA)** the energy change that occurs when one mole of electrons combines with one mole of atoms or ions in the gas phase.

## 3.12 Electron Affinities

In the preceding section, we examined the periodic nature of the energy required to remove electrons from atoms. Now we look at a complementary process and examine the change in energy when electrons are *added* to atoms to form monatomic anions. The energies involved are called **electron affinities (EAs)**. They are the energy changes that occur when one mole of electrons is added to one mole of atoms or ions in the gas phase.

For example, the energy associated with adding one mole of electrons to one mole of chlorine atoms in the gas phase is

$$\text{Cl} + \text{e}^- \rightarrow \text{Cl}^- \qquad \text{EA} = -349 \text{ kJ/mol}$$

The negative sign tells us that energy is lost, or released, when chlorine forms chloride ions. The same is true for most main group elements, as shown in **Figure 3.39**, because the association of a negatively charged electron with a positively charged nucleus usually produces an ion that has less energy than the free atom and electron had before they came together.

| 1 | 2 | 13 | 14 | 15 | 16 | 17 | 18 |
|---|---|---|---|---|---|---|---|
| H −72.6 | | | | | | | He (0.0)[a] |
| Li −59.6 | Be >0 | B −26.7 | C −122 | N +7 | O −141 | F −328 | Ne (+29)[a] |
| Na −52.9 | Mg >0 | Al −42.5 | Si −134 | P −72.0 | S −200 | Cl −349 | Ar (+35)[a] |
| K −48.4 | Ca −2.4 | Ga −28.9 | Ge −119 | As −78.2 | Se −195 | Br −325 | Kr (+39)[a] |
| Rb −46.9 | Sr −5.0 | In −28.9 | Sn −107 | Sb −103 | Te −190 | I −295 | Xe (+41)[a] |
| Cs −45.5 | Ba −14 | Tl −19.2 | Pb −35.2 | Bi −91.3 | Po −183.3 | At −270[a] | Rn (+41)[a] |

[a]Calculated values.

**FIGURE 3.39** Electron affinity (EA) values of main group elements are expressed in kilojoules per mole. The more negative the value, the more energy is released when one mole of atoms combines with one mole of electrons to form one mole of anions with a 1− charge. A greater release of energy reflects more attraction between the atoms of the elements and free electrons.

As evident from the EA values in Figure 3.39, the trends in EA are not as regular as the trends in atomic size and ionization energy. Electron affinities generally increase (become less negative) with increasing atomic number among the group 1 and group 17 elements (except for F and Cl), but other groups do not display a clear trend. In general, EA becomes more negative with increasing atomic number across a row, but that trend has exceptions, too. The halogens of group 17 have the most negative EA values of the elements in each of their rows because of the relatively high $Z_{\text{eff}}$ experienced by electrons in their valence-shell $p$ orbitals (including the electron each of their atoms acquires in becoming an anion) and because the anions they form have the stable electron configurations of noble gases.

By contrast, adding an electron to a group 18 atom requires adding energy because noble gas atoms have stable electron configurations already. Beryllium and magnesium have positive electron affinities because the added electrons have to occupy outer-shell $p$ orbitals that are significantly higher in energy than the outer-shell $s$ orbitals. We can also rationalize the positive electron affinity of nitrogen by noting that adding an electron to a nitrogen atom means that the atom loses the stability associated with a half-filled set of $2p$ orbitals:

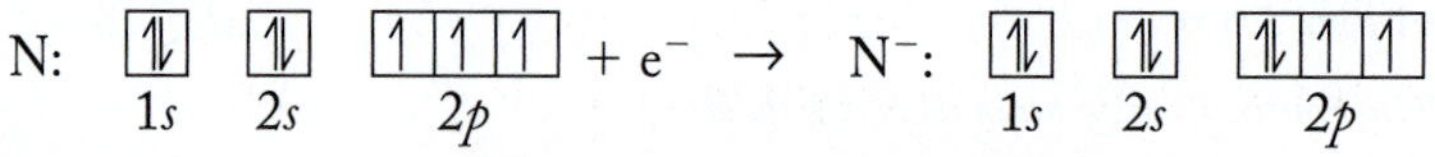

$$\text{N: } \underset{1s}{[\uparrow\downarrow]}\ \underset{2s}{[\uparrow\downarrow]}\ \underset{2p}{[\uparrow|\uparrow|\uparrow]} + \text{e}^- \rightarrow \text{N}^-\text{: } \underset{1s}{[\uparrow\downarrow]}\ \underset{2s}{[\uparrow\downarrow]}\ \underset{2p}{[\uparrow\downarrow|\uparrow|\uparrow]}$$

### CONCEPT **TEST**

Describe at least one similarity and one difference in the periodic trends in first ionization energies and electron affinities among the main group elements.

*(Answers to Concept Tests are in the back of the book.)*

To end this chapter, let's return to the first three decades of the 20th century, which saw remarkable advances in our understanding of the structure of atoms and how nature works at the atomic and subatomic levels. In this chapter we have tried to connect the advances, showing how one led to another (**Figure 3.40**), while also conveying a sense of how profoundly unsettling those new ideas about the laws of nature were to the leading scientists of the time.

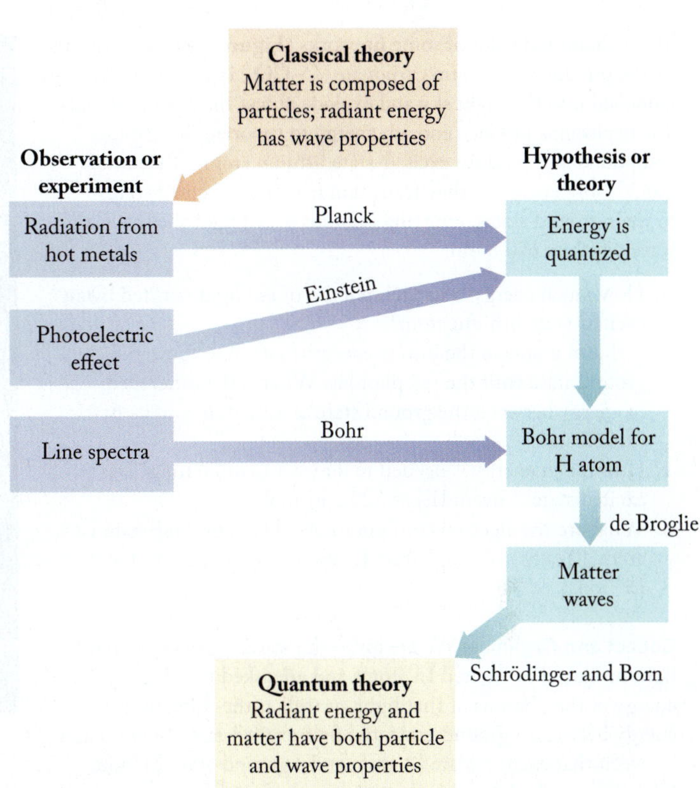


**FIGURE 3.40** During the first three decades of the 20th century, quantum theory evolved from classical 19th-century theories of the nature of matter and energy. The arrows trace the development of modern quantum theory, which assumes that radiant energy has both wave properties and particle properties and that mass (matter) also has both wave properties and particle properties.

Consensus in the scientific community on the ideas of quantum theory did not come easily. For example, we have seen how excited-state sodium atoms emit photons of yellow-orange light as they return to the ground state. Einstein puzzled over that phenomenon for several years before proposing that neither the exact moment when emission occurs nor the path of the emitted photon could be predicted exactly. He concluded that quantum theory allows us to calculate only the probability of a spontaneous electron transition; the details of the event are left to chance. In other words, no force of nature causes an excited-state sodium atom to emit a photon and return to a lower energy level at a particular instant.

That probabilistic view of the interaction of matter and energy was very different from the familiar cause-and-effect interactions involving energy exchange between large objects. For example, a marble rolling off a table drops immediately to the floor, yet an electron remains in an excited state for an indeterminate time before falling to a lower energy level. That lack of determinacy bothered Einstein and many of his colleagues. Had they discovered an underlying theme of nature—that some processes cannot be described or known with certainty? Do fundamental limits exist on how well we can know and understand our world and the events that change it?

Many scientists in the early decades of the 20th century did not agree with that uncertain view of nature. They preferred the Newtonian view, in which events occur for a reason and where clearly understood causes and effects are at work. Those scientists believed that the more they studied nature with ever more sophisticated tools, the more they would understand why things happen the way they do. Soon after Max Born published his probabilistic interpretation of Schrödinger's wave functions in 1926, Einstein wrote Born a letter in which he contrasted the new theories with the certainties many people find in religious beliefs:

> Quantum mechanics is very impressive. But an inner voice tells me that it is not yet the real thing. The theory produces a great deal but hardly brings us closer to the secret of the Old One. I am at all events convinced that He does not play dice.[2]

[2]Letter to Max Born, 12 December 1926; quoted in R. W. Clark, *Einstein: The Life and Times* (New York: HarperCollins, 1984), p. 880.

## SAMPLE EXERCISE 3.16 Integrating Concepts: Red Fireworks

The brilliant red color of some fireworks (**Figure 3.41**) is produced by the presence of lithium carbonate ($Li_2CO_3$) in shells that are launched into the night sky and explode at just the right moment. The explosions produce enough energy to vaporize the lithium compound and produce excited-state lithium atoms. The atoms quickly lose energy as they transition from their excited states to their ground states, emitting photons of red light that have a wavelength of 670.8 nm.

a. How much energy is in each photon of red light emitted by an excited-state lithium atom?
b. Lithium atoms in the lowest energy (first) state above the ground state emit the red photons. What is the difference in energy between the ground state of a Li atom and its first excited state?
c. How much energy is needed to ionize a Li atom in its first excited state? Given: $IE_1 = 520.2$ kJ/mol.
d. What are the electron configurations of (1) a ground-state Li atom, (2) a first-excited-state Li atom, and (3) a ground-state $Li^+$ ion?

**FIGURE 3.41** The red color of some fireworks is produced by the transition of electrons in lithium atoms from their first excited state to their ground state.

**Collect and Organize** We are given the wavelength of the red light produced by excited Li atoms and are asked to calculate the energy of the photons of that light, as well as the differences in energy between a ground-state and first-excited-state Li atom and between that excited-state Li atom and a ground-state $Li^+$ ion. We also are asked to write the electron configurations of the three states. The energy of a photon is related to its wavelength ($\lambda$) by Equation 3.4:

$$E = \frac{hc}{\lambda}$$

**Analyze** The energy of a photon ($E$) emitted by an excited-state atom is the same as the difference in energies ($\Delta E$) between the excited state and the lower-energy state (here, the ground state) of the transition that produced the photon. Ionizing one mole ($6.0221 \times 10^{23}$ atoms) of Li takes 520.2 kJ, so dividing 520.2 kJ by $6.0221 \times 10^{23}$ gives us the energy required to ionize one Li atom. That value should be greater than the energy needed to raise a Li atom to its first excited state, and the difference between the energies is that required to ionize the excited-state atom. We have seen in other calculations in this chapter that photons of visible light have energies in the $10^{-19}$ J range.

**Solve**

a. The energy of a photon with a wavelength of 670.8 nm is

$$E = \frac{hc}{\lambda} = \frac{(6.626 \times 10^{-34}\ \text{J}\cdot\text{s})(2.998 \times 10^{8}\ \cancel{\text{m}}/\text{s})}{670.8\ \cancel{\text{nm}}\left(\dfrac{10^{-9}\ \cancel{\text{m}}}{1\ \cancel{\text{nm}}}\right)}$$
$$= 2.961 \times 10^{-19}\ \text{J}$$

b. The difference in energies between the ground state of a Li atom and its lowest-energy excited state must match the energy of the photon emitted; therefore,

$$\Delta E = 2.961 \times 10^{-19}\ \text{J}$$

c. The energy required to ionize a ground-state Li atom is

$$\left(520.2\ \frac{\cancel{\text{kJ}}}{\cancel{\text{mol}}}\right)\left(\frac{1000\ \text{J}}{1\ \cancel{\text{kJ}}}\right)\left(\frac{1\ \cancel{\text{mol}}}{6.0221 \times 10^{23}\ \text{atoms}}\right)$$
$$= 8.638 \times 10^{-19}\ \text{J}$$

The energy required to ionize a Li atom in the first excited state ($2.961 \times 10^{-19}$ J above the ground state) is

$$\begin{array}{r} 8.638 \times 10^{-19}\ \text{J} \\ -\ 2.961 \times 10^{-19}\ \text{J} \\ \hline 5.677 \times 10^{-19}\ \text{J} \end{array}$$

d. Li is the first atom in the second row of the periodic table, which means it has one electron in its $2s$ orbital. Therefore,
1. Ground-state Li atoms have the electron configuration $1s^22s^1$.
2. The first subshell (and electron energy level) above $2s$ is $2p$, so the valence electron in a Li atom in the lowest excited state is in a $2p$ orbital, giving it the electron configuration $1s^22p^1$.
3. The valence electron of a Li atom is lost when it forms a $Li^+$ ion, which means that the electron configuration of a ground-state $Li^+$ ion is simply $1s^2$.

**Think About It** You might expect that the colors that lithium and other elements make in fireworks' explosions are the same ones they produce in Bunsen burner flames. You would be right. Many of those colors are produced by transitions from first excited states to ground states because the lowest-energy excited state is the one most easily populated by the thermal energy available in a gas flame or an exploding fireworks shell.

## SUMMARY

**LO1** Visible light is one form of **electromagnetic radiation**. Like the other forms, it has wave properties described by characteristic **wavelengths ($\lambda$)** and **frequencies ($\nu$)**. The speed of light is a constant in a vacuum, $c = 2.998 \times 10^8$ m/s, but slows as it passes through liquid and solids. (Section 3.1)

**LO2** According to quantum theory, discrete energy levels exist in atoms, which means they absorb or emit discrete amounts of energy called **photons**. Planck used those energy **quanta** to explain the radiation emitted by incandescent objects, and Einstein used them to explain the **photoelectric effect**. The radiant energy required to dislodge a photoelectron from a metal surface is called the **work function ($\Phi$)** of the metal. (Section 3.3)

**LO3** **Electron transitions** between energy levels in an atom produce **atomic emission spectra** and **atomic absorption spectra** consisting of narrow lines at wavelengths corresponding to the change in energy of the electrons. (Sections 3.2 and 3.4)

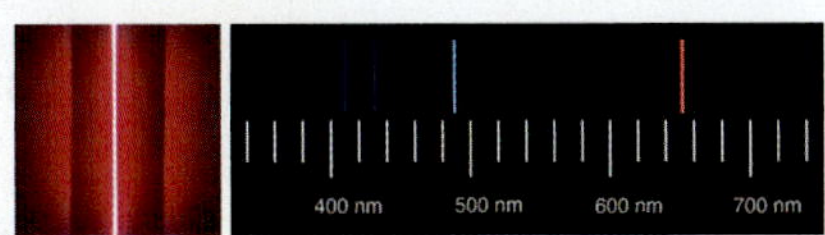


**LO4** De Broglie proposed that electrons in atoms, as well as all other moving particles, have wave properties and can be treated as **matter waves**. He explained the stability of the electron orbits in the Bohr hydrogen atom in terms of those waves. The **Heisenberg uncertainty principle** states that both the position and momentum of an electron cannot be precisely known at the same time. (Section 3.5)

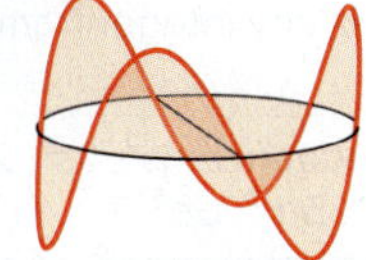

**LO5** The solutions to the **Schrödinger wave equation** are mathematical expressions called **wave functions ($\psi$)**, where $\psi^2$ defines the regions within an atom, called **orbitals**, that describe the probability of finding an electron at a given distance from the nucleus. Orbitals have characteristic three-dimensional sizes, shapes, and orientations that are depicted by boundary–surface representations. Each orbital has a unique set of three **quantum numbers**: **principal quantum number $n$**, which defines orbital size and energy level; **angular momentum quantum number $\ell$**, which defines orbital shape; and **magnetic quantum number $m_\ell$**, which defines orbital orientation in space. A fourth quantum number, **spin quantum number $m_s$**, allows two opposite-spin electrons to share an orbital. The **Pauli exclusion principle** states that no two electrons in an atom can have the same four values of $n$, $\ell$, $m_\ell$, and $m_s$. (Sections 3.6 and 3.7)

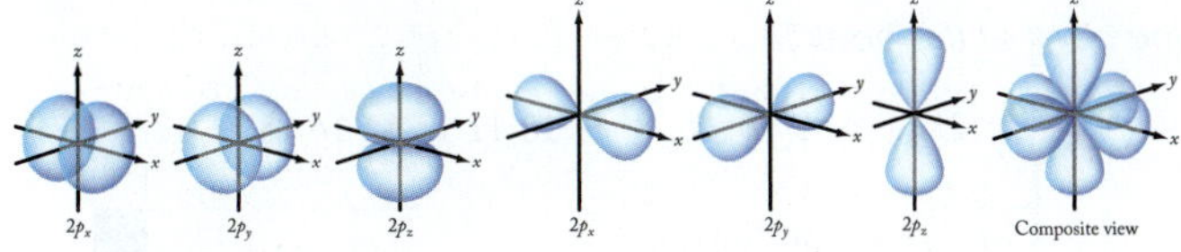


**LO6** According to the **aufbau principle**, electrons fill the lowest-energy atomic orbitals of a ground-state atom first. An **electron configuration** is a set of numbers and letters expressing the number of electrons that occupy each orbital in an atom. The electrons in the outermost occupied shell of an atom are **valence electrons**; the electrons in the filled inner shells are **core electrons**. All the orbitals of a given $p$, $d$, or $f$ subshell are said to be degenerate, which means they all have the same energy. **Hund's rule** states that in any set of degenerate orbitals, one electron must occupy each orbital before a second electron occupies any orbital in the set, and that in the lowest-energy configuration, all unpaired electrons have the same spin. (Sections 3.8 and 3.9)

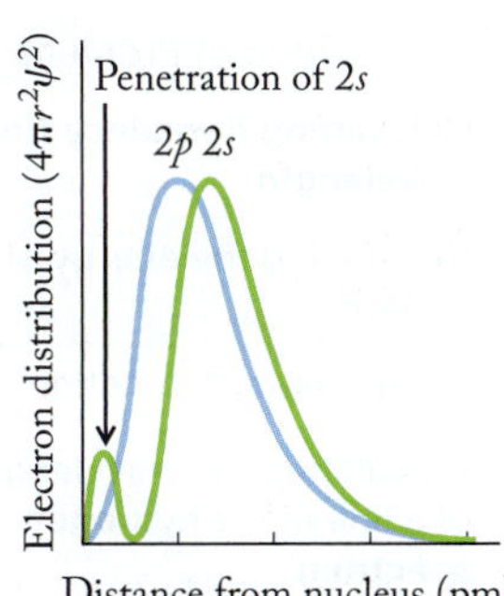


**LO7** **Effective nuclear charge ($Z_{eff}$)** is the net nuclear charge experienced by outer-shell electrons when they are shielded by inner-shell electrons. The sizes of atoms increase with increasing atomic number down a group of elements because valence-shell electrons with higher $n$ values are, on average, farther from the nucleus. However, the sizes of atoms decrease with increasing atomic number across a row of elements because the valence electrons experience higher effective nuclear charges. Anions are larger than their parent atoms because of additional electron–electron repulsion, but cations are smaller than their parent atoms. **Ionization energy (IE)** is the energy needed to remove one mole of electrons from one mole of atoms or ions in the gas phase. IE values generally increase with increasing effective nuclear charge across a row and decrease with increasing atomic number down a group in the periodic table. **Photoelectron spectra**, which plot the relative number of electrons in an element and their binding energies, can be used to identify metals and determine their electron configurations. (Sections 3.8, 3.10, and 3.11)

**LO8** **Electron affinity (EA)** is the energy change that occurs when one mole of electrons combines with one mole of atoms or ions in the gas phase. The EA values of many main group elements are negative, indicating that energy is released when they acquire electrons. (Section 3.12)

## PARTICULATE **PREVIEW WRAP-UP**

1. Model (a) shows electrons moving around the nucleus of an atom, in keeping with the Rutherford model.
2. Concentric circles suggest that electrons orbit the nucleus at different distances from it.
3. Model (a) suggests that electrons orbit the nuclei of atoms in concentric circular orbits; model (b) shows a more probabilistic picture of electron distribution, with regions of high electron densities at different distances from the nucleus.

## PROBLEM-SOLVING SUMMARY

| Type of Problem | Concepts and Equations | Sample Exercises |
|---|---|---|
| **Calculating frequency from wavelength** | $\nu = \frac{c}{\lambda}$ (3.2) | **3.1** |
| **Calculating the energy of a photon** | $E = \frac{hc}{\lambda}$ (3.4) | **3.2** |
| **Using the work function** | $\Phi = h\nu_0 = h\nu - \text{KE}_{\text{electron}}$ (3.5, 3.6) | **3.3** |
| **Calculating the wavelength of a line in the hydrogen spectrum** | $\lambda = 364.56 \text{ nm}\left(\frac{m^2}{m^2 - n^2}\right)$ (3.7)<br>$\frac{1}{\lambda} = R_{\text{H}}\left(\frac{1}{n_1{}^2} - \frac{1}{n_2{}^2}\right)$ (3.8) | **3.4** |
| **Calculating the energy of a transition in a hydrogen atom** | $\Delta E = -2.178 \times 10^{-18} \text{ J}\left(\frac{1}{n_{\text{final}}{}^2} - \frac{1}{n_{\text{initial}}{}^2}\right)$ (3.11) | **3.5** |
| **Calculating the wavelength of a particle in motion** | $\lambda = \frac{h}{mu}$ (3.12) | **3.6** |
| **Calculating Heisenberg uncertainties** | $\Delta x \cdot m\Delta u \geq \frac{h}{4\pi}$ (3.14) | **3.7** |
| **Identifying the subshells and orbitals in an energy level and valid quantum number sets** | $n$ is the shell number; $\ell$ defines both the subshell and the type of orbital; an orbital has a unique combination of allowed $n$, $\ell$, and $m_\ell$ values; $\ell$ is any integer from 0 to $(n - 1)$; $m_\ell$ is any integer from $-\ell$ to $+\ell$, including zero; $m_s$ equals $+\frac{1}{2}$ or $-\frac{1}{2}$. | **3.8, 3.9** |
| **Writing electron configurations of atoms and ions** | Orbitals fill up in the following sequence: $1s^2$, $2s^2$, $2p^6$, $3s^2$, $3p^6$, $4s^2$, $3d^{10}$, $4p^6$, $5s^2$, $4d^{10}$, $5p^6$, $6s^2$, $4f^{14}$, $5d^{10}$, $6p^6$, $7s^2$, $5f^{14}$, $6d^{10}$, $7p^6$, where superscripts represent the maximum numbers of electrons. Orbitals in electron configurations are arranged based on increasing value of $n$. In forming transition metal ions, electrons in orbitals with the highest $n$ value are removed first; stability is enhanced in half-filled and filled $d$ subshells. | **3.10–3.12** |
| **Ordering atoms and ions by size** | In general, sizes decrease left to right across a row and increase down a column of the periodic table; cations are smaller and anions are larger than their parent atoms. | **3.13** |
| **Recognizing trends in ionization energies** | First ionization energies increase across a row and decrease down a column. | **3.14** |
| **Determining electron configurations from PES data** | The binding energy of electrons in a metal can be determined by using X-rays to eject electrons. Photoelectron spectra plot the number of electrons versus the binding energy and can be used to identify unknown metals and determine their electron configurations. | **3.15** |

## VISUAL PROBLEMS

*(Answers to boldface end-of-chapter questions and problems are in the back of the book.)*

**3.1.** Which of the elements highlighted in Figure P3.1 consist of atoms with

a. a single $s$ electron in their outermost shells?
b. filled sets of $s$ and $p$ orbitals in their outermost shells?
c. filled sets of $d$ orbitals?
d. half-filled sets of $d$ orbitals?
e. two $s$ electrons in their outermost shells?
(More than one answer is possible.)

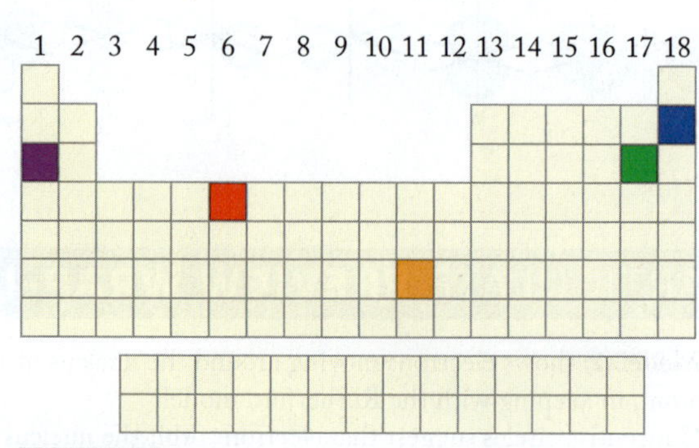


FIGURE P3.1

3.2. Which of the highlighted elements in Figure P3.1
a. forms a common monatomic ion that is larger than its parent atom?
b. has the most unpaired electrons per ground-state atom?
c. has a complete outer shell?
d. forms a common monatomic ion that is isoelectronic with another highlighted element?

**3.3.** Which of the highlighted elements in Figure P3.1 form(s) common monatomic ions that are (a) larger than their parent atoms and (b) smaller than their parent atoms?

3.4. Which of the highlighted elements in Figure P3.4 forms monatomic ions by
a. losing an *s* electron?
b. losing two *s* electrons?
c. losing two *s* electrons and a *d* electron?
d. adding an electron to a *p* orbital?
e. adding electrons to two *p* orbitals?

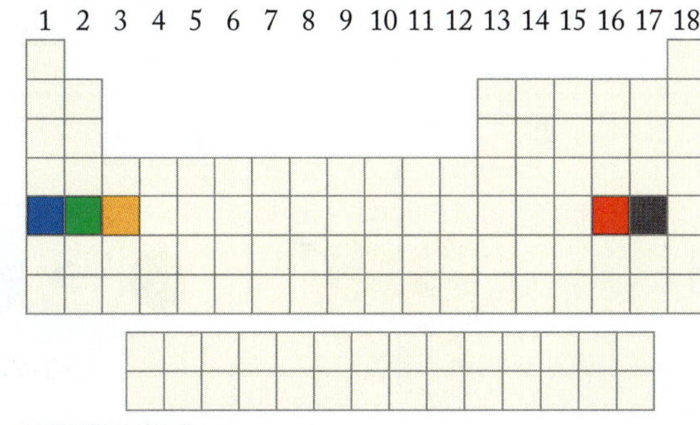


**FIGURE P3.4**

**3.5.** Which of the highlighted elements in Figure P3.4 form(s) common monatomic ions that are smaller than their parent atoms?

3.6. Rank the highlighted elements in Figure P3.4 in order of (a) increasing size of their atoms and (b) increasing first ionization energy.

**3.7.** Suppose three beams of radiation are focused on a negatively charged metallic surface. The beam represented by the A waves in Figure P3.7 causes photoelectrons to be emitted from the surface. Which of the following statements accurately describes the abilities of the beams represented by waves B and C to also produce photoelectrons from that surface?
a. Neither the B nor the C wave can produce photoelectrons.
b. Both the B and C waves should produce photoelectrons if the sources of the waves are bright enough.
c. The B wave may or may not produce photoelectrons, but the C wave surely will.
d. The C wave may or may not produce photoelectrons, but the B wave surely will.

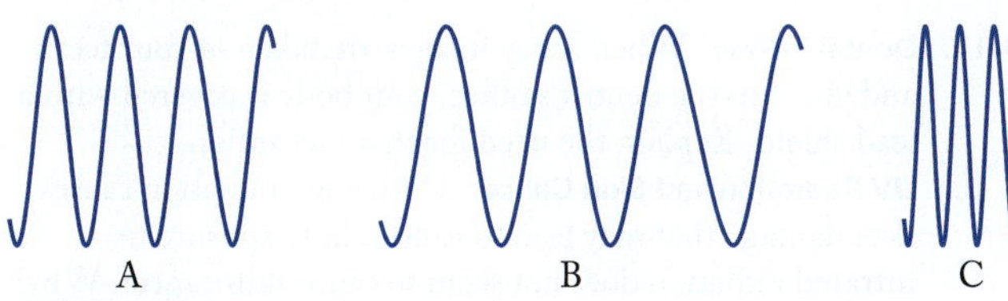


**FIGURE P3.7**

3.8. A group 2 element (M) and a group 17 element (X) have nearly the same atomic radius, as shown by the relative sizes of the blue and red spheres representing their atoms in Figure P3.8. Are the two elements in the same row of the periodic table? If not, which one is in the higher row?

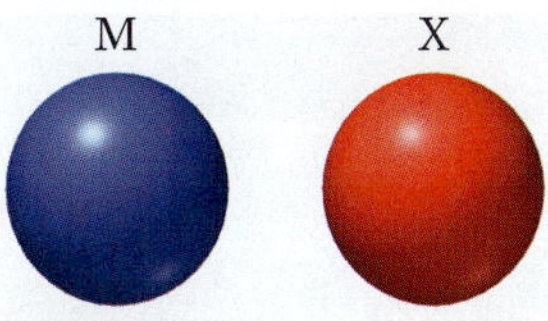


**FIGURE P3.8**

**3.9.** Which of the pairs of spheres in Figure P3.9 best depicts the relative sizes of the cation and anion formed by atoms of the elements M and X from Figure P3.8?

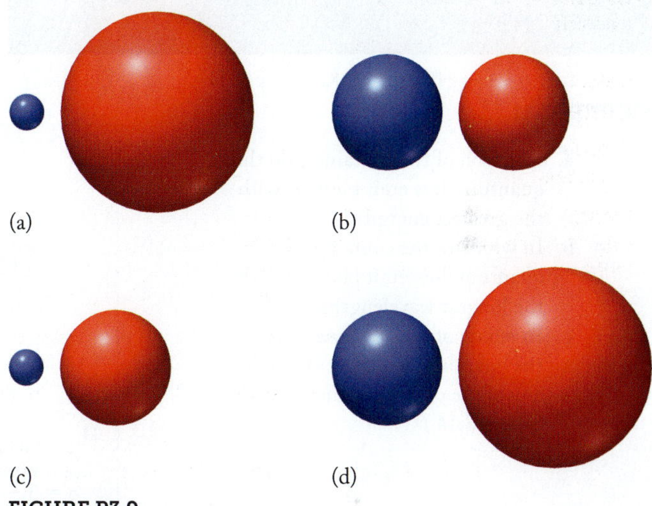


**FIGURE P3.9**

3.10. The arrows A, B, and C in Figure P3.10 show three transitions among the $n = 3$, 4, and 5 energy levels in a single-electron ion.
a. If we assume that the transitions are accompanied by the loss or gain of electromagnetic energy, do the arrows depict absorption or emission of radiation?
b. The spectral lines of the transitions are, in order of increasing wavelength, 80, 117, and 253 nm. Which wavelength goes with which arrow?

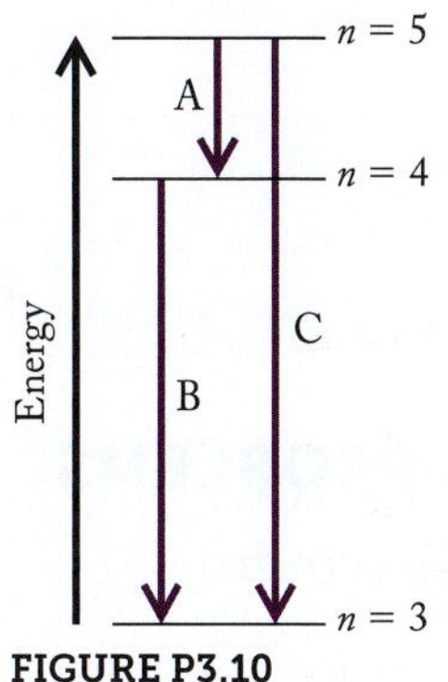


**FIGURE P3.10**

**3.11.** **Quantum Dot Televisions** Some flat-panel displays and TV sets use quantum dots (nanocrystals of cadmium selenide and cadmium sulfide) to generate colors. The colors of the quantum dots in Figure P3.11 are determined by particle size and composition.

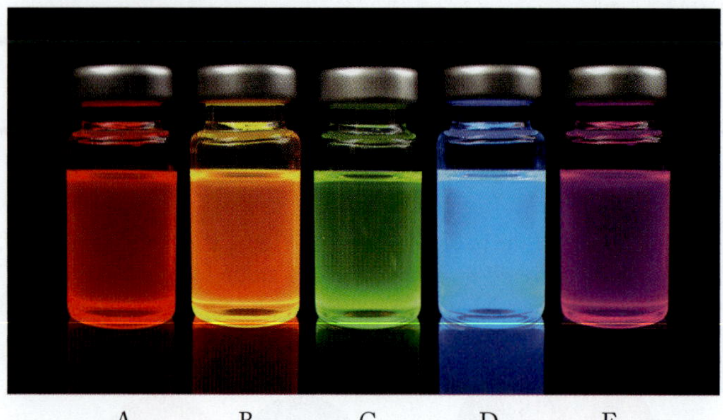


**FIGURE P3.11**

a. In which of the containers do the quantum dots emit photons with the greatest energies?
b. In which of the containers do the quantum dots emit photons with the longest wavelengths?
c. In which of the containers are the quantum dots larger than in any of the others? (Hint: see Figure 3.14.)

3.12. Use representations [A] through [I] in Figure P3.12 to answer questions a–f.

a. Between [A] and [I], which valence-shell $s$ electron experiences the stronger attraction to its nucleus?
b. Which representations depict the loss of an electron and which depict the gain of an electron?
c. Which representation depicts quantized processes?
d. Which representations depict the probabilistic nature of quantum mechanics?
e. What processes are represented in [C], [D], [F], and [G]?
f. Which representation is consistent with zero probability of an electron's being in the nucleus of an atom?

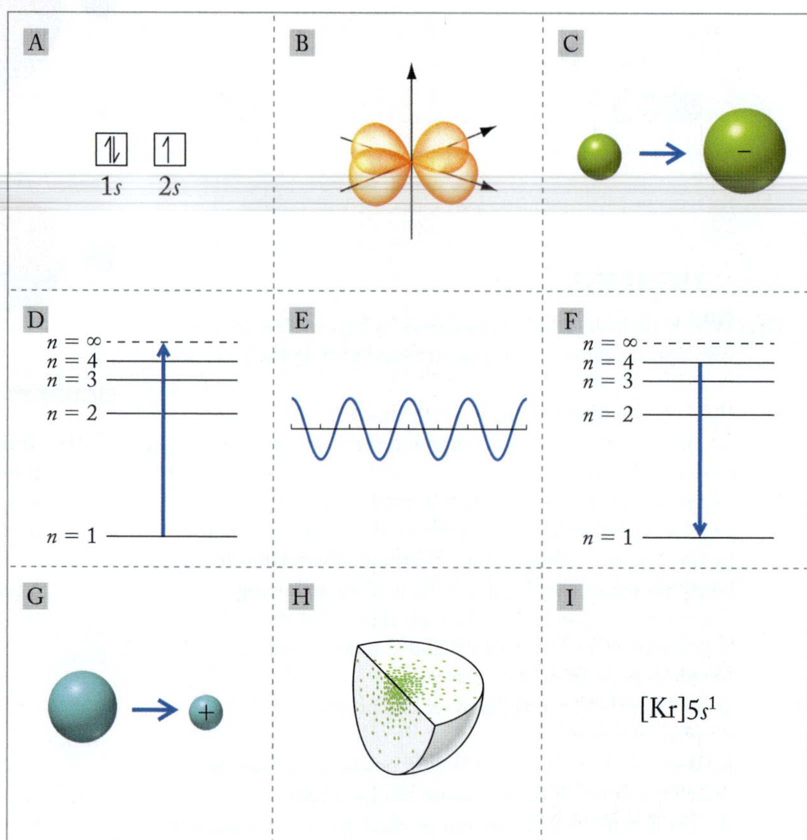


**FIGURE P3.12**

# QUESTIONS AND PROBLEMS

## *The Electromagnetic Spectrum*

### Concept Review

**3.13.** Why are the various forms of radiant energy called *electromagnetic* radiation?

3.14. Explain with a sketch why long-wavelength waves of electromagnetic radiation have lower frequencies than short-wavelength waves.

**3.15.** **Dental X-Rays** When X-ray images are taken of your teeth and gums in the dentist's office, your body is covered with a lead shield. Explain the need for that precaution.

3.16. **UV Radiation and Skin Cancer** Ultraviolet radiation causes skin damage that may lead to cancer, but exposure to infrared radiation does not seem to cause skin cancer. Why?

**3.17.** **Lava** As hot molten lava cools, it begins to solidify and no longer glows in the dark. Does that mean it no longer emits any kind of electromagnetic radiation? If not, what kind of radiation is it likely to emit once it is no longer "red" hot?

*3.18. If light consists of waves, why don't objects look "wavy" to us?

### Problems

**3.19.** **Digital Screens** Tablets and cell phones emit blue light (Figure P3.19) that disrupts sleep patterns when viewed before going to bed. The longest wavelength that corresponds to blue light is 495 nm. What is the frequency of that radiation?

**FIGURE P3.19**

3.20. **Submarine Communications** The Russian and U.S. navies developed extremely low-frequency communications networks to send messages to submerged submarines. The frequency of the carrier wave of the Russian network was 82 Hz, whereas the United States used 76 Hz.

a. What was the ratio of the wavelengths of the Russian network to the wavelengths of the U.S. network?
b. To calculate the actual underwater wavelength of the transmissions in either network, what additional information would you need?

---

**3.21.** **Drones and Phones** Remote control aerial vehicles such as the drone in Figure P3.21 typically operate at gigahertz (GHz) frequencies, whereas cellular phones operate using megahertz (MHz) frequencies. Calculate the wavelengths corresponding to these frequencies:

a. quadcopter drone, 2.4 GHz
b. Apple iPhone, 848 MHz

**FIGURE P3.21**

3.22. **Remote Keyless Systems** The doors of most automobiles can be unlocked by pushing a button on a key chain (Figure P3.22). Those devices operate on frequencies in the megahertz range. Calculate the wavelength of (a) the 315 MHz radiation emitted by remote keyless systems in North American cars and (b) the 434 MHz radiation used to unlock the doors of European and Asian cars. Where do those frequencies lie in the electromagnetic spectrum?

**FIGURE P3.22**

---

**3.23.** In 1895, German physicist Wilhelm Röntgen discovered X-rays. He also discovered that X-rays emitted by different metals have different wavelengths. Which X-rays have the higher frequency, those emitted by (a) Cu ($\lambda = 0.154$ nm) or (b) iron ($\lambda = 194$ pm)?

3.24. **Garage Door Openers** The remote control units for garage door openers transmit electromagnetic radiation. Before 2005, they operated on a frequency of 390 MHz, but since 2005, the operating wavelength has been 952 mm. Which radiation has the lower frequency, that emitted by the pre-2005 or the post-2005 devices?

---

**3.25.** **Speed of Light** How long does light from the sun take to reach Earth when the distance between them is 149.6 million kilometers?

3.26. **Exploration of the Solar System** The *Voyager 1* spacecraft was launched in 1977 to explore the outer solar system and interstellar space. By November 2018, the probe was $2.16 \times 10^{10}$ km from Earth and still sending and receiving data. How long did a signal from Earth take to reach *Voyager 1* over that distance?

## Atomic Spectra

### Concept Review

**3.27.** Describe the similarities and differences in the atomic emission and absorption spectra of an element.

3.28. Are the Fraunhofer lines the result of atomic emission or atomic absorption?

**3.29.** How did the study of the atomic emission spectra of the elements lead to identifying the origins of the Fraunhofer lines in sunlight?

*3.30. How would the appearance of the Fraunhofer lines in the solar spectrum be changed if sunlight were passed through a flame containing high-temperature sodium atoms?

## Particles of Light: Quantum Theory

### Concept Review

**3.31.** What is a quantum?

3.32. What is a photon?

**3.33.** If a piece of tungsten metal were heated to 1000 K, would it emit light in the dark? If so, what color?

3.34. **Incandescent Lightbulbs** A variable power supply is connected to an incandescent lightbulb. At the lowest power setting, the bulb feels warm to the touch but produces no light. At medium power, the lightbulb filament emits a red glow. At the highest power, the lightbulb emits white light. Explain that emission pattern.

### Problems

**3.35.** **Tanning Booths** Prolonged exposure to ultraviolet radiation in tanning booths significantly increases the risk of skin cancer. What is the energy of a photon of UV light with a wavelength of $3.00 \times 10^{-7}$ m?

3.36. **Remote Control** Most remote controllers for audiovisual equipment emit radiation with the wavelength profile shown in Figure P3.36. What type of radiation is shown, and what is the energy of a photon at the peak wavelength?

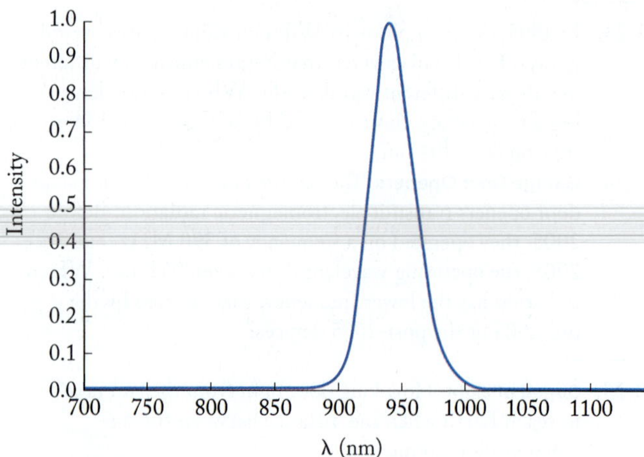


**FIGURE P3.36**

**3.37.** Which of the following have quantized values? Explain your selections.

a. The elevation of the treads of a moving escalator
b. The elevations at which the doors of an elevator open
c. The speed of an automobile

3.38. Which of the following have quantized values? Explain your selections.

a. The pitch of a note played on a slide trombone
b. The pitch of a note played on a flute
c. The wavelengths of light produced by the heating elements in a toaster
d. The wind speed at the top of Mt. Everest

**3.39.** When a piece of gold foil is irradiated with UV radiation ($\lambda = 132$ nm), electrons are ejected with a kinetic energy of $7.34 \times 10^{-19}$ J. What is the work function of gold?

3.40. The first ionization energy of a gas-phase atom of a particular element is $6.24 \times 10^{-19}$ J. What is the maximum wavelength of electromagnetic radiation that could ionize that atom?

**3.41.** **Solar Power** Photovoltaic cells convert solar energy into electricity. Could germanium ($\Phi = 7.21 \times 10^{-19}$ J) be used to convert visible sunlight to electricity? Assume that most of the electromagnetic energy from the sun in the visible region is at wavelengths shorter than 600 nm.

3.42. With reference to Problem 3.41, could tin ($\Phi = 6.20 \times 10^{-19}$ J) be used to construct solar cells?

***3.43.** Pieces of potassium ($\Phi = 3.68 \times 10^{-19}$ J) and sodium ($\Phi = 4.41 \times 10^{-19}$ J) metal are exposed to radiation with a wavelength of 300 nm. Which metal emits electrons with the greater velocity? What is the velocity of the electrons?

3.44. Titanium ($\Phi = 6.94 \times 10^{-19}$ J) and silicon ($\Phi = 7.24 \times 10^{-19}$ J) surfaces are irradiated with UV radiation with a wavelength of 250 nm. Which surface emits electrons with the longer wavelength? What is the wavelength of the electrons emitted by the titanium surface?

**3.45.** **Red Lasers** The power of a red laser ($\lambda = 630$ nm) is 1.00 watt (abbreviated W, where 1 W = 1 J/s). How many photons per second does the laser emit?

*3.46. **Starlight** The energy density of starlight in interstellar space is $10^{-15}$ J/m$^3$. If the average wavelength of starlight is 500 nm, what is the corresponding density of photons per cubic meter of space?

## The Hydrogen Spectrum and the Bohr Model

### Concept Review

**3.47.** Why is the wavelength of light required to ionize a hydrogen atom shorter than the wavelength corresponding to any of the other electron transitions in Figure 3.18?

3.48. How does the value of $n$ of an orbit in the Bohr model of hydrogen relate to the energy of an electron in that orbit?

**3.49.** Does the electromagnetic energy emitted by an excited-state H atom depend on the individual values of $n_1$ and $n_2$ or only on the difference between them ($n_1 - n_2$)?

3.50. Explain the difference between a ground-state H atom and an excited-state H atom.

**3.51.** Without calculating any wavelength values, identify which of the four electron transitions listed below for the hydrogen atom is associated with emitting a photon of the shortest wavelength.

a. $n = 1 \rightarrow n = 2$
b. $n = 2 \rightarrow n = 1$
c. $n = 1 \rightarrow n = 4$
d. $n = 4 \rightarrow n = 1$

3.52. Without calculating any frequency values, identify which of the four electron transitions listed below for the hydrogen atom is associated with absorbing a photon of the highest frequency.

a. $n = 1 \rightarrow n = 3$
b. $n = 3 \rightarrow n = 1$
c. $n = 3 \rightarrow n = 6$
d. $n = 6 \rightarrow n = 3$

### Problems

**3.53.** Calculate the wavelength of the photons emitted by hydrogen atoms when they undergo $n = 4 \rightarrow n = 3$ transitions. In which region of the electromagnetic spectrum does that radiation occur?

3.54. Calculate the frequency of the photons emitted by hydrogen atoms when they undergo $n = 5 \rightarrow n = 3$ transitions. In which region of the electromagnetic spectrum does that radiation occur?

---

***3.55.** The energies of photons emitted by one-electron atoms and ions fit the equation

$$E = (2.18 \times 10^{-18}\,\text{J})Z^2\left(\frac{1}{n_1^{\,2}} - \frac{1}{n_2^{\,2}}\right)$$

where $Z$ is the atomic number, $n_1$ and $n_2$ are positive integers, and $n_2 > n_1$. Is the emission associated with the $n = 2 \rightarrow n = 1$ transition in a one-electron ion ever in the visible region? Why or why not?

*3.56. Can transitions from higher-energy states to the $n = 2$ state in $He^+$ ever produce visible light? If so, for what values of $n_2$? Refer to the equation in Problem 3.55.

---

***3.57.** By absorbing different wavelengths of light, an electron in a hydrogen atom undergoes a transition from $n = 2$ to $n = 3$ and then from $n = 3$ to $n = 4$.

a. Are the wavelengths for the two transitions additive—that is, does $\lambda_{2\rightarrow4} = \lambda_{2\rightarrow3} + \lambda_{3\rightarrow4}$?

b. Are the energies of the two transitions additive—that is, does $E_{2\rightarrow4} = E_{2\rightarrow3} + E_{3\rightarrow4}$?

*3.58. The hydrogen atomic emission spectrum includes a UV line with a wavelength of 92.3 nm.

a. Is that line associated with a transition between different excited states or between an excited state and the ground state?

b. What is the energy of the longest-wavelength photon that a ground-state hydrogen atom can absorb?

## Electrons as Waves

### Concept Review

**3.59.** Why do matter waves *not* add significantly to the challenge of hitting a baseball thrown at 44 m/s (98 mph)?

3.60. How does de Broglie's hypothesis that electrons behave like waves explain the stability of the electron orbits in a hydrogen atom?

**3.61.** Two objects are moving at the same speed. Which (if any) of the following statements about them are true?

a. The de Broglie wavelength of the heavier object is longer than that of the lighter one.

b. If one object has twice as much mass as the other, its wavelength is one-half the wavelength of the other.

c. Doubling the speed of one of the objects will have the same effect on its wavelength as doubling its mass.

3.62. Which (if any) of the following statements about the frequency of a particle are true?

a. Heavy, fast-moving objects have lower frequencies than those of lighter, faster-moving objects.

b. Only very light particles can have high frequencies.

c. Doubling the mass of an object and halving its speed result in no change in its frequency.

### Problems

**3.63.** Calculate the wavelengths of the following objects:

a. A muon (a subatomic particle with a mass of $1.884 \times 10^{-28}$ kg) traveling at 325 m/s

b. Electrons ($m_e = 9.10938 \times 10^{-31}$ kg) moving at $4.50 \times 10^6$ m/s in an electron microscope

c. An 82-kg sprinter running at 9.9 m/s

d. Earth (mass $= 6.0 \times 10^{24}$ kg) moving through space at $3.0 \times 10^4$ m/s

3.64. How rapidly would each of the following particles be moving if they all had the same wavelength as a photon of red light ($\lambda = 750$ nm)?

a. An electron of mass $9.10938 \times 10^{-28}$ g

b. A proton of mass $1.67262 \times 10^{-24}$ g

c. A neutron of mass $1.67493 \times 10^{-24}$ g

d. An $\alpha$ particle of mass $6.64 \times 10^{-24}$ g

---

**3.65.** **Particles in a Cyclotron** The first cyclotron was built in 1930 at the University of California, Berkeley, and was used to accelerate molecular ions of hydrogen, $H_2^+$, to a velocity of $4 \times 10^6$ m/s. (Modern cyclotrons can accelerate particles to nearly the speed of light.) If the relative uncertainty in the velocity of the $H_2^+$ ion was 3%, what was the uncertainty of its position?

3.66. **Radiation Therapy** An effective treatment for some cancerous tumors involves irradiation with "fast" neutrons. The neutrons from one treatment source have an average velocity of $3.1 \times 10^7$ m/s. If the velocities of individual neutrons are known to within 2% of that value, what is the uncertainty in the position of one of them?

## Quantum Numbers and the Sizes and Shapes of Atomic Orbitals

### Concept Review

**3.67.** How does the concept of an orbit in the Bohr model of the hydrogen atom differ from the concept of an orbital in quantum theory?

3.68. What properties of an orbital are defined by each of the three quantum numbers $n$, $\ell$, and $m_\ell$?

**3.69.** Why does the size of an *s* orbital increase as the value of quantum number $n$ increases?

3.70. Why do we need three quantum numbers to define an orbital when four are needed to describe an electron in an orbital?

### Problems

**3.71.** How many orbitals are in an atom with each of the following principal quantum numbers? (a) 1; (b) 2; (c) 3; (d) 4; (e) 5

3.72. How many orbitals are described by each combination of quantum numbers?

a. $n = 3, \ell = 2$

b. $n = 3, \ell = 1$

c. $n = 4, \ell = 2, m_\ell = 2$

---

**3.73.** What are the possible values of quantum number $\ell$ when $n = 4$?

3.74. What are the possible values of quantum number $m_\ell$ when $\ell = 2$?

**3.75.** Which subshell corresponds to each of the following sets of quantum numbers?
a. $n = 6, \ell = 0$
b. $n = 3, \ell = 2$
c. $n = 2, \ell = 1$
d. $n = 5, \ell = 4$

3.76. Which subshell corresponds to each of the following sets of quantum numbers?
a. $n = 7, \ell = 1$
b. $n = 4, \ell = 2$
c. $n = 3, \ell = 0$
d. $n = 6, \ell = 5$

---

**3.77.** How many electrons could occupy orbitals with the following quantum numbers?
a. $n = 2, \ell = 0$
b. $n = 3, \ell = 1, m_\ell = 0$
c. $n = 4, \ell = 2$
d. $n = 1, \ell = 0, m_\ell = 0$

3.78. How many electrons could occupy orbitals with the following quantum numbers?
a. $n = 3, \ell = 2$
b. $n = 5, \ell = 4$
c. $n = 3, \ell = 0$
d. $n = 4, \ell = 1, m_\ell = 1$

---

**3.79.** Which of the following combinations of quantum numbers are allowed?
a. $n = 1, \ell = 1, m_\ell = 0, m_s = +\frac{1}{2}$
b. $n = 3, \ell = 0, m_\ell = 0, m_s = -\frac{1}{2}$
c. $n = 1, \ell = 0, m_\ell = 1, m_s = -\frac{1}{2}$
d. $n = 2, \ell = 1, m_\ell = 2, m_s = +\frac{1}{2}$

3.80. Which of the following combinations of quantum numbers are allowed?
a. $n = 3, \ell = 2, m_\ell = 0, m_s = -\frac{1}{2}$
b. $n = 5, \ell = 4, m_\ell = 4, m_s = +\frac{1}{2}$
c. $n = 3, \ell = 0, m_\ell = 1, m_s = +\frac{1}{2}$
d. $n = 4, \ell = 4, m_\ell = 1, m_s = -\frac{1}{2}$

## The Periodic Table and Filling Orbitals; Electron Configurations of Ions

### Concept Review

**3.81.** What is meant when two or more orbitals are said to be degenerate?

3.82. Explain how the electron configurations of the group 2 elements are linked to their location in the periodic table.

**3.83.** How do we know from examining the periodic table that the 4*s* orbital is filled before the 3*d* orbitals?

3.84. Why do so many transition metals form ions with a 2+ charge?

### Problems

**3.85.** Identify the subshells with the following combinations of quantum numbers and arrange them in order of increasing energy in a multielectron atom:
a. $n = 3, \ell = 2$
b. $n = 7, \ell = 3$
c. $n = 3, \ell = 0$
d. $n = 4, \ell = 1$

3.86. Identify the subshells with the following combinations of quantum numbers and arrange them in order of increasing energy in an atom of gold:
a. $n = 2, \ell = 1$
b. $n = 5, \ell = 0$
c. $n = 3, \ell = 2$
d. $n = 4, \ell = 3$

---

**3.87.** What are the electron configurations of $Li^+$, Ca, $F^-$, $Mg^{2+}$, and $Al^{3+}$?

3.88. Which species in Problem 3.87 are isoelectronic with Ne?

---

**3.89.** Which of the orbital diagrams in Figure P3.89 best describes the ground-state electron configuration of Mn? Which one is the orbital diagram for $Mn^{2+}$?

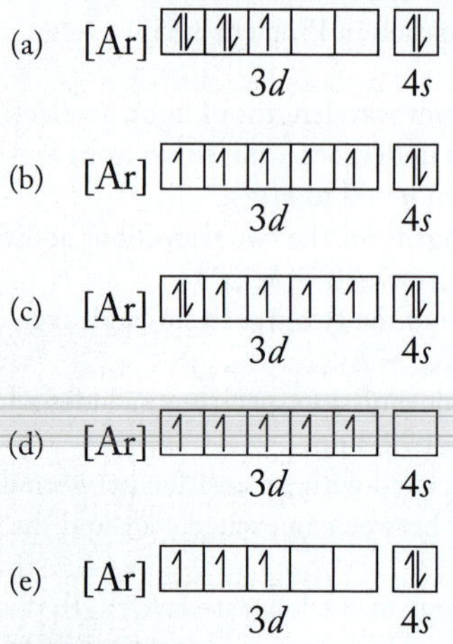

(a) [Ar] 3*d*: [↑↓][↑↓][↑][ ][ ] 4*s*: [↑↓]
(b) [Ar] 3*d*: [↑][↑][↑][↑][↑] 4*s*: [↑↓]
(c) [Ar] 3*d*: [↑↓][↑][↑][↑][↑] 4*s*: [↑↓]
(d) [Ar] 3*d*: [↑][↑][↑][↑][↑] 4*s*: [ ]
(e) [Ar] 3*d*: [↑][↑][↑][ ][ ] 4*s*: [↑↓]

**FIGURE P3.89**

3.90. Which of the orbital diagrams in Figure P3.90 best describe the ground-state electron configurations of Pb, $Pb^{2+}$, and $Pb^{4+}$?

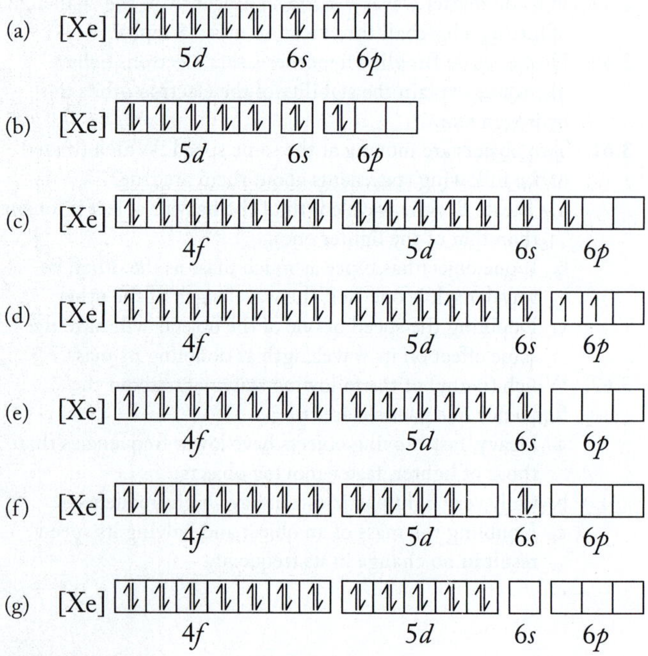

(a) [Xe] 5*d*: [↑↓][↑↓][↑↓][↑↓][↑↓] 6*s*: [↑↓] 6*p*: [↑][↑][ ]
(b) [Xe] 5*d*: [↑↓][↑↓][↑↓][↑↓][↑↓] 6*s*: [↑↓] 6*p*: [↑↓][ ][ ]
(c) [Xe] 4*f*: [↑↓][↑↓][↑↓][↑↓][↑↓][↑↓][↑↓] 5*d*: [↑↓][↑↓][↑↓][↑↓][↑↓] 6*s*: [↑↓] 6*p*: [↑↓][ ][ ]
(d) [Xe] 4*f*: [↑↓][↑↓][↑↓][↑↓][↑↓][↑↓][↑↓] 5*d*: [↑↓][↑↓][↑↓][↑↓][↑↓] 6*s*: [↑↓] 6*p*: [↑][↑][ ]
(e) [Xe] 4*f*: [↑↓][↑↓][↑↓][↑↓][↑↓][↑↓][↑↓] 5*d*: [↑↓][↑↓][↑↓][↑↓][↑↓] 6*s*: [↑↓] 6*p*: [ ][ ][ ]
(f) [Xe] 4*f*: [↑↓][↑↓][↑↓][↑↓][↑↓][↑↓][↑↓] 5*d*: [↑↓][↑↓][↑↓][↑↓][ ] 6*s*: [↑↓] 6*p*: [ ][ ][ ]
(g) [Xe] 4*f*: [↑↓][↑↓][↑↓][↑↓][↑↓][↑↓][↑↓] 5*d*: [↑↓][↑↓][↑↓][↑↓][↑↓] 6*s*: [ ] 6*p*: [ ][ ][ ]

**FIGURE P3.90**

**3.91.** What are the condensed electron configurations of the following ground-state atoms and ions? How many unpaired electrons are in each? (a) K; (b) $K^+$; (c) $S^{2-}$; (d) N; (e) Ba; (f) $Ti^{4+}$; (g) Al

3.92. What are the condensed electron configurations of the following ground-state atoms and ions? How many unpaired electrons are in each? (a) O; (b) $P^{3-}$; (c) $Na^+$; (d) Sc; (e) $Ag^+$; (f) $Cd^{2+}$; (g) $Zr^{2+}$

---

**3.93.** Identify the element and the number of unpaired electrons in each of the following condensed electron configurations of ground-state gas-phase atoms:

a. $[Ar]3d^24s^2$
b. $[Ar]3d^54s^1$
c. $[Ar]3d^{10}4s^1$
*d. $[Kr]4d^{10}$

3.94. Identify the ion and the number of unpaired electrons in each of the following condensed electron configurations of ground-state monatomic cations:

a. [Ne]
b. $[Kr]4d^{10}5s^2$
c. $[Ar]3d^{10}$
d. $1s^22s^22p^6$

---

**3.95.** Which monatomic ion has a charge of 1− and the condensed electron configuration $[Ne]3s^23p^6$? How many unpaired electrons are in the ground state of that ion?

3.96. Which monatomic ion has a charge of 1+ and the electron configuration $[Kr]4d^{10}5s^2$? How many unpaired electrons are in the ground state of that ion?

---

**3.97.** Identify the element with the lowest atomic number that has two unpaired electrons in its ground state.

3.98. Identify the first-row transition metal that has no unpaired electrons in its ground state.

---

**3.99.** Which of the following condensed electron configurations represent an excited state?

a. $[He]2s^12p^5$
b. $[Kr]4d^{10}5s^25p^1$
c. $[Ar]3d^{10}4s^24p^5$
d. $[Ne]3s^23p^24s^1$

3.100. Which of the following condensed electron configurations represent an excited state? Could any represent *ground-state* electron configurations of 2+ ions?

a. $[Ar]4s^24p^1$
b. $[Ar]3d^{10}$
c. $[Kr]4d^{10}5s^1$
d. $[Ar]3s^23p^63d^1$

---

**3.101.** In which subshell are the highest-energy electrons in a ground-state atom of the isotope $^{131}I$? Are the electron configurations of $^{131}I$ and $^{127}I$ the same?

*3.102. No known element contains electrons in an $\ell = 4$ subshell in the ground state. If such an element were synthesized, what minimum atomic number would it have to have?

---

***3.103.** Appendix 3 lists the ground state of gas-phase palladium atoms as $[Kr]4d^{10}$. Which of the following is the *best* explanation for that observation?

a. The aufbau principle applies only to transition metal atoms in period 4.
b. Gas-phase palladium atoms are always in an excited state.
c. The energy of the 5*s* orbital must be higher than that of the 4*d* orbital for Pd.
d. Palladium is always present as a $Pd^{2+}$ cation because of the photoelectric effect.

*3.104. The ground-state electron configuration in Appendix 3 for gas-phase silver atoms is $[Kr]4d^{10}5s^1$. Which of the following statements best explains the electron configuration of silver *and* is consistent with the answer to Problem 3.103?

a. *nd* orbitals always fill before $(n + 1)s$ orbitals.
b. The energy of the 4*d* orbital is still lower in energy than the 5*s* orbital in Ag atoms.
c. The repulsions between electrons in the 4*d* orbital are greater than those in the 5*s* orbital.
d. Gas-phase silver atoms are always in an excited state.

---

***3.105.** Removing a 1*s* electron from beryllium requires X-rays with wavelength 110.7 nm. Table 3.2 lists the first ionization energy of Be as 900 kJ/mol.

a. Why don't those data represent the same process?
b. Does the energy to remove the 1*s* electron correspond to the second ionization energy of Be (1758 kJ/mol)?

*3.106. The first ionization energies of group 1 elements decrease as we go down the group. The energies to move an electron from $n = 1$ to $n = \infty$, however, increase from Li to Cs. Why are those two trends opposite?

## The Sizes of Atoms and Ions

### Concept Review

**3.107.** Sodium atoms are much larger than chlorine atoms, but sodium ions are much smaller than chloride ions. Why?

3.108. Why does atomic size tend to decrease with increasing atomic number across a row of the periodic table?

### Problems

**3.109.** Using only the periodic table as a guide, arrange each set of particles by size, largest to smallest:

a. Al, P, Cl, Ar
b. C, Si, Ge, Sn
c. $Li^+$, Li, Na, K
d. F, Ne, Cl, $Cl^-$

3.110. Using only the periodic table as a guide, arrange each set of particles by size, largest to smallest:

a. Li, B, N, Ne
b. Mg, K, Ca, Sr
c. $Rb^+$, $Sr^{2+}$, Cs, Fr
d. $S^{2-}$, $Cl^-$, Ar, $K^+$

## Ionization Energies and Photoelectron Spectroscopy

### Concept Review

**3.111.** How do ionization energies change with increasing atomic number (a) down a group of elements in the periodic table and (b) from left to right across a row?

3.112. Explain the differences in ionization energy between (a) He and Li; (b) Li and Be; (c) Be and B; (d) N and O.

**3.113.** How does the wavelength of light required to ionize a gas-phase atom change with increasing atomic number down a group in the periodic table?

3.114. Why is the first ionization energy of Al less than that of Mg *and* less than that of Si?

### Problems

**3.115.** Without referring to Figure 3.37, arrange the following groups of elements in order of increasing first ionization energy.

a. F, Cl, Br, I
b. Li, Be, Na, Mg
c. N, O, F, Ne

3.116. Without referring to Figure 3.37, arrange the following groups of elements in order of increasing first ionization energy.

a. Mg, Ca, Sr, Ba
b. He, Ne, Ar, Kr
c. P, S, Cl, Ar

---

**3.117.** Assign each peak in the photoelectron spectrum shown in Figure P3.117 to the orbital(s) whose electrons have that binding energy. What element has that photoelectron spectrum?

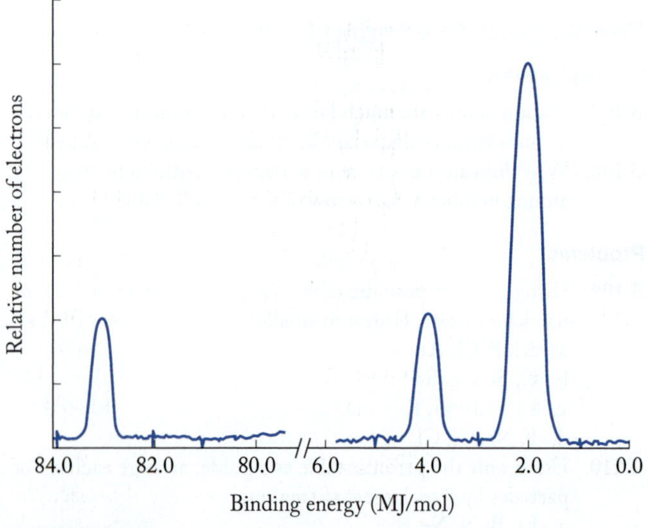


**FIGURE P3.117**

3.118. Assign each peak in the photoelectron spectrum shown in Figure P3.118 to the orbital(s) whose electrons have that binding energy. What element has that photoelectron spectrum?

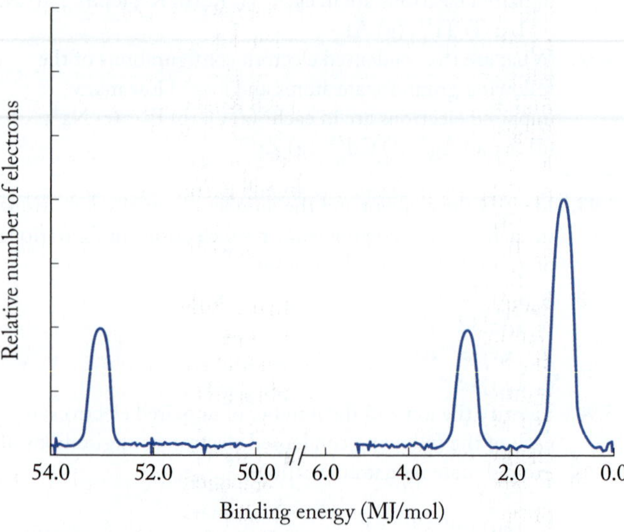


**FIGURE P3.118**

## Electron Affinities

### Concept Review

**3.119.** An electron affinity (EA) value that is negative indicates that the free atoms of an element are higher in energy than the 1− anions they form by acquiring electrons. Does that mean that all the elements with negative EA values exist in nature as anions? Give some examples to support your answer.

3.120. The electron affinities of the group 17 elements are all negative values, but the EA values of the group 18 noble gases are all positive. Explain that difference.

**3.121.** The electron affinities of the group 17 elements increase with increasing atomic number. Suggest a reason for that trend.

3.122. Ionization energies generally increase with increasing atomic number across the second row of the periodic table, but electron affinities generally decrease. Explain those opposing trends.

## Additional Problems

**3.123.** **Searching for Extraterrestrial Life** Project Ozma was an early attempt to find advanced civilizations elsewhere in our universe. Scientists involved with the project decided to search space for signals being broadcast at 1240 MHz, a frequency naturally emitted by H atoms. The scientists reasoned that other intelligent life might come to the same conclusion about a "universal" communication frequency.

a. Calculate the wavelength and energy of that light.
b. Does that frequency correspond to Balmer, Lyman, or Paschen lines in the emission spectrum of H?

3.124. **X-ray Absorption Spectroscopy** Electrons in multielectron atoms absorb X-rays at characteristic energies, leading to ionization. The characteristic energies for each element allow scientists to identify the element. For magnesium, X-rays with $\lambda = 952$ pm are required to selectively eject an electron from the $n = 1$ energy level, and X-rays with $\lambda = 197$ nm remove an electron from the $n = 2$ energy level.
a. Calculate the frequency and energy of those X-rays.
b. Why are the wavelengths and energies different for the two electrons?
c. In terms of Bohr's values of $n$, what transitions do those represent?

**3.125.** **Satellite Radio and Weather Radio** Sirius XM radio transmits at 2.3 GHz by using a network of satellites. The National Oceanic and Atmospheric Administration transmits warnings for tornadoes over weather radios at 162 MHz. Which of those signals uses photons of (a) higher energy? (b) shorter wavelength? (c) higher frequency?

*3.126. **Colors of Fireworks** Barium compounds are a source of the green colors in many fireworks displays.
a. What is the ground-state electron configuration for Ba?
b. The lowest-energy excited state of Ba has the electron configuration [Xe]$5d^16s^1$. What are the possible quantum numbers $n$, $\ell$, and $m_\ell$ of a $5d$ electron?
c. The excited state in part b is $1.79 \times 10^{-19}$ J above the ground state. Could emission from that excited state account for the green color in fireworks?
d. Another Ba excited state has the electron configuration [Xe]$6s^16p^1$. Its energy is $3.59 \times 10^{-19}$ J above the ground state. Could transitions from that state to the ground state account for the green color in fireworks?

***3.127.** When an atom absorbs an X-ray of sufficient energy, one of its $2s$ electrons may be ejected, creating a hole that can be spontaneously filled when an electron in a higher-energy orbital—$2p$, for example—falls into it. A photon of electromagnetic radiation with an energy that matches the energy lost in the $2p \rightarrow 2s$ transition is emitted. Predict how the wavelengths of $2p \rightarrow 2s$ photons would differ between (a) different elements in the fourth row of the periodic table and (b) different elements in the same column (for example, between the noble gases from Ne to Rn).

*3.128. Two helium ions ($He^+$) in the $n = 3$ excited state emit photons of radiation as they return to the ground state. One ion does so in a single transition from $n = 3$ to $n = 1$. The other does so in two steps: $n = 3$ to $n = 2$ and then $n = 2$ to $n = 1$. Which of the following statements about the two pathways is true?
a. The sum of the energies lost in the two-step process is the same as the energy lost in the single transition from $n = 3$ to $n = 1$.
b. The sum of the wavelengths of the two photons emitted in the two-step process is equal to the wavelength of the single photon emitted in the transition from $n = 3$ to $n = 1$.
c. The sum of the frequencies of the two photons emitted in the two-step process is equal to the frequency of the single photon emitted in the transition from $n = 3$ to $n = 1$.
d. The wavelength of the photon emitted by the $He^+$ ion in the $n = 3$ to $n = 1$ transition is shorter than the wavelength of a photon emitted by a H atom in an $n = 3$ to $n = 1$ transition.

**3.129.** Use your knowledge of electron configurations to explain the following observations:
a. Silver tends to form ions with a charge of 1+, but the elements to the left and right of silver in the periodic table tend to form ions with 2+ charges.
b. The heavier group 13 elements (Ga, In, and Tl) tend to form ions with charges of 1+ or 3+, but not 2+.
c. The heavier elements of group 14 (Sn and Pb) and group 4 (Ti, Zr, and Hf) tend to form ions with charges of 2+ or 4+.

3.130. **Chemistry of Photo-Gray Glasses** "Photo-gray" lenses for eyeglasses darken in bright sunshine because the lenses contain tiny, transparent AgCl crystals. Exposure to light removes electrons from $Cl^-$ ions, forming a chlorine atom in an excited state (indicated here by the asterisk):

$$Cl^- \xrightarrow{h\nu} Cl^* + e^-$$

The electrons are transferred to $Ag^-$ ions, forming silver metal:

$$Ag^+ + e^- \rightarrow Ag$$

Silver metal is reflective, producing the photo-gray color. How might substituting AgBr for AgCl affect the light sensitivity of photo-gray lenses? In answering that question, consider whether more energy or less is needed to remove an electron from a $Br^-$ ion than from a $Cl^-$ ion.

**3.131.** Tin (in group 14) forms both $Sn^{2+}$ and $Sn^{4+}$ ions, but magnesium (in group 2) forms only $Mg^{2+}$ ions.
a. Write condensed ground-state electron configurations for the ions $Sn^{2+}$, $Sn^{4+}$, and $Mg^{2+}$.
b. Which neutral atoms have ground-state electron configurations identical to $Sn^{2+}$ and $Mg^{2+}$?
c. Which 2+ ion is isoelectronic with $Sn^{4+}$?

*3.132. Effective nuclear charge ($Z_{eff}$) is related to atomic number ($Z$) by a factor called the shielding parameter ($\sigma$) according to the equation $Z_{eff} = Z - \sigma$.
a. Calculate $Z_{eff}$ for the outermost $s$ electrons of Ne and Ar, given that $\sigma = 4.24$ for Ne and $\sigma = 11.24$ for Ar.
b. Explain why the shielding parameter is much greater for Ar than for Ne.

**3.133.** **Millikan's Experiment** In his oil-drop experiment, Millikan used X-rays to ionize $N_2$ gas. Electrons lost by $N_2$ were absorbed by oil droplets. If the wavelength of the X-ray was 154 pm, could he instead have filled his device with argon gas, which has an ionization energy of 1521 kJ/mol?

*3.134. How can an electron get from one lobe of a $p$ orbital to the other without going through the point of zero electron density between them?

# 4

# Chemical Bonding

## Understanding Climate Change

**LIVESTOCK AND CLIMATE CHANGE** Human demand for meat and milk is driving an increase in the number of cows on Earth—which, in turn, increases emissions of methane, a powerful greenhouse gas.

PARTICULATE **REVIEW**

### *Molecules and Valence Electrons*

In Chapter 4, we learn how the electronic structure of atoms is related to how atoms bond, forming molecules. Consider these models of three molecules abundant in Earth's atmosphere.

- What are the molecular formulas of these three atmospheric gases?
- What is the total number of electrons in each molecule?
- How many of these electrons come from the valence shells of the atoms in each molecule?

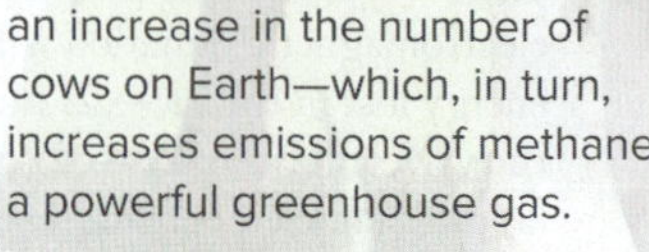


(Review Section 3.8 if you need help.)

*(Answers to Particulate Review questions are in the back of the book.)*

## *Bonds and Bonding Capacity*

The figure shows molecules of two gases that contribute to climate change. As you read Chapter 4, look for information that will help you answer these questions:

- One of the molecules contains only single bonds, whereas the other contains only double bonds. Which bonds are present in which molecules?
- How many bonds does the carbon atom in each molecule form?
- What structural information do space-filling models provide that Lewis structures do not, and what information do Lewis structures provide that space-filling models do not?

$CH_4$ (a) $CO_2$ (b)

## Learning Outcomes

**LO1** Describe how covalent, ionic, and metallic bonds are alike and how they differ

**LO2** Calculate and compare the relative strengths of ion–ion attractions
**Sample Exercise 4.1**

**LO3** Predict the polarity of covalent bonds on the basis of differences in the electronegativity between the bonded atoms
**Sample Exercise 4.2**

**LO4** Name molecular and ionic compounds and write their formulas
**Sample Exercises 4.3–4.8**

**LO5** Draw Lewis structures of ionic compounds, molecular compounds, and polyatomic ions
**Sample Exercises 4.9–4.13**

**LO6** Draw resonance structures and use formal charges to evaluate their relative importance
**Sample Exercises 4.14, 4.16–4.18**

**LO7** Describe how bond order, bond energy, and bond length are related
**Sample Exercise 4.15**

**LO8** Explain how molecules of some atmospheric gases absorb infrared radiation and contribute to the greenhouse effect

## 4.1 Chemical Bonds and Greenhouse Gases

In Chapter 3 we learned how scientific advances in the early decades of the 20th century produced a radically different, probabilistic view of the structure of atoms and their interactions with electromagnetic radiation. Here we continue our exploration of matter and energy, but we expand our focus to include the structure and behavior of the particles that make up ionic and molecular compounds and metallic solids, and we explore how molecules interact with electromagnetic radiation.

Let's consider one example of how radiant energy—here, infrared radiation—interacts with molecules in Earth's atmosphere, particularly molecules of carbon dioxide ($CO_2$). Scientists can explain the connection between the rate at which the average temperature of Earth's surface is increasing (nearly one Celsius degree in the last half century) and the increasing atmospheric concentrations of $CO_2$ and other greenhouse gases, such as methane, $CH_4$. Those gases trap Earth's heat much like panes of glass trap heat in a greenhouse. Greenhouse gases are transparent to visible radiation, allowing sunlight to pass through the atmosphere and warm Earth's surface, but they trap heat that would otherwise flow from Earth's surface back into space. Without greenhouse gases, Earth would be too cold to be habitable.

The amount of some greenhouse gases in the atmosphere may surprise you. Oil drilling and fracking operations, livestock animals, landfills, and wetlands introduce more than 0.5 teragram of $CH_4$ (1 teragram = 1 Tg = 1 million metric tons = $10^{12}$ g) into the atmosphere *every* day. Increases in atmospheric concentrations of other greenhouse gases have been linked to human activity, particularly to increasing rates of fossil fuel combustion and the destruction of forests that would otherwise consume $CO_2$ during photosynthesis. During the last 50 years, atmospheric concentrations of $CO_2$ have increased from about 315 parts per million (ppm) to more than 400 ppm now. The atmosphere hasn't contained so much $CO_2$ in more than half a million years, long before our species evolved. In this chapter we do not address the social consequences of climate change, but we do explore why $CO_2$ in the atmosphere traps heat, whereas the two most abundant atmospheric gases, $N_2$ and $O_2$, do not.

Carbon dioxide is good at trapping heat because it absorbs infrared radiation. To understand why it does, we need to understand the properties of chemical bonds and the role of valence electrons in forming those bonds. We also need to understand why some pairs of atoms share bonding electrons equally, whereas others do not, and how unequal sharing allows some compounds, such as $CO_2$, to absorb infrared radiation and contribute to climate change.

We begin this chapter by exploring how atoms combine, forming chemical bonds that link them in molecular and ionic compounds. Fewer than 100 stable (nonradioactive) elements make up all the matter in the universe, yet those elements make up more than 150 million compounds (with the number growing daily as chemists synthesize new compounds). How can such a relatively small number of elements combine to make so many compounds? It turns out that most compounds are more stable than the elements from which they are made. Stable compounds exist because chemical bonds form when atoms or ions are attracted to each other.

We learned in Chapter 3 that an atom's outermost, or *valence*, electrons are involved in forming chemical bonds. Actually, chemists use the term **valence** all by itself to describe the capacity of the atoms of a particular element to form chemical bonds. Later in this chapter we will see how the atoms of some elements may form different numbers of chemical bonds and thus exhibit more than one *valency*. Before we explore how and why some elements display multiple valencies, let's first examine the three major categories of chemical bonds—namely, ionic bonds, covalent bonds, and metallic bonds.

**valence** the capacity of the atoms of an element to form chemical bonds.

**electrostatic potential energy ($E_{el}$)** the energy a charged particle has because of its position relative to another charged particle; $E_{el}$ is directly proportional to the product of the charges of the particles and inversely proportional to the distance between them; also called *coulombic attraction*.

**CHEMTOUR**

Bonding

## Ionic Bonds

Recall from Chapter 2 that binary (two-element) ionic compounds consist of cations formed from metallic elements and anions formed from nonmetals. Those ionic compounds are held together by electrostatic attractions between ions of opposite charge. The strength of those attractions is a form of potential energy called **electrostatic potential energy ($E_{el}$)**. It is also called *coulombic attraction*.

The value of $E_{el}$ between a pair of ions is directly proportional to the product of their charges ($Q_1$ and $Q_2$) and inversely proportional to the distance ($d$) between their nuclei:

$$E_{el} \propto \frac{Q_1 \times Q_2}{d}$$

We can turn that expression into an equation by replacing the "proportional to" symbol ($\propto$) with a constant of proportionality that gives us energy in joules when we express distance in nanometers and when the values of $Q$ are the whole-number charges on the ions, such as 2+ for $Ca^{2+}$ and 1− for $Cl^-$:

$$E_{el} = 2.31 \times 10^{-19}\ \text{J} \cdot \text{nm}\left(\frac{Q_1 \times Q_2}{d}\right) \tag{4.1}$$

Two ions with the same charge—either two positive ions or two negative ions—repel each other. According to Equation 4.1, that repulsion generates a positive $E_{el}$ because multiplying together either two positive or two negative values yields a positive value. Conversely, the attraction between a positive ion and a negative ion produces a negative $E_{el}$. Moreover, the greater the attraction (that is, the larger the charges or the shorter the distance between them), the more negative the $E_{el}$ value.

**CONNECTION** In Chapter 2 we learned that ions (charged particles) are further classified as cations (positively charged) and anions (negatively charged).

**CONNECTION** The coulomb is the SI unit of electrical charge. We used that unit when we explored the charge on the electron in Section 2.1.

**CONNECTION** In Chapter 1 we defined potential energy as the energy of position or composition. The electrostatic potential energy between two ions depends on both the distance between them (their position) and their electrical charges (their composition).

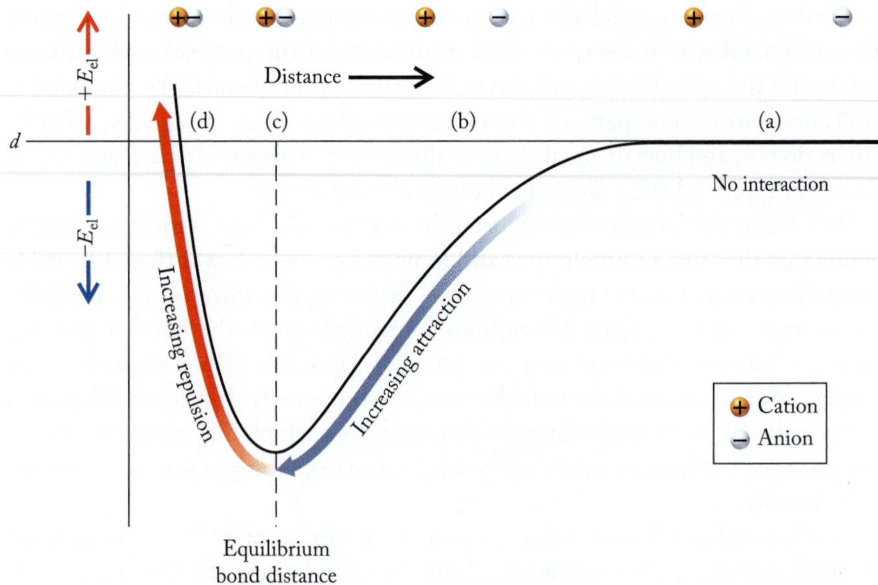


**FIGURE 4.1** Changes in electrostatic potential energy ($E_{el}$) as an ionic bond forms. A pair of positive and negative ions far apart from each other do not interact ($E_{el} = 0$). As the ions move closer together, the electrostatic potential energy between them becomes more negative, reaching a minimum that corresponds to the energy of the ionic bond between the two. If the ions are forced even closer together, $E_{el}$ increases due to increasing repulsion.

The change in $E_{el}$ that two oppositely charged particles experience as they approach each other is illustrated in **Figure 4.1**. When the ions are far apart (the right end of the curve), no attraction occurs between them, and $E_{el} = 0$. As the ions approach each other, $E_{el}$ decreases as electrostatic attractions increase, reaching an energy minimum that corresponds to maximum stability and the formation of an **ionic bond** between the ions. The absolute value of $E_{el}$ at the energy minimum represents the amount of energy required to pull the ions apart and break the ionic bond. If the ions are pushed even closer together (to the left of the energy minimum position in Figure 4.1), $E_{el}$ rises and the ion pair becomes less stable as the decreasing distance between their positively charged nuclei creates a growing electrostatic repulsion between the ions.

When the ions involved in Figure 4.1 are $Na^+$ and $Cl^-$ ions, the distance between their nuclei at the energy minimum is 0.282 nm. That distance of maximum stability matches the distance between the nuclei of adjacent $Na^+$ and $Cl^-$ ions in a crystal of table salt, sodium chloride.

**CONNECTION** We discussed how to predict the charges of monatomic ions from the positions of their parent elements in the periodic table in Chapter 2.

We can use Equation 4.1 to calculate the strength of the electrostatic attraction between a pair of oppositely charged ions if we know their charges and the distance between their nuclei. The latter value can be calculated for pairs of ions of the main group elements by summing their ionic radii (see Figure 3.36), as illustrated in Sample Exercise 4.1.

### SAMPLE EXERCISE 4.1 Calculating the Electrostatic Potential Energy of an Ionic Bond — LO2

What is the electrostatic potential energy ($E_{el}$) of the ionic bond between a potassium ion and a chloride ion?

**Collect and Organize** We are asked to calculate the $E_{el}$ of an ionic bond. Equation 4.1 relates $E_{el}$ to the charges on a pair of ions and the distance between them. An ionic bond forms when $E_{el}$ reaches a minimum, and that happens when the distance between the nuclei of the ions is close to the sum of their ionic radii. Ionic radii of main group elements are given in Figure 3.36.

**Analyze** Potassium is in group 1, so it forms 1+ ions. Chlorine is in group 17, so it forms 1− ions. According to Figure 3.36, the ionic radii of $K^+$ and $Cl^-$ ions are 138 and 181 pm, respectively.

**Solve** The distance ($d$) between the nuclei of a pair of $K^+$ and $Cl^-$ ions is

$$d = (K^+ \text{ ionic radius}) + (Cl^- \text{ ionic radius}) = (138 \text{ pm} + 181 \text{ pm}) = 319 \text{ pm}$$

The distance unit in Equation 4.1 is nanometers, so we need to convert that value for $d$ to nanometers as part of the calculation of $E_{el}$:

$$E_{el} = 2.31 \times 10^{-19} \text{ J} \cdot \text{nm}\left(\frac{Q_1 \times Q_2}{d}\right)$$

$$= 2.31 \times 10^{-19} \text{ J} \cdot \cancel{\text{nm}}\left(\frac{(1+) \times (1-)}{319 \cancel{\text{pm}} \times (1 \cancel{\text{nm}}/1000 \cancel{\text{pm}})}\right)$$

$$= -7.24 \times 10^{-19} \text{ J}$$

**Think About It** $E_{el}$ has a negative value because $K^+$ and $Cl^-$ are at a lower energy when they are connected by an ionic bond than when they are apart. To break the bond and separate the ions would take $7.24 \times 10^{-19}$ J of energy.

**Practice Exercise** What is the electrostatic potential energy ($E_{el}$) of the ionic bond between $Ca^{2+}$ and $S^{2-}$? Before doing the calculation, predict whether it will be less than (more negative) or greater than the $E_{el}$ value for KCl.

*(Answers to Practice Exercises are in the back of the book.)*

## CONCEPT TEST

Without doing any calculations, rank the pairs of ions in each group in order of increasing coulombic attraction between their ions: (a) NaCl, KCl, RbCl; (b) KF, MgO, CaO.

*(Answers to Concept Tests are in the back of the book.)*

In Sample Exercise 4.1 we calculated the electrostatic potential energy in the ionic bond between one $K^+$ ion and one $Cl^-$ ion. However, the energy associated with bonding is more conventionally expressed per *mole* of ions. If we multiply the result from Sample Exercise 4.1 by the Avogadro constant, we obtain the following:

$$\frac{6.022 \times 10^{23} \cancel{\text{ion pairs}}}{1 \text{ mol}} \times \frac{-7.24 \times 10^{-19} \text{ J}}{\cancel{\text{ion pairs}}} = -4.36 \times 10^5 \text{ J/mol, or } -436 \text{ kJ/mol}$$

That answer is close to the true value, but it's not exactly right. The actual change in energy that occurs when one mole of $K^+$ ions in the gas phase combines with one mole of gaseous $Cl^-$ ions to form one mole of solid KCl is −720 kJ/mol, not −436 kJ/mol. What accounts for the difference? The answer is contained in the name we give the real value: the **lattice energy ($U$)** of KCl. Lattice energy is the energy released when free, gas-phase ions combine to form one mole of a crystalline solid. A crystalline solid consists of an ordered three-dimensional array of particles (atoms, ions, or molecules) called a **crystal lattice**. The left column in **Table 4.1** shows a photo of solid KCl and a drawing of its crystal lattice. Note the tightly packed array of $K^+$ ions (purple spheres) and $Cl^-$ ions (green spheres).

In that array, every $K^+$ ion touches six $Cl^-$ ions, not just one, and every $Cl^-$ ion touches six $K^+$ ions. All the additional interactions between ions of opposite charge make the ionic bonds stronger, which makes $E_{el}$ in solid KCl more negative. The tight packing also produces some repulsions between ions of the same charge that are close to one another but do not touch. The net effect of the interactions is an overall decrease in $E_{el}$, corresponding to stronger ionic bonding than our calculation predicted on the basis of one pair of ions.

**ionic bond** a bond resulting from the electrostatic attraction of a cation to an anion.

**lattice energy ($U$)** the energy released when one mole of an ionic compound forms from its free ions in the gas phase.

**crystal lattice** an ordered three-dimensional array of particles.

**TABLE 4.1** Types of Chemical Bonds

| | Ionic | Covalent | Metallic |
|---|---|---|---|
| Elements Involved | Metals and nonmetals | Nonmetals and metalloids | Metals |
| Electron Distribution | Transferred | Shared | Delocalized |
| Particulate View | $Cl^-$ $K^+$ | Br Br | Cu |
| Macroscopic View | | | |

**TABLE 4.2** Lattice Energies ($U$) of Some Common Ionic Compounds

| Compound | $U$ (kJ/mol) |
|---|---|
| LiF | −1049 |
| LiCl | −864 |
| NaF | −930 |
| NaCl | −786 |
| NaBr | −754 |
| KCl | −720 |
| KBr | −691 |
| MgO | −3791 |
| $MgCl_2$ | −2540 |

**Table 4.2** lists lattice energy values for several common ionic compounds. Those values show trends consistent with the strengths of the ion-pair interactions predicted by Equation 4.1. For example, the only major difference between the crystal lattices of NaF, NaCl, and NaBr is the size of the anion ($F^- < Cl^- < Br^-$). Differences in ionic radii correspond to differences in the distance, $d$, between the nuclei of the pairs of ions. A large value of $d$ in the denominator of Equation 4.1 means that the attractions between $Na^+$ ions and $Br^-$ ions are weaker than the attractions between $Na^+$ ions and $F^-$ ions. Thus, NaF has the most negative lattice energy of the three compounds, and NaBr has the least negative $U$ value.

The large difference between the lattice energies of NaF and MgO cannot be explained solely by differences in $d$ because the sums of the radii of the ions involved are nearly the same. However, the crystal lattice of MgO consists of ions with charges of 2+ and 2−, whereas NaF is made of 1+ and 1− ions. Inserting those charges into the numerator of Equation 4.1 gives us $Q_1Q_2$ values of −4 for MgO and −1 for NaF. As a result, the lattice energy of MgO is about four times more negative than that of NaF.

## Covalent Bonds

**Figure 4.2** shows how the potential energy of two hydrogen atoms changes as they approach each other. The two are far enough apart on the right end of the curve that they do not interact with each other and have zero potential energy of interaction. As the two atoms approach each other, the proton in the nucleus of one is attracted to the electron of the other and vice versa. That mutual attraction of both electrons for both nuclei is accompanied by a drop in potential energy that reaches a minimum at $d = 74$ pm. At that distance the strength of the attractions between one hydrogen nucleus (which is a proton) and each of the two electrons equals the strength of the attractions between the second hydrogen nucleus and the electrons. We say that the atoms "share" the two electrons equally; that is, they have formed a single **covalent bond**. They are no longer two separate H atoms but rather a more stable single molecule of $H_2$. If the atoms come any closer together, their potential energy rises as the repulsion between their two positive nuclei increases.

The distance of minimum energy at 74 pm represents the **bond length** of the H—H bond. Covalent bond lengths have values close to the sum of the atomic radii of the atoms involved in making the bonds. The minimum potential energy of −436 kJ/mol in Figure 4.2 is the energy released when two moles of H atoms form one mole of $H_2$ molecules. That value also is the amount of energy required to break one mole of $H_2$ molecules into two moles of free H atoms. It is called the **bond energy** or *bond strength* of a H—H bond. The significance of bond lengths and bond energies is discussed further in Section 4.6.

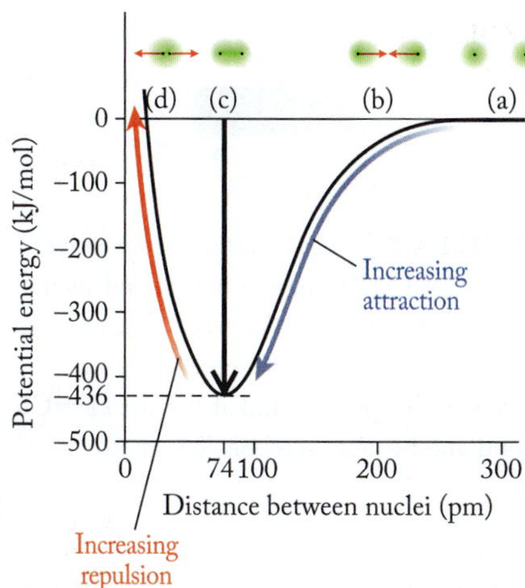


**FIGURE 4.2** Changes in potential energy when an H—H covalent bond forms. (a) Two hydrogen atoms far apart do not interact, but (b) as the atoms move closer to each other, the positive nuclei and negative electrons on nearby atoms are mutually attracted, producing decreased electrostatic potential energy. At (c) an internuclear distance of 74 pm, potential energy reaches a minimum value, corresponding to a covalent bond between the atoms. If the atoms are even closer together (d), repulsion between their nuclei causes potential energy to increase.

## Metallic Bonds

A third type of chemical bond holds the atoms together in metallic solids. As with atoms in molecules, the positive nucleus of each atom in a metallic solid is attracted to the electrons of the atoms that surround it. Those attractions, coupled with overlapping valence-shell orbitals, result in the formation of **metallic bonds**. Those bonds are different from covalent bonds in that they are not pairs of electrons shared by pairs of atoms. Instead, the shared electrons in metallic bonds form a "sea" of mobile electrons that flows freely among all the atoms in a piece of metal.

To understand the mobility of metallic bonds, let's focus on the bonding in copper. The electron configuration of Cu ($Z = 29$) is $[Ar]3d^{10}4s^1$. If two Cu atoms are close enough to form a bond, we might expect them to share their 4*s* electrons and form a molecule of $Cu_2$—in much the way two hydrogen atoms form a covalent bond and a molecule of $H_2$ by sharing their 1*s* electrons. However, no experimental evidence indicates the existence of $Cu_2$ molecules in copper metal. Instead, a crystal lattice of copper atoms is present, as shown in the right column of Table 4.1. Inside that lattice, each Cu atom is surrounded by 12 other Cu atoms. That means that each Cu atom shares its 4*s* electron with 12 other atoms, not just one, and those 12 share their 4*s* electrons with 12 others, and so on. As a result, bonding electrons are not confined to particular pairs of Cu atoms but instead can easily move from one atom to another throughout a piece of copper. Because of that electron mobility, copper (and metals in general) are excellent conductors of electricity.

Table 4.1 summarizes some similarities and differences across the three types of chemical bonds. Keep in mind, though, that many bonds do not fall exclusively into just one category—they may instead have some covalent, ionic, and even metallic character simultaneously. In the next section, we will explore how to determine the extent to which a given set of atoms will form ionic bonds or covalent bonds.

**covalent bond** a bond created by two atoms sharing one or more pairs of electrons.

**bond length** the distance between the nuclei of two atoms joined in a bond.

**bond energy** the energy needed to break one mole of a specific covalent bond in the gas phase; also called *bond strength*.

**metallic bond** a bond consisting of the nuclei of metal atoms surrounded by a "sea" of shared electrons.

# 4.2 Electronegativity, Unequal Sharing, and Polar Bonds

Covalent bonds are pairs of shared valence electrons. However, that pair of electrons is not necessarily shared *equally* by the two atoms in the bond. For example, the pair of electrons in a molecule of HCl is closer to the chlorine end of the bond

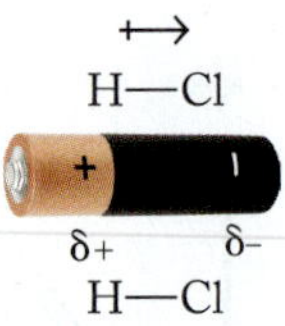


**FIGURE 4.3** Just as a battery has positive and negative terminals, a polar bond such as the one in a molecule of HCl has positive and negative ends, represented here by the arrow with a positive tail above the H—Cl bond and by delta symbols (δ+ and δ−).

than the hydrogen end. That unequal sharing makes the H—Cl bond a **polar covalent bond**.

The polarity of the bond in HCl means that the bond functions as a tiny electric **dipole**, a pair of opposite charges separated by a distance. A slightly positive pole is at the H end of the bond, with a slightly negative pole at the Cl end—analogous to the positive and negative ends of the battery shown in **Figure 4.3**. That figure also shows two representations we use to depict unequal sharing of pairs of electrons. One of them incorporates an arrow with a plus sign embedded in its tail. The arrow points toward the more negative, electron-rich end of the bond, whereas the plus sign represents the more positive, electron-poor end. Another representation makes use of the lowercase Greek delta, δ, followed by a + or − sign. The deltas represent *partial* electrical charges, as opposed to the full electrical charges that designate the complete transfer of one or more electrons when atoms become cations or anions.

**Figure 4.4** shows examples of equal and unequal sharing of pairs of electrons. In Figure 4.4(a), the Cl atoms of $Cl_2$ are connected by a **nonpolar covalent bond** because the bonding electrons are shared equally between two identical atoms. Figure 4.4(b), by contrast, depicts the polar covalent bond in HCl, and Figure 4.4(c) depicts the ionic bond between a pair of $Na^+$ and $Cl^-$ ions in NaCl. In NaCl, the valence electron of sodium has been completely transferred to the chlorine atom, creating a $Na^+$ ion and a $Cl^-$ ion and resulting in *total*, rather than *partial*, separation of electrical charge: 1+ on Na and 1− on Cl. The degree of charge separation may be represented using color, as shown in the calibration bar at the top of Figure 4.4, where yellow-green represents equal sharing (nonpolar covalent bonding) and no partial charges. The red and violet ends represent full charge separation (ionic bonding), whereas the colors between represent various degrees of partial charge separation (polar covalent bonding).

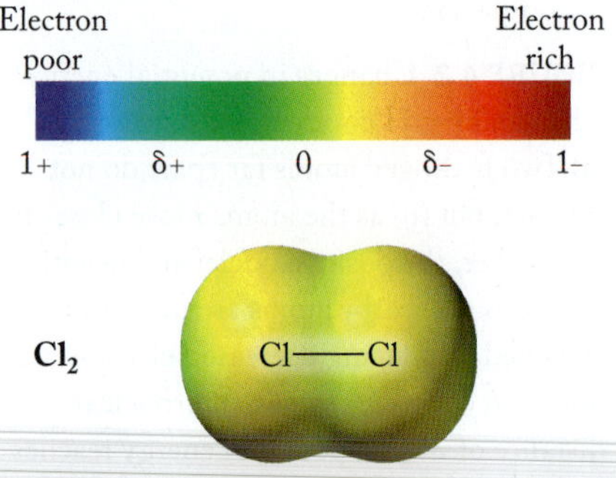


(a) Nonpolar covalent: even charge distribution

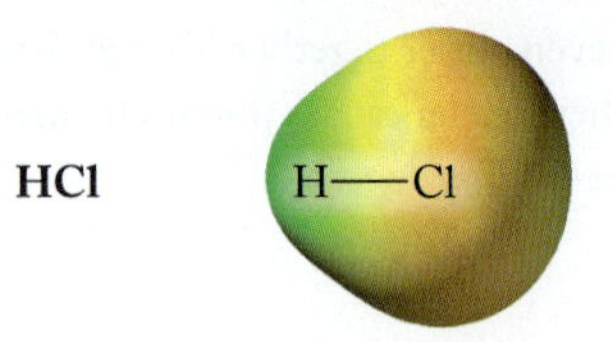


(b) Polar covalent: uneven charge distribution

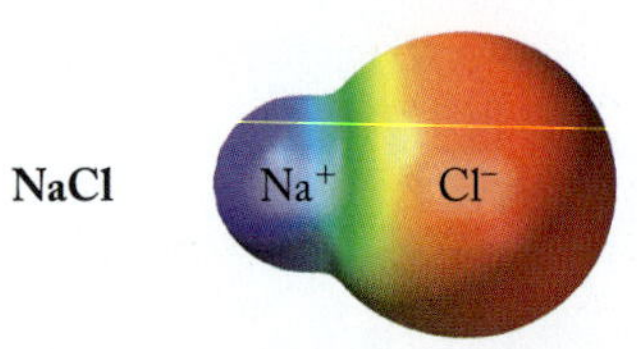


(c) Ionic: complete transfer of electron

**FIGURE 4.4** Variations in valence electron distribution are represented using colored surfaces in these molecular models. (a) In the covalent bond in $Cl_2$, the same color pattern on both ends of the bond means that the two atoms share their pair of electrons equally. (b) Unequal sharing of the pair of electrons in HCl is shown by the partial negative charge and orange-red color of the Cl atom and the partial positive charge and blue-green color around the H atom. (c) In ionic NaCl, the violet color of the sodium ion indicates that it has a 1+ charge, and the deep red of the chloride ion reflects its charge of 1−.

American chemist Linus Pauling (1901–1994) explained why some covalent bonds are polar by using the concept of **electronegativity**, represented by the Greek letter $\chi$ (chi) and defined as an atom's tendency to attract electrons toward itself within a chemical bond. Central to that concept is the assumption that a bond between two atoms of the same element is 100% covalent, whereas the bonds between atoms of two different elements are less than 100% covalent. Pauling developed electronegativity values for the most common elements (**Figure 4.5**). The polarity of a bond depends on the difference ($\Delta\chi$) in the electronegativity values of the two elements the bond connects.

When $\Delta\chi$ is less than or equal to 0.4, a covalent bond is considered essentially nonpolar—for example, the bond between Cl ($\chi = 3.0$) and Br ($\chi = 2.8$). When $\Delta\chi$ is between 0.4 and 2.0, bonds are considered polar covalent, and when $\Delta\chi$ is equal to or greater than 2.0, bonds are considered ionic. Thus, H (2.1), F (4.0), and Cl (3.0) form polar covalent bonds in HF and HCl, but the H—F bond ($\Delta\chi = 1.9$) is *more* polar than the H—Cl bond ($\Delta\chi = 0.9$). Calcium oxide is considered an ionic compound because the $\Delta\chi$ between Ca (1.0) and O (3.5) is 2.5. Keep in mind that those cutoff values are more like guidelines than strict limits. We revisit the characterization of bonds as either ionic or polar covalent in Chapter 5, in which we discuss the polarity of entire molecules.

The data in Figure 4.5 show that electronegativity is a periodic property of the elements, with values generally increasing from left to right across a row in the periodic table and decreasing from top to bottom down a group. Those trends arise for essentially the same reasons that produce the similar trends in first ionization

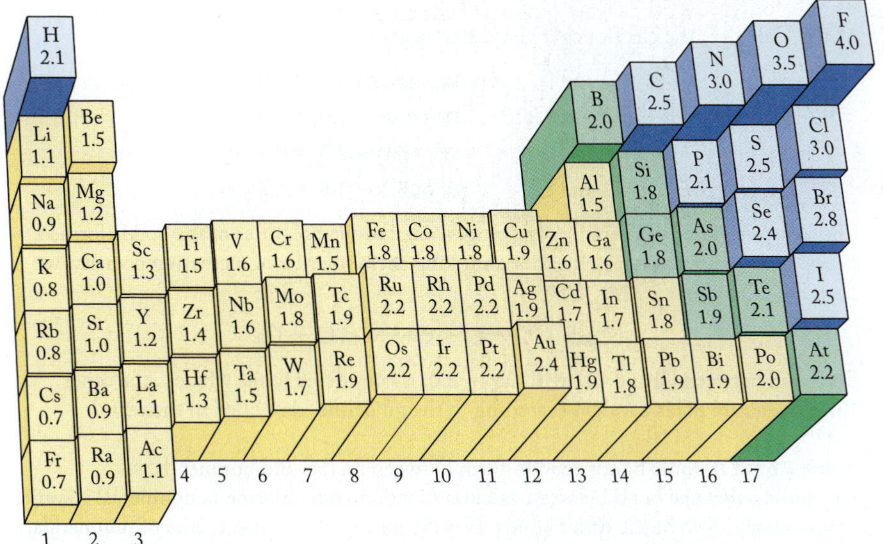


**FIGURE 4.5** The electronegativity values of the elements increase from left to right across a period and decrease from top to bottom down a group. The greater an element's electronegativity, the greater an atom's ability to attract electrons toward itself within a chemical bond.

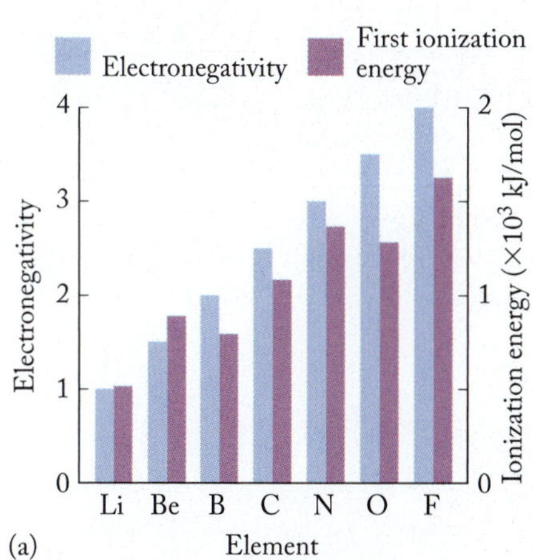


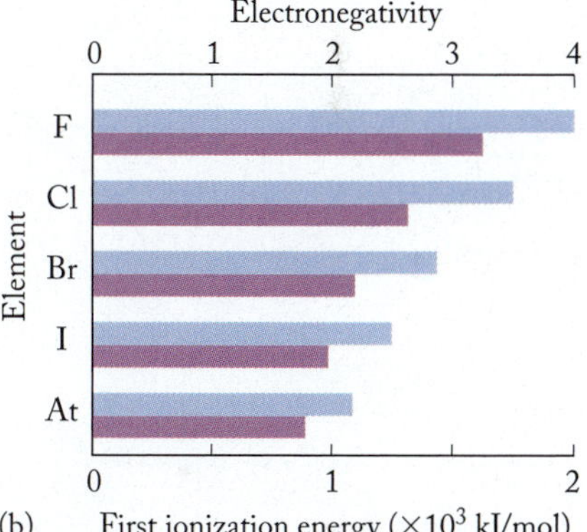


**FIGURE 4.6** The trends in the electronegativities of the main group elements follow those of first ionization energies: both tend to (a) increase with increasing atomic number across a row and (b) decrease with increasing atomic number within a group.

energies (see Figure 3.37). Greater attraction between the nuclei of atoms and their outer-shell electrons produces both higher ionization energies and greater electronegativities across a row (**Figure 4.6a**). Down a group of elements, the weaker attraction between nuclei and valence-shell electrons as atomic number increases leads to lower ionization energies and smaller electronegativities (**Figure 4.6b**). For those two reasons, the most electronegative elements—fluorine, oxygen, nitrogen, and chlorine—are in the upper right corner of the periodic table, whereas the least electronegative elements—francium, cesium, rubidium, and potassium—are in the lower left corner.

## CONCEPT TEST

Draw an arrow like the one in Figure 4.3 to indicate the polarity of a carbon–oxygen bond.

*(Answers to Concept Tests are in the back of the book.)*

**polar covalent bond** a bond resulting from unequal sharing of pairs of electrons between atoms.

**dipole** a pair of oppositely charged poles separated by a distance.

**nonpolar covalent bond** a bond characterized by an even distribution of charge; the two atoms equally share the electrons in the bond.

**electronegativity** a relative measure of an atom's ability to attract electrons to itself within a bond.

### SAMPLE EXERCISE 4.2 Comparing Bond Polarities LO3

Rank the bonds formed between atoms of O and C, Cl and Ca, N and S, and O and Si in order of increasing polarity. Categorize each bond as nonpolar covalent, polar covalent, or ionic.

**Collect, Organize, and Analyze** To rank the polarities of the bonds between different pairs of atoms, we need to calculate the differences in the parent elements' electronegativities ($\Delta\chi$; see Figure 4.5).

**CHEMTOUR**
Bond Polarity and Polar Molecules

**Solve** Calculate the electronegativity differences:

| | |
|---|---|
| O and C: | $\Delta\chi = 3.5 - 2.5 = 1.0$ |
| Cl and Ca: | $\Delta\chi = 3.0 - 1.0 = 2.0$ |
| N and S: | $\Delta\chi = 3.0 - 2.5 = 0.5$ |
| O and Si: | $\Delta\chi = 3.5 - 1.8 = 1.7$ |

The electronegativity differences are proportional to the polarities of the bonds formed between the pairs of atoms. Therefore, ranking them in order of increasing polarity, we have the following:

$$\text{N—S} < \text{O—C} < \text{O—Si} < \text{Cl—Ca}$$

The bond between Cl and Ca, with $\Delta\chi = 2.0$, is considered ionic, whereas the other three bonds are polar covalent according to the guidelines described in the text.

**Think About It** Ionic bonds tend to form between metals and nonmetals, so concluding that the bond between calcium (a metal) and chlorine (a nonmetal) is ionic is reasonable. Two of the other bonds, N—S and O—C, connect pairs of nonmetals, and the O—Si bond connects a nonmetal with a metalloid. We expect those three bonds to be covalent.

**Practice Exercise** Which pair forms the most polar bond: O and S, Be and Cl, N and H, or C and Br? Is the bond between that pair ionic?

*(Answers to Practice Exercises are in the back of the book.)*

## 4.3 Naming Compounds and Writing Formulas

Most of the matter on Earth consists of molecular and ionic compounds. As we explore the particulate nature of some of those compounds, we need to establish some rules for naming compounds and writing their chemical formulas. We focus on simple compounds, starting with binary (two-element) molecular compounds formed by combinations of main group elements.

### Binary Molecular Compounds

When two nonmetals react, they form binary molecular compounds. To translate the molecular formula of such a compound into its two-word name, follow these three steps:

1. The first word is the name of the first element in the formula.
2. For the second word, change the ending of the name of the second element to *-ide*.
3. Use prefixes (**Table 4.3**) to indicate the number of atoms of each type in the molecule. (Exception: do not use the prefix *mono-* with the first element in a name.)

For example, $SO_2$ is sulfur dioxide (not *mono*sulfur dioxide), and $N_2O$ is dinitrogen monoxide. When prefixes ending in *o-* or *a-* (such as *mono-* and *tetra-*) precede a name that begins with a vowel (such as *oxide*), the *o* or *a* at the end of the prefix is deleted to make the combination of prefix and name easier to pronounce. For example, the name of $N_2O$ is dinitrogen monoxide, not dinitrogen *mono*oxide.

**TABLE 4.3 Naming Prefixes for Molecular Compounds**

| | |
|---|---|
| one | *mono-* |
| two | *di-* |
| three | *tri-* |
| four | *tetra-* |
| five | *penta-* |
| six | *hexa-* |
| seven | *hepta-* |
| eight | *octa-* |
| nine | *nona-* |
| ten | *deca-* |

The order in which the elements are written (and named) in formulas corresponds to their relative positions in the periodic table: the element with the smaller group number appears first. If a compound contains two elements from the same group—for example, sulfur and oxygen—the element with the higher atomic number goes first.

**SAMPLE EXERCISE 4.3** Naming and Writing Formulas of Binary Molecular Compounds **LO4**

What are the names of the compounds with the following chemical formulas? (a) $NO_2$; (b) $N_2O_4$; (c) $N_2O_5$? What are the molecular formulas of the following compounds? (d) sulfur trioxide; (e) sulfur monoxide; (f) diphosphorus pentoxide?

**Collect and Organize** We are asked to translate the formulas of three molecular compounds into names and the names of three compounds into molecular formulas. The prefixes in Table 4.3 are used in the names of compounds to indicate the number of atoms of each element in a molecule: *mono-* means one atom, *di-* means two atoms, *tri-* means three, *tetra-* means four, and *penta-* means five.

**Analyze** In the first question, the first element in all three compounds is nitrogen, so the first word in each name is *nitrogen* with the appropriate prefix. Because only one nitrogen atom is in the first compound, no prefix is necessary. The second element in all three compounds is oxygen, so the second word in each name is *oxide* with the appropriate prefix. In the second question, all the molecules contain oxygen because their names end in *oxide*.

**Solve** (a) nitrogen dioxide; (b) dinitrogen tetroxide; (c) dinitrogen pentoxide; (d) $SO_3$; (e) SO; (f) $P_2O_5$.

**Think About It** To avoid awkward *ao* vowel combinations in the middle of the second word in (b) and (c), we deleted the last letter *a* of the prefixes *tetra-* and *penta-* before *oxide*.

**Practice Exercise** Name these compounds: (a) $P_4O_{10}$; (b) CO; (c) $NCl_3$. Write the formulas for these compounds: (d) sulfur hexafluoride; (e) iodine monochloride; (f) dibromine monoxide.

*(Answers to Practice Exercises are in the back of the book.)*

## Binary Ionic Compounds of Main Group Elements

NaCl Reaction

Ionic compounds also have two-word names. To name a binary ionic compound, follow these steps:

1. The first word is the name of the cation, which is simply the name of its parent element.
2. The second word is the name of the anion, which is the name of its parent element with the ending changed to *-ide*.

Prefixes are not used in naming binary ionic compounds of representative elements because metals in groups 1 and 2 and aluminum in group 13 all have characteristic positive charges, as do the monatomic anions formed by the elements in groups 16 and 17. Ionic compounds are electrically neutral, so the negative and positive charges in an ionic compound must balance, which dictates the number of each ion in the formula. Therefore, the name *magnesium fluoride*, for example, is unambiguous: it can mean only $MgF_2$.

**SAMPLE EXERCISE 4.4** Naming and Writing Formulas of Binary Ionic Compounds **LO4**

What are the names and formulas of the compounds formed by ions of these elements? (a) potassium and bromine; (b) calcium and oxygen; (c) sodium and sulfur; (d) magnesium and chlorine; (e) aluminum and oxygen?

**Collect, Organize, and Analyze** We need to write the formulas of five binary ionic compounds. Checking the names of the compounds against the positions of the elements in the periodic table and the charges on the monatomic ions they form (Figure 2.11), we see that all five are made up of main group elements that form these ions: $K^+$, $Br^-$, $Ca^{2+}$, $O^{2-}$, $Na^+$, $S^{2-}$, $Mg^{2+}$, $Cl^-$, and $Al^{3+}$. When you are writing the name and the formula, the cation always comes before the anion.

**Solve** The names of the compounds are listed in the table below, followed by the formulas of the ions in each compound from Figure 2.11. The ratios of the 1+ and 1− ions in (a) and the 2+ and 2− ions in (b) must both be 1:1 to balance their charges. However, twice as many $Na^+$ as $S^{2-}$ ions must be present in compound (c), and twice as many $Cl^-$ ions as $Mg^{2+}$ ions must be in compound (d). Two $Al^{3+}$ ions must be present for every three $O^{2-}$ ions in compound (e). Therefore, the names of the compounds and their formulas are

| | Name | Ions | Formula |
|---|---|---|---|
| (a) | Potassium bromide | $K^+$, $Br^-$ | KBr |
| (b) | Calcium oxide | $Ca^{2+}$, $O^{2-}$ | CaO |
| (c) | Sodium sulfide | $Na^+$, $S^{2-}$ | $Na_2S$ |
| (d) | Magnesium chloride | $Mg^{2+}$, $Cl^-$ | $MgCl_2$ |
| (e) | Aluminum oxide | $Al^{3+}$, $O^{2-}$ | $Al_2O_3$ |

**Think About It** The positive charge on the cations and the negative charge on the anions must balance so that they sum to a net charge of zero. Subscripts indicate the ratio of the cations to anions in the neutral ionic compound. When no subscript is written for a particular ion, it is assumed to be 1. Here we used Figure 2.11 to identify the charges of the ions. Try using only the periodic table inside the front cover as you complete the Practice Exercise.

**Practice Exercise** Without referring to Figure 2.11, write the names and formulas of the compounds formed by ions of these elements: (a) strontium and chlorine; (b) magnesium and oxygen; (c) sodium and fluorine; (d) calcium and bromine.

*(Answers to Practice Exercises are in the back of the book.)*

## Binary Ionic Compounds of Transition Metals

Many metallic elements, including most of the transition metals, form multiple cations with different charges. For example, most of the copper found in nature is present as $Cu^{2+}$; however, some copper compounds contain $Cu^+$ ions. Therefore, the name *copper oxide* is ambiguous because it does not distinguish between CuO and $Cu_2O$ (**Figure 4.7**). To name those compounds, chemists historically used different names for $Cu^+$ and $Cu^{2+}$ ions: they were called *cuprous* and *cupric* ions, respectively. Similarly, $Fe^{2+}$ and $Fe^{3+}$ ions were called *ferrous* and *ferric* ions. Note that, in both pairs of ions, the name with the lower charge ends in *-ous* and the name with the higher charge ends in *-ic*.

More recently, chemists have adopted a more systematic way of distinguishing between differently charged ions of the same transition metal. The new system, called the Stock system after the German chemist Alfred Stock (1876–1946), uses a Roman numeral to indicate the charge on the transition metal ion in a compound. Thus, the modern name of CuO is copper(II) oxide (pronounced "copper-two oxide"), and $Cu_2O$ is copper(I) oxide. Roman numerals are used to indicate the charge of transition metal cations unless the metal forms just one cation, as with $Ag^+$, $Cd^{2+}$, and $Zn^{2+}$. For example, the compounds they form with $O^{2-}$ are simply called silver oxide ($Ag_2O$), cadmium oxide (CdO), and zinc oxide (ZnO), respectively.

**FIGURE 4.7** Two oxides of copper: (a) copper(II) oxide, CuO, and (b) copper(I) oxide, $Cu_2O$.

**SAMPLE EXERCISE 4.5** Naming and Writing Formulas of Transition Metal Compounds **LO4**

What are the chemical formulas of (a) iron(II) sulfide and (b) iron(III) oxide? What are the names of (c) $V_2O_5$ and (d) NiS?

**Collect, Organize, and Analyze** We are asked to write the chemical formulas for two binary ionic compounds of iron and to write the names of two other transition metal compounds from their formulas. The Roman numerals in the names of the compounds indicate the charges in their iron ions: (II) means 2+ and (III) means 3+. Oxygen and sulfur are group 16 elements, which form 2− monatomic ions. As with all ionic compounds, the sum of the positive and negative charges of their ions must be zero.

**Solve**

a. A charge balance in iron(II) sulfide is achieved with a 1:1 ratio of $Fe^{2+}$ to $S^{2-}$ ions, so the chemical formula is FeS.
b. To balance the charges on $Fe^{3+}$ and $O^{2-}$, we need three $O^{2-}$ ions for every two $Fe^{3+}$ ions because $[2 \times (3+)] + [3 \times (2-)] = 0$. Thus, the formula of iron(III) oxide is $Fe_2O_3$.
c. The compound with the formula $V_2O_5$ has five $O^{2-}$ ions for every two vanadium ions. That means the charge on the two vanadium ions must sum to 10+, which means that each vanadium ion is 5+. Expressing that 5+ charge with the Roman numeral V gives us the compound name vanadium(V) oxide.
d. NiS has one $S^{2-}$ ion for each nickel ion. That means the charge on the nickel ion must be 2+. Expressing that 2+ charge with the Roman numeral II gives us the compound name nickel(II) sulfide.

**Think About It** The Roman numeral system for conveying the charges on transition metal ions simplifies relating the names and formulas of the compounds they form. Writing the formulas of the two compounds on the basis of their historical names would require memorizing the charges of ferrous and ferric ions.

**Practice Exercise** Write the formulas of manganese(II) chloride and manganese(IV) oxide. What are the names of $Cr_2O_3$ and PbS?

*(Answers to Practice Exercises are in the back of the book.)*

**TABLE 4.4** Some Common Polyatomic Ions

| Chemical Formula | Name |
|---|---|
| $CO_3^{2-}$ | Carbonate |
| $HCO_3^-$ | Hydrogen carbonate or bicarbonate |
| $CH_3COO^-$ | Acetate |
| $CN^-$ | Cyanide |
| $SCN^-$ | Thiocyanate |
| $ClO^-$ | Hypochlorite |
| $ClO_2^-$ | Chlorite |
| $ClO_3^-$ | Chlorate |
| $ClO_4^-$ | Perchlorate |
| $CrO_4^{2-}$ | Chromate |
| $Cr_2O_7^{2-}$ | Dichromate |
| $MnO_4^-$ | Permanganate |
| $N_3^-$ | Azide |
| $NH_4^+$ | Ammonium |
| $NO_2^-$ | Nitrite |
| $NO_3^-$ | Nitrate |
| $OH^-$ | Hydroxide |
| $O_2^{2-}$ | Peroxide |
| $PO_4^{3-}$ | Phosphate |
| $HPO_4^{2-}$ | Hydrogen phosphate |
| $H_2PO_4^-$ | Dihydrogen phosphate |
| $S_2^{2-}$ | Disulfide |
| $SO_3^{2-}$ | Sulfite |
| $SO_4^{2-}$ | Sulfate |
| $HSO_3^-$ | Hydrogen sulfite or bisulfite |
| $HSO_4^-$ | Hydrogen sulfate or bisulfate |

## Polyatomic Ions

**Table 4.4** provides the formulas, charges, and names of common **polyatomic ions**, which means that the ions consist of two or more atoms joined by covalent bonds. The ammonium ion ($NH_4^+$) is the most common cation among the polyatomic ions; all others in the table are anions.

Polyatomic ions containing oxygen and one or more other elements are called **oxoanions**. Most oxoanions have a name based on the name of the element that appears first in the formula, with its ending changed to either *-ite* or *-ate*. The *-ate* oxoanion of an element has more oxygen atoms than its *-ite* counterpart; for example, $SO_4^{2-}$ is the sulfate ion and $SO_3^{2-}$ is the sulfite ion.

If an element forms more than two oxoanions (for example, chlorine, bromine, and iodine each form four oxoanions), we use prefixes to distinguish among them (**Table 4.5**). The oxoanion with the most oxygen atoms has the prefix *per-* and ends in *-ate*, and the oxoanion with the fewest oxygen atoms has the prefix *hypo-* and ends in *-ite*. The four oxoanions of chlorine are therefore *per*chlor*ate* ($ClO_4^-$), chlor*ate* ($ClO_3^-$), chlor*ite* ($ClO_2^-$), and *hypo*chlor*ite* ($ClO^-$).

Other groups of oxoanions differ by the presence of one or more H atoms in their formulas. For example, Table 4.4 includes phosphate ($PO_4^{3-}$), hydrogen phosphate ($HPO_4^{2-}$), and dihydrogen phosphate ($H_2PO_4^-$). Note how the negative charges of those ions decrease by 1 with each added H atom in the formula. Also note that some of the oxoanions in Table 4.4 have historical names that begin with the prefix *bi-*. Sometimes those names are more commonly used than the modern ones, as with $HCO_3^-$, which is often called bicarbonate rather than hydrogen carbonate.

Although the names and formulas of polyatomic ions show several trends, a name by itself does not tell you the chemical formula or the charge. For example, nitrate and sulfate both end in *-ate*, but nitrate is $NO_3^-$, whereas sulfate is $SO_4^{2-}$. Therefore, you need to memorize the formulas, charges, and names of the common polyatomic ions in Table 4.4. Two important polyatomic ions that contain oxygen but are not considered oxoanions are the hydroxide ($OH^-$) and acetate ($CH_3COO^-$) ions. You will encounter them often in this book.

**TABLE 4.5** Oxoanions of Chlorine and Their Corresponding Acids

| OXOANION | | OXOACID | |
|---|---|---|---|
| Formula | Name | Formula | Name |
| $ClO^-$ | Hypochlorite | HClO | Hypochlorous acid |
| $ClO_2^-$ | Chlorite | $HClO_2$ | Chlorous acid |
| $ClO_3^-$ | Chlorate | $HClO_3$ | Chloric acid |
| $ClO_4^-$ | Perchlorate | $HClO_4$ | Perchloric acid |

**SAMPLE EXERCISE 4.6** Writing Formulas of Compounds Containing Oxoanions **LO4**

What are the chemical formulas of (a) sodium sulfite; (b) magnesium phosphate; (c) ammonium nitrite?

**Collect, Organize, and Analyze** To write the formulas, we need to know the formulas and charges of the cations and anions that make up those ionic compounds. Then we

**polyatomic ion** a charged group of atoms joined by covalent bonds.

**oxoanion** a polyatomic anion that contains at least one nonoxygen central atom bonded to one or more oxygen atoms.

need to identify a ratio of cations to anions that balances the positive and negative charges. The cations are $Na^+$ (a group 1 element), $Mg^{2+}$ (a group 2 element), and $NH_4^+$, which is listed in Table 4.4 along with the formulas and charges of the anions in those compounds: sulfite ($SO_3^{2-}$), phosphate ($PO_4^{3-}$), and nitrite ($NO_2^-$).

**Solve**

a. A 2:1 ratio of $Na^+$ ions to $SO_3^{2-}$ ions balances the charges in sodium sulfite, so the formula is $Na_2SO_3$.
b. We need three $Mg^{2+}$ ions for every two $PO_4^{3-}$ ions, so the formula of magnesium phosphate is $Mg_3(PO_4)_2$.
c. A 1:1 ratio of $NH_4^+$ ions to $NO_2^-$ ions balances the charges in ammonium nitrite, so the formula is $NH_4NO_2$.

**Think About It** In writing the formula of magnesium phosphate, we used parentheses around the phosphate ion to clarify that the subscript 2 applies to all the atoms in $PO_4^{3-}$. We do not need parentheses in the formula of ammonium nitrite because only one of each ion appears in the formula.

 **Practice Exercise** What are the chemical formulas of (a) strontium nitrate; (b) potassium bicarbonate; (c) barium chromate?

*(Answers to Practice Exercises are in the back of the book.)*

**SAMPLE EXERCISE 4.7 Naming Compounds Containing Oxoanions LO4**

What are the names of the compounds with these chemical formulas: (a) $CaCO_3$; (b) $LiNO_3$; (c) $Mg(ClO_4)_2$; (d) $(NH_4)_2SO_4$; (e) $KClO_3$; (f) $NaHCO_3$?

**Collect, Organize, and Analyze** We are given the formulas of six compounds, each containing at least one oxoanion, and asked to name them. The names of ionic compounds start with the name of the cation and end with the name of the anion. The cations in five of the compounds are those formed by atoms of these elements: (a) calcium, (b) lithium, (c) magnesium, (e) potassium, and (f) sodium. According to the list of polyatomic ion names in Table 4.4, the cation in formula (d) is the ammonium ion, and the oxoanions in the six compounds are (a) carbonate, (b) nitrate, (c) perchlorate, (d) sulfate, (e) chlorate, and (f) hydrogen carbonate.

**Solve** Combining the names of the cations and anions, we get (a) calcium carbonate, (b) lithium nitrate, (c) magnesium perchlorate, (d) ammonium sulfate, (e) potassium chlorate, and (f) sodium hydrogen carbonate.

**Think About It** As noted in the text, hydrogen carbonate is commonly called bicarbonate. The charges on the most common monatomic ions can be inferred from the positions of their parent elements in the periodic table, but knowing the names and formulas of common polyatomic ions requires at least some memorization.

 **Practice Exercise** Name the following compounds: (a) $Ca_3(PO_4)_2$; (b) $MgSO_3$; (c) $LiNO_2$; (d) $NaClO$; (e) $KMnO_4$.

*(Answers to Practice Exercises are in the back of the book.)*

## Binary Acids

Some compounds have special names because of their particular chemical properties. Among those are acids. We will discuss the properties of acids further in later chapters, but for now it is enough to say that acids are compounds that release

**oxoacid** a compound composed of oxoanions bonded to $H^+$ ions.

**octet rule** the tendency of atoms of main group elements to make bonds by gaining, losing, or sharing electrons to achieve a valence shell containing eight electrons, or four electron pairs.

**Lewis symbol** the chemical symbol for an element surrounded by one or more dots representing valence electrons; also called a *Lewis dot symbol*.

**bonding capacity** the number of covalent bonds an atom forms to have an octet of electrons in its valence shell.

hydrogen ions ($H^+$) when they dissolve in water. For example, when the molecular compound hydrogen chloride (HCl) dissolves in water, it produces the solution we call *hydrochloric acid*. In that aqueous solution, every molecule of HCl has separated into a $H^+$ ion and a $Cl^-$ ion. To name the aqueous solution of acids such as HCl, follow these steps:

1. Add the prefix *hydro-* to the name of the second element in the formula.
2. Replace the last syllable in the second element's name with the suffix *-ic* and add the word *acid*.

To distinguish between a hydrogen halide (a molecular compound with covalent bonds between the hydrogen atom and the halogen atom) and the aqueous form ($H^+$ ions and $X^-$ ions dissolved in water), we add the symbol (*aq*) for "aqueous" to the latter. For example, HBr is hydrogen bromide, but HBr(*aq*) is hydrobromic acid.

**CONCEPT TEST**

What is the name of the solution produced when HF dissolves in water?

*(Answers to Concept Tests are in the back of the book.)*

## Oxoacids

The scheme for naming the acids of oxoanions, called **oxoacids**, is illustrated in Table 4.5. If the oxoanion name ends in *-ate*, the name of the corresponding oxoacid ends in *-ic*. Thus, $SO_4^{2-}$ is the sulf*ate* ion and $H_2SO_4$ is sulfur*ic* acid. If the oxoanion name ends in *-ite*, the name of the oxoacid ends in *-ous*. Thus, $NO_2^-$ is the nitr*ite* ion and $HNO_2$ is nitr*ous* acid.

**SAMPLE EXERCISE 4.8** Naming and Writing Formulas of Oxoacids **LO4**

What are the names and formulas of the oxoacids formed by these oxoanions? (a) $SO_3^{2-}$; (b) $ClO_4^-$; (c) $PO_4^{3-}$?

**Collect, Organize, and Analyze** We are given the formulas of three oxoanions and are asked to name the oxoacids they form when they bond with $H^+$ ions. According to Table 4.4, the names of the oxoanions are (a) sulfite, (b) perchlorate, and (c) phosphate. When the oxoanion name ends in *-ite*, the corresponding oxoacid name ends in *-ous acid*. When the oxoanion name ends in *-ate*, the oxoacid name ends in *-ic acid*.

**Solve** Making the appropriate changes to the endings of the oxoanion names and adding the word *acid*, we get (a) sulfurous acid, $H_2SO_3$; (b) perchloric acid, $HClO_4$; and (c) phosphoric acid, $H_3PO_4$.

**Think About It** Once we know the names of the common oxoanions, naming the corresponding oxoacids is simply a matter of applying the correct ending (*-ous* or *-ic*) and then adding the word *acid*.

**Practice Exercise** Name the following acids: (a) HBrO; (b) $HBrO_2$; (c) $H_2CO_3$.

*(Answers to Practice Exercises are in the back of the book.)*

# 4.4 Lewis Symbols and Lewis Structures

In 1916, American chemist Gilbert N. Lewis (1875–1946) proposed that atoms form chemical bonds by sharing pairs of electrons. He further suggested that, through that sharing, each atom acquires enough valence electrons to mimic the outer-shell electron configuration of a noble gas. Today we associate those configurations with a filled 1*s* orbital (corresponding to a helium atom) or with a full set of *s* and *p* orbitals, which is the valence-shell configuration of the other noble gases. Although Lewis's view of chemical bonding predated quantum mechanics and the notion of atomic orbitals, it was consistent with what he called the **octet rule**: all atoms except the very smallest (for example, hydrogen) tend to lose, gain, or share electrons so that each atom has eight valence electrons, or an *octet* of them. However, a hydrogen atom forms only one bond and, by doing so, mimics the electron configuration of a helium atom.

**CONNECTION** In Chapter 3 we learned that up to two electrons can occupy an *s* orbital, whereas up to six electrons can occupy the three *p* orbitals.

## Lewis Symbols

Lewis developed a system of symbols, called **Lewis symbols** or *Lewis dot symbols*, to depict an atom's **bonding capacity**—that is, the number of chemical bonds that each atom of an element typically forms to complete its octet. A Lewis symbol consists of the symbol of an element surrounded by dots representing its valence electrons. The dots are placed on the four sides of the symbol (top, bottom, right, and left). The order in which they are placed does not matter as long as one dot is placed on each side before any dots are paired. The number of *unpaired* dots in a Lewis symbol indicates the typical bonding capacity for atoms of that element.

**Figure 4.8** shows the Lewis symbols of the main group elements. Because all elements in a family have the same number of valence electrons, they all have the same arrangement of dots in their Lewis symbols. For example, the Lewis symbols of carbon and all other group 14 elements have four unpaired dots, representing four unpaired electrons. Four unpaired electrons means that carbon atoms each tend to form four chemical bonds. When they do, their four valence electrons plus four additional electrons from the atoms they bond with yield an octet of eight valence electrons. Similarly, the Lewis symbols of nitrogen and all other group 15 elements have three unpaired dots, representing three unpaired electrons. A nitrogen atom has a bonding capacity of three and typically forms three bonds to complete its octet.

Before we start to use Lewis symbols to explore how elements form compounds, we need to keep in mind that the number of unpaired dots in the symbols indicates the number of bonds that the atoms of an element *typically* form. We will see in this chapter that atoms may exceed their bonding capacities in some molecules and polyatomic ions and fail to reach them in others. Bonding capacity is a useful concept, but we will treat it more like a guideline than a strict rule.

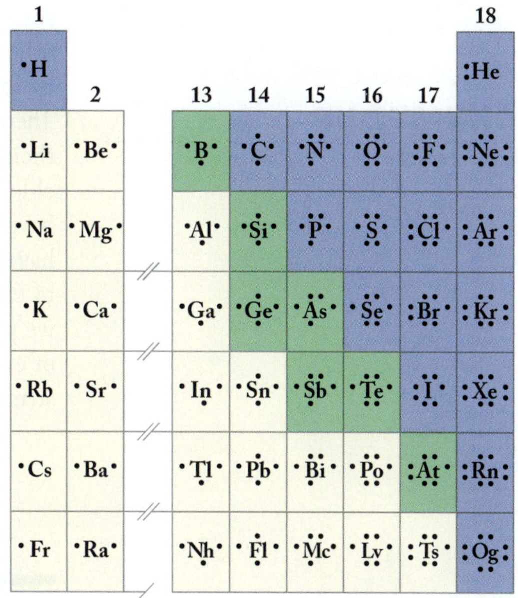


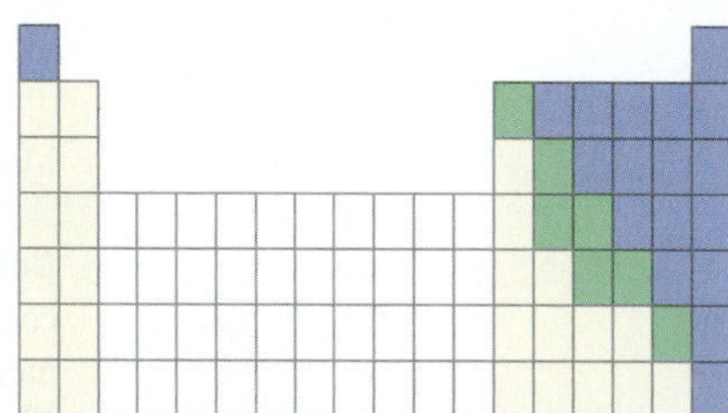

**FIGURE 4.8** Lewis symbols of the main group elements. Elements in the same group have the same number of valence electrons and the same arrangement of dots representing the valence electrons in their Lewis symbols.

**CONCEPT TEST**

Devise a formula that relates the bonding capacity of the atoms of the elements in groups 14–17 to their group number.

*(Answers to Concept Tests are in the back of the book.)*

**Lewis structure** a two-dimensional representation of the bonds and lone pairs of valence electrons in an ionic or molecular compound.

**bonding pair** a pair of electrons that two atoms share.

**single bond** a bond that results when two atoms share one pair of electrons.

**lone pair** a pair of electrons not shared.

**CHEMTOUR**

Lewis Structures

## Lewis Structures of Ionic Compounds

A **Lewis structure** is a two-dimensional representation of a compound showing how its atoms are connected. Though Lewis structures are most often used to show the bonding in molecular compounds, they also offer insights into the composition of polyatomic ions and ionic compounds. The Lewis structures of binary ionic compounds such as NaCl are some of the simplest to draw, so we start with them.

Crystals of sodium chloride are held together by the attraction between oppositely charged $Na^+$ and $Cl^-$ ions. We discussed in Chapter 3 how sodium and the other group 1 and 2 elements achieve the electron configurations of noble gases by losing their valence-shell *s* electrons and forming positively charged cations. Therefore, the ions of group 1 and 2 elements have no valence-shell electrons, and their Lewis structures have no dots around them. However, nonmetals such as chlorine, which acquire electrons to achieve noble gas electron configurations, have filled *s* and *p* orbitals in their valence shells. Therefore, their monatomic ions have Lewis structures with four pairs of dots. We place brackets around the dots to emphasize that all eight valence electrons are associated with the anion, and we place the charge of the ion outside the bracket to indicate the overall charge of everything inside. Applying that notation to NaCl gives the following Lewis structure:

$$\mathrm{Na}\cdot + \cdot\ddot{\underset{..}{\mathrm{Cl}}}: \rightarrow \mathrm{Na}^+[:\ddot{\underset{..}{\mathrm{Cl}}}:]^-$$

### SAMPLE EXERCISE 4.9 Drawing the Lewis Structure of a Binary Ionic Compound LO5

Draw the Lewis structure of calcium fluoride.

**Collect, Organize, and Analyze** We are asked to draw the Lewis structure of the ionic compound formed by calcium and fluoride ions. Calcium is a group 2 element, which means its atoms form ions with 2+ charges by losing both their valence (3*s*) electrons. Fluorine is a group 17 element, which means it forms 1− fluoride ions by gaining one valence-shell electron to give it a complete octet.

**Solve** A $Ca^{2+}$ ion has no valence electrons left, so its Lewis structure is simply

$$\mathrm{Ca}^{2+}$$

After acquiring an electron, a F ion has a complete octet and the following Lewis structure:

$$[:\ddot{\underset{..}{\mathrm{F}}}:]^-$$

Two fluoride ions must be present for every one calcium ion in $CaF_2$ to have a neutral compound. A Lewis structure reflecting that 2:1 ratio is

$$[:\ddot{\underset{..}{\mathrm{F}}}:]^- \ \mathrm{Ca}^{2+} \ [:\ddot{\underset{..}{\mathrm{F}}}:]^-$$

**Think About It** The lack of dots in the structure of $Ca^{2+}$ reinforces the fact that Ca atoms, like those of the other main group metals, lose all their valence-shell electrons when they form monatomic cations. By contrast, atoms of the nonmetals form monatomic anions with complete octets.

**Practice Exercise**
Draw the Lewis structure of sodium oxide.

*(Answers to Practice Exercises are in the back of the book.)*

## Lewis Structures of Molecular Compounds

Molecular compounds are held together by covalent bonds. Because covalent bonds are *pairs* of shared valence electrons, the Lewis structures of molecular compounds focus on how the shared pairs, also called **bonding pairs**, are distributed among the atoms in their molecules. A single bonding pair of electrons is called a **single bond**, often represented by a dash in Lewis structures, as in the structural formulas we first saw in Chapter 1. Electron pairs not involved in bonds appear as pairs of dots on one atom. The unshared electron pairs are called **lone pairs**.

A five-step process for drawing Lewis structures follows. It is particularly useful for drawing the structures of small molecules and polyatomic ions that have a single central atom bonded to atoms with lower bonding capacities.

## Five Steps for Drawing Lewis Structures

1. *Determine the total number of valence electrons.* For a neutral molecule, count the valence electrons in all the atoms in the molecule. For a polyatomic ion, count the valence electrons and then add (for an anion) or subtract (for a cation) the number of electrons needed to account for the charge on the ion.
2. *Arrange the symbols of the elements to show how their atoms are bonded (a skeletal structure) and then connect them with single bonds (single pairs of bonding electrons).* Identify the element with the greatest bonding capacity and position it as the central atom. If two elements have the same bonding capacity, choose the one that is less *electronegative*. Place the remaining atoms around the central atom, connecting each atom with a single bond. Some helpful hints: A hydrogen atom will never be a central atom because it has a bonding capacity of one (it can form only one bond). Sometimes the formulas of compounds and polyatomic ions suggest a skeletal structure. For example, the skeletal structures of HNCO and $SCN^-$ are H—N—C—O and S—C—N, respectively.
3. *Complete the octets of all the atoms (except hydrogen, which can form only one bond) bonded to the central atom by adding lone pairs of electrons.* Place lone pairs on all the outer atoms until each has an octet (including the two electrons used to connect it to the central atom in the skeletal structure).
4. *Compare the number of valence electrons in the Lewis structure with the number determined in step 1.* If valence electrons remain unused in the structure, place lone pairs of electrons on the central atom (even if doing so means giving it more than an octet of valence electrons), until the Lewis structure includes all the valence electrons counted in step 1.
5. *Complete the octet on the central atom.* If the central atom has an octet, the structure is complete. If the central atom has less than an octet, create additional bonds to it by converting one or more lone pairs of electrons on the outer atoms into bonding pairs.

## SAMPLE EXERCISE 4.10 Drawing the Lewis Structure of a Small Molecule I — LO5

Chloroform is a volatile liquid once used as an anesthetic in surgery. It has the molecular formula $CHCl_3$. Draw its Lewis structure.

**Collect, Organize, and Analyze** The formula $CHCl_3$ tells us that a chloroform molecule contains one carbon atom, one hydrogen atom, and three chlorine atoms. Carbon is a group 14 element, so the carbon atom has four valence electrons, and four more will complete its octet. Hydrogen, in group 1, has one valence electron and can form only one bond that consists of two electrons. Chlorine, in group 17, has seven valence electrons, so one more will complete its octet. We can use the five-step process described previously to generate the Lewis structure.

**Solve**

Step 1. Summing the data in the last column of the following table gives us the total number of valence electrons in a molecule of $CHCl_3$: 26.

| ELEMENT | | | VALENCE ELECTRONS | | |
|---|---|---|---|---|---|
| Symbol | No. of atoms | | No. of valence $e^-$ per atom | | Total |
| C | 1 | × | 4 | = | 4 |
| H | 1 | × | 1 | = | 1 |
| Cl | 3 | × | 7 | = | 21 |
| **Valence electrons in $CHCl_3$** | | | | | **26** |

Step 2. Carbon has the greatest bonding capacity of the three elements, so we put a carbon atom at the center of the skeletal structure of $CHCl_3$:

```
      H
      |
  Cl—C—Cl
      |
      Cl
```

Step 3. The four single bonds give the carbon atom a complete octet but each chlorine atom needs three lone pairs of electrons to complete its octet:

```
        H
        |
  :C̈l̤—C—C̈l̤:
        |
       :C̈l̤:
```

Step 4. That structure contains four pairs of bonding electrons and nine lone pairs, for a total of $(4 \times 2) + (9 \times 2) = 26$ electrons, which matches the total number of valence electrons calculated in step 1.

Step 5. The carbon atom is surrounded by four bonds, which means it has eight valence electrons and a full octet. The structure is complete.

**Think About It** Carbon has a bonding capacity of four and is bonded to four atoms in that structure, so the carbon atom has no lone pairs of electrons.

**Practice Exercise**

Draw the Lewis structure of methane, $CH_4$.

*(Answers to Practice Exercises are in the back of the book.)*

**SAMPLE EXERCISE 4.11** Drawing the Lewis Structure of a Small Molecule II **LO5**

Draw the Lewis structure of ammonia, $NH_3$.

**Collect, Organize, and Analyze** The chemical formula tells us that $NH_3$ contains one atom of nitrogen and three atoms of hydrogen per molecule. Nitrogen is a group 15 element with five valence electrons and a bonding capacity of three. Hydrogen atoms have one valence electron each and a bonding capacity of one. We can use the five-step process described previously to generate the Lewis structure.

**Solve**

Step 1. Summing the data in the last column of the following table gives us the total number of valence electrons in a molecule of $NH_3$: eight.

| ELEMENT | | | VALENCE ELECTRONS | | |
|---|---|---|---|---|---|
| Symbol | No. of atoms | | No. of valence $e^-$ per atom | | Total |
| N | 1 | × | 5 | = | 5 |
| H | 3 | × | 1 | = | 3 |
| **Valence electrons in $NH_3$** | | | | | **8** |

Step 2. Nitrogen has the greatest bonding capacity of the three elements, so we put a nitrogen atom at the center of the skeletal structure:

```
H—N—H
  |
  H
```

Step 3. The single bond to each hydrogen atom completes its valence shell.

Step 4. The three bonds in the structure represent $3 \times 2 = 6$ valence electrons, but we started with a total of eight valence electrons in step 1. Adding a lone pair of electrons to nitrogen gives us eight valence electrons and completes the octet on nitrogen:

```
  ··
H—N—H
  |
  H
```

Step 5. The nitrogen atom has a complete octet, so the Lewis structure is complete.

**Think About It** The structure makes sense because the nitrogen atom exhibits the bonding capacity predicted by its Lewis symbol. However, we will soon encounter a polyatomic ion, $NH_4^+$, in which the central nitrogen atom bonds to *four* hydrogen atoms. Keep in mind that Lewis bonding capacities are guidelines rather than strictly observed rules.

**Practice Exercise** Draw the Lewis structure of phosphorus trichloride.

*(Answers to Practice Exercises are in the back of the book.)*

**SAMPLE EXERCISE 4.12** Drawing the Lewis Structure of a Polyatomic Ion **LO5**

Draw the Lewis structure of the hydroxide ($OH^-$) ion.

**Collect, Organize, and Analyze** Hydroxide ions each contain one atom of oxygen, a group 16 element with six valence electrons and a typical bonding capacity of two, and a hydrogen atom that has a bonding capacity of one. As before, we use the five-step process to draw its Lewis structure.

**Solve**

Step 1. The charge of 1− means that one more valence electron is in the structure in addition to the six from the oxygen atom and one from the hydrogen atom, for a total of 1 + 6 + 1 = 8 valence electrons.

Step 2. Connecting the hydrogen atom to the oxygen atom with a covalent bond yields:

$$\text{O—H}$$

Step 3. The bonded H atom has a complete valence shell.

Step 4. The one bond in the structure represents two valence electrons, but eight are in the ion. So we add six more by placing three lone pairs of electrons around the O atom, which completes its octet:

$$:\ddot{\underset{..}{\text{O}}}\text{—H}$$

Step 5. The Lewis structure represents an ion, so square brackets are drawn around the structure and the ion's charge is placed outside the brackets. The Lewis structure is now complete.

$$\left[:\ddot{\underset{..}{\text{O}}}\text{—H}\right]^-$$

**Think About It** The bonding capacity of oxygen is two, yet only one bond is present in a $OH^-$ ion. However, that ion also has the capacity to form an ionic bond with a positively charged cation, such as $Na^+$, producing a neutral ionic compound (NaOH) with equal numbers of ionic and covalent bonds.

**Practice Exercise** Draw the Lewis structure of the ammonium ion, $NH_4^+$.

*(Answers to Practice Exercises are in the back of the book.)*

To draw the Lewis structure of a molecule with more than one "central" atom, drawing pieces of the molecule, each containing one of the central atoms, is often useful—as shown for hydrogen peroxide ($H_2O_2$):

$$\text{H—}\ddot{\underset{..}{\text{O}}}\text{—}\ddot{\underset{..}{\text{O}}}\text{—H}$$

Here each oxygen atom is the "central atom" of a three-atom subunit, and the bond between the two O atoms contributes two valence electrons toward completing the octets of both:

$$\text{—}\ddot{\underset{..}{\text{O}}}\text{—}\ddot{\underset{..}{\text{O}}}\text{—H}$$

$$\text{H—}\ddot{\underset{..}{\text{O}}}\text{—}\ddot{\underset{..}{\text{O}}}\text{—}$$

## Lewis Structures of Molecules with Double and Triple Bonds

**double bond** a bond formed when two atoms share two pairs of electrons.

**triple bond** a bond formed when two atoms share three pairs of electrons.

Lewis structures also can be used to show the bonding in molecules in which two atoms share more than one pair of bonding electrons. A bond in which two atoms share two pairs of electrons is called a **double bond**. For example, the two oxygen atoms in a molecule of $O_2$ share two pairs of electrons, forming an O═O double bond. When the two nitrogen atoms in a molecule of $N_2$ share three pairs of electrons, they form a N≡N **triple bond**. In Section 4.6 we discuss the characteristics of multiple bonds and compare them with those of single bonds.

How do we know when a Lewis structure has a double or triple bond? Typically we realize it when we apply steps 3–5 in the guidelines. Suppose we complete the octets of all the atoms attached to the central atom in step 3, and in doing so we use all the available valence electrons. If the central atom does not yet have an octet, we cannot just add valence electrons to the structure. But we can complete the central atom's octet in step 5 by converting one or more lone pairs of electrons from outer atom(s) into bonding pairs. The following example—drawing the Lewis structure of formaldehyde ($CH_2O$)—illustrates how that is done.

Step 1. The total number of valence electrons in a $CH_2O$ molecule is

| ELEMENT | | | VALENCE ELECTRONS | | |
|---|---|---|---|---|---|
| Symbol | No. of atoms | | No. of valence $e^-$ per atom | | Total |
| C | 1 | × | 4 | = | 4 |
| H | 2 | × | 1 | = | 2 |
| O | 1 | × | 6 | = | 6 |
| **Valence electrons in $CH_2O$** | | | | | **12** |

Step 2. Carbon has the greatest bonding capacity (four) and is the central atom. Connecting it with single bonds to the other three atoms produces this skeletal structure:

H—C—H
  |
  O

Step 3. Each H atom has a single covalent bond (two electrons), thus completing its valence shell. Adding three lone pairs of electrons completes the octet for oxygen:

H—C—H
  |
 :Ö:

Step 4. The structure has 12 valence electrons, which matches the number in the molecule. No additional electrons (dots) can be added to the structure.

Step 5. The central C atom has only six electrons. To give the carbon atom the two additional electrons it needs to have an octet—without removing any electrons from oxygen, which already has an octet—we convert one of the lone pairs on the oxygen atom into a bonding pair between C and O:

H—C—H → H₂C=O (H\C/H over double bond to O with two lone pairs)

Which of the three lone pairs is converted into a bonding pair does not matter because all three are equivalent. The central carbon atom now has a complete octet, as does the oxygen atom. The structure is reasonable because the four covalent bonds around carbon—two single bonds and one double bond—match its bonding capacity. The double bond to oxygen is reasonable because oxygen is a group 16 element with a bonding capacity of two, and it has two bonds in the structure.

Notice that we have drawn the double bond and the two single bonds around the central carbon atom and the double bond and lone pairs of electrons on the oxygen atom in a way that maximizes the separation between the pairs. We will learn in Chapter 5 that such arrangements more closely represent the actual bonding and lone pair orientations in molecules. For now, be assured that nothing is wrong with drawing a Lewis structure for formaldehyde in which the bond angles are, for example, 90° instead of 120°. Lewis structures show how atoms are bonded to each other in molecules, not necessarily how the bonds are oriented in space.

**SAMPLE EXERCISE 4.13** Drawing Lewis Structures with Double and Triple Bonds **LO5**

Draw the Lewis structure of acetylene, $C_2H_2$, the fuel used in oxyacetylene torches for cutting steel and other metals.

**Collect, Organize, and Analyze** One molecule of acetylene contains two carbon atoms and two hydrogen atoms. Carbon is a group 14 element with a bonding capacity of four, and hydrogen has a bonding capacity of one. We follow the steps in the guidelines, using double or triple bonds as needed.

**Solve**

Step 1. The two carbon atoms contribute four valence electrons each, and the two hydrogen atoms each contribute one, for a total of $(2 \times 4) + (2 \times 1) = 10$.

Step 2. The two C atoms serve as central atoms in three-atom subunits, each bonded to the other and to one of the H atoms, which gives us the skeletal structure:

$$\text{H—C—C—H}$$

Step 3. Each H atom has a single covalent bond (consisting of two electrons) and a complete valence shell.

Step 4. Our structure has six valence electrons, but the molecule has 10, so we have to add four more. No additional electrons can be placed on the H atoms.

Step 5. The four remaining valence electrons must be placed on the two central carbon atoms. One way to do that is to add two more bonding pairs between the carbon atoms, for a total of six shared electrons between the two carbon atoms. Doing so gives the structure the right number of valence electrons, and it completes the octets of both carbon atoms:

$$\text{H—C}\equiv\text{C—H}$$

**Think About It** The carbon atoms each form four bonds, as carbon atoms usually do. Here the four consist of one single bond and one triple bond.

**Practice Exercise** Draw the Lewis structure of carbon dioxide.

*(Answers to Practice Exercises are in the back of the book.)*

**allotropes** different molecular forms of the same element, such as oxygen ($O_2$) and ozone ($O_3$).

**resonance** a characteristic of electron distributions in which two or more equivalent Lewis structures can be drawn for one compound.

**resonance structure** one of two or more Lewis structures with the same arrangement of atoms but different arrangements of bonding pairs of electrons.

## 4.5 Resonance

The atmosphere contains two kinds of oxygen molecules. Most are $O_2$, but a tiny fraction are $O_3$, a form of oxygen called ozone. Different molecular forms of the same element are called **allotropes**, and they have different chemical and physical properties. Ozone, for example, is an acrid, pale blue gas that is toxic even at low concentrations, whereas $O_2$ is a colorless, odorless gas essential for most life-forms. Ozone is produced naturally by lightning (**Figure 4.9**) and is the source of the pungent odor you may have smelled after a severe thunderstorm. Ozone in the lower atmosphere is sometimes referred to as "bad ozone" because high levels in polluted air can damage crops, harm trees, and cause human health problems. Ozone in the upper atmosphere, however, is considered "good ozone" because it shields life on Earth from potentially harmful ultraviolet radiation from the sun.

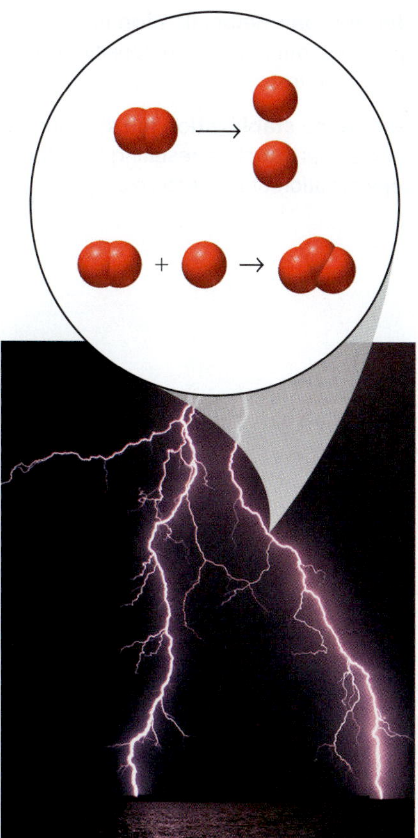

**FIGURE 4.9** Lightning strikes contain enough energy to break oxygen–oxygen double bonds. The O atoms formed in that fashion collide with other $O_2$ molecules, forming ozone ($O_3$), an allotrope of oxygen.

**CHEMTOUR**
Resonance

Let's draw the Lewis structure of ozone by following our five-step process from Section 4.4. Oxygen is a group 16 element, so it has six valence electrons. The total number of valence electrons in an ozone molecule, then, is 3 × 6 = 18 (step 1). Connecting the three O atoms with single bonds (step 2) gives

$$\mathrm{O{-}O{-}O}$$

Using 12 of the 14 remaining electrons to complete the octets of the atoms on the ends (step 3) gives

$$:\ddot{\underset{\cdot\cdot}{\mathrm{O}}}{-}\mathrm{O}{-}\ddot{\underset{\cdot\cdot}{\mathrm{O}}}:$$

The last two electrons are added as a lone pair to the central oxygen atom (step 4):

$$:\ddot{\underset{\cdot\cdot}{\mathrm{O}}}{-}\underset{\cdot\cdot}{\mathrm{O}}{-}\ddot{\underset{\cdot\cdot}{\mathrm{O}}}:$$

We have used all 18 valence electrons, but that structure leaves the central atom two electrons shy of an octet, so we convert one of the lone pairs on the leftmost O atom into a bonding pair (step 5):

$$\cdot\underset{\cdot\cdot}{\mathrm{O}}{=}\ddot{\mathrm{O}}{-}\ddot{\underset{\cdot\cdot}{\mathrm{O}}}:$$

We also could have formed the double bond with a lone pair from the rightmost O atom:

$$:\ddot{\underset{\cdot\cdot}{\mathrm{O}}}{-}\ddot{\mathrm{O}}{=}\underset{\cdot\cdot}{\mathrm{O}}\cdot$$

Those two structures illustrate an important concept in Lewis theory called **resonance**: the existence of multiple Lewis structures, called **resonance structures**, which have the same arrangement of atoms but different arrangements of bonding electrons and lone pairs.

Having seen that drawing two equivalent Lewis structures for the same molecules is possible, you may be wondering whether either one accurately describes the bonding in ozone molecules. Experimental evidence indicates that, technically, neither structure is correct. Scientists have determined that the two bonds in ozone have the same length, 128 pm. As we will see in Section 4.6, that value is about halfway between the length of an O—O single bond (148 pm) and an O═O double bond (121 pm). One way to explain that result is to assume that the actual bonding in an $O_3$ molecule is the average of the two resonance structures, which means identical bonds between the O atoms that are intermediate in length and strength between O—O single and O═O double bonds. It is as though two bonding pairs of electrons connect the center atom to the two others and that a third bonding pair is spread out across all three atoms.

**electron-pair delocalization** the spreading out of electron density over several atoms.

**resonance stabilization** the stability of a molecular structure resulting from the delocalization of its electrons.

That spreading out, or **delocalization**, of the third bonding pair can be explained by assuming that bonding and lone pairs of electrons can be rearranged as shown by the red arrows in the following images.

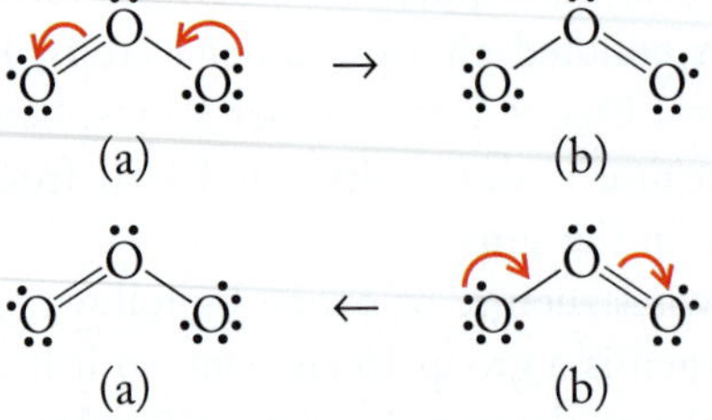

(a) → (b)

(a) ← (b)

That means that the bonding in a molecule of ozone is described by neither resonance structure (a) nor (b) but by an average of the two. A double-headed arrow is used between resonance forms to symbolize the averaging effect of bonding-pair delocalization:

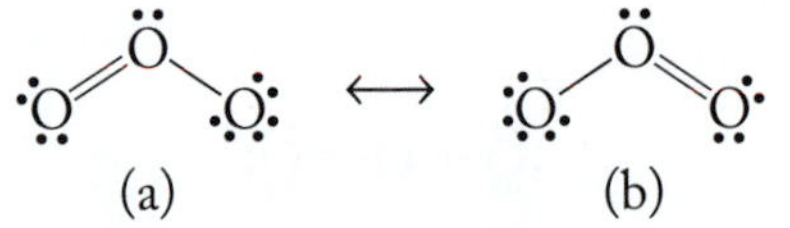

(a) ⟷ (b)

A key point about delocalization is that it reduces the electrons' potential energy and lowers the energy of the molecule, a phenomenon called **resonance stabilization**. We will see in the chapters ahead that resonance can strongly influence the chemical properties of molecular substances. Resonance may also occur in polyatomic ions, as illustrated in Sample Exercise 4.14.

**SAMPLE EXERCISE 4.14** Drawing Resonance Structures **LO6**

Draw all the resonance structures of the nitrate ion, $NO_3^-$.

**Collect, Organize, and Analyze** We are asked to draw the resonance structures of the $NO_3^-$ ion, which contains one atom of nitrogen (a group 15 element) and three atoms of oxygen (a group 16 element). The charge of the ion (1−) means it contains an additional valence electron.

**Solve**

Step 1. The number of valence electrons in a $NO_3^-$ ion is

| ELEMENT | | | VALENCE ELECTRONS | | |
|---|---|---|---|---|---|
| Symbol | No. of atoms | | No. of valence $e^-$ per atom | | Total |
| N | 1 | × | 5 | = | 5 |
| O | 3 | × | 6 | = | 18 |
| Plus one electron for the 1− charge | | | | | 1 |
| **Valence electrons in $NO_3^-$** | | | | | **24** |

Step 2. Nitrogen has the higher bonding capacity, so N is the central atom. Connecting it with single bonds to the three O atoms gives us the following skeletal structure:

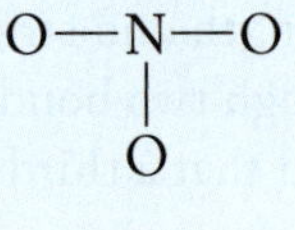

```
O—N—O
  |
  O
```

Step 3. Each O atom needs three lone pairs of electrons to complete its octet:

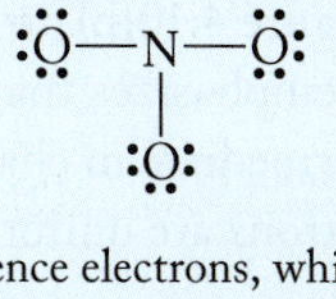

Step 4. The structure has 24 valence electrons, which matches the number determined in step 1.

Step 5. The central N atom has only six electrons, two short of an octet, so we convert a lone pair on one of the oxygen atoms into a bonding pair:

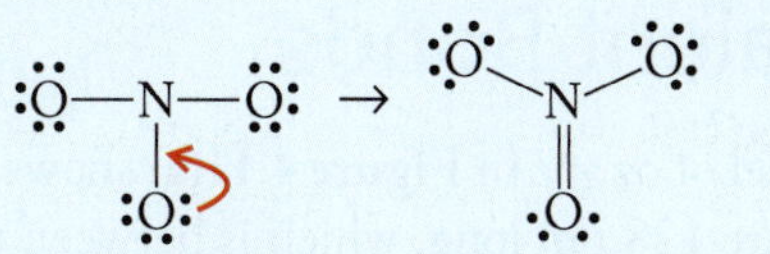

The nitrogen atom now has a complete octet. Adding brackets and the ionic charge, we have a complete Lewis structure:

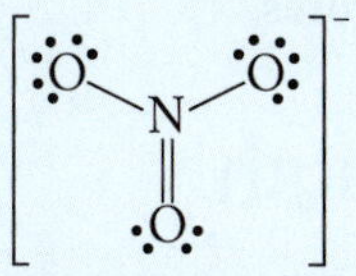

Using the O atom to the left or right of the central N atom to form the double bond creates two additional resonance forms, or three in all:

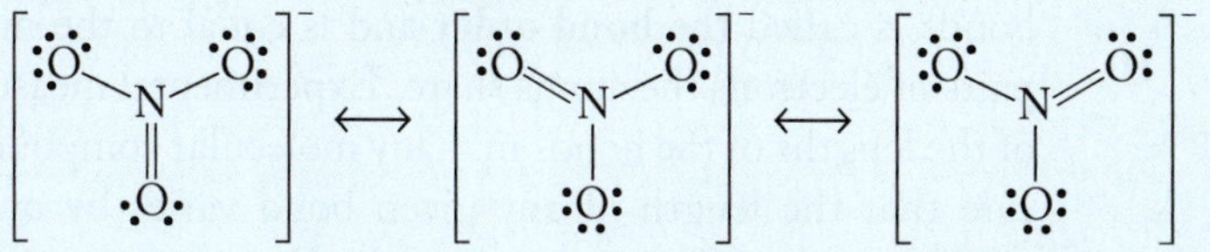

**Think About It** None of the three resonance structures accurately represents the nitrate ion; the true structure is an average of those three—a molecular ion with three equivalent bonds rather than a double bond and two single bonds.

**Practice Exercise** Draw all the resonance forms of the azide ion, $N_3^-$, and the nitronium ion, $NO_2^+$.

*(Answers to Practice Exercises are in the back of the book.)*

A test of whether a compound has multiple equivalent resonance forms—and therefore exhibits resonance stabilization—is the presence of one or more atoms having both single and double bonds to two or more atoms of the same element, as in $O_3$ and $NO_3^-$. A molecule of benzene ($C_6H_6$) contains a ring of six carbon atoms with alternating single and double bonds (**Figure 4.10**), so it has that

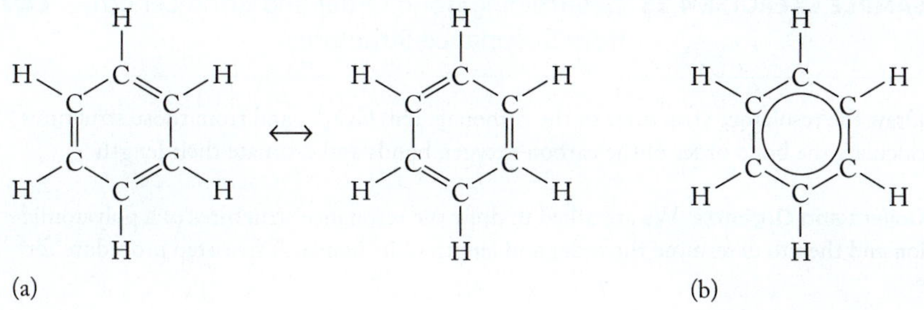

(a)

(b)

**FIGURE 4.10** (a) The molecular structure of benzene is an average of two equivalent structures. (b) The average is often represented by a circle inside a hexagonal ring of single bonds to indicate the uniform distribution of the electrons in the bonds around the ring.

**bond order** the number of bonds between atoms: 1 for a single bond, 2 for a double bond, and 3 for a triple bond.

property, too. Those single and double bonds can be drawn in two equivalent ways (Figure 4.10a), but chemists often draw benzene molecules with a circle in the center of the ring, as shown in Figure 4.10(b), to represent an averaging of the two resonance structures. The circle emphasizes that the six carbon–carbon bonds in the ring are all identical and intermediate in character between single and double bonds and that the bonding electrons are uniformly distributed around the ring.

## 4.6 The Lengths and Strengths of Covalent Bonds

The space-filling model of ozone in **Figure 4.11(a)** shows two equivalent oxygen–oxygen bonds. Both are 128 pm long, which is between the length of the typical O═O double bond (121 pm) and O—O single bond (148 pm) shown in **Figure 4.11(b)**. In this section we use bond length and bond strength to rationalize and validate molecular structures.

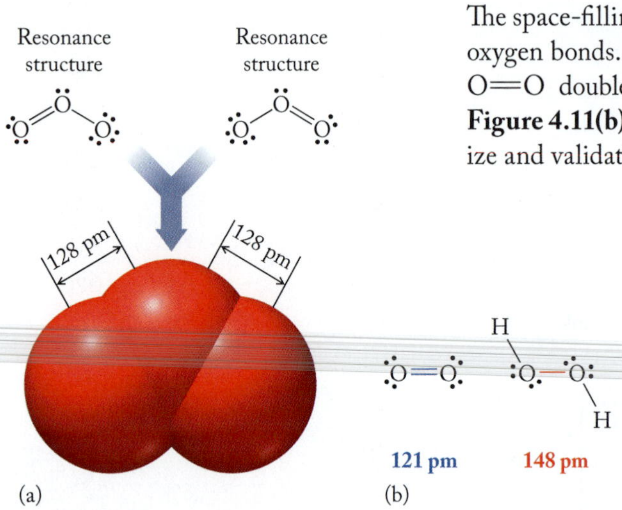

Resonance structure

Resonance structure

128 pm

128 pm

(a)

121 pm

148 pm

(b)

**FIGURE 4.11** (a) The molecular structure of ozone is an average of the two resonance structures shown at the top. Both bonds in ozone are 128 pm long. (b) The ozone bond length falls between the average length of an O═O double bond (121 pm) and the average length of an O—O single bond (148 pm). That intermediate value indicates that the bonds in ozone molecules are neither single bonds nor double bonds, but something between.

### Bond Length

The distance between atoms bonded to each other—that is, the length of the bond they form—depends on the identities of the two atoms and the number of bonds they form. The number of bonds is called the **bond order** and is equal to the number of pairs of electrons the atoms share. Experimental measurements of the lengths of the bonds in many molecular compounds indicate that the length of any given bond varies by only a few picometers in different compounds. For example, the C—H bond length in formaldehyde, $CH_2O$, is nearly the same as the C—H bond length in methane, $CH_4$. Similarly, the lengths of the C═O double bonds in formaldehyde and $CO_2$ are nearly identical (**Figure 4.12**).

As the bond order between two atoms increases, bond length decreases. We see that trend in the lengths of the O—O bond in hydrogen peroxide, $H_2O_2$, and the O═O bond in $O_2$ (Figure 4.11b); in the C═O bonds in $CO_2$ and the C≡O bond in CO (Figure 4.12); and in the average lengths of several bonds listed in **Table 4.6**. Bond length, then, can be used to determine bond order, and vice versa, even when bond order is not a simple whole number. For example, the actual bond order in $O_3$ is neither 1 nor 2 but rather 1.5, with each pair of O atoms sharing three electrons.

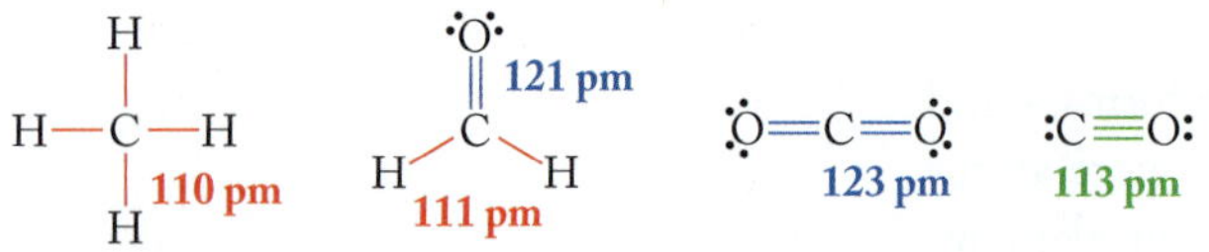

110 pm

121 pm

111 pm

123 pm

113 pm

**FIGURE 4.12** Bond lengths depend on the identity of the two atoms forming the bond and decrease with increasing bond order. The lengths of the C—H single bonds in $CH_4$ and $CH_2O$ are nearly the same, as are the lengths of the C═O double bonds in $CO_2$ and $CH_2O$. However, the C≡O triple bond in CO is much shorter than the C═O double bonds in $CH_2O$ and $CO_2$.

**SAMPLE EXERCISE 4.15** Determining Bond Order and Bond Length from Resonance Structures **LO7**

Draw the resonance structures of the carbonate ion, $CO_3^{2-}$, and from those structures calculate the bond order of the carbon–oxygen bonds and estimate their length.

**Collect and Organize** We are asked to draw the resonance structures of a polyatomic ion and then to determine the order and length of its bonds. A five-step procedure for drawing Lewis structures is described in Section 4.4. Table 4.6 contains average bond lengths, which vary inversely with bond order.

**Analyze** We drew the resonance structures of the nitrate ion, $NO_3^-$, in Sample Exercise 4.14. Carbon atoms (group 14) have one fewer valence electron than nitrogen atoms (group 15), but carbonate ions have one more negative charge than nitrate ions. Therefore, $CO_3^{2-}$ and $NO_3^-$ are isoelectronic and should have similar Lewis structures.

**Solve**

Step 1. The number of valence electrons in a $CO_3^{2-}$ ion is

| ELEMENT | | | VALENCE ELECTRONS | | |
|---|---|---|---|---|---|
| Symbol | No. of atoms | | No. of valence $e^-$ per atom | | Total |
| C | 1 | × | 4 | = | 4 |
| O | 3 | × | 6 | = | 18 |
| Plus two electrons for the 2− charge | | | | | 2 |
| **Valence electrons in $CO_3^{2-}$** | | | | | **24** |

Step 2. Connecting the carbon atom with single bonds to the three O atoms gives the following skeletal structure:

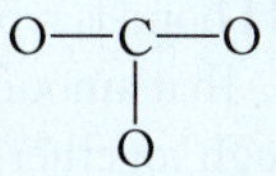

```
O—C—O
  |
  O
```

Step 3. Adding lone pairs of electrons to complete the octets of the three O atoms:

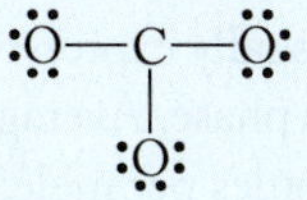

Step 4. Twenty-four valence electrons are in the structure, which matches the number determined in step 1.

Step 5. The central C atom has only six electrons, so we convert a lone pair on one of the oxygen atoms into a bonding pair:

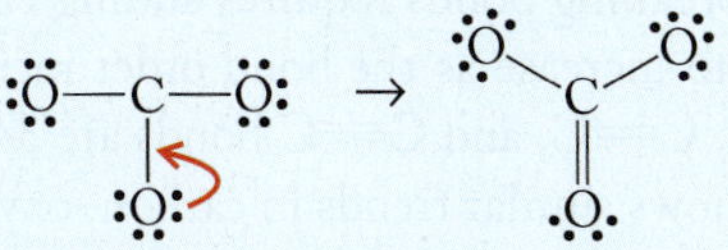

The carbon atom now has a complete octet. Adding brackets and the ionic charge, we have a complete Lewis structure:

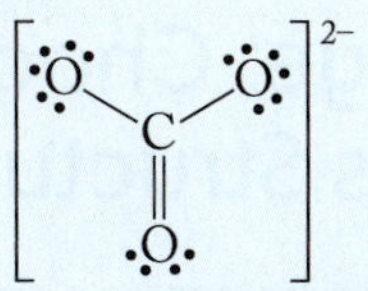

Using the O atom to the left or right of the central C atom to form the double bond creates two additional resonance forms, or three in all:

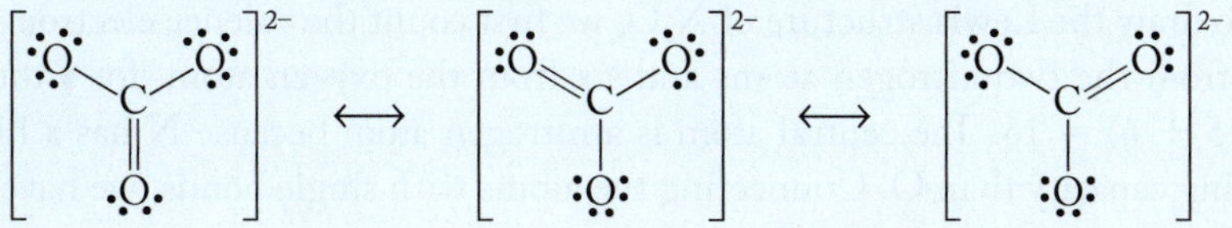

**TABLE 4.6 Average Lengths and Energies of Selected Covalent Bonds**

| Bond | Bond Length (pm) | Bond Energy (kJ/mol) |
|---|---|---|
| C—C | 154 | 348 |
| C=C | 134 | 614 |
| C≡C | 120 | 839 |
| C—N | 147 | 293 |
| C=N | 127 | 615 |
| C≡N | 116 | 891 |
| C—O | 143 | 358 |
| C=O | 123 | 743[a] |
| C≡O | 113 | 1072 |
| C—H | 110 | 413 |
| C—F | 133 | 485 |
| C—Cl | 177 | 328 |
| N—H | 104 | 391 |
| N—N | 147 | 163 |
| N=N | 124 | 418 |
| N≡N | 110 | 945 |
| N—O | 136 | 201 |
| N=O | 122 | 607 |
| N≡O | 106 | 678 |
| O—O | 148 | 146 |
| O=O | 121 | 498 |
| O—H | 96 | 463 |
| S—O | 151 | 265 |
| S=O | 143 | 523 |
| S—S | 204 | 266 |
| S—H | 134 | 347 |
| H—H | 74 | 436 |
| H—F | 92 | 567 |
| H—Cl | 127 | 431 |
| H—Br | 141 | 366 |
| H—I | 161 | 299 |
| F—F | 143 | 155 |
| Cl—Cl | 200 | 243 |
| Br—Br | 228 | 193 |
| I—I | 266 | 151 |

[a]The bond energy of the C—O bond in $CO_2$ is 799 kJ/mol.

Each resonance structure contains two C—O bonds and one C=O bond, which means a total of four bonding pairs of electrons is distributed evenly among three pairs of bonded atoms. Therefore, each bond consists of 4/3 = 1.33 bonding pairs, giving it a bond order of 1.33.

According to Table 4.6, the average length of a C—O bond is 143 pm and the average length of a C=O bond is 123 pm. The length of the bonds in $CO_3^{2-}$ ions should be between those values.

**Think About It** Here we used resonance structures to determine bond order and, in turn, to estimate bond length. The experimentally determined value of the bond lengths in $CO_3^{2-}$ is 129 pm, which confirms our estimate. In the text we used an experimentally determined bond length in ozone to determine bond order and confirm the validity of the resonance structures we had drawn. The process seems to work well in both directions.

**Practice Exercise** Draw the resonance structures of $HNO_3$. Are all the N—O bonds the same length?

*(Answers to Practice Exercises are in the back of the book.)*

## Bond Energies

The energy needed to break a H—H bond is represented by the depth of the potential energy minimum in Figure 4.2. That amount of energy (436 kJ/mol) is released when two H atoms come close enough together to form a bond. That same amount also must be used to break a single H—H bond—that is, to move the atoms so far apart that they are no longer electrostatically attracted to each other. Bond energy (or *bond strength*) for any bond is usually expressed in terms of the energy needed to break one mole of bonds in the gas phase. Average bond energies for some common covalent bonds, expressed in kilojoules per mole, are given in Table 4.6.

Bond energies (like bond lengths) are average values because they vary depending on the structure of the rest of the molecule. For example, the bond energy of a C=O bond in carbon dioxide is 799 kJ/mol, whereas the C=O bond energy in formaldehyde is only 743 kJ/mol. Bond energies are always positive quantities because breaking bonds requires adding energy.

Bond energies tend to increase as the bond order increases. For example, the bond energies of C—C, C=C, and C≡C bonds are 348, 614, and 839 kJ/mol, respectively. Table 4.6 shows similar trends in carbon–oxygen, nitrogen–nitrogen, and nitrogen–oxygen bonds.

# 4.7 Formal Charge: Choosing among Lewis Structures

Dinitrogen monoxide ($N_2O$), also known as nitrous oxide, is an atmospheric gas used as an anesthetic in dentistry and medicine. Its common name—laughing gas—is derived from the euphoria people feel when they inhale it.

To draw the Lewis structure of $N_2O$, we first count the valence electrons: five each from the two nitrogen atoms and six from the oxygen atom, for a total of $(2 \times 5 + 6) = 16$. The central atom is a nitrogen atom because N has a higher bonding capacity than O. Connecting the atoms with single bonds, we have

$$N—N—O$$

Completing the octets of the N and O atoms on the ends gives us a structure with 16 valence electrons, the number determined in step 1:

$$:\ddot{\underset{..}{N}}—N—\ddot{\underset{..}{O}}:$$

However, the central N atom has only two single bonds and thus only four valence electrons. To give it four more electrons and complete its octet, we need to convert lone pairs on the end atoms to bonding pairs. Which lone pairs should we choose? We have three choices. We could use two lone pairs from the N atom on the left to form a N≡N triple bond:

$$:N{\equiv}N—\ddot{\underset{..}{O}}:$$

Or we could use two lone pairs from the O atom to make a N≡O triple bond:

$$:\ddot{\underset{..}{N}}—N{\equiv}O:$$

Finally, we could use one lone pair from each end atom to make two double bonds:

$$\ddot{N}{=}N{=}\ddot{O}$$

Are those three resonance structures equivalent, as we saw with the carbonate ion $CO_3^{2-}$ in Sample Exercise 4.15? So far, in the sets of resonance structures we have considered, such as those of $O_3$ and $NO_3^-$, all the structures have been equivalent, so no one of them has been more important than another in representing the actual bonding in the molecule. However, the three resonance structures for $N_2O$ are not equivalent to one another because one contains a N≡N triple bond, one contains a N≡O triple bond, and one contains both a N=N double bond and a N=O double bond. To help us decide which resonance form in a nonequivalent set is the most important in representing the actual bonding pattern in a molecule, we apply the concept of formal charge.

A **formal charge (FC)** is not a real charge, but rather an accounting system for the number of electrons *formally assigned* to an atom in a structure of a molecule or polyatomic ion. To understand what formal charge means, let's calculate it for each of the three atoms in $N_2O$.

## Calculating Formal Charge

1. Determine the number of valence electrons in the free atom (the number of dots in its Lewis symbol).
2. Count the lone-pair electrons on the atom in the structure.
3. Count the electrons in bonds to the atom and divide that number by 2.
4. Sum the results of steps 2 and 3, and subtract that sum from the number determined in step 1.

Summarizing those steps in the form of an equation, we have

$$FC = \begin{pmatrix}\text{number of}\\ \text{valence e}^-\end{pmatrix} - \left[\begin{matrix}\text{number of e}^-\\ \text{in lone pairs}\end{matrix} + \frac{1}{2}\begin{pmatrix}\text{number of}\\ \text{shared e}^-\end{pmatrix}\right] \quad (4.2)$$

The calculation of formal charge assumes that each atom is *formally assigned* all its lone-pair electrons and half the electrons shared in bonding pairs. If that number matches the number of valence electrons in a single atom of the element (as reflected in its Lewis symbol), the formal charge on the atom is zero.

**formal charge (FC)** the value calculated for an atom in a molecule or polyatomic ion by determining the difference between the number of valence electrons in the free atom and the sum of the lone-pair electrons plus half the electrons in the atom's bonding pairs.

We can confirm that the three resonance forms of $N_2O$ we have drawn are not equivalent by calculating the formal charges on the three atoms in each structure. The steps in that calculation are highlighted in the table that follows, in which we have colored the lone pairs of electrons red and the shared pairs green to make the quantities of the electrons easier to track in the formal charge calculations. The numbers of valence electrons in free atoms of N (5) and O (6) are shown in blue.

**FORMAL CHARGE CALCULATIONS FOR THE RESONANCE STRUCTURES OF $N_2O$**

| STEP | :N≡N—Ö: A | | | :N̈=N=Ö: B | | | :N̈—N≡O: C | | |
|---|---|---|---|---|---|---|---|---|---|
| 1 Number of valence electrons | 5 | 5 | 6 | 5 | 5 | 6 | 5 | 5 | 6 |
| 2 Number of lone-pair electrons | 2 | 0 | 6 | 4 | 0 | 4 | 6 | 0 | 2 |
| 3 Number of shared electrons | 6 | 8 | 2 | 4 | 8 | 4 | 2 | 8 | 6 |
| 4 FC = valence $e^-$ – [lone pair $e^-$ + ½ (shared $e^-$)] | 0 | +1 | –1 | –1 | +1 | 0 | –2 | +1 | +1 |

To illustrate one of the formal charge calculations in the table, consider the N atom at the end in structure A. It has two electrons in a lone pair and six electrons in three shared (bonding) pairs. Using Equation 4.2 to calculate the formal charge on this N atom,

$$FC = 5 - [2 + \tfrac{1}{2}(6)] = 0$$

The results of similar FC calculations for all the other atoms in the three resonance structures complete the bottom row in the table. Note that the formal charges of the three atoms in each of the three structures sum to zero, as they should for a neutral molecule. When we analyze the formal charges of the atoms in a polyatomic ion, the formal charges on its atoms must add up to the charge on the ion.

Having calculated the formal charges, we now need to use the results to decide which of the three $N_2O$ structures best represents the bonding in those molecules. To decide, we use three criteria:

1. The best structure is the one in which the formal charge on each atom is zero.
2. If no such structure can be drawn, or if the structure is that of a polyatomic ion, the best structure is the one in which most atoms have formal charges equal to zero or as close to zero as possible.
3. Any negative formal charges should be on the atom(s) of the more/most electronegative element.

Let's apply those criteria to the nonequivalent resonance structures of $N_2O$. First of all, none of the structures meets criterion (1) because each structure has at least two nonzero FC values. So we proceed to criterion (2) to find the structure with the most FC values closest to zero, such as −1 or +1. When we do, we have a tie between structure A (0, +1, −1) and structure B (−1, +1, 0).

To break the tie, we invoke criterion (3) by asking the question, "In which structure is the negative formal charge on the more electronegative atom?" Oxygen is more electronegative than nitrogen, so structure A, in which the formal charge on the O atom is −1, best represents the actual bonding in $N_2O$.

How does calculating formal charges and choosing a best structure from among the nonequivalent possible structures compare with reality? The bond between the two nitrogen atoms is shorter than a N=N bond but longer than a N≡N bond, so the true structure is something between the left structure and the middle structure. As a result, we say that the middle structure "contributes to" the bonding in $N_2O$. That reality check is important as we interpret the results of formal charge analyses. Just because a resonance structure scores the best in an FC analysis does not mean that it matches experimental data. The true structure is often intermediate between two contributing structures.

## CONCEPT TEST

What is the formal charge on a sulfur atom that has three lone pairs of electrons and one bonding pair?

*(Answers to Concept Tests are in the back of the book.)*

Does a link exist between the calculated formal charge on an atom in a resonance structure and the bonding capacity of that atom? The Lewis symbol of nitrogen has three unpaired electrons, so a nitrogen atom can complete its octet by forming three bonds. The Lewis symbol of oxygen has two unpaired electrons, so the bonding capacity of an oxygen atom is two. In the $N_2O$ calculations, the nitrogen atom with three bonds has a formal charge of zero (structure A), and the oxygen atom with two bonds has a formal charge of zero (O in structure B). In general—and if the octet rule is obeyed—atoms have formal charges of zero in resonance structures in which the numbers of bonds they form match their bonding capacities. If the number of bonds an atom forms is one more than its bonding capacity, such as for an oxygen atom with three bonds (O in structure C), the formal charge is +1. If the number of bonds it forms is one fewer than the bonding capacity, such as for an oxygen atom with one bond (O in structure A), the formal charge is −1.

## SAMPLE EXERCISE 4.16 Selecting a Resonance Structure on the Basis of Formal Charges — LO6

Which of these resonance forms best describes the actual bonding in a molecule of $CO_2$?

$$:\ddot{\underset{\cdot\cdot}{O}}-C\equiv O: \longleftrightarrow \ddot{O}=C=\ddot{O} \longleftrightarrow :O\equiv C-\ddot{\underset{\cdot\cdot}{O}}:$$

**Collect, Organize, and Analyze** We are given three nonequivalent resonance structures for $CO_2$. Formal charges can be used to select the structure most likely to match the experimentally determined bond lengths. The preferred structure is one in which the formal charges are closest to zero and any negative formal charges are on the more electronegative atom (O). Here, oxygen is a group 16 element with six valence electrons, and carbon is a group 14 element with four valence electrons.

**Solve** We use Equation 4.2 to calculate the formal charge on each atom. The results are tallied in the following table, in which we have applied the same color-coding scheme used for the $N_2O$ calculations:

**FORMAL CHARGE CALCULATIONS FOR THE RESONANCE STRUCTURES OF $CO_2$**

| STEP | $:\ddot{O}-C\equiv O:$ | | | $\ddot{O}=C=\ddot{O}$ | | | $:O\equiv C-\ddot{O}:$ | | |
|---|---|---|---|---|---|---|---|---|---|
| 1 Number of valence electrons | 6 | 4 | 6 | 6 | 4 | 6 | 6 | 4 | 6 |
| 2 Number of lone pair electrons | 6 | 0 | 2 | 4 | 0 | 4 | 2 | 0 | 6 |
| 3 Number of shared electrons | 2 | 8 | 6 | 4 | 8 | 4 | 6 | 8 | 2 |
| 4 FC = valence $e^-$ − [lone pair $e^-$ + ½ (shared $e^-$)] | −1 | 0 | +1 | 0 | 0 | 0 | +1 | 0 | −1 |

The formal charges are all zero on the atoms in the structure with two double bonds. Therefore, that structure best represents the actual bonding in $CO_2$.

**Think About It** The sum of the formal charges is zero in all three resonance structures, as it should be for a neutral molecule.

**Practice Exercise** Which resonance structure(s) of the nitronium ion, $NO_2^+$ (see Practice Exercise 4.14), contribute(s) the most to the actual bonding in the ion?

*(Answers to Practice Exercises are in the back of the book.)*

## 4.8 Exceptions to the Octet Rule

Not all atoms achieve complete octets when forming covalent bonds. The Lewis symbol of beryllium, •Be•, indicates that a Be atom has the capacity to form two bonds. However, adding two more electrons to the two it already has results in only four electrons in its valence shell, not eight. Similarly, the three dots in the Lewis symbols of boron and aluminum indicate that their atoms can form three bonds, but doing so results in only six valence-shell electrons. Thus, according to Lewis theory, Be, B, and Al tend to form *electron-deficient* molecules. Other examples include molecules that have odd numbers of electrons.

### Odd-Electron Molecules

Traces of nitrogen monoxide (NO) enter the atmosphere from automobile exhaust and then react with $O_2$ to form nitrogen dioxide, $NO_2$, which plays a key role in forming the air pollution known as photochemical smog. Molecules of NO and $NO_2$ have odd numbers of valence electrons: $5 + 6 = 11$ in NO and $5 + 2 \times 6 = 17$ in $NO_2$. To understand what that means, let's draw the Lewis structure of NO. Because the molecule has only two atoms, it has no *central* atom. Therefore, we start with a single bond between N and O and then complete the octet around O, the more electronegative element:

$$\text{N—}\ddot{\underset{\cdot\cdot}{\text{O}}}\text{:}$$

Placing the remaining three electrons around the N atom

$$\dot{\underset{\cdot\cdot}{\text{N}}}\text{—}\ddot{\underset{\cdot\cdot}{\text{O}}}\text{:}$$

leaves the N atom with only five valence electrons. To increase the number, we convert a lone pair on the O atom into a bonding pair:

$$.\dot{\text{N}}\text{=}\ddot{\text{O}}\text{:}$$

That change has the added advantage of creating a double-bonded O atom, which gives it a formal charge of zero. The formal charge on nitrogen also is zero.

The only problem with that structure is that nitrogen has only seven electrons. Do we have a way to give it eight electrons? Not really. We could use another of the O atom lone pairs to make a bonding pair,

$$.\dot{\text{N}}\equiv\text{O:}$$ [structure crossed out]

but the N atom in that new structure now has *nine* valence-shell electrons, which is impossible for an atom with only four orbitals in its valence shell. Moving only one electron from the O atom to the N atom gives the N atom eight valence electrons, but now the O atom has only seven valence electrons:

:N=O.

That makes matters worse because now the more electronegative element (O), which attracts bonding electrons more strongly than the other (N), has an incomplete octet. As a result, choosing the structure in which oxygen has an octet but nitrogen does not is reasonable when not enough electrons are available to complete the octets of both. Compounds that contain unpaired valence electrons are called **free radicals**. They typically are very reactive species because acquiring or sharing an electron from another molecule or ion is often energetically favorable.

**free radical** an atom, ion, or molecule with unpaired electrons.

**SAMPLE EXERCISE 4.17** Drawing the Lewis Structures of an Odd-Electron Molecule **LO6**

Draw the resonance structures of nitrogen dioxide ($NO_2$) that have formal charges closest to zero.

**Collect, Organize, and Analyze** We need to draw the resonance structures of $NO_2$ that best describe the actual bonding in its molecules. Each molecule of $NO_2$ contains one atom of nitrogen (bonding capacity of three) and two atoms of oxygen (bonding capacity of two).

**Solve** The number of valence electrons is

| ELEMENT | | | VALENCE ELECTRONS | | |
|---|---|---|---|---|---|
| Symbol | No. of atoms | | No. of valence $e^-$ per atom | | Total |
| N | 1 | × | 5 | = | 5 |
| O | 2 | × | 6 | = | 12 |
| **Valence electrons in $NO_2$** | | | | | **17** |

Nitrogen has the greater bonding capacity, so it is the central atom:

O—N—O

One molecule of $NO_2$ contains an odd number (17) of valence electrons, which means one of its atoms has an incomplete octet. Completing the octets on the O atoms gives

:Ö—N—Ö:

The structure has 16 valence electrons, so we add the one remaining valence electron to the N atom:

:Ö—Ṅ—Ö:

Only five valence electrons are around the N atom, leaving it three short of an octet. We increase the number by converting a lone pair on one of the O atoms to a bonding pair.

:Ö—N—Ö:

Two equivalent resonance structures can be drawn with the formal charges shown in red:

O(0)=N(+1)—O(−1) ⟷ O(−1)—N(+1)=O(0)

**Think About It** The two resonance structures are equivalent because each contains one N—O single bond and one N═O double bond. Neither satisfies the octet rule, but the formal charges of the atoms are close to zero, and the negative formal charge is on the more electronegative element (oxygen). Both O atoms have complete octets, leaving the less electronegative N atom one electron short in the odd-electron molecule. The odd electron on the N atom in $NO_2$ makes it chemically reactive. For example, two $NO_2$ molecules may combine by sharing the odd electrons on their N atoms, forming a N—N bond and a molecule of $N_2O_4$:

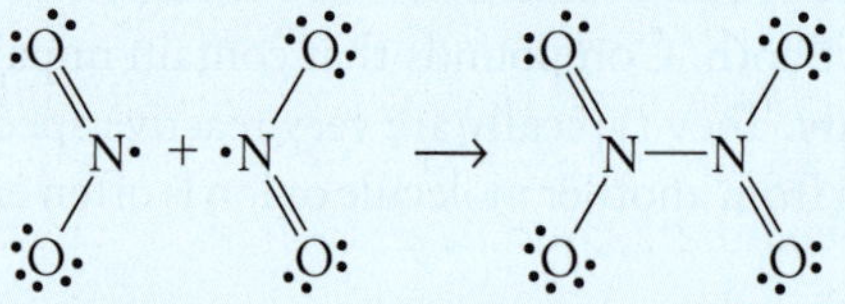

**Practice Exercise** Nitrogen trioxide ($NO_3$) may form in polluted air when $NO_2$ reacts with $O_3$. Draw its Lewis structure(s).

*(Answers to Practice Exercises are in the back of the book.)*

## Atoms with More than an Octet

In some molecules, atoms of nonmetals in the third row and below in the periodic table ($Z > 12$) appear to have more than an octet of valence electrons. Consider the Lewis structures of $PCl_5$ and $SF_6$:

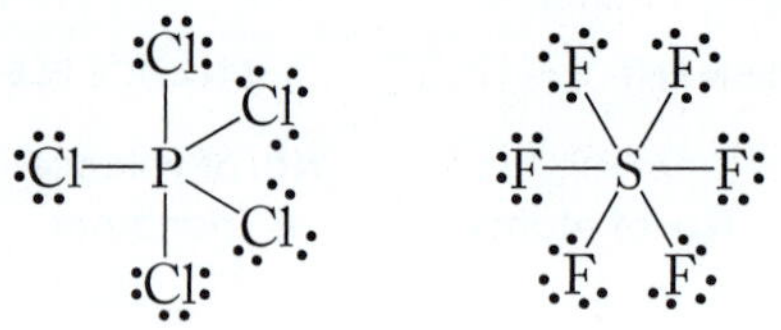

(a) (b)

The five covalent bonds in $PCl_5$ and the six in $SF_6$ indicate a total of 10 and 12 valence electrons around the central P and S atoms, respectively. All the atoms in both structures have formal charges of zero, so the two structures seem to be perfectly acceptable representations of the bonding in those molecules. But how can P and S atoms have more than eight valence electrons?

Over the years, chemists have debated several explanations for what many people have called the *hypervalency* of phosphorus, sulfur, and other nonmetals in the third row and below in the periodic table. Because those elements have empty valence-shell *d* orbitals, you might think that perhaps they use the *d* orbitals to form additional bonds. However, recent studies used computer models that allowed chemists to predict how different atomic orbitals contribute to bond formation. Those studies have shown that *d* orbitals contribute little to the bonding in molecules such as $PCl_5$ and $SF_6$. We will explore a theory of covalent bond formation in Chapter 5 that explains the bonding in those molecules without incorporating valence-shell *d* orbitals.

The fact that some atoms may have more than eight valence electrons within a compound does not necessarily mean they always do. Rather, that tends to happen when

1. They bond with strongly electronegative elements—particularly F, O, and Cl.
2. Doing so results in a structure whose atoms' formal charges are closer to zero.

Consider the S atom in a $SO_4^{2-}$ ion. Following the usual steps to draw its Lewis structure and assign formal charges, we get

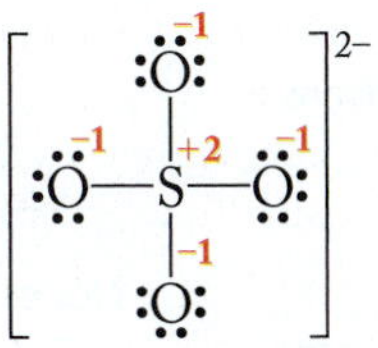

The sum of the formal charges on atoms in the ion is (+2) + [4 × (−1)] = −2, which is equal to the overall ionic charge, as it should be. However, we need to consider how that Lewis structure could be redrawn so that at least some formal charges are zero. We could draw such a structure by converting two lone pairs of electrons into bonding pairs:

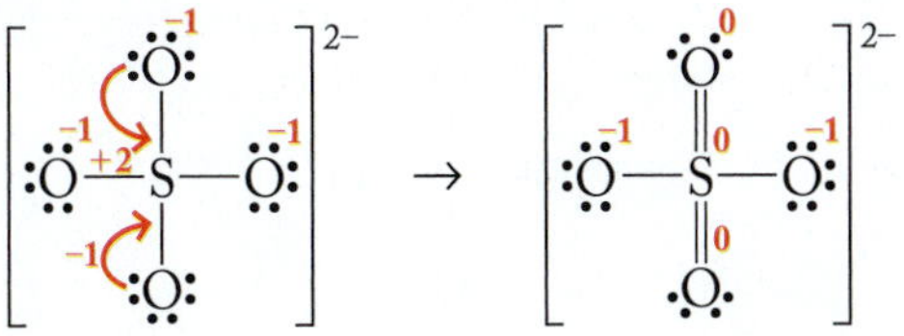

Each O atom still has a complete octet, but the S atom accommodates 12 electrons. In that structure the formal charge on the S atom is zero, as are the formal charges on two of the four O atoms. A formal charge of −1 is assigned to one of the other two oxygen atoms, which sums to an overall 2− charge of the ion. We could draw the two double bonds to any two of the O atoms, which means the structure is stabilized by resonance.

We can draw the Lewis structure of $H_2SO_4$ by combining two hydrogen ions ($H^+$) with the two oxygen atoms with 1− charges:

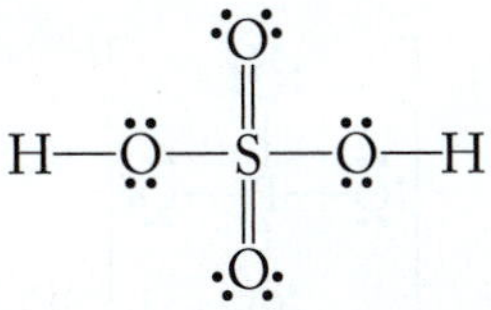

Each oxygen atom has an octet, and every atom has a formal charge of zero.

From the preceding analyses of formal charge, we might conclude that the preferred bonding pattern in $SO_4^{2-}$ ions and molecules of $H_2SO_4$ includes two S═O double bonds and zero formal charges all around. However, experimental evidence suggests that although the structures having zero formal charges do contribute to the bonding of $SO_4^{2-}$ and $H_2SO_4$, other structures that obey the octet rule and have no S═O double bonds contribute as well. Thus, the actual bonding in those particles is an average of the structures with an octet and the structures with more than an octet of valence electrons.

**SAMPLE EXERCISE 4.18** Drawing a Lewis Structure Containing an Atom with More than an Octet of Valence Electrons **LO6**

Draw a Lewis structure for the phosphate ion ($PO_4^{3-}$) that minimizes the formal charges on its atoms.

**Collect, Organize, and Analyze** Each ion contains one atom of phosphorus and four atoms of oxygen and has an overall charge of 3−. Phosphorus and oxygen are in groups

15 and 16 and have bonding capacities of three and two, respectively. Phosphorus is in row 3 ($Z = 15$), so it may exhibit hypervalency.

**Solve** The number of valence electrons is

| ELEMENT | | | VALENCE ELECTRONS | | |
|---|---|---|---|---|---|
| Symbol | No. of atoms | | No. of valence $e^-$ per atom | | Total |
| P | 1 | × | 5 | = | 5 |
| O | 4 | × | 6 | = | 24 |
| Plus three electrons for the 3− charge | | | | | 3 |
| **Valence electrons in $PO_4^{3-}$** | | | | | **32** |

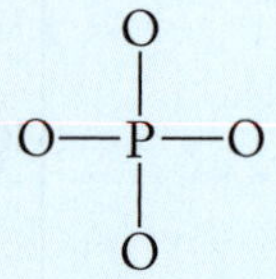

Each O atom needs three lone pairs of electrons to complete its octet:

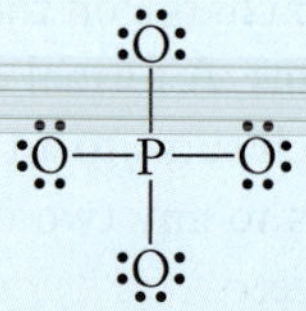

The structure has 32 valence electrons, which matches the number in the ion. To complete the Lewis structure, we add brackets and its electrical charge:

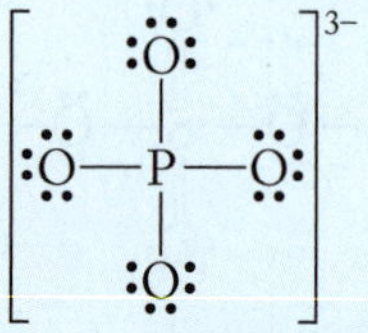

Each O has a single bond and a formal charge of −1; the four bonds around the P atom are one more than its bonding capacity, so its formal charge is +1. The sum of the formal charges, $+1 + 4(-1) = -3$, matches the charge on the ion, 3−.

We can reduce the formal charge on P by increasing the number of bonds to it, and we can do that by converting a lone pair on one of the O atoms into a bonding pair:

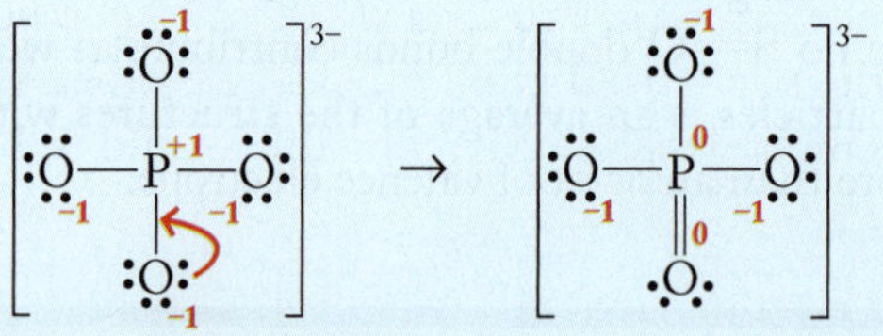

At the same time, we change a single-bonded O atom into a double-bonded O atom and thereby make its formal charge zero. Therefore, the structure on the right, in which the P atom has more than an octet of valence electrons, is the one that minimizes the formal charges on the phosphate ion.

**Think About It** The phosphorus atom in the final structure has more than an octet of valence electrons. It is also stabilized by resonance: we can draw the P═O double bond between any of the four O atoms and the central P atom.

**Practice Exercise** Draw the resonance structures of the selenite ion ($SeO_3^{2-}$) that minimize the formal charges on the atoms.

*(Answers to Practice Exercises are in the back of the book.)*

# 4.9 Vibrating Bonds and the Greenhouse Effect

**CHEMTOUR** Vibrational Modes

**CHEMTOUR** Greenhouse Effect

Covalent bonds are not rigid. They all vibrate a little, stretching and bending like tiny atomic-sized springs (**Figure 4.13**). As polar bonds vibrate, the strengths of tiny electrical fields produced by the partial separations of charge in the bonds fluctuate at the same frequencies as their vibrations. The natural frequencies of the vibrations correspond to frequencies of infrared radiation. As we described in Chapter 3, all forms of radiant energy, including infrared rays, travel through space as oscillating electrical and magnetic fields. Now, suppose a photon of infrared radiation traveling through Earth's atmosphere strikes a molecule containing a polar bond vibrating at the same frequency as the photon. The fluctuating fields of the photon and the vibrating bond may interact. The molecule might absorb that photon, temporarily increasing the molecule's internal energy, and later emit a photon of the same energy as it returns to its ground state. That molecule–photon interaction is at the heart of the greenhouse effect.

To understand the connection between vibrating bonds and a warming atmosphere, recall that infrared radiation is the part of the electromagnetic spectrum that we cannot see but that we feel as heat. Any warm object, including Earth's surface, emits infrared radiation. When infrared photons strike atmospheric molecules that contain polar bonds, such as $CO_2$, the photons may be absorbed. When they are reemitted, they are just as likely to move back toward Earth's surface as they are to go upward toward space. In that process of absorption and reemission, a significant fraction of the heat flowing from Earth's surface is trapped in the atmosphere, much in the way that the glass covering a greenhouse traps heat inside it.

Not all polar bond vibrations result in the absorption and emission of infrared radiation. For example, two kinds of stretching vibrations can occur in a molecule of $CO_2$, which has two C=O bonds on opposite sides of the central C atom. One is a *symmetric* stretching vibration (Figure 4.13a), in which the two C=O bonds stretch and then compress at the same time. The two fluctuating electrical fields produced by the two C=O bonds cancel each other out, and no infrared absorption or emission is possible. That vibration is said to be *infrared inactive.*

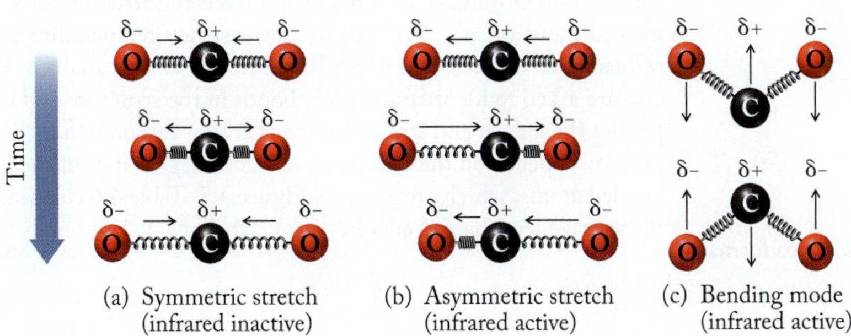


**FIGURE 4.13** Three modes of bond vibration in a molecule of $CO_2$ include (a) symmetric stretching of the C=O bonds, which produces no overall change in the polarity of the molecule; (b) asymmetric stretching, which produces side-to-side fluctuations in polarity that may result in absorption of infrared (IR) radiation; and (c) a bending mode, which produces up-and-down fluctuations that may also absorb IR radiation.

However, when the bonds stretch such that one C═O bond gets shorter as the other gets longer (Figure 4.13b), the changes in charge separation do not cancel. That *asymmetric* stretch produces a fluctuating electrical field that enables $CO_2$ to absorb infrared radiation, so the vibration is *infrared active*. Molecules can also bend (Figure 4.13c) to produce fluctuating electrical fields that match the frequencies of other photons of infrared radiation. Because the frequencies of the asymmetric stretching and bending of the bonds in $CO_2$ are in the same range as the frequencies of infrared radiation emitted from Earth's surface, carbon dioxide is a potent greenhouse gas.

**CONNECTION** The average temperature of Earth's surface is 287 K, which means that it emits its peak intensity of electromagnetic radiation in the infrared region (see Figure 3.10).

## CONCEPT TEST

Nitrogen and oxygen make up about 99% of the gases in the atmosphere. Is the stretching of the N≡N and O═O bonds in the molecules infrared active? Why or why not?

*(Answers to Concept Tests are in the back of the book.)*

In this chapter we have explored the electrostatic potential energy between oppositely charged ions that leads to ionic bond formation. We also have explored the nature of the covalent bonds that hold together molecules and polyatomic ions, observing that the bonds owe their strength to the pairs of electrons shared between nuclei of atoms. Sharing does not necessarily mean equal sharing, and unequal sharing coupled with bond vibration accounts for the ability of some atmospheric gases to absorb and emit infrared radiation. In so doing, the molecules function as potent greenhouse gases.

Early in the chapter we noted that moderate concentrations of greenhouse gases are required for climate stability and to make our planet habitable. The escalating concern of many scientists is that Earth's climate is being destabilized by too much of a good thing. Policies being made by the world's governments today will significantly affect the problem of climate change, one way or the other. As an informed member of the world community, you have the opportunity to influence how those policy decisions are made. We hope that you will make the most of that opportunity.

## SAMPLE EXERCISE 4.19 Integrating Concepts: Mothballs

A compound often referred to by the acronym PDB is the active ingredient in most mothballs. It is also used to control mold and mildew, as a deodorant, and as a disinfectant. Tablets containing it are often stuck under the lids of garbage cans or placed in the urinals in public restrooms, producing a distinctive aroma. Molecules of PDB have the following skeletal structure:

```
       H     H
        \   /
         C—C
        /   \
  Cl—C       C—Cl
        \   /
         C—C
        /   \
       H     H
```

a. Draw the Lewis structure of PDB and note any nonzero formal charges.

b. Is the structure stabilized by resonance? If so, draw all resonance structures.

c. In the structure you drew, which bonds, if any, are polar?

d. Predict the average carbon–carbon bond length and bond strength in the structure you drew.

**Collect and Organize** We are given the skeletal structure of a molecule and are asked to draw its Lewis structure, including all resonance structures, and to perform a formal charge analysis. We also are asked to identify any polar bonds in the structure and to predict the length and strength of the carbon–carbon bonds. Bond polarity depends on the difference in electronegativities of the bonded atoms, which are given in Figure 4.5. Table 4.6 contains the average lengths and energies (strengths) of covalent bonds.

**Analyze** The five-step procedure used in Sample Exercises 4.10–4.14 to draw the Lewis structures of other small molecules should be useful in drawing the Lewis structure of PDB.

**Solve**

a and b. The number of valence electrons is

| ELEMENT | | | VALENCE ELECTRONS | | |
|---|---|---|---|---|---|
| Symbol | No. of atoms | | No. of valence e⁻ per atom | | Total |
| C | 6 | × | 4 | = | 24 |
| H | 4 | × | 1 | = | 4 |
| Cl | 2 | × | 7 | = | 14 |
| **Valence electrons in molecule of PDB** | | | | | **42** |

Completing the octets on the Cl atoms

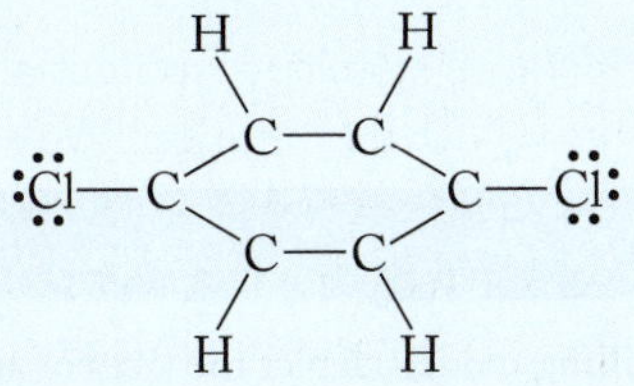

gives us a structure with 36 valence electrons (12 bonding pairs and 6 lone pairs). We need six more. We add six by adding three more bonds between carbon atoms, turning three C—C single bonds into C═C double bonds. We have to distribute them evenly around the ring to avoid any C atoms with five bonds. Two equivalent resonance structures, analogous to those for benzene (Figure 4.10), can be drawn to show the bonding pattern:

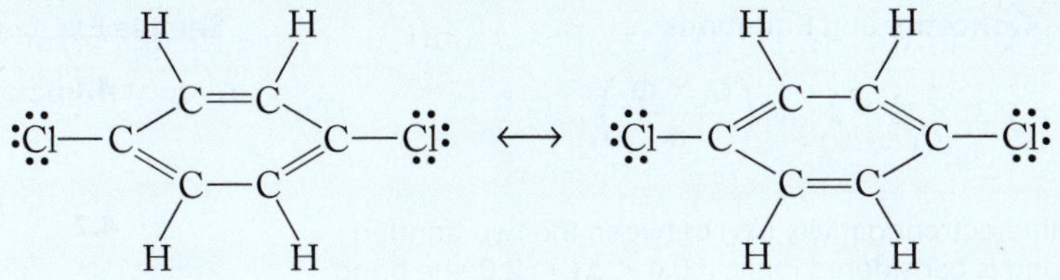

Resonance stabilizes the structure of PDB. Each C atom has four bonds, and each H and Cl atom has one bond, so every atom has the number of bonds that matches its bond capacity. That means that all formal charges are zero.

c. The differences in electronegativities for the bonded pairs of atoms are

C—C $\Delta\chi = 0$

Cl—C $\Delta\chi = 3.0 - 2.5 = 0.5$

C—H $\Delta\chi = 2.5 - 2.1 = 0.4$

Of the three pairs, only the Cl—C bond meets our polar bond guidelines ($0.4 < \Delta\chi < 2.0$).

d. The even distribution of a total of nine bonding pairs of electrons among six C atoms means that, on average, each pair shares 1.5 pairs of bonding electrons. The corresponding bond length and bond strength should be about halfway between those of the C—C single and C═C double bonds given in Table 4.6:

Approximate bond length:

$$[(154 + 134)/2] \text{ pm} = 144 \text{ pm}$$

Approximate bond strength:

$$[(348 + 614)/2] \text{ kJ/mol} = 481 \text{ kJ/mol}$$

**Think About It** The resonance structures resemble those of benzene, which is reflected in the common name of PDB, *para*-dichlorobenzene. Note that the two polar C—Cl bonds in PDB are oriented in opposite directions. Thus, the unequal sharing of the bonding pair of electrons in the Cl—C bond on the left side of the molecule is offset by the unequal sharing of the bonding pair of electrons in the C—Cl bond on the right side. In Chapter 5 we will explain that offsetting bond polarities in symmetrical molecules such as PDB explains why those substances are nonpolar overall.

## SUMMARY

**LO1** A chemical bond results from two ions being attracted to each other (an **ionic bond**) or from two atoms sharing electrons (a **covalent bond**). The atoms in metallic solids pool their electrons to form **metallic bonds**. (Section 4.1)

**LO2 Electrostatic potential energy ($E_{el}$)** is a measure of the strength of the attractions between cations and anions in an ionic compound. $E_{el}$ is directly proportional to the product of the ion charges and inversely proportional to the distance between the nuclei of the ions. (Section 4.1)

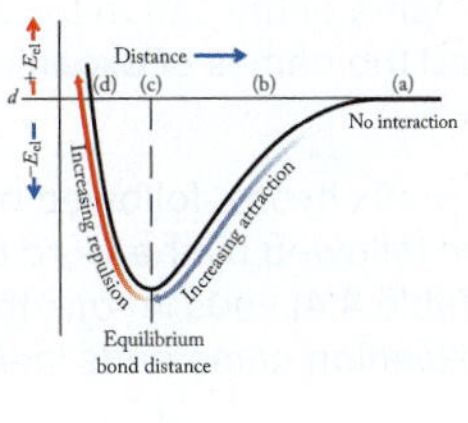

**LO3** Unequal electron sharing between atoms of different elements results in **polar covalent bonds**. Bond polarity is a measure of how unequally the electrons in covalent bonds are shared. More polarity results from greater differences between the **electronegativities** of the bonded atoms. Electronegativity generally increases with increasing ionization energy. (Section 4.2)

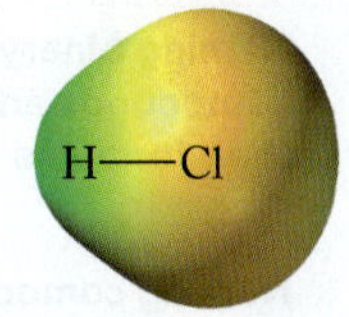

**LO4** Standardized naming conventions allow us to translate the names of molecular and ionic compounds into their chemical formulas and vice versa. (Section 4.3)

**LO5** **Lewis symbols** use dots to represent paired and unpaired electrons in the ground states of atoms. The number of unpaired electrons indicates how many bonds the element is likely to form—that is, its **bonding capacity**. Chemical stability is achieved when atoms have eight electrons in their valence *s* and *p* orbitals, which are then said to be following the **octet rule**. A **Lewis structure** shows the bonding pattern in molecules and polyatomic ions; pairs of dots represent **lone pairs** of electrons that do not contribute to bonding. A **single bond** consists of a single pair of electrons shared between two atoms; a **double bond** has two shared pairs, and a **triple bond** has three shared pairs. (Section 4.4)

**LO6** Two or more equivalent Lewis structures—called **resonance structures**—can sometimes be drawn for one molecule or polyatomic ion. The actual bonding pattern in a molecule is an average of equivalent resonance structures. The preferred resonance structure of a molecule is one in which the **formal charge (FC)** on each atom is zero or as close to zero as possible, and any negative formal charges are on the more electronegative atoms. **Free radicals** include reactive molecules that have an odd number of valence electrons and contain atoms with incomplete octets. Atoms of elements in the third row of the periodic table with $Z > 12$ and beyond may accommodate more than an octet of electrons. (Sections 4.5, 4.7, and 4.8)

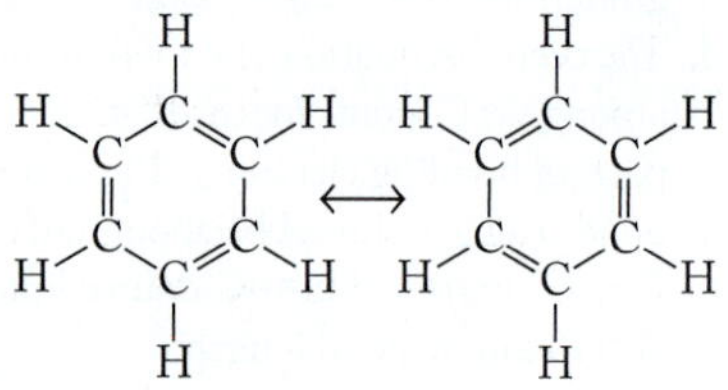

**LO7** **Bond order** is the number of bonding pairs in a covalent bond. **Bond energy** is the energy change that accompanies the breaking of one mole of a particular covalent bond in the gas phase. As the bond order between two atoms increases, the bond length decreases and the bond energy increases. (Section 4.6)

**LO8** Covalent bonds behave more like flexible springs than rigid rods. The bonds can undergo a variety of vibrations. The vibrations of polar bonds may create fluctuating electrical fields that allow molecules to absorb infrared (IR) electromagnetic radiation. When atmospheric gases absorb IR radiation, they contribute to the greenhouse effect. (Section 4.9)

## PARTICULATE **PREVIEW WRAP-UP**

1. Four single (C—H) bonds are in the molecule of $CH_4$ in image (a); two double (C═O) bonds are in the molecule of $CO_2$ in image (b).
2. Carbon atoms form four single bonds in $CH_4$ and two double bonds in $CO_2$.
3. Space-filling models depict the sizes of atoms and the shapes of molecules, but they don't show the types of bonds (single, double, or triple) or lone pairs of electrons.

## PROBLEM-SOLVING SUMMARY

| Type of Problem | Concepts and Equations | Sample Exercises |
|---|---|---|
| **Calculating the electrostatic potential energy of ionic bonds** | $E_{el} = 2.31 \times 10^{-19}\ \text{J} \cdot \text{nm}\left(\frac{Q_1 \times Q_2}{d}\right)$ (4.1) | **4.1** |
| **Comparing bond polarities** | Calculate the difference in electronegativity ($\Delta\chi$) between the two bonded atoms. If $\Delta\chi \geq 2.0$, the bond is considered ionic; if $0.4 < \Delta\chi < 2.0$, the bond is considered polar covalent; if $\Delta\chi \leq 0.4$, the bond is considered nonpolar covalent. | **4.2** |
| **Naming binary molecular compounds and writing their formulas** | First write the name of the element to the left of or, if the elements are in the same group, below the other one in the periodic table. Then write the name of the other element, changing its ending to *-ide*. Use the prefixes in Table 4.3 to indicate the number of atoms of each element. | **4.3** |
| **Naming binary ionic compounds and writing their formulas** | First write the name of the cation's parent element; if it is a transition metal that forms ions with different charges, use a Roman numeral to represent the charge. Then write the name of the anion's parent element with its ending changed to *-ide*. | **4.4, 4.5** |
| **Naming compounds of polyatomic ions and writing their formulas** | As with a binary compound, write the name of the cation followed by the name of the anion. Use Table 4.4 to find the names of oxoanions (which end in *-ate* or *-ite*) and other polyatomic ions. | **4.6, 4.7** |
| **Naming acids and writing their formulas** | For a binary acid (HX), begin with the prefix *hydro-* followed by the name of element X, but change its ending to *-ic* followed by the word *acid*. For an oxoacid, if the name of its oxoanion (Table 4.4) ends in *-ate*, the corresponding oxoacid name ends in *-ic acid*. If the oxoanion name ends in *-ite*, the oxoacid name ends in *-ous acid*. | **4.8** |

| Type of Problem | Concepts and Equations | Sample Exercises |
|---|---|---|
| **Drawing Lewis structures** | Connect the atoms with single covalent bonds, distributing the valence electrons to give each outer atom eight valence electrons (except two for H); use multiple bonds where necessary to complete the central atom's octet. | **4.9–4.13** |
| **Drawing resonance structures** | Include all possible arrangements of covalent bonds in the molecule if more than one equivalent structure can be drawn. | **4.14** |
| **Determining bond order and bond length from resonance structures** | Draw resonance structures to determine the average bond order for the equivalent bonds. Relate bond order to bond length by using Table 4.6. | **4.15** |
| **Selecting resonance structures on the basis of formal charges** | Calculate formal charge on each atom by using $$\text{FC} = \begin{pmatrix}\text{number of}\\ \text{valence e}^-\end{pmatrix} - \left[\begin{matrix}\text{number of e}^-\\ \text{in lone pairs}\end{matrix} + \frac{1}{2}\begin{pmatrix}\text{number of}\\ \text{shared e}^-\end{pmatrix}\right] \quad (4.2)$$ Select structures with formal charges closest to zero and with negative formal charges on the most electronegative atoms. | **4.16** |
| **Drawing Lewis structures of odd-electron molecules** | Distribute the valence electrons in the Lewis structure to leave the most electronegative atom(s) with eight valence electrons and the least electronegative atom with the odd number of electrons. | **4.17** |
| **Drawing Lewis structures containing atoms with more than an octet of valence electrons** | Distribute the valence electrons in the Lewis structure, allowing atoms of elements in period 3 and beyond to have more than eight valence electrons if more than four bonds are needed or if doing so results in a structure with formal charges closer to zero. | **4.18** |

## VISUAL PROBLEMS

*(Answers to boldface end-of-chapter questions and problems are in the back of the book.)*

**4.1.** Which Lewis symbol in Figure P4.1 correctly portrays the most stable ion of aluminum?

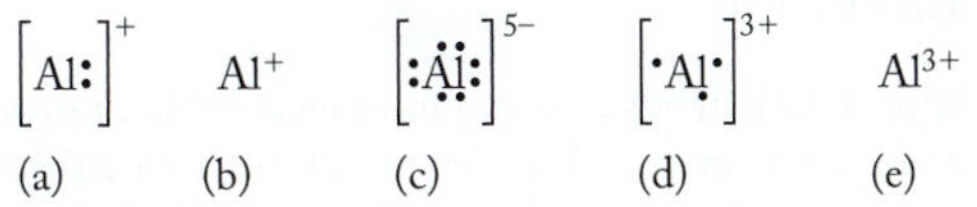

$[\text{Al:}]^{+}$ (a) $\quad \text{Al}^{+}$ (b) $\quad [\text{:}\ddot{\text{Al}}\text{:}]^{5-}$ (c) $\quad [\cdot\text{Al}\cdot]^{3+}$ (d) $\quad \text{Al}^{3+}$ (e)

**FIGURE P4.1**

4.2. Which Lewis symbols in Figure P4.2 are correct?

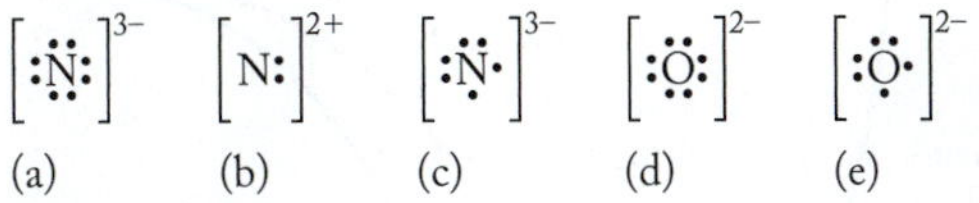

$[\text{:}\ddot{\text{N}}\text{:}]^{3-}$ (a) $\quad [\text{N:}]^{2+}$ (b) $\quad [\text{:}\dot{\text{N}}\cdot]^{3-}$ (c) $\quad [\text{:}\ddot{\text{O}}\text{:}]^{2-}$ (d) $\quad [\text{:}\ddot{\text{O}}\cdot]^{2-}$ (e)

**FIGURE P4.2**

*4.3. Which, if any, of the Lewis structures in Figure P4.3 are resonance structures for the thiocyanate ion ($SCN^-$)? Explain your selection(s).

$[\text{:}\ddot{\text{S}}\text{—N}\equiv\text{C:}]^-$ $\quad [\text{:N}\equiv\text{C—}\ddot{\text{S}}\text{:}]^-$ $\quad [\text{:}\ddot{\text{N}}\text{—S}\equiv\text{C:}]^-$

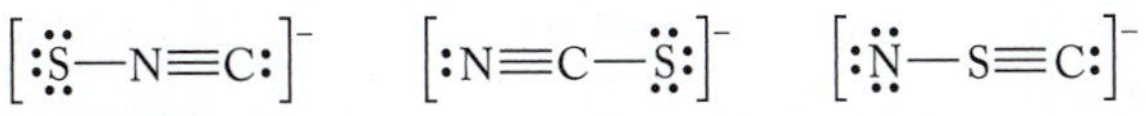

**FIGURE P4.3**

*4.4. Which of the Lewis structures in Figure P4.4 are resonance forms of the molecule $S_2O$? Explain your selections.

$\text{:}\ddot{\text{S}}\text{=}\ddot{\text{S}}\text{—}\ddot{\text{O}}\text{:}$ $\quad \text{:}\ddot{\text{S}}\text{—}\ddot{\text{O}}\text{=}\ddot{\text{S}}\text{:}$

S=S—O (bent) $\quad$ S—O=S (bent)

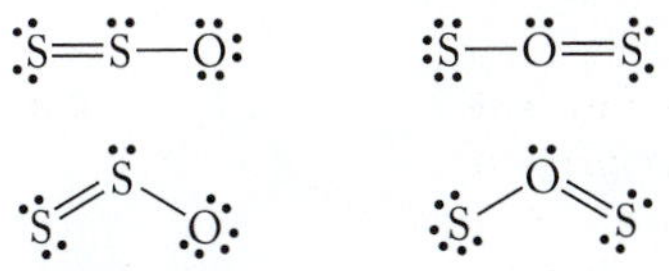

**FIGURE P4.4**

*Note*: The color scale used to indicate electron density in Problems 4.5–4.8 and 4.12 is the same as in Figure 4.4, where violet represents a charge of 1+, dark red is 1−, and yellow-green is 0.

**4.5.** Which image in Figure P4.5 best describes the distribution of electron density in BrCl?

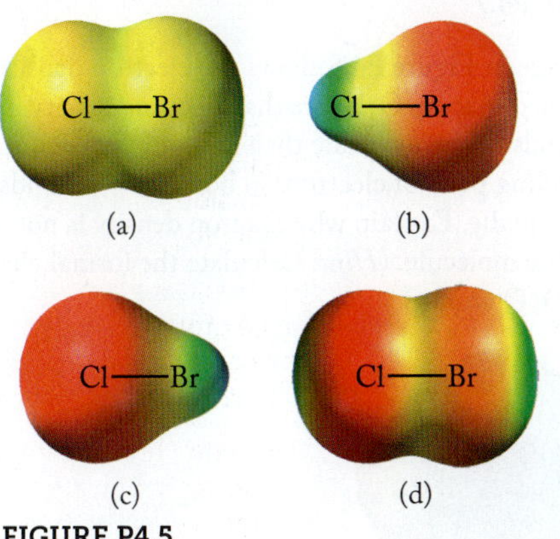


**FIGURE P4.5**

**4.6.** Which image in Figure P4.6 best describes the distribution of electron density in CsI?

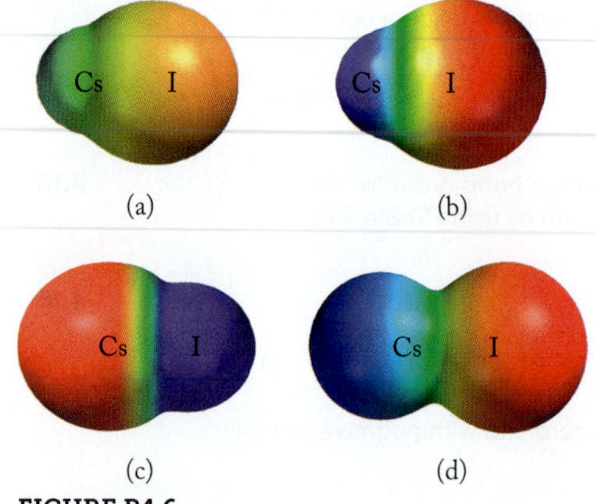


**FIGURE P4.6**

**4.7.** Which image in Figure P4.7 most accurately describes the distribution of electron density in $SO_2$? Explain your answer.

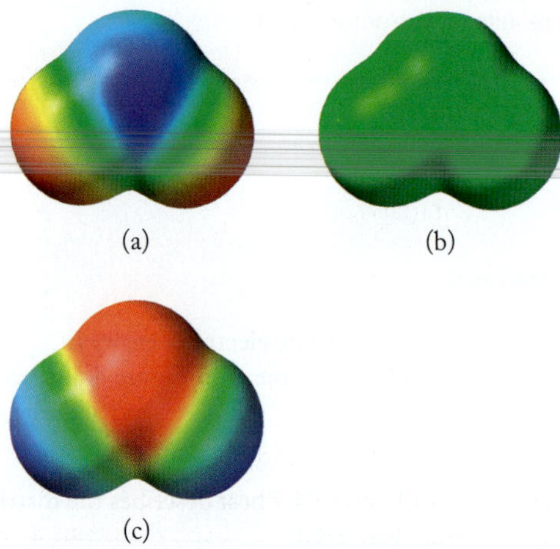


**FIGURE P4.7**

***4.8.** The image in Figure P4.8 shows the electron density in a molecule of ozone, $O_3$. Note that electron density is higher at the ends of the molecule than in its center even though the bonding pairs of electrons in both O—O bonds are shared equally. Explain why electron density is not uniform across the molecule. (*Hint*: Calculate the formal charges on the three O atoms.)

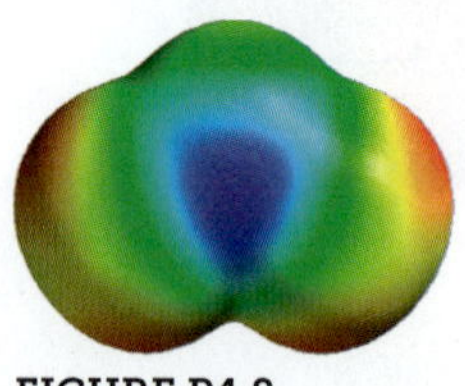

**FIGURE P4.8**

**4.9.** Water in the atmosphere is a greenhouse gas, which means its molecules are transparent to visible light but may absorb photons of infrared radiation. Which of the three modes of bond vibration shown in Figure P4.9 are infrared active? Note that molecules of $H_2O$ are not linear.

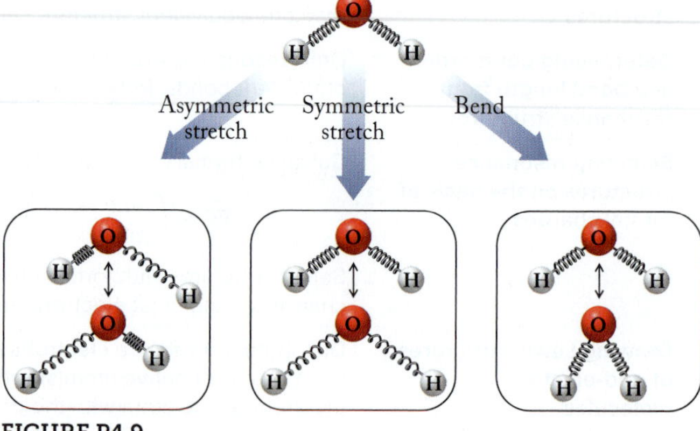


**FIGURE P4.9**

**4.10.** Which highlighted elements in Figure P4.10 may have more than eight valence electrons when bonding to a highly electronegative element?

**FIGURE P4.10**

**4.11.** Figure P4.11 shows two graphs of electrostatic potential energy versus internuclear distance. One is for a pair of potassium and chloride ions, and the other is for a pair of potassium and fluoride ions. Which is which?

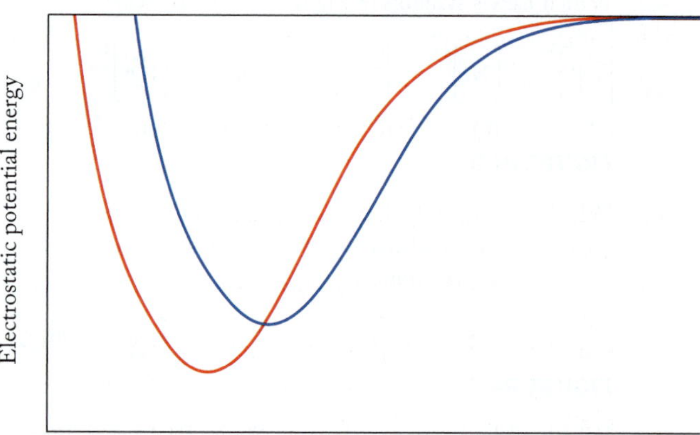


**FIGURE P4.11**

4.12. Use images [A] through [I] in Figure P4.12 to answer questions a–f.
 a. Which processes require energy?
 b. Which processes release energy?
 c. Which process contributes to the greenhouse effect?
 d. Which representations depict ionic bonding?
 e. Which representations depict covalent bonding?
 f. Which representation depicts both covalent bonding *and* ionic bonding?

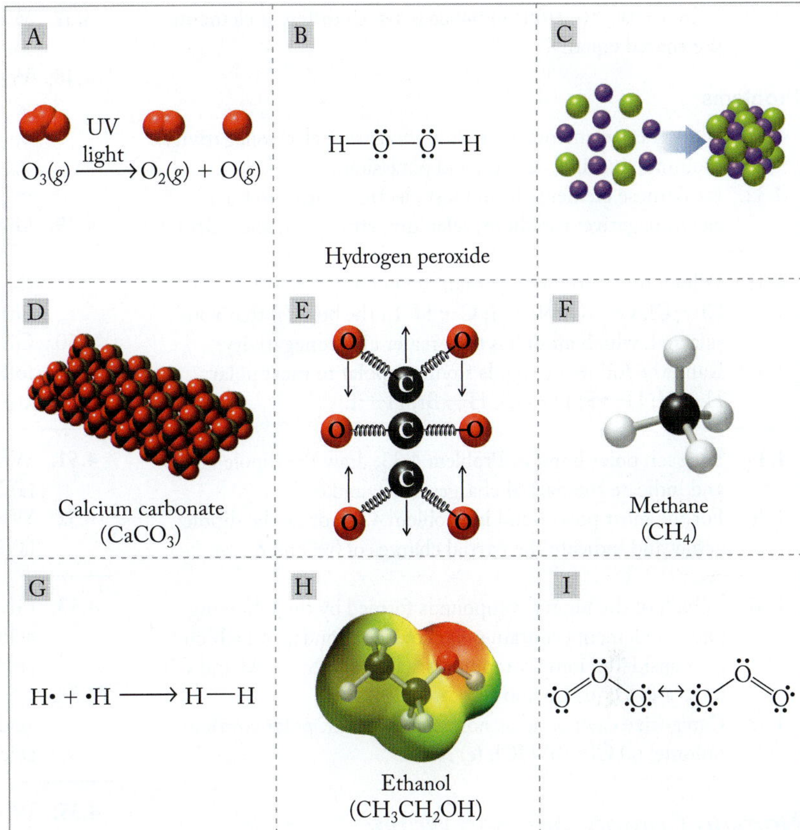


**FIGURE P4.12**

# QUESTIONS AND PROBLEMS

## Chemical Bonds and Greenhouse Gases

### Concept Review

**4.13.** Describe the differences in bonding between *covalent* and *ionic* compounds.

4.14. How can a compound contain both covalent and ionic bonds?

**4.15.** In a hydrogen molecule, what particles are electrostatically attracted to each other? What particles electrostatically repel each other?

4.16. What name is given to the distance at which the potential energy of electrostatic attractions and repulsions is minimized in a molecule?

**4.17.** In an ionic compound, what particles are electrostatically attracted to each other? Name two factors that affect the strength of those attractions.

4.18. What structural feature of metal atoms enables the positively charged nuclei to be electrostatically attracted to electrons that can move throughout the lattice of atoms?

### Problems

**4.19.** Calculate the electrostatic potential energy between a pair of potassium and bromide ions in solid KBr. (*Hint*: See Figure 3.36.)

4.20. Calculate the electrostatic potential energy between a pair of aluminum and oxide ions in solid $Al_2O_3$.

**4.21.** Which of these substances has the most negative lattice energy? (a) KCl; (b) $TiO_2$; (c) $BaCl_2$; (d) KI

4.22. Rank the following ionic compounds, which have the same crystal structure, from least negative to most negative lattice energy: CsCl, CsBr, and CsI.

**4.23.** Rank the following ionic compounds in order of increasing coulombic attraction between their ions: KBr, $SrBr_2$, and CsBr.

4.24. Rank the following ionic compounds in order of increasing coulombic attraction between their ions: BaO, $BaCl_2$, and CaO.

## Electronegativity, Unequal Sharing, and Polar Bonds

### Concept Review

**4.25.** How can we use electronegativity to predict whether a bond between two atoms is likely to be covalent or ionic?

4.26. How do the electronegativities of the elements change across a row and down a group in the periodic table?

**4.27.** How are trends in electronegativity related to trends in atomic size?

4.28. Is the element with the most valence electrons in a row of the periodic table also the most electronegative?

**4.29.** What does the term *polar covalent bond* mean? How does a polar covalent bond differ from an ionic bond?

4.30. Why are the electrons in bonds between different elements not shared equally?

### Problems

**4.31.** Rank these elements in order of increasing electronegativity: bromine, fluorine, cesium, and potassium.

4.32. Rank these elements from most electronegative to least electronegative: rubidium, selenium, strontium, and sulfur.

---

**4.33.** Which of the following bonds are polar? C—Se; C—O; Cl—Cl; O=O; N—H; C—H. In the bond(s) that you selected, which atom has the greater electronegativity?

4.34. Rank the following bonds from nonpolar to most polar: H—H; H—F; H—Cl; H—Br; H—I.

---

**4.35.** For each polar bond in Problem 4.33, draw the dipole arrow and indicate the partial charges of $\delta+$ and $\delta-$.

4.36. For the most polar bond in Problem 4.34, draw the dipole arrow and indicate the partial charges of $\delta+$ and $\delta-$.

---

**4.37.** Which of the binary compounds formed by the following pairs of elements contain polar covalent bonds, and which are considered ionic compounds? (a) C and S; (b) Al and Cl; (c) C and O; (d) Ca and O

4.38. Categorize each bond as nonpolar covalent, polar covalent, or ionic: (a) $Cl_2$; (b) HCl; (c) NaCl.

## Naming Compounds and Writing Formulas

### Concept Review

**4.39.** What is the role of Roman numerals in the names of the compounds formed by transition metals?

4.40. Why does the name of a binary ionic compound in which the cation is from a group 1 or group 2 element not need a Roman numeral after the element's name?

**4.41.** Consider a mythical element X, which forms two oxoanions: $XO_2^{2-}$ and $XO_3^{2-}$. Which has a name that ends in *-ite*?

4.42. Concerning the oxoanions in Problem 4.41, would either name require a prefix such as *hypo-* or *per-*? Explain why or why not.

### Problems

**4.43.** What are the names of these compounds of nitrogen and oxygen? (a) $NO_3$; (b) $N_2O_5$; (c) $N_2O_4$; (d) $NO_2$; (e) $N_2O_3$; (f) NO; (g) $N_2O$; (h) $N_4O$

4.44. More than a dozen neutral compounds containing only sulfur and oxygen have been identified. What are the chemical formulas of the following six? (a) sulfur monoxide; (b) sulfur dioxide; (c) sulfur trioxide; (d) disulfur monoxide; (e) hexasulfur monoxide; (f) heptasulfur dioxide

---

**4.45.** Predict the formula and give the name of the ionic compound formed by the following pairs of elements: (a) sodium and sulfur; (b) strontium and fluorine; (c) aluminum and oxygen; (d) lithium and hydrogen.

4.46. Predict the formula and give the name of the ionic compound formed by the following pairs of elements: (a) potassium and bromine; (b) calcium and hydrogen; (c) lithium and nitrogen; (d) aluminum and chlorine.

---

**4.47.** What are the names of the cobalt oxides that have the following formulas? (a) CoO; (b) $Co_2O_3$; (c) $CoO_2$

4.48. What are the formulas of the following copper minerals?
a. cuprite, copper(I) oxide
b. chalcocite, copper(I) sulfide
c. covellite, copper(II) sulfide

---

**4.49.** Give the formula and charge of the oxoanion in each of the following compounds: (a) sodium hypobromite; (b) potassium sulfate; (c) lithium iodate; (d) magnesium nitrite.

*4.50. Give the formula and charge of the oxoanion in each of the following compounds: (a) sodium tellurite; (b) potassium arsenate; (c) barium selenate; (d) potassium bromate.

---

**4.51.** What are the names of the following ionic compounds? (a) $NiCO_3$; (b) NaCN; (c) $LiHCO_3$; (d) $Ca(ClO)_2$

4.52. What are the names of the following ionic compounds? (a) $Mg(ClO_4)_2$; (b) $NH_4NO_3$; (c) $Cu(CH_3COO)_2$; (d) $K_2SO_3$

---

**4.53.** Give the name or chemical formula of each of the following acids: (a) HF; (b) $H_2SO_3$; (c) phosphoric acid; (d) nitrous acid.

*4.54. Give the name or chemical formula of each of the following acids: (a) HBr; (b) $HIO_4$; (c) selenous acid; (d) hydrocyanic acid.

---

**4.55.** Write the chemical formulas of the following compounds: (a) potassium sulfide; (b) potassium selenide; (c) rubidium sulfate; (d) rubidium nitrite; (e) magnesium sulfate.

4.56. Write the chemical formulas of the following compounds: (a) aluminum nitride; (b) ammonium sulfite; (c) rubidium chromate; (d) ammonium nitrate; (e) aluminum selenite.

---

**4.57.** What are the names of the following compounds? (a) MnS; (b) $V_3N_2$; (c) $Cr_2(SO_4)_3$; (d) $Co(NO_3)_2$; (e) $Fe_2O_3$

4.58. What are the names of the following compounds? (a) RuS; (b) $PdCl_2$; (c) $Ag_2O$; (d) $WO_3$; (e) $PtO_2$

---

**4.59.** Which is the formula of sodium sulfite? (a) $Na_2S$; (b) $Na_2SO_3$; (c) $Na_2SO_4$; (d) NaSH

4.60. Which is the formula of barium nitrate? (a) $Ba_3N_2$; (b) $Ba_2NO_3$; (c) $Ba_2(NO_3)_2$; (d) $Ba(NO_3)_2$

---

**4.61.** Provide the name or formula: (a) $H_2SO_4$; (b) $Na_2SO_4$; (c) $Cu_2SO_4$; (d) $SO_3$; (d) sulfurous acid; (e) sulfur dioxide; (f) sulfite.

4.62. Provide the name or formula: (a) $HNO_2$; (b) $N_3^-$; (c) $Mn(NO_3)_2$; (d) NO; (d) nitric acid; (e) dinitrogen trioxide; (f) titanium(IV) nitride.

## Lewis Symbols and Lewis Structures

### Concept Review

**4.63.** How does the number of valence electrons in the neutral atom of an element relate to the element's group number?

4.64. Does the octet rule mean that a diatomic molecule must have 16 valence electrons?

**4.65.** Why is the bonding pattern in water H—O—H and not H—H—O?

4.66. Describe the relationship between the bonding capacity of an atom and its number of valence electrons.

**Problems**

**4.67.** Draw Lewis symbols of potassium, magnesium, and phosphorus.

4.68. Draw Lewis symbols of gallium, tellurium, and iodine.

**4.69.** Draw Lewis symbols for $K^+$, $Al^{3+}$, $N^{3-}$, and $I^-$.

4.70. Draw Lewis symbols for the most stable ions formed by lithium, magnesium, aluminum, and fluorine.

**4.71.** How many valence electrons does each of the following species contain? (a) $N_2$; (b) HCl; (c) $NH_4^+$; (d) $CN^-$

4.72. How many valence electrons does each of the following species contain? (a) $H^+$; (b) $H_3O^+$; (c) $CO_2$; (d) $CH_4$

**4.73.** Draw Lewis structures for the following diatomic molecules and ions: (a) CO; (b) $O_2$; (c) $ClO^-$; (d) $CN^-$.

4.74. Draw Lewis structures for the following molecules and ions: (a) $Br_2$; (b) $H_3O^+$; (c) $N_2$; (d) HF.

**4.75.** Draw Lewis structures for the following molecular compounds and ions: (a) $CCl_4$; (b) $BH_3$; (c) $SiF_4$; (d) $BH_4^-$; (e) $PH_4^+$.

4.76. Draw Lewis structures for the following molecular compounds and ions: (a) $AlCl_3$; (b) $PH_3$; (c) $H_2Se$; (d) $NO_2^-$; (e) $AlH_4^-$.

**4.77.** **Greenhouse Gases** Chlorofluorocarbons (CFCs) are compounds linked to the depletion of stratospheric ozone. They also are greenhouse gases. Draw Lewis structures for the following CFCs:

a. $CCl_3F$ (Freon 11)
b. $CCl_2F_2$ (Freon 12)
c. $CClF_3$ (Freon 13)
d. $Cl_2FC—CClF_2$ (Freon 113)
e. $ClF_2C—CClF_2$ (Freon 114)

4.78. Replacing a halogen atom in a CFC molecule with a hydrogen atom makes the compound more environmentally "friendly." Draw Lewis structures for the following compounds:

a. $CHCl_2F$ (Freon 21)
b. $CHF_2Cl$ (Freon 22)
c. $CH_2ClF$ (Freon 31)
d. $F_3C—CHBrCl$ (Halon 2311)
e. $Cl_2FC—CH_3$ (HCFC 141b)

**4.79.** Draw Lewis structures for the following oxoanions: (a) $ClO_2^-$; (b) $SO_3^{2-}$; (c) $HCO_3^-$.

4.80. Draw Lewis structures for the following oxoanions: (a) $BrO_4^-$; (b) $SeO_4^{2-}$; (c) $HPO_4^{2-}$.

**4.81.** **Skunks and Rotten Eggs** Many sulfur-containing organic compounds have characteristically foul odors: butanethiol ($CH_3CH_2CH_2CH_2SH$) is responsible for the odor of skunks, and rotten eggs smell the way they do because they produce tiny amounts of pungent hydrogen sulfide, $H_2S$. Draw the Lewis structures for $CH_3CH_2CH_2CH_2SH$ and $H_2S$.

4.82. Draw the Lewis structure of the molecule whose atoms are connected as shown in Figure P4.82.

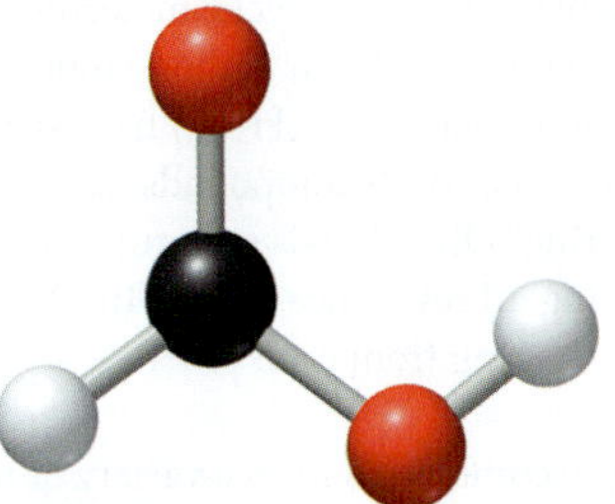

FIGURE P4.82

**4.83.** **Chlorine Bleach** Chlorine combines with oxygen in several proportions. Dichlorine monoxide ($Cl_2O$) is used in manufacturing bleaching agents. Potassium chlorate ($KClO_3$) is used in oxygen generators aboard aircraft. Draw the Lewis structures for $Cl_2O$ and $ClO_3^-$ (Cl is the central atom).

4.84. **Dangers of Mixing Cleansers** Labels on household cleansers caution against mixing bleach with ammonia (Figure P4.84) because they react with each other to produce monochloramine ($NH_2Cl$) and hydrazine ($N_2H_4$), both of which are toxic. Draw the Lewis structures for monochloramine and hydrazine.

FIGURE P4.84

***4.85.** Methanol is a fuel for race cars with the molecular formula $CH_4O$. Draw the Lewis structure for methanol.

4.86. Carbon disulfide, $CS_2$, is a flammable, low-boiling liquid. Draw the Lewis structure for $CS_2$.

## Resonance

**Concept Review**

**4.87.** How does resonance influence the stability of a molecule or an ion?

4.88. What structural features do all the resonance forms of a molecule or ion have in common?

**4.89.** Explain why $NO_2$ is more likely to exhibit resonance than $CO_2$.

*4.90. Three equivalent resonance structures can be drawn for a nitrate ion. How much of the time does the bonding in a nitrate ion match any one of them?

**Problems**

**4.91.** Draw two Lewis structures showing the resonance that occurs in cyclobutadiene ($C_4H_4$), a cyclic molecule with a structure that includes a ring of four carbon atoms.

*4.92. Pyridine ($C_5H_5N$) and pyrazine ($C_4H_4N_2$) have structures similar to that of benzene. Both compounds have structures with six atoms in a ring. Draw Lewis structures for pyridine and pyrazine, showing all resonance forms. The N atoms in pyrazine are across the ring from each other.

---

***4.93.** Oxygen and nitrogen combine to form a variety of nitrogen oxides, including the following two unstable compounds that each have two nitrogen atoms per molecule: $N_2O_2$ and $N_2O_3$. Draw Lewis structures for the molecules and show all resonance forms.

*4.94. Oxygen and sulfur combine to form a variety of sulfur oxides. Some are stable molecules, and some, including $S_2O_2$ and $S_2O_3$, decompose when heated. Draw Lewis structures for those two compounds and show all resonance forms.

---

**4.95.** Draw Lewis structures for fulminic acid (HCNO) that show all resonance forms.

4.96. Draw Lewis structures for hydrazoic acid ($HN_3$) that show all resonance forms.

---

**4.97.** Draw Lewis structures that show the resonance that occurs in dinitrogen pentoxide. (*Hint*: $N_2O_5$ has an O atom at its center.)

4.98. **Bacteria Make Nitrites** Nitrogen-fixing bacteria convert urea [$H_2NC(O)NH_2$] into nitrite ions. Draw Lewis structures for the two species. Include all resonance forms. (*Hint*: Urea has a C=O bond.)

## The Lengths and Strengths of Covalent Bonds

**Concept Review**

**4.99.** Why is the oxygen–oxygen bond length in $O_3$ different from the one in $O_2$?

4.100. Do you expect the sulfur–oxygen bond lengths in sulfite ($SO_3^{2-}$) and sulfate ($SO_4^{2-}$) ions to be about the same? Why?

**4.101.** Rank the following ions in order of increasing nitrogen–oxygen bond lengths: $NO_2^-$, $NO^+$, and $NO_3^-$.

4.102. Rank the following substances in order of increasing carbon–oxygen bond lengths: CO, $CO_2$, and $CO_3^{2-}$.

**4.103.** Rank the following ions in order of increasing nitrogen–oxygen bond energy: $NO_2^-$, $NO^+$, and $NO_3^-$.

4.104. Rank the following substances in order of increasing carbon–oxygen bond energy: CO, $CO_2$, and $CO_3^{2-}$.

**4.105.** Which has the longer carbon–carbon bond: acetylene ($C_2H_2$) or ethane ($C_2H_6$)?

4.106. Which has the stronger carbon–carbon bond: acetylene ($C_2H_2$) or ethane ($C_2H_6$)?

## Formal Charge: Choosing among Lewis Structures

**Concept Review**

**4.107.** Describe how formal charges are used to choose between possible molecular structures.

4.108. How do the electronegativities of elements influence the selection of which Lewis structure is favored?

**4.109.** In a molecule containing S and O atoms, is a structure with a negative formal charge on sulfur more likely to contribute to bonding than an alternative structure with a negative formal charge on oxygen?

4.110. In a cation containing N and O, why do Lewis structures with a positive formal charge on nitrogen contribute more to the actual bonding in the molecule than those structures with a positive formal charge on oxygen?

**Problems**

**4.111.** Hydrogen isocyanide (HNC) has the same elemental composition as hydrogen cyanide (HCN), but the H in HNC is bonded to the nitrogen atom. Draw a Lewis structure for HNC, and assign formal charges to each atom. How do the formal charges on the atoms differ in the Lewis structures for HCN and HNC?

4.112. **Molecules in Interstellar Space** Hydrogen cyanide (HCN) and cyanoacetylene ($HC_3N$) have been detected in the interstellar regions of space. Draw Lewis structures for the molecules, and assign formal charges to each atom. The hydrogen atom is bonded to carbon in both cases.

---

**4.113.** **Origins of Life** The discovery of polyatomic organic molecules such as cyanamide ($H_2NCN$) in interstellar space has led some scientists to believe that the molecules from which life began on Earth may have come from space. Draw Lewis structures for cyanamide and select the preferred structure on the basis of formal charges.

4.114. Complete the Lewis structures for and assign formal charges to the atoms in five of the resonance forms of thionitrosyl azide ($SN_4$). Indicate which of your structures should be most stable. The molecule is linear with S at one end.

---

***4.115.** Nitrogen is the central atom in molecules of nitrous oxide ($N_2O$). Draw Lewis structures for another possible arrangement: N—O—N. Assign formal charges and suggest a reason why the structure is not likely to be stable.

4.116. **More Molecules in Space** Formamide ($HCONH_2$) and methyl formate ($HCO_2CH_3$) have been detected in space. Draw the Lewis structures of the compounds, on the basis of the skeletal structures in Figure P4.116, and assign formal charges.

```
    O    H            O         H
    |   /             |         |
H—C—N              H—C—O—C—H
        \                       |
         H                      H
```

**FIGURE P4.116**

---

***4.117.** Nitromethane ($CH_3NO_2$) reacts with hydrogen cyanide (HCN) to produce $CNNO_2$ and $CH_4$.

a. Draw Lewis structures for $CH_3NO_2$ and show all resonance forms.

b. Draw Lewis structures for $CNNO_2$, showing all resonance forms, on the basis of the two possible skeletal structures for it in Figure P4.117. Assign formal charges and predict which structure is more likely to exist.

c. Are the two structures of $CNNO_2$ resonance forms of each other?

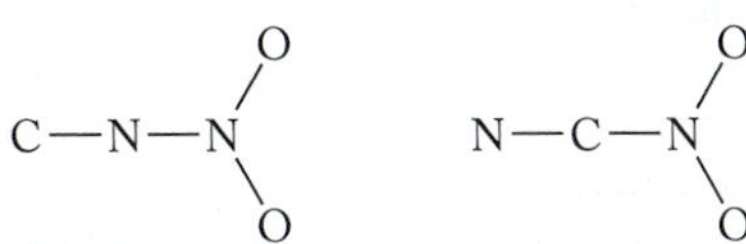


**FIGURE P4.117**

4.118. Use formal charges to determine which resonance form of each of the following ions is preferred: $CNO^-$, $NCO^-$, and $CON^-$.

## *Exceptions to the Octet Rule*

### Concept Review

**4.119.** Are all odd-electron molecules exceptions to the octet rule?

4.120. Do atoms in rows 3 and below always have more than eight valence electrons? Explain your answer.

### Problems

**4.121.** In which of the following molecules does the sulfur atom have more than eight valence electrons? (a) $SF_6$; (b) $SF_5$; (c) $SF_4$; (d) $SF_2$

4.122. In which of the following molecules does the phosphorus atom have more than eight valence electrons? (a) $POCl_3$; (b) $PF_5$; (c) $PF_3$; (d) $P_2F_4$ (which has a P—P bond)

**4.123.** How many electrons are present in the covalent bonds surrounding the sulfur atom in the following species? (a) $SF_4O$; (b) $SOF_2$; (c) $SO_3$; (d) $SF_5^-$

4.124. How many electrons are in the covalent bonds surrounding the phosphorus atom in the following species? (a) $POCl_3$; (b) $H_3PO_4$; (c) $H_3PO_3$; (d) $PF_6^-$

**4.125.** Dissolving NaF in selenium tetrafluoride ($SeF_4$) produces $NaSeF_5$. Draw the Lewis structures of $SeF_4$ and $SeF_5^-$. In which structure does Se have more than eight valence electrons?

4.126. The reaction between $NF_3$, $F_2$, and $SbF_3$ at 200°C and 100 atm pressure produces the ionic compound $NF_4SbF_6$. Draw the Lewis structures of the ions in the product.

**4.127.** **Ozone Depletion** The compound $Cl_2O_2$ may play a role in ozone depletion in the stratosphere. Draw the Lewis structure of $Cl_2O_2$ on the basis of the arrangement of atoms in Figure P4.127. Does either chlorine atom in the structure have more than eight valence electrons?

**FIGURE P4.127**

4.128. $Cl_2O_2$ decomposes to $Cl_2$ and $ClO_2$. Draw the Lewis structure of $ClO_2$.

**4.129.** Which of the following chlorine oxides are odd-electron molecules? (a) $Cl_2O_7$; (b) $Cl_2O_6$; (c) $ClO_4$; (d) $ClO_3$; (e) $ClO_2$

4.130. Which of the following nitrogen oxides are most likely to have an unpaired electron? (a) NO; (b) $NO_2$; (c) $NO_3$; (d) $N_2O_4$; (e) $N_2O_5$

**4.131.** Which of the Lewis structures in Figure P4.131 contributes most to the bonding in CNO?

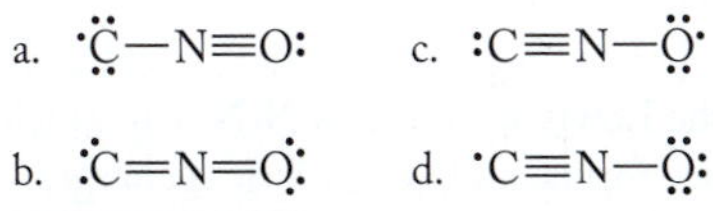

a. ·C̤̈—N≡O: c. :C≡N—Ö̤·

b. :C̤=N=O̤: d. ·C≡N—Ö̤:

**FIGURE P4.131**

4.132. Why is the Lewis structure in Figure P4.132 unlikely to contribute much to the bonding in NCO?

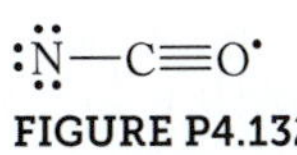

:N̤̈—C≡O·

**FIGURE P4.132**

## *Vibrating Bonds and the Greenhouse Effect*

### Concept Review

**4.133.** Describe how atmospheric greenhouse gases act like the panes of glass in a greenhouse.

*4.134. Water vapor in the atmosphere contributes more to the greenhouse effect than carbon dioxide, yet water vapor is not considered an important factor in climate change. Propose a reason why.

***4.135.** Increasing concentrations of nitrous oxide in the atmosphere may be contributing to climate change. Is the ability of $N_2O$ to absorb IR radiation due to nitrogen–nitrogen bond stretching, nitrogen–oxygen bond stretching, or both? Explain your answer.

4.136. Is the ability of $H_2O$ molecules to absorb photons of IR radiation due to symmetrical stretching or asymmetrical stretching of its O—H bonds, or both? Explain your answer. (*Hint*: The angle between the two O—H bonds in $H_2O$ is 104.5°.)

**4.137.** Can molecules of carbon monoxide in the atmosphere absorb photons of IR radiation? Explain why or why not.

*4.138. How does the high-temperature conversion of limestone ($CaCO_3$) to lime (CaO) during the production of cement contribute to climate change?

**4.139.** Why does infrared radiation cause bonds to vibrate but not break (as UV radiation can)?

4.140. Argon is the third most abundant species in the atmosphere. Why isn't it a greenhouse gas?

***4.141.** Which C—O bond has a higher stretching frequency: the one in CO or the one in $CH_2O$? Explain your selection.

*4.142. Which compound, NO or $NO_2$, absorbs IR radiation of a longer wavelength?

## *Additional Problems*

**4.143.** On the basis of the Lewis symbols in Figure P4.143, predict to which group in the periodic table element X belongs.

a. ·Ẋ· b. ·Ẋ: c. :Ẋ: d. :Ẋ:

**FIGURE P4.143**

4.144. Use formal charges to predict the order of the atoms in each compound.
 a. Are the atoms in carbon disulfide arranged CSS or SCS?
 b. Are the atoms in hypochlorous acid arranged HOCl or HClO?

**4.145.** **Chemical Weapons** Draw the Lewis structure of phosgene, $COCl_2$, a poisonous gas used in chemical warfare during World War I.

4.146. The dinitramide anion $[N(NO_2)_2^-]$ was first isolated in 1996. The arrangement of atoms in $N(NO_2)_2^-$ is shown in Figure P4.146.
 a. Complete the Lewis structure of $N(NO_2)_2^-$, including any resonance forms, and assign formal charges.
 b. Explain why the nitrogen–oxygen bond lengths in $N(NO_2)_2^-$ and $N_2O$ should (or should not) be similar.
 c. $N(NO_2)_2^-$ was isolated as $[NH_4^+][N(NO_2)_2^-]$. Draw the Lewis structure of $NH_4^+$.

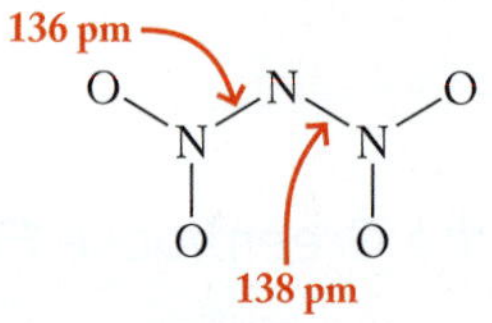

136 pm
138 pm

**FIGURE P4.146**

***4.147.** Silver cyanate (AgOCN) is a source of the cyanate ion $(OCN^-)$. Under certain conditions, the species OCN is an anion with a charge of 1−; under others, it is a neutral, odd-electron molecule, OCN.
 a. Two molecules of OCN combine to form OCNNCO. Draw the Lewis structure of the molecule, including all resonance forms.
 b. The $OCN^-$ ion reacts with BrNO, forming the unstable molecule OCNNO. Draw the Lewis structures of BrNO and OCNNO, including all resonance forms.
 c. The $OCN^-$ ion reacts with $Br_2$ and $NO_2$ to produce $N_2O$, $CO_2$, BrNCO, and OCN(CO)NCO. Draw the resonance structures of OCN(CO)NCO, which has the arrangement of atoms shown in Figure P4.147.

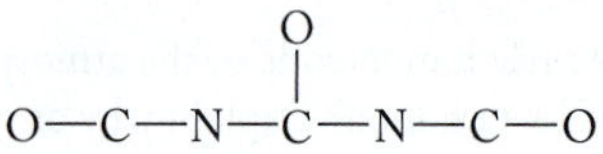

**FIGURE P4.147**

*4.148. During the reaction of the cyanate ion $(OCN^-)$ with $Br_2$ and $NO_2$, a very unstable substance called an *intermediate* forms and then quickly falls apart. Its formula is $O_2NNCO$.
 a. Draw three of the resonance forms for $O_2NNCO$, assign formal charges, and predict which of the three contributes most to the bonding in $O_2NNCO$. Its skeletal structure is shown in Figure P4.148(a).

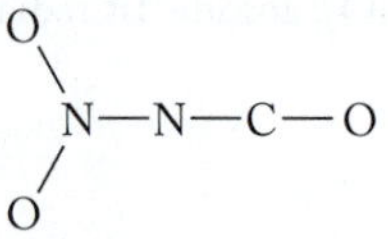

**FIGURE P4.148(a)**

 b. Draw Lewis structures for the different arrangement of the N, C, and O atoms in $O_2NNCO$ shown in Figure P4.148(b).

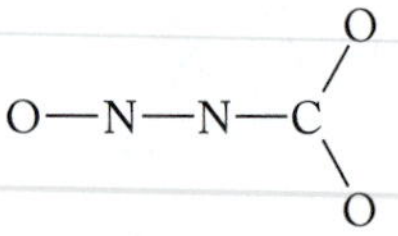

**FIGURE P4.148(b)**

**4.149.** A compound with the formula $Cl_2O_6$ decomposes to a mixture of $ClO_2$ and $ClO_4$. Draw two Lewis structures for $Cl_2O_6$: one with a chlorine–chlorine bond and one with a Cl—O—Cl arrangement of atoms.

*4.150. A compound consisting of chlorine and oxygen, $Cl_2O_7$, decomposes to $ClO_4$ and $ClO_3$.
 a. Draw two Lewis structures of $Cl_2O_7$: one with a chlorine–chlorine bond and one with a Cl—O—Cl arrangement of atoms.
 b. Draw the Lewis structure of $ClO_3$.

***4.151.** The odd-electron molecule CN reacts with itself to form cyanogen, $C_2N_2$.
 a. Draw the Lewis structure of CN, and predict which arrangement for cyanogen is more likely: NCCN or CNNC.
 b. Cyanogen reacts slowly with water to produce oxalic acid $(H_2C_2O_4)$ and ammonia; the Lewis structure of oxalic acid is shown in Figure P4.151. Compare the structure with your answer in part (a). Do you still believe the structure you selected in part (a) is the better one?

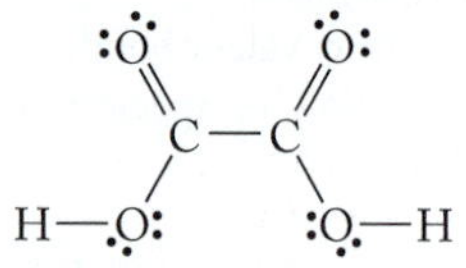

**FIGURE P4.151**

4.152. The odd-electron molecule SN forms $S_2N_2$, which has a cyclic structure (the atoms form a ring).
 a. Draw a Lewis structure of SN and complete the possible Lewis structures for $S_2N_2$ in Figure P4.152.
 b. Which is the preferred structure for $S_2N_2$?

**FIGURE P4.152**

***4.153.** The molecular structure of sulfur cyanide trifluoride $(SF_3CN)$ has the arrangement of atoms with the bond lengths indicated in Figure P4.153. Using the observed bond lengths as a guide, complete the Lewis structure of $SF_3CN$ and assign formal charges.

116 pm
174 pm
160 pm

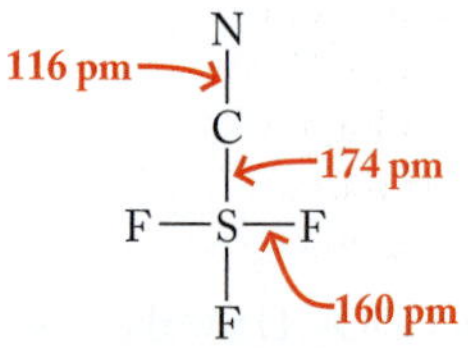

**FIGURE P4.153**

4.154. **Strike-Anywhere Matches** Heating phosphorus with sulfur produces $P_4S_3$, a solid used in the heads of strike-anywhere matches. $P_4S_3$ has the skeletal structure shown in Figure P4.154. Complete its Lewis structure.

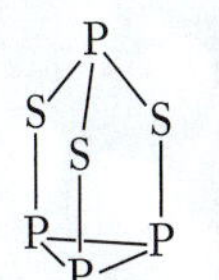

**FIGURE P4.154**

***4.155.** The $TeOF_6^{2-}$ anion was first synthesized in 1993. Draw its Lewis structure.

*4.156. **Sulfur in the Environment** Sulfur is cycled in the environment through compounds such as dimethyl sulfide ($CH_3SCH_3$), zinc sulfide, and sulfite and sulfate ions. Draw Lewis structures for those four species.

**4.157.** **Antacid Tablets** Antacids commonly contain calcium carbonate, magnesium hydroxide, or both. Draw the Lewis structures for calcium carbonate and magnesium hydroxide.

4.158. How many pairs of electrons does xenon share in the following molecules and ions? (a) $XeF_2$; (b) $XeOF_2$; (c) $XeF^+$; (d) $XeF_5^+$; (e) $XeO_4$

***4.159.** A short-lived allotrope of nitrogen, $N_4$, was reported in 2002.

a. Draw the Lewis structures of all the resonance forms of linear $N_4$ (N—N—N—N).
b. Assign formal charges and determine which resonance structure best depicts $N_4$.
c. Draw a Lewis structure of a ring (cyclic) form of $N_4$ and assign formal charges.

*4.160. Scientists have predicted the existence of $O_4$, even though the molecule has never been observed. However, $O_4^{2-}$ has been detected. Draw the Lewis structures for $O_4$ and $O_4^{2-}$.

**4.161.** Which of the following molecules and ions contains an atom with more than eight valence electrons? (a) $Cl_2$; (b) $ClF_3$; (c) $ClI_3$; (d) $ClO^-$

4.162. Which of the following molecules contains an atom with more than eight valence electrons? (a) $XeF_2$; (b) $GaCl_3$; (c) $ONF_3$; (d) $SeO_2F_2$

***4.163.** A linear nitrogen anion, $N_5^-$, was isolated for the first time in 1999.

a. Draw the Lewis structures for four resonance forms of linear $N_5^-$.
b. Assign formal charges to the atoms in the structures in part (a), and identify the structures that contribute most to the bonding in $N_5^-$.
c. Compare the Lewis structures for $N_5^-$ and $N_3^-$. In which ion do the nitrogen–nitrogen bonds have the higher average bond order?

*4.164. Carbon tetroxide ($CO_4$) was discovered in 2003.

a. Draw the Lewis structure of $CO_4$ on the basis of the skeletal structure shown in Figure P4.164.
b. Do any resonance forms of the structure you drew have zero formal charges on all atoms?
c. Can you draw a structure in which all four oxygen atoms in $CO_4$ are bonded to carbon?

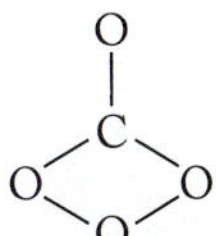


**FIGURE P4.164**

**4.165.** Plot the electronegativities of elements with $Z = 3$ to 9 ($y$-axis) versus their first ionization energy ($x$-axis). Is the plot linear? Use your graph to predict the electronegativity of neon, whose first ionization energy is 2081 kJ/mol.

*4.166. In the typical Lewis structure of $BF_3$, only six valence electrons are on the boron atom and each B—F bond is a single bond. However, the length and strength of those bonds indicate that they have a small measure of double-bond character—that is, their bond order is slightly greater than 1.

a. Draw a Lewis structure of $BF_3$, including all resonance structures, with one B=F double bond.
b. What is the formal charge on the B atom, and what is the average formal charge on each F atom?
c. On the basis of formal charges alone, what should be the bond order of each B—F bond in $BF_3$?
d. What factor might support a bond order slightly greater than 1?

**4.167.** The cation $N_2F^+$ is isoelectronic with $N_2O$.

a. What does it mean to be isoelectronic?
b. Draw the Lewis structure of $N_2F^+$. (*Hint*: The molecule contains a nitrogen–nitrogen bond.)
c. Which atom has the +1 formal charge in the structure you drew in part (b)?
d. Does $N_2F^+$ have resonance forms?
e. Could the middle atom in the $N_2F^+$ ion be a fluorine atom? Explain your answer.

4.168. **Ozone Depletion** Methyl bromide ($CH_3Br$) is produced naturally by fungi. Methyl bromide has also been used in agriculture as a fumigant, but its use is being phased out because the compound has been linked to ozone depletion in the upper atmosphere.

a. Draw the Lewis structure of $CH_3Br$.
b. Which bond in $CH_3Br$ is more polar, carbon–hydrogen or carbon–bromine?

**4.169.** Draw the Lewis structure for dimethyl ether, $C_2H_6O$, given that the structure contains an oxygen atom bonded to two carbons: C—O—C.

*4.170. Draw another Lewis structure for $C_2H_6O$ that has a different connectivity from that in Problem 4.169. (*Hint*: Remember that the bonding capacity of hydrogen is one.)

**4.171.** Draw the Lewis structure for butane, $C_4H_{10}$, given that the structure contains four carbon atoms bonded in a row: C—C—C—C.

*4.172. Draw another Lewis structure for $C_4H_{10}$ that has a different connectivity from that in Problem 4.171. (*Hint*: Given that the bonding capacity of hydrogen is one, how else might the carbon atoms be connected?)

# 5

# Bonding Theories

## Explaining Molecular Geometry

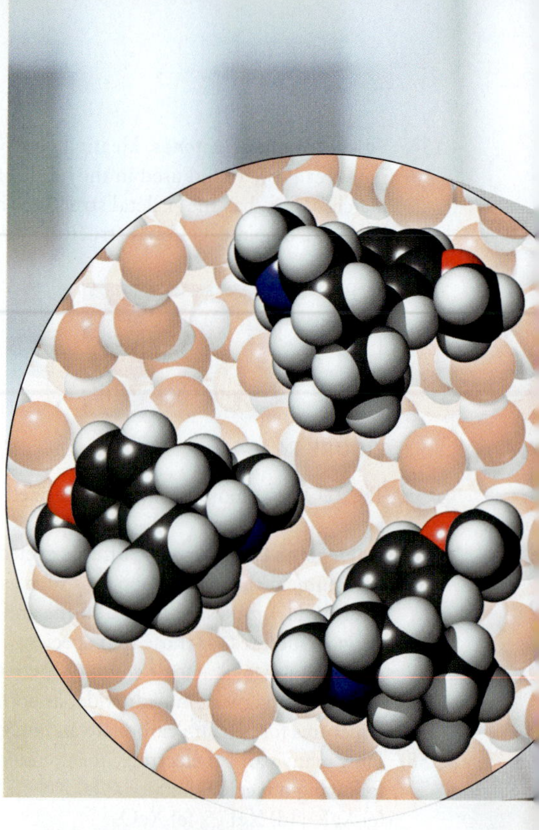

COUGH MEDICINE
Dextromethorphan has the molecular formula $C_{18}H_{25}NO$ and is a common ingredient in cough syrup. The bonds in dextromethorphan, and the direction those bonds point in space, are responsible for the shape of the molecule and its cough-suppressant properties. Another molecule, levomethorphan, with the same atoms and bonds as dextromethorphan, has bonds oriented differently, giving levomethorphan a different shape and very different properties: the latter compound is an opioid narcotic.

PARTICULATE **REVIEW**

### *Lewis Structure and Polar Bonds*

In Chapter 5, we explore the shapes of molecules. Consider this ball-and-stick representation of a molecule of acetaminophen, the active ingredient in Tylenol:

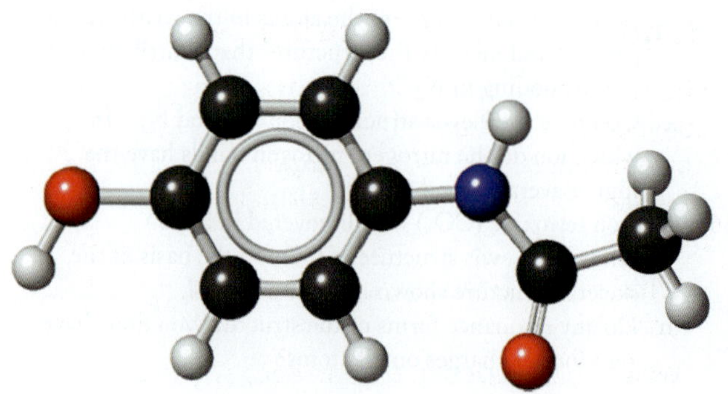

- Draw the Lewis structure of acetaminophen.
- What does the circle inside the hexagon of carbon atoms represent?
- Identify the three most polar bonds in acetaminophen.

(Review Sections 4.3, 4.4, and 4.6 if you need help.)

*(Answers to Particulate Review questions are in the back of the book.)*

PARTICULATE **PREVIEW**

## *Bond Angles and Bonding Orbitals*

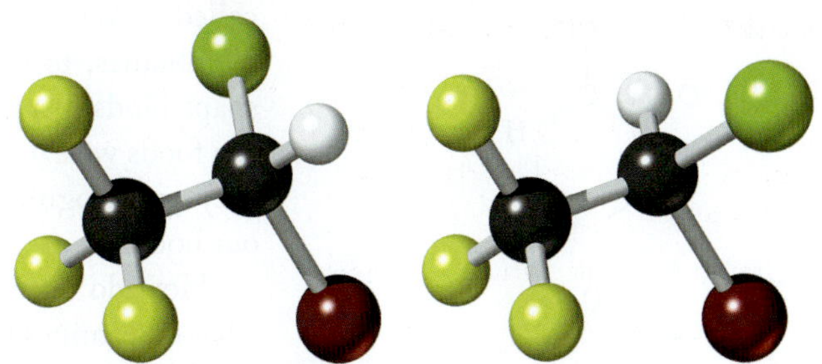

Shown here are two molecules of a compound commonly called halothane, $CF_3CHBrCl$. It is a powerful anesthetic used in surgery. As you read Chapter 5, look for ideas that will help you answer these questions:

- What are the bond angles between the F—C—F bonds?
- What orbitals overlap to form each C—H bond?
- Do the two ball-and-stick representations of halothane depict the same molecule or two different molecules? If different, how are they related?

## Learning Outcomes

**LO1** Use the theory of valence-shell electron-pair repulsion (VSEPR) and the concept of steric number to predict the bond angles in molecules and the shapes of molecules with one central atom
**Sample Exercises 5.1–5.3**

**LO2** Predict whether a substance is polar or nonpolar on the basis of its molecular structure
**Sample Exercise 5.4**

**LO3** Use atomic hybridization and valence bond theory to explain orbital overlap, bond angles, and molecular shape
**Sample Exercise 5.5**

**LO4** Recognize chiral molecules
**Sample Exercise 5.6**

**LO5** Draw molecular orbital (MO) diagrams of small molecules and use MO theory to predict bond order and explain the magnetic properties and UV/visible spectra of molecular compounds
**Sample Exercises 5.7–5.9**

# 5.1 Biological Activity and Molecular Shape

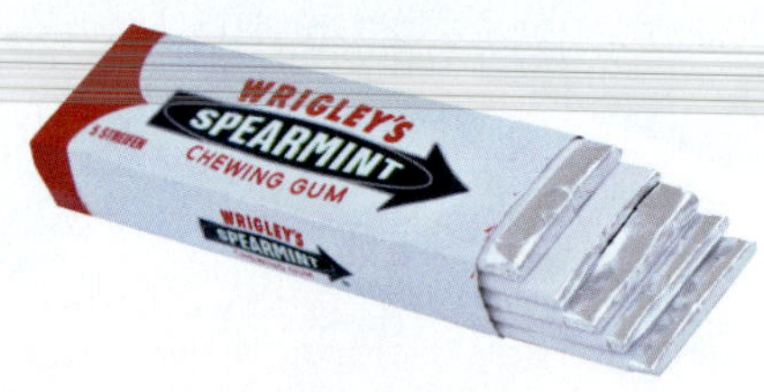

(a)

(b)

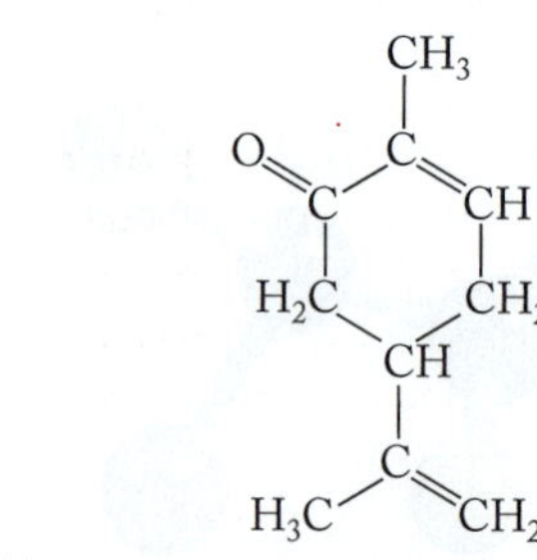

(c)

**FIGURE 5.1** The smell and taste of (a) spearmint and (b) caraway result from two different arrangements in space of the atoms in $C_{10}H_{14}O$, whose condensed molecular structure is drawn in (c).

Hold your hands out in front of you, palms up, fingers extended. Now rotate your wrists inward so that your thumbs point straight up. Your right hand looks the same as the image your left hand makes in a mirror. Does that mean that your two hands have the same shape? If you have ever tried to put your right hand in a glove made for your left, you know that they do *not* have the same shape—they are very similar but not the same. Many other objects in our world have a "handedness" about them, too, including scissors, golf clubs, and the cough suppressants in many cold medicines.

This chapter focuses on the importance of shape at the molecular level. For example, the compound that produces the refreshing aroma of spearmint has the molecular formula $C_{10}H_{14}O$. The compound responsible for the musty aroma of caraway seeds in rye bread has that same molecular formula *and* the same Lewis structure (**Figure 5.1**). To understand how two compounds could be so much alike and still have different properties, we have to consider their structures in three dimensions.

We perceive a difference in their aromas in part because each molecule has a unique site at which it attaches to our nasal membranes. Just as a left glove fits only a left hand, the spearmint molecule fits only the spearmint-shaped site, and the caraway molecule fits only the caraway-shaped site. That phenomenon is called *molecular recognition*, and it enables biomolecular structures, such as nasal membranes, to recognize and react when a particular molecule with a specific shape binds to a part of the structure known as an *active site*. Many substances in the foods we eat and in the medicines we take exert physiological effects because they are recognized by, and bind to, specific sites in the molecules that make up our bodies.

How do chemists describe the three-dimensional shapes of molecules, and what determines those shapes? Can we predict shapes if we know how atoms are bonded in molecules? In this chapter, we examine several theories of bonding that explain molecular shapes, and we begin to explore how molecular shape affects the physical, chemical, and biological properties of compounds.

The shape of a molecule can affect many properties of the substance, including its physical state at room temperature, its aroma, its biological activity, and its distribution in the environment. In Chapter 4 we drew Lewis structures to account for the bonding in molecules and polyatomic ions, but Lewis structures are only two-dimensional representations of how atoms and the electron pairs that surround them are arranged in molecules. Lewis structures show what atoms are *connected* to one another in molecules and polyatomic ions, but they don't show how the atoms are *oriented* in three dimensions, nor do they necessarily reveal the overall shape of the molecule.

To illustrate that point, look at the Lewis structures and ball-and-stick models of carbon dioxide and methane shown in **Figure 5.2**. The linear array of atoms and bonding electrons in the Lewis structure of $CO_2$ corresponds to the actual linear shape of the molecule as represented by the ball-and-stick model. The angle between the two C═O bonds is 180°, just as in the Lewis structure. By contrast, the Lewis structure of methane shows 90° angles between the four C—H bonds, whereas the experimentally measured H—C—H **bond angles**, depicted in the ball-and-stick model, are 109.5°.

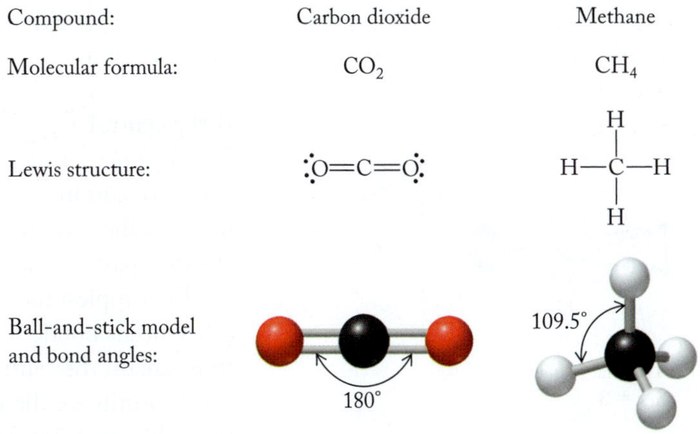

| Compound: | Carbon dioxide | Methane |
|---|---|---|
| Molecular formula: | $CO_2$ | $CH_4$ |
| Lewis structure: | :Ö═C═Ö: | H—C—H (with H above and below C) |
| Ball-and-stick model and bond angles: | 180° | 109.5° |

**FIGURE 5.2** The Lewis structure of $CO_2$ matches its molecular shape, but the Lewis structure of $CH_4$ does not because the C—H bonds in $CH_4$ actually extend in three dimensions.

In this chapter, we explore theories of covalent bonding that account for and predict the shapes of molecules. During that exploration, you may wonder which bonding theory is the best and ask which you should use. Different models and theories are well suited to answering some bonding questions but not others. For example, the Lewis structures we drew in Chapter 4 do a good job of explaining the bonding in molecules in which the atoms obey the octet rule, but the structures do not describe the three-dimensional shapes of those molecules or the angles between their bonds. As different theories of covalent bonding are described in this chapter, you will see which theory is best suited to answer a particular question about bonding and molecular structure.

## 5.2 Valence-Shell Electron-Pair Repulsion Theory

The theory of **valence-shell electron-pair repulsion (VSEPR)** is based on the fundamental chemical principle that electrons have negative charges and repel each other. VSEPR theory applies that principle by assuming that pairs of valence electrons are arranged about central atoms in ways that minimize repulsions between the pairs. To use VSEPR to predict the shape of a molecule, we must consider its **electron-pair geometry**, which describes the relative positions in three-dimensional space of all the bonding pairs and lone pairs of valence electrons on the central atom, and its **molecular geometry**, which describes the relative positions of the atoms in a molecule. To accurately predict the molecular geometry, we first need to know the electron-pair geometry. If no lone pairs of electrons are present, the process is simplified because the electron-pair geometry *is* the molecular geometry. Let's begin with that simpler case and consider the shapes of molecules that have different numbers of atoms bonded to a single central atom that has no lone pairs of electrons. To make our exploration even simpler, we will initially focus on molecules in which the same kind of atom is bonded to each central atom, so that all the bonds in each molecule are identical.

**bond angle** the angle (in degrees) defined by lines joining the centers of two atoms to a third atom to which they are chemically bonded.

**valence-shell electron-pair repulsion theory (VSEPR)** a model predicting the arrangement of valence electron pairs around a central atom that minimizes their mutual repulsion to produce the lowest energy orientations.

**electron-pair geometry** the three-dimensional arrangement of bonding pairs and lone pairs of electrons about a central atom.

**molecular geometry** the three-dimensional arrangement of the atoms in a molecule.

## Central Atoms with No Lone Pairs

To determine the geometry of a molecule with a single central atom, we start by drawing its Lewis structure. From the Lewis structure, we determine a parameter called the **steric number (SN)** of the central atom, which is the sum of the number of atoms bonded to that atom and the number of lone pairs on it:

$$SN = \begin{pmatrix} \text{number of atoms} \\ \text{bonded to central atom} \end{pmatrix} + \begin{pmatrix} \text{number of lone pairs} \\ \text{on central atom} \end{pmatrix} \quad (5.1)$$

If the central atom has no lone pairs, the *steric number equals the number of atoms bonded to the central atom*. In evaluating the shapes of the molecules, we will generate five common shapes—shown in **Figure 5.3** and called linear, trigonal planar, tetrahedral, trigonal bipyramidal, and octahedral—which describe both electron-pair geometries and molecular geometries.

The simplest molecular structure with a central atom is one that has only two other atoms bonded to the central atom. As long as no lone pairs of valence electrons are on the central atom, its steric number is 2. The electron pairs in the two bonds minimize their mutual repulsion by being on opposite sides of the central atom (Figure 5.3a). That gives a **linear** electron-pair geometry and a linear molecular geometry. The three atoms in the molecule are arranged in a straight line, and the angle between the two bonds is 180°.

If the central atom is bonded to three other atoms and has no lone pairs of valence electrons, SN = 3. The three bonding atoms are as far apart as possible when located at the three corners of an equilateral triangle. The angle between each bond is 120° (Figure 5.3b). That electron-pair and molecular geometry is called **trigonal planar**.

If the central atom is bonded to four atoms and has no lone pairs, SN = 4. The atoms bonded to the central atom occupy the four vertices of a *tetrahedron*, a four-sided pyramid (*tetra* is Greek, meaning "four"). The bonding pairs form bond angles of 109.5° with each other, as shown in Figure 5.3(c). The electron-pair geometry and molecular geometry are both **tetrahedral**.

When a central atom is bonded to five atoms and has no lone pairs, SN = 5 and the five other atoms occupy the five corners of two triangular pyramids that share the same base. The central atom of the molecule is at the common center of the two bases, as shown in Figure 5.3(d). One bonding pair points to the top of the upper pyramid, one points to the bottom of the lower pyramid, and the other three point to the three vertices of the shared triangular base. Thinking of the three vertices as three points along a circle at the equator of a sphere may be helpful. In that model, the atoms that occupy the three sites and the bonds that connect them to the central atom are called *equatorial* atoms and bonds. The bond angles between the three equatorial bonds are 120° (just as in the trigonal planar geometry in Figure 5.3b). The bond angle between an equatorial bond and either vertical, or *axial*, bond is 90°, and the angle between the two axial bonds is 180°. A molecule in which the atoms are arranged that way is said to have a **trigonal bipyramidal** electron-pair and molecular geometry.

For SN = 6, picture two pyramids that have a square base (Figure 5.3e). Put them together, base to base, and you form a shape in which all six corner positions are equivalent. We can think of the six bonding pairs of electrons as three sets of two electron pairs each. The two pairs in each set are oriented at 180° to each other and at 90° to the other two pairs, just like the axes of an *xyz* coordinate system. In

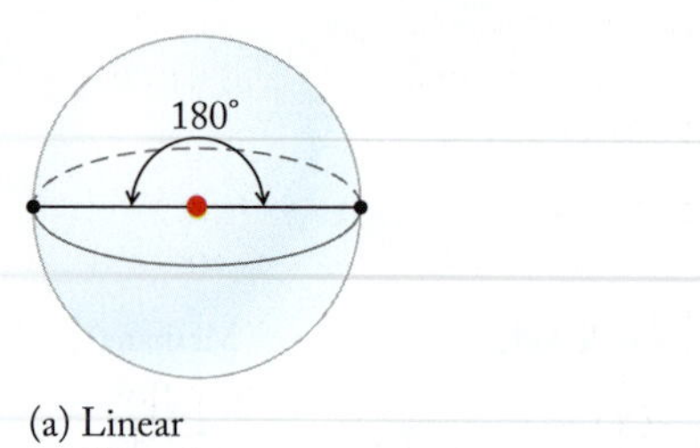


(a) Linear
SN = 2

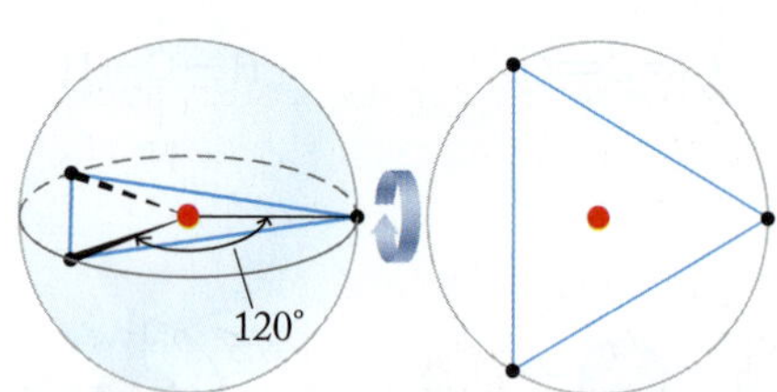


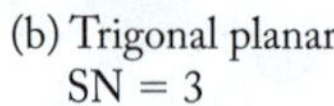

(b) Trigonal planar
SN = 3

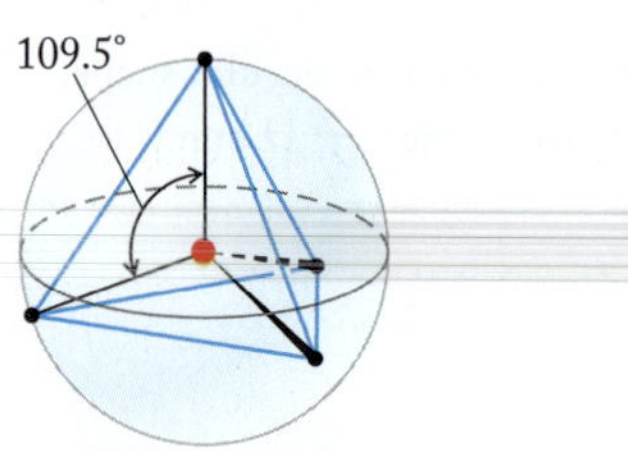


(c) Tetrahedral
SN = 4

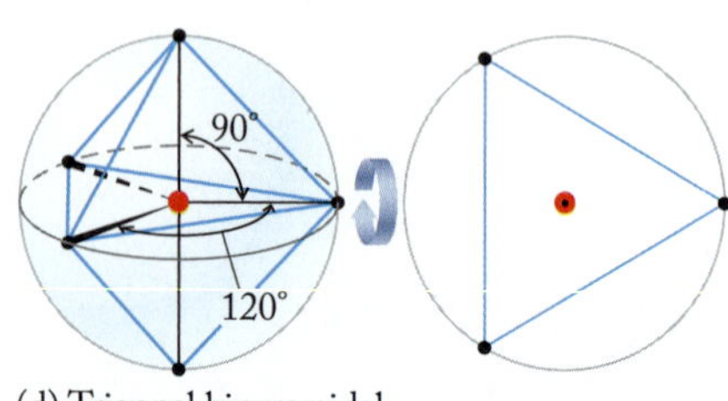


(d) Trigonal bipyramidal
SN = 5

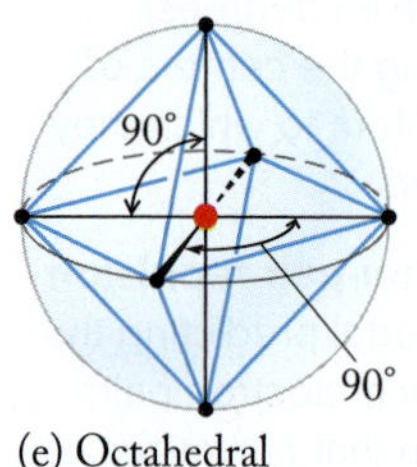


(e) Octahedral
SN = 6

**FIGURE 5.3** Electron-pair geometries depend on the steric number (SN) of the central atom in a molecule. In these images, the central atoms (red dots) have no lone pairs of electrons, so the SN is equal to the number of atoms bonded to the central atom, and the molecular geometry is the same as the electron-pair geometry. The black lines represent covalent bonds; the blue lines outline the geometric forms that give these shapes their names. Parts (b) and (d) show spheres rotated 90° to display the equilateral triangles in the equatorial plane.

our spherical model, four atoms are at the four vertices of the common square base. The bonds to them are 90° apart. A fifth atom lies at the top of the upper pyramid, and a sixth lies at the bottom of the lower one. That arrangement defines **octahedral** electron-pair and molecular geometries.

**steric number (SN)** the sum of the number of atoms bonded to a central atom plus the number of lone pairs of electrons on the central atom.

**linear** molecular geometry about a central atom with a steric number of 2 and no lone pairs of electrons.

**trigonal planar** molecular geometry about a central atom with a steric number of 3 and no lone pairs of electrons.

**tetrahedral** molecular geometry about a central atom with a steric number of 4 and no lone pairs of electrons.

**trigonal bipyramidal** molecular geometry about a central atom with a steric number of 5 and no lone pairs of electrons, in which three atoms occupy equatorial sites and two other atoms occupy axial sites above and below the equatorial plane.

**octahedral** molecular geometry about a central atom with a steric number of 6 and no lone pairs of electrons, in which all six sites are equivalent.

### CONCEPT TEST

Assume that all the bonds in molecules with the five shapes described previously are single bonds and that none of the central atoms have lone pairs of electrons. Which SN values correspond to central atoms with less than an octet of valence electrons, which SN values correspond to central atoms that may have more than an octet of valence electrons, and which SN value has a central atom with eight valence electrons?

*(Answers to Concept Tests are in the back of the book.)*

Now let's consider a real-world analogy to those bond orientations. Beginning with fully inflated balloons, we tie together clusters of two, three, four, five, and six of them (**Figure 5.4**) so tightly that they push against each other. If the tie points of the clusters represent the central atom in our balloon model, the opposite ends represent the atoms bonded to the central atom. Note how our clusters of balloons produce the same orientations that resulted from selecting points as far apart as possible. The long axes of the balloons in Figure 5.4 accurately represent the bond directions in Figure 5.3.

**FIGURE 5.4** Balloons tied tightly together push against each other and orient themselves as far away from each other as possible. In doing so, they mimic the locations of different numbers of electron pairs about a central atom. Each balloon represents one electron pair. Note the similarities between the patterns of the balloons and the geometric shapes in Figure 5.3.

Now let's look at five simple molecules that have no lone pairs about the central atom and apply VSEPR and the concept of steric number to predict their electron-pair and molecular geometries. To do so, we follow three steps:

1. Draw the Lewis structure for the molecule.
2. Determine the steric number of the central atom.
3. Use the steric number and Figure 5.3 to predict the electron-pair and molecular geometries.

#### Example I: Carbon Dioxide, $CO_2$

1. Lewis structure:

$$:\ddot{O}=C=\ddot{O}:$$

2. The central carbon atom has two atoms bonded to it and no lone pairs, so the steric number is 2.
3. Because SN = 2, the O=C=O bond angle is 180° (see Figure 5.3a) and the electron-pair and molecular geometries are both linear.

**CONNECTION** We learned in Chapter 4 that some molecules have central atoms with incomplete octets of electrons.

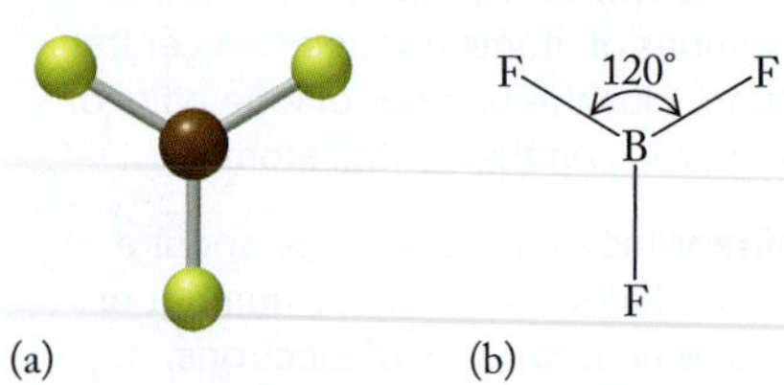


**FIGURE 5.5** The ball-and-stick model shows the orientation of the atoms in boron trifluoride. All F—B—F bond angles are 120° in this trigonal planar molecular geometry.

### Example II: Boron Trifluoride, $BF_3$

1. The Lewis structure of $BF_3$ that results in zero formal charges on all atoms is

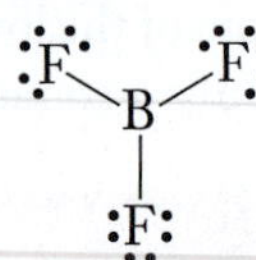


2. Three fluorine atoms are bonded to the central boron atom, and no lone pairs are present on boron, so SN = 3.
3. If SN = 3, all four atoms lie in the same plane. The three F atoms form an equilateral triangle, and the F—B—F bond angles are 120°. The electron-pair and molecular geometries are trigonal planar (Figure 5.3b and **Figure 5.5**).

### Example III: Carbon Tetrachloride, $CCl_4$

1. Lewis structure:
2. Four chlorine atoms are bonded to the central carbon atom, which gives carbon a full octet, so SN = 4.
3. If SN = 4, the chlorine atoms are located at the vertices of a tetrahedron and all four Cl—C—Cl bond angles are 109.5°, producing tetrahedral electron-pair and molecular geometries (Figure 5.3c and **Figure 5.6**).

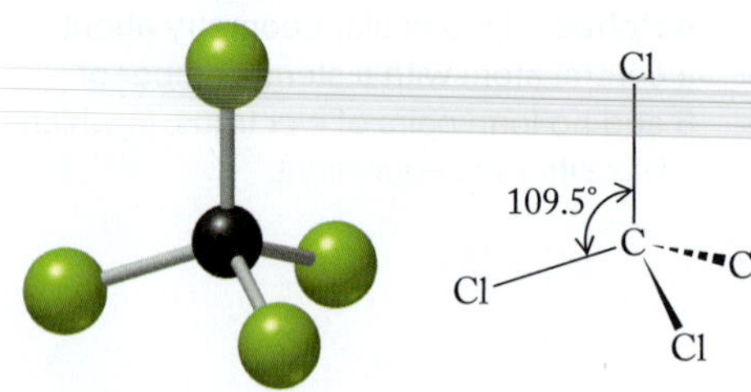


**FIGURE 5.6** The ball-and-stick model shows the orientation of the atoms in carbon tetrachloride. All Cl—C—Cl bond angles are 109.5° in this tetrahedral molecular geometry.

Figure 5.6 illustrates one of the conventions chemists use to draw three-dimensional structures on a two-dimensional surface such as a textbook page. A solid wedge (◄) is used to indicate a bond that comes out of the page toward the viewer. Thus, the solid wedge in Figure 5.6 means that the chlorine atom in that position points toward the viewer at a downward angle. A dashed wedge (····), by contrast, indicates a bond that goes into the page away from the viewer. Solid lines indicate bonds that lie in the plane of the page.

### Example IV: Phosphorus Pentafluoride, $PF_5$

1. Lewis structure:
2. Five fluorine atoms are bonded to the central phosphorus atom, which has no lone pairs, so SN = 5.
3. If SN =5, the fluorine atoms are located at the vertices of a trigonal bipyramid, and the electron-pair geometry and molecular geometry are trigonal bipyramidal (Figure 5.3d and **Figure 5.7**).

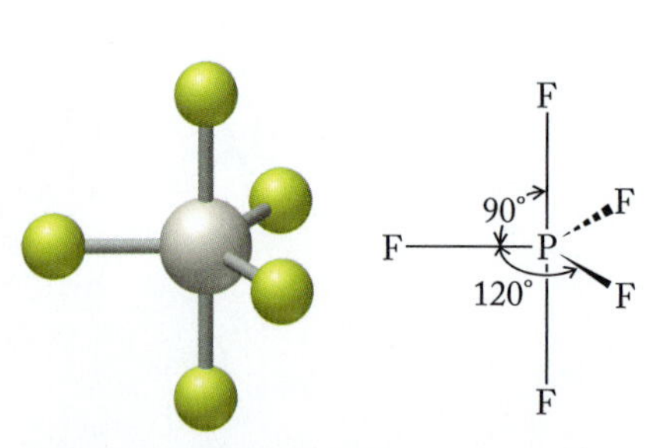


**FIGURE 5.7** The ball-and-stick model shows the orientation of the atoms in phosphorus pentafluoride. The equatorial P—F bonds are 120° from each other. Each of them is 90° from the two bonds connecting the central phosphorus atom to the fluorine atoms in the axial positions. This molecular geometry is trigonal bipyramidal.

### Example V: Sulfur Hexafluoride, $SF_6$

1. Lewis structure:

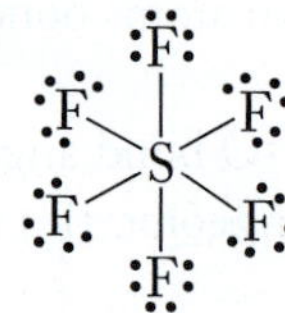

2. Six fluorine atoms are bonded to the central sulfur atom, which has no lone pairs, so SN = 6.
3. If SN = 6, the fluorine atoms are located at the vertices of an octahedron and the electron-pair and molecular geometries are octahedral (Figure 5.3e and **Figure 5.8**).

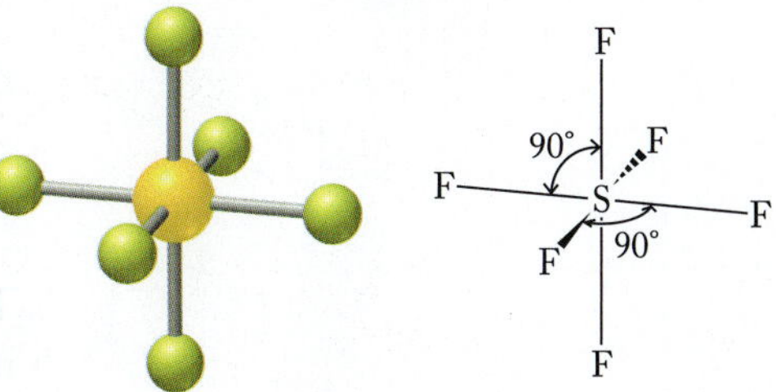


**FIGURE 5.8** The ball-and-stick model shows how the six fluorine atoms are oriented in three dimensions about the central sulfur atom in this octahedral molecular geometry.

**SAMPLE EXERCISE 5.1** Using VSEPR to Predict Molecular Geometry I **LO1**

Formaldehyde, $CH_2O$, is a gas at room temperature and an ingredient in solutions used to preserve biological samples. Use VSEPR to predict the molecular geometry of formaldehyde.

**Collect, Organize, and Analyze** To predict the geometry, we need to (1) draw the Lewis structure, (2) determine the steric number of the central atom by using Equation 5.1, and (3) identify the electron-pair and molecular geometries by using Figure 5.3. Carbon has the greatest bonding capacity of the three component elements, so it is the central atom.

**Solve** Using the procedure developed in Chapter 4 for drawing Lewis structures, we obtain the following one for formaldehyde:

:O:
||
C
H H

Three atoms are bonded to the central atom, which has no lone pairs on it, so SN = 3 and the electron-pair and molecular geometries are both trigonal planar.

**Think About It** The key to predicting the correct molecular geometry of a molecule with no lone pairs on its central atom is to determine the steric number, which is the number of atoms bonded to the central atom.

**Practice Exercise** Use VSEPR to determine the molecular geometry of the chloroform molecule, $CHCl_3$, and draw the molecule by using the solid-wedge, dashed-wedge convention.

*(Answers to Practice Exercises are in the back of the book.)*

Measurements of the bond angles in formaldehyde show that the H—C—H bond angle is slightly smaller than the 120° predicted for a trigonal planar geometry (Figure 5.3b), and the H—C═O bond angles are about 1° larger. The C═O double bond consists of two pairs of bonding electrons that exert greater repulsion on adjacent bonds than a single bonding pair would. That greater repulsion increases the H—C═O bond angles and decreases the H—C—H bond angle. VSEPR does not enable us to predict the actual values of those bond angles, but it does allow us to correctly predict that bond angles deviate from the ideal values of an idealized trigonal planar molecule.

## Central Atoms with Lone Pairs

If SN = 2 and one of the electron pairs on the central atom is a lone pair, only one other atom is in the molecule (Equation 5.1). Because two points define a straight line, two-atom molecules are all linear and have no bond angles.

Moving on to molecules with three or more atoms, recall from Figure 4.11 that we drew ozone molecules with an angular shape:

$$:\ddot{O}=\ddot{O}-\ddot{O}: \longleftrightarrow :\ddot{O}-\ddot{O}=\ddot{O}:$$

To calculate the steric number of the central O atom in both resonance structures, notice that it is bonded to two other atoms and has one lone pair of electrons. Therefore, its SN value in both resonance structures is 2 + 1 = 3. When SN = 3, the arrangement of atoms and lone pairs about the central atom is trigonal planar (Figure 5.3b). That means that the electron-pair geometry, which takes into account both the atoms and the nonbonding electron pairs around the central atom, is trigonal planar (**Figure 5.9a**). The molecular geometry, however, describes the relative positions of the *atoms* in the molecule only. With SN = 3 and two atoms attached to the central O atom, the ozone molecule has the **angular** (or **bent**) molecular geometry shown in **Figure 5.9(b)**.

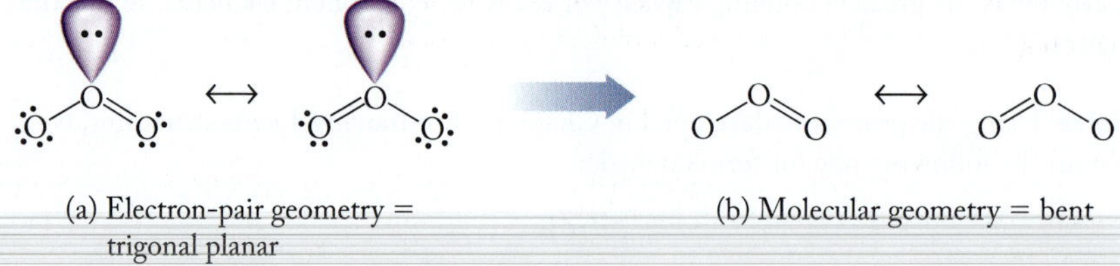


**FIGURE 5.9** (a) The electron-pair geometry of $O_3$ is trigonal planar because the steric number of the central oxygen atom is 3 (it is bonded to two atoms and has one lone pair of electrons). (b) The molecular geometry is bent because the central oxygen atom in the structure has no bonded atom, only a lone pair of electrons.

Experimental measurements confirm that $O_3$ is indeed a bent molecule with a bond angle of 117°. The angle is smaller than the 120° we would predict for a symmetrical trigonal planar electron-pair geometry. Using VSEPR, we can explain the smaller angle by comparing the amount of space near atoms occupied by bonding electrons with the amount of space occupied by electrons in a lone pair. Because the bonding electrons are attracted to two nuclei, they have a high probability of being located between the two atomic centers that share them. That puts the lone pair of electrons closer to the bonding pairs and produces greater repulsion (**Figure 5.10**). As a result, the lone pair pushes the bonding pairs closer together, thereby reducing the bond angle. In general,

- repulsion between lone pairs and bonding pairs is greater than repulsion between bonding pairs,
- repulsion caused by a lone pair is greater than repulsion caused by a double bond,
- repulsion caused by a double bond is greater than repulsion caused by a single bond, and
- two lone pairs of electrons on a central atom exert a greater repulsive force on the atom's bonding pairs than does one lone pair.

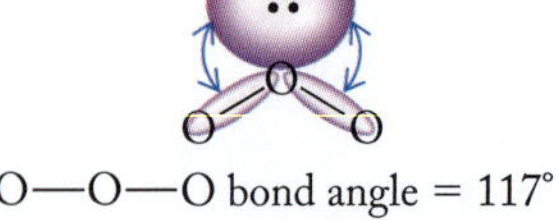

O—O—O bond angle = 117°

**FIGURE 5.10** The lone pair of electrons on the central oxygen atom in $O_3$ occupies more space (larger purple region) than the electron pairs in the O—O bonds (smaller purple regions). The increased repulsion (represented by the double-headed arrows) resulting from the larger volume occupied by the lone pair forces the noncentral oxygen atoms closer together, making the bond angle 117° instead of 120°.

**SAMPLE EXERCISE 5.2** Predicting Relative Sizes of Bond Angles **LO1**

Rank $NH_3$, $CH_4$, and $H_2O$ in order of decreasing bond angles in their molecular structures.

**Collect, Organize, and Analyze** The bond angles surrounding a central atom are linked to the steric number of the central atom, as shown in Figure 5.3, and to the number of lone pairs of electrons on the central atom. To find the steric numbers, we need to draw the Lewis structures of the three molecules.

**Solve** Using the method for drawing Lewis structures from Chapter 4, we obtain the following results:

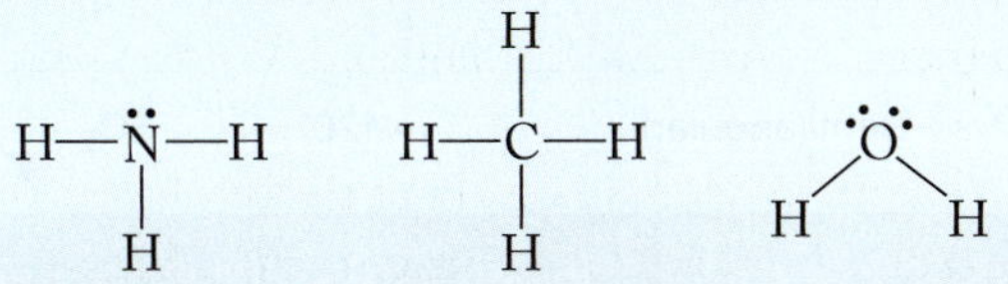

H—N—H (with H below N)   H—C—H (with H above and below C)   H—O—H (bent, with two lone pairs on O)

Because a total of four atoms or lone pairs of electrons surround each central atom in the three molecules, SN = 4 for all three. That means they all have tetrahedral electron-pair geometries. In $CH_4$, all four tetrahedral electron pairs are equivalent bonding pairs, so all four bond angles are 109.5°. In $NH_3$ and $H_2O$, however, repulsion from the lone pair(s) of electrons pushes the bonds closer together, reducing the bond angles. The two lone pairs on the O atom in $H_2O$ exert a greater repulsive force on its bonding pairs than the single lone pair on the N atom exerts on the bonding pairs in $NH_3$. Therefore, the bond angle in $H_2O$ should be less than the bond angles in $NH_3$. Ranking the three molecules in order of decreasing bond angle, we have

$$CH_4 > NH_3 > H_2O$$

**Think About It** The logic used in answering this problem is supported by experimental evidence: the bond angles in molecules of $CH_4$, $NH_3$, and $H_2O$ are 109.5°, 107.0°, and 104.5°, respectively.

**Practice Exercise** Determine which of the following species has the largest bond angle and which has the smallest: $NO_2$, $N_2O$, and $NO_2^-$.

*(Answers to Practice Exercises are in the back of the book.)*

**angular** or **bent** molecular geometry about a central atom with a steric number of 3 and one lone pair or a steric number of 4 and two lone pairs.

**trigonal pyramidal** molecular geometry about a central atom with a steric number of 4 and one lone pair of electrons.

As we saw in Sample Exercise 5.2, three combinations of atoms and lone pairs are possible for a central atom with SN = 4: four atoms and no lone pairs, three atoms and one lone pair, or two atoms and two lone pairs (**Table 5.1**). The first case is illustrated by the molecular structure of methane, $CH_4$, in which a tetrahedral electron-pair geometry translates into a tetrahedral molecular geometry. In ammonia, $NH_3$—which also has a tetrahedral electron-pair geometry by virtue of its three bonding pairs and one lone pair—one of the four vertices is the lone pair on the N atom (**Figure 5.11**). The resulting molecular geometry is called **trigonal pyramidal.** As we discussed in Sample Exercise 5.2, the strong repulsion produced by the diffuse lone pair of electrons on the N atom pushes the three N—H bonds closer together in $NH_3$ and reduces the angles between them from 109.5° (as in $CH_4$) to 107.0°.

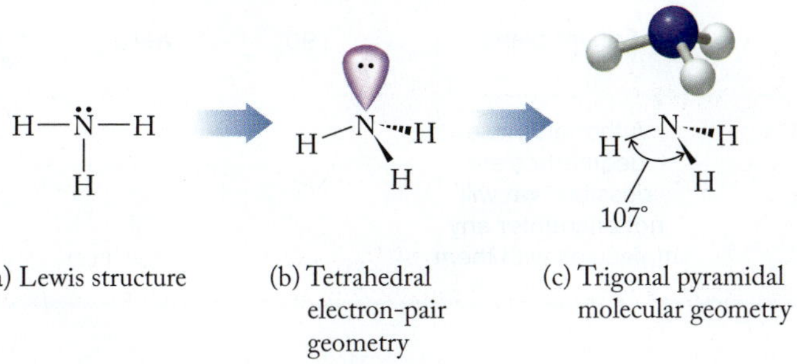

(a) Lewis structure

(b) Tetrahedral electron-pair geometry

(c) Trigonal pyramidal molecular geometry — 107°

**FIGURE 5.11** The steric number of the N atom in $NH_3$ is 4 (it is bonded to three atoms and has one lone pair of electrons), so its electron-pair geometry is tetrahedral. However, one vertex of the tetrahedron is occupied by a lone pair of electrons, not an atom, so the molecular geometry is trigonal pyramidal.

**TABLE 5.1** Electron-Pair Geometries and Molecular Geometries

| | Electron-Pair Geometry | No. of Bonded Atoms | No. of Lone Pairs | Molecular Geometry | Theoretical Bond Angles | Example |
|---|---|---|---|---|---|---|
| SN = 3 | | | | | | |
| | Trigonal planar | 3 | 0 | Trigonal planar | 120° | $CH_2O$ |
| | Trigonal planar | 2 | 1 | Bent (angular) | <120° | $SO_2$ |
| SN = 4 | | | | | | |
| | Tetrahedral | 4 | 0 | Tetrahedral | 109.5° | $CH_2Cl_2$ |
| | Tetrahedral | 3 | 1 | Trigonal pyramidal | <109.5° | $NH_3$ |
| | Tetrahedral | 2 | 2 | Bent (angular) | <109.5° | $H_2O$ |
| SN = 5 | | | | | | |
| | Trigonal bipyramidal | 5 | 0 | Trigonal bipyramidal | 90°, 120° | $PF_3Cl_2$ |
| | Trigonal bipyramidal | 4 | 1 | Seesaw | <90°, <120° | $SCl_4$ |
| | Trigonal bipyramidal | 3 | 2 | T-shaped | <90° | $BrF_3$ |
| | Trigonal bipyramidal | 2 | 3 | Linear | 180° | $XeF_2$ |
| SN = 6 | | | | | | |
| | Octahedral | 6 | 0 | Octahedral | 90° | $SF_6$ |
| | Octahedral | 5 | 1 | Square pyramidal | <90° | $IF_5$ |
| | Octahedral | 4 | 2 | Square planar | 90° | $XeF_4$ |
| | Octahedral | 3 | 3 | Although these geometries are possible, we will not encounter any molecules with them | | |
| | Octahedral | 2 | 4 | | | |

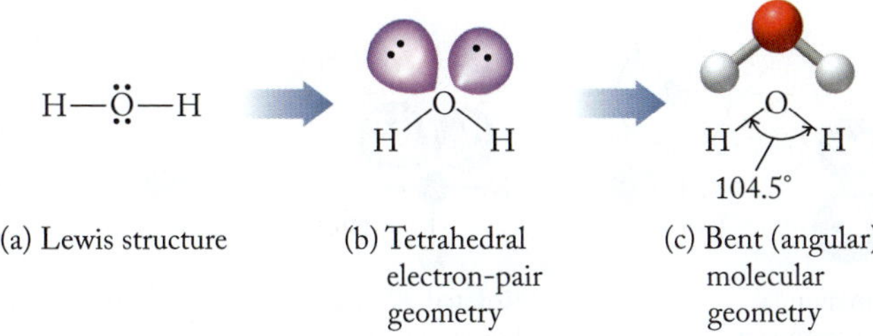


**FIGURE 5.12** Two vertices of the tetrahedron are occupied by lone pairs of electrons, so only two atoms define the molecular geometry of $H_2O$, which is bent.

The central O atom in a molecule of $H_2O$ has SN = 4 and a tetrahedral electron-pair geometry because it is bonded to two H atoms and has two lone pairs of electrons (**Figure 5.12**). The molecular geometry of $H_2O$, however, is bent (or angular) because molecular geometries ignore lone pairs and take into account only the atoms bonded to the central atom. As we also discussed in Sample Exercise 5.2, the H—O—H bond angle in water is smaller than 109.5° as a result of repulsion between the two lone pairs and each bonding pair. The actual bond angle is 104.5°.

Molecules with trigonal bipyramidal electron-pair geometry (SN = 5) have four possible molecular geometries depending on the number of lone pairs per molecule (see Table 5.1). Those options arise because a trigonal bipyramid has axial and equatorial vertices (Figure 5.3d). VSEPR enables us to predict which vertices are occupied by bonded atoms and which by lone pairs. The key to those predictions is that the repulsions between pairs of electrons increase as the angle between them decreases: two electron pairs at 90° experience a greater mutual repulsion than two at 120°, which in turn have a greater repulsion than two at 180°. To minimize repulsions involving lone pairs, VSEPR predicts that they preferentially occupy equatorial, rather than axial, vertices. Why? Because an equatorial lone pair has *two* 90° repulsions with the two axial electron pairs, as shown in **Figure 5.13(a)**, but an axial lone pair has *three* 90° repulsions with three equatorial electron pairs, as shown in **Figure 5.13(b)**.

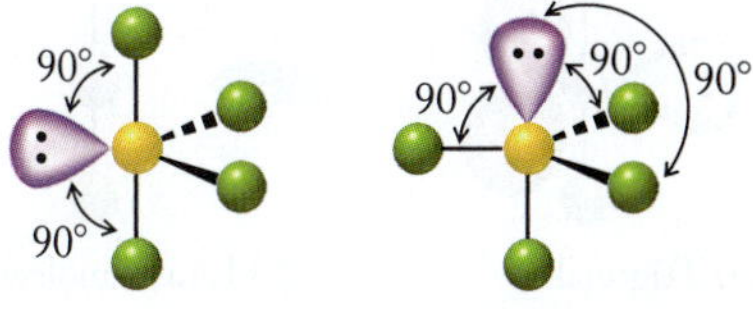


(a) Equatorial lone pair (b) Axial lone pair DOES NOT EXIST

**FIGURE 5.13** A lone pair of electrons in an equatorial position of a molecule with trigonal bipyramidal electron-pair geometry interacts through 90° with two other electron pairs. A lone pair in an axial position interacts through 90° with *three* other electron pairs. Fewer 90° interactions reduce internal electron-pair repulsion and lead to greater stability.

When we assign one, two, or three lone pairs of valence electrons to equatorial vertices, we get three of the molecular geometries for SN = 5 in Table 5.1. When a single lone pair occupies an equatorial site, we get a molecular geometry called **seesaw** (**Figure 5.14**) because its shape, when rotated 90° clockwise, resembles a playground seesaw. (That shape's formal name is *disphenoidal*.)

When two lone pairs occupy equatorial sites, the molecular geometry that results is called **T-shaped** (**Figure 5.15**). That designation becomes more apparent when the structure in Figure 5.15(a) is rotated 90° counterclockwise. Lone-pair repulsions result in bond angles slightly less than the 90° and 180° we would expect from a perfectly shaped "T" geometry.

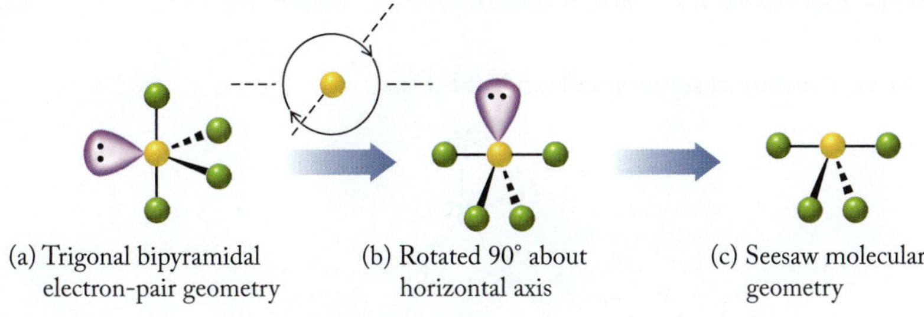


**FIGURE 5.14** A single lone pair of electrons in an equatorial position of a trigonal bipyramidal electron-pair geometry produces a seesaw molecular geometry.

**seesaw** molecular geometry about a central atom with a steric number of 5 and one lone pair of electrons in an equatorial position; the atoms occupy two axial sites and two equatorial sites.

**T-shaped** molecular geometry about a central atom with a steric number of 5 and two lone pairs of electrons that occupy equatorial positions; the atoms occupy two axial sites and one equatorial site.

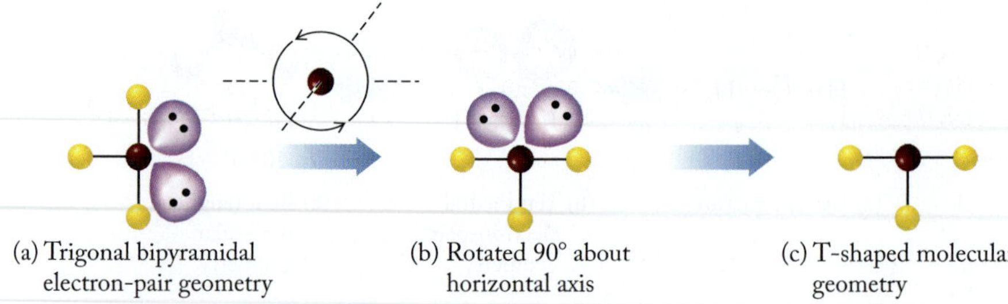

(a) Trigonal bipyramidal electron-pair geometry (b) Rotated 90° about horizontal axis (c) T-shaped molecular geometry

**FIGURE 5.15** Two lone pairs of electrons in equatorial positions of a trigonal bipyramidal electron-pair geometry produce a T-shaped molecular geometry.

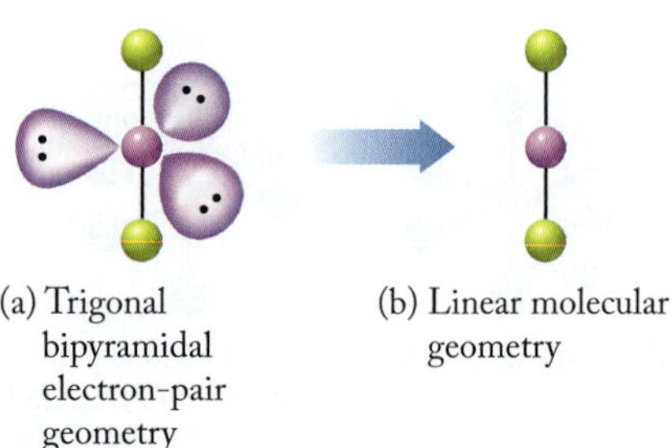

(a) Trigonal bipyramidal electron-pair geometry (b) Linear molecular geometry

**FIGURE 5.16** Three lone pairs of electrons in equatorial positions of a trigonal bipyramidal electron-pair geometry produce a linear molecular geometry.

Finally, a SN = 5 molecule with three lone pairs and two bonded atoms has a linear geometry because all three lone pairs occupy equatorial sites and the bonding pairs are in the two axial positions (**Figure 5.16**).

## CONCEPT TEST

The bond angles in a trigonal bipyramidal molecular structure are 90°, 120°, or 180°. Are the corresponding bond angles in a seesaw structure likely to be larger than, the same as, or smaller than those values?

*(Answers to Concept Tests are in the back of the book.)*

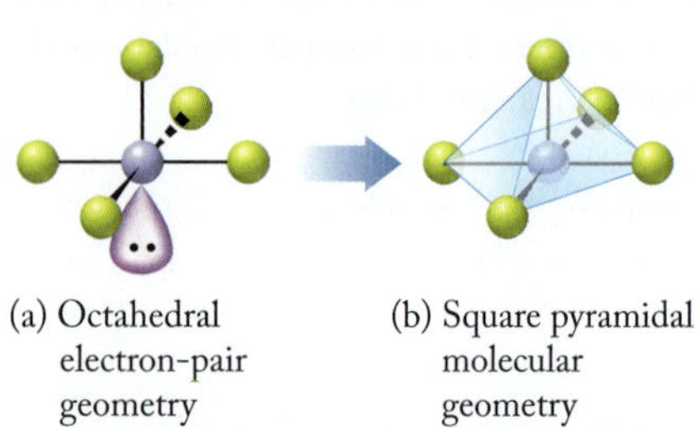

(a) Octahedral electron-pair geometry (b) Square pyramidal molecular geometry

**FIGURE 5.17** A lone pair of electrons in an octahedral electron-pair geometry produces a square pyramidal molecular geometry.

Molecules that have a central atom with SN = 6 have an octahedral electron-pair geometry (Figure 5.3e) and the molecular geometries listed in Table 5.1. With one lone pair of electrons, only one molecular geometry is possible because all the sites in an octahedron are equivalent (**Figure 5.17**). That one molecular geometry is called **square pyramidal** and consists of a pyramid with a square base, four triangular sides, and the central atom "embedded" in the base.

Because of stronger repulsion between lone pairs and bonding pairs, the bond angles in a square pyramidal molecule should be slightly less than the ideal angle of 90°. In $BrF_5$, for example, the angles between its equatorial and axial bonds are 85°.

When two lone pairs are present at the central atom in a molecule with octahedral electron-pair geometry, they occupy vertices on opposite sides of the octahedron to minimize the repulsions between them. The resultant molecular geometry is called **square planar** because the molecule is shaped like a square and all five atoms reside in the same plane (**Figure 5.18**). Because the two lone pairs are on opposite sides of the bonding pairs, the presence of the lone pairs does not distort the bond angles: they are all 90° or 180°.

(a) Octahedral electron-pair geometry (b) Square planar molecular geometry

**FIGURE 5.18** Two lone pairs of electrons on opposite sides of an octahedral electron-pair geometry produce a square planar molecular geometry.

### SAMPLE EXERCISE 5.3 Using VSEPR to Predict Molecular Geometry II LO1

The Lewis structure of sulfur tetrafluoride ($SF_4$) is

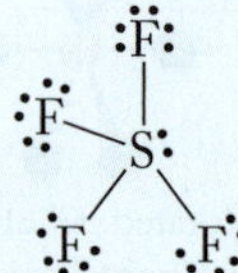

What is its molecular geometry and what are the angles between the S—F bonds?

**Collect, Organize, and Analyze** The Lewis structure of $SF_4$ shows that the central S atom is bonded to four F atoms and has one lone pair of electrons, which means that the S atom has SN = 4 + 1 = 5. Table 5.1 contains molecular geometry options for molecules with different SN values.

**Solve** A steric number of 5 means that the S atom has a trigonal bipyramidal electron-pair geometry, but the presence of one lone pair of electrons on the S atom means that its molecular geometry (from Table 5.1) is seesaw:

The proximity of the equatorial lone pair to the center of the S atom increases the electron-pair repulsion experienced by the bonding pairs, which decreases the bond angles between them from their normal 90° between the axial and equatorial bonds, 120° between the two equatorial bonds, and 180° between the two axial bonds.

**Think About It** Any molecule with SN = 5 and one lone pair should have a seesaw molecular geometry like that of $SF_4$.

**Practice Exercise** Determine the molecular geometries and the bond angles of (a) $SO_2Cl_2$, (b) $ClF_3$, and (c) $XeF_4$.

*(Answers to Practice Exercises are in the back of the book.)*

## 5.3 Polar Bonds and Polar Molecules

We learned in Chapter 4 that an unequal distribution of bonding electrons between two atoms produces a partial negative charge on one end of the bond and a partial positive charge on the other. That charge separation, caused by the difference in electronegativity between the two atoms, results in a **bond dipole**. Now that we know about the three-dimensional geometry of molecules, we can explore how bond dipoles contribute to a molecule's *overall* polarity.

The polarity of a molecule can be determined by summing the polarities of all its individual bond dipoles. That summing, which can be done using vector addition (see Appendix 1), must take into account both the strengths of the individual bond dipoles and their orientations with respect to one another. $CO_2$, for example, has two C=O bonds. Oxygen is more electronegative than carbon, so each bond has a dipole, which we indicate by drawing an arrow pointing toward the oxygen atom. The two dipoles are equivalent in strength because they involve the same atoms and the same kind of bond. The linear shape of the molecule means that the direction of one C=O bond dipole is opposite the direction of the other:

$$\ddot{O}=C=\ddot{O}$$

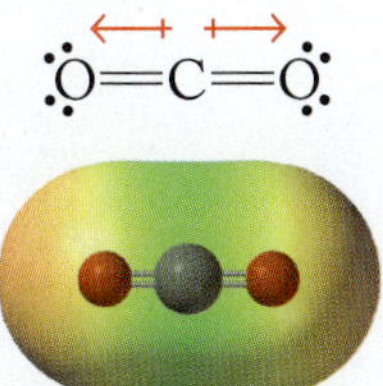

Because the two bond dipoles are equal in magnitude but point in opposite directions, $CO_2$ is a nonpolar molecule. Even though $CO_2$ has polar bonds, $CO_2$ has

**CONNECTION** See Figure 4.3 to review using arrows to indicate bond polarity and Figure 4.4 to review using electron-density surfaces to indicate bond polarity.

**CONNECTION** In Chapter 4 we introduced electronegativity and the unequal distribution of electrons in polar covalent bonds. See Figure 4.5 to review the periodic trends in electronegativity values.

**square pyramidal** molecular geometry about a central atom with a steric number of 6 and one lone pair of electrons; as typically drawn, the atoms occupy four equatorial sites and one axial site.

**square planar** molecular geometry about a central atom with a steric number of 6 and two lone pairs of electrons that occupy axial sites; the atoms occupy four equatorial positions.

**bond dipole** separation of electrical charge created when atoms with different electronegativities form a covalent bond.

**permanent dipole** permanent separation of electrical charge in a molecule resulting from unequal distributions of bonding and/or lone pairs of electrons.

**dipole moment ($\mu$)** a measure of the degree to which a molecule aligns itself in an applied electric field; a quantitative expression of the polarity of a molecule.

no overall **permanent dipole**. We use the adjective *permanent* here because, as we saw in Chapter 4, the vibration of the bonds in molecules such as $CO_2$ can create fluctuating electric fields that allow $CO_2$ to absorb infrared electromagnetic radiation. The fluctuations create high-frequency *temporary* dipoles, even in many nonpolar molecules.

Carbon tetrafluoride, $CF_4$, also is a nonpolar molecule, even though its C—F bonds are polar. The four C—F bond dipoles all have the same strength because the same atoms are involved. $CF_4$ has tetrahedral molecular geometry, however, so the bond dipoles offset one another and the $CF_4$ molecule is nonpolar overall:

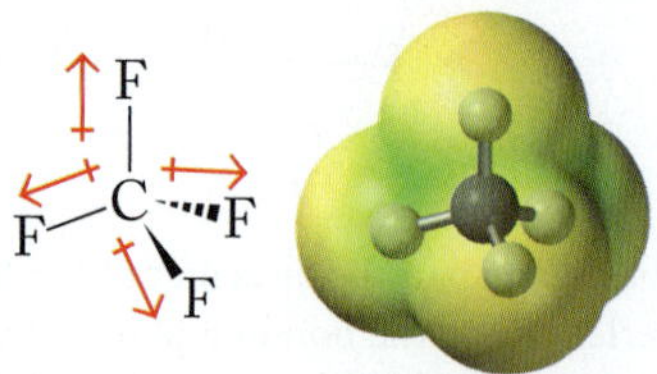


Water molecules are bent, not linear as molecules of $CO_2$ are. Therefore, the dipoles of the two O—H bonds in a molecule of $H_2O$ do *not* offset each other. Instead, the molecule has an overall permanent dipole, with the negative end directed toward the O atom and the positive end directed toward the hydrogen atoms:

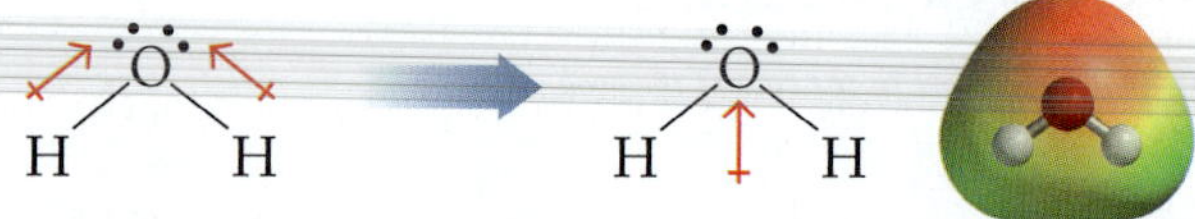


The presence of that overall dipole means that water is a polar molecule. Water's polarity leads to interactions between H and O atoms on adjacent $H_2O$ molecules in liquid water or solid ice. Other polar molecules exhibit similar interactions. We discuss those interactions and their consequences in Chapter 6.

A molecule's polarity can be determined experimentally by measuring its permanent **dipole moment ($\mu$)**. The value of $\mu$ expresses the extent of the overall separation of positive and negative charge in the molecule, and it is determined by measuring the degree to which the molecule aligns with a strong electric field, as shown in **Figure 5.19**. Note how the negative (F) ends of the HF molecules in the

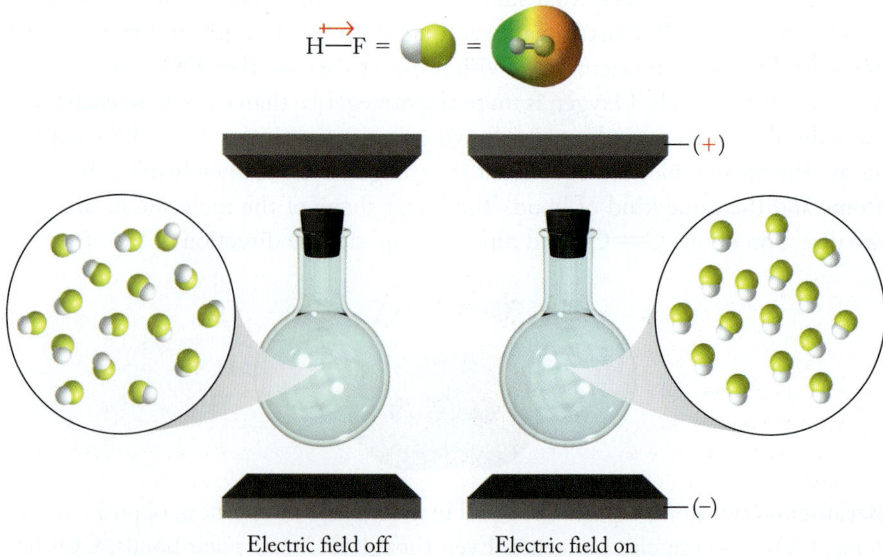


**FIGURE 5.19** Gaseous HF molecules are oriented randomly in the absence of an electric field but align when an electric field is applied to two metal plates. The negative (fluorine) end of each molecule is directed toward the positively charged plate; the positive (hydrogen) end is directed toward the negative plate.

**TABLE 5.2** Permanent Dipole Moments of Several Polar Molecules

| Formula | Structure with Bond Dipole(s) | Direction of Permanent Dipole | Dipole Moment (debyes) |
|---|---|---|---|
| HF | H—F | ⟶ | 1.82 |
| $H_2O$ | H–O–H | ↑ | 1.85 |
| $NH_3$ | H–N(–H)–H | ↑ | 1.47 |
| $CHCl_3$ | H–C(–Cl)(–Cl)–Cl | ↓ | 1.01 |
| $CCl_3F$ | F–C(–Cl)(–Cl)–Cl | ↑ | 0.45 |

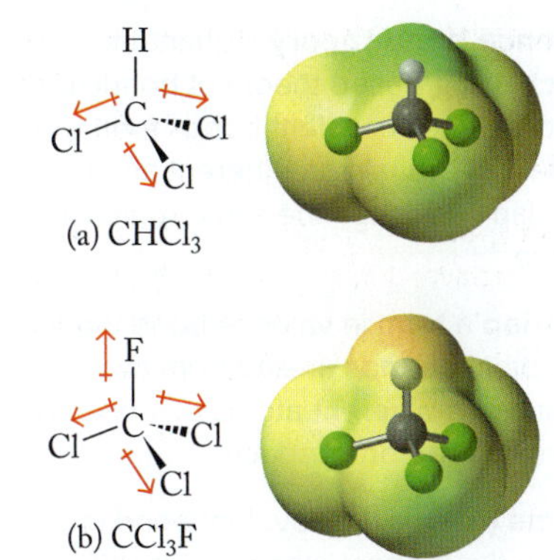


**FIGURE 5.20** (a) The four bonds in chloroform ($CHCl_3$) are not equivalent, which causes an asymmetrical distribution of bonding electrons in the molecule, resulting in a permanent dipole directed toward the bottom of $CHCl_3$ as drawn. (b) The C—F bond dipole in $CCl_3F$ is stronger than the C—Cl bond dipoles and is not completely offset by them, which makes $CCl_3F$ a slightly polar substance with its dipole directed toward the fluorine atom.

figure are oriented toward the positive plate and how the positive (H) ends are oriented toward the negative plate. The more polar a molecule, the more strongly it aligns with an electric field.

Dipole moments are usually expressed in units of *debyes* (D), where 1 D = $3.34 \times 10^{-30}$ coulomb-meter. The dipole moments of several polar substances are listed in **Table 5.2**. Chloroform ($CHCl_3$) has a relatively strong dipole moment (1.01 D) because chlorine is more electronegative than carbon, and the two electrons in each C—Cl bond are pulled away from C and toward Cl (**Figure 5.20a**). Because hydrogen is only slightly less electronegative than carbon ($\Delta\chi = 0.4$), the H—C bond is considered essentially nonpolar. Consequently, the electron distribution is away from the top part of the molecule as drawn and toward the bottom.

In trichlorofluoromethane ($CCl_3F$), all four bond dipoles point away from the central carbon atom because fluorine ($\chi = 4.0$) and chlorine ($\chi = 3.0$) are both more electronegative than carbon ($\chi = 2.5$; **Figure 5.20b**). However, a C—F bond dipole is stronger than a C—Cl bond dipole because fluorine is more electronegative than chlorine. As a result, bonding electrons are pulled more toward the C—F (top) part of the molecule than toward the C—Cl (bottom) part. The net direction of the summed bond dipoles is upward.

**SAMPLE EXERCISE 5.4** Predicting Whether a Substance Is Polar **LO2**

Does either of these molecules have a permanent dipole? (a) $CH_2O$ (formaldehyde); (b) $CH_2Cl_2$ (dichloromethane, an organic solvent)

**Collect, Organize, and Analyze** To predict whether a molecular compound has a permanent dipole, we need to determine whether its molecules contain polar bonds and whether the bond dipoles offset each other. A bond's polarity depends on the difference in electronegativity between the bonded pair of atoms. From Figure 4.5, we know that the electronegativities of the atoms in formaldehyde are H = 2.1, C = 2.5, and O = 3.5, and the electronegativity of the chlorine atoms in dichloromethane is Cl = 3.0.

**valence bond theory** a quantum mechanics–based theory of bonding that assumes covalent bonds form when half-filled orbitals on different atoms overlap or occupy the same region in space.

**overlap** a term in valence bond theory describing bonds arising from two orbitals on different atoms that occupy the same region of space.

**sigma (σ) bond** a covalent bond in which the highest electron density lies between the two atoms along the internuclear axis.

**hybridization** in valence bond theory, the mixing of atomic orbitals to generate new sets of orbitals that may form sigma bonds with other atoms.

**hybrid atomic orbital** in valence bond theory, one of a set of equivalent orbitals about an atom created when specific atomic orbitals are mixed.

**Solve**

a. In Sample Exercise 5.1, we determined that formaldehyde has a trigonal planar structure:

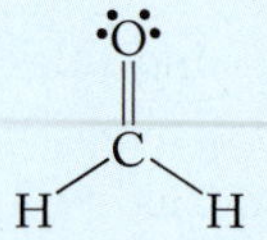

The large difference between the electronegativities of C and O and the negligible difference between the electronegativities of C and H mean that the polar C═O bond gives formaldehyde a permanent dipole:

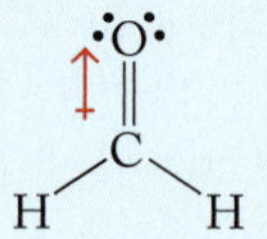

b. Using the procedure to create Lewis structures that we learned in Chapter 4, we can draw the Lewis structure for $CH_2Cl_2$ with the two hydrogen atoms next to each other or in an alternating fashion with the chlorine atoms:

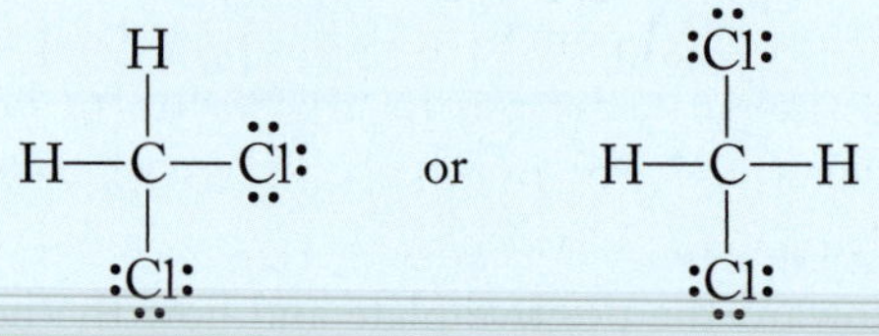

Remember, however, that a Lewis structure simply depicts what bonds are present, not how those bonds are oriented in space. SN = 4 for the carbon atom in each Lewis structure drawn above for $CH_2Cl_2$, and no lone pairs are on the carbon atom. Regardless of how the Lewis structure is drawn, the electron-pair geometry and the molecular geometry are both tetrahedral for $CH_2Cl_2$:

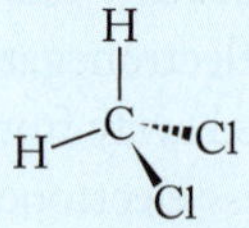

The large difference in the electronegativities between C and Cl means that both the C—Cl bonds are polar, with a negligible difference in electronegativity for the C—H bonds. Therefore, $CH_2Cl_2$ has a permanent dipole:

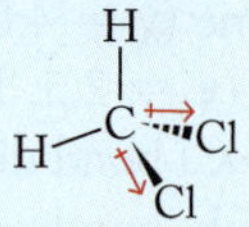

**Think About It** When more than one kind of atom is bonded to a central atom, the molecule is highly likely to have a permanent dipole because the orientation and/or magnitude of the individual bond dipoles do not cancel out.

**Practice Exercise** (a) Does carbon disulfide ($CS_2$), a gas present in small amounts in crude petroleum, have a dipole moment? (b) Does the ozone molecule ($O_3$) have a permanent dipole moment?

*(Answers to Practice Exercises are in the back of the book.)*

## CONCEPT TEST

Water and hydrogen sulfide both have a bent molecular geometry with dipole moments of 1.85 D and 0.97 D, respectively. Why is the dipole moment of $H_2S$ smaller than that of $H_2O$?

*(Answers to Concept Tests are in the back of the book.)*

# 5.4 Valence Bond Theory and Hybrid Orbitals

Drawing Lewis structures and applying VSEPR theory enable us to reliably predict the geometry of many molecules. However, we have yet to make a connection between molecular geometry and the idea, first advanced in Chapter 3, that electrons exist in atomic orbitals. The absence of a connection between the electronic structure of atoms and the electronic structure of molecules led to the development of bonding theories that use atomic orbitals to account for the molecular geometries predicted by VSEPR. We explore one of those theories now.

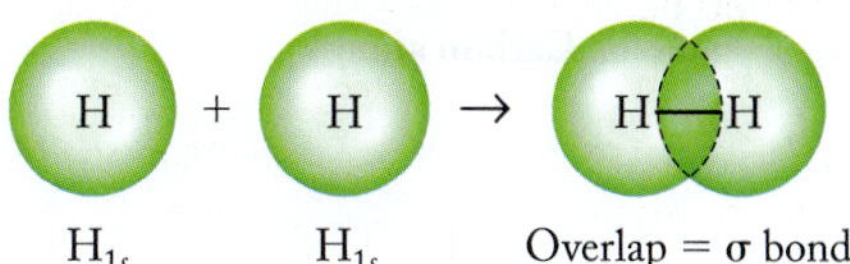


**FIGURE 5.21** The overlap of the 1*s* orbitals on two hydrogen atoms results in a single σ bond between the two hydrogen atoms in a $H_2$ molecule.

**CHEMTOUR**
Hybridization

**Valence bond theory** evolved in the late 1920s when Linus Pauling merged quantum mechanics with Lewis's model of shared electron pairs to develop a theory to explain molecular bonding. According to valence bond theory, a chemical bond forms when the atomic orbitals of two atoms **overlap**. The electrons in overlapping orbitals are attracted to the nuclei of both bonded atoms. That attraction leads to lower potential energy and greater chemical stability than if the atoms were completely free (review Figure 4.2). Valence bond theory was an important advance over Lewis's model because it could be used to explain the molecular geometries we have been exploring in this chapter. Let's see how it works.

We begin by applying Pauling's valence bond theory to $H_2$, the simplest diatomic molecule. A free H atom has a single electron in a 1*s* atomic orbital. According to valence bond theory, overlap of two half-filled 1*s* orbitals leads to increased electron density between the nuclei of the bonding atoms as the two nuclei share the two 1*s* electrons. As a result, a H—H covalent bond forms (**Figure 5.21**). The region of highest electron density lies along the internuclear axis between the two atoms, which makes the H—H bond an example of a **sigma (σ) bond**.

## $sp^3$ Hybrid Orbitals

Now let's consider the bonding in methane, $CH_4$. Recall from Chapter 3 that an individual carbon atom has the electron configuration $[He]2s^22p^2$, that *s* orbitals are spherical, and that *p* orbitals are usually drawn with teardrop shapes (actually, they're shaped more like mushroom caps—review Figure 3.24). The lobes of the three *p* orbitals in a subshell are oriented at 90° to one another, as shown in **Figure 5.22**. Further, we learned in Chapter 4 that carbon atoms form four covalent bonds to complete their octets.

We saw in Section 5.2 that the central carbon atom in methane has a steric number of 4 (four bonds, no lone pairs) and a tetrahedral molecular geometry. How can a carbon atom with a filled, spherical 2*s* orbital and two half-filled 2*p* orbitals oriented 90° from each other result in a tetrahedral molecule with bond angles of 109.5°? VSEPR cannot answer that question, but valence bond theory can. In valence bond theory, the 2*s* and 2*p* orbitals can be reconfigured in a process called **hybridization**.

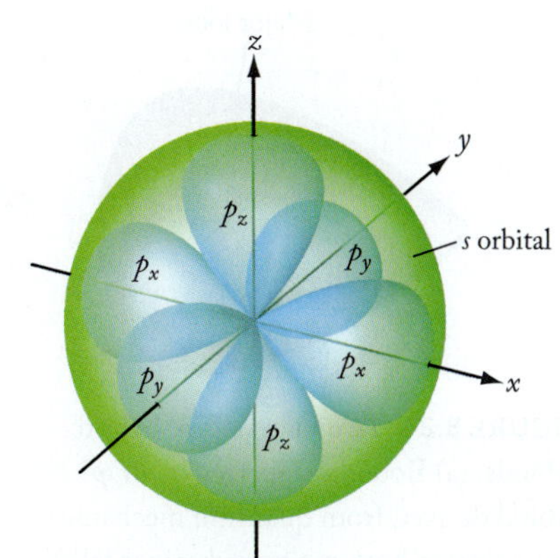


**FIGURE 5.22** Relative orientation in space of one 2*s* orbital (spherical) and three 2*p* orbitals (teardrop-shaped) oriented at 90° to each other along the *x*-, *y*-, and *z*-axes of a coordinate system.

Theoretically, hybridization of the carbon atom in a molecule of methane occurs as the valence-shell atomic orbitals of the C atom begin to overlap with the 1*s* orbitals of the four H atoms. The 2*s*, $2p_x$, $2p_y$, and $2p_z$ orbitals of the C atom no longer represent the lowest-energy orientations of the atom's valence orbitals. What happens next can be viewed as a mathematical mixing of the wave functions defining the atom's 2*s* and 2*p* orbitals. The product of mixing is a new set of wave functions that defines four equivalent **hybrid atomic orbitals** with shapes and orientations unlike those of 2*s* and 2*p* orbitals (**Figure 5.23**). Because a single *s* orbital and three *p* orbitals are involved in that mixing process, we call

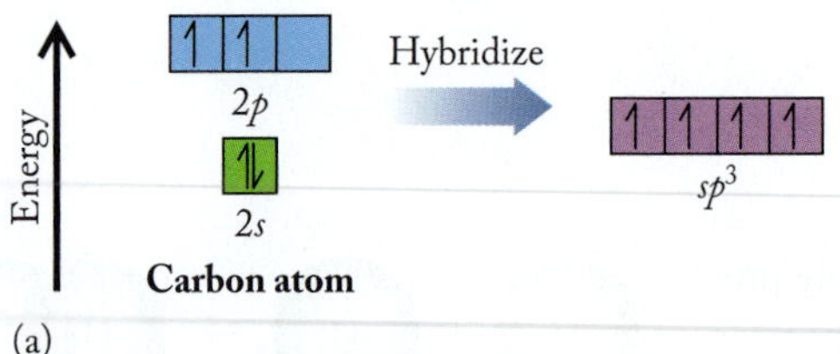


**FIGURE 5.23** (a) When the 2*s* and 2*p* atomic orbitals of carbon are hybridized, the four orbitals are mixed to create four $sp^3$ hybrid orbitals, each containing one electron. (b) The hybrid orbitals are oriented 109.5° from each other, which is the orientation of the bonds in a molecule of methane and in other molecules with a tetrahedral molecular geometry.

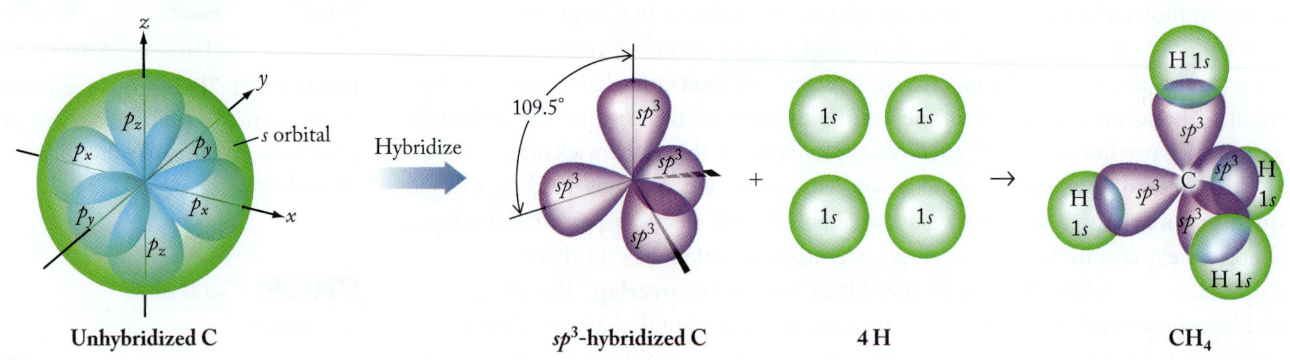


**CONNECTION** Recall from Chapter 3 that the shapes of the *p* orbitals we use in this book are longer and narrower than their true shapes to make their orientation easier to see (see Figure 3.24).

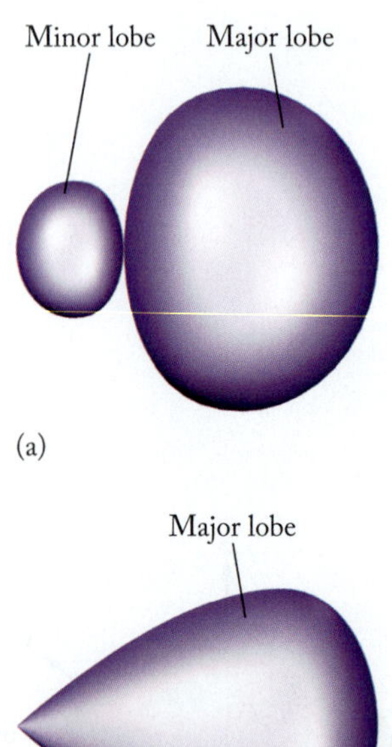


**FIGURE 5.24** Two views of $sp^3$ hybrid orbitals. (a) Boundary surface of an $sp^3$ orbital derived from quantum mechanics, consisting of both major and minor lobes. (b) To better show the orientation of bonds formed by hybrid orbitals, we elongate the major lobe and ignore the minor one (which is not involved in bond formation).

the result a set of **$sp^3$ hybrid orbitals**, as shown in Figure 5.23(a). The lobes are all 109.5° from each other because the distance between them is *maximized*, which means that the repulsion between the electrons in them is *minimized*. Overlap of the lobes with the 1*s* orbitals of four hydrogen atoms results in the formation of four σ bonds and one molecule of methane that has the tetrahedral molecular geometry shown in Figure 5.23(b).

Those $sp^3$ hybrid orbitals appear to have one lobe each, but in fact every $sp^3$ hybrid orbital consists of a major lobe and a minor lobe, as shown in **Figure 5.24(a)**. Because our current focus is on bonding and the minor lobe is not involved in bonding, we ignore it in this chapter and omit it from the illustrations. Also, the shapes of the hybrid orbitals that we use in this book are more elongated than the boundary surfaces calculated from quantum mechanics, as shown in **Figure 5.24(b)**. As with our *p* orbital images, we stretch the lobes of hybrid orbitals a bit to better show their orientation.

According to valence bond theory, *any* atom with *s* and *p* orbitals in its valence shell can form a set of $sp^3$ hybrid orbitals when it is the central atom in a molecule with a tetrahedral electron-pair geometry. The hybrid orbitals may contain single electrons or pairs of electrons. For example, $sp^3$ hybridization of the nitrogen atom, which has five valence electrons, produces three hybrid orbitals that are half-filled and one hybrid orbital that is full. In a molecule of ammonia ($NH_3$; **Figure 5.25a**), the three half-filled orbitals form three σ bonds by overlapping with the 1*s* orbitals of three hydrogen atoms, and the filled hybrid orbital contains the N atom's lone pair of electrons.

Similarly, the oxygen atom in water forms four $sp^3$ hybrid orbitals: two contain lone pairs, and two are half-filled and available for bond formation (**Figure 5.25b**). The half-filled orbitals overlap with 1*s* orbitals from hydrogen atoms and form two σ bonds, whereas the two filled orbitals represent the two lone pairs on O. Thus, a carbon atom that forms four σ bonds, a nitrogen atom that forms three σ bonds and has one lone pair of electrons, and an oxygen atom that forms two σ bonds and has two lone pairs of electrons all have the same steric number (4) and are all $sp^3$-hybridized atoms. That consistency is not accidental. As we will see, the SN of an atom with an octet of valence electrons is directly linked to both its hybridization and its molecular geometry.

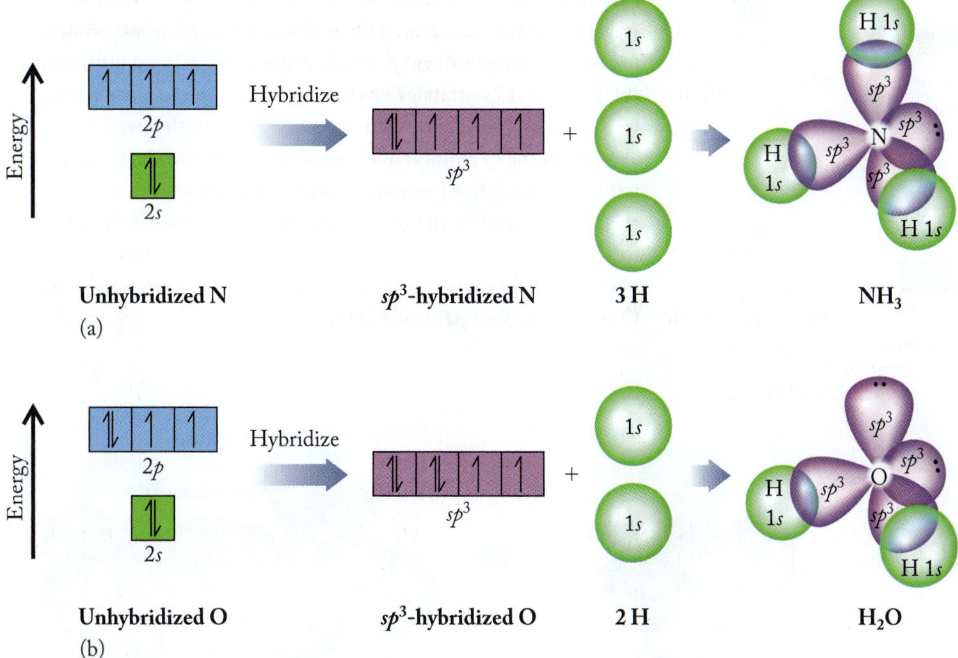


**FIGURE 5.25** A set of four $sp^3$ hybrid orbitals points toward the vertices of a tetrahedron. (a) The nonbonding electron pair of the N in $NH_3$ occupies one of the four hybrid orbitals, giving the molecule a trigonal pyramidal molecular geometry. (b) The O atom in $H_2O$ is also $sp^3$ hybridized. Only two of the four orbitals contain bonding pairs of electrons, giving $H_2O$ a bent molecular geometry.

## $sp^2$ Hybrid Orbitals

Recall from Sample Exercise 5.1 that formaldehyde has a trigonal planar molecular geometry. According to the VSEPR model, the SN of the central carbon atom is 3, so three atomic orbitals must be mixed to make three equivalent hybrid orbitals. Instead of mixing all four atomic orbitals, as in $sp^3$ hybridization, we mix only three: the $2s$ orbital and two of the $2p$ orbitals. That leaves the third $2p$ orbital unhybridized (**Figure 5.26a**).

Mixing one $s$ and two $p$ orbitals generates three hybrid orbitals called **$sp^2$ hybrid orbitals.** The energy levels of $sp^2$ orbitals are slightly lower than those of $sp^3$ orbitals because only two $p$ orbitals are mixed with the $s$ orbital in a set of $sp^2$ orbitals. The orbitals in an $sp^2$-hybridized atom all lie in the same plane and are 120° apart. The two lobes of the unhybridized $p$ orbital lie above and below the plane of the triangle defined by the $sp^2$ hybrid orbitals. An $sp^2$-hybridized atom can form up to three σ bonds with its three hybridized orbitals, and the carbon atom in formaldehyde does so—that is, it forms two σ bonds to the two H atoms and one σ bond to the O atom.

For its part, the O atom also has a steric number of 3 (it's bonded to one atom and has two lone pairs of electrons), so it is also $sp^2$ hybridized (**Figure 5.26b**). The O atom's unhybridized $p_x$ orbital is parallel to the C atom's unhybridized $p_x$ orbital, producing a **pi (π) bond** in which electron densities are highest above and below the internuclear axis, as shown in the last image in **Figure 5.26(c)**. Note that the π bond is *one* bond formed by the two atoms' $p$ orbitals through *two* points of contact—it is *not* two bonds. That π bond and the σ bond between C and O together make up the C═O double bond in $CH_2O$ (**Figure 5.27**). Note how the trigonal planar geometries around the

**$sp^3$ hybrid orbitals** four hybrid orbitals with a tetrahedral orientation produced by mixing one $s$ and three $p$ orbitals.

**$sp^2$ hybrid orbitals** three hybrid orbitals in a trigonal planar orientation formed by mixing one $s$ and two $p$ orbitals.

**pi (π) bond** a covalent bond in which electron density is greatest above and below the internuclear axis.

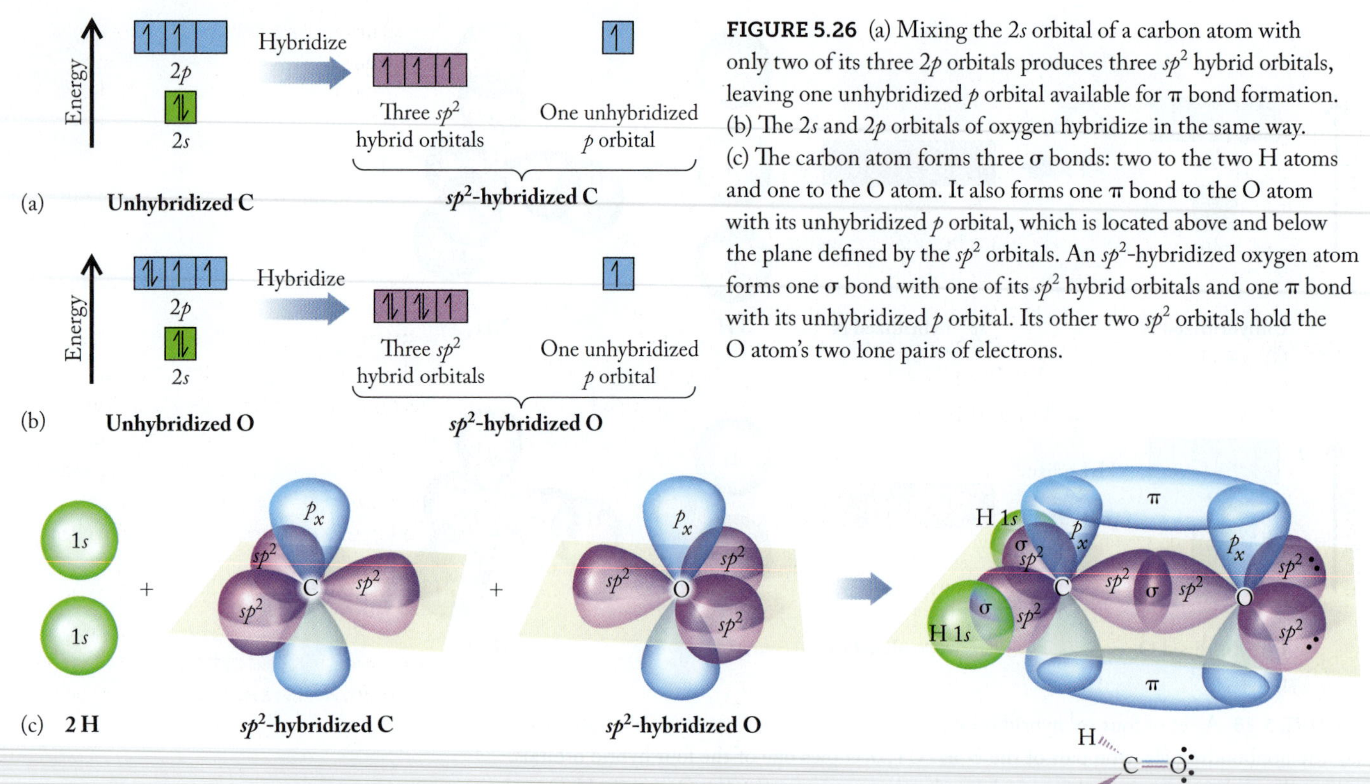


**FIGURE 5.26** (a) Mixing the 2*s* orbital of a carbon atom with only two of its three 2*p* orbitals produces three $sp^2$ hybrid orbitals, leaving one unhybridized *p* orbital available for π bond formation. (b) The 2*s* and 2*p* orbitals of oxygen hybridize in the same way. (c) The carbon atom forms three σ bonds: two to the two H atoms and one to the O atom. It also forms one π bond to the O atom with its unhybridized *p* orbital, which is located above and below the plane defined by the $sp^2$ orbitals. An $sp^2$-hybridized oxygen atom forms one σ bond with one of its $sp^2$ hybrid orbitals and one π bond with its unhybridized *p* orbital. Its other two $sp^2$ orbitals hold the O atom's two lone pairs of electrons.

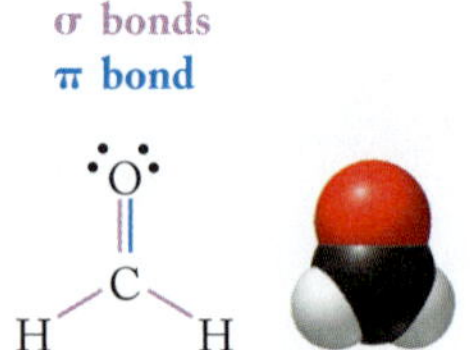


**FIGURE 5.27** Formaldehyde has three σ bonds and one π bond.

C and O atoms align with each other. Also, the widths of the *p* orbitals' lobes and the distances between atoms are not drawn to scale in Figure 5.26(c). The lobes are actually much wider than drawn (see Figure 3.24), and therefore they overlap.

A nitrogen atom also can form $sp^2$ hybrid orbitals, as shown in **Figure 5.28**. When it does, its lone pair occupies one of the three hybrid orbitals; the other two $sp^2$ hybrid orbitals are half-filled and can form two σ bonds, whereas the unhybridized half-filled *p* orbital can form one π bond. That combination of one lone pair plus the capacity to form two σ bonds to two other atoms gives N a steric number of 3, the same SN value as $sp^2$-hybridized C and O atoms. Thus, central atoms with trigonal planar electron-pair geometries are often $sp^2$ hybridized.

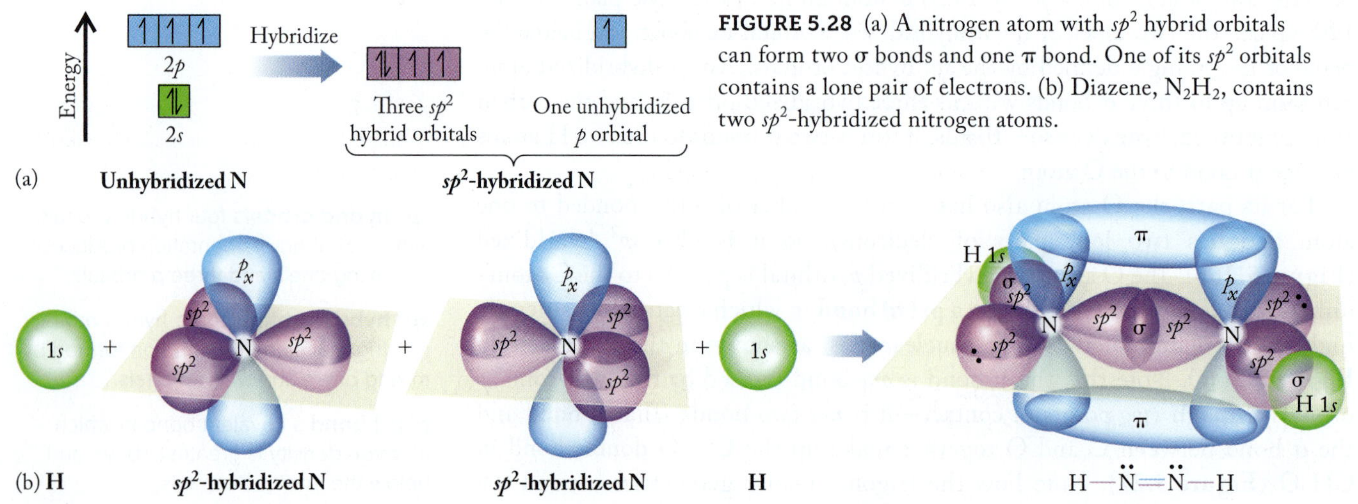


**FIGURE 5.28** (a) A nitrogen atom with $sp^2$ hybrid orbitals can form two σ bonds and one π bond. One of its $sp^2$ orbitals contains a lone pair of electrons. (b) Diazene, $N_2H_2$, contains two $sp^2$-hybridized nitrogen atoms.

## *sp* Hybrid Orbitals

**sp hybrid orbitals** two hybrid orbitals on opposite sides of the hybridized atom formed by mixing one s and one p orbital.

Hybrid orbitals also can be generated by mixing one *s* orbital with one *p* orbital, forming two *sp* orbitals.

In the linear molecule acetylene, $C_2H_2$, both carbon atoms have steric numbers equal to 2 according to VSEPR. Because steric number always indicates the number of hybrid orbitals that must be created, we must mix two atomic orbitals to make two hybrid orbitals on each carbon atom. Mixing one 2*s* and one 2*p* orbital on a carbon atom and leaving the other two 2*p* orbitals unhybridized produces two ***sp* hybrid orbitals** with major lobes on opposite sides of the carbon atom, as shown in **Figure 5.29**. The lobes of the two unhybridized *p* orbitals are above and below and in front of and behind the axis of the *sp* hybrid orbitals.

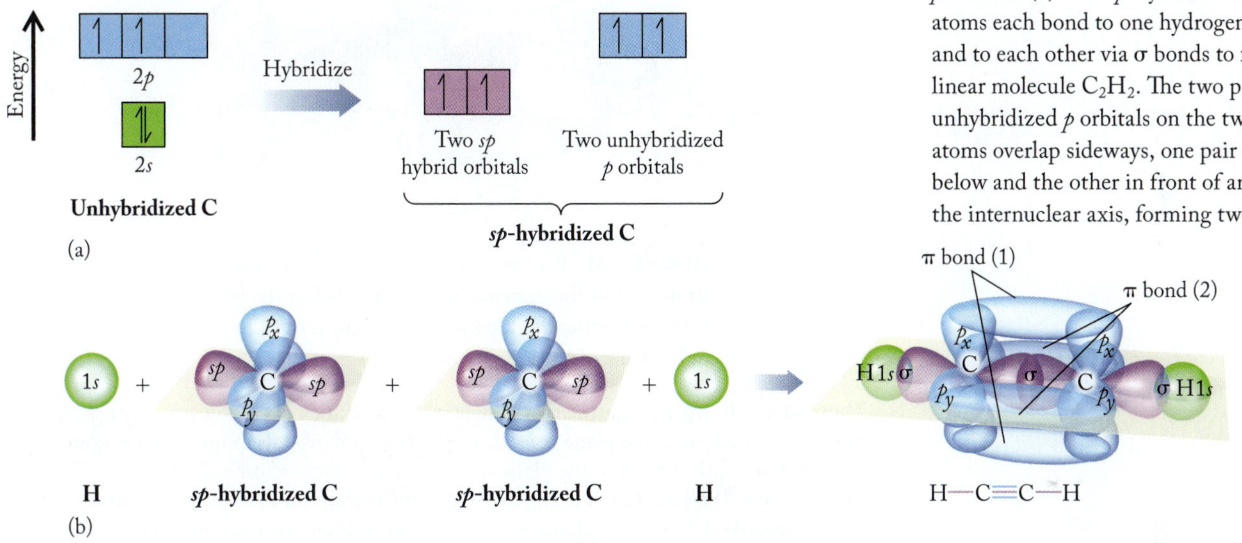

**FIGURE 5.29** (a) Mixing one 2*s* orbital and one 2*p* orbital on a carbon atom creates two *sp* hybrid orbitals and leaves the carbon atom with two unhybridized *p* orbitals. (b) Two *sp*-hybridized carbon atoms each bond to one hydrogen atom and to each other via σ bonds to form the linear molecule $C_2H_2$. The two pairs of unhybridized *p* orbitals on the two carbon atoms overlap sideways, one pair above and below and the other in front of and behind the internuclear axis, forming two π bonds.

The valence bond view of $C_2H_2$ shows a σ-bonding framework in which a hydrogen 1*s* orbital overlaps with one *sp* hybrid orbital on each carbon atom. The remaining *sp* hybrid orbitals, one on each carbon atom, overlap (Figure 5.29b). That arrangement brings the two sets of unhybridized *p* orbitals on the two carbon atoms into parallel alignment. They then overlap to form two π bonds between the two carbon atoms, so that the one σ bond and the two π bonds form the triple bond of HC≡CH. The steric number of each *sp*-hybridized carbon atom in acetylene is 2 (bonds to two atoms with no lone pairs). The link between SN = 2 and *sp* hybridization also applies to nitrogen atoms in $N_2$. A lone pair occupies one of the two *sp* orbitals on each nitrogen atom, and the other *sp* hybrid orbital forms the σ bond, as in **Figure 5.30**. The shapes associated with all the hybridization schemes discussed so far are summarized in **Table 5.3**.

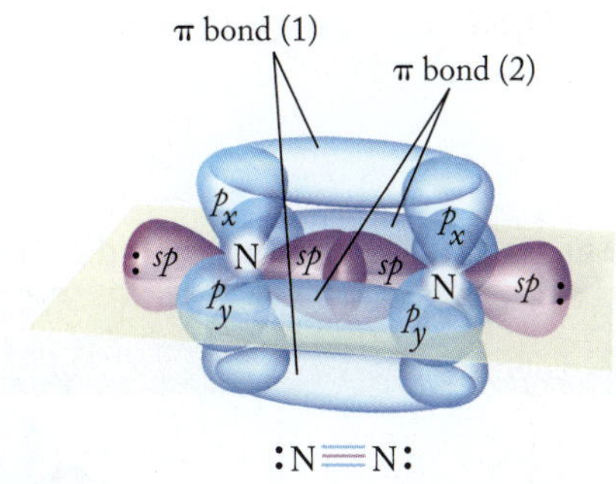


**FIGURE 5.30** The triple bond between *sp*-hybridized nitrogen atoms in $N_2$ consists of one σ bond and two π bonds.

**SAMPLE EXERCISE 5.5** Using Valence Bond Theory to Explain Molecular Shape **LO3**

Use valence bond theory to explain the linear molecular geometry of $CO_2$. Determine the hybridization of carbon and oxygen in this molecule and describe the orbitals that overlap to form the bonds.

**Collect, Organize, and Analyze** We know that $CO_2$ is linear and that the Lewis structure of the molecule is

$$:\ddot{O}=C=\ddot{O}:$$

**Solve** The central carbon atom has no lone pairs, so SN = 2. Both of its double bonds are composed of one σ bond and one π bond. Therefore, the carbon atom must be *sp* hybridized because its two half-filled *sp* hybrid orbitals have the capacity to form two σ bonds, and its two half-filled unhybridized *p* orbitals are available to form the two π bonds (**Figure 5.31**). The oxygen atoms must be $sp^2$ hybridized because that hybridization leaves the O atoms with one unhybridized *p* orbital to form a π bond. The σ bonds form when each *sp* orbital on the carbon atom overlaps with one $sp^2$ orbital on an oxygen atom. The lone pairs on oxygen occupy $sp^2$ orbitals.

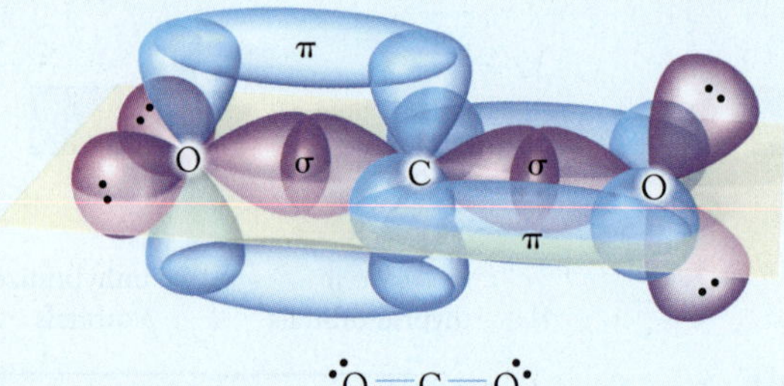


$$:\ddot{O}=C=\ddot{O}:$$

**FIGURE 5.31** The bonding pattern in $CO_2$. The π bond on the left in the drawing is above and below the plane of the molecule; the π bond on the right is in front of and in back of the internuclear axis.

**Think About It** Notice that the two unhybridized *p* orbitals of carbon are oriented 90° to each other and that the plane containing the three $sp^2$ orbitals of one oxygen atom is rotated 90° with respect to the plane containing the three $sp^2$ orbitals of the other oxygen atom. The orientation of the lone pairs of electrons on each oxygen atom and its σ bond to carbon is trigonal planar, which minimizes their mutual repulsion.

**Practice Exercise** In which of these molecules does the central atom have $sp^3$ hybrid orbitals? (a) $CCl_4$; (b) HCN; (c) $SO_2$; (d) $PH_3$

*(Answers to Practice Exercises are in the back of the book.)*

**TABLE 5.3 Summary of Hybridization Schemes and Orbital Orientations**

| Steric Number | Hybridization | Orientation of Hybrid Orbitals | Numbers of σ Bonds | Molecular Geometries | Angles between Hybrid Orbitals |
|---|---|---|---|---|---|
| 2 | *sp* |  | 2 | Linear | 180° |
| 3 | $sp^2$ |  | 3<br>2 | Trigonal planar<br>Bent | 120° |
| 4 | $sp^3$ | 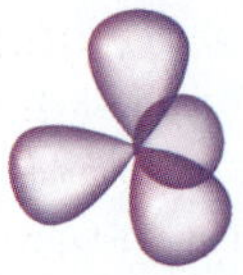 | 4<br>3<br>2 | Tetrahedral<br>Trigonal pyramidal<br>Bent | 109.5° |

CONCEPT **TEST**

Why can't a carbon atom with $sp^3$ hybrid orbitals form $\pi$ bonds?

*(Answers to Concept Tests are in the back of the book.)*

**molecular recognition** the process by which molecules interact with other molecules to produce a biological effect.

## 5.5 Molecules with Multiple "Central" Atoms

We began this chapter by pointing out that two smells—rye bread and spearmint—actually come from molecules with the same composition (atoms) and structure (bonds). When you smell any odor, the process by which the molecules interact with the *receptors* or active sites in your tissues is known as **molecular recognition**. Molecular recognition happens for a variety of processes in living things, such as molecules interacting with receptors in your nose or medicines interacting with proteins on the surface of bacteria. Recognition does not usually involve forming covalent bonds. Instead, interactions between molecules require that the biologically active molecules and the receptors that respond to them fit tightly together, which means that they must have complementary three-dimensional shapes (**Figure 5.32**). An example of a biological effect caused by molecular recognition is the process by which green tomatoes and other types of produce ripen. Tomatoes ripen faster when stored in a paper or plastic bag instead of sitting exposed on a kitchen counter. Why? Because tomatoes give off ethylene gas ($C_2H_4$) as they ripen, and they have receptors that respond to molecules of ethylene by accelerating the ripening process (**Figure 5.33**). A bag traps that gas.

Both carbon atoms in ethylene are bonded to three other atoms and have no lone pairs of electrons (**Figure 5.34a**), so SN = 3. That means that the geometry around each carbon atom is trigonal planar and that the carbon atoms are both $sp^2$ hybridized (see Table 5.3). For each carbon atom, two of the $sp^2$ orbitals form $\sigma$ bonds with hydrogen 1$s$ orbitals; the third forms the C=C $\sigma$ bond shown in **Figure 5.34(b)**. The C=C $\pi$ bond is formed by overlap of the unhybridized 2$p$ orbitals on the carbon atoms. Taken together, the two trigonal planar carbon atoms produce an overall geometry for ethylene in which all six atoms lie in the same plane. The $\pi$ bond in $C_2H_4$ molecules contributes to an overall distribution of electrons with maximum densities above and below the plane, as shown by the boundary surfaces in **Figure 5.34(c)**.

Now let's consider a biologically active molecule with three "central" atoms. Acrolein is one component of barbeque smoke that contributes to the distinctive odor of a cookout—and is a possible cancer-causing compound. The Lewis

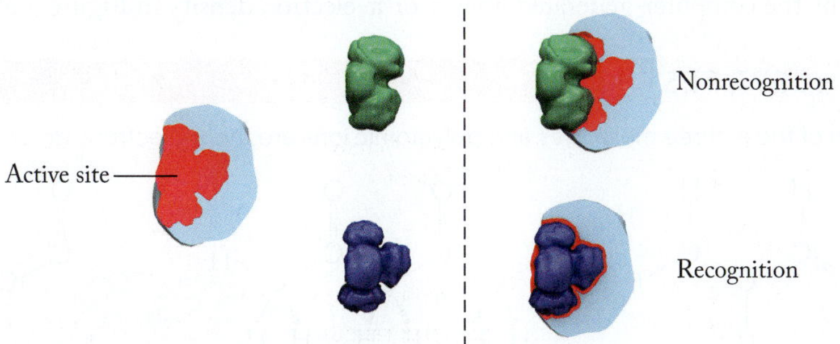


**FIGURE 5.32** Molecular recognition in biological systems. Only the blue molecule matches the shape of the active site and interacts with it.

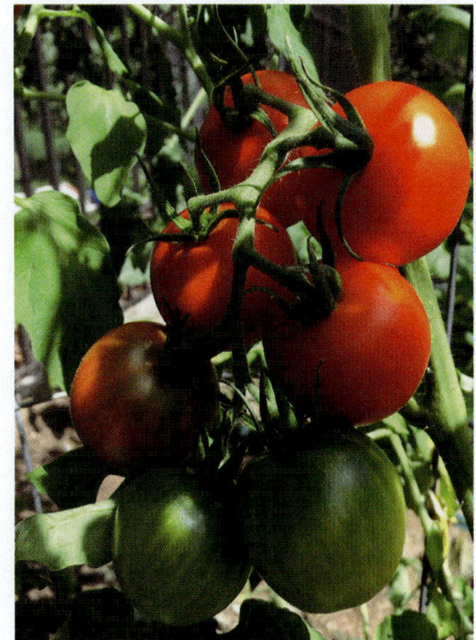

**FIGURE 5.33** Ripening tomatoes give off ethylene gas, which speeds up the ripening process. Ethylene molecules have a unique shape that interacts with the receptors that control ripening.

SN = 3 SN = 3

Trigonal planar Trigonal planar

(a) (b) (c)

**FIGURE 5.34** Ethylene molecules contain two carbon atoms that are (a) at the centers of two overlapping triangular planes that are also coplanar. That combination means that all the atoms are in the same plane. (b) The carbon atoms are both $sp^2$ hybridized. (c) The $\pi$ bond in ethylene results in electron density above and below the plane of the molecule.

structure of acrolein shows that the molecule contains three carbon atoms, each with a steric number of 3 and $sp^2$ hybridization:

That network of three carbon atoms, each representing a trigonal planar "center," means that *all* the atoms in a molecule of acrolein could be in the same plane.

Another component of barbeque smoke, benzene, also consists of molecules in which all the atoms are in the same plane as a result of $sp^2$ hybridization and trigonal planar geometries on, here, six carbon atoms (**Figure 5.35**). As we saw in Chapter 4, a benzene molecule features a hexagon of carbon atoms, each bonded to two other carbon atoms and one hydrogen atom. That geometry is a perfect match for $sp^2$ hybridization because the 120° bond angles those orbitals produce are the same as the 120° inside angles of a regular hexagon.

As drawn in Figure 5.35(a), bonding around the benzene ring consists of alternating single and double carbon–carbon bonds. That pattern is an example of *conjugation*—a term that applies to alternating single and multiple bonds in molecular compounds in which adjacent atoms have unhybridized $p$ orbitals. The resonance structures in Figure 5.35(a) correspond to the two ways of arranging the $\pi$ bonds formed by overlapping $2p$ orbitals on the six carbon atoms. All the carbon $2p$ orbitals are aligned perpendicular to the plane of the ring and are the same distance apart, so all are equally likely to overlap with the $2p$ orbitals on either of their neighbors around the ring. As a result, the electrons in the $\pi$ bonds are delocalized over *all six* carbon atoms, forming two clouds of electrons above and below the hexagonal rings, as shown by the computer-generated images of $\pi$-electron density in Figure 5.35(b).

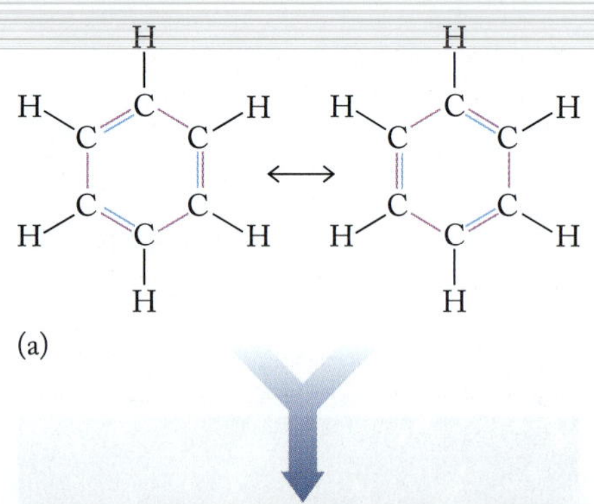

(a)

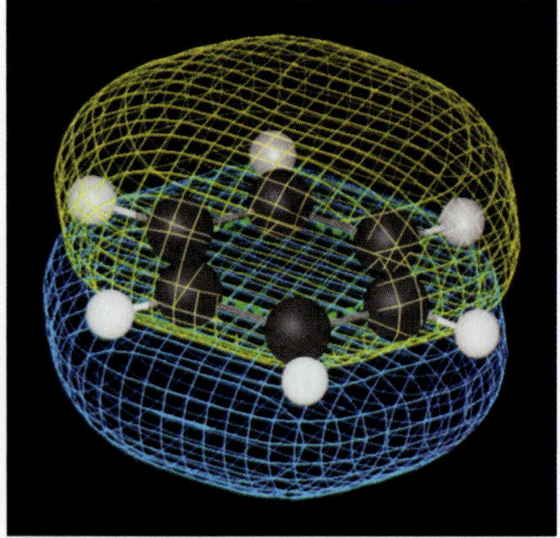

(b)

**FIGURE 5.35** Two views of the resonance in benzene. (a) The Lewis resonance structures of benzene can be explained using valence bond theory by showing localized $\pi$ bonds between adjacent carbon atoms. (b) Resonance leads to complete delocalization of the electrons in the $\pi$ bonds around the benzene ring.

## CONCEPT TEST

In which of these three molecules and polyatomic ions are the $\pi$ electrons delocalized?

(a) (b) (c)

*(Answers to Concept Tests are in the back of the book.)*

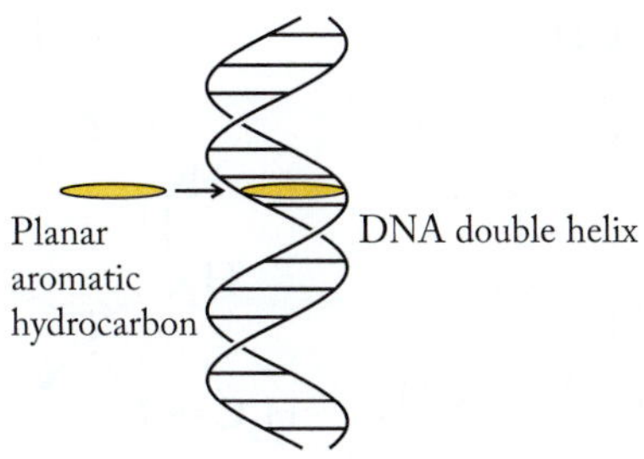


**FIGURE 5.36** The molecules of these polycyclic aromatic hydrocarbons (PAHs) consist of fused benzene rings whose $\pi$ bonds are delocalized over all the rings in each molecule. Molecules with this planar shape can slip between the strands of DNA and disrupt cell replication, which can lead to cell death or induce malignancy.

Naphthalene

Phenanthrene

Anthracene

Benzo[*a*]pyrene

The delocalization of bonding pairs is an important phenomenon because it spreads out their electrons, thereby reducing their mutual repulsion and lowering the potential energy of a molecule or polyatomic ion, making it more stable. In general, the greater the degree of delocalization, the greater the stability. That trend is evident in a class of compounds called **aromatic compounds** that includes benzene and other molecules with planar ring structures. An important subclass of aromatic compounds known as *polycyclic aromatic hydrocarbons* (PAHs) consists of molecules containing multiple fused six-carbon rings (**Figure 5.36**). PAHs are formed in fires fueled by coal, oil, or natural gas, and they are found in cigarette smoke. In 2004, they were discovered in interstellar space.

The chemical stability of PAHs means that they persist in the environment. If we inhale or ingest these compounds, they may bind to the DNA in our cells in a process called *intercalation*. Because PAHs are flat, they can slide into the double helix that DNA forms, as shown in Figure 5.36. Once there, they may alter or prevent DNA replication and thereby damage or kill cells. Intercalation in DNA is one step in the process by which PAHs may induce cancer.

**CONNECTION** We discussed electron delocalization when learning to draw resonance structures in Chapter 4.

**CONNECTION** Drawing molecular structures of benzene with a circle in the middle of the hexagon of carbon atoms (see Figure 4.10) is another way to represent the clouds of $\pi$ electrons above and below the plane of the molecule.

**CHEMTOUR**
Structure of Benzene

## 5.6 Chirality and Molecular Recognition

Before ending our study of molecular shape, let's return to the subject of handedness introduced at the beginning of the chapter. There we described how two molecules with the same molecular formula and Lewis structure interact differently with receptors in our nasal membranes. As a result, one produces the smell of caraway seeds and the other, spearmint leaves (see Figure 5.1).

You may find the different odors surprising when you consider the molecular structures of the two compounds (**Figure 5.37**). At first glance they may seem identical. In fact, they have the same common chemical name, carvone. But look closely at the bonding pattern around the carbon atoms at the bottom of the rings highlighted with red circles. The hydrogen atom bonded to the bottom carbon atom is on the back side of the ring in the caraway compound, but it is on the front side

**aromatic compound** a cyclic, planar compound with delocalized $\pi$ electrons above and below the plane of the molecule.

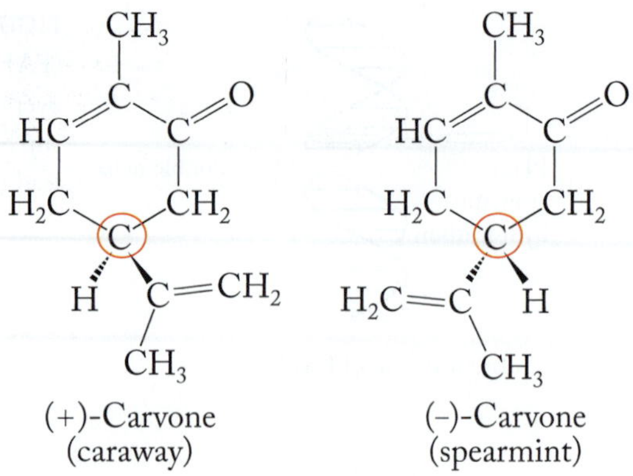

**FIGURE 5.37** The only difference between the two forms of carvone is the orientation of the two groups attached to the carbon atoms highlighted with the red circles.

(+)-Carvone (caraway)

(−)-Carvone (spearmint)

of the ring in the spearmint compound. That minor difference in bond orientation creates a difference in molecular shape that triggers different receptors in our noses.

Compounds that have the same chemical formula but different molecular structures are called **isomers**. The term is derived from *iso* (Greek for "same") and *mer* (Greek for "unit" or "part"). In this and later chapters we will encounter several kinds of isomerism. The carvone molecules in Figure 5.37 represent one kind: they have structures with the same atoms bonded in the same ways—that is, they have identical Lewis structures, but some of the bonds are oriented in three-dimensional space in ways that are not the same. Isomers of that kind are called **stereoisomers**.

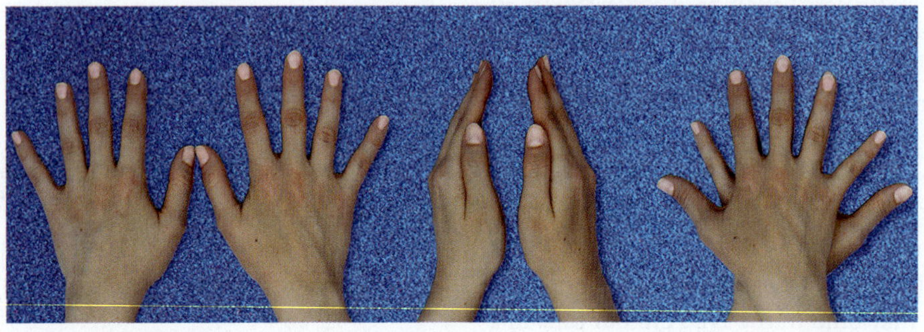

**FIGURE 5.38** The left hand and right hand are mirror images of each other but cannot be superimposed. Molecules having that relationship are known as enantiomers.

**CHEMTOUR**

Optical Activity

**isomer** one of a group of compounds having the same chemical formula but different molecular structures.

**stereoisomers** molecules with the same formula and bonding order, but with different spatial arrangements of their atoms.

**enantiomer** one of a pair of optical isomers of a compound.

**chiral** describes a molecule that is not superimposable on its mirror image.

The two forms of carvone represent a particular kind of stereoisomerism—they are mirror images of each other (just as your right hand and left hand are) but cannot be superimposed (**Figure 5.38**). Compounds whose molecular structures are nonsuperimposable mirror images—that is, they have the same composition and same bonds but different three-dimensional shapes—are called **enantiomers**.

Evidence for the existence of enantiomers comes from the fact that they interact differently with a kind of light called *plane-polarized* light. In a normal, unpolarized beam of light, the electric fields oscillate in all directions as the beam travels through space, as shown on the left in **Figure 5.39**. In plane-polarized light, however, the electric fields oscillate in only one plane. To distinguish between a pair of enantiomers, (+) and (−) signs are added to their names to refer to the specific effect each isomer has on plane-polarized light. When a beam of plane-polarized light passes through a solution containing (+)-carvone, or any (+) stereoisomer, the plane of oscillation rotates clockwise; if the beam passes through a solution of a (−) stereoisomer, such as (−)-carvone, the plane rotates counterclockwise. Stereoisomers that can rotate plane-polarized electromagnetic radiation in that way are said to be *optically active* and are called *optical* isomers.

Observing how a compound interacts with plane-polarized light is one way to determine whether the compound is optically active. Another way is to examine its molecular structure and, in particular, to determine whether it has any $sp^3$ carbon atoms bonded to four different atoms or groups of atoms. Such a carbon atom is called a **chiral** carbon atom, or *stereocenter*. A chiral compound

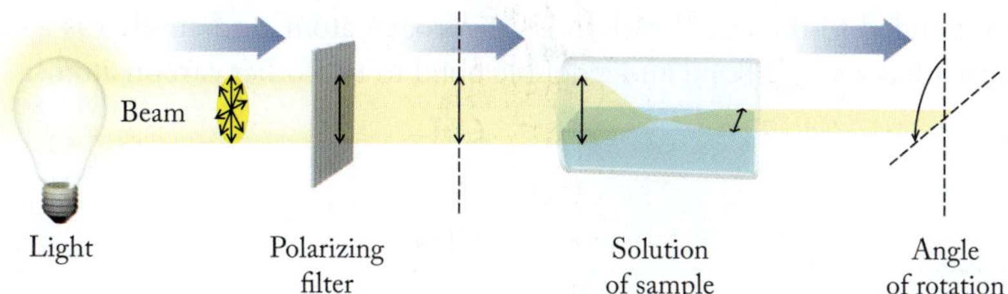


**FIGURE 5.39** A beam of plane-polarized light consists of electric field vectors that oscillate in only one direction. The plane of oscillation rotates if the beam passes through a solution of one enantiomer of an optically active compound. The (+) enantiomer causes the beam to rotate clockwise; the (−) enantiomer causes the beam to rotate counterclockwise.

contains one or more chiral atoms, which explains why the compound is optically active. Many compounds of biological importance, including the proteins and carbohydrates that we consume each day, are chiral compounds. "Chirality" comes from the Greek word *cheir*, or "hand," and is correctly called "handedness." Although several features within molecules can lead to chirality, the most common is the presence of a chiral carbon atom.

To see how chirality works, let's look at a compound that contains a central carbon atom bonded to four different atoms. The compound is bromochlorofluoromethane (CHBrClF) (**Figure 5.40a**). It is used in fire extinguishers on airplanes. We will compare its molecular structure with that of dibromochloromethane ($CHBr_2Cl$), a compound that may form during the purification of municipal water supplies with chlorine and that has only *three* different atoms bonded to its central carbon atom (**Figure 5.40b**). In the first step of our comparison, we generate mirror images of both molecules. Then we rotate the mirror images 180° to try to superimpose each mirror image on its original image. The reflected, rotated images of CHBrClF cannot be superimposed. For example, when we superimpose the F, C, and H atoms of the two structures in Figure 5.40(a), the Br and Cl atoms are not aligned. That means that CHBrClF is a chiral compound. However, the molecular images of $CHBr_2Cl$ *can* be superimposed in that way, so $CHBr_2Cl$ is not a chiral compound—it is an *achiral* (meaning *not* chiral) compound.

Where are the chiral carbon atoms in (+)- and (−)-carvone? If you guessed the circled carbon atoms in Figure 5.37, you were right. One of the four different

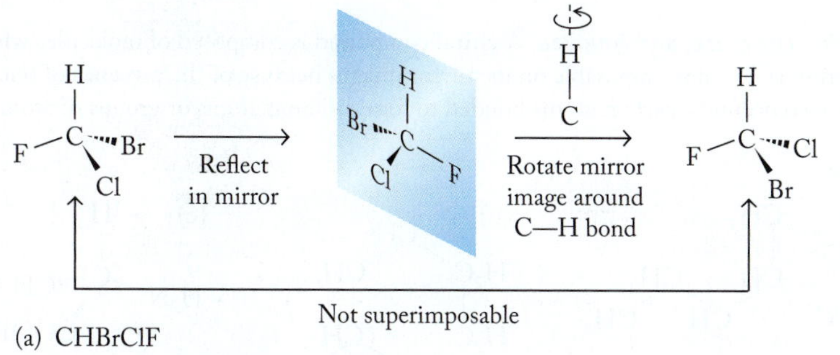

Reflect in mirror
Rotate mirror image around C—H bond
Not superimposable
(a) CHBrClF

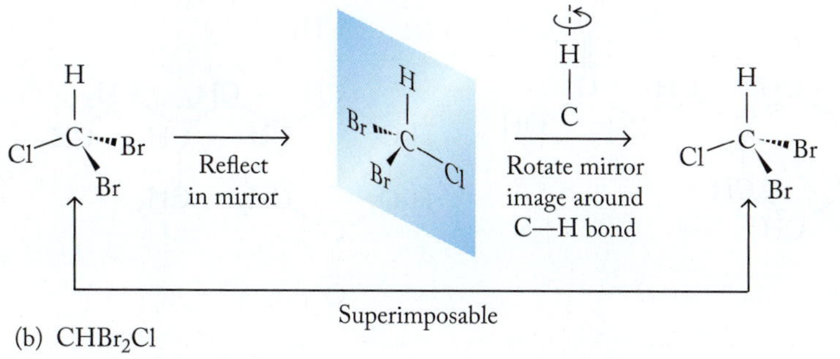

Reflect in mirror
Rotate mirror image around C—H bond
Superimposable
(b) $CHBr_2Cl$

**FIGURE 5.40** (a) A molecule of a chiral compound, such as CHBrClF, is not superimposable on its mirror image. (b) A molecule of a compound that is not chiral, such as $CHBr_2Cl$, is superimposable on its mirror image.

groups bonded to the chiral carbon is a hydrogen atom, and another is a carbon atom that has a single bond and a double bond to two other carbon atoms:

$$-C(=CH_2)-CH_3$$

The other two bonds from the chiral carbon are part of the six-carbon ring. They both connect to $-CH_2-$ groups, so how are they different? The answer lies in other groups to which the $-CH_2-$ groups are also bonded. Note that the left side of the ring contains a C═C double bond, whereas the right side contains a C═O double bond. That asymmetry in the ring is the reason that carvone is not superimposable on its mirror image.

**CHEMTOUR**
Chiral Centers

## SAMPLE EXERCISE 5.6 Recognizing Chiral Structures **LO4**

Identify which of the following molecules are chiral and circle their stereocenters. Some molecules may have more than one chiral center.

(a) $CH_3$, CH, $CH_2$, $H_3C$, CH, $CH_3$, $CH_3$

(b) CH, $H_2C$, CH, $H_2C$, CH, $CH_2$, $CH_3$

(c) H, $H_2N$, C, $CH_2CH_3$, COOH

(d) O, CH, CH, C, HC, C, CH, OH, HC, CH, CH

(e) $CH_3$, CH, $CH_2$, $CH_2$, $H_2C$, CH, CH, $CH_3$, $H_2C$, $CH_2$, $CH_3$, C, O

**Collect, Organize, and Analyze** A chiral compound is composed of molecules whose structure is not superimposable on its mirror images because of the presence of one or more stereocenters: carbon atoms bonded to four different atoms or groups of atoms.

**Solve**

(a) $CH_3$, CH, $CH_2$, $H_3C$, CH, $CH_3$, $CH_3$

(b) CH, $H_2C$, CH, $H_2C$, CH, $CH_2$, $CH_3$

(c) H, $H_2N$, C, $CH_2CH_3$, COOH

(d) O, CH, CH, C, HC, C, CH, OH, HC, CH, CH

(e) $CH_3$, CH, $CH_2$, $CH_2$, $H_2C$, CH, CH, $CH_3$, $H_2C$, $CH_2$, $CH_3$, C, O

The circled carbon atom in compound (a) is bonded to four different groups: the $CH_3$ group below it, the H atom above it, a group to its right that contains two carbon and five hydrogen atoms, and a group to its left that contains three carbon and seven hydrogen atoms. Therefore, the circled carbon atom is a stereocenter and the compound is chiral.

In compound (b), the circled carbon atom is part of a six-carbon ring and bonded to a $CH_3$ group and a H atom. The ring is not symmetrical: one side has a C═C double bond and the other side does not. Therefore, the circled atom is chiral.

The circled carbon atom in (c) is bonded to four different groups of atoms: –H, $–NH_2$, –COOH, and $–CH_2CH_3$, so that compound is chiral. It also is an example of a class of compounds called amino acids.

All the carbon atoms in compound (d) are $sp^2$ hybridized, giving them planar molecular geometries. Planar molecules cannot be chiral because they have superimposable mirror images. Therefore, compound (d) is achiral.

Compound (e) has three chiral centers. As we work from right to left, the first one is similar to the chiral center in compound (a). The other two are in the six-carbon ring. Each is bonded to different groups outside the ring, and the ring itself is not symmetrical: a $CH_3$ group is bonded to the top carbon atom in the ring, and the bottom atom is double bonded to an oxygen atom. That asymmetry means that both circled atoms are bonded to four different groups of atoms and are chiral centers. The presence of these stereocenters makes molecule (e) chiral.

**Think About It** To decide whether an $sp^3$-hybridized carbon atom is a stereocenter, we often have to look beyond the atoms bonded directly to it. When the atom is in a ring of atoms, encountering different groups as we work our way around the ring may identify an asymmetry that ensures that the atom is bonded to at least two different groups. If it is also bonded to two additional different groups, the atom is a stereocenter.

**Practice Exercise** Identify which of the following molecules are chiral. Circle the stereocenters in each structure.

(a) $CH_2$, $CH_2$, $H_2C$, CH, $CH_2$, $H_2C$, CH, $CH_2$, $CH_2$, $CH_2$

(b) $CH_3$, $CH_3$, CH, CH, $H_3C$, CH, $CH_3$, $CH_3$

(c) O, $CH_3$, $CH_2$, C, CH, $CH_3$, $H_3C$, $CH_2$, O, $CH_2$

(d) $NH_2$, H, C, COOH, C, HC, CH, HC, CH, C, OH

(e) $CH_3$, $H_2C$, O, C, OH, $H_3C$, C, $CH_2$, $CH_2$, OH

*(Answers to Practice Exercises are in the back of the book.)*

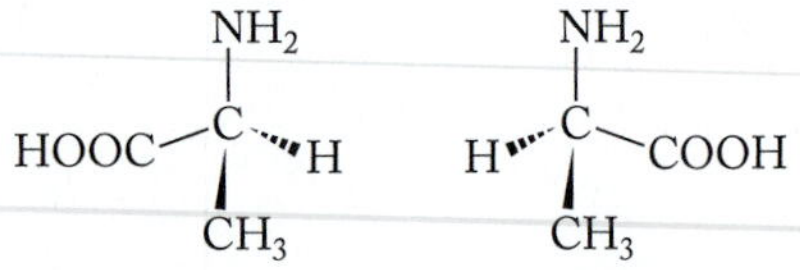

NH$_2$ | HOOC—C—H | CH$_3$   (+)-Alanine

NH$_2$ | H—C—COOH | CH$_3$   (−)-Alanine

**FIGURE 5.41** The two enantiomers of the amino acid alanine.

## Chirality in Nature

Many compounds formed by living systems are chiral. For example, the proteins in our tissues and in the foods we consume are made of amino acids that are nearly all chiral compounds. The two enantiomers of the amino acid alanine are shown in **Figure 5.41**. Our bodies can make proteins only from the (+) enantiomers of the 20 common amino acids. The origins of that biological preference for one enantiomer over another are unknown, though the amino acids in meteorites striking Earth have slightly more of the enantiomers found in living things in Earth, which may indicate that the chiral preferences observed in Earth's biosphere are not limited to our planet.

Whatever the origin of the chirality, processes requiring molecular recognition in living systems often depend on the selectivity conveyed by chirality. The human body is a chiral environment, so the handedness of molecules matters, including those we take into our bodies for therapeutic purposes. As many as half of today's prescription drugs are chiral and owe their function to recognition by a receptor that favors one enantiomer over the other. Typically, only one enantiomer of a chiral drug is active. A classic example is the drug albuterol (**Figure 5.42**), used to treat asthma and other respiratory disorders. One isomer causes bronchodilation (a widening of the air passages of the lungs, which eases breathing), whereas the other causes bronchial constriction and may actually be detrimental to the patient. The compounds in this chapter's opening photo, dextromethorphan and levomethorphan, are also enantiomers with different biochemical properties.

As noted previously, living organisms typically produce only one stereoisomer of a chiral compound. When a chiral compound such as albuterol is synthesized in a laboratory, however, both enantiomers are often produced in equal amounts. The resulting 50:50 mixture is called a **racemic mixture**. Because one isomer rotates the plane of polarized light in one direction, and the other to the same extent in the opposite direction, a racemic mixture does not rotate plane-polarized light at all. Because the two isomers interact differently with receptors, the pharmaceutical industry routinely faces two choices: devise a special synthetic procedure that yields only the isomer of interest or separate the two isomers during the manufacturing process. Both approaches are used in producing chiral pharmaceuticals, and both contribute to the cost of those drugs.

**racemic mixture** a sample containing equal amounts of both enantiomers of a compound.

**molecular orbital (MO) theory** a bonding theory based on the mixing of atomic orbitals of similar shapes and energies to form molecular orbitals that belong to the molecule as a whole.

**molecular orbital** a region of characteristic shape and energy where electrons in a molecule are located.

**bonding orbital** a term in MO theory describing regions of increased electron density between nuclear centers that serve to hold atoms together in molecules.

**antibonding orbital** a term in MO theory describing regions of electron density in a molecule that destabilize the molecule because they do not increase the electron density between nuclear centers.

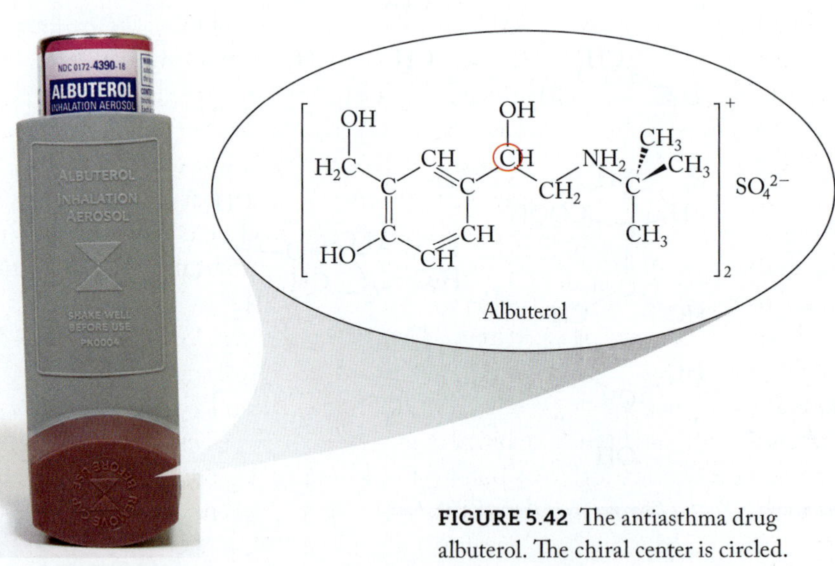


**FIGURE 5.42** The antiasthma drug albuterol. The chiral center is circled.

**CONCEPT TEST**

*The Documents in the Case*, a mystery written by Dorothy L. Sayers and Robert Eustace in 1930, involves an authority's suspicious death caused by wild, edible mushrooms. Allegedly, the victim ate a stew made of poisonous mushrooms that contained a toxic natural product called muscarine (**Figure 5.43**). A forensic specialist evaluated the contents of the victim's stomach and found that they contained muscarine but that they did not rotate plane-polarized light. From those findings, the coroner concluded that the victim had been murdered. How did the coroner reach that conclusion?

*(Answers to Concept Tests are in the back of the book.)*

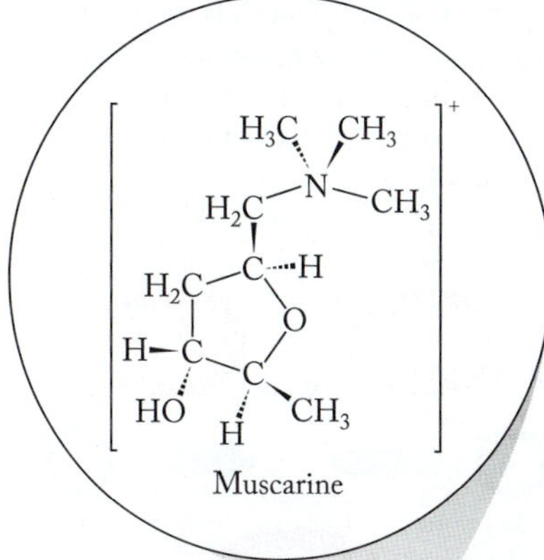

**FIGURE 5.43** Several varieties of mushrooms, including *Amanita muscaria*, contain the toxic substance muscarine.

## 5.7 Molecular Orbital Theory

Lewis structures and valence bond theory help us understand the bonding capacities of individual atoms and the bonding patterns in molecules and ions, whereas VSEPR and valence bond theory account for their molecular shapes. However, some phenomena cannot be explained by any of those models. For example, none explain why $O_2$ is attracted to a magnetic field or how molecules and ions in Earth's upper atmosphere produce the shimmering, colorful display known as an *aurora* (**Figure 5.44**). To explain those phenomena, we need another model that describes covalent bonding: **molecular orbital (MO) theory**.

MO theory is based on quantum mechanics. Like atomic orbitals, **molecular orbitals** are wave functions that represent discrete energy states inside molecules. As with atomic orbitals, the lowest-energy MOs always fill first. Electrons in them move to higher-energy MOs when molecules absorb quanta of electromagnetic radiation. When the electrons return to lower-energy MOs, distinctive wavelengths of UV and visible radiation may be emitted—producing, for example, some of the colors in an aurora. Note how those processes of molecular absorption and emission of electromagnetic radiation parallel the absorption and emission of radiation by atoms, discussed in Chapter 3.

Unlike atomic orbitals, including the hybrid atomic orbitals of valence bond theory, MOs are not linked to single atoms but rather belong to *all* the atoms in a molecule. That delocalized view of covalent bonding is particularly effective at describing the bonding in molecules such as benzene, in which conjugation allows $\pi$ bonds to spread out over more than one pair of atoms.

In MO theory, molecular orbitals are formed by combining atomic orbitals from each atom in the molecule. In this book, we limit our discussion of MO theory to simple molecules in which MOs are formed from the atomic orbitals of only a few atoms. Some MOs have lobes of high electron density that lie *between* bonded pairs of atoms; those are called **bonding orbitals**. The energies of bonding MOs are lower than the energies of the atomic orbitals that combined to form them, so populating them with electrons (each MO can hold two electrons, just like an atomic orbital) stabilizes the molecule and contributes to the strength of the bonds holding its atoms together.

Other MOs have lobes of high electron density not located between the bonded atoms. They are **antibonding orbitals** and have energies higher than those of the atomic orbitals that combined to form them. When electrons are in antibonding orbitals, they destabilize the molecule. If a molecule were to have the same number of electrons in its bonding and antibonding orbitals, no net energy would be holding the molecule together, and it would fall apart (actually, it would never have formed in the first place). Another key point in MO theory is that the

**FIGURE 5.44** Auroras are spectacular displays of color produced by excited-state atoms, ions, and molecules in Earth's upper atmosphere.

**CONNECTION** Molecular orbital theory also explains the "sea of electrons" model for metallic bonding (Section 4.1) and the electrical conductivity of metals. That topic is revisited in Chapter 18.

*total number of molecular orbitals must match the number of atomic orbitals involved in forming them*. For example, combining two atomic orbitals, one from each of two atoms, produces one low-energy bonding orbital and one high-energy antibonding orbital. To better understand how atomic orbitals combine to form molecular orbitals, let's apply MO theory to the bond that holds together $H_2$.

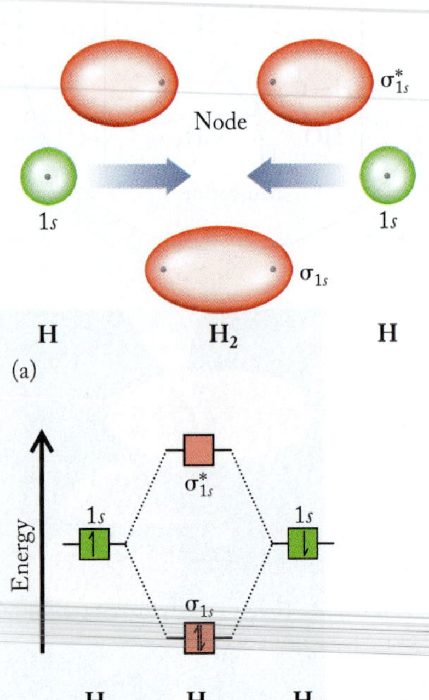


**FIGURE 5.45** Mixing the 1s orbitals of two hydrogen atoms creates two molecular orbitals: a bonding $\sigma_{1s}$ orbital containing two electrons and an empty antibonding $\sigma^*_{1s}$ orbital. (a) The lower red oval is the bonding orbital. Dots show the locations of the two hydrogen nuclei. The two red ovals at the top together make up the antibonding orbital. *Note*: The two top ovals represent only *one* molecular orbital with a node of zero electron density between. (b) A molecular orbital diagram shows the relative energies of the bonding and antibonding molecular orbitals and of the atomic orbitals that formed them.

## Molecular Orbitals of $H_2$

According to MO theory, a hydrogen molecule is formed when the 1*s* atomic orbitals on two hydrogen atoms combine to form the two molecular orbitals shown in **Figure 5.45(a)**. MO theory stipulates that mixing two atomic orbitals creates two molecular orbitals. The lower-energy, bonding molecular orbital is oval shaped and spans the two atomic centers. The high density of electrons between the two atoms makes that bonding MO a **sigma (σ) molecular orbital**. When two electrons occupy it, a single σ bond is formed. The σ bonding molecular orbital in $H_2$ is labeled $\sigma_{1s}$ because it is formed by mixing two 1*s* atomic orbitals.

The higher-energy, antibonding molecular orbital formed from two hydrogen atomic orbitals is designated $\sigma^*_{1s}$ (pronounced "sigma star"). That antibonding orbital has two separate lobes of electron density oriented away from the internuclear axis and a region of zero electron density (a node) between the atoms, as shown in Figure 5.45(a).

**Figure 5.45(b)** is a **molecular orbital diagram**, analogous to the *atomic* orbital diagrams we discussed in Chapter 3. Note that the $\sigma_{1s}$ MO is lower in energy and therefore more stable than the 1*s* atomic orbitals by about the same amount that the $\sigma^*_{1s}$ MO is higher in energy than the 1*s* atomic orbitals. Therefore, the formation of the two MOs does not significantly change the total energy of the system. A hydrogen molecule has two valence electrons, one from each H atom, both residing in the lower-energy $\sigma_{1s}$ orbital. As in atomic orbitals, those two $\sigma_{1s}$ electrons must have opposite spins. The electron configuration that corresponds to the MO diagram in Figure 5.45(b) is written $(\sigma_{1s})^2$, where the superscript indicates that two electrons are in the $\sigma_{1s}$ molecular orbital. Because the energy of the electrons in a $(\sigma_{1s})^2$ configuration is lower than the energy of the electrons in two isolated hydrogen atoms, $H_2$ molecules are lower in energy and therefore more stable than a pair of free H atoms.

Hydrogen is a diatomic gas, but helium exists as free atoms and not as molecular $He_2$. MO theory explains why. Each He atom has two valence electrons in its 1*s* atomic orbital. Mixing two He 1*s* orbitals yields the same set of molecular orbitals (**Figure 5.46**) as those of $H_2$. Adding four electrons (two from each He atom) fills both MOs. The presence of two electrons in the $\sigma^*_{1s}$ orbital cancels the stability gained from having two electrons in the $\sigma_{1s}$ orbital. Because no net gain in stability occurs, $He_2$, $(\sigma_{1s})^2 (\sigma^*_{1s})^2$, does not exist.

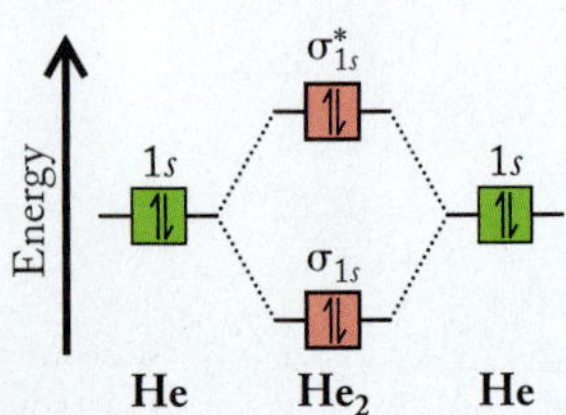


**FIGURE 5.46** The molecular orbital diagram for nonexistent $He_2$ shows that the same number of electrons occupy the antibonding orbital and the bonding orbital. Therefore, the bond order is zero and the molecule is not stable.

Another way to compare the bonding in $H_2$ and $He_2$ is to consider the *bond order* in each. We previously defined bond order as the number of bonds between two atoms: a bond order of 1 for Cl—Cl, 2 for O=O, and 3 for N≡N. In MO theory, we define bond order as follows:

$$\text{Bond order} = \tfrac{1}{2}\left[\begin{pmatrix}\text{number of}\\ \text{bonding electrons}\end{pmatrix} - \begin{pmatrix}\text{number of}\\ \text{antibonding electrons}\end{pmatrix}\right] \quad (5.2)$$

In a molecule of $H_2$, two electrons are in the bonding MO and none are in the antibonding MO, so

$$\text{Bond order in } H_2 = \tfrac{1}{2}(2 - 0) = 1$$

In nonexistent $He_2$, the bond order is 0 because the number of electrons in bonding and antibonding orbitals is equal:

$$\text{Bond order in } He_2 = \tfrac{1}{2}(2 - 2) = 0$$

In general, the strength of the bond and the stability of the molecule both increase as the bond order increases.

**CONNECTION** We first defined bond order in Chapter 4 when we related the lengths and strengths of bonds to the number of electron pairs shared by two atoms.

**CHEMTOUR**
Molecular Orbitals

**SAMPLE EXERCISE 5.7** Using MO Theory to Predict Bond Order I **LO5**

Draw the MO diagram of $H_2^-$, determine the bond order in the ion, and predict whether the ion is more stable than $H_2$.

**Collect and Organize** We are to apply MO theory to draw an orbital diagram for $H_2^-$ and then determine the bond order by using Equation 5.2. If the bond order is greater than zero, the ion may be stable.

**Analyze** We should be able to base the MO diagram for $H_2^-$ on the MO diagram for $H_2$ (Figure 5.45b) because the $H_2^-$ ion has only one more electron than $H_2$ and the empty $\sigma_{1s}^*$ orbital in $H_2$ can accommodate up to two electrons.

**Solve** The $\sigma_{1s}$ orbital is filled in $H_2$. The third electron of $H_2^-$ goes into the $\sigma_{1s}^*$ orbital, so the MO diagram is as shown in **Figure 5.47**. The notation for that electron configuration is $(\sigma_{1s})^2\,(\sigma_{1s}^*)^1$ (listing the molecular orbitals in order of increasing energy). The bond order is

$$\text{Bond order in } H_2^- = \tfrac{1}{2}(2 - 1) = \tfrac{1}{2}$$

The $H_2^-$ ion exists, but it is not as stable as $H_2$.

**Think About It** We encountered fractional bond orders in Chapter 4 in discussing resonance, and we encounter them again here in the MO treatment of $H_2^-$. A bond order of $\tfrac{1}{2}$ in MO theory means that the bond between the two atoms in $H_2^-$ is weaker than the single bond in $H_2$, making $H_2^-$ less stable than $H_2$.

**Practice Exercise**
Use MO theory to predict the bond order in a $H_2^+$ ion.

*(Answers to Practice Exercises are in the back of the book.)*

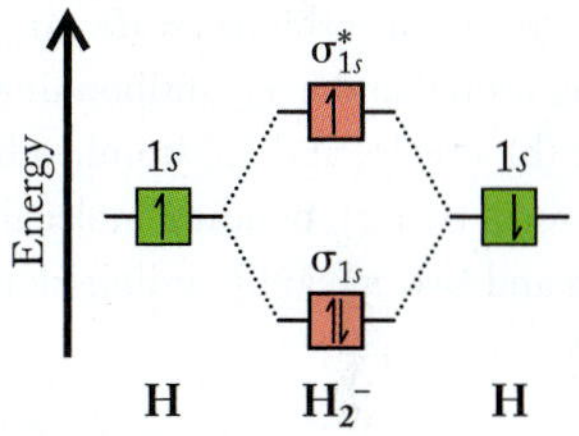

**FIGURE 5.47** The molecular orbital diagram for $H_2^-$.

## Molecular Orbitals of Other Homonuclear Diatomic Molecules

Molecular orbital diagrams for homonuclear (same-atom) diatomic molecules such as $N_2$ and $O_2$ are more complex than the diagram for $H_2$ because of the greater number and variety of atomic orbitals in $N_2$ and $O_2$. Not all combinations of atomic orbitals result in bonding, but some general guidelines exist for constructing the MO diagram for any molecule:

1. The number of molecular orbitals equals the number of atomic orbitals used to create them.
2. Atomic orbitals with similar energy and orientation mix more effectively than do those with different energies and orientations.
3. Better mixing leads to a larger energy difference between bonding and antibonding orbitals and thus greater stabilization of the bonding MOs.

**sigma (σ) molecular orbital** in MO theory, the orbital that results in the highest electron density between the two bonded atoms.

**molecular orbital diagram** in MO theory, an energy-level diagram showing the relative energies and electron occupancy of the molecular orbitals for a molecule.

4. A molecular orbital can accommodate a maximum of two electrons; two electrons in the same MO have opposite spins.
5. Electrons in ground-state molecules occupy the lowest-energy molecular orbitals available, following the aufbau principle and Hund's rule.

In mixing atomic orbitals to create molecular orbitals, we focus on the atom's valence electrons because core electrons do not participate in bonding. Focusing on $N_2$ and $O_2$ as examples, we first mix their 2*s* orbitals. The mixing process is analogous to the one we used for $H_2$, except that the resulting MOs are designated $\sigma_{2s}$ and $\sigma^*_{2s}$.

Next, we mix the three pairs of 2*p* orbitals, producing a total of six MOs. The different spatial orientations of the $2p_x$, $2p_y$, and $2p_z$ atomic orbitals result in different kinds of MOs (**Figure 5.48**). The $2p_z$ atomic orbitals point toward each other. When they mix, two molecular orbitals form: a $\sigma_{2p}$ bonding orbital and a $\sigma^*_{2p}$ antibonding orbital. The lobes of the $2p_x$ and $2p_y$ atomic orbitals are oriented at 90° both to the internuclear axis and to each other. When the $2p_x$ orbitals mix, and when the $2p_y$ orbitals mix, they do so around the internuclear axis instead of

**FIGURE 5.48** Two sets of three *p* atomic orbitals mix to form six molecular orbitals. (a) The $2p_z$ atomic orbitals create a $\sigma_{2p}$ bonding orbital and a $\sigma^*_{2p}$ antibonding orbital. (b) The $2p_x$ and $2p_y$ atomic orbitals mix to form two $\pi_{2p}$ bonding molecular orbitals and two $\pi^*_{2p}$ antibonding molecular orbitals.

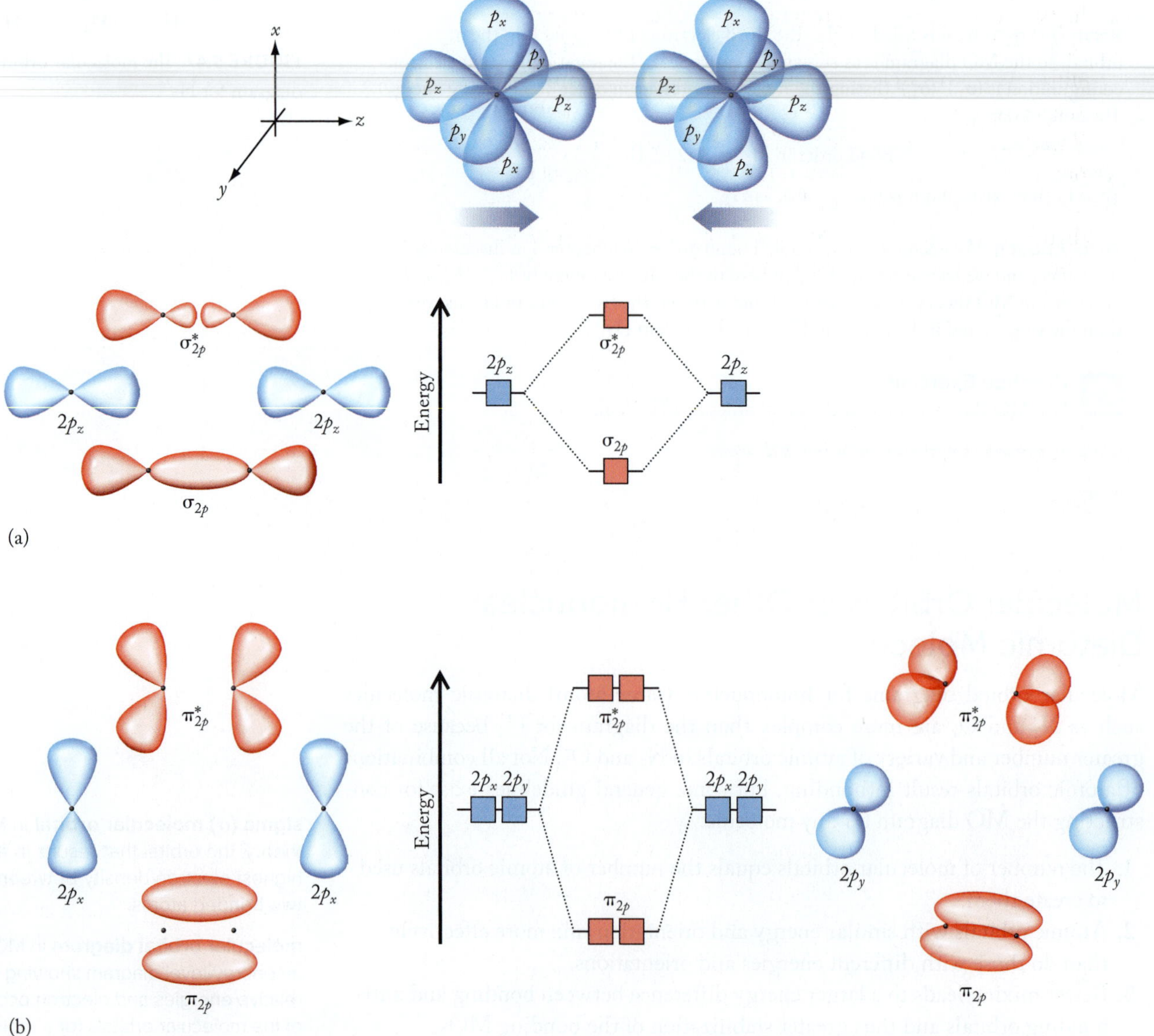

along it. That mixing produces **pi (π) molecular orbitals**. As with σ molecular orbitals, the numbers of low-energy π and high-energy π* molecular orbitals are equal. Electrons that occupy π orbitals contribute to forming π bonds; electrons that populate π* orbitals detract from π bond formation.

**pi (π) molecular orbital** in MO theory, an orbital formed by the mixing of atomic orbitals oriented above and below, or in front of and behind, the internuclear axis.

The relative energies of the σ and π molecular orbitals for $N_2$ and $O_2$ are shown in **Figure 5.49**. In each molecule, the energies of the $\sigma_{2s}$ and $\sigma^*_{2s}$ MOs are lower than the energy of the $\sigma_{2p}$ MO because they are formed from the mixing of 2*s* atomic orbitals, which are lower in energy than 2*p* atomic orbitals. However, the relative energies of the MOs formed by mixing the 2*p* orbitals in $N_2$ and $O_2$ are not the same. Let's begin with $O_2$ (Figure 5.49b) because its diagram is representative of most homonuclear diatomic molecules, including all the halogens. In order of increasing energy, the MOs are $\sigma_{2s}$, $\sigma^*_{2s}$, $\sigma_{2p}$, $\pi_{2p}$, $\pi^*_{2p}$, and $\sigma^*_{2p}$. Keep in mind that groups of two $\pi_{2p}$ and two $\pi^*_{2p}$ orbitals are present, each of which can hold up to four electrons. Adding 12 valence electrons (six from each O atom in the $O_2$ molecule) to the MOs, starting with the lowest-energy MO and working our way up, produces the following electron configuration:

$$O_2\text{: } (\sigma_{2s})^2(\sigma^*_{2s})^2(\sigma_{2p})^2(\pi_{2p})^4(\pi^*_{2p})^2$$

The distribution of electrons in the MO diagram reflects that sequence. Note that the two $\pi^*_{2p}$ orbitals are degenerate (equivalent in energy), so each contains a single electron, in accordance with Hund's rule. From that MO diagram, we can determine that a molecule of $O_2$ has two unpaired electrons. That finding contradicts the representation of bonding obtained from the Lewis structure of $O_2$, from VSEPR, and from valence bond theory, all of which predict that the valence electrons are paired. We will return to that point shortly, but for now let's compare the MO diagrams for $N_2$ and for $O_2$ in Figure 5.49.

When we do, notice the difference in the relative energies of the $\pi_{2p}$ and $\sigma_{2p}$ MOs. In $N_2$, the $\pi_{2p}$ molecular orbitals are lower in energy than the $\sigma_{2p}$ orbital,

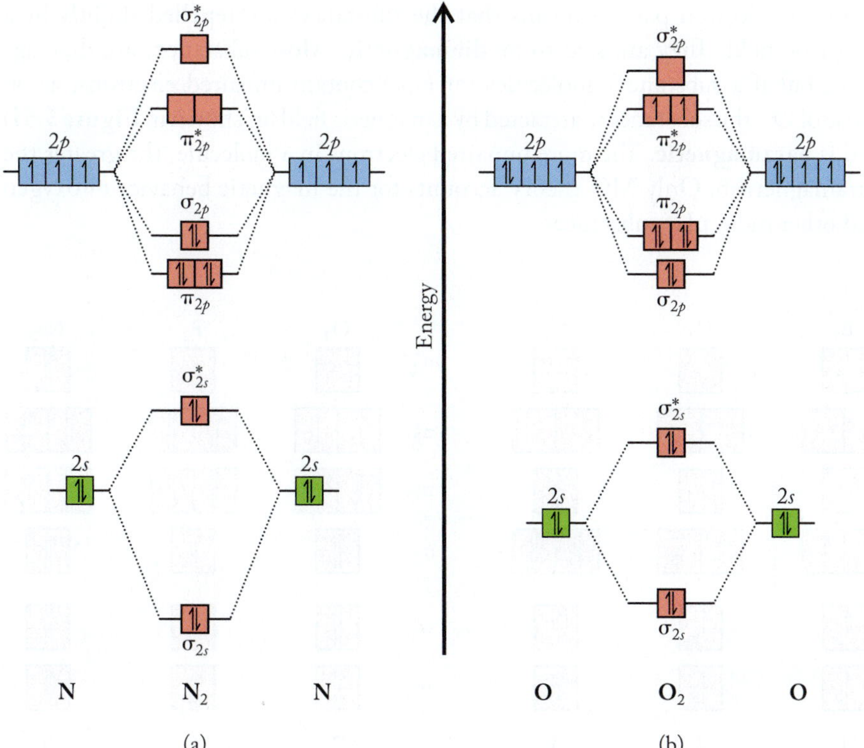


**FIGURE 5.49** Molecular orbital diagrams of (a) $N_2$ and (b) $O_2$. The vertical sequence of orbitals in $N_2$ applies to homonuclear diatomic molecules of elements with $Z \leq 7$. The $O_2$ sequence applies to homonuclear diatomic molecules of elements beyond oxygen ($Z \geq 8$).

**diamagnetic** describes a substance with no unpaired electrons that is weakly repelled by a magnetic field.

**paramagnetic** describes a substance with unpaired electrons that is attracted to a magnetic field.

whereas in $O_2$ they are reversed. That difference is related to the energy differences between the 2*s* and 2*p* orbitals, which increase as the atomic number increases. The greater the difference, the more likely the MOs are to have the relative energies we see in the $O_2$ orbital diagram. Lesser differences result in some mixing of the 2*s* and 2*p* orbitals during MO formation, which lowers the energy of the $\sigma_{2s}$ orbital and raises the energy of the $\sigma_{2p}$ orbital—enough to put it above the $\pi_{2p}$ orbital, as in the $N_2$ orbital diagram. Adding 10 valence electrons to that stack of MOs in $N_2$, lowest energy first, produces the following electron configuration:

$$N_2: (\sigma_{2s})^2(\sigma_{2s}^*)^2(\pi_{2p})^4(\sigma_{2p})^2$$

The distribution of electrons in the MO diagram in Figure 5.49(a) reflects that electron configuration. A total of eight electrons are in bonding MOs and two are in antibonding MOs. Using Equation 5.2 to calculate the bond order for $N_2$, we get

$$\text{Bond order in } N_2 = \tfrac{1}{2}(8 - 2) = 3$$

In $O_2$, eight electrons occupy bonding orbitals and four electrons occupy antibonding orbitals, so

$$\text{Bond order in } O_2 = \tfrac{1}{2}(8 - 4) = 2$$

On the basis of their Lewis structures (Section 4.4), we predicted a triple bond in $N_2$ and a double bond in $O_2$, and MO theory leads us to the same conclusions.

One strength of MO theory is that it enables us to explain properties of molecular compounds that other bonding theories cannot. The magnetic behavior of homonuclear diatomic molecules is a case in point. Electrons in atoms have two possible spin orientations depending on the value of their spin quantum number, $m_s$. **Figure 5.50** shows the MO-based electron configurations in diatomic molecules of the period 2 elements. In most of the molecules, all the electrons are paired; the molecules that make up most substances contain only paired electrons. Complete electron pairing means that the substances are repelled slightly by a magnetic field. They are said to be **diamagnetic**. Most substances are diamagnetic, but if a substance's molecules (or ions) contain unpaired electrons, as do those of $O_2$, the substance is attracted by a magnetic field (as shown in **Figure 5.51**) and is **paramagnetic**. The more unpaired electrons in a molecule, the greater the paramagnetism. Only MO theory accounts for the magnetic behavior of oxygen and other molecular substances.

**CONNECTION** In Chapter 3 we introduced the spin quantum number ($m_s$), which has a value of either $+\frac{1}{2}$ or $-\frac{1}{2}$ and defines the orientation of an electron in a magnetic field.

**FIGURE 5.50** Valence-shell molecular orbital diagrams and bond orders of the homonuclear diatomic molecules of the second-row elements.

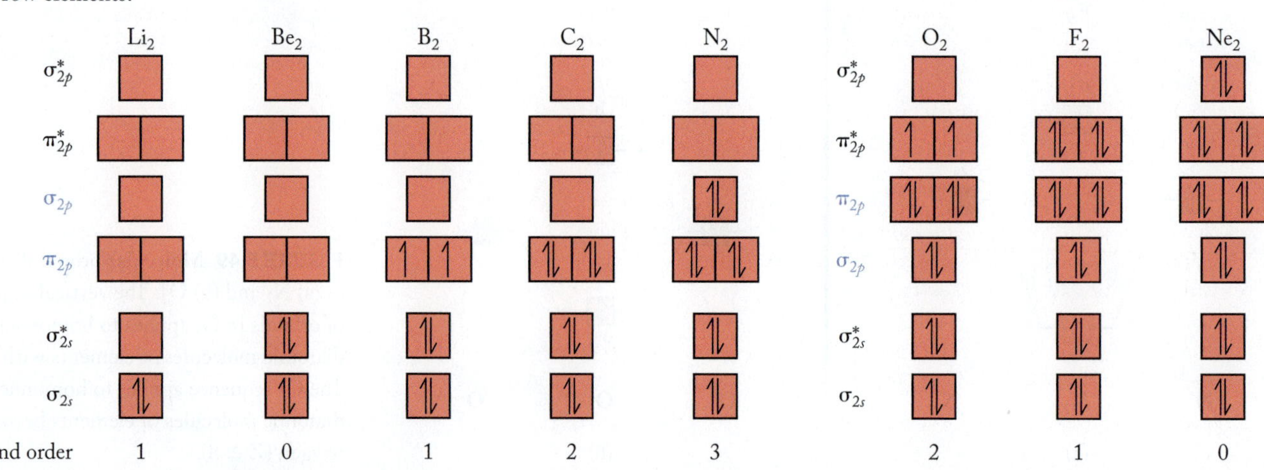

## CONCEPT TEST

a. Which diatomic molecules in Figure 5.50 do not exist?

b. Which one is paramagnetic (in addition to $O_2$)?

*(Answers to Concept Tests are in the back of the book.)*

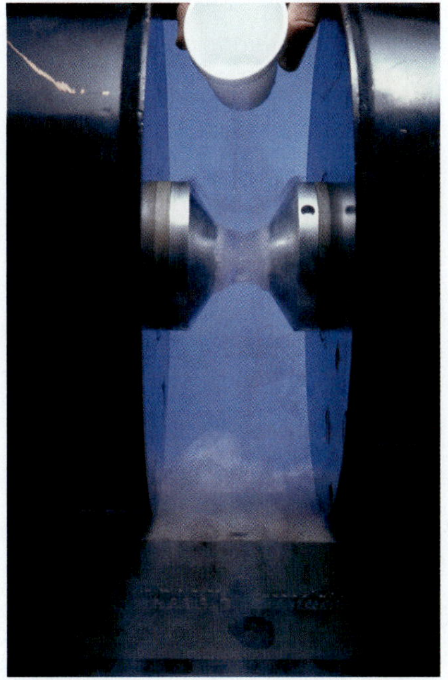

**FIGURE 5.51** Liquid $O_2$ poured from a cup is suspended in the space between the poles of this magnet because the unpaired electrons in its molecules make $O_2$ paramagnetic. Paramagnetic substances are attracted to magnetic fields.

### SAMPLE EXERCISE 5.8 Using MO Theory to Predict Bond Order II — LO5

For which molecules in Figure 5.50 would bond order increase if they lost one electron each, forming singly charged diatomic cations ($X_2^+$)?

**Collect and Organize** We are asked to determine which of the ions $Li_2^+$, $Be_2^+$, $B_2^+$, $C_2^+$, $N_2^+$, $O_2^+$, $F_2^+$, and $Ne_2^+$ have a greater bond order than that of their parent molecules. The MO electron configurations of the parent molecules are given in Figure 5.50. Bond order can be calculated using Equation 5.2:

$$\text{Bond order} = \frac{1}{2}\left[\begin{pmatrix}\text{number of}\\ \text{bonding electrons}\end{pmatrix} - \begin{pmatrix}\text{number of}\\ \text{antibonding electrons}\end{pmatrix}\right] \quad (5.2)$$

**Analyze** The electron configurations of the ions of interest can be written by reducing by 1 the number of electrons in the highest-energy occupied MOs (see the following table) in the parent molecules. Bond order will increase if the electron is lost from an antibonding orbital and will decrease if an electron is lost from a bonding orbital.

**Solve** Removing one electron from each MO diagram in Figure 5.50 yields the results in the following table.

| Ion | Electron Configuration | Bond Order | Bond Order in Parent Molecule |
|---|---|---|---|
| $Li_2^+$ | $(\sigma_{2s})^1$ | $\frac{1}{2}(1-0)=0.5$ | $\frac{1}{2}(2-0)=1$ |
| $Be_2^+$ | $(\sigma_{2s})^2(\sigma_{2s}^*)^1$ | $\frac{1}{2}(2-1)=0.5$ | $\frac{1}{2}(2-2)=0$ |
| $B_2^+$ | $(\sigma_{2s})^2(\sigma_{2s}^*)^2(\pi_{2p})^1$ | $\frac{1}{2}(3-2)=0.5$ | $\frac{1}{2}(4-2)=1$ |
| $C_2^+$ | $(\sigma_{2s})^2(\sigma_{2s}^*)^2(\pi_{2p})^3$ | $\frac{1}{2}(5-2)=1.5$ | $\frac{1}{2}(6-2)=2$ |
| $N_2^+$ | $(\sigma_{2s})^2(\sigma_{2s}^*)^2(\pi_{2p})^4(\sigma_{2p})^1$ | $\frac{1}{2}(7-2)=2.5$ | $\frac{1}{2}(8-2)=3$ |
| $O_2^+$ | $(\sigma_{2s})^2(\sigma_{2s}^*)^2(\sigma_{2p})^2(\pi_{2p})^4(\pi_{2p}^*)^1$ | $\frac{1}{2}(8-3)=2.5$ | $\frac{1}{2}(8-4)=2$ |
| $F_2^+$ | $(\sigma_{2s})^2(\sigma_{2s}^*)^2(\sigma_{2p})^2(\pi_{2p})^4(\pi_{2p}^*)^3$ | $\frac{1}{2}(8-5)=1.5$ | $\frac{1}{2}(8-6)=1$ |
| $Ne_2^+$ | $(\sigma_{2s})^2(\sigma_{2s}^*)^2(\sigma_{2p})^2(\pi_{2p})^4(\pi_{2p}^*)^4(\sigma_{2p}^*)^1$ | $\frac{1}{2}(8-7)=0.5$ | $\frac{1}{2}(8-8)=0$ |

Comparing the last two columns, we find that the bonds in $Be_2^+$, $O_2^+$, $F_2^+$, and $Ne_2^+$ have a higher bond order than the bond in their parent molecules, although the bond order of zero in $Be_2$ and $Ne_2$ means that those molecules don't exist.

**Think About It** Ionizing $O_2$, $F_2$, and (if they existed) $Be_2$ and $Ne_2$ reduces the number of electrons in antibonding molecular orbitals, which increases bond order. However, ionizing $Li_2$, $B_2$, $C_2$, and $N_2$ reduces the number of electrons in bonding molecular orbitals, which decreases bond order.

 **Practice Exercise** Which molecules in Figure 5.50 experience an increase in bond order when they form anions with 1− charges?

*(Answers to Practice Exercises are in the back of the book.)*

## Molecular Orbitals of Heteronuclear Diatomic Molecules

Molecular orbital theory also enables us to account for the bonding in *heteronuclear* diatomic molecules, which are molecules made of two different atoms. The bonding in some of those molecules is difficult to explain with other bonding theories. For example, drawing a single Lewis structure for an odd-electron molecule such as nitrogen monoxide, NO, is often difficult. In Chapter 4, we considered several arrangements of the valence electrons in NO, such as

$$\dot{\ddot{N}}=\dot{\ddot{O}}: \qquad :\dot{N}=\dot{\ddot{O}}:$$

We predicted that oxygen was more likely to have a complete octet of valence electrons because it is the more electronegative element. In addition, experimental evidence rules out structures with unpaired electrons on the oxygen atom. Therefore, the preferred structure is the one shown in red. However, the length of the NO bond (115 pm) is considerably shorter than the value in Table 4.6 for an average N═O double bond (122 pm).

Molecular orbital theory helps explain both the bonding in NO and the shorter-than-expected bond length. Let's look at the bonding first. The MO diagram for nitrogen monoxide is different from the diagrams of homonuclear diatomic gases. Nitrogen and oxygen atoms have different numbers of protons and electrons, and the difference in effective nuclear charge between N and O atoms means that their atomic orbitals have different energies, as **Figure 5.52** shows for the $2s$ and $2p$ orbitals.

In constructing the MO diagram for NO, the guidelines described previously still apply. The number of MOs formed must equal the number of atomic orbitals combined, and the energy and orientation of the atomic orbitals being mixed must be considered. One additional factor influences the energies of the MOs in heteronuclear diatomic molecules: *bonding* MOs tend to be closer in energy to the atomic orbitals of the more electronegative atom (here, O), and *antibonding* MOs tend to be closer in energy to the atomic orbitals of the less electronegative atom (N). Note in Figure 5.52 how the energy of the bonding $\sigma_{2s}$ orbital is closer to that of the $2s$ orbital of the oxygen atom, and the energy of the antibonding $\sigma^*_{2s}$ orbital is closer to that of the $2s$ orbital of the nitrogen atom. Similarly, the $\pi_{2p}$ MOs in NO are closer in energy to the $2p$ orbitals of oxygen, and the $\pi^*_{2p}$ MOs are closer in energy to the $2p$ orbitals of nitrogen. The proximity of the nitrogen $2p$ atomic orbitals to the $\pi^*_{2p}$ MOs is consistent with the single electron in the $\pi^*_{2p}$ MOs being on nitrogen rather than on oxygen. That prediction is consistent with our Lewis structure in which the odd electron in NO is on the nitrogen atom.

MO theory also enables us to rationalize the relatively short bond length in NO. Using Equation 5.2, we find that the bond order is $\frac{1}{2}(8 - 3) = 2.5$, halfway between the bond orders for N═O and N≡O and consistent with a bond length of 115 pm, which is halfway between the lengths of the N═O bond (122 pm) and the N≡O bond (106 pm).

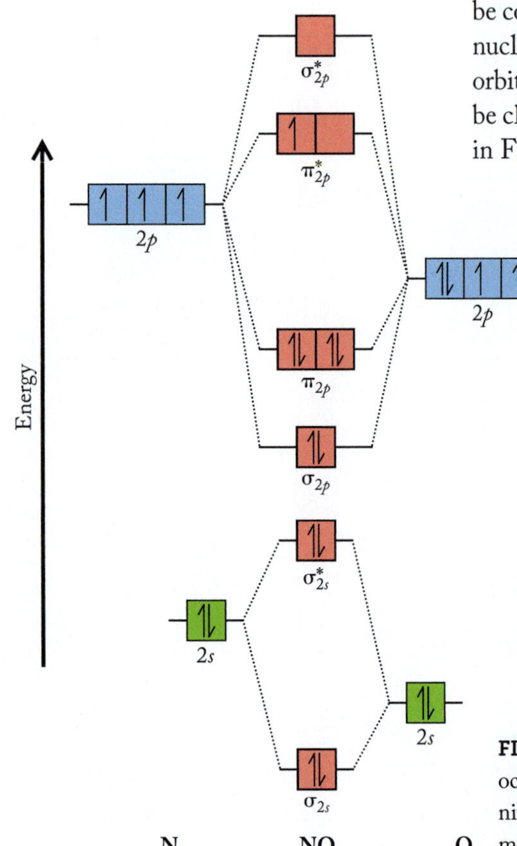


**FIGURE 5.52** The molecular orbital diagram for NO shows that the unpaired electron occupies a $\pi^*_{2p}$ antibonding orbital, which is closer in energy to the $2p$ atomic orbitals of nitrogen than to the $2p$ atomic orbitals of oxygen. Because of that proximity, the electron has more nitrogen character and is more likely to be on the nitrogen atom than the oxygen atom.

**SAMPLE EXERCISE 5.9** Using MO Theory to Determine the Bond Order of Heteronuclear Diatomic Molecules **LO5**

Nitrogen monoxide reacts with many transition metal compounds including the hemoglobin in our blood. In those compounds, NO may be $NO^+$ or $NO^-$. Use Figure 5.52 to predict the bond order of $NO^+$ and $NO^-$.

**Collect and Organize** We need to determine the bond order of the bonds in two diatomic ions according to the MO diagram of their parent molecule, NO, in Figure 5.52. Equation 5.2 relates bond order to the number of electrons in bonding and antibonding orbitals.

**Analyze** A single electron is in the $\pi^*_{2p}$ MO, which is the highest-energy occupied MO in the molecule. That electron is lost when NO forms $NO^+$, and a second electron occupies that orbital in $NO^-$.

**Solve** The electron configuration of $NO^+$ is

$$NO^+: (\sigma_{2s})^2(\sigma^*_{2s})^2(\sigma_{2p})^2(\pi_{2p})^4$$

Adding a second electron to the $\pi^*_{2p}$ orbital in NO yields the following electron configuration for $NO^-$:

$$NO^-: (\sigma_{2s})^2(\sigma^*_{2s})^2(\sigma_{2s})^2(\pi_{2p})^4(\pi^*_{2p})^2$$

The bond orders of the two ions are

$$NO^+: \text{Bond order} = \tfrac{1}{2}(8 - 2) = 3$$

$$NO^-: \text{Bond order} = \tfrac{1}{2}(8 - 4) = 2$$

**Think About It** The bond orders in $N_2$ and $O_2$ are 3 and 2, respectively. The cation $NO^+$ is isoelectronic with $N_2$, so our calculated bond order makes sense. The anion $NO^-$ is isoelectronic with $O_2$, and so our calculated bond order for it is reasonable, too.

**Practice Exercise** Using Figure 5.52 as a guide, draw the MO diagram for carbon monoxide and determine the bond order for the carbon–oxygen bond.

*(Answers to Practice Exercises are in the back of the book.)*

## Molecular Orbitals of $N_2^+$ and the Colors of Auroras

In addition to predicting the magnetic properties of molecules, MO theory is particularly useful for predicting their spectroscopic properties. Recall from Section 3.4 that the light emitted by excited free atoms is quantized because it is linked to the movement of electrons between atomic orbitals with discrete energies. Broadly speaking, the same is true in molecules: electrons can move from one molecular orbital to another when molecules absorb or emit light.

We can use that information to look again at how the colors of the auroras are produced. The principal chemical species involved are listed in **Table 5.4**. An asterisk indicates a molecule or molecular ion in an excited state. Excited $N_2^+$ ions produce blue-violet (391–470 nm) light, and excited $N_2$ molecules produce deep crimson red (650–680 nm) light. The MO diagrams of $N_2^*$ and $N_2$, compared in **Figure 5.53(a)**, show that one of the two electrons originally in the $\sigma_{2p}$ orbital in $N_2$ has been raised to a $\pi^*_{2p}$ orbital in $N_2^*$, leaving an unpaired $\sigma_{2p}$ electron behind. **Figure 5.53(b)**, which compares the MO diagrams of $N_2^{+*}$ and $N_2^+$, shows us that $N_2^{+*}$ also has one electron in a $\pi^*_{2p}$ orbital, but its $\sigma_{2p}$ orbital is empty because

**CONNECTION** We used orbital diagrams in Chapter 3 to show electron transitions between energy levels as atoms absorb and emit electromagnetic radiation.

**TABLE 5.4** Origins of Colors in the Aurora

| Wavelength (nm) | Color | Chemical Species |
|---|---|---|
| 650–680 | Deep red | $N_2^*$ |
| 630 | Red | $O^*$ |
| 558 | Green | $O^*$ |
| 391–470 | Blue-violet | $N_2^{+*}$ |

*In an excited state.

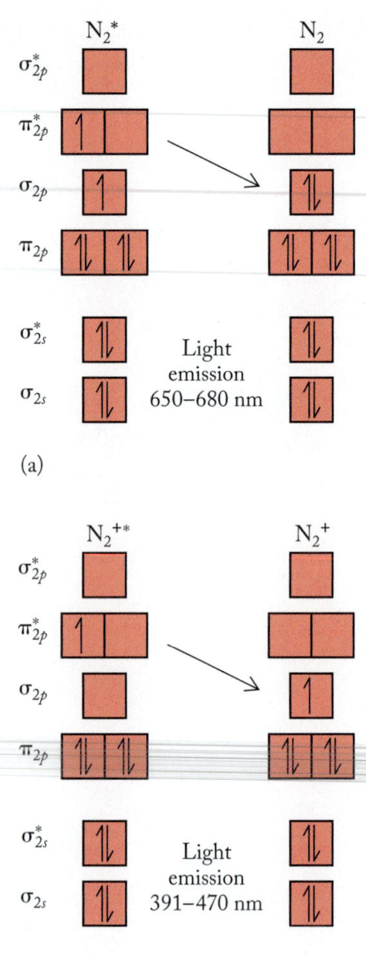


**FIGURE 5.53** Molecular orbital diagrams for (a) $N_2$ and (b) $N_2^+$ show electronic transitions that result in the emission of visible light. High-energy collisions between particles in the upper atmosphere excite electrons from the $\sigma_{2p}$ orbital in $N_2$ molecules to the $\pi^*_{2p}$ orbital and the formation of $N_2^{+*}$ molecular ions that also have electrons in $\pi^*_{2p}$ orbitals. When the electrons return to the ground state, red and blue-violet light is emitted.

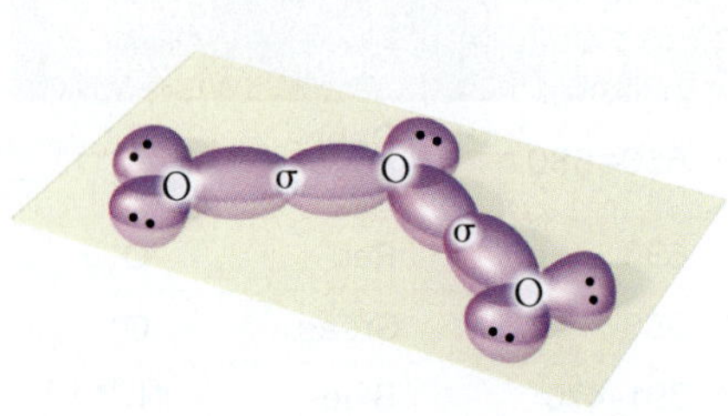


**FIGURE 5.54** Two sigma bonds and five lone pairs of electrons in a molecule of $O_3$.

the other $\pi^*_{2p}$ electron originally in the $N_2$ molecule was lost when the molecule was ionized. As the $\pi^*_{2p}$ electrons return from their antibonding orbital in the excited state to the bonding $\sigma_{2p}$ orbital in the ground state, the distinctive celestial emissions of $N_2^+$ and $N_2$ are produced.

### CONCEPT TEST

Are the bond orders of the excited-state species in Figure 5.53 the same as those of the ground-state species?

*(Answers to Concept Tests are in the back of the book.)*

## Using MO Theory to Explain Fractional Bond Orders and Resonance

In Chapter 4, our attempts to draw the Lewis structure of ozone led to the conclusion that molecules of $O_3$ are held together by two O—O single bonds plus a shared pair of electrons distributed equally across both bonds. The concept of resonance was used to describe the delocalization of the second bonding pair and the equivalency of the bonds in $O_3$, which are shorter and stronger than O—O single bonds but not as short or strong as O═O double bonds. In this chapter we have learned how VSEPR explains the bent shape of $O_3$ molecules and how valence bond theory and the formation of $sp^2$ hybrid orbitals help us visualize the trigonal planar orientation of bonding and lone pairs of electrons that account for ozone's molecular shape.

Now let's explore how MO theory offers another perspective on the delocalized bonding in $O_3$—one that does not require drawing resonance structures that, as we have admitted, do not accurately convey where all the bonding and lone pairs of electrons really are. MO theory is great at explaining bonding pair delocalization because MOs are not isolated between pairs of atoms but rather are spread across several atoms or even entire molecules. To simplify our exploration, we start with three O atoms connected by two single (σ) bonds and that have a total of five lone pairs of valence electrons (**Figure 5.54**). That trigonal planar pattern of electron pairs is what we would expect if each O atom had $sp^2$-hybridized atomic orbitals. Figure 5.54 shows 14 valence electrons, leaving $(3 \times 6) - 14 = 4$ electrons unaccounted for. Let's distribute the four electrons among a set of molecular orbitals formed by mixing the three unhybridized *p* orbitals of the three O atoms.

In MO theory, mixing three *p* orbitals with lobes above and below the bonding plane produces three molecular orbitals. One of them is a low-energy π MO and another is a high-energy π* MO, but what about the third? The third orbital is neither bonding nor antibonding. Instead, it is a *nonbonding* (*n*) molecular orbital. Electrons in nonbonding MOs neither lower the energy of a molecular structure, which would stabilize it, nor raise its energy, which would destabilize it. Rather, nonbonding MOs have the same energy as the atomic orbitals from which they formed, as shown in the partial MO diagram of ozone in **Figure 5.55**. Adding four electrons to those three molecular orbitals puts two electrons in the π orbital, two in the nonbonding orbital, and zero in π*. Therefore, $\frac{1}{2}(2 - 0) = 1$ π bond is distributed across the entire molecule. The overall bond order of 2 σ bonds + 1 π bond = 3 is divided equally between the two pairs of atoms, giving an average bond order for ozone of 1.5. The pair of nonbonding electrons is also delocalized. Each of the two noncentral O atoms gets half ownership of that nonbonding pair, for a total of 2.5 lone pairs of electrons each, which is the

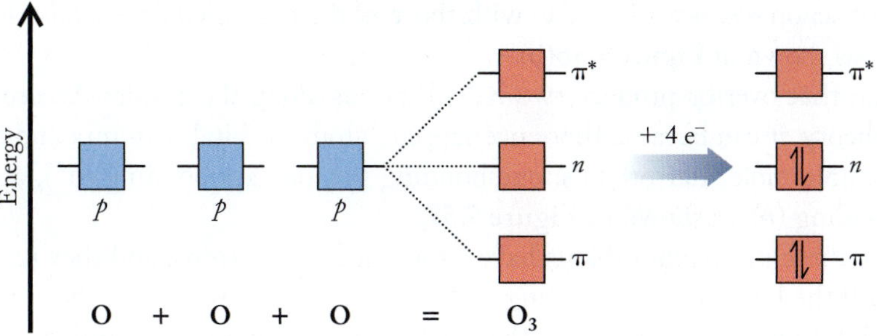


**FIGURE 5.55** Forming molecular orbitals from unhybridized $p$ orbitals in $O_3$.

average of two lone pairs on one noncentral O atom and three on the other in the resonance structures of $O_3$:

$$\ddot{O}\text{–}\ddot{O}\text{=}\ddot{O} \longleftrightarrow \ddot{O}\text{=}\ddot{O}\text{–}\ddot{O}$$

## A Bonding Theory for SN > 4

In Chapter 4 we noted that the central atoms in some molecules form more than four bonds, as shown in these Lewis structures of $PCl_5$ and $SF_6$:

(a) (b)

Until recently, a popular explanation for how atoms such as phosphorus and sulfur can form more than four bonds was based on their using valence-shell $d$ atomic orbitals to form hybrid orbitals that expanded their bonding capacity and allowed them to have more than an octet of valence electrons. Here we explore another way to explain the capacity of atoms to form more than four bonds, or, in general, to have more than four bonds and lone pairs of electrons in their valence shells—but without expanding their octets.

Let's begin by examining the bonding in molecules of a noble gas compound, $XeF_2$, whose synthesis was first reported in 1962. The Lewis symbol of xenon, which contains eight dots like all noble gases except helium (see Figure 4.8), tells us that the atom has a full octet in its valence shell. It has no unpaired electrons, and, theoretically, no capacity to form covalent bonds. One way to explain the existence of $XeF_2$ begins by hybridizing the 5$s$ orbital and two of the 5$p$ orbitals in xenon's valence shell, which yields a set of three $sp^2$ hybrid orbitals and one unhybridized $p$ orbital, as shown in **Figure 5.56(a)**. Now we consider the possibility that the two

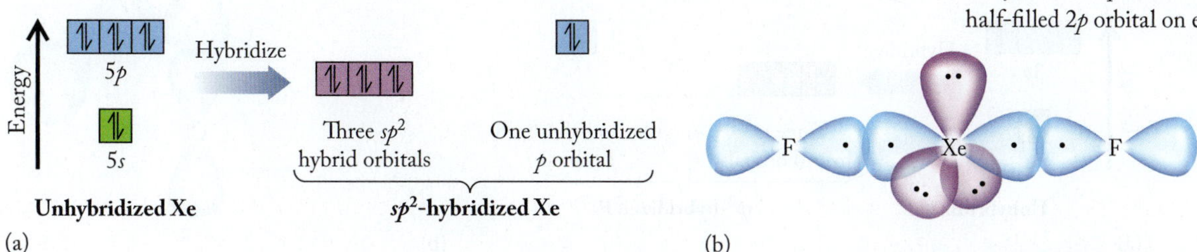


**FIGURE 5.56** (a) Mixing one 5$s$ orbital and two 5$p$ orbitals on a xenon atom creates three $sp^2$ hybrid orbitals and leaves the xenon atom with one unhybridized $p$ orbital. (b) The two lobes of the xenon atom's unhybridized 5$p$ orbital overlap with the half-filled 2$p$ orbital on each fluorine atom.

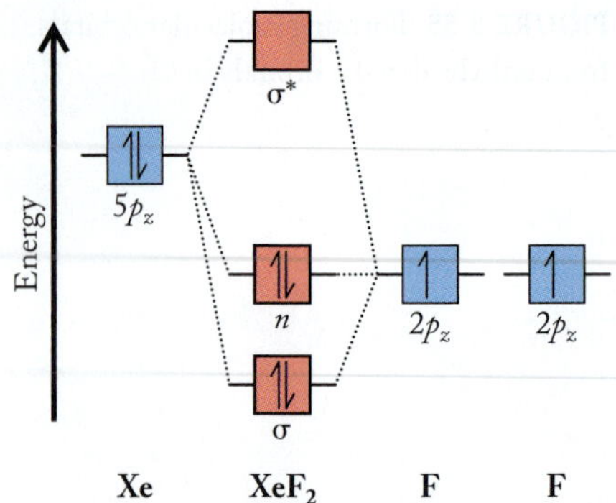


**FIGURE 5.57** Partial molecular orbital diagram of $XeF_2$. These are the only MOs that contribute to bonding.

lobes of xenon's $5p$ orbital overlap with those of the half-filled $2p$ orbitals of two F atoms, as shown in **Figure 5.56(b)**.

Can that overlap produce two Xe—F bonds along the $z$-axis? According to MO theory, it can because those overlapping atomic orbitals can mix and form a set of three molecular orbitals: one bonding ($\sigma$), one antibonding ($\sigma^*$), and one nonbonding ($n$), as shown in **Figure 5.57**.

The three molecular orbitals have a total of four electrons, and they preferentially fill the lower-energy bonding and nonbonding MOs, leaving the antibonding orbital empty. Electrons in the nonbonding orbital do not influence bond order, so the presence of two electrons in the bonding orbital together with the absence of any electrons in the antibonding orbital produces a bond order of $\frac{1}{2}(2 - 0) = 1$. That bond order of 1 is delocalized over both Xe—F bonds, leaving each one with an effective bond order of only $\frac{1}{2}$ (the two nonbonding electrons are located on the F atoms).

Thus, MO theory predicts that $XeF_2$ should exist, and it does. In addition, using xenon's unhybridized $p$ orbital to form the two Xe—F bonds means that the molecule should be linear, which it is. That molecular shape is reinforced by offsetting interactions between the bonding electrons and the three lone pairs of electrons in xenon's $sp^2$ hybrid orbitals, which are evenly distributed in lobes that are all 120° from each other in the equatorial plane of the molecule and 90° from the axial bonding orbitals. That arrangement means that the electron-pair geometry of the molecule is trigonal bipyramidal, as we would expect from a molecule with three lone pairs of electrons and two bonding pairs of electrons.

The bonding pattern in a molecule of $PCl_5$ resembles the pattern in $XeF_2$ because $PCl_5$ has the same trigonal bipyramidal electron-pair geometry. As with the Xe atom in $XeF_2$, the valence-shell orbitals of the P atom in $PCl_5$ are $sp^2$ hybridized—but here, the valence shell contains three half-filled $sp^2$ hybrid orbitals and one full $p$ orbital, as shown in **Figure 5.58(a)**.

The two lobes of the P atom's full $3p$ orbital (shown in blue in **Figure 5.58b**) overlap with those of the half-filled $3p$ orbitals of two Cl atoms, leading to the formation of two axial P—Cl bonds that each have a bond order of $\frac{1}{2}$. The P atom's half-filled $sp^2$ hybrid orbitals (shown in purple) overlap with the half-filled $3p$ orbitals of three Cl atoms to produce three equatorial P—Cl bonds that each have a bond order of 1. Because of their lower bond order, you might expect the axial bonds to be weaker and longer than the equatorial single bonds, and you

**FIGURE 5.58** (a) The hybridization of the $3s$ orbital and two $3p$ orbitals in a phosphorus atom creates three $sp^2$ hybrid orbitals and leaves the atom with one unhybridized $p$ orbital. (b) The two lobes of the phosphorus atom's unhybridized $3p$ orbital each overlap with the half-filled $3p$ orbital of a chlorine atom, as does each of the phosphorus atom's $sp^2$ hybrid orbitals.

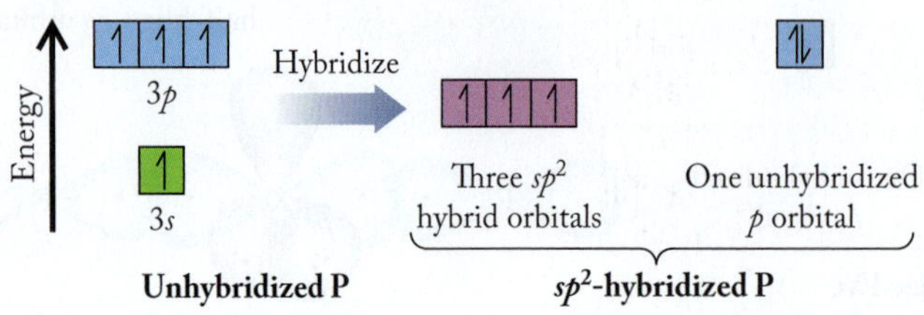


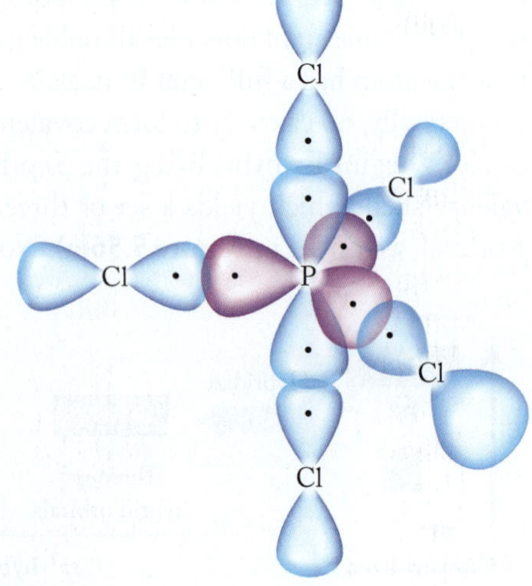

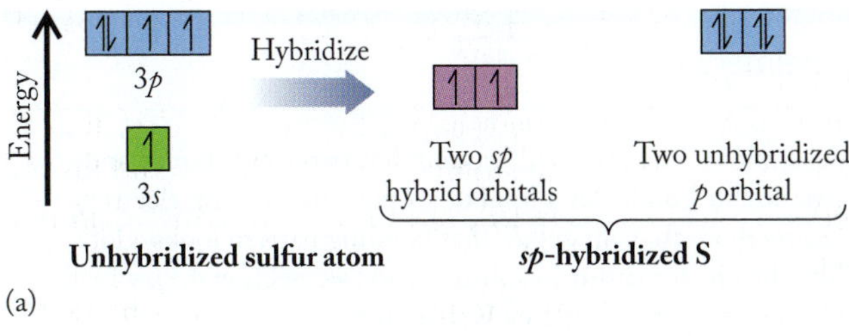


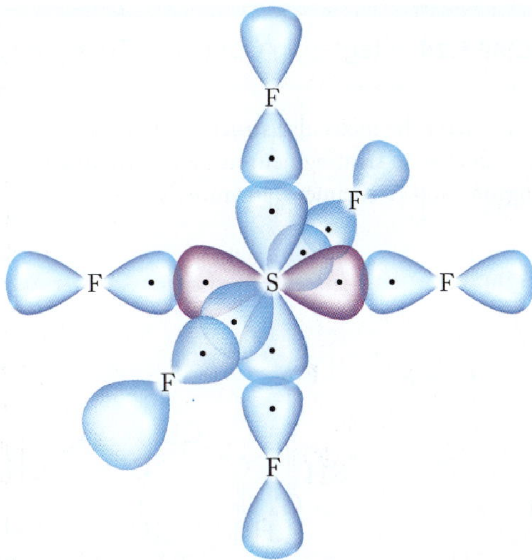


**FIGURE 5.59** (a) The hybridization of the 3*s* orbital and one 3*p* orbital in a sulfur atom creates two *sp* hybrid orbitals and leaves the atom with two unhybridized *p* orbitals. (b) The four lobes of the sulfur atom's unhybridized 3*p* orbitals each overlap with the half-filled 2*p* orbital of a fluorine atom, as does each *sp* hybrid orbital of the sulfur atom.

would be correct: the equatorial bonds are 202 pm long, whereas the axial bonds are 214 pm. No lone pairs of electrons are in the valence shell of the P atom, so the molecular geometry of $PCl_5$ matches its electron-pair geometry: trigonal bipyramidal, as predicted by VSEPR for SN = 5.

Finally, let's consider the bonding in a molecule of $SF_6$. How might we model the bonding capacity of a sulfur atom expanding from 2 to 6 by using only the four orbitals in its 3*s* and 3*p* subshells? One such model uses two of sulfur's 3*p* orbitals to make a total of four S—F bonds. We form *sp*-hybridized orbitals for the valence-shell electrons, resulting in two half-filled *sp* hybrid orbitals and two full 3*p* orbitals, as shown in **Figure 5.59(a)**.

The two lobes of each of the S atom's two full 3*p* orbitals (shown in blue in **Figure 5.59b**) overlap with the half-filled 2*p* orbitals of four F atoms to form four S—F bonds: two oriented above and below the S atom and two oriented behind and in front of the S atom. Each of those four bonds, according to MO theory, has a bond order of $\frac{1}{2}$. The S atom's two half-filled *sp* hybrid orbitals (shown in purple) overlap with the half-filled 2*p* orbitals of two F atoms to produce the two S—F bonds to the left and right of the S atom. Those bonds each have a bond order of 1. The six S—F bonds are all oriented 90° from each other and give the molecule an octahedral shape, as predicted for SN = 6.

Because of their different bond orders, you might expect two of the S—F bonds to be shorter than the other four, but actually they are all the same length. The reason why is that all six bonds are directed toward equivalent fluorine atoms at the corners of an octahedron and all the bonding electrons are evenly distributed over the entire molecule. That uniformity means that any differences between hybridized *sp* and unhybridized *p* orbitals in the S atom's valence shell disappear once the molecule forms. That even distribution of bonding electrons means that each equivalent S—F bond in $SF_6$ has a bond order of:

$$[(4 \text{ bonds} \times \tfrac{1}{2}) + (2 \text{ bonds} \times 1)]/6 = \tfrac{2}{3}$$

Before ending this chapter, we should recall the several theories of chemical bonding that have been presented and applied in Chapters 4 and 5. Each theory has strengths and limitations. The best one to apply to a given situation depends on the question asked. If our focus is on how the atoms in the second row of elements in the periodic table bond in molecules, Lewis structures usually suffice. However, Lewis structures cannot explain the experimentally measured bond angles within molecules. If we are interested in visualizing and explaining the shapes of molecules, VSEPR and valence bond theory are more useful models. However, valence bond theory cannot adequately explain the magnetic properties of molecular compounds and their ability to absorb and emit UV radiation and visible light, so we turn to molecular orbital theory as our most powerful model.

## SAMPLE EXERCISE 5.10 Integrating Concepts: Treatment for Alzheimer's Disease

In 2011, a compound with the molecular structure shown here was reported to be effective in treating neurodegenerative diseases, including Huntington's and Alzheimer's, in animals:

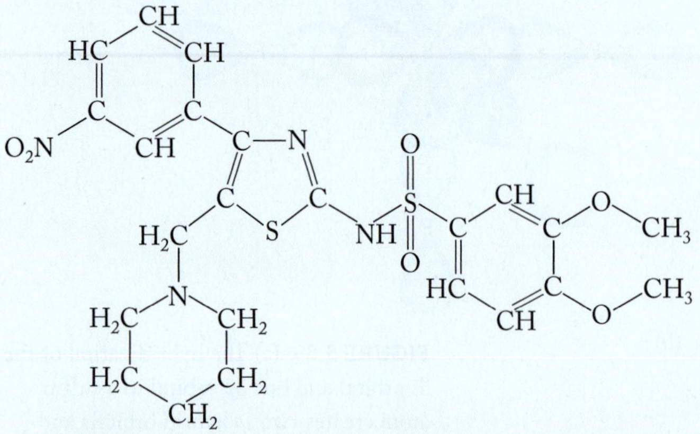

The structure has two S atoms. One has two bonds, consistent with its Lewis bonding capacity, but the other has six. Answer the following questions about this drug:

a. Many pharmaceuticals are chiral. Are any carbon atoms in the compound stereocenters?
b. What are the steric numbers and electron-pair geometries of the two sulfur atoms?
c. What is the bond angle between the two S═O double bonds?
d. The structure has two six-carbon rings. Are either of them aromatic rings? Which one(s)?
e. Does the compound have a permanent dipole?

**Collect and Organize** We are given the molecular structure of a biologically active compound and are asked questions about it. A chiral carbon atom is one bonded to four different atoms or groups of atoms. The steric number of an atom is the sum of its lone pairs of valence electrons and the number of atoms bonded to it. Steric numbers allow us to predict electron-pair geometries and bond angles and, indirectly, molecular geometries. Aromatic rings have six $sp^2$-hybridized C atoms and are stabilized by the delocalization of the electrons in three $\pi$ bonds around the ring. Molecules in which bond dipoles do not cancel have permanent dipoles—that is, molecules with asymmetrical distributions of valence electrons.

**Analyze**

a. Only carbon atoms with four single ($\sigma$) bonds can be stereocenters, which eliminates all those in double-bonded rings. Also, the C atoms in $-CH_2-$ groups can't be chiral because two of the four atoms or groups bonded to them are the same (H atoms).
b. and c. Sulfur atoms complete their octets by forming two covalent bonds, giving them two bonding pairs and two lone pairs. The sulfur atom in the five-atom ring fits that description. The other sulfur atom has two single bonds and two double bonds, for a total of six pairs of bonding electrons—clearly more than an octet. That bonding pattern looks a lot like the one for sulfur in sulfuric acid (see Section 4.8), which has no lone pairs in its Lewis structure.
c. The atoms in both six-carbon rings are bonded to each other with alternating single and double bonds, which allows the $\pi$ electrons in the three double bonds to be completely delocalized around each ring.
d. The molecule consists mostly of C and H atoms, which form bonds that are essentially nonpolar, but it also contains C—O, N═O, N—H, and S═O bonds, which are polar. Asymmetry is also present in the overall structure of the molecule.

**Solve**

a. None of the $sp^3$ carbon atoms are bonded to four different atoms or groups of atoms, so none are chiral.
b. One of the two sulfur atoms has two atoms bonded to it and two lone pairs of valence electrons, which adds up to SN = 4. The corresponding electron-pair geometry is tetrahedral. The other S atom has six bonds to four atoms and no lone pairs, which means that its SN also is 4 and it also has a tetrahedral electron-pair geometry.
c. Given the tetrahedral electron-pair geometry around S and the lack of lone pairs, we expect tetrahedral molecular geometry and an O═S═O bond angle of about 109°.
d. Because of the presence of delocalized $\pi$ electrons, both six-carbon rings are aromatic.
e. Given its generally asymmetric structure and the presence of polar bonds, we can conclude that the molecule has a permanent dipole.

**Think About It** We established that the compound has no chiral carbon atoms, but does it have no stereocenters at all? Look carefully at the nitrogen atom in the middle of the structure. It is $sp^3$ hybridized and surrounded by (1) a H atom; (2) the right side of the molecule, which is different from (3) the left side; and (4) a lone pair of electrons. Do the four different groups (counting the lone pair as a "group") make that N atom a stereocenter? They would, except for a process known as *inversion*. If we label the right side of the molecule R and the left side L, the structure surrounding the N atom could have two enantiomers:

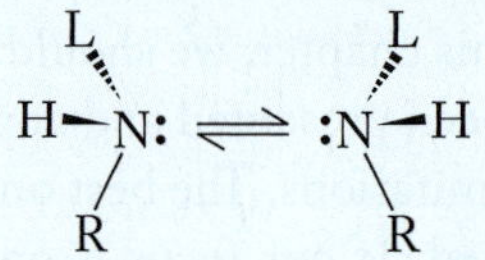

The double-arrow symbol between them means that the bonds of one enantiomer can reversibly flip around the N atom to form the other enantiomer. Because the bonds interconvert, we can't produce a sample of just one enantiomer.

## SUMMARY

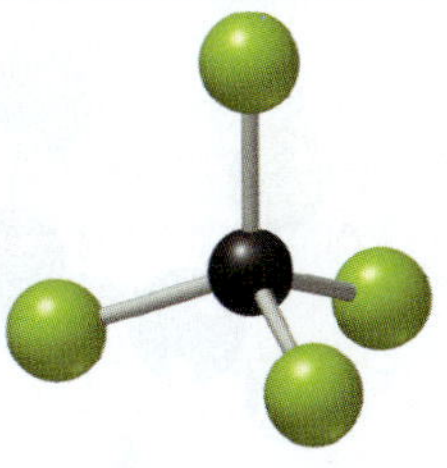

**LO1** The shape of a molecule reflects the arrangement of the atoms in three-dimensional space and is determined largely by characteristic **bond angles**. Minimizing repulsion between pairs of valence electrons (the **VSEPR** model) results in the lowest-energy orientations of bonding and nonbonding electron pairs and accounts for the observed **molecular geometries** of molecules. The shape of a molecule can be determined from its **steric number** (the sum of the number of bonded atoms and lone pairs around a central atom) and the **electron-pair geometry**, or arrangement of atoms and lone pairs. The observed bond angles in molecules deviate from the ideal values as a result of unequal repulsions between lone pairs and bonding pairs of electrons. (Sections 5.1 and 5.2)

**LO2** Two covalently bonded atoms with different electronegativities have partial electrical charges of opposite sign, creating a **bond dipole**. If the individual bond dipoles in a molecule do not offset each other, the molecule is polar. If they do offset each other, the molecule is nonpolar. **Dipole moment ($\mu$)** is a quantitative measure of the polarity of a molecule. (Section 5.3)

Electric field on

**LO3** According to **valence bond theory**, the **overlap** of half-filled orbitals results in covalent bonds between pairs of atoms in molecules. Molecular geometry is explained by the mixing, or **hybridization**, of *s* and *p* atomic orbitals to create $sp^3$, $sp^2$, or $sp$ **hybrid atomic orbitals**. Covalent bonds in which the electron density is greatest along the internuclear axis are **sigma ($\sigma$) bonds**. Covalent bonds in which the electron density is greatest either above and below or in front of and behind the internuclear axis are **pi ($\pi$) bonds**. The molecules of **aromatic compounds** contain planar rings of six $sp^2$-hybridized carbon atoms with alternating $\pi$ bonds whose electrons are delocalized over the entire ring system. (Sections 5.4 and 5.5)

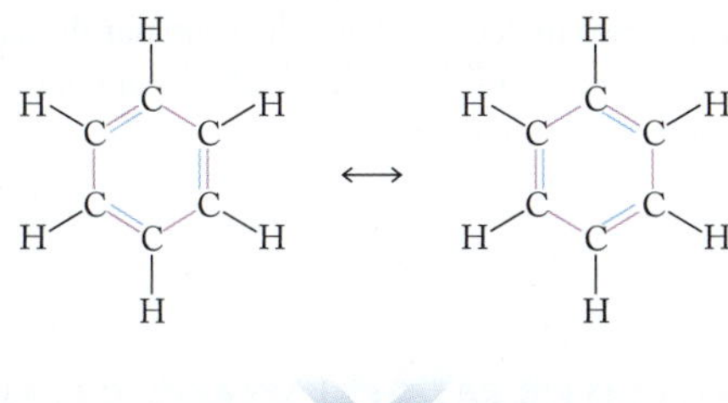

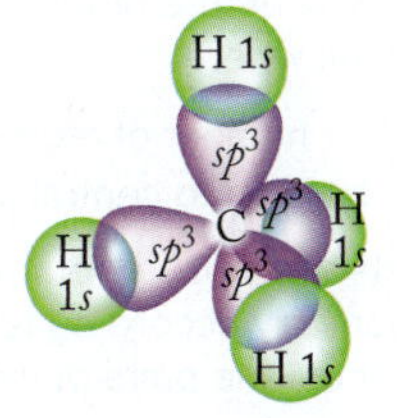


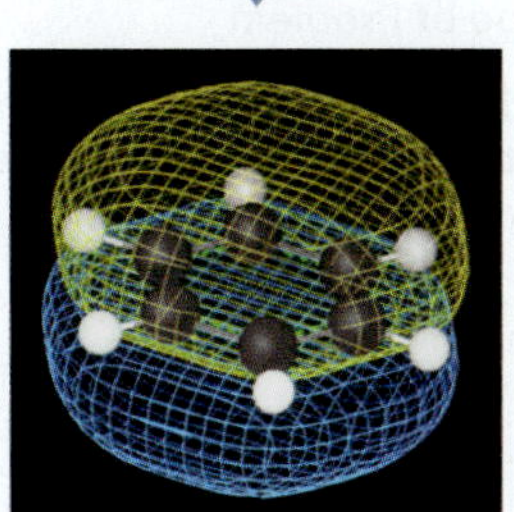

**LO4** **Isomers** are compounds that have the same chemical formula but different molecular structures. **Chiral** molecules have isomers with nonsuperimposable mirror images and have different properties. Many contain an $sp^3$-hybridized carbon atom bonded to four different atoms or groups of atoms. (Section 5.6)

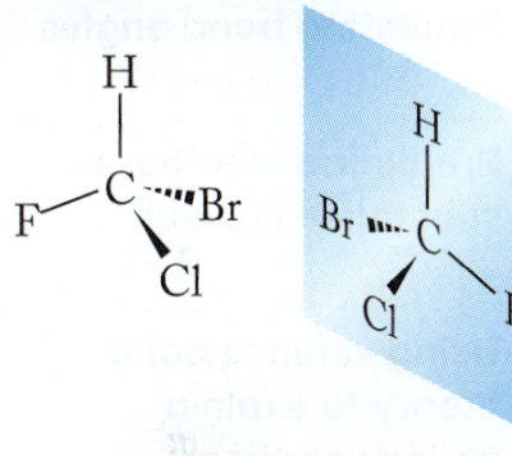


**LO5** **Molecular orbital (MO) theory** is based on the formation of **molecular orbitals**, which are orbitals belonging to an entire molecule. MO theory describes the magnetic and spectroscopic properties of molecular compounds in ways that valence bond theory cannot. Mixing two atomic orbitals creates one **bonding orbital** and one **antibonding orbital**. The region of highest electron density in **sigma ($\sigma$) molecular orbitals** lies along the internuclear axis. The regions of highest electron density in **pi ($\pi$) molecular orbitals** are found above and below or behind and in front of the internuclear axis. A **molecular orbital diagram** shows the relative energies of the molecular orbitals in a molecule. (Section 5.7)

## PARTICULATE PREVIEW WRAP-UP

1. Both carbon atoms in halothane have a tetrahedral molecular geometry, so the F—C—F bond angles are 109.5°.
2. The C—H bond is formed from the overlap of the C $sp^3$ hybrid orbital and the H 1*s* orbital.
3. The two molecules shown here are not the same molecule; they are enantiomers of each other (note the orientation of the H, Cl, and Br atoms around carbon).

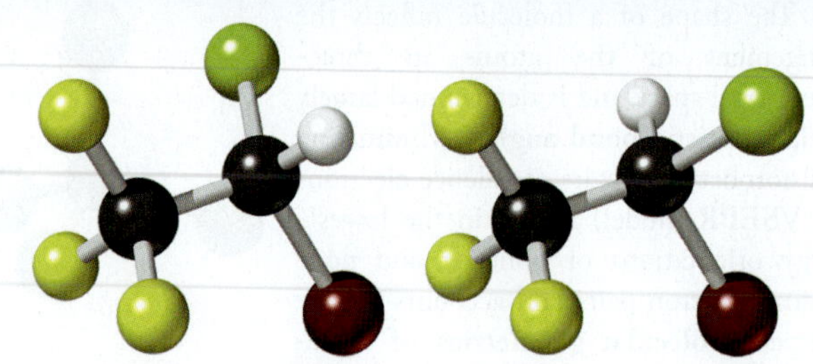

## PROBLEM-SOLVING SUMMARY

| Type of Problem | Concepts and Equations | Sample Exercises |
|---|---|---|
| **Predicting molecular geometry** | Draw the Lewis structure of the molecule. Determine the steric number (SN) of the central atom, where $$\text{SN} = \left(\begin{matrix}\text{number of atoms}\\ \text{bonded to central atom}\end{matrix}\right) + \left(\begin{matrix}\text{number of lone pairs}\\ \text{on central atom}\end{matrix}\right) \quad (5.1)$$ Choose the electron-pair geometry that corresponds to the SN. Choose a molecular geometry according to the electron-pair geometry that accounts for the number of lone pairs of valence electrons on the central atom (see Table 5.1). | **5.1, 5.3** |
| **Predicting bond angles** | Repulsion from lone pairs on a central atom pushes bonding pairs closer together and decreases bond angles. | **5.2** |
| **Predicting whether a substance is polar** | A substance is polar if a molecule of it has an overall permanent dipole—that is, if its bond dipoles and the locations of its lone pairs do not offset each other. | **5.4** |
| **Using valence bond theory to explain molecular shape** | Translate the observed electron-pair geometry into the appropriate central-atom hybridization by following these guidelines: (see table below) | **5.5** |
| **Recognizing chiral molecules** | Look for an $sp^3$ carbon atom bonded to four different atoms or groups of atoms. | **5.6** |
| **Using MO theory to predict bond order** | Draw the molecular orbital diagram and count the electrons in bonding and antibonding orbitals. $$\text{Bond order} = \tfrac{1}{2}\left[\left(\begin{matrix}\text{number of}\\ \text{bonding electrons}\end{matrix}\right) - \left(\begin{matrix}\text{number of}\\ \text{antibonding electrons}\end{matrix}\right)\right] \quad (5.2)$$ | **5.7–5.9** |

| Steric Number | Electron-Pair Geometry | Hybridization |
|---|---|---|
| 2 | Linear | *sp* |
| 3 | Trigonal planar | $sp^2$ |
| 4 | Tetrahedral | $sp^3$ |

## VISUAL PROBLEMS

*(Answers to boldface end-of-chapter questions and problems are in the back of the book.)*

**5.1.** The two compounds with the molecular structures shown in Figure P5.1 have the same molecular formula: $C_2H_3F_3$. Which molecule in Figure P5.1 has the greater dipole moment?

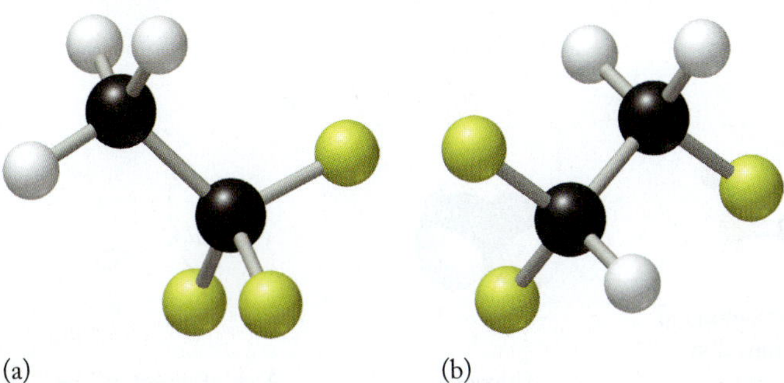

(a) (b)

**FIGURE P5.1**

5.2. Could you distinguish between the two structures of $N_2H_2$ shown in Figure P5.2 by the magnitude of their dipole moments?

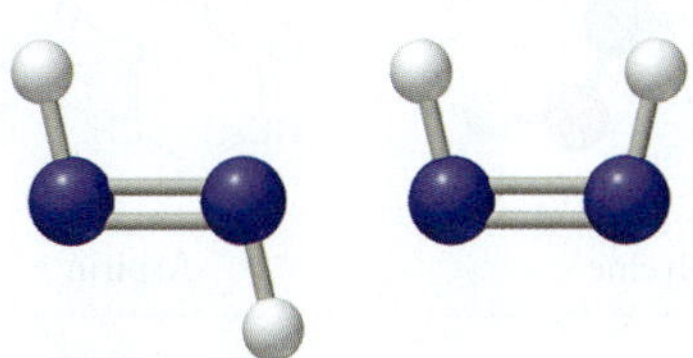

**FIGURE P5.2**

**5.3.** In which molecules shown in Figure P5.3 are all the atoms in a single plane? Do any of the molecules have delocalized $\pi$ electrons?

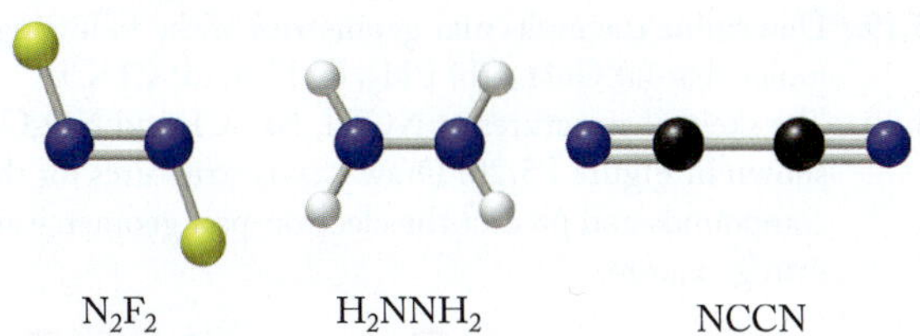


**FIGURE P5.3**

5.4. The ball-and-stick model in Figure P5.4 shows the skeletal structure of $C_4H_4$.

a. Draw the Lewis structure(s) of $C_4H_4$.
b. Are all the atoms in a single plane?
c. Are delocalized $\pi$ electrons present in a molecule of $C_4H_4$?

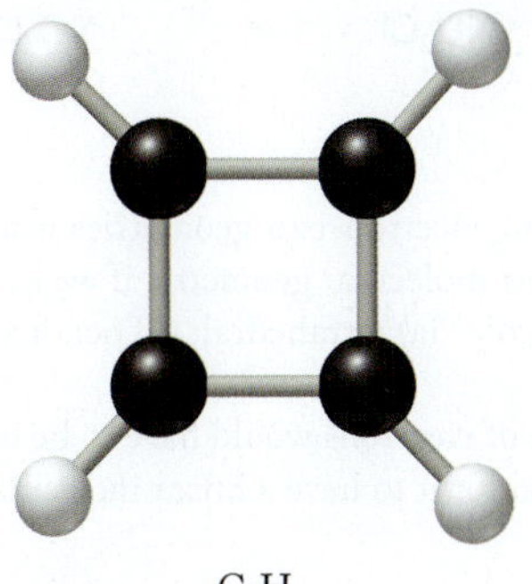


**FIGURE P5.4**

**5.5.** Use the MO diagram in Figure P5.5 to predict whether $O_2^+$ has more or fewer electrons in antibonding molecular orbitals than $O_2^{2+}$.

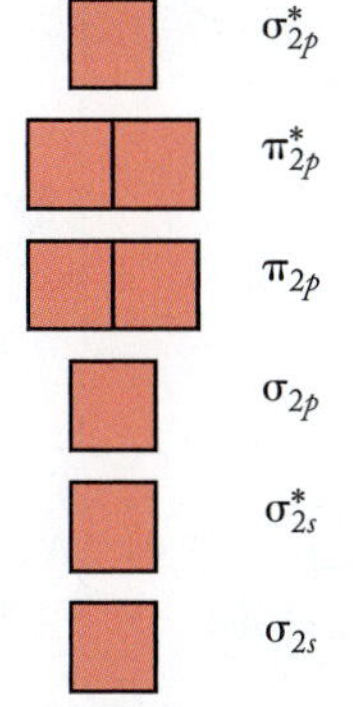


**FIGURE P5.5**

5.6. Under appropriate reaction conditions, diatomic molecules of iodine are ionized, forming $I_2^+$ cations. The corresponding anion, $I_2^-$, is unknown. Use the molecular orbital diagram in Figure P5.6 to explain why $I_2^+$ is more stable than $I_2^-$.

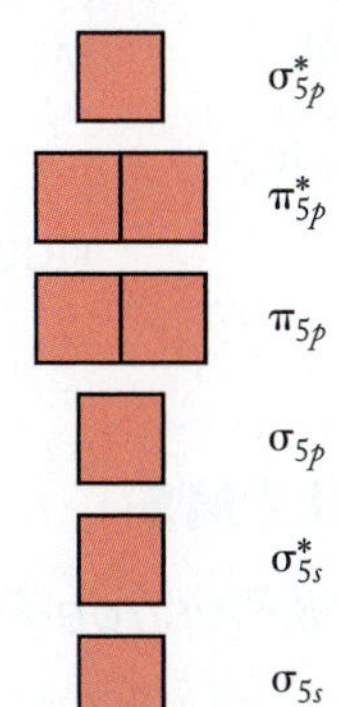


**FIGURE P5.6**

**5.7.** Figure P5.7 shows the molecular structure of a constituent of pine oil that contributes to its characteristic aroma. Is the compound chiral? Explain your answer.

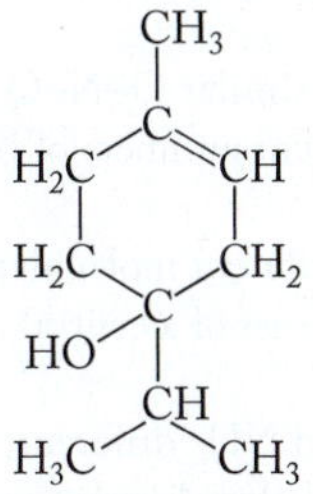

$CH_3$
C
$H_2C$ CH
$H_2C$ $CH_2$
HO C
CH
$H_3C$ $CH_3$

**FIGURE P5.7**

5.8. Figure P5.8 shows the molecular structure of menthol, a chiral compound that gives mint leaves their characteristic aroma. Locate the stereocenter(s) in the structure.

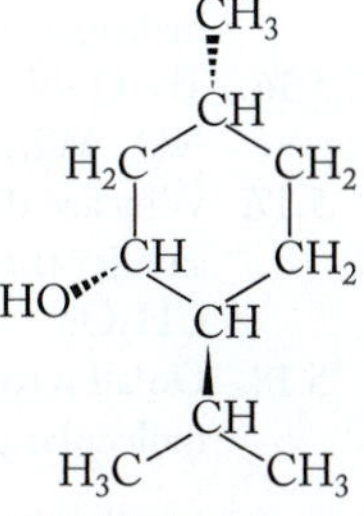


**FIGURE P5.8**

**5.9.** The molecular geometry of $ReF_7$ is an uncommon structure called a pentagonal bipyramid, shown in Figure P5.9. What are the bond angles in a pentagonal bipyramid?

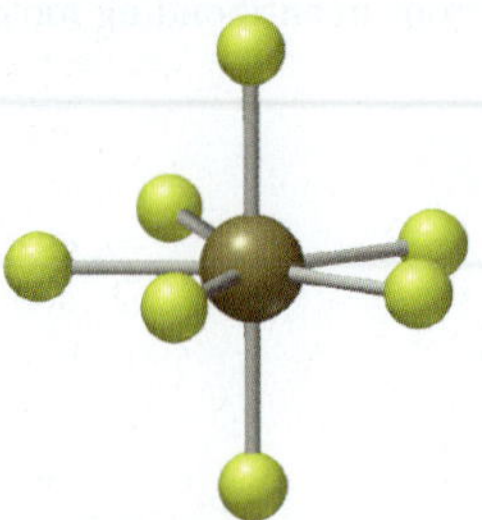

**FIGURE P5.9**

5.10. Use representations [A] through [I] in Figure P5.10 to answer questions a–f.

a. Which compounds contain delocalized $\pi$ electrons?

b. Which compounds contain a $\pi$ bond but not delocalized $\pi$ electrons?

c. Of [B], [E], and [H], which is/are planar?

d. Locate the chiral carbon atom(s).

e. Compare the bonding angles around the sulfur atoms in [D] and [G].

f. [C] is a CFC (chlorofluorocarbon) responsible for the destruction of the ozone layer. How is [F] related to [C]?

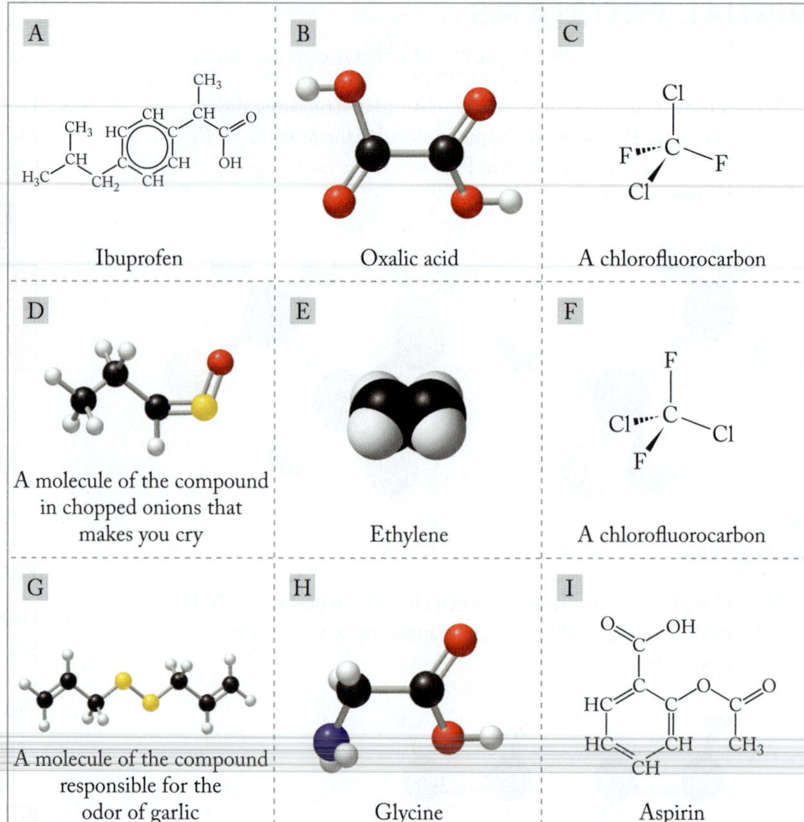


**FIGURE P5.10**

# QUESTIONS AND PROBLEMS

## Molecular Shape; Valence-Shell Electron-Pair Repulsion Theory (VSEPR)

### Concept Review

**5.11.** Why is a molecule's shape determined by repulsions between electron pairs and not by repulsions between nuclei?

5.12. Why do we need to draw the Lewis structure of a molecule before predicting its geometry?

**5.13.** Why do $NO_3^-$ and $NO_2^-$ ions have similar O–N–O bond angles, even though they have different numbers of N—O bonds?

5.14. Why do $CF_4$, $SF_4$, and $XeF_4$ have different molecular geometries, even though they all consist of a central atom bonded to four fluorine atoms?

**5.15.** Why are the bond angles in $BH_3$ and $NH_3$ different, even though they both consist of a central atom bonded to three hydrogen atoms?

5.16. The O–N–O bond angle is larger in $NO_2^+$ than it is in $NO_2$. Why?

**5.17.** Why are the H–C–H bond angles in molecules of $CH_4$ smaller than the H–C–H bond angles in molecules of $CH_2O$?

*5.18. Do all resonance forms of a molecule have the same molecular geometry? Explain your answer.

### Problems

**5.19.** Determine the molecular geometries of the following molecules: (a) $GeH_4$; (b) $PH_3$; (c) $H_2S$; (d) $CHCl_3$.

5.20. The skeletal structures of NOCl, $NO_2Cl$, and $NO_3Cl$ are shown in Figure P5.20. Draw Lewis structures for the three compounds and predict the electron-pair geometry at each nitrogen atom.

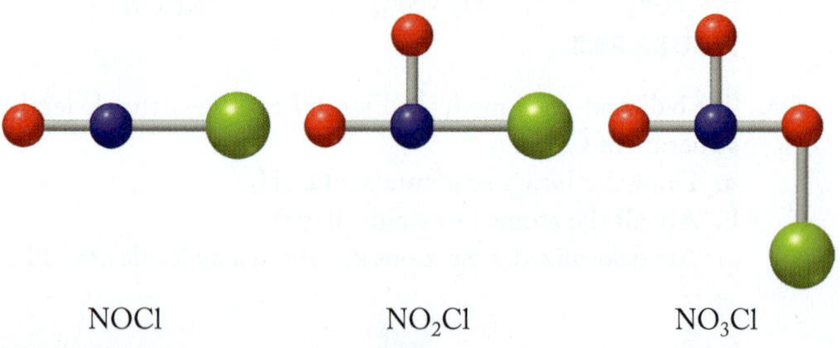


**FIGURE P5.20**

**5.21.** Which of the following electron-pair geometries is not consistent with a linear molecular geometry, if we assume three atoms per molecule? (a) tetrahedral; (b) octahedral; (c) trigonal planar

5.22. How many lone pairs of electrons would have to be on a SN = 6 central atom for it to have a linear molecular geometry?

**5.23.** Determine the molecular geometries of the following ions: (a) $NH_4^+$; (b) $CO_3^{2-}$; (c) $NO_2^-$; (d) $XeF_5^+$.

5.24. Determine the molecular geometries of the following ions: (a) $SCN^-$; (b) $CH_3PCl_3^+$ (P is the central atom and is bonded to the C atom); (c) $ICl_2^-$; (d) $PO_3^{3-}$.

---

**5.25.** Determine the molecular geometries of the following ions: (a) $S_2O_3^{2-}$; (b) $PO_4^{3-}$; (c) $NO_3^-$; (d) $NCO^-$.

5.26. Determine the molecular geometries of the following molecules: (a) $ClO_2$; (b) $ClO_3$; (c) $IF_3$; (d) $SF_4$.

---

**5.27.** Which two of the triatomic molecules $O_3$, $SO_2$, and $CO_2$ have the same molecular geometry?

5.28. Which two of the species $N_3^-$, $O_3$, and $CO_2$ have the same molecular geometry?

---

**5.29.** Which two of the ions $SCN^-$, $CNO^-$, and $NO_2^-$ have the same molecular geometry?

5.30. Which two of the molecules $N_2O$, $Se_2O$, and $CO_2$ have the same molecular geometry?

---

**5.31.** Rank the following molecules in order of increasing bond angles: (a) $NH_2Cl$; (b) $CCl_4$; (c) $H_2S$.

5.32. Rank the following molecular geometries in order of increasing bond angles: (a) trigonal pyramidal; (b) trigonal planar; (c) square planar.

---

**5.33. The Venusian Atmosphere** Several sulfur oxides not found in Earth's atmosphere have been detected in the atmosphere of Venus (Figure P5.33), including $S_2O$ and $S_2O_2$. Draw Lewis structures for $S_2O$ and $S_2O_2$, and determine their molecular geometries.

FIGURE P5.33

5.34. Determine the molecular geometries of the following molecules and ions: (a) $NO_3^-$; (b) $NO_4^{3-}$; (c) $NCN^{2-}$; (d) $NF_3$.

---

***5.35.** For many years, it was believed that the noble gases could not form covalently bonded compounds. However, xenon does react with fluorine. One of the products is the pentafluoroxenate anion, $XeF_5^-$. Draw the Lewis structure of $XeF_5^-$ and predict its geometry.

*5.36. The first compound containing a xenon–sulfur bond was isolated in 1998. Draw a Lewis structure for HXeSH and determine its molecular geometry.

---

***5.37. Chemical Warfare I** Sarin is a colorless, odorless gas that disrupts the nervous system, and just one drop can lead to death within minutes. It was created by German scientists in the 1930s who were trying to develop new pesticides. Like all chemical weapons, sarin is banned under international law. The structure in Figure P5.37 shows the connectivity of the atoms in the sarin molecule. Complete the Lewis structure by adding bonds and lone pairs as necessary. Assign formal charges to the P and O atoms, and determine the molecular geometry around P.

```
                 H
                 |
   H  O   H—C—H
   |   |        |
H—C—P—O—C—H
   |   |        |
   H  F   H—C—H
                 |
                 H
```

FIGURE P5.37

5.38. Determine the electron-pair geometries around the nitrogen atoms in the following unstable nitrogen oxides: (a) $N_2O_2$; (b) $N_2O_5$; (c) $N_2O_3$. ($N_2O_2$ and $N_2O_3$ have N—N bonds; $N_2O_5$ does not.)

## Polar Bonds and Polar Molecules

### Concept Review

**5.39.** Explain the difference between a polar bond and a polar molecule.

5.40. Must a polar molecule contain polar covalent bonds? Why?

**5.41.** Can a nonpolar molecule contain polar covalent bonds?

5.42. What does a dipole moment measure?

### Problems

**5.43.** Vinyl chloride (Figure P5.43) is used to make plastics. Which bond in vinyl chloride is most polar?

```
H
  \
   C=CH2
  /
:Cl:
```

FIGURE P5.43

5.44. Which bond in Figure P5.43 has no dipole moment?

---

**5.45. Kidney Function** Urea (Figure P5.45) is a waste product of human metabolism that the kidneys remove. Urea levels are measured to monitor healthy kidney function. Which bond in urea is most polar?

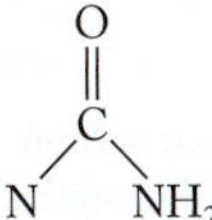


FIGURE P5.45

5.46. Does the molecule of urea (Figure P5.45) have $\mu = 0$?

**5.47.** The following molecules contain polar covalent bonds. Which are polar molecules and which have no permanent dipoles? (a) $CCl_4$; (b) $CHCl_3$; (c) $CS_2$; (d) $H_2S$; (e) $SCl_2$

5.48. Which of the following molecules has a permanent dipole? (a) $C{\equiv}O$; (b) $N{\equiv}N$; (c) $H{-}C{\equiv}N$; (d) $H{-}C{\equiv}C{-}H$

**5.49.** Compounds containing carbon, chlorine, and fluorine are known as chlorofluorocarbons (CFCs). Which of the following CFCs are polar and which are nonpolar? (a) $CFCl_3$; (b) $CF_2Cl_2$; (c) $Cl_2FCCF_2Cl$

5.50. Which of the following molecules has a permanent dipole? (a) $C_4F_8$ (cyclic structure); (b) $ClFCCF_2$; (c) $Cl_2HCCClF_2$

**5.51.** Predict which molecule in each of the following pairs is more polar: (a) $CClF_3$ or $CBrF_3$; (b) $CF_2Cl_2$ or $CHF_2Cl$.

5.52. Which molecule in each of the following pairs is more polar? (a) $NH_3$ or $PH_3$; (b) $CCl_2F_2$ or $CBr_2F_2$

**5.53.** **Volcanic Gas Emissions** In 2018 the Kilauea volcano on Hawaii erupted (Figure P5.53) and sent large amounts of four gases into the atmosphere: $H_2O$, $H_2S$, $SO_2$, and $CO_2$. (a) Determine whether each of those molecules is polar or nonpolar. (b) The structure of each of those four compounds can be described as a central atom bonded to two other identical atoms. Why aren't all the molecules nonpolar?

**FIGURE P5.53**

5.54. **Chemical Warfare II** A compound with the formula $COCl_2$ has been used as a chemical warfare agent. It and two similar compounds, $COBr_2$ and $COI_2$, are all eye irritants and cause skin to blister. The severity of the skin reactions is influenced by the polarity of the compounds. Rank the compounds in order of increasing polarity of their C—X bonds (where X is a halogen atom).

## Valence Bond Theory and Hybrid Orbitals

### Concept Review

**5.55.** What must atomic orbitals have in common to mix and form hybrid orbitals?

5.56. Why do atomic orbitals form hybrid orbitals?

### Problems

**5.57.** What is the hybridization of the underlined carbon atom in each of these condensed structural formulas? (a) $CH_3{-}\underline{C}H_2{-}CH_3$; (b) $CH_2{=}\underline{C}H{-}CH_3$; (c) $CH_3{-}\underline{C}H(CH_3){-}CH_3$; (d) $\underline{C}H{\equiv}C{-}CH_3$; (e) $CH{\equiv}\underline{C}{-}CH_3$

5.58. What is the hybridization of the nitrogen atom in each of the following ions and molecules? (a) $NO_2^+$; (b) $NO_2^-$; (c) $N_2O$; (d) $N_2O_5$; (e) $N_2O_3$

**5.59.** $N_2F_2$ has the two possible structures shown in Figure P5.59. Are the differences between the structures related to differences in the hybridization of nitrogen in $N_2F_2$? Identify the hybrid orbitals that account for the bonding in $N_2F_2$. Are they the same as those in acetylene, $C_2H_2$?

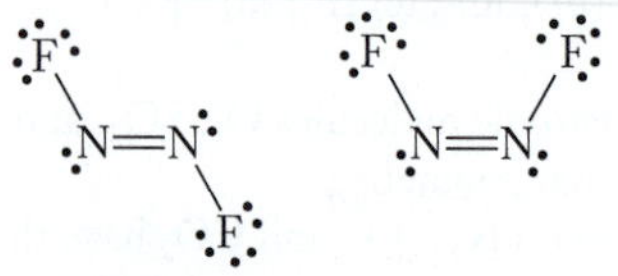

**FIGURE P5.59**

5.60. **Air Bag Chemistry** Azides such as sodium azide, $NaN_3$, are used in automobile air bags as a source of nitrogen gas. Another compound with three bonded nitrogen atoms is $N_3F$. How are the arrangements of the electrons around the nitrogen atoms different between the azide ion ($N_3^-$) and $N_3F$? Is anything different in the hybridization of the central nitrogen atom?

**5.61.** How does the hybridization of the central atom change in the series $CO_2$, $NO_2$, $O_3$?

5.62. How does the hybridization of the sulfur atom change when we compare $SO_2$ with $SO_2^{2-}$?

**5.63.** Draw a Lewis structure for $Cl_3^+$. Determine its molecular geometry and the hybridization of the central Cl atom.

5.64. Do all resonance forms of $N_2O$ have the same hybridization at the central N atom?

**5.65.** The Lewis structure for $N_4O$, with the skeletal structure O–N–N–N–N, contains one N—N single bond, one N=N double bond, and one N≡N triple bond. Is the hybridization of all the nitrogen atoms the same?

*5.66. Several resonance forms can be drawn for the anion $[C(CN)_3]^-$, including the two structures shown in Figure P5.66. Do they have the same geometry about the central carbon? What is the hybridization of each carbon atom?

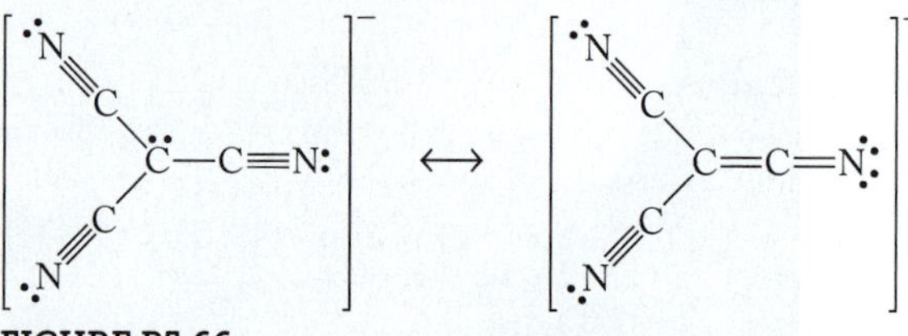

**FIGURE P5.66**

## Molecules with Multiple "Central" Atoms

### Concept Review

**5.67.** Can molecules with more than one central atom have resonance forms?

5.68. Why is assigning a single geometry to a molecule with more than one central atom difficult?

***5.69.** Are resonance structures examples of electron delocalization? Explain your answer.

*5.70. Can hybrid orbitals be associated with more than one atom?

## Problems

**5.71.** The two nitrogen atoms in nitramide are connected, with two oxygen atoms on one nitrogen atom and two hydrogen atoms on the other (Figure P5.71). What is the molecular geometry of each nitrogen atom in nitramide? Is the hybridization of both nitrogen atoms the same?

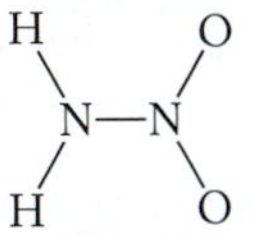

**FIGURE P5.71**

5.72. Cyclic structures exist for many compounds of carbon and hydrogen. Describe the molecular geometry and hybridization around each carbon atom in cyclopropene ($C_3H_4$), cyclobutane ($C_4H_8$), cyclobutene ($C_4H_6$), and cyclobutadiene ($C_4H_4$) (Figure P5.72).

Cyclopropene Cyclobutane Cyclobutene Cyclobutadiene

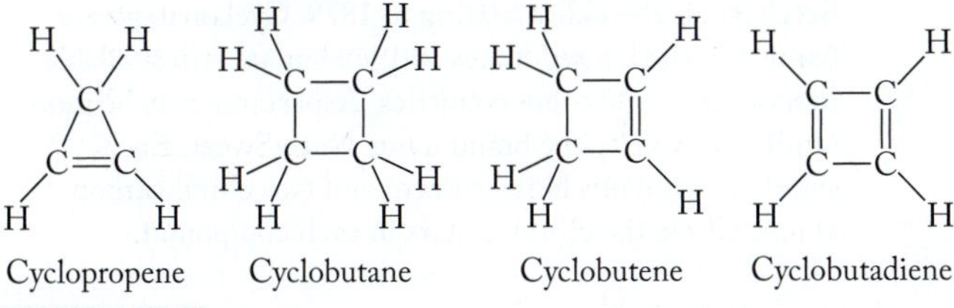

**FIGURE P5.72**

---

**5.73. Butter** The molecular structure of a compound that contributes to the flavor of butter is shown in Figure P5.73. What is the molecular geometry and bond angle at each carbon atom?

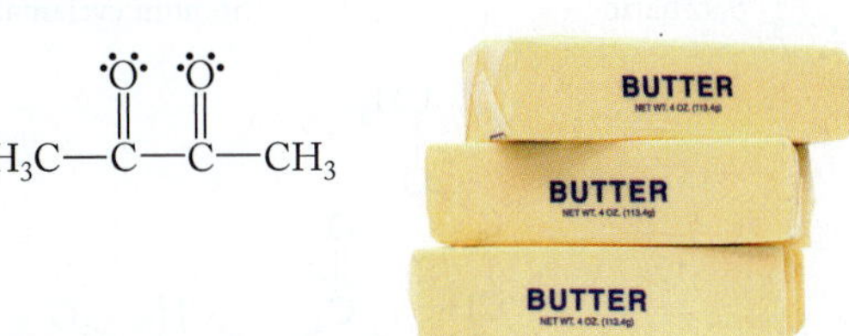

**FIGURE P5.73**

5.74. What is the hybridization of each carbon atom in Figure P5.73?

---

**5.75. Sex Hormones** Progesterone (Figure P5.75a) is a hormone involved in regulating menstrual cycles and pregnancy, and testosterone (Figure 5.75b) is the primary male sex hormone. Identify the $sp^2$-hybridized carbon atoms in both progesterone and testosterone.

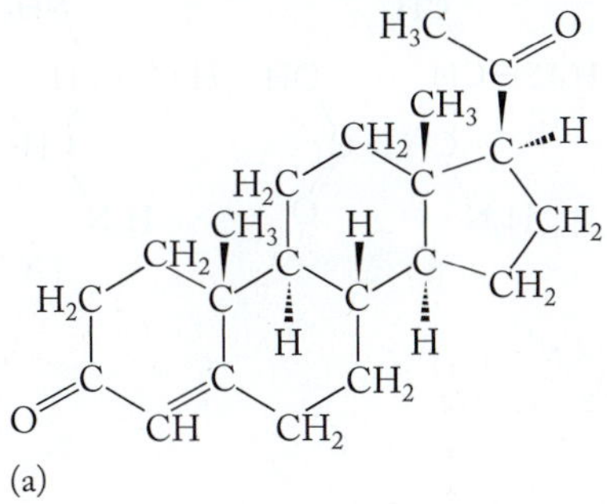

(a)

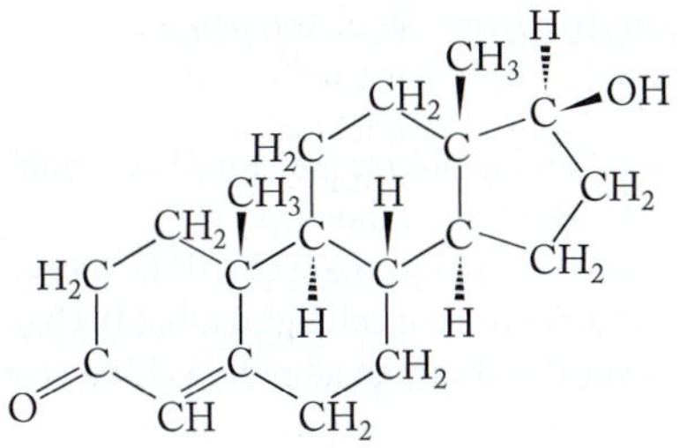

(b)

**FIGURE P5.75**

5.76. **Steroid Hormones** Cortisone (Figure P5.76) is a steroid produced by the body in response to stress. The hormone suppresses the immune system and can be administered as a drug to reduce inflammation, pain, and swelling. Identify the $sp^2$-hybridized carbon atoms in cortisone.

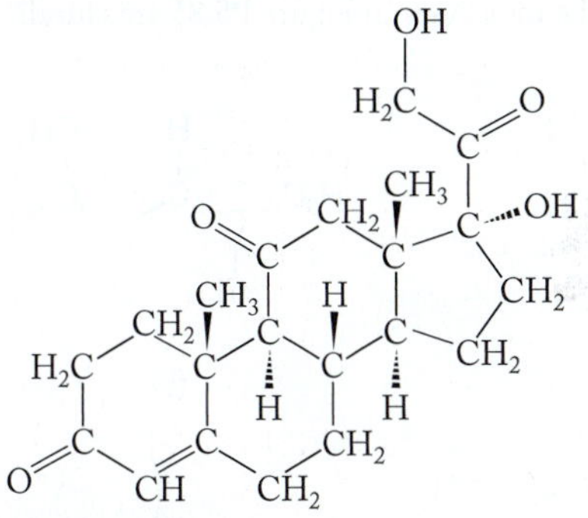

**FIGURE P5.76**

---

**5.77. Pheromones I** Bombykol is the compound synthesized by female silkworm moths to attract mates. Locate the conjugated double bonds in Figure P5.77 and identify the hybridization of the carbon atoms in those bonds.

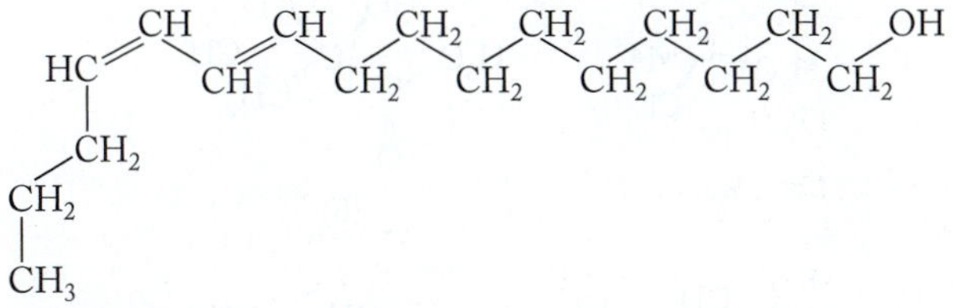

**FIGURE P5.77**

5.78. **Pheromones II** Fucoserratene is the compound synthesized by a brown alga to reproduce. Locate the conjugated double bonds in Figure P5.78 and identify the hybridization of the carbon atoms in those bonds.

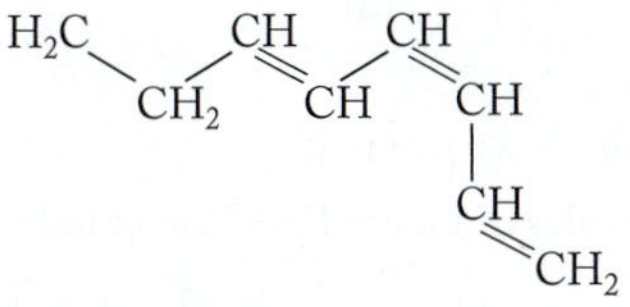

**FIGURE P5.78**

## Chirality and Molecular Recognition

### Concept Review

**5.79.** Which of the following objects are chiral? (a) a golf club; (b) a spoon; (c) a glove; (d) a shoe

5.80. Which of the following objects are chiral? (a) a key; (b) a screwdriver; (c) a fluorescent coil lightbulb; (d) a baseball

**5.81.** Could an *sp*-hybridized carbon atom be a chiral center? Explain your answer.

5.82. Two compounds have the same Lewis structure and the same optical activity. Are they enantiomers or the same compound?

**5.83.** Are racemic mixtures homogeneous or heterogeneous?

5.84. Can a mixture of enantiomers rotate plane-polarized light? Explain your answer.

### Problems

**5.85.** Which of the molecules in Figure P5.85 are chiral?

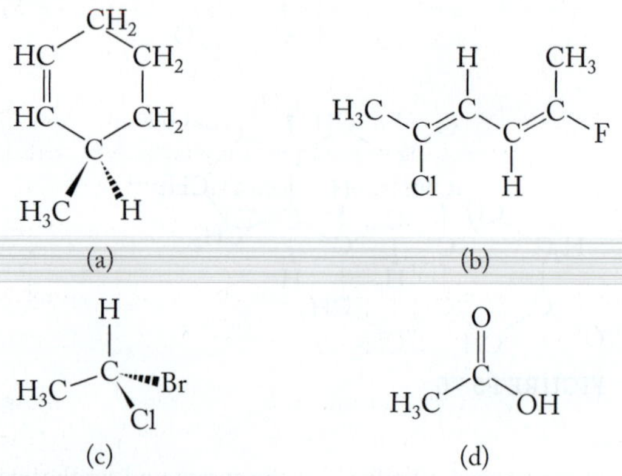

(a) (b) (c) (d)

**FIGURE P5.85**

5.86. Which of the molecules in Figure P5.86 are chiral?

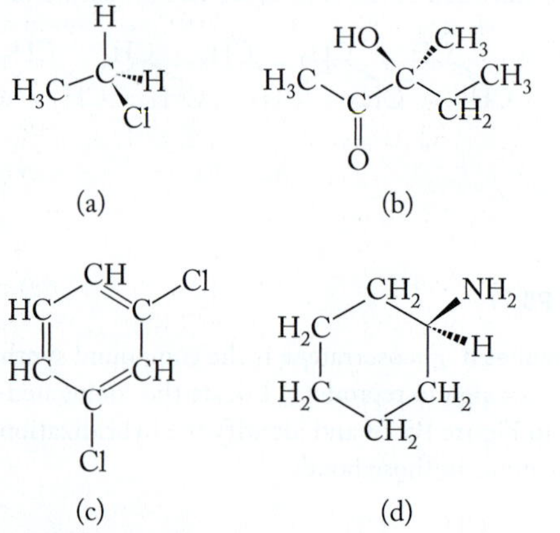


**FIGURE P5.86**

**5.87.** Which of the molecules in Figure P5.87 are chiral?

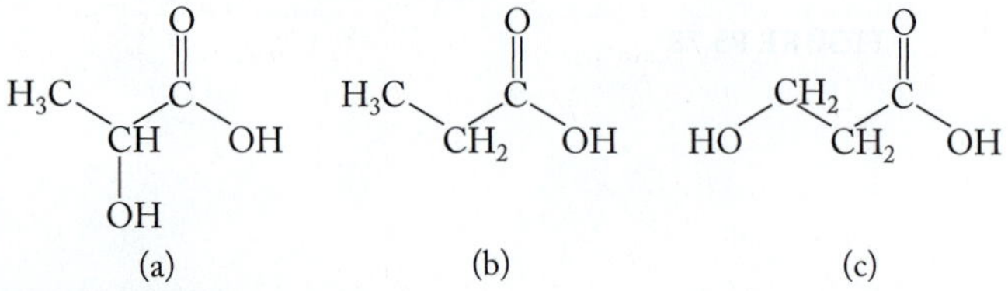

(a) (b) (c)

**FIGURE P5.87**

5.88. Which of the molecules in Figure P5.88 are chiral?

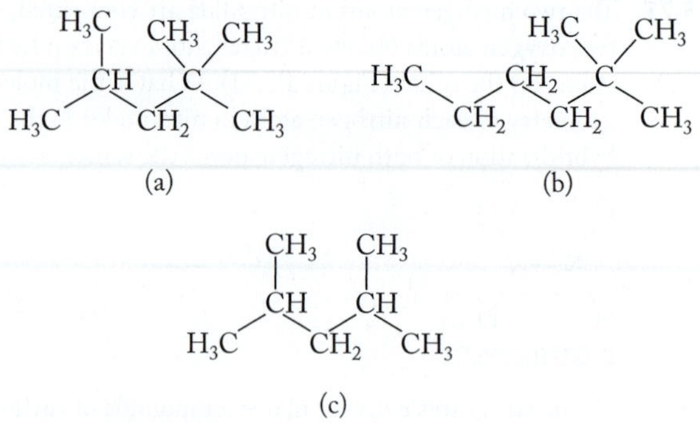

(a) (b) (c)

**FIGURE P5.88**

**5.89. Artificial Sweeteners** Figure P5.89 shows three artificial sweeteners that have been used in food and beverages. Saccharin is the oldest, dating to 1879. Cyclamates were banned in the United States in 1969 but are still available in more than 100 other countries. Aspartame may be more familiar to you by the brand name NutraSweet. Each sweetener contains between zero and two chiral carbon atoms. Circle the chiral centers in each compound.

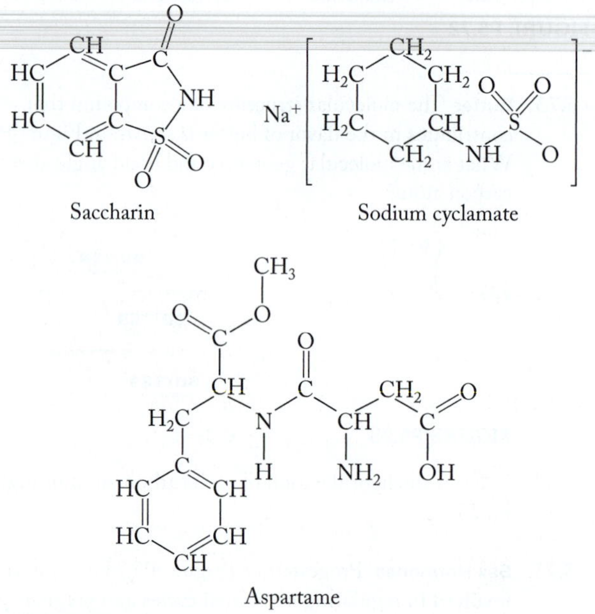

Saccharin Sodium cyclamate

Aspartame

**FIGURE P5.89**

5.90. Identify the chiral centers in each molecule in Figure P5.90.

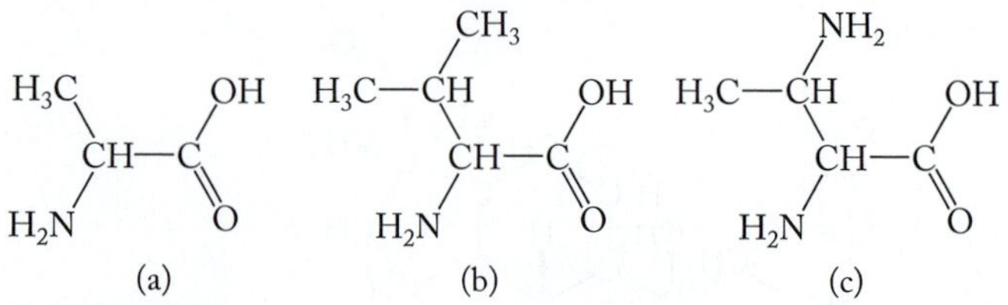

(a) (b) (c)

**FIGURE P5.90**

**5.91. The Smell of Raspberries** The compound with the structure in Figure P5.91 is a major contributor to the aroma of ripe raspberries. Identify any chiral center(s).

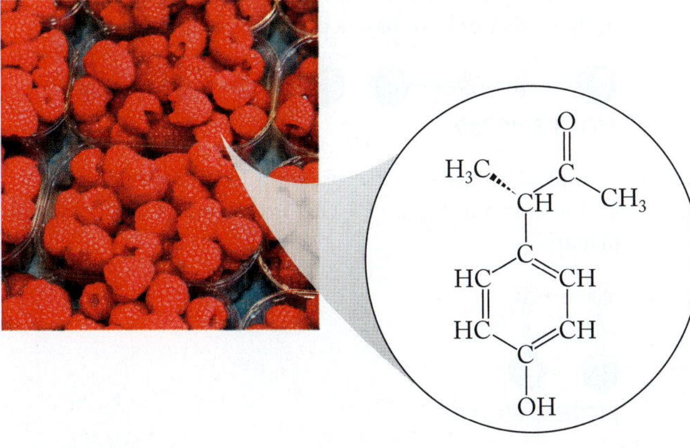


**FIGURE P5.91**

5.92. **Antidepressants** Figure P5.92 shows the structure of the drug bupropion, used to treat depression. Identify any chiral center(s).

$H_3C$ $CH_3$
C
$H_3C$ NH
H
O C
C $CH_3$
C
HC CH
HC C
CH Cl

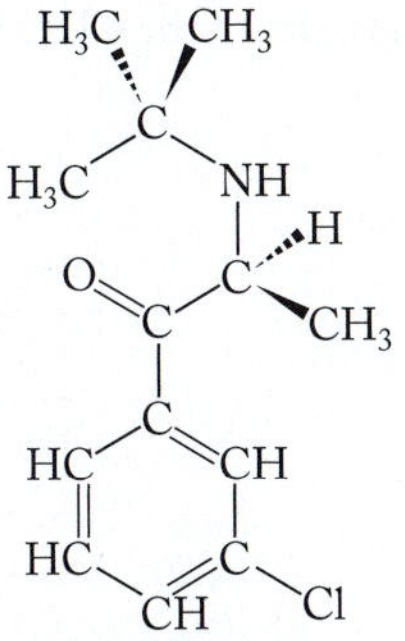

**FIGURE P5.92**

## Molecular Orbital Theory

### Concept Review

**5.93.** Which better explains the visible emission spectra of molecular substances: valence bond theory or molecular orbital theory?

5.94. Which better explains the magnetic properties of molecular substances: valence bond theory or molecular orbital theory?

**5.95.** Do all $\sigma$ molecular orbitals result from the overlap of *s* atomic orbitals?

5.96. Do all $\pi$ molecular orbitals result from the overlap of *p* atomic orbitals?

### Problems

**5.97.** Make a sketch showing how two 1*s* orbitals overlap to form a $\sigma_{1s}$ bonding molecular orbital and a $\sigma^*_{1s}$ antibonding molecular orbital.

5.98. Make a sketch showing how two $2p_y$ orbitals overlap "sideways" to form a $\pi_{2p}$ bonding molecular orbital and a $\pi^*_{2p}$ antibonding molecular orbital.

---

**5.99.** Use MO theory to predict the bond orders of the following molecular ions: $N_2^+$, $O_2^+$, $C_2^+$, and $Br_2^{2-}$. Do you expect any of the species to exist?

5.100. Diatomic noble gas molecules, such as $He_2$ and $Ne_2$, do not exist. Would removing an electron create molecular ions, such as $He_2^+$ and $Ne_2^+$, that are more stable than $He_2$ and $Ne_2$?

---

**5.101.** Which of the following molecular ions are paramagnetic? (a) $N_2^+$; (b) $O_2^+$; (c) $C_2^{2+}$; (d) $Br_2^{2-}$

5.102. Which of the following molecular ions are diamagnetic? (a) $O_2^-$; (b) $O_2^{2-}$; (c) $N_2^{2-}$; (d) $F_2^+$

---

**5.103.** Which of the following molecular anions have electrons in $\pi$ antibonding orbitals? (a) $C_2^{2-}$; (b) $N_2^{2-}$; (c) $O_2^{2-}$; (d) $Br_2^{2-}$

5.104. Which of the following molecular cations have electrons in $\pi$ antibonding orbitals? (a) $N_2^+$; (b) $O_2^+$; (c) $C_2^{2+}$; (d) $Br_2^{2+}$

---

**5.105.** For which of the following diatomic molecules does the bond order increase with the gain of two electrons, forming the corresponding 2− anion? (a) $B_2$; (b) $C_2$; (c) $N_2$; (d) $O_2$

5.106. For which of the following diatomic molecules does the bond order increase with the loss of two electrons, forming the corresponding 2+ cation? (a) $B_2$; (b) $C_2$; (c) $N_2$; (d) $O_2$

---

**5.107.** Do the 1+ cations of homonuclear diatomic molecules of the second-row elements always have shorter bonds than those of the corresponding neutral molecules?

5.108. Do any of the anions of the homonuclear diatomic molecules formed by B, C, N, O, and F have shorter bonds than those of the corresponding neutral molecules? Consider only the 1− and 2− anions.

---

***5.109. Perchlorate Ion and Human Health** Perchlorate compounds adversely affect human health by interfering with the uptake of iodine in the thyroid gland. Because of that behavior, though, they are also used to treat hyperthyroidism, or overactive thyroid. Draw the Lewis structure(s) of the perchlorate ion, $ClO_4^-$. What is the shape of the ion? What hybridization scheme for the central chlorine atom accounts for that shape?

*5.110. **Bleaching Agents** Draw the Lewis structure of the chlorite ion, $ClO_2^-$, which is used as a bleaching agent. What is the shape of the ion? What hybridization scheme for the central chlorine atom accounts for the structure you drew?

---

**5.111.** What is the molecular geometry around sulfur and nitrogen in the sulfamate anion shown in Figure P5.111? Which atomic or hybrid orbitals overlap to form the S—O and S—N bonds in the sulfamate anion?

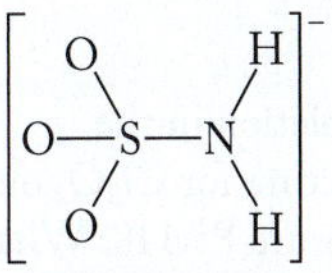

[O H; O—S—N; O H]$^-$

**FIGURE P5.111**

5.112. What is the geometry around each sulfur atom in the disulfate anion shown in Figure P5.112? What is the hybridization of the central oxygen atom?

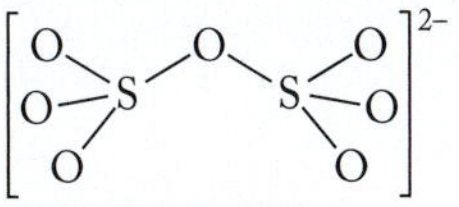

[O O O; O—S S—O; O O]$^{2-}$

**FIGURE P5.112**

## Additional Problems

**5.113.** Figure P5.113 depicts the principal nitrogen-containing molecule in mammalian urine. Elevated concentrations of that compound in mammalian blood may indicate kidney failure. Draw the Lewis structure that corresponds to Figure P5.113. Is that compound a planar molecule?

**FIGURE P5.113**

5.114. **Ant Venom** Formic acid occurs naturally in the venom of stinging ants. Draw the Lewis structure of formic acid that corresponds to Figure P5.114. Is formic acid a planar molecule?

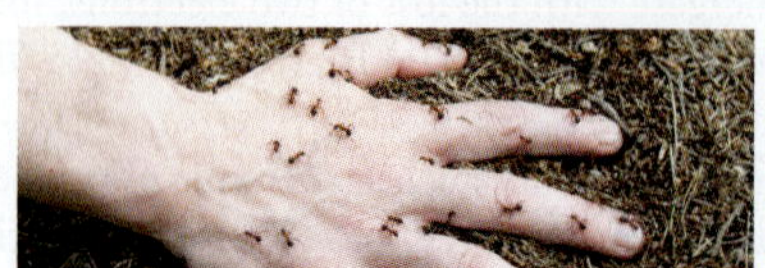

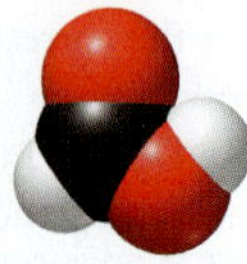

**FIGURE P5.114**

**5.115.** **Rocket Propellants** Draw the Lewis structure for the two ions in ammonium perchlorate ($NH_4ClO_4$), which is used as a propellant in solid fuel rockets, and determine the molecular geometries of the two polyatomic ions.

5.116. **Pressure-Treated Lumber** By December 31, 2003, concerns over arsenic contamination had prompted the manufacturers of pressure-treated lumber to voluntarily cease producing lumber treated with chromated copper arsenate (CCA) for residential use. CCA-treated lumber has a light greenish color and was widely used to build decks, sandboxes, and playground structures. Draw the Lewis structure for the arsenate ion ($AsO_4^{3-}$). Predict the angles between the arsenic–oxygen bonds in the arsenate anion.

**5.117.** Consider the molecular structure in Figure P5.117. What is the angle formed by the N—C—C bonds in the structure? What are the O═C—O and C—O—H bond angles?

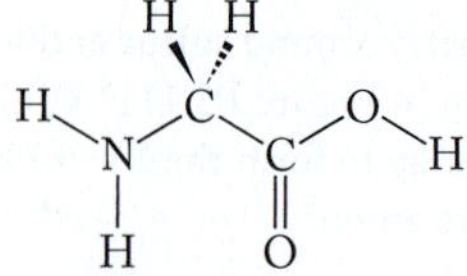


**FIGURE P5.117**

5.118. $Cl_2O_2$ may play a role in ozone depletion in the stratosphere. Draw the Lewis structure for $Cl_2O_2$ on the basis of the skeletal structure in Figure P5.118. What is the geometry about the central chlorine atom?

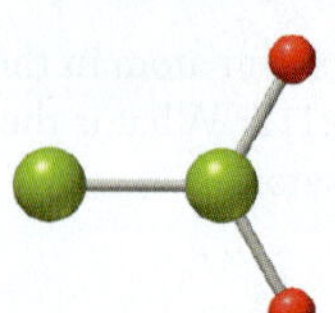

**FIGURE P5.118**

**5.119.** Bombardment of $Cl_2O_2$ molecules with intense UV radiation is thought to produce the two compounds with the skeletal structures shown in Figure P5.119.

a. Are either or both of those molecules linear?

b. Do either or both have a permanent dipole?

**FIGURE P5.119**

*5.120. Complete the Lewis structure for the cyclic structure of $Cl_2O_2$ shown in Figure P5.120. Is the cyclic $Cl_2O_2$ molecule planar?

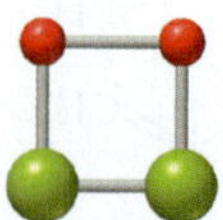

**FIGURE P5.120**

**5.121.** **Ozone Depletion** In 1999, the $ClO^+$ ion, a potential contributor to stratospheric ozone depletion, was isolated in the laboratory.

a. Draw the Lewis structure for $ClO^+$.

b. Using the molecular orbital diagram for $ClO^+$ in Figure P5.121, determine the order of the Cl–O bond in $ClO^+$.

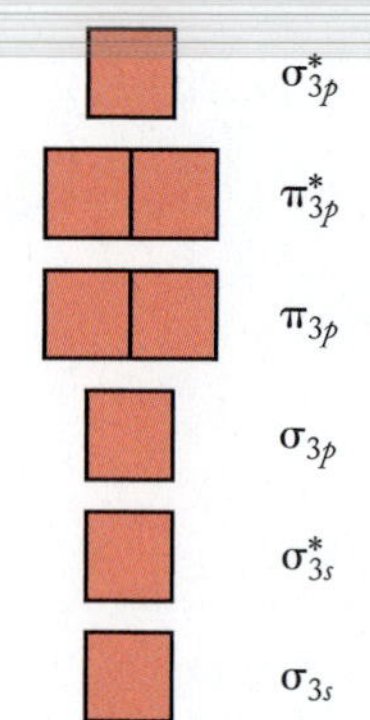


**FIGURE P5.121**

5.122. **Early Earth** Some scientists believe that an anion with the skeletal structure shown in Figure P5.122 may have played a role in forming nucleic acids before life existed on Earth.

a. Complete the Lewis structure of that anion.

b. Predict the C–P–O bond angle in the anion.

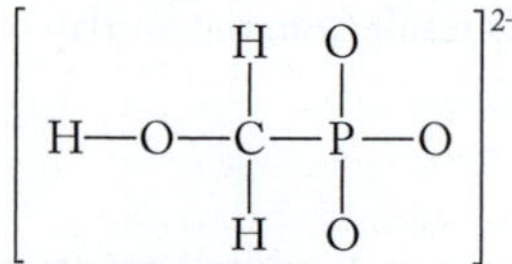


**FIGURE P5.122**

**5.123.** **Cola Beverages** Phosphoric acid imparts a tart flavor to cola beverages. The structure of phosphoric acid is shown in Figure P5.123. Complete the Lewis structure of phosphoric acid and predict the hybridization of the P and O atoms. What is the molecular geometry around the phosphorus atom in your structure?

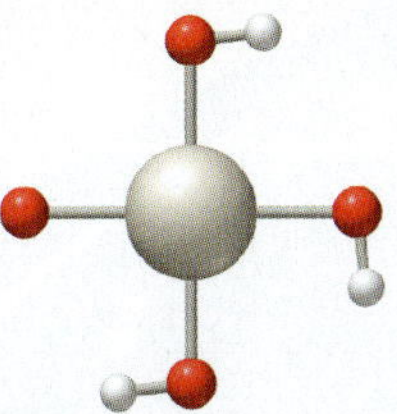

**FIGURE P5.123**

*5.124. Ozone ($O_3$) has a small dipole moment (0.54 D). How can a molecule with only one kind of atom have a dipole moment?

**5.125.** **Rocket Fuel** Hydrazine, $NH_2NH_2$, fueled the spacecraft that delivered the rover *Curiosity* to the surface of Mars in 2012.

a. Draw the Lewis structure of hydrazine.
b. What is the hybridization of its N atoms?
c. Is hydrazine a polar compound? Explain why or why not.

5.126. Two compounds formed by the reaction of boron with carbon monoxide have the following skeletal structures: B—B—C—O and O—C—B—B—C—O.

a. Draw the Lewis structures of both compounds that minimize formal charges.
b. What are the B–B–C bond angles in the molecules?

***5.127.** Boron reacts with NO, forming a compound with the formula BNO.

a. Draw the Lewis structure for BNO, including any resonance forms.
b. Assign formal charges and predict which structure best describes the bonding in that molecule.
c. Predict the molecular geometry of BNO.

5.128. Borazine ($B_3N_3H_6$, a cyclic compound with alternating B and N atoms in the ring) is isoelectronic with benzene ($C_6H_6$). Are delocalized $\pi$ electrons present in borazine?

***5.129.** **Compounds That May Help Prevent Cancer** Broccoli, cabbage, and kale contain compounds that break down in the human body to form isothiocyanates, whose presence may reduce the risk of getting certain cancers. The simplest isothiocyanate is methyl isothiocyanate, $CH_3NCS$.

a. Draw the Lewis structure for methyl isothiocyanate, including all resonance forms. *Hint*: The nitrogen atom is bonded to the methyl (—$CH_3$) group.
b. Assign formal charges and determine which structure is likely to contribute the most to bonding.
c. Predict the molecular geometry of the molecule at both carbon atoms.

*5.130. **Toxic to Insects and People** Methyl thiocyanate ($CH_3SCN$) is used as an agricultural pesticide and fumigant. It is slightly water soluble and is readily absorbed through the skin, but it is highly toxic if ingested. It is toxic in part because it metabolizes to produce the cyanide ion.

a. Draw the Lewis structure for methyl thiocyanate, including all resonance forms.
b. Assign formal charges and predict which structure contributes the most to bonding.
c. Predict the molecular geometry of the molecule at both carbon atoms.

**5.131.** Some chemists think that HArF consists of $H^+$ ions and $ArF^-$ ions. Using an appropriate MO diagram, determine the bond order of the Ar–F bond in $ArF^-$.

*5.132. To model the bonding in $SF_6$ gas, some chemists assume the existence of $SF_4^{2+}$ cations surrounded by two $F^-$ ions.

a. Draw the Lewis structure of a $SF_4^{2+}$ ion.
b. What are the formal charges on S and F in the structure you drew?
c. What is the ion's shape?

***5.133.** Which of the unstable nitrogen oxides $N_2O_2$, $N_2O_5$, and $N_2O_3$ are polar molecules? ($N_2O_2$ and $N_2O_3$ have N—N bonds; $N_2O_5$ does not.)

5.134. Hydrogen atoms have one electron. Does that mean that hydrogen gas is paramagnetic? Why or why not?

**5.135.** Draw the molecular orbital diagram of the valence shell of a $F_2^+$ ion, and use it to determine the bond order in the ion.

5.136. **Antioxidant Protection for Cells** The superoxide ion is a metabolism by-product that would cause cellular damage in the absence of important enzymes that convert the superoxide ion into oxygen, $O_2$, or hydrogen peroxide, $H_2O_2$. Use molecular orbital diagrams to determine the bond order of the peroxide ($O_2^{2-}$) and superoxide ($O_2^-$) ions. Are the bond order values consistent with those predicted from Lewis structures?

***5.137.** Trimethylamine, $(CH_3)_3N$, has a trigonal *pyramidal* structure, whereas trisilylamine, $(SiH_3)_3N$, has a trigonal *planar* geometry. Draw Lewis structures for both compounds consistent with the observed geometries and explain your reasoning.

5.138. Elemental sulfur has several allotropic forms, including cyclic $S_8$ molecules. What is the orbital hybridization of sulfur atoms in the $S_8$ allotrope? The bond angles are about 108°.

**5.139.** Using an appropriate molecular orbital diagram, show that the bond order in the disulfide anion, $S_2^{2-}$, is equal to 1. Is $S_2^{2-}$ diamagnetic or paramagnetic?

# 6

# Intermolecular Forces

## Attractions between Particles

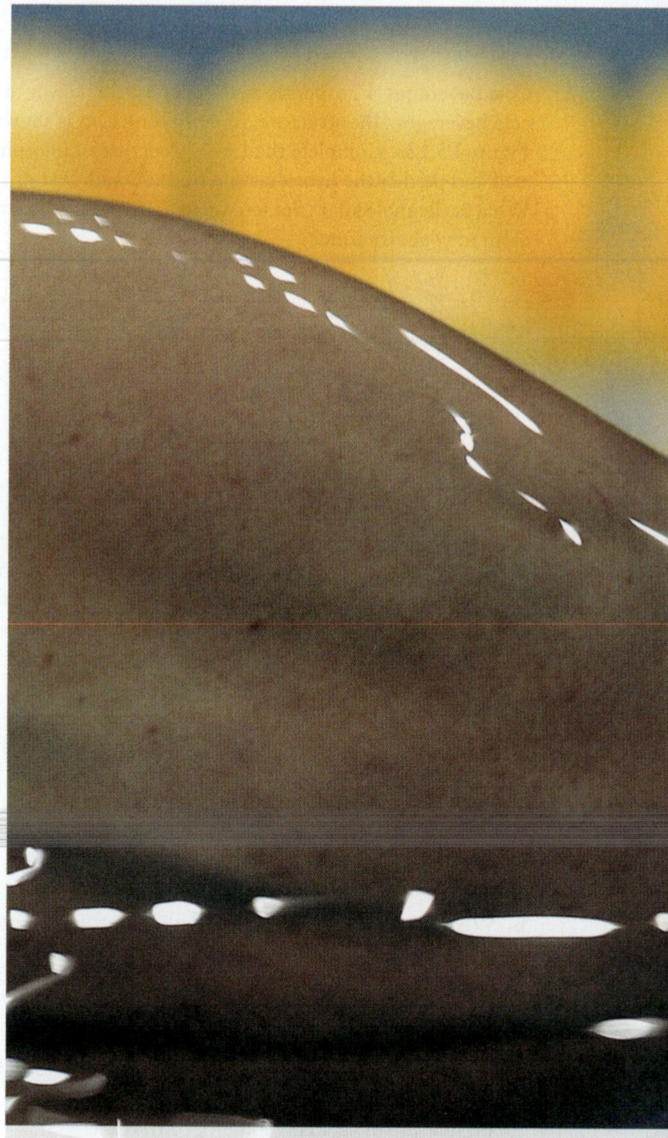

**SURFACE TENSION** A swimmer must break the surface tension of water while doing the backstroke.

### PARTICULATE REVIEW

#### *Polar Bonds versus Polar Molecules*

In Chapter 6, we explore the connections between the structure of molecules and how they interact with each other. Here are representations of four molecules: carbon dioxide, oxygen, water, and ozone.

- Which molecule is polar and contains polar bonds?
- Which molecule is nonpolar despite containing polar bonds?
- Which molecule is polar despite containing nonpolar bonds?

(Review Section 5.3 if you need help.)

*(Answers to Particulate Review questions are in the back of the book.)*

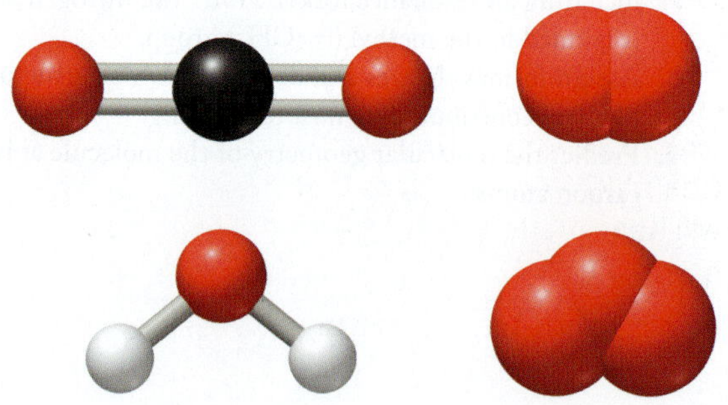

PARTICULATE **PREVIEW**

### *Bonds and Functional Groups*

Each molecule shown here contains one or two carbon–oxygen bonds. As you read Chapter 6, look for ideas that will help you answer these questions:

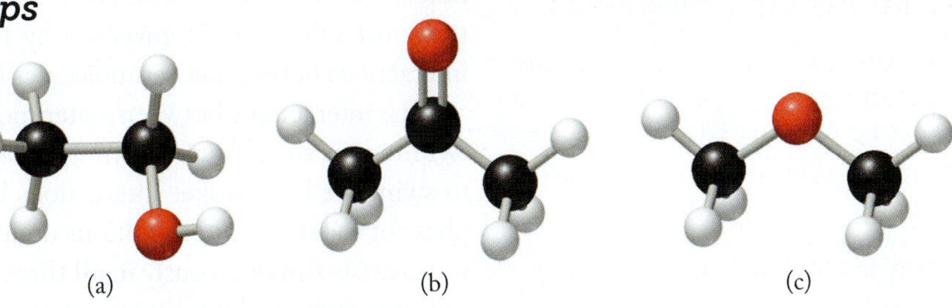


- What is the functional group in each molecule?
- What intermolecular forces exist between molecules of (a)? Between molecules of (b)? Between molecules of (c)?
- Rank these three compounds from lowest to highest boiling point.

## Learning Outcomes

**LO1** Describe the types of intermolecular forces and how they and their strengths are related to the structures of molecules and to the sizes and charges of ions
**Sample Exercise 6.1**

**LO2** Explain how intermolecular forces affect the boiling points of compounds
**Sample Exercise 6.2**

**LO3** Explain why ionic compounds and polar molecular compounds dissolve in polar solvents and why nonpolar molecular compounds dissolve in nonpolar solvents
**Sample Exercises 6.3, 6.4**

**LO4** Identify the regions of a phase diagram and explain how temperature and pressure affect phase changes
**Sample Exercise 6.5**

**LO5** Describe water's unusual properties and how they relate to the hydrogen bonds that water molecules form

# 6.1 Intramolecular Forces versus Intermolecular Forces

In the past two chapters, we have examined bonding in molecules and have explored how the combination of attractive and repulsive electrical forces between pairs of electrons determines molecular geometry. We refer to these interactions *within* a molecule as *intramolecular forces*. In this chapter, we begin the study of the forces that act *between* molecules and *between* molecules and ions. These *intermolecular forces* also are electrostatic, but they are weaker than intramolecular forces and typically act over longer distances. Intermolecular forces have considerable influence on the physical properties of all substances. For example, interactions between solvent and solute particles in solutions affect the solubility of one substance in another.

We often take water for granted because it is such a familiar substance. However, water's physical properties are truly remarkable. They are so essential to life as we know it that astronomers searching for life on other planets look first for the presence of liquid water. Water molecules are polar, which makes water capable of dissolving many substances because of favorable interactions between water molecules and the particles of these substances. Seawater contains both dissolved ionic compounds, such as sodium chloride, and dissolved molecular species, such as oxygen gas. Aquatic life—indeed, all life on Earth—relies on the presence of these and many other substances in salt water, freshwater, and water-based biological fluids, such as blood in animals and sap in plants. Water's ability to dissolve and transport substances is governed by the strength and number of intermolecular interactions between water molecules and other ions and molecules.

The interactions between water molecules are strong enough that the Olympic swimmer in the chapter-opening photo must "break through" the water's surface to swim the backstroke. Interactions between water molecules also influence the phase of water and the conditions under which phase transitions occur. Sometimes water exists simultaneously in all three phases, such as on an early spring day when solid ice melts to liquid water in the sunlight while water vapor condenses into white clouds that dot the sky. The polarity of water influences its physical properties, such as its boiling point, which is the temperature at which liquid vaporizes to gas, and its melting point, the temperature at which solid melts to liquid.

More generally, let's think about the three common states of matter (**Figure 6.1**) and how they differ based on the kinetic energy of the particles in them and their

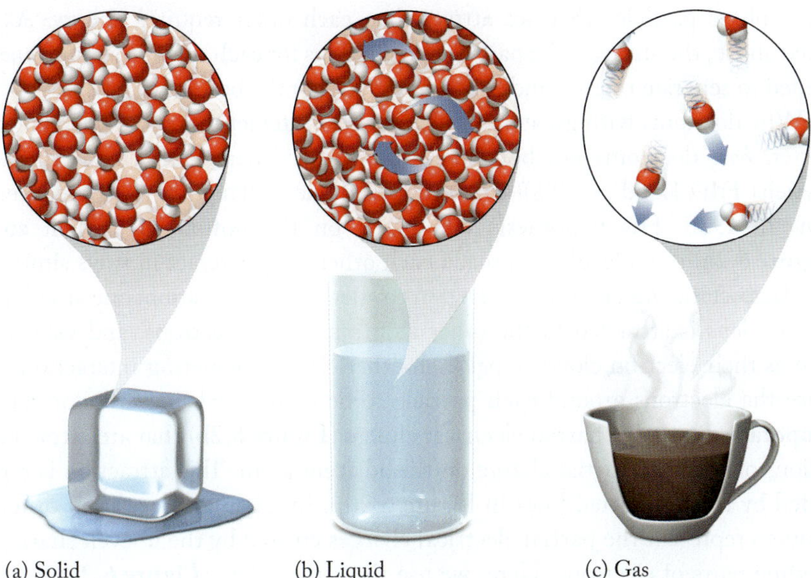


**FIGURE 6.1** (a) The molecules of $H_2O$ in solid ice are locked in place by the strength of intermolecular attractions. (b) In liquid water, molecules of $H_2O$ have more kinetic energy and flow past one another. (c) In the water vapor above a hot cup of coffee, the molecules have enough kinetic energy to overcome nearly all intermolecular attractions and move freely throughout the space they occupy.

ability to overcome the attractive forces between particles. In a solid, the average kinetic energy of the particles is not enough to overcome these forces of attraction. Consequently, these particles have the same nearest neighbors over time and do not move much. In a liquid, the average kinetic energy of the particles is enough to overcome some of the attractive forces; particles in a liquid experience more freedom of motion and can move past one another. In gases, the average kinetic energy of the particles is enough to overcome essentially all the attractive forces between them, imparting nearly complete freedom of motion to the widely separated particles.

**CONNECTION** We defined kinetic energy in Section 1.5 as the energy that an object in motion has because of its mass and its speed.

The stronger the attractive forces among the particles in a solid, the more energy needed to overcome those forces to cause melting or vaporization. Thus, a substance made of particles that interact relatively strongly has high melting and boiling points, which means the substance is likely to be a solid at room temperature and normal pressure. Under the same conditions, a substance with weaker particle–particle interactions has a lower melting point and is more likely to be a liquid. A substance with very weak particle–particle interactions has even lower melting and boiling points and is more likely to be a gas. Let's begin our study of intermolecular forces by examining the types of interactions we observe between atoms and nonpolar molecules.

**TABLE 6.1** Boiling Points of Noble Gases

| Noble Gas | Atomic View | Z | Boiling Point (K) |
|---|---|---|---|
| He | | 2 | 4 |
| Ne | | 10 | 27 |
| Ar | | 18 | 87 |
| Kr | | 36 | 120 |
| Xe | | 54 | 165 |
| Rn | | 86 | 211 |

## 6.2 London Dispersion Forces

The melting and boiling points of substances, among other observable and measurable macroscopic properties, arise from the electrostatic interactions between the particles that make up those substances. Let's see how this happens by looking at a group of elements that exist as single atoms: the noble gases. **Table 6.1** lists their atomic numbers and boiling points. Note the correlation: their boiling points increase as their atomic numbers increase. To understand why this correlation exists, let's think about what happens at the particle level when a liquid vaporizes. The particles in liquids (and solids) are in direct contact with each other (as shown in Figure 6.1). However, when a liquid vaporizes, those contacts are broken: the gas-phase particles become essentially independent. Separating

**CONNECTION** Polar bond formation and the color scales and symbols used to represent it were introduced in Chapter 4.

liquid-phase particles that are attracted to each other requires energy. As we noted above, the stronger the particles' attractions for each other, the more energy needed to separate them—and, in turn, the higher the boiling points.

Why do atoms with greater atomic numbers interact more strongly? For that matter, *how* do atoms not bonded to each other interact? German-American physicist Fritz London (1900–1954) proposed one explanation for these interactions in 1930. His hypothesis was based on the notion that when atoms (**Figure 6.2a**) or molecules approach each other, they interact in ways similar to the electrostatic interactions involved in covalent bond formation: one atom's positive nucleus is attracted to the other atom's negative electrons, and vice versa, even as their electron clouds repel each other. These competing interactions can cause the electrons around each particle to be distributed unevenly, producing **temporary dipoles** of partial electrical charge (**Figure 6.2b**) that are attracted to regions of opposite partial charge on the adjacent atom. This attraction is represented by the red dotted lines in Figure 6.2(b). In Chapter 4, we used different colors to represent the partial electrical charges created by the uneven sharing of bonding pairs of electrons. Here, we use the same colors (**Figure 6.2c**) to show partial electrical charges caused by the uneven distributions of electrons in neutral atoms and over entire molecules.

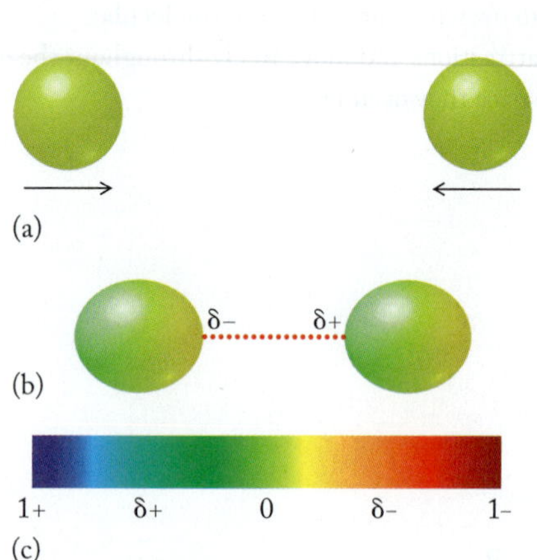


**FIGURE 6.2** (a) Two atoms, each with a symmetrical distribution of electrons, approach each other and (b) create two temporary dipoles as their nuclei and electron clouds interact. (c) The strengths of temporary dipoles are shown with the same color scale used in Chapter 4 to represent the strengths of permanent dipoles in molecules.

The presence of temporary dipoles in atoms and molecules creates a way for them to interact with other atoms and molecules. In a molecule, though, the partial charges in temporary dipoles are likely to be distributed over considerably greater distances than in single atoms because the atomic nuclei and the clouds of electrons shared by groups of atoms within the molecules (or even the entire molecule) interact with the electrons and nuclei in neighboring atoms or molecules.

Interactions based on the presence of temporary dipoles are called **London dispersion forces** (represented by the red dots in Figure 6.2) in honor of Fritz London's pioneering work. The strengths of the interactions increase as the number of electrons in atoms and molecules increases. Because all atoms and molecules have electrons, all atoms and molecules experience London dispersion forces to some degree. The larger the cloud of electrons surrounding a nucleus in an atom or the multiple nuclei in a molecule, the more likely those electrons are to be distributed unevenly, or *polarized*. Greater **polarizability** leads to stronger temporary dipoles and stronger intermolecular interactions, so London dispersion forces become stronger as atoms and molecules become larger.

Polarizability also explains the boiling points of the halogens (**Table 6.2**). These elements exist as diatomic molecules in which equal sharing of the bonding pairs of electrons by identical atoms means that the molecules have no permanent dipoles. Comparing the boiling points of the halogens with their molar masses (as a measure of particle size) reveals a trend that mimics the one we saw with the noble gases: boiling point increases as particle size increases.

A similar trend is also observed in the boiling points of a class of compounds called straight-chain **hydrocarbons**, so named because their molecules are composed of atoms of hydrogen and carbon (**Figure 6.3**). These particular hydrocarbons are called **alkanes** because they are composed of molecules in which each carbon atom is bonded to four other carbon or hydrogen atoms. This means that all the

**TABLE 6.2** Boiling Points of the Halogens

| Halogen | Molecular View | Molar Mass (g/mol) | Boiling Point (K) |
|---|---|---|---|
| $F_2$ | | 38 | 85 |
| $Cl_2$ | | 71 | 239 |
| $Br_2$ | | 160 | 332 |
| $I_2$ | | 254 | 457 |
| $At_2$ | | 420 | 610* |

*Estimated.

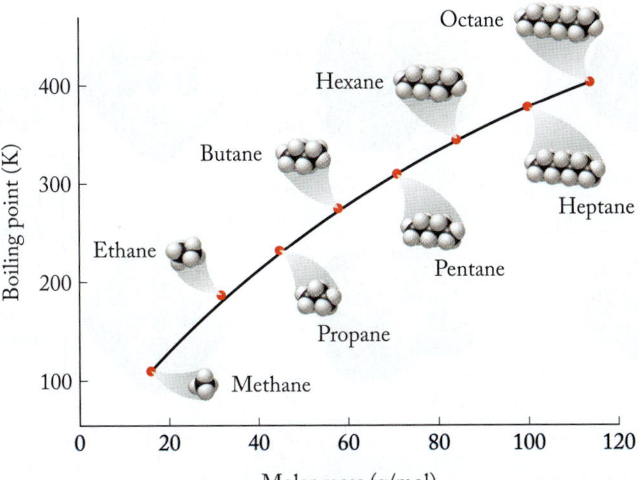


**FIGURE 6.3** Boiling points of $C_1$ to $C_8$ alkanes.

bonds in their molecules are single bonds. The alkanes in Figure 6.3 contain carbon atoms bonded to no more than two other carbon atoms, which gives their molecules a chainlike appearance when each molecule has four or more carbon atoms. London's explanation of the trend in Figure 6.3 was the same as for the noble gases and the halogens: larger clouds of increasing numbers of electrons per molecule are more polarizable. Greater polarizability means that these molecules are more likely to form temporary dipoles that attract other molecules in the liquid phase. Again, vaporizing a liquid requires separating liquid-phase particles that are attracted to each other, which requires energy, and the stronger the particles' attractions for each other, the more energy needed to separate them. Greater energy requirements means higher boiling points.

CONCEPT **TEST**

Explain why, at room temperature, $CF_4$ is a gas but $CCl_4$ is a liquid.

*(Answers to Concept Tests are in the back of the book.)*

## The Importance of Shape

Molecular shape, as well as molecular mass, helps determine the strength of London dispersion forces. The three hydrocarbons depicted in **Figure 6.4** all have the same molecular formula, $C_5H_{12}$, and molar mass, 72 g/mol, but the locations of their bonds and their molecular shapes are different. Compounds such as these, with the same molecular formula but different connections between the atoms in their molecules, are called **constitutional isomers**, or *structural* isomers. Constitutional isomers have different physical and chemical properties despite their shared molecular formulas.

For example, all three of the $C_5$ (five carbon atoms per molecule) hydrocarbons in Figure 6.4 are nonpolar, which means that none of them has a permanent dipole. Therefore, the only intermolecular forces they experience are London dispersion forces (the red dotted lines in Figure 6.4) caused by temporary dipoles. Both pentane and 2,2-dimethylpropane have the same number of electrons. However, the carbon atoms in pentane form a straight chain, which gives the molecule the overall shape of a stubby piece of chalk. As a result of this shape, the pentane molecule has a larger surface area than that of 2,2-dimethylpropane and thus experiences more interactions with other molecules of the same shape.

**temporary dipole** the separation of charge produced in an atom or molecule by a momentary uneven distribution of electrons; also called *induced dipole*.

**London dispersion force** an intermolecular force between atoms or molecules caused by the presence of temporary dipoles in the molecules.

**polarizability** the relative ease with which the electron cloud in a molecule, ion, or atom can be distorted, inducing a temporary dipole.

**hydrocarbon** an organic compound whose molecules contain only carbon and hydrogen atoms.

**alkane** a hydrocarbon in which each carbon atom is bonded to four other carbon or hydrogen atoms.

**constitutional isomer** one of a set of compounds with the same molecular formula but different connections between the atoms in their molecules; also called *structural isomer*.

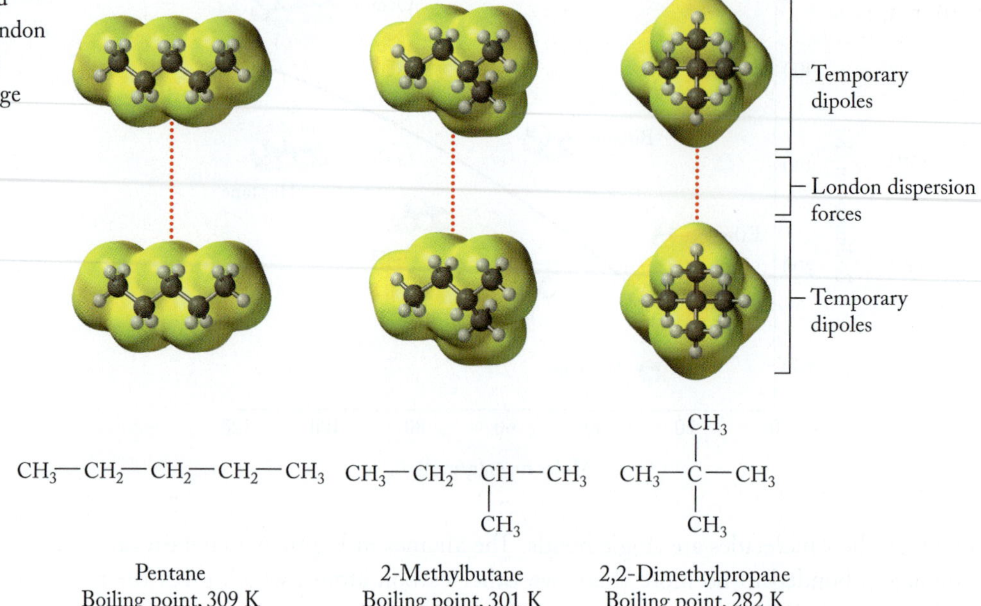


**FIGURE 6.4** Molecular structures and boiling points of three $C_5$ isomers. London dispersion forces (red dotted lines) are strongest for pentane because of its large surface area.

**CONNECTION** Isomers are introduced and defined in Section 5.6.

Molecules of 2,2-dimethylpropane are shaped more like lumpy spheres, have a smaller surface-area-to-volume ratio, and interact less strongly with adjacent molecules. Weaker London dispersion forces explain why 2,2-dimethylpropane has a lower boiling point than that of pentane. The remaining molecule, 2-methylbutane, is neither as straight as pentane nor as spherical as 2,2-dimethylpropane. Therefore, 2-methylbutane boils at a temperature between the boiling points of the other two. In general, when molecules with similar molar masses and chemical features are compared, the ones with more branching in their structures have lower boiling points.

**CONCEPT TEST**

Predict which compound has the higher boiling point in each of the following pairs: (a) $CH_3Cl$ or $CH_3Br$; (b) $CH_3CH_2OH$ or $CH_3OH$; (c) $(CH_3CH_2)_3N$ or $(CH_3)_3N$.

## Viscosity

In addition to boiling point, another property of liquids that is influenced by London dispersion forces is viscosity. The **viscosity** of a fluid (a fluid is a liquid or a gas—anything that can flow) is a measure of its resistance to flow. Viscous, or "thick," fluids such as honey, molasses, and the oil used to lubricate engines have high resistances to flow. Low-viscosity substances such as water and gasoline flow or pour easily. Gases have ultralow viscosities several orders of magnitude below those of the lowest-viscosity liquids.

**FIGURE 6.5** Using a Zahn cup to determine viscosity.

**viscosity** a measure of a fluid's resistance to flow.

Many ways exist to measure viscosity and to express the results of those measurements. One of the simplest measurements uses a device called a Zahn cup (**Figure 6.5**), a small metal cup with a hole in the bottom. The cup is dipped into a liquid sample and then lifted out. The time the liquid takes to drain through the hole is measured and converted into units of dynamic viscosity. (Although other types exist, *dynamic* viscosity is the one most widely used in chemistry.) A common unit for expressing viscosity is the *centipoise* (cP), where 1.00 cP is the viscosity of pure water at 20°C.

**TABLE 6.3** Viscosities of Some Liquid Alkanes

| Compound | Molecular Structure | Molar Mass (g/mol) | Viscosity at 20°C (cP) |
|---|---|---|---|
| Hexane ($C_6H_{14}$) | | 86 | 0.29 |
| Octane ($C_8H_{18}$) | | 114 | 0.54 |
| Decane ($C_{10}H_{22}$) | | 142 | 0.92 |
| Dodecane ($C_{12}H_{26}$) | | 170 | 1.34 |
| Hexadecane ($C_{16}H_{34}$) | | 226 | 3.34 |

**Table 6.3** lists the viscosities of a group of liquid alkanes with between 6 (hexane) and 16 (hexadecane) carbon atoms per molecule. Notice how the viscosities increase with increasing molecular size. Molecules of hexadecane experience the greatest London dispersion forces in this group because they are the largest molecules. Stronger intermolecular attractions mean, in turn, that they do not slip past each other as easily as those experiencing weaker intermolecular attractions. More resistance to flow means higher viscosity.

# 6.3 Interactions Involving Polar Molecules

In Section 5.3, we saw how the unequal distributions of bonding pairs of electrons result in partial negative charges on some bonded atoms and partial positive charges on others. When bond dipoles are arranged asymmetrically within molecules, the molecules themselves have permanent dipoles. Molecules with permanent dipoles can interact with each other and with ions in ionic compounds. Such interactions are much weaker than the strengths of ionic or covalent bonds, but they are strong enough that they warrant consideration in addition to the London dispersion forces that all molecules experience.

**STEPWISE ANIMATION**
Interactions Involving Polar Molecules

## Dipole–Dipole Interactions

Let's consider two compounds having the same molar masses (58 g/mol) and having electron clouds with similar shapes: methylpropane and 2-propanone (commonly called acetone). Both molecules are shown in **Figure 6.6**. They should experience similar London dispersion forces, yet the boiling point of acetone (329 K) is almost 70 K higher than that of methylpropane (261 K). Why?

The reason has to do with the differences in the polarities of these two molecules. Methylpropane is a nonpolar hydrocarbon, but acetone has an overall dipole moment of 2.9 D because of its double bond between the highly

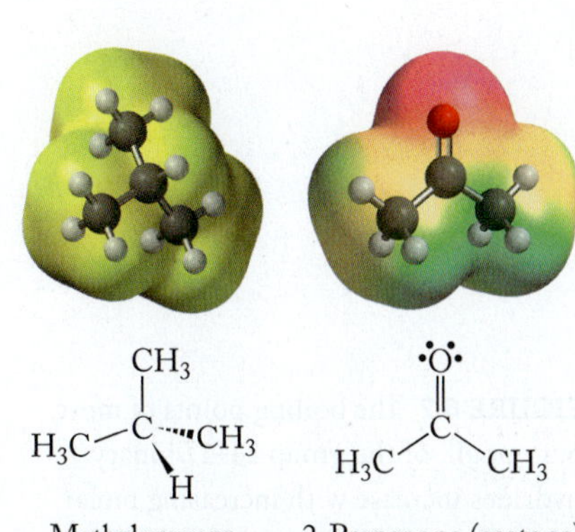


Methylpropane
Boiling point, 261 K

2-Propanone (acetone)
Boiling point, 329 K

**FIGURE 6.6** Molecular structures and boiling points of methylpropane and acetone.

**dipole–dipole interaction** an attraction between regions of polar molecules that have partial charges of opposite sign.

**organic compound** A molecule containing carbon atoms whose structure typically consists of carbon–carbon bonds and carbon–hydrogen bonds and that may include one or more heteroatoms such as oxygen, nitrogen, sulfur, phosphorus, or the halogens.

**functional group** a group of atoms in an organic compound's molecular structure that imparts characteristic physical and chemical properties.

**carbonyl group** a functional group that consists of a carbon atom with a double bond to an oxygen atom.

**ketone** an organic compound that contains a carbonyl group bonded to two other carbon atoms.

**hydrogen bond** a strong dipole–dipole interaction, which occurs between a hydrogen atom bonded to a N, O, or F atom and another N, O, or F atom.

electronegative ($\chi$ = 3.5) oxygen atom and the less electronegative ($\chi$ = 2.5) carbon atom. Molecules with permanent dipole moments are polar, and they can interact with each other through a second kind of intermolecular force called **dipole–dipole interactions**. These interactions occur between acetone molecules when the partial negative charges on the oxygen atoms of some molecules are attracted to the partial positive charges on other acetone molecules. The strength of these interactions must be overcome to vaporize molecules of liquid acetone, and these interactions are why acetone's boiling point is higher than methylpropane's.

The C═O bond constitutes one of the most common functional groups in **organic compounds**—molecules that always contain carbon atoms, almost always contain hydrogen atoms, and often contain oxygen, nitrogen, sulfur, phosphorus, and/or the halogens. A **functional group** is a particular group of atoms in a compound's molecular structure that significantly affects the compound's physical and chemical properties. The C═O functional group in an organic compound such as acetone is called a **carbonyl group**. When the carbon atom of a carbonyl group in a molecule is bonded to two other carbon atoms, the compound is called a **ketone**. Acetone is the ketone with the simplest molecular structure and smallest molar mass.

## Hydrogen Bonds

The graph in **Figure 6.7** compares the boiling points of compounds whose molecules each contain one atom of a group 14, 15, 16, or 17 element bonded to enough hydrogen atoms to have a complete valence-shell octet. Most of the boiling points increase with increasing molar masses, as we saw with nonpolar hydrocarbons in Figure 6.3. However, the boiling points of the lowest-molar-mass compounds of groups 15, 16, and 17—namely, $NH_3$, $H_2O$, and HF—are unusually high in comparison with the others in their series. To understand why, we need to focus on the polar bonds formed between H atoms and N, O, or F atoms. The H atoms share just two electrons, and when they are shared with a highly electronegative atom such as N, O, or F, the electron density on the surface of the H atom is reduced so much that the H atoms interact strongly with electronegative N, O, and F atoms on neighboring molecules.

**CONNECTION** In Section 5.3, we learned that permanent dipole moments are experimentally measured values that can be expressed in debyes (D).

The resulting high-strength dipole–dipole interactions between molecules—involving a H atom on one molecule and a N, O, or F atom on another

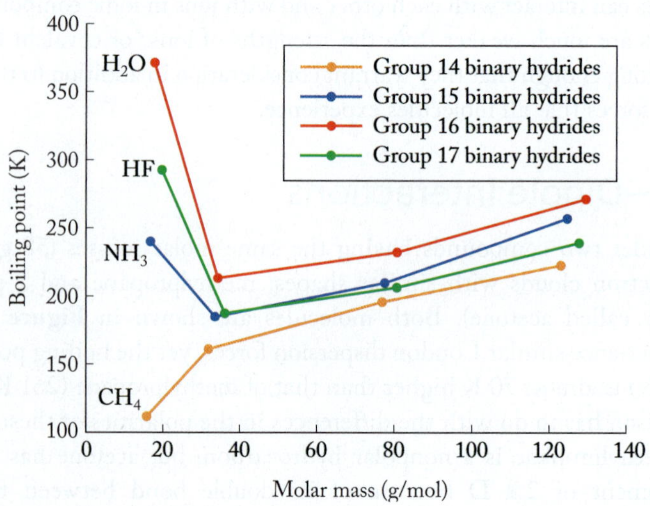


**FIGURE 6.7** The boiling points of most, but not all, of the group 14–17 binary hydrides increase with increasing molar mass. The boiling points of $H_2O$, HF, and $NH_3$ are much higher than expected because of hydrogen bonding between their molecules.

molecule—are called **hydrogen bonds** and are represented by the blue dotted lines in **Figure 6.8**. Hydrogen bonds are the strongest dipole–dipole interactions and can be nearly one-tenth the strength of some covalent bonds.

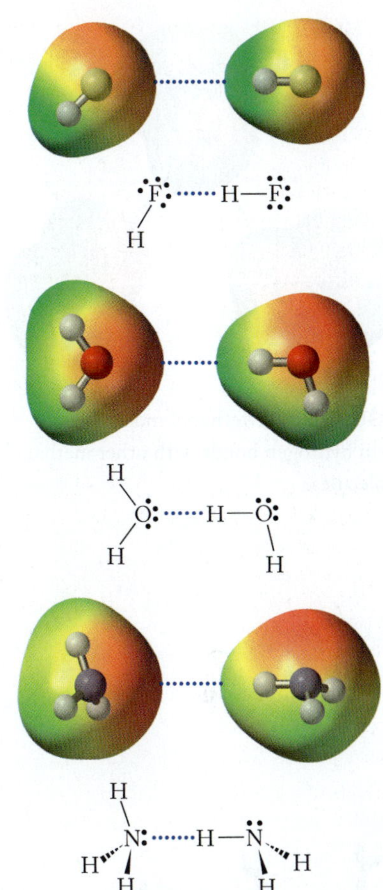

**FIGURE 6.8** Hydrogen bonds (blue dotted lines) occur between hydrogen atoms bonded to F, O, or N atoms in one molecule and F, O, or N atoms in adjacent molecules.

### SAMPLE EXERCISE 6.1 Identifying Intermolecular Forces — LO1

Many indoor swimming pools use some form of chlorine as the principal disinfectant to maintain the health and safety of their facility. Chemical reactions between the chlorine and urea (in urine) lead to the formation of several compounds, including chloramine ($NH_2Cl$) and trichloramine ($NCl_3$). The latter compound accounts for the distinctive odor of most indoor pools. Identify the intermolecular forces between molecules of (a) diatomic chlorine, $Cl_2$; (b) $NH_2Cl$; and (c) $NCl_3$.

**Collect, Organize and Analyze** We need to identify the intermolecular forces acting between molecules of three compounds. All molecules experience London dispersion forces, but we will need their molecular structures and geometries to determine whether dipole–dipole interactions are present. Finally, we need to consider whether the structure of any of the molecules would result in the presence of hydrogen bonding.

**Solve** The Lewis structures for $Cl_2$, $NH_2Cl$, and $NCl_3$ are:

Cl—Cl    H—N(—H)—Cl    Cl—N(—Cl)—Cl

Diatomic chlorine has a linear geometry and is nonpolar, so molecules of $Cl_2$ experience only London dispersion forces. Both $NH_2Cl$ and $NCl_3$ have trigonal pyramidal geometries with a lone pair of electrons on the nitrogen atom. Because they are polar molecules, both compounds experience dipole–dipole forces in addition to London dispersion forces. In addition, chloramine has N—H bonds, so it can form hydrogen bonds between the hydrogen atoms on one molecule of $NH_2Cl$ and the partial negative charge on the nitrogen atom of a neighboring molecule.

**Think About It** A molecule's geometry contributes to its polarity and hence to the presence or absence of dipole–dipole forces. Hydrogen bonding can occur between molecules of the same compound if N—H, O—H, or F—H bonds are present in the molecule.

**Practice Exercise** The chemistry of pool water can be complex. Some pool operators use bromine in addition to chlorine as part of their water treatment protocol, which leads to the formation of small amounts of chloroform ($CHCl_3$), bromoform ($CHBr_3$), dichlorobromomethane ($CHCl_2Br$), and dibromochloromethane ($CHClBr_2$). Identify the intermolecular forces present between molecules of each compound.

*(Answers to Practice Exercises are in the back of the book.)*

### CONCEPT TEST

Both $CH_3F$ and HF contain one fluorine atom and at least one hydrogen atom. Explain why only one of these compounds forms hydrogen bonds with molecules of itself.

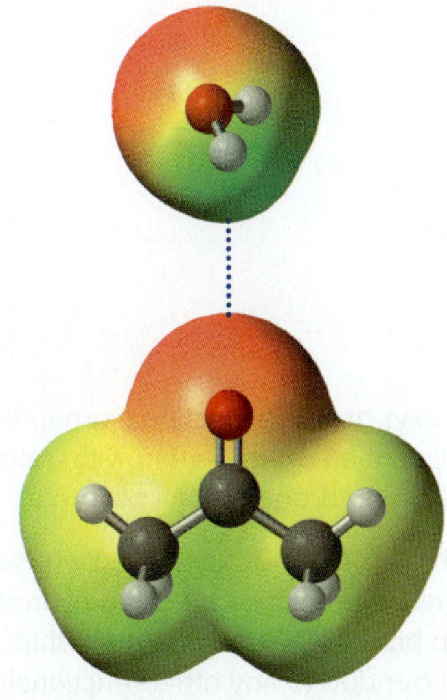

**FIGURE 6.9** In solutions of acetone in water, hydrogen bonds form between the hydrogen atoms of water molecules and the oxygen atoms in acetone molecules.

Hydrogen bonds also can form between molecules of different substances—even when one of them has no H atom bonded to a N, O, or F atom. For example, when acetone dissolves in water, hydrogen bonds form between the H atoms of water molecules and the O atoms of acetone molecules, as shown in **Figure 6.9**.

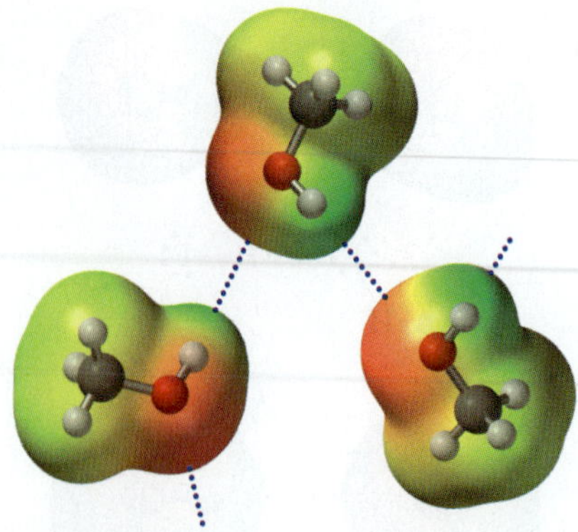

**FIGURE 6.10** Methanol molecules can form hydrogen bonds with other methanol molecules.

These interactions happen even though the H atoms in acetone are bonded to C atoms and no hydrogen bonds are present between molecules of acetone. The key to hydrogen bonds forming between acetone and water molecules is the negative partial charge on the electron-rich O atoms in molecules of acetone, as indicated by the red color on the electrostatic potential map of acetone in Figure 6.9, and the positive partial charge indicated by the blue-green color on the electron-poor H atoms in molecules of $H_2O$.

To consider another perspective on the importance of hydrogen bonding, let's examine the boiling points and intermolecular forces between molecules of ethane, formaldehyde, and methanol. The molar masses of these three organic compounds are nearly the same, so we might expect the strengths of their London dispersion forces and their boiling points to be similar. **Table 6.4** shows electrostatic potential maps of their molecular structures and lists some of their properties. Note that the boiling point of formaldehyde is 70 K higher than that of ethane. We can explain this because the carbonyl functional group in formaldehyde gives the molecule a dipole moment of 2.33 D. As a result, dipole–dipole interactions occur between formaldehyde molecules—interactions that do not exist between nonpolar ethane molecules. However, the difference between the boiling points of ethane and methanol is even greater, 154 K, even though the dipole moment of methanol is less than that of formaldehyde. Methanol's boiling point is so much higher because the –OH group leads to the formation of hydrogen bonds between methanol molecules, as represented by the blue dotted lines in **Figure 6.10**. The presence of the –OH functional group, called a **hydroxyl group**, indicates that methanol is a member of the class of organic compounds called **alcohols**. The physical and chemical properties of alcohols are closely linked to the hydroxyl groups in their structures and their capacity to form hydrogen bonds. Sample Exercise 6.2 introduces another functional group called an **ether** group that consists of an oxygen atom with two single bonds to two carbon atoms.

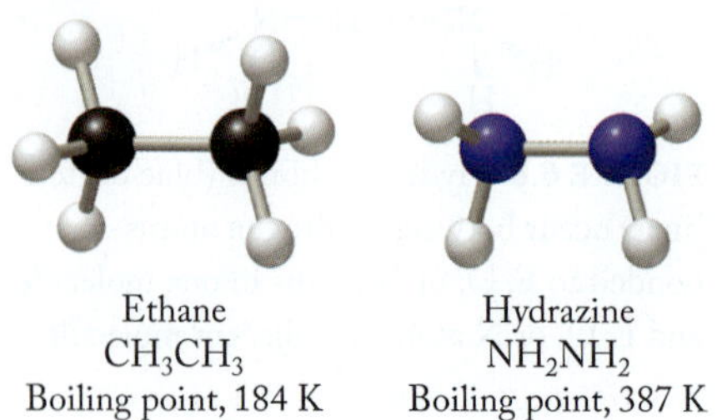


**FIGURE 6.11** Ethane and hydrazine have similar structures but different boiling points.

## CONCEPT TEST

Ethane and hydrazine have similar molecular structures (**Figure 6.11**), yet the boiling point of hydrazine (387 K) is more than 200 K greater than that of ethane (184 K). What intermolecular interactions account for this huge difference in boiling points?

**TABLE 6.4** Some Properties of Ethane, Formaldehyde, and Methanol

| | Ethane | Formaldehyde | Methanol |
|---|---|---|---|
| Formula | $CH_3CH_3$ | $CH_2O$ | $CH_3OH$ |
| Structure | | | |
| $\mathcal{M}$ (g/mol) | 30.0 | 30.0 | 32.0 |
| Dipole Moment (D) | 0.00 | 2.33 | 1.69 |
| Boiling Point (K) | 184 | 254 | 338 |

**hydroxyl group** a functional group that consists of an oxygen atom with a single bond to a hydrogen atom.

**alcohol** an organic compound whose molecular structure includes a hydroxyl group bonded to a carbon atom that is not bonded to any other functional group(s).

**ether** an organic compound that contains an oxygen atom with single bonds to two carbon atoms.

## SAMPLE EXERCISE 6.2 Explaining Differences in Boiling Point LO2

Dimethyl ether ($C_2H_6O$) has a molar mass of 46 g/mol and a boiling point of 248 K. Ethanol has the same chemical formula and molar mass but a boiling point of 351 K. Explain the difference in boiling point. Their structures are shown in **Figure 6.12**.

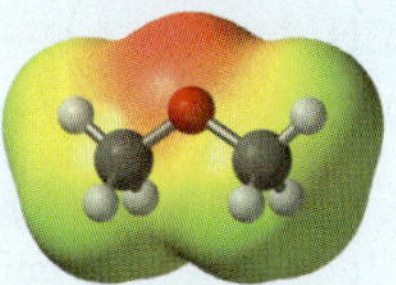

Dimethyl ether
$CH_3OCH_3$
Boiling point, 248 K

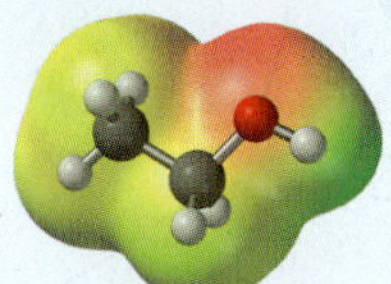

Ethanol
$CH_3CH_2OH$
Boiling point, 351 K

**FIGURE 6.12** Dimethyl ether and ethanol have the same molar mass but different boiling points.

**Collect, Organize and Analyze** We need to explain the large difference between the boiling points of two compounds that have the same molar mass. Molecules of both compounds contain polar groups. The –OH group in the ethanol molecule is polar because of unequal sharing of the bonding pair of electrons between O and H. The C—O—C group in dimethyl ether is polar not only because of the bond dipoles between C and O but also because the bond angle between the three atoms should be about 109°, given the $sp^3$ hybridization of the O atom. The resulting asymmetry in each molecule gives both dimethyl ether and ethanol permanent dipoles. The electrostatic potential maps of the compounds indicate that both have overall dipole moments. In addition, hydrogen bonds can form between the –OH groups on adjacent ethanol molecules, as shown in **Figure 6.13**.

**Solve** The same molar masses and similar shapes suggest that molecules of both compounds should experience similar London dispersion forces. The difference in their boiling points results from hydrogen bonds—particularly strong dipole–dipole interactions—between molecules of ethanol. These interactions would be stronger than the dipole–dipole interactions between molecules of dimethyl ether. Because overcoming those stronger interactions requires more energy, ethanol has a higher boiling point.

**Think About It** Dimethyl ether and ethanol are constitutional isomers, yet their boiling points differ by more than 100 K—a difference due to the hydrogen bonding between the –OH groups present in molecules of ethanol but not present in molecules of dimethyl ether.

**Practice Exercise** Isopropanol is the compound commonly known as rubbing alcohol. Its boiling point is 355 K. Ethylene glycol, the principal ingredient in automotive antifreeze, has about the same molar mass and boils at 469 K. Why do these substances (**Figure 6.14**) have such different boiling points?

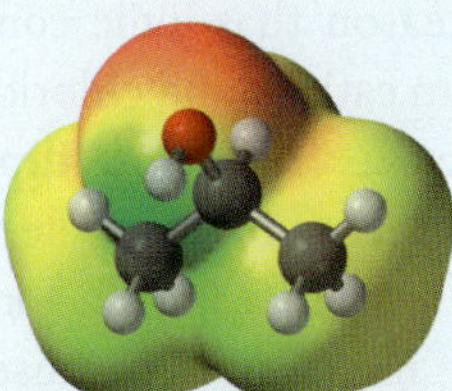

Isopropanol
$CH_3CH(OH)CH_3$
Boiling point, 355 K

Ethylene glycol
$HOCH_2CH_2OH$
Boiling point, 469 K

**FIGURE 6.14** Isopropanol and ethylene glycol have similar molar masses but different boiling points.

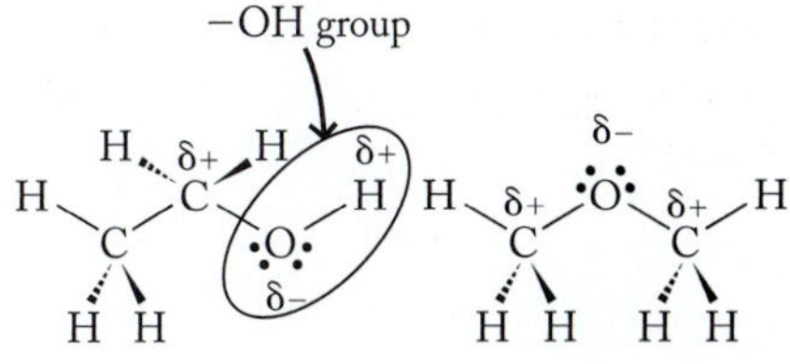


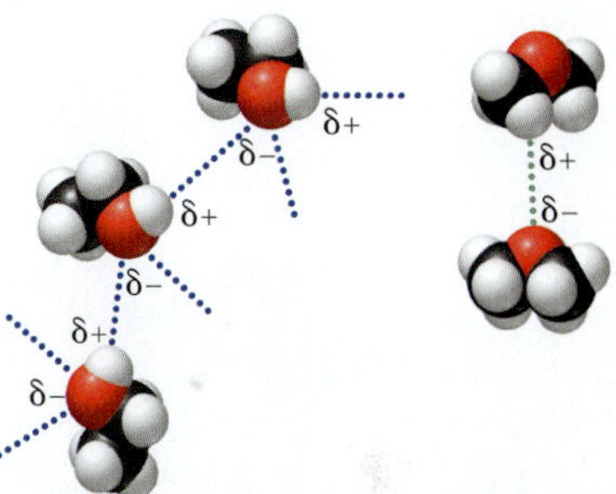


**FIGURE 6.13** The polar –OH group in ethanol leads to hydrogen bonding (blue dotted lines) between its molecules. Dimethyl ether does not form hydrogen bonds but does experience dipole–dipole interactions, which are represented by the green dots in Figure 6.13.

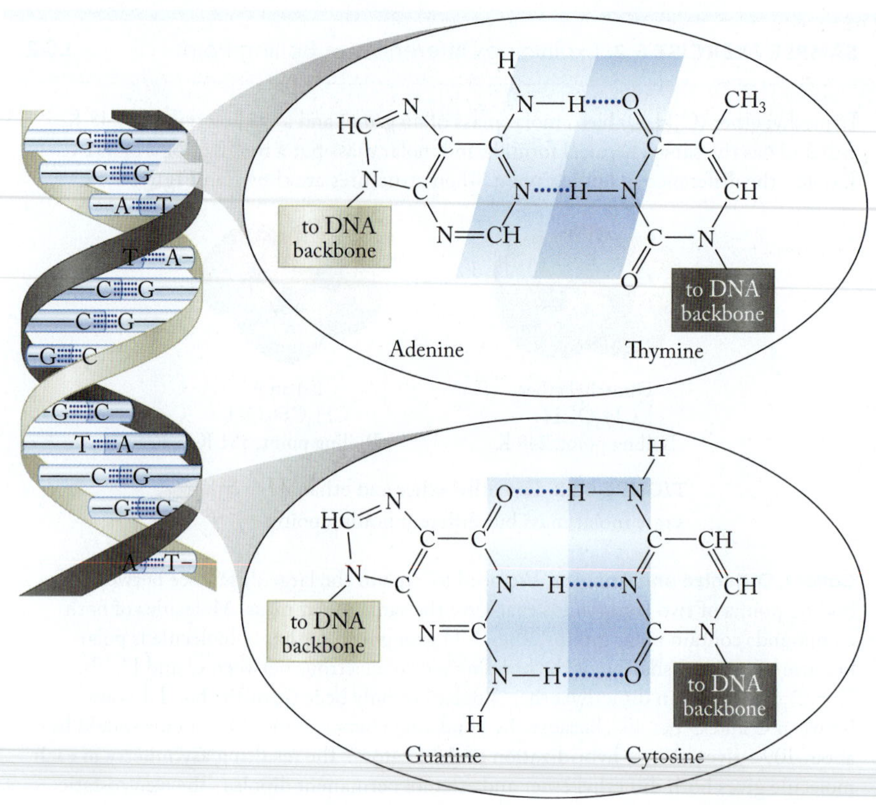


**FIGURE 6.15** Hydrogen bonds (blue dotted lines) that occur between hydrogen atoms and either nitrogen or oxygen atoms in adjacent strands of DNA stabilize the double-helix structure. The two detailed views show the four building blocks of DNA: adenine (A), thymine (T), guanine (G), and cytosine (C).

Hydrogen bonding also plays a role in defining the shapes of very large molecules, especially those of biological interest such as proteins and DNA. Proteins are so large that their long chains of atoms fold back and wrap around themselves, allowing atoms in the protein to hydrogen-bond with other atoms in the protein in much the same way that atoms can hydrogen-bond in adjacent molecules. The double strands of DNA form a three-dimensional shape called a double helix, in which pairs of the molecular building blocks of DNA, called nucleotides, form hydrogen bonds that keep the strands linked together. Pairs of nucleotides named adenine (A) and thymine (T) on adjacent DNA strands form two hydrogen bonds, whereas guanine (G) and cytosine (C) form three hydrogen bonds (**Figure 6.15**).

## Ion–Dipole Interactions

Water is sometimes called "*nature's solvent*" because many ionic solids and polar molecular compounds dissolve in it. One reason that ionic compounds such as NaCl are so soluble in water is that the sodium cations and chloride anions interact with the permanent dipoles of water molecules. These **ion–dipole interactions** (represented by the purple dots in **Figure 6.16**) compete with the ion–ion (coulombic) interactions holding the ions in a lattice. Ion–dipole interactions are stronger than dipole–dipole interactions because an ion involved in an ion–dipole interaction has completely lost or gained one or more electrons and has a charge of at least 1+ or 1−, whereas dipole–dipole interactions involve atoms with only partial charges. **Table 6.5** summarizes the relative strengths of ion–dipole interactions and the other intermolecular forces discussed in this chapter (including *dipole–induced dipole interactions*, which we discuss in the next section).

**FIGURE 6.16** Ion–dipole interactions. The hydrogen atoms (positive poles) of $H_2O$ molecules are attracted to the $Cl^-$ ions of NaCl. Similarly, the oxygen atoms (negative poles) of $H_2O$ are attracted to the $Na^+$ cations.

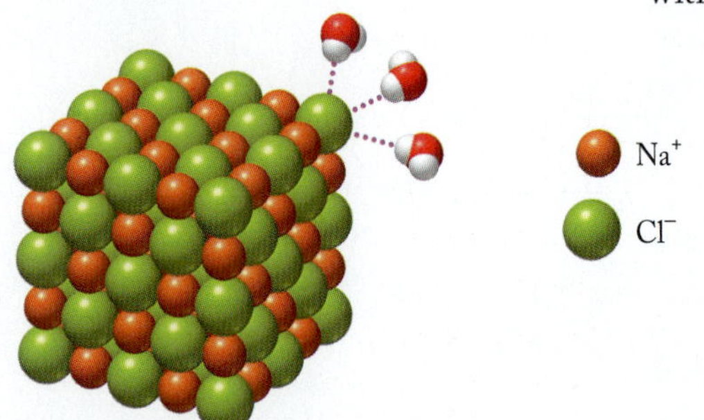

**TABLE 6.5** Relative Strengths of Intermolecular Forces and Some Phenomena They Explain

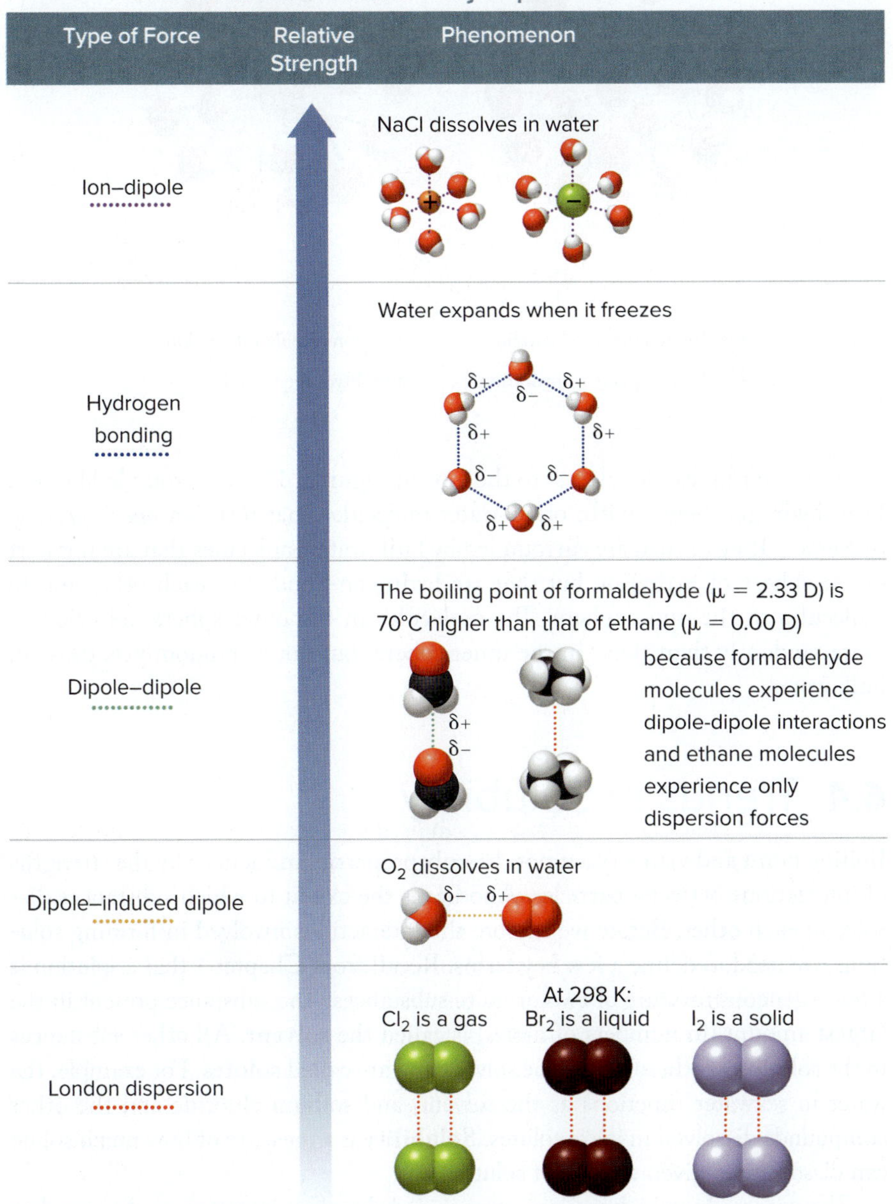

| Type of Force | Relative Strength | Phenomenon |
|---|---|---|
| Ion–dipole | (strongest) | NaCl dissolves in water |
| Hydrogen bonding | | Water expands when it freezes |
| Dipole–dipole | | The boiling point of formaldehyde ($\mu$ = 2.33 D) is 70°C higher than that of ethane ($\mu$ = 0.00 D) because formaldehyde molecules experience dipole-dipole interactions and ethane molecules experience only dispersion forces |
| Dipole–induced dipole | | $O_2$ dissolves in water |
| London dispersion | (weakest) | At 298 K: $Cl_2$ is a gas; $Br_2$ is a liquid; $I_2$ is a solid |

Intermolecular Forces

As an ion is removed from its neighbors in the lattice, it becomes surrounded by a cluster of water molecules that forms a **sphere of hydration** (**Figure 6.17**). The dissolved ions are said to be *hydrated*. The strengths of these multiple ion–dipole interactions (represented by the green dashed lines in Figures 6.16 and 6.17) help overcome the lattice energy—that is, the energy of the ionic bonds that hold ions together in crystalline ionic solids. Within a sphere of hydration, the water molecules closest to the ion are oriented so that their oxygen atoms (negative poles) are directed toward cations and their hydrogen atoms (positive poles) are directed toward anions, as shown for $Na^+$ and $Cl^-$ ions in Figure 6.17. The number of water molecules in the *inner sphere of hydration* can vary; six water molecules hydrate most ions, but the number can range from four to nine. When a substance dissolves in a liquid other than water, the cluster of host molecules is called a *sphere of solvation*, and the dissolved particles are said to be *solvated*.

**CONNECTION** In Section 4.1, we defined lattice energy as the energy released when gas-phase cations and anions form one mole of an ionic solid.

**ion–dipole interaction** an attractive force between an ion and a molecule that has a permanent dipole.

**sphere of hydration** the cluster of water molecules surrounding an ion in an aqueous solution.

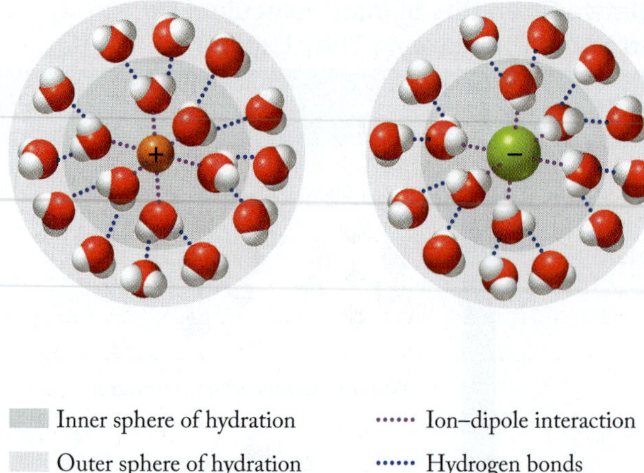


**FIGURE 6.17** Each hydrated $Na^+$ and $Cl^-$ ion in a solution of NaCl is surrounded by six water molecules oriented toward the center ion as a result of ion–dipole interactions (purple dotted lines). Those water molecules make up an inner sphere of hydration. Water molecules in an outer sphere of hydration are oriented as a result of hydrogen bonds (blue dotted lines) with molecules in the inner sphere.

The water molecules closest to the ions in Figure 6.17 are surrounded by, and form hydrogen bonds with, other water molecules that form an *outer sphere of hydration*. They in turn are surrounded by bulk water molecules that are not part of the sphere of hydration but that are hydrogen-bonded to each other and to molecules in the outer sphere. The molecules in the outer sphere are oriented more randomly than those in the inner sphere, but not as randomly as those in bulk water.

## 6.4 Trends in Solubility

Boiling point and viscosity are not the only properties influenced by the strengths of interactions between particles. Another is the extent to which substances dissolve in each other. Before we explore the interactions involved in forming solutions, we need to define a few key terms. Recall from Chapter 1 that a solution is a homogeneous mixture of two or more substances. The substance present in the largest amount (in numbers of moles) is called the **solvent**. All other substances in the solution are dissolved in the solvent and are called **solutes**. For example, the water in seawater functions as the solvent, and sodium chloride and the other compounds dissolved in it are solutes. **Solubility** is a measure of how much solute can dissolve in a given volume of solution.

The process by which an ionic compound dissolves in water, as discussed in Section 6.3, involves a competition between the ion–dipole interactions and the lattice energy of the ionic compound. Other processes influenced by the strengths of intermolecular forces include dissolving a gas in a liquid or dissolving one liquid in another. For example, low-molar-mass alcohols such as methanol, isopropanol, and ethylene glycol dissolve in water because of the hydrogen bonds that form between the –OH groups in the alcohols and in water. The strengths and numbers of the hydrogen bonds, as shown for ethylene glycol in **Figure 6.18**, help these alcohols to dissolve in water and help water to dissolve them in all proportions. When two liquid compounds have unlimited solubility in each other, chemists say that they are **miscible** with each other. However, when two liquid compounds have only limited solubility in each other, they are said to be immiscible. Gasoline is immiscible with water.

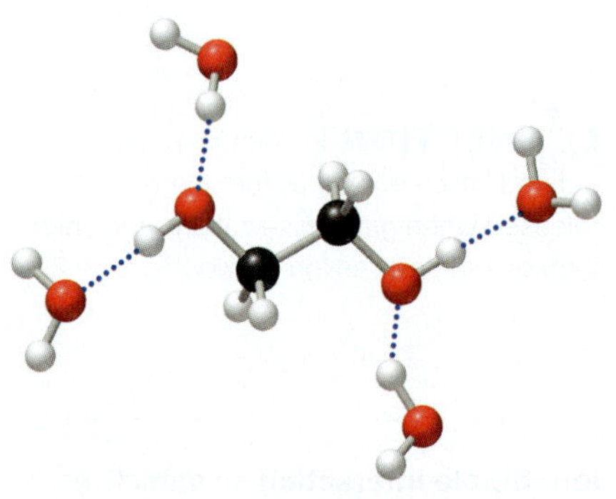

**FIGURE 6.18** A single molecule of ethylene glycol ($HOCH_2CH_2OH$) can form multiple hydrogen bonds. The strength of these solute–solvent interactions makes ethylene glycol miscible with water.

To predict whether a given solute is likely to be highly soluble in a given solvent or only slightly soluble, we need to consider the strengths and numbers of the interactions among solute particles and the strengths of the interactions among solvent particles. The stronger those interactions are, the harder it will be for the solute particles to separate from one another and dissolve in the solvent, and the harder it will be for the solvent particles to separate from one another to accommodate solute particles. Conversely, the stronger the interactions are between solute and solvent molecules, the easier it will be for the solute to dissolve.

Let's discuss the solute–solvent interactions first because they help drive the dissolution process. Polar solutes tend to dissolve in polar solvents due to strong dipole–dipole interactions between molecules of solute and solvent—especially when hydrogen bonds form between the solute and solvent molecules (as when ethylene glycol or methanol dissolves in water). By contrast, nonpolar solutes do not dissolve in polar solvents, or dissolve only a little, because the solute–solvent interactions that promote dissolution are much weaker than the interactions that keep solute molecules attracted to other solute molecules and the interactions that keep solvent molecules attracted to other solvent molecules. For example, octane does not dissolve in water because hydrogen bonding is the predominant attraction between $H_2O$ molecules, whereas London dispersion forces are the predominant attractions between octane molecules. The result is two separate liquid phases that neither mix with nor dissolve in each other when octane and water are combined (**Figure 6.19**).

What little solubility octane or any nonpolar solute has in water (or any polar solvent) is promoted by **dipole–induced dipole interactions** (represented by the orange dots in **Figure 6.20**). We saw in Section 6.2 how temporary dipoles can form between nonpolar molecules through temporary shifts in the distribution of electrons in molecules. An even stronger shift may occur when a polar molecule such as $H_2O$ collides with a nonpolar molecule such as $O_2$ (Figure 6.20). When the O atom in $H_2O$ is near one end of an $O_2$ molecule, the partial negative charge on the O atom in $H_2O$ repels the electrons in $O_2$, pushing them toward the other end of that molecule and creating, or *inducing*, a temporary dipole within the normally nonpolar $O_2$. The end of the $O_2$ molecule closest to the O atom in $H_2O$

**solvent** the component of a solution for which the largest number of moles is present.

**solute** any component in a solution other than the solvent. A solution may contain one or more solutes.

**solubility** the maximum quantity of a substance that can dissolve in a given volume of solution.

**miscible** capable of being mixed in any proportion (without reacting chemically).

**dipole–induced dipole interaction** an attraction between a polar molecule and the oppositely charged pole it temporarily induces in another molecule.

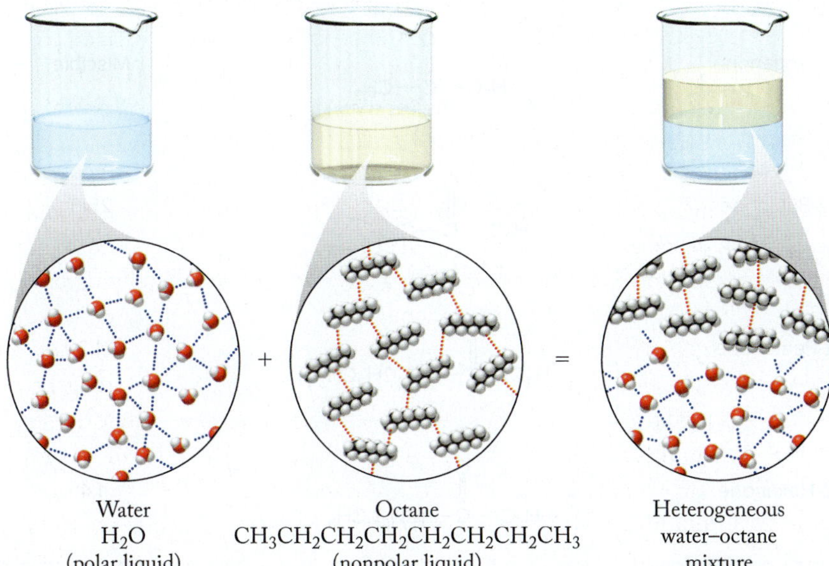


**FIGURE 6.19** The strength of London dispersion forces, represented by the red dotted lines, holds nonpolar molecules of octane together in the liquid phase. Water molecules are attracted to one another by hydrogen bonding (blue dotted lines). The lack of strong solute–solvent interactions needed to overcome these forces explains why octane and water are immiscible.

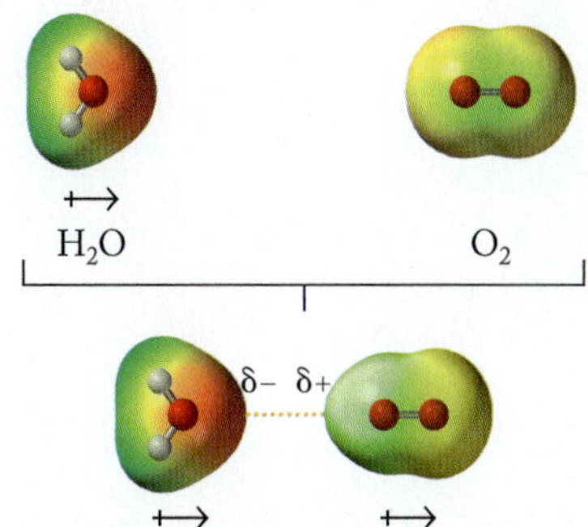


**FIGURE 6.20** Dipole–induced dipole interactions between molecules of $H_2O$ and $O_2$. The permanent dipole of $H_2O$ induces a temporary dipole in a normally nonpolar molecule of $O_2$.

**hydrophobic** describes a "water-fearing," or repulsive, interaction between a solute and water that decreases water solubility.

**hydrophilic** describes a "water-loving," or attractive, interaction between a solute and water that increases water solubility.

temporarily acquires a partial positive charge while the opposite end has a temporary partial negative charge. When the two molecules move apart, the partial charges in the $O_2$ molecule disappear. The temporary induced dipole is not as strong as the permanent dipole that induced it, but it is strong enough that $O_2$ is slightly soluble in water—soluble enough that the amount of dissolved oxygen sustains a multitude of aquatic life-forms.

## Competing Intermolecular Forces

If polar solutes readily dissolve in polar solvents, and nonpolar solutes do not, what is the solubility of solutes whose molecules contain both polar and nonpolar groups in a polar solvent such as water? To answer that, let's consider the solubilities of several ketones (**Table 6.6**) in water. Note how their solubilities decrease as the number of nonpolar $-CH_2-$ groups per molecule increases. The additional nonpolar groups strengthen the London dispersion forces between the ketone molecules, but they contribute little to the interactions between the ketone molecules and water molecules, which occur mainly through hydrogen bonding. As we have seen, strong interactions between solute molecules that are not offset by strong solute–solvent interactions tend to hold the solute molecules together in a single phase and inhibit their mixing with, and dissolving in, solvents.

Nonpolar interactions such as the London dispersion forces between the hydrocarbon chains of the ketones in Table 6.6 are called **hydrophobic** ("water fearing") interactions because they tend to keep the compounds from dissolving in water. However, dipole–dipole interactions, especially hydrogen bonding, promote solubility in water and are called **hydrophilic** ("water loving")

**TABLE 6.6** Solubilities of Some Ketones in Water at 20°C

| Compound | Condensed Molecular Structure | Solubility in Water (g/100 mL) |
|---|---|---|
| 2-Propanone | $H_3C-C(=O)-CH_3$ | Miscible |
| 2-Butanone | $H_3C-C(=O)-CH_2CH_3$ | 25.6 |
| 2-Pentanone | $H_3C-C(=O)-CH_2CH_2CH_3$ | 4.3 |
| 2-Hexanone | $H_3C-C(=O)-(CH_2)_3CH_3$ | 1.4 |
| 2-Heptanone | $H_3C-C(=O)-(CH_2)_4CH_3$ | 0.4 |

interactions. The solubility in water of compounds whose molecules contain both polar and nonpolar groups, as ketones do, is the net result of offsetting interactions. Hydrophilic interactions between molecules of solute and water promote solubility, whereas hydrophobic interactions between solute molecules inhibit it. As the sizes of the nonpolar regions of solute molecules increase, the strengths of solute–solute hydrophobic interactions increase and their solubilities in water decrease.

**CONCEPT TEST**

Rank these four alcohols from most soluble to least soluble in water:

| $CH_3(CH_2)_3OH$ | $CH_3(CH_2)_5OH$ | $CH_3(CH_2)_4OH$ | $CH_3(CH_2)_2OH$ |
|---|---|---|---|
| Butanol | Hexanol | Pentanol | Propanol |

We have established that polar solutes tend to dissolve in polar solvents and that nonpolar solutes do not, at least not much. Instead, nonpolar solutes tend to dissolve in nonpolar solvents, which leads to the following useful and general solubility guideline: *like dissolves like.* For example, the principal ingredient in some household cleaners (**Figure 6.21**) that remove grease, crayon wax, label adhesives, and other nonpolar materials from surfaces and fabrics is often labeled "petroleum distillate" or "petroleum naphtha." Those are common names for a mixture of hydrocarbons derived from crude oil, most of which have between 5 and 11 carbon atoms per molecule. They are effective at dissolving nonpolar substances because hydrocarbons also are nonpolar.

**FIGURE 6.21** Household cleaners such as this one are effective at removing grease stains and other nonpolar materials from surfaces and fabrics because the principal ingredient is a mixture of nonpolar hydrocarbons.

**SAMPLE EXERCISE 6.3** Predicting Solubility in Water **LO3**

Which of these four compounds have limited solubility in water: carbon tetrachloride ($CCl_4$), ammonia ($NH_3$), hydrogen fluoride (HF), and nitrogen ($N_2$)?

**Collect, Organize, and Analyze** Water is polar, so nonpolar compounds should have limited solubility in it. To judge which compounds are polar and which are nonpolar, we need to draw all their Lewis structures and determine which molecules have polar bonds and whether the arrangement of these bonds gives these molecules permanent dipoles.

**Solve** The Lewis structures of the four compounds are shown in **Figure 6.22**. Molecules of $CCl_4$ are perfectly tetrahedral and symmetrical, so this compound is nonpolar and should have limited solubility in water. Because molecules of $N_2$ have nonpolar bonds, this compound is nonpolar as well and also should have limited solubility. However, molecules of both $NH_3$ and HF are asymmetrical and have polar bonds. Moreover, both have the capacity to form hydrogen bonds with water molecules, so these compounds should be very soluble in water.

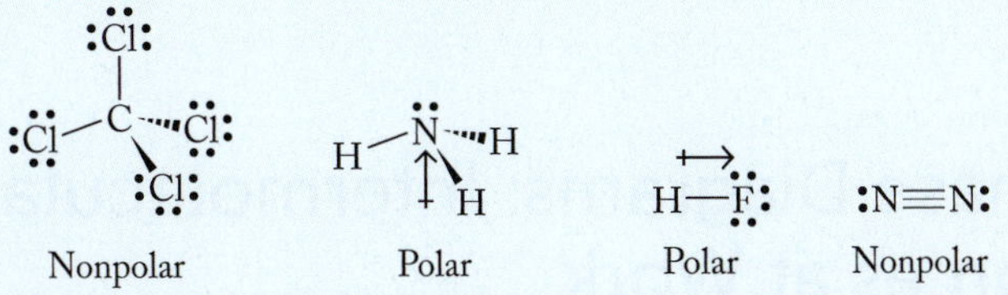

Nonpolar    Polar    Polar    Nonpolar

**FIGURE 6.22** Molecular geometries for carbon tetrachloride, ammonia, hydrogen fluoride, and diatomic nitrogen, showing the directions of the permanent dipoles for ammonia and hydrogen fluoride.

**Think About It** Nonpolar solutes dissolve in nonpolar solvents, but they have limited solubility in a polar solvent such as water, in accordance with the "like dissolves like" principle.

**Practice Exercise** Deep-sea divers breathe a mixture of gases rich in helium because the solubility of helium gas in blood is lower than the solubility of nitrogen gas in blood. Explain why helium is less soluble.

**SAMPLE EXERCISE 6.4** Explaining Solubility Trends for Ionic Compounds **LO3**

The solubilities of magnesium chloride ($MgCl_2$) and calcium chloride ($CaCl_2$) in water at 20°C are 54.5 and 74.5 g/100 mL, respectively. Identify some of the factors that account for the lower solubility of magnesium chloride.

**Collect, Organize, and Analyze** Any compound's solubility in a solvent depends on the relative strengths of the solute–solute, solvent–solvent, and solute–solvent interactions. To explain the solubility of ionic compounds in water, we first need to consider the ion–ion attractions and the hydrogen bonding in the water. We then need to compare these forces with the ion–dipole attractions between the ions and the water molecules in the solution. The strength of ion–ion attractions (ionic bonds) is described by Equation 4.1, in which the attraction is proportional to $(Q_1 \times Q_2)/d$, where $Q_1$ and $Q_2$ represent the charges on the ions and $d$ is the distance between them. Because Ca is below Mg in group 2 of the periodic table, we know that $Ca^{2+}$ has a larger radius than $Mg^{2+}$; their radii are also given in Figure 3.36.

**Solve** When we compare the solubilities of $MgCl_2$ and $CaCl_2$ in water, the strength of the hydrogen bonds between water molecules and the strength of the ion–dipole interactions between water molecules and chloride ions will be the same for both compounds. The difference between the solubilities of the two compounds depends on the relative strengths of the ion–ion forces and the ion–dipole forces between the cations and water molecules. The charges on the ions in $MgCl_2$ and $CaCl_2$ ($Q_1$ and $Q_2$) are the same. However, $Ca^{2+}$ has a larger radius than that of $Mg^{2+}$, so the distance between the cations and anions, $d$, will be greater for $CaCl_2$ than for $MgCl_2$. Therefore, Equation 4.1 predicts that the ion–ion forces of $MgCl_2$ will be stronger than those of $CaCl_2$, reducing the solubility of $MgCl_2$.

**Think About It** The balance between intermolecular forces can help us understand the differences in solubility of ionic compounds that have similarly charged ions, even if both compounds have appreciable solubility in water.

**Practice Exercise** The solubilities of $MgBr_2$ and $CaBr_2$ in water at 20°C are 101.5 and 142 g/100 mL, respectively. Identify some of the factors that account for why these bromide compounds have greater solubility than that of the chloride compounds $MgCl_2$ and $CaCl_2$.

## 6.5 Phase Diagrams: Intermolecular Forces at Work

**pressure (*P*)** the ratio of a force to the surface area over which the force is applied.

**standard atmosphere (atm)** the average pressure at sea level on Earth.

**phase diagram** a graphical representation of the dependence of the stabilities of the physical states of a substance on temperature and pressure.

The strength of the attractive forces between particles determines whether a substance is a solid, liquid, or gas at a given temperature. Heating substances to higher temperatures supplies the particles in them with more energy, until the kinetic energy of the particles eventually exceeds the energy of the intermolecular forces that hold them rigidly in place in a solid or keep them close to each other in a liquid. At this point, the solid should melt or the liquid should vaporize. It turns out, however, that pressure, in addition to temperature and the strength of intermolecular forces, influences the physical state of a substance.

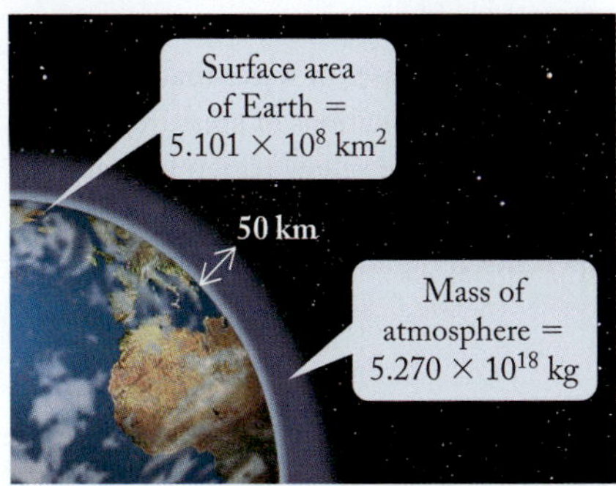


**FIGURE 6.23** Atmospheric pressure results from the force that the gases of Earth's atmosphere exert on the surface.

## Pressure

Why does **pressure (*P*)**, the ratio of force to surface area, influence the physical states of substances? For most substances (with the notable exception of $H_2O$), the solid phases are the most dense and the vapor phases are the least dense, so most substances occupy a slightly larger volume when they melt, and all substances occupy a significantly larger volume after vaporization (or sublimation). Increasing the pressure on substances keeps them from expanding, thereby raising the temperatures at which most substances melt and raising the temperatures at which *all* substances vaporize and sublime.

The pressure that most substances of interest to us experience is *atmospheric* pressure, the weight of Earth's atmosphere pressing down on its surface divided by its surface area (**Figure 6.23**). We can calculate how strong this pressure is by multiplying the mass of all the gases in the atmosphere by the acceleration due to gravity that all objects on Earth's surface experience, 9.807 m/s$^2$. The product is the force ($F$) that the mass of atmospheric gases exerts on Earth's surface because of gravity:

$$F = (5.270 \times 10^{18}\ \text{kg}) \times \left(9.807\ \frac{\text{m}}{\text{s}^2}\right) = 5.168 \times 10^{19}\ \frac{\text{kg} \cdot \text{m}}{\text{s}^2}$$

This combination of units, kg • m/s$^2$, is defined as a newton (N) in honor of Sir Isaac Newton and is the SI unit of force. If we divide the force by Earth's surface area ($A$), we obtain the following average atmospheric pressure at sea level:

$$P = \frac{F}{A} \tag{6.1}$$

$$= \frac{(5.168 \times 10^{19}\ \text{N})}{5.101 \times 10^{8}\ \cancel{\text{km}^2} \times \left(\frac{1000\ \text{m}}{\cancel{\text{km}}}\right)^2} = 1.013 \times 10^5\ \frac{\text{N}}{\text{m}^2}$$

One newton per square meter is defined as one pascal (Pa), the SI unit of pressure. The average pressure that the atmosphere exerts on Earth's surface (at sea level) is, therefore, $1.013 \times 10^5$ Pa. This average also defines a quantity of pressure called the **standard atmosphere (atm)**.

To avoid the need for exponents, scientists often express atmospheric pressure in units such as the kilopascal (kPa), the bar, or the millibar (mb):

$$1\ \text{atm} = 101.3\ \text{kPa} = 1.013\ \text{bar} = 1013\ \text{mb}$$

The bar is not an official SI unit, but it is based on one: 1 bar = $10^5$ Pa, and it is widely used in science. Meteorologists tend to use millibars to express atmospheric pressures, as shown in **Figure 6.24**. Doing so simplifies pressure values on weather maps by avoiding the need for decimal points.

**FIGURE 6.24** The lines on this weather map mark different atmospheric pressures.

## Phase Diagrams

**Phase diagrams** are graphs of pressure (on the $y$-axis) versus temperature (on the $x$-axis). Scientists use these graphs to represent which phases of a substance are the most stable at different combinations of temperature and pressure. To depict a range of pressures, the scale of the $y$-axis is usually logarithmic rather than linear.

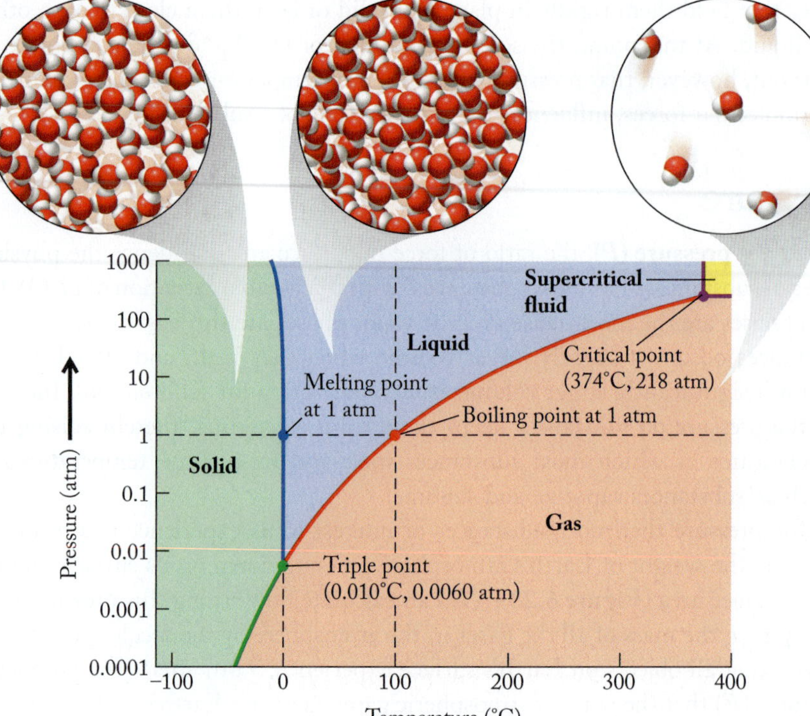


**FIGURE 6.25** The phase diagram for water indicates the phases of water that exist at various combinations of pressure and temperature. The yellow region indicates the conditions under which water exists as a supercritical liquid.

**CHEMTOUR**
Phase Diagrams

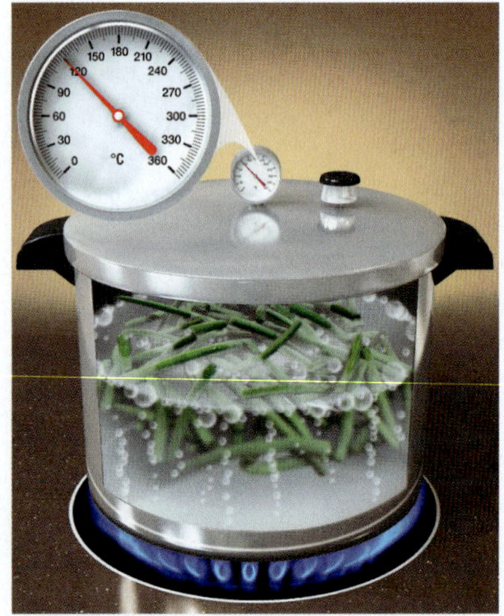

**FIGURE 6.26** Steam is restricted from escaping a pressure cooker as the water inside it starts to boil. The higher resulting pressure causes the water to boil at a higher temperature, so the food inside cooks faster.

Phase diagrams such as that of water (**Figure 6.25**) have three regions corresponding to the three states of matter. The lines separating the regions represent combinations of temperature and pressure at which the two phases on both sides of each line coexist in equilibrium. Thus, the blue line separating the solid and liquid regions in Figure 6.25 represents a collection of melting (and freezing) points, the red line separating the liquid and gas regions represents a series of boiling (and condensation) points, and the green line separating the solid and gas regions represents a series of sublimation (and deposition) points.

The red line curves from lower left to upper right because when the pressure above a liquid is increased, as in a pressure cooker (**Figure 6.26**), more energy is necessary to overcome that pressure and convert liquid molecules into vapor, where they take up much more space. Heating the liquid to an even higher temperature gives it that additional energy. The solid–gas (green) line curves similarly because higher pressures make sublimation into water vapor more difficult for molecules of ice. In both cases, a phase transition from a dense, condensed (liquid or solid) phase to a much less dense vapor phase occurs at higher temperatures under increasing pressure.

The blue melting–freezing equilibrium line for water is nearly vertical at low to moderate pressures and then bends to the left at high pressures. Thus, the temperature at which ice melts decreases as pressure increases. The trend is opposite that observed for almost all other substances (for example, compare it with the curve of the blue line in the phase diagram of $CO_2$ in **Figure 6.27**). The reason for water's unusual melting–freezing behavior is that water expands as it freezes, whereas almost all other substances contract—that is, the solid phases of nearly all substances are more dense than their liquid states. Why does water expand and become less dense? As it freezes, more hydrogen bonds form between its molecules as they cease to flow past each other and instead take up positions in a solid structure where they are surrounded by, and hydrogen-bonded to, an extended array of other water molecules. These additional hydrogen bonds create a slightly

**CONNECTION** See Figure 1.10 to review the names of phase changes and the physical states involved.

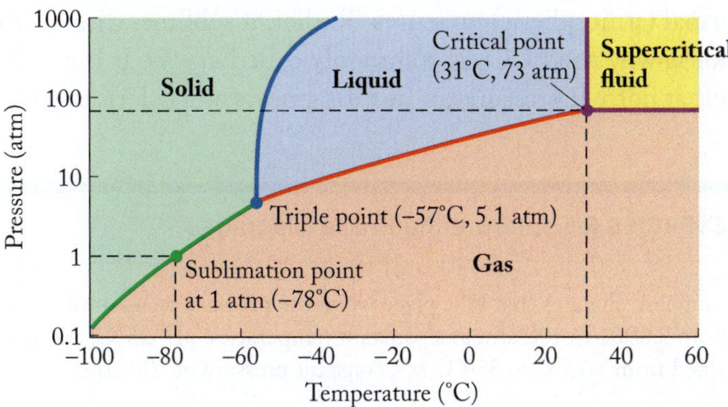


**FIGURE 6.27** Phase diagram for carbon dioxide.

more open structure in ice than in liquid water, which makes ice less dense. If we apply enough pressure to ice, we can force it to melt.

## CONCEPT TEST

A length of wire with heavy weights on each end can pass downward through a block of ice without cutting it in two (**Figure 6.28**). The temperature of the ice stays below its melting point the whole time, and the ice is still a single block after the wire has passed all the way through it. Explain how the wire can pass through a frozen block of ice without cutting it in two.

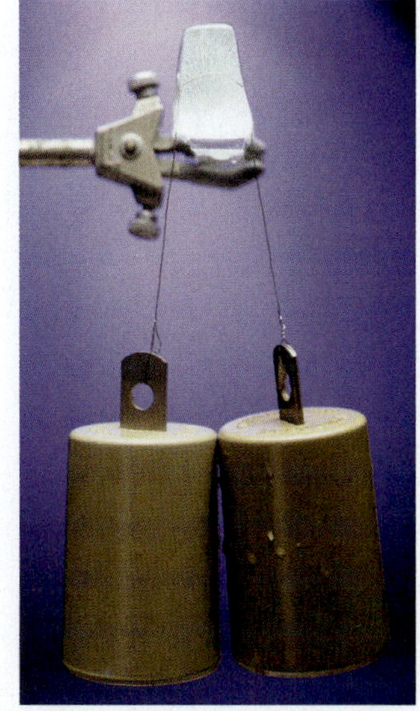

**FIGURE 6.28** A weighted wire passes downward through a block of ice without cutting the ice in two.

The point on a phase diagram at which all three phase transition lines meet is known as the **triple point** (see Figures 6.25 and 6.27). This point represents the combination of temperature and pressure at which the liquid, solid, and vapor states of a substance exist simultaneously. The triple point of water is at a temperature of 0.010°C—just above its normal (1 standard atmosphere) freezing point of 0°C—but at a pressure of only 0.0060 atm.

The **critical point**, however, is where the liquid–gas equilibrium line ends and the two states are indistinguishable. The critical point is reached because thermal expansion at high temperature decreases the density of the liquid state while high pressure compresses the gas into a small volume, increasing its density. At the critical point, the densities of the liquid and vapor states are equal. At temperature–pressure combinations above the critical point of a substance, it exists as a **supercritical fluid** (see the yellow regions in Figures 6.25 and 6.27).

A supercritical fluid has the ultralow viscosity of a gas and can easily diffuse through many solid materials, yet it can dissolve substances in those materials as though it were a liquid. Supercritical $CO_2$ is used in the food processing industry to decaffeinate coffee beans and to remove fat from potato chips. Supercritical $CO_2$ also is used to extract essential oils from plants to make perfumes and to dry-clean clothes. Because supercritical $CO_2$ can dissolve nonpolar substances, it offers an alternative to the once widely used organic solvents. That is one example of the **green chemistry** principle that the use of hazardous substances derived from nonrenewable sources should be reduced or eliminated. In these applications, supercritical $CO_2$ has the advantage of being nontoxic and nonflammable and of forming at a relatively low temperature (31°C), which is an advantage in extracting pharmaceuticals and other compounds that may decompose at high temperature.

Notice in Figure 6.27 that the blue region representing liquid $CO_2$ does not exist below a pressure of 5.1 atm. This means that solid $CO_2$ does not melt into a

**triple point** the temperature and pressure at which all three phases of a substance coexist. Freezing and melting, boiling and liquefaction, and sublimation and deposition all proceed at the same rate, so no net change takes place in the system.

**critical point** a specific temperature and pressure at which the liquid and gas phases of a substance have the same density and are indistinguishable from each other.

**supercritical fluid** a substance in a state that is above the temperature and pressure at the critical point, at which the liquid and vapor phases are indistinguishable.

**green chemistry** laboratory practices that reduce or eliminate the use or generation of hazardous substances.

**surface tension** the ability of the surface of a liquid to resist an external force.

**meniscus** the concave or convex surface of a liquid.

liquid at normal (atmospheric) pressures. Rather, it sublimes directly to $CO_2$ gas. This behavior is why solid $CO_2$ is commonly called *dry ice*. It is a cold solid that does not melt at normal pressures; it just disappears as a colorless gas.

## SAMPLE EXERCISE 6.5 Interpreting Phase Diagrams LO4

Describe the phase changes that take place when the pressure on a sample of water is increased from 0.0001 to 100 atm at a constant temperature of −25°C, and the sample is then warmed from −25°C to 350°C at a constant pressure of 100 atm.

**Collect and Organize** We are asked to describe the phase changes that a sample of water undergoes as its pressure increases at constant temperature and its temperature increases at constant pressure. Figure 6.25 shows which phases of water are stable at various combinations of temperature and pressure.

**Analyze** The change in pressure at constant temperature defines two points on the phase diagram of water with the coordinates (−25°C, 0.0001 atm) and (−25°C, 100 atm). Connecting those points will give us a vertical line. If the line crosses a phase boundary, a change in physical state will occur. The change in temperature at constant pressure defines a third point (350°C, 100 atm) on the phase diagram, which will connect to the second point with a horizontal line.

**Solve** **Figure 6.29** shows a plot of our sample's changes in pressure and temperature. At the bottom end of vertical line 1, water is a gas (vapor). As pressure increases along line 1, it crosses the boundary between gas and solid (point A), which means that vapor turns directly into solid ice, which is stable above 100 atm at −25°C (point B). As the temperature of the ice increases along horizontal line 2, it intersects the solid–liquid boundary (point C) and the ice melts. At even higher temperatures, just below 350°C, the line intersects the liquid–gas boundary (point D) and the liquid water vaporizes.

**Think About It** The solid-to-liquid and liquid-to-gas transitions with increasing temperature along line 2 are what we would expect when a solid substance is warmed to its melting point and then the liquid is heated to its boiling point at a given pressure. The transition along line 1 is less familiar because it is caused by increasing the pressure on a gas at a temperature below the triple point temperature, so the vapor never condenses. Instead, it is deposited as a solid.

**Practice Exercise** Describe the phase changes that occur when the temperature of $CO_2$ is increased from −100°C to 50°C at a pressure of 25 atm and the pressure is then increased to 100 atm.

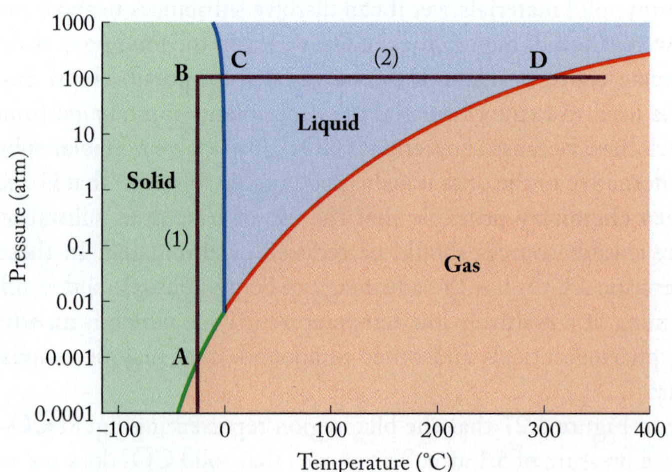


**FIGURE 6.29** Phase diagram for water.

# 6.6 Some Remarkable Properties of Water

Water covers more than 70% of Earth's surface, our muscle tissue is about 75% (by mass) water, and our blood is about 95% water. Water's omnipresence and familiarity may lead us to take its remarkable properties for granted. For example, although nearly all substances are more dense as solids than as liquids, water is not. Every other substance of comparable molar mass is a gas at room temperature and pressure, but water is a liquid. We have seen how hydrogen bonding between water molecules can explain its unusually high boiling point, and here we explore how hydrogen bonding is linked to these other remarkable properties. Hydrogen bonding allows a carefully placed steel needle to float on water and enables insects to walk on water. Hydrogen bonding also enables liquid water to overcome the force of gravity and climb to the top of the world's tallest trees.

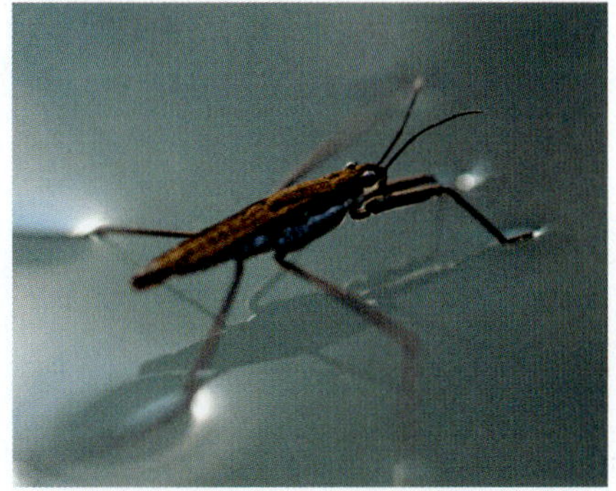

**FIGURE 6.30** Surface tension allows a water strider to walk on water without sinking.

**Surface tension** is the ability of the surface of a liquid to resist an external force, such as the weight of an object. We usually think of surface tension as a property of liquids, which enables a water strider to literally walk on water (**Figure 6.30**), but the property also applies to solids, in which surface tension is called *surface stress*, the energy needed to stretch a surface so that its area increases by a unit amount.

**STEPWISE ANIMATION**
Surface Tension

To understand how surface tension depends on intermolecular forces, consider the microscopic view of a steel needle suspended on the surface of a dish of water (**Figure 6.31**). Water molecules in the interior of the water are surrounded by other water molecules and form hydrogen bonds to them in all directions. However, no molecules of liquid water exist above those at the surface, so those water molecules form fewer hydrogen bonds per molecule than those in the bulk liquid. Gravity pulls the steel needle downward, but to move downward the needle must penetrate the surface. This means pushing aside surface water molecules, increasing the area of the surface, and turning what were once interior water molecules into surface molecules. This transformation requires breaking some of the hydrogen bonds experienced by the interior molecules, which takes 23 kJ of energy per mole of hydrogen bonds. The downward pressure of the needle on the surface is not enough to overcome this energy, so the needle floats.

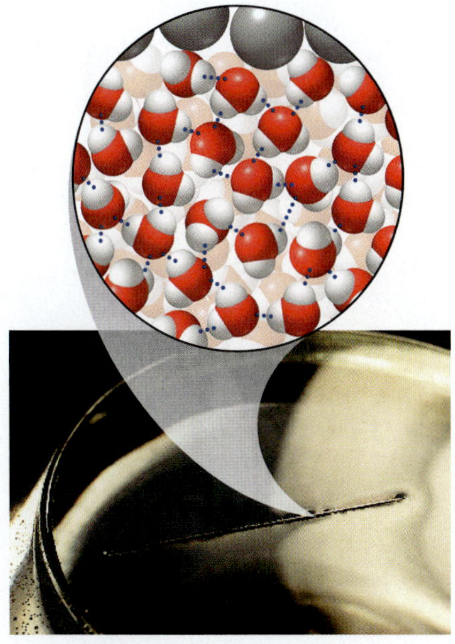

**FIGURE 6.31** Surface water molecules form fewer hydrogen bonds than interior water molecules. To fall through the surface, a needle must briefly increase the area of the surface, which means turning some interior water molecules into surface water molecules, which means breaking some of the hydrogen bonds around those interior molecules. The strength of these hydrogen bonds stops the needle from penetrating the surface.

Another phenomenon linked to the strength of intermolecular forces is the shape of a liquid's surface in a graduated cylinder or other small-diameter container made of glass, which is mostly $SiO_2$. As shown in **Figure 6.32**, the surface of water in a partially filled glass test tube is concave, whereas the surface of liquid mercury is convex. Either curved surface is called a **meniscus**. In both liquids, the meniscus is the result of two competing forces: *cohesive forces*, interactions between like particles (such as the hydrogen bonds between water molecules), and *adhesive forces*, interactions between materials made of different substances. In the water sample in Figure 6.32(a), the adhesive forces are dipole–dipole interactions between water molecules and polar Si—O—Si groups on the surface of the glass and even hydrogen bonding between water molecules and Si—O—H groups on the surface. These adhesive forces are strong enough to cause the water to climb up the glass. Cohesive interactions with other water molecules pull them up to nearly the same height as the surface molecules. Water molecules farther from the inner wall of the tube are pulled up progressively less, creating a concave meniscus.

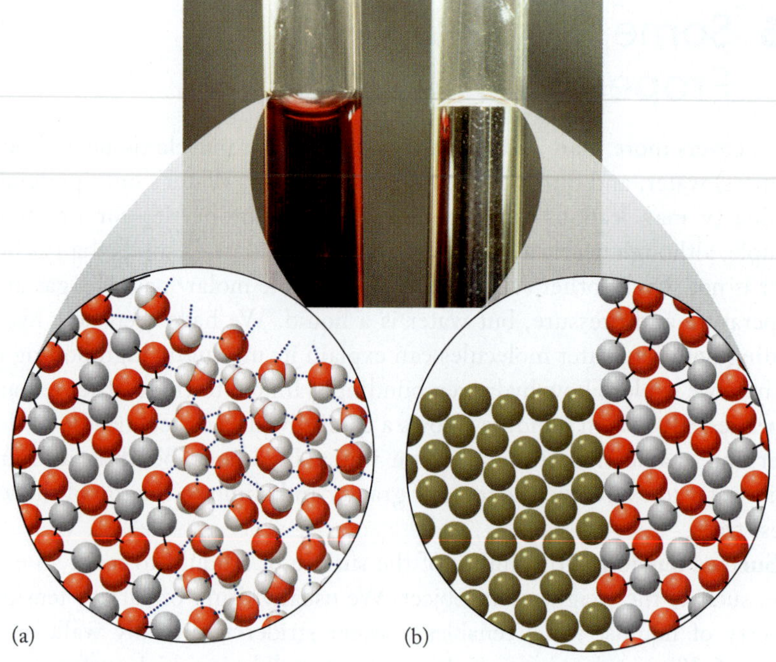


**FIGURE 6.32** (a) A combination of adhesive and cohesive forces causes the water (dyed red here) to form a concave meniscus in a test tube made of silica glass, which contains polar surface groups. (b) Atoms of liquid mercury stick more strongly to other atoms of mercury than they adhere to glass, so mercury forms a convex meniscus.

**CHEMTOUR** Capillary Action

When mercury partially fills a glass test tube (Figure 6.32b), strong cohesive forces exist in the form of metallic bonds between mercury atoms. The only adhesive forces are relatively weak interactions between induced dipoles in surface mercury atoms and the polar groups on the glass surface. Here, mercury atoms are more strongly attracted to one another than to the glass. As a result, mercury pulls away from the surface, mounding up in a way that reduces its interaction with the test tube and increases cohesive interactions. The result is a convex meniscus.

The principle behind water's concave meniscus is taken to the extreme when water enters very narrow tubes called capillaries. The tiny inner diameter of capillaries means that all water molecules inside a capillary either experience adhesive intermolecular forces directly or are a few hydrogen bonds away from molecules that do. As a result, adhesion pulls the water molecules along the wall of the capillary and cohesion pulls all the others. The result of this combination of intermolecular forces is a phenomenon called **capillary action**, which is a liquid's ability to flow against gravity, spontaneously rising in a narrow tube or in structures made up of narrow pores, such as a carnation (**Figure 6.33**) or the trunks of trees.

**FIGURE 6.33** Water (containing blue dye) rises in a carnation as a result of capillary action.

Capillary action also is responsible for wicking, the movement of a fluid away from its source through a porous material. Wicking is the process by which paper towels absorb spills and by which some fabrics, such as those made of polyester microfibers, draw perspiration away from the skin. Undergarments made of such fabrics serve as a base layer of clothing below other layers that provide thermal insulation. This combination keeps athletes warm while allowing them to practice and play in clothing that is not waterlogged with perspiration (**Figure 6.34**).

**capillary action** the rise of a liquid in a narrow tube as a result of adhesive forces between the liquid and the tube and cohesive forces within the liquid.

CONCEPT **TEST**

Is mercury spontaneously drawn up into a glass capillary tube? Why or why not?

Another unusual property of water is the way its density changes as its temperature drops to near its freezing point. As happens for nearly all substances,

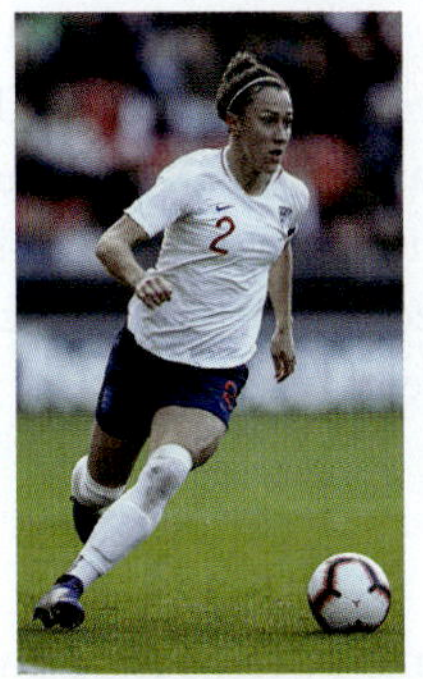

**FIGURE 6.34** Athletes wear clothing made with fabrics that can wick away moisture.

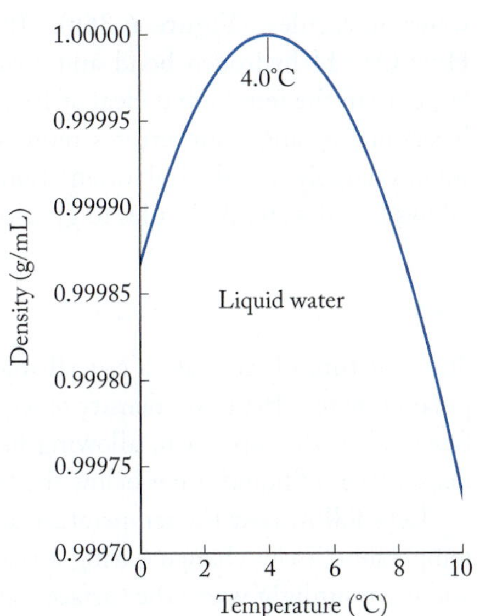


**FIGURE 6.35** As water is cooled, its density increases until the temperature reaches 4°C. At that temperature the density has its maximum value, 1.000 g/mL. As the water cools from 4°C to its freezing point at 0°C, the density decreases.

water's density increases as its temperature decreases. However, water's density reaches a maximum at 4°C (**Figure 6.35**) and actually *decreases* as it cools toward its freezing point. When water begins to freeze, its density drops even more, to about 0.92 g/mL, as the oxygen atom in each water molecule becomes the center of an array of two covalent bonds (to the H atoms within the molecule) and two hydrogen bonds (to two other $H_2O$ molecules), as shown in **Figure 6.36(a)**. The hydrogen bonds are more than twice as long as those of the covalent bonds in $H_2O$, but the larger distance is not the reason that ice is less dense than water; rather, the reason is the *directionality* of the hydrogen bonds in ice. Recall from Chapter 5 that the electron-group geometry of the valence electrons in the O atoms (SN = 4) is tetrahedral. However, ice crystals feature *hexagonal* arrays of

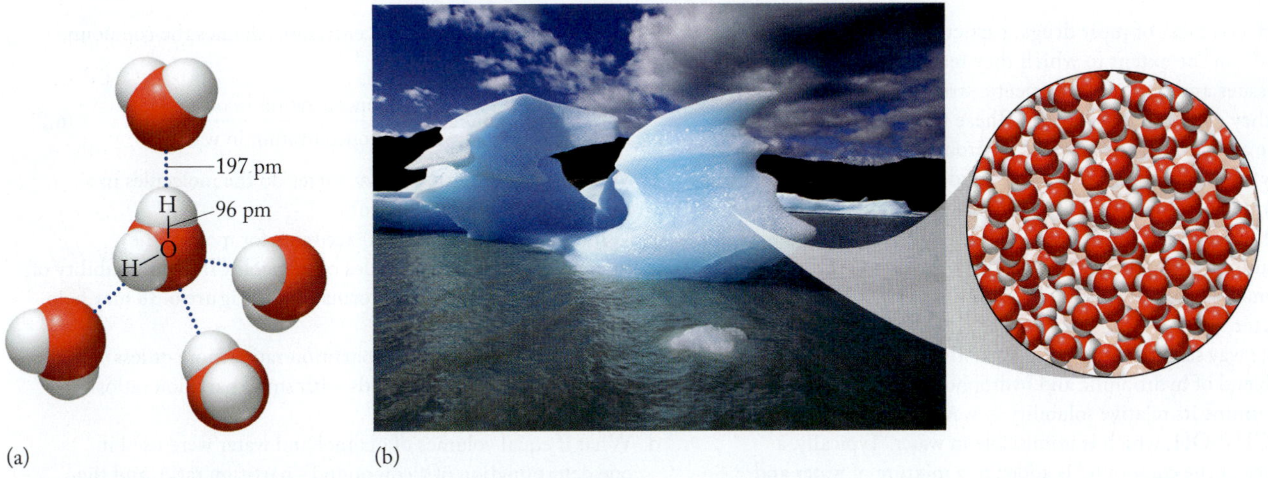


**FIGURE 6.36** (a) The oxygen atoms in ice form two covalent bonds to H atoms and two longer hydrogen bonds to the H atoms in nearby molecules. (b) Ice is less dense than water because of its three-dimensional molecular structure created by the covalent and hydrogen bonds within and between its molecules.

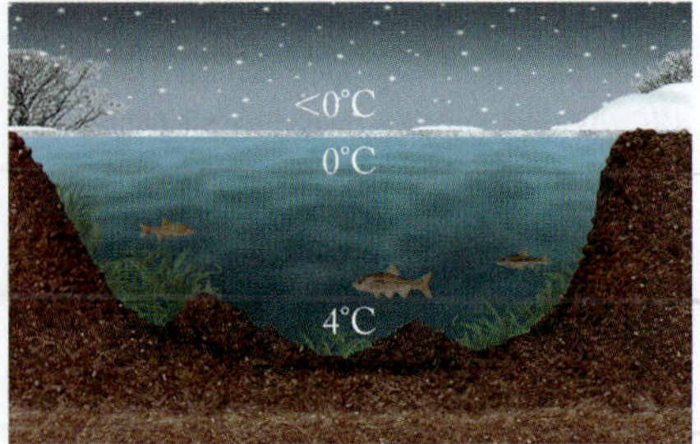


**FIGURE 6.37** Wintering over. As the surface water of a lake in a temperate climate cools to 0°C and then freezes during the winter, the deepest, densest water in the lake remains a relatively warm 4°C.

water molecules (**Figure 6.36b**). This means that the angle between the H⋯⋯O—H hydrogen bond and covalent bond at each corner is about 120°, larger than the tetrahedral ideal of 109.5°. The larger corner angles mean that ice's hexagonal arrangement creates more space between the molecules than liquid water's largely tetrahedral orientation of hydrogen-bonded molecules, which allows liquid water molecules to get closer to their nearest neighbors.

## Water and Aquatic Life

The structure of ice plays a crucial role in aquatic ecosystems in temperate and polar climates. The lower density of ice means that lakes, rivers, and polar oceans freeze from the top down, allowing fish and other aquatic life to survive in the denser layer of liquid water below the frozen surface.

Let's follow how the temperature and density of the water in a deep lake in a temperate climate change during a year. In early spring, rising air temperatures and more sunlight warm the surface waters of the lake. Continued heating creates an upper layer of warm, low-density water separated from the colder, denser water below by a *thermocline,* a sharp change in temperature between the two layers. Little mixing takes place between the two layers over the summer and into the fall. Then declining air temperatures cool the surface waters until they are colder and denser than the water below. The cold surface water sinks, forcing warmer water to the surface, where it also cools and sinks. The process continues until all the water in the lake reaches 4°C and is uniformly mixed.

As the surface water cools further with the approach of winter, it becomes less dense and ice eventually forms. The layer of ice floats above and insulates the water below, including the 4°C water that fills the deepest parts of the lake because it is the densest (**Figure 6.37**). This water serves as a liquid haven for aquatic life-forms and allows them to survive the subfreezing air temperatures of winter.

### SAMPLE EXERCISE 6.6 Integrating Concepts: Drug Efficacy and Partition Ratios

The effectiveness of most drugs, particularly those taken orally, depends on the extent to which they reach the target organs and tissues and on their therapeutic strength and persistence once they arrive at their targets. These factors are linked to how soluble a drug is in the aqueous environments of blood serum and cytosol (the liquid inside cells) and to how well the drug can penetrate nonaqueous, nonpolar parts of the body, such as the epithelium surrounding the small intestine and the membranes surrounding individual cells. In other words, to be effective, a drug must have at least some hydrophilic *and* some hydrophobic character.

One way scientists can predict whether a molecular compound has a blend of hydrophilic and hydrophobic properties is to determine its relative solubility in water and in octanol, $CH_3(CH_2)_7OH$, which is immiscible in water. Typically, a quantity of the compound is added to a mixture of water and octanol. The mixture is shaken vigorously to allow the compound to dissolve in either liquid, the liquids are allowed to separate, and then the concentration of the compound in each liquid is measured. The ratio of the concentrations defines the compound's partition ratio:

$$\text{Partition ratio} = \frac{\text{concentration in octanol}}{\text{concentration in water}} \tag{6.2}$$

a. What types of intermolecular forces do the molecules in a sample of octanol experience?
b. Which force cited in part (a) accounts for most of the interaction between molecules of octanol? (The immiscibility of octanol in water and the information in **Figure 6.38** may help you decide.)
c. Are compounds with large partition ratios more or less hydrophobic than compounds with small partition ratios? Why?
d. What if equal volumes of octanol and water were used in one determination of a compound's partition ratio, and then twice as much octanol was used in another determination? How would this difference in volumes influence the two experimentally determined partition ratio values?

$CH_3(CH_2)_7OH$, $\mathcal{M}$ = 130 g/mol
Boiling point, 468 K

Octanol

$CH_3(CH_2)_7CH_3$, $\mathcal{M}$ = 128 g/mol
Boiling point, 424 K

Nonane

**FIGURE 6.38** Octanol and nonane have similar boiling points despite the polar –OH group in octanol.

**Collect and Organize** We know the molecular structure of octanol and that it is immiscible in water. We are asked what types of intermolecular forces octanol molecules experience, which force is the strongest, and how the partitioning of a compound between octanol and water relates to its hydrophobicity. We also are asked to predict how changing the relative volumes of octanol and water affects the ratio of the concentrations of a solute in the two liquids.

**Analyze** Octanol is an alcohol with one –OH group at the end of a chain of eight carbon atoms. The O and H atoms in –OH groups can form hydrogen bonds to the H and O atoms on neighboring molecules. All molecules experience London dispersion forces that get stronger with increasing molar mass. The immiscibility of octanol in water means that it is, on balance, hydrophobic and should therefore be a good solvent for hydrophobic solutes. If the partition ratio of a compound is a constant, the ratio of concentrations of the compound in octanol and water must be a constant that does not change with changes in the volumes of either or both solvents.

**Solve**

a. Molecules of octanol interact with one another through London dispersion forces, as all molecules do, and through hydrogen bonding between the OH groups on adjacent molecules.

b. The hydrophobicity of octanol indicates that its molecules experience strong London dispersion forces because of their long, nonpolar chains of carbon and hydrogen atoms. The importance of these interactions is reinforced by the relatively small difference between the boiling points of octanol and nonpolar nonane, $CH_3(CH_2)_7CH_3$, a compound with nearly the same molar mass and molecular shape as octanol (Figure 6.38). Therefore, London dispersion forces are the principal intermolecular force among octanol molecules.

c. The long nonpolar chains in molecules of octanol make it an effective solvent for nonpolar, or hydrophobic, solutes. Those same solutes should have little solubility in water, which would give a relatively high concentration ratio in Equation 6.2 during a partition ratio determination. Thus, a compound with a large partition ratio is more hydrophobic than one with a smaller partition ratio.

d. Because the concentration ratio in Equation 6.2 is a constant, changing the relative volumes of octanol and water in a partition ratio determination should not alter the calculated value. However, doubling the volume of octanol would mean that twice as much of the total *quantity* of solute in the mixture would end up in the octanol phase.

**Think About It** Even though octanol contains an alcohol functional group, hydrogen bonding is not the predominant intermolecular force because a larger proportion of each molecule—the hydrocarbon chain—experiences London dispersion forces with other molecules of octanol. Boiling points can be explained by considering not only what kind of intermolecular forces may exist but also the relative magnitude of competing intermolecular forces.

## SUMMARY

**LO1** All atoms and molecules experience **London dispersion forces** with other atoms and molecules as a result of their **polarizability** and the existence of **temporary dipoles** even in particles that have no permanent dipole. Stronger London dispersion forces lead to higher boiling points and greater **viscosities.** Ions in aqueous solution interact with water molecules through **ion–dipole interactions**, forming a **sphere of hydration** around the ion. Molecules with permanent dipoles interact through a combination of London dispersion forces and **dipole–dipole interactions**. The strongest dipole–dipole interactions are **hydrogen bonds**, which form between H atoms bonded to N, O, and F atoms and other N, O, and F atoms. Polar **organic compounds** contain polar **functional groups**. Among them are **carbonyl groups**, which contain C═O bonds and are found in **ketones**. Other polar functional groups include the **hydroxyl group** (–OH) in **alcohols**. In contrast, nonpolar **hydrocarbons** such as **alkanes** contain only hydrogen and carbon atoms. (Sections 6.2 and 6.3)

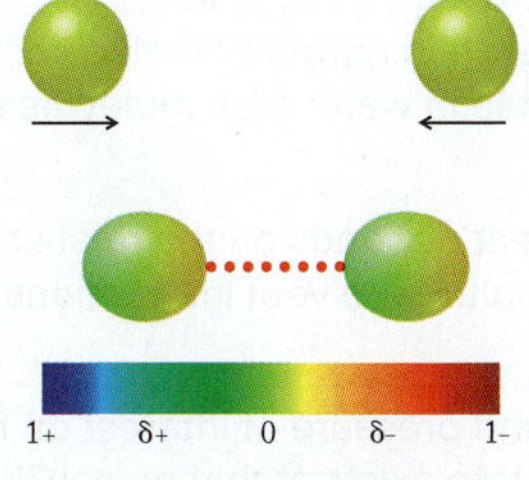


**LO2** The melting and boiling points of compounds are related to the intermolecular forces they experience. Stronger intermolecular attractions between molecules correlate with higher melting points and higher boiling points. (Section 6.3)

**LO3** In a solution, the substance present in the largest amount (in number of moles) is the **solvent**. All other substances are **solutes**. Polar solutes dissolve in polar solvents when the dipole–dipole interactions between solute and solvent molecules offset the interactions that keep either solute molecules or solvent molecules together. The limited **solubility** of nonpolar solutes in polar solvents is a result of **dipole–induced dipole interactions**. **Hydrophilic** substances are more soluble in water than are **hydrophobic** substances, which are more soluble in nonpolar solvents. In general, polar solutes dissolve in polar solvents and nonpolar solutes dissolve in nonpolar solvents—that is, like dissolves like. (Section 6.4)

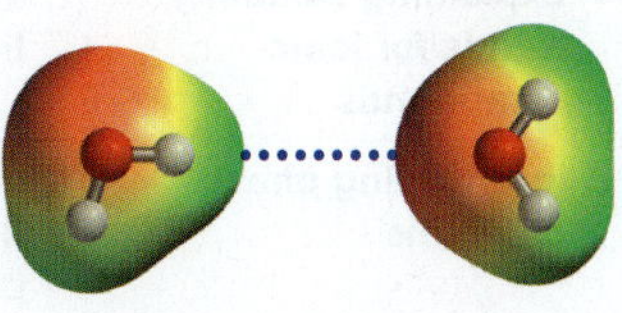

**LO4** The ratio of force to surface area is defined as **pressure (*P*)**, which is often expressed in **standard atmospheres (atm)**, a unit equal to Earth's average atmospheric pressure at sea level. The **phase diagram** of a substance indicates its state at a particular pressure and temperature. The three states of solid, liquid, and gas exist in equilibrium at the **triple point**. Above the temperature and pressure of its **critical point**, a substance exists as a **supercritical fluid** with many of the physical properties of a gas but with the ability to dissolve other substances as though it were a liquid. (Section 6.5)

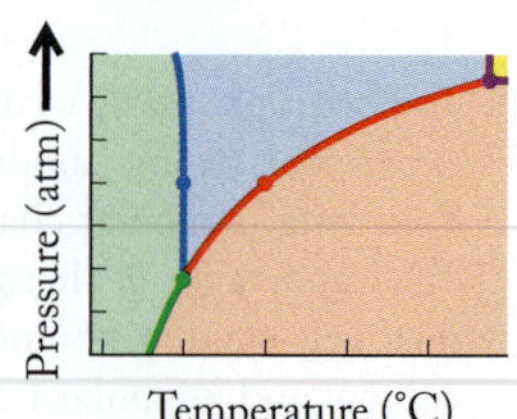


**LO5** The remarkable behavior of water, including its unusually high **surface tension**, results from the strength of hydrogen bonding between its molecules. Those interactions also cause water to expand when it freezes and contribute to **capillary action**. (Section 6.6)

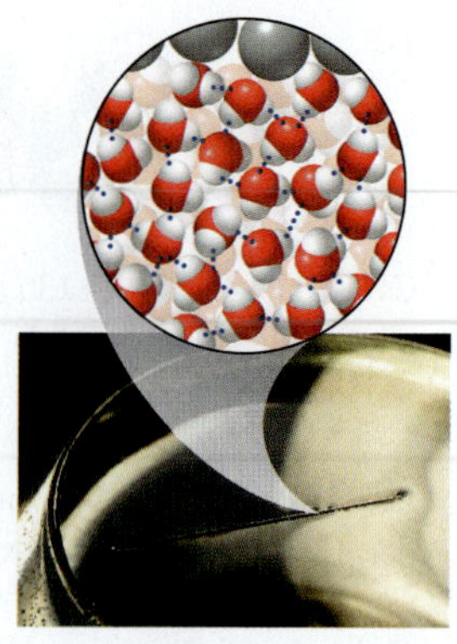

## PARTICULATE **PREVIEW WRAP-UP**

The functional group in each molecule is (a) an alcohol functional group, (b) a ketone functional group, and (c) an ether functional group. The intermolecular forces present between molecules of (a) are London dispersion forces, dipole–dipole forces, and hydrogen bonding. The intermolecular forces present between molecules of (b) are London dispersion forces and dipole–dipole forces. The intermolecular forces present between molecules of (c) are London dispersion forces. Compound (c) has the lowest boiling point, and compound (a) has the highest boiling point.

## PROBLEM-SOLVING SUMMARY

| Type of Problem | Concepts and Equations | Sample Exercises |
|---|---|---|
| **Identifying intermolecular forces** | All atoms and molecules experience London dispersion forces. Ions in aqueous solution interact with water molecules through ion–dipole interactions. Molecules with permanent dipoles interact through a combination of London dispersion forces and dipole–dipole interactions. The strongest dipole–dipole interactions are hydrogen bonds, which form between H atoms bonded to N, O, and F atoms and other N, O, and F atoms. | 6.1 |
| **Explaining differences and trends in boiling points of liquids** | Because of London dispersion forces, substances made of large molecules usually have higher boiling points than those with smaller molecules. Because of dipole–dipole interactions, polar compounds have higher boiling points than nonpolar compounds of similar molar mass. Compounds whose molecules form hydrogen bonds have even higher boiling points because H bonds are especially strong dipole–dipole interactions. | 6.2 |
| **Predicting solubility in water** | Polar molecules are more soluble in water than nonpolar molecules. Molecules that form hydrogen bonds are more soluble in water than molecules that cannot form these bonds. Like dissolves like. | 6.3 |
| **Explaining solubility trends for ionic compounds** | The solubility of any compound in a solvent depends on the relative strengths of the solute–solute, solvent–solvent, and solute–solvent interactions. | 6.4 |
| **Interpreting phase diagrams** | Locate the combination of temperature and pressure of interest on the phase diagram, and determine which physical state exists at that point. Changes in pressure at constant temperature are represented by vertical paths, and changes in temperature at constant pressure are represented by horizontal paths. If a path crosses a phase boundary line, a change in phase occurs. | 6.5 |

## VISUAL PROBLEMS

*(Answers to boldface end-of-chapter questions and problems are in the back of the book.)*

**6.1.** Figure P6.1 shows molecular structures of four constitutional isomers of heptane ($C_7H_{16}$). Which has the highest boiling point?

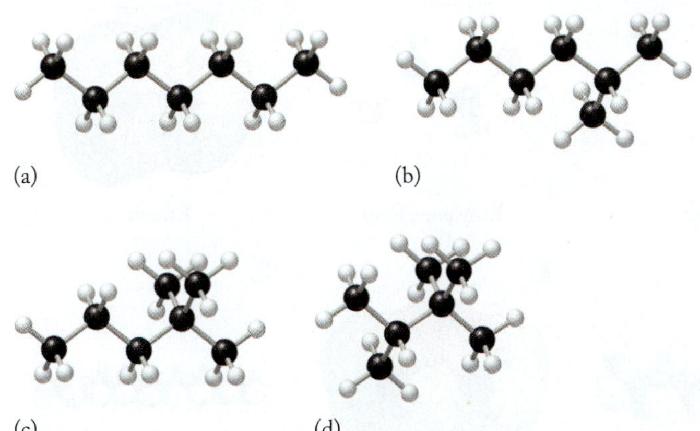


**FIGURE P6.1**

6.2. Which constitutional isomer of heptane in Figure P6.1 is the most viscous at 20°C?

**6.3.** Figure P6.3 contains the Lewis structures of ammonia ($NH_3$) and phosphine ($PH_3$). The boiling points of these compounds at $P$ = 1.00 atm are 185 K and 240 K. Which one boils at the higher temperature? Explain your selection.

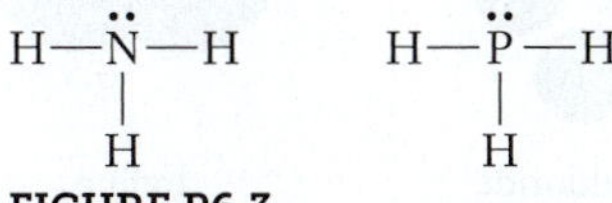


**FIGURE P6.3**

6.4. Figure P6.4 shows space-filling models of pentane ($C_5H_{12}$) and decane ($C_{10}H_{22}$). Which substance has the lower freezing point? Explain your answer.

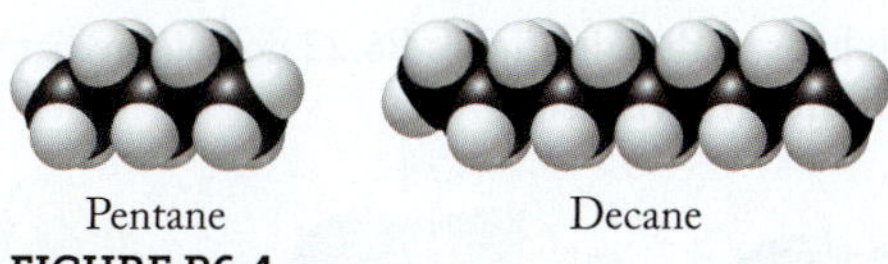


**FIGURE P6.4**

**6.5.** Identify the functional group in each molecule in Figure P6.5.

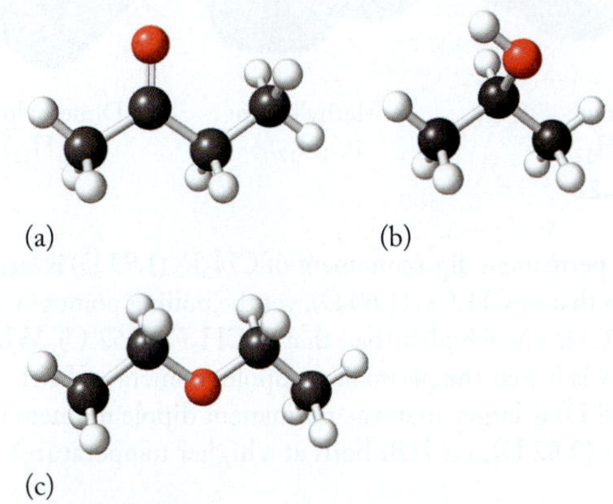


**FIGURE P6.5**

6.6. Which molecules in Figure P6.5 are constitutional isomers?

**6.7.** Figure P6.7 shows the phase diagram of imaginary molecular compound X. If a sample of X is left outside in a sealed container on a summer day, will the X in the container be a solid, liquid, or gas at $P$ = 1.00 atm?

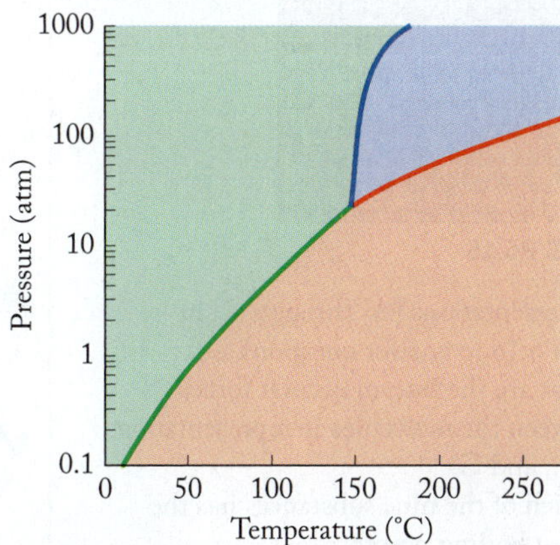


**FIGURE P6.7**

6.8. Suppose you bring the sample of X from Problem 6.7 inside and place it in an uncovered pot of boiling water on your kitchen stove. What phase changes, if any, will occur?

**6.9.** Another sample of X (Figure P6.7) is stored in a pressurized container ($P$ = 100 atm) at 0°C. If the sample is then transferred to an oven and slowly warmed to 250°C, what phase changes, if any, will X undergo?

6.10. Does the green line in Figure P6.7 represent a series of (a) freezing points, (b) sublimation points, (c) boiling points, or (d) critical points?

**6.11.** Does the solid form of compound X (Figure P6.7) float on the liquid as the liquid begins to freeze at $P$ = 300 atm?

6.12. The graph in Figure P6.12 is an expanded view of part of the phase diagram of water. In this version the pressure scale is linear instead of logarithmic. Which phases are represented by the blue and pink colors?

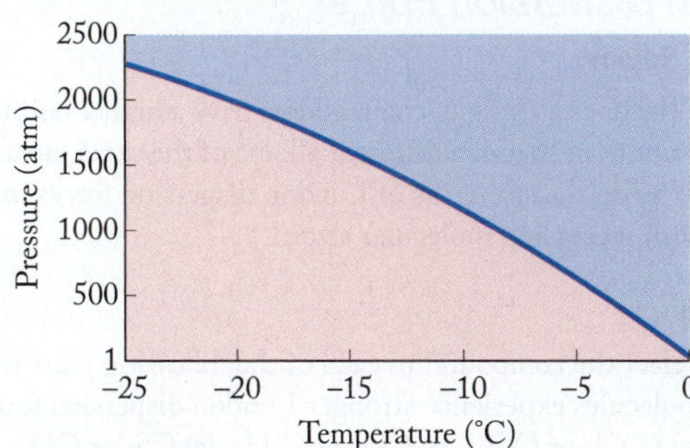


**FIGURE P6.12**

**6.13.** In Figure P6.12, what phase change, if any, takes place if the temperature of a sample of water is increased from −25°C to −15°C while the pressure on it is decreased from 2500 to 1000 atm? Assume that the changes in temperature and pressure occur at constant rates over the same time interval.

*6.14. Suppose an ice skater weighing 50.0 kg and wearing newly sharpened ice skates stands on ice at a temperature of −10°C. If the surface area of the edges of the skates pressing down on the ice is 0.033 $cm^2$, will the ice under the skate edges melt? Figure P6.12 may help you make your prediction.

**6.15.** In Figure P6.15, identify the physical state (solid, liquid, or gas) of xenon and classify the attractive forces between the xenon atoms.

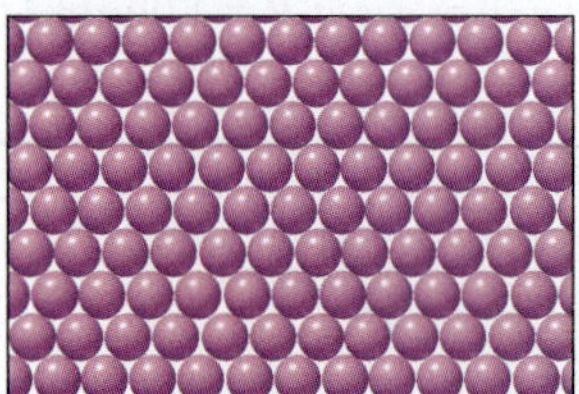

**FIGURE P6.15**

6.16. Use representations [A] through [I] in Figure P6.16 to answer questions a–f.

a. What are the intermolecular forces between the molecules in representations C, E, and G?
b. Which of the nine substances has the lowest boiling point?
c. Which representations are isomers of one another?
d. Which is more soluble in water: ethylene glycol or iodine? Which is more soluble in carbon tetrachloride? Explain your answers in terms of the intermolecular forces involved.
e. Which hydrocarbon has the higher boiling point?
f. Which substance does not form hydrogen bonds between its molecules but can form hydrogen bonds to another molecule?

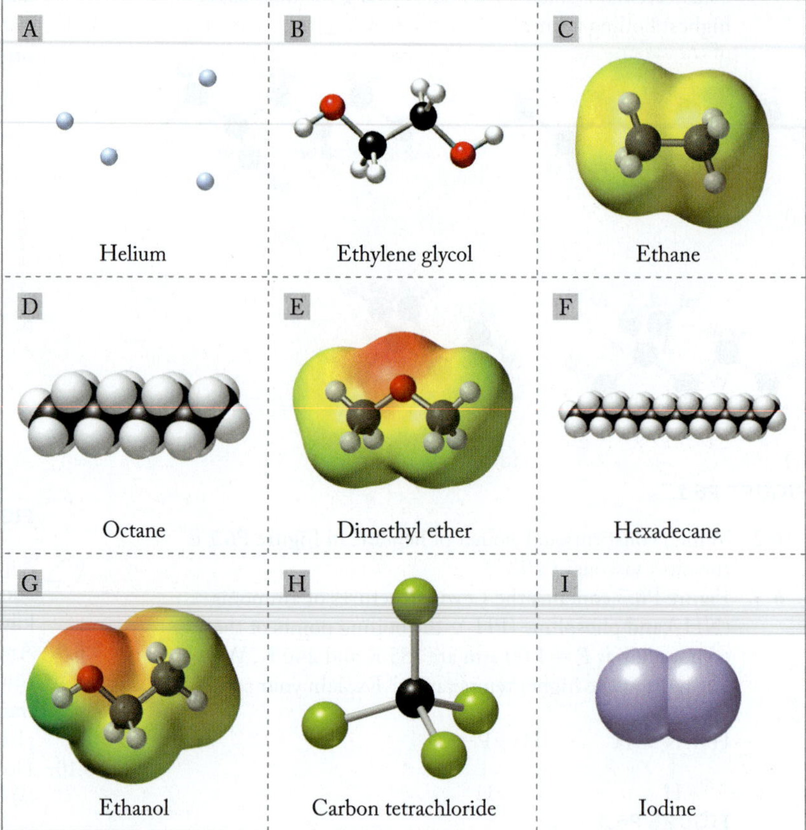


**FIGURE P6.16**

# QUESTIONS AND PROBLEMS

## London Dispersion Forces

### Concept Review

**6.17.** Why does a straight-chain alkane have a higher boiling point than that of a different alkane of the same molar mass?

6.18. Why do the strengths of London dispersion forces increase with increasing molecular size?

### Problems

**6.19.** Select the compound in each of the following pairs whose molecules experience stronger London dispersion forces. (a) $C_2Cl_6$ or $C_2F_6$; (b) $CH_4$ or $C_3H_8$; (c) $CS_2$ or $CO_2$

6.20. The three most abundant gases in air are $N_2$, $O_2$, and Ar. Which has the highest boiling point, and which has the lowest boiling point?

---

**6.21. Fuels from Crude Oil** Petroleum (crude oil) is a complex mixture of mostly hydrocarbons that can be separated into useful fuels by distillation. Common petroleum-based fuels in order of increasing boiling point are gasoline, jet fuel, kerosene, diesel oil, and fuel oil. Which fuel is the most viscous at 20°C?

6.22. Which hydrocarbon in Figure P6.22 do you predict will have the greatest viscosity?

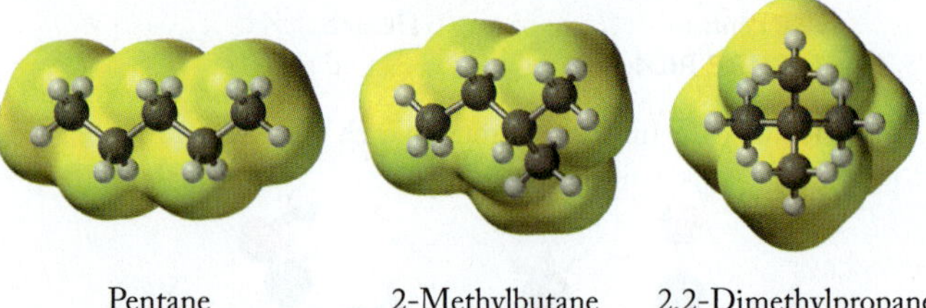

Pentane ($C_5H_{12}$) 2-Methylbutane ($C_5H_{12}$) 2,2-Dimethylpropane ($C_5H_{12}$)

**FIGURE P6.22**

---

**6.23.** The permanent dipole moment of $CH_2F_2$ (1.93 D) is larger than that of $CH_2Cl_2$ (1.60 D), yet the boiling point of $CH_2Cl_2$ (40°C) is much higher than that of $CH_2F_2$ (−52°C). Why?

6.24. How is it that the permanent dipole moment of HCl (1.08 D) is larger than the permanent dipole moment of HBr (0.82 D), yet HBr boils at a higher temperature?

---

**6.25.** In the following pairs of molecules, which experiences the stronger London dispersion forces between molecules of the same compound? (a) $CCl_4$ or $CF_4$; (b) $CH_4$ or $C_3H_8$

6.26. What kinds of intermolecular forces must be overcome as (a) solid $CO_2$ sublimes, (b) $CHCl_3$ boils, and (c) ice melts?

***6.27.** The compound with the molecular structure in Figure P6.27 (b) melts at a higher temperature than the compound in (a). Explain why, using the different types and strengths of the intermolecular forces experienced by the molecules.

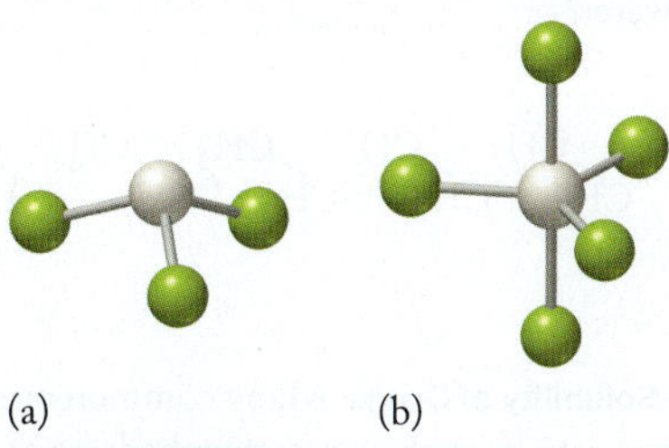

(a) (b)

**FIGURE P6.27**

*6.28. Which of the two compounds with the condensed molecular structures and formulas shown in Figure P6.28 has the higher boiling point?

$ClCH_2CH_2Cl$ $CHCl_2CH_3$

**FIGURE P6.28**

## Interactions Involving Polar Molecules

### Concept Review

**6.29.** How are water molecules oriented around anions in aqueous solutions?

6.30. How are water molecules oriented around cations in aqueous solutions?

**6.31.** Why are dipole–dipole interactions generally weaker than ion–dipole interactions?

6.32. Two liquids—one polar, one nonpolar—have the same molar mass. Which one is likely to have the higher boiling point? Which one is likely to have the greater viscosity?

**6.33.** Why are hydrogen bonds considered a special class of dipole–dipole interactions?

6.34. Can all polar hydrogen-containing molecules form hydrogen bonds?

### Problems

**6.35. Sulfur Compounds in Nature** Cows produce methanethiol, $CH_3SH$, whereas skunks spray a mixture containing butanethiol, $C_4H_9SH$ (Figure P6.35). Both compounds have distinct, unpleasant odors. Why are the boiling points of methane thiol and butane thiol lower than the boiling points of methanol and butanol, the corresponding alcohols?

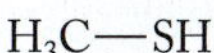

Methane thiol
Boiling point = 279 K

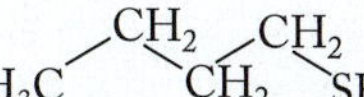

Butane thiol
Boiling point = 371 K

$H_3C$—OH

Methanol
Boiling point = 338 K



Butanol
Boiling point = 391 K

**FIGURE P6.35**

*6.36. **The Smell of the Sea** The distinct odor of the seashore at low tide is due in part to the presence of dimethyl sulfide, $CH_3SCH_3$ (boiling point, 310 K), a molecule with the same structure as dimethyl ether, $CH_3OCH_3$ (boiling point, 250 K). Why is the boiling point of dimethyl ether lower than the boiling point of dimethyl sulfide?

**6.37.** Why do molecules of methanol ($CH_3OH$) form hydrogen bonds, but molecules of methane ($CH_4$) do not?

6.38. Why does hydrogen bonding in $(CH_3CH_2)_2EH$ (where E is a group 15 element) decrease down the group?

**6.39.** In which of the following compounds do the molecules experience dipole–dipole interactions? (a) $CF_4$; (b) $CF_2Cl_2$; (c) $CCl_4$; (d) $CFCl_3$

6.40. In which of the following compounds do molecules experience dipole–dipole interactions? (a) $CO_2$; (b) $NO_2$; (c) $SO_2$; (d) $H_2S$

**6.41.** Which of the following molecules can hydrogen-bond among themselves in pure samples of bulk material? (a) methanol ($CH_3OH$); (b) ethane ($CH_3CH_3$); (c) dimethyl ether ($CH_3OCH_3$); (d) acetic acid ($CH_3COOH$)

6.42. Which of the following molecules can hydrogen-bond with molecules of water? (a) methanol ($CH_3OH$); (b) ethane ($CH_3CH_3$); (c) dimethyl ether ($CH_3OCH_3$); (d) acetic acid ($CH_3COOH$)

(Before solving the following problems, you may find it useful to review Section 4.1 on the strengths of ionic bonds.)

**6.43.** In an aqueous solution containing chloride, bromide, and iodide ions, which anion would you expect to experience the strongest ion–dipole interactions with surrounding water molecules?

6.44. In an aqueous solution containing $Na^+$, $Mg^{2+}$, $K^+$, and $Ca^{2+}$ ions, which cation would you expect to experience the strongest ion–dipole interactions?

## Trends in Solubility

### Concept Review

**6.45.** What is the difference, if any, between the terms *miscible* and *soluble*?

6.46. Which of the following substances have little solubility in water? (a) $CH_4$; (b) benzene, $C_6H_6$; (c) KBr; (d) Ar

**6.47.** In what context do the terms *hydrophobic* and *hydrophilic* relate to the solubilities of substances in water?

6.48. A series of alcohols has the generic molecular formula $C_nH_{(2n+1)}OH$. How does the value of $n$ affect the solubility of these alcohols in water?

**6.49.** Which of the two compounds in Figure P6.27 is more soluble in water? Explain your reasoning.

*6.50. Are the two compounds in Figure P6.28 miscible with each other? Why do you think they are or are not?

## Problems

**6.51.** Which of the following compounds is likely to be the most soluble in water? (a) NaCl; (b) KI; (c) $Ca(OH)_2$; (d) CaO?

6.52. Which of the following compounds is likely to be the most soluble in water?
a. $CH_3CH_2OCH_2CH_3$
b. $CS_2$
c. $CH_3CH_2CH_2CH_3$
d. $CH_3CH_2OH$

---

**6.53.** Which of the following pairs of substances are likely to be miscible?
a. $Br_2$ and benzene ($C_6H_6$)
b. $CH_3CH_2OCH_2CH_3$ (diethyl ether) and $CH_3COOH$ (acetic acid)
c. $C_6H_{12}$ (cyclohexane) and hexane ($CH_3CH_2CH_2CH_2CH_2CH_3$)
d. $CS_2$ (carbon disulfide) and $CCl_4$ (carbon tetrachloride)

6.54. Which of the following pairs of substances is likely to be miscible?
a. $CH_3CH_2OH$ (ethanol) and $CH_3CH_2OCH_2CH_3$ (diethyl ether)
b. $CH_3OH$ (methanol) and methyl amine ($CH_3NH_2$)
c. $CH_3CN$ (acetonitrile) and acetone ($CH_3COCH_3$)
d. $CF_3CHF_2$ (a Freon replacement) and $CH_3CH_2CH_2CH_2CH_3$ (pentane)

---

**6.55. Flex-Fuel Vehicles** Many automobile manufacturers offer vehicles that can burn a range of fuels from gasoline to alcohols, including various blends of hydrocarbons with alcohols. Ethanol ($CH_3CH_2OH$) from corn is the most frequent choice, but industrial processes for making butanol ($CH_3CH_2CH_2CH_2OH$) have also been developed.
a. Why does ethanol dissolve readily in hydrocarbons containing between 5 and 10 carbon atoms?
b. Diesel fuels contain hydrocarbons ranging in size from $C_{14}H_{30}$ to $C_{20}H_{42}$. Is ethanol or butanol more likely to have greater solubility in diesel fuels?

*6.56. **Shampoos** Many shampoos contain ionic compounds such as sodium lauryl sulfate (Figure P6.56) as the first ingredient after water. The function of this compound is to dissolve dirt particles, which are largely nonpolar.
a. Which part of the sodium lauryl sulfate molecule is likely to dissolve dirt particles?
b. How does the sulfate group make sodium lauryl sulfate soluble in water?

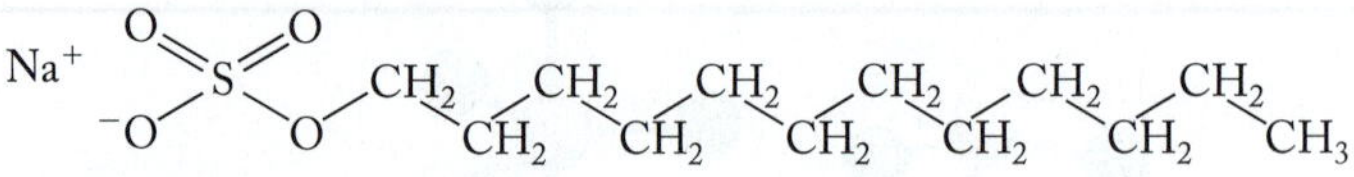

**FIGURE P6.56**

---

***6.57. Improving the Solubility of Drugs** Many common over-the-counter medications such as pseudoephedrine and acetylsalicylic acid are formulated with either sodium cations or chloride anions to improve their water solubility. Why would the structures in Figure P6.57 have improved solubility in water?

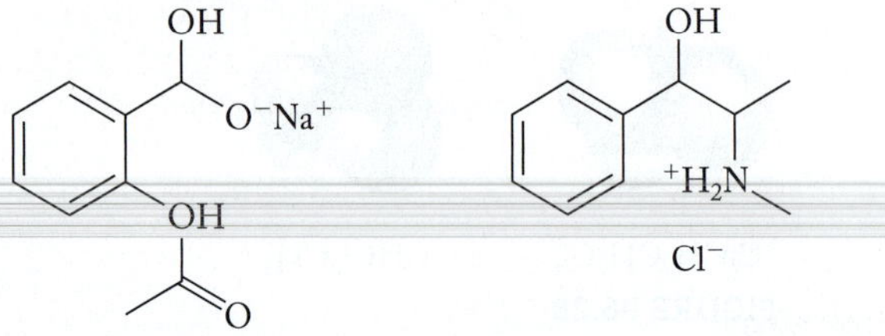

Sodium acetylsalicyclate    Pseudoephedrine hydrochloride

**FIGURE P6.57**

6.58. Rank the ketones in Figure P6.58 from least soluble to most soluble in water.

(a) (b)

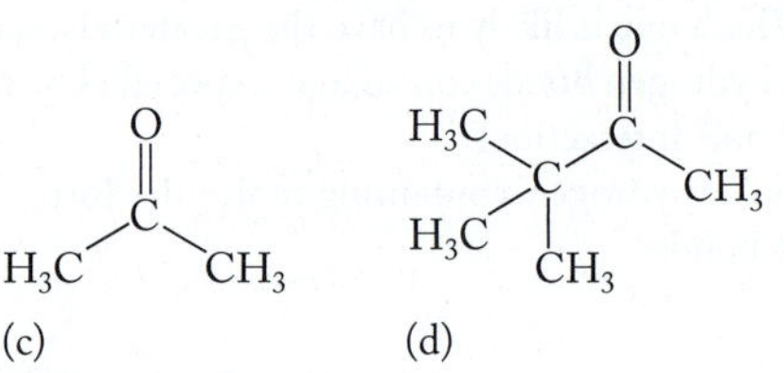

**FIGURE P6.58**

## Phase Diagrams: Intermolecular Forces at Work

### Concept Review

**6.59.** Explain the difference between sublimation and evaporation.

6.60. Can ice be melted merely by applying pressure? How about dry ice? Explain your answers.

**6.61.** Explain the meaning of the term *equilibrium line* as it is used in Section 6.5.

6.62. Explain how the solid–liquid line in the phase diagram of water (Figure 6.25) differs in character from the solid–liquid line in the phase diagrams of most other substances, such as $CO_2$ (Figure 6.27).

**6.63.** Which phase of a substance (gas, liquid, or solid) is most likely to be the stable phase: (a) at low temperatures and high pressures; (b) at high temperatures and low pressures?

6.64. At what temperatures and pressures does a substance behave as a supercritical fluid?

**6.65.** **Preserving Food** Freeze-drying is used to preserve food at low temperature with minimal loss of flavor. Freeze-drying works by freezing the food and then lowering the pressure with a vacuum pump to sublime the ice. Must the pressure be lower than the pressure at the triple point of $H_2O$?

6.66. Solid helium cannot be converted directly into the vapor phase. Does the phase diagram of He have a triple point?

### Problems

(To solve Problems 6.67–6.76, consult Figures 6.25 and 6.27.)

**6.67.** What is the boiling point of water at a pressure of 200 atm?

6.68. What is the freezing point of water at a pressure of 1000 atm?

---

**6.69.** What phase changes, if any, does liquid water at 100°C undergo if the initial pressure of 5.0 atm is reduced to 0.5 atm at constant temperature?

6.70. What phase changes, if any, occur if a sample of $CO_2$ initially at −80°C and 5.0 atm is warmed to −25°C at constant pressure?

---

**6.71.** Below what temperature can solid $CO_2$ (dry ice) be converted into $CO_2$ gas simply by lowering the pressure?

6.72. What is the maximum pressure at which solid $CO_2$ (dry ice) can be converted into $CO_2$ gas without melting?

---

**6.73.** Predict the phase of water at 100°C and 5.0 atm. What phase changes, if any, does water undergo when the pressure is reduced to 0.5 atm at constant temperature?

6.74. Predict the phase of $CO_2$ at −80°C and 8.0 atm. What phase changes, if any, does $CO_2$ undergo when allowed to warm to −25°C at 8.0 atm?

---

**6.75.** **Glaciers** Melting glaciers in Greenland could significantly raise the level of the ocean. An article on the Jakobshavn glacier contained the following passage: "Because of the extreme pressures involved, the freezing point of the ice itself changes, making it more susceptible to melting."

a. Is this statement valid?

b. Use the phase diagram of water to estimate the pressure required to reduce the melting point of ice by 0.5°C.

6.76. **Sequestration of Carbon Dioxide** One solution to combat the increasing concentration of $CO_2$ in Earth's atmosphere is to capture and store it under high pressure.

a. What pressures are required to convert $CO_2$ gas to solid $CO_2$ (dry ice) and to liquid $CO_2$ at −50°C?

b. What volume would $1.0 \times 10^6$ g of $CO_2$ vapor ($d = 1.977$ g/L at 0°C) occupy when cooled and condensed to liquid $CO_2$ ($d = 1.101$ g/mL at −37°C)?

## Some Remarkable Properties of Water

### Concept Review

**6.77.** Explain why a needle floats on water but sinks in a container of methanol.

6.78. Explain why different liquids do not reach the same height in capillary tubes of the same diameter.

**6.79.** Explain why the plumbing in a house may burst if the temperature in the house drops below 0°C.

*6.80. A needle sinks when placed on hot (90°C) water but floats when placed on cold (5°C) water. Why?

**6.81.** The meniscus of mercury in a mercury thermometer (Figure P6.81) is convex. Why?

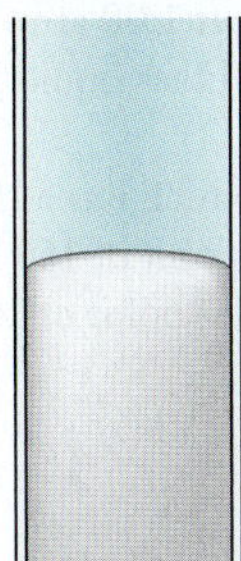

**FIGURE P6.81**

6.82. The mercury level in a capillary tube inserted into a dish of mercury is below the surface of the mercury in the dish. Why?

**6.83.** Describe the origin of surface tension in terms of intermolecular interactions.

6.84. Why does the density of water increase when it is cooled from 6°C to 4°C but then increase when it is cooled from 4°C to 2°C?

**6.85.** One of two glass capillary tubes of the same diameter is placed in a dish of water and the other is placed in a dish of ethanol ($CH_3CH_2OH$). Which liquid will rise higher in its tube?

6.86. Would you expect water to rise to the same height in a tube made of a polyethylene plastic as it does in a glass capillary tube of the same diameter? The molecular structure of a repeating unit in polyethylene is shown in Figure P6.86.

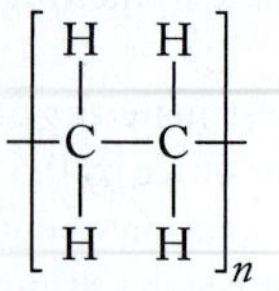

**FIGURE P6.86**

## Additional Problems

**6.87.** Which type of intermolecular force exists in all substances?

6.88. Why does water climb higher in a capillary tube than in a test tube?

**6.89.** Methanol has a larger molar mass than that of water but boils at a lower temperature. Suggest a reason why.

6.90. What kinds of intermolecular forces must be overcome as (a) solid $CO_2$ sublimes; (b) $CHCl_3$ boils; (c) ice melts?

***6.91.** Which of these compounds has the lowest boiling point? (a) $CH_4$; (b) $CH_3Cl$; (c) $CH_2Cl_2$; (d) $CHCl_3$; (e) $CCl_4$

*6.92. A partially filled bottle contains an equal number of moles of methanol ($CH_3OH$), ethanol ($CH_3CH_2OH$), and propanol ($CH_3CH_2CH_2OH$). Which of these alcohols would be present in the largest amount in the vapor above the liquid mixture in the bottle?

**6.93.** Does the sublimation point of ice increase or decrease with increasing pressure?

6.94. Why is methanol miscible with water but methane is not?

***6.95.** The melting point of hydrogen is 15.0 K at 1.00 atm. The temperature of its triple point is 13.8 K. Does liquid $H_2$ expand or contract when it freezes?

6.96. Sketch a phase diagram for element Z, which has a triple point at 152 K and 0.371 atm, a boiling point of 166 K at a pressure of 1.00 bar, and a normal melting point of 161 K.

***6.97.** From among the compounds with the molecular structures shown in Figure P6.97, pick the ones that you think should be soluble in both water and octanol (see Sample Exercise 6.6).

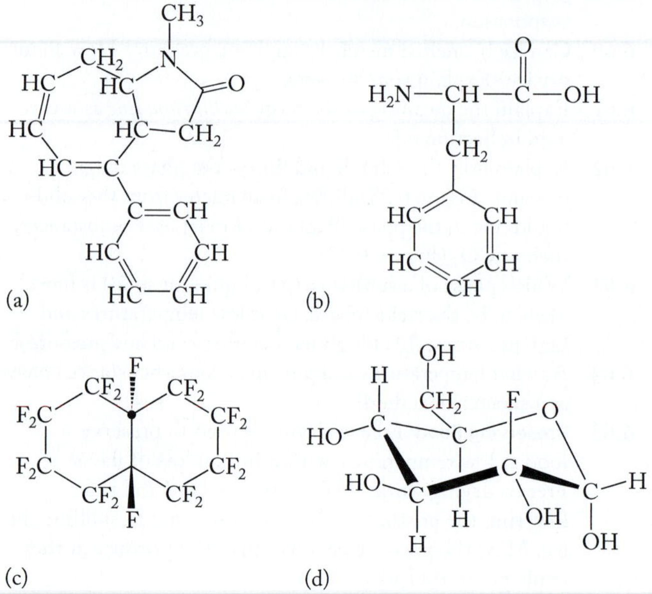

(a) (b) (c) (d)

**FIGURE P6.97**

*6.98. **First Aid for Bruises** Compounds with low boiling points may be sprayed on the skin as a topical anesthetic. Such compounds chill the skin as they evaporate and provide short-term relief from injuries. Predict which compound among those in **Figure P6.98** has the lowest boiling point.

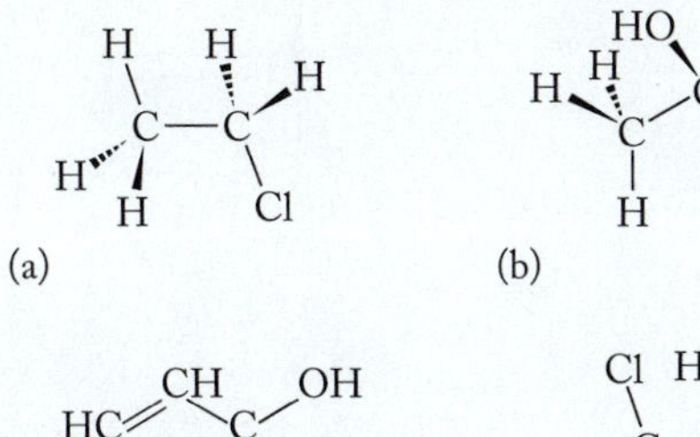


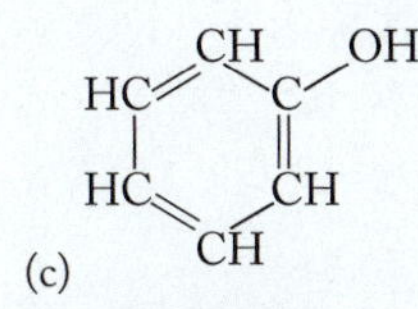


(d)

**FIGURE P6.98**

***6.99.** The graph in Figure P6.99 shows the boiling points of three compounds—$CH_3CH_2OCH_2CH_3$, $CH_3CH_2CH_2CH_2OH$, and $CH_3CH_2SCH_2CH_3$—plotted against their dipole moments. Which compound corresponds to each of the points labeled a, b, and c?

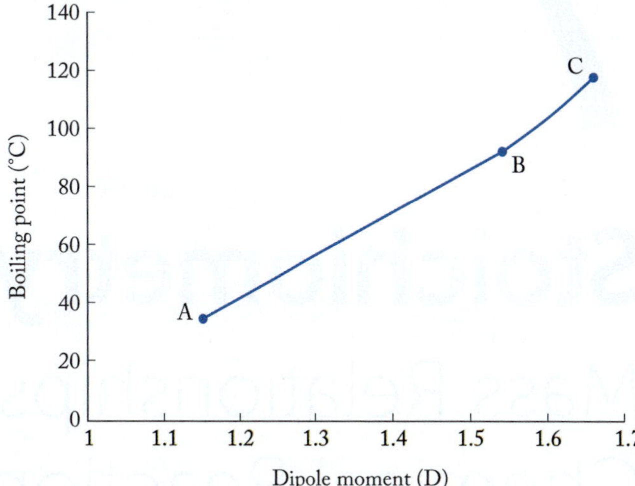


**FIGURE P6.99**

6.100. For each of the following pairs of substances, (a) identify all the intermolecular forces present between particles in a sample of each substance, and (b) predict which substance is likely to be more soluble in $CCl_4$.

a. $Br_2$ or NaBr
b. $CH_3CH_2OH$ or $CH_3OCH_3$
c. $CS_2$ or KOH
d. $I_2$ or $CaF_2$

# 7

# Stoichiometry

## Mass Relationships and Chemical Reactions

**PHOTOSYNTHESIS BY MICROORGANISMS** Intense blooms of cyanobacteria turn a slow-moving river bright green in Puy-de-Dôme, France.

PARTICULATE **REVIEW**

### *Molecular Formulas and Ratios*

In Chapter 7, we explore the composition of pure substances and the relationship between the quantities of reactants and products in a chemical reaction. The molecular structures of glucose and lactic acid—two components of carbohydrate metabolism—are shown here.

- Write each compound's molecular formula.
- What is each compound's ratio of carbon to hydrogen to oxygen?
- What functional groups are present in these compounds?

(Review Sections 1.6 and 6.3 if you need help.)

*(Answers to Particulate Review questions are in the back of the book.)*

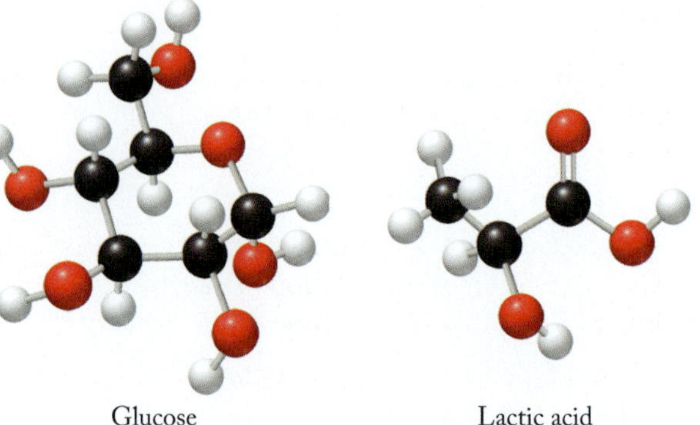

PARTICULATE **PREVIEW**

### *Unreacted Reactants*

Nitromethane, $CH_3NO_2$, is a fuel used in drag racing. As you read Chapter 7, look for ideas that will help you answer these questions:

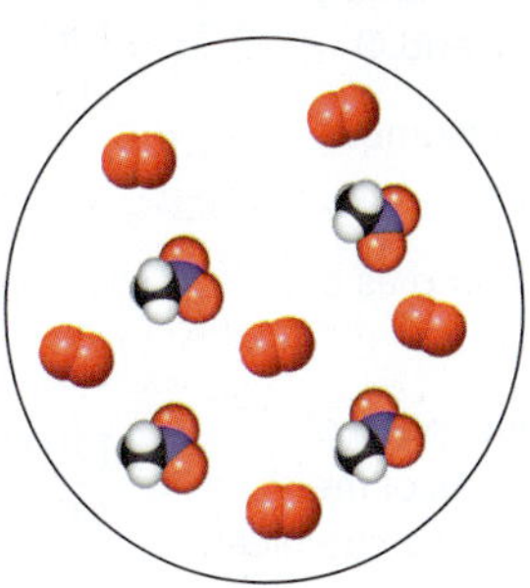

- What products are formed when a mixture of $CH_3NO_2$ and $O_2$ reacts? *Hint*: The nitrogen in the reactant is converted into $N_2(g)$ during the reaction.
- Write a balanced chemical equation describing the combustion of nitromethane.
- When the mixture shown in the circle reacts, how many $CH_3NO_2$ molecules remain unreacted? How many $O_2$ molecules remain unreacted?

## Learning Outcomes

**LO1** Write balanced chemical equations to describe chemical reactions
**Sample Exercises 7.1–7.3**

**LO2** Use mole ratios from balanced chemical equations to relate the masses of substances reacted and the masses of products formed
**Sample Exercises 7.4, 7.5**

**LO3** Identify the limiting reactant in a chemical reaction
**Sample Exercise 7.6**

**LO4** Calculate the theoretical and percent yields in a chemical reaction
**Sample Exercises 7.7, 7.8**

**LO5** Interconvert the chemical formula and percent composition of a compound
**Sample Exercises 7.9–7.11**

**LO6** Determine the molecular formula of a compound from its percent composition and molecular mass
**Sample Exercise 7.12**

**LO7** Use data from combustion analysis to determine a substance's empirical formula
**Sample Exercise 7.13**

## 7.1 Chemical Reactions and the Carbon Cycle

**CHEMTOUR** Carbon Cycle

Elements essential to the chemistry of life, including carbon, nitrogen, phosphorus, and sulfur, circulate through the environment by way of intricate chemical webs, or cycles, that operate both on a global scale and locally within individual ecosystems. We begin this chapter by examining some of the chemical reactions in the carbon cycle, which includes the six pathways shown in **Figure 7.1**. Five of these pathways have been operating for a very long time—ever since a phylum of bacteria called cyanobacteria evolved about 2.5 billion years ago. These organisms are still an abundant component of Earth's biosphere, as seen in this chapter's opening photo.

**CONNECTION** In Section 3.3, we saw that the energy of a photon can be calculated by multiplying the Planck constant ($h$) by the frequency ($\nu$) of the photon.

Cyanobacteria played a key role in launching the carbon cycle because they were among the first organisms capable of **photosynthesis**. Photosynthesis is a set of chemical reactions driven by energy from sunlight that, in its entirety, consumes carbon dioxide and water and turns these starting materials, or **reactants**, into different **products** depending on the organism involved. The photosynthesis that occurs in cyanobacteria and in green plants today produces $O_2$ gas and glucose ($C_6H_{12}O_6$), which serves as an energy source and as a building block for much of the biomass in all living things (see pathways ① and ② in Figure 7.1).

Equation 7.1, an example of a **chemical equation**, represents photosynthesis with symbols. Note that the formulas of the two reactants appear on the left, the formulas of the products are on the right, and a reaction arrow between them shows the direction of the reaction.

$$6\,CO_2(g) + 6\,H_2O(\ell) \xrightarrow{h\nu} C_6H_{12}O_6(aq) + 6\,O_2(g) \qquad (7.1)$$

The letters $h\nu$ over the reaction arrow indicate that the reaction requires photons of electromagnetic energy (sunlight). The letters in parentheses indicate the physical states of the reactants and products: ($g$) for gases, ($\ell$) for liquids, and ($aq$) for substances dissolved in water, which means that they are in an *aqueous* solution. Had a solid reactant or product been present, its formula would have been followed by ($s$).

Equation 7.1, like all properly written chemical equations, is *balanced*, which means that the total number of atoms of each element is the same on the left and right sides of the reaction arrow. These atom counts must be equal because atoms do not change their identities in chemical reactions. This need for balance is the

**photosynthesis** a set of chemical reactions driven by the energy of sunlight that convert carbon dioxide and water into oxygen and a variety of molecules containing C, H, and O.

**reactants** substance(s) consumed during a chemical reaction.

**products** substance(s) formed as a result of a chemical reaction.

**chemical equation** a description of the identities and proportions of the reactants and the products in a chemical reaction.

**law of conservation of mass** the principle that the sum of the masses of the reactants in a chemical reaction is equal to the sum of the masses of the products.

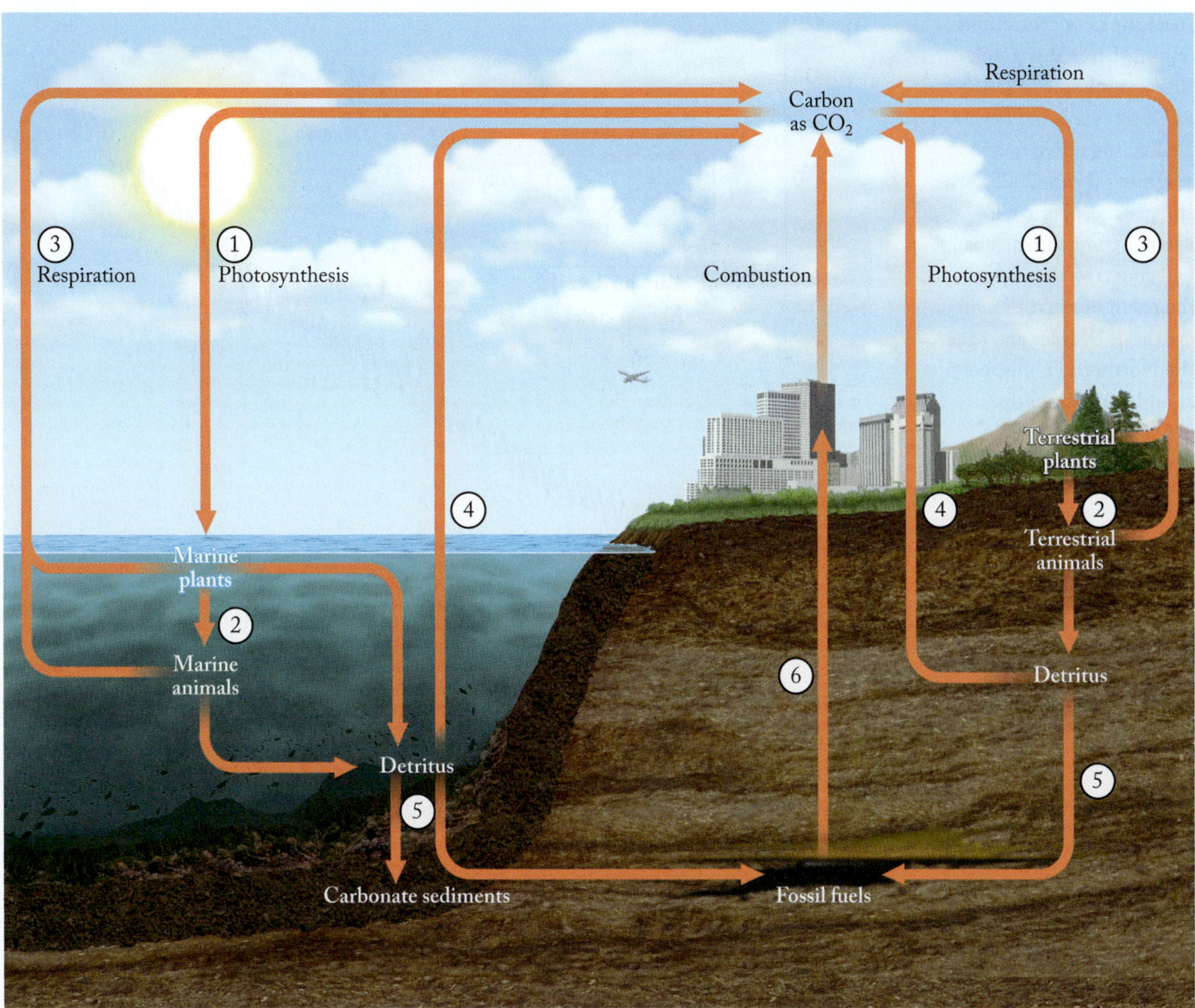


**FIGURE 7.1** The carbon cycle. ① Terrestrial plants and marine plants incorporate $CO_2$ into their biomass. ② Some of the plant biomass becomes the biomass of animals. ③ As plants and animals respire, they release $CO_2$ back into the environment. ④ When they die, the decay of their tissues releases most of their carbon content as $CO_2$, but about 0.01% is incorporated into ⑤ carbonate minerals and deposits of coal, petroleum, and natural gas (fossil fuels). ⑥ Mining and the combustion of fossil fuels for human use are returning this carbon to the atmosphere.

reason that the molecular formulas $CO_2$, $H_2O$, and $O_2$ are each preceded by the number 6, which means that six molecules of $CO_2$ and six molecules of $H_2O$ are needed to produce one molecule of glucose and six molecules of $O_2$.

Equal numbers of atoms also means that the sum of the masses of the reactants always equals the sum of the masses of the products in a chemical equation. This equality is known as the **law of conservation of mass**, and it applies to all chemical reactions.

To harvest the chemical energy in glucose, plants and animals rely on a series of reactions called *respiration*, in which photosynthesis runs in reverse (pathway ③ in Figure 7.1):

$$C_6H_{12}O_6(aq) + 6\ O_2(g) \rightarrow 6\ CO_2(g) + 6\ H_2O(\ell) \quad (7.2)$$

On a global scale, photosynthesis and respiration proceed at nearly, but not exactly, the same rates. Each year, about 0.01% of the decaying mass of dead plants and animals (called *detritus*) is incorporated into sediments and soils. Shielded from atmospheric $O_2$, this buried carbon is not converted back into $CO_2$, at least not right away. Although 0.01% may not seem like much, over more than 2 billion years the process has removed about $10^{20}$ kg of carbon dioxide from the atmosphere. About $10^{15}$ kg of the buried carbon is in the form of fossil fuels: coal, petroleum, and natural gas.

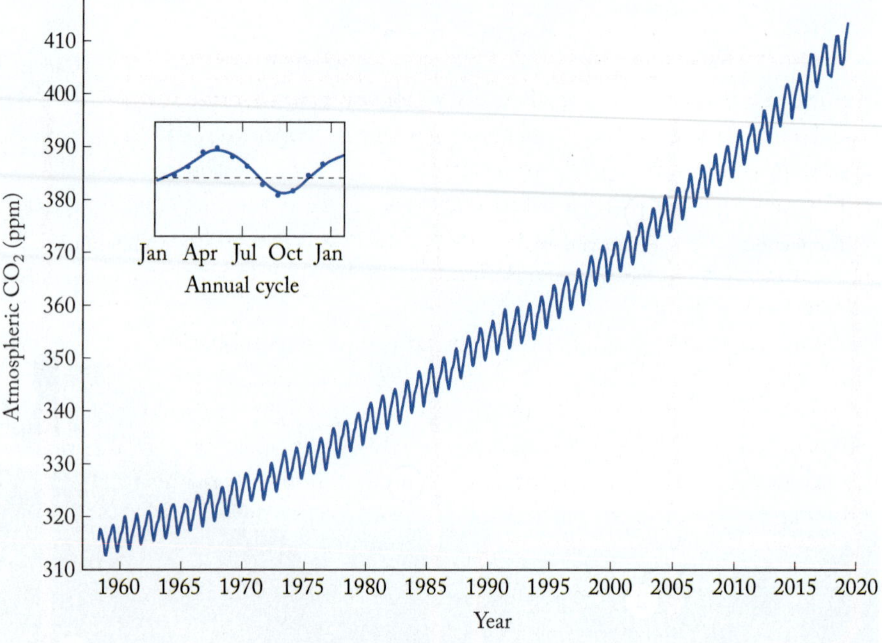


**FIGURE 7.2** Increasing concentrations of $CO_2$ in the atmosphere. This graph is based on analyses done since 1958 at the Mauna Loa Observatory in Hawaii and is called the Keeling curve to honor Charles David Keeling (1928–2005), who began determining atmospheric $CO_2$ concentrations long before climate change was a widespread concern. The annual cycle in the data is the result of the different rates of photosynthesis by trees and other vegetation in the Northern Hemisphere during summer and winter months.

Fossil fuels have served as the world's primary energy resource for more than a century, with the rate of consumption of these fuels increasing rapidly since 1950. In 2018, the combustion of fossil fuels and a variety of industrial processes converted 10.1 trillion ($1.01 \times 10^{13}$) kg of fossil carbon into fossil carbon dioxide gas, which was released into the atmosphere.[1] As a result, the atmospheric concentration of $CO_2$ has increased sharply, especially over the past 60 years (**Figure 7.2**). It is now more than 400 parts per million (ppm). To put this value in perspective: atmospheric $CO_2$ concentrations oscillated between 190 and 280 ppm for the previous 800,000 years and probably longer (**Figure 7.3**). The concentration of $CO_2$ in the atmosphere today is more than 40% higher than at any time between when our species evolved and the beginning of the Industrial Revolution.

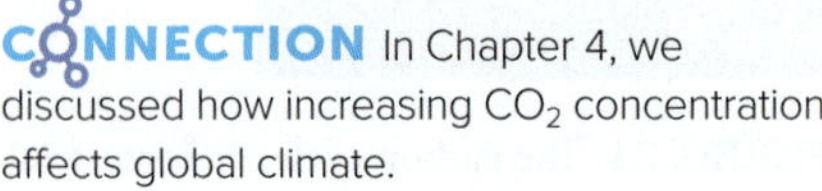
**CONNECTION** In Chapter 4, we discussed how increasing $CO_2$ concentration affects global climate.

[1]*Carbon Budget 2018* report, Global Carbon Project (www.globalcarbonproject.org).

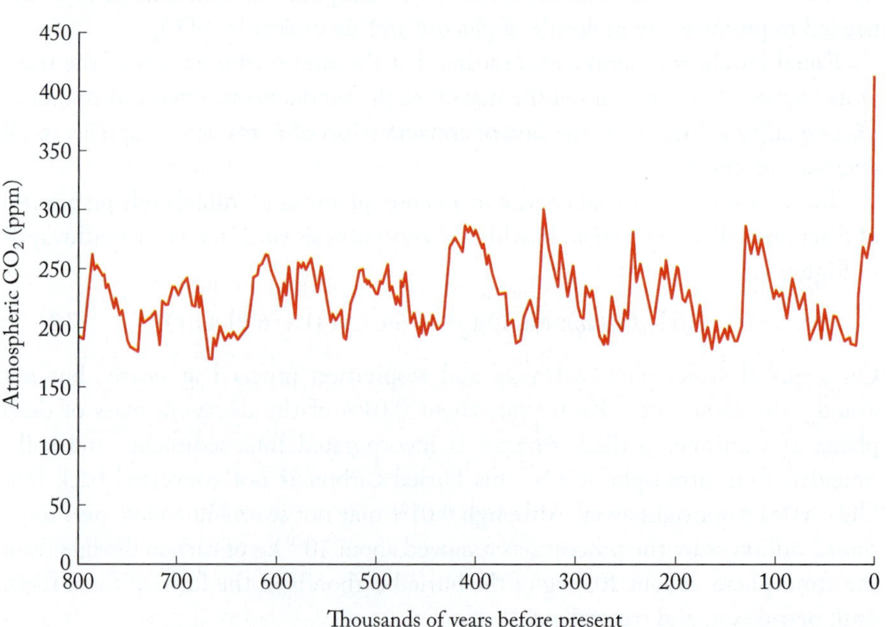


**FIGURE 7.3** History of the atmospheric concentration of $CO_2$. Analyses of bubbles of air trapped in Antarctic glaciers have enabled scientists to create a history of atmospheric $CO_2$ levels over the past 800,000 years. The data show that atmospheric concentrations of $CO_2$ oscillated between about 190 and 280 ppm. Low $CO_2$ concentrations coincide with ice ages in the Northern Hemisphere. High $CO_2$ concentrations correspond to relatively warm periods called interglacials, including the one we are now experiencing. The most recent data are based on the Keeling curve in Figure 7.2.

# 7.2 Writing Balanced Chemical Equations

In this chapter, we use balanced chemical equations to describe several reactions linked to the carbon cycle and energy production as well as to relate the quantities of substances consumed and produced in these reactions. You must be able to write complete, balanced chemical equations if given the identities of at least some of the reactants and products involved. To help you develop this ability, let's start with simple **combination reactions** in which two or more reactants combine to form a single product.

Some important combination reactions are also **combustion** reactions, in which fuel rapidly reacts with $O_2$ and in the process releases energy. One of the simplest combustion reactions involves the complete combustion of carbon, the main component of coal. The product of this reaction is carbon dioxide. At the atomic level, forming one molecule of carbon dioxide takes just one atom of carbon and one molecule (two atoms) of oxygen:

CHEMTOUR Balancing Equations

Translating this atom-scale view of the reaction into a chemical equation, we get

$$C(s) + O_2(g) \rightarrow CO_2(g) \quad (7.3)$$

Note that Equation 7.3 is already balanced when written with one atom or one molecule of each substance.

## Moles and Chemical Equations

During the combustion of coal that contains sulfur impurities, a combination reaction between sulfur and oxygen occurs:

$$S(s) + O_2(g) \rightarrow SO_2(g) \quad (7.4)$$

The lack of any coefficients in Equation 7.4 tells us that *one atom* of sulfur reacts with *one molecule* of oxygen, producing *one molecule* of sulfur dioxide. Recall that in Chapter 2, we developed the concepts of the mole and molar mass to describe macroscopic quantities of substances in terms of the number of particles they contain.

**CONNECTION** In Section 2.5, we learned that one mole of a substance contains a number of particles equal to the Avogadro constant ($N_A = 6.022 \times 10^{23}$) and has a mass known as its molar mass.

The concepts of the mole and molar mass enable us to interpret Equation 7.4 in three more ways:

1. The coefficients in the chemical equation tell us that one *mole* of sulfur reacts with one *mole* of oxygen, producing one *mole* of sulfur dioxide.
2. The Avogadro constant tells us that $6.022 \times 10^{23}$ *atoms* of S react with $6.022 \times 10^{23}$ *molecules* of $O_2$, forming $6.022 \times 10^{23}$ *molecules* of $SO_2$.
3. The molar masses of the reactants and products allow us to say that 32.06 g of S reacts with 32.00 g of $O_2$ to produce 64.06 g of $SO_2$.

The several interpretations of this equation, however, do not limit the amount of S that can react to only 32.06 g. Reacting twice that amount, or 64.12 g of S, is possible as long as twice the molar mass of oxygen, 64.00 g of $O_2$, is available. That is, 2 moles of S can react with 2 moles of $O_2$ to produce 2 moles of $SO_2$. In

**combination reaction** a reaction in which two or more substances form a single product.

**combustion** a rapid reaction between fuel and oxygen that produces energy.

**TABLE 7.1** Conservation of Mass in the Combustion of Sulfur to Form $SO_2$

| Variable | S(*s*) | + | $O_2(g)$ | → | $SO_2(g)$ |
|---|---|---|---|---|---|
| **Particle Ratio** | 1 atom | + | 1 molecule | → | 1 molecule |
| **Mole Ratio** | 1 mol | + | 1 mol | → | 1 mol |
| **Mass Ratio** | 32.06 g | + | 32.00 g | → | 64.06 g |
| **General Case (moles)** | *x* mol | + | *x* mol | → | *x* mol |
| **General Case (masses)** | *x*(32.06 g) | + | *x*(32.00 g) | → | *x*(64.06 g) |

fact, the equation tells us that *any* number of moles (*x* mol) of S can react with an equal number of moles (*x* mol) of $O_2$ to produce the same quantity (*x* mol) of $SO_2$. This interpretation is valid because the mole ratio (the ratio of the coefficients) of S to $O_2$ to $SO_2$ in the equation is 1:1:1. From Equation 7.4, we can determine how much of one reactant is needed to completely react with any quantity of the other reactant. We can also calculate how much product can be made from any quantity of reactants by multiplying *x* by the appropriate molar mass. We summarize these interpretations in **Table 7.1**.

In other words, balanced chemical equations are consistent with the law of conservation of mass because (1) the total number of atoms (and hence moles) of each element to the left of the reaction arrow must equal the total number of atoms (and moles) of that element to the right of the reaction arrow, and (2) the identity of atoms does not change in a chemical reaction. Note that the mole ratio differs when different substances combine and is therefore specific for a given reaction. For example, in the reaction depicted in Equation 7.2 between glucose and oxygen:

$$C_6H_{12}O_6(aq) + 6\ O_2(g) \rightarrow 6\ CO_2(g) + 6\ H_2O(\ell)$$

the mole ratio of $C_6H_{12}O_6$ to $O_2$ is 1:6.

**SAMPLE EXERCISE 7.1** Visualizing Chemical Reactions **LO1**

Using blue spheres to represent nitrogen atoms and red spheres to represent oxygen atoms, sketch a picture describing the reaction between two molecules of $N_2$ and four molecules of $O_2$ to form $NO_2$.

**Collect, Organize, and Analyze** We are given the chemical formulas for two reactants and will use appropriately colored spheres to sketch their conversion to a product. We will use the atoms contained in two molecules of nitrogen and four molecules of oxygen to construct the appropriate number of $NO_2$ molecules.

**Solve** From a total of four nitrogen atoms and eight oxygen atoms, we can make four molecules of $NO_2$:

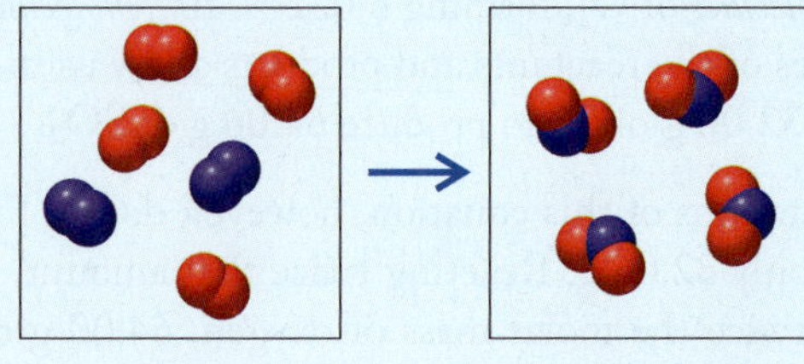

**Think About It** A chemical reaction involves rearrangements of atoms and molecules, resulting in new molecules. The number of each type of atom present before the reaction equals the number of each type of atom present after the reaction.

**Practice Exercise** Using colored spheres to represent nitrogen atoms (blue) and hydrogen atoms (white), sketch a picture describing the reaction between two molecules of $N_2$ and six molecules of $H_2$ to form ammonia, $NH_3$.

*(Answers to Practice Exercises are in the back of the book.)*

Let's practice writing a balanced chemical equation for another reaction. When the $SO_2$ produced in Equation 7.4 enters the atmosphere as coal is burned, it may form sulfur trioxide in another combination reaction:

$$SO_2(g) + O_2(g) \rightarrow SO_3(g) \qquad (7.5)$$

We know the formulas of the reactants and products, but we don't know the mole ratios of the two reactants and the product in the balanced chemical equation. Several approaches exist for writing chemical equations. Let's apply the following four-step method, which works well if we have correctly identified all the reactants and products.

1. *Write a preliminary expression containing a single particle (atom, molecule, or formula unit) of each reactant and product with a reaction arrow separating reactants from products. Include phase symbols indicating physical states.*

Below is such an expression for the reaction of interest, including particulate models of the molecules involved.

$$SO_2(g) + O_2(g) \rightarrow SO_3(g)$$

2. *Check whether the expression is balanced by counting the atoms of each element on each side of the reaction arrow.*

Adding up the atoms of each element in the reactants and products, we find that the expression above is balanced in terms of the numbers of S atoms but not in terms of O atoms: four oxygen atoms are on the reactant side, but only three are on the product side.

| Element | Reactant Side | Product Side | Balanced? |
|---|---|---|---|
| S | 1 | 1 | ✔ |
| O | 2 + 2 = 4 | 3 | ✗ |

3. *Choose an element that appears in only one reactant and one product and balance it first.*

In this example, the only element that occurs only once on each side of the reaction arrow is S, but it is already balanced, so we can skip this step.

4. *Choose coefficients for the other substances so that the number of atoms for each element is the same on both sides of the reaction arrow.*

We need one more O atom on the product side to have a balanced equation. Changing the coefficient in front of $SO_3$ from 1 to 2 increases the number of O atoms from three for one molecule of $SO_3$ to six for two molecules of $SO_3$:

$$SO_2(g) + O_2(g) \rightarrow 2\ SO_3(g)$$

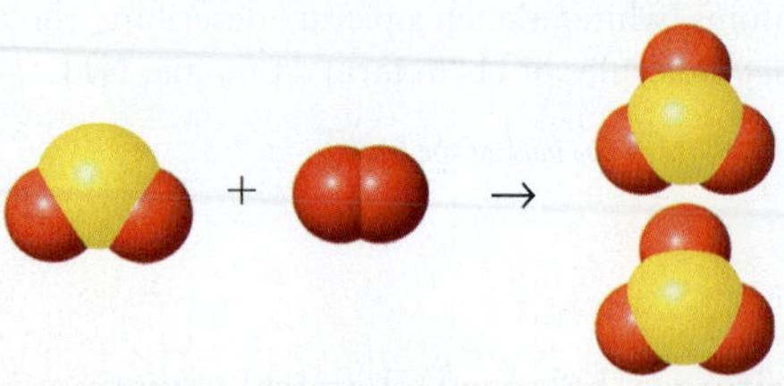

| Element | Reactant Side | Product Side | Balanced? |
|---|---|---|---|
| S | 1 | 2 × 1 = 2 | ✗ |
| O | 2 + 2 = 4 | 2 × 3 = 6 | ✗ |

However, by increasing the number of oxygen atoms on the product side of the equation, we have also increased the number of sulfur atoms, and thus the number of sulfur atoms on the product side is now two while the number on the reactant side is still just one. We need to restore the balance of the sulfur atoms by changing the coefficient for $SO_2$ from 1 to 2:

$$2\ SO_2(g) + O_2(g) \rightarrow 2\ SO_3(g)$$

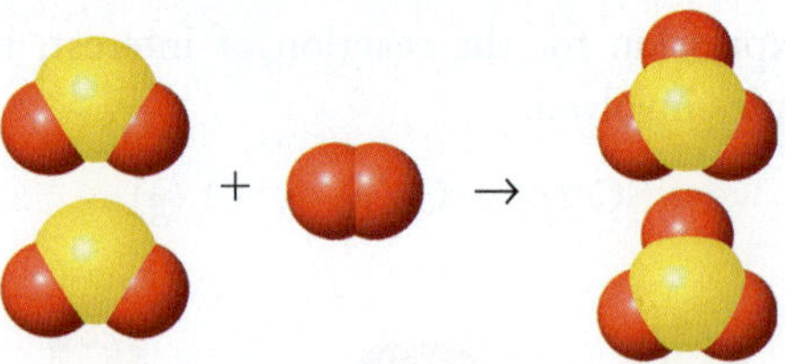

Doing so also changes the number of oxygen atoms on the reactant side because we now have two molecules of $SO_2$, each with two oxygen atoms, plus one molecule of oxygen (two more oxygen atoms). A final check of the number of atoms of each element on both sides of the reaction now confirms that the reaction is balanced:

| Element | Reactant Side | Product Side | Balanced? |
|---|---|---|---|
| S | 2 × 1 = 2 | 2 × 1 = 2 | ✔ |
| O | (2 × 2) + 2 = 6 | 2 × 3 = 6 | ✔ |

Balancing two elements at once is not unusual in writing chemical equations. Note that a balanced equation was achieved only by changing coefficients, *not* by changing any subscripts in chemical formulas. Changing subscripts would change the identities of the substances themselves, and that would change the nature of the reaction. Only coefficients may be changed when you are writing balanced chemical equations.

When $SO_3$ forms in the atmosphere as a result of the reaction in Equation 7.5, it can react with water vapor to produce liquid sulfuric acid, $H_2SO_4$:

$$SO_3(g) + H_2O(g) \rightarrow H_2SO_4(\ell) \qquad (7.6)$$

Atmospheric $H_2SO_4$ dissolves in rain, producing an aqueous solution of sulfuric acid known as acid rain. If this rain lands on objects made of limestone or marble,

**FIGURE 7.4** Atmospheric sulfuric acid, made when $SO_3(g)$ dissolves in rainwater, attacks marble statues by converting the calcium carbonate to calcium sulfate, which is slowly washed away by rain and melting snow. The photograph on the left was taken in 1908; the one on the right, in 1968.

which are minerals composed of calcium carbonate ($CaCO_3$), the following chemical reaction occurs:

$$CaCO_3(s) + H_2SO_4(aq) \rightarrow CaSO_4(aq) + H_2O(\ell) + CO_2(g) \quad (7.7)$$

The above is not a combination reaction because it has multiple products. One of them, calcium sulfate ($CaSO_4$), is more soluble in water than $CaCO_3$. Thus, the reaction in Equation 7.7 can result in the loss of distinctive features in statues and other structures made of limestone or marble (**Figure 7.4**). In these contexts, the reaction is an example of *chemical weathering*, a process that is continually changing minerals on Earth's surface—including, sadly, objects sculpted from those minerals.

## CONCEPT TEST

Why can't we write a balanced equation for the reaction between $SO_2$ and $O_2$ in Equation 7.5 this way?

$$SO_2(g) + O(g) \rightarrow SO_3(g)$$

*(Answers to Concept Tests are in the back of the book.)*

### SAMPLE EXERCISE 7.2 Writing a Balanced Chemical Equation I — LO1

Write a balanced chemical equation describing the chemical weathering reaction between acid rain that contains sulfuric acid and the iron-containing mineral hematite ($Fe_2O_3$), also known as iron(III) oxide. One product of the reaction is water-soluble $Fe_2(SO_4)_3$; the other is water itself.

**Collect and Organize** We are given the identities of the reactants and products. We also know their physical states: $Fe_2(SO_4)_3$ is soluble in water, so its symbol is (*aq*); $H_2SO_4$ is dissolved in rain, so its symbol is also (*aq*). A mineral is a solid substance of limited solubility in liquid water, so the symbol after $Fe_2O_3$ is (*s*).

**Analyze** To write a balanced chemical equation, we start with a preliminary expression that contains one molecule or formula unit of each reactant and product. Then we count the number of atoms of each element in the reaction mixture and we balance them, starting with those that appear in only one reactant and product.

**Solve**

1. The preliminary expression relating reactants and products is

$$Fe_2O_3(s) + H_2SO_4(aq) \rightarrow Fe_2(SO_4)_3(aq) + H_2O(\ell)$$

2. When we count the atoms on both sides, we get

| Element | Reactant Side | Product Side | Balanced? |
|---|---|---|---|
| Fe | 2 | 2 | ✔ |
| H | 2 | 2 | ✔ |
| S | 1 | 3 × 1 = 3 | ✗ |
| O | 3 + 4 = 7 | (3 × 4) + 1 = 13 | ✗ |

3. The iron and hydrogen atoms are already balanced. We turn to balancing the sulfur atoms because sulfur is present in one substance on each side of the equation. To balance sulfur, we place a coefficient of 3 in front of $H_2SO_4$:

$$Fe_2O_3(s) + \underline{3}\ H_2SO_4(aq) \rightarrow Fe_2(SO_4)_3(aq) + H_2O(\ell)$$

| Element | Reactant Side | Product Side | Balanced? |
|---|---|---|---|
| Fe | 2 | 2 | ✔ |
| H | 3 × 2 = 6 | 2 | ✗ |
| S | 3 × 1 = 3 | 3 × 1 = 3 | ✔ |
| O | 3 + (3 × 4) = 15 | (3 × 4) + 1 = 13 | ✗ |

4. Note that changing the coefficient for $H_2SO_4$ balanced the S atoms, but it undid the balance of H atoms. We can rebalance the H atoms by putting a coefficient of 3 in front of $H_2O$:

$$Fe_2O_3(s) + \underline{3}\ H_2SO_4(aq) \rightarrow Fe_2(SO_4)_3(aq) + \underline{3}\ H_2O(\ell)$$

| Element | Reactant Side | Product Side | Balanced? |
|---|---|---|---|
| Fe | 2 | 2 | ✔ |
| H | 3 × 2 = 6 | 3 × 2 = 6 | ✔ |
| S | 3 × 1 = 3 | 3 × 1 = 3 | ✔ |
| O | 3 + (3 × 4) = 15 | (3 × 4) + (3 × 1) = 15 | ✔ |

Note that balancing the H atoms with a coefficient of 3 in front of water also balanced the O atoms.

**Think About It** Balancing the equation required balancing the numbers of atoms of four elements. Balancing the three that were part of only one reactant and one product also balanced the fourth (oxygen), which was part of all four reactants and products.

**Practice Exercise** Write a balanced chemical equation for the reaction between $P_4O_{10}(s)$ and liquid water that produces phosphoric acid, $H_3PO_4(\ell)$.

## Combustion of Hydrocarbons

In recent years, the use of coal as a fuel for heating and generating electricity has been declining, replaced largely by natural gas, a cleaner-burning fuel. Natural gas is mostly methane, $CH_4$. The simplest of all hydrocarbons, $CH_4$ produces more energy in the United States and in many other industrialized countries than any other single substance.

The complete combustion of methane (or any hydrocarbon) rapidly consumes oxygen and produces carbon dioxide and water vapor. Let's get more practice with writing balanced equations by considering the combustion of methane.

1. *Write a preliminary expression containing a single particle of each reactant and product.* Methane has the formula $CH_4$, oxygen is $O_2$, carbon dioxide is $CO_2$, and water is $H_2O$.

$$CH_4(g) + O_2(g) \rightarrow CO_2(g) + H_2O(g)$$

2. *Check whether the expression is balanced.*

| Element | Reactant Side | Product Side | Balanced? |
|---|---|---|---|
| C | 1 | 1 | ✔ |
| H | 4 | 2 | ✗ |
| O | 2 | 2 + 1 = 3 | ✗ |

The equation as written is not yet balanced.

3. *Choose an element that appears in only one reactant and one product to balance first.*

Here we could choose either carbon or hydrogen. Carbon is already balanced, so we focus on hydrogen. With 4 H atoms on the reactant side and 2 H atoms on the product side, we can make the number of H atoms equal on both sides of the equation by placing a coefficient of 2 in front of $H_2O$:

$$CH_4(g) + O_2(g) \rightarrow CO_2(g) + \underline{2}\ H_2O(g)$$

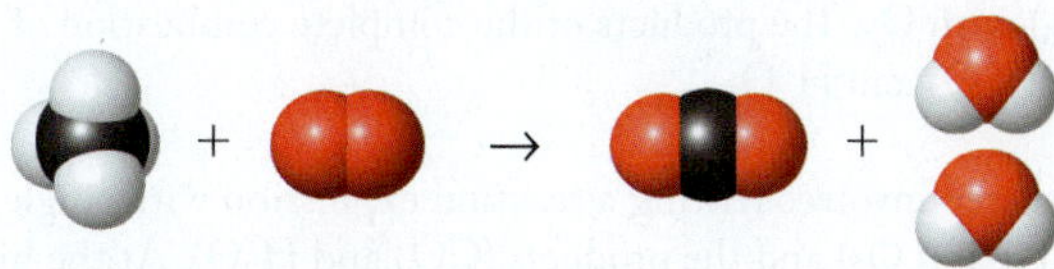

Assigning this coefficient balances the number of hydrogen atoms but not the number of oxygen atoms.

| Element | Reactant Side | Product Side | Balanced? |
|---|---|---|---|
| C | 1 | 1 | ✔ |
| H | 4 | 2 × 2 = 4 | ✔ |
| O | 2 | 2 + (2 × 1) = 4 | ✗ |

4. *Choose coefficients for the other substances so that the numbers of atoms of each element are the same on both sides of the equation.*

We can balance the O atoms by placing a coefficient of 2 in front of $O_2$, which finishes the balancing process:

$$CH_4(g) + \underline{2}\ O_2(g) \rightarrow CO_2(g) + \underline{2}\ H_2O(g) \qquad (7.8)$$

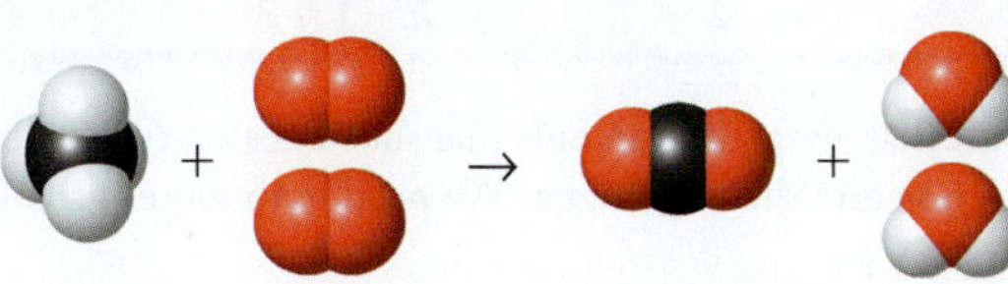

| Element | Reactant Side | Product Side | Balanced? |
|---|---|---|---|
| C | 1 | 1 | ✔ |
| H | 4 | 2 × 2 = 4 | ✔ |
| O | 2 × 2 = 4 | 2 + (2 × 1) = 4 | ✔ |

In the hot flame produced by the combustion of methane, the water would exist as water vapor. However, if the reaction mixture cooled to room temperature, the complete reaction equation would be

$$CH_4(g) + 2\,O_2(g) \rightarrow CO_2(g) + 2\,H_2O(\ell) \tag{7.9}$$

This may seem like an unimportant detail, but the physical states of reactants and products influence how much energy is released or absorbed in a chemical reaction. We will study this phenomenon in detail in Chapter 10. For now, be assured that expressing the right physical states in chemical equations is important.

## SAMPLE EXERCISE 7.3 Writing a Balanced Chemical Equation II — LO1

Methane is the principal ingredient in natural gas, but ethane, $C_2H_6$, is also present. Write a balanced chemical equation for the complete combustion of $C_2H_6$.

**Collect and Organize** We need to write a balanced chemical equation describing the complete combustion of ethane. Combustion reactions involve the rapid reaction of fuels (here, $C_2H_6$) with $O_2$. The products of the complete combustion of hydrocarbons such as ethane are $CO_2$ and $H_2O$.

**Analyze** The first step involves writing a reaction expression with single molecules of the reactants ($C_2H_6$ and $O_2$) and the products ($CO_2$ and $H_2O$). At the high temperature of a combustion flame, all four species in the reaction are gases. Next, we balance the number of C and H atoms because they are each present in only one reactant and one product, and we finish by balancing the total number of O atoms on each side of the reaction arrow.

**Solve**

1. Write a preliminary expression containing a single particle of each reactant and product:

$$C_2H_6(g) + O_2(g) \rightarrow CO_2(g) + H_2O(g)$$

2. Counting the atoms on both the reactant and the product sides indicates that the chemical equation is not balanced:

| Element | Reactant Side | Product Side | Balanced? |
|---|---|---|---|
| C | 2 | 1 | ✗ |
| H | 6 | 2 | ✗ |
| O | 2 | 2 + 1 = 3 | ✗ |

3. To balance atoms that are present in only one substance on each side of the equation, we start with carbon and then hydrogen. We need both more carbon atoms and

more hydrogen atoms on the product side, so we increase the coefficient of $CO_2$ to 2 and we increase the coefficient of $H_2O$ to 3:

$$C_2H_6(g) + O_2(g) \rightarrow \underline{2}\ CO_2(g) + \underline{3}\ H_2O(g)$$

| Element | Reactant Side | Product Side | Balanced? |
|---|---|---|---|
| C | 2 | 2 × 1 = 2 | ✔ |
| H | 6 | 3 × 2 = 6 | ✔ |
| O | 2 | (2 × 2) + (3 × 1) = 7 | ✗ |

Note that although changing these coefficients balances the number of carbon atoms and the number of hydrogen atoms, it also changes the number of oxygen atoms.

Balancing the oxygen atoms is all that remains. With 2 O atoms on the reactant side and 7 O atoms on the product side, we cannot place a whole-number coefficient in front of $O_2$ that will make the number of O atoms equal on both sides of the equation. A coefficient of $\frac{7}{2}$ in front of $O_2$ would balance the number of O atoms, but we need to use whole numbers. Because $\frac{7}{2}$ times 2 is the whole number 7, one way to balance the number of oxygen atoms (without disrupting the balance of the C atoms or the H atoms) is to multiply all the coefficients in the expression by 2:

$$2\ C_2H_6(g) + 7\ O_2(g) \rightarrow 4\ CO_2(g) + 6\ H_2O(g)$$

| Element | Reactant Side | Product Side | Balanced? |
|---|---|---|---|
| C | 2 × 2 = 4 | 4 × 1 = 4 | ✔ |
| H | 2 × 6 = 12 | 6 × 2 = 12 | ✔ |
| O | 7 × 2 = 14 | (4 × 2) + (6 × 1) = 14 | ✔ |

**Think About It** Writing a balanced chemical equation for the combustion of $C_2H_6$ required one more step (to get an even number of O atoms on both sides of the reaction arrow) than writing the chemical equation for the combustion of $CH_4$. This step was needed because we had an odd number of molecules of $H_2O$ on the right side of the reaction arrow after balancing the number of H atoms. Can you see how this will happen every time the number of H atoms per molecule of a hydrocarbon fuel *is not divisible by 4*? For another illustration, try the following Practice Exercise.

**Practice Exercise** Write the chemical equation describing the complete combustion of butane ($C_4H_{10}$), the fuel in disposable lighters.

## Progression of Combustion Reactions

If we could peer inside the blue flame (**Figure 7.5**) produced by the complete combustion of methane and observe the molecules formed during the reaction, we would discover that the reactants are not transformed into final products all at once. Instead, the reaction happens in stages. Both the speed at which a reaction takes place and the molecular-level view of the steps in the reaction are topics in chemical kinetics, the subject of Chapter 13. For now, let us note that, in an early stage of the reaction, $CH_4$ molecules split into molecular fragments that combine with oxygen, forming a mixture of carbon monoxide, hydrogen gas, and water vapor:

$$CH_4(g) + O_2(g) \rightarrow CO(g) + H_2(g) + H_2O(g) \qquad (7.10)$$

**FIGURE 7.5** Complete combustion of natural gas (mostly $CH_4$) produces a characteristic blue flame as a result of emission from excited-state molecular fragments, which are formed early in the reaction.

Note that the equation is balanced as written. Next, the hydrogen formed in the reaction combines with more oxygen to form more water vapor. Writing single molecules of these reactants and products in a preliminary reaction expression gives us this unbalanced expression:

$$H_2(g) + O_2(g) \rightarrow H_2O(g)$$

We can balance it by placing a coefficient of 2 before both $H_2$ and $H_2O$:

$$2\,H_2(g) + O_2(g) \rightarrow 2\,H_2O(g) \qquad (7.11)$$

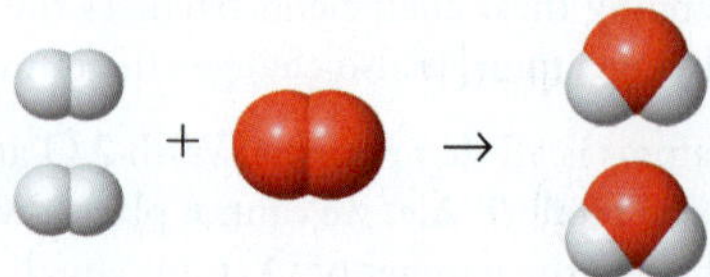

Next, the carbon monoxide formed in the first step of the process (Equation 7.10) combines with more $O_2$ to form $CO_2$:

$$CO(g) + O_2(g) \rightarrow CO_2(g)$$

To balance this preliminary expression, we place a coefficient of 2 in front of CO and $CO_2$:

$$2\,CO(g) + O_2(g) \rightarrow 2\,CO_2(g) \qquad (7.12)$$

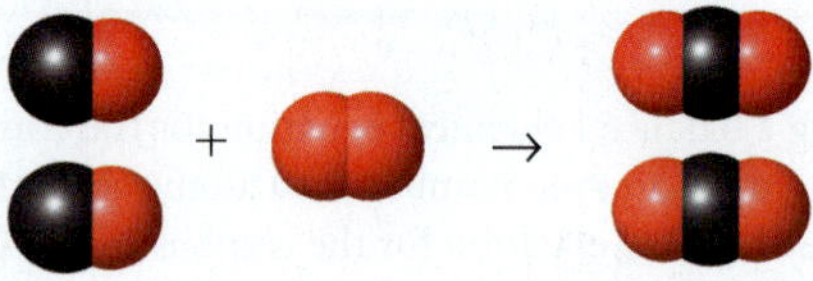

We now have three chemical equations describing the stepwise combustion of methane:

$$CH_4(g) + O_2(g) \rightarrow CO(g) + H_2(g) + H_2O(g) \qquad (7.10)$$

$$2\,H_2(g) + O_2(g) \rightarrow 2\,H_2O(g) \qquad (7.11)$$

$$2\,CO(g) + O_2(g) \rightarrow 2\,CO_2(g) \qquad (7.12)$$

All three reactions occur rapidly in a flame, although the rate of the last one, the conversion of CO to $CO_2$, is not as rapid as the other two. Inevitably, some CO escapes from the reaction mixture (the flame) before the third reaction is complete. This situation is hazardous because CO is toxic, producing nausea and headaches at concentrations between 10 and 100 ppm in the air, and CO can be fatal at concentrations over 100 ppm (**Figure 7.6**). The formation of CO is a concern in the combustion of many hydrocarbons, including the gasoline in automobile engines.

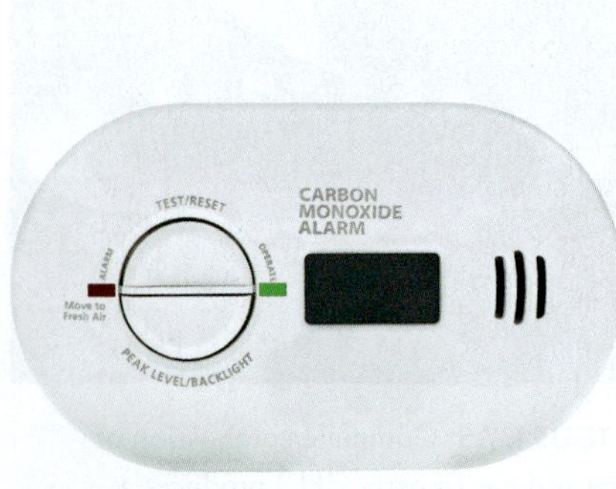


**FIGURE 7.6** Detectors like this one are widely used because CO is a toxic, but colorless and odorless, gas.

If we combine Equations 7.10–7.12, we should produce an equation describing the overall combustion reactions, that is, Equation 7.8. Note that CO is a product in Equation 7.10, and it is a reactant in Equation 7.12. Likewise, $H_2$ is a product in Equation 7.10 and a reactant in Equation 7.11. To combine these equations, we need to balance the numbers of molecules of $H_2$ and CO on the product side of Equation 7.10 with the numbers of molecules of $H_2$ and CO on the reactant sides of Equation 7.11 and Equation 7.12. We can do this by multiplying all the coefficients in Equation 7.10 by a factor of 2:

$$2\,CH_4(g) + 2\,O_2(g) \rightarrow 2\,CO_2(g) + 2\,H_2(g) + 2\,H_2O(g) \quad (7.13)$$

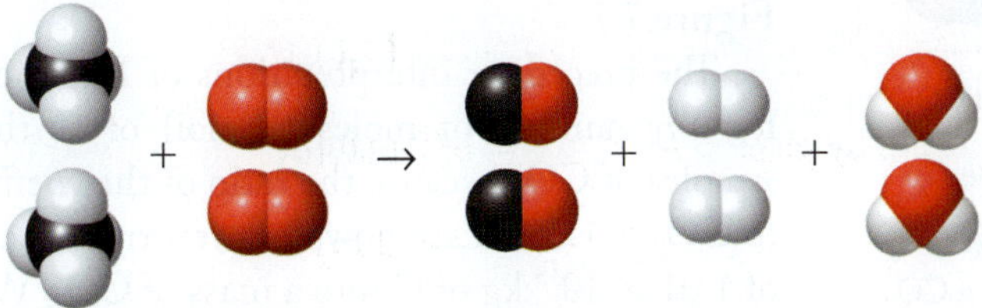

Adding Equation 7.13 to Equations 7.11 and 7.12:

Equation 7.13: $2\,CH_4(g) + 2\,O_2(g) \rightarrow 2\,CO(g) + 2\,H_2(g) + 2\,H_2O(g)$

Equation 7.11: $2\,H_2(g) + O_2(g) \rightarrow 2\,H_2O(g)$

Equation 7.12: $2\,CO(g) + O_2(g) \rightarrow 2\,CO_2(g)$

$$2\,CH_4(g) + 2\,H_2(g) + 2\,CO(g) + 4\,O_2(g) \rightarrow 2\,CO(g) + 2\,H_2(g) + 2\,CO_2(g) + 4\,H_2O(g)$$

and canceling out the substances that appear on both sides of the reaction arrow in equal amounts:

$$2\,CH_4(g) + \cancel{2\,H_2(g)} + \cancel{2\,CO(g)} + 4\,O_2(g) \rightarrow \cancel{2\,CO(g)} + \cancel{2\,H_2(g)} + 2\,CO_2(g) + 4\,H_2O(g)$$

gives us

$$2\,CH_4(g) + 4\,O_2(g) \rightarrow 2\,CO_2(g) + 4\,H_2O(g)$$

By convention, a balanced equation is reported with the smallest whole-number coefficients, so we simplify this chemical equation by dividing all coefficients by 2:

$$CH_4(g) + 2\,O_2(g) \rightarrow CO_2(g) + 2\,H_2O(g)$$

This is Equation 7.8, which describes the complete combustion of methane.

## 7.3 Stoichiometric Calculations

In Section 7.1, we noted that human activity added $1.01 \times 10^{13}$ kg of carbon to the atmosphere in 2018 in the form of $CO_2$. What is the mass of this much $CO_2$? The answer must be more than $1.01 \times 10^{13}$ kg because this value represents only the mass due to carbon. The balanced chemical equation for the complete combustion of carbon is

$$C(s) + O_2(g) \rightarrow CO_2(g) \quad (7.3)$$

Equation 7.3 tells us that the complete combustion of one atom of C produces one molecule of $CO_2$. It also tells us that one *mole* of C atoms produces one *mole* of $CO_2$ molecules (see Table 7.1). This mole ratio of reactant to product is a part of the **stoichiometry** of the reaction—that is, the mole ratios of all the

**CONNECTION** In Chapter 2, we learned that the molecular mass of a compound is the sum of the average atomic masses of the atoms in one of its molecules and that its molar mass ($\mathcal{M}$) has the same numerical value as its molecular mass but is expressed in grams per mole.

**stoichiometry** the mole ratios among the reactants and products in a chemical reaction.

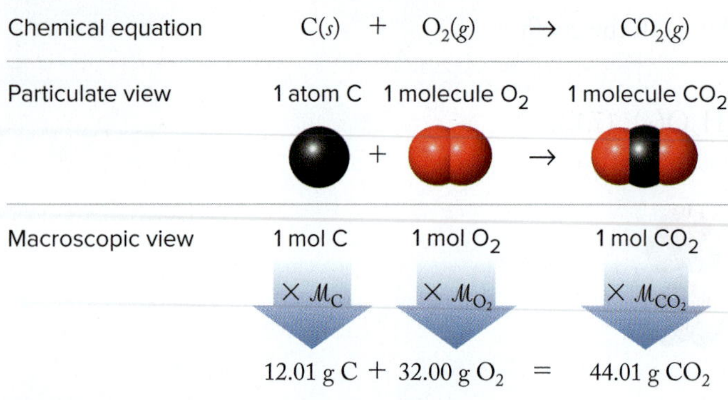


**FIGURE 7.7** Particulate- and macroscopic-scale views of the combustion of carbon.

reactants and products in a chemical reaction. Invoking the concept of molar mass, we can also state that one *molar mass* (12.01 g) of C will produce one *molar mass* (44.01 g) of $CO_2$. These equivalencies are illustrated in **Figure 7.7**.

The preceding interpretations of Equation 7.3 work for *any* number of moles ($x$ mol) of C that produce $x$ moles of $CO_2$ because the ratio of the coefficients of C to $CO_2$ is 1:1. To see how, let's return to the conversion of $1.01 \times 10^{13}$ kg of C into a mass of $CO_2$. We first convert the mass of C into moles of C:

$$1.01 \times 10^{13} \cancel{\text{kg C}} \times \frac{10^3 \cancel{\text{g C}}}{1 \cancel{\text{kg C}}} \times \frac{1 \text{ mol C}}{12.01 \cancel{\text{g C}}} = 8.410 \times 10^{14} \text{ mol C}$$

**CONNECTION** A summary of how to convert the number of moles of an element into the number of moles of a compound with that element in it and to relate the number of moles of a compound to the mass of the compound as shown in Figure 2.18.

Note that the value of "moles of C" is expressed with four significant figures, even though the mass of C is known to only three significant figures. We carry an additional digit because our result here is an intermediate value in the calculation. A danger always exists that rounding off an intermediate value in a multistep calculation can introduce a "rounding error." To avoid such errors, round off only the final calculated value to the appropriate number of significant figures. We know from the stoichiometry of the reaction that one mole of $CO_2$ is produced for every mole of C consumed; therefore,

$$8.410 \times 10^{14} \cancel{\text{mol C}} \times \frac{1 \text{ mol } CO_2}{1 \cancel{\text{mol C}}} = 8.410 \times 10^{14} \text{ mol } CO_2$$

To convert moles of $CO_2$ to mass, we first calculate the molar mass of $CO_2$:

$$\mathcal{M}_{CO_2} = 12.01 \text{ g/mol} + 2(16.00 \text{ g/mol}) = 44.01 \text{ g/mol } CO_2$$

Multiplying the moles of $CO_2$ by the molar mass of $CO_2$ and converting grams to kilograms gives us the mass of $CO_2$ added to the atmosphere:

$$8.410 \times 10^{14} \cancel{\text{mol } CO_2} \times \frac{44.01 \cancel{\text{g}} \text{ } CO_2}{1 \cancel{\text{mol } CO_2}} \times \frac{1 \text{ kg}}{10^3 \cancel{\text{g}}} = 3.70 \times 10^{13} \text{ kg } CO_2$$

Because the answer in each of these steps is the starting point for the next step, we could have combined the three calculations:

$$1.01 \times 10^{13} \cancel{\text{kg C}} \times \frac{10^3 \cancel{\text{g}}}{1 \cancel{\text{kg}}} \times \frac{1 \cancel{\text{mol C}}}{12.01 \cancel{\text{g C}}} \times \frac{1 \cancel{\text{mol } CO_2}}{1 \cancel{\text{mol C}}} \times \frac{44.01 \cancel{\text{g}} \text{ } CO_2}{1 \cancel{\text{mol } CO_2}} \times \frac{1 \text{ kg}}{10^3 \cancel{\text{g}}}$$
$$= 3.70 \times 10^{13} \text{ kg } CO_2$$

Note that in the second step of the calculation we converted kilograms into grams, and in the last step we converted grams into kilograms. These two steps effectively multiplied and then divided by 1000, so they had no overall effect on the final answer. As you hone your skills in solving multistep calculations such as this one, check for offsetting conversion factors and simply cancel them out. Doing so will save you unnecessary calculation steps and reduce the chance of data entry errors.

The steps we followed in the preceding calculation can be used to determine the mass of any substance (reactant or product) involved in any chemical reaction, provided that we know the quantity of another component of the reaction mixture and the stoichiometric relationship between the two substances—that is, their mole ratio in the balanced chemical equation describing the reaction.

## CONCEPT TEST

The bright white patterns of light produced during fireworks displays are sometimes produced by the combustion of magnesium metal:

$$2\ Mg(s) + O_2(g) \rightarrow 2\ MgO(s) \qquad (7.14)$$

Is the total mass of Mg and $O_2$ consumed equal to the mass of MgO produced? Is the mass of Mg consumed equal to the mass of $O_2$ consumed? Explain both answers.

---

### SAMPLE EXERCISE 7.4 Calculating the Mass of a Product from the Mass of a Reactant — LO2

In 2017, U.S. power plants consumed more than $1.45 \times 10^{11}$ kg of natural gas. How many kilograms of $CO_2$ ($\mathcal{M}$ = 44.01 g/mol) were released into the atmosphere from these power plants? Natural gas is mostly methane ($CH_4$; $\mathcal{M}$ = 16.04 g/mol), so base your solution on its combustion reaction (Equation 7.8):

$$CH_4(g) + 2\ O_2(g) \rightarrow CO_2(g) + 2\ H_2O(g)$$

**Collect and Organize** We are asked to calculate the mass of $CO_2$ released by the combustion of $1.45 \times 10^{11}$ kg of $CH_4$ (our model compound for natural gas). We are given the molar masses of the two compounds and the balanced chemical equation relating the moles of $CO_2$ produced to the moles of $CH_4$ consumed.

**Analyze** To use Equation 7.8, we must convert the mass of $CH_4$ into moles of $CH_4$ by first converting kilograms of $CH_4$ into grams of $CH_4$ and then dividing by the molar mass of $CH_4$. The number of moles of $CH_4$ consumed equals the number of moles of $CO_2$ produced. When we multiply this value by the molar mass of $CO_2$, we get the number of grams of $CO_2$, which we then convert to kilograms. The molar mass of $CH_4$ is about 16 g/mol, and the molar mass of $CO_2$ is about 44 g/mol, or about 2.5 times as large. Therefore, our final answer should be about 2.5 times the mass of natural gas given in the problem, or about $3.6 \times 10^{11}$ kg of $CO_2$.

**Solve** Converting kilograms of $CH_4$ into moles of $CH_4$ (and carrying an extra significant figure to avoid a rounding error):

$$1.45 \times 10^{11}\ \cancel{\text{kg CH}_4} \times \frac{10^3\ \text{g}}{1\ \cancel{\text{kg}}} \times \frac{1\ \text{mol CH}_4}{16.04\ \cancel{\text{g CH}_4}} = 9.040 \times 10^{12}\ \text{mol CH}_4$$

Converting moles of $CH_4$ into moles of $CO_2$:

$$9.040 \times 10^{12}\ \cancel{\text{mol CH}_4} \times \frac{1\ \text{mol CO}_2}{1\ \cancel{\text{mol CH}_4}} = 9.040 \times 10^{12}\ \text{mol CO}_2$$

Converting moles of $CO_2$ into kilograms of $CO_2$:

$$9.040 \times 10^{12}\ \cancel{\text{mol CO}_2} \times \frac{44.01\ \cancel{\text{g}}\ \text{CO}_2}{1\ \cancel{\text{mol CO}_2}} \times \frac{1\ \text{kg}}{10^3\ \cancel{\text{g}}} = 3.978 \times 10^{11}\ \text{kg CO}_2$$

We can combine the three calculations, and in later problems we will do this routinely and not show the individual steps:

$$1.45 \times 10^{11}\ \cancel{\text{kg CH}_4} \times \frac{10^3\ \cancel{\text{g}}}{1\ \cancel{\text{kg}}} \times \frac{1\ \cancel{\text{mol CH}_4}}{16.04\ \cancel{\text{g CH}_4}} \times \frac{1\ \cancel{\text{mol CO}_2}}{1\ \cancel{\text{mol CH}_4}} \times \frac{44.01\ \cancel{\text{g}}\ \text{CO}_2}{1\ \cancel{\text{mol CO}_2}} \times \frac{1\ \text{kg}}{10^3\ \cancel{\text{g}}}$$

$$= 3.98 \times 10^{11}\ \text{kg CO}_2$$

**Think About It** The mass of $CO_2$ released into the atmosphere by these power plants is close to our estimated value. Globally, emissions of $CO_2$ and other greenhouse gases such as $CH_4$ are increasing every year as a result of human activity. In Chapter 4, we learned how these gases contribute to global climate change.

**Practice Exercise** How many grams of $O_2$ are consumed during the complete combustion of 1.00 g of butane, $C_4H_{10}$?

**SAMPLE EXERCISE 7.5** Calculating the Masses of Reactants **LO2**

Copper was among the first metals to be refined from minerals collected by ancient metalworkers. The production of copper was already an industry by 3500 BCE. Cuprite is a copper mineral commonly found near Earth's surface, making it a likely resource for Bronze Age coppersmiths. Cuprite has the formula $Cu_2O$ and can be converted to copper metal by reacting it with charcoal:

$$2\ Cu_2O(s) + C(s) \rightarrow 4\ Cu(s) + CO_2(g)$$

How much cuprite and how much carbon are needed to prepare a 256 g copper bracelet such as the one shown in **Figure 7.8**?

**FIGURE 7.8** A copper bracelet.

**Collect and Organize** We are given a balanced chemical equation and the mass of product. We are asked to find the masses of the reactants. Because the chemical equation relates moles, not masses, of reactants and products, we need to find the molar mass of each substance in the reaction.

**Analyze** The balanced chemical equation tells us that for every 4 moles of copper we produce, 2 moles of $Cu_2O$ and 1 mole of C must react. To use these mole ratios, we first need to convert the mass of copper to the equivalent number of moles of Cu by using the molar mass of Cu. We then work a separate conversion for each reactant: the first conversion uses the mole ratio of $Cu_2O$ to Cu (2:4) from the balanced equation, and the second uses the mole ratio of C to Cu (1:4). Finally, we use the molar masses of $Cu_2O$ and C to find the mass of each required for the reaction.

**Solve** Carrying out the steps described in the analysis of the problem:

$$256\ \cancel{\text{g Cu}} \times \frac{1\ \cancel{\text{mol Cu}}}{63.55\ \cancel{\text{g Cu}}} \times \frac{2\ \cancel{\text{mol Cu}_2\text{O}}}{4\ \cancel{\text{mol Cu}}} \times \frac{143.09\ \text{g Cu}_2\text{O}}{1\ \cancel{\text{mol Cu}_2\text{O}}} = 288\ \text{g Cu}_2\text{O}$$

$$256\ \cancel{\text{g Cu}} \times \frac{1\ \cancel{\text{mol Cu}}}{63.55\ \cancel{\text{g Cu}}} \times \frac{1\ \cancel{\text{mol C}}}{4\ \cancel{\text{mol Cu}}} \times \frac{12.01\ \text{g C}}{1\ \cancel{\text{mol C}}} = 12.1\ \text{g C}$$

**Think About It** To prepare a given mass of copper, we must have enough copper ore and enough charcoal. Usually, the supply of ore is limited, but more than enough charcoal is available.

**Practice Exercise** The copper mineral chalcocite, $Cu_2S$, can be converted to copper simply by heating in air: $Cu_2S(s) + O_2(g) \rightarrow 2\ Cu(s) + SO_2(g)$. How much $Cu_2S$ is needed to make 256 g of Cu? How many grams of $SO_2$ are produced?

## 7.4 Limiting Reactants and Percent Yield

**limiting reactant** a reactant that is consumed completely in a chemical reaction. The amount of product formed depends on the amount of the limiting reactant available.

Did you notice that Sample Exercise 7.4 assumed that we had an unlimited supply of oxygen? For our solution to be correct, more than enough $O_2$ had to be in the air to convert all the carbon in natural gas (methane) to $CO_2$. In the balanced

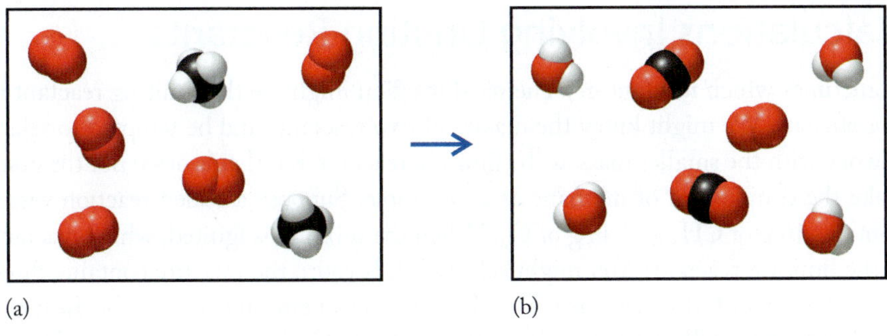


**FIGURE 7.9** (a) A combustion reaction mixture containing five $O_2$ molecules and two $CH_4$ molecules. (b) The reaction produces two molecules of $CO_2$ and four molecules of $H_2O$, leaving one molecule of $O_2$ left over. Methane is the limiting reactant.

**FIGURE 7.10** A macroscopic illustration of the principle of the limiting reactant (or ingredient): how many modest salami and cheese sandwiches (with only one slice of salami and one slice of cheese per sandwich) can be prepared from eight slices of bread, four slices of cheese, and three slices of salami? Only three: salami is the limiting ingredient, and two slices of bread and one slice of cheese are left over.

equation for this reaction (Equation 7.8), the stoichiometry requires 2 molecules of $O_2$ to combust every 1 molecule of $CH_4$:

$$CH_4(g) + 2\,O_2(g) \rightarrow CO_2(g) + 2\,H_2O(g) \qquad (7.8)$$

Now consider the mixture in **Figure 7.9(a)**, which consists of 2 molecules of $CH_4$ and 5 molecules of $O_2$. When this mixture reacts, the products (**Figure 7.9b**) are 2 molecules of $CO_2$ and 4 molecules of $H_2O$. But 1 molecule of unreacted $O_2$ also appears in Figure 7.9(b). In the initial reaction mixture, $CH_4$ is considered the **limiting reactant**, which means that the amount of product is limited by the quantity of $CH_4$ initially available. Similarly, $O_2$ is described as being “in excess” because more $O_2$ is initially available than can react with the amount of $CH_4$. **Figure 7.10** provides another—perhaps more familiar—macroscopic example of the concept of a limiting reactant in the context of making sandwiches. Examine Figure 7.10 and convince yourself that the salami is the limiting reactant, whereas the slices of bread and cheese are present “in excess.”

**CHEMTOUR**
Limiting Reactant

**CONCEPT TEST**

Hydrogen and oxygen can react to produce water. Examine **Figure 7.11**. If this mixture reacts to produce water, which substance will be the limiting reactant? Which substance will be present in excess?

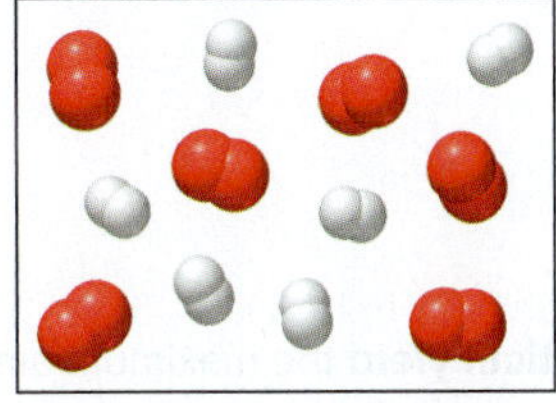

**FIGURE 7.11** A reaction mixture of hydrogen molecules and oxygen molecules.

## Calculations Involving Limiting Reactants

Sometimes which reactant in a chemical reaction might be the limiting reactant is not obvious. We might know the masses of two reactants and be tempted to select the one with the smaller mass as the limiting reactant, but that is often not the case. Take the combustion of methane as an example. Suppose a sealed reaction vessel contains 16 g of $CH_4$ and 48 g of $O_2$. When the mixture is ignited, which reactant is the limiting reactant? You might select $CH_4$ because the mixture contains three times as much $O_2$ (by mass) as $CH_4$. However, the chemical equation for the combustion reaction tells us that we need two moles of $O_2$ for every one mole of $CH_4$, and the molar mass of $O_2$ (32.00 g/mol) is about twice that of $CH_4$ (16.04 g/mol). Twice as many moles of a reactant that has twice the molar mass means that we need four times as much of it by mass. However, the mass of $O_2$ in the reaction vessel is only three times the mass of $CH_4$, so $O_2$ is the limiting reactant.

Let's create a procedure for identifying the limiting reactant for any reaction in which we know the quantities of two or more reactants. Such a problem can be solved in several ways. We explore one of them here, starting with a chemical equation having the general form A + B → C. First, we calculate how much product would be formed if reactant A were completely consumed in the reaction. Then we repeat the calculation for reactant B.

Let's try this approach with the reaction that has produced electricity in spacecraft since the dawn of crewed space flight in the 1960s. The reaction uses fuel cells according to the following chemical reaction:

$$2\,H_2(g) + O_2(g) \rightarrow 2\,H_2O(\ell) \qquad (7.15)$$

Suppose a gas-delivery system provides hydrogen and oxygen to a fuel cell so that the mixture in the cell consists of 0.055 g of $H_2$ and 0.48 g of $O_2$. Which gas is the limiting reactant?

Equation 7.15 tells us that two moles of $H_2$ combines with one mole of $O_2$ to produce two moles of water, so let's calculate the amount of water each reactant would produce if it were completely consumed.

$$0.055\ \cancel{g\ H_2} \times \frac{1\ \cancel{mol\ H_2}}{2.016\ \cancel{g\ H_2}} \times \frac{2\ \cancel{mol\ H_2O}}{2\ \cancel{mol\ H_2}} \times \frac{18.02\ g\ H_2O}{1\ \cancel{mol\ H_2O}} = 0.49\ g\ H_2O$$

Now we calculate how much water would be produced if $O_2$ were completely consumed:

$$0.48\ \cancel{g\ O_2} \times \frac{1\ \cancel{mol\ O_2}}{32.00\ \cancel{g\ O_2}} \times \frac{2\ \cancel{mol\ H_2O}}{1\ \cancel{mol\ O_2}} \times \frac{18.02\ g\ H_2O}{1\ \cancel{mol\ H_2O}} = 0.54\ g\ H_2O$$

Less $H_2O$ is produced from the $H_2$ supply than from $O_2$, so $H_2$ must be the limiting reactant. Some $O_2$ will be left unreacted. We can determine how much by calculating the mass of $O_2$ consumed for the given amount of $H_2$. We use the 2:1 stoichiometric ratio between $H_2$ and $O_2$ and the molar mass of $O_2$:

$$0.055\ \cancel{g\ H_2} \times \frac{1\ \cancel{mol\ H_2}}{2.016\ \cancel{g\ H_2}} \times \frac{1\ \cancel{mol\ O_2}}{2\ \cancel{mol\ H_2}} \times \frac{32.00\ g\ O_2}{1\ \cancel{mol\ O_2}} = 0.44\ g\ O_2$$

The difference between the mass of $O_2$ supplied (0.48 g) and the mass consumed (0.44 g) means that 0.04 g of $O_2$ is left over.

This method allows us to not only identify the limiting reactant in a process but also determine the **theoretical yield** of the product—the maximum amount of product that could be obtained from the given quantities of reactants. Theoretical yield is sometimes called *stoichiometric yield* or 100% yield. Here, the theoretical yield of water from the given amounts of reactants is 0.49 g of $H_2O$.

**theoretical yield** the maximum amount of product possible in a chemical reaction for the given quantities of reactants; also called the stoichiometric yield.

## CONCEPT TEST

Suppose the reaction A + 2B + 3C yields a desired product, D. If a reaction mixture contains 1 mole of A, 2.5 moles of B, and 1.5 moles of C, (a) which is the limiting reactant, and (b) how much of the other reactants remain after the limiting reactant is consumed?

### SAMPLE EXERCISE 7.6 Identifying the Limiting Reactant in a Reaction Mixture LO3

The flame in an oxyacetylene torch (**Figure 7.12**) reaches temperatures as high as 3500°C as a result of the combustion of a mixture of acetylene ($C_2H_2$) and pure $O_2$. If these two gases flow from high-pressure tanks at the rates of 52.0 g of $C_2H_2$ and 188 g of $O_2$ per minute, which reactant is the limiting reactant, or is the mixture stoichiometric?

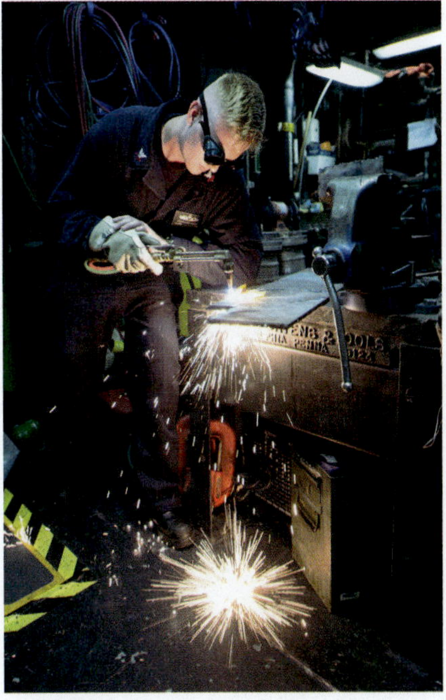

**FIGURE 7.12** An oxyacetylene torch is used to create art from sheet metal.

**Collect and Organize** We are given the rates (in grams per minute) at which two reactants are introduced to a reaction system (the torch) and asked to determine whether one of them is limiting and, if so, which one. Complete combustion of hydrocarbons produces $CO_2$ and $H_2O$.

**Analyze** We need to write a balanced chemical equation describing the combustion reaction to determine the required mole ratio of $C_2H_2$ to $O_2$. We can directly convert the rates at which the reactants flow to the torch into masses in grams by assuming that the torch is operated for exactly 1 min. These masses of the two reactants can then be converted into moles of product by using the stoichiometry of the reaction. The lesser of the two quantities is the answer we seek.

**Solve** The unbalanced chemical equation for the reaction is

$$C_2H_2(g) + O_2(g) \rightarrow CO_2(g) + H_2O(g)$$

Although the number of hydrogen atoms is balanced, the numbers of carbon and oxygen atoms are not. Using the four-step procedure outlined in Section 7.2, we can determine that the balanced equation for this reaction is

$$2\ C_2H_2(g) + 5\ O_2(g) \rightarrow 4\ CO_2(g) + 2\ H_2O(g)$$

Calculating the quantity of $CO_2$ that could be produced by each reactant if it were the limiting reactant:

$$188\ \cancel{\text{g O}_2} \times \frac{1\ \cancel{\text{mol O}_2}}{32.00\ \cancel{\text{g O}_2}} \times \frac{4\ \text{mol CO}_2}{5\ \cancel{\text{mol O}_2}} = 4.70\ \text{mol CO}_2$$

$$52.0\ \cancel{\text{g C}_2\text{H}_2} \times \frac{1\ \cancel{\text{mol C}_2\text{H}_2}}{26.04\ \cancel{\text{g C}_2\text{H}_2}} \times \frac{4\ \text{mol CO}_2}{2\ \cancel{\text{mol C}_2\text{H}_2}} = 3.99\ \text{mol CO}_2$$

The smaller quantity of $CO_2$ that could be produced by acetylene tells us that (1) it is the limiting reactant and (2) the acetylene : oxygen ratio entering the torch was not a stoichiometric one.

**Think About It** We compared the amount of $CO_2$ produced in the reaction to determine the limiting reactant. We could also have compared the amount of water produced. We also could have calculated the amount of one of the reactants required for the complete consumption of the other.

**Practice Exercise** Any fuel–oxygen mixture that contains less fuel than oxygen needed to burn the fuel completely is called a *fuel-lean* (or simply *lean*) mixture, and a combustion mixture containing too little oxygen is called a *fuel-rich* mixture. A high-performance torch that burns propane, $C_3H_8$, is adjusted so that 88 g of $O_2$ flows into its flame for every 23 g of $C_3H_8$. Is the mixture fuel-rich, fuel-lean, or stoichiometric?

**FIGURE 7.13** Ibuprofen is the active ingredient in the pain relievers Advil and Motrin.

## SAMPLE EXERCISE 7.7 Calculating the Theoretical Yield — LO4

Ibuprofen (**Figure 7.13**) is an over-the-counter pharmaceutical (for example, sold under the brand names Advil and Motrin) for pain relief. Like aspirin, ibuprofen is an NSAID (nonsteroidal anti-inflammatory drug). The original six-step synthesis patented in the 1960s resulted in significant waste, so a green chemistry alternative synthesis is now used that is cheaper and more environmentally friendly. The first step in the modern industrial synthesis involves the reaction of isobutylbenzene, or IBB ($C_{10}H_{14}$; $\mathcal{M}$ = 134.21 g/mol) with acetic anhydride, or AA ($C_4H_6O_3$; $\mathcal{M}$ = 102.09 g/mol), which produces 4-isobutylacetophenone, or IBA ($C_{12}H_{16}O$; $\mathcal{M}$ = 176.25 g/mol). Acetic acid ($CH_3COOH$) also is produced (**Figure 7.14**). Suppose that this reaction is studied on a laboratory scale in which 265 g of IBB is reacted with 198 g of AA.

a. Which is the limiting reactant?
b. How many grams of IBA are produced if the limiting reactant is completely consumed?
c. How much of the other reactant will be left over?

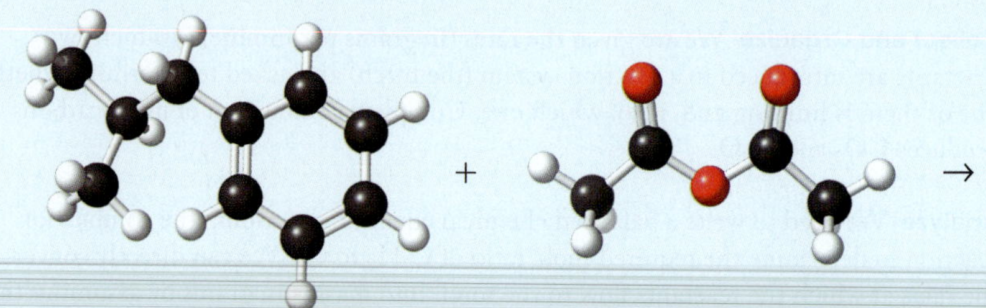

Isobutylbenzene (IBB) + Acetic anhydride (AA) →

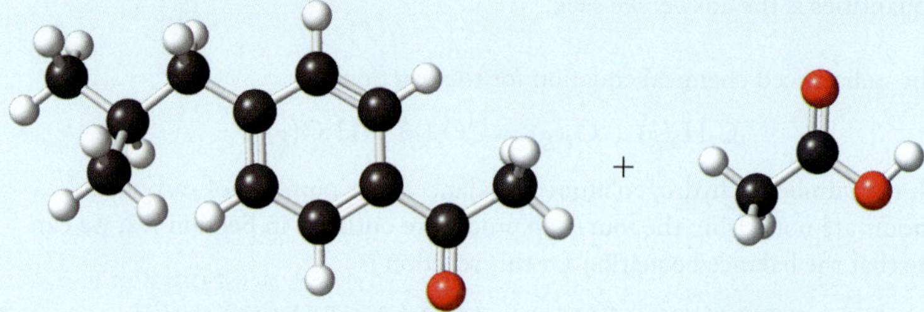

4-Isobutylacetophenone (IBA) + Acetic acid

**FIGURE 7.14** Reaction for the first step in synthesizing ibuprofen.

**Collect and Organize** We are given the quantities of two reactants, the molar masses of the reactants and the desired product, and a balanced equation describing their reaction. We are asked to determine which is the limiting reactant, how much product could be made, and how much of the nonlimiting reactant will be left over.

**Analyze** Start with a balanced equation for the reaction in Figure 7.14. To identify the limiting reactant, determine the quantity of product that could be made by first assuming that IBB is the limiting reactant and then assuming that AA is the limiting reactant. The limiting reactant will produce the smaller mass of IBA, which is also the answer to part (b). To calculate the quantity of the nonlimiting reactant that will be left over, we use the stoichiometry of the reaction to calculate how much of it is consumed during the reaction and then subtract that value from the initial amount present in the reaction mixture.

**Solve**

a. The chemical equation for the reaction of IBB with AA is balanced when every molecule has a coefficient of 1:

$$C_{10}H_{14} + C_4H_6O_3 \rightarrow C_{12}H_{16}O + C_2H_4O_2$$

Calculating the mass of IBA that could be produced if IBB is the limiting reactant:

$$265\ \cancel{\text{g IBB}} \times \frac{1\ \cancel{\text{mol IBB}}}{134.21\ \cancel{\text{g IBB}}} \times \frac{1\ \cancel{\text{mol IBA}}}{1\ \cancel{\text{mol IBB}}} \times \frac{176.25\ \text{g IBA}}{1\ \cancel{\text{mol IBA}}} = 348\ \text{g IBA}$$

Repeating the preceding calculation with 198 g of AA:

$$198\ \cancel{\text{g AA}} \times \frac{1\ \cancel{\text{mol AA}}}{102.09\ \cancel{\text{g AA}}} \times \frac{1\ \cancel{\text{mol IBA}}}{1\ \cancel{\text{mol AA}}} \times \frac{176.25\ \text{g IBA}}{1\ \cancel{\text{mol IBA}}} = 342\ \text{g IBA}$$

AA produces the smaller amount of product; therefore, it is the limiting reactant.

b. The maximum amount of IBA that can be made is 342 g.

c. The mass of IBB consumed in the reaction is

$$198\ \cancel{\text{g AA}} \times \frac{1\ \cancel{\text{mol AA}}}{102.09\ \cancel{\text{g AA}}} \times \frac{1\ \cancel{\text{mol IBB}}}{1\ \cancel{\text{mol AA}}} \times \frac{134.21\ \text{g IBB}}{1\ \cancel{\text{mol IBB}}} = 260\ \text{g IBB}$$

The mass of IBB left over is (265 − 260) g = 5 g.

**Think About It** For various scientific and economic reasons, industrial processes are often run with nonstoichiometric amounts of reactant. We will encounter some of these reasons as we explore chemical kinetics and equilibrium in later chapters.

**Practice Exercise** Sodium borohydride ($NaBH_4$; $\mathcal{M}$ = 37.83 g/mol) is a solid material that can be used to store hydrogen gas. One way to prepare it is to react trimethylborate [$B(OCH_3)_3$; $\mathcal{M}$ = 103.91 g/mol] with sodium hydride (NaH; $\mathcal{M}$ = 24.00 g/mol):

$$B(OCH_3)_3(\ell) + 4\ NaH(s) \rightarrow NaBH_4(s) + 3\ NaOCH_3(s)$$

If 71.5 g of trimethylborate is reacted with 56.2 g of sodium hydride, how many grams of sodium borohydride can be produced?

## Percent Yield: Actual versus Theoretical

Recall that the *theoretical* yield is the maximum quantity of product possible from the given quantities of reactants. In nature, industry, or the laboratory, the *actual* yield is often less than the theoretical yield for several reasons. Sometimes reactants combine to produce products other than the ones desired. Other reactions are so slow that some reactants remain unreacted even after long reaction times. Still other reactions do not go to completion no matter how long they are allowed to run. That can happen when a reaction also runs in reverse, turning products back into reactants. When this happens, the result can be a mixture of reactants and products in which the composition does not change with time. These reactions have reached a state of *chemical equilibrium*, a concept we explore in considerable detail in Chapter 14. So, for various reasons, distinguishing between the theoretical and actual yields of a chemical reaction as well as calculating the **percent yield** is useful:

$$\text{Percent yield} = \frac{\text{actual yield}}{\text{theoretical yield}} \times 100\% \qquad (7.16)$$

**percent yield** the ratio, expressed as a percentage, of the actual yield of a chemical reaction to the theoretical yield.

**SAMPLE EXERCISE 7.8** Calculating Percent Yield **LO4**

The industrial process for making the ammonia used in fertilizer, explosives, and many other products is based on the reaction between nitrogen and hydrogen at high temperature and pressure:

$$N_2(g) + 3\ H_2(g) \rightarrow 2\ NH_3(g)$$

If 18.2 kg of $NH_3$ ($\mathcal{M}$ = 17.03 g/mol) is produced by a reaction mixture that initially contains 6.00 kg of $H_2$ ($\mathcal{M}$ = 2.016 g/mol) and an excess of $N_2$, what is the percent yield of the reaction?

**Collect and Organize** We know that the actual yield of $NH_3$ is 18.2 kg. We also know that $H_2$ must be the limiting reactant because an excess of $N_2$ is present and that the reaction mixture initially contained 6.00 kg of $H_2$. The mole ratio of $NH_3$ to $H_2$ in the chemical equation is 2:3.

**Analyze** One way to calculate percent yield involves the following steps:

1. Convert the mass of $H_2$ into moles of $H_2$.
2. Convert the moles of $H_2$ consumed into moles of $NH_3$ produced by using the 3:2 stoichiometric ratio in the balanced chemical equation.
3. Convert moles of $NH_3$ into grams of $NH_3$ by multiplying by the molar mass of $NH_3$.
4. Convert grams of $NH_3$ into kilograms of $NH_3$ to calculate the theoretical yield of the reaction.
5. Divide the actual yield (18.2 kg of $NH_3$) by the theoretical yield from step 4 to obtain the percent yield.

**Solve** Calculating the theoretical yield (combining steps 1–4):

$$6.00\ \cancel{\text{kg H}_2} \times \frac{10^3\ \text{g}}{1\ \cancel{\text{kg}}} \times \frac{1\ \cancel{\text{mol H}_2}}{2.016\ \cancel{\text{g H}_2}} \times \frac{2\ \cancel{\text{mol NH}_3}}{3\ \cancel{\text{mol H}_2}} \times \frac{17.03\ \text{g NH}_3}{1\ \cancel{\text{mol NH}_3}} \times \frac{1\ \text{kg}}{10^3\ \text{g}}$$
$$= 33.8\ \text{kg NH}_3$$

Dividing the actual yield by this theoretical yield (step 5):

$$\frac{18.2\ \cancel{\text{kg NH}_3}}{33.8\ \cancel{\text{kg NH}_3}} \times 100\% = 53.8\%$$

**Think About It** Although a yield of about 54% may seem low, it is not unusual for a reaction such as this one, which reaches chemical equilibrium before the limiting reactant has been completely consumed. In Chapter 14, we will discuss ways that chemists and chemical engineers manipulate reaction conditions to try to improve the percent yields of reactions that reach equilibrium.

 **Practice Exercise** The industrial synthesis of $H_2$ begins with the steam–methane reforming reaction, in which methane reacts with high-temperature steam:

$$CH_4(g) + H_2O(g) \rightarrow CO(g) + 3\ H_2(g)$$

What is the percent yield when a reaction vessel that initially contains 64 kg of $CH_4$ and excess steam yields 18.1 kg of $H_2$?

## 7.5 Percent Composition and Empirical Formulas

**CHEMTOUR** Percent Composition

**percent composition** the composition of a compound expressed in terms of the percentage by mass of each element in the compound.

During the Industrial Revolution, access to geological deposits of iron oxides was a key to economic development. The more accessible an iron-containing mineral was and the higher its iron content, the more valuable it was. Several ways exist to express the elemental content of substances in mixtures. A popular one is **percent composition**: the percentage by mass of the constituent elements of a compound with respect to its total mass. Given the iron-containing minerals wüstite (FeO) and hematite ($Fe_2O_3$), which has the higher iron content, and what is the iron content value expressed as a percent composition?

One way to answer the first question is to examine the mole ratios of Fe to O, which are 1:1 in FeO and 2:3 (or 1:1.5) in $Fe_2O_3$. In hematite, 1.5 atoms of O are present for every atom of Fe, but only 1 atom of O per atom of Fe in wüstite, which means a higher percentage of O and, hence, a lower percentage of Fe in $Fe_2O_3$. Therefore, FeO must have the higher iron content.

To find the percent composition of FeO, let's assume that we have 1 mole of it. If we do, we have 1 mole of Fe and 1 mole of O. We can convert the number of moles to equivalent masses by using the molar masses of Fe (55.85 g/mol) and O (16.00 g/mol):

$$\left(1 \cancel{\text{mol}} \text{ Fe} \times \frac{55.85 \text{ g}}{1 \cancel{\text{mol}}}\right) + \left(1 \cancel{\text{mol}} \text{ O} \times \frac{16.00 \text{ g}}{1 \cancel{\text{mol}}}\right) = 71.85 \text{ g FeO}$$

Of this 71.85 g, Fe accounts for 55.85 g, so the Fe content of FeO is

$$\frac{\text{mass of Fe}}{\text{total mass}} = \frac{55.85 \cancel{\text{g}} \text{ Fe}}{71.85 \cancel{\text{g}} \text{ FeO}} = 0.7773, \text{ or } 77.73\% \text{ Fe}$$

It follows that the oxygen content is

$$\frac{\text{mass of O}}{\text{total mass}} = \frac{16.00 \cancel{\text{g}} \text{ O}}{71.85 \cancel{\text{g}} \text{ FeO}} = 0.2227, \text{ or } 22.27\% \text{ O}$$

Because FeO contains only Fe and O, we also could have determined the percentage of O by subtracting the percentage of Fe from 100%:

$$100.00\% - 77.73\% = 22.27\%$$

**SAMPLE EXERCISE 7.9** Calculating Percent Composition from a Chemical Formula **LO5**

What is the percent composition of the mineral forsterite, $Mg_2SiO_4$ (**Figure 7.15**)?

**Collect, Organize, and Analyze** The chemical formula of forsterite tells us that one mole of it contains 2 moles of Mg, 1 mole of Si, and 4 moles of O. To convert these mole quantities into percent composition values based on mass, we must first convert the mole values to mass values by multiplying each by the appropriate molar mass. Summing these element masses, then dividing each one by the total mass, and then expressing the results as percentages gives the percent composition of the mineral.

**Solve** Calculating the mass of each element in one mole of $Mg_2SiO_4$ and the molar mass of $Mg_2SiO_4$:

$$\begin{aligned} 2 \cancel{\text{mol}} \text{ Mg} \times 24.31 \text{ g}/\cancel{\text{mol}} &= 48.62 \text{ g Mg} \\ + 1 \cancel{\text{mol}} \text{ Si} \times 28.09 \text{ g}/\cancel{\text{mol}} &= 28.09 \text{ g Si} \\ + 4 \cancel{\text{mol}} \text{ O} \times 16.00 \text{ g}/\cancel{\text{mol}} &= 64.00 \text{ g O} \\ \hline \text{Molar mass of } Mg_2SiO_4 &= 140.71 \text{ g/mol} \end{aligned}$$

Calculating percent composition:

$$\%\text{Mg} = \frac{48.62 \cancel{\text{g}} \text{ Mg}}{140.71 \cancel{\text{g}}} \times 100\% = 34.55\% \text{ Mg}$$

$$\%\text{Si} = \frac{28.09 \cancel{\text{g}} \text{ Si}}{140.71 \cancel{\text{g}}} \times 100\% = 19.96\% \text{ Si}$$

$$\%\text{O} = \frac{64.00 \cancel{\text{g}} \text{ O}}{140.71 \cancel{\text{g}}} \times 100\% = 45.48\% \text{ O}$$

**FIGURE 7.15** Crystals of pure $Mg_2SiO_4$ are colorless, but sometimes they contain impurities, such as $Fe^{2+}$ ions (in place of $Mg^{2+}$ ions) that make the crystals green.

**Think About It** The molar masses of the three elements are similar, so it makes sense that the largest mass percent value belongs to O, followed by Mg and then Si because the mole ratios of these elements in forsterite are 4:2:1, respectively. Also, note that the three percentages sum to 99.99%. Percent composition values should always sum to 100%, or very close to it, if we have accounted for all the elements in the compound and done the math correctly.

**Practice Exercise** What is the percent composition of the antibiotic tetracycline, which has the molecular formula $C_{22}H_{24}N_2O_8$?

**CONNECTION** We defined empirical formulas in Section 1.6 when we introduced ionic compounds. We introduced the concept of formula units in Section 2.4.

We typically determine the percent composition of a substance in the laboratory by measuring the amount of each element in a given mass of the substance. We can use these data to determine the substance's *empirical formula*, a formula based on the lowest whole-number ratio of its component elements. The formulas of nearly all ionic compounds are empirical because they are based on formula units, which contain the fewest numbers of cations and anions that together have no overall charge. The word *empirical* is synonymous with *experimental*, meaning "derived from experimental data." The term's use here is appropriate because empirical formulas are often derived from percent composition data obtained through chemical analyses.

To derive an empirical formula from percent composition data, we first assume that we have exactly 100 g of the compound. This way, the percent composition values become the masses of each element expressed in grams. Suppose that the results of an elemental analysis of a sample of an iron oxide show that it is 69.94% Fe and 30.06% O. Therefore, exactly 100 g of the sample would have 69.94 g of Fe and 30.06 g of O. Next, we convert these masses into equivalent numbers of moles by dividing by the appropriate molar masses:

$$69.94\ \cancel{\text{g Fe}} \times \frac{1\text{ mol Fe}}{55.85\ \cancel{\text{g Fe}}} = 1.252\text{ mol Fe}$$

and

$$30.06\ \cancel{\text{g O}} \times \frac{1\text{ mol O}}{16.00\ \cancel{\text{g O}}} = 1.879\text{ mol O}$$

To convert these values to a ratio of small whole numbers, we divide both by the smaller value:

$$\frac{1.879\text{ mol O}}{1.252} = 1.501\text{ mol O} \approx \tfrac{3}{2}\text{ mol O}$$

$$\frac{1.252\text{ mol Fe}}{1.252} = 1.000\text{ mol Fe}$$

To transform this mole ratio into whole numbers that can be the subscripts in an empirical formula, we need to recognize that 1.501 is really close to $1\frac{1}{2}$, or $\frac{3}{2}$. If we multiply each mole value by 2, we then have 2 moles of Fe and 3 moles of O, which translates into the empirical formula $Fe_2O_3$.

## CONCEPT TEST

Which pairs of the following compounds have the same percent composition?

a. ethylene ($C_2H_4$), a gas used to ripen green tomatoes

b. eicosene ($C_{20}H_{40}$), a compound used to trap Japanese beetles

c. acetylene ($C_2H_2$), a gas used in torches to cut steel and other metals

d. benzene ($C_6H_6$), a carcinogenic compound found in petroleum

### SAMPLE EXERCISE 7.10 Deriving an Empirical Formula from Percent Composition LO5

Structures containing the mineral magnetite (**Figure 7.16**) are found in birds' beaks and may help them navigate. Magnetite is an oxide of iron that is 72.36% iron by mass. What is its empirical formula?

**Collect and Organize** We are given the iron content of magnetite and asked to determine its empirical formula—that is, the simplest ratio of iron ions to oxide ions in the mineral.

**Analyze** Because the only elements in the mineral are iron and oxygen, the percent composition by mass of the oxygen may be obtained by subtracting the percentage of Fe from 100%. We can use the steps described in the preceding discussion to:

1. Convert percentage values into mass values in grams by assuming that we have exactly 100 g of magnetite.
2. Convert mass values into mole values by dividing by the molar masses of Fe and O.
3. Calculate a mole ratio by dividing both by the smaller of the two values.
4. Convert the mole ratio from step 3 into small whole numbers, if necessary.

**Solve**

1. A 100 g sample has 72.36 g of Fe, and

$$(100.00 - 72.36)\text{ g Fe} = 27.64\text{ g O}$$

2. Converting grams into moles:

$$72.36\ \cancel{\text{g Fe}} \times \frac{1\text{ mol Fe}}{55.85\ \cancel{\text{g Fe}}} = 1.296\text{ mol Fe}$$

and

$$27.64\ \cancel{\text{g O}} \times \frac{1\text{ mol O}}{16.00\ \cancel{\text{g O}}} = 1.728\text{ mol O}$$

3. Calculating a mole ratio of O to Fe:

$$\frac{1.728\text{ mol O}}{1.296\text{ mol Fe}} = 1.333\text{ mol O/mol Fe}$$

4. To simplify this ratio, we need to recognize that 1.333 is the decimal equivalent of the improper fraction 4/3, which means that the mole ratio of O to Fe is $\frac{4}{3}$:1, or simply 4:3. The corresponding empirical formula is $Fe_3O_4$.

**Think About It** As we have seen again in this Sample Exercise, a key to deriving an empirical formula from percent composition data is to translate mole ratios that are decimal values greater than 1 into equivalent improper fractions. Although we said in Chapter 2 that iron occurs in nature as $Fe^{2+}$ and $Fe^{3+}$ ions, the iron in magnetite *appears* to have a fractional charge of $\frac{8}{3}$, or 2.67. Actually, the formula unit of $Fe_3O_4$ contains two $Fe^{3+}$ ions and one $Fe^{2+}$ ion.

**Practice Exercise** For thousands of years, the mineral chalcocite (**Figure 7.17**) has been a valuable source of copper. Its chemical composition is 79.85% Cu and 20.15% S. What is its empirical formula?

**FIGURE 7.16** The distinctive gray crystals in this mineral sample are composed of magnetite.

**FIGURE 7.17** Chalcocite is an important source of copper. The blue and blue-green colors in the sample are common for copper-containing compounds.

### SAMPLE EXERCISE 7.11 Deriving an Empirical Formula of a Compound Containing More than Two Elements LO5

A sample of dolomite, a carbonate mineral, is 21.73% Ca, 13.18% Mg, 13.03% C, and 52.06% O. What is its empirical formula?

**Collect and Organize** We are given the percent composition by mass of dolomite and asked to determine its empirical formula. The percent composition data should add up to 100% (which they do).

**Analyze** We follow the same steps in deriving the empirical formula of dolomite as we did in Sample Exercise 7.10, except that here we must calculate more than one mole ratio because the mineral has four elements.

**Solve** Converting percentage values into grams and then into moles:

$$21.73\ \cancel{\text{g Ca}} \times \frac{1\ \text{mol Ca}}{40.08\ \cancel{\text{g Ca}}} = 0.5422\ \text{mol Ca}$$

$$13.18\ \cancel{\text{g Mg}} \times \frac{1\ \text{mol Mg}}{24.31\ \cancel{\text{g Mg}}} = 0.5422\ \text{mol Mg}$$

$$13.03\ \cancel{\text{g C}} \times \frac{1\ \text{mol C}}{12.01\ \cancel{\text{g C}}} = 1.085\ \text{mol C}$$

$$52.06\ \cancel{\text{g O}} \times \frac{1\ \text{mol O}}{16.00\ \cancel{\text{g O}}} = 3.254\ \text{mol O}$$

Dividing each mole value by the smallest one (0.5422):

$$\frac{0.5422\ \text{mol Ca}}{0.5422} = 1.000\ \text{mol Ca} \qquad \frac{0.5422\ \text{mol Mg}}{0.5422} = 1.000\ \text{mol Mg}$$

$$\frac{1.085\ \text{mol C}}{0.5422} = 2.001\ \text{mol C} \qquad \frac{3.254\ \text{mol O}}{0.5422} = 6.001\ \text{mol O}$$

All results are either whole numbers or very close to whole numbers. Rounding them to single digits gives us a mole ratio of 1 Ca:1 Mg:2 C:6 O, which corresponds to the empirical formula $CaMgC_2O_6$.

**Think About It** The problem states that dolomite is a carbonate mineral, which means that it contains $CO_3^{2-}$ ions. Thus, we can rewrite the empirical formula to indicate that the mineral is actually a blend of calcium carbonate and magnesium carbonate: $CaCO_3 \cdot MgCO_3$.

**Practice Exercise** Determine the empirical formula of a mineral that is 24.95% Fe, 46.46% Cr, and 28.59% O by mass.

## 7.6 Comparing Empirical and Molecular Formulas

An empirical formula provides the simplest mole ratios of the elements in a compound. It represents one formula unit of an ionic compound. It may—or may not—provide the number of atoms of each element in one molecule of a molecular compound. To explore these two possibilities, consider the molecular structures in **Figure 7.18**. The structure shown in Figure 7.18(a) is formaldehyde, with the molecular formula $CH_2O$, which is also the simplest mole ratio of its component elements. Thus, formaldehyde's molecular and empirical formulas are the same. However, the structure shown in Figure 7.18(b) is glycolaldehyde, with the molecular formula $C_2H_4O_2$. This molecular formula is not the simplest ratio of the moles of each element in the compound because it can be further simplified by dividing each subscript by 2. Doing so produces the same empirical formula, $CH_2O$, as formaldehyde as well as a whole family of compounds known as simple sugars, including the glucose ($C_6H_{12}O_6$) produced during photosynthesis in green plants. For simple sugars and

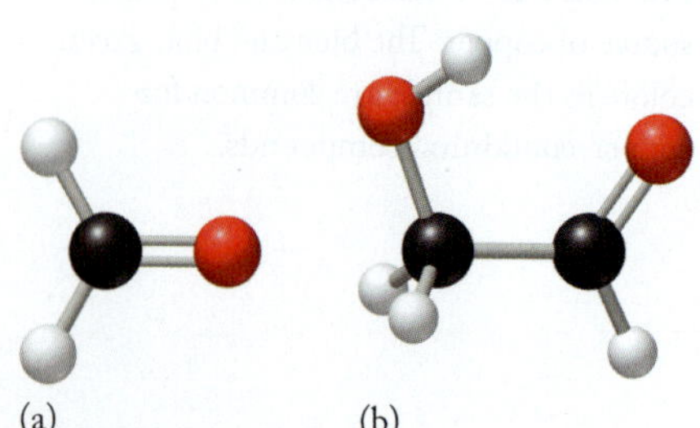


**FIGURE 7.18** Molecular structures of (a) formaldehyde and (b) glycolaldehyde: two compounds with the same empirical formula, $CH_2O$.

many other molecular compounds, we need to know the molar masses to determine their molecular formulas from percent composition data.

**CONCEPT TEST**

Which of the following compounds have empirical formulas that match their molecular formulas?

a. ethylene glycol, $C_2H_6O_2$

b. isopropanol, $C_3H_8O$

c. glucose, $C_6H_{12}O_6$

To see how to use the empirical formula of a compound and its molecular mass to determine its molecular formula, let's reconsider glycolaldehyde and glucose. Their molecular formulas can be expressed in terms of their common empirical formula ($CH_2O$) as follows:

$$C_2H_4O_2 = (CH_2O)_2 \qquad C_6H_{12}O_6 = (CH_2O)_6$$

Extending these two equations to any molecular compound, we have the following general relation between molecular and empirical formulas:

$$(\text{Molecular formula}) = (\text{empirical formula})_n \tag{7.17}$$

where $n$ is a positive integer equal to or greater than 1.

The key to deriving a molecular formula from an empirical formula is to determine the value of $n$. That's where molecular mass comes in. Using glucose as an example, let's first calculate the mass of one of its molecules:

$$\left(6\ \cancel{\text{atoms C}} \times \frac{12.01\ \text{u}}{1\ \cancel{\text{atom C}}}\right) + \left(12\ \cancel{\text{atoms H}} \times \frac{1.008\ \text{u}}{1\ \cancel{\text{atom H}}}\right) + \left(6\ \cancel{\text{atoms O}} \times \frac{16.00\ \text{u}}{1\ \cancel{\text{atom O}}}\right) = 180.16\ \text{u}$$

Now we calculate the mass of its empirical formula unit ($CH_2O$):

$$\left(1\ \cancel{\text{atom C}} \times \frac{12.01\ \text{u}}{1\ \cancel{\text{atom C}}}\right) + \left(2\ \cancel{\text{atoms H}} \times \frac{1.008\ \text{u}}{1\ \cancel{\text{atom H}}}\right) + \left(1\ \cancel{\text{atom O}} \times \frac{16.00\ \text{u}}{1\ \cancel{\text{atom O}}}\right) = 30.03\ \text{u}$$

Pretend that we did not already know the molecular formula of glucose, but we did know its empirical formula and could calculate the mass of one empirical formula unit (30.03 u). If we also knew that its molecular mass was 180.16 u, we could divide the molecular mass by the mass of one empirical formula unit to obtain the value of $n$ to use in Equation 7.17:

$$n = \frac{\text{molecular mass}}{\text{mass of 1 empirical formula unit}} \tag{7.18}$$

$$= \frac{180.16\ \cancel{\text{u}}}{30.03\ \cancel{\text{u}}} = 5.999 \approx 6$$

Inserting this value of $n$ into Equation 7.17 yields the actual molecular formula of glucose:

$$(CH_2O)_6 = C_6H_{12}O_6$$

Determining the molecular formula is possible once you know both the empirical formula and the molecular mass, but how do chemists experimentally determine the molecular mass? In Section 2.6, we introduced *mass spectrometry* as a technique to determine the molecular mass of compounds and measure isotopic abundances. The *mass spectrum* shows the fragments of an ionized molecule according to their $m/z$ (mass/charge) ratio, and the highest $m/z$ value in the spectrum often represents the molecular ion, $M^+$. If we assume that the molecular fragments are singly charged ions ($z = 1+$), the $m/z$ value of that molecular-ion peak is simply $m$, the molecular mass of the compound.

**CONNECTION** We examined the mass spectra of several compounds in Section 2.6.

## CONCEPT TEST

Does the molecular-ion peak in a mass spectrum correspond to the mass associated with the empirical formula or the molecular formula of a compound?

## SAMPLE EXERCISE 7.12 Using Percent Composition and Molecular Mass to Derive a Molecular Formula **LO6**

Pheromones are chemical substances that members of a species secrete to stimulate a response in other individuals of the same species. The percent composition of eicosene, a compound similar to the mating pheromone secreted by the Japanese beetle, is 85.63% C and 14.37% H. Its molecular mass, as determined by mass spectrometry, is 280 u. What is the molecular formula of eicosene?

**Collect and Organize** We are given the percent composition of eicosene and its molecular mass. The percent composition data can be used to derive the empirical formula. Equation 7.17 relates the empirical and molecular formulas of compounds. Equation 7.18 can be used to calculate the value of $n$ from the mass of an empirical formula unit.

**Analyze** We can follow the procedure used in Sample Exercise 7.10 to determine the empirical formula of eicosene. Next we calculate the mass of one formula unit. Dividing that mass into the molecular mass (280 u) given in the problem will give us the value of $n$ for Equation 7.17 and will allow us to convert the empirical formula into a molecular formula. Given the large molecular mass of eicosene and the low atomic masses of C (12.01 u) and H (1.008 u), $n$ may be a large number.

**Solve** Converting percentage values into grams and then into moles:

$$85.63\ \cancel{\text{g C}} \times \frac{1\ \text{mol C}}{12.01\ \cancel{\text{g C}}} = 7.130\ \text{mol C}$$

$$14.37\ \cancel{\text{g H}} \times \frac{1\ \text{mol H}}{1.008\ \cancel{\text{g H}}} = 14.26\ \text{mol H}$$

Dividing moles of H by moles of C (the smaller value of the two) gives us the simplified mole ratio of 1 C:2 H and the empirical formula $CH_2$.

The mass of one empirical formula unit is

$$\left(1\ \cancel{\text{atom C}} \times \frac{12.01\ \text{u}}{1\ \cancel{\text{atom C}}}\right) + \left(2\ \cancel{\text{atoms H}} \times \frac{1.008\ \text{u}}{1\ \cancel{\text{atom H}}}\right) = 14.03\ \text{u}$$

Using Equation 7.18 to calculate the formula multiplier $n$:

$$n = \frac{\text{molecular mass}}{\text{mass of 1 empirical formula unit}} = \frac{280\ \cancel{\text{u}}}{14.03\ \cancel{\text{u}}} = 19.96 \approx 20$$

Using Equation 7.17 to determine the molecular formula:

$$\text{Molecular formula} = (CH_2)_n = (CH_2)_{20} = C_{20}H_{40}$$

**Think About It** The molecular formula of eicosene represents the number of C and H atoms in one molecule of eicosene. The compound is one of many hydrocarbons with the empirical formula $CH_2$.

**Practice Exercise** Determine the empirical and molecular formulas of a compound that contains 43.64% P and 56.36% O and has a molar mass of 284 g/mol.

## 7.7 Combustion Analysis

**STEPWISE ANIMATION**
Combustion Analysis

Recall from Section 7.2 that combustion reactions involve burning substances in oxygen. In **combustion analysis**, the complete combustion of a compound followed by an analysis of the products enables chemists to determine the chemical composition of that compound. To ensure that combustion is complete, the process is carried out in an atmosphere of pure oxygen. This type of elemental analysis relies on complete combustion so that all the carbon in the sample is converted to $CO_2$ and all the hydrogen to $H_2O$. For a generic hydrocarbon $C_aH_b$, the overall reaction is

$$C_aH_b + \text{excess } O_2(g) \rightarrow a\ CO_2(g) + \frac{b}{2}\ H_2O(g) \tag{7.19}$$

As shown in **Figure 7.19**, the products of combustion in an elemental analyzer flow through a tube packed with $Mg(ClO_4)_2(s)$, which selectively absorbs the $H_2O(g)$, and then through a tube containing $NaOH(s)$, which absorbs the $CO_2(g)$. The masses of the tubes are measured before and after combustion. The differences equal the masses of $CO_2$ and $H_2O$ produced by the combustion of the

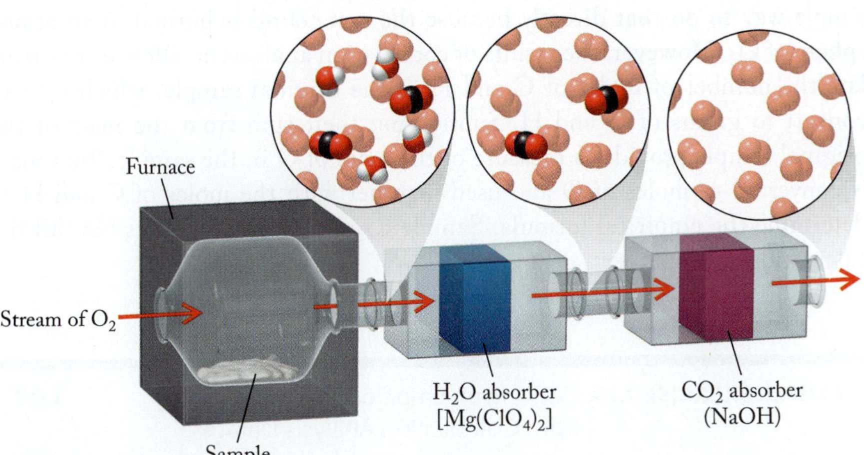


**FIGURE 7.19** A carbon–hydrogen elemental analyzer relies on the complete combustion of organic compounds. The products are $H_2O$ vapor and $CO_2$. Water vapor is absorbed by a cartridge packed with solid $Mg(ClO_4)_2$, whereas $CO_2$ is absorbed by solid NaOH. The empirical formula of the compound is calculated from the masses of $H_2O$ and $CO_2$ absorbed.

**combustion analysis** a laboratory procedure in which a substance is burned completely in oxygen to produce known compounds whose masses are used to determine the composition of the original material.

sample. Suppose that complete combustion of a sample of a hydrocarbon results in a 1.320 g increase in the mass of the tube that traps $CO_2(g)$ and a 0.541 g increase in the mass of the tube that traps $H_2O(g)$. How can we use these results to determine the empirical formula of the hydrocarbon?

Empirical formulas express the mole ratios of the elements in a compound, so we start by converting the masses of $CO_2$ and $H_2O$ caught in the traps into moles of $CO_2$ and $H_2O$ and then to equivalent numbers of moles of C and H in the original sample:

$$1.320 \cancel{\text{g CO}_2} \times \frac{1 \cancel{\text{mol CO}_2}}{44.01 \cancel{\text{g CO}_2}} \times \frac{1 \text{ mol C}}{1 \cancel{\text{mol CO}_2}} = 0.02999 \text{ mol C} \approx 0.0300 \text{ mol C}$$

$$0.541 \cancel{\text{g H}_2\text{O}} \times \frac{1 \cancel{\text{mol H}_2\text{O}}}{18.02 \cancel{\text{g H}_2\text{O}}} \times \frac{2 \text{ mol H}}{1 \cancel{\text{mol H}_2\text{O}}} = 0.0600 \text{ mol H}$$

The mole ratio of H to C is (0.0600 mol/0.0300 mol) = 2:1, so the empirical formula is $CH_2$.

If we want to extend this method to determining a molecular formula, we need to know the molecular mass. Suppose the mass spectrum of the hydrocarbon of interest has a molecular ion at 84 u. The formula mass of $CH_2$ is 14 u. Dividing the molecular mass by the formula mass, we get the multiplier, $n$:

$$n = \frac{84 \cancel{\text{u}}}{14 \cancel{\text{u}}} = 6$$

The molecular formula is therefore

$$(CH_2)_6 = C_6H_{12}$$

Note that here we did not need to know the initial mass of the sample to determine its empirical formula. We needed to know only that the sample was a hydrocarbon and that it was completely converted into the stated amounts of $CO_2$ and $H_2O$.

What if we knew that our sample was not a hydrocarbon? What if we had isolated a pharmacologically promising compound from a tropical plant, and we knew that its molecules contained atoms of carbon, hydrogen, and oxygen? We would need to determine the percentage of oxygen in it, but we have no simple way to do that directly because the compound is burned in an atmosphere of $O_2$. However, the results of combustion analysis do allow us to calculate the number of moles of C and H in the original sample, which we can convert to grams of C and H. Subtracting their sum from the mass of the original sample provides a measure of the mass of O in the sample. That mass is converted to moles of O and used together with the moles of C and H to determine the empirical formula. Sample Exercise 7.13 illustrates how all the calculations fit together.

**SAMPLE EXERCISE 7.13** Deriving an Empirical Formula from Combustion Analysis Data **LO7**

Combustion of 1.000 g of an organic compound known to contain only carbon, hydrogen, and oxygen produces 2.360 g of $CO_2$ and 0.640 g of $H_2O$. What is the empirical formula of the compound?

**Collect and Organize** We are given the initial mass of a sample of a compound that consists of C, H, and O, and we also know the masses of $CO_2$ and $H_2O$ produced during its combustion. We are asked to determine the empirical formula of the compound.

**Analyze** We assume complete combustion of the sample, which means that its entire carbon content is converted into $CO_2$, and all its hydrogen content becomes $H_2O$. We need to determine the number of moles of C and H in 2.360 g of $CO_2$ and 0.640 g of $H_2O$, respectively. We also need to calculate the masses of C and H because the difference between the sum of the masses of C and H and the mass of the original sample is equal to the O content of the original sample. We then determine the C:H:O mole ratio and convert it into a ratio of small whole numbers.

**Solve** The numbers of moles of C and H in the $CO_2$ and $H_2O$ collected during combustion are

$$2.360\ \cancel{\text{g CO}_2} \times \frac{1\ \cancel{\text{mol CO}_2}}{44.01\ \cancel{\text{g CO}_2}} \times \frac{1\ \text{mol C}}{1\ \cancel{\text{mol CO}_2}} = 0.05362\ \text{mol C}$$

$$0.640\ \cancel{\text{g H}_2\text{O}} \times \frac{1\ \cancel{\text{mol H}_2\text{O}}}{18.02\ \cancel{\text{g H}_2\text{O}}} \times \frac{2\ \text{mol H}}{1\ \cancel{\text{mol H}_2\text{O}}} = 0.07103\ \text{mol H}$$

The masses of C and H are

$$0.05362\ \cancel{\text{mol C}} \times \frac{12.01\ \text{g C}}{1\ \cancel{\text{mol C}}} = 0.6440\ \text{g C}$$

$$0.07103\ \cancel{\text{mol H}} \times \frac{1.008\ \text{g H}}{1\ \cancel{\text{mol H}}} = 0.07160\ \text{g H}$$

The difference between the original sample mass (1.000 g) and the sum of the masses of C and H (0.6440 g + 0.07160 g = 0.7156 g) is the mass of the oxygen in the sample:

$$\text{Mass of oxygen} = 1.000\ \text{g} - 0.7156\ \text{g} = 0.2844\ \text{g O}$$

The number of moles of O atoms in the sample is

$$0.2844\ \cancel{\text{g O}} \times \frac{1\ \text{mol O}}{16.00\ \cancel{\text{g O}}} = 0.01778\ \text{mol O}$$

Combining the above results, we get the mole ratio of the three elements in the sample:

$$0.05362\ \text{mol C} : 0.07103\ \text{mol H} : 0.01778\ \text{mol O}$$

To simplify the ratios, we divide them by the smallest of their values (0.01778):

$$\frac{0.05362\ \text{mol C}}{0.01778} = 3.016 \approx 3\ \text{mol C}$$

$$\frac{0.07103\ \text{mol H}}{0.01778} = 3.995 \approx 4\ \text{mol H}$$

$$\frac{0.01778\ \text{mol O}}{0.01778} = 1\ \text{mol O}$$

The whole-number mole ratio that most closely matches the calculated values is 3:4:1, making the empirical formula of the sample $C_3H_4O$.

**Think About It** The results of our mole ratio calculations for C and H are very close to small whole numbers, which means that our calculations and the formula based on them are probably correct.

**Practice Exercise** Vanillin is the compound containing carbon, hydrogen, and oxygen that gives vanilla beans their distinctive flavor. The combustion of 30.4 mg of vanillin produces 70.4 mg of $CO_2$ and 14.4 mg of $H_2O$. The mass spectrum of vanillin shows a molecular-ion peak at 152 u. Use this information to determine the molecular formula of vanillin.

## SAMPLE EXERCISE 7.14 Integrating Concepts: Synthesizing Hydrogen Gas

Practice Exercise 7.8 describes the first stage in the steam–methane reforming reaction for synthesizing $H_2$ gas. The carbon monoxide by-product of the reaction can be reacted with more steam in a second stage to produce more hydrogen and carbon dioxide.

a. Write a balanced chemical equation for the reaction between CO and steam that produces $H_2$ and $CO_2$.
b. Combine your answer from part (a) with the chemical equation for the first stage of the steam–methane reforming reaction from Practice Exercise 7.8 to write an overall reaction in which methane and steam react to form hydrogen gas and carbon dioxide.
c. Suppose a reaction vessel for the two-stage process initially contains 48.0 kg of $CH_4$ and 118 kg of $H_2O(g)$ and that the process yields 16 kg of $H_2$. Is the initial reaction mixture stoichiometric? If not, which is the limiting reactant?
d. What is the percent yield of the reaction mixture in part (c)?

**Collect and Organize** We need to write a balanced chemical equation for the reaction between CO and $H_2O$ that produces $H_2$ and $CO_2$ and then combine it with the steam–methane reforming reaction to write an overall equation describing the formation of $H_2$ and $CO_2$ from $CH_4$ and $H_2O$. We must then determine whether a reaction mixture of $CH_4$ and $H_2O$ is stoichiometric and calculate the percent yield of the reaction. We know the masses of $CH_4$ and $H_2O$ in the reaction mixture and the mass of $H_2$ they produce. We also know the chemical equation of the steam–methane reforming reaction, the first step in the two-step process. From previous calculations in this chapter, we know the following molar masses: $H_2O$, 18.02 g/mol; $CH_4$, 16.04 g/mol; and $H_2$, 2.016 g/mol.

**Analyze** To combine two chemical equations, we need to make sure that the coefficients of products in the first equation that are reactants in the second equation are the same, so that they cancel out when the equations are combined. One way to determine whether a reaction mixture is stoichiometric is to calculate the quantities of product that each reactant could produce if it were completely consumed. Using this approach makes sense because we need to determine the percent yield of the reaction, which involves calculating the theoretical yield and comparing it with the actual yield.

**Solve**

a. First we write the chemical equation for the reaction between CO and steam that produces $H_2$ and $CO_2$. We start with one molecule of each reactant and product,

$$CO(g) + H_2O(g) \rightarrow H_2(g) + CO_2(g)$$

and we find that the equation is balanced already.

b. Comparing this equation with the one for the steam–methane reforming reaction,

$$CH_4(g) + H_2O(g) \rightarrow CO(g) + 3\,H_2(g)$$

we find that CO is a product of that reaction as well as a reactant in the second-stage reaction. Its coefficient is 1 in both reactions, so the two can be combined without modification:

$$\begin{array}{r} CH_4(g) + H_2O(g) \rightarrow \cancel{CO(g)} + 3\,H_2(g) \\ + \cancel{CO(g)} + H_2O(g) \rightarrow H_2(g) + CO_2(g) \\ \hline CH_4(g) + 2\,H_2O(g) \rightarrow 4\,H_2(g) + CO_2(g) \end{array}$$

c. To determine whether the reaction mixture is stoichiometric, we calculate the theoretical yields of $H_2$ produced by each reactant:

$$48.0\ \cancel{\text{kg CH}_4} \times \frac{10^3\ \cancel{\text{g}}}{1\ \cancel{\text{kg}}} \times \frac{1\ \cancel{\text{mol CH}_4}}{16.04\ \cancel{\text{g CH}_4}} \times \frac{4\ \cancel{\text{mol H}_2}}{1\ \cancel{\text{mol CH}_4}} \times \frac{2.016\ \cancel{\text{g}}\ \text{H}_2}{1\ \cancel{\text{mol H}_2}} \times \frac{1\ \text{kg}}{10^3\ \cancel{\text{g}}} = 24.13\ \text{kg H}_2$$

$$118\ \cancel{\text{kg H}_2\text{O}} \times \frac{10^3\ \cancel{\text{g}}}{1\ \cancel{\text{kg}}} \times \frac{1\ \cancel{\text{mol H}_2\text{O}}}{18.02\ \cancel{\text{g H}_2\text{O}}} \times \frac{4\ \cancel{\text{mol H}_2}}{2\ \cancel{\text{mol H}_2\text{O}}} \times \frac{2.016\ \cancel{\text{g}}\ \text{H}_2}{1\ \cancel{\text{mol H}_2}} \times \frac{1\ \text{kg}}{10^3\ \cancel{\text{g}}} = 26.40\ \text{kg H}_2$$

Methane produces less $H_2$, so the reaction mixture is not stoichiometric and $CH_4$ is the limiting reactant.

d. The percent yield calculation is based on the lesser theoretical yield (24.13 kg of $H_2$):

$$\frac{16\ \cancel{\text{kg H}_2}}{24.13\ \cancel{\text{kg H}_2}} \times 100\% = 66\%\ \text{yield}$$

**Think About It** The second-stage reaction in the synthesis of $H_2$ is called the *water–gas shift reaction*. Many important industrial processes rely on chemical reactions with yields less than 100%, and the production of hydrogen gas is one of them. As in other multistep calculations, we retained extra significant figures for intermediate values and did not round off to two digits until the end. Another way to avoid rounding errors is to use values in calculator memory for each new step in a calculation rather than reenter the results of previous calculations.

## SUMMARY

**LO1** **Chemical equations**, which contain the chemical formulas of the **reactants** on the left and the **products** on the right, provide concise, symbolic descriptions of chemical reactions. The proportions of the reactants and products are expressed by coefficients preceding their formulas. Writing balanced chemical equations requires adjusting these coefficients until the number of atoms of each element is the same on the reactant and product sides of the equation. The **law of conservation of mass** states that the sum of the masses of the reactants in a chemical reaction is equal to the sum of the masses of the products. (Sections 7.1 and 7.2)

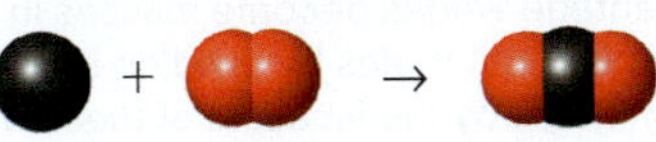

**LO2** The mole ratios of reactants and products in a chemical reaction are called the **stoichiometry** of the reaction. These ratios can be used to relate the masses of the reactants consumed and the products made in a chemical reaction. (Section 7.3)

**LO3** Chemical reactions are not always run with exact stoichiometric amounts of material. The **limiting reactant** in a reaction mixture is the reactant that limits how much product can be made. (Section 7.4)

**LO4** The maximum amount of product that can form in a chemical reaction is the **theoretical yield**. The measured amount of product formed in a reaction is the actual yield, and the ratio of actual yield to theoretical yield, expressed as a percentage, is the **percent yield** for the reaction. (Section 7.4)

**LO5** The **percent composition** of a compound is the percentage by mass of each element in the compound. The empirical formula of a compound represents the lowest whole-number ratio of its elements. (Section 7.5)

**LO6** A compound's empirical formula may or may not be the same as its molecular formula, which indicates the number of each type of atom in one molecule. Converting the empirical formula of a compound into a molecular formula requires knowing the molecular mass of the compound. (Section 7.6)

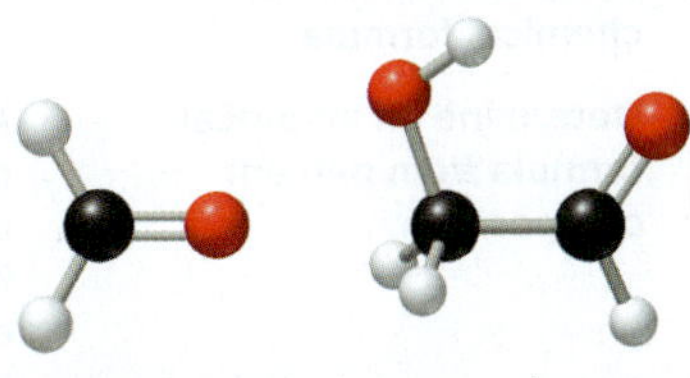

**LO7** In **combustion analysis**, a known mass of an organic compound is burned in a stream of oxygen gas. The carbon in the sample is converted into $CO_2$ and the hydrogen is converted into $H_2O$. The masses of $CO_2$ and $H_2O$ are measured and used to determine the masses of C and H in the organic compound and then the compound's empirical formula. (Section 7.7)

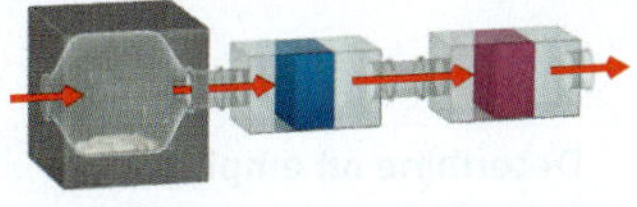

## PARTICULATE PREVIEW WRAP-UP

Nitromethane burns in oxygen to produce carbon dioxide, nitrogen gas, and water:

$$4\ CH_3NO_2(\ell) + 3\ O_2(g) \rightarrow 4\ CO_2(g) + 2\ N_2(g) + 6\ H_2O(g).$$

If the mixture shown were to react, every molecule of $CH_3NO_2$ would react, but 3 molecules of $O_2(g)$ would remain unreacted.

## PROBLEM-SOLVING SUMMARY

| Type of Problem | Concepts and Equations | Sample Exercises |
|---|---|---|
| **Write and balance a chemical equation** | Adjust coefficients to balance number of atoms of each element on both sides of the reaction arrow. Start with elements that are in only one reactant and product on each side. | **7.1–7.3** |
| **Write and balance a chemical equation for a combustion reaction** | C and H in organic compounds react with $O_2$ to form $CO_2$ and $H_2O$; for example: $CH_4(g) + 2\ O_2(g) \rightarrow CO_2(g) + 2\ H_2O(g)$ Balance the atoms of C first, then H, and then O. | **7.3** |
| **Calculate the mass of a product from the mass of a reactant** | Convert mass of reactant into moles of reactant; use reaction stoichiometry to calculate moles of product, and then convert moles of product into mass of product. | **7.4, 7.5** |
| **Identify the limiting reactant in a reaction mixture** | Calculate how much product each reactant could make; the reactant making the least amount of product is the limiting reactant. | **7.6, 7.7** |

| Type of Problem | Concepts and Equations | Sample Exercises |
|---|---|---|
| **Calculate theoretical yield and percent yield** | Use the quantity of the limiting reactant to calculate the theoretical yield of product. Divide actual yield (given or measured) by theoretical yield: $$\text{Percent yield} = \frac{\text{actual yield}}{\text{theoretical yield}} \times 100\% \quad (7.16)$$ | **7.7, 7.8** |
| **Calculate percent composition from a chemical formula** | Calculate the mass of each element in one mole of the compound. Divide each element's mass by the molar mass; express the results as percentages. | **7.9** |
| **Determine an empirical formula from percent composition** | Assume a 100 g sample so that percentage values become masses in grams. Convert the mass of each element into moles by dividing by its molar mass. Divide these numbers of moles by the smallest of their values. If necessary, multiply by an appropriate factor to remove any fractions and obtain whole-number mole values, which become the subscript numbers in the empirical formula of the compound. | **7.10, 7.11** |
| **Relate empirical and molecular formulas** | Derive an empirical formula from percent composition data. Calculate the multiplier *n* by dividing the molecular mass by the mass of one empirical formula unit. Multiply subscripts in the empirical formula by *n* to obtain the molecular formula. | **7.12** |
| **Determine an empirical formula from combustion analysis data** | For hydrocarbons, convert given masses of $CO_2$ and $H_2O$ into moles of $CO_2$ and $H_2O$ and then to moles of C and H. For compounds also containing O, convert moles of C and H into masses of C and H and subtract the sum of these values from the sample mass to calculate the mass of O. Convert mass of O to moles of O. Simplify mole ratios of C to H to O. | **7.13** |

## VISUAL PROBLEMS

*(Answers to boldface end-of-chapter questions and problems are in the back of the book.)*

**7.1.** The molecular models in Figure P7.1 represent five oxides of nitrogen. Write the empirical and molecular formulas of the one(s) for which the two formulas are not the same.

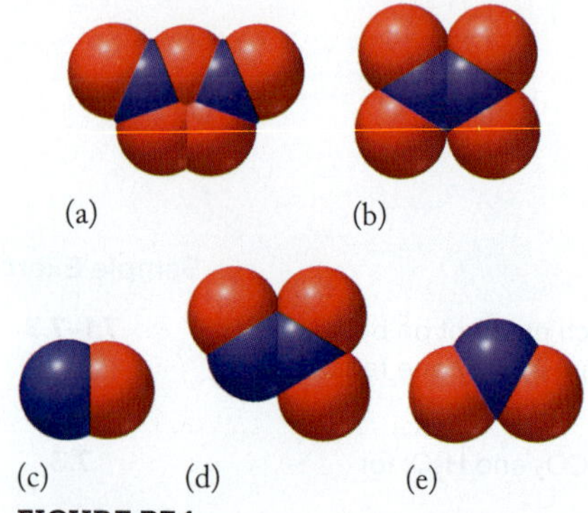


**FIGURE P7.1**

7.2. Which of the $C_2$ hydrocarbons in Figure P7.2 have different empirical and molecular formulas? What are those formulas?

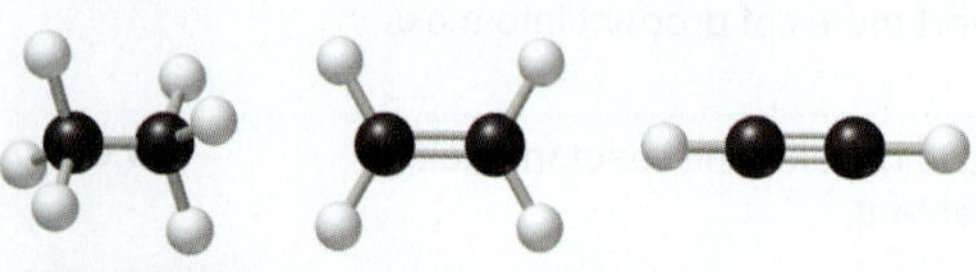

**FIGURE P7.2**

**7.3.** Which two hydrocarbons with the molecular structures shown in Figure P7.3 have the same percent composition?

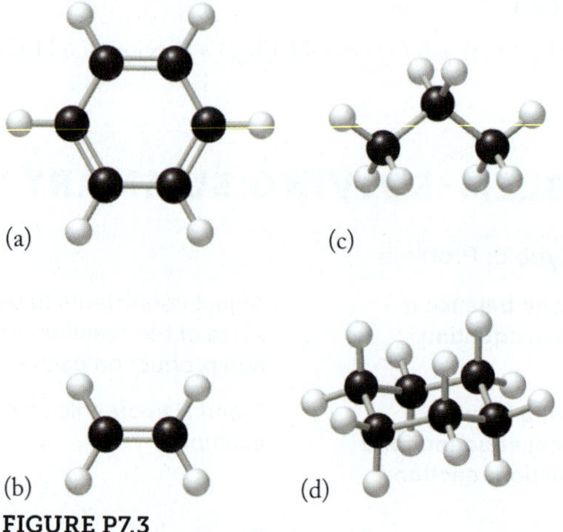


**FIGURE P7.3**

7.4. Each pair of images in Figure P7.4 contains substances composed of two elements: X (red spheres) and Y (blue spheres). Write a balanced chemical equation for the reaction taking place in each pair of images. Be sure to indicate the physical states of the reactants and products by using the appropriate symbols in parentheses.

**7.5.** Identify the limiting reactant in each pair of containers pictured in Figure P7.5. The red spheres represent atoms of element X, whereas the blue spheres are atoms of element Y. Each question mark indicates that unreacted reactant is left over.

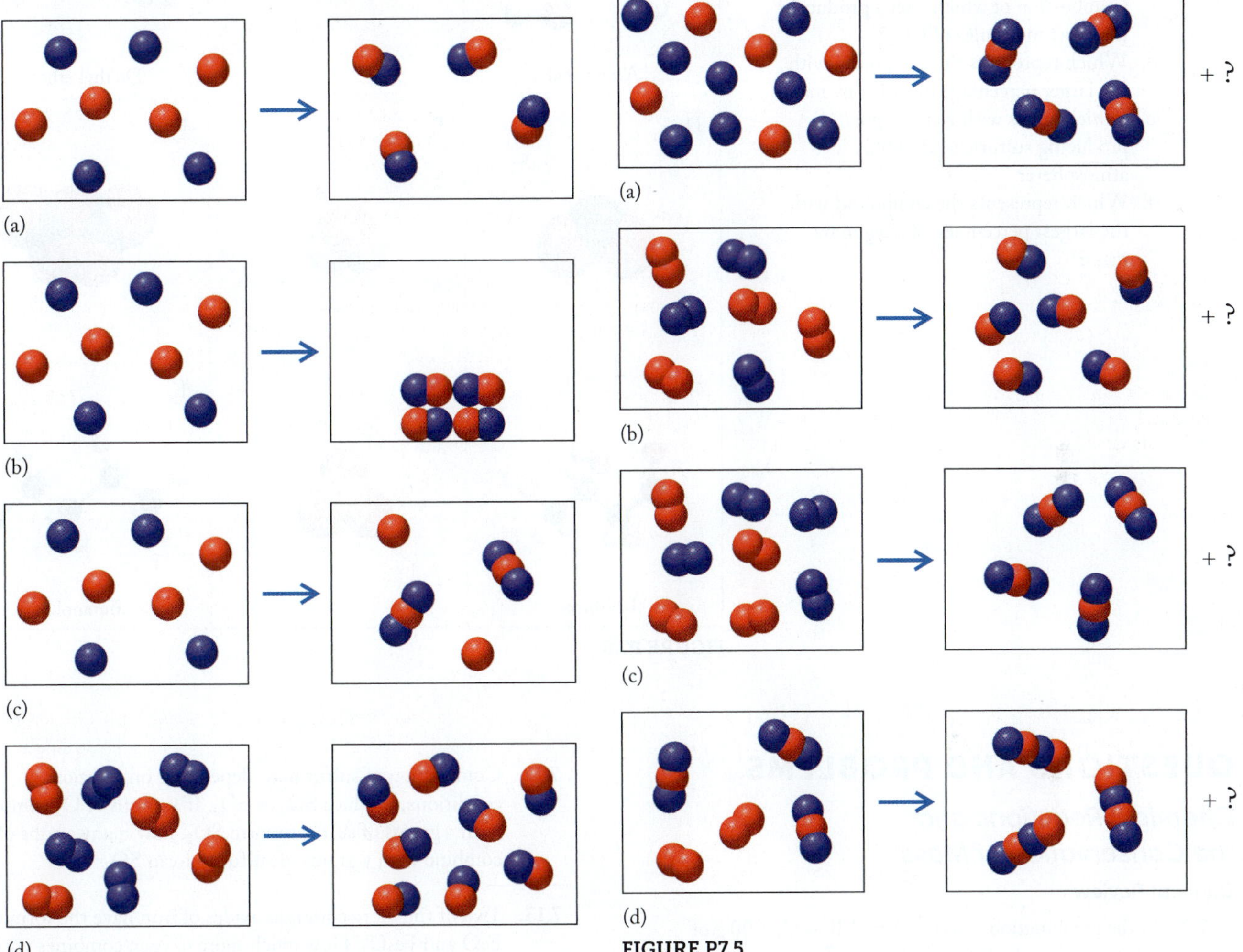


**FIGURE P7.4**

**FIGURE P7.5**

7.6. Use representations [A] through [I] in Figure P7.6 to answer questions a–f.
a. Which represent compounds with the same empirical formula?
b. Which represent compounds with the same percent composition?
c. Combustion of which one(s) produce(s) the most molecules of $CO_2$?
d. Which represents the compound with the larger percentage of sulfur by mass?
e. Which reacts with water vapor, producing sulfuric acid in the atmosphere?
f. Which represents the compound with the largest percentage of oxygen by mass?

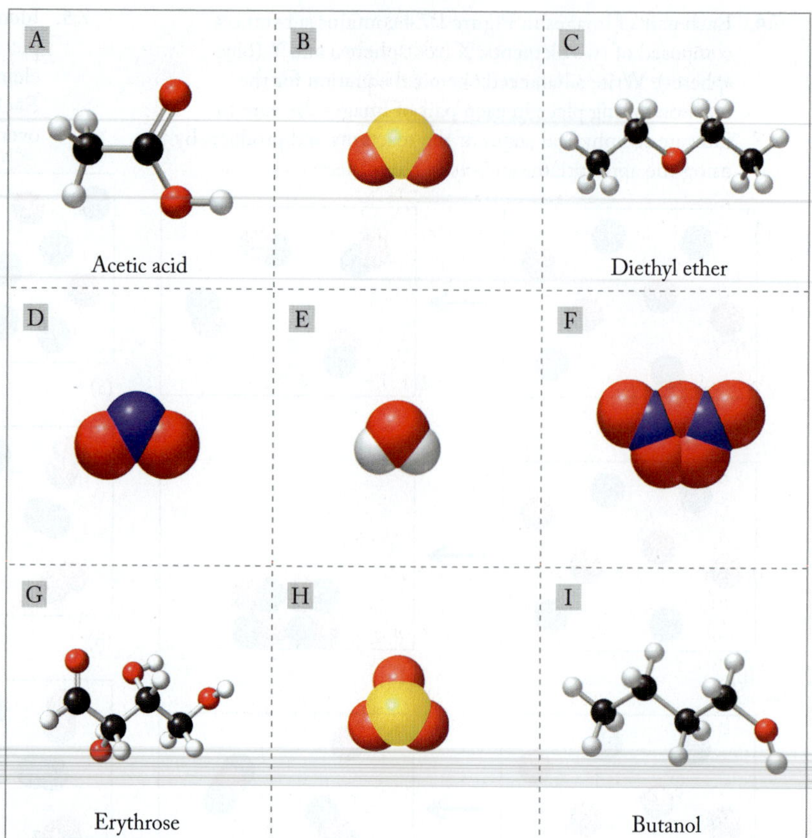

**FIGURE P7.6**

## QUESTIONS AND PROBLEMS

### Chemical Reactions and the Conservation of Mass

**Concept Review**

**7.7.** In the combination reaction $A + 2\,B \rightarrow C$, 1.00 g of substance A and 4.00 g of substance B are consumed. How many grams of substance C are formed?

7.8. In the reaction $A + B \rightarrow C + D$, 3.00 g of substance C and 3.00 g of substance D are produced as 2.00 g of substance A are consumed. How many grams of substance B are also consumed?

**7.9.** Among the four reactants and products in Question 7.8, which one has the largest molar mass?

*7.10. Among the three substances in Question 7.7 and the four substances in Question 7.8, which ones *could* be elements?

**Problems**

**7.11.** Depending on reaction conditions, $O_2$ may combine with $N_2$ to form NO or $NO_2$. If *x* grams of $O_2$ combine with *y* grams of $N_2$ to form NO, how many grams of $O_2$ combine with *y* grams of $N_2$ to form $NO_2$?

7.12. Combustion of sulfur may, depending on reaction conditions, produce $SO_2$ or $SO_3$. If *x* grams of $O_2$ combine with *y* grams of sulfur to form $SO_2$, how many grams of $O_2$ combine with *y* grams of sulfur to form $SO_3$?

**7.13.** Two of the more common oxides of iron have the formulas FeO and $Fe_2O_3$. How much more oxygen combines with a given mass of iron to form $Fe_2O_3$ than combines with the same mass of iron to form FeO?

7.14. Tin(IV) chloride is prepared by the following combination reaction:

$$Sn(s) + 2\,Cl_2(g) \rightarrow SnCl_4(\ell)$$

If *x* grams of chlorine combine with *y* grams of tin to form tin(IV) chloride, how many grams of chlorine are present in a sample of tin(II) chloride that contains *y* grams of tin?

### Writing Balanced Chemical Equations; Combustion Reactions

**Concept Review**

**7.15.** In a balanced chemical equation, does the number of atoms in the reactants always equal the number of atoms in the products?

7.16. In a balanced chemical equation for the complete combustion of a hydrocarbon, what is the ratio of atoms of C in the hydrocarbon to molecules of $CO_2$ produced?

**7.17.** How many moles of water vapor are produced for every mole of methane consumed in the combustion reaction $CH_4(g) + 2\,O_2(g) \rightarrow CO_2(g) + 2\,H_2O(g)$?

7.18. Why is CO produced during the combustion of hydrocarbons even when enough $O_2$ is present for complete combustion?

## Problems

**7.19.** Balance the following reactions for the formation of nitrogen compounds:
a. $N_2(g) + O_2(g) \rightarrow NO(g)$
b. $N_2(g) + O_2(g) \rightarrow N_2O(g)$
c. $NO(g) + NO_3(g) \rightarrow NO_2(g)$
d. $NO(g) + O_2(g) + H_2O(\ell) \rightarrow HNO_2(\ell)$

7.20. **Chemistry of Geothermal Vents** Some scientists believe that life on Earth originated near geothermal vents. Balance the following reactions, which are among those taking place near such vents:
a. $CH_3SH(aq) + CO(aq) \rightarrow CH_3COSCH_3(aq) + H_2S(aq)$
b. $H_2S(aq) + CO(aq) \rightarrow CH_3CO_2H(aq) + S_8(s)$

---

***7.21.** Write a balanced chemical equation for each of the following reactions:
a. Dinitrogen pentoxide reacts with sodium metal to produce sodium nitrate and nitrogen dioxide.
b. A mixture of nitric acid and nitrous acid is formed when water reacts with dinitrogen tetroxide.
c. At high pressure, nitrogen monoxide decomposes to dinitrogen monoxide and nitrogen dioxide.

7.22. Write a balanced chemical equation for each of the following reactions:
a. Hydrazine, $N_2H_4$, reacts with oxygen to produce water and the element nitrogen.
b. Ammonia ($NH_3$) burns in oxygen to produce elemental nitrogen and water.
c. Silicon dioxide reacts with carbon to produce the element silicon and carbon monoxide.

---

**7.23.** Complete and balance the following chemical equations for the complete combustion of several hydrocarbons.
a. $C_3H_8(g) + O_2(g) \rightarrow$
b. $C_4H_{10}(g) + O_2(g) \rightarrow$
c. $C_6H_6(\ell) + O_2(g) \rightarrow$
d. $C_8H_{18}(\ell) + O_2(g) \rightarrow$

7.24. Complete and balance the following chemical equations for the complete combustion of several hydrocarbons.
a. $C_5H_{10}(\ell) + O_2(g) \rightarrow$
b. $C_6H_{14}(\ell) + O_2(g) \rightarrow$
c. $C_8H_{10}(\ell) + O_2(g) \rightarrow$
d. $C_9H_{12}(\ell) + O_2(g) \rightarrow$

---

**7.25.** Write balanced chemical equations for the complete combustion of the gaseous hydrocarbons with the molecular structures in Figure P7.25.

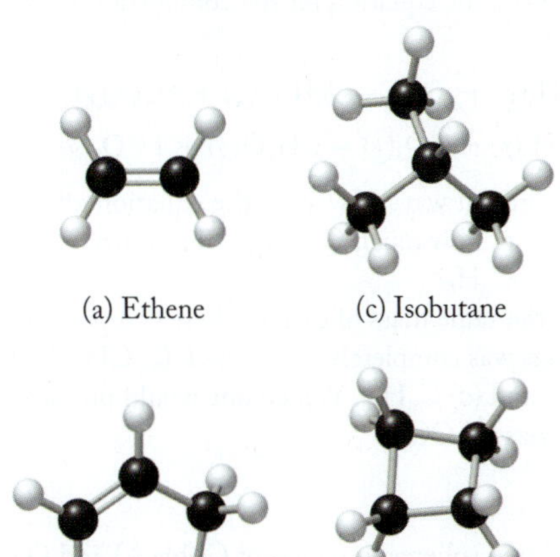


**FIGURE P7.25**

7.26. Write balanced chemical equations for the complete combustion of the liquid hydrocarbons with the molecular structures in Figure P7.26.

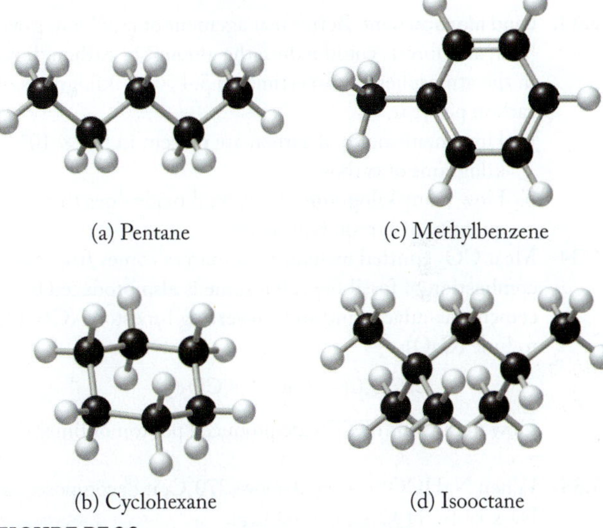


**FIGURE P7.26**

---

**7.27.** **Chemistry of Volcanic Gases** Balance the following reactions that occur during volcanic eruptions:
a. $SO_2(g) + O_2(g) \rightarrow SO_3(g)$
b. $H_2S(g) + O_2(g) \rightarrow SO_2(g) + H_2O(g)$
*c. $H_2S(g) + SO_2(g) \rightarrow S_8(s) + H_2O(g)$

*7.28. Copper was one of the first metals used by humans because it can be recovered from several copper minerals, including cuprite ($Cu_2O$), chalcocite ($Cu_2S$), and malachite [$Cu_2CO_3(OH)_2$]. Balance the following reactions for converting these minerals into copper metal:
a. $Cu_2O(s) + C(s) \rightarrow Cu(s) + CO_2(g)$
b. $Cu_2O(s) + Cu_2S(s) \rightarrow Cu(s) + SO_2(g)$
c. $Cu_2CO_3(OH)_2(s) + C(s) \rightarrow Cu(s) + CO_2(g) + H_2O(g)$

## Stoichiometric Calculations

### Concept Review

**7.29.** We can write the equation for the combustion of ethane in two ways:

$$C_2H_6(g) + \frac{7}{2}\,O_2(g) \rightarrow 3\,H_2O(g) + 2\,CO_2(g)$$

$$2\,C_2H_6(g) + 7\,O_2(g) \rightarrow 6\,H_2O(g) + 4\,CO_2(g)$$

Do the different ways of writing the equation affect the calculation of how much $CO_2$ is produced from a known quantity of $C_2H_6$?

7.30. Suppose the same mass of each of these components of natural gas was completely combusted: (a) $CH_4$, (b) $C_2H_6$, (c) $C_3H_8$, and (d) $C_4H_{10}$. Which one would produce the greatest mass of $CO_2$?

### Problems

**7.31.** Using colored spheres to represent C (black) and O (red), sketch the reaction between five C atoms and the necessary number of $O_2$ molecules to produce a 50% mixture of CO and $CO_2$.

7.32. Using colored spheres to represent N (blue) and O (red), sketch the reaction between three molecules of $N_2$ and enough $O_2$ to produce a mixture containing 50% $NO_2$ and 50% $N_2O_4$.

---

**7.33.** **Land Management** Better management of cropland, grazing land, and forests could reduce the amount of carbon dioxide in the atmosphere by an estimated $5.4 \times 10^9$ kilograms of carbon per year.

a. How many moles of carbon are present in $5.4 \times 10^9$ kilograms of carbon?

b. How many kilograms of carbon dioxide does this quantity of carbon represent?

7.34. Most $CO_2$ emitted by industrial sources comes from the combustion of fossil fuels, but some is also produced by cement manufacturing and converting limestone ($CaCO_3$) to lime (CaO):

$$CaCO_3(s) \rightarrow CaO(s) + CO_2(g)$$

How many tons of $CO_2$ are produced per ton of lime?

---

**7.35.** When $NaHCO_3$ is heated above 270°C, it decomposes to $Na_2CO_3(s)$, $H_2O(g)$, and $CO_2(g)$.

a. Write a balanced chemical equation for this reaction.

b. Calculate the mass of $CO_2$ produced when 25.0 g of $NaHCO_3$ decomposes.

7.36. **Pigments for Stoplights** Cadmium yellow (cadmium sulfide) is a lemon-yellow pigment used in the lenses of stoplights. Its formula is CdS, and it is very insoluble in water. The recommended recipe for cadmium yellow is to mix cadmium nitrate with sodium sulfide in water. The cadmium yellow forms as a solid, whereas the other product, sodium nitrate, remains dissolved in the water.

a. Write a balanced chemical equation for the reaction.

b. Calculate the mass of cadmium nitrate you must start with to make 125 g of CdS.

---

**7.37.** **Oxygen for First Responders** In self-contained breathing devices (Figure P7.37) used by first responders, potassium superoxide, $KO_2$, reacts with exhaled carbon dioxide to produce potassium carbonate and oxygen:

$$4\,KO_2(s) + 2\,CO_2(g) \rightarrow 2\,K_2CO_3(s) + 3\,O_2(g)$$

How much $O_2$ could be produced from 85 g of $KO_2$?

**FIGURE P7.37**

7.38. **Capturing $CO_2$** Carbon dioxide can be removed from a gas stream by reacting it with potassium carbonate in the presence of water:

$$CO_2(g) + K_2CO_3(s) + H_2O(\ell) \rightarrow 2\,KHCO_3(s)$$

If a resting human exhales 36 mg of $CO_2$ in one breath, how much potassium carbonate would be required to capture it all?

---

***7.39.** The uranium minerals found in nature must be refined and enriched in $^{235}U$ before the uranium can be used as a fuel in nuclear reactors. One procedure for enriching uranium relies on reacting $UO_2$ with HF at high temperatures to form $UF_4$, which is then converted into $UF_6$ in another high-temperature reaction with fluorine gas:

$$(1)\; UO_2(g) + 4\,HF(aq) \rightarrow UF_4(g) + 2\,H_2O(\ell)$$

$$(2)\; UF_4(g) + F_2(g) \rightarrow UF_6(g)$$

a. How many kilograms of HF are needed to completely react with 5.00 kg of $UO_2$?

b. How much $UF_6$ can be produced from 850.0 g of $UO_2$?

7.40. The mineral bauxite, which is mostly $Al_2O_3$, is the principal industrial source of aluminum metal. How much aluminum can be produced from 1.00 metric ton ($1.00 \times 10^3$ kg) of $Al_2O_3$?

---

***7.41.** Chalcopyrite ($CuFeS_2$) is an abundant copper mineral that can be converted into elemental copper. How much Cu is present in 1.00 kg of $CuFeS_2$?

*7.42. **Mining for Gold** Unlike most metals, gold occurs in nature as the pure element. Miners in California in 1849 searched for gold nuggets and gold dust in streambeds, where the denser gold could be easily separated from sand and gravel. However, larger deposits of gold are found in veins of rock and can be separated chemically in the following two-step process:

$$(1)\; 4\,Au(s) + 8\,NaCN(aq) + O_2(g) + 2\,H_2O(\ell) \rightarrow 4\,NaAu(CN)_2(aq) + 4\,NaOH(aq)$$

$$(2)\; 2\,NaAu(CN)_2(aq) + Zn(s) \rightarrow 2\,Au(s) + Na_2[Zn(CN)_4](aq)$$

If 23 kg of ore is 0.19% gold by mass, how much Zn is needed to react with the gold in the ore? Assume that reactions 1 and 2 are 100% efficient.

## *Limiting Reactants and Percent Yield*

### Concept Review

**7.43.** If fewer moles of A are present in a reaction between A and B, A must be the limiting reactant. What is wrong with this statement?

7.44. A reaction vessel contains equal masses of magnesium metal and oxygen gas. The mixture is ignited, forming MgO. After the reaction has gone to completion, the mass of the MgO is less than the mass of the reactants. Does this result violate the law of conservation of mass? Explain your answer.

**7.45.** Explain how the parameters of theoretical yield and percent yield differ.

7.46. Can the percent yield of a chemical reaction ever exceed 100%?

**7.47.** **Laboratory Errors** A student goes to the balance to measure out 2.50 g of a reactant onto a piece of filter paper. While carrying the filter paper back to the lab bench, the student spills some of the solid onto the floor. The student carries out a reaction using the remaining solid without redetermining its mass. Using the original mass from the balance, the student calculates a theoretical yield for the reaction. How does this laboratory error affect the actual yield? How does it affect the percent yield?

7.48. Consider the reaction of 10 g of A with 10 g of B in the combination reaction $A + B \rightarrow C$. How do the theoretical yield, actual yield, and percent yield change (or not change) if 20 g of A are reacted with 20 g of B?

### Problems

**7.49.** **S'mores** The campfire treat known as a s'more is made from putting two squares of chocolate and one marshmallow between two square graham crackers (Figure P7.49). One chocolate bar can be broken into 10 squares, one bag of marshmallows contains 52 marshmallows, and one box of graham crackers contains 70 large rectangular crackers, each of which can be broken into two s'more-sized square crackers. How many s'mores can be made from eight chocolate bars, one bag of marshmallows, and one box of graham crackers?

**FIGURE P7.49**

7.50. A factory making toy wagons (Figure P7.50) has 13,466 wheels, 3360 handles, and 2400 wagon beds in stock. How many wagons with four wheels can the factory make from these components?

**FIGURE P7.50**

**7.51.** Given the amounts of reactants shown, calculate the theoretical yield in grams of the product of each of these *unbalanced* chemical equations.

a. $Li(s) + N_2(g) \rightarrow Li_3N(s)$
   5.0 g   2.0 g   ? g

b. $P_2O_5(s) + H_2O(\ell) \rightarrow H_3PO_4(aq)$
   25.0 g   36.0 g   ? g

c. $SO_2(g) + O_2(g) \rightarrow SO_3(g)$
   6.4 g   4.0 g   ? g

7.52. Given the amounts of reactants shown, calculate the theoretical yield in grams of the product indicated by the question mark for each of these *unbalanced* chemical reactions.

a. $Cu_2O(s) + H_2(g) \rightarrow Cu(s) + H_2O(\ell)$
   30.0 g   12.0 g   ? g

b. $Mg(s) + HCl(g) \rightarrow MgCl_2(s) + H_2(g)$
   24.3 g   10.0 g   ? g

c. $CuCl_2(aq) + Zn(s) \rightarrow Cu(s) + ZnCl_2(aq)$
   11.36 g   10.0 g   ? g

***7.53.** **Sulfur in Coal** Suppose 75 metric tons of coal that is 3.0% sulfur by mass is burned at a power plant. During combustion, the sulfur is converted into $SO_2$. Antipollution scrubbers installed in the smokestacks of the power plant capture 3.9 metric tons of $SO_2$. How efficient are the scrubbers in capturing $SO_2$? How many metric tons of $SO_2$ escape?

7.54. One reaction in the production of sulfuric acid involves converting sulfur dioxide to sulfur trioxide. In the presence of excess $O_2$, 88 kg of $SO_2$ produces 106 kg of $SO_3$. What is the percent yield?

***7.55.** An important industrial use of chloroform [$CHCl_3(\ell)$] is a reaction with $HF(g)$ to produce $CHClF_2(g)$:

i. $CHCl_3(\ell) + 2\,HF(g) \rightarrow CHClF_2(g) + 2\,HCl(g)$

The product $CHClF_2$ is then heated to a high temperature in the absence of oxygen to produce $C_2F_4$, which is used to make the polymer known as Teflon.

ii. $2\,CHClF_2(g) \rightarrow C_2F_4(g) + 2\,HCl(g)$

a. In a reaction mixture of 775 g of $CHCl_3$ and 775 g of HF, which is the limiting reactant?

b. Assume that the yield in reaction (i) is stoichiometric and in reaction (ii) is 95%. How many grams of $C_2F_4$ could be prepared in reaction (ii) from the product formed in reaction (i)?

c. How much of which reactant is left over in the reaction mixture in part (i)?

7.56. A reaction vessel contains 10.0 g of CO and 10.0 g of $O_2$, which combine to form $CO_2$:

$$2\,CO(g) + O_2(g) \rightarrow 2\,CO_2(g)$$

a. Which reactant is limiting?
b. How many grams of $CO_2$ could be produced?
c. How many grams of the nonlimiting reactant are left over?

**7.57.** **Chemistry of Fermentation** Yeast converts glucose ($C_6H_{12}O_6$) in aqueous solution into ethanol ($CH_3CH_2OH$; $d$ = 0.789 g/mL) in a process called fermentation. Carbon dioxide also is produced.
a. Write a balanced chemical equation for the fermentation reaction.
b. If 100.0 g of glucose yields 50.0 mL of ethanol, what is the percent yield for the reaction?

*7.58. **Black Powder** Gunpowder, also known as black powder, is generally thought to have been invented by the Chinese in the 9th century. The first description in English was given by Roger Bacon in the 13th century. The basic formulation has not changed since then: 40% potassium nitrate, 30% carbon, and 30% sulfur by weight. The products of the reaction that provide the explosive force when the powder is ignited are three gases: carbon monoxide, carbon dioxide, and nitrogen. An additional product is potassium sulfite. Consider a sample of 100 g of black powder.
a. Which reactant is limiting?
b. How much of each reactant in excess is left over after the explosion?

## Percent Composition and Empirical and Molecular Formulas

### Concept Review

**7.59.** What is the difference between an empirical formula and a molecular formula?

7.60. Do the empirical and molecular formulas of a compound have the same percent composition values? Explain your answer.

**7.61.** Is the element with the largest atomic mass always the element present in the highest percentage by mass in a compound? Explain your answer.

7.62. Why is the empirical formula of an ionic compound represented by a formula unit? Why is the term *molecular formula* not appropriate to use when describing ionic compounds?

### Problems

**7.63.** Calculate the percent composition of (a) $Na_2O$, (b) NaOH, (c) $NaHCO_3$, and (d) $Na_2CO_3$.

7.64. Calculate the percent composition of (a) sodium sulfate, (b) dinitrogen tetroxide, (c) strontium nitrate, and (d) aluminum sulfide.

**7.65.** **Organic Compounds in Space** The following compounds have been detected in space. Which contains the greatest percentage of carbon by mass?
a. naphthalene, $C_{10}H_8$
b. chrysene, $C_{18}H_{12}$
c. pentacene, $C_{22}H_{14}$
d. pyrene, $C_{16}H_{10}$

7.66. **Lead Compounds as Pigments** Ancient Egyptians used lead compounds, including PbS, $PbCO_3$, and $Pb_2Cl_2CO_3$, as pigments in cosmetics, and many people suffered from chronic lead poisoning as a result. Calculate the percentage of lead in each of those compounds.

**7.67.** Do any two of the following compounds, which have been detected in outer space, have the same empirical formula?
a. naphthalene, $C_{10}H_8$
b. chrysene, $C_{18}H_{12}$
c. anthracene, $C_{14}H_{10}$
d. pyrene, $C_{16}H_{10}$
e. benzoperylene, $C_{22}H_{12}$
f. coronene, $C_{24}H_{12}$

7.68. Which of the following nitrogen oxides have the same empirical formulas? (a) $N_2O$; (b) NO; (c) $NO_2$; (d) $N_2O_2$; (e) $N_2O_4$

**7.69.** **Surgical-Grade Titanium** Medical implants (Figure P7.69) and high-quality jewelry items for body piercings are often made of a material known as G23Ti, or surgical-grade titanium. The percent composition of the material is 64.39% titanium, 24.19% aluminum, and 11.42% vanadium. What is the empirical formula of surgical-grade titanium?

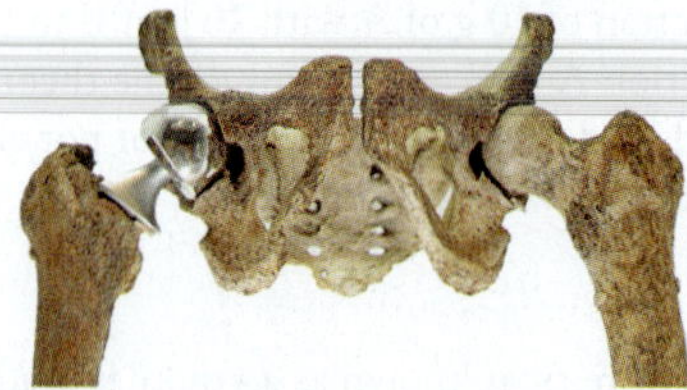

**FIGURE P7.69**

7.70. A sample of an iron-containing compound is 22.0% iron, 50.2% oxygen, and 27.8% chlorine by mass. What is the empirical formula of this compound?

**7.71.** **Asbestos and Lung Disease** Inhaling asbestos fibers may lead to a lung disease known as asbestosis and to a form of lung cancer called mesothelioma. One form of asbestos, chrysotile, is 26.31% magnesium, 20.27% silicon, and 1.45% hydrogen by mass, with the rest of the mass as oxygen. What is the empirical formula of chrysotile?

7.72. **Chemistry of Soot** A piece of glass held over a candle flame becomes coated with soot, the result of the incomplete combustion of candle wax. Elemental analysis of a compound extracted from a sample of soot gave the following results: 7.74% H and 92.26% C by mass. Calculate the empirical formula of the compound.

**7.73.** What is the empirical formula of the compound that is 24.2% Cu, 27.0% Cl, and 48.8% O by mass?

7.74. A chlorine oxide used to kill anthrax spores in contaminated buildings is 52.6% Cl by mass. What is its empirical formula?

**7.75.** **Sour Candy** Tartaric acid, $C_4H_6O_6$, and citric acid, $C_6H_8O_7$, are both used commercially to give sour candies (Figure P7.75) their characteristic sour taste. Which compound has the larger percentage of C by mass?

**FIGURE P7.75**

7.76. **Chlorofluorocarbons** CFCs (chlorofluorocarbons) are molecules used as refrigerants, but they also contribute to the destruction of the ozone layer. One CFC known as Freon consists of two carbon atoms, two fluorine atoms, and four chlorine atoms. What is the empirical formula of Freon? What is its molecular formula?

**7.77.** **Making the DNA Bases** Adenine (135.14 g/mol; 44.44% C, 3.73% H, and 51.84% N) was detected in mixtures of HCN, ammonia, and water under conditions that simulate those on early Earth. This observation suggests a possible origin for one of the bases found in DNA. What are the empirical and molecular formulas for adenine?

7.78. **Making Sugars for RNA** Ribose, the sugar found in RNA, has been detected in experiments designed to mimic the conditions of early Earth. If ribose contains 40.00% C, 6.71% H, and 53.28% O, with a molar mass of 150.13 g/mol, what are the empirical and molecular formulas for ribose?

## Combustion Analysis

### Concept Review

**7.79.** Explain why combustion analysis must be carried out in an excess of oxygen.

7.80. Why is the quantity of $CO_2$ obtained in a combustion analysis not a direct measure of the oxygen content of the starting compound?

**7.81.** Can the results of a combustion analysis ever give the true molecular formula of a compound?

7.82. A chemical reaction used to analyze a compound is expected to have a 100% yield. However, when a compound is synthesized, a 100% yield is almost never expected. Explain this difference.

### Problems

**7.83.** A 0.100 g sample of a compound containing C, H, and O is burned in oxygen, producing 0.1783 g of $CO_2$ and 0.0734 g of $H_2O$. What is the empirical formula of the compound?

7.84. **GRAS List for Food Additives** The alcohol geraniol is on the U.S. Food and Drug Administration's GRAS (generally recognized as safe) list and can be used in foods and personal care products. By itself, geraniol has a roselike odor, but it is often blended with other scents to produce the fruity fragrances of some personal care products. Complete combustion of 175 mg of geraniol produces 499 mg of $CO_2$ and 184 mg of $H_2O$. What is the empirical formula of geraniol?

**7.85.** Combustion of 135.0 mg of a hydrocarbon produces 440.0 mg of $CO_2$ and 135.0 mg of $H_2O$. The molar mass of the hydrocarbon is 270 g/mol. What are the empirical and molecular formulas of this compound?

7.86. The combustion of 40.5 mg of a compound extracted from the bark of the sassafras tree—and known to contain only C, H, and O—produces 110.0 mg of $CO_2$ and 22.5 mg of $H_2O$. The molar mass of the compound is 162 g/mol. What are its empirical and molecular formulas?

**7.87.** **Ethnobotany** One ingredient in a Native American stomachache remedy derived from common chokecherry is caffeic acid, which contains only carbon, hydrogen, and oxygen. Combustion of $1.00 \times 10^2$ mg of caffeic acid yields 220 mg of $CO_2$ and 40.3 mg of $H_2O$. Determine the empirical formula of caffeic acid.

7.88. Coniine, a substance isolated from poison hemlock, contains only carbon, hydrogen, and nitrogen. Combustion of 5.024 mg of coniine yields 13.90 mg of $CO_2$ and 6.048 mg of $H_2O$. What is the empirical formula of coniine?

## Additional Problems

***7.89.** **Artificial Bones for Medical Implants** The material often used to make artificial bones is the same material that gives natural bones their strength. Its common name is hydroxyapatite, and its formula is $Ca_5(PO_4)_3OH$.

a. Propose a systematic name for this compound.
b. What is the mass percentage of calcium in it?
c. When treated with hydrogen fluoride, hydroxyapatite becomes fluorapatite [$Ca_5(PO_4)_3F$], an even stronger substance. Does the percent mass of Ca increase or decrease as a result of this substitution?

*7.90. As a solution of copper sulfate slowly evaporates, beautiful blue crystals form. Their chemical formula is $CuSO_4 \cdot 5\,H_2O$.

a. What is the percentage of water in this compound?
b. At high temperatures, the water is driven off as steam. What fraction of the original sample's mass is lost as a result?

**7.91.** **Fiber in the Diet** Dietary fiber is a mixture of many compounds, including xylose ($C_5H_{10}O_5$) and methyl galacturonate ($C_7H_{12}O_7$).

a. Do these compounds have the same empirical formula?
b. Write balanced chemical equations for the complete combustion of xylose and methyl galacturonate.

7.92. **Chemistry of Copper Production** The mineral chalcopyrite ($CuFeS_2$) is an important source of copper metal, though recovering the metal requires several chemical reactions that transform $CuFeS_2$ into CuS, then $Cu_2S$, and finally Cu metal. The pennies minted in the United States between 1909 and 1982 weighed 3.11 g and were 95% by mass copper.

a. How much chalcopyrite had to be mined to produce one dollar's worth of these pennies?
b. How much chalcopyrite had to be mined to produce one dollar's worth of the pennies if the first reaction had a percent yield of 85% and the second and third reactions had yields of essentially 100%?
c. How much chalcopyrite had to be mined to produce one dollar's worth of the pennies if each reaction proceeded with 85% yield?

***7.93.** **Mining for Gold** Gold can be extracted from the surrounding rock by using a solution of sodium cyanide. Though effective for isolating gold, toxic cyanide finds its way into watersheds, causing environmental damage and harming human health.

$$4\,Au(s) + 8\,NaCN(aq) + O_2(g) + 2\,H_2O(\ell) \rightarrow 4\,NaAu(CN)_2(aq) + 4\,NaOH(aq)$$

$$2\,NaAu(CN)_2(aq) + Zn(s) \rightarrow 2\,Au(s) + Na_2[Zn(CN)_4](aq)$$

a. If a sample of rock contains 0.009% gold by mass, how much NaCN is needed to extract the gold as $NaAu(CN)_2$ from 1 metric ton ($10^3$ kg) of rock?
b. How much zinc is needed to convert the $NaAu(CN)_2$ from part (a) to metallic gold?
c. The gold recovered in part (b) is manufactured into a gold ingot in the shape of a cube. The density of gold is 19.3 g/cm$^3$. How big is the block of gold in cubic centimeters?

*7.94. Phosgenite, a lead compound with the formula $Pb_2Cl_2CO_3$, is found in Egyptian cosmetics. Phosgenite was prepared by the reaction of PbO, NaCl, and $CO_2$. An unbalanced expression of the reactant mixture is

$$PbO(s) + NaCl(aq) + H_2O(\ell) + CO_2(g) \rightarrow Pb_2Cl_2CO_3(s) + NaOH(aq)$$

a. Balance the equation.
b. How many grams of phosgenite can be obtained from 10.0 g of PbO and 10.0 g of NaCl in the presence of excess water and $CO_2$?
c. Phosgenite can be considered a mixture of two lead compounds. Which compounds appear to be combined to make phosgenite?

***7.95.** Uranium oxides used in preparing fuel for nuclear reactors are separated from other metals in minerals by converting the uranium to $UO_x(NO_3)_y(H_2O)_z$, where uranium has a positive charge ranging from 3+ to 6+.

a. Roasting $UO_x(NO_3)_y(H_2O)_z$ at 400°C leads to loss of water and produces nitrogen oxides, leaving behind a product with the formula $U_aO_b$ that is 83.22% U by mass. What are the values of *a* and *b*? What is the charge on U in $U_aO_b$?
b. Higher temperatures produce a different uranium oxide, $U_cO_d$, which is 84.8% U by mass. What are the values of *c* and *d*? What is the charge on U in $U_cO_d$?
c. The values of *x*, *y*, and *z* in $UO_x(NO_3)_y(H_2O)_z$ are found by gently heating the compound to remove all the water. In a laboratory experiment, 1.328 g of $UO_x(NO_3)_y(H_2O)_z$ produced 1.042 g of $UO_x(NO_3)_y$. Continued heating generated 0.742 g of $U_nO_m$. Using the information in parts (a) and (b), calculate *x*, *y*, and *z*.

*7.96. Corn farmers in the midwestern United States typically use $5.0 \times 10^3$ kg of ammonium nitrate fertilizer per square kilometer of cornfield per year. Some of the fertilizer washes into the Mississippi River and eventually flows into the Gulf of Mexico, promoting the growth of algae and endangering other aquatic life.

a. Ammonium nitrate can be prepared by the following reaction:

$$NH_3(g) + HNO_3(aq) \rightarrow NH_4NO_3(aq)$$

How much nitric acid would be required to make the fertilizer needed for 1 km$^2$ of cornfield per year?

b. Ammonium ions dissolved in groundwater may be converted into $NO_3$ ions by bacterial action:

$$NH_4^+(aq) + 2\,O_2(g) \rightarrow NO_3^-(aq) + H_2O(\ell) + 2\,H^+(aq)$$

If 10% of the ammonium component of $5.0 \times 10^3$ kilograms of fertilizer ends up as nitrate ions, how much oxygen would be consumed?

***7.97.** Several chemical reactions have been proposed for the formation of organic compounds from inorganic precursors, including the following:

$$H_2S(g) + FeS(s) + CO_2(g) \rightarrow FeS_2(s) + HCO_2H(\ell)$$

a. Identify the ions in FeS and $FeS_2$.
b. What are the names of FeS and $FeS_2$?
c. How much $HCO_2H$ is obtained by reacting 1.00 g of FeS, 0.50 g of $H_2S$, and 0.50 g of $CO_2$ if the reaction results in a 50.0% yield?

*7.98. Organic compounds called *carbohydrates* may be formed in reactions between iron(II) sulfide and carbonic acid:

$$2\,FeS + H_2CO_3 \rightarrow 2\,FeO + \tfrac{1}{n}(CH_2O)_n + 2\,S$$

a. What is the empirical formula of these carbohydrates?
b. How much carbohydrate is produced from a reaction mixture that initially contains 211 g of FeS and excess $H_2CO_3$ if the reaction results in a 78.5% yield?
c. If the carbohydrate product has a molecular mass of 300 u, what is its molecular formula?

***7.99.** **Marine Chemistry of Iron** On the seafloor, solid iron(II) oxide may react with water to form solid $Fe_3O_4$ and hydrogen gas.

a. Write a balanced chemical equation for the reaction.
b. When $CO_2$ is also present, the product of the reaction is methane, not hydrogen. Write a balanced chemical equation for this reaction.

7.100. Titanium dioxide and zinc oxide are common names of two of the active ingredients approved by the U.S. FDA for use in sunscreens.

a. What are the chemical formulas of these compounds?
b. What are the proper names of the compounds according to the rules for naming described in Chapter 4?
c. Which compound contains the higher percentage of oxygen by mass?

***7.101.** **Ethanol in Fuel** E-85 is an alternative fuel for automobiles and light trucks that consists of 85% (by volume) ethanol ($CH_3CH_2OH$) and 15% gasoline. The density of ethanol is 0.79 g/mL.

a. How many moles of ethanol are in a gallon of E-85?

b. How many moles of carbon dioxide are produced by the complete combustion of the ethanol in a gallon of E-85 fuel?

7.102. **Military Balloons** The ferrosilicon process is used in the military to produce hydrogen quickly to inflate balloons. A small truck-mounted generator contains a steel vessel charged with a source of silicon (usually an iron–silicon mixture, from which the process gets its name) and sodium hydroxide. When water is added, a reaction ensues between Si, NaOH, and water that produces sodium silicate ($Na_2SiO_3$) and hydrogen gas. How many grams of silicon and sodium hydroxide would you need to produce $1.55 \times 10^4$ g of hydrogen gas, enough to inflate a reconnaissance balloon?

***7.103.** You are given a 0.6240 g sample of a substance with the generic formula $MCl_2 \cdot 2\,H_2O$. After completely drying the sample (which means removing the 2 moles of $H_2O$ per mole of $MCl_2$), the sample has a mass of 0.5471 g. What is the identity of element M?

7.104. A compound found in crude oil is 93.71% C and 6.29% H by mass. The molar mass of the compound is 128 g/mol. What is its molecular formula?

**7.105.** A reaction vessel for synthesizing ammonia by reacting nitrogen and hydrogen is charged with 6.04 kg of $H_2$ and excess $N_2$. If 28.0 kg of $NH_3$ is produced, what is the percent yield of the reaction?

7.106. If a cube of table sugar, which is made of sucrose, $C_{12}H_{22}O_{11}$, is added to concentrated sulfuric acid, the acid "dehydrates" the sugar: it removes the hydrogen and oxygen from it, leaving behind a lump of carbon. What percentage of the initial mass of sugar is carbon?

**7.107.** **Reducing $SO_2$ Emissions** One way to remove $SO_2$ from the "stack" gases of coal-burning power plants is to spray the gases with fine particles of solid calcium oxide suspended in $O_2$ gas. The product of the reaction of $SO_2$, CaO, and $O_2$ is calcium sulfate.

a. Write a balanced chemical equation for this reaction.

b. How many metric tons of calcium sulfate would be produced from each ton of $SO_2$ that is trapped?

7.108. **Gas Grill Reaction** The burner in a gas grill mixes 24 volumes of air for every one volume of propane ($C_3H_8$) fuel. Like all gases, the volume that propane occupies is directly proportional to the number of moles of it at a given temperature and pressure. Air is 21% (by volume) $O_2$. Is the flame produced by the burner fuel-rich (excess propane in the reaction mixture), fuel-lean (not enough propane), or stoichiometric (just right)?

**7.109.** A common mineral in Earth's crust has the chemical composition 34.55% Mg, 19.96% Si, and 45.49% O. What is its empirical formula?

7.110. **Ozone Generators** Some indoor air-purification systems work by converting a little of the oxygen in the air to ozone, which kills mold and mildew spores and other biological air pollutants. The chemical equation for the ozone generation reaction is

$$3\,O_2(g) \rightarrow 2\,O_3(g)$$

It is claimed that one such system generates 4.0 g of $O_3$ per hour from dry air passing through the purifier at a flow rate of 5.0 L/min. If exactly 1 liter of indoor air contains 0.28 g of $O_2$, what percentage of the $O_2$ is converted to $O_3$ by the air purifier?

# 8

# Reactions in Aqueous Solutions

## Chemistry of the Hydrosphere

**THE BLUE PLANET** Life exists on Earth because liquid water exists here. Most of it is seawater, which is principally a solution of $Na^+$ ions (the orange spheres) and $Cl^-$ ions (the green spheres).

PARTICULATE **REVIEW**

### *Ions and Ionic Compounds*

In Chapter 8, we explore the solubility of ionic compounds and the reactions that can occur when aqueous solutions of ionic compounds are mixed.

- Which particles shown here are present in solid $Fe(NO_3)_2$?
- What is the charge on the cation? What is the charge on the anion?
- What is the ratio of cations to anions in this compound?

(Review Sections 4.1 and 4.2 if you need help.)

*(Answers to Particulate Review questions are in the back of the book.)*

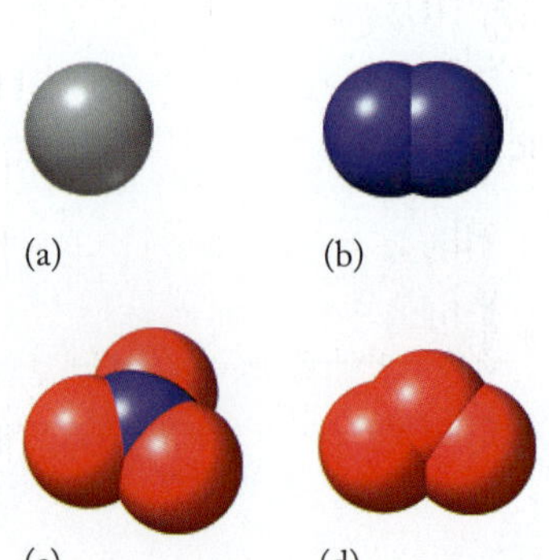

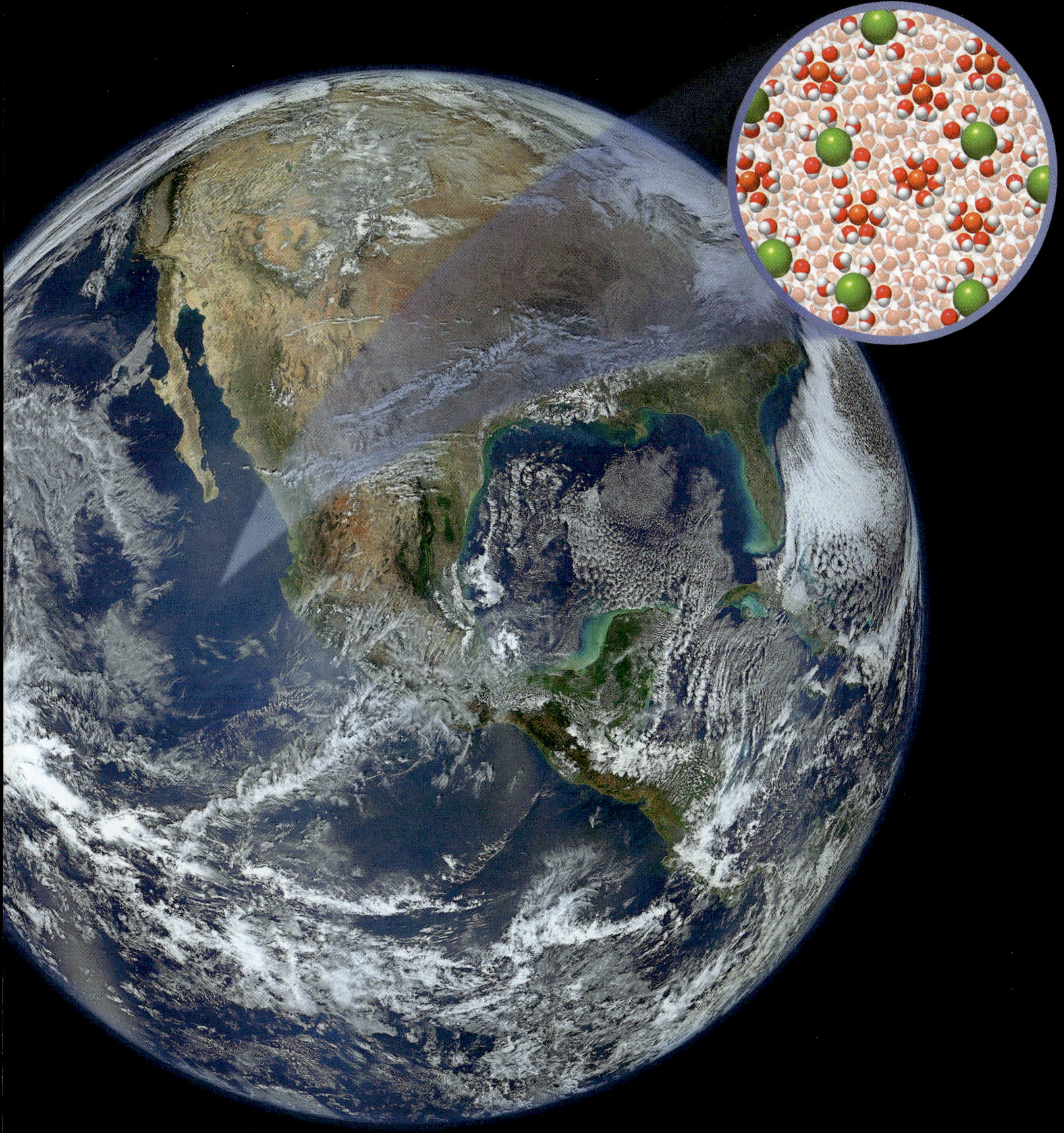

PARTICULATE **PREVIEW**

### *Hydrogen Atom versus Acidic Proton*

Hydrogen atoms are present in many molecules and ions. As you read Chapter 8, look for ideas that will help you answer these questions:

- Why is a hydrogen ion referred to as a proton?
- Which of the molecules shown here can ionize when it dissolves in water, releasing protons?
- In that molecule, which hydrogen atom is released as a proton?

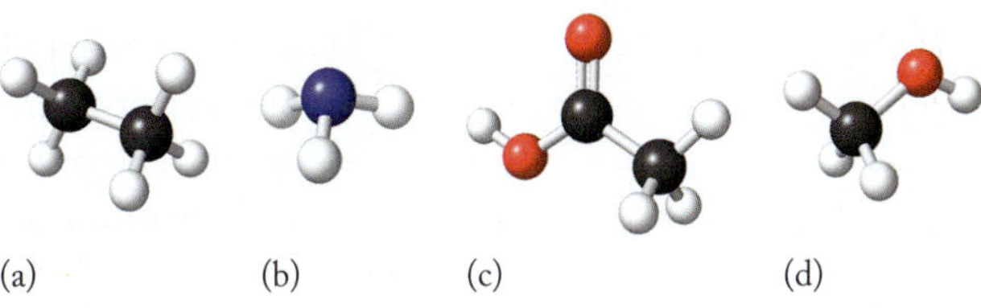

## Learning Outcomes

**LO1** Express the concentrations of solutions in different units, including molarity
**Sample Exercises 8.1, 8.2**

**LO2** Calculate the mass of a solute or the volume of a concentrated solution required to make a dilute solution of specified volume and concentration
**Sample Exercises 8.3, 8.4**

**LO3** Use Beer's Law and absorbance data to determine the concentration of an unknown solution
**Sample Exercise 8.5**

**LO4** Write molecular, overall ionic, and net ionic equations for reactions in aqueous solution
**Sample Exercises 8.6, 8.7, 8.11, 8.18**

**LO5** Distinguish among strong electrolytes, weak electrolytes, and nonelectrolytes
**Sample Exercise 8.8**

**LO6** Identify Brønsted–Lowry acids and bases and determine the concentration of a solution from titration data
**Sample Exercises 8.9, 8.10**

**LO7** Use solubility guidelines to predict the products of precipitation reactions and to quantify results from precipitation reactions
**Sample Exercises 8.11–8.13**

**LO8** Determine oxidation numbers and use them to identify redox reactions, oxidizing agents, and reducing agents
**Sample Exercises 8.14, 8.15**

**LO9** Use half-reactions to write balanced equations for redox reactions, and use the activity series to predict redox reactions
**Sample Exercises 8.16–8.19**

## 8.1 Solutions and Their Concentrations

Depressions in Earth's crust contain more than $10^{21}$ liters of freshwater and seawater—enough to cover 70% of the planet. All this water contains dissolved ionic and molecular compounds. The concentrations of these solutes vary considerably from one body of freshwater to the next, but the proportions of the major ions in seawater are essentially the same all over the world. Perhaps more remarkably, their concentrations appear to have changed little over the last billion years.

The constant composition of the oceans may come as a surprise, given that rivers deliver about $4 \times 10^{12}$ kg of dissolved ionic compounds to the sea every year. However, because such a large volume of freshwater ($4 \times 10^{16}$ L per year) flows to the sea, the total concentration of all ionic compounds in river water is typically less than 1/100 the salinity of seawater. In addition, the proportions of the ions in river water don't match those of seawater.

Given these differences in composition, how did the sea become so salty in the first place, and how has it stayed that way? Not all the salt got there through physical erosion and chemical weathering of the land. Much of the $Cl^-$ content, for example, probably came from HCl released by underwater volcanoes billions of years ago, and most of the $Na^+$ ions probably leached out of the ocean's floor as it first filled with water. Today, an elaborate system of physical and chemical processes operates within and above Earth's oceans to maintain their composition. About 90% of the water vapor in the atmosphere—the source of all that river water flowing into the sea—is evaporated seawater. Therefore, river water does not really dilute the saltiness of seawater because the evaporation process that created river water actually made the sea a little saltier in the first place.

By contrast, $Ca^{2+}$ ions flowing into the sea don't remain in seawater very long. They are taken up by corals, shellfish, and other sea creatures to grow the hard parts of their bodies (made mostly of $CaCO_3$). Other ions, such as $Fe^{3+}$ and $Mn^{2+}$, are soluble in freshwater—which is, on average, slightly acidic—but become insoluble when they reach the sea, which is slightly basic.

The composition of the sea and the survival of the creatures in it are linked to chemical and biochemical reactions that depend on acid–base balance, the formation of insoluble ionic compounds, and reactions in which some elements are oxidized as others are reduced. In this chapter, we explore each of these types of reactions. As we examine reactions that take place in solution, we need to be able to quantify the amounts of substances present so that we can carry out calculations based on stoichiometry. To do so, we must understand the various modes of expressing the concentration of dissolved substances.

## Concentration Units

In Chapter 1, we defined a solution as a homogeneous mixture of two or more substances. Though we usually think of solutions as liquids, homogeneous mixtures of solids and gases also exist. For example, familiar metals such as brass (a mixture of copper and zinc), bronze (copper and tin), and stainless steel (chromium and other metals mixed with iron) are solutions; so is filtered air. The substance present in the greatest proportion in a solution is called the solvent, and all the substances dissolved in it are solutes. Solutions in which water is the solvent are called *aqueous* solutions and are the focus of this chapter.

**CONNECTION** Solvents and solutes were introduced in Chapter 6, in which we used intermolecular forces to predict the solubility of molecular substances.

The concentration of an aqueous solution describes the quantity of solute dissolved in a specific quantity of either solvent or total solution. This ratio may be expressed as the mass of the solute in a given mass or volume of solvent or in a given mass or volume of solution. Concentration units based on mass-to-mass ratios, such as milligrams of solute per kilogram of solvent (mg/kg), are commonly used in disciplines requiring exact measurements over wide ranges of temperature and pressure because volumes change under these conditions, but masses do not. In daily life, however, concentration units are often based on mass-to-volume ratios, such as milligrams of solute per liter of solution (mg/L). Actually, units of milligrams per kilogram (mg/kg) and milligrams per liter (mg/L) are equivalent for dilute aqueous solutions near room temperature because the densities of these solutions are nearly the same as that of water, which is 1.00 kg/L at 20°C.

Another concentration unit synonymous with milligrams per kilogram (mg/kg) is parts per million (ppm). These units are equivalent because $1 \text{ mg} = 10^{-3}$ g and $1 \text{ kg} = 10^3$ g. Therefore,

$$\frac{1 \text{ mg}}{\text{kg}} = \frac{10^{-3}\text{ g}}{10^{3}\text{ g}} = \frac{1\text{ g}}{10^{6}\text{ g}} = 1 \text{ ppm}$$

One-millionth of a gram per gram is the same as one gram per million grams, or simply one part per million. Even smaller concentrations may be expressed in micrograms per kilogram (μg/kg) or micrograms per liter (μg/L), which are also equivalent for dilute aqueous solutions and are the same as parts per billion (ppb). Environmental pollutants reported in the press are often expressed in milligrams per liter (mg/L) or parts per million (ppm). For example, the maximum contaminant level for arsenic in drinking water is 0.010 mg/L, or 0.010 ppm. Results from clinical laboratories, such as those for blood tests, also are expressed as mass-to-volume ratios. For example, the acceptable range for fasting blood sugar is 70–99 mg/dL, where "dL" is deciliter (1/10 of a liter, or 100 mL).

**molarity ($M$)** the number of moles of solute per liter of solution: $M = n/V$; also called *molar concentration*.

**CHEMTOUR**
Molarity

Because we are interested in describing reactions in solutions, we will work with quantities of dissolved reactants and products that are related by how many *moles* of each of them are in reaction mixtures. Consequently, we need to express their concentrations in units based on moles of solute per mass or volume of solution. One such unit is **molarity ($M$)**, the number of moles ($n$) of solute in a volume ($V$) of 1 L of solution.[1] In equation form:

$$M = \frac{\text{moles of solute}}{\text{liter of solution}} = \frac{n}{V} \tag{8.1}$$

Often we know the volume and concentration of a solution and need to calculate the number of moles of solute in it. Equation 8.1 can be rearranged to solve for $n$:

$$n = V \times M = \cancel{\text{L}} \times \frac{\text{mol}}{\cancel{\text{L}}} \tag{8.2}$$

To calculate the mass ($m$) in grams of the solute in a solution of a known volume and molar concentration, we can multiply the number of moles of solute obtained using Equation 8.2 by the molar mass ($\mathcal{M}$) of the solute:

$$m = (V \times M) \times \mathcal{M} = \left(\cancel{\text{L}} \times \frac{\cancel{\text{mol}}}{\cancel{\text{L}}}\right) \times \frac{\text{g}}{\cancel{\text{mol}}} \tag{8.3}$$

Equation 8.3 is particularly useful when we need to calculate the mass of solute required to prepare a solution of a desired volume and concentration.

In many environmental and biological systems, solute concentrations are much less than 1.0 $M$. In **Table 8.1**, for instance, the average concentrations of most of the major ions in seawater and in human serum are more conveniently expressed in *milli*moles per liter (mmol/L), or millimolarity (m$M$), where 1 m$M$ = $10^{-3}$ $M$. Even smaller concentrations of minor and trace substances in seawater and in biological samples such as blood and urine are expressed in micromolarity (μ$M$; 1 μ$M$ = $10^{-6}$ $M$), nanomolarity (n$M$; 1 n$M$ = $10^{-9}$ $M$), and even picomolarity (p$M$; 1 p$M$ = $10^{-12}$ $M$).

The values in the first column of concentrations in Table 8.1 are expressed in millimoles of solute per kilogram of seawater. Environmental scientists and especially oceanographers prefer to use concentration units based on the masses of water samples rather than their volumes because the volume of a given mass of water varies with temperature and pressure, whereas its mass and the quantities of solutes dissolved in that mass remain constant.

**TABLE 8.1** Average Concentrations of 11 Major Constituents of Seawater and Human Serum

| | SEAWATER | | HUMAN SERUM |
|---|---|---|---|
| Constituent | mmol/kg | m$M$ | m$M$ |
| $Na^+$ | 468.96 | 480.57 | 135–145 |
| $K^+$ | 10.21 | 10.46 | 3.5–5.0 |
| $Mg^{2+}$ | 52.83 | 54.14 | 0.08–0.12 |
| $Ca^{2+}$ | 10.28 | 10.53 | 0.2–0.3 |
| $Sr^{2+}$ | 0.0906 | 0.0928 | $<3 \times 10^{-4}$ |
| $Cl^-$ | 545.88 | 559.40 | 98–108 |
| $SO_4^{2-}$ | 28.23 | 28.93 | 0.3 |
| $HCO_3^-$ | 2.06 | 2.11 | 22–30 |
| $Br^-$ | 0.844 | 0.865 | 0.04–0.06 |
| $B(OH)_3$ | 0.416 | 0.426 | $<8 \times 10^{-4}$ |
| $F^-$ | 0.068 | 0.070 | 5–6 |

CONCEPT **TEST**

Rank the following solutions from most concentrated to least concentrated: (a) 0.0053 $M$ NaCl; (b) 54 m$M$ NaCl; (c) 550 μ$M$ NaCl; (d) 56,000 n$M$ NaCl.

*(Answers to Concept Tests are in the back of the book.)*

[1]Note that molarity is symbolized by an italic $M$, whereas molar mass is a script $\mathcal{M}$ throughout this text.

### SAMPLE EXERCISE 8.1 Converting Mass-per-Volume Concentrations into Molarity LO1

Vinyl chloride (**Figure 8.1**) is a widely used industrial chemical (about $10^{10}$ kg are produced each year, mostly for making polyvinyl chloride plastic). It also is highly toxic and a carcinogen. In the United States, the maximum concentration of vinyl chloride allowed in drinking water is 0.002 mg/L. What is that concentration in moles per liter?

**Collect, Organize, and Analyze** Because the solute mass is given in milligrams, we need to convert it to grams and then to moles. Given the two C atoms and one Cl atom in its molecular structure, the molar mass of vinyl chloride is likely to be about 60 g/mol, so 0.002 mg/L should translate into 0.00003 mmol/L or $3 \times 10^{-8}$ *M*.

**Solve** The molar mass of vinyl chloride ($C_2H_3Cl$) is

$$\mathcal{M} = 2(12.01 \text{ g/mol}) + 3(1.008 \text{ g/mol}) + 35.45 \text{ g/mol} = 62.49 \text{ g/mol}$$

The molarity equivalent of 0.002 mg/L of vinyl chloride is

$$\frac{0.002 \cancel{\text{mg}}}{1 \text{ L}} \times \frac{1 \text{ g}}{10^3 \cancel{\text{mg}}} \times \frac{1 \text{ mol}}{62.49 \cancel{\text{g}}} = \frac{3 \times 10^{-8} \text{ mol}}{1 \text{ L}} = 3 \times 10^{-8} M$$

**Think About It** This concentration matches our estimated value. Expressing the concentration with micromolarity units to one significant figure is appropriate because the initial concentration value was known to only one significant figure. Given the hazards that vinyl chloride poses to human health, having a very small allowed concentration of it in drinking water is reasonable.

**Practice Exercise** When a 1.00 L sample of water from the surface of the Dead Sea is evaporated, 179 g of $MgCl_2$ are recovered. What was the molarity of $MgCl_2$ in the sample?

*(Answers to Practice Exercises are in the back of the book.)*

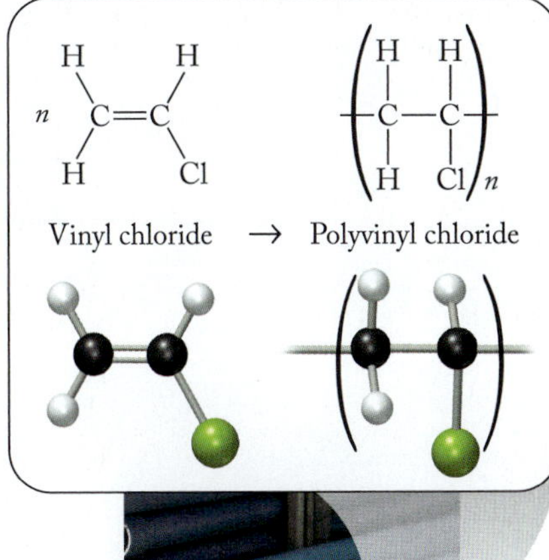

**FIGURE 8.1** Vinyl chloride is the common name of a small molecule, called a monomer (Greek for "one unit"), from which very large molecules, or polymers ("many units"), called polyvinyl chloride (PVC) are made. PVC is used to make many products, including the pipes for drainage and sewer systems.

### SAMPLE EXERCISE 8.2 Converting Mass-per-Mass Concentrations into Molarity LO1

A water sample from the Great Salt Lake in Utah contains 83.6 mg of $Na^+$ per gram of lake water. What is the molar concentration of $Na^+$ ions if the density of the lake water is 1.160 g/mL?

**Collect and Organize** Our task is to convert concentration units from milligrams of $Na^+$ per gram of solution to moles of $Na^+$ per liter of solution. We know the lake water's density, which we will use to convert a concentration based on mass of solution to molarity, which is based on volume of solution. The molar mass of Na is 22.99 g/mol.

**Analyze** The initial concentration value of 83.6 mg $Na^+$/g is the same as 83.6 g/kg. This mass of $Na^+$ corresponds to about 4 moles of $Na^+$, and the volume of a kilogram of lake water will be only a little less than a liter, given a density of about 1.2 kg/L, so the answer should be close to 4 *M*.

**Solve** Calculating the moles of $Na^+$ in 83.6 mg of $Na^+$:

$$83.6 \cancel{\text{mg}} \text{ Na}^+ \times \frac{1 \text{ g}}{1000 \cancel{\text{mg}}} \times \frac{1 \text{ mol}}{22.99 \cancel{\text{g}}} = 3.636 \times 10^{-3} \text{ mol Na}^+$$

The volume of exactly 1 g of lake water is

$$1\ \text{g} \times \frac{1\ \text{mL}}{1.160\ \text{g}} \times \frac{1\ \text{L}}{1000\ \text{mL}} = 8.621 \times 10^{-4}\ \text{L}$$

and the molarity of $Na^+$ ions in the lake is

$$M = \frac{3.636 \times 10^{-3}\ \text{mol Na}^+}{8.621 \times 10^{-4}\ \text{L}} = 4.22\ M$$

**Think About It** According to Table 8.1, the average concentration of $Na^+$ in seawater is 480.57 m$M$, or 0.48057 $M$. The concentration of $Na^+$ in the Great Salt Lake is nearly 10 times greater.

**Practice Exercise** If the density of seawater at a depth of 10,000 m is 1.071 g/mL, and a 25.0 g sample of water from that depth contains 99.7 mg of $K^+$, what is the molarity of potassium ions in the sample?

## SAMPLE EXERCISE 8.3 Calculating the Quantity of Solute Needed to Prepare a Solution — LO2

An aqueous solution called phosphate-buffered saline (PBS) is used in biological research to wash and store living cells. Its ingredients include 10.0 m$M$ $Na_2HPO_4$. How many grams of this solute would you need to prepare 10.0 L of PBS?

**Collect and Organize** We are asked to calculate the mass of solute needed to prepare a solution of specified volume and concentration. We know the formula of the solute. Equation 8.3 relates the mass of a solute in a solution to the volume and concentration of the solution and the molar mass of the solute.

**Analyze** We need to use the formula of the solute to calculate its molar mass and then use that value and the volume and concentration of the solution in Equation 8.3 to calculate the mass we need. We need enough solute to make 10.0 L of 10.0 m$M$ (or 0.0100 $M$) PBS solution, which translates into 10.0 L × 0.01 mol/L, or 0.1 mole. From its formula, the molar mass of the solute is about 200 g/mol, so 0.1 mole should have a mass of about 20 g.

**Solve** The molar mass of $Na_2HPO_4$ is

$$\begin{aligned}\mathcal{M} &= 2(22.99\ \text{g/mol}) + 1.008\ \text{g/mol} + 30.97\ \text{g/mol} + 4(16.00\ \text{g/mol}) \\ &= 141.96\ \text{g/mol}\end{aligned}$$

Calculating the mass of $Na_2HPO_4$ with Equation 8.3:

$$\begin{aligned}m &= (V \times M) \times \mathcal{M} \\ &= \left(10.0\ \text{L} \times 10.0\ \frac{\text{mmol}}{\text{L}} \times \frac{1\ \text{mol}}{1000\ \text{mmol}}\right) \times \frac{141.96\ \text{g}}{1\ \text{mol}} = 14.2\ \text{g}\end{aligned}$$

**Think About It** In solving the problem, we converted "mmol/L" to "mol/L" so that the product ($V \times M$) yielded moles of solute, which we then converted into a mass in grams. The result of the calculation is close to our estimate.

**Practice Exercise** An aqueous solution known as Ringer's lactate is administered intravenously to trauma victims suffering from blood loss or severe burns. The solution contains the chloride salts of sodium, potassium, and calcium and is 4.00 m$M$ in sodium lactate, $CH_3CH(OH)COONa$. How many grams of sodium lactate are needed to prepare 10.0 L of Ringer's lactate?

**standard solution** a solution of accurately known concentration that is used in chemical analysis.

**dilution** the process of lowering the concentration of a solution by adding more solvent.

## 8.2 Dilutions

Scientists and technicians who routinely analyze environmental, biological, or clinical samples often use commercially available **standard solutions**. Standard solutions contain accurately known concentrations of one or more solutes. These concentrations are often higher than required for laboratory work, so they must be diluted before use. **Dilution** is the process of reducing solute concentration by adding more solvent to a solution. Adding solvent does not change the number of solute particles in the solution, only the volume they occupy, as shown in **Figure 8.2**.

In chemistry laboratories, dilution often involves transferring a precisely measured volume of a concentrated solution to a volumetric flask and then filling the flask to its calibration mark with solvent (**Figure 8.3**) and thoroughly mixing its contents. The number of particles of solute transferred from the concentrated solution to the flask is the same number of particles present in the entire volume in the flask after dilution. Figure 8.3 shows how to prepare 250.0 mL of 0.100 *M* $CuSO_4$ starting with a 1.00 *M* standard solution. Note that a 25 mL volumetric pipette is used to transfer that volume of the more concentrated solution to the

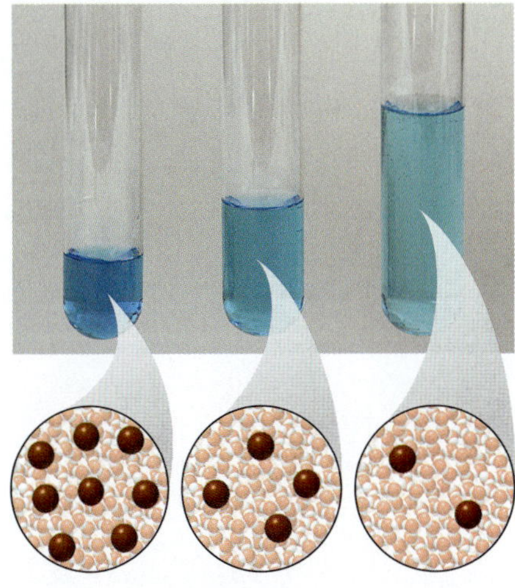

**FIGURE 8.2** Adding more solvent to the solution in the test tube on the left does not change the amount of solute, but it does dilute the concentration of the solution and the intensity of the color.

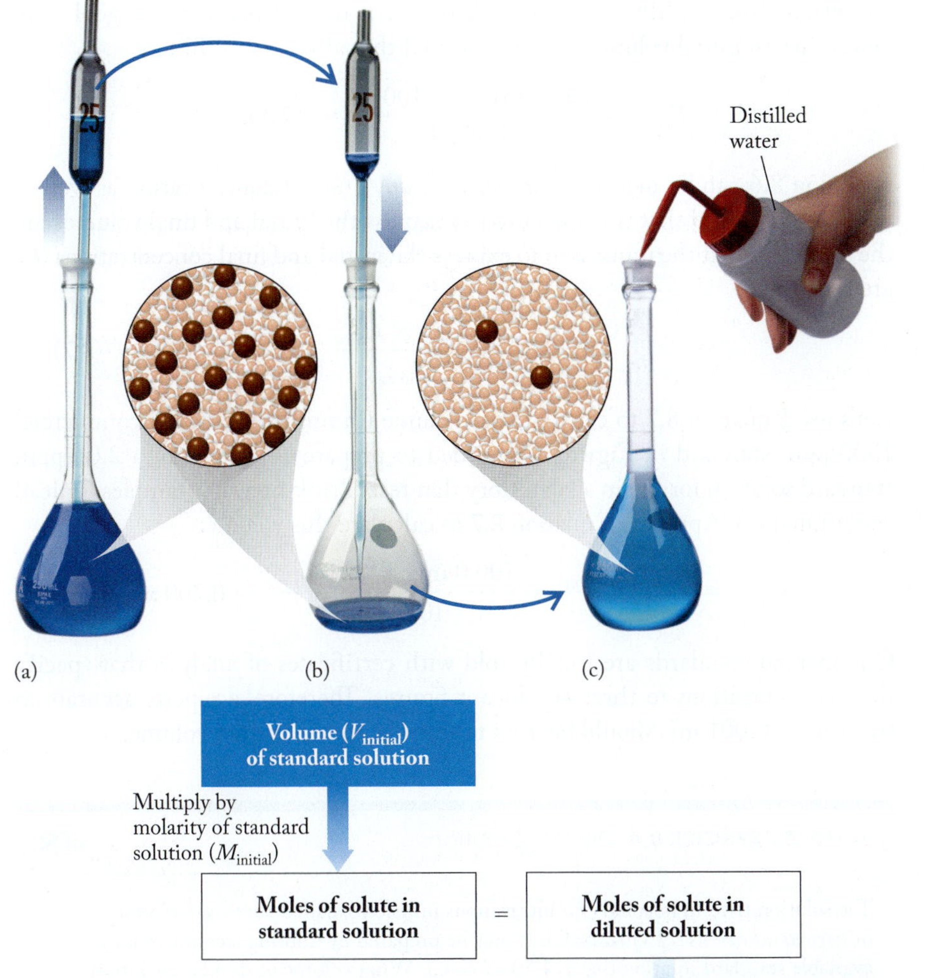


**CHEMTOUR**
Dilutions

**FIGURE 8.3** To prepare 250.0 mL of 0.100 *M* $CuSO_4$, (a) a pipette is used to withdraw 25.0 mL of 1.00 *M* $CuSO_4$. (b) This volume of standard solution is transferred to a 250.0 mL volumetric flask. (c) Distilled water is added to bring the volume of the dilute solution to 250.0 mL. The dilute solution is a lighter color than the standard solution. The number of colored spheres in each molecular view represents the relative concentrations of $Cu^{2+}$ ions. (The sulfate ions are not shown for simplicity.)

flask in which it will be diluted to obtain the desired solution. To show how this volume was calculated, we start with a variation of Equation 8.2:

$$n_{\text{initial}} = V_{\text{initial}} \times M_{\text{initial}} \tag{8.4}$$

where the subscripts refer to volume, molarity, and number of moles of $Cu^{2+}$ ions in the *initial* (concentrated) solution used in the dilution. The number of moles of $Cu^{2+}$ ions in the *final* (dilute) solution:

$$n_{\text{final}} = V_{\text{final}} \times M_{\text{final}} \tag{8.5}$$

is the same as the number transferred. Because the two $n$ values are equal, the right-hand sides of Equations 8.4 and 8.5 also are equal:

$$V_{\text{initial}} \times M_{\text{initial}} = V_{\text{final}} \times M_{\text{final}}$$

Rearranging the terms in this equation to solve for $V_{\text{initial}}$, we obtain

$$V_{\text{initial}} = \frac{V_{\text{final}} \times M_{\text{final}}}{M_{\text{initial}}} \tag{8.6}$$

To use Equation 8.6 to calculate the volume of 1.00 $M$ $CuSO_4$ needed to prepare 250.0 mL of 0.100 $M$ $CuSO_4$, you might think that $V_{\text{final}}$ would first have to be converted from milliliters to liters. Actually, it doesn't because our goal is to determine an initial volume to be transferred that also is in milliliters:

$$V_{\text{initial}} = \frac{250.0\ \text{mL} \times 0.100\ \cancel{M}}{1.00\ \cancel{M}} = 25.0\ \text{mL}$$

Equation 8.6 can be modified for use with *any* units of concentration as well as volume—provided that the units used to express the initial and final volumes are the same and that the units used to express the initial and final concentrations ($C$) are the same:

$$V_{\text{initial}} = \frac{V_{\text{final}} \times C_{\text{final}}}{C_{\text{initial}}} \tag{8.7}$$

Let's use Equation 8.7 to calculate the volume (in milliliters) of the commercial 1000 ppm standard in **Figure 8.4** needed to prepare 100.0 mL of a 2.00 ppm standard solution for use in a laboratory that tests drinking water samples for lead contamination. Applying Equation 8.7 to calculate this volume:

$$V_{\text{initial}} = \frac{V_{\text{final}} \times C_{\text{final}}}{C_{\text{initial}}} = \frac{100.0\ \text{mL} \times 2.00\ \cancel{\text{ppm}}}{1000\ \cancel{\text{ppm}}} = 0.200\ \text{mL}$$

Commercial standards are usually sold with certificates of analysis that specify their concentrations to three significant figures. Therefore, a pipette accurate to the nearest 0.001 mL should be used to transfer the calculated volume.

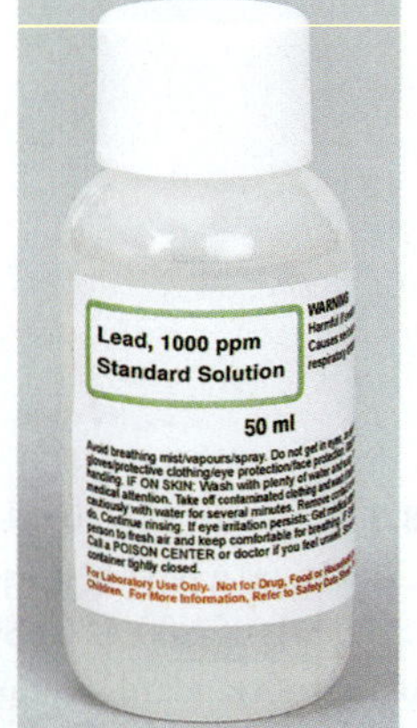


**FIGURE 8.4** Commercial standard solutions such as this 1000 ppm lead standard are widely used in testing laboratories that analyze many environmental, clinical, and other samples. Their solute concentrations are often higher than those required for laboratory work, so they must be diluted before use.

**SAMPLE EXERCISE 8.4** Diluting Solutions **LO2**

The solution used in hospitals for intravenous infusion—called *physiological saline* or *normal saline*—is 0.155 $M$ NaCl. It may be prepared by diluting a commercially available standard solution that is 1.50 $M$ NaCl. What volume of standard solution is required to prepare 10.0 L of physiological saline?

**Collect, Organize, and Analyze** We know the volume and concentration of the dilute solution (normal saline) to be prepared and the concentration of the standard solution to be diluted. Calculating the volume of the standard solution to be transferred is a matter of inserting the three quantities given into Equation 8.7, where $V_{final} = 10.0$ L, $C_{final} = 0.155\ M$, and $C_{initial} = 1.50\ M$. The final concentration is about 1/10 the initial concentration, so the volume to be transferred should be about 1/10 the final volume, or about 1 L.

**Solve**

$$V_{initial} = \frac{V_{final} \times C_{final}}{C_{initial}} = \frac{10.0 \text{ L} \times 0.155\ \cancel{M}}{1.50\ \cancel{M}} = 1.03 \text{ L}$$

**Think About It** Our calculated volume is close to our estimate. How would we measure out 1.03 L of the standard solution with acceptable precision and accuracy? If we knew (or could accurately determine) the density of the initial solution, we might use an electronic balance, such as the one shown in **Figure 8.5**, which could be used to dispense up to 10 kg of solution with a precision of 0.01 g.

**Practice Exercise** The concentration of $Ca^{2+}$ in a standard solution is 10.00 mg/mL. What volume of the standard solution should be used to prepare 0.500 L of a solution in which the $Ca^{2+}$ concentration is 50.0 mg/L?

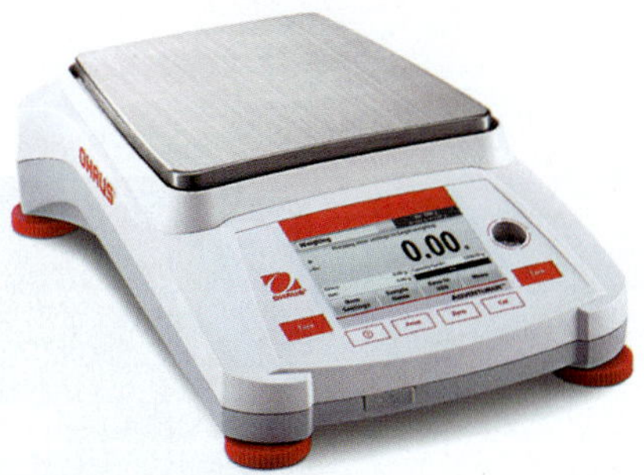



**FIGURE 8.5** This electronic balance can be used to measure masses up to 10 kg to the nearest 0.01 g.

CONCEPT **TEST**

**Figure 8.6** shows several solutions of a red cough syrup dissolved in water. Order these solutions from the most dilute to the most concentrated.

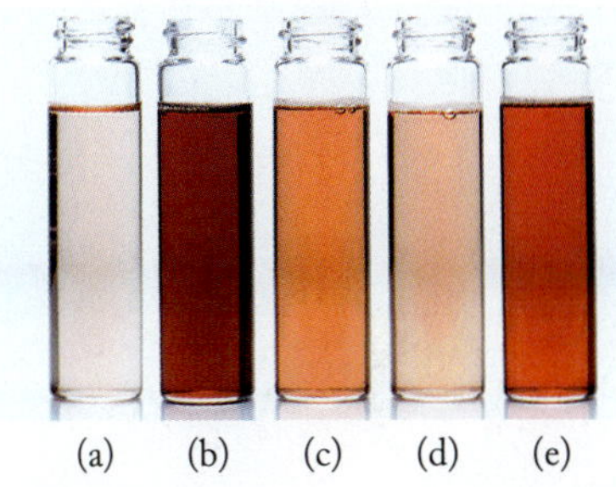



**FIGURE 8.6** Aqueous solutions of cough syrup diluted to various concentrations.

## Determining Concentration

The intensity of the blue color of the $Cu^{2+}$ solution in Figure 8.2 decreases as we dilute the solution. Can we use the intensity of the color to quantify the concentration? The color of the $Cu^{2+}$ solution arises from the absorption of visible light, a topic we will explore in more detail in Chapter 23. For now, knowing that a solution's **absorbance (*A*)** is a measure of the intensity of the color is enough. **Beer's law** (Equation 8.8) relates absorbance to three quantities: concentration (*c*), the path length that the light travels through the solution (*b*), and a constant (called the **molar absorptivity**, or **ε**) characteristic of the dissolved solute.

$$A = \varepsilon bc \qquad (8.8)$$

Absorbance is measured with a spectrophotometer like that shown in **Figure 8.7(a)**, in which a solution is placed in a sample cell of fixed length *b*, typically 1.00 cm. The cell is placed in a beam of light. The value of *A* represents how much light the solution absorbs. If we express the concentration of the solution, *c*, in molarity (*M*), we can calculate ε. A graph of *A* versus *c* for solutions of different $Cu^{2+}$ concentration yields a straight line because the amount of light that a sample absorbs is *directly proportional* to the concentration of that sample. Such a graph (**Figure 8.7b**) is called a **calibration curve.**

Once a calibration curve is generated, the concentration of any solution of $Cu^{2+}$ can be determined by measuring its absorbance, finding that value on the *y*-axis, reading across to where it intersects the calibration curve, and then reading the value on the *x*-axis below, which is the concentration. We also can use a calibration curve to determine molar absorptivity. If $b = 1.00$ cm, the slope of the graph is equal to ε in units of $M^{-1}$ cm$^{-1}$. This method of analysis is called spectrophotometry. Let's apply Beer's law to find the concentration of a solution of iron ions in Sample Exercise 8.5.

**absorbance (*A*)** a measure of the quantity of light that a sample absorbs.

**Beer's law** the relation of the absorbance of a solution (*A*) to concentration (*c*), the light's path length (*b*), and the solute's molar absorptivity (ε) by the equation $A = \varepsilon bc$.

**molar absorptivity (ε)** a measure of how well a compound or ion absorbs light.

**calibration curve** a graph showing how a measurable property, such as absorbance, varies for a set of standard samples of known concentration that can later be used to identify an unknown concentration from a measured absorbance.

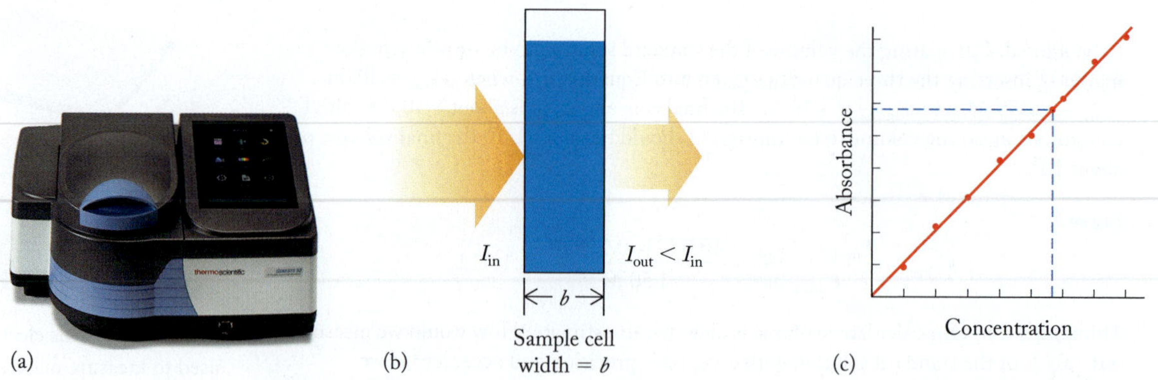


**FIGURE 8.7** (a) A spectrophotometer, used to measure the absorbance of solutions. (b) The intensity of light leaving the solution ($I_{out}$) is less than the intensity of light entering the solution ($I_{in}$). (c) The red line represents a calibration curve for Beer's law, plotting absorbance versus concentration for a series of standard samples. The horizontal dashed blue line represents the absorbance of a solution of unknown concentration. By extending the line to where it meets the calibration curve and reading down, we can determine the concentration of the unknown solution.

**FIGURE 8.8** Adding a colorless solution of phenanthroline to a colorless solution of $Fe^{2+}$ forms an orange-red solution when the phenanthroline molecules bond to the $Fe^{2+}$ ions.

### SAMPLE EXERCISE 8.5 Applying Beer's Law — LO3

One way to determine the concentration of $Fe^{2+}$ in an environmental sample is to add a solution of phenanthroline molecules that will bond to the metal ions, creating an intensely orange-red solution (**Figure 8.8**). The absorbance of a $5.00 \times 10^{-5}$ *M* solution is measured as 0.55 in a 1.00 cm cell. (a) Calculate the molar absorptivity, ε, and (b) determine the concentration of such a solution whose absorbance is 0.36.

**Collect and Organize** Given the concentration and absorbance of a solution, we are asked to calculate the molar absorptivity (ε).

**Analyze** (a) We can rearrange Beer's law to solve for ε and then use this value to find the concentration of the unknown solution. (b) Concentration and absorbance are linearly related by Beer's law, so we predict that a smaller absorbance for the unknown solution should correspond to a lower concentration. (And in fact, a concentration of zero would correspond to zero absorbance.)

**Solve**

a. Rearranging Equation 8.8 and substituting for *A*, *b*, and *c*:

$$\varepsilon = \frac{A}{bc} = \frac{0.55}{(1.00\ \text{cm})(5.00 \times 10^{-5}\ M)} = 1.1 \times 10^4\ M^{-1}\ \text{cm}^{-1}$$

b. Solving Beer's law for concentration and substituting for *A*, *b*, and ε:

$$c = \frac{A}{\varepsilon b} = \frac{0.36}{(1.1 \times 10^4\ M^{-1} \cdot \cancel{\text{cm}^{-1}})(1.00\ \cancel{\text{cm}})} = 3.3 \times 10^{-5}\ M$$

**Think About It** As predicted, the concentration of the unknown solution is lower than $5.00 \times 10^{-5}$ *M*.

 **Practice Exercise** One analytical method to determine the concentration of copper ions in a sample uses a solution of cuproine molecules to bond to

$Cu^{2+}$ ions, similar to the bonds between $Fe^{2+}$ ions and phenanthroline molecules. The following data were collected for four standard solutions of $Cu^{2+}$ with cuproine:

| Concentration of Copper(II) ($M$) | Absorbance ($A$) |
|---|---|
| $1.00 \times 10^{-4}$ | 0.64 |
| $1.25 \times 10^{-4}$ | 0.80 |
| $1.50 \times 10^{-4}$ | 0.96 |
| $1.75 \times 10^{-4}$ | 1.12 |

Construct a calibration curve based on these data, and use it to determine the concentration of a copper solution with $A = 0.74$.

## 8.3 Electrolytes and Nonelectrolytes

The images in **Figure 8.9** show an apparatus in which a lightbulb and two graphite rods are connected in series to a source of electricity (not shown) so that the bulb lights up only when enough electric current flows between the rods. As you can see, this happens in some but not all cases in Figure 8.9. It does *not* happen in Figure 8.9(a) when the two rods are immersed in distilled water, nor does the bulb light in Figure 8.9(b) when the rods are immersed in a beaker of table salt. The lack of electric current means that both distilled

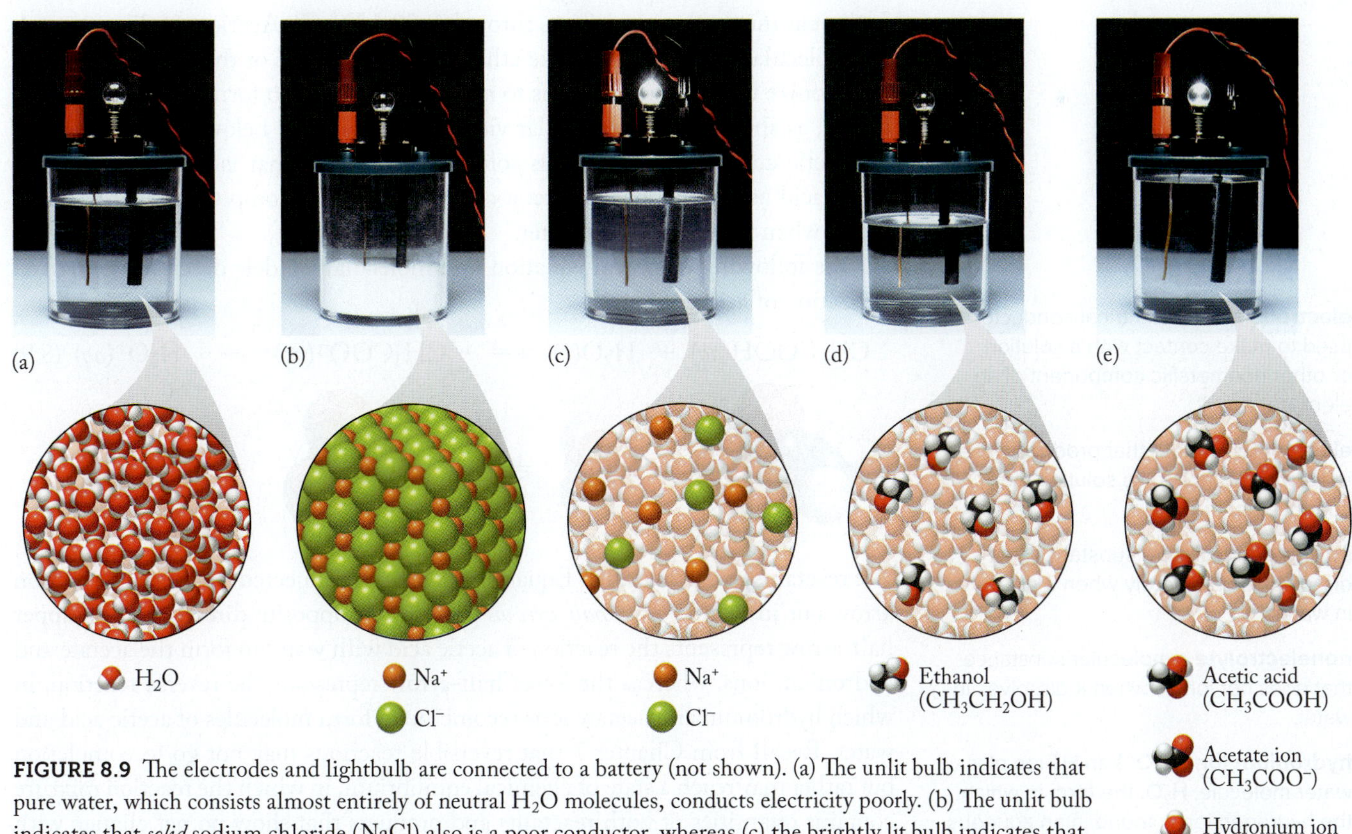


**FIGURE 8.9** The electrodes and lightbulb are connected to a battery (not shown). (a) The unlit bulb indicates that pure water, which consists almost entirely of neutral $H_2O$ molecules, conducts electricity poorly. (b) The unlit bulb indicates that *solid* sodium chloride (NaCl) also is a poor conductor, whereas (c) the brightly lit bulb indicates that a 0.50 $M$ solution of NaCl conducts electricity well. (d) The unlit bulb indicates that a 0.50 $M$ solution of ethanol ($CH_3CH_2OH$) is no better a conductor than pure water. (e) The dimly lit bulb indicates that a 0.50 $M$ solution of acetic acid ($CH_3COOH$) is a better conductor than pure water but not as good as 0.50 $M$ NaCl.

water and solid NaCl are poor conductors of electricity. The bulb does light, however, when the rods are immersed in 0.50 *M* NaCl (Figure 8.9c), so this solution is a *good* conductor of electricity. How does adding NaCl to water result in a good conductor?

Ions in Solution

When NaCl dissolves, its $Na^+$ and $Cl^-$ ions are liberated from their crystal lattice in Figure 8.9(b), and they can migrate independently through the solvent, as shown in the particulate view of Figure 8.9(c). As the ions move, so do the electric charges. The $Na^+$ ions are attracted to and move toward the graphite rod, or **electrode**, connected to the battery's *negative* terminal, whereas the $Cl^-$ ions move toward the *positive* graphite electrode. This ion movement allows electricity to flow through the solution.

Any ionic solute (such as NaCl) that produces a solution containing ions is called an **electrolyte** because movement of the dissolved ions allows the solution to conduct electricity. NaCl dissociates completely into $Na^+$ and $Cl^-$ ions when it dissolves in water, so it is called a **strong electrolyte**, even though solid NaCl alone is not a good conductor of electricity. However, if NaCl is heated above its melting point, the molten NaCl produced *is* a good conductor of electricity because the ions in liquid NaCl can move.

In Figure 8.9(d), the graphite electrodes are immersed in a 0.50 *M* solution of ethanol ($CH_3CH_2OH$). The unlit bulb indicates that ethanol conducts electricity no better than pure water. Thus, when molecules of $CH_3CH_2OH$ dissolve in water, they remain intact as whole molecules—they do not dissociate into ions. Solutes that do not form ions when they dissolve are called **nonelectrolytes**.

Finally, the lightbulb in Figure 8.9(e) produces a faint glow, which means that some current is flowing through the 0.50 *M* solution of acetic acid in the beaker—just not as much current as flows through 0.50 *M* NaCl. Acetic acid, like ethanol, is a molecular compound. Unlike ethanol, though, some of the acetic acid molecules ionize by donating $H^+$ ions to molecules of $H_2O$ to form **hydronium ions, $H_3O^+$**, as shown in the molecular view of Equation 8.9 below. Less than 1% of the acetic acid molecules in this solution ionize, but that is enough to qualify acetic acid as a **weak electrolyte**: a class of molecular compounds that partially ionize when they dissolve in water.

The following chemical equation and molecular models describe the partial ionization of acetic acid:

$$CH_3COOH(aq) + H_2O(\ell) \rightleftharpoons CH_3COO^-(aq) + H_3O^+(aq) \quad (8.9)$$

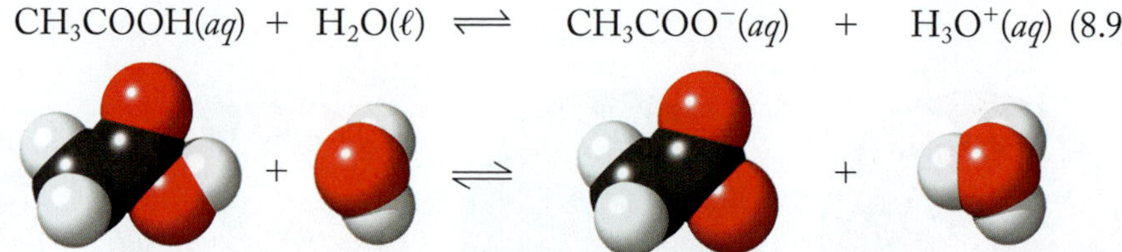

The reactants and products in Equation 8.9 are not connected by the usual reaction arrow but instead by two *half-arrows* pointing in opposite directions. The upper half-arrow represents the reaction of acetic acid with water to form the acetate and hydronium ions, whereas the lower half-arrow represents the reverse reaction, in which hydronium and acetate ions recombine to form molecules of acetic acid and water. Recall from Chapter 7 that reversible reactions may not go to completion but rather may reach a state of chemical equilibrium, in which the reaction mixture contains quantities of both reactants and products that show no net change with time. An aqueous solution of acetic acid, such as the one in Figure 8.9(e), achieves chemical equilibrium after relatively few of its molecules have ionized. Chemical equilibria are discussed in much greater detail in Chapters 14–16.

**electrode** a solid electrical conductor used to make contact with a solution or other nonmetallic component of an electrical circuit.

**electrolyte** a solute that produces ions in solution, enabling its solutions to conduct electricity.

**strong electrolyte** a substance that dissociates completely when it dissolves in water.

**nonelectrolyte** a molecular substance that does not ionize when it dissolves in water.

**hydronium ion ($H_3O^+$)** an $H^+$ ion plus a water molecule, $H_2O$; the form in which the hydrogen ion is found in an aqueous solution.

**weak electrolyte** a substance that only partly ionizes when it dissolves in water.

**acid (Brønsted–Lowry acid)** a proton donor.

**base (Brønsted–Lowry base)** a proton acceptor.

CONCEPT **TEST**

The conductivity test in Figure 8.9 is carried out with three new solutions: 0.1 *M* $CH_3OH$ (methanol), 0.1 *M* HF, and 0.1 *M* LiF. The lightbulb glows brightly when immersed in the 0.1 *M* LiF solution, has a faint glow in the 0.1 *M* HF solution, and does not glow at all in the 0.1 *M* $CH_3OH$ solution. Classify each of the three solutes as a strong electrolyte, weak electrolyte, or nonelectrolyte. Determine whether the solute dissociates in water, and if it does, identify the dissociated particles.

## 8.4 Acid–Base Reactions: Proton Transfer

Although the behavior of acetic acid molecules when dissolved in water is best described as a dynamic equilibrium, not all acids are weak electrolytes. Let's look at the behavior of another common acid: hydrochloric acid (HCl). Pure HCl is a molecular compound and a gas. However, when HCl dissolves in water, it ionizes completely, producing $H_3O^+(aq)$ and $Cl^-(aq)$ (**Figure 8.10**).

**CONNECTION** We discussed the naming of acids in Section 4.3.

As a molecule of HCl dissolves, it donates a proton to a water molecule:

$$\underset{\substack{\text{proton donor}\\\text{(acid)}}}{HCl(g)} + \underset{\substack{\text{proton acceptor}\\\text{(base)}}}{H_2O(\ell)} \rightarrow H_3O^+(aq) + Cl^-(aq) \quad (8.10)$$

This behavior is consistent with our definition of an **acid** as a *proton donor*. Because the water molecule accepts the proton, water here is a *proton acceptor*, the corresponding definition of a **base**. These definitions were originally proposed by chemists Johannes Brønsted (1879–1947) and Thomas Lowry (1874–1936), so they are called **Brønsted–Lowry acids** and **Brønsted–Lowry bases**, respectively. We will develop this concept of proton transfer more completely later, but for now it offers a convenient way to define acids and bases in aqueous solutions.

Chemists often simplify the chemical equations describing the behavior of acids in aqueous solution by leaving $H_2O$ out of the reactants. Instead, we use (*aq*) symbols to indicate the physical state of all solutes, and we write $H^+(aq)$ instead of $H_3O^+(aq)$ as a product. Thus, Equation 8.10 is sometimes written

$$HCl(g) \xrightarrow{H_2O(\ell)} H^+(aq) + Cl^-(aq) \quad (8.11)$$

When you see $H^+(aq)$ used as in Equation 8.11, always remember that hydrogen ions do not have an independent existence in water. Each proton reacts with a water molecule to form a hydronium ion ($H_3O^+$):

$$H^+(aq) + H_2O(\ell) \rightarrow H_3O^+(aq) \quad (8.12)$$

In aqueous acid–base chemistry, the terms *hydrogen ion*, *proton*, and *hydronium ion* all refer to the same species. We will encounter many other chemical equations in the paragraphs and chapters ahead that have been simplified by leaving out water molecules as individual molecules or parts of hydronium ions.

In the reaction between aqueous solutions of HCl and NaOH,

$$HCl(aq) + NaOH(aq) \rightarrow NaCl(aq) + H_2O(\ell) \quad (8.13)$$

the HCl functions as an acid by donating a proton ($H^+$) to the hydroxide ion ($OH^-$), and the hydroxide ion functions as a base by accepting the proton; the $H^+$ and $OH^-$ ions react to form a molecule of water. The remaining two ions—the anion $Cl^-$ and the cation $Na^+$—form a salt, sodium chloride, which remains

**FIGURE 8.10** Hydrochloric acid is a strong acid because all the molecules of HCl react with molecules of $H_2O$ to form $H_3O^+$ and $Cl^-$ ions.

**neutralization reaction** a reaction that takes place when an acid reacts with a base and produces a solution of a salt in water.

**salt** the product of a neutralization reaction, made up of the cation of the base in the reaction plus the anion of the acid.

**molecular equation** a balanced equation describing a reaction in solution in which the reactants and products are written as neutral compounds.

**overall ionic equation** a balanced equation that shows all the species, both ionic and molecular, present in a reaction occurring in aqueous solution.

**net ionic equation** a balanced equation that describes the actual reaction taking place in aqueous solution; it is obtained by eliminating the spectator ions from the overall ionic equation.

**spectator ion** an ion present in a reaction vessel when a chemical reaction takes place but is unchanged by the reaction; spectator ions appear in an overall ionic equation but not in a net ionic equation.

**strong acid** an acid that completely ionizes in aqueous solution.

**weak acid** a weak electrolyte that only partially ionizes in aqueous solution.

dissolved and dissociated in the aqueous solution. This reaction is called a **neutralization reaction** because the acid and the base no longer exist as a result of the reaction. The products of the neutralization of an acid such as HCl with a base such as NaOH in solution are always referred to as water and a salt. In fact, a **salt** is defined as a substance formed along with water as a product of a neutralization reaction.

Equation 8.13 is written as a **molecular equation**, meaning that each reactant and product is written as a neutral compound. Molecular equations are sometimes the easiest equations to balance, so reactions involving substances that dissociate in solution are often written first as molecular equations to indicate each reactant and product and to simplify balancing the equation.

Another representation of the reaction in Equation 8.13 shows the reactants and products as the particles that exist in solution, which means that any substance that dissociates completely is written as individual ions:

$$\overbrace{H^+(aq) + Cl^-(aq)}^{HCl(aq)} + \overbrace{Na^+(aq) + OH^-(aq)}^{NaOH(aq)} \rightarrow \overbrace{Na^+(aq) + Cl^-(aq)}^{NaCl(aq)} + H_2O(\ell) \quad (8.14)$$

This form, called an **overall ionic equation**, distinguishes ionic substances from molecular substances in a chemical reaction taking place in solution. Any nonelectrolytes or weak electrolytes are written as neutral molecules.

Notice that in Equation 8.14, chloride ions and sodium ions appear on both sides of the reaction arrow. If this were an algebraic equation, we would remove terms that are the same on both sides. When we do the same thing in chemistry, the result is a **net ionic equation**, as written in Equation 8.15. The ions removed (here, sodium ion and chloride ion) are **spectator ions**, and the resulting equation shows only the species taking part in the reaction:

$$H^+(aq) + OH^-(aq) \rightarrow H_2O(\ell) \quad (8.15)$$

The spectator ions are unchanged by the reaction and remain in solution. The chemical change in a neutralization reaction occurs when a hydrogen ion reacts with a hydroxide ion to produce a molecule of water. The net ionic equation helps us focus on those species that actually react. The three ways of depicting the reaction between HCl and NaOH—molecular, overall ionic, and net ionic equations—are summarized in **Figure 8.11**.

(a) Molecular equation $HCl(aq) + NaOH(aq) \rightarrow NaCl(aq) + H_2O(\ell)$

(b) Overall ionic equation
$H^+(aq) + Cl^-(aq) + Na^+(aq) + OH^-(aq) \rightarrow Na^+(aq) + Cl^-(aq) + H_2O(\ell)$

(c) Net ionic equation = overall ionic equation − spectator ions
$H^+(aq) + \cancel{Cl^-(aq)} + \cancel{Na^+(aq)} + OH^-(aq) \rightarrow \cancel{Na^+(aq)} + \cancel{Cl^-(aq)} + H_2O(\ell)$
$H^+(aq) + OH^-(aq) \rightarrow H_2O(\ell)$

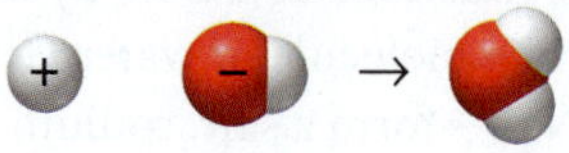

**FIGURE 8.11** The neutralization reaction between aqueous hydrochloric acid and sodium hydroxide that produces sodium chloride and water can be written in three ways: (a) a molecular equation, (b) an overall ionic equation, and (c) a net ionic equation.

## Strong Acids and Weak Acids

**Table 8.2** lists the names and formulas of the most common **strong acids**, so named because they completely dissociate in water. Hydrobromic acid and hydroiodic acid form when hydrogen bromide and hydrogen iodide, respectively, dissolve in water, just as hydrogen chloride dissolves in water to form hydrochloric acid. Sulfuric acid and nitric acid are among the most important industrial chemicals in the world.

Sulfuric acid is made by combining sulfur trioxide and water vapor, as discussed in Chapter 7:

$$SO_3(g) + H_2O(\ell) \rightarrow H_2SO_4(\ell) \qquad (7.6)$$

Nitric acid can be made from nitrogen dioxide gas and water vapor:

$$3\,NO_2(g) + H_2O(g) \rightarrow NO(g) + 2\,HNO_3(\ell) \qquad (8.16)$$

When $HNO_3(\ell)$ and $H_2SO_4(\ell)$ dissolve in water, they completely ionize:

$$HNO_3(\ell) + H_2O(\ell) \rightarrow H_3O^+(aq) + NO_3^-(aq) \qquad (8.17)$$

$$H_2SO_4(\ell) + H_2O(\ell) \rightarrow H_3O^+(aq) + HSO_4^-(aq) \qquad (8.18)$$

Strong acids are strong electrolytes. The arrows in the ball-and-stick figures of the acids in Table 8.2 point to the bonds that dissociate in solution to produce protons. Because one molecule of nitric acid donates one proton, nitric acid is called a *monoprotic acid*. Hydrochloric acid also is a strong monoprotic acid. Sulfuric acid can donate up to two protons per molecule, making it a *diprotic acid*. However, in most $H_2SO_4$ solutions, only one of the two H atoms in each molecule ionizes. This behavior means that sulfuric acid is a strong acid in terms of donating the first proton. This ionization results in the formation of the $HSO_4^-$ ion (hydrogen sulfate), which also is an acid because some $HSO_4^-$ ions in solution dissociate into $H^+$ and $SO_4^{2-}$ ions. However, because not all $HSO_4^-$ ions dissociate, the hydrogen sulfate anion is a **weak acid**. Other common weak acids are shown in **Table 8.3**.

**TABLE 8.2 Structures of Some Common Strong Acids**

| Acid | Molecular Formula | Bonds Ionizing in Solution |
|---|---|---|
| Hydrochloric acid | $HCl$ | |
| Hydrobromic acid | $HBr$ | |
| Hydroiodic acid | $HI$ | |
| Nitric acid | $HNO_3$ | |
| Sulfuric acid | $H_2SO_4$ | |
| Perchloric acid | $HClO_4$ | |

The two equations for the ionization of sulfuric acid are written differently to reflect these different degrees of ionization. First, the complete ionization of the strong acid $H_2SO_4$:

$$H_2SO_4(\ell) + H_2O(\ell) \rightarrow H_3O^+(aq) + HSO_4^-(aq) \qquad (8.18)$$

Second, the dynamic equilibrium and partial ionization of the weak acid $HSO_4^-$:

$$HSO_4^-(aq) + H_2O(\ell) \rightleftharpoons H_3O^+(aq) + SO_4^{2-}(aq) \qquad (8.19)$$

For nearly all diprotic acids, the singly charged anion that forms when a molecule of the acid donates one $H^+$ ion is a weaker acid than the original one. The same is true for *triprotic acids*. For example, phosphoric acid ($H_3PO_4$; Table 8.3) is a stronger acid than the dihydrogen phosphate ion, $H_2PO_4^-$, which is a stronger acid than the hydrogen phosphate ion, $HPO_4^{2-}$.

**STEPWISE ANIMATION**
Strong versus Weak Acids

**TABLE 8.3 Structures of Some Common Weak Acids**

| Name and Formula | Molecular Structure | Origin or Use |
|---|---|---|
| Hydrofluoric acid $HF$ | | HF gas dissolved in water |
| Acetic acid $CH_3COOH$ | | Vinegar is a dilute aqueous solution of acetic acid |
| Carbonic acid $H_2CO_3$ | | $CO_2$ gas dissolved in water |
| Hydrosulfuric acid $H_2S$ | | Hydrogen sulfide gas dissolved in water |
| Phosphoric acid $H_3PO_4$ | | $P_4O_{10}(s)$ dissolved in water |

**SAMPLE EXERCISE 8.6 Writing the Net Ionic Equation for a Neutralization Reaction I** **LO4**

Write the balanced (a) molecular, (b) overall ionic, and (c) net ionic equations that describe the reaction that takes place when an aqueous solution of nitric acid is neutralized by an aqueous solution of calcium hydroxide.

**Collect and Organize** Solutions of nitric acid and calcium hydroxide take part in a neutralization reaction. Our task is to write three equations describing the neutralization reaction.

**Analyze** We know that calcium hydroxide [$Ca(OH)_2$] is the base and nitric acid ($HNO_3$) is the acid. The products of a neutralization reaction are water and a salt. (a) All species are written as neutral compounds in the molecular equation. The products are water and calcium nitrate, the salt made from the cation of the base and the anion of the acid. (b) We can write each substance in the balanced molecular equation in ionic form to generate the overall ionic equation. (c) Removing spectator ions from the overall ionic equation gives us the net ionic equation.

**Solve**

a. The molecular equation is

$$HNO_3\,(aq) + Ca(OH)_2(aq) \rightarrow Ca(NO_3)_2(aq) + H_2O(\ell) \quad \text{(unbalanced)}$$

$$2\,HNO_3\,(aq) + Ca(OH)_2(aq) \rightarrow Ca(NO_3)_2(aq) + 2\,H_2O(\ell) \quad \text{(balanced)}$$

b. The overall ionic equation is

$$2\,H^+(aq) + 2\,NO_3^-(aq) + Ca^{2+}(aq) + 2\,OH^-(aq) \rightarrow Ca^{2+}(aq) + 2\,NO_3^-(aq) + 2\,H_2O(\ell)$$

c. The spectator ions are $Ca^{2+}$ and $NO_3^-$. Removing them gives us the net ionic equation:

$$2\,H^+(aq) + 2\,OH^-(aq) \rightarrow 2\,H_2O(\ell)$$

or, simplified to show the smallest coefficients,

$$H^+(aq) + OH^-(aq) \rightarrow H_2O(\ell)$$

**Think About It** We treat chemical equations just like algebraic equations, canceling out spectator ions that are unchanged in the reaction and appear on both sides of the reaction arrow in the *overall* ionic equation. After canceling the spectator ions, we are left with the net ionic equation, which focuses on the species that were changed as a result of the reaction. Also remember that nitrate is a polyatomic ion (see Table 4.4) that retains its identity in solution and does not separate into atoms and ions. Note that the net ionic equation is the same as in the reaction of an aqueous solution of HCl with aqueous NaOH.

**Practice Exercise** Write balanced (a) molecular, (b) overall ionic, and (c) net ionic equations for the reaction between an aqueous solution of sodium hydroxide and an aqueous solution of acetic acid, $CH_3COOH(aq)$, the weak acid shown in Equation 8.9. The products are sodium acetate and water.

## SAMPLE EXERCISE 8.7 Writing the Net Ionic Equation for a Neutralization Reaction II — LO4

Write the net ionic equation for the reaction that takes place when acid rain containing sulfuric acid reacts with the surface of a marble statue, as depicted in Figure 7.4 and as described by Equation 7.7:

$$CaCO_3(s) + H_2SO_4(aq) \rightarrow CaSO_4(aq) + H_2O(\ell) + CO_2(g) \quad (7.7)$$

**Collect, Organize, and Analyze** To write the net ionic equation for the reaction, we must first write the overall ionic equation, which contains the formulas of all the individual particles in the reaction mixture. Sulfuric acid ($H_2SO_4$) is the $H^+$ ion donor in the reaction, making $CaCO_3$ the $H^+$ ion acceptor. Each mole of $H_2SO_4$ can donate 2 moles of $H^+$ ions. The products of the reaction include $CaSO_4$ and two others, $H_2O$ and $CO_2$, which must have formed when 1 mole of $CO_3^{2-}$ ions in $CaCO_3$ accepted the 2 moles of $H^+$ ions as the statue dissolved:

$$2\,H^+(aq) + CaCO_3(s) \rightarrow Ca^{2+}(aq) + H_2O(\ell) + CO_2(g)$$

If this equation is correct, carbonate ions are the species acting as the base in the reaction.

$H_2SO_4$ is a strong acid and completely ionizes, forming $H^+$ and $HSO_4^-$ ions (Equation 8.18). However, $HSO_4^-$ is a weak acid and only partially separates into $H^+$ and $SO_4^{2-}$ ions (Equation 8.19), so it remains as $HSO_4^-$ in the overall ionic equation. Calcium sulfate is a soluble ionic compound and completely separates into $Ca^{2+}$ and $SO_4^{2-}$ ions in solution.

**Solve** Building an overall ionic equation from the preceding analysis of the reaction in Equation 7.7:

$$CaCO_3(s) + H^+(aq) + HSO_4^-(aq) \rightarrow Ca^{2+}(aq) + SO_4^{2-}(aq) + H_2O(\ell) + CO_2(g)$$

To write a net ionic equation, we eliminate spectator ions from the overall ionic equation. However, no spectator ions are present here: none of the ions on the reactant side of the reaction arrow appear on the product side. Therefore, the overall ionic equation *is* the net ionic equation:

$$CaCO_3(s) + H^+(aq) + HSO_4^-(aq) \rightarrow Ca^{2+}(aq) + SO_4^{2-}(aq) + H_2O(\ell) + CO_2(g)$$

**Think About It** Because one reactant was a solid and the products included a gas and liquid water, the number of possible spectator ions in the overall ionic equation was not large to begin with. That fact, coupled with the partial ionization of $HSO_4^-$ ions, led to the absence of any spectator ions.

**Practice Exercise** Write balanced (a) molecular, (b) overall ionic, and (c) net ionic equations for the reaction between aqueous solutions of phosphoric acid ($H_3PO_4$) and sodium hydroxide. The products are sodium phosphate and water.

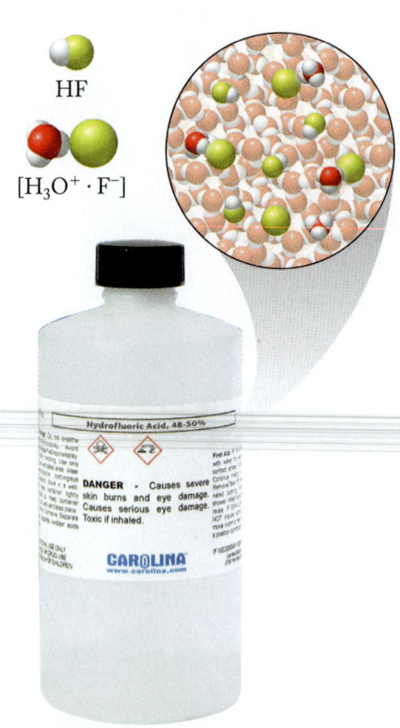


**FIGURE 8.12** Hydrofluoric acid is a weak acid because only some of the HF molecules react with molecules of $H_2O$, and even when they do, they form mostly $[H_3O^+ \cdot F^-]$ ion pairs rather than free $H_3O^+$ ions.

Hydrofluoric acid (formed from hydrogen fluoride, HF) is a common weak acid shown in Table 8.3. However, we know from conductivity tests such as those pictured in Figure 8.9 that an aqueous solution of HF contains far fewer $H_3O^+$ ions than the number of HF molecules dissolved in solution (**Figure 8.12**). How do chemists explain this observation? One explanation is that the covalent bonds that hold HF molecules together are much stronger than the bonds in the other hydrogen halides and therefore don't break as easily. Another explanation is that even if the bonds do break and HF ionizes,

$$HF(g) + H_2O(\ell) \rightarrow F^-(aq) + H_3O^+(aq)$$

the ions that form may not move about freely in solution. Dissolved fluoride ions are much smaller than those of the other halogens, so they can get closer to and are more strongly attracted to hydronium ions. This electrostatic attraction is so strong that the ions form stable ion pairs:

$$F^-(aq) + H_3O^+(aq) \rightleftharpoons [H_3O^+ \cdot F^-](aq) \tag{8.20}$$

The reaction in Equation 8.20 is reversible and does not consume *all* the $H_3O^+$ ions in solution. The relatively few that are still free give hydrofluoric acid its weakly acidic properties.

Another weak acid listed in Table 8.3 is acetic acid, $CH_3COOH$, our example of a weak electrolyte in Section 8.3. Acetic acid is a member of a large group of organic compounds called **carboxylic acids** because their structures all contain the –COOH group (that is, the *carboxyl group*). The H atom in this group partially dissociates when it dissolves in water; hence, carboxylic acids are weak acids. Also in Table 8.3 are carbonic acid, $H_2CO_3$, which forms when carbon dioxide dissolves in water, and hydrosulfuric acid, $H_2S$, which forms when highly toxic and intensely foul-smelling hydrogen sulfide gas dissolves in water.

**CONNECTION** We learned about the size of ions in Section 3.10.

CONCEPT **TEST**

A molecule called EDTA has four carboxylic acid groups and hence four acidic protons. We can symbolize it as $H_4E$ to highlight this property. Pick the more acidic member of each pair listed: (a) $H_2E^{2-}$ or $HE^{3-}$; (b) $H_4E$ or $H_3E^-$.

## Strong Bases, Weak Bases, and Amphiprotic Substances

In a classification scheme similar to that used for acids, bases are classified as **strong bases** or **weak bases**, depending on the extent to which they dissociate in aqueous solution. Strong bases are strong electrolytes; weak bases are weak electrolytes. Important sources of $OH^-$ ions include the alkali metal hydroxides, solid ionic compounds that have the generic formula MOH (where M represents any alkali metal). They all dissociate completely into $M^+$ and $OH^-$ ions when they dissolve in water. The hydroxides of $Ca^{2+}$, $Sr^{2+}$, and $Ba^{2+}$ (group 2) also dissociate completely when they dissolve in water, but they are not as soluble as the group 1 hydroxides. Because of the limited solubility of the group 2 hydroxides, the concentrations of dissolved $OH^-$ ions in their solutions cannot equal those of concentrated solutions of the group 1 hydroxides.

Although substances such as $NaOH(aq)$ and $Ca(OH)_2(aq)$ can be considered bases because they produce hydroxide ions in aqueous solution, it is important to remember the Brønsted–Lowry definition of a base as a proton acceptor. Hydroxide ions are bases because they can accept protons. This definition permits other substances that do not consist of hydroxide ions to be categorized as bases as well.

One such example is ammonia ($NH_3$), which is a weak base even though it does not contain hydroxide ions. Ammonia in its molecular form is a gas at room temperature. When $NH_3$ dissolves in water, it produces a solution that conducts electricity weakly, which means that ammonia is a weak electrolyte:

$$\underset{\substack{\text{proton}\\\text{acceptor}\\\text{(base)}}}{NH_3(g)} + \underset{\substack{\text{proton}\\\text{donor}\\\text{(acid)}}}{H_2O(\ell)} \rightleftharpoons NH_4^+(aq) + OH^-(aq) \quad (8.21)$$

At equilibrium, this solution contains mostly aqueous ammonia molecules, with smaller concentrations of ammonium and hydroxide ions. If the apparatus in Figure 8.9 were immersed in 0.50 $M$ $NH_3$, the bulb would light about as brightly as it does in 0.50 $M$ acetic acid (Figure 8.9e). Compounds such as $NH_3$ that contain an $-NH_2$ group typically function as proton acceptors—that is, Brønsted–Lowry bases.

Some compounds can act as either acids or bases because they have the capacity to accept $H^+$ ions in the presence of acids and donate them in the presence of bases. One compound that exhibits this behavior is water:

$$HCl(g) + H_2O(\ell) \rightarrow Cl^-(aq) + H_3O^+(aq) \quad (8.10)$$

$$NH_3(g) + H_2O(\ell) \rightleftharpoons NH_4^+(aq) + OH^-(aq) \quad (8.21)$$

In a solution of hydrochloric acid, molecules of $H_2O$ accept $H^+$ ions created by the ionization of molecules of HCl. In an aqueous solution of ammonia, molecules of $H_2O$ ionize, donating $H^+$ ions to molecules of $NH_3$. Substances that can function as either a Brønsted–Lowry base or a Brønsted–Lowry acid (such as water) are said to be **amphiprotic**.

Another common amphiprotic substance is sodium hydrogen carbonate, $NaHCO_3$—more commonly called sodium bicarbonate—the principal ingredient in baking soda. In a solution that also contains a strong acid, bicarbonate ions act like bases, accepting $H^+$ ions from the acid and forming the weak acid $H_2CO_3$, which decomposes to water and $CO_2$.

$$H^+(aq) + HCO_3^-(aq) \rightarrow H_2CO_3(aq) \rightarrow H_2O(\ell) + CO_2(g) \quad (8.22)$$

In a solution that also contains a strong base, bicarbonate acts like an acid, donating $H^+$ ions to the $OH^-$ ions from the base:

$$OH^-(aq) + HCO_3^-(aq) \rightarrow H_2O(\ell) + CO_3^{2-}(aq) \quad (8.23)$$

Note that adding acid to compounds that contain either $HCO_3^-$ or $CO_3^{2-}$ ions produces the same two products that we saw in Sample Exercise 8.7: liquid $H_2O$ and $CO_2$ gas.

**carboxylic acid** a compound containing the –COOH functional group.

**strong base** a base that completely dissociates into ions in aqueous solution.

**weak base** a base that is a weak electrolyte and so has a limited capacity to accept protons.

**amphiprotic** describes a substance that can behave as either a proton acceptor or a proton donor.

## CONCEPT TEST

Identify each of the following compounds in aqueous solution as an acid, a base, or neither. If the compound is an acid or base, identify it as weak or strong. (a) $CH_3CH_2COOH$; (b) $CH_3OH$; (c) $Mg(OH)_2$; (d) $HNO_3$

## SAMPLE EXERCISE 8.8 Comparing Electrolytes, Acids, and Bases LO5

Classify each of the following compounds in aqueous solution as a strong electrolyte, a weak electrolyte, or a nonelectrolyte, and then again as an acid, a base, or neither: (a) NaCl; (b) HCl; (c) NaOH; (d) $CH_3COOH$; (e) $CH_3CH_2NH_2$.

**Collect and Organize** We are given the formulas of five compounds and are asked to classify them according to their behavior in an aqueous solution. We will work with the definitions of the given terms.

**Analyze** Strong electrolytes are substances that dissociate completely into ions when dissolved in water. Weak electrolytes dissociate only partially in water. Nonelectrolytes do not dissociate at all. A Brønsted–Lowry acid is defined as a proton donor, and a Brønsted–Lowry base is a proton acceptor. Structures with a –COOH group function as weak acids, whereas compounds containing the $–NH_2$ group are weak bases.

**Solve**

a. NaCl is an ionic compound that dissociates completely into ions in aqueous solution, as shown in Figure 8.9(c). Therefore, NaCl is a strong electrolyte. Sodium chloride has no protons to donate, and neither $Na^+$ nor $Cl^-$ ions readily accept protons. Thus, NaCl is neither an acid nor a base.
b. HCl dissociates completely into $H^+$ and $Cl^-$ ions when dissolved in water, making HCl a strong electrolyte. As a proton donor, aqueous HCl is also a Brønsted–Lowry acid.
c. NaOH is an ionic compound that dissociates completely into $Na^+$ and $OH^-$ ions in aqueous solution. NaOH is a strong electrolyte. The $OH^-$ ion formed upon dissolution of NaOH readily accepts a proton to form water, making NaOH a Brønsted–Lowry base.
d. Some molecules of acetic acid, $CH_3COOH$, when dissolved in water, dissociate into $H^+$ and $CH_3COO^-$ ions (Figure 8.9e), creating a weakly conducting solution; thus, it is a weak electrolyte. Because acetic acid is a proton donor, it is considered a Brønsted–Lowry acid.
e. Some molecules of $CH_3CH_2NH_2$ function as proton acceptors when dissolved in water (which functions as a proton donor) to form $CH_3CH_2NH_3{}^+$ ions, creating a weakly conducting solution. Therefore, $CH_3CH_2NH_2$ is both a weak electrolyte and a Brønsted–Lowry base.

**Think About It** You should not be surprised that some strong electrolytes are also acids or bases. Acidity and basicity have different definitions from those for electrolytes, and substances can be classified as either an acid or a base while also being either a strong or weak electrolyte. A weak electrolyte that donates a proton is a weak acid.

**Practice Exercise** Classify aqueous solutions of each of the following compounds as a strong electrolyte, weak electrolyte, or nonelectrolyte, and as an acid, a base, or neither: (a) potassium sulfate; (b) HBr; (c) $NH_3$; (d) ethanol ($CH_3CH_2OH$).

## 8.5 Titrations

Neutralization reactions such as the one in Sample Exercise 8.6 highlight an important point about the behavior of acids and bases: even though weak acids and bases are only partially ionized in their aqueous solutions, they become completely ionized during neutralization reactions. For example, completely neutralizing one mole of $H_2SO_4$ takes two moles of a strong base such as NaOH because one molecule of $H_2SO_4$ has two *ionizable* hydrogen atoms. That both H atoms may not be ionized in a solution of sulfuric acid does not matter; they both *do* ionize during a neutralization reaction. Similarly, completely neutralizing one mole of acetic acid takes one mole of NaOH, even though less than 1% of the molecules of acetic acid in solution are ionized before the reaction begins.

The same principle applies to weak bases. For example, one mole of hydrochloric acid (a strong acid) is needed to neutralize one mole of ammonia in solution, even though most of the molecules of $NH_3$ have not reacted with $H_2O$ to form $NH_4^+$ and $OH^-$ ions before HCl is added. As the neutralization reaction proceeds, more and more $NH_3$ reacts with $H_2O$ to form $NH_4^+$ and $OH^-$ ions, and the $OH^-$ ions are consumed by combining with $H^+$ ions from the acid to form $H_2O$.

Aqueous-phase reactions can be used to determine the concentrations of dissolved substances with excellent precision and accuracy. The analytical technique used is called **titration**. It is classified as a *volumetric* method of analysis because it is based on measuring the volumes of both a standard solution and a sample solution. The standard solution, called the **titrant**, contains a known concentration of a substance that reacts with a substance in the sample. The goal is to precisely determine the concentration of this second substance, called the **analyte**. The chemistries of titration reactions are often based on acid–base neutralization reactions.

To get a feel for how titrations work, let's consider an application from environmental science: analyzing acidic water draining from abandoned coal mines. Sulfide-containing minerals called pyrites are often present in and around seams of coal. Mining the coal exposes the pyrite to chemical and biochemical weathering. Those processes produce sulfuric acid, which dissolves in groundwater seeping through abandoned mines. To assess and control the environmental impact of this acidic water, scientists and engineers need to know how much sulfuric acid is dissolved in it. That's where titrations come in.

Suppose we have a sample of mine drainage containing an unknown concentration of sulfuric acid. We can use an acid–base titration in which the sulfuric acid in the water sample is neutralized by reacting it with a standard solution of strong base, such as NaOH. The neutralization reaction is

$$H_2SO_4(aq) + 2\,NaOH(aq) \rightarrow Na_2SO_4(aq) + 2\,H_2O(\ell)$$

Note that both ionizable hydrogen atoms on each molecule of $H_2SO_4$ are neutralized in the reaction. Neutralizing both takes two moles of NaOH for every mole of $H_2SO_4$.

To determine the concentration of analyte (sulfuric acid) in the sample, we need to measure the volume of titrant (standard solution of NaOH) needed to neutralize it. We start by carefully transferring a known volume of the sample into a reaction vessel such as an Erlenmeyer flask. A few drops of a color indicator are added to detect when just enough titrant has been added to neutralize

**titration** an analytical method of determining the concentration of a solute in a sample by reacting the solute with a solution of known concentration.

**titrant** the standard solution added to the sample in a titration.

**analyte** the substance whose concentration is to be determined in a chemical analysis.

**equivalence point** the point in a titration at which just enough titrant has been added to react with all the analyte in the sample.

**end point** the point in a titration at which a color change or other signal indicates that enough titrant has been added to react with all the analyte in the sample.

the sulfuric acid (**Figure 8.13**). The titrant is added slowly by using a buret, a narrow glass cylinder with volume markings and fitted with a stopcock to quickly stop the flow when just enough titrant has been added. This stage of a titration, when the reaction is just complete, is called the **equivalence point** because a stoichiometric quantity of titrant has been added that is *equivalent* to the quantity of the analyte.

The volume of titrant required to make the indicator change color may not be *exactly* the same as the volume of titrant needed to consume all the analyte in the sample. The volume needed to make the indicator change color is actually the **end point** of the titration. It is the volume we record and use in calculating how much analyte was in the sample. If the correct indicator has been chosen, the equivalence point is very close to the end point.

To detect the end point in a sulfuric acid–sodium hydroxide titration, we might use the indicator *phenolphthalein*, which is colorless in acidic solutions but pink in basic solutions (Figure 8.13). The part of the titration that requires the most skill is adding just enough NaOH solution to reach the end point, the point at which a pink color first persists in the solution being titrated. To catch the end point precisely, the standard solution must be added no faster than one drop at a time, with thorough mixing between drops.

Suppose that 22.40 mL of 0.00100 $M$ NaOH is necessary to react completely with the $H_2SO_4$ in a 100.0 mL sample of acidic mine drainage. We know the volume and molarity of the NaOH standard solution, so we can adapt Equation 8.2 to calculate the number of moles of NaOH consumed after first converting the volume into liters:

$$n_{\text{NaOH}} = V_{\text{NaOH}} \times M_{\text{NaOH}}$$

$$= 0.02240\ \cancel{\text{L}} \times \frac{1.00 \times 10^{-3}\ \text{mol NaOH}}{\cancel{\text{L}}} = 2.240 \times 10^{-5}\ \text{mol NaOH}$$

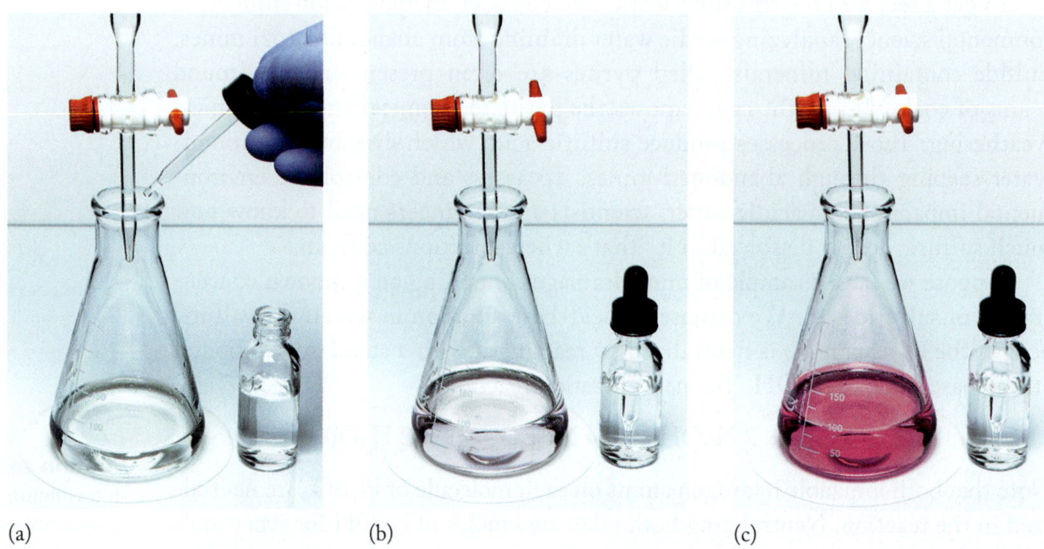

(a) (b) (c)

**FIGURE 8.13** Determining a sulfuric acid concentration. (a) A known volume of a sample containing $H_2SO_4$ is placed in the flask. The buret is filled with the titrant, an aqueous NaOH solution of known concentration. A few drops of phenolphthalein indicator solution are added to the flask. (b) Titrant is carefully added to the flask until the indicator changes from colorless to faint pink, signaling that the acid has been neutralized and the end point has been reached. (c) If more titrant is added, the color becomes darker, signaling that excess titrant was added and the end point has been passed.

We also know from the stoichiometry of the reaction that two moles of NaOH is required to neutralize one mole of $H_2SO_4$, so the number of moles of $H_2SO_4$ in the 100.0 mL sample must have been

$$n_{H_2SO_4} = 2.240 \times 10^{-5} \text{ mol NaOH} \times \frac{1 \text{ mol } H_2SO_4}{2 \text{ mol NaOH}} = 1.120 \times 10^{-5} \text{ mol } H_2SO_4$$

Dividing this number of moles of $H_2SO_4$ by the volume of the sample ($V_{sample}$) in liters gives us the molar concentration of $H_2SO_4$:

$$M_{H_2SO_4} = \frac{n_{H_2SO_4}}{V_{sample}}$$

$$= \frac{1.120 \times 10^{-5} \text{ mol } H_2SO_4}{0.1000 \text{ L}} = \frac{1.120 \times 10^{-4} \text{ mol } H_2SO_4}{\text{L}}$$

$$= 1.12 \times 10^{-4}\, M\, H_2SO_4$$

**SAMPLE EXERCISE 8.9** Determining Solute Concentration from Titration Data **LO6**

Apple cider vinegar is an aqueous solution of acetic acid made from fermented apple juice. Commercial vinegar must contain at least 4 g of acetic acid per 100 mL of vinegar. Suppose that the titration of a 25.00 mL sample of vinegar requires 12.15 mL of 1.885 *M* NaOH. What is the molarity of acetic acid in the vinegar sample? Does it meet the 4 g/100 mL standard?

**Collect and Organize** We have the results of a titration of acetic acid in vinegar with a strongly basic titrant. We know that acetic acid is $CH_3COOH$ and has a molar mass of 60.05 g/mol. We know the volume and concentration of the titrant and the volume of the sample, and we need to calculate the concentration of acetic acid in the sample.

**Analyze** We can use Equation 8.2 to calculate the number of moles of basic titrant consumed from its volume and concentration. We need a balanced chemical equation to relate the number of moles of titrant (base) to the number of moles of acid in the sample during the titration reaction. One ionizable H atom exists in a molecule of $CH_3COOH$ (highlighted in red), and one $OH^-$ ion is present per formula unit of NaOH. Therefore, the mole ratio of $CH_3COOH$ to NaOH in the balanced chemical equation for the titration reaction is 1:1. The volume of titrant consumed is about half the sample volume, which means that the concentration of acetic acid in the sample should be about half the concentration of NaOH in the titrant, or about 1 *M*.

**Solve** The balanced chemical equation for the neutralization reaction is

$$CH_3COOH(aq) + NaOH(aq) \rightarrow H_2O(\ell) + CH_3COONa(aq)$$

The number of moles of NaOH required to reach the end point is

$$n_{NaOH} = V_{NaOH} \times M_{NaOH}$$

$$= \left(12.15 \text{ mL} \times \frac{1 \text{ L}}{1000 \text{ mL}}\right) \times \frac{1.885 \text{ mol}}{\text{L}} = 0.02290 \text{ mol NaOH}$$

Converting moles of NaOH into moles of $CH_3COOH$:

$$n_{CH_3COOH} = 0.02290 \text{ mol NaOH} \times \frac{1 \text{ mol } CH_3COOH}{1 \text{ mol NaOH}} = 0.02290 \text{ mol } CH_3COOH$$

Dividing moles of $CH_3COOH$ by the sample volume (in liters), we obtain the molar concentration of $CH_3COOH$ in the sample:

$$M_{CH_3COOH} = \frac{n_{CH_3COOH}}{V_{sample}}$$

$$= \frac{0.02290 \text{ mol } CH_3COOH}{25.00 \text{ mL}} \times \frac{1000 \text{ mL}}{1 \text{ L}} = 0.9160 \text{ mol/L} = 0.9160\ M$$

Converting this molar concentration into grams per 100 mL:

$$\frac{0.9160 \text{ mol}}{\text{L}} \times \frac{60.05 \text{ g}}{\text{mol}} \times \frac{1 \text{ L}}{1000 \text{ mL}} = \frac{0.05501 \text{ g}}{\text{mL}}, \text{ which equals } \frac{5.501 \text{ g}}{100 \text{ mL}}$$

The sample meets the 4 g/100 mL standard for apple cider vinegar.

**Think About It** The calculated molar concentration of acetic acid in the vinegar sample was close to our 1 *M* estimate, and the equivalent concentration expressed in grams per 100 mL is just above the standard for apple cider vinegar, so our results are entirely reasonable.

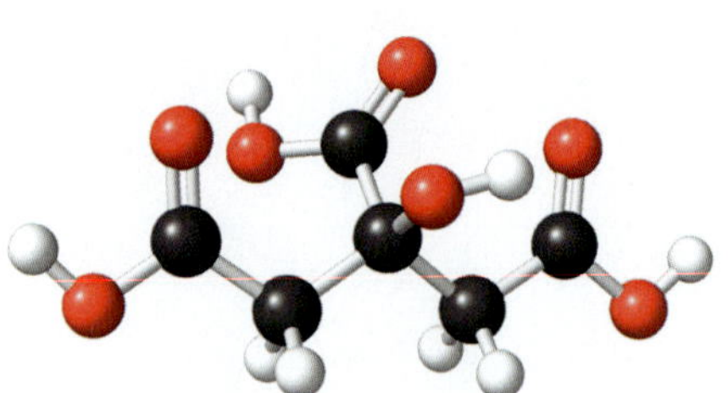

**FIGURE 8.14** Citric acid, $C_6H_8O_7$.

**Practice Exercise** Citric acid is a triprotic acid with a molar mass of 192.12 g/mol. It has three carboxylic acid groups, as shown in **Figure 8.14**. It is found in lemon juice and other citrus products and is widely used as a flavoring in beverages. What is the molarity of citric acid in lemon juice if 24.26 mL of 0.904 *M* NaOH is required in a titration to neutralize 25.00 mL of juice? How many grams of citric acid are in 0.100 L of juice?

**SAMPLE EXERCISE 8.10** Calculating Molarity in a Gas-Forming Neutralization Reaction **LO6**

Many over-the-counter antacids, such as the one shown in **Figure 8.15**, contain sodium bicarbonate. Sodium bicarbonate reacts with excess stomach acid (aqueous HCl) in a neutralization reaction. A 650 mg tablet is dissolved in 25.00 mL of water. Titrating this solution requires 25.27 mL of a 0.3000 *M* solution of HCl. (a) Write a net ionic equation for the reaction between $NaHCO_3$ and HCl. (b) What is the molarity of the antacid solution? (c) Is the tablet pure $NaHCO_3$?

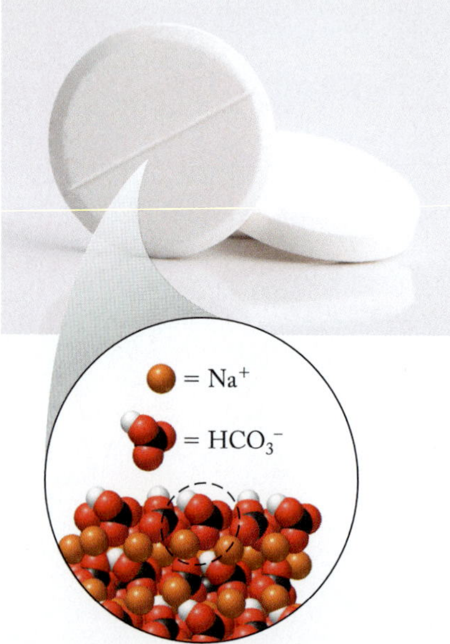


**FIGURE 8.15** Sodium bicarbonate antacid tablets.

**Collect and Organize** We are assigned three tasks. We are given the names of the reactants and are asked to write a balanced net ionic equation. We are given the volume and concentration of the titrant and the volume of the sample, and we are asked to calculate the concentration of the solution. Finally, using the results of our calculations, we are asked to assess the purity of the $NaHCO_3$ present in the original sample.

**Analyze** (a) The products of the reaction between $NaHCO_3$ and HCl are sodium chloride, water, and carbon dioxide. We will first balance the molecular equation and then generate the overall ionic equation and the net ionic equation. (b) To calculate the molarity of the sodium bicarbonate solution, we will use the balanced molecular equation to determine the ratio of the moles of titrant (acid) to the moles of sample (base). (c) We can determine the purity of the tablet by converting the number of moles of $NaHCO_3$ calculated in part (b) into an equivalent number of grams and then comparing the result with the mass of the tablet (650 mg).

**Solve**

a. The balanced molecular equation for the neutralization reaction is

$$NaHCO_3(aq) + HCl(aq) \rightarrow H_2O(\ell) + CO_2(g) + NaCl(aq)$$

The balanced ionic equation is

$$Na^+(aq) + HCO_3^-(aq) + H^+(aq) + Cl^-(aq) \rightarrow H_2O(\ell) + CO_2(g) + Na^+(aq) + Cl^-(aq)$$

and the net ionic equation is

$$\cancel{Na^+(aq)} + HCO_3^-(aq) + H^+(aq) + \cancel{Cl^-(aq)} \rightarrow H_2O(\ell) + CO_2(g) + \cancel{Na^+(aq)} + \cancel{Cl^-(aq)}$$

$$HCO_3^-(aq) + H^+(aq) \rightarrow H_2O(\ell) + CO_2(g)$$

b. The stoichiometry of the reaction shows that 1 mole of $NaHCO_3$ reacts with 1 mole of HCl. The number of moles of $NaHCO_3$ titrated is

$$0.02527\ \cancel{\text{L HCl}} \times \frac{0.3000\ \cancel{\text{mol HCl}}}{1\ \cancel{\text{L HCl}}} \times \frac{1\ \text{mol NaHCO}_3}{1\ \cancel{\text{mol HCl}}} = 0.007581\ \text{mol NaHCO}_3$$

The molarity of the $HCO_3^-$ is

$$\frac{0.007581\ \text{mol NaHCO}_3}{0.02500\ \text{L solution}} = 0.3032\ M\ \text{NaHCO}_3$$

c. To calculate the number of grams of sodium bicarbonate in the sample, we first need to calculate the molar mass of $NaHCO_3$:

$$\mathcal{M}_{\text{NaHCO}_3} = 22.99\ \text{g/mol} + 1.008\ \text{g/mol} + 12.01\ \text{g/mol} + 3(16.00\ \text{g/mol})$$
$$= 84.01\ \text{g/mol NaHCO}_3$$

Therefore, the mass of $NaHCO_3$ in the tablet is

$$0.007581\ \cancel{\text{mol NaHCO}_3} \times \frac{84.01\ \text{g NaHCO}_3}{1\ \cancel{\text{mol NaHCO}_3}} = 0.6369\ \text{g NaHCO}_3$$

The mass of the tablet was 650 mg. On the basis of the titration, the sample contained only 637 mg of $NaHCO_3$ and hence is not pure $NaHCO_3$.

**Think About It** The stoichiometry of the net ionic equation indicates a 1:1 ratio of acid to base. Because a nearly equivalent volume of aqueous HCl was needed to titrate 25 mL of antacid solution, it makes sense that the concentration of the base is similar to that of the acid.

**Practice Exercise** What is the molarity of calcium bicarbonate if 9.870 mL of 1.000 $M$ $HNO_3$ is required in a titration to neutralize 50.00 mL of a solution of $Ca(HCO_3)_2$?

## 8.6 Precipitation Reactions

Some of the most abundant elements in Earth's crust, including silicon (Si), aluminum (Al), and iron (Fe), are not abundant in seawater because the minerals that contain these elements have limited solubility in water. Many minerals contain ionic compounds, so why aren't they soluble? Recall from Chapter 6 that compounds are *not* soluble when solute–solvent interactions (ion–dipole here) are not strong enough to offset solute–solute interactions (ion–ion here) and solvent–solvent interactions such as hydrogen bonding between water molecules.

The strengths of ion–ion interactions increase with increasing charge and decrease with increasing ion size (see Equation 4.1). These trends are reflected in the solubility guidelines listed in **Table 8.4** and in the corresponding solubility matrix in **Table 8.5** for some common ionic compounds. For example, all salts that contain alkali metal cations are soluble—as we would expect, given their 1+ charges. In addition, all nitrate and acetate compounds are soluble because both ions are relatively large polyatomic ions and have charges of only 1−. However, nearly all ionic compounds in which the cations' charges are 3+ or 4+ and the anions' charges are 2− or 3− (many transition metal oxides and sulfides fall into this category) have limited solubility because the

**TABLE 8.4** Solubility Guidelines for Common Ionic Compounds in Water

| | |
|---|---|
| (1) | All compounds containing the following ions are *soluble*:<br>• Cations: group 1 ions (alkali metals) and $NH_4^+$<br>• Anions: $NO_3^-$, $ClO_4^-$, and $CH_3COO^-$ (acetate) |
| (2) | Compounds containing the following anions are *soluble* except as noted:<br>• $Cl^-$, $Br^-$, and $I^-$, except those of $Ag^+$, $Cu^+$, $Hg_2^{2+}$, and $Pb^{2+}$<br>• $SO_4^{2-}$, except those of $Ba^{2+}$, $Ca^{2+}$, $Hg_2^{2+}$, $Pb^{2+}$, and $Sr^{2+}$ |
| (3) | Compounds containing the following anions are considered to have *limited solubility*:<br>• $F^-$, $O^{2-}$, $CO_3^{2-}$, $PO_4^{3-}$, $AsO_4^{3-}$, $CrO_4^{2-}$, $C_2O_4^{2-}$ (oxalates), except those of group 1 cations and $NH_4^+$, which are soluble<br>• $OH^-$, except those of group 1 cations and $NH_4^+$, $Ca^{2+}$, $Sr^{2+}$, and $Ba^{2+}$<br>• $S^{2-}$, except those of group 1 cations and $NH_4^+$, $Ca^{2+}$, $Sr^{2+}$, and $Ba^{2+}$<br>[Note: The solubilities of the group 2 hydroxides and sulfides increase with increasing cation atomic number. MgS decomposes in water, forming $H_2S$ and $Mg(OH)_2$.] |

interactions between ions with multiple charges are strong. However, other factors influence the solubilities of ionic compounds, too, thereby complicating solubility trends based only on ionic charges. As a result, some compounds composed of 1+ and 2+ cations and 1− anions have limited solubility in water, such as the halides of silver(I), mercury(I), lead(II), and many transition metal hydroxides.

**CONNECTION** The role of ion–dipole forces in the dissolution of ionic compounds was discussed in Chapter 6.

## CONCEPT TEST

Write the chemical formulas of the following compounds and then use the solubility guidelines in Tables 8.4 and 8.5 to classify them as either soluble or only slightly soluble in water: (a) calcium chloride; (b) potassium phosphate; (c) iron(III) hydroxide; (d) sodium sulfide; (e) zinc oxide.

When two aqueous solutions combine, a solid product called a **precipitate** may form. Such reactions are called **precipitation reactions.** The solubility guidelines in Tables 8.4 and 8.5 allow us to predict whether a precipitate will form

**TABLE 8.5** Solubility Matrix for Common Ionic Compounds in Water

| | | CATION | | | |
|---|---|---|---|---|---|
| | | Group 1 or $NH_4^+$ | $Ca^{2+}$ or $Sr^{2+}$ or $Ba^{2+}$ | $Cu^+$ or $Ag^+$ | $Hg_2^{2+}$ or $Pb^{2+}$ |
| ANION | $NO_3^-$ or $ClO_4^-$ or $CH_3COO^-$ | Soluble | Soluble | Soluble | Soluble |
| | $Cl^-$ or $Br^-$ or $I^-$ | Soluble | Soluble | Limited solubility | Limited solubility |
| | $OH^-$ or $S^{2-}$ | Soluble | Soluble | Limited solubility | Limited solubility |
| | $SO_4^{2-}$ | Soluble | Limited solubility | Soluble | Limited solubility |
| | $F^-$ or $O^{2-}$ or $CO_3^{2-}$ or $PO_4^{3-}$ or $AsO_4^{3-}$ or $CrO_4^{2-}$ or $C_2O_4^{2-}$ | Soluble | Limited solubility | Limited solubility | Limited solubility |

when two aqueous solutions containing soluble ionic compounds are mixed. For example, does a precipitate form when a solution of sodium iodide (NaI) is mixed with a solution of lead(II) nitrate, $Pb(NO_3)_2$? Put another way, will either cation in the two solutions combine with the anion from the other solution to form a compound that has limited solubility? To answer this question, we need to do the following:

**precipitate** a solid product formed from a reaction in solution.

**precipitation reaction** a reaction in which soluble reactants form a product that has limited solubility.

1. Identify the ions dissolved in the two solutions after the two ionic compounds dissolve and separate into their component ions. Here, one solution contains $Na^+$ ions and $I^-$ ions, whereas the other contains $Pb^{2+}$ ions and $NO_3^-$ ions.
2. Determine whether any collisions between cations and anions create a product of limited solubility. For example, the $Na^+$ ions can collide with $I^-$ ions or with $NO_3^-$ ions. The $Pb^{2+}$ ions also can collide with $I^-$ ions or with $NO_3^-$ ions. NaI and $Pb(NO_3)_2$ are the reactants, so we need to evaluate $NaNO_3$ and $PbI_2$ as products. According to the guidelines in Tables 8.4 and 8.5, sodium nitrate is soluble because it consists of a group 1 cation (all of which are soluble) and because it is a nitrate compound (all of which are soluble, too). However, $PbI_2$ has limited solubility in water because it is a lead(II) halide, so it precipitates when the two solutions mix (**Figure 8.16**).

Let's write balanced molecular and ionic equations for this precipitation reaction. We start with single formula units of reactants and products:

$$Pb(NO_3)_2(aq) + NaI(aq) \rightarrow PbI_2(s) + NaNO_3(aq)$$

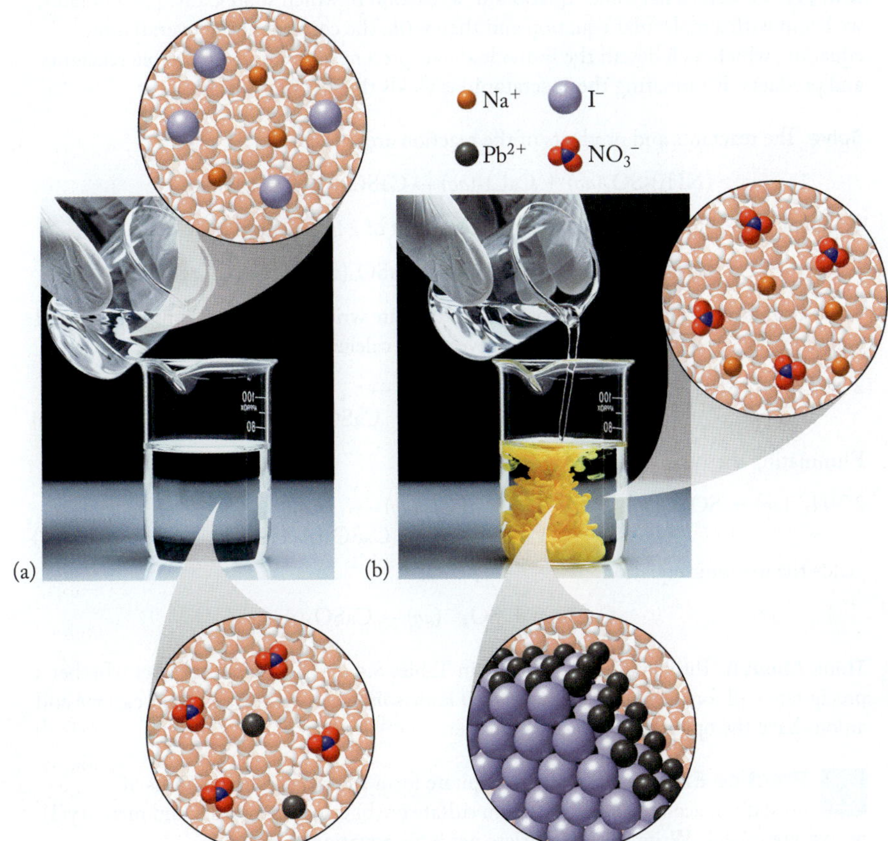


**FIGURE 8.16** (a) One beaker contains 0.1 *M* $Pb(NO_3)_2$, and the other contains 0.2 *M* NaI. Both solutions are colorless. (b) As the NaI solution is poured into the $Pb(NO_3)_2$ solution, a yellow precipitate of $PbI_2$ forms. Sodium ions and nitrate ions remain in solution.

To produce a balanced molecular equation, we need to balance the number of iodide and nitrate ions on both sides of the reaction arrow. We do this by placing coefficients of 2 in front of NaI and $NaNO_3$:

$$Pb(NO_3)_2(aq) + 2\ NaI(aq) \rightarrow PbI_2(s) + 2\ NaNO_3(aq)$$

Next, we write an overall ionic equation in which we separate all the soluble salts into their component ions:

$$Pb^{2+}(aq) + 2\ NO_3^-(aq) + 2\ Na^+(aq) + 2\ I^-(aq) \rightarrow PbI_2(s) + 2\ Na^+(aq) + 2\ NO_3^-(aq)$$

Note how the $Na^+$ and $NO_3^-$ ions are in solution on both sides of the reaction arrow. They are spectator ions in the reaction, so we delete them to write the net ionic equation:

$$Pb^{2+}(aq) + 2\ I^-(aq) \rightarrow PbI_2(s)$$

**SAMPLE EXERCISE 8.11** Writing the Net Ionic Equation for a Precipitation Reaction **LO4, LO7**

A precipitate forms when aqueous solutions of ammonium sulfate and calcium chloride are mixed. Write the net ionic equation for the reaction.

**Collect and Organize** We know that mixing solutions of ammonium sulfate, $(NH_4)_2SO_4$, and calcium chloride, $CaCl_2$, yields a product with limited solubility. We need to identify it to be able to write the net ionic equation. According to the solubility guidelines in Tables 8.4 and 8.5, ammonium salts and most sulfate salts are soluble in water; however, calcium sulfate is only slightly soluble.

**Analyze** To write a net ionic equation for a reaction in which solid $CaSO_4$ precipitates, we begin with a molecular equation and then write the corresponding overall ionic equation, which includes all the individual ions present in solutions of soluble reactants and products. Eliminating the spectator ions yields the net ionic equation.

**Solve** The reactants and products of the reaction are

$$(NH_4)_2SO_4(aq) + CaCl_2(aq) \rightarrow CaSO_4(s) + NH_4Cl(aq)$$

To balance the equation, we just need a coefficient of 2 in front of $NH_4Cl$:

$$(NH_4)_2SO_4(aq) + CaCl_2(aq) \rightarrow CaSO_4(s) + 2\ NH_4Cl(aq)$$

The reactants are soluble ionic compounds, so they are written as separate ions in an overall ionic equation, as is soluble $NH_4Cl$. However, solid calcium sulfate is written as $CaSO_4(s)$:

$$2\ NH_4^+(aq) + SO_4^{2-}(aq) + Ca^{2+}(aq) + 2\ Cl^-(aq) \rightarrow CaSO_4(s) + 2\ NH_4^+(aq) + 2\ Cl^-(aq)$$

Eliminating the $NH_4^+$ and $Cl^-$ spectator ions

$$\cancel{2\ NH_4^+(aq)} + SO_4^{2-}(aq) + Ca^{2+}(aq) + \cancel{2\ Cl^-(aq)} \rightarrow CaSO_4(s) + \cancel{2\ NH_4^+(aq)} + \cancel{2\ Cl^-(aq)}$$

yields the net ionic equation

$$Ca^{2+}(aq) + SO_4^{2-}(aq) \rightarrow CaSO_4(s)$$

**Think About It** The solubility guidelines in Tables 8.4 and 8.5 help us predict whether a precipitate will form when solutions of two ionic solutes are mixed and their cations and anions have the opportunity to swap partners.

**Practice Exercise** Does a precipitate form when aqueous solutions of (a) sodium acetate and ammonium sulfate or (b) calcium chloride and mercury(I) nitrate are mixed? Write the appropriate net ionic equation(s).

## CONCEPT TEST

Lead(II) dichromate ($PbCr_2O_7$; the dichromate ion is $Cr_2O_7^{2-}$) is the solid pigment called school bus yellow once used to paint lines on highways. Propose a synthesis of school bus yellow that uses a precipitation reaction.

Precipitation reactions may be used to determine the concentrations of ions in solution. For example, NaCl (in the form of rock salt) is used to melt ice and snow on roads during winter. Some of this NaCl may find its way into nearby drinking water supplies. To determine whether sources of drinking water have been contaminated with road salt, an analyst can determine the concentration of chloride ion by reacting a sample of the water with a solution containing more than enough silver nitrate, $AgNO_3$, to precipitate the $Cl^-$ ions as AgCl:

$$NaCl(aq) + AgNO_3(aq) \rightarrow AgCl(s) + NaNO_3(aq)$$

The precipitate can be filtered, dried, and weighed. The mass of the AgCl precipitate is divided by its molar mass to calculate the number of moles of AgCl on the filter. This value is the same as the number of moles of $Cl^-$ ions in the original water sample. Precipitation reactions used in analytical chemistry sometimes include indicators that provide a visual signal (such as a color change) when an excess of the precipitation agent is present to ensure that all the ions of interest have precipitated. That way, the analyst can be sure that $Cl^-$ ions are the limiting reactant in the precipitation of AgCl.

**CONNECTION** The concept of a limiting reactant was introduced in Chapter 7.

### SAMPLE EXERCISE 8.12 Calculating Solute Concentration from Mass of a Precipitate — LO7

To determine the concentration of chloride ion in a 100.0 mL sample of groundwater, a chemist adds a large enough volume of a solution of $AgNO_3$ to precipitate all the $Cl^-$ ions as AgCl. The mass of the resulting precipitate is 71.7 mg. What is the $Cl^-$ concentration in the sample in milligrams per liter?

**Collect and Organize** We are given the sample volume and mass of AgCl formed. Our task is to determine the chloride concentration in milligrams of $Cl^-$ per liter of groundwater.

**Analyze** We need to calculate the mass of the $Cl^-$ ions in the weighed mass of AgCl and then divide that mass of $Cl^-$ ions by the volume of the water sample to calculate a concentration in milligrams per liter. The molar mass of chlorine is about 1/3 that of silver and about 1/4 the formula mass of AgCl. So, 71.7 mg of AgCl should contain a little less than 20 mg of $Cl^-$ ions, which were originally dissolved in a 100 mL sample. This mass-to-volume ratio is a little less than 200 mg/1000 mL, or 200 mg/L.

**Solve** The molar masses of Cl and Ag are 35.45 g/mol and 107.87 g/mol, respectively, so the molar mass of AgCl is 143.32 g/mol. The mass of $Cl^-$ ions in 71.7 mg of AgCl is

$$71.7\ \cancel{\text{mg AgCl}} \times \frac{1\ \cancel{\text{g}}}{1000\ \cancel{\text{mg}}} \times \frac{1\ \cancel{\text{mol AgCl}}}{143.32\ \cancel{\text{g AgCl}}} \times \frac{1\ \cancel{\text{mol Cl}^-}}{1\ \cancel{\text{mol AgCl}}} \times \frac{35.45\ \text{g Cl}^-}{1\ \cancel{\text{mol Cl}^-}} \times \frac{1000\ \text{mg}}{1\ \text{g}}$$

$$= 17.7\ \text{mg Cl}^-$$

This 17.7 mg mass of $Cl^-$ ions was originally in a 100.0 mL sample. Converting this mass-to-volume ratio to milligrams per liter yields

$$\frac{17.7\ \text{mg Cl}^-}{100.0\ \cancel{\text{mL}}} \times \frac{1000\ \cancel{\text{mL}}}{1\ \text{L}} = 177\ \text{mg Cl}^-/\text{L}$$

**Think About It** Did you notice that the first step in the first calculation converted milligrams to grams (of AgCl) and that the last step converted grams to milligrams (of $Cl^-$ ions)? Numerically, the two steps involved dividing by 1000 and then multiplying by 1000. Had we recognized that the initial and final quantities were both expressed in milligrams, we could have eliminated those two steps and carried the *milli-* prefix through all the steps from first to last, thereby simplifying the calculation without affecting the result.

**Practice Exercise** The concentration of $SO_4^{2-}$ ions in a 50.0 mL sample of coastal seawater is determined by adding a solution of $BaCl_2$ to the sample and precipitating the $SO_4^{2-}$ as $BaSO_4$. The precipitate is removed from the sample by filtration and then dried and weighed. If the mass of $BaSO_4$ recovered from the sample is 0.311 g, what is the sulfate concentration of the sample expressed in millimoles per liter?

## SAMPLE EXERCISE 8.13 Predicting the Mass of a Precipitate LO7

Barium sulfate is used to enhance X-ray imaging of the upper and lower gastrointestinal (GI) tracts. In upper GI imaging, patients drink a suspension of solid $BaSO_4$ in water, which has the consistency of a dense, chalky milkshake and, even with flavoring added, tastes awful. The compound is not toxic because of its low solubility. To make pure $BaSO_4$, a precipitation reaction is employed: aqueous solutions of soluble barium nitrate and sodium sulfate are mixed, and solid $BaSO_4$ is separated from the reaction mixture by filtration. If a 1 L volumetric flask is used to prepare a 1.55 *M* $Ba(NO_3)_2$ solution, how many grams of $BaSO_4$ ($\mathcal{M}$ = 233.40 g/mol) will be produced if the entire solution is reacted with excess $Na_2SO_4$?

**Collect and Organize** We know the soluble reactants [$Ba(NO_3)_2$ and $Na_2SO_4$] and the insoluble product ($BaSO_4$) of a precipitation reaction and the volume and molar concentration of a solution containing the limiting reactant. We are asked to calculate the mass of $BaSO_4$ that will be produced.

**Analyze** This stoichiometry problem is like the ones we solved in Chapter 7, except that this one is based on quantities of dissolved reactants [$Ba(NO_3)_2$ and $Na_2SO_4$] and a solid product ($BaSO_4$):

$$Ba(NO_3)_2(aq) + Na_2SO_4(aq) \rightarrow BaSO_4(s) + 2\ NaNO_3(aq)$$

One liter of a 1.55 *M* barium nitrate solution has about 1.5 moles of $Ba(NO_3)_2$, which translates into about 1.5 moles of $Ba^{2+}$ ions and 1.5 moles of $BaSO_4$. Multiplying 1.5 by a molar mass of about 233 g/mol should give us an answer of about 350 g.

**Solve** We first multiply the volume and concentration of the $Ba(NO_3)_2$ solution to obtain an equivalent number of moles of $Ba(NO_3)_2$. Two more conversion factors are required to convert moles of $Ba(NO_3)_2$ to moles of $BaSO_4$ and finally into a mass of $BaSO_4$:

$$1.00\ \cancel{L} \times \frac{1.55\ \cancel{mol\ Ba(NO_3)_2}}{\cancel{L}} \times \frac{1\ \cancel{mol\ BaSO_4}}{1\ \cancel{mol\ Ba(NO_3)_2}} \times \frac{233.40\ g\ BaSO_4}{1\ \cancel{mol\ BaSO_4}} = 362\ g\ BaSO_4$$

**Think About It** As in many calculations based on reactions in solution, using "mol/L" instead of "*M*" helps us properly convert units and obtain a correct answer, which is close to our estimate.

**Practice Exercise** Vermilion, also known as Chinese red, is a rare and expensive solid natural pigment used to print the artist's signature on works of art. The pigment is mercury(II) sulfide, and it is insoluble in water. What mass of vermilion can be produced when 50.00 mL of 0.0150 *M* mercury(II) nitrate is mixed with a solution containing excess sodium sulfide?

**saturated solution** a solution that contains the maximum concentration of a solute possible at a given temperature.

**unsaturated solution** a solution that contains less than the maximum quantity of solute predicted to be soluble in a given volume of solution at a given temperature.

**supersaturated solution** a solution that contains more than the maximum quantity of solute predicted to be soluble in a given volume of solution at a given temperature.

**FIGURE 8.17** (a) A seed crystal is added to a supersaturated solution of sodium acetate. (b) The seed crystal becomes a site for rapid growth of sodium acetate crystals. (c) Crystal growth continues until the solution is no longer supersaturated. Instead, it is merely saturated with sodium acetate.

**FIGURE 8.18** Gypsum ($CaSO_4 \cdot 2\,H_2O$) crystals in this cave in Naica, Mexico, grew to enormous size over half a million years as gypsum slowly precipitated.

## Saturated Solutions and Supersaturation

The aqueous solubilities of many ionic and molecular solids increase with increasing temperature. For example, more table sugar (or sucrose, $C_{12}H_{22}O_{11}$) dissolves in hot water than in cold—a phenomenon familiar to anyone who has tried to sweeten a glass of iced tea. When as much sugar as possible has dissolved in hot water, the resulting solution is said to be **saturated** with sugar because it can't hold any more. If we continue to heat the solution without adding more sugar, it becomes **unsaturated** because more sugar would dissolve in it if more were available. If the solution is then cooled to a temperature at which less sugar is soluble, but no sugar precipitates, then the solution is said to be **supersaturated**.

Supersaturated solutions can remain so for a long time if left undisturbed, but eventually mechanical shock or the addition of a *seed crystal* (a small crystal of the solute that serves as a site for crystallization) causes the solute to rapidly precipitate (**Figure 8.17**). More dramatic examples of crystal growth from supersaturated solutions were discovered in 2000 by miners in Naica, Mexico (**Figure 8.18**). They happened upon a cave containing enormous crystals of the mineral gypsum (calcium sulfate), some nearly 12 m long and 2 m across. Apparently the crystals formed from $Ca^{2+}$ and $SO_4^{2-}$ ions dissolved in groundwater that had been geothermally heated to nearly 60°C. As this water cooled in the cave to about 54°C, it became slightly supersaturated, and crystals of gypsum began to grow very slowly for more than half a million years. When the supersaturated gypsum solution was pumped out of the cave as part of current mining operations, the crystals were revealed.

Solutions also may become supersaturated when solvent evaporates. For example, groundwater leaching through porous limestone becomes saturated with $CaCO_3$. If the water then seeps into a cave where it simultaneously drips from the ceiling and evaporates, $CaCO_3$ precipitates, forming deposits (**Figure 8.19**) called stalactites when attached to the ceiling and called stalagmites when built up from the floor.

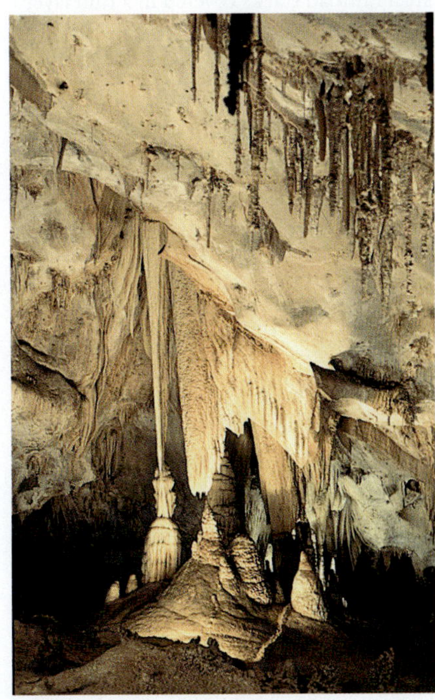

**FIGURE 8.19** In Carlsbad Caverns in New Mexico, stalagmites of limestone grow up from the cavern floor, and stalactites grow downward from the ceiling.

# 8.7 Oxidation–Reduction Reactions: Electron Transfer

Oxygen is one of the most abundant elements on Earth; it makes up 50% of the crust and 89% of the water covering Earth's surface by mass. Its molecules also make up 21% of all those in Earth's atmosphere. Oxygen is an essential element for nearly all creatures, even aquatic organisms, who must obtain it from the small concentrations that dissolve in water. Oxygen is at the core of many chemical reactions that release energy and sustain life, too: *oxidation–reduction* reactions, or *redox* reactions.

Although this chapter focuses on reactions in solution, we begin our coverage of redox chemistry with oxygen reactions that occur in the air because many are both familiar and simple to describe with balanced chemical equations. Oxygen combines with nonmetals such as carbon, sulfur, and nitrogen to produce volatile oxides. It also combines with metals and semimetals, producing solid oxides. All these reactions are examples of redox reactions. Indeed, *oxidation* was once defined as a reaction that increased the oxygen content of a substance. According to this definition, combustion of a hydrocarbon such as methane, which we discussed in Chapter 7, involves oxidation because the products, $CO_2$ and $H_2O$, contain more oxygen than $CH_4$:

$$CH_4(g) + 2\ O_2(g) \rightarrow CO_2(g) + 2\ H_2O(\ell) \qquad (7.9)$$

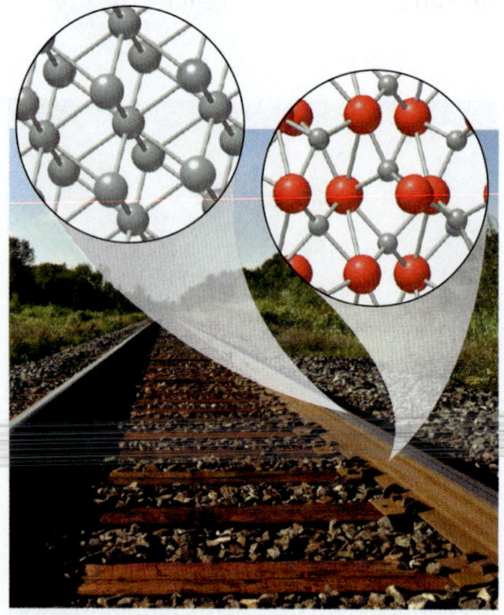

Similarly, the corrosion of metals is oxidation because the products of corrosion, such as rusted iron (**Figure 8.20**), have a higher oxygen content than that of the original metal (which had none):

$$4\ Fe(s) + 3\ O_2(g) \rightarrow 2\ Fe_2O_3(s) \qquad (8.24)$$

**FIGURE 8.20** Corrosion (oxidation) leaves steel railroad tracks covered in a thin layer of orange $Fe_2O_3$. Any rust that forms on the top of the tracks is worn away by the wheels of trains passing over them, leaving shiny metallic surfaces of mostly iron that is available to be oxidized further.

*Reduction* reactions were originally defined as reactions in which the oxygen content of a substance was reduced. A classic example is the reduction of iron ores such as $Fe_2O_3$ to metallic Fe:

$$Fe_2O_3(s) + 3\ CO(g) \rightarrow 2\ Fe(s) + 3\ CO_2(g) \qquad (8.25)$$

The reaction in Equation 8.25 describes the reduction of $Fe_2O_3$, but it is more involved than that. Oxidations and reductions don't happen in isolation; they happen together. This means that as $Fe_2O_3$ is reduced, an element in the other reactant must be oxidized. That other element is the carbon in CO, which is oxidized from CO to a compound with an even higher oxygen content, $CO_2$.

**CONCEPT TEST**

The following equation describes the conversion of one iron mineral called magnetite ($Fe_3O_4$) into another called hematite ($Fe_2O_3$):

$$4\ Fe_3O_4(s) + O_2(g) \rightarrow 6\ Fe_2O_3(s)$$

Is iron oxidized or reduced in the reaction?

## Oxidation Numbers

Today we use *redox* to describe other reactions in addition to those that increase or decrease the oxygen content of a substance. An element is said to undergo **oxidation** when it *loses* electrons and **reduction** when it *gains* them.[2] These definitions

[2]The mnemonic *OIL RIG* may be helpful for remembering these expanded definitions: "oxidation is loss; reduction is gain."

are more general than the older ones, but they are consistent with them, too. To see how, let's reconsider the process of rust ($Fe_2O_3$) forming on an iron object.

$$4\ Fe(s) + 3\ O_2(g) \rightarrow 2\ Fe_2O_3(s) \tag{8.24}$$

Fe is oxidized according to the old definition—its oxygen content increases when it forms $Fe_2O_3$—but it does so by *losing* electrons. The Fe atoms in elemental iron have their normal free-atom complement of 26 electrons each, but the $Fe^{3+}$ ions in $Fe_2O_3$ have only 23 electrons. This loss of three electrons means that iron is oxidized under the new definition, too. At the same time, oxygen is reduced from its elemental state (in which the O atoms have their normal complement of eight electrons) to oxide ($O^{2-}$) ions with 10 electrons each. The simultaneous loss and gain of electrons is the key feature of all redox reactions: if one element loses electrons, another must gain them so that the losses and gains are in balance. In Equation 8.24, this balance is achieved because four Fe atoms each lose three electrons, for a total loss of 12 electrons, while six O atoms gain two electrons each, for a total gain of 12. Thus, Equation 8.24 is balanced because the numbers of Fe and O atoms on both sides of the reaction arrow are the same *and* because the number of electrons gained and lost is the same.

**oxidation** a chemical change in which an element loses electrons; the oxidation number of the element increases.

**reduction** a chemical change in which an element gains electrons; the oxidation number of the element decreases.

**oxidation number (O.N.)** or **oxidation state** a numerical value (which can be positive, negative, or zero) based on the number of electrons that an atom gains or loses when it forms an ion or that it shares when it forms a covalent bond with an atom of another element.

To help track electron losses and gains in redox reactions, chemists follow changes in the oxidation states of the atoms in reactants and products. The **oxidation state**, or **oxidation number (O.N.)**, of an atom in a molecule or ion is a measure of the number of electrons it has compared to the number it would have if it were a free atom. The O.N. values of monatomic ions are the same as their electrical charges. In molecules and polyatomic ions, atoms have O.N. values that are related to the number of covalent bonds they form, but their O.N. values also depend on the elements they are bonded to and the O.N. values of those elements. The rules for assigning O.N. values are summarized in **Table 8.6**.

In assigning O.N. values to the atoms in molecules, a hierarchy exists that is based on electronegativity. Fluorine is always in a −1 oxidation state in molecular compounds because it is the most electronegative element and it forms one bond per atom. Other halogen atoms also have O.N. values of −1 in their compounds, but not when they are bonded to the more electronegative fluorine or oxygen atoms. As illustrated in **Table 8.7**, chlorine and the higher *Z* halogens can be in oxidation states up to +7 depending on the number of O atoms they are bonded to.

Oxygen nearly always has an O.N. of −2 in its compounds because it is the next-most-electronegative element after fluorine and tends to form two bonds per atom. However, oxygen's O.N. is +2 in $OF_2$ because of rule 3 in Table 8.6

**TABLE 8.6 Rules for Assigning Oxidation Numbers (O.N.) to Atoms**

1. O.N. = 0 for atoms in pure elements.
2. O.N. = the charge of single-atom ions.
3. O.N. of fluorine atoms = −1 in all compounds.
4. O.N. of oxygen atoms = −2 in *nearly* all compounds. Exceptions: when oxygen atoms are bonded to fluorine atoms (for example, $OF_2$), where its O.N. = +2, or when O is bonded to another O atom in a peroxide (for example, $H_2O_2$, where its O.N. = −1) or in a superoxide (for example, $KO_2$, where its O.N. = $-\frac{1}{2}$).
5. O.N. of hydrogen atoms = +1 in *nearly* all compounds. The exception is hydrogen in metal hydrides (for example, LiH), where H is the more electronegative element and its O.N. = −1.
6. O.N. values of the atoms in a neutral compound sum to zero.
7. O.N. values of the atoms in a polyatomic ion sum to the charge on the ion.

**TABLE 8.7 Some Oxidation States of Chlorine**

| Compound | Chlorine O.N. |
|---|---|
| HCl | −1 |
| HClO | +1 |
| $HClO_2$ | +3 |
| $ClO_2$ | +4 |
| $HClO_3$ | +5 |
| $HClO_4$ | +7 |

(fluorine's O.N. is always −1 in its compounds) and rule 6 (the O.N. values of the elements in a compound must sum to zero). The oxidation state of O is also different when it bonds to itself in compounds called peroxides. These compounds have –O–O– functional groups in which oxygen's O.N. equals −1.

**SAMPLE EXERCISE 8.14** Determining Oxidation Numbers **LO8**

What are the oxidation states of sulfur in (a) $S_8$, (b) $SO_2$, (c) $Na_2S$, and (d) $CaSO_4$?

**Collect and Organize** We are asked to assign oxidation numbers to sulfur in $S_8$ and in three sulfur compounds. The rules for assigning oxidation numbers are given in Table 8.6.

**Analyze**

a. $S_8$ is a molecular form of the element sulfur.
b. In $SO_2$, sulfur is bonded to oxygen, which is a more electronegative element and has an O.N. of −2 (rule 4).
c. $Na_2S$ is a binary ionic compound; the charge of the sulfide ion is 2−.
d. Sulfur in $CaSO_4$ is part of the sulfate ion, which means that its oxidation number added to those of the four O atoms must add up to the charge of the ion, which is 2− (rule 7).

**Solve**

a. The oxidation state of sulfur in elemental sulfur ($S_8$) is 0 (rule 1).
b. The sum of the oxidation numbers for S and the two O atoms in this neutral molecule must be zero (rule 6). Assigning an O.N. of −2 to each oxygen atom, and letting the oxidation state of sulfur be $x$, we can set up the following equation:

$$x + 2(-2) = 0$$
$$x = +4$$

So, the oxidation state of sulfur in $SO_2$ is +4.
c. The oxidation state of the atom in a monatomic ion is the same as the charge on the ion (rule 2). Therefore, the oxidation state of sulfur in the sulfide ion is −2.
d. The charge on the $SO_4^{2-}$ ion is 2−. Each O atom has an O.N. of −2. Letting $x$ be the O.N. of sulfur and applying rule 7:

$$x + 4(-2) = -2$$
$$x = +6$$

So, the oxidation state of S in $CaSO_4$ is +6.

**Think About It** We found that sulfur, like chlorine in the compounds in Table 8.7, can exist in oxidation states that range from negative to positive values. The most positive values occur when single sulfur atoms are bonded to multiple oxygen atoms: the more O atoms, the higher the oxidation state.

 **Practice Exercise** Determine the oxidation states of nitrogen in (a) $NO_2$; (b) $N_2O$; (c) $HNO_3$.

## Considering Changes in Oxidation Number in Redox Reactions

Redox reactions involve the transfer of electrons between atoms, and these transfers result in changes in the oxidation states of the atoms. We know from rules 1 and 2 in Table 8.6 that when an atom acquires an electron, forming a 1− anion, its oxidation number is reduced from 0 to −1 (ΔO.N. = −1). But when an atom loses an electron and forms a 1+ cation, its oxidation number increases from 0 to +1 (ΔO.N. = +1). We can generalize this pattern to describe the gain or loss of any number (*n*) of electrons:

| | | |
|---|---|---|
| Oxidation | Lose *n* electrons | ΔO.N. = +*n* |
| Reduction | Gain *n* electrons | ΔO.N. = −*n* |

This general pattern can help us track the flow of electrons from the elements losing them (that is, being oxidized) to the elements gaining them (that is, being reduced) in a reaction. The tracking system involves calculating the changes in oxidation number of the elements that are oxidized and reduced.

To see how the system works, let's examine the redox properties of two neighbors in the periodic table: copper ($Z = 29$) and zinc ($Z = 30$). Elemental Zn has considerable electron-donating power, and $Cu^{2+}$ ions tend to accept electrons. These complementary properties are illustrated in **Figure 8.21**.

When a strip of Zn metal is placed in a solution of $CuSO_4$—that is, a solution of $Cu^{2+}$ ions and $SO_4^{2-}$ ions—Zn atoms spontaneously donate electrons to $Cu^{2+}$ ions, forming $Zn^{2+}$ ions and Cu atoms. We can see this redox reaction happen: the shiny zinc surface turns dark brown as a textured layer of copper metal accumulates on it, and the distinctive blue color of $Cu^{2+}(aq)$ ions fades as they acquire electrons and become atoms of copper metal. The chemical equation describing this reaction is

$$Zn(s) + CuSO_4(aq) \rightarrow ZnSO_4(aq) + Cu(s)$$

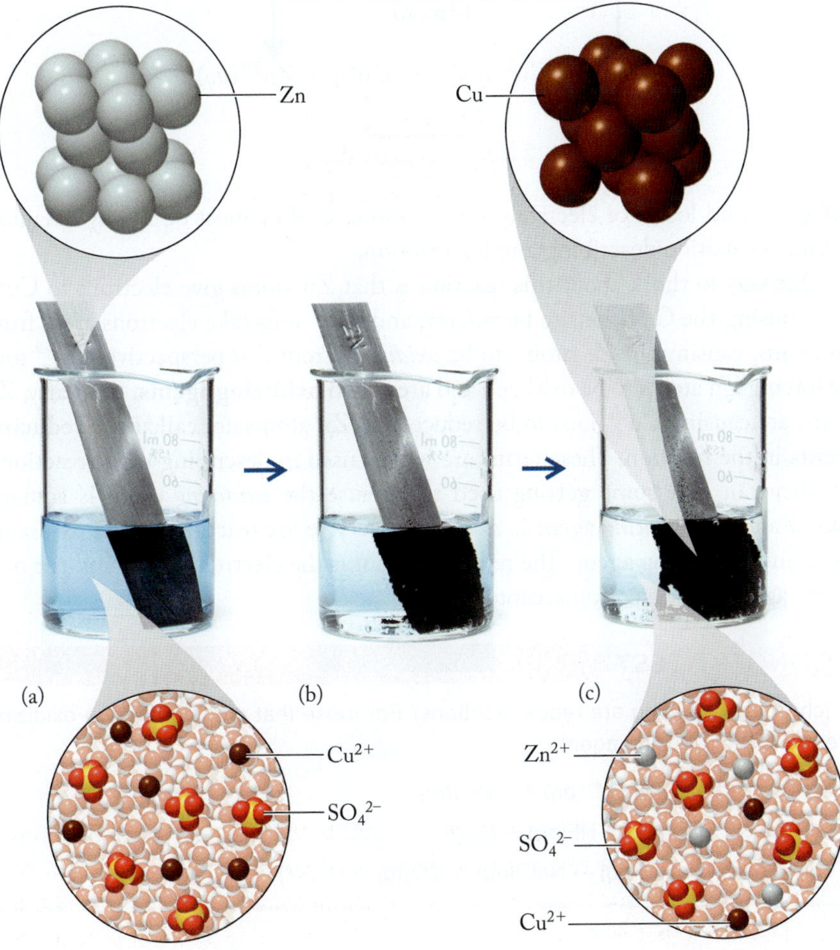


**FIGURE 8.21** Redox reaction of Zn metal with $Cu^{2+}$ ions. (a) A strip of Zn is placed in a solution of $CuSO_4$. (b) Over time, the surface of the Zn strip darkens as it is coated by a layer of Cu atoms. (c) Eventually the blue color of $Cu^{2+}$ ions in solution fades as Cu atoms form. They accumulate on the Zn strip and then fall to the bottom of the beaker.

and the overall ionic equation including all ions in solution is

$$Zn(s) + Cu^{2+}(aq) + SO_4{}^{2-}(aq) \rightarrow Zn^{2+}(aq) + SO_4{}^{2-}(aq) + Cu(s)$$

Eliminating $SO_4{}^{2-}$ ions, which are unchanged (spectator ions) in the reaction, produces the following net ionic equation:

$$Zn(s) + Cu^{2+}(aq) \rightarrow Zn^{2+}(aq) + Cu(s) \qquad (8.26)$$

Now let's assign oxidation numbers to the Cu and Zn atoms and ions in Equation 8.26. According to rules 1 and 2 in Table 8.6, the oxidation states of the atoms in Zn and Cu metal are 0, and the oxidation states of the atoms in both their ions are +2. Therefore, the changes in oxidation number (ΔO.N.) are +2 for Zn atoms and −2 for $Cu^{2+}$ ions:

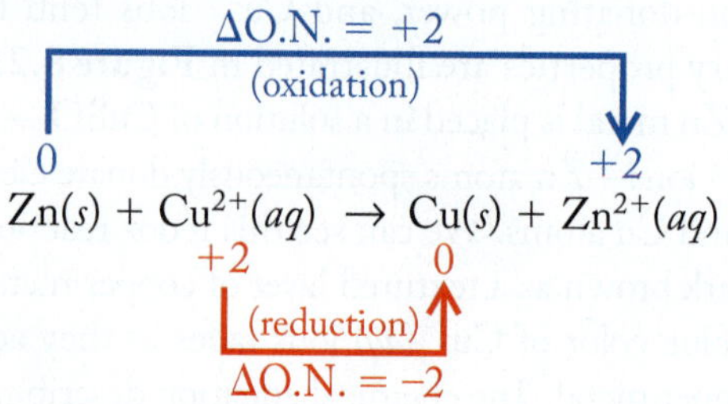

ΔO.N. = +2
(oxidation)
0 → +2

$$Zn(s) + Cu^{2+}(aq) \rightarrow Cu(s) + Zn^{2+}(aq)$$

+2 → 0
(reduction)
ΔO.N. = −2

These changes mean that each Zn atom *loses* two electrons and each $Cu^{2+}$ ion *gains* two electrons.

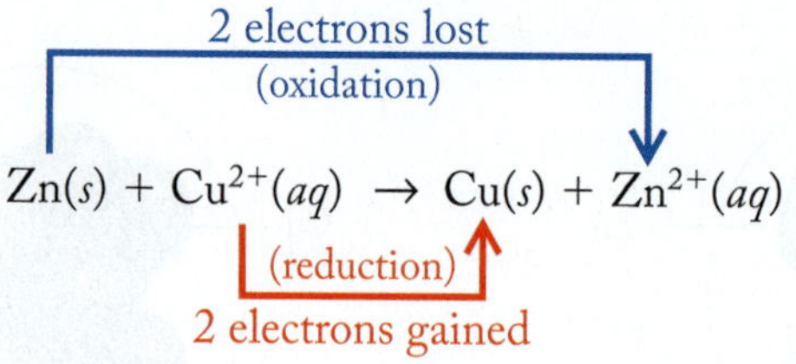

2 electrons lost
(oxidation)

$$Zn(s) + Cu^{2+}(aq) \rightarrow Cu(s) + Zn^{2+}(aq)$$

(reduction)
2 electrons gained

The gains and losses of electrons are the same, as they must be in any balanced chemical equation describing a redox reaction.

One way to think about this reaction is that Zn atoms give electrons to $Cu^{2+}$ ions, causing the $Cu^{2+}$ ions to be *reduced*, and $Cu^{2+}$ ions take electrons away from Zn atoms, causing the Zn atoms to be *oxidized*. From this perspective, $Cu^{2+}$ ions are forcing Zn atoms to be oxidized and are called **oxidizing agents**. Similarly, Zn atoms are causing $Cu^{2+}$ ions to be reduced, so Zn atoms are called the **reducing agents** in the reaction. These terms are widely used in describing redox reactions, but they can take some getting used to because the *oxidizing* agent is actually *reduced* and the *reducing* agent is *oxidized*. Every redox reaction has an oxidizing agent and a reducing agent. The reducing agent is the electron donor, and the oxidizing agent is the electron acceptor.

**oxidizing agent** a reactant that accepts electrons from another in a redox reaction, thereby oxidizing the other reactant; the oxidizing agent is reduced in the reaction.

**reducing agent** a reactant that donates electrons to another in a redox reaction, thereby reducing the other reactant; the reducing agent is oxidized in the reaction.

## CONCEPT **TEST**

Which of the following are redox reactions? For those that are, identify the oxidizing agent and the reducing agent:

a. $Sn^{2+}(aq) + Br_2(\ell) \rightarrow Sn^{4+}(aq) + 2\,Br^-(aq)$

b. $2\,F_2(g) + 2\,H_2O(\ell) \rightarrow 4\,HF(aq) + O_2(g)$

c. $NaHCO_3(aq) + HCl(aq) \rightarrow NaCl(aq) + CO_2(g) + H_2O(\ell)$

**SAMPLE EXERCISE 8.15** Identifying Oxidizing and Reducing Agents and Their Changes in Oxidation State **LO8**

Energy released by the reaction of hydrazine and dinitrogen tetroxide:

$$2\,N_2H_4(\ell) + N_2O_4(g) \rightarrow 3\,N_2(g) + 4\,H_2O(g)$$

is used to orient and maneuver spacecraft and to propel rockets into space (**Figure 8.22**). Identify the elements that are oxidized and reduced and their changes in oxidation state. Identify the oxidizing agent and the reducing agent.

**FIGURE 8.22** The NASA *Juno* spacecraft went into orbit around Jupiter in July 2016 by firing its main engine, which is powered by the redox reaction between hydrazine ($N_2H_4$) and dinitrogen tetroxide ($N_2O_4$).

**Collect, Organize, and Analyze** The reactants are hydrazine, $N_2H_4$, and dinitrogen tetroxide, $N_2O_4$, and the products are $N_2$ and $H_2O$. According to the rules in Table 8.6, the oxidation state of nitrogen in molecules of $N_2$ (a pure element) is 0, and the O.N. values of O and H in the three reactant and product compounds are −2 and +1, respectively. Because the O.N. values of O and H do not change during the reaction, the only element oxidized and reduced in the reaction is nitrogen. Rule 6 in Table 8.6 can be used to calculate the oxidation states of nitrogen in $N_2H_4$ and $N_2O_4$.

**Solve** The oxidation numbers of N and H in $N_2H_4$ must sum to 0 (rule 6). The O.N. value of hydrogen is +1, so the O.N. of nitrogen is

$$2x + 4(+1) = 0$$
$$x = -2$$

Applying rule 6 to $N_2O_4$, we find that the O.N. of nitrogen in the molecule is

$$2x + 4(-2) = 0$$
$$x = +4$$

The O.N. of each N atom in $N_2H_4$ changes from −2 to 0 for a gain of +2. Each molecule of $N_2H_4$ has two of these N atoms, and two $N_2H_4$ molecules are in the chemical equation, so the overall ΔO.N. is $2 \times 2 \times (+2) = +8$. The O.N. of each N atom in $N_2O_4$ changes from +4 to 0, for a ΔO.N. of −4. Two N atoms are in the chemical equation's single $N_2O_4$ molecule, so the overall ΔO.N. is $2 \times (-4) = -8$.

These overall changes in oxidation state are shown in the chemical equation below:

$$\Delta\text{O.N.} = 2 \times 2 \times (+2) = +8 \text{ (oxidation): } -2 \rightarrow 0$$

$$2\,N_2H_4(g) + N_2O_4(g) \rightarrow 3\,N_2(g) + 4\,H_2O(g)$$

$$\Delta\text{O.N.} = 2 \times (-4) = -8 \text{ (reduction): } +4 \rightarrow 0$$

In the reaction, $N_2H_4$ is the reducing agent because the N atoms in it are oxidized, and $N_2O_4$ is the oxidizing agent because the N atoms in it are reduced.

**Think About It** We tend to identify the specific atoms that are oxidized or reduced, whereas whole molecules or ions are identified as the oxidizing agents or reducing agents. In the equation in this exercise, nitrogen is oxidized but $N_2H_4$ is the reducing agent.

**Practice Exercise** In the reaction between $O_2(g)$ and $SO_2(g)$ to make $SO_3(g)$, what is the element oxidized, the element reduced, the change in oxidation number of each element, the oxidizing agent, and the reducing agent?

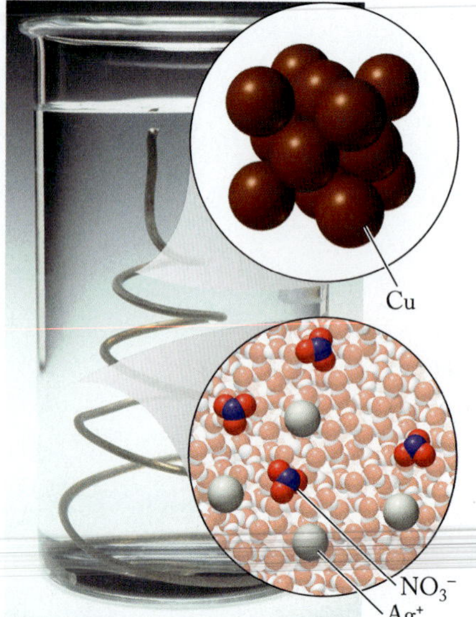


(a)

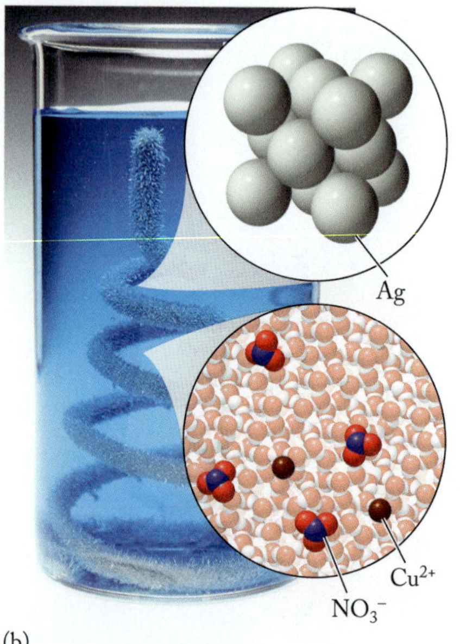


(b)

**FIGURE 8.23** (a) When a Cu wire is immersed in a solution of $AgNO_3$, Cu metal oxidizes to $Cu^{2+}$ ions as $Ag^+$ ions are reduced to Ag metal. (b) A day later, the solution has the blue color of $Cu^{2+}$ ions in solution, and the wire is coated with Ag metal.

## Balancing Redox Reactions by Using Half-Reactions

The losses and gains of electrons must balance in any redox reaction. To apply this principle, let's examine the reaction that occurs when a piece of copper wire [elemental copper, Cu(*s*)] is placed in a colorless solution of silver nitrate (containing $Ag^+$ and $NO_3^-$ ions). The solution gradually takes on the blue color of a solution of $Cu^{2+}$ ions, and branchlike structures of solid Ag form on the copper wire (**Figure 8.23**). In this chemical reaction, Cu metal is oxidized to $Cu^{2+}$ ions and $Ag^+$ ions are reduced to Ag metal. If we ignore the nitrate ions, which are only spectator ions in the reaction, the reactants and products that should appear in the net ionic equation are

$$Ag^+(aq) + Cu(s) \rightarrow Ag(s) + Cu^{2+}(aq)$$

At first glance, the equation for the redox reaction between $Ag^+(aq)$ and Cu(*s*) may seem balanced—and in a material sense it is: the number of silver and copper atoms or ions is the same on both sides. However, the charges are not balanced: the sum of the charges is 1+ for the reactants but 2+ for the products. The reason for this imbalance is that one electron is involved in the reduction reaction, but two are involved in the oxidation reaction. Equal numbers of electrons must be lost and gained, so to balance equations describing redox reactions such as this one, we must develop a method that accounts for electron transfer.

The change in oxidation number reflects the change in the number of electrons associated with an atom or ion. This change in the number of electrons happens because electrons are transferred from one atom or ion to another during a redox reaction. By tracking changes in oxidation numbers, we can gain additional insight into the transfer of electrons as the redox reaction takes place.

Think of the reaction between Cu(*s*) and $AgNO_3(aq)$ as consisting of one oxidation and one reduction. Each of these reactions represents *half* of the overall reaction; hence, each is called a **half-reaction**. We balance the equation by following these five steps:

**half-reaction** one of the halves of an oxidation–reduction reaction; one half-reaction is the oxidation component, and the other is the reduction component.

**STEPWISE ANIMATION**

Balancing Redox Reactions

1. *Write one equation for the oxidation half-reaction and a separate equation for the reduction half-reaction*:

   Oxidation: $Cu(s) \rightarrow Cu^{2+}(aq)$
   Reduction: $Ag^{+}(aq) \rightarrow Ag(s)$

2. *Balance the number of atoms of each element in each half-reaction*. This example requires no changes at this point.
3. *Balance the charge in each half-reaction* by adding electrons to the appropriate side. Always *add* electrons; never subtract them. We add two electrons to the product side of the Cu half-reaction:

   Oxidation: $Cu(s) \rightarrow Cu^{2+}(aq) + 2\ e^{-}$

   and one electron to the reactant side of the Ag half-reaction:

   Reduction: $1\ e^{-} + Ag^{+}(aq) \rightarrow Ag(s)$

   This gives us a total charge of zero on both sides of both half-reactions. Note that electrons are added on the product side of the oxidation equation (oxidation is the loss of electrons) and on the reactant side of the reduction equation (reduction is the gain of electrons). This is always the case.
4. *Multiply each half-reaction by the appropriate whole number* to make the number of electrons lost in the oxidation half-reaction equal the number of electrons gained in the reduction half-reaction:

   Oxidation: $Cu(s) \rightarrow Cu^{2+}(aq) + 2\ e^{-}$
   Reduction: $2 \times [1\ e^{-} + Ag^{+}(aq) \rightarrow Ag(s)]$

   As a result, we have the following half-reactions:

   Oxidation: $Cu(s) \rightarrow Cu^{2+}(aq) + 2\ e^{-}$
   Reduction: $2\ e^{-} + 2\ Ag^{+}(aq) \rightarrow 2\ Ag(s)$

5. *Add the two half-reactions to generate the equation representing the overall redox reaction*:

   Oxidation: $Cu(s) \rightarrow Cu^{2+}(aq) + \cancel{2\ e^{-}}$
   Reduction: $\cancel{2\ e^{-}} + 2\ Ag^{+}(aq) \rightarrow 2\ Ag(s)$

   Overall equation: $2\ Ag^{+}(aq) + Cu(s) \rightarrow 2\ Ag(s) + Cu^{2+}(aq)$

The equation is now balanced in terms of mass and charge. Note that the overall equation is actually a net ionic equation because only the species involved in the redox reaction are shown. The nitrate ions from the silver nitrate solution (Figure 8.23) are spectator ions and are not shown in the final net ionic equation or in any intermediate equations we used to generate it.

We will review the methods for balancing chemical equations as described here when we discuss applications of oxidation–reduction reactions in Chapter 17.

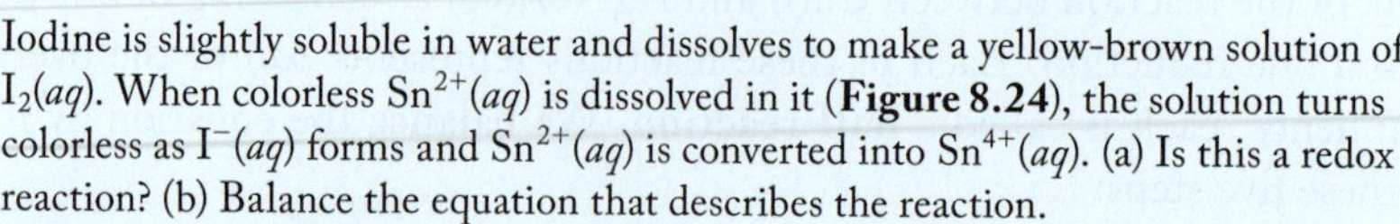

**SAMPLE EXERCISE 8.16** Writing Balanced Equations for Redox Reactions Using Half-Reactions **LO9**

Iodine is slightly soluble in water and dissolves to make a yellow-brown solution of $I_2(aq)$. When colorless $Sn^{2+}(aq)$ is dissolved in it (**Figure 8.24**), the solution turns colorless as $I^-(aq)$ forms and $Sn^{2+}(aq)$ is converted into $Sn^{4+}(aq)$. (a) Is this a redox reaction? (b) Balance the equation that describes the reaction.

**Collect and Organize** The reactants in this process are $I_2(aq)$ and $Sn^{2+}(aq)$; the products are $I^-(aq)$ and $Sn^{4+}(aq)$.

**Analyze** The oxidation number of tin changes from +2 in $Sn^{2+}(aq)$ to +4 in $Sn^{4+}(aq)$. The oxidation number of iodine changes from 0 in $I_2(aq)$ to −1 in $I^-(aq)$. We write tin and iodine in separate equations when we write the respective half-reactions.

**Solve**

a. Because the oxidation numbers change, this is a redox reaction.

b. Linking the reactants and products with a reaction arrow produces this starting point in writing a balanced equation:

$$Sn^{2+}(aq) + I_2(aq) \rightarrow Sn^{4+}(aq) + I^-(aq)$$

We balance it with our five-step procedure:

1. Separate the half-reactions.

Oxidation: $Sn^{2+}(aq) \rightarrow Sn^{4+}(aq)$
Reduction: $I_2(aq) \rightarrow I^-(aq)$

2. Balance the number of atoms of each element.

Oxidation: $Sn^{2+}(aq) \rightarrow Sn^{4+}(aq)$
Reduction: $I_2(aq) \rightarrow 2\,I^-(aq)$

3. Balance the charges by adding electrons.

Oxidation: $Sn^{2+}(aq) \rightarrow Sn^{4+}(aq) + 2\,e^-$
Reduction: $I_2(aq) + 2\,e^- \rightarrow 2\,I^-(aq)$

4. Balance the numbers of electrons. This step is not needed here because the numbers of electrons are the same in the two half-reactions.

5. Add the two half-reactions.

Oxidation: $Sn^{2+}(aq) \rightarrow Sn^{4+}(aq) + \cancel{2\,e^-}$
Reduction $I_2(aq) + \cancel{2\,e^-} \rightarrow 2\,I^-(aq)$

Overall equation: $Sn^{2+}(aq) + I_2(aq) \rightarrow Sn^{4+}(aq) + 2\,I^-(aq)$

Check the overall equation for mass balance: 1 Sn + 2 I = 1 Sn + 2 I. Check the charge balance: reactant side, 2+; product side, (4+) + 2(1−) = 2+.

**Think About It** Writing and combining the two half-reactions produced a balanced net ionic equation describing the redox reaction and showed that two moles of electrons are transferred to iodine for each mole of $Sn^{2+}$ ions oxidized.

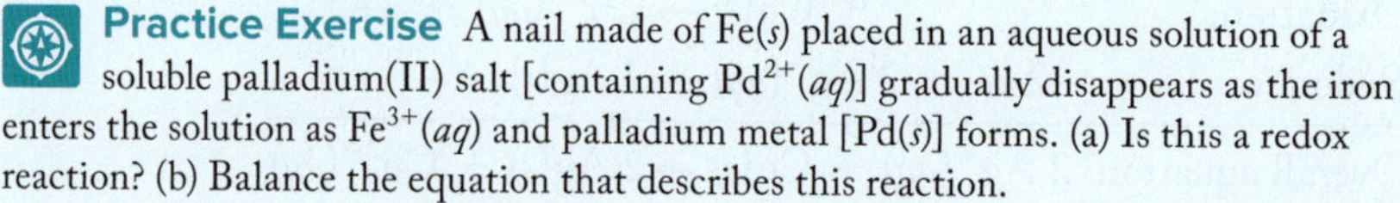

**Practice Exercise** A nail made of Fe(*s*) placed in an aqueous solution of a soluble palladium(II) salt [containing $Pd^{2+}(aq)$] gradually disappears as the iron enters the solution as $Fe^{3+}(aq)$ and palladium metal [Pd(*s*)] forms. (a) Is this a redox reaction? (b) Balance the equation that describes this reaction.

(a)

(b)

(c)

**FIGURE 8.24** The beaker in this photo contains an aqueous solution of iodine. (a) When drops of a solution of $SnCl_2$ are added to the beaker, the yellow-brown color of the iodine solution begins to fade. (b) $Sn^{2+}$ ions reduce $I_2$ to colorless $I^-$ ions. (c) The color disappears when enough $Sn^{2+}$ has been added to reduce all the $I_2$.

## The Activity Series of Metals

If we replaced the $AgNO_3(aq)$ solution in Figure 8.23a with a $Zn(NO_3)_2$ solution, no reaction between it and Cu metal would be observed. By immersing copper wire in solutions of other metal cations, we would find, for example, that Cu

**TABLE 8.8** An Activity Series for Metals in Aqueous Solution

| Metal | Oxidation Reaction |
|---|---|
| Lithium | $Li(s) \rightarrow Li^{+}(aq) + e^{-}$ |
| Potassium | $K(s) \rightarrow K^{+}(aq) + e^{-}$ |
| Barium | $Ba(s) \rightarrow Ba^{2+}(aq) + 2\ e^{-}$ |
| Calcium | $Ca(s) \rightarrow Ca^{2+}(aq) + 2\ e^{-}$ |
| Sodium | $Na(s) \rightarrow Na^{+}(aq) + e^{-}$ |
| Magnesium | $Mg(s) \rightarrow Mg^{2+}(aq) + 2\ e^{-}$ |
| Aluminum | $Al(s) \rightarrow Al^{3+}(aq) + 3\ e^{-}$ |
| Manganese | $Mn(s) \rightarrow Mn^{2+}(aq) + 2\ e^{-}$ |
| Zinc | $Zn(s) \rightarrow Zn^{2+}(aq) + 2\ e^{-}$ |
| Chromium | $Cr(s) \rightarrow Cr^{3+}(aq) + 3\ e^{-}$ |
| Iron | $Fe(s) \rightarrow Fe^{2+}(aq) + 2\ e^{-}$ |
| Cobalt | $Co(s) \rightarrow Co^{2+}(aq) + 2\ e^{-}$ |
| Nickel | $Ni(s) \rightarrow Ni^{2+}(aq) + 2\ e^{-}$ |
| Tin | $Sn(s) \rightarrow Sn^{2+}(aq) + 2\ e^{-}$ |
| Lead | $Pb(s) \rightarrow Pb^{2+}(aq) + 2\ e^{-}$ |
| Hydrogen | $H_2(g) \rightarrow 2\ H^{+}(aq) + 2\ e^{-}$ |
| Copper | $Cu(s) \rightarrow Cu^{2+}(aq) + 2\ e^{-}$ |
| Silver | $Ag(s) \rightarrow Ag^{+}(aq) + e^{-}$ |
| Mercury | $Hg(\ell) \rightarrow Hg^{2+}(aq) + 2\ e^{-}$ |
| Platinum | $Pt(s) \rightarrow Pt^{2+}(aq) + 2\ e^{-}$ |
| Gold | $Au(s) \rightarrow Au^{3+}(aq) + 3\ e^{-}$ |

metal reduces $Ag^{+}$ and $Au^{3+}$ ions to Ag and Au metal, but it does not reduce $Zn^{2+}$, $Ni^{2+}$, and $Al^{3+}$ ions to their metals. Similarly, we could explore the ability of zinc metal to reduce these same cations and find that all except $Al^{3+}$ (and $Zn^{2+}$) are reduced by Zn metal.

Why do particular metals reduce some metal cations but not others? A detailed explanation awaits in Chapter 17, but for now we can use the **activity series** in **Table 8.8** to rank the strengths of metals as reducing agents: from the highly reactive group 1 and group 2 metals at the top of the table to the largely inert metals, such as platinum and gold, at the bottom. A metal at the top of the table can reduce the cation of a metal below it, as the top metal's oxidation half-reaction proceeds as written in the table, whereas the lower metal's cations are reduced to metal atoms as its half-reaction proceeds in reverse—as a reduction half-reaction.

**activity series** a high-to-low ranking of metals on the basis of their strengths as reducing agents.

**SAMPLE EXERCISE 8.17** Using the Activity Series **LO9**

An aluminum wire is dipped into a solution containing $Na^{+}$, $Sn^{2+}$, and $Cu^{2+}$ ions. Which ions will be present in the solution after the reactions are complete?

**Collect and Organize** We are asked to predict which metal ions in solution will be reduced by aluminum. We use the activity series in Table 8.8 to guide our answer.

**Analyze** The activity series is arranged based on metals' decreasing strengths as reducing agents. Alternatively, the ease of oxidation of the corresponding metals at the top of the series can reduce the cations of the metals below them.

**Solve** Aluminum is above copper and tin but below sodium in Table 8.8. Therefore, Al reduces both $Cu^{2+}$ and $Sn^{2+}$ ions, removing them from the solution, but $Na^+$ ions remain. Aluminum is oxidized as $Cu^{2+}$ and $Sn^{2+}$ ions are reduced, so $Al^{3+}$ ions are also present in the solution when the reaction is over.

**Think About It** The arrangement of metals in the activity series makes sense from what we know about the chemical reactivity of groups of metals: the highly reactive group 1 and group 2 metals are clustered at the top and many of those that are stable as free metals such as gold, silver, and copper, which are used to make coins, are near the bottom.

**Practice Exercise** Which substance listed in Table 8.8 is most easily oxidized? Which substance is most easily reduced?

Why is hydrogen gas listed in Table 8.8? What does its position in the activity series tell us? Table 8.8 allows us to predict which metals will react with strong aqueous acids such as HCl. Any metal below $H_2$ in Table 8.8 will not react with $H^+$ and hence will be resistant to strong acids. One reason that coins and jewelry made from silver, gold, and platinum last is that they do not react with most acids.

**SAMPLE EXERCISE 8.18** Writing Net Ionic Equations for Redox Reactions **LO4, LO9**

Write balanced molecular and net ionic equations for the reaction between zinc metal and aqueous hydrochloric acid. Identify the oxidation and reduction half-reactions.

**Collect and Organize** The reactants in this process are Zn(*s*) and HCl(*aq*). The products are determined by the activity series.

**Analyze** Zinc metal is listed above $H^+$ in Table 8.8; thus, it is oxidized by $H^+$ from hydrochloric acid, producing $Zn^{2+}$ and hydrogen gas. After writing a molecular equation for the reaction, we will convert it to a net ionic equation. After assigning oxidation numbers, we can identify the half-reactions.

**Solve** The known reactants and products are

$$Zn(s) + HCl(aq) \rightarrow ZnCl_2(aq) + H_2(g)$$

Adding a coefficient of 2 in front of HCl balances all the elements:

$$Zn(s) + 2\,HCl(aq) \rightarrow ZnCl_2(aq) + H_2(g)$$

The overall ionic equation is

$$Zn(s) + 2\,H^+(aq) + 2\,Cl^-(aq) \rightarrow Zn^{2+}(aq) + 2\,Cl^-(aq) + H_2(g)$$

The net ionic equation becomes

$$Zn(s) + 2\,H^+(aq) + \cancel{2\,Cl^-(aq)} \rightarrow Zn^{2+}(aq) + \cancel{2\,Cl^-(aq)} + H_2(g)$$

$$Zn(s) + 2\,H^+(aq) \rightarrow Zn^{2+}(aq) + H_2(g)$$

The oxidation numbers of Zn, $Zn^{2+}$, and $H^+$ are 0, +2, and +1, respectively. The oxidation number of $H_2$ gas is zero because it is the elemental form of hydrogen. Zinc is oxidized and hydrogen ion is reduced. The two half-reactions are

Oxidation: $Zn(s) \rightarrow Zn^{2+}(aq) + 2\,e^-$

Reduction: $2\,H^+(aq) + 2\,e^- \rightarrow H_2(g)$

**Think About It** All metals above $H^+(aq)$ in the activity series will reduce a strong acid such as HCl(*aq*) or HBr(*aq*) to produce $H_2(g)$.

**Practice Exercise** Predict the products of the reaction between aluminum metal and silver nitrate. Write balanced molecular and net ionic equations for the reaction. Identify the oxidation and reduction half-reactions.

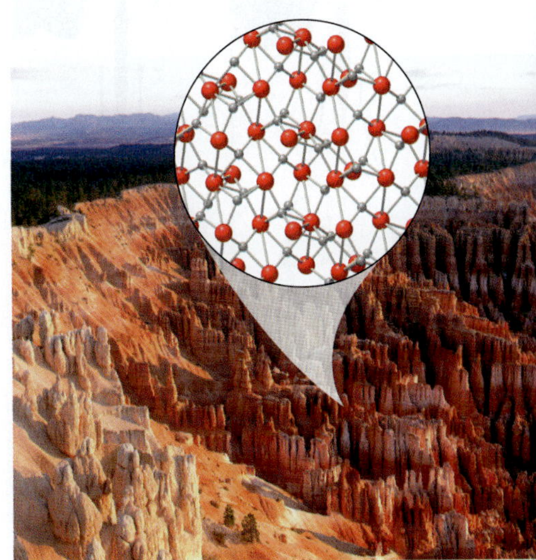

**FIGURE 8.25** Red-orange rock formations called hoodoos in Bryce Canyon, Utah, owe their color to high concentrations of iron(III) oxide.

## Redox Reactions in Acidic or Basic Solutions

Redox processes play a major role in determining the character of Earth's rocks and soils. For example, the orange-red spires called hoodoos in Bryce Canyon National Park in Utah owe their distinctive color to an iron(III) oxide mineral called hematite (**Figure 8.25**). These rocks are also relatively fragile and susceptible to physical and chemical weathering, which has resulted in the fascinating shapes of the hoodoos for which Bryce is famous. Iron(III) oxide also is the form of iron known as rust, the crumbly, red-brown solid that forms on iron objects exposed to water and oxygen.

Soil color is influenced by mineral content and by the availability of molecular oxygen. Wetlands are areas either saturated with water year-round or flooded during much of the year. Even though wetland soils have considerable iron content, they do not have the distinctive red-orange color of iron(III) compounds because the waterlogged soils have so little $O_2$ that the dominant oxidation state of Fe is +2 instead of +3. In fact, soil color is one way that environmental scientists can tell whether an area is a wetland (**Figure 8.26**).

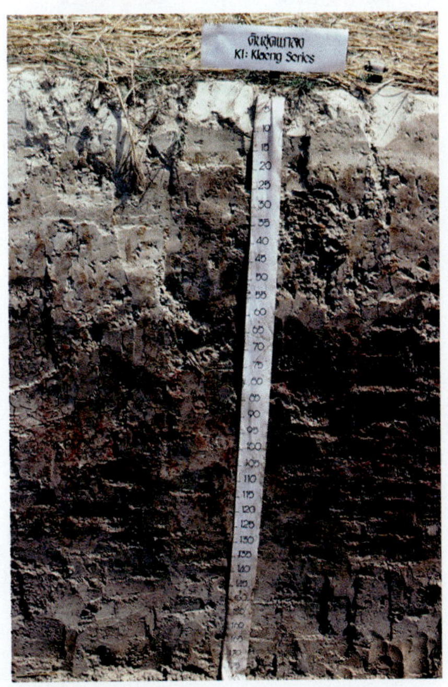


**FIGURE 8.26** Wetland soil has a blue-gray color because of the presence of iron(II) compounds, but red-orange mottling indicates the presence of iron(III) oxides formed as a result of $O_2$ permeation through channels sometimes made by plant roots.

Many redox reactions in nature take place either in acidic solutions or in basic solutions, and the $H^+(aq)$, $OH^-(aq)$, or even the water may play a role in the reaction. Let's look at what we must do to balance and write net ionic equations for such reactions.

A method to determine the concentration of $Fe^{2+}(aq)$ in an acidic solution involves its oxidation to $Fe^{3+}(aq)$ by the intensely purple permanganate ion, $MnO_4^-(aq)$ (**Figure 8.27**). In the reaction, the permanganate ion is reduced to $Mn^{2+}(aq)$:

| $Fe^{2+}(aq)$ | + | $MnO_4^-(aq)$ | → | $Fe^{3+}(aq)$ | + | $Mn^{2+}(aq)$ |
|---|---|---|---|---|---|---|
| appears colorless | | deep purple | | yellow-orange | | pale pink |

This reaction occurs in an acidic solution, which suggests that it involves protons, and it is clearly redox because $Fe^{2+}(aq)$ is oxidized to $Fe^{3+}(aq)$. The permanganate

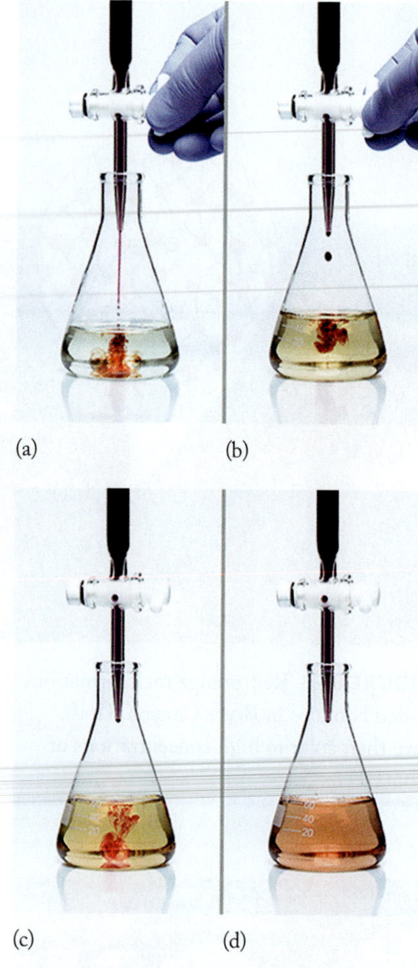

**FIGURE 8.27** In this colorful redox titration, (a) a dark purple standard solution of permanganate, $MnO_4^-$, ions is slowly added to an acidic sample containing nearly colorless $Fe^{2+}$ ions. (b) As the titration reaction proceeds, $Fe^{2+}$ ions are oxidized to $Fe^{3+}$ ions while $MnO_4^-$ ions are reduced to $Mn^{2+}$ ions. (c) Increasing concentrations of pale yellow $Fe^{3+}$ ions and faintly pink $Mn^{2+}$ ions combine to give the sample a pale amber color. (d) When all the $Fe^{2+}$ ions have been consumed, the next drop of unreacted purple $MnO_4^-$ ions gives the sample a peach color that signals the endpoint of the titration.

ion must be the oxidizing agent, so it must be reduced. Indeed, the O.N. of the Mn atom in $MnO_4^-$ is +7, and it is reduced to +2 in $Mn^{2+}$.

A chemical equation describing the reaction will be balanced when (a) the number of atoms of each element on both sides of the reaction is the same and (b) the total charges on each side of the reaction arrow are the same. We proceed in the same way as for the previous examples in neutral solutions, but we must add a few new steps to address the role of the protons in the reaction.

1. Separate iron and manganese in their respective half-reactions.

   Oxidation: $Fe^{2+}(aq) \rightarrow Fe^{3+}(aq)$
   Reduction: $MnO_4^-(aq) \rightarrow Mn^{2+}(aq)$

2. Balance the number of atoms of each element in three steps. First, we balance all the elements *except hydrogen and oxygen*:

   Oxidation: $Fe^{2+}(aq) \rightarrow Fe^{3+}(aq)$ (atoms are balanced)
   Reduction: $MnO_4^-(aq) \rightarrow Mn^{2+}(aq)$ (O atoms are not balanced)

   Next we balance oxygen by adding $H_2O(\ell)$.

   Oxidation: $Fe^{2+}(aq) \rightarrow Fe^{3+}(aq)$ (atoms are balanced)
   Reduction: $MnO_4^-(aq) \rightarrow Mn^{2+}(aq) + 4\,H_2O(\ell)$
   (O atoms are balanced, H atoms are not balanced)

   In acidic solutions, we balance hydrogen by adding $H^+(aq)$:

   Oxidation: $Fe^{2+}(aq) \rightarrow Fe^{3+}(aq)$ (atoms are balanced)
   Reduction: $8\,H^+(aq) + MnO_4^-(aq) \rightarrow Mn^{2+}(aq) + 4\,H_2O(\ell)$
   (atoms are balanced)

3. Balance charge by adding electrons.

   Oxidation: $Fe^{2+}(aq) \rightarrow Fe^{3+}(aq) + 1\,e^-$
   Reduction: $5\,e^- + 8\,H^+(aq) + MnO_4^-(aq) \rightarrow Mn^{2+}(aq) + 4\,H_2O(\ell)$

   Note that the number of electrons added to the reduction half-reaction also accounts for the oxidation state of Mn changing from +7 to +2.

4. Balance numbers of electrons lost and gained.

   Oxidation: $5 \times [Fe^{2+}(aq) \rightarrow Fe^{3+}(aq) + 1\,e^-]$
   Reduction: $1 \times [5\,e^- + 8\,H^+(aq) + MnO_4^-(aq) \rightarrow Mn^{2+}(aq) + 4\,H_2O(\ell)]$

5. Add the two equations.

   Oxidation: $5\,Fe^{2+}(aq) \rightarrow 5\,Fe^{3+}(aq) + \cancel{5\,e^-}$
   Reduction: $\cancel{5\,e^-} + 8\,H^+(aq) + MnO_4^-(aq) \rightarrow Mn^{2+}(aq) + 4\,H_2O(\ell)]$

   $$8\,H^+(aq) + MnO_4^-(aq) + 5\,Fe^{2+}(aq) \rightarrow 5\,Fe^{3+}(aq) + Mn^{2+}(aq) + 4\,H_2O(\ell)$$

Always check the atom and charge balance of the final net ionic equation. The reactant side contains 8 H, 1 Mn, 4 O, and 5 Fe atoms, and the product side contains 8 H, 1 Mn, 4 O, and 5 Fe atoms, so the atom count for each element is balanced. On the reactant side, the charge is (8+) + (1−) + (10+) = 17+, and on the product side it is (15+) + (2+) = 17+, so charge is balanced. This method works for writing the net ionic equation for any redox reaction taking place in an aqueous acid.

Now that we've detailed the step-by-step procedure for balancing a redox reaction equation in acidic solution, we turn to Sample Exercise 8.19 to examine how to balance a redox reaction equation in basic solution.

### SAMPLE EXERCISE 8.19 Writing Balanced Net Ionic Equations for Redox Reactions in Basic Solutions — LO9

Some wastewater treatment processes involve removing the permanganate ion ($MnO_4^-$) by reacting it with oxalate, $C_2O_4^{2-}$, a substance commonly found in plants. Write the balanced net ionic equation for this reaction in a basic solution, starting with these reactants and products:

$$MnO_4^-(aq) + C_2O_4^{2-}(aq) \rightarrow MnO_2(s) + CO_3^{2-}(aq)$$

**Collect and Organize** We are given some of the reactants and products in a redox reaction that occurs in a basic solution. Our task is to write a balanced net ionic equation describing the reaction.

**Analyze** We use oxidation numbers to determine which element is oxidized and which is reduced. The O.N. of manganese changes from +7 in $MnO_4^-(aq)$ to +4 in $MnO_2(s)$, so manganese is reduced. The O.N. of carbon changes from +3 in $C_2O_4^{2-}(aq)$ to +4 in $CO_3^{2-}(aq)$, so carbon is oxidized.

**Solve**

1. Separate the elements undergoing oxidation and reduction.

   Oxidation: $MnO_4^-(aq) \rightarrow MnO_2(s)$
   Reduction: $C_2O_4^{2-}(aq) \rightarrow CO_3^{2-}(aq)$

2. a. Balance the masses in three steps to account for any role played by the aqueous base. First, we balance all the elements *except hydrogen and oxygen*:

   Oxidation: $MnO_4^-(aq) \rightarrow MnO_2(s)$ (atoms are balanced except for O)
   Reduction: $C_2O_4^{2-}(aq) \rightarrow 2\,CO_3^{2-}(aq)$ (atoms are balanced except for O)

   b. Balance O by adding water as needed.

   Oxidation: $MnO_4^-(aq) \rightarrow MnO_2(s) + 2\,H_2O(\ell)$
   (atoms are balanced except for H)
   Reduction: $2\,H_2O(\ell) + C_2O_4^{2-}(aq) \rightarrow 2\,CO_3^{2-}(aq)$
   (atoms are balanced except for H)

   c. Balance H by adding $H^+(aq)$ as needed. Adding acidic protons may seem strange when the reaction occurs in basic solution, but we will remove $H^+(aq)$ from the equation in a later step while still keeping it balanced.

   Oxidation: $4\,H^+(aq) + MnO_4^-(aq) \rightarrow MnO_2(s) + 2\,H_2O(\ell)$
   Reduction: $2\,H_2O(\ell) + C_2O_4^{2-}(aq) \rightarrow 2\,CO_3^{2-}(aq) + 4\,H^+(aq)$

3. Balance the charges by adding electrons.

   Oxidation: $3\,e^- + 4\,H^+(aq) + MnO_4^-(aq) \rightarrow MnO_2(s) + 2\,H_2O(\ell)$
   Reduction: $2\,H_2O(\ell) + C_2O_4^{2-}(aq) \rightarrow 2\,CO_3^{2-}(aq) + 4\,H^+(aq) + 2\,e^-$

4. Balance the numbers of electrons lost and gained.

   Oxidation: $2 \times [3\,e^- + 4\,H^+(aq) + MnO_4^-(aq) \rightarrow MnO_2(s) + 2\,H_2O(\ell)]$
   Reduction: $3 \times [2\,H_2O(\ell) + C_2O_4^{2-}(aq) \rightarrow 2\,CO_3^{2-}(aq) + 4\,H^+(aq) + 2\,e^-]$

5. Add the two equations.

   Oxidation: $6\,e^- + 8\,H^+(aq) + 2\,MnO_4^-(aq) \rightarrow 2\,MnO_2(s) + 4\,H_2O(\ell)]$
   Reduction: $6\,H_2O(\ell) + 3\,C_2O_4^{2-}(aq) \rightarrow 6\,CO_3^{2-}(aq) + 12\,H^+(aq) + 6\,e^-]$

   This gives us the balanced equation:

$$2\,MnO_4^-(aq) + 2\,H_2O(\ell) + 3\,C_2O_4^{2-}(aq) \rightarrow 2\,MnO_2(s) + 6\,CO_3^{2-}(aq) + 4\,H^+(aq)$$

6. Now we can address the issue that we have acidic protons present in our balanced equation for a process that occurs in basic solution. To resolve this, we can add $OH^-(aq)$ ions to both sides of our balanced equation:

$$4\,OH^-(aq) + 2\,MnO_4^-(aq) + 2\,H_2O(\ell) + 3\,C_2O_4^{2-}(aq) \rightarrow$$
$$2\,MnO_2(s) + 6\,CO_3^{2-}(aq) + 4\,H^+(aq) + 4\,OH^-(aq)$$

which can be simplified to:

$$4\,OH^-(aq) + 2\,MnO_4^-(aq) + \cancel{2\,H_2O(\ell)} + 3\,C_2O_4^{2-}(aq) \rightarrow 2\,MnO_2(s) + 6\,CO_3^{2-}(aq) + \cancel{2}\,\cancel{4}\,H_2O(\ell)$$

or

$$4\,OH^-(aq) + 2\,MnO_4^-(aq) + 3\,C_2O_4^{2-}(aq) \rightarrow 2\,MnO_2(s) + 6\,CO_3^{2-}(aq) + 2\,H_2O(\ell)$$

**Think About It** This reaction occurs in basic solution, so the balanced chemical equation should not include acidic protons, but it should include hydroxide ions.

**Practice Exercise** Wetland soil is blue-gray as a result of $Fe(OH)_2(s)$, whereas well-aerated soils are often orange-red because of the presence of $Fe(OH)_3(s)$, as we saw in Figure 8.26. Write the balanced equation for the reaction of $O_2(g)$ with $Fe(OH)_2(s)$ in soil that produces $Fe(OH)_3(s)$ under basic conditions.

We can observe the characteristic colors associated with soils rich in iron(II) or iron(III) compounds in the laboratory. **Figure 8.28(a)** shows a solution of iron(II) ammonium sulfate, $(NH_4)_2Fe(SO_4)_2$, as a solution of NaOH is added to it. Adding base causes $Fe(OH)_2$ to precipitate as a blue-gray solid, similar in color to the wetland soil shown in Figure 8.26. When this mixture is filtered, the iron(II) hydroxide residue immediately starts to darken (**Figure 8.28b**). After about 20 minutes of exposure to oxygen in the air, the precipitate has turned the orange-red color of iron(III) hydroxide (**Figure 8.28c**). The moist iron(II) hydroxide has been oxidized to iron(III) hydroxide—in much the way that the iron(II) compounds in a wetland soil are oxidized when exposed to oxygen.

Reactions in aqueous solutions are an integral part of our daily lives. On a large scale, in oceans, rivers, and rain, they shape our physical world. Many reactions that produce the substances that are part of modern life—as varied as paint pigments and drugs—are run in water, and many analytical procedures rely on

(a)

(b)

(c)

**FIGURE 8.28** (a) A solution of iron(II) ammonium sulfate forms a blue-green precipitate of iron(II) hydroxide when NaOH is added. (b) When the precipitate is isolated on a paper filter, it rapidly turns a darker color. (c) After being exposed to air for about 20 minutes, the precipitate has changed to the red-orange color of iron(III) oxide.

reactions in water to determine the content of aqueous solutions that we drink, swim in, and use in countless consumer products such as car batteries and shampoos. On a small scale, reactions in the water within the cells of our bodies and in all living organisms make possible the chemical processes essential to life. On Earth, water may exist without life, but no life exists without water.

## SAMPLE EXERCISE 8.20 Integrating Concepts: Selecting an Antacid

A 5.0 mL dose of the liquid antacid in **Figure 8.29(a)** contains 200 mg of $Mg(OH)_2$ and 200 mg of $Al(OH)_3$. A tablet of the antacid in **Figure 8.29(b)** contains 500 mg of $CaCO_3$. According to their labels, the bottle of liquid antacid holds 12 fluid ounces (355 mL), and the other bottle contains 150 tablets. Assume that the masses of the ingredients are known to two significant figures. In many drugstores, the two bottles sell for about the same price.

(a)

(b)

**FIGURE 8.29** Over-the-counter antacids: (a) this liquid contains suspended $Mg(OH)_2$ and $Al(OH)_3$; (b) Tums tablets are composed mostly of $CaCO_3$.

a. If the maximum dissolved concentration (solubility) of $Mg(OH)_2$ is $1.2 \times 10^{-4}\ M$, what percentage of the $Mg(OH)_2$ in the liquid antacid is dissolved and not solid particles suspended in the viscous liquid?
b. Write net ionic equations for the neutralization of excess stomach acid, $HCl(aq)$, by the three active ingredients in the two antacids. Assume that all three are solids.
c. How many moles of HCl could be neutralized by one 5.0 mL dose of the liquid antacid?
d. How many moles of HCl could be neutralized by one tablet of the solid antacid containing 500 mg of the active ingredient?
e. Which of the two antacids is a better buy in terms of acid-neutralizing capacity per bottle?

**Collect and Organize** We know the quantities and identities of the active ingredients in the doses of two antacids: 200 mg of $Mg(OH)_2$ and 200 mg of $Al(OH)_3$ in 5.0 mL of the liquid, and 500 mg of $CaCO_3$ per tablet of the solid. The bottle of solid antacid has 150 tablets, and one bottle of the liquid has $\frac{355}{5.0} = 71$ doses. Our tasks include calculating how much of the $Mg(OH)_2$ is dissolved in the liquid and not just suspended in it, writing balanced net ionic equations describing the neutralization reactions, calculating the quantity of stomach acid that one dose of each antacid can neutralize, and determining which is the better buy.

**Analyze** To answer part (a), we could use the solubility of $Mg(OH)_2$ to calculate the mass of $Mg(OH)_2$ that dissolves in 5.0 mL and compare this quantity to 200 mg. Because molar masses are given in grams per mole and molarity is the same as moles per liter, it might be simpler to scale up all the mass and volume quantities by 1000 and calculate the mass of $Mg(OH)_2$ in grams that dissolves in 5.0 L and then compare this value to 200 g. All the active ingredients in the antacids are solids, so they should appear as whole formula units and not individual ions in the net ionic equations in part (b). The balanced equations in part (b) and the calculations in part (c) will be based on the following stoichiometric considerations: (1) 1 mole of $Mg(OH)_2$ contains 2 moles of $OH^-$ ions, so it reacts with 2 moles of HCl; (2) 1 mole of $Al(OH)_3$ contains 3 moles of $OH^-$ ions, so it can neutralize 3 moles of HCl; (3) the 1 mole of $CO_3^{2-}$ ions in a mole of $CaCO_3$ can accept 2 moles of $H^+$ ions, which means that it can neutralize 2 moles of HCl. The molar masses of the three reactants are 58.32 g of $Mg(OH)_2$/mol, 78.00 g of $Al(OH)_3$/mol, and 100.09 g of $CaCO_3$/mol.

**Solve**

a. If 5.0 mL of the liquid antacid has 200 mg of $Mg(OH)_2$, 5.0 L of the liquid has 200 g of $Mg(OH)_2$. Therefore, we convert the solubility of $Mg(OH)_2$ from molarity to the number of grams of it that will dissolve in 5.0 L:

$$5.0\ \cancel{L} \times \frac{1.2 \times 10^{-4}\ \cancel{mol\ Mg(OH)_2}}{\cancel{L}} \times \frac{58.32\ g\ Mg(OH)_2}{1\ \cancel{mol\ Mg(OH)_2}}$$
$$= 0.0350\ g\ Mg(OH)_2$$

We divide this result by 200 g, expressing the quotient as the percentage of $Mg(OH)_2$ that is dissolved (the rest is tiny particles in suspension):

$$\frac{0.0350\ g}{200\ g} \times 100\% = 0.017\%\ \text{dissolved}$$

b. According to the stoichiometric ratios developed in the Analyze section, the molecular equations for the three neutralization reactions with the appropriate salts and moles of water are

$$Mg(OH)_2(s) + 2\ HCl(aq) \rightarrow MgCl_2(aq) + 2\ H_2O(\ell)$$
$$Al(OH)_3(s) + 3\ HCl(aq) \rightarrow AlCl_3(aq) + 3\ H_2O(\ell)$$
$$CaCO_3(s) + 2\ HCl(aq) \rightarrow CaCl_2(aq) + H_2O(\ell) + CO_2(g)$$

Separating the soluble ionic compounds into their free ions, we obtain overall ionic equations:

$$Mg(OH)_2(s) + 2\ H^+(aq) + 2\ Cl^-(aq) \rightarrow Mg^{2+}(aq) + 2\ Cl^-(aq) + 2\ H_2O(\ell)$$

$$Al(OH)_3(s) + 3\ H^+(aq) + 3\ Cl^-(aq) \rightarrow Al^{3+}(aq) + 3\ Cl^-(aq) + 3\ H_2O(\ell)$$

$$CaCO_3(s) + 2\ H^+(aq) + 2\ Cl^-(aq) \rightarrow Ca^{2+}(aq) + 2\ Cl^-(aq) + H_2O(\ell) + CO_2(g)$$

In each of the preceding equations, $Cl^-$ ions are the only spectator ions. Deleting them yields the following net ionic equations:

$$Mg(OH)_2(s) + 2\,H^+(aq) \rightarrow Mg^{2+}(aq) + 2\,H_2O(\ell)$$

$$Al(OH)_3(s) + 3\,H^+(aq) \rightarrow Al^{3+}(aq) + 3\,H_2O(\ell)$$

$$CaCO_3(s) + 2\,H^+(aq) \rightarrow Ca^{2+}(aq) + H_2O(\ell) + CO_2(g)$$

c. Calculating the number of moles of HCl neutralized by one dose (5.0 mL) of the liquid antacid (containing 200 mg of each active ingredient):

$$200\ \text{mg Mg(OH)}_2 \times \frac{1\ \text{g}}{1000\ \text{mg}} \times \frac{1\ \text{mol Mg(OH)}_2}{58.32\ \text{g Mg(OH)}_2} \times \frac{2\ \text{mol HCl}}{1\ \text{mol Mg(OH)}_2} = 0.00686\ \text{mol HCl}$$

$$+\ 200\ \text{mg Al(OH)}_3 \times \frac{1\ \text{g}}{1000\ \text{mg}} \times \frac{1\ \text{mol Al(OH)}_3}{78.00\ \text{g Al(OH)}_3} \times \frac{3\ \text{mol HCl}}{1\ \text{mol Al(OH)}_3} = 0.00769\ \text{mol HCl}$$

$$= 0.01455\ \text{mol HCl}$$

d. Calculating the number of moles of HCl neutralized by one tablet (500 mg) of the solid antacid:

$$500\ \text{mg CaCO}_3 \times \frac{1\ \text{g}}{1000\ \text{mg}} \times \frac{1\ \text{mol CaCO}_3}{100.09\ \text{g CaCO}_3} \times \frac{2\ \text{mol HCl}}{1\ \text{mol CaCO}_3} = 0.00999\ \text{mol HCl}$$

The 5.0 mL dose of the liquid contains $0.01455/0.00999 \approx 1.5$ times more acid-neutralizing power than one tablet of the solid antacid.

e. One bottle of the liquid antacid has 71 doses, which can neutralize (71 doses × 0.01455 mol HCl/dose) = 1.033 mol HCl. A bottle of the solid antacid has 150 tablets, which can neutralize (150 tablets × 0.00999 mol HCl/tablet) = 1.50 mol HCl. Therefore, a bottle of the solid antacid has more neutralizing capacity and is the better buy.

**Think About It** All three active ingredients are solids with limited solubility in water. However, they react with—and dissolve in—acid solutions such as stomach acid. The solid antacid provides about half again as much acid-neutralizing capacity as the liquid for the same (or at least a comparable) price. In fairness to this particular liquid product, it also contains an ingredient that relieves gas.

## SUMMARY

**LO1** The concentration of solute in a solution can be expressed as mass of solute per mass of solution, such as milligrams of solute per kilogram of solution, which is the same as parts per million (ppm). Concentrations also can be expressed as mass of solute per volume of solution or as moles of solute per liter of solution (mol/L), which is called **molarity (*M*)**. (Section 8.1)

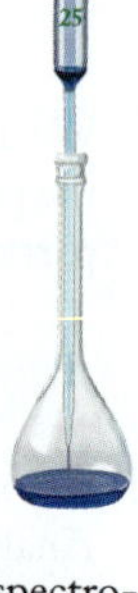

**LO2** During **dilution**, the quantity of solute in a sample of a concentrated solution does not change, but adding solvent increases the volume of the sample and decreases the concentration of solute. (Section 8.2)

**LO3** We can determine the concentration of a solution by spectrophotometry and by applying **Beer's law**, which relates the **absorbance** of a solution ($A$) to concentration ($c$), path length ($b$), and **molar absorptivity (ε)** by the equation $A = \varepsilon bc$. (Section 8.2)

**LO4** Chemical equations can be written in three ways. **Molecular equations** may be the easiest to balance; **overall ionic equations** show all the species as they exist in solution; **net ionic equations** eliminate **spectator ions** and show only the species involved in the reaction. (Section 8.4)

**LO5** A solute that dissociates into ions in aqueous solution is called an **electrolyte**. Mobility of these ions makes the solution a better electrical conductor than pure water. **Strong electrolytes** dissociate completely in water, **weak electrolytes** dissociate partially, and **nonelectrolytes** do not dissociate at all. (Section 8.3)

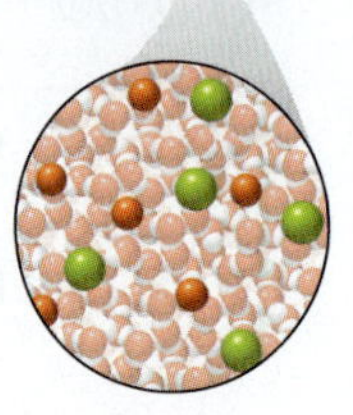


**LO6 Brønsted–Lowry acids** are proton ($H^+$) donors, and **Brønsted–Lowry bases** are proton acceptors. In a **neutralization reaction**, $H^+$ ions from the acid combine with $OH^-$ ions from the base, forming $H_2O$ and a **salt. Titrations** are volumetric methods of chemical analysis in which the concentrations of solutes called **analytes** are determined by reacting them with known volumes and concentrations of standard solutions called **titrants.** (Sections 8.4, 8.5)

**LO7 Solubility** guidelines define substances that dissolve readily in water and those that have very limited solubility. In a **precipitation reaction**, mixing dissolved reactants produces an insoluble **precipitate** as cations and anions switch partners. Balanced chemical equations and concentrations can be used to calculate the amount of precipitate formed when solutions are mixed. (Section 8.6)

**LO8** In a redox reaction, substances either gain electrons, and thereby undergo **reduction**, or they lose electrons and undergo **oxidation**. A reaction is a redox reaction if the **oxidation numbers (O.N.)**, or **oxidation states**, of elements in the reactants change during the reaction. (Section 8.7)

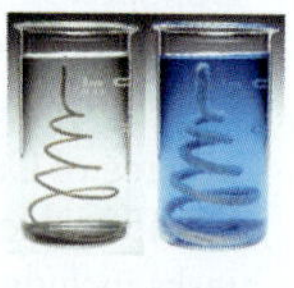

**LO9** Oxidation and reduction are complementary processes. The **activity series** allows us to predict whether a particular metal ion will oxidize a different metal. When we are balancing equations, the number of electrons lost by the substance being oxidized must equal the number of electrons gained by the substance being reduced. (Section 8.7)

## PARTICULATE PREVIEW WRAP-UP

Most hydrogen atoms consist of one proton and one electron, so $H^+$ is literally a proton. $CH_3COOH$ [molecule (c)] is a weak acid, meaning that a small fraction of these molecules ionize in water. Only the hydrogen atoms in –COOH groups ionize; those in $CH_3$ groups do not ionize. None of the H atoms in molecules (a), (b), and (d) ionize in water.

## PROBLEM-SOLVING SUMMARY

| Type of Problem | Concepts and Equations | Sample Exercises |
|---|---|---|
| **Converting concentration units** | Use the solute/solvent ratios in concentration units as conversion factors. Use units of moles per liter to represent molarity. | **8.1, 8.2** |
| **Calculating quantity of solute to prepare a solution** | $m = (V \times M) \times \mathcal{M} = \left(\cancel{\text{L}} \times \frac{\cancel{\text{mol}}}{\cancel{\text{L}}}\right) \times \frac{\text{g}}{\cancel{\text{mol}}}$ (8.3) | **8.3** |
| **Preparing dilute solutions** | $V_{\text{initial}} = \frac{V_{\text{final}} \times C_{\text{final}}}{C_{\text{initial}}}$ (8.7) | **8.4** |
| **Applying Beer's law** | Use Equation 8.8 and solve for $\varepsilon$ for a solution of known concentration, or solve for $c$ for a solution of unknown concentration: $A = \varepsilon bc$ (8.8) | **8.5** |
| **Writing neutralization reaction equations** | Write a balanced molecular equation in which the number of acidic hydrogen atoms matches the number of $H^+$ ions that can be accepted by the base. Write an overall ionic equation in which strong acids are ionized and all soluble ionic compounds are represented by their component ions. All other compounds are unchanged from the molecular equation. Delete spectator ions to write the net ionic equation. | **8.6, 8.7** |
| **Comparing electrolytes, acids, and bases** | Use the definitions of a strong electrolyte, a weak electrolyte, a nonelectrolyte, an acid, and a base to classify compounds into one or more categories. | **8.8** |
| **Determining solute concentration from titration data** | Calculate moles of titrant from its concentration and the volume needed to reach the end point. Use the stoichiometry of the titration reaction to calculate moles of analyte and then divide by sample volume to calculate analyte concentration. | **8.9, 8.10** |
| **Writing a net ionic equation for a precipitation reaction** | Balance the overall ionic equation, writing the precipitates as solids. Delete spectator ions to write the net ionic equation. | **8.11** |
| **Calculating solute concentration from the mass of a precipitate** | Convert precipitate mass into moles by dividing by its molar mass. Convert moles of precipitate into moles of solute. Calculate molarity of the solute in the sample by dividing by the volume of the sample in liters. | **8.12** |
| **Predicting the mass of a precipitate** | Identify the limiting reactant if necessary. Use the stoichiometry of the net ionic equation to calculate moles of precipitate, and then convert moles to mass. | **8.13** |
| **Determining oxidation numbers (O.N.)** | Follow the rules in Table 8.7. In general, the O.N. of a pure element is 0, the O.N. of a monatomic ion is the charge on the ion, and the O.N. values of O and H in most compounds are −2 and +1, respectively. For polyatomic ions, the sum of the O.N. values must equal the charge on the ion. | **8.14** |
| **Identifying oxidizing and reducing agents and determining the number of electrons transferred** | The oxidizing agent contains an atom whose O.N. decreases during the reaction; the reducing agent contains an atom whose O.N. increases; the changes in O.N. determine the number of electrons transferred. | **8.15** |
| **Writing net ionic equations for redox reactions with half-reactions** | Follow the five steps described on pages 373–374 for writing half-reactions and combining them to write net ionic equations describing a redox reaction. | **8.16** |
| **Using the activity series** | Any metal in Table 8.8 will reduce the cation of an element listed below it. | **8.17, 8.18** |
| **Writing net ionic equations for redox reactions in acidic or basic solutions** | Follow the steps described on page 378 and in Sample Exercise 8.19 for balancing the numbers of H and O atoms in the two half reactions before combining them to write a net ionic equation. | **8.19** |

## VISUAL PROBLEMS

*(Answers to boldface end-of-chapter questions and problems are in the back of the book.)*

**8.1.** Figure P8.1 shows a solution containing three binary (HX) acids. One of them is a weak acid and the other two are strong acids. Which color sphere is formed by ionization of the weak acid?

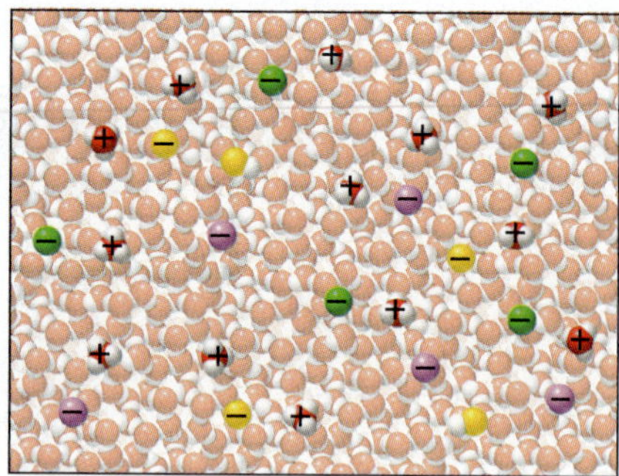

**FIGURE P8.1**

8.2. Figure P8.2 represents the products of the reaction that occurs when solutions of NaCl and $AgNO_3$ are mixed. The silver spheres represent $Ag^+$ ions. Which spheres represent (a) $Na^+$ ions; (b) $Cl^-$ ions; (c) $NO_3^-$ ions?

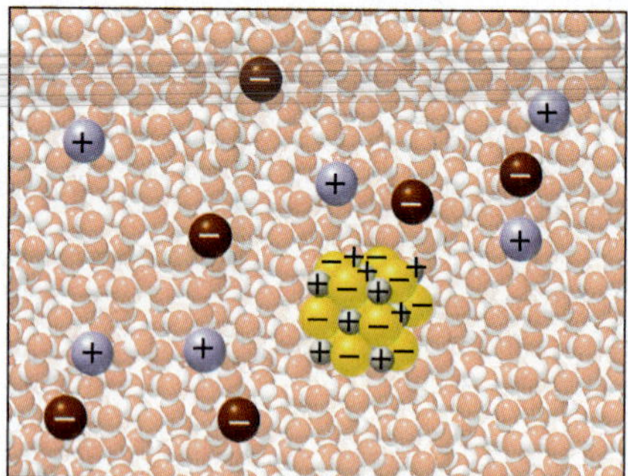

**FIGURE P8.2**

**8.3.** Figure P8.3 represents the reaction mixture at the equivalence point in a titration of a sample of battery acid ($H_2SO_4$) carried out by using a standard solution of NaOH as the titrant. Which ion is represented by the green spheres and which by the blue spheres?

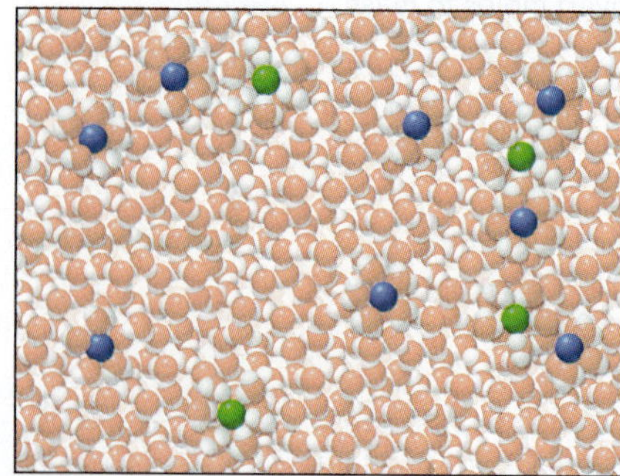

**FIGURE P8.3**

8.4. Suppose that two antacid tablets, each containing 0.50 g of $CaCO_3$ ($\mathcal{M}$ = 100 g/mol), react with 0.02 mole of HCl. Which image in Figure P8.4, in which green spheres represent cations and blue spheres represent anions, accurately reflects the population of ions in solution after the reaction is over?

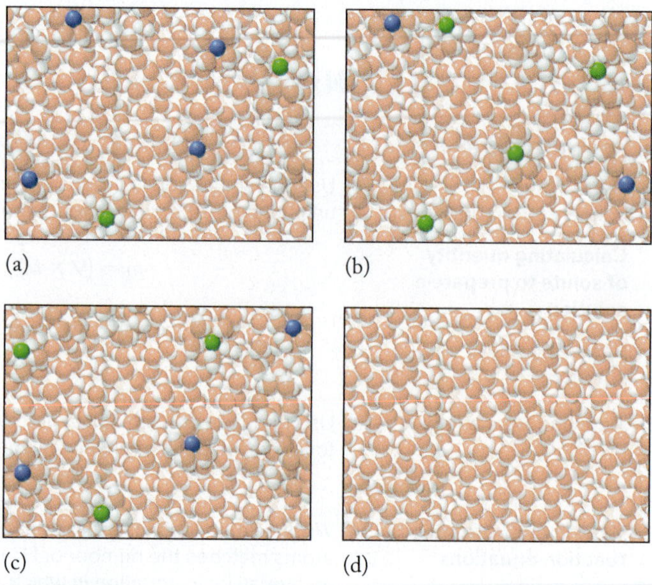

**FIGURE P8.4**

**8.5.** What is the change in oxidation state of nitrogen in the reaction described by the molecular models in Figure P8.5? (Blue spheres are nitrogen; red spheres are oxygen.)

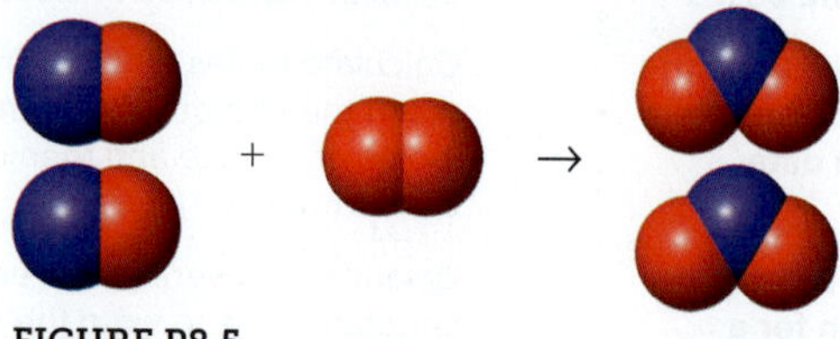

**FIGURE P8.5**

8.6. Identify the oxidizing and reducing agents in the reaction described by the molecular models in Figure P8.5.

**8.7.** Rank the nitrogen oxide molecules depicted by the models in Figure P8.7 in order of decreasing oxidation number of the nitrogen (blue spheres) in them. (The red spheres are oxygen atoms.)

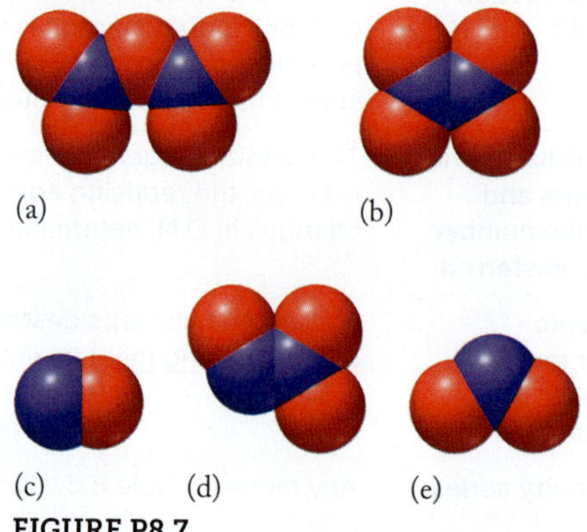

**FIGURE P8.7**

8.8. Rank the hydrocarbons depicted by the molecular models in Figure P8.8 according to decreasing oxidation state of the carbon (black spheres) in them.

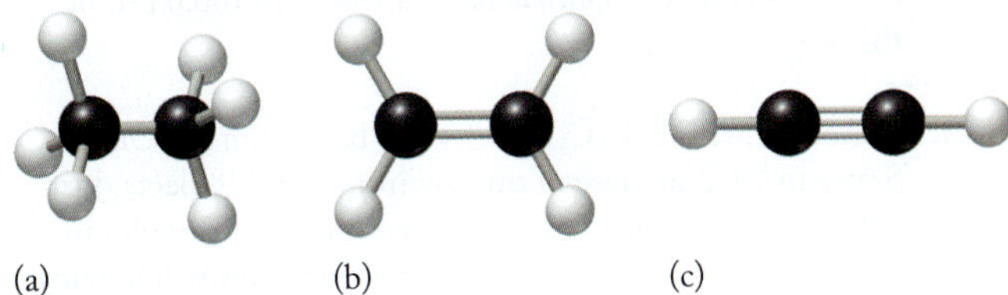


**FIGURE P8.8**

***8.9.** One way to follow the progress of a titration and detect its equivalence point is to monitor the conductivity of the titration reaction mixture. For example, consider how the conductivity of a sample of sulfuric acid changes as it is titrated with a standard solution of barium hydroxide before and then after the equivalence point.

a. Write the overall ionic equation for the titration reaction.
b. Which of the four graphs in Figure P8.9 comes closest to representing the changes in conductivity during the titration? (The zero point on the $y$-axis of these graphs represents the conductivity of pure water; the break points on the $x$-axis represent the equivalence point in the titration.)

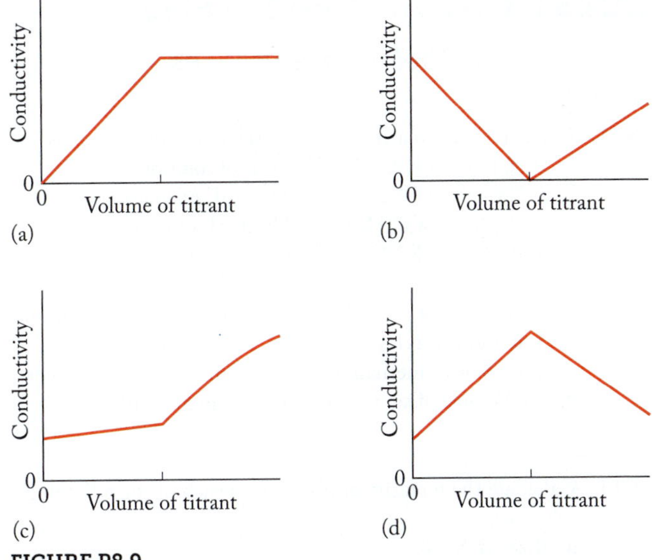


**FIGURE P8.9**

8.10. Use representations [A] through [I] in Figure P8.10 to answer questions a–f.

a. Which solids dissolve in water?
b. Which solids dissolve to create a solution that conducts electricity?
c. Which solutions can neutralize barium hydroxide? What would the spectator ions be?
d. Which solutions will form a precipitate upon mixing?
e. Name an aqueous solution that could be added to [A] to form the precipitate in [E].
f. Which combination of [A], [B], and [C] will cause a redox reaction?

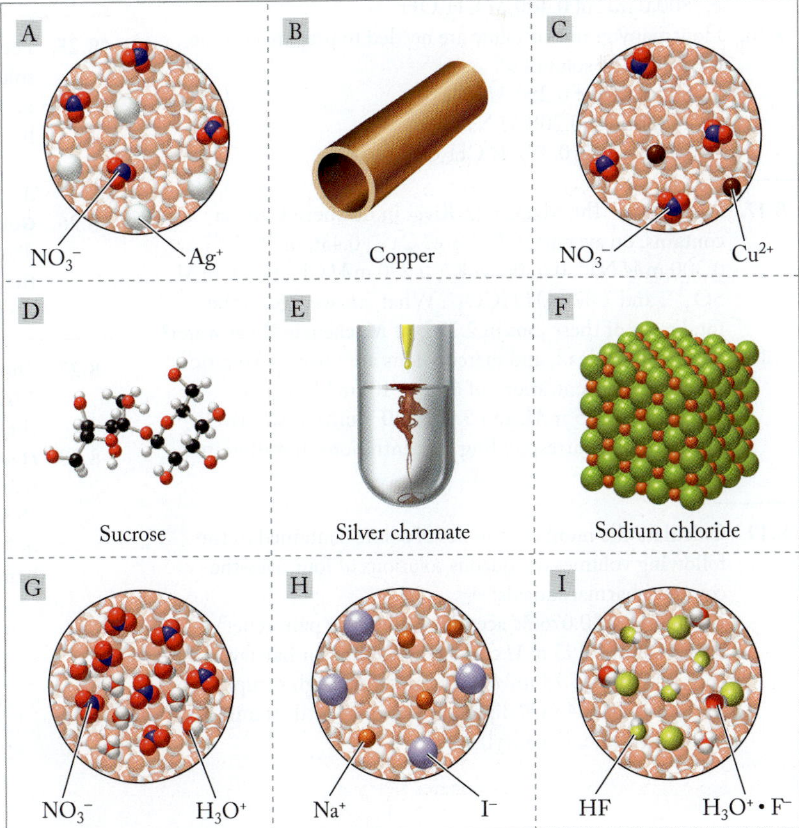


**FIGURE P8.10**

# QUESTIONS AND PROBLEMS

## Solutions and Their Concentrations

### Problems

**8.11.** Calculate the molarity of each of the following solutions:
a. 0.56 mole of $BaCl_2$ in 100.0 mL of solution
b. 0.200 mole of $Na_2CO_3$ in 200.0 mL of solution
c. 0.325 mole of $C_6H_{12}O_6$ in 250.0 mL of solution
d. 1.48 mole of $KNO_3$ in 250.0 mL of solution

8.12. Calculate the molarity of each of the following solutions:
a. 0.150 mole of urea $[(NH_2)_2CO]$ in 250.0 mL of solution
b. 1.46 mole of $NH_4CH_3CO_2$ in 1.000 L of solution
c. 1.94 mole of methanol ($CH_3OH$) in 5.000 L of solution
d. 0.045 mole of sucrose ($C_{12}H_{22}O_{11}$) in 50.0 mL of solution

**8.13.** Calculate the molarity of $Na^+$ ions in each of the following solutions:
a. 0.29 $M$ $NaNO_3$
b. 0.33 g of NaCl in 25 mL of solution
c. 0.88 $M$ $Na_2SO_4$
d. 0.46 g of $Na_3PO_4$ in 100 mL of solution

8.14. Calculate the molarity of each of the following solutions:
a. 31.76 g of $LiClO_4$ in 475.0 mL of solution
b. 6.37 g of $(NH_4)_2SO_4$ in 250.0 mL of solution
c. 2.97 g of KBr in 75.0 mL of solution
d. 0.773 g of $Pb(NO_3)_2$ in 100.0 mL of solution

**8.15.** How many grams of solute are needed to prepare each of the following solutions?
a. 1.000 L of 0.200 $M$ NaCl
b. 250.0 mL of 0.125 $M$ $CuSO_4$
c. 500.0 mL of 0.400 $M$ $CH_3OH$

8.16. How many grams of solute are needed to prepare each of the following solutions?
a. 500.0 mL of 0.250 $M$ KBr
b. 25.0 mL of 0.200 $M$ $NaNO_3$
c. 100.0 mL of 0.375 $M$ $CH_3OH$

**8.17.** **River Water** The Mackenzie River in northern Canada contains, on average, 0.820 m$M$ $Ca^{2+}$, 0.430 m$M$ $Mg^{2+}$, 0.300 m$M$ $Na^+$, 0.0200 $M$ $K^+$, 0.250 m$M$ $Cl^-$, 0.380 m$M$ $SO_4^{2-}$, and 1.82 m$M$ $HCO_3^-$. What, on average, is the total mass of these ions in 2.75 L of Mackenzie River water?

8.18. Zinc, copper, lead, and mercury ions are toxic to Atlantic salmon at concentrations of $6.42 \times 10^2$ m$M$, $7.16 \times 10^{-3}$ m$M$, 0.965 m$M$, and $5.00 \times 10^{-2}$ m$M$, respectively. What are the corresponding concentrations in milligrams per liter?

**8.19.** Calculate the number of moles of solute contained in the following volumes of aqueous solutions of four over-the-counter pharmaceuticals:
a. 0.250 L of 0.076 $M$ acetaminophen (for pain relief)
b. 2.11 L of 0.193 m$M$ cromolyn sodium (for hay fever)
c. 0.0475 L of 5.73 m$M$ benzocaine (in cough syrup)
d. 14.6 L of 27.4 m$M$ diphenhydramine (antihistamine; brand name, Benadryl)

8.20. A sample of crude oil contains 3.13 m$M$ naphthalene, 12.0 m$M$ methylnaphthalene, 23.8 m$M$ dimethylnaphthalene, and 14.1 m$M$ trimethylnaphthalene. What is the total number of moles of all the naphthalene compounds in 100.0 mL of the oil?

**8.21.** The pesticide DDT ($C_{14}H_9Cl_5$) was banned in the United States in 1972 because of dire environmental impacts. Just before it was banned, analyses of groundwater samples in Pennsylvania between 1969 and 1971 yielded the following results:

| Location | Sample | Mass of DDT |
|---|---|---|
| Orchard | 250.0 mL | 0.030 mg |
| Residential | 1.750 L | 0.035 mg |
| Residential after a storm | 50.0 mL | 0.57 mg |

Express these data in units of millimoles of DDT per liter.

8.22. Between 1969 and 1975, pesticide concentrations in the portion of the Rhine River that flows between Germany and France averaged 0.55 mg/L of hexachlorobenzene ($C_6Cl_6$), 0.06 mg/L of dieldrin ($C_{12}H_8Cl_6O$), and 1.02 mg/L of hexachlorocyclohexane ($C_6H_6Cl_6$). Express these concentrations in millimoles per liter.

**8.23.** Tap water in North America from groundwater sources contains an average of 48 mg/L of $Ca^{2+}$ ion. What is the molarity of calcium ion in this water?

*8.24. The concentration of manganese in one brand of soluble plant fertilizer is 0.05% by mass. If a 20 g sample of the fertilizer is dissolved in 2.0 L of solution, what is the molarity of dissolved manganese in the solution?

***8.25.** For which of the following compounds is making a 1.0 $M$ solution at 0°C possible?
a. $CuSO_4 \cdot 5\,H_2O$, solubility = 23.1 g/100 mL
b. $AgNO_3$, solubility = 122 g/100 mL
c. $Fe(NO_3)_2 \cdot 6\,H_2O$, solubility = 113 g/100 mL
d. $Ca(OH)_2$, solubility = 0.185 g/100 mL

8.26. **Gold in the Ocean** About $6 \times 10^9$ g of gold is thought to be dissolved in the oceans of the world. If the total volume of the oceans is $1.5 \times 10^{21}$ L, what is the average molarity of gold in seawater?

**8.27.** The concentration of $Mg^{2+}$ in a sample of coastal seawater is 1.09 g/kg. What is the molarity of $Mg^{2+}$ in this seawater if the seawater has a density of 1.02 g/mL?

8.28. **Hemoglobin in Blood** A typical adult body contains 6.0 L of blood. The hemoglobin content of blood is about 15.5 g/100.0 mL. The approximate molar mass of hemoglobin is 64,500 g/mol. How many moles of hemoglobin are present in a typical adult?

## Dilutions

### Problems

**8.29.** Calculate the final concentrations of the following aqueous solutions after each has been diluted to a final volume of 25.0 mL:

a. 1.00 mL of 0.452 $M$ $Na^+$
b. 2.00 mL of 3.4 m$M$ LiCl
c. 5.00 mL of $6.42 \times 10^{-2}$ m$M$ $Zn^{2+}$

8.30. Chemists who analyze samples for dissolved trace elements may buy standard solutions that contain 1.000 g/L concentrations of the elements. If a chemist needs to prepare 0.500 L of a working standard that has a concentration of 5.00 mg/L, what volume of the 1.000 g/L standard is needed?

---

***8.31.** A puddle of coastal seawater, caught in a depression formed by some coastal rocks at high tide, begins to evaporate on a hot summer day as the tide goes out. If the volume of the puddle decreases to 23% of its initial volume, what is the concentration of $Na^+$ after evaporation if it was 0.449 $M$ initially?

8.32. What volume of 2.5 $M$ $SrCl_2$ is needed to prepare 500.0 mL of 5.0 m$M$ solution?

---

***8.33.** **Dilution of Adult-Strength Cough Syrup** A standard dose of an over-the-counter cough suppressant for adults is 20.0 mL. A portion this size contains 35 mg of the active pharmaceutical ingredient (API). A pediatrician prescribes the drug for a 6-year-old child, but the child may take only 10.0 mL at a time and receive a maximum of 4.00 mg of the API. What is the concentration in milligrams per milliliter of the adult-strength medication, and how many milliliters of it should be diluted to make 100.0 mL of child-strength cough syrup?

*8.34. **Mixing Fertilizer** The label on a bottle of "organic" liquid fertilizer concentrate states that it contains 8 g of phosphate per 100.0 mL and that 16 fluid ounces should be diluted with water to make 32 gallons of fertilizer to be applied to growing plants. What is the phosphate concentration in grams per liter in the diluted fertilizer? (1 gallon = 128 fluid ounces.)

---

**8.35.** A 1.00 $M$ solution of $CuSO_4$ has an initial volume of 15.0 mL. If the solution's absorbance due to $Cu^{2+}(aq)$ ions decreases by 45% upon dilution, how much water has been added to it?

8.36. By what percentage does the absorbance due to $Cr^{3+}(aq)$ ions decrease if 12.25 mL of water is added to a 16.75 mL sample of 0.500 $M$ $Cr(NO_3)_3$?

---

***8.37.** The reaction of $SnCl_2(aq)$ with $Pt^{4+}(aq)$ in aqueous HCl yields a yellow-orange solution of a 1:1 Pt–Sn compound with a molar absorptivity ($\varepsilon$) of $1.3 \times 10^4\ M^{-1}\ cm^{-1}$. What is the absorbance in a cell with a path length of 1.00 cm of a solution prepared by adding 100 mL of an aqueous solution of 5.2 mg of $(NH_4)_2PtCl_6$ to 100 mL of an aqueous solution of 2.2 mg of $SnCl_2$?

*8.38. The reaction of $SnCl_2(aq)$ with $RhCl_3(aq)$ in aqueous HCl yields a red solution of a 1:1 Rh–Sn compound. If a solution prepared by adding 150 mL of a 0.272 m$M$ aqueous solution of $SnCl_2$ to 50 mL of an aqueous solution of 8.5 mg of $RhCl_3$ has an absorbance of 0.85, as measured in a 1.00 cm cell, what is the molar absorptivity of the red compound?

## Electrolytes and Nonelectrolytes

### Concept Review

**8.39.** A solution of table salt is a good conductor of electricity, but a solution containing an equal molar concentration of table sugar is not. Why?

8.40. **Fuel Cells** The electrolyte in an electricity-generating device called *a fuel cell* consists of a mixture of $Li_2CO_3$ and $K_2CO_3$ heated to 650°C. At this temperature, these ionic solids melt. Explain how the mixture of molten carbonates can conduct electricity.

***8.41.** Rank the following solutions on the basis of their ability to conduct electricity, starting with the most conductive: (a) 1.0 $M$ NaCl; (b) 1.2 $M$ KCl; (c) 1.0 $M$ $Na_2SO_4$; (d) 0.75 $M$ LiCl.

8.42. Rank the conductivities of 1 $M$ aqueous solutions of each of the following solutes, starting with the most conductive: (a) acetic acid; (b) methanol; (c) sucrose (table sugar); (d) hydrochloric acid.

### Problems

**8.43.** What is the molarity of $Na^+$ ions in a 0.025 $M$ aqueous solution of (a) NaBr; (b) $Na_2SO_4$; (c) $Na_3PO_4$?

8.44. What is the molarity of each ion in a 0.035 $M$ aqueous solution of (a) $NH_4Cl$; (b) $Li_2SO_4$; (c) $MgBr_2$?

## Acid–Base Reactions: Proton Transfer

### Concept Review

**8.45.** What is the difference between a strong acid and a weak acid?

8.46. Give the formulas of two strong acids and two weak acids.

**8.47.** What chemical property of a base makes it a base?

8.48. What is the difference between a strong base and a weak base?

**8.49.** Give the formulas of two strong bases and two weak bases.

8.50. Write the net ionic equation for the neutralization of a strong acid by a strong base.

### Problems

**8.51.** For each of the following acid–base reactions, identify the acid and the base and then write the net ionic equation:

a. $H_2SO_4(aq) + Ca(OH)_2(s) \rightarrow CaSO_4(aq) + 2\ H_2O(\ell)$
b. $PbCO_3(s) + H_2SO_4(aq) \rightarrow PbSO_4(s) + CO_2(g) + H_2O(\ell)$
c. $Ca(OH)_2(s) + 2\ CH_3COOH(aq) \rightarrow$ $Ca(CH_3COO)_2(aq) + 2\ H_2O(\ell)$

8.52. Complete and balance each of the following neutralization reactions, name the products, and write the net ionic equations.

a. $HI(aq) + LiOH(aq) \rightarrow$
b. $H_3PO_4(aq) + KOH(aq) \rightarrow$
c. $Al(OH)_3(s) + CH_3COOH(aq) \rightarrow$
d. $HNO_3(aq) + Ba(OH)_2(aq) \rightarrow$

**8.53.** Write a balanced molecular equation and a net ionic equation for the following reactions:
a. Solid magnesium hydroxide reacts with a solution of sulfuric acid.
b. Solid magnesium carbonate reacts with a solution of hydrochloric acid.
c. Ammonia gas reacts with hydrogen chloride gas.

8.54. Write a balanced molecular equation and a net ionic equation for the following reactions:
a. Solid aluminum hydroxide reacts with a solution of hydrobromic acid.
b. A solution of sulfuric acid reacts with solid sodium carbonate.
c. A solution of calcium hydroxide reacts with a solution of nitric acid.

---

**8.55.** **Toxicity of Lead Pigments** The use of lead(II) carbonate and lead(II) hydroxide as white pigments in paint was discontinued in the United States in 1978 because these compounds dissolved in the stomachs of young children who ingested paint chips. The $Pb^{2+}$ ions released when the compounds dissolve interfere with neurotransmissions in the brain, causing neurological disorders. Using net ionic equations, show why lead(II) carbonate and lead(II) hydroxide dissolve in acidic solutions.

8.56. **Lawn Care** Many homeowners treat their lawns with $CaCO_3(s)$ to make the soil less acidic. Write a net ionic equation for the reaction of $CaCO_3(s)$ with a strong acid.

## *Titrations*

### Problems

**8.57.** How many milliliters of 0.100 *M* NaOH is required to neutralize the following solutions?
a. 10.0 mL of 0.0500 *M* HCl
b. 25.0 mL of 0.126 *M* $HNO_3$
c. 50.0 mL of 0.215 *M* $H_2SO_4$

8.58. How many milliliters of 0.100 *M* $HNO_3$ is needed to neutralize the following solutions?
a. 45.0 mL of 0.667 *M* KOH
b. 58.5 mL of 0.0100 *M* $Al(OH)_3$
c. 34.7 mL of 0.775 *M* NaOH

---

***8.59.** The solubility of slaked lime, $Ca(OH)_2$, in water at 20°C is 0.185 g/100.0 mL. What volume of 0.00100 *M* HCl is needed to neutralize 10.0 mL of a saturated $Ca(OH)_2$ solution?

8.60. The solubility of magnesium hydroxide, $Mg(OH)_2$, in water is $9.0 \times 10^{-4}$ g/100.0 mL. What volume of 0.00100 *M* $HNO_3$ is required to neutralize 1.00 L of a saturated $Mg(OH)_2$ solution?

---

**8.61.** **Chlorinity of Seawater** Scientists can precisely determine the chloride ion concentration, or *chlorinity*, of seawater samples by using a titration called the Mohr method. The titrant is a solution of $AgNO_3$. The indicator is a few drops of $K_2CrO_4$ solution, which imparts a light-yellow color to the titration mixture before the equivalence point (Figure P8.61a). However, after all the $Cl^-$ ions have been precipitated as AgCl, the first excess of $Ag^+$ ions combines with $CrO_4^{2-}$ ions to form $Ag_2CrO_4$, which makes the milky suspension of AgCl look pink (Figure P8.61b). If 27.80 mL of 0.5000 *M* $AgNO_3$ is needed to titrate a 25.00 mL sample of seawater, what is the concentration of $Cl^-$ in the sample? Express your answer in millimolar concentration and in grams per kilogram (the density of the sample is 1.025 g/mL).

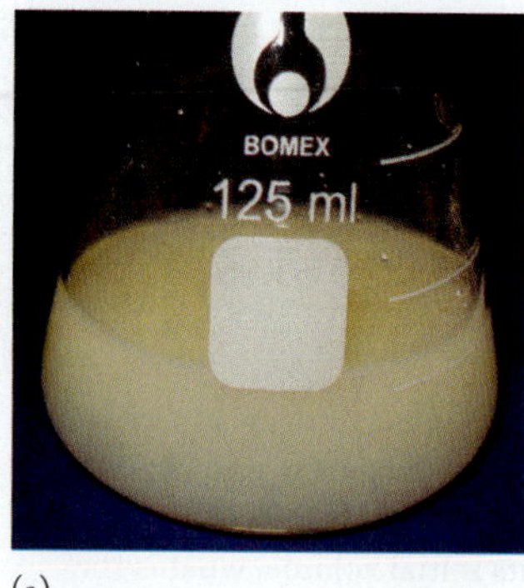


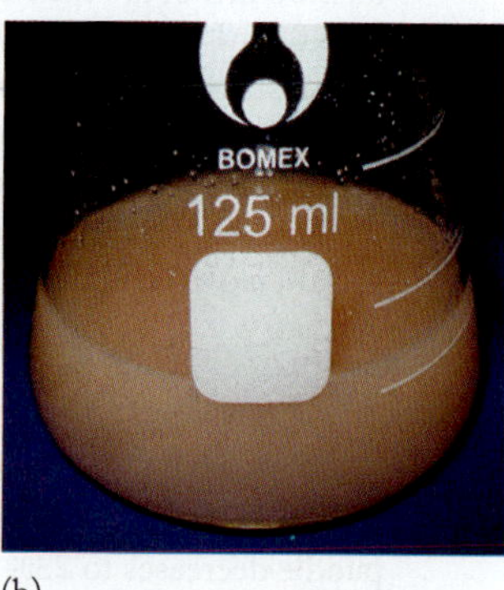


(a) (b)

**FIGURE P8.61**

8.62. **Exercise Physiology** Lactic acid accumulates in muscle when glucose is metabolized under conditions in which oxygen is the limiting reagent. Lactic acid has one carboxylic acid functional group per molecule (Figure P8.62). To determine the concentration of a solution of lactic acid, a chemist titrates a 20.00 mL sample of it with 0.1010 *M* NaOH and finds that 12.77 mL of titrant is required to reach the equivalence point. What is the molarity of the lactic acid solution?

**FIGURE P8.62**

## *Precipitation Reactions*

### Concept Review

**8.63.** What is the difference between a saturated solution and a supersaturated solution?

8.64. Rain and snow are commonly called "precipitation." How are they like a solid that forms in an aqueous solution?

**8.65.** Is a saturated solution always a concentrated solution? Explain.

8.66. **Behavior of Honey** Honey is a concentrated solution of sugar molecules in water. Clear, viscous honey becomes cloudy after being stored for long periods. Explain how this transition illustrates supersaturation.

### Problems

**8.67.** According to the solubility guidelines in Tables 8.4 and 8.5, which of the following compounds have limited solubility in water? (a) barium sulfate; (b) barium hydroxide; (c) lanthanum nitrate; (d) sodium acetate; (e) lead hydroxide; (f) calcium phosphate

8.68. **Ocean Vents** The black "smoke" that flows out of deep ocean hydrothermal vents (Figure P8.68) is made of insoluble metal sulfides suspended in seawater. Of the following cations present in the water flowing up through these vents, which ones could contribute to the formation of the black smoke? $Na^+$, $Li^+$, $Mn^{2+}$, $Fe^{2+}$, $Ca^{2+}$, $Mg^{2+}$, $Zn^{2+}$, $Pb^{2+}$, $Cu^{2+}$

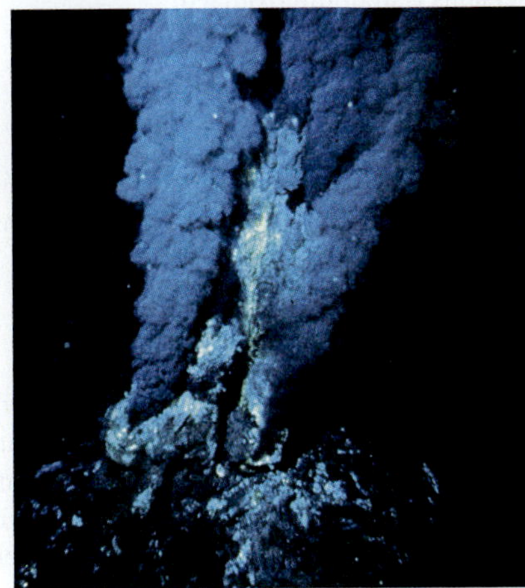

**FIGURE P8.68**

---

**8.69.** Complete and balance the chemical equations for the precipitation reactions, between the following pairs of reactants, and write the net ionic equations:

a. $Pb(NO_3)_2(aq) + Na_2SO_4(aq) \rightarrow$
b. $NiCl_2(aq) + NH_4NO_3(aq) \rightarrow$
c. $FeCl_2(aq) + Na_2S(aq) \rightarrow$
d. $MgSO_4(aq) + BaCl_2(aq) \rightarrow$

*8.70. **Wastewater Treatment** Show, with appropriate net ionic reactions, how treating wastewater with solutions of sodium hydroxide can remove $Cr^{3+}$ and $Cd^{2+}$.

---

***8.71.** An aqueous solution containing $Ca^{2+}$, $Cl^-$, $CO_3^{2-}$, and $NO_3^-$ is allowed to evaporate. Which compound will precipitate first?

8.72. The solubility of $CaCl_2$ in water at 25°C is 81.1 g/100 mL; at 0°C, its solubility decreases to 59.5 g/100 mL. Answer the following questions about an aqueous solution of 26.4 g of $CaCl_2$ in a volume of 37.5 mL:

a. At 25°C, is this a saturated solution?
b. If the solution is cooled to 0°C, do you expect a precipitate to form?
c. If the solution is slowly cooled to 0°C and no precipitate forms, what kind of solution is it?

---

**8.73.** Calculate the mass of $MgCO_3$ precipitated by mixing 10.0 mL of a 0.200 $M$ $Na_2CO_3$ solution with 5.00 mL of a 0.0500 $M$ $Mg(NO_3)_2$ solution.

*8.74. **Phosphates in Sewage** Eutrophication, the rapid growth of algae and the death of fish, may be caused by the presence of excess phosphates in water. Treatment plants that process sewage may add $Ca(OH)_2$ (slaked lime) to water to remove phosphates before returning the water to the environment. Although the phosphates may be present in several forms, we can use $HPO_4^-$ as a representative phosphate in the net ionic equation:

$$5\,Ca^{2+}(aq) + 3\,HPO_4^-(aq) + 4\,OH^-(aq) \rightarrow Ca_5(OH)(PO_4)_3(s) + 3\,H_2O(\ell)$$

If phosphates (as $HPO_4^-$) are present at a level of 15.7 mg/L in wastewater, how much $Ca(OH)_2$ would need to be added to $1.00 \times 10^5$ L of water to precipitate 95% of the phosphate ion present?

---

**8.75.** Iron(II) can be precipitated from a slightly basic aqueous solution by bubbling oxygen through the solution, which converts $Fe^{2+}$ to insoluble $Fe^{3+}$:

$$4\,Fe(OH)^+(aq) + 4\,OH^-(aq) + O_2(g) + 2\,H_2O(\ell) \rightarrow 4\,Fe(OH)_3(s)$$

How many grams of $O_2$ are consumed to precipitate all the iron in 75 mL of 0.090 $M$ $Fe^{2+}$?

8.76. Given the following equation, how many grams of $PbCO_3$ will dissolve when 1.00 L of 1.00 $M$ $H^+$ is added to 5.00 g of $PbCO_3$?

$$PbCO_3(s) + 2\,H^+(aq) \rightarrow Pb^{2+}(aq) + H_2O(\ell) + CO_2(g)$$

## Oxidation–Reduction Reactions

### Concept Review

**8.77.** How are the gains or losses of electrons related to changes in oxidation numbers?

8.78. What is the sum of the oxidation numbers of the atoms in a molecule?

**8.79.** What is the sum of the oxidation numbers of all the atoms in each of the following polyatomic ions? (a) $OH^-$; (b) $NH_4^+$; (c) $SO_4^{2-}$; (d) $PO_4^{3-}$

8.80. Gold does not dissolve in concentrated $H_2SO_4$ but readily dissolves in $H_2SeO_4$ (selenic acid). Which acid is the stronger oxidizing agent?

**8.81.** Silver dissolves in sulfuric acid to form silver sulfate and $H_2$, but gold does not dissolve in sulfuric acid to form gold sulfate. Which of the two metals is the better reducing agent?

8.82. In which one of the following fluorides is chromium in the highest oxidation state? (a) $CrF_3$, (b) $CrF_6$, (c) $CrF_5$, (d) $CrF_2$

**8.83.** What is meant by a half-reaction?

8.84. Why is the oxidizing agent in a redox reaction reduced and the reducing agent oxidized?

### Problems

**8.85.** What is the oxidation number of chlorine in each of the following oxoacids? (a) hypochlorous acid (HClO); (b) chloric acid ($HClO_3$); (c) perchloric acid ($HClO_4$)

8.86. What is the oxidation number of nitrogen in each of the following species? (a) elemental nitrogen ($N_2$); (b) hydrazine ($N_2H_4$); (c) ammonium ion ($NH_4^+$)

---

**8.87.** What is the change (if any) in the oxidation state of carbon in this reaction?

$$C_{12}H_{22}O_{11}(s) \rightarrow 12\,C(s) + 11\,H_2O(\ell)$$

8.88. **Nitrogen Cycle** In one stage of the nitrogen cycle, *Nitrosomonas* bacteria convert ammonia and oxygen into nitrite ions.
a. What is the change in oxidation state of nitrogen during the reaction?
b. Write a balanced net ionic equation for the reaction in acidic groundwater.

---

**8.89.** Complete the following half-reactions by adding the appropriate number of electrons. Identify the oxidation half-reactions and the reduction half-reactions.
a. $Br_2(\ell) \rightarrow 2\,Br^-(aq)$
b. $Pb(s) + 2\,Cl^-(aq) \rightarrow PbCl_2(s)$
c. $O_3(g) + 2\,H^+(aq) \rightarrow O_2(g) + H_2O(\ell)$
d. $2\,H_2SO_3(aq) + H^+(aq) \rightarrow HS_2O_4^-(aq) + 2\,H_2O(\ell)$

8.90. Complete the following half-reactions by adding the appropriate number of electrons. Which are oxidation half-reactions and which are reduction half-reactions?
a. $Fe^{2+}(aq) \rightarrow Fe^{3+}(aq)$
b. $AgI(s) \rightarrow Ag(s) + I^-(aq)$
c. $VO_2^+(aq) + 2\,H^+(aq) \rightarrow VO^{2+}(aq) + H_2O(\ell)$
d. $I_2(s) + 6\,H_2O(\ell) \rightarrow 2\,IO_3^-(aq) + 12\,H^+(aq)$

---

**8.91.** Complete the following net ionic equations. Identify the oxidizing agent and the reducing agent. Identify which elements are oxidized and which are reduced:
a. $MnO_2(s) + HCl(aq) \rightarrow Mn^{2+}(aq) + Cl_2(g)$
b. $I_2(s) + S_2O_3^{2-}(aq) \rightarrow S_4O_6^{2-}(aq) + I^-(aq)$
c. $MnO_4^-(aq) + Fe^{2+}(aq) \rightarrow Mn^{2+}(aq) + Fe^{3+}(aq)$

8.92. Complete the following net ionic equations. Identify the oxidizing agent and the reducing agent. Identify which elements are oxidized and which are reduced:
a. $MnO_4^-(aq) + S^{2-}(aq) \rightarrow MnO_2(s) + S(s)$
b. $IO_3^-(aq) + I^-(aq) \rightarrow I_2(s)$
c. $Mn^{2+}(aq) + BiO_3^-(aq) \rightarrow MnO_4^-(aq) + Bi^{3+}(aq)$

---

**8.93.** **Natural Weathering of Ores** Iron is oxidized in several chemical weathering processes. How many moles of $O_2$ are consumed when one mole of magnetite ($Fe_3O_4$) is converted into hematite ($Fe_2O_3$)?

*8.94. The mineral rhodochrosite [manganese(II) carbonate; $MnCO_3$] is a commercially important source of manganese. How many moles of $O_2$ are consumed when one mole of $MnCO_3$ is converted into $MnO_2$ and $CO_2$?

---

**8.95.** **Earth's Crust** The following chemical reactions have helped shape Earth's crust. Determine the oxidation numbers of all the elements in the reactants and products, and identify which elements are oxidized and which are reduced:
a. $3\,SiO_2(s) + 2\,Fe_3O_4(s) \rightarrow 3\,Fe_2SiO_4(s) + O_2(g)$
b. $SiO_2(s) + 2\,Fe(s) + O_2(g) \rightarrow Fe_2SiO_4(s)$
c. $4\,FeO(s) + O_2(g) + 6\,H_2O(\ell) \rightarrow 4\,Fe(OH)_3(s)$

8.96. Determine the oxidation numbers of each of the elements in the following reactions, and identify which of them, if any, are oxidized or reduced:
a. $SiO_2(s) + 2\,H_2O(\ell) \rightarrow H_4SiO_4(aq)$
b. $2\,MnCO_3(s) + O_2(g) \rightarrow 2\,MnO_2(s) + 2\,CO_2(g)$
c. $3\,NO_2(g) + H_2O(\ell) \rightarrow 2\,NO_3^-(aq) + NO(g) + 2\,H^+(aq)$

---

**8.97.** How many moles of $O_2$ are consumed in the conversion of one mole of $FeCO_3$ to each of the following compounds? Assume that $CO_2$ also is produced. (a) $Fe_2O_3$; (b) $Fe_3O_4$

8.98. Uranium is found in Earth's crust as $UO_2$ and in an assortment of compounds containing $UO_2^{n+}$ cations. How many moles of electrons are transferred in the conversion of one mole of $UO_2$ to each of the following species? In which of the conversions is uranium oxidized?

(a) $UO_2(CO_3)_3^{4-}(aq)$; (b) $UO_2(HPO_4)_2^{2-}(aq)$

---

**8.99.** Nitrogen in the hydrosphere is found primarily as ammonium ions and nitrate ions. Complete and balance the following chemical equation for the oxidation of ammonium ions to nitrate ions in acidic solution:

$$NH_4^+(aq) + O_2(g) \rightarrow NO_3^-(aq)$$

8.100. In sediments and waterlogged soil, dissolved $O_2$ concentrations are so low that the microorganisms living there must rely on other sources of oxygen for respiration. Some bacteria can extract the oxygen from sulfate ions, reducing the sulfur in them to hydrogen sulfide gas and giving the sediments or soil a distinctive rotten-egg odor.
a. What is the change in oxidation state of sulfur as a result of this reaction?
b. Write the balanced net ionic equation for the reaction under acidic conditions, which releases $O_2$ from sulfate and forms hydrogen sulfide gas.

---

**8.101.** The solubilities of Fe and Mn compounds in freshwater streams are affected by changes in the oxidation states of these metals. Complete and balance the following net ionic equation describing a redox reaction occurring in slightly acidic freshwater:

$$Fe(OH)_2^+(aq) + Mn^{2+}(aq) \rightarrow MnO_2(s) + Fe^{2+}(aq)$$

8.102. A method to determine the quantity of dissolved oxygen in natural waters requires a series of redox reactions. Complete the following chemical equations in that series under the conditions indicated:
a. $Mn^{2+}(aq) + O_2(g) \rightarrow MnO_2(s)$ (basic solution)
b. $MnO_2(s) + I^-(aq) \rightarrow Mn^{2+}(aq) + I_2(s)$ (acidic solution)
c. $I_2(s) + S_2O_3^{2-}(aq) \rightarrow I^-(aq) + S_4O_6^{2-}(aq)$ (neutral solution)

**8.103.** Silver can be extracted from rocks by using cyanide ion. Complete and balance the following reaction for this process:

$Ag(s) + CN^-(aq) + O_2(g) \rightarrow Ag(CN)_2^-(aq)$ (basic solution)

8.104. Permanganate ion ($MnO_4^-$) is used in water purification to remove oxidizable substances. Complete and balance the following equations describing the removal of sulfide, cyanide, and sulfite. Assume that reaction conditions are basic:
a. $MnO_4^-(aq) + S^{2-}(aq) \rightarrow MnS(s) + S_8(s)$
b. $MnO_4^-(aq) + CN^-(aq) \rightarrow MnO_2(s) + CNO^-(aq)$
c. $MnO_4^-(aq) + SO_3^{2-}(aq) \rightarrow MnO_2(s) + SO_4^{2-}(aq)$

---

**8.105.** **Biocide Chemistry** The water-soluble gas $ClO_2$ is known as an oxidative biocide. It destroys bacteria by oxidizing their cell walls and viruses by attacking their viral envelopes.

$ClO_2$ may be prepared for use as a decontaminating agent from several different starting materials in slightly acidic solutions. Complete and balance the following chemical equations describing the synthesis of $ClO_2$.

a. $ClO_3^-(aq) + SO_2(g) \rightarrow ClO_2(g) + SO_4^{2-}(aq)$
b. $ClO_3^-(aq) + Cl^-(aq) \rightarrow ClO_2(g) + Cl_2(g)$
c. $ClO_3^-(aq) + Cl_2(g) \rightarrow ClO_2(g) + O_2(g)$

*8.106. Toxic cyanide ions can be removed from wastewater by adding hypochlorite:

$$2\,CN^-(aq) + 5\,OCl^-(aq) + H_2O(\ell) \rightarrow N_2(g) + 2\,HCO_3^-(aq) + 5\,Cl^-(aq)$$

a. Identify the oxidizing agent in this reaction.
b. How many liters of 0.125 *M* $OCl^-$ are required to remove the $CN^-$ in $3.4 \times 10^6$ L of wastewater in which the $CN^-$ concentration is 0.58 mg/L?

---

**8.107.** Refer to Table 8.8 to determine which of the following metals will reduce aqueous $Fe^{2+}$ to iron metal: lead, copper, zinc, or aluminum.

8.108. Which of the following ions will oxidize aluminum? $Li^+$; $Ca^{2+}$; $Ag^+$; $Sn^{2+}$

---

**8.109.** Through appropriate experiments, we could expand the activity series in Table 8.8 to include additional metals. If aluminum is oxidized by $V^{3+}$, but aluminum does not reduce $Sc^{3+}$, where would you place vanadium and scandium in the activity series?

8.110. If iron is oxidized by $Cd^{2+}$, but iron does not reduce $Ga^{3+}$, where would you place cadmium and gallium in the activity series?

## Additional Problems

**8.111.** **Rhubarb Leaves** Rhubarb leaves contain 0.520 g of oxalic acid ($H_2C_2O_4$) per 100.0 g of leaves. The oxalic acid can react with calcium ion to make insoluble calcium oxalate, a major constituent of kidney stones. To study this reaction in the laboratory, 375 mL of a 0.866 *M* solution of oxalic acid is treated with excess 0.133 *M* calcium hydroxide solution.

a. How much calcium oxalate is formed in this reaction?
*b. What volume of calcium hydroxide solution is required if it must be present in 20% excess to ensure complete reaction?

*8.112. **Antifreeze** Ethylene glycol is the common name for the liquid used to keep the coolant in automobile cooling systems from freezing. Ethylene glycol is 38.7% carbon, 9.7% hydrogen, and 51.6% oxygen by mass. Its molar mass is 62.07 g/mol, and its density is 1.106 g/mL at 20°C.

a. What is the molecular formula of ethylene glycol?
b. In a solution prepared by mixing equal volumes of water and ethylene glycol, which ingredient is the solute and which is the solvent?

***8.113.** According to the label on a bottle of concentrated hydrochloric acid, the contents are 36.0% HCl by mass and have a density of 1.18 g/mL.

a. What is the molarity of this concentrated HCl?
b. What volume of it would you need to prepare 0.250 L of 2.00 *M* HCl?
c. What mass of sodium hydrogen carbonate would be needed to neutralize the spill if a bottle containing 1.75 L of this concentrated HCl dropped on a lab floor and broke open?

8.114. **Synthesis and Toxicity of Chlorine** Chlorine was first prepared in 1774 by heating a mixture of NaCl and $MnO_2$ in sulfuric acid:

$$NaCl(aq) + H_2SO_4(aq) + MnO_2(s) \rightarrow Na_2SO_4(aq) + MnCl_2(aq) + H_2O(\ell) + Cl_2(g)$$

a. Assign oxidation numbers to the elements in each compound, and balance the redox reaction in acid solution.
b. Write a net ionic equation describing the reaction for the formation of chlorine.
c. If chlorine gas is inhaled, it causes pulmonary edema (fluid in the lungs) because it reacts with water in the alveolar sacs of the lungs to produce the strong acid HCl and the weaker acid HOCl. Balance the equation for the conversion of $Cl_2$ to HCl and HOCl.

***8.115.** When a solution of dithionite ions ($S_2O_4^{2-}$) is added to a solution of chromate ions ($CrO_4^{2-}$), the products of the ensuing chemical reaction that occurs under basic conditions include soluble sulfite ions and solid chromium(III) hydroxide. This reaction is used to remove $Cr^{6+}$ from wastewater generated by factories that make chrome-plated metals.

a. Write the net ionic equation for this redox reaction.
b. Which element is oxidized and which is reduced?
c. Identify the oxidizing and reducing agents in the reaction.
d. How many grams of sodium dithionite would be needed to remove the $Cr^{6+}$ in 100.0 L of wastewater that contains 0.00148 *M* chromate ion?

*8.116. A prototype battery based on iron compounds with large, positive oxidation numbers was developed in 1999. In the following reactions, assign oxidation numbers to the elements in each compound and balance the redox reactions in basic solution:

a. $FeO_4^{2-}(aq) + H_2O(\ell) \rightarrow FeOOH(s) + O_2(g) + OH^-(aq)$
b. $FeO_4^{2-}(aq) + H_2O(\ell) \rightarrow Fe_2O_3(s) + O_2(g) + OH^-(aq)$

***8.117.** **Polishing Silver** Silver tarnish is the result of silver metal reacting with sulfur compounds, such as $H_2S$, and $O_2$ in the air. The tarnish on silverware ($Ag_2S$) can be removed by soaking the silverware in a slightly basic solution of $NaHCO_3$ (baking soda) in a basin lined with aluminum foil.

a. Write a balanced chemical equation for the tarnish formation reaction.
b. Write a balanced net ionic equation for the tarnish removal process in which $Ag_2S$ reacts with Al metal, forming $Al(OH)_3(s)$, Ag metal, and $HS^-$ ions.

8.118. Give the formulas of the acids formed in the following chemical reactions of chlorine oxides.

a. $ClO + H_2O \rightarrow ? + ?$
b. $Cl_2O + H_2O \rightarrow HCl + ?$
c. $Cl_2O_6 + H_2O \rightarrow ? + ?$

**8.119.** Many nonmetal oxides react with water to form acidic solutions. Give the formulas of the acids produced in the following reactions:

a. $P_4O_{10} + 6\ H_2O \rightarrow ?$
b. $SeO_2 + H_2O \rightarrow ?$
c. $B_2O_3 + 3\ H_2O \rightarrow ?$

8.120. Write net ionic equations for the reactions that occur when

a. a sample of acetic acid is titrated with a solution of KOH.
b. a solution of sodium carbonate is mixed with a solution of calcium chloride.
c. calcium oxide dissolves in water.

**8.121.** One way to determine the concentration of hypochlorite ions ($ClO^-$) in solution is to first react them with $I^-$ ions. Under acidic conditions, the products of the reaction are $I_2$ and $Cl^-$ ions. Then the $I_2$ produced in the first reaction is titrated with a solution of thiosulfate ions ($S_2O_3^{2-}$). The products of the titration reaction are $S_4O_6^{2-}$ and $I^-$ ions. Write net ionic equations for the two reactions.

*8.122. **Fluoride Ion in Drinking Water** Sodium fluoride is added to drinking water in many municipalities to protect teeth against cavities. The target of the fluoridation is hydroxyapatite, $Ca_{10}(PO_4)_6(OH)_2$, a compound in tooth enamel. However, concern has arisen that fluoride ions in water may contribute to skeletal fluorosis, an arthritis-like disease.

a. Write a net ionic equation for the reaction between hydroxyapatite and sodium fluoride that produces fluorapatite, $Ca_{10}(PO_4)_6F_2$.
b. The U.S. Environmental Protection Agency now restricts the concentration of $F^-$ in drinking water to 4 mg/L. Express this concentration of $F^-$ in molarity.
c. The findings of one study of skeletal fluorosis suggest that drinking water with a fluoride concentration of 4 mg/L for 20 years raises the fluoride content in bone to 6 mg/g, a level at which a patient may experience stiff joints and other symptoms. How much fluoride (in milligrams) is present in a 100 mg sample of bone with this fluoride concentration?

***8.123. Rocket Fuel in Drinking Water** Near Las Vegas, Nevada, improper disposal of perchlorates used to manufacture rocket fuel contaminated a stream flowing into Lake Mead, the largest artificial lake in the United States and a major supply of drinking and irrigation water for the southwestern United States. The U.S. EPA has proposed an advisory range for perchlorate concentrations in drinking water of 4 to 18 μg/L. The perchlorate concentration in the stream averages 700.0 μg/L, and the stream flows at an average rate of 42.5 million liters per day.

a. What are the formulas of sodium perchlorate and ammonium perchlorate?
b. How many kilograms of perchlorate flow from the Las Vegas stream into Lake Mead each day?
c. What volume of perchlorate-free lake water would have to mix with the stream water each day to dilute the stream's perchlorate concentration from 700.0 to 4 μg/L?
d. Since 2003, the states of Maryland, Massachusetts, and New Mexico have limited perchlorate concentrations in drinking water to 0.1 μg/L. Five replicate samples were analyzed for perchlorates by laboratories in each state, and the following data (micrograms per liter) were collected:

| Maryland | Massachusetts | New Mexico |
|---|---|---|
| 1.1 | 0.90 | 1.2 |
| 1.1 | 0.95 | 1.2 |
| 1.4 | 0.92 | 1.3 |
| 1.3 | 0.90 | 1.4 |
| 0.9 | 0.93 | 1.1 |

Which lab produced the most precise analytical results?

*8.124. **Analyzing Fruit Juice for Vitamin C** The amount of ascorbic acid (vitamin C) in fruit juice is determined by a titration using a redox reaction (Figure P8.124). Iodine is the titrant, but because iodine solutions in water are unstable, the iodine is generated by a reaction in which a titrant containing iodate ($IO_3^-$) ions is added to a solution containing the juice sample, iodide ions ($I^-$), and a few drops of a starch solution that turns dark blue in the presence of iodine. Iodate ions are reduced to iodine ($I_2$) while iodide ions are oxidized to iodine:

i. $IO_3^-(aq) + I^-(aq) \rightarrow I_2(aq)$ (not balanced)

However, any iodine formed by reaction (i) is immediately reduced back to iodide by ascorbic acid from the juice sample:

ii. $C_6H_8O_6(aq) + I_2(aq) \rightarrow C_6H_6O_6(aq) + 2\ I^-(aq) + 2\ H^+(aq)$

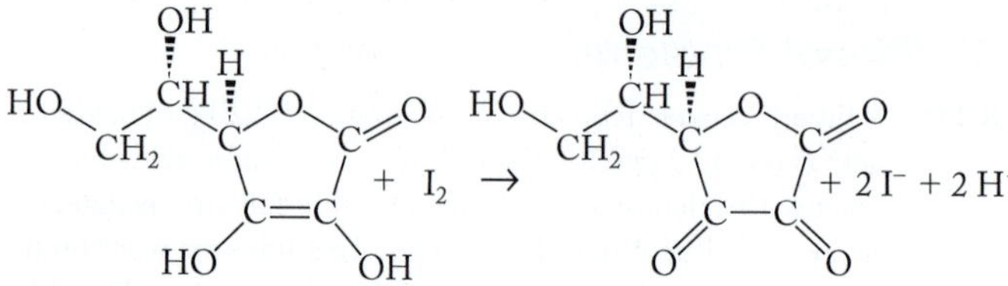

**FIGURE P8.124**

As long as ascorbic acid is present in the sample, no iodine can last in the solution and the starch indicator does not turn blue. As more iodate titrant is added, more iodine is generated by reaction (i) and consumed by reaction (ii), which also consumes the ascorbic acid in the sample. When enough iodate has been added and all the ascorbic acid has been consumed, the next drop of iodate titrant generates iodine that is not consumed, and the starch indicator in the test solution turns blue. This is the end point of the titration.

a. Write a balanced chemical equation based on the reaction described in (i).
b. Combine equations (i) and (ii) to show the overall reaction in the titration.
c. A 25 mL portion of lemon juice is mixed with iodide solution, HCl, and water to a total volume of 100.0 mL. A 0.002455 *M* solution of $KIO_3$ is used as a titrant. A volume of 10.47 mL of titrant is required to reach the end point of the titration. What is the concentration in milligrams per liter of ascorbic acid in the lemon juice?

*8.125. **Making Apple Cider Vinegar** Some people who prefer natural foods make their own apple cider vinegar. They start with freshly squeezed apple juice that contains about 6% natural sugars. These sugars, which all have nearly the same empirical formula, $CH_2O$, are fermented with yeast in a chemical reaction that produces equal numbers of moles of ethanol ($CH_3CH_2OH$) and carbon dioxide. The product of fermentation, called hard cider, undergoes an acid fermentation step in which ethanol and dissolved oxygen gas react to form acetic acid ($CH_3COOH$) and water. This acetic acid is the principal solute in vinegar.

a. Write a balanced chemical equation for the fermentation of natural sugars to ethanol and carbon dioxide. You may use in the equation the empirical formula given in the preceding paragraph.
b. Write a balanced chemical equation for the acid fermentation of ethanol to acetic acid.
c. What are the oxidation states of carbon in the reactants and products of the two fermentation reactions?
d. If a sample of apple juice contains $1.00 \times 10^2$ g of natural sugar, what is the maximum quantity of acetic acid that could be produced by the two fermentation reactions?

8.126. A food chemist determines the concentration of acetic acid in a sample of apple cider vinegar (see Problem 8.125) by acid–base titration. What is the concentration of acetic acid in the vinegar if the density of the sample is 1.01 g/mL, the titrant is 1.002 *M* NaOH, and the average volume of titrant required to titrate 25.00 mL subsamples of the vinegar is 20.78 mL? Express your answer the way a food chemist probably would: as percent by mass.

**8.127. Cave Formations** The stalactites and stalagmites in most caves are made of calcium carbonate. In the Lower Kane Cave in Wyoming, however, they are made of gypsum (calcium sulfate; see Figure 8.18). The presence of $CaSO_4$ is explained by the following sequence of reactions:

$$H_2S(aq) + 2\,O_2(g) \rightarrow H_2SO_4(aq)$$

$$H_2SO_4(aq) + CaCO_3(s) \rightarrow CaSO_4(s) + H_2O(\ell) + CO_2(g)$$

a. Which (if either) of these reactions is a redox reaction?
b. Write a net ionic equation for the reaction of $H_2SO_4$ with $CaCO_3$.
c. How would the net ionic equation be different if the reaction were written as follows?

$$H_2SO_4(aq) + CaCO_3(s) \rightarrow CaSO_4(s) + H_2CO_3(aq)$$

8.128. **Gardening Chemistry** Dolomite is a mixed carbonate mineral (Figure P8.128) with the formula $MgCa(CO_3)_2$. Gardeners add dolomite granules to soil and potting mixes to reduce acidity and provide a source of $Mg^{2+}$ ions, which plants need to grow. Would a 50-pound bag of dolomite granules neutralize more acid than a 50-pound bag of limestone ($CaCO_3$) granules? How much more? Express your answer as a percentage.

**FIGURE P8.128**

**8.129.** Which of the following reactions of calcium compounds is or are redox reactions?

a. $CaCO_3(s) \rightarrow CaO(s) + CO_2(g)$
b. $CaO(s) + SO_2(g) \rightarrow CaSO_3(s)$
c. $CaCl_2(s) \rightarrow Ca(s) + Cl_2(g)$
d. $3\,Ca(s) + N_2(g) \rightarrow Ca_3N_2(s)$

8.130. HF is prepared by reacting $CaF_2$ with $H_2SO_4$:

$$CaF_2(s) + H_2SO_4(\ell) \rightarrow 2\,HF(g) + CaSO_4(s)$$

HF can be electrolyzed, in turn, when dissolved in molten KF to produce fluorine gas:

$$2\,HF(\ell) \rightarrow F_2(g) + H_2(g)$$

Fluorine is extremely reactive, so it is typically sold as a 5% mixture by volume in an inert gas such as helium. How much $CaF_2$ is required to produce 500.0 L of 5% $F_2$ in helium? Assume that the density of $F_2$ gas is 1.70 g/L.

# 9

# Properties of Gases

## The Air We Breathe

**SCUBA DIVER AND SEA GOLDIES** The gas exhaled by the diver forms bubbles that increase in size as they rise to the surface.

PARTICULATE **REVIEW**

### *How Many Moles of Gas?*

In Chapter 9, we explore the physical and chemical properties of gases. The reactants for the combustion of methane are depicted in the molecular view here.

- Write a balanced equation for the combustion reaction of $CH_4$.
- Identify the limiting reactant in the reaction mixture in the image.
- If each molecule represents one mole of gas, how many moles of gas are present after the reaction is complete?

(Review Sections 7.3 and 7.4 if you need help.)

*(Answers to Particulate Review questions are in the back of the book.)*

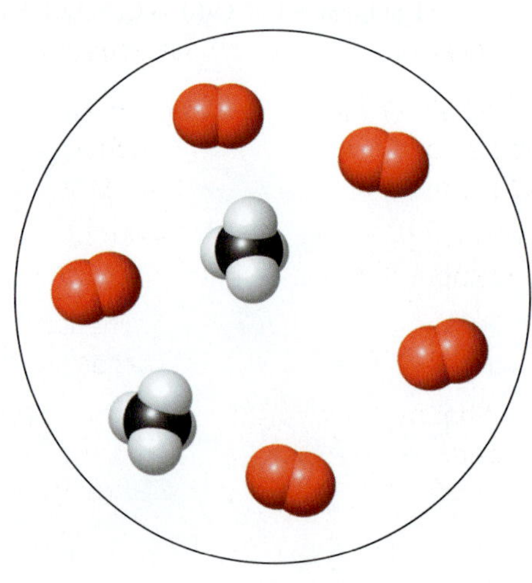

PARTICULATE **PREVIEW**

### *Expanding or Contracting Gases?*

The balloons in the drawing are filled with an equal number of moles of nitrogen and helium. As you read Chapter 9, look for ideas that will help you answer these questions:

- How does the volume of each balloon change when the temperature increases?
- How does the volume of each balloon change when the atmospheric pressure increases?
- What happens to the frequency of collisions between the particles when the temperature of the gas increases? When the atmospheric pressure increases?

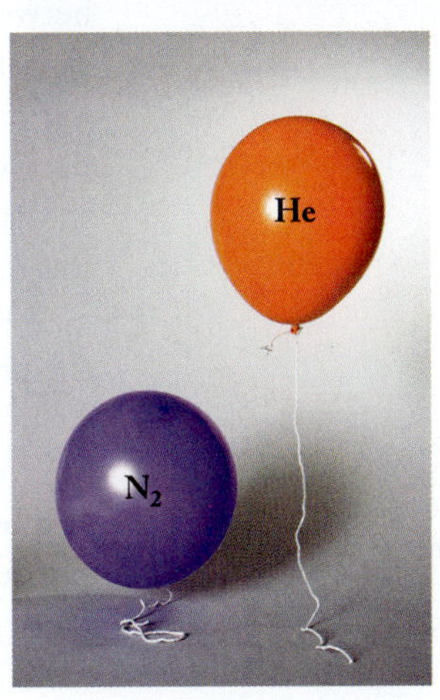

## Learning Outcomes

**LO1** Use kinetic molecular theory to explain the properties of gases

**LO2** Use Graham's law to establish relative rates of effusion of gases
**Sample Exercise 9.1**

**LO3** Relate the root-mean-square speed of a gas to its mass and absolute temperature
**Sample Exercise 9.2**

**LO4** Describe how a manometer measures gas pressure and convert between different units used to express pressure
**Sample Exercise 9.3**

**LO5** Use the individual, combined, and ideal gas laws to relate changes in the volume, temperature, pressure, and number of moles of a gas, and express these changes quantitatively
**Sample Exercises 9.4–9.8**

**LO6** Use the ideal gas law to relate density and molar mass of an ideal gas
**Sample Exercises 9.9, 9.10**

**LO7** Use balanced chemical equations to relate the volumes of gas-phase reactants and products by using the stoichiometries of the reactions and the ideal gas law
**Sample Exercise 9.11**

**LO8** Relate the mole fraction, partial pressure, and quantity of a gas in a mixture
**Sample Exercises 9.12, 9.13**

**LO9** Use intermolecular forces to explain the nonideal behavior of gases and use the van der Waals equation to correct for this behavior
**Sample Exercise 9.14**

## 9.1 An Invisible Necessity: The Properties of Gases

How often do you think about the composition of the air you breathe? Chances are it is not often. Breathing is something you do unconsciously, if you are in good health and if your senses are not telling you that something about the air around you might be unhealthy. Air is supposed to be colorless, tasteless, and odorless. If it's not, you have good reason to quickly search for cleaner air to breathe. Another assumption is that you are not engaged in an activity, such as fighting a fire or diving underwater, that limits your access to clean air and the precious oxygen in it.

Let's consider some situations for which the preceding assumptions may not be true. For example, people suffering from impaired lung function may require supplemental oxygen. Often this $O_2$ flows from a tank of pure oxygen at a rate between 1 and 2 L per minute. Even small, portable tanks contain enough oxygen to last several hours because the $O_2$ gas inside the tank has been compressed. A highly pressurized tank with an internal volume of less than 2.5 L can deliver hundreds of liters of $O_2$ at normal atmospheric pressure. Similarly, the air inside the tanks that scuba divers often wear can last for half an hour or more because these tanks are filled with air at pressures as high as 300 atm.

Gases are compressible, but liquids and solids are not. This is why increasing the pressure on a gas allows more gas to be compressed in the fixed volume of a small, solid cylinder of air carried by the firefighter in **Figure 9.1(a)**. By contrast, the volume of the gas in a weather balloon *increases* from its size at launch (**Figure 9.1b**) as it ascends into the atmosphere because the pressure of the air surrounding it *decreases*. Both observations illustrate a well-known property of most gases: The volume a gas occupies is inversely proportional to its pressure ($V \propto 1/P$).

**CONNECTION** Average atmospheric pressure at sea level is, by definition, 1 atmosphere (atm), as described in Section 6.5.

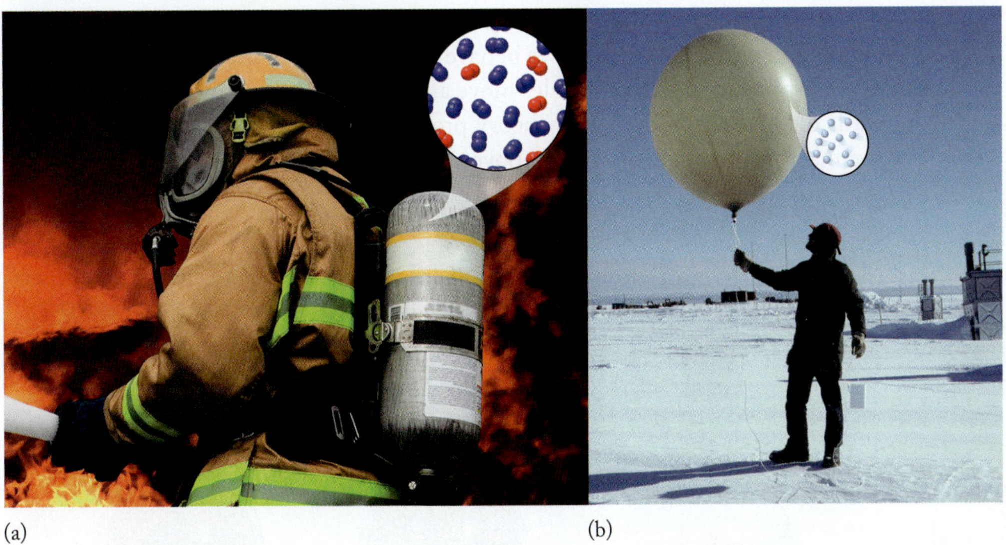

(a) (b)

**FIGURE 9.1** The volume of a quantity of gas is inversely proportional to the pressure around it. (a) The tank of air for a firefighter is referred to as compressed air because the particles are under high pressure in the small tank. (b) The volume of this weather balloon will expand as it ascends into the sky and experiences decreasing atmospheric pressure with increasing altitude.

Following is a list of other properties of gases with some examples that you may have observed:

- The pressure of a quantity of gas in a rigid container is directly proportional to its absolute temperature ($P \propto T$). This property is one reason why a big drop in temperature in early winter may set off low-tire-pressure sensors in automobiles even though no air has leaked out of the tires.
- The gas pressure inside a rigid container at constant temperature is directly proportional to the quantity (number of moles) of gas in the container ($P \propto n$). For example, a pressure gauge attached to a scuba tank (**Figure 9.2**) not only measures the internal pressure of the tank but also provides a diver with a measure of how much air is left in the tank.
- The volume of a gas is inversely proportional to the pressure we apply ($P \propto 1/V$). For example, a sealed bag of potato chips in the hold of an airliner during flight may appear to be ready to burst your baggage because the atmospheric pressure at 35,000 feet is much lower than at sea level.
- All gases are miscible with all other gases—that is, gas mixtures are homogeneous (unless the gases react with each other). For example, when a diver takes a breath of air from a scuba tank, the proportion of $O_2$ in the air is always the same, whether it's the first breath taken at the beginning of a dive or the last one before reaching the surface.
- At a given temperature and pressure, the densities of gaseous substances are directly proportional to their molar masses and are much lower than their densities as solids or liquids. For example, the density of liquid butane (the fuel in disposable lighters) at its boiling point of 0°C is 0.60 g/mL, which is more than 200 times greater than the density of butane vapor at 0°C (0.0026 g/mL).
- Gases expand to occupy the entire volume of their containers. For example, the shape of a party balloon is essentially the same whether it is partially or fully inflated. The gas in the balloon uniformly occupies all the available space, pushing outward with the same pressure in all directions.
- Gases in inflated party balloons escape, or *effuse*, through the walls of the balloons at rates that are inversely proportional to the molar masses of the gases. Those with the smallest molar masses, such as helium, escape the most rapidly.

**FIGURE 9.2** The internal pressure of a scuba tank is a direct measure of the amount of air inside it.

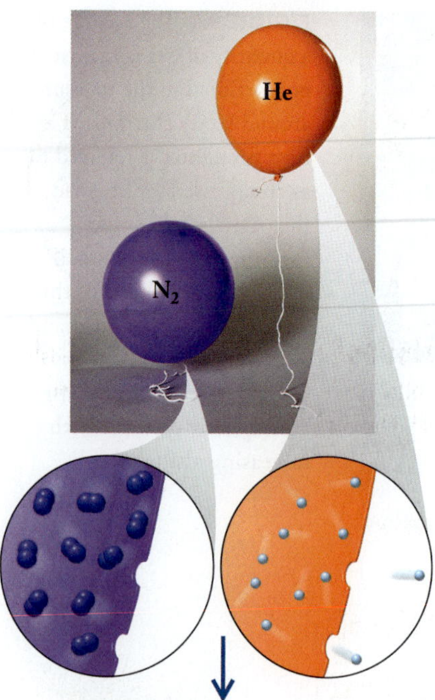


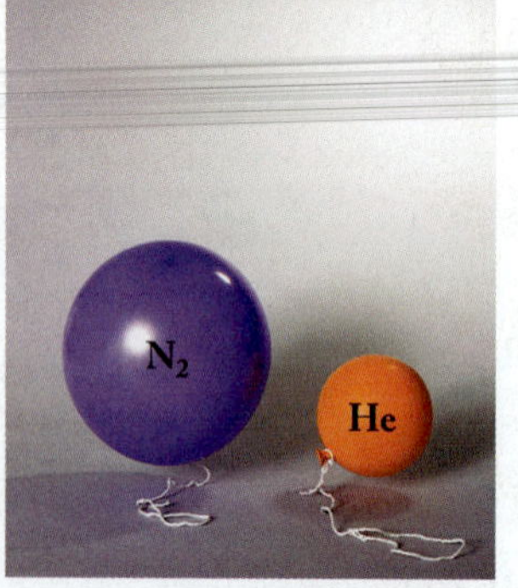


**FIGURE 9.4** Two balloons at the same temperature and pressure, one filled with nitrogen gas and the other filled with an equal volume of helium gas. Over time, the volume of the helium balloon decreases much faster.

**effusion** the process by which a gas escapes from its container through a tiny hole into a region of lower pressure.

**Graham's law of effusion** the principle that the rate of effusion of a gas is inversely proportional to the square root of its molar mass.

**kinetic molecular theory (KMT)** a model that explains the behavior of gases on the basis of the motion of the particles that make them up.

**root-mean-square speed ($u_{rms}$)** the square root of the average of the squared speeds of all the particles in a population of gas particles.

In this chapter, we investigate these and other properties of gases in more detail, and we explain why gases behave the way they do on the basis of a theory of how their atoms and molecules move and interact with each other and with the walls of their containers. This theory provides particle-based explanations of all the macroscale properties listed above.

**CONCEPT TEST**

Nitrogen dioxide, $NO_2$, is a red-brown gas that contributes to the formation of photochemical smog. Which of the images in **Figure 9.3** best depicts a sealed container of $NO_2$?

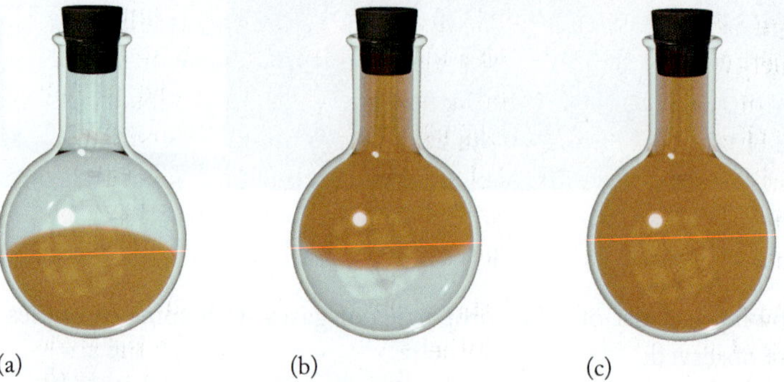


**FIGURE 9.3** Three proposed depictions of a sealed container of gas.

*(Answers to Concept Tests are in the back of the book.)*

## 9.2 Effusion, Diffusion, and the Kinetic Molecular Theory of Gases

Let's begin our exploration of how and why gases behave the way they do by examining the last property of gases listed in Section 9.1. Why is it that helium-filled party balloons made of latex rubber deflate and are no longer buoyant after a day or two, whereas the same balloons filled with air remain inflated for many days? The answer to this question involves the structure of party balloons and the motion of the atoms and molecules inside them.

Tiny imperfections (**Figure 9.4**) in latex rubber balloons allow gas particles to slowly escape in a process called **effusion**. This phenomenon was investigated in the early 19th century by Scottish chemist Thomas Graham (1805–1869), who discovered that the rate of effusion of different gases was inversely proportional to the square root of their densities. As noted in Section 9.1, the densities of gases are proportional to their molar masses, so the modern interpretation of **Graham's law of effusion** states that the effusion rate of a gas is inversely proportional to the square root of its molar mass and its density:

$$\text{Effusion rate} \propto \sqrt{\frac{1}{\text{density}}} \propto \sqrt{\frac{1}{\text{molar mass}}}$$

Why do gases with lower molar masses effuse faster? This behavior can be explained using the **kinetic molecular theory (KMT)** of gases. KMT is based on the following set of assumptions about the particulate nature of gases:

- Gas molecules have tiny volumes compared with the volume they occupy. Their individual volumes are so small as to be considered negligible, allowing particles in a gas to be treated as *point masses*—masses with essentially no volume. Gas molecules are separated by large distances; hence, a gas is mostly empty space. This assumption is consistent with the low density of gases when compared to the densities of liquids and solids. For example, because a given volume of liquid butane contains 223 times as many molecules as the same volume of butane vapor at 0°C, the butane molecules must be spaced much farther apart in the vapor phase.

- Gas molecules move constantly and randomly throughout the space they occupy, continually colliding with one another and with their container walls (**Figure 9.5**). These collisions are *elastic*; that is, they result in no net transfer of energy to the walls. To get a mental picture of what an elastic collision is like, imagine a cue ball colliding with one or more stationary balls on a pool table (**Figure 9.6**). Most of the balls exit the collision in different directions at different speeds. Some of the kinetic energy of the cue ball has transferred to the others, but the *average* kinetic energy of all the balls is the same immediately after the collision as it was before.
- Because gas particles are so spread out, we assume their intermolecular forces of attraction to one another are negligible, unlike the attractive forces between particles in liquids and solids. In fact, KMT assumes the gas particles don't interact *at all*.
- The average kinetic energy of the molecules in a gas is proportional to the absolute temperature of the gas. All populations of gas molecules at the same temperature have the same average kinetic energy.

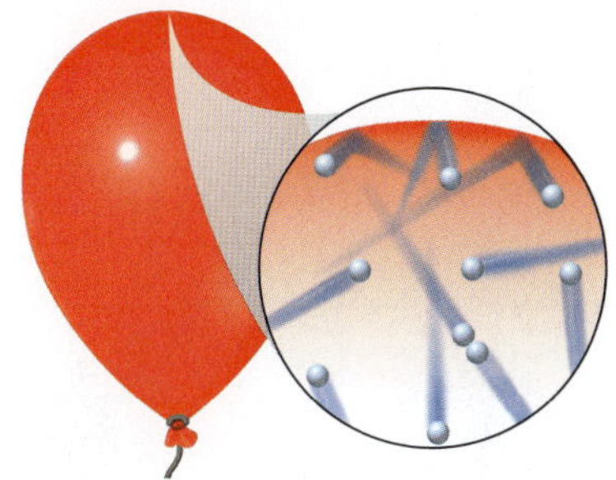

**FIGURE 9.5** Helium atoms continuously collide with one another and with their container walls.

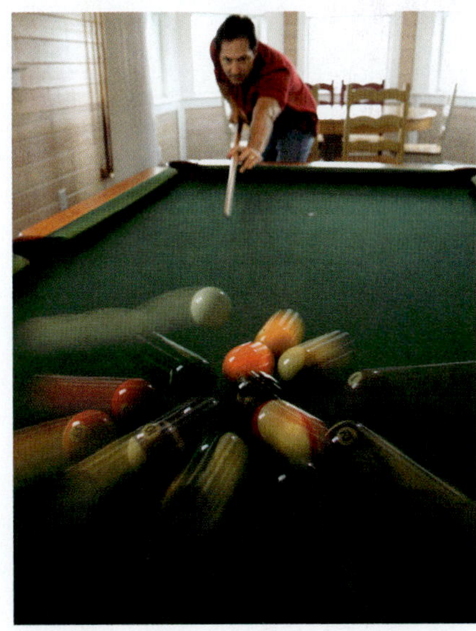

**FIGURE 9.6** The collisions between the balls on a pool table are a good model for elastic collisions.

Recall from Chapter 1 that the kinetic energy of any object in motion—including a gas-phase atom or molecule—is proportional to its mass ($m$) and the square of its speed ($u$):

$$\text{KE} = \tfrac{1}{2}mu^2 \qquad (1.3)$$

We use the term *average* kinetic energy in describing the behavior of molecules of a gas because the speeds of particles with the same mass and at the same temperature are not all the same. Instead, they are distributed over a range of values, as shown in the graph in **Figure 9.7**. The $x$-axis of the graph is particle speed; the $y$-axis represents the relative number of particles at each speed. The peak in the speed distribution curve represents the *most probable speed* ($u_m$), the speed of the largest number of particles. Because the distribution of speeds is not symmetrical, the *average speed* ($u_{avg}$), which is simply the arithmetic average of the individual speeds of all the particles, is a little higher than the most probable speed. A third kind of average speed, called the **root-mean-square speed ($u_{rms}$)**, is larger still. Its name comes from its theoretical definition: the square root of the average of the squared speeds of all the particles of a gas. It is important to us because $u_{rms}$ is the speed of a particle whose kinetic energy is the same as the *average* kinetic energy of all the particles in a gas. The equation form of this definition is

$$\text{KE}_{avg} = \tfrac{1}{2}mu_{rms}^2 \qquad (9.1)$$

We can use Equation 9.1 to compare the $u_{rms}$ speeds of gases at the same temperature. For example, suppose we have a mixture of $O_2$ ($\mathcal{M}$ = 32.00 g/mol), $N_2$ ($\mathcal{M}$ = 28.01 g/mol), He ($\mathcal{M}$ = 4.003 g/mol), and $H_2$ ($\mathcal{M}$ = 2.016 g/mol). Because all four gases are at the same temperature, the average kinetic energy of the particles in each gas is also the same. This is true because $\text{KE}_{avg}$ is directly

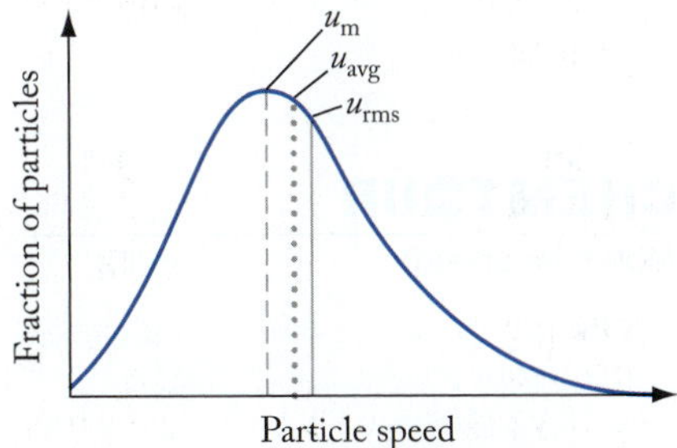


**FIGURE 9.7** At any given temperature, the speeds of the particles that make up a gas cover a range of values. The dashed line represents the most probable particle speed ($u_m$); the dotted line is the arithmetic average speed ($u_{avg}$); and the solid line is the root-mean-square speed ($u_{rms}$), which is directly proportional to the average kinetic energy of the particles.

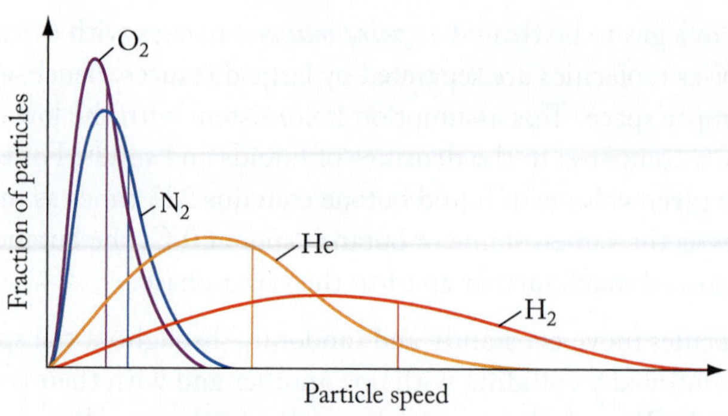


**FIGURE 9.8** The speeds of gas particles at a given temperature are inversely related to their masses: the most massive (here, molecules of $O_2$) move, on average, the slowest, and the least massive (molecules of $H_2$) move the fastest and have the most widely distributed speeds. The vertical lines represent the root mean square speeds.

proportional to the temperature of a gas according to the following equation, where $k_B$ is a constant known as the Boltzmann constant:

$$KE_{avg} = \frac{3}{2}k_BT \tag{9.2}$$

Therefore, the $\frac{1}{2}mu^2_{rms}$ term in Equation 9.1 is also the same for each gas. However, the masses of their particles are not the same, which means their $u_{rms}$ speeds cannot be the same: the least massive particles (the molecules of $H_2$) should have the highest $u_{rms}$ speeds, whereas the most massive particles (the molecules of $O_2$) should have the slowest $u_{rms}$ speeds. This prediction is supported by the speed distribution curves for these four elements in **Figure 9.8**. Note how particle speeds are inversely related to particle masses and that the least massive particles (the atoms of He and molecules of $H_2$) have not only the highest average speeds but also the most widely distributed speed profiles.

Let's use Equation 9.1 to make a more quantitative connection between particle speed and particle mass. We start with a mixture of a heavier gas ($N_2$) and a lighter gas (He) from Figure 9.8. The mixture has a single temperature, so the two gases have the same average kinetic energies:

$$KE_{avg,He} = KE_{avg,N_2} \tag{9.3}$$

Substituting the right side of Equation 9.1 into both sides of Equation 9.3:

$$\frac{1}{2}m_{He}u^2_{rms,He} = \frac{1}{2}m_{N_2}u^2_{rms,N_2} \tag{9.4}$$

Molecular Speed

Rearranging the terms in Equation 9.4 so that the masses are on one side of the equation and the $u_{rms}$ speeds are on the other:

$$\frac{u^2_{rms,He}}{u^2_{rms,N_2}} = \frac{m_{N_2}}{m_{He}}$$

Taking the square root of both sides:

$$\frac{u_{rms,He}}{u_{rms,N_2}} = \sqrt{\frac{m_{N_2}}{m_{He}}} \tag{9.5}$$

Now let's scale up the mass values on the right side of Equation 9.5 by multiplying both of them by the Avogadro constant. Doing so converts molecular masses into molar masses without changing the value of their ratio:

$$\frac{u_{rms,He}}{u_{rms,N_2}} = \sqrt{\frac{\mathcal{M}_{N_2}}{\mathcal{M}_{He}}} = \sqrt{\frac{28.02 \cancel{\text{g/mol}}}{4.003 \cancel{\text{g/mol}}}} = 2.646 \tag{9.6}$$

Equation 9.6 tells us that the root-mean-square speeds of the two gases are inversely proportional to the square roots of their molar masses. Therefore, atoms of He collide with the walls of their container (such as the green party balloon in Figure 9.4) 2.646 times more frequently than molecules of $N_2$ at the same pressure and temperature. This means He atoms encounter and pass through atomic-scale escape routes in their balloon 2.646 times more frequently than $N_2$ molecules do. As a result, He effuses 2.646 times faster than $N_2$, and a helium-filled balloon deflates much faster than one filled with air (mostly $N_2$).

Equation 9.6 can be used with any pair of gases that behave as predicted by KMT. We can write a generic form of it for any two gases A and B:

$$\frac{u_{\text{rms,A}}}{u_{\text{rms,B}}} = \sqrt{\frac{\mathcal{M}_B}{\mathcal{M}_A}} \qquad (9.7)$$

If particles of gas A have higher speeds than particles of gas B, then gas A particles collide more frequently with the walls of their container, which increases the frequency with which they encounter and pass through microscopic holes in the wall and undergo effusion. We express this in equation form as

$$\frac{\text{effusion rate}_A}{\text{effusion rate}_B} = \sqrt{\frac{\mathcal{M}_B}{\mathcal{M}_A}} \qquad (9.8)$$

Equation 9.8 is a mathematical representation of Graham's law that can be used to calculate the relative rates of effusion of pairs of gases.

**SAMPLE EXERCISE 9.1** Calculating Relative Rates of Effusion **LO2**

An odorous gas emitted by a hot spring was found to effuse at 0.342 times the rate at which helium effuses. What is the molar mass of the emitted gas?

**Collect and Organize** We are asked to determine the molar mass of an unknown gas on the basis of its rate of effusion relative to the rate for helium ($\mathcal{M}_{He}$ = 4.003 g/mol). Equation 9.8 provides the mathematical relationship between the effusion rates and molar masses of a pair of gases.

**Analyze** We can estimate an answer by considering Equation 9.8. Because the unknown gas effuses about 1/3 as fast as He does, the molar mass should be nearly $3^2$ or 9 times the mass of He, or about 9 × 4 = 36 g/mol.

**Solve** Using the symbol X for the unknown gas in Equation 9.8:

$$\frac{\text{effusion rate}_X}{\text{effusion rate}_{He}} = \sqrt{\frac{\mathcal{M}_{He}}{\mathcal{M}_X}} = 0.342$$

Squaring both sides of the rightmost equality yields

$$\frac{\mathcal{M}_{He}}{\mathcal{M}_X} = (0.342)^2$$

$$\mathcal{M}_X = \frac{4.003 \text{ g/mol}}{(0.342)^2} = 34.2 \text{ g/mol}$$

**Think About It** The molar mass of the unidentified gas is 34.2 g/mol, which is consistent with our prediction. One possibility for the identity of this gas is $H_2S(g)$, $\mathcal{M}$ = 34.08 g/mol, a toxic gas with a foul smell associated with rotten eggs or low tide on a seaweed-covered shoreline.

 **Practice Exercise**
Helium effuses 3.16 times as fast as which other noble gas?

*(Answers to Practice Exercises are in the back of the book.)*

Equation 9.7 allows us to compare $u_{rms}$ values for any two gases, but it does not provide actual $u_{rms}$ values. We can relate $u_{rms}$ to the molar mass and temperature of a gas by using Equations 9.1 and 9.2. Setting their right sides equal to each other:

$$\frac{1}{2}mu_{rms}^2 = \frac{3}{2}k_B T \quad (9.9)$$

Solving Equation 9.9 for $u_{rms}$, we obtain:

$$u_{rms} = \sqrt{\frac{3k_B T}{m}} \quad (9.10)$$

The Boltzmann constant, $k_B$, is related to another constant $R$ and to the Avogadro constant: $k_B = R/N_A$. Substituting this equality into Equation 9.10 and recognizing that $\mathcal{M} = mN_A$, we arrive at an equation that relates $u_{rms}$ to $\mathcal{M}$ and $T$:

$$u_{rms} = \sqrt{\frac{3RT}{\mathcal{M}}} \quad (9.11)$$

$R$ is called the **universal gas constant**, and we explore its origin in Section 9.6. For now, we will simply treat it as a constant with the following value:

$$R = 8.314 \frac{\text{kg} \cdot \text{m}^2}{\text{s}^2 \cdot \text{mol} \cdot \text{K}}$$

**CONNECTION** We learned in Chapter 1 that Kelvin and Celsius temperatures are related by the equation $T(\text{K}) = T(°\text{C}) + 273.15$.

In using Equation 9.11, we must express molar masses in kilograms (not grams) per mole to be consistent with the units on $R$, and the temperature must be expressed in kelvin. The units on $u_{rms}$ are meters per second.

**universal gas constant** the constant $R$ in the ideal gas equation; its value and units depend on the units used for the variables in the equation.

**diffusion** the spread of one substance (usually a gas or liquid) through another.

**mean free path** the average distance that a particle can travel through air or any gas before colliding with another particle.

### SAMPLE EXERCISE 9.2 Calculating Root-Mean-Square Speeds — LO3

Calculate the root-mean-square speed of nitrogen molecules at 25°C.

**Collect and Organize** We are asked to calculate the root-mean-square speed ($u_{rms}$) of $N_2$ molecules at 25°C (298 K). Nitrogen gas has a molar mass of 28.01 g/mol. Equation 9.11 relates the $u_{rms}$ speed of the particles in a gas to its absolute temperature and molar mass.

**Analyze** We can estimate the value of $u_{rms}$ from the approximate values of $R$, $T$, and $\mathcal{M}$ on the right side of Equation 9.11: $R$ is about 10 and $T$ is about 300 K, so the value of the numerator is approximately $3 \times 10 \times 300$, or nearly 10,000. Expressing the molar mass of $N_2$ in kilograms per mole (to be compatible with the units on $R$) makes the denominator about 0.03, and the fraction inside the square root sign is about 10,000/0.03, or 333,000, the square root of which is between 500 and 600 m/s.

**Solve**

$$u_{rms} = \sqrt{\frac{3RT}{\mathcal{M}}}$$

$$= \sqrt{\frac{(3)\left(8.314 \frac{\text{kg} \cdot \text{m}^2}{\text{s}^2 \cdot \cancel{\text{mol}} \cdot \cancel{\text{K}}}\right)(273 + 25)\ \cancel{\text{K}}}{\left(\frac{28.01\ \cancel{\text{g}}}{1\ \cancel{\text{mol}}}\right)\left(\frac{1\ \cancel{\text{kg}}}{1000\ \cancel{\text{g}}}\right)}} = 515\ \text{m/s}$$

**Think About It** The calculated result is within our estimated range, and it suggests that the molecules in the air surrounding us are moving *very* rapidly. In fact, 515 m/s corresponds to 1150 miles per hour. An airplane flying that fast would be supersonic.

**Practice Exercise** Calculate the root-mean-square speed of helium at 25°C in meters per second, and compare your result with the root-mean-square speed of nitrogen calculated in Sample Exercise 9.2.

Equation 9.11 and the foundations of kinetic molecular theory tell us that the root-mean-square speed of the particles of a gas is proportional to the square root of the absolute temperature of the gas. This increase in $u_{rms}$ with increasing $T$ is illustrated in **Figure 9.9**. Note that increasing $u_{rms}$ values are accompanied by wider distributions in particle speeds with increasing temperature. The profiles in Figure 9.9 also show that no matter how hot a gas is, a finite possibility exists that at least some particles in it are hardly moving.

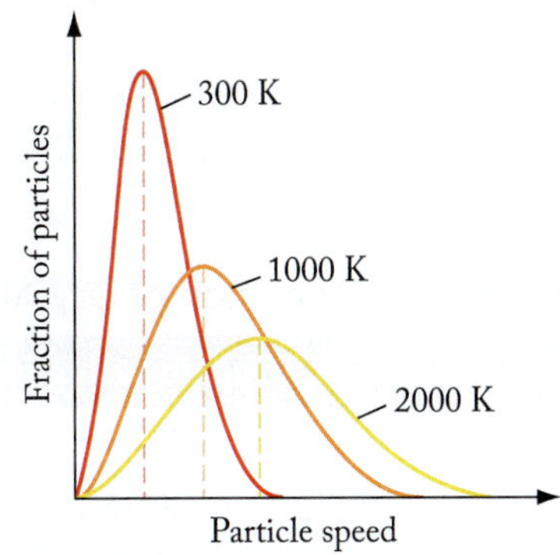


**FIGURE 9.9** The most probable speeds (dashed lines) of the particles of a gas increase with increasing temperature. Notice that the distributions of particle speeds also broaden as the temperature increases.

The effusion of gases is closely related to their **diffusion**, which is defined as the spread of one substance through another. Gas-phase diffusion occurs when there are differences in composition of a mixture of gases within the space they occupy. Diffusion is one process by which odors spread from their source through the air, such as the smell of perfume from the person wearing it or the smell of baking bread from the kitchen. Given the high speeds at which gas particles move, you might expect diffusion in air to be very rapid. It's not. If the air is still, the scent of an open bottle of perfume may take several minutes to spread throughout a space the size of a bedroom. It takes this long because the molecules of scent keep colliding with molecules of $N_2$, $O_2$, and other components of air (**Figure 9.10**). Even though ambient air is mostly empty space, these molecules still take up enough space to limit how far any molecule can travel before it collides with one of them.

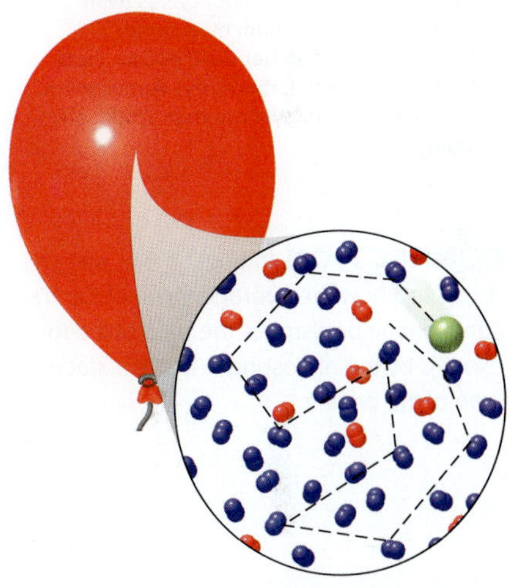

**FIGURE 9.10** Diffusion rates of gases are slowed by frequent collisions between gas-phase particles. The dashed line shows the path followed by one particle as it collides with others. The average distance between collisions is called the mean free path.

The average distance that a particle can travel through air or any gas before colliding with another one is called the **mean free path** of the particle. Mean free paths depend on how densely the air particles are packed, and that depends on their pressure. At an atmospheric pressure of 1 atm, the mean free path is about $6.8 \times 10^{-8}$ m. For particles traveling at hundreds of meters per second, such a tiny distance translates into about $10^{10}$ collisions per second. Such a high collision frequency translates into a lot of bouncing around with little forward progress—and a slow rate of diffusion.

Thomas Graham also studied the diffusion of gases. One of his favorite experimental designs involved measuring the time that different gases took to diffuse through porous plugs made of plaster of Paris. The results of the experiments were essentially the same as those for gas effusion—namely, gases diffuse at rates that are inversely proportional to the square root of their densities or their molar masses. Thus, Graham's law of effusion and *Graham's law of diffusion* are interchangeable, and Equation 9.8 applies to both the effusion and diffusion of gases.

## CONCEPT TEST

Which graph in **Figure 9.11** best describes the ratio of the rates of diffusion of two gases (X and Y) as a function of temperature?

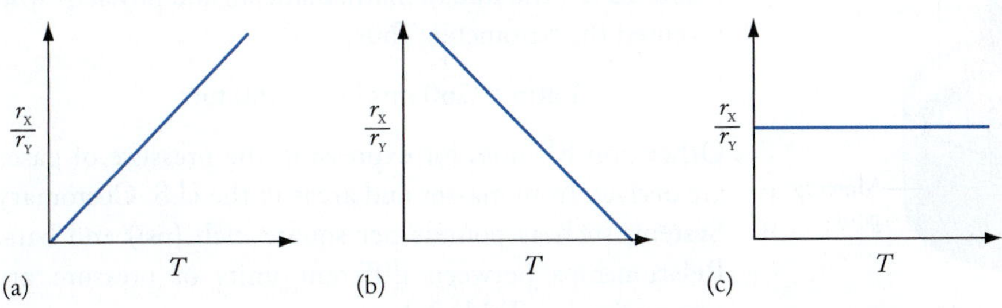


**FIGURE 9.11** Three proposed graphs of temperature versus ratio of rates of diffusion.

## 9.3 Atmospheric Pressure

We have established that gas molecules are in constant rapid motion and have molecular speeds defined by their molar masses and their temperature. What happens when these gas molecules collide with other molecules or with the walls of a container such as the firefighter's air tank or the weather balloon from Figure 9.1? The frequency and force of these collisions results in pressure, a concept we first encountered in Chapter 6. Simply put, more frequent or more forceful collisions involving gas molecules lead to higher pressures.

In this section, we begin our exploration of the pressures of gases by focusing on the ones we draw into our lungs with each breath. Earth is surrounded by an atmosphere that is about 50 km thick and is composed primarily of nitrogen (78% by volume), oxygen (21%), and lesser proportions of other gases (**Table 9.1**). To put the thickness of the atmosphere in perspective, if Earth were the size of an apple, the atmosphere would be about as thick as the apple's skin.

**TABLE 9.1 Composition of Dry Air**[a]

| Compound | % (by volume) |
|---|---|
| Nitrogen | 78.08 |
| Oxygen | 20.95 |
| Argon | 0.934 |
| Carbon dioxide | 0.0408[b] |
| Neon | 0.0018 |
| Helium | 0.00052 |
| Methane | 0.00018 |
| Krypton | 0.00011 |

[a]Includes major and minor gases (with concentrations >1 ppm by volume).
[b]Value as of November 2018 (NOAA Earth System Research Laboratory). Atmospheric $CO_2$ is increasing by more than 2 ppm each year.

Earth's atmosphere is pulled downward by gravity and exerts a force spread across the surface of the planet. The ratio of force, $F$, to surface area, $A$, defines pressure, $P$:

$$P = \frac{F}{A} \tag{6.1}$$

For Earth's atmosphere, the $P$ in Equation 6.1 is atmospheric pressure, which can be measured with an instrument called a **barometer**. A simple but effective barometer design consists of a narrow tube about 1 m long, sealed at one end and filled with mercury. The tube is inverted, and its open end is placed into a pool of mercury that is open to the atmosphere (**Figure 9.12**). Gravity pulls downward on the mercury in the tube, creating a vacuum at the top of the tube. Meanwhile, atmospheric pressure pushes downward on the mercury in the pool, which pushes it up into the tube. The opposing forces on the mercury column in the tube create a stable column height that provides a measure of the atmospheric pressure.

**CONNECTION** In Section 6.5, we discussed how the average value of Earth's atmospheric pressure is the ratio of the force exerted by the atmosphere to the surface area of the planet.

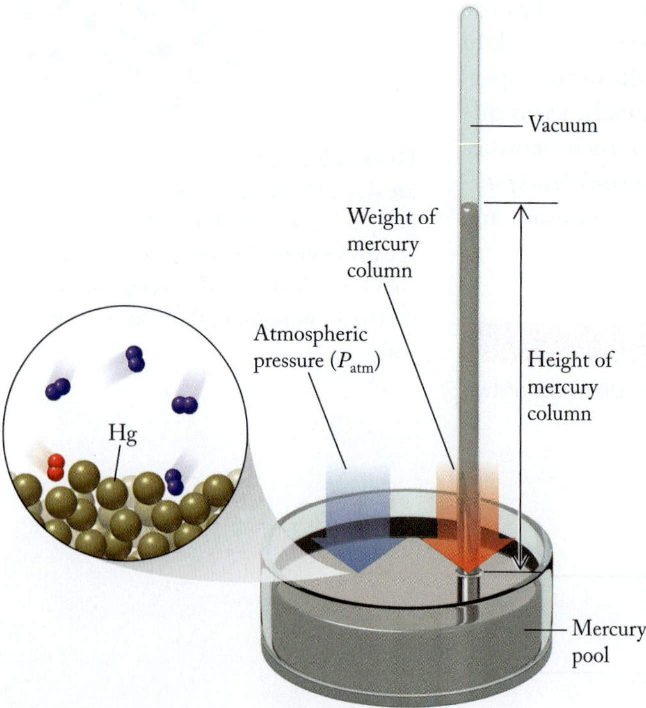


**FIGURE 9.12** The height of the mercury column in this simple barometer designed by Evangelista Torricelli is proportional to atmospheric pressure.

Atmospheric pressure varies from place to place and with changing weather conditions. As we noted in Chapter 6, the average atmospheric pressure at sea level was used to define the standard atmosphere (1 atm), a non-SI unit of pressure. This pressure can support a column of mercury 760 mm high, which is the basis for another non-SI unit of pressure: millimeters of mercury (mmHg). Pressure in millimeters of mercury is also expressed in a unit called the torr in honor of Evangelista Torricelli (1608–1647), the Italian mathematician and physicist who invented the barometer. Thus,

$$1 \text{ atm} = 760 \text{ mmHg} = 760 \text{ torr}$$

Other non-SI units for expressing the pressure of gases are derived from masses and areas in the U.S. Customary System, such as pounds per square inch (psi) and bars. Relationships between different units of pressure are summarized in **Table 9.2**.

The SI unit of pressure is the pascal (Pa), named in honor of French mathematician and physicist Blaise Pascal

**TABLE 9.2** Units for Expressing Pressure

| Unit | Value |
|---|---|
| Standard atmosphere (atm) | 1 atm |
| Pascal (Pa) | 1 atm = $1.01325 \times 10^5$ Pa |
| Kilopascal (kPa) | 1 atm = 101.325 kPa |
| Millimeter of mercury (mmHg) | 1 atm = 760 mmHg |
| Torr | 1 atm = 760 torr |
| Bar | 1 atm = 1.01325 bar |
| Millibar (mbar or mb) | 1 atm = 1013.25 mbar |
| Pounds per square inch (psi) | 1 atm = 14.7 psi |
| Inches of mercury | 1 atm = 29.92 inches of Hg |

**barometer** an instrument that measures atmospheric pressure.

**manometer** an instrument for measuring the pressure exerted by a gas.

(1623–1662), the first to propose that atmospheric pressure decreases with increasing altitude. We can explain this phenomenon by noting that the atmospheric pressure anywhere on Earth's surface is related to the mass of the column of air *above* that location. As altitude increases, the mass of the column of air *above* that altitude decreases. Less mass means a smaller force exerted downward by the air at higher altitude, which means less pressure (**Figure 9.13**).

**STEPWISE ANIMATION**
Manometer

Scientists conducting experiments with gases usually need to know the pressures of gases in closed systems. A **manometer** is an instrument that was once widely used to measure these pressures. One common type of manometer is illustrated in **Figure 9.14**. It is a U-shaped tube filled with mercury (or another dense liquid). One end of the tube is connected with a valve to a sample vessel containing the gas of interest; the other end is open to the atmosphere.

**STEPWISE ANIMATION**
Measuring Gas Pressure

This type of manometer is particularly useful for determining gas pressures greater than atmospheric pressure. Initially, the mercury levels are the same in both arms of the tube ($\Delta h = 0$; Figure 9.14a). After the manometer is connected to the sample vessel and the valve is opened (Figure 9.14b), the greater pressure in

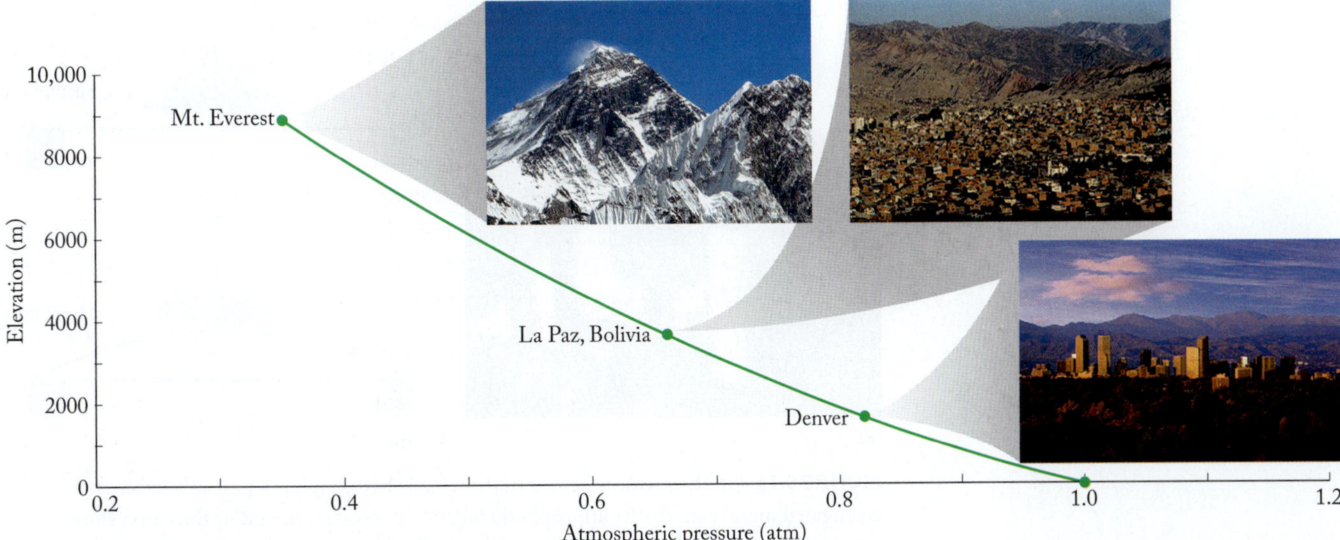


**FIGURE 9.13** Atmospheric pressure decreases with increasing altitude because the mass of the column of air above a given area decreases with increasing altitude.

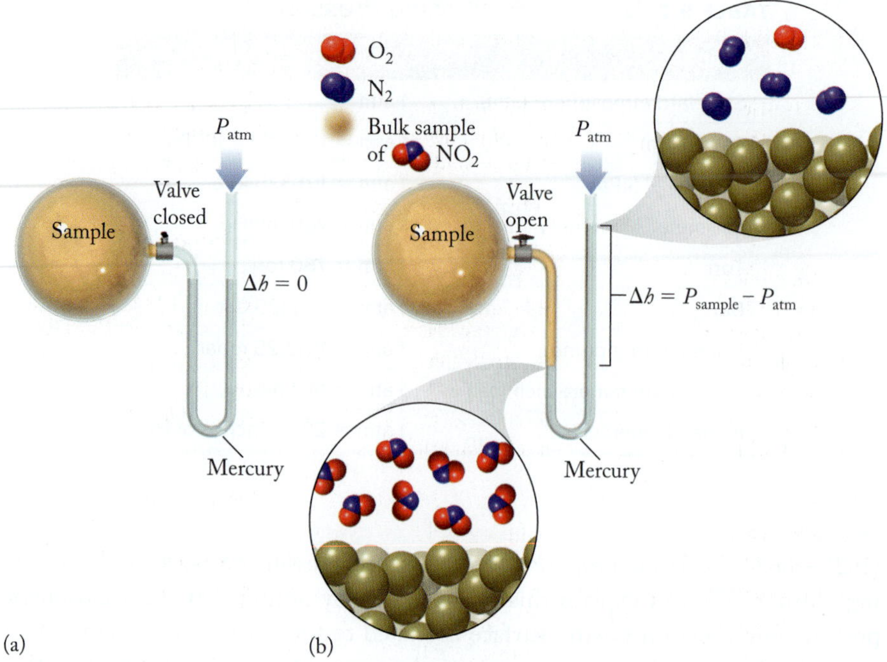


**FIGURE 9.14** A manometer for measuring sample pressures that are greater than atmospheric pressure. (a) When the valve is closed, the mercury levels are the same. (b) When the valve is opened, the pressure of the gas in the bulb pushes down on the mercury, making the two mercury levels different.

the sample flask pushes down on the mercury in the left arm of the tube, creating a $\Delta h$ that is a measure of the difference between the sample pressure and atmospheric pressure.

Today, manometers have been largely replaced by electronic pressure sensors based on flexible metallic or ceramic diaphragms. As the pressure on one side of the diaphragm increases, it distorts away from that side. This is the same mechanism used to sense changes in atmospheric pressure in most barometers, including the recording barometer, or *barograph*, shown in **Figure 9.15**.

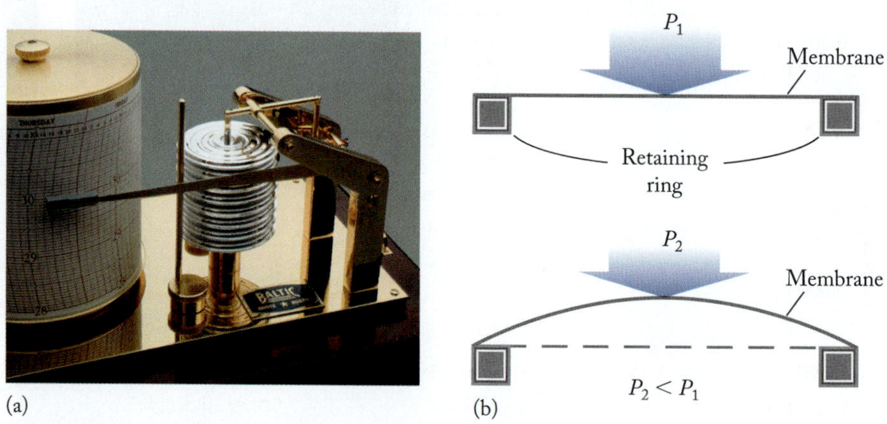


**FIGURE 9.15** (a) The pressure sensor in this classic barograph is a partially evacuated corrugated metal can. (b) As atmospheric pressure decreases, the lid of the can distorts outward. This motion is amplified by a series of levers and transmitted via the barograph's horizontal arm to a pen tip that records the pressure on the graph paper as the drum slowly turns. A week's worth of barometric data can be recorded in this way.

**SAMPLE EXERCISE 9.3** Measuring Gas Pressure with a Manometer **LO4**

For centuries, high-temperature kilns such as the one shown in **Figure 9.16** have been used to convert limestone ($CaCO_3$) into quicklime (CaO), which is used to "sweeten" (reduce the acidity of) soil in agriculture and to make mortar and concrete:

$$CaCO_3(s) \rightarrow CaO(s) + CO_2(g)$$

The same experiment can be carried out in the apparatus shown in **Figure 9.17**. Figure 9.17(a) shows a sample of $CaCO_3$ before heating. All the air has been removed from the flask with a vacuum pump, and the flask has been connected to a manometer. Then the flask is heated, decomposing the $CaCO_3$ to CaO and $CO_2$ gas. After the flask has cooled back to its initial temperature, the valve is opened and the difference in levels of mercury ($\Delta h$) in the arms of the manometer is 144 mm (Figure 9.17b). Calculate the pressure of the $CO_2$ in (a) millimeters of mercury, (b) atmospheres, and (c) kilopascals on a day when atmospheric pressure is 756 mmHg.

**FIGURE 9.16** A historic lime kiln.

**Collect, Organize, and Analyze** We are given the difference in height of the mercury column in a manometer, which is a measure of the pressure of $CO_2$ produced by the decomposition reaction above that of atmospheric pressure. We are asked to express this pressure in units of millimeters of mercury, atmospheres, and kilopascals. The appropriate conversion factors, found in Table 9.2, are 1 atm = 760 mmHg = 101.325 kPa. The measured pressure of 144 mmHg is about 1/5 of 760, so we can estimate that the total pressure in the flask will be about 1.2 atm. About 100 kPa are in 1 atmosphere, so 1.2 atm is equivalent to about 120 kPa.

**Solve**

a. The total pressure in millimeters of mercury is

$$756 + 144 = 900 \text{ mmHg} = 9.00 \times 10^2 \text{ mmHg}$$

b. Converting millimeters of mercury to atmospheres:

$$9.00 \times 10^2 \cancel{\text{mmHg}} \times \frac{1 \text{ atm}}{760 \cancel{\text{mmHg}}} = 1.18 \text{ atm}$$

c. Converting millimeters of mercury to kilopascals:

$$9.00 \times 10^2 \cancel{\text{mmHg}} \times \frac{101.325 \text{ kPa}}{760 \cancel{\text{mmHg}}} = 1.20 \times 10^2 \text{ kPa}$$

**Think About It** The calculated values are very close to our estimated values. All are expressed to three significant figures because that many were in the initial sum of the manometer and ambient pressures.

**Practice Exercise** Suppose the measurements in Sample Exercise 9.3 were repeated, but the mass of the $CaCO_3$ sample was only 88.2% of the sample in the original measurement. What would be the value of $\Delta h$ in mmHg in the final measurement?

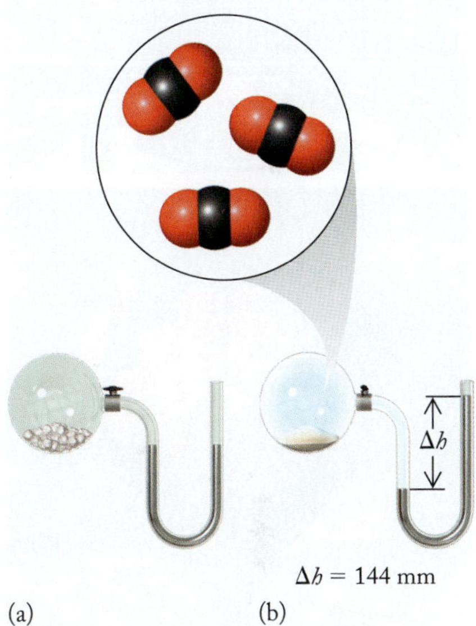


**FIGURE 9.17** Apparatus for monitoring the thermal decomposition of $CaCO_3$ by using a manometer. (a) A flask containing chunks of $CaCO_3$. (b) The same setup after a decomposition reaction that releases $CO_2$ into the flask and manometer.

## 9.4 Relating *P*, *T*, and *V*: The Gas Laws

In Section 9.1, we summarized some properties of gases and described how pressure and temperature affect volume, mostly in qualitative terms. Our knowledge of the quantitative relationships among *P*, *T*, and *V* goes back more than three centuries to a time before the field of chemistry as we know it even existed. Some experiments that led to our understanding of how gases behave were inspired by the development in the 18th century of the first hot-air balloons big enough (and safe enough) to carry passengers (**Figure 9.18**).

**FIGURE 9.18** Hot-air balloons then and now. Fascination with hot-air ballooning in the late 18th and early 19th centuries led to important discoveries about the properties of gases.

**Boyle's law** the principle that the volume of a fixed quantity of gas at constant temperature is inversely proportional to its pressure.

## Boyle's Law: Relating Pressure and Volume

In Section 9.1, we explored the compressibility of gases and the inverse relation between the pressure and volume of a fixed quantity of gas at constant temperature. This relationship was discovered by British chemist Robert Boyle (1627–1691) and is known as **Boyle's law**:

$$P \propto \frac{1}{V} \quad \text{(at constant } n \text{ and } T\text{)} \tag{9.12}$$

In his experiments, Boyle used a J-shaped tube similar in construction to a manometer that is closed at one end and open at the other. With no mercury in the tube (**Figure 9.19a**), the pressure on the air inside the J-tube is simply atmospheric pressure (let's assume it to be 760 mmHg). When just enough mercury is added to fill the bottom of the J-tube (**Figure 9.19b**), the pressure on the trapped air is still 760 mmHg. However, adding more mercury increases the pressure on the trapped air by an amount equal to the difference in height ($\Delta h$) of the mercury in the two sides of the tube (**Figure 9.19c**). The total pressure on the trapped air is 760 mmHg plus $\Delta h$. As the total pressure increases, the volume of the trapped air decreases as predicted by Equation 9.12.

We can turn Equation 9.12 from a proportionality into an equality by multiplying its right side by a constant:

$$P = (\text{constant})\,\frac{1}{V}$$

or

$$PV = \text{constant} \tag{9.13}$$

where the value of the constant depends on how much air is trapped ($n$) and its temperature ($T$). If $PV$ is a constant, then the product of the pressure and volume of a given quantity of gas under one set of conditions—that is, $P_1 \times V_1$—has the same value under any other set of conditions—say, $P_2 \times V_2$—as long as the temperature remains the same. Putting this equality in equation form gives us a handy mathematical expression of Boyle's law:

$$P_1V_1 = P_2V_2 \tag{9.14}$$

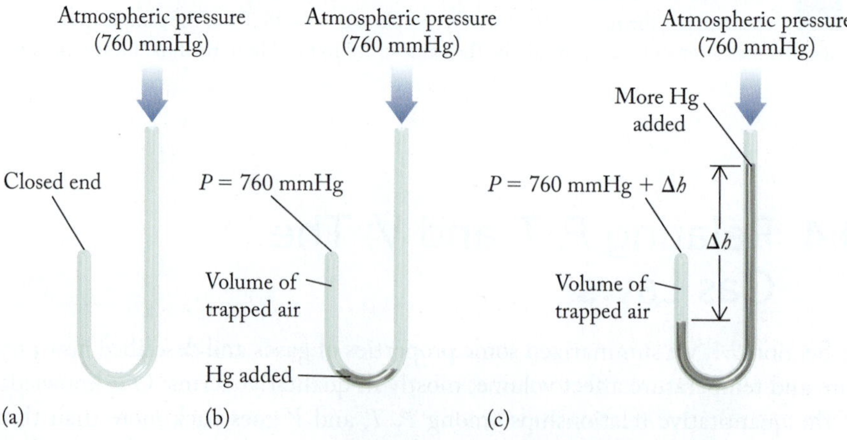


**FIGURE 9.19** Boyle used a J-shaped tube for his experiments on the relationship between pressure and volume. The volume of the air trapped in the closed end of the tube decreased as more mercury was added, and the difference in the height of mercury in the two sides of the tube—a measure of the pressure on the trapped gas—increased.

CONCEPT **TEST**

Which graph in **Figure 9.20** correctly describes the relationship between the product of pressure and volume ($PV$) as a function of pressure ($P$) for a given quantity of gas at constant temperature?

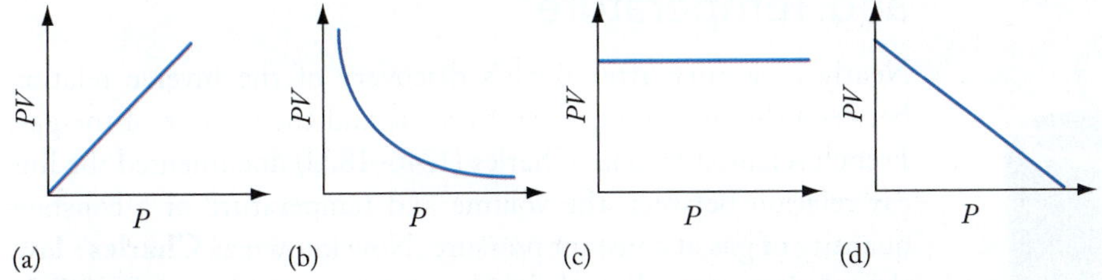


**FIGURE 9.20** Four proposed relationships between $P$ and $PV$.

**SAMPLE EXERCISE 9.4** Applying Boyle's Law **LO5**

A balloon is partly inflated with 5.00 L of helium at sea level, where the atmospheric pressure is 1.00 atm. The balloon ascends to an altitude of 1600 m, where the pressure is 0.83 atm. What is the volume of the balloon at the higher altitude if the temperature of the helium does not change during the ascent?

**Collect, Organize, and Analyze** We are given the initial volume ($V_1 = 5.00$ L) of a gas and its initial pressure ($P_1 = 1.00$ atm) and we are asked to find the volume ($V_2$) of the gas when the pressure changes ($P_2 = 0.83$ atm). The balloon contains a fixed amount of gas and its temperature is constant, so pressure and volume are related by Boyle's law (Equation 9.14). Therefore, its volume should increase because pressure decreases.

**Solve**

Rearranging Equation 9.14 to solve for $V_2$ and inserting the given values of $P_1$, $P_2$, and $V_1$ gives

$$V_2 = \frac{P_1 V_1}{P_2} = \frac{(1.00\ \cancel{\text{atm}})(5.00\ \text{L})}{0.83\ \cancel{\text{atm}}} = 6.0\ \text{L}$$

**Think About It** The prediction we made that the volume would increase as the pressure decreased is confirmed. Equation 9.14 can also be used to calculate the change in pressure that takes place at constant temperature when the volume of a quantity of gas changes.

**Practice Exercise** A scuba diver exhales 3.50 L of air while swimming at a depth of 20.0 m, where the sum of atmospheric pressure and water pressure is 3.00 atm. By the time the exhaled air rises to the surface, where the pressure is 1.00 atm, what is the change in volume?

We can explain Boyle's law by using kinetic molecular theory, which asserts that particles of a gas are in constant random motion, colliding with one another and with the walls of their container, producing the pressure exerted by the gas. If the particles are squeezed into a smaller volume of space (**Figure 9.21**), then the density of the particles is greater and the frequency with which they collide with the walls of their container is greater. The more frequent the collisions, the greater the force exerted by the particles per unit of interior surface area and the greater the pressure. The converse also is true: as the volume of a container of gas

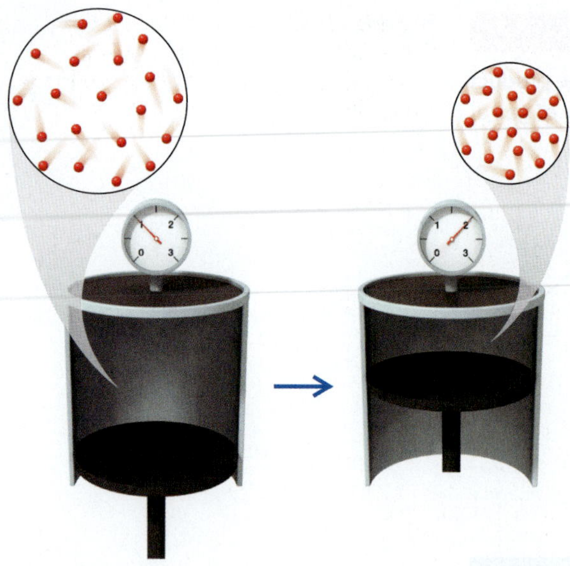

**FIGURE 9.21** Gas particles are in constant motion and exert pressure through collisions with the interior surface of their container. When a quantity of gas is squeezed into half its original volume, the particles collide more frequently with their container, causing the pressure to double.

(such as the weather balloon in Figure 9.1b) increases, collision frequency decreases and pressure drops.

## Charles's Law: Relating Volume and Temperature

Nearly a century after Boyle's discovery of the inverse relation between the pressure exerted by a gas and the volume of the gas, French scientist Jacques Charles (1746–1823) documented the linear relation between the volume and temperature of a constant quantity of gas at constant pressure. Now known as **Charles's law**, the relation states that when the pressure exerted on a gas is held constant, the volume of a fixed quantity of gas is directly proportional to the *absolute* temperature (the temperature expressed in kelvin) of the gas:

$$V \propto T \text{ (at constant } P \text{ and } n) \tag{9.15}$$

The effects of Charles's law can be seen in **Figure 9.22**, in which a balloon has been attached to a flask, trapping a fixed amount of gas in the apparatus. Heating the flask causes the gas to expand, inflating the balloon. As with Boyle's law, an equals sign can replace the proportionality symbol in Equation 9.15 if we include a constant:

$$V = (\text{constant}) \times T$$

or

$$\frac{V}{T} = \text{constant} \tag{9.16}$$

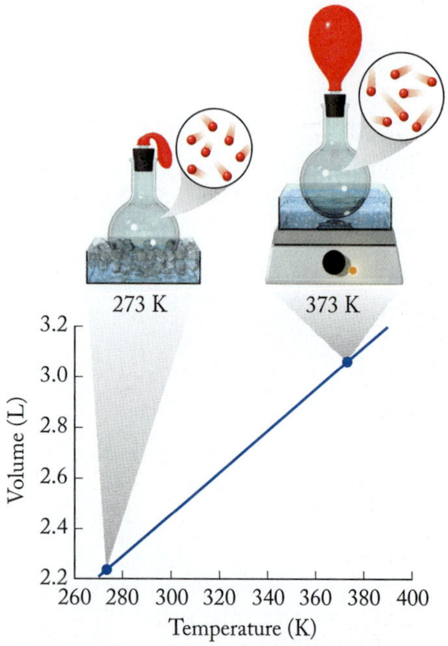

**FIGURE 9.22** A balloon attached to a flask inflates as the temperature of the gas inside the flask increases from 273 K to 373 K at constant pressure. This behavior is described by Charles's law.

The value of the constant depends on the number of moles of gas in the sample ($n$) and on the pressure of the gas ($P$). When these two parameters are held constant, the ratio $V/T$ does not change, and any two combinations of volume and temperature are related as follows:

$$\frac{V_1}{T_1} = \frac{V_2}{T_2} \tag{9.17}$$

We can rearrange the terms in Equation 9.17 to obtain another expression of the proportionality between the volume of a gas and its absolute temperature:

$$\frac{V_1}{V_2} = \frac{T_1}{T_2} \tag{9.18}$$

**SAMPLE EXERCISE 9.5** Applying Charles's Law **LO5**

Several students at a northern New England college are hosting a party celebrating the mid-January start of spring semester classes. They decide to decorate the front door of their apartment building with party balloons. The air in the inflated balloons is initially

21°C (70°F). After an hour outside, the temperature of the balloons is −24°C (−12°F). If we assume that no air leaks from the balloons and that the pressure in them does not change significantly, how much does their volume change? Express your answer as a percentage of the initial volume.

**Collect and Organize** We are asked to calculate the change in volume that accompanies a temperature change from 21°C to −24°C. According to Charles's law, the volume of a fixed quantity of a gas at constant pressure is proportional to the *absolute* temperature of the gas.

**Analyze** More than one way exists to express Charles's law mathematically. Because we are given temperature values ($T_1$ and $T_2$) and are asked to calculate the corresponding ratio of volumes ($V_1$ and $V_2$), Equation 9.18 is particularly well suited to our calculation:

$$\frac{V_1}{V_2} = \frac{T_1}{T_2}$$

The given temperatures are about 293 and 253 K. The ratio of the smaller to the larger temperature (253/293) is a little more than 5/6, or about 85%. Therefore, the volume should *decrease* by about 15%.

**Solve** If we let $T_1$ be the lower temperature, the value of the right side of Equation 9.18 is

$$\frac{T_1}{T_2} = \frac{(-24 + 273)\cancel{\text{K}}}{(21 + 273)\cancel{\text{K}}} = \frac{249\ \cancel{\text{K}}}{294\ \cancel{\text{K}}} = 0.8469 = 84.7\%$$

This is the same value as the ratio of the volumes ($V_1/V_2$) on the left side, so the answer we seek is the difference between 100% and 84.7%, or a 15.3% decrease.

**Think About It** A substantial decrease in temperature (from a human's comfort perspective) translates into a relatively small decrease in volume, as we predicted. The small change makes sense because the difference in temperature on the Kelvin scale is the difference between two large values close to each other in value, which means their ratio is not very far from 1.

**Practice Exercise** If the temperature of a gas confined to the cylinder in **Figure 9.23** is raised from 245°C to 560°C, what is the ratio of the final volume to the initial volume if the pressure inside the cylinder remains constant?

**Charles's law** the principle that the volume of a fixed quantity of gas at constant pressure is directly proportional to its absolute temperature.

How does kinetic molecular theory explain Charles's law? Temperature is directly related to molecular motion. The average speed at which gas-phase particles move increases with increasing temperature. For a given number of particles, increasing temperature increases their motion and therefore increases the frequency and average kinetic energy of their collisions with the walls of their container. More frequent, higher-energy collisions would produce an increase in pressure if volume stayed the same. However, if pressure is to remain constant—as required by Charles's law—the volume must expand so that the faster-moving particles have to travel farther to collide with the walls of their container. Thus, a higher temperature means more energetic particle–wall collisions, but a larger volume means that these collisions happen less frequently so that, overall, pressure does not change.

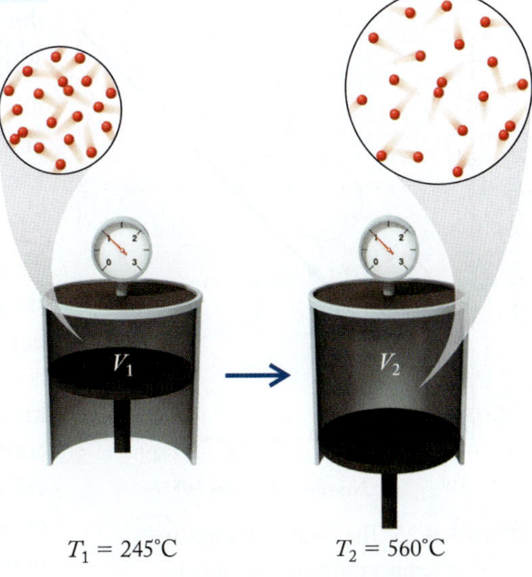


**FIGURE 9.23** A molecular view of Charles's law. The volume of a gas increases when its temperature increases and pressure stays the same.

Charles's law made possible the experimental determination that absolute zero (0 K) is equal to −273.15°C. The graph in **Figure 9.24** shows what happens when we plot the volume of fixed quantities of helium at three constant pressures as a

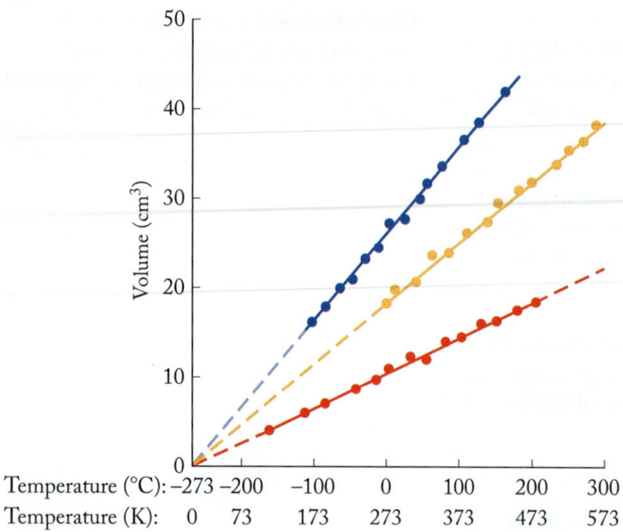


**FIGURE 9.24** A plot of the volumes of a constant mass of helium gas at constant pressure versus temperature. Each line represents a different gas pressure. When the curves are extrapolated to $V = 0$, the corresponding temperatures are all −273.15°C, or 0 K (absolute zero).

function of temperature. All three curves show a linear relationship between volume and temperature. The dashed lines in Figure 9.24 all cross the temperature axis ($V = 0$) at −273°C, which is 0 K, or absolute zero. Thus, the volume of any gas that obeys Charles's law is directly proportional to its absolute temperature at a given pressure.

## Avogadro's Law: Relating Volume and Quantity of Gas

Boyle's and Charles's laws apply to isolated systems that contain constant quantities of gases. However, many physical and chemical systems are open systems in which the quantities of gases *do* change. An example of an open system is the air in a scuba tank. As we noted in Section 9.1, the pressure of a scuba tank during a dive is a reliable indicator of how much air is left in the tank because when temperature and volume are constant, the pressure ($P$) of a gas, or of a mixture of gases, is proportional to the quantity (moles, $n$) of gas in its container:

$$P \propto n \text{ (at constant } V \text{ and } T)$$

or

$$\frac{P}{n} = \text{constant} \tag{9.19}$$

**CONNECTION** The number of particles in a mole is called the Avogadro constant in recognition of his early work with gases that led to the determination of atomic masses, as described in Chapter 2.

According to kinetic molecular theory, increasing the number of particles of a gas increases the frequency with which the particles collide with the walls of the container, so long as the volume and temperature are constant. If the number of particles doubles, for example, then the number of collisions doubles, and more collisions means higher pressure.

Some containers (balloons, for example) are not rigid. When gas is added to them, their volumes increase. If temperature and pressure remain constant, then the increase in volume ($V$) is directly proportional to the quantity (moles, $n$) of gas added. Expressing this relationship mathematically:

$$V \propto n \text{ (at constant } P \text{ and } T)$$

or

$$\frac{V}{n} = \text{constant} \tag{9.20}$$

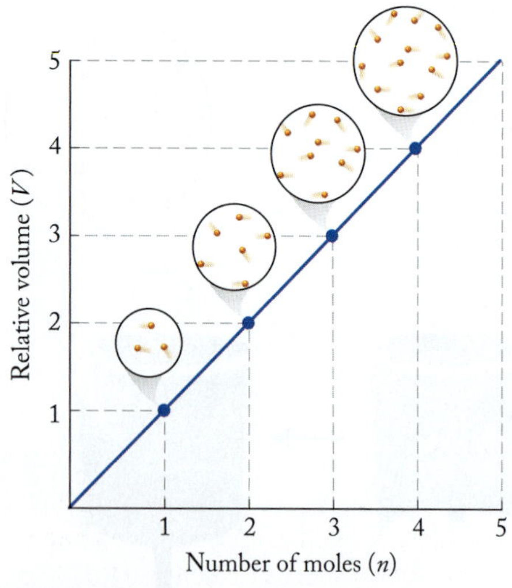


**FIGURE 9.25** The volume of a gas at constant temperature and pressure is directly related to the quantity (number of moles) of the gas, a relationship known as Avogadro's law.

Amedeo Avogadro (1776–1856), whom we know from the Avogadro constant ($N_A$), first recognized that the volume of a gas is proportional to the number of particles in it at constant temperature and pressure, and this relationship is called **Avogadro's law** in his honor. Avogadro's law can also be explained by kinetic molecular theory. As we just mentioned, adding particles of gas to a container increases their concentration, which increases the frequency of their collisions with the walls of the container, thus increasing the pressure. However, if the volume of the container increases enough to maintain the same density of particles and the same particle collision frequency, then the pressure will remain constant. Thus, pressure remains constant as long as the volume of the system is proportional to the number of particles of gas in it, as shown in **Figure 9.25**.

## CONCEPT TEST

Which graph in **Figure 9.26** correctly describes the relationship between $n$ and the value of $V/n$ (at constant $P$ and $T$)?

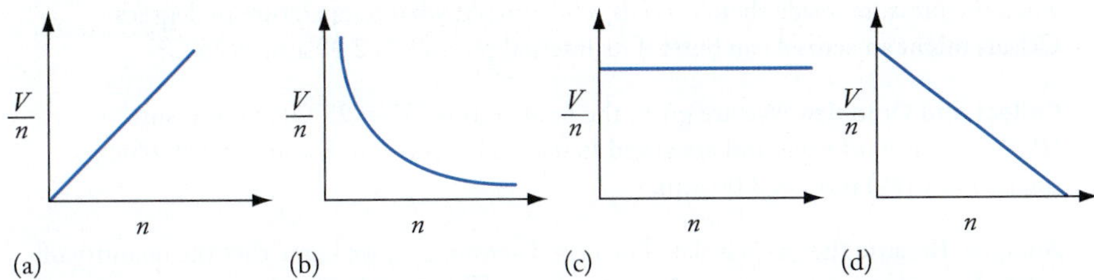


**FIGURE 9.26** Four proposed relationships between $n$ and $V/n$.

## Amontons's Law: Relating Pressure and Temperature

According to Charles's law, the volume of a gas is proportional to its temperature at constant pressure. An increase in temperature increases the speed of gas particles, which in turn increases both the frequency and force of their collisions with the walls of their container. These increases produce an increase in pressure if the volume remains constant. However, an increase in volume relieves the increase in pressure, which is the essence of Charles's law. However, if the gas is confined to a rigid container, then the volume remains constant and the pressure increases. In general, the pressure of any quantity of gas is directly proportional to its absolute temperature if its volume does not change:

$$P \propto T \quad \text{or} \quad \frac{P}{T} = \text{constant} \quad (\text{at constant } V \text{ and } n) \tag{9.21}$$

This relationship between $P$ and $T$ is called **Amontons's law** in honor of French physicist Guillaume Amontons (1663–1705), a contemporary of Robert Boyle who constructed a thermometer on the basis of the observation that the pressure of a gas is directly proportional to its temperature.[1]

Where do we see evidence of Amontons's law? Suppose we inflate a bicycle tire to a prescribed pressure of 6.8 atm (100 psi) on a warm afternoon in late autumn when the temperature is 25°C. The next morning, after an early freeze in which the temperature drops to 0°C, the pressure in the tire drops to about 6.2 atm (91 psi). Did the tire leak? Probably not. Instead, the decrease in pressure can be explained by Amontons's law. We can express the law in the form of an equation relating the two pressures and temperatures:

$$\frac{P_1}{T_1} = \frac{P_2}{T_2} \tag{9.22}$$

Solving for $P_2$ (the pressure at the lower temperature), we get the expected result:

$$P_2 = \frac{P_1 T_2}{T_1} = \frac{(6.8 \text{ atm})(273 \cancel{\text{K}})}{298 \cancel{\text{K}}} = 6.2 \text{ atm}$$

[1]Amontons's law is also referred to as Gay-Lussac's law, but Amontons was first to recognize the relationship between pressure and temperature, working some 50–60 years before French chemist Joseph Louis Gay-Lussac (1778–1850).

**Avogadro's law** the principle that the volume of a gas at constant temperature and pressure is proportional to the quantity (number of moles) of the gas.

**Amontons's law** the principle that the pressure of a fixed quantity of gas is proportional to its absolute temperature if its volume does not change.

**SAMPLE EXERCISE 9.6** Applying Amontons's Law **LO5**

Labels on aerosol cans caution against incinerating them because the cans may explode when the pressure inside them exceeds 3.00 atm. At what temperature in degrees Celsius might an aerosol can burst if its internal pressure is 2.00 atm at 25°C?

**Collect and Organize** We are given the temperature ($T_1$ = 25°C) and pressure ($P_1$ = 2.00 atm) of a gas and are asked to determine the temperature ($T_2$) at which the pressure ($P_2$) reaches 3.00 atm.

**Analyze** Because the gas is isolated in a rigid aerosol can, we know that the quantity of gas and its volume are constant. Amontons's law (Equation 9.22) relates the pressures of a confined quantity of gas at two temperatures. To estimate our answer, we note that the pressure in the can must increase by 50% to reach 3.00 atm. Pressure is directly proportional to absolute temperature, so the temperature also must increase by 50%. The initial temperature of 25°C is nearly 300 K, so a 50% increase in absolute temperature corresponds to a final temperature near 450 K, or about 175°C.

**Solve** Rearranging Equation 9.22 to solve for $T_2$:

$$T_2 = \frac{T_1 P_2}{P_1}$$

and inserting the given $T$ and $P$ values:

$$T_2 = \frac{[(25 + 273)\text{K}](3.00 \cancel{\text{atm}})}{2.00 \cancel{\text{atm}}} = 447 \text{ K}$$

Converting $T_2$ to degrees Celsius:

$$T_2 = 447 \text{ K} - 273 = 174°\text{C}$$

**Think About It** This temperature is close to our estimated value, and it is well below the temperature of 850°C that solid-waste incinerators are supposed to achieve to decompose hazardous organic material. Such a high temperature makes the warning label on the can all the more important.

**Practice Exercise** Air pressure in each tire of an automobile is adjusted to 32 psi (221 kPa) at a gas station in San Diego, California, where the air temperature is 20°C (68°F) and atmospheric pressure is 101.0 kPa. After a 3-hour drive along Interstate Highway 8, the car and driver are in Yuma, Arizona, where the temperature is 43°C (110°F). What is the pressure in the tires, if atmospheric pressure is still 100.7 kPa? *Note*: The measured tire pressures are *gauge* pressures, meaning that they indicate the pressures in the tires *above atmospheric pressure*.

## 9.5 The Combined Gas Law

Boyle's law:

$$P_1V_1 = P_2V_2 \tag{9.14}$$

can be combined with Amontons's law:

$$\frac{P_1}{T_1} = \frac{P_2}{T_2} \tag{9.22}$$

to obtain a single equation that relates the pressure, volume, and temperature of a quantity of gas changing from one set of conditions ($P_1$, $V_1$, $T_1$) to another ($P_2$, $V_2$, $T_2$):

$$\frac{P_1V_1}{T_1} = \frac{P_2V_2}{T_2} \text{ (at constant } n) \tag{9.23}$$

Equation 9.23 is known as the **combined gas law.** It is extremely useful for calculating the impact of changing temperature and pressure on the volume of a gaseous system.

**SAMPLE EXERCISE 9.7** Applying the Combined Gas Law **LO5**

The pressure inside a weather balloon as it is released is 798 mmHg. If the volume and temperature of the balloon are 131 L and 20°C, what is the volume of the balloon when it reaches an altitude where its internal pressure is 235 mmHg and $T = -52°C$?

**Collect and Organize** We are given the initial temperature, pressure, and volume of a gas, and we are asked to determine the final volume after the pressure and temperature have both decreased.

**Analyze** The quantity of gas is a constant, so $PV/T$ is also a constant according to the combined gas law (Equation 9.23) as long as we convert the Celsius temperatures to Kelvin temperatures. We expect the balloon's volume to decrease as its temperature decreases but to increase as its pressure decreases. The final pressure is only about 1/3 of the initial pressure, which would triple the balloon's volume were it not for the simultaneous decrease in temperature. However, the relative change in absolute temperature is not as great as the decrease in pressure, so we can predict that the volume of the balloon should increase by almost, but not quite, a factor of 3.

**Solve** First, we convert the given Celsius temperatures to Kelvin temperatures:

$$T_1 = 20°C + 273 = 293 \text{ K}$$
$$T_2 = -52°C + 273 = 221 \text{ K}$$

We then solve Equation 9.23 for $V_2$:

$$\frac{P_1V_1}{T_1} = \frac{P_2V_2}{T_2}$$

$$V_2 = V_1 \times \frac{P_1}{P_2} \times \frac{T_2}{T_1} = 131 \text{ L} \times \frac{798 \cancel{\text{mmHg}}}{235 \cancel{\text{mmHg}}} \times \frac{221 \cancel{\text{K}}}{293 \cancel{\text{K}}} = 336 \text{ L}$$

**Think About It** The volume increases by a factor of 336 L/131 L, or 2.56 times, which is in good agreement with our estimate of a nearly threefold increase.

**Practice Exercise** The balloon in Sample Exercise 9.7 is designed to continue its ascent to an altitude of 30 km, where it bursts, releasing a package of meteorological instruments that parachutes back to Earth. If the pressure inside the balloon at 30 km is 33 mmHg and the temperature is −45°C, what is the volume of the balloon when it bursts?

## 9.6 Ideal Gases and the Ideal Gas Law

Gases that behave in accordance with the combined gas law are called **ideal gases.** Most gases exhibit ideal behavior at the pressures and temperatures we typically encounter in nature and while doing general chemistry lab experiments. Under these conditions, gases behave ideally because the assumptions about their composition that we described in Section 9.2 are valid. This means that the volumes of the individual gas particles are insignificant in comparison to the overall volume occupied by the gas, and the particles do not interact with one another. Instead, they move independently with speeds that are related to their masses and to the temperature of the gas.

**combined gas law** the principle that the ratio $PV/T$ for a given quantity of gas is a constant.

**ideal gas** a gas whose behavior is predicted by the linear relations defined by the combined gas law.

In Section 9.4, we explored how the pressure of an ideal gas is

- proportional to the moles of particles in it, $P \propto n$;
- proportional to its absolute temperature, $P \propto T$; and
- inversely proportional to its volume, $P \propto \frac{1}{V}$.

Combining these three expressions, we obtain

$$P \propto n \times T \times \frac{1}{V}$$

As in Section 9.4, we can convert the proportionality to an equality by introducing a constant:

**CHEMTOUR**
Ideal Gas Law

$$P = (\text{constant}) \times n \times T \times \frac{1}{V}$$

The constant in this expression is the same *universal gas constant*, $R$, that we encountered in Section 9.2:

$$P = R \times n \times T \times \frac{1}{V}$$

Rearranging the terms and simplifying, we get

$$PV = nRT \tag{9.24}$$

Equation 9.24 is probably the most important one in this chapter. It is called the **ideal gas equation**—a mathematical expression of the **ideal gas law**. The value of $R$, however, depends on the units used for pressure and volume. (We always express the quantity of gas in moles, and we use the Kelvin scale for temperature.) If we use SI units for volume (cubic meters) and pressure (pascals), then

$$R = 8.314 \frac{\text{m}^3 \cdot \text{Pa}}{\text{mol} \cdot \text{K}}$$

(Note that the numerical value of $R$ here is the same as in Section 9.2 because $1\ \text{Pa} = 1\ \text{kg/m} \cdot \text{s}^2$.)

As a practical matter, it is often more convenient to express volumes using units smaller than cubic meters, such as liters ($1\ \text{L} = 10^{-3}\ \text{m}^3$) and to express pressures in larger units such as kilopascals ($1\ \text{kPa} = 10^3\ \text{Pa}$). Because these units are 1/1000 and 1000 times the size of the first set of units, respectively, the two factors cancel out and $R$ has the same value:

$$R = 8.314 \frac{\text{L} \cdot \text{kPa}}{\text{mol} \cdot \text{K}}$$

In many calculations in this book, we express pressure in standard atmospheres (atm), so we will also use an $R$ value based on the equality $1\ \text{atm} = 101.325\ \text{kPa}$. To avoid a rounding error, we will also add one more significant figure to the initial value of $R$:

$$R = 8.3145 \frac{\text{L} \cdot \cancel{\text{kPa}}}{\text{mol} \cdot \text{K}} \times \frac{1\ \text{atm}}{101.325\ \cancel{\text{kPa}}} = 0.08206 \frac{\text{L} \cdot \text{atm}}{\text{mol} \cdot \text{K}}$$

Note that both values for $R$ have a numerator with units of volume × pressure. In Chapter 10, we will learn that the product of volume and pressure represents

**ideal gas equation** (also called the **ideal gas law**) the principle relating the pressure, volume, number of moles, and temperature of an ideal gas, expressed by the equation $PV = nRT$, where $R$ is the universal gas constant.

**standard temperature and pressure (STP)** 0°C and 1 bar as defined by IUPAC; 0°C and 1 atm are commonly used in the United States.

the *work* that a gas does when it expands against an opposing pressure. *Energy* is the ability to do work, so the numerator of $R$ can also be expressed with units of energy. The SI unit of energy is the joule, where 1 J is equivalent to 1 $m^3 \cdot Pa$, which gives another set of units for $R$:

$$R = 8.314 \frac{\text{J}}{\text{mol} \cdot \text{K}}$$

The various values for $R$ are summarized in **Table 9.3**.

**TABLE 9.3** Values of the Universal Gas Constant ($R$)

| Value of $R$ | Units |
|---|---|
| 0.08206 | $L \cdot atm/(mol \cdot K)$ |
| 8.314 | $kg \cdot m^2/(s^2 \cdot mol \cdot K)$ |
| 8.314 | $J/(mol \cdot K)$ |
| 8.314 | $m^3 \cdot Pa/(mol \cdot K)$ |
| 62.37 | $L \cdot torr/(mol \cdot K)$ |

Before closing this section, you need to know some useful reference conditions in the properties of gases. One of them is **standard temperature and pressure (STP)**. Since 1982, the International Union of Pure and Applied Chemistry (IUPAC) has defined STP as 0°C and 1 bar of pressure. Before that, STP was defined as 0°C and 1 standard atmosphere of pressure, and that definition has lived on, especially among some scientists and science teachers in the United States. One reason for the persistence of the 1-atm definition is that the standard atmosphere is equal to 1.01325 bar, so the difference between the two pressures is relatively small.

One value that depends on how we define STP is *molar volume*, which is the volume that one mole of an ideal gas occupies at STP (**Figure 9.27**). If we define STP pressure as 1 atm, then the volume of one mole of an ideal gas at 0°C (273 K) is

$$V = \frac{nRT}{P} = \frac{(1\ \cancel{\text{mol}})\left(0.08206 \frac{\text{L} \cdot \cancel{\text{atm}}}{\cancel{\text{mol}} \cdot \cancel{\text{K}}}\right)(273\ \cancel{\text{K}})}{1\ \cancel{\text{atm}}} = 22.4\ \text{L}$$

Molar volume based on the modern definition of STP ($P = 1$ bar, or 100 kPa) is slightly larger:

$$V = \frac{(1\ \cancel{\text{mol}})\left(8.314 \frac{\text{L} \cdot \cancel{\text{kPa}}}{\cancel{\text{mol}} \cdot \cancel{\text{K}}}\right)(273\ \cancel{\text{K}})}{100\ \cancel{\text{kPa}}} = 22.7\ \text{L}$$

At the level of accuracy of calculations in this text, substituting 1 atm for 1 bar makes little difference, so we consider STP to be 0 °C and 1 atm.

Many chemical and biochemical processes (including those that sustain us) take place at pressures near 1 atm and at temperatures between 0°C and 40°C. Under these conditions, the volume of one mole of a gas is no more than about 15% greater than the molar volume. Therefore, the volume of a gas can be estimated if the number of moles of the gas is known. An important feature of the molar volume of a gas at ambient temperatures and pressures is that it applies to essentially all gases, independently of their chemical composition, because they all behave like ideal gases.

**FIGURE 9.27** This box represents the volume of one mole of gas at STP. A basketball would fit loosely inside the box.

**SAMPLE EXERCISE 9.8** Applying the Ideal Gas Law **LO5**

Bottles of compressed $O_2$ carried by climbers ascending Mt. Everest contain 1.00 kg of the gas. What volume of $O_2$ can one bottle deliver to a climber at an altitude where the temperature is −38°C and the atmospheric pressure is 0.35 atm?

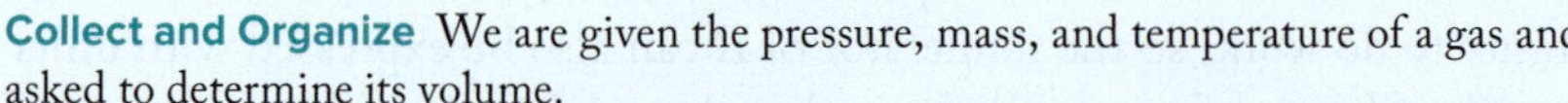

**Collect and Organize** We are given the pressure, mass, and temperature of a gas and asked to determine its volume.

**Analyze** The ideal gas equation enables us to relate $P$, $V$, $T$, and $n$, the number of moles of $O_2$. We are not given $n$, but we do know the mass of $O_2$, and we can use its molar mass ($\mathcal{M}$ = 32.00 g/mol) to calculate $n$ and then use the ideal gas equation to calculate $V$. To estimate our answer, we note that a kilogram of $O_2$ contains 1000 g × (1 mol/32 g), or about 30 mol, of the gas. This quantity would occupy 30 × 22.4 L, or about 660 L, at STP. However, conditions on the top of Mt. Everest are far from STP. The temperature on an absolute scale is lower, which would tend to reduce the volume of gas, but the pressure is only about 1/3 of what it is at STP, which means the volume should be nearly three times 660 L, or nearly 2000 L.

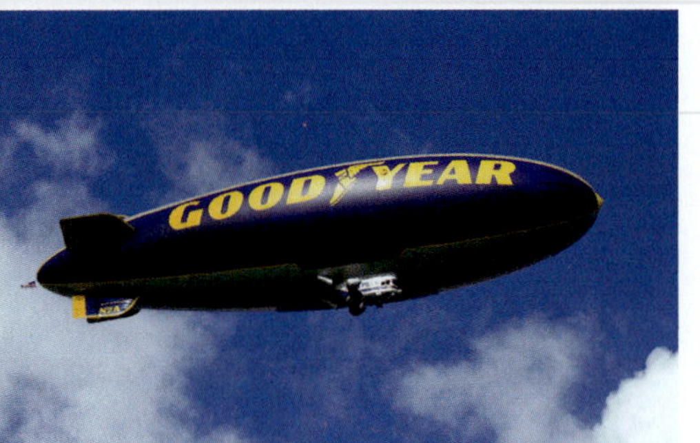

FIGURE 9.28 The Goodyear blimp *Spirit of Innovation*.

**Solve** Let's start by converting the mass of $O_2$ into moles:

$$1.00\ \cancel{\text{kg O}_2} \times \frac{1000\ \text{g}}{1\ \cancel{\text{kg}}} \times \frac{1\ \text{mol O}_2}{32.00\ \cancel{\text{g O}_2}} = 31.3\ \text{mol O}_2$$

Solving the ideal gas equation for $V$ and inserting the values known:

$$PV = nRT$$

$$V = \frac{nRT}{P} = \frac{(31.3\ \cancel{\text{mol}})\left(0.08206\ \dfrac{\text{L} \cdot \cancel{\text{atm}}}{\cancel{\text{mol}} \cdot \cancel{\text{K}}}\right)(273 - 38)\ \cancel{\text{K}}}{0.35\ \cancel{\text{atm}}} = 1.7 \times 10^3\ \text{L}$$

**Think About It** Our answer is certainly reasonable according to our estimate. Most climbers require several of these bottles to climb Mt. Everest and return. We used the ideal gas law to solve this problem because all four gas law variables—$P$, $V$, $T$, and $n$—factored into the calculation.

**Practice Exercise** Buoyancy for the Goodyear blimp *Spirit of Innovation* comes from 5.75 × $10^3$ cubic meters (2.03 × $10^5$ cubic feet) of helium (**Figure 9.28**).

a. What is the mass of this much helium at 25°C and 1.00 atm of pressure?
b. What is the buoyancy of the balloon—that is, the difference between the mass of the He and the mass of the same volume of dry air at the same temperature and pressure?

(a)

(b)

FIGURE 9.29 More than 1800 people were asphyxiated by a dense cloud of $CO_2$ released by Lake Nyos in northwestern Cameroon on August 21, 1986. (a) The lake normally has a deep blue color. (b) The $CO_2$ event stirred up sediment from the lake floor, turning the water brown.

## 9.7 Densities of Gases

Carbon dioxide is a relatively minor component of our atmosphere, but it has received considerable attention in recent years because of its increasing concentration in the atmosphere and the resulting impact on global climate. Although this increase has been linked to human activity, many natural sources such as volcanoes also add $CO_2$ to the atmosphere. The quantities of $CO_2$ involved in volcanic eruptions are not normally a threat to human health, but when this gas collects in the waters of a deep crater lake and is then released all at once, the results can be catastrophic. On August 21, 1986, volcanic Lake Nyos in Cameroon, Africa, suddenly released about a cubic kilometer of $CO_2$ into the air. The gas emerged from the lake in a frothy mist an estimated 100 m high, which flowed out of the crater and into surrounding valleys, killing more than 1800 people and their farm animals (**Figure 9.29**). Because $CO_2$ is denser than air, the deadly mist formed a ground-hugging layer that displaced the air and the oxygen necessary for life.

The density of any gas at STP can be calculated by dividing its molar mass by the molar volume. Carbon dioxide, for example, has a molar mass of 44.0 g/mol, so the density of $CO_2$ at 0°C and 1 atm is

$$\frac{44.0\ \text{g}/\cancel{\text{mol}}}{22.4\ \text{L}/\cancel{\text{mol}}} = 1.96\ \text{g/L}$$

The density of air, which is mostly $N_2$ ($\mathcal{M}$ = 28.01 g/mol) and $O_2$ ($\mathcal{M}$ = 32.00 g/mol), is only 1.21 g/L at STP, so 1 km$^3$ of $CO_2$, a colorless, odorless gas, could easily displace enough air to become a silent killer. The greater density of $CO_2$ and the fact that it does not support combustion also makes it effective in fighting small fires. The gas from a $CO_2$ fire extinguisher effectively blankets burning fuel, depriving it of $O_2$ and extinguishing the fire.

We can use the ideal gas equation to calculate the density of a gas at any temperature and pressure. Density is expressed in units of mass per unit volume, and the number of moles of a pure substance is related to mass by its molar mass. We can rearrange the ideal gas law to bring the $n$ and $V$ terms onto one side of the equation:

$$\frac{n}{V} = \frac{P}{RT}$$

If we multiply both sides of the equation by the molar mass in grams per mole, $\mathcal{M}$, the left-hand side of the equation has units of grams per liter, or density. The resulting equation relates the density of any gas to its molar mass and pressure:

$$\frac{\mathcal{M}n}{V} = \frac{\left(\frac{\text{g}}{\cancel{\text{mol}}}\right)(\cancel{\text{mol}})}{\text{L}} = d = \frac{\mathcal{M}P}{RT} \qquad (9.25)$$

We can determine the density of a gas experimentally by using an apparatus such as the one shown in **Figure 9.30**. In this apparatus, a glass bulb of known volume is attached to a vacuum pump and the air is removed. The mass of the bulb is determined twice: first while it is evacuated and again after the bulb is filled with the test gas. The difference in the masses divided by the volume of the bulb is the density of the gas.

If we also knew the temperature and pressure of the gas in the bulb in Figure 9.30, we could use the values of $V$, $T$, and $P$ to calculate the number of moles of gas in the bulb:

$$n = \frac{PV}{RT}$$

Dividing the mass ($m$) of the gas in the bulb (in grams) by the number of moles of gas ($n$) gives the molar mass of the gas:

$$\mathcal{M} = \frac{m}{n}$$

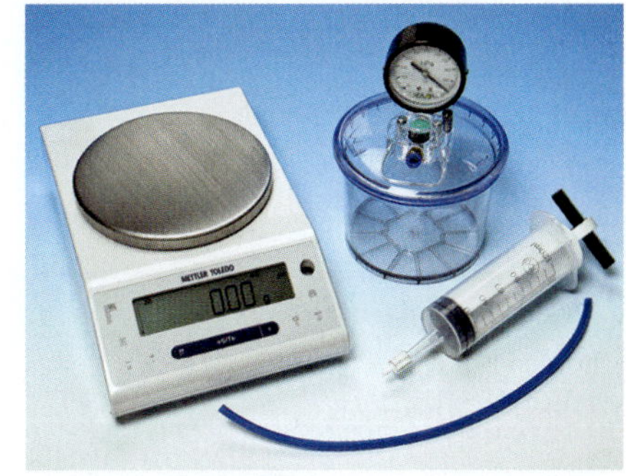

(a)

(b)

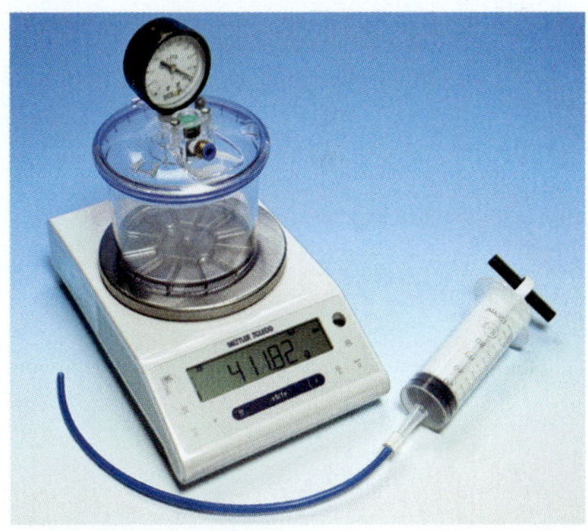

(c)

**FIGURE 9.30** Apparatus for determining the density of a gas. (a) A syringe is used to remove the atmosphere from this chamber with a volume of 880 mL. (b) The mass of the empty chamber is measured. (c) The chamber is then opened to the atmosphere to be refilled with air. The difference in mass between the filled and evacuated tube (1.02 g or 1020 mg) is the mass of the air inside it. The density of the air sample is 1020 mg/880 mL = 1.16 mg/mL = 1.16 g/L.

## CONCEPT TEST

Which graphs in **Figure 9.31** best approximate the following for an ideal gas?

a. The relationship between density and pressure ($n$ and $T$ constant)

b. The relationship between density and temperature ($n$ and $P$ constant)

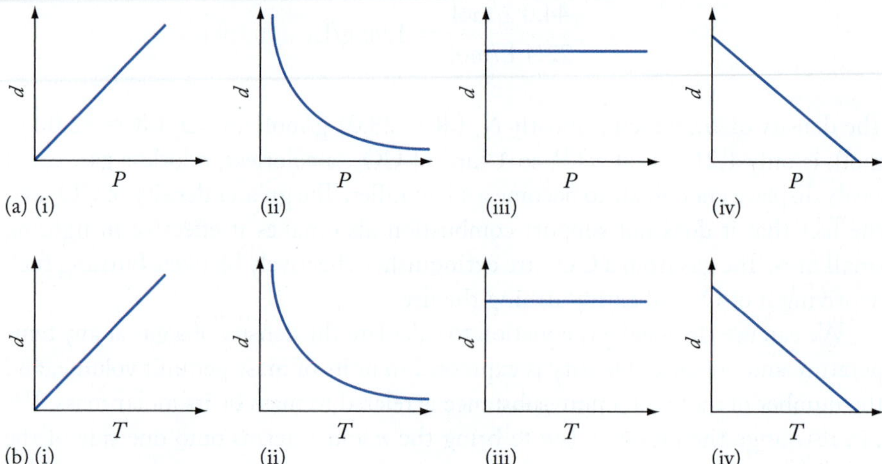


**FIGURE 9.31** Four proposed relationships (a) between $P$ and $d$, and (b) between $T$ and $d$.

### SAMPLE EXERCISE 9.9 Calculating the Density of a Gas — LO6

According to the U.S. National Weather Service, the air temperature in Phoenix, Arizona, reached 26°C (78°F) on January 1, 2012, when atmospheric pressure was 1024 mbar. What was the density of the air? Assume that the average molar mass of air is 28.8 g/mol, which is the weighted average of the molar masses of the various gases in dry air (see Table 9.1).

**Collect and Organize** We are given the average molar mass, temperature, and atmospheric pressure of an air mass, and we are asked to calculate its density, where $d = m/V$.

**Analyze** We may assume that the air behaves like an ideal gas and obeys the ideal gas law. One way to calculate the density of air (or any ideal gas) is to divide the mass ($m$) of one mole of it (28.8 g here) by the volume ($V$) that one mole occupies. This volume can be calculated using the ideal gas equation and the temperature and pressure values provided. We will need to convert $T$ and $P$ into units that are compatible with one of our $R$ values. Because the density of air at STP is about 1.3 g/L, we can estimate that the air density at Phoenix's warm temperature is about 10% lower than at STP, or about 1.2 g/L.

**Solve** Several options are available for converting millibars to another pressure unit. Let's use the equality 1 atm = 1013.25 mbar and calculate the volume of one mole of air, starting with the ideal gas equation: $PV = nRT$. Solving for $V$:

$$V = \frac{nRT}{P} = \frac{(1\ \text{mol})\left(0.08206\ \frac{\text{L}\cdot\text{atm}}{\text{mol}\cdot\text{K}}\right)(26 + 273)\text{K}}{(1024\ \text{mbar})\left(\frac{1\ \text{atm}}{1013.25\ \text{mbar}}\right)} = 24.28\ \text{L}$$

Calculating density:

$$d = \frac{m}{V} = \frac{28.8\ \text{g}}{24.28\ \text{L}} = 1.19\ \text{g/L}$$

**Think About It** The calculated result is consistent with our estimate and with the general trend that gases are about 10% less dense at comfortable ambient temperatures than at STP (0°C). We could also have used Equation 9.25 to calculate the density, but the approach we chose follows the method demonstrated by the apparatus in Figure 9.30.

**Practice Exercise** Air is a mixture of mostly nitrogen and oxygen. A balloon is filled with pure oxygen and released in a room full of air. Will it sink to the floor or float to the ceiling?

**SAMPLE EXERCISE 9.10** Calculating Molar Mass from Density **LO6**

Vent pipes at solid-waste landfills emit gases released by decomposing material. A sample of such a gas has a density of 0.65 g/L at 25.0°C and 757 mmHg. What is its molar mass?

**Collect and Organize** We are given the density (in grams per liter), temperature (in degrees Celsius), and pressure (in millimeters of mercury) of a gaseous sample, and we are asked to calculate its molar mass.

**Analyze** Equation 9.25 relates the density, pressure, and molar mass of a gas. We need to convert the temperature to kelvin, and we need to convert the pressure to atmospheres (1 atm = 760 mmHg). We can use the concept of molar volume from Section 9.6 to estimate the molar mass of the gas. The volume of one mole of an ideal gas is about 24 L at 25°C and 1 atm. If the mass of 1 L is 0.65 g, or about 2/3 of a gram, then the mass of 24 L is about 2/3 of 24, or 16 g/mol.

**Solve** Solving Equation 9.25 for $\mathcal{M}$ and substituting the values given for $d$, $R$, $T$, and $P$:

$$\mathcal{M} = \frac{dRT}{P} = \frac{\left(\frac{0.65\ \text{g}}{\cancel{\text{L}}}\right)\left(0.08206\ \frac{\cancel{\text{L}} \cdot \cancel{\text{atm}}}{\text{mol} \cdot \cancel{\text{K}}}\right)(25 + 273)\ \cancel{\text{K}}}{(757\ \cancel{\text{mmHg}})\left(\frac{1\ \text{atm}}{760\ \cancel{\text{mmHg}}}\right)} = 16\ \text{g/mol}$$

**Think About It** Equation 9.25 provides a clear route to solving problems such as this one. However, if you hadn't memorized the route or been given it, you could still solve the problem by using the central equation in this chapter: $PV = nRT$. The density of the gas tells us the mass of 1 L of it. If we assumed that $V =$ exactly 1 L, then we could use the given $T$ and $P$ values and $PV = nRT$ to calculate the number of moles ($n$) in 1 L. Dividing 0.65 g/L by the calculated $n$ per liter yields the molar mass in grams per mole.

**Practice Exercise** When HCl(*aq*) and $NaHCO_3$(*aq*) are mixed, a chemical reaction takes place in which a gas is one of the products. A sample of the gas has a density of 1.81 g/L at 1.00 atm and 23.0°C. What is the molar mass of the gas? What might be the identity of the gas?

## 9.8 Gases in Chemical Reactions

Many important chemical reactions, including those that supply us with most of our energy needs, involve gaseous reactants or products. Even when the fuel is a solid or a liquid, the actual combustion reaction takes place in the gas phase, such as when gasoline burns in the combustion chamber of a car engine or the flames of a charcoal fire cook dinner at a backyard barbecue.

In a chemical reaction involving a gas as either reactant or product, the quantity of it (in moles) in the reaction mixture can be determined using the ideal gas law if we know the volume of the gas and the temperature and pressure at which the reaction proceeds. Once we know the value of $n$, we can calculate the quantities of the other reactants or products, including the quantity of energy released or consumed, from the stoichiometry of the reaction. Alternatively, we may know the quantity of another reactant and need to calculate the volume of a gaseous reactant needed for a complete reaction. The next Sample Exercise provides an example of such a calculation from the world of backyard barbecues.

**SAMPLE EXERCISE 9.11** Combining Stoichiometry and the Ideal Gas Law **LO7**

The maximum flow rate of propane fuel through one of the burners in a gas grill is 5.5 g of $C_3H_8$ per minute. What volume of air must mix with the propane ($\mathcal{M}$ = 44.1 g/mol) each minute to supply enough $O_2$ for complete combustion? Assume that $P$ = 1.00 atm and $T$ = 25°C. The balanced chemical equation for the reaction is:

$$C_3H_8(g) + 5\,O_2(s) \rightarrow 3\,CO_2(g) + 4\,H_2O(g)$$

**Collect and Organize** We know that each minute 5.5 g of $C_3H_8$ is burned, and we are asked to calculate the volume of air required for complete combustion. According to the data in Table 9.1, air is 20.95% $O_2$. The volume of an ideal gas is related to its amount, temperature, and pressure by the ideal gas law: $PV = nRT$.

**Analyze** The balanced chemical equation tells us we need 5 mol $O_2$ to completely combust 1 mol $C_3H_8$, so we need to convert 5.5 g of $C_3H_8$ to moles of $C_3H_8$ and then to moles of $O_2$ by using the 5 : 1 mole ratio. Moles of $O_2$ can be converted to a volume of $O_2$ by using the ideal gas law and then to a corresponding volume of air. To estimate our answer, we note that (5.5 g of $C_3H_8$)/(44 g/mol) is a little more than 0.1 mol of $C_3H_8$. We will need about five times that number of moles of $O_2$ to convert 3 mol of C and 8 mol of H into $CO_2$ and $H_2O$. If we assume that we need about 0.5 mol of $O_2$, the volume of $O_2$ will be a little more than half the molar volume, or about 12 L. Air is about 1/5 $O_2$, so we will need about 12 × 5, or 60, L of air.

**Solve**

The number of moles of $C_3H_8$ is

$$5.5\ \cancel{\text{g } C_3H_8} \times \frac{1 \text{ mol } C_3H_8}{44.1\ \cancel{\text{g } C_3H_8}} = 0.125 \text{ mol } C_3H_8$$

Applying the 5 : 1 stoichiometric ratio of $O_2$ to $C_3H_8$, we obtain

$$0.125\ \cancel{\text{mol } C_3H_8} \times \frac{5 \text{ mol } O_2}{1\ \cancel{\text{mol } C_3H_8}} = 0.625 \text{ mol } O_2$$

The pressure is given as 1.00 atm, so we use the $R$ value with atmospheres in the numerator in the ideal gas law to convert moles of $O_2$ to liters of $O_2$:

$$V = \frac{(0.625\ \cancel{\text{mol}}\ O_2)\left(0.08206\ \frac{\text{L} \cdot \cancel{\text{atm}}}{\cancel{\text{mol}} \cdot \cancel{\text{K}}}\right)(25 + 273)\ \cancel{\text{K}}}{1.00\ \cancel{\text{atm}}} = 15.3 \text{ L } O_2$$

Air is 20.95% $O_2$, so the volume of air that contains 15.3 L of $O_2$ is

$$V_{\text{air}} = 15.3 \text{ L } \cancel{O_2} \times \frac{100.00\ \%\ \text{air}}{20.95\ \cancel{\%\ O_2}} = 73 \text{ L air}$$

**Think About It** Our calculated result, expressed with two significant digits, is close to the predicted value. It is reasonable, given the size of a molar volume and the fact that air is only about 20% $O_2$ by volume.

**Practice Exercise** Automobile airbags (**Figure 9.32**) inflate during a crash or sudden stop by the rapid generation of nitrogen gas from sodium azide. The first step of the airbag reaction may be written as

$$2\ NaN_3(s) \rightarrow 2\ Na(s) + 3\ N_2(g)$$

How many grams of sodium azide are needed to supply enough nitrogen gas to fill a 45 × 45 × 25 cm bag to a pressure of 1.20 atm at 15°C?

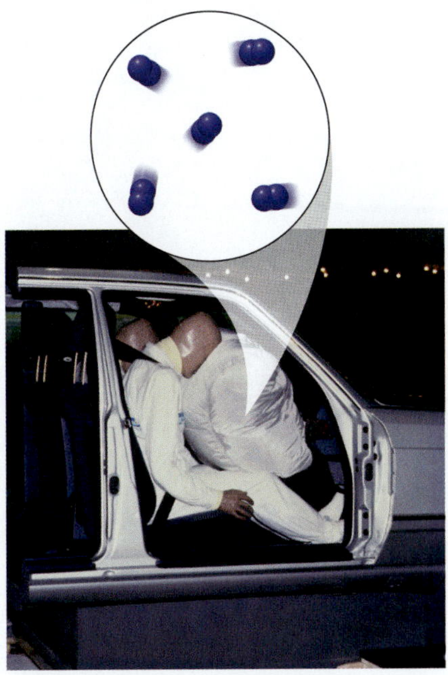

**FIGURE 9.32** An automobile airbag inflates when solid $NaN_3$ rapidly decomposes, producing $N_2$ gas.

## 9.9 Mixtures of Gases

As noted in Table 9.1, dry air is mostly $N_2$ and $O_2$ plus much smaller concentrations of other gases, such as argon and carbon dioxide. Each gas in air, or in any other mixture of gases, exerts its own pressure, called its **partial pressure**. Atmospheric pressure is the sum of the partial pressures of all the gases in the air:

$$P_{atm} = P_{N_2} + P_{O_2} + P_{Ar} + P_{CO_2} + \cdots$$

A similar expression can be written for any other mixture of gases:

$$P_{total} = P_1 + P_2 + P_3 + \cdots \quad (9.26)$$

Equation 9.26 is a mathematical expression of **Dalton's law of partial pressures**: the total pressure of any mixture of gases is the sum of the partial pressures of the gases in the mixture.

The most abundant gases in a mixture (such as $N_2$ and $O_2$ in air) have the greatest partial pressures and contribute the most to the total pressure of the mixture. The mathematical term used to express the abundance of a specific component $i$ in a mixture of gases is its **mole fraction ($x_i$)**,

$$x_i = \frac{n_i}{n_{total}} \quad (9.27)$$

where $n_i$ is the number of moles of component $i$ and $n_{total}$ is the total number of moles of gas in the mixture. The partial pressure of each component is the product of its mole fraction times the total pressure of the mixture:

$$P_i = x_i P_{total} \quad (9.28)$$

Let's apply Equation 9.28 to an air sample at a total pressure of 998 mbar in which the mole fractions of $N_2$ and $O_2$ are 0.7808 and 0.2095. The partial pressures of $N_2$ and $O_2$ are:

$$P_{N_2} = x_{N_2} P_{total} = (0.7808)(998\text{ mbar}) = 779\text{ mbar}$$
$$P_{O_2} = x_{O_2} P_{total} = (0.2095)(998\text{ mbar}) = 209\text{ mbar}$$

Note that the partial pressure of each component gas depends only on the component's mole fraction and the total pressure of the gas mixture. It does not depend on the identity of the component gas or of the other gases in the mixture. Note, too, that mole fractions have no units. This means that they can be used when describing the behavior of any kind of homogeneous mixture (solid, liquid, or gas). Finally, the sum of the mole fractions of all the components in a mixture should always add up to 1 (or, because of rounding, very close to 1).

The kinetic molecular theory of gases explains the additive nature of partial pressures and the dependence of partial pressure on mole fraction. According to the theory, the particles of a pure gas, or of a mixture of gases, collide with the

**CONNECTION** Dalton's law of partial pressures was articulated by John Dalton, the same English chemist who developed the law of multiple proportions and the scientific theory that matter is composed of atoms. Both are described in Chapter 1.

**partial pressure** the contribution to the total pressure made by a component in a mixture of gases.

**Dalton's law of partial pressures** the principle that the total pressure of a mixture of gases is the sum of the partial pressures of all the gases in the mixture.

**mole fraction ($x_i$)** the ratio of the number of moles of a particular component $i$ in a mixture to the total number of moles in the mixture.

walls of the mixture's container. The force of these collisions, when averaged over the surface area of the container, produces a total pressure. The more particles that are present, the more frequent the collisions and the greater the pressure. The theory assumes that all particles have, on average, the same kinetic energy, which means that the frequency and force of their collisions, and the contributions these collisions make to overall pressure, depend only on how many of the particles are present. The theory further assumes that the particles do not interact with each other, so their identities are irrelevant—indeed, the particles of all the components of a mixture are equal-opportunity colliders. The particles of each gas contribute to the total pressure of the mixture in exact proportion to how many of them are in the container as compared to the total number of particles.

**CHEMTOUR**
Dalton's Law

**SAMPLE EXERCISE 9.12** Calculating Mole Fractions and Partial Pressures **LO8**

Scuba divers who dive to depths below 50 m may breathe a gas mixture called trimix during the deepest parts of their dives. One formulation of the mixture, called *trimix 10/70*, is 10% oxygen, 70% helium, and 20% nitrogen by volume. What is the mole fraction of each gas in this mixture, and what is the partial pressure of oxygen in the lungs of a diver at a depth of 60 m (where the ambient pressure is 7.0 atm)?

**Collect and Organize** We are given the composition of a scuba diver's breathing mixture expressed in percent by volume of its three components: $O_2$, He, and $N_2$. We are asked to determine the mole fractions of the gases and the partial pressure of $O_2$ in the lungs of a diver breathing the mixture at a depth at which the pressure is 7.0 atm.

**Analyze** According to Avogadro's law, equal volumes of ideal gases at the same temperature and pressure contain equal numbers of moles of gas. Therefore, the mole fraction of each gas in a mixture is equal to the volume of that gas divided by the sum of the volumes of all the gases that were mixed. Thus, the mole fractions of the gases are the same as their percent-by-volume values. The partial pressure of a component of a gas mixture may be calculated using Equation 9.28, $P_i = x_i P_{total}$.

**Solve** Expressing the percent by volume values as mole fractions:

| Gas | % by Volume | Mole Fraction |
|---|---|---|
| $O_2$ | 10 | 0.10 |
| He | 70 | 0.70 |
| $N_2$ | 20 | 0.20 |

Calculating $P_{O_2}$:

$$P_{O_2} = x_{O_2} P_{total} = 0.10 \times 7.0 \text{ atm} = 0.70 \text{ atm}$$

**Think About It** This exercise introduces a useful equality: the concentrations of the gases in a mixture, such as those listed in Table 9.1 for dry air, are equivalent to the mole fractions of the gases, though decimal points may need to be moved—for example, two places to the left to convert percent values into mole fractions, or six places to the right to convert into parts-per-million values.

**Practice Exercise** What is the partial pressure, in atmospheres, of $O_2$ in the air outside an airliner cruising at an altitude of 10 km, where the atmospheric pressure is 190 mmHg? How much must the outside air be compressed to produce a cabin pressure in which $P_{O_2} = 0.200$ atm?

**CONCEPT TEST**

Does each gas in an equimolar mixture of four gases have the same mole fraction?

Sample Exercise 9.12 and its Practice Exercise illustrate how our health and well-being rely on access to the appropriate breathing gases when we venture a few kilometers above sea level or a few meters below it. At more dweller-friendly altitudes, Dalton's law of partial pressures is used to determine the quantities of gaseous products in chemical reactions. For example, heating potassium chlorate ($KClO_3$) in the presence of $MnO_2$ causes it to decompose into KCl(*s*) and $O_2(g)$. The oxygen gas produced by this reaction can be collected by bubbling the gas into an inverted bottle that is initially filled with water (**Figure 9.33**). As the reaction proceeds, $O_2(g)$ displaces the water in the bottle. When the reaction is complete, the volume of water displaced provides a measure of the volume of $O_2$ produced. If the temperature of the water and atmospheric pressure are known, the ideal gas law can be used to determine the number of moles of $O_2$ produced.

Water displacement can be used to collect and measure the volume of any gas that neither reacts with nor dissolves appreciably in water. However, one more step is needed to apply the ideal gas law to calculating the number of moles of gas produced. At room temperature (nominally 20–25°C), any enclosed space above a pool of liquid water contains water vapor in addition to the gas produced by the reaction. Therefore, the gas collected in the $KClO_3$ reaction in Figure 9.33 is a mixture of $O_2$ and $H_2O$ vapor. Each gas exerts its own partial pressure, and Dalton's law of partial pressures tells us the total pressure of the mixture when the water level inside the collection vessel matches the water level outside the vessel:

$$P_{total} = P_{atm} = P_{O_2} + P_{H_2O} \quad (9.29)$$

To calculate the quantity of oxygen produced using the ideal gas law, we need to know $P_{O_2}$ which we get by subtracting $P_{H_2O}$ (**Table 9.4**) from $P_{atm}$. If we know the values of $P_{O_2}$, $T$, and $V$, we can calculate the number of moles or number of grams of oxygen produced.

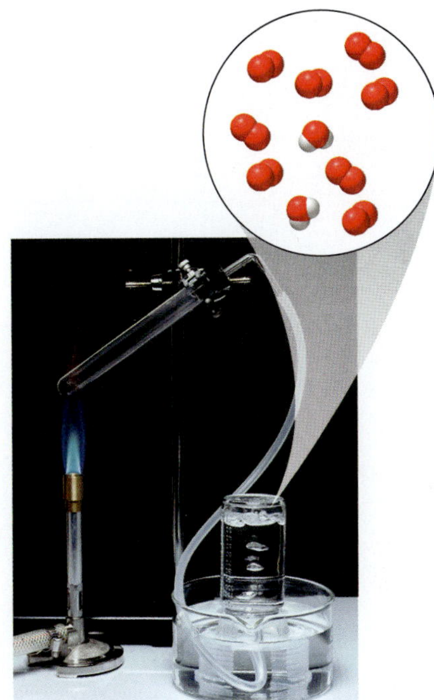

**FIGURE 9.33** Collecting $O_2$ gas by water displacement. Oxygen is generated by the thermal decomposition of $KClO_3$. The delivery tube is removed when the height of the water in the collection jar matches the level of water around it, ensuring that the pressure in the jar is equal to atmospheric pressure.

**SAMPLE EXERCISE 9.13** Calculating the Quantity of a Gas Collected by Water Displacement **LO8**

During the decomposition of $KClO_3$, 92.0 mL of gas is collected by the displacement of water at 25.0°C. If atmospheric pressure is 756 mmHg, what mass of $O_2$ is collected?

**Collect and Organize** We are given values for the total pressure (756 mmHg), volume (92.0 mL), and temperature (25.0°C) of the gas collected by water displacement during the decomposition of $KClO_3$, and we are asked to calculate the mass of $O_2$ in the gas. Table 9.4 contains values for the partial pressure of water vapor at several temperatures. The ideal gas law relates the number of moles of an ideal gas to its pressure, volume, and absolute temperature.

**Analyze** Oxygen is collected over water, so the collection vessel contains both $O_2$ and water vapor. To calculate $P_{O_2}$, we subtract the partial pressure of water vapor at 25°C (23.8 mmHg, according to Table 9.4) from atmospheric (total) pressure. To calculate the number of moles of $O_2$ by using the ideal gas equation, we need to convert the pressure value to atmospheres and the temperature to kelvin. Multiplying that number of moles by the molar mass of $O_2$ will give us the mass of $O_2$ collected. To estimate the result, let's round the volume of the gas to 100 mL, or 0.1 L. At 25°C,

**TABLE 9.4** Partial Pressure of Water Vapor at Selected Temperatures

| Temperature (°C) | Pressure (mmHg) |
|---|---|
| 5 | 6.5 |
| 10 | 9.2 |
| 15 | 12.8 |
| 20 | 17.5 |
| 25 | 23.8 |
| 30 | 31.8 |
| 35 | 42.2 |
| 40 | 55.3 |
| 45 | 71.9 |
| 50 | 92.5 |

the molar volume of an ideal gas is nearly 25 L, so the volume collected corresponds to 0.1/25, or about 0.004 mol of gas. The partial pressure of water vapor is smaller than atmospheric pressure, so most of the gas collected is $O_2$. The mass of this quantity of $O_2$ is 0.004 mol × 32 g/mol, or about 0.13 g.

**Solve** The value of $P_{O_2}$ is

$$P_{O_2} = P_{atm} - P_{H_2O} = (756 - 23.8)\ \text{mmHg} = 732\ \text{mmHg}$$

The number of moles of $O_2$ is

$$n = \frac{\left(732\ \text{mmHg} \times \dfrac{1\ \text{atm}}{760\ \text{mmHg}}\right)\left(92.0\ \text{mL} \times \dfrac{1\ \text{L}}{1000\ \text{mL}}\right)}{\left(0.08206\ \dfrac{\text{L} \cdot \text{atm}}{\text{mol} \cdot \text{K}}\right)(25 + 273)\ \text{K}} = 0.00362\ \text{mol}$$

and the mass of $O_2$ is

$$m = 0.00362\ \text{mol}\left(\frac{32.00\ \text{g}}{1\ \text{mol}}\right) = 0.116\ \text{g}$$

**Think About It** The calculated mass of $O_2$ is about what was expected. The value of $P_{O_2}$ is 732/756, or 96.8%, of $P_{atm}$—close enough to justify our ignoring the contribution of $P_{H_2O}$ in our estimate, but not so close that it could be ignored in the actual calculation.

**Practice Exercise** Electrical energy can be used to separate water into $O_2$ and $H_2$. In one demonstration of this reaction, 27 mL of $H_2$ is collected over water at 25°C. Atmospheric pressure is 761 mmHg. How many milligrams of $H_2$ are collected?

## 9.10 Real Gases

Up to now, all our calculations of *P*, *V*, *T*, and *n* have assumed ideal gas behavior, and this is acceptable because, under typical atmospheric pressures and temperatures, most gases *do* behave ideally. We have assumed, according to kinetic molecular theory, that the volume occupied by individual gas particles is negligible compared with the total volume occupied by the gas. In addition, we have assumed that no interactions other than random elastic collisions occur between particles. These assumptions are not valid, however, when gases are subjected to extremely high pressures or to temperatures so low that they approach the temperatures at which the gases condense.

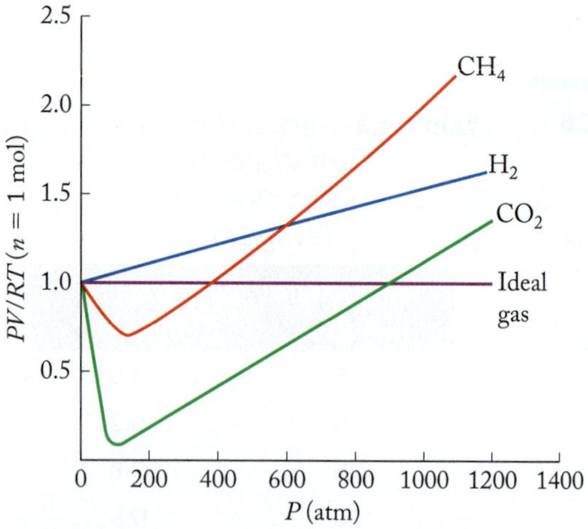

**FIGURE 9.34** The effect of very high pressure on the behavior of one mole of real and ideal gases. The curves for real gases diverge from ideal behavior in a manner specific to each gas and due to two competing factors: interactions between particles (which lower the value of $PV/RT$) and the incompressibility of the particles themselves (which raises the value of $PV/RT$).

### Deviations from Ideality

Let's start by considering the behavior of one mole of a gas as we increase the pressure on the gas. From the ideal gas law, we know that $PV/RT = n$, so for one mole of gas, $PV/RT$ should remain equal to 1 regardless of how we change the pressure. This relationship between $PV/RT$ and $P$ for an ideal gas is shown by the horizontal purple line in **Figure 9.34**. However, the curves for $CH_4$, $H_2$, and $CO_2$ at pressures above 10 atm are not horizontal lines. Not only do their curves diverge from the ideal, but the shapes of the curves also differ for each gas, indicating that deviations from ideal behavior depend on the identity of the gas.

Why don't real gases behave like ideal gases at high pressure? One reason is that particles of ideal gases do not interact with each other, but real atoms and molecules *do*, in the various ways discussed in Chapter 6.

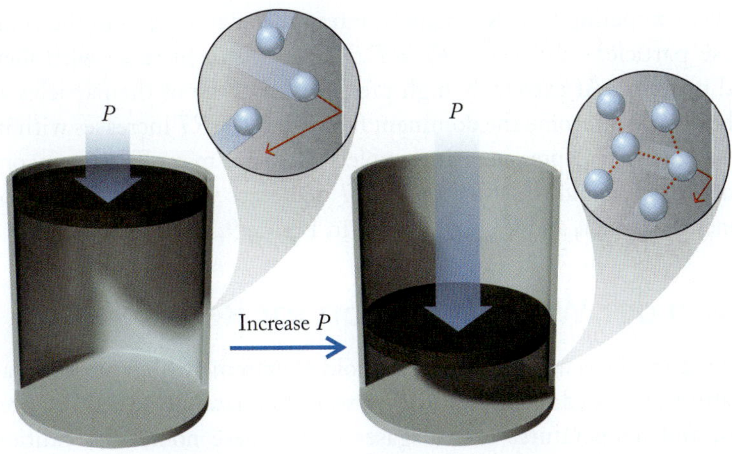


**FIGURE 9.35** At very high pressures, more intermolecular forces exist between particles (represented by the dotted red lines in the expanded view). These interactions reduce the frequency and force of particle collisions with the walls of the container (thin red arrow), thereby reducing the pressure.

Gas-phase particles experience more intermolecular forces when extremely high pressures push them closer together or when lower temperatures force them to move more slowly and they take longer to pass by one another. Both conditions—high pressure and low temperature—favor intermolecular interactions (represented by the dotted red lines in the right-side particulate view in **Figure 9.35**) that the ideal gas law does not take into account. The more often that particles interact with each other, the less often and forcefully they collide with the walls of their container and the lower the pressure. Likewise, lowering the temperature of a gas causes molecules to move more slowly, leading to more attractions between molecules and more deviations from ideal gas behavior.

The different intermolecular forces experienced by different compounds affect how significantly a particular gas will deviate from ideal behavior under high-pressure and/or low-temperature conditions. For example, polar molecules such as chloromethane ($CH_3Cl$) will experience dipole–dipole forces that result in attractions between the molecules. These forces are typically stronger than the attractive London dispersion forces experienced by nonpolar molecules such as $CH_4$, and we can predict greater deviation from ideality for chloromethane than for methane.

**CONNECTION** The nature and relative strengths of intermolecular forces were discussed in Chapter 6.

### CONCEPT **TEST**

Which of the following gases would you expect to deviate from ideal behavior: $H_2$ or HBr?

Another reason for nonideal behavior is that atoms and molecules of gases do have some volume. Under extremely high pressure, the volume of the particles themselves becomes a significant fraction of the total volume of the gas as the particles are squeezed more tightly together. However, only the empty space between particles makes gases compressible. The particles themselves are not compressible: they have finite dimensions and do not shrink with increasing pressure. Therefore, when the pressure doubles, the volume of a real gas does not shrink by 50% (as it does in an ideal gas) because a significant part of that volume can't shrink at all. If the volume shrinks by less than 50% when pressure doubles, then the product $P \times V$ in the numerator of $PV/RT$ is no longer a constant. Instead, $P \times V$ and the $PV/RT$ ratio increase as $P$ increases. Because all atoms and molecules have a finite volume, this positive deviation in $PV/RT$ is observed for all gases, including the three in Figure 9.34, if $P$ is high enough.

So two competing factors—namely, intermolecular forces and the volumes of gas-phase particles—can cause $PV/RT$ to decrease or increase with increasing external pressure. At extremely high pressures, the size of the particles and the space they occupy become the dominant factor and $PV/RT$ increases with increasing $P$. At less extreme pressures, intermolecular forces may offset the incompressibility of particles, and the value of $PV/RT$ decreases with increasing $P$ for some gases, including $CH_4$ and $CO_2$, as shown in Figure 9.34.

## The van der Waals Equation for Real Gases

Because the ideal gas equation does not hold at extremely high pressures and low temperatures, we need to modify it if we want to relate the quantities, pressures, volumes, and temperatures of real gases under these nonideal conditions. An equation developed in 1873 by Dutch scientist Johannes Diderik van der Waals (1837–1923) incorporates the needed modifications:

$$\left[P + a\left(\frac{n}{V}\right)^2\right](V - nb) = nRT \tag{9.30}$$

Equation 9.30 is called the **van der Waals equation** and includes two terms to correct for (1) intermolecular forces that lower the number of independent particles and the pressure they create by colliding with container walls [the $a(n/V)^2$ term] and (2) the volume taken up by the particles of a gas, which are not compressible (the $nb$ term). The values of $a$ and $b$ in these terms, called *van der Waals constants*, have been determined experimentally for many gases (**Table 9.5**). The values of both constants increase with increasing molar mass and with the number of atoms per molecule, which makes sense because larger molecules both take up more space and experience stronger London dispersion forces, which lead to stronger intermolecular forces.

Note that the pressure correction term, $a(n/V)^2$, increases with the square of the concentration $(n/V)$ of particles. This dependence makes sense if we consider that the chance that two particles will get close enough to interact depends on the concentration of both particles—that is, $(n/V)^2$. In Chapter 6, we explored the many ways that molecules can interact: London dispersion, dipole–dipole, hydrogen bonding, and dipole–induced dipole interactions. All these interactions may contribute to the value of the $a$ constant of a gaseous substance, depending on its molecular structure, so they are collectively called **van der Waals forces**.

**TABLE 9.5** van der Waals Constants of Selected Gases

| Substance | $a$ ($L^2 \cdot atm/mol^2$) | $b$ (L/mol) |
|---|---|---|
| He | 0.0341 | 0.02370 |
| Ar | 1.34 | 0.0322 |
| $H_2$ | 0.244 | 0.0266 |
| $N_2$ | 1.39 | 0.0391 |
| $O_2$ | 1.36 | 0.0318 |
| $CH_4$ | 2.25 | 0.0428 |
| $CO_2$ | 3.59 | 0.0427 |
| CO | 1.45 | 0.0395 |
| $H_2O$ | 5.46 | 0.0305 |
| NO | 1.34 | 0.02789 |
| $NO_2$ | 5.28 | 0.04424 |
| HCl | 3.67 | 0.04081 |
| $SO_2$ | 6.71 | 0.05636 |

**CONCEPT TEST**

The van der Waals constant $a$ for $SO_2$, 6.71 $L^2 \cdot atm/mol^2$, is nearly twice the value of $a$ for $CO_2$, 3.59 $L^2 \cdot atm/mol^2$. Suggest a reason for this difference.

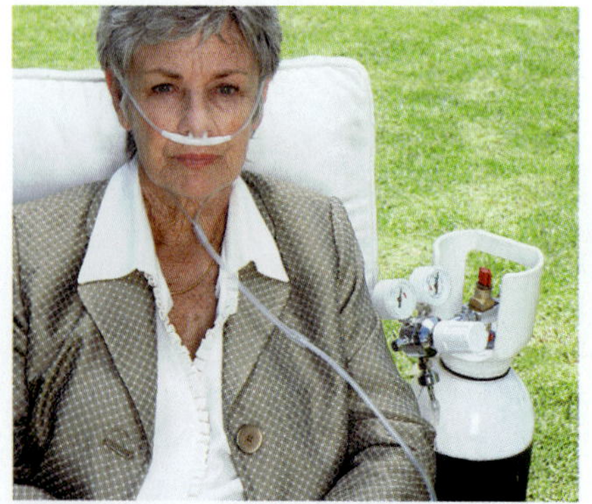

**FIGURE 9.36** A patient breathing supplemental oxygen from a portable tank.

**SAMPLE EXERCISE 9.14** Using the van der Waals Equation **LO9**

When they are out and about, patients suffering from chronic lung disease often rely on portable tanks of compressed $O_2$ (**Figure 9.36**). The tanks have an internal volume of 2.24 L and typically contain 0.500 kg of $O_2$ when filled to their recommended pressure. What is that pressure at 20°C? Calculate your answer by using both the ideal gas equation and the van der Waals equation.

**Collect, Organize, and Analyze** We are given the mass, volume, and temperature of a quantity of $O_2$, and we are asked to calculate its pressure by using both the van der Waals equation and the ideal gas equation.

The van der Waals constants for $O_2$, listed in Table 9.5, are $a = 1.36\ L^2 \cdot atm/mol^2$ and $b = 0.0318$ L/mol. These values are a little more than half the $a$ and $b$ values for $CH_4$, which exhibits the nonideal behavior shown in Figure 9.34—namely, intermolecular forces produce a decrease in $PV/RT$ below 400 atm. $O_2$ molecules should experience weaker, but still significant, intermolecular forces that lead to pressures below those predicted by the ideal gas law.

**Solve** Let's start by converting the mass of $O_2$ into the corresponding number of moles because $n$ appears several times in both the ideal gas and van der Waals equations.

$$0.500\ \text{kg O}_2\left(\frac{1000\ \text{g}}{1\ \text{kg}}\right)\left(\frac{1\ \text{mol O}_2}{32.00\ \text{g O}_2}\right) = 15.62\ \text{mol O}_2$$

Assuming ideal behavior and solving the ideal gas equation for $P$ yields

$$P = \frac{nRT}{V} = \frac{(15.62\ \text{mol})\left(0.08206\ \frac{\text{L}\cdot\text{atm}}{\text{mol}\cdot\text{K}}\right)(273 + 20)\ \text{K}}{(2.24\ \text{L})} = 168\ \text{atm}$$

Repeating this calculation by using the van der Waals equation, we obtain

$$P = \frac{nRT}{V - nb} - a\left(\frac{n}{V}\right)^2$$

$$= \frac{(15.62\ \text{mol})\left(0.08206\ \frac{\text{L}\cdot\text{atm}}{\text{mol}\cdot\text{K}}\right)(273 + 20)\ \text{K}}{2.24\ \text{L} - (15.62\ \text{mol})\left(0.0318\ \frac{\text{L}}{\text{mol}}\right)} - \left(1.36\ \frac{\text{L}^2\cdot\text{atm}}{\text{mol}^2}\right)\left(\frac{15.62\ \text{mol}}{2.24\ \text{L}}\right)^2$$

$$= 215.4\ \text{atm} - 66.1\ \text{atm}$$

$$= 149\ \text{atm}$$

**Think About It** As predicted, the "real" pressure is less than the "ideal" pressure as a result of intermolecular forces between $O_2$ molecules. The last step in the calculation shows that the pressure in the tank would have been much higher (215 atm) than in an ideal gas (168 atm) because of the volume of the gas occupied by $O_2$ molecules—the $nb$ correction term. However, this increase was more than offset by the decrease in pressure (−66 atm) because of interactions between molecules—the $a(n/V)^2$ correction term.

**Practice Exercise** Thousands of buses in the United States and other countries run on compressed natural gas (CNG) to reduce pollution and save fuel costs (**Figure 9.37**). CNG, which is mostly methane, is stored on the roofs of the buses in large tanks in which the density of $CH_4$ is as high as 165 g/L. Use both the ideal gas equation and the van der Waals equation to calculate the pressure in one of these tanks at 25°C.

**van der Waals equation** an equation describing how the pressure, volume, and temperature of a quantity of a real gas are related; it includes terms that account for the incompressibility of gas particles and interactions between them.

**van der Waals force** any interaction between neutral atoms and molecules, including hydrogen bonds, other dipole–dipole interactions, and London dispersion forces; the term does not apply to interactions involving ions.

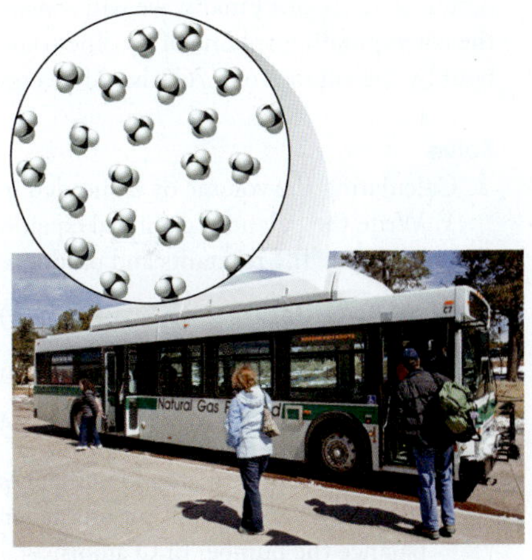


**FIGURE 9.37** A bus fueled by compressed natural gas.

**SAMPLE EXERCISE 9.15** Integrating Concepts: Air for a Jet Engine

The Boeing 767 (**Figure 9.38**) has been a popular wide-body commercial airliner (more than 1100 have been built) and one of the most fuel efficient. While cruising at 853 km/h (530 mph) at an altitude of 11,000 m (36,000 ft), a 767-200ER (the extended-range version of the plane) consumes about 6510 L (1720 U.S. gallons) of jet fuel per hour.

Use the following facts to solve this problem: (1) the density of jet fuel is 0.804 g/mL; (2) at an altitude of 11,000 m, $P_{atm}$ = 211 mmHg and $T = -56$°C; and (3) dodecane, $C_{12}H_{26}$, is considered an appropriate model hydrocarbon for jet fuel.

a. What volume of air, in liters, does a cruising 767 need so that it can completely burn an hour's worth of jet fuel?

b. Fuel efficiency is often based on the number of passengers times distance traveled per volume of fuel consumed. In the United States, this value is typically expressed in units of passenger-miles per gallon. However, in much of the rest of the world, efficiency units are inverted and are typically expressed in liters per 100 km per passenger. Express the fuel efficiency of a full 767-200ER (which holds 224 passengers) in both sets of units.

**FIGURE 9.38** A Boeing 767-200ER.

**Collect and Organize** We know the quantity of dodecane to be combusted, and we are asked to calculate the volume of air needed for complete combustion—that is, to convert its C and H content into $CO_2$ and $H_2O$. We know the pressure and temperature of the air. According to Table 9.1, dry air is 20.95% $O_2$. The molar mass of dodecane is 170.33 g/mol.

**Analyze** This exercise involves a chemical reaction, so writing a balanced chemical equation for it is a good place to start. Next, we need to convert the volume of fuel into an equivalent number of moles of fuel and then to use the stoichiometry of the reaction to convert that value to moles of $O_2$. We will then calculate the equivalent volume of $O_2$ by using the ideal gas equation. (The pressure is below 1 atm, so we should not need to correct for nonideal behavior.) Finally, we will convert the volume of $O_2$ to the corresponding volume of air. The volume of air needed each hour by the engines of a 767 should be enormous.

**Solve**

a. Calculating the volume of air needed in 1 hour:

1. Write the balanced chemical equation for the combustion reaction. The reactants and products are

$$C_{12}H_{26}(\ell) + O_2(g) \rightarrow CO_2(g) + H_2O(g)$$

We first balance the numbers of C and H atoms:

$$C_{12}H_{26}(\ell) + O_2(g) \rightarrow 12\ CO_2(g) + 13\ H_2O(g)$$

This leaves us with an odd number of O atoms on the right, requiring that we multiply all the terms by 2 and then balance the number of O atoms:

$$2\ C_{12}H_{26}(\ell) + 37\ O_2(g) \rightarrow 24\ CO_2(g) + 26\ H_2O(g)$$

2. Converting the volume of fuel consumed in 1 hour into an equivalent number of moles of $C_{12}H_{26}$ (and carrying an extra digit in intermediate values):

$$6510\ \cancel{\text{L}}\left(\frac{1000\ \cancel{\text{mL}}}{1\ \cancel{\text{L}}}\right)\left(\frac{0.804\ \cancel{\text{g}}}{1\ \cancel{\text{mL}}}\right)\left(\frac{1\ \text{mol}}{170.33\ \cancel{\text{g}}}\right)$$

$$= 3.073 \times 10^4\ \text{mol}\ C_{12}H_{26}$$

3. Converting moles of $C_{12}H_{26}$ into moles of $O_2$:

$$3.073 \times 10^4\ \cancel{\text{mol}\ C_{12}H_{26}}\left(\frac{37\ \text{mol}\ O_2}{2\ \cancel{\text{mol}\ C_{12}H_{26}}}\right) = 5.685 \times 10^5\ \text{mol}\ O_2$$

4. In converting moles of $O_2$ into a volume of air by using the ideal gas law, we need to convert pressure units from millimeters of mercury to atmospheres and temperature to the Kelvin scale:

$$V = \frac{nRT}{P} = \frac{(5.685 \times 10^5\ \cancel{\text{mol}})\left(0.08206\ \frac{\text{L} \cdot \cancel{\text{atm}}}{\cancel{\text{mol}} \cdot \cancel{\text{K}}}\right)(273 - 56)\ \cancel{\text{K}}}{211\ \cancel{\text{mmHg}}\left(\frac{1\ \text{atm}}{760\ \cancel{\text{mmHg}}}\right)}$$

$$= 3.664 \times 10^7\ \text{L}\ O_2$$

5. Air is 20.95% $O_2$ by volume, so the volume of air that the engines must take in each hour is

$$3.664 \times 10^7\ \cancel{\text{L}\ O_2}\left(\frac{100\ \text{L air}}{20.95\ \cancel{\text{L}\ O_2}}\right) = 1.75 \times 10^8\ \text{L air}$$

b. Calculating fuel efficiency in customary U.S. units, on the basis of the distance traveled and the fuel consumed in 1 hour:

$$\frac{224\ \text{passengers} \times 530\ \text{miles}}{1720\ \text{gal}} = 69.0\ \text{passenger-miles/gallon}$$

The corresponding efficiency value in liters per 100 km per passenger is

$$\frac{6510\ \text{L}}{224\ \text{passengers} \times 853\ \text{km}} \times 100 = 3.41\ \text{L/100 km/passenger}$$

**Think About It** The calculated volume of air is 175 million liters. That is a lot of air, but then 6510 L of jet fuel is a lot of fuel. Perhaps a more interesting value is the fuel efficiency of 69.0 passenger-miles per gallon, which is much greater than that of most automobiles with two occupants—and gets you to your destination a lot faster.

## SUMMARY

**LO1** The behavior of gases is explained by **kinetic molecular theory (KMT)**, which assumes that gases are composed of particles in constant random motion, that the volume of the particles is insignificant compared to the volume occupied by the gas, and that the collisions of the particles are elastic. (Section 9.1 and 9.2)

**LO2** According to **Graham's law of effusion**, the rate of **effusion** (escape through a pinhole) of a gas at a fixed temperature is inversely proportional to the square root of its molar mass. Gas **diffusion** is the spread of one gas through another. Graham's law of effusion describes diffusion as well: at a fixed temperature, the rate at which a gas spreads is inversely proportional to the square root of its molar mass. (Section 9.2)

**LO3** Gas particles move at **root-mean-square speeds** ($u_{rms}$) that are inversely proportional to the square root of their molar masses. (Section 9.2)

**LO4** The mass of the gases in Earth's atmosphere combined with the force of gravity results in an atmospheric pressure on the surface of the planet. Pressure is defined as the ratio of force to surface area and is measured with **barometers** and **manometers**. (Section 9.3)

**LO5** The laws describing how the pressure, volume, and temperature of a gas are related bear the names of the scientists who discovered them: **Boyle's law**, **Charles's law**, **Avogadro's law**, and **Amontons's law**. The **combined gas law** relates the pressure, volume, and temperature of a fixed quantity of gas. The **ideal gas law** and the **ideal gas equation** ($PV = nRT$, where $R$ is the **universal gas constant**) describe the behavior of gases under normal conditions. At **standard temperature and pressure (STP)**, which in the United States is traditionally defined as 0°C and 1 atm, the *molar volume* of an **ideal gas** is 22.4 L. (Sections 9.4–9.6)

**LO6** The density of a gas is proportional to its molar mass and pressure and inversely proportional to its absolute temperature. (Section 9.7)

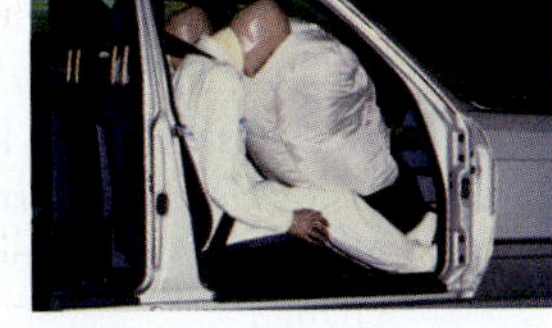

**LO7** The ideal gas equation and the stoichiometry of a chemical reaction can be used to calculate the volumes of gases required or produced in the reaction. (Section 9.8)

**LO8** In a gas mixture, the contribution that each component gas makes to the total gas pressure is called the **partial pressure** of that gas. **Dalton's law of partial pressures** states that the total pressure of a gas mixture is the sum of the partial pressures of its components. Dalton's law allows us to calculate the partial pressure ($P_i$) of any constituent gas $i$ in a gas mixture if we know its **mole fraction ($x_i$)** and the total pressure. The total pressure of a gas sample collected over water is the sum of the partial pressure of water vapor and the partial pressure of the gas. (Section 9.9)

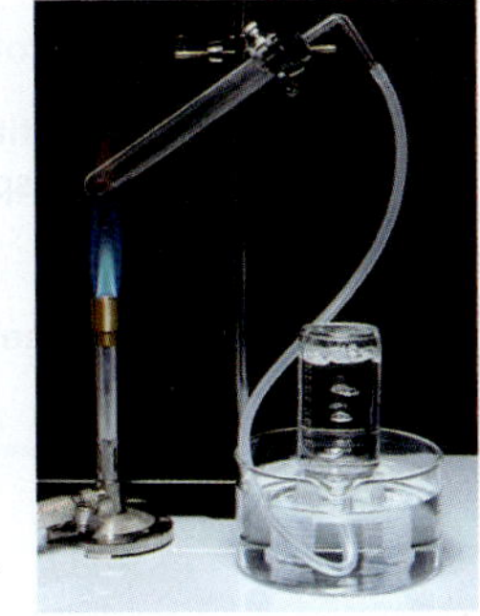

**LO9** At high pressures or very low temperatures, real gases deviate from the predictions of the ideal gas law because of intermolecular interactions between gas phase particles and the incompressibility of these particles. The **van der Waals equation**, a modified form of the ideal gas equation, accounts for the behavior of real gases. (Section 9.10)

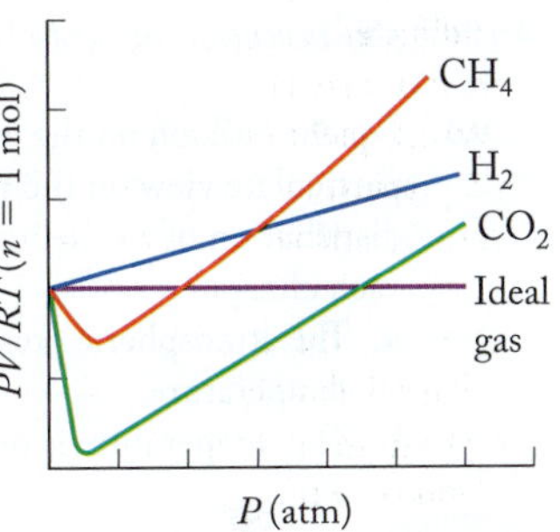


## PARTICULATE PREVIEW WRAP-UP

As the temperature increases, the balloon expands, and the number of collisions between particles (helium atoms or nitrogen molecules) decreases. As the atmospheric pressure increases, the balloon shrinks, and the number of collisions between the particles increases.

## PROBLEM-SOLVING SUMMARY

| Type of Problem | Concepts and Equations | | Sample Exercises |
|---|---|---|---|
| **Calculating relative rates of effusion** | $\frac{\text{effusion rate}_A}{\text{effusion rate}_B} = \sqrt{\frac{\mathcal{M}_B}{\mathcal{M}_A}}$ | (9.8) | **9.1** |
| **Calculating root-mean-square speeds** | $u_{rms} = \sqrt{\frac{3RT}{\mathcal{M}}}$ | (9.11) | **9.2** |
| **Measuring gas pressure with a manometer; converting pressure units** | See Table 9.2 and the conversion factors inside the back cover. | | **9.3** |
| **Applying Boyle's law** | $P_1V_1 = P_2V_2$ | (9.14) | **9.4** |
| **Applying Charles's law** | $\frac{V_1}{T_1} = \frac{V_2}{T_2}$ | (9.17) | **9.5** |
| **Applying Amontons's law** | $\frac{P_1}{T_1} = \frac{P_2}{T_2}$ | (9.22) | **9.6** |
| **Applying the combined gas law** | $\frac{P_1V_1}{T_1} = \frac{P_2V_2}{T_2}$ | (9.23) | **9.7** |
| **Applying the ideal gas law** | $PV = nRT$ | (9.24) | **9.8–9.11** |
| **Calculating mole fractions and partial pressures** | $x_i = \frac{n_i}{n_{total}}$ | (9.27) | **9.12** |
| | $P_i = x_iP_{total}$ | (9.28) | |

| Type of Problem | Concepts and Equations | Sample Exercises |
|---|---|---|
| **Calculating the quantity of a gas collected by water displacement** | Calculate the partial pressure ($P_i$) of the collected gas by using the equation $$P_i = P_{atm} - P_{H_2O}$$ | **9.13** |
| **Using the van der Waals equation** | $$\left[P + a\left(\frac{n}{V}\right)^2\right](V - nb) = nRT \quad (9.30)$$ | **9.14** |

## VISUAL PROBLEMS

*(Answers to boldface end-of-chapter questions and problems are in the back of the book.)*

**9.1.** For the balloon on the left in Figure P9.1, which particulate view on the right (i, ii, or iii) best represents the distribution of the helium atoms inside the balloon after each change?

a. The atmospheric pressure is increased at constant temperature.

b. The temperature is decreased at constant pressure.

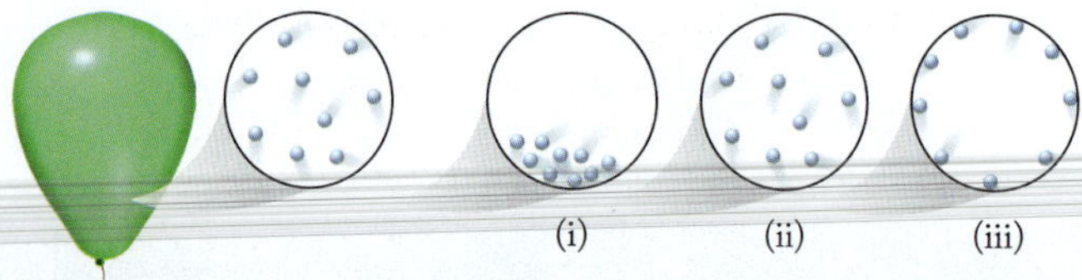


**FIGURE P9.1**

9.2. How does the downward motion of the piston in Figure P9.2(a) affect the pressure of the gas in the cylinder? Assume that the temperature of the gas does not change significantly but that the volume is reduced by half. Draw a particulate representation for the gas in each cylinder.

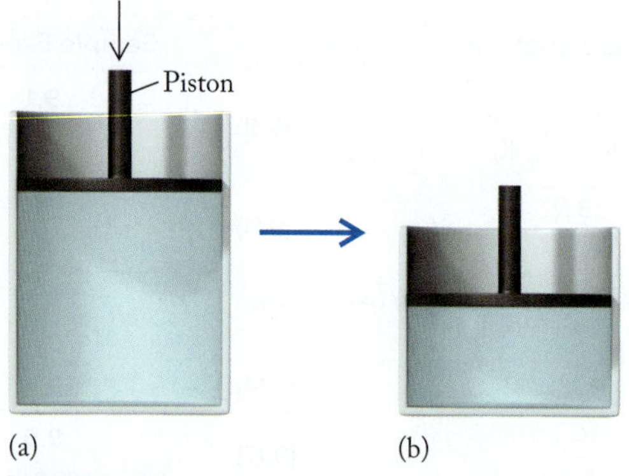


**FIGURE P9.2**

**9.3.** How does the motion of the piston in Problem 9.2 affect the root-mean-square speed of the gas particles in the cylinder?

9.4. Suppose the temperature of the gas in Figure P9.2(b) increases from 200 K to 400 K but that the pressure on the piston remains constant. Does the position of the piston change? If so, by how much does the volume of the gas change?

**9.5.** Enough gas is added to the cylinder in Figure P9.5(a) to double the number of particles inside the cylinder while the temperature remains constant.

a. If we assume that the position of the piston does not change, by how much does the pressure inside the cylinder change?

b. By how much does the frequency of the collisions between gas particles and the inner walls of the cylinder change?

c. By how much does the most probable speed of the gas particles change?

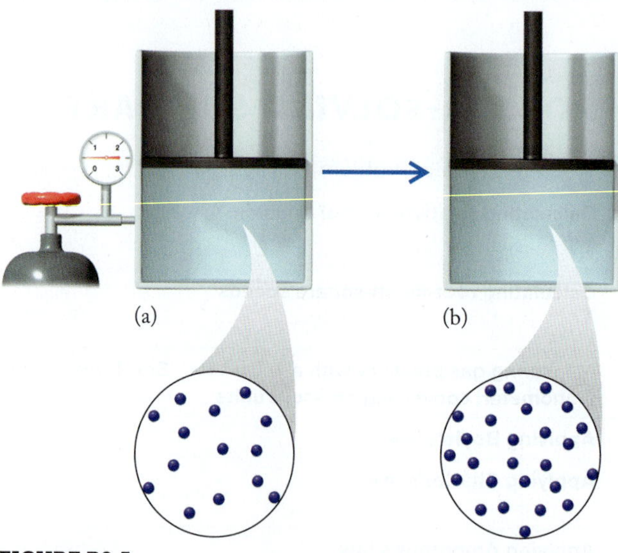


**FIGURE P9.5**

9.6. Suppose the same quantity of gas from Problem 9.5 was added to the cylinder in Figure P9.5(a) at constant temperature *and* constant pressure.

a. In which direction would the piston move?

b. By how much would the volume of the gas in the cylinder change?

**9.7.** Consider the two gas mixtures in Figure P9.7. Assume that their volumes and temperatures are the same. Molecules of gas A are represented by red spheres.

a. In which mixture is the mole fraction of A greater?
b. In which mixture is the partial pressure of A greater?
c. In which mixture is the total pressure greater?

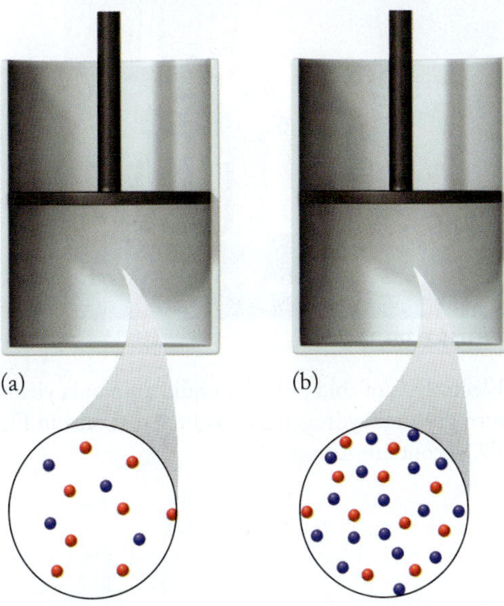


**FIGURE P9.7**

9.8. Suppose that the piston in Figure P9.7(a) was pushed downward so that the volume of the gas mixture was about 2/3 its initial value. Which of the following values would increase, which would decrease, and which would remain the same, if we assume that no change in temperature or in the numbers of red and blue spheres occurs?

a. The mole fraction of A (whose molecules are represented by red spheres)
b. The partial pressure of A
c. The total pressure of the mixture
d. The most probable speed of the molecules of A

**9.9.** Figure P9.9 shows the distribution of molecular speeds of $CO_2$ and $SO_2$ molecules at 25°C. Which curve is the profile for $SO_2$? Which profile should also match that of propane ($C_3H_8$), a common fuel in portable grills?

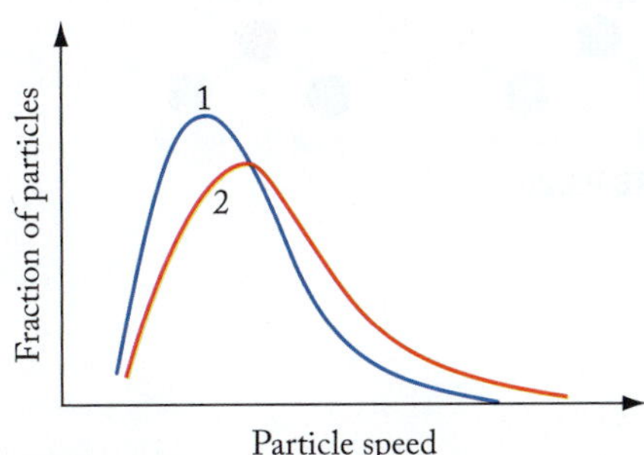


**FIGURE P9.9**

9.10. How would a graph showing the distribution of molecular speeds of $CO_2$ at −60°C differ from the curve for $CO_2$ shown in Figure P9.9?

**9.11.** Figure P9.11 describes the relationship between the volume of a gas and temperature at constant pressure (Charles's law).

a. If each line represents 1.00 mol of gas, which line represents the highest pressure?
b. Estimate the slope of this line on the graph. What is the pressure of the gas?

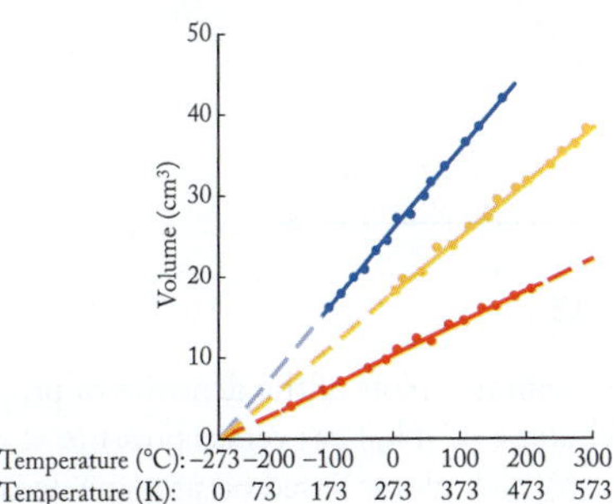


**FIGURE P9.11**

9.12. Use the graph in Figure P9.11 to answer the following questions.

a. Is the slope of the graph independent of the identity of the gas (for example, He vs. $N_2$) if the number of moles of gas and the pressure are constant?
b. If each line in Figure P9.11 is at the *same* pressure (for example, $P = 1.00$ atm), which line represents the greatest number of moles of gas?

**9.13.** The two plots of volume versus reciprocal pressure in Figure P9.13 correspond to the same quantity of gas at two temperatures. Which line corresponds to the higher temperature?

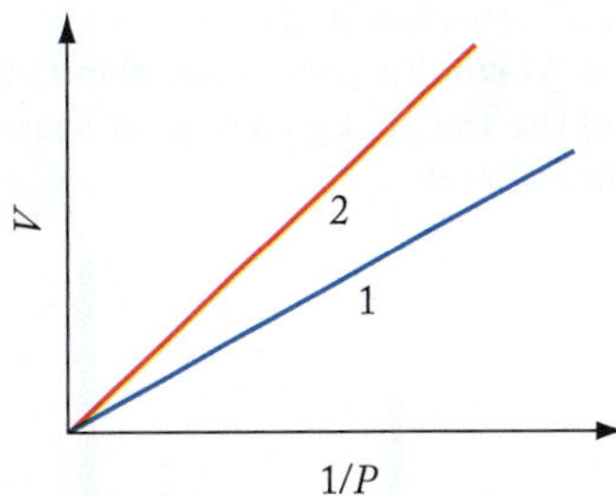


**FIGURE P9.13**

9.14. In Figure P9.14, which of the two plots of volume versus pressure at constant temperature is consistent with the ideal gas law?

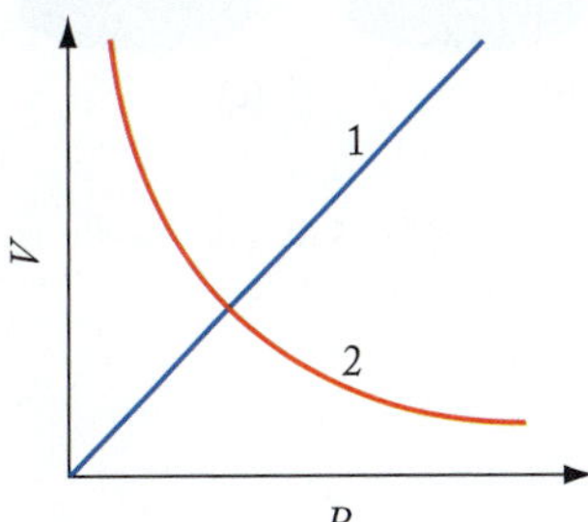


**FIGURE P9.14**

**9.15.** In Figure P9.15, which of the two plots of volume versus temperature at constant pressure is *not* consistent with the ideal gas law?

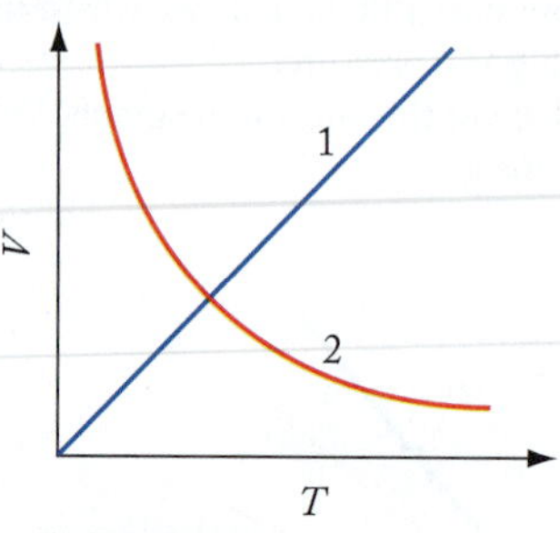


FIGURE P9.15

9.16. Figure P9.16 contains plots of the densities of propane ($C_3H_8$) and butane ($C_4H_{10}$) gas versus pressure at the same temperature. Which plot is based on propane densities?

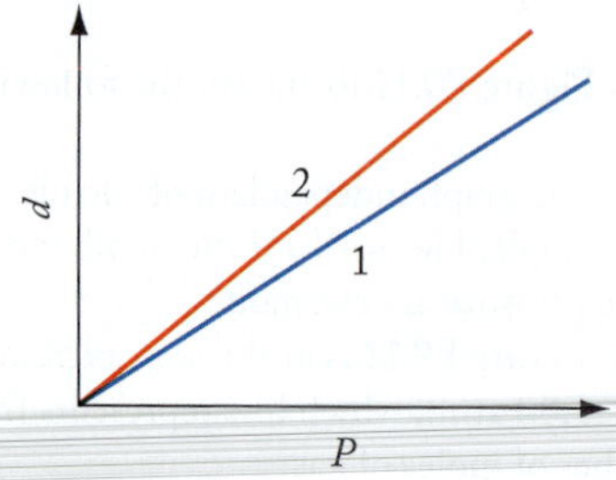


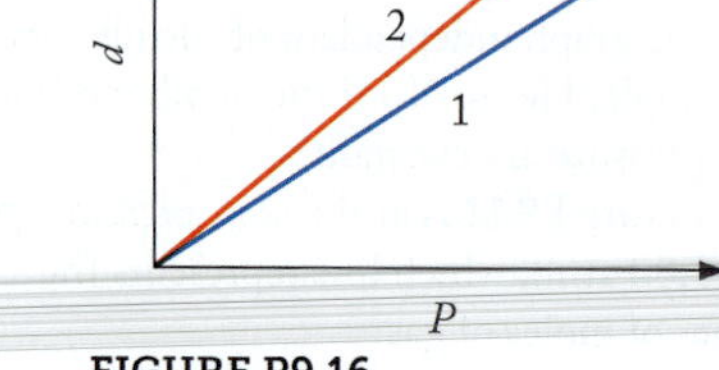

FIGURE P9.16

**9.17.** Consider the three Torricelli barometers in Figure P9.17. Which one is most likely sensing atmospheric pressure at each of the following locations?

1. San Diego, California (sea level)
2. The summit of Mount Everest (8.8 km above sea level)
3. The bottom of the Tau Tona gold mine in South Africa (3.9 km below sea level)

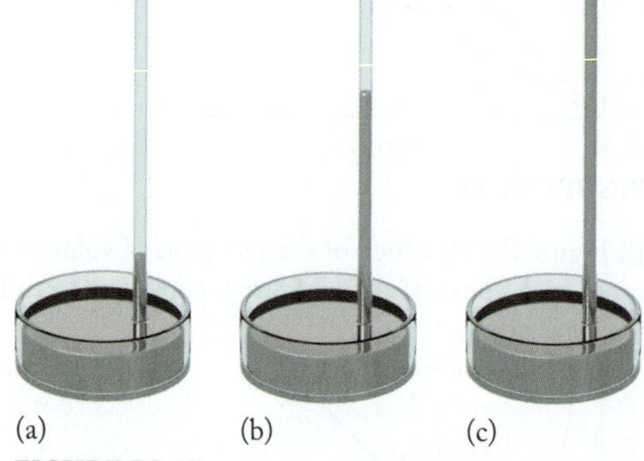


FIGURE P9.17

9.18. Electrolysis of water produces hydrogen and oxygen. A simple apparatus for this reaction is shown in Figure P9.18. Which test tube contains hydrogen?

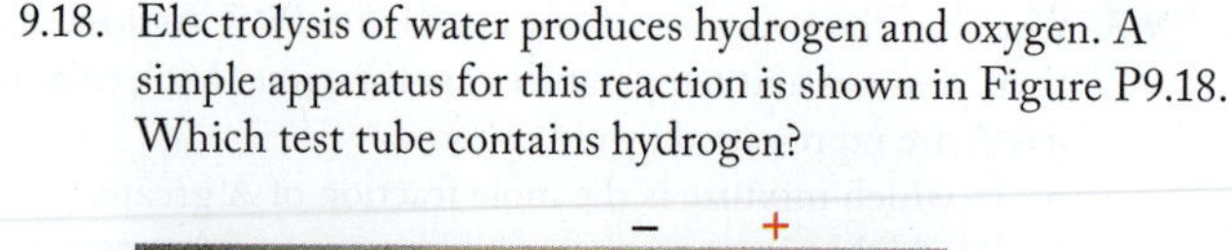


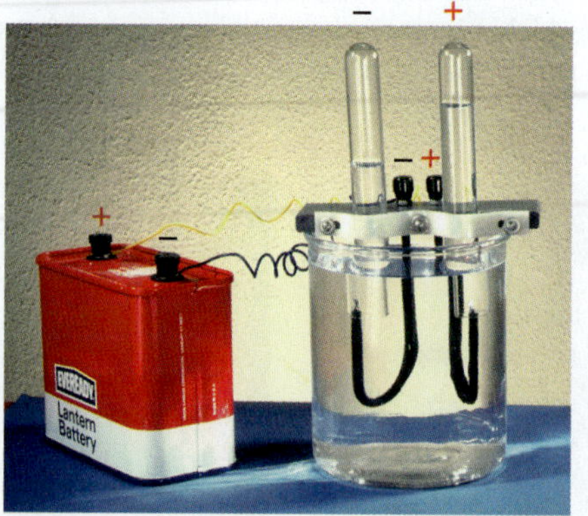


FIGURE P9.18

**9.19.** Electrolysis of cold (−37°C) liquid ammonia yields nitrogen and hydrogen gas. Which test tube in Figure P9.19 contains nitrogen?

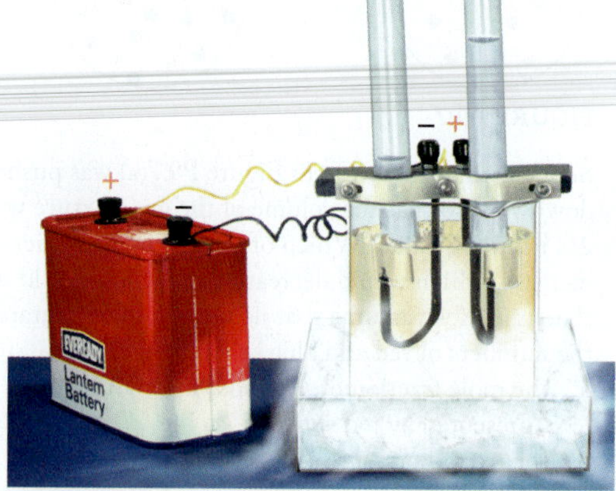


FIGURE P9.19

9.20. Which molecule in Figure P9.20 is likely to have the greater value of $a$ in the van der Waals equation? Do you expect them to have similar values of $b$?

FIGURE P9.20

*9.21. Figure P9.21 represents a plot of $PV/RT$ as a function of $P$ for methane and propane. Which line on the graph in Figure P9.21 represents propane?

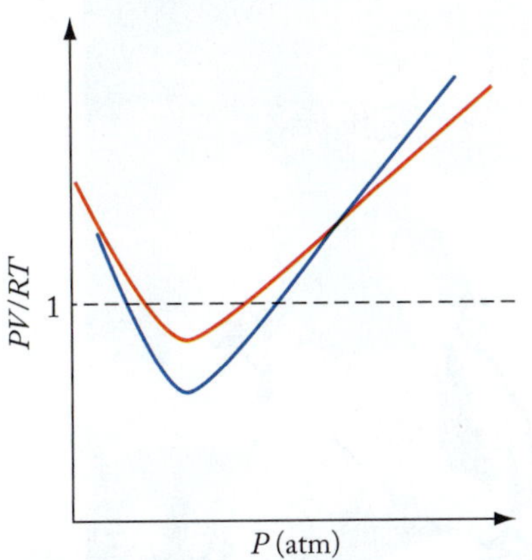


**FIGURE P9.21**

9.22. Use representations [A] through [I] in Figure P9.22 to answer questions a–f. The pink balloons contain hydrogen, the yellow balloons contain nitrogen, and the gray balloons contain oxygen.

a. Identify three changes that could be responsible for the change in size of the pink balloon from [A] to [C].
b. If the smaller pink balloon in [A] corresponds to the particulate view in [B], which of the changes identified in the previous question would result in the larger pink balloon in [C] also corresponding to the particulate view in [B]?
c. If the smaller yellow balloon in [D] corresponds to the particulate view in [E], which particulate view corresponds to the larger yellow balloon in [F] if no additional nitrogen has been added?
d. If the larger gray balloon in [I] corresponds to the particulate view in [E], which particulate view represents the gas at a lower temperature?
e. If the gray balloons in [G] and [I] each contain 1 mol of gas at 25°C, in which gray balloon are the collisions between the oxygen molecules and the inside of the balloon more frequent?
f. Which balloon contains the gas with the shortest mean free path?

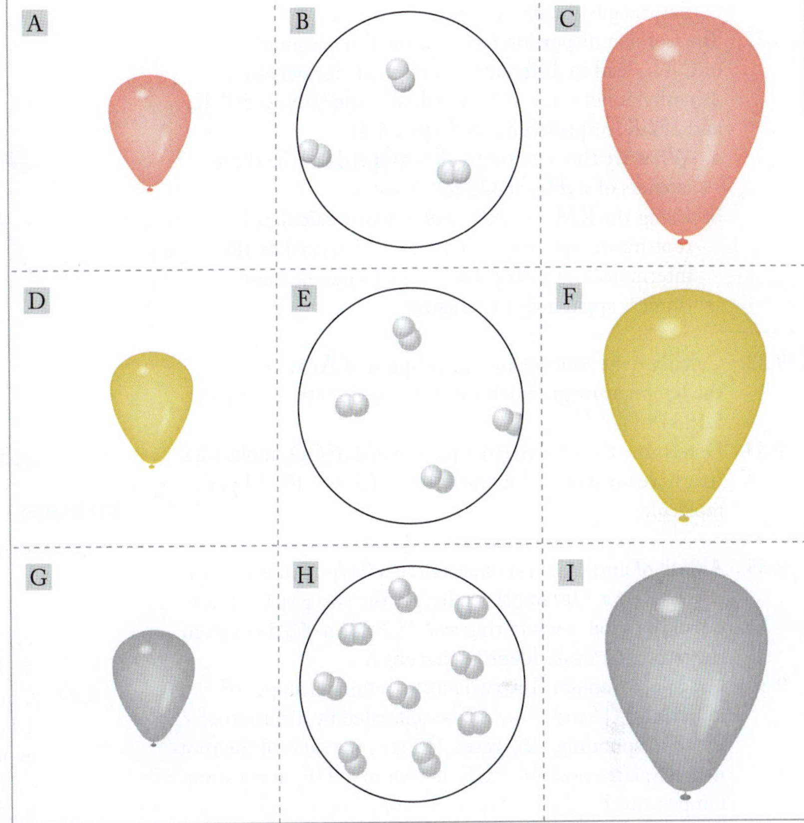


**FIGURE P9.22**

# QUESTIONS AND PROBLEMS

## Effusion, Diffusion, and the Kinetic Molecular Theory of Gases

### Concept Review

**9.23.** What is meant by the *root-mean-square* speed of gas particles?

*9.24. Why is the root-mean-square speed of gas particles greater than the simple average of their speeds?

**9.25.** Does pressure affect the root-mean-square speed of the particles in a gas? Why or why not?

9.26. How is the rate of effusion of a gas related to each of the following?

a. Molar mass
b. Root-mean-square speed
c. Temperature
d. Density

### Problems

**9.27.** Rank the gases NO, $NO_2$, $N_2O_4$, and $N_2O_5$ in order of increasing root-mean-square speed at 0°C.

9.28. **Chlorofluorocarbons** Chlorofluorocarbons with the general formula $CF_nCl_{4-n}$ were used for decades in refrigerators and air conditioners until they were found to degrade Earth's ozone layer. For which value of $n$ is the rate of effusion the slowest? Which compound will effuse at the fastest rate?

**9.29.** Molecular hydrogen effuses 4.0 times as fast as gas X at the same temperature. What is the molar mass of gas X?

9.30. Which noble gas effuses 3.2 times faster than argon at 0°C?

**9.31.** Methane ($CH_4$) and propane ($C_3H_8$) are both used as fuel for cooking.
a. What are the root-mean-square speeds of $CH_4$, ethane ($C_2H_6$), and $C_3H_8$ at 298 K?
b. Plot the root-mean-square speeds of these gases as a function of their molar mass and use it to predict the root-mean-square speed of butane ($C_4H_{10}$) and pentane ($C_5H_{12}$).
c. Of the major components of natural gas—$CH_4$, $C_2H_6$, $C_3H_8$, and $C_4H_{10}$—which gas effuses from a sample of natural gas the most rapidly?

9.32. The London dispersion forces in the three isomers of $C_5H_{12}$ lead to different boiling points for pentane, 2-methylbutane, and 2,2-dimethylbutane (309 K, 301 K, and 282 K, respectively; see Figure 6.4).
a. What are the root-mean-square speeds of the three isomers of $C_5H_{12}$ at 325 K?
b. Using the KMT model, describe why calculated root-mean-square speeds are not influenced by the intermolecular forces that the molecules of these isomers apparently experience.

---

**9.33.** Calculate the root-mean-square speed of Ar atoms at the temperature at which their average kinetic energy is 5.18 kJ/mol.

9.34. Determine the root-mean-square speed of $CO_2$ molecules that have an average kinetic energy of $3.2 \times 10^{-21}$ J per molecule.

---

**9.35.** A flask of ammonia is connected to a flask of an unknown acid HX by a 1.00 m glass tube. As the two gases diffuse down the tube, a white ring of $NH_4X$ forms 68.5 cm from the ammonia flask. Identify element X.

9.36. **Enriching Uranium** The two most common isotopes of uranium, $^{238}U$ and $^{235}U$, can be separated by diffusion of the corresponding $UF_6$ gases. What is the ratio of the root-mean-square speed of $^{238}UF_6$ to that of $^{235}UF_6$ at constant temperature?

## *Atmospheric Pressure*

### Concept Review

**9.37.** Describe the difference between force and pressure.

9.38. Why does atmospheric pressure decrease with increasing elevation?

**9.39.** Three barometers based on Torricelli's design are constructed using water ($d = 1.00$ g/mL), ethanol ($d = 0.789$ g/mL), and mercury ($d = 13.546$ g/mL). Which barometer contains the tallest column of liquid?

9.40. In constructing a barometer based on Torricelli's design, what advantage is there in choosing a dense liquid?

**9.41.** Why does an ice skater exert more pressure on ice when wearing newly sharpened skates than when wearing skates with dull blades?

9.42. Why is traveling over deep snow easier when wearing boots and snowshoes (Figure P9.42) rather than just boots?

**FIGURE P9.42**

### Problems

**9.43.** Which quantity represents the larger pressure?
a. 10 torr or 10 atm
b. 10 torr or 100 mmHg
c. 100 torr or 1.0 atm
d. 2.0 kPa or 562 mmHg

9.44. Consider a balloon filled with helium to a volume of 1 L. Under which pressure condition do the He atoms experience more collisions at room temperature?
a. 0.1 atm or 0.1 torr
b. 1.0 atm or 1000 mmHg
c. 390 torr or 390 mmHg
d. 0.541 atm or 2.8 kPa

---

**9.45.** **Record High Atmospheric Pressure** The highest atmospheric pressure ever recorded on Earth was 108.6 kPa at Tosontsengel, Mongolia, on December 19, 2001. Express this pressure in (a) millimeters of mercury, (b) atmospheres, and (c) millibars.

9.46. **Record Low Atmospheric Pressure** Despite the destruction from Hurricane Katrina in August 2005, the lowest pressure for a hurricane in the Atlantic Ocean was measured several weeks after Katrina. Hurricane Wilma registered an atmospheric pressure of 88.2 kPa on October 19, 2005, about 2 kPa lower than that of Hurricane Katrina. What was the *difference* in pressure between the two hurricanes in (a) millimeters of mercury, (b) atmospheres, and (c) millibars?

## *Relating P, T, and V: The Gas Laws*

### Concept Review

**9.47.** How does kinetic molecular theory explain why pressure is directly proportional to temperature at fixed volume (Amontons's law)?

9.48. Explain Boyle's law by using kinetic molecular theory.

**9.49.** A hot-air balloonist is rising too fast for her liking. Should she increase or decrease the temperature of the gas in the balloon?

9.50. Could the pilot of the balloon in Problem 9.49 reduce her rate of ascent by allowing some gas to leak out of the balloon? Explain your answer.

### Problems

**9.51.** The volume of a quantity of gas at 1.00 atm is compressed from 3.25 L to 2.24 L. What is the final pressure of the gas if no change in temperature occurs?

9.52. The pressure on a sample of an ideal gas is increased from 715 mmHg to 3.55 atm at constant temperature. If the initial volume of the gas is 485 mL, what is the final volume of the gas?

---

**9.53.** **Temperature Effects on Bicycle Tires** A bicycle racer inflates his tires to 7.1 atm on a warm autumn afternoon when temperatures reach 27°C. By the next morning, the temperature has dropped to 5.0°C. What is the pressure in the tires if we assume that the volume of the tire does not change significantly?

9.54. Balloons for a New Year's Eve party in Potsdam, New York, are filled to a volume of 5.0 L at a temperature of 20°C and then hung outside, where the temperature is −25°C. What is the volume of the balloons after they have cooled to the outside temperature? Assume that atmospheric pressure inside and outside the house is the same.

---

**9.55.** Glassblowers use a torch fueled by methane and oxygen. In the morning, a cylinder of oxygen gas contains 136 atm (2000 psi) of oxygen. By the end of the glassblower's workday, the pressure in the cylinder has decreased to 102 atm. What percentage of the oxygen in the tank was consumed if the temperature of the cylinder did not change?

9.56. Effusion causes the volume of a helium balloon to decrease by 25.0% after several days at constant temperature and atmospheric pressure.

a. What fraction of the helium was lost?

*b. What fraction of gas would be lost over the same period if the balloon was instead filled with CO, if we assume the same constant temperature and atmospheric pressure?

---

**9.57.** **Underwater Archaeology** A scuba diver releases a balloon containing 115 L of air attached to a tray of artifacts at an underwater archaeological site (Figure P9.57). When the balloon reaches the surface, it has expanded to a volume of 352 L. The pressure at the surface is 1.00 atm.

a. What is the pressure at the underwater site? Assume that the water temperature is constant.

b. If the pressure increases by 1.0 atm for every 10 m of depth, at what depth was the diver working?

**FIGURE P9.57**

*9.58. **Scuba Diving** A popular scuba tank for sport diving has an internal volume of 12.0 L and can be filled with air up to a pressure of 232 bar. Suppose that a diver consumes air at the rate of 21 L/min while diving on a coral reef where the sum of atmospheric pressure and water pressure averages 2.2 bar. How long will the diver take to use up a full tank of air if the temperatures of the air and water on the reef are both 28°C?

---

**9.59.** Which of the following actions would produce the greater increase in the volume of a gas sample?

a. Lowering the pressure from 760 mmHg to 700 mmHg at constant temperature

b. Raising the temperature from 10°C to 35°C at constant pressure

9.60. Which of the following actions would produce the greater increase in the volume of a gas sample?

a. Doubling the amount of gas in the sample at constant temperature and pressure

b. Raising the temperature from 244°C to 1100°C

---

**9.61.** What happens to the volume of gas in a cylinder with a movable piston under the following conditions?

a. Both the absolute temperature and the external pressure on the piston double.

b. The absolute temperature is halved, and the external pressure on the piston doubles.

c. The absolute temperature increases by 75%, and the external pressure on the piston increases by 50%.

9.62. What happens to the pressure of a gas under the following conditions?

a. The absolute temperature is halved and the volume doubles.

b. Both the absolute temperature and the volume double.

c. The absolute temperature increases by 75%, and the volume decreases by 50%.

---

**9.63.** A weather balloon containing 6.1 mol of He has a volume of 200.0 L and an internal pressure of 1.17 atm when it is filled at 20°C.

a. The weather balloon loses He at a rate of 10 mmol/h through effusion. What is the volume of the balloon after 24 hours?

b. Do the molecules of He that effuse in the first 12 hours move faster than those that effuse in the second 12 hours?

c. The balloon is released after 24 hours and rises to an altitude in the stratosphere where its internal pressure is 63 mmHg and the temperature is 210 K. What is the volume of the balloon at this altitude?

d. How much larger would the balloon have been if it had been released immediately after filling?

*9.64. A 200.0 L weather balloon identical to the one in Problem 9.63 is filled with 6.1 mol of $H_2$ at an internal pressure of 1.17 atm and a temperature of 20°C.

a. If the balloon loses $H_2$ entirely from effusion, what is the volume of this balloon after 24 hours?

b. Why doesn't this balloon have the same volume after 24 hours as the one in Problem 9.63?

c. This balloon is also released after 24 hours and rises to the same altitude as the He balloon in Problem 9.63. What is the volume of the balloon at this altitude?

d. Why would the two balloons have the same volume immediately after filling?

---

***9.65.** The temperature of a sealed sample of $N_2$ gas is raised so that the $u_{rms}$ of the gas increases by 22.5%. What is the final pressure of the gas if the initial pressure was 745 mmHg?

*9.66. The temperature of a sample of He gas in the piston in Figure P9.66 is raised so that its $u_{rms}$ increases by 33.3%. By what percentage does the volume of the gas increase if the piston is allowed to move to maintain a constant pressure?

**FIGURE P9.66**

## *Ideal Gases and the Ideal Gas Law*

### Concept Review

**9.67.** What is meant by standard temperature and pressure (STP)? What is the volume of one mole of an ideal gas at STP?

9.68. Which of the following are *not* characteristics of an ideal gas?

a. The gas atoms or molecules have insignificant volume compared with the volume that they occupy.

b. The volume of the gas is independent of temperature.

c. The density of the gas is the same for all ideal gases.

d. The gas atoms or molecules do not interact with one another.

### Problems

**9.69.** How many moles of air must be in a racing bicycle tire with a volume of 2.36 L if it has an internal pressure of 6.8 atm at 17.0°C?

9.70. The fuel canister for a portable camping stove contains 190 g of pressurized butane ($C_4H_{10}$). What volume of butane does this represent at 25°C if it escapes into the air at 750 mmHg?

---

**9.71.** **Hyperbaric Oxygen Therapy** Hyperbaric oxygen chambers are used to treat divers suffering from decompression sickness (the "bends") with pure oxygen at greater than atmospheric pressure. Other clinical uses include treatment of patients with thermal burns and CO poisoning. What is the pressure in a chamber with a volume of $4.85 \times 10^3$ L that contains 15.00 kg of $O_2$ gas at a temperature of 298 K?

9.72. **The *Hindenburg*** The German airship *Hindenburg* (Figure P9.72) made several transatlantic crossings in the early 1930s before a fiery crash in Lakehurst, New Jersey, in 1937, sealed the fate of "lighter-than-air" dirigibles. The *Hindenburg* held $1.6 \times 10^4$ kg of hydrogen gas. What was the volume of the hydrogen in the *Hindenburg* at 18°C and 0.98 atm?

FIGURE P9.72

**9.73.** **CNG-Powered Buses** In some cities, public buses are fueled by compressed natural gas (CNG; mostly methane, $CH_4$) How many grams of $CH_4$ gas are in a 250 L fuel tank at a pressure of 255 bar at 20°C?

9.74. **Liquid Nitrogen–Powered Car** Students at the University of North Texas and the University of Washington built a car propelled by compressed nitrogen gas. The gas was obtained by boiling liquid nitrogen stored in a 182 L tank. What volume of $N_2$ is released at 0.927 atm of pressure and 25°C from a tank full of liquid $N_2$ ($d$ = 0.808 g/mL)?

***9.75.** Suppose that the atmospheric temperature and pressure at the top of a ski run are −5°C and 713 mmHg, respectively. At the bottom of the run, the temperature and pressure are 0°C and 734 mmHg, respectively. How many more moles of oxygen does a skier take in with a lungful of air at the bottom of the run than at the top? Express your answer as a percentage.

*9.76. A balloon vendor at a street fair is using a tank of helium to fill her balloons. The tank has an internal volume of 45.0 L and a pressure of 195 atm at 22°C. After a while, she notices that the valve has not been closed properly and the pressure has dropped to 115 atm. How many moles of He have been lost?

**9.77.** The volume of a 4.0 g sample of a gaseous substance was measured at the temperatures listed in the following table. The pressure of the gas was 1.00 atm at all times.

a. How many moles of a gas were in the sample?
b. What was the probable identity of the gas?

| *V* (L) | *T* (K) |
|---|---|
| 7.88 | 96 |
| 3.94 | 48 |
| 1.97 | 24 |
| 0.79 | 9.6 |
| 0.39 | 4.8 |

9.78. The volume of 0.50 mol of a noble gas was measured at the temperatures listed in the following table. The pressure of the gas was 1.00 atm at all times. Use the data in the table to calculate the value of *R*, the universal gas constant.

| *V* (L) | *T* (K) |
|---|---|
| 3.94 | 96 |
| 1.97 | 48 |
| 0.79 | 24 |
| 0.39 | 9.6 |
| 0.20 | 4.8 |

## *Densities of Gases*

### Concept Review

**9.79.** Do all gases at the same pressure and temperature have the same density? Explain your answer.

9.80. Birds and sailplanes take advantage of thermals (rising columns of warm air) to gain altitude with less effort than usual. Why does warm air rise?

**9.81.** How does the density of a gas sample change when (a) its pressure is increased and (b) its temperature is decreased?

9.82. Rank the noble gases in order of decreasing density.

### Problems

**9.83.** **Biological Effects of Radon Exposure** Radon is a naturally occurring radioactive gas found in the ground and in building materials. Cumulative radon exposure is a significant risk factor for lung cancer.

a. Calculate the density of radon at 298 K and 1 atm of pressure.
b. Are radon concentrations likely to be greater in the basement or on the top floor of a building?

*9.84. Four empty balloons, each with a mass of 10.0 g, are inflated to a volume of 20.0 L. The first balloon contains He; the second, Ne; the third, $CO_2$; and the fourth, CO. If the density of air at 25°C and 1.00 atm is 1.17 g/L, how many of the balloons float in it?

**9.85.** A 30.0 mL flask contains 0.078 g of a volatile oxide of sulfur. The pressure in the flask is 750 mmHg, and the temperature is 22°C.

a. What is the density of the gas?
b. Is the gas $SO_2$ or $SO_3$?

9.86. A 100.0 mL flask contains 0.193 g of a volatile oxide of nitrogen. The pressure in the flask is 760 mmHg at 17°C.

a. Calculate the density of the gas.
b. Is the gas NO, $NO_2$, or $N_2O_5$?

**9.87.** The density of an unknown gas is 1.107 g/L at 300 K and 740 mmHg.

a. What is the molar mass of the gas?
b. Could this gas be CO or $CO_2$?

9.88. A gas containing chlorine and oxygen has a density of 2.875 g/L at 756 mmHg and 11°C.

a. Calculate the molar mass of the gas.
b. What is the most likely molecular formula of the gas?

## Gases in Chemical Reactions

### Problems

**9.89.** **Rescue Breathing Devices** Self-contained self-rescue breathing devices, like the one shown in Figure P9.89, convert $CO_2$ into $O_2$ according to the following reaction:

$$4\ KO_2(s) + 2\ CO_2(g) \rightarrow 2\ K_2CO_3(s) + 3\ O_2(g)$$

How many grams of $KO_2$ are needed to produce 100.0 L of $O_2$ at 20°C and 1.00 atm?

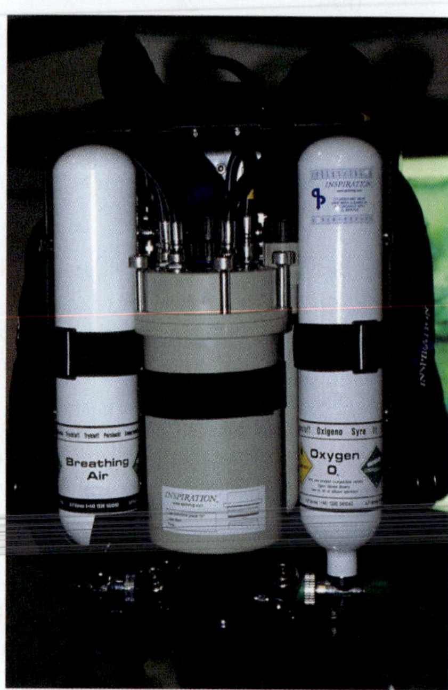


**FIGURE P9.89**

9.90. Acid precipitation dripping on limestone produces carbon dioxide by the following reaction:

$$CaCO_3(s) + 2\ H^+(aq) \rightarrow Ca^{2+}(aq) + CO_2(g) + H_2O(\ell)$$

Suppose that 8.6 mL of $CO_2$ are produced at 15°C and 760 mmHg.

a. How many moles of $CO_2$ are produced?
b. How many milligrams of $CaCO_3$ are consumed?

---

**9.91.** Oxygen is generated by the thermal decomposition of potassium chlorate:

$$2\ KClO_3(s) \rightarrow 2\ KCl(s) + 3\ O_2(g)$$

How many grams of $KClO_3$ are needed to generate 200.0 L of oxygen at 0.85 atm and 273 K?

9.92. Nitrogen gas is a product of the thermal decomposition of ammonium dichromate, $(NH_4)_2Cr_2O_7$:

$$(NH_4)_2Cr_2O_7(s) \rightarrow N_2(g) + Cr_2O_3(s) + 4\ H_2O(g)$$

How many liters of $N_2$ are produced during the decomposition of 100.0 g of ammonium dichromate ($\mathcal{M}$ = 252.07 g/mol) at 22°C and a pressure of 757 torr?

---

**9.93.** **Healthy Air for Submariners** The $CO_2$ that builds up in the air of a submerged submarine can be removed by reacting it with an aqueous solution of 2-aminoethanol:

$$CO_2(g) + 2\ HOCH_2CH_2NH_2(aq) \rightarrow HOCH_2CH_2NH_3^+(aq) + HOCH_2CH_2NHCO_2^-(aq)$$

If a sailor exhales 125 mL of $CO_2$ per minute at 23°C and 1.02 atm, what volume of 4.0 *M* 2-aminoethanol is needed per sailor in a 24-hour period?

9.94. **Miners' Lamps** Before the development of reliable batteries, miners' lamps burned acetylene produced by the reaction of calcium carbide with water:

$$CaC_2(s) + H_2O(\ell) \rightarrow C_2H_2(g) + CaO(s)$$

Suppose that a lamp uses 4.8 L of acetylene per hour at 1.02 atm pressure and 25°C.

a. How many moles of $C_2H_2$ are used per hour?
b. How many grams of calcium carbide are consumed for a 4-hour shift?

## Mixtures of Gases

### Concept Review

**9.95.** What is meant by the *partial pressure* of a gas?

9.96. Can a barometer be used to measure just the partial pressure of oxygen in the atmosphere? Why or why not?

### Problems

**9.97.** A gas mixture with a total volume of 1.0 L contains 0.70 mol of $N_2$, 0.20 mol of $H_2$, and 0.10 mol of $CH_4$.

a. What is the mole fraction of $H_2$ in the mixture?
b. What are the partial pressures of each gas at 10°C?

9.98. A gas mixture with a total volume of 2.0 L contains 7.0 g of $N_2$, 2.0 g of $H_2$, and 16.0 g of $CH_4$.

a. What is the mole fraction of $H_2$ in the mixture?
b. What are the partial pressures of each gas at 10°C?

---

**9.99.** All three gases effuse from a balloon containing the mixture in Problem 9.97. Which gas will see the greatest *decrease* in its mole fraction?

*9.100. All three gases effuse from a balloon containing the mixture in Problem 9.98. Which gas will see the largest *increase* in its partial pressure?

---

**9.101.** A sample of oxygen is collected over water at 25°C and 1.00 atm. If the total sample volume is 0.480 L, how many moles of $O_2$ are collected?

9.102. Water vapor is removed from the $O_2$ sample in Problem 9.101. What is the volume of the dry $O_2$ at 25°C and 1.00 atm?

---

**9.103.** The following reactions are carried out in sealed containers. Will the total pressure after each reaction is complete be greater than, less than, or equal to the total pressure before the reaction? Assume that all reactants and products are gases at the same temperature.

a. $N_2O_5(g) + NO_2(g) \rightarrow 3\ NO(g) + 2\ O_2(g)$
b. $2\ SO_2(g) + O_2(g) \rightarrow 2\ SO_3(g)$
c. $C_3H_8(g) + 5\ O_2(g) \rightarrow 3\ CO_2(g) + 4\ H_2O(g)$

9.104. In each of the following gas-phase reactions, determine whether the total pressure at the end of the reaction (carried out in a sealed, rigid vessel) will be greater than, less than, or equal to the total pressure at the beginning. Assume that all reactants and products are gases at the same temperature.

a. $H_2(g) + Cl_2(g) \rightarrow 2\ HCl(g)$

b. $4\ NH_3(g) + 5\ O_2(g) \rightarrow 4\ NO(g) + 6\ H_2O(g)$

c. $2\ NO(g) + O_2(g) \rightarrow 2\ NO_2(g)$

***9.105.** **High-Altitude Mountaineering** Most alpine climbers breathe pure oxygen near the summits of the world's highest mountains. How much more $O_2$ is in a lungful of pure $O_2$ at an elevation at which atmospheric pressure is 266 mmHg than in a lungful of air at sea level? Express your answer as a percentage.

9.106. **Scuba Diving** A scuba diver is at a depth of 50 m, where the pressure is 5.0 atm. What should be the mole fraction of $O_2$ in the gas mixture the diver breathes to achieve the same $P_{O_2}$ as at sea level?

***9.107.** Ammonia is produced industrially from the reaction of hydrogen with nitrogen under pressure in a sealed reactor. What is the percent decrease in total pressure of a sealed reaction vessel during the reaction between $H_2$ at a partial pressure of 2.4 atm and $N_2$ at a partial pressure of 3.6 atm if half of the $H_2$ is consumed? No other gases are present initially in the reactor.

*9.108. In a sealed 10.0 L vessel, 0.156 mol of C is reacted with 0.117 mol of $O_2$ at 500 K, producing a mixture of CO and $CO_2$. The total pressure is 0.640 atm. What is the partial pressure of CO?

## Real Gases

### Concept Review

**9.109.** Why do real gases behave nonideally at very low temperatures and very high pressures?

9.110. Under what conditions is the pressure exerted by a real gas *less* than that predicted for an ideal gas?

**9.111.** Why do the values of the van der Waals constant $b$ of the noble gas elements increase with atomic number?

9.112. Why does the value of the constant $a$ in the van der Waals equation generally increase with the molar mass of the gas?

**9.113.** Explain why the van der Waals constant $a$ is greater for Ar than it is for He.

9.114. The van der Waals constant $a$ for $CO_2$ is 3.59 $L^2 \cdot atm/mol^2$. Would you expect the value of $a$ for $CS_2$ to be larger or smaller than that?

### Problems

**9.115.** The graphs of $PV/RT$ versus $P$ (see Figure 9.34) for one mole of $CH_4$ and one mole of $H_2$ differ in how they deviate from ideal behavior. For which gas is the effect of the volume occupied by the gas molecules more important than the attractive forces between molecules at $P = 200$ atm?

9.116. Which noble gas is expected to deviate the most from ideal behavior in a graph of $PV/RT$ versus $P$?

**9.117.** Silane ($SiH_4$), methane ($CH_4$), and methyl chloride ($CH_3Cl$) are all gases at 298 K and 1 atm.

a. Draw Lewis structures for each compound and determine their molecular geometries.

b. Match these three compounds with the following values of the van der Waals constant $a$: 2.253, 4.320, and 7.471 $L^2 \cdot atm/mol^2$.

9.118. Silicon tetrafluoride, sulfur tetrafluoride, and nitrogen trifluoride are all gases at 298 K and 1 atm.

a. Draw Lewis structures of each molecule, and determine whether they are polar or nonpolar.

b. Which gas do you predict to have the smallest van der Waals constant $a$?

**9.119.** At high pressures, real gases do not behave ideally.

a. Use the van der Waals equation and data in the text to calculate the pressure exerted by 50.0 g of $H_2$ at 20°C in a 1.00 L container.

b. Repeat the calculation, assuming that the gas behaves like an ideal gas.

c. Explain the difference between your answers to parts (a) and (b) by using kinetic molecular theory.

9.120. Calculate the pressure exerted by 5.00 mol of $CO_2$ in a 1.00 L vessel at 300 K (a) assuming that the gas behaves ideally and (b) using the van der Waals equation. (c) Explain the difference between your answers to parts (a) and (b) by using kinetic molecular theory.

## Additional Problems

**9.121.** Each cylinder in the engine of the sports car in Figure P9.121 contains 633 mL of air and gasoline vapor before the motion of a piston increases the pressure inside the cylinder by 9.0 times.

a. What is the volume of the air and gasoline vapor mixture after the compression, if we assume that the temperature of the mixture does not change significantly?

b. Is the assumption of no temperature change realistic? Why or why not?

**FIGURE P9.121**

9.122. **Early Earth Atmosphere** Life as we know it on Earth depends on a constant atmospheric composition of 21% $O_2$ and 78% $N_2$ by volume. Over the past 3 billion years, the amount of oxygen in the atmosphere has ranged from as low as 15% to as high as 35% by volume.

a. What partial pressures of oxygen does this range correspond to?

b. How much does the density of air change over this range of oxygen concentrations if the balance of the atmosphere is assumed to be nitrogen?

**9.123.** Scientists have used laser light to slow atoms to speeds corresponding to temperatures below 0.00010 K. At this temperature, what is the root-mean-square speed of argon atoms?

9.124. **Blood Pressure** A typical blood pressure in a resting adult is "120 over 80," meaning 120 mmHg of pressure with each beat of the heart and 80 mmHg between heartbeats. Express these pressures in the following units: (a) torr; (b) atmospheres; (c) bars; (d) kilopascals.

***9.125.** A popular style of scuba tank is called the "aluminum 80" because it can deliver 80 cubic feet of air at "normal" temperature (22°C) and pressure (1.00 atm) when filled with air at a pressure of 200 atm. A particular aluminum 80 tank has a mass of 15 kg empty. What is its mass when filled with air at 200 atm?

9.126. The flame produced by the burner of a gas (propane; $C_3H_8$) grill is a pale blue color when enough air mixes with the propane to burn it completely. For every gram of propane that flows through the burner, what volume of air is needed to burn it completely? Assume that the temperature of the burner is 200°C, the pressure is 1.00 atm, and the mole fraction of $O_2$ in air is 0.21.

***9.127.** The apparatus in Figure P9.127 is used in experiments in which a cotton ball stuffed in the right end of a 1-meter-long glass tube is soaked in either hydrochloric acid (HCl) or acetic acid ($CH_3COOH$). A cotton ball at the left end of the tube is soaked in one of three amines (a class of basic organic compounds): $CH_3NH_2$, $(CH_3)_2NH$, or $(CH_3)_3N$. Vapors from the acids and the basic amines diffuse down the tube toward each other. When they meet, they neutralize each other, producing solid salts that form white rings in the tube.

a. In one combination of acid and amine, a white ring was observed halfway between the two ends. Which acid and which amine were used?

b. Which combination of acid and amine would produce a ring closest to the amine end of the tube?

c. Would two of the six possible combinations produce rings in the same position? Assume that measurements can be made to the nearest centimeter.

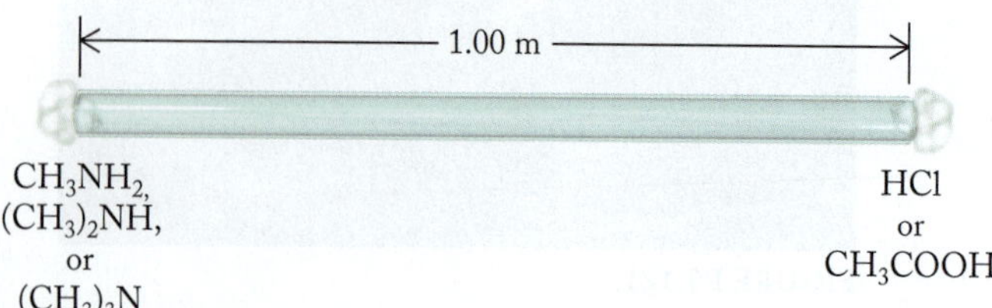


**FIGURE P9.127**

9.128. The pressure in an aerosol can is 1.2 atm at 27°C. The can will withstand a pressure of 3.0 atm. Will it burst if heated in a campfire to 450°C?

**9.129.** **Denitrification in the Environment** In some aquatic ecosystems, nitrate ($NO_3^-$) is converted to nitrite ($NO_2^-$), which then decomposes to nitrogen and water. As an example of this second reaction, consider the decomposition of ammonium nitrite:

$$NH_4NO_2(aq) \rightarrow N_2(g) + 2\,H_2O(\ell)$$

What is the change in pressure in a sealed 10.0 L vessel resulting from the formation of $N_2$ gas when the ammonium nitrite in 1.00 L of 1.0 $M$ $NH_4NO_2$ decomposes at 25°C?

*9.130. When sulfur dioxide bubbles through a solution containing nitrite, chemical reactions that produce gaseous $N_2O$ and NO may occur.

a. How much faster on average are NO molecules moving than $N_2O$ molecules in such a reaction mixture?

b. If these two nitrogen oxides are separated on the basis of differences in their rates of effusion, will unreacted $SO_2$ interfere with the separation? Explain your answer.

***9.131.** **Using Wetlands to Treat Agricultural Waste** Wetlands can play a significant role in removing fertilizer residues from rain runoff and groundwater. One way they do this is through denitrification, which converts nitrate ions to nitrogen gas:

$$2\,NO_3^-(aq) + 5\,CO(g) + 2\,H^+(aq) \rightarrow N_2(g) + H_2O(\ell) + 5\,CO_2(g)$$

Suppose that 200.0 g of $NO_3^-$ flows into a swamp each day.

a. What volume of $N_2$ would be produced at 17°C and 1.00 atm if the denitrification process were complete?

b. What volume of $CO_2$ would be produced?

c. Suppose that the gas mixture produced by the decomposition reaction is trapped in a container at 17°C; what is the density of the mixture, if we assume that $P_{total} = 1.00$ atm?

9.132. Ammonium nitrate decomposes on heating. The products depend on the reaction temperature:

$$NH_4NO_3(s) \xrightarrow{300^\circ C} N_2(g) + \tfrac{1}{2}\,O_2(g) + 2\,H_2O(g)$$

$$NH_4NO_3(s) \xrightarrow{200^\circ C\text{–}260^\circ C} N_2O(g) + 2\,H_2O(g)$$

A sample of $NH_4NO_3$ decomposes at an unspecified temperature, and the resulting gases are collected over water at 20°C.

a. Without completing a calculation, predict whether the volume of gases collected can be used to distinguish between the two reaction pathways. Explain your answer.

b. The gas produced during the thermal decomposition of 0.256 g of $NH_4NO_3$ displaces 79 mL of water at 20°C and 760 mmHg of atmospheric pressure. Is the gas $N_2O$ or a mixture of $N_2$ and $O_2$?

**9.133.** **Airbag Chemistry** The following chemical equation describes the overall reaction in an automobile airbag:

$$20\ NaN_3(s) + 6\ SiO_2(s) + 4\ KNO_3(s) \rightarrow 32\ N_2(g) + 5\ Na_4SiO_4(s) + K_4SiO_4(s)$$

Calculate how many grams of sodium azide ($NaN_3$) are needed to inflate a 40 × 40 × 20 cm airbag to a pressure of 1.25 atm at a temperature of 20°C. How much more sodium azide is needed if the airbag must produce the same pressure at 10°C?

9.134. **Decay Products of Uranium Minerals** Radon and helium are both by-products of the radioactive decay of uranium minerals. A fresh sample of carnotite, $K_2(UO_2)_2(VO_4)_2 \cdot 3\ H_2O$, is put on display in a museum. Calculate the relative rates of diffusion of helium and radon under fixed conditions of pressure and temperature. Which gas diffuses more rapidly through the display case?

**9.135.** **Rocket Fuel** During the Second World War, German scientists and engineers developed the first operational rocket-propelled warplane (Figure P9.135). The engine was powered by the reaction of hydrogen peroxide and hydrazine, producing nitrogen and water vapor.

$$2\ H_2O_2(\ell) + N_2H_4(\ell) \rightarrow N_2(g) + 4\ H_2O(g)$$

What is the total pressure of the gases released by the reaction of 8.50 g of $H_2O_2$ and 4.00 g of $N_2H_4$ in a 125 L reaction vessel (rocket engine) at 1110°C?

**FIGURE P9.135**

9.136. On October 26, 2014, Alan Eustace set a record for the highest parachute jump when he dropped from a balloon at an altitude of 41,419 m, where the atmospheric pressure is only 5.6 mmHg.

a. What is the density of air at this height?
b. Is the mean free path of a gas molecule longer or shorter at this altitude than at sea level?

# 10

# Thermochemistry

## Energy Changes in Chemical Reactions

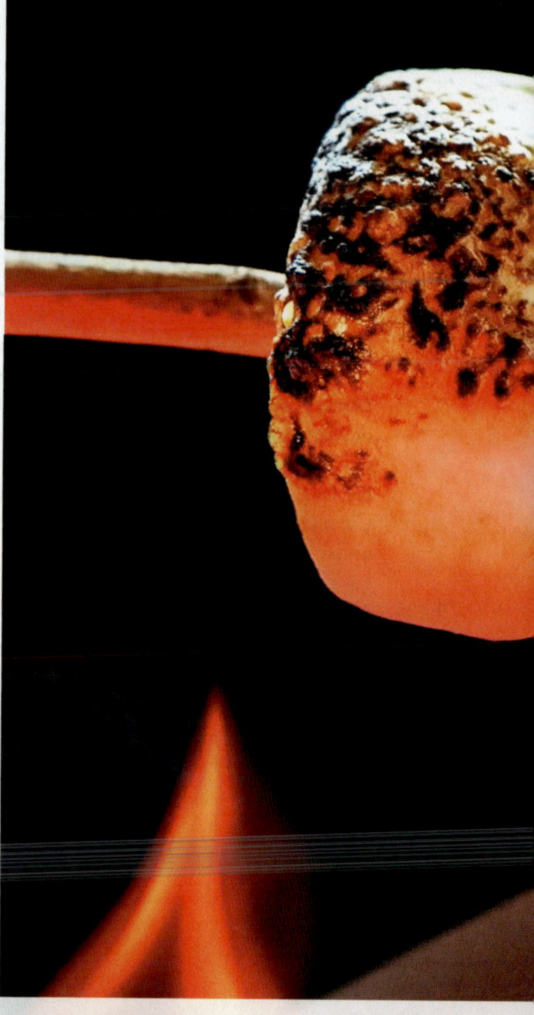

**CAMPFIRE** Burning wood involves the combustion of cellulose and other large molecules rich in carbon and hydrogen. The products are $CO_2$, $H_2O$, and a considerable quantity of energy that can be used to roast marshmallows.

PARTICULATE **REVIEW**

### *Phase Changes and Energy*

In Chapter 10, we explore the energy changes that accompany phase changes. Particulate representations of the three phases of water are shown here.

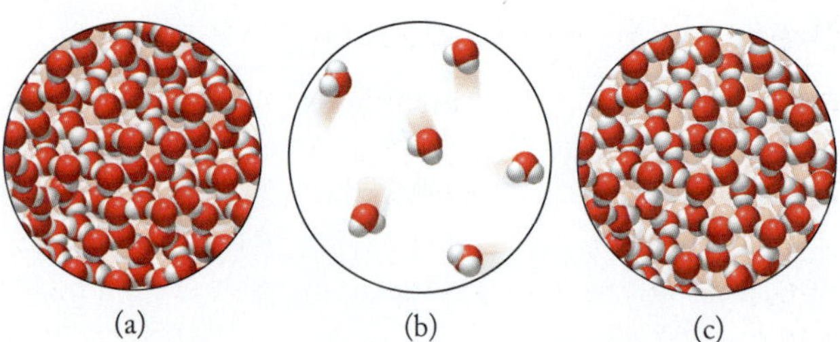


- Which representation depicts the solid phase of water? The liquid? The gaseous?
- Is energy added or released during the phase change from (a) to (b)? What intermolecular forces are involved?
- Describe the energy changes that accompany the phase changes from (a) to (c) and from (c) to (a).

(Review Section 1.4 and Section 6.2 if you need help.)

*(Answers to Particulate Review questions are in the back of the book.)*

PARTICULATE **PREVIEW**

### *Breaking Bonds and Energy Changes*

Calcium chloride, shown in the accompanying figure, is used to melt ice on sidewalks. As you read Chapter 10, look for ideas that will help you answer these questions.

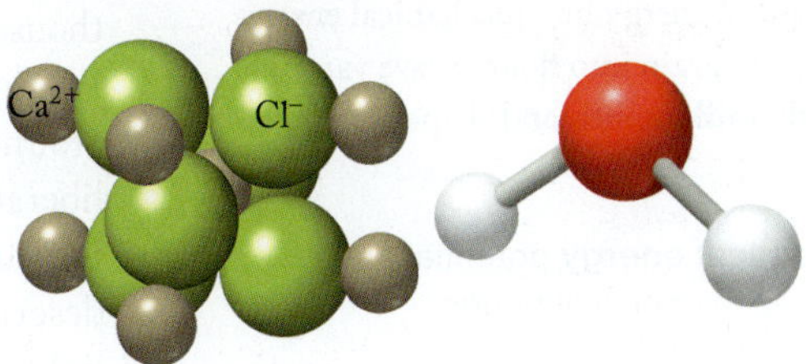


- What kind of bonds must be broken for calcium chloride to dissolve in water? Is energy absorbed or released in order to break these bonds?
- Label the polar covalent bonds in water by using $\delta+$ and $\delta-$ symbols.
- What intermolecular interactions form as the salt dissolves? Is energy absorbed or released as these attractions form?

## Learning Outcomes

**LO1** Distinguish between systems and surroundings and between endothermic and exothermic processes
**Sample Exercise 10.1**

**LO2** Relate changes in the internal energies of thermodynamic systems to heat flows and work done
**Sample Exercises 10.2, 10.3**

**LO3** Calculate the energy gained or lost during changes in temperature and physical state
**Sample Exercises 10.4, 10.5**

**LO4** Use calorimetry data to calculate enthalpies of reaction and heat capacities of calorimeters
**Sample Exercises 10.6, 10.7**

**LO5** Calculate enthalpies of reaction by using Hess's law and enthalpies of formation
**Sample Exercises 10.8–10.10**

**LO6** Estimate enthalpies of reaction by using average bond energies
**Sample Exercise 10.11**

**LO7** Estimate enthalpies of solution and lattice energies by using the Born–Haber cycle and Hess's law
**Sample Exercise 10.12**

**LO8** Calculate and compare fuel values and fuel densities
**Sample Exercises 10.13, 10.14**

## 10.1 Energy as a Reactant or Product

FIGURE 10.1 Water about to go over a waterwheel has more potential energy than water at the bottom of the wheel. Waterwheels were developed by the ancient Greeks and widely used in medieval Europe and colonial America to convert this potential energy into mechanical energy to grind grain into flour, weave yarn into cloth, mill lumber, and shape metals.

Combustion reactions produce most of the energy we consume, but where does this energy come from? The combustion of coal (which is mostly carbon) generates much of the electricity in China, India, and the United States; the combustion of natural gas (mostly $CH_4$) heats water, warms homes, and is also used to generate electricity; and then there is the combustion of gasoline, which many of us rely on to get us where we need to go. All the fuels in these combustion reactions are fossil fuels: the decomposed remains of plants and animals that lived millions of years ago. To trace the origin of the energy in fossil fuels, we need to analyze the food chains that supplied nutrition to those plants and animals. Nearly all those food chains started with green plants that harvested the energy of the sun through photosynthesis.

The connection to the sun is perhaps more obvious when we consider renewable sources of energy. In many parts of the world, wood is the principal fuel for heating homes and for cooking food. Dry wood is mostly cellulose, the most common organic compound in the world, consisting of long chains of glucose molecules. As we discussed in Chapter 7, green plants harness the energy of the sun to synthesize glucose from carbon dioxide and water via photosynthesis. Trees require hundreds or even thousands of molecules of glucose to synthesize one long chain of cellulose.

Burning wood in a campfire (or combusting food inside us) turns this process around, converting cellulose (or glucose) and $O_2$ back into $CO_2$ and $H_2O$ and liberating the energy of the sun that was originally harvested by green plants. R. Buckminster Fuller (1895–1983), an architect, inventor, and futurist, once described the release of energy that happens when wood burns as "all that sunlight unwinding."

**chemical energy** potential energy stored in chemical bonds.

**internal energy (*E*)** the sum of all the kinetic and potential energies of all the components of a system.

**thermochemistry** the study of the changes in energy that accompany chemical reactions.

**thermodynamics** the study of energy and its transformations.

### Forms of Energy

We learned in Chapter 1 that there are two broad categories of energy: *kinetic energy*, which is the energy of an object in motion, and *potential energy*, which can be the energy that an object has because of its position, as illustrated by the water in a mill pond in **Figure 10.1**. Another form of potential energy is the energy that a substance has because of its composition—that is, the way in which the atoms inside it are bonded together. This kind of potential energy is called **chemical energy**. The sum of all the kinetic and potential (including chemical) energies of an object, or a collection of many objects, is the **internal energy ($E$)** of the object or collection. When the collective internal energies of the reactants in a chemical reaction are different from those of the products—and they nearly always are—the difference constitutes the *change in energy* ($\Delta E$) that accompanies the reaction.

As a practical matter, the *absolute* value of a system's internal energy is extremely difficult to determine, but *changes* in energy are fairly easy to measure because they correspond to changes in the system's physical state or temperature. The branch of chemistry that explores these changes in energy is called **thermochemistry**, our primary focus in this chapter. Thermochemistry is related to a more expansive scientific domain called **thermodynamics**, which is the study of energy and its transformations.

In our exploration of the energy changes that accompany chemical reactions, we will often encounter reactions in which the collective internal energy of the reactants is greater than that of the products, such as the combustion of hydrogen (**Figure 10.2**). The result is a release of energy, which has to go somewhere. It can't just disappear because that would violate the principle articulated by the law of conservation of energy and the **first law of thermodynamics**: energy cannot be created or destroyed, though it can change from one form to another. Expressed another way, the total energy of the universe is a constant. The impact of this principle may be clearer if we consider the universe to consist of only two parts: the part we are interested in, which we call the **system**, and everything else, which we call the **surroundings**. We can express these definitions in the form of an equation:

$$\text{Universe} = \text{system} + \text{surroundings} \quad (10.1)$$

Let's use Equation 10.1 to connect changes in the internal energy of a system and its surroundings to the principle of energy conservation. If energy cannot be created or destroyed, then the energy of the universe ($E_{\text{univ}}$) is a constant, which means that $\Delta E_{\text{univ}}$ during any process is always equal to zero. If so, then

$$\Delta E_{\text{univ}} = \Delta E_{\text{sys}} + \Delta E_{\text{surr}} = 0$$

and

$$\Delta E_{\text{sys}} = -\Delta E_{\text{surr}} \quad (10.2)$$

When a thermodynamic system releases energy, as when molecules of $H_2$ and $O_2$ combine to form molecules of $H_2O$, the energy flows into its surroundings. Whatever energy the system loses ($\Delta E_{\text{sys}} < 0$), its surroundings gain ($\Delta E_{\text{surr}} > 0$).

What effect does this increase in energy have on the surroundings? To get one answer to this question, let's consider the combustion of hydrogen that powers the engines of a Delta IV rocket (**Figure 10.3**). If the combustion gases (the reactants and products) are the system, then everything else in the universe—including the rocket, its launchpad, and the air through which the rocket travels on its way into space—makes up the surroundings. The most significant energy transfer from the system to the surroundings occurs when the rocket and its cargo are launched into space. This involves exerting a large[1] force ($F$) through a considerable distance ($d$), which meets our definition from Chapter 1 of doing work ($w$):

$$w = F \times d \quad (1.2)$$

[1]At full throttle, one of the Delta IV engines produces up to $3.2 \times 10^6$ newtons of thrust (force), which is about 15 times that of a Boeing 747 engine at takeoff.

**first law of thermodynamics** the principle that the energy gained or lost by a system must equal the energy lost or gained by the surroundings. Energy cannot be created or destroyed.

**system** the part of the universe that is the focus of a thermochemical study.

**surroundings** everything in a thermochemical study that is not part of the system.

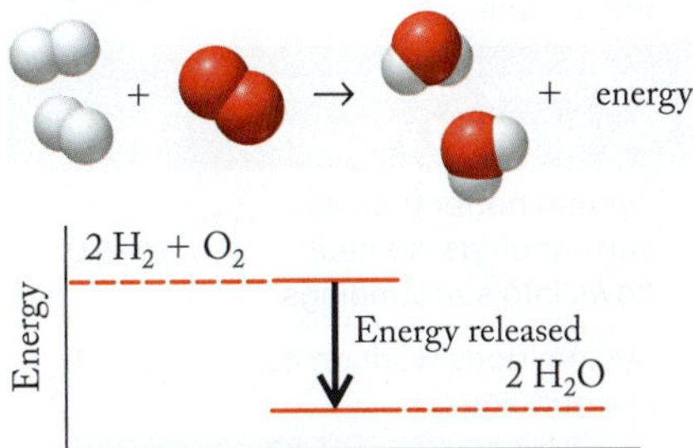


**FIGURE 10.2** Energy is released during chemical reactions when higher-energy reactants, such as $H_2$ and $O_2$, form lower-energy products, such as $H_2O$.

**CONNECTION** In Chapter 1, we defined *energy* as the ability to do work. We also introduced the *law of conservation of energy* and the concept that energy cannot be created or destroyed but can be changed from one form of energy to another.

**FIGURE 10.3** Up to three hydrogen-fueled engines make up the first stage of a Delta IV Heavy rocket such as this one, which can send a satellite weighing 23 metric tons into low Earth orbit.

**TABLE 10.1** Flows of Heat and Work and Their Impact on the Energy of a System

| Processes That Increase the Energy of a System | Result |
|---|---|
| Surroundings hotter than the system, so heat flows into the system | $q > 0$ |
| Surroundings do work on the system | $w > 0$ |
| **Processes That Decrease the Energy of a System** | **Result** |
| System hotter than its surroundings, so heat flows into surroundings | $q < 0$ |
| System does work on its surroundings | $w < 0$ |

In addition to doing work, the hot gases of the system heat their surroundings. These two processes, doing work and transferring heat, represent the two ways in which energy moves between a system and its surroundings. In general, if heat flows *out from* a system, or if work is done *by* the system on its surroundings, the internal energy $E$ of the system decreases. However, if heat flows *into* a system or if work is done *on* a system, the internal energy of the system increases. We can express this last relationship in an equation, using the symbol $q$ to represent heat:

$$\Delta E = q + w \tag{10.3}$$

Heat flowing into a system means that $q$ has a positive value. Doing work on a system means that $w$ has a positive value. If both $q$ and $w$ are positive, then so is $\Delta E$—that is, $\Delta E > 0$. When a system does work *on* its surroundings, like the reaction mixture in the rocket engines in Figure 10.3, then $w$ is negative, and if heat flows out from the system (see again Figure 10.3), then $q$ is also negative. Taken together, these changes produce a decrease in the internal energy of the system ($\Delta E < 0$). The various combinations of heat flow and work and their impact on $\Delta E$ are summarized in **Table 10.1** and in **Figure 10.4**. Note that Equation 10.3 applies to all the changes of $\Delta E$, $q$, and $w$ in the figure and is essentially a statement of the first law of thermodynamics: any change in energy experienced by the system is balanced by a change of equal magnitude but opposite sign in the energy of its surroundings, as expressed by the quantity of heat that flowed between them and the quantity of work that one did on the other.

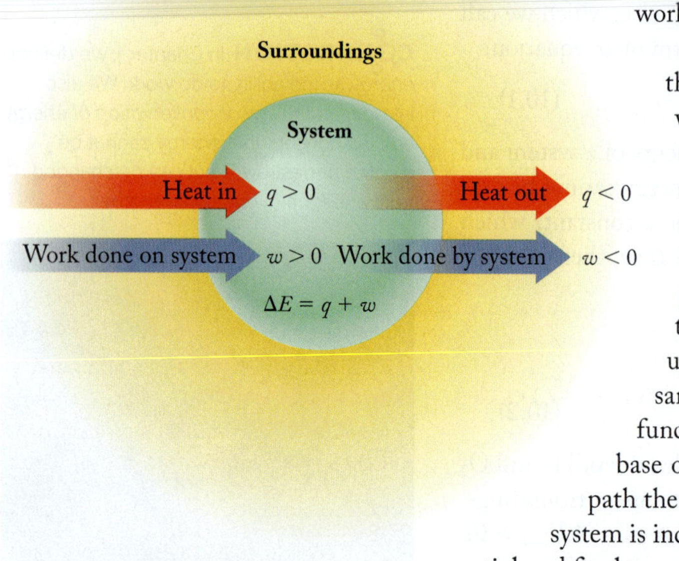

**FIGURE 10.4** The change in internal energy ($\Delta E$) of a system is positive when energy flows into it ($q > 0$) or work is done on it ($w > 0$). A system loses internal energy ($\Delta E < 0$) when energy flows out from it ($q < 0$) or the system does work on its surroundings ($w < 0$).

No matter how the internal energy of a system is changed, the final internal energy of the system is *independent* of how it was acquired. It makes no difference whether the system lost or gained heat, or whether it did work or had work done on it, because internal energy is a **state function**: it depends only on how much potential and kinetic energy the system has and not at all on the pathway or sequence of events that produced those levels of energy. Think of skiers ascending to the top of a mountain for their run: whether they ride straight up on a ski lift or hike up a winding path, they arrive at the same height in the end (**Figure 10.5**). Their altitude is a state function, as is the change in altitude they experienced from the base of the mountain to its summit: both are independent of the path the skiers take. Similarly, any change in the internal energy of a system is independent of the processes causing the change. Only the initial and final energy levels of the system define $\Delta E$:

$$\Delta E = E_{\text{final}} - E_{\text{initial}}$$

We will encounter other examples of system properties that are state functions in later sections of this chapter and in later chapters of this book.

At this point we need to be clear on what we mean by *heat*. Heat is energy that is in the process of being transferred from a higher-temperature object to a lower-temperature one. An operating engine of a Delta IV rocket gets very hot. The quantity of **thermal energy** in the hot engine is determined not only by its temperature but also by the composition of the materials of which it is made and how massive it is (a concept called *heat capacity*, which we explore in Section 10.4). By *thermal* energy, we mean that part of the total internal energy of a system that is linked to how hot it is.

**CONNECTION** In Chapter 1, we defined *heat* as the spontaneous transfer of energy from a warmer object to a cooler one.

**FIGURE 10.5** Whether a skier at the base of a mountain rides a ski lift (1) or hikes up a mountain trail (2), the skier's increase in altitude and potential energy is the same.

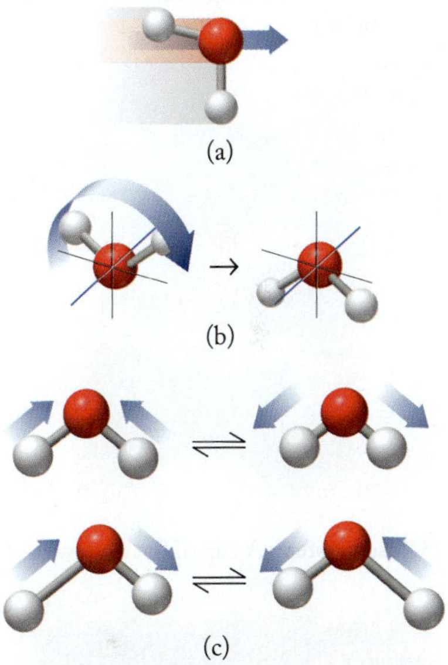

At the atomic level, the thermal energy of a system is the sum of the kinetic energies of the particles that make up the system. We learned in Chapter 1 that the kinetic energy (KE) of any object in motion, whether it is a rocket or an atom, is a function of its mass ($m$) and its speed ($u$):

$$\mathrm{KE} = \tfrac{1}{2}mu^2 \qquad (1.3)$$

The motion of particles of gases as they collide with one another and the walls of the container they occupy is called *translational* kinetic energy (**Figure 10.6a**). Remember that we learned when studying kinetic molecular theory in Section 9.2 that the average kinetic energy of the molecules in a gas *is directly proportional* to its *absolute* (kelvin) temperature. As the particles within a gas collide with one another, they can transfer energy from one particle to another.

Molecules also have nontranslational, or *internal,* forms of kinetic energy. These include *rotational* kinetic energy (**Figure 10.6b**) and *vibrational* kinetic energy (**Figure 10.6c**). Bond vibrations also provide additional forms of potential energy in molecules in much the way that stretching, compressing, or bending a spring increases its potential energy. We noted in Chapter 4 that there are several kinds of bond vibrations in molecules with more than two atoms, including symmetric and asymmetric stretching and bending as the angles between bonds increase and decrease.

(a)

(b)

(c)

**FIGURE 10.6** Some of the types of molecular motion that contribute to the internal energy of a system are (a) translational motion, the motion from place to place along a path; (b) rotational motion, the motion about a fixed axis; and (c) vibrational motion, the movement back and forth from some central position.

## 10.2 Transferring Energy and Doing Work

Systems come in different sizes. They can be as large as galaxies or as small as single living cells. Most of the systems we discuss in this chapter fit on a laboratory bench, and we usually limit our view of their surroundings to that part of the universe that can exchange energy and matter with the system. In evaluating how much energy is gained or lost in a chemical reaction, the system may be just the particles involved in the reaction, but it may also include the vessel that contains them.

**state function** a property of an entity based solely on its chemical or physical state or both but not on how it achieved that state.

**thermal energy** the portion of the total internal energy of a system that is proportional to its absolute temperature.

(a) **Isolated system:** A thermos bottle containing hot soup with the lid screwed on tightly

(b) **Closed system:** A cup of hot soup with a lid

(c) **Open system:** An open cup of hot soup

**FIGURE 10.7** Illustrations of isolated, closed, and open systems: (a) Hot soup in a tightly sealed thermos bottle approximates an isolated system: no vapor escapes, no matter is added or removed, and no energy is transferred to the surroundings. (b) Hot soup in a cup with a lid is a closed system; the soup transfers energy to the surroundings as it cools, but no matter escapes and none is added. (c) Hot soup in a cup with no lid is an open system; it transfers both matter (steam) and energy to the surroundings as it cools. Matter in the form of pepper, grated cheese, or other flavor enhancers might also be added to the system.

## Isolated, Closed, and Open Systems

As we explore a system's energy production and consumption, we need to be clear on how the system interacts with its surroundings. For example, can energy be transferred from the system to its surroundings? Can matter be exchanged between the system and its surroundings? The answers to these questions define the type of system we have. If both matter and energy can be transferred between a system and its surroundings, we have an *open* system. If only energy can flow between them, we have a *closed* system. And if a system is completely cut off from its surroundings so that neither matter nor energy can be transferred, we have an *isolated* system.

The containers of hot soup in **Figure 10.7** illustrate the three types of systems. Figure 10.7(a) shows an isolated system: a serving of soup inside a thermos bottle that has a perfect seal and that is perfectly insulated, allowing neither thermal energy nor matter to escape or enter. The mass of the system never changes, nor does its temperature or its energy content. No such thermos bottle actually exists because even the best ones can't keep soup hot forever. For short times, however, substances in sealed, well-insulated containers approximate isolated systems.

Hot soup in a cup with a lid on it represents a *closed* thermodynamic system (Figure 10.7b). The lid keeps the soup and any vapor it produces from escaping, and it keeps any matter from getting into the soup. Only energy is exchanged between the soup and its surroundings, as thermal energy is transferred from the soup through the cup and into the air and the tabletop. Gradually the soup cools to room temperature. Many real thermodynamic systems are closed systems.

Soup in an open cup represents an *open* thermodynamic system because it can exchange both energy and matter with its surroundings (Figure 10.7c). Thermal energy flowing from the soup into its surroundings and matter—in the form of water vapor, for example—is free to leave the system. Matter can also enter the system—perhaps a little seasoning to enhance the flavor.

### CONCEPT TEST

Suppose two identical pots of water are warmed on a stove until the water inside them begins to boil. Both pots are then removed from the stove. One of the two is covered with a tight lid, the other is not, and both are allowed to cool.

a. What type of thermodynamic system—open, closed, or isolated—describes each of the cooling pots?
b. Which pot cools faster? Why?

*(Answers to Concept Tests are in the back of the book.)*

## Exothermic and Endothermic Processes

Chemists classify reactions on the basis of whether they give off or absorb energy. A chemical reaction that results in the transfer of energy—usually in the form of thermal energy—from the reaction mixture (the system) to its surroundings is an **exothermic** reaction. This energy flow can usually be detected by an increase in the temperature of the surroundings. Combustion reactions are good examples of exothermic reactions. Chemical reactions that absorb energy from their surroundings are **endothermic** reactions.

Nonchemical processes that give off or take in energy can also be described as exothermic or endothermic. For example, your body cools itself during strenuous exercise by sweating. Cooling happens as the sweat on your skin evaporates

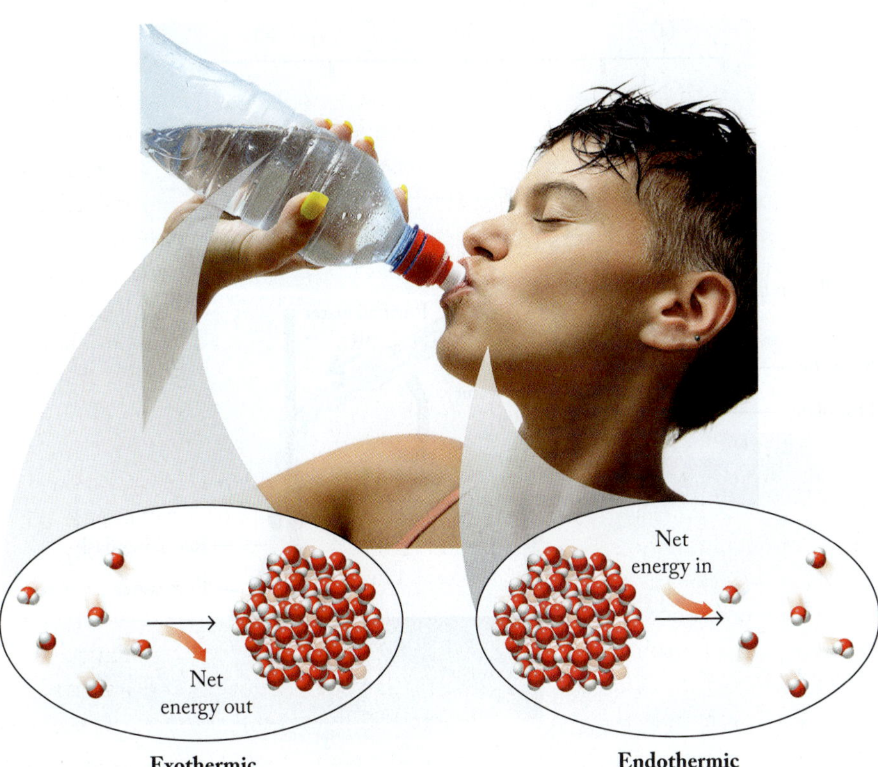


**FIGURE 10.8** (a) The condensation of atmospheric water vapor on the outside of a cold water bottle is exothermic, whereas (b) the evaporation of perspiration from an athlete's skin and clothing is endothermic.

because the conversion of a liquid to a gas is an endothermic process, requiring the flow of energy into the system (sweat) from its surroundings (your body), as shown in **Figure 10.8**. However, the reverse process, water vapor condensing, as on the outside of a bottle of cold water on a humid summer day, is exothermic. Energy is transferred from the system (condensing water vapor) to its surroundings, including the water in the bottle in Figure 10.8.

### SAMPLE EXERCISE 10.1 Identifying Exothermic and Endothermic Processes — LO1

Describe the flow of thermal energy (heat) during the desalination of seawater by distillation (**Figure 10.9**), identify the steps as either endothermic or exothermic, and give the sign of $q$ associated with each step. Consider the water being purified to be the system.

**Collect and Organize** The water is the system, so we need to evaluate how the water gains or loses thermal energy during distillation. A positive value of $q$ represents a gain in thermal energy.

**Analyze** During distillation, energy flows into the seawater, raising its temperature by increasing the kinetic energy of water molecules and particles of dissolved sea salts. Additional heating vaporizes the water by providing its molecules with enough kinetic energy to overcome the hydrogen bonds and other intermolecular forces that keep them together in the liquid phase. As water vapor passes through the cold condenser, energy flows from the vapor into the condenser. As a result, the kinetic energy of the vaporized molecules decreases, which allows their attractions for each other to pull them together. These molecules form a layer of liquid water on the inside wall of the condenser.

**Solve** Energy flows from the surroundings (hot plate) to heat the system (seawater) to its boiling point. That process is endothermic and the sign of $q$ is positive. Additional

**exothermic process** a process in which energy in the form of heat flows from a system into its surroundings.

**endothermic process** a process in which energy in the form of heat flows from the surroundings into the system.

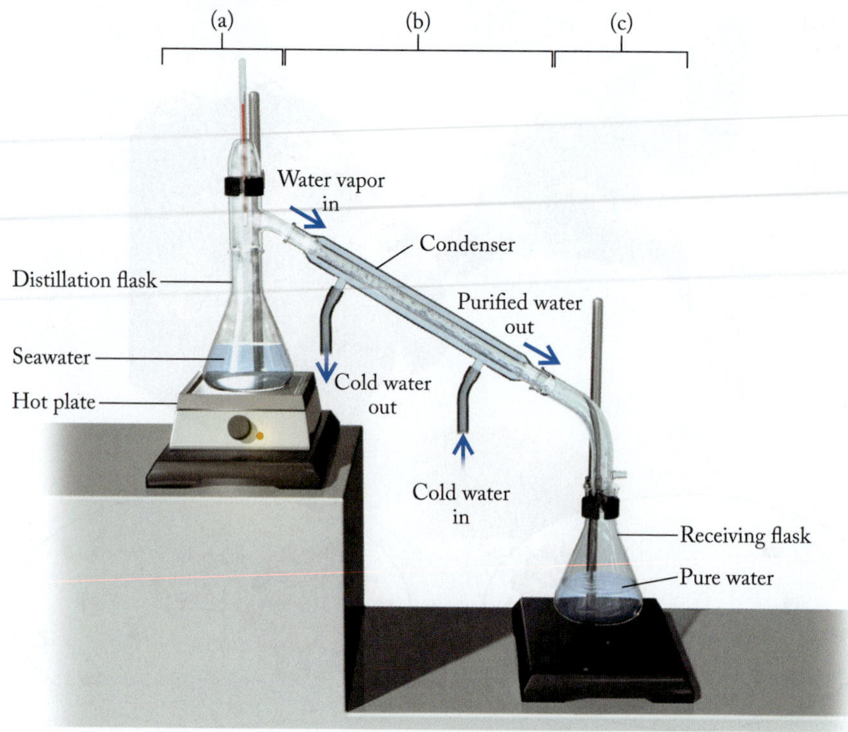


**FIGURE 10.9** A laboratory setup for desalinating seawater. (a) Seawater is heated to the boiling point in the distillation flask. (b) Water vapor rises and enters the condenser, where it is liquefied. (c) The purified liquid is collected in the receiving flask.

energy flows from the surroundings into the system as the water vaporizes—this is another endothermic process for which $q$ is positive. However, energy then flows from the system (in the form of hot water vapor) into the surroundings (cold condenser walls) as the water vapor condenses and further cools: these are both exothermic processes, and the sign of $q$ is negative for both of them.

**Think About It** As any liquid is heated to its boiling point and vaporized, the particles in it gain enough kinetic energy to overcome the intermolecular forces that kept them together in the liquid state. As any vapor cools and condenses, its particles lose the kinetic energy that allowed them to overcome their attraction for each other, and they come together in the liquid state.

**Practice Exercise** What is the sign of $q$ as (a) a match burns, (b) drops of molten candle wax solidify, and (c) a piece of dry ice disappears at room temperature? In each case, define the system and indicate whether the process is endothermic or exothermic.

*(Answers to Practice Exercises are in the back of the book.)*

## Pressure–Volume Work

We have discussed how the internal energy $E$ of a system increases ($\Delta E > 0$) when heat flows into it ($q > 0$) or work is done on it by its surroundings ($w > 0$), and how the internal energy of a system decreases ($\Delta E < 0$) when heat flows out from it ($q < 0$) or it does work on its surroundings ($w < 0$). Let's explore a system that rapidly cycles between having its surroundings do work on it and then it doing work on its surroundings.

The system is the gases in the combustion chamber of one of the cylinders in a diesel engine (**Figure 10.10**). First, the downward motion of the piston during the intake stroke (Figure 10.10a) allows air to enter the cylinder, where it mixes

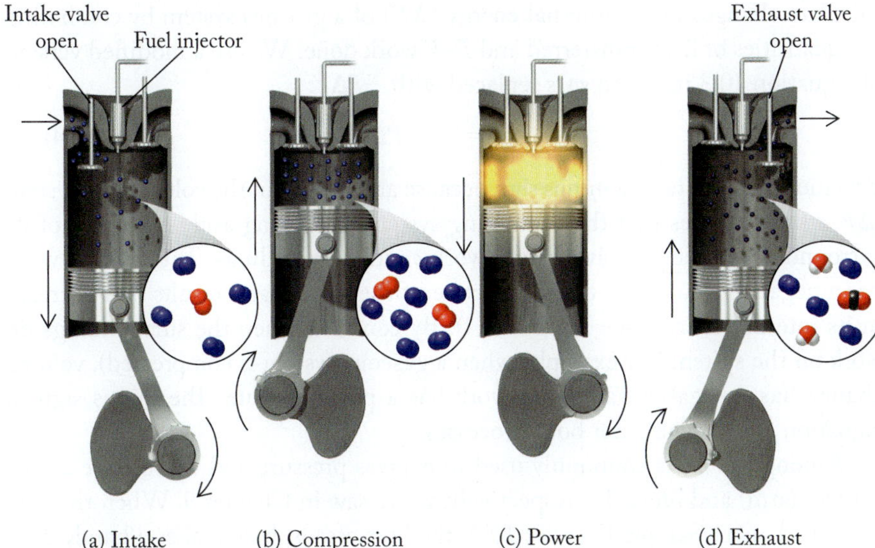


**FIGURE 10.10** Diesel engines are based on a four-stroke cycle in which (a) fuel is injected and air (represented by $N_2$ and $O_2$) is drawn into the combustion cylinder as the piston goes down, (b) the mixture is compressed as the piston moves up, (c) compression heating causes the mixture to ignite spontaneously and release energy that pushes the piston down, and (d) the products of combustion ($CO_2$ and $H_2O$) are pushed out as the piston moves up again.

with an injection of a few microliters of diesel fuel. Then the intake valve closes, and the upward motion of the piston (Figure 10.10b) squeezes the air and fuel vapors (the system) in the cylinder into less than one-tenth of their initial volume. We learned in Section 9.4 that decreasing the volume of a gas increases its pressure. The compression heats up the reaction mixture (a clear sign its internal energy has increased) until, at the point of maximum compression, the mixture gets so hot it spontaneously ignites. The energy released by combustion rapidly raises the temperature of the reaction mixture even further, which increases the pressure it exerts on the top of the piston. This pressure drives the piston downward during the power stroke (Figure 10.10c) as the volume of the combustion mixture rapidly expands. Compressing the reaction mixture increased the mixture's internal energy because its surroundings (the piston) did work on the system. Then expansion of the hot products of combustion did work on their surroundings (the piston again), releasing energy they had gained through compression and combustion by pushing the piston downward (Figure 10.10d).

The work done *on* the gases in an engine cylinder as they are compressed and the work done *by* the gases as they expand are two examples of **pressure–volume ($P$–$V$) work**. The work done by the gases as they expand is the work that any gaseous system does when its volume increases by an amount $\Delta V$ (where $\Delta V = V_f - V_i$) against an opposing pressure $P$. The amount of work done is the product $P \times \Delta V$. This pressure–volume work is equivalent to force × distance ($F \times d$) work. To understand why, remember from Section 6.5 that pressure is force per unit of surface area:

$$P = \frac{F}{A} \tag{6.1}$$

Area has dimensions of distance squared ($d^2$), and volume is often expressed in units of distance cubed ($d^3$). If we substitute $d^2$ and $d^3$ for area and volume change, then

$$P \times \Delta V = \frac{F}{d^2} \times d^3 = F \times d$$

Now that we have established the equivalency of pressure–volume work and force–distance work, we need to know how to convert the units used to express $P$–$V$ work into units of energy (for example, joules). Doing so enables us to

**CHEMTOUR**
Pressure–Volume Work

**pressure–volume ($P$–$V$) work** the work associated with the expansion or compression of a gas.

calculate changes in the internal energy ($\Delta E$) of a gaseous system by combining the quantities of heat transferred and $P$–$V$ work done. We use a modified version of Equation 10.3 in which $w$ is replaced with $-P\Delta V$:

$$\Delta E = q - P\Delta V \tag{10.4}$$

Equation 10.4 contains a minus sign because an increase in the volume of a system ($\Delta V > 0$) indicates that the expanding system is pushing aside a portion of its surroundings, which involves doing work *on* its surroundings. This kind of work has a negative value from the perspective of the system and results in a decrease in its internal energy (see Figure 10.4). By contrast, when the surroundings do work on the system (for example, when a gaseous system is compressed), volume change has a negative value, but work has a positive value. The minus sign in Equation 10.4 accounts for both processes.

Among the units commonly used to express pressure and volume are atmospheres (atm) and liters (L), respectively, as we saw in Chapter 9. When they are combined, as in Sample Exercise 10.2, the liter-atmosphere makes a handy conversion factor:

$$1 \text{ liter-atmosphere (L} \cdot \text{atm)} = 101.325 \text{ J}$$

### SAMPLE EXERCISE 10.2 Calculating $P$–$V$ Work — LO2

A tank of compressed helium is used to inflate 100 balloons for sale at a carnival on a day when the atmospheric pressure is 1.01 atm. If each balloon is inflated with 4.8 L of He, how much $P$–$V$ work is done by the compressed helium? Express your answer in joules.

**Collect and Organize** Our task is to determine how much $P$–$V$ work is done by inflating the 100 balloons, each with 4.8 L of helium, and each expanding against an opposing pressure of 1.01 atm.

**Analyze** We focus on the work done on the atmosphere (the surroundings) by the helium (the system) as the volume of the balloons increases. Inflating each of 100 balloons with about 5 L of helium against a pressure of about 1 atm means that $P\Delta V$ will be about 500 liter-atmospheres and about 100 times that number of joules, or 50,000 J.

**Solve** The work ($w$) done by the compressed helium is

$$\begin{aligned} w &= -P\Delta V \\ &= -1.01 \cancel{\text{atm}} \times \frac{4.8 \cancel{\text{L}}}{1 \cancel{\text{balloon}}} \times 100 \cancel{\text{balloons}} \times \frac{101.325 \text{ J}}{1 \cancel{\text{L}} \cdot \cancel{\text{atm}}} \\ &= -4.9 \times 10^4 \text{ J or } -49 \text{ kJ} \end{aligned}$$

**Think About It** The internal energy of the compressed helium (the system) decreases when some of it expands to inflate the balloons because work is done *by* the system on its surroundings. Lower internal energy in this case corresponds to the lower kinetic energy (and lower temperature) of both the helium atoms in the balloons and those remaining in the tank.

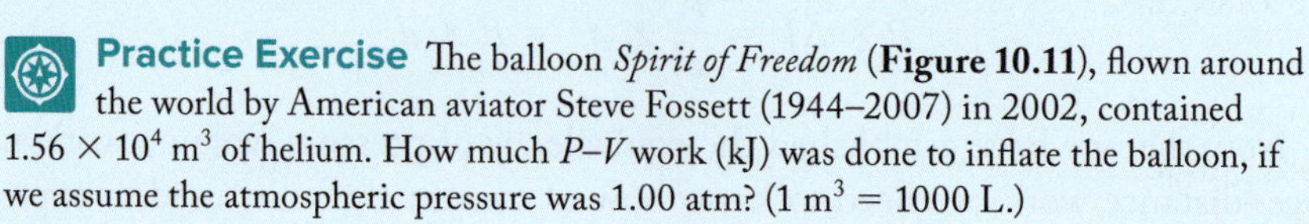

**Practice Exercise** The balloon *Spirit of Freedom* (**Figure 10.11**), flown around the world by American aviator Steve Fossett (1944–2007) in 2002, contained $1.56 \times 10^4 \text{ m}^3$ of helium. How much $P$–$V$ work (kJ) was done to inflate the balloon, if we assume the atmospheric pressure was 1.00 atm? ($1 \text{ m}^3 = 1000$ L.)

**FIGURE 10.11** The *Spirit of Freedom* was the first balloon to be flown nonstop around the world.

**SAMPLE EXERCISE 10.3** Relating Energy and Heat to *P–V* Work **LO2**

The racing cars in **Figure 10.12** are powered by V8 engines in which the motion of each piston in its cylinder displaces a volume of 0.733 L. If combustion of the mixture of gasoline vapor and air in one cylinder releases 1.68 kJ of energy, and if 33% of the energy does *P–V* work, how much energy is transferred as heat from the reaction mixture to its surroundings? How much pressure, on average, does the combustion reaction mixture exert on each piston?

**Collect, Organize, and Analyze** We know the quantity of energy (1.68 kJ) released by a combustion reaction and that 33% of this energy does *P–V* work on its surroundings. The rest (100 − 33 = 67%) must be lost as heat, as described by Equation 10.4: $\Delta E = q - P\Delta V$. The system loses energy; therefore, $\Delta E = -1.68$ kJ, and the value of $q$ will be about 2/3 of −1.68 kJ, or about −1 kJ.

**Solve** Using the information given and Equation 10.4 to solve for $q$ and $P$:

$$q = (0.67) \times (-1.68 \text{ kJ}) = -1.13 \text{ kJ}$$

$$P = -\frac{0.33 \times \Delta E}{\Delta V} = -\frac{0.33 \times (-1.68 \text{ kJ})}{0.733 \text{ L}} \times \frac{1 \text{ L} \cdot \text{atm}}{101.325 \text{ J}} \times \frac{1000 \text{ J}}{1 \text{ kJ}} = 7.5 \text{ atm}$$

**Think About It** We used a negative sign in the last equation because $P\Delta V$ represents work done *by* the system and is part of the energy lost by the system (−1.68 kJ). The value of $q$ is close to our estimate.

**Practice Exercise** The air compressor used with a paint sprayer does 64 J of work pumping air into a tank. Warmed by the process, air in the tank gives off 32 J of energy to its surroundings. What is the change in the internal energy of the air in the tank as a result of the two processes?

**FIGURE 10.12** These racing cars are powered by internal combustion engines in which hot gases do *P–V* work that is harnessed by the engine and drivetrain and propels the cars to speeds over 200 miles per hour (320 km/hr).

## 10.3 Enthalpy and Enthalpy Changes

Many phase changes and chemical reactions take place at constant pressure. These include chemical and biochemical reactions that occur in organisms living on Earth's surface and the chemical reactions done in open beakers and flasks on laboratory benches. We use a thermodynamic parameter called *enthalpy* to track the changes in energy and the flow of heat into or out from these and other constant-pressure systems. **Enthalpy ($H$)** is a measure of the total energy of a system. By *total*, we mean the system's internal energy ($E$) *and* the energy expended to push aside its surroundings to make space for the system at a given pressure. In equation form, this definition is

$$H = E + PV \quad (10.5)$$

In this chapter, our focus is not so much on the enthalpy of systems as on the *change* in enthalpy that accompanies, for example, a chemical reaction or a phase change. **Enthalpy change ($\Delta H$)** is equal to the quantity of heat transferred into or out from a system during a chemical reaction or phase change that occurs at constant pressure—that is, $q_P$. To see how $\Delta H$ and $q_P$ are equivalent, we can rewrite Equation 10.5 to describe the changes in the three terms:

$$\Delta H = \Delta E + \Delta(PV)$$

During processes occurring at constant pressure, the $\Delta(PV)$ term becomes $P\Delta V$:

$$\Delta H = \Delta E + P\Delta V \quad (10.6)$$

**CHEMTOUR** Internal Energy

**CHEMTOUR** Estimating Enthalpy Changes

**enthalpy ($H$)** a measure of the total energy of a system; the sum of the internal energy and the pressure–volume product of a system: $H = E + PV$.

**enthalpy change ($\Delta H$)** the energy absorbed by an endothermic process or given off by an exothermic process that takes place at constant pressure.

Next, we apply Equation 10.4 ($\Delta E = q - P\Delta V$) to a process occurring at constant pressure so that $q = q_P$:

$$\Delta E = q_P - P\Delta V$$

If we rearrange the terms to isolate $q_P$ on the left side, we get an equation with the same right side as Equation 10.6:

$$q_P = \Delta E + P\Delta V$$

In other words:

$$\Delta H = q_P \tag{10.7}$$

Like $\Delta E$, $\Delta H$ is a state function, so the value of $\Delta H$ depends only on the difference in enthalpy between the initial and final states, not on the pathway followed in going from one to the other.

Recall that the terms *exothermic* and *endothermic* describe processes that release or absorb energy, respectively, typically in the form of heat. These labels also describe processes in which the changes in enthalpy of the system are less than or greater than zero:

| Description | Enthalpy Change |
|---|---|
| Exothermic | $\Delta H < 0$ |
| Endothermic | $\Delta H > 0$ |

When an exothermic process runs in reverse, it becomes an endothermic process, and vice versa. For example, 6.01 kJ of thermal energy must be added to melt one mole of ice at 0°C, which is an endothermic process, but 6.01 kJ of thermal energy must be removed from a mole of liquid water at 0°C to freeze it, which is an exothermic process. This enthalpy change of 6.01 kJ/mol is called the **enthalpy of fusion ($\Delta H_{fus}$)** of water, where the subscript "fus" indicates melting (or *fusion*) of the solid. Sometimes $\Delta H_{fus}$ is referred to as the *heat of fusion*. The enthalpy changes associated with other paired phase changes—vaporization and condensation, and sublimation and deposition—also have complementary values. For example, it takes 40.7 kJ of thermal energy to vaporize a mole of liquid water at 100°C, and the same 40.7 kJ of thermal energy is released when one mole of water vapor condenses. As with $\Delta H_{fus}$, the **enthalpy of vaporization ($\Delta H_{vap}$)** of water is often referred to as the *heat of vaporization*.

**Table 10.2** lists $\Delta H_{fus}$ and $\Delta H_{vap}$ values for several common compounds. The values for the alkanes containing one to four carbons increase with increasing molar mass because these molecules interact via London dispersion forces that increase with increasing molecular size. Thus, the thermal energy required to convert them from solids into liquids and from liquids into gases also increases with increasing molecular size.

Most of the enthalpy of vaporization values in Table 10.2 are based on the energy required to vaporize substances at their normal boiling points. However, the $\Delta H_{vap}$ values for water include additional values at temperatures below the normal boiling point of 100°C. These $\Delta H_{vap}$ values decrease as temperature increases because the kinetic energy of the $H_2O$ molecules in liquid water increases as temperature increases. More kinetic energy means greater molecular motion, which also means less interaction between molecules: the average number of hydrogen bonds per water molecule decreases from an estimated 3.6 at 25°C to only 3.2 at 100°C. As the interactions between molecules decrease, the energy required to separate them into independent, gas-phase particles decreases as well.

**enthalpy of fusion ($\Delta H_{fus}$)** the energy required to convert one mole of a solid substance at its melting point into the liquid state; also called the *heat of fusion*.

**enthalpy of vaporization ($\Delta H_{vap}$)** the energy required to convert one mole of a liquid substance at its boiling point into the vapor state; also called the *heat of vaporization*.

**TABLE 10.2** Enthalpies of Fusion and Vaporization

| Compound and Molecular Structure | $\mathcal{M}$ (g/mol) | $\Delta H_{fus}$ (kJ/mol) | $\Delta H_{vap}$ (kJ/mol) |
|---|---|---|---|
| Methane | 16.04 | 0.94 | 8.2 |
| Ethane | 30.07 | 2.86 | 14.7 |
| Propane | 44.10 | 3.53 | 15.7 |
| Butane | 58.12 | 4.66 | 21.0 |
| Methanol | 32.04 | 3.18 | 35.3 |
| Ethanol | 46.07 | 5.02 | 38.6 |
| Acetone | 58.08 | 5.69 | 31.3 |
| Ammonia | 17.03 | 5.65 | 23.4 |
| Water | 18.02 | 6.01 | 44.0 at 25°C<br>43.5 at 40°C<br>42.5 at 60°C<br>41.6 at 80°C<br>40.7 at 100°C |

**CONCEPT TEST**

Using intermolecular forces, explain why water has the highest $\Delta H_{vap}$ value of all the compounds in Table 10.2 (and the highest boiling point, too).

## 10.4 Heating Curves and Heat Capacity

Heating Curves

Hikers and cross-country skiers use portable stoves fueled by propane or butane to prepare hot meals. In the winter, the only source of water may be ice or snow. What changes in temperature and physical state does the water undergo as the energy from a portable stove turns a saucepan filled with snow initially at −18.0°C into boiling water? We can track the changes by using a type of graph called a heating curve (**Figure 10.13**), which plots the increasing temperature of a system while it is heated at a constant rate.

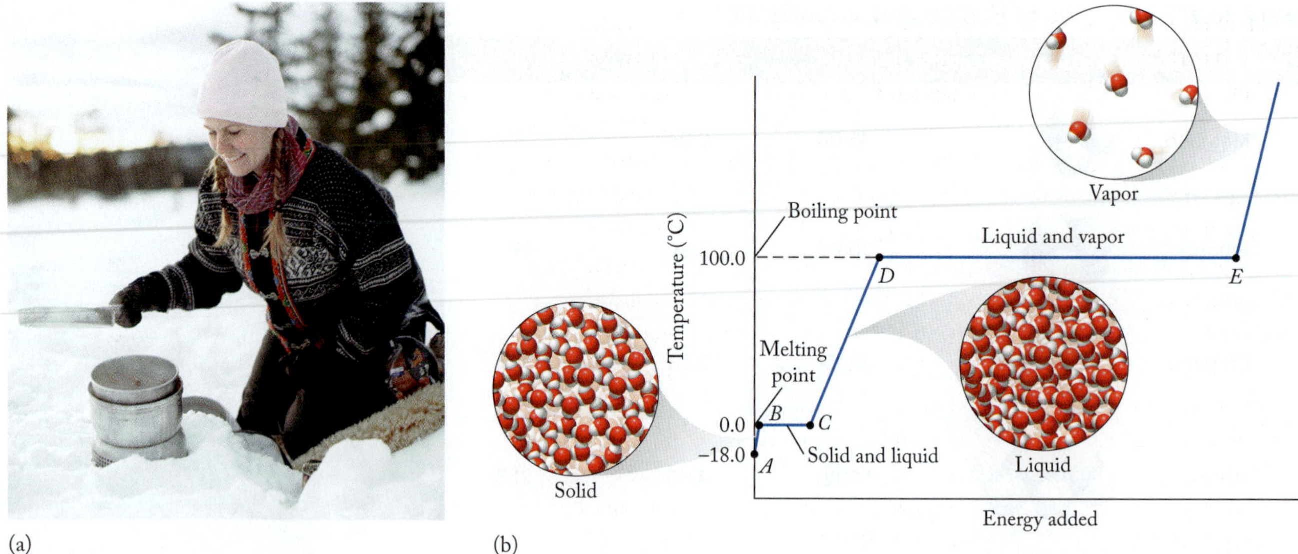


**FIGURE 10.13** (a) The hiker is using a small cooking stove protected by a windscreen as a heat source. (b) The energy required to melt snow and boil the resultant water is illustrated by the four line segments on the heating curve of water: heating snow to its melting point ($\overline{AB}$); melting the snow to form liquid water ($\overline{BC}$); heating the water to its boiling point ($\overline{CD}$); and boiling the water to convert it to vapor ($\overline{DE}$).

At first, the added energy raises the temperature of the snow as shown by the $\overline{AB}$ segment of the heating curve in Figure 10.13(b). The temperature stops rising when the snow reaches its melting point at 0.0°C, and it stays there (horizontal segment $\overline{BC}$) until all the snow has melted. Continued heating causes the temperature of the pot's contents, now liquid water, to rise again until the liquid reaches its boiling point at 100.0°C (segment $\overline{CD}$). At 100.0°C, the temperature of the water remains steady again as liquid water becomes water vapor (horizontal segment $\overline{DE}$). Only if all the liquid water in the pan vaporizes can the temperature of the water vapor in the pan rise above 100.0°C, as indicated by the line segment beyond point $E$ in Figure 10.13(b).

The difference between the $x$-axis coordinates at the beginning and end of each line segment in Figure 10.13(b) indicates how much energy is required during each step of the overall process. In the first step (line segment $\overline{AB}$), the energy required to raise the temperature of the snow from −18.0°C to 0.0°C is related to the snow's **heat capacity ($C_P$)**, a factor representing how much thermal energy the snow can store. Mathematically, $C_P$ is the quantity of energy that must flow into an object at constant pressure (hence the "P" subscript) to raise its temperature by 1°C. Thus, the units of $C_P$ are J/°C. If we know the value of $C_P$, we can calculate how much thermal energy ($q$) must be transferred to an object to raise its temperature by an amount $\Delta T$:

$$q = C_P \Delta T \tag{10.8}$$

When an object is a pure substance (like the snow in the hiker's pot), its heat capacity can then be calculated using that substance's *specific heat capacity*, or **specific heat ($c_s$)**, which is the quantity of thermal energy required to raise the temperature of one gram of a substance by 1°C at constant pressure. Thus, the units of $c_s$ are J/(g · °C). The product of the mass ($m$) in grams of an object composed of a single substance times the specific heat of the substance ($c_s$) is the same as the heat capacity of the object:

$$C_P(\text{J/°C}) = m(\text{g}) \times c_s[\text{J/(g} \cdot \text{°C)}]$$

Substituting ($mc_s$) for $C_P$ in Equation 10.8 gives us an equation that relates energy flow to the change in temperature of a mass $m$ of a pure substance:

$$q = mc_s\Delta T \tag{10.9}$$

**heat capacity ($C_P$)** the energy required to raise the temperature of an object by 1°C at constant pressure.

**specific heat ($c_s$)** the energy required to raise the temperature of 1 g of a substance by 1°C at constant pressure.

**molar heat capacity ($c_P$)** the energy required to raise the temperature of one mole of a substance by 1°C at constant pressure.

Heat capacity is an extensive property that depends on how massive an object is as well as what it's made of, whereas specific heat is an intensive property that is characteristic of a particular substance in a particular physical state (**Table 10.3**). For example, the $c_s$ of liquid water, 4.18 J/(g · °C), is about twice that of ice. Why are they different? The answer has to do with the interactions between water molecules and their freedom of motion. Water molecules in solid ice have little freedom of motion, though they can vibrate in place. Some of these vibrations stretch and compress the O—H bonds in the molecules or change the angle between the bonds. Still others stretch and compress the hydrogen bonds between water molecules. These various modes of vibration provide ways for ice to disperse any energy flowing into it. The more ways there are, the greater the specific heat. The molecules in liquid water enjoy even more freedom of motion and even more ways to disperse absorbed thermal energy; therefore, liquid water has a higher specific heat than ice.

Another intensive property related to specific heat is **molar heat capacity ($c_P$)**, which is the quantity of energy required to raise the temperature of one mole of a substance by 1°C at constant pressure. If we know the number of moles ($n$) of a substance in a sample and also the temperature change in degrees Celsius ($\Delta T$) that the substance experiences, we can calculate $q$ for the process:

$$q = nc_P\Delta T \tag{10.10}$$

**TABLE 10.3 Specific Heat and Molar Heat Capacity Values**

| Substance | Phase | $c_s$ [J/(g · °C)] | $c_P$ [J/(mol · °C)] |
|---|---|---|---|
| **Elements** | | | |
| Aluminum | (s) | 0.897 | 24.2 |
| Carbon (graphite) | (s) | 0.71 | 8.5 |
| Carbon (diamond) | (s) | 0.127 | 1.5 |
| Chromium | (s) | 0.449 | 23.3 |
| Copper | (s) | 0.385 | 24.5 |
| Gold | (s) | 0.129 | 25.4 |
| Iron | (s) | 0.45 | 25.1 |
| Lead | (s) | 0.129 | 26.7 |
| Silicon | (s) | 0.71 | 19.9 |
| Silver | (s) | 0.233 | 25.1 |
| Tin | (s) | 0.227 | 26.9 |
| Titanium | (s) | 0.523 | 25.0 |
| Zinc | (s) | 0.387 | 25.3 |
| **Compounds** | | | |
| Ammonia | (ℓ) | 4.75 | 80.9 |
| Silicon dioxide | (s) | 0.703 | 42.2 |
| Water (−10°C) | (s) | 2.11 | 38.0 |
| Water (25°C) | (ℓ) | 4.18 | 75.3 |
| Water (102°C) | (g) | 1.89 | 34.1 |
| **Mixture** | | | |
| Air[a] | (g) | 1.003 | 29.1 |

[a]Dry air at 0°C and 1 atmosphere of pressure.

## Hot Soup on a Cold Day

Suppose the hiker in Figure 10.13(a) decides to make soup starting with 275 g of snow. She puts the snow, which has an initial temperature ($T_i$) of −18.0°C, into the pan and starts to heat it up. At first, energy is consumed to warm the snow to its melting point, which makes the final temperature of the snow ($T_f$) 0.0°C. We can use Equation 10.9 to calculate how many joules of thermal energy must be added to do this. The value of $m$ is 275 g, $c_s$ is the specific heat of ice [2.11 J/(g · °C)], and the change in temperature is the difference between the final and initial temperatures, or

$$\Delta T = T_f - T_i = [0.0 - (-18.0)]°C = 18.0°C$$

We insert these values into Equation 10.9:

$$q = mc_s\Delta T$$

$$= 275\ \cancel{g} \times \frac{2.11\ J}{\cancel{g} \cdot \cancel{°C}} \times 18.0\ \cancel{°C} = 1.04 \times 10^4\ J = 10.4\ kJ$$

Notice that this $q$ value is positive, which means that the system (the snow) gains energy as it warms up. Also note that line segment $\overline{AB}$ on the heating curve in Figure 10.13(b) has a positive slope because the temperature of the snow increases as the energy is added.

The thermal energy absorbed in the next step is that required to melt (or *fuse*) the snow, which is the process that occurs along line segment $\overline{BC}$ on the heating curve. The value of $q$ for this step can be calculated using the enthalpy of fusion ($\Delta H_{fus}$) of water and the number of moles ($n$) of water in 275 g of snow:

$$q = n\Delta H_{fus} \qquad (10.11)$$

$$= 275\ \cancel{g} \times \frac{1\ \cancel{mol}}{18.02\ \cancel{g}} \times \frac{6.01\ kJ}{\cancel{mol}} = 91.7\ kJ$$

This value is also positive because energy flows into the system even though its temperature doesn't change. Thermal energy breaks some of the hydrogen bonds that hold molecules of water in a rigid three-dimensional array in the solid phase. Disrupting these intermolecular forces increases the *potential* energy of the water molecules because they are no longer confined to a single location—they can flow past their nearest molecular neighbors. However, this does not result in actual motion of the $H_2O$ molecules, and a corresponding increase in their average *kinetic* energies or temperature, until all the ice has melted.

As energy continues to flow into the pot of 0.0°C water, its temperature rises. If enough energy is added, the temperature of the water reaches 100.0°C. The relation between this increase in temperature ($\Delta T$) of 100.0°C and the quantity of energy that was absorbed by the water can again be calculated using Equation 10.9, but this time we use the specific heat of liquid water:

$$q = mc_s\Delta T$$

$$= 275\ \cancel{g} \times \frac{4.18\ J}{\cancel{g} \cdot \cancel{°C}} \times 100.0\ \cancel{°C} = 1.15 \times 10^5\ J = 115\ kJ$$

Note that line segment $\overline{CD}$ on the heating curve has a positive slope because the temperature of the water increases as energy is added.

At this point in our story, we assume that our hiker–chef accidentally leaves the boiling water unattended and it vaporizes completely. As line segment $\overline{DE}$ on the heating curve shows, the temperature of the water remains at 100.0°C until all of it is vaporized. The enthalpy change and the energy transferred are the product of the number of moles of liquid water and water's enthalpy of vaporization ($\Delta H_{vap}$):

$$q = n\Delta H_{vap} \tag{10.12}$$

$$= 275 \text{ g} \times \frac{1 \text{ mol}}{18.02 \text{ g}} \times \frac{40.7 \text{ kJ}}{\text{mol}} = 621 \text{ kJ}$$

As happened when the snow sample melted, the energy absorbed by molecules of liquid $H_2O$ during vaporization is converted into greater *potential* energy of the molecules, as essentially all the hydrogen bonds between them are broken. However, their average kinetic energy and temperature do not increase. Only after all the liquid water has completely vaporized does the temperature of the water vapor increase above 100.0°C.

Notice in Figure 10.13(b) that line segment $\overline{DE}$, representing the phase change from liquid water to water vapor, is much longer than line segment $\overline{BC}$, which represents the change from solid snow to liquid water. The relative lengths of the lines reflect the fact that the enthalpy of vaporization of water (40.7 kJ/mol) is much larger than the enthalpy of fusion of ice (6.01 kJ/mol). It takes more energy to boil one mole of water than to melt one mole of ice because only a fraction of the hydrogen bonds in ice are broken upon melting, whereas essentially all the hydrogen bonds in liquid water are broken when it vaporizes.

Heating curves and their corresponding calculations represent changes in energy and temperature as a substance is heated. If we were to reverse this process and cool a sample of a gas to form a liquid and then cool it further to form a solid, the resultant graph would be called a cooling curve. Just like heating curves, cooling curves have constant-temperature horizontal segments where the state of matter changes. However, unlike heating curves, the line segments that show temperature changes have negative slopes in cooling curves because temperature decreases as energy is lost.

**FIGURE 10.14** An ice-resurfacing machine uses hot water to melt the ice, which then rapidly refreezes to create a smooth rink for skating.

**SAMPLE EXERCISE 10.4** Calculating Energy Transfer **LO3**

Between periods of a hockey game, an ice-resurfacing machine (**Figure 10.14**) spreads $3.00 \times 10^2$ liters of hot (40.0°C) water across a skating rink. (a) Identify the segments on the cooling curve (**Figure 10.15**) in terms of the change occurring on the part of the water. (b) How much energy must the water lose as it cools to its freezing point, freezes, and further cools to −10.0°C? Assume that the water is the system and that its density is 0.992 g/mL at 40.0°C.

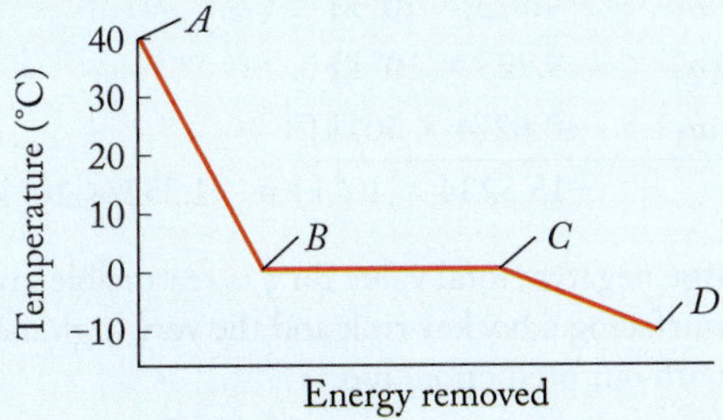


**FIGURE 10.15** A cooling curve for water.

**Collect and Organize** We are given a cooling curve, and we know the temperature, volume, and density of water used to resurface a skating rink. We are asked to identify the changes that correlate to the segments on the curve and then to calculate how much energy the water loses as it cools from 40.0°C to −10.0°C. Table 10.3 lists specific heat and molar heat capacity values for liquid water and solid ice. The enthalpy of fusion ($\Delta H_{fus}$) of water is 6.01 kJ/mol.

**Analyze** Multiplying the volume of 40.0°C water applied to the ice by its density will give us the mass of water in grams. Dividing that number by the molar mass of $H_2O$ (18.02 g/mol) will give us the number of moles, which is a value we can use with the molar heat capacities of liquid water and solid ice to calculate the energy losses during the cooling stages, using Equation 10.10. The number of moles multiplied by the enthalpy of fusion (Equation 10.11) gives us how much energy must be removed to freeze the water. Because the water is the system and it transfers energy to the surroundings, the enthalpy of fusion will be a negative number. Given the large number of moles of $H_2O$ in 300 L (~300 kg) of water and the high molar heat capacity of liquid water, we expect a large, negative value for the total energy removed.

**Solve** (a) First, we identify the line segments on the cooling curve with respect to the changes occurring: segment $\overline{AB}$, water cooled from 40°C to freezing point (0°C); segment $\overline{BC}$, liquid water converted to solid ice; segment $\overline{CD}$, ice at 0°C cooled to −10°C. (b) Next, we convert $3.00 \times 10^2$ L of water into moles:

$$3.00 \times 10^2\ \cancel{\text{L}} \times \frac{1000\ \cancel{\text{mL}}}{1\ \cancel{\text{L}}} \times \frac{0.992\ \cancel{\text{g}}}{1\ \cancel{\text{mL}}} \times \frac{1\ \text{mol}}{18.02\ \cancel{\text{g}}} = 1.651 \times 10^4\ \text{mol}$$

Then we calculate the energy lost in each stage of the cooling/freezing process.

1. Cooling 40.0°C water to its freezing point:

$$\begin{aligned} q_1 &= nc_P\Delta T \\ &= 1.651 \times 10^4\ \cancel{\text{mol}} \times \frac{75.3\ \text{J}}{\cancel{\text{mol}} \cdot \cancel{^\circ\text{C}}} \times (0.0 - 40.0)\cancel{^\circ\text{C}} \\ &= -4.973 \times 10^7\ \text{J} = -4.973 \times 10^4\ \text{kJ} \end{aligned}$$

2. Freezing the water:

$$\begin{aligned} q_2 &= n(-\Delta H_{fus}) \\ &= 1.651 \times 10^4\ \cancel{\text{mol}} \times \frac{-6.01\ \text{kJ}}{\cancel{\text{mol}}} = -9.923 \times 10^4\ \text{kJ} \end{aligned}$$

3. Cooling the ice:

$$\begin{aligned} q_3 &= nc_P\Delta T \\ &= 1.651 \times 10^4\ \cancel{\text{mol}} \times \frac{38.0\ \text{J}}{\cancel{\text{mol}} \cdot \cancel{^\circ\text{C}}} \times (-10.0 - 0.0)\cancel{^\circ\text{C}} \\ &= -6.274 \times 10^6\ \text{J} = -6.274 \times 10^3\ \text{kJ} \end{aligned}$$

Summing these three $q$ values and rounding off the total to the correct number of significant figures:

$$\begin{array}{rl} q_1 & -4.973 \times 10^4\ \text{kJ} \\ +\ q_2 & -9.923 \times 10^4\ \text{kJ} \\ +\ q_3 & -0.6274 \times 10^4\ \text{kJ} \\ \hline & -15.5234 \times 10^4\ \text{kJ} = -1.552 \times 10^5\ \text{kJ} \end{array}$$

**Think About It** The large negative total value for $q$ is reasonable given the large volume of water involved in resurfacing a hockey rink and the very high molar heat capacity of water. It is consistent with our prediction, too.

**Practice Exercise** The flame in a torch used to cut metal is produced by burning acetylene ($C_2H_2$) in pure oxygen. If the combustion of one mole of acetylene releases 1251 kJ of energy, what mass of acetylene is needed to cut through a piece of steel if the process requires $5.42 \times 10^4$ kJ of energy?

Water is an extraordinary substance for many reasons, and its high specific heat value (the second-highest one in Table 10.3) is one of them. The ability of water to absorb large quantities of thermal energy is why it is used as a *heat sink* in both automobile radiators and our bodies. The term *heat sink* is often used to identify matter that can absorb energy without changing phase or significantly changing its temperature. Weather and climate changes are largely driven and regulated by cycles involving the retention of energy by Earth's oceans, which serve as enormous heat sinks.

Why does liquid water have such an unusually high specific heat? One of the principal reasons is hydrogen bonding. As we discussed in Chapter 6, molecules of $H_2O$ can form up to four hydrogen bonds each. These strong intermolecular interactions contribute to a lower potential energy. As water absorbs thermal energy, its internal energy increases. Part of the increase goes toward increasing the kinetic energies of the water molecules and raising the temperature of the water. However, a significant part goes toward increasing the potential energy of the water by breaking some of the hydrogen bonds between its molecules via increased molecular motion. As we have seen in this section, increasing potential energy produces no increase in temperature. As a result, even more energy must be absorbed to produce a temperature change of, say, 1°C. In this way hydrogen bonding contributes to the rather high molar heat capacity of water (compared to the other compounds in Table 10.3). When we factor in its small molar mass compared to most other compounds, the extraordinarily high specific heat of liquid water seems reasonable.

**CONNECTION** As shown in Figure 6.36, a molecule of $H_2O$ in ice or liquid water can form up to four hydrogen bonds: each of its H atoms can hydrogen-bond to the O atom of a nearby water molecule, and its O atom, with its two lone pairs of electrons, can form two hydrogen bonds to nearby H atoms.

## Cold Drinks on a Hot Day

Let's consider another application of energy transfer. Suppose we throw a party and plan to chill three cases of beverages by placing the cans (72 aluminum cans, each containing 355 mL), which are initially at a temperature of 25.0°C, in an insulated cooler and covering them with ice cubes that are initially at −18.0°C. If we assume that (1) the ice is sold in 4.5-kilogram (10-pound) bags; (2) the mass of aluminum in each can is 12.5 grams; (3) the cans contain mostly water, which has a density of 1.00 g/mL; and (4) the other ingredients are present in such small concentrations that they will not affect our energy transfer calculation, then how many bags of ice do we need to chill the cans and their contents to 0.0°C (that is, "ice cold")?

We can use the energy transfer relationships we have defined to predict how much ice is required. In doing so, we assume that whatever energy is absorbed by the ice is lost by the cans and the beverages in them. In equation form, this assumption is

$$q_{\text{lost}} = -q_{\text{gained}} \tag{10.13}$$

and it applies to energy transfer processes that occur between warm and cold objects in any isolated system.

Let's first consider the energy lost by the cans and their contents. By substance, their masses are

$$72\ \cancel{\text{cans}} \times \frac{12.5\text{ g Al}}{\cancel{\text{can}}} = 9.00 \times 10^2\text{ g Al}$$

plus

$$72\ \cancel{\text{cans}} \times \frac{355\ \cancel{\text{mL}}\ H_2O}{\cancel{\text{can}}} \times \frac{1.00\text{ g}}{\cancel{\text{mL}}} = 2.56 \times 10^4\text{ g }H_2O$$

We can use Equation 10.9 and the specific heat values for aluminum and liquid water in Table 10.3 to calculate the quantity of energy that must be transferred to lower the temperatures of the beverages by −25.0°C:

$$q = mc_s\Delta T$$

$$= 9.00 \times 10^2\ \text{g Al} \times \frac{0.897\ \text{J}}{\text{g Al} \cdot {}^\circ\text{C}} \times (-25.0^\circ\text{C})$$

$$+ 2.56 \times 10^4\ \text{g H}_2\text{O} \times \frac{4.18\ \text{J}}{\text{g H}_2\text{O} \cdot {}^\circ\text{C}} \times (-25.0^\circ\text{C})$$

$$= -2.02 \times 10^4\ \text{J} + (-2.68 \times 10^6\ \text{J}) = -2.70 \times 10^6\ \text{J} = -2.70 \times 10^3\ \text{kJ}$$

If the beverages transfer $2.70 \times 10^3$ kJ, then according to Equation 10.13 (and the law of conservation of energy), the ice gains $2.70 \times 10^3$ kJ of energy. This energy serves two purposes: warming ice cubes to 0.0°C and then melting them. To calculate how many bags of ice melt, we'll need to solve for $n$, the number of moles of ice that we need. We focus on moles rather than grams because the enthalpy of fusion is expressed in kJ/mol, and we can calculate the energy required to warm the ice by using its molar heat capacity [$c_P$ = 38.0 J/(mol · °C)] from Table 10.3 and Equation 10.10. To keep all energy units the same, let's convert $c_P$ to 0.0380 kJ/(mol · °C) before combining the energy absorbed by (1) warming and (2) melting the ice:

$$q_1 + q_2 = nc_P\Delta T + n\Delta H_{fus} = 2.70 \times 10^3\ \text{kJ}$$

$$\left(n \times \frac{0.0380\ \text{kJ}}{\text{mol} \cdot {}^\circ\text{C}} \times 18.0^\circ\text{C}\right) + \left(n \times \frac{6.01\ \text{kJ}}{\text{mol}}\right) = 2.70 \times 10^3\ \text{kJ}$$

$$n \times \left(\frac{0.684\ \text{kJ}}{\text{mol}} + \frac{6.01\ \text{kJ}}{\text{mol}}\right) = 2.70 \times 10^3\ \text{kJ}$$

$$n = 403\ \text{mol}$$

Converting moles into grams, then kilograms, and then 4.5-kilogram bags:

$$403\ \text{mol} \times \frac{18.02\ \text{g}}{\text{mol}} \times \frac{1\ \text{kg}}{1000\ \text{g}} \times \frac{1\ \text{bag}}{4.5\ \text{kg}} = 1.6\ \text{bags}$$

We can't buy a fraction of a bag, so we'll need to buy two bags of ice for our party.

## SAMPLE EXERCISE 10.5 Calculating a Final Temperature from Energy Gain and Loss — LO3

Suppose you have exactly 237 g (1 cup) of hot (100.0°C) brewed tea in an insulated mug and that you add to it $2.50 \times 10^2$ g of ice initially at −18.0°C. If all the ice melts, what is the final temperature of the tea? Assume that tea has the same thermal properties as water.

**Collect and Organize** We know the mass of the tea, its initial temperature, and its specific heat (the $c_s$ value for liquid water from Table 10.3). We also know the mass of the ice and its initial temperature, and from Tables 10.2 and 10.3 we know the specific heats and molar heat capacities of ice and liquid water and the enthalpy of fusion of ice. Our task is to find the final temperature of the tea.

**Analyze** The energy lost by the tea has the same absolute value as, but the opposite sign of, the energy gained by the ice cubes. In equation form:

$$q_{\text{ice}} = -q_{\text{tea}}$$

We can assume that once all the ice has melted, the 250 g of 0.0°C water produced will mix with and be warmed by the still warmer 237 g of tea. Together they reach the same final temperature. Therefore, there are three parts to $q_{\text{ice}}$: the energy gained when the ice warms from −18.0°C to 0.0°C, the energy gained when the ice melts, and the energy gained when the temperature of the melted ice increases from 0.0°C to the final temperature (which we'll call $x$°C). We have to calculate the number of moles of $H_2O$ in the ice to use $\Delta H_{\text{fus}}$, so we'll use the molar heat capacities of ice and water to calculate $q_{\text{ice}}$. The tea undergoes no phase change, so its only loss of energy is due to its temperature change from 100.0°C to $x$°C.

**Solve**

1. Converting grams of ice to moles:

$$2.50 \times 10^2 \text{ g} \times \frac{1 \text{ mol}}{18.02 \text{ g}} = 13.87 \text{ mol}$$

2. The ice gains energy as it
   a. Warms to 0.0°C:

$$q_{\text{ice}} = nc_{\text{P}}\Delta T = 13.87 \text{ mol} \times \frac{38.0 \text{ J}}{\text{mol} \cdot °\text{C}} \times (18.0°\text{C}) \times \frac{1 \text{ kJ}}{1000 \text{ J}} = 9.49 \text{ kJ}$$

   b. Melts:

$$q_{\text{ice}} = n\Delta H_{\text{fus}} = 13.87 \text{ mol} \times \frac{6.01 \text{ kJ}}{\text{mol}} = 83.4 \text{ kJ}$$

   c. Warms to final temperature $x$:

$$q_{\text{ice}} = nc_{\text{P}}\Delta T = 13.87 \text{ mol} \times \frac{75.3 \text{ J}}{\text{mol} \cdot °\text{C}} \times \frac{1 \text{ kJ}}{1000 \text{ J}} \times (x - 0.0)°\text{C} = 1.044(x)\text{kJ}$$

3. The energy lost by the tea ($q_{\text{tea}}$) as it cools to $x$°C is

$$q_{\text{tea}} = mc_s\Delta T = 237 \text{ g} \times \frac{4.18 \text{ J}}{\text{g} \cdot °\text{C}} \times (x - 100.0)°\text{C} \times \frac{1 \text{ kJ}}{1000 \text{ J}} = 0.991(x - 100.0)\text{kJ}$$

4. Balancing the loss and gain of energy:

$$q_{\text{ice}} = -q_{\text{tea}}$$

$$(9.49 + 83.4 + 1.044x) \text{ kJ} = -(0.991x - 99.1) \text{ kJ}$$

$$x = 3.1°\text{C}$$

**Think About It** The answer is consistent with the assumption that after all the ice melted, the temperature of the liquid water thus produced would increase a little to reach the same temperature as the tea. There were no temperature units in the final values used to calculate $x$, but remember that $x$ was assumed to have units of °C when we inserted it into the $\Delta T$ terms in the first steps of the calculation.

**Practice Exercise** Calculate the final temperature of a mixture of 0.350 kg of ice initially at −18°C and 437 g of water initially at 100.0°C.

## Determining Specific Heat

Suppose your laboratory instructor gives you some metal pellets and asks you to determine whether the metal is aluminum, titanium, zinc, tin, or lead. You are not allowed to chemically alter the metal in order to determine its identity. In addition to the glassware in your lab drawer, you have access to a Bunsen burner, a balance, and Styrofoam coffee cups. How can you decide which metal it is?

According to Table 10.3, the five possible metals have quite different specific heat values, ranging from 0.129 to 0.897 J/(g · °C). How can we determine the specific heat of the unknown metal? We need to link its mass (we have a balance) and a change in its temperature (easily measured with a thermometer) to the quantity of energy transferred during a temperature change. That is the challenge. One way to meet it is to couple the energy gained or lost by the metal to the energy lost or gained by a known substance. A good choice for this substance is water.

We can use an experimental setup like the one in **Figure 10.16**. It includes a Bunsen burner that we can use to heat a beaker of water to its boiling point: 100.0°C. Then we measure the mass of the pellets—let's say it is 25.0 g—and place them in a test tube to keep them dry. The test tube is immersed in the boiling water. After a while, the temperature of the metal should also be 100.0°C. Meanwhile, we add 50.0 g of water to two nested Styrofoam coffee cups. We use the coffee cups because they have good insulating capability and negligible mass compared to the water.

The critical part of the experiment comes when we measure the temperature of the water in the nested cups—it's 19.3°C—and then quickly remove the test tube from the boiling water bath and pour the metal out of it and into the water in the cups. Then we measure the temperature of that water until it reaches a maximum value, which is 27.1°C. At this temperature we may assume that the energy gained by the water and the energy lost by the metal are in balance:

$$q_{\text{water}} = -q_{\text{metal}}$$

The left side of the equation is a quantity we can calculate using Equation 10.9:

$$q_{\text{water}} = mc_s\Delta T$$
$$= 50.0\ \cancel{\text{g}} \times \frac{4.18\ \text{J}}{\cancel{\text{g}} \cdot \cancel{°\text{C}}} \times (27.1 - 19.3)\cancel{°\text{C}} = 1.63 \times 10^3\ \text{J}$$

We apply this value to what we know about the right side of the equation:

$$1.63 \times 10^3\ \text{J} = -q_{\text{metal}} = -mc_s\Delta T$$
$$= -25.0\ \text{g} \times c_s \times (27.1 - 100.0)°\text{C} = (1823\ \text{g} \cdot °\text{C})\ c_s$$
$$c_s = 0.894\ \text{J/(g} \cdot °\text{C)}$$

According to the $c_s$ values in Table 10.3, the mystery metal is aluminum.

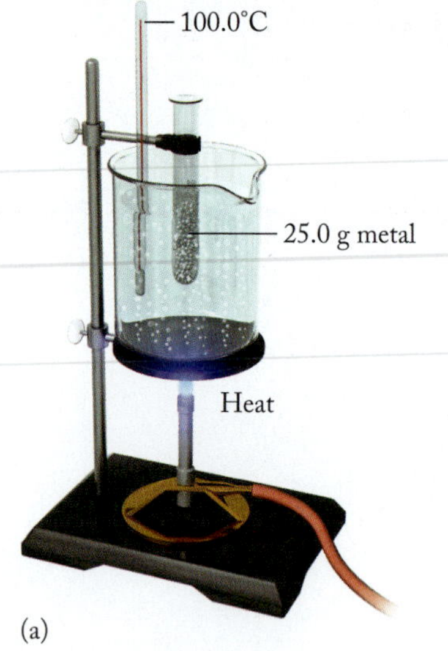

**FIGURE 10.16** Experimental setup to determine the specific heat of a metal. (a) Heat 25.0 grams of the metal to 100.0°C in boiling water. (b) Add 50.0 grams of water at 19.3°C to nested Styrofoam cups. (c) Transfer the hot metal to the water, and record the peak temperature of the metal–water mixture, which reaches 27.1°C in this experiment.

## CONCEPT **TEST**

In the preceding determination of the specific heat of a metal, how would the following factors have affected the experimental $c_s$ value?

a. Slowly transferring the metal from the hot-water bath to the Styrofoam cups so that the metal's temperature was less than 100.0°C when it hit the water.

b. Drops of hot water adhering to the metal when it was transferred to the Styrofoam cups.

c. Ignoring the heat capacity of the thermometer.

d. Energy transfer through the Styrofoam cups or the open top. Assume the lab temperature was 22°C.

# 10.5 Enthalpies of Reaction and Calorimetry

**CHEMTOUR**
Calorimetry

In previous sections, we have used specific heat, molar heat capacity, and enthalpies of fusion and vaporization to track flows of thermal energy as substances were warmed or cooled or changed physical state, but we have not explained how we know the values of those parameters. One way to obtain them is through experiments based on calorimetry. **Calorimetry** is an analytical technique in which changes in the temperature of a device with a known heat capacity, called a **calorimeter**, are used to determine the energy released or absorbed by a process occurring inside the calorimeter. The process may be a physical exchange of thermal energy, or it can be a chemical reaction in which energy flows either from or into a reaction mixture.

The magnitude of the energy absorbed or released during a chemical reaction is proportional to the quantity of the reactants consumed and to a property of the reaction called its **enthalpy of reaction ($\Delta H_{rxn}$)**, often referred to as its *heat of reaction*. The subscript may be changed to reflect a specific type of reaction. For example, $\Delta H_{comb}$ is sometimes used to represent the enthalpy change that accompanies a combustion reaction.

The value of $\Delta H_{rxn}$ of a particular reaction depends on the difference in enthalpy between its products and reactants. It also depends on how we choose to represent the reaction in a balanced chemical equation. For example, suppose we write the balanced equation for the complete combustion of butane, the fuel in disposable lighters, in the following way:

$$2\ C_4H_{10}(g) + 13\ O_2(g) \rightarrow 8\ CO_2(g) + 10\ H_2O(\ell)$$

We can turn this chemical equation into a **thermochemical equation** by adding the enthalpy change that accompanies the combustion of 2 moles of $C_4H_{10}$ by 13 moles of $O_2$ to form 8 moles of $CO_2$ and 10 moles of liquid $H_2O$:

$$2\ C_4H_{10}(g) + 13\ O_2(g) \rightarrow 8\ CO_2(g) + 10\ H_2O(\ell) \qquad \Delta H_{rxn} = -5754\ \text{kJ}$$

However, we might want to express the enthalpy change that accompanies the combustion of only 1 mole of butane. As you might expect, burning half the fuel releases half the energy and is accompanied by only half the change in enthalpy, as represented by the following thermochemical equation:

$$C_4H_{10}(g) + 13/2\ O_2(g) \rightarrow 4\ CO_2(g) + 5\ H_2O(\ell) \qquad \Delta H_{rxn} = -2877\ \text{kJ}$$

**CONCEPT TEST**

The enthalpy change accompanying the combustion of one mole of hydrogen gas, producing one mole of liquid water, is −286 kJ. What is the value of $\Delta H_{rxn}$ for the following reaction?

$$2\ H_2(g) + O_2(g) \rightarrow 2\ H_2O(\ell) \qquad \Delta H_{rxn} = ?$$

Absolute enthalpy values are difficult to determine, but we can determine *changes* in enthalpy experimentally. The apparatus in Figure 10.16 is a kind of calorimeter called a *coffee-cup calorimeter*, which is particularly useful for determining $\Delta H_{rxn}$ for reactions in aqueous solutions. Because the reactions happen at constant (ambient) pressure, the energy absorbed or released by the reaction ($q_{rxn}$) is equal to the enthalpy change that accompanies the reaction (see

**calorimetry** the experimental determination of the quantity of energy transferred during a phase change or chemical process.

**calorimeter** a device used to measure the absorption or release of energy by a phase change or chemical process.

**enthalpy of reaction ($\Delta H_{rxn}$)** the enthalpy change that accompanies a chemical reaction; also called the *heat of reaction*.

**thermochemical equation** the chemical equation of a reaction that includes the change in enthalpy that accompanies the reaction.

Equation 10.7). If all this energy is transferred to or from the contents of the calorimeter, then

$$q_{rxn} = -q_{calorimeter} \quad (10.14)$$

The value of $q_{calorimeter}$ can be calculated from the measured change in temperature ($\Delta T$) and the heat capacity of the calorimeter ($C_{calorimeter}$):

$$q_{calorimeter} = C_{calorimeter}\,\Delta T \quad (10.15)$$

In reactions involving dilute aqueous solutions, nearly all the mass ($m$) of the calorimeter comes from the mass of the water in it. This fact and the very high specific heat of water [$c_{s,H_2O} = 4.18$ J/(g · °C)] allow us to assume that the heat capacity of the calorimeter is essentially the same as the heat capacity of the water in it:

$$q_{calorimeter} = \left(\frac{4.18\ \text{J}}{\text{g} \cdot {}^\circ\text{C}}\right) m\Delta T \quad (10.16)$$

**SAMPLE EXERCISE 10.6** Calculating $\Delta H_{rxn}$ from Calorimetry Data **LO4**

When 0.200 L of 0.200 $M$ HCl is mixed with 0.200 L of 0.200 $M$ NaOH in a coffee-cup calorimeter, the temperature of the mixture increases from 22.15°C to 23.48°C. If the densities of the two solutions are 1.00 g/mL, what is the value of $\Delta H_{rxn}$ for the following reaction?

$$HCl(aq) + NaOH(aq) \rightarrow NaCl(aq) + H_2O(\ell) \qquad \Delta H_{rxn} = ?$$

**Collect and Organize** We know the (a) concentrations, (b) volumes, (c) densities, and (d) initial and final temperatures of two solutions: a strong acid and a strong base. We are asked to calculate the value of $\Delta H_{rxn}$ for the reaction in which they neutralize each other.

**Analyze** The densities of both aqueous solutions are nearly the same as that of water (1.00 g/mL), which confirms that they are dilute solutions with heat capacities that are essentially the same as that of water. Therefore, Equation 10.16 can be used to calculate $q_{calorimeter}$. Their common density also means that the total mass of the solutions in grams is the same as their total volume in milliliters: 400. Using this value in Equation 10.16 and a $\Delta T$ value of about 1.3 gives a $q_{calorimeter}$ of about $4 \times 400 \times 1.3 \approx 2000$ J, or 2 kJ. The energy gained by the solution represents about −2 kJ released by the exothermic neutralization reaction ($q_{rxn}$). To calculate $\Delta H_{rxn}$ for the reaction as written (one mole of each reactant), we will need to calculate the number of moles of each reactant in the reaction mixture. The volumes of each solution are 200 mL = 1/5 L and their concentrations are 1/5 molar, so there is only 1/5 × 1/5 = 1/25 of a mole of each reactant. Therefore, the value of $\Delta H_{rxn}$ for a reaction in which one mole of each reactant is consumed should be about −2 kJ × 25 = −50 kJ.

**Solve** Calculating the value of $q_{calorimeter}$ ($-q_{rxn}$) with Equation 10.16:

$$q_{calorimeter} = \left(\frac{4.18\ \text{J}}{\text{g} \cdot {}^\circ\text{C}}\right) m\Delta T$$

$$= \frac{4.18\ \text{J}}{(\cancel{\text{g}} \cdot \cancel{{}^\circ\text{C}})} \times (4.00 \times 10^2\ \cancel{\text{g}}) \times (23.48 - 22.15)\cancel{{}^\circ\text{C}} = 2224\ \text{J} = 2.224\ \text{kJ}$$

Therefore, $q_{rxn} = -2.224$ kJ. Calculating the value of $\Delta H_{rxn}$:

$$\Delta H_{rxn} = \frac{-2.224\ \text{kJ}}{0.200\ \cancel{\text{L}} \times \dfrac{0.200\ \text{mol}}{\cancel{\text{L}}}} = -55.6\ \text{kJ/mol reactant, or } -55.6\ \text{kJ}$$

**Think About It** In the second step, we converted the energy released by the reaction mixture ($q_{rxn}$) into the energy that would have been released had there been one mole of each reactant, which is the quantity of each of them in the balanced chemical equation.

**Practice Exercise** Addition of 1.31 g of zinc metal to 100.0 mL of 0.200 $M$ HCl in a coffee-cup calorimeter causes the temperature to increase from 11.96°C to 15.21°C. If the density of the HCl solution is 1.00 g/mL, what is the value of $\Delta H_{rxn}$ for the following reaction?

$$Zn(s) + 2\ HCl(aq) \rightarrow ZnCl_2(aq) + H_2(g)$$

## Bomb Calorimetry

One of the most important categories of chemical reactions in terms of their transfer of thermal energy is combustion reactions. The quantities of energy they produce can be determined with devices called **bomb calorimeters** (**Figure 10.17**). To use these instruments, a combustible sample is placed in a sealed vessel (called a *bomb*) capable of withstanding high pressures; the bomb, in turn, is submerged in a large volume of water in a heavily insulated container. Oxygen is introduced into the bomb, and the mixture is ignited with an electric spark. As combustion occurs, thermal energy generated by the reaction (the system) flows into the walls of the bomb and then into the water in the calorimeter (the surroundings).

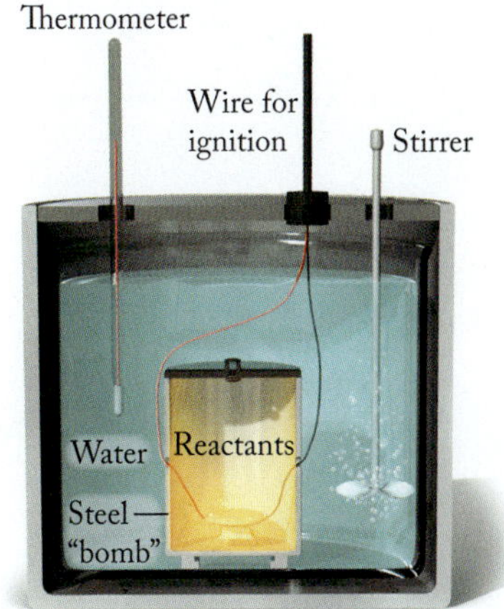


**FIGURE 10.17** A bomb calorimeter.

A well-designed bomb calorimeter keeps the system contained within the bomb and ensures that all energy released by the reaction stays inside the calorimeter. Specifically, the calorimeter consists of the bomb, the water, the insulated container, and minor components (stirrer, thermometer, and any other materials). The energy produced by the reaction is determined by measuring the temperature of the water before and after the reaction. The water is at the same temperature as all the other insulated parts, so the temperature change of the water tracks the temperature change of the entire calorimeter.

Measuring the change in temperature of the water is not the whole story. We also need to know the heat capacity of the water and all the other insulated components of the calorimeter—that is, its **calorimeter constant,** $C_{calorimeter}$. Why? Unlike in our coffee-cup calorimeter, where essentially all the energy warms a mass of water, the energy in a bomb calorimeter is absorbed by its many components, each of which has a heat capacity of its own. If we know the value of $C_{calorimeter}$ and if we can measure the change in water temperature, then we can calculate the quantity of energy that flows from a reaction mixture into a calorimeter.

How is $C_{calorimeter}$ determined? One way is to burn a known quantity of a material with a known enthalpy of combustion. The quantity of energy released by the reaction is then known, and the value of $C_{calorimeter}$ can be calculated from the change in temperature of the calorimeter produced by that quantity of energy. Benzoic acid ($C_6H_5COOH$) is often used for this purpose because very pure samples of it can be obtained. Complete combustion of exactly one gram of benzoic acid is known to release 26.38 kJ of thermal energy. Once $C_{calorimeter}$ has been determined, the calorimeter can be used to determine the quantities of energy produced by other combustion reactions on a per-gram or per-mole basis.

Because there is no change in the volume of the reaction mixture in a bomb calorimeter, this technique is referred to as *constant-volume calorimetry*. No $P$–$V$ work is done, so Equation 10.4

$$\Delta E = q - P\Delta V$$

simplifies to

$$q = \Delta E$$

**bomb calorimeter** a constant-volume device used to measure the energy released during a combustion reaction.

**calorimeter constant ($C_{calorimeter}$)** the heat capacity of a calorimeter.

The energy lost by the reaction mixture in a bomb calorimeter is the energy gained by the calorimeter (including the water in it), so an increase in the temperature of the surroundings also provides a measure of the energy released by the reaction happening inside the apparatus. Perhaps you are wondering how the $\Delta E$ measured in a bomb calorimeter is related to the $\Delta H$ of a reaction. For many reactions $\Delta E$ and $\Delta H$ are nearly the same, so we do not take into account the small differences between their $\Delta E$ and $\Delta H$ values.

**SAMPLE EXERCISE 10.7** Determining the Heat Capacity of a Calorimeter **LO4**

Before we can determine the energy change of a reaction run in a calorimeter, we must determine the heat capacity of the calorimeter, $C_{\text{calorimeter}}$. What is the value of $C_{\text{calorimeter}}$ if burning 1.000 g of benzoic acid increases the temperature of a calorimeter by 7.248°C? Combustion of benzoic acid releases 26.38 kJ of energy per gram of benzoic acid.

**Collect, Organize, and Analyze** We have been asked to find the calorimeter constant, which is the energy required to increase the temperature of a calorimeter by 1°C. We know that a reaction that releases 26.38 kJ of energy increases the temperature of the calorimeter by 7.248°C. Therefore, the calorimeter constant should be about 1/7 of 26.38, or a little less than 4 kJ/°C.

**Solve** Applying Equation 10.15 and solving for $C_{\text{calorimeter}}$:

$$q_{\text{calorimeter}} = C_{\text{calorimeter}}\Delta T$$

$$C_{\text{calorimeter}} = \frac{q_{\text{calorimeter}}}{\Delta T}$$

$$= \frac{26.38 \text{ kJ}}{7.248°\text{C}} = 3.640 \text{ kJ/°C}$$

**Think About It** The calorimeter constant is determined for a specific calorimeter like the one in Figure 10.17. Once $C_{\text{calorimeter}}$ is known, the calorimeter can be used to determine the enthalpy of combustion of any combustible material. If any of the internal parts of the calorimeter change or are replaced, a new constant must be determined.

**Practice Exercise** When a 0.500 g sample of biodiesel fuel prepared from waste vegetable oil is burned in the bomb calorimeter from Sample Exercise 10.7, its temperature rises by 4.86°C. How much energy (in kilojoules) was released during combustion of the biodiesel sample?

## 10.6 Hess's Law and Standard Enthalpies of Reaction

As noted in the previous section, calorimetry can be used to determine the energy and enthalpy changes that accompany chemical reactions. However, there may be times when determining $\Delta H_{\text{rxn}}$ directly is not possible. For example, $CO_2$ is the principal product of the combustion of carbon in the form of charcoal:

$$(1) \qquad C(s) + O_2(g) \rightarrow CO_2(g) \qquad \Delta H_1$$

When the oxygen supply is limited, however, the products may include carbon monoxide:

$$(2) \qquad 2\,C(s) + O_2(g) \rightarrow 2\,CO(g) \qquad \Delta H_2$$

It is difficult to directly determine the enthalpy change that accompanies reaction (2) because as long as any oxygen is present, some of the CO formed may combine with $O_2$ to form $CO_2$, yielding a mixture of CO and $CO_2$. However, we can calculate $\Delta H$ for reaction (2) *indirectly* by starting with $\Delta H_{rxn}$ values that we *can* determine. For example, we can determine the enthalpy changes that accompany both the reaction in equation (1) and the combustion of a sample of pure CO gas:

$$(3) \qquad 2\,CO(g) + O_2(g) \rightarrow 2\,CO_2(g) \qquad \Delta H_3$$

Look closely at the reactants and products of chemical equations (1), (2), and (3). Equation (1) represents the complete combustion of carbon to $CO_2$, whereas equations (2) and (3) represent the stepwise combustion of carbon: first to CO and then to $CO_2$. As a result, the enthalpy change that accompanies the overall reaction (1) is related to the sum of the enthalpy changes associated with the stepwise reactions (2) and (3). Why? Because, as noted in Section 10.3, enthalpy change is a *state* function. This means that the $\Delta H_{rxn}$ value of an overall reaction that may occur in two or more steps, such as reaction (1), is the sum of the $\Delta H_{rxn}$ values of those steps. This holds true because the various steps consume the same reactants and eventually form the same products as the overall reaction.

**CHEMTOUR**
State Functions and Path Functions

Combining enthalpies of reaction in this way is in accordance with **Hess's law**, also known as *Hess's law of constant heat of summation,* which states that the change in enthalpy that accompanies a process that occurs in more than one step is the sum of the enthalpy changes that occur in each of those steps. Let's put Hess's law to work by deriving an expression for enthalpy change that accompanies the incomplete combustion of C to CO as described in reaction (2) by using the measurable $\Delta H_{rxn}$ values of reactions (1) and (3).

**CHEMTOUR**
Hess's Law

Combining both the equations that describe chemical reactions and their $\Delta H_{rxn}$ values is an exercise in pattern recognition. The key is to look for the reactants and products of the reaction whose $\Delta H_{rxn}$ we want to find in chemical equations describing the reactions whose $\Delta H_{rxn}$ values we know. In this example, we need to combine the chemical equations describing reactions (1) and (3) in a way that gives us the equation for reaction (2). Carbon and $O_2$ are reactants in equation (2), and CO is the only product. Inspecting the other two equations, we find that CO is a reactant in equation (3). Because CO is on the product side of equation (2), we flip equation (3) so that the reaction is written in reverse. Recall from Section 10.3 that when an exothermic process runs in reverse, it becomes an endothermic process, and vice versa. Therefore, reversing a reaction *changes the sign* of its $\Delta H$ value. Applying this principle to the reaction in equation (3), we get

$$(4) \qquad 2\,CO_2(g) \rightarrow 2\,CO(g) + O_2(g) \qquad \Delta H_4 = -\Delta H_3$$

Flipping equation (3) puts $O_2$ on the product side of equation (4), which is not where we want it. $O_2$ is on the reactant side of equation (1), so combining equations (1) and (4) may result in canceling out $O_2$ from the product side. However, we can't combine equations (1) and (4) just yet because there are two molecules of $CO_2$ on the reactant side of equation (4) but only one on the product side of equation (1). We'd prefer that these values be the same so that they cancel out when we combine equations (1) and (4) because there are no $CO_2$ terms at all in equation (2). To get two molecules of $CO_2$ on the product side of equation (1), we multiply all the terms in the equation, *including* $\Delta H_1$, by 2:

**Hess's law** the principle that the enthalpy of reaction ($\Delta H_{rxn}$) for a process that is the sum of two or more reactions is equal to the sum of the $\Delta H_{rxn}$ values of the constituent reactions; also called *Hess's law of constant heat of summation.*

$$(5) \qquad 2\,C(s) + 2\,O_2(g) \rightarrow 2\,CO_2(g) \qquad \Delta H_5 = 2\Delta H_1$$

Now we combine equations (4) and (5), *which includes summing their* $\Delta H_{rxn}$ *values:*

$$\begin{array}{lll} (4) & 2\ CO_2(g) \rightarrow 2\ CO(g) + O_2(g) & \Delta H_4 = -\Delta H_3 \\ +\ (5) & 2\ C(s) + 2\ O_2(g) \rightarrow 2\ CO_2(g) & \Delta H_5 = 2\Delta H_1 \\ \hline \end{array}$$

$$\cancel{2\ CO_2(g)} + 2\ C(s) + 2\ O_2(g) \rightarrow 2\ CO(g) + \cancel{O_2(g)} + \cancel{2\ CO_2(g)}$$

or

$$(2) \qquad 2\ C(s) + O_2(g) \rightarrow 2\ CO(g) \qquad \Delta H_2 = 2\Delta H_1 - \Delta H_3$$

Why did we multiply the enthalpy change that accompanies reaction (1) by 2 when we multiplied the coefficients of each of the reactants and product by 2? After all, why should changing the way we write the chemical equation impact the enthalpy change that accompanies the formation of the same product from the same reactants? Recall that the coefficients in chemical equations can represent the numbers of moles of reactants and products (as well as the numbers of atoms and molecules). Thus, reaction (1) describes the incomplete combustion of *one mole* of carbon, but reaction (5) describes the incomplete combustion of *two moles* of carbon. Burning twice as much fuel should generate twice as much heat—and be accompanied by twice the change in enthalpy.

## Standard Enthalpy of Reaction ($\Delta H^\circ_{rxn}$)

Now let's put some values on the enthalpy changes that accompany chemical reactions. Our focus will be the reactions that make up one industrial process for producing hydrogen gas (see Sample Exercise 7.14). In the first step of hydrogen production, methane reacts with a limited supply of high-temperature steam to produce carbon monoxide and hydrogen gas:

$$CH_4(g) + H_2O(g) \rightarrow CO(g) + 3\ H_2(g)$$

**CONNECTION** We introduced the conversion factor 1 atm = 1.01325 bar in Table 9.2.

This reaction is called the steam–methane reforming reaction and is quite endothermic, requiring a flow of thermal energy into the reaction mixture to turn reactants into products. To describe *how* endothermic the reaction is, we use a thermodynamic property called the **standard enthalpy of reaction ($\Delta H^\circ_{rxn}$)**, which is often called the *standard heat of reaction*. The adjective *standard*, indicated by the symbol °, describes the enthalpy change that accompanies a reaction under **standard conditions**, which means gases at a constant pressure of 1 bar (which is approximately the same as 1 atm) and solutions at a concentration of 1 *M*. There is no universal standard temperature, though many tables of thermodynamic data, including those in Appendix 4 of this book, apply to processes occurring at 25°C.

**standard enthalpy of reaction ($\Delta H^\circ_{rxn}$)** the enthalpy change associated with a reaction that takes place under standard conditions; also called the *standard heat of reaction*.

**standard conditions** in thermodynamics: a pressure of 1 bar (~1 atm) and some specified temperature, assumed to be 25°C unless otherwise stated; for solutions, a concentration of 1 *M* is specified.

**standard state** the most stable form of a substance under a pressure of 1 bar (~1 atm) and some specified temperature (usually 25°C).

Implied in our notion of standard conditions is the assumption that parameters such as $\Delta H$ change with temperature and pressure. That assumption is correct, as we saw with enthalpies of vaporization in Section 10.3, but the changes are so small that we ignore them in the calculations in this chapter and those that follow. We also use the term **standard state** to describe the most stable physical state of a substance under standard conditions. In their standard states at $P$ = 1 bar and $T$ = 25°C, oxygen is a gas, water is a liquid, and carbon is solid graphite. Additional examples of standard states include that most metals and metalloids are solids, mercury and bromine are liquids, and $H_2$, $N_2$, $F_2$, $Cl_2$, and the group 18 elements are gases.

Returning to the steam–methane reforming reaction and including its standard enthalpy of reaction value yields the following thermochemical equation:

$$(1)\quad CH_4(g) + H_2O(g) \rightarrow CO(g) + 3\,H_2(g) \qquad \Delta H_1^\circ = 206\ \text{kJ}$$

In the second step in hydrogen production, CO from the first step reacts with more steam to produce $CO_2$ and more $H_2$ gas in a reaction called the *water–gas shift reaction*:

$$(2)\quad CO(g) + H_2O(g) \rightarrow CO_2(g) + H_2(g) \qquad \Delta H_2^\circ = -41\ \text{kJ}$$

In accordance with Hess's law, we can write an overall thermochemical equation for the process by adding reactions (1) and (2) and their $\Delta H^\circ$ values:

$$\begin{array}{lll} (1) & CH_4(g) + H_2O(g) \rightarrow \cancel{CO(g)} + 3\,H_2(g) & \Delta H_1^\circ = 206\ \text{kJ} \\ +\,(2) & \cancel{CO(g)} + H_2O(g) \rightarrow CO_2(g) + H_2(g) & \Delta H_2^\circ = -41\ \text{kJ} \\ \hline (3) & CH_4(g) + 2\,H_2O(g) \rightarrow CO_2(g) + 4\,H_2(g) & \Delta H_3^\circ = 165\ \text{kJ} \end{array}$$

This result is illustrated graphically in **Figure 10.18**.

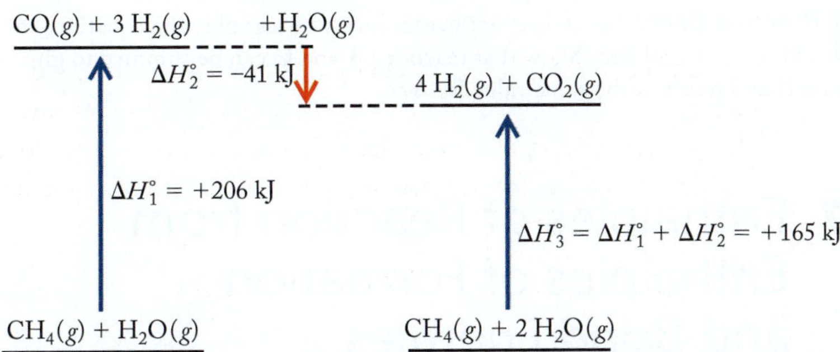


**FIGURE 10.18** Hess's law predicts that the enthalpy change ($\Delta H_3^\circ$) for the reaction in which 1 mole of $CH_4$ and 2 moles of $H_2O$ vapor produce 4 moles of $H_2$ and 1 mole of $CO_2$ is the sum of the enthalpy changes that accompany each of the two steps in the overall reaction: (1) 1 mole of $CH_4$ reacts with 1 mole of $H_2O$ vapor to produce 1 mole of CO and 3 moles of $H_2$ ($\Delta H_1^\circ$), and (2) the reaction of the CO produced in the first step with another mole of $H_2O$ vapor to form a fourth mole of $H_2$ and 1 mole of $CO_2$ ($\Delta H_2^\circ$).

### SAMPLE EXERCISE 10.8 Calculating $\Delta H^\circ_{rxn}$ by Using Hess's Law — LO5

One reason furnaces and water heaters fueled by natural gas need to be vented is that incomplete combustion can produce toxic carbon monoxide:

$$\text{Equation A:}\quad 2\,CH_4(g) + 3\,O_2(g) \rightarrow 2\,CO(g) + 4\,H_2O(g) \qquad \Delta H_A^\circ = ?$$

Use thermochemical equations B and C to calculate $\Delta H_A^\circ$:

$$\text{Equation B:}\quad CH_4(g) + 2\,O_2(g) \rightarrow CO_2(g) + 2\,H_2O(g) \qquad \Delta H_B^\circ = -802\ \text{kJ}$$

$$\text{Equation C:}\quad 2\,CO(g) + O_2(g) \rightarrow 2\,CO_2(g) \qquad \Delta H_C^\circ = -566\ \text{kJ}$$

**Collect and Organize** We are given two equations (B and C) with thermochemical data and a third (A) for which we are asked to find $\Delta H^\circ$. All the reactants and products in equation A are present in B and/or C.

**Analyze** We can manipulate equations B and C algebraically so that they sum to give the equation for which $\Delta H^\circ$ is unknown. Then we can calculate the unknown value by applying Hess's law. Equations A and B both contain methane as a reactant, so we will use B in the direction written. CO is a product in A but a reactant in C, so we have to reverse C to get CO on the product side. Reversing C means that we must also change the sign of $\Delta H_C^\circ$. If the coefficients in B and the reverse of C do not allow us to sum the two equations to obtain equation A, we will need to multiply one or both by appropriate factors.

**Solve** We start with equation B as written and add the reverse of equation C. We must also remember to change the sign of $\Delta H_C$ for the latter:

| | | |
|---|---|---|
| (B) | $CH_4(g) + 2\,O_2(g) \rightarrow CO_2(g) + 2\,H_2O(g)$ | $\Delta H^\circ_B = -802$ kJ |
| (C, reversed) | $2\,CO_2(g) \rightarrow 2\,CO(g) + O_2(g)$ | $-\Delta H^\circ_C = +566$ kJ |

However, these reactions do not sum to equation A because the coefficient of $CH_4$ is 2 in equation A but only 1 in equation B. To fix this, we need to multiply all the terms in B by 2, including $\Delta H^\circ_B$:

| | | |
|---|---|---|
| (2B) | $2\,CH_4(g) + 4\,O_2(g) \rightarrow 2\,CO_2(g) + 4\,H_2O(g)$ | $2\,\Delta H^\circ_B = -1604$ kJ |

When we sum C (reversed) and 2B, the $CO_2$ terms cancel out and we obtain equation A:

| | | |
|---|---|---|
| (C, reversed) | $\cancel{2\,CO_2(g)} \rightarrow 2\,CO(g) + \cancel{O_2(g)}$ | $-\Delta H^\circ_C = +566$ kJ |
| + (2B) | $2\,CH_4(g) + \overset{3}{\cancel{4}}\,O_2(g) \rightarrow \cancel{2\,CO_2(g)} + 4\,H_2O(g)$ | $2\,\Delta H^\circ_B = -1604$ kJ |
| (A) | $2\,CH_4(g) + 3\,O_2(g) \rightarrow 2\,CO(g) + 4\,H_2O(g)$ | $\Delta H^\circ_A = -1038$ kJ |

**Think About It** Our calculation shows that incomplete combustion of two moles of methane is less exothermic ($\Delta H^\circ_A = -1038$ kJ) than their complete combustion ($2\,\Delta H^\circ_B = -1604$ kJ), which makes sense because the CO produced in incomplete combustion reacts exothermically with more $O_2$ to form $CO_2$. In fact, the value of $\Delta H^\circ_C$ for the reaction $2\,CO(g) + O_2(g) \rightarrow 2\,CO_2(g)$ is the difference between $-1604$ kJ and $-1038$ kJ.

**Practice Exercise** It does not matter how you assemble the equations in a Hess's law problem. Show that reactions A and C can be summed to give reaction B and result in the same value for $\Delta H^\circ_B$.

## 10.7 Enthalpies of Reaction from Enthalpies of Formation and Bond Energies

As noted in Section 10.1, it is extremely difficult to measure the *absolute* value of the internal energy of a substance, and the same is true for the enthalpy of a substance. However, we can establish *relative* enthalpy values that are referenced to a convenient standard: the **standard enthalpy of formation ($\Delta H^\circ_f$)** for a substance, which is defined as the enthalpy change that takes place at a constant pressure of 1 bar when one mole of a substance is formed from its constituent elements in their standard states. A reaction that fits this description is known as a **formation reaction.**

The standard enthalpy of formation of any pure element in its standard state is, by definition, zero. This is the zero point of all other enthalpy values (like using the freezing point of water as the zero point on the Celsius temperature scale). Standard enthalpy of formation values for several compounds are given in **Table 10.4**, and a more complete list can be found in Appendix 4. Because the definition of a formation reaction specifies one mole of product, writing balanced equations for formation reactions may require the use of something we have typically tried to avoid until now: fractional coefficients in balanced equations. For example, the balanced thermochemical equation describing the formation of nitrogen monoxide is usually written

$$N_2(g) + O_2(g) \rightarrow 2\,NO(g) \qquad \Delta H^\circ_{rxn} = 180.6\ \text{kJ}$$

Although all reactants in the equation are in their standard states, it is not a formation reaction because two moles of product are formed. Therefore, $\Delta H^\circ_{rxn}$ for

**standard enthalpy of formation ($\Delta H^\circ_f$)** the enthalpy change of a formation reaction; also called the *standard heat of formation.*

**formation reaction** a reaction in which one mole of a substance is formed from its component elements in their standard states.

**TABLE 10.4** Standard Enthalpies of Formation of Selected Substances at 25°C

| Substance | $\Delta H_f^\circ$ (kJ/mol) | Substance | $\Delta H_f^\circ$ (kJ/mol) |
|---|---|---|---|
| $Br_2(\ell)$ | 0 | $O_2(g)$ | 0 |
| $H_2(g)$ | 0 | $N_2(g)$ | 0 |
| $H_2O(g)$ | −241.8 | $NO(g)$ | 90.3 |
| $H_2O(\ell)$ | −285.8 | $NH_3(g)$, ammonia | −46.1 |
| C(*s*, graphite) | 0 | $N_2H_4(g)$, hydrazine | 95.35 |
| $CH_4(g)$, methane | −74.8 | $N_2H_4(\ell)$ | 50.63 |
| $C_2H_2(g)$, acetylene | 226.7 | $CO(g)$ | −110.5 |
| $C_2H_4(g)$, ethylene | 52.4 | $CO_2(g)$ | −393.5 |
| $C_2H_6(g)$, ethane | −84.67 | $CH_3OH(\ell)$, methanol | −238.7 |
| $C_3H_8(g)$, propane | −103.8 | $CH_3CH_2OH(\ell)$, ethanol | −277.7 |
| $C_4H_{10}(g)$, butane | −125.6 | $CH_3COOH(\ell)$, acetic acid | −484.5 |

this reaction does not equal $\Delta H_f^\circ$. In fact, $\Delta H_{rxn}^\circ$ is twice $\Delta H_f^\circ$ because the equation as written describes the formation of *two* moles of NO. To write an equation describing the formation of one mole, we divide each coefficient (and $\Delta H_{rxn}^\circ$) in the preceding equation by 2:

$$\tfrac{1}{2} N_2(g) + \tfrac{1}{2} O_2(g) \rightarrow NO(g) \qquad \Delta H_f^\circ = 90.3 \text{ kJ}$$

**SAMPLE EXERCISE 10.9** Recognizing Formation Reactions **LO5**

Which of the following $\Delta H_{rxn}^\circ$ values are $\Delta H_f^\circ$ values, if we assume that each reaction takes place at 25°C? For those that are not formation reactions, explain why not.

a. $H_2(g) + \tfrac{1}{2} O_2(g) \rightarrow H_2O(g)$ $\quad \Delta H_{rxn}^\circ = -241.8$ kJ
b. $C(s, \text{graphite}) + 2\, H_2(g) + \tfrac{1}{2} O_2(g) \rightarrow CH_3OH(\ell)$ $\quad \Delta H_{rxn}^\circ = -238.7$ kJ
c. $CH_4(g) + 2\, O_2(g) \rightarrow CO_2(g) + 2\, H_2O(g)$ $\quad \Delta H_{rxn}^\circ = -802.3$ kJ
d. $P_4(s, \text{white}) + 6\, Cl_2(g) \rightarrow 4\, PCl_3(\ell)$ $\quad \Delta H_{rxn}^\circ = -1278$ kJ

**Collect, Organize, and Analyze** We are given four balanced thermochemical equations and are asked to determine which of the $\Delta H_{rxn}^\circ$ values are also $\Delta H_f^\circ$ values. The standard enthalpy of formation of a substance is the enthalpy of a reaction in which one mole of the substance is formed from its constituent elements, each in their standard state.

**Solve**

a. One mole of water vapor is formed from its constituent elements in their standard states. Therefore, this is the formation reaction for $H_2O(g)$, and its enthalpy of reaction is the $\Delta H_f^\circ$ of water vapor.
b. One mole of liquid methanol is formed from its constituent elements in their standard states. Therefore, the equation describes a formation reaction, and the enthalpy of reaction is $\Delta H_f^\circ$.
c. The reactants are not elements in their standard states and more than one mole of product is formed, so the enthalpy of reaction is not $\Delta H_f^\circ$.
d. According to Table A4.3 in Appendix 4, the standard form of phosphorus is white phosphorus, which has the molecular formula $P_4$, so both reactants are elements in their standard states. However, the reaction produces *four* moles of $PCl_3$, so the enthalpy of reaction is not the $\Delta H_f^\circ$ of $PCl_3$. Actually, it is four times the value of $\Delta H_f^\circ$.

**Think About It** Just because we can write formation reactions for substances like methanol does not mean that anyone would ever use that reaction to make methanol. Formation reactions are defined to provide a standard so that the flows of thermal energy in reactions can be evaluated.

 **Practice Exercise** Write formation reactions for (a) $CaCO_3(s)$; (b) $CH_3COOH(\ell)$ (acetic acid); (c) $KMnO_4(s)$.

Standard enthalpies of formation can be used to predict standard enthalpies of reaction. To do so, we break down each reaction into a series of formation reactions for the reactants and products. Then we apply Hess's law to combine the $\Delta H^\circ_f$ values of the reactants and products into an overall $\Delta H^\circ_{rxn}$ value. Let's use the combustion of methane to illustrate how this is done. As we saw in Chapter 7, the overall combustion reaction can be written

$$CH_4(g) + 2\,O_2(g) \rightarrow CO_2(g) + 2\,H_2O(g) \qquad (7.8)$$

The formation reactions for three of the four reactants and products ($O_2$ is a pure element in its standard state, so its $\Delta H^\circ_f = 0$) are the following:

$$\text{(A)} \quad C(s,\text{ graphite}) + 2\,H_2(g) \rightarrow CH_4(g) \qquad \Delta H^\circ_f = -74.8\text{ kJ}$$

$$\text{(B)} \quad C(s,\text{ graphite}) + O_2(g) \rightarrow CO_2(g) \qquad \Delta H^\circ_f = -393.5\text{ kJ}$$

$$\text{(C)} \quad H_2(g) + \tfrac{1}{2}\,O_2(g) \rightarrow H_2O(g) \qquad \Delta H^\circ_f = -241.8\text{ kJ}$$

To combine equations A, B, and C to end up with Equation 7.8, we need to first reverse equation A to put $CH_4$ on the reactant side. We also need to multiply equation C by 2 because we need a coefficient of 2 in front of $H_2O$ in Equation 7.8, and also because we need a coefficient of 2 in front of $H_2$ to cancel out the $H_2$ term in the reverse of equation A. Making these changes and summing the three equations that result:

$$\text{(A, reversed)} \quad CH_4(g) \rightarrow C(s,\text{ graphite}) + 2\,H_2(g) \qquad \Delta H^\circ_f = 74.8\text{ kJ}$$

$$+\text{(B)} \quad C(s,\text{ graphite}) + O_2(g) \rightarrow CO_2(g) \qquad \Delta H^\circ_f = -393.5\text{ kJ}$$

$$+\text{(2C)} \quad 2\,H_2(g) + O_2(g) \rightarrow 2\,H_2O(g) \qquad \Delta H^\circ_f = -483.6\text{ kJ}$$

$$CH_4(g) + \cancel{C(s,\text{ graphite})} + 2\,O_2(g) + \cancel{2\,H_2(g)} \rightarrow \cancel{C(s,\text{ graphite})} + \cancel{2\,H_2(g)} + CO_2(g) + 2\,H_2O(g)$$

or

$$CH_4(g) + 2\,O_2(g) \rightarrow CO_2(g) + 2\,H_2O(g) \qquad \Delta H^\circ_{rxn} = -802.3\text{ kJ}$$

where $\Delta H^\circ_{rxn} = (74.8 - 393.5 - 483.6)\text{ kJ} = -802.3\text{ kJ}$.

Note how the reactant in the overall reaction is separated into its component elements in the first step in the preceding sequence. In the second and third steps, the elements formed in the first step recombine to form the products. Thus, this sequence of steps describes the enthalpy changes that take place as the elemental building blocks of the reactant become the building blocks (with oxygen) of the products. We can create an analogous sequence for any chemical reaction by taking the following approach: sum the $\Delta H^\circ_f$ values of the products, each multiplied by its coefficient in a balanced chemical equation describing the reaction, and then subtract the sum of the $\Delta H^\circ_f$ values of the reactants, each multiplied by its coefficient. Expressing this sequence in equation form:

$$\Delta H^\circ_{rxn} = \sum n_{\text{products}}\,\Delta H^\circ_{f,\text{products}} - \sum n_{\text{reactants}}\,\Delta H^\circ_{f,\text{reactants}} \qquad (10.17)$$

where $n_{products}$ is the number of moles of each product in the balanced equation and $n_{reactants}$ is the number of moles of each reactant. Applying Equation 10.17 to the combustion of methane:

$$\begin{aligned}\Delta H^\circ_{rxn} &= [(1 \text{ mol } CO_2)(-393.5 \text{ kJ/mol}) + (2 \text{ mol } H_2O)(-241.8 \text{ kJ/mol})] \\ &\quad - [(1 \text{ mol } CH_4)(-74.8 \text{ kJ/mol}) + (2 \text{ mol } O_2)(0.0 \text{ kJ/mol})] \\ &= [(-393.5 \text{ kJ}) + (-483.6 \text{ kJ})] - [(-74.8 \text{ kJ}) + (0.0 \text{ kJ})] \\ &= -802.3 \text{ kJ}\end{aligned}$$

**CONCEPT TEST**

Explain why the $\Delta H^\circ_f$ value for $CO_2(g)$ is the same as the $\Delta H^\circ_{comb}$ value for C(*s*, graphite).

**SAMPLE EXERCISE 10.10** Calculating Standard Enthalpies of Reaction from Standard Enthalpies of Formation **LO5**

Use the appropriate standard enthalpy of formation values to calculate $\Delta H^\circ_{rxn}$ for the complete combustion of propane:

$$C_3H_8(g) + 5\,O_2(g) \rightarrow 3\,CO_2(g) + 4\,H_2O(g)$$

**Collect and Organize** The standard enthalpies of formation of all the reactants and products are listed in Table 10.4. Our task is to use the balanced chemical equation and the $\Delta H^\circ_f$ data from Table 10.4 to calculate the enthalpy of combustion.

**Analyze** Equation 10.17 defines the relation between (1) standard enthalpies of formation of reactants and products and (2) the standard enthalpy of a reaction. Because $O_2$ gas is a pure element in its standard state, its $\Delta H^\circ_f$ value is zero. We expect the reaction to be very exothermic ($\Delta H^\circ_{rxn} = 0$) because it represents combustion.

**Solve** Inserting $\Delta H^\circ_f$ values for the products and reactants and their coefficients from the balanced chemical equation into Equation 10.17:

$$\begin{aligned}\Delta H^\circ_{rxn} &= \left[(3 \text{ mol } CO_2)\left(-393.5\,\frac{\text{kJ}}{\text{mol}}\right) + (4 \text{ mol } H_2O)\left(-241.8\,\frac{\text{kJ}}{\text{mol}}\right)\right] \\ &\quad - \left[(1 \text{ mol } C_3H_8)\left(-103.8\,\frac{\text{kJ}}{\text{mol}}\right) + (5 \text{ mol } O_2)\left(0.0\,\frac{\text{kJ}}{\text{mol}}\right)\right] \\ &= -2043.9 \text{ kJ}\end{aligned}$$

**Think About It** The large negative value is expected for the combustion of a hydrocarbon fuel. Propane is used in many backyard grills. Remember that the enthalpy value calculated in this problem corresponds only to the specific quantities of reactants and products described in the balanced chemical equation.

**Practice Exercise** Use standard enthalpy of reaction values to calculate $\Delta H^\circ_{rxn}$ for the water–gas shift reaction:

$$CO(g) + H_2O(g) \rightarrow CO_2(g) + H_2(g)$$

## Enthalpies of Reaction and Bond Energies

The energy changes associated with chemical reactions depend on how much energy is required to break the bonds in the reactants and how much is released as the bonds in products form. For example, in the methane combustion reaction

(Equation 7.8), the C—H bonds in $CH_4$ and the O═O bonds in $O_2$ must be broken before the C═O bonds in $CO_2$ and the O—H bonds in $H_2O$ can form. Breaking bonds is endothermic (the "Energy in" arrow in **Figure 10.19**), and forming bonds is exothermic (the "Energy out" arrows). If a chemical reaction is exothermic, as methane combustion is, more energy is released in forming the bonds in molecules of products than is consumed in breaking the bonds in molecules of reactants. If a chemical reaction is endothermic, however, more energy is consumed in breaking bonds in the reactants than is released in forming the bonds in the products.

Bond energy (or bond strength) is the enthalpy change ($\Delta H$) that occurs when one mole of bonds in the gas phase is broken. The quantity of energy needed to break a particular bond is equal in magnitude but opposite in sign to the quantity of energy released when that same bond forms. In other words, breaking a bond requires an investment of energy and is a highly endothermic process ($\Delta H > 0$), whereas energy is released when atoms come together to make a covalent bond (and a more stable molecular structure). Thus, bond formation is a highly exothermic process ($\Delta H < 0$).

**CONNECTION** We first encountered bond energy in Chapter 4. Bond energies for some common covalent bonds are listed in Table 4.6, and a more complete list is found in Appendix 4.

During the combustion of one mole of $CH_4$ (Figure 10.19), we assume that four moles of C—H bonds and two moles of O═O bonds are broken. The formation of one mole of $CO_2$ and two moles of $H_2O$ requires the formation of two moles of C═O bonds and four moles of O—H bonds. The net change in energy resulting from breaking bonds in the reactants and forming bonds in the products can be estimated from the average bond energies. We start by taking an inventory of the bond energies involved:

| | Bond | Number of Bonds (mol) | Bond Energy (kJ/mol) | $\Delta H$ |
|---|---|---|---|---|
| BONDS BROKEN | C—H | 4 | 413 | 4 mol × 413 kJ/mol |
| | O═O | 2 | 498 | 2 mol × 498 kJ/mol |
| BONDS FORMED | O—H | 4 | 463 | −(4 mol × 463 kJ/mol) |
| | C═O | 2 | 799 | −(2 mol × 799 kJ/mol) |

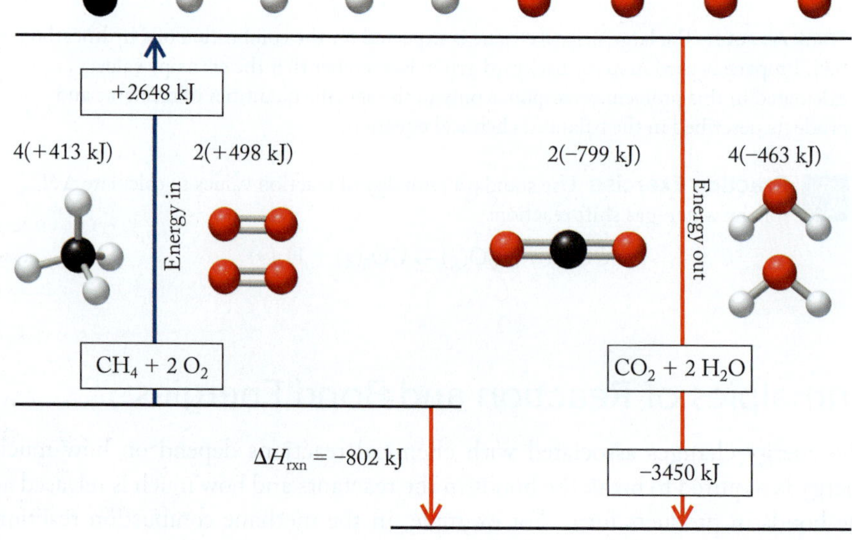


**FIGURE 10.19** The complete combustion of 1 mole of methane requires that 4 moles of C—H bonds and 2 moles of O═O bonds be broken. Breaking bonds requires energy and is accompanied by an increase in enthalpy. Formation of 2 moles of C═O bonds and 4 moles of O—H bonds is accompanied by an even greater decrease in enthalpy, so the overall reaction is accompanied by a decrease in enthalpy and is exothermic.

Next we sum the positive $\Delta H$ values associated with breaking bonds and the negative $\Delta H$ values associated with forming them to obtain an estimate of the enthalpy change of the overall reaction:

$$\begin{aligned}\Delta H_{rxn} &= (4\ \cancel{mol} \times 413\ kJ/\cancel{mol}) \\ &\quad + (2\ \cancel{mol} \times 498\ kJ/\cancel{mol}) \\ &\quad - (4\ \cancel{mol} \times 463\ kJ/\cancel{mol}) \\ &\quad - (2\ \cancel{mol} \times 799\ kJ/\cancel{mol}) \\ &= -802\ kJ\end{aligned}$$

This estimate is essentially the same as the $\Delta H^\circ_{rxn}$ value, −802.3 kJ, which we derived from the standard enthalpies of formation of the reactants and products. We need to keep in mind that the energy of a particular type of bond, such as a C—H bond, can vary from one molecule to another or even within a molecule. However, bond energies are usually close enough to their average values that the averages can be used to calculate reliable estimates of the $\Delta H_{rxn}$ values of reactions that occur *in the gas phase*, where the impact of intermolecular forces on the energies of reactants and products is negligible. The procedure we follow in estimating $\Delta H_{rxn}$ values for gas-phase reactions is expressed in equation form as follows:

$$\Delta H_{rxn} = \sum \Delta H_{\text{bond breaking}} - \sum \Delta H_{\text{bond forming}} \tag{10.18}$$

Here the negative sign is a reminder that we combine positive $\Delta H$ values (bond energies) for bonds that are broken with *negative* $\Delta H$ values (negative bond energies) for bonds that are formed.

**SAMPLE EXERCISE 10.11** Estimating $\Delta H_{rxn}$ Values of Gas-Phase Reactions from Average Bond Energies **LO6**

Use average bond energies to estimate $\Delta H_{rxn}$ for the industrial synthesis of ammonia:

$$N_2(g) + 3\ H_2(g) \rightarrow 2\ NH_3(g)$$

**Collect and Organize** We need to estimate the value of $\Delta H_{rxn}$ of a gas-phase reaction from the bond energies of its reactants and products. Table A4.1 in Appendix 4 lists average values for the energies of covalent bonds. Equation 10.18 relates the average bond energies of the reactants and products to the enthalpy of reaction.

**Analyze** First, we need to determine which types of bonds hold together molecules of $N_2$, $H_2$, and $NH_3$. This will require drawing their Lewis structures, as we learned to do in Chapter 4. Once we have the structures, we can take an inventory of the number of each type of bond that is broken or formed, look up the average bond energies for the bonds, and combine them using Equation 10.18. The reaction involves the formation of two moles of a compound from its component elements in their standard states. Therefore, the result of this calculation should be close to two times the standard enthalpy of formation ($\Delta H^\circ_f$) of $NH_3$, which is −46.1 kJ/mol.

**Solve** Let's use Lewis structures of the reactants and products to count all the bonds involved:

:N≡N: + 3 H—H → 2 H—N̈—H (with a third H bonded below N)

We then construct a table, using the appropriate bond energies from Appendix 4:

| | Bond | Number of Bonds (mol) | Bond Energy (kJ/mol) | $\Delta H$ |
|---|---|---|---|---|
| BONDS BROKEN | N≡N | 1 | 945 | 1 mol × 945 kJ/mol |
| | H—H | 3 | 436 | 3 mol × 436 kJ/mol |
| BONDS FORMED | N—H | 6 | 391 | −(6 mol × 391 kJ/mol) |

Summing the values in the right-hand column to estimate $\Delta H_{rxn}$:

$$\Delta H_{rxn} = (1 \text{ mol} \times 945 \text{ kJ/mol}) + (3 \text{ mol} \times 436 \text{ kJ/mol}) - (6 \text{ mol} \times 391 \text{ kJ/mol})$$
$$= -93 \text{ kJ}$$

**Think About It** An enthalpy change of −93 kJ means that the reaction is exothermic. The value for $\Delta H_f^\circ$ of $NH_3$ (see Table 10.4 or Appendix 4) is −46.1 kJ/mol. Two moles of $NH_3$ are produced in the reaction, and multiplying the $\Delta H_f^\circ$ value by 2 gives us −92.2 kJ, which is very close to our calculated value. Remember that the enthalpies of reaction calculated using bond energies are estimates because they use *average* bond energies. They generally vary from actual experimentally measured values.

**Practice Exercise** Use average bond energies to estimate $\Delta H_{rxn}$ for the water–gas shift reaction, which is part of the industrial process for making hydrogen gas:

$$CO(g) + H_2O(g) \rightarrow CO_2(g) + H_2(g)$$

## 10.8 Energy Changes When Substances Dissolve

**FIGURE 10.20** Sand and salt (NaCl) help keep streets and highways clear of snow and ice during the winter.

In many cities and towns, winter brings subfreezing temperatures and snow—sometimes a lot of it. To keep walkways clear after plowing or shoveling, pellets of calcium chloride are often spread to melt packed snow and ice and to prevent new ice from forming. Meanwhile, road crews apply mixtures of sand and sodium chloride to keep streets and highways navigable (**Figure 10.20**).

In Chapter 11, we will explore why $CaCl_2$ and NaCl effectively melt ice, but for now, we explain why $CaCl_2$, which costs much more than NaCl, is so widely used to keep sidewalks free of ice. There are several reasons, but we focus here on a thermochemical one: when NaCl dissolves in melting ice, there is little change in the temperature of the ice or the meltwater, but when $CaCl_2$ dissolves, the water heats up, which helps melt more ice. Why is dissolving $CaCl_2$ a decidedly exothermic process, but dissolving NaCl is slightly endothermic (**Figure 10.21**)? To understand why, we need to analyze, step by step, how salts dissolve in water and to track the changes in enthalpy that accompany each of these steps.

**enthalpy of solution ($\Delta H_{solution}$)** the overall change in enthalpy that occurs when a solute is dissolved in a solvent; also called the *heat of solution*.

In Section 10.7, we used covalent bond energies to estimate enthalpies of reaction. In this section, we consider energies associated with interactions between the ions in salts and molecules of water. As we have discussed in prior chapters, ionic compounds are held together by ionic bonds that have bond energies of many hundreds of kilojoules per mole. Molecules of $H_2O$ interact with each other principally through hydrogen bonding. When an ionic solid dissolves in water, the ionic bonds between its ions break. Similarly, many hydrogen bonds break as molecules of $H_2O$ cluster around the dissolved ions. These endothermic processes

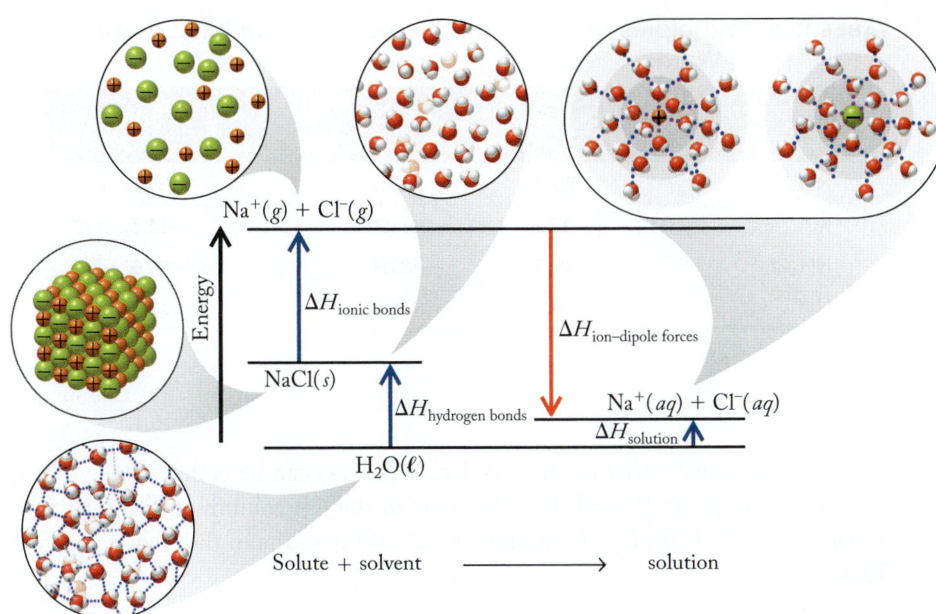


**FIGURE 10.21** The enthalpy of solution ($\Delta H_{\text{solution}}$) of NaCl is the sum of the enthalpy changes that accompany breaking the hydrogen bonds between molecules of water, separating the $Na^+$ and $Cl^-$ ions in solid NaCl, and forming ion–dipole interactions in solution.

require investments of energy, which we can calculate by summing the enthalpy changes that accompany each process:

$$\text{Thermal energy invested} = \Delta H_{\text{ionic bonds}} + \Delta H_{\text{hydrogen bonds}}$$

**CONNECTION** In Section 6.4, we discussed the intermolecular forces involved in determining solubility.

When ionic compounds dissolve in water, their ions interact with the permanent dipoles of the water molecules to form spheres of hydration around the ions. We use the symbol $\Delta H_{\text{ion–dipole}}$ to represent the change in enthalpy that accompanies these interactions. It is the exothermic part of the overall change in enthalpy, called the **enthalpy of solution ($\Delta H_{\text{solution}}$)**, or the *heat of solution*, that accompanies the dissolution process (see Figure 10.21):

$$\Delta H_{\text{solution}} = \underbrace{[\Delta H_{\text{ionic bonds}} + \Delta H_{\text{hydrogen bonds}}]}_{\text{Endothermic}} + \underbrace{\Delta H_{\text{ion-dipole forces}}}_{\text{Exothermic}} \quad (10.19)$$

Keep in mind that the energy required to break the ionic bonds in one mole of an ionic compound is equal in magnitude but opposite in sign to the lattice energy ($U$) of the compound:

$$\Delta H_{\text{ionic bonds}} = -U \quad (10.20)$$

We can determine the value of $\Delta H_{\text{solution}}$ experimentally by using calorimetric methods such as the one shown in **Figure 10.22.** The enthalpy of solution can be

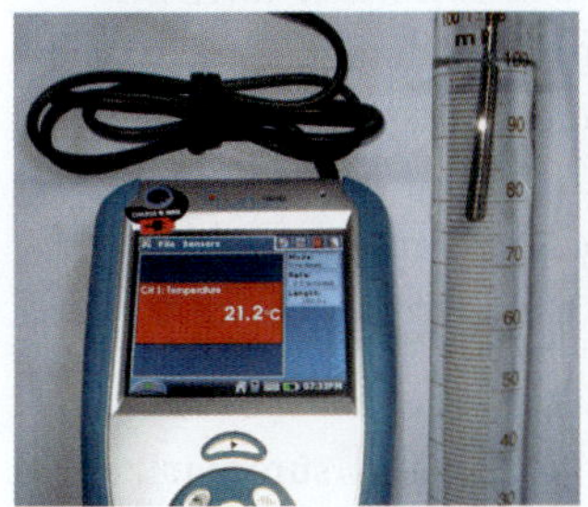


Initial water temperature = 21.2°C

Adding $CaCl_2$ to water

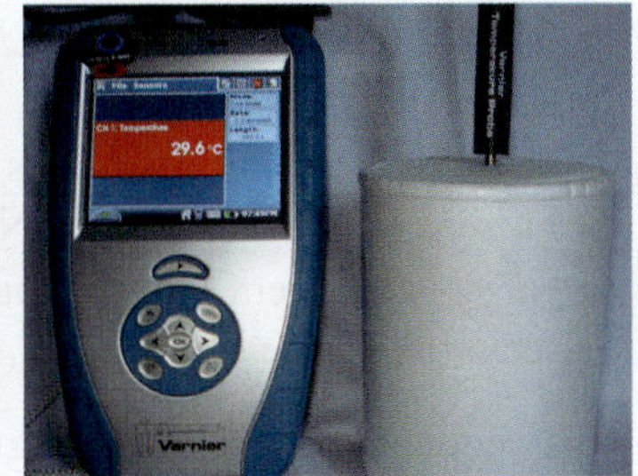


Final temperature of $CaCl_2$ solution = 29.6°C

**FIGURE 10.22** Determining the enthalpy of solution. The temperature of 100 mL of water increases from 21.2°C to 29.6°C when 5.0 g of $CaCl_2$ dissolves in it. This temperature change and the heat capacity of water can be used to calculate the enthalpy of solution of $CaCl_2$.

**TABLE 10.5** Enthalpies of Solution of Some Common Ionic Compounds in Water

| Endothermic Compound | $\Delta H_{solution}$ (kJ/mol) | Exothermic Compound | $\Delta H_{solution}$ (kJ/mol) |
|---|---|---|---|
| $NH_4NO_3$ | 25.7 | LiCl | −37.1 |
| KCl | 17.2 | NaOH | −44.5 |
| $NH_4Cl$ | 15.2 | KOH | −57.6 |
| NaCl | 4.0 | $CaCl_2$ | −82.2 |

**CONNECTION** Lattice energy is the energy released when gas-phase ions come together to form one mole of an ionic compound (see Section 4.1). This quantity of energy is consumed in separating a mole of the compound into individual gas-phase ions.

positive or negative—that is, the dissolution process can be endothermic, as it is for NaCl (Figure 10.21) and the other salts in the left column of **Table 10.5**, or exothermic, as it is for $CaCl_2$ (Figure 10.22) and the salts in the right column of Table 10.5.

## Calculating Lattice Energies by Using the Born–Haber Cycle

**STEPWISE ANIMATION**
Born–Haber Cycle

Lattice energies are difficult to determine experimentally, but the lattice energy of a binary ionic compound can be calculated from its standard enthalpy of formation and a judicious application of Hess's law (Section 10.7). To illustrate how this works, let's calculate the lattice energy of NaCl. We start with its standard enthalpy of formation:

$$\text{Na}(s) + \tfrac{1}{2}\,\text{Cl}_2(g) \rightarrow \text{NaCl}(s) \qquad \Delta H^\circ_f = -411 \text{ kJ}$$

The value of $\Delta H^\circ_f$ can be determined experimentally with a calorimeter containing known quantities of elemental Na and $Cl_2$. Think of the formation reaction as consisting of five steps, as shown in **Figure 10.23** and described as follows:

1. Subliming one mole of Na metal, producing one mole of gas-phase Na atoms. The enthalpy change, $\Delta H_{\text{sub,Na}}$, that accompanies this step is the enthalpy of sublimation of sodium. Sublimation is an endothermic process (arrow points up in Figure 10.23).
2. Breaking the covalent bonds in 0.5 mole of $Cl_2$ molecules, producing one mole of gas-phase Cl atoms. The enthalpy change here is half the bond energy (BE) of one mole of Cl–Cl bonds, or $\tfrac{1}{2}\,\text{BE}_{\text{Cl}_2}$ (also endothermic).
3. Ionizing one mole of Na atoms to one mole of $Na^+$ ions in the gas phase, which requires adding a quantity of thermal energy equal to the first ionization energy of sodium, $\text{IE}_{1,\text{Na}}$ (also endothermic).
4. Combining one mole of Cl atoms with one mole of electrons to form one mole of $Cl^-$ ions, which is accompanied by an enthalpy change equal to the electron affinity of chlorine, $\text{EA}_{\text{Cl}}$ (an exothermic process for Cl and many other elements; arrow points down in Figure 10.23).
5. Forming one mole of solid NaCl from one mole each of gas-phase $Na^+$ and $Cl^-$ ions. The enthalpy change that accompanies this step is the quantity we seek: the lattice energy ($U_{\text{NaCl}}$) of NaCl.

This sequence of steps is diagrammed in Figure 10.23 and summarized in **Table 10.6**. It is called the **Born–Haber cycle**. The enthalpy changes that

**Born–Haber cycle** a series of steps with corresponding enthalpy changes that describes the formation of an ionic solid from its constituent elements.

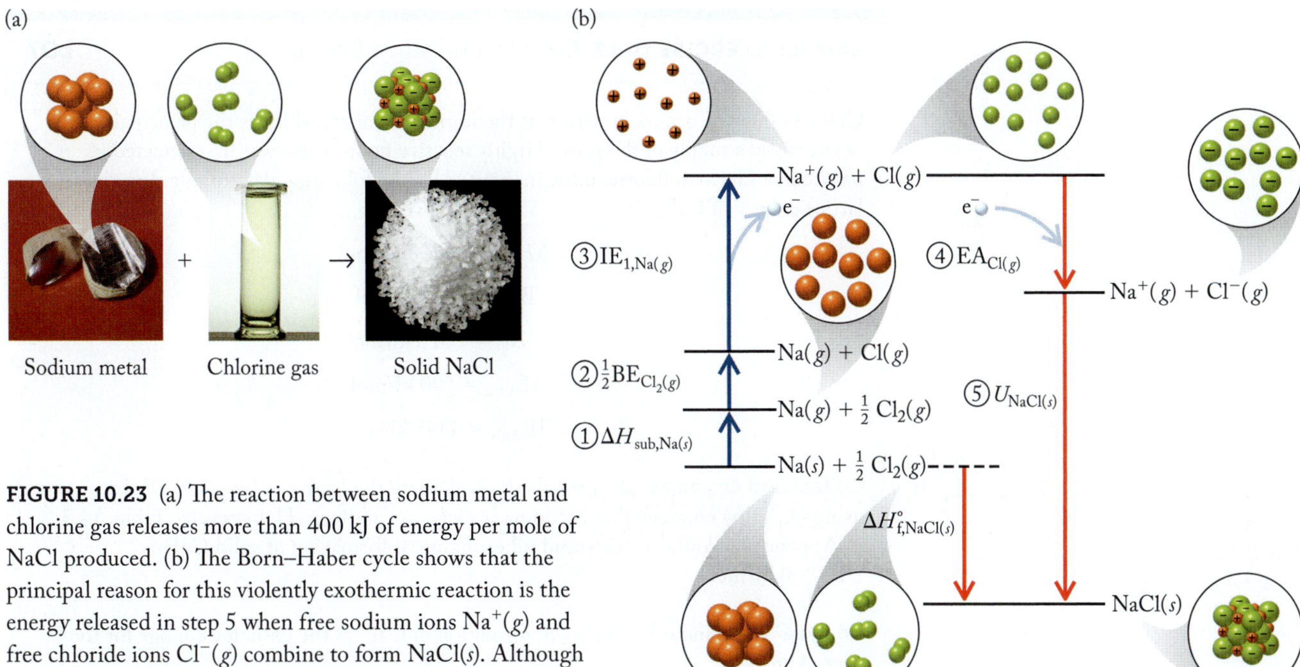


**FIGURE 10.23** (a) The reaction between sodium metal and chlorine gas releases more than 400 kJ of energy per mole of NaCl produced. (b) The Born–Haber cycle shows that the principal reason for this violently exothermic reaction is the energy released in step 5 when free sodium ions $Na^+(g)$ and free chloride ions $Cl^-(g)$ combine to form NaCl(*s*). Although the particulate representations indicate whether each element or compound exists as atoms, molecules, or ions, they are not intended to represent the exact number of particles involved.

accompany the five steps add up to the standard enthalpy of formation of NaCl, which is −411 kJ/mol:

$$\Delta H^\circ_{f,NaCl} = \Delta H_{sub,Na} + \tfrac{1}{2}\,BE_{Cl_2} + IE_{1,Na} + EA_{Cl} + U_{NaCl}$$

We insert the values from Table 10.6 and solve for $U$:

$$-411\text{ kJ} = (108\text{ kJ}) + \tfrac{1}{2}(243\text{ kJ}) + (495\text{ kJ}) + (-349\text{ kJ}) + U_{NaCl}$$

$$U_{NaCl} = (-411\text{ kJ}) - (108\text{ kJ}) - (121.5\text{ kJ}) - (495\text{ kJ}) - (-349\text{ kJ}) = -786\text{ kJ}$$

The Born–Haber cycle can also be used to calculate enthalpy changes in other steps in the cycle that are difficult to determine experimentally, such as electron affinities. To do so, though, we need to know the enthalpy changes of all the other steps and the value of $\Delta H^\circ_f$.

**CONNECTION** First ionization energy and electron affinity values are given in Figures 3.36 and 3.37, respectively.

**TABLE 10.6** Born–Haber Cycle for the Formation of NaCl(*s*)

| Step | Description | Chemical Equation | Enthalpy Change (kJ) |
|---|---|---|---|
| 1 | Sublime 1 mol Na(*s*) | $Na(s) \rightarrow Na(g)$ | $\Delta H_{sub,Na} = 108$ |
| 2 | Break $\frac{1}{2}$ mol of Cl—Cl bonds | $\frac{1}{2}\,Cl_2(g) \rightarrow Cl(g)$ | $\frac{1}{2}\,BE_{Cl_2} = \frac{1}{2}(243) = 121.5$ |
| 3 | Ionize 1 mol of Na(*g*) atoms to form 1 mol of $Na^+$ ions | $Na(g) \rightarrow Na^+(g) + e^-$ | $IE_{1,Na} = 495$ |
| 4 | 1 mol of Cl atoms acquires 1 mol of electrons to form 1 mol of $Cl^-$ ions | $Cl(g) + e^- \rightarrow Cl^-(g)$ | $EA_{Cl} = -349$ |
| 5 | 1 mol of $Na^+$ ions combine with 1 mol of $Cl^-$ ions to form 1 mol of NaCl(*s*) | $Na^+(g) + Cl^-(g) \rightarrow NaCl(s)$ | $U_{NaCl} = ?$ |

**SAMPLE EXERCISE 10.12** Calculating Lattice Energy **LO7**

Calcium fluoride occurs in nature as the mineral fluorite, which is the principal source of the world's supply of fluorine. (Highly reactive fluorine gas was first detected in nature in a German fluorite mine in 2012.) Use the following data to calculate the lattice energy of $CaF_2$.

$$\Delta H_{sub,Ca} = 168 \text{ kJ/mol}$$
$$BE_{F_2} = 155 \text{ kJ/mol}$$
$$EA_F = -328 \text{ kJ/mol}$$
$$IE_{1,Ca} = 590 \text{ kJ/mol}$$
$$IE_{2,Ca} = 1145 \text{ kJ/mol}$$

**Collect and Organize** We are asked to calculate the lattice energy of $CaF_2$ by using enthalpy changes that accompany steps in the Born–Haber cycle. Table A4.3 in Appendix 4 contains the standard enthalpy of formation of solid $CaF_2$: −1228.0 kJ/mol.

**Analyze** The standard enthalpy of formation of $CaF_2$ is the enthalpy change for the overall process:

$$Ca(s) + F_2(g) \rightarrow CaF_2(s) \qquad \Delta H_f^\circ = -1228 \text{ kJ/mol}$$

When we break down this reaction into the steps of the Born–Haber cycle, we need to include both the first and second ionization energies of Ca because its atoms lose two electrons when forming $Ca^{2+}$ ions. Two moles of fluorine atoms are needed to react with one mole of calcium atoms, so we need to break the F—F bonds in one mole of $F_2$, and we need to multiply the electron affinity of F by 2 to calculate the enthalpy change accompanying the formation of *two* moles of $F^-$ ions. **Figure 10.24** summarizes the Born–Haber cycle for calculating the lattice energy of $CaF_2$.

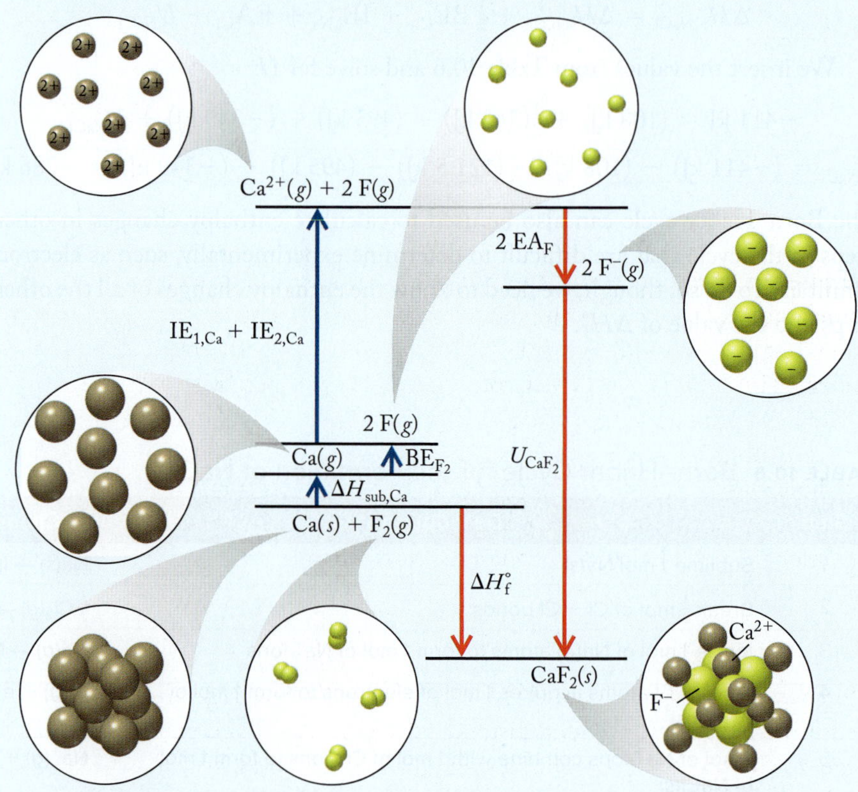


**FIGURE 10.24** Born–Haber cycle for the formation of $CaF_2$.

To estimate what a reasonable lattice energy value is, refer to Table 4.2, which lists the lattice energy of $MgCl_2$, another halide of an alkaline earth metal: $U_{MgCl_2} = -2540$ kJ/mol. The charges on the ions in $CaF_2$ and $MgCl_2$ are the same. Even though $Ca^{2+}$ ions are larger than $Mg^{2+}$ ions, $F^-$ ions are smaller than $Cl^-$ ions, so the overall differences in ionic radii may offset, which would lead to comparable *d* values in coulombic attraction expressions (Equation 4.1) for the two compounds. Therefore, a $U_{CaF_2}$ value near −2540 kJ/mol would be reasonable.

**Solve** The Born–Haber cycle for forming $CaF_2$ is

$$\Delta H_f^\circ = \Delta H_{sub,Ca} + BE_{F_2} + IE_{1,Ca} + IE_{2,Ca} + 2\ EA_F + U_{CaF_2}$$

Substituting in the values:

$$-1228\ \text{kJ} = 168\ \text{kJ} + 155\ \text{kJ} + 590\ \text{kJ} + 1145\ \text{kJ} + 2(-328\ \text{kJ}) + U_{CaF_2}$$

Solving for the lattice energy:

$$U_{CaF_2} = -2630\ \text{kJ/mol}\ CaF_2$$

**Think About It** The lattice energies of $CaF_2$ and $MgCl_2$ are similar in strength, just as we estimated.

 **Practice Exercise** Burning magnesium metal in air produces MgO and a very bright white light, making the reaction popular in fireworks and signaling devices:

$$Mg(s) + \tfrac{1}{2}O_2(g) \rightarrow MgO(s) \qquad \Delta H_f^\circ = -602\ \text{kJ}$$

Calculate the lattice energy of MgO from the following endothermic changes in enthalpy:

| Process | Enthalpy Change (kJ/mol) | Process | Enthalpy Change (kJ/mol) |
|---|---|---|---|
| $Mg(s) \rightarrow Mg(g)$ | 148 | $Mg(g) \rightarrow Mg^{2+}(g) + 2\ e^-$ | 2188 |
| $O_2(g) \rightarrow 2\ O(g)$ | 498 | $O(g) + 2\ e^- \rightarrow O^{2-}(g)$ | 605 |

Because the lattice energy of an ionic compound must be overcome if the compound is to dissolve, ionic compounds with very large lattice energies often have limited solubility in water. This result is consistent with the solubility rules described in Chapter 8—namely, ionic compounds formed by the group 1 cations (all 1+) and ammonium ions (also 1+) are all soluble in water, as are all nitrates (1−), acetates (1−), and perchlorates (1−). We would expect compounds containing these ions to require less energy to separate into their component ions because of their small charges, which result in relatively small values in the numerator of Equation 4.1, which describes coulombic attraction:

$$E_{el} = 2.31 \times 10^{-19}\ \text{J} \cdot \text{nm}\left(\frac{Q_1 \times Q_2}{d}\right) \qquad (4.1)$$

Moreover, $NH_4^+$, $NO_3^-$, and $CH_3COO^-$ have relatively large ionic radii, which lead directly to relatively long distances between anion and cation centers and large *d* values in the denominator of Equation 4.1. Thus, NaF ($U = -910$ kJ/mol) is more soluble in water than MgO ($U = -3792$ kJ/mol) due to the greater

**TABLE 10.7 Enthalpies of Solution of Some Common Molecular Compounds in Water**

| Compound | $\Delta H_{solution}$ (kJ/mol) |
|---|---|
| HCl(*g*) | −74.8 |
| $NH_3$(*g*) | −30.5 |
| $CH_3CH_2OH$(ℓ) | −10.6 |
| $CH_3OH$(ℓ) | −3.0 |
| $CH_3COOH$(ℓ) | −1.5 |

attraction of the 2+ and 2− charges of the ions in MgO, as compared to the 1+ and 1− charges of the ions of comparable size in NaF. The solubility of NaF in water is 12 g/100 mL at 25°C, whereas the solubility of MgO is only 0.000062 g/100 mL.

## Molecular Solutes

Until now, our focus has been on enthalpy changes that accompany the dissolution of ionic solutes in water. However, many polar molecular compounds, especially those that can form hydrogen bonds with water molecules, are also soluble in water. In Chapter 6, we learned that low-molar-mass alcohols such as methanol ($CH_3OH$), ethanol ($CH_3CH_2OH$), and ethylene glycol ($HOCH_2CH_2OH$) are *miscible* with water—that is, they dissolve in water, and water dissolves in them, in all proportions.

The enthalpies of solution of several molecular compounds are listed in **Table 10.7**. Like all $\Delta H_{solution}$ values, the ones in Table 10.7 represent the net sum of the energy investments needed to overcome solvent–solvent and solute–solute interactions and the energy released when particles of solute and solvent interact. Equation 10.21 presents this relationship for the case of methanol dissolving in water.

$$\Delta H_{solution} = \Delta H_{H_2O-H_2O} + \Delta H_{CH_3OH-CH_3OH} + \Delta H_{CH_3OH-H_2O} \qquad (10.21)$$

As we have seen for ionic solutes, the value of $\Delta H_{solution}$ can be determined using calorimetric methods, but independently determining the three terms on the right side of Equation 10.21 is difficult. When enthalpies of solution have values near zero, as in the case of methanol, at least we know that the energies required to disrupt solvent–solvent and solute–solute interactions are nearly balanced by the energy released when new interactions form between molecules of solute and solvent. In the case of methanol, there are negligible differences between the hydrogen bonds between water molecules that are broken, the hydrogen bonds between methanol molecules that are broken, and the hydrogen bonds that form between molecules of water and methanol.

The negative $\Delta H_{solution}$ values in Table 10.7 mean that the solute–solvent intermolecular interactions are stronger than the solute–solute and solvent–solvent intermolecular interactions. However, not all molecular solutes have exothermic enthalpies of solution in water. Among those that don't are alcohols with four or more carbon atoms per molecule. Stronger London dispersion forces between the nonpolar hydrocarbon portions of these molecules lead to stronger solute–solute interactions and contribute to endothermic enthalpies of solution. The increasing nonpolar (hydrophobic) nature of these alcohols also causes them to be less soluble in water.

**CONNECTION** The lower solubility in water of alcohols with greater molecular masses (and more nonpolar —$CH_2$— groups per molecule) was discussed in Section 6.4.

## 10.9 More Applications of Thermochemistry

**fuel value** the quantity of energy released during the complete combustion of 1 g of a substance.

**fuel density** the quantity of energy released during the complete combustion of a particular volume of a liquid fuel.

In Section 10.7, we calculated the standard enthalpies of reaction for the combustion of one mole of methane (−802.3 kJ) and one mole of propane (−2043.9 kJ). Does the much more negative (exothermic) value for propane make it an inherently better (higher-energy) fuel? Not necessarily. Expressing $\Delta H^\circ_{rxn}$ values on a per-mole basis is the only way to ensure that we are talking about the same number of molecules. However, we do not purchase fuels, or anything else for that matter, in units of moles. Depending on the fuel, we buy it either by mass (coal by the ton) or by volume (gasoline by the liter or the gallon).

To better compare methane and propane as fuels, let's calculate the enthalpy change that takes place when 1 g of each burns in air to produce $CO_2$ and water vapor. We divide the value of $\Delta H^\circ_{rxn}$ (in kilojoules per mole) for each reaction by the molar mass of the hydrocarbon to determine the number of kilojoules of energy released per gram of substance:

$$CH_4: \quad \frac{-802.3 \text{ kJ}}{\cancel{\text{mol}}} \times \frac{1 \cancel{\text{mol}}}{16.04 \text{ g}} = -50.02 \text{ kJ/g}$$

$$C_3H_8: \quad \frac{-2043.9 \text{ kJ}}{\cancel{\text{mol}}} \times \frac{1 \cancel{\text{mol}}}{44.10 \text{ g}} = -46.35 \text{ kJ/g}$$

From the perspective of the surroundings that absorb this energy, such as the air in a gas furnace, the water in a water heater, or the food on a stove or backyard grill, equal masses of the two fuels provide comparable quantities of energy. Energies like these, calculated on a per-gram basis, are called **fuel values**.

Fuel values of the $C_1$ to $C_{10}$ alkanes are compared in **Figure 10.25**. Note how the fuel values *decrease* as the carbon numbers (and molar masses) *increase*. What is different about these compounds (besides the sizes of their molecules) that could explain this trend? One difference is the hydrogen-to-carbon ratios. As the number of carbon atoms per molecule increases, the hydrogen-to-carbon ratio decreases: from four atoms of H per atom of C in methane to about two atoms of H per atom of C in higher-molar-mass alkanes. To understand why this ratio is important, remember that it takes only 2.0 g of hydrogen atoms to make one mole of water vapor and release 241.8 kJ of thermal energy. However, it takes 12.0 g of carbon to form one mole of $CO_2$ and release 393.5 kJ. Thus, it takes six times as much carbon (by mass) to produce only about 60% more thermal energy than hydrogen. This is why pure hydrogen has the highest fuel value of all fuels.

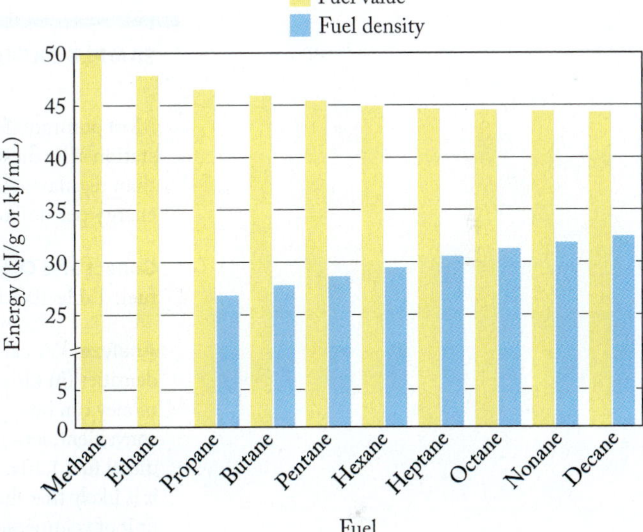


**FIGURE 10.25** Fuel values and fuel densities of the $C_1$ to $C_{10}$ alkanes according to their complete combustion to $CO_2$ and $H_2O$ vapor.

Fuel values of the major fossil and renewable fuels are listed in **Table 10.8**. These data are based on combustion reactions that produce $CO_2$ and *liquid* $H_2O$, and they are somewhat higher than those plotted in Figure 10.25 for the same compounds (propane, for example) on the basis of their combustion to $CO_2$ and $H_2O$ *vapor*. Why are their fuel values different? Because it takes energy (water's enthalpy of vaporization) to vaporize a mole of liquid water. Therefore, liquid water is a lower-energy product than water vapor, and a combustion process that produces liquid water, such as the combustion of methane (**Figure 10.26**), is accompanied by a greater decrease in enthalpy than if water vapor had been produced.

Another term used to compare the energy content of liquid fuels is **fuel density**. Fuel density is the quantity of energy released per unit volume of a liquid fuel. Many liquid fuels are priced based on volume (gallons in the United States, liters in the rest of the world), so fuel density can be more useful than fuel value in comparing them as sources of thermal energy, as we do in Sample Exercise 10.13.

**TABLE 10.8** Fuel Values of Some Common Fuels

| Fuel | Fuel Value[a] (kJ/g) |
|---|---|
| Hydrogen | 141.8 |
| Natural gas | 54.0 |
| Propane | 50.3 |
| Butane | 49.5 |
| Gasoline | 47.4 |
| Diesel | 44.8 |
| Ethanol | 29.8 |
| Coal (anthracite) | 27 |
| Wood | 17 |

[a]Based on complete combustion to $CO_2$ and liquid $H_2O$.

## CONCEPT TEST

The data in Figure 10.25 show that the fuel densities of the $C_3$–$C_{10}$ alkanes increase with increasing carbon number (and molar mass) even though their fuel values decrease. Why?

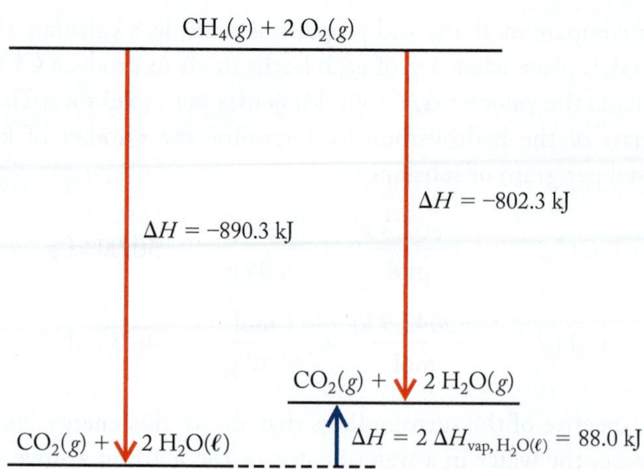


**FIGURE 10.26** Methane's enthalpies of combustion have different values depending on whether water vapor or liquid water is produced. The difference is due to the enthalpy of vaporization of water.

**SAMPLE EXERCISE 10.13** Comparing Fuel Values and Fuel Densities **LO8**

Most automobiles run on either gasoline or diesel fuel. Suppose a particular service station sells diesel fuel (density = 0.83 g/mL) for exactly 10% more per unit volume than regular-grade gasoline (density = 0.75 g/mL). Which fuel contains more thermal energy per dollar?

**Collect and Organize** We know the densities and relative prices of gasoline and diesel fuel. Table 10.8 lists their fuel values: 47.4 kJ/g for gasoline and 44.8 kJ/g for diesel fuel.

**Analyze** We can use the given density values to convert fuel values (in kJ/g) into fuel densities (in kJ/mL) and then compare the relative volumes that the same amount of money can buy. We are not given the individual prices of the fuels, so to make the units convenient, let's make $x$ the price of 1 mL of gasoline. That makes the price of 1 mL of diesel fuel $1.10x$. Gasoline has a slightly higher fuel value but significantly lower density, so it is likely that diesel fuel will have the greater fuel density. However, diesel costs more per unit of volume, so the difference in energy content per $x$ dollars (or cents) may be small.

**Solve**

$$\text{Gasoline:}\quad \frac{47.4\ \text{kJ}}{\text{g}} \times \frac{0.75\ \text{g}}{\text{mL}} \times \frac{1\ \text{mL}}{x} = 36\ \text{kJ}/x$$

$$\text{Diesel fuel:}\quad \frac{44.8\ \text{kJ}}{\text{g}} \times \frac{0.83\ \text{g}}{\text{mL}} \times \frac{1\ \text{mL}}{1.10\ x} = 34\ \text{kJ}/x$$

Therefore, gasoline provides slightly more energy content (about 6%) for the same amount of money $x$.

**Think About It** Comparable energy content for the money is only part of the energy story in selecting fuels (or buying cars). Another part is the efficiencies with which internal combustion (gasoline) engines and diesel engines convert the chemical energy in their fuels to mechanical work. On this score, diesel engines are considerably better, which translates into much greater distances traveled on the same fuel budget. Fuels with different fuel values may also produce different amounts of carbon dioxide, further complicating the optimum choice of fuel.

**Practice Exercise** In November 2011, an Atlas V rocket launched NASA's Mars Science Laboratory, including the Mars rover *Curiosity*, into space. The first-stage engine of the Atlas V (**Figure 10.27**) is fueled by kerosene, a hydrocarbon mixture with a density of 0.79 g/mL and a fuel value of 46 kJ/g. At full throttle, the Atlas V engine generates 20 gigawatts ($2.0 \times 10^{10}$ J/s) of power. If the engine is 50% efficient at converting chemical energy into mechanical energy (the thrust lifting the rocket into space), how rapidly (in liters per second) must kerosene be supplied to and completely burned in the engine?

**FIGURE 10.27** Launch of the Mars Science Lab in 2011.

## Energy from Food

Food is the fuel of living systems. The overall biochemical processes that convert foods into energy resemble combustion reactions, though each process involves many more steps to convert the carbon and hydrogen content of foods, such as carbohydrates, into carbon dioxide and water. The energy content of most foods can be determined using the same instruments used to evaluate the energy released by other combustible substances. For example, we can burn dry grain or dehydrated fruit in bomb calorimeters to determine the energy they would provide if we ate them.

As an illustration, let's determine the fuel value of peanuts. Suppose our sample is a single peanut with a mass of 1.89 g. We put it in a calorimeter for which $C_{\text{calorimeter}} = 19.31$ kJ/°C and burn it completely in excess oxygen. The resulting temperature increase of the calorimeter is 2.334°C. We determine the fuel value of the peanut much as we would determine the enthalpy of combustion of any combustible material: by first using Equation 10.15 to calculate the energy required to raise the temperature of the calorimeter by 2.334°C:

$$\begin{aligned} q_{\text{calorimeter}} &= C_{\text{calorimeter}}\,\Delta T \\ &= (19.31\ \text{kJ/}\cancel{°\text{C}})(2.334\ \cancel{°\text{C}}) \\ &= 45.07\ \text{kJ} \end{aligned}$$

This quantity of thermal energy is generated by the combustion of 1.89 g, so the fuel value of the peanut is

$$\frac{45.07\ \text{kJ}}{1.89\ \text{g}} = 23.8\ \text{kJ/g}$$

The energy content of food is often expressed in Calories (Cal), where 1 Calorie = $10^3$ cal = 4.184 kJ. The fuel value of the peanut in Calories is

$$\frac{23.8\ \cancel{\text{kJ}}}{1\ \text{g}} \times \frac{1\ \text{Cal}}{4.184\ \cancel{\text{kJ}}} = 5.69\ \text{Cal/g}$$

**SAMPLE EXERCISE 10.14** Calculating the Fuel Value of Food **LO8**

Glucose ($C_6H_{12}O_6$) is a simple sugar formed by photosynthesis in plants. The complete combustion of 0.5763 g of glucose in a calorimeter ($C_{\text{calorimeter}} = 6.20$ kJ/°C) raises the temperature of the calorimeter by 1.45°C. What is the fuel value of glucose in Calories per gram?

**Collect and Organize** We are asked to determine the fuel value of glucose, which means the energy given off when 1 g is completely converted to $CO_2$ and liquid $H_2O$. We know the mass of the glucose sample that is completely combusted and the temperature change produced by the energy released during combustion in a calorimeter with a known calorimeter constant. Equation 10.15 relates these quantities.

**Analyze** The energy released during combustion of the sample and gained by the calorimeter is the product of the calorimeter constant and the change in temperature. We can convert this quantity into Calories by using the conversion factor 1 Cal = 4.184 kJ. We anticipate a fuel value similar to that of the peanut.

**Solve**

$$q_{\text{calorimeter}} = C_{\text{calorimeter}}\,\Delta T = (6.20\ \text{kJ/}\cancel{°\text{C}})(1.45\cancel{°\text{C}}) = 8.99\ \text{kJ}$$

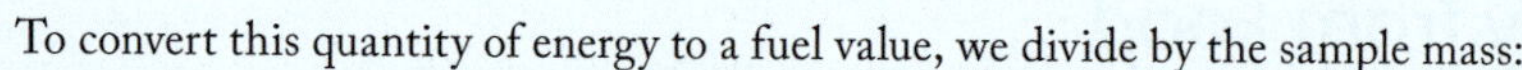

To convert this quantity of energy to a fuel value, we divide by the sample mass:

$$\frac{8.99 \text{ kJ}}{0.5763 \text{ g}} = 15.6 \text{ kJ/g}$$

and then convert into Calories:

$$(15.6 \text{ kJ/g})\left(\frac{1 \text{ Cal}}{4.184 \text{ kJ}}\right) = 3.73 \text{ Cal/g}$$

**Think About It** Food and nutrition scientists persist in using Calories to describe the energy content of foods even though the SI unit of energy, which is widely used in the physical sciences, is the joule. To add to the confusion, there are calories (lowercase "c") and there are Calories (uppercase "C"). They are not the same: 1 Cal = 1000 cal = 1 kcal. All three units are widely used in the health sciences and medicine.

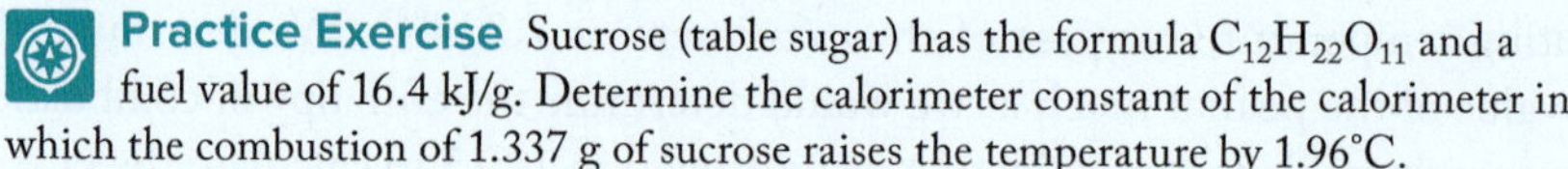

**Practice Exercise** Sucrose (table sugar) has the formula $C_{12}H_{22}O_{11}$ and a fuel value of 16.4 kJ/g. Determine the calorimeter constant of the calorimeter in which the combustion of 1.337 g of sucrose raises the temperature by 1.96°C.

**FIGURE 10.28** Charles M. Hall (top) and Paul Louis-Toussaint Héroult (middle) independently developed the same electrolytic process for producing aluminum metal from aluminum ore. Hall's sister Julia (bottom), also a chemistry major at Oberlin College, assisted her brother in the lab, and her business skills made their aluminum production company a financial success. The company became the Aluminum Company of America, shortened to Alcoa.

**CONCEPT TEST**

Combustion of one gram of glucose ($C_6H_{12}O_6$) produces less energy than combustion of one gram of sucrose ($C_{12}H_{22}O_{11}$). Suggest a reason why.

## Recycling Aluminum

Analyses of the energy required to carry out industrial procedures are frequently done to assess costs of operations and support new approaches to producing materials. The following evaluation of the energy requirements associated with the production and recycling of aluminum illustrates a practical application of thermochemical concepts.

Over the last century, aluminum, both alone and in combination with other metals, replaced steel for building structures in which high strength-to-weight ratios and corrosion resistance are paramount. These structures include major components of airplanes, motor vehicles, and the facades of buildings.

The industrial process for converting aluminum ore ($Al_2O_3$) into aluminum metal was developed by two 23-year-old chemists, Charles Hall and Paul Louis-Toussaint Héroult (both 1863–1914), working independently in the United States (Hall) and France (Héroult) (**Figure 10.28**). Their process is based on passing an electric current through a solution of aluminum oxide dissolved in molten cryolite ($Na_3AlF_6$). As electricity passes through the solution, $Al^{3+}$ ions are reduced to aluminum metal while a positively charged carbon electrode is oxidized to carbon dioxide. The process is described by the following reaction:

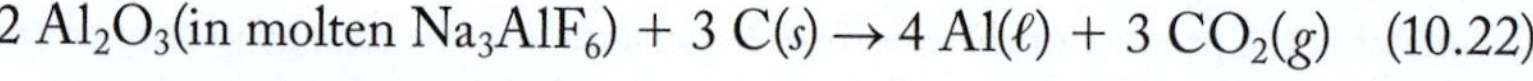

$$2\ Al_2O_3(\text{in molten } Na_3AlF_6) + 3\ C(s) \rightarrow 4\ Al(\ell) + 3\ CO_2(g) \quad (10.22)$$

The principal energy cost of the Hall–Héroult process is the electricity needed to reduce $Al^{3+}$ to Al. The major cost in recycling aluminum is the energy required to melt aluminum metal so that it can be reshaped into new products. We can use thermochemical principles to estimate the energy required for the Hall–Héroult process and compare it to the energy needed for recycling.

We can estimate the energy involved in producing aluminum metal from aluminum ore from the standard enthalpies of formation of the reactants and products in Equation 10.22:

$$\Delta H^\circ_{rxn} = [3(\Delta H^\circ_{f,CO_2} + 4(\Delta H^\circ_{f,Al(\ell)})] - [2(\Delta H^\circ_{f,Al_2O_3}) + 3(\Delta H^\circ_{f,C(s)})]$$

$$= \left[(3 \cancel{\text{mol CO}_2})\left(\frac{-393.5\text{ kJ}}{1 \cancel{\text{mol CO}_2}}\right) + (4 \cancel{\text{mol Al}})\left(\frac{10.79\text{ kJ}}{1 \cancel{\text{mol Al}}}\right)\right]$$

$$-\left[(2 \cancel{\text{mol Al}_2\text{O}_3})\left(\frac{-1675.7\text{ kJ}}{1 \cancel{\text{mol Al}_2\text{O}_3}}\right) + (3 \cancel{\text{mol C}})\left(\frac{0.0\text{ kJ}}{1 \cancel{\text{mol C}}}\right)\right]$$

$$= +2214.1\text{ kJ}$$

Dividing this value by the four moles of aluminum produced in the reaction as written, we get

$$\frac{2214.1\text{ kJ}}{4\text{ mol Al}} = 553.52\,\frac{\text{kJ}}{\text{mol Al}}$$

In this calculation we used a nonzero standard enthalpy of formation value for aluminum metal. This may seem odd because $\Delta H^\circ_f$ values of pure elements are normally zero. However, the zero value applies only to an element in its standard state, which for aluminum is solid, not the molten aluminum produced in the Hall–Héroult process. Therefore, we used the enthalpy of fusion of aluminum (10.79 kJ/mol) as the standard enthalpy of formation for liquid aluminum.

Our estimate of the energy requirements to reduce aluminum ore to aluminum metal did not include the energy needed to melt the cryolite in which the ore dissolves (its melting point is 1012°C), nor did we consider the inefficiency of using electricity to drive a chemical reaction. These and other factors make the energy requirements for producing one mole of Al from aluminum ore in a modern refinery closer to 1000 kJ.

We can estimate the energy required to recycle one mole of aluminum by calculating the energy required to raise its temperature from 25°C to its melting point (660°C) and then melt it. The molar heat capacity of Al is 24.2 J/(mol · °C), so the energy needed to warm it to its melting point is

$$q = nc_P\Delta T$$

$$= 1.00 \cancel{\text{mol}} \times \frac{24.2 \cancel{\text{J}}}{\cancel{\text{mol}} \cdot \cancel{^\circ\text{C}}} \times (660 - 25)\cancel{^\circ\text{C}} \times \frac{1\text{ kJ}}{1000 \cancel{\text{J}}}$$

$$= 15.4\text{ kJ}$$

Once aluminum reaches its melting point, the energy required to melt it ($\Delta H_{fus}$ = 10.79 kJ/mol) is

$$\frac{10.79\text{ kJ}}{\cancel{\text{mol}}} \times 1.00 \cancel{\text{mol}} = 10.8\text{ kJ}$$

The estimated total energy to heat and melt 1.00 mole of aluminum is then

$$15.4\text{ kJ} + 10.8\text{ kJ} = 26.2\text{ kJ}$$

This value represents only a small percentage of the energy needed to produce one mole of aluminum from its ore.

High energy costs make recycling aluminum economically attractive in addition to environmentally sound. The production and recycling of aluminum both incur energy costs, but detailed analyses of both processes have shown that recycling saves aluminum manufacturers about 95% of the energy required to produce the metal from ore. This energy saving has inspired the rapid growth of a global aluminum recycling industry. In the United States alone, aluminum recycling is a $1 billion-per-year business. In most industrialized countries, nearly all the aluminum in motor vehicles and building materials is recycled, as is about half the aluminum in food and beverage containers.

**SAMPLE EXERCISE 10.15** Integrating Concepts: Selecting a Heating System

Suppose some friends who happen to be avid skiers decide to pool their resources and buy a vacation home near their favorite ski resort. They find one that meets their needs but then discover that it needs a new furnace, though it does have an operational wood stove. They collect information about replacement furnace options, focusing on two fuels: home heating oil, which is derived from crude oil and has a hydrocarbon distribution similar to that of diesel fuel, and propane. Their research on the two fuels yields the results summarized in the following table:

| Fuel | Price of Fuel (\$/L) | Fuel Value (kJ/g) | Density (g/mL) | Furnace Efficiency |
|---|---|---|---|---|
| Heating oil | 0.72 | 44.9 | 0.83 | 83% |
| Propane | 0.75 | 50.3 | 0.62 | 95% |

a. According to these data, which of the two fuels is the better buy?
b. The brochure for the propane furnace they are considering describes it as a "condensing" furnace, noting that it is 10% more efficient than a noncondensing model. What does "condensing" mean in this context, and how much does it contribute to the higher efficiency of the furnace?
c. The most frugal of the friends suggests that they use the wood stove to save fuel costs. A local firewood dealer sells one cord of seasoned hardwood, which weighs about one metric ton, for $250. How much money would the friends save burning a cord of wood instead of the better fuel identified in part (a)? Assume the fuel value of the wood is 17 kJ/g and that the wood stove is 50% efficient.
d. Which of the fuels under consideration produces the most "fossil" $CO_2$ per kJ delivered to the home's living space?

**Collect and Organize** We know the price per liter of two fossil fuels and the price of wood, a renewable alternative. We also know the densities of the fossil fuels and the mass of a cord of wood, and we know the fuel values of all three fuels. We need to select which of the two fossil fuels is the better energy buy and then compare its cost-effectiveness with that of wood. The efficiencies with which stoves and furnaces convert the chemical energy of fuels into heat for the living space are also a factor, and we have to explain why a condensing furnace for burning propane is so efficient.

**Analyze** One way to compare the costs of the three fuels is to start with the fuel values and convert them from kJ/g into kJ/\$. The prices are based on volumes, so we will need to use the densities of the fuels as one of the conversion factors. Fuel values such as those given in Table 10.8 and in this problem are based on combustion reactions that convert the carbon and hydrogen content of the fuel into $CO_2$ and liquid $H_2O$. To compare fuel costs, we need to convert price-per-volume values into price per mass and then price per unit of energy produced. Then we use the efficiencies of the furnaces to calculate the price per unit of heat delivered to the vacation home's living space.

The $CO_2$ produced by the combustion of propane and heating oil is considered fossil $CO_2$ because the fuels are derived from natural gas and crude oil, making them fossil fuels. The fossil fuel with the higher carbon-to-hydrogen ratio will be the one that produces more fossil $CO_2$ per kilojoule of energy released.

**Solve**

a. Converting the fuel value of heating oil into kilojoules for the home:

$$\frac{44.9\text{ kJ}}{\text{g}} \times \frac{0.83\text{ g}}{\cancel{\text{mL}}} \times \frac{1000\ \cancel{\text{mL}}}{\cancel{\text{L}}} \times \frac{1\ \cancel{\text{L}}}{\$0.72} \times \frac{83\%}{100\%} = 4.3 \times 10^4\text{ kJ/\$}$$

Doing the same conversion for propane:

$$\frac{50.3\text{ kJ}}{\text{g}} \times \frac{0.62\text{ g}}{\cancel{\text{mL}}} \times \frac{1000\ \cancel{\text{mL}}}{\cancel{\text{L}}} \times \frac{1\ \cancel{\text{L}}}{\$0.75} \times \frac{95\%}{100\%} = 4.0 \times 10^4\text{ kJ/\$}$$

Therefore, heating oil provides more energy per dollar.

b. The higher efficiency of the *condensing* propane furnace is the result of extracting additional heat from the reaction mixture as it cools and the water vapor in it *condenses*, giving the following overall combustion reaction:

$$C_3H_8(g) + 5\,O_2(g) \rightarrow 3\,CO_2(g) + 4\,H_2O(\ell)$$

instead of the one from Sample Exercise 10.10:

$$C_3H_8(g) + 5\,O_2(g) \rightarrow 3\,CO_2(g) + 4\,H_2O(g) \qquad \Delta H^\circ_{rxn} = -2043.9\text{ kJ}$$

The difference between the $\Delta H^\circ_{rxn}$ values of these two reactions is four times the enthalpy of vaporization of one mole of water at 298 K:

$$H_2O(\ell) \rightarrow H_2O(g) \qquad \Delta H_{vap} = 44.0\text{ kJ}$$

According to Hess's law, we can reverse this vaporization equation and multiply it by 4 to match the four moles of $H_2O$ in the combustion reactions. We add it to the thermochemical equation from Sample Exercise 10.10 to get our desired equation:

$$C_3H_8(g) + 5\,O_2(g) \rightarrow 3\,CO_2(g) + \cancel{4\,H_2O(g)} \qquad \Delta H^\circ_{rxn} = -2043.9\text{ kJ}$$

$$+\cancel{4\,H_2O(g)} \rightarrow 4\,H_2O(\ell) \qquad \Delta H^\circ_{rxn} = -176.0\text{ kJ}$$

$$C_3H_8(g) + 5\,O_2(g) \rightarrow 3\,CO_2(g) + 4\,H_2O(\ell) \qquad \Delta H^\circ_{rxn} = -2219.9\text{ kJ}$$

The ratio of $\Delta H^\circ_{rxn}$ values is

$$\frac{-2219.9\text{ kJ}}{-2043.9\text{ kJ}} = 1.0861$$

This nearly 9% increase in $\Delta H^\circ_{rxn}$ accounts for most of the greater efficiency of the condensing propane furnace as compared to the noncondensing model.

c. Calculating the energy obtained by burning a cord of hardwood in the wood stove:

$$\frac{17\text{ kJ}}{\text{g}} \times \frac{1000\text{ g}}{\text{kg}} \times \frac{1000\text{ kg}}{\text{ton}} \times \frac{1\text{ ton}}{\text{cord}} \times 1\text{ cord} \times \frac{50\%}{100\%}$$

$$= 8.5 \times 10^6\text{ kJ}$$

The cost of the heating oil needed to produce this much heat is

$$8.5 \times 10^6\text{ kJ} \times \frac{\$1}{4.3 \times 10^4\text{ kJ}} = \$198 \approx \$200$$

This cost is below that of a cord of hardwood, so burning wood would achieve no cost savings over heating oil.

d. We know from its formula that the ratio of C to H atoms in propane is 3:8, but we are not given this information for the hydrocarbons in heating oil. However, we do know that heating oil has a lower fuel value than propane, which means that it is composed of hydrocarbons with higher molar masses that have higher carbon-to-hydrogen ratios than propane. Therefore, more of heating oil's fuel value comes from the conversion of C to $CO_2$, which means it produces more $CO_2$ per kilojoule than propane. On top of that, the heating oil furnace is less efficient than the propane furnace, so heating oil produces even more fossil $CO_2$ to heat the home than propane. Wood is a renewable biofuel and produces no fossil $CO_2$ during combustion.

**Think About It** These calculations of energy costs considered the price and fuel values of three fuels and the efficiencies with which their chemical energies could be used to heat a living space. All three factors are important, though there are others we did not consider, such as the cost of installing a new furnace and connecting it to the home's existing heating system. Environmental concerns regarding climate change might also lead the skiing friends to use the wood stove to heat the home on ski weekends.

## SUMMARY

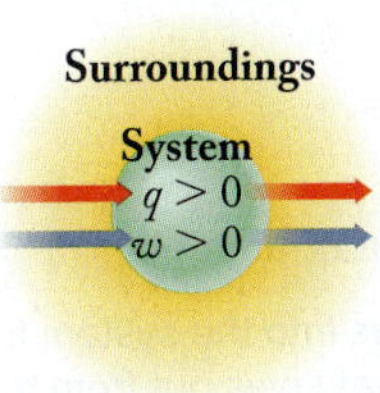


**LO1** **Thermodynamics** is the study of energy and its transformations, whereas **thermochemistry** is the study of the changes in energy that accompany chemical reactions. The **first law of thermodynamics** states that energy is neither created nor destroyed, so the energy gained or lost by a **system** equals the energy lost or gained by the **surroundings**. We use the symbol $q$ to represent the *quantity* of thermal energy transferred between a system and its surroundings. In an **exothermic process**, the system loses energy by heating its surroundings ($q < 0$); in an **endothermic process**, the system absorbs energy from its warmer surroundings ($q > 0$). (Sections 10.1 and 10.2)

**LO2** The sum of the kinetic and potential energies of a system is its **internal energy ($E$)**. The internal energy of a system is increased ($\Delta E = E_{final} - E_{initial}$ is positive) when it is heated ($q > 0$) or if work is done on it ($w > 0$). (Section 10.2)

**LO3** The **enthalpy ($H$)** of a system is given by $H = E + PV$. The **enthalpy change ($\Delta H$)** of a system is equal to the energy ($q_P$) added to or removed from the system at constant pressure: $\Delta H > 0$ for endothermic reactions and $\Delta H < 0$ for exothermic reactions. The **enthalpy of fusion ($\Delta H_{fus}$)** is the change in enthalpy when one mole of the solid substance melts at its melting point. The **enthalpy of vaporization ($\Delta H_{vap}$)** is the enthalpy change that occurs when one mole of the liquid substance is vaporized at its boiling point. Characteristic values like **specific heat ($c_s$)** and **molar heat capacity ($c_P$)** can be used to quantify thermodynamic changes in a system. (Sections 10.3 and 10.4)

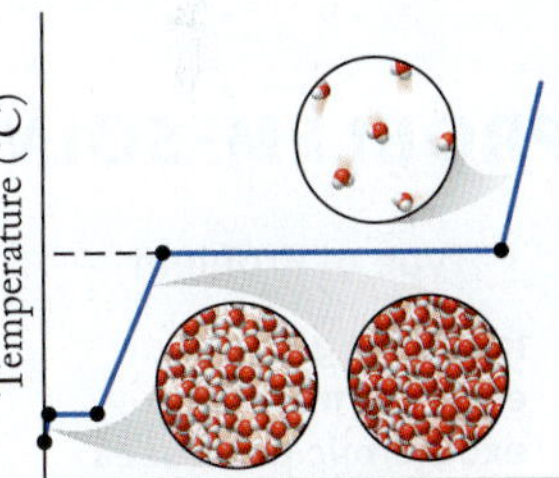


**LO4** A **calorimeter** is a device for determining how much energy is transferred from a system (often a reaction mixture) to its surroundings. A coffee-cup calorimeter operates at constant (atmospheric) pressure and is useful for determining the enthalpy changes or **enthalpies of reaction ($\Delta H_{rxn}$)** of aqueous-phase chemical reactions. A **bomb calorimeter** is a constant-volume device used to determine the energy released during combustion reactions. A **thermochemical equation** includes the change in enthalpy that accompanies a chemical reaction. (Section 10.5)

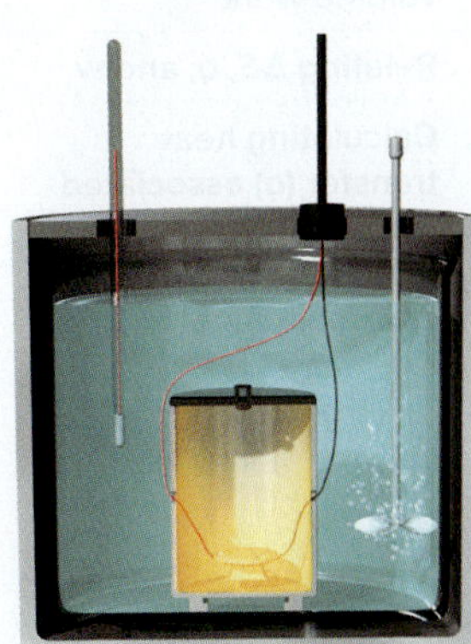

**LO5** A **standard enthalpy of reaction ($\Delta H^\circ_{rxn}$)** is the enthalpy change associated with a reaction occurring under **standard conditions**: a constant pressure of 1 bar and a specified temperature. **Hess's law** states that the enthalpy of a reaction that occurs in more than one step is the sum of the enthalpies of reaction of the steps. Hess's law can be used to calculate enthalpy changes in reactions that are hard or impossible to measure directly. The **standard enthalpy of formation ($\Delta H^\circ_f$)** of a substance is the enthalpy change that accompanies a **formation reaction,** in which one mole of the substance is made from its constituent elements in their standard states. Enthalpies of reaction can be calculated from the enthalpies of formation of reactants and products. (Sections 10.6 and 10.7)

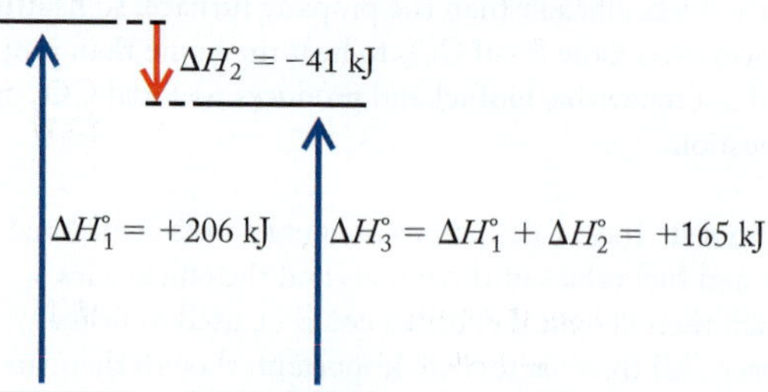


**LO6** Enthalpies of reaction for gas-phase reactions can be estimated by considering the energy required to break bonds in reactants and the energy released when bonds in products are formed. (Section 10.7)

**LO7** The **enthalpy of solution ($\Delta H_{solution}$)** for an ionic compound dissolving in water is the sum of the energies consumed to overcome the compound's lattice energy and to separate water molecules to make space for the ions and the energy released due to the formation of ion–dipole interactions between solute ions and the water molecules surrounding them. Lattice energies can be calculated with a **Born–Haber cycle** in an application of Hess's law. (Section 10.8)

**LO8** **Fuel value** is the amount of energy released during complete combustion of 1 g of a fuel, whereas **fuel density** is the amount of energy released per unit volume of a liquid fuel. Enthalpies of combustion can be used to evaluate the fuel values of dried foods because the products ($CO_2 + H_2O$) are the same, so the enthalpy change that accompanies respiration in organisms is the same as that of combustion. (Section 10.9)

## PARTICULATE **PREVIEW WRAP-UP**

- The ionic bonds in calcium chloride must be broken by the addition of energy to overcome the electrostatic attractions between the calcium cations and the chloride anions.
- Ion–dipole forces form between the $Ca^{2+}$ ions and the $\delta-$ oxygen atoms in water, and between the $Cl^-$ ions and the $\delta+$ hydrogen atoms in water. The formation of these ion–dipole forces is accompanied by a release of energy.

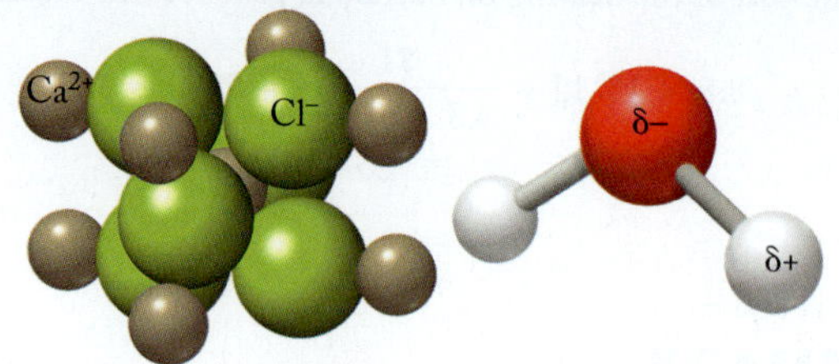


## PROBLEM-SOLVING SUMMARY

| Type of Problem | Concepts and Equations | | Sample Exercises |
|---|---|---|---|
| **Identifying endothermic and exothermic processes** | During an endothermic process, heat flows into the system from its surroundings ($q > 0$). During an exothermic process, heat flows out from the system into its surroundings ($q < 0$). | | **10.1** |
| **Calculating pressure–volume work** | $w = -P\Delta V$ | | **10.2** |
| **Relating $\Delta E$, $q$, and $w$** | $\Delta E = q + w = q - P\Delta V$ | (10.3, 10.4) | **10.3** |
| **Calculating heat transfer ($q$) associated with a change of temperature or state of a substance** | Heating either an object: | | **10.4, 10.5** |
| | $q = C_P\Delta T$ | (10.8) | |
| | or a mass ($m$) of a pure substance: | | |
| | $q = mc_s\Delta T$ | (10.9) | |
| | or a quantity of a pure substance in moles ($n$): | | |
| | $q = nc_P\,\Delta T$ | (10.10) | |
| | Melting a solid at its melting point: | | |
| | $q = n\Delta H_{fus}$ | (10.11) | |
| | Vaporizing a liquid at its boiling point: | | |
| | $q = n\Delta H_{vap}$ | (10.12) | |

| Type of Problem | Concepts and Equations | | Sample Exercises |
|---|---|---|---|
| **Calculating $C_{calorimeter}$ and $\Delta H_{rxn}$ from calorimetry data** | $q_{rxn} = -q_{calorimeter}$<br>$q_{calorimeter} = C_{calorimeter}\,\Delta T$ | (10.14)<br>(10.15) | **10.6, 10.7** |
| **Calculating $\Delta H_{rxn}$ by using Hess's law** | Reorganize the information so that the reactions add together as desired. Reversing a reaction changes the sign of the reaction's $\Delta H_{rxn}$ value. Multiplying the coefficients in a reaction by a factor means the reaction's $\Delta H_{rxn}$ value has to be multiplied by the same factor. | | **10.8** |
| **Recognizing formation reactions** | The reactants must be elements in their standard states, and the product must be one mole of a single compound. | | **10.9** |
| **Calculating $\Delta H^\circ_{rxn}$ from standard enthalpies of formation** | $\Delta H^\circ_{rxn} = \sum n_{products}\,\Delta H^\circ_{f,products} - \sum n_{reactants}\,\Delta H^\circ_{f,reactants}$ | (10.17) | **10.10** |
| **Estimating enthalpies of gas-phase reactions from average bond energies** | $\Delta H_{rxn} = \sum \Delta H_{bond\ breaking} - \sum \Delta H_{bond\ forming}$ | (10.18) | **10.11** |
| **Calculating lattice energy** | The lattice energy of an ionic compound can be calculated from the enthalpy of formation of the compound by applying Hess's law to the steps involved in converting free elements into an ionic solid (the Born–Haber cycle). | | **10.12** |
| **Calculating fuel values and fuel densities** | The fuel value of a substance is the energy released by the complete combustion of 1 g of the substance. Fuel density is the energy released during combustion of a unit volume (such as 1 mL) of a substance. | | **10.13, 10.14** |

## VISUAL PROBLEMS

*(Answers to boldface end-of-chapter questions and problems are in the back of the book.)*

**10.1.** Figure P10.1 shows the compression stroke of a diesel engine. How does upward motion of the piston alter the internal energy of the gases trapped in the cylinder?

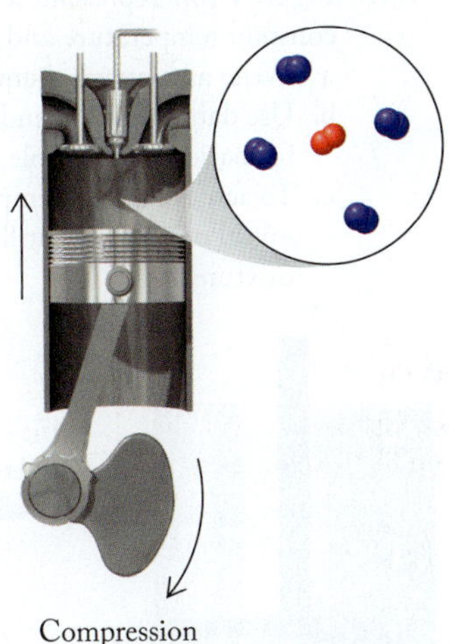

Compression

**FIGURE P10.1**

10.2. Figure P10.2 shows the power stroke of a diesel engine as energy released by the rapid combustion of the air and fuel vapor inside the cylinder pushes the cylinder downward. If the gases inside the cylinder are a thermodynamic system, how does the internal energy of the system change as a result of the combustion reaction and downward motion of the piston? In your description, indicate the signs on $\Delta E$, $q$, and $w$.

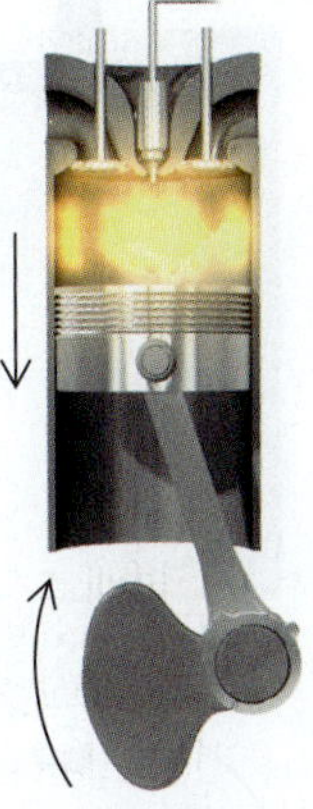

Power

**FIGURE P10.2**

**10.3.** From their molecular structures, predict which of the four hydrocarbons in Figure P10.3 has the highest fuel value and which has the lowest.

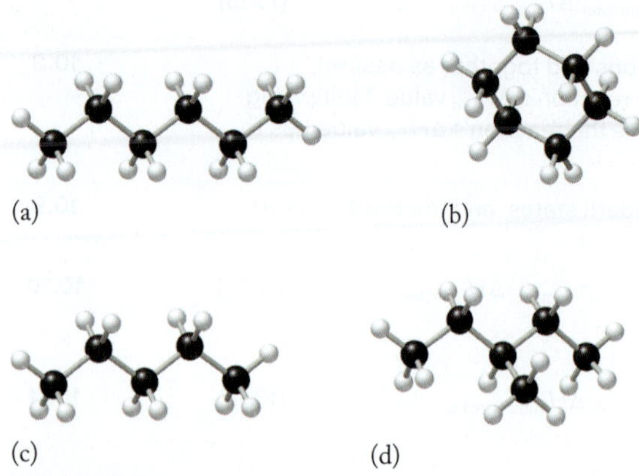


**FIGURE P10.3**

10.4. The diagram in Figure P10.4 shows how the volume of a reaction mixture at constant pressure and temperature changes as $N_2$ and $H_2$ combine, forming $NH_3$.

a. In this reaction, does the reaction mixture do work on the surroundings, or vice versa?
b. Use data from Appendix 4 to calculate $\Delta H^\circ_{rxn}$ for the formation of one mole of product.
c. To achieve a final temperature that is the same as the initial one, does heat flow out from or into the reaction mixture?

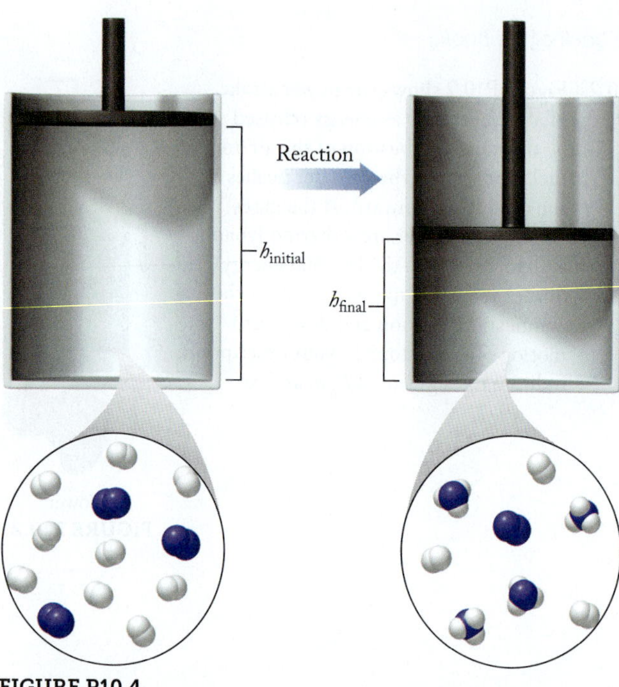


**FIGURE P10.4**

**10.5.** Assuming the reaction mixture in Figure P10.4 is a thermodynamic system, what are the signs on $q$, $w$, and $\Delta E$? If each molecule in the figure represents one mole of reactant or product, what is the percent yield of the reaction?

10.6. The diagram in Figure P10.6 is based on the standard enthalpies of formation ($\Delta H^\circ_f$) values for four compounds made from the elements listed on the "zero" line of the vertical axis.

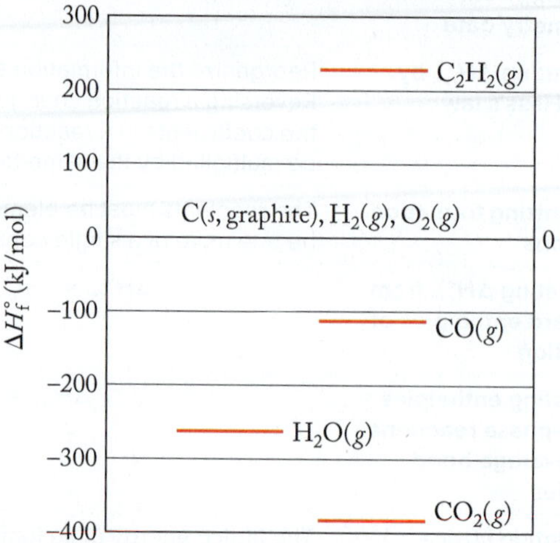


**FIGURE P10.6**

a. Why are the elements all on the same horizontal line?
b. Why is $C_2H_2(g)$ sometimes called an "endothermic" compound?
c. From these $\Delta H^\circ_f$ values, predict which of these two reactions has the more negative enthalpy change:

$$2\,C_2H_2(g) + 3\,O_2(g) \rightarrow 4\,CO(g) + 2\,H_2O(g)$$

or

$$2\,C_2H_2(g) + 5\,O_2(g) \rightarrow 4\,CO_2(g) + 2\,H_2O(g)$$

**10.7.** Figure P10.7 represents a chemical reaction taking place at constant temperature and pressure.

a. Write a balanced chemical equation for the reaction.
b. Use data from Appendix 4 to calculate $\Delta H^\circ_{rxn}$ for the formation of one mole of CO(g).
c. To achieve a final temperature that is the same as the initial one, does heat flow out from or into the reaction mixture?

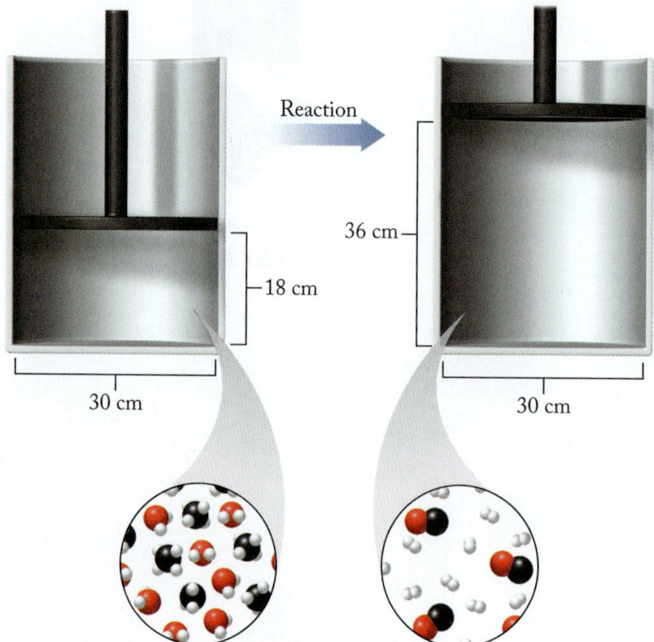


**FIGURE P10.7**

10.8. Rubbing alcohol (2-propanol) quickly evaporates when applied to a patient's skin before drawing a blood sample. Figure P10.8 illustrates the heating curve for the conversion of 1.00 moles of liquid 2-propanol to the gas phase. Use Figure P10.8 to estimate the following values:

a. The melting point of 2-propanol
b. $\Delta H_{vap}$ for 2-propanol
*c. The specific heat for liquid 2-propanol
*d. The specific heat for gaseous 2-propanol

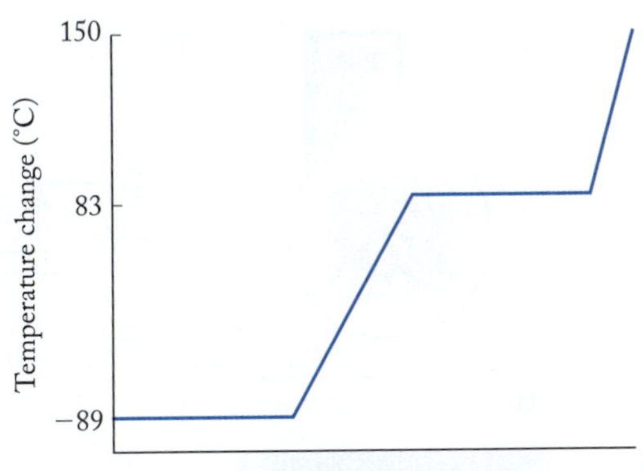


**FIGURE P10.8**

**10.9.** Eicosane, $C_{20}H_{42}$, is a hydrocarbon detected in cell walls. It is a solid at room temperature (20°C) but melts at 37°C. Use the heating curve for eicosane in Figure P10.9 to estimate:

a. $\Delta H_{fus}$ for $C_{20}H_{42}$
b. The boiling point of eicosane
*c. The specific heat for solid $C_{20}H_{42}$
*d. The specific heat for liquid $C_{20}H_{42}$

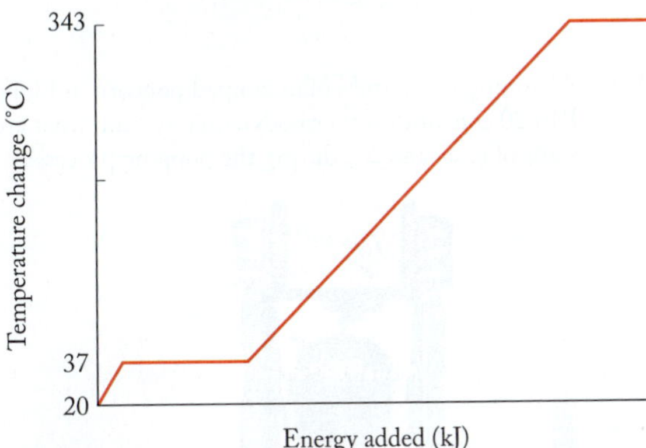


**FIGURE P10.9**

10.10. Which of the reactions shown in Figure P10.10 will have $\Delta H_{rxn} < 0$?

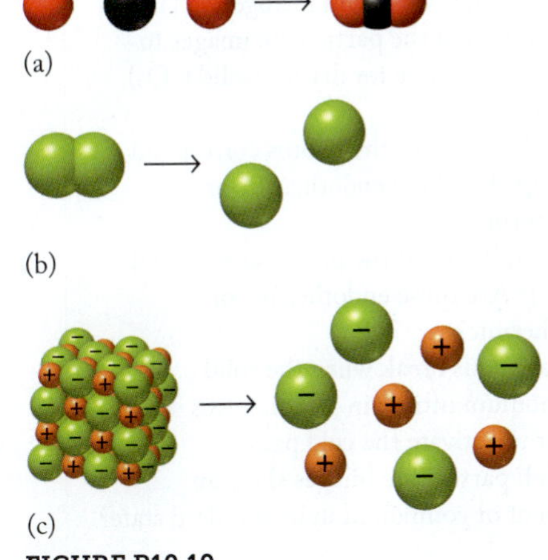


**FIGURE P10.10**

**10.11.** Use the data on bond enthalpies in Table 4.6 to determine $\Delta H_{rxn}$ for each of the reactions in Figure P10.11.

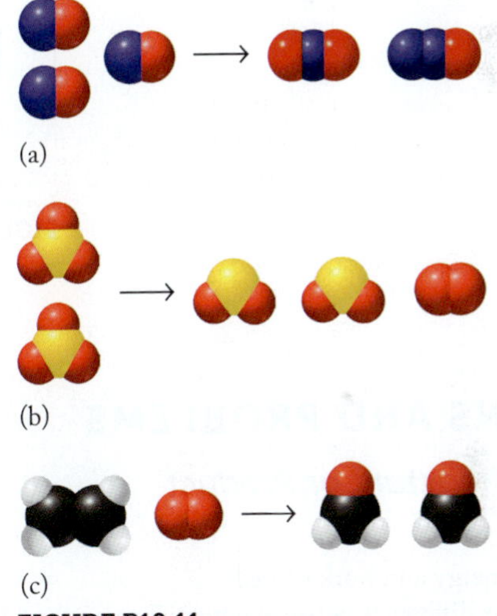


**FIGURE P10.11**

10.12. Use representations [A] through [I] in Figure P10.12 to answer questions a–f.
a. Match two of the particulate images to the phase change for liquid nitrogen in [B].
b. Match two of the particulate images to the phase change for dry ice (solid $CO_2$) in [H].
c. Which, if any, of the photos correspond to [D]? Are these endothermic or exothermic?
d. Which, if any, of the photos correspond to [F]? Are these endothermic or exothermic?
e. What bonds break when the solid ammonium nitrate in [E] dissolves in water to activate the cold pack?
f. Which particulate images show an element or compound in its standard state?

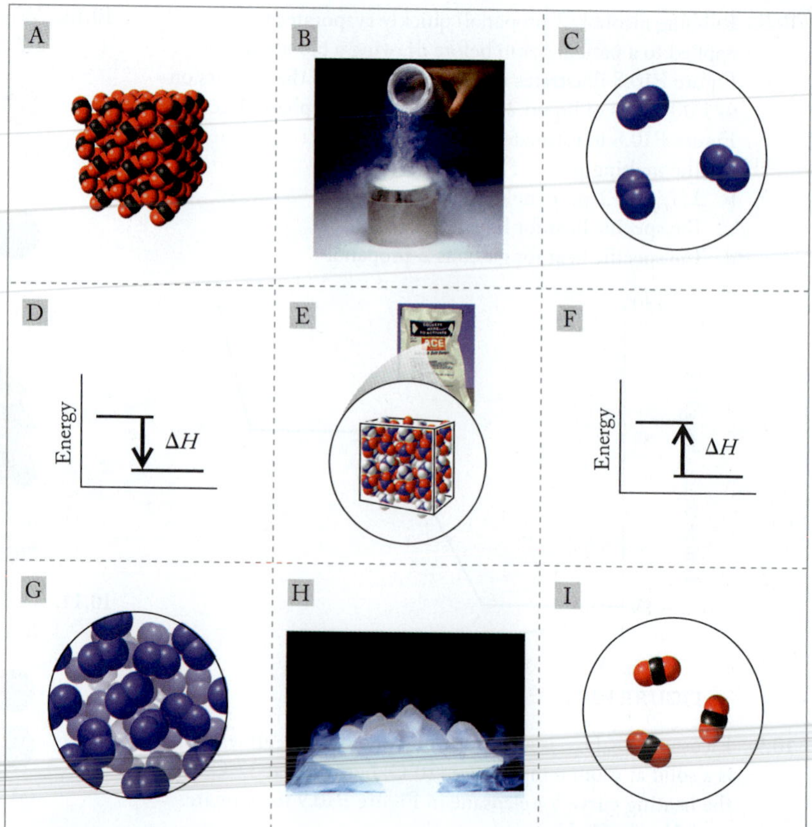


FIGURE P10.12

## QUESTIONS AND PROBLEMS

### Energy as a Reactant or Product

#### Concept Review

**10.13.** How are energy and work related?

10.14. Explain the difference between potential energy and kinetic energy in molecules.

**10.15.** Explain what is meant by a state function.

10.16. Are kinetic energy and potential energy both state functions?

**10.17.** If the potential energy of two particles increases as they move away from each other, do the two particles attract or repel each other?

10.18. Explain how there can be kinetic energy in a stationary ice cube.

### Transferring Energy and Doing Work

#### Concept Review

**10.19.** Describe two ways to increase the internal energy of a gas sample.

10.20. Assuming the kernels of unpopped popcorn in Figure P10.20 constitute a thermodynamic system, what are the signs of $q$, $w$, and $\Delta E$ during the popping process?

FIGURE P10.20

***10.21.** Does the amount of $P$–$V$ work done by a gas on its surroundings depend on the identity of the gas?

10.22. Why is there a negative sign in front of the $P\Delta V$ term in $\Delta E = q - P\Delta V$?

### Problems

**10.23.** Which of the following processes are exothermic, and which are endothermic?

a. Molten aluminum solidifies.
b. Rubbing alcohol evaporates from the skin.
c. Fog forms over San Francisco Bay.

10.24. Which of the following processes are exothermic, and which are endothermic?

a. Ice cubes solidify in the freezer.
b. Ice cubes in a frost-free freezer slowly lose mass.
c. Dew forms on a lawn overnight.

---

**10.25.** What happens to the internal energy of a liquid at its boiling point when it vaporizes?

10.26. What happens to the internal energy of a gas when it expands (with no heat flow)?

---

**10.27.** **Atmospheric Research** Atmospheric scientists often use balloons to carry their instruments aloft. How much $P$–$V$ work does a gas in a balloon do on its surroundings at a constant pressure of 1.00 atm if the volume of gas triples from $1.250 \times 10^5$ L to $1.050 \times 10^6$ L? Express your answer in L · atm and in joules (J).

10.28. An expanding gas does 150.0 J of work on its surroundings at a constant pressure of 1.01 atm. If the gas initially occupied 68 mL, what is the final volume of the gas?

---

**10.29.** Calculate $\Delta E$ when

a. $q = 100.0$ J; $w = 50.0$ J
b. $q = 6.2$ kJ; $w = 0.70$ L · atm
c. $q = -615$ kJ; $w = -3.25$ kilowatt-hours (1 kWh = 3600 kJ)

10.30. Calculate $\Delta E$ for a system that absorbs 726 kJ of heat from its surroundings and does 526 kJ of work on its surroundings.

---

**10.31.** Calculate $\Delta E$ for the combustion of a gas that releases 210.0 kJ of heat to its surroundings and does 65.5 kJ of work on its surroundings.

10.32. Calculate $\Delta E$ for a chemical reaction that releases 90.7 kJ of heat to its surroundings but does no work on them.

---

***10.33.** The following reactions take place in a cylinder equipped with a movable piston at atmospheric pressure. Which reactions will result in work being done on the surroundings? What is the sign of $w$? Assume the system returns to an initial temperature of 110°C. *Hint*: The volume of a gas is proportional to the number of moles ($n$) at constant temperature and pressure.

a. $CH_4(g) + 2\,O_2(g) \rightarrow CO_2(g) + 2\,H_2O(g)$
b. $C_3H_8(g) + 5\,O_2(g) \rightarrow 3\,CO_2(g) + 4\,H_2O(g)$
c. $N_2(g) + 2\,O_2(g) \rightarrow 2\,NO_2(g)$

*10.34. In which direction will the piston described in Problem 10.33 move when the following reactions are carried out at atmospheric pressure inside the cylinder and after the system has returned to its initial temperature of 110°C? What is the sign on $w$? *Hint*: The volume of a gas is proportional to the number of moles ($n$) at constant temperature and pressure.

a. $N_2(g) + 3\,H_2(g) \rightarrow 2\,NH_3(g)$
b. $C(s) + O_2(g) \rightarrow CO_2(g)$
c. $CH_3CH_2OH(g) + 3\,O_2(g) \rightarrow 2\,CO_2(g) + 3\,H_2O(g)$

## *Enthalpy and Enthalpy Changes*

### Concept Review

**10.35.** What is meant by an *enthalpy change*?

10.36. Describe the difference between an internal energy change ($\Delta E$) and an enthalpy change ($\Delta H$).

**10.37.** Why is the sign of $\Delta H$ negative for an exothermic process?

10.38. What happens to the magnitude and sign of the enthalpy change when a process is reversed?

### Problems

**10.39.** Adding Drano to a clogged sink causes the drainpipe to get warm. What is the sign of $\Delta H$ when Drano dissolves in water?

10.40. **Instant Cold-Pack Chemistry** Breaking the small pouch of water inside a chemical cold pack containing ammonium nitrate activates the pack, which is used by sports trainers for injured athletes. What is the sign of $\Delta H$ for the process taking place in the cold pack?

---

**10.41.** **Diatomic Elements** The stable forms of hydrogen and oxygen at room temperature and pressure are the gas-phase diatomic molecules $H_2$ and $O_2$. What is the sign of $\Delta H$ for the following processes?

a. A solid with metallic properties is formed when hydrogen gas is compressed under extremely high pressure: $H_2(g) \rightarrow H_2(s)$.
b. High-energy light shines on oxygen gas in the upper reaches of the atmosphere, converting oxygen gas to oxygen atoms: $O_2(g) \rightarrow 2\,O(g)$.

10.42. **Plaster of Paris** Gypsum is the common name of calcium sulfate dihydrate, which has the formula $CaSO_4 \cdot 2\,H_2O$. When gypsum is heated to 150°C, it loses most of the water in its formula and forms plaster of Paris ($CaSO_4 \cdot 0.5\,H_2O$):

$$2\,CaSO_4 \cdot 2\,H_2O(s) \rightarrow 2\,CaSO_4 \cdot 0.5\,H_2O(s) + 3\,H_2O(g)$$

What is the sign of $\Delta H$ for making plaster of Paris from gypsum?

## *Heating Curves and Heat Capacity*

### Concept Review

**10.43.** What is the difference between *specific heat* and *heat capacity*?

10.44. Which has a larger heat capacity: one liter of water or one cubic meter of water? Which has a larger molar heat capacity?

**10.45.** Why is the enthalpy of vaporization of water so much larger than its enthalpy of fusion?

10.46. How does a cooling curve for water differ from its heating curve?

**10.47.** Figure P10.47 shows portions of the heating curves for two moles each of three liquids: chloroform (boiling point, 142°C), water (boiling point, 100°C), and ethanol (boiling point, 78°C). Rank the liquids in order of their molar enthalpies of vaporization from lowest to highest.

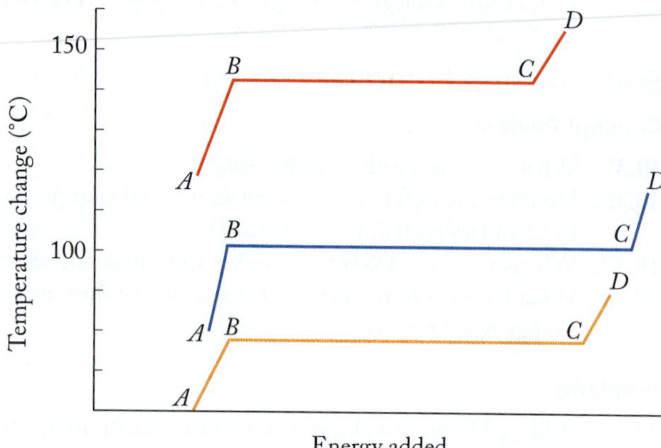


**FIGURE P10.47**

10.48. In Figure P10.47, why do the line segments $\overline{AB}$ have slightly different slopes in each heating curve?

**10.49.** On a cooling curve for water, what do the slopes of the lines indicate?

10.50. The same quantity of heat warms equal masses of metals A and B. Does the metal with the larger specific heat reach the higher temperature?

***10.51.** **Cooling an Automobile Engine** Most automobile engines are cooled by water circulating through them and a radiator. However, the original Volkswagen Beetle had an air-cooled engine. Why might car designers choose water cooling over air cooling?

*10.52. **Nuclear Reactor Coolants** The reactor-core cooling systems in some nuclear power plants use liquid sodium as the coolant. Sodium has a thermal conductivity of 1.42 J/(cm · s · K), which is quite high compared with that of water [$6.1 \times 10^{-3}$ J/(cm · s · K)]. The respective molar heat capacities are 28.3 J/(mol · K) and 75.3 kJ/(mol · K). What is the advantage of using liquid sodium over water in this application?

### Problems

**10.53.** How much energy must be absorbed by 100.0 g of water to raise its temperature from 30.0°C to 100.0°C?

10.54. At an elevation where the boiling point of water is 93°C, 1.33 kg of water at 30°C absorbs 290.0 kJ from a mountain climber's stove. Is this amount of thermal energy sufficient to heat the water to its boiling point?

**10.55.** Use the following data to sketch a heating curve for one mole of methanol. Start the curve at −100°C and end it at 100°C.

| | |
|---|---|
| **Boiling point** | 65°C |
| **Melting point** | −94°C |
| **Enthalpy of vaporization ($\Delta H_{vap}$)** | 35.3 kJ/mol |
| **Enthalpy of fusion ($\Delta H_{fus}$)** | 3.18 kJ/mol |
| **Molar heat capacity (ℓ)** | 81.1 J/(mol · °C) |
| **Molar heat capacity (g)** | 43.9 J/(mol · °C) |
| **Molar heat capacity (s)** | 48.7 J/(mol · °C) |

10.56. Use the following data to sketch a heating curve for 1.5 moles of acetic acid. Start the curve at +16°C and end it at 130°C.

| | |
|---|---|
| **Boiling point** | 118°C |
| **Melting point** | 16°C |
| **Enthalpy of vaporization ($\Delta H_{vap}$)** | 23.7 kJ/mol |
| **Enthalpy of fusion ($\Delta H_{fus}$)** | 11.7 kJ/mol |
| **Molar heat capacity (ℓ)** | 123.1 J/(mol · °C) |
| **Molar heat capacity (g)** | 63.4 J/(mol · °C) |

**10.57.** **Keeping an Athlete Cool** During a strenuous workout, an athlete generates 233 kJ of thermal energy. What mass of water would have to evaporate from the athlete's skin to dissipate this energy?

10.58. The same quantity of thermal energy is added to equal masses of titanium, iron, silver, and lead. All of them are initially at the same temperature. Which metal has the highest final temperature?

**10.59.** Exactly 10.0 mL of water at 25.0°C is added to a hot iron skillet. All the water is converted into steam at 100.0°C. The mass of the pan is 1.20 kg. What is the change in temperature of the skillet?

*10.60. A 20.0 g piece of iron and a 20.0 g piece of gold at 100.0°C are dropped into 1.00 L of water at 20.0°C. What is the final temperature of the water and the pieces of metal?

## *Enthalpies of Reaction and Calorimetry*

### Concept Review

**10.61.** Why is it necessary to know the heat capacity of a calorimeter?

10.62. Could an endothermic reaction be used to determine the heat capacity of a calorimeter?

**10.63.** If we replace the water in a bomb calorimeter with another liquid, do we need to determine a new calorimeter constant?

10.64. When measuring the enthalpy of combustion of a very small amount of material, would you prefer to use a calorimeter having a heat capacity that is small or large?

**Problems**

**10.65.** Calculate the heat capacity of a calorimeter if the combustion of 5.000 g of benzoic acid produces a temperature increase of 16.397°C.

10.66. Calculate the heat capacity of a calorimeter if the combustion of 4.663 g of benzoic acid produces an increase in temperature of 7.149°C.

---

**10.67.** The complete combustion of 1.200 g of cinnamaldehyde ($C_9H_8O$, one of the compounds in cinnamon) in a bomb calorimeter ($C_{calorimeter}$ = 3.460 kJ/°C) produced an increase in temperature of 12.79°C. How much thermal energy is produced during the complete combustion of one mole of cinnamaldehyde?

10.68. **Spices** The aromatic hydrocarbon cymene ($C_{10}H_{14}$) is found in nearly 100 spices and fragrances, including coriander, anise, and thyme. The complete combustion of 1.608 g of cymene in a bomb calorimeter ($C_{calorimeter}$ = 3.460 kJ/°C) produced an increase in temperature of 19.35°C. How much thermal energy is produced during the complete combustion of one mole of cymene?

---

**10.69.** **Hormone Mimics** Phthalates that are used to make plastics flexible are among the most abundant industrial contaminants in the environment. Several phthalates have been shown to act as hormone mimics in humans by activating the receptors for estrogen, a female sex hormone. Combustion of one mole of one of these compounds, dimethyl phthalate ($C_{10}H_{10}O_4$), produces 4685 kJ of thermal energy. If 1.00 g of dimethyl phthalate is combusted in a bomb calorimeter whose heat capacity ($C_{calorimeter}$) is 7.854 kJ/°C, what is the change in temperature of the calorimeter?

10.70. Diborane, $B_2H_6$, was once considered for use as a rocket fuel. It is a highly reactive gas that produces a green flame when burned in air:

$$B_2H_6(g) + 3\,O_2(g) \rightarrow B_2O_3(s) + 3\,H_2O(g)$$

The $\Delta H_{comb}$ value for diborane is –1958 kJ/mol. Assume 0.368 g of diborane is combusted in a calorimeter whose heat capacity ($C_{calorimeter}$) is 7.854 kJ/°C at 20.611°C. What is the final temperature of the calorimeter?

---

**10.71.** A coffee-cup calorimeter contains 100.0 mL of 1.00 *M* HCl at 22.4°C. When 0.243 g of Mg metal is added to the acid, the ensuing reaction:

$$Mg(s) + 2\,HCl(aq) \rightarrow MgCl_2(aq) + H_2(g) \qquad \Delta H_{rxn} = ?$$

causes the temperature of the solution to increase to 33.4°C. What is the value of $\Delta H_{rxn}$ of the reaction? Assume the density of the solution is 1.01 g/mL and that its specific heat is 4.18 J/(g · °C).

10.72. **Chemical Cold Pack** When 4.00 g $NH_4NO_3$ ($\mathcal{M}$ = 80.04 g/mol)—the active ingredient in some chemical cold packs—is dissolved in 96.0 g of $H_2O$, the temperature of the resulting solution is 3.07°C colder than the water and ammonium nitrate were before they were mixed. What is the value of $\Delta H_{rxn}$ for the following dissolution process?

$$NH_4NO_3(s) \rightarrow NH_4NO_3(aq) \qquad \Delta H = ?$$

## Hess's Law and Standard Enthalpies of Reaction

**Concept Review**

**10.73.** How is Hess's law consistent with the law of conservation of energy?

10.74. Would Hess's law be valid if enthalpy were not a state function? Why or why not?

**Problems**

**10.75.** Use the $\Delta H^\circ_{rxn}$ values of the following reactions:

$$2\,SO_2(g) + O_2(g) \rightarrow 2\,SO_3(g) \qquad \Delta H^\circ_{rxn} = -196\text{ kJ}$$
$$\tfrac{1}{4}\,S_8(s) + 3\,O_2(g) \rightarrow 2\,SO_3(g) \qquad \Delta H^\circ_{rxn} = -790\text{ kJ}$$

to calculate the $\Delta H^\circ_{rxn}$ value of this reaction:

$$\tfrac{1}{8}\,S_8(s) + O_2(g) \rightarrow SO_2(g) \qquad \Delta H^\circ_{rxn} = ?$$

10.76. **Ozone Layer** The destruction of the ozone layer by chlorofluorocarbons (CFCs) can be described by the following reactions:

$$ClO(g) + O_3(g) \rightarrow Cl(g) + 2\,O_2(g) \qquad \Delta H^\circ_{rxn} = -29.90\text{ kJ}$$
$$2\,O_3(g) \rightarrow 3\,O_2(g) \qquad \Delta H^\circ_{rxn} = +24.18\text{ kJ}$$

Use the preceding $\Delta H^\circ_{rxn}$ values to determine the value of the standard enthalpy of reaction for this reaction:

$$Cl(g) + O_3(g) \rightarrow ClO(g) + O_2(g) \qquad \Delta H^\circ_{rxn} = ?$$

---

**10.77.** The mineral spodumene ($LiAlSi_2O_6$) exists in two crystalline forms called α and β. Use Hess's law and the following information to calculate $\Delta H^\circ_{rxn}$ for the conversion of α-spodumene into β-spodumene:

$$Li_2O(s) + 2\,Al(s) + 4\,SiO_2(s) + \tfrac{3}{2}\,O_2(g) \rightarrow 2\,\alpha\text{-}LiAlSi_2O_6(s) \qquad \Delta H^\circ_{rxn} = -1870.6\text{ kJ}$$

$$Li_2O(s) + 2\,Al(s) + 4\,SiO_2(s) + \tfrac{3}{2}\,O_2(g) \rightarrow 2\,\beta\text{-}LiAlSi_2O_6(s) \qquad \Delta H^\circ_{rxn} = -1814.6\text{ kJ}$$

10.78. **Gas from Coal or Oil Shale** Synthetic natural gas (SNG), sometimes called substitute natural gas, is a methane-containing mixture produced from the gasification of coal or oil shale directly at the site of the mine or oil field. One reaction for the production of SNG is:

$$4\,CO(g) + 8\,H_2(g) \rightarrow 3\,CH_4(g) + CO_2(g) + 2\,H_2O(g)$$

Use the following thermochemical equations to determine $\Delta H^\circ$ for the reaction as written.

$$C(s,\text{ graphite}) + 2\,H_2(g) \rightarrow CH_4(g) \qquad \Delta H^\circ = -74.8\text{ kJ}$$
$$C(s,\text{ graphite}) + \tfrac{1}{2}\,O_2(g) \rightarrow CO(g) \qquad \Delta H^\circ = -110.5\text{ kJ}$$
$$CO(g) + \tfrac{1}{2}\,O_2(g) \rightarrow CO_2(g) \qquad \Delta H^\circ = -283.0\text{ kJ}$$
$$H_2(g) + \tfrac{1}{2}\,O_2(g) \rightarrow H_2O(g) \qquad \Delta H^\circ = -285.8\text{ kJ}$$

---

**10.79.** **Burning Phosphorus** White phosphorus, $P_4$, ignites in air to form $P_4O_{10}$ in a rapid reaction used in incendiary devices. Red phosphorus, an allotrope of element 15, also reacts with $O_2$ in air to form the same product, but at a much slower rate.

a. Describe how to calculate $\Delta H_{rxn}$ for the conversion of white phosphorus to red phosphorus (reaction i) using reactions ii and iii and their corresponding $\Delta H_{rxn}$ values.

i. $P_4(s, \text{white}) \rightarrow 4\ P(s, \text{red})$

ii. $P_4(s, \text{white}) + 5\ O_2\ (g) \rightarrow P_4O_{10}(s)$

iii. $4\ P(s, \text{red}) + 5\ O_2(g) \rightarrow P_4O_{10}(s)$

b. Does $\Delta H_{rxn}$ for reaction (i) in part a represent the enthalpy of formation of red phosphorus?

10.80. **Low-Emission Vehicles** Adding ammonia to the exhaust gases from motor vehicles can reduce the emissions of nitrogen oxide pollutants through selective catalytic reduction (SCR):

i. $6\ NO(g) + 4\ NH_3(g) \rightarrow 5\ N_2(g) + 6\ H_2O(g)$

ii. $6\ NO_2(g) + 8\ NH_3(g) \rightarrow 7\ N_2(g) + 12\ H_2O(g)$

Describe how to use reactions i and ii and their corresponding $\Delta H_{rxn}$ values to determine the value of the enthalpy of reaction for the following reaction:

iii. $3\ NO(g) + N_2(g) + 3\ H_2O(g) \rightarrow 3\ NO_2(g) + 7\ NH_3(g)$

## Enthalpies of Reaction from Enthalpies of Formation and Bond Energies

### Concept Review

**10.81.** Why is the standard enthalpy of formation of CO(*g*) difficult to measure experimentally?

10.82. Explain how the use of $\Delta H^\circ_f$ to calculate $\Delta H^\circ_{rxn}$ is an example of Hess's law.

**10.83.** Are the standard enthalpies of formation of all allotropes of an element the same? Explain.

10.84. Why are the standard enthalpies of formation of elements in their standard states assigned a value of zero?

**10.85.** Why must the stoichiometry of a reaction be known in order to estimate the enthalpy change from bond energies?

10.86. Why must the structures of the reactants and products be known in order to estimate the enthalpy change of a reaction from bond energies?

***10.87.** When calculating the enthalpy change for a chemical reaction by using bond energies, why is it important that the reactants and products all be gases?

*10.88. If the energy needed to break two moles of C=O bonds is greater than the sum of the energies needed to break the O=O bonds in one mole of $O_2$ and vaporize one mole of carbon, why does the combustion of pure carbon release heat?

### Problems

**10.89.** For which of the following reactions does $\Delta H^\circ_{rxn}$ also represent an enthalpy of formation?

a. $C(s) + \frac{1}{2} O_2(g) \rightarrow CO(g)$

b. $6\ MnO(s) + O_2(g) \rightarrow 2\ Mn_3O_4(s)$

c. $2\ Na(s) + C(s, \text{graphite}) + 3/2\ O_2(g) \rightarrow Na_2CO_3(s)$

d. $4\ H_2(g) + 2\ C(s) \rightarrow 2\ CH_4(g)$

10.90. For which of the following reactions does $\Delta H^\circ_{rxn}$ also represent an enthalpy of formation?

a. $2\ N_2(g) + 3\ O_2(g) \rightarrow 2\ NO_2(g) + 2\ NO(g)$

b. $N_2(g) + O_2(g) \rightarrow 2\ NO(g)$

c. $2\ NO_2(g) \rightarrow N_2O_4(g)$

d. $N_2(g) + 2\ O_2(g) \rightarrow 2\ NO_2(g)$

**10.91.** **Rocket Plane Fuel** The X-37B surveillance plane under development by the U.S. Air Force uses a mixture of hydrazine ($N_2H_4$) and dinitrogen tetroxide ($N_2O_4$) as fuel.

a. The standard enthalpies of formation of $N_2H_4(g)$, $N_2O_4(g)$, and $H_2O(g)$ are 95.35 kJ/mol, 9.2 kJ/mol, and −241.8 kJ/mol, respectively. Use this information to calculate the enthalpy change for the reaction:

$$2\ N_2H_4(g) + N_2O_4(g) \rightarrow 3\ N_2(g) + 4\ H_2O(g)$$

b. Would more energy be released if the hydrazine was in the liquid state, given $\Delta H^\circ_f$ of $N_2H_4(\ell) = 50.63$ kJ/mol?

10.92. **Methanogenesis** Methanogenic bacteria use carbon dioxide and hydrogen to make methane.

a. Write a balanced chemical reaction for the synthesis of methane and liquid water from hydrogen and carbon dioxide.

b. Calculate the standard enthalpy of the reaction in part a by using the appropriate enthalpies of formation from Appendix 4.

**10.93.** Ammonium nitrate decomposes to $N_2O$ and water vapor at temperatures between 250°C and 300°C. Write a balanced chemical reaction describing the decomposition of ammonium nitrate, and calculate its $\Delta H^\circ_{rxn}$ by using the appropriate $\Delta H^\circ_f$ values from Appendix 4.

*10.94. **Chemistry of TNT** Trinitrotoluene (TNT) is a highly explosive compound. The thermal decomposition of TNT is described by the following chemical equation:

$$2\ C_7H_5N_3O_6(s) \rightarrow 12\ CO(g) + 5\ H_2(g) + 3\ N_2(g) + 2\ C(s)$$

If $\Delta H^\circ_{rxn}$ for the reaction is −10,153 kJ, how much TNT is needed to equal the explosive power of one mole of ammonium nitrate (see Problem 10.93)?

**10.95.** A 1.252 g sample of ethanol, $C_2H_6O$, is combusted in a bomb calorimeter with a calorimeter constant $C_{calorimeter}$ = 4.368 kJ/°C, causing the temperature to rise by 6.804 °C.

a. Calculate $q$ for the combustion reaction.

b. Calculate $\Delta H_{comb}$ for $C_2H_6O$.

c. Write a balanced chemical equation for the combustion of $C_2H_6O$.

d. Use $\Delta H_{comb}$ for $C_2H_6O$ from part b, and the values of $\Delta H^\circ_f$ for $O_2$ and the gaseous combustion products from Appendix 4, to calculate $\Delta H^\circ_{f,ethanol}$.

10.96. A 0.8691 g sample of butanol, $C_4H_{10}O$, is combusted in a bomb calorimeter with a calorimeter constant $C_{calorimeter}$ = 4.368 kJ/°C, causing the temperature to rise by 8.279 °C.

a. Calculate $q$ for the combustion reaction.

b. Calculate $\Delta H_{comb}$ for $C_4H_{10}O$.

c. Write a balanced chemical equation for the combustion of $C_4H_{10}O$.

d. Use $\Delta H_{comb}$ for $C_4H_{10}O$ from part b, and the values of $\Delta H^\circ_f$ for $O_2$ and the gaseous combustion products from Appendix 4, to calculate $\Delta H^\circ_{f,butanol}$.

*Note*: Use the average bond energy values in Table A4.1 of Appendix 4 to answer Problems 10.97–10.102.

**10.97.** Use average bond energies to estimate the enthalpy changes of the following reactions under standard conditions:

a. $N_2(g) + 3\ H_2(g) \rightarrow 2\ NH_3(g)$

b. $N_2(g) + 2\ H_2(g) \rightarrow H_2NNH_2(g)$

c. $2\ N_2(g) + O_2(g) \rightarrow 2\ N_2O(g)$

10.98. Use average bond energies to estimate the enthalpy changes of the following reactions under standard conditions:

a. $CO_2(g) + H_2(g) \rightarrow H_2O(g) + CO(g)$

b. $N_2(g) + O_2(g) \rightarrow 2\ NO(g)$

*c. $C(s, \text{graphite}) + CO_2(g) \rightarrow 2\ CO(g)$

*Hint*: The enthalpy of sublimation of C(*s*, graphite), is 719 kJ/mol.

**10.99.** Use average bond energies to estimate the difference in $\Delta H^\circ_{rxn}$ values between the incomplete combustion of one mole of ethane to carbon monoxide and water vapor and the complete combustion of ethane to carbon dioxide and water vapor.

10.100. Use average bond energies to estimate the difference in $\Delta H^\circ_{rxn}$ values between these two reactions:

$$C(s) + O_2(g) \rightarrow CO_2(g)$$

$$C(s) + \tfrac{1}{2}O_2(g) \rightarrow CO(g)$$

***10.101.** Use average bond energies to estimate $\Delta H^\circ_{rxn}$ for the following reaction:

$$4\ NH_3(g) + 7\ O_2(g) \rightarrow 4\ NO_2(g) + 6\ H_2O(g)$$

*10.102. Estimate the energy of the bonds in $SO_2$ given that $\Delta H^\circ_{rxn} = -1036$ kJ for this reaction:

$$2\ H_2S(g) + 3\ O_2(g) \rightarrow 2\ SO_2(g) + 2\ H_2O(g)$$

## Energy Changes When Substances Dissolve; More Applications of Thermochemistry

### Concept Review

**10.103.** How do ion–dipole interactions influence whether an ionic compound's enthalpy of solution is exothermic or endothermic?

10.104. Sodium hydroxide is more soluble in hot water than in cold water, but dissolving sodium hydroxide in water is an exothermic process. How can this be the case?

**10.105.** What is meant by *fuel value*?

10.106. What are the units of fuel values?

**10.107.** Without doing any calculations, predict which compound in each pair releases more energy during combustion:

a. 1 mole of $CH_4$ or 1 mole of $H_2$

b. 1 g of $CH_4$ or 1 g of $H_2$

10.108. Propane is a gas at 1 bar of pressure above −42°C, though it is transported as a liquid under pressure. How does the physical state of propane affect its fuel value and its fuel density?

### Problems

**10.109.** Use a Born–Haber cycle to calculate the lattice energy of potassium chloride (KCl) from the following data:

Ionization energy of K(*g*) = 419 kJ/mol
Electron affinity of Cl(*g*) = −349 kJ/mol
Energy to sublime K(*s*) = 89 kJ/mol
Bond energy of $Cl_2(g)$ = 243 kJ/mol
Standard enthalpy of formation of KCl = −436.5 kJ/mol

10.110. Electron affinities cannot be measured, but they can be estimated from other thermodynamic values if a value for the lattice energy is known. Calculate the electron affinity of sodium oxide ($Na_2O$) from the following data:

Ionization energy of Na(*g*) = 495 kJ/mol
Lattice energy of $Na_2O$ = 2480 kJ/mol
Energy to sublime Na(*s*) = 109 kJ/mol
Bond energy of $O_2(g)$ = 498 kJ/mol
$\Delta H_{rxn}$ for $2\ Na(s) + \tfrac{1}{2}O_2(g) \rightarrow Na_2O(s) = -416$ kJ

**10.111. Flex-Fuel Vehicles** An increasing number of vehicles in the United States can run on either gasoline [a typical component of gasoline has the formula $C_9H_{20}(\ell)$; $\Delta H_{comb} = -6160$ kJ/mol] or ethanol [$CH_3CH_2OH(\ell)$; $\Delta H_{comb} = -1367$ kJ/mol].

a. Which fuel has the greater fuel value?

b. The densities of $C_9H_{20}$ and $CH_3CH_2OH$ are 0.718 g/mL and 0.794 g/mL, respectively. Which has the greater fuel value, as expressed in kJ/L?

c. How many grams of each fuel are needed to heat 1.00 kg of water from 25.0°C to 85.0°C? Assume that all the energy released during combustion is used to heat the water.

10.112. **Biofuels** The microorganism *Clostridium acetobutylicum* is capable of converting glucose into butanol, $C_4H_{10}O$. Butanol is touted as a better renewable biofuel than ethanol.

a. Calculate the fuel value of $C_4H_{10}O$, given $\Delta H^\circ_{comb} = -2676$ kJ/mol.

b. How much energy is released during the combustion of 1.00 kg of $C_4H_{10}O$?

c. How many grams of $C_4H_{10}O$ must be burned to heat 1.00 kg of water from 20.0°C to 90.0°C? Assume that all the energy released during combustion is used to heat the water.

## Additional Problems

***10.113. Production of HCl** The industrial production of hydrogen chloride gas is most frequently carried out by direct synthesis from hydrogen and chlorine:

$$H_2(g) + Cl_2(g) \rightarrow 2\ HCl(g)$$

Smaller quantities of HCl(*g*) may be produced on the laboratory scale by the reaction of sodium chloride and sulfuric acid:

$$2\ NaCl(s) + H_2SO_4(\ell) \rightarrow 2\ HCl(g) + Na_2SO_4(s)$$

Apply concepts discussed in this chapter and data from Appendix 4 to determine whether either heating or cooling is required when these reactions are carried out.

10.114. **Burning Off Calories** A typical double-patty hamburger from a fast-food establishment contains about 563 Calories. (Remember that the dietary "Calorie" is actually a kilocalorie.) Walking at a brisk pace burns about 4.70 Calories per minute. How many minutes would you need to walk to "burn off" the Calories in one double burger?

**10.115.** A 100.0 mL sample of 1.0 *M* NaOH is mixed with 50.0 mL of 1.0 *M* $H_2SO_4$ in a large Styrofoam coffee cup; a thermometer is mounted in the lid of the cup to measure the temperature of the contents. The temperature of each solution before mixing is 22.3°C. After mixing, their temperature reaches 31.4°C. Assume that (1) the density of the mixed solutions is 1.00 g/mL, (2) the specific heat of the mixed solutions is 4.18 J/(g · °C), and (3) no heat is lost to the surroundings.

a. Write a balanced chemical equation for the reaction that takes place in the cup.
b. Is any NaOH or $H_2SO_4$ left in the cup when the reaction is over?
c. Calculate the enthalpy change per mole of $H_2O$ produced in the reaction.

10.116. With reference to Problem 10.115, what if 65.0 mL of 1.0 *M* $H_2SO_4$ is mixed with 100.0 mL of 1.0 *M* NaOH? Will the increase in temperature be less than, more than, or the same as that measured in Problem 10.115?

***10.117.** An insulated container holds 50.0 g of water at 25.0°C. A cube of copper with dimensions 0.933 × 0.933 × 0.933 cm that has been heated to 100.1°C is dropped into the water. What is the final shared temperature of the copper and the water? (The density of copper is 8.92 g/cm$^3$.)

10.118. If all the energy obtained from burning 0.454 kg (1.00 pound) of propane is used to heat water, how many kilograms of water can be heated from 20.0°C to 45.0°C?

**10.119.** Endothermic compounds have positive standard enthalpies of formation. An example is acetylene, $C_2H_2$($\Delta H^\circ_f$ = 226.7 kJ/mol). Combustion of acetylene in pure oxygen produces a flame hot enough to cut and weld steel.

a. What is the standard enthalpy of combustion of acetylene?
b. What is the fuel value of acetylene, assuming the products are $CO_2$ and $H_2O$ vapor?

10.120. Baking soda ($NaHCO_3$) thermally decomposes to soda ash ($Na_2CO_3$), $CO_2$, and $H_2O$:

$$2\,NaHCO_3(s) \rightarrow Na_2CO_3(s) + CO_2(g) + H_2O(g)$$

Use appropriate standard enthalpies of formation to calculate the enthalpy change that accompanies the thermal decomposition of one mole of $NaHCO_3$.

***10.121.** **Polymer Chemistry** Use appropriate bond energies from Table A4.1 of Appendix 4 to predict whether the reaction in which ethylene forms polyethylene plastic is exothermic, endothermic, or involves no change in enthalpy. The reaction can be written:

$$n\,CH_2{=}CH_2 \rightarrow [-CH_2-CH_2-]_n$$

where the structure in the brackets is the *repeating unit* of polyethylene and the value of *n* is typically in the thousands.

10.122. **Metabolism of Methanol** In December 2011, over 100 people died in West Bengal, India, after drinking bootleg (illegal) liquor spiked with methanol. Methanol is toxic because it is metabolized in a two-step process that produces formic acid (HCOOH):

$$O_2(g) + 2\,CH_3OH(aq) \rightarrow 2\,HCOOH(aq) + 2\,H_2O(\ell) \qquad \Delta H^\circ_{rxn} = -1019.5\text{ kJ}$$

a. Is the reaction endothermic or exothermic?
b. What change in enthalpy accompanies the metabolism of 60.0 g of methanol?
c. In the first step of the process, methanol is converted into formaldehyde ($CH_2O$), which is then converted into formic acid. Would you expect $\Delta H^\circ_{rxn}$ for the metabolism of one mole of methanol to formaldehyde to be larger or smaller than the conversion of one mole of methanol to formic acid?

**10.123.** In a high-temperature gas-phase reaction, methanol ($CH_3OH$) reacts with $N_2$ to produce HCN, $O_2$, and $NH_3$. The reaction is endothermic, requiring 164 kJ of thermal energy per mole of methanol under standard conditions.

a. Write a balanced chemical equation for this reaction.
b. What is the change in enthalpy under standard conditions if 60.0 g of $CH_3OH(g)$ reacts with excess $N_2(g)$, forming $HCN(g)$, $O_2(g)$, and $NH_3(g)$?

10.124. Calculate $\Delta H^\circ_{rxn}$ for the reaction

$$2\,Ni(s) + \tfrac{1}{4}\,S_8(s) + 3\,O_2(g) \rightarrow 2\,NiSO_3(s) \qquad \Delta H^\circ_{rxn} = ?$$

from the following data:

$$(1)\ NiSO_3(s) \rightarrow NiO(s) + SO_2(g) \qquad \Delta H^\circ_{rxn} = 156\text{ kJ}$$
$$(2)\ \tfrac{1}{8}\,S_8(s) + O_2(g) \rightarrow SO_2(g) \qquad \Delta H^\circ_{rxn} = -297\text{ kJ}$$
$$(3)\ Ni(s) + \tfrac{1}{2}\,O_2(g) \rightarrow NiO_2(s) \qquad \Delta H^\circ_{rxn} = -241\text{ kJ}$$

**10.125.** Use the information in thermochemical equations (1) through (3) to calculate the value of $\Delta H^\circ_{rxn}$ for the reaction in equation (4).

$$(1)\ Pb(s) + \tfrac{1}{2}\,O_2(g) \rightarrow PbO(s) \qquad \Delta H^\circ_{rxn} = -219\text{ kJ}$$
$$(2)\ C(s) + O_2(g) \rightarrow CO_2(g) \qquad \Delta H^\circ_{rxn} = -394\text{ kJ}$$
$$(3)\ PbCO_3(s) \rightarrow PbO(s) + CO_2(g) \qquad \Delta H^\circ_{rxn} = 86\text{ kJ}$$
$$(4)\ 2\,Pb(s) + 2\,C(s) + 3\,O_2(g) \rightarrow 2\,PbCO_3(s) \qquad \Delta H^\circ_{rxn} = ?$$

*10.126. **Specific Heats of Metals** In 1819, Pierre Dulong and Alexis Petit reported that the product of the atomic mass of a metal times its specific heat is approximately constant, an observation called the *law of Dulong and Petit.*

a. Explain how the specific heat and molar heat capacity data for the metals in Table 10.3 support this law.
b. Use these data to predict the specific heat values of nickel and platinum.

**10.127.** **Odor of Urine** Urine odor gets worse with time because it contains the metabolic product urea, $CO(NH_2)_2$, a compound that is slowly converted to carbon dioxide and ammonia, which has a sharp, unpleasant odor:

$$CO(NH_2)_2(aq) + H_2O(\ell) \rightarrow CO_2(aq) + 2\,NH_3(aq)$$

This reaction is much too slow for $\Delta H^\circ_{rxn}$ to be determined experimentally by measuring a change in temperature, but it can be calculated from the appropriate $\Delta H^\circ_f$ values. Calculate $\Delta H^\circ_{rxn}$. [The value of $\Delta H^\circ_f$ of $CO(NH_2)_2(aq)$ is −319.2 kJ/mol.]

10.128. Calculate the lattice energy of magnesium chloride ($\Delta H^\circ_f = -641$ kJ/mol) by (a) a Born–Haber cycle and (b) the enthalpy of solution of $MgCl_2$ ($\Delta H^\circ_{solution} = -158$ kJ/mol.) Compare the values of *U* obtained from these two methods. The following data are needed:

Ionization energy of Mg to $Mg^{2+}$ = 2188 kJ/mol
Electron affinity of Cl = −349 kJ/mol
Energy to sublime Mg = 150 kJ/mol
Bond energy of $Cl_2$ = 240 kJ/mol
Enthalpy of hydration of $Mg^{2+}(aq)$ = −1903 kJ/mol
Enthalpy of hydration of $Cl^-(aq)$ = −368 kJ/mol

***10.129.** ***The Martian*, Part 1** In the novel and movie *The Martian*, astronaut Mark Watney is stranded on Mars and must figure out a way to produce water to survive. He decides to decompose rocket fuel (hydrazine, $N_2H_4$) to a mixture of hydrogen and nitrogen and then react the hydrogen with oxygen to make water.

a. Write a balanced chemical equation for the decomposition of hydrazine to nitrogen and hydrogen. What is the relationship between $H^\circ_{rxn}$ and $\Delta H^\circ_f$ for this reaction?
b. Use $\Delta H^\circ_f$ for $N_2H_4(\ell)$ and $H_2O(\ell)$ to find $\Delta H_{comb}$ for hydrazine:

$$N_2(g) + 2\,H_2(g) \rightarrow N_2H_4(\ell) \qquad \Delta H^\circ_f = 50.6 \text{ kJ/mol}$$
$$H_2(g) + \tfrac{1}{2}\,O_2(g) \rightarrow H_2O(\ell) \qquad \Delta H^\circ_f = -285.8 \text{ kJ/mol}$$
$$N_2H_4(\ell) + O_2(g) \rightarrow N_2(g) + 2\,H_2O(\ell) \qquad \Delta H_{comb} = ?$$

c. What is the fuel value for liquid hydrazine?
d. What is the fuel density (in kJ/L) for liquid hydrazine?
e. Would you expect the $\Delta H^\circ_f$ for hydrazine *gas* to be greater than or less than 50.6 kJ/mol?

*10.130. ***The Martian*, Part 2** In Problem 10.129 we looked at the thermochemistry of hydrazine in the context of the novel and movie *The Martian*.

a. Use the average bond energies listed in Table A4.1 in Appendix 4 to calculate $H^\circ_{rxn}$ for the decomposition of hydrazine to nitrogen and oxygen.
b. Does your answer for part a match the liquid or gas-phase value of $\Delta H^\circ_f$ for hydrazine in Table 10.4?
c. Is the vaporization of liquid hydrazine an endothermic or an exothermic process?
d. Vaporization represents the disruption of intermolecular forces. What types of intermolecular forces are present between molecules of liquid hydrazine?

# 11

# Properties of Solutions

## Their Concentrations and Colligative Properties

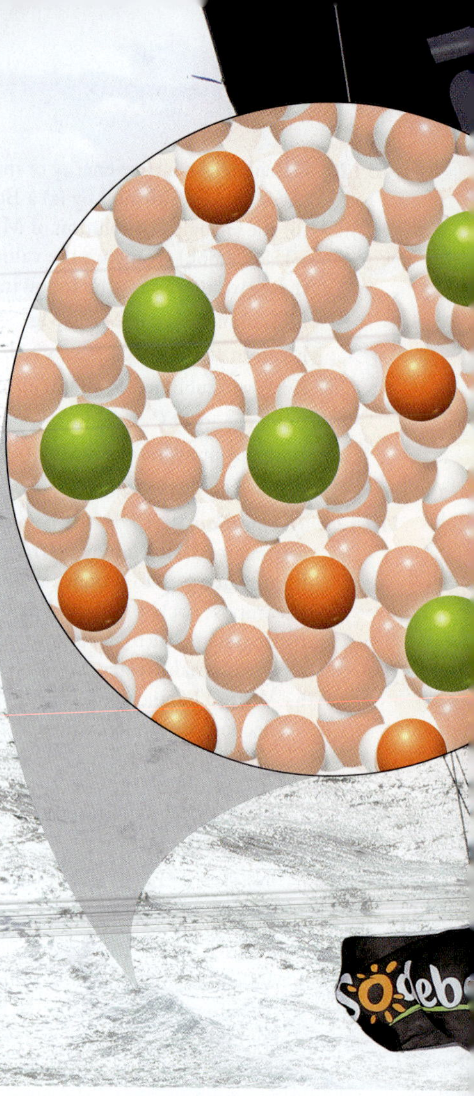

**MAKING POTABLE WATER AT SEA** Sailboats designed to race around the world nonstop have reverse osmosis systems onboard that turn seawater into potable drinking water.

PARTICULATE **REVIEW**

### *Same Composition, Different Properties*

In Chapter 11, we explore how solutes change the properties of the solvent when forming a solution. Consider these two possible solutes, both of which have the molecular formula $C_2H_6O$.

- What intermolecular forces exist between molecules of ethanol?
- What intermolecular forces exist between molecules of dimethyl ether?
- Which compound is more soluble in water?

(Review Sections 6.2–6.4 if you need help.)

*(Answers to Particulate Review questions are in the back of the book.)*

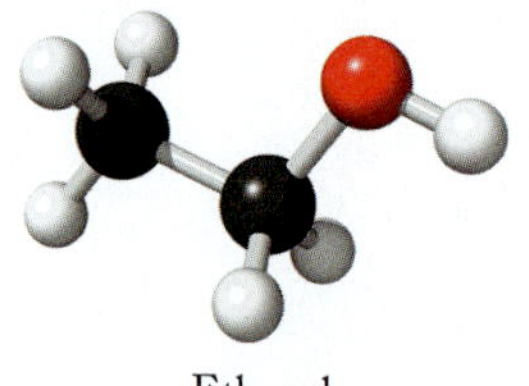

Ethanol

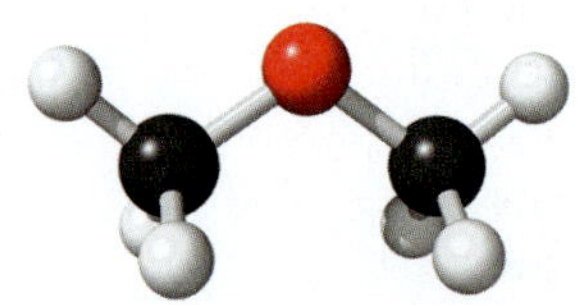

Dimethyl ether

PARTICULATE **PREVIEW**

### *Solvent versus Solution*

The covered containers shown here hold pure water (on the left) and an aqueous solution of NaCl (on the right). As you read Chapter 11, look for ideas that will help you answer these questions.

- When NaCl(*s*) is added to water, what intermolecular forces produce the solution on the right?
- What happens to the amount of water in the vapor phase when salt is added?
- How does the presence of a dissolved solute like salt affect the boiling point of water?

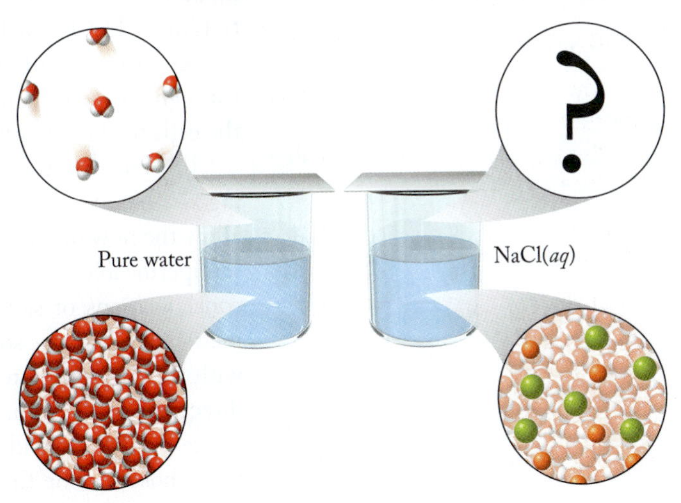

## Learning Outcomes

**LO1** Describe osmosis and relate the direction of osmotic flow and osmotic pressure to solute concentration
**Sample Exercises 11.1, 11.2**

**LO2** Use the van 't Hoff factor to account for the colligative properties of solutions
**Sample Exercises 11.2, 11.10**

**LO3** Use osmotic pressure to determine molar mass
**Sample Exercise 11.3**

**LO4** Use the Clausius–Clapeyron equation to calculate vapor pressure
**Sample Exercise 11.4**

**LO5** Calculate vapor pressures of solutions by using Raoult's law
**Sample Exercises 11.5, 11.6**

**LO6** Express concentrations in molality
**Sample Exercise 11.7**

**LO7** Calculate the freezing points and boiling points of solutions and use them to determine molar mass
**Sample Exercises 11.8, 11.9, 11.11**

**LO8** Explain how temperature, pressure, and intermolecular forces influence the solubilities of gases in liquids and use Henry's law to calculate those solubilities
**Sample Exercise 11.12**

## 11.1 Osmosis: "Water, Water, Everywhere"

In "Rime of the Ancient Mariner," British poet Samuel Taylor Coleridge (1772–1834) wrote: "Water, water, everywhere, and all the boards did shrink. Water, water, everywhere, nor any drop to drink." The water in Coleridge's verse is seawater, which is mostly an aqueous solution of NaCl and other salts. Even though all life-forms need water, sailors who drink seawater suffer from dehydration, and drinking enough of it can be fatal.

In this chapter, we explore the science behind Coleridge's poetry—namely, how and why the properties of solutions depend on the concentration of the solutes in them. We also explore a process by which modern mariners, such as the sailor in this chapter's opening photo, are able to make seawater fit to drink.

Why is drinking seawater a really bad idea? The problem is the difference in salinity (salt concentration) between seawater and the fluid in most living cells (even those in the tissues of marine organisms). If human blood cells are immersed in seawater, they dehydrate, shrivel up, and die because the concentration of ions—mostly $Na^+$ and $Cl^-$—in seawater is much greater than in the aqueous medium inside the cell. Too high a concentration of ions means too low a concentration of $H_2O$: much lower than in most living cells. The difference in $H_2O$ concentrations results in a net movement of water molecules from inside the cell, through the cell membrane, into the seawater surrounding the cell. This causes the cells to shrivel and collapse, as shown in **Figure 11.1(a)**.

This outward movement of water happens because the membranes surrounding red blood cells are *semipermeable*, which means they contain channels that allow the flow of molecules of water but not particles of hydrated solutes. When a semipermeable membrane separates two aqueous solutions with different total concentrations of salts (or any solutes), water spontaneously flows through the membrane from the solution with the lower solute concentration into the solution with the higher solute concentration. Water molecules can and do flow in both directions through the cell membrane, but because of the higher concentration of solute particles outside the cell, the water molecules leave the cell at a faster rate than those outside the cell flow into it, so the *net* flow of water is outward. This

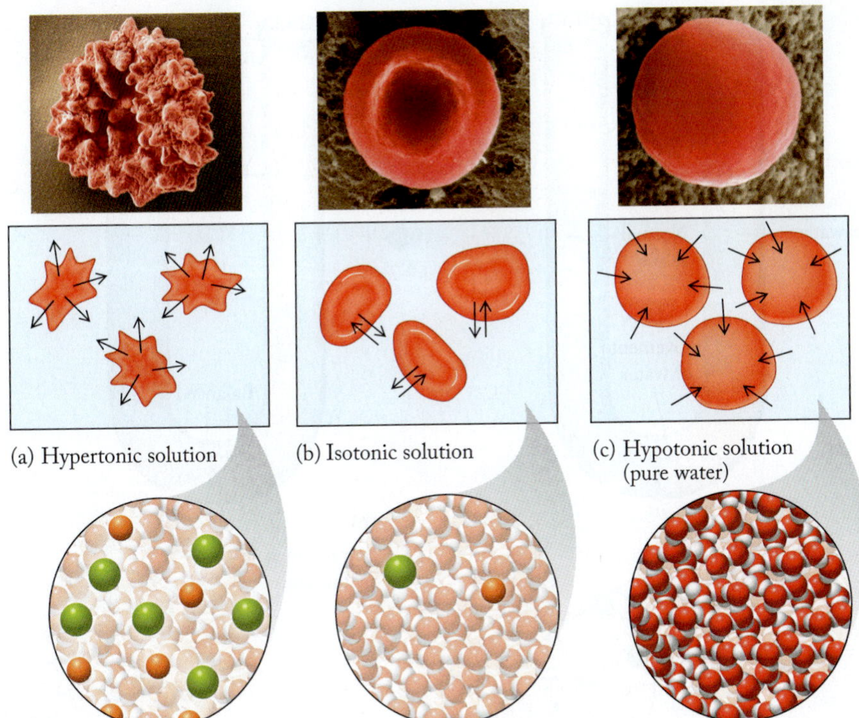


**FIGURE 11.1** The membrane of a red blood cell is semipermeable, which means that water flows by osmosis into and out of the cell to equalize solute concentrations inside and out. (a) When a cell is bathed in a high-salinity *hypertonic* solution, there is a net osmotic flow out of the cell, and the cell shrinks. (b) When a cell is bathed in an *isotonic* solution, in which the solute concentration matches that inside the cell, there is no net osmotic flow into or out of the cell, and its size does not change. (c) When the cell is bathed in pure water or a *hypotonic* solution, there is a net osmotic flow into the cell, and it expands.

spontaneous movement of water through cell membranes (or any solvent through any semipermeable membrane) is called **osmosis**.

The outward flow of water by osmosis happens when red blood cells are immersed in seawater or any *hypertonic* solution, that is, a solution containing a higher concentration of solutes than inside the cell. Human cells contain some dissolved electrolytes, which is why fluids administered by intravenous (IV) injection in hospitals to deliver medications or to rehydrate patients contain the equivalent of about 0.155 *M* NaCl (0.90% NaCl). This value matches the concentration of the electrolytes in most of our cells and is called *normal saline*. Such solutions are said to be *isotonic* solutions (**Figure 11.1b**).

Just as too high a total concentration of solute particles can be dangerous, too low a concentration can be, too. An IV injection without the needed electrolytes (a *hypotonic* solution; **Figure 11.1c**) would cause a patient to retain too much water, which could also have life-threatening consequences.

Although drinking salt water can be fatal to humans, saltwater fish thrive in the oceans because their kidneys have evolved to handle the higher salinity of the seawater they absorb. When placed in fresh water, though, saltwater fish find themselves immersed in a hypotonic solution and do not survive. Similarly, seawater is a lethal hypertonic solution for freshwater fish.

**Figure 11.2** provides a molecular view of why the water in a red blood cell immersed in seawater flows out of the cell and into the seawater. The apparatus in the figure consists of two chambers: one containing seawater and the other normal saline. The two chambers are separated by a semipermeable membrane that allows water molecules, but not the hydrated ions, to move through it. Initially, the volumes and liquid levels in the two compartments are the same. Over time, however, the volume of the normal saline decreases and the volume of the seawater increases due to a net flow of water molecules through the membrane from the normal saline side (lower concentration of solute) to the seawater side (higher concentration of solute).

**osmosis** the spontaneous flow of solvent molecules through a semipermeable membrane from a more dilute solution into a more concentrated one.

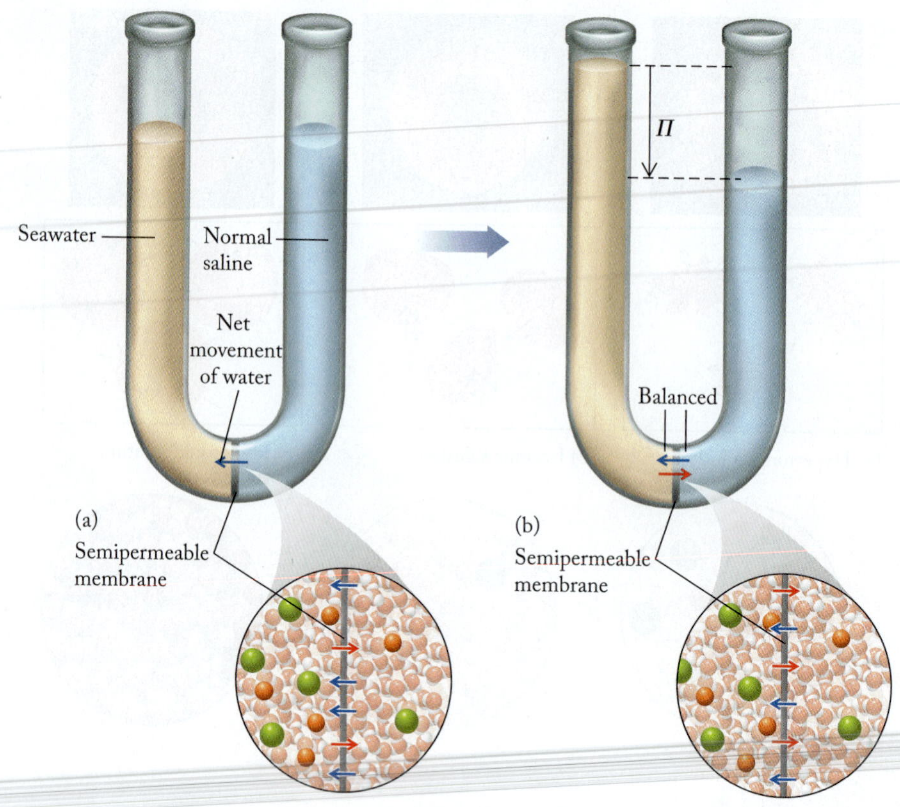


**FIGURE 11.2** (a) When equal volumes of seawater and normal saline are separated by a semipermeable membrane, there is a net osmotic flow of water from the normal saline chamber to the seawater chamber. (b) Net osmotic flow stops when the difference between the heights of the two liquids creates a sufficiently large opposing osmotic pressure ($\Pi$) and additional flow from the seawater side to the normal saline side to balance osmotic flow.

Osmosis happens because the concentration of solute particles on the normal saline side of the membrane is lower than it is on the seawater side. This lower concentration means that the water molecules on the normal saline side of the membrane pass through it at a faster rate than those on the seawater side pass through in the opposite direction, as symbolized by the greater number of blue arrows than red ones in Figure 11.2(a).

## 11.2 Osmotic Pressure and the van 't Hoff Factor

The osmotic flow of water shown in Figure 11.2 eventually reaches a steady state, as shown in Figure 11.2(b). The net flow of water molecules through the membrane into the seawater compartment continues until it is balanced by an opposing flow created by pressure resulting from the difference in the liquid levels in the two compartments. This opposing pressure is called the **osmotic pressure ($\Pi$)** of the solution. If we had filled the right-hand compartment with pure water instead of normal saline, then the difference in solute concentrations would have been even greater and the osmotic pressure would have been even higher. Equation 11.1 can be used to calculate this pressure:

$$\Pi = MRT \tag{11.1}$$

Here the Greek capital letter pi ($\Pi$) is used as the symbol for osmotic pressure to distinguish it from the pressure ($P$) exerted by gases. The value of $\Pi$ is equal to the total molar ($M$) concentration of all the solute particles in seawater (or any solution) multiplied by the universal gas constant, $R$, and the absolute temperature, $T$. When we use an $R$ value of 0.08206 L · atm/(mol · K), we get osmotic pressure values in atmospheres.

**CONNECTION** Molarity—concentration in units of moles of solute per liter of solution—was introduced in Chapter 8. The universal gas constant was introduced in Chapter 9.

**SAMPLE EXERCISE 11.1** Calculating Osmotic Pressure **LO1**

If the total concentration of solute particles in the fluid inside a red blood cell is 0.310 *M*, what is the osmotic pressure of this fluid at body temperature (37°C)? The membrane surrounding the cell is semipermeable.

**Collect, Organize, and Analyze** We are asked to calculate the osmotic pressure of a solution with a total particle concentration of 0.310 *M* at 37°C. Equation 11.1 relates the osmotic pressure of solutions to their solute particle concentrations and absolute temperatures.

**Solve**

$$\Pi = MRT = 0.310\ \frac{\cancel{\text{mol}}}{\cancel{\text{L}}} \times \frac{0.08206\ \cancel{\text{L}} \cdot \text{atm}}{\cancel{\text{mol}} \cdot \cancel{\text{K}}} \times 310\ \cancel{\text{K}} = 7.89\ \text{atm}$$

**Think About It** The calculated osmotic pressure is nearly 8 atm. That is a lot of pressure—comparable to that in high-performance bicycle tires. In the absence of this opposing pressure, water would migrate across the semipermeable membrane of the red blood cell if it were immersed in distilled water, causing the cell to swell up (Figure 11.1c) and eventually burst.

**Practice Exercise** Calculate the osmotic pressure across a semipermeable membrane separating pure water from a sample of seawater at 20°C. Assume the total concentration of all the ions in the sample is 1.148 *M*.

*(Answers to Practice Exercises are in the back of the book.)*

Is there a connection between osmotic pressure ($\Pi$) and the pressure exerted by ideal gases, as described in Chapter 9? Actually, the two kinds of pressure are analogous, as shown when we solve the ideal gas law $PV = nRT$ for pressure and write it this way:

$$P = \left(\frac{n}{V}\right)RT = MRT$$

The $n/V$ part of the equation is the same as molarity (assuming $V$ is expressed in liters).

Osmotic pressure is a **colligative property** of a solution, which means that it depends only on the total concentration of solute particles and not on their identities. This independence from solute identity arises because the key factor in defining the relative rates of migration of water molecules across a semipermeable membrane is the difference in the concentrations of solute on both sides of the membrane. The more moles per liter of solute particles there are on one side, the greater the osmotic pressure.

When a semipermeable membrane separates two solutions that have different solute concentrations, such as seawater and normal saline, we can calculate the osmotic pressure between them by calculating the difference in solute concentrations on opposite sides of the membrane. In the case of seawater surrounding a red blood cell, this difference is 1.148 − 0.310 = 0.838 *M*. Inserting this value into Equation 11.1 and assuming $T = 20°C$:

$$\Pi = MRT = 0.838\ \frac{\cancel{\text{mol}}}{\cancel{\text{L}}} \times 0.08206\ \frac{\cancel{\text{L}} \cdot \text{atm}}{\cancel{\text{mol}} \cdot \cancel{\text{K}}} \times 293\ \cancel{\text{K}} = 20.1\ \text{atm}$$

In the absence of this opposing pressure, the water inside a red blood cell immersed in seawater spontaneously flows out through the cell membrane into the seawater, causing the cell to collapse as illustrated in Figure 11.1(a).

**CHEMTOUR** Osmotic Pressure

**osmotic pressure ($\Pi$)** the pressure applied across a semipermeable membrane to stop the flow of water from the compartment containing pure solvent or a less concentrated solution to the compartment containing a more concentrated solution.

**colligative properties** characteristics of solutions that depend on the concentration, not the identity, of particles dissolved in the solvent.

## van 't Hoff Factors

When we use Equation 11.1 to calculate the osmotic pressure of a solution, the molarity term $M$ represents the total concentration of solute particles in solution. If a solute is an electrolyte, then one mole of it produces more than one mole of particles in solution, and the value of $M$ in Equation 11.1 is greater than the molar concentration of the solute itself.

Frequently we know the concentration of the solute in a solution (because we know the quantity of it used to prepare the solution) better than we know the concentration of particles present. This is the case for weak electrolytes that are only partially ionized in aqueous solutions, and it is even true for solutions of strong electrolytes such as NaCl at especially high concentrations. For example, 1 mL of a 0.01 $M$ solution of NaCl contains 0.02 mmol of ions (0.01 mmol of $Na^+$ and 0.01 mmol of $Cl^-$). However, 1 mL of a 1.0 $M$ NaCl solution contains less than 2.0 mmol of ions. Why does this happen? The answer is that cations and anions do not behave as completely independent particles in solution as their concentrations increase. Instead, they cluster together as intermolecular forces (ion–ion forces) between ions contribute to the properties of the solution. The simplest and most common cluster is an **ion pair**: one cation and one anion acting as a single particle. This ion pairing reduces the total concentration of $Na^+$ and $Cl^-$ ions in more concentrated solutions of electrolytes.

**CONNECTION** Electrolytes were introduced in Section 8.3.

The degree of ion pairing affects the colligative properties of a solution. A Dutch chemist, Jacobus van 't Hoff (1852–1911), addressed this problem by defining a term $i$, now called the **van 't Hoff factor** (or ***i* factor**), which is the ratio of the concentration of solute particles in a solution to the concentration of particles that would be there if the solute did not dissociate. Consider the data plotted in **Figure 11.3** for several solutes, all 0.10 $M$. The height of each of the bars represents the factor by which the concentration of each solute should be multiplied to calculate the total concentration of solute particles (ions and/or molecules) that are present in each solution. Each red bar represents a theoretical van 't Hoff factor. For example, the theoretical value for NaCl is $i = 2$ because one mole of this salt should produce two moles of ions when it dissolves in water. The theoretical $i$ factor for $MgSO_4$ is also 2, but it is 3 for $CaCl_2$ and 4 for $Na_3PO_4$ because a mole of these last two salts should produce three and four moles of ions, respectively, when they dissolve.

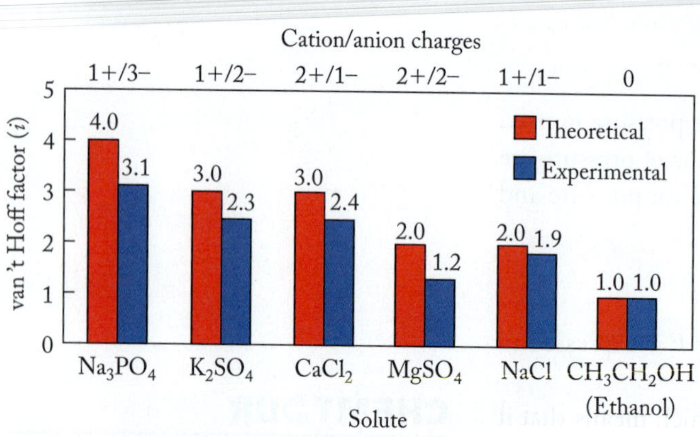


**FIGURE 11.3** Theoretical and experimentally measured values for the van 't Hoff factors for 0.10 $M$ solutions of several electrolytes and the nonelectrolyte ethanol. The larger the charges on the ions, the greater the difference between the theoretical and experimentally measured values.

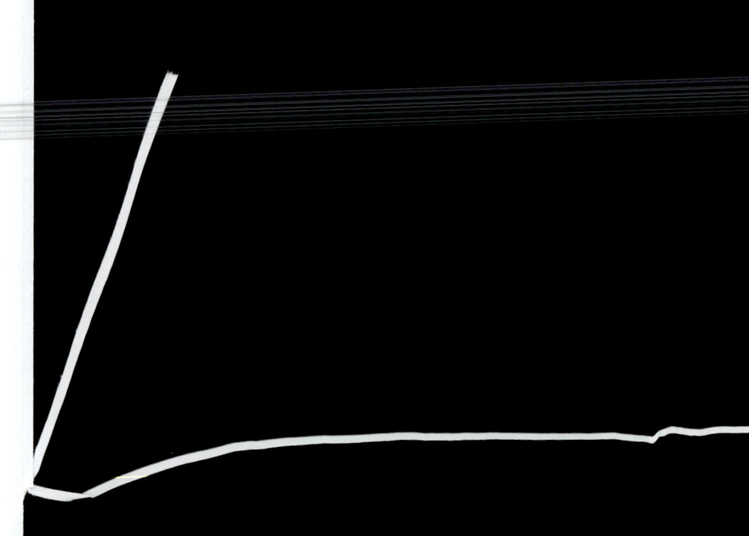

The blue bars in Figure 11.3 represent the actual ratios of particle concentrations to solute concentrations, which are determined experimentally by measuring the colligative properties of these solutions. For nonelectrolytes such as ethanol, the theoretical value and the actual value are both 1.0 because one mole of dissolved solute produces one mole of dissolved molecules.

In solutions of electrolytes with concentrations less than 0.01 $M$, the experimentally measured van 't Hoff factors are essentially the same as the theoretical values. However, at concentrations above 0.01 $M$, the experimental values decrease with increasing concentration, which is reflected in the shorter blue bars for the electrolytes in Figure 11.3. For example, ion pairing reduces the total concentration of $Na^+$ and $Cl^-$ ions in 0.10 $M$ NaCl from 0.20 $M$ to 0.19 $M$, so the experimental van 't Hoff factor for 0.10 $M$ NaCl is 1.9.

Note how the difference between theoretical and experimental values in Figure 11.3, reflecting greater ion pairing, increases with increasing ionic charges.

This behavior is consistent with the stronger attraction between cations and anions of higher charges. Knowledge of the van 't Hoff factor ($i$) of a solution allows us to predict the osmotic pressure of the solution from the molar concentration ($M$) of the solute:

$$\Pi = iMRT \tag{11.2}$$

For example, the osmotic pressure of 0.10 $M$ $K_2SO_4$ at 25°C is

$$\Pi = iMRT = 2.3 \times 0.10 \frac{\text{mol}}{\cancel{\text{L}}} \times 0.08206 \frac{\cancel{\text{L}} \cdot \text{atm}}{\cancel{\text{mol} \cdot \text{K}}} \times 298\ \cancel{\text{K}} = 5.6\ \text{atm}$$

**ion pair** a cluster formed when a cation and an anion associate with each other in solution.

**van 't Hoff factor (*i*)** the ratio of the concentration of solute particles in a solution to the concentration of particles that would be there if the solute did not dissociate.

**reverse osmosis (RO)** a purification process in which solvent is forced through semipermeable membranes, leaving dissolved impurities behind.

### CONCEPT TEST

If a 1.0 $M$ glucose solution is inside a semipermeable membrane and 1.0 $M$ KCl is on the outside, in which direction do water molecules migrate through the membrane?

*(Answers to Concept Tests are in the back of the book.)*

The preceding osmotic pressure calculations show that relatively small differences in solute concentration can produce osmotic pressures large enough to distort and even destroy living cells. Fluids administered intravenously to hospital patients must have nearly the same concentration of solute particles as the patients' blood and blood cells.

Two common IV solutions are physiological saline, which contains 9.0 g/L NaCl, and D5W, which is $5.0 \times 10^1$ g/L dextrose (another name for glucose; $\mathcal{M}$ = 180.16 g/mol). Let's convert these concentrations to molarity values:

$$\text{Physiological saline:}\quad \frac{9.0\ \cancel{\text{g NaCl}}}{1\ \text{L}} \times \frac{1\ \text{mol NaCl}}{58.44\ \cancel{\text{g NaCl}}} = 0.15\ M\ \text{NaCl}$$

$$\text{D5W:}\quad \frac{5.0 \times 10^1\ \cancel{\text{g dextrose}}}{1\ \text{L}} \times \frac{1\ \text{mol dextrose}}{180.16\ \cancel{\text{g dextrose}}} = 0.28\ M\ \text{dextrose}$$

Although the *molarity* of D5W is nearly twice that of physiological saline, the van 't Hoff factor of a 0.15 $M$ solution of NaCl is nearly 2. Dextrose, however, is a molecular compound that does not ionize in water and has a van 't Hoff factor of 1. Thus, the concentrations of dissolved particles in the two solutions are essentially the same.

## Reverse Osmosis and Ion Exchange: Making Drinkable Water

We have developed a molecular-level explanation based on osmosis for why people shouldn't drink seawater. However, applying a technique based on osmosis can make seawater drinkable. The process known as *desalination* involves removing most of the ions from seawater. Distillation is one desalination method that is widely used in the desert countries surrounding the Persian Gulf; however, a less energy-intensive approach involves pumping seawater through semipermeable membranes. Molecules of water pass through the membranes, and the ions dissolved in the seawater are left behind. This process, called **reverse osmosis (RO)**, works because very-high-pressure pumps overcome the osmotic pressure of seawater and reverse the normally spontaneous flow of $H_2O$ molecules from pure water into seawater.

**CONNECTION** Simple distillation, such as the desalination of seawater as described in Chapter 1, is used to separate volatile solvents from nonvolatile solutes.

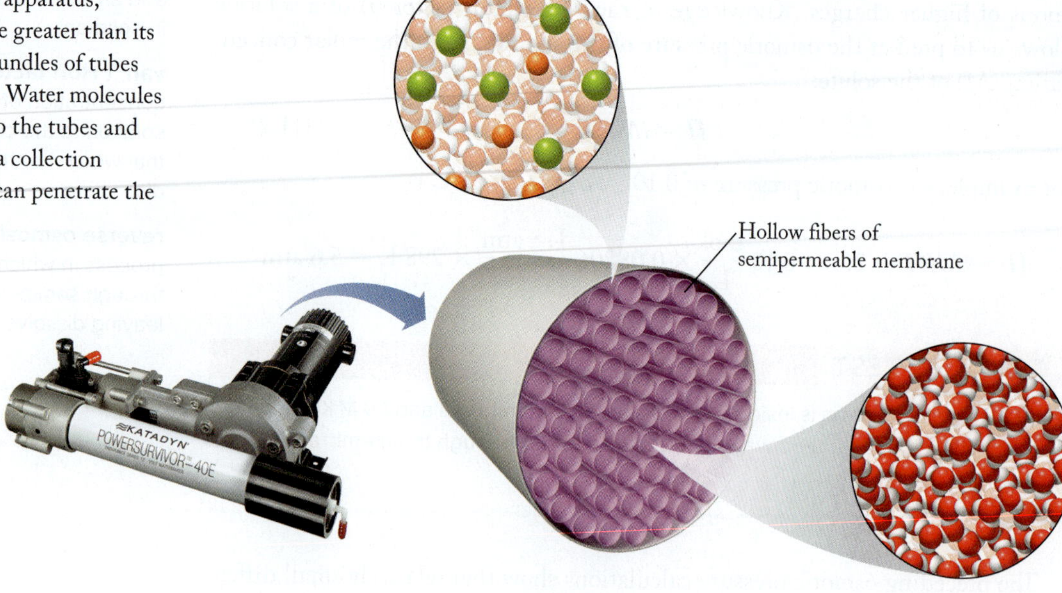


**FIGURE 11.4** Seawater can be desalinated by reverse osmosis. In this apparatus, seawater flows at a pressure greater than its osmotic pressure around bundles of tubes with semipermeable walls. Water molecules pass from the seawater into the tubes and flow through the tubes to a collection vessel. Few hydrated ions can penetrate the walls of the tubes.

The RO system shown in **Figure 11.4** consists of a cylindrical chamber containing many narrow tubes, each made of a semipermeable membrane. Seawater is forced through the chamber so that it washes over the exterior of the tubes, which are filled with flowing pure water. Ordinarily, the direction of osmosis would be from the inner tubes into the surrounding seawater. However, when a pressure greater than the osmotic pressure of seawater is exerted on the seawater in the outer chamber, water flows in the other direction: from the chamber, through the membranes, into the tubes, and out of the apparatus to a pure-water collector.

Some municipal water-supply systems use reverse osmosis to make slightly saline (brackish) water fit to drink. Some industries use this method to purify conventional tap water, and it provides a renewable supply of drinking water for sailors on ocean-crossing yachts. Strong pumps and fittings and rugged membranes are required because reverse osmosis systems must operate at very high pressures (see Sample Exercise 11.2). Moreover, semipermeable membranes are not 100% efficient, so RO systems cannot turn seawater into completely deionized water. Some solvated $Na^+$ and $Cl^-$ ions inevitably pass through the membrane, but their concentrations in the product water are sufficiently low so that the water is drinkable.

**SAMPLE EXERCISE 11.2** Calculating the Pressure Needed for Reverse Osmosis **LO1, 2**

What is the reverse osmotic pressure required at 20°C to purify well water that is equivalent to 0.177 *M* NaCl if the resulting water must meet the drinking water standard that the concentration of $Na^+$ ions be no greater than 20 mg/L? Assume the *i* value of the well water electrolytes is 1.85.

**Collect and Organize** We are asked to calculate the pressure needed to purify water by reverse osmosis. We are given the temperature, the concentration of NaCl in the water to be purified, and the target concentration of $Na^+$ ions in the water to be produced. We can adapt Equation 11.2 to calculate the difference in osmotic pressure between the source and the product water.

**Analyze** To apply Equation 11.2, we need to convert the concentration of $Na^+$ ions in the purified water to molarity. We may assume that this value also represents the concentration of $Cl^-$ ions in the drinking water, which must be there to balance ionic charges. This concentration value will probably be smaller than the salt concentration of the well water, so we can estimate the osmotic pressure that must be overcome because the NaCl concentration in the well water is close to that of physiological saline, which has an osmotic pressure of about 8 atm (see Sample Exercise 11.1).

**Solve** Convert 20 mg/L $Na^+$ to molarity:

$$\frac{20 \text{ mg Na}^+}{\text{L}} \times \frac{1 \text{ g}}{1000 \text{ mg}} \times \frac{1 \text{ mol Na}^+}{22.99 \text{ g Na}^+} = \frac{8.7 \times 10^{-4} \text{ mol Na}^+}{\text{L}}$$

The difference in NaCl concentration between the well water and target drinking water is

$$(0.177 - 0.00087) = 0.176\ M \text{ NaCl}$$

The corresponding osmotic pressure is

$$\begin{aligned} \Pi &= iMRT \\ &= 1.85 \times 0.176 \frac{\text{mol}}{\text{L}} \times 0.08206 \frac{\text{L} \cdot \text{atm}}{\text{mol} \cdot \text{K}} \times 293 \text{ K} \\ &= 7.8 \text{ atm} \end{aligned}$$

**Think About It** In the absence of any external pressure, solvent will flow from the product water to the well water. However, an external pressure greater than 7.8 atm will force the spontaneous flow to reverse, forming drinkable product water from well water.

**Practice Exercise** Calculate the minimum external pressure that must be applied in a reverse osmosis system at 20°C to reduce the salinity of a seawater sample from 3.5% by mass NaCl ($d$ = 1.020 kg/L; $i$ = 1.84) to 48 mg/L.

For many households, a different approach is used to remove certain ions from water. Water containing certain metal ions—principally $Ca^{2+}$ and $Mg^{2+}$—is called *hard water*, and it causes problems in industrial processes and in homes. Because hard water combines with soap to form a gray scum, clothes washed in hard water appear gray and dingy. Hard water forms scale (an incrustation) in boilers, pipes, and kettles, diminishing their ability to conduct heat and carry water, and hard water sometimes has an unpleasant taste. In one common method of *water softening* (removing the ions responsible for the undesirable properties of hard water), hard water is passed through a system that exchanges calcium and magnesium ions for sodium ions (**Figure 11.5**) in a process called **ion exchange**.

To *soften* hard water, it is pumped through tanks filled with small plastic beads called resin, labeled "R" in Figure 11.5. In the tanks, $Ca^{2+}$ and $Mg^{2+}$ ions in the water are exchanged for $Na^+$ ions that are loosely bound to ion-exchange sites on the resin. In many resins, these sites consist of carboxylate ($RCOO^-$) groups. As hard water flows through the resin, $Ca^{2+}$ and $Mg^{2+}$ ions exchange places with $Na^+$ ions on the resin because cations with 2+ and 3+ charges bind more strongly to the $RCOO^-$ groups than $Na^+$ ions. The following chemical equation illustrates this ion-exchange process:

$$2(\text{RCOO}^-)\text{Na}^+(s) + \text{Ca}^{2+}(aq) \rightarrow (\text{RCOO}^-)_2\text{Ca}^{2+}(s) + 2\ \text{Na}^+(aq)$$

Hard water that has been softened in this way contains increased concentrations of sodium ion. Although this may not be a problem for healthy children and

**ion exchange** a process by which one ion is displaced by another.

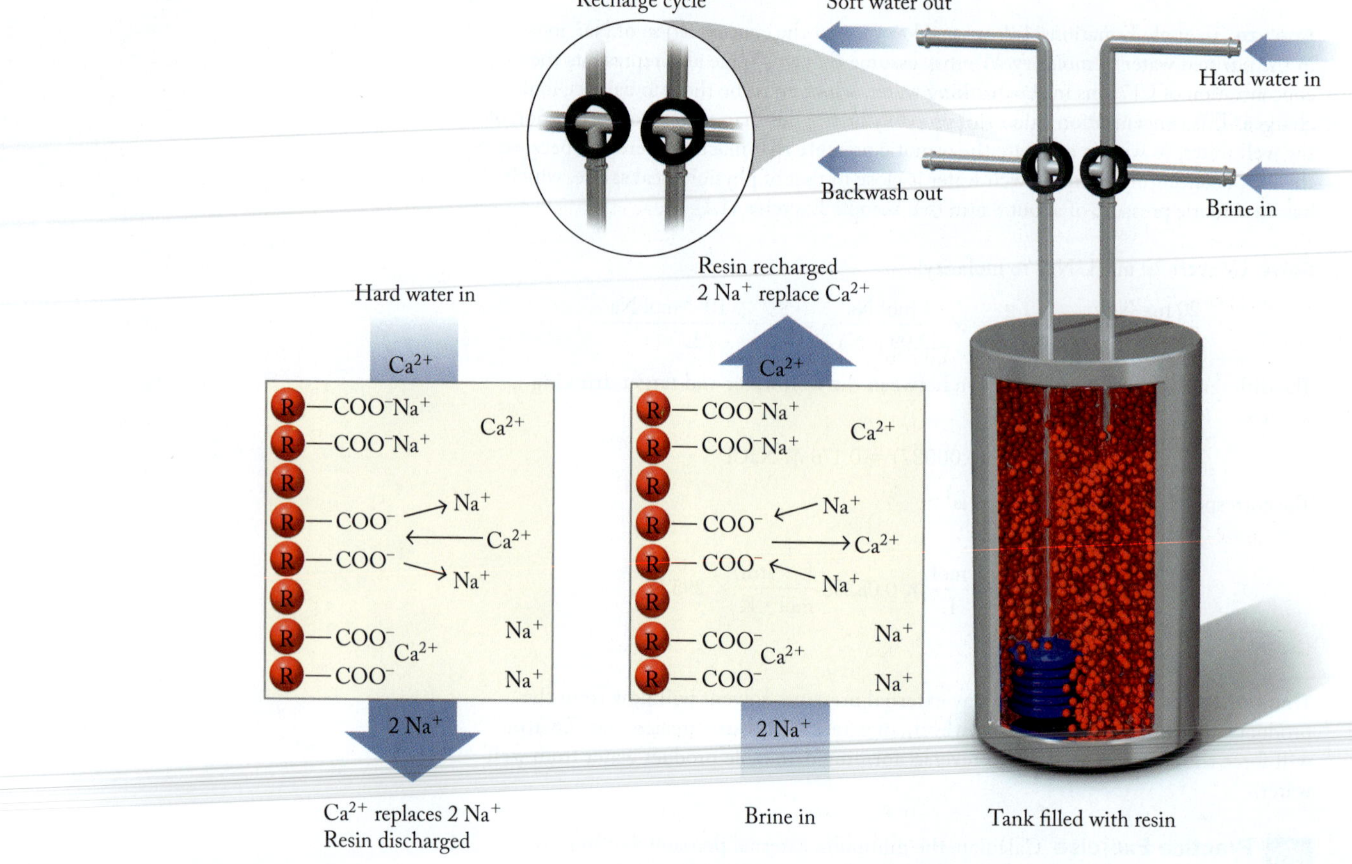


**FIGURE 11.5** Residential water softeners use ion exchange to remove 2+ ions (such as $Ca^{2+}$) that make water hard. The ion-exchange resin contains cation-exchange sites that are initially occupied by $Na^+$ ions. These ions are replaced by 2+ "hardness" ions as water flows through the resin. Eventually, most of the ion-exchange sites are occupied by 2+ ions, and the system loses its water-softening ability. The resin is then backwashed with a saturated solution of NaCl (brine), which displaces the hardness ions, restoring the resin to its $Na^+$ form.

adults, people suffering from high blood pressure often must limit their intake of $Na^+$ and so should not drink water softened by this kind of ion-exchange reaction. Note that whereas reverse osmosis results in the net reduction in the concentration of dissolved species, ion exchange merely exchanges certain dissolved ions for others.

## Using Osmotic Pressure to Determine Molar Mass

Osmotic pressure measurements have the advantage of producing large, easily measured pressures, even from dilute solutions. By measuring the osmotic pressure of a solution containing a dissolved nonelectrolyte, we can determine the molar mass of the solute, even for small samples of biologically important substances that often have large molar masses. For example, the molar mass of the albumin in the serum of humans, horses, and cows (~66,000 g/mol) was determined from osmotic pressure measurements as early as the 1920s.

**CONNECTION** In Chapter 9, we used measurements of density and the ideal gas law to calculate the molar masses of gases.

**SAMPLE EXERCISE 11.3** Using Osmotic Pressure to Determine Molar Mass **LO3**

A 47 mg sample of a water-soluble nonelectrolyte is isolated from a South African tree. The sample is dissolved in enough water to make 2.50 mL of solution. The osmotic pressure of the solution is 0.489 atm at 25°C. What is the molar mass of the compound?

**Collect and Organize** We are asked to determine the molar mass of a compound. We know the concentration of a solution (in mg/mL) of the unknown substance and the osmotic pressure of the solution at a given temperature. Equation 11.2 relates the concentration in molarity to osmotic pressure. Because the solute is a nonelectrolyte, its $i$ factor is 1.

**Analyze** Equation 11.2 can be used to calculate the molarity ($M$) of the solution from the known values of $\Pi$, $i$, and $T$. We can then divide the calculated mol/L concentration value into the given mg/mL (which is the same as g/L) concentration value to obtain g/mol, or molar mass, of the solute.

**Solve** Rearranging Equation 11.2 to solve for $M$ and substituting the relevant known values:

$$M = \frac{\Pi}{iRT} = \frac{0.489\ \text{atm}}{1 \times 0.08206\ \dfrac{\text{L} \cdot \text{atm}}{\text{mol} \cdot \text{K}} \times 298\ \text{K}} = 2.00 \times 10^{-2}\ \text{mol/L}$$

The sample's concentration in g/L is:

$$\frac{47\ \text{mg}}{2.50\ \text{mL}} = 18.8\ \text{mg/mL} = 18.8\ \text{g/L}$$

To calculate molar mass, we divide the g/L concentration value by the mol/L concentration value, which gives us g/mol:

$$\text{Molar mass} = \frac{18.8\ \text{g/L}}{2.00 \times 10^{-2}\ \text{mol/L}} = 9.4 \times 10^{2}\ \text{g/mol}$$

**Think About It** One of the advantages of determining molar mass by osmotic pressure is that only a small quantity of sample is required to produce an osmotic pressure sufficiently large (0.489 atm) to allow for an accurate determination of a molar mass of nearly 1000 g/mol. In most laboratories, the most popular technique for determining molar mass is mass spectrometry, as we discussed in Chapter 2. It requires even less sample than does an osmotic pressure determination of molar mass.

**Practice Exercise** A solution is made by dissolving 5.00 mg of a polysaccharide in water to give a final volume of 1.00 mL. The osmotic pressure of this nonelectrolyte solution is $1.91 \times 10^{-3}$ atm at 25°C. What is the molar mass of the polysaccharide?

**CONNECTION** In Chapter 7, we used molar masses determined by mass spectrometry to convert empirical formulas derived from elemental analyses into molecular formulas.

## 11.3 Vapor Pressure

Ask yourself what happens if you leave a glass of water on a kitchen counter for several days. If the glass is left undisturbed, the water level drops as the water slowly evaporates. At the particle level, some of the molecules of $H_2O$ on the surface escape from the liquid and enter the air above it. They can escape because the water molecules at the surface of the liquid have a distribution of kinetic energies (as we saw with gas-phase molecules in Chapter 9), and some of them have sufficient energy to overcome the intermolecular forces holding them on the surface. Once these high-energy molecules have left the liquid phase, taking their kinetic energies with them, the remaining water molecules redistribute their energies, drawing energy from their surroundings to do so, and the evaporation process continues. Eventually, the glass goes dry without ever being above room temperature.

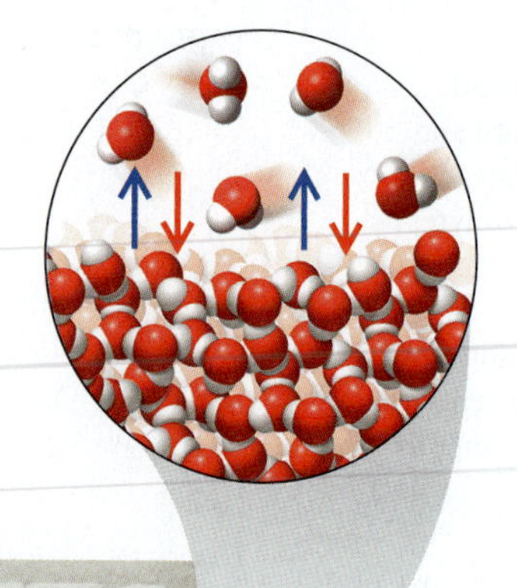

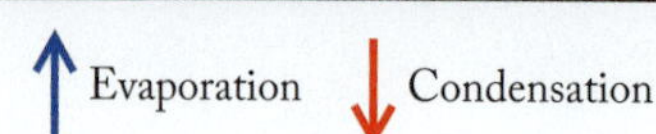


**FIGURE 11.6** A covered glass of water achieves a dynamic equilibrium where the rate at which liquid water is lost to evaporation equals the rate at which liquid water is gained by condensation.

If an identical glass of water is left *covered* on the same counter (**Figure 11.6**), water molecules still evaporate at the same rate, but they are confined to the space above the water and below the cover. As the number of water molecules in this space increases, the rate at which they condense onto the cover and the walls of the glass and, more significantly, into the liquid surface also increases. Soon, the rates of evaporation and condensation equalize as water molecules in the liquid and gas phases achieve a state of dynamic equilibrium. The number of water molecules in the gas phase remains constant as long as the temperature does not change and the cover stays on.

The water vapor in the space above the liquid water in the covered glass exerts a partial pressure that is equal to the **vapor pressure ($P_{vap}$)** of the liquid water, which is defined as the partial pressure exerted by a gas in equilibrium with its liquid state at a given temperature. A liquid is said to be **volatile** when enough of its molecules vaporize to produce a significant vapor pressure at a given temperature: the higher its vapor pressure, the more volatile the liquid is.

The higher the temperature of a volatile liquid, the greater the fraction of its molecules having enough energy to break away from its surface and enter the gas phase, and the higher its vapor pressure. This trend is shown in **Figure 11.7**. The least volatile liquid in the figure, ethylene glycol, requires the highest temperatures to exert significant vapor pressures; the most volatile liquid, diethyl ether, vaporizes at much lower temperatures. If the temperature of a liquid is high enough, its vapor pressure reaches ambient atmospheric pressure (the horizontal line in Figure 11.7). The liquid vaporizes readily under these conditions because it has reached its boiling point. When this ambient pressure is a standard atmosphere (1 atm), the liquid is at its *normal* boiling point, such as 100.0°C for water.

Let's examine why the vapor pressures of the four substances in Figure 11.7 span such a broad range. The answer lies in the different types and strengths of

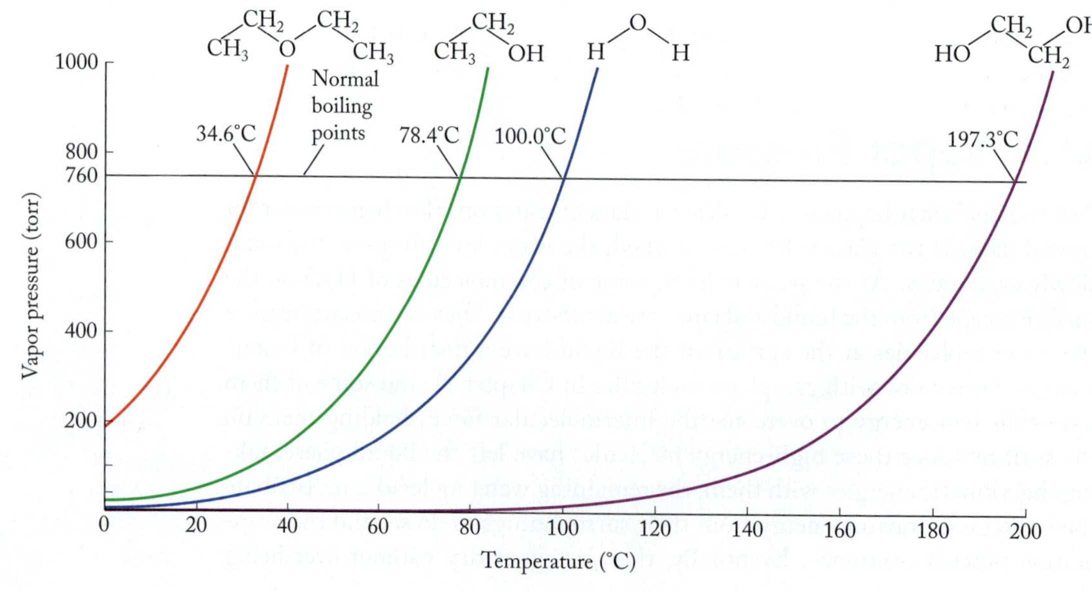


**FIGURE 11.7** A graph of vapor pressure versus temperature for four liquids shows that vapor pressure increases with increasing temperature. The temperature at which the vapor pressure equals 1 atm is the normal boiling point of the liquid.

the intermolecular forces acting between the molecules of each substance. As we saw in Section 6.6, water forms hydrogen bonds (strong dipole–dipole forces) between its molecules. Molecules of both ethylene glycol and ethanol also form hydrogen bonds, and the presence of two –OH groups in the former accounts for the lower vapor pressure of ethylene glycol relative to ethanol, which has only one –OH group. Furthermore, the London dispersion forces that form between all molecules contribute to the lower vapor pressure for ethylene glycol relative to water. Diethyl ether is a polar molecule, and its molecules interact mainly through dipole–dipole forces, but the lack of hydrogen bonding between its molecules results in it being more volatile than, and therefore having a greater vapor pressure than, the other three substances.

**vapor pressure ($P_{vap}$)** the pressure exerted by a gas in equilibrium with its liquid phase at a given temperature.

**volatile** having a significant vapor pressure at a given temperature.

**Clausius–Clapeyron equation** a relationship between the vapor pressure of a substance at different temperatures and its enthalpy of vaporization.

## CONCEPT TEST

Rank the compounds with these condensed molecular structures in order of increasing vapor pressure at 0°C:

(a) $H_3C{-}CH_2{-}CH_2{-}O{-}H$

(b) $H_3C{-}C({=}O){-}CH_3$

(c) $H_3C{-}CH_2{-}CH_2{-}CH_3$

**STEPWISE ANIMATION**
Vapor Pressure

## The Clausius–Clapeyron Equation

The vapor pressure of a volatile liquid increases with increasing temperature, and as Figure 11.7 shows, the increase is not linear. However, if we graph the natural logarithm of the vapor pressure versus $1/T$, where $T$ is the absolute temperature, we do get a straight line (**Figure 11.8**). The slope of this line depends on the enthalpy of vaporization ($\Delta H_{vap}$) of the liquid:

$$\ln(P_{vap}) = -\frac{\Delta H_{vap}}{R}\left(\frac{1}{T}\right) + C \qquad (11.3)$$

where $R$ is the universal gas constant [8.314 J/(mol · K)] and $C$ is a constant that depends on the identity of the liquid. We can use Equation 11.3 to relate the vapor pressures at two temperatures ($T_1$ and $T_2$) to $\Delta H_{vap}$:

$$\ln(P_{vap,T_1}) + \frac{\Delta H_{vap}}{R}\left(\frac{1}{T_1}\right) = C = \ln(P_{vap,T_2}) + \frac{\Delta H_{vap}}{R}\left(\frac{1}{T_2}\right)$$

or

$$\ln\left(\frac{P_{vap,T_2}}{P_{vap,T_1}}\right) = -\frac{\Delta H_{vap}}{R}\left(\frac{1}{T_2} - \frac{1}{T_1}\right) \qquad (11.4)$$

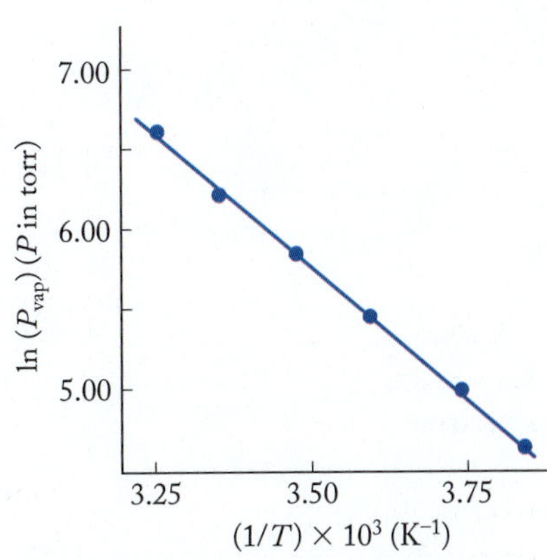


**FIGURE 11.8** Plotting the natural logarithm of the vapor pressure versus the reciprocal of the absolute temperature gives a straight line described by the Clausius–Clapeyron equation. The graph shows the plot for pentane.

Equation 11.4 is called the **Clausius–Clapeyron equation**. We can use it to calculate the vapor pressure of a liquid at any temperature if we know its normal boiling point and $\Delta H_{vap}$ value, as shown in Sample Exercise 11.4, or to determine the value of $\Delta H_{vap}$ from measurements of vapor pressure at different temperatures.

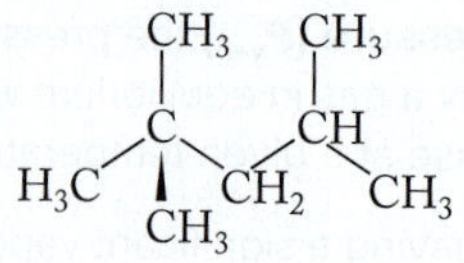

$$\begin{array}{c} CH_3 \quad CH_3 \\ | \qquad\quad | \\ H_3C - C - CH_2 - CH - CH_3 \\ | \\ CH_3 \end{array}$$

**FIGURE 11.9** Structural formula of isooctane.

**FIGURE 11.10** Octane ratings for gasoline.

## SAMPLE EXERCISE 11.4 Calculating Vapor Pressure by Using the Clausius–Clapeyron Equation — LO4

The compound with the common name *isooctane* (its official name is 2,2,4-trimethylpentane) has the structure shown in **Figure 11.9**. It defines the "100" value on the octane rating scale used to grade gasoline (**Figure 11.10**). Its normal boiling point is 99°C, and its enthalpy of vaporization is 35.2 kJ/mol. What is the vapor pressure of isooctane at 25°C in torr?

**Collect and Organize** We are given the normal boiling point and enthalpy of vaporization of isooctane. We are asked to calculate its vapor pressure at 25°C. The Clausius–Clapeyron equation relates the vapor pressure values of a liquid at two temperatures to its $\Delta H_{vap}$ value.

**Analyze** To use the Clausius–Clapeyron equation, we must express temperatures on the Kelvin scale and convert $\Delta H_{vap}$ to joules per mole to be compatible with the units on $R$, J/(mol · K). Vapor pressure decreases sharply (Figure 11.7) as temperatures decrease below the normal boiling point of a liquid, so the vapor pressure of isooctane at 25°C should be only a fraction of the ambient atmospheric pressure, 760 torr.

**Solve**

$\Delta H_{vap}$ in joules per mole is

$$35.2\,\frac{\text{kJ}}{\text{mol}} \times \frac{1000\text{ J}}{1\text{ kJ}} = 3.52 \times 10^4\,\frac{\text{J}}{\text{mol}}$$

Entering these values in Equation 11.4 and solving for $P_{vap,T_2}$ yields

$$\ln\left(\frac{P_{vap,T_2}}{P_{vap,T_1}}\right) = -\frac{\Delta H_{vap}}{R}\left(\frac{1}{T_2} - \frac{1}{T_1}\right)$$

$$\ln\left(\frac{P_{vap,T_2}}{760\text{ torr}}\right) = -\frac{3.52 \times 10^4\,\frac{\text{J}}{\text{mol}}}{8.314\,\frac{\text{J}}{\text{mol}\cdot\text{K}}}\left(\frac{1}{298\text{ K}} - \frac{1}{372\text{ K}}\right) = -2.826$$

$$\frac{P_{vap,T_2}}{760\text{ torr}} = e^{-2.826} = 5.925 \times 10^{-2}$$

$$P_{vap,T_2} = 45.0\text{ torr}$$

**Think About It** We expected isooctane to have a relatively low vapor pressure at a temperature well below its normal boiling point, so this number is reasonable. Its value and the value of other hydrocarbons in gasoline at summerlike temperatures are well above the vapor pressure of water (23.8 torr), which means that their evaporation rates are high enough to impact the quality of the air near gas stations and to make it economically feasible to use gas pump nozzles that trap some of these vapors.

**Practice Exercise** Pentane ($C_5H_{12}$) gas is used to blow the bubbles in molten polystyrene, which is used in coffee cups and other products that have good thermal insulation properties. The normal boiling point of pentane is 36°C; its vapor pressure at 25°C is 505 torr. What is the enthalpy of vaporization of pentane?

## CONCEPT TEST

Diesel fuel is made of hydrocarbons with an average of 13 carbon atoms per molecule, and gasoline is made of hydrocarbons with an average of 7 carbon atoms per molecule. Which fuel has the higher vapor pressure at room temperature?

## 11.4 Solutions of Volatile Substances

Pharmaceutical companies often use large amounts of volatile solvents during the manufacture of drugs. To reduce cost and practice better environmental stewardship, the companies try to purify and reuse these solvents. Separating them from the mixtures requires understanding their chemistry.

The behavior of volatile liquids was extensively studied by French chemist François Marie Raoult (1830–1901), who published the following succinct description of it in 1882: *The total vapor pressure of an ideal solution depends on the vapor pressure of each component in the solution and its mole fraction in the liquid mixture.* In equation form, this description is known as **Raoult's law**:

$$P_{\text{total}} = x_1P_1^\circ + x_2P_2^\circ + x_3P_3^\circ + \cdots \tag{11.5}$$

where $x_i$ is the mole fraction of each volatile component of a solution and $P_i^\circ$ is the vapor pressure of the pure component. Each term in the series represents the contribution that each volatile component of a solution makes to the total vapor pressure of the solution. Thus, Raoult's law is analogous to Dalton's law of partial pressures, but for volatile liquids.

Although simple distillation can separate mixtures of volatile substances where the boiling point differences between the components are large—for example, a difference of more than 70°C—this technique will not separate substances with much smaller differences in boiling point. Instead, volatile liquids with small differences in their vapor pressures are separated using a process called **fractional distillation** (**Figure 11.11**). Like simple distillation, fractional distillation separates the components of mixtures according to differences in their volatilities through selective evaporation and condensation. However, unlike simple distillation, fractional distillation employs a column above the distillation flask in which many evaporation–condensation cycles separate substances that have only slightly different volatilities. Such a column, called a fractionating column, contains surfaces on which hot vapors rising up from the boiling flask can condense.

To accomplish fractional distillation and recover expensive solvents in pharmaceutical manufacturing processes requires a much larger and more elaborate system than the bench-top system in Figure 11.11. An industrial-scale fractional distillation system is illustrated in **Figure 11.12**. As the vapors percolate upward in the tower, they encounter cooler surfaces where the least volatile of them condense, collecting in trays. As heat rises in the tower, the substances collected in each tray reach temperatures at which they revaporize. However, the vapor produced by the liquid in each tray has a different composition from that of the liquid that condensed into the tray: the vapor-phase mixture is richer in the more volatile components of the liquid. When this enriched vapor condenses in another, cooler tray and then revaporizes, the vapor produced this time is even richer in the more volatile components. These cycles of vaporization, condensation, and revaporization continue to repeat, eventually resulting in the separation of volatile substances with only slightly different vapor pressures.

To explore how fractional distillation works in more detail, let's discuss the separation of a 50:50 (by moles) mixture of two volatile hydrocarbons: heptane ($C_7H_{16}$; boiling point, 98°C) and octane ($C_8H_{18}$; boiling point, 126°C). Octane

**Raoult's law** the principle that the vapor pressure of a solution is the sum of the vapor pressures of the volatile components of the solution, which are each the product of the vapor pressure of the pure component and its mole fraction in the solution.

**fractional distillation** a method of separating a mixture of compounds on the basis of their different boiling points.

**CHEMTOUR** Raoult's Law

**CONNECTION** According to Dalton's law of partial pressures (Section 9.9), each gas in a mixture contributes a partial pressure equal to the product of the mole fraction of each gas times the total pressure of the mixture.

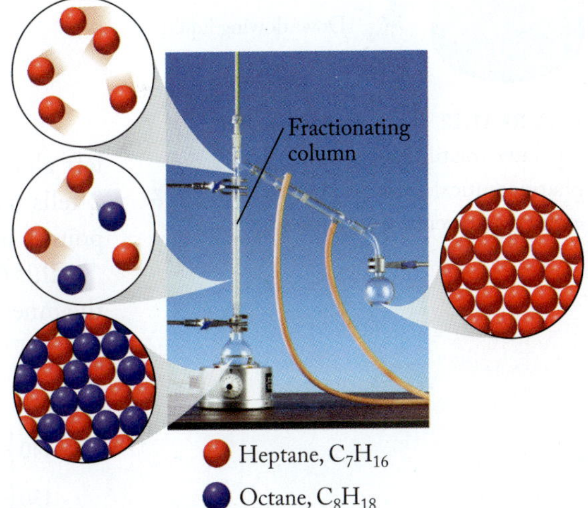


**FIGURE 11.11** Fractional distillation apparatus. Vapors rise through a fractionating column, where they repeatedly condense and revaporize. The most volatile component distills first into the collecting flask. Increasingly less volatile, higher-boiling-point components are distilled in turn. The progress of the distillation process is monitored using the thermometer at the top of the fractionating column.

**CHEMTOUR** Fractional Distillation

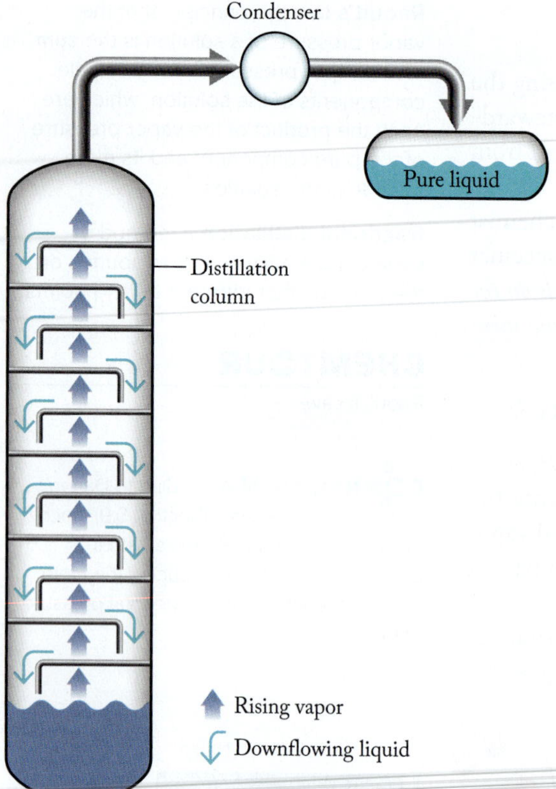


**FIGURE 11.12** Fractional distillation separates volatile substances during pharmaceutical production while recovering expensive solvents to be reused.

has the higher boiling point because its molecules are larger and experience stronger London dispersion forces. These forces inhibit the vaporization of octane molecules until they acquire greater kinetic energies at higher temperatures.

Suppose a sample of our 50:50 mixture is heated in a simple distillation apparatus. The mixture starts to boil at a temperature at which the sum of their vapor pressures reaches ambient pressure (760 torr). This temperature turns out to be 108°C, which is between the normal boiling points of the two components of the mixture. At this temperature, heptane has the greater vapor pressure because it is more volatile, but octane has a significant vapor pressure, too, so the vapors that condense in the distillation apparatus are enriched in heptane but still have a significant octane component. This one-step, simple distillation does not separate them completely, so we turn to a fractional distillation.

As heptane-enriched vapors from a boiling 50:50 mixture reach the bottom of a fractionating column, they condense on the inner surfaces, forming a liquid that is enriched in heptane. The process is shown graphically in **Figure 11.13**. The blue curve represents temperatures at which different mixtures of the two liquids boil. The temperature at point 1 on the curve confirms that the initial boiling point of a 50:50 solution is about 108°C. The red line on the graph shows the composition of the vapor that is formed as these solutions boil. To find the composition of the vapor produced by a 50:50 mixture of the two liquids boiling at 108°C, we move horizontally to the left along the dashed line between points 1 and 2. The $x$-coordinate of point 2 tells us that the composition of the vapor produced by the liquid boiling at point 1 is about 65% heptane and only 35% octane.

This 65:35 vapor rises higher in the fractionating distillation column, cools on the inner surfaces, and condenses as a 65:35 liquid in a process represented by the

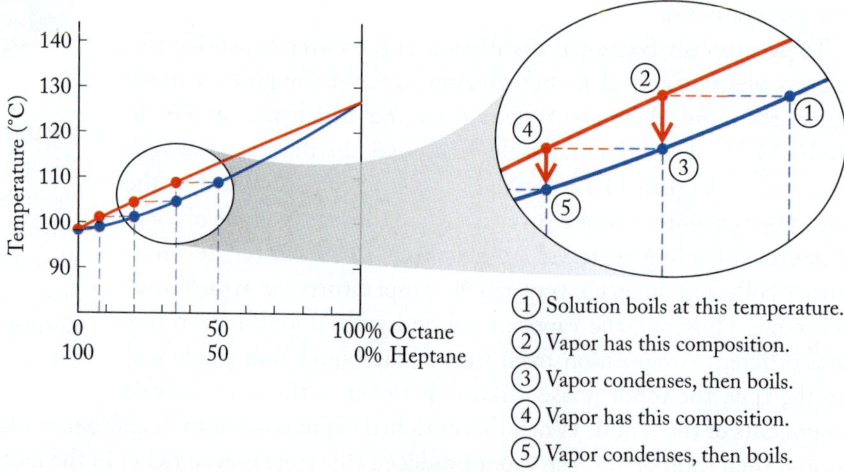


**FIGURE 11.13** Stages in the fractional distillation of a 50:50 mixture of heptane and octane. The blue line tracks the boiling points of the original sample and the liquids that condense in the fraction column. Points on the red line show the composition of the vapor produced as the sample and the condensed liquids vaporize. With each vaporization and condensation cycle, the composition of the liquid in the column comes closer to being pure heptane.

red arrow from point 2 to point 3. Continued heating of the column warms this liquid, and it vaporizes at about 104°C (the *y*-coordinate of point 3). To find the composition of the vapor above this boiling liquid, we again move left on the temperature axis until we intersect the red curve (point 4). Reading down from point 4 to the concentration axis, we see that the vapor concentration is now about 80% heptane and only 20% octane. This 80:20 vapor rises still higher in the fractionating column where it cools and condenses, and the distillation cycle repeats once again.

If we continue this process of redistilling mixtures with increasing concentrations of heptane and then cooling and condensing the vapors, we eventually obtain a condensate that is pure heptane. If we monitor the temperature at which vapors condense at the very top of our fractionating distillation column, we will see a profile of temperature versus volume of distillate produced that looks like **Figure 11.14**. The first liquid to be produced is nearly pure heptane, which has a boiling point of 98°C. Ideally, all the heptane in the original sample is recovered before the octane is collected as the temperature at the top rises to 126°C.

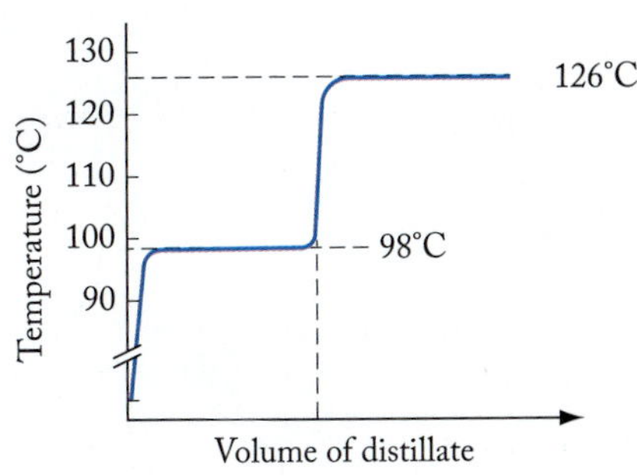


**FIGURE 11.14** Temperature profile of the fractional distillation of a 50:50 mixture of heptane and octane.

## CONCEPT TEST

The condensed structures and dipole moments of dimethyl ether and acetone are shown in **Figure 11.15**. Which of the two compounds would you expect to be present in higher concentration in the vapors of a boiling 50:50 (mol/mol) mixture? Explain your selection.

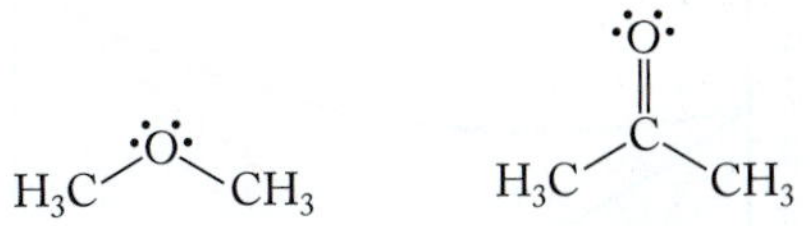

$H_3C$–O–$CH_3$
Dimethyl ether
$\mu = 1.30$ D

$H_3C$–C(=O)–$CH_3$
Acetone
$\mu = 2.88$ D

**FIGURE 11.15** Condensed structures and dipole moments of dimethyl ether and acetone.

### SAMPLE EXERCISE 11.5 Calculating the Vapor Pressure of a Solution of Volatile Solutes — LO5

What is the vapor pressure of a solution prepared by dissolving 13 g of heptane ($C_7H_{16}$) in 87 g of octane ($C_8H_{18}$) at 25°C? How much higher is the mole ratio of heptane to octane in the vapor above the solution than in the solution itself? The vapor pressures of heptane and octane at 25°C are 31 torr and 11 torr, respectively.

**Collect and Organize** We are asked to calculate the vapor pressure of a solution of a volatile solute (heptane) dissolved in a volatile solvent (octane) and to predict the composition of the vapor produced by the solution at 25°C. We know the vapor pressures of the pure solute and pure solvent at this temperature. Raoult's law (Equation 11.5) relates the vapor pressure of a mixture of volatile substances to the individual vapor pressures of the components of the mixture.

**Analyze** To calculate the vapor pressures of the volatile components of a mixture (the $x_iP_i$ terms in Equation 11.5), we need to first calculate their mole fractions in the mixture. The relative concentrations of the vapors trapped above a solution of volatile components should be proportional to the ratios of their vapor pressures—assuming the vapors behave like ideal gases. If they do, then they obey the ideal gas law, $PV = nRT$, and the partial pressure of each is proportional to its concentration (moles per liter) in the gas phase:

$$P \propto \frac{n}{V} \text{ at constant } T$$

The vapor pressure of the mixture should be between that of pure heptane and pure octane—and closer to the octane value because there is more of it. The vapor pressure of heptane is nearly three times that of octane, so the vapor should be enriched by about a factor of 3 in heptane compared to the composition of the liquid.

**Solve** The number of moles of each component is

$$13 \text{ g } C_7H_{16} \times \frac{1 \text{ mol } C_7H_{16}}{100.20 \text{ g } C_7H_{16}} = 0.130 \text{ mol } C_7H_{16}$$

$$87 \text{ g } C_8H_{18} \times \frac{1 \text{ mol } C_8H_{18}}{114.23 \text{ g } C_8H_{18}} = 0.762 \text{ mol } C_8H_{18}$$

The mole fraction of each component in the mixture is

$$x_{\text{heptane}} = \frac{0.130 \text{ mol}}{(0.130 + 0.762) \text{ mol}} = 0.146$$

$$x_{\text{octane}} = 1 - x_{\text{heptane}} = 0.854$$

Using these mole fraction values and the vapor pressures of the two hydrocarbons in Equation 11.5, we have

$$\begin{aligned} P_{\text{total}} &= x_{\text{heptane}}P_{\text{heptane}} + x_{\text{octane}}P_{\text{octane}} \\ &= 0.146(31 \text{ torr}) + 0.854(11 \text{ torr}) \\ &= 4.5 \text{ torr} + 9.4 \text{ torr} = 13.9 \text{ torr} \end{aligned}$$

As discussed above, the heptane/octane concentration ratio in the gas phase should be the same as the ratio of their vapor pressures:

$$\frac{4.5 \text{ torr}}{9.4 \text{ torr}} = 0.48$$

The mole ratio of heptane to octane in the liquid mixture is

$$\frac{0.13 \text{ mol}}{0.76 \text{ mol}} = 0.17$$

Therefore, the vapor phase is enriched in heptane by a factor of

$$\frac{0.48}{0.17} = 2.8$$

**Think About It** As expected, the vapor pressure of the mixture is between the vapor pressures of the separate components, and the vapors are enriched in the more volatile component, heptane, by about a factor of 3 compared to the liquid.

**Practice Exercise** Benzene ($C_6H_6$) is a trace component of gasoline. What is the mole ratio of benzene to octane in the vapor above a solution of 11% benzene and 89% octane by mass at 25°C? The vapor pressures of benzene and octane at 25°C are 95 torr and 11 torr, respectively.

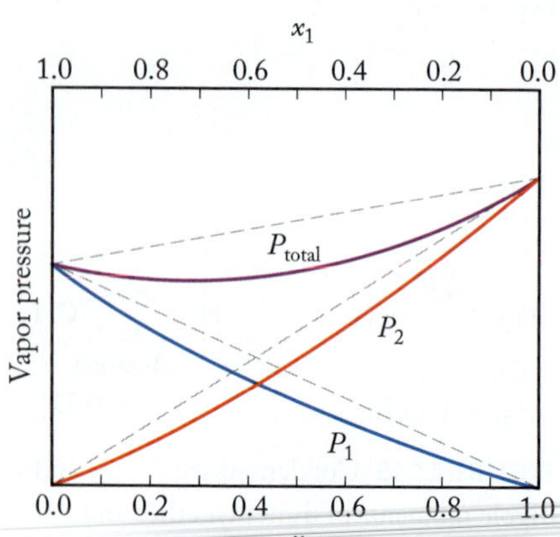


(a)

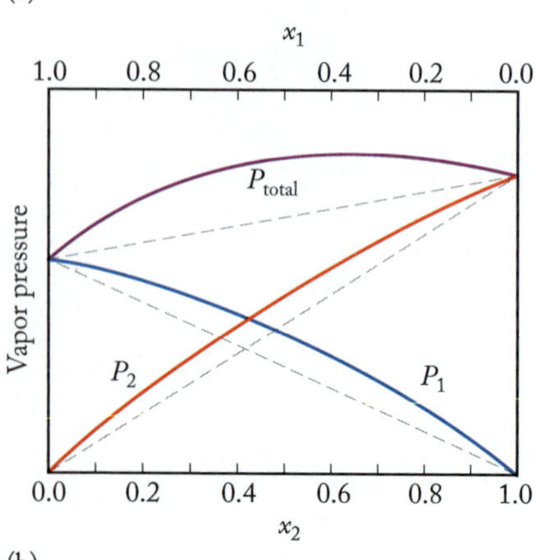


(b)

**FIGURE 11.16** In a mixture of two volatile substances, the vapor pressures $P_1$ and $P_2$ may deviate from the ideal behavior predicted by Raoult's law and described by the dashed lines. (a) If solute–solvent interactions are *stronger* than solvent–solvent or solute–solute interactions, the deviations from Raoult's law are *negative*. (b) If solute–solvent interactions are *weaker* than solvent–solvent or solute–solute interactions, the deviations are *positive*.

**ideal solution** a solution that obeys Raoult's law.

Raoult's law applies to **ideal solutions**. What is an ideal solution? Solutions such as the mixture of heptane and octane in Sample Exercise 11.5 obey Raoult's law when the intermolecular interactions between all the molecules in the mixture have comparable strengths. In the case of a binary mixture in which the more abundant component (in terms of number of moles) is designated the *solvent* and the other component is the *solute*, an ideal solution is one in which the strengths of the solvent–solvent, solute–solute, and solute–solvent interactions are nearly the same.

Solute–solvent interactions, though, are sometimes stronger than solvent–solvent or solute–solute interactions. When this happens, the *adhesive* forces between solute and solvent molecules are greater than the *cohesive* forces between solute molecules and between solvent molecules. As a result, vaporization of both the solute and solvent is inhibited by their strong attraction for each other in the liquid phase, which produces *negative* deviations from the vapor pressures predicted by Raoult's law, as shown in **Figure 11.16(a)**. Solutions of chloroform and

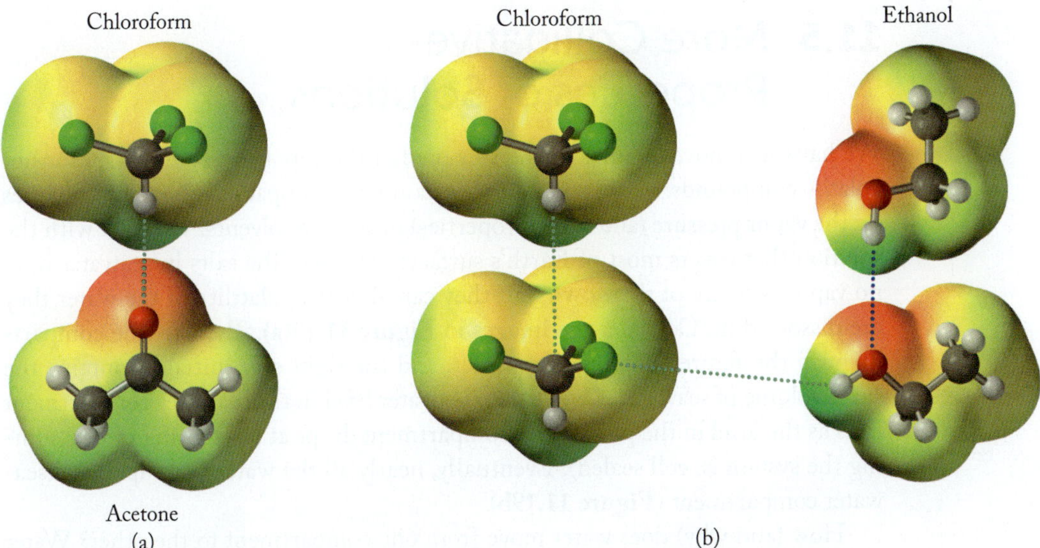


**FIGURE 11.17** (a) In a solution of chloroform and acetone, solute–solvent interactions are stronger than solute–solute or solvent–solvent interactions. (b) In a solution of chloroform and ethanol, solute–solvent interactions are weaker than solute–solute or solvent–solvent interactions.

acetone (**Figure 11.17a**) exhibit this behavior, in part due to the strong dipole–dipole interactions that form between acetone and chloroform molecules.

*Positive* deviations from Raoult's law (**Figure 11.16b**) are observed when the molecules of solute–solvent pairs experience *weaker* adhesive interactions than the cohesive forces between solute molecules and between solvent molecules. Solutions of chloroform and ethanol (**Figure 11.17b**) exhibit this behavior because ethanol molecules hydrogen-bond more strongly with each other than with molecules of $CHCl_3$, thereby reducing the likelihood of strong solute–solvent interactions.

**CONNECTION** In Chapter 6, we learned that strong *adhesive* forces between the molecules of water and the molecular structure of a capillary coupled with strong *cohesive* forces between molecules of water produce capillary action.

## CONCEPT TEST

Which of the following solutions is least likely to follow Raoult's law: (a) acetone and ethanol; (b) pentane and hexane; (c) methanol and water? The molecular structures of these molecules are shown in **Figure 11.18**.

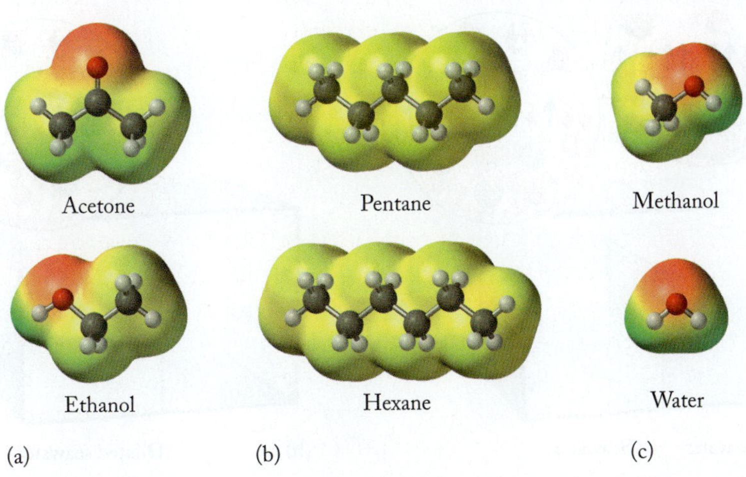


**FIGURE 11.18** The molecular components of three binary mixtures.

## 11.5 More Colligative Properties of Solutions

We have seen how Raoult's law lets us predict the vapor pressure of solutions of volatile compounds. In this section, we examine the impact of *non*volatile solutes on the vapor pressure (and other properties) of volatile solvents. We start with the solution that covers most of Earth's surface: seawater. The salts in seawater have no vapor pressure of their own, but they can alter the volatility of the water they are dissolved in. Consider the images in **Figure 11.19(a).** The left-side compartment in the figure contains pure water, and the right side initially contains the same volume of seawater. Over time, the water level in the seawater compartment rises as the level in the pure-water compartment drops at a matching rate (assuming the system is well sealed). Eventually, nearly all the water ends up in the seawater compartment (**Figure 11.19b**).

How (and why) does water move from one compartment to the other? Water in both compartments evaporates, producing water vapor in the headspace above them. Eventually the amount of water vapor stabilizes, exerting a partial pressure that matches the vapor pressure of water at the temperature inside the chamber. At this point, the overall rate of evaporation from the compartments matches the rate of condensation so that $P_{H_2O}$ remains constant.

Because both compartments share a common headspace, the water vapor is equally distributed throughout it. As a result, the rate of condensation into both compartments should be equal. If the rates of evaporation were also equal, the two liquid levels would not change over time. However, the levels clearly do change, so the rates of evaporation must not be equal. The rate of evaporation of $H_2O$ from seawater must be lower because water is transported from the pure-water to the seawater compartment. Thus, the presence of solute particles lowers the vapor pressure of the water in seawater. Let's explore why.

The lower vapor pressure of seawater can be understood in the context of Raoult's law. In this case the vapor pressure of a solution, $P_{\text{solution}}$, that includes a volatile solvent is related to the mole fraction of the solvent times the vapor pressure of the pure solvent, $P^\circ_{\text{solvent}}$, according to the following variation of Raoult's law:

$$P_{\text{solution}} = x_{\text{solvent}} P^\circ_{\text{solvent}} \quad (11.6)$$

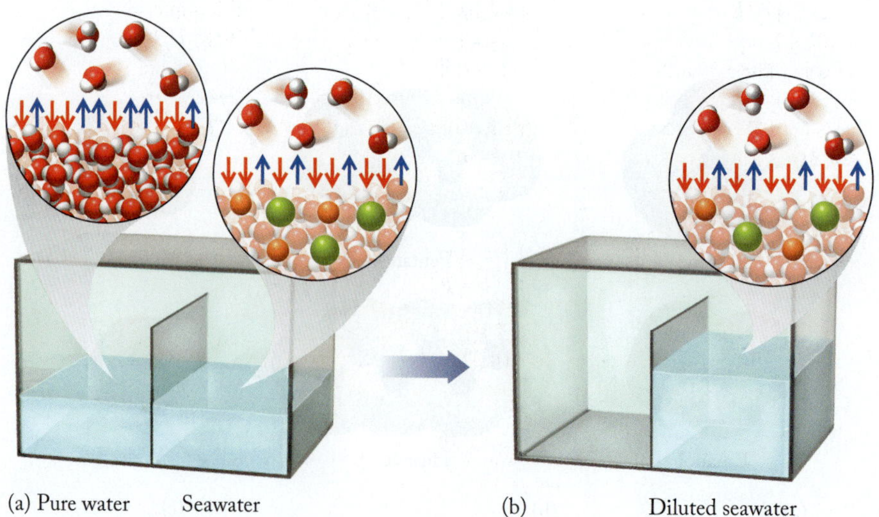


**FIGURE 11.19** (a) Adjoining compartments are partially filled with pure water and seawater. (b) The slightly higher vapor pressure of the pure water leads to a net transfer of water from the pure-water compartment to the seawater compartment.

According to Equation 11.6, the more nonvolatile solute particles there are in a solution, the lower the mole fraction of volatile solvent particles, and the lower the vapor pressure of the solution. In addition, the value of $P_{\text{solution}}$ depends only on the number of solute particles and not on their nature. In other words, its lower vapor pressure is a colligative property of the solution.

Let's use Raoult's law to calculate the vapor pressure of a sample of seawater that contains 53.6 moles of water and 1.120 moles of ions. The mole fraction of water in seawater is

$$x_{H_2O} = \frac{53.6\ \cancel{\text{mol}}}{(53.6 + 1.120)\ \cancel{\text{mol}}} = 0.980$$

Thus, the vapor pressure of seawater is 98.0% of the vapor pressure of pure water at any given temperature. This slightly lower vapor pressure explains the slow migration of water from the pure-water compartment to the seawater compartment in Figure 11.19.

## CONCEPT TEST

Is the vapor pressure of a liquid substance an intensive or extensive property of the substance?

### SAMPLE EXERCISE 11.6 Calculating the Vapor Pressure of a Solution of One or More Nonvolatile Solutes LO5

Most of the world's supply of maple syrup is produced in Quebec, Canada, where the sap from maple trees is evaporated (**Figure 11.20**) until the concentration of sugar (mostly sucrose) in the sap reaches at least 66% by mass. What is the vapor pressure in atm of a 66% aqueous solution of sucrose ($\mathcal{M}$ = 342.30 g/mol) at 100°C? Assume that the solution obeys Raoult's law.

**Collect and Organize** We are asked to calculate the vapor pressure of a 66% solution by mass of sucrose. The vapor pressure of the solution is a colligative property that depends on the mole fraction of solvent ($x_{H_2O}$):

$$P_{\text{solution}} = x_{H_2O} P^{\circ}_{H_2O}$$

The vapor pressure of pure water ($P^{\circ}_{H_2O}$) at 100°C (its normal boiling point) is 1.00 atm. Its molar mass is 18.02 g/mol.

**Analyze** A concentration of 66% by mass means that a 100 g sample of the solution contains 66 g of sucrose and 34 g of water. To calculate the mole fraction of water in a 100 g sample, we need to convert these two masses to moles of sucrose and moles of water and then calculate the mole fraction of water. A mass of 66 g of sucrose ($\mathcal{M}$ = 342 g/mol) represents about 0.2 mole. There are about 2 moles of water in 34 g of water ($\mathcal{M}$ = 18 g/mol), so the mole fraction of water is about 2/2.2, or about 0.9, and the vapor pressure should be about 0.9 atm.

**Solve** In 100 g of syrup there are

$$66\ \cancel{\text{g sucrose}} \times \frac{1\ \text{mol sucrose}}{342.30\ \cancel{\text{g sucrose}}} = 0.193\ \text{mol sucrose}$$

$$34\ \cancel{\text{g water}} \times \frac{1\ \text{mol water}}{18.02\ \cancel{\text{g water}}} = 1.89\ \text{mol water}$$

The mole fraction of water is

$$x_{H_2O} = \frac{1.89\ \cancel{\text{mol}}}{(1.89 + 0.193)\ \cancel{\text{mol}}} = 0.91$$

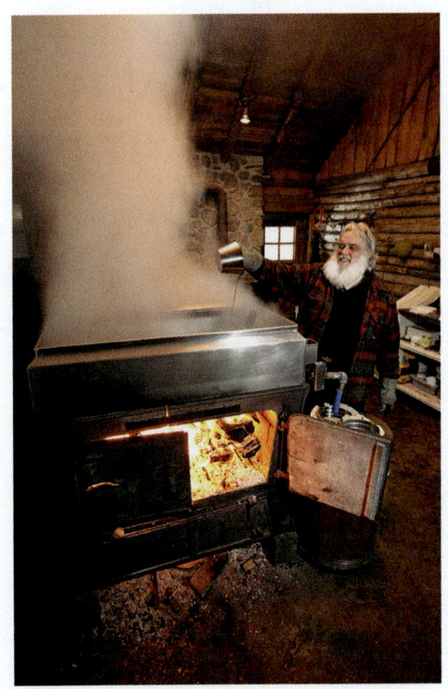

**FIGURE 11.20** Nonvolatile solutes in the sap from maple trees lower the vapor pressure of the solution. Evaporation of water as the sap is converted to maple syrup requires a considerable amount of time and energy.

**boiling point elevation** an increase in the normal boiling point of solutions relative to that of their pure solvents.

**freezing point depression** a decrease in the normal freezing point of solutions relative to that of their pure solvents.

**molality (*m*)** concentration expressed as the number of moles of solute per kilogram of solvent.

Vapor pressure of the solution is

$$P_{\text{solution}} = x_{H_2O}P^{\circ}_{H_2O} = 0.91 \times 1.00 \text{ atm} = 0.91 \text{ atm}$$

**Think About It** The calculated value is close to the estimated one and less than the vapor pressure of pure water, as we expect in a concentrated solution. Because its vapor pressure is less than atmospheric pressure, maple syrup does not boil at 100°C. Its boiling point at $P = 1.00$ atm, like those of other aqueous solutions containing nonvolatile solutes, is higher than 100°C.

**Practice Exercise** The liquid used in automobile cooling systems is prepared by dissolving ethylene glycol ($HOCH_2CH_2OH$; $\mathcal{M} = 62.07$ g/mol) in water. What is the vapor pressure at 50°C of a solution prepared by mixing equal volumes of ethylene glycol (density, 1.114 g/mL) and water (density, 1.000 g/mL)? Express your answer to the nearest torr. The vapor pressure of pure water at 50°C is 92 torr. The vapor pressure of ethylene glycol is less than 1 torr at 50°C.

The impact of solute particles on the properties of a concentrated aqueous solution can be seen in the phase diagram in **Figure 11.21**. The orange line represents the boiling points of pure water at different pressures and temperatures. Keep in mind that the pressure values along this line, which represent a series of boiling points, are actually vapor pressure values at different temperatures. This is because water boils when its vapor pressure reaches ambient pressure. The dark red line in Figure 11.21 represents vapor pressures of a concentrated aqueous solution, such as antifreeze, over the same range of temperatures. Notice that the vapor pressure of the solution is lower than that of pure water at all temperatures and that the solution reaches a vapor pressure of 1 atm at a temperature well above 100°C. This phenomenon is called **boiling point elevation.** The difference between the normal boiling point and the boiling point of the solution is labeled $\Delta T_b$ and is proportional to the total concentration of all the solute particles in the solution—it is, in fact, a colligative property.

At its lower end, the orange line intersects a faint blue line separating solid ice and pure liquid water (representing melting points) and a faint green line separating ice and steam (representing sublimation points). This point of intersection is

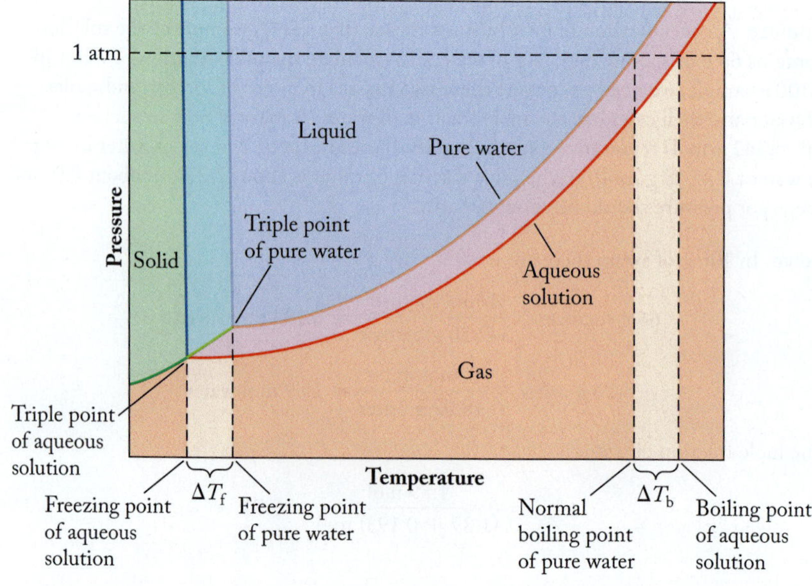


**FIGURE 11.21** Combined phase diagram for pure water and a solution of a nonvolatile solute in water. The solution's boiling point is higher than the boiling point of the water, and the solution's freezing point is lower than the freezing point of the water.

the triple point of water. There is an analogous triple point for the solution, and there are analogous lines for melting point (blue) and sublimation point (green). The blue lines are separated because solutions freeze at lower temperatures than their pure solvents—a phenomenon called **freezing point depression**. The symbol $\Delta T_f$ represents the difference in freezing points. As with boiling point elevation, $\Delta T_f$ is a colligative property; it is proportional to the total concentration of all the solute particles in solution. Before we can use this proportionality to calculate the values of $\Delta T_b$ and $\Delta T_f$, or to measure these temperature differences and use them to determine the concentrations of solutions, we need to become acquainted with the concentration unit known as *molality*.

**CONNECTION** We introduced phase diagrams in Section 6.5 to show how the physical states of substances change with changing temperatures and pressures.

## Molality

Molarity, or moles of solute per liter of solution, is probably the most widely used unit of concentration in chemistry. Mol<u>al</u>ity is closely related to molarity. It, too, is based on the number of moles of solute in solution, but instead of moles per liter of solution, **molality (*m*)** is based on moles of solute *per kilogram of solvent*:

$$m = \frac{n_{\text{solute}}}{\text{kg solvent}} \tag{11.7}$$

Molality, not molarity, is used to express solute concentrations when calculating boiling point elevation or freezing point depression because liquids, like all forms of matter, tend to expand as their temperatures increase and contract as their temperatures decrease. Exactly 1 kilogram of water has a volume of 1.0000 L at 4.0°C, but it expands to occupy 1.0434 L when heated to 100°C. This increase in volume means that the molarity of an aqueous solution is about 4% lower near its boiling point than it is at 4.0°C. However, the mole fractions of solute and solvent (which define shifts in vapor pressure, boiling point, and freezing point) remain unchanged. Molalities, by contrast, do not change with temperature because the mass of solvent in which a given quantity of solute has dissolved does not change with temperature.

**Figure 11.22** summarizes the pathways that can be followed when calculating concentrations in molality. All lead to a ratio of the number of moles of solute to the mass of solvent. Let's follow the pathway that starts at the bottom of Figure 11.22 to calculate the molality of an aqueous solution of sodium chloride that is 0.558 *M* in NaCl (approximately the NaCl concentration in seawater) and that has a density of 1.022 g/mL. We start by assuming a solution volume of one liter (the bottom box in Figure 11.22). Next, we convert this volume (1 L = 1000 mL) into a mass of solution by multiplying by its density:

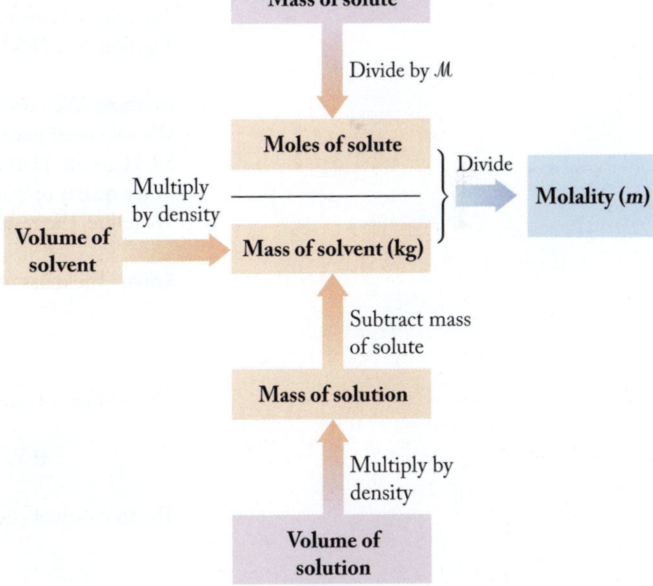


**FIGURE 11.22** Flow diagram for calculating molality.

$$1000\ \cancel{\text{mL}} \times \frac{1.022\ \text{g}}{\cancel{\text{mL}}} = 1022\ \text{g}$$

To calculate the solvent portion of this mass, we need to first calculate the mass of NaCl in it:

$$\frac{0.558\ \cancel{\text{mol NaCl}}}{1\ \text{L solution}} \times \frac{58.44\ \text{g NaCl}}{1\ \cancel{\text{mol NaCl}}} = 32.6\ \text{g NaCl in 1 L of solution}$$

We subtract this mass of solute from the total mass of solution (1022 g) to obtain the mass of solvent: (1022 g − 32.6 g) = 989 g = 0.989 kg. The molality of the solution (number of moles of solute per kilogram of solvent) is then

$$\frac{0.558 \text{ mol NaCl}}{0.989 \text{ kg solvent}} = 0.564\ m \text{ NaCl}$$

The molarity (0.558 $M$) and molality (0.564 $m$) are nearly the same, but the molality is greater than the molarity, which is true for all aqueous solutions.

### CONCEPT TEST

The difference between the molar concentration and molal concentration of any dilute aqueous solution is small. Why?

### SAMPLE EXERCISE 11.7 Calculating the Molality of a Solution LO6

A popular recipe[1] for preparing corned beef (a St. Patrick's Day favorite in many places) calls for preparing a seasoned brine (salt solution) that contains 0.50 pound of kosher salt (NaCl) dissolved in 3.0 quarts of water. Assuming the density of water is 1.00 g/mL, what is the molality of NaCl in this solution for "corning" beef?

**Collect and Organize** We are asked to calculate the molality of a salt solution prepared from a known mass of salt and volume of water. Molality is moles of solute per kilogram of solvent. We know the density of water is 1.00 g/mL, which is the same as 1.00 kg/L; the latter value should be more useful in this calculation given the volume of the brine. The table of conversion factors at the back of this book includes 1 pound = 453.59 g and 1 gallon = 3.785 L. There are four quarts in one gallon.

**Analyze** We need to convert 3.0 quarts of water into an equivalent mass in kilograms. We also need to calculate the number of moles of NaCl in 0.50 pound. Its molar mass is 58.44 g/mol. Half a pound of NaCl contains (453.59 g/2)/(58.44 g) moles, or about 4 mol. Three quarts of water is a little less than three liters, so its mass is a little less than 3 kg. Therefore, the results of our molality calculation should be about 4 mol/3 kg, or 1.3 $m$.

**Solve** The mass of water in the brine is

$$3.0 \text{ qt} \times \frac{1 \text{ gal}}{4 \text{ qt}} \times \frac{3.785 \text{ L}}{1 \text{ gal}} \times \frac{1.00 \text{ kg}}{1 \text{ L}} = 2.84 \text{ kg}$$

The number of moles of NaCl is

$$0.50 \text{ lb NaCl} \times \frac{453.59 \text{ g}}{1 \text{ lb}} \times \frac{1 \text{ mol NaCl}}{58.44 \text{ g NaCl}} = 3.88 \text{ mol NaCl}$$

The molality of the brine is

$$\frac{3.88 \text{ mol NaCl}}{2.84 \text{ kg}} = 1.4\ m \text{ NaCl}$$

**Think About It** The calculated molality is consistent with our prediction. It is rounded off to only two significant figures, which may seem stingy, but it's appropriate given the small number of significant figures in the amounts of ingredients in the recipe. Actually, the quantities in this and most recipes are typically expressed with only one significant figure.

**Practice Exercise** What is the molality of a solution prepared by dissolving 78.2 g of ethylene glycol, $HOCH_2CH_2OH$, in 1.50 L of water? Assume the density of water is 1.00 g/mL.

[1]A. Brown, *Good Eats 3*, 2011, p. 32.

## Boiling Point Elevation and Freezing Point Depression

**CHEMTOUR**
Boiling and Freezing Points of Solutions

Boiling point elevation is a colligative property, so it does not depend on the identity of the solute particles—only on their number. However, each solvent has its own sensitivity to the presence of solute particles among its own molecules, so each solvent has its own *boiling-point-elevation constant*, $K_b$. This constant is used to relate boiling point elevation ($\Delta T_b$) to the concentration of solute particles expressed in molality, or $i \cdot m$, where $i$ is the van 't Hoff factor of the solute in a solution in which its molal concentration is $m$:

$$\Delta T_b = K_b\, i\, m \tag{11.8}$$

The $K_b$ of water is 0.52°C/$m$, which means that dissolving one mole of solute particles in 1 kg of water ($i\, m = 1$) increases the boiling point of the water by 0.52°C.

**SAMPLE EXERCISE 11.8** Calculating the Boiling Point Elevation of an Aqueous Solution **LO7**

What is the boiling point of seawater in which the concentration of sea salts is equivalent to 0.560 mol of NaCl/kg of seawater, and the van 't Hoff factor of NaCl is 1.85?

**Collect and Organize** We are asked to calculate the boiling point of 0.560 mol/kg NaCl. The $K_b$ of water is 0.52°C/$m$, and its normal boiling point is 100.0°C. Boiling point elevation can be calculated using Equation 11.8: $\Delta T_b = K_b\, i\, m$.

**Analyze** To use Equation 11.8, we need to express the salt concentration in moles of dissolved particles per kilogram of *solvent*. The value we are given is in moles of NaCl per kilogram of *solution* (seawater). Therefore, we need to calculate the mass of $H_2O$ in one kilogram of seawater by subtracting the mass of NaCl and then use the mass of $H_2O$ to calculate molality. If we assume that the molal concentration we calculate is similar to the mol/kg concentration we started with, the value of ($i\, m$) will be about 2 × 0.56 $m$ or 1.12 $m$. This concentration will elevate the boiling point of water by (1.12 × 0.52)°C, or about 0.6°C.

**Solve** Mass of $H_2O$ in one kilogram of seawater:

$$\left(1.000\ \cancel{\text{kg}}\ \text{solution} \times \frac{1000\ \text{g}}{1\ \cancel{\text{kg}}}\right) - \left(0.560\ \cancel{\text{mol NaCl}} \times \frac{58.44\ \text{g NaCl}}{1\ \cancel{\text{mol NaCl}}}\right)$$

$$= 967\ \text{g} = 0.967\ \text{kg}$$

Concentration of NaCl:

$$\frac{0.560\ \text{mol NaCl}}{0.967\ \text{kg}\ H_2O} = 0.579\ m\ \text{NaCl}$$

Boiling point elevation:

$$\Delta T_b = K_b\, i\, m = 0.52\ \frac{°\text{C}}{\cancel{m}} \times 1.85 \times 0.579\ \cancel{m} = 0.56°\text{C}$$

To calculate the boiling point of seawater, we add $\Delta T_b$ to the boiling point of pure water

$$T_{b,\text{seawater}} = T_{b,\text{solvent}} + \Delta T_b = (100.0 + 0.56)°\text{C} = 100.6°\text{C}$$

**Think About It** As predicted, the boiling point of seawater is about 0.6°C higher than the boiling point of pure water. The change seems small given the saltiness of seawater, but consider that the value of $K_b$ for water is itself small.

**Practice Exercise** Crude oil pumped out of the ground may be accompanied by saline *formation water*. If the boiling point of a sample of formation water is 2.3°C above the normal boiling point of pure water, what is the molality of dissolved particles in the sample? Assume that $i = 1.85$.

Freezing point depression is another colligative property that is directly proportional to the molal concentration of dissolved solute particles:

$$\Delta T_f = K_f\, i\, m \tag{11.9}$$

Here, $\Delta T_f$ is the change in the freezing temperature of the solvent, $K_f$ is the *freezing-point-depression constant* of the solvent, and $i$ is the van 't Hoff factor of the solute in a solution in which its molal concentration is $m$.

### SAMPLE EXERCISE 11.9 Calculating the Freezing Point of a Solution LO7

What is the freezing point of automobile radiator fluid prepared by mixing equal volumes of ethylene glycol ($\mathcal{M}$ = 62.07 g/mol) and water at a temperature where the density of ethylene glycol is 1.114 g/mL and the density of water is 1.000 g/mL? The freezing-point-depression constant of water, $K_f$, is 1.86°C/$m$.

**Collect and Organize** We are asked to determine the freezing point of a 50:50 (by volume) solution of ethylene glycol in water. We are given the densities of the two liquids and the value of $K_f$, which relates the molality of ethylene glycol to freezing point depression (Equation 11.9). Ethylene glycol is a nonelectrolyte, so its $i$ value is exactly 1.

**Analyze** To use Equation 11.9, we need to convert the 50:50 volume ratio into a solute concentration expressed in moles of ethylene glycol per kilogram of water. To simplify the calculation, let's assume that we mix 1.000 liter of each of the two liquids. This is a handy volume because one liter of water has a mass of one kilogram (because a density of 1.000 g/mL is the same as 1.000 kg/L). The mass of one liter of ethylene glycol is 1114 g. It contains 1114/62.07, or about 1200/60 = 20 moles of ethylene glycol. The freezing point depression of a 20 $m$ aqueous solution of ethylene glycol is (1.86 × 1 × 20)°C, or nearly 40°C. Therefore, the freezing point of the radiator fluid should be a bit above −40°C.

**Solve** The solvent mass is

$$1.000\ \cancel{\text{L}}\ \text{H}_2\text{O} \times \frac{1000\ \cancel{\text{mL}}}{1\ \cancel{\text{L}}} \times \frac{1.000\ \cancel{\text{g}}}{\cancel{\text{mL}}} \times \frac{0.001\ \text{kg}}{1\ \cancel{\text{g}}} = 1.000\ \text{kg H}_2\text{O}$$

Moles of ethylene glycol in 1.000 L:

$$1.000\ \cancel{\text{L ethylene glycol}} \times \frac{1000\ \cancel{\text{mL}}}{1\ \cancel{\text{L}}} \times \frac{1.114\ \cancel{\text{g}}}{1\ \cancel{\text{mL}}} \times \frac{1\ \text{mol ethylene glycol}}{62.07\ \cancel{\text{g ethylene glycol}}}$$

$$= 17.95\ \text{mol ethylene glycol}$$

Concentration of ethylene glycol:

$$m = \frac{17.95\ \text{mol ethylene glycol}}{1.000\ \text{kg H}_2\text{O}} = 17.95\ m$$

Using Equation 11.9 gives us

$$\Delta T_f = K_f\, i\, m = 1.86\ \frac{°\text{C}}{\cancel{m}} \times 1 \times 17.95\ \cancel{m} = 33.4°\text{C}$$

To calculate the freezing point of the solution, we subtract $\Delta T_f$ from the freezing point of the pure solvent:

$$\text{Freezing point of radiator fluid} = T_{f,\text{solvent}} - \Delta T_f = 0.0°\text{C} - 33.4°\text{C} = -33.4°\text{C}$$

**Think About It** The answer is about what we expected. The freezing point of radiator fluid should be below the coldest expected temperatures, and −33°C (−27°F) is colder than most places in the United States ever get. In much of Alaska and Canada, however, 60% to 70% by volume solutions of ethylene glycol would be advisable to prevent engines from freezing up.

**Practice Exercise** Some Thanksgiving dinner chefs "brine" their turkeys before roasting them to help the birds retain moisture and to season the meat. Brining involves fully immersing a turkey for about 6 hours in a brine (salt solution) prepared by dissolving 2.0 pounds of salt (NaCl; $\mathcal{M}$ = 58.44 g/mol) in 2.0 gallons of water (a recipe from an American cookbook). Suppose a turkey soaking in such a solution is left for 6 hours on an unheated porch. At what temperature in °C is the brine in danger of freezing? The $K_f$ of water is 1.86°C/$m$; assume that $i$ = 1.82 for this NaCl solution.

## SAMPLE EXERCISE 11.10 Assessing Particle Interactions in Solution LO2

The experimentally measured freezing point of a 1.90 $m$ aqueous solution of NaCl is −6.57°C. What is the value of the van 't Hoff factor for this solution? Is the solution behaving ideally, or is there evidence that solute particles are interacting with one another? The freezing-point-depression constant of water is $K_f$ = 1.86°C/$m$. Assume the freezing point of pure water is 0.00°C.

**Collect and Organize** We are asked to calculate the $i$ factor of a solution of known molality on the basis of a measured freezing point and the $K_f$ value. Equation 11.8 relates $\Delta T_f$, $K_f$, the van 't Hoff factor ($i$) for this solution, and its molality ($m$). We are also asked whether the solution behaves ideally.

**Analyze** Ideally, $i$ = 2 for solutions of NaCl. However, given the high concentration of NaCl (1.90 $m$), a value less than 2 is likely.

**Solve** Rearranging Equation 11.9 and solving for $i$:

$$i = \Delta T_f / K_f m = 6.57\cancel{°C} / \left(1.86\ \frac{\cancel{°C}}{\cancel{m}} \times 1.90\ \cancel{m}\right) = 1.86$$

The value of $i$ is less than 2 (the theoretical value for NaCl), so the solution is not behaving ideally. Ion pairs must be forming in solution.

**Think About It** As predicted, the value for $i$ for this solution is less than the theoretical value of 2.

**Practice Exercise** The freezing point of a 1.12 $M$ solution of $MgSO_4$ is 2.31°C below the freezing point of pure water. If the density of the solution is 1.126 g/mL, what is the value of $i$?

In Section 11.2 we used osmotic pressure to determine the molar mass of compounds. In principle, any colligative property can be used to determine the molar mass of a substance, including measuring the boiling point of a solution containing a known quantity of a substance or measuring the melting point/freezing point of a solid solution formed from two well-mixed solids. These methods work best for solvents with large values of $K_b$ or $K_f$ and for nonelectrolyte solutes for which the van 't Hoff factor $i$ = 1. Water, for example, has fairly small $K_b$ and $K_f$ values.

Measurements of molar mass using $K_b$ or $K_f$ suffer from several disadvantages compared to osmotic pressure measurements. First, to create a solution that gives a measurable boiling point or freezing point change for a solution, the concentration of the solution has to be much higher than may be readily achievable. Second, biomaterials such as proteins and carbohydrates are nearly always available only in small quantities, and often they are not very soluble in nonaqueous solvents that have larger $K_f$ or $K_b$ values. Furthermore, these biomaterials often have high molar masses, which means that large quantities are needed to give high enough molal concentrations for reliable $\Delta T_f$ or $\Delta T_b$ measurements. Third, a solute might need to be recovered unchanged for other uses, which rules out boiling-point-elevation measurements for heat-sensitive solutes.

### SAMPLE EXERCISE 11.11 Using Freezing Point Depression to Determine Molar Mass — LO7

Eicosene is a molecular compound and nonelectrolyte with the empirical formula $CH_2$. It is used in water-resistant sunscreens. The freezing point of a solution prepared by dissolving 100 mg of eicosene in 1.00 g of benzene was 1.75°C lower than the freezing point of pure benzene. What is the molar mass of eicosene? ($K_f$ for benzene is 4.90°C/*m*.)

**Collect and Organize** We are asked to determine the molar mass of a compound. We know the mass of the compound, the mass of the solvent, the change in freezing point, and the $K_f$ of the solvent. Because the solute is a nonelectrolyte, $i = 1$. Equation 11.9 relates concentration in molality (moles of solute per kilogram of solvent) to the change in freezing point.

**Analyze** Our sample consists of 100 mg, or 0.100 g. To find the molar mass, we need to determine how many moles of eicosene are contained in the sample. We can first use the experimental data to calculate molality, the concentration term in Equation 11.9. Once we know the molality of the eicosene, we can calculate the number of moles of eicosene in the sample by using the mass of the solvent benzene. Then we can calculate the molar mass of eicosene. The empirical formula of eicosene enables us to check the validity of our answer because the molar mass of eicosene must be a whole-number multiple of the mass of $CH_2$ (14 g/mol).

**Solve** The molality of the eicosene solution is

$$m = \frac{\Delta T_f}{iK_f} = \frac{1.75\cancel{°C}}{(1 \times 4.90\cancel{°C}/m)} = 0.357\ m \text{ eicosene}$$

This means that 0.357 mole of eicosene is dissolved per kilogram of solvent. Only 1.00 g ($1.00 \times 10^{-3}$ kg) was used in this sample. Calculating the moles of eicosene in the sample:

$$0.357\ m = \frac{\text{moles of eicosene}}{1.00 \times 10^{-3}\ \text{kg benzene}}$$

$$\text{Moles of eicosene} = \frac{0.357\ \text{mol eicosene}}{1\ \cancel{\text{kg benzene}}} \times 1.00 \times 10^{-3}\ \cancel{\text{kg benzene}}$$
$$= 3.57 \times 10^{-4}\ \text{mol eicosene}$$

Because the molar mass is the mass of 1 mole of eicosene, and 100 mg of eicosene was used to prepare the solution,

$$\text{Molar mass} = \frac{0.100\ \text{g eicosene}}{3.57 \times 10^{-4}\ \text{mol}} = 280\ \text{g/mol}$$

**Think About It** The molar mass of eicosene, 280 g/mol, is reasonable because it corresponds to $(CH_2)_n$, where $n = 20$. The molecular formula of eicosene is therefore $C_{20}H_{40}$.

**Practice Exercise** A solution prepared by dissolving 360 mg of a sugar (a molecular compound and a nonelectrolyte) in 1.00 g of water froze at −3.72°C. What is the molar mass of this sugar? The value of $K_f$ of water is 1.86°C/*m*.

## 11.6 Henry's Law and the Solubility of Gases

**CHEMTOUR**
Henry's Law

Finally, we turn to solutions of gases dissolved in liquids. Recall from Chapter 6 that the relatively weak dipole–induced dipole forces between polar water molecules and nonpolar oxygen molecules lead to a concentration of dissolved oxygen sufficient to sustain aquatic plants and fish. The solubility of $O_2$ (or any sparingly soluble gas) in a liquid such as water is directly proportional to the partial pressure of the gas above the surface of the liquid. This relationship is known as **Henry's law**, in honor of William Henry (1775–1836), a British physician who first proposed the relationship. In equation form, the relationship is:

$$C_{gas} = k_H P_{gas} \tag{11.10}$$

where $C_{gas}$ represents the maximum concentration (solubility) of a gas in a particular solvent, $k_H$ is the Henry's law constant for the gas in that solvent, and $P_{gas}$ is the partial pressure of the gas in the environment surrounding the solvent. When $C_{gas}$ is expressed in molarity (mol/L), the units of the Henry's law constant are moles per liter-atmosphere, mol/(L · atm). **Table 11.1** lists $k_H$ values for several common gases in water. The magnitude of $k_H$ reflects both differences in the intermolecular forces in the dissolved gases and deviations from ideal gas behavior. For example, small, weakly polarizable noble gas atoms such as He have smaller Henry's law constants than larger, more polarizable molecules like $O_2$ or $CH_4$. Polar molecules like $H_2S$ and $CH_3Cl$ have even larger values of $k_H$.

**TABLE 11.1** Henry's Law Constants for Several Gases in Water at 20°C

| Gas | $k_H$ [mol/(L · atm)] |
|---|---|
| He | $3.5 \times 10^{-4}$ |
| $O_2$ | $1.3 \times 10^{-3}$ |
| $N_2$ | $6.7 \times 10^{-4}$ |
| $CO_2$ | $3.5 \times 10^{-2}$ |
| $H_2S$ | $8.7 \times 10^{-2}$ |
| $CH_4$ | $1.5 \times 10^{-3}$ |
| $CH_3Cl$ | $1.3 \times 10^{-1}$ |

According to Henry's law, the concentration of dissolved oxygen in blood is proportional to the partial pressure of oxygen in the air we inhale and thus proportional to atmospheric pressure. This is an accurate statement of Henry's law, but it does not tell the whole story. It does not mean, for example, that residents of Denver, Colorado (average $P_{O_2} = 0.178$ atm), live with less blood oxygen than residents of New York City or anywhere else near sea level (where $P_{O_2}$ averages 0.209 atm).

The concentration of $O_2$ in the blood also depends on the concentration of hemoglobin, an oxygen-transporting protein that attaches to molecules of $O_2$ as blood passes through the lungs. The hemoglobin binding sites in most people can be saturated with $O_2$ even when the $P_{O_2}$ in their lungs is as low as 0.110 atm. If $P_{O_2}$ decreases to about 0.066 atm (as it does on high mountains), only about 80% of the hemoglobin molecules pick up molecules of $O_2$ passing through the lungs. Over several weeks, the body responds to low $P_{O_2}$ by producing more hemoglobin so that the same amount of $O_2$ is delivered to tissues. Some endurance athletes choose to train at high altitudes so that, when they compete at lower altitudes, their bodies transport $O_2$ more efficiently.

**Henry's law** the principle that the concentration of a sparingly soluble gas in a liquid is proportional to the partial pressure of the gas.

**FIGURE 11.23** Opening a warm can of soda can be risky because dissolved $CO_2$ gas escapes from the fluid as the pressure is released.

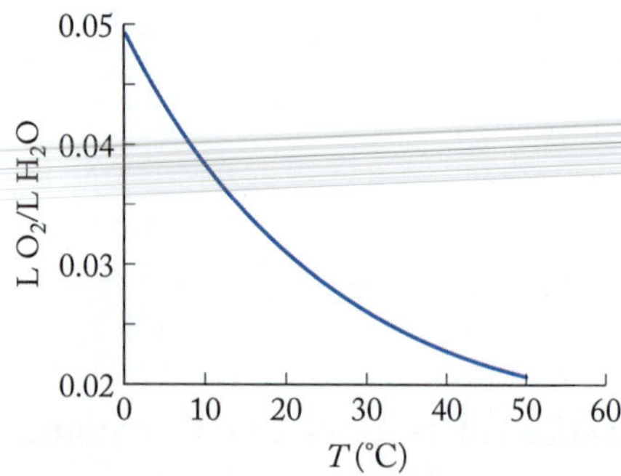


**FIGURE 11.24** The solubility of $O_2$ in water decreases as the temperature increases (data are based on $P_{O_2}$ = 1.00 atm).

**TABLE 11.2** Henry's Law Constants for $O_2$ Gas in Water

| Temperature (°C) | $k_H$ [mol/(L · atm)] |
|---|---|
| 0 | $2.18 \times 10^{-3}$ |
| 5 | $1.90 \times 10^{-3}$ |
| 10 | $1.68 \times 10^{-3}$ |
| 15 | $1.50 \times 10^{-3}$ |
| 20 | $1.35 \times 10^{-3}$ |
| 25 | $1.23 \times 10^{-3}$ |
| 30 | $1.12 \times 10^{-3}$ |
| 35 | $1.04 \times 10^{-3}$ |
| 40 | $0.96 \times 10^{-3}$ |
| 45 | $0.90 \times 10^{-3}$ |
| 50 | $0.84 \times 10^{-3}$ |

The solubility of gases in water and other liquids decreases as the temperature of the liquid increases. Opening a can of warm soda leads to an explosive release of $CO_2$ (**Figure 11.23**) because the solubility of gas is lower in warm soda than in cold soda, but the increase in pressure caused by an increased amount in gas in the can forces more gas into solution than would normally dissolve at the higher temperature. When the can is opened, the pressure suddenly drops and the $CO_2$ gas escapes from the fluid, resulting in bubbles and fizz. The graph in **Figure 11.24** shows that the solubility of $O_2$ in water decreases as the temperature rises from 0°C to 50°C. The $k_H$ value for $O_2$, which decreases as water temperature increases (**Table 11.2**), is consistent with this trend, too.

**SAMPLE EXERCISE 11.12** Calculating Gas Solubility by Using Henry's Law **LO8**

Lake Titicaca is located high in the Andes Mountains between Peru and Bolivia. Its surface is 3811 m above sea level, where the average atmospheric pressure is 0.636 atm. During the summer, the average temperature of the water's surface rarely exceeds 15°C. What is the solubility of oxygen in Lake Titicaca at that temperature? Express your answer in molarity and in mg/L.

**Collect and Organize** We are asked to determine the solubility (the maximum concentration) of oxygen gas in water at 0.636 atm and 15°C. According to Table 9.1, dry air is 20.95% $O_2$. Henry's law (Equation 11.10) relates the solubility of gases to their partial pressures. The Henry's law constants for $O_2$ in water at different temperatures are listed in Table 11.2.

**Analyze** We need the partial pressure of oxygen to calculate its molar solubility by using Equation 11.10. According to Equation 9.28, partial pressure is the product of the mole fraction of $O_2$ in air (20.95% expressed as a decimal value) and atmospheric pressure (0.636 atm). To express solubility in mg/L will require converting moles to grams by multiplying by the molar mass of $O_2$. To estimate our answer, we note that all the Henry's law constants for $O_2$ in water are about $10^{-3}$ mol/(L · atm), and the value of $P_{O_2}$ is about $0.2 \times 0.6 = 0.12$ atm, so the product of $k_H \times P_{O_2}$ (the solubility according to Equation 11.10) will be around $10^{-4}$ mol/L. Multiplying this value by the molar mass of $O_2$ (32 g/mol) should give us a solubility near $3 \times 10^{-3}$ g/L, or 3 mg/L.

**Solve** The partial pressure of oxygen is calculated using Equation 9.28:

$$P_{O_2} = x_{O_2}P_{\text{total}} = (0.2095)(0.636 \text{ atm}) = 0.133 \text{ atm}$$

We use this value for $P_{O_2}$ in Equation 11.10 and the $k_H$ value for $O_2$ in water at 15°C from Table 11.2 to calculate the solubility ($C_{O_2}$):

$$C_{O_2} = k_H P_{O_2} = \left(1.50 \times 10^{-3} \frac{\text{mol}}{\text{L} \cdot \cancel{\text{atm}}}\right)(0.133 \cancel{\text{atm}}) = 2.00 \times 10^{-4} \text{ mol/L}$$

$$(2.00 \times 10^{-4} \cancel{\text{mol}}/\text{L})(32.00 \text{ g}/\cancel{\text{mol}}) = 6.40 \times 10^{-3} \text{ g/L} = 6.40 \text{ mg/L}$$

**Think About It** The calculated solubility values in mol/L and mg/L are in the same ballpark as those we predicted and illustrate the limited solubility of $O_2$ in water. Very little $O_2$ dissolves because it is a nonpolar solute and water is a polar solvent. As a result, the solute–solvent intermolecular forces that promote solubility are weak in this case.

**Practice Exercise** If the pressure of $CO_2$ inside a 1-liter bottle of seltzer is 3.0 atm at 20°C, how much $CO_2$ is dissolved in the seltzer? Express your answer in grams.

## SAMPLE EXERCISE 11.13 Integrating Concepts: Antifreeze in Car Batteries

The aqueous electrolyte inside a fully charged lead–acid battery (the kind used to start automobile engines) is 35% (by volume) sulfuric acid, $H_2SO_4$. Pure sulfuric acid is a dense (1.84 g/mL), nonvolatile liquid (its vapor pressure at 25°C is <0.001 torr). It is also a strong acid that is completely ionized in concentrated aqueous solutions:

$$H_2SO_4(aq) \rightarrow H^+(aq) + HSO_4^-(aq)$$

The power to start the engine comes from the following chemical reaction as the battery is discharged:

$$Pb(s) + PbO_2(s) + 2\,H_2SO_4(aq) \rightarrow 2\,PbSO_4(s) + 2\,H_2O(\ell)$$

a. What is the vapor pressure of the solution at 25°C?
b. What is the freezing point of the solution?
c. How do vapor pressure and freezing point change as the battery is discharged?

**Collect and Organize** We know the concentration of a solution expressed as the volume ratio of a nonvolatile liquid solute ($H_2SO_4$, $\mathcal{M}$ = 98.08 g/mol) to solvent ($H_2O$). We are asked to calculate the freezing point of the solution and its vapor pressure at 25°C. We also need to describe how these freezing point and vapor pressure values change during a reaction in which $H_2SO_4$ is a reactant and $H_2O$ is a product. The vapor pressure of such a solution can be calculated using Equation 11.6: $P_{solution} = x_{solvent}P°_{solvent}$. Its freezing point can be calculated using Equation 11.9: $\Delta T_f = K_f\, i\, m$. The vapor pressure of pure water at 25°C is 23.8 torr, and its $K_f$ value is 1.86°C/*m*.

**Analyze** Solutions of nonvolatile solutes in volatile solvents have lower vapor pressures and freezing points than their pure solvents. To find out how much lower our solution's values are, we must first calculate the mole fraction of the solvent in solution (needed in Equation 11.6) and the molal concentration of solute (needed in Equation 11.9). As the battery is discharged, the concentration of $H_2SO_4$ decreases. Assume the $H_2SO_4$ completely dissociates into $H^+$ and $HSO_4^-$ ions, so the van 't Hoff factor of sulfuric acid is 2.

**Solve**

a. If we assume a sample volume of 100 mL, then 35% $H_2SO_4$ consists of 35 mL of $H_2SO_4$ and (100 − 35) = 65 mL of $H_2O$. Converting these volumes into masses and then moles,

$$35\ \text{mL}\ H_2SO_4 \times 1.84\ \frac{\text{g}}{\text{mL}} \times \frac{1\ \text{mol}\ H_2SO_4}{98.08\ \text{g}\ H_2SO_4} \times \frac{2\ \text{mol ions}}{1\ \text{mol}\ H_2SO_4}$$

$$= 1.31\ \text{mol ions}$$

$$65\ \text{mL}\ H_2O \times \frac{1.00\ \text{g}}{\text{mL}} \times \frac{1\ \text{mol}\ H_2O}{18.02\ \text{g}\ H_2O}$$

$$= 3.61\ \text{mol}\ H_2O$$

The mole fraction of $H_2O$ is then

$$3.61\ \text{mol}/(3.61 + 1.31)\ \text{mol} = 0.734$$

And the vapor pressure of the battery acid is

$$P_{solution} = x_{solvent}P°_{solvent} = 0.734 \times 23.8\ \text{torr} = 17.5\ \text{torr}$$

b. To calculate freezing point depression, we first need to calculate the molality of solute particles:

$$\frac{1.31\ \text{mol ions}}{65\ \text{mL}\ H_2O \times 1.00\ \text{g/mL}} \times \frac{1000\ \text{g}}{\text{kg}} = 20.2\ \text{mol/kg} = 20.2\ m$$

Calculating freezing point depression ($\Delta T_f$):

$$\Delta T_f = K_f m = 1.86°\text{C}/m \times 20.2\ m = 38°\text{C}$$

Therefore, the freezing point of battery acid is −38°C.

c. The chemical reaction that provides the battery's energy consumes solute, converting it into more solvent. The result is a less concentrated solution of sulfuric acid and a greater mole fraction of $H_2O$. Such a solution has a higher freezing point (closer to that of pure water) and a greater vapor pressure (again, closer to that of pure water).

**Think About It** We did not include the van 't Hoff factor of 2 when using Equation 11.9 in part b because it had already been used to calculate the number of moles of solute particles in part a. Thus, the *m* value we used in part b was actually the product of *i* and *m*.

## SUMMARY

**LO1** In **osmosis**, solvent flows through a semipermeable membrane from a solution of lower solute concentration into a solution of higher solute concentration. **Osmotic pressure (*Π*)** is defined as the pressure required to halt the net flow of solvent through a semipermeable membrane. It is a **colligative property** of a solution that depends only on the concentration of dissolved particles and not their identities. In **reverse osmosis (RO)**, water is pumped through a semipermeable membrane with a pressure that reverses osmotic flow. The technique is used to purify water. (Sections 11.1 and 11.2)

**LO2** The **van 't Hoff factor (*i*)** accounts for the colligative properties of solutions of electrolytes, which may form **ion pairs** that reduce the number of particles in solution. (Section 11.2)

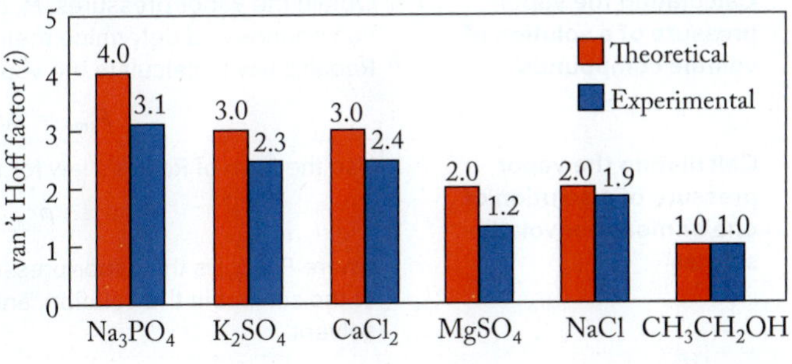

**LO3** Colligative properties, such as osmotic pressure, of solutions of nonelectrolytes can be used to determine the molar mass of nonelectrolytes. (Section 11.2)

**LO4** Molecules at the surface of a liquid evaporate by breaking intermolecular interactions with neighboring molecules and entering the gas phase. The **vapor pressure** of a **volatile** liquid is proportional to the fraction of its molecules that enter the gas phase and increases with increasing temperature as determined by the **Clausius–Clapeyron equation**. (Section 11.3)

**LO5** Mixtures of volatile liquids can be separated by **fractional distillation**. The vapor pressure of an **ideal solution** of volatile compounds follows **Raoult's law**. (Section 11.4)

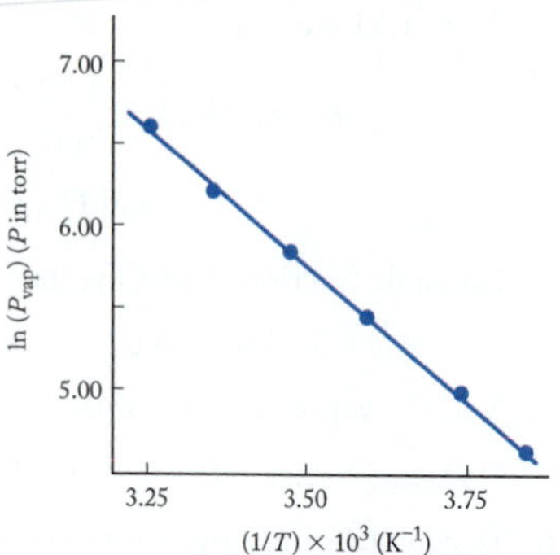


**LO6** The units used for describing concentration include molarity and **molality** (*m*), which is the number of moles of solute per kilogram of solvent. (Section 11.5)

**LO7** Nonvolatile solutes in solution decrease the solvent's vapor pressure, elevate the solvent's boiling point, and depress its freezing point. **Boiling point elevation** and **freezing point depression** are colligative properties and can be used to determine the molar mass of the solute. (Section 11.5)

**LO8** According to **Henry's law**, the solubilities of gases increase with increasing partial pressure. Solubilities decrease with increasing temperature. (Section 11.6)

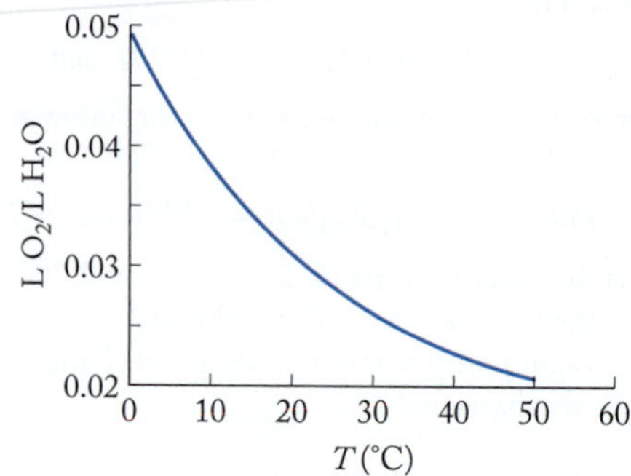


## PARTICULATE PREVIEW WRAP-UP

Ion–dipole forces form between the sodium cations and the δ− oxygen atoms and between the chloride anions and the δ+ hydrogen atoms.

The amount of vapor is reduced above a solution when compared to pure solvent, and consequently the boiling point increases.

## PROBLEM-SOLVING SUMMARY

| Type of Problem | Concepts and Equations | Sample Exercises |
|---|---|---|
| **Calculating osmotic pressure and reverse osmotic pressure** | $\Pi = iMRT$ (11.2)<br>where $\Pi$ is the osmotic pressure, $i$ is the van 't Hoff factor, $M$ is the molar concentration of the solution, and $T$ is the absolute temperature of the solution. | **11.1, 11.2** |
| **Determining molar mass from osmotic pressure** | Use $\Pi = iMRT$ to calculate the molarity, $M$, of a solution containing a known mass of solute. Then use $M$ to determine the molar mass of the solute. | **11.3** |
| **Calculating vapor pressure ($\Delta H_{vap}$) by using the Clausius–Clapeyron equation** | $\ln\left(\frac{P_{vap,T_2}}{P_{vap,T_1}}\right) = -\frac{\Delta H_{vap}}{R}\left(\frac{1}{T_2} - \frac{1}{T_1}\right)$ (11.4) | **11.4** |
| **Calculating the vapor pressure of a solution of volatile compounds** | Obtain the vapor pressures, $P_i$, of the components of a solution as pure compounds and determine their mole fractions ($x_i$) in the solution. Use Raoult's law to calculate the vapor pressure of the solution:<br>$P_{total} = x_1P_1^\circ + x_2P_2^\circ + x_3P_3^\circ + \cdots$ (11.5) | **11.5** |
| **Calculating the vapor pressure of a solution of one or more nonvolatile solutes** | Use the form of Raoult's law for nonvolatile solutes:<br>$P_{solution} = x_{solvent}P^\circ_{solvent}$ (11.6)<br>where $P_{solution}$ is the vapor pressure of the solution, $x_{solvent}$ is the mole fraction of the solvent in the solution, and $P^\circ_{solvent}$ is the vapor pressure of the pure solvent. | **11.6** |
| **Calculating the molality of a solution** | $m = \frac{n_{solute}}{\text{kg solvent}}$ (11.7) | **11.7** |

| Type of Problem | Concepts and Equations | Sample Exercises |
|---|---|---|
| **Calculating boiling points and freezing points of solutions** | $\Delta T_b = K_b i\, m$ (11.8)<br>where $\Delta T_b$ is the elevation in the boiling point of the solvent and<br>$\Delta T_f = K_f i\, m$ (11.9)<br>where $\Delta T_f$ is the depression in the freezing point of the solvent. | **11.8, 11.9** |
| **Assessing ion clustering in electrolyte solutions** | Use measured freezing point depression or boiling point elevation values and Equation 11.8 or 11.9 to determine the value of the van 't Hoff factor. | **11.10** |
| **Determining molar mass from boiling point elevation or freezing point depression** | Use $\Delta T_b = K_b i\, m$ or $\Delta T_f = K_f i\, m$ to calculate the molality, $m$, of a solution containing a known mass of solute. Then use $m$ to determine the molar mass of the solute. | **11.11** |
| **Calculating gas solubility by using Henry's law** | $C_{gas} = k_H P_{gas}$ (11.10)<br>where $k_H$ is the Henry's law constant for the gas in that solvent at the given temperature. | **11.12** |

## VISUAL PROBLEMS

*(Answers to boldface end-of-chapter questions and problems are in the back of the book.)*

**11.1.** Figure P11.1 provides a particle-level view of a sealed container partially filled with a solution that has two components: X (blue spheres) and Y (red spheres). Which of the following statements about substances X and Y are true?

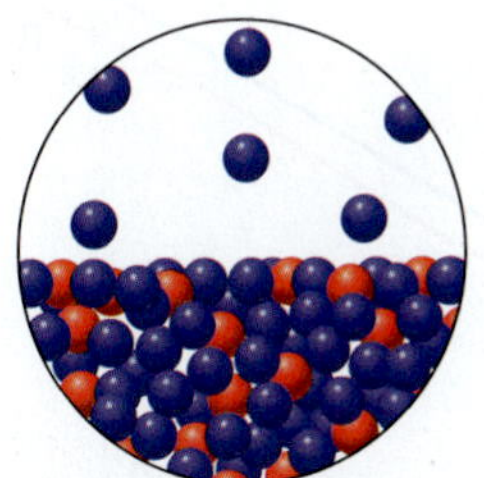

**FIGURE P11.1**

a. X is the solvent in this solution.
b. Pure Y is a volatile liquid.
c. If Y were not present, there would be fewer X particles in the gas above the liquid solution.
d. The presence of Y increases the vapor pressure of X.

11.2. Figure P11.2 provides a particle-level view of a sealed container partially filled with a solution of two miscible liquids: X (blue spheres) and Y (red spheres). Which of the following statements about substances X and Y are true?

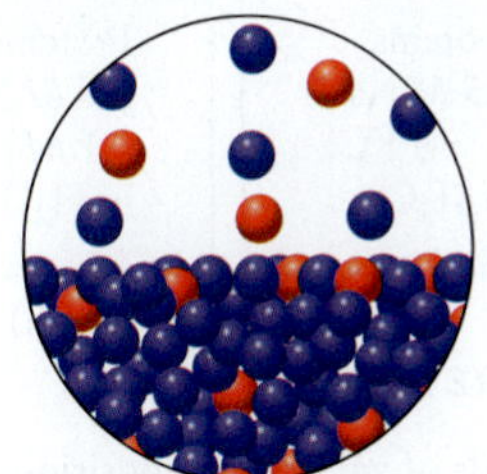

**FIGURE P11.2**

a. Y is the solvent in this solution.
b. Pure Y has a higher vapor pressure than pure X.
c. The presence of Y in the solution lowers the vapor pressure of X.
d. If Y were not present, there would be fewer total particles in the gas above the liquid solution.

**11.3.** Figure P11.3 provides particle-level views of 0.001 *M* aqueous solutions of the following four solutes: $C_6H_{12}O_6$, NaCl, $MgCl_2$, and $K_3PO_4$. The blue spheres represent particles of solute. Which compounds are represented in images (a)–(d)?

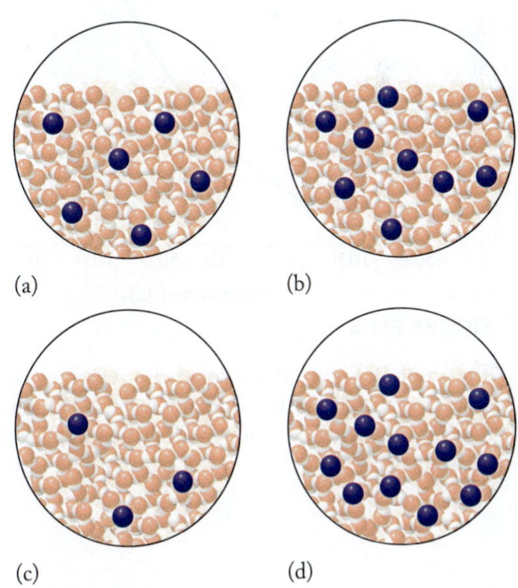


**FIGURE P11.3**

11.4. Which of the four solutions in Figure P11.4 has the highest (i) vapor pressure; (ii) boiling point; (iii) freezing point; (iv) osmotic pressure?

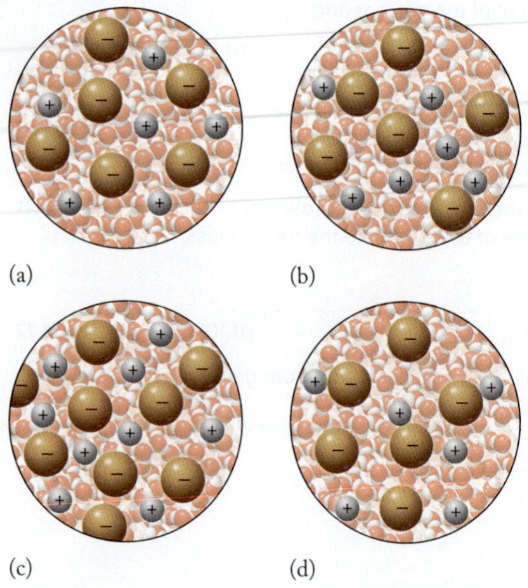


**FIGURE P11.4**

**11.5.** Use the graph in Figure P11.5 to estimate the normal boiling points of substances X and Y. Molecules of which substance experience the stronger intermolecular forces?

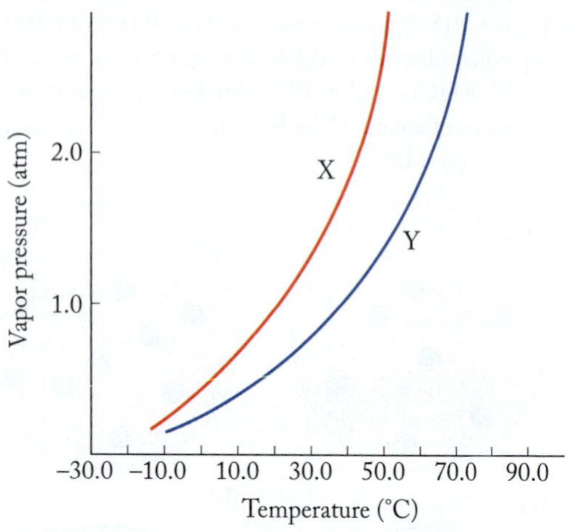


**FIGURE P11.5**

11.6. The graph in Figure P11.5 could also represent the effect of adding a nonvolatile solute to a solvent. Which curve, X or Y, represents the solution after the solute has been added?

***11.7.** Which of the three curves of osmotic pressure versus temperature in Figure P11.7 represents a strong electrolyte if the concentrations of solutions A, B, and C are all $1.21 \times 10^{-5}$ *M*?

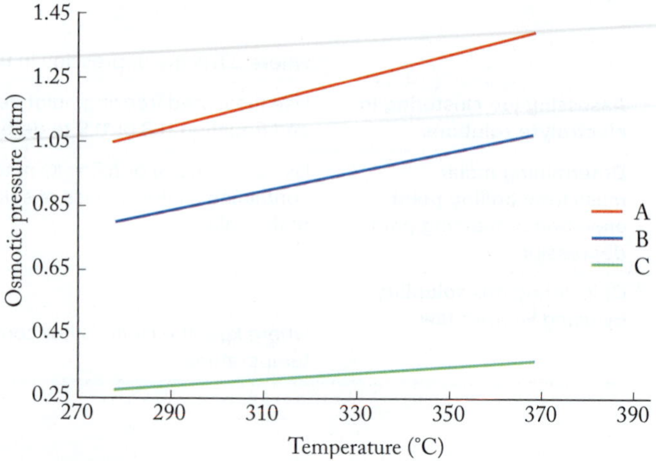


**FIGURE P11.7**

11.8. Which curve of freezing point depression versus molality in Figure P11.8 represents the behavior of an ideal solution of a 1:1 electrolyte? What might the other curve represent?

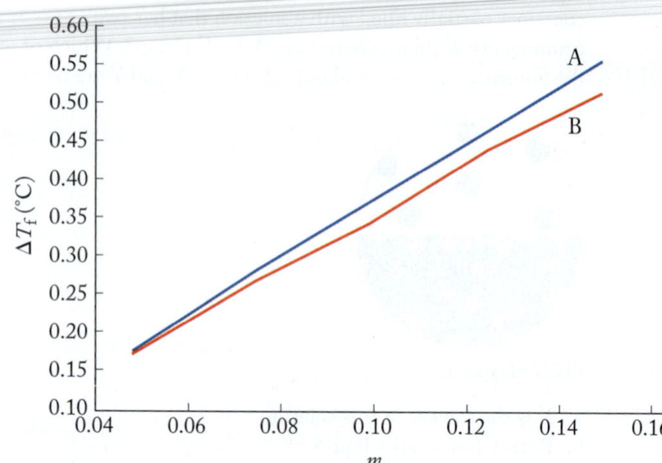


**FIGURE P11.8**

***11.9. Kidney Dialysis** Semipermeable membranes of the sort used in kidney dialysis do not allow large molecules and cells to pass but do allow small ions and water to pass. Figure P11.9 shows such a membrane separating fluids of various compositions.

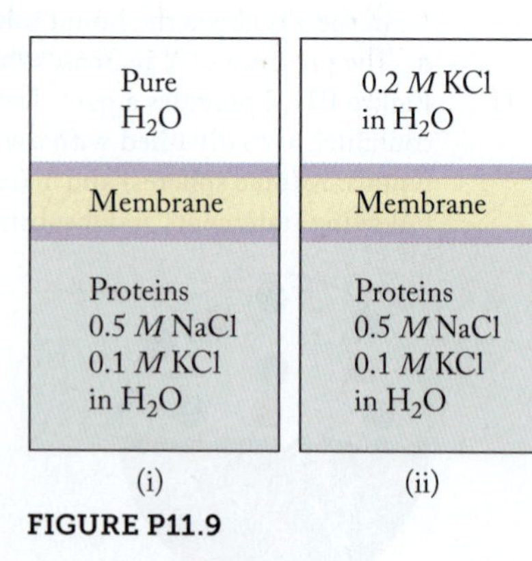


**FIGURE P11.9**

a. In which direction does the water flow in each solution?
b. In which direction do the sodium ions flow in each solution?
c. In which direction do the potassium ions flow in each solution?

11.10. Which of the images in Figure P11.10 represents the gas with the larger Henry's law constant, $k_H$, if the atmospheric pressure is the same in each case?

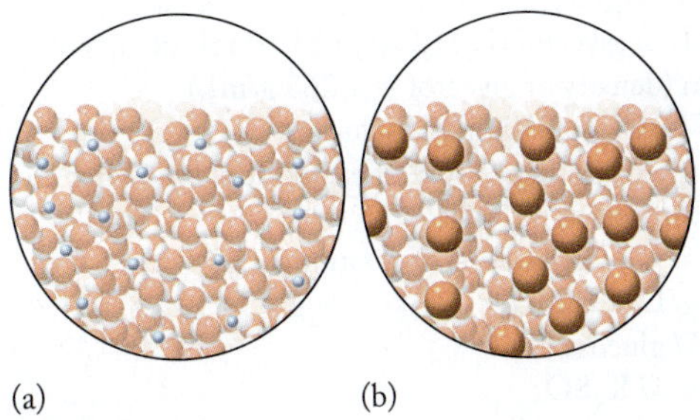


**FIGURE P11.10**

**11.11.** Which of the images in Figure P11.11 best describes the effect of pressure on the solubility of a gas?

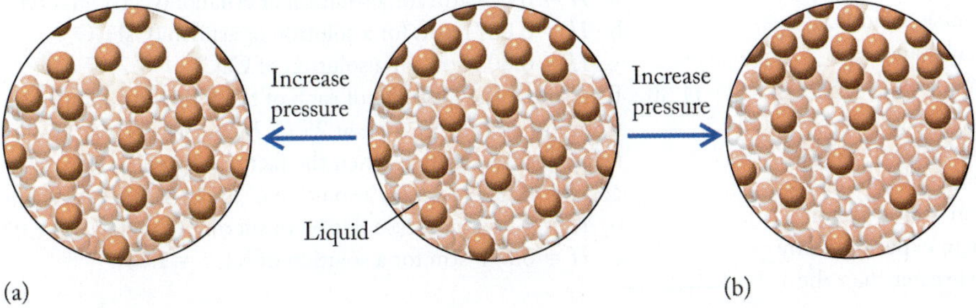


**FIGURE P11.11**

11.12. Use representations [A] through [I] in Figure P11.12 to answer questions a–f.

a. Which compound, [B], [C], or [F] has the highest boiling point?
b. Between [A] and [I], which has the higher vapor pressure?
c. Between [A] and [I], which has the lower freezing point?
d. If solutions [A] and [I] were separated by a semipermeable membrane, as shown in [E], what molecules or ions would flow across the membrane and in what direction?
e. Name two variables that could be changed to increase the solubility of the gas in [D].
f. Which solution, [G] or [H], is at a lower temperature?

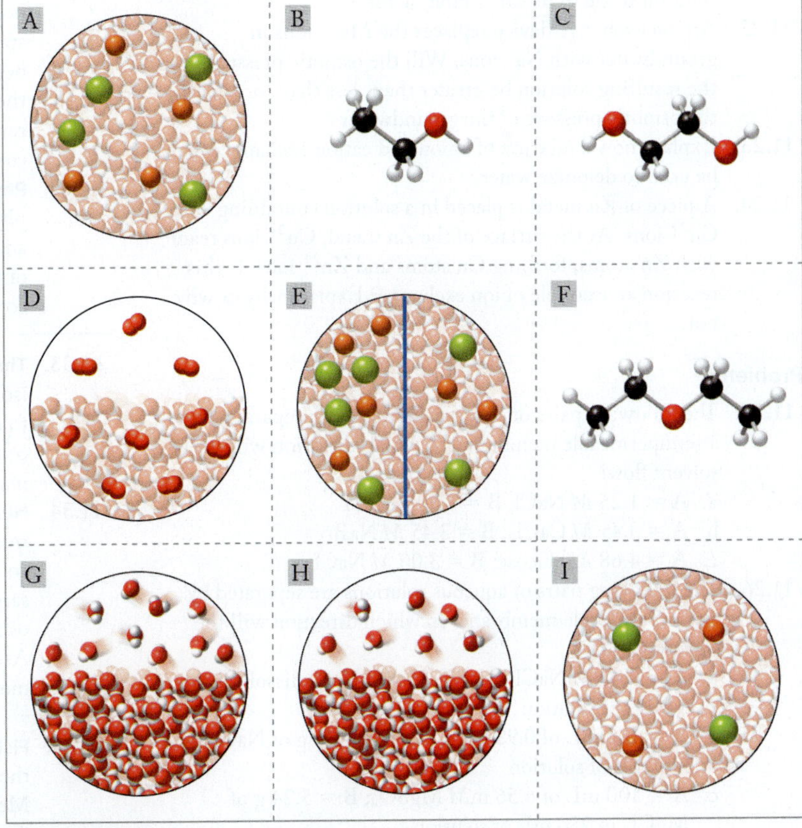


**FIGURE P11.12**

# QUESTIONS AND PROBLEMS

## Osmosis, Osmotic Pressure, and the van 't Hoff Factor

### Concept Review

**11.13.** What is a semipermeable membrane?

11.14. A pure solvent is separated from a solution containing the same solvent by a semipermeable membrane. In which direction is there a net flow of solvent across the membrane, and why?

**11.15.** A dilute solution is separated from a more concentrated solution containing the same solvent by a semipermeable membrane. In which direction is there a net flow of solvent across the membrane, and why?

11.16. How is the osmotic pressure of a solution related to its molar concentration and its temperature?

**11.17.** What is reverse osmosis? List the basic components of equipment used to purify seawater by reverse osmosis.

11.18. Why is it important to know whether a substance is a strong electrolyte before predicting its effect on the osmotic pressure of a solution?

**11.19.** Why is the van 't Hoff factor for solutions of acetic acid, $CH_3COOH$, slightly greater than 1 but far short of $i = 2$?

11.20. Why are theoretical and experimental van 't Hoff factors of electrolytes sometimes different? Can an experimentally measured value of a van 't Hoff factor be greater than the theoretical value? Why or why not?

**11.21.** How might the electrical conductivity of an aqueous solution relate to its van 't Hoff factor?

*11.22. An ion exchange device replaces the $Mg^{2+}$ ions in groundwater with $Na^+$ ions. Will the osmotic pressure of the resulting solution be greater than, less than, or equal to the osmotic pressure of the groundwater?

**11.23.** Explain how a mixture of anion and cation exchangers can be used to deionize water.

11.24. A piece of Zn metal is placed in a solution containing $Cu^{2+}$ ions. At the surface of the Zn metal, $Cu^{2+}$ ions react with Zn atoms, forming Cu atoms and $Zn^{2+}$ ions. Is this reaction an example of ion exchange? Explain why or why not.

### Problems

**11.25.** The following pairs of aqueous solutions are separated by a semipermeable membrane. In which direction will the solvent flow?

a. A = 1.25 *M* NaCl; B = 1.50 *M* KCl
b. A = 3.45 *M* $CaCl_2$; B = 3.45 *M* NaBr
c. A = 4.68 *M* glucose; B = 3.00 *M* NaCl

11.26. The following pairs of aqueous solutions are separated by a semipermeable membrane. In which direction will the solvent flow?

a. A = 0.48 *M* NaCl; B = 55.85 g of NaCl dissolved in 1.00 L of solution
b. A = 100 mL of 0.982 *M* $CaCl_2$; B = 16 g of NaCl in 100 mL of solution
c. A = 100 mL of 6.56 m*M* $MgSO_4$; B = 5.24 g of $MgCl_2$ in 250 mL of solution

---

**11.27.** Calculate the osmotic pressure of each of the following aqueous solutions at 20°C:

a. 2.39 *M* methanol ($CH_3OH$)
b. 9.45 m*M* $MgCl_2$
c. 40.0 mL of glycerol ($C_3H_8O_3$) in 250.0 mL of aqueous solution (density of glycerol = 1.265 g/mL)
d. 25 g of $CaCl_2$ in 350 mL of solution

11.28. Calculate the osmotic pressure of each of the following aqueous solutions at 27°C:

a. 10.0 g of NaCl in 1.50 L of solution
b. 10.0 mg/L of $LiNO_3$
c. 0.222 *M* glucose
d. 0.00764 *M* $K_2SO_4$

---

**11.29.** Determine the molarity of each of the following solutions from its osmotic pressure at 25°C. Include the van 't Hoff factor for the solution when the factor is given.

a. $\Pi = 0.674$ atm for a solution of ethanol ($CH_3CH_2OH$)
b. $\Pi = 0.0271$ atm for a solution of aspirin ($C_9H_8O_4$)
c. $\Pi = 0.605$ atm for a solution of $CaCl_2$, $i = 2.47$

11.30. Determine the molarity of each of the following solutions from its osmotic pressure at 25°C. Include the van 't Hoff factor for the solution when the factor is given.

a. $\Pi = 0.0259$ atm for a solution of urea [$H_2NC(O)NH_2$]
b. $\Pi = 1.56$ atm for a solution of sucrose ($C_{12}H_{22}O_{11}$)
c. $\Pi = 0.697$ atm for a solution of KI, $i = 1.90$

---

**11.31. Making Maple Syrup I** Traditional methods for making maple syrup concentrate the sap from maple trees by boiling away much of the water. Some manufacturers have started to use reverse osmosis to reduce the concentration before heating. If the average concentration of sugars in the sap is 2% sucrose ($C_{12}H_{11}O_{12}$) by mass, how much reverse osmotic pressure must be applied to double the concentration by reverse osmosis?

11.32. **Physiological Saline** After 100.0 mL of a solution of physiological saline (0.90% NaCl by mass) is diluted by the addition of 250.0 mL of water, what is the osmotic pressure of the final solution at 37°C? Assume that NaCl dissociates completely into $Na^+(aq)$ and $Cl^+(aq)$.

---

**11.33. Throat Lozenges** A 188 mg sample of a nonelectrolyte isolated from throat lozenges is dissolved in enough water to make 10.0 mL of solution at 25°C. The osmotic pressure of the resulting solution is 4.89 atm. Calculate the molar mass of the compound.

11.34. **Healing Herbs** Alpetin is a compound found in *Alpinia speciosa*, a tropical evergreen used historically for treating cold, flu, fever, flatulence, and indigestion. A 54.1 mg sample of alpetin is dissolved in 75.0 mL of water. The osmotic pressure of the solution is 0.0657 atm at 27°C. Assuming that alpetin is a nonelectrolyte, calculate the molar mass of alpetin.

---

**11.35. Fish Proteins** Northern cod produce proteins that protect their cells from damage caused by subzero temperatures. Measurements of the osmotic pressure for two "antifreeze" proteins at 18°C yielded the following data: 50.4 mg of protein A in 1.5 mL of water had an osmotic pressure of

0.299 atm, whereas an osmotic pressure of 0.218 atm was measured for a solution of 52.6 mg of protein B in 1.75 mL of water. What are the molar masses of proteins A and B?

*11.36. An unknown compound (152 mg) is dissolved in water to make 75.0 mL of solution. The solution does not conduct electricity and has an osmotic pressure of 0.328 atm at 27°C. Elemental analysis reveals the substance to be 78.90% C, 10.59% H, and 10.51% O. Determine the molecular formula of this compound.

## *Vapor Pressure*

### Concept Review

**11.37.** Use kinetic molecular theory to explain why the vapor pressure of a liquid increases with increasing temperature. Is vapor pressure an intensive or extensive property of a volatile substance?

11.38. Why is the vapor pressure of neopentane (its condensed molecular structure is shown in Figure P11.38) higher than that of pentane at the same temperature? In general, how is the vapor pressure of a liquid affected by the strength of intermolecular forces?

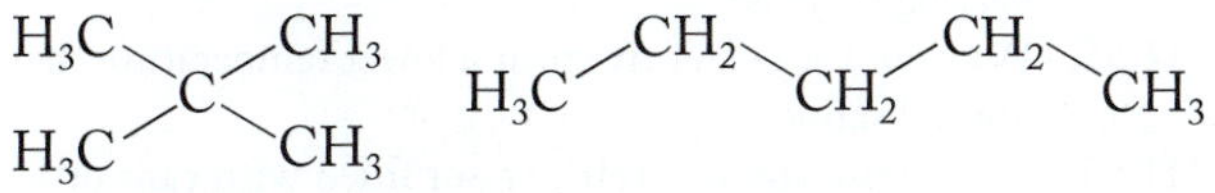

Neopentane Pentane

**FIGURE P11.38**

**11.39.** Rank the compounds in Figure P11.39 in order of increasing vapor pressure at 298 K.

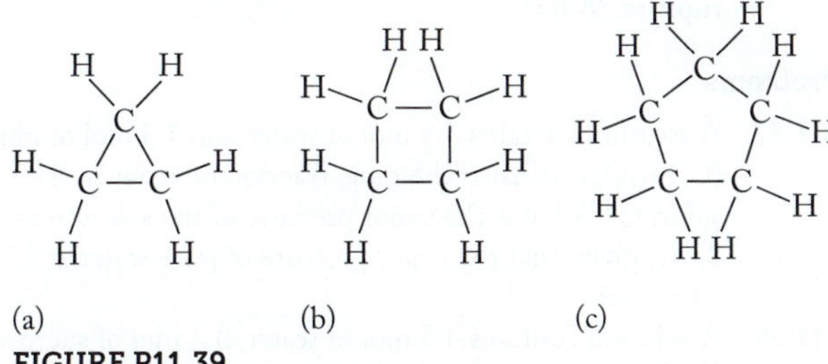


**FIGURE P11.39**

11.40. Figure P11.40 plots the vapor pressure of three substances—$CH_3OH$, $CH_3OCH_3$, and $CH_3CH_2CH_3$—as a function of temperature. Which curve corresponds to which substance?

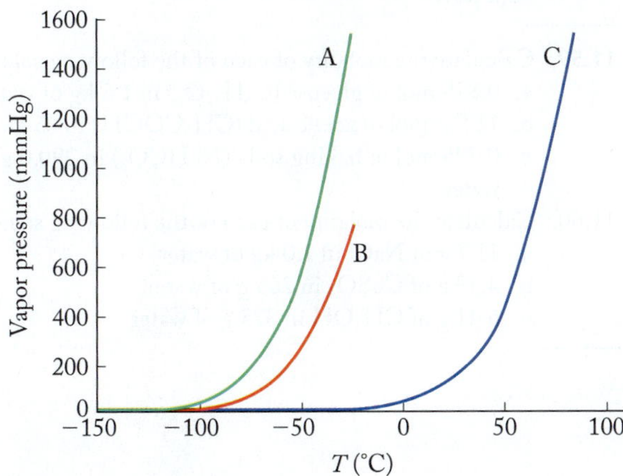


**FIGURE P11.40**

### Problems

**11.41. Pine Oil** The smell of fresh-cut pine is due in part to the cyclic alkene pinene, whose structure is shown in Figure P11.41. Use the data in the table to calculate the heat of vaporization, $\Delta H_{vap}$, of pinene.

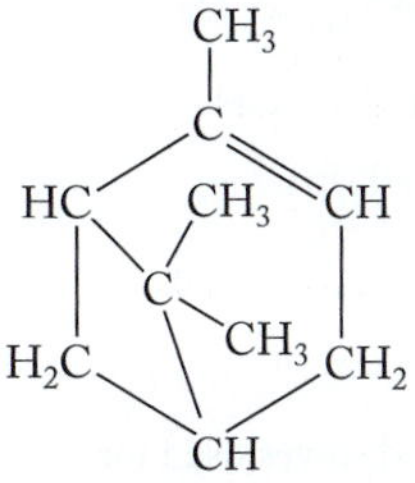

Pinene

**FIGURE P11.41**

| Vapor Pressure (torr) | Temperature (K) |
|---|---|
| 760 | 429 |
| 515 | 415 |
| 340 | 401 |
| 218 | 387 |
| 135 | 373 |

11.42. **Almonds and Cherries** Almonds and almond extracts are common ingredients in baked goods. Almonds contain the compound benzaldehyde (shown in Figure P11.42), which accounts for the odor of the nut. Benzaldehyde is also partly responsible for the aroma of cherries. Use the data in the table to calculate the enthalpy of vaporization, $\Delta H_{vap}$, of benzaldehyde.

| Vapor Pressure (torr) | Temperature (K) |
|---|---|
| 50 | 373 |
| 111 | 393 |
| 230 | 413 |
| 442 | 433 |
| 805 | 453 |

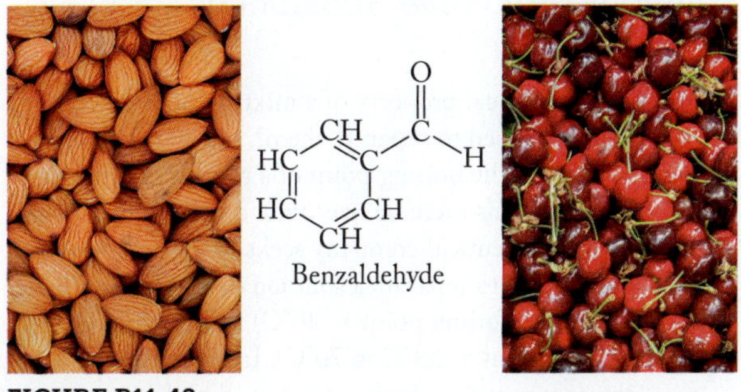


**FIGURE P11.42**

**11.43. High-Octane Gasoline** Gasoline is a complex mixture of hydrocarbons. It is sold with a variety of octane ratings that are based on the comparison of the gasoline with the combustion properties of isooctane, a compound with the molecular formula $C_8H_{18}$. The structures of isooctane and another compound with molecular formula $C_8H_{18}$ are shown in Figure P11.43, along with their normal boiling points and enthalpies of vaporization. Determine the vapor pressure of each isomer on a day when the temperature is 38°C.

Isooctane
bp = 98.2°C
$\Delta H_{vap}$ = 35.8 kJ/mol

Tetramethylbutane
bp = 106.5°C
$\Delta H_{vap}$ = 43.3 kJ/mol

**FIGURE P11.43**

11.44. **Camping Fuel** Portable lanterns and stoves used for camping and backpacking often use a mixture of $C_5$ and $C_6$ hydrocarbons known as "white gas." Figure P11.44 shows the structure of pentane, $C_5H_{12}$, along with its normal boiling point and enthalpy of vaporization. Determine the vapor pressure of pentane on a morning when the temperature is 5.0°C.

Pentane
bp = 36.0°C
$\Delta H_{vap}$ = 27.6 kJ/mol

**FIGURE P11.44**

## Solutions of Volatile Substances

### Concept Review

**11.45.** What physical property of a mixture of hydrocarbons would be used to separate them?

11.46. Why does the boiling point of a mixture of volatile hydrocarbons increase over time during a distillation?

**11.47.** A pharmaceutical company seeks to use fractional distillation to separate a solution of dichloromethane ($CH_2Cl_2$; boiling point = 40°C) and hexane ($C_6H_{14}$; boiling point = 68°C to 70°C). In an equimolar mixture of the compounds, which is present in higher concentration in the vapor above the solution?

*11.48. Could a mixture of two volatile solids be separated by fractional sublimation? Explain.

### Problems

**11.49.** **Greener Fuel** Most gasoline sold in the United States contains ethanol, $C_2H_5OH$, in addition to a mixture of hydrocarbons such as octane, $C_8H_{18}$. At 20°C, the vapor pressure of ethanol is 45 torr and the vapor pressure of octane is 10 torr. What is the vapor pressure at 20°C of a solution prepared by mixing 8.16 g of ethanol and 63.3 g of octane?

11.50. **Alcohol Fuels** Using butanol, $C_4H_9OH$, rather than ethanol as a gasoline additive has an advantage in providing a higher fuel value when mixed with octane. At 20°C, the vapor pressure of butanol is 5 torr and the vapor pressure of octane is 10 torr. What is the vapor pressure at 20°C of a solution prepared by mixing 16.2 g of butanol and 127 g of octane?

## More Colligative Properties of Solutions

### Concept Review

**11.51.** Explain how nearly all the water ends up in one compartment in Figure 11.19.

11.52. A dam separates the fresh water of the Charles River from the seawater of Boston Harbor. If the two bodies of water are at the same temperature, which one evaporates faster on a hot summer day?

**11.53.** What is the difference between molarity and molality?

11.54. **Making Maple Syrup II** Heating maple tree sap converts the dilute sugar solution into the viscous amber syrup we use as a topping on pancakes and waffles. How do the boiling points and vapor pressures of the sap change during the heating process?

**11.55.** Why does seawater freeze at a lower temperature than normal saline?

*11.56. The thermostat in a refrigerator filled with cans of soft drinks malfunctions and the temperature of the refrigerator drops below 0°C. The contents of the cans of diet soft drinks freeze, rupturing many of the cans and causing an awful mess. However, none of the cans containing regular, nondiet soft drinks freeze and rupture. Why?

### Problems

**11.57.** A solution contains 3.5 mol of water and 1.5 mol of glucose ($C_6H_{12}O_6$). What is the mole fraction of water in the solution? What is the vapor pressure of the solution at 25°C, given that the vapor pressure of pure water at 25°C is 23.8 torr?

11.58. A solution contains 4.5 mol of water, 0.3 mol of sucrose ($C_{12}H_{22}O_{11}$), and 0.2 mol of glucose. Sucrose and glucose are nonvolatile. What is the mole fraction of water in the solution? What is the vapor pressure of the solution at 35°C, given that the vapor pressure of pure water at 35°C is 42.2 torr?

---

**11.59.** Calculate the molality of each of the following solutions:
a. 0.875 mol of glucose ($C_6H_{12}O_6$) in 1.5 kg of water
b. 11.5 mmol of acetic acid ($CH_3COOH$) in 65 g of water
c. 0.325 mol of baking soda ($NaHCO_3$) in 290.0 g of water

11.60. Calculate the molality of each of the following solutions:
a. 11.7 g of NaCl in 1.0 kg of water
b. 4.99 g of $CuSO_4$ in 265 g of water
c. 6.41 g of $CH_3OH$ in 375 g of water

---

**11.61.** Calculate the molality of each of the following aqueous solutions:
a. 1.30 $M$ $CaCl_2$ ($d$ = 1.113 g/mL)
b. 2.02 $M$ fructose ($C_6H_{12}O_6$; $d$ = 1.139 g/mL)
c. 8.94 $M$ ethylene glycol (antifreeze, $HOCH_2CH_2OH$; $d$ = 1.069 g/mL)
d. 1.97 $M$ LiCl ($d$ = 1.046 g/mL)

11.62. Table 8.1 lists molar concentrations of the major ions in seawater. Using a density of 1.025 g/mL for seawater, convert each concentration into a molality.

---

**11.63.** What mass of the following solutions contains 0.100 mol of solute? (a) 0.334 $m$ $NH_4NO_3$; (b) 1.24 $m$ ethylene glycol, $HOCH_2CH_2OH$; (c) 5.65 $m$ $CaCl_2$

11.64. How many moles of solute are there in the following solutions?
a. 0.150 $m$ glucose solution made by dissolving the glucose in 100.0 kg of water
b. 0.028 $m$ $Na_2CrO_4$ solution made by dissolving the $Na_2CrO_4$ in 1000.0 g of water
c. 0.100 $m$ urea solution made by dissolving the urea in 500.0 g of water

---

**11.65.** **Fish Kills** High concentrations of ammonia ($NH_3$), nitrite ion, and nitrate ion in water can kill fish. Lethal concentrations of these substances for rainbow trout are 1.1 mg/L, 0.40 mg/L, and 1361 mg/L, respectively. Express these concentrations in molality units, assuming a solution density of 1.00 g/mL.

11.66. **Mineral Content of River Water** The concentrations of six important elements in a sample of river water are 0.050 mg/kg of $Al^{3+}$, 0.040 mg/kg of $Fe^{3+}$, 13.4 mg/kg of $Ca^{2+}$, 5.2 mg/kg of $Na^+$, 1.3 mg/kg of $K^+$, and 3.4 mg/kg of $Mg^{2+}$. Express each of these concentrations in molality units.

---

**11.67.** **Cinnamon** Cinnamon owes its flavor and odor to cinnamaldehyde ($C_9H_8O$). Determine the freezing point of a solution of 75 mg of cinnamaldehyde dissolved in 1.00 g of benzene ($K_f$ = 4.90°C/$m$; normal freezing point = 5.5°C).

11.68. **Spearmint** Determine the boiling point elevation of a solution of 125 mg of carvone ($C_{10}H_{14}O$; oil of spearmint) dissolved in 1.50 g of carbon disulfide ($K_b$ = 2.34°C/$m$).

---

**11.69.** What molality of a nonvolatile, nonelectrolyte solute is needed to lower the melting point of camphor by 1.000°C ($K_f$ = 39.7°C/$m$)?

11.70. What molality of a nonvolatile, nonelectrolyte solute is needed to raise the boiling point of water by 7.60°C ($K_b$ = 0.52°C/$m$)?

---

**11.71.** **Saccharin** Determine the melting point of an aqueous solution made by adding 186 mg of saccharin ($C_7H_5O_3NS$) to 1.00 mL of water ($d$ = 1.00 g/mL; $K_f$ = 1.86°C/$m$).

11.72. Determine the boiling point of an aqueous solution that is 2.50 $m$ ethylene glycol ($HOCH_2CH_2OH$); $K_b$ for water is 0.52°C/$m$. Assume that the boiling point of pure water is 100.00°C.

---

**11.73.** Which aqueous solution has the lowest freezing point:
a. 0.5 $m$ glucose, 0.5 $m$ NaCl, or 0.5 $m$ $CaCl_2$?
b. 0.0500 $m$ $C_6H_{12}O_6$, 0.0450 $m$ NaI, or 0.0125 $m$ $K_2SO_4$?

11.74. Arrange the following solutions in order of increasing freezing point depression:
a. 0.10 $m$ $MgCl_2$ in water, $i$ = 2.7, $K_f$ = 1.86°C/$m$
b. 0.20 $m$ toluene in diethyl ether, $i$ = 1.00, $K_f$ = 1.79°C/$m$
c. 0.20 $m$ ethylene glycol in ethanol, $i$ = 1.00, $K_f$ = 1.99°C/$m$

---

**11.75.** Which aqueous solution has the highest boiling point:
a. 0.5 $m$ glucose, 0.5 $m$ NaCl, or 0.5 $m$ $CaCl_2$?
b. 0.0400 $m$ $NH_4NO_3$, 0.0165 $m$ LiCl, or 0.0105 $m$ $Cu(NO_3)_2$?

11.76. Arrange the following aqueous solutions in order of increasing boiling point:
a. 0.06 $m$ $FeCl_3$ ($i$ = 3.4)
b. 0.10 $m$ $MgCl_2$ ($i$ = 2.7)
c. 0.20 $m$ KCl ($i$ = 1.9)

---

***11.77.** **Caffeine** The freezing point of a solution prepared by dissolving 150 mg of caffeine in 10.0 g of camphor is lower by 3.07°C than that of pure camphor ($K_f$ = 39.7°C/$m$). What is the molar mass of caffeine? Elemental analysis of caffeine yields the following results: 49.49% C, 5.15% H, 28.87% N, and the remainder is O. What is the molecular formula of caffeine?

*11.78. **Ethnobotany** One of the ingredients in the Native American stomachache remedy derived from common chokecherry is caffeic acid.
a. Combustion of 100 mg of caffeic acid yielded 220 mg of $CO_2$ and 40.3 mg of $H_2O$. Determine the empirical formula of caffeic acid.
b. A solution of 0.272 g of caffeic acid in 10.0 g of carbon tetrachloride causes the freezing point of the solution to decrease by 4.47°C. Given that $K_f$ for $CCl_4$ is 29.8°C/$m$ and that caffeic acid is a nonelectrolyte, calculate the molar mass and molecular formula of caffeic acid.

## *Henry's Law and the Solubility of Gases*

### Concept Review

**11.79.** Why is the Henry's law constant for $CO_2$ so much larger than those for $N_2$ and $O_2$ at the same temperature? *Hint*: Does $CO_2$ react with water?

11.80. The values of $k_H$ for NO and CO gas in water at 20°C are $1.4 \times 10^{-3}$ $M$/atm and $7.4 \times 10^{-3}$ $M$/atm, respectively. Why is CO more soluble in water?

**11.81.** As water in a beaker is heated, bubbles form inside the beaker at temperatures well below the boiling point of water. What gas is in the bubbles?

*11.82. Air is primarily a mixture of nitrogen and oxygen. Is the Henry's law constant for the solubility of air in water the sum of $k_H$ for $N_2$ and $k_H$ for $O_2$? Explain why or why not.

**Problems**

*11.83. **Arterial Blood** Arterial blood contains about 0.25 g of oxygen per liter at 37°C and standard atmospheric pressure. What is the Henry's law constant, in mol/(L · atm), for $O_2$ dissolution in blood at 37°C?

11.84. **Laughing Gas** Nitrous oxide, $N_2O$, is used in dental clinics as an anesthetic. The solubility of $N_2O$ in water is 1.1 g/L at an atmospheric pressure of 1 atm and a temperature of 20°C.

a. Calculate the Henry's law constant of $N_2O$ at 20°C.
b. Compare the value for $k_H$ for $N_2O$ with the value for $O_2$ in Table 11.1. Why are they different?

---

*11.85. **Oxygen for Climbers and Divers** Use the Henry's law constant for $O_2$ dissolved in arterial blood from Problem 11.83 to calculate the solubility of $O_2$ in the blood of (a) a climber on Mt. Everest ($P_{atm}$ = 0.35 atm) and (b) a scuba diver breathing air at a depth of 20 meters ($P \approx 3.0$ atm).

11.86. **Fracking and Air Quality** Hydraulic fracturing, or "fracking," allows methane gas to be extracted from otherwise inaccessible shale formations. Scientists estimate that about 2.3% of the methane produced is lost to the atmosphere, where it behaves as a greenhouse gas.

a. Calculate the solubility of methane in water under a pure methane atmosphere of 1.0 atm. (The Henry's law constant for methane in water is approximately $1.5 \times 10^{-3}$ mol/(L · atm) at 20°C and 1.0 atm.)
b. Calculate the maximum solubility of the methane found in a sample of air that contains $1.89 \times 10^{-4}$% methane gas ($CH_4$).
*c. What fraction of the methane released by fracking is likely to end up dissolved in water?

## Additional Problems

*11.87. **Melting Ice** $CaCl_2$ is used to melt ice on sidewalks. Could $CaCl_2$ melt ice at −20°C? Assume that the solubility of $CaCl_2$ at this temperature is 70.1 g of $CaCl_2$/100.0 g of $H_2O$ and that the van 't Hoff factor for a saturated solution of $CaCl_2$ is 2.5.

*11.88. **Making Ice Cream** A mixture of table salt and ice is used to chill the contents of hand-operated ice-cream makers. What is the melting point of a mixture of 2.00 kg of NaCl and 12.00 kg of ice if exactly half the ice melts? Assume that all the NaCl dissolves in the melted ice and that the van 't Hoff factor for the resulting solution is 1.44.

**11.89.** The freezing points of 0.0935 *m* ammonium chloride and 0.0378 *m* ammonium sulfate in water were found to be −0.322°C and −0.173°C, respectively. What are the values of the van 't Hoff factors for these salts?

11.90. The following data were collected for three compounds in aqueous solution. Determine the value of the van 't Hoff factor for each salt ($K_f$ for water = 1.86°C/*m*).

| Compound | Concentration (g/kg) | Experimentally Measured $\Delta T_f$(°C) |
|---|---|---|
| LiCl | 5.0 | 0.410 |
| HCl | 5.0 | 0.486 |
| NaCl | 5.0 | 0.299 |

**11.91.** **BPA** Bisphenol A, or BPA, is used in the manufacture of water bottles and other plastic containers. BPA is suspected to have adverse health effects, and efforts are underway to find a suitable replacement. BPA is 78.92% C, 7.06% H, and 14.02% O by mass.

a. A 224 mg sample of BPA is dissolved in 1.00 g of chloroform ($K_b$ = 3.63°C/*m*), increasing the boiling point of chloroform by 3.56°C. Calculate the molar mass of BPA.
b. What is the empirical formula for BPA?
c. What is the molecular formula of BPA?

11.92. **Human Hemoglobin** The hemoglobin found in human red blood cells has a molar mass of about $6.2 \times 10^4$ g/mol. What concentration range of hemoglobin will result in an osmotic pressure between 0.35 atm and 0.55 atm at 37°C in aqueous solution?

**11.93.** A solution of 7.50 mg of a small protein in 5.00 mL of aqueous solution has an osmotic pressure of 6.50 torr at 23.1°C. What is the molar mass of the protein?

11.94. **Kidney Dialysis** Hemodialysis, a method of removing waste products from the blood if the kidneys have failed, uses a tube made of a cellulose membrane that is immersed in a large volume of aqueous solution. Blood is pumped through the tube and is then returned to the patient's vein. The membrane allows small ions, urea, and water, but not large protein molecules and cells, to pass through it. Assume that a physician wants to decrease the concentration of sodium ion and urea in a patient's blood while maintaining the concentration of potassium ion and chloride ion in the blood. What materials must be dissolved in the aqueous solution in which the dialysis tube is immersed? How must the concentrations of ions in the immersion fluid compare to those in blood?

*11.95. The Henry's law constant for the solubility of a gas and the van der Waals constants $a$ and $b$ in Equation 9.30 all depend on intermolecular forces.

a. Plot the values of $a$ and $b$ from Table 9.5 as a function of $k_H$ (Table 11.1) for He, $O_2$, $N_2$, $CH_4$, and $CO_2$. Which gas seems to be an outlier?
b. Why might this gas be an outlier?

11.96. **Psoriasis Treatment** Xanthotoxin is a natural product extracted from the fruits of *Heracleum persicum* that seems to have some success in treating psoriasis, a skin disease.

a. Combustion of 0.0100 g of xanthotoxin in excess oxygen yielded 0.0244 g of $CO_2$ and 0.00333 g of $H_2O$. Calculate the empirical formula of xanthotoxin.
b. A solution of 0.250 g of xanthotoxin in 10.0 g of carbon tetrachloride causes the freezing point of the solution to decrease by 3.46°C. Given that $K_f$ for $CCl_4$ is 29.8°C/$m$ and that xanthotoxin is a nonelectrolyte, calculate the molar mass and molecular formula of xanthotoxin.

# 12

# Thermodynamics

## Why Chemical Reactions Happen

**CORROSION AT SEA** Most metal objects in contact with seawater, including the remains of this ship, corrode as a result of spontaneous chemical reactions. Here, iron metal has turned into iron(III) oxide.

PARTICULATE **REVIEW**

### *Phase Changes and Energy Changes*

In Chapter 12, we will investigate how energy changes affect the spontaneity of chemical reactions and phase changes.

- What two phase changes are represented in these particulate images?
- Which phase change occurs due to a flow of energy from the system into its surroundings on a cold day (−5°C)?
- Label each phase change as either endothermic or exothermic.

(Review Sections 10.1 and 10.2 if you need help.)

*(Answers to Particulate Review questions are in the back of the book.)*

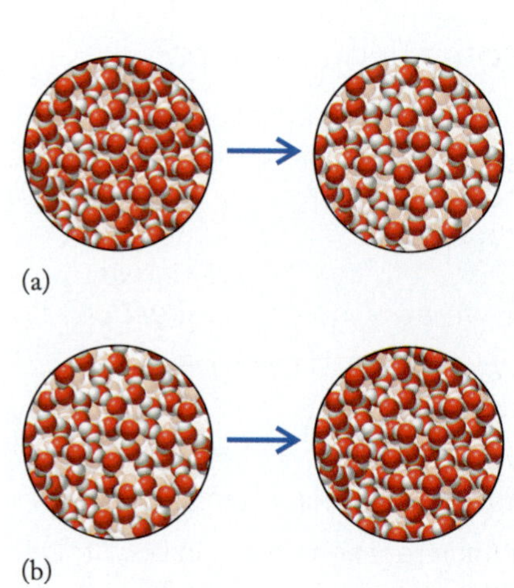

PARTICULATE **PREVIEW**

### *Spontaneous or Not?*

When we exercise, we rely on the oxidation of glucose ($C_6H_{12}O_6$) to carbon dioxide and water for energy. When photosynthesis takes place in green plants, carbon dioxide and water react to produce glucose and oxygen. As you read Chapter 12, look for ideas that will help you answer these questions.

- Write balanced chemical equations describing these two reactions.
- Which reaction results in breaking larger molecules into smaller molecules? Does it occur spontaneously?
- Which reaction results in building larger molecules from smaller molecules? Does it occur spontaneously?

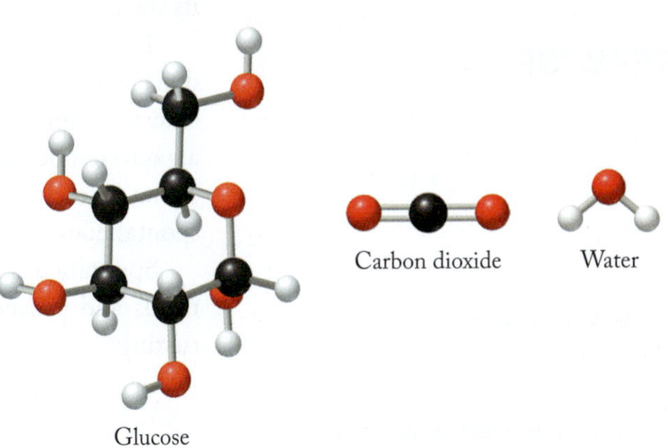

## Learning Outcomes

**LO1** Predict the signs of entropy changes for spontaneous and nonspontaneous chemical reactions and physical processes
**Sample Exercise 12.1**

**LO2** Use microstates to explain why a perfect crystalline solid has zero entropy

**LO3** Predict the relative entropies of substances on the basis of their molecular structures
**Sample Exercise 12.2**

**LO4** Calculate entropy changes in chemical reactions by using standard molar entropies
**Sample Exercise 12.3**

**LO5** Calculate standard free-energy changes in chemical reactions and physical processes
**Sample Exercises 12.4–12.6**

**LO6** Predict the spontaneity of a chemical reaction or physical process as a function of temperature and in terms of changes in its enthalpy and entropy
**Sample Exercise 12.7**

**LO7** Calculate the net change in free energy of coupled spontaneous and nonspontaneous reactions
**Sample Exercise 12.8**

## 12.1 Spontaneous Processes

Some processes are so familiar that we don't question why they happen. If a car tire is punctured, the air inside rushes out and the tire goes flat; the air does *not* rush back into a punctured tire and reinflate it. Discarded objects made of iron slowly turn into clumps of rust; they don't turn back into shiny metal all by themselves. A tray of ice cubes left on a kitchen counter turns into a tray of liquid water; as long as the kitchen stays warm, the water in the tray will not turn back into ice.

All these processes are **spontaneous** because they happen without any ongoing intervention. In each case, the reverse process is nonspontaneous because it cannot happen on its own. So, why are some processes spontaneous and others nonspontaneous?

Let's answer this question by considering the spontaneous chemical reaction in 19th-century streetlights. Before Thomas Edison invented the lightbulb, city streets were often illuminated at night by gas-fueled lamps. Once lit, they burned through the night until their fuel was cut off as dawn approached. Chemical reactions such as the combustion of gas-lamp fuel are examples of spontaneous reactions: once started, they proceed without outside assistance as long as all the reactants are available. The reverse reaction—here, converting carbon dioxide and water back into fuel and oxygen—is **nonspontaneous**: it cannot happen on its own.

**STEPWISE ANIMATION**
Spontaneous Processes

The word *spontaneous* can be misleading because many spontaneous reactions don't start all by themselves: they need a little energy boost, such as a spark or an external flame that ignites the flame in a gas lamp or the gasoline–air mixture in an automobile engine. However, once initiated, a spontaneous reaction continues on its own. It is the self-sustaining nature of these reactions that makes them spontaneous.

**spontaneous process** a process that proceeds without continuous outside intervention.

**nonspontaneous process** a process that requires continuous outside intervention to proceed.

Spontaneous does not mean fast. Though combustion reactions are spontaneous and proceed rapidly, other spontaneous reactions do not, including iron rusting:

$$4\ Fe(s) + 3\ O_2(g) \rightarrow 2\ Fe_2O_3(s) \qquad \Delta H^\circ_{rxn} = -1648.4\ kJ$$

Spontaneous is also *not* a synonym for exothermic, though many scientists once thought so. In the mid-19th century, many of them believed that exothermic

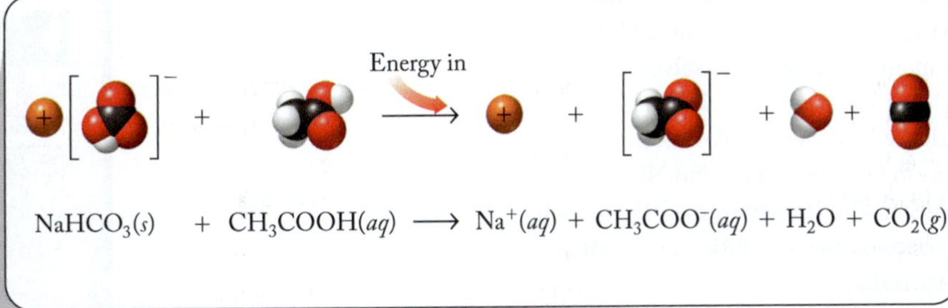


**FIGURE 12.1** Adding vinegar to baking soda produces sodium ions, acetate ions, water, and carbon dioxide in an endothermic reaction ($\Delta H^\circ_{rxn}$ = 48.5 kJ/mol) that is also spontaneous.

reactions were always spontaneous, and it is true that most are. However, others are not spontaneous, and some endothermic reactions are spontaneous. For example, when room-temperature vinegar (dilute acetic acid) is added to baking soda (sodium bicarbonate) as shown in **Figure 12.1**, the foaming reaction mixture tells us that a chemical change is taking place, and a decrease in the temperature of the reaction mixture tells us that the reaction is endothermic ($\Delta H_{rxn} > 0$).

**CONNECTION** We learned in Chapter 10 that enthalpy change ($\Delta H$) is a measure of how much energy flows into ($\Delta H > 0$) or out from ($\Delta H < 0$) a system during a process occurring at constant pressure.

Why is the vinegar and baking soda reaction spontaneous? The particles that make up the products are more spread out and/or have more freedom of motion than the particles in the reactants. For example, the reactants are liquid and solid, but the products include a gas ($CO_2$). The particles that make up these three states of matter have different freedoms of motion, as illustrated for the molecules of water in ice, liquid water, and steam in **Figure 12.2**. The particles that make up ice or any crystalline solid occupy fixed positions, and their motion is limited to vibrating in place without going anywhere (unless the whole crystal moves). The particles in liquids are more mobile: they are able to flow past and tumble over each other, which means they have some *translational* and *rotational* freedom of motion, respectively. Particles in the gas phase are much more spread out—which is why gases are compressible, whereas liquids and solids are not—and they have much more translational and rotational freedom of motion. Therefore, when either solid or liquid reactants form one or more products that are gases, the particles are more spread out and have more freedom to move about.

An increase in particle motion also takes place in the process that makes instant cold packs cold (**Figure 12.3**). There are two compartments in an instant cold pack. One is filled with water, and the other contains a water-soluble compound such as ammonium nitrate that has a positive enthalpy of solution ($\Delta H_{solution} > 0$). When the membrane separating the two compartments is ruptured, the solid compound mixes with and dissolves in water. The resulting solution gets very cold, becoming an effective anti-inflammatory treatment for bruises and muscle sprains. The spontaneity of this process is due to the production of particles (dissolved $NH_4^+$ ions and $NO_3^-$ ions) that are more spread out and have more freedom of motion than they had in the solid phase. This concept applies to the

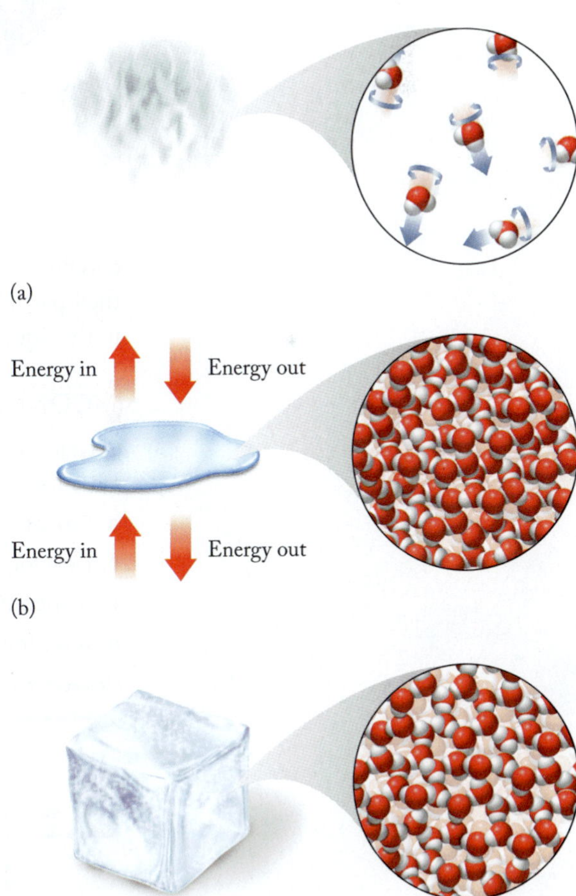


**FIGURE 12.2** Molecular motion in the three phases of water. (a) Molecules of water vapor have much more translational and rotational motion than particles in either condensed phase. (b) Molecules in water can flow past and tumble over each other. (c) Molecules in ice can only vibrate in place.

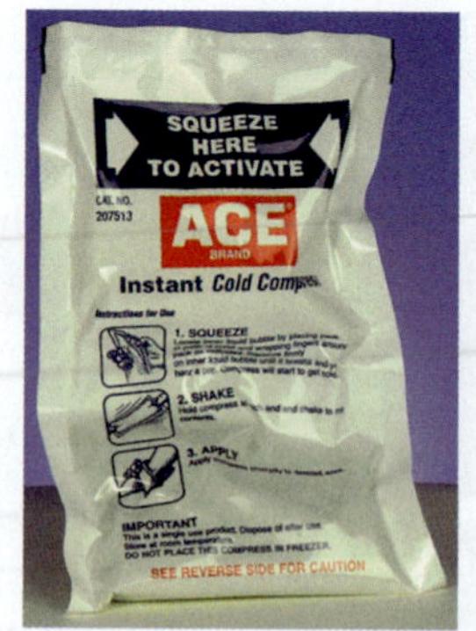


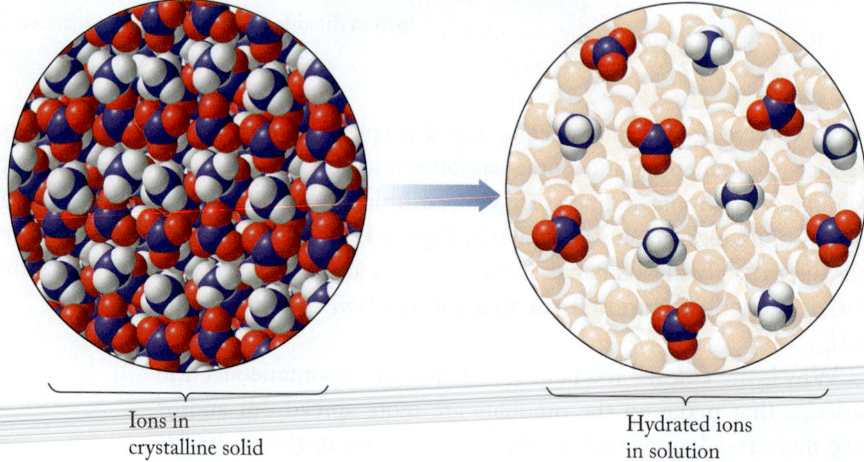


**FIGURE 12.3** An instant cold pack gets cold when a membrane inside it is ruptured, allowing a water-soluble compound such as ammonium nitrate to dissolve. The $NH_4^+$ and $NO_3^-$ ions in solid $NH_4NO_3$ experience increased freedom of motion as they form hydrated $NH_4^+$ and $NO_3^-$ ions in solution. The temperature of the solution drops because the dissolution process is endothermic.

Dissolution of Ammonium Nitrate

dissolution of all solids in liquid solvents. It also applies to melting solids because their particles are usually more spread out and always have more freedom of motion in the liquid state, as shown for ice and water in Figure 12.2.

**CONCEPT TEST**

In which of the following processes do particles become more spread out and gain more freedom of motion?

a. Perspiration evaporates.

b. Dew forms overnight on blades of grass.

c. A spoonful of sugar dissolves in a cup of coffee.

d. A wooden log burns in a fireplace.

*(Answers to Concept Tests are in the back of the book.)*

## 12.2 Entropy and the Second Law of Thermodynamics

**CONNECTION** The first law of thermodynamics is described in Section 10.1.

The first law of thermodynamics tells us that energy cannot be created or destroyed. This means we cannot do more work than is allowed by the available energy. The **second law of thermodynamics** says, in effect, that not all the energy released by a spontaneous reaction, such as burning gasoline in a car engine, is available to do useful work. These two principles of energy explain why so-called perpetual motion machines are impossible. In every process that releases energy, some of what is released cannot be used to do anything useful, like propelling a car down the road.

If energy cannot be destroyed, what happens to the energy that is unavailable to do useful work? Essentially, this energy simply spreads out, becoming less concentrated over time. For both the automobile engine and solar energy, the low-grade energy flows uselessly into the surroundings in a much less concentrated form.

Whenever an isolated thermodynamic system undergoes a change in which the particles in the system disperse into a larger volume and/or gain freedom of motion, the process is *always* spontaneous. The system is said to experience an increase in **entropy (*S*)**. Entropy is a measure of how dispersed the energy of a system is. The formal description of the second law of thermodynamics states that the entropy of an isolated thermodynamic system always increases during a spontaneous process. What exactly is entropy?

**second law of thermodynamics** the principle stating that the total entropy of the universe increases in any spontaneous process.

**entropy (*S*)** a measure of how dispersed the energy in a system is at a specific temperature.

**CHEMTOUR**
Entropy

The real meaning of entropy is linked to how energy is distributed in a thermodynamic system. To illustrate what we mean by energy dispersion of a system, let's revisit the experiment in Chapter 10 in which we determined the specific heat of a metal (see Figure 10.16). Recall how the metal pellets were first heated to 100°C in a bath of boiling water and then transferred to a small volume of cool water in an insulated container. The temperature of the water in the container increased, and we used the magnitude of that change to calculate how much energy flowed from the metal into the water as the metal and water (a closed thermodynamic system) came to the same temperature. Energy initially confined to the metal flowed spontaneously from the hot metal into the cool water and was dispersed throughout the metal–water system.

**CONNECTION** In Chapter 9, we learned how kinetic molecular theory explains many of the physical properties of gases, including the rates at which they effuse and diffuse as described by Graham's law.

Energy dispersion can happen even when all the components of a system, or the system and its surroundings, are at the same temperature. Let's revisit a spontaneous, isothermal process from Chapter 9: the diffusion of ideal gases. The apparatus in **Figure 12.4** contains two equal-volume chambers. Suppose the one on the left contains 10 neon atoms, and the one on the right contains 10 argon atoms. We assume that the gases behave as ideal gases, which means their atoms are essentially tiny points of mass that do not interact (except for frequent elastic collisions). We also assume the temperatures of both chambers are the same, which means that their pressures are also the same.

When the partition between the two chambers is removed, atoms of each gas spontaneously diffuse into the space originally occupied by the other gas. This behavior is explained by kinetic molecular theory and the random motion of the particles that make up ideal gases. According to Graham's law, the less massive neon atoms will diffuse faster into the space occupied by the more massive argon atoms, but, given enough time, the atoms of both gases will end up evenly distributed throughout the combined volume.

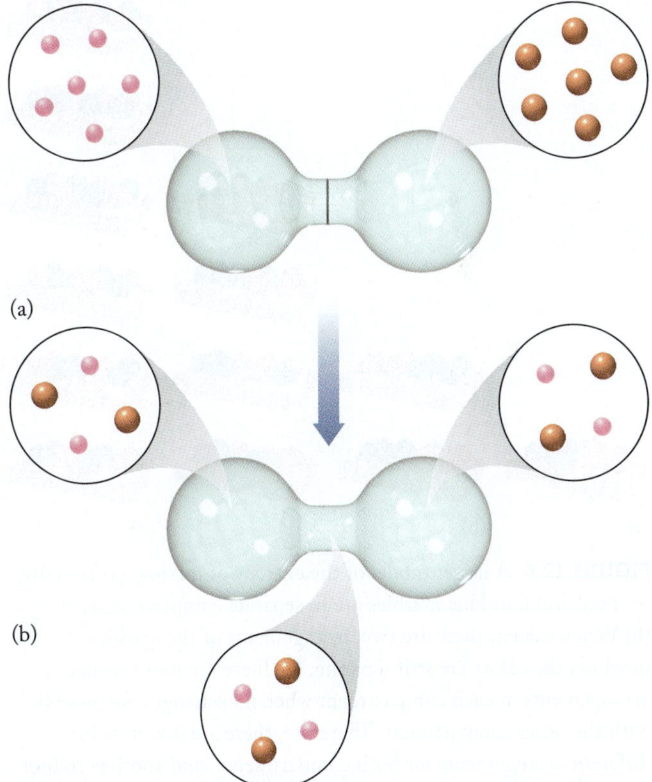


**FIGURE 12.4** Entropy of mixing. (a) When the partition separating two containers with different noble gases is removed, (b) the two gases spontaneously diffuse into each other's compartment, resulting in a uniform mixture of randomly distributed atoms of the two gases.

Though kinetic molecular theory is a powerful tool for predicting that gases mix spontaneously, there is another way to account for the spontaneity of mixing—one based on a statistical view of energy dispersion and entropy. To

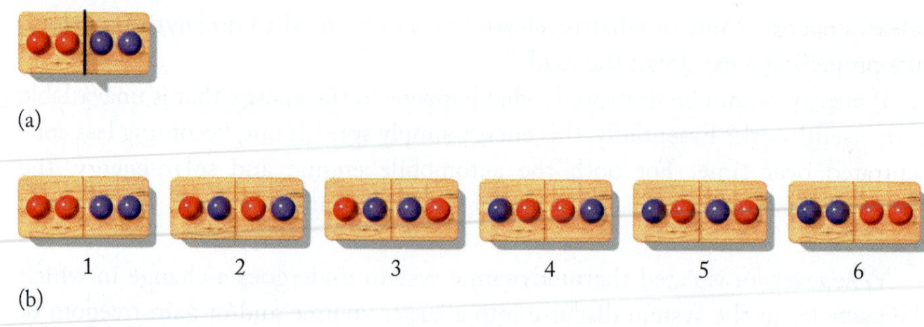


**FIGURE 12.5** A simple model of the entropy of mixing. (a) Initially, pairs of red and blue marbles are in separate compartments. (b) When mixed, the marbles can form six different patterns. In two of the arrangements (1 and 6), the marbles are still in separate compartments, but in four of them (2–5), they are mixed: one of each color marble in both compartments. Thus, there are twice as many arrangements with red–blue pairs than with single-color pairs.

illustrate how it works, let's start with a vastly simplified model of two gases mixing (**Figure 12.5**). We start with two red marbles representing atoms of one of the gases and two blue marbles representing atoms of the other. Initially, they are separated into two compartments as shown in Figure 12.5(a). Then the partition separating them is removed and the marbles are able to occupy any of the sites in either compartment. There are six ways to arrange the two red and two blue marbles among the four sites, as shown in Figure 12.5(b). In only two of these (numbered 1 and 6) are the same color marbles in the same compartment. In the other four, they are mixed. If there is an equal chance any marble could be in any one of the positions, then it is twice as likely that they will be mixed together (arrangements 2–5) rather than separate (arrangements 1 and 6).

Now let's double the number of marbles and marble-holding sites, as shown in **Figure 12.6(a)**. If we randomly arrange all eight marbles among the eight available sites, there are again only two possible arrangements in which each compartment contains marbles of only one color (**Figure 12.6b**). There are several mixing options: one or two or three of the red marbles could end up on the side that originally contained only blue marbles, and vice versa. The exchange of one marble produces four options for arranging the blue one and the remaining three red ones in the originally all-red compartment, as shown in **Figure 12.6(c)**. For each of these four arrangements, there are four ways to arrange the one red marble and three blue ones in the other compartment. This means that there are $4 \times 4 = 16$ different ways to arrange the marbles if one red and one blue marble are exchanged between compartments. Can you see how there are also 16 different ways to arrange the marbles if three red and three blue ones are exchanged between compartments? The third option is equal mixing so that each compartment has two red and two blue marbles. There are six ways to arrange these marbles in each compartment (**Figure 12.6d**) for a total of $6 \times 6 = 36$ arrangements. If all arrangements are equally probable, then out of a total of $2 + 16 + 16 + 36 = 70$ arrangements, only two (or $2/70 = 0.029 = 2.9\%$) of them produce no mixing of the marbles. The remaining 68 arrangements (97.1%) produce at least some mixing, and the most probable mixing pattern (36 out of 70) produces a uniform distribution of marbles between the two compartments.

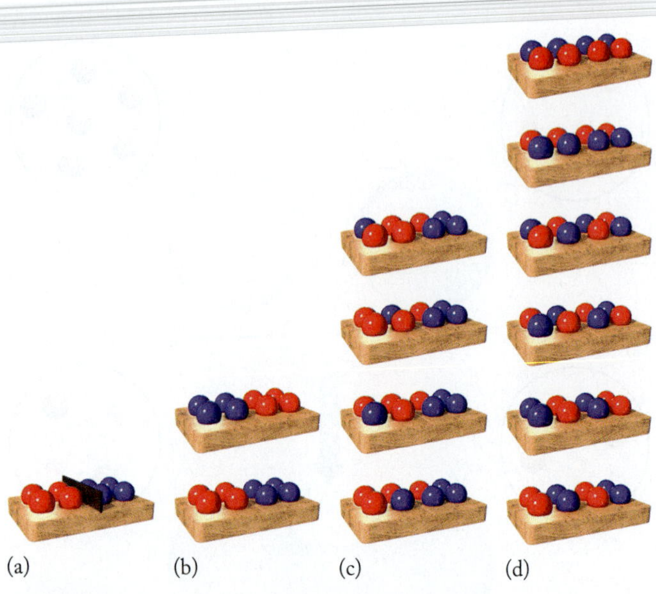


**FIGURE 12.6** A larger model of the entropy of mixing. (a) Initially, four red and four blue marbles are in separate compartments. (b) When mixed, there are two arrangements of the marbles in which the colors are still separate. (c) There are four possible arrangements in each compartment when it exchanges one marble with the other compartment. Therefore, there are $4 \times 4 = 16$ different arrangements for both compartments, and another 16 (not shown) when the compartments exchange three marbles. (d) The greatest number of arrangements—six for each compartment, or $6 \times 6 = 36$ for both—occurs when each compartment contains two red and two blue marbles. Therefore, the most probable arrangement is even mixing, the next most probable is uneven mixing, and by far the least probable is no mixing at all.

If we extend the message contained in mixing the small numbers of marbles in Figures 12.5 and 12.6 to extremely large numbers of particles (think of the Avogadro constant), we come to the following conclusions:

1. Substances (such as gases) that are miscible will mix with each other spontaneously because there is a higher probability of a mixed distribution.
2. The most probable mixing pattern produces a uniform distribution of particles throughout the volume they occupy.

This probabilistic view of particle behavior was first connected to the entropy of thermodynamic systems in the 1870s by German physicist Ludwig Boltzmann (1844–1906), after his pioneering work on the kinetic molecular theory of gases. Boltzmann proposed that the entropy of a system was related to the number of different ways the particles in it could be arranged that was consistent with the system's total internal energy. This number of energy-equivalent arrangements of particles, in which each particle is in a particular position and has a particular momentum (the product of its mass and speed), defined a variable that Boltzmann called *Wahrscheinlichkeit* ($W$)—the German word for probability. He then linked the entropy ($S$) of the system to the number of probable arrangements ($W$) of its particles and particle energies:

$$S = k_B \ln W \tag{12.1}$$

where $k_B$ is the Boltzmann constant, which is equal to the universal gas constant ($R$) divided by the Avogadro constant ($N_A$):

$$k_B = \frac{R}{N_A} = \frac{8.314 \dfrac{\text{J}}{\cancel{\text{mol}} \cdot \text{K}}}{6.0221 \times 10^{23}/\cancel{\text{mol}}} = 1.381 \times 10^{-23}\ \text{J/K}$$

**STEPWISE ANIMATION**
Microstates

The probable number of arrangements of the particles in a system at a given temperature is called the number of **accessible microstates**. To get a feel for what this phrase means, let's return to Figure 12.5 and the arrangements numbered 2–5. For now, think of the marbles in these arrangements as stationary atoms in an isolated thermodynamic system. Each of the four arrangements represents an accessible microstate—an equivalent arrangement of the *macro*state in which two atoms of two elements are mixed so that each compartment has one atom of each element. Arrangements 1 and 6 are equivalent microstates of the macrostate in which the four atoms are not mixed. The fact that the mixed macrostate is represented by more microstates means that it has greater entropy. This means that the mixing process is accompanied by an increase in the number of accessible microstates and an increase in entropy and, according to the second law, is spontaneous.

Boltzmann's probabilistic $W$ also incorporates particle motion and position. In addressing particle motion, we need to think of all the ways that particles can move. Put another way, we need to consider the kinds of *motional* energy they have. As we discussed in Chapter 10 (see Figure 10.6), molecules (and atoms) have translational energy, rotational energy, and (in molecules) bond vibrational energy. Recall from Chapter 3 that energy on the atomic scale is quantized, not continuous. Thus, the translational and rotational energies of atoms and molecules and the vibrational energies of the bonds in molecules are all quantized. The differences between translational and rotational energy states of gas-phase atoms and molecules are so small that the particles have access to enormous numbers of energy states at room temperature. To illustrate how many, let's use Equation 12.1 to calculate the number of microstates that are accessible to a single neon atom at 25°C. To calculate the number of accessible microstates, we rearrange the terms in Equation 12.1 to solve for $W$ and then insert the entropy value of one mole of

**accessible microstate** a unique arrangement of the positions and momenta of the particles in a thermodynamic system.

Ne under standard conditions at 25°C (146.3 J/mol · K; see Section 12.3 and Appendix 4) and the quantity of Ne (one atom):

$$S = k_B \ln W$$

$$\ln W = \frac{S}{k_B}$$

$$\ln W = \frac{\left(146.3 \frac{\text{J}}{\text{mol} \cdot \text{K}}\right)\left(\frac{1 \text{ atom}}{6.0221 \times 10^{23} \frac{\text{atoms}}{\text{mol}}}\right)}{1.381 \times 10^{-23} \frac{\text{J}}{\text{K}}}$$

$$\ln W = 17.59$$

$$W = e^{17.59} = 4.358 \times 10^7$$

This value illustrates the enormous number of ways that the energy of a single gas-phase atom can be dispersed. The motion of an atom in a solid or a liquid is much more restricted, so it has access to many fewer microstates. An atom of carbon in the rigid crystalline structure of a diamond, for example, is restricted to vibrating within a very confined space. There is little uncertainty in the position of the atom, and it has little motional energy. These limitations are reflected in the small entropy value of diamond, 2.4 J/(mol · K) and in the number of microstates accessible to one of its carbon atoms:

$$S = k_B \ln W$$

$$\ln W = \frac{S}{k_B} = \frac{\left(2.4 \frac{\text{J}}{\text{mol} \cdot \text{K}}\right)\left(\frac{1 \text{ atom}}{6.0221 \times 10^{23} \frac{\text{atoms}}{\text{mol}}}\right)}{1.381 \times 10^{-23} \frac{\text{J}}{\text{K}}} = 0.29$$

$$W = e^{0.29} = 1.3$$

Thus, there is on average little more than one microstate available to each atom in a diamond at 25°C and 1 bar, but that value may be misleading. When we attempt to calculate the number of accessible microstates in one *mole* of diamond at 25°C and 1 bar, the result is so enormous that most calculators can't process it and will generate an error message:

$$\ln W = \frac{\left(2.4 \frac{\text{J}}{\text{mol} \cdot \text{K}}\right) \times 1 \text{ mol}}{1.381 \times 10^{-23} \frac{\text{J}}{\text{K}}} = 1.7 \times 10^{23}$$

$$W = e^{1.7 \times 10^{23}} = [\text{Error:overflow}]$$

As a result, the number of microstates accessible to a mole of particles is enormous, even when they are locked in place in a crystalline solid.

The statistical definition of entropy based on accessible microstates is conceptually compatible with the thermodynamic, macroscopic view based on energy dispersion. After all, each accessible microstate represents one of the multitude of ways that energy can be dispersed in a thermodynamic system. Actually, some facets of entropy are easier to understand using the microstate model. One of them is the concept of absolute entropy, and the notion embedded in the **third law of thermodynamics**, which states that a perfect crystalline solid has zero entropy at a temperature of absolute zero. If the particles of a crystalline solid are uniformly arranged in three-dimensional space (**Figure 12.7**) and have zero translational kinetic energy, the crystal has only one microstate, and according to Equation 12.1 its entropy is zero:

$$S = k_B \ln W = k_B \ln 1 = k_B \times 0 = 0$$

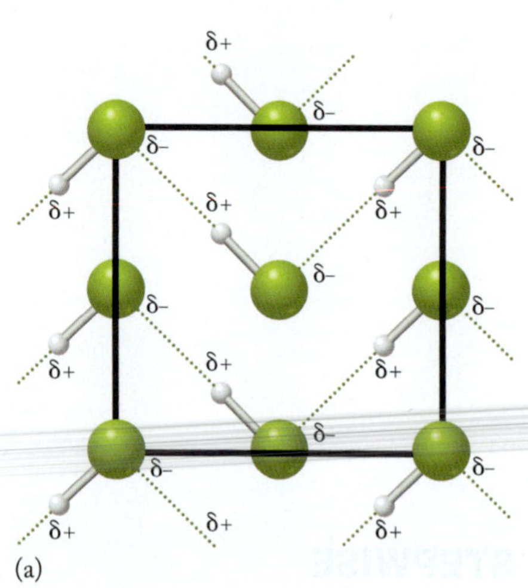


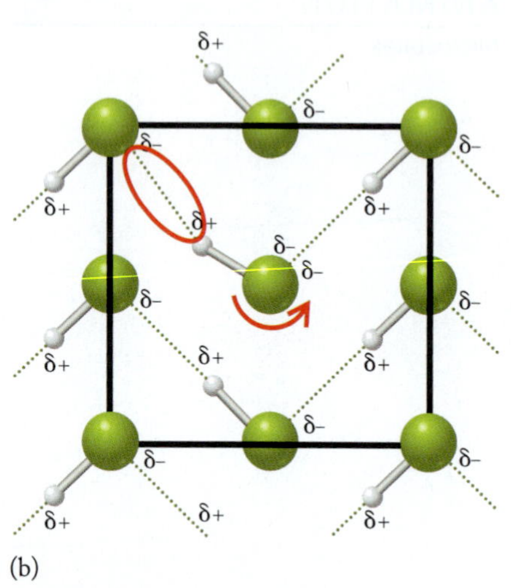


**FIGURE 12.7** Two views of the layers of molecules in solid HCl. (a) A perfect crystal of HCl at $T = 0$ K has zero entropy because all the molecules have zero translational kinetic energy and are arranged uniformly, which means they all experience equivalent dipole–dipole interactions. (b) In this imperfect crystal at $T > 0$ K, the center HCl molecule is rotated a little, creating a slightly different dipole–dipole interaction and a disordered structure. The entropy of this crystal is greater than 0.

## 12.3 Absolute Entropy and Molecular Structure

The third law of thermodynamics provides a vital reference point for the absolute entropy scale: a perfect crystal at 0 K has zero entropy. From this starting point and by carefully measuring the rise in temperature of substances as known quantities of thermal energy are added to them, scientists have been able to establish absolute entropy values for many substances at many temperatures. The most common quantity for expressing these values is **standard molar entropy ($S°$)**, which is the entropy of one mole of a substance in its standard state (see **Table 12.1**) at a pressure of 1 bar (~1 atm). Some sample $S°$ values at 298 K are listed in **Table 12.2**.

**TABLE 12.1** Standard States of Pure Substances and Solutions

| Physical State | Standard State | Pressure[a] | Temperature[b] |
|---|---|---|---|
| Solid | Pure solid, most stable allotrope of an element | 1 bar | 25°C |
| Liquid | Pure liquid | 1 bar | 25°C |
| Gas | Pure gas | 1 bar | 25°C |
| Solution | 1 *M* | 1 bar | 25°C |

[a]Since 1982, 1 bar has been the standard pressure for tabulating all thermodynamic data. Before 1982, standard pressure was 1 atmosphere (atm) = 1.01325 bar.

[b]The thermodynamic data in Appendix 4 and used elsewhere in this book are based on a temperature of 25°C (298 K). This temperature is *not* the same as the STP value we use for ideal gases, which is 273 K.

The $S°$ values for liquid water and water vapor in Table 12.2 illustrate a difference between the entropies of liquids and gases that we discussed in the previous section: the molecules in a gas under standard conditions are much more widely dispersed than the molecules in a liquid, and the entropies of the different phases of a substance at a given temperature follow the order $S_{solid} < S_{liquid} < S_{gas}$.

**TABLE 12.2** Selected Standard Molar Entropy Values[a]

| Formula | $S°$ [J/(mol · K)] | Formula | Name | $S°$ [J/(mol · K)] |
|---|---|---|---|---|
| $Br_2(g)$ | 245.5 | $CH_4(g)$ | Methane | 186.2 |
| $Br_2(\ell)$ | 152.2 | $CH_3CH_3(g)$ | Ethane | 229.5 |
| $C_{diamond}(s)$ | 2.4 | $CH_3CH_2CH_3(g)$ | Propane | 269.9 |
| $C_{graphite}(s)$ | 5.7 | $CH_3(CH_2)_2CH_3(g)$ | Butane | 310.0 |
| $CO(g)$ | 197.7 | $CH_3(CH_2)_2CH_3(\ell)$ | | 231.0 |
| $CO_2(g)$ | 213.8 | $CH_3OH(g)$ | Methanol | 239.9 |
| $H_2(g)$ | 130.6 | $CH_3OH(\ell)$ | | 126.8 |
| $N_2(g)$ | 191.5 | $CH_3CH_2OH(g)$ | Ethanol | 282.6 |
| $O_2(g)$ | 205.0 | $CH_3CH_2OH(\ell)$ | | 160.7 |
| $H_2O(g)$ | 188.8 | $C_6H_6(g)$ | Benzene | 269.2 |
| $H_2O(\ell)$ | 69.9 | $C_6H_6(\ell)$ | | 172.9 |
| $NH_3(g)$ | 192.5 | $C_{12}H_{22}O_{11}(s)$ | Sucrose | 360.2 |

[a]Values for additional substances are given in Appendix 4.

**third law of thermodynamics** the principle stating that the entropy of a perfect crystal is zero at absolute zero.

**standard molar entropy ($S°$)** the absolute entropy of one mole of a substance in its standard state.

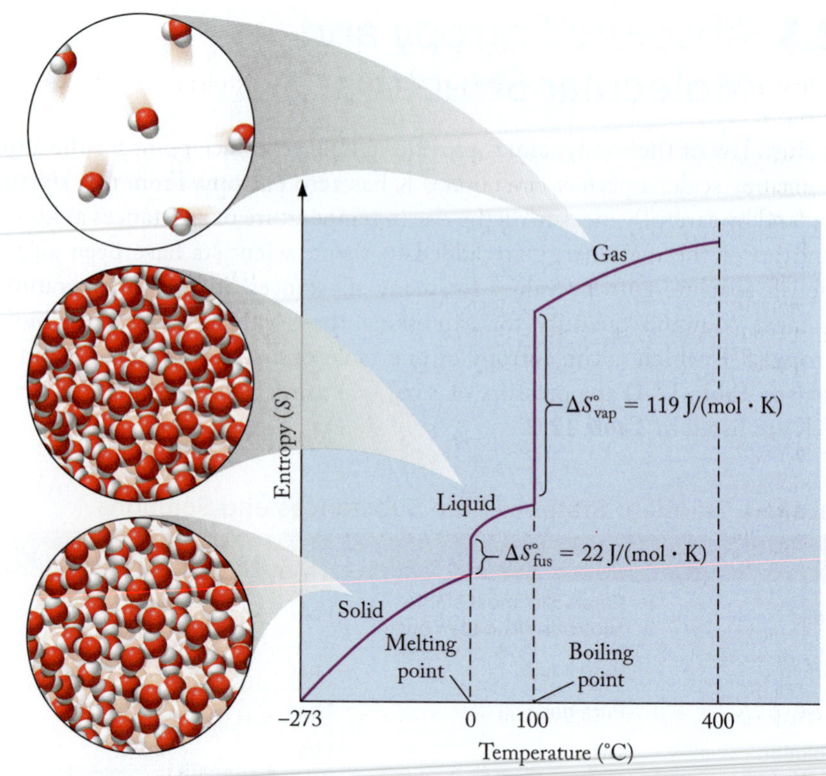


**FIGURE 12.8** Abrupt increases in entropy accompany changes of state, as shown for water, with the greater increase occurring during the transition from liquid to gas. In both phase changes, entropy increases because each change increases the dispersion of the energy of a system's particles. In other words, the particles have access to more microstates. We can often make qualitative predictions about entropy changes that accompany chemical reactions based on these factors, even if we have no thermodynamic data about the reactants and products.

## CONCEPT TEST

Predict which one of the following values comes closest to the number of microstates accessible to one molecule of liquid water at 25°C: (a) 1, (b) $10^3$, (c) $10^6$, (d) $10^9$, (e) $10^{12}$. Use the appropriate $S°$ value from Table 12.2 and Equation 12.1 to determine whether your selection was the best one.

The entropy changes that occur as one mole of ice at 0 K is heated are shown in **Figure 12.8**. Note the jump in entropy as the ice melts at 0°C and the even bigger jump as the liquid water vaporizes at 100°C. Also note that the lines between the phase changes are curved. The change in entropy, $\Delta S$, is not linear with temperature because heating a substance at a higher temperature produces a smaller entropy increase than adding the same quantity of thermal energy to the same substance at a lower temperature.

The factors that affect entropy change can be summarized as follows:

1. Entropy increases when temperature increases.
2. Entropy increases when volume increases.
3. Entropy increases when the number of independent particles increases.

(a) $NH_3(g) + HCl(g) \rightarrow NH_4Cl(s)$

(b) $C_{12}H_{22}O_{11}(s) \rightarrow C_{12}H_{22}O_{11}(aq)$

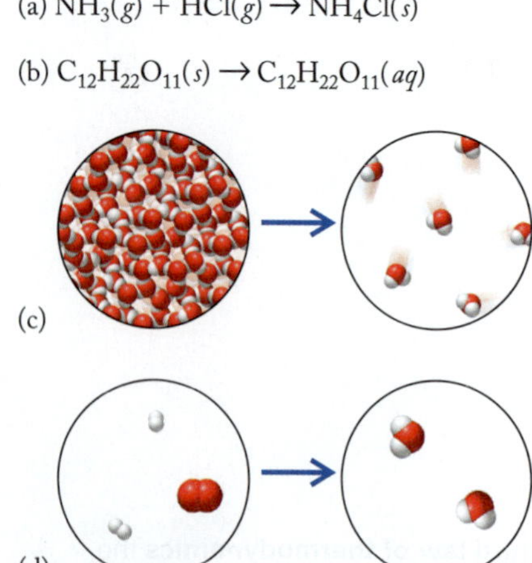


**FIGURE 12.9** These four processes occur at constant temperature (Sample Exercise 12.1).

### SAMPLE EXERCISE 12.1 Predicting the Sign of an Entropy Change LO1

Predict whether an increase or decrease in entropy accompanies each of the processes in **Figure 12.9** when they occur at constant temperature.

**Collect, Organize, and Analyze** We are given a chemical equation or a visual depiction of four processes, and we are asked to predict the signs of the accompanying entropy changes.

a. Two gaseous compounds form a single solid compound, which means the particles of product occupy much less space and have much less freedom of motion.
b. A solid dissolves in water, forming molecules dispersed in an aqueous solution that have more freedom of motion.
c. Molecules of liquid water become molecules of water vapor, which means the molecules and their kinetic energies are more dispersed.
d. A total of three moles of gaseous reactants combine to form two moles of a gaseous product. Having fewer gas-phase particles means less freedom of motion.

**Solve**

a. $\Delta S < 0$.
b. $\Delta S > 0$.
c. $\Delta S > 0$.
d. $\Delta S < 0$.

**Think About It** Entropy increases when solids melt and liquids vaporize because their particles experience increased freedom of motion and greater dispersion of their kinetic energies. When solids dissolve in liquids, they also gain freedom of motion and experience an increase in entropy. When gases combine to form liquids, solids, or fewer particles of gases, they lose freedom of motion and entropy decreases.

**Practice Exercise** Does each process in **Figure 12.10** result in an increase or decrease in the entropy of the system? Assume the reactants and products are at the same temperature.

*(Answers to Practice Exercises are in the back of the book.)*

(a) $CaCO_3(s) + 2\,HCl(aq) \rightarrow CaCl_2(aq) + CO_2(g) + H_2O(\ell)$

(b) $NH_3(g) + BF_3(g) \rightarrow NH_3BF_3(s)$

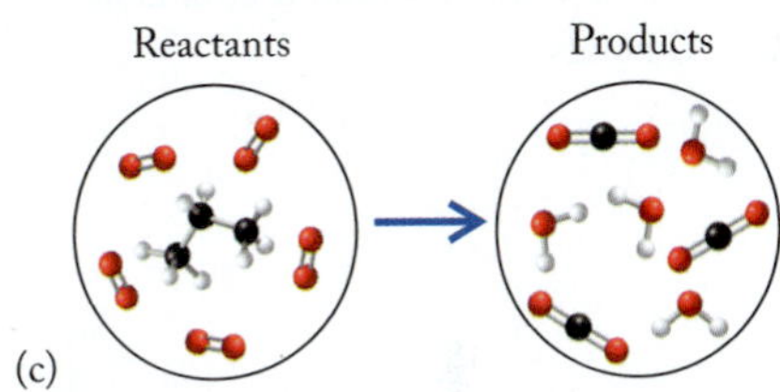


(c)

**FIGURE 12.10** These three processes occur at constant temperature (Practice Exercise 12.1).

The data in Table 12.2 contain an important message about the standard molar entropies of molecular substances: they are linked to the masses of their molecules and their molecular structures. To see the result of this influence, consider the standard molar entropies of the $C_1$ to $C_4$ alkanes in natural gas in Table 12.2. Note how $S°$ values increase with increasing molar mass. This trend is due to the different ways that these molecules move in addition to simple translational motion in three-dimensional space. For example, we have noted that molecules in the gas phase are free to rotate, which adds to the number of accessible microstates. Even liquid molecules tumble over each other as they flow past their nearest neighbors, adding to the ways that energy can be dispersed among them.

The quantities of energy involved in rotational motion depend on how massive the molecules are and on how their masses are distributed. Consider, for example, the $S°$ values of octane and tetramethylbutane in **Figure 12.11**. The isomers have the same molar masses but different molecular shapes. The longer, narrower shape of octane means that more of its mass is farther from the center of the molecule and farther from the axis of rotation when the molecule tumbles end over end. When the mass of a molecule is more spread out, the spacing between its rotational energy states is smaller and more microstates are accessible. Therefore, molecules of octane have greater entropy than the same number of molecules of tetramethylbutane at the same temperature.

Another way to think about the distribution of mass and entropy is to consider how many different orientations can be drawn for the molecule due to free

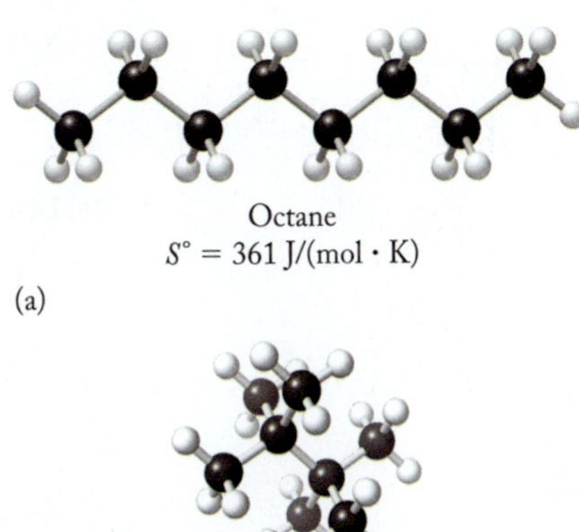


**FIGURE 12.11** Ball-and-stick models and $S°$ values of (a) octane and (b) tetramethylbutane.

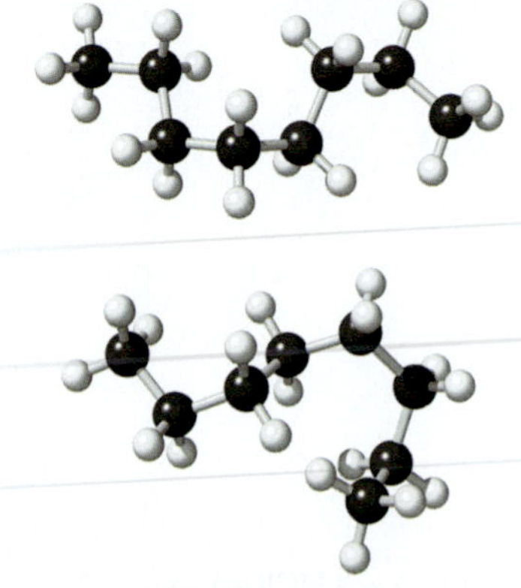

**FIGURE 12.12** Linear alkanes don't have to be linear. These models represent a few of the many orientations of the atoms in octane as a result of rotations around its C—C bonds.

rotation about C—C single bonds. Octane is linear and has many more possible conformations than tetramethylbutane. Only a few of the possible orientations are shown in **Figure 12.12**.

Another structural feature that influences entropy is rigidity. The two most common forms of carbon—namely, diamond and graphite—both consist of extended arrays of C—C bonds. However, diamond is an extremely hard substance because it has a rigid three-dimensional atomic structure (**Figure 12.13a**). Its atoms have access to fewer microstates and it has less entropy [$S° = 2.4$ J/(mol · K)] than soft, easily deformed graphite [$S° = 5.7$ J/(mol · K)], which consists of two-dimensional sheets (**Figure 12.13b**) that are separated by more than the length of a C—C bond. This means that they are attracted to each other by only London dispersion forces, which allows them to slide past each other when a modest shear force is applied.

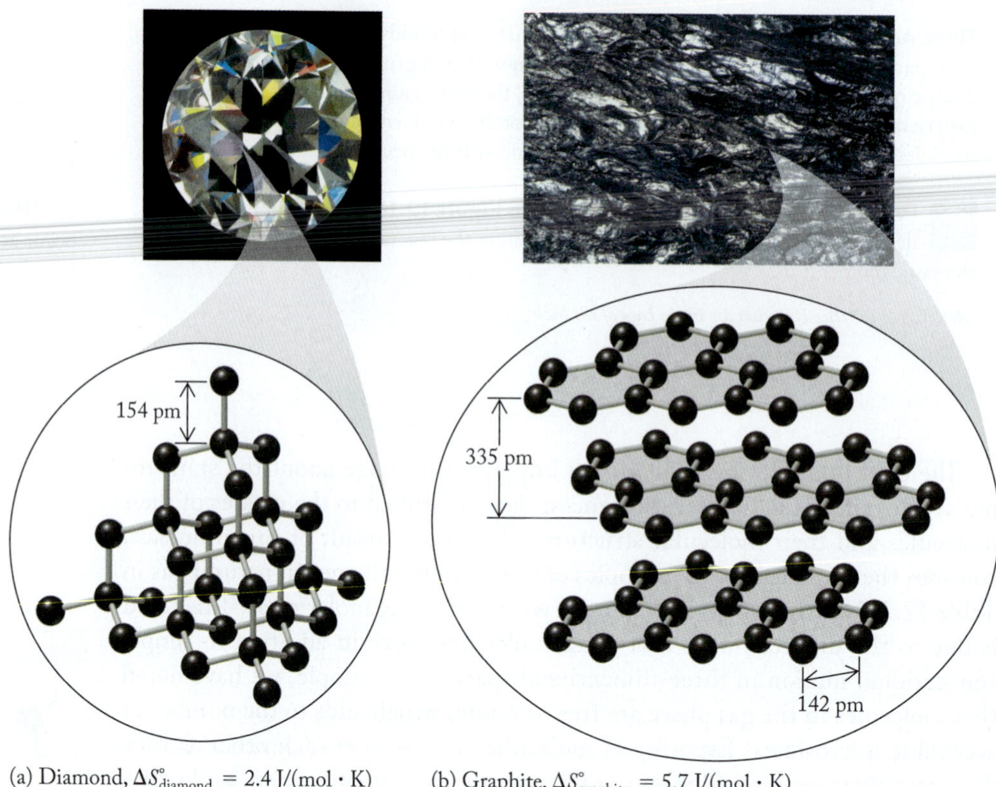


**FIGURE 12.13** (a) Diamonds have a rigid three-dimensional structure, which correlates with a small standard molar entropy. (b) The larger $S°$ for graphite reflects the ease with which layers of carbon atoms can slide past each other.

**SAMPLE EXERCISE 12.2** Comparing Standard Molar Entropy Values **LO3**

Without consulting any reference sources, select the substance in each pair in **Figure 12.14** that has the greater standard molar entropy at 298 K. Assume there is one mole of each substance in its standard state (that is, the pressure of each gas is 1 bar and the concentration of each solution is 1 *M*).

(a) $HCl(g)$ or $HCl(aq)$

(b) $CH_3OH(\ell)$ or $CH_3CH_2OH(\ell)$

(c)

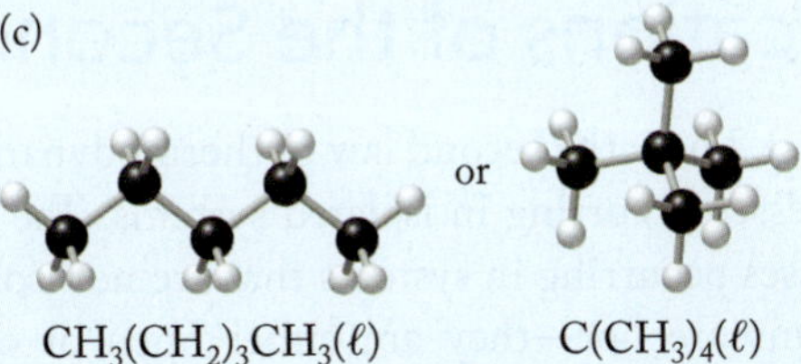


**FIGURE 12.14** Three pairs of substances (Sample Exercise 12.2).

**Collect and Organize** We are given chemical formulas or molecular models of pairs of substances, and we are asked to select which substance in each pair has the greater entropy per mole under standard conditions—that is, the greater standard molar entropy ($S°$) at 298 K.

**Analyze** Particles in the vapor state of a substance are more dispersed and have more entropy than they do in the liquid state at the same temperature, which in turn have more entropy than particles in the solid state. Substances with larger molar masses tend to have greater $S°$ values than those in the same physical state that are composed of less massive particles. Substances composed of compact molecules have lower standard molar entropies than those composed of elongated molecules of the same mass. The volume of one mole of an ideal gas at $P = 1$ bar and $T = 298$ K is about 24.8 liters; the volume of one mole of solute in a 1 *M* solution is only 1 liter.

**Solve**

a. One mole of HCl dissolved in one liter of water occupies much less volume and its particles have much less freedom of motion than one mole of HCl gas. Therefore, $HCl(g)$ has a greater $S°$ value than $HCl(aq)$.
b. Both compounds are liquid alcohols, but molecules of ethanol ($CH_3CH_2OH$) are more massive than those of methanol ($CH_3OH$). Therefore, ethanol has a greater $S°$ value than methanol.
c. The two compounds are isomers with the same molar mass. However, molecules of $CH_3(CH_2)_3CH_3$ are more elongated and have more opportunities than $C(CH_3)_4$ to disperse their kinetic energy. Thus, there are more ways to draw molecules of $CH_3(CH_2)_3CH_3$ because of free rotation about the C—C single bonds (**Figure 12.15**). The greater number of possible orientations of C—C bonds in $CH_3(CH_2)_3CH_3$ means that linear $CH_3(CH_2)_3CH_3$ has more opportunities to disperse its kinetic energy and therefore has a greater $S°$ value than that of $C(CH_3)_4$.

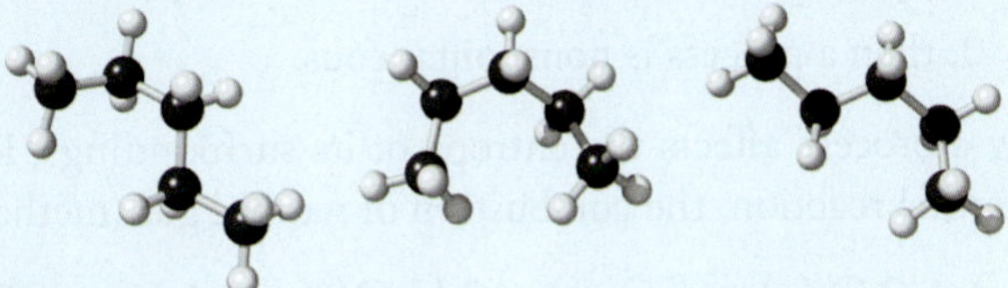

**FIGURE 12.15** Three possible conformations of $CH_3(CH_2)_3CH_3$.

**Think About It** The comparison in part (a) is complicated by the fact that HCl is a strong acid, which means that each molecule produces two ions in solution. However, they are liquid-phase ions and have less total entropy than half as many gas-phase HCl molecules. [Compare the $S°$ values of $HCl(g)$, $H^+(aq)$, and $Cl^-(aq)$ in Appendix 4 to see for yourself.]

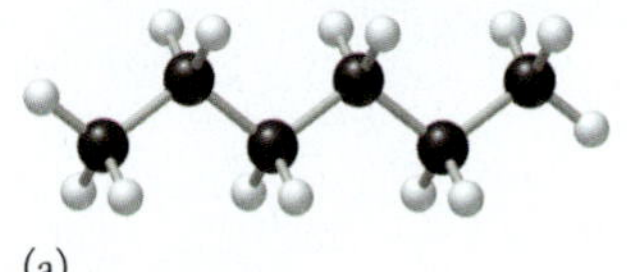

(a)

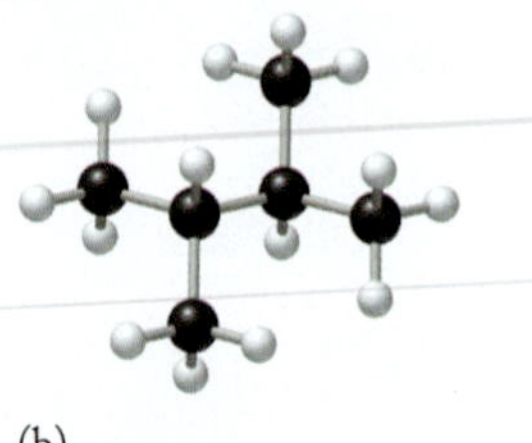

(b)

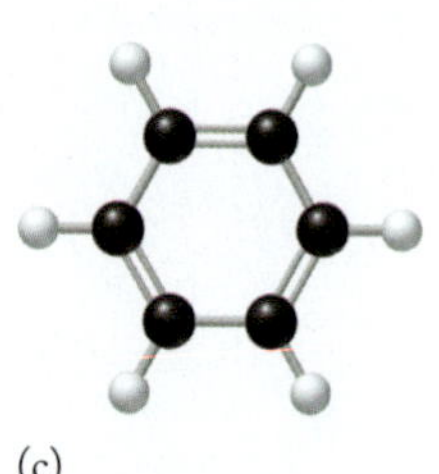

(c)

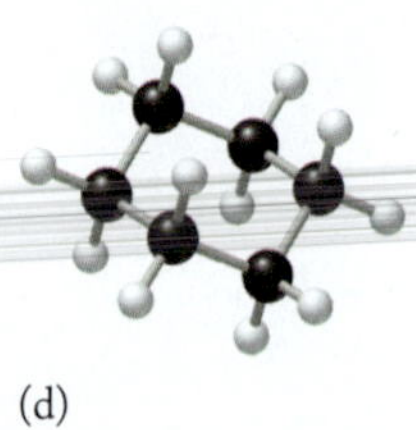

(d)

**FIGURE 12.16** Four hydrocarbons, each of which contains six carbon atoms.

**CONNECTION** In Section 10.2, we defined an isolated thermodynamic system as one that exchanges neither energy nor matter with its surroundings. A closed system exchanges energy but not matter, and an open system exchanges both.

**Practice Exercise** **Figure 12.16** shows ball-and-stick models of four hydrocarbons that each contain six carbon atoms per molecule. Rank these compounds in order of decreasing $S°$ value.

## 12.4 Applications of the Second Law

We saw in Section 12.2 how the second law of thermodynamics accounts for the spontaneity of processes occurring in isolated systems. The second law can also be applied to processes occurring in systems that are not isolated (which is good because most systems are not—they are either open or closed). Recall from Chapter 10 that in thermodynamics we divide the universe into two parts: the part we are interested in (the system) and everything else (the surroundings). Mathematically,

$$\text{Universe} = \text{system} + \text{surroundings} \quad (10.1)$$

By analogy, the overall change in the entropy of the universe as a result of a physical or chemical change is the sum of the entropy changes experienced by the system and its surroundings:

$$\Delta S_{\text{univ}} = \Delta S_{\text{sys}} + \Delta S_{\text{surr}} \quad (12.2)$$

When a spontaneous process occurs in an isolated system, $\Delta S_{\text{sys}} > 0$. Because the system is isolated, the process has no impact on its surroundings, so $\Delta S_{\text{surr}} = 0$. Therefore, according to Equation 12.2, $\Delta S_{\text{univ}}$ must be greater than zero because $\Delta S_{\text{sys}} > 0$. The positive value of $\Delta S_{\text{univ}}$ is the basis for another way of expressing the second law of thermodynamics that applies to all systems: *a spontaneous process produces an increase in the entropy of the universe.*

The latter version of the second law assumes that a physical or chemical change in a closed or open thermodynamic system can alter the entropies of both the system and its surroundings. The second law says that a process is spontaneous when $\Delta S_{\text{univ}} > 0$. The second law also provides a thermodynamic requirement for nonspontaneity: a process that produces a decrease in the entropy of the universe will not occur spontaneously. Therefore, the reverse of any spontaneous process has to be nonspontaneous because it can be spontaneous in only one direction—the one for which $\Delta S_{\text{univ}} > 0$. Reversing a spontaneous process reverses the sign of $\Delta S_{\text{univ}}$, making it less than zero, which means the process is nonspontaneous.

- If $\Delta S_{\text{univ}} > 0$, then a process is spontaneous.
- If $\Delta S_{\text{univ}} < 0$, then a process is nonspontaneous.

To see how a process affects the entropy of its surroundings, let's focus on a familiar exothermic reaction, the combustion of natural gas (methane):

$$CH_4(g) + 2\,O_2(g) \rightarrow CO_2(g) + 2\,H_2O(\ell) \qquad \Delta H° = -890 \text{ kJ}$$

As written, the reaction consumes three moles of gases, and it produces two moles of a liquid product and one mole of $CO_2$ gas, so there are fewer moles of gases on the product side of the reaction equation. Given the much greater freedom of motion of particles in the gas phase, we predict that there will be a decrease in entropy of the reaction mixture, which is our thermodynamic system:

$$\Delta S_{\text{sys}} < 0$$

The reaction is spontaneous, however, so

$$\Delta S_{univ} > 0$$

How do we reconcile these opposing inequalities? Equation 12.2 supplies an explanation. If $\Delta S_{univ} > 0$, then the sum of $\Delta S_{sys}$ and $\Delta S_{surr}$ must also be greater than zero. The fact that $\Delta S_{sys} < 0$ simply means that $\Delta S_{surr}$ is not only greater than zero but also must be a large enough positive value to more than offset the negative value of $\Delta S_{sys}$. Expressing this relationship in terms of the absolute values of $\Delta S_{surr}$ and $\Delta S_{sys}$:

$$|\Delta S_{surr}| > |\Delta S_{sys}|$$

Is the combustion of methane likely to produce a large, positive $\Delta S_{surr}$? Absolutely it will—because the reaction is highly exothermic: combustion of only one mole (16 grams) of methane releases 890 kJ of thermal energy. As energy flows from the system into its surroundings, dispersion of this energy produces a positive $\Delta S_{surr}$ that more than compensates for the unfavorable (negative) value of $\Delta S_{sys}$.

All exothermic reactions have the capacity to raise the entropy of their surroundings. The more energy that flows into the surroundings, or into any collection of particles, the greater the dispersion of energy among the particles and the greater the increase in their entropy ($\Delta S$). However, adding thermal energy to particles at a higher temperature produces a smaller gain in entropy than heating the same particles at a lower temperature. This inverse relationship between entropy gain and temperature, which we discussed with Figure 12.8, can be mathematically expressed as

$$\Delta S = \frac{q_{rev}}{T} \quad (12.3)$$

where $q_{rev}$ represents the reversible flow of energy in the form of heat caused by the difference in temperature between the system and its surroundings.

**STEPWISE ANIMATION**
Reversible Processes

Theoretically, a **reversible process** is one that happens so slowly that, after an incremental change in the system has occurred, the process can be reversed by a tiny change in reaction conditions that restores the original state of the system with no net flow of energy into or out of the system. In other words, the system and its surroundings finish exactly as they were before the process began. At best, real chemical reactions and physical changes are only approximately reversible, but many are close enough that we can adapt Equation 12.3 to calculate $\Delta S_{sys}$:

$$\Delta S_{sys} = \frac{q_{sys}}{T} \quad (12.4)$$

Let's apply Equation 12.4 to the melting of ice. It takes 6.01 kJ (or $6.01 \times 10^3$ J) of energy to melt 1.00 mol of ice at 0°C. If we assume that the process occurs reversibly, then

$$\Delta S_{sys} = \frac{q_{sys}}{T} = \frac{(1.00 \ \cancel{\text{mol}})(6.01 \times 10^3 \text{ J}/\cancel{\text{mol}})}{273 \text{ K}} = 22.0 \text{ J/K}$$

Because energy flows into the ice, $q_{sys}$ is positive. This also makes $\Delta S_{sys}$ positive, as it should be, given the greater freedom of motion of particles in the liquid phase. Note that the units of entropy in this calculation are joules per kelvin.

Now suppose that the melting process occurs as thermal energy flows into one mole of ice from room-temperature (22°C, or 295 K) surroundings. The

**reversible process** a process that happens so slowly that an incremental change can be reversed by another tiny change, restoring the original state of the system with no net flow of energy between the system and its surroundings.

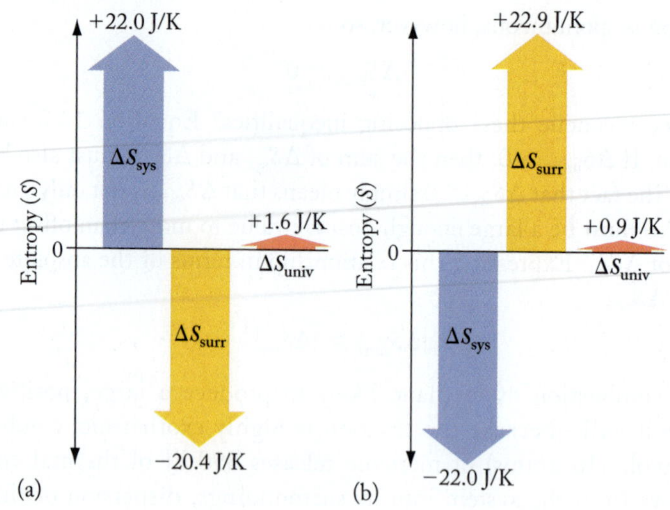


**FIGURE 12.17** (a) If the surroundings experience a decrease in entropy, and the system (ice) experiences an increase in entropy that more than offsets the decrease, then the process (melting) is spontaneous. (b) If the surroundings are at a lower temperature than the system, then energy spontaneously flows into the surroundings due to a decrease in entropy for the system and an increase in entropy for the surroundings.

change in entropy of the surroundings can be calculated using another adaptation of Equation 12.3:

$$\Delta S_{surr} = \frac{q_{surr}}{T} = \frac{(1.00\ \cancel{mol})(-6.01 \times 10^3\ J/\cancel{mol})}{295\ K} = -20.4\ J/K$$

Note that the sign of $q_{surr}$ is negative because energy flows from the surroundings into the system (ice cube). Note, too, that (1) $\Delta S_{surr} < 0$, and (2) the magnitude of the decrease in $\Delta S_{surr}$ is less than the increase in $\Delta S_{sys}$, because the same absolute value of $q$ is divided by a higher temperature to calculate $\Delta S_{surr}$. Therefore, when we sum $\Delta S_{sys}$ and $\Delta S_{surr}$ (**Figure 12.17a**), we get a positive value for $\Delta S_{univ}$:

$$\Delta S_{univ} = \Delta S_{sys} + \Delta S_{surr} = (22.0 - 20.4)\ J/K = 1.6\ J/K$$

The result of this calculation is one example of a general truth about the flow of thermal energy—namely, it flows spontaneously into a system when the system is cooler than its surroundings. Even more generally, we say that thermal energy flows spontaneously from a warm object to an adjacent cooler object.

Why did we ignore the change in temperature of the surroundings as thermal energy flowed from it into the melting ice? The answer lies in the sheer size of the surroundings (the universe minus a small cube of ice). The temperature of such a gigantic thermal mass does not change significantly.

Now let's consider how entropy changes when the temperature of the surroundings is *lower* than the temperature of the system. Suppose 1.00 mol of liquid water at 0°C is placed in a freezer at −10°C. Because the temperature of the surroundings (the freezer) is lower than the temperature of the liquid water, energy spontaneously flows from the water into its surroundings. The net entropy change for this process is

$$\begin{aligned}\Delta S_{univ} &= \Delta S_{sys} + \Delta S_{surr}\\ &= \frac{(1.00\ \cancel{mol})(-6.01 \times 10^3\ J/\cancel{mol})}{273\ K} + \frac{(1.00\ \cancel{mol})(+6.01 \times 10^3\ J/\cancel{mol})}{263\ K}\\ &= (-22.0\ J/K) + (+22.9\ J/K)\\ &= 0.9\ J/K\end{aligned}$$

Once again, there is an increase in the entropy of the universe (**Figure 12.17b**) as energy flows spontaneously from the warmer object (liquid water at 0°C) into the colder surroundings (the freezer at −10°C).

**FIGURE 12.18** Water vapor exhaled by this Inuit hunter in the Northwest Territories of Canada was spontaneously deposited on his facial hair as frost—a testament to how cold it was when this photo was taken.

**CONCEPT TEST**

Is $\Delta S_{univ}$ greater than, less than, or equal to zero when water vapor exhaled by the Inuit hunter in **Figure 12.18** is deposited as crystals of ice on his beard?

Entropy-change calculations based on Equation 12.3 assume process reversibility, which as we have discussed is an idealized, theoretical concept. In reality, the $\Delta S$ values calculated in this way are *minimum* $\Delta S$ values. When processes take place in the real world, the accompanying changes in entropy are inevitably greater than those calculated using Equation 12.3.

## 12.5 Calculating Entropy Changes

The entropy of a system (like its enthalpy and internal energy) is a state function, which means that the change in entropy that accompanies a process depends only on the initial and final states of the system, not on the pathway of the process. Therefore, the change in entropy experienced by the system is simply the difference between its initial and final absolute entropy levels:

**CONNECTION** We defined state functions in Section 10.1.

$$\Delta S_{sys} = \Delta S_{final} - \Delta S_{initial} \qquad (12.5)$$

We can adapt Equation 12.5 to calculate the change in entropy that accompanies a chemical reaction under standard conditions, $\Delta S^\circ_{rxn}$, from the difference between the standard molar entropies of $n_{reactants}$ moles of reactants (the equivalent of $S_{initial}$ in Equation 12.5) and the standard molar entropies of $n_{products}$ moles of products (that is, $S_{final}$):

$$\Delta S^\circ_{rxn} = \sum n_{products} S^\circ_{products} - \sum n_{reactants} S^\circ_{reactants} \qquad (12.6)$$

Each $S^\circ$ value for a product or reactant is multiplied by the appropriate number of moles from the balanced chemical equation. In other words, $\Delta S^\circ_{rxn}$, like $\Delta H^\circ_{rxn}$, is an extensive thermodynamic property that depends on the amounts of substances consumed or produced in a reaction. Standard molar entropies of selected substances are listed in Table 12.2 and in Appendix 4.

**SAMPLE EXERCISE 12.3** Calculating Entropy Changes **LO4**

A process called methanol reforming is used to generate hydrogen for fuel cells from methanol and steam. Use the appropriate standard molar entropy values to calculate the value of $\Delta S^\circ_{rxn}$ at 298 K for the steam–methane reforming reaction:

$$CH_3OH(g) + H_2O(g) \rightarrow CO_2(g) + 3\,H_2(g)$$

**Collect, Organize, and Analyze** We are given a balanced chemical equation describing a reaction and are asked to use the appropriate standard molar entropy values to calculate the change in entropy that occurs during the reaction under standard conditions. Entropy changes associated with chemical reactions depend on the entropies of the reactants and products, as described in Equation 12.6. Table 12.2 contains the following $S°$ values:

| **Substance** | $CH_3OH(g)$ | $H_2O(g)$ | $CO_2(g)$ | $H_2(g)$ |
|---|---|---|---|---|
| **$S°$ (J/mol·K)** | 239.9 | 188.8 | 213.8 | 130.6 |

During the reaction, two moles of gaseous reactants produce four moles of gaseous products, so it is likely the value of $\Delta S°_{rxn}$ will be greater than zero.

**Solve**

$$\Delta S°_{rxn} = \sum n_{products} S°_{products} - \sum n_{reactants} S°_{reactants}$$

$$= \left[1\ \text{mol CO}_2 \times 213.8\ \frac{\text{J}}{\text{mol} \cdot \text{K}} + 3\ \text{mol H}_2 \times 130.6\ \frac{\text{J}}{\text{mol} \cdot \text{K}}\right]$$

$$-\left[1\ \text{mol CH}_3\text{OH} \times 239.9\ \frac{\text{J}}{\text{mol} \cdot \text{K}} + 1\ \text{mol H}_2\text{O} \times 188.8\ \frac{\text{J}}{\text{mol} \cdot \text{K}}\right]$$

$$= (605.6 - 428.7)\ \text{J/K} = 176.9\ \text{J/K}$$

**Think About It** As predicted, $\Delta S°_{rxn} > 0$ because of the greater freedom of motion and dispersion of energy associated with four moles of gas than with two moles.

**Practice Exercise** Calculate the standard molar entropy change for the combustion of methane gas by using $S°$ values from Table 12.2. Before carrying out the calculation, predict whether the entropy of the system increases or decreases. Assume that liquid water is one of the products.

## 12.6 Free Energy

In Sample Exercise 12.3, we used standard molar entropy values to calculate $\Delta S°_{rxn}$. The reactants and products together constituted a closed thermodynamic system, which meant that thermal energy could flow into or out from it but that matter was not exchanged with its surroundings. Therefore, the $\Delta S°$ value that was calculated applied only to the system. There was no evaluation of the change in the entropy of the system's surroundings accompanying the reaction. If the reaction were endothermic, thermal energy flowing from the surroundings into the system would lower the temperature of the surroundings and produce a decrease in $S_{surr}$. If the reaction were exothermic, thermal energy would flow from the system into its surroundings and $S_{surr}$ would increase.

**CONNECTION** In Section 10.3, we defined a change in enthalpy ($\Delta H$) as the energy gained or lost as heat during a process taking place at constant pressure.

Thus, the entropy change experienced by the surroundings of any chemical system depends on whether the process occurring in the system is exothermic or endothermic. When energy in the form of heat flows from an exothermic process occurring at constant pressure into a system's surroundings, the quantity of energy is equal in magnitude but opposite in sign to the enthalpy change of the system:

$$q_{surr} = -\Delta H_{sys} \tag{12.7}$$

When a system undergoes an endothermic process, the direction of energy flow is reversed but Equation 12.7 still applies. Assuming that these transfers of heat occur reversibly, we can calculate the value of $\Delta S_{surr}$ by substituting $-\Delta H_{sys}$ for $q_{surr}$ in Equation 12.3:

$$\Delta S_{surr} = \frac{-\Delta H_{sys}}{T}$$

Combining this equation with Equation 12.2 ($\Delta S_{univ} = \Delta S_{sys} + S_{surr}$) gives

$$\Delta S_{univ} = \Delta S_{sys} - \frac{\Delta H_{sys}}{T} \quad (12.8)$$

The beauty of Equation 12.8 is that it allows us to predict whether a process is spontaneous at a particular temperature once we calculate the enthalpy and entropy changes accompanying the process. The downside of Equation 12.8 is that spontaneity relies on the value of a parameter ($\Delta S_{univ}$) that is impossible to determine directly and that has little physical meaning. It would be useful if we could substitute a single thermodynamic parameter for $\Delta S_{univ}$ that is based only on the system and not the entire universe. Such a parameter exists, and it is called the change in the system's **free energy**. Note that *free* in this context does not mean "at no cost"; it means energy that is *freed* during a process to do useful work.

**free energy** a measure of the maximum amount of work that a thermodynamic system can perform.

**Gibbs free energy (*G*)** the maximum amount of energy released by a process occurring at constant temperature and pressure that is available to do useful work.

In chemistry we focus on a particular kind of free energy, called **Gibbs free energy (*G*)** in honor of American scientist J. Willard Gibbs (1839–1903). Gibbs free energy is the maximum amount of energy available to do useful work in processes happening at constant temperature and pressure or once the temperatures and pressures of reaction mixtures have returned to their initial values. These conditions are met in the chemical reactions and physical changes we discuss in the following sections. As a result, the changes in free energy that accompany these processes will be changes in Gibbs free energy, and we will use the symbol $\Delta G$ to represent these changes. Gibbs free energy, like enthalpy and entropy, is a state function.

**CHEMTOUR**
Gibbs Free Energy

Like many of the thermodynamic properties we have examined, absolute free-energy values of substances are often of less interest than the *changes* in free energy that accompany chemical reactions and other processes. Gibbs proposed that the change in free energy ($\Delta G_{sys}$) of a process occurring at constant temperature and pressure is linked directly to that temperature and $\Delta S_{univ}$:

$$\Delta G_{sys} = -T\Delta S_{univ}$$

Because of the ($-T$) multiplier, *negative* values of $\Delta G_{sys}$ correspond to *positive* values of $\Delta S_{univ}$. Therefore,

- If $\Delta G_{sys} < 0$, then $\Delta S_{univ} > 0$, and the reaction is spontaneous.
- If $\Delta G_{sys} > 0$, then $\Delta S_{univ} < 0$, and the reaction is nonspontaneous. Instead, the reaction running in reverse of the process is spontaneous.
- If $\Delta G_{sys} = 0$, then $\Delta S_{univ} = 0$, and the composition of the reaction mixture does not change with time. Instead, the reaction has reached equilibrium.

These points are summarized in **Table 12.3**.

The Gibbs equation ($\Delta G_{sys} = -T\Delta S_{univ}$) can be combined with Equation 12.8 once we multiply all the terms in Equation 12.8 by ($-T$):

$$-T\Delta S_{univ} = -T\Delta S_{sys} + \Delta H_{sys} \quad (12.9)$$

The left side of Equation 12.9 is equal to $\Delta G_{sys}$. Making that substitution and rearranging the terms on the right side gives

$$\Delta G_{sys} = \Delta H_{sys} - T\Delta S_{sys}$$

Because all the parameters in this equation apply to the system, we typically simplify the equation by eliminating the subscripts:

$$\Delta G = \Delta H - T\Delta S \quad (12.10)$$

**TABLE 12.3** Links between $\Delta G^\circ_{sys}$, $\Delta S^\circ_{univ}$, and Spontaneity

| $\Delta G^\circ_{sys}$ | $\Delta S^\circ_{univ}$ | Spontaneity |
|---|---|---|
| $< 0$ | $> 0$ | Spontaneous |
| $> 0$ | $< 0$ | Nonspontaneous |
| 0 | 0 | No change with time; at equilibrium |

Equation 12.10 highlights the two thermodynamic factors that contribute to a decrease in free energy and to making a process spontaneous at constant temperature and pressure:

1. The system experiences an increase in entropy ($\Delta S > 0$).
2. The process is exothermic ($\Delta H < 0$).

If a reaction is spontaneous, at least one of these conditions must be true.

The change in free energy of a process can be calculated using Equation 12.10 if we first calculate the values of $\Delta H$ and $\Delta S$. For a chemical reaction occurring under standard conditions, the *standard* change in free energy $\Delta G^\circ_{rxn}$ can be calculated using a modified version of Equation 12.10:

$$\Delta G^\circ_{rxn} = \Delta H^\circ_{rxn} - T\Delta S^\circ_{rxn} \tag{12.11}$$

In Chapter 10, we calculated $\Delta H^\circ_{rxn}$ values from the difference in the standard enthalpies of formation ($\Delta H^\circ_f$) of products and reactants:

$$\Delta H^\circ_{rxn} = \sum n_{products}\Delta H^\circ_{f,products} - \sum n_{reactants}\Delta H^\circ_{f,reactants} \tag{10.17}$$

The value of $\Delta S^\circ_{rxn}$ can be calculated using Equation 12.6:

$$\Delta S^\circ_{rxn} = \sum n_{products}S^\circ_{products} - \sum n_{reactants}S^\circ_{reactants} \tag{12.6}$$

The results of these two calculations are used in Equation 12.11 to calculate $\Delta G^\circ_{rxn}$, as illustrated in Sample Exercises 12.4 and 12.5.

### SAMPLE EXERCISE 12.4 Calculating $\Delta G^\circ$ for a Chemical Reaction **LO5**

Nitrogen dioxide ($NO_2$) is the substance responsible for the brown haze visible in the atmosphere in some cities. In automobile engines, nitrogen and oxygen react to form nitrogen monoxide (NO), which then combines with additional oxygen in the atmosphere to form $NO_2$ in the following reaction:

$$2\ NO(g) + O_2(g) \rightarrow 2\ NO_2(g)$$

What is the value of $\Delta G^\circ_{rxn}$ for this reaction at 298 K if $\Delta H^\circ_{rxn} = -114.2$ kJ and $\Delta S^\circ_{rxn} = -146.5$ J/K?

**Collect, Organize, and Analyze** We are given the $\Delta H^\circ_{rxn}$ and $\Delta S^\circ_{rxn}$ values for a reaction and the temperature, and we are asked to calculate the value of $\Delta G^\circ_{rxn}$. Equation 12.11 connects these three thermodynamic parameters. The change in entropy is given in J/K, so we must convert that value to kJ/K before we can subtract $T\Delta S^\circ_{rxn}$ from $\Delta H^\circ_{rxn}$.

**Solve**

$$\begin{aligned}\Delta G^\circ_{rxn} &= \Delta H^\circ_{rxn} - T\Delta S^\circ_{rxn}\\ &= (-114.2\text{ kJ}) - \left[298\text{ K} \times (-146.5\text{ J/K}) \times \frac{1\text{ kJ}}{1000\text{ J}}\right]\\ &= -70.5\text{ kJ}\end{aligned}$$

**Think About It** The negative value of $\Delta G^\circ_{rxn}$ means that the reaction is spontaneous under standard conditions and $T = 298$ K.

 **Practice Exercise** We opened the chapter by discussing one of the reactions responsible for iron rusting:

$$4\ Fe(s) + 3\ O_2(g) \rightarrow 2\ Fe_2O_3(s)$$

Calculate $\Delta G^\circ_{rxn}$ for this reaction at $T = 298$ K if $\Delta H^\circ_{rxn} = -1648.4$ kJ and $\Delta S^\circ_{rxn} = -549.3$ J/K.

**SAMPLE EXERCISE 12.5** Predicting Reaction Spontaneity under Standard Conditions **LO5**

Consider the reaction of nitrogen gas and hydrogen gas (**Figure 12.19**) at 298 K to make ammonia at the same temperature:

$$N_2(g) + 3\,H_2(g) \rightarrow 2\,NH_3(g)$$

a. Before doing any calculations, predict the sign of $\Delta S^\circ_{rxn}$.
b. What is the actual value of $\Delta S^\circ_{rxn}$?
c. What is the value of $\Delta H^\circ_{rxn}$?
d. What is the value of $\Delta G^\circ_{rxn}$ at 298 K?
e. Is the reaction spontaneous at 298 K and 1 bar of pressure?

**Collect and Organize** We are asked to predict the sign and calculate the value of the standard entropy change of a reaction. We are also asked to calculate the enthalpy change of the reaction and its change in free energy, and then to predict its spontaneity under standard conditions. Standard molar entropies and standard enthalpies of formation of the reactants and product are listed in Table 10.4, Table 12.2, and Appendix 4.

**Analyze** Figure 12.19 reinforces the point that there are more moles of gaseous reactants than products in the reaction. Equation 12.6 can be used to calculate entropy changes under standard conditions:

$$\Delta S^\circ_{rxn} = \sum n_{products}S^\circ_{products} - \sum n_{reactants}S^\circ_{reactants} \qquad (12.6)$$

We learned in Chapter 10 how to calculate $\Delta H^\circ_{rxn}$ from standard enthalpies of formation (in Appendix 4) by using Equation 10.17:

$$\Delta H^\circ_{rxn} = \sum n_{products}\Delta H^\circ_{f,products} - \sum n_{products}\Delta H^\circ_{f,reactants} \qquad (10.17)$$

The calculated values of $\Delta S^\circ_{rxn}$ and $\Delta H^\circ_{rxn}$ can be used in Equation 12.11 to calculate $\Delta G^\circ_{rxn}$:

$$\Delta G^\circ_{rxn} = \Delta H^\circ_{rxn} - T\Delta S^\circ_{rxn} \qquad (12.11)$$

Whether the reaction is spontaneous ($\Delta G^\circ_{rxn} < 0$) or nonspontaneous ($\Delta G^\circ_{rxn} > 0$) under standard conditions (and $T = 298$ K) is difficult to predict without knowing the magnitudes of $\Delta H^\circ_{rxn}$ and $T\Delta S^\circ_{rxn}$.

**Solve**

a. The number of gas-phase molecules decreases as the reaction proceeds, so the sign of $\Delta S^\circ_{rxn}$ is probably negative.
b. Using data from Table 12.2 in Equation 12.6 to calculate $\Delta S^\circ_{rxn}$,

$$\begin{aligned} S^\circ_{rxn} &= \sum n_{products}S^\circ_{products} - \sum n_{reactants}S^\circ_{reactants} \\ &= \left[2\ \text{mol NH}_3 \times \left(192.5\ \frac{\text{J}}{\text{mol}\cdot\text{K}}\right)\right] \\ &\quad - \left[1\ \text{mol N}_2 \times \left(191.5\ \frac{\text{J}}{\text{mol}\cdot\text{K}}\right) + 3\ \text{mol H}_2 \times \left(130.6\ \frac{\text{J}}{\text{mol}\cdot\text{K}}\right)\right] \\ &= -198.3\ \text{J/K} \end{aligned}$$

The entropy change is negative, as predicted.

c. The change in enthalpy that accompanies the reaction under standard conditions is

$$\begin{aligned} \Delta H^\circ_{rxn} &= \sum n_{products}\Delta H^\circ_{f,products} - \sum n_{reactants}\Delta H^\circ_{f,reactants} \\ &= \left[2\ \text{mol NH}_3 \times \left(-46.1\ \frac{\text{kJ}}{\text{mol}}\right)\right] \\ &\quad - \left[1\ \text{mol N}_2 \times \left(0.0\ \frac{\text{kJ}}{\text{mol}}\right) + 3\ \text{mol H}_2 \times \left(0.0\ \frac{\text{kJ}}{\text{mol}}\right)\right] \\ &= -92.2\ \text{kJ} \end{aligned}$$

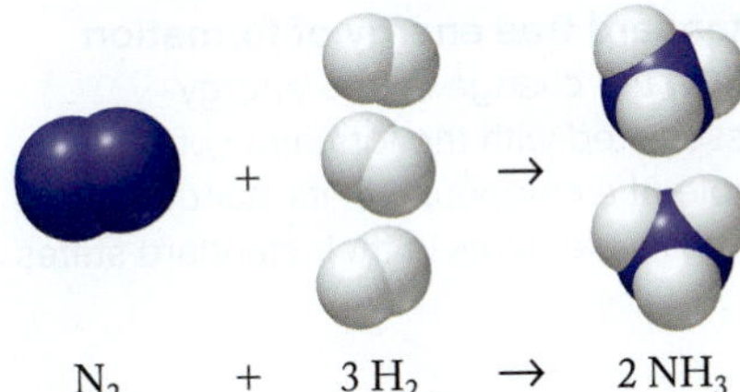


$N_2 \quad + \quad 3\,H_2 \quad \rightarrow \quad 2\,NH_3$

**FIGURE 12.19** In the synthesis of ammonia, one molecule of nitrogen reacts with three molecules of hydrogen to yield two molecules of ammonia. All substances are gases.

**standard free energy of formation ($\Delta G^\circ_f$)** the change in free energy associated with the formation of one mole of a compound in its standard state from its elements in their standard states.

d. Using the values of $\Delta S^\circ_{rxn}$ and $\Delta H^\circ_{rxn}$ calculated in parts (b) and (c) in Equation 12.11,

$$\Delta G^\circ_{rxn} = \Delta H^\circ_{rxn} - T\Delta S^\circ_{rxn}$$

$$= -92.2 \text{ kJ} - (298 \text{ K})\left(-198.3 \frac{\text{J}}{\text{K}}\right)\left(\frac{1 \text{ kJ}}{1000 \text{ J}}\right)$$

$$= -33.1 \text{ kJ}$$

e. $\Delta G^\circ_{rxn} < 0$, so the reaction is spontaneous under standard conditions and $T = 298$ K.

**Think About It** We could not predict the spontaneity of the reaction because we didn't have a feel for the magnitudes of $\Delta H^\circ_{rxn}$ and $\Delta S^\circ_{rxn}$. Here, an unfavorable entropy change ($\Delta S^\circ_{rxn} < 0$) is more than offset by a favorable enthalpy change ($\Delta H^\circ_{rxn} < 0$) so that, overall, $\Delta G^\circ_{rxn} < 0$.

**Practice Exercise**

For the reaction $2\ H_2(g) + O_2(g) \rightarrow 2\ H_2O(\ell)$,

a. Predict the sign of the entropy change for the reaction.
b. What is the value of $\Delta S^\circ_{rxn}$?
c. What is the value of $\Delta H^\circ_{rxn}$?
d. Is the reaction spontaneous at 298 K and 1 bar pressure?

## CONCEPT TEST

The preparation of ammonia from nitrogen and hydrogen in Sample Exercise 12.5 is predicted to be spontaneous under standard conditions, yet if we mix the two gases at 298 K, no reaction is observed. Suggest a reason why.

There is another way to calculate the change in free energy of a reaction under standard conditions. It is based on another thermodynamic property of substances listed in Appendix 4: the **standard free energy of formation ($\Delta G^\circ_f$)**. A compound's $\Delta G^\circ_f$ value is the change in free energy associated with the formation of one mole of it in its standard state from its elements in their standard states.

In much the way we calculated standard enthalpies of reaction ($\Delta H^\circ_{rxn}$) from the difference in the standard enthalpies of formation ($\Delta H^\circ_f$) of their products and reactants in Chapter 10, so too can we calculate the change in standard free energy of a reaction under standard conditions from the difference in the standard free energies of formation of its products and reactants. As with standard enthalpies of formation, the standard free energy of formation of the most stable forms of elements in their standard states is defined to be zero. The similarities between the two calculations can be seen from the similar formats of the equations used to calculate $\Delta G^\circ_{rxn}$,

$$\Delta G^\circ_{rxn} = \sum n_{products}\Delta G^\circ_{f,products} - \sum n_{reactants}\Delta G^\circ_{f,reactants} \quad (12.12)$$

and $\Delta H^\circ_{rxn}$,

$$\Delta H^\circ_{rxn} = \sum n_{products}\Delta H^\circ_{f,products} - \sum n_{reactants}\Delta H^\circ_{f,reactants} \quad (10.17)$$

Sample Exercise 12.6 illustrates just how similar the two calculations are.

## CONCEPT TEST

Molecular structures and standard free energies of formation for three structural isomers with the molecular formula $C_8H_{18}$ are shown in **Figure 12.20**. All three isomers burn in air, as described by the same chemical equation:

$$2\ C_8H_{18}(\ell) + 25\ O_2(g) \rightarrow 16\ CO_2(g) + 18\ H_2O(g)$$

Are the $\Delta G^\circ_{rxn}$ values for the three combustion reactions also the same? Why or why not?

Octane
$CH_3(CH_2)_6CH_3$
$\Delta G^\circ_f = 16.3$ kJ/mol

(a)

2-Methylheptane
$(CH_3)_2CH(CH_2)_4CH_3$
$\Delta G^\circ_f = 11.7$ kJ/mol

(b)

3,3-Dimethylhexane
$CH_3CH_2C(CH_3)_2CH_2CH_2CH_3$
$\Delta G^\circ_f = 12.6$ kJ/mol

(c)

**FIGURE 12.20** Molecular structures and $\Delta G^\circ_f$ values of three $C_8H_{18}$ isomers.

**SAMPLE EXERCISE 12.6** Calculating $\Delta G^\circ_{rxn}$ by Using Appropriate $\Delta G^\circ_f$ Values **LO5**

Use the appropriate standard free energy of formation values in Appendix 4 to calculate the change in free energy as ethanol burns under standard conditions. Assume that the reaction proceeds as described by the following chemical equation:

$$CH_3CH_2OH(\ell) + 3\,O_2(g) \rightarrow 2\,CO_2(g) + 3\,H_2O(\ell)$$

**Collect and Organize** We are asked to calculate the value of $\Delta G^\circ_{rxn}$ for the combustion of ethanol by using the "appropriate" $\Delta G^\circ_f$ values, which are, according to Equation 12.12, the $\Delta G^\circ_f$ values of the reactants and products in the combustion reaction:

| Substance | $CH_3CH_2OH(\ell)$ | $O_2(g)$ | $CO_2(g)$ | $H_2O(\ell)$ |
|---|---|---|---|---|
| $\Delta G^\circ_f$ (kJ/mol) | −174.9 | 0 | −394.4 | −237.2 |

**Analyze** The reaction consumes one mole of liquid ethanol and three moles of oxygen gas and produces two moles of $CO_2$ gas and three moles of liquid $H_2O$. Multiplying the $\Delta G^\circ_f$ values by the appropriate numbers of moles and subtracting the sum of the reactant values from the sum of the product values, as described in Equation 12.12, will yield the value of $\Delta G^\circ_{rxn}$. The combustion of ethanol, a common additive in gasoline in the United States, is spontaneous, so $\Delta G^\circ_{rxn}$ should be less than zero.

**Solve** Inserting the appropriate numbers of moles and $\Delta G^\circ_f$ values in Equation 12.12,

$$\begin{aligned}\Delta G^\circ_{rxn} &= \sum n_{products}\Delta G^\circ_{f,products} - \sum n_{reactants}\Delta G^\circ_{f,reactants} \\ &= [2\,\text{mol CO}_2 \times (-394.4\text{ kJ/mol}) + 3\,\text{mol H}_2\text{O} \times (-237.2\text{ kJ/mol})] \\ &\quad -[1\,\text{mol CH}_3\text{CH}_2\text{OH} \times (-174.9\text{ kJ/mol}) + 3\,\text{mol O}_2 \times (0\text{ kJ/mol})] \\ &= -1325.5\text{ kJ/mol}\end{aligned}$$

**Think About It** The negative value of $\Delta G^\circ_{rxn}$ confirms our prediction that the reaction is spontaneous. The calculated value represents the maximum amount of the total energy released by the combustion of one mole of ethanol under standard conditions that is available to do useful work.

**Practice Exercise** Use the appropriate standard free energy of formation values in Appendix 4 to calculate the value of $\Delta G^\circ_{rxn}$ for the steam–methane reforming reaction, which is used commercially to produce $H_2$ gas:

$$CH_4(g) + H_2O(g) \rightarrow CO(g) + 3\,H_2(g)$$

**CONCEPT TEST**

a. Given the thermodynamic data in Table A4.3, is the conversion of diamond to graphite spontaneous? Explain your answer.

b. If yes, does knowing that the conversion is spontaneous tell you how rapidly the conversion occurs?

## The Meaning of *Free* Energy

What exactly does "energy available to do useful work" mean? We answer this question by considering the internal combustion (gasoline) engines used to power most automobiles. A combustion reaction mixture is a thermodynamic system that experiences a decrease in internal energy ($\Delta E$) as heat ($q$) flows out from it

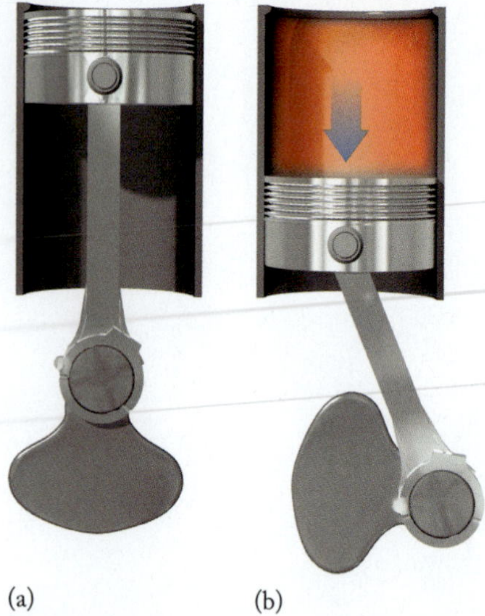


FIGURE 12.21 (a) In a car engine, thermal expansion causes the gases in a cylinder to (b) push down on a piston with a pressure, $P$, represented by the blue arrow. The product of $P$ and the change in volume of the gases, $\Delta V$, is the work done by the expanding gases, which propels the car.

into its surroundings and as it does work ($w$) on its surroundings. These three variables are related by Equation 10.3,

$$\Delta E = q + w \tag{10.3}$$

and all three are less than zero from the perspective of the system. Internal combustion engines have cooling systems to manage the dissipation of heat, which is wasted energy that does nothing to power the car. The energy that does move the car is derived from the rapid expansion of the gaseous products of combustion in the cylinders of its engine. As **Figure 12.21** shows, this expansion pushes down on the piston of a cylinder, increasing the volume of the reaction mixture. The product of the pressure exerted by the reacting gases and their products on the piston and the resulting volume change is $P$–$V$ work (discussed in Section 10.2), which propels the car. At constant pressure:

$$w = -P\Delta V$$

Free energy is a measure of the *maximum* amount of work that can be done by the energy released during combustion. To see how this theoretical quantity of work compares with the total energy released, we can rearrange Equation 12.10 to isolate the $\Delta H$ term:

$$\Delta H = \Delta G + T\Delta S \tag{12.13}$$

Equation 12.13 tells us that the enthalpy change that accompanies the making and breaking of chemical bonds during a chemical reaction may be divided into two parts, a $\Delta G$ part and a $T\Delta S$ part. The $\Delta G$ part is the energy that can theoretically be converted into motion and other useful work (such as generating electricity for the car's electrical system). The $T\Delta S$ part, however, is unusable: it is the portion of energy that disperses when, for example, hot gases flow out of the end of an automobile exhaust pipe. This part of $\Delta H$ is wasted. Consequently, the conversion of chemical energy into useful mechanical energy ($\Delta G$) is never 100% efficient. In addition, some portion of $\Delta G$ is also wasted because the combustion and energy conversion happen quickly and, therefore, irreversibly. Maximum efficiency comes with very slow, reversible reactions and energy conversion rates, but that is not how automobile engines operate. It turns out that gasoline engines convert only about 30% of the energy produced during combustion into useful work.

## 12.7 Temperature and Spontaneity

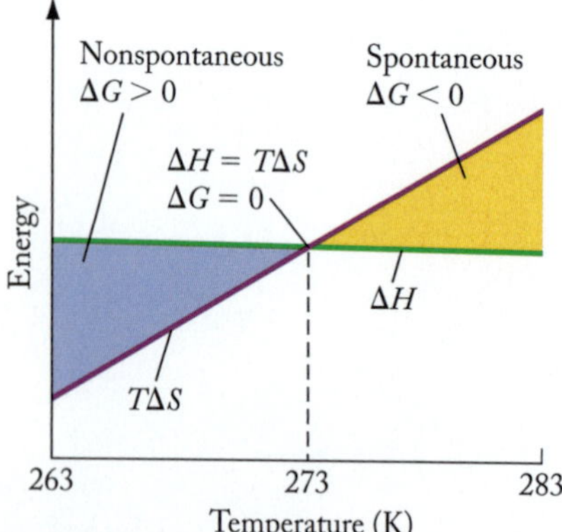


FIGURE 12.22 Changes in the values of $\Delta H$, the quantity $T\Delta S$, and $\Delta G$ for ice melting as the temperature changes from −10°C (263 K) to +10°C (283 K).

Let's revisit the process of ice melting (Figure 12.2), this time focusing on how the values of $\Delta H$, $T\Delta S$, and $\Delta G$ change as the temperature of a mixture of ice and water increases from 263 K to 283 K (**Figure 12.22**). As temperatures rise over this range, there is little impact on the enthalpy of fusion, $\Delta H$, as shown by the nearly flat green line in Figure 12.22. Increasing the temperature, though, increases the value of $T\Delta S$ (as shown by the upward slope of the purple line). The $\Delta S$ of ice (or any solid) melting has a positive value, so $T\Delta S$ must increase as $T$ increases. The green and purple lines intersect at 273 K, or 0°C, which means $\Delta H$ is equal to $T\Delta S$ at that temperature. Put another way, the difference between $\Delta H$ and $T\Delta S$ is zero at 0°C—and so is $\Delta G$ (because $\Delta G = \Delta H - T\Delta S$). When $\Delta G = 0$ for a process, then that process is at equilibrium, such as when ice and water coexist at 0°C.

Ice melts spontaneously above 0°C, so $\Delta G$ for the melting process must be less than zero. The graph in Figure 12.22 shows that indeed it is. Above 0°C, the value of $T\Delta S$ is greater than $\Delta H$. Therefore, the difference between them ($\Delta H - T\Delta S$)

is less than zero and becomes more negative with increasing temperature, as illustrated by the distance from the purple line to the green line in Figure 12.22.

Below 0°C, ice does not melt spontaneously, which means that $\Delta G > 0$. The graph shows why this is true: at $T < 273$ K (0°C), $T\Delta S$ values (the purple line) are less than the $\Delta H$ values (the green line), which means that $\Delta H - T\Delta S$ (and $\Delta G$) is greater than zero. Positive $\Delta G$ values mean that the process is nonspontaneous below 273 K and that ice does not melt below its freezing point. However, the opposite process—liquid water freezing—*is* spontaneous because reversing a process keeps the absolute values but switches the signs of $\Delta H$, $\Delta S$, and $\Delta G$. Therefore, if the melting process is nonspontaneous, then the exothermic freezing process is spontaneous at low temperatures.

**CONNECTION** In Section 6.5, we discussed the equilibrium between phases of a substance and used phase diagrams to illustrate which physical states are stable at various combinations of temperature and pressure.

### CONCEPT **TEST**

A process is spontaneous at lower temperatures and nonspontaneous at higher temperatures. Sketch a graph similar to that in Figure 12.22 for this process.

The temperature at the point in Figure 12.22 where $\Delta G = 0$ defines the melting (or freezing) point of water, which is the temperature at which solid ice and liquid water coexist in a state of thermodynamic equilibrium. Similarly, at 100°C and 1 atm, liquid water and water vapor coexist because the free-energy changes that accompany vaporization and condensation both equal zero.

The temperature at which $\Delta G = 0$ for a process can be calculated from Equation 12.10 if we know the values of $\Delta H$ and $\Delta S$. These values for the fusion of ice are

$$H_2O(s) \rightarrow H_2O(\ell) \qquad \Delta H^\circ = 6.01 \times 10^3 \text{ J/mol}; \Delta S^\circ = 22.0 \text{ J/(mol} \cdot \text{K)}$$

Assuming that $\Delta H^\circ$ and $\Delta S^\circ$ do not change significantly with small changes in temperature, we can use these values in calculations at temperatures other than 298 K. This turns out to be a reasonable assumption, so that

$$\Delta G = \Delta H - T\Delta S \approx \Delta H^\circ - T\Delta S^\circ \tag{12.14}$$

Inserting the values of $\Delta H^\circ$ and $\Delta S^\circ$ and using the fact that $\Delta G = 0$ for a process at equilibrium, we get

$$\Delta G = (6.01 \times 10^3 \text{ J/mol}) - T[22.0 \text{ J/(mol} \cdot \text{K)}] = 0$$

$$T = \frac{6.01 \times 10^3 \text{ J/mol}}{22.0 \text{ J/(mol} \cdot \text{K)}} = 273 \text{ K} = 0°\text{C}$$

which is the familiar value for the melting point of ice.

Like physical processes, chemical reactions under standard conditions may be spontaneous below or above a characteristic temperature ($T$) at which $\Delta G^\circ = 0$, and

$$T = \frac{\Delta H^\circ}{\Delta S^\circ}$$

As summarized in **Table 12.4**, an exothermic process ($\Delta H^\circ < 0$) for which $\Delta S^\circ$ is also less than zero—for example, the freezing of liquid water—is spontaneous only below a characteristic temperature. An endothermic process ($\Delta H^\circ > 0$) for which $\Delta S^\circ$ is also greater than zero—for example, the melting of ice into liquid water—is spontaneous only above a characteristic temperature. This pattern makes sense because lower temperatures make the value of the $T\Delta S^\circ$ term in Equation 12.11 less negative if $\Delta S^\circ < 0$. This helps make $\Delta G^\circ$ more negative

**TABLE 12.4** Effects of $\Delta H$, $\Delta S$, and $T$ on $\Delta G$ and Spontaneity

| $\Delta H$ | $\Delta S$ | $\Delta G$ | Spontaneity |
|---|---|---|---|
| − | + | Always $<0$ | Always spontaneous |
| − | − | $<0$ at lower temperature | Spontaneous at lower temperature |
| + | + | $<0$ at higher temperature | Spontaneous at higher temperature |
| + | − | Always $>0$ | Never spontaneous |

because of the minus sign in front of the $T\Delta S^\circ$ term. Higher temperatures make the value of $T\Delta S^\circ$ more positive if $\Delta S^\circ > 0$, which again helps make $\Delta G^\circ$ more negative.

Chemical reactions can also have zero change in free energy. As a spontaneous reaction ($\Delta G_{rxn} < 0$) proceeds, reactants become products and $\Delta G_{rxn}$ becomes less negative, as we will see in the next section. In fact, $\Delta G_{rxn}$ may reach zero before all the reactants are consumed. Here, there is no further net increase in the quantities of products and no further net loss in the quantities of reactants in the reaction mixture. Rather, reactants and products coexist in equilibrium. Reactants still react, and products are still formed, but the reverse reaction, in which products become reactants, proceeds at the same rate as the forward reaction. This is the state we know as chemical equilibrium. No net change in the amounts of reactants and products occurs once equilibrium is reached.

**SAMPLE EXERCISE 12.7** Relating Reaction Spontaneity to $\Delta H_{rxn}$ and $\Delta S_{rxn}$ **LO6**

The gas-phase reaction between methanol and water that produces carbon dioxide and three moles of hydrogen is accompanied by an increase in entropy, $\Delta S^\circ_{rxn} = 176.9$ J/K, under standard conditions (see Sample Exercise 12.3).

$$CH_3OH(g) + H_2O(g) \rightarrow CO_2(g) + 3\ H_2(g)$$

At what temperatures is this reaction spontaneous?

**Collect and Organize** We are asked to determine the temperatures at which a reaction with $\Delta S^\circ > 0$ is spontaneous. Equation 12.14 relates spontaneity ($\Delta G^\circ < 0$) to $\Delta H^\circ_{rxn}$, $\Delta S^\circ_{rxn}$, and $T$.

$$\Delta G_{rxn} \approx \Delta H^\circ_{rxn} - T\Delta S^\circ_{rxn} < 0$$

**Analyze** According to Table 12.4, a reaction in which entropy increases ($\Delta S^\circ > 0$) will be spontaneous at all temperatures if it is also exothermic ($\Delta H^\circ_{rxn} < 0$). However, if the reaction is accompanied by a positive enthalpy change, $\Delta H^\circ_{rxn} > 0$ (endothermic), then the reaction will be spontaneous only above a certain temperature. We need to calculate $\Delta H^\circ_{rxn}$ by using standard enthalpies of formation from Appendix 4 and then solve the relationship:

$$T > \Delta H^\circ_{rxn}/\Delta S^\circ_{rxn}$$

**Solve** The enthalpy change for the reaction is calculated as follows:

$$\begin{aligned}\Delta H^\circ_{rxn} &= [1\ \cancel{\text{mol}\ CO_2} \times (-393.5\ \text{kJ}/\cancel{\text{mol}}) + 3\ \cancel{\text{mol}\ H_2} \times (0\ \text{kJ}/\cancel{\text{mol}})] \\ &\quad - [1\ \cancel{\text{mol}\ CH_3OH} \times (-200.7\ \text{kJ}/\cancel{\text{mol}}) + 1\ \cancel{\text{mol}\ H_2O} \times (-241.8\ \text{kJ}/\cancel{\text{mol}})] \\ &= 49.0\ \text{kJ}\end{aligned}$$

Substituting into the relationship $T > \Delta H^\circ_{rxn}/\Delta S^\circ_{rxn}$:

$$\frac{\Delta H^\circ_{rxn}}{\Delta S^\circ_{rxn}} = \frac{49.0 \text{ kJ}}{0.1769 \text{ kJ/K}} = 277 \text{ K}$$

$$T > 277 \text{ K}$$

Thus, the conversion of methanol vapor and steam to hydrogen and carbon dioxide should be spontaneous above 277 K, or 4°C.

**Think About It** The enthalpy change for the reaction is a relatively small value, so the reaction does not require an enormously high temperature to be spontaneous.

**Practice Exercise** At high temperatures, ammonia decomposes into nitrogen and hydrogen gases:

$$2\,NH_3(g) \rightarrow N_2(g) + 3\,H_2(g)$$

Using the information in Appendix 4, determine the temperatures at which this reaction will be spontaneous.

## 12.8 Driving the Human Engine: Coupled Reactions

The laws of thermodynamics that govern chemical reactions in the laboratory also govern all the chemical reactions that take place in living systems (**Figure 12.23**). Organisms carry out reactions that release energy by oxidizing compounds derived from the nutrients they consume, and then use that energy to do work and to sustain an array of other essential biological functions. Just like mechanical engines, however, humans and other life-forms are far from 100% efficient, which means that life requires a continuous input of energy. Thus, we must continually absorb energy in the form of the caloric content of the food we eat and then release heat and waste products into our surroundings. According to the USDA, adult women

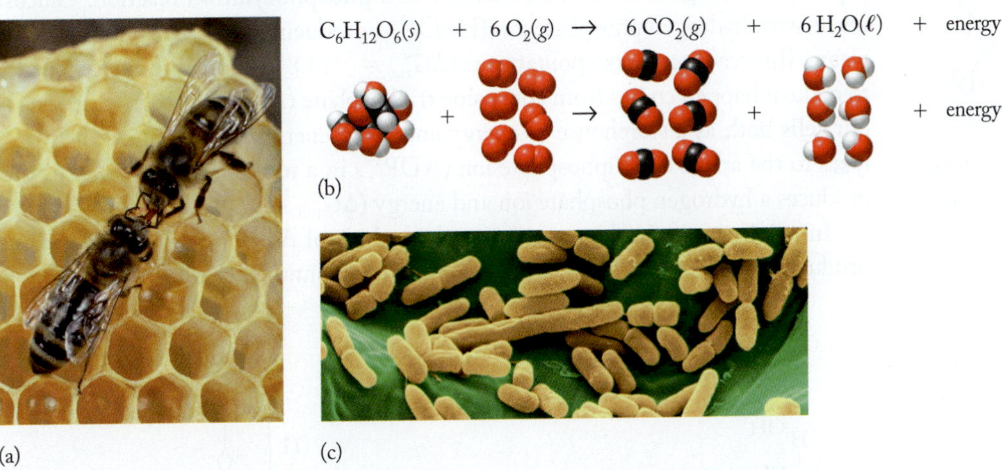


**FIGURE 12.23** The rules of thermodynamics apply to all living systems. (a) Honeybees extract energy from nutrients to support life. As bees collect and process plant nectars, they store the energy as honey, which is then consumed by the bees, by humans, or by other animals. (b) When the sugar in honey is digested, carbon dioxide, water, and energy are released, increasing the entropy of the universe. (c) The spontaneous life processes of all organisms—including the *E. coli* bacteria and many others that live in our gastrointestinal tracts—increase the entropy of the universe.

**glycolysis** a series of reactions that converts glucose into pyruvate; a major anaerobic (no oxygen required) pathway for the metabolism of glucose in the cells of almost all living organisms.

**phosphorylation** a reaction resulting in the addition of a phosphate group to an organic molecule.

**CONNECTION** In Section 10.9, we defined 1 Calorie, the "calorie" used in discussing food, as being equivalent to 1 kcal.

**CONNECTION** Photosynthesis and the carbon cycle were introduced in Section 7.1.

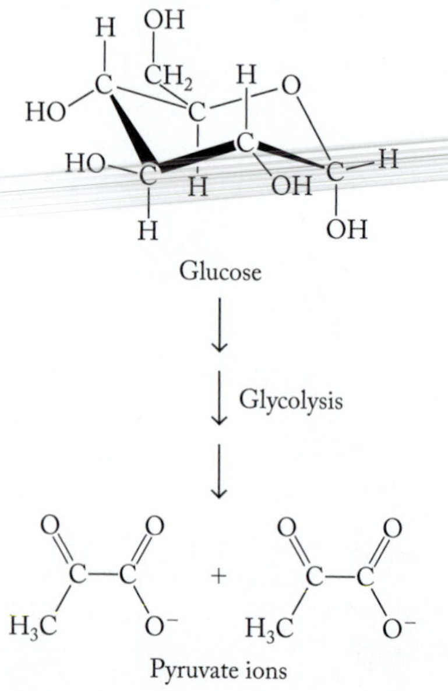

need an estimated 1600–2400 Calories per day, whereas adult men need 2000–3000 Calories. This level of caloric intake provides the energy we need to function at all levels, from thinking to getting out of bed in the morning.

In living systems, spontaneous reactions ($\Delta G_{rxn} < 0$) typically involve breaking food down, while nonspontaneous reactions ($\Delta G_{rxn} > 0$) involve building molecules needed by the body. Living systems use the energy from spontaneous reactions to run nonspontaneous reactions; we say that the spontaneous reactions are *coupled* to the nonspontaneous reactions. Part of the study of biochemistry involves deciphering the molecular mechanisms that enable reaction coupling. This topic is covered in Chapter 20, but for now it is enough to know that elegant molecular processes have evolved to enable living systems to couple chemical reactions so that the energy obtained from spontaneous reactions can be used to drive the nonspontaneous reactions that maintain life. The metabolic chemical reactions we look at here are not presented for you to memorize. Rather, the intent is to aid your understanding of how changes in free energy allow spontaneous reactions to drive nonspontaneous reactions and to illustrate operationally what the phrase *coupled reactions* actually means.

As noted in Chapter 10, the energy contained in the food we eat comes indirectly from solar energy. Green plants store energy from sunlight by producing glucose ($C_6H_{12}O_6$) from $CO_2$ and $H_2O$ during photosynthesis. This is a nonspontaneous process, which is why the plants require the energy of sunlight to carry out this reaction. Animals that consume plants use the energy stored in glucose and other molecules, and they release $CO_2$ and $H_2O$ back into the environment:

$$C_6H_{12}O_6(aq) + 6\ O_2(g) \rightarrow 6\ CO_2(g) + 6\ H_2O(\ell)$$

This reaction increases the entropy of the universe and is therefore spontaneous. However, the reaction takes place in living organisms in many steps, some of which are nonspontaneous. **Figure 12.24** summarizes one portion of glucose metabolism—**glycolysis**—that involves a series of spontaneous and nonspontaneous reactions.

Glucose → Glycolysis → Pyruvate ions

**FIGURE 12.24** In glycolysis, each molecule of glucose is converted into two pyruvate ions.

In glycolysis, each mole of glucose is converted into two moles of pyruvate ion ($CH_3C(O)COO^-$). An early step is the conversion of glucose into glucose 6-phosphate (**Figure 12.25**), an example of a **phosphorylation** reaction. Glucose reacts with hydrogen phosphate ion ($HPO_4^{2-}$), producing glucose 6-phosphate and water. This reaction is nonspontaneous ($\Delta G^\circ_{rxn} = +13.8$ kJ/mol). The energy needed to make it happen comes from adenosine triphosphate ($ATP^{4-}$), which functions in our cells both as a storehouse of energy and as an energy-transfer agent: it hydrolyzes to the adenosine diphosphate ion ($ADP^{3-}$) in a reaction (**Figure 12.26**) that produces a hydrogen phosphate ion and energy ($\Delta G^\circ_{rxn} = -30.5$ kJ/mol).

In a living system, the spontaneous hydrolysis of $ATP^{4-}$ consumes water and produces $HPO_4^{2-}$ and $H^+$, whereas the nonspontaneous phosphorylation of

$$\text{Glucose}(aq) + HPO_4^{2-}(aq) \rightarrow \text{Glucose 6-phosphate}(aq) + H_2O(\ell)$$

**FIGURE 12.25** The conversion of glucose into glucose 6-phosphate is an early step in glycolysis. This reaction is nonspontaneous, which means that energy must be added to make the reaction go: $\Delta G^\circ_{rxn} = 13.8$ kJ/mol.

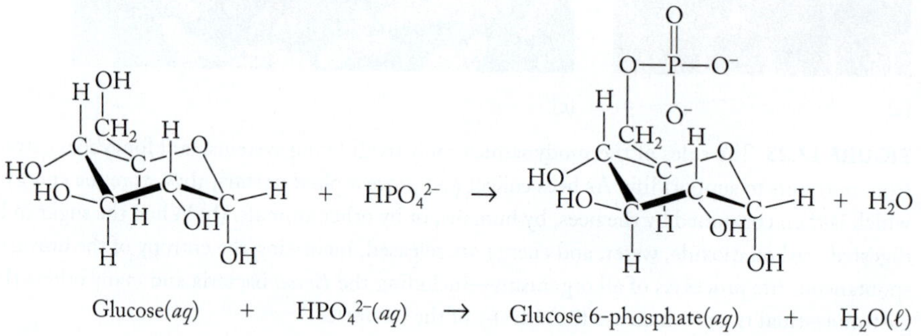

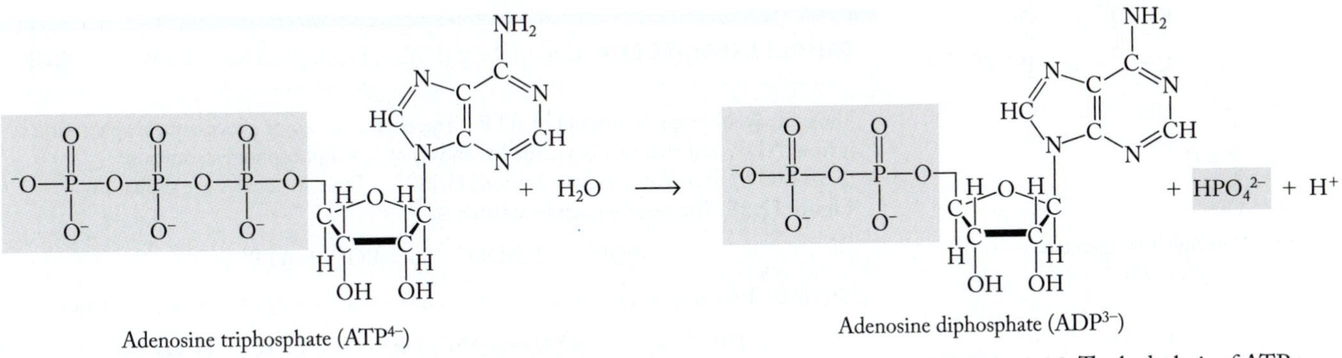

Adenosine triphosphate ($ATP^{4-}$) + $H_2O$ → Adenosine diphosphate ($ADP^{3-}$) + $HPO_4^{2-}$ + $H^+$

**FIGURE 12.26** The hydrolysis of ATP to ADP is a spontaneous reaction: $\Delta G° = -30.5$ kJ/mol. The body couples this reaction to nonspontaneous reactions so that the energy released can drive them (see Figure 12.25).

glucose consumes $HPO_4^{2-}$ and produces water. Their complementary reactants and products allow the two reactions to couple: the spontaneous $ATP^{4-} \rightarrow ADP^{3-}$ reaction supplies the energy that drives the nonspontaneous formation of glucose 6-phosphate (**Figure 12.27**).

This example illustrates another important general point about reactions and $\Delta G°_{rxn}$ values: those for coupled reactions (and for sequential reactions) are additive. This is true for any set of reactions, not just those occurring in living systems.

The first two steps in glycolysis are as follows:

$$(1) \quad ATP^{4-}(aq) + H_2O(\ell) \rightarrow ADP^{3-}(aq) + HPO_4^{2-}(aq) + H^+(aq) \qquad \Delta G°_{rxn} = -30.5 \text{ kJ}$$

$$(2) \quad \underset{\text{Glucose}}{C_6H_{12}O_6(aq)} + \underset{\text{Hydrogen phosphate}}{HPO_4^{2-}(aq)} \rightarrow \underset{\text{Glucose 6-phosphate}}{C_6H_{11}O_6PO_3^{2-}(aq)} + H_2O(\ell) \qquad \Delta G°_{rxn} = 13.8 \text{ kJ}$$

If we add these reactions and their free energies, we get

$$ATP^{4-}(aq) + \cancel{H_2O(\ell)} + C_6H_{12}O_6(aq) + \cancel{HPO_4^{2-}(aq)} \rightarrow$$
$$ADP^{3-}(aq) + \cancel{HPO_4^{2-}(aq)} + H^+(aq) + C_6H_{11}O_6PO_3^{2-}(aq) + \cancel{H_2O(\ell)}$$
$$\Delta G°_{rxn} = (-30.5 + 13.8) \text{ kJ} = -16.7 \text{ kJ}$$

Because equal quantities of $H_2O$ and $HPO_4^{2-}$ appear on both sides of the combined equation, they cancel out, leaving the following net ionic equation:

$$C_6H_{12}O_6(aq) + ATP^{4-}(aq) \rightarrow ADP^{3-}(aq) + C_6H_{11}O_6PO_3^{2-}(aq) + H^+(aq) \qquad \Delta G°_{rxn} = -16.7 \text{ kJ}$$

Because $\Delta G°_{rxn}$ for the net reaction is negative, the reaction is spontaneous.

**CONNECTION** In Chapter 10, we used Hess's law (the enthalpy change of a reaction that is the sum of two or more reactions equals the sum of the enthalpy changes of the constituent reactions) to calculate $\Delta H°_{rxn}$. We apply a similar principle here when adding $\Delta G°_{rxn}$ values of coupled reactions.

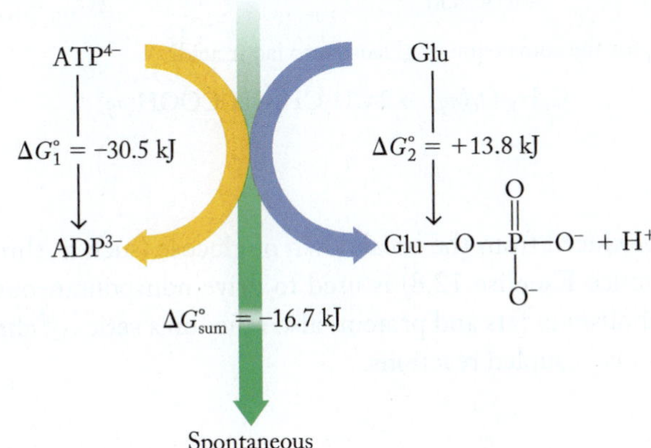


**FIGURE 12.27** The spontaneous hydrolysis of ATP is coupled to the nonspontaneous phosphorylation of glucose (Glu). The overall reaction—the sum of the two individual reactions—is spontaneous.

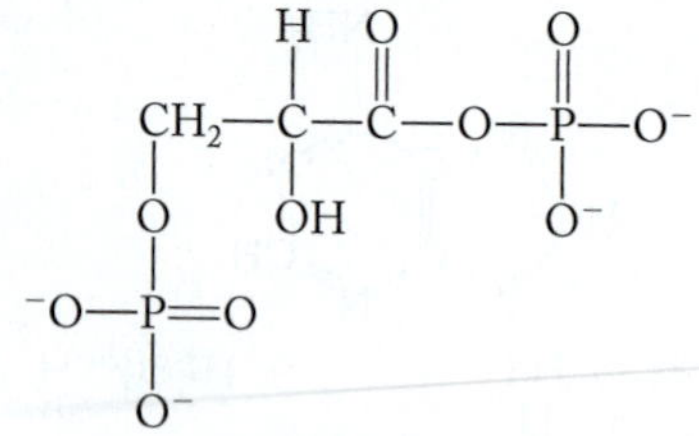

1,3-Diphosphoglycerate ($1,3\text{-}DPG^{4-}$)

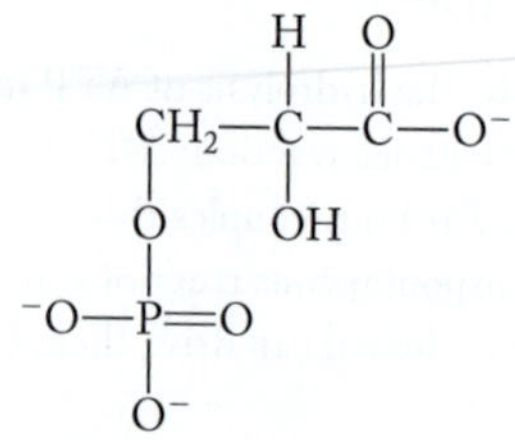

3-Phosphoglycerate ($3\text{-}PG^{3-}$)

**FIGURE 12.28** Structures of 1,3-diphosphoglycerate and 3-phosphoglycerate.

## SAMPLE EXERCISE 12.8 Calculating $\Delta G^\circ_{rxn}$ of Coupled Reactions — LO7

The body would rapidly run out of ATP if there were not some process for regenerating it from ADP, and that process is the hydrolysis of 1,3-diphosphoglycerate ions ($1,3\text{-}DPG^{4-}$) to 3-phosphoglycerate ions ($3\text{-}PG^{3-}$). Their structures are shown in **Figure 12.28**. The reaction can be written as

$$ADP^{3-} + 1,3\text{-}DPG^{4-} \rightarrow 3\text{-}PG^{3-} + ATP^{4-}$$

This hydrolysis is spontaneous. Calculate its $\Delta G^\circ$ value from the following $\Delta G^\circ$ values:

(1) $1,3\text{-}DPG^{4-}(aq) + H_2O(\ell) \rightarrow 3\text{-}PG^{3-}(aq) + HPO_4^{\ 2-}(aq) + H^+(aq)$ $\qquad \Delta G^\circ_{rxn} = -49.0\ \text{kJ}$

(2) $ADP^{3-}(aq) + HPO_4^{\ 2-}(aq) + H^+(aq) \rightarrow ATP^{4-}(aq) + H_2O(\ell)$ $\qquad \Delta G^\circ_{rxn} = 30.5\ \text{kJ}$

**Collect and Organize** We need to calculate $\Delta G^\circ_{rxn}$ for a reaction that is the sum of two reactions. If the reactions in equations 1 and 2 add up to the overall reaction, then the overall $\Delta G^\circ_{rxn}$ is the sum of the $\Delta G^\circ_{rxn}$ values for the individual reactions.

**Analyze** First, we add the reactions described by equations 1 and 2. Assuming that the overall reaction between $ADP^{3-}$ and $1,3\text{-}DPG^{4-}$ is the sum of the reactions describing the hydrolysis of $1,3\text{-}DPG^{4-}$ and the phosphorylation of $ADP^{3-}$, we know that the sum of $\Delta G^\circ_1$ and $\Delta G^\circ_2$ will be less than zero because the overall reaction is spontaneous.

**Solve** Summing the reactions in equations 1 and 2 confirms that they equal the overall reaction:

(1) $1,3\text{-}DPG^{4-}(aq) + H_2O(\ell) \rightarrow 3\text{-}PG^{3-}(aq) + HPO_4^{\ 2-}(aq) + H^+(aq)$

(2) $ADP^{3-}(aq) + HPO_4^{\ 2-}(aq) + H^+(aq) \rightarrow ATP^{4-}(aq) + H_2O(\ell)$

$$1,3\text{-}DPG^{4-}(aq) + \cancel{H_2O(\ell)} + ADP^{3-}(aq) + \cancel{HPO_4^{\ 2-}(aq)} + \cancel{H^+(aq)} \rightarrow$$
$$3\text{-}PG^{3-}(aq) + \cancel{HPO_4^{\ 2-}(aq)} + \cancel{H^+(aq)} + ATP^{4-}(aq) + \cancel{H_2O(\ell)}$$

We sum the $\Delta G^\circ_{rxn}$ values for steps 1 and 2 to determine $\Delta G^\circ_{rxn}$ for the overall reaction:

$$\Delta G^\circ_{overall} = \Delta G^\circ_1 + \Delta G^\circ_2 = (-49.0 + 30.5)\ \text{kJ} = -18.5\ \text{kJ}$$

**Think About It** As predicted, the overall $\Delta G^\circ_{rxn}$ value is less than zero for this spontaneous reaction. Hydrolysis of $1,3\text{-}DPG^{4-}$ provides more than enough free energy to convert $ADP^{3-}$ into $ATP^{4-}$.

**Practice Exercise** Conversion of glucose into lactic acid (**Figure 12.29**) drives the phosphorylation of two moles of ADP to ATP:

$$\underset{\text{Glucose}}{C_6H_{12}O_6(aq)} + 2\ HPO_4^{\ 2-}(aq) + 2\ ADP^{3-}(aq) + 2H^+(aq) \rightarrow$$
$$2\ \underset{\text{Lactic acid}}{CH_3CH(OH)COOH(aq)} + 2\ ATP^{4-}(aq) + 2\ H_2O(\ell) \qquad \Delta G^\circ_{rxn} = -135\ \text{kJ/mol}$$

What is $\Delta G^\circ_{rxn}$ for the conversion of glucose into lactic acid?

$$C_6H_{12}O_6(aq) \rightarrow 2\ CH_3CH(OH)COOH(aq)$$

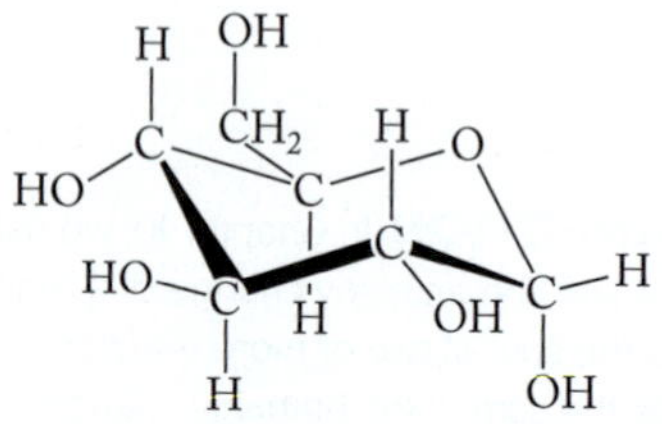

Glucose

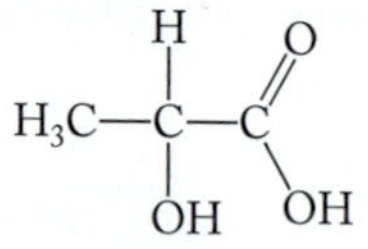

Lactic acid

**FIGURE 12.29** Structures of glucose and lactic acid.

The ATP produced from the breakdown of glucose (such as through the first reaction in Practice Exercise 12.8) is used to drive nonspontaneous reactions in cells. The metabolism of fats and proteins also relies on a series of chemical cycles, all of which involve coupled reactions.

## SAMPLE EXERCISE 12.9 Integrating Concepts: Trouton's Rule

In 1883–1884, while still an undergraduate student at Trinity College in Dublin, Ireland, Frederick Trouton (1863–1922) published two short papers describing what we now call Trouton's rule. Trouton's rule states that the ratio of the enthalpy of vaporization ($\Delta H_{vap}$) of a liquid to its normal boiling point (in K) is approximately constant:

$$\frac{\Delta H_{vap}}{T_b} \approx 88 \frac{J}{mol \cdot K}$$

This ratio is also known as the entropy of vaporization ($\Delta S_{vap}$) and may be used to estimate $\Delta H_{vap}$ values of liquids whose boiling points are known.

a. Suggest why $\Delta H_{vap}/T$ should be approximately constant.
b. Using Trouton's rule, calculate the values of $\Delta H_{vap}$ for the substances given in the table that follows (whose structures are shown in **Figure 12.30**), and compare them to the experimentally determined values. Also calculate $\Delta S_{vap}$ for each substance by dividing the experimentally determined $\Delta H_{vap}$ values by the normal boiling points of the liquids. Identify those compounds whose actual $\Delta H_{vap}$ values deviate by more than 10% from those estimated using Trouton's rule.

| Substance | Boiling Point $T_b$ (°C) | Calculated $\Delta H_{vap}$ (J/mol) | Experimentally Determined $\Delta H_{vap}$ (J/mol) | Calculated Value of $\Delta S_{vap}$ [J/(mol · K)] |
|---|---|---|---|---|
| Ethanol | 78.3 | | 38,600 | |
| Acetone | 56.0 | | 29,100 | |
| Benzene | 80.0 | | 30,700 | |
| Chloroform | 61.1 | | 29,200 | |
| Formic acid | 100.8 | | 22,700 | |
| Hexane | 68.7 | | 28,900 | |
| Water | 100.0 | | 40,660 | |

c. On the basis of the results in part (b) and the molecular structures of the seven compounds, explain why some of them deviate from Trouton's rule.
d. Predict which four of the compounds whose molecular structures are shown in **Figure 12.31** obey Trouton's rule.
e. What is the sign of the change in standard free energy of vaporization ($\Delta G^\circ_{vap}$) of each of the seven liquids in the table at 298 K?

**Collect and Organize** We are given the boiling points of seven liquids. Trouton's rule relates the boiling points of liquids to their $\Delta H_{vap}$ values. In part (a), we are asked to explain why this relationship exists. In part (b), we use Trouton's rule to estimate the $\Delta H_{vap}$ values of seven liquids, calculate actual $\Delta S_{vap}$ values, and compare these results to their actual $\Delta H_{vap}$ values, explaining any deviations in part (c). In part (d), we predict how well Trouton's rule will apply to six other liquids, and in part (e), we are asked to predict the sign of $\Delta G^\circ_{vap}$ of the seven liquids at 298 K.

**Analyze** The molecular structures of compounds influence how they interact, and the strengths of their intermolecular interactions help define physical properties such as boiling points and enthalpies of vaporization. Our task is to look for structural features that are associated with particularly strong intermolecular interactions. One of these is hydrogen bonds, which form in three of the liquids: ethanol, formic acid, and water. These three may be among those that do not obey Trouton's rule.

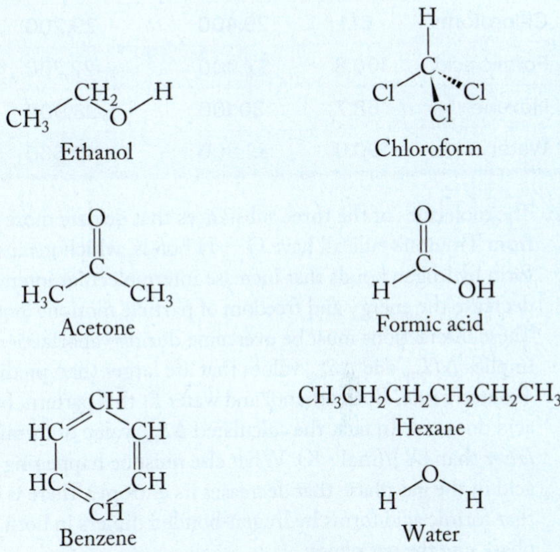


**FIGURE 12.30** Seven molecular structures.

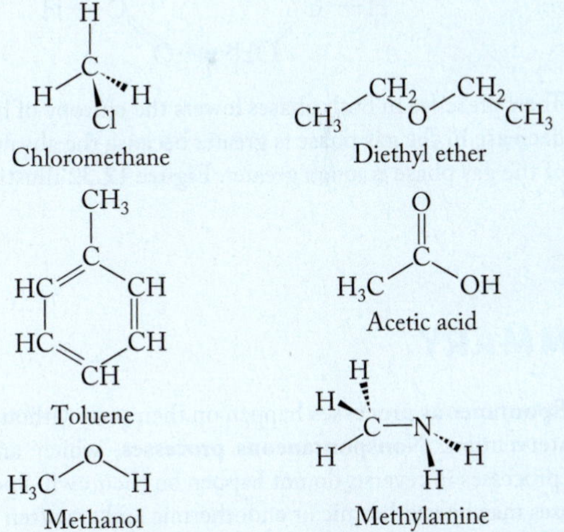


**FIGURE 12.31** Six molecular structures.

**Solve**

a. The boiling points and $\Delta H_{vap}$ values of molecular compounds both depend on the strengths of interactions between molecules in the liquid phase. Given this mutual dependence, higher boiling points should be associated with higher $\Delta H_{vap}$ values so that the ratio between the two is (nearly) constant.

b. The calculated values of $\Delta H_{vap}$ for ethanol, formic acid, and water all vary by more than 10% from the measured values.

| Substance | Boiling Point $T_b$ (°C) | Calculated $\Delta H_{vap}$ (J/mol) | Experimentally Determined $\Delta H_{vap}$ (J/mol) | Calculated Value of $\Delta S_{vap}$ [J/(mol · K)] |
|---|---|---|---|---|
| Ethanol | 78.3 | 30,900 | 38,600 | 110 |
| Acetone | 56.0 | 29,000 | 29,100 | 88 |
| Benzene | 80.0 | 31,100 | 30,700 | 87 |
| Chloroform | 61.1 | 29,400 | 29,200 | 87 |
| Formic acid | 100.8 | 32,900 | 22,700 | 61 |
| Hexane | 68.7 | 30,100 | 28,900 | 85 |
| Water | 100.0 | 32,800 | 40,660 | 109 |

c. The molecules of the three substances that deviate more than 10% from Trouton's rule all have O—H bonds, which means they can form hydrogen bonds that increase intermolecular interactions and decrease the energy and freedom of particle motions (entropy). These interactions must be overcome during vaporization, which implies $\Delta H_{vap}$ and $\Delta S_{vap}$ values that are larger than predicted by Trouton's rule. Both ethanol and water fit this pattern, but formic acid does not. In fact, the calculated $\Delta S_{vap}$ value of formic acid is *lower* than 88 J/(mol · K). What else must be happening to formic acid in the gas phase that decreases its entropy? There is evidence that formic acid forms hydrogen-bonded dimers in both the liquid phase *and* the gas phase:

H—C(=O······HO)(—OH······O=)C—H

Their presence in both phases lowers the entropy of both, but the decrease in the gas phase is greater because the absolute entropy of the gas phase is much greater. **Figure 12.32** illustrates the relationships between the entropies of the liquid and gas phases of these compounds.

d. The $\Delta H_{vap}$ values of chloromethane, toluene, and diethyl ether that are calculated using Trouton's rule should be close to the experimental $\Delta H_{vap}$ values because these compounds do not form hydrogen bonds.

e. The boiling points of all seven liquids are above 25°C (298 K). Therefore, vaporization of their liquids under standard conditions and at $T = 298$ K is nonspontaneous, and the $\Delta G°_{vap}$ values of all seven should be greater than zero.

**Think About It** Trouton's rule was developed empirically, but our answer to part (a) explains why it should be true. This exercise also shows that compounds experiencing intermolecular forces, such as hydrogen bonds that add significantly to London dispersion forces in the liquid phase, have $\Delta H_{vap}$ values that are significantly greater than those predicted by Trouton's rule.

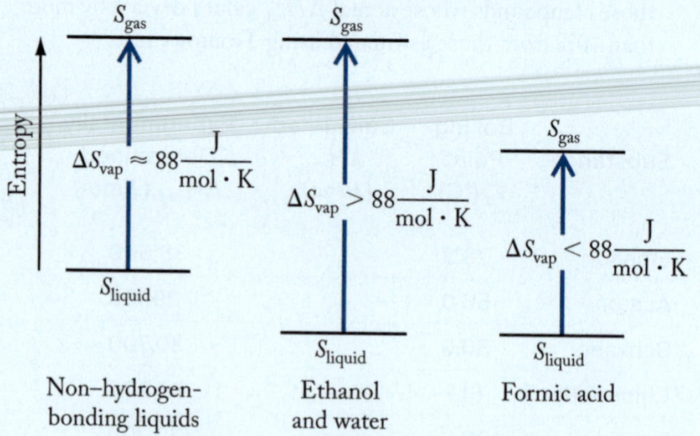


**FIGURE 12.32** $S_{liquid}$ and $S_{gas}$ relationships.

## SUMMARY

**LO1** **Spontaneous processes** happen on their own without continuing intervention. **Nonspontaneous processes**, which are spontaneous processes in reverse, do not happen on their own. Spontaneous processes may be exothermic or endothermic and are often accompanied by an increase in the freedom of motion of the particles involved in the process. Spontaneous reactions are not necessarily rapid. (Section 12.1)

**LO2** **Entropy (*S*)** is a thermodynamic property that provides a measure of how dispersed the energy is in a system at a given temperature. As temperature increases, the translational, rotational, and vibrational motion of molecules increases. The entropy of a system increases as the number of probable arrangements of its particles, called **accessible microstates**, increases. A perfect crystal of a pure substance is defined to have zero entropy at absolute zero. (Section 12.2)

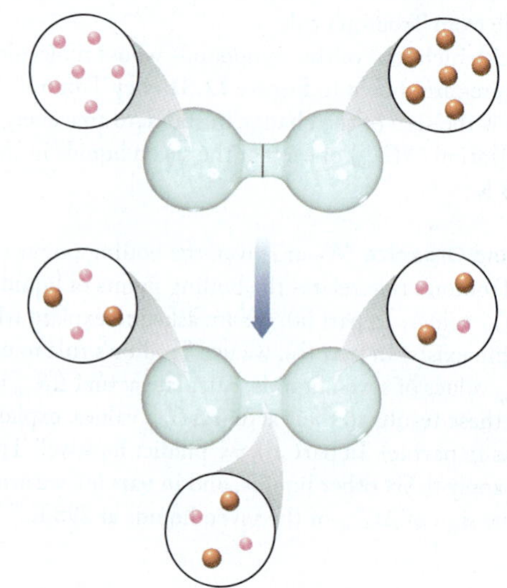

**LO3** According to the **second law of thermodynamics**, a spontaneous process is accompanied by an increase in the entropy of an isolated system or of the universe for a process occurring in any system. A **reversible process** takes place in very small steps and very slowly so that the system can be restored to its initial state with no net flow of energy to or from its surroundings. All substances have positive entropies at temperatures above absolute zero. **Standard molar entropies ($S°$)** are entropy values for substances in their standard states. The entropy of a system increases with increasing molecular complexity and with increasing temperature. (Sections 12.2, 12.3, and 12.4)

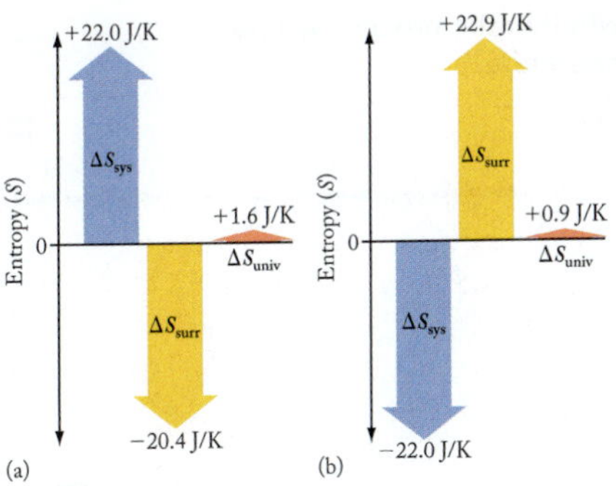


**LO4** Any process is spontaneous if it produces an increase in the entropy of the universe. The entropy change in a reaction under standard conditions can be calculated from the standard entropies of the products and reactants and their coefficients in the balanced chemical equation. (Section 12.5)

**LO5 Gibbs free energy ($G$)** is the maximum energy available to do useful work for a process occurring at constant temperature and pressure. When a process results in a decrease in **free energy** of a system ($\Delta G < 0$), the process is spontaneous; when $\Delta G > 0$, the process is nonspontaneous. The change in free energy of a reaction under standard conditions can be calculated either from the **standard free energies of formation ($\Delta G°_f$)** of the products and reactants or from the values of $\Delta H°_{rxn}$ and $\Delta S°_{rxn}$. (Section 12.6)

**LO6** The temperature range over which a process is spontaneous depends on the signs and the relative magnitudes of $\Delta H$ and $\Delta S$. (Section 12.7)

$\Delta H = T\Delta S$

$\Delta G = 0$

Energy

$\Delta H$

$T\Delta S$

Temperature (K)

**LO7** Many important biochemical processes, including **glycolysis** and **phosphorylation**, are made possible by the coupling of spontaneous and nonspontaneous reactions. The free energy released in the spontaneous processes going on in the body is used to drive nonspontaneous processes. (Section 12.8)

## PARTICULATE **PREVIEW WRAP-UP**

The chemical equations describing (a) respiration and (b) photosynthesis are

$$\text{(a) } C_6H_{12}O_6(aq) + 6\ O_2(g) \rightarrow 6\ CO_2(g) + 6\ H_2O(\ell)$$

$$\text{(b) } 6\ CO_2(g) + 6\ H_2O(\ell) \rightarrow C_6H_{12}O_6(aq) + 6\ O_2(g)$$

During energy production, relatively large $C_6H_{12}O_6$ molecules are converted into much smaller molecules of $CO_2$ and $H_2O$. This reaction is spontaneous. Photosynthesis is the reverse reaction: building larger $C_6H_{12}O_6$ molecules from much smaller molecules of $CO_2$ and $H_2O$ in a reaction that is not spontaneous and relies on the energy in sunlight to happen.

## PROBLEM-SOLVING SUMMARY

| Type of Problem | Concepts and Equations | Sample Exercises |
|---|---|---|
| **Predicting the sign of an entropy change** | Look for fewer moles of gas as products than as reactants or for the precipitation of a solute from solution ($\Delta S < 0$ for both). For the reverse processes, $\Delta S > 0$. | **12.1** |
| **Comparing standard molar entropy values** | Substances composed of larger, less rigid molecules have more entropy. Among substances with similar molar masses, gases have more entropy than liquids, which have more than solids. | **12.2** |
| **Calculating the entropy change of a chemical reaction** | Use $\Delta S°_{rxn} = \sum n_{products}S°_{products} - \sum n_{reactants}S°_{reactants}$ (12.6) where $S°_{products}$ and $S°_{reactants}$ are standard molar entropies of the products and reactants, and $n_{products}$ and $n_{reactants}$ are their stoichiometric coefficients in the balanced chemical equation describing the reaction. | **12.3** |
| **Predicting reaction spontaneity under standard conditions** | A reaction is spontaneous if $\Delta G°_{rxn} < 0$, where $\Delta G°_{rxn} = \Delta H°_{rxn} - T\Delta S°_{rxn}$ (12.11) | **12.4, 12.5** |

| Type of Problem | Concepts and Equations | Sample Exercises |
|---|---|---|
| **Calculating $\Delta G^\circ_{rxn}$ by using appropriate $\Delta G^\circ_f$ values** | Use $$\Delta G^\circ_{rxn} = \sum n_{products}\Delta G^\circ_{f,products} - \sum n_{reactants}\Delta G^\circ_{f,reactants} \quad (12.12)$$ where $\Delta G^\circ_{f,products}$ and $\Delta G^\circ_{f,reactants}$ are the standard molar free energies of formation of the products and reactants, and $n_{products}$ and $n_{reactants}$ are their stoichiometric coefficients in the balanced chemical equation describing the reaction. | **12.6** |
| **Relating reaction spontaneity to temperature, $\Delta H^\circ_{rxn}$, and $\Delta S^\circ_{rxn}$** | The temperature at which a process is at equilibrium can be calculated using $$\Delta G_{rxn} \approx \Delta H^\circ_{rxn} - T\Delta S^\circ_{rxn} \quad (12.14)$$ and setting $\Delta G_{rxn} = 0$. Use Table 12.4 to determine whether the process will be spontaneous above or below that characteristic temperature. | **12.7** |
| **Calculating $\Delta G^\circ_{rxn}$ of coupled reactions** | Free-energy changes of coupled reactions are additive. | **12.8** |

## VISUAL PROBLEMS

*(Answers to boldface end-of-chapter questions and problems are in the back of the book.)*

**12.1.** The marbles in Figure P12.1 occupy two of the three depressions in each of the blocks. After the black divider is removed, the marbles in each block can occupy any of the sites in that block. How many arrangements of the marbles are possible before and after removal of the divider in the (a) block and in the (b) block?

**FIGURE P12.1**

12.2. The marbles in Figure P12.2 occupy two of the four depressions in each of the blocks. After the black divider is removed, the marbles in each block can occupy any of the sites in that block. How many arrangements of the marbles are possible before and after removal of the divider in the (a) block and in the (b) block?

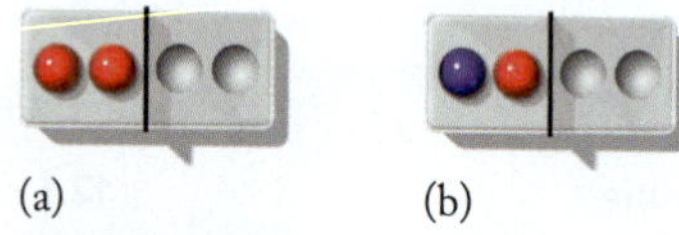


**FIGURE P12.2**

**12.3.** Two tires shown in Figure P12.3 are inflated at the same temperature to the same volume, though more air is used to inflate the tire on the right. In which tire is the gas under greater internal pressure, and in which does the gas have greater total entropy?

**FIGURE P12.3**

12.4. Two cubic containers (Figure P12.4) contain the same number of moles of gas at the same temperature.

a. Which cube contains the gas with more entropy?
b. If the sample in cube (b) is left unchanged but the sample in cube (a) is cooled so that it condenses, which sample has the higher entropy?

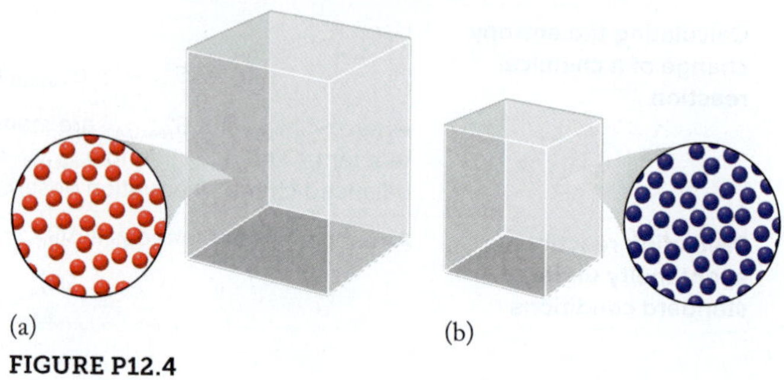


**FIGURE P12.4**

**12.5.** Figure P12.5 shows a tank that has just been filled with a mixture of two ideal gases: A (red spheres) and B (blue spheres). If the molar mass of A is twice that of B, will the atoms of A eventually fill the bottom of the tank and the atoms of B fill the top of the tank? Why or why not? Does the tank represent an isolated system?

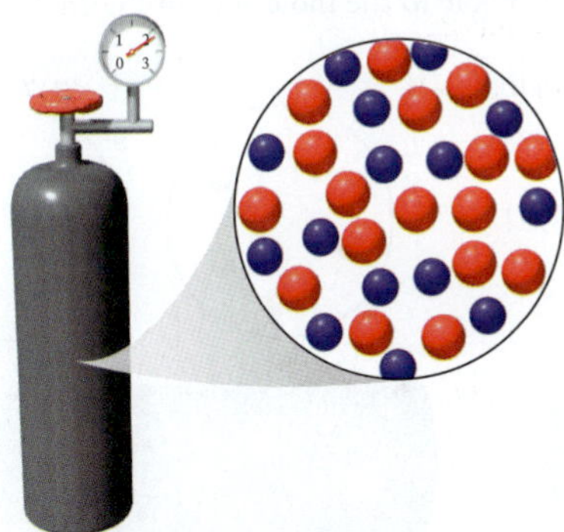

**FIGURE P12.5**

12.6. The box on the left in Figure P12.6 represents a mixture of two diatomic gases: $A_2$ (red spheres) and $B_2$ (blue spheres). How do the entropies of $A_2$ and $B_2$ change as a result of the process depicted by the arrow?

**FIGURE P12.6**

**12.7.** Is the process in Figure P12.6 more likely to be spontaneous at high temperature or low temperature, or is it unaffected by changing temperature?

12.8. Figure P12.8 shows the plots of $\Delta H$ and $T\Delta S$ for a phase change as a function of temperature.

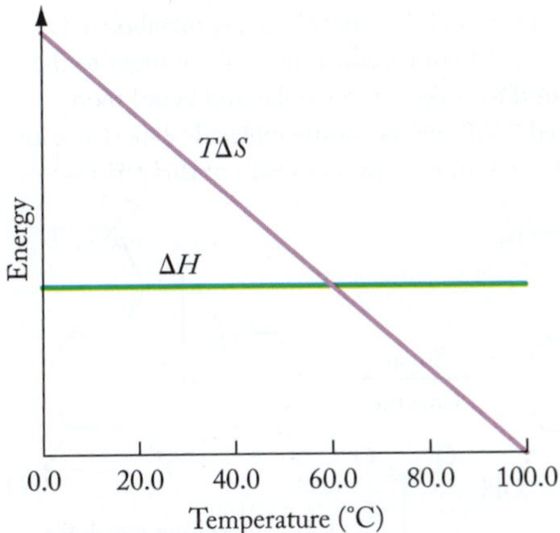


**FIGURE P12.8**

a. What is the status of the process at the point where the two lines intersect?
b. Over what temperature range is the process spontaneous?

**12.9.** Of the six phase changes—melting, vaporization, condensation, freezing, sublimation, and deposition—which ones have thermodynamic profiles that fit the pattern in Figure P12.8?

12.10. Use representations [A] through [I] in Figure P12.10 to answer questions a–f.

a. What is the sign of $\Delta S_{sys}$ for the formation of frost on a window pane in image [B]? Under what conditions would this process *not* be spontaneous?
b. Image [H] shows a neutralization reaction between antacid tablets (sodium bicarbonate) and stomach acid, $HCl(aq)$. What is the sign of $\Delta S_{rxn}$?
c. Image [D] depicts the mixing of two colorless solutions that results in image [F]. Is $\Delta S_{rxn}$ greater than, less than, or equal to zero?
d. If you mix the particulate substances in [A] and [C] to create the mixture in [E], does the entropy increase, decrease, or remain unchanged?
e. When you mix the particulate substances in [G] and [I] to form mixture [E], how does the change in volume affect the entropy?
f. Does all mixing lead to an increase in entropy of the system?

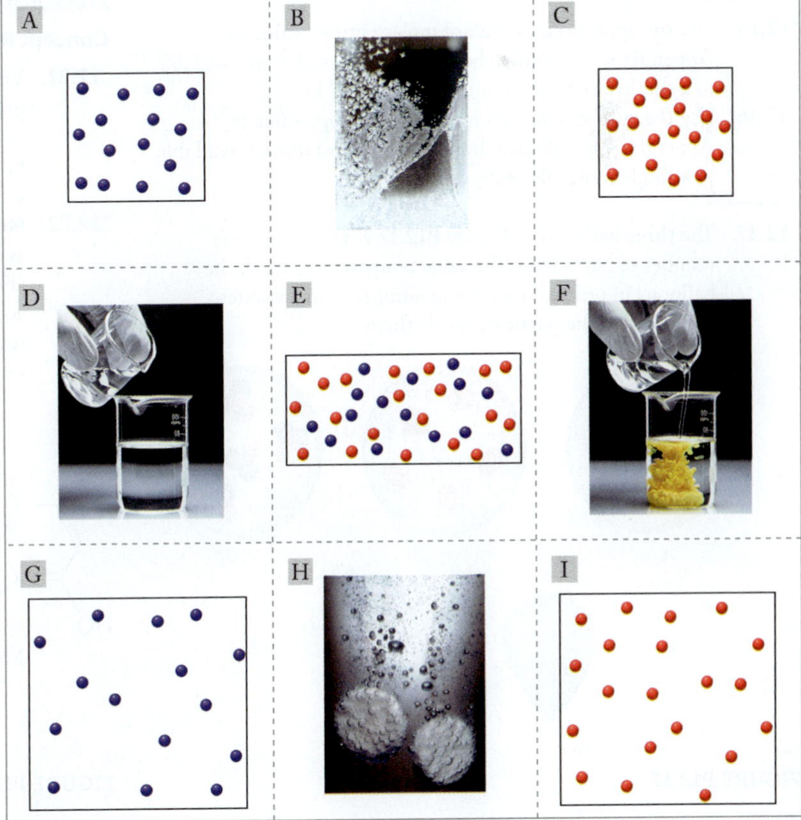


**FIGURE P12.10**

# QUESTIONS AND PROBLEMS

## Spontaneous Processes and Entropy

### Concept Review

**12.11.** How does the entropy change that accompanies a reaction relate to the entropy change that happens when the reaction runs in reverse?

12.12. Identify the following processes as spontaneous or nonspontaneous, and explain your choice.

a. A photovoltaic cell in a portable device charges your cell phone.
b. Dry ice (solid $CO_2$) sublimes at room temperature.
c. A radiopharmaceutical imaging agent containing technetium emits gamma rays.

**12.13.** You flip three coins, assigning the values +1 for heads (H) and −1 for tails (T). Each outcome of the three flips constitutes a microstate. How many different microstates are possible from flipping the three coins? Which value or values for the sums in the microstates are most likely? *Hint*: The sequence HHT (+1 +1 −1) is one possible outcome, or microstate. Note, however, that this outcome differs from THH (−1 +1 +1), even though the two sequences sum to the same value.

12.14. Imagine you have four identical chairs to arrange on four steps leading up to a stage, one chair on each step. The chairs have numbers on their backs: 1, 2, 3, and 4. How many different microstates for the chairs are possible? (When viewed from the front, all the microstates look the same. When viewed from the back, you can identify the different microstates because you can distinguish the chairs by their numbers.)

### Problems

**12.15.** Use the appropriate standard molar entropy value in Appendix 4 to calculate how many microstates are available to a single molecule of liquid $H_2O$ at 298 K.

12.16. Use the appropriate standard molar entropy value in Appendix 4 to calculate how many microstates are available to a single molecule of $N_2$ at 298 K.

---

**12.17.** The three balloons in Figure P12.17 contain the same number of particles at the same temperature. Rank the balloons in order of increasing number of microstates accessible to the particles inside them.

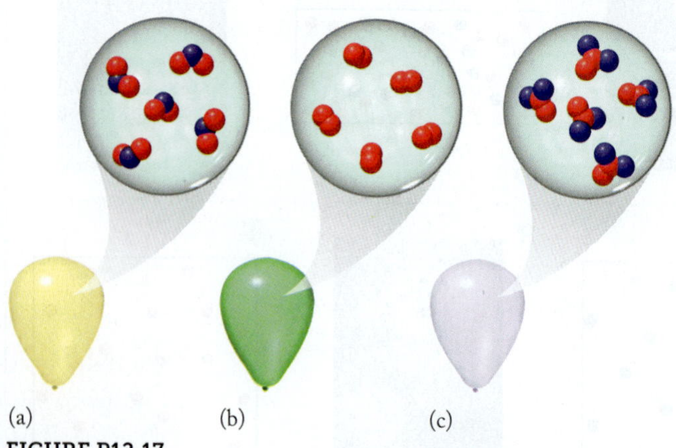


**FIGURE P12.17**

12.18. Figure P12.18(a) shows a cylinder within a cylinder that contains a population of gaseous molecules. The volume occupied by the molecules can be increased by pulling the inside cylinder out (Figure P12.18b), much like a telescope. The number of molecules within the cylinder remains the same during this operation. Compare the number of microstates available to the molecules in Figure P12.18(a) and (b).

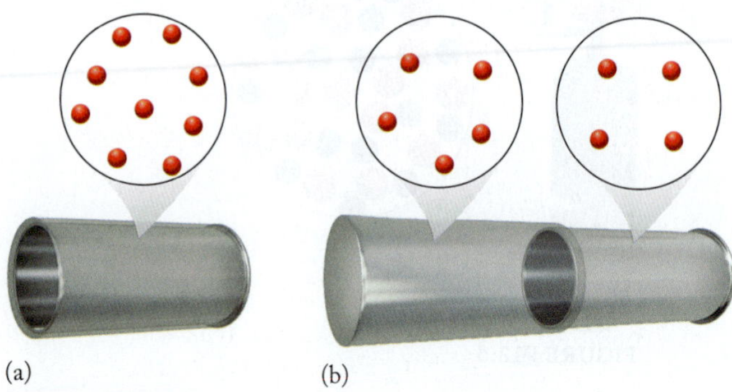


**FIGURE P12.18**

---

**12.19.** Which of the following ionic solutes experiences the greatest increase in entropy when 0.0100 mol of it dissolves in 1.00 liter of water? (a) $CaCl_2$; (b) NaBr; (c) KCl; (d) $Cr(NO_3)_3$; (e) LiOH

12.20. Which of the following molecular solutes experience an increase in entropy when dissolved in water? (a) $CO_2(g)$; (b) $HF(g)$; (c) $CH_3OH(\ell)$; (d) $CH_3COOH(\ell)$; (e) $C_{12}H_{22}O_{11}(s)$

## Absolute Entropy and Molecular Structure

### Concept Review

**12.21.** Which component in each of the following pairs has the greater entropy?

a. 1 mole of $S_2(g)$ or 1 mole of $S_8(g)$
b. 1 mole of $S_2(g)$ or 1 mole of $S_8(s)$
c. 1 mole of $O_2(g)$ or 1 mole of $O_3(g)$

*12.22. **Metabolism of Pharmaceuticals** When some pharmaceutical agents such as the pain killer morphine are metabolized (Figure P12.22), our bodies add a molecule of sugar to the active compound to make it more polar and hence more readily excreted. Will the morphine molecule experience an increase or decrease in entropy as a result of this process?

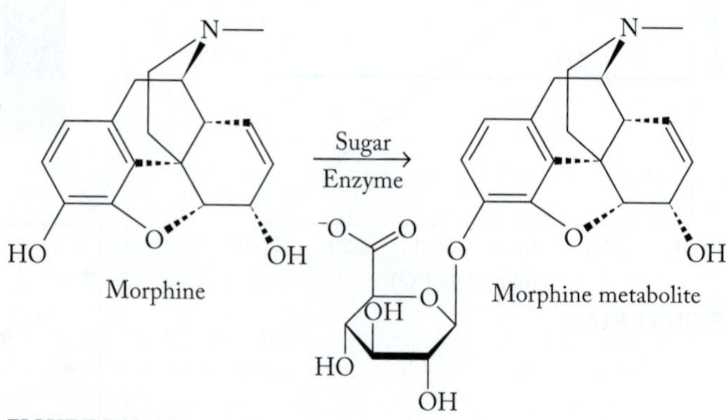


**FIGURE P12.22**

***12.23.** Diamond and the fullerenes are allotropes of carbon. On the basis of their different structures and properties, predict which has the higher standard molar entropy.

12.24. **Superfluids** The 1996 Nobel Prize in Physics was awarded to Douglas Osheroff, Robert Richardson, and David Lee for discovering *superfluidity* (apparently frictionless flow) in $^3$He. When $^3$He is cooled to 2.7 mK, the liquid settles into an *ordered* superfluid state. Predict the sign of the entropy change for the conversion of liquid $^3$He into its superfluid state.

**Problems**

**12.25.** Rank the compounds in each of the following groups in order of increasing standard molar entropy ($S°$):

a. $CH_4(g)$, $CF_4(g)$, and $CCl_4(g)$
b. $CH_2O(g)$, $CH_3CHO(g)$, and $CH_3CH_2CHO(g)$
c. $HF(g)$, $H_2O(g)$, and $NH_3(g)$

12.26. Rank the compounds in each of the following groups in order of increasing standard molar entropy ($S°$):

a. $CH_4(g)$, $CH_3CH_3(g)$, and $CH_3CH_2CH_3(g)$
b. $CCl_4(\ell)$, $CHCl_3(\ell)$, and $CH_2Cl_2(\ell)$
c. $CO_2(\ell)$, $CO_2(g)$, and $CS_2(g)$

---

**12.27.** Figure P12.27 shows a chunk of dry ice at room temperature. Write a balanced equation that represents the change occurring for the dry ice, and determine whether the entropy change for this process is greater than, equal to, or less than zero.

**FIGURE P12.27**

12.28. In Chapter 8, we studied redox reactions. Does the entropy increase or decrease in the redox reaction shown in Figure P12.28 between hydrazine, $N_2H_4$, and dinitrogen tetroxide, $N_2O_4$, to yield a mixture of nitrogen and water?

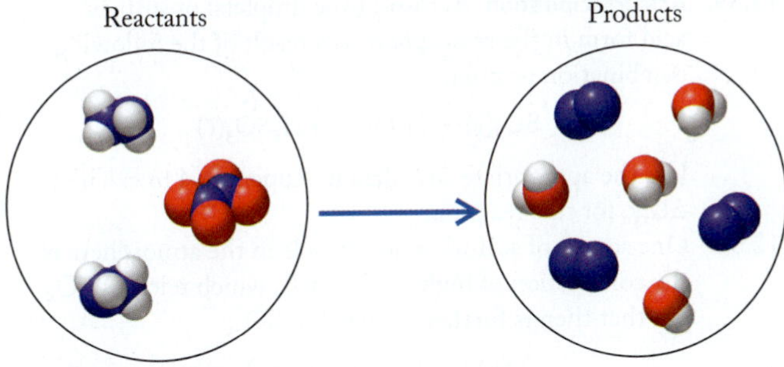


**FIGURE P12.28**

## Applications of the Second Law

**Concept Review**

**12.29.** Ice cubes melt in a glass of lemonade, cooling the lemonade from 10.0°C to 0.0°C. If the ice cubes are the system, what are the signs of $\Delta S_{sys}$ and $\Delta S_{surr}$?

12.30. Adding sidewalk deicer (calcium chloride) to water lowers the freezing point of the water. If solid $CaCl_2$ is the system, what are the signs of $\Delta S_{sys}$ and $\Delta S_{surr}$?

**Problems**

**12.31.** Which of the following combinations of entropy changes for a process are mathematically possible?

a. $\Delta S_{sys} > 0, \Delta S_{surr} > 0, \Delta S_{univ} > 0$
b. $\Delta S_{sys} > 0, \Delta S_{surr} < 0, \Delta S_{univ} > 0$
c. $\Delta S_{sys} > 0, \Delta S_{surr} > 0, \Delta S_{univ} < 0$

12.32. Which of the following combinations of entropy changes for a process are mathematically possible?

a. $\Delta S_{sys} < 0, \Delta S_{surr} > 0, \Delta S_{univ} > 0$
b. $\Delta S_{sys} < 0, \Delta S_{surr} < 0, \Delta S_{univ} > 0$
c. $\Delta S_{sys} < 0, \Delta S_{surr} > 0, \Delta S_{univ} < 0$

---

**12.33.** **Cleaning Natural Gas** Predict whether the entropy of the system increases or decreases for the following reaction, which describes the process used to remove hydrogen sulfide from natural gas:

$$8\,H_2S(g) + 4\,O_2(g) \rightarrow S_8(s) + 8\,H_2O(g)$$

12.34. **Making Sodium Hydroxide** Predict whether the entropy of the system increases or decreases for the following reaction, which describes the process used industrially to produce sodium hydroxide and chlorine by passing electric current through brine:

$$2\,NaCl(aq) + 2\,H_2O(\ell) \rightarrow 2\,NaOH(aq) + H_2(g) + Cl_2(g)$$

---

**12.35.** If the value of $\Delta S_{rxn}$ of the nonspontaneous reaction $A + B \rightarrow C$ is −66.0 J/K, what is the maximum entropy change in the reaction's surroundings?

12.36. The value of $\Delta S_{rxn}$ of the spontaneous reaction $D + E \rightarrow F$ is 72.0 J/K. What is the minimum value of the entropy change in the reaction's surroundings?

## Calculating Entropy Changes

**Concept Review**

**12.37.** Under standard conditions, the products of a reaction have, overall, greater entropy than the reactants. What is the sign of $\Delta S°_{rxn}$?

12.38. Do decomposition reactions tend to have $\Delta S°_{rxn}$ values that are greater than zero or less than zero? Why?

**12.39.** Do precipitation reactions tend to have $\Delta S°_{rxn}$ values that are greater than zero or less than zero? Why?

12.40. For each of the reactions given, indicate whether $\Delta S$ should have a positive sign or a negative sign. If it is not possible to judge the sign of $\Delta S$ from the information provided, indicate why that is the case.

a. $2\,Na(s) + Cl_2(g) \rightarrow 2\,NaCl(s)$
b. $4\,H_3PO_3(\ell) \rightarrow PH_3(g) + 3\,H_3PO_4(\ell)$
c. $CO(g) + H_2O(g) \rightarrow CO_2(g) + H_2(g)$
d. $Ca(OH)_2(s) + CO_2(g) \rightarrow CaCO_3(s) + H_2O(g)$

### Problems

**12.41.** **Smog** Use the standard molar entropies in Appendix 4 to calculate the $\Delta S°$ value for each of the following atmospheric reactions that contribute to the formation of photochemical smog.
a. $N_2(g) + O_2(g) \rightarrow 2\,NO(g)$
b. $2\,NO(g) + O_2(g) \rightarrow 2\,NO_2(g)$
c. $NO(g) + \frac{1}{2}O_2(g) \rightarrow NO_2(g)$
d. $2\,NO_2(g) \rightarrow N_2O_4(g)$

12.42. Use the standard molar entropies in Appendix 4 to calculate the $\Delta S°$ value for each of the following reactions of sulfur compounds.
a. $H_2S(g) + \frac{3}{2}O_2(g) \rightarrow H_2O(g) + SO_2(g)$
b. $2\,SO_2(g) + O_2(g) \rightarrow 2\,SO_3(g)$
c. $SO_3(g) + H_2O(\ell) \rightarrow H_2SO_4(aq)$
d. $S(g) + O_2(g) \rightarrow SO_2(g)$

**12.43.** **Ozone Layer** The following reaction plays a key role in the destruction of ozone in the atmosphere:

$$Cl(g) + O_3(g) \rightarrow ClO(g) + O_2(g)$$

The standard entropy change ($\Delta S°_{rxn}$) is 19.9 J/(mol · K). Use the standard molar entropies in Appendix 4 to calculate the $S°$ value of ClO($g$).

12.44. Calculate the $\Delta S°$ value for the conversion of ozone to oxygen,

$$2\,O_3(g) \rightarrow 3\,O_2(g)$$

in the absence of Cl atoms, and compare it with the $\Delta S°$ value for the reaction in Problem 12.43.

## Free Energy

### Concept Review

**12.45.** What does the sign of $\Delta G$ tell you about the spontaneity of a process?

12.46. What does the sign of $\Delta G$ tell you about the rate of a reaction?

***12.47.** Many 19th-century scientists believed that all exothermic reactions were spontaneous. Why did so many of them share this belief?

12.48. In which direction does a reaction proceed when (a) $\Delta G_{rxn} < 0$; (b) $\Delta G_{rxn} = 0$; (c) $\Delta G_{rxn} > 0$?

**12.49.** What are the signs of $\Delta S$, $\Delta H$, and $\Delta G$ for the sublimation of dry ice (solid $CO_2$) at 25°C?

12.50. What are the signs of $\Delta S$, $\Delta H$, and $\Delta G$ for the formation of dew on a cool night?

**12.51.** Which of the following processes is/are spontaneous?
a. A tornado forms.
b. A broken touchscreen fixes itself.
c. You get an A in this course.
d. Hot soup gets cold before it is served.

12.52. Which of the following processes is/are spontaneous?
a. Wood burns in air.
b. Water vapor condenses on the sides of a glass of iced tea.
c. Salt dissolves in water.
d. Photosynthesis occurs.

### Problems

**12.53.** Calculate the free-energy change for the dissolution in water of one mole of NaBr and one mole of NaI at 298 K from the values in the following table.

| | $\Delta H°_{solution}$ (kJ/mol) | $\Delta S°_{solution}$ [J/(mol · K)] |
|---|---|---|
| NaBr | −0.60 | 57 |
| NaI | −7.5 | 74 |

12.54. The values of $\Delta H°_{rxn}$ and $\Delta S°_{rxn}$ for the reaction

$$2\,NO(g) + O_2(g) \rightarrow 2\,NO_2(g)$$

are −12 kJ and −146 J/K, respectively.
a. Use these values to calculate $\Delta G°_{rxn}$ at 298 K.
b. Explain why the value of $\Delta S°_{rxn}$ is negative.

**12.55.** A mixture of CO($g$) and $H_2(g)$ is produced by passing steam over hot charcoal:

$$H_2O(g) + C(s) \rightarrow H_2(g) + CO(g)$$

Calculate the $\Delta G°_{rxn}$ value for the reaction from the appropriate $\Delta G°_f$ data in Appendix 4.

12.56. Use the appropriate $\Delta G°_f$ data in Appendix 4 to calculate $\Delta G°_{rxn}$ for the complete combustion of methanol:

$$2\,CH_3OH(g) + 3\,O_2(g) \rightarrow 2\,CO_2(g) + 4\,H_2O(g)$$

**12.57.** **Photochemical Smog** Use the appropriate $\Delta G°_f$ data in Appendix 4 to calculate $\Delta G°_{rxn}$ for the oxidation of NO to $NO_2$—a key reaction in the formation of photochemical smog:

$$NO(g) + \tfrac{1}{2}O_2(g) \rightarrow NO_2(g)$$

12.58. Use the free energies of formation from Appendix 4 to calculate the standard free-energy change for the decomposition of ammonia in the following reaction:

$$2\,NH_3(g) \rightarrow N_2(g) + 3\,H_2(g)$$

Is the reaction spontaneous under standard conditions?

**12.59.** **Acid Precipitation** Aerosols (fine droplets) of sulfuric acid form in the atmosphere as a result of the following combination reaction:

$$SO_3(g) + H_2O(g) \rightarrow H_2SO_4(\ell)$$

Use the appropriate $\Delta G°_f$ data in Appendix 4 to calculate $\Delta G°_{rxn}$ for this reaction.

12.60. One source of sulfuric acid aerosols in the atmosphere is the combustion of high-sulfur fuels, which releases $SO_2$ gas that then is further oxidized to $SO_3$:

$$2\,SO_2(g) + O_2(g) \rightarrow 2\,SO_3(g)$$

Use the appropriate $\Delta G°_f$ data in Appendix 4 to calculate $\Delta G°_{rxn}$ for this combination reaction at 25°C. Is it spontaneous under standard conditions?

## Temperature and Spontaneity

### Concept Review

**12.61.** Are exothermic reactions spontaneous only at low temperature? Explain your answer.

12.62. Are endothermic reactions never spontaneous at low temperature? Explain your answer.

### Problems

**12.63.** What is the lowest temperature at which the following reaction (see Problem 12.55) is spontaneous?

$$H_2O(g) + C(s) \rightarrow H_2(g) + CO(g)$$

12.64. Above what temperature does nitrogen monoxide form from nitrogen and oxygen?

$$N_2(g) + O_2(g) \rightarrow 2\ NO(g)$$

Assume that the values of $\Delta H^\circ_{rxn}$ and $\Delta S^\circ_{rxn}$ do not change appreciably with temperature.

---

**12.65.** Use the data in Appendix 4 to calculate $\Delta H^\circ_{rxn}$ and $\Delta S^\circ_{rxn}$ for the vaporization of hydrogen peroxide:

$$H_2O_2(\ell) \rightarrow H_2O_2(g)$$

If we assume that the calculated values are independent of temperature, what is the boiling point of hydrogen peroxide at $P = 1.00$ atm?

12.66. **Volcanoes** Deposits of elemental sulfur are often seen near active volcanoes. Their presence there may be due to the following reaction of $SO_2$ with $H_2S$:

$$SO_2(g) + 2\ H_2S(g) \rightarrow \tfrac{3}{8}\ S_8(s) + 2\ H_2O(g)$$

If we assume that the values of $\Delta H^\circ_{rxn}$ and $\Delta S^\circ_{rxn}$ do not change appreciably with temperature, over what temperature range is the reaction spontaneous?

---

**12.67.** Which of the following reactions is spontaneous (i) only at low temperatures; (ii) only at high temperatures; (iii) at all temperatures?

a. $2\ NO(g) + O_2(g) \rightarrow 2\ NO_2(g)$

b. $2\ NH_3(g) + 2\ O_2(g) \rightarrow N_2O(g) + 3\ H_2O(g)$

c. $NH_4NO_3(s) \rightarrow 2\ H_2O(g) + N_2O(g)$

12.68. Which of the following reactions is spontaneous (i) only at low temperatures; (ii) only at high temperatures; (iii) at all temperatures?

a. $2\ Mg(s) + O_2(g) \rightarrow 2\ MgO(s)$

b. $2\ CH_3OH(\ell) + 3\ O_2(g) \rightarrow 2\ CO_2(g) + 4\ H_2O(\ell)$

c. $N_2(g) + O_2(g) \rightarrow 2\ NO(g)$

---

**12.69.** One method for the industrial production of methanol uses the following reaction:

$$CO(g) + 2\ H_2(g) \rightarrow CH_3OH(\ell)$$

a. Use the data in Appendix 4 to calculate $\Delta G^\circ$ for this reaction at 298 K.

b. The reaction is normally run at a minimum temperature of 475 K. What is the value of $\Delta G$ at that temperature? Is the reaction spontaneous at that temperature?

12.70. Gas streams containing $CO_2$ are often passed through absorption tubes filled with $CaO(s)$, where the following reaction takes place to remove the $CO_2$ from the stream:

$$CaO(s) + CO_2(g) \rightarrow CaCO_3(s)$$

a. Use the data in Appendix 4 to calculate $\Delta G^\circ$ at 298 K for this reaction.

b. Is the reaction spontaneous at 298 K?

c. Calculate $\Delta G$ for this reaction at 1500 K, a typical temperature for a lime kiln. (Assume that $\Delta H$ and $\Delta S$ do not change with temperature.)

d. Is the reaction as written spontaneous at 1500 K?

e. In a lime kiln, calcium carbonate (in the form of oyster shells) is roasted to produce CaO and $CO_2$. Is this process spontaneous at the temperature of a kiln?

## Driving the Human Engine: Coupled Reactions

### Concept Review

**12.71.** Describe the ways in which two chemical reactions must complement each other so that the decrease in free energy of the spontaneous one can drive the nonspontaneous one.

12.72. How do you calculate the value of $\Delta G^\circ$ for a reaction that is the result of coupling a spontaneous reaction ($\Delta G^\circ_{spon} < 0$) and a nonspontaneous reaction ($\Delta G^\circ_{nonspon} > 0$)?

**12.73.** Why is it important that at least some of the spontaneous steps in glycolysis convert ADP to ATP?

12.74. The second step in glycolysis converts glucose 6-phosphate into fructose 6-phosphate (Figure P12.74). Suggest a reason why $\Delta G^\circ$ for this reaction is close to zero.

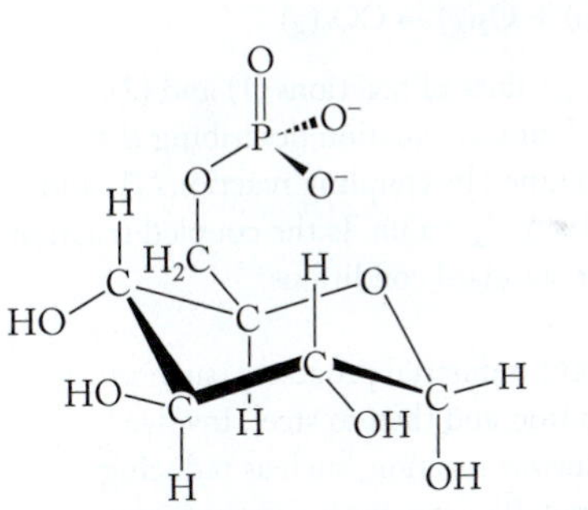


**FIGURE P12.74**

**Problems**

**12.75.** The methane in natural gas is an important starting material, or feedstock, for producing industrial chemicals, including $H_2$ gas.

a. Use the appropriate $\Delta G_f^\circ$ value(s) from Appendix 4 to calculate $\Delta G_{rxn}^\circ$ for the reaction known as *steam–methane reforming*:

$$CH_4(g) + H_2O(g) \rightarrow CO(g) + 3\,H_2(g)$$

b. To help drive this nonspontaneous reaction, the CO that is produced can be oxidized to $CO_2$ by using more steam:

$$CO(g) + H_2O(g) \rightarrow CO_2(g) + H_2(g)$$

Use the appropriate $\Delta G_f^\circ$ value(s) from Appendix 4 to calculate $\Delta G_{rxn}^\circ$ for this reaction, which is known as the *water–gas shift reaction*.

c. Combine these two reactions and write the chemical equation of the overall reaction in which methane and steam combine to produce hydrogen gas and carbon dioxide.

d. Calculate the $\Delta G_{rxn}^\circ$ value of the overall reaction. Is it spontaneous under standard conditions?

12.76. In addition to the reactions described in Problem 12.75, methane can, in theory, be used to produce hydrogen gas by a process in which it decomposes into elemental carbon and hydrogen:

$$(1) \quad CH_4(g) \rightarrow C(s) + 2\,H_2(g)$$

and the carbon produced in the first step is then oxidized to $CO_2$:

$$(2) \quad C(s) + O_2(g) \rightarrow CO_2(g)$$

a. Calculate the $\Delta G_{rxn}^\circ$ values of reactions (1) and (2).

b. Write a balanced chemical equation describing the overall reaction obtained by coupling reactions (1) and (2), and calculate its $\Delta G_{rxn}^\circ$ value. Is the coupled reaction spontaneous under standard conditions?

**12.77.** **Making Steel** Important industrial processes, such as converting iron ore to iron and then to steel, involve coupling a nonspontaneous reaction, such as reducing the iron in $Fe_2O_3$ to metallic iron, with a spontaneous one, such as the oxidation of the carbon in CO to $CO_2$. Use the appropriate thermodynamic data in Appendix 4 to calculate the $\Delta G_{rxn}^\circ$ value of the following reaction at 1450°C:

$$Fe_2O_3(s) + 3\,CO(g) \rightarrow 2\,Fe(s) + 3\,CO_2(g)$$

12.78. One source of the carbon monoxide reactant in Problem 12.77 is pure hot carbon, called *coke*, which is derived from coal. What is the overall $\Delta G_{rxn}^\circ$ value of the iron reduction reaction at 1450°C starting with carbon as the reducing agent instead of carbon monoxide? Assume that coke has the thermodynamic properties of graphite.

## Additional Problems

**12.79.** Chlorofluorocarbons (CFCs) are no longer used as refrigerants because they catalyze the decomposition of stratospheric ozone. Trichlorofluoromethane ($CCl_3F$) boils at 23.8°C and its molar enthalpy of vaporization is 24.8 kJ/mol. What is the molar entropy of vaporization of $CCl_3F(\ell)$?

12.80. **Methane-Producing Bacteria** Methanogenic bacteria convert liquid acetic acid ($CH_3COOH$) into $CO_2(g)$ and $CH_4(g)$.

a. Is this process endothermic or exothermic under standard conditions?

b. Is the reaction spontaneous under standard conditions?

***12.81.** At what temperature is the standard free-energy change for the following reaction equal to zero?

$$NH_4Cl(s) \rightarrow NH_3(g) + HCl(g)$$

12.82. Consider the precipitation reactions described by the following net ionic equations:

$$Mg^{2+}(aq) + 2\,OH^-(aq) \rightarrow Mg(OH)_2(s)$$

$$Ag^+(aq) + Cl^-(aq) \rightarrow AgCl(s)$$

a. Predict the sign of $\Delta S_{rxn}^\circ$ for the reactions.

b. Using the values for $S^\circ$ from Appendix 4, calculate $\Delta S^\circ$ for these reactions.

c. Do your calculations support your prediction?

***12.83.** Calculate the standard free-energy change of the following reaction. Is it spontaneous?

$$2\,NO(g) + 2\,H_2(g) \rightarrow N_2(g) + 2\,H_2O(g)$$

12.84. The absolute entropy ($S$) of a perfect, defect-free solid equals zero and has one accessible microstate. It is impossible to make such a material, but silicon chip manufacturers strive for as few defects as possible in their products. Calculate the absolute molar entropy for a piece of silicon with several probable arrangements of (a) $W = 16$, (b) $W = 625$, and (c) $W = 2500$ per atom of Si.

**12.85.** Determine the value of $\Delta G^\circ$ for the reduction of iron ore with hydrogen gas:

$$Fe_2O_3(s) + 3\,H_2(g) \rightarrow 2\,Fe(s) + 3\,H_2O(g)$$

given the following thermodynamic properties at 25°C:

| | $\Delta H_f^\circ$ (kJ/mol) | $S^\circ$ [J/(mol · K)] |
|---|---|---|
| $Fe_2O_3(s)$ | −824.2 | 87.4 |
| $H_2(g)$ | 0 | 130.6 |
| $Fe(s)$ | 0 | 27.3 |
| $H_2O(g)$ | −241.8 | 188.8 |

*12.86. Show that hydrogen cyanide (HCN) is a gas at 25°C by estimating its normal boiling point from the following data:

| | $\Delta H^\circ_f$ (kJ/mol) | $S^\circ$ [J/(mol · K)] |
|---|---|---|
| HCN($\ell$) | 108.9 | 113 |
| HCN($g$) | 135.1 | 202 |

**12.87.** Figure P12.87 presents the $\Delta G^\circ_f$ values of several elements and compounds selected from Appendix 4. Which of the following conversions are spontaneous? (a) $C_6H_6(\ell)$ to $CO_2(g)$ and $H_2O(\ell)$; (b) $CO_2(g)$ to $C_2H_2(g)$; (c) $H_2(g)$ and $O_2(g)$ to $H_2O(\ell)$. Explain your reasoning. *Hint*: Begin by writing a balanced equation describing the conversion. Add other substances as needed to complete the equation.

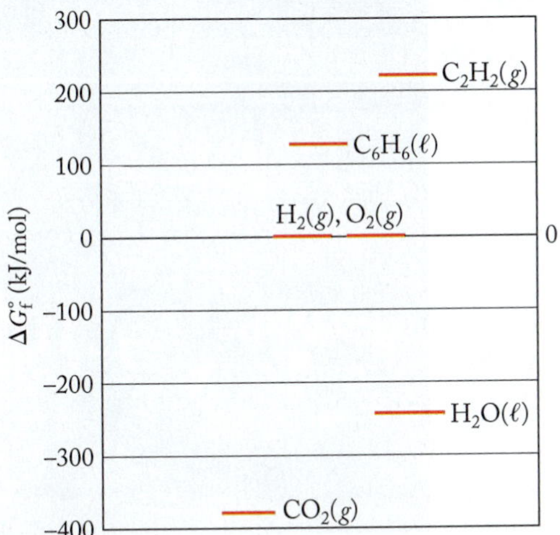


**FIGURE P12.87**

12.88. Write two equations for the complete combustion of one mole of acetylene, $C_2H_2(g)$, in oxygen at 298 K: in the first equation, the water produced as a product is a liquid; in the second equation, the water is in the gas phase.

a. Determine $\Delta G^\circ$ for each reaction.

*b. Suggest a way you could determine the difference between the two $\Delta G^\circ$ values without having to solve for $\Delta G^\circ$ for both reactions.

**12.89. Lightbulb Filaments** Tungsten (W) is the favored metal for lightbulb filaments, in part because of its high melting point (3422°C). The enthalpy of fusion of tungsten is 35.4 kJ/mol. What is its entropy of fusion?

*12.90. Over what temperature range is the reduction of tungsten(VI) oxide by hydrogen to give metallic tungsten and water spontaneous? The standard enthalpy of formation of $WO_3(s)$ is −843 kJ/mol, and its standard molar entropy is 76 J/(mol · K).

***12.91. Melting Organic Compounds** When dicarboxylic acids (compounds with two –COOH groups in their structures) melt, they often decompose to produce one mole of $CO_2$ gas for every mole of dicarboxylic acid melted (see Figure P12.91).

a. What are the signs of $\Delta H$ and $\Delta S$ for the process as written?

b. Do you think the dicarboxylic acid will re-form when the melted material cools? Why or why not?

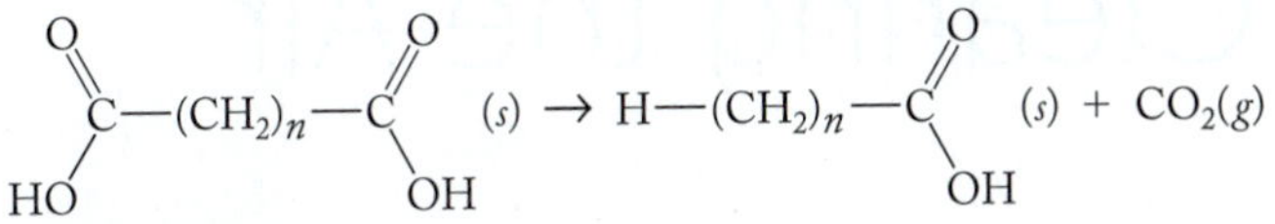

$$\text{HO–C(=O)–(CH}_2)_n\text{–C(=O)–OH}\,(s) \rightarrow \text{H–(CH}_2)_n\text{–C(=O)–OH}\,(s) + CO_2(g)$$

**FIGURE P12.91**

*12.92. **Melting DNA** When a solution of DNA in water is heated, the DNA double helix separates into two single strands.

a. What is the sign of $\Delta S$ for the separation process?

b. The DNA double helix re-forms as the system cools. What is the sign of $\Delta S$ for the process by which two single strands re-form the double helix?

c. The melting point of DNA is defined as the temperature at which $\Delta G = 0$. At that temperature, the melting reaction produces two single strands as fast as two single strands recombine to form the double helix. Write an equation that defines the melting temperature ($T$) of DNA in terms of $\Delta H$ and $\Delta S$.

# 13

# Chemical Kinetics

## Clearing the Air

PHOTOCHEMICAL SMOG The polluted air in Beijing is a mixture of fine particulate matter and the oxides of both sulfur and nitrogen, produced by the combustion of fossil fuels.

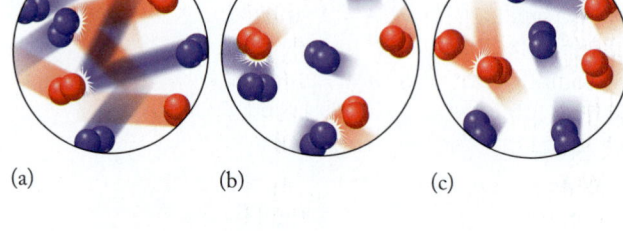

PARTICULATE **REVIEW**

### *Temperature and Collisions*

In Chapter 13, we examine the factors that influence the rates of reactions. Some of the air pollution visible in the photo above is due to a reaction between nitrogen and oxygen to form nitrogen monoxide.

- Write a balanced chemical equation for this reaction.
- Pictured are three mixtures of nitrogen and oxygen gases. Which mixture is at the highest temperature? In which mixture are the average velocities of the $N_2$ and $O_2$ molecules the highest?
- In which mixture do the molecules collide most frequently?

(a) (b) (c)

(Review Section 9.2 if you need help.)

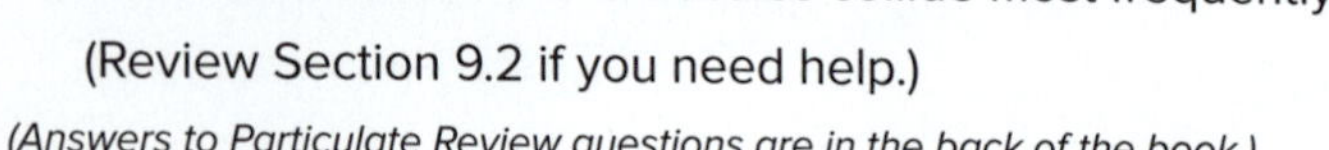

*(Answers to Particulate Review questions are in the back of the book.)*

PARTICULATE **PREVIEW**

## *Reaction Steps*

Many reactions take place in multiple steps. These images provide a molecular view of how the decomposition of $NO_2$ takes place in two steps at high temperatures. As you read Chapter 13, look for ideas that will help you answer these questions:

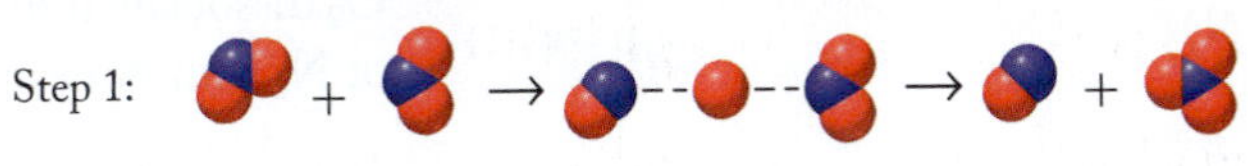

- Write a balanced chemical equation for each step in the reaction and add the two reactions together to obtain an overall reaction.
- What two processes do the dashed lines represent?
- Explain why one of the products in step 1 is *not* a product of the overall reaction.

## Learning Outcomes

**LO1** Use collision theory to explain the dependence of reaction rate on reactant concentration and temperature

**LO2** Use the stoichiometry of a reaction to relate the rates at which the concentrations of its reactants and products change as the reaction proceeds
**Sample Exercises 13.1, 13.2**

**LO3** Determine average and instantaneous reaction rates
**Sample Exercise 13.3**

**LO4** Use initial reaction rate data for a chemical reaction to determine the rate law, overall reaction order, and rate constant value and units
**Sample Exercise 13.4**

**LO5** Use integrated rate laws to identify the orders of reactions and determine their rate constants
**Sample Exercises 13.5, 13.7, 13.9**

**LO6** Calculate the half-life of a reactant and its concentration during a reaction
**Sample Exercises 13.6, 13.8**

**LO7** Relate rates of reactions to temperature and their activation energies
**Sample Exercises 13.10, 13.11**

**LO8** Use rate laws to assess the validity of a reaction mechanism
**Sample Exercises 13.12, 13.13**

**LO9** Identify catalysts and describe their impact on reaction rates and mechanisms
**Sample Exercise 13.14**

## 13.1 Cars and Air Quality

In one of the photographs at the beginning of this chapter, a thick haze obscures the buildings surrounding Tiananmen Square in Beijing. A major component of this haze is called **photochemical smog**. Its photochemical origins are linked to chemical reactions initiated by radiant energy from the sun interacting with nitrogen oxides emitted by vehicles and factories. The term *smog* was originally used to describe air pollution that occurred when *sm*oke mixed with f*og*.

How do the combustion reactions in car and truck engines, for example, produce these atmospheric nitrogen oxides? A key first step takes place at the high temperatures inside these engines, which promote a highly endothermic combination reaction between $N_2$ and $O_2$:

$$N_2(g) + O_2(g) \rightarrow 2\,NO(g) \qquad \Delta H^\circ_{rxn} = 180.6 \text{ kJ} \qquad (13.1)$$

The production of NO in the combustion chambers of engines is a little more complicated than gas-phase $N_2$ and $O_2$ molecules simply colliding to form molecules of NO. At the high temperatures of operating engines, some molecules of $O_2$ dissociate into individual O atoms. Collisions between O atoms and molecules of $N_2$ sometimes produce this reaction:

$$O(g) + N_2(g) \rightarrow NO(g) + N(g)$$

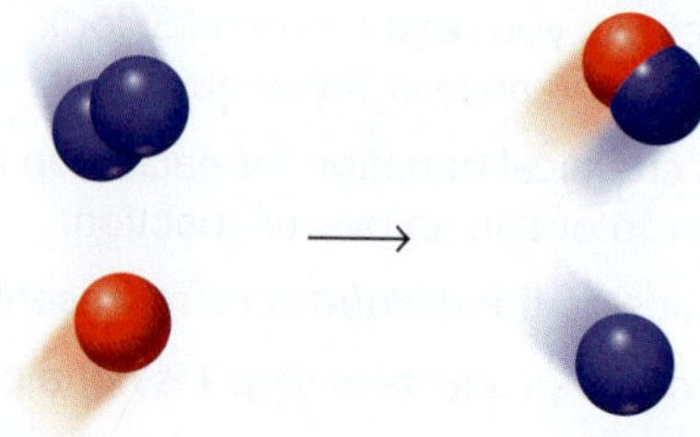

**photochemical smog** a mixture of gases formed in the lower atmosphere when sunlight interacts with compounds produced in internal combustion engines and other pollution sources.

Collisions between these N atoms and molecules of $O_2$ may then result in this reaction:

$$N(g) + O_2(g) \rightarrow NO(g) + O(g)$$

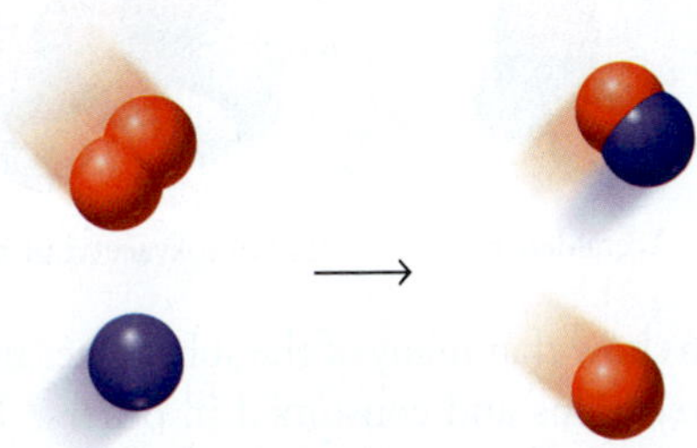

Summing these two reactions and simplifying, we obtain Equation 13.1:

$$\begin{array}{r} O(g) + N_2(g) \rightarrow NO(g) + N(g) \\ + N(g) + O_2(g) \rightarrow NO(g) + O(g) \\ \hline \cancel{O(g)} + N_2(g) + \cancel{N(g)} + O_2(g) \rightarrow NO(g) + \cancel{N(g)} + NO(g) + \cancel{O(g)} \end{array}$$

or $$N_2(g) + O_2(g) \rightarrow 2\,NO(g)$$

## CONCEPT TEST

The reaction in Equation 13.1 requires extremely high temperatures to become spontaneous.

a. What is the sign of $\Delta S^\circ_{rxn}$?

b. Why is the sign of $\Delta S^\circ_{rxn}$ difficult to predict only from the information in Equation 13.1?

*(Answers to Concept Tests are in the back of the book.)*

When molecules of NO enter the atmosphere, they may react with molecules of $O_2$, producing brown $NO_2$ gas:

$$2\,NO(g) + O_2(g) \rightarrow 2\,NO_2(g) \qquad \Delta H^\circ_{rxn} = -114.2\text{ kJ} \quad (13.2)$$

Sunlight provides enough energy to break the bonds in $NO_2$, re-forming NO and reactive oxygen atoms:

$$NO_2(g) \xrightarrow{\text{sunlight}} NO(g) + O(g) \qquad \Delta H^\circ_{rxn} = 306.3\text{ kJ} \quad (13.3)$$

These photochemically generated O atoms combine with molecules of $O_2$, producing ozone ($O_3$),

$$O_2(g) + O(g) \rightarrow O_3(g) \qquad \Delta H^\circ_{rxn} = -106.5\text{ kJ} \quad (13.4)$$

Oxygen atoms can also react with molecules of water vapor, producing hydroxyl radicals:

$$H_2O(g) + O(g) \rightarrow 2\,OH(g) \qquad \Delta H^\circ_{rxn} = 70.6\text{ kJ} \quad (13.5)$$

In the atmosphere, $O_3$ and OH radicals react with volatile organic compounds (VOCs) to form a variety of noxious substances. In one such reaction,

**CONNECTION** In Section 4.8, we learned that odd-electron species such as OH belong to a highly reactive group of substances called free radicals.

acetaldehyde is transformed into peroxyacetyl nitrate, which is highly irritating to one's eyes, nose, and throat and can cause respiratory distress:

$$N_2(g) + 3\,O_2(g) + OH(g) + CH_3CHO(g) \rightarrow CH_3C(O)O_2NO_2(g) + H_2O(g) + NO_2(g) \quad (13.6)$$

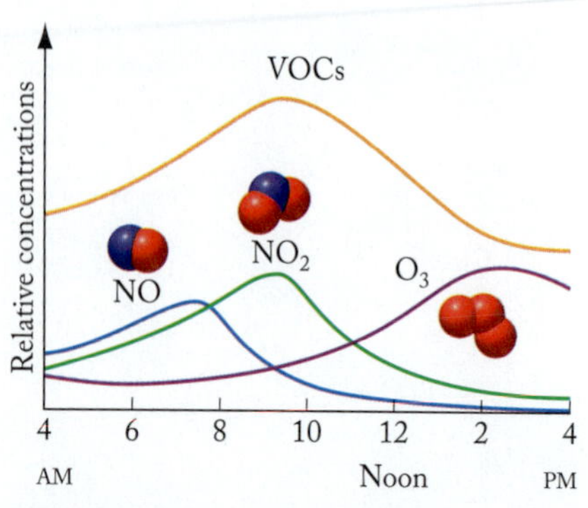


**FIGURE 13.1** In photochemical smog, NO from engine exhaust builds up in the early morning and then decreases as it reacts with atmospheric $O_2$ to form $NO_2$, whose concentration is highest in late morning. Photodecomposition of $NO_2$ leads to the formation of high levels of $O_3$ in the afternoon.

Equations 13.1–13.6 show that many of the substances in photochemical smog are produced in some reactions and consumed in others. The rates of these and other atmospheric reactions influence when smog happens, how intense it is, and how long it persists. The graphs in **Figure 13.1** show that the maximum NO concentration occurs during the morning rush hour. Later in the morning, the concentration of $NO_2$ reaches a maximum. This sequence makes sense because the NO produced in Equation 13.1 is a reactant in Equation 13.2, which produces $NO_2$. The highest ozone concentrations are reached in the middle of the afternoon because the reactions in Equations 13.2 and 13.3 produce a supply of free O atoms that are reactants in the formation of ozone (Equation 13.4), hydroxyl radicals (Equation 13.5), and peroxyacetyl nitrate (Equation 13.6).

Ozone also reacts with NO, forming $NO_2$ and $O_2$:

$$O_3(g) + NO(g) \rightarrow O_2(g) + NO_2(g) \qquad \Delta H^\circ_{rxn} = -199.8\text{ kJ} \quad (13.7)$$

Concentrations of $NO_2$ drop during the afternoons of smoggy days because $NO_2$, ozone, and hydrocarbons react to form an array of other (mostly unpleasant) compounds. A principal source of VOCs is unburnt hydrocarbons from car and truck fossil fuels. Their peak concentration also tracks with the morning rush hour. If we were to extend the graph in Figure 13.1, we would see a second maximum in VOC concentration after the evening rush hour.

Photochemical smog became a significant environmental problem in many urban areas during the last half of the 20th century as the use of private automobiles expanded rapidly. By 1975, smog was so widespread that the U.S. Environmental Protection Agency required that pollutant emissions from cars and light trucks be dramatically reduced. Automotive engineers met these standards by designing engines that burned fuel more efficiently and cleanly and by developing devices called catalytic converters to further reduce pollutant concentrations in engine exhaust. In the years since 1975, these devices have removed billions of tons of pollution from urban air in industrialized countries around the world.

Designing catalytic converters required extensive study of the by-products of the combustion reactions in automobile engines and the rates of the reactions that convert these substances into less noxious gases. The target compounds include unburned hydrocarbons and carbon monoxide in addition to various oxides of nitrogen. Catalytic converters speed up the oxidation of hydrocarbons and CO to $CO_2$ and $H_2O$ and the conversion of nitrogen oxides into $N_2$ and $O_2$. These reactions happen during the fraction of a second that molecules spend in contact with the catalytic surfaces. These devices and cleaner-burning engines don't remove all the pollutants from engine exhaust, but they do remove most of them.

In this chapter, we will examine how the rates of smog formation and other chemical reactions are measured and explore the factors that influence those rates, which vary over a wide range (**Figure 13.2**), from so slow that we barely perceive they are occurring to explosively fast. We will also see how scientists use the study

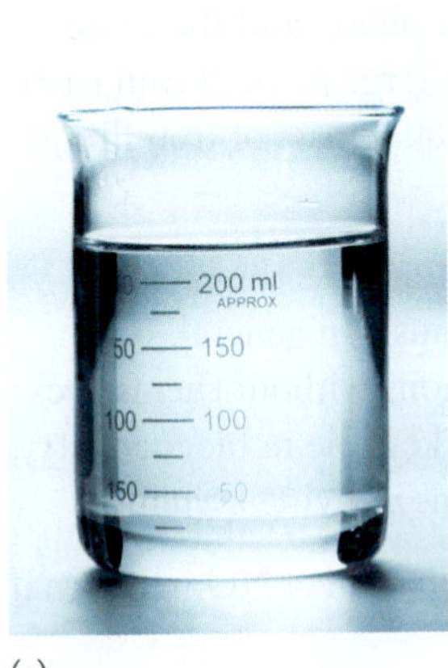

(a)

(b)

(c)

**FIGURE 13.2** (a) A 30% aqueous solution of $H_2O_2$ decomposes to water and oxygen gas so slowly that no change is observable. (b) Pouring the solution on a wedge of potato, which contains the enzyme catalase, causes the reaction to proceed at a faster rate, visibly producing bubbles of $O_2$ gas. (c) Adding a small amount of $MnO_2$ causes the reaction to proceed so rapidly that the energy liberated causes the solution to boil.

of reaction rates, **chemical kinetics**, to better understand how reactions happen on a molecular level and to manipulate the rates of chemical reactions to improve the quality of our environment and our lives.

## 13.2 Reaction Rates

Although chemical equations represent the overall changes that occur in a chemical reaction, they do not necessarily explain *how* those changes happen in terms of successful collisions between reactants to form products. As we will learn in this chapter, though, studies of the rates of chemical reactions *do* provide insights into how chemical reactions happen.

Chapters 9 and 10 provide a foundation on which we can build our understanding of chemical kinetics. In Chapter 9, for example, we explored the kinetic molecular theory of gases and the relationship between temperature and the average speeds of molecules in the gas phase. Faster speeds mean more frequent and forceful collisions between molecules, which can result in more rapid reactions.

In this chapter, we focus on the possibility that molecular collisions having enough kinetic energy may cause the bonds inside the molecules to break, allowing their atoms to recombine into different molecular structures. In other words, chemical reactions can occur. This view of how chemistry happens in the gas phase, which evolved in the early 20th century, is still a useful model. It is called **collision theory**. In this chapter, we will examine five factors that influence reaction rates:

1. *Physical states of the reactants.* The more frequently that particles interact—that is, the more often they collide with one another—the faster they can react with each other. Particles in solids have the same nearest neighbors over time, whereas the particles in liquids move in relation to one another. The particles in gases are widely separated and move very rapidly. These differences in motion are one reason why reactions in the solid phase tend to be much slower than reactions in liquids and gases.
2. *Concentration of reactants.* As the concentrations of reactants increase, the rates of chemical reactions tend to increase because reactant particles collide more frequently.
3. *Temperature.* As temperature increases, the average kinetic energy (KE) of particles increases which leads to an increase in reaction rate due to more frequent and forceful collisions between particles.

**CONNECTION** We discussed kinetic molecular theory, a model that describes the behavior of gases, in Section 9.2.

**CHEMTOUR**
Reaction Rate

**CHEMTOUR**
Collision Theory

**CONNECTION** We discussed translational (and vibrational and rotational) motion in Section 10.1.

**chemical kinetics** the study of the rates of chemical reactions and the factors that influence reaction rates.

**collision theory** a model that describes how collisions between gas-phase atoms or molecules cause chemical reactions to occur.

**reaction rate** a measure of how rapidly a reaction occurs; it is related to rates of change in the concentrations of reactants and products over time.

4. *Surface area.* The larger the surface area of a substance, the more frequent the collisions between the reactants and those in another phase, and therefore the faster the reaction between them. For example, a large piece of iron rusts slowly as it reacts with atmospheric oxygen, but an equal mass of iron filings react so rapidly that they can actually burn.
5. *Catalysts.* Catalytic converters contain materials that increase the rates of reactions that consume pollutants in automobile exhaust. In general, catalysts are materials or substances that accelerate reactions without themselves being consumed in the process. Catalysts also play a key role in biochemistry, where proteins called enzymes catalyze most reactions in living systems.

With these five factors in mind, let's return to the formation of NO in internal combustion engines:

$$N_2(g) + O_2(g) \rightarrow 2\ NO(g) \tag{13.1}$$

How much NO is in the exhaust from an automobile engine? Combustion reaction gases spend very little time inside the engine, so it depends on how rapidly the reaction in Equation 13.1 proceeds. We express **reaction rate** as the change in the concentration of a product, such as NO, or a reactant, such as $N_2$ or $O_2$, that occurs over some interval of time.

The rate of change in the concentration of a reactant ($\Delta$[reactant]/$\Delta t$) has a negative value because reactant concentrations decrease as a reaction proceeds. Therefore, reactant concentration at the end of the interval ($[\text{reactant}]_{\text{final}}$) is always less than it was at the beginning ($[\text{reactant}]_{\text{initial}}$). Therefore, $\Delta[\text{reactant}] = [\text{reactant}]_{\text{final}} - [\text{reactant}]_{\text{initial}}$ must have a negative value, and so must $\Delta$[reactant]/$\Delta t$. However, the rate of any reaction is defined as a positive quantity because it describes the rate at which reactants form products, so a minus sign is used with $\Delta$[reactant]/$\Delta t$ values to obtain an overall positive value for reaction rate. Thus, the rate at which $N_2$ and $O_2$ combine to form NO can be expressed as follows:

$$\text{Reaction rate} = (-)\ \text{the rate of change in}\ [N_2] = -\frac{\Delta[N_2]}{\Delta t} \tag{13.8}$$

We use brackets around $N_2$ to represent its concentration in moles per liter—a practice we will follow in the remaining chapters of this book. Reaction rates are usually expressed in units of concentration per time, as in the fraction on the right in Equation 13.8. A commonly encountered example is molarity per second (*M*/s).

Nitrogen and oxygen gas are consumed at the same rate in the balanced reaction in Equation 13.1 because their stoichiometric coefficients are both 1. However, two moles of NO form for every one mole of $N_2$ or $O_2$ consumed. Therefore, the rate of increase in the concentration of NO is twice the rate of decrease in the concentrations of $N_2$ and $O_2$. These relative rates of change can be expressed using the following equation:

$$\frac{\Delta[NO]}{\Delta t} = -2\frac{\Delta[N_2]}{\Delta t} = -2\frac{\Delta[O_2]}{\Delta t} \tag{13.9}$$

The negative signs are needed to convert the rates at which $[N_2]$ and $[O_2]$ are decreasing into positive values. When these values are multiplied by 2, they equal the rate at which [NO] is increasing. Now let's divide all the terms in Equation 13.9 by 2 and rearrange them so that they match the sequence of reactants and products in Equation 13.1:

$$-\frac{\Delta[N_2]}{\Delta t} = -\frac{\Delta[O_2]}{\Delta t} = \frac{1}{2}\frac{\Delta[NO]}{\Delta t} \tag{13.10}$$

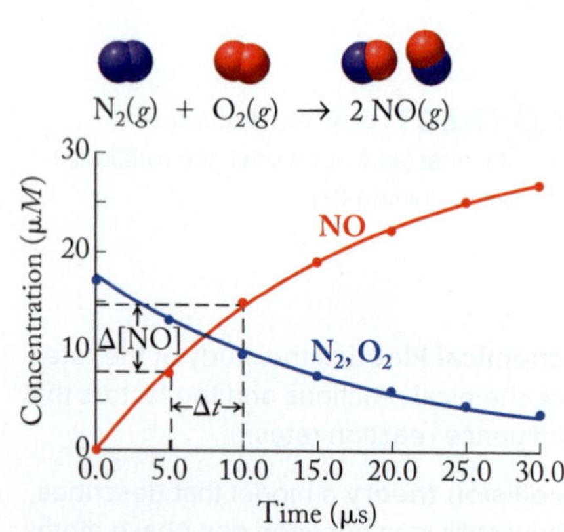


**FIGURE 13.3** Concentrations of $N_2$, $O_2$, and NO over 30.0 μs for the reaction $N_2(g) + O_2(g) \rightarrow 2\ NO(g)$.

Thus, the concentrations of $N_2$ and $O_2$ decrease at half the rate at which the concentration of NO increases. The curves in **Figure 13.3** depict the changing concentrations of these compounds as the reaction proceeds.

Note how the coefficient of 2 in front of NO in the chemical equation has become its reciprocal, $\frac{1}{2}$, in Equation 13.10. This pattern holds for all reactions: the coefficients in an equation expressing the relative rates of change of reactants and products are the reciprocals of the coefficients in a balanced chemical equation describing the reaction. To give all the terms positive values, the rates representing decreasing reactant concentrations receive minus signs. Sample Exercise 13.1 provides practice in applying this approach to the changing concentrations of reactants and product in an important industrial process: the synthesis of ammonia.

**SAMPLE EXERCISE 13.1** Predicting Relative Rates of Concentration Change in a Chemical Reaction **LO2**

Each year, about 130 million metric tons ($1.3 \times 10^{11}$ kg) of ammonia is produced worldwide, mostly for use as fertilizer in agriculture. The process for making ammonia is based on the following reaction:

$$N_2(g) + 3\ H_2(g) \rightarrow 2\ NH_3(g)$$

How is the rate of formation of $NH_3$ related to the rates of consumption of $N_2$ and $H_2$?

**Collect, Organize, and Analyze** The coefficients in an equation relating the rates of change in the concentrations of $N_2$, $H_2$, and $NH_3$ are the reciprocals of their coefficients (1, 3, and 2, respectively) in the balanced chemical equation. Nitrogen and hydrogen are consumed in the reaction, so their rate terms are preceded by minus signs.

**Solve** We insert the reciprocals of 1, 3, and 2 in front of the rates of change in the concentrations of $N_2$, $H_2$, and $NH_3$, and we place minus signs before the two reactant terms. This gives us the relative rates:

$$-\frac{\Delta[N_2]}{\Delta t} = -\frac{1}{3}\frac{\Delta[H_2]}{\Delta t} = \frac{1}{2}\frac{\Delta[NH_3]}{\Delta t}$$

**Think About It** The balanced equation indicates that two moles of $NH_3$ are formed for each mole of $N_2$ consumed. This means that the rate of consumption of $N_2$ should be half the rate of formation of ammonia, which is what the first and last terms in the answer tell us. Similarly, we know that three moles of $H_2$ are consumed for each mole of $N_2$ consumed. Therefore, the rate at which $N_2$ is consumed is one-third the rate at which $H_2$ is consumed, as described in the first and second terms of the answer.

**Practice Exercise**
In the reaction $2\ CO(g) + O_2(g) \rightarrow 2\ CO_2(g)$,

a. Which reactant (CO or $O_2$) is consumed at the higher rate during the oxidation of carbon monoxide?
b. How is the rate of change in the concentration of $CO_2$ related to the rate of change in the concentration of $O_2$?

*(Answers to Practice Exercises are in the back of the book.)*

## Reaction Rate Values

Up to this point, our focus has been on the relative rates at which reactants are consumed and products are formed in chemical reactions. We have seen how these rates differ depending on the values of the coefficients in balanced chemical equations—that is, depending on reaction stoichiometry. The questions to answer now are the following:

- How do we know the actual rates at which concentrations change in a chemical reaction?
- Which of these rates do we use to express the rate of the reaction?

The answer to the first question is that rates of change are determined experimentally. Suppose, for example, that analysis of the gas mixture in the ammonia synthesis in Sample Exercise 13.1 reveals that, over a 10.0-second time interval, the concentration of ammonia increases from 0.133 *M* to 0.605 *M*. These data mean that the average rate of change in ammonia concentration over the interval is

$$\frac{\Delta[NH_3]}{\Delta t} = \frac{(0.605 - 0.133)M}{10.0 \text{ s}} = 0.0472 \; M/\text{s}$$

To calculate the rate of change in the concentration of $N_2$, we use the equation we derived in the exercise:

$$-\frac{\Delta[N_2]}{\Delta t} = \frac{1}{2}\frac{\Delta[NH_3]}{\Delta t} = \frac{1}{2}(0.0472 \; M/\text{s})$$

$$\frac{\Delta[N_2]}{\Delta t} = -0.0236 \; M/\text{s}$$

Which of these two values should we use to express the rate of the reaction? Both values work if we have referred to the compound in the chemical equation whose concentration was monitored as a function of time. However, it is conventional to use the rate of change of the reactant or product with a coefficient of 1 in the chemical equation—here, $N_2$—as the basis for expressing the rate of the overall reaction:

$$\text{Rate} = -\frac{\Delta[N_2]}{\Delta t} = 0.0236 \; M/\text{s}$$

If all the coefficients in a chemical equation are greater than 1, then we can represent the rate of the reaction in terms of the rate of change in the concentration of any chosen reactant or product. To do so, we multiply the rate by the reciprocal of that substance's coefficient in the balanced chemical equation describing the reaction. For example, the rate of the reaction $2\,O_3(g) \rightarrow 3\,O_2(g)$ can be expressed as the rate of change of either $[O_3]$ or $[O_2]$:

$$\text{Rate} = \frac{1}{3}\frac{\Delta[O_2]}{\Delta t} = -\frac{1}{2}\frac{\Delta[O_3]}{\Delta t}$$

**SAMPLE EXERCISE 13.2** Calculating Rates of Change in Reactant and Product Concentrations by Using Stoichiometric Ratios **LO2**

In a high-temperature reaction between NO and $H_2$:

$$2\,NO(g) + 2\,H_2(g) \rightarrow N_2(g) + 2\,H_2O(g)$$

the average rate of change in the concentration of NO is $-6.0 \times 10^{-5}\; M\,s^{-1}$. What are the average rates of change in the concentrations of the other three substances in the reaction mixture?

**Collect, Organize, and Analyze** We have the balanced chemical equation for a reaction and the average rate of change in concentration of one reactant. The stoichiometry enables us to relate the consumption of reactants (NO and $H_2$) to the formation of products ($N_2$ and $H_2O$).

**Solve** For every two moles of NO consumed in the reaction, two moles of $H_2$ are also consumed, and one mole of $N_2$ and two moles of $H_2O$ are generated. Inserting the reciprocals of 2, 2, 1, and 2 in front of the rates of change in the concentrations of NO, $H_2$, $N_2$, and $H_2O$ and placing minus signs before the two reactant terms, we obtain the relative rates:

$$\text{Rate} = -\frac{1}{2}\frac{\Delta[\text{NO}]}{\Delta t} = -\frac{1}{2}\frac{\Delta[\text{H}_2]}{\Delta t} = \frac{\Delta[\text{N}_2]}{\Delta t} = \frac{1}{2}\frac{\Delta[\text{H}_2\text{O}]}{\Delta t}$$

The rates of change in the concentrations of all four species are

$$\frac{\Delta[\text{NO}]}{\Delta t} = -6.0 \times 10^{-5}\ M\,\text{s}^{-1}$$

$$-\frac{1}{2}\frac{\Delta[\text{H}_2]}{\Delta t} = -\frac{1}{2}\frac{\Delta[\text{NO}]}{\Delta t} = -6.0 \times 10^{-5}\ M\,\text{s}^{-1}$$

$$\frac{\Delta[\text{N}_2]}{\Delta t} = -\frac{1}{2}\frac{\Delta[\text{NO}]}{\Delta t} = -\frac{1}{2} \times (-6.0 \times 10^{-5}\ M\,\text{s}^{-1}) = 3.0 \times 10^{-5}\ M\,\text{s}^{-1}$$

$$\frac{1}{2}\frac{\Delta[\text{H}_2\text{O}]}{\Delta t} = -\frac{1}{2}\frac{\Delta[\text{NO}]}{\Delta t} = -(-6.0 \times 10^{-5}\ M\,\text{s}^{-1}) = 6.0 \times 10^{-5}\ M\,\text{s}^{-1}$$

**Think About It** Reactants are consumed during a reaction, so the rates of change in their concentrations have negative signs. However, product concentrations increase, so their rates of change are positive. Three of the rates have the same magnitude and the fourth is half that magnitude, which is consistent with the stoichiometry of the balanced equation.

**Practice Exercise** Consider the following reaction:

$$8\,\text{A}(g) + 5\,\text{B}(g) \rightarrow 8\,\text{C}(g) + 6\,\text{D}(g)$$

If [C] increases at an average rate of 4.0 $M\,\text{s}^{-1}$, what are the average rates at which the concentrations of the other species change?

## Average and Instantaneous Reaction Rates

Suppose we run an experiment to determine the rate of formation of NO in an automobile engine. In the laboratory, we use a reaction vessel as hot as the combustion chambers in the engine and we monitor the concentrations of reactants and products over time, obtaining the data in **Table 13.1**, which correspond to the graph in Figure 13.3. Let's calculate the average rate of formation of NO between 5.0 μs and 10.0 μs:

$$\frac{\Delta[\text{NO}]}{\Delta t} = \frac{[\text{NO}]_{10.0\ \mu s} - [\text{NO}]_{5.0\ \mu s}}{t_{10.0\ \mu s} - t_{5.0\ \mu s}} = \frac{(14.8 - 7.8)\ \mu M}{(10.0 - 5.0)\ \mu s} = 1.4\ M/\text{s}$$

During the same interval, the average rate of change in the concentration of $N_2$ (and $O_2$) is

$$\frac{\Delta[\text{N}_2]}{\Delta t} = \frac{[\text{N}_2]_{10.0\ \mu s} - [\text{N}_2]_{5.0\ \mu s}}{t_{10.0\ \mu s} - t_{5.0\ \mu s}} = \frac{(9.6 - 13.1)\ \mu M}{(10.0 - 5.0)\ \mu s} = -0.70\ M/\text{s}$$

These results give us two values on which to base the average rate of the reaction. In this example, we select the rate of change in $[N_2]$ or $[O_2]$ because they both have a coefficient of 1 in the balanced chemical equation. Therefore, the average rate of this reaction between 5.0 μs and 10.0 μs is

$$\text{Rate} = -\frac{\Delta[\text{N}_2]}{\Delta t} = -(-0.70\ M/\text{s}) = 0.70\ M/\text{s}$$

**TABLE 13.1** Changing Concentrations of Reactants and Products for the Reaction $N_2(g) + O_2(g) \rightarrow 2\,NO(g)$

| Time (μs) | $[N_2]$, $[O_2]$ (μ*M*) | [NO] (μ*M*) |
|---|---|---|
| 0 | 17.0 | 0.0 |
| 5.0 | 13.1 | 7.8 |
| 10.0 | 9.6 | 14.8 |
| 15.0 | 7.6 | 18.6 |
| 20.0 | 5.8 | 22.2 |
| 25.0 | 4.5 | 24.8 |
| 30.0 | 3.6 | 26.7 |

The curvature of the lines in Figure 13.3 tells us that this average rate value applies only to that 5.0-μs time interval. Any other time interval has a different average rate. For instance, from $t = 25.0$ μs to $t = 30.0$ μs, the rate of the reaction is

$$-\frac{\Delta[N_2]}{\Delta t} = -\frac{[N_2]_{30.0\ \mu s} - [N_2]_{25.0\ \mu s}}{t_{30.0\ \mu s} - t_{25.0\ \mu s}} = -\frac{(3.6 - 4.5)\ \mu M}{(30.0 - 25.0)\ \mu s} = 0.18\ M/s$$

We can also determine the *instantaneous* rate of a reaction—that is, the rate of the reaction at a time after it began. The difference between average and instantaneous reaction rates is analogous to the difference between the average and instantaneous speeds of a runner. If a competitor in a marathon runs from mile 10 to mile 20 in 1 hour, her average speed over that distance was 10 mph. At a given instant during the run, however, her instantaneous speed could have been 12 mph while going downhill or 8 mph while going uphill.

Let's again consider the exothermic conversion of NO into $NO_2$ in the atmosphere:

$$2\ NO(g) + O_2(g) \rightarrow 2\ NO_2(g) \qquad \Delta H^\circ_{rxn} = -114.2\ kJ \quad (13.2)$$

The rate of this reaction based on the rate of consumption of $O_2$ is described by the data in **Table 13.2**, which are plotted in **Figure 13.4**. One way to estimate the instantaneous rate of the reaction is to draw a tangent to the $[O_2]$ curve at a point in time and determine the slope of the tangent. Figure 13.4 illustrates how to do this at $t = 2000.0$ s. In this example, we select two convenient points along the tangent, at $t = 1000.0$ and 3000.0 s; determine the corresponding $[O_2]$ values at those times; and then use those time and concentration values to calculate the instantaneous reaction rate:

$$\text{Slope (at } t = 2000.0\text{ s)} = \frac{\Delta[O_2]}{\Delta t} = \frac{(0.0072 - 0.0084)\ M}{(3000.0 - 1000.0)\ s}$$
$$= -6.0 \times 10^{-7}\ M\ s^{-1}$$

$$\text{Instantaneous rate (at } t = 2000.0\text{ s)} = -\frac{\Delta[O_2]}{\Delta t} = -(-6.0 \times 10^{-7}\ M\ s^{-1})$$
$$= 6.0 \times 10^{-7}\ M\ s^{-1}$$

**TABLE 13.2** Concentrations of NO, $O_2$, and $NO_2$ as a Function of Time

| REACTION: $2\ NO(g) + O_2(g) \rightarrow 2\ NO_2(g)$ AT 25°C | | | |
|---|---|---|---|
| Time (s) | [NO] (*M*) | $[O_2]$ (*M*) | $[NO_2]$ (*M*) |
| 0.0 | 0.0100 | 0.0100 | 0.0000 |
| 285.0 | 0.0090 | 0.0095 | 0.0010 |
| 660.0 | 0.0080 | 0.0090 | 0.0020 |
| 1175.0 | 0.0070 | 0.0085 | 0.0030 |
| 1895.0 | 0.0060 | 0.0080 | 0.0040 |
| 2975.0 | 0.0050 | 0.0075 | 0.0050 |
| 4700.0 | 0.0040 | 0.0070 | 0.0060 |
| 7800.0 | 0.0030 | 0.0065 | 0.0070 |

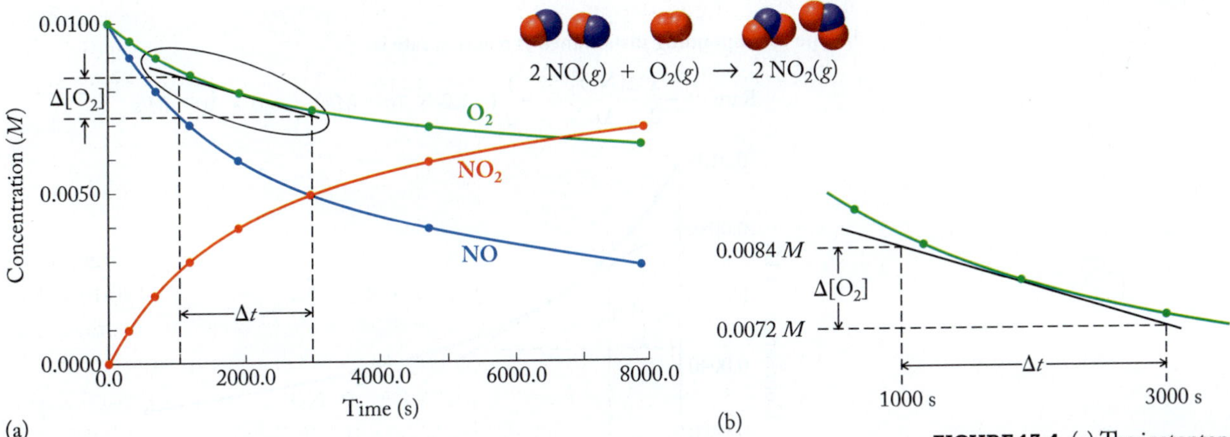


**FIGURE 13.4** (a) The instantaneous rate of change in [$O_2$] in the reaction 2 NO(*g*) + $O_2$(*g*) → 2 $NO_2$(*g*) is equal to the slope of a tangent to the curve of [$O_2$] versus time. (b) An expanded view of the instantaneous rate of change in [$O_2$] at $t = 2000$ s. Data from Table 13.2.

Note that we use data points along the tangent line, not along the curve, to calculate instantaneous reaction rate.[1]

## CONCEPT TEST

Which of the following statements is true about the instantaneous rate of the chemical reaction A → B as the reaction progresses? Assume that the rate of the reaction decreases as [A] decreases.

a. $-\Delta[A]/\Delta t$ increases, $\Delta[B]/\Delta t$ decreases

b. $-\Delta[A]/\Delta t$ decreases, $\Delta[B]/\Delta t$ increases

c. $-\Delta[A]/\Delta t$ and $\Delta[B]/\Delta t$ both increase

d. $-\Delta[A]/\Delta t$ and $\Delta[B]/\Delta t$ both decrease

## SAMPLE EXERCISE 13.3 Determining an Instantaneous Reaction Rate LO3

a. What is the instantaneous rate of change of [NO] at $t = 2000.0$ s in the experiment that produced the data in Table 13.2?

b. What is the rate of the reaction on the basis of your result in part (a)?

**Collect, Organize, and Analyze** We are asked to determine the instantaneous rate of change of [NO] and the corresponding reaction rate at $t = 2000.0$ s by using the data in Table 13.2. The coefficient of NO is 2 in the balanced chemical equation: 2 NO(*g*) + $O_2$(*g*) → 2 $NO_2$(*g*).

The corresponding reaction rate will be (−1/2) the value of the instantaneous rate of change of [NO].

**Solve**

a. First, we plot [NO] versus time (**Figure 13.5**) and draw a tangent to the curve at $t = 2000.0$ s. Choosing two points along the tangent, $t = 1000.0$ s and 3000.0 s, we determine the concentrations corresponding to those times along the vertical axis and use those values to calculate the slope of the line:

$$\frac{\Delta[\text{NO}]}{\Delta t} = \frac{(0.0046 - 0.0070)\ M}{(3000.0 - 1000.0)\ \text{s}} = -1.2 \times 10^{-6}\ M/\text{s}$$

[1]If you have studied calculus, you may recall that the average rate of a reaction approaches the instantaneous rate as $\Delta t$ approaches zero. Using calculus, we would say $\lim_{\Delta t \to 0} \Delta[O_2]/\Delta t = d[O_2]/dt$. The slope of a tangent to a curve is the derivative of the curve at a given point and can be calculated using many scientific calculators.

b. The corresponding instantaneous reaction rate is

$$\text{Rate} = -\frac{1}{2}\frac{\Delta[\text{NO}]}{\Delta t} = -\frac{1}{2}(-1.2 \times 10^{-6}\ M/\text{s}) = 6.0 \times 10^{-7}\ M/\text{s}$$

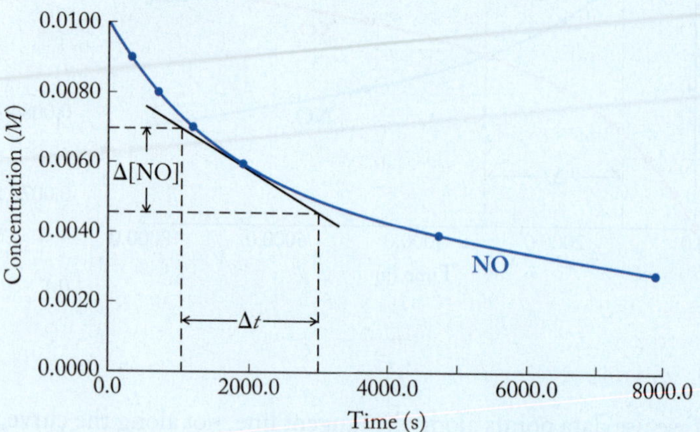


**FIGURE 13.5** Plot of concentration versus time, using data from Table 13.2.

**Think About It** Note that the instantaneous reaction rate calculated in this Sample Exercise is the same as the rate we calculated earlier based on the rate of change of $[O_2]$. This helps confirm the accuracy of the two determinations of reaction rate.

**Practice Exercise** What is the instantaneous rate of change in $[NO_2]$ at $t = 2000.0$ s in the experiment that produced the data in Table 13.2?

## 13.3 Effect of Concentration on Reaction Rate

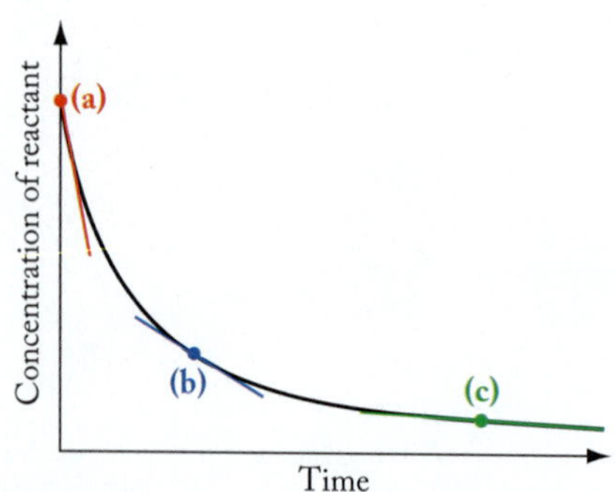


**FIGURE 13.6** Typical plot of reactant concentration as a function of time: (a) tangent at $t = 0$; (b) tangent at the midpoint of the reaction; (c) tangent close to the end of the reaction.

**Figure 13.6** shows a typical plot of reactant concentration versus time. Tangents are drawn to the line at three points: (a), at the instant the reaction begins; (b), at an instant when the reaction is about halfway to completion; and (c), when the reaction is nearly over. The slope of the tangent at point (a) defines the *initial* rate of the reaction, which is the rate observed at the instant the reactants are mixed at $t = 0$. The slopes of the other tangents become less negative as the reaction proceeds, which means that the reaction is slowing down. Eventually the slopes of the tangents and the rate of the reaction approach zero. This behavior is typical of most reactions.

Collision theory explains this trend of decreasing reaction rates with time. If gas-phase reactions take place because of collisions between reactant molecules, then the more reactant molecules there are in a given reaction space, the more collisions per unit time and the more opportunities there are for reactants to turn into products. As reactants are consumed and their concentrations decrease, the frequency of collisions between reactant molecules decreases, and the reaction slows down.

### Reaction Order and Rate Constants

Reaction Order

Experimental observations and theoretical considerations tell us that reaction rates depend on reactant concentrations. However, they do not tell us to what extent rates depend on reactant concentrations. For example, if the concentration of a reactant decreases by half, does the reaction rate also decrease by half? The answer to this question is linked to a parameter called **reaction order,** which tells us how reaction rates change with changing reactant concentrations. A key point here is that reaction order must be determined experimentally. There is no guarantee that the reaction order is linked to the coefficient of the reactant in the balanced chemical equation. The balanced chemical equation tells us which substances react and are formed, and in what ratios, but it does not tell us which molecules collide in order for bonds to break and new bonds to form.

To understand what reaction order means quantitatively, let's examine the chemical reaction that occurs between two components of smog, $O_3$ and NO:

$$O_3(g) + NO(g) \rightarrow O_2(g) + NO_2(g)$$

Suppose different concentrations of $O_3$ and NO are injected into a reaction vessel at 298 K and the initial rates of reaction for each mixture are determined based on the instantaneous rate at $t = 0$. **Table 13.3** contains the results of three determinations of initial reaction rate for this reaction. Note that we use a subscript 0 to denote the initial concentrations of the reactants.

In experiment 1, the initial concentrations of $O_3$ and NO are both 0.0010 *M* and the initial rate is 11 *M*/s. In experiment 2, the value $[O_3]_0$ is double that in experiment 1, but $[NO]_0$ is the same as in experiment 1. Comparing the reaction rates in experiments 2 and 1, we see that doubling $[O_3]_0$ doubles the initial reaction rate from 11 *M*/s to 22 *M*/s. In experiment 3, the value of $[O_3]_0$ is the same as in experiment 1, but $[NO]_0$ has doubled and so has the initial reaction rate.

These results tell us that the rate of the reaction is directly proportional to the concentration of $O_3$ and that it is directly proportional to the concentration of NO. Expressed another way, reaction rate is a function of the reactants' concentrations each raised to the first power. In the language of chemical kinetics, this dependence means that the reaction is first order in $O_3$ and first order in NO. These two concentration dependencies of reaction rate can be expressed mathematically using Equation 13.11:

$$\text{Rate} = k[O_3][NO] \qquad (13.11)$$

where $k$ is called the **rate constant** of the reaction and the entire equation is called the **rate law** of the reaction. Also, the first-order dependence on two reactants means that this reaction is second order overall. The **overall reaction order** is the sum of the powers in the rate law, which here is $1 + 1 = 2$.

We can calculate the value of $k$ in Equation 13.11 by inserting the concentration and reaction rate data from any of the three experiments into the equation. Using the data from experiment 1:

$$\text{Rate} = k[O_3][NO] = 11\ M/\text{s} = k(0.0010\ M)^2$$

$$k = \frac{11\ \cancel{M}/\text{s}}{(0.0010\ \cancel{M})(0.0010\ M)} = 1.1 \times 10^7\ M^{-1}\,\text{s}^{-1}$$

**reaction order** an experimentally determined number defining the dependence of the reaction rate on the concentration of a reactant.

**rate constant** the proportionality constant that relates the rate of a reaction to the concentrations of reactants.

**rate law** an equation that defines the experimentally determined relation between the concentration of reactants in a chemical reaction and the rate of that reaction.

**overall reaction order** the sum of the exponents of the concentration terms in the rate law.

**TABLE 13.3** Effect of Reactant Concentrations on Initial Reaction Rates for $O_3(g) + NO(g) \rightarrow O_2(g) + NO_2(g)$

| Experiment | $[O_3]_0$ (*M*) | $[NO]_0$ (*M*) | Initial Reaction Rate (*M*/s) |
|---|---|---|---|
| 1 | 0.0010 | 0.0010 | 11 |
| 2 | 0.0020 | 0.0010 | 22 |
| 3 | 0.0010 | 0.0020 | 22 |

Using this value of $k$ in Equation 13.11 gives us a rate law for the reaction at 298 K:

$$\text{Rate} = 1.1 \times 10^7\ M^{-1}\ \text{s}^{-1}\ [O_3][NO]$$

Now let's write a generic rate law expression for any chemical reaction with two reactants, A and B:

$$A + B \rightarrow C$$

The rate law for this reaction may be written

$$\text{Rate} = k[A]^m[B]^n \qquad (13.12)$$

where $m$ is the reaction order with respect to A, and $n$ is the reaction order with respect to B.

Perhaps you are wondering why the two concentration terms in Equation 13.12 are multiplied together and not added together, given that the chemical equation is written as A + B. A molecular view of the reason is provided in **Figure 13.7,** which shows how different numbers of NO and $O_3$ molecules in a reaction vessel might collide with each other during the reaction we explored above. Kinetics studies of this reaction have shown that it happens due to collisions between NO and $O_3$ molecules (unlike the high-temperature reaction between $N_2$ and $O_2$ in Equation 13.1, which we noted is *not* due to collisions between $N_2$ and $O_2$ molecules). Look closely at how increasing the numbers of molecules in Figure 13.7(a)–(e) produces increasing numbers of collisions that are proportional to the *product* of the number of molecules of each reactant in the reaction mixture. Increasing the number of molecules of each type is equivalent to increasing the concentrations of the two gases. Therefore, the rate of the reaction should depend on the product of the concentrations of NO and $O_3$.

## CONCEPT TEST

If there had been data from a fourth experiment in Table 13.3 in which the initial concentrations of $O_3$ and NO had both been 0.0020 *M*, what would the initial rate of the reaction have been?

The data in Table 13.3 made it relatively easy to see the dependence of reaction rate on reactant concentrations. This is not always the case. Consider, for example, the results in **Table 13.4** from a kinetics study of Equation 13.2, another important reaction in atmospheric chemistry:

$$2\ NO(g) + O_2(g) \rightarrow 2\ NO_2(g)$$

The relationships between reaction rate and the reactants' concentrations in these data are not as evident as they were in the Table 13.3 data. So, to determine

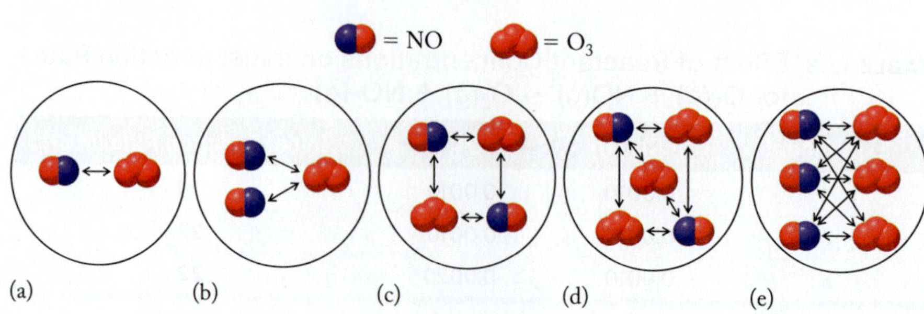


**FIGURE 13.7** Increasing the concentration increases the number of possible effective collisions (double-headed arrows) and therefore the number of potential reaction events. Reaction rate depends on the number of collisions between molecules, which are shown for the reaction between NO and ozone ($O_3$) that produces $NO_2$ and $O_2$. (a) With only one NO molecule and one $O_3$ molecule, each molecule can collide only with the other, giving a relative reaction rate of $1 \times 1 = 1$. (e) Three molecules of NO can collide with three molecules of $O_3$ for a relative reaction rate of $3 \times 3 = 9$ times the relative rate in (a).

**TABLE 13.4** **Effect of Reactant Concentrations on Initial Reaction Rates for $2\ NO(g) + O_2(g) \rightarrow 2\ NO_2(g)$**

| Experiment | $[NO]_0$ (*M*) | $[O_2]_0$ (*M*) | Initial Reaction Rate (*M*/s) |
|---|---|---|---|
| 1 | 0.025 | 0.025 | $1.6 \times 10^{-5}$ |
| 2 | 0.025 | 0.066 | $4.1 \times 10^{-5}$ |
| 3 | 0.050 | 0.025 | $6.2 \times 10^{-5}$ |

reaction order, we take a more mathematical approach, inserting the concentrations of the reactants into Equation 13.12:

$$\text{Rate} = k[NO]^m[O_2]^n \qquad (13.13)$$

Our task is to determine the values of $m$ and $n$ from the data in Table 13.4. We start by comparing the results of experiments 1 and 2, in which $[NO]_0$ is unchanged between experiments but $[O_2]_0$ and the initial reaction rate are both higher in experiment 2. The ratio of the initial reaction rates in experiments 2 and 1, $\text{Rate}_2/\text{Rate}_1$, is related to the ratio of the concentrations of $O_2$ in experiments 2 and 1, $[O_2]_2/[O_2]_1$, and to the dependence of reaction rate on $[O_2]$, that is, on the value of $n$. Expressing this relationship in equation form and then making $n$ a coefficient by taking the logarithm of both sides:

$$\frac{\text{Rate}_2}{\text{Rate}_1} = \left(\frac{[O_2]_2}{[O_2]_1}\right)^n \quad \text{or} \quad \log\left(\frac{\text{Rate}_2}{\text{Rate}_1}\right) = n \log\left(\frac{[O_2]_2}{[O_2]_1}\right)$$

Rearranging the terms and inserting the appropriate values from experiments 1 and 2:

$$n = \frac{\log\left(\frac{\text{Rate}_2}{\text{Rate}_1}\right)}{\log\left(\frac{[O_2]_2}{[O_2]_1}\right)} = \frac{\log \frac{4.1 \times 10^{-5}\ \cancel{M/s}}{1.6 \times 10^{-5}\ \cancel{M/s}}}{\log \frac{0.066\ \cancel{M}}{0.025\ \cancel{M}}} = \frac{0.409}{0.422} = 0.97, \text{ or about } 1$$

The key to interpreting this value of $n$ is to recognize that it is very close to 1, which means that the reaction is first order in $O_2$.

Taking a similar approach with the data from experiments 1 and 3, in which the initial concentrations of $O_2$ are the same but those of NO differ, we calculate the value of $m$:

$$m = \frac{\log\left(\frac{\text{Rate}_3}{\text{Rate}_1}\right)}{\log\left(\frac{[NO]_3}{[NO]_1}\right)} = \frac{\log \frac{6.2 \times 10^{-5}\ \cancel{M/s}}{1.6 \times 10^{-5}\ \cancel{M/s}}}{\log \frac{0.050\ \cancel{M}}{0.025\ \cancel{M}}} = \frac{0.588}{0.301} = 1.95, \text{ or about } 2$$

Therefore, the reaction is second order in NO, which gives us the following rate law expression:

$$\text{Rate} = k[NO]^2[O_2]$$

Using the data from experiment 3 to solve for $k$:

$$\text{Rate} = k[NO]^2[O_2] = 6.2 \times 10^{-5}\ M/\text{s} = k(0.050\ M)^2(0.025\ M)$$

$$k = \frac{6.2 \times 10^{-5}\ \cancel{M}/\text{s}}{(0.050\ M)^2(0.025\ \cancel{M})} = 0.99\ M^{-2}\ \text{s}^{-1}$$

Therefore, the rate law for the reaction at 298 K is

$$\text{Rate} = 0.99\ M^{-2}\ \text{s}^{-1}\ [NO]^2[O_2]$$

Thus, the reaction of NO with $O_2$ is second order in NO, first order in $O_2$, and third order overall. The procedure we followed here can be used to calculate the order ($m$) of any reaction with respect to any reactant (A) whose concentration differs in a pair of reaction rate experiments:

$$m = \frac{\log\left(\frac{\text{Rate}_2}{\text{Rate}_1}\right)}{\log\left(\frac{[\text{A}]_2}{[\text{A}]_1}\right)} \quad (13.14)$$

In the preceding exercise, calculated exponent values that were very close to 1 and 2 were rounded to these values given the uncertainty in the data used to calculate them. However, this does not mean that the values of concentration exponents in rate law equations are always whole numbers, as we will see in Sample Exercise 13.4. They may be fractions, zero, or, in rare cases, negative. It is important to remember that rate laws and reaction orders are different from relative rates in that they must be determined experimentally. They cannot be predicted from the coefficients in a balanced chemical equation. The significance of reaction orders in describing how reactions take place is explored further in Section 13.5.

The values of the rate constants calculated using the data in Tables 13.3 and 13.4 are unique to these reactions and to the temperature at which the experiments were run. Keep in mind that reaction rates depend on the concentrations of reactants, but rate constants do not. However, rate constants do change with changing temperature. The units of reaction rates are usually expressed in *M*/s, but the units on a rate constant depend on the overall order of the reaction. The $k$ value we calculated for the data in Table 13.4 had units of $M^{-2}\ s^{-1}$ for a reaction that was third order overall, and the $k$ value for the data in Table 13.3 had units of $M^{-1}\ s^{-1}$ for a reaction that was second order overall. Do you see a pattern in these units and reaction order? If so, what are the units of $k$ for a first-order reaction? If you said $s^{-1}$, you are right. The dependence of the units of $k$ on overall reaction order are summarized in **Table 13.5**.

**TABLE 13.5** Relationship between Overall Reaction Order and Units of the Rate Constant $k$

| Overall Reaction Order | Units of $k$ |
|---|---|
| 1 | $s^{-1}$ |
| 2 | $M^{-1}\ s^{-1}$ |
| 3 | $M^{-2}\ s^{-1}$ |
| 4 | $M^{-3}\ s^{-1}$ |

**SAMPLE EXERCISE 13.4** Deriving a Rate Law from Initial Reaction Rate Data — **LO4**

Write the rate law for Equation 13.1, the reaction of $N_2$ with $O_2$,

$$N_2(g) + O_2(g) \rightarrow 2\ NO(g)$$

using the data in **Table 13.6**. Determine the overall reaction order and the value of the rate constant.

**TABLE 13.6** Initial Reaction Rates for $N_2(g) + O_2(g) \rightarrow 2\ NO(g)$

| Experiment | $[N_2]_0$ (*M*) | $[O_2]_0$ (*M*) | Initial Reaction Rate (*M*/s) |
|---|---|---|---|
| 1 | 0.040 | 0.020 | 701 |
| 2 | 0.040 | 0.010 | 496 |
| 3 | 0.010 | 0.010 | 124 |

**Collect and Organize** We are asked to determine the rate law and rate constant for a reaction from initial reaction rate data (note the subscript zero on the concentration terms in Table 13.6) for each of three sets of initial reactant concentrations. Equation 13.14 can be used to derive the reaction order with respect to each of the reactants from initial reaction rate data.

**Analyze** The generic rate law expression (Equation 13.12) for this reaction is

$$\text{Rate} = k[N_2]^m[O_2]^n$$

The overall order of the reaction is the sum of the orders with respect to the individual reactants. Once we have established the rate law, the rate constant can be calculated using concentrations of reactants from any single row in Table 13.6. The rate constant for the reaction must have units that express the reaction rate in $M\,s^{-1}$.

**Solve** To determine the value of $m$, we select a pair of experiments with the same $O_2$ and different $N_2$ concentrations, such as experiments 2 and 3, and insert rate and $[N_2]$ values into Equation 13.14:

$$m = \frac{\log\left(\frac{\text{Rate}_2}{\text{Rate}_3}\right)}{\log\left(\frac{[N_2]_2}{[N_2]_3}\right)} = \frac{\log\left(\frac{496\ \cancel{M}/s}{124\ \cancel{M}/s}\right)}{\log\left(\frac{0.040\ \cancel{M}}{0.010\ \cancel{M}}\right)} = \frac{0.602}{0.602} = 1.0$$

To calculate $n$, we use experiments 1 and 2 because they have different $[O_2]$ values but the same $[N_2]$ values.

$$n = \frac{\log\left(\frac{\text{Rate}_1}{\text{Rate}_2}\right)}{\log\left(\frac{[O_2]_1}{[O_2]_2}\right)} = \frac{\log\left(\frac{701\ \cancel{M}/s}{496\ \cancel{M}/s}\right)}{\log\left(\frac{0.020\ \cancel{M}}{0.010\ \cancel{M}}\right)} = \frac{0.150}{0.301} = 0.50$$

Thus, the reaction is one-half order with respect to $O_2$, first order with respect to $N_2$, and $(\frac{1}{2} + 1) = \frac{3}{2}$ order overall:

$$\text{Rate} = k[N_2][O_2]^{1/2}$$

We can use the data from any one of the experiments to obtain the value of $k$. Let's use experiment 1:

$$\text{Rate} = k[N_2][O_2]^{1/2} = 701\ M/s = k(0.040\ M)(0.020\ M)^{1/2}$$

$$k = \frac{701\ \cancel{M}/s}{(0.040\ \cancel{M})(0.020\ M)^{1/2}} = 1.2 \times 10^5\ M^{-1/2}\,s^{-1}$$

**Think About It** The units on the rate constant, $M^{-1/2}\,s^{-1}$, are unusual and not included in Table 13.5, but they are appropriate given the fractional order of the reaction with respect to $O_2$. The reason for the fractional order of the reaction with respect to $O_2$ will be discussed in the next section.

**Practice Exercise** Nitric oxide reacts rapidly with unstable nitrogen trioxide ($NO_3$), forming $NO_2$:

$$NO(g) + NO_3(g) \rightarrow 2\ NO_2(g)$$

Determine the rate law for the reaction and calculate the rate constant from the data in **Table 13.7.**

**TABLE 13.7** Initial Reaction Rates for the Formation of $NO_2$ from the Reaction $NO(g) + NO_3(g) \rightarrow 2\ NO_2(g)$ at 25°C

| Experiment | $[NO]_0$ ($M$) | $[NO_3]_0$ ($M$) | Initial Reaction Rate ($M$/s) |
|---|---|---|---|
| 1 | $1.25 \times 10^{-3}$ | $1.25 \times 10^{-3}$ | $2.45 \times 10^4$ |
| 2 | $2.50 \times 10^{-3}$ | $1.25 \times 10^{-3}$ | $4.90 \times 10^4$ |
| 3 | $2.50 \times 10^{-3}$ | $2.50 \times 10^{-3}$ | $9.80 \times 10^4$ |

## Integrated Rate Laws: First-Order Reactions

Determining a rate law by using initial reaction rate data has some disadvantages. The method requires several experiments with different concentrations of reactants that must be varied in a systematic fashion. We also must accurately determine the reaction rate at the instant the reaction begins. It would be much easier if we could determine the rate law and calculate the rate constant for a reaction from the plot of concentration versus time in a single experiment.

In fact, this can be accomplished for reactions in which the reaction rate depends on the concentration of only one reactant. The photochemical decomposition of ozone is such a reaction:

$$O_3(g) \xrightarrow{\text{sunlight}} O_2(g) + O(g)$$

**TABLE 13.8** Rate of Photochemical Decomposition of Ozone

| Time (s) | $[O_3]$ ($M$) | $\ln[O_3]$ |
|---|---|---|
| 0.0 | $1.000 \times 10^{-4}$ | −9.21 |
| 300.0 | $7.19 \times 10^{-5}$ | −9.64 |
| 600.0 | $5.17 \times 10^{-5}$ | −9.87 |
| 900.0 | $3.72 \times 10^{-5}$ | −10.20 |
| 1200.0 | $2.67 \times 10^{-5}$ | −10.53 |
| 1500.0 | $1.92 \times 10^{-5}$ | −10.86 |

This reaction can be studied in the laboratory by using high-intensity ultraviolet light to simulate sunlight. One such study yielded the results listed in **Table 13.8** and plotted in **Figure 13.8(a)** with the graphing feature of a spreadsheet software program. Because ozone is the only reactant, the rate law for the reaction should depend only on the ozone concentration. Let's start by assuming that the reaction is first order in ozone and has the following rate law:

$$\text{Rate} = k[O_3]$$

Because the coefficient of $O_3$ in the balanced equation is 1, the $O_3$ consumption rate, $-\Delta[O_3]/\Delta t$, is equal to the reaction rate, and we can write

$$\text{Rate} = -\frac{\Delta[O_3]}{\Delta t} = k[O_3]$$

This rate law can be transformed using calculus into an expression that relates the concentration of ozone $[O_3]_t$ at any time ($t$) during the reaction to the initial concentration $[O_3]_0$:

$$\ln\frac{[O_3]_t}{[O_3]_0} = -kt \quad (13.15)$$

This version of the rate law is called an **integrated rate law** because integral calculus is used to derive it. It describes the change in reactant concentration

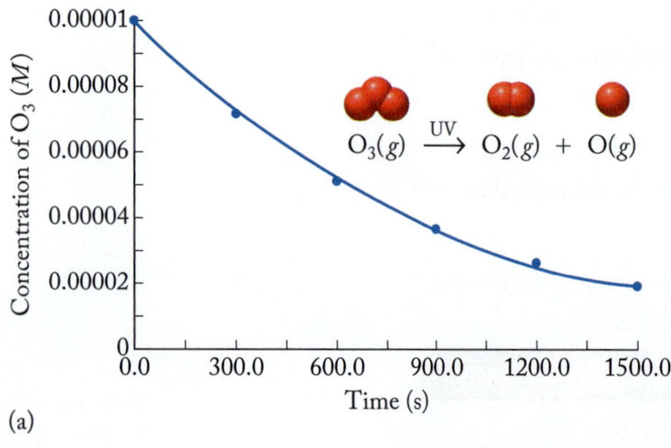


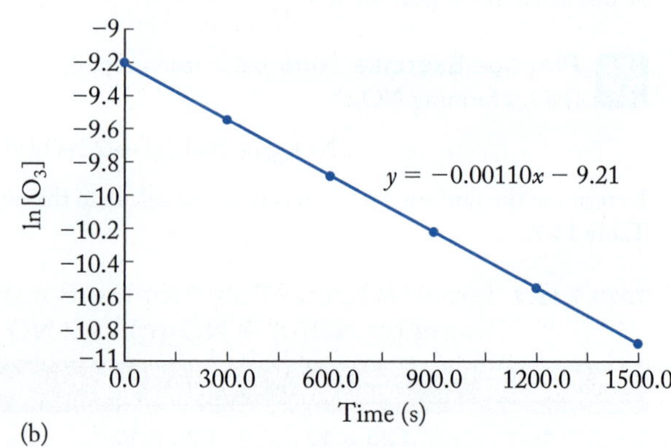


**FIGURE 13.8** The decomposition of $O_3$ is plotted as (a) $[O_3]$ versus time and (b) $\ln[O_3]$ versus time, using data from Table 13.8. The line in (b) is straight, indicating that the decomposition reaction is first order in $O_3$.

with time. The general integrated rate law for any reaction that is first order in reactant X is

$$\ln \frac{[X]_t}{[X]_0} = -kt \quad (13.16)$$

**integrated rate law** a mathematical expression that describes the change in concentration of a reactant in a chemical reaction with time.

Using the identity $\ln(a/b) = \ln a - \ln b$, we can rearrange this equation to

$$\ln[X]_t = -kt + \ln[X]_0 \quad (13.17)$$

which is the equation of a straight line of the form

$$y = mx + b$$

where $\ln[X]_t$ is the $y$ variable and $t$ is the $x$ variable. The slope of the line ($m$) is $-k$, and the $y$-intercept ($b$) is $\ln[X]_0$. Rearranging Equation 13.15 to fit the format of Equation 13.17 gives

$$\ln[O_3] = -kt + \ln[O_3]_0 \quad (13.18)$$

A graph of the natural logarithm of $[O_3]$ versus time does indeed produce a straight line, as shown in **Figure 13.8(b)**. This linearity means that our assumption was correct and the reaction is first order in $O_3$.

The spreadsheet software used to plot the data can also be used to calculate the equation of the line that best fits all the data points, which is shown in Figure 13.8(b). If you have a graphing calculator, you can use its curve-fitting function to obtain the slope, $\Delta(\ln[O_3])/\Delta t$, of the straight line that best fits the data points. The spreadsheet software does not display the units of the slope, but we know that the slope of the line represents $\Delta\ln[O_3]/\Delta t$. Log terms are unitless, and the time values in this data set are expressed in seconds, so the units on the slope value are 1/s, or $s^{-1}$. Therefore, the value of $m$ is $-0.00110\ s^{-1}$, and the value of $k$ is $-(-0.00110\ s^{-1})$, or $1.10 \times 10^{-3}\ s^{-1}$.

### SAMPLE EXERCISE 13.5 Using an Integrated Law to Determine a Rate Constant — LO5

Dinitrogen pentoxide is one of the least abundant oxides of nitrogen in the atmosphere because, if it forms, it rapidly decomposes:

$$2\ N_2O_5(g) \rightarrow 4\ NO_2(g) + O_2(g)$$

A kinetic study of the decomposition of $N_2O_5$ yields the data in the first two columns of **Table 13.9**.

a. Test the validity of the assumption that the decomposition of $N_2O_5$ is first order in $N_2O_5$.
b. Determine the value of the rate constant.

**Collect and Organize** We are given experimental data showing the concentration of a single reactant as a function of time and told to assume a first-order reaction. We can verify this assumption by using the integrated rate law for a first-order reaction (Equation 13.17) and then calculate the rate constant.

**Analyze** Equation 13.17 has the form $y = mx + b$, which means that a plot of $\ln[N_2O_5]$ ($y$) versus time ($x$) should be linear if the decomposition of $N_2O_5$ is first order. The slope ($m$) of the graph corresponds to $-k$, the negative of the rate constant.

**TABLE 13.9** Rate of Decomposition of $N_2O_5$

| Time (s) | $[N_2O_5]$ ($M$) | $\ln[N_2O_5]$ |
|---|---|---|
| 0.0 | 0.1000 | −2.303 |
| 50.0 | 0.0649 | −2.735 |
| 100.0 | 0.0561 | −2.881 |
| 200.0 | 0.0250 | −3.689 |
| 300.0 | 0.0136 | −4.298 |
| 400.0 | 0.0059 | −5.133 |

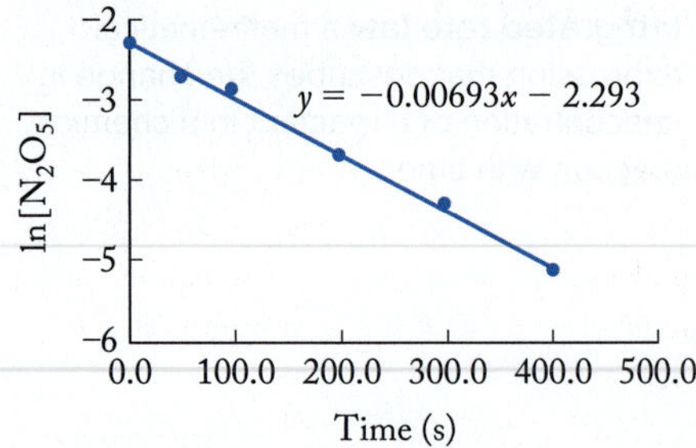


**FIGURE 13.9** Plot of $\ln[N_2O_5]$ as a function of time, using data from Table 13.9.

**Solve**

a. The first step is to convert the $[N_2O_5]$ values into $\ln[N_2O_5]$ values, which are listed in the third column of Table 13.9. Plotting the time and $\ln[N_2O_5]$ values as $x$ and $y$ variables produces the graph shown in **Figure 13.9**. The spreadsheet software that plots the data also calculates the equation of the straight line that best fits the data: $y = -0.00693x - 2.293$. The fact that all the points lie near a straight line confirms that the reaction is first order in $N_2O_5$.

b. The value of the rate constant $k$ is $-m$, where $m$ is the value of the slope of the line, and it has units of $s^{-1}$, which makes it $-(-0.00693\ s^{-1})$, or $6.93 \times 10^{-3}\ s^{-1}$.

**Think About It** Using a calculator or spreadsheet software to plot the data and determine the slope of the line that best fits it tends to generate the most accurate slope and rate constant values because these tools use all the points in the data set. An alternative method based on calculating $\Delta y/\Delta x$ by using the coordinates of two of the plotted points is likely to be less accurate because there is inherent uncertainty in experimental results that can cause some of the plotted points to be slightly above or below the line that best fits them, as you can see in Figure 13.9. For example, had we estimated the slope by using the coordinates for the points at $t = 100.0$ seconds and 400.0 seconds, we would have obtained a value equal to $(-5.133 - (-2.881))/(400.0-100.0)\ s = -7.51 \times 10^{-3}\ s^{-1}$, which is 8.4% greater in magnitude than the slope calculated using all the data points.

**Practice Exercise** Hydrogen peroxide ($H_2O_2$) decomposes into water and oxygen:

$$H_2O_2(\ell) \rightarrow H_2O(\ell) + \tfrac{1}{2}\,O_2(g)$$

Use the data in **Table 13.10** to determine whether the decomposition of $H_2O_2$ is first order in $H_2O_2$, and calculate the value of the rate constant at the temperature of the experiment that produced the data.

**TABLE 13.10** Rate of Decomposition of Hydrogen Peroxide

| Time (s) | $[H_2O_2]$ (*M*) |
|---|---|
| 0.0 | 0.500 |
| 100.0 | 0.444 |
| 200.0 | 0.434 |
| 500.0 | 0.347 |
| 1000.0 | 0.205 |
| 1500.0 | 0.144 |

## Half-Lives

The **half-life ($t_{1/2}$)** of a reactant is the interval during which the concentration of a reactant decreases by half, as shown for the decomposition of $N_2O$ (laughing gas) in **Figure 13.10**:

$$2\,N_2O(g) \rightarrow 2\,N_2(g) + O_2(g)$$

Over each half-life interval, one-half of the quantity of reactant present at the beginning of the interval is consumed; the other half remains. Half-life is inversely related to the rate constant of a reaction: the higher the reaction rate, the shorter the half-life.

For first-order reactions, such as the decomposition of $N_2O$, we can derive a mathematical relation between the half-life $t_{1/2}$ and rate constant $k$ by starting with Equation 13.16:

$$\ln \frac{[X]_t}{[X]_0} = -kt \qquad (13.16)$$

After one half-life has passed ($t = t_{1/2}$), the concentration of X is half its original value: $[X]_t = \frac{1}{2}[X]_0$. Inserting these values for [X] and $t$ into the equation yields

$$\ln \frac{\frac{1}{2}[X]_0}{[X]_0} = -kt_{1/2}$$

$$\ln \tfrac{1}{2} = -kt_{1/2}$$

**half-life ($t_{1/2}$)** the reaction time interval during which the concentration of a reactant decreases by half.

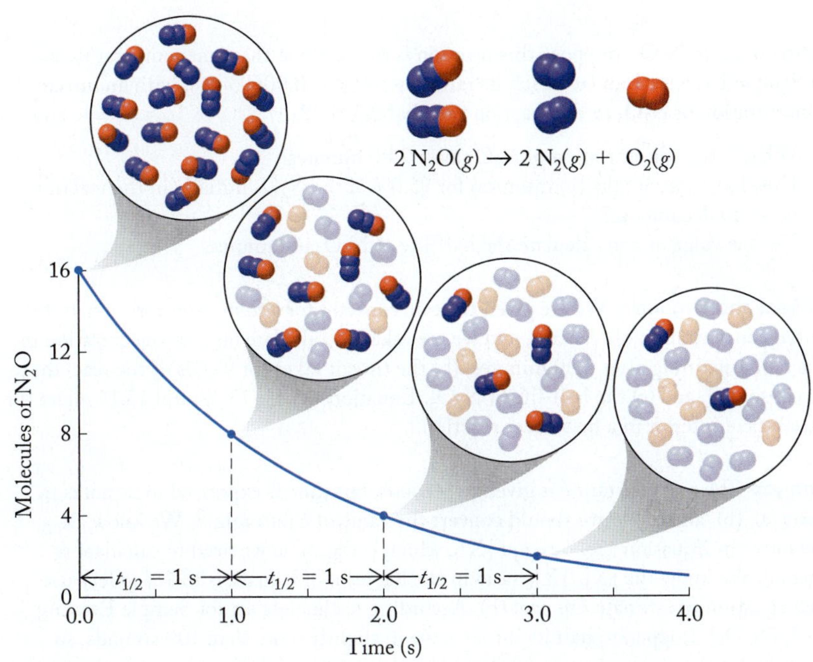


**FIGURE 13.10** The decomposition of $N_2O(g)$ is first order in $N_2O$. At a particular temperature the half-life of the reaction is 1.0 s, which means that, on average, half of a population of 16 $N_2O$ molecules decomposes in 1.0 s, half of the remaining 8 molecules decompose in the next 1.0 s, and so on.

The natural log of $\frac{1}{2}$ is $-0.693$, so

$$-0.693 = -kt_{1/2}$$

$$t_{1/2} = \frac{0.693}{k} \tag{13.19}$$

Thus, the half-life of a first-order reaction is inversely proportional to the rate constant, as noted at the beginning of this discussion. The absence of any concentration term in Equation 13.19 means that no matter what the initial concentration of the reactant in a first-order reaction is, half of it is consumed in one half-life. In other words, the half-life of a first-order chemical reaction (and only a first-order reaction) is constant at a given temperature.

Equations based on integrated rate laws can also be used to determine how much time it takes for a reactant concentration to drop to a target value or how much of it remains after a particular reaction time. Calculations such as these are widely used in chemistry (see Sample Exercise 13.6) and especially in nuclear chemistry. They are used to determine, for example, the time it will take for the level of radioactivity in material removed from a nuclear reactor to reach acceptably low levels or to calculate the age of an ancient wooden sculpture on the basis of its concentration of a radioactive isotope that occurs naturally in all plant material. We explore these and other applications of radioactive decay rates, which always follow first-order kinetics, in Chapter 21.

**SAMPLE EXERCISE 13.6** Relating Reactant Concentration, Reaction Time, and Half-Life in a First-Order Reaction **LO6**

The results of the experiment described in Sample Exercise 13.5 showed that decomposition of dinitrogen pentoxide

$$2\,N_2O_5(g) \rightarrow 2\,N_2O_4(g) + O_2(g)$$

is first order in $N_2O_5$. Suppose this reaction is run again at the same temperature as in Sample Exercise 13.5 (at which its rate constant $k = 0.00693\ s^{-1}$), with an initial concentration of $N_2O_5$ in the reaction vessel of 0.375 *M*.

a. What is the concentration of $N_2O_5$ after 3.00 minutes?
b. How long does it take (in minutes) for 95.0% of the $N_2O_5$ initially in the reaction vessel to decompose?
c. Use the value of *k* to calculate the half-life of $N_2O_5$ in minutes.

**Collect and Organize** We are given the rate constant for a first-order reaction and the initial concentration of reactant, and we are asked to calculate (a) the concentration of reactant remaining after 3.00 minutes, (b) the time it takes for 95.0% of the reactant to be consumed, and (c) the half-life of $N_2O_5$. Equations 13.16, 13.17, and 13.19 relate the variables of interest in a first-order reaction.

**Analyze** The rate constant *k* is given in seconds, but time is expressed in minutes in parts (a), (b), and (c), so we should convert the value of *k* into $min^{-1}$. We know all the terms in Equation 13.17 except $[X]_t$, which is the value we need to calculate in part (a). We know the $[X]_t/[X]_0$ ratio in the log term of Equation 13.16, so let's use that equation to estimate *t* in part (b). According to the data set for Sample Exercise 13.5, $[N_2O_5]$ dropped to half its initial value in slightly more than 100 seconds, so the calculated $t_{1/2}$ value in part (c) should be about $100\ s \times \frac{1\ min}{60\ s} = \frac{5}{3}$, or $1\frac{2}{3}$ minutes. It would take four half-lives to consume all but $(\frac{1}{2})^4 = 1/16$ or about 6%, of the initial concentration of $N_2O_5$ in the sample. Therefore, it should take a little more than four half-lives, or $4 \times \frac{5}{3} = \frac{20}{3} \approx 7$ minutes, to consume about 95% of the initial quantity of $N_2O_5$ in the sample.

**Solve**

a. Converting *k* to $min^{-1}$, and then using the given concentration and time values in Equation 13.17:

$$k = 0.00693\ s^{-1} \times (60\ s\ min^{-1}) = 0.416\ min^{-1}$$

$$\ln[N_2O_5]_t = -(0.416\ \cancel{min^{-1}}) \times 3.00\ \cancel{min} + \ln(0.375\ M)$$

$$= -2.229;\ \text{therefore, } [N_2O_5]_t = e^{-2.229} = 0.108\ M$$

b. If 95.0% of $[N_2O_5]_0$ is consumed, then $[N_2O_5]_t = (100.0 - 95.0) = 5.0\%$ or 0.050 $[N_2O_5]_0$. Using this value in Equation 13.16:

$$\ln \frac{0.050\ \cancel{[N_2O_5]_0}}{\cancel{[N_2O_5]_0}} = -(0.416\ min^{-1})t$$

$$t = \frac{-2.996}{-0.416\ min^{-1}} = 7.2\ min$$

c. Using the calculated *k* value from part (a) in Equation 13.19 to calculate $t_{1/2}$:

$$t_{1/2} = \frac{0.693}{k} = \frac{0.693}{0.416\ min^{-1}} = 1.67\ min$$

**Think About It** The calculated answers in parts (b) and (c) agree with the predicted results, and the answer in (a) is reasonable given that the reaction time of 3.00 minutes represents nearly two half-lives. Therefore, a $[N_2O_5]_t$ value that is between $\frac{1}{3}$ and $\frac{1}{4}$ of $[N_2O_5]_0$ makes sense.

**Practice Exercise** A wristwatch made in 1920 has a dial that glows in the dark because it contains a radioactive isotope of radium. The half-life of this isotope is 1600 years.

a. What is the rate constant of this first order radioactive decay process that makes the watch dial glow?

b. What fraction of the radioactive radium in the watch when it was made will still be there in 2020? (Express your answer as a percentage.)
c. How long will it take for 99.0% of the radioactive radium in the watch to decay?

## Integrated Rate Laws: Second-Order Reactions

In Section 13.1, we described how $NO_2$ exposed to sunlight decomposes to NO and O (Equation 13.3). Nitrogen dioxide may also undergo thermal decomposition, producing NO and $O_2$:

$$2\ NO_2(g) \rightarrow 2\ NO(g) + O_2(g) \qquad (13.20)$$

Because Equation 13.20 has only one reactant, like the decomposition reactions of $N_2O_5$ and $H_2O_2$ in Sample Exercise 13.5 and its Practice Exercise, we might expect this reaction to be first order. However, when we use the data in **Table 13.11** to evaluate the reaction order and rate constant, we find that the plot of $\ln[NO_2]$ versus time (**Figure 13.11a**) is curved, not linear, which means that the thermal decomposition of $NO_2$ is *not* first order.

What is the reaction order? The answer to this question lies in understanding how the reaction takes place. If each $NO_2$ molecule simply fell apart, the reaction would be first order, much like the decomposition of $N_2O_5$. However, if the reaction happens because of collisions between pairs of $NO_2$ molecules, it would be first order with respect to each $NO_2$ molecule and second order overall. In other words, if the reaction were second order, the decomposition described by Equation 13.20 would depend on collisions between pairs of molecules that just happen to be molecules of the same substance, $NO_2$. The rate law expression is

$$\text{Rate} = k[NO_2]^2 \qquad (13.21)$$

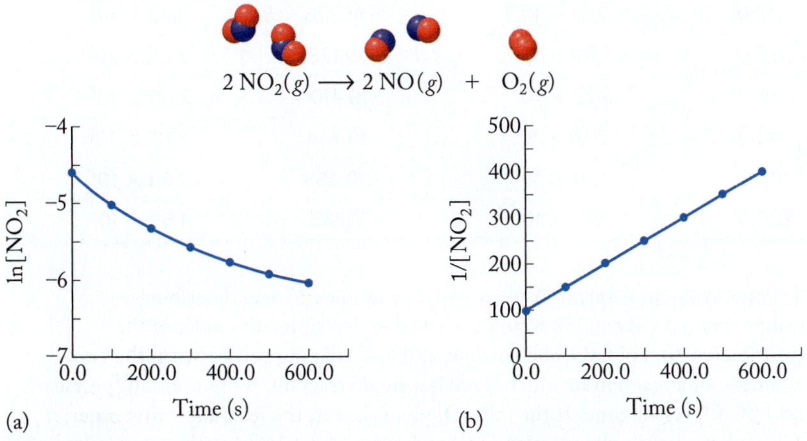


**FIGURE 13.11** At high temperatures, $NO_2$ slowly decomposes into NO and $O_2$. (a) The plot of $\ln[NO_2]$ versus time is not linear, indicating that the reaction is not first order. (b) The plot of $1/[NO_2]$ versus time is linear, indicating that the reaction is second order in $NO_2$. The slope of the line in this graph equals the rate constant. Data from Table 13.11.

**TABLE 13.11** Rate of Decomposition of $NO_2$ to NO and $O_2$

| Time (s) | $[NO_2]$ (*M*) | $\ln[NO_2]$ | $1/[NO_2]$ ($M^{-1}$) |
|---|---|---|---|
| 0.0 | $1.00 \times 10^{-2}$ | −4.605 | 100 |
| 100.0 | $6.48 \times 10^{-3}$ | −5.039 | 154 |
| 200.0 | $4.79 \times 10^{-3}$ | −5.341 | 209 |
| 300.0 | $3.80 \times 10^{-3}$ | −5.573 | 263 |
| 400.0 | $3.15 \times 10^{-3}$ | −5.760 | 317 |
| 500.0 | $2.69 \times 10^{-3}$ | −5.918 | 372 |
| 600.0 | $2.35 \times 10^{-3}$ | −6.057 | 426 |

How can we determine whether this decomposition is really second order? One way is to assume that it is and then test that assumption. The test entails transforming the rate law in Equation 13.21 into the integrated rate law for a second-order reaction, again using calculus. The result of the transformation is

$$\frac{1}{[NO_2]} = kt + \frac{1}{[NO_2]_0} \quad (13.22)$$

which, like Equation 13.18, has the form $y = mx + b$ and is the equation of a straight line, this time with $1/[NO_2]$ as the $y$ variable and $t$ as the $x$ variable.

The graph obtained using data from columns 1 and 4 of Table 13.11 is shown in **Figure 13.11(b)**. The plot is linear, so this $NO_2$ decomposition reaction is second order overall. The slope of the line provides a direct measure of $k$, which is $0.544\ M^{-1}\ s^{-1}$.

A general form of Equation 13.22 that applies to any reaction that is second order in a single reactant (X) is

$$\frac{1}{[X]} = kt + \frac{1}{[X]_0} \quad (13.23)$$

**SAMPLE EXERCISE 13.7** Distinguishing between First- and Second-Order Reactions **LO5**

Chlorine monoxide accumulates in the stratosphere above Antarctica each winter and plays a key role in forming the ozone hole above the South Pole each spring. Eventually, ClO decomposes according to the equation

$$2\ ClO(g) \rightarrow Cl_2(g) + O_2(g)$$

The kinetics of this reaction were studied in a laboratory experiment at 298 K; the results are listed in the first two columns of **Table 13.12**. Determine whether the reaction is first or second order in ClO and the value of $k$ at 298 K.

**TABLE 13.12** Rate of Decomposition of Chlorine Monoxide

| Time (ms) | [ClO] ($M$) | ln[ClO] | 1/[ClO] ($M^{-1}$) |
|---|---|---|---|
| 0.0 | $1.50 \times 10^{-8}$ | −18.015 | $6.67 \times 10^{7}$ |
| 10.0 | $7.01 \times 10^{-9}$ | −18.776 | $1.43 \times 10^{8}$ |
| 20.0 | $4.88 \times 10^{-9}$ | −19.138 | $2.05 \times 10^{8}$ |
| 30.0 | $3.72 \times 10^{-9}$ | −19.410 | $2.69 \times 10^{8}$ |
| 40.0 | $2.61 \times 10^{-9}$ | −19.764 | $3.83 \times 10^{8}$ |
| 100.0 | $1.31 \times 10^{-9}$ | −20.453 | $7.63 \times 10^{8}$ |
| 200.0 | $6.3 \times 10^{-10}$ | −21.185 | $1.59 \times 10^{9}$ |

**Collect, Organize, and Analyze** We are given experimental data describing the decomposition of ClO at 298 K and are asked to determine the order of the decomposition reaction of ClO, the rate law, and $k$. To distinguish between the two reaction orders for a reaction in which there is a single reactant, we plot ln[ClO] versus time and 1/[ClO] versus time. If the ln[ClO] plot is linear, the reaction is first order; if the 1/[ClO] plot is linear, the reaction is second order. We determine the rate constant from the slope of whichever plot is linear.

**Solve** To evaluate the two possibilities, we need to calculate ln[ClO] and 1/[ClO] values for each value of [ClO] provided. The results of these calculations are listed in the third and fourth columns of Table 13.12. The plots of ln[ClO] and 1/[ClO] versus time are shown in **Figure 13.12**. The ln[ClO] plot is not linear, but the 1/[ClO] plot is, from which we may conclude that the reaction is second order in ClO and second order overall. The line that best fits the data has a slope of $7.55 \times 10^9\ M^{-1}\ s^{-1}$, which is the value of $k$.

**Think About It** Sample Exercises 13.5 and 13.7 demonstrate that we can use integrated rate laws to distinguish between first- and second-order reactions in a single reactant.

**Practice Exercise** Experimental evidence shows that the rate of the following reaction depends only on the concentration of $NO_2$:

$$NO_2(g) + CO(g) \rightarrow NO(g) + CO_2(g)$$

Determine whether the reaction is first or second order in $NO_2$, and calculate the rate constant from the data in **Table 13.13**, which were obtained at 488 K.

**TABLE 13.13** Concentration of $NO_2$ as a Function of Time

| Time (h) | $[NO_2]$ ($M$) |
|---|---|
| 0.00 | 0.250 |
| 1.39 | 0.198 |
| 3.06 | 0.159 |
| 4.72 | 0.132 |
| 6.39 | 0.114 |
| 8.06 | 0.099 |
| 9.72 | 0.088 |
| 11.39 | 0.080 |

Second-order reactions have half-lives, too. The relation between $k$ and $t_{1/2}$ for the decomposition of $NO_2$ can be derived from Equation 13.23 if we first rearrange the terms to solve for $kt$:

$$kt = \frac{1}{[X]} - \frac{1}{[X]_0}$$

After one half-life has elapsed ($t = t_{1/2}$), [X] has decreased to half its initial concentration. Substituting this information into the preceding equation, we have

$$kt_{1/2} = \frac{1}{\frac{1}{2}[X]_0} - \frac{1}{[X]_0}$$

$$= \frac{2}{[X]_0} - \frac{1}{[X]_0} = \frac{1}{[X]_0}$$

or

$$t_{1/2} = \frac{1}{k[X]_0} \qquad (13.24)$$

Thus, the value of $t_{1/2}$ is inversely proportional to the initial concentration of X in a second-order reaction. This dependence on concentration contrasts with the $t_{1/2}$ values of first-order reactions, which are independent of concentration.

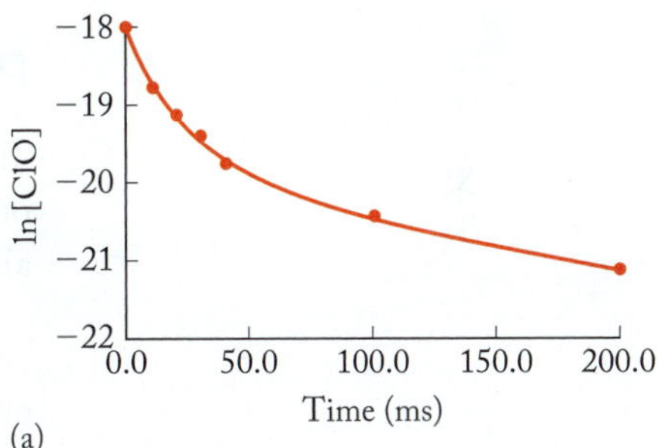


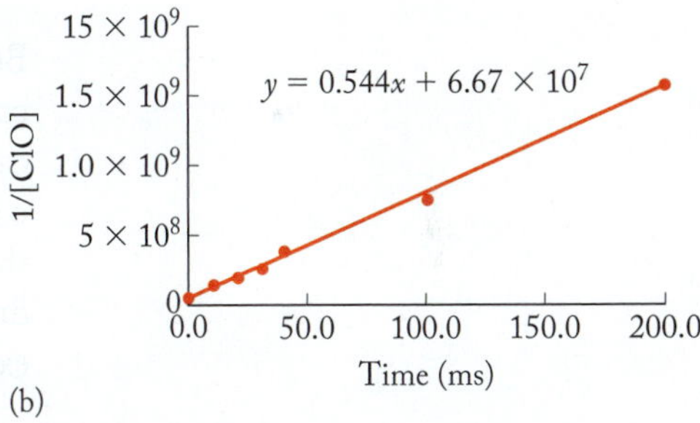


**FIGURE 13.12** Plots of (a) ln[ClO] versus time and (b) 1/[ClO] versus time, using data from Table 13.12.

**SAMPLE EXERCISE 13.8** Calculating the Half-Life of a Second-Order Reaction **LO6**

Calculate the half-life of the second-order decomposition of $NO_2$ (Equation 13.20) if $k = 0.544\ M^{-1}\ s^{-1}$ and the initial concentration of $NO_2$ is 0.0100 $M$.

**Collect and Organize** We are asked to find the half-life of $NO_2$, the time required for the concentration of $NO_2$ to decrease by one-half. We are given the rate constant and initial concentration of $NO_2$ and told that the reaction is second order. Equation 13.24 gives $t_{1/2}$ of a second-order reaction in which there is only one reactant.

**Analyze** The half-life is inversely proportional to the rate constant $k$ and to the initial concentration of $NO_2$. The value of $k$ is about $5 \times 10^{-1}\ M^{-1}\ s^{-1}$ at an initial concentration of $10^{-2}\ M$, so $t_{1/2}$ is inversely proportional to about $5 \times 10^{-3}\ s^{-1}$. Thus, we predict $t_{1/2} \approx 200$ s.

**Solve**

$$t_{1/2} = \frac{1}{k[NO_2]_0} = \frac{1}{(0.544\ \cancel{M}^{-1}\ s^{-1})(1.00 \times 10^{-2}\ \cancel{M})} = 184\ s$$

**Think About It** The calculated value of $t_{1/2}$ is very close to the predicted value of 200 s. As in all second-order reactions with one reactant, we need to know its initial concentration to calculate its half-life.

**Practice Exercise** The rate constant for the decomposition of ClO is $7.55 \times 10^9\ M^{-1}\ s^{-1}$ at 298 K. Determine $t_{1/2}$ of ClO when $[ClO]_0 = 1.50 \times 10^{-8}\ M$.

## Pseudo–First-Order Reactions

The integrated rate law in Equation 13.23 applies only to reactions that are second order in a single reactant. It does not apply to reactions that are second order overall but are first order in two reactants, such as

$$NO(g) + O_3(g) \rightarrow NO_2(g) + O_2(g)$$

or

$$NO_2(g) + O_3(g) \rightarrow NO_3(g) + O_2(g)$$

Because the integrated rate law for a reaction that is first order in two reactants is complicated, chemists often adjust reaction conditions so that a simpler rate law can be used. One approach is to have one of the reactants present at a much higher concentration than the other. This condition is common for many components in the urban atmosphere where, for example, ozone concentrations are often hundreds to thousands of times greater than NO concentrations. With such a large excess, the ozone concentration remains virtually constant during the reaction

$$NO(g) + O_3(g) \rightarrow NO_2(g) + O_2(g)$$

Thus, the rate law,

$$\text{Rate} = k[NO][O_3]$$

may be simplified to

$$\text{Rate} = k'[NO] \qquad (13.25)$$

where

$$k' = k[O_3]_0 \qquad (13.26)$$

and where $[O_3]_0$ is the initial concentration of ozone, which remains virtually constant throughout the reaction.

Equation 13.25 looks like the rate law for a first-order reaction (Rate = $k$[X]), but it is really a **pseudo–first-order** rate law because it only appears to obey first-order kinetics. A pseudo–first-order reaction has the same integrated rate law as a first-order reaction, but the rate of the pseudo–first-order reaction depends on the concentration of more than one reactant.

**pseudo–first order** describes a reaction in which all the reactants but one are present at such high concentrations that they do not decrease significantly during the reaction, so that the reaction rate is controlled by the concentration of the limiting reactant.

## SAMPLE EXERCISE 13.9 Deriving a Pseudo–First-Order Rate Law LO5

The data in the first two columns of **Table 13.14** were obtained in a study of the oxidation of trace levels of NO in the presence of a large excess of ozone at 298 K:

$$NO(g) + O_3(g) \rightarrow NO_2(g) + O_2(g)$$

(Note the NO concentration units in Table 13.14. When the molar concentrations of reactants are very low, as in the case of many atmospheric pollutants, it is easier to express them in units of molecules/cm$^3$.)

a. Verify that the reaction is pseudo–first order and determine the pseudo–first-order rate constant $k'$.
b. If the ozone concentration is 100 times the NO concentration at $t = 0$, what is the second-order rate constant $k$? Express this $k$ in $M^{-1}\ s^{-1}$.

**Collect and Organize** Under pseudo–first-order conditions, a large excess of one reactant allows us to use the integrated rate law for a reaction that is first order in the other reactant. We are asked to verify that the reaction is pseudo–first order and to determine the pseudo–first-order rate constant. We are given experimental data showing the change of concentration of NO with time. We are also given the initial concentration of $O_3$ and asked to calculate the second-order rate constant for the reaction.

**Analyze** We can verify our assumption of pseudo–first order by treating the data as we would for a reaction that is first order in a single reactant (NO) and plotting ln[NO] versus time. The plot should be a straight line with a slope equal to $-k'$. We can solve for the second-order rate constant $k$ by using Equation 13.26. The second-order rate constant is probably large because the changes in [NO] were determined during a total reaction time of only one millisecond.

**Solve**

a. The plot of ln[NO] values (third column of Table 13.14) versus time is linear, as shown in **Figure 13.13**, which indicates that the reaction is indeed pseudo–first order. The slope of the line that best fits the data is $-1.81 \times 10^{-3}\ \mu s^{-1}$. Changing the sign of the slope and converting the units from $\mu s^{-1}$ to $s^{-1}$ gives us $1.81 \times 10^3\ s^{-1}$ for the value of the pseudo–first-order rate constant $k'$.
b. We calculate the second-order rate constant $k$ by using Equation 13.26, $k' = k[O_3]_0$, where

$$[O_3]_0 = 100[NO]_0 = 100(1.00 \times 10^9 \text{ molecules/cm}^3)$$

Solving for $k$:

$$k = \frac{k'}{[O_3]_0} = \frac{1.81 \times 10^3\ s^{-1}}{1.00 \times 10^{11} \text{ molecules/cm}^3} = \frac{1.81 \times 10^{-8}\ \text{cm}^3}{\text{molecules} \cdot \text{s}}$$

To convert the units to $M^{-1}\ s^{-1}$, we need to convert the reciprocal concentration units of cm$^3$/molecule into $M^{-1}$:

$$k = \frac{1.81 \times 10^{-8}\ \text{cm}^3\ s^{-1}}{\text{molecules}} \times \frac{6.022 \times 10^{23} \text{ molecules}}{1 \text{ mol}} \times \frac{1 \text{ L}}{1000 \text{ cm}^3}$$

$$= 1.09 \times 10^{13} \frac{\text{L} \cdot s^{-1}}{\text{mol}} = 1.09 \times 10^{13}\ M^{-1}\ s^{-1}$$

**Think About It** As predicted, the second-order rate constant for the reaction is large. The pseudo–first-order ($k'$) and second-order ($k$) rate constants are very different because $k'$ contains a term for the initial concentration of ozone, which is very small when expressed in moles per liter. The value of $k'$ will be different for every initial concentration of $O_3$, but the value of $k$ will be independent of $[O_3]_0$.

**TABLE 13.14** Concentration of NO and ln[NO] as a Function of Time

| Time ($\mu$s) | [NO] (molecules/cm$^3$) | ln[NO] |
|---|---|---|
| 0.0 | $1.00 \times 10^9$ | 20.723 |
| 100.0 | $8.36 \times 10^8$ | 20.544 |
| 200.0 | $6.98 \times 10^8$ | 20.364 |
| 300.0 | $5.83 \times 10^8$ | 20.184 |
| 400.0 | $4.87 \times 10^8$ | 20.004 |
| 500.0 | $4.07 \times 10^8$ | 19.824 |
| 1000.0 | $1.65 \times 10^8$ | 18.915 |

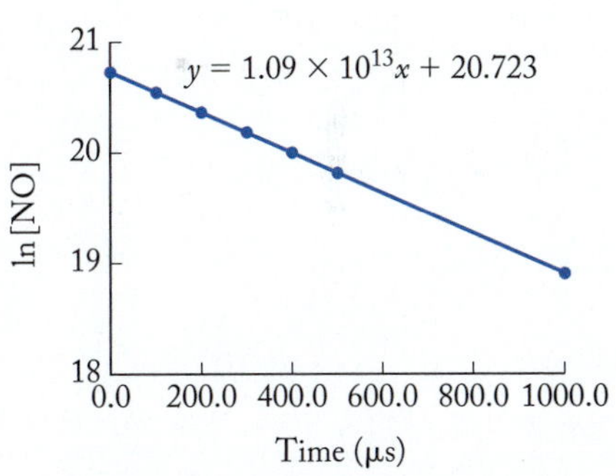


**FIGURE 13.13** Plot of ln[NO] as a function of time, using data from Table 13.14.

**TABLE 13.15** Concentration of Chlorine Atoms as a Function of Time

| Time (μs) | [Cl] (*M*) |
|---|---|
| 0.0 | $5.60 \times 10^{-14}$ |
| 100.0 | $5.27 \times 10^{-14}$ |
| 600.0 | $3.89 \times 10^{-14}$ |
| 1200.0 | $2.69 \times 10^{-14}$ |
| 1850.0 | $1.81 \times 10^{-14}$ |

**Practice Exercise** The reaction

$$Cl(g) + O_3(g) \rightarrow ClO(g) + O_2(g)$$

is first order in both reactants. Determine the pseudo–first-order and second-order rate constants for the reaction from the data in **Table 13.15** if $[O_3]_0 = 8.5 \times 10^{-11}$ *M*.

CONCEPT **TEST**

A student measured the pseudo–first-order rate constant for the reaction of NO with $O_3$ in Sample Exercise 13.9 at four initial concentrations of ozone. If all four $[O_3]_0$ values are much greater than [NO], how could the student determine the second-order rate constant graphically?

## Zero-Order Reactions

We introduced the following reaction in the Practice Exercise accompanying Sample Exercise 13.7:

$$NO_2(g) + CO(g) \rightarrow NO(g) + CO_2(g)$$

The rate law for this reaction is

$$\text{Rate} = k[NO_2]^2 \quad (13.27)$$

The rate of the reaction does not change when [CO] is changed, even when the concentrations of CO and $NO_2$ are comparable. This situation is not the same as in pseudo–first-order reactions, in which the rate does not change as one of the reactants is consumed because that reactant is present in large excess.

One interpretation of Equation 13.27 is that it contains a [CO] term to the zeroth power, making the reaction *zero order* in that reactant. Because any value raised to the zeroth power is 1, we have

$$\text{Rate} = k[NO_2]^2[CO]^0 = k[NO_2]^2(1) = k[NO_2]^2$$

Reactions with a true zero-order rate law are rare, but let's consider a generic reaction involving a single reactant X, with the reaction zero order in reactant X and zero order overall. The rate law is

$$\text{Rate} = -\Delta[X]/\Delta t = k[X]^0 = k$$

and the integrated rate law is

$$[X] = -kt + [X]_0$$

The slope of a plot of reactant concentration versus time (**Figure 13.14**) equals the negative of the zero-order rate constant *k*.

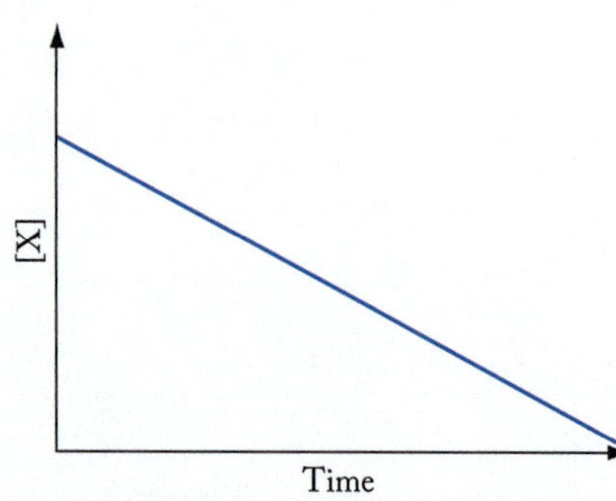


**FIGURE 13.14** The change in concentration of the reactant X in the zero-order reaction $X \rightarrow Y$ is constant over time.

We can calculate the half-life of a zero-order reaction by substituting $t = t_{1/2}$ and $[X] = [X]_0/2$ into the integrated rate law:

$$\frac{[X]_0}{2} = -kt_{1/2} + [X]_0$$

$$kt_{1/2} = [X]_0 - \frac{[X]_0}{2} = \frac{[X]_0}{2}$$

$$t_{1/2} = \frac{[X]_0}{2k}$$

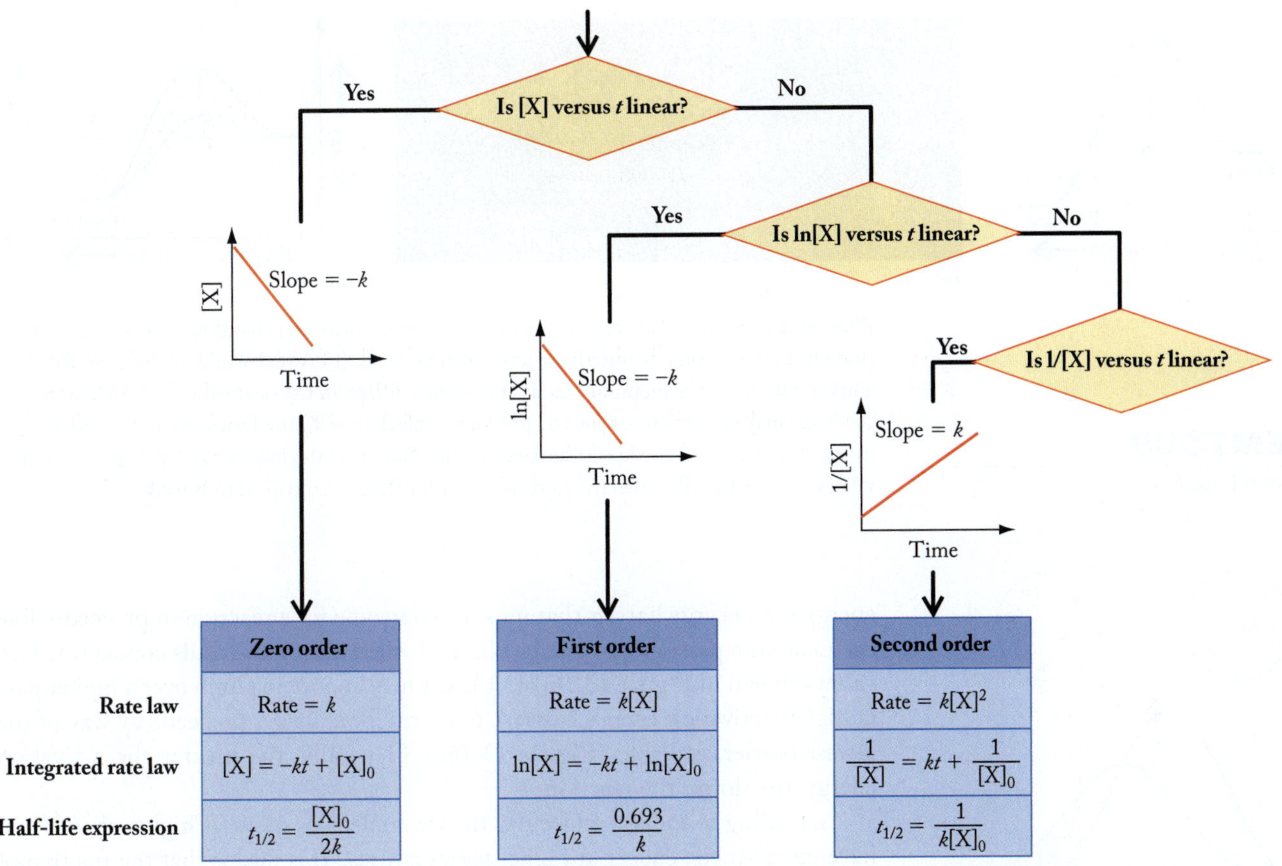


**FIGURE 13.15** Summary of how to distinguish between zero-order, first-order, and second-order kinetics for reactions involving a single reactant (X).

For now, we leave the discussion of zero-order reactions with this purely mathematical treatment. We return to them and examine their meaning at the molecular level in Sections 13.5 and 13.6.

**Figure 13.15** summarizes how to determine whether a reaction with a single reactant is zero, first, or second order. The figure also lists the rate laws and integrated rate laws for these reactions and the equations used to calculate $t_{1/2}$ of X. The decision-making process shown in Figure 13.15 can also be used to determine the kinetics of reactions involving two reactants (X and Y) if the reaction is zero order in Y. In such cases, we focus on how the concentration of X changes with time. Finally, if a reaction is first order in both X and Y, which means second order overall, and Y is present in the reaction mixture at a much higher concentration than X, then we can calculate the pseudo–first-order rate constant $k'$ from a plot of ln[X] versus $t$ and then calculate the second-order rate constant by dividing $k'$ by [Y].

## 13.4 Reaction Rates, Temperature, and the Arrhenius Equation

Why do rate constants for different reactions have different values? According to collision theory, chemical reactions take place in the gas phase when molecules collide with enough energy to break bonds in reactants and allow bonds in products to form. The minimum amount of energy that enables a chemical reaction to happen is called the **activation energy ($E_a$) (Figure 13.16a)**, and every reaction has a characteristic activation energy, usually expressed in kJ/mol. Activation

**activation energy ($E_a$)** the minimum energy that molecules need to react when they collide.

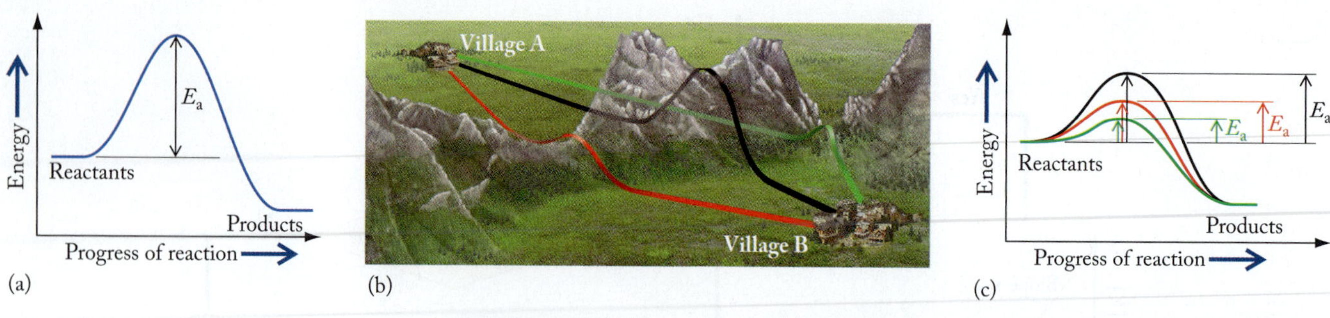


**FIGURE 13.16** (a) The energy profile of a reaction includes an activation energy barrier, $E_a$, that must be overcome before the reaction can proceed. (b) A real-world analogy confronts a hiker climbing over mountain passes to get to a village in the next valley. (c) Although the hiker may choose one of the steeper routes (black or red), the fastest route to products for most molecules is the lowest barrier (green). Note that the lowest barrier in going from village A to village B is also the lowest barrier for the return trip from B to A.

**CHEMTOUR**
Arrhenius Equation

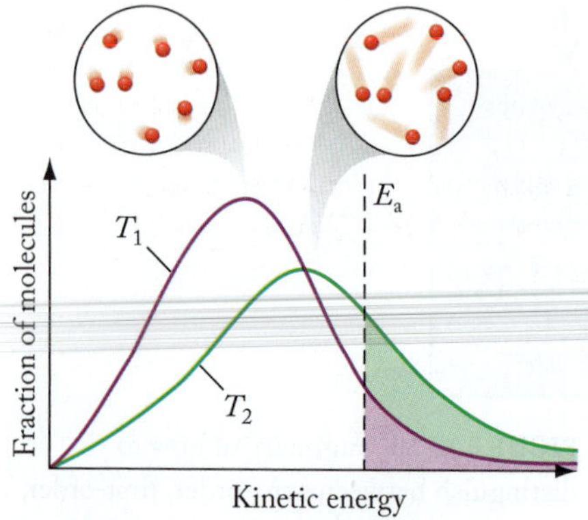


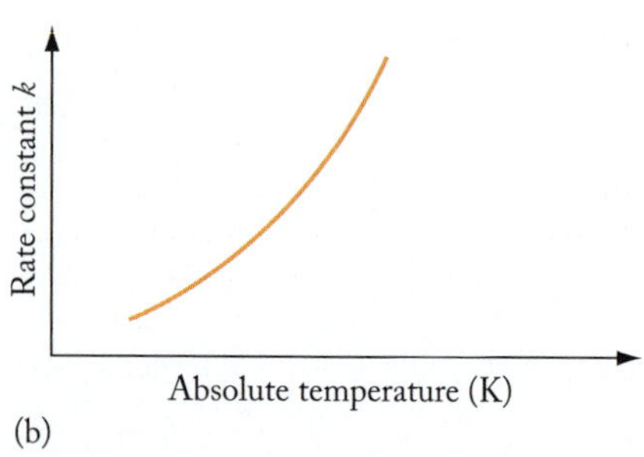


**FIGURE 13.17** (a) According to kinetic molecular theory, a fraction of reactant molecules have kinetic energies equal to or greater than the activation energy ($E_a$) of the reaction. As temperature increases from $T_1$ to $T_2$, the number of molecules with energies exceeding $E_a$ increases, leading to an increase in reaction rate. (b) The rate constant for any reaction increases with increasing temperature.

energy is an energy barrier that must be overcome for a reaction to proceed—like the mountain passes that must be climbed when hiking the trails connecting two valleys shown in **Figure 13.16(b)**. A hiker may choose to climb over a higher pass (a higher activation energy barrier), but a reaction always proceeds by way of the lowest barrier available (**Figure 13.16c**). Generally, the greater the activation energy, the slower the reaction.

According to kinetic molecular theory, molecules move at higher speeds and have more kinetic energy at higher temperatures. This means that the fraction of molecules with kinetic energies greater than a given activation energy increases with increasing temperature, as shown in **Figure 13.17(a)**. Therefore, the rates of chemical reactions should increase with increasing temperature, and indeed they do (**Figure 13.17b**).

In the late 19th century, experiments carried out in the laboratories of Jacobus van 't Hoff and Svante Arrhenius led to a fundamental advance in understanding how temperature affects rates of chemical reactions. The mathematical connection between temperature, the rate constant $k$ for a reaction, and its activation energy is given by the **Arrhenius equation**:

$$k = Ae^{-E_a/RT} \tag{13.28}$$

where $R$ is the universal gas constant in J/(mol · K) and $T$ is the reaction temperature in kelvins. The factor $A$, called the **frequency factor**, is the product of collision frequency and a term that corrects for the fact that some collisions do not lead to products because the colliding molecules are not oriented in the right way with respect to each other.

To examine the importance of molecular orientation during collisions, let's revisit the reaction between $O_3$ and NO:

$$O_3(g) + NO(g) \rightarrow O_2(g) + NO_2(g) \tag{13.7}$$

**Figure 13.18** shows two ways that ozone and nitric oxide molecules might approach each other. Only one of these orientations, the one in which an $O_3$ molecule approaches the nitrogen atom of NO, leads to a chemical reaction between the two molecules.

A collision between $O_3$ and NO molecules with the correct orientation and enough kinetic energy may result in the formation of the **activated complex**

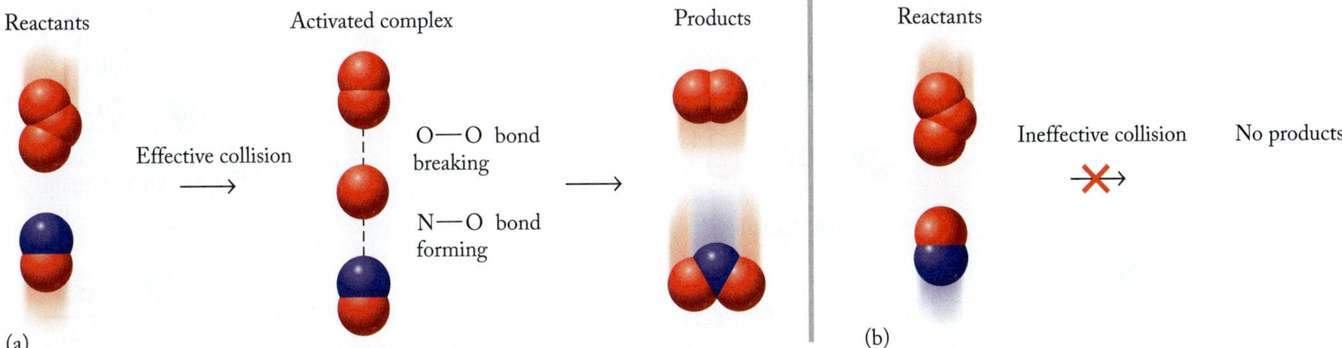


**FIGURE 13.18** The effect of molecular orientation on reaction rate. (a) When the $O_3$ and NO molecules are oriented such that the collision is between an $O_3$ oxygen and the NO nitrogen, the collision is effective and an activated complex forms, yielding the two product molecules $O_2$ and $NO_2$. (b) When the reactant molecules are oriented such that the collision is between an $O_3$ oxygen and the NO oxygen, no activated complex forms and no reaction occurs.

shown in Figure 13.18(a). In this species, one of the O—O bonds in the $O_3$ molecule has started to break and the new N—O bond is beginning to form. Activated complexes represent midway points in chemical reactions. They have extremely brief lifetimes and fall apart rapidly, either forming products or re-forming reactants. Activated complexes are formed by reacting species that have acquired enough energy to react with each other. The internal energy of an activated complex represents a high-energy **transition state** of a reaction and defines the height of an activation energy barrier in the chemical reaction. The magnitudes of activation energies for most reactions can vary from a few kilojoules to hundreds of kilojoules per mole.

**CONNECTION** The relationship between temperature and speeds of particles in the gas phase was introduced in Section 9.2 (see Figure 9.9).

We can draw an **energy profile** for a chemical reaction that shows the changes in potential energy for the reaction as a function of the progress of the reaction from reactants to products. We have already seen one energy profile in Figure 13.16(a). Now consider the energy profile for the reaction between NO and $O_3$ that produces $NO_2$ and $O_2$, as shown in **Figure 13.19(a)**. The $x$-axis represents the progress of the reaction and the $y$-axis represents chemical energy. Activation energy represents the difference in energy between the reactants and the transition state they must pass through to become products. The size of the activation energy barrier depends on whether the reaction is proceeding in the forward direction ($E_a$ = 10.5 kJ/mol), as shown in Figure 13.19(a), or in the reverse direction ($E_a$ = 210 kJ/mol), as shown in **Figure 13.19(b)**. The smaller activation energy barrier in the forward direction means that the forward reaction proceeds at a higher rate than the reverse reaction, assuming equal concentrations of reactants and products.

One of the many uses of the Arrhenius equation is to calculate the value of $E_a$ for a chemical reaction. When we take the natural logarithm of both sides of Equation 13.28,

$$\ln k = -\frac{E_a}{R}\left(\frac{1}{T}\right) + \ln A \qquad (13.29)$$

the result fits the general equation of a straight line ($y = mx + b$) if we make ($\ln k$) the $y$ variable and ($1/T$) the $x$ variable. We can calculate $E_a$ by determining the rate constant $k$ for the reaction at several temperatures. Plotting ln $k$

**Arrhenius equation** an equation relating the rate constant of a reaction to the absolute temperature ($T$), the activation energy of the reaction ($E_a$), and the frequency factor ($A$).

**frequency factor ($A$)** the product of the frequency of molecular collisions and a factor that expresses the probability that the relative orientation of the molecules is appropriate for a reaction to occur.

**activated complex** a short-lived species formed in a chemical reaction in which molecules have both the proper orientation and enough energy to react with each other.

**transition state** a high-energy state between reactants and products in a chemical reaction.

**energy profile** a graph showing the changes in chemical energy during a reaction.

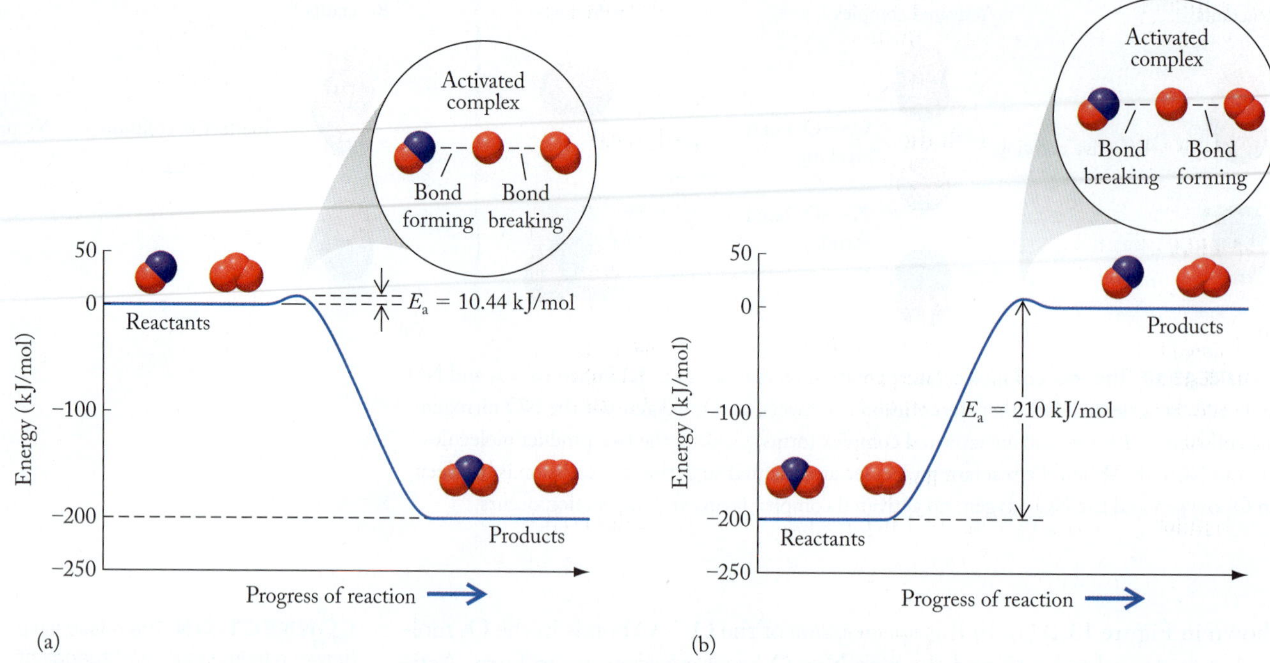


**FIGURE 13.19** (a) The energy profile for the reaction $NO(g) + O_3(g) \rightarrow NO_2(g) + O_2(g)$ includes an activation energy barrier of 10.5 kJ/mol. (b) The reverse reaction has a much larger activation energy of 210 kJ/mol.

versus $1/T$ should give a straight line, the slope of which is $-E_a/R$. **Table 13.16** and **Figure 13.20** show data for the reaction between NO and $O_3$ at six temperatures. The straight line that best fits the points in the figure has a slope of $-1.256 \times 10^3$ K, and the activation energy of the reaction is

$$E_a = -\text{slope} \times R$$

$$= -(-1.256 \times 10^3 \text{ K}) \times \left(8.314 \frac{\text{J}}{\text{mol} \cdot \text{K}}\right) = 1.044 \times 10^4 \text{ J/mol}$$

$$= 10.44 \text{ kJ/mol}$$

**TABLE 13.16** Temperature Dependence of the Rate of the Reaction $NO(g) + O_3(g) \rightarrow NO_2(g) + O_2(g)$

| $T$ (K) | $k$ ($M^{-1}$ $s^{-1}$) | $1/T$ ($K^{-1}$) | ln $k$ |
|---|---|---|---|
| 300 | $1.21 \times 10^{10}$ | $3.33 \times 10^{-3}$ | 23.216 |
| 325 | $1.67 \times 10^{10}$ | $3.08 \times 10^{-3}$ | 23.539 |
| 350 | $2.20 \times 10^{10}$ | $2.86 \times 10^{-3}$ | 23.814 |
| 375 | $2.79 \times 10^{10}$ | $2.67 \times 10^{-3}$ | 24.052 |
| 400 | $3.45 \times 10^{10}$ | $2.50 \times 10^{-3}$ | 24.264 |
| 425 | $4.15 \times 10^{10}$ | $2.35 \times 10^{-3}$ | 24.449 |

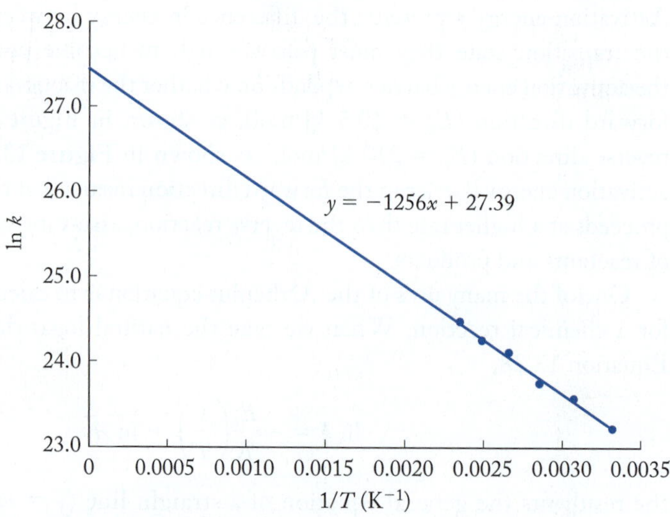


**FIGURE 13.20** A graph of ln $k$ versus $1/T$ yields a straight line with a slope equal to $-E_a/R$ and a $y$-intercept equal to ln $A$, the natural logarithm of the frequency factor. Data from Table 13.16.

The $y$-intercept ($1/T = 0$) in Figure 13.20 is 27.4. According to Equation 13.29, this value represents ln $A$, which means that

$$A = e^{27.39} = 7.9 \times 10^{11}\ M^{-1}\ s^{-1}$$

Note that the units on $A$ match those of the $k$ values used to calculate it.

**SAMPLE EXERCISE 13.10** Calculating an Activation Energy and Frequency Factor **LO7**

As noted in Sample Exercise 13.7, ClO is a highly reactive gas that plays a key role in the destruction of ozone in the stratosphere. The data in **Table 13.17(a)** were collected in a study of the effect of temperature on the rate at which ClO decomposes into $Cl_2$ and $O_2$:

$$2\ ClO(g) \rightarrow Cl_2(g) + O_2(g)$$

Determine the activation energy ($E_a$) and frequency factor ($A$) for this reaction.

**TABLE 13.17(a)** Rate Constant as a Function of Temperature for the Decomposition of ClO

| $T$ (K) | $k$ ($M^{-1}$ s$^{-1}$) |
|---|---|
| 238 | $1.9 \times 10^9$ |
| 258 | $3.1 \times 10^9$ |
| 278 | $4.9 \times 10^9$ |
| 298 | $7.2 \times 10^9$ |

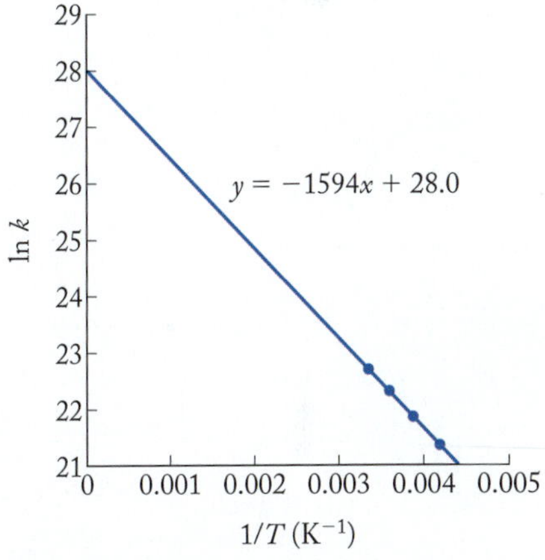


**FIGURE 13.21** Plot of ln $k$ versus $1/T$ for the decomposition of ClO($g$), using data from Table 13.17(b).

**Collect and Organize** We are given rate constant values of a reaction at different temperatures, and we are asked to find the reaction's $A$ and $E_a$ values. According to the Arrhenius equation, the slope of a plot of ln $k$ versus $1/T$ is equal to $-E_a/R$, and the $y$-intercept is equal to ln $A$.

**Analyze** We need to convert the $T$ and $k$ values from Table 13.17(a) to $1/T$ and ln $k$, respectively, to use them in the Arrhenius equation. The rate constants for the reaction are large, about $10^9\ M^{-1}\ s^{-1}$, near room temperature, so the activation energy barrier should be relatively low.

**Solve** Expanding the data table to include columns for ln $k$ and $1/T$ yields **Table 13.17(b)**. A plot of ln $k$ versus $1/T$ gives us a straight line with a slope of −1594 K (**Figure 13.21**) and its $y$-intercept at 28.0. Therefore, the values of $A$ and $E_a$ are

$$y\text{-intercept} = 28.0 = \ln A, \qquad A = e^{28.0} = 1.45 \times 10^{12}\ M^{-1}\ s^{-1}$$

$$E_a = -\text{slope} \times R$$

$$= -(-1594\ \cancel{K}) \times 8.314 \frac{J}{mol \cdot \cancel{K}} = 1.33 \times 10^4\ J/mol = 13.3\ kJ/mol$$

**TABLE 13.17(b)** Summary of $T$, $1/T$, $k$, and ln $k$ for the Decomposition of ClO

| $T$ (K) | $1/T$ (K$^{-1}$) | $k$ ($M^{-1}$ s$^{-1}$) | ln $k$ |
|---|---|---|---|
| 238 | $4.20 \times 10^{-3}$ | $1.9 \times 10^9$ | 21.365 |
| 258 | $3.88 \times 10^{-3}$ | $3.1 \times 10^9$ | 21.855 |
| 278 | $3.60 \times 10^{-3}$ | $4.9 \times 10^9$ | 22.313 |
| 298 | $3.36 \times 10^{-3}$ | $7.2 \times 10^9$ | 22.697 |

**TABLE 13.18** Rate Constants as a Function of Temperature for the Reaction of Br with $O_3$

| $T$ (K) | $k$ [cm$^3$/(molecule · s)] |
|---|---|
| 238 | $5.9 \times 10^{-13}$ |
| 258 | $7.7 \times 10^{-13}$ |
| 278 | $9.6 \times 10^{-13}$ |
| 298 | $1.2 \times 10^{-12}$ |

**Think About It** The activation energy is the height of a barrier to be overcome before a reaction can proceed. As predicted, the activation energy for ClO decomposition is relatively small.

**Practice Exercise** The rate constant for the reaction

$$Br(g) + O_3(g) \rightarrow BrO(g) + O_2(g)$$

was determined at the four temperatures shown in **Table 13.18**. Calculate the values of $E_a$ and $A$ for this reaction.

Calculating activation energies by using the graphical method generally requires measuring the rate constant at a minimum of three temperatures to verify that the plot of ln $k$ versus $1/T$ is a straight line. Once we know the value of $E_a$, we can use it and the value of the rate constant ($k_1$) of a reaction at one temperature ($T_1$) to calculate the value of the rate constant ($k_2$) at another temperature ($T_2$) without having to determine the frequency factor, $A$. We start by substituting $k_1$, $k_2$, $T_1$, and $T_2$ into Equation 13.29:

$$\ln k_1 = -\frac{E_a}{R}\left(\frac{1}{T_1}\right) + \ln A \qquad \ln k_2 = -\frac{E_a}{R}\left(\frac{1}{T_2}\right) + \ln A$$

Subtracting these two expressions, $\ln k_2 - \ln k_1$, gives

$$\ln k_2 - \ln k_1 = \left[-\frac{E_a}{R}\left(\frac{1}{T_2}\right) + \ln A\right] - \left[-\frac{E_a}{R}\left(\frac{1}{T_1}\right) + \ln A\right]$$

Using the mathematical properties of logarithms, we can rearrange this equation as follows:

$$\ln \frac{k_2}{k_1} = -\frac{E_a}{R}\left(\frac{1}{T_2} - \frac{1}{T_1}\right) \qquad (13.30)$$

**SAMPLE EXERCISE 13.11** Calculating the Value of a Rate Constant at a Second Temperature **LO7**

The rate constant ($k$) for the reaction $2\ ClO(g) \rightarrow Cl_2(g) + O_2(g)$ is $7.2 \times 10^9\ M^{-1}\ s^{-1}$ at 298 K. In Sample Exercise 13.10, we determined that the activation energy of this reaction is 12.8 kJ/mol. What is the rate constant of this reaction in the lower stratosphere, where $T = 217$ K?

**Collect, Organize, and Analyze** We are given a rate constant value at one temperature and asked to calculate its value at a lower temperature. We know the activation energy of the reaction, so we can use Equation 13.30 to solve for $k$ at the lower temperature. According to the rate constant data for this reaction in Table 13.17(a), the value of $k$ at 217 K should be near $1 \times 10^9\ M^{-1}\ s^{-1}$. As part of the calculation, we need to convert $E_a$ from kilojoules to joules to be compatible with the value of $R$: 8.314 J/(mol · K).

**Solve** Letting $T_2$ be 217 K and $T_1$ be 298 K in Equation 13.30 and solving for $k_2$:

$$\ln\frac{k_2}{k_1} = -\frac{E_a}{R}\left(\frac{1}{T_2} - \frac{1}{T_1}\right) = \left(\frac{-12.8\,\frac{\text{kJ}}{\text{mol}} \times \frac{1000\text{ J}}{\text{kJ}}}{8.314\,\frac{\text{J}}{\text{mol}\cdot\text{K}}}\right)\left(\frac{1}{217\text{ K}} - \frac{1}{298\text{ K}}\right) = -1.928$$

$$\frac{k_2}{k_1} = \frac{k_2}{7.2 \times 10^9\ M^{-1}\,\text{s}^{-1}} = e^{-1.928} = 0.145$$

$$k_2 = 1.0 \times 10^9\ M^{-1}\,\text{s}^{-1}$$

**Think About It** The calculated rate constant corresponds to the predicted value. Equation 13.30 comes in handy when we know the rate constant of a reaction at a standard reference temperature, such as 298 K (25°C), but we need to know it at a much higher or lower temperature.

**Practice Exercise** The activation energy for the reaction $O_3(g) + NO(g) \rightarrow O_2(g) + NO_2(g)$ is 10.2 kJ/mol. How many times larger is the rate constant for this reaction at 400 K than at 200 K?

### CONCEPT TEST

Which of the following statements is/are true about activation energies?

a. Fast reactions have large rate constants *and* large activation energies.

b. The forward reaction always has a lower activation energy than that of the reverse reaction.

c. Raising the temperature of a reaction mixture lowers the activation energy of the reaction.

## 13.5 Reaction Mechanisms

We ended the previous section with an equation that enables us to calculate the rate constant of a reaction at any temperature from the value of $k$ at a reference temperature and the value of $E_a$. This calculation assumes, however, that the reaction proceeds in the same way no matter what the temperature is.[2] In other words, Equation 13.30 holds if the reaction happens through the same set of collisions and involves the same sequence of bonds breaking in reactants and bonds forming in products at both $T_1$ and $T_2$. These molecular steps define its **reaction mechanism**.

In this section, we explore how kinetics data help chemists understand how chemical reactions actually happen. To obtain such data, they can test for the presence of products formed in preliminary steps. Technology enables chemists to follow many chemical reactions through the transformation of reactants to products in the time that it takes for individual covalent bonds to break and new ones to form. These processes occur as rapidly as the bonds vibrate, that is, in femtoseconds ($10^{-15}$ s), so the area of research based on monitoring these ultrafast processes is called *femtochemistry*.[3]

**CHEMTOUR** Reaction Mechanisms

**reaction mechanism** a set of steps that describes how a reaction occurs at the molecular level; the mechanism must be consistent with the rate law for the reaction.

[2]This assumption is not always valid, especially when $T_1$ and $T_2$ are far apart. Under these circumstances, Equation 13.30 may not hold.

[3]In 1999, Ahmed H. Zewail (1946–2016) received the Nobel Prize in Chemistry for his pioneering research in the development of femtochemistry.

## Elementary Steps

Let's start by revisiting the thermal decomposition of $NO_2$:

$$2\ NO_2(g) \rightarrow 2\ NO(g) + O_2(g) \quad (13.20)$$

We noted in Section 13.3 that this reaction is second order in $NO_2$ because it results from collisions between pairs of $NO_2$ molecules. How do the atoms in two colliding $NO_2$ molecules rearrange themselves to form two NO molecules and one $O_2$ molecule?

One answer to this question is contained in the two-step reaction pathway shown in **Figure 13.22**. In the first step, a collision between two $NO_2$ molecules produces a molecule of NO and a molecule of $NO_3$. In a second step, the $NO_3$ decomposes to NO and $O_2$. Both steps in the mechanism involve transient activated complexes. In the activated complex of the first step, two molecules share an oxygen atom. The bonds in the activated complex of the second step rearrange so that two oxygen atoms become bonded, forming a molecule of $O_2$ and leaving behind a molecule of NO. The molecule $NO_3$ is an **intermediate** in this mechanism because it is produced in one step and consumed in a later step. Intermediates are not considered reactants or products and do not appear in the equation describing a reaction. Intermediates in chemical reactions are sometimes long-lived enough to be isolated. In contrast, activated complexes have never been isolated, although they have been detected.

This reaction mechanism is a combination of two **elementary steps**. An elementary step that involves a single molecule is **unimolecular**, one that involves a collision between two molecules is **bimolecular**, and one that involves a collision between three molecules is **termolecular**. Bimolecular elementary steps are much more common than termolecular elementary steps because the chance of three molecules colliding at the same time is much smaller. The terms uni-, bi-, and termolecular are used by chemists to describe the **molecularity** of an elementary step, which refers to the number of reacting atoms, ions, or molecules involved in that step. The molecularity of an elementary step is the same as its reaction order.

A valid reaction mechanism must be consistent with the stoichiometry of the reaction. In other words, the elementary steps in Figure 13.22,

$$\text{Elementary step 1} \quad 2\ NO_2(g) \rightarrow NO(g) + NO_3(g)$$

$$\text{Elementary step 2} \quad NO_3(g) \rightarrow NO(g) + O_2(g)$$

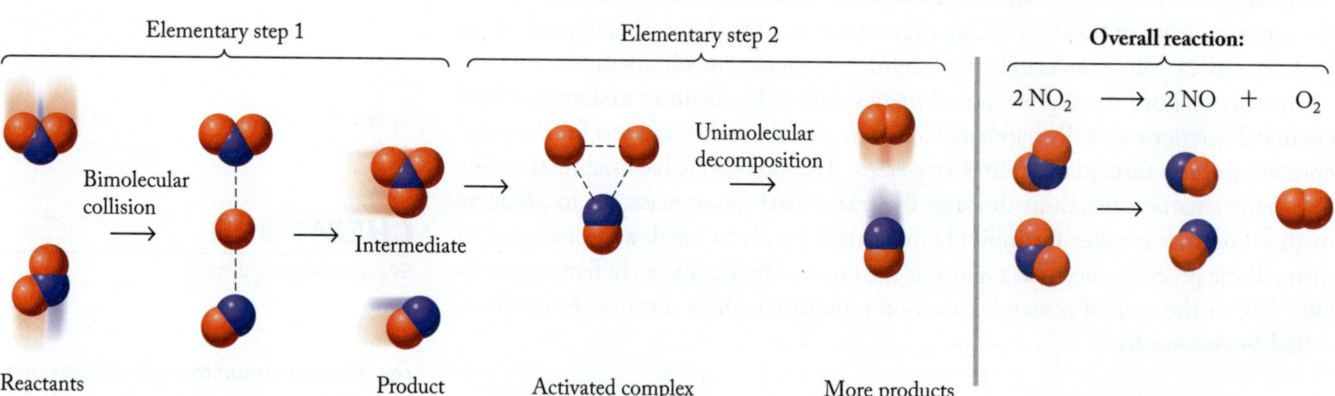


**FIGURE 13.22** The decomposition of $NO_2$ begins when two $NO_2$ molecules collide, producing NO and $NO_3$ (elementary step 1). The $NO_3$ then rapidly decomposes into NO and $O_2$ (elementary step 2).

must sum to yield the balanced chemical equation, which they do once the intermediate ($NO_3$) is canceled out:

$$2\,NO_2(g) + \cancel{NO_3(g)} \rightarrow 2\,NO(g) + \cancel{NO_3(g)} + O_2(g)$$
$$2\,NO_2(g) \rightarrow 2\,NO(g) + O_2(g)$$

How does the concept of activation energy apply to a two-step reaction such as this one? The two elementary steps produce an energy profile with two maxima. In elementary step 1, collisions between pairs of $NO_2$ molecules result in the formation of an activated complex associated with the first transition state in **Figure 13.23**. As this activated complex transforms into NO and $NO_3$, the energy of the system drops to the bottom of the trough between the two maxima. In elementary step 2, $NO_3$ forms the activated complex associated with the second transition state. As these complexes transform into final products, the energy of the system drops to its final level.

Figure 13.23 shows that the energy barrier for elementary step 1 is much greater than that for elementary step 2. This difference is consistent with the relative rates of the two steps: step 1 is slower than step 2. If the reaction were to proceed in the reverse direction (as NO and $O_2$ react, forming $NO_2$), the energy barrier on the right would be the smaller of the two, and that elementary step would be the more rapid one. Experimental evidence supports these expectations.

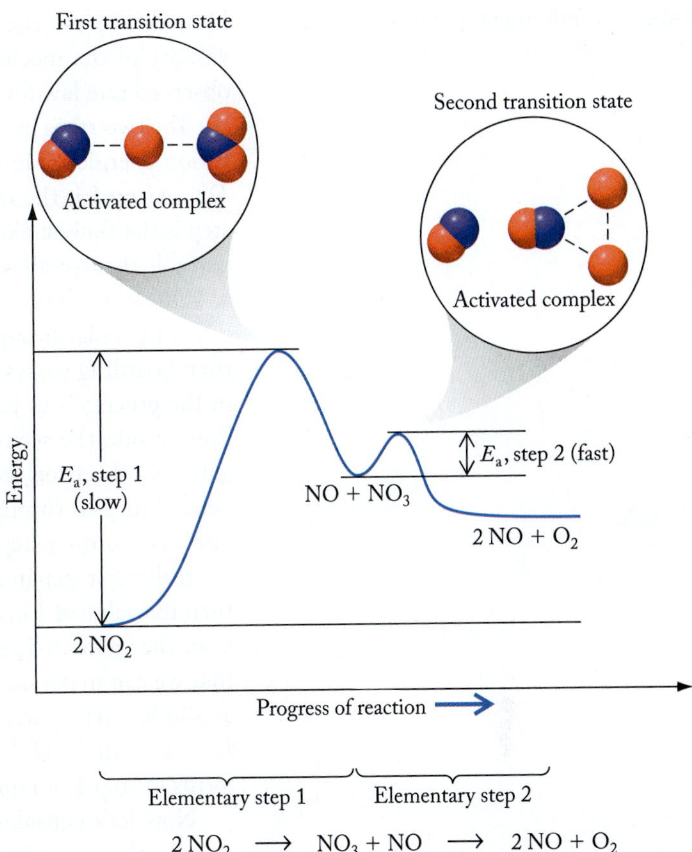


**FIGURE 13.23** The energy profile for the decomposition of $NO_2$ to NO and $O_2$ shows activation energy barriers for both elementary steps. The activation energy of the first step is larger than that of the second step, so the first step is the slower of the two.

**CONCEPT TEST**

In a reaction mechanism with *n* elementary steps, how many activation energies does the overall reaction have?

## Rate Laws and Reaction Mechanisms

Any mechanism proposed for a reaction must be consistent with the rate law derived from experimental data. As noted in Section 13.3, the rate law for the decomposition of $NO_2$ is second order in $NO_2$:

$$\text{Rate} = k[NO_2]^2 \tag{13.21}$$

The balanced chemical equation for this (or any) reaction does not provide enough information to write the overall rate law for the reaction. However, we can use the balanced chemical equation for any elementary step to write a rate law for that step. For the decomposition of $NO_2$, the rate law for the first elementary step, $2\,NO_2(g) \rightarrow NO_3(g) + NO(g)$, is

$$\text{Rate}_1 = k_1[NO_2]^2 \tag{13.31}$$

which we obtain by writing the concentration for each reactant raised to a power equal to the coefficient of that reactant in the balanced equation for that step. For the second elementary step in the $NO_2$ decomposition, $NO_3(g) \rightarrow NO(g) + O_2(g)$, the sole reactant, $NO_3$, has a coefficient of 1, so the rate law is

$$\text{Rate}_2 = k_2[NO_3] \tag{13.32}$$

**intermediate** a species produced in one step of a reaction and consumed in a later step.

**elementary step** a molecular-level view of a single process taking place in a chemical reaction.

**unimolecular step** a step in a reaction mechanism involving only one molecule (or atom or ion) on the reactant side.

**bimolecular step** a step in a reaction mechanism involving a collision between two molecules (or atoms or ions).

**termolecular step** a step in a reaction mechanism involving a collision among three molecules (or atoms or ions).

**molecularity** the number of molecules (or atoms or ions) involved in an elementary step in a reaction.

**rate-determining step** the slowest step in a multistep chemical reaction.

How do we use the rate laws in Equations 13.31 and 13.32 to determine the validity of the mechanism—that is, how do we show that they conform to the observed rate law for the decomposition of $NO_2$ in Equation 13.21?

The two steps in the mechanism proceed at different rates, with different activation energies (Figure 13.23). Step 1 is slower (larger $E_a$) than step 2 (smaller $E_a$), so step 1 is the **rate-determining step** in the reaction. The rate-determining step is the slowest elementary step in a chemical reaction, and the rate of this step controls the overall reaction rate.

One way of visualizing the concept of a rate-determining step is to analyze the flow of people through a busy airport. Many travelers arrive at the airport with their boarding passes in hand or print them from convenient kiosks. The next step in the process is to pass through security. Typically, the number of people in the line outside the security point greatly exceeds the number of available security gates, so the time required to make it to a flight depends mostly on the time needed to pass through security. For many passengers, security screening is the rate-determining step on the way to their boarding gate.

If the first step in the mechanism in Figure 13.22 is the rate-determining step, then the value of $k$ in Equation 13.21 is equal to $k_1$ from Equation 13.31. In addition, the value of $k_1$ must be smaller than the value of $k_2$. $NO_3$ is stable enough that we can make small amounts of it and test whether $k_1 < k_2$. Experiments run at 300 K starting with $NO_2$ or $NO_3$ yield the vastly different rate constant values $k_1 \approx 1 \times 10^{-10}\ M^{-1}\ s^{-1}$ and $k_2 \approx 6.3 \times 10^4\ s^{-1}$. Therefore, as soon as any $NO_3$ forms in step 1, it rapidly falls apart to NO and $O_2$ in step 2.

Now let's consider the reaction that is the reverse of $NO_2$ decomposition—namely, the formation of $NO_2$ from NO and $O_2$:

$$\text{Overall reaction} \qquad 2\,NO(g) + O_2(g) \rightarrow 2\,NO_2(g)$$

We determined in Section 13.3 that this reaction is second order in NO and first order in $O_2$:

$$\text{Rate} = k[NO]^2[O_2] \qquad (13.33)$$

One proposed mechanism is shown in **Figure 13.24**. It has two elementary steps:

$$\text{Step 1} \quad NO(g) + O_2(g) \rightleftharpoons NO_3(g) \qquad \text{Rate} = k_1[NO][O_2] \quad (13.34)$$

$$\text{Step 2} \quad NO_3(g) + NO(g) \rightarrow 2\,NO_2(g) \qquad \text{Rate} = k_2[NO_3][NO] \quad (13.35)$$

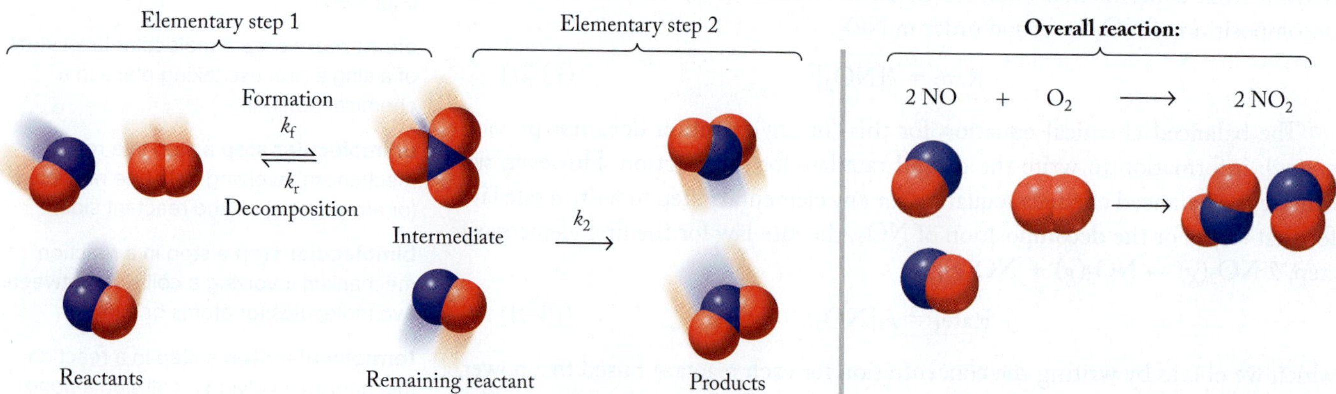


**FIGURE 13.24** A mechanism for the formation of $NO_2$ from NO and $O_2$ has two elementary steps: (1) a fast, reversible bimolecular reaction in which NO and $O_2$ form $NO_3$ and (2) a slower, rate-determining bimolecular reaction in which $NO_3$ reacts with a molecule of NO to form two molecules of $NO_2$.

The rate laws in Equations 13.34 and 13.35 were obtained by writing the concentration for each reactant raised to a power equal to the reactant's coefficient in the balanced equation. If step 1 were the rate-determining step, the reaction would be first order in NO and $O_2$, but that is not what the experimentally determined rate law indicates. If step 2 were the rate-determining step, the reaction would be first order in NO and $NO_3$, but that is inconsistent with the rate law, too. How can we account for the experimental rate law?

Consider what happens if step 2 is slow while step 1 is fast and reversible, which means that $NO_3$ forms rapidly from NO and $O_2$ but decomposes just as rapidly back into NO and $O_2$. We indicate reversibility with a double arrow in Figure 13.24. Expressing this situation in equation form:

$$\text{Rate of forward reaction} = k_f[\text{NO}][\text{O}_2] = \text{fast}$$

$$\text{Rate of reverse reaction} = k_r[\text{NO}_3] = \text{equally fast}$$

The subscripts "f" and "r" refer to the "forward" and "reverse" reactions.

Combining these two expressions, we have

$$k_f[\text{NO}][\text{O}_2] = k_r[\text{NO}_3]$$

$$[\text{NO}_3] = \frac{k_f}{k_r}[\text{NO}][\text{O}_2] \qquad (13.36)$$

Now, if we replace the $[NO_3]$ term in Equation 13.35 (which we select because step 2 is the rate-determining step) with the right side of Equation 13.36, we get

$$\text{Rate} = k_2 \frac{k_f}{k_r}[\text{NO}]^2[\text{O}_2]$$

The three rate constants can be combined,

$$k_{\text{overall}} = k_2 \frac{k_f}{k_r}$$

and the rate law for the overall reaction becomes

$$\text{Rate} = k_{\text{overall}}[\text{NO}]^2[\text{O}_2]$$

This expression matches the overall rate law in Equation 13.33, so the proposed mechanism, a fast and reversible step 1 followed by a slow step 2, *may* be valid.

Even though the proposed reaction mechanism is consistent with the overall stoichiometry of the reaction and with the experimentally derived rate law, these consistencies do not prove that the proposed mechanism is correct. We need more experimental evidence, such as detecting the presence of an intermediate in the reaction mixture. By contrast, not finding a reactive (and transient) intermediate would not necessarily disprove a reaction mechanism.

## SAMPLE EXERCISE 13.12 Linking Reaction Mechanisms to an Experimental Rate Law

**LO8**

The experimentally determined rate law for the reduction reaction,

$$2\,\text{H}_2(g) + 2\,\text{NO}(g) \rightarrow \text{N}_2(g) + 2\,\text{H}_2\text{O}(g)$$

which occurs at high temperatures, is

$$\text{Rate} = k[\text{NO}]^2[\text{H}_2]$$

A proposed mechanism for the reaction, shown in **Figure 13.25**, consists of the following steps:

Elementary step 1 $H_2(g) + 2\,NO(g) \rightarrow N_2O(g) + H_2O(g)$

Elementary step 2 $H_2(g) + N_2O(g) \rightarrow N_2(g) + H_2O(g)$

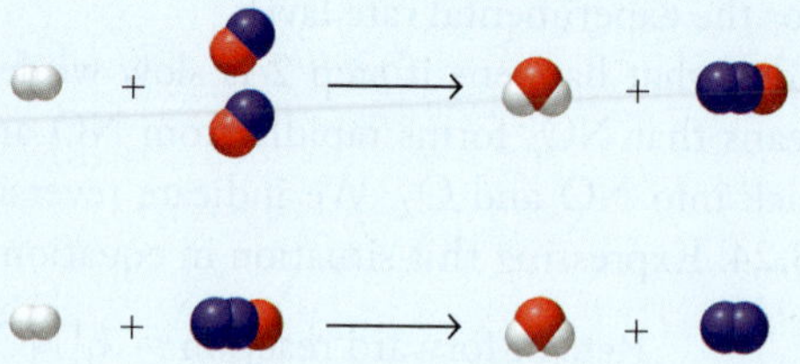

Elementary step 1 + ⟶ +

Elementary step 2 + ⟶ +

**FIGURE 13.25** A proposed mechanism for the reaction between $NO(g)$ and $H_2(g)$.

Is this reaction mechanism consistent with the stoichiometry of the reaction and with the rate law? If so, which is the rate-determining step?

**Collect, Organize, and Analyze** To answer the questions posed in this exercise, we need to do the following:

1. Determine whether the chemical equations of the elementary steps add up to the overall reaction equation.
2. Write rate laws for each elementary step.
3. Compare the rate laws of the elementary steps to the experimental rate law of the overall reaction to assess the validity of the mechanism.
4. Determine which step is rate determining by matching its rate law to the observed rate law.

**Solve** Let's test whether the elementary steps add up to the overall reaction:

$$(1) \quad H_2(g) + 2\,NO(g) \rightarrow \cancel{N_2O(g)} + H_2O(g)$$
$$(2) \quad H_2(g) + \cancel{N_2O(g)} \rightarrow N_2(g) + H_2O(g)$$
$$2\,H_2(g) + 2\,NO(g) \rightarrow N_2(g) + 2\,H_2O(g)$$

This is indeed the equation of the overall reaction, so the elementary steps are consistent with the stoichiometry of the overall reaction.

Next, we focus on the reaction mechanism. Elementary step 1 involves two molecules of NO colliding with one molecule of $H_2$ in a rare termolecular reaction. Elementary step 2 is a bimolecular reaction between the $N_2O$ produced in step 1 and another molecule of $H_2$. The rate law for any elementary step can be written directly from the balanced equations, using the stoichiometric coefficients as exponents in the rate law:

Step 1 $H_2(g) + 2\,NO(g) \rightarrow N_2O(g) + H_2O(g)$ $\quad \text{Rate} = k_1[H_2][NO]^2$

Step 2 $H_2(g) + N_2O(g) \rightarrow N_2(g) + H_2O(g)$ $\quad \text{Rate} = k_2[H_2][N_2O]$

The rate law of step 1 matches the observed rate law of the overall reaction. Therefore, the proposed two-step mechanism is consistent with the experimental rate law, and step 1 is the rate-determining step.

**Think About It** We do not have direct proof that this is the correct mechanism, though it is consistent with the available data. A plausible mechanism for a reaction must yield a balanced equation whose stoichiometry matches the overall reaction and must be consistent with the rate law for the overall process. Both conditions are met here in Sample Exercise 13.12.

 **Practice Exercise** The following mechanism is proposed for a reaction between compounds A and B:

| | | |
|---|---|---|
| Step 1 | $2\,A(g) + B(g) \rightleftharpoons C(g)$ | fast and reversible |
| Step 2 | $B(g) + C(g) \rightarrow D(g)$ | slow |
| Overall | $2\,A(g) + 2\,B(g) \rightarrow D(g)$ | |

What is the rate law for the overall reaction based on the proposed mechanism?

**SAMPLE EXERCISE 13.13** Testing a Proposed Reaction Mechanism **LO8**

One proposed mechanism for the decomposition of $N_2O_5$ to $NO_2$ involves the following three elementary steps:

| | | |
|---|---|---|
| Step 1 | $2\,N_2O_5(g) \rightleftharpoons N_4O_{10}(g)$ | fast and reversible |
| Step 2 | $N_4O_{10}(g) \rightarrow N_2O_3(g) + 2\,NO_2(g) + O_3(g)$ | slow |
| Step 3 | $N_2O_3(g) + O_3(g) \rightarrow 2\,NO_2(g) + O_2(g)$ | fast |
| Overall: | $2\,N_2O_5(g) \rightarrow 4\,NO_2(g) + O_2(g)$ | |

What is the rate law for the overall reaction based on the proposed mechanism?

**Collect and Organize** We are asked to find the rate law for the overall reaction, which will reflect the rate laws for the elementary reactions preceding and including the rate-determining step. We are given three elementary steps and their relative rates.

**Analyze** The rate law for an elementary step can be written directly from the balanced chemical equation for that step. We are told that step 1 is fast *and reversible*, which means that as the reaction proceeds and $[N_4O_{10}]$ increases, the rate of step 1 in the forward direction is eventually matched by the rate of step 1 in the reverse direction. Only elementary steps 1 and 2 will contribute to the rate law for the overall reaction because step 2, the only slow step, must be the rate-determining step.

**Solve** The rate laws for step 1 in the forward and reverse directions are

$$\text{Rate of forward step 1} = k_f[N_2O_5]^2$$

$$\text{Rate of reverse step 1} = k_r[N_4O_{10}]$$

The rate law for step 2 is

$$\text{Rate of step 2} = k_2[N_4O_{10}]$$

$N_4O_{10}$ is an intermediate in this reaction, so it should not appear in the overall rate law. To find the overall rate law, we need to express $[N_4O_{10}]$ in terms of $[N_2O_5]$ by setting the step 1 rates equal to each other and rearranging the terms:

$$[N_4O_{10}] = \frac{k_f}{k_r}[N_2O_5]^2$$

We then substitute this expression for $[N_4O_{10}]$ in the rate law for step 2, which is the rate-determining step, and then combine all the rate constants into one overall $k$:

$$\text{Rate} = k_2[N_4O_{10}] = k_2\frac{k_f}{k_r}[N_2O_5]^2 = k[N_2O_5]^2$$

**Think About It** The rate law for the overall reaction must depend on the concentration of $N_2O_5$. The order of the reaction in $N_2O_5$ depends on the proposed mechanism for the reaction. Notice that step 3 has no impact on the rate law for this mechanism because it comes after the slowest step. In addition, note that the rate law for the mechanism here in Sample Exercise 13.13 does not match the experimentally determined rate law from Sample Exercise 13.5: Rate = $k[N_2O_5]$. Therefore, the mechanism that starts with the dimerization of $N_2O_5$ cannot be correct.

**Practice Exercise** Here is another proposed mechanism for the reaction of NO with $H_2$ (Sample Exercise 13.12):

| | |
|---|---|
| Elementary step 1 | $H_2(g) + NO(g) \rightarrow N(g) + H_2O(g)$ |
| Elementary step 2 | $N(g) + NO(g) \rightarrow N_2(g) + O(g)$ |
| Elementary step 3 | $H_2(g) + O(g) \rightarrow H_2O(g)$ |

Is this a valid mechanism?

## Mechanisms and One Meaning of Zero Order

Let's revisit the reaction between $NO_2$ and CO, which has an experimentally determined rate law that is zero order in CO, second order in $NO_2$, and second order overall:

$$NO_2(g) + CO(g) \rightarrow NO(g) + CO_2(g) \qquad \text{Rate} = k[NO_2]^2$$

What does this overall rate law tell us about the reaction? Remember that the overall rate law depends on the concentration of reactants in the rate-determining step, which means that CO is not a reactant in the rate-determining step. Carbon monoxide is clearly involved in the reaction—it is converted into $CO_2$—but whatever step involves CO must be more rapid than the rate-determining step. This leads us to conclude that the reaction must have at least two elementary steps.

It has been proposed that the reaction mechanism is the following:

$$(1) \qquad 2\,NO_2(g) \rightarrow NO_3(g) + NO(g) \qquad \text{Rate} = k_1[NO_2]^2$$

$$(2) \qquad NO_3(g) + CO(g) \rightarrow NO_2(g) + CO_2(g) \qquad \text{Rate} = k_2[NO_3][CO]$$

The experimentally determined overall rate law matches the rate law for the first step, which must be the slower, rate-determining step. The overall reaction is zero order in CO because CO is not a reactant in that step.

**CONCEPT TEST**

The rate law for the reaction $XO_2(g) + M(g) \rightarrow XO(g) + MO(g)$, where X and M represent two different elements, is second order in $[XO_2]$ and independent of [M] for a wide variety of compounds and elements. Why can we conclude that these reactions probably proceed by the same mechanism?

Before closing this discussion on reaction mechanisms, let's revisit the first reaction we introduced in this chapter:

$$N_2(g) + O_2(g) \rightarrow 2\,NO(g) \qquad (13.1)$$

We noted in Section 13.1 that the reaction does not happen in one step due to collisions between molecules of $N_2$ and $O_2$, but rather in two steps. The first step, involving collisions of O atoms and $N_2$ molecules, is followed by a second step, involving collisions between N atoms and molecules of $O_2$:

$$\text{(1)} \quad O(g) + N_2(g) \rightarrow NO(g) + N(g)$$

$$\text{(2)} \quad N(g) + O_2(g) \rightarrow NO(g) + O(g)$$

These are the elementary steps that result in the formation of NO at the high temperatures in internal combustion engines. Experiments have indicated that the first reaction is slower than the second, which makes the first reaction the rate-determining step. The molecularity of this step should therefore define the rate constant expression for the overall reaction:

$$\text{Rate} = k_1[O][N_2]$$

Unfortunately, this equation includes a concentration term for O atoms, but atomic oxygen is not a reactant in the overall equation describing the reaction. Moreover, analysis of kinetics data led to our deriving the following rate law for the same reaction in Sample Exercise 13.4:

$$\text{Rate} = k[O_2]^{1/2}[N_2]$$

Is there a way to link these two rate laws? The answer is yes because the concentrations of O atoms inside automobile engines' combustion chambers are the result of a separate reaction not directly involved in NO formation. This reaction involves collisions between very stable molecules (M) and molecules of $O_2$, and it results in the dissociation of $O_2$ into atomic O:

$$\text{(3)} \quad M(g) + O_2(g) \rightleftharpoons M(g) + 2\,O(g)$$

Reaction (3) is rapid and reversible, and the rates of the forward and reverse reactions become equal as molecules of $O_2$ break up into atoms of O as rapidly as O atoms recombine into molecules of $O_2$. Assigning rate constant symbols $k_f$ and $k_r$ to the forward and reverse reactions, we can set up the following rate equality:

$$k_f[M][O_2] = k_r[M][O]^2$$

Rearranging the terms and solving for [O]:

$$[O]^2 = \frac{k_f}{k_r}\frac{\cancel{[M]}[O_2]}{\cancel{[M]}} = \frac{k_f}{k_r}[O_2] \quad \text{or} \quad [O] = \left(\frac{k_f}{k_r}\right)^{1/2}[O_2]^{1/2}$$

Replacing the [O] term in the rate law for reaction (1) with the right side of the above equation and combining the three $k$ terms into an overall $k$:

$$\text{Rate} = k_1[O][N_2] = k_1\left(\frac{k_f}{k_r}\right)^{1/2}[O_2]^{1/2}[N_2] = k[O_2]^{1/2}[N_2]$$

As you can see, a mechanism based on the molecular collisions we described in Section 13.1 is consistent with the rate law we developed from initial reaction rate data in Sample Exercise 13.4.

## 13.6 Catalysts

In reaction (3) in the preceding discussion, you may have noticed that the stable molecule M was unchanged in the reaction even though it played an important role in the dissociation of $O_2$ to O. A substance such as M, which increases the rate of a reaction but is not consumed in the process, is called a **catalyst.** Let's conclude this chapter by exploring two kinds of catalysts that affect our everyday lives—one that causes air pollution and one that helps keep the air clean.

**catalyst** a substance added to a reaction that increases the rate of the reaction but is not consumed in the process.

## Catalysts and the Ozone Layer

Although the ozone in the troposphere (between 6 km and 10 km above Earth's surface) is a pollutant and health hazard, the ozone in the stratosphere (between 10 km and 40 km above Earth's surface) is necessary to protect us from solar UV radiation. In this section, we discuss the role of catalysts in the breakdown of stratospheric ozone that has led to the annual formation of ozone holes over Antarctica during early spring in the Southern Hemisphere (September and October).

The natural photodecomposition of ozone in the stratosphere,

$$2\ O_3(g) \xrightarrow{\text{sunlight}} 3\ O_2(g) \qquad (13.37)$$

begins with the absorption of UV radiation from the sun and the generation of atomic oxygen:

$$(1) \qquad O_3(g) \rightarrow O_2(g) + O(g)$$

The oxygen atom can react with an ozone molecule to form two more molecules of oxygen:

$$(2) \qquad O_3(g) + O(g) \rightarrow 2\ O_2(g)$$

The second elementary step is slow because its activation energy is relatively high: 17.7 kJ/mol.

In 1974, two American scientists, Sherwood Rowland (1927–2012) and Mario Molina (b. 1943), predicted significant depletion of stratospheric ozone because of the release of a class of volatile compounds called chlorofluorocarbons (CFCs) into the atmosphere at ground level that ultimately enter the stratosphere. This prediction was later supported by experimental evidence of a thinning of the ozone layer and the formation of annual ozone holes over Antarctica. In September 2000, stratospheric ozone concentrations over Antarctica were the lowest ever observed—less than half of what they were in 1980—and the ozone hole covered nearly all of Antarctica and the tip of South America (**Figure 13.26**).

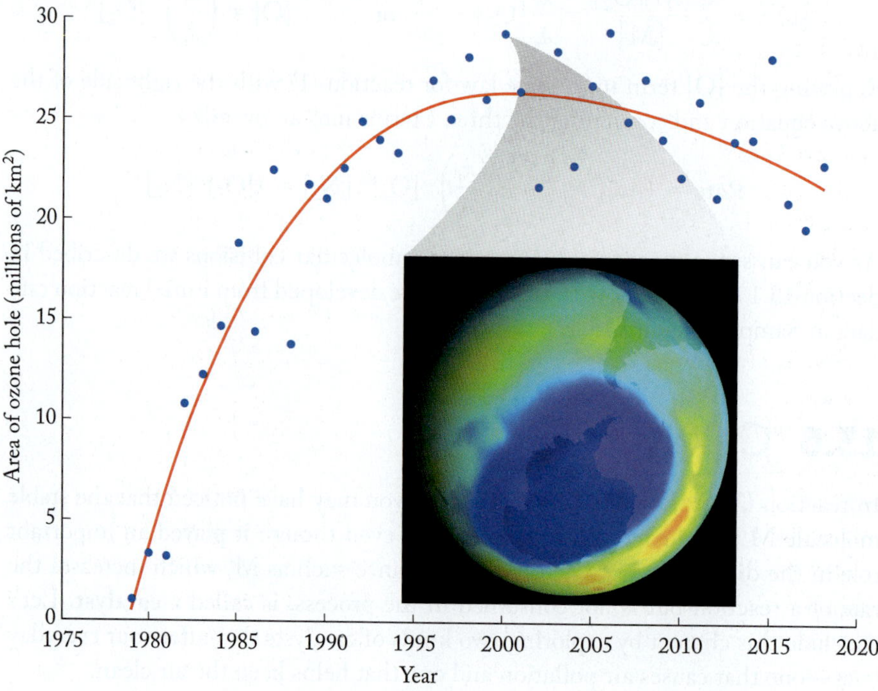


**FIGURE 13.26** Trends in stratospheric ozone depletion over the South Pole. The data points represent the maximum area of the ozone hole observed during each spring in the Southern Hemisphere (usually in September or October). The satellite image shows the largest hole ever recorded, which occurred in 2000. The blue-violet color over Antarctica represents ozone concentrations that are less than half their normal values.

In 1987, an international agreement known as the Montreal Protocol called for an end to the production of ozone-depleting CFCs, which were in common use as refrigerants and aerosol propellants. The Montreal Protocol has had a dramatic effect on CFC production and emission into the atmosphere, and as the trend line in Figure 13.26 indicates, the ozone layer over Antarctica may be slowly recovering. In 2018, the area of the ozone hole was 22.9 million km$^2$, down from the 2000 maximum of 29.4 million km$^2$.

**homogeneous catalyst** a catalyst in the same phase as the reactants.

How do CFCs contribute to the destruction of ozone? Three of the more widely used CFCs were $CCl_2F_2$, $CCl_3F$, and $CClF_3$. In the stratosphere, they encounter UV radiation with enough energy to break C—Cl bonds, releasing chlorine atoms:

$$CCl_3F(g) \rightarrow CCl_2F(g) + Cl(g) \quad (13.38)$$

Free chlorine atoms react with ozone, forming chlorine monoxide:

$$Cl(g) + O_3(g) \rightarrow ClO(g) + O_2(g) \quad (13.39)$$

Chlorine monoxide then reacts with ozone, producing oxygen and regenerating atomic chlorine:

$$ClO(g) + O_3(g) \rightarrow Cl(g) + 2\,O_2(g) \quad (13.40)$$

If we add Equations 13.39 and 13.40 and cancel species as needed, we get

$$2\,O_3(g) \rightarrow 3\,O_2(g)$$

The overall reaction is the same as that for the natural photodecomposition of ozone in Equation 13.37. The difference is the presence of chlorine atoms in Equations 13.39 and 13.40. Chlorine atoms act as a catalyst for the destruction of ozone because they speed up the reaction but are not consumed by it. Rather, they are consumed in one elementary step but then regenerated in a later elementary step. Chlorine is a catalyst, not an intermediate, because a catalyst is consumed in an early step of a reaction mechanism and then regenerated in a later step, whereas an intermediate is produced before it is consumed. A single chlorine atom can catalyze the destruction of hundreds to thousands of stratospheric $O_3$ molecules before it combines with other atoms and forms a less reactive molecule.

The activation energy for the Cl-catalyzed decomposition of $O_3$ is only 2.2 kJ/mol, whereas the activation energy for the uncatalyzed reaction is 17.7 kJ/mol (**Figure 13.27**). The smaller $E_a$ value of the catalyzed destruction of ozone means that it occurs faster than the natural photodecomposition process. In the reaction describing the destruction of ozone, the catalyst, Cl($g$), and reactant, $O_3(g)$, exist in the same physical phase. When a catalyst and the reacting species are in the same phase, the catalyst is a **homogeneous catalyst**.

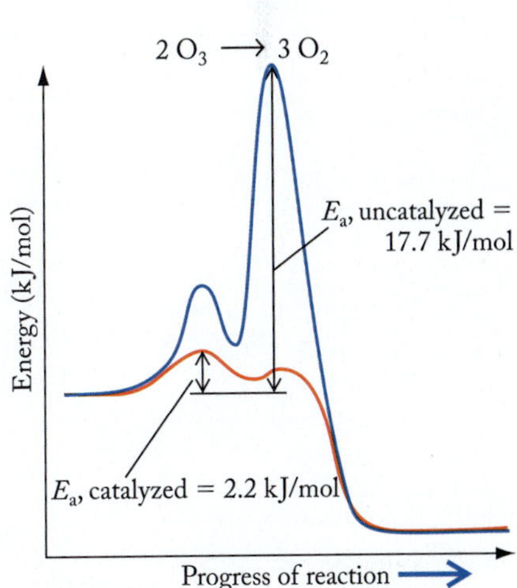


**FIGURE 13.27** The decomposition of $O_3$ in the presence of chlorine atoms has a lower activation energy (2.2 kJ/mol) than that of the naturally occurring photodecomposition of $O_3$ to $O_2$ (17.7 kJ/mol). The catalytic effect of chlorine is a key factor in the depletion of stratospheric ozone and the formation of an ozone hole over the South Pole.

The rate of ozone depletion accelerates over the South Pole toward the end of the Southern Hemisphere winter because of the clouds that cover much of Antarctica. These clouds form from crystals of ice and tiny drops of nitric acid in the winter, when temperatures in the lower stratosphere dip to −80°C. The clouds become collection sites for chlorine compounds formed by the photodecomposition of CFCs, and they catalyze the transformation of these compounds into molecular chlorine, $Cl_2$. During late winter and early spring, the sun rises over Antarctica and its UV rays decompose $Cl_2$ into free Cl atoms. Because atomic chlorine is a powerful catalyst for ozone destruction, high concentrations of newly formed atomic Cl mean rapid and widespread loss of stratospheric $O_3$ over Antarctica and the Southern Ocean.

### SAMPLE EXERCISE 13.14 Identifying a Catalyst in a Reaction Mechanism LO9

A reaction mechanism proposed for the decomposition of ozone in the presence of NO at high temperatures consists of three elementary steps:

$$(1) \quad O_3(g) + NO(g) \rightarrow O_2(g) + NO_2(g)$$

$$(2) \quad NO_2(g) \rightarrow NO(g) + O(g)$$

$$(3) \quad O(g) + O_3(g) \rightarrow 2\,O_2(g)$$

If the rate of the overall reaction is higher than the rate of the uncatalyzed decomposition of ozone to oxygen (Equation 13.37), is NO a catalyst in the reaction?

**Collect and Organize** We are asked to determine whether NO is a catalyst in a reaction. A catalyst increases the rate of a reaction and is not consumed by the overall reaction. We are given the elementary steps of the reaction and are told that the reaction is more rapid in the presence of NO.

**Analyze** We can sum the reactions to determine the overall reaction. If NO is consumed in an early step before it is regenerated in a later one and it is not consumed in the overall process, it is a catalyst.

**Solve** Summing the three elementary steps gives the overall reaction:

$$O_3(g) + \cancel{NO(g)} + \cancel{NO_2(g)} + \cancel{O(g)} + O_3(g) \rightarrow$$

$$O_2(g) + \cancel{NO_2(g)} + \cancel{NO(g)} + \cancel{O(g)} + 2\,O_2(g)$$

$$2\,O_3(g) \rightarrow 3\,O_2(g)$$

This equation does not include NO, and the rate of the reaction is higher when NO is present. Thus, NO fulfills both requirements for being a catalyst.

**Think About It** NO behaves much like the Cl atoms in Equations 13.39 and 13.40. NO is not an intermediate because it is consumed *before* it is produced.

**Practice Exercise** The combustion of fossil fuels results in the release of $SO_2$ into the atmosphere, where it reacts with oxygen to form $SO_3$:

$$2\,SO_2(g) + O_2(g) \rightarrow 2\,SO_3(g)$$

In the atmosphere, $SO_2$ may react with $NO_2$, forming $SO_3$ and NO:

$$NO_2(g) + SO_2(g) \rightarrow NO(g) + SO_3(g)$$

The rate of reaction of $SO_2$ with $NO_2$ is faster than the rate of reaction of $SO_2$ with oxygen. If the NO produced in the reaction of $NO_2$ and $SO_2$ is then oxidized to $NO_2$,

$$2\,NO(g) + O_2(g) \rightarrow 2\,NO_2(g)$$

is $NO_2$ a catalyst in the reaction of $SO_2$ with $O_2$?

## Catalytic Converters

In Section 13.1 we discussed the air pollution caused by vehicles and the technology that has been developed to clean the air. **Figure 13.28** shows a catalytic converter in a car's exhaust system and illustrates how it provides a surface on which NO in engine exhaust decomposes to $N_2$ and $O_2$. A catalytic converter is an example of a **heterogeneous catalyst**, where the reactants and the catalyst have different phases. To maximize the surface area inside a catalytic converter,

**heterogeneous catalyst** a catalyst in a different phase from that of the reactants.

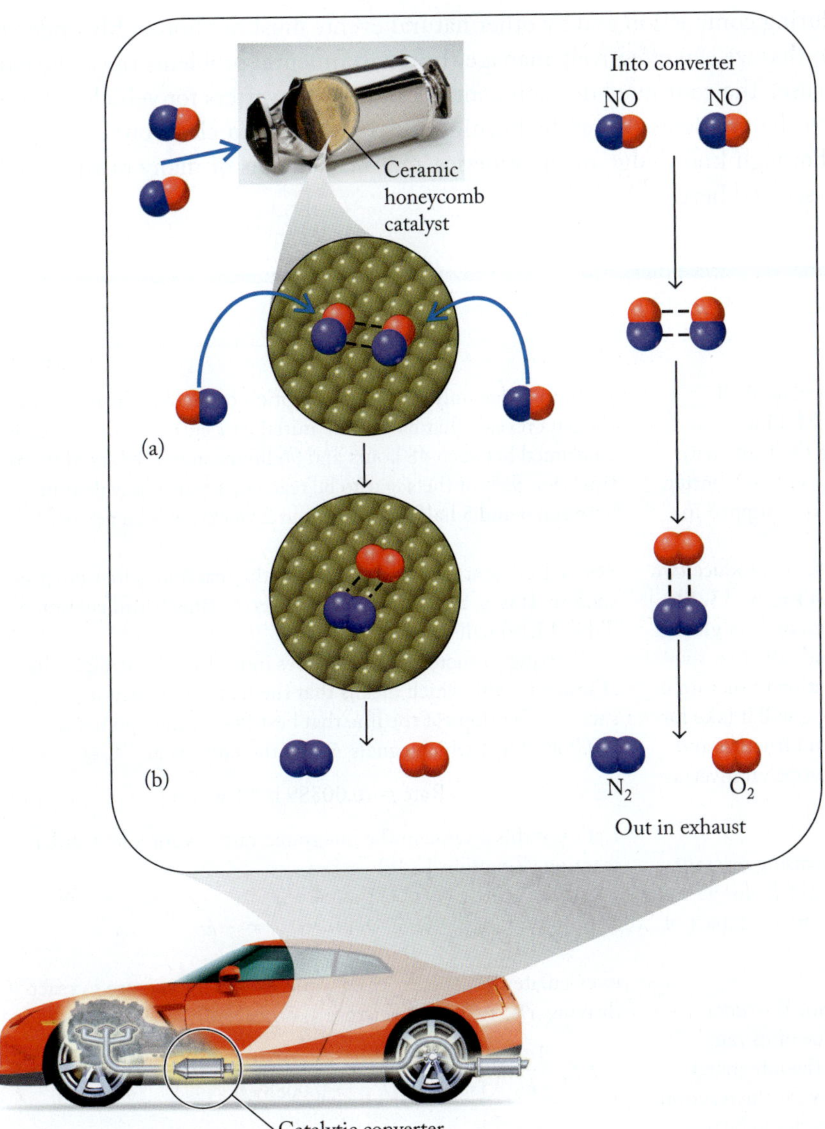


**FIGURE 13.28** Catalytic converters in automobiles reduce emissions of NO by lowering the activation energy of its decomposition into $N_2$ and $O_2$. Metal catalysts are supported on a porous ceramic honeycomb. At the molecular level, (a) NO molecules are adsorbed onto the surface of metal clusters, where their NO bonds are broken, and (b) pairs of O atoms and N atoms form $O_2$ and $N_2$. The $O_2$ and $N_2$ desorb from the surface and are released to the atmosphere.

it is filled with a fine honeycomb mesh coated with transition metals such as palladium, platinum, and rhodium. The metals that make up catalytic coatings are first dissolved as metal salts and dispersed on the mesh, where they are reduced to metallic clusters 2 nm to 10 nm in diameter. The small size of these clusters and the large number of them add to the total surface area available to promote the removal of pollutants. Some of these clusters specifically speed up the decomposition of NO, whereas others promote the oxidation of carbon monoxide to $CO_2$ and unburned hydrocarbons to $CO_2$ and water vapor.

Many gas-phase reactions that occur on solid heterogeneous catalysts exhibit zero-order reaction kinetics. This happens when there are more reactant molecules in the system than there are catalytic surface sites available to promote their decomposition. Under these conditions, adding more reactant has no effect on the catalyzed reaction rate, and the reaction appears to be zero order.

In the reactions responsible for photochemical smog, we saw that determining the rates of chemical reactions is crucial for us to comprehend processes on a molecular level. The mechanisms of the reactions of volatile oxides produced during combustion and by other natural events must be thoroughly understood so that we can effectively manage the environmental problems these substances cause. The continued development of catalytic converters for vehicles to diminish the problems caused by burning fossil fuels and to clean our air requires a thorough knowledge of the kinetics and mechanisms of many of the reactions discussed here.

### SAMPLE EXERCISE 13.15 Integrating Concepts: Chocolate-Covered Cherries

Chocolate-covered cherries consist of a solid chocolate shell enclosing a sweet liquid that surrounds a cherry. Making chocolate-covered cherries involves first covering the fruits with fondant, a doughlike mixture of powdered sugar (sucrose), butter, cherry juice, and the enzyme invertase. They are then dipped in chocolate and stored at 5°C.

During storage, sucrose (table sugar) hydrolyzes to produce two isomeric sugars, glucose and fructose, as shown in **Figure 13.29**. Invertase acts as a catalyst for this reaction. The mixture of glucose and fructose is called invert sugar, and it is a liquid. The first two columns of **Table 13.19** list data on the concentration of sucrose in the fondant as a function of time at 5°C. How long will it take for 95% of the sucrose to be converted into glucose and fructose and for the cherries to be ready to eat? What would happen if invertase were left out of the recipe?

**Collect and Organize** We are given data on decreasing reactant concentration with time and asked how long it will take for 95% of the reactant to be consumed. We must also describe the impact of leaving out a catalyst from the reaction mixture.

**Analyze** The sucrose data can be used to determine the order of the hydrolysis reaction and to calculate the value of its rate constant at 5°C. This $k$ value can then be used in the integrated rate law with the available data to derive a rate law for the reaction and to estimate the time we must wait for the cherries to be ready. The reaction takes place in an aqueous solution, so we may assume that the concentration of water is very large and does not change during the reaction. Therefore, any change in the rate of the reaction is due only to the consumption of sucrose. Inspection of the data reveals that half of the initial concentration of sucrose is consumed between 48 hours and 96 hours, about 3 days of reaction time. For 95% of the sugar to be reacted, a time equivalent to between 4 and 5 half-lives, or about 2 weeks, will be needed.

**Solve** Let's start by testing whether the reaction is first order in sucrose. If it is, a plot of ln[sucrose] versus time (third column of Table 13.19) will be linear.

Plotting ln[sucrose] versus $t$ does indeed yield a straight line (**Figure 13.30**), which means that the reaction is first order in sucrose. The slope of the line that best fits the data points is $-0.00889\ h^{-1}$, which equals $-k$, so the rate law at 5°C is

$$\text{Rate} = (0.00889\ h^{-1})[\text{sucrose}]$$

Let's use this $k$ value in the integrated rate law for a first-order reaction (Equation 13.16)

$$\ln\left(\frac{[\text{sucrose}]_t}{[\text{sucrose}]_0}\right) = -kt$$

to calculate the time $t$ necessary for 95% of the sucrose to react (leaving 5%, or $0.05 \times [\text{sucrose}]_0$, unreacted):

$$t = \frac{1}{-k}\ln\left(\frac{[\text{sucrose}]_t}{[\text{sucrose}]_0}\right) = \frac{1}{-0.00889\ h^{-1}}\ln\left(\frac{0.05\ \cancel{[\text{sucrose}]_0}}{\cancel{[\text{sucrose}]_0}}\right)$$

$$= 337\ h$$

$$\left(337\ \cancel{h} \times \frac{1\ d}{24\ \cancel{h}}\right) = 14\ d\ (\text{or 2 weeks})$$

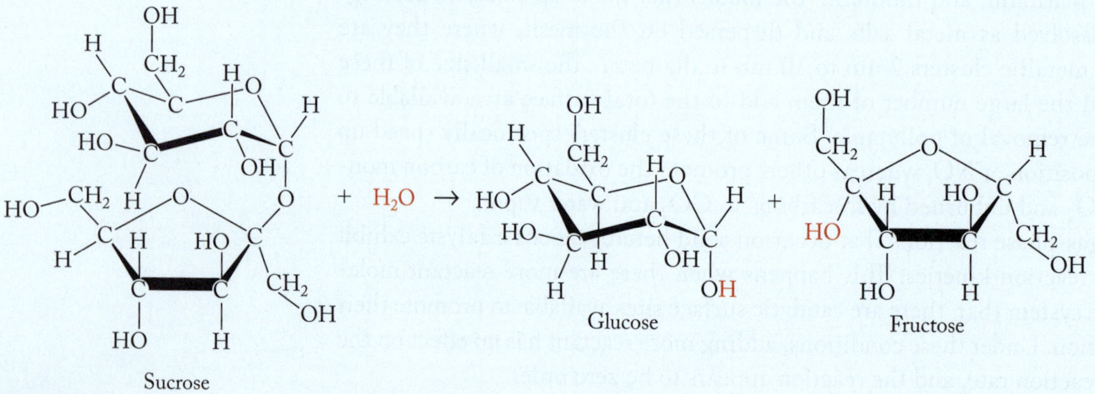

**FIGURE 13.29** Hydrolysis of sucrose produces glucose and fructose.

**TABLE 13.19** Concentration of Sucrose and ln[Sucrose] as a Function of Time

| Time (h) | [Sucrose] in Fondant (g/cm$^3$) | ln[Sucrose] |
|---|---|---|
| 0.0 | 0.844 | −0.170 |
| 24.0 | 0.681 | −0.384 |
| 48.0 | 0.550 | −0.598 |
| 96.0 | 0.357 | −1.030 |
| 192.0 | 0.153 | −1.877 |

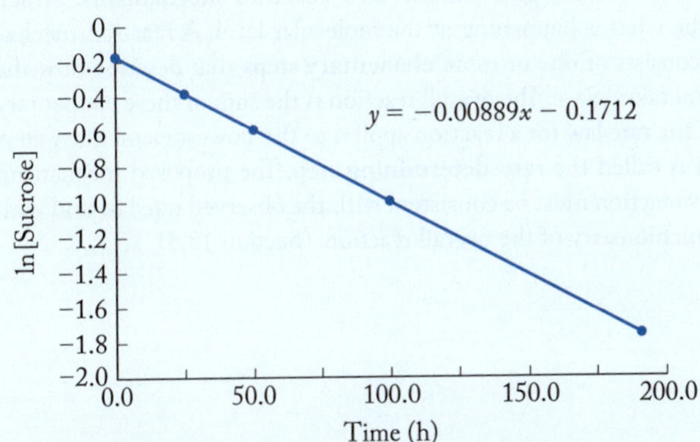


**FIGURE 13.30** Plot of ln[sucrose] as a function of time, using data from Table 13.19.

If the catalyst had not been included, the reaction rate would have been slower, and the time to reach 95% conversion of sucrose to glucose and fructose would have been a lot longer.

**Think About It** The calculated reaction time agrees with our original estimate based on the approximate number of half-lives required to reach 95% conversion. The reaction was run at 5 °C, probably to preserve the cherries and keep the chocolate solid, but that probably also slowed the rate of hydrolysis. Running the reaction at higher temperatures might have sped up the reaction but resulted in melted chocolate and spoiled cherries.

## SUMMARY

**LO1** **Collision theory** is a model that explains the rates of gas-phase reactions in terms of the frequency and energy of collision between the free atoms and molecules of reactants. Higher temperatures and higher concentrations produce more frequent collisions and more rapid reaction rates. (Section 13.2)

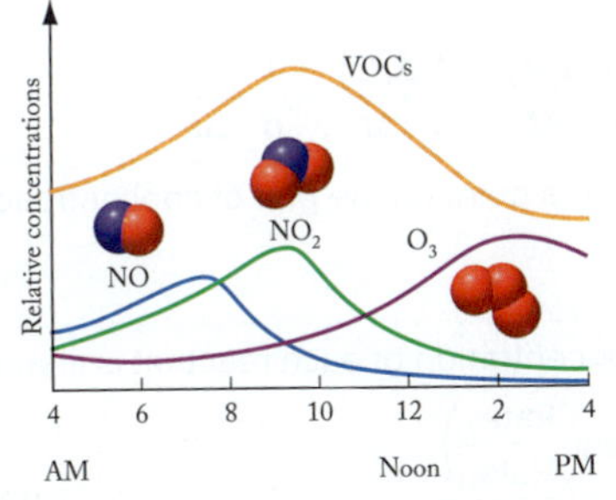


**LO2** The relative rates of consumption of reactants and formation of products are related by the stoichiometry of the reaction. (Section 13.2)

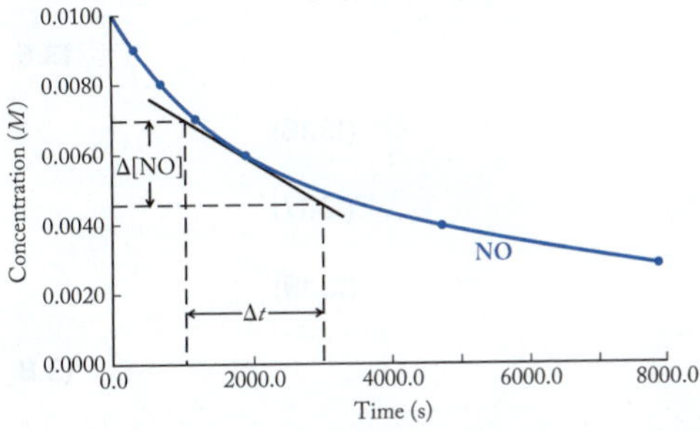


**LO3** The overall rate of a reaction can be expressed as either an average rate or an instantaneous rate. **Reaction rates** are typically determined from experimental measurements. In most reactions, the reaction rate decreases as the reaction proceeds. (Section 13.2)

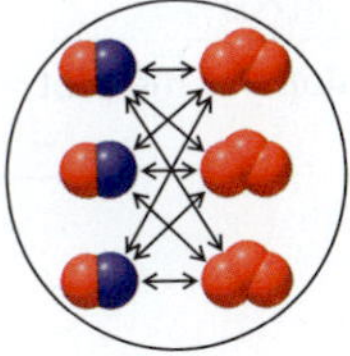

**LO4** The dependence of the rate of a reaction X + Y → Z on reactant concentrations is expressed in the **rate law** for the reaction: Rate = $k[A]^m[B]^n$, where $m$ and $n$ are the **reaction order** in reactants A and B, respectively, and $k$ is the **rate constant**. The units of a rate constant depend on the **overall reaction order**, which is the sum of the reaction orders with respect to individual reactants. (Section 13.3)

**LO5** The order of a reaction and the rate law for the reaction can be determined from differences in the initial rates of reaction observed with different concentrations of reactants or from single kinetics experiments by using **integrated rate laws**. (Section 13.3)

**LO6** The **half-life** ($t_{1/2}$) of a reactant is the time required for the concentration of a reactant to decrease to one-half its starting concentration. (Section 13.3)

**LO7** Increasing the temperature of a chemical reaction increases its rate and rate constant. The **activation energy** ($E_a$) of a reaction is the minimum energy that molecules need to react when they collide. Reactions with large activation energies are usually slow. Measuring the rate constant of a reaction at different temperatures allows the calculation of activation energies by using the **Arrhenius equation**. (Section 13.4)

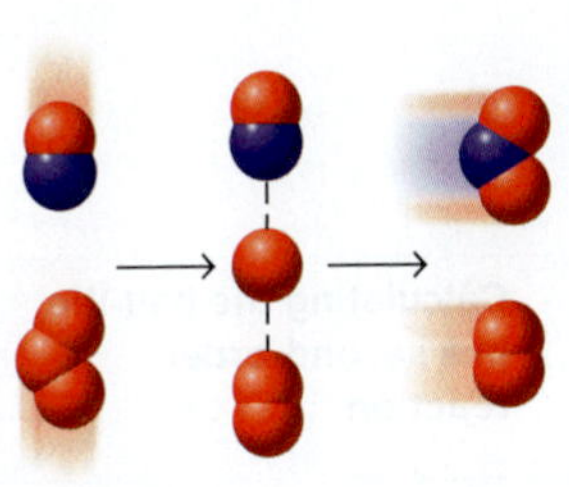

**LO8** Rate studies give insight into **reaction mechanisms,** which describe what is happening at the molecular level. A reaction mechanism consists of one or more **elementary steps** that describe how the reaction takes place. The overall reaction is the sum of these elementary steps. The rate law for a reaction applies to the slowest elementary step, which is called the **rate-determining step.** The proposed mechanism for any reaction must be consistent with the observed rate law and with the stoichiometry of the overall reaction. (Section 13.5)

**LO9** **Catalysts** increase reaction rates by changing the mechanism of a reaction and by decreasing the activation energy. A **homogeneous catalyst** is in the same phase as the reactants in the reaction being catalyzed. A **heterogeneous catalyst** is in a phase different from the phase of the reactant. (Section 13.6)

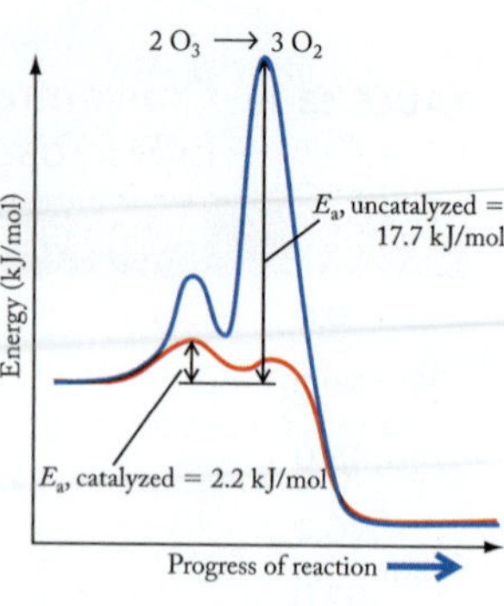


## PARTICULATE PREVIEW WRAP-UP

$$\begin{array}{ll} \text{Step 1:} & 2\,NO_2(g) \rightarrow \cancel{NO_3(g)} + NO(g) \\ \text{Step 2:} & \cancel{NO_3(g)} \rightarrow NO(g) + O_2(g) \\ \hline & 2\,NO_2(g) \rightarrow 2\,NO(g) + O_2(g) \end{array}$$

The dashed lines represent the simultaneous breaking of bonds in the reactants and forming of bonds in the intermediate and/or products. The $NO_3$ molecule produced in step 1 is an intermediate that decomposes in step 2.

## PROBLEM-SOLVING SUMMARY

| Type of Problem | Concepts and Equations | Sample Exercises |
|---|---|---|
| **Predicting relative reaction rates** | Determine relative rates from the stoichiometry of the balanced chemical equation. For $a\text{A} + b\text{B} \rightarrow c\text{C} + d\text{D}$, $$\text{Rate} = -\frac{1}{a}\frac{\Delta[\text{A}]}{\Delta t} = -\frac{1}{b}\frac{\Delta[\text{B}]}{\Delta t} = \frac{1}{c}\frac{\Delta[\text{C}]}{\Delta t} = \frac{1}{d}\frac{\Delta[\text{D}]}{\Delta t}$$ | **13.1, 13.2** |
| **Determining an instantaneous reaction rate** | Determine the slope of a line tangent to a point on the plot of concentration versus time. | **13.3** |
| **Deriving a rate law from initial reaction rate data** | Compare the change in rate as the concentration of each reactant is changed. $$m = \frac{\log\left(\frac{\text{Rate}_2}{\text{Rate}_1}\right)}{\log\left(\frac{[\text{A}]_2}{[\text{A}]_1}\right)} \quad (13.14)$$ | **13.4** |
| **Distinguishing between first- and second-order reactions** | A linear plot of the natural logarithm of reactant concentration (ln[X]) versus time indicates a first-order reaction, whereas a linear plot of the reciprocal of reactant concentration (1/[X]) versus time indicates a second-order reaction. | **13.5, 13.7** |
| **Relating reactant concentration, reaction time, and half-life** | For first-order reactions: $$\ln\frac{[\text{X}]_t}{[\text{X}]_0} = -kt \quad (13.16)$$ $$\ln[\text{X}]_t = -kt + \ln[\text{X}]_0 \quad (13.17)$$ $$t_{1/2} = \frac{0.693}{k} \quad (13.19)$$ | **13.6** |
| **Calculating the half-life in a second-order reaction** | In a second-order reaction, $$t_{1/2} = \frac{1}{k[\text{X}]_0} \quad (13.24)$$ | **13.8** |
| **Deriving pseudo–first-order rate laws** | Plot ln [limiting reactant] versus $t$. The slope of the plot = [excess reactant] × $k$. | **13.9** |

| Type of Problem | Concepts and Equations | Sample Exercises |
|---|---|---|
| **Calculating activation energies and frequency factors** | Determine $k$ values at different temperatures. Then plot ln $k$ versus $1/T$ and solve for $E_a$ and $A$: $$\ln k = -\frac{E_a}{R}\left(\frac{1}{T}\right) + \ln A \quad (13.29)$$ | **13.10** |
| **Calculating rate constants for a reaction at different temperatures** | Solve for $k_2$ in $$\ln \frac{k_2}{k_1} = -\frac{E_a}{R}\left(\frac{1}{T_2} - \frac{1}{T_1}\right) \quad (13.30)$$ | **13.11** |
| **Linking reaction mechanisms to experimental rate laws** | The chemical equations describing the steps in the mechanism must sum to the equation that describes the overall reaction; the rate law of the rate-determining step must be consistent with the rate law of the overall reaction. | **13.12, 13.13** |
| **Identifying catalysts in reaction mechanisms** | A homogeneous catalyst is consumed in an early step and regenerated in a later one. | **13.14** |

## VISUAL PROBLEMS

*(Answers to boldface end-of-chapter questions and problems are in the back of the book.)*

**13.1.** Nitrous oxide decomposes to nitrogen and oxygen in the following reaction:

$$2\ N_2O(g) \rightarrow 2\ N_2(g) + O_2(g)$$

In Figure P13.1, which curve represents $[N_2O]$ and which curve represents $[O_2]$?

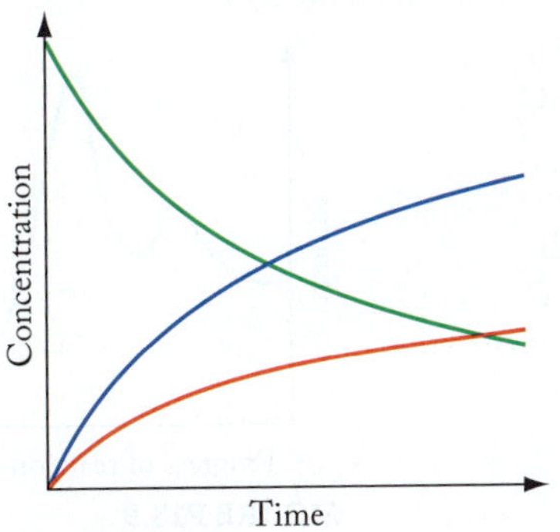


**FIGURE P13.1**

13.2. Sulfur trioxide is formed in the reaction

$$SO_2(g) + \tfrac{1}{2}\ O_2(g) \rightarrow SO_3(g)$$

In Figure P13.2, which curve represents $[SO_2]$ and which curve represents $[O_2]$? All three gases are present initially.

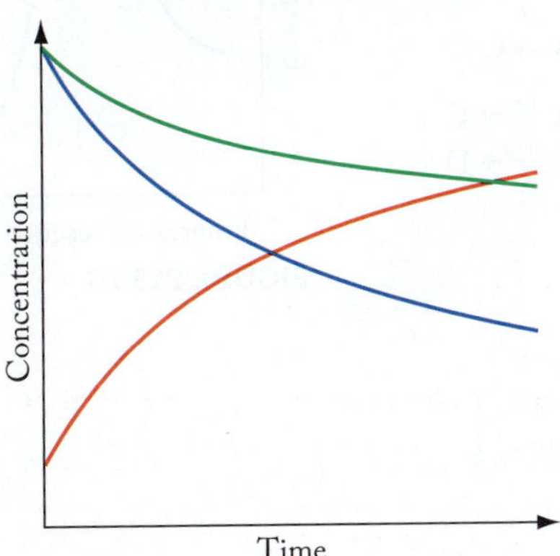


**FIGURE P13.2**

**13.3.** The rate law for the reaction $2\ A \rightarrow B$ is second order in A. Figure P13.3 represents samples with different concentrations of A, represented by red spheres. In which sample will the reaction $A \rightarrow B$ proceed most rapidly?

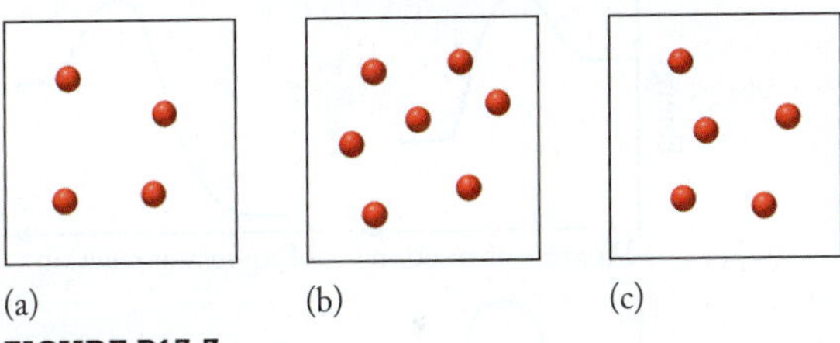


**FIGURE P13.3**

13.4. The rate law for the reaction $A + B \rightarrow C$ is first order in both A and B. Figure P13.4 represents samples with different concentrations of A (red spheres) and B (blue spheres). In which sample will the reaction $A + B \rightarrow C$ proceed most rapidly?

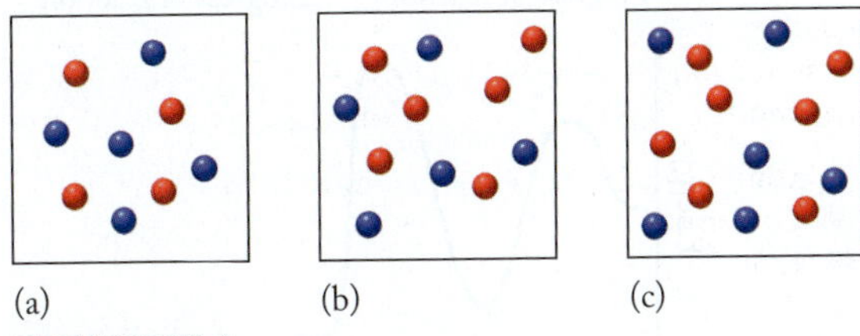


**FIGURE P13.4**

**13.5.** Using the numbers 1–5 on the reaction profile (Figure P13.5), identify the following:

a. The energy of the reactants
b. The energy of the products
c. The activation energy of the forward reaction
d. The activation energy of the reverse reaction
e. The energy change of the reaction

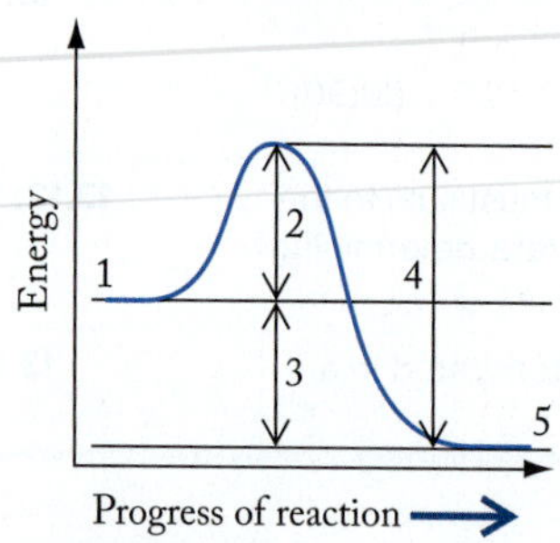


**FIGURE P13.5**

13.6. Select the energy profile from those given in Figure P13.6 that best corresponds to the following:

a. A highly exothermic reaction with a large activation energy
b. A highly endothermic reaction with a large activation energy
c. A reaction involving a stable intermediate

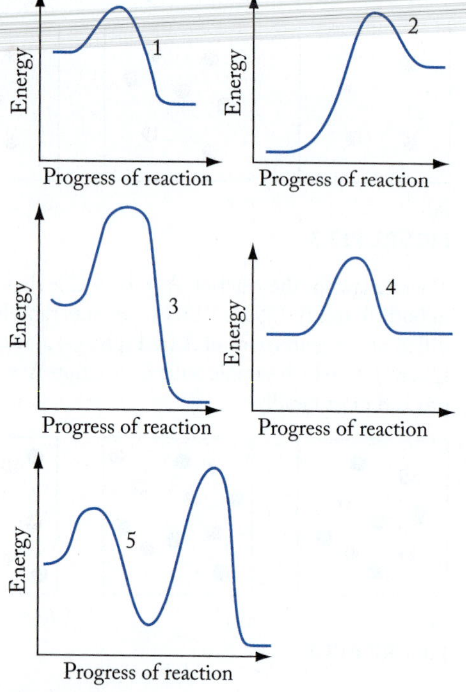


**FIGURE P13.6**

**13.7.** Which of the energy profiles in Figure P13.7 represents the reaction with the smallest rate constant at constant $T$?

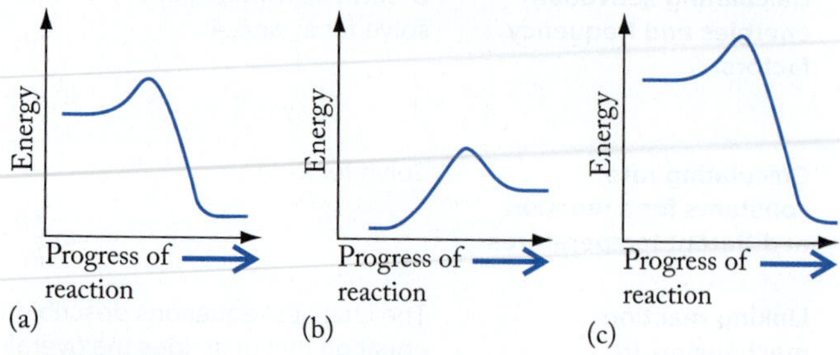


**FIGURE P13.7**

13.8. Which of the energy profiles in Figure P13.8 represents the reaction with the largest rate constant at constant $T$?

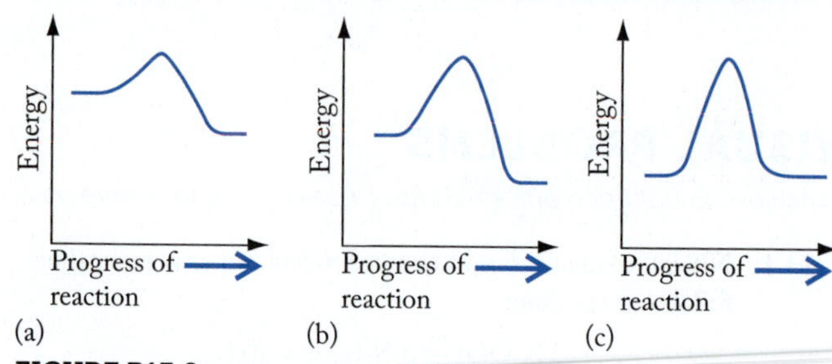


**FIGURE P13.8**

**13.9.** Which of the following mechanisms is consistent with the energy profile shown in Figure P13.9?

a. $2\,A \xrightarrow{\text{slow}} B$
$B \xrightarrow{\text{fast}} C$

b. $A + B \rightarrow C$

c. $2\,A \underset{}{\overset{\text{fast}}{\rightleftharpoons}} B$
$B \xrightarrow{\text{slow}} C$

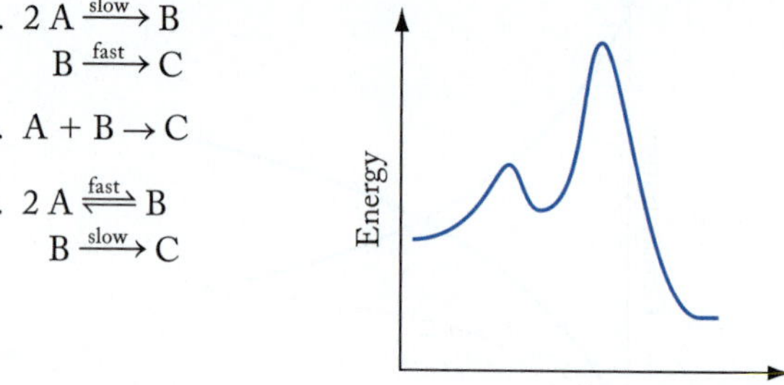


**FIGURE P13.9**

13.10. Which of the following mechanisms is consistent with the energy profile shown in Figure P13.10?

a. $A + B \xrightarrow{\text{slow}} C$
$C \xrightarrow{\text{fast}} D$

b. $A + B \rightarrow C$

c. $2\,A \xrightarrow{\text{fast}} C$
$B + C \xrightarrow{\text{slow}} D$

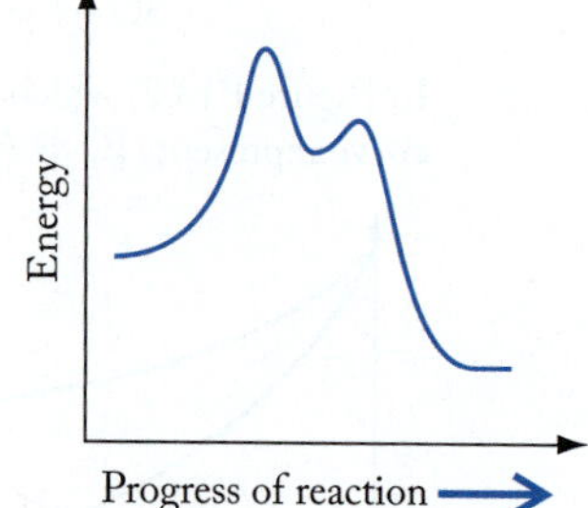


**FIGURE P13.10**

**13.11.** Which of the energy profiles in Figure P13.11 represents the effect of a catalyst on the rate of a reaction?

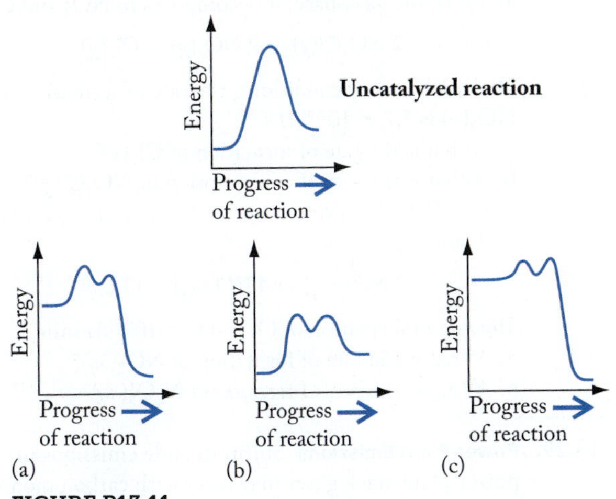


**FIGURE P13.11**

13.12. Explain why the energy profiles in Figure P13.12(a) and (b) do not represent the effect of a catalyst on the rate of the reaction.

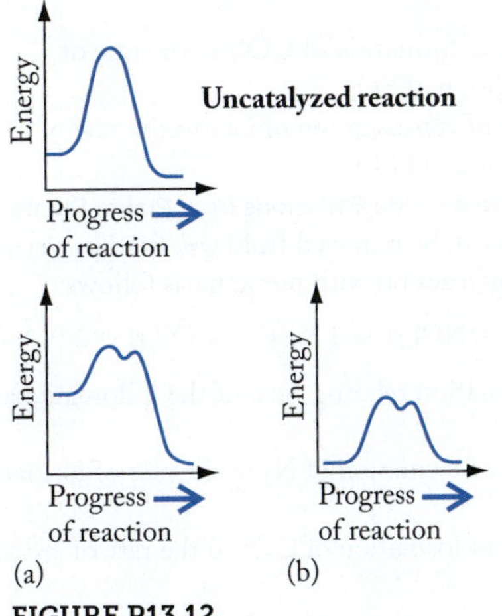


**FIGURE P13.12**

**13.13.** If we assume that the uncatalyzed reaction profiled in Figure P13.12 is reversible, is the value of $k_f$ at a given temperature (a) equal to, (b) greater than, or (c) less than $k_r$?

13.14. Use representations [A] through [I] in Figure P13.14 to answer questions a–f.

a. Which two elementary steps, when combined, would involve an intermediate of nitrogen trioxide?
b. What is the overall reaction for the two elementary steps in Problem 13.14(a)?
c. Which representation shows the photodecomposition of a chlorofluorocarbon?
d. Which two elementary steps, when combined, depict the chlorine-catalyzed destruction of ozone?
e. Which representation is the overall reaction for the chlorine-catalyzed destruction of ozone?
f. What happens to the number of collisions between chlorine radicals and ozone between representations [C], [F], and [I]?

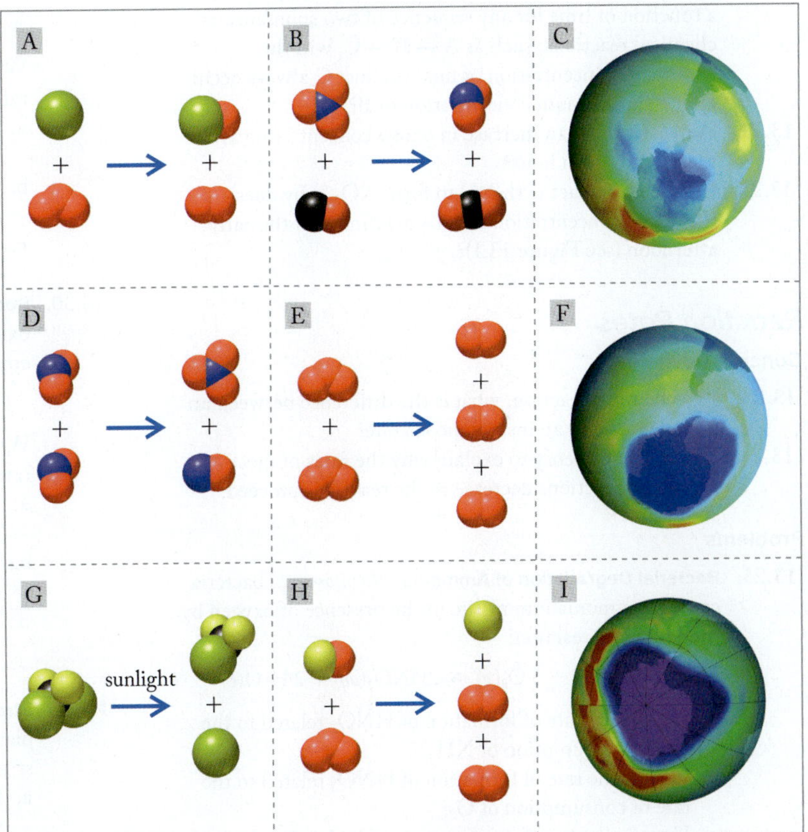


**FIGURE P13.14**

# QUESTIONS AND PROBLEMS

## Colliding Particles and the Chemistry of Smog Formation

### Concept Review

**13.15.** Give two reasons why increasing the temperature of a gas-phase reaction mixture increases the rate of the reaction.

13.16. Will compressing a reaction mixture into half its initial volume with no change in temperature affect the rate of the reaction? If so, how and why?

**13.17.** Of the atmospheric gases NO, $NO_2$, $O_3$, CO, and Ar, which one has the highest root-mean-square speed at 30°C? Which one has the highest average kinetic energy?

*13.18. Why do collisions between molecules of $N_2$ and other atmospheric gases rarely result in chemical reactions?

**13.19.** Why does the maximum concentration of ozone in Figure 13.1 occur later than the maximum concentrations of NO and $NO_2$?

13.20. In a plot of the concentration of reactants and products as a function of time for any sequence of two spontaneous chemical reactions, such as $A \rightarrow B \rightarrow C$, will the maximum concentration of final product C always occur after the maximum concentration of B?

**13.21.** Why isn't there an increase in ozone concentration after the evening rush hour?

13.22. If ozone can react with NO to form $NO_2$, why does the ozone concentration reach a maximum in the early afternoon (see Figure 13.1)?

## Reaction Rates

### Concept Review

**13.23.** In a chemical reaction, what is the difference between an average rate and an instantaneous rate?

13.24. Use collision theory to explain why the rates of most gas-phase reactions decrease as the reactions proceed.

### Problems

**13.25.** **Bacterial Degradation of Ammonia** *Nitrosomonas* bacteria convert ammonia into nitrite in the presence of oxygen by the following reaction:

$$2\,NH_3(aq) + 3\,O_2(g) \rightarrow 2\,HNO_2(aq) + 2\,H_2O(\ell)$$

a. How is the rate of formation of $HNO_2$ related to the rate of consumption of $NH_3$?
b. How is the rate of formation of $HNO_2$ related to the rate of consumption of $O_2$?
c. How is the rate of consumption of $NH_3$ related to the rate of consumption of $O_2$?

13.26. **Catalytic Converters and Combustion** Catalytic converters in automobiles combat air pollution by converting NO and CO into $N_2$ and $CO_2$:

$$2\,CO(g) + 2\,NO(g) \rightarrow N_2(g) + 2\,CO_2(g)$$

a. How is the rate of formation of $N_2$ related to the rate of consumption of CO?
b. How is the rate of formation of $CO_2$ related to the rate of consumption of NO?
c. How is the rate of consumption of CO related to the rate of consumption of NO?

**13.27.** Nitryl chloride, $NO_2Cl$, is a reactive chlorine-containing species sometimes found in marine sediments in industrial areas. In the gas phase, it decomposes to $NO_2$ and $Cl_2$:

$$2\,NO_2Cl(g) \rightarrow 2\,NO_2(g) + Cl_2(g)$$

Under laboratory conditions, the rate of formation of $NO_2(g)$ is $5.7 \times 10^{-6}\,M\,s^{-1}$.

a. What is the rate of formation of $Cl_2(g)$?
b. What is the rate of consumption of $NO_2Cl(g)$?

13.28. $N_2O_5$, when dissolved in water, decomposes to produce $NO_2$ and $O_2$:

$$2\,N_2O_5(aq) \rightarrow 4\,NO_2(aq) + O_2(aq)$$

The rate of formation of $O_2$ is $4.0 \times 10^{-3}\,M\,min^{-1}$.

a. What is the rate of formation of $NO_2(aq)$?
b. What is the rate of reaction of $N_2O_5(aq)$?

**13.29.** **Power-Plant Emissions** Sulfur dioxide emissions in power-plant stack gases may react with carbon monoxide as follows:

$$SO_2(g) + 3\,CO(g) \rightarrow 2\,CO_2(g) + COS(g)$$

Write an equation relating each of the following pairs of rates:

a. The rate of formation of $CO_2$ to the rate of consumption of CO
b. The rate of formation of COS to the rate of consumption of $SO_2$
c. The rate of consumption of CO to the rate of consumption of $SO_2$

13.30. **Reducing Nitric Oxide Emissions from Power Plants** Nitric oxide (NO) can be removed from gas-fired power-plant emissions by reaction with methane as follows:

$$CH_4(g) + 4\,NO(g) \rightarrow 2\,N_2(g) + CO_2(g) + 2\,H_2O(g)$$

Write an equation relating each of the following pairs of rates:

a. The rate of formation of $N_2$ to the rate of formation of $CO_2$
b. The rate of formation of $CO_2$ to the rate of consumption of NO
c. The rate of consumption of $CH_4$ to the rate of formation of $H_2O$

**13.31.** **Stratospheric Ozone Depletion** Chlorine monoxide (ClO) plays a major role in creating the ozone holes in the stratosphere over Earth's polar regions.

a. If $\Delta[ClO]/\Delta t = -2.3 \times 10^7\,M/s$ at 298 K, what is the rate of change in $[Cl_2]$ and $[O_2]$ in the following reaction?

$$2\,ClO(g) \rightarrow Cl_2(g) + O_2(g)$$

b. If $\Delta[ClO]/\Delta t = -2.9 \times 10^4\,M/s$, what is the rate of formation of oxygen and $ClO_2$ in the following reaction?

$$ClO(g) + O_3(g) \rightarrow O_2(g) + ClO_2(g)$$

13.32. The chemistry of smog formation includes $NO_3$ as an intermediate in several reactions.

a. If $\Delta[NO_3]/\Delta t = -2.2 \times 10^5$ m$M$/min in the following reaction, what is the rate of formation of $NO_2$?

$$NO_3(g) + NO(g) \rightarrow 2\ NO_2(g)$$

b. What is the rate of change of $[NO_2]$ in the following reaction if $\Delta[NO_3]/\Delta t = -2.3$ m$M$/min?

$$2\ NO_3(g) \rightarrow 2\ NO_2(g) + O_2(g)$$

**13.33.** Nitrite ion reacts with ozone in aqueous solution, producing nitrate ion and oxygen:

$$NO_2^-(aq) + O_3(g) \rightarrow NO_3^-(aq) + O_2(g)$$

The following data were collected for this reaction at 298 K. Calculate the average reaction rate between 0.0 μs and 100.0 μs (microseconds) and between 200.0 μs and 300.0 μs.

| Time (μs) | $[O_3]$ ($M$) |
|---|---|
| 0.0 | $1.13 \times 10^{-2}$ |
| 100.0 | $9.93 \times 10^{-3}$ |
| 200.0 | $8.70 \times 10^{-3}$ |
| 300.0 | $8.15 \times 10^{-3}$ |

13.34. At room temperature in the gas phase, dinitrogen pentoxide ($N_2O_5$) decomposes to dinitrogen tetroxide and oxygen:

$$2\ N_2O_5(g) \rightarrow 2\ N_2O_4(g) + O_2(g)$$

Calculate the average rate of this reaction between consecutive measurements listed in the following table.

| Time (s) | $[N_2O_5]$ ($M$) |
|---|---|
| 0.0 | 0.200 |
| 300.0 | 0.180 |
| 600.0 | 0.161 |
| 900.0 | 0.144 |
| 1200.0 | 0.130 |

**13.35.** The following data were collected for the dimerization of ClO to $Cl_2O_2$ at 298 K.

| Time (s) | [ClO] (molecules/cm$^3$) |
|---|---|
| 0.0 | $2.60 \times 10^{11}$ |
| 1.0 | $1.08 \times 10^{11}$ |
| 2.0 | $6.83 \times 10^{10}$ |
| 3.0 | $4.99 \times 10^{10}$ |
| 4.0 | $3.93 \times 10^{10}$ |
| 5.0 | $3.24 \times 10^{10}$ |
| 6.0 | $2.76 \times 10^{10}$ |

Plot [ClO] and $[Cl_2O_2]$ as a function of time and determine the instantaneous rates of change in both at 1.0 s.

13.36. **Tropospheric Ozone** Tropospheric (lower atmosphere) ozone is rapidly consumed in many reactions, including

$$O_3(g) + NO(g) \rightarrow NO_2(g) + O_2(g)$$

Use the following data to calculate the instantaneous rate of the reaction at $t = 0.000$ s and $t = 0.052$ s.

| Time (s) | [NO] ($M$) |
|---|---|
| 0.000 | $2.0 \times 10^{-8}$ |
| 0.011 | $1.8 \times 10^{-8}$ |
| 0.027 | $1.6 \times 10^{-8}$ |
| 0.052 | $1.4 \times 10^{-8}$ |
| 0.102 | $1.2 \times 10^{-8}$ |

## Effect of Concentration on Reaction Rate

### Concept Review

**13.37.** The decomposition of A is second order in A. What effect does doubling the concentration of A have on the rate constant?

13.38. The decomposition of A is second order in A. What effect does doubling the concentration of A have on the half-life of A?

**13.39.** In a study of the rate of decomposition of compound A, a plot of [A] versus time is not linear, but a plot of ln [A] versus time is linear. What is the order of the reaction?

13.40. In a study of the rate of decomposition of compound B, plots of [B] versus time and ln [B] versus time are not linear, but a plot of 1/[B] versus time is linear. What is the order of the reaction?

**13.41.** What effect does doubling the initial concentration of a reactant have on the half-life in a reaction that is first order in the reactant? Explain this effect in terms of collisions between reactant molecules.

13.42. Two first-order decomposition reactions of the form $A \rightarrow B + C$ have the same rate constant at a given temperature. Do the reactants in the two reactions have the same half-lives at this temperature?

### Problems

**13.43.** For each of the following rate laws, determine the order with respect to each reactant and the overall reaction order.

a. Rate = $k[A]$
b. Rate = $k[A][B]$
c. Rate = $k[A][B]^2$

13.44. Determine the overall order of the following rate laws and the order with respect to each reactant.

a. Rate = $k[A]^{1/2}[B]^2$
b. Rate = $k[A]^2[B]$
c. Rate = $k[A][B][C]$

**13.45.** Write rate laws and determine the units (using *M* for concentration and seconds for time) of the rate constant for the following reactions:
a. The reaction of oxygen atoms with $NO_2$ is first order in both reactants.
b. The reaction between NO and $Cl_2$ is second order in NO and first order in $Cl_2$.
c. The reaction between $Cl_2$ and chloroform ($CHCl_3$) is first order in $CHCl_3$ and one-half order in $Cl_2$.
*d. The decomposition of ozone ($O_3$) to $O_2$ is second order in $O_3$ and an order of −1 in O atoms.

13.46. Compounds A and B react to give a single product, C. Write the rate law for each of the following cases and determine the units of the rate constant by using the units *M* for concentration and seconds for time:
a. The reaction is first order in A and first order in B.
b. The reaction is first order in A and first order overall.
c. The reaction is independent of the concentration of A and second order overall.
d. The reaction is second order in A and third order overall.

**13.47.** Predict the rate law for the reaction $2\,XO(g) \rightarrow X_2(g) + O_2(g)$ under each of the following conditions:
a. The rate doubles when [XO] doubles.
b. The rate quadruples when [XO] doubles.
c. The rate is halved when [XO] is halved.
d. The rate is unchanged when [XO] is doubled.

13.48. Predict the order with respect to NO for the reaction $NO(g) + Br_2(g) \rightarrow NOBr_2(g)$ under each of the following conditions:
a. The rate doubles when [NO] is doubled and $[Br_2]$ remains constant.
b. The rate increases by 1.56 times when [NO] is increased 1.25 times and $[Br_2]$ remains constant.
c. The rate is halved when [NO] is doubled and $[Br_2]$ remains constant.

**13.49.** In the reaction of NO with ClO,

$$NO(g) + ClO(g) \rightarrow NO_2(g) + Cl(g)$$

the initial rate of reaction quadruples when the concentrations of both reactants are doubled. What additional information do we need to determine whether the reaction is first order in each reactant?

13.50. If the reaction in Problem 13.49 is not first order in each reactant, what other order(s) are possible?

**13.51. Rate Laws for the Destruction of Tropospheric Ozone** The reaction of $NO_2$ with ozone produces $NO_3$ in a second-order reaction overall:

$$NO_2(g) + O_3(g) \rightarrow NO_3(g) + O_2(g)$$

a. Write the rate law for the reaction if the reaction is first order in each reactant.
b. The rate constant for the reaction is $1.93 \times 10^4\ M^{-1}\,s^{-1}$ at 298 K. What is the rate of the reaction when $[NO_2] = 1.8 \times 10^{-8}\ M$ and $[O_3] = 1.4 \times 10^{-7}\ M$?
c. What is the rate of formation of $NO_3$ under these conditions?
d. What happens to the rate of the reaction if the concentration of $O_3(g)$ is doubled?

13.52. **Sources of Nitric Acid in the Atmosphere** The reaction between $N_2O_5$ and water is a source of nitric acid in the atmosphere:

$$N_2O_5(g) + H_2O(g) \rightarrow 2\,HNO_3(g)$$

a. The reaction is first order in each reactant. Write the rate law for the reaction.
b. When $[N_2O_5] = 0.132$ m*M* and $[H_2O] = 230$ m*M*, the rate of the reaction is $4.55 \times 10^{-4}$ m*M*/min. What is the rate constant for the reaction?

**13.53.** Each of the following reactions is first order in each reactant and second order overall. Which reaction is the most rapid if the initial concentrations of all the reactants are the same?
a. $ClO_2(g) + O_3(g) \rightarrow ClO_3(g) + O_2(g)$
$k = 3.0 \times 10^{-19}\ cm^3/(molecule \cdot s)$
b. $ClO_2(g) + NO(g) \rightarrow NO_2(g) + ClO(g)$
$k = 3.4 \times 10^{-13}\ cm^3/(molecule \cdot s)$
c. $ClO(g) + NO(g) \rightarrow Cl(g) + NO_2(g)$
$k = 1.7 \times 10^{-11}\ cm^3/(molecule \cdot s)$
d. $ClO(g) + O_3(g) \rightarrow ClO_2(g) + O_2(g)$
$k = 1.5 \times 10^{-17}\ cm^3/(molecule \cdot s)$

13.54. Two reactions in which there is a single reactant have nearly the same magnitude rate constant. One is first order; the other is second order.
a. If the initial concentrations of the reactants are both 0.010 *M*, which reaction will proceed at the higher rate?
b. If the initial concentrations of the reactants are both 2.0 *M*, which reaction will proceed at the higher rate?

**13.55.** In the presence of water, NO and $NO_2$ react to form nitrous acid ($HNO_2$) by the following reaction:

$$NO(g) + NO_2(g) + H_2O(\ell) \rightarrow 2\,HNO_2(aq)$$

When the concentration of NO or $NO_2$ is doubled, the initial rate of reaction doubles. If the rate of the reaction does not depend on $[H_2O]$, what is the rate law for this reaction?

13.56. **Hydroperoxyl Radicals in the Atmosphere** During a smog event, trace amounts of many highly reactive substances are present in the atmosphere. One of these is the hydroperoxyl radical, $HO_2$. The rate constant for the reaction between $HO_2$ and another atmospheric pollutant, $SO_3$, is

$$2\,HO_2(g) + SO_3(g) \rightarrow H_2SO_3(g) + 2\,O_2(g)$$

$$k = 2.6 \times 10^{11}\ M^{-1}\,s^{-1} \text{ at } 298\text{ K}$$

The initial rate of the reaction doubles when the concentration of either $SO_3$ or $HO_2$ is doubled. What is the rate law for the reaction?

**13.57.** **Disinfecting Municipal Water Supplies** Chlorine dioxide ($ClO_2$) is a disinfectant used in municipal water-treatment plants (Figure P13.57). It dissolves in basic solution, producing $ClO_3^-$ and $ClO_2^-$:

$$2\,ClO_2(g) + 2\,OH^-(aq) \rightarrow ClO_3^-(aq) + ClO_2^-(aq) + H_2O(\ell)$$

**FIGURE P13.57**

The following reaction rate data were obtained at 298 K:

| Experiment | $[ClO_2]_0$ | $[OH^-]_0$ (*M*) | Initial Rate (*M*/s) |
|---|---|---|---|
| 1 | 0.060 | 0.030 | 0.0248 |
| 2 | 0.020 | 0.030 | 0.00827 |
| 3 | 0.020 | 0.090 | 0.0247 |

Determine the rate law and the rate constant for this reaction at 298 K.

13.58. The following data were collected for the reaction:

$$H_2(g) + NO(g) \rightarrow N_2O(g) + H_2O(g)$$

| Experiment | $[H_2]_0$ (*M*) | $[NO]_0$ (*M*) | Initial Rate (*M*/s) |
|---|---|---|---|
| 1 | 0.35 | 0.30 | $2.835 \times 10^{-3}$ |
| 2 | 0.35 | 0.60 | $1.134 \times 10^{-2}$ |
| 3 | 0.70 | 0.60 | $2.268 \times 10^{-2}$ |

Determine the rate law and the rate constant for the reaction at the temperature for which the data were collected.

**13.59.** Hydrogen gas reduces NO to $N_2$ in the following reaction:

$$2\,H_2(g) + 2\,NO(g) \rightarrow 2\,H_2O(g) + N_2(g)$$

The initial reaction rates of four mixtures of $H_2$ and NO were measured at 900°C with the following results:

| Experiment | $[H_2]_0$ (*M*) | $[NO]_0$ (*M*) | Initial Rate (*M*/s) |
|---|---|---|---|
| 1 | 0.212 | 0.136 | 0.0248 |
| 2 | 0.212 | 0.272 | 0.0991 |
| 3 | 0.424 | 0.544 | 0.793 |
| 4 | 0.848 | 0.544 | 1.59 |

Determine the rate law and the rate constant for the reaction at 900°C.

13.60. The rate of the reaction

$$NO_2(g) + CO(g) \rightarrow NO(g) + CO_2(g)$$

was determined in three experiments at 225°C. The results are given in the following table:

| Experiment | $[NO_2]_0$ (*M*) | $[CO]_0$ (*M*) | Initial Rate (*M*/s) |
|---|---|---|---|
| 1 | 0.263 | 0.826 | $1.44 \times 10^{-5}$ |
| 2 | 0.263 | 0.413 | $1.44 \times 10^{-5}$ |
| 3 | 0.526 | 0.413 | $5.76 \times 10^{-5}$ |

a. Determine the rate law for the reaction.
b. Calculate the value of the rate constant at 225°C.
c. Calculate the rate of formation of $CO_2$ when $[NO_2] = [CO] = 0.500\ M$.

**13.61.** In addition to being studied in the gas phase, the decomposition of $N_2O_5$ has been evaluated in solution. In carbon tetrachloride ($CCl_4$) at 45°C,

$$2\,N_2O_5 \rightarrow 4\,NO_2 + O_2$$

is a first-order reaction and $k = 6.32 \times 10^{-4}\ s^{-1}$. How much $N_2O_5$ remains in solution after 1 h if the initial concentration of $N_2O_5$ was 0.50 mol/L? What percent of the $N_2O_5$ has reacted at that point?

13.62. Because the units of concentration in the term $\ln([X]/[X]_0)$ cancel out in the integrated rate law for first-order reactions (Equation 13.16), molar concentration can be replaced by any concentration term. With gases, for example, partial pressures may be used. The decomposition of phosphine gas ($PH_3$) at 600°C is first order in $PH_3$ with $k = 0.023\ s^{-1}$.

$$4\,PH_3(g) \rightarrow P_4(g) + 6\,H_2(g)$$

If the initial partial pressure of $PH_3$ is 375 torr, what percent of $PH_3$ reacts in 1 min?

**13.63. Laughing Gas in the Atmosphere** Nitrous oxide ($N_2O$) is used as an anesthetic (laughing gas) and in aerosol cans to produce whipped cream. It is a potent greenhouse gas and decomposes slowly to $N_2$ and $O_2$:

$$2\ N_2O(g) \rightarrow 2\ N_2(g) + O_2(g)$$

a. If the plot of $\ln[N_2O]$ as a function of time is linear, what is the rate law for the reaction?
b. How many half-lives will it take for the concentration of $N_2O$ to reach 6.25% of its original concentration?

13.64. The unsaturated hydrocarbon butadiene ($C_4H_6$) in Figure P13.64 dimerizes to 4-vinylcyclohexene ($C_8H_{12}$). When data collected in studies of the kinetics of this reaction were plotted against reaction time, plots of $[C_4H_6]$ or $\ln[C_4H_6]$ produced curved lines, but the plot of $1/[C_4H_6]$ was linear.

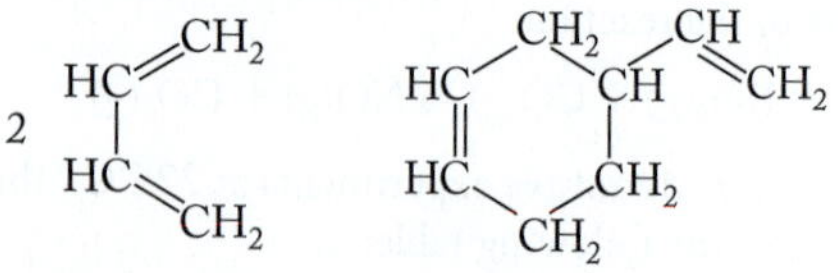

2 Butadiene → Vinylcyclohexene

**FIGURE P13.64**

a. What is the rate law for the reaction?
b. How many half-lives will it take for the $[C_4H_6]$ to decrease to 3.1% of its original concentration?

***NOTE: Problems 13.65–13.74 involve graphing experimental data. Using spreadsheet software or a graphing calculator will make solving them much easier.***

**13.65.** Nitrogen trioxide decomposes to $NO_2$ and $O_2$ in the following reaction:

$$2\ NO_3(g) \rightarrow 2\ NO_2(g) + O_2(g)$$

The following data were collected at 298 K:

| Time (min) | $[NO_3]$ ($\mu M$) |
|---|---|
| 0.0 | $1.470 \times 10^{-3}$ |
| 10.0 | $1.463 \times 10^{-3}$ |
| 100.0 | $1.404 \times 10^{-3}$ |
| 200.0 | $1.344 \times 10^{-3}$ |
| 300.0 | $1.288 \times 10^{-3}$ |
| 400.0 | $1.237 \times 10^{-3}$ |
| 500.0 | $1.190 \times 10^{-3}$ |

Calculate the value of the second-order rate constant at 298 K.

13.66. Two structural isomers of $ClO_2$ are shown in Figure P13.66.

**FIGURE P13.66**

The isomer with the Cl—O—O skeletal arrangement is unstable and rapidly decomposes according to the reaction $2\ ClOO(g) \rightarrow Cl_2(g) + 2\ O_2(g)$. The following data were collected for the decomposition of ClOO at 298 K:

| Time (μs) | [ClOO] ($M$) |
|---|---|
| 0.0 | $1.76 \times 10^{-6}$ |
| 0.7 | $2.36 \times 10^{-7}$ |
| 1.3 | $3.56 \times 10^{-8}$ |
| 2.1 | $3.23 \times 10^{-9}$ |
| 2.8 | $3.96 \times 10^{-10}$ |

Determine the rate law and the rate constant for the reaction at 298 K.

**13.67.** At high temperatures, ammonia spontaneously decomposes into $N_2$ and $H_2$. The following data were collected at one such temperature:

| Time (s) | $[NH_3]$ ($M$) |
|---|---|
| 0 | $2.56 \times 10^{-2}$ |
| 12 | $2.47 \times 10^{-2}$ |
| 56 | $2.16 \times 10^{-2}$ |
| 224 | $1.31 \times 10^{-2}$ |
| 532 | $5.19 \times 10^{-3}$ |
| 746 | $2.73 \times 10^{-3}$ |

Determine the rate law for the decomposition of ammonia and the value of the rate constant at the temperature of the experiment.

13.68. **Atmospheric Chemistry of Hydroperoxyl Radicals** Atmospheric chemistry involves highly reactive, odd-electron molecules such as the hydroperoxyl radical $HO_2$, which decomposes into $H_2O_2$ and $O_2$. Determine the rate law for the reaction and the value of the rate constant at 298 K by using the following data obtained at 298 K.

| Time (μs) | $[HO_2]$ ($\mu M$) |
|---|---|
| 0.0 | 8.5 |
| 0.6 | 5.1 |
| 1.0 | 3.6 |
| 1.4 | 2.6 |
| 1.8 | 1.8 |
| 2.4 | 1.1 |

**13.69. Tracing Phosphorus in Organisms** Radioactive isotopes such as $^{32}P$ are used to follow biological processes. The following radioactivity data (in relative radioactivity values) were collected for a sample containing $^{32}P$:

| Time (days) | Relative Radioactivity |
|---|---|
| 0 | 10.00 |
| 1 | 9.53 |
| 2 | 9.08 |
| 5 | 7.85 |
| 10 | 6.16 |
| 20 | 3.79 |

a. Write the rate law for the decay of $^{32}P$.
b. Determine the value of the rate constant.
c. Determine the half-life of $^{32}P$.
d. How many days does it take for 99% of a sample of $^{32}P$ to decay?

13.70. Nitrous acid slowly decomposes to NO, $NO_2$, and water in the following second-order reaction:

$$2\,HNO_2(aq) \rightarrow NO(g) + NO_2(g) + H_2O(\ell)$$

a. Use the following data to determine the rate constant for this reaction at 298 K:

| Time (min) | $[HNO_2]$ ($\mu M$) |
|---|---|
| 0.0 | 0.1560 |
| 1000.0 | 0.1466 |
| 1500.0 | 0.1424 |
| 2000.0 | 0.1383 |
| 2500.0 | 0.1345 |
| 3000.0 | 0.1309 |

b. Determine the half-life for the decomposition of $HNO_2$.
c. If the experiment that yielded the results in the preceding table had been continued for 3000.0 minutes more, what would the concentration of $HNO_2$ have been?

**13.71.** The dimerization of ClO,

$$2\,ClO(g) \rightarrow Cl_2O_2(g)$$

is second order in ClO.

a. Use the following data to determine the value of $k$ at 298 K.

| Time (s) | [ClO] (molecules/cm$^3$) |
|---|---|
| 0 | $2.60 \times 10^{11}$ |
| 1.00 | $1.08 \times 10^{11}$ |
| 2.00 | $6.83 \times 10^{10}$ |
| 3.00 | $4.99 \times 10^{10}$ |
| 4.00 | $3.93 \times 10^{10}$ |

b. Determine the half-life for the dimerization of ClO in this experiment.

13.72. Kinetics data for the reaction $Cl_2O_2(g) \rightarrow 2\,ClO(g)$ at a given temperature are summarized in the following table.

| Time ($\mu$s) | $[Cl_2O_2]$ ($M$) |
|---|---|
| 0 | $6.60 \times 10^{-8}$ |
| 172 | $5.68 \times 10^{-8}$ |
| 345 | $4.89 \times 10^{-8}$ |
| 517 | $4.21 \times 10^{-8}$ |
| 690 | $3.62 \times 10^{-8}$ |
| 862 | $3.12 \times 10^{-8}$ |

a. Determine the value of the rate constant at this temperature.
b. Determine $t_{1/2}$ for the decomposition of $Cl_2O_2$ at 298 K.

**13.73. Kinetics of Sucrose Hydrolysis** The metabolism of table sugar (sucrose; $C_{12}H_{22}O_{11}$) begins with the hydrolysis of the disaccharide to glucose and fructose (both $C_6H_{12}O_6$):

$$C_{12}H_{22}O_{11}(aq) + H_2O(\ell) \rightarrow 2\,C_6H_{12}O_6(aq)$$

The kinetics of the reaction were studied at 24°C in a reaction system with a large excess of water, so the reaction was pseudo–first order in sucrose. Determine the rate law and the pseudo–first-order rate constant for the reaction from the following data:

| Time (s) | $[C_{12}H_{22}O_{11}]$ ($M$) |
|---|---|
| 0.0 | 0.562 |
| 612.0 | 0.541 |
| 1600.0 | 0.509 |
| 2420.0 | 0.484 |
| 3160.0 | 0.462 |
| 4800.0 | 0.442 |

13.74. Hydroperoxyl radicals react rapidly with ozone to produce oxygen and •OH radicals:

$$HO_2(g) + O_3(g) \rightarrow OH(g) + 2\,O_2(g)$$

The rate of this reaction was studied in the presence of a large excess of ozone. Determine the pseudo–first-order rate constant and the second-order rate constant for the reaction from the following data:

| Time (ms) | $[HO_2]$ ($M$) | $[O_3]$ ($M$) |
|---|---|---|
| 0.0 | $3.2 \times 10^{-6}$ | $1.0 \times 10^{-3}$ |
| 10.0 | $2.9 \times 10^{-6}$ | $1.0 \times 10^{-3}$ |
| 20.0 | $2.6 \times 10^{-6}$ | $1.0 \times 10^{-3}$ |
| 30.0 | $2.4 \times 10^{-6}$ | $1.0 \times 10^{-3}$ |
| 80.0 | $1.4 \times 10^{-6}$ | $1.0 \times 10^{-3}$ |

## Reaction Rates, Temperature, and the Arrhenius Equation

### Concept Review

**13.75.** How does the magnitude of a reaction's activation energy influence the rate of a reaction?

13.76. Do all spontaneous reactions happen instantaneously at room temperature?

**13.77.** Under what circumstances is the activation energy of a reaction proceeding in the forward direction less than the activation energy if it proceeds in reverse?

13.78. How does increasing the temperature of a reaction affect its activation energy?

**13.79.** Does reducing the activation energy of a reaction by $\frac{1}{2}$ increase its rate constant by a factor of 2?

*13.80. Two first-order reactions have activation energies of 15 kJ/mol and 150 kJ/mol. Which reaction will show the larger percent increase in rate as temperature is increased?

### Problems

***NOTE: Problems 13.81–13.84 involve graphing experimental data. Using spreadsheet software or a graphing calculator will make solving them much easier.***

**13.81.** **Activation Energy of Smog-Forming Reactions** The initial step in the formation of smog is the reaction between nitrogen and oxygen. At the temperatures indicated, values of the rate constant of the reaction

$$N_2(g) + O_2(g) \rightarrow 2\ NO(g)$$

are as follows:

| $T$ (K) | $k$ ($M^{-1/2}$ s$^{-1}$) |
|---|---|
| 2000 | 318 |
| 2100 | 782 |
| 2200 | 1770 |
| 2300 | 3733 |
| 2400 | 7396 |

a. Calculate the activation energy of the reaction.
b. Calculate the frequency factor for the reaction.
c. Calculate the value of the rate constant at ambient temperature, $T$ = 300 K.

13.82. Values of the rate constant for the decomposition of $N_2O_5$ gas at four temperatures are as follows:

| $T$ (K) | $k$ (s$^{-1}$) |
|---|---|
| 658 | $2.14 \times 10^5$ |
| 673 | $3.23 \times 10^5$ |
| 688 | $4.81 \times 10^5$ |
| 703 | $7.03 \times 10^5$ |

a. Determine the activation energy of the decomposition reaction.
b. Calculate the value of the rate constant at 300 K.

**13.83.** **Activation Energy of Stratospheric Ozone Destruction Reactions** The kinetics of the reaction between chlorine dioxide and ozone are relevant to the study of atmospheric ozone destruction. The value of the rate constant for the reaction between chlorine dioxide and ozone was measured at four temperatures between 193 K and 208 K. The results were as follows:

| $T$ (K) | $k$ ($M^{-1}$ s$^{-1}$) |
|---|---|
| 193 | 34.0 |
| 198 | 62.8 |
| 203 | 112.8 |
| 208 | 196.7 |

a. Calculate the values of the activation energy and the frequency factor for the reaction.
b. What is the value of the rate constant in the upper stratosphere, where $T$ = 258 K?

13.84. Chlorine atoms react with methane, forming HCl and $CH_3$. The rate constant for the reaction is $6.0 \times 10^7\ M^{-1}\ s^{-1}$ at 298 K. When the experiment was run at three other temperatures, the following data were collected:

| $T$ (K) | $k$ ($M^{-1}$ s$^{-1}$) |
|---|---|
| 303 | $6.5 \times 10^7$ |
| 308 | $7.0 \times 10^7$ |
| 313 | $7.5 \times 10^7$ |

a. Calculate the values of the activation energy and the frequency factor for the reaction.
b. What is the value of the rate constant in the lower stratosphere, where $T$ = 218 K?

**13.85.** The rates of many reactions nearly double for every 10°C rise in temperature. This is not always the case, but it is a convenient concept to apply when making estimations. Azomethane ($CH_3NNCH_3$) is a heat-sensitive explosive that decomposes to yield nitrogen gas and methyl radicals:

$$CH_3NNCH_3(g) \rightarrow N_2(g) + 2\ CH_3(g)$$

At 600°C, the reaction has an activation energy of $2.14 \times 10^4$ J/mol and $k = 2.00 \times 10^8\ s^{-1}$. Does the rule of thumb hold for this reaction?

13.86. The compound 1,1-difluoroethane decomposes at elevated temperatures to give fluoroethylene and hydrogen fluoride:

$$CH_3CHF_2(g) \rightarrow CH_2CHF(g) + HF(g)$$

At 460°C, $k = 5.8 \times 10^{-6}\ s^{-1}$ and $E_a$ = 265 kJ/mol. To what temperature would you have to raise the reaction to make it go four times as fast?

## Reaction Mechanisms

### Concept Review

**13.87.** Under what reaction conditions does a bimolecular reaction obey pseudo–first-order reaction kinetics?

*13.88. If a reaction is zero order in a reactant, does that mean that the reactant is never involved in collisions with other reactants? Explain your answer.

### Problems

**13.89.** The hypothetical reaction $A \rightarrow B$ has an activation energy of 50.0 kJ/mol. Draw a reaction profile for each of the following mechanisms:

a. A single elementary step
b. A two-step reaction in which the activation energy of the second step is 15 kJ/mol
c. A two-step reaction in which the activation energy of the second step is the rate-determining barrier

13.90. For the spontaneous reaction $A + B \rightarrow C \rightarrow D + E$, draw three reaction profiles, one for each of the following mechanisms:

a. C is an activated complex.
b. The reaction has two elementary steps; the first step is rate determining and C is an intermediate.
c. The reaction has two elementary steps; the second step is rate determining and C is an intermediate.

**13.91.** Write the rate laws for the following elementary steps and identify them as uni-, bi-, or termolecular steps:

a. $SO_2Cl_2(g) \rightarrow SO_2(g) + Cl_2(g)$
b. $NO_2(g) + CO(g) \rightarrow NO(g) + CO_2(g)$
c. $2\ NO_2(g) \rightarrow NO_3(g) + NO(g)$

13.92. Write the rate laws for the following elementary steps and identify them as uni-, bi-, or termolecular steps:

a. $Cl(g) + O_3(g) \rightarrow ClO(g) + O_2(g)$
b. $2\ NO_2(g) \rightarrow N_2O_4(g)$
*c. ${}^{14}_{6}C \rightarrow {}^{14}_{7}N + {}^{0}_{-1}\beta$

**13.93.** A common classroom demonstration involves mixing 30% hydrogen peroxide with a solution of potassium iodide. The following mechanism has been proposed for the reaction:

Step 1 $\quad H_2O_2(aq) + I^-(aq) \rightarrow H_2O(\ell) + IO^-(aq)$ slow

Step 2 $\quad H_2O_2(aq) + IO^-(aq) \rightarrow H_2O(\ell) + O_2(g) + I^-(aq)$ fast

a. Write the equation for the overall reaction.
b. Write the rate law predicted by the mechanism for the overall reaction.
c. Which species is a catalyst?
d. Identify any intermediates in the reaction.

13.94. A proposed mechanism for the reaction of $NO_2(g)$ and $CO(g)$ is

Step 1 $\quad 2\ NO_2(g) \rightarrow NO(g) + NO_3(g)$ slow

Step 2 $\quad NO_3(g) + CO(g) \rightarrow NO_2(g) + CO_2(g)$ fast

a. Write the equation for the overall reaction.
b. Write the rate law predicted by the mechanism for the overall reaction.
c. Identify the reactants and products of the reaction.
d. Identify any intermediates in the reaction.

***13.95.** In Sample Exercise 13.4, we determined that the formation of NO from $N_2$ and $O_2$ at high temperatures is first order in $N_2$ and $\frac{1}{2}$ order in $O_2$. Is the following reaction mechanism consistent with these concentration dependencies? If so, which is the rate-determining step?

(1) $\quad N_2(g) \rightleftharpoons 2\ N(g)$

(2) $\quad N(g) + O_2(g) \rightarrow NO(g) + O(g)$

(3) $\quad N(g) + O(g) \rightarrow NO(g)$

Overall: $\quad N_2(g) + O_2(g) \rightarrow 2\ NO(g)$

13.96. A proposed mechanism for the gas-phase decomposition of hydrogen peroxide at an elevated temperature consists of three elementary steps:

$$H_2O_2(g) \rightarrow 2\ OH(g)$$

$$H_2O_2(g) + OH(g) \rightarrow H_2O(g) + HO_2(g)$$

$$HO_2(g) + OH(g) \rightarrow H_2O(g) + O_2(g)$$

If the rate law for the reaction is first order in $H_2O_2$, which step in the mechanism is the rate-determining step?

**13.97.** At a given temperature, the rate of the reaction between NO and $Cl_2$ is proportional to the product of the concentrations of the two gases: $[NO][Cl_2]$. The following two-step mechanism has been proposed for the reaction:

(1) $\quad NO(g) + Cl_2(g) \rightarrow NOCl_2(g)$

(2) $\quad NOCl_2(g) + NO(g) \rightarrow 2\ NOCl(g)$

Overall: $\quad 2\ NO(g) + Cl_2(g) \rightarrow 2\ NOCl(g)$

Which step must be the rate-determining step if this mechanism is correct?

13.98. **Mechanism of Ozone Destruction** Ozone decomposes thermally to oxygen in the following reaction:

$$2\ O_3(g) \rightarrow 3\ O_2(g)$$

The following mechanism has been proposed:

$$O_3(g) \rightarrow O(g) + O_2(g)$$

$$O(g) + O_3(g) \rightarrow 2\ O_2(g)$$

The reaction is second order in ozone. What properties of the two elementary steps (specifically, relative rate and reversibility) are consistent with this mechanism?

**13.99. Mechanism of $NO_2$ Destruction** The rate laws for the thermal and photochemical decomposition of $NO_2$ are different. Which of the following mechanisms are possible for the thermal decomposition of $NO_2$, and which are possible for the photochemical decomposition of $NO_2$? For the thermal decomposition, Rate $= k[NO_2]^2$, and for the photochemical decomposition, Rate $= k[NO_2]$.

a. $NO_2(g) \xrightarrow{\text{slow}} NO(g) + O(g)$
$O(g) + NO_2(g) \xrightarrow{\text{fast}} NO(g) + O_2(g)$

b. $NO_2(g) + NO_2(g) \xrightarrow{\text{fast}} N_2O_4(g)$
$N_2O_4(g) \xrightarrow{\text{slow}} NO(g) + NO_3(g)$
$NO_3(g) \xrightarrow{\text{fast}} NO(g) + O_2(g)$

c. $NO_2(g) + NO_2(g) \xrightarrow{\text{slow}} NO(g) + NO_3(g)$
$NO_3(g) \xrightarrow{\text{fast}} NO(g) + O_2(g)$

13.100. The rate laws for the thermal and photochemical decomposition of $NO_2$ are different. Which of the following mechanisms are possible for the thermal decomposition of $NO_2$, and which are possible for the photochemical decomposition of $NO_2$? For the thermal decomposition, Rate = $k[NO_2]^2$, and for the photochemical decomposition, Rate = $k[NO_2]$.

a. $NO_2(g) + NO_2(g) \xrightarrow{\text{slow}} N_2O_4(g)$
$N_2O_4(g) \xrightarrow{\text{fast}} N_2O_3(g) + O(g)$
$N_2O_3(g) + O(g) \xrightarrow{\text{fast}} N_2O_2(g) + O_2(g)$
$N_2O_2(g) \xrightarrow{\text{fast}} 2\ NO(g)$

b. $NO_2(g) + NO_2(g) \xrightarrow{\text{slow}} NO(g) + NO_3(g)$
$NO_3(g) \xrightarrow{\text{fast}} NO(g) + O_2(g)$

c. $NO_2(g) \xrightarrow{\text{slow}} N(g) + O_2(g)$
$N(g) + NO_2(g) \xrightarrow{\text{fast}} N_2O_2(g)$
$N_2O_2(g) \xrightarrow{\text{fast}} 2\ NO(g)$

## Catalysts

### Concept Review

**13.101.** Does a catalyst affect both the rate and the rate constant of a reaction?

*13.102. Is the rate law for a catalyzed reaction the same as that for the uncatalyzed reaction?

**13.103.** Does a substance that increases the rate of a reaction also increase the rate of the reverse reaction?

13.104. The rate of the reaction between $NO_2$ and CO is independent of [CO]. Does this mean that CO is a catalyst for the reaction?

***13.105.** Can the concentration of a homogeneous catalyst appear in the rate law for the reaction it catalyzes?

*13.106. The rate of a chemical reaction is too slow to measure at room temperature. We could either raise the temperature or add a catalyst. Which would be a better solution to accurately determine the rate constant?

### Problems

**13.107.** Is NO a catalyst for the decomposition of $N_2O$ in the following two-step reaction mechanism, or is $N_2O$ a catalyst for the conversion of NO to $NO_2$?

(1) $NO(g) + N_2O(g) \rightarrow N_2(g) + NO_2(g)$

(2) $2\ NO_2(g) \rightarrow 2\ NO(g) + O_2(g)$

13.108. **NO as a Catalyst for Ozone Destruction** Explain why NO is a catalyst in the following two-step process that results in the depletion of ozone in the stratosphere:

(1) $NO(g) + O_3(g) \rightarrow NO_2(g) + O_2(g)$

(2) $O(g) + NO_2(g) \rightarrow NO(g) + O_2(g)$

Overall: $O(g) + O_3(g) \rightarrow 2\ O_2(g)$

**13.109.** On the basis of the frequency factors and activation energy values of the following two reactions, determine which one will have the larger rate constant at room temperature (298 K).

$$O_3(g) + O(g) \rightarrow O_2(g) + O_2(g)$$
$A = 8.0 \times 10^{-12}\ \text{cm}^3/(\text{molecules} \cdot \text{s})$ $E_a = 17.7$ kJ/mol

$$O_3(g) + Cl(g) \rightarrow ClO(g) + O_2(g)$$
$A = 2.9 \times 10^{-11}\ \text{cm}^3/(\text{molecules} \cdot \text{s})$ $E_a = 2.16$ kJ/mol

13.110. On the basis of the frequency factors and activation energy values of the following two reactions, determine which one will have the larger rate constant at room temperature (298 K).

$$O_3(g) + Cl(g) \rightarrow ClO(g) + O_2(g)$$
$A = 2.9 \times 10^{-11}\ \text{cm}^3/(\text{molecules} \cdot \text{s})$ $E_a = 2.16$ kJ/mol

$$O_3(g) + NO(g) \rightarrow NO_2(g) + O_2(g)$$
$A = 2.0 \times 10^{-12}\ \text{cm}^3/(\text{molecules} \cdot \text{s})$ $E_a = 11.6$ kJ/mol

## Additional Problems

**13.111.** A student inserts a glowing wood splint into a test tube filled with $O_2$. The splint quickly catches on fire. Use collision theory to explain why the splint burns so much faster in pure $O_2$ than in air.

*13.112. A backyard chef turns on the propane gas to a barbecue grill. Even though the reaction between propane and oxygen is spontaneous, the gas does not begin to burn until the chef pushes an igniter button to produce a spark. Why is the spark needed?

**13.113.** On average, someone who falls through the ice covering a frozen lake is less likely to experience anoxia (lack of oxygen) than someone who falls into a warm pool and is underwater for the same length of time. Why?

13.114. The rate at which drugs are metabolized depends on age: children metabolize some drugs more rapidly than adults, whereas the elderly metabolize drugs more slowly. Diazepam is used to treat anxiety disorders and seizures in patients in all age groups. Its half-life in hours is estimated to be equal to the patient's age in years; in a 50-year-old, for example, diazepam would have a 50-hour half-life. How long will it take for 95% of a dose of diazepam to be metabolized in a 5-year-old child compared to a 50-year-old adult, if we assume a first order process?

**13.115.** If the rate of the reverse reaction is much slower than the rate of the forward reaction, does the method used to determine a rate law from initial concentrations and initial rates also work at some other time $t$?

13.116. A teaching assistant is designing a synthesis experiment for use in a 3-hour laboratory. The literature preparation specifies that 125 mL of a 1.0 $M$ solution of reactant A should be mixed with 125 mL of a 1.0 $M$ solution of B at room temperature in a 500-mL flask. The rate of the reaction was such that, after 6 hours of sitting undisturbed on the lab bench, only 50% of the stoichiometric yield of the desired product C was produced. Suggest three reasonable changes in the procedure the assistant could try to improve the yield of product over a 3-hour period.

**13.117.** How can a plot of $1/[X] - 1/[X]_0$ as a function of $t$ be used to determine the value of a second-order rate constant?

13.118. Why can't an elementary step in a mechanism have a rate law that is zero order in a reactant?

**13.119.** In the reaction between nitrogen dioxide and ozone,

$$2\ NO_2(g) + O_3(g) \rightarrow N_2O_5(g) + O_2(g)$$

how are the rates of change in the concentrations of the reactants and products related?

13.120. The following table contains kinetics data for the reaction

$$2\ NO(g) + Cl_2(g) \rightarrow 2\ NOCl(g)$$

| Experiment | $[NO]_0$ ($M$) | $[Cl_2]_0$ ($M$) | Initial Rate ($M$/s) |
|---|---|---|---|
| 1 | 0.20 | 0.10 | 0.63 |
| 2 | 0.20 | 0.30 | 5.70 |
| 3 | 0.80 | 0.10 | 2.58 |
| 4 | 0.40 | 0.20 | ? |

Predict the initial rate of reaction in experiment 4.

**13.121.** The following is an important reaction in the formation of photochemical smog:

$$NO(g) + O_3(g) \rightarrow NO_2(g) + O_2(g)$$

The reaction is first order in NO and $O_3$. The rate constant of the reaction is 80 $M^{-1}\ s^{-1}$ at 25°C and 3000 $M^{-1}\ s^{-1}$ at 75°C.

a. If this reaction were to occur in a single step, would the rate law be consistent with the observed order of the reaction for NO and $O_3$?

b. What is the value of the activation energy of the reaction?

c. What is the rate of the reaction at 25°C when $[NO] = 3 \times 10^{-6}\ M$ and $[O_3] = 5 \times 10^{-9}\ M$?

d. Predict the values of the rate constant at 10°C and 35°C.

13.122. Ammonia reacts with nitrous acid to form an intermediate, ammonium nitrite ($NH_4NO_2$), which decomposes to $N_2$ and $H_2O$:

$$NH_3(g) + HNO_2(aq) \rightarrow NH_4NO_2(aq) \rightarrow N_2(g) + 2\ H_2O(\ell)$$

a. The reaction is first order in ammonia and second order in nitrous acid. What is the rate law for the reaction? What are the units on the rate constant if concentrations are expressed in molarity and time, in seconds?

b. The rate law for the reaction has also been written as

$$\text{Rate} = k[NH_4^+][NO_2^-][HNO_2]$$

Is this expression equivalent to the one you wrote in part (a)?

c. With the data in Appendix 4, calculate the value of $\Delta H^\circ_{rxn}$ for the overall reaction $\Delta H^\circ_{f,HNO_2(aq)} = -128.9$ kJ/mol.

d. Draw an energy profile for the process with the assumption that the $E_a$ of the first step is lower than the $E_a$ of the second step.

**13.123.** When ionic compounds such as NaCl dissolve in water, the sodium ions are surrounded by six water molecules. The bound water molecules exchange with those in bulk solution as described by the reaction involving $^{18}$O-enriched water:

$$Na(H_2O)_6^+(aq) + H_2{}^{18}O(\ell) \rightarrow Na(H_2O)_5(H_2{}^{18}O)^+(aq) + H_2O(\ell)$$

a. The following reaction mechanism has been proposed:

(1) $Na(H_2O)_6^+(aq) \rightarrow Na(H_2O)_5^+(aq) + H_2O(\ell)$

(2) $Na(H_2O)_5^+(aq) + H_2{}^{18}O(\ell) \rightarrow Na(H_2O)_5(H_2{}^{18}O)^+(aq)$

What is the rate law if the first step is the rate-determining step?

b. Sketch the reaction's energy profile that is consistent with this two-step mechanism.

13.124. **Lachrymators in Smog** The combination of ozone, volatile hydrocarbons, nitrogen oxide, and sunlight in urban environments produces peroxyacetyl nitrate (PAN), a potent lachrymator (a substance that causes tears to form in the eyes). PAN decomposes to acetyl radicals and nitrogen dioxide in a process that is second order in PAN, as shown in Figure P13.124:

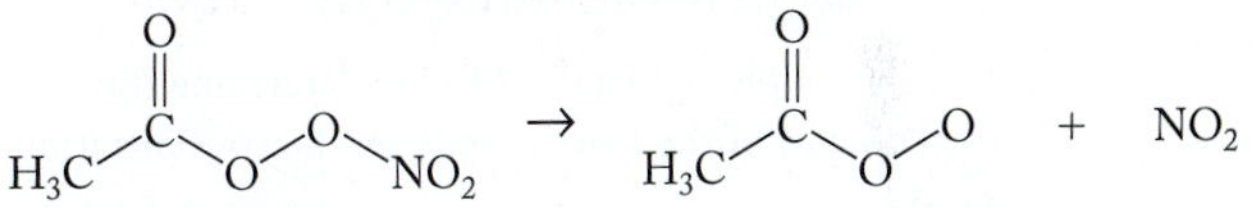

$$H_3C{-}C({=}O){-}O{-}O{-}NO_2 \rightarrow H_3C{-}C({=}O){-}O{-}O + NO_2$$

**FIGURE P13.124**

a. The half-life of the reaction, at 23°C and $P_{CH_3CO_3NO_2} = 10.5$ torr, is 100 h. Calculate the rate constant for the reaction.

b. Determine the rate of the reaction at 23°C and $P_{CH_3CO_3NO_2} = 10.5$ torr.

c. Draw a graph showing $P_{PAN}$ as a function of time from 0 h to 200 h starting with $P_{CH_3CO_3NO_2} = 10.5$ torr.

**13.125.** **Nitric Oxide in the Human Body** Nitric oxide (NO) is a gaseous free radical that plays many biological roles, including regulating neurotransmission and the human immune system. One of its many reactions involves the peroxynitrite ion ($ONOO^-$):

$$NO(g) + ONOO^-(aq) \rightarrow NO_2(g) + NO_2^-(aq)$$

a. Use the following data to determine the rate law and rate constant of the reaction at the experimental temperature at which these data were generated.

| Experiment | $[NO]_0$ ($M$) | $[ONOO^-]_0$ ($M$) | Rate ($M$/s) |
|---|---|---|---|
| 1 | $1.25 \times 10^{-4}$ | $1.25 \times 10^{-4}$ | $2.03 \times 10^{-11}$ |
| 2 | $1.25 \times 10^{-4}$ | $0.625 \times 10^{-4}$ | $1.02 \times 10^{-11}$ |
| 3 | $0.625 \times 10^{-4}$ | $2.50 \times 10^{-4}$ | $2.03 \times 10^{-11}$ |
| 4 | $0.625 \times 10^{-4}$ | $3.75 \times 10^{-4}$ | $3.05 \times 10^{-11}$ |

b. Draw the Lewis structure of peroxynitrite ion (including all resonance forms) and assign formal charges. Note which form is preferred.

c. Use the average bond energies in Appendix Table A4.1 to estimate the value of $\Delta H^\circ_{rxn}$ by using the preferred structure from part (b).

13.126. **Kinetics of Protein Chemistry** In the presence of $O_2$, NO reacts with sulfur-containing proteins to form *S*-nitrosothiols, such as $C_6H_{13}SNO$. This compound decomposes according to this reaction:

$$2\,C_6H_{13}SNO(aq) \rightarrow 2\,NO(g) + C_{12}H_{26}S_2(aq)$$

The following data were collected for the decomposition reaction at 69°C.

| Time (min) | $[C_6H_{13}SNO]$ (*M*) |
|---|---|
| 0.0 | $1.05 \times 10^{-3}$ |
| 10.0 | $9.84 \times 10^{-4}$ |
| 20.0 | $9.22 \times 10^{-4}$ |
| 30.0 | $8.64 \times 10^{-4}$ |
| 60.0 | $7.11 \times 10^{-4}$ |

Calculate the value of the first-order rate constant for the reaction.

**13.127.** Solutions of nitrous acid ($HNO_2$) in $^{18}O$-labeled water undergo isotope exchange:

$$HNO_2(aq) + H_2{}^{18}O(\ell) \rightarrow HN^{18}O_2(aq) + H_2O(\ell)$$

a. Use the following data at 24°C to determine the dependence of the reaction rate on the concentration of $HNO_2$.

| Time (min) | $[HNO_2]$ (*M*) |
|---|---|
| 0.0 | $5.4 \times 10^{-2}$ |
| 20.0 | $1.5 \times 10^{-3}$ |
| 40.0 | $7.7 \times 10^{-4}$ |
| 60.0 | $5.2 \times 10^{-4}$ |

b. Does the reaction rate depend on the concentration of $H_2{}^{18}O$?

13.128. Ethylene ($C_2H_4$) reacts with ozone to form 2 moles of formaldehyde (a probable human carcinogen) per mole of ethylene, as shown in Figure P13.128. The following data were collected at 298 K.

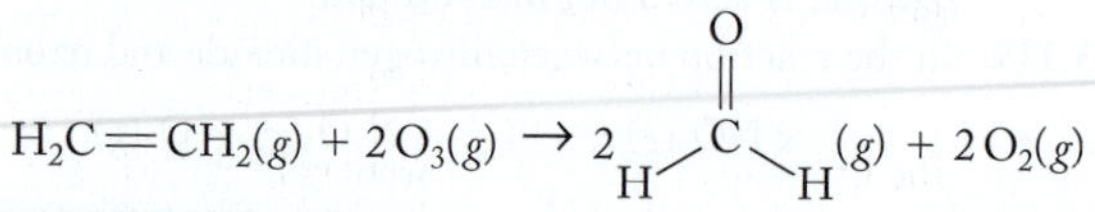

$$H_2C{=}CH_2(g) + 2\,O_3(g) \rightarrow 2\,H_2C{=}O(g) + 2\,O_2(g)$$

**FIGURE P13.128**

| Experiment | $[O_3]_0$ (*M*) | $[C_2H_4]_0$ (*M*) | Rate (*M*/s) |
|---|---|---|---|
| 1 | $0.86 \times 10^{-2}$ | $1.00 \times 10^{-2}$ | 0.0877 |
| 2 | $0.43 \times 10^{-2}$ | $1.00 \times 10^{-2}$ | 0.0439 |
| 3 | $0.22 \times 10^{-2}$ | $0.50 \times 10^{-2}$ | 0.0110 |

a. Determine the rate law and the value of the rate constant of the reaction at 298 K.

b. The rate constant was determined at several additional temperatures. Calculate the activation energy of the reaction from the following data.

| *T* (K) | *k* ($M^{-1}\,s^{-1}$) |
|---|---|
| 263 | $3.28 \times 10^2$ |
| 273 | $4.73 \times 10^2$ |
| 283 | $6.65 \times 10^2$ |
| 293 | $9.13 \times 10^2$ |

**13.129. Reducing NO Emissions** Adding $NH_3$ to the stack gases at an electric power–generating plant can reduce $NO_x$ emissions. This selective noncatalytic reduction (SNR) process depends on the reaction between $NH_2$ (an odd-electron compound) and NO:

$$NH_2(g) + NO(g) \rightarrow N_2(g) + H_2O(g)$$

The following kinetics data were collected at 1200 K.

| Experiment | $[NH_2]_0$ (*M*) | $[NO]_0$ (*M*) | Rate (*M*/s) |
|---|---|---|---|
| 1 | $1.00 \times 10^{-5}$ | $1.00 \times 10^{-5}$ | 0.12 |
| 2 | $2.00 \times 10^{-5}$ | $1.00 \times 10^{-5}$ | 0.24 |
| 3 | $2.00 \times 10^{-5}$ | $1.50 \times 10^{-5}$ | 0.36 |
| 4 | $2.50 \times 10^{-5}$ | $1.50 \times 10^{-5}$ | 0.45 |

a. What is the rate law for the reaction?
b. What is the value of the rate constant at 1200 K?

13.130. **Testing for a Banned Herbicide** Sodium chlorate is used in some weed-control products, though its sale has been banned in EU countries since 2009. A simple colorimetric test for the presence of the chlorate ion in a solution of herbicide relies on the following reaction:

$$2\,MnO_4^-(aq) + 5\,ClO_3^-(aq) + 6\,H^+(aq) \rightarrow 2\,Mn^{2+}(aq) + 5\,ClO_4^-(aq) + 3\,H_2O(\ell)$$

The table contains rate data for this reaction at a given temperature.

| Experiment | $[MnO_4^-]_0$ (*M*) | $[ClO_3^-]_0$ (*M*) | $[H^+]_0$ (*M*) | Initial Rate (*M*/s) |
|---|---|---|---|---|
| 1 | 0.10 | 0.10 | 0.10 | $5.2 \times 10^{-3}$ |
| 2 | 0.25 | 0.10 | 0.10 | $3.3 \times 10^{-2}$ |
| 3 | 0.10 | 0.30 | 0.10 | $1.6 \times 10^{-2}$ |
| 4 | 0.10 | 0.10 | 0.20 | $7.4 \times 10^{-3}$ |

Determine the rate law and the rate constant for this reaction at this temperature.

# 14

# Chemical Equilibrium

## Equal but Opposite Reaction Rates

**CHEMISTRY AND THE FOOD SUPPLY** Most crops require nitrogen fertilizers, and the source of most of this nitrogen is ammonia, which is synthesized by the reaction $N_2(g) + 3\ H_2(g) \rightleftharpoons 2\ NH_3(g)$.

PARTICULATE **REVIEW**

### *Forward Rate versus Reverse Rate*

In Chapter 14, we investigate reversible reactions that proceed until the rate of the reverse reaction matches the rate of the forward reaction.

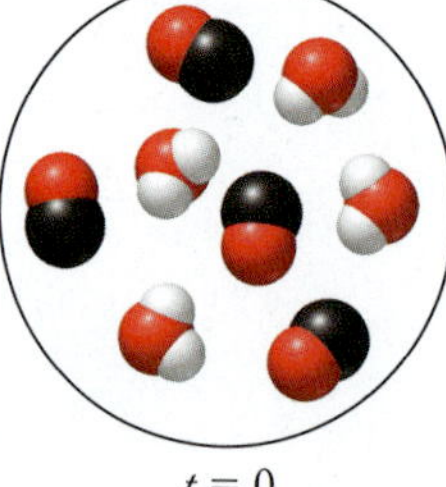

$t = 0$

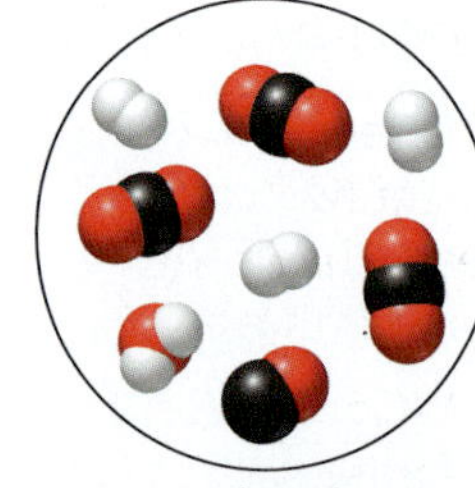

$t = 1$ hour later

- Write a balanced equation that includes both a forward reaction and a reverse reaction for this system that initially contains only carbon monoxide and water.
- If more carbon monoxide were added to the system at $t = 1$ hour, how would this affect the frequency of collisions between the reactants? How would this affect the rate of the forward reaction?

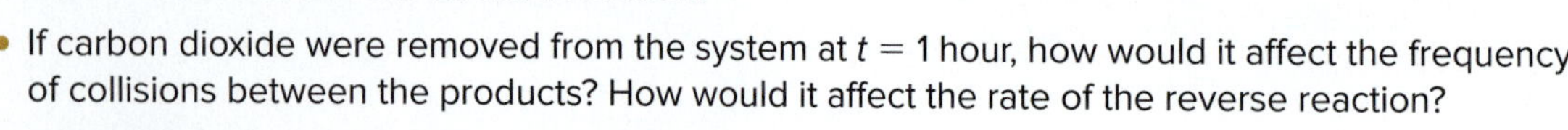

- If carbon dioxide were removed from the system at $t = 1$ hour, how would it affect the frequency of collisions between the products? How would it affect the rate of the reverse reaction?

(Review Section 13.3 if you need help.)

*(Answers to Particulate Review questions are in the back of the book.)*

PARTICULATE **PREVIEW**

## *Spontaneous Shifts*

The reaction $2\ NO_2(g) \rightleftharpoons N_2O_4(g)$ is depicted here under two different sets of initial conditions. As you read Chapter 14, look for ideas that will help you answer these questions:

- Which initial condition results in the spontaneous formation of products?
- Which initial condition results in the spontaneous formation of reactants?
- What is the sign of $\Delta G_{rxn}$ under initial condition A? under initial condition B? Describe the composition of a reaction mixture for which $\Delta G_{rxn} = 0$.

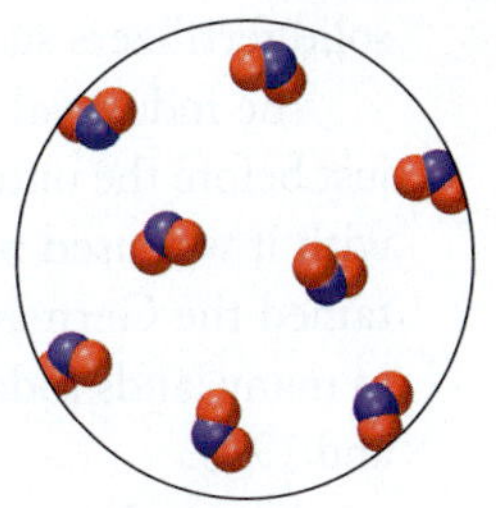

Initial condition A

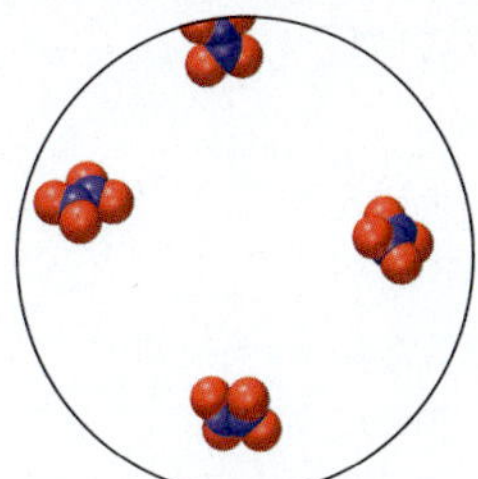

Initial condition B

## Learning Outcomes

**LO1** Describe the dynamic nature of chemical equilibria

**LO2** Write equilibrium constant expressions for reversible reactions, including those involving heterogeneous equilibria
**Sample Exercises 14.1, 14.8**

**LO3** Use equilibrium concentration or partial-pressure data to calculate the value of $K_c$ or $K_p$
**Sample Exercises 14.2, 14.3**

**LO4** Interconvert $K_c$ and $K_p$ values of gas-phase reactions
**Sample Exercise 14.4**

**LO5** Calculate *K* values of related chemical reactions
**Sample Exercises 14.5, 14.6**

**LO6** Calculate the value of a reaction quotient and use it to predict the direction of a reversible chemical reaction
**Sample Exercise 14.7**

**LO7** Predict how a reaction at equilibrium responds to changes in reaction conditions
**Sample Exercises 14.9–14.11**

**LO8** Calculate the concentrations or partial pressures of reactants and products in a reaction mixture at equilibrium from their initial values and the value of *K*
**Sample Exercises 14.12, 14.13**

**LO9** Relate the standard free-energy change of a reversible reaction to its equilibrium constant
**Sample Exercise 14.14**

**LO10** Predict the value of *K* at any temperature from thermodynamic data
**Sample Exercise 14.15**

## 14.1 The Dynamics of Chemical Equilibrium

For centuries, farmers have applied fertilizer to their fields, providing their crops with nitrogen, potassium, and phosphorus compounds that are essential to plant growth. In the past, farmers used mostly biological fertilizers: manure and pulverized bones from farm animals and the droppings of wild ones, including guano collected from bat caves. Today, large-scale production of corn, wheat, and other grains has created a demand for fertilizer that is too large to be met by traditional biological sources. Instead, most of the potassium and phosphorus comes from mined minerals, and most of the nitrogen comes from a synthetic chemistry process in which atmospheric $N_2$ is combined with $H_2$ to form ammonia, $NH_3$. Each year about 200 million tons of ammonia and nitrogen compounds derived from it are used to enhance the world's food supply. Ammonia can be injected directly into the ground, as shown in this chapter's opening photograph, or used to make solid fertilizers such as urea and ammonium nitrate.

The industrial process for synthesizing ammonia was developed in Germany just before the outbreak of World War I, and much of the ammonia first produced with it was used not to grow food but rather to manufacture explosives that sustained the German war effort. It is ironic that a process that sustains human life in many lands today contributed to the deaths of millions of people between 1914 and 1918.

Ammonia synthesis is based on the following chemical reaction:

$$N_2(g) + 3\,H_2(g) \rightleftharpoons 2\,NH_3(g)$$

The reaction is exothermic ($\Delta H^\circ_{rxn} < 0$) and spontaneous ($\Delta G^\circ_{rxn} < 0$) under standard conditions, but it is also very slow at 25°C. Catalysts and elevated temperatures increase the rate of the reaction but lower its yield. In other words, less ammonia is synthesized at higher temperatures, but what is produced is made more rapidly. Why does yield decrease as temperatures rise? A clue is contained in the fact that it takes four moles of reactant gases to make two moles of ammonia gas. This decrease in the number of gas-phase molecules indicates that the reaction mixture experiences a decrease in entropy ($\Delta S^\circ_{rxn} < 0$), which means that the value of $\Delta G^\circ_{rxn}$ in this key equation from Chapter 12,

$$\Delta G^\circ_{rxn} = \Delta H^\circ_{rxn} - T\Delta S^\circ_{rxn} \quad (12.11)$$

increases as $T$ increases and eventually becomes greater than zero. In Chapter 12, we associated positive values of $\Delta G^\circ_{rxn}$ with nonspontaneous reactions. In this chapter, we discover that reaction spontaneity is not a yes-or-no proposition but rather a measure of how far a reaction proceeds until it reaches *chemical equilibrium* and no further increase occurs in product concentration. The more positive the value of $\Delta G^\circ_{rxn}$, the less product forms before equilibrium is achieved.

How, then, does the chemical industry produce millions of tons of ammonia each year? Scientists and engineers have developed ways to manipulate the equilibrium state in the reactors in which ammonia is synthesized. For example, the equilibrium reaction mixture may be pumped through condensers that liquefy and remove $NH_3$ but leave $N_2$ and $H_2$ in the gas phase, where these reactants combine to produce more ammonia. In this chapter, we explore why this technique works, causing the reaction to make more ammonia and providing the fertilizer that helps feed billions of people.

**chemical equilibrium** a dynamic process in which the concentrations of reactants and products remain constant over time and the rate of a reaction in the forward direction matches its rate in the reverse direction.

**Chemical equilibrium** is a *dynamic* process because reactants form products and products form reactants at equal rates. Equilibrium is represented by a pair of single-headed arrows pointing in opposite directions:

**CHEMTOUR**
Equilibrium

$$\text{Reactants} \rightleftharpoons \text{Products}$$

and by the following equality in reaction rates at equilibrium:

$$\text{Rate}_{\text{forward}} = \text{Rate}_{\text{reverse}}$$

Some reactions reach equilibrium only after nearly all the reactants have formed products. We say that these equilibria *lie far to the right* (in the direction of the forward reaction arrow). In other reactions, little product is formed. These equilibria are said to favor reactants and *lie far to the left* (the direction of the reverse reaction arrow). Nothing can be inferred from the position of the equilibrium regarding how much time the reaction takes to reach equilibrium. Studies of equilibrium reveal only the extent to which a reaction proceeds, not how rapidly it proceeds.

To explore the dynamics of chemical equilibrium, let's look at a two-step industrial process for making $H_2$ gas. The reactants are methane and steam. In the first step, methane reacts with steam in the presence of a nickel or iron oxide catalyst at temperatures near 1000°C:

$$CH_4(g) + H_2O(g) \rightleftharpoons CO(g) + 3\,H_2(g) \quad (14.1)$$

This reaction, called the steam–methane reforming reaction, is followed by another, called the water–gas shift reaction, in which the carbon monoxide formed in the first step is reacted with more steam in the presence of a Cu/ZnO catalyst at around 200°C:

$$H_2O(g) + CO(g) \rightleftharpoons H_2(g) + CO_2(g) \quad (14.2)$$

Let's explore the dynamics of Equation 14.2. If we put an equal number of moles of water vapor and carbon monoxide in a closed chamber and they react, the concentrations of CO and $H_2O$ initially fall as the concentrations of $H_2$ and $CO_2$ increase, as shown in **Figure 14.1**. Because $H_2O$ and CO react in a

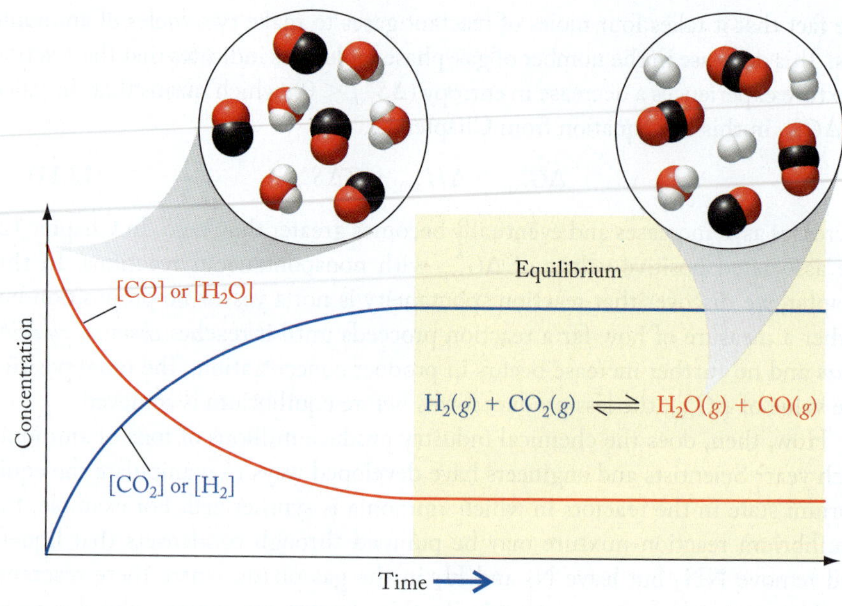


**FIGURE 14.1** Concentrations of reactants and products in the water–gas shift reaction change over time until equilibrium is reached. At equilibrium in the water–gas shift reaction, $H_2$ and $CO_2$ are being formed at the same rate at which they are reacting to re-form $H_2O$ and CO.

1:1 stoichiometric ratio, their concentrations decrease at the same rate and are always equivalent in the reaction chamber. Similarly, $H_2$ and $CO_2$ are formed in a 1:1 ratio, so their concentrations increase at the same rate and are always equal.

**CONNECTION** The method for calculating the percent yield of a reaction was described in Section 7.4.

**CONNECTION** In Chapter 13, we discussed how reaction rate depends on molecular collisions; the higher the concentrations of reactants, the more frequently molecules collide, and the faster a reaction proceeds.

The water–gas shift reaction is reversible. Reversible reactions can proceed in both the forward and the reverse directions. In fact, they proceed in both directions at the same time. One requirement for reversibility is that the products must remain in contact with each other. If one or more of the products is a gas, then we need to run the reaction in a sealed chamber.

The graph in Figure 14.1 shows that eventually the concentrations of reactants and products in the water–gas shift reaction no longer change with time, at which point the reaction has reached chemical equilibrium. Note, however, that the concentrations of the reactants (CO and $H_2O$) do not go to zero. The presence of these reactants in the reaction mixture when the reaction has reached equilibrium means that the yield of the reaction never reaches 100%.

How do the rates of the forward and reverse reactions change during the water–gas shift reaction? When the CO and $H_2O$ are initially mixed, their concentrations are at a maximum and the rate of the forward reaction, as indicated by the slope of the [CO] and [$H_2O$] curve at $t = 0$ in Figure 14.1, is also at its maximum. As the reaction proceeds, reactant concentrations decrease, and the rate of the forward reaction also decreases, because the likelihood of collisions between reactant molecules decreases.

The concentrations of products are initially zero, and so is the rate of the reverse reaction. As product molecules form, the likelihood that they will collide with one another to re-form reactant molecules increases, and so does the rate of the reverse reaction. Ultimately, the rate of the reverse reaction equals the rate of the forward reaction, as shown in **Figure 14.2**. At this point equilibrium has been achieved.

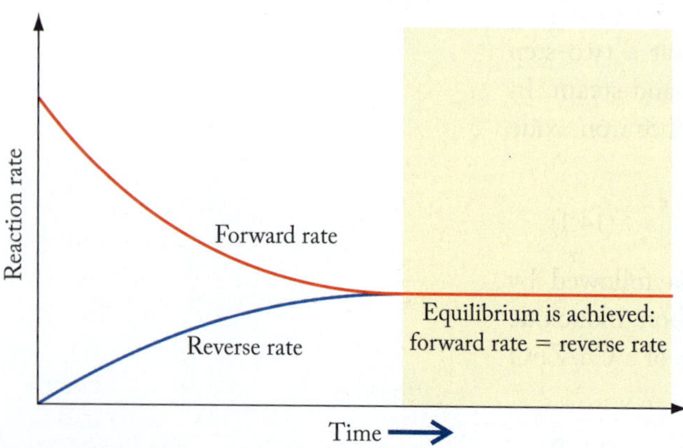


**FIGURE 14.2** The rates of forward and reverse reactions decrease and increase, respectively, until they are the same. Once equilibrium is achieved, the composition of the reaction mixture remains constant.

Let's explore a chemical reaction of $NO_2$ in which we can watch a system reach equilibrium. Recall from Chapter 13 that nitrogen dioxide, $NO_2$, is the gas that gives photochemical smog its distinctive brown color (**Figure 14.3**). Pure nitrogen dioxide dimerizes, forming $N_2O_4$, which is a colorless gas with a boiling point of 21°C:

$$\underset{\text{(brown)}}{2\,NO_2(g)} \rightleftharpoons \underset{\text{(colorless)}}{N_2O_4(g)} \tag{14.3}$$

The rate of the forward reaction ($\text{Rate}_f$) is second order in $NO_2$:

$$\text{Rate}_f = k_f[NO_2]^2 \tag{14.4}$$

At equilibrium, $\text{Rate}_f$ equals $\text{Rate}_r$, the rate of the reverse reaction, which is first order in $N_2O_4$:

$$\text{Rate}_r = k_r[N_2O_4] \tag{14.5}$$

As the concentration of $NO_2$ gas decreases, the rate of the forward reaction also decreases until it matches the rate of the reverse reaction,

$$\text{Rate}_f = \text{Rate}_r \tag{14.6}$$

and equilibrium is established. Replacing the terms in Equation 14.6 with the right sides of Equations 14.4 and 14.5 yields

$$k_f[NO_2]^2 = k_r[N_2O_4]$$

which we can rearrange to

$$\frac{k_f}{k_r} = \frac{[N_2O_4]}{[NO_2]^2} \tag{14.7}$$

This ratio $k_f/k_r$ is the ratio of two constants, which is simply another constant. This constant is called an **equilibrium constant (*K*)**. Equating it to the ratio of product concentrations to reactant concentrations on the right side of Equation 14.7 gives us the **equilibrium constant expression** for this reaction:

$$K = \frac{[N_2O_4]}{[NO_2]^2} \tag{14.8}$$

We just derived the $K$ expression for the dimerization of $NO_2$ from the ratio of its forward and reverse rate laws. However, prior knowledge of the kinetics of reactions is not needed to write their equilibrium constant expressions. As we will see in the next section, all we need are balanced chemical equations. This is because the composition of an equilibrium mixture of reactants and products does not depend on the terms and exponents of the forward and reverse rate laws or on how rapidly the reaction proceeds. Some reactions achieve equilibrium quickly, whereas others take a very long time. The value of an equilibrium constant $K$ says nothing about how rapidly a reaction proceeds toward equilibrium; it tells us only the extent to which the reaction proceeds before it reaches equilibrium.

**FIGURE 14.3** The brown layer in the atmosphere seen here is caused by high concentrations of $NO_2$, a key ingredient in photochemical smog.

**CONNECTION** We discussed the rates of reactions of nitrogen oxides in the atmosphere in Chapter 13 in our discussion of the chemistry of photochemical smog.

**equilibrium constant (*K*)** the value of the ratio of concentration (or partial pressure) terms in the equilibrium constant expression at a specific temperature.

**equilibrium constant expression** the ratio of the equilibrium concentrations or partial pressures of products to reactants, with each term raised to a power equal to the coefficient of that substance in the balanced chemical equation for the reaction.

## CONCEPT TEST

In the reversible reaction $A(g) \rightleftharpoons B(g)$, the rate constant of the forward reaction at a particular temperature is 3.0 times the value of the rate constant of the reverse reaction. What is the value of the equilibrium constant at that temperature?

*(Answers to Concept Tests are in the back of the book.)*

## 14.2 Writing Equilibrium Constant Expressions

Let's revisit the water–gas shift reaction:

$$H_2O(g) + CO(g) \rightleftharpoons H_2(g) + CO_2(g) \tag{14.2}$$

The data in **Table 14.1** describe the results of four experiments in which different quantities of $H_2O$ vapor, CO, $H_2$, and $CO_2$ are injected into a sealed reaction vessel heated to 500 K. The gases react, and the final concentrations of all four gases are determined once the reaction has reached equilibrium.

In experiment 1, the reaction vessel initially contains equimolar concentrations of $H_2O$ and CO but no $H_2$ or $CO_2$. In experiment 2, the vessel initially contains equimolar concentrations of $H_2$ and $CO_2$ but no $H_2O$ or CO. The data in Table 14.1 indicate that when the reaction mixtures in experiments 1 and 2 achieve chemical equilibrium, the concentrations of $H_2O$ and CO are the same (0.0034 *M*) in both experiments, and so too are the concentrations of $H_2$ and $CO_2$ (0.0166 *M*). (The equilibrium concentrations of both reactants are equal here because their initial concentrations and their coefficients in the balanced equation are the same. The same statement can be made about the equilibrium concentrations of the products.) Experiments 1 and 2 show that the composition of a reaction mixture at equilibrium is independent of the direction in which a particular reaction proceeds to achieve equilibrium.

The data from experiment 3 in Table 14.1 show that the concentrations of products increase and the concentrations of reactants decrease even when the concentrations of the products are initially *greater than* the concentrations of the reactants. The data from experiment 4 show that the concentrations of products increase and the concentrations of reactants also decrease when the concentrations of the products are initially the same as the concentrations of the reactants.

The significance of these observations, taken with those from experiments 1 and 2, can be appreciated if we do the following math: multiply the equilibrium concentrations of the products ($H_2$ and $CO_2$) together and divide that value by the equilibrium concentrations of the reactants ($H_2O$ and CO) multiplied together:

$$\text{Experiments 1 and 2} \qquad \frac{[H_2][CO_2]}{[H_2O][CO]} = \frac{(0.0166)(0.0166)}{(0.0034)(0.0034)} = 24$$

$$\text{Experiment 3} \qquad \frac{[H_2][CO_2]}{[H_2O][CO]} = \frac{(0.0354)(0.0454)}{(0.0046)(0.0146)} = 24$$

$$\text{Experiment 4} \qquad \frac{[H_2][CO_2]}{[H_2O][CO]} = \frac{(0.0282)(0.0182)}{(0.0118)(0.0018)} = 24$$

In every experiment, this calculation yields the same result.

In fact, we would get the same ratio of product to reactant concentrations at equilibrium from any combination of initial concentrations of these four gases at 500 K. This constancy applies to other reaction mixtures and has been known since the mid-19th century, when Norwegian chemists Cato Guldberg (1836–1902) and Peter Waage (1833–1900) discovered that any reversible reaction eventually reaches a state in which the ratio of the concentrations of products to reactants, with each value raised to a power corresponding to the coefficient for that substance in the balanced chemical equation for the reaction, has a characteristic value at a given temperature. They called this phenomenon the **law of mass action**. This ratio of concentration terms is the equilibrium constant

**TABLE 14.1** Initial and Equilibrium Concentrations of the Reactants and Products in the Water–Gas Shift Reaction $H_2O(g) + CO(g) \rightleftharpoons H_2(g) + CO_2(g)$ at 500 K

| Experiment | INITIAL CONCENTRATION (*M*) | | | | EQUILIBRIUM CONCENTRATION (*M*) | | | |
|---|---|---|---|---|---|---|---|---|
| | $[H_2O]$ | [CO] | $[H_2]$ | $[CO_2]$ | $[H_2O]$ | [CO] | $[H_2]$ | $[CO_2]$ |
| 1 | 0.0200 | 0.0200 | 0 | 0 | 0.0034 | 0.0034 | 0.0166 | 0.0166 |
| 2 | 0 | 0 | 0.0200 | 0.0200 | 0.0034 | 0.0034 | 0.0166 | 0.0166 |
| 3 | 0.0100 | 0.0200 | 0.0300 | 0.0400 | 0.0046 | 0.0146 | 0.0354 | 0.0454 |
| 4 | 0.0200 | 0.0100 | 0.0200 | 0.0100 | 0.0118 | 0.0018 | 0.0282 | 0.0182 |

expression for the reaction. It is also called the **mass action expression**, but we will limit the use of that term to reactions that are not at equilibrium. For the water–gas shift reaction at 500 K, the equilibrium constant expression is

$$K = \frac{[H_2][CO_2]}{[H_2O][CO]} = 24$$

The exponents in the equilibrium constant expression are the same as the coefficients in the balanced equation. This is in contrast to rate law expressions, in which the exponents are often not the same as the coefficients in the balanced equation, because rate law exponents reflect the stoichiometry of the rate-determining step, not necessarily the stoichiometry of the overall reaction.

## CONCEPT TEST

To run the water–gas shift reaction (Equation 14.2), equal numbers of moles of water vapor, carbon dioxide, carbon monoxide, and hydrogen gas are injected into a rigid, sealed reaction vessel and heated to 500 K. Which expression about the composition of the equilibrium reaction mixture is true?

a. $[CO] = [H_2] = [CO_2] = [H_2O]$

b. $[CO_2] = [H_2] > [CO] = [H_2O]$

c. $[CO_2] = [H_2] < [CO] = [H_2O]$

d. $[H_2O] = [H_2] > [CO_2] = [CO]$

e. $[H_2O] = [H_2] < [CO_2] = [CO]$

For the generic reaction in which $a$ moles of reactant A react with $b$ moles of B to form $c$ moles of substance C and $d$ moles of D,

$$a\,A + b\,B \rightleftharpoons c\,C + d\,D$$

the equilibrium constant expression is

$$K_c = \frac{[C]^c[D]^d}{[A]^a[B]^b} \quad (14.9)$$

where the subscript "c" of the equilibrium constant represents concentration. If substances A, B, C, and D are gases, then the equilibrium constant may also be expressed in terms of their partial pressures:

$$K_p = \frac{(P_C)^c(P_D)^d}{(P_A)^a(P_B)^b} \quad (14.10)$$

As we'll see later in this chapter, the values of $K_c$ and $K_p$ for a given reaction and temperature may or may not be the same. It depends on whether the

**law of mass action** the principle relating the balanced chemical equation of a reversible reaction to its mass action expression (or equilibrium constant expression).

**mass action expression** an expression equivalent in form to the equilibrium constant expression, but applied to reaction mixtures that may or may not be at equilibrium.

number of moles of gaseous reactants is the same as the number of moles of gaseous products.

Equilibrium constants never have units. This is true even when different numbers of moles of reactants and products are present because the concentrations and partial pressures in $K_c$ or $K_p$ expressions are actually ratios of the concentrations or pressures of substances to an ideal standard concentration (1.000 *M*) or partial pressure (1.000 atm). The units in such ratios cancel, so the terms in the equilibrium constant expression have no units. In practice, we use concentration and partial pressure values directly in equilibrium calculations and simply report the value of *K* as a number without units.

### SAMPLE EXERCISE 14.1 Writing Equilibrium Constant Expressions — LO2

A key reaction in the formation of acid rain involves the reversible combination of $SO_2$ and $O_2$ in the atmosphere, which produces $SO_3$:

$$2\,SO_2(g) + O_2(g) \rightleftharpoons 2\,SO_3(g)$$

Write the $K_c$ and $K_p$ expressions for this reaction.

**Collect, Organize, and Analyze** We are given the balanced chemical equation for a reaction and asked to write $K_c$ and $K_p$ expressions for the reaction. Equilibrium constant expressions are ratios of the concentrations ($K_c$) or partial pressures ($K_p$) of products to reactants, with each term raised to the power equal to its coefficient in the balanced chemical equation. In the reaction of interest, the coefficients of $SO_2$ and $SO_3$ are both 2, so the $SO_2$ and $SO_3$ terms in the $K_c$ and $K_p$ expressions will be squared.

**Solve**

$$K_c = \frac{[SO_3]^2}{[SO_2]^2[O_2]}$$

$$K_p = \frac{(P_{SO_3})^2}{(P_{SO_2})^2(P_{O_2})}$$

**Think About It** The $K_c$ and $K_p$ expressions have the same form: their numerators and denominators contain terms for the same products and reactants, each raised to the same power. The difference between them is the nature of the terms: molar concentrations in the $K_c$ expression and pressures in the $K_p$ expression.

**Practice Exercise** Write the equilibrium constant expressions $K_c$ and $K_p$ for the following reaction:

$$CH_4(g) + H_2O(g) \rightleftharpoons CO(g) + 3\,H_2(g)$$

*(Answers to Practice Exercises are in the back of the book.)*

### SAMPLE EXERCISE 14.2 Calculating the Value of $K_c$ — LO3

**Table 14.2** contains data from four experiments on the dimerization of $NO_2$:

$$2\,NO_2(g) \rightleftharpoons N_2O_4(g)$$

The experiments were run at 100°C in a rigid, closed container. Use the data from each experiment to calculate a value of the equilibrium constant $K_c$ for the dimerization reaction.

**TABLE 14.2** Data for the Reaction $2\ NO_2(g) \rightleftharpoons N_2O_4(g)$ at 100°C

| Experiment | INITIAL CONCENTRATION (*M*) | | EQUILIBRIUM CONCENTRATION (*M*) | |
|---|---|---|---|---|
| | $[NO_2]$ | $[N_2O_4]$ | $[NO_2]$ | $[N_2O_4]$ |
| 1 | 0.0200 | 0.0000 | 0.0172 | 0.00139 |
| 2 | 0.0300 | 0.0000 | 0.0244 | 0.00280 |
| 3 | 0.0400 | 0.0000 | 0.0310 | 0.00452 |
| 4 | 0.0000 | 0.0200 | 0.0310 | 0.00452 |

**Collect and Organize** We are given four sets of data that contain initial and equilibrium concentrations of a reactant and product. We are asked to determine the value of the equilibrium constant $K_c$ in each experiment. The equilibrium constant expression (Equation 14.8) for this reaction is

$$K_c = \frac{[N_2O_4]}{[NO_2]^2}$$

**Analyze** In each of the four experiments, the concentration of $NO_2$ is nearly 10 times the concentration of $N_2O_4$ at equilibrium. However, both values in each experiment are much less than 1, and the $[NO_2]$ term is squared. These two factors taken together mean that the values of the numerators in the equilibrium constant expressions will probably be greater than the denominators, so the calculated values of $K_c$ will be greater than 1.

**Solve**

Experiment 1: $$K_c = \frac{[N_2O_4]}{[NO_2]^2} = \frac{0.00139}{(0.0172)^2} = 4.70$$

Experiment 2: $$K_c = \frac{0.00280}{(0.0244)^2} = 4.70$$

Experiments 3 and 4: $$K_c = \frac{0.00452}{(0.0310)^2} = 4.70$$

**Think About It** The values calculated for $K_c$ are the same, as they should be for the same reaction at the same temperature. They are greater than 1, too, as predicted.

**Practice Exercise** A mixture of gaseous CO and $H_2$, called *synthesis gas*, is used commercially to prepare methanol ($CH_3OH$), a compound considered to be an alternative fuel to gasoline. Under equilibrium conditions at 700 K, $[H_2] = 0.074\ M$, $[CO] = 0.025\ M$, and $[CH_3OH] = 0.040\ M$. What is the value of $K_c$ for this reaction at 700 K?

## SAMPLE EXERCISE 14.3 Calculating the Value of $K_p$ LO3

A sealed chamber contains an equilibrium mixture of $NO_2(g)$ and $N_2O_4(g)$ at 15°C. Their partial pressures are $P_{NO_2} = 0.34$ atm and $P_{N_2O_4} = 0.80$ atm. What is the value of $K_p$ for the following reaction under these conditions?

$$2\ NO_2(g) \rightleftharpoons N_2O_4(g)$$

**Collect and Organize** We are given the partial pressure values of reactant and product gases at equilibrium and asked to determine the value of the equilibrium constant $K_p$ for the dimerization reaction of $NO_2$ to form $N_2O_4$. $K_p$ expressions are like $K_c$

expressions except that the concentration terms in the $K_c$ expression are replaced with partial-pressure terms (Equation 14.10).

**Analyze** The $K_p$ expression for this reaction is:

$$K_p = \frac{P_{N_2O_4}}{(P_{NO_2})^2}$$

The partial pressure of $NO_2$ is slightly greater than the partial pressure of $N_2O_4$ at equilibrium, but both values are less than 1. Moreover, the $P_{NO_2}$ term is squared. These two factors taken together mean that the value of the numerator in the $K_p$ expression will probably be greater than that of the denominator, so the calculated value of $K_p$ will be greater than 1.

**Solve**

$$K_p = \frac{0.80}{(0.34)^2} = 6.9$$

**Think About It** As we predicted, $K_p > 1$, even though the equilibrium partial pressure of the product is actually less than that of the reactant. Relatively little $N_2O_4$ forms here because $P_{NO_2}$ is so low, and two moles of $NO_2$ are required to make one mole of $N_2O_4$.

**Practice Exercise** A reaction vessel contains an equilibrium mixture of $SO_2$, $O_2$, and $SO_3$. If the partial pressures are $P_{SO_2} = 0.0018$ atm, $P_{O_2} = 0.0032$ atm, and $P_{SO_3} = 0.0166$ atm, what is the value of $K_p$ for the following reaction?

$$2\ SO_2(g) + O_2(g) \rightleftharpoons 2\ SO_3(g)$$

The value of $K$ indicates how far a reaction proceeds at a given temperature. The values we have seen thus far—$K_c = 24$ for the water–gas shift reaction at 500 K and $K_c = 4.7$ for the dimerization of $NO_2$ at 100°C—are considered intermediate values. Because the range of $K$ values is so large ($0 < K < \infty$), both of these values are considered close to 1, which means that comparable concentrations of reactants and products are likely to be present at equilibrium.

In contrast, the reaction between $H_2$ and $O_2$ to form water proceeds until one of the reactants is almost completely consumed. This observation is consistent with a large value of $K$ at 25°C:

$$2\ H_2(g) + O_2(g) \rightleftharpoons 2\ H_2O(g) \qquad K_c = 3 \times 10^{81}$$

The decomposition of $CO_2$ to CO and $O_2$ at 25°C proceeds hardly at all and is consistent with a very small value of $K$ and virtually no product being formed:

$$2\ CO_2(g) \rightleftharpoons 2\ CO(g) + O_2(g) \qquad K_c = 3 \times 10^{-92}$$

## 14.3 Relationships between $K_c$ and $K_p$ Values

**CHEMTOUR**
Equilibrium in the Gas Phase

As we noted in Section 14.2, the values of $K_c$ and $K_p$ for a given reaction and temperature may or may not be the same, depending on the numbers of moles of gaseous reactants and products. To better understand this relationship, we begin with the ideal gas law:

$$PV = nRT \qquad (9.24)$$

If we solve for $P$ and express volume in liters, then $n/V$ has units of moles per liter, which is the same as molarity ($M$):

$$P = \frac{n}{V}RT = MRT \qquad (14.11)$$

Applying Equation 14.11 to the gases in the $NO_2/N_2O_4$ equilibrium from Sample Exercise 14.3 gives

$$P_{NO_2} = \frac{n_{NO_2}}{V}RT = [NO_2]RT$$

$$P_{N_2O_4} = \frac{n_{N_2O_4}}{V}RT = [N_2O_4]RT$$

Substituting these values into the expression for $K_p$ from Sample Exercise 14.3, we get

$$K_p = \frac{P_{N_2O_4}}{(P_{NO_2})^2} = \frac{[N_2O_4]RT}{([NO_2]RT)^2} = \frac{[N_2O_4]\cancel{RT}}{[NO_2]^2(RT)^2}$$

The ratio of concentration terms in the expression on the right, $[N_2O_4]/[NO_2]^2$, is the same as the $K_c$ expression for this reaction. Substituting $K_c$ for those terms and simplifying the $RT$ terms gives

$$K_p = K_c\frac{1}{RT}$$

This last equation defines the specific relationship between $K_c$ and $K_p$ for this reaction.

A more general expression can be derived for the generic reaction of gases A and B to form gases C and D:

$$a\,A + b\,B \rightleftharpoons c\,C + d\,D$$

The $K_p$ expression for this reaction is

$$K_p = \frac{(P_C)^c(P_D)^d}{(P_A)^a(P_B)^b} \qquad (14.10)$$

Replacing each partial pressure term in Equation 14.10 with the corresponding molar concentration term $\times$ $RT$ (from Equation 14.11) gives us the general expression

$$K_p = \frac{([C]RT)^c([D]RT)^d}{([A]RT)^a([B]RT)^b} \qquad (14.12)$$

Combining the $RT$ terms, we get

$$K_p = \frac{[C]^c[D]^d}{[A]^a[B]^b} \times (RT)^{[(c+d)-(a+b)]} \qquad (14.13)$$

The concentration ratio on the right side of this equation matches the $K_c$ expression for this generic reaction (see Equation 14.9). Substituting this equality into Equation 14.13 gives us

$$K_p = K_c(RT)^{[(c+d)-(a+b)]} \qquad (14.14)$$

To simplify Equation 14.14, consider this: $(c + d)$ represents the sum of the coefficients of the gaseous products in the reaction—that is, the sum of the number of moles of gases. Similarly, $(a + b)$ represents the sum of the number of moles of gaseous reactants. The difference between the two sums, $(c + d) - (a + b)$, represents the change in the number of moles of gases from the reactant side to the product side of the balanced chemical equation. We use the symbol $\Delta n$ to represent this change. Substituting $\Delta n$ for $(c + d) - (a + b)$ in Equation 14.14 gives us

$$K_p = K_c(RT)^{\Delta n} \quad (14.15)$$

Equation 14.15 provides a quantitative interpretation of the opening statement of this section: the relationship between the $K_p$ and $K_c$ values of a chemical reaction involving gases depends on the number of moles of gaseous reactants and products. In reactions such as the water–gas shift reaction,

$$H_2O(g) + CO(g) \rightleftharpoons H_2(g) + CO_2(g) \quad (14.2)$$

in which the number of moles of gas on both sides of the reaction arrow is the same, $\Delta n = 0$ and $K_p = K_c$. However, in the steam–methane reforming reaction,

$$CH_4(g) + H_2O(g) \rightleftharpoons CO(g) + 3\ H_2(g) \quad (14.1)$$

two moles of gaseous reactants form four moles of gaseous products. Therefore,

$$\Delta n = 4\text{ mol} - 2\text{ mol} = 2\text{ mol}$$

Inserting this value for $\Delta n$ in Equation 14.15 gives us the following relationship between $K_p$ and $K_c$ for this reaction:

$$K_p = K_c(RT)^2$$

A final point about the relative sizes of $K_p$ and $K_c$ values: one mole of an ideal gas at STP (273 K and 1 atm) occupies a volume of 22.4 L. Therefore, its molar concentration is 1 mol/22.4 L = 0.0446 *M*. Thus, the pressure of a pure gas at STP is 22.4 times its molar concentration. The *RT* term in Equation 14.15 is essentially a conversion factor for changing molar concentrations into partial pressures. The value of *R* that we use depends on the units used to express concentration, pressure, and temperature. If we stick with molarity, atmospheres, and *T* in kelvin, the value of *R* is 0.08206 (L · atm)/(mol · K).

**CONNECTION** As stated in Section 9.6, the *molar volume* of an ideal gas is 22.4 L at $T = 273$ K and $P = 1$ atm. These values are considered *standard* temperature and pressure (STP).

**SAMPLE EXERCISE 14.4** Calculating $K_c$ from $K_p$ **LO4**

In Sample Exercise 14.3, we calculated the value of $K_p$ (6.9) for the dimerization of $NO_2$ to $N_2O_4$ at 15°C. What is the value of $K_c$ for this reaction at 15°C?

**Collect and Organize** We are given a $K_p$ value and asked to calculate the corresponding $K_c$ value for the same reaction at the same temperature. Equation 14.15 relates $K_p$ and $K_c$ values:

$$K_p = K_c(RT)^{\Delta n}$$

where $\Delta n$ represents the change in the number of moles of gas when going from reactant to product.

**Analyze** We need a balanced chemical equation to determine the value of $\Delta n$. From Sample Exercise 14.3 we know that the equation is

$$2\ NO_2(g) \rightleftharpoons N_2O_4(g)$$

Because the number of moles of gas is not the same on both sides of the reaction arrow, the values of $K_c$ and $K_p$ will not be the same.

**Solve** According to the balanced equation, two moles of gaseous reactants yield one mole of gaseous product. Therefore:

$$\Delta n = 1 \text{ mol} - 2 \text{ mol} = -1 \text{ mol}$$

Inserting this value, the known value of $K_p$, and the temperature into Equation 14.15,

$$K_p = K_c(RT)^{\Delta n}$$

$$6.9 = K_c[0.08206 \times (15 + 273)]^{-1}$$

$$K_c = (6.9)(0.08206)(288) = 1.6 \times 10^2$$

**Think About It** The value of $K_c$ is nearly 25 times larger than that of $K_p$ because two moles of gaseous reactants produce only one mole of gaseous product, and the reaction was carried out at a high temperature, which makes the $RT$ term large.

**Practice Exercise** What is the value of $K_p$ for the industrial synthesis of ammonia at 500°C?

$$N_2(g) + 3\,H_2(g) \rightleftharpoons 2\,NH_3(g) \qquad K_c = 5.8 \times 10^{-2} \text{ at } 500°\text{C}$$

# 14.4 Manipulating Equilibrium Constant Expressions

If $K_c = 100$ for the following hypothetical combination reaction,

$$A + B \rightleftharpoons 2\,C \qquad K_c = 1.00 \times 10^2$$

then what is $K_c$ for the reverse reaction?

$$2\,C \rightleftharpoons A + B \qquad K_{c,\text{reverse}} = ?$$

Moreover, what is $K_c$ for the reaction in which only one mole of C is made—that is, in which all the coefficients are half what they were in the original equation?

$$\tfrac{1}{2}A + \tfrac{1}{2}B \rightleftharpoons C \qquad K_{c,1/2} = ?$$

To answer questions like these, we need to think about how the equilibrium constant expressions for these reactions are related. We explore these relationships in this section.

## *K* for Reverse Reactions

The $K_c$ expression for our combination reaction, $A + B \rightleftharpoons 2\,C$, is

$$K_c = \frac{[C]^2}{[A][B]} = 1.00 \times 10^2$$

The $K_c$ expression for the reverse reaction, $2\,C \rightleftharpoons A + B$, is

$$K_{c,\text{reverse}} = \frac{[A][B]}{[C]^2}$$

Comparing the $K_c$ expressions for the forward and reverse reactions, we see that $K_{c,\text{reverse}}$ is the reciprocal of $K_c$, which means that its value must be $1/1.00 \times 10^2 = 0.0100$:

$$2\,C \rightleftharpoons A + B \qquad K_{c,\text{reverse}} = 0.0100$$

Generalizing this result for any chemical reaction, we have

$$K_{\text{forward}} = \frac{1}{K_{\text{reverse}}} \tag{14.16}$$

This inverse relationship between $K_{\text{forward}}$ and $K_{\text{reverse}}$ makes sense when we apply it to our hypothetical reaction. The relatively large value of $K_{\text{forward}}$ ($1.00 \times 10^2$) means that the concentration of product C in a reaction mixture at equilibrium is likely to be much higher than the concentrations of reactants A and B. The corresponding small value of $K_{\text{reverse}}$ (0.0100) means the same thing in terms of the relative values of [A], [B], and [C]. The difference is that the equilibrium mixture in the reverse reaction contains mostly reactant (C) and little of the products (A and B), as we would expect given the small value of $K_{\text{reverse}}$.

## *K* for an Equation Multiplied by a Number

The $K_c$ expression for our combination reaction representing the preparation of only one mole of C, $\frac{1}{2}\text{ A} + \frac{1}{2}\text{ B} \rightleftharpoons \text{C}$, is

$$K_c = \frac{[\text{C}]}{[\text{A}]^{1/2}[\text{B}]^{1/2}}$$

Comparing this expression with the $K_c$ expression for the forward reaction,

$$K_c = \frac{[\text{C}]^2}{[\text{A}][\text{B}]}$$

we see that the $K_c$ expression for the preparation of only one mole of C is equal to the square root of the $K_c$ expression for the forward reaction that produces two moles of C. The C term is squared, whereas the A and B terms are raised to the first power in the latter; but the C term is raised only to the first power, whereas the A and B terms are raised to the $\frac{1}{2}$ power in the former. Therefore, the value of $K_{c,1/2}$ is the square root of the value of $K_c$, or $\sqrt{1.00 \times 10^2} = 10.0$. We can extend this pattern to all chemical equilibria. If the balanced chemical equation of a reaction is multiplied by some factor $n$, then the value of $K$ is raised to the $n$th power ($K^n$).

**SAMPLE EXERCISE 14.5** Calculating the *K* Values of Related Chemical Reactions **LO5**

Recall from Chapter 13 that a key reaction in the formation of photochemical smog is the reaction between NO and $O_2$:

$$(1) \qquad 2\text{ NO}(g) + \text{O}_2(g) \rightleftharpoons 2\text{ NO}_2(g)$$

The $K_p$ value of this reaction is $2.4 \times 10^{12}$ at 25°C.

a. What is the $K_p$ value of the decomposition of $NO_2(g)$?

$$(2) \qquad 2\text{ NO}_2(g) \rightleftharpoons 2\text{ NO}(g) + \text{O}_2(g)$$

b. What is the $K_p$ value of the reaction in which NO and $O_2$ combine to form only one mole of $NO_2$?

$$(3) \qquad \text{NO}(g) + \tfrac{1}{2}\text{ O}_2(g) \rightleftharpoons \text{NO}_2(g)$$

**Collect, Organize, and Analyze** We know the $K_p$ value of reaction 1 but not of reactions 2 and 3. Reaction 2 is the reverse of reaction 1. Therefore, the $K_p$ value of reaction 2 is the reciprocal of the $K_p$ value of reaction 1. Reaction 3 has the same reactants and product as reaction 1, but the stoichiometric coefficients in its chemical equation are one-half of the values used in the chemical equation for reaction 1. When the coefficients of a chemical equation are multiplied by a factor $n$ ($\frac{1}{2}$ here), the value of the equilibrium constant is raised to the $n$th power, which here means that the $K_p$ value of reaction 3 is the square root of the $K_p$ value of reaction 1. The value of $K_{p,rxn1}$ is huge—a bit more than $10^{12}$—which means that the value of $K_{p,rxn2}$ for the reverse reaction should be very small—somewhat less than $10^{-12}$. The value of $K_{p,rxn3}$ is equal to $\sqrt{K_{p,rxn1}}$, or about $10^6$.

**Solve**

a. $K_{p,rxn2} = \dfrac{1}{K_{p,rxn1}} = \dfrac{1}{2.4 \times 10^{12}} = 4.2 \times 10^{-13}$

b. $K_{p,rxn3} = \sqrt{K_{p,rxn1}} = \sqrt{2.4 \times 10^{12}} = 1.5 \times 10^{6}$

**Think About It** As predicted, the value of $K_{p,rxn2}$ is very small and the value of $K_{p,rxn3}$ is quite large. Both values carry the same message: NO should (eventually) be oxidized to $NO_2$ in the atmosphere. How long the reaction takes, as we saw in Chapter 13, is another matter.

 **Practice Exercise** $K_c = 9.6$ for the synthesis of ammonia at 300°C:

$$N_2(g) + 3\,H_2(g) \rightleftharpoons 2\,NH_3(g)$$

What is the value of $K_c$ for the following decomposition of ammonia at 300°C?

$$NH_3(g) \rightleftharpoons \tfrac{1}{2}\,N_2(g) + \tfrac{3}{2}\,H_2(g)$$

Given two correct equilibrium constant expressions for the same reaction, you may wonder how the same reaction with the same reactants and products can have two or more equilibrium constant values. Surely the same ingredients should be present in the same proportions at equilibrium no matter how we choose to write a balanced equation describing their reaction. In fact, they are. The difference in $K$ values is not chemical; it's mathematical. It has to do with the two ways we wrote balanced chemical equations describing the reaction, which resulted in two different sets of exponents on the concentration terms in the two $K$ expressions. Thus, the same concentrations of reactants and products, raised to different powers, produce different values of $K$.

## Combining $K$ Values

In Chapter 10, we applied Hess's law to calculate the enthalpies of combined reactions. We carry out a similar process here to determine the overall $K$ for a reaction that is the sum of two or more other reactions.

Consider two reactions from Chapter 13 involved in the formation of photochemical smog, wherein the NO produced in a car's engine at high temperatures is oxidized to $NO_2$ in the atmosphere:

$$\begin{array}{ll} (1) & N_2(g) + O_2(g) \rightleftharpoons \cancel{2\,NO(g)} \\ (2) & \cancel{2\,NO(g)} + O_2(g) \rightleftharpoons 2\,NO_2(g) \\ \hline \text{Overall:} & N_2(g) + 2\,O_2(g) \rightleftharpoons 2\,NO_2(g) \end{array}$$

The equilibrium constant expression for the overall reaction is

$$K_c = \frac{[NO_2]^2}{[N_2][O_2]^2}$$

We can derive this expression from the equilibrium constant expressions for reactions 1 and 2,

$$K_1 = \frac{[NO]^2}{[N_2][O_2]} \quad \text{and} \quad K_2 = \frac{[NO_2]^2}{[NO]^2[O_2]}$$

if we multiply $K_1$ by $K_2$:

$$K_1 \times K_2 = \frac{\cancel{[NO]^2}}{[N_2][O_2]} \times \frac{[NO_2]^2}{\cancel{[NO]^2}[O_2]} = \frac{[NO_2]^2}{[N_2][O_2]^2} = K_{overall}$$

This approach works for all series of reactions, and as a general rule:

$$K_{overall} = K_1 \times K_2 \times K_3 \times \cdots \times K_n \tag{14.17}$$

The overall equilibrium constant for a sum of two or more reactions is the product of the equilibrium constants of the individual reactions. Thus, the value of $K_c$ for the overall reaction for the formation of $NO_2$ from $N_2$ and $O_2$ at 1000 K is the product of the equilibrium constants for reactions 1 and 2:

$$K_1 = \frac{[NO]^2}{[N_2][O_2]} = 7.2 \times 10^{-9}$$

$$K_2 = \frac{[NO_2]^2}{[NO]^2[O_2]} = 0.020$$

$$K_{overall} = K_1 \times K_2 = (7.2 \times 10^{-9})(0.020) = 1.4 \times 10^{-10}$$

Remember that the equilibrium constant expression for the overall reaction must contain the appropriate terms for the products and reactants of that reaction. Just as with Hess's law in thermochemical calculations, we may need to reverse an equation or multiply an equation by a factor when we combine it with another to create the equation of interest. If we reverse a reaction, we must take the reciprocal of its $K$. If we multiply a reaction by a constant, we must raise its $K$ to that power.

## SAMPLE EXERCISE 14.6 Calculating Overall $K$ Values of Combined Reactions — LO5

At 1000 K, the $K_c$ values of the following reactions are as given:

(1) $N_2O_4(g) \rightleftharpoons 2\ NO_2(g)$ $\quad K_c = 1.5 \times 10^6$

(2) $N_2(g) + 2\ O_2(g) \rightleftharpoons 2\ NO_2(g)$ $\quad K_c = 1.4 \times 10^{-10}$

What is the $K_c$ value at 1000 K of the following reaction?

$$N_2(g) + 2\ O_2(g) \rightleftharpoons N_2O_4(g)$$

**Collect and Organize** We are given two reactions and their $K_c$ values. We need to combine the two reactions in such a way that $N_2$ and $O_2$ are on the reactant side of the overall equation and $N_2O_4$ is on the product side, and then calculate the $K_c$ value of the overall reaction.

**Analyze** The overall reaction is the sum of the reverse of reaction 1 and reaction 2 as written:

| | |
|---|---|
| Reaction 1 reversed | $\cancel{2\ NO_2(g)} \rightleftharpoons N_2O_4(g)$ |
| Reaction 2 | $N_2(g) + 2\ O_2(g) \rightleftharpoons \cancel{2\ NO_2(g)}$ |
| Overall | $N_2(g) + 2\ O_2(g) \rightleftharpoons N_2O_4(g)$ |

Reversing a chemical reaction requires taking the reciprocal of its $K_c$ value, and when two reactions are added, the value of $K_c$ of the overall reaction is the product of the $K_c$ values of the two reactions. The $K_c$ value of reaction 1 is large ($>10^6$), which means that its reciprocal is small ($<10^{-6}$). The product of the latter value and the even smaller $K_c$ value of reaction 2 ($\sim 10^{-10}$) should be a very small value—about $10^{-16}$.

**Solve**

$$K_{overall} = \frac{1}{K_1} \times K_2 = \left(\frac{1}{1.5 \times 10^6}\right)(1.4 \times 10^{-10}) = 9.3 \times 10^{-17}$$

**Think About It** The result of the $K_{overall}$ calculation is reassuringly close to the estimated value. Combining reactions (1) and (2) after reversing reaction (1) produced the same result as if we were calculating $K_{overall}$ for a two-step reaction in which the product of the first step ($NO_2$) was the reactant in the second:

$$N_2(g) + 2\,O_2(g) \rightleftharpoons 2\,NO_2(g) \rightleftharpoons N_2O_4(g)$$

**Practice Exercise** Calculate the value of $K_c$ for the hypothetical reaction

$$Q(g) + X(g) \rightleftharpoons M(g)$$

given the following information:

$$2\,M(g) \rightleftharpoons Z(g) \qquad K_c = 6.2 \times 10^{-4}$$
$$Z(g) \rightleftharpoons 2\,Q(g) + 2\,X(g) \qquad K_c = 5.6 \times 10^{-2}$$

To summarize the key points for manipulating equilibrium constants:

- The $K$ value of a reaction running in reverse is the reciprocal of $K$ for the forward reaction.
- If the original chemical equation describing a reversible reaction is multiplied by a factor $n$, the value of $K$ of the new equilibrium constant expression is the value of the original $K$ raised to the $n$th power.
- If an overall chemical reaction is the sum of two or more other reactions, the overall value of $K$ is the product of the $K$ values of the other reactions.

## 14.5 Equilibrium Constants and Reaction Quotients

In Sections 14.1 and 14.2, we introduced two key terms: *equilibrium constant expression* and *mass action expression*. Until now, we have used the first term almost exclusively because we have been dealing with chemical reactions that have achieved equilibrium. Now it is time to reintroduce the concept of a mass action expression because we can apply it not only to the concentrations (or partial pressures) of products and reactants in reaction mixtures that have reached equilibrium but also to reaction mixtures that are not yet at equilibrium but will eventually be.

Even if a reversible chemical reaction has not reached equilibrium, we can still insert reactant and product concentrations (or partial pressures) into its mass action expression. The mathematical result is not a $K$ value because the reaction is not yet at equilibrium. Instead, it is a $Q$ value, where $Q$ stands for **reaction quotient**.

**reaction quotient (*Q*)** the numerical value of the mass action expression for any values of the concentrations (or partial pressures) of reactants and products; at equilibrium, $Q = K$.

The value of $Q$ provides us with a kind of status report on how a reaction is proceeding. To see how, let's revisit the water–gas shift reaction,

$$H_2O(g) + CO(g) \rightleftharpoons H_2(g) + CO_2(g) \qquad K_c = 24 \text{ at } 500\text{ K}$$

and the data from experiment 3 in Table 14.1, where the initial concentrations of reactants and products are as follows:

| $[H_2O]$ (*M*) | [CO] (*M*) | $[H_2]$ (*M*) | $[CO_2]$ (*M*) |
|---|---|---|---|
| 0.0100 | 0.0200 | 0.0300 | 0.0400 |

Inserting these values into the mass action expression for the reaction yields a $Q$ value based on concentration—that is, $Q_c$:

$$Q_c = \frac{[H_2][CO_2]}{[H_2O][CO]} = \frac{(0.0300)(0.0400)}{(0.0100)(0.0200)} = 6.00$$

$K_c = 24$ for this reaction, so $Q_c < K_c$, which means that proportionally smaller concentrations of products and larger concentrations of reactants are in the initial reaction mixture than will be present at equilibrium. To achieve equilibrium, some of the reactants must form products, increasing the value of $Q_c$ until it matches the value of $K_c$ and the following equilibrium concentrations of reactants and products are present in the reaction vessel:

| $[H_2O]$ (*M*) | [CO] (*M*) | $[H_2]$ (*M*) | $[CO_2]$ (*M*) |
|---|---|---|---|
| 0.0046 | 0.0146 | 0.0354 | 0.0454 |

To put the results from the data in Table 14.1 in context, let's consider the curves in **Figure 14.4**. Starting at the left end of the graph (zone a), we have the initial conditions of experiment 1 from Table 14.1—namely, equal concentrations of reactants are present, but no products. Over time, reactant concentrations (the red curve) decrease as product concentrations (the blue curve) increase. At the

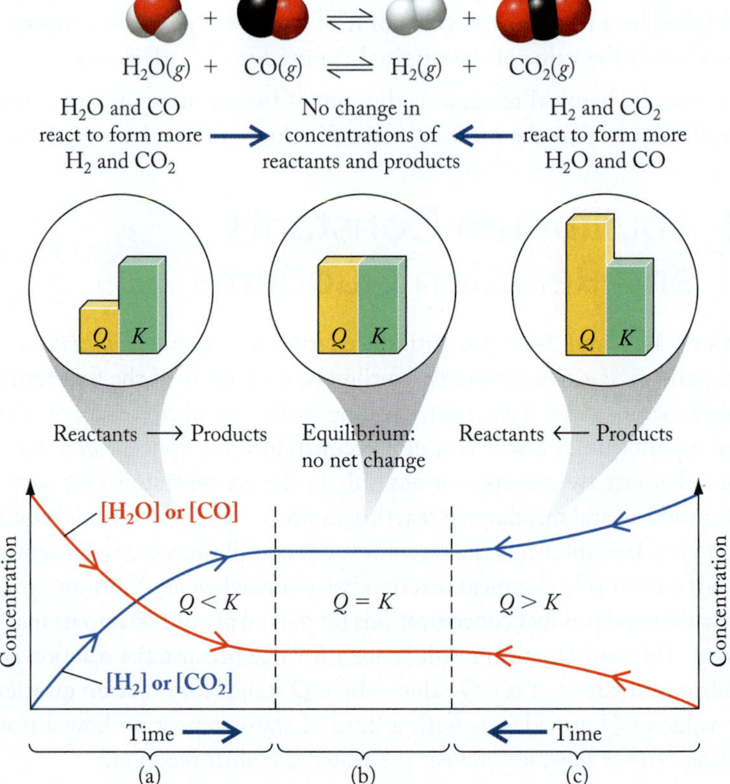


**FIGURE 14.4** The value of the reaction quotient $Q$ relative to the equilibrium constant $K$ for the water–gas shift reaction. (a) Reactant concentrations (red) are higher than their equilibrium concentrations, and product concentrations (blue) are lower than at equilibrium; $Q < K$, and the reactants form more products. (b) Equilibrium concentrations are achieved; $Q = K$, and no net change in concentrations takes place. (c) Product concentrations are higher than at equilibrium, and reactant concentrations are lower than at equilibrium; $Q > K$, and products form more reactants as the reaction runs in reverse.

right end of the graph (zone c), we have the initial conditions of experiment 2—namely, products are present, but no reactants. Over time, reactant concentrations increase as product concentrations decrease. In the middle of the graph (zone b), no net change occurs in the composition because the reaction is at equilibrium.

We can also characterize the three zones on the basis of the value of $Q$ compared to $K$. In zone (a), $Q < K$ and a net conversion of reactants into products occurs as the forward reaction dominates. In zone (c), $Q > K$ and a net conversion of products into reactants takes place as the reverse reaction dominates. In the middle, zone (b), $Q = K$ and no net change in the composition of the reaction mixture occurs over time. The relative values of $Q$ and $K$ and their consequences are summarized in **Table 14.3.**

**TABLE 14.3** Interpreting Relative Values of *Q* and *K*

| Value of Q Relative to K | Meaning |
|---|---|
| $Q < K$ | Reaction as written proceeds in forward direction (→). |
| $Q = K$ | Reaction is at equilibrium (⇌). |
| $Q > K$ | Reaction as written proceeds in reverse direction (←). |

### CONCEPT TEST

Do the initial reaction conditions in experiments 3 and 4 in Table 14.1 belong in zone (a), (b), or (c) in Figure 14.4?

### SAMPLE EXERCISE 14.7 Using Q and K Values to Predict the Direction of a Reaction — LO6

At 2300 K, the value of $K_c$ of the following reaction is $1.5 \times 10^{-3}$:

$$N_2(g) + O_2(g) \rightleftharpoons 2\,NO(g)$$

At the instant when a reaction vessel at 2300 K contains 0.50 $M$ $N_2$, 0.25 $M$ $O_2$, and 0.0042 $M$ NO, is the reaction mixture at equilibrium? If not, in which direction will the reaction proceed to reach equilibrium?

**Collect and Organize** We are asked whether a reaction mixture is at equilibrium. We are given the value of $K_c$ and the concentrations of reactants and product.

**Analyze** The mass action expression for this reaction based on concentrations is

$$Q_c = \frac{[NO]^2}{[N_2][O_2]}$$

If we insert the given concentration values into this expression, we can determine the value of $Q_c$. Comparing $Q_c$ with $K_c$ enables us to determine (1) whether the reaction is at equilibrium and (2) if it is not at equilibrium, in which direction the reaction will proceed. The value of [NO] is about $10^{-2}$ times the concentrations of the reactants, and the [NO] term is squared in the mass action expression. These two factors together make the value of the numerator about $10^{-4}$ that of the denominator. Therefore, the value of $Q_c$ may be less than the value of $K_c$.

**Solve**

$$Q_c = \frac{(0.0042)^2}{(0.50)(0.25)} = 1.4 \times 10^{-4}$$

$Q_c < K_c$, so the reaction mixture is not at equilibrium. To achieve equilibrium, more reactants must form products so that the numerator of the mass action expression increases and the denominator decreases. This happens if the reaction proceeds in the forward direction.

**Think About It** Our prediction that $Q_c$ would be less than $K_c$ was correct. The value of $K_c$ is small, but the value of $Q_c$ is even smaller.

**Practice Exercise** The value of $K_c$ for the reaction

$$2\ NO_2(g) \rightleftharpoons N_2O_4(g)$$

is 4.7 at 373 K. Is a mixture of the two gases in which $[NO_2] = 0.025\ M$ and $[N_2O_4] = 0.0014\ M$ in chemical equilibrium? If not, in which direction does the reaction proceed to achieve equilibrium?

## 14.6 Heterogeneous Equilibria

Thus far in this chapter, we have focused on reactions in the gas phase. However, the principles of chemical equilibrium also apply to reactions in the liquid phase, particularly reactions in solution. Equilibria in which products and reactants are all in the same phase are **homogeneous equilibria**. Equilibria in which reactants and products are in different phases are **heterogeneous equilibria**.

In Sample Exercise 14.1, we considered the equilibrium associated with the oxidation of $SO_2$ to $SO_3$, a key step in forming aerosols of $H_2SO_4$ in the atmosphere. One way to prevent this reaction from happening is to "scrub" $SO_2$ from the exhaust gases emitted by factories where sulfur-containing fuels are burned. Solid lime, CaO, is a widely used scrubbing agent. Sprayed into the exhaust gases, it combines with $SO_2$ to form calcium sulfite:

$$CaO(s) + SO_2(g) \rightleftharpoons CaSO_3(s)$$

The large quantities of lime needed for this reaction and for many other industrial and agricultural uses come from heating pulverized limestone, which is mostly $CaCO_3$, in kilns (**Figure 14.5**) operated at temperatures near 1000°C. At these temperatures, $CaCO_3$ decomposes into lime (CaO) and $CO_2$ gas:

$$CaCO_3(s) \rightleftharpoons CaO(s) + CO_2(g) \qquad \Delta H^\circ_{rxn} = 178.1\ kJ$$

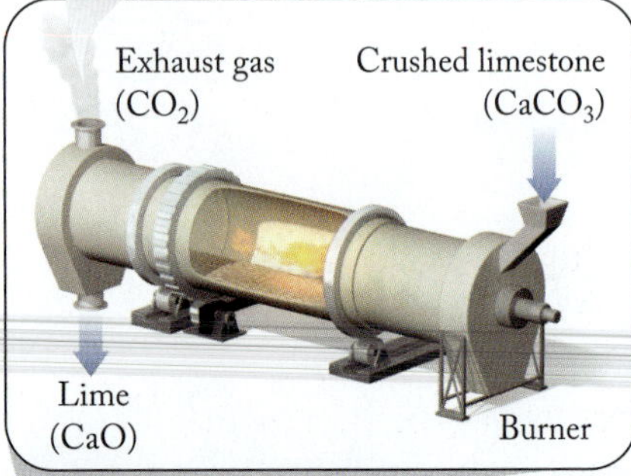

**FIGURE 14.5** Rotary kilns operating at 1000°C are used to convert limestone into lime. Rotation of these large cylindrical kilns ensures that crushed limestone ($CaCO_3$) is uniformly heated and has thermally decomposed to lime (CaO) and $CO_2$ gas by the time it passes through the kiln.

According to what we have learned so far, the concentration-based equilibrium constant expression for this reaction would be as follows:

$$K_c = \frac{[CaO][CO_2]}{[CaCO_3]}$$

This expression contains concentration terms for two solids: CaO and $CaCO_3$. But what do we mean by the concentration of a solid? Any pure solid has a constant "concentration" because its mass (and number of moles) per unit volume is always the same. As long as any CaO or $CaCO_3$ is present, the effective "concentration" of both is 1, and [CaO] and $[CaCO_3]$ terms are not included in the equilibrium constant expression. This leaves us with

$$K_c = [CO_2]$$

This expression means that, as long as some CaO or $CaCO_3$ is present, the equilibrium concentration of $CO_2$ gas does not vary at a given temperature, as shown in **Figure 14.6**. Instead, the concentration of $CO_2$ is the same as the value of $K_c$ at that temperature.

The same concept of constant concentration applies to pure liquids that are involved in reversible chemical reactions. As long as the liquid is present, its "concentration" is considered constant during the reaction and does not appear in the equilibrium constant expression. Similarly, $K_c$ expressions for most reactions in aqueous solutions do not include a term for $[H_2O]$, even when water is a reactant

**homogeneous equilibrium** an equilibrium in which reactants and products are in the same phase.

**heterogeneous equilibrium** an equilibrium in which reactants and products are in more than one phase.

or product, because its concentration does not change significantly. In writing equilibrium constant expressions for heterogeneous equilibria, we follow the rules we learned earlier, with the additional rule that pure liquids and solids do not appear in the expression.

(a)

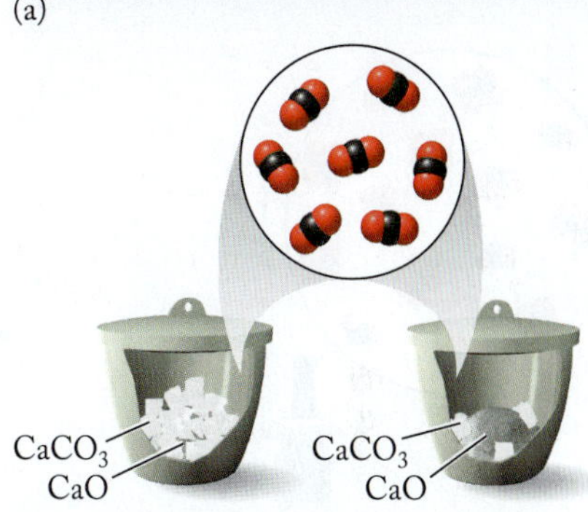


(b)

**FIGURE 14.6** The position of a heterogeneous equilibrium between $CaCO_3$, CaO, and $CO_2$ at constant temperature depends only on the concentration of $CO_2$ gas present. As long as some solid is in the system, the equilibrium concentration of $CO_2$ remains the same. (a) Muffle furnace for heating crucibles containing $CaCO_3$. (b) Two crucibles containing different amounts of solid material at the same temperature have the same concentration of $CO_2$ gas.

### SAMPLE EXERCISE 14.8 Relating Equilibrium Constant Expressions for Heterogeneous Equilibria LO2

Pulverized lime (CaO) is used to remove $SO_2$ from the stack gases of coal-fired power plants as described by this reaction:

$$CaO(s) + SO_2(g) \rightleftharpoons CaSO_3(s)$$

a. Write $K_c$ and $K_p$ expressions for the reaction described above.
b. If the value of $K_p$ for the reaction is $2.2 \times 10^6$ at 325°C, what is the value of $K_c$ at this temperature?

**Collect and Organize** We are given a reversible reaction, and we are asked to write $K_c$ and $K_p$ equilibrium constant expressions for it. We need to identify any pure liquids or solids involved in the reaction and, because they are considered to have constant concentrations, to omit their terms from the equilibrium constant expressions. $K_c$ and $K_p$ values are related by Equation 14.15: $K_p = K_c(RT)^{\Delta n}$.

**Analyze** The reaction involves two solids, CaO and $CaSO_3$, and one reactant gas, $SO_2$. The equilibrium constant expressions will, therefore, have only a term for $SO_2$ in their denominators. Calculating a $K_c$ value from a $K_p$ value will require rearranging the terms in Equation 14.15. The chemical equation describes a reaction having one mole of gaseous reactant and no gaseous products, so the value of $\Delta n$ is −1.

**Solve**

a. The equilibrium constant expressions are

$$K_c = \frac{1}{[SO_2]} \text{ and } K_p = \frac{1}{P_{SO_2}}$$

b. Rearranging the terms in Equation 14.15 to solve for $K_c$:

$$K_c = K_p/(RT)^{\Delta n} = 2.2 \times 10^6/(0.08206 \times (325 + 273)) = 1.1 \times 10^8$$

**Think About It** The calculated $K_c$ value is 50 times larger than the $K_p$ value, which is more than the $K_c/K_p$ ratio would have been at 298 K. The larger difference at the higher temperature makes sense because the concentration of $SO_2$ (or any ideal gas) is much lower at the higher temperature and constant pressure.

**Practice Exercise** At temperatures above 100°C, baking soda decomposes as described by the chemical equation below. However, the small value of $K_p$ for this reaction at 298 K indicates that baking soda is thermally stable at room temperature:

$$2\,NaHCO_3(s) \rightleftharpoons Na_2CO_3(s) + CO_2(g) + H_2O(g) \qquad K_p = 8.6 \times 10^{-7}$$

Write $K_p$ and $K_c$ expressions for this reaction, and calculate the value of $K_c$ at 298 K.

### CONCEPT TEST

Explain why $K_c = [H_2O(g)]$ is the mass action expression for the equilibrium $H_2O(\ell) \rightleftharpoons H_2O(g)$.

## 14.7 Le Châtelier's Principle

Once a chemical reaction has reached equilibrium, the composition of the system remains unchanged as long as no external forces perturb it. What happens when a system at equilibrium is subjected to some external stress? To explore that, let's revisit the reaction between $NO_2$, a brown gas, and $N_2O_4$, its colorless dimer:

$$\underset{\text{(brown)}}{2\ NO_2(g)} \rightleftharpoons \underset{\text{(colorless)}}{N_2O_4(g)} \tag{14.3}$$

The flask on the left in **Figure 14.7** contains an equilibrium mixture of $NO_2$ and $N_2O_4$ at room temperature. When an identical flask (on the right in Figure 14.7) is placed in an ice bath, most of the $N_2O_4$ in the mixture condenses, dramatically lowering its gas-phase concentration. A lower concentration of $N_2O_4$ gas means that the rate of the reverse reaction decreases. As a result, the forward reaction dominates as more $NO_2$ dimerizes to form more of the colorless $N_2O_4$, most of which condenses, and the cycle repeats itself. Eventually very little of either compound is left in the gas phase, as shown in the molecular balloon on the right in Figure 14.7, and the brown color of the reaction mixture has nearly faded away. **Figure 14.8** shows that as a mixture at 0°C is warmed, the equilibrium shifts to a higher concentration of $NO_2$, consistent with the darker brown color as the mixture returns to room temperature (23°C) and then warms to 35°C.

**FIGURE 14.7** Shifting equilibrium. The flask on the left contains an equilibrium reaction mixture of brown $NO_2$ gas and colorless $N_2O_4$ gas at room temperature. When an identical flask is placed in an ice bath, $N_2O_4$ condenses, which shifts the equilibrium to produce $N_2O_4$, lowering the concentration of $NO_2$ gas and causing the brown color of the gas mixture to fade.

From this observation, we may conclude that heating an equilibrium mixture of these gases perturbs this equilibrium, causing the rate of the reverse reaction to be higher than the rate of the forward reaction and resulting in a net increase in the partial pressure of $NO_2$. Likewise, cooling this reaction mixture causes its brown color to fade, so we can conclude that the rate of the forward reaction is higher than the rate of the reverse reaction at lower temperatures, which results in a net decrease in $P_{NO_2}$.

In this section, we explore how and why reaction mixtures at equilibrium respond not only to changes in temperature but also to other changes, such as increasing or decreasing the partial pressure (or concentration) of a reactant or product. We will see, for example, how adding or removing an ingredient in an equilibrium mixture alters the value of the reaction quotient $Q$ so that it is no longer equal to the value of $K$. We must also remember from Chapter 13 that changing the temperature of a system also changes the value of $K$. When a system at equilibrium undergoes one of these changes, the perturbed system is no longer at equilibrium, and the composition of the system will change to restore equilibrium.

**Le Châtelier's principle** the principle that a system at equilibrium responds to a stress in such a way that it relieves that stress.

One of the first scientists to study and then successfully predict how chemical equilibria respond to such perturbations was French chemist Henri Louis Le Châtelier (1850–1936). He articulated what is now known as **Le Châtelier's principle**, which states that if a system at equilibrium is perturbed (or stressed), the position of the equilibrium shifts in the direction that relieves the stress. Through the years, chemists have used Le Châtelier's principle to increase the yields of chemical reactions that would otherwise have produced very little of a desired compound.

**FIGURE 14.8** Equilibrium mixtures of brown $NO_2$ and colorless $N_2O_4$ at different temperatures.

## Effects of Adding or Removing Reactants or Products

When a reactant or product is added or removed, a system at chemical equilibrium is perturbed. Following Le Châtelier's principle, the system responds to restore equilibrium.

To explore how industrial chemists exploit Le Châtelier's principle, let's return to the water–gas shift reaction for making hydrogen:

$$H_2O(g) + CO(g) \rightleftharpoons H_2(g) + CO_2(g) \quad (14.2)$$

To shift the equilibrium toward the production of more $H_2$, chemists pass the reaction mixture through a scrubber containing a concentrated aqueous solution of $K_2CO_3$. Doing this removes $CO_2$ (**Figure 14.9**) from the gaseous mixture as a result of the following reaction:

$$CO_2(g) + H_2O(\ell) + K_2CO_3(aq) \rightleftharpoons 2\,KHCO_3(s)$$

Removing $CO_2$ means that fewer molecules of it are available to collide with molecules of $H_2$ to drive the reverse reaction in Equation 14.2. As a result, the rate of the reverse reaction becomes slower than the rate of the forward reaction. This means that the system is no longer in equilibrium. To return to equilibrium, the reaction proceeds in the forward direction (we say that the reaction *shifts to the right*), making more $CO_2$ gas to restore some of what was removed, and in the process making more $H_2$ gas, until a new equilibrium is achieved. The new equilibrium is like the old one in that the value of the mass action expression,

$$\frac{[H_2][CO_2]}{[H_2O][CO]}$$

is equal to the value of *K*. This is true even though the concentrations of the individual reactants and products have changed; the overall ratio in the *K* expression is restored.

**CHEMTOUR**
Le Châtelier's Principle

$$CO(g) + H_2O(g) \rightleftharpoons CO_2(g) + H_2(g)$$

At equilibrium
$Q = K$
$Q < K$
New equilibrium established
$Q = K$
$H_2$
$CO_2, H_2$
$CO_2$ removed
$CO_2$
Partial pressure
$CO, H_2O$
$CO, H_2O$
Time

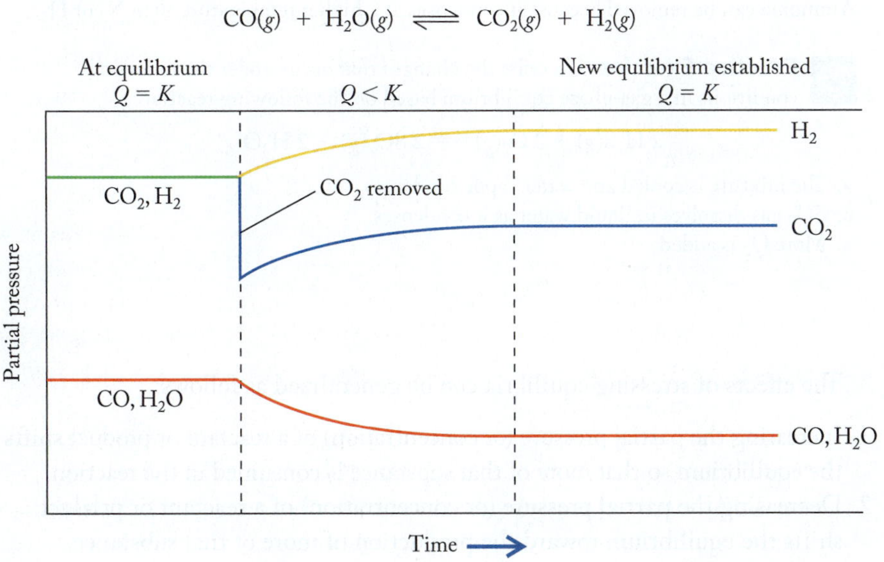

**FIGURE 14.9** Removing $CO_2$ from an equilibrium mixture of the water–gas shift reaction results in fewer collisions between $CO_2$ and $H_2$ molecules, slowing the rate of the reverse reaction. Meanwhile, the forward reaction continues at the same rate and generates additional $CO_2$ and $H_2$ until the mixture establishes a new equilibrium.

Another way to shift a reaction mixture at equilibrium is to add more reactant or product. If the goal is to form more products, then adding more reactants increases their concentration in the reaction mixture, which increases the rate of the forward reaction. Some of the added reactants are converted into additional products. This approach works even when only one reactant is added in a multireactant reaction if some of the other reactant(s) is still available. To understand why, consider what happens when just one of the reactant concentration terms in the equilibrium constant expression is increased. The result is a reaction quotient $Q$ that is less than $K$, which means the reaction will proceed in the forward direction, forming more products, until equilibrium is restored (that is, until $Q = K$ again).

### SAMPLE EXERCISE 14.9 Stressing an Equilibrium by Adding or Removing Reactants or Products — LO7

Suggest three ways in which the production of ammonia through the reaction

$$N_2(g) + 3\,H_2(g) \rightleftharpoons 2\,NH_3(g)$$

could be increased without changing the reaction temperature.

**Collect, Organize, and Analyze** We are given the balanced chemical equation of a reversible reaction and are asked to suggest three ways to increase its yield—that is, to shift its equilibrium to the right. We need to use the mass action expression for the reaction:

$$Q_p = \frac{(P_{NH_3})^2}{(P_{H_2})^3(P_{N_2})}$$

**Solve** This reaction's equilibrium can be shifted to the right by removing $NH_3$, which reduces the value of the numerator of the mass action expression.

The equilibrium can also be shifted to the right by increasing the partial pressure of $N_2$ or $H_2$, which increases the value of the denominator. Either approach will produce a smaller reaction quotient $Q_p$. If $Q_p < K_p$, the reaction system will respond by consuming $N_2$ and $H_2$ and forming more $NH_3$.

Therefore, if we (1) increase the partial pressure of $N_2$, (2) increase the partial pressure of $H_2$, or (3) remove $NH_3$ from the system, the equilibrium will shift to the right.

**Think About It** The industrial synthesis of ammonia relies on shifting the equilibrium to the right by running the reaction at high partial pressures of the reactants and by removing the product $NH_3$ by passing the reaction mixture through chilled condensers. Ammonia can be removed because it condenses at a higher temperature than $N_2$ or $H_2$.

**Practice Exercise** Describe the changes that occur under the following conditions in a gas-phase equilibrium based on the following reaction:

$$2\,H_2S(g) + 3\,O_2(g) \rightleftharpoons 2\,SO_2(g) + 2\,H_2O(g)$$

a. The mixture is cooled and water vapor condenses.
b. $SO_2$ gas dissolves in liquid water as it condenses.
c. More $O_2$ is added.

The effects of stressing equilibria can be generalized as follows:

1. Increasing the partial pressure (or concentration) of a reactant or product shifts the equilibrium so that more of that substance is consumed in the reaction.
2. Decreasing the partial pressure (or concentration) of a reactant or product shifts the equilibrium toward the production of more of that substance.

## Effects of Changes in Pressure and Volume

A reaction involving gaseous reactants or products may be perturbed by altering the partial pressures of the reactants and products. A simple way to do that is by changing the volume of the reaction mixture while keeping the temperature constant—that is, through an *isothermal* process.

**CONNECTION** According to Boyle's law (Chapter 9), the pressure of an ideal gas is inversely proportional to its volume at constant temperature.

To see how such a change in volume perturbs a gas-phase equilibrium, let's again return to the equilibrium between $NO_2$ and its dimer, $N_2O_4$. Suppose a gas-tight syringe contains 50 mL of an equilibrium mixture of the two gases (**Figure 14.10a**). This mixture is then rapidly compressed at constant temperature into a volume of 25 mL (**Figure 14.10b**). The darker brown color indicates an increase in the partial pressures (and concentrations) of the gases.

Let's assume that the equilibrium partial pressures of the two gases in Figure 14.10(a) are $P_{NO_2} = X$ and $P_{N_2O_4} = Y$. This means that

$$K_p = \frac{P_{N_2O_4}}{(P_{NO_2})^2} = \frac{Y}{X^2}$$

When the mixture is compressed into half its original volume, the partial pressure of each gas should double so that $P_{NO_2} = 2X$ and $P_{N_2O_4} = 2Y$. Inserting these new values into the mass action expression gives a $Q_p$ value that is half the value of $K_p$:

$$Q_p = \frac{2Y}{(2X)^2} = \frac{2Y}{4X^2} = \frac{1}{2}K_p$$

When $Q_p < K_p$, the reaction proceeds in the forward direction, consuming reactants and forming products. Here, $NO_2$ is consumed and $N_2O_4$ forms, which causes the dark brown in Figure 14.10(b) to fade a little, as shown in Figure **14.10(c)**.

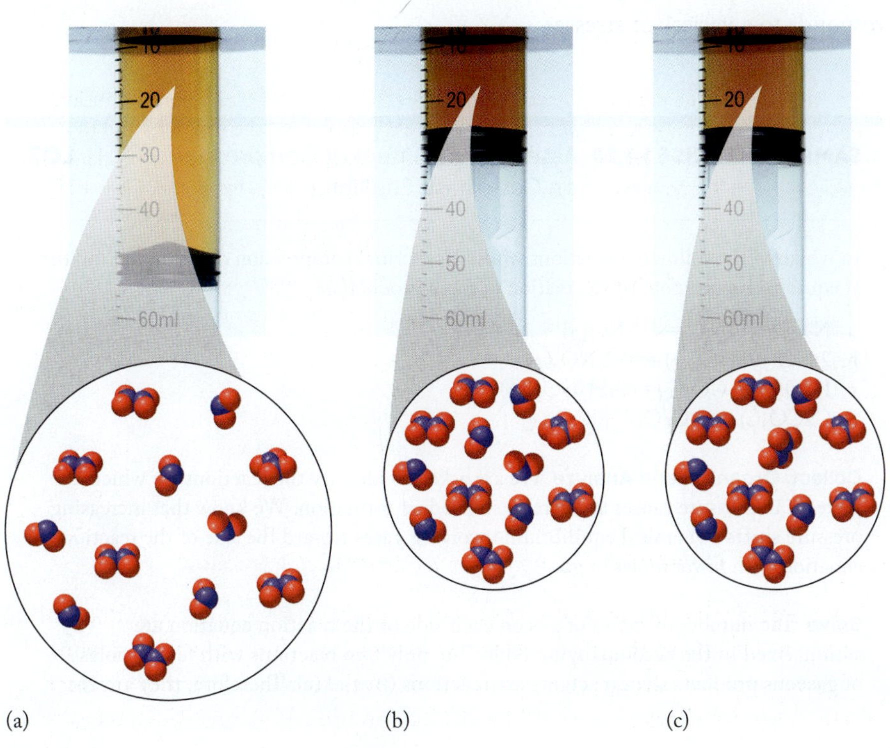


**FIGURE 14.10** An equilibrium mixture of brown $NO_2$ and colorless $N_2O_4$ (a) before compression, (b) immediately after compression, and (c) 10 seconds after compression.

**TABLE 14.4** Responses to Different Kinds of Stress for the Reaction $2\ A(g) \rightleftharpoons B(g)$ at Equilibrium

| Kind of Stress | How System Responds | Direction of Shift |
|---|---|---|
| Add A | Consume A | Right |
| Remove A | Produce A | Left |
| Add B | Consume B | Left |
| Remove B | Produce B | Right |
| Compress the reaction mixture (decrease volume, increase pressure) | Reduce moles of gas | Right |
| Expand the reaction mixture (increase volume, decrease pressure) | Increase moles of gas | Left |

Notice that converting $NO_2$ into $N_2O_4$ reduced the total number of moles of gas in the reaction mixture (because two moles of $NO_2$ are consumed for every one mole of $N_2O_4$ produced). In fact, any time a reaction mixture at equilibrium is compressed—and the partial pressures of its gas-phase reactants and products increase—the equilibrium shifts toward the side of the reaction equation with fewer moles of gases. Remember that as the pressure increased, so did the number of collisions, resulting in an increase in the forward rate of the reaction. However, allowing a reaction mixture to expand at constant temperature lowers the partial pressures of all the gaseous reactants and products and shifts the equilibrium toward the side of the reaction equation with more moles of gases. These shifts occur in all gas-phase reactions with different numbers of moles of reactant and product gases in which chemical equilibrium is perturbed by isothermal compression or expansion. **Table 14.4** summarizes how a generic gas-phase reversible reaction at equilibrium in which 2 moles of reactants form 1 mole of product responds to a variety of stresses.

**SAMPLE EXERCISE 14.10** Assessing the Effect of Compression on Gas-Phase Equilibria **LO7**

In which of the following reactions would isothermal compression of a reaction mixture at equilibrium promote the formation of more product(s)?

a. $N_2(g) + O_2(g) \rightleftharpoons 2\ NO(g)$
b. $2\ NO(g) + O_2(g) \rightleftharpoons 2\ NO_2(g)$
c. $H_2O(\ell) + CO_2(g) \rightleftharpoons H_2CO_3(aq)$
d. $CaCO_3(s) \rightleftharpoons CaO(s) + CO_2(g)$

**Collect, Organize, and Analyze** We are asked to identify the reactions for which an increase in pressure causes an increase in product formation. We know that increasing pressure shifts a chemical equilibrium involving gases toward the side of the reaction equation with fewer moles of gas.

**Solve** The number of moles of gas on each side of the reaction equation are summarized in the accompanying table. The only two reactions with fewer moles of gaseous products than reactants are reactions (b) and (c). Therefore, they are the only two in which an increase in the total pressure of the reacting gases increases product formation.

| Reaction | Moles of Gaseous Reactants | Moles of Gaseous Products |
|---|---|---|
| a | 2 | 2 |
| b | 3 | 2 |
| c | 1 | 0 |
| d | 0 | 1 |

**Think About It** We could have used reaction quotients to predict the impacts of isothermal compression. For example, the mass action expression $Q_p$ for reaction (c) is $1/P_{CO_2}$. If $P_{CO_2} = X$ at equilibrium, then $Q_p = K_p = \frac{1}{X}$. If the reaction mixture is compressed, then $P_{CO_2} > X$, which decreases the value of $\frac{1}{X}$, making $Q_p < K_p$. To restore equilibrium, the reaction proceeds in the forward direction, increasing product formation by consuming $CO_2$ until its partial pressure is once again equal to $X$.

**Practice Exercise** How does isothermal compression impact the composition of an equilibrium reaction mixture in the gas-phase reaction depicted here?

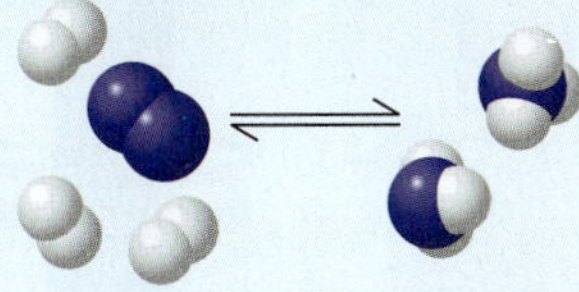

## Effects of Temperature Changes

We saw in Figure 14.7 that increasing the temperature of an equilibrium mixture of $NO_2$ and $N_2O_4$ shifts the equilibrium in favor of the formation of more $NO_2$, which imparts a darker brown color to the reaction mixture. This shift makes sense because the dimerization reaction is exothermic:

$$2\,NO_2(g) \rightleftharpoons N_2O_4(g) \qquad \Delta H^\circ_{rxn} = -57.2\text{ kJ/mol}$$

Because thermal energy flows from the reaction mixture to its surroundings, you might be tempted to think of energy as a "product" of the forward reaction and then reason that raising the temperature of the reaction mixture is similar to adding more "product," which would favor the reverse reaction and the formation of more $NO_2$. Although this prediction is consistent with what we observe, it does not explain *why* a decrease in temperature favors the reactants in an exothermic reaction. We need to be clear that no "energy" term is in the reaction quotient for this or for any chemical reaction, so changing the temperature does not really perturb an equilibrium by changing the ratio of product to reactant concentrations or partial pressures—that is, by changing the value of $Q$. Rather, the better explanation is that changing temperature *changes the value of* $K$. As a result, $Q$ and $K$ are no longer equal, and the reaction proceeds in the direction that makes them equal again.

We examine the influence of temperature on $K$ values in more detail in Section 14.10. For now, it is enough to note that raising the temperature of an endothermic reaction mixture promotes the forward reaction and raises its $K$ value, whereas heating an exothermic reaction mixture promotes the reverse reaction and lowers the value of $K$. **Table 14.5** summarizes how a system at equilibrium responds to changes in temperature.

**TABLE 14.5** Responses to Temperature Changes for the Reaction $A(g) \rightleftharpoons B(g)$ at Equilibrium

| Type of Reaction | Kind of Stress | How System Responds |
|---|---|---|
| Endothermic | Increase temperature | Shift to form more B(*g*) |
| | Decrease temperature | Shift to form more A(*g*) |
| Exothermic | Increase temperature | Shift to form more A(*g*) |
| | Decrease temperature | Shift to form more B(*g*) |

**SAMPLE EXERCISE 14.11** Predicting How Temperature Affects Chemical Equilibria **LO7**

The color of an aqueous acidic solution of cobalt(II) chloride depends on the temperature (**Figure 14.11**). In aqueous HCl, the solution is pink at 0°C, magenta at 25°C, and dark blue at 75°C. Is the reaction that produces the pink-to-blue color change exothermic or endothermic?

Temperature = 5°C Temperature = 75°C

$$\underbrace{Co(H_2O)_6^{2+}(aq)}_{\text{Pink}} + 4\,Cl^-(aq) \rightleftharpoons \underbrace{CoCl_4^{2-}(aq)}_{\text{Royal blue}} + 6\,H_2O(\ell)$$

**FIGURE 14.11** Two forms of cobalt(II) ion, one pink and one blue, are in equilibrium in aqueous hydrochloric acid solution. The position of equilibrium shifts to the right as temperature increases from 5°C to 75°C, causing the color of the solution to change from pink to blue as the concentration of $CoCl_4^{2-}$ increases. Chapter 16 describes the formation of complex ions such as these.

**Collect and Organize** We are given a reversible reaction and asked to determine whether it is exothermic or endothermic. We know that the solution changes color from pink to blue when the temperature is increased.

**Analyze** If the reaction is endothermic, then increasing the temperature should shift the reaction toward the formation of the blue product. If the reaction is exothermic, then increasing the temperature should shift the reaction toward the formation of the pink solution of reactants.

**Solve** The blue product is favored at higher temperatures, so the reaction as written must be endothermic.

**Think About It** This exercise shows how determining the effect of changing the temperature of an equilibrium reaction mixture can tell us whether the reaction is exothermic or endothermic. The fact that the reaction mixture in this exercise is magenta at room temperature means that both the pink and blue forms are present and that the value of $K$ at room temperature is close to 1.

**Practice Exercise** Predict how the value of the equilibrium constant of the reaction

$$N_2(g) + O_2(g) \rightleftharpoons 2\,NO(g) \qquad \Delta H^\circ_{rxn} = 181\text{ kJ}$$

changes with increasing temperature.

**CONCEPT TEST**

Consider this gas-phase system at equilibrium:

$$N_2(g) + 3\,H_2(g) \rightleftharpoons 2\,NH_3(g)$$

Identify five stresses that could be applied to this exothermic reaction that would result in forming more ammonia.

## Catalysts and Equilibrium

The industrial production of ammonia (see Sample Exercise 14.9) was developed by the German chemists Fritz Haber (1868–1934) and Carl Bosch (1874–1940) in the early 20th century and is still widely referred to as the Haber–Bosch process. What makes the process commercially feasible is the use of catalysts. As discussed in Chapter 13, a catalyst increases the rate of a chemical reaction by lowering its activation energy. If a catalyst increases the rate of a reaction, does that catalyst affect the equilibrium constant of the reaction?

To answer this question, consider the energy profiles of the catalyzed and uncatalyzed reactions in **Figure 14.12**. The catalyst increases the rate of the reaction by changing the reaction mechanism and, as a result, decreasing the height of the activation energy barrier. However, the barrier height is reduced by the same amount whether the reaction as written proceeds in the forward or reverse direction, so the increase in reaction rate produced by the catalyst is the same in both directions. Therefore, a catalyst has no effect on the equilibrium constant of a reaction or on the composition of an equilibrium reaction mixture. A catalyst does, however, decrease the amount of time needed for a reaction to reach equilibrium.

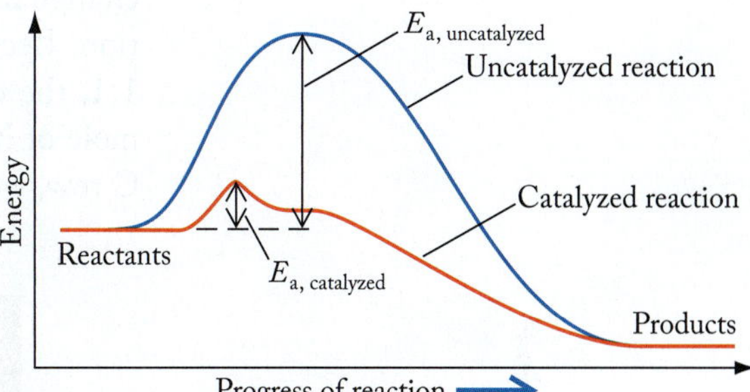


**FIGURE 14.12** The effect of a catalyst on a reaction. A catalyst lowers the activation energy barrier, so the rate of the reaction increases. However, because both the forward reaction and the reverse reaction occur more rapidly, the position of equilibrium (that is, the value of $K$) does not change. The system reaches equilibrium more rapidly, but the relative amounts of product and reactant present at equilibrium do not change.

# 14.8 Calculations Based on $K$

The values of equilibrium constants for chemical reactions are used in several kinds of calculations, including those in which

1. We want to determine whether a reaction mixture has reached equilibrium (Sample Exercise 14.7).
2. We know the value of $K$ and the starting concentrations or partial pressures of reactants and/or products, and we want to calculate their equilibrium concentrations or pressures.

**CHEMTOUR**
Solving Equilibrium Problems

In this section, we focus on the second type of calculation and introduce a useful way of handling such problems: a table of reactant and product concentration (or partial pressure) values called a *RICE table*. "RICE" is an acronym: the table starts with the balanced chemical equation describing the ***R***eaction followed by rows that contain ***I***nitial concentration values, ***C***hanges in those initial values as the reaction proceeds toward equilibrium, and ***E***quilibrium values.

In our first example, we calculate how much nitrogen monoxide forms in a sample of air heated to a temperature at which $K_p = 1.00 \times 10^{-5}$ for the reaction

$$N_2(g) + O_2(g) \rightleftharpoons 2\,NO(g)$$

The initial partial pressures are $P_{N_2} = 0.79$ atm and $P_{O_2} = 0.21$ atm, and we assume that no NO is present.

We start by writing the given (initial) information in our RICE table:

| REACTION | $N_2(g)$ | + | $O_2(g)$ | $\rightleftharpoons$ | $2\ NO(g)$ |
|---|---|---|---|---|---|
| | $P_{N_2}$ (atm) | | $P_{O_2}$ (atm) | | $P_{NO}$ (atm) |
| Initial | 0.79 | | 0.21 | | 0 |
| Change | | | | | |
| Equilibrium | | | | | |

The reaction will proceed in the forward direction because no product is initially present, so $Q_p = 0 < K_p$.

We need to use algebra to fill in rows C and E. We don't know how much $N_2$ or $O_2$ will be consumed or how much NO will be formed. We can define the change in partial pressure of $N_2$ as $-x$ because $N_2$ is consumed during the reaction. Because the mole ratio of $N_2$ to $O_2$ in the balanced chemical equation is 1:1, the change in $O_2$ is also $-x$. Two moles of NO are produced from each mole of $N_2$ and $O_2$, so the change in $P_{NO}$ is $+2x$. Inserting these values in the C row, we have

| REACTION | $N_2(g)$ | + | $O_2(g)$ | $\rightleftharpoons$ | $2\ NO(g)$ |
|---|---|---|---|---|---|
| | $P_{N_2}$ (atm) | | $P_{O_2}$ (atm) | | $P_{NO}$ (atm) |
| Initial | 0.79 | | 0.21 | | 0 |
| Change | $-x$ | | $-x$ | | $+2x$ |
| Equilibrium | | | | | |

Combining the I and C rows, we obtain expressions for the three partial pressures at equilibrium:

| REACTION | $N_2(g)$ | + | $O_2(g)$ | $\rightleftharpoons$ | $2\ NO(g)$ |
|---|---|---|---|---|---|
| | $P_{N_2}$ (atm) | | $P_{O_2}$ (atm) | | $P_{NO}$ (atm) |
| Initial | 0.79 | | 0.21 | | 0 |
| Change | $-x$ | | $-x$ | | $+2x$ |
| Equilibrium | $0.79 - x$ | | $0.21 - x$ | | $2x$ |

The next step is to substitute the terms from the E row into the $K_p$ expression for the reaction:

$$K_p = \frac{(P_{NO})^2}{(P_{N_2})(P_{O_2})}$$

$$= \frac{(2x)^2}{(0.79 - x)(0.21 - x)} \quad (14.18)$$

Multiplying the terms in the denominator of Equation 14.18 gives

$$K_p = \frac{4x^2}{0.1659 - 1.00x + x^2} = 1.00 \times 10^{-5}$$

Cross-multiplying, we get

$$1.659 \times 10^{-6} - (1.00 \times 10^{-5})x + (1.00 \times 10^{-5})x^2 = 4x^2$$

Combining the $x^2$ terms and rearranging, we have

$$3.99999\ x^2 + (1.00 \times 10^{-5})x - 1.659 \times 10^{-6} = 0$$

This equation fits the general form of a quadratic equation:

$$ax^2 + bx + c = 0$$

which can be solved for $x$ using a variety of tools, including the problem-solving functions in scientific calculators or spreadsheet programs, or by pencil and paper through using the quadratic formula:

$$x = \frac{-b \pm \sqrt{b^2 - 4ac}}{2a}$$

Taking this low-tech approach:

$$x = \frac{-1.00 \times 10^{-5} \pm \sqrt{(1.00 \times 10^{-5})^2 - 4(3.99999)(-1.659 \times 10^{-6})}}{2(3.99999)}$$

Solving this equation yields two $x$ values: $6.43 \times 10^{-4}$ and $-6.45 \times 10^{-4}$. We focus on the positive value because a gas cannot have a negative partial pressure. Calculating equilibrium partial pressures to two significant figures:

$$\begin{aligned} P_{O_2} &= 0.21 - x \\ &= 0.21 - (6.43 \times 10^{-4}) = 0.21 \text{ atm} \\ P_{N_2} &= 0.79 - x \\ &= 0.79 - (6.43 \times 10^{-4}) = 0.79 \text{ atm} \\ P_{NO} &= 2x \\ &= 2(6.43 \times 10^{-4}) = 1.286 \times 10^{-3} = 0.0013 \text{ atm} \end{aligned}$$

The small quantity of NO produced by the reaction means that no change occurs in the partial pressures of $N_2$ or $O_2$ to two significant figures.

Because the $x$ terms in the denominator of Equation 14.18 are very much smaller than the initial partial pressures, we can simplify the calculation of $P_{NO}$ by ignoring the $x$ terms in the denominator and using the initial values for $P_{N_2}$ and $P_{O_2}$ instead:

$$\begin{aligned} K_p = 1.0 \times 10^{-5} &= \frac{(P_{NO})^2}{(P_{N_2})(P_{O_2})} \\ &= \frac{4x^2}{(0.79 - x)(0.21 - x)} \approx \frac{4x^2}{(0.79)(0.21)} \\ 4x^2 &= (0.79)(0.21)(1.0 \times 10^{-5}) \\ &= 1.659 \times 10^{-6} \\ x^2 &= 4.148 \times 10^{-7} \\ x &= 6.44 \times 10^{-4} \text{ atm} \end{aligned}$$

This value of $x$ is nearly the same as that obtained by solving the quadratic equation ($6.43 \times 10^{-4}$ atm) and is identical to two significant figures. Generally speaking, we can ignore the $-x$ or $+x$ component of an equilibrium concentration or partial pressure term if the value of $x$ is less than 5% of the initial value. This often happens when $K$ values are small ($<\sim 3 \times 10^{-5}$) and initial concentration or partial pressure values are greater than about 0.03.

Equilibrium calculations can also be simplified when initial reactant concentrations are the same. For example, suppose we have a vessel containing 0.100 *M* $N_2$ and 0.100 *M* $O_2$ at a temperature at which $K_c$ for the NO formation reaction is 0.100. What is the equilibrium concentration of NO? The RICE table in this case is

| REACTION | $N_2(g)$ | + | $O_2(g)$ | $\rightleftharpoons$ | $2\ NO(g)$ |
|---|---|---|---|---|---|
| | $[N_2]$ ($M$) | | $[O_2]$ ($M$) | | [NO] ($M$) |
| Initial | 0.100 | | 0.100 | | 0 |
| Change | $-x$ | | $-x$ | | $+2x$ |
| Equilibrium | $0.100 - x$ | | $0.100 - x$ | | $2x$ |

Inserting the values from the E row into the expression for $K_c$ gives

$$K_c = \frac{(2x)^2}{(0.100 - x)(0.100 - x)} = 0.100$$

Taking the square root of each side:

$$\frac{2x}{0.100 - x} = 0.316$$

To solve for $x$, we first cross-multiply and then combine the two $x$ terms:

$$2x = (0.316)\ (0.100 - x) = 0.0316 - 0.316\ x$$

$$2x + 0.316\ x = 2.316\ x = 0.0316$$

$$x = 0.0136\ M$$

The equilibrium concentrations are $[N_2] = [O_2] = 0.100\ M - 0.0316\ M = 0.086\ M$ and $[NO] = 2x = 0.0272\ M$.

It's a good idea to check that these concentrations are consistent with the known value of $K_c$. Substituting into the $K_c$ expression, we get

$$K_c = \frac{(0.0272)^2}{(0.086)^2} = 0.10$$

which is the value of $K_c$ given for the reaction in question.

**SAMPLE EXERCISE 14.12** Calculating an Equilibrium Partial Pressure I **LO8**

Much of the $H_2$ used in the Haber–Bosch process is produced by the water–gas shift reaction:

$$CO(g) + H_2O(g) \rightleftharpoons CO_2(g) + H_2(g) \quad (14.2)$$

If a reaction vessel at 400°C is filled with an equimolar mixture of CO and steam such that $P_{CO} = P_{H_2O} = 2.00$ atm, what is the partial pressure of $H_2$ at equilibrium? The equilibrium constant $K_p = 10.0$ at 400°C.

**Collect and Organize** We are asked to find the partial pressure of a gaseous product in an equilibrium mixture given the initial partial pressures of reactants and the value of $K_p$. One way to approach this problem involves (1) setting up a RICE table, (2) using the partial pressures from the E row in the equilibrium constant expression, and (3) solving for $P_{H_2}$.

**Analyze** The system initially contains no product. This means that the reaction quotient $Q_p$ is equal to zero and is thus less than $K$. Therefore, the reaction proceeds in the forward direction, decreasing the partial pressures of the reactants while increasing those of the products. A $K_p$ value of 10.0 means that most of the reactants should be converted into products, but there should be significant partial pressures of both reactants and products at equilibrium.

**Solve** Let $x$ be the increase in partial pressure of $H_2$ as a result of the reaction. The stoichiometry of the reaction tells us that the change in $P_{CO_2}$ is also $x$ and that the changes in both $P_{CO}$ and $P_{H_2}$ are $-x$:

| REACTION | $CO(g)$ | + | $H_2O(g)$ | $\rightleftharpoons$ | $CO_2(g)$ | + | $H_2(g)$ |
|---|---|---|---|---|---|---|---|
| | $P_{CO}$ (atm) | | $P_{H_2O}$ (atm) | | $P_{CO_2}$ (atm) | | $P_{H_2}$ (atm) |
| Initial | 2.00 | | 2.00 | | 0.00 | | 0.00 |
| Change | $-x$ | | $-x$ | | $+x$ | | $+x$ |
| Equilibrium | $2.00 - x$ | | $2.00 - x$ | | $x$ | | $x$ |

Inserting these equilibrium terms into the equilibrium constant expression for the reaction gives

$$K_p = \frac{(P_{CO_2})(P_{H_2})}{(P_{CO})(P_{H_2O})} = \frac{(x)(x)}{(2.00 - x)(2.00 - x)} = 10.0$$

This equation can be simplified by taking the square root of both sides:

$$\frac{x}{2.00 - x} = \sqrt{10.0} = 3.16$$

Solving for $x$ gives $x = 1.52$ atm, which is the equilibrium partial pressure of $H_2$ (and $CO_2$).

**Think About It** Our prediction that most, but far from all, of the reactants would form products was correct. It is a good idea to substitute the results into the equilibrium constant expression as a check on the validity of the solution. The calculated partial pressures are $P_{H_2} = P_{CO_2} = 1.52$ atm and $P_{H_2O} = P_{CO} = 2.00 - 1.52 = 0.48$ atm. Inserting these values in the equilibrium constant expression gives $K_p = (1.52)^2/(0.48)^2 = 10.03$, which is not significantly different from the given value of 10.0 and which confirms that our calculation is correct.

 **Practice Exercise** The chemical equation for the formation of hydrogen iodide from $H_2$ and $I_2$ is

$$H_2(g) + I_2(g) \rightleftharpoons 2\,HI(g)$$

$K_p = 50.0$ for this reaction at 450°C. What is the partial pressure of HI in a sealed reaction vessel at 450°C if the initial partial pressures of $H_2$ and $I_2$ are both 0.100 atm and no HI is initially present?

**SAMPLE EXERCISE 14.13** Calculating an Equilibrium Partial Pressure II **LO8**

Suppose that in a reaction vessel running the water–gas shift reaction at 400°C,

$$CO(g) + H_2O(g) \rightleftharpoons CO_2(g) + H_2(g)$$

the initial partial pressures are $P_{CO} = 2.00$ atm, $P_{H_2O} = 2.00$ atm, $P_{H_2} = 0.15$ atm, and $P_{CO_2} = 0.00$ atm. What is the partial pressure of $H_2$ at equilibrium, given $K_p = 10.0$ at 400°C?

**Collect and Organize** We are asked to calculate the partial pressure of a product at equilibrium. We know the initial partial pressures of all reactants and of both products as well as the value of $K_p$. The difference between this problem and Sample Exercise 14.12 is that here we have product present before the reaction starts.

**Analyze** Comparing the reaction quotient $Q$ with the value of $K$ lets us know in which direction the reaction proceeds to attain equilibrium. No $CO_2$ is initially present, so $Q = 0$. Therefore, the reaction as written proceeds in the forward direction. Our strategy is to solve the problem by setting up a RICE table.

**Solve** Let $x$ be the increase in $P_{H_2}$. The change in $P_{CO_2}$ is also $x$, and the changes in $P_{CO}$ and $P_{H_2O}$ are $-x$:

| REACTION | $CO(g)$ | + | $H_2O(g)$ | $\rightleftharpoons$ | $CO_2(g)$ | + | $H_2(g)$ |
|---|---|---|---|---|---|---|---|
| | $P_{CO}$ (atm) | | $P_{H_2O}$ (atm) | | $P_{CO_2}$ (atm) | | $P_{H_2}$ (atm) |
| Initial | 2.00 | | 2.00 | | 0.00 | | 0.15 |
| Change | $-x$ | | $-x$ | | $+x$ | | $+x$ |
| Equilibrium | $2.00 - x$ | | $2.00 - x$ | | $x$ | | $0.15 + x$ |

$$K_p = \frac{(P_{CO_2})(P_{H_2})}{(P_{CO})(P_{H_2O})} = \frac{(x)(0.15 + x)}{(2.00 - x)(2.00 - x)} = 10.0$$

Solving for $x$ gives two values: 1.50 and 2.96. The 2.96 value is not useful because it produces negative partial pressures of CO and $H_2O$: $(2.00 - 2.96)$ atm $= -0.96$ atm, which is impossible. Therefore, $x = 1.50$, and at equilibrium $P_{H_2} = 0.15 + 1.50 = 1.65$ atm.

**Think About It** Using $x = 1.50$ in the terms in the E row of the RICE table, we find that the equilibrium partial pressures of CO, $H_2O$, and $CO_2$ are 0.50, 0.50, and 1.50, respectively. These values result in a $K_p$ value of $(1.65)(1.50)/(0.50)(0.50) = 9.9$, which is acceptably close to the given value of 10.0. Comparing the results of this Sample Exercise to the previous one, we find that having an initial $P_{H_2}$ of 0.15 atm results in a higher final $P_{H_2}$ value (1.65 vs. 1.52 atm); however, the presence of $H_2$ at the start of the reaction results in slightly less conversion of reactants to products than when no $H_2$ is present initially. This result makes sense because the presence of some product before the reaction starts means that less product must form before $Q_p = K_p$.

 **Practice Exercise** The value of $K_c$ for the reaction

$$N_2O_4(g) \rightleftharpoons 2\,NO_2(g)$$

is 0.21 at 373 K. If a reaction vessel at that temperature initially contains 0.030 $M$ $NO_2$ and 0.030 $M$ $N_2O_4$, what are the concentrations of the two gases at equilibrium?

## 14.9 Equilibrium and Thermodynamics

**CHEMTOUR** Equilibrium and Thermodynamics

**CONNECTION** In Chapter 12, we defined the change in free energy of a reaction, $\Delta G_{rxn}$, as the energy available to do useful work at a particular temperature and pressure.

In Section 12.6, we explored how spontaneous reactions increase the entropy of the universe and experience a decrease in their own free energy: $\Delta G_{rxn}$ is negative, and the reaction as written proceeds in the forward direction. Alternatively, if $\Delta G_{rxn}$ is positive, the reaction is nonspontaneous; instead, the reverse reaction is spontaneous, and the reaction as written proceeds in the reverse direction. As a spontaneous reaction under constant temperature and pressure proceeds, the concentrations of reactants and products change, and the free energy of the system changes as well. Eventually, $\Delta G_{rxn}$ reaches zero. When it does, no free energy is left to do useful work. The reaction has achieved chemical equilibrium.

The magnitude of $\Delta G_{rxn}$—how far it is from zero in either a negative or positive direction—indicates how far a system is from its equilibrium position. Similarly, when $Q$ is much larger or smaller than $K$, we know that a chemical system is far from equilibrium. It is reasonable, then, to think that the separation between the values of $Q$ and $K$ and the sign and magnitude of $\Delta G_{rxn}$ are somehow related. Indeed they are, and their mathematical relationship is one of the most important connections in chemistry because it enables us to relate the thermodynamics of a chemical reaction to the composition of a reaction mixture at equilibrium.

The thermodynamic view of equilibrium and the relationship between $\Delta G_{rxn}$ and $Q$ are described by this equation,

$$\Delta G_{rxn} = \Delta G^\circ_{rxn} + RT \ln Q \tag{14.19}$$

where $\Delta G^\circ_{rxn}$ is the change in free energy under standard conditions. To better understand what this means, let's return once more to the dimerization of $NO_2$:

$$2\, NO_2(g) \rightleftharpoons N_2O_4(g) \tag{14.3}$$

We can calculate the change in standard free energy for the reaction ($\Delta G^\circ_{rxn}$) as we did in Chapter 12, using standard free energy of formation ($\Delta G^\circ_f$) values from Table A4.3 in Appendix 4 and the formula

$$\Delta G^\circ_{rxn} = \sum n_{products}\Delta G^\circ_{f,products} - \sum n_{reactants}\Delta G^\circ_{f,reactants} \tag{12.12}$$

$$= 1\ \cancel{mol}\ (99.8\ kJ/\cancel{mol}) - 2\ \cancel{mol}\ (51.3\ kJ/\cancel{mol}) = -2.8\ kJ$$

The negative value of $\Delta G^\circ_{rxn}$ indicates that the reaction as written should, under standard conditions, proceed spontaneously in the forward direction at 298 K. However, this value of $\Delta G^\circ_{rxn}$ applies only when $P_{NO_2} = P_{N_2O_4} = 1$ bar, which rarely happens.

**CONNECTION** Recall from Section 10.6 and Table 12.1 that standard conditions for thermodynamic values are at a pressure of 1 bar and a temperature of 298 K.

Let's consider what happens in a reaction vessel at 298 K if it initially contains pure $NO_2$ at a partial pressure of 2.0 bar with no (or hardly any) $N_2O_4$ present. Under these conditions, $Q_p = 0$:

$$Q_p = \frac{P_{N_2O_4}}{(P_{NO_2})^2} = \frac{0}{4.0} = 0$$

The natural log (ln) of zero is undefined, so we can't use Equation 14.19 to calculate the value of $\Delta G_{rxn}$. However, if the reaction mixture contained a trace of $N_2O_4$ so that, for example, its partial pressure was 1.0 mbar, then $Q_p$ would be $0.0010/4.0 = 0.00025$, and $\Delta G_{rxn}$ would be:

$$\Delta G_{rxn} = \Delta G^\circ_{rxn} + RT \ln Q = -2.8\ kJ/mol + [8.314\ \cancel{J}/(mol \cdot \cancel{K})$$

$$\times \frac{1\ kJ}{1000\ \cancel{J}} \times 298\ \cancel{K} \times \ln 0.00025] = -23.3\ kJ/mol$$

This value of $\Delta G_{rxn}$ is plotted as point ① on the graph in **Figure 14.13**. Its negative value tells us that the reaction will proceed spontaneously, consuming molecules of $NO_2$ at twice the rate that it produces molecules of $N_2O_4$. As the reaction proceeds, $P_{NO_2}$ decreases and $P_{N_2O_4}$ increases, as does the value of $Q$, which makes the $RT \ln Q$ term in Equation 14.19 greater and the value of $\Delta G_{rxn}$ less negative. Eventually, $P_{NO_2}$ falls to 0.49 bar, $P_{N_2O_4}$ increases to 0.75 bar, and the value of $Q_p$ reaches $0.75/(0.49)^2 = 3.1$. At this point ② on the graph, the reaction has come to chemical equilibrium, which means $K_p = Q_p = 3.1$.

At equilibrium, Equation 14.19 becomes

$$\Delta G_{rxn} = \Delta G^\circ_{rxn} + RT \ln K = 0$$

or

$$\Delta G^\circ_{rxn} = -RT \ln K \tag{14.20}$$

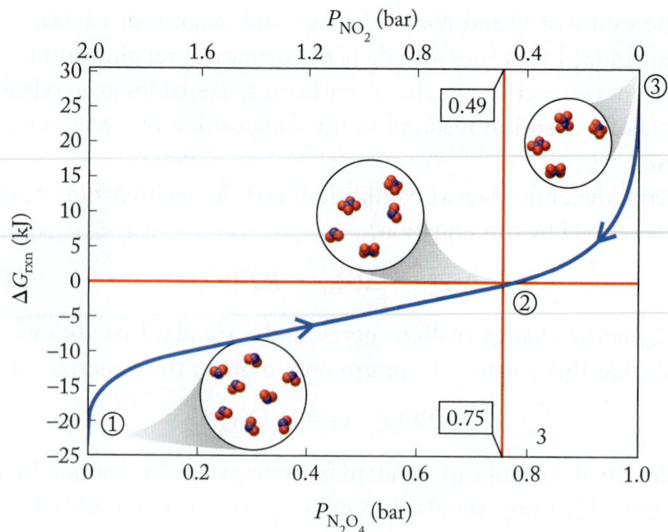


**FIGURE 14.13** Changes in the value of $\Delta G_{rxn}$ during the dimerization reaction $2\ NO_2(g) \rightleftharpoons N_2O_4(g)$. At point ① the system consists of nearly pure reactant, $NO_2$ gas; $\Delta G_{rxn}$ has a large negative value, and the dimerization reaction proceeds spontaneously. As it does, $P_{NO_2}$ decreases and $P_{N_2O_4}$ increases, which means $Q_p$ and $\Delta G_{rxn}$ also increase until $\Delta G_{rxn}$ reaches zero at point ②. Here $Q_p = K_p$; the reaction has reached equilibrium, and no further change in the composition of the reaction mixture occurs. At point ③ the system consists of nearly pure $N_2O_4$; $\Delta G_{rxn}$ has a large positive value, which means the dimerization reaction is nonspontaneous but the reverse reaction, the decomposition of $N_2O_4$, proceeds spontaneously. As it does, $P_{N_2O_4}$ decreases and $P_{NO_2}$ increases, which means $Q_p$ and $\Delta G_{rxn}$ both decrease until $\Delta G_{rxn}$ reaches zero at point ②, and the system is again at equilibrium.

We can use Equation 14.20 to confirm that the above $K_p$ value is consistent with the value of $\Delta G^\circ_{rxn}$ for the dimerization reaction:

$$\Delta G^\circ_{rxn} = -RT \ln K = -8.314\ \text{J/(mol} \cdot \text{K)} \times \frac{1\ \text{kJ}}{1000\ \text{J}} \times 298\ \text{K} \times \ln(3.1) = -2.8\ \text{kJ/mol}$$

Starting with a reaction mixture that is essentially pure reactant is a common way to run a reaction, but Figure 14.13 shows the results of another approach: starting with (nearly) pure product. Consider what happens if a reaction vessel at 298 K contains a reaction mixture in which the partial pressure of $N_2O_4$ is 1.0 bar with a trace amount (2.0 mbar) of $NO_2$. Using these values in Equation 14.19 yields a $\Delta G_{rxn}$ value of +28.0 kJ for the reaction. This positive value, which is plotted at point ③ in Figure 14.13, indicates that the forward reaction is not spontaneous. However, the reverse reaction is spontaneous: the dimer spontaneously dissociates into molecules of $NO_2$, causing $P_{N_2O_4}$ to decrease, $P_{NO_2}$ to increase, and the values of $Q_p$ (of the dimerization reaction) and $RT \ln Q_p$ to decrease. When $RT \ln Q_p$ falls to +2.8 kJ/mol, the value of $\Delta G_{rxn}$ again reaches $(-2.8 + 2.8) = 0$ at point ② on the graph, but this time as $N_2O_4$ decomposes into $NO_2$:

$$N_2O_4(g) \rightleftharpoons 2\ NO_2(g)$$

## CONCEPT TEST

Suppose that $\Delta G^\circ_{rxn} = -10.0$ kJ/mol for the hypothetical chemical reaction $A \rightleftharpoons B$. Which of the following statements about an equilibrium mixture of A and B at 298 K is true?

a. Only A is present.
b. Only B is present.
c. An equimolar mixture of A and B is present.
d. More A than B is present.
e. More B than A is present.

Rearranging Equation 14.20 allows us to calculate the $K$ value for a reaction from its change in standard free energy and absolute temperature. First, we rearrange the terms,

$$\ln K = \frac{-\Delta G^\circ_{rxn}}{RT} \tag{14.21}$$

and then take the antilogarithm of both sides:

$$K = e^{-\Delta G^\circ_{\text{rxn}}/RT} \tag{14.22}$$

Equation 14.22 provides the following interpretation of reaction spontaneity under standard conditions. Whenever $\Delta G^\circ_{\text{rxn}}$ is negative, the exponent $-\Delta G^\circ_{\text{rxn}}/RT$ in Equation 14.22 is positive, and $e^{-\Delta G^\circ_{\text{rxn}}/RT} > 1$, making $K > 1$. Therefore, any reversible reaction with an equilibrium constant greater than 1 proceeds spontaneously in the forward direction under standard conditions, as shown in **Figure 14.14(a)**. This spontaneity has its limits. As reactants are consumed and products are formed, the value of the reaction quotient increases, making the value

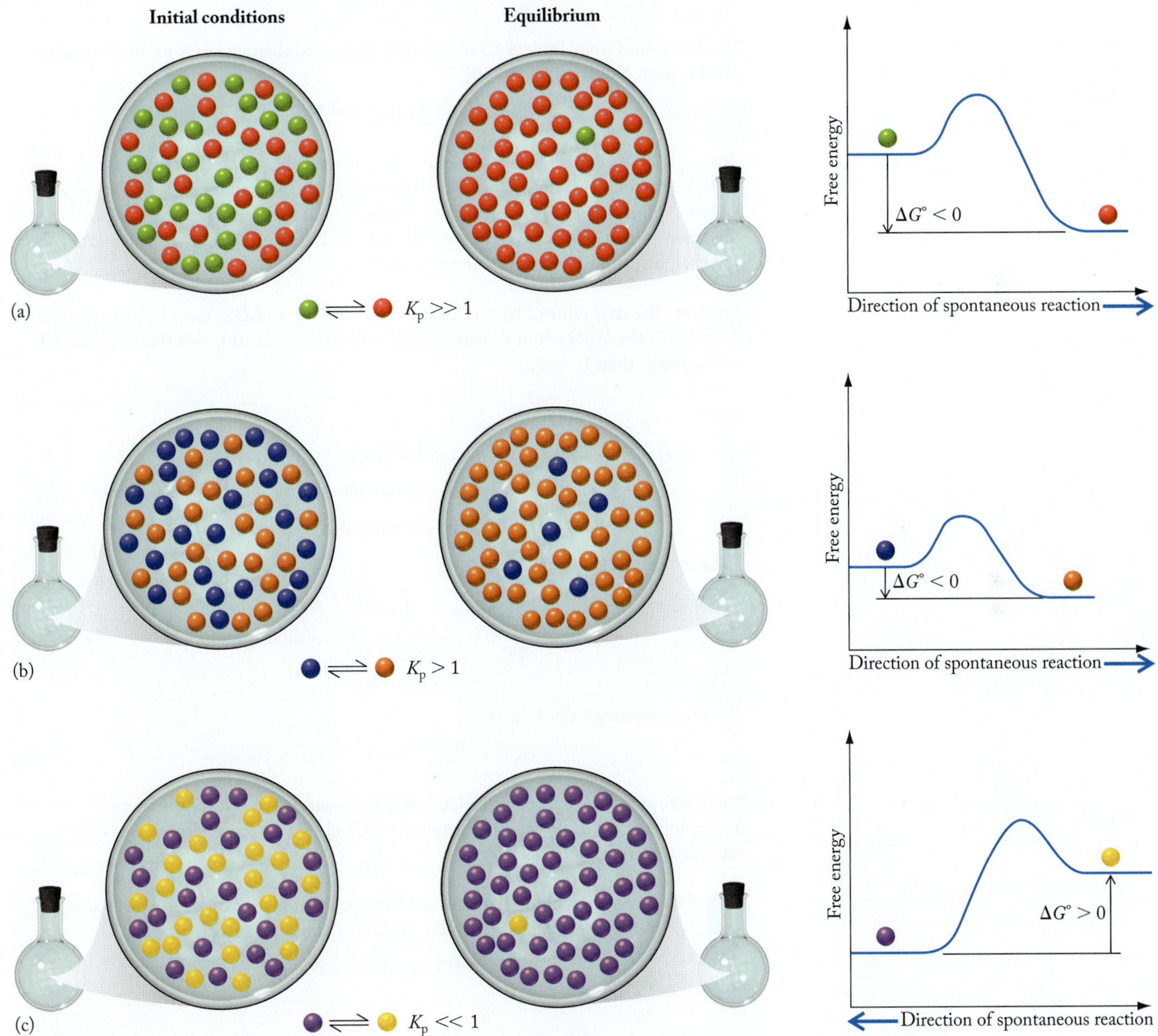


**FIGURE 14.14** The equilibrium constant of a chemical reaction is linked to its $\Delta G^\circ$ value. The initial conditions for the three flasks on the left each contain equimolar mixtures of different pairs of gases. The initial partial pressure of each gas is 1 bar. The three images on the right represent the mixtures in these same flasks at equilibrium. (a) The value of $\Delta G^\circ$ for the formation of "red" gas from "green" gas has a large negative value, which makes $K$ much greater than 1. The reaction proceeds in the forward direction, leaving little green gas unreacted. (b) If the value of $\Delta G^\circ$ for the formation of "orange" gas from "blue" gas has a less negative value than in part (a), the value of $K$ is smaller, though still greater than 1, and more orange gas is present at equilibrium than blue gas. (c) If the value of $\Delta G^\circ$ for the formation of "yellow" gas from "purple" gas is positive, $K$ is less than 1 and the reaction runs in the reverse direction, forming purple gas from yellow gas.

of $\Delta G_{rxn}$ less negative. When it reaches zero, the composition of the reaction mixture does not change further because chemical equilibrium has been achieved.

It follows that a reversible reaction with a less negative value of $\Delta G^\circ_{rxn}$ (**Figure 14.14b**) is still spontaneous, but it has a smaller equilibrium constant, so less reactant is consumed and less product is formed before $\Delta G_{rxn}$ reaches zero and equilibrium is reached. Finally, a reaction that has a positive value of $\Delta G^\circ_{rxn}$ (**Figure 14.14c**) has an equilibrium constant that is less than 1, and it is not spontaneous under standard conditions, but the reverse of the reaction is.

**SAMPLE EXERCISE 14.14** Relating $K$ and $\Delta G^\circ_{rxn}$ **LO9**

Use $\Delta G^\circ_f$ values from Table A4.3 to calculate $\Delta G^\circ_{rxn}$ and the value of $K$ for the formation of $NO_2$ from NO and $O_2$ at 298 K:

$$NO(g) + \tfrac{1}{2}\,O_2(g) \rightleftharpoons NO_2(g)$$

**Collect and Organize** We are asked to calculate the values of $\Delta G^\circ_{rxn}$ and $K$ for a reaction starting with $\Delta G^\circ_f$ values from Table A4.3: 51.3 kJ/mol for $NO_2$ and 86.6 kJ/mol for NO. Because $O_2$ gas is the most stable form of the element under standard conditions, its $\Delta G^\circ_f$ value is 0.0 kJ/mol. We can use Equation 12.12 to calculate $\Delta G^\circ_{rxn}$ and then Equation 14.22 to calculate the value of $K$.

**Analyze** The $\Delta G^\circ_f$ value of $NO_2$ is less than that of NO, so $\Delta G^\circ_{rxn}$ must be negative, and the value of the exponent in Equation 14.22 will be positive. This will translate into a $K$ value greater than 1.

**Solve**

$$\begin{aligned}\Delta G^\circ_{rxn} &= [\Delta G^\circ_{f,NO_2}] - [\Delta G^\circ_{f,NO} + \tfrac{1}{2}\,\Delta G^\circ_{f,O_2}] \\ &= [1\ \cancel{mol}\ (51.3\ kJ/\cancel{mol})] - [1\ \cancel{mol}\ (86.6\ kJ/\cancel{mol}) + \tfrac{1}{2}\ \cancel{mol}\ (0.0\ kJ/\cancel{mol})] \\ &= -35.3\ kJ \text{ or } -35{,}300\ J \text{ per mole of } NO_2 \text{ produced}\end{aligned}$$

The exponent in Equation 14.22 is

$$\frac{-\Delta G^\circ_{rxn}}{RT} = -\frac{-\left(\dfrac{-35{,}300\ \cancel{J}}{\cancel{mol}}\right)}{\left(\dfrac{8.314\ \cancel{J}}{\cancel{mol}\cdot\cancel{K}}\right)(298\ \cancel{K})} = 14.2$$

The corresponding value of $K$ is

$$K = e^{-\Delta G^\circ_{rxn}/RT} = e^{14.2} = 1.5 \times 10^6$$

**Think About It** The exponential relationship between $\Delta G^\circ_f$ and $K$ means that a moderately negative free-energy change of −35.3 kJ/mol corresponds to a very large value of $K$: here, greater than $10^6$.

**Practice Exercise** The standard free energy of formation of ammonia at 298 K is −16.5 kJ/mol. What is the value of $K$ for the reaction at 298 K?

$$N_2(g) + 3\,H_2(g) \rightleftharpoons 2\,NH_3(g)$$

Which equilibrium constant, $K_c$ or $K_p$, is related to $\Delta G^\circ$ by Equation 14.22? $\Delta G^\circ$ represents a change in free energy under standard conditions, and the standard state for a gaseous reactant or product is one in which its partial pressure is 1 bar. Thus, the $\Delta G^\circ$ of a reaction *in the gas phase* is linked by Equation 14.22 to its $K_p$ value. However, standard conditions for reactions in solution (the focus of

Chapter 15) mean that all dissolved reactants and products are present at a concentration of 1.00 *M*. Therefore, the $\Delta G°$ of a reaction *in solution* is related by Equation 14.22 to its $K_c$ value.

The values of $\Delta G°$ and $K$ are also linked by their dependence on how we choose to write the chemical equation describing a reaction. For example, multiplying the coefficients in an equation by $n$ means multiplying its $\Delta G°$ value by $n$. However, in this situation, $K$ is not multiplied by $n$ but rather is raised to the $n$th power: $K$ becomes $K^n$. This difference makes sense given the logarithmic relationship between $\Delta G°$ and $K$ and the fact that $\ln (X^n) = n \ln X$. This logarithmic relationship also explains why reversing a reaction changes the sign of $\Delta G°$ but produces a $K$ that is the reciprocal of the original: $\ln (1/X) = -\ln X$.

## 14.10 Changing *K* with Changing Temperature

We have noted often in this chapter that the value of $K$ changes with changing temperature. In this section, we look more closely at that relationship, developing several important equations that link $K$ and $T$.

### Temperature, *K*, and $\Delta G°_{rxn}$

Let's begin by combining a key equation from Chapter 12,

$$\Delta G°_{rxn} = \Delta H°_{rxn} - T\Delta S°_{rxn} \quad (12.11)$$

with one from this chapter,

$$\ln K = \frac{-\Delta G°_{rxn}}{RT} \quad (14.21)$$

to derive an equation that relates $K$ to $\Delta H°_{rxn}$ and $\Delta S°_{rxn}$,

$$\ln K = \frac{-\Delta G°_{rxn}}{RT} = \frac{-\Delta H°_{rxn}}{RT} + \frac{T\Delta S°_{rxn}}{RT} = \frac{-\Delta H°_{rxn}}{RT} + \frac{\Delta S°_{rxn}}{R} \quad (14.23)$$

Note how a negative value of $\Delta H°_{rxn}$ or a positive value of $\Delta S°_{rxn}$ contributes to a large value of $K$. These dependencies are predicted because negative values of $\Delta H°_{rxn}$ and positive values of $\Delta S°_{rxn}$ are the two factors that contribute to making reactions spontaneous.

Because we are discussing the influence of temperature on $K$, let's identify the factors affected by changes in $T$ in Equation 14.23. First, the values of $\Delta H°_{rxn}$ and $\Delta S°_{rxn}$, such as those in Appendix 4 for $T = 298$ K, change very little with changing temperature, so we will treat $\Delta H°_{rxn}$ and $\Delta S°_{rxn}$ as constants in this discussion. However, the value of the first term on the right side of Equation 14.23 is inversely proportional to temperature, which means that the influence of $\Delta H°_{rxn}$ on $K$ decreases as temperature increases. Thus, a favorable (negative) $\Delta H°_{rxn}$ contributes less to increasing the value of $K$ as temperature increases. This temperature dependence makes sense if we invoke Le Châtelier's principle and the notion that energy is absorbed from the surroundings in endothermic ($\Delta H°_{rxn} > 0$) reactions and, conversely, energy is released to the surroundings during exothermic ($\Delta H°_{rxn} < 0$) reactions. Increasing the temperature shifts an equilibrium in the direction of the endothermic process and inhibits the exothermic process.

If $\Delta H°_{rxn}$ and $\Delta S°_{rxn}$ do not vary much with temperature, then Equation 14.23 predicts that $\ln K$ will be a linear function of $1/T$. Furthermore, we expect a graph

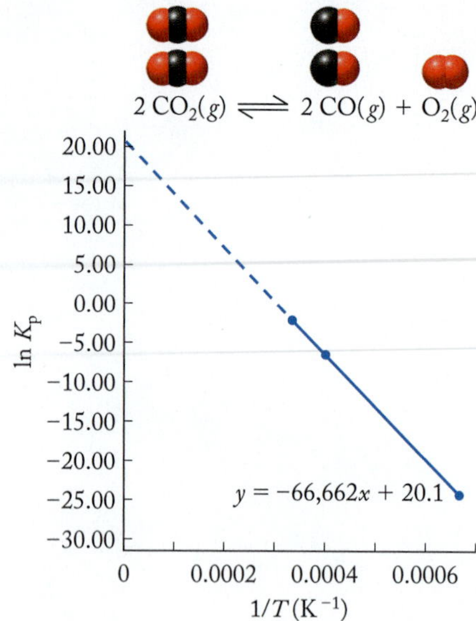


**FIGURE 14.15** The graph of ln $K_p$ versus $1/T$ is a straight line. The slope of the line is $-\Delta H°/R$ and the $y$-intercept is $\Delta S°/R$. The negative slope indicates an endothermic process.

of ln $K$ versus $1/T$ to have a positive slope for an exothermic process and a negative slope for an endothermic process. We can determine $\Delta H°$ from the slope of the line and $\Delta S°_{rxn}$ from its $y$-intercept. Thus, we can calculate fundamental thermodynamic values of a reaction at equilibrium by determining its equilibrium constant at different temperatures. For example, let's determine the values of $\Delta H°_{rxn}$ and $\Delta S°_{rxn}$ for the reaction

$$2\,CO_2(g) \rightleftharpoons 2\,CO(g) + O_2(g)$$

starting with its $K_p$ values of $2.57 \times 10^{-11}$ at 1500 K, $1.42 \times 10^{-3}$ at 2500 K, and 0.112 at 3000 K. $K_p$ increases as temperature increases, so the reaction must be endothermic. We can use these data to determine the values of $\Delta H°_{rxn}$ and $\Delta S°_{rxn}$ by plotting ln $K_p$ values versus $1/T$. To see how this graphical method works, let's rewrite Equation 14.23 so that it fits the form of the equation for a straight line ($y = mx + b$):

$$\ln K = \frac{-\Delta H°_{rxn}}{R}\left(\frac{1}{T}\right) + \frac{\Delta S°_{rxn}}{R} \tag{14.24}$$

The graph of ln $K_p$ versus $1/T$ (**Figure 14.15**) is indeed a straight line. The slope of this line is −66,662 K, which is equal to $-\Delta H°_{rxn}/R$. The corresponding value of $\Delta H°_{rxn}$ is

$$\Delta H°_{rxn} = -\text{slope} \times R$$
$$= -(-66{,}662\ \text{K})\left(\frac{8.314\ \text{J}}{\text{mol} \cdot \text{K}}\right)$$
$$= 554{,}228\ \text{J/mol} = 554.2\ \text{kJ/mol}$$

The $y$-intercept ($\Delta S°_{rxn}/R$) of the graph is 20.1. The corresponding value of $\Delta S°_{rxn}$ is

$$\Delta S°_{rxn} = (20.1)\left(\frac{8.314\ \text{J}}{\text{mol} \cdot \text{K}}\right) = 167\ \frac{\text{J}}{\text{mol} \cdot \text{K}}$$

The entropy change is positive because the forward reaction converts two moles of gaseous $CO_2$ into three moles of gaseous products.

We can use Equation 14.24 to compare the $K$ values of a reaction at two temperatures. In the two-temperature version of Equation 14.24, the $\Delta S°_{rxn}/R$ terms drop out and we have

$$\ln\left(\frac{K_2}{K_1}\right) = \frac{-\Delta H°_{rxn}}{R}\left(\frac{1}{T_2} - \frac{1}{T_1}\right) \tag{14.25}$$

This equation looks very much like the *Clausius–Clapeyron equation* we developed in Chapter 11, which describes the relationship between the vapor pressures of a liquid at two temperatures:

$$\ln\left(\frac{P_{vap,T_2}}{P_{vap,T_1}}\right) = \frac{-\Delta H_{vap}}{R}\left(\frac{1}{T_2} - \frac{1}{T_1}\right) \tag{11.4}$$

Indeed, the Clausius–Clapeyron equation is a special case of Equation 14.25 applied to a change in physical state (liquid → gas) instead of a change in chemical composition. The pressures in Equation 11.4 are actually equilibrium vapor pressures, and the enthalpy of reaction is the enthalpy of vaporization ($\Delta H_{vap}$). Equation 14.25 is called the *van 't Hoff equation* because it was first derived by Jacobus van 't Hoff. It is particularly useful for calculating the value of $K$ at a very high or very low temperature if we already know $K$ at a standard reference temperature, such as 298 K. We use such a calculation in Sample Exercise 14.15.

**SAMPLE EXERCISE 14.15** Calculating an Equilibrium Constant Value at a Specific Temperature **LO10**

Use data from Appendix 4 to calculate the equilibrium constant $K_p$ for the exothermic reaction

$$N_2(g) + 3\,H_2(g) \rightleftharpoons 2\,NH_3(g)$$

at 298 K and at 773 K, a typical temperature used in the Haber–Bosch process for synthesizing ammonia.

**Collect and Organize** We are asked to calculate $K_p$ for a reaction at two temperatures. One of the temperatures is 298 K, which is the reference temperature for the standard thermodynamic data in Appendix 4. The $\Delta G_f^\circ$ value of $NH_3$ is −16.5 kJ/mol. Note that $\Delta G_f^\circ$ values are valid only for $T$ = 298 K but that $\Delta H_f^\circ$ and $S^\circ$ values may be used at other temperatures.

**Analyze** We can use Equation 14.22 to calculate the value of $K_p$ at 298 K from the value of $\Delta G^\circ_{rxn}$. Then we can use Equation 14.25 to calculate the value of $K_p$ at 773 K from the value of $K_p$ at 298 K and the value of $\Delta H^\circ_{rxn}$. The $\Delta H^\circ_{rxn}$ value can be calculated from the enthalpy of formation of $NH_3$ gas (−46.1 kJ/mol). The negative value of $\Delta H_f^\circ$ means that $\Delta H^\circ_{rxn}$ is also negative, so the reaction is exothermic and we predict that the value of $K_p$ will be smaller at 773 K than at 298 K.

**Solve** The value of $\Delta G^\circ_{rxn}$ for the reaction is

$$\Delta G^\circ_{rxn} = \frac{2\ \text{mol NH}_3}{1\ \text{mol N}_2} \times \frac{-16.5\ \text{kJ}}{1\ \text{mol NH}_3} \times \frac{1000\ \text{J}}{1\ \text{kJ}} = -33{,}000\ \frac{\text{J}}{\text{mol N}_2} = -3.30 \times 10^4\ \text{J/mol}$$

Using this value in Equation 14.22 to calculate the value of $K_p$,

$$K_p = e^{-\Delta G^\circ_{rxn}/RT} = \exp\left(\frac{-\left(-33{,}000\ \frac{\text{J}}{\text{mol}}\right)}{8.314\ \frac{\text{J}}{\text{mol}\cdot\text{K}} \times 298\ \text{K}}\right) = 6.09 \times 10^5$$

Once we know the value of $K_p$ at 298 K, we can calculate $K_p$ at 773 K by using Equation 14.25. We must first calculate the value of $\Delta H^\circ_{rxn}$, which is twice the $\Delta H_f^\circ$ of $NH_3$:

$$\Delta H^\circ_{rxn} = \frac{2\ \text{mol NH}_3}{1\ \text{mol N}_2} \times \frac{-46.1\ \text{kJ}}{1\ \text{mol NH}_3} \times \frac{1000\ \text{J}}{1\ \text{kJ}} = -92{,}200\ \frac{\text{J}}{\text{mol N}_2}$$

After substituting $K_1 = 6.09 \times 10^5$, $T_1$ = 298 K, and $T_2$ = 773 K into the equation, we solve for $K_2$:

$$\ln\left(\frac{K_2}{6.09 \times 10^5}\right) = -\frac{\left(-92{,}200\ \frac{\text{J}}{\text{mol}}\right)}{8.314\ \frac{\text{J}}{\text{mol}\cdot\text{K}}}\left(\frac{1}{773\ \text{K}} - \frac{1}{298\ \text{K}}\right)$$

$$= -22.87$$

$$\frac{K_2}{6.09 \times 10^5} = e^{-22.87} = 1.17 \times 10^{-10}$$

$$K_2 = 7.12 \times 10^{-5}\ \text{at 773 K}$$

**Think About It** The equilibrium constant decreases markedly when the temperature of the reaction is raised from 298 K to 773 K. This decrease agrees with our prediction for an exothermic reaction.

 **Practice Exercise** Use data from Appendix 4 to calculate the value of $K_p$ for the following reaction at both 298 K and 2000 K:

$$2\,N_2(g) + O_2(g) \rightleftharpoons 2\,N_2O(g)$$

The equilibrium constant determined in Sample Exercise 14.15 for the ammonia synthesis reaction is greater than $10^5$ at room temperature but about $10^{-5}$ at 773 K. That's 10 orders of magnitude smaller. We now have a more complete picture of this reaction, which was the basis for Sample Exercise 14.9. Ammonia is usually synthesized from $N_2$ and $H_2$ at temperatures near 400°C. These high temperatures increase the rate of this exothermic reaction, even though they significantly decrease $K_p$ and decrease the amount of ammonia produced at equilibrium. Here, the practical benefit of a more favorable reaction rate outweighs the less favorable thermodynamics of running the reaction at high temperature. In practice, the ammonia produced in the reaction is continuously removed from the reaction vessel by condensation, shifting the equilibrium toward the formation of more product as predicted by Le Châtelier's principle.

### SAMPLE EXERCISE 14.16 Integrating Concepts: Fire Extinguishers

Dry chemical fire extinguishers are the most common type purchased for use in the home. They are also used to fight fires at airport crash rescue sites. A dry chemical extinguisher sprays a fine powder of sodium bicarbonate ($NaHCO_3$, baking soda) or potassium bicarbonate ($KHCO_3$). Both substances decompose at the elevated temperatures of a fire and release carbon dioxide to extinguish it. $K_p = 0.25$ for the baking soda decomposition reaction at 125°C:

$$2\,NaHCO_3(s) \rightleftharpoons Na_2CO_3(s) + CO_2(g) + H_2O(g)$$

a. Neither $NaHCO_3$ nor $KHCO_3$ decomposes at room temperature, but both decompose rapidly at high temperatures. Are their decomposition reactions exothermic or endothermic? What does this mean about the stability of these compounds in fire extinguishers?

b. $KHCO_3$ is marketed as being about two times as effective on fires involving flammable liquids or gases (class B fires) as $NaHCO_3$. Use thermochemical data from Appendix 4 to assess this claim by comparing the amounts of $CO_2$ generated when each material decomposes. Assume that the temperature of the fire is 1200°C.

**Collect and Organize** We are asked how long fire extinguishers containing $NaHCO_3$ or $KHCO_3$ can be kept at room temperature without losing their efficacy and whether manufacturers' claims about fighting fires with these compounds are valid. We are given an equilibrium constant for the $NaHCO_3$ reaction at 125°C (398 K), the fact that the compounds do not decompose at room temperature but do at elevated temperatures, and tables in Appendix 4 that contain thermochemical data on both compounds as well as the reaction products. To assess the claim of greater efficacy of $KHCO_3$, we need to determine $K_p$ values for reactions of both substances at the fire temperature. We can then see whether one produces more $CO_2$ than the other.

**Analyze** We can use $\Delta G^\circ_f$ values for the reactants and products to calculate $\Delta G^\circ_{rxn}$ at 298 K for the decomposition of $KHCO_3$ and then use that value to calculate $K_p$ for the reaction at 298 K. Then we can use $\Delta H^\circ_f$ values for the reactants and products to calculate $\Delta H^\circ_{rxn}$ values for both reactions. These values and their known $K_p$ values at lower temperatures can be used with Equation 14.25 to calculate their $K_p$ values at 1200°C (1473 K). Comparing these two equilibrium constant values will help us determine whether one or the other bicarbonate compound releases more $CO_2$ at flame temperatures.

**Solve**

a. $NaHCO_3$ and $KHCO_3$ do not decompose at room temperature but do at flame temperatures, so we predict that their decomposition reactions are endothermic. Energy is absorbed by the system in endothermic reactions. These compounds should be stable in fire extinguishers stored at room temperature.

b. Let's first calculate the value of $K_p$ for the decomposition of $NaHCO_3$ at 1200°C by using Equation 14.25:

$$\ln\left(\frac{K_2}{K_1}\right) = \frac{-\Delta H^\circ_{rxn}}{R}\left(\frac{1}{T_2} - \frac{1}{T_1}\right)$$

We need $\Delta H^\circ_{rxn}$ for the reaction of $NaHCO_3$, which we can calculate from data given in Appendix 4 and methods we learned in Chapter 10.

$$\Delta H^\circ_{rxn} = \Sigma n_{products}\,\Delta H^\circ_{f,products} - \Sigma n_{reactants}\,\Delta H^\circ_{f,reactants} \qquad (10.17)$$

$$= (\Delta H^\circ_{f,Na_2CO_3} + \Delta H^\circ_{f,H_2O} + \Delta H^\circ_{f,CO_2}) - 2(\Delta H^\circ_{f,NaHCO_3})$$

$$= [(-1130.7) + (-241.8) + (-393.5) - (2)(-950.8)]\ \text{kJ/mol} = 135.6\ \text{kJ/mol}$$

$$= 135{,}600\ \text{J/mol}$$

Then

$$\ln\frac{K_2}{0.25} = \frac{-135{,}600\ \text{J/mol}}{8.314\ \dfrac{\text{J}}{\text{mol}\cdot\text{K}}}\left(\frac{1}{1473\ \text{K}} - \frac{1}{398\ \text{K}}\right)$$

$$\ln K_2 - \ln(0.25) = 29.91$$

$$\ln K_2 = 28.52$$

$$K_2 = 2.43 \times 10^{12} = K_{p,NaHCO_3} \text{ at } 1200^\circ\text{C}$$

Now we need to calculate $K_p$ for $KHCO_3$ at 1200°C by using Equation 14.22:

$$K = e^{-\Delta G^\circ_{rxn}/RT}$$

We can use $\Delta G^\circ_f$ values from the Appendix to calculate $\Delta G^\circ_{rxn}$, as we learned in Chapter 12, and from that we can determine $K_p$ at 25°C. Then we can use Equation 14.25 and $\Delta H^\circ_f$ values from the Appendix to estimate $K_{p,KHCO_3}$ at 1200°C. The reaction of interest is

$$2\ KHCO_3(s) \rightleftharpoons K_2CO_3(s) + CO_2(g) + H_2O(g)$$

$$\Delta G^\circ_{rxn} = \Sigma n_{products}\,\Delta G^\circ_{f,products} - \Sigma n_{reactants}\,\Delta G^\circ_{f,reactants} \quad (12.12)$$

$$= (\Delta G^\circ_{f,K_2CO_3} + \Delta G^\circ_{f,H_2O} + \Delta G^\circ_{f,CO_2}) - 2(\Delta G^\circ_{f,KHCO_3})$$

$$= [(-1063.5) + (-228.6) + (-394.4) - (2)(-863.6)]\ kJ/mol$$

$$= 40.7\ kJ/mol = 4.07 \times 10^4\ J/mol$$

Then

$$K = \exp\left(\frac{-4.07 \times 10^4 \frac{J}{mol}}{8.314 \frac{J}{mol \cdot K} \times 298\ K}\right) = e^{-16.427} = 7.34 \times 10^{-8}$$

We can now use $\Delta H^\circ_f$ values from the Appendix to find $\Delta H^\circ_{rxn}$ and apply Equation 14.25 again to calculate the value of $K_{p,KHCO_3}$ at 1200°C:

$$\Delta H^\circ_{rxn} = \Sigma n_{products}\,\Delta H^\circ_{f,products} - \Sigma n_{reactants}\,\Delta H^\circ_{f,reactants}$$

$$= (\Delta H^\circ_{f,K_2CO_3} + \Delta H^\circ_{f,H_2O} + \Delta H^\circ_{f,CO_2}) - 2(\Delta H^\circ_{f,KHCO_3})$$

$$= [(-1151.0) + (-241.8) + (-393.5) - (2)(-963.2)]\ kJ/mol = 140.1\ kJ/mol$$

$$= 1.401 \times 10^5\ J/mol$$

Then

$$\ln \frac{K_2}{7.34 \times 10^{-8}} = \frac{-1.401 \times 10^5\ J/mol}{8.314 \frac{J}{mol \cdot K}}\left(\frac{1}{1473\ K} - \frac{1}{298\ K}\right)$$

$$\ln K_2 = 28.68 \qquad K_{p,KHCO_3} = 2.85 \times 10^{12}$$

The expressions for both equilibrium constants are the same,

$$K_{p,NaHCO_3} = [P_{CO_2}][P_{H_2O}]$$

$$K_{p,KHCO_3} = [P_{CO_2}][P_{H_2O}]$$

because the other ingredients are solids. We may assume the initial partial pressures of $CO_2$ and water vapor are insignificant compared to their equilibrium partial pressures. Their 1:1 mole ratio in the two reactions allows us to represent the equilibrium partial pressure of each with the same symbol. Let's use $x$ in the $NaHCO_3$ reaction and $y$ in the $KHCO_3$ reaction:

$$K_{p,NaHCO_3} = (P_{CO_2})(P_{H_2O}) = (x)(x) = 2.43 \times 10^{12}$$

$$K_{p,KHCO_3} = (P_{CO_2})(P_{H_2O}) = (y)(y) = 2.85 \times 10^{12}$$

Solving for $x$ and $y$, we find that the partial pressure of $CO_2$ ($P_{CO_2}$) in a fire at 1200°C is $1.56 \times 10^6$ atm from the decomposition of $NaHCO_3$ and $1.69 \times 10^6$ atm from the decomposition of $KHCO_3$. These values are nearly the same, so the advantage of using $KHCO_3$ is apparently not based on its ability to produce more $CO_2$.

**Think About It** The $\Delta H^\circ_{rxn}$ values for both compounds are positive, which agrees with our prediction that the reactions are endothermic. The similar $P_{CO_2}$ values mean that $KHCO_3$ must do something else chemically that makes it a better fire-extinguishing agent than $NaHCO_3$. In fact, it does do something different: $KHCO_3$ better inhibits reactions involving free radical intermediates formed in a fire.

## SUMMARY

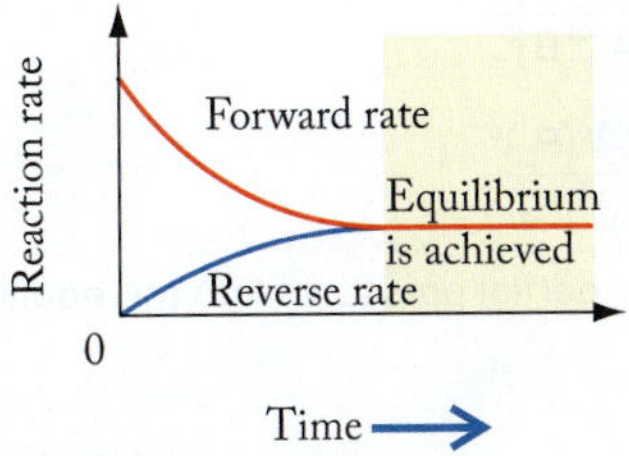


**LO1 Chemical equilibrium** can be approached from either reaction direction and is achieved when the forward and reverse reaction rates are the same. At equilibrium, a reaction vessel may contain comparable amounts of reactants and products, may contain mostly reactants (the equilibrium *lies to the left*), or may contain mostly products (the equilibrium *lies to the right*). The value of the reaction's **equilibrium constant (*K*)** tells us whether products or reactants are favored at equilibrium. (Section 14.1)

**LO2** According to the **law of mass action**, the **equilibrium constant expression** (or **mass action expression**) for $K_c$ is the ratio of the equilibrium molar concentrations of the products divided by the equilibrium molar concentrations of the reactants, each raised to the respective stoichiometric coefficient in the balanced equation. If the substances involved in an equilibrium are gases, then an equilibrium constant $K_p$ may be written based on the partial pressures of the gases. **Heterogeneous equilibria** involve more than one phase. The concentrations of pure liquids and solids do not change during a reaction and are omitted from equilibrium constant expressions. Equilibrium constants have no units. (Sections 14.2, 14.6)

**LO3** The value of $K_c$ or $K_p$ may be calculated using the equilibrium constant expression and concentrations or partial pressures of reactants and products at equilibrium. (Section 14.2)

**LO4** The relationship between $K_c$ and $K_p$ depends on the number of moles of gaseous reactants and products in the balanced chemical equation. (Section 14.3)

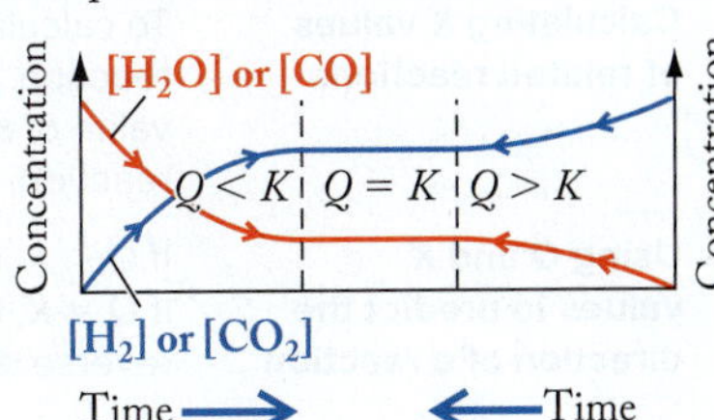


**LO5** The reverse of a reaction has an equilibrium constant that is the reciprocal of $K$ for the forward reaction. If the balanced equation for a reaction is

multiplied by some factor $n$, the value of $K$ for that reaction is raised to the $n$th power. If reactions are summed to give an overall reaction, their equilibrium constants are multiplied together to obtain an overall equilibrium constant. (Section 14.4)

**LO6** The **reaction quotient ($Q$)** is the value of the mass action expression at any instant during a reaction. At equilibrium, $Q = K$, but for nonequilibrium conditions, the value of $Q$ indicates whether the concentrations of products ($Q < K$) or reactants ($Q > K$) will increase. (Section 14.5)

**LO7** According to **Le Châtelier's principle**, chemical reactions at equilibrium respond to stress, such as removing or adding a reactant or product, by shifting position to relieve the stress. Changing the temperature also stresses an equilibrium mixture because it changes the value of $K$. A catalyst decreases the time that a system takes to achieve equilibrium but does not change the value of the equilibrium constant. (Section 14.7)

**LO8** Equilibrium concentrations or partial pressures of reactants and products can be calculated from initial concentrations or pressures, the reaction stoichiometry, and the value of the equilibrium constant. (Section 14.8)

**LO9** A negative value of $\Delta G^\circ_{rxn}$ corresponds to $K > 1$ (products favored), and a positive value of $\Delta G^\circ_{rxn}$ corresponds to $K < 1$ (reactants favored). Concentrations of reactants and products at equilibrium are similar if the value of $\Delta G^\circ_{rxn}$ is near 0 ($K$ is near 1). (Section 14.9)

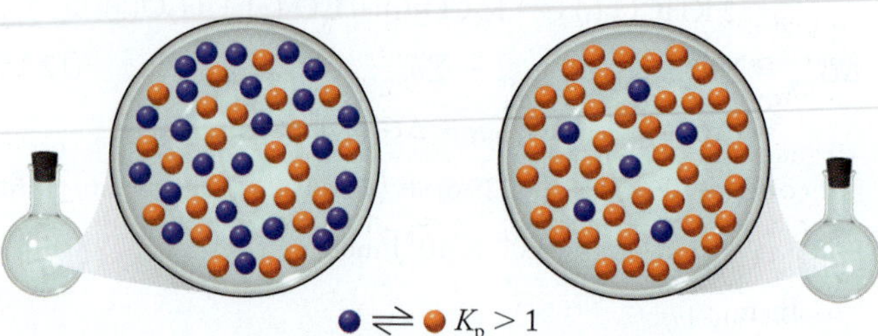


**LO10** Higher reaction temperatures increase the equilibrium constant of an endothermic reaction, but decrease the equilibrium constant of an exothermic reaction. The slope of a plot of ln $K$ versus $1/T$ for an equilibrium system is used to determine $\Delta H^\circ_{rxn}$, and the $y$-intercept of the plot is used to determine the value of $\Delta S^\circ_{rxn}$. (Section 14.10)

## PARTICULATE **PREVIEW WRAP-UP**

Initial condition A spontaneously forms product ($N_2O_4$). Initial condition B spontaneously forms reactant ($NO_2$). Both conditions result in spontaneous reactions, although in opposite directions. Therefore, a decrease in free energy ($\Delta G_{rxn} < 0$) occurs starting with either condition. The forward and reverse reactions both proceed until $\Delta G_{rxn} = 0$, at which point the two reactions have produced the same equilibrium mixture of the two gases.

## PROBLEM-SOLVING SUMMARY

| Type of Problem | Concepts and Equations | Sample Exercises |
|---|---|---|
| **Writing equilibrium constant expressions** | For the reaction $a\text{A} + b\text{B} \rightleftharpoons c\text{C} + d\text{D}$ <br> $K_c = \frac{[\text{C}]^c[\text{D}]^d}{[\text{A}]^a[\text{B}]^b}$ (14.9) <br> and <br> $K_p = \frac{(P_C)^c(P_D)^d}{(P_A)^a(P_B)^b}$ (14.10) | **14.1** |
| **Calculating $K_c$ or $K_p$** | Insert equilibrium molar concentrations or partial pressures into the equilibrium constant expression. | **14.2, 14.3** |
| **Interconverting $K_c$ and $K_p$** | Use the relation <br> $K_p = K_c(RT)^{\Delta n}$ (14.15) <br> where $\Delta n$ is the number of moles of product gas minus the number of moles of reactant gas in the balanced chemical equation. | **14.4** |
| **Calculating $K$ values of related reactions** | To calculate $K$ for the reverse of a reaction, take the reciprocal of $K$ for the forward reaction. If all the coefficients in a chemical equation are multiplied by $n$, the value of $K$ increases by the power of $n$. If reactions are summed to give an overall reaction, their equilibrium constants are multiplied together to obtain an overall $K$. | **14.5, 14.6** |
| **Using Q and K values to predict the direction of a reaction** | If $Q < K$, the reaction proceeds in the forward direction to make more products; if $Q = K$, the reaction is at equilibrium; if $Q > K$, the reaction proceeds in the reverse direction to make more reactants. | **14.7** |

| Type of Problem | Concepts and Equations | Sample Exercises |
|---|---|---|
| **Writing equilibrium constant expressions for heterogeneous equilibria** | Molar concentrations of pure liquids and pure solids are omitted from equilibrium constant expressions because these concentrations are constant. | **14.8** |
| **Stressing an equilibrium by adding or removing reactants or products** | Decreasing the concentration of a substance involved in an equilibrium shifts the equilibrium toward the production of more of that substance. Increasing the concentration of a substance shifts the equilibrium so that some of that substance is consumed in the reaction. | **14.9** |
| **Predicting the effect of compression or expansion on gas-phase equilibria** | Equilibria involving different numbers of gaseous reactants and products shift in response to an increase (or decrease) in volume toward the side of the reaction equation with more (or fewer) moles of gases. | **14.10** |
| **Predicting how temperature changes impact chemical equilibrium** | The value of $K$ for an endothermic reaction increases with increasing temperature; the value of $K$ for an exothermic reaction decreases with increasing temperature. | **14.11** |
| **Calculating concentrations or partial pressures of reactants and products at equilibrium** | Use a RICE table to develop algebraic terms for each reactant's and product's partial pressure or concentration at equilibrium. Let $x$ be the change in partial pressure or concentration of one component of the reaction. Express the changes in the other components in terms of $x$. Substitute these terms into the expression for $K$ and solve for $x$. | **14.12, 14.13** |
| **Relating $K$ and $\Delta G^\circ_{rxn}$** | Calculate $\Delta G^\circ_{rxn}$ from $\Delta G^\circ_{rxn} = \sum n_{products}\,\Delta G^\circ_{f,products} - \sum n_{reactants}\,\Delta G^\circ_{f,reactants}$ (12.12) Then use $K = e^{-\Delta G^\circ_{rxn}/RT}$ (14.22) Be sure to convert $\Delta G^\circ_{rxn}$ to joules per mole and use $R = 8.314$ J/(mol · K). | **14.14** |
| **Calculating equilibrium constant values at specific temperatures** | Use $\ln\left(\frac{K_2}{K_1}\right) = \frac{-\Delta H^\circ_{rxn}}{R}\left(\frac{1}{T_2} - \frac{1}{T_1}\right)$ (14.25) Convert $\Delta H^\circ_{rxn}$ from kilojoules per mole to joules per mole to match the units of $R$. | **14.15** |

## VISUAL PROBLEMS

*(Answers to boldface end-of-chapter questions and problems are in the back of the book.)*

**14.1.** Figure P14.1 shows the energy profiles of reactions $A \rightleftharpoons B$ and $C \rightleftharpoons D$, respectively. Which reaction has the larger forward rate constant? Which reaction has the smaller reverse rate constant? Which reaction has the larger equilibrium constant $K_c$?

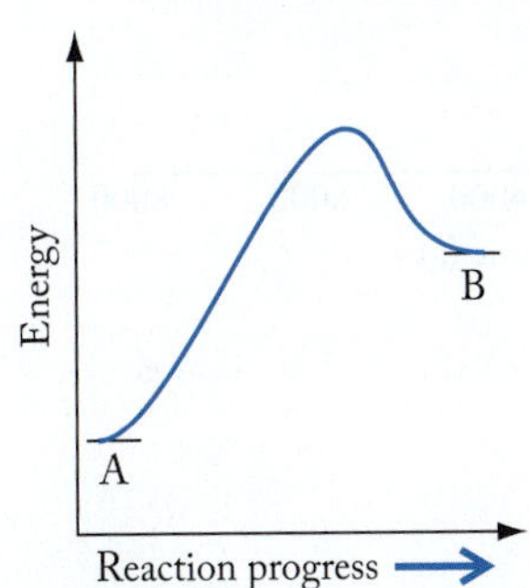


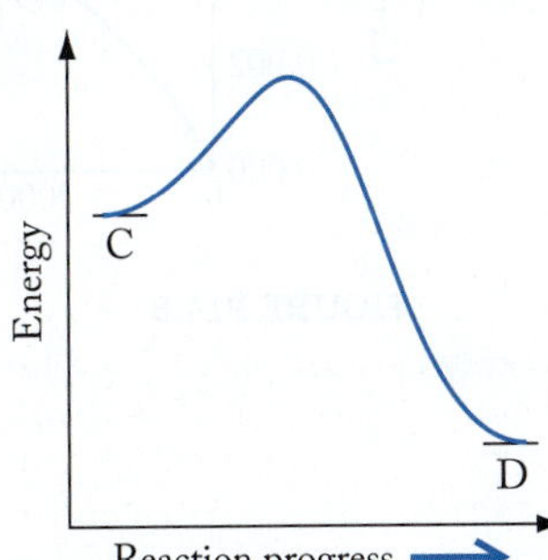


**FIGURE P14.1**

14.2. A particulate view of the progress of the reaction A (red spheres) $\rightleftharpoons$ B (blue spheres) is shown in Figure P14.2.

a. Has the system reached equilibrium?

b. If the system has not yet reached equilibrium, in which direction will the reaction continue to proceed? If the system has reached equilibrium, what is the value of the equilibrium constant ($K$)?

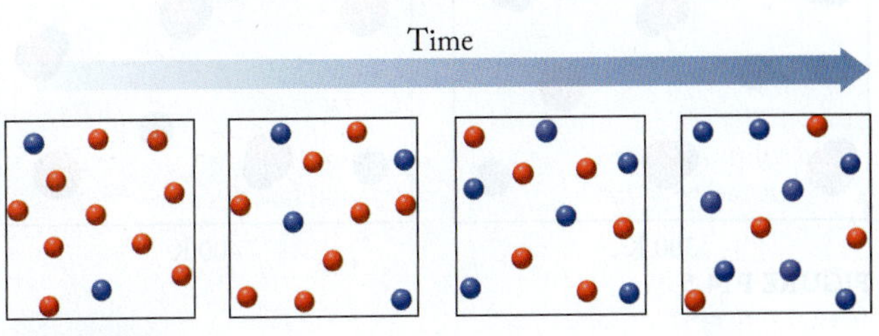


**FIGURE P14.2**

**14.3.** Figure P14.3 provides a particulate view of an equilibrium reaction mixture of A (red spheres) and B (blue spheres). What are the $K_c$ and $K_p$ values of the reaction $A(g) \rightleftharpoons B(g)$?

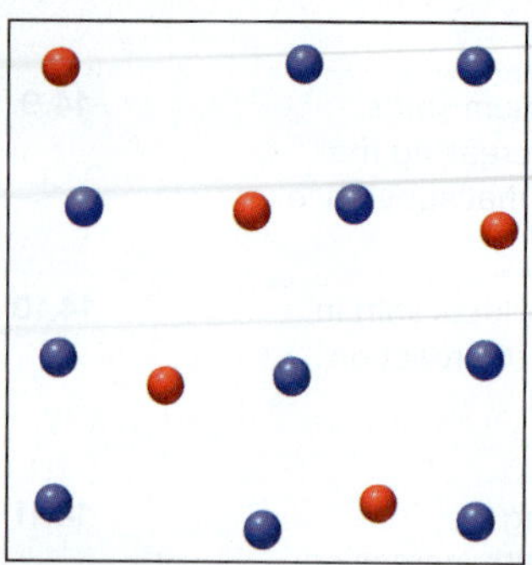

**FIGURE P14.3**

14.4. Figure P14.4 represents a mixture of reactants A (red spheres) and B (blue spheres) and product AB. Each sphere represents a partial pressure of 0.1 atm for that reactant or product. If the value of $K_p$ for the reaction $A(g) + B(g) \rightleftharpoons AB(g)$ is 2.0, is the mixture at equilibrium? If not, in what direction (to the left or to the right) will the system shift to attain equilibrium?

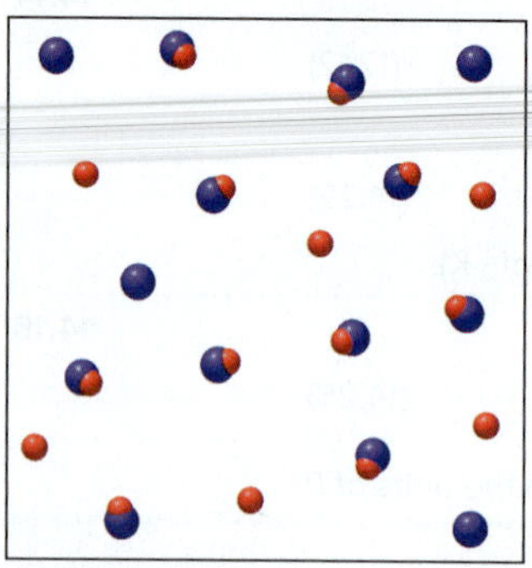

**FIGURE P14.4**

**14.5.** The left and right diagrams in Figure P14.5 represent equilibrium states of the gas-phase reaction

$$A \text{ (red spheres)} + B \text{ (blue spheres)} \rightleftharpoons AB$$

at 300 K and 400 K, respectively. Is this reaction endothermic or exothermic? Explain.

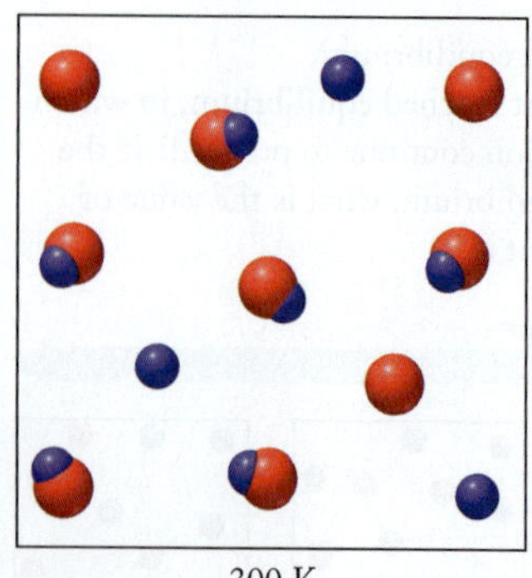


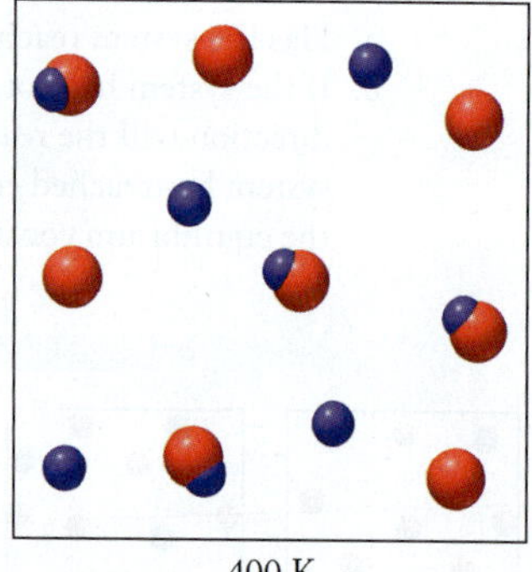


**FIGURE P14.5**

14.6. Figure P14.6 shows a plot of $\ln K_c$ versus $1/T$ for the reaction $A(g) + 2\,B(g) \rightleftharpoons AB_2(g)$. Is the reaction endothermic or exothermic?

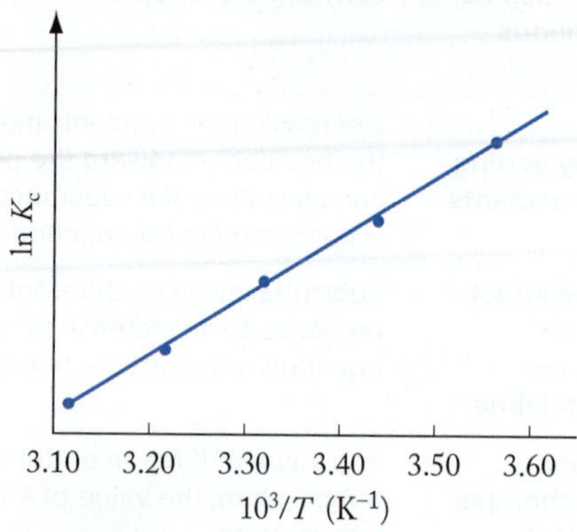


**FIGURE P14.6**

**14.7.** Use the graph in Figure P14.7 to estimate the value of $K_c$ for the following reaction:

$$N_2O(g) \rightleftharpoons N_2(g) + \tfrac{1}{2}O_2(g)$$

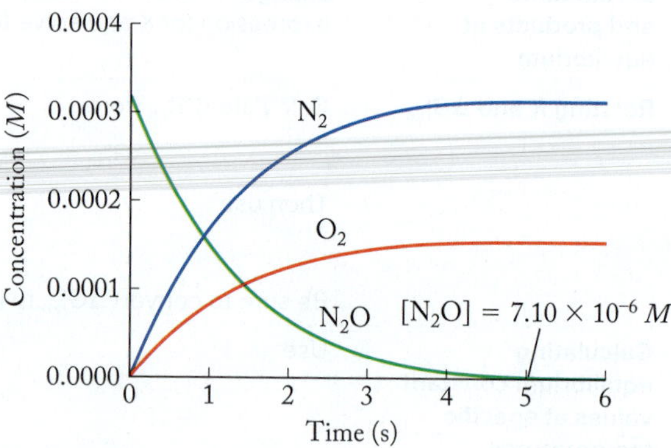


**FIGURE P14.7**

14.8. Use the data in Figure P14.8 to estimate the value of $K_c$ for the following reaction:

$$2\,NO(g) + O_2(g) \rightleftharpoons 2\,NO_2(g)$$

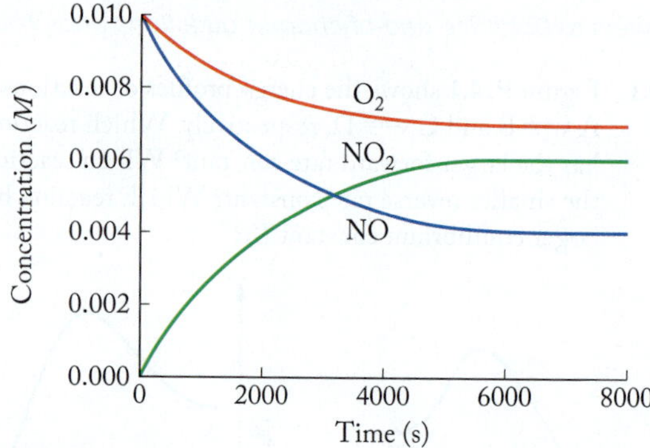


**FIGURE P14.8**

**14.9.** The organic molecule on the left in Figure P14.9 is a β-diketone (A). It exists in equilibrium with its isomer (B), called an enol (alk**ene**-alcoh**ol**). Molecules of A are represented by red spheres and molecules of B by blue spheres. The boxes to the left of the reaction arrows in (a) and (b) contain 26 and 28 molecules of A, respectively.

a. Fill in the empty box with the appropriate number of molecules of each type at equilibrium if the equilibrium constant for the isomerization is $K = 0.3$.

b. Fill in the empty box with the appropriate number of molecules of each type at equilibrium if the equilibrium constant for the isomerization is $K = 3$.

A ⇌ B

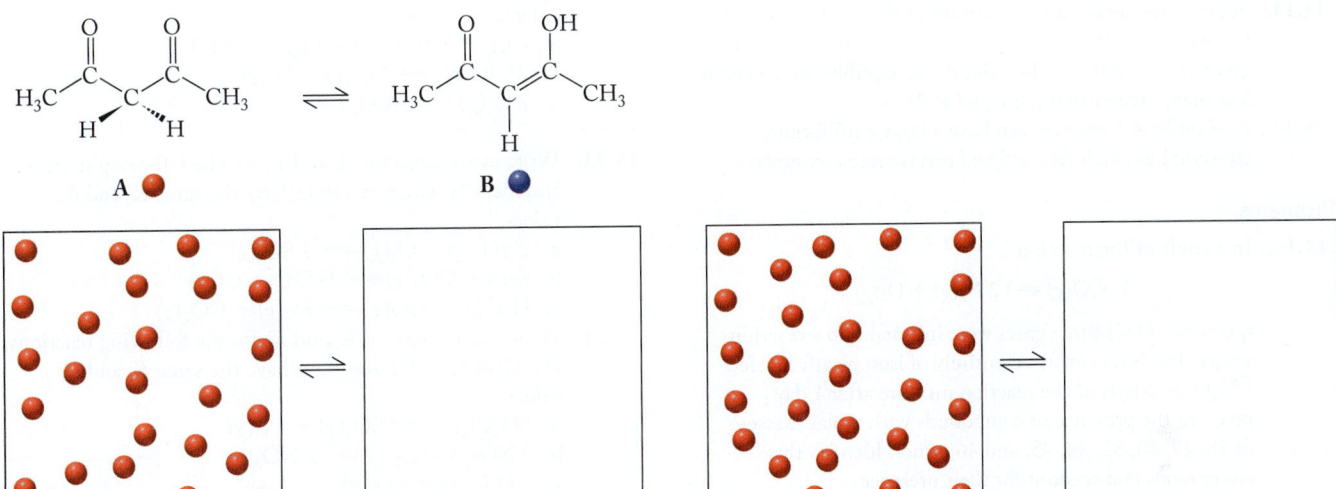

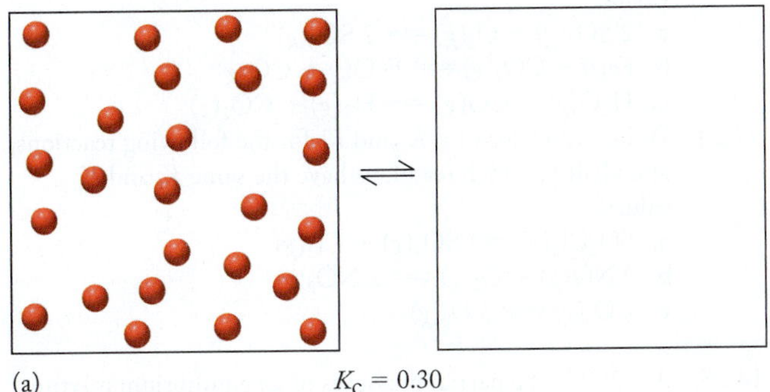


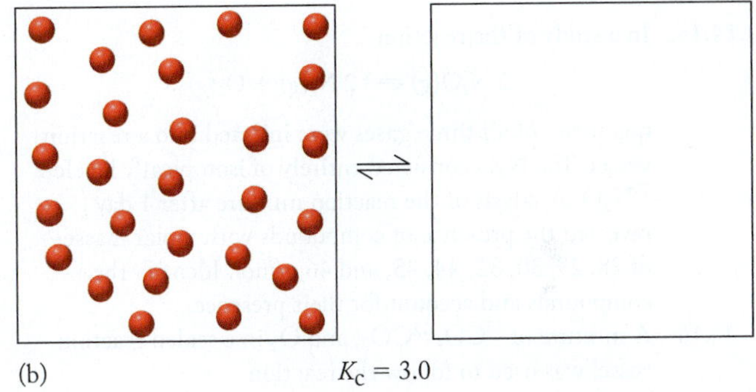


**FIGURE P14.9**

14.10. Use representations [A] through [I] in Figure P14.10 to answer questions a–f.

a. Write a balanced chemical equation describing the reversible reaction in [E], assuming that $NO_2$ is the reactant.

b. Identify one stress to the system that would spontaneously transform [E] to [C].

c. Whereas [E] is an equilibrium mixture, the reaction mixture in [C] is not yet at equilibrium. Which representation depicts the composition of [C] after equilibrium is reestablished?

d. Are [G] and [I] at equilibrium? How can you tell?

e. Given that $N_2O_4$ is colorless and $NO_2$ is brown, match the pictures of the syringes to their respective particulate images.

f. Does the syringe in [H] contain any $N_2O_4$?

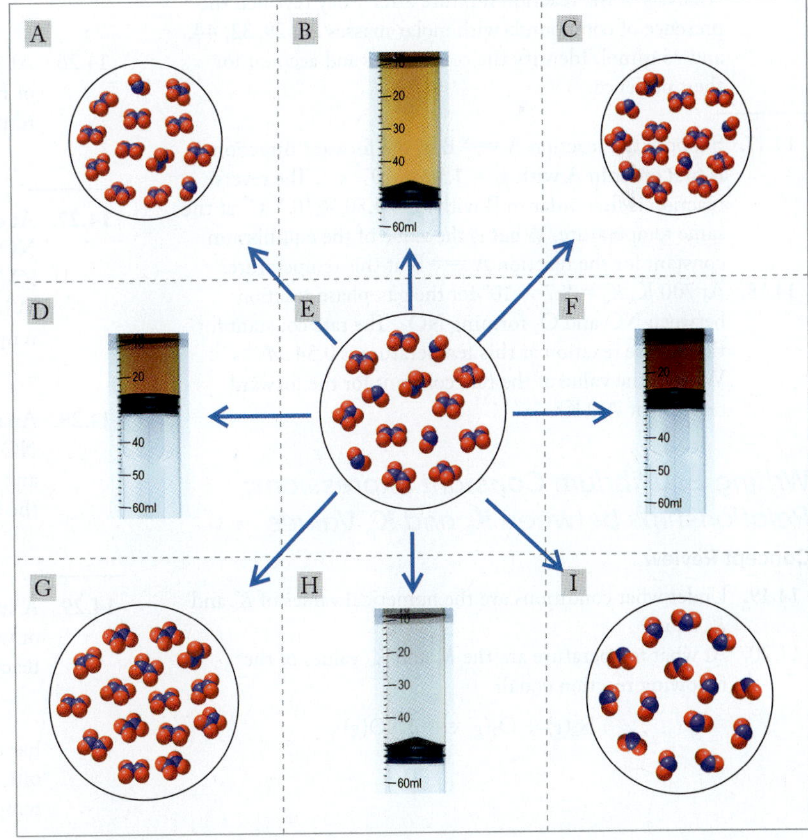


**FIGURE P14.10**

# QUESTIONS AND PROBLEMS

## The Dynamics of Chemical Equilibrium

### Concept Review

**14.11.** Can a reaction that is not reversible achieve chemical equilibrium?

14.12. At equilibrium, is the sum of the concentrations of all the reactants always equal to the sum of the concentrations of the products? Explain.

**14.13.** Suppose the forward rate constant of the reaction $A \rightleftharpoons B$ is greater than the rate constant of the reverse reaction at a given temperature. Is the value of the equilibrium constant less than, greater than, or equal to 1?

14.14. Explain how a reaction can have a large equilibrium constant but small forward and reverse rate constants.

### Problems

**14.15.** In a study of the reaction

$$2\ N_2O(g) \rightleftharpoons 2\ N_2(g) + O_2(g)$$

quantities of all three gases were injected into a reaction vessel. The $N_2O$ consisted entirely of isotopically labeled $^{15}N_2O$. Analysis of the reaction mixture after 1 day revealed the presence of compounds with molar masses of 28, 29, 30, 32, 44, 45, and 46 g/mol. Identify the compounds and account for their presence.

14.16. A mixture of $^{13}CO$, $^{12}CO_2$, and $O_2$ in a sealed reaction vessel was used to follow the reaction

$$2\ CO(g) + O_2(g) \rightleftharpoons 2\ CO_2(g)$$

Analysis of the reaction mixture after 1 day revealed the presence of compounds with molar masses 28, 29, 32, 44, and 45 g/mol. Identify the compounds and account for their presence.

**14.17.** Suppose the reaction $A \rightleftharpoons B$ in the forward direction is first order in A with $k_f = 1.50 \times 10^{-2}\ s^{-1}$. The reverse reaction is first order in B with $k_r = 4.50 \times 10^{-2}\ s^{-1}$ at the same temperature. What is the value of the equilibrium constant for the reaction $A \rightleftharpoons B$ at this temperature?

14.18. At 700 K, $K_c = 8.7 \times 10^6$ for the gas-phase reaction between NO and $O_2$ forming $NO_2$. The rate constant for the reverse reaction at this temperature is 0.54 $M^{-1}\ s^{-1}$. What is the value of the rate constant for the forward reaction at 700 K?

## Writing Equilibrium Constant Expressions; Relationships between $K_c$ and $K_p$ Values

### Concept Review

**14.19.** Under what conditions are the numerical values of $K_c$ and $K_p$ equal?

14.20. At what temperature are the $K_c$ and $K_p$ values of the following reaction equal?

$$N_2(g) + O_2(g) \rightleftharpoons 2\ NO(g)$$

### Problems

**14.21.** Write expressions for $K_c$ and $K_p$ for the following reactions, and identify which reactions have the same $K_c$ and $K_p$ values.

a. $N_2(g) + 2\ O_2(g) \rightleftharpoons N_2O_4(g)$
b. $3\ NO(g) \rightleftharpoons NO_2(g) + N_2O(g)$
c. $2\ N_2O(g) \rightleftharpoons 2\ N_2(g) + O_2(g)$

14.22. Write expressions for $K_c$ and $K_p$ for the following reactions, and identify which reactions have the same $K_c$ and $K_p$ values.

a. $Cl(g) + O_3(g) \rightleftharpoons ClO(g) + O_2(g)$
b. $2\ ClO(g) \rightleftharpoons 2\ Cl(g) + O_2(g)$
c. $2\ O_3(g) \rightleftharpoons 3\ O_2(g)$

**14.23.** Write expressions for $K_c$ and $K_p$ for the following reactions, and identify which reactions have the same $K_c$ and $K_p$ values.

a. $2\ SO_2(g) + O_2(g) \rightleftharpoons 2\ SO_3(g)$
b. $Fe(s) + CO_2(g) \rightleftharpoons FeO(s) + CO(g)$
c. $H_2O(g) + CO(g) \rightleftharpoons H_2(g) + CO_2(g)$

14.24. Write expressions for $K_c$ and $K_p$ for the following reactions, and identify which reactions have the same $K_c$ and $K_p$ values.

a. $SO_2Cl_2(g) \rightleftharpoons SO_2(g) + Cl_2(g)$
b. $2\ NO(g) + O_2(g) \rightleftharpoons 2\ NO_2(g)$
c. $2\ O_3(g) \rightleftharpoons 3\ O_2(g)$

**14.25.** At 1200 K the partial pressures of an equilibrium mixture of $H_2S$, $H_2$, and S are 0.020, 0.045, and 0.030 atm, respectively. Calculate the value of $K_p$ at 1200 K.

$$H_2S(g) \rightleftharpoons H_2(g) + S(g)$$

14.26. At 1045 K the partial pressures of an equilibrium mixture of $H_2O$, $H_2$, and $O_2$ are 0.040, 0.0045, and 0.0030 atm, respectively. Calculate the value of $K_p$ at 1045 K.

$$2\ H_2O(g) \rightleftharpoons 2\ H_2(g) + O_2(g)$$

**14.27.** At equilibrium, the concentrations of gaseous $N_2$, $O_2$, and NO in a sealed reaction vessel are $[N_2] = 3.3 \times 10^{-3}\ M$, $[O_2] = 5.8 \times 10^{-3}\ M$, and $[NO] = 3.1 \times 10^{-3}\ M$. What is the value of $K_c$ for the following reaction at the temperature of the mixture?

$$N_2(g) + O_2(g) \rightleftharpoons 2\ NO(g)$$

14.28. Analyses of an equilibrium mixture of gaseous $N_2O_4$ and $NO_2$ gave the following results: $[NO_2] = 4.2 \times 10^{-3}\ M$ and $[N_2O_4] = 2.9 \times 10^{-3}\ M$. What is the value of $K_c$ for the following reaction at the temperature of the mixture?

$$2\ NO_2(g) \rightleftharpoons N_2O_4(g)$$

**14.29.** A sealed reaction vessel initially contains $1.50 \times 10^{-2}$ mol of water vapor and $1.50 \times 10^{-2}$ mol of CO. After the reaction

$$H_2O(g) + CO(g) \rightleftharpoons H_2(g) + CO_2(g)$$

has come to equilibrium, the vessel contains $8.3 \times 10^{-3}$ mol of $CO_2$. What is the value of $K_c$ of the reaction at the temperature of the vessel?

*14.30. A 100 mL reaction vessel initially contains $2.60 \times 10^{-2}$ mol of NO and $1.30 \times 10^{-2}$ mol of $H_2$. At equilibrium, the concentration of NO in the vessel is 0.161 *M*. What is the value of $K_c$ for the following reaction?

$$2\,H_2(g) + 2\,NO(g) \rightleftharpoons 2\,H_2O(g) + N_2(g)$$

---

**14.31.** $K_p = 2.40$ for the following equilibrium at 100°C.

$$2\,NO(g) + Br_2(g) \rightleftharpoons 2\,NOBr(g)$$

What is the value of $K_c$ at the same temperature?

14.32. $K_p = 0.176$ for the following equilibrium at 1773 K.

$$CO(g) + 3\,H_2(g) \rightleftharpoons CH_4(g) + H_2O(g)$$

What is $K_c$ for this reaction at the same temperature?

---

**14.33.** The value of $K_p$ for the reaction

$$4\,HCl(g) + O_2(g) \rightleftharpoons 2\,Cl_2(g) + 2\,H_2O(g)$$

is 13.3 at 480°C. What is the value of $K_c$ at this temperature?

14.34. The value of $K_c$ for the synthesis of ammonia,

$$N_2(g) + 3\,H_2(g) \rightleftharpoons 2\,NH_3(g)$$

is 0.50 at 400°C. What is the value of $K_p$ at this temperature?

---

**14.35. Bulletproof Glass** Phosgene ($COCl_2$) is used in manufacturing foam rubber and bulletproof glass. Phosgene is formed from carbon monoxide and chlorine in the following reaction:

$$Cl_2(g) + CO(g) \rightleftharpoons COCl_2(g)$$

The value of $K_c$ for the reaction is 5.0 at 327°C. What is the value of $K_p$ at 327°C?

14.36. If the value of $K_p$ for the following reaction

$$SO_2(g) + NO_2(g) \rightleftharpoons NO(g) + SO_3(g)$$

is 3.45 at 298 K, what is the value of $K_c$ for the reaction at this temperature?

## Manipulating Equilibrium Constant Expressions

### Concept Review

**14.37.** How is the value of the equilibrium constant affected by scaling up or down the coefficients of the reactants and products in the chemical equation describing the reaction?

14.38. Is the numerical value of $K_p$ for the reaction

$$H_2(g) + I_2(g) \rightleftharpoons 2\,HI(g)$$

greater than, equal to, or less than the value of the equilibrium constant for the following reaction?

$$\tfrac{1}{2}\,H_2(g) + \tfrac{1}{2}\,I_2(g) \rightleftharpoons HI(g)$$

### Problems

**14.39.** $K_c = 120$ for the following reaction at 425 K:

$$I_2(g) + Br_2(g) \rightleftharpoons 2\,IBr(g)$$

What is the value of $K_c$ for the following reaction, also at 425 K?

$$\tfrac{1}{2}\,I_2(g) + \tfrac{1}{2}\,Br_2(g) \rightleftharpoons IBr(g)$$

14.40. The equilibrium constant $K_p$ for the synthesis of ammonia,

$$N_2(g) + 3\,H_2(g) \rightleftharpoons 2\,NH_3(g)$$

is $4.3 \times 10^{-3}$ at 300°C. What is the value of $K_p$ for the following equilibrium, also at 300°C?

$$\tfrac{1}{2}\,N_2(g) + \tfrac{3}{2}\,H_2(g) \rightleftharpoons NH_3(g)$$

---

**14.41.** The following reaction is one of the elementary steps in the oxidation of NO:

$$NO(g) + NO_3(g) \rightleftharpoons 2\,NO_2(g)$$

Write an expression for the equilibrium constant $K_c$ for this reaction and for the reverse reaction:

$$2\,NO_2(g) \rightleftharpoons NO(g) + NO_3(g)$$

How are the two $K_c$ expressions related?

14.42. At a given temperature, the equilibrium constant $K_c$ for the reaction

$$2\,NO(g) + 2\,H_2(g) \rightleftharpoons N_2(g) + 2\,H_2O(g)$$

is 0.11. What is the equilibrium constant $K_p$ for the following reaction?

$$NO(g) + H_2(g) \rightleftharpoons \tfrac{1}{2}\,N_2(g) + H_2O(g)$$

---

**14.43. Air Pollutants** Sulfur oxides are major air pollutants. The reaction between sulfur dioxide and oxygen can be written in two ways:

$$SO_2(g) + \tfrac{1}{2}\,O_2(g) \rightleftharpoons SO_3(g)$$

and

$$2\,SO_2(g) + O_2(g) \rightleftharpoons 2\,SO_3(g)$$

Write an equation relating the value of $K_c$ of the second reaction to the value of $K_p$ of the first reaction.

14.44. At a given temperature, $K_c$ for the reaction

$$2\,NO(g) + 2\,H_2(g) \rightleftharpoons N_2(g) + 2\,H_2O(g)$$

is 0.11. What is the value of $K_p$ for the following reaction at the same temperature?

$$N_2(g) + 2\,H_2O(g) \rightleftharpoons 2\,NO(g) + 2\,H_2(g)$$

---

**14.45.** At a given temperature, the value of $K_c$ for the reaction

$$2\,SO_2(g) + O_2(g) \rightleftharpoons 2\,SO_3(g)$$

is $2.4 \times 10^{-3}$. What is the value of $K_p$ for each of the following reactions at the same temperature?

a. $SO_2(g) + \tfrac{1}{2}\,O_2(g) \rightleftharpoons SO_3(g)$
b. $2\,SO_3(g) \rightleftharpoons 2\,SO_2(g) + O_2(g)$
c. $SO_3(g) \rightleftharpoons SO_2(g) + \tfrac{1}{2}\,O_2(g)$

14.46. If $K_c = 5 \times 10^{12}$ at a given temperature for the reaction

$$2\ NO(g) + O_2(g) \rightleftharpoons 2\ NO_2(g)$$

what is the value of $K_p$ for each of the following reactions at the same temperature?
a. $NO(g) + \frac{1}{2} O_2(g) \rightleftharpoons NO_2(g)$
b. $2\ NO_2(g) \rightleftharpoons 2\ NO(g) + O_2(g)$
c. $NO_2(g) \rightleftharpoons NO(g) + \frac{1}{2} O_2(g)$

**14.47.** Calculate the value of the equilibrium constant $K_c$ for the reaction

$$2\ D \rightleftharpoons A + 2\ B$$

given the following information:

$$A + 2\ B \rightleftharpoons C \qquad K_c = 3.3$$

$$C \rightleftharpoons 2\ D \qquad K_c = 0.041$$

14.48. **Reactions in Photochemical Smog** The brown gas responsible for the color in photochemical smog, $NO_2$, exists in equilibrium with its colorless dimer, $N_2O_4(g)$. In one of several possible decomposition reactions, $N_2O_4$ produces $N_2O$ and oxygen. Calculate the value of $K_c$ for the reaction

$$2\ N_2O_4(g) \rightleftharpoons 2\ N_2O(g) + 3\ O_2(g)$$

from the following information:

$$N_2(g) + \tfrac{1}{2} O_2(g) \rightleftharpoons N_2O(g) \qquad K_c = 2.7 \times 10^{-18}$$

$$N_2O_4(g) \rightleftharpoons 2\ NO_2(g) \qquad K_c = 4.6 \times 10^{-3}$$

$$N_2(g) + 2\ O_2(g) \rightleftharpoons 2\ NO_2(g) \qquad K_c = 1.7 \times 10^{-17}$$

## Equilibrium Constants and Reaction Quotients

### Concept Review

**14.49.** What is a reaction quotient?

14.50. As a reaction proceeds in the forward direction, how does the value of $Q$ change?

**14.51.** What does it mean when the reaction quotient $Q$ is numerically equal to the equilibrium constant $K$?

14.52. Explain how knowing $Q$ and $K$ for an equilibrium system enables you to say whether it is at equilibrium or whether it will shift in one direction or another.

### Problems

**14.53.** If $K_c = 22$ for the hypothetical reaction $A(g) \rightleftharpoons B(g)$ at a given temperature, and if $[A] = 0.10\ M$ and $[B] = 2.0\ M$ in a reaction mixture at that temperature, is the reaction at chemical equilibrium? If not, in which direction will the reaction proceed to reach equilibrium?

14.54. $K_c = 1.2 \times 10^3$ at 400°C for the reaction to produce the highly toxic nerve gas phosgene:

$$CO(g) + Cl_2(g) \rightleftharpoons COCl_2(g)$$

If equimolar amounts of CO, $Cl_2$, and $COCl_2$ are present at some point in the reaction in a 10.0 L reaction vessel, is the reaction at equilibrium? If not, in which direction will the reaction proceed to reach equilibrium?

**14.55.** Suppose that the value of the equilibrium constant $K_p$ of the following hypothetical reaction

$$A(g) + B(g) \rightleftharpoons C(g)$$

is 2.0 at a particular temperature. Which of the following reaction mixtures is at chemical equilibrium at that temperature?
a. $P_A = P_B = P_C = 2.0$ atm
b. $P_A = P_B = P_C = 0.5$ atm
c. $P_A = P_B = 1.0$ atm, $P_C = 2.0$ atm
d. $P_A = P_B = P_C = 0.0005$ atm

14.56. In which direction will the following hypothetical reaction proceed to reach equilibrium under each of the conditions given?

$$A(g) + B(g) \rightleftharpoons C(g) \qquad K_p = 1.00$$

a. $[A] = [B] = [C] = 1.0\ M$
b. $[A] = [B] = [C] = 2.0\ M$
c. $[A] = [B] = [C] = 0.5\ M$
d. $[A] = [B] = [C] = 1.0\ mM$

**14.57.** If $K_c = 1.5 \times 10^{-3}$ for the reaction

$$N_2(g) + O_2(g) \rightleftharpoons 2\ NO(g)$$

in which direction will the reaction proceed if the partial pressures of the three gases are all $1.00 \times 10^{-3}$ atm?

14.58. At 650 K, the value of $K_p$ for the ammonia synthesis reaction,

$$N_2(g) + 3\ H_2(g) \rightleftharpoons 2\ NH_3(g)$$

is $4.3 \times 10^{-4}$. If a vessel contains a reaction mixture in which $[N_2] = 0.010\ M$, $[H_2] = 0.030\ M$, and $[NH_3] = 0.00020\ M$, will more ammonia form?

**14.59.** The hypothetical equilibrium $X(aq) + Y(aq) \rightleftharpoons Z(aq)$ has $K_c = 1.00$ at 350 K. If the initial molar concentrations of X, Y, and Z in a solution are all 0.2 $M$, in which direction will the reaction shift to reach equilibrium?
a. To the left, making more X and Y
b. To the right, making more Z
c. The system is at equilibrium and the concentrations will not change.

*14.60. Reactions between carboxylic acids and alcohols to produce esters typically do not go to completion. Ethyl acetate, a compound used industrially to decaffeinate coffee and tea, is made in the following reaction, for which $K_c = 3.87$ at 75°C:

$$CH_3CH_2OH(\ell) + CH_3COOH(\ell) \rightleftharpoons CH_3COOCH_2CH_3(\ell) + H_2O(\ell)$$

a. Is a mixture of 125 g of ethanol ($CH_3CH_2OH$), 125 g of acetic acid ($CH_3COOH$), 125 g of ethyl acetate ($CH_3COOCH_2CH_3$), and 125 g of water in a 5 L reactor at equilibrium?
b. If not, in which direction will the reaction shift to reach equilibrium?

## Heterogeneous Equilibria

### Concept Review

**14.61.** Why doesn't the $K_c$ expression for the reaction

$$CaCO_3(s) \rightleftharpoons CaO(s) + CO_2(g)$$

contain terms for the concentrations of $CaCO_3$ and CaO?

14.62. How many partial pressure terms are in the $K_p$ expression for the thermal decomposition of sodium bicarbonate?

$$2\ NaHCO_3(s) \rightleftharpoons Na_2CO_3(s) + CO_2(g) + H_2O(g)$$

### Problems

**14.63.** Write $K_c$ expressions for the following reactions:

a. $CuS(s) \rightleftharpoons Cu^{2+}(aq) + S^{2-}(aq)$

b. $HCl(g) + H_2O(\ell) \rightleftharpoons H_3O^+(aq) + Cl^-(aq)$

c. $Al_2O_3(s) + 3\ H_2O(\ell) \rightleftharpoons 2\ Al^{3+}(aq) + 6\ OH^-(aq)$

d. $NH_3(g) + H_2O(\ell) \rightleftharpoons NH_4{}^+(aq) + OH^-(aq)$

14.64. Write $K_p$ expressions for the following reactions:

a. $CO_2(g) + C(s) \rightleftharpoons 2\ CO(g)$

b. $CO_2(g) + MgO(s) \rightleftharpoons MgCO_3(s)$

c. $HCl(g) + NH_3(g) \rightleftharpoons NH_4Cl(s)$

d. $Ca(HCO_3)_2(s) \rightleftharpoons CaCO_3(s) + CO_2(g) + H_2O(g)$

---

**14.65.** The value of $K_p$ for the reaction

$$CO_2(g) + C(s) \rightleftharpoons 2\ CO(g)$$

is 1.5 at 700°C. Calculate the equilibrium partial pressures of CO and $CO_2$ at 700°C if initially $P_{CO_2} = 5.0$ atm and $P_{CO} = 0.0$ atm. Pure graphite is present in excess.

14.66. **Jupiter's Atmosphere** Ammonium hydrogen sulfide ($NH_4SH$) has been detected in the atmosphere of Jupiter, where it probably exists in equilibrium with ammonia and hydrogen sulfide:

$$NH_4SH(s) \rightleftharpoons NH_3(g) + H_2S(g)$$

The value of $K_p$ for the reaction at 24°C is 0.126. Suppose a sealed flask at this temperature contains an equilibrium mixture of $NH_4SH$, $NH_3$, and $H_2S$. At equilibrium, the partial pressure of $H_2S$ is 0.355 atm. What is the partial pressure of $NH_3$?

---

**14.67.** **Pottery Glaze** Cobalt(II) oxide has been used for centuries as a glaze for pottery because of its deep blue color, which is known as cobalt blue. The oxide can be decomposed into cobalt metal by reduction with CO:

$$CoO(s) + CO(g) \rightleftharpoons Co(s) + CO_2(g)$$

The value of $K_c$ for this reaction at 500°C is $4.9 \times 10^2$.

a. What is the value of $K_p$ at 500°C?

b. If the total pressure in the reactor in which the oxide is being reduced is 6.50 atm, what are the partial pressures of CO and $CO_2$?

14.68. The value of $K_p$ for the reaction

$$CO_2(g) + H_2(g) \rightleftharpoons CO(g) + H_2O(\ell)$$

is $3.2 \times 10^{-4}$ at 25°C.

a. What is the value of $K_c$ at 25°C?

b. If the partial pressures of $CO_2$ and $H_2$ are the same and the total pressure of the reaction mixture is 5.00 atm, what are the partial pressures of each of the gases in a reaction mixture at equilibrium at 25°C?

## Le Châtelier's Principle

### Concept Review

**14.69.** How does manipulating the concentrations of the reactants and products in a reaction mixture affect the value of the equilibrium constant?

14.70. How does adding a reactant affect the rates of the forward and reverse reactions?

**14.71.** **Carbon Monoxide Poisoning** Patients suffering from carbon monoxide poisoning are treated with pure oxygen to remove CO from the hemoglobin (Hb) in their blood. The two relevant equilibria are

$$Hb + 4\ CO(g) \rightleftharpoons Hb(CO)_4$$

$$Hb + 4\ O_2(g) \rightleftharpoons Hb(O_2)_4$$

The value of the equilibrium constant for CO binding to Hb is greater than that for $O_2$. How, then, does this treatment work?

14.72. Is the equilibrium constant $K_p$ for the reaction

$$2\ NO_2(g) \rightleftharpoons N_2O_4(g)$$

in air the same in Los Angeles as in Denver if the atmospheric pressure in Denver is lower but the temperature is the same?

**14.73.** Henry's law (Chapter 11) predicts that the solubility of a gas in a liquid increases with its partial pressure. Explain Henry's law in relation to Le Châtelier's principle.

*14.74. For the reaction

$$2\ CO(g) + O_2(g) \rightleftharpoons 2\ CO_2(g)$$

why does adding an inert gas such as argon to an equilibrium mixture of CO, $O_2$, and $CO_2$ in a sealed vessel increase the total pressure of the system but not affect the position of the equilibrium?

### Problems

**14.75.** Which of the following equilibria will shift toward formation of more products if an equilibrium mixture is compressed into half its volume?

a. $2\ N_2O(g) \rightleftharpoons 2\ N_2(g) + O_2(g)$

b. $2\ CO(g) + O_2(g) \rightleftharpoons 2\ CO_2(g)$

c. $N_2(g) + O_2(g) \rightleftharpoons 2\ NO(g)$

d. $2\ NO(g) + O_2(g) \rightleftharpoons 2\ NO_2(g)$

14.76. Which of the following equilibria will shift toward formation of more products if the volume of a reaction mixture at equilibrium increases by a factor of 2?

a. $2\ SO_2(g) + O_2(g) \rightleftharpoons 2\ SO_3(g)$

b. $NO(g) + O_3(g) \rightleftharpoons NO_2(g) + O_2(g)$

c. $2\ N_2O_5(g) \rightleftharpoons 4\ NO_2(g) + O_2(g)$

d. $N_2O_4(g) \rightleftharpoons 2\ NO_2(g)$

**14.77.** What would be the effect of the changes listed on the equilibrium concentrations of reactants and products in the following reaction?

$$2\ O_3(g) \rightleftharpoons 3\ O_2(g)$$

a. $O_3$ is added to the system.

b. $O_2$ is added to the system.

c. The mixture is compressed to 1/10 its initial volume.

14.78. How will the changes listed affect the position of the following equilibrium?

$$2\ NO_2(g) \rightleftharpoons NO(g) + NO_3(g)$$

a. The concentration of NO is increased.

b. The concentration of $NO_2$ is increased.

c. The volume of the system is allowed to expand to 5 times its initial value.

**14.79.** How would reducing the partial pressure of $O_2(g)$ affect the position of the equilibrium in the following reaction?

$$2\ SO_2(g) + O_2(g) \rightleftharpoons 2\ SO_3(g)$$

*14.80. Ammonia is added to a gaseous reaction mixture containing $H_2$, $Cl_2$, and HCl that is at chemical equilibrium. How will the addition of ammonia affect the relative concentrations of $H_2$, $Cl_2$, and HCl if the equilibrium constant of reaction 2 is much greater than the equilibrium constant of reaction 1?

(1) $H_2(g) + Cl_2(g) \rightleftharpoons 2\ HCl(g)$

(2) $HCl(g) + NH_3(g) \rightleftharpoons NH_4Cl(s)$

**14.81.** In which of the following equilibria should the forward reaction be favored by an increase in temperature?

a. $ClO(g) + O_3(g) \rightleftharpoons Cl(g) + 2\ O_2(g)$ $\quad \Delta H^\circ_{rxn} = -29.9$ kJ

b. $2\ NO_2(g) \rightleftharpoons 2\ NO(g) + O_2(g)$ $\quad \Delta H^\circ_{rxn} = 112$ kJ

c. $2\ SO_2(g) + O_2(g) \rightleftharpoons 2\ SO_3(g)$ $\quad \Delta H^\circ_{rxn} = -196$ kJ

14.82. In which of the following equilibria should the forward reaction be favored by a decrease in temperature?

a. $CaO(g) + CO_2(g) \rightleftharpoons CaCO_3(s)$ $\quad \Delta H^\circ_{rxn} = -227.3$ kJ

b. $2\ NO_2(g) \rightleftharpoons N_2O_4\ (g)$ $\quad \Delta H^\circ_{rxn} = -57.2$ kJ

c. $2\ NOCl(g) \rightleftharpoons 2\ NO(g) + Cl_2(g)$ $\quad \Delta H^\circ_{rxn} = 77.2$ kJ

## Calculations Based on K

### Problems

**14.83.** Consider the following reaction:

$$PCl_5(g) \rightleftharpoons PCl_3(g) + Cl_2(g) \qquad K_p = 23.6 \text{ at } 500\ K$$

a. Calculate the equilibrium partial pressures of the reactants and products at 500 K if the initial pressures are $P_{PCl_5} = 0.560$ atm and $P_{PCl_3} = 0.500$ atm.

b. If more chlorine is added after equilibrium is reached, how will the concentrations of $PCl_5$ and $PCl_3$ change?

14.84. Enough $NO_2$ gas is injected into a cylindrical vessel to produce a partial pressure, $P_{NO_2}$, of 1.00 atm at 298 K. Calculate the equilibrium partial pressures of $NO_2$ and $N_2O_4$, given

$$2\ NO_2(g) \rightleftharpoons N_2O_4(g) \qquad K_p = 3.1 \text{ at } 298\ K$$

**14.85.** The value of $K_c$ for the reaction between water vapor and dichlorine monoxide

$$H_2O(g) + Cl_2O(g) \rightleftharpoons 2\ HOCl(g)$$

is 0.0900 at 25°C. Determine the equilibrium concentrations of all three compounds at 25°C if the starting concentrations of both reactants are 0.00432 *M* and no HOCl is present.

14.86. The value of $K_p$ for the reaction

$$3\ H_2(g) + N_2(g) \rightleftharpoons 2\ NH_3(g)$$

is $4.3 \times 10^{-4}$ at 648 K. Determine the equilibrium partial pressure of $NH_3$ in a reaction vessel that initially contained 0.900 atm $N_2$ and 0.500 atm $H_2$ at 648 K.

**14.87.** The value of $K_p$ for the reaction

$$NO(g) + \tfrac{1}{2}O_2(g) \rightleftharpoons NO_2(g)$$

is $2.3 \times 10^6$ at 25°C. At equilibrium, what is the ratio of $P_{NO_2}$ to $P_{NO}$ in air at 25°C? Assume that $P_{O_2} = 0.21$ atm and does not change.

*14.88. **Making Hydrogen Gas** Passing steam over hot carbon produces a mixture of carbon monoxide and hydrogen:

$$H_2O(g) + C(s) \rightleftharpoons CO(g) + H_2(g)$$

The value of $K_c$ for the reaction at 1000°C is $3.0 \times 10^{-2}$.

a. Calculate the equilibrium partial pressures of the products and reactants at 1000°C if $P_{H_2O} = 0.442$ atm and $P_{CO} = 5.0$ atm at the start of the reaction. Assume that the carbon is in excess.

b. Determine the equilibrium partial pressures of the reactants and products after enough CO and $H_2$ are added to the equilibrium mixture in part (a) to initially increase the partial pressures of both gases by 0.075 atm.

***14.89.** A flask containing pure $NO_2$ is heated to 1000 K, a temperature at which $K_p = 158$ for the decomposition of $NO_2$:

$$2\ NO_2(g) \rightleftharpoons 2\ NO(g) + O_2(g)$$

The partial pressure of $O_2$ at equilibrium is 0.136 atm.

a. Calculate the partial pressures of NO and $NO_2$.

b. Calculate the total pressure in the flask at equilibrium.

14.90. The equilibrium constant $K_p$ of the reaction

$$2\ SO_3(g) \rightleftharpoons 2\ SO_2(g) + O_2(g)$$

is 7.69 at 830°C. If a vessel at this temperature initially contains pure $SO_3$ and if the partial pressure of $SO_3$ at equilibrium is 0.100 atm, what is the partial pressure of $O_2$ in the flask at equilibrium?

*14.91. **$NO_x$ Pollution** In a study of the formation of $NO_x$ air pollution, a chamber heated to 2200°C was filled with air (0.79 atm $N_2$, 0.21 atm $O_2$). What are the equilibrium partial pressures of $N_2$, $O_2$, and NO if $K_p = 0.050$ for the following reaction at 2200°C?

$$N_2(g) + O_2(g) \rightleftharpoons 2\,NO(g)$$

*14.92. The equilibrium constant $K_p$ for the thermal decomposition of $NO_2$

$$2\,NO_2(g) \rightleftharpoons 2\,NO(g) + O_2(g)$$

is $6.5 \times 10^{-6}$ at 450°C. If a reaction vessel at this temperature initially contains 0.500 atm $NO_2$, what will the partial pressures of $NO_2$, NO, and $O_2$ be in the vessel when equilibrium has been attained?

**14.93.** The value of $K_c$ for the thermal decomposition of hydrogen sulfide

$$2\,H_2S(g) \rightleftharpoons 2\,H_2(g) + S_2(g)$$

is $2.2 \times 10^{-4}$ at 1400 K. A sample of gas in which $[H_2S] = 6.00\,M$ is heated to 1400 K in a sealed high-pressure vessel. After chemical equilibrium has been achieved, what is the value of $[H_2S]$? Assume that no $H_2$ or $S_2$ was present in the original sample.

14.94. **Urban Air** On a smoggy day, the equilibrium concentration of $NO_2$ in the air over an urban area reaches $2.2 \times 10^{-7}$ $M$. If the temperature of the air is 25°C, what is the concentration of the dimer $N_2O_4$ in the air?

$$N_2O_4(g) \rightleftharpoons 2\,NO_2(g) \qquad K_c = 6.1 \times 10^{-3} \qquad \text{at } 25°C$$

***14.95.** **Chemical Weapon** Phosgene, $COCl_2$, gained notoriety as a chemical weapon in World War I. Phosgene is produced by the reaction of carbon monoxide with chlorine:

$$CO(g) + Cl_2(g) \rightleftharpoons COCl_2(g)$$

$K_c = 5.0$ for this reaction at 600 K. What are the equilibrium partial pressures of the three gases if a reaction vessel at 600 K initially contains a mixture of the reactants in which $P_{CO} = P_{Cl_2} = 0.265$ atm and $P_{COCl_2} = 0.000$ atm?

*14.96. At 2000°C, $K_c = 1.0$ for the following reaction:

$$2\,CO(g) + O_2(g) \rightleftharpoons 2\,CO_2(g)$$

What is the ratio of [CO] to $[CO_2]$ in an atmosphere at 2000°C in which $[O_2] = 0.0045\,M$?

***14.97.** The water–gas shift reaction is an important source of hydrogen. The value of $K_c$ for the reaction

$$CO(g) + H_2O(g) \rightleftharpoons CO_2(g) + H_2(g)$$

at 700 K is 5.1. Calculate the equilibrium concentrations of the four gases if the initial concentration of each of them is 0.050 $M$ at 700 K.

*14.98. Sulfur dioxide reacts with $NO_2$, forming $SO_3$ and NO:

$$SO_2(g) + NO_2(g) \rightleftharpoons SO_3(g) + NO(g)$$

If $K_c = 2.50$ for this reaction at its temperature, what are the equilibrium concentrations of the products if the reaction mixture was initially 0.50 $M$ $SO_2$, 0.50 $M$ $NO_2$, 0.0050 $M$ $SO_3$, and 0.0050 $M$ NO?

**14.99.** Nitrogen dioxide and nitric oxide react to form dinitrogen trioxide. $K_p = 0.535$ for this reaction at 25°C:

$$NO(g) + NO_2(g) \rightleftharpoons N_2O_3(g)$$

If a 4.0 L flask at 25°C initially contains 15 g of NO and 69 g of $NO_2$, what are the concentrations of all species at equilibrium?

14.100. For the decomposition of HI($g$) into $H_2(g)$ and $I_2(g)$ at 400°C, $K_c = 0.0183$:

$$2\,HI(g) \rightleftharpoons H_2(g) + I_2(g)$$

If 80.0 g of HI($g$) is placed in a 2.5 L chamber at 400°C, what are the concentrations of all species when the system comes to equilibrium?

## Equilibrium and Thermodynamics

### Concept Review

**14.101.** Do all reactions with equilibrium constants <1 have values of $\Delta G^\circ_{rxn} > 0$?

*14.102. The equation $\Delta G^\circ_{rxn} = -RT \ln K$ relates the value of $K_p$, not $K_c$, to the change in standard free energy for a reaction in the gas phase. Explain why.

**14.103.** Starting with pure reactants, in which direction will an equilibrium shift if $\Delta G^\circ_{rxn} < 0$?

14.104. Starting with pure products, in which direction will an equilibrium shift if $\Delta G^\circ_{rxn} < 0$?

### Problems

**14.105.** Which of the following reactions has the largest equilibrium constant at 25°C?

a. $Cl_2(g) + F_2(g) \rightleftharpoons 2\,ClF(g)$ $\quad \Delta G^\circ_{rxn} = 115.4$ kJ
b. $Cl_2(g) + Br_2(g) \rightleftharpoons 2\,ClBr(g)$ $\quad \Delta G^\circ_{rxn} = -2.0$ kJ
c. $Cl_2(g) + I_2(g) \rightleftharpoons 2\,ICl(g)$ $\quad \Delta G^\circ_{rxn} = -27.9$ kJ

14.106. Which of the following reactions has the smallest equilibrium constant at 25°C?

a. $N_2(g) + 3\,H_2(g) \rightleftharpoons 2\,NH_3(g)$ $\quad \Delta G^\circ_{rxn} = -106.0$ kJ
b. $P_4(s) + 6\,H_2(g) \rightleftharpoons 4\,PH_3(g)$ $\quad \Delta G^\circ_{rxn} = 53.6$ kJ
c. $2\,As(s) + 3\,H_2(g) \rightleftharpoons 2\,AsH_3(g)$ $\quad \Delta G^\circ_{rxn} = 137.8$ kJ

**14.107.** Use the $\Delta G^\circ$ data given here to calculate the value of $K_p$ at 298 K for the following reaction:

$$N_2(g) + 2\,O_2(g) \rightleftharpoons 2\,NO_2(g)$$

$$N_2(g) + O_2(g) \rightleftharpoons 2\,NO(g) \qquad \Delta G^\circ_{rxn} = 173.2 \text{ kJ}$$

$$2\,NO(g) + O_2(g) \rightleftharpoons 2\,NO_2(g) \qquad \Delta G^\circ_{rxn} = -69.7 \text{ kJ}$$

14.108. Under the appropriate conditions, NO forms $N_2O$ and $NO_2$:

$$3\,NO(g) \rightleftharpoons N_2O(g) + NO_2(g)$$

Use the given $\Delta G^\circ$ values for the following reactions to calculate the value of $K_p$ for the preceding reaction at 500°C.

$$2\,NO(g) + O_2(g) \rightleftharpoons 2\,NO_2(g) \qquad \Delta G^\circ_{rxn} = -69.7 \text{ kJ}$$

$$2\,N_2O(g) \rightleftharpoons 2\,NO(g) + N_2(g) \qquad \Delta G^\circ_{rxn} = -33.8 \text{ kJ}$$

$$N_2(g) + O_2(g) \rightleftharpoons 2\,NO(g) \qquad \Delta G^\circ_{rxn} = 173.2 \text{ kJ}$$

*14.109. At a temperature of 1000 K, $SO_2(g)$ combines with oxygen to make $SO_3(g)$:

$$2\,SO_2(g) + O_2(g) \rightleftharpoons 2\,SO_3(g)$$

Under these conditions, $K_p = 3.4$.

a. Use the appropriate thermodynamic data in Appendix 4 to calculate the value of $\Delta H^\circ_{rxn}$ for this reaction.
b. What is the value of $K_p$ for this reaction at 298 K?
c. Use the answer from part (b) to calculate the value of $\Delta G^\circ_{rxn}$ at 298 K, and compare it to the value you obtained using the $\Delta G^\circ_{rxn}$ values in Appendix 4.

*14.110. The value of the equilibrium constant $K_p$ for the reaction

$$H_2(g) + CO_2(g) \rightleftharpoons H_2O(g) + CO(g)$$

is 0.534 at 700°C.

a. What is the value of $K_p$ for this reaction at 298 K? (*Hint*: To perform this $K_p$ calculation, you will need to calculate the value of $\Delta H^\circ_{rxn}$ by using the appropriate data from Appendix 4.)
b. Calculate the value of $\Delta G^\circ_{rxn}$ at 298 K by using your answer from part (a). Then compare it to the $\Delta G^\circ_{rxn}$ value you obtained using the $\Delta G^\circ_{rxn}$ values in Appendix 4.

## Changing K with Changing Temperature

### Concept Review

**14.111.** The value of the equilibrium constant of a reaction decreases with increasing temperature. Is this reaction endothermic or exothermic?

14.112. The reaction

$$2\,CO(g) + O_2(g) \rightleftharpoons 2\,CO_2(g)$$

is exothermic. Does the value of $K_p$ increase or decrease with increasing temperature?

**14.113.** The value of $K_p$ for the water–gas shift reaction

$$CO(g) + H_2O(g) \rightleftharpoons H_2(g) + CO_2(g)$$

increases as the temperature decreases. Is the reaction exothermic or endothermic?

14.114. Does the value of $K_p$ for the reaction

$$CH_4(g) + H_2O(g) \rightleftharpoons 3\,H_2(g) + CO(g) \qquad \Delta H^\circ_{rxn} = 206\text{ kJ}$$

increase, decrease, or remain unchanged as the temperature increases?

### Problems

**14.115. Air Pollution** Automobiles and trucks pollute the air with NO. At 2000°C, $K_c$ for the reaction

$$N_2(g) + O_2(g) \rightleftharpoons 2\,NO(g)$$

is $4.10 \times 10^{-4}$, and $\Delta H^\circ_{rxn} = 180.6$ kJ. What is the value of $K_c$ at 25°C?

14.116. At 400 K, the value of $K_p$ for the reaction

$$N_2(g) + 3\,H_2(g) \rightleftharpoons 2\,NH_3(g)$$

is 41, and $\Delta H^\circ = -92.2$ kJ. What is the value of $K_p$ at 700 K?

---

**14.117.** The equilibrium constant for the reaction

$$2\,NO(g) + O_2(g) \rightleftharpoons 2\,NO_2(g)$$

decreases from $1.5 \times 10^5$ at 430°C to 23 at 1000°C. From these data, calculate the value of $\Delta H^\circ_{rxn}$.

14.118. The value of $K_c$ for the reaction $A \rightleftharpoons B$ is 0.455 at 50°C and 0.655 at 100°C. Calculate $\Delta H^\circ_{rxn}$.

## Additional Problems

***14.119. CO as a Fuel** Is carbon dioxide a viable source of the fuel CO? Pure $CO_2$ decomposes into CO and $O_2$ at high temperatures:

$$2\,CO_2(g) \rightleftharpoons 2\,CO(g) + O_2(g)$$

The yield of the reaction increases with increasing temperature:

| Temperature (K) | Yield (%) |
|---|---|
| 1500 | 0.048 |
| 2500 | 17.6 |
| 3000 | 54.8 |

Is the reaction endothermic? Calculate the value of $K_p$ at each temperature and discuss the results. Is the decomposition of $CO_2$ an antidote for climate change?

14.120. Ammonia decomposes at high temperatures. In an experiment to explore this behavior, 2.00 mol of gaseous $NH_3$ is sealed in a rigid 1.00 L vessel. The vessel is heated to 785 K and some of the $NH_3$ decomposes in the following reaction:

$$2\,NH_3(g) \rightleftharpoons N_2(g) + 3\,H_2(g)$$

The system eventually reaches equilibrium and is found to contain 0.0040 mol of $NH_3$. What are the values of $K_p$ and $K_c$ for the decomposition reaction at 785 K?

***14.121.** Elements of group 16 form hydrides with the generic formula $H_2X$. When gaseous $H_2X$ is bubbled through a solution containing 0.3 $M$ hydrochloric acid, the solution becomes saturated and $[H_2X] = 0.1\ M$. The following equilibria exist in this solution:

$$H_2X(aq) + H_2O(\ell) \rightleftharpoons HX^-(aq) + H_3O^+(aq) \qquad K_1 = 8.3 \times 10^{-8}$$

$$HX^-(aq) + H_2O(\ell) \rightleftharpoons X^{2-}(aq) + H_3O^+(aq) \qquad K_2 = 1 \times 10^{-14}$$

Calculate the concentration of $X^{2-}$ in the solution.

*14.122. **Making Hydrogen** Debate continues on the practicality of using $H_2$ gas as a fuel for cars. The equilibrium constant $K_c$ for the reaction

$$CO(g) + H_2O(g) \rightleftharpoons CO_2(g) + H_2(g)$$

is $1.0 \times 10^5$ at 25°C. Starting with this value, calculate the value of $\Delta G^\circ_{rxn}$ at 25°C and, without doing any calculations, guess the sign of $\Delta H^\circ_{rxn}$.

***14.123.** **Air Pollution Control** Calcium oxide is used to remove the pollutant $SO_2$ from smokestack gases. The $\Delta G°$ of the overall reaction

$$CaO(s) + SO_2(g) + \tfrac{1}{2}O_2(g) \rightleftharpoons CaSO_4(s)$$

is −418.6 kJ. What is $P_{SO_2}$ in equilibrium with air ($P_{O_2} = 0.21$ atm) and solid CaO?

14.124. **Hydrogen Production** The steam–methane reforming reaction plays a key role in producing hydrogen gas for use as a fuel and as a reactant in ammonia production. The equilibrium constant ($K_p$) of the reaction $CH_4(g) + H_2O(g) \rightleftharpoons 3\,H_2(g) + CO(g)$ is 13.0 at 700°C.

a. Describe two advantages in running this endothermic ($\Delta H^\circ_{rxn} = 206$ kJ) reaction at 700°C instead of 100°C.

b. If the initial partial pressures of the two reactants are each 5.00 atm at 700°C and no products are present, what is the partial pressure of $H_2$ gas after equilibrium is achieved?

# 15

# Acid–Base Equilibria

## Proton Transfer in Biological Systems

**SHADES OF PINK AND BLUE AND IN BETWEEN** The colors of hydrangea blossoms depend on the acidity of the soil they grow in.

PARTICULATE **REVIEW**

### *Strong Acids versus Weak Acids*

In Chapter 15, we explore the properties of weak acids and bases. Two molecules, HCl and $CH_3COOH$, are shown here. Let's compare what happens when molecules of HCl(*g*) dissolve in water with what happens when molecules of $CH_3COOH(\ell)$ dissolve in water.

- What covalent bonds break as these molecules dissolve? What covalent bonds form?
- What intermolecular forces form between which particles?
- What are the primary particles in the two aqueous solutions?

(Review Sections 8.3 and 8.4 if you need help.)

*(Answers to Particulate Review questions are in the back of the book.)*

PARTICULATE **PREVIEW**

### *Two Acids and Two Bases*

The equilibrium present in an aqueous solution of ammonia is depicted here. As you read Chapter 15, look for ideas that will help you answer these questions:

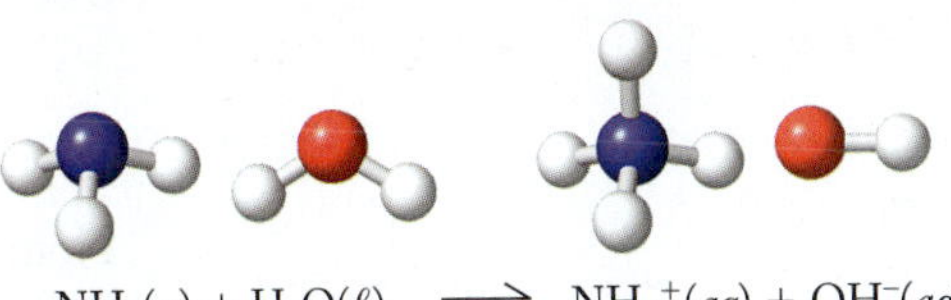


- Identify both the Brønsted–Lowry base and the Brønsted–Lowry acid in the forward reaction.
- What species functions as a Brønsted–Lowry base in the reverse reaction? Is a Brønsted–Lowry acid also present in the reverse reaction?
- Which species is the most abundant acid and which is the most abundant base present in the solution if $K << 1$ for the reaction as written?

## Learning Outcomes

**LO1** Relate the strengths of acids and bases to their $K_a$ and $K_b$ values
**Sample Exercise 15.1**

**LO2** Relate the strengths of acids to their molecular structures
**Sample Exercises 15.2, 15.3**

**LO3** Identify the structural features of strong and weak bases that facilitate proton transfer

**LO4** Recognize conjugate acid–base pairs and their complementary acid or base strengths
**Sample Exercises 15.4, 15.5**

**LO5** Interconvert $[H_3O^+]$, pH, pOH, and $[OH^-]$
**Sample Exercises 15.6–15.8**

**LO6** Relate the pH, percent ionization, and the principal solute particles present in solution to the $K_a$ values of weak acids and the $K_b$ values of weak bases
**Sample Exercises 15.9, 15.10**

**LO7** Relate the pH of solutions of strong and weak acids and bases to their $K$ values and the concentrations of their solutions
**Sample Exercises 15.11–15.14**

**LO8** Calculate the pH of solutions of polyprotic acids
**Sample Exercises 15.15, 15.16**

**LO9** Predict whether a salt is acidic, basic, or neutral, and calculate the pH of a salt solution
**Sample Exercises 15.17–15.19**

## 15.1 Acids and Bases: A Balancing Act

The pink and blue colors of the blossoms in the chapter-opening photograph are produced by the same species of plant (a hydrangea) grown in the same garden but in soils with different composition. The colors of hydrangea blossoms are controlled by the availability of aluminum ions ($Al^{3+}$) in the soil they grow in. Aluminum ions are soluble in acidic soils, and hydrangeas grown in acidic soils form blue blossoms. In neutral or slightly basic soils, however, aluminum ions precipitate as aluminum hydroxide and are unavailable. As a result, hydrangea blossoms grown in those soils lack blue pigment and are pink.

For many biological systems, including the human body, acid–base balance is vital to normal function and good health. Our blood, for example, is normally slightly basic, regulated by the proportions of $CO_2$ gas and bicarbonate ($HCO_3^-$) ions dissolved in it. This regulation happens because of the following reversible chemical reactions:

$$CO_2(g) + H_2O(\ell) \rightleftharpoons H_2CO_3(aq) \qquad (15.1)$$

$$H_2CO_3(aq) + H_2O(\ell) \rightleftharpoons HCO_3^-(aq) + H_3O^+(aq) \qquad (15.2)$$

Note how molecules of $CO_2$ and $H_2O$ combine to form molecules of carbonic acid ($H_2CO_3$), which donate $H^+$ ions to more water molecules, thereby forming bicarbonate and hydronium ($H_3O^+$) ions.

As we learned in Chapter 8, compounds that produce $H_3O^+$ ions when dissolved in water are acids, whereas compounds that produce hydroxide ions ($OH^-$) in aqueous solution are bases. More precisely, these compounds are referred to as **Arrhenius acids** and **Arrhenius bases**, respectively, in honor of Svante Arrhenius, whose research on the behavior of electrolytic solutions was recognized with the 1903 Nobel Prize in Chemistry.

**Arrhenius acid** a compound that produces $H_3O^+$ ions in aqueous solution.

**Arrhenius base** a compound that produces $OH^-$ ions in aqueous solution.

If we apply Le Châtelier's principle to Equation 15.2, the concentration of $H_3O^+$ ions will increase (as does the acidity of the solution) when the concentration of $H_2CO_3$ increases. In addition, the equilibrium described in Equation 15.1 tells us that the concentration of $H_2CO_3$ increases when the concentration of dissolved $CO_2$ increases. Therefore, an increase in the concentration of dissolved $CO_2$ shifts the equilibria in *both* Equations 15.1 and 15.2 to the right, making a solution of carbonic acid more acidic. By contrast, an increase in the concentration of $HCO_3^-$ ions shifts both equilibria to the left, decreasing the concentration of $H_3O^+$ ions and reducing acidity.

**CONNECTION** In Chapter 8, we learned that an aqueous solution's acidity is proportional to the concentration of $H_3O^+$ ions in it.

The equilibria in Equations 15.1 and 15.2 play a key role in human health. Normally, the concentration of $HCO_3^-$ ions in our blood is about 20 times the dissolved concentration of $CO_2$ because the dissociation of carbonic acid is not our only source of $HCO_3^-$ ions. The abundance of $HCO_3^-$ ions in blood drives down the concentration of $H_3O^+$ ions to below $10^{-7}$ *M*. As we discuss later in this chapter, a $[H_3O^+]$ value slightly below $10^{-7}$ *M* means that our blood is slightly basic.

Unfortunately, some people suffer from medical conditions that alter the normal balance of $CO_2$ and $HCO_3^-$ in their blood. For example, chronic lung disease impairs the ability not only to inhale $O_2$ but also to exhale $CO_2$. As a result, $CO_2$ builds up in the blood and in the tissues where it is produced. The result is a condition called *respiratory acidosis*. Its symptoms include fatigue, lethargy, and shortness of breath. In severe cases, the disease can be fatal.

**CONCEPT TEST**

When people hyperventilate, they breathe too rapidly, resulting in carbon dioxide being removed from the bloodstream more quickly than the body produces it through metabolism. How does this process affect the equilibria in Equations 15.1 and 15.2?

*(Answers to Concept Tests are in the back of the book.)*

In this chapter, we examine acid–base balance in environmental waters and in our bodies. We study why changes in this balance occur, some of the consequences of these changes, and what we can do to control them.

## 15.2 Acid Strength and Molecular Structure

The Arrhenius model of acids and bases explains many, but not all, of the properties of acidic and basic compounds in aqueous solutions. For example, Arrhenius acids produce $H_3O^+$ ions in aqueous solution, but these compounds do not contain $H_3O^+$ ions (or even $H^+$ ions) in their molecular structures. Arrhenius bases produce $OH^-$ ions in aqueous solution even though few of them include $OH^-$ ions in their structures. The desire to explain the macroscopic properties of acids and bases on the basis of their molecular structures led to the development of the *Brønsted–Lowry* model of acids and bases, which defines an acid as a substance that *donates* $H^+$ ions and a base as a substance that *accepts* $H^+$ ions.

**CONNECTION** We introduced acids, bases, and neutralization reactions in Section 8.4, where we defined Brønsted–Lowry acids as substances that donate $H^+$ ions and bases as substances that accept $H^+$ ions.

We introduced the Brønsted–Lowry model in Chapter 8, where we also defined *strong acids* as those that ionize completely in aqueous solution and *weak acids* as those that only partially ionize. Later in this chapter, we will discover that the degree of ionization of an acid depends on its concentration in solution: the more dilute it is, the higher its degree of ionization. Thus, even weak acids may ionize completely in very dilute solutions. This behavior requires us to refine our definitions of weak and strong acids.

Throughout this chapter and those that follow, an acid (HA) will be considered a strong acid if the equilibrium constant for the reaction

$$HA(aq) + H_2O(\ell) \rightleftharpoons A^-(aq) + H_3O^+(aq) \quad (15.3)$$

is much greater than 1:

$$K_a = \frac{[A^-][H_3O^+]}{[HA]} >> 1 \quad (15.4)$$

Here the subscript "a" identifies the equilibrium constant $K_a$ as an *a*cid ionization constant. Note that this $K_a$ expression has no $[H_2O]$ term. We do this with all aqueous equilibrium constants because the concentration of water in aqueous solutions is very large and rarely changes significantly due to reactions involving particles of solutes. Therefore, the value of $[H_2O]$ is a constant that is folded into equilibrium constant values for reactions in aqueous solutions. **Table 15.1** contains the names of the common strong acids, and their molecular structures,

**CONNECTION** The omission of $H_2O(\ell)$ from equilibrium expressions was discussed in Section 14.6.

**CONNECTION** The use of molecular models with colored surfaces to show variations in the distribution of valence electrons was introduced in Chapter 4 (see Figure 4.4).

**CHEMTOUR** Acid–Base Ionization

**CHEMTOUR** Acid Strength and Molecular Structure

**TABLE 15.1** The Structures and Acidic Properties of the Common Strong Acids

| Acid | Molecular Structure | Reaction in Water |
|---|---|---|
| Hydrochloric | | $HCl(aq) + H_2O(\ell) \rightarrow Cl^-(aq) + H_3O^+(aq)$ |
| Hydrobromic | | $HBr(aq) + H_2O(\ell) \rightarrow Br^-(aq) + H_3O^+(aq)$ |
| Hydroiodic | | $HI(aq) + H_2O(\ell) \rightarrow I^-(aq) + H_3O^+(aq)$ |
| Perchloric | | $HClO_4(aq) + H_2O(\ell) \rightarrow ClO_4^-(aq) + H_3O^+(aq)$ |
| Sulfuric | | $H_2SO_4(aq) + H_2O(\ell) \rightarrow HSO_4^-(aq) + H_3O^+(aq)$ |
| Nitric | | $HNO_3(aq) + H_2O(\ell) \rightarrow NO_3^-(aq) + H_3O^+(aq)$ |

**TABLE 15.2** The Molecular Structures and $K_a$ Values of Some Common Weak Acids

| Acid | Molecular Structure | $K_a$ |
|---|---|---|
| Acetic | $H_3C$—C(=O)—O—H | $1.76 \times 10^{-5}$ |
| Formic | H—C(=O)—O—H | $1.77 \times 10^{-4}$ |
| Hydrofluoric | H—F | $6.8 \times 10^{-4}$ |
| Hypochlorous | H—O—Cl | $2.9 \times 10^{-8}$ |
| Nitrous | H—O—N=O | $4.0 \times 10^{-4}$ |

including colored surfaces that show where valence electron densities are particularly high (red regions) or low (blue regions).

Weak acids have $K_a$ values that are smaller, often much smaller, than 1. The smaller a $K_a$ value, the weaker the acid. **Table 15.2** contains the molecular structures and $K_a$ values at 25°C of several common weak acids, and ionizable H atoms are shown in red. In this table and throughout this chapter, values of $K$ are determined at 25°C unless otherwise specified.

**SAMPLE EXERCISE 15.1** Relating $K_a$ and Acid Strength **LO1**

Suppose 0.100 $M$ aqueous solutions of each of the weak acids in Table 15.2 are prepared. Rank them from most acidic to least acidic.

**Collect, Organize, and Analyze** The acidity of an aqueous solution is proportional to the concentration of $H_3O^+$ ions in it, and the higher the value of $K_a$, the higher the concentration of $H_3O^+$ ions produced by a given concentration of acid.

**Solve** Ranking the acids in order of decreasing $K_a$ values, we have hydrofluoric acid > nitrous acid > formic acid > acetic acid > hypochlorous acid.

**Think About It** The list of weak acids includes four in which the ionizable H atom is bonded to an O atom and one in which it is bonded to an F atom. The fact that all these H atoms are bonded to atoms of two of the most electronegative elements in the periodic table is not a coincidence, as we will see later in this section.

**Practice Exercise** Rank the following acids in order of decreasing acid strength.

| Weak Acid | $K_a$ |
|---|---|
| $HN_3$ | $1.9 \times 10^{-5}$ |
| $CH_2ClCOOH$ | $1.4 \times 10^{-3}$ |
| HCN | $6.2 \times 10^{-10}$ |
| $HClO_2$ | $1.1 \times 10^{-2}$ |
| $C_6H_5COOH$ | $6.5 \times 10^{-5}$ |

*(Answers to Practice Exercises are in the back of the book.)*

## Strengths of Binary Acids

Let's examine the molecular structures of the strong acids in Table 15.1 to explore what makes them strong. Two types of acids appear in the table. The first group includes three binary acids with the generic formula HX, where X is a group 17 element: Cl, Br, or I. Hydrofluoric acid is not in the table because HF is a weak acid, which is reflected in the small value of its acid ionization constant ($K_a = 6.8 \times 10^{-4}$).

The binary acids in Table 15.1 are listed in order of increasing atomic number of the group 17 element as well as increasing acid strength: HCl < HBr < HI. Why is this trend observed? We can begin to answer this question by investigating the polarity of the H—X bond, as shown for HCl in **Figure 15.1**. Unequal sharing of each bonding pair of electrons leaves the H atoms with partial positive charges and the Cl atoms with partial negative charges. This polarity leads to strong dipole–dipole interactions between the H atoms and the lone pairs of electrons on the O atoms in $H_2O$ molecules (represented by the dotted blue line in Figure 15.1). This interaction facilitates ionization of the H—Cl bond, producing a $Cl^-$ anion and a $H^+$ cation that bonds to a molecule of $H_2O$ to form a hydronium ($H_3O^+$) ion.

If bond polarity and the dipole–dipole interactions in Figure 15.1 were the only factors that determined the strengths of these binary acids, then HF should be strongest, followed by HCl, HBr, and HI, because the electronegativities of the halogens follow the trend F > Cl > Br > I. In other words, because the H—F bond is more polar than the H—I bond, the H—F bond should ionize more easily than the H—I bond. However, HF is a weak acid, and the relative strengths of the other acids trend in the opposite direction: HI is a stronger acid than HBr, which is in turn stronger than HCl. To explain this trend, we need to consider another important factor: the bond energies of H—X bonds, which *decrease* with increasing atomic number (**Table 15.3**). Weaker H—X bonds are

**TABLE 15.3** Average H—X Bond Dissociation Energies and $X^-$ Ionic Radii for Four Binary Acids

| Acid | Bond Energy (kJ/mol) | Ionic Radius (pm) |
|---|---|---|
| HF | 567 | 133 |
| HCl | 431 | 181 |
| HBr | 366 | 195 |
| HI | 299 | 216 |

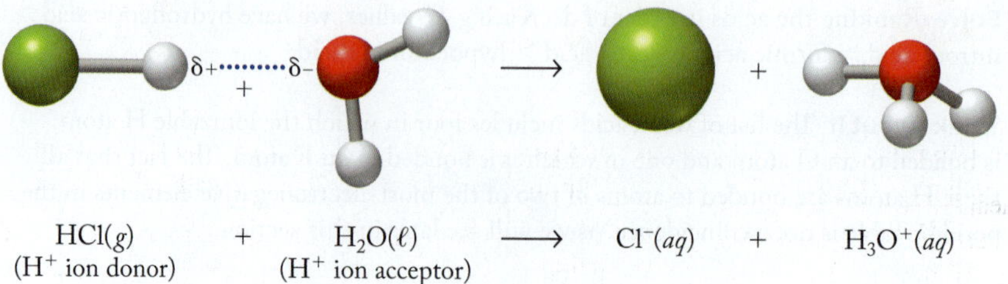


**FIGURE 15.1** When hydrogen chloride gas dissolves in water, each molecule of HCl ionizes by reacting with a molecule of $H_2O$, forming $Cl^-$ and $H_3O^+$ ions.

easier to break into $H^+$ and $X^-$ ions and therefore more readily form $H_3O^+$ ions, as shown in Figure 15.1. Thus, the observed strengths of the binary acids of the halogens result from two competing factors: the polarity of the H—X bond and the bond dissociation energy of the H—X bond.

**CONNECTION** The magnitude of the electrostatic potential energy between pairs of ions is inversely proportional to the distance between their ion centers (Section 4.1).

CONCEPT **TEST**

The strengths of binary acids increase with increasing atomic number among the elements in the same group of nonmetals, such as those in group 17 (**Figure 15.2**). Acid strength also increases with increasing atomic number across a row, such as the hydrides of the nonmetals in row 3. Explain why.

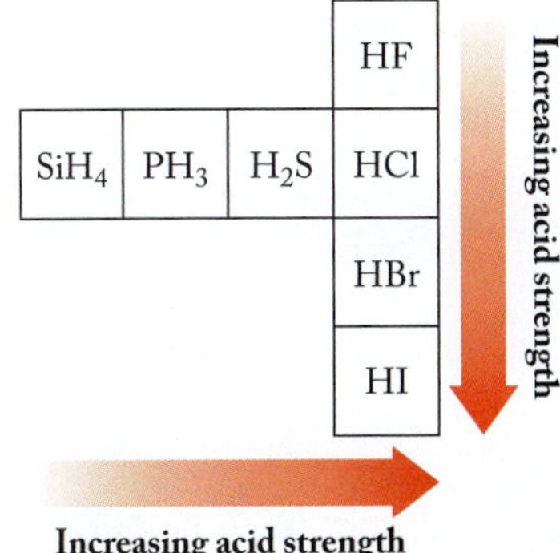


**FIGURE 15.2** Trends in increasing acid strength.

## Oxoacids

The second group of strong acids in Table 15.1 is composed of oxoacids. Most oxoacids are not strong acids, but those in Table 15.1 are. To understand why only a few oxoacids are strong and most are weak, we need to expand the preceding discussion of factors that influence the strengths of acids.

Let's begin by considering two oxoacids (**Figure 15.3**) that differ only in their number of O atoms per molecule: $HNO_3$, a strong acid, and $HNO_2$, a weak acid. The greater strength of $HNO_3$ is reflected in the bar graph in Figure 15.3(a), which shows that essentially all the $HNO_3$ molecules in a 0.100 *M* aqueous solution are ionized:

$$HNO_3(\ell) + H_2O(\ell) \rightarrow NO_3^-(aq) + H_3O^+(aq) \quad (15.5)$$

**CONNECTION** The electronegativities of the elements are given in Figure 4.5.

In contrast, only 6.1% of the nitrous acid molecules in 0.100 *M* $HNO_2$ are ionized (Figure 15.3b). This behavior is consistent with the $K_a$ value of this weak acid:

**CONNECTION** The rules for naming oxoacids are described in Section 4.3.

$$HNO_2(\ell) + H_2O(\ell) \rightleftharpoons NO_2^-(aq) + H_3O^+(aq) \qquad K_a = 4.0 \times 10^{-4} \quad (15.6)$$

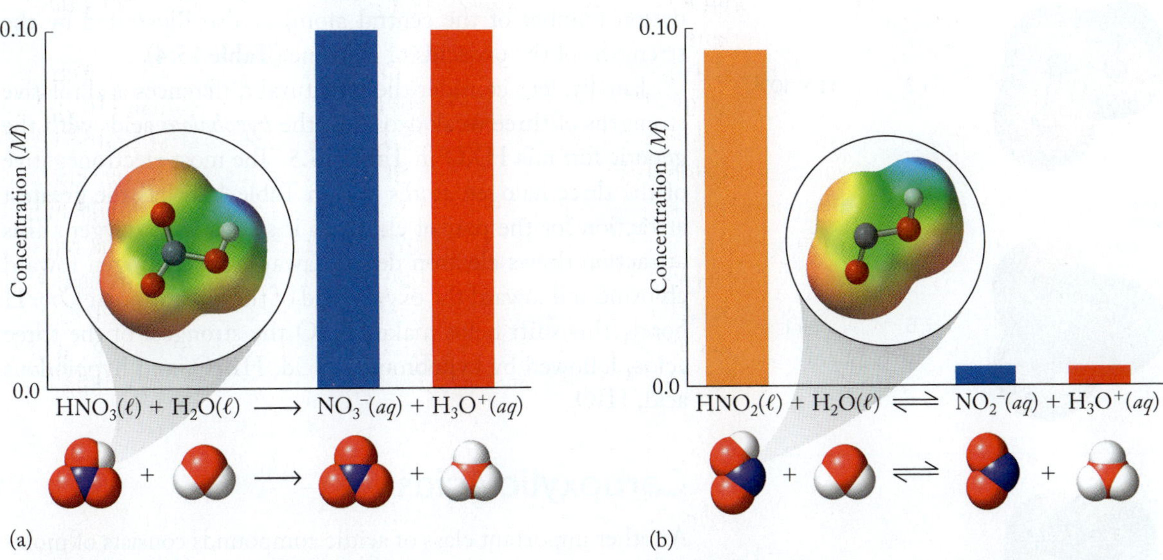


**FIGURE 15.3** Degrees of ionization of a strong acid versus a weak acid. (a) Essentially all the molecules of $HNO_3$ are ionized in a 0.100 *M* solution of the acid. (b) Only 6.1% of the nitrous acid molecules are ionized in 0.100 *M* $HNO_2$. The molecular structures show that the greater number of O atoms in molecules of $HNO_3$ than in $HNO_2$ draws more electron density away from the N atoms, stabilizing the $NO_3^-$ anion more than the $NO_2^-$ anion in aqueous solutions.

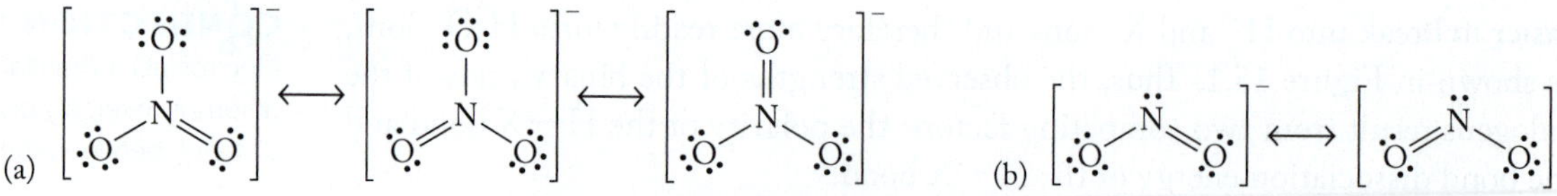

**FIGURE 15.4** The greater number of O atoms in a $NO_3^-$ ion (a) delocalize its negative charge more than in a $NO_2^-$ ion (b), which provides the $NO_3^-$ ion with more resonance stability and helps make $HNO_3$ a stronger acid.

Why is $HNO_3$ a stronger acid than $HNO_2$? The key is the electronegativity of oxygen and the presence of one more O atom in each molecule of $HNO_3$. The more O atoms bonded to the central atom of an oxoacid, the more that the electron density is pulled toward these atoms and away from the hydrogen end of the O—H bond. These shifts in electron density are depicted in the electron density maps surrounding the molecular structures in Figure 15.3. The red regions surrounding the three O atoms in $HNO_3$ indicate a greater shift in electron density away from the O—H bond than in in $HNO_2$, and that means stronger dipole–dipole interactions between the H atoms in molecules of $HNO_3$ and the O atoms in water molecules, and that means greater acid strength.

**CONNECTION** Delocalization and resonance stabilization are described in Section 4.5.

Another factor is the greater stability of the $NO_3^-$ ions, compared to the $NO_2^-$ ions, that are formed when the acids donate protons. Nitrate ions are more stable because they have more O atoms and therefore more ways to spread out, or delocalize, their negative charges. Greater delocalization can be visualized by the greater number of resonance Lewis structures that can be drawn for nitrate ions compared to nitrite ions, as shown in **Figure 15.4**. In Chapter 4, we explored resonance stabilization, which is achieved by delocalizing $\pi$ bonds over multiple pairs of atoms. In this chapter, the negative charges on oxoanions are similarly delocalized and with the same result: greater chemical stability. This trend of increasing acid strength with increasing numbers of oxygen atoms bonded to the central atom (that is, with increasing oxidation number of the central atom) is also illustrated by the strengths of the oxoacids of chlorine (**Table 15.4**).

**TABLE 15.4** Structures and $K_a$ Values of the Oxoacids of Chlorine

| Acid | Molecular Structure | Cl Oxidation State | $K_a$ |
|---|---|---|---|
| Hypochlorous HClO | 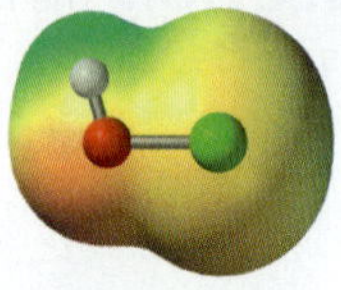 | +1 | $2.9 \times 10^{-8}$ |
| Chlorous $HClO_2$ | 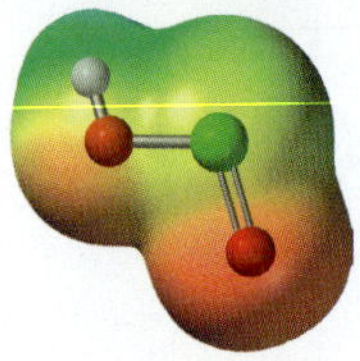 | +3 | $1.1 \times 10^{-2}$ |
| Chloric $HClO_3$ | 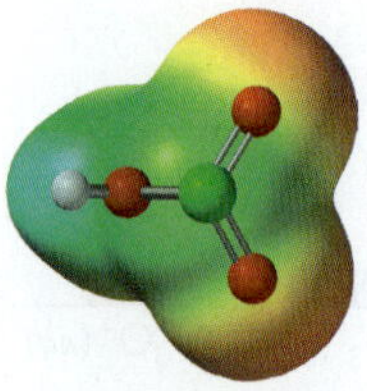 | +5 | ~1 |
| Perchloric $HClO_4$ | 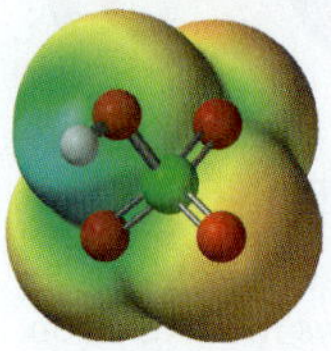 | +7 | Strong Acid |

Finally, let's consider the structural differences and relative strengths of three weak oxoacids, the *hypohalous* acids with the generic formula HXO in **Table 15.5**. The most electronegative of the three halogen atoms (Cl) in Table 15.5 has the greatest attraction for the pair of electrons it shares with oxygen. This attraction draws electron density away from hydrogen toward chlorine and toward the oxygen end of the already polar O—H bond. This shift helps make HClO the strongest of the three acids, followed by hypobromous acid, HBrO, and hypoiodous acid, HIO.

## Carboxylic Acids

Another important class of acidic compounds consists of molecules containing the carboxylic acid functional group, and the most common of these carboxylic acids is acetic acid. Lewis structures provide a view of the ionization of acetic acid (**Figure 15.5**) that illustrates how the strength of the acid is

**TABLE 15.5** Strengths of the Hypohalous Acids

| Acid | Molecular Structure | Electronegativity of the Halogen Atom | $K_a$ |
|---|---|---|---|
| Hypochlorous HClO | 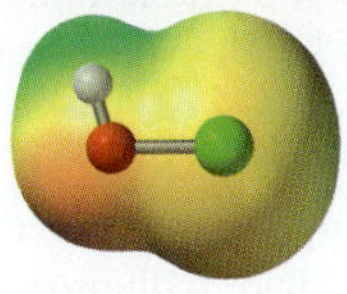 | 3.0 | $2.9 \times 10^{-8}$ |
| Hypobromous HBrO |  | 2.8 | $2.3 \times 10^{-9}$ |
| Hypoiodous HIO | 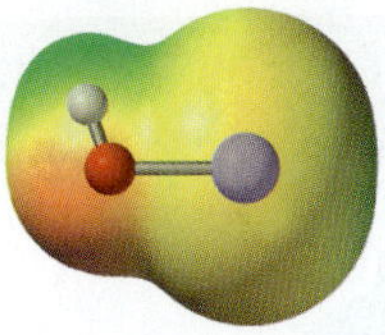 | 2.5 | $2.3 \times 10^{-11}$ |

$$CH_3COOH(aq) + H_2O(\ell) \rightleftharpoons CH_3COO^-(aq) + H_3O^+(aq)$$

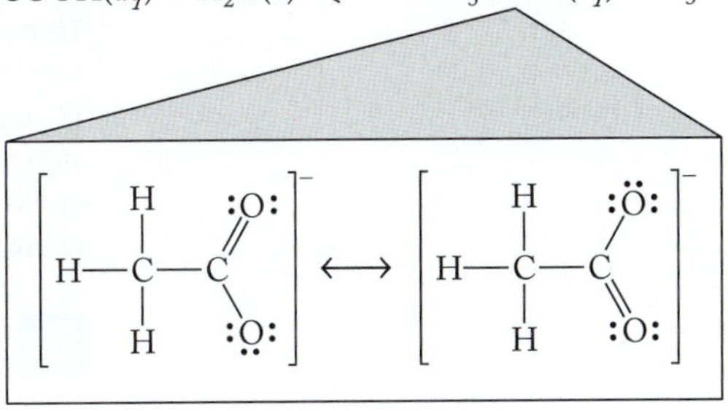

**FIGURE 15.5** Ionization of acetic acid.

linked to delocalization of the negative charge on the acetate ions it forms when it donates $H^+$ ions. The acid strengths of other carboxylate ions are enhanced by incorporating electronegative elements in their molecular structures, as shown in **Table 15.6**. Compare the $K_a$ value of acetic acid with those of the three acids beneath it in the table. Increasing the number of chlorine atoms pulls electron density away from the O—H bond, as represented by the red color on the O atoms in the carboxylic acid groups of these molecules. At the bottom of Table 15.6, trifluoroacetic acid is even stronger than trichloroacetic acid because fluorine is more electronegative than chlorine. Its atoms attract additional electron density toward their end of the molecule (note the yellow color surrounding the F atoms), away from the O—H bond.

**CONNECTION** The structure of the carboxylic acid functional group was introduced in Section 8.4.

**SAMPLE EXERCISE 15.2** Relating Acid Strength to Molecular Structure I — LO2

Rank the following compounds in order of decreasing acid strength: HBrO, $HBrO_2$, $HBrO_3$, $HBrO_4$, and $HClO_4$.

**Collect, Organize, and Analyze** All the compounds are oxoacids of group 17 elements. The strengths of these acids increase with increasing electronegativity of the central atom and with increasing numbers of oxygen atoms bonded to it, and chlorine is slightly more electronegative than bromine.

**Solve** The strengths of the four bromine oxoacids decrease in the order $HBrO_4 >$ $HBrO_3 > HBrO_2 >$ HBrO on the basis of the number of O atoms bonded to the central Br atoms. Because Cl is more electronegative than Br, $HClO_4$ should be stronger than $HBrO_4$. Therefore, the overall rank-ordered list is $HClO_4 > HBrO_4 >$ $HBrO_3 > HBrO_2 >$ HBrO.

**Think About It** Had we compared the strength of HCl and HBr instead of $HClO_4$ and $HBrO_4$, the ranking would have been reversed: HBr > HCl because the ionizable H atoms in these binary acids are bonded directly to the halogen atoms instead of O atoms. The bond dissociation energy of the H—Cl bond (Table 15.3) is 431 kJ/mol, or 431 − 366 = 65 kJ/mol stronger than the H—Br bond, making it much harder to break.

**Practice Exercise** Rank the following compounds in order of decreasing acid strength: fluoroacetic acid, difluoroacetic acid, trifluoroacetic acid, chloroacetic acid, and bromoacetic acid. Explain why you ranked the acids in the order that you did.

**TABLE 15.6** Molecular Structures and $K_a$ Values of Five Carboxylic Acids

| Acid | Molecular Structure | $K_a$ |
|---|---|---|
| Acetic $CH_3COOH$ | | $1.76 \times 10^{-5}$ |
| Chloroacetic $CH_2ClCOOH$ | | $1.4 \times 10^{-3}$ |
| Dichloroacetic $CHCl_2COOH$ | | $5.5 \times 10^{-2}$ |
| Trichloroacetic $CCl_3COOH$ | | $2.3 \times 10^{-1}$ |
| Trifluoroacetic $CF_3COOH$ | | $5.9 \times 10^{-1}$ |

**SAMPLE EXERCISE 15.3** Relating Acid Strength to Molecular Structure II **LO2**

Rank the following compounds in order of decreasing acid strength: $H_3PO_4$, $H_3AsO_4$, and $H_3SbO_4$.

**Collect and Organize** Each of these three oxoacids contains a different group 15 element. In each acid, the central atom is bonded to the same number of O atoms. The electronegativities of the elements decrease with increasing atomic number down a column in the periodic table.

**Analyze** In oxoacids, a more electronegative central atom draws electron density away from the H end of the O—H bond and helps disperse the negative charge that results when an oxoacid releases $H^+$ to form an oxoanion.

**Solve** Ranking the oxoacids in order of decreasing acid strength is a matter of ranking them in order of decreasing electronegativity of the central atom—that is, in order of increasing atomic number (and row number) in the periodic table: $H_3PO_4 > H_3AsO_4 > H_3SbO_4$.

**Think About It** This trend of decreasing oxoacid strength with increasing atomic number holds for other groups of nonmetals due to the same trend in electronegativity.

**Practice Exercise** Consider the following compounds: $H_2SeO_4$, $H_2SO_4$, $H_2SeO_3$, and $H_2SO_3$. Which is the strongest acid? Which is the weakest acid?

## 15.3 Strong and Weak Bases

Like acids, bases are also classified as either weak or strong. Among the most common strong bases are the hydroxides of the elements of groups 1 and 2. **Table 15.7** lists the ions produced by some of these compounds when they dissolve in water and dissociate completely. Note that one mole of a group 1 hydroxide produces one mole of $OH^-$ ions when it dissolves in water and that one mole of a group 2 hydroxide produces two moles of $OH^-$ ions. Some group 2 hydroxides have limited solubility, which limits their strength as bases and reduces the hazards associated with handling them compared to the group 1 bases. Indeed, many people ingest solid $Mg(OH)_2$ as an antacid.

Many bases don't contain hydroxide ions in their structures but produce them by reacting with water in solution. These compounds are strong bases if their

**TABLE 15.7** Some Strong Bases and Their Dissociation in Water

| Strong Base | Solid | Solvated Ions |
|---|---|---|
| Lithium hydroxide | $LiOH(s)$ | $Li^+(aq) + OH^-(aq)$ |
| Sodium hydroxide | $NaOH(s)$ | $Na^+(aq) + OH^-(aq)$ |
| Potassium hydroxide | $KOH(s)$ | $K^+(aq) + OH^-(aq)$ |
| Calcium hydroxide | $Ca(OH)_2(s)$ | $Ca^{2+}(aq) + 2\ OH^-(aq)$ |
| Barium hydroxide | $Ba(OH)_2(s)$ | $Ba^{2+}(aq) + 2\ OH^-(aq)$ |
| Strontium hydroxide | $Sr(OH)_2(s)$ | $Sr^{2+}(aq) + 2\ OH^-(aq)$ |

reactions with water are complete. They include soluble ionic oxides such as calcium oxide, CaO, which has the common name quicklime:

$$CaO(s) + H_2O(\ell) \rightarrow Ca^{2+}(aq) + 2\ OH^-(aq) \qquad (15.7)$$

The water-soluble group 1 and group 2 oxides are all strong bases.

Weak bases also produce $OH^-$ ions when they dissolve in and react with water, but their reactions reach equilibrium before all the base reacts. As we saw in Chapter 8, ammonia is a widely used weak base. Its basic properties are linked to the hydrogen bond that forms between the nitrogen atom in a molecule of $NH_3$ and a hydrogen atom in a molecule of $H_2O$. The strength of this interaction can deprotonate the water molecule and form a fourth N—H bond, transforming $NH_3$ into a **protonated** ammonium ($NH_4^+$) ion (**Figure 15.6**).

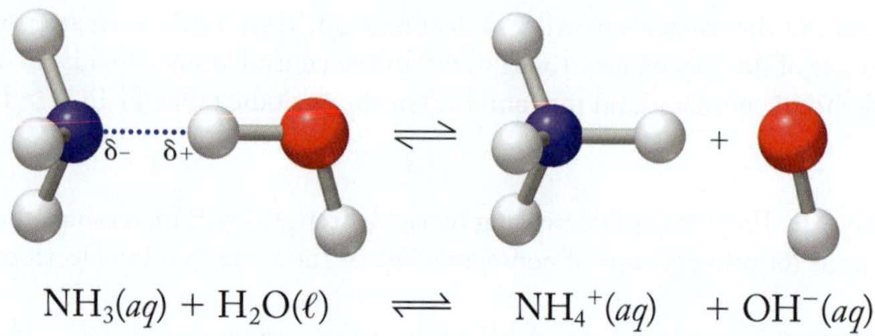


**FIGURE 15.6** Ammonia and water react to form ammonium ions and hydroxide ions.

The degree to which a protonation reaction proceeds once it reaches chemical equilibrium is reflected in the value of its equilibrium constant, $K_b$, where the "b" subscript describes a *b*ase ionization reaction. The $K_b$ expression and value for ammonia in water is

$$K_b = \frac{[NH_4^+][OH^-]}{[NH_3]} = 1.76 \times 10^{-5} \qquad (15.8)$$

As with $K_a$ expressions, we simplify $K_b$ expressions by omitting the $[H_2O]$ term because of its large and nearly constant value in most reactions in aqueous solution.

## Amines

Amines are organic compounds with functional groups that resemble $NH_3$, except that N atoms in amines are bonded to one or more carbon atoms. **Table 15.8** lists the $K_b$ values for several amines. Note that the $K_b$ value for the simplest amine, methylamine ($CH_3NH_2$), is more than 20 times larger than that of ammonia, $NH_3$. The N atom within $CH_3NH_2$ is more electronegative than the C and H atoms in the methyl group, which means that the N atom draws electron density away from the $-CH_3$ group and toward itself. It also draws pairs of electrons from the three N—H bonds in ammonia toward itself, but a methyl group has many more electrons than a H atom. The greater electron density around the N atom in $CH_3NH_2$ than around the N in $NH_3$ means that the lone pair of electrons on the N in $CH_3NH_2$ forms a stronger bond with a hydrogen ion, which makes $CH_3NH_2$ a stronger Brønsted–Lowry base. Ethylamine and dimethyl amine are even stronger bases than $CH_3NH_2$ because of the still larger electron densities around their N atoms.

**protonated** describes the molecular structure of a base that has accepted a proton ($H^+$).

Not all amines are stronger bases than ammonia. Some amines are weaker bases than ammonia because the atoms and bonds in their molecular structures draw electron density *away* from their nitrogen atoms. For example, the lone pair

**TABLE 15.8** Molecular Structures and $K_b$ Values of Five Weak Bases

| Base | Molecular Structure | $K_b$ |
|---|---|---|
| Ammonia $NH_3$ | | $1.76 \times 10^{-5}$ |
| Methylamine $CH_3NH_2$ | | $4.4 \times 10^{-4}$ |
| Dimethylamine $(CH_3)_2NH$ | | $5.9 \times 10^{-4}$ |
| Ethylamine $CH_3CH_2NH_2$ | | $5.6 \times 10^{-4}$ |
| Aniline (Phenylamine) $C_6H_5NH_2$ | | $4.0 \times 10^{-10}$ |

of electrons on the nitrogen atom in a molecule of phenylamine (Table 15.8) is drawn into the cloud of delocalized π electrons around the aromatic ring, which pulls the electrons away from the nitrogen atom and decreases its ability to accept hydrogen ions.

**CONCEPT TEST**

Is chloramine, $NH_2Cl$, a stronger or weaker base than ammonia? Explain your reasoning.

## 15.4 Conjugate Pairs

Let's revisit the ionization of nitrous acid, in which $HNO_2$ functions as a Brønsted–Lowry acid by donating $H^+$ ions to molecules of $H_2O$ (**Figure 15.7a**). In the reverse reaction, $NO_2^-$ ions accept $H^+$ ions and function as a Brønsted–Lowry base. Structurally, the difference between $HNO_2$ and $NO_2^-$ is the $H^+$ ion that a molecule of $HNO_2$ donates, forming a $NO_2^-$ ion, and that a $NO_2^-$

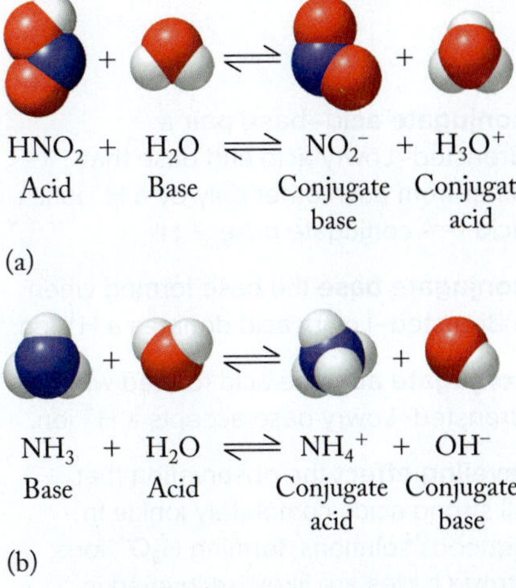


**FIGURE 15.7** Acid–base conjugate pairs.

**STEPWISE ANIMATION**

Conjugate Acids and Bases

ion accepts in the reverse reaction to re-form a molecule of $HNO_2$. An acid and a base related in this way are called a **conjugate acid–base pair**, or *conjugate pair.* An acid forms its **conjugate base** when it donates a $H^+$ ion, and a base forms its **conjugate acid** when it accepts a $H^+$ ion.

**Figure 15.7(b)** shows an ammonia molecule acting as a Brønsted–Lowry base, accepting a $H^+$ ion from a molecule of $H_2O$ and producing an ammonium ion and a hydroxide ion. When $NH_4^+$ ions donate $H^+$ ions, they function as Brønsted–Lowry acids. The structures of $NH_3$ molecules and $NH_4^+$ ions differ only by the $H^+$ ion that a molecule of $NH_3$ accepts and that an $NH_4^+$ ion donates, so $NH_4^+$ and $NH_3$ are another conjugate acid–base pair.

Is water an acid or a base? In Figure 15.7(a), $H_2O$ functions as a base because it accepts a $H^+$ ion from nitrous acid and forms $H_3O^+$, its conjugate acid. Thus, $H_2O$ and $H_3O^+$ are a conjugate pair. In Figure 15.7(b), however, $H_2O$ functions as an acid by donating a $H^+$ ion to ammonia to form the $OH^-$ ion, its conjugate base. Here, $H_2O$ and $OH^-$ are a conjugate pair. The two reactions in Figure 15.7 demonstrate that water can act as an acid or a base, depending on the acid–base properties of the substance dissolved in it.

The following generic chemical equations summarize the relationships between conjugate acid–base pairs in aqueous solutions:

$$\text{Acid}(aq) + H_2O(\ell) \rightleftharpoons \text{conjugate base}(aq) + H_3O^+(aq)$$

$$\text{Base}(aq) + H_2O(\ell) \rightleftharpoons \text{conjugate acid}(aq) + OH^-(aq)$$

The difference within each conjugate acid–base pair is the $H^+$ ion donated by the acid to form its conjugate base and that a base accepts to form its conjugate acid.

**SAMPLE EXERCISE 15.4** Identifying Conjugate Acid–Base Pairs **LO4**

Identify the conjugate acid–base pairs in the ionization reactions that occur when (a) perchloric acid ($HClO_4$) and (b) formic acid (HCOOH) dissolve in water.

**Collect, Organize, and Analyze** Brønsted–Lowry acids form their conjugate bases when they donate $H^+$ ions to molecules of $H_2O$. Therefore, the formulas of their conjugate bases are the formulas of the original acids minus a $H^+$ ion.

**Solve**

a. Perchloric acid, $HClO_4$, has only one H atom per molecule, so it must lose that one as a $H^+$ ion:

$$\underset{\text{Acid}}{HClO_4(aq)} + H_2O(\ell) \rightarrow \underset{\text{Conjugate base}}{ClO_4^-(aq)} + H_3O^+(aq)$$

b. Formic acid, HCOOH, has two H atoms per molecule, but only the one bonded to an O atom in the carboxylic acid group (see Table 15.2) is ionizable in water:

$$\underset{\text{Acid}}{HCOOH(aq)} + H_2O(\ell) \rightleftharpoons \underset{\text{Conjugate base}}{HCOO^-(aq)} + H_3O^+(aq)$$

In both reactions, $H_3O^+$ and $H_2O$ are also a conjugate acid–base pair: $H_2O$ is the base and $H_3O^+$ is its conjugate acid.

**Think About It** Perchloric acid is a strong acid and ionizes completely in aqueous solution. As a result, $ClO_4^-$ is such a weak base that the reverse reaction, in which $ClO_4^-$ adds a $H^+$ ion to re-form $HClO_4$, essentially never occurs. Therefore, a single arrow is used to depict the ionization reaction of $HClO_4$.

 **Practice Exercise** Identify the conjugate acid–base pairs in the reaction that occurs when acetic acid ($CH_3COOH$) dissolves in water.

**conjugate acid–base pair** a Brønsted–Lowry acid and base that differ from each other only by a $H^+$ ion: acid $\rightleftharpoons$ conjugate base + $H^+$.

**conjugate base** the base formed when a Brønsted–Lowry acid donates a $H^+$ ion.

**conjugate acid** the acid formed when a Brønsted–Lowry base accepts a $H^+$ ion.

**leveling effect** the observation that all strong acids completely ionize in aqueous solutions, forming $H_3O^+$ ions; strong bases are likewise leveled in water and are completely converted into their conjugate acids and $OH^-$ ions.

## Relative Strengths of Conjugate Acids and Bases

HCl is a strong acid and ionizes completely:

$$HCl(g) + H_2O(\ell) \rightarrow Cl^-(aq) + H_3O^+(aq) \quad (15.9)$$

Essentially, the reverse reaction does not happen, which means that the $Cl^-$ ion (the conjugate base of HCl) is an extremely weak base whose presence in an aqueous solution does not significantly affect acidity. This contrast in relative strengths applies to all conjugate pairs: strong acids have very weak conjugate bases and strong bases have very weak conjugate acids, as shown in **Figure 15.8**.

Between these extremes exist many weak acids with weak conjugate bases. For example, $HNO_2$ is a weak acid ($K_a = 4.0 \times 10^{-4}$), which means that its conjugate base, $NO_2^-$, is a weak base. This pairing of weak acids and weak bases applies to most of the conjugate acids and bases.

All the strong acids in Figure 15.8 ionize completely in water: essentially every molecule of acid transfers a $H^+$ ion to a molecule of $H_2O$. In this context, water is said to *level* the strengths of these acids when they dissolve in water: they all are equally strong because they cannot be more than 100% ionized. This **leveling effect** means that $H_3O^+$, the conjugate acid of $H_2O$, is the strongest acid that can exist in water. Any acid stronger than $H_3O^+$ simply donates all its ionizable H atoms to water molecules, forming $H_3O^+$ ions, when it dissolves in water. However, weak acids are differentiated by the fact that only a small fraction of the molecules donate their ionizable H atoms to water molecules. The weak acids higher on the list in Figure 15.8 form more acidic aqueous solutions than acids lower on the list because a greater fraction of their molecules ionize in water.

A similar leveling effect occurs for strong bases. The strongest base that can exist in water is the $OH^-$ ion, the conjugate base of $H_2O$. Any base stronger than $OH^-$, such as an oxide ion ($O^{2-}$), hydrolyzes in water, producing two $OH^-$ ions (**Figure 15.9**). The strengths of bases weaker than $OH^-$ ions can be differentiated by the fraction of their molecules that accept $H^+$ ions from water molecules in aqueous solutions. Weaker bases are higher on the list in Figure 15.8; stronger bases are lower on the list.

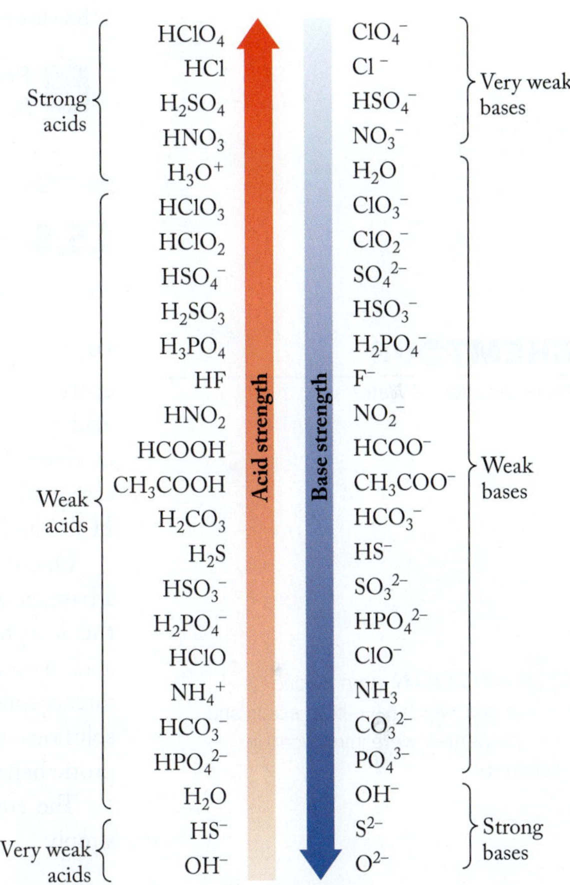


**FIGURE 15.8** Opposing trends characterize the relative strengths of acids and their conjugate bases: the stronger the acid, the weaker its conjugate base. The same is true for bases: the stronger the base, the weaker its conjugate acid.

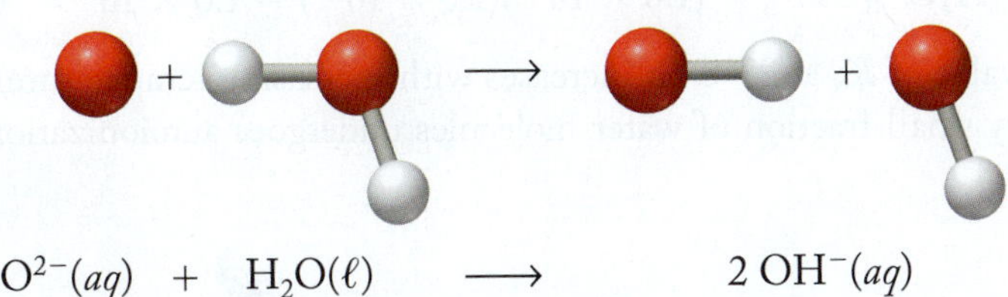


**FIGURE 15.9** The oxide ion hydrolyzes in water, producing two hydroxide ions.

**SAMPLE EXERCISE 15.5** Relating the Strengths of Conjugate Pairs **LO4**

List the following anions in order of decreasing strength as Brønsted–Lowry bases: $F^-$, $Cl^-$, $OH^-$, $HCOO^-$, and $NO_2^-$.

**Collect, Organize, and Analyze** All five anions are listed among the bases in Figure 15.8. Therefore, we can rank them according to their location in the figure: the strongest base will be the closest to the bottom, and the weakest will be the closest to the top.

**Solve** The anions in decreasing order of strength as bases ($H^+$ ion acceptors) are $OH^-$, $HCOO^-$, $NO_2^-$, $F^-$, and $Cl^-$.

**Think About It** According to Figure 15.8, we have also sorted the anions in increasing order for the strength of their conjugate acids. The complementary sorting makes sense because the stronger the acid, the weaker its conjugate base, and vice versa.

**Practice Exercise** List the following anions in order of decreasing strength as Brønsted–Lowry bases: $Br^-$, $S^{2-}$, $ClO^-$, $CH_3COO^-$, and $HSO_3^-$.

## 15.5 pH and the Autoionization of Water

**CHEMTOUR**
Autoionization of Water

We have seen that the acidity of an aqueous solution is directly related to its concentration of $H_3O^+$ ions. In this section, we examine another way to quantify acidity. First, though, we need to understand what is meant by the **autoionization** of water. It is a process that occurs between water molecules and results in their essentially ionizing each other, forming equal but very small concentrations of $H_3O^+$ and $OH^-$ ions in pure water (**Figure 15.10**).

One water molecule, acting as an acid, donates a $H^+$ ion; the other, acting as a base, accepts a $H^+$ ion. The donor molecule produces its conjugate base ($OH^-$); the acceptor molecule produces its conjugate acid ($H_3O^+$). We encountered this acid–base duality of water in Figure 15.7, in which molecules of $H_2O$ acted as $H^+$ ion acceptors (bases) in solutions of acidic solutes and as $H^+$ ion donors (acids) in solutions of basic solutes. The autoionization of water is an example of its amphiprotic behavior.

**CONNECTION** *Amphiprotic* compounds, which have both acidic and basic properties, were introduced in Section 8.4.

The equilibrium constant expression for the autoionization of water ($K_w$) is simply

$$K_w = [H_3O^+][OH^-] \quad (15.10)$$

because water concentration terms are not included in aqueous equilibria. In pure water at 25°C, $[H_3O^+] = [OH^-] = 1.0 \times 10^{-7}$ *M*. Inserting these values into Equation 15.10 gives

$$K_w = [H_3O^+][OH^-] = (1.0 \times 10^{-7})(1.0 \times 10^{-7}) = 1.0 \times 10^{-14} \quad (15.11)$$

The tiny value of $K_w$ at 25°C (it increases with increasing temperature) confirms that a very small fraction of water molecules undergoes autoionization at room

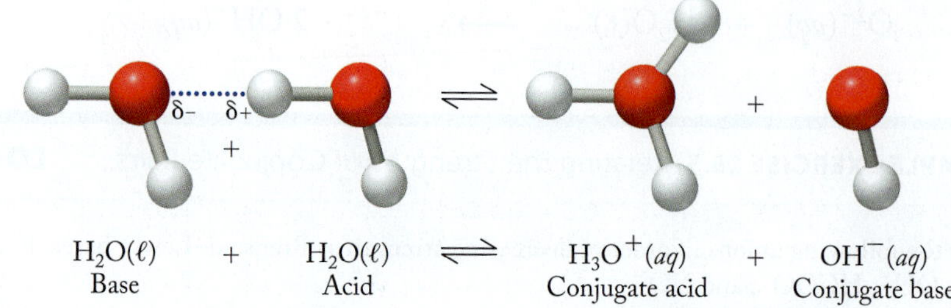


**FIGURE 15.10** The autoionization of water takes place when a proton is transferred from one water molecule to another.

temperature. The reverse of autoionization—the reaction between $H_3O^+$ and $OH^-$ to produce $H_2O$—has an equilibrium constant of $1/K_w = 1.0 \times 10^{14}$ at 25°C and essentially goes to completion:

$$H_3O^+(aq) + OH^-(aq) \rightleftharpoons 2\,H_2O(\ell) \qquad K = 1/K_w = 1.0 \times 10^{14}$$

Equation 15.11 means that an inverse relationship exists between $[H_3O^+]$ and $[OH^-]$ in any aqueous sample: as the concentration of one increases, the concentration of the other must decrease so that the product of the two is always $1.0 \times 10^{-14}$. A solution in which $[H_3O^+] > [OH^-]$ is acidic, a solution in which $[H_3O^+] < [OH^-]$ is basic, and a solution in which $[H_3O^+] = [OH^-] = 1.0 \times 10^{-7}\ M$ is neutral (neither acidic nor basic).

The tiny value of $K_w$ means that the autoionization of water does not contribute significantly to $[H_3O^+]$ in solutions of most acids or to $[OH^-]$ in solutions of most bases, so we can ignore the contribution of autoionization in most calculations of acid or base strength. However, if acids or bases are extremely weak, or if their concentrations are extremely low, $H_2O$ autoionization may need to be considered.

**autoionization** the process that produces equal and very small concentrations of $H_3O^+$ and $OH^-$ ions in pure water.

**pH** the negative logarithm of the hydronium ion concentration in an aqueous solution.

## The pH Scale

In the early 1900s, scientists developed a device called the *hydrogen electrode* to determine the concentrations of hydronium ions in solutions. The electrical voltage, or *potential*, produced by the hydrogen electrode is a linear function of the logarithm of $[H_3O^+]$. This relationship led Danish biochemist Søren Sørensen (1868–1939) to propose a scale for expressing acidity and basicity on the basis of what he called "the potential of the hydrogen ion," abbreviated **pH**. Mathematically, we define pH as the negative logarithm of $[H_3O^+]$:

$$\text{pH} = -\log[H_3O^+] \qquad (15.12)$$

For example, the pH of a solution in which $[H_3O^+] = 5.0 \times 10^{-3}\ M$ is

$$\text{pH} = -\log(5.0 \times 10^{-3}) = -(-2.30) = 2.30$$

Sørensen's pH scale has several attractive features. Because it is logarithmic, it uses no exponents, as are commonly encountered in values of $[H_3O^+]$. The logarithmic scale also means that a change of one pH unit corresponds to a 10-fold change in $[H_3O^+]$, so that a solution with a pH of 5.0 has 10 times the concentration of $H_3O^+$ ions and is 10 times as acidic as a solution with a pH of 6.0. Similarly, the concentration of $H_3O^+$ ions in a solution with a pH of 12.0 is 1/10 that of a solution with a pH of 11.0. Conversely, the concentration of $OH^-$ ions in a solution with a pH of 12.0 is 10 times that of a solution with a pH of 11.0.

The negative sign in front of the logarithmic term in Equation 15.12 means that pH values of most aqueous solutions are positive numbers between 0 and 14. It also means that *large* pH values correspond to *small* values of $[H_3O^+]$. Acidic solutions have pH values less than 7.00 ($[H_3O^+] > 1.0 \times 10^{-7}\ M$), and basic solutions have pH values greater than 7.00 ($[H_3O^+] < 1.0 \times 10^{-7}\ M$). A solution with a pH of exactly 7.00 is neutral. The pH values for some common aqueous solutions are shown in **Figure 15.11**.

**CHEMTOUR**
pH Scale

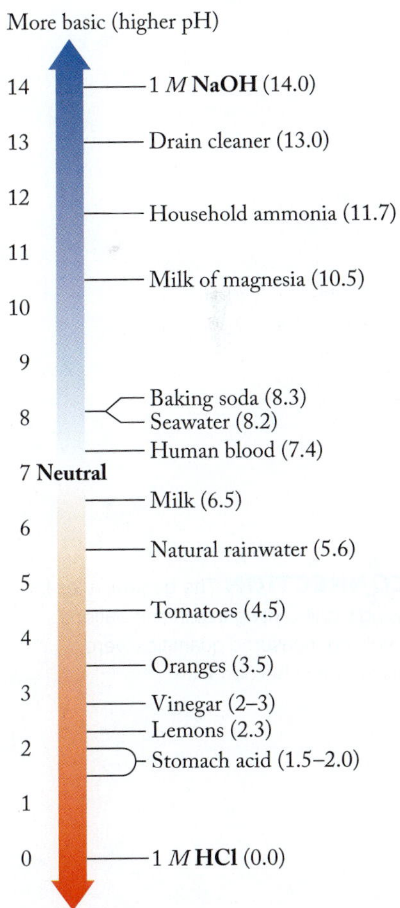


**FIGURE 15.11** The pH scale is a convenient way to express the range of acidic or basic properties of some common materials.

**CONCEPT TEST**

Identify each of these five pH values as strongly acidic, weakly acidic, weakly basic, strongly basic or neutral: a. 13.77 b. 10.03 c. 7.00 d. 4.37 e. 0.22.

**SAMPLE EXERCISE 15.6** Relating pH and $[H_3O^+]$ **LO5**

Suppose that the pH of solution A is 8.0 and the pH of solution B is 4.0. Which of the following statements about the two solutions are true, and which are false?

a. Solution A is 2 times as acidic as solution B.
b. Solution B is 10,000 times as acidic as solution A.
c. The concentration of $OH^-$ ions in solution A is half the concentration in solution B.
d. The values of $[OH^-]$ and $[H_3O^+]$ are closer to each other in solution A than in solution B.

**Collect, Organize, and Analyze** The negative sign in the pH formula ($pH = -\log[H_3O^+]$) means that a higher pH value corresponds to a (much) lower $[H_3O^+]$. The pH scale is logarithmic, so a decrease of one pH unit corresponds to a 10-fold increase in $[H_3O^+]$.

**Solve**

a. False, because pH is a log scale, and because higher pH values mean lower $[H_3O^+]$.
b. True, because a decrease of four pH units corresponds to a $10^4$ increase in $[H_3O^+]$ and acidity.
c. False, because pH is a log scale. Actually, $[OH^-]$ in solution A is $10^4$ times that of solution B.
d. True, because a pH of 8.0 is only one unit away from the pH (7.0) where $[H_3O^+] = [OH^-]$. A pH of 4.0 is three units below 7.0, so $[H_3O^+] = 1000 \times [OH^-]$ in solution B.

**Think About It** You may find at first that calculating a logarithm in your head is challenging. It helps to understand that an increase of one pH unit corresponds to a decrease in $[H_3O^+]$ to only 1/10, or 10%, of its initial value.

**Practice Exercise** Which of the following changes in pH corresponds to the greatest percent increase in $[H_3O^+]$? Which change corresponds to the greatest percent decrease in $[H_3O^+]$? a. pH 1 → pH 3; b. pH 7 → pH 4; c. pH 12 → pH 14; d. pH 5 → pH 9; e. pH 2 → pH 0.

When interconverting $[H_3O^+]$ and pH values, we need to know how to express pH values with the appropriate number of significant figures. Because a pH value is the negative logarithm of a hydronium ion concentration, the digit or digits before the decimal point in the pH value define the location of the decimal point in the $[H_3O^+]$ value. They do not define how precisely we know the concentration value, so they are not considered significant digits. Suppose, for example, the pH of a solution of a weak acid is determined to be 2.57. The concentration of $H_3O^+$ ions in the solution is therefore $10^{-2.57} = 2.69 \times 10^{-3}$ *M* if expressed to three significant figures. However, we are not allowed to use three because the "2" before the decimal place in the pH value tells us only that the value of $[H_3O^+]$ is somewhere between $10^{-2}$ *M* and $10^{-3}$ *M*. It does not tell us precisely where it is within that range; the two digits after the decimal point in the pH value tell us that. Therefore, the value of $[H_3O^+]$ that corresponds to a pH of 2.57 is $2.7 \times 10^{-3}$ *M*.

**CONNECTION** The general rules for using significant figures in calculations involving measured quantities were discussed in Section 1.7.

**SAMPLE EXERCISE 15.7** Interconverting pH and $[H_3O^+]$ **LO5**

Cells located in the upper region of the human stomach, called the fundus, secrete gastric acid (about 0.16 *M* HCl) during digestion. After this acid mixes with the food being digested, the pH of the contents of the stomach is often between 3 and 4.

a. How much more acidic is 0.16 *M* HCl than stomach contents with a pH of 3.20?
b. Which of the two samples in part a has the lower $[H_3O^+]$?

**Collect and Organize** We are asked to compare the acidities (the ratio of the $[H_3O^+]$ values) of two solutions that both contain hydrochloric acid. We know the pH of one and the HCl concentration of the other and must determine which sample has the lower $[H_3O^+]$. Equation 15.12 relates the pH and $[H_3O^+]$ values of solutions.

**Analyze** Hydrochloric acid is a strong acid, so the $[H_3O^+]$ value of a solution of HCl is the same as the concentration of the acid. To use Equation 15.12 to convert pH 3.20 into its corresponding $[H_3O^+]$ value requires setting both sides of the equation as exponents of 10, which removes the log function preceding the $[H_3O^+]$ term and turns the pH value into an exponent. The results of this conversion, when divided into 0.16 *M*, will give the desired ratio of acidities.

**Solve**

a. Converting pH 3.20 to $[H_3O^+]$:

$$\text{pH} = -\log[H_3O^+]$$

or

$$\log[H_3O^+] = -\text{pH}$$

Taking the antilog ($10^x$) of both sides: $[H_3O^+] = 10^{-\text{pH}}$

Inserting pH = 3.20: $[H_3O^+] = 10^{-3.20} = 6.3 \times 10^{-4}\ M$

Calculating the ratio of acidities ($[H_3O^+]$ values):

$$\frac{[H_3O^+](\text{gastric acid})}{[H_3O^+](\text{stomach contents})} = \frac{0.16\ \cancel{M}}{6.3 \times 10^{-4}\ \cancel{M}} = 2.5 \times 10^2$$

Thus, the acidity of secreted gastric acid is $2.5 \times 10^2$ times the acidity of the digestion mixture in the stomach.

b. Using the values calculated in part a, the stomach contents have a lower $[H_3O^+]$.

**Think About It** Converting pH 3.20 into its corresponding $[H_3O^+]$ value yields a number with only two significant figures. The "3" in 3.20 tells us only that the $[H_3O^+]$ value is somewhere between $10^{-3}$ *M* and $10^{-4}$ *M*, but not precisely where.

**Practice Exercise** The results of recent oceanographic studies indicate that the average pH of Earth's oceans is now about 8.07. The average hydrogen ion concentration of the ocean at the start of the industrial revolution (around 1750) was $6.61 \times 10^{-9}$ *M*. How much more acidic is the ocean now?

**CONCEPT TEST**

Is the pH of a 1.00 *M* solution of a weak acid higher or lower than the pH of a 1.00 *M* solution of a strong acid?

## pOH, p$K_a$, and p$K_b$ Values

The letter *p* as used in pH is also used with other symbols to mean *the negative logarithm* of the variable that follows it. For example, just as every aqueous solution has a pH value, it also has a **pOH** value, defined as

$$\text{pOH} = -\log[OH^-] \tag{15.13}$$

We can use the $K_w$ expression, Equation 15.11, to relate pOH to pH. We start by taking the negative logarithm of both sides of the equation and then rearrange the terms:

$$K_w = [H_3O^+][OH^-] = 1.00 \times 10^{-14}$$

$$-\log K_w = -\log([H_3O^+][OH^-]) = -\log(1.0 \times 10^{-14})$$

$$pK_w = -\log[H_3O^+] - \log[OH^-] = -(-14.00)$$

$$pK_w = \text{pH} + \text{pOH} = 14.00 \tag{15.14}$$

**pOH** the negative logarithm of the hydroxide ion concentration in an aqueous solution.

Many tables of equilibrium constants list p$K$ values rather than $K$ values because doing so does not require using exponential notation, thus saving space. The tables in Appendix 5 of this book contain both sets of values. Use them whenever you need a $K$ or p$K$ value that is not given in a problem.

**SAMPLE EXERCISE 15.8** Relating $[H_3O^+]$, $[OH^-]$, pH, and pOH **LO5**

Digested food leaving your stomach and entering your small intestine undergoes an increase in pH from about 4.50 to 7.50. Convert these pH values to pOH, $[H_3O^+]$, and $[OH^-]$ values.

**Collect, Organize, and Analyze** We are given two pH values, and we want to determine the corresponding pOH, $[H_3O^+]$, and $[OH^-]$ values.

- Use Equation 15.12 to convert from pH to $[H_3O^+]$, as we did in Sample Exercise 15.7.
- Use Equation 15.14 to convert from pH to pOH.
- Use Equation 15.13 to convert from pOH to $[OH^-]$.

The $[H_3O^+]$ value we calculate for pH 4.50 should be $10^3$, or 1000, times the $[H_3O^+]$ value we calculate for pH 7.50 because $7.50 - 4.50 = 3.00$. Correspondingly, the $[OH^-]$ value we calculate for pH 7.50 should be $10^3$, or 1000, times the $[OH^-]$ value we calculate for pH 4.50. The inverse relationship between pH and pOH means that the pOH value we calculate for pH 4.50 should be three units larger than the one we get for pH 7.50.

**Solve**

1. Converting pH to $[H_3O^+]$: $$\text{pH} = -\log[H_3O^+] \quad (15.12)$$

or $$[H_3O^+] = 10^{-\text{pH}}$$

Stomach: $$[H_3O^+] = 10^{-4.50} = 3.2 \times 10^{-5}\ M$$

Small intestine: $$[H_3O^+] = 10^{-7.50} = 3.2 \times 10^{-8}\ M$$

2. Converting pH to pOH: $$\text{pH} + \text{pOH} = 14.00 \quad (15.14)$$

$$\text{pOH} = 14.00 - \text{pH}$$

Stomach: $$\text{pOH} = 14.00 - 4.50 = 9.50$$

Small intestine: $$\text{pOH} = 14.00 - 7.50 = 6.50$$

3. Converting pOH to [pOH⁻]: $$\text{pOH} = -\log[OH^-] \quad (15.13)$$

or $$[OH^-] = 10^{-\text{pOH}}$$

Stomach: $$[OH^-] = 10^{-9.50} = 3.2 \times 10^{-10}\ M$$

Small intestine: $$[OH^-] = 10^{-6.50} = 3.2 \times 10^{-7}\ M$$

**Think About It** As predicted, the value of $[H_3O^+]$ in the stomach is 1000 times that of the small intestine. The differences in pOH values are also as predicted, as are the differences in $[OH^-]$. All calculated concentration values were rounded to two significant digits because each starting pH value had only two digits after the decimal point.

**Practice Exercise** What are the values of $[H_3O^+]$ and $[OH^-]$ in a sample of household ammonia, an aqueous solution of $NH_3$, that has a pH of 11.70?

**Figure 15.12** summarizes the ways we can interconvert $[H^+]$, $[OH^-]$, pH, and pOH. The diagram may leave you wondering which path to follow to make conversions, such as $[OH^-]$ to pH. Two equivalent sets of calculations are shown: (i) $[OH^-] \rightarrow \text{pOH} \rightarrow \text{pH}$ or (ii) $[OH^-] \rightarrow [H_3O^+] \rightarrow \text{pH}$. Using either results in the same answer.

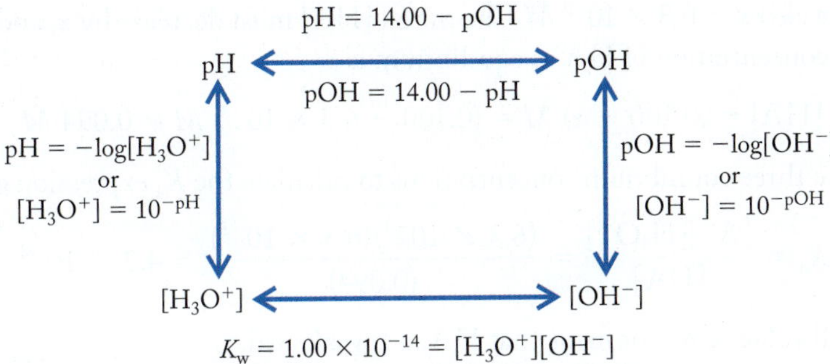


**FIGURE 15.12** Pathways for interconverting $[H_3O^+]$, $[OH^-]$, pH, and pOH values.

# 15.6 $K_a$, $K_b$, and the Ionization of Weak Acids and Bases

Most of the acids and bases in nature are weak acids and bases with $K_a$ and $K_b$ values that are much less than 1. In this section and the next, we explore how to quantify the degree to which these molecules react with molecules of water to form either $H_3O^+$ or $OH^-$ ions.

## Weak Acids

Let's begin with the ionization of a generic weak acid (HA), as expressed by the chemical equation

$$HA(aq) + H_2O(\ell) \rightleftharpoons A^-(aq) + H_3O^+(aq) \qquad (15.3)$$

The corresponding $K_a$ expression is

$$K_a = \frac{[A^-][H_3O^+]}{[HA]} \qquad (15.15)$$

We can use Equation 15.15 to calculate the $K_a$ value of an unknown weak acid if we know both the concentration of a solution of the acid and its pH. Suppose we determine that a 0.100 *M* solution of HA has a pH of 2.20. To calculate $K_a$, we first convert the pH value to $[H_3O^+]$:

$$[H_3O^+] = 10^{-2.20} = 6.3 \times 10^{-3}\ M$$

We used RICE tables in Chapter 14 to calculate the concentrations of reactants and products in reaction mixtures at equilibrium given the *K* of the reactions. Let's use one here to organize what we know so far, if we assume no hydronium ions are initially present and that the only source of $H_3O^+$ ions is the ionization of HA:

**CONNECTION** RICE tables were introduced in Section 14.8 to help us calculate the changes (C) from initial (I) to equilibrium (E) concentrations for a given reaction (R).

| REACTION (R) | $HA(aq) + H_2O(\ell)$ | $\rightleftharpoons$ $A^-(aq)$ | + $H_3O^+(aq)$ |
|---|---|---|---|
| | [HA] (*M*) | $[A^-]$ (*M*) | $[H_3O^+]$ (*M*) |
| Initial (I) | 0.100 | 0 | 0 |
| Change (C) | $-x$ | $+x$ | $+x$ |
| Equilibrium (E) | $0.100 - x$ | $x$ | $6.3 \times 10^{-3}$ |

We can see that $+x$ for $[H_3O^+]$ equals $6.3 \times 10^{-3}$ *M*. Because one mole of HA ionizes to produce one mole of $H_3O^+$ and one mole of $A^-$, the equilibrium value

**degree of ionization** the ratio of the quantity of a substance that is ionized to the concentration of the substance before ionization; when expressed as a percentage, called **percent ionization**.

of $[A^-]$ is also $x = 6.3 \times 10^{-3}\ M$. The initial [HA] must decrease by $x$, and therefore the concentration of HA at equilibrium is

$$[HA] = (0.100 - x)\ M = (0.100 - 6.3 \times 10^{-3})\ M = 0.094\ M$$

Using the three equilibrium concentrations to calculate the $K_a$ expression gives us

$$K_a = \frac{[A^-][H_3O^+]}{[HA]} = \frac{(6.3 \times 10^{-3})(6.3 \times 10^{-3})}{(0.094)} = 4.2 \times 10^{-4}$$

The small value of $K_a$ confirms that HA is a weak acid.

The ratio of the equilibrium $[H_3O^+]$ to the initial [HA] represents the **degree of ionization** of HA, usually expressed as a percentage of the initial acid concentration. For this reason it is also called **percent ionization**. In equation form, this relationship is

$$\text{Percent ionization} = \frac{[H_3O^+]_{\text{equilibrium}}}{[HA]_{\text{initial}}} \times 100\% \quad (15.16)$$

Inserting the data from the previous calculation into Equation 15.16:

$$\text{Percent ionization} = \frac{6.3 \times 10^{-3}\ \cancel{M}}{0.100\ \cancel{M}} \times 100\% = 6.3\%$$

**CONCEPT TEST**

Describe how the percent ionization of two different weak acids, each in a .001 *M* solution, are related to their $K_a$ values.

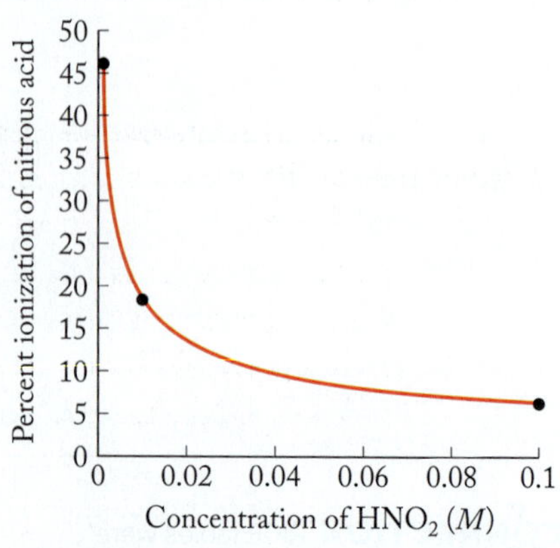


**FIGURE 15.13** The degree of ionization of a weak acid increases with decreasing acid concentration. Here, the degree of ionization of nitrous acid increases from about 6% in a 0.100 *M* solution to 18% in a 0.010 *M* solution to 46% in a 0.001 *M* solution.

As noted in Section 15.2, the degree of ionization increases with increasing acid strength, as indicated by larger $K_a$ values. Now let's examine how the degree of ionization of a weak acid depends on its concentration in solution. A plot of percent ionization of nitrous acid as a function of the acid's initial concentration is shown in **Figure 15.13**. Note that the degree of ionization increases as the concentration of the acid decreases. This trend is observed for all weak acids.

Suppose that we have a 1.0 *M* solution of a generic weak acid, HA, in which the concentrations of $A^-$ and $H_3O^+$ ions at equilibrium are both 0.0010 *M*. Using these concentrations in Equation 15.15 to solve for $K_a$, we have

$$K_a = \frac{[A^-][H_3O^+]}{[HA]} = \frac{(0.0010)(0.0010)}{(1.0 - 0.0010)} = 1.0 \times 10^{-6}$$

Now suppose that more water is added to the HA solution, increasing its volume by a factor of 10. If no change in the degree of ionization occurred, then [HA] would be 0.10 *M*, and the concentrations of $A^-$ and $H_3O^+$ ions would both be $1.0 \times 10^{-4}\ M$. However, if we insert these values into the equilibrium constant expression, we get a reaction quotient ($Q$) value of

$$Q = \frac{(1.0 \times 10^{-4})(1.0 \times 10^{-4})}{(0.10 - 1.0 \times 10^{-4})} = 1.0 \times 10^{-7}$$

This $Q$ value is only 1/10 the value of $K_a$, which means that the acid ionization reaction is no longer at equilibrium. Because $Q < K_a$, the reaction proceeds in the forward direction. This means that the concentrations of $A^-$ and $H_3O^+$ ions increase, as does the percentage of HA molecules that ionize. Further dilutions would produce even greater percent ionization values, following the trend we saw for nitrous acid in Figure 15.13.

**CONNECTION** We introduced the reaction quotient, $Q$, in Section 14.5.

## SAMPLE EXERCISE 15.9 Relating pH, $K_a$, and Percent Ionization of a Weak Acid LO6

The pH of a 1.00 *M* solution of formic acid (HCOOH) at 25 °C is 1.879.

a. What is the percent ionization of 1.00 *M* HCOOH?
b. What is the $K_a$ value of formic acid?

**Collect and Organize** We are asked to determine the $K_a$ value of formic acid and its percent ionization in a solution of known concentration and pH. A molecule of formic acid has two H atoms, but only one is in a carboxylic acid group that can ionize in an aqueous solution. As a result, the acid ionization reaction is

$$HCOOH(aq) + H_2O(\ell) \rightleftharpoons HCOO^-(aq) + H_3O^+(aq)$$

**Analyze** We can use a RICE table to derive a $K_a$ value from the composition of a reaction mixture at equilibrium. First, we need to convert the given pH value to $[H_3O^+]$ by using a version of Equation 15.12: $[H_3O^+] = 10^{-pH}$. The pH of the 1.00 *M* solution is close to 2, which corresponds to $[H_3O^+] = 10^{-2}\ M$. Therefore, the percent ionization of formic acid in this solution should be about 1%.

**Solve**

a. The pH of the solution is 1.879. The corresponding $[H_3O^+]$ is

$$[H_3O^+] = 10^{-1.879} = 1.32 \times 10^{-2}\ M$$

If we assume that the ionization of formic acid is the only source of $H_3O^+$ ions, then $[H_3O^+] = [HCOO^-] = 1.32 \times 10^{-2}\ M$. Inserting this value and the initial concentration of HCOOH into Equation 15.16:

$$\text{Percent ionization} = \frac{[HCOO^-]_{equilibrium}}{[HCOOH]_{initial}} \times 100\%$$

$$= \frac{1.32 \times 10^{-2}\ \cancel{M}}{1.00\ \cancel{M}} \times 100\% = 1.32\%$$

b. On the basis of the above assumption, the RICE table for the reaction should look like this:

| Reaction (R) | HCOOH(*aq*) + $H_2O(\ell)$ | $\rightleftharpoons$ | $HCOO^-(aq)$ | + | $H_3O^+(aq)$ |
|---|---|---|---|---|---|
| | [HCOOH] (*M*) | | $[HCOO^-]$ (*M*) | | $[H_3O^+]$ (*M*) |
| Initial (I) | 1.00 | | 0 | | 0 |
| Change (C) | $-1.32 \times 10^{-2}$ | | $+1.32 \times 10^{-2}$ | | $+1.32 \times 10^{-2}$ |
| Equilibrium (E) | 0.987 | | $1.32 \times 10^{-2}$ | | $1.32 \times 10^{-2}$ |

Inserting the equilibrium terms into the $K_a$ expression for the acid ionization reaction:

$$K_a = \frac{[HCOO^-][H_3O^+]}{[HCOOH]} = \frac{(1.32 \times 10^{-2})(1.32 \times 10^{-2})}{(0.987)} = 1.77 \times 10^{-4}$$

**Think About It** The result of the percent ionization calculation is close to our estimate of about 1%, and the calculated $K_a$ value is the same as the $K_a$ value for formic acid listed in Table 15.2.

**Practice Exercise** The value of $[H_3O^+]$ in a 0.050 *M* solution of an organic acid at 25 °C is $5.9 \times 10^{-3}\ M$. What is the pH of the solution, the percent ionization of the acid, and its $K_a$ value?

CONCEPT **TEST**

Three weak acids have the following $K_a$ values:

| Acid | A | B | C |
|---|---|---|---|
| $K_a$ | $3.6 \times 10^{-5}$ | $4.9 \times 10^{-4}$ | $9.2 \times 10^{-4}$ |

a. Which of the three acids is the most extensively ionized in a 0.10 *M* solution of the acid?

b. Which of the three acids has the lowest percent ionization in a 1.00 *M* solution of the acid?

## Weak Bases

The concept of percent ionization also applies to Brønsted–Lowry bases and their acceptance of protons in aqueous solutions. Consider the reaction of generic base B in an aqueous solution:

$$B(aq) + H_2O(\ell) \rightleftharpoons BH^+(aq) + OH^-(aq)$$

The corresponding equilibrium constant expression is

$$K_b = \frac{[BH^+][OH^-]}{[B]} \tag{15.17}$$

As with weak acids, we can calculate the degree of ionization of a weak base and its $K_b$ value if we know the pH and the initial concentration of the base in the solution:

$$\text{Percent ionization} = \frac{[OH^-]_{\text{equilibrium}}}{[B]_{\text{initial}}} \times 100\% \tag{15.18}$$

Doing so takes more steps than in the corresponding calculation for a weak acid, as shown in Sample Exercise 15.10.

**SAMPLE EXERCISE 15.10** Relating pH, $K_b$, and Percent Ionization of a Weak Base **LO6**

Trimethylamine, $(CH_3)_3N$, is a particularly foul-smelling volatile organic compound that forms when plant and animal matter decay. It is soluble in water, and it reacts with water molecules according to the following chemical equation:

$$(CH_3)_3N(aq) + H_2O(\ell) \rightleftharpoons (CH_3)_3NH^+(aq) + OH^-(aq)$$

The pH of a 0.500 *M* solution of trimethylamine is 11.75 at 25°C.

a. What is the percent ionization of trimethylamine in a 0.500 *M* solution?
b. What is the $K_b$ value of trimethylamine?

**Collect, Organize, and Analyze** We are asked to determine the $K_b$ value of trimethylamine and its percent ionization in a solution of known concentration and pH. According to the stoichiometry of the reaction, $[(CH_3)_3NH^+] = [OH^-]$ in an aqueous solution of $(CH_3)_3N$ at equilibrium. In addition, $[(CH_3)_3N]$ is equal to the initial concentration of the base minus the portion of it that reacts with water to form $OH^-$ ions:

$$[(CH_3)_3N]_{\text{equilibrium}} = [(CH_3)_3N]_{\text{initial}} - [OH^-]_{\text{equilibrium}}$$

Converting pH to pOH and then calculating $[OH^-]$ will give us $[OH^-]$ and $[(CH_3)_3NH^+]$ at equilibrium. We can then calculate the percent ionization by using Equation 15.18. As in Sample Exercise 15.9, using a RICE table should be a useful strategy for deriving the value of the equilibrium constant from the composition of the reaction mixture at equilibrium.

The pH of the 0.500 *M* solution is a little less than 12, which corresponds to a pOH that is a little more than 2 and a value for $[OH^-]$ that is a little less than $10^{-2}$ *M*. Therefore, the percent ionization of trimethylamine in this solution should be 1% to 2%.

**Solve**

a. The pH of the 0.500 *M* solution is 11.75. Calculating the corresponding pOH:

$$\text{pH} + \text{pOH} = 14.00$$

or

$$\text{pOH} = 14.00 - \text{pH} = 14.00 - 11.75 = 2.25$$

Calculating $[OH^-]$: $$[OH^-] = 10^{-\text{pOH}} = 10^{-2.25} = 5.6 \times 10^{-3} M$$

This value is the same as $[(CH_3)_3NH^+]$ at equilibrium. Using it and the initial concentration of trimethylamine to calculate percent ionization:

$$\text{Percent ionization} = \frac{[BH^+]_{\text{equilibrium}}}{[B]_{\text{initial}}} \times 100\%$$

$$= \frac{5.6 \times 10^{-3}\ M}{0.500\ M} \times 100\% = 1.1\%$$

Using a RICE table to determine the equilibrium concentrations of the reaction components:

| Reaction (R) | $(CH_3)_3N(aq) + H_2O(\ell)$ $\rightleftharpoons$ | $(CH_3)_3NH^+(aq)$ + | $OH^-(aq)$ |
|---|---|---|---|
| | $[(CH_3)_3N]$ (*M*) | $[(CH_3)_3NH^+]$ (*M*) | $[OH^-]$ (*M*) |
| Initial (I) | 0.500 | 0 | 0 |
| Change (C) | $-5.6 \times 10^{-3}$ | $+5.6 \times 10^{-3}$ | $+5.6 \times 10^{-3}$ |
| Equilibrium (E) | 0.494 | $5.6 \times 10^{-3}$ | $5.6 \times 10^{-3}$ |

Inserting these equilibrium values in the $K_b$ expression for the reaction:

$$K_b = \frac{[(CH_3)_3NH^+][OH^-]}{[(CH_3)_3N]} = \frac{(5.6 \times 10^{-3})(5.6 \times 10^{-3})}{0.494}$$

$$= 6.3 \times 10^{-5}$$

**Think About It** The result of the percent ionization calculation is in our estimated range of 1% to 2%, which reflects the fact that although trimethylamine may have a strong odor, it is only a weak base. The latter point is reinforced by the small $K_b$ value.

**Practice Exercise** The pH of a 0.100 *M* aqueous solution of ethylamine, $C_2H_5NH_2$, is 11.86 at 25°C. What percentage of the $C_2H_5NH_2$ molecules in this solution ionizes as described in the chemical equation that follows, and what is the $K_b$ value of ethylamine?

$$C_2H_5NH_2(aq) + H_2O(\ell) \rightleftharpoons C_2H_5NH_3^+(aq) + OH^-(aq)$$

## 15.7 Calculating the pH of Acidic and Basic Solutions

In this section, we calculate the pH values of solutions of acids and bases. We begin with solutions of strong acids and strong bases.

## Strong Acids and Strong Bases

We have seen that strong acids ionize completely in aqueous solutions. This means that in nearly all their solutions, the concentration of $H_3O^+$ ions is the same as the initial concentration of the strong acid. For example, the pH of muriatic acid (7.4 *M* HCl), which is sold in hardware and building supply stores to clean concrete surfaces, is

$$pH = -\log[H_3O^+] = -\log[HCl] = -\log(7.4) = -0.87$$

To calculate the pH of a solution of a strong base, we need to calculate the concentration of $OH^-$ ions from the initial concentration of the base and then convert $[OH^-]$ to pOH and convert pOH to pH. Sample Exercise 15.11 illustrates the steps involved.

FIGURE 15.14 Cleaners used to remove grease and hair clogs from drain pipes typically contain high concentrations of a strong base, such as sodium hydroxide.

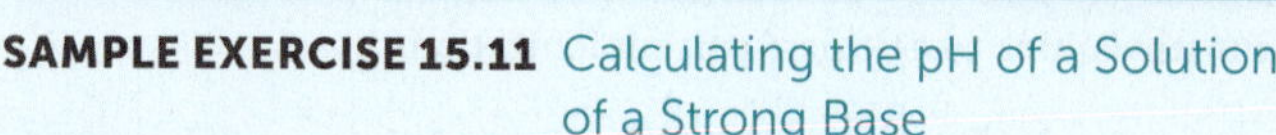

**SAMPLE EXERCISE 15.11** Calculating the pH of a Solution of a Strong Base **LO7**

Liquid cleaners such as the one shown in **Figure 15.14** are used to remove clogs from bathroom and kitchen drains. They contain sodium hydroxide at concentrations as high as 1.2 *M*. What is the pH of 1.2 *M* NaOH?

**Collect, Organize, and Analyze** NaOH is a strong base that produces one mole of hydroxide ions per mole of NaOH in solution, so $[OH^-]$ is the same as the initial concentration of the base. Therefore, $[OH^-] = 1.2\ M$. We can convert this value to pOH and then to pH by following the steps shown in Figure 15.12.

**Solve**

Calculating pOH: $pOH = -\log[OH^-] = -\log(1.2) = -0.079$

Calculating pH: $pH = 14.00 - pOH = 14.00 - (-0.079) = 14.08$

**Think About It** The calculated pH is slightly above 14 because $[OH^-]$ is slightly greater than one molar.

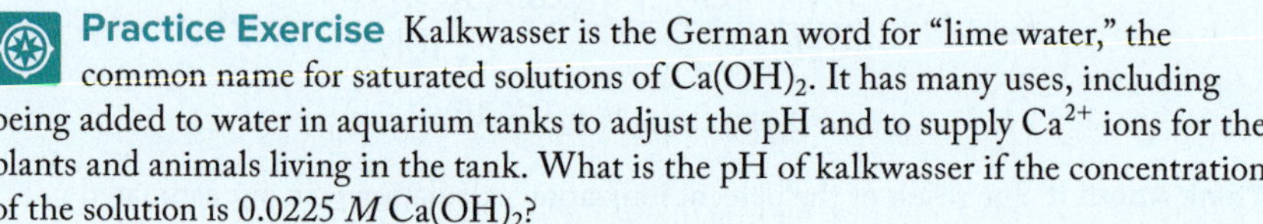

**Practice Exercise** Kalkwasser is the German word for "lime water," the common name for saturated solutions of $Ca(OH)_2$. It has many uses, including being added to water in aquarium tanks to adjust the pH and to supply $Ca^{2+}$ ions for the plants and animals living in the tank. What is the pH of kalkwasser if the concentration of the solution is 0.0225 *M* $Ca(OH)_2$?

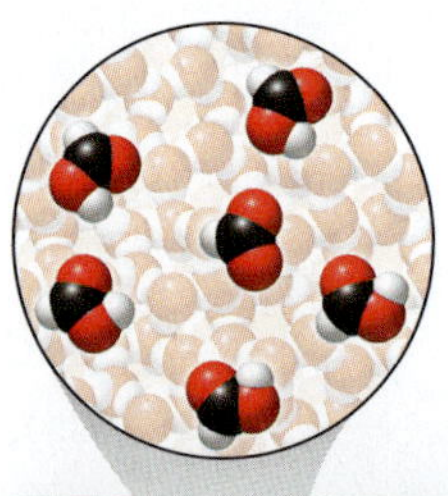

FIGURE 15.15 Ants secrete concentrated formic acid as a defense against predators.

## Weak Acids and Weak Bases

In Sample Exercises 15.12 and 15.13, we calculate the pH values of aqueous solutions of a weak acid and a weak base. We use the initial concentrations of the solutions to set up RICE tables like those used in Chapter 14.

**SAMPLE EXERCISE 15.12** Calculating the pH of a Solution of a Weak Acid **LO7**

Formic acid is a common carboxylic acid and the one with the simplest molecular structure. A diluted sample of the secretions of the ant shown in **Figure 15.15** contains 0.040 *M* formic acid, HCOOH. What is the pH of 0.040 *M* formic acid? The $K_a$ value of the acid is $1.77 \times 10^{-4}$.

**Collect and Organize** We are asked to determine the pH of a known concentration of a weak acid. We also know its $K_a$ value. The molecular structure of formic acid in Table 15.2 indicates that only one H atom per molecule is ionizable in aqueous solution.

**Analyze** The acid ionization reaction is

$$HCOOH(aq) + H_2O(\ell) \rightleftharpoons HCOO^-(aq) + H_3O^+(aq)$$

and the corresponding equilibrium constant expression is

$$K_a = \frac{[HCOO^-][H_3O^+]}{[HCOOH]} = 1.77 \times 10^{-4}$$

Calculating pH involves first solving for $[H_3O^+]$, which we can do by setting up a RICE table based on the above reaction.

**Solve** The objective of the RICE table is to solve for $[H_3O^+]$ at equilibrium, so we give it the symbol $x$. Filling in the other cells in the table on the basis of the 1:1:1 stoichiometry of the reaction gives us

| Reaction (R) | $HCOOH(aq) + H_2O(\ell) \rightleftharpoons$ | $HCOO^-(aq)$ + | $H_3O^+(aq)$ |
|---|---|---|---|
| | $[HCOOH]$ ($M$) | $[HCOO^-]$ ($M$) | $[H_3O^+]$ ($M$) |
| Initial (I) | 0.040 | 0 | 0 |
| Change (C) | $-x$ | $+x$ | $+x$ |
| Equilibrium (E) | $0.040 - x$ | $x$ | $x$ |

Inserting the equilibrium terms into the $K_a$ expression,

$$K_a = \frac{[HCOO^-][H_3O^+]}{[HCOOH]} = \frac{(x)(x)}{(0.040 - x)} = 1.77 \times 10^{-4}$$

Solving for $x$ by first cross-multiplying,

$$x^2 = 7.08 \times 10^{-6} - (1.77 \times 10^{-4})x$$

and rearranging the terms,

$$x^2 + (1.77 \times 10^{-4})x - 7.08 \times 10^{-6} = 0$$

gives us a quadratic equation with two solutions:

$$x = -0.00275\ M \quad \text{and} \quad x = 0.00257\ M$$

The negative $x$ value has no physical meaning because it results in negative concentration values, so $[H_3O^+] = 0.00257\ M$.

Solving for pH: $\text{pH} = -\log[H_3O^+] = -\log(0.00257) = 2.59$

**Think About It** Why didn't we try to simplify our $K_a$ expression by ignoring the $-x$ term instead of solving the quadratic equation? We learned in Section 14.8 that we could simplify the algebra of our problem solving by ignoring $-x$ when $x < 5\%$ of the initial concentration of the weak acid—that is, when less than 5% of the acid ionizes. Here the value of $x$ compared to the initial concentration is 0.00257/0.040, meaning that the acid is 6.4% ionized. This finding represents a significant decrease in [HCOOH] due to acid ionization.

**Practice Exercise** Acetic acid, the main ingredient in vinegar, is also a carboxylic acid. What is the pH of a 0.035 $M$ solution of acetic acid given that its $K_a = 1.76 \times 10^{-5}$?

**SAMPLE EXERCISE 15.13** Calculating the pH of a Solution of a Weak Base **LO7**

The concentration of $NH_3$ in the household ammonia used to clean windows ranges between 50 g/L and 100 g/L, or from about 3 $M$ to almost 6 $M$. What is the pH of 3.0 $M$ $NH_3$? The $K_b$ value for ammonia is $1.76 \times 10^{-5}$.

**Collect, Organize, and Analyze** We are asked to determine the pH of 3.0 $M$ $NH_3$, which reacts with water to form $NH_4^+$ and $OH^-$ ions:

$$NH_3(aq) + H_2O(\ell) \rightleftharpoons NH_4^+(aq) + OH^-(aq) \qquad K_b = 1.76 \times 10^{-5}$$

To calculate pH, we must first determine the equilibrium concentration of $OH^-$ ions and then convert $[OH^-]$ to pOH and finally to pH. Given the 1:1:1 stoichiometry of $NH_3$, $NH_4^+$, and $OH^-$, we know that $[NH_4^+] = [OH^-]$ at equilibrium. If we let these concentrations be $x$, the change in $[NH_3]$ as the reaction proceeds is $-x$. The pH value we obtain for a concentrated solution of a weak base should be well above 7 but below 14.

**Solve** We begin by setting up a RICE table in which we let $[NH_4^+] = [OH^-] = x$ at equilibrium:

| Reaction (R) | $NH_3(aq) + H_2O(\ell)$ | $\rightleftharpoons$ | $NH_4^+(aq)$ | + | $OH^-(aq)$ |
|---|---|---|---|---|---|
| | $[NH_3]$ ($M$) | | $[NH_4^+]$ ($M$) | | $[OH^-]$ ($M$) |
| Initial (I) | 3.0 | | 0 | | 0 |
| Change (C) | $-x$ | | $+x$ | | $+x$ |
| Equilibrium (E) | $3.0 - x$ | | $x$ | | $x$ |

Because $K_b$ is much smaller ($1.76 \times 10^{-5}$) than the initial concentration of base (3.0 $M$), we can make the simplifying assumption that $x$ will be smaller than 3.0 $M$, and so $3.0 - x \approx 3.0$. With this assumption, our equilibrium constant expression is

$$K_b = \frac{[NH_4^+][OH^-]}{[NH_3]} = \frac{(x)(x)}{3.0} = \frac{x^2}{3.0} = 1.76 \times 10^{-5}$$

so

$$x^2 = 5.28 \times 10^{-5}$$

Solving for $x$ gives us

$$x = [OH^-] = \sqrt{5.28 \times 10^{-5}} = 7.3 \times 10^{-3}\ M$$

Taking the negative logarithm of $[OH^-]$ to calculate pOH:

$$pOH = -\log[OH^-] = -\log(7.3 \times 10^{-3}\ M) = 2.14$$

Then we subtract this value from 14.00 to obtain the pH:

$$pH = 14.00 - pOH = 14.00 - 2.14 = 11.86$$

**Think About It** The calculated pH value falls in the range we predicted, given the small $K_b$ value but the relatively high initial concentration of ammonia. To check our simplifying assumption, let's compare the value of $x$ to the initial $[NH_3]$ value of 3.0 $M$:

$$\frac{7.3 \times 10^{-3}\ M}{3.0\ M} = 0.0024 \times 100\% = 0.24\%$$

This small percentage change in $[NH_3]$ confirms that our simplifying assumption was justified.

 **Practice Exercise** What is the pH of a 0.200 $M$ solution of methylamine ($CH_3NH_2$, $K_b = 4.4 \times 10^{-4}$)?

## pH of Very Dilute Solutions of Strong Acids

Recall that the autoionization of water contributes very little to the equilibrium concentrations of $H_3O^+$ or $OH^-$ ions due to the small value of $K_w$. Now that we have developed the concept of pH and explored how it helps us quantify the acidity of a solution, let's consider a situation where the concentration of hydronium ions produced by the autoionization of water is greater than the concentration of hydronium ions produced by an acid.

### SAMPLE EXERCISE 15.14 pH Calculations Involving the Autoionization of Water — LO7

What is the pH of $1.0 \times 10^{-8}$ $M$ HCl at 25°C?

**Collect and Organize** We are asked to calculate the pH of a very dilute solution of HCl, which is a strong acid and ionizes completely. The autoionization of pure water at 25°C produces an equilibrium at $[H_3O^+] = 1.0 \times 10^{-7}$ $M$.

**Analyze** In a $1.0 \times 10^{-8}$ $M$ HCl solution, $[H_3O^+] = 1.0 \times 10^{-8}$ $M$, and the pH should be

$$pH = -\log[H_3O^+] = -\log(1.0 \times 10^{-8}) = 8.00$$

This answer is not reasonable because a solution of a strong acid, no matter how dilute it is, cannot be basic (pH > 7). To calculate pH, we must also consider the autoionization of water. Let's set up a RICE table based on the autoionization equilibrium in which $x$ represents the increase in $[H_3O^+]$ resulting from autoionization and in which the initial value of $[H_3O^+]$ is $1.0 \times 10^{-8}$ $M$.

**Solve** Setting up the RICE table and filling in the rows as described above:

| Reaction (R) | $H_2O(\ell) + H_2O(\ell)$ | $\rightleftharpoons$ | $H_3O^+(aq)$ | + | $OH^-(aq)$ |
|---|---|---|---|---|---|
| | | | $[H_3O^+]$ ($M$) | | $[OH^-]$ ($M$) |
| Initial (I) | | | $1.0 \times 10^{-8}$ | | 0 |
| Change (C) | | | $+x$ | | $+x$ |
| Equilibrium (E) | | | $(1.0 \times 10^{-8}) + x$ | | $x$ |

Substituting equilibrium values into Equation 15.11 and solving for $x$:

$$K_w = [H_3O^+][OH^-] = 1.0 \times 10^{-14}$$

or,

$$(1.0 \times 10^{-8} + x)(x) = 1.0 \times 10^{-14}$$

Rearranging this equation to solve for $x$ gives the quadratic equation

$$x^2 + (1.0 \times 10^{-8})x - 1.0 \times 10^{-14} = 0$$

which has two solutions:

$$x = 9.5 \times 10^{-8}\ M \text{ and } x = -1.1 \times 10^{-7}\ M$$

Only the positive value makes physical sense; a negative value for $x$ would mean a negative concentration of hydroxide ions at equilibrium. Thus, the total concentration of hydrogen ions in the solution is

$$[H_3O^+] = (1.0 \times 10^{-8}\ M) + (9.5 \times 10^{-8}\ M) = 10.5 \times 10^{-8}\ M = 1.05 \times 10^{-7}\ M$$

and the pH is

$$pH = -\log(1.05 \times 10^{-7}\ M) = 6.98$$

**Think About It** This value agrees with our prediction that the solution should be (very) slightly acidic.

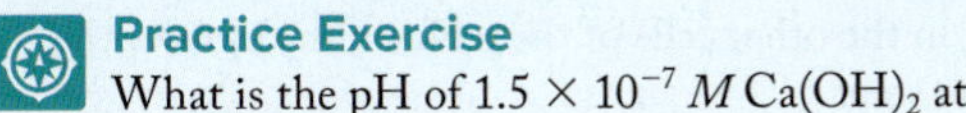

**Practice Exercise**
What is the pH of $1.5 \times 10^{-7}$ $M$ $Ca(OH)_2$ at 25°C?

**monoprotic acid** an acid that has one ionizable hydrogen atom per molecule.

**polyprotic acid** an acid that has two or more ionizable hydrogen atoms per molecule.

## 15.8 Polyprotic Acids

Until now, we have dealt with **monoprotic acids**, which have only one ionizable hydrogen atom per molecule. Acids that contain more than one ionizable hydrogen atom—such as sulfuric acid ($H_2SO_4$) and phosphoric acid ($H_3PO_4$)—are called **polyprotic acids**. For molecules with two and three ionizable hydrogen atoms, we use the more specific terms *diprotic acids* and *triprotic acids*, respectively. Let's consider the acidic properties of sulfuric acid, a strong, diprotic acid.

**CHEMTOUR**
Acid Rain

### Acid Rain

Coal naturally contains sulfur impurities. When coal is burned to produce electricity, these impurities are released into the atmosphere as $SO_2$. Some of this $SO_2$ is oxidized to $SO_3$, which combines with water vapor to form particles of liquid sulfuric acid ($H_2SO_4$), a principal component of acid rain in many parts of the world. Sulfuric acid is a strong acid (see Table 15.1) that is essentially completely ionized in aqueous solutions:

$$H_2SO_4(aq) + H_2O(\ell) \rightarrow HSO_4^-(aq) + H_3O^+(aq)$$

However, the second ionization step often does not result in complete ionization, depending on the concentration of the acid:

$$HSO_4^-(aq) + H_2O(\ell) \rightleftharpoons SO_4^{2-}(aq) + H_3O^+(aq) \qquad K_{a_2} = 0.012$$

Note that $K_{a_2}$ corresponds to the donation of the second $H^+$ ion per molecule.

The combination of one complete and one incomplete ionization reaction means that many aqueous solutions of $H_2SO_4$ contain more than one mole but less than two moles of $H_3O^+$ ions for every mole of $H_2SO_4$ dissolved.

**SAMPLE EXERCISE 15.15** Calculating the pH of a Solution of a Strong Diprotic Acid **LO8**

What is the pH of 0.100 $M$ $H_2SO_4$ at 25°C?

**Collect and Organize** We are given the concentration of a solution of sulfuric acid and asked to calculate its pH. Sulfuric acid is a strong diprotic acid in that one hydrogen atom per molecule ionizes completely, but the second hydrogen atom may not be ionized for every molecule ($K_{a_2} = 0.012$).

**Analyze** We start with the first ionization reaction that goes to completion:

$$H_2SO_4(aq) + H_2O(\ell) \rightarrow HSO_4^-(aq) + H_3O^+(aq)$$

Therefore, as the second ionization step begins, $[HSO_4^-] = [H_3O^+] = 0.100\ M$. Ionization of $HSO_4^-$ then produces additional $H_3O^+$ ions:

$$HSO_4^-(aq) + H_2O(\ell) \rightleftharpoons SO_4^{2-}(aq) + H_3O^+(aq)$$

The increase in $[H_3O^+]$ due to the second ionization reaction will be between 0 $M$ and 0.1 $M$, which means that the total $[H_3O^+]$ value will be between 0.1 $M$ and 0.2 $M$. The pH of the solution at equilibrium should therefore be a little less than 1.

**Solve** We begin by setting up a RICE table based on the second ionization reaction. Initially, $[HSO_4^-] = [H_3O^+] = 0.100\ M$. We let the change in $[H_3O^+]$ during the second ionization step be $+x$. Filling in the other cells of the table on the basis of the stoichiometry of the second ionization step:

| Reaction (R) | $HSO_4^-(aq) + H_2O(\ell)$ | $\rightleftharpoons$ | $SO_4^{2-}(aq)$ | + | $H_3O^+(aq)$ |
|---|---|---|---|---|---|
| | $[HSO_4^-]$ (*M*) | | $[SO_4^{2-}]$ (*M*) | | $[H_3O^+]$ (*M*) |
| Initial (I) | 0.100 | | 0 | | 0.100 |
| Change(C) | $-x$ | | $+x$ | | $+x$ |
| Equilibrium (E) | $0.100 - x$ | | $x$ | | $0.100 + x$ |

Inserting the equilibrium concentrations in the equilibrium constant expression for $K_{a_2}$:

$$K_{a_2} = \frac{[H_3O^+][SO_4^{2-}]}{[HSO_4^-]} = \frac{(0.100 + x)(x)}{(0.100 - x)} = 1.2 \times 10^{-2}$$

Cross-multiplying and rearranging the terms:

$$x^2 + 0.112x - 1.2 \times 10^{-3} = 0$$

Solving this quadratic equation for $x$ yields a positive value and a negative value:

$$x = 0.00985 \quad \text{and} \quad x = -0.122$$

The negative value for $x$ has no physical meaning because it gives us a negative $[SO_4^{2-}]$ value. Therefore,

$$[H_3O^+] = (0.100 + x) = (0.100 + 0.00985) = 0.10985 = 0.110\ M$$

The corresponding pH is

$$pH = -\log[H_3O^+] = -\log(0.110) = 0.96$$

**Think About It** As predicted, the value of $[H_3O^+]$ is between 0.1 *M* and 0.2 *M* and the pH of the solution is a little less than 1. The degree of ionization of $HSO_4^-$ is

$$\frac{[SO_4^{2-}]_{\text{equilibrium}}}{[HSO_4^-]_{\text{initial}}} = \frac{0.00985\ M}{0.100\ M} = 0.0985 = 9.8\%$$

This means that ignoring the decrease in $HSO_4^-$ to avoid solving a quadratic equation would have been a bad idea.

**Practice Exercise** What is the pH of a 0.200 *M* solution of $H_2SO_4$? How should doubling the concentration of $H_2SO_4$ affect the pH when compared to the 0.100 *M* solution of $H_2SO_4$ in Sample Exercise 15.15?

**CONCEPT TEST**

Identify the most abundant phosphorus-containing species present in an aqueous solution of phosphoric acid, $H_3PO_4$.

## Normal Rain

Did you know that rainwater falling from the sky is naturally acidic? The fourth-most-abundant gas in the atmosphere is $CO_2$, which dissolves in water to form a small amount of carbonic acid, as we discussed in Section 15.1:

$$CO_2(g) + H_2O(\ell) \rightleftharpoons H_2CO_3(aq) \qquad (15.1)$$

The equilibrium constant for this reaction is only about $10^{-3}$, so most of the dissolved carbon dioxide remains in the form of hydrated molecules of $CO_2$. However, to simplify equilibrium calculations involving solutions of carbon dioxide, we routinely represent the total concentration of $CO_2(aq)$ and $H_2CO_3(aq)$ as $[H_2CO_3]$, even though carbonic acid is not the principal species in solution. Thus, the acidic properties of dissolved $CO_2$ are represented by these chemical equations:

$$H_2CO_3(aq) + H_2O(\ell) \rightleftharpoons HCO_3^-(aq) + H_3O^+(aq) \qquad K_{a_1} = 4.3 \times 10^{-7}$$

$$HCO_3^-(aq) + H_2O(\ell) \rightleftharpoons CO_3^{2-}(aq) + H_3O^+(aq) \qquad K_{a_2} = 4.7 \times 10^{-11}$$

Calculating the pH of a solution of carbonic acid is simpler than the pH calculation for sulfuric acid. Because the $K_{a_2}$ value of $H_2CO_3$ is so small, it does not contribute significantly to the concentration of $H_3O^+$ ions in a solution of the acid and can be ignored in calculating pH. We can explain why $K_{a_2} << K_{a_1}$ on the basis of electrostatic attractions between oppositely charged ions. The first ionization step produces a negatively charged oxoanion, $HCO_3^-$. The second requires that a $H^+$ ion dissociate from $HCO_3^-$ to produce an even more negative oxoanion, $CO_3^{2-}$. Separating oppositely charged ions that are naturally attracted to each other is not a favored process, which is confirmed by the much smaller value of $K_{a_2}$.

## SAMPLE EXERCISE 15.16 Calculating the pH of a Solution of a Weak Diprotic Acid — LO8

What is the pH of rainwater at 25°C in equilibrium with atmospheric $CO_2$, which gives the water a constant dissolved $CO_2$ concentration of $1.4 \times 10^{-5}$ *M*?

**Collect and Organize** We are asked to determine the pH of a dilute solution of dissolved $CO_2$. There are two ionizable H atoms in $H_2CO_3$. The $K_{a_1}$ and $K_{a_2}$ values are given in the text. Any $H_2CO_3$ that dissociates to produce bicarbonate and hydronium ions will be replaced by more $CO_2$ dissolving from the atmosphere, so the $[H_2CO_3]$ value in the denominator of the $K_{a_1}$ expression will be constant at $1.4 \times 10^{-5}$ *M*.

**Analyze** The large difference between the $K_{a_1}$ and $K_{a_2}$ values means that the pH of the solution is controlled by the first ionization equilibrium:

$$H_2CO_3(aq) + H_2O(\ell) \rightleftharpoons HCO_3^-(aq) + H_3O^+(aq) \qquad K_{a_1} = 4.3 \times 10^{-7}$$

Because of the small value of $K_{a_1}$ and the small concentration of dissolved $CO_2$, we should obtain a pH value that is less than 7 but a lot closer to 7 than 0.

**Solve** First we set up a RICE table in which $x = [H_3O^+] = [HCO_3^-]$ at equilibrium and the value of $[H_2CO_3]$ is a constant $1.4 \times 10^{-5}$ *M*.

| Reaction (R) | $H_2CO_3(aq) + H_2O(\ell)$ | $\rightleftharpoons$ | $HCO_3^-(aq)$ | + | $H_3O^+(aq)$ |
|---|---|---|---|---|---|
| | $[H_2CO_3]$ (*M*) | | $[HCO_3^-]$ (*M*) | | $[H_3O^+]$ (*M*) |
| Initial (I) | $1.4 \times 10^{-5}$ | | 0 | | 0 |
| Change (C) | 0 | | $+x$ | | $+x$ |
| Equilibrium (E) | $1.4 \times 10^{-5}$ | | $x$ | | $x$ |

$$K_{a_1} = \frac{[HCO_3^-][H_3O^+]}{[H_2CO_3]} = \frac{(x)(x)}{1.4 \times 10^{-5}} = 4.3 \times 10^{-7}$$

so

$$x^2 = 6.02 \times 10^{-12}$$

Therefore,

$$x = [H_3O^+] = 2.45 \times 10^{-6}\ M$$

Taking the negative logarithm of $[H_3O^+]$ to calculate pH:

$$pH = -\log[H_3O^+] = -\log(2.45 \times 10^{-6}\ M) = 5.61$$

**Think About It** Carbonic acid is a weak acid, and its concentration here is small, so obtaining a pH value that is only about 1.4 units below neutral pH is reasonable. The calculated pH value is expressed with only two significant figures after the decimal point because the concentration of the acid and its $K_{a_1}$ value were known to only two.

**Practice Exercise** The proximity of the calculated pH value to 7.00 raises the question of whether the autoionization of water contributes significantly to $[H_3O^+]$ in the rainwater sample. Recalculate the pH of the rainwater sample in Sample Exercise 15.16, assuming that the initial concentration of $H_3O^+ = 1.00 \times 10^{-7}$ *M*.

Some acids have three ionizable H atoms per molecule. Two important triprotic acids are phosphoric acid ($H_3PO_4$) and citric acid, the acid responsible for the tart flavor of citrus fruits. Note in **Table 15.9** that $K_{a_1} > K_{a_2} > K_{a_3}$ for both acids. This pattern is much like that for the $K_{a_1} > K_{a_2}$ values of diprotic acids and for the same reason: removing a second $H^+$ ion from the negatively charged ion formed after the first $H^+$ ion is removed is difficult, and removing a third $H^+$ ion from an ion with a 2− charge is even more difficult.

**TABLE 15.9** Ionization Equilibria for Two Triprotic Acids

| Reaction | $K_a$ |
|---|---|
| **Phosphoric Acid** | |
| (1) $HO{-}P({=}O)(OH){-}OH + H_2O \rightleftharpoons HO{-}P({=}O)(OH){-}O^- + H_3O^+$ | $K_{a_1} = 7.11 \times 10^{-3}$ |
| (2) $HO{-}P({=}O)(OH){-}O^- + H_2O \rightleftharpoons {}^-O{-}P({=}O)(OH){-}O^- + H_3O^+$ | $K_{a_2} = 6.32 \times 10^{-8}$ |
| (3) ${}^-O{-}P({=}O)(OH){-}O^- + H_2O \rightleftharpoons {}^-O{-}P({=}O)(O^-){-}O^- + H_3O^+$ | $K_{a_3} = 4.5 \times 10^{-13}$ |
| **Citric Acid** | |
| (1) $HO{-}C(CH_2COOH)(CH_2COOH){-}COOH + H_2O \rightleftharpoons HO{-}C(CH_2COO^-)(CH_2COOH){-}COOH + H_3O^+$ | $K_{a_1} = 7.44 \times 10^{-4}$ |
| (2) $HO{-}C(CH_2COO^-)(CH_2COOH){-}COOH + H_2O \rightleftharpoons HO{-}C(CH_2COO^-)(CH_2COOH){-}COO^- + H_3O^+$ | $K_{a_2} = 1.73 \times 10^{-5}$ |
| (3) $HO{-}C(CH_2COO^-)(CH_2COOH){-}COO^- + H_2O \rightleftharpoons HO{-}C(CH_2COO^-)(CH_2COO^-){-}COO^- + H_3O^+$ | $K_{a_3} = 4.02 \times 10^{-7}$ |

**CONCEPT TEST**

Do you expect the second or third acid ionization step in phosphoric acid and citric acid to influence the pH of 0.100 *M* solutions of either acid? Why or why not?

## 15.9 Acidic and Basic Salts

Seawater and the freshwater in many rivers and lakes have pH values that range from weakly basic to weakly acidic. How can these waters be more basic than the acidic rainwater (pH $\leq$ 5.6) that serves, directly or indirectly, as their water supply? When rain soaks into the ground, its pH changes as it flows through soils that contain basic components. To understand the chemical processes that produce neutral or slightly basic groundwater, we first need to examine the acid–base properties of some common ionic compounds present in these waters.

As discussed in Chapter 8, soluble ionic compounds separate into their component ions when they dissolve in water. For example, a 0.01 *M* solution of NaCl contains 0.01 *M* $Na^+$ ions and 0.01 *M* $Cl^-$ ions. It is also a neutral solution. Neither $Na^+$ ions nor $Cl^-$ ions hydrolyze (react with water) to form $H_3O^+$ or $OH^-$ ions when they dissolve in water. The $Cl^-$ ion is the conjugate base of a strong acid (HCl), so it is such a weak base that its base strength can be considered negligible.

When NaF dissolves in water, however, it produces $F^-$ ions, which are the conjugate base of HF, a weak acid. Therefore, $F^-$ ions are weakly basic, producing at least some $OH^-$ ions when they dissolve in water (**Figure 15.16**). The sodium ions don't hydrolyze and are merely spectator ions in the reaction, as with NaCl.

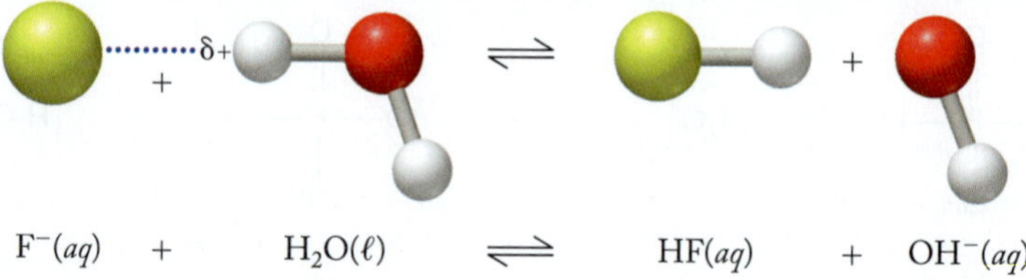


$$F^-(aq) + H_2O(\ell) \rightleftharpoons HF(aq) + OH^-(aq)$$

**FIGURE 15.16** In aqueous solution, $F^-$ ions are weakly basic.

If salts that contain the conjugate bases of weak acids are basic, then salts that contain the conjugate acids of weak bases are acidic. One example of such a salt is $NH_4Cl$. Its $Cl^-$ ions don't hydrolyze and do not affect pH, but the $NH_4{}^+$ ions are the conjugate acid of $NH_3$, a weak base. As a result, $NH_4{}^+$ ions are weakly acidic, producing at least some $H_3O^+$ ions when they dissolve in water (**Figure 15.17**). Consequently, aqueous solutions of $NH_4Cl$ are weakly acidic.

**Table 15.10** summarizes how salts can be acidic, basic, or neutral depending on whether they include cations that are the conjugate acids of weak bases, anions that are the conjugate bases of weak acids, or both. Note that salts that contain

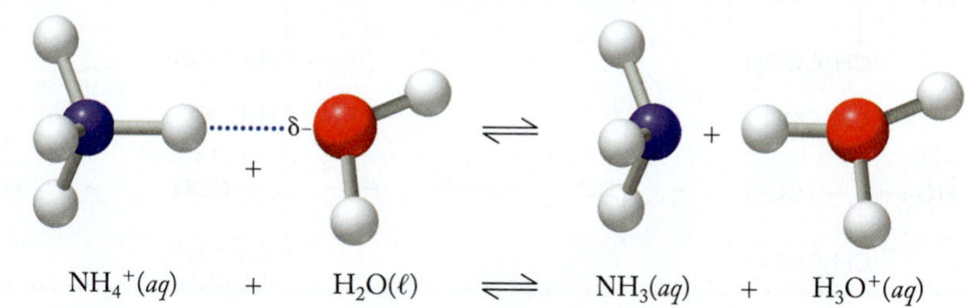


$$NH_4{}^+(aq) + H_2O(\ell) \rightleftharpoons NH_3(aq) + H_3O^+(aq)$$

**FIGURE 15.17** In aqueous solution, $NH_4{}^+$ ions are weakly acidic.

**TABLE 15.10** Acid–Base Properties of Salts

| Anion of a | Cation of a | Aqueous Solutions Are | Example |
|---|---|---|---|
| Strong acid | Strong base | Neutral | $NaCl$ |
| Strong acid | Weak base | Acidic | $NH_4Cl$ |
| Weak acid | Strong base | Basic | $NaF$ |
| Weak acid | Weak base | Neutral,[a]<br>Acidic,[b]<br>or Basic[c] | $CH_3COONH_4$<br>$NH_4F$<br>$NH_4HCO_3$ |

[a]If $K_a$ (of weak acid) $= K_b$ (of weak base)
[b]If $K_a$ (of weak acid) $> K_b$ (of weak base)
[c]If $K_a$ (of weak acid) $< K_b$ (of weak base)

both the conjugate base of a weak acid and the conjugate acid of a weak base may be acidic, basic, or neutral, depending on the relative strengths of the acid and base. Ammonium acetate represents the rare example of a salt in which the strengths of the conjugate acid (acetic acid) and the base (ammonia) happen to be the same: $K_a$ of acetic acid $= K_b$ of ammonia $= 1.76 \times 10^{-5}$. As a result, ammonium acetate is a neutral salt.

## CONCEPT TEST

Pair one of three cations ($K^+$, $NH_4^+$, and $Ca^{2+}$) with one of three anions ($F^-$, $CH_3COO^-$, $NO_3^-$) to form (a) an acidic salt, (b) a basic salt, and (c) a neutral salt.

### SAMPLE EXERCISE 15.17 Predicting Whether a Salt Is Acidic, Basic, or Neutral — LO9

NaClO is the active ingredient in chlorine bleach. Is an aqueous solution of NaClO acidic, basic, or neutral? Write a chemical equation to explain your answer.

**Collect, Organize, and Analyze** Sodium ions do not hydrolyze and do not affect the pH of aqueous solutions. However, $ClO^-$ ions are the conjugate base of HClO, which is a weak acid (Table 15.2). Therefore, $ClO^-$ ions should be weak bases that partially hydrolyze in water and generate $OH^-$ ions:

$$ClO^-(aq) + H_2O(\ell) \rightleftharpoons HClO(aq) + OH^-(aq)$$

**Solve** Because the hydrolysis of $ClO^-$ ions produces $OH^-$ ions, solutions of NaClO are weakly basic.

**Think About It** Any sodium salt that contains an anion that is the conjugate base of a weak acid produces weakly basic aqueous solutions.

**Practice Exercise** Write a chemical equation for the hydrolysis reaction that explains why an aqueous solution of $K_2SO_4$ is basic.

Salts can be acidic, basic, or neutral, but how do we calculate the pH of their aqueous solutions? Our approach is much like the one we used to calculate the pH of solutions of weak acids and bases, but an extra step is involved. For example, if we want to calculate the pH of a solution of ammonium chloride, we need to know the equilibrium constant of the reaction in which the $NH_4^+$ ion functions as a Brønsted–Lowry acid:

$$NH_4^+(aq) + H_2O(\ell) \rightleftharpoons NH_3(aq) + H_3O^+(aq)$$

$K_a$ values are typically not listed for conjugate acids of weak bases. Instead, the $K_b$ values for the weak bases are listed, as for ammonia in Appendix 5:

$$NH_3(aq) + H_2O(\ell) \rightleftharpoons NH_4^+(aq) + OH^-(aq) \qquad K_b = 1.76 \times 10^{-5}$$

The strengths of conjugate acids and bases are complementary: the stronger one is, the weaker the other, so we can derive the $K_a$ value for ammonium ions from the $K_b$ value for ammonia. To see how, let's write the $K_a$ and $K_b$ equilibrium constant expressions for $NH_4^+$ and $NH_3$:

$$K_a = \frac{[NH_3][H_3O^+]}{[NH_4^+]} \qquad K_b = \frac{[NH_4^+][OH^-]}{[NH_3]}$$

These expressions are similar, although any shared terms that appear in the numerator of one expression are found in the denominator of the other. When we multiply the two expressions together,

$$K_a \times K_b = \frac{\cancel{[NH_3]}[H_3O^+]}{\cancel{[NH_4^+]}} \times \frac{\cancel{[NH_4^+]}[OH^-]}{\cancel{[NH_3]}} = [H_3O^+][OH^-]$$

we get the $K_w$ expression for the autoionization of water:

$$K_a \times K_b = K_w \qquad (15.19)$$

Equation 15.19 is useful because (1) it works for all conjugate acid–base pairs, and (2) it allows us to calculate either the $K_b$ of the anion in a basic salt from the $K_a$ of its conjugate acid or the $K_a$ of the cation in an acidic salt from the $K_b$ of its conjugate base.

**FIGURE 15.18** The active ingredient in chlorine bleach is sodium hypochlorite.

**SAMPLE EXERCISE 15.18** Calculating the pH of a Solution of a Basic Salt **LO9**

The bottle of chlorine bleach shown in **Figure 15.18** holds an aqueous solution that contains 82.5 g/L of NaClO. What is the pH of this solution at 25°C?

**Collect and Organize** We know the concentration of an aqueous solution of hypochlorite, NaClO, and we are asked to calculate its pH. We determined in Sample Exercise 15.17 that NaClO is a basic salt because the $ClO^-$ ion is the conjugate base of HClO, a weak acid. Sodium ions do not hydrolyze and are spectator ions as $ClO^-$ ions react with molecules of $H_2O$, forming molecules of HClO and $OH^-$ ions. This reaction is like the one shown in Figure 15.16 for the hydrolysis of $F^-$ ions. The $K_a$ of HClO is $2.9 \times 10^{-8}$.

**Analyze** First we need to convert the $K_a$ value for HClO into the $K_b$ value for the $ClO^-$ ion by using Equation 15.19: $K_a \times K_b = K_w$. We also need to convert the concentration of NaClO from g/L to molarity. That value will be the initial value of the reactant in a RICE table based on its hydrolysis reaction: $ClO^-(aq) + H_2O(\ell) \rightleftharpoons HClO(aq) + OH^-(aq)$. We will solve for $[OH^-]$, then pOH, and then pH. The concentration of the solution should be about 1 *M*, and $K_b$ should be a little more than $10^{-7}$. Therefore, $[OH^-]$ of a ~1 *M* solution should be near the square root of that value, or $\sim 10^{-3}$, making pOH ≈ 3 and pH ≈ 11.

**Solve**

Calculating $K_b$: $$K_b = \frac{K_w}{K_a} = \frac{1.0 \times 10^{-14}}{2.9 \times 10^{-8}} = 3.45 \times 10^{-7}$$

and the molar concentration of NaClO to use in the RICE table:

$$\frac{82.5\ \text{g}}{\text{L}} \times \frac{1\ \text{mol}}{74.44\ \text{g}} = 1.108\ \text{mol/L}$$

| Reaction (R) | $ClO^-(aq) + H_2O(\ell)$ | $\rightleftharpoons$ | $HClO(aq)$ | + | $OH^-(aq)$ |
|---|---|---|---|---|---|
| | $[ClO^-]$ (*M*) | | $[HClO]$ (*M*) | | $[OH^-]$ (*M*) |
| Initial (I) | 1.108 | | 0 | | 0 |
| Change (C) | $-x$ | | $+x$ | | $+x$ |
| Equilibrium (E) | $1.108 - x$ | | $x$ | | $x$ |

To solve for $x$, we make the simplifying assumption that $x$ will be smaller than 1.108 *M* because the value of $K_b$ is so small ($<< 10^{-5}$):

$$K_b = \frac{[HClO][OH^-]}{[ClO^-]} = \frac{(x)(x)}{1.108 - x} \approx \frac{x^2}{1.108} = 3.45 \times 10^{-7}$$

$$x = [OH^-] = 6.18 \times 10^{-4}$$

Calculating pOH: $\text{pOH} = -\log[OH^-] = -\log(6.18 \times 10^{-4}) = 3.21$

and pH: $\text{pH} = 14.00 - \text{pOH} = 14.00 - 3.21 = 10.79$

**Think About It** The calculated pH (10.79) is close to our estimate of 11 and confirms that NaClO is indeed a basic salt. Also, the calculated $[OH^-]$ value is much less than 5% of the initial $[ClO^-]$ value, so our assumption that we could ignore the $-x$ term in the denominator of the $K_b$ expression is justified.

**Practice Exercise** The pH of swimming pools is made slightly basic by spreading solid $Na_2CO_3$ across the surface. Another approach involves preparing concentrated solutions of $Na_2CO_3$ and slowly adding them to the water circulating through the pool pump and filter. What is the pH of an aqueous solution of 0.100 *M* $Na_2CO_3$?

## SAMPLE EXERCISE 15.19 Calculating the pH of a Solution of an Acidic Salt — LO9

Aqueous solutions of ammonium chloride are sometimes used in organic chemistry when weakly acidic reaction conditions are needed. What is the pH of 0.25 *M* $NH_4Cl$ at 25°C?

**Collect and Organize** We are asked to calculate the pH of a solution of $NH_4Cl$. When $NH_4Cl$ dissolves in water, $NH_4^+$ and $Cl^-$ ions are released into solution. The $NH_4^+$ ion is the conjugate acid of $NH_3$, a weak base. Chloride ions do not hydrolyze and are only spectator ions as $NH_4^+$ ions react with water molecules, forming molecules of $NH_3$ and $H_3O^+$ ions as shown in Figure 15.17.

**Analyze** The $K_a$ value of $NH_4^+$ can be calculated by dividing $K_w$ by the $K_b$ of ammonia. The $K_b$ value of ammonia is close to $10^{-5}$, so the $K_a$ value of the ammonium ion will be close to $10^{-9}$. Therefore, $[H_3O^+]$ of a 0.25 *M* solution should be the square root of $\sim 10^{-10}$, or $\sim 10^{-5}$, giving a pH $\approx 5$.

**Solve** The $K_a$ expression for the $NH_4^+$ ion is

$$K_a = \frac{[NH_3][H_3O^+]}{[NH_4^+]}$$

Rearranging Equation 15.19 to solve for $K_a$:

$$K_a = \frac{K_w}{K_b} = \frac{1.00 \times 10^{-14}}{1.76 \times 10^{-5}} = 5.68 \times 10^{-10} = \frac{[NH_3][H_3O^+]}{[NH_4^+]}$$

We set up a RICE table in which we make the usual assumptions that the reaction is the only significant source of $H^+$ and that $x = [H_3O^+] = [NH_3]$ at equilibrium:

| Reaction (R) | $NH_4^+(aq) + H_2O(\ell)$ | $\rightleftharpoons$ | $NH_3(aq)$ | + | $H_3O^+(aq)$ |
|---|---|---|---|---|---|
| | $[NH_4^+]$ (*M*) | | $[NH_3]$ (*M*) | | $[H_3O^+]$ (*M*) |
| Initial (I) | 0.25 | | 0 | | 0 |
| Change (C) | $-x$ | | $+x$ | | $+x$ |
| Equilibrium (E) | $0.25 - x$ | | $x$ | | $x$ |

Given the very small value of $K_a$, we can make the simplifying assumption that $(0.25\ M - x) \approx 0.25\ M$, which gives us

$$K_a = \frac{[NH_3][H_3O^+]}{[NH_4^+]} = \frac{(x)(x)}{0.25 - x} \approx \frac{x^2}{0.25} = 5.68 \times 10^{-10}$$

$$x = [H_3O^+] = 1.19 \times 10^{-5}$$

$$pH = -\log[H_3O^+] = -\log(1.19 \times 10^{-5}) = 4.92$$

**Think About It** This result matches our prediction closely and confirms that $NH_4Cl$ is an acidic salt. The calculated $[H_3O^+]$ is much less than 5% of the initial concentration of $NH_4^+$, so our simplifying assumption is valid.

 **Practice Exercise** What is the difference in pH between a 0.25 *M* solution of dimethyl ammonium chloride, $(CH_3)_2NH_2Cl$, and a 0.25 *M* $NH_4Cl$ solution?

## SAMPLE EXERCISE 15.20 Integrating Concepts: The pH of Human Blood

We began this chapter by noting how essential it is that our circulation and respiration systems efficiently remove from our bodies the $CO_2$ produced in our cells. An enzyme, carbonic anhydrase, plays a key role in this process by speeding up the rate at which $CO_2$ hydrolyzes,

$$CO_2(g) + H_2O(\ell) \rightleftharpoons H_2CO_3(aq)$$

and the rate at which $H_2CO_3$ ionizes,

$$H_2CO_3(aq) + H_2O(\ell) \rightleftharpoons HCO_3^-(aq) + H_3O^+(aq)$$

a. Does carbonic anhydrase increase the acid strength of dissolved $CO_2$—that is, the $K_{a_1}$ of carbonic acid on the basis of the total concentration of $CO_2$ in solution?

b. Suppose that, during strenuous exercise, the total concentration of dissolved $CO_2$ in the blood flowing through muscle tissues is $2.7 \times 10^{-3}$ *M*. If the concentration of $HCO_3^-$ ions in the blood is 0.028 *M*, what is the pH of the blood?

**Collect and Organize** We are asked whether the presence of an enzyme that increases the rates at which $CO_2$ hydrolyzes and undergoes acid ionization makes carbonic acid a stronger acid. We then need to calculate the pH of blood in which the initial $[HCO_3^-]$ and $[CO_2]$ values are known. Carbonic acid is a weak diprotic acid; its $K_{a_1}$ is $4.3 \times 10^{-7}$.

**Analyze**

a. We learned in Chapter 14 that catalysts speed up reactions but do not alter their equilibrium constants.

b. Calculating the pH of a carbonic acid solution will require a RICE table based on $K_{a_1}$ in which $[H_3O^+] = x$ at equilibrium but in which the initial concentration of $HCO_3^-$ is 0.028 *M*, not zero. The normal pH of human blood is close to 7.4, so the calculated pH should be close to 7.4.

**Solve**

a. Carbonic anhydrase should not affect the value of $K_{a_1}$, although it would produce increases in both $[H_2CO_3]$ and $[HCO_3^-]$. These increases in the numerator and denominator of the $K_{a_1}$ expression offset each other and do not affect pH.

b. Set up a RICE table based on the $K_{a_1}$ reaction, where the values in the $[H_2CO_3]$ column represent the total concentration of dissolved $CO_2$ ($[CO_2(aq)] + [H_2CO_3(aq)]$):

| Reaction (R) | $H_2CO_3(aq)$ + $H_2O(\ell)$ ⇌ | $HCO_3^-(aq)$ + | $H_3O^+(aq)$ |
|---|---|---|---|
| | $[H_2CO_3]$ (*M*) | $[HCO_3^-]$ (*M*) | $[H_3O^+]$ (*M*) |
| Initial (I) | $2.7 \times 10^{-3}$ | 0.028 | 0 |
| Change (C) | $-x$ | $+x$ | $+x$ |
| Equilibrium (E) | $2.7 \times 10^{-3} - x$ | $0.028 + x$ | $x$ |

In solving for $x$, we make the simplifying assumption that its value will be much less than 0.028 *M* (as it must be if the calculated pH is close to 7.4, which means that $[H_3O^+] \approx 10^{-7}$), and we ignore the very small $x^2$ term in the numerator of the calculation:

$$K_{a_1} = \frac{[HCO_3^-][H_3O^+]}{[H_2CO_3]} = \frac{(0.028 + x)(x)}{2.7 \times 10^{-3}} \approx \frac{0.028x}{2.7 \times 10^{-3}}$$

$$= 4.3 \times 10^{-7}$$

$$x = [H_3O^+] = 4.15 \times 10^{-8}$$

Therefore, the pH of the blood is

$$pH = -\log[H_3O^+] = -\log(4.15 \times 10^{-8}) = 7.38$$

**Think About It** As expected, the calculated pH is slightly basic. As discussed in Section 15.1, the high concentration of bicarbonate ions in the blood shifts the carbonic acid ionization equilibrium to the left, lowering $[H_3O^+]$ and raising pH. We explore the impact of having a second source of product ions on other equilibria in Chapter 16.

## SUMMARY

**LO1** The strengths of acids and bases are related to the values of their acid and base ionization constants, $K_a$ and $K_b$. Most acids are weak, which means that their $K_a$ values are much less than 1 and that they ionize only partially in water. (Section 15.2)

**LO2** The strengths of acids are related to the strengths of dipole–dipole interactions between their ionizable H atoms and the O atoms of water molecules, and to the stability of the anions they form when they release $H^+$ ions. The stability of oxoanions is enhanced by multiple oxygen atoms bonded to their central atoms, an arrangement that disperses the negative charge(s) over the anion. (Section 15.2)

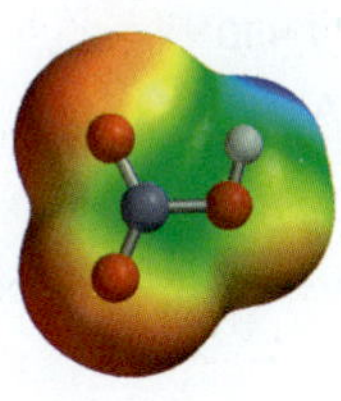

**LO3** The soluble hydroxides of the group 1 and 2 elements are strong bases. Nitrogen-containing bases are weak bases whose strengths are related to their N atoms' ability to use their lone pairs of electrons to bond with $H^+$ ions. (Section 15.3)

**LO4** When an acid (HA) ionizes, it forms its **conjugate base**, $A^-$. When a base (B) reacts with a $H^+$ ion, it forms its **conjugate acid**, $HB^+$. (Section 15.4)

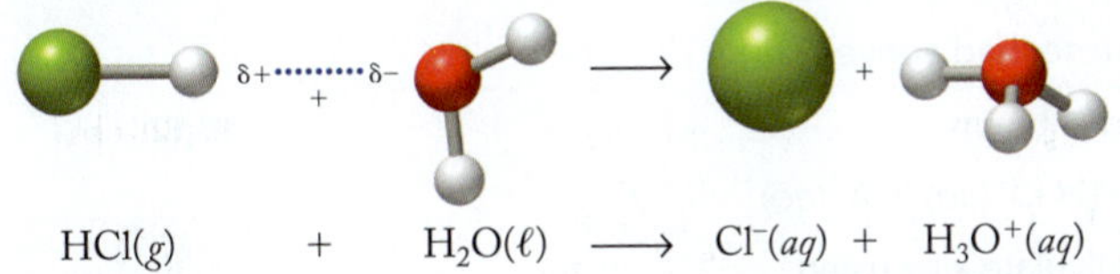


**LO5** In a neutral solution, $[H_3O^+] = [OH^-] = 1.0 \times 10^{-7}$. The **pH** scale is a logarithmic scale for expressing the acidic or basic strength of solutions. Acidic solutions have pH values less than 7; basic solutions have pH values greater than 7. Because pH is the negative logarithm of the $H_3O^+$ concentration, the higher the pH, the lower the $H_3O^+$ concentration. An increase in one pH unit represents a decrease in $[H_3O^+]$ to 1/10 of its initial value. Likewise, **pOH** is the negative logarithm of the $OH^-$ concentration. The sum of pH and pOH equals 14 in an aqueous solution at 25°C. (Section 15.5)

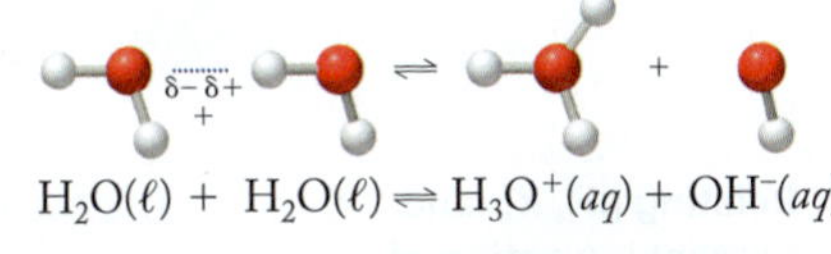


**LO6** The $K_a$ and $K_b$ values of acids and bases can be used to identify the primary species present in solution and to calculate the extent to which their solutions are ionized—that is, their **degree of ionization** or **percent ionization**—and vice versa. (Section 15.6)

**LO7** To calculate the pH of a weak acid or base, use a RICE table based on the acid or base ionization reaction to determine the equilibrium value of $[H_3O^+]$ or $[OH^-]$ in a solution of a weak acid or base. (Section 15.7)

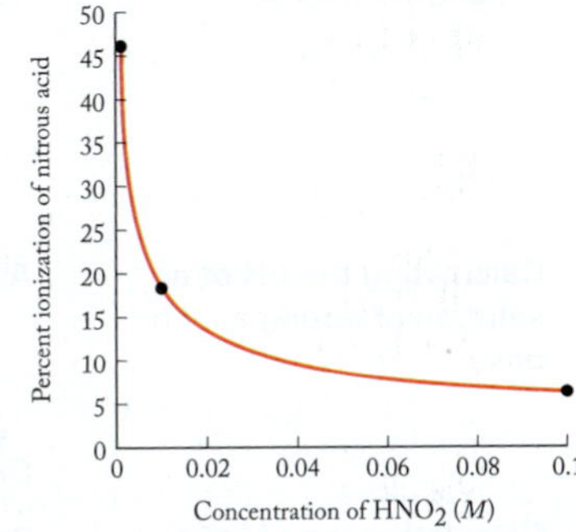


**LO8 Polyprotic acids** can undergo more than one acid ionization reaction, but for most, the first ionization reaction is the one that controls pH. (Section 15.8)

**LO9** A salt solution is acidic if the cation in the salt is the conjugate acid of a weak base and the anion is the conjugate base of a strong acid. A salt solution is basic if the anion in the salt is the conjugate base of a weak acid and the cation is the conjugate acid of a strong base. (Section 15.9)

## PARTICULATE PREVIEW WRAP-UP

$NH_3$ is the Brønsted–Lowry base (proton acceptor) and $H_2O$ is the Brønsted–Lowry acid (proton donor) in the forward reaction. In the reverse reaction, the ammonium ion donates a proton (Brønsted–Lowry acid), while the hydroxide ion accepts a proton (Brønsted–Lowry base). If $K << 1$, then the equilibrium lies far to the left and the most abundant acid and most abundant base in solution are $H_2O$ and $NH_3$, respectively.

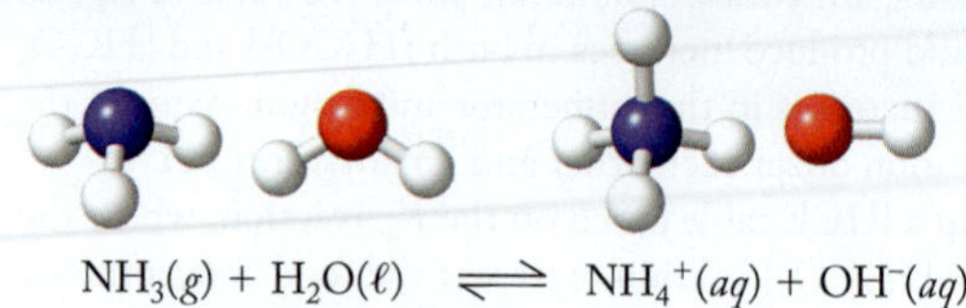

$$NH_3(g) + H_2O(\ell) \rightleftharpoons NH_4^+(aq) + OH^-(aq)$$

## PROBLEM-SOLVING SUMMARY

| Type of Problem | Concepts and Equations | Sample Exercises |
|---|---|---|
| **Ranking acids and bases from strongest to weakest** | The strengths of acids and bases are proportional to their $K_a$ and $K_b$ values.<br>The strengths of oxoacids increase with increasing electronegativity of the central atom or with increasing numbers of oxygen atoms bonded to it. | **15.1–15.3** |
| **Identifying acid–base conjugate pairs and predicting their relative strengths** | The formula of the base in a conjugate pair is the formula of the acid with one fewer $H^+$ ion.<br>The stronger an acid or base, the weaker its conjugate base or acid. | **15.4, 15.5** |
| **Interconverting $[H_3O^+]$, $[OH^-]$, pH, and pOH** | Use the following:<br>$pH = -\log[H_3O^+]$ (15.12)<br>$pOH = -\log[OH^-]$ (15.13)<br>$pK_w = pH + pOH = 14.00$ (15.14) | **15.6–15.8** |
| **Relating pH, $K_a$, and percent ionization of a weak acid** | Use the following equations:<br>$[H_3O^+] = 10^{-pH}$<br>$K_a = \frac{[A^-][H_3O^+]}{[HA]}$ (15.15)<br>$\text{Percent ionization} = \frac{[H_3O^+]_{equilibrium}}{[HA]_{initial}} \times 100\%$ (15.16) | **15.9** |
| **Determining percent ionization and $K_b$ given the pH of a weak base** | Use the following:<br>$[OH^-] = 10^{-pOH}$<br>$K_b = \frac{[BH^+][OH^-]}{[B]}$ (15.17)<br>$\text{Percent ionization} = \frac{[OH^-]_{equilibrium}}{[B]_{initial}} \times 100\%$ (15.18) | **15.10** |
| **Calculating the pH of a solution of strong acid or base** | Assume 100% ionization so that<br>$[H_3O^+] = [HX]$ and<br>$[OH^-] = [MOH]$ or $[OH^-] = 2\,[M(OH)_2]$<br>Calculate pH from $[H_3O^+]$ or $[OH^-]$ as described above. | **15.11** |
| **Calculating the pH of a solution of weak acid HA** | Set up a RICE table based on the $K_a$ equilibrium<br>$HA(aq) + H_2O(\ell) \rightleftharpoons H_3O^+(aq) + A^-(aq)$<br>Let $x = [H_3O^+] = [A^-]$ at equilibrium. Calculate $x$ by using<br>$K_a = \frac{[A^-][H_3O^+]}{[HA]}$<br>Then calculate $pH = -\log[H_3O^+]$. | **15.12** |

| Type of Problem | Concepts and Equations | Sample Exercises |
|---|---|---|
| **Calculating the pH of a solution of weak base B** | Set up a RICE table based on the equilibrium $B(aq) + H_2O(\ell) \rightleftharpoons BH^+(aq) + OH^-(aq)$ Let $x = [OH^-] = [HB^+]$ at equilibrium. Calculate $x$ by using $K_b = \frac{[BH^+][OH^-]}{[B]}$ Then use $pOH = -\log[OH^-]$ and $pH = 14.00 - pOH$ | 15.13 |
| **Calculating the pH of a solution of very dilute acidic (or very dilute basic) solution while considering the autoionization of water** | Set up a RICE table based on the equilibrium $H_2O(\ell) + H_2O(\ell) \rightleftharpoons H_3O^+(aq) + OH^-(aq)$ Let $x = [H_3O^+] = [OH^-]$ due to autoionization. Add $x$ to the $[H_3O^+]$ due to the dilute acid (or to the $[OH^-]$ due to the weak base). Solve for $x$ by using $K_w = [H_3O^+][OH^-]$ (15.11) Then convert $[H_3O^+]$ to pH. | 15.14 |
| **Calculating the pH of a solution of a strong diprotic acid** | Assume that $H_2A(aq) + H_2O(\ell) \rightarrow H_3O^+(aq) + HA^-(aq)$ is complete. Set up a RICE table based on the $K_{a_2}$ equilibrium $HA^-(aq) + H_2O(\ell) \rightleftharpoons H_3O^+(aq) + A^{2-}(aq)$ Let $x$ = additional $[H_3O^+]$ from the second ionization step; $[HA^-]_{\text{initial (2nd step)}} = [H_2A]_{\text{initial}}$. Calculate $x$ by using $K_{a_2} = \frac{[H_3O^+][A^{2-}]}{[HA^-]}$ Combine initial and additional $[H_3O^+]$ values and convert to pH. | 15.15 |
| **Calculating the pH of a solution of a weak diprotic acid** | Set up a RICE table based on the $K_{a_1}$ equilibrium $H_2A(aq) + H_2O(\ell) \rightleftharpoons H_3O^+(aq) + HA^-(aq)$ Let $x = [H_3O^+] = [HA^-]$ at equilibrium. Calculate $x$ by using $K_{a_1} = \frac{x^2}{[H_2A] - x}$ Then convert $x = [H_3O^+]$ to pH. | 15.16 |
| **Distinguishing acidic, basic, and neutral salts** | The cations in acidic salts are the conjugate acids of weak bases. The anions in basic salts are the conjugate bases of weak acids. | 15.17 |
| **Calculating the pH of a solution of a basic salt** | Assume that the salt (MA) completely dissociates into $M^+$ and $A^-$. Set up a RICE table for the equilibrium $A^-(aq) + H_2O(\ell) \rightleftharpoons HA(aq) + OH^-(aq)$ Let $x = [HA] = [OH^-] =$ at equilibrium. Calculate $x$ using $K_b = K_w/K_{a\,\text{(of conjugate acid, HA)}}$ $K_b = \frac{K_w}{K_a} = \frac{x^2}{[A^-] - x}$ Then convert $[OH^-]$ to pOH and then pOH to pH. | 15.18 |
| **Calculating the pH of a solution of an acidic salt** | Assume that the salt (BHX) completely dissociates into $BH^+$ and $X^-$. Set up a RICE table for the equilibrium $BH^+(aq) + H_2O(\ell) \rightleftharpoons B(aq) + H_3O^+(aq)$ Let $x = [H_3O^+] = [B]$ at equilibrium. Calculate $x$ by using $K_a = K_w/K_{b\,\text{(of conjugate base, B)}}$ $K_a = \frac{K_w}{K_b} = \frac{x^2}{[BH^+] - x}$ Then convert $x = [H_3O^+]$ to pH. | 15.19 |

## VISUAL PROBLEMS

*(Answers to boldface end-of-chapter questions and problems are in the back of the book.)*

**15.1.** Which of the lines in Figure P15.1 best represents the relation between the percent ionization of acetic acid and its concentration in aqueous solution?

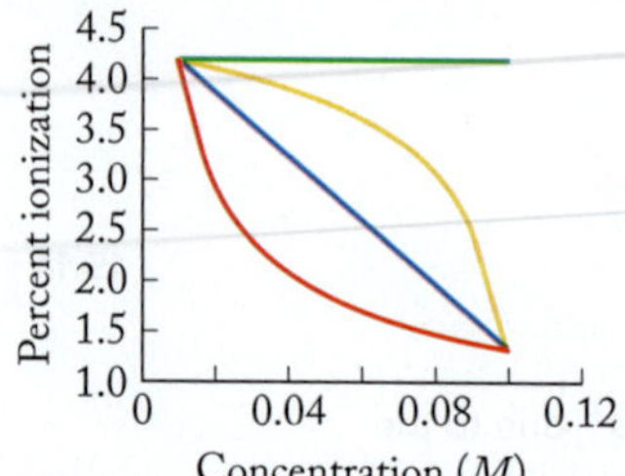


**FIGURE P15.1**

15.2. The graph in Figure P15.2 shows the percent ionization of two acids as a function of concentration in water. Which line describes the behavior of $HNO_3$, and which line describes the behavior of acetic acid ($CH_3COOH$)?

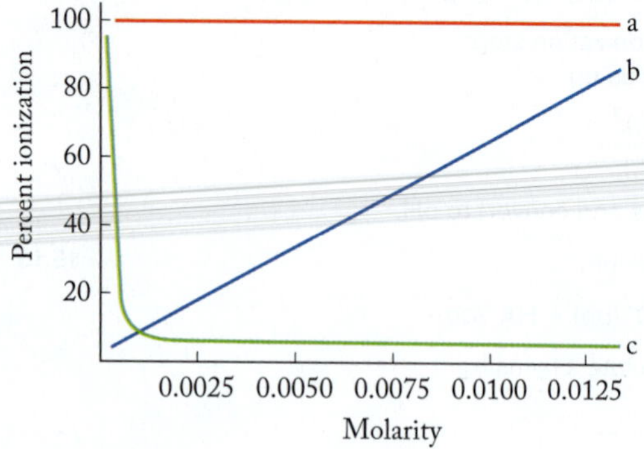


**FIGURE P15.2**

**15.3.** The bar graph in Figure P15.3 shows the percent ionization of $1 \times 10^{-3}$ *M* solutions of three hypohalous acids: HClO, HBrO, and HIO. Which bar corresponds to HIO?

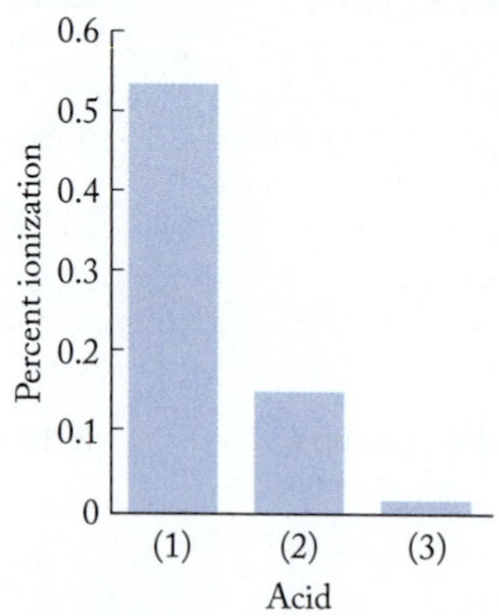


**FIGURE P15.3**

15.4. Figure P15.4 shows the molecular structure of alanine. Is an aqueous solution of alanine acidic, basic, or neutral? Explain your selection or describe what additional information you need to make a more informed selection.

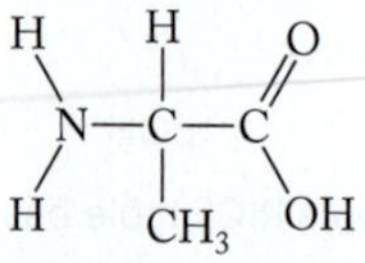

H H O
N—C—C
H $CH_3$ OH

**FIGURE P15.4**

***15.5.** Figure P15.5 shows the condensed molecular structures of piperidine and morpholine. Which is the stronger base? Explain your selection.

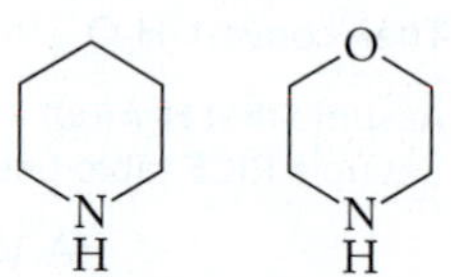


**FIGURE P15.5**

15.6. On the basis of the skeletal structures shown in Figure P15.6, predict whether ethanolamine or ethylamine is the stronger base. Explain your prediction.

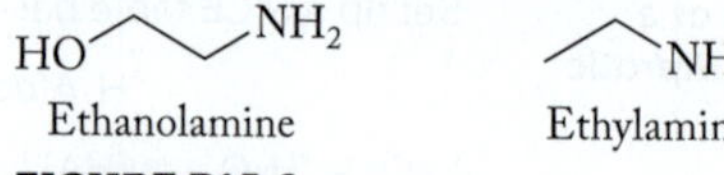


**FIGURE P15.6**

**15.7.** Figure P15.7 shows the structure of pseudoephedrine, a widely used decongestant and stimulant.
a. Is pseudoephedrine an acidic, basic, or neutral compound?
b. Which functional groups in its structure gives the property you selected in part a?

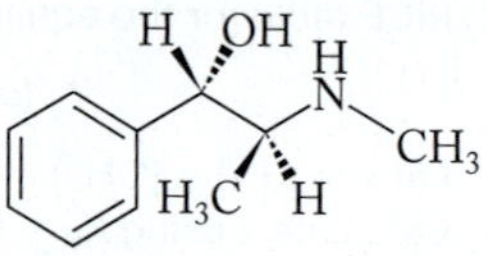

H OH H N $CH_3$ $H_3C$ H

**FIGURE P15.7**

*15.8. The p$K_a$ values of the ions shown in Figure P15.8 increase as more methyl ($-CH_3$) groups are added. Do more methyl groups increase or decrease the strength of the neutral conjugate bases of these ions?

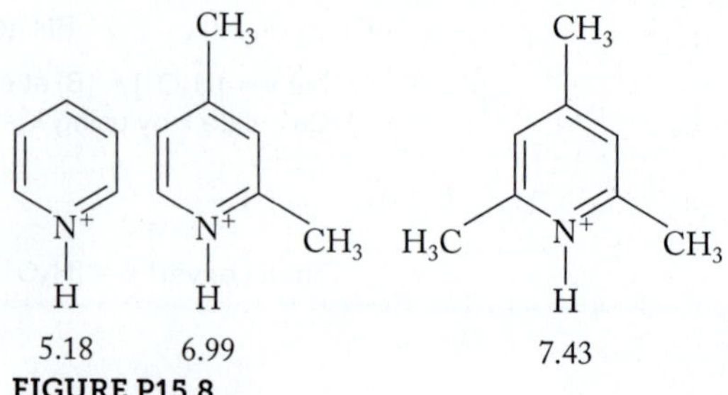


**FIGURE P15.8**

**15.9.** The value of $K_{a_1}$ of phosphorous acid, $H_3PO_3$, is nearly the same as that of phosphoric acid, $H_3PO_4$. Identify the ionizable hydrogen atoms in Figure P15.9.

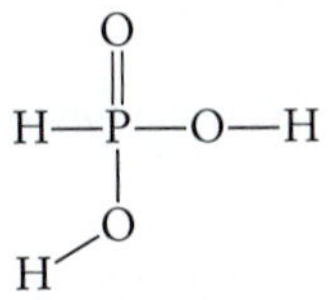

$$\mathrm{H{-}P({=}O)(OH){-}O{-}H}$$

**FIGURE P15.9**

15.10. Use representations [A] through [I] in Figure P15.10 to answer questions a–f.
a. Identify the strong acid(s).
b. Identify the weak acid(s).
c. Identify the strong base(s).
d. Identify the weak base(s).
e. Which compound can function as both a weak acid and a weak base?
f. Which compound contains hydrogen atoms but is neither an acid nor a base?

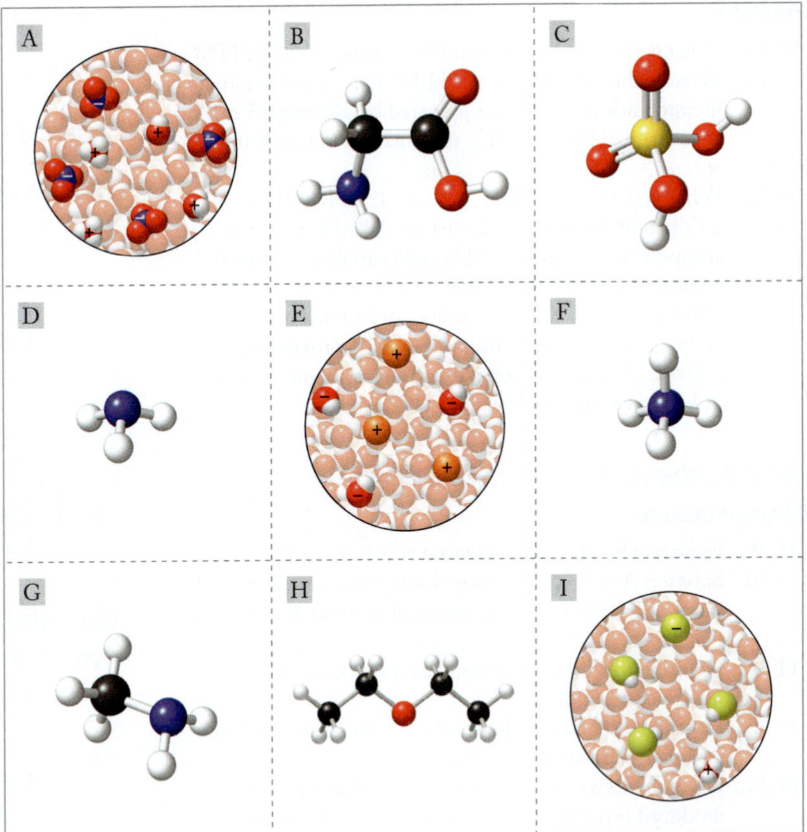


**FIGURE P15.10**

# QUESTIONS AND PROBLEMS

## Strong and Weak Acids and Bases

### Concept Review

**15.11.** In an aqueous solution of HF, which compound acts as a Brønsted–Lowry acid and which is the Brønsted–Lowry base?

15.12. In an aqueous solution of $HNO_3$, which compound acts as a Brønsted–Lowry acid and which is the Brønsted–Lowry base?

**15.13.** In an aqueous solution of $NH_3$, which species acts as a Brønsted–Lowry acid and which is the Brønsted–Lowry base?

15.14. Both KOH and $Ba(OH)_2$ are strong bases. Does this mean that solutions of the two compounds with the same molarity have the same capacity to accept hydrogen ions? Why or why not?

**15.15.** Identify the acids and bases in the following reactions:
a. $HCl(aq) + NaOH(aq) \rightarrow NaCl(aq) + H_2O(\ell)$
b. $MgCO_3(s) + 2\,HCl(aq) \rightarrow MgCl_2(aq) + CO_2(g) + H_2O(\ell)$
c. $2\,NH_3(aq) + H_2SO_4(aq) \rightarrow (NH_4)_2SO_4(aq)$

15.16. Identify the acids and bases in the following reactions:
a. $(CH_3)_3N(aq) + H_2O(\ell) \rightleftharpoons (CH_3)_3NH^+(aq) + OH^-(aq)$
b. $H_2CO_3(aq) + H_2O(\ell) \rightleftharpoons HCO_3^-(aq) + H_3O^+(aq)$
c. $(CH_3)_3COH(aq) + H_3O^+(aq) \rightleftharpoons (CH_3)_3COH_2^+(aq) + H_2O(\ell)$

**15.17.** Identify the conjugate base of each of the following substances: $HNO_2$, HClO, $H_3PO_4$, and $NH_4^+$.

15.18. Identify the conjugate acid of each of the following species: $(CH_3)_3N$, $CH_3COO^-$, $HSO_4^-$, and $OH^-$.

**15.19.** What is the conjugate acid of the bisulfate ion, $HSO_4^-$, and what is its conjugate base?

*15.20. Compounds that do not ionize in water have been known to ionize in nonaqueous solvents. In such a solvent, what would be the conjugate acid and conjugate base of methanol, $CH_3OH$?

**15.21.** Explain why $H_2SO_4$ is a stronger acid than $H_2SeO_4$.

15.22. Explain why $H_2SO_4$ is a stronger acid than $H_2SO_3$.

**15.23.** Predict which acid in the following pairs of acids is stronger: (a) $H_2SO_3$ or $H_2SeO_3$; (b) $H_2SeO_4$ or $H_2SeO_3$.

15.24. Trifluoroacetic acid, $CF_3COOH$, is more than $10^4$ times as strong as acetic acid, $CH_3COOH$. Explain why.

**Problems**

**15.25.** What is the concentration of $H_3O^+$ ions in 0.65 $M$ $HNO_3$?

15.26. What is the concentration of $H_3O^+$ ions in a solution of hydrochloric acid that was prepared by diluting 7.5 mL of concentrated (11.6 $M$) HCl to a final volume of 100.0 L?

---

**15.27.** What is the value of $[OH^-]$ in 0.0205 $M$ $Ba(OH)_2$?

15.28. Calcium hydroxide, also known as slaked lime, is the cheapest strong base available and is used in industrial processes in which low concentrations of base are required. Only 0.16 g of $Ca(OH)_2$ dissolves in 100 mL of water at 25°C. What is the concentration of hydroxide ions in 250 mL of a solution containing the maximum amount of dissolved calcium hydroxide?

## pH and the Autoionization of Water

**Concept Review**

**15.29.** Explain why pH values decrease as acidity increases.

15.30. Solution A is 100 times more acidic than solution B. What is the difference in the pH values of solution A and solution B?

**15.31.** Describe a solution (solute and concentration) that has a negative pH value.

15.32. Describe a solution (solute and concentration) that has a negative pOH value.

***15.33.** Draw the Lewis structures of the ions that would be produced if pure ethanol underwent autoionization.

*15.34. Liquid ammonia at a temperature of 223 K undergoes autoionization. The value of the equilibrium constant for the autoionization of ammonia is considerably less than that of water. Write an equation for the autoionization of ammonia and suggest a reason why the value of $K$ for the process is less than that of water.

**Problems**

**15.35.** Calculate the pH and pOH of solutions with the following $[H_3O^+]$ or $[OH^-]$ values. Indicate which solutions are acidic, basic, or neutral.
a. $[H_3O^+] = 5.3 \times 10^{-3}$ $M$
b. $[H_3O^+] = 3.8 \times 10^{-9}$ $M$
c. $[H_3O^+] = 7.2 \times 10^{-6}$ $M$
d. $[OH^-] = 1.0 \times 10^{-14}$ $M$

15.36. Calculate the pH and pOH of the solutions with the following hydronium ion or hydroxide ion concentrations. Indicate which solutions are acidic, basic, or neutral.
a. $[OH^-] = 8.2 \times 10^{-11}$ $M$
b. $[OH^-] = 7.7 \times 10^{-6}$ $M$
c. $[H_3O^+] = 3.2 \times 10^{-4}$ $M$
d. $[H_3O^+] = 1.0 \times 10^{-7}$ $M$

---

**15.37.** Calculate the concentration of the following ions in the solution described:
a. $[H_3O^+]$ in $8.4 \times 10^{-4}$ $M$ NaOH
b. $[H_3O^+]$ in $6.6 \times 10^{-5}$ $M$ $Ca(OH)_2$
c. $[OH^-]$ in $4.5 \times 10^{-3}$ $M$ HCl
d. $[OH^-]$ in $2.9 \times 10^{-5}$ $M$ HCl

15.38. Determine the indicated pH or pOH values:
a. pH of a solution whose pOH = 5.5
b. pH of a solution whose pOH = 6.8
c. pOH of a solution whose pH = 9.7
d. pOH of a solution whose pH = 4.4

---

**15.39.** Calculate the pH and pOH of the following solutions:
a. stomach acid in which [HCl] = 0.155 $M$
b. 0.00500 $M$ $HNO_3$
c. a 2:1 mixture of 0.0125 $M$ HCl and 0.0125 $M$ NaOH
d. a 3:1 mixture of 0.0125 $M$ $H_2SO_4$ and 0.0125 $M$ KOH

15.40. Calculate the pH and pOH of the following solutions:
a. 0.0450 $M$ NaOH
b. 0.160 $M$ $Ca(OH)_2$
c. a 1:1 mixture of 0.0125 $M$ HCl and 0.0125 $M$ $Ca(OH)_2$
d. a 2:3 mixture of 0.0125 $M$ $HNO_3$ and 0.0125 $M$ KOH

---

**15.41.** Calculate the pH of $1.33 \times 10^{-9}$ $M$ LiOH.

15.42. Calculate the pH of $6.9 \times 10^{-8}$ $M$ HBr.

## Calculations Involving pH, $K_a$, and $K_b$

**Concept Review**

**15.43.** A researcher prepares 1 $M$ solutions of the following acids: $CH_3COOH$, $HNO_2$, HClO, and HCl.
a. Rank them in order of decreasing $[H_3O^+]$.
b. Rank them in order of increasing strength as acids (weakest to strongest).

15.44. On the basis of the following degree-of-ionization data for 0.100 $M$ solutions, select which acid has the largest $K_a$.

| Acid | $C_6H_5COOH$ | HF | $HN_3$ | $CH_3COOH$ |
|---|---|---|---|---|
| Degree of Ionization (%) | 2.5 | 8.5 | 1.4 | 1.3 |

**15.45.** A 1.0 $M$ aqueous solution of $HNO_3$ is a much better conductor of electricity than is a 1.0 $M$ solution of $HNO_2$. Explain why.

15.46. Hydrogen chloride and water are molecular compounds, yet a solution of HCl dissolved in $H_2O$ is an excellent conductor of electricity. Explain why.

**15.47.** Hydrofluoric acid is a weak acid. Write the mass action expression for its acid ionization reaction.

15.48. **Early Antiseptic** The use of phenol, also known as carbolic acid, was pioneered in the 19th century by Sir Joseph Lister (after whom Listerine was named) as an antiseptic in surgery. Its formula is $C_6H_5OH$ (the red hydrogen atom is ionizable). Write the mass action expression for the acid ionization equilibrium of phenol.

***15.49.** The $K_a$ values of weak acids depend on the solvent in which they dissolve. For example, the $K_a$ of alanine in aqueous ethanol is less than its $K_a$ in water.
a. In which solvent does alanine ionize more?
b. Which is the stronger Brønsted–Lowry base: water or ethanol?

*15.50. The $K_a$ of proline is $2.5 \times 10^{-11}$ in water, $2.8 \times 10^{-11}$ in an aqueous solution that is 28% ethanol, and $1.66 \times 10^{-8}$ in aqueous formaldehyde at 25°C.
a. In which solvent is proline the strongest acid?
b. Rank these compounds in order of their strengths as Brønsted–Lowry bases: water, ethanol, and formaldehyde.

**15.51.** When methylamine, $CH_3NH_2$, dissolves in water, the resulting solution is slightly basic. Which compound is the Brønsted–Lowry acid and which is the base?

*15.52. When 1,2-diaminoethane, $H_2NCH_2CH_2NH_2$, dissolves in water, the resulting solution is basic. Write the formula of the ionic compound formed when hydrochloric acid is added to a solution of 1,2-diaminoethane.

### Problems

**15.53.** **Muscle Physiology** During strenuous exercise, lactic acid, $H_3CCH(OH)COOH$, builds up in muscle tissues. In a 1.00 *M* aqueous solution, 2.94% of lactic acid is ionized. What is the value of its $K_a$?

15.54. **Rancid Butter** The odor of spoiled butter is due in part to butanoic acid, $H_3C(CH_2)_2COOH$, which results from the chemical breakdown of butterfat. A 0.100 *M* solution of butanoic acid is 1.23% ionized. Calculate the value of $K_a$ for butanoic acid.

---

**15.55.** At equilibrium, the value of $[H_3O^+]$ in 0.125 *M* of an unknown acid is $4.07 \times 10^{-3}$ *M*. Determine the percent ionization and the $K_a$ of this acid.

15.56. Nitric acid ($HNO_3$) is a strong acid that is essentially completely ionized in aqueous solutions of concentrations ranging from 1% to 10% (1.5 *M*). However, in more concentrated solutions, part of the nitric acid is present as un-ionized molecules of $HNO_3$. For example, in a 50% solution (7.5 *M*) at 25°C, only 33% of the molecules of $HNO_3$ dissociate into $H^+$ and $NO_3^-$. What is the $K_a$ value of $HNO_3$ at 25°C?

---

**15.57.** What is the percent ionization of a 0.145 *M* hypobromous solution?

15.58. What is the percent ionization of a 0.0290 *M* HIO(*aq*) solution?

---

**15.59.** **Acid Rain I** A weather system moving through the American Midwest produced rain with an average pH of 5.02. By the time the system reached New England, the rain it produced had an average pH of 4.66. How much more acidic was the rain falling in New England?

15.60. **Acid Rain II** A newspaper reported that the "level of acidity" in a sample taken from an extensively studied watershed in New Hampshire in 1998 was "an astounding 200 times lower than the worst measurement" taken in the preceding 23 years. What is this difference expressed in units of pH?

---

**15.61.** The $K_b$ of aminoethanol, $HOCH_2CH_2NH_2$, is $3.1 \times 10^{-5}$.
a. Is aminoethanol a stronger or weaker base than ethylamine, $pK_b = 3.36$?
b. Calculate the pH of $1.67 \times 10^{-2}$ *M* aminoethanol.
c. Calculate the $[OH^-]$ concentration of $4.25 \times 10^{-4}$ *M* aminoethanol.

15.62. **Food Dye** Quinoline is a weakly basic liquid used in manufacturing quinoline yellow, a greenish-yellow dye for foods, and in producing niacin. Its $pK_b$ is 9.15.
a. What is the pH of 0.0752 *M* quinoline?
b. What is the hydroxide ion concentration of the solution in part a?

---

**15.63.** **Painkillers** Morphine is an effective painkiller but is also highly addictive. Codeine is a popular prescription painkiller because it is much less addictive than morphine. Codeine contains a basic nitrogen atom that can be protonated to give the conjugate acid of codeine.
a. Calculate the pH of $1.8 \times 10^{-3}$ *M* morphine if its $pK_b = 5.79$.
b. Calculate the pH of $2.7 \times 10^{-4}$ *M* codeine if the $pK_a$ of the conjugate acid is 8.21.

15.64. The awful odor of dead fish is due mostly to trimethylamine, $(CH_3)_3N$, one of three compounds related to ammonia in which methyl groups replace one, two, or all three H atoms in ammonia.
a. The $K_b$ of trimethylamine $[(CH_3)_3N]$ is $6.5 \times 10^{-5}$ at 25°C. Calculate the pH of $3.00 \times 10^{-4}$ *M* trimethylamine.
b. The $K_b$ of methylamine $[(CH_3)NH_2]$ is $4.4 \times 10^{-4}$ at 25°C. Calculate the pH of $2.88 \times 10^{-3}$ *M* methylamine.
*c. The $K_b$ of dimethylamine $[(CH_3)_2NH]$ is $5.9 \times 10^{-4}$ at 25°C. What concentration of dimethylamine is needed for the solution to have the same pH as the solution in part b?

## Polyprotic Acids

### Concept Review

**15.65.** Why is the $K_{a_2}$ value of phosphoric acid less than its $K_{a_1}$ value but greater than its $K_{a_3}$ value?

15.66. In calculating the pH of 1.0 *M* $H_2SO_3$, we can ignore the $H^+$ ions produced by the ionization of the bisulfite ($HSO_3^-$) ion; however, in calculating the pH of 1.0 *M* sulfuric acid, we cannot ignore the $H^+$ ions produced by the ionization of the bisulfate ion. Why?

**15.67.** Carbonic acid, $H_2CO_3$, is a very weak diprotic acid ($K_{a_1} = 4.3 \times 10^{-7}$), but germanic acid, $H_2GeO_3$, is even weaker ($K_{a_1} = 9.8 \times 10^{-10}$). Suggest a reason why.

**15.68.** Figure P15.68 shows skeletal structures of the dicarboxylic acids malonic acid (left) and oxalic acid (right). The $K_{a_1}$ of malonic acid is about $10^4$ times as large as its $K_{a_2}$, whereas the $K_{a_1}$ of oxalic acid is about $10^3$ times as large as its $K_{a_2}$. Suggest a reason why the separation in $K_a$ values is greater for malonic acid.

Malonic acid Oxalic acid

**FIGURE P15.68**

### Problems

**15.69.** What is the pH of 0.75 $M$ $H_2SO_4$?

15.70. What is the pH of $5.00 \times 10^{-4}$ $M$ $H_2SO_4$?

---

**15.71.** Ascorbic acid (vitamin C) is a weak diprotic acid. What is the pH of 0.250 $M$ ascorbic acid?

15.72. **Rhubarb Pie** The leaves of the rhubarb plant contain high concentrations of diprotic oxalic acid (HOOCCOOH) and must be removed before the stems are used to make rhubarb pie. What is the pH of 0.0288 $M$ oxalic acid?

---

**15.73. Malaria Treatment** Quinine occurs naturally in the bark of the cinchona tree. For centuries, quinine was the only treatment for malaria. Calculate the pH of 0.01050 $M$ quinine.

15.74. Dozens of pharmaceuticals as varied as cyclizine for motion sickness and Viagra for impotence are derived from the organic compound piperazine, whose structure is shown in Figure P15.74.

a. Solutions of piperazine are basic ($K_{b_1} = 5.38 \times 10^{-5}$; $K_{b_2} = 2.15 \times 10^{-9}$). What is the pH of 0.0125 $M$ piperazine?

*b. Draw the structure of the ionic form of piperazine that would be present in stomach acid (about 0.16 $M$ HCl).

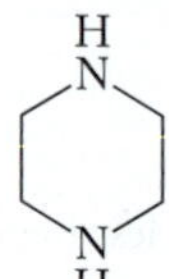

**FIGURE P15.74**

## Acidic and Basic Salts

### Concept Review

**15.75.** How is it that aqueous solutions of NaF are basic, but solutions of NaCl are neutral?

15.76. Why is publishing tables of $K_b$ values of the conjugate bases of weak acids whose $K_a$ values are known unnecessary?

**15.77.** Which of the following salts produces an acidic solution in water: ammonium acetate, ammonium nitrate, or sodium formate?

15.78. Which of the following salts produces a basic solution in water: $NaNO_2$, $KNO_3$, or $NH_4Cl$?

**15.79. Neutralizing the Smell of Fish** Trimethylamine, $(CH_3)_3N$ ($K_b = 6.5 \times 10^{-5}$ at 25°C), contributes to the "fishy" odor of not-so-fresh seafood. Some people squeeze fresh lemon juice (which contains a high concentration of citric acid) on cooked fish to reduce the fishy odor. Why is this practice effective?

*15.80. **Nutritional Value of Beets** Beets contain high concentrations of the calcium salt of malonic acid (see Figure P15.68). Could the presence of the calcium salt of malonic acid affect the pH balance of beets? If so, in which direction? Explain.

### Problems

**15.81.** The $K_a$ of the conjugate acid of the artificial sweetener saccharin is $2.1 \times 10^{-11}$. What is the p$K_b$ of saccharin?

15.82. The $K_{a_1}$ value for oxalic acid (HOOCCOOH) is $5.9 \times 10^{-2}$, and the $K_{a_2}$ value is $6.4 \times 10^{-5}$. What are the values of $K_{b_1}$ and $K_{b_2}$ of the oxalate anion ($^-OOCCOO^-$)?

---

**15.83. Decongestants I** Calculate the pH of a $1.25 \times 10^{-2}$ $M$ solution of the decongestant ephedrine hydrochloride. The p$K_b$ of its parent compound and conjugate base, ephedrine, is 3.86.

15.84. **Decongestants II** Pseudoephedrine hydrochloride (Figure P15.84) is a common ingredient in cough syrups and decongestants. The p$K_b$ of its parent compound, pseudoephedrine, is 4.78. What is the pH of 0.0295 $M$ pseudoephedrine hydrochloride?

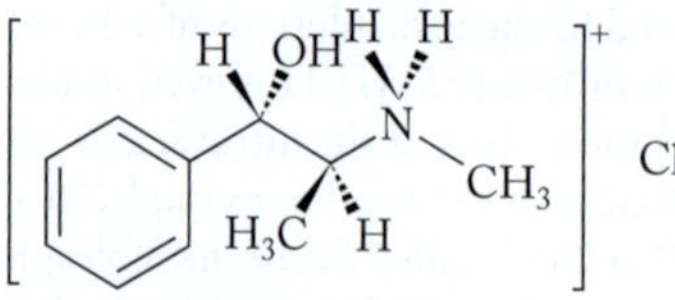

**FIGURE P15.84**

---

**15.85. Potato Chips** Salt and vinegar potato chips get their distinctive flavor from a combination of acetic acid and sodium acetate. The $K_a$ of acetic acid is $1.76 \times 10^{-5}$ at 25°C. Calculate the pH of 0.255 $M$ sodium acetate.

15.86. **Poisonous Plant** The gifblaar plant is a small South African shrub and one of the most poisonous plants known because it contains fluoroacetate ions. If the $K_a$ of fluoroacetic acid is $3.0 \times 10^{-3}$ at 25°C, what is the pH of a 0.125 $M$ solution of sodium fluoroacetate?

## Additional Problems

**15.87.** Consider the following compounds: $CH_3NH_2$, $CH_3COOH$, $Ca(OH)_2$, and $HClO_4$.

a. Identify the Arrhenius acid(s).

b. Identify the Arrhenius base(s).

c. Identify the Brønsted–Lowry acid(s).

d. Identify the Brønsted–Lowry base(s).

15.88. Are all Arrhenius acids also Brønsted–Lowry acids? Are all Brønsted–Lowry acids also Arrhenius acids? If yes, explain why. If not, give an example to demonstrate the difference.

**15.89.** Are all Arrhenius bases also Brønsted–Lowry bases? Are all Brønsted–Lowry bases also Arrhenius bases? If yes, explain why. If not, give an example to demonstrate the difference.

15.90. Describe the intermolecular forces and changes in bonding that lead to the formation of a basic solution when methylamine ($CH_3NH_2$) dissolves in water.

***15.91.** Describe the chemical reactions of sulfur that begin with the burning of high-sulfur fossil fuel and that end with the reaction between acid rain and building exteriors made of marble ($CaCO_3$).

15.92. The $K_{a_1}$ of phosphorous acid, $H_3PO_3$, is nearly the same as that of phosphoric acid, $H_3PO_4$.

a. Draw the Lewis structure of phosphorous acid.
b. Identify the ionizable hydrogen atoms in the structure.
c. Explain why the $K_{a_1}$ values of phosphoric acid and phosphorous acid are similar.

***15.93. pH of Natural Waters** In a 1985 study of Little Rock Lake in Wisconsin, 400 gallons of 18 *M* sulfuric acid were added to the lake over 6 years. The initial pH of the lake was 6.1 and the final pH was 4.7. If none of the acid was consumed in chemical reactions, estimate the volume of the lake.

15.94. **Acid–Base Properties of Pharmaceuticals I** Zoloft is a prescription drug for the treatment of depression. It is sold as its hydrochloride salt, which is produced as shown in Figure P15.94. When the hydrochloride dissolves in water, will the resulting solution be acidic or basic?

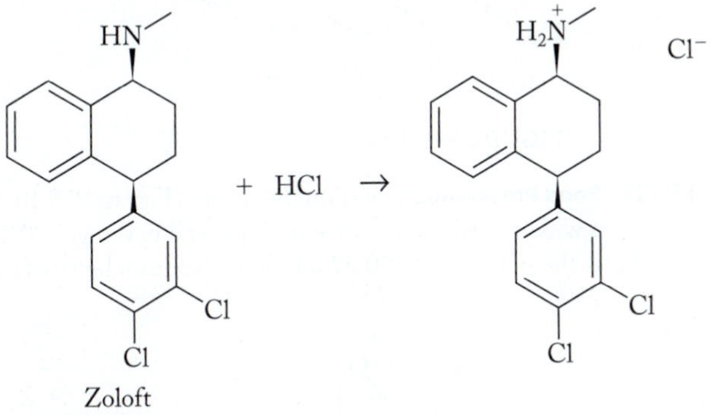

Zoloft + HCl → ($Cl^-$ salt)

Zoloft

**FIGURE P15.94**

**15.95. Acid–Base Properties of Pharmaceuticals II** Prozac, an antidepressant drug, has the structure given in Figure P15.95.

a. Is a solution of Prozac in water likely to be acidic or basic? Explain your answer.
b. Prozac is also sold as a hydrochloride salt. Which functional group is more likely to react with HCl?
c. Prozac is sold as its hydrochloride salt because the salt is more soluble in water than unreacted Prozac. Why is the salt more soluble?

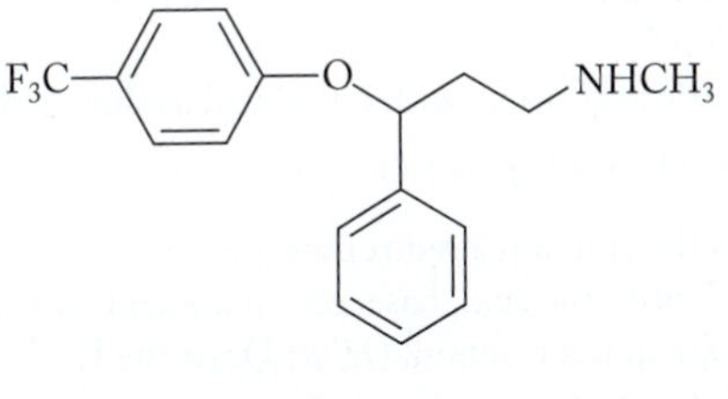

Prozac

**FIGURE P15.95**

15.96. Naproxen (sold as Aleve) is an anti-inflammatory drug used to reduce pain, fever, inflammation, and stiffness caused by conditions such as osteoarthritis and rheumatoid arthritis. Naproxen is an organic acid; its structure is shown in Figure P15.96. Naproxen has limited solubility in water, so it is sold as its sodium salt.

a. Draw the molecular structure of the sodium salt.
b. Is an aqueous solution of the salt acidic or basic? Explain why.
c. Explain why the salt is more soluble in water than naproxen itself.

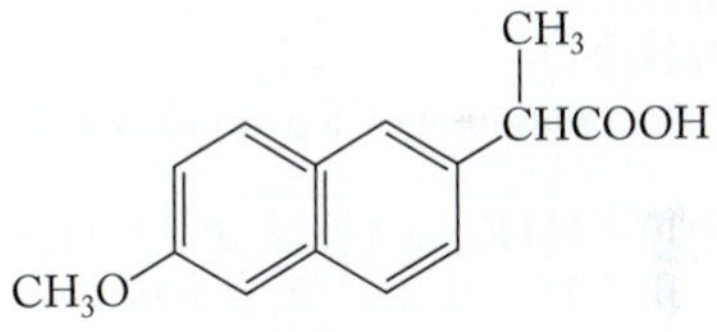

Naproxen

**FIGURE P15.96**

***15.97.** Pentafluorocyclopentadiene, which has the structure shown in Figure P15.97, is a strong acid.

a. Draw the conjugate base of $C_5F_5H$.
b. Why is the compound so acidic when most organic acids are weak?

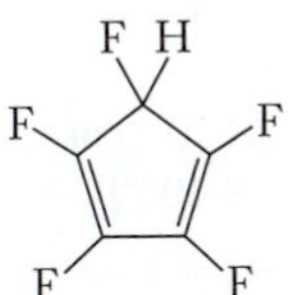


**FIGURE P15.97**

15.98. **Ocean Acidification** Some climate models predict a decrease in the pH of the oceans of 0.3 pH unit to 0.5 pH unit by 2100 because of increases in atmospheric carbon dioxide.
  a. Explain, using the appropriate chemical reactions and equilibria, how an increase in atmospheric $CO_2$ could produce a decrease in oceanic pH.
  b. How much more acidic would the oceans be if their pH dropped this much?
  c. Oceanographers are concerned about the how a drop in oceanic pH would affect the survival of coral reefs. Why?

***15.99.** Sulfuric acid reacts with nitric acid as shown below:

$$HNO_3(aq) + 2\,H_2SO_4(aq) \rightarrow NO_2^+(aq) + H_3O^+(aq) + 2\,HSO_4^-(aq)$$

  a. Is the reaction a redox process?
  b. Identify the acid, base, conjugate acid, and conjugate base in the reaction. (*Hint*: Draw the Lewis structures for each.)

15.100. Thiosulfuric acid, $H_2S_2O_3$, can be synthesized by reacting $H_2S$ with $HSO_3Cl$:

$$HSO_3Cl(\ell) + H_2S(g) \rightarrow HCl(g) + H_2S_2O_3(\ell)$$

  a. Draw a Lewis structure for $H_2S_2O_3$. (*Hint*: it is isostructural with $H_2SO_4$.)
  b. Do you expect $H_2S_2O_3$ to be a stronger or weaker acid than $H_2SO_4$? Explain your answer.

**15.101.** Which of these solutions is the most acidic? Which is the most basic?
  a. 1.0 $M$ $H_2SO_3$
  b. 0.10 $M$ $H_2SO_4$
  c. 0.30 $M$ $NaHSO_4$
  d. 0.30 $M$ $Na_2SO_4$
  e. 0.30 $M$ $Na_2SO_3$

15.102. Predict which solution in each pair below will have the lower pH.
  a. $2.56 \times 10^{-2}$ $M$ HCl or $4.96 \times 10^{-2}$ $M$ HBr
  b. $1.00 \times 10^{-5}$ $M$ acetic acid ($K_a = 1.76 \times 10^{-5}$) or $1.00 \times 10^{-5}$ $M$ formic acid ($K_a = 1.77 \times 10^{-4}$)
  c. 22 m$M$ $CH_3NH_2$ (p$K_b$ = 3.36) or 22 m$M$ $(CH_3)_2NH$ ($K_b = 5.9 \times 10^{-4}$)
  d. 158 m$M$ $NH_3$ (p$K_b$ = 4.75) or 158 m$M$ acetic acid (p$K_a$ = 4.75)
  e. 0.00395 $M$ $HNO_3$ or 0.00145 $M$ $HClO_4$
  f. $2.05 \times 10^{-1}$ $M$ propionic acid ($K_a = 1.4 \times 10^{-5}$) or $2.05 \times 10^{-1}$ $M$ fluoroacetic acid ($K_a = 2.6 \times 10^{-3}$)
  g. 375 m$M$ pyridine (p$K_b$ = 8.77) or 375 m$M$ aniline (p$K_b$ = 9.40)
  h. 0.555 $M$ $Fe(H_2O)_6^{3+}$ ($K_a = 3 \times 10^{-3}$) or 0.355 $M$ $Cr(H_2O)_6^{3+}$ ($K_a = 1 \times 10^{-4}$)

***15.103.** Nicotine, $C_{10}H_{14}N_2$, is a highly addictive drug that occurs naturally in tobacco and is added to other products, including the liquids in e-cigarettes. Nicotine molecules (see Figure P15.103) contain two nitrogen atoms, making the compound dibasic.
  a. Explain why the N atom in the 5-member ring in a molecule of nicotine is a stronger base (p$K_{b_1}$ = 6.0) than the N atom in the 6-member ring (p$K_{b_2}$ = 10.9).
  b. What is the pH of a 0.100 $M$ solution of nicotine hydrochloride, $C_{10}H_{15}N_2 \cdot Cl$?

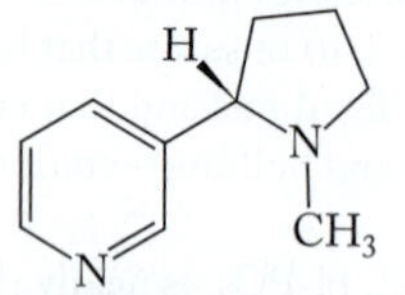

**FIGURE P15.103**

*15.104. Quinine, $C_{20}H_{24}N_2O_2$, is derived from the bark of the cinchona tree, and it has been used to treat malaria and other diseases for nearly 400 years. Its molecules contain two basic nitrogen atoms, highlighted in the molecular structure in Figure P15.104.
  a. Which N atom is the stronger base? Explain why you think so.
  b. Quinine's p$K_b$ values are 5.48 and 9.90. What is the pH of a 0.200 $M$ solution of quinine sulfate, $C_{20}H_{26}N_2O_2 \cdot SO_4$?

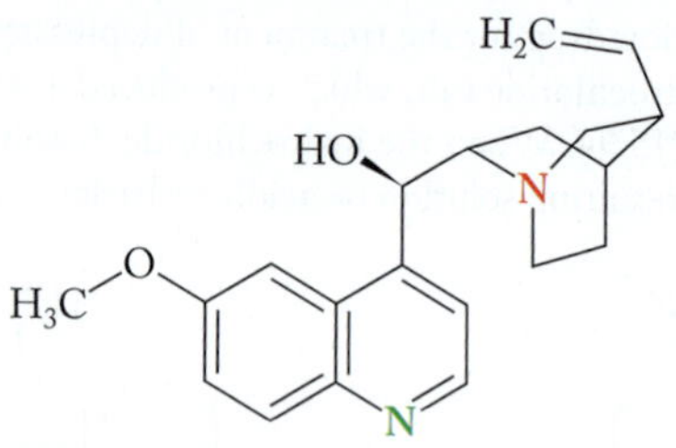

**FIGURE P15.104**

15.105. **Food Preservation** Sodium benzoate (Figure P15.105) is a widely used preservative in foods and beverages. What is the pH of a 0.0500 $M$ solution of sodium benzoate at 25°C?

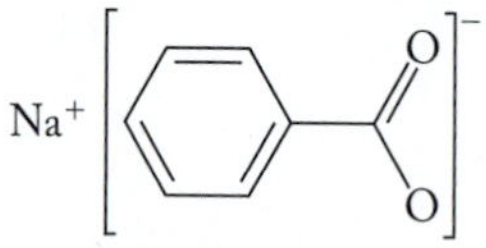

**FIGURE P15.105**

15.106. The value of $K_w$ increases as temperature increases.
a. If $pK_w = 13.017$ at 60°C, what is the value of $[H^+]$?
b. What is the pH of water at 60°C?

**15.107.** Identify the conjugate base for the following weak acids and calculate $K_b$ for each. Which conjugate base is the strongest?
a. $ClCH_2COOH$
b. $NH_4^+$
c. HCN
d. $CH_3CH_2OH$

15.108. Identify the conjugate acid for the following weak bases and calculate $K_a$ for each. Which conjugate acid is the strongest?
a. $CH_3NH_2$
b. $HPO_4^{2-}$
c. $H_2NC(O)NH_2$
d. $NO_2^-$

**15.109.** For each molecular equation, write the net ionic equation and identify the Brønsted–Lowry acids and bases:
a. $Na_2CO_3(aq) + H_2SO_4(aq) \rightarrow Na_2SO_4(aq) + CO_2(g) + H_2O(\ell)$
b. $2\,CH_3COOH(aq) + Mg(OH)_2(s) \rightarrow (CH_3COO)_2Mg(aq) + 2\,H_2O(\ell)$
c. $CaO(s) + H_2O(\ell) \rightarrow Ca(OH)_2(s)$
d. $NaSH(aq) + HNO_3(aq) \rightarrow NaNO_3(aq) + H_2S(g)$

15.110. Write both the molecular equation and the net ionic equation for the reaction that occurs when each pair of compounds is mixed in aqueous solution. For each reaction, label the Brønsted–Lowry acids and bases.
a. HCl and $Ca(OH)_2$
b. $H_3PO_4$ and KOH
c. $HNO_3$ and $Na_2CO_3$
d. $H_3PO_4$ and $Ca(OH)_2$

# 16

# Additional Aqueous Equilibria

## Chemistry and the Oceans

**OCEAN ACIDIFICATION AFFECTS CORAL REEFS** Increasing concentrations of $CO_2$ in the atmosphere are making seawater more acidic, which threatens corals and other marine life that form exoskeletons made of calcium carbonate.

## PARTICULATE **REVIEW**

### *Ionic Compounds and Solubility*

In Chapter 16, we revisit the solubility of ionic compounds. Ionic lattices for calcium carbonate, lithium oxide, and potassium chloride are shown here.

- Write the empirical formulas for these compounds.
- Match each of the ionic lattices to the name of the compound.
- If you added one mole of each compound to one liter of water, which ion(s) would be present in the highest concentration?

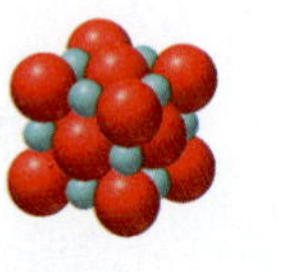

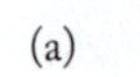

(a)

(b)

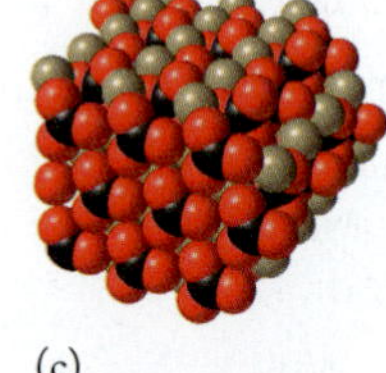

(c)

(Review Section 8.6 if you need help.)

*(Answers to Particulate Review questions are in the back of the book.)*

PARTICULATE **PREVIEW**

### *Donating Protons versus Donating Electron Pairs*

The molecules and ions shown here can function as acids or bases in the Brønsted–Lowry model of proton transfer and the Lewis model of electron-pair donation. As you read Chapter 16, look for ideas that will help you answer these questions:

- Which of these particles can donate a proton? Which can accept a proton?
- Which of these particles has one or more lone pairs of electrons on the central atom?
- Which of these particles can donate a lone pair of electrons to another particle?

## Learning Outcomes

**LO1** Explain the common ion effect and use it to calculate the pH of a solution containing a weak acid or base and its conjugate base or acid
**Sample Exercises 16.1, 16.2**

**LO2** Describe pH buffering and how to prepare a buffer with a desired pH
**Sample Exercises 16.3–16.5**

**LO3** Evaluate a buffer's capacity to resist changes in its pH due to the addition of acid or base
**Sample Exercises 16.6–16.8**

**LO4** Predict and interpret the results of an acid–base titration
**Sample Exercises 16.9–16.12**

**LO5** Identify a compound in a reaction as a Lewis acid or Lewis base
**Sample Exercise 16.13**

**LO6** Use formation constants to calculate the concentrations of free or complexed metal ions in solution
**Sample Exercise 16.14**

**LO7** Relate the acid strength of hydrated metal ions and their limited solubility in basic solutions to the charges on their ions

**LO8** Relate the solubility of an ionic compound to its solubility product and to solution pH
**Sample Exercises 16.15–16.18**

**LO9** Separate mixtures of ionic compounds by selective precipitation reactions
**Sample Exercise 16.19**

## 16.1 Ocean Acidification: Equilibrium under Stress

In Chapter 7, we reviewed data that showed a significant increase in the concentration of carbon dioxide in the atmosphere since the Industrial Revolution began around 1750 (see Figures 7.2 and 7.3). The impact of an increasing concentration of atmospheric $CO_2$ on climate change is a widely discussed topic these days, but another impact also worthy of our attention is ocean acidification.

As we saw at the beginning of Chapter 15, $CO_2$ gas dissolves in water to form carbonic acid:

$$CO_2(g) + H_2O(\ell) \rightleftharpoons H_2CO_3(aq) \quad (16.1)$$

Although $H_2CO_3$ is present in only low concentrations in aqueous solutions, we discussed in Section 15.7 why we use the symbol $[H_2CO_3]$ to represent the total amount of $CO_2(g)$ dissolved in water and that when $CO_2(g)$ dissolves in water, it forms an acidic solution:

$$H_2CO_3(aq) + H_2O(\ell) \rightleftharpoons HCO_3^-(aq) + H_3O^+(aq) \quad K_{a_1} = 4.3 \times 10^{-7} \quad (16.2)$$

And we saw that the bicarbonate ions also donate $H^+$ ions in aqueous solutions:

$$HCO_3^-(aq) + H_2O(\ell) \rightleftharpoons CO_3^{2-}(aq) + H_3O^+(aq) \quad K_{a_2} = 4.7 \times 10^{-11} \quad (16.3)$$

Remember, however, that because $K_{a_2} << K_{a_1}$, the concentration of $CO_3^{2-}$ ions in slightly basic seawater is much smaller than the concentration of $HCO_3^-$ ions.

The solubility of atmospheric $CO_2$ increases as its partial pressure increases, so higher concentrations of $CO_2$ in the atmosphere mean higher concentrations of dissolved $CO_2$ in the sea. Scientists estimate that about 30% of the "fossil" $CO_2$ added to the atmosphere by burning coal, natural gas, and petroleum-based fuels dissolves in the sea. Increasing concentrations of dissolved $CO_2$ have shifted the equilibrium in Equation 16.1 to the right—which, in turn, has shifted the equilibria in Equation 16.2 to the right, increasing the acidity of the sea. As a result, the average pH of seawater has dropped by about 0.11 pH units since the 18th century.

**CONNECTION** Henry's law, discussed in Section 11.6, states that the solubility of a gas is proportional to its partial pressure.

This is an enormous problem—especially if oceanic pH continues to drop—because many marine organisms such as plankton, shellfish, and corals like those in the chapter-opening photo have exoskeletons made of $CaCO_3$. To make $CaCO_3$, these organisms need a supply of dissolved $Ca^{2+}$ and $CO_3^{2-}$ ions so that the following precipitation reaction can take place:

$$Ca^{2+}(aq) + CO_3^{2-}(aq) \rightleftharpoons CaCO_3(s)$$

Unfortunately, increasing concentrations of $H_3O^+$ ions in the sea reduce the already low concentrations of $CO_3^{2-}$ ions by shifting the equilibrium in Equation 16.3 to the left. Concern is growing that the oceanic $[CO_3^{2-}]$ may become too small for marine organisms to form their $CaCO_3$ skeletons, and that they—and the aquatic systems that rely on them, such as tropical coral reefs—will not survive.

In this chapter, we explore the equilibria that control the pH of the sea and of other environmental and biological systems. We also examine how changes in pH can affect other equilibria, including those key to the survival of marine and other species.

## 16.2 The Common-Ion Effect

The chemical equilibria that control acid–base balance in natural systems—such as seawater and the fluids in all living cells—depend on the presence of both acidic and basic compounds. We encountered one example in Sample Exercise 15.20, where a sample of blood plasma that initially contained $2.7 \times 10^{-3}$ *M* dissolved $CO_2$ and 0.028 *M* $HCO_3^-$ had a pH of 7.38. Now let's calculate the pH of a solution that initially contains $2.7 \times 10^{-3}$ *M* dissolved $CO_2$ but no bicarbonate ions.

We can create a RICE table to track the molar concentrations of the reactants and products, continuing to use $[H_2CO_3]$ to represent the total amount of dissolved $CO_2(g)$ in solution:

| Reaction (R) | $H_2CO_3(aq) + H_2O(\ell)$ | $\rightleftharpoons$ | $HCO_3^-(aq)$ | + | $H_3O^+(aq)$ |
|---|---|---|---|---|---|
| | $[H_2CO_3]$ (*M*) | | $[HCO_3^-]$ (*M*) | | $[H_3O^+]$ (*M*) |
| Initial (I) | $2.7 \times 10^{-3}$ | | 0 | | 0 |
| Change (C) | $-x$ | | $+x$ | | $+x$ |
| Equilibrium (E) | $(2.7 \times 10^{-3}) - x$ | | $x$ | | $x$ |

We can then use the $K_{a_1}$ value for carbonic acid ($4.3 \times 10^{-7}$) from Equation 16.2 to solve for $x$, making the assumption to ignore the $-x$ term in the denominator of our expression as being less than 5% of $2.7 \times 10^{-3}$ (and, of course, checking the assumption):

$$K_{a_1} = \frac{[HCO_3^-][H_3O^+]}{[CO_2]} = \frac{(x)(x)}{(2.7 \times 10^{-3}) - x} \approx \frac{x^2}{2.7 \times 10^{-3}} = 4.3 \times 10^{-7}$$

$$x = [H_3O^+] = 3.4 \times 10^{-5}\ M$$

Calculating pH: $\quad \text{pH} = -\log[H_3O^+] = -\log(3.4 \times 10^{-5}) = 4.47$

Note that the pH value of this solution is nearly three units lower than the plasma sample in Sample Exercise 15.20, corresponding to a $[H_3O^+]$ value nearly $10^3$ larger. This difference exists because the bicarbonate ions in blood plasma inhibit the forward reaction in Equation 16.2. This result is an example of the general principle that the ionization of any weak acid (HA) is inhibited by the presence of a second source of its conjugate base ($A^-$). To see this, let's consider the general equation for this process:

$$HA(aq) + H_2O(\ell) \rightleftharpoons A^-(aq) + H_3O^+(aq)$$

**CONNECTION** Le Châtelier's principle was introduced in Section 14.7 to describe how a system at equilibrium responds to an external stress.

Le Châtelier's principle predicts that adding a product to a reaction mixture at equilibrium will react to re-form reactants until equilibrium is reestablished.

This application of Le Châtelier's principle to a weak acid equilibrium illustrates a phenomenon known as the **common-ion effect**: in any equilibrium involving ions, a reaction that produces an ion is suppressed when more of that ion is added to the system. One result of this suppression for a weak acid solution is that the initial concentrations of the conjugate acid–base pair change very little before equilibrium is achieved. Therefore, $[HA]_{initial} \approx [HA]_{equilibrium}$ and $[A^-]_{initial} \approx [A^-]_{equilibrium}$, and we can write the $K_a$ expression for this solution as

**STEPWISE ANIMATION**
Common-Ion Effect

$$K_a = \frac{[H_3O^+][A^-]_{initial}}{[HA]_{initial}}$$

If we take the negative logarithm of both sides of this expression, $[H_3O^+]$ becomes pH and $K_a$ becomes $pK_a$:

$$pK_a = pH - \log\frac{[A^-]_{initial}}{[HA]_{initial}}$$

We can write a more general form of this equation by replacing $[HA]_{initial}$ and $[A^-]_{initial}$ with [acid] and [base], respectively:

$$pK_a = pH - \log\frac{[base]}{[acid]}$$

or

$$pH = pK_a + \log\frac{[base]}{[acid]} \quad (16.4)$$

Equation 16.4 is called the **Henderson–Hasselbalch equation**. It can be used to calculate the pH of a solution in which the ionization of a weak acid is suppressed by an additional source of its conjugate base, so the concentrations of both do not change significantly before equilibrium is achieved.

**CONNECTION** In Chapter 15, we introduced pH and $pK_a$ using logarithmic scales. The smaller the $[H_3O^+]$ or $K_a$ value, the higher the pH or $pK_a$, respectively.

What happens to the logarithm term in the Henderson–Hasselbalch equation when the concentrations of acid and base are equal? The value of the fraction is 1, and $\log(1) = 0$, so $pH = pK_a$. This equality serves as a handy reference point in working with solutions of conjugate acid–base pairs. If the concentration of the base is greater than that of the acid, the logarithmic term is greater than zero and $pH > pK_a$. If the concentration of the base is less than that of the acid, the logarithmic term is less than zero, and $pH < pK_a$.

Suppose, for example, the concentration of the base is 10 times the concentration of the acid—that is, [base] = 10[acid]. Substituting this equality into Equation 16.4, we have

$$pH = pK_a + \log\frac{10[acid]}{[acid]}$$

$$= pK_a + \log 10 = pK_a + 1$$

Thus, a 10-fold higher concentration of base produces a pH that is one unit above the $pK_a$ value. Similarly, if the concentration of the acid component is 10 times that of the base, then $pH = pK_a - 1$.

**common-ion effect** the shift in the position of an equilibrium caused by the presence or addition of an ion taking part in the reaction.

**Henderson–Hasselbalch equation** an equation used to calculate the pH of a solution in which the concentrations of acid and conjugate base are known.

Throughout this chapter, we assume that the Henderson–Hasselbalch equation can be used to calculate the pH of solutions containing weak acids and their conjugate bases, or weak bases and their conjugate acids, as long as the concentrations of the two components do not differ from one another by more than a factor of 10. Sample Exercises 16.1 and 16.2 illustrate how the Henderson–Hasselbalch equation simplifies pH calculations when we know the concentrations of both components of a conjugate pair.

**SAMPLE EXERCISE 16.1** Calculating the pH of a Solution of a Weak Acid and Its Conjugate Base **LO1**

What is the pH of a sample of river water at 25°C in which $[HCO_3^-] = 1.0 \times 10^{-4}\ M$ and in which $[H_2CO_3]$ in equilibrium with atmospheric $CO_2$ is $1.4 \times 10^{-5}\ M$?

**Collect and Organize** We are given the initial concentrations of a weak acid and its conjugate base. The value of $[H_2CO_3]$ is constant because it is linked to the solubility of atmospheric $CO_2$ gas. If we assume that little change in $[HCO_3^-]$ occurs before equilibrium is achieved, the pH of the river water can be calculated using the Henderson–Hasselbalch equation.

**Analyze** The conjugate acid–base pair relationship in this reaction is described by Equation 16.2:

$$H_2CO_3(aq) + H_2O(\ell) \rightleftharpoons HCO_3^-(aq) + H_3O^+(aq)$$

This reaction has a $pK_{a_1}$ value of 6.37 (see Table A5.1). The concentration of the base is nearly 10 times that of the acid, so the log term in the Henderson–Hasselbalch equation should be almost 1, and the calculated pH should be a little greater than 7.

**Solve** Inserting the concentration and $pK_{a_1}$ values into the Henderson–Hasselbalch equation:

$$pH = pK_{a_1} + \log\frac{[\text{base}]}{[\text{acid}]} = 6.37 + \log\frac{1.0 \times 10^{-4}}{1.4 \times 10^{-5}}$$

$$= 6.37 + 0.85 = 7.22$$

**Think About It** This result is a little greater than 7 as we predicted. It contains two significant figures (the "2"s in the tenths and hundredths places) because two significant figures are present in all three values used to calculate it.

**Practice Exercise** What is the pH of a formic acid solution in which $[HCOOH] = 2.5 \times 10^{-2}\ M$ and $[HCOO^-] = 7.8 \times 10^{-3}\ M$?

*(Answers to Practice Exercises are in the back of the book.)*

We can also use the Henderson–Hasselbalch equation to calculate the pH of a solution of a weak base and its conjugate acid. An extra step is usually involved because we may know the $pK_b$ value of the base but not the $pK_a$ value of its conjugate acid, and only $pK_a$ values are used in the Henderson–Hasselbalch equation. However, we learned in Chapter 15 that the equilibrium constants for conjugate acid–base pairs are related by Equation 15.19 ($K_a \times K_b = K_w$). To recast Equation 15.19 in p$K$ terms, we can insert the numerical value of $K_w$ at 25°C ($1.0 \times 10^{-14}$):

$$K_a \times K_b = K_w = 1.0 \times 10^{-14}$$

and take the −log values of all three terms:

$$pK_a + pK_b = 14.00 \quad (16.5)$$

Thus, we can convert the $pK_b$ of a base into the $pK_a$ of its conjugate acid by subtracting the $pK_b$ value from 14.00.

**SAMPLE EXERCISE 16.2** Calculating the pH of a Solution of a Weak Base and Its Conjugate Acid **LO1**

What is the pH of a solution that is 0.200 *M* in $NH_3$ and 0.300 *M* in $NH_4Cl$?

**Collect, Organize, and Analyze** We are asked to calculate the pH of a solution containing known concentrations of a weak base ($NH_3$) and a salt of its conjugate acid ($NH_4^+$). Assuming the initial concentrations of $NH_3$ and $NH_4^+$ do not change significantly as equilibrium is reached, we can use the Henderson–Hasselbalch equation to calculate the pH of such a solution from the concentrations of the two components and the $pK_a$ of the acid. Table A5.3 contains $K_b$ and $pK_b$ values of common bases. The $pK_a$ and $pK_b$ values of a conjugate acid–base pair are related by Equation 16.5:

$$pK_a + pK_b = 14.00$$

Our approach involves converting the $pK_b$ value for $NH_3$ from Table A5.3 into a $pK_a$ value. Adding ammonium ion to a solution of ammonia should produce a solution that is still basic, but not as basic as a solution that contains only ammonia.

**Solve** Inserting the $pK_b$ value for $NH_3$ (4.75) into Equation 16.5 and solving for $pK_a$:

$$pK_a = 14.00 - pK_b = 14.00 - 4.75 = 9.25$$

We can then use this value and the given concentrations of $NH_3$ and $NH_4^+$ in Equation 16.4:

$$\text{pH} = pK_a + \log\frac{[\text{base}]}{[\text{acid}]}$$

$$= 9.25 + \log\frac{0.200}{0.300} = 9.07$$

**Think About It** We predicted that the solution would be basic, but not as basic as a solution of ammonia alone. To check whether this prediction is true, we can calculate the pH of 0.200 *M* $NH_3$ by using the approach we followed in Sample Exercise 15.13. The result is a pH of 11.27—more than two pH units higher (more basic) than the solution of ammonia and ammonium chloride in this exercise.

**Practice Exercise** Calculate the pH of a solution that is 0.25 *M* in methylamine, $CH_3NH_2$, and 0.75 *M* in methylamine hydrochloride, $CH_3NH_3Cl$.

**CHEMTOUR**
Buffers

## 16.3 pH Buffers

**pH buffer** a solution that resists changes in pH when acids or bases are added to it; typically, a solution of a weak acid and its conjugate base.

The common-ion effect plays a key role in controlling the pH of aqueous solutions (including those inside our bodies and covering 70% of Earth's surface). These solutions have the capacity to withstand additions of acidic or basic substances with little change in their pH because they contain **pH buffers**. These buffers are composed of either a weak acid and its conjugate base or of a weak base and its conjugate acid. A buffer's acid component gives it the capacity to neutralize small additions of basic substances, which convert some of the acid

(HA) into its conjugate base ($A^-$), and the buffer's basic component gives it the capacity to neutralize small additions of acid, converting some of the conjugate base $A^-$ back into HA. As long as the addition of acid or base does not react most of either component of the buffer, the resulting changes in pH are likely to be small.

To illustrate the capacity of a buffer to minimize changes in pH, let's examine 1.00 L of a buffer solution that initially contains 0.0100 mol of HA and 0.0100 mol of $A^-$, and to which 0.0015 mol of strong base such as sodium hydroxide (represented by $OH^-$) is accidentally added. This added base neutralizes 0.0015 mol of HA and produces an additional 0.0015 mol of [$A^-$], as shown in the table below:

**STEPWISE ANIMATION**
The Buffer System

| Reaction | HA(*aq*) (mol) | + | $OH^-$(*aq*) (mol) | → | $A^-$(*aq*) (mol) | + $H_2O(\ell)$ |
|---|---|---|---|---|---|---|
| Initial | 0.0100 | | 0.0015 | | 0.0100 | |
| Change | −0.0015 | | −0.0015 | | +0.0015 | |
| After | 0.0085 | | 0 | | 0.0115 | |

Note that this is not a RICE table because it does not contain an equilibrium reaction. The strong base reacts completely, so the reaction arrow in this table is a single forward arrow, not a pair of equilibrium arrows. The numbers of moles in the final row of the table are easily converted back to concentrations expressed in mol/L, or molarity (*M*), by using the solution volume of 1.00 L.

According to the Henderson–Hasselbalch equation, the initial pH of the buffer is equal to the $pK_a$ value of HA:

$$\text{pH} = \text{p}K_a + \log\frac{[A^-]_{\text{initial}}}{[HA]_{\text{initial}}} = \text{p}K_a + \log\frac{0.0100\ \cancel{M}}{0.0100\ \cancel{M}} = \text{p}K_a + \log(1) = \text{p}K_a$$

After addition of the strong base, the final pH of the buffer is

$$\text{pH} = \text{p}K_a + \log\frac{[A^-]_{\text{final}}}{[HA]_{\text{final}}} = \text{p}K_a + \log\frac{0.0115\ \cancel{M}}{0.0085\ \cancel{M}}$$
$$= \text{p}K_a + \log(1.35) = \text{p}K_a + 0.13$$

For many reaction systems, including biochemical systems, an increase in pH of 0.13 units is tolerable. By contrast, consider the change in pH if 0.0015 mol of strong base had been added to a liter of pure water at pH = 7.00. The final [$OH^-$] would be 0.0015 *M*, so pOH = −log(0.0015) = 2.82, and pH = 14.00 − 2.82 = 11.18. Thus, the increase in pH would have been 4.18 units instead of 0.13.

Most buffers contain similar concentrations of the two components and can neutralize small additions of either strong acids or strong bases equally well, as shown in **Figure 16.1**. As we have seen, when the ratio of the conjugate pair is exactly 1, the pH = the $pK_a$ of the weak acid. Buffers can be prepared using different proportions of acid and conjugate base that effectively control pH over a range of values from about one unit below the $pK_a$ value of the acid to one pH unit above it. In the paragraphs that follow, we explore how to select an appropriate conjugate acid–base pair and how to combine them in just the right proportion to prepare a buffer with a desired pH.

Suppose we need to run a biochemical reaction at a pH of exactly 5.00. What combination of weak acid and conjugate base would be a good choice, and in what ratio of acid to base? To select an appropriate conjugate pair, we must find a weak acid that has a $pK_a$ value within 1 unit of 5.00—the closer the better. A search of Table A5.1 of Appendix 5 discloses a familiar candidate: acetic acid, which has a $pK_a$ of 4.75, or only 0.25 units below the target pH of 5.00. We also need to

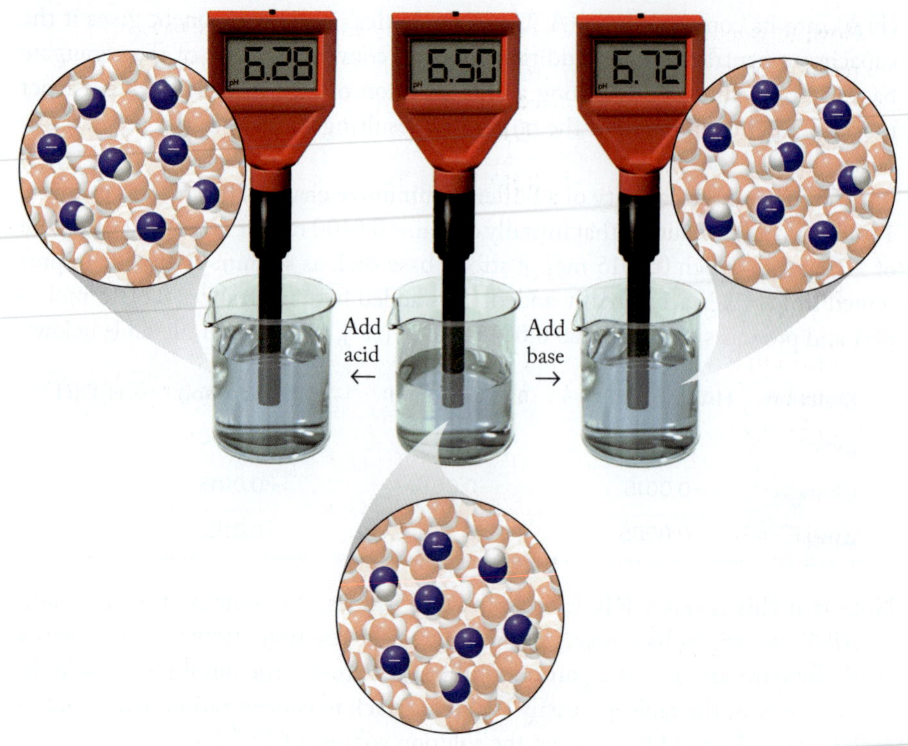


**FIGURE 16.1** A buffer with a pH = 6.50 (the middle beaker) contains an equal number of moles of a weak acid (HA) and its conjugate base ($A^-$). If enough strong acid were added to the buffer to neutralize 25% of the buffer's conjugate base, converting it into weak acid through the reaction $H_3O^+(aq) + A^-(aq) \rightarrow HA(aq) + H_2O(\ell)$, the pH of the buffer would fall, but only by 0.22. If a similar quantity of strong base were added to the original buffer solution in the middle beaker, it would react 25% of the buffer's weak acid through the reaction $OH^-(aq) + HA(aq) \rightarrow A^-(aq) + H_2O(\ell)$; pH would rise, but again only by 0.22 unit.

choose a salt of the conjugate base of acetic acid, for example, sodium acetate. Next, we can use the Henderson–Hasselbalch equation to determine the ratio of acid to conjugate base to use in preparing the buffer. We can solve for the ratio of acid to base by using Equation 16.4:

$$\log \frac{[\text{base}]}{[\text{acid}]} = \text{pH} - \text{p}K_a = 5.00 - 4.75 = 0.25$$

$$\frac{[\text{base}]}{[\text{acid}]} = 10^{0.25} = 1.78$$

Therefore, the buffer solution should contain 1.00 mol of acetic acid for every 1.78 mol of acetate ions. This ratio makes sense because a 1:1 mole ratio of acid to base would produce a pH 4.75 buffer. Preparing a slightly more basic buffer (pH = 5.00) should require proportionally more base and less acid.

Knowing the ratio of acidic to basic buffer components does not yet let us specify their actual concentrations. For that, we need some specification of overall buffer concentration, which is linked to the capacity of the buffer to control pH within an acceptable range when other acidic or basic substances are added. Suppose we need to prepare 5.00 L of a pH 5.00 buffer in which the combined concentrations of acid and base equal 0.100 *M*. What masses of these ingredients should we weigh out?

We can use algebra to calculate the desired concentrations of the two components. If we let $x$ equal the molarity of the acid, then the concentration of base is $(0.100 - x)$ *M*. We use the ratio of the two, 1.78, to solve for $x$:

$$\frac{[\text{base}]}{[\text{acid}]} = 1.78 = \frac{0.100 - x}{x},\ x = 0.036$$

Therefore, the concentration of acid in the buffer is 0.036 *M*, and the concentration of base is $0.100 - 0.036 = 0.064$ *M*.

Now, to calculate the necessary quantities of these two substances, we need the molar masses of acetic acid ($CH_3COOH$; $\mathcal{M}$ = 60.0 g/mol) and a salt of its conjugate base, such as sodium acetate ($CH_3COONa$; $\mathcal{M}$ = 82.0 g/mol). The quantities needed to prepare 5.00 L of pH 5.00 buffer are:

$$5.00\ \cancel{\text{L}} \times \frac{0.036\ \cancel{\text{mol}}\ \text{CH}_3\text{COOH}}{\cancel{\text{L}}} \times \frac{60.05\ \text{g}}{\cancel{\text{mol}}} = 10.8\ \text{g CH}_3\text{COOH}$$

$$5.00\ \cancel{\text{L}} \times \frac{0.064\ \cancel{\text{mol}}\ \text{CH}_3\text{COONa}}{\cancel{\text{L}}} \times \frac{82.05\ \text{g}}{\cancel{\text{mol}}} = 26.3\ \text{g CH}_3\text{COONa}$$

The following Sample Exercises provide additional practice in recognizing combinations of acids and bases that should serve as effective pH buffers and in preparing pH buffers.

### SAMPLE EXERCISE 16.3 Recognizing an Effective pH Buffer LO2

An effective pH buffer has the capacity to reduce changes in pH caused by additions of other acidic and basic substances. Which of the following pairs of substances could form such a solution: (a) hydrochloric acid and sodium chloride; (b) formic acid and sodium formate; (c) trifluoroacetic acid and sodium trifluoroacetate; (d) acetic acid and sodium formate; (e) ammonia and ammonium nitrate?

**Collect and Organize** A pH buffer contains either a weak acid and its conjugate base or a weak base and its conjugate acid. These combinations of acidic and basic components give a buffer the capacity to neutralize additions of other acidic or basic substances and reduce changes in pH.

**Analyze** The solution formed by mixing the substances in (a) contains a strong acid and its conjugate base. The substances in (b) and (c) form solutions of weak acids and their conjugate bases. The substances in (d) form a solution of a weak acid and a salt whose anion is not the conjugate base of the acid. The substances in (e) form a solution of a weak base and its conjugate acid.

**Solve** The pairs of substances in (b), (c), and (e) satisfy the criteria for preparing pH buffers. Pair (a) does not because it has no capacity to reduce a change in pH due to adding more strong acid. Pair (d) does not because adding strong acid converts sodium formate to formic acid, which is a stronger acid than acetic acid, and adding base produces acetate ions, which are a stronger base than formate ions.

**Think About It** Another criterion for selecting weak acids for aqueous buffers is that they must be soluble in water. All the acids selected are carboxylic acids with relatively small molar masses, and they are soluble in water. However, organic acids with large hydrocarbon (hydrophobic) regions in their molecular structures, such as octanoic acid and benzoic acid (**Figure 16.2**), are only slightly soluble in water.

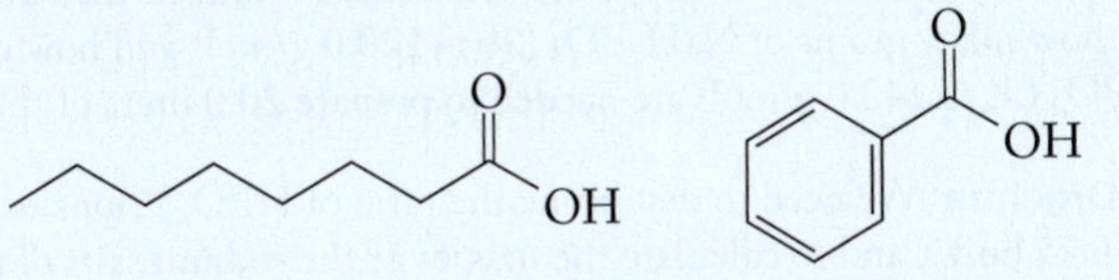

O
OH
Octanoic acid

O
OH
Benzoic acid

**FIGURE 16.2** Two slightly soluble organic acids.

**Practice Exercise** Which of the following mixtures could form effective buffers? (a) $KH_2PO_4$ and $K_2HPO_4$; (b) $NaHCO_3$ and $Na_2CO_3$; (c) $NaHCO_3$ alone; (d) $H_2SO_4$ and $KHSO_4$; (e) HClO and $NaClO_2$.

**SAMPLE EXERCISE 16.4** Preparing a Basic Buffer **LO2**

Select a weak base in Table A5.3 that, when mixed with the chloride salt of its conjugate acid in approximately equimolar proportions, produces a buffer with a pH of 9.20. What will be the ratio of base to conjugate acid in the buffer?

**Collect, Organize, and Analyze** When we prepared a pH 5.00 buffer (see page 770), we searched Table A5.1 in Appendix 5 for a weak acid that had a $pK_a$ value close to the target pH. To prepare a pH 9.20 buffer, we need to search Table A5.3 for a weak base whose *conjugate acid* has $pK_a$ close to the target pH of 9.20. This means we are looking for a base with a $pK_b$ value near $14.00 - 9.20 = 4.80$. We can then use the conjugate acid's $pK_a$ value in the Henderson–Hasselbalch equation to find the molar concentration ratio of base to acid.

**Solve** According to Table A5.3, $NH_3$ has a $pK_b$ of 4.75, which is very close to the pH value of the buffer. Using the Henderson–Hasselbalch equation to calculate the ratio of base ($NH_3$) to conjugate acid ($NH_4^+$):

$$\log \frac{[\text{base}]}{[\text{acid}]} = \text{pH} - pK_a$$

$$\log \frac{[NH_3]}{[NH_4^+]} = 9.20 - 9.25 = -0.05$$

$$\frac{[NH_3]}{[NH_4^+]} = 10^{-0.05} = 0.89$$

**Think About It** The ratio of $[NH_3]$ to $[NH_4^+]$ is slightly less than 1, which makes sense because the pH 9.20 buffer is less basic than a solution that contains a 1:1 ratio of $[NH_3]$ to $[NH_4^+]$, whose pH = 9.25. Also, ammonia and ammonium salts such as $NH_4Cl$ are soluble in water, on the assumption that they do not chemically react with other solutes in the solution whose pH we wish to control.

**Practice Exercise** Select a weak base in Table A5.3 that, when mixed with the chloride salt of its conjugate acid in approximately equimolar proportions, produces a buffer with a pH of 10.75. Calculate the ratio of base to conjugate acid in the buffer solution.

**SAMPLE EXERCISE 16.5** Preparing a Buffer Solution with a Desired pH and Total Buffer Concentration **LO2**

A buffer system containing dihydrogen phosphate ($H_2PO_4^-$) and hydrogen phosphate ($HPO_4^{2-}$) helps regulate the pH of cytoplasm in living cells.

a. What is the ratio of $HPO_4^{2-}$ ions to $H_2PO_4^-$ ions in a buffer with a pH of 6.75?
b. If the combined concentration of $H_2PO_4^-$ and $HPO_4^{2-}$ ions in the buffer is to be 0.200 *M*, how many grams of $NaH_2PO_4$ ($\mathcal{M}$ = 120.0 g/mol) and how many grams of $Na_2HPO_4$ ($\mathcal{M}$ = 142.0 g/mol) are needed to prepare 20.0 liters of the buffer?

**Collect and Organize** We need to determine the ratio of $HPO_4^{2-}$ ions to $H_2PO_4^-$ ions in a pH 6.75 buffer and to calculate the masses of the sodium salts of these ions that are needed to make 20.0 L of 0.200 *M* buffer. According to Table A5.1, the $pK_a$ of $H_2PO_4^-$ (the $pK_{a_2}$ of $H_3PO_4$) is 7.19. The Henderson–Hasselbalch equation relates the pH of a buffer to the $pK_a$ of its acid component and the concentrations of that acid and its conjugate base.

**Analyze** The pH of this buffer is controlled by the acid ionization equilibrium of dihydrogen phosphate ions:

$$H_2PO_4^-(aq) + H_2O(\ell) \rightleftharpoons HPO_4^{2-}(aq) + H_3O^+(aq)$$

The target pH (6.75) is less than the p$K_a$ (7.19), so the buffer must contain a higher concentration of $H_2PO_4^-$ ions than of $HPO_4^{2-}$ ions. The sum of $[H_2PO_4^-]$ and $[HPO_4^{2-}]$ is 0.200 $M$, so if we let $x = [HPO_4^{2-}]$, then $[H_2PO_4^-] = (0.200 - x)\ M$.

**Solve**

a. Using the Henderson–Hasselbalch equation to solve for the ratio of base to acid:

$$\log \frac{[HPO_4^{2-}]}{[H_2PO_4^-]} = \text{pH} - \text{p}K_a = 6.75 - 7.19 = -0.44$$

$$\frac{[HPO_4^{2-}]}{[H_2PO_4^-]} = 10^{-0.44} = 0.36$$

b. Calculating the dissolved concentrations of $HPO_4^{2-}$ ($x$) and $H_2PO_4^-$ ($0.200 - x$):

$$\frac{x}{0.200 - x} = 0.36$$

$$x = [HPO_4^{2-}] = 0.053\ M$$

$$(0.200 - x) = [H_2PO_4^-] = 0.147\ M$$

The masses of their sodium salts in 20.0 liters of buffer are

$$Na_2HPO_4\text{: } 20.0\ \cancel{L} \times 0.053\ \frac{\cancel{mol}}{\cancel{L}} \times 142.0\ \frac{g}{\cancel{mol}} = 150\ g$$

$$NaH_2PO_4\text{: } 20.0\ \cancel{L} \times 0.147\ \frac{\cancel{mol}}{\cancel{L}} \times 120.0\ \frac{g}{\cancel{mol}} = 353\ g$$

**Think About It** The buffer contains a higher concentration of $H_2PO_4^-$ ions than of $HPO_4^{2-}$ ions, which is consistent with our prediction and with the ratio of acid to conjugate base of any buffer that has a pH below the p$K_a$ of the acid used to make it.

**Practice Exercise** How many kilograms of sodium ascorbate ($NaC_6H_7O_6$) and ascorbic acid ($C_6H_8O_6$) are needed to make 10.0 liters of pH 4.25 buffer if the total concentration of the two buffer components is 0.500 $M$?

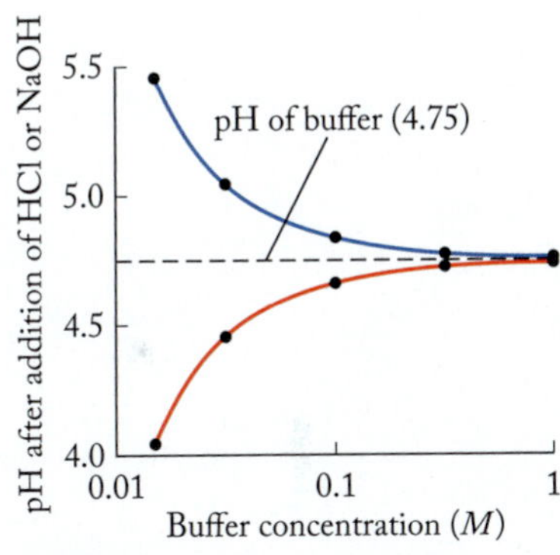


**FIGURE 16.3** When strong acid (red line) or strong base (blue line) is added to a buffer solution, the extent to which the pH changes is inversely proportional to buffer concentration: the higher the concentrations of the buffer components, the smaller the change in pH. Here, 100 mL samples of five solutions that are 0.015 $M$, 0.030 $M$, 0.100 $M$, 0.300 $M$, and 1.000 $M$ acetic acid and sodium acetate all have an initial pH of 4.75 (dashed line). The graph shows the pH values of these solutions after 1.00 mL of 1.00 $M$ HCl or 1.00 $M$ NaOH has been added.

## Buffer Capacity

The greater the concentrations of the conjugate pair in a buffer, the greater its ability to withstand additions of acid or base without significant changes in pH (**Figure 16.3**)—that is, the greater its **buffer capacity**. We have already examined the response of a hypothetical buffer system to an addition of strong base. The following Sample Exercise addresses the response of a natural buffer system to an addition of a strong acid.

**buffer capacity** the quantity of acid or base that a pH buffer can neutralize while keeping its pH within a desired range.

### SAMPLE EXERCISE 16.6 Calculating Buffer Response to the Addition of Acid or Base — LO3

a. What is the change in pH of a 1.00 L sample of river water, in which $[H_2CO_3] = 1.39 \times 10^{-5}\ M$ and $[HCO_3^-] = 1.05 \times 10^{-4}\ M$, when 10.0 mL of $1.0 \times 10^{-3}\ M$ $HNO_3$ is added to it?

b. Compare the pH change in part a to the pH change when the same quantity of strong acid is added to 1.00 L of pure (pH 7.00) water.

**Collect and Organize** We know the volume of a river water sample and the concentrations of a weak acid ($H_2CO_3$) and its conjugate base ($HCO_3^-$) that are dissolved in it. The presence of these species enables the sample to function as a pH buffer. We can assume that the sample's $[H_2CO_3]$ is linked to the concentration of $CO_2$ in the atmosphere, as described in Section 16.2, and is unaffected by the addition of acid. We are asked to determine the pH change when a known volume and concentration of a strong acid ($HNO_3$) is added to the sample, as well as to compare that result to the change that occurs when the same quantity of $HNO_3$ is added to pure water. The Henderson–Hasselbalch equation is useful for calculating the pH of buffer solutions with comparable concentrations of the conjugate acid–base pair. The $pK_{a_1}$ of carbonic acid is 6.37.

**Analyze** (a) When strong acid is added to a $H_2CO_3/HCO_3^-$ buffer, the acid reacts with $HCO_3^-$ ions, as described by this chemical equation:

$$HCO_3^-(aq) + H_3O^+(aq) \rightarrow H_2CO_3(aq) + H_2O(\ell)$$

To predict the impact of added acid on the composition of the buffer and on sample pH, we first need to calculate the initial pH of the river water sample. Then we will calculate the number of moles of $HCO_3^-$ ions in the initial sample and the number of moles of $H_3O^+$ ions added, which will tell us how many moles of $HCO_3^-$ ions react. Then we can then determine the remaining number of moles of $HCO_3^-$ ions and their concentration. The change in pH of the river water sample is likely to be small: a few tenths of a pH unit at most. The final pH of the pure water sample can be calculated from the number of moles of $H_3O^+$ ions divided by the total volume of the sample.

**Solve**

a. Calculating the initial pH of the river water sample:

$$pH_{initial} = pK_a + \log\frac{[base]}{[acid]} = pK_{a_1} + \log\frac{[HCO_3^-]_{initial}}{[H_2CO_3]}$$

$$= 6.37 + \log(1.05 \times 10^{-4}\,M/1.39 \times 10^{-5}\,M) = 7.25$$

Calculating the number of moles ($n$) of $HCO_3^-$ ions in the original sample:

$$n_{HCO_3^-,\,initial} = 1.00\,\text{L} \times \left(\frac{1.05 \times 10^{-4}\text{ mol } HCO_3^-}{\text{L}}\right) = 1.05 \times 10^{-4}\text{ mol } HCO_3^-$$

Calculating the number of moles of $H_3O^+$ ions added and $HCO_3^-$ ions that react with the strong acid:

$$n = 10.0\text{ mL} \times \left(\frac{1.00 \times 10^{-3}\text{ mol } H_3O^+}{1\text{ L}}\right) \times \left(\frac{1\text{ L}}{1000\text{ mL}}\right)$$

$$= 1.00 \times 10^{-5}\text{ mol } H_3O^+\text{ added} = 1.00 \times 10^{-5}\text{ mol } HCO_3^-\text{ reacted}$$

The difference between the initial number of moles and the number of moles reacted equals the number of moles of $HCO_3^-$ ions that remain:

$$1.05 \times 10^{-4}\text{ mol } HCO_3^- - 1.00 \times 10^{-5}\text{ mol } HCO_3^- = 9.5 \times 10^{-5}\text{ mol } HCO_3^-\text{ remaining}$$

To determine the concentration of the remaining $HCO_3^-$ ions, we divide the number of moles by the new total volume of 1.00 L of sample + 10.0 mL of $HNO_3$ solution:

$$M_{HCO_3^-,\,final} = \frac{9.5 \times 10^{-5}\text{ mol } HCO_3^-}{(1.00 + 0.0100)\text{ L}} = 9.4 \times 10^{-5}\,M\,HCO_3^-$$

Now we can calculate the final pH of the river water sample:

$$pH_{final} = pK_a + \log\frac{[HCO_3^-]}{[H_2CO_3]} = 6.37 + \log\frac{9.4 \times 10^{-5}\,M}{1.39 \times 10^{-5}\,M} = 7.20$$

and the change in pH = $pH_{final} - pH_{initial} = 7.20 - 7.25 = -0.05$ pH units.

b. To calculate the pH of 1.00 L of pure water after the addition of acid:

$$[H_3O^+]_{final} = \frac{\text{mol } H_3O^+\text{ added}}{\text{total volume (L)}} = \frac{1.00 \times 10^{-5}\text{ mol } H_3O^+}{1.01\text{ L}} = 9.9 \times 10^{-6}\,M$$

$$pH_{final} = -\log[H_3O^+]_{final} = -\log(9.9 \times 10^{-6}) = 5.00$$

and the change in pH = $pH_{final} - pH_{initial} = 5.00 - 7.00 = -2.00$ pH units.

**Think About It** Adding strong acid to the river water lowered the pH by only 0.05 pH units because it reacted with only about 10% of the basic component of the buffer system. This decrease in pH is a tiny fraction of the change in pH (−2.00 pH units) that happens when the same quantity of acid is added to pure water.

**Practice Exercise** What is the pH of a buffer that contains 0.225 *M* acetic acid and 0.375 *M* sodium acetate? What is the pH of 100.0 mL of the buffer after 10.0 mL of 0.318 *M* NaOH is added?

## SAMPLE EXERCISE 16.7 Effect of Concentration on Buffer Capacity LO3

Calculate the final pH after 0.0100 mol of $H_3O^+$ ions are added to (a) 0.100 L of Buffer A, containing 1.00 *M* acetic acid and 1.00 *M* sodium acetate, and (b) 0.100 L of Buffer B, containing 0.150 *M* acetic acid and 0.150 *M* sodium acetate.

**Collect and Organize** We are asked to calculate the final pH of equal volumes of two buffer solutions after adding the same quantity of acid (0.0100 mol of $H_3O^+$). From Table A5.1, we know that the $pK_a$ of $CH_3COOH$ is 4.75. The pH of a solution with known concentrations of a conjugate acid–base pair can be calculated using the Henderson–Hasselbalch equation.

**Analyze** The pH of both buffers is controlled by the acid ionization equilibrium of acetic acid:

$$CH_3COOH(aq) + H_2O(\ell) \rightleftharpoons CH_3COO^-(aq) + H_3O^+(aq)$$

Addition of 0.0100 mol of $H_3O^+$ to the two buffers will react 0.0100 mol of $CH_3COO^-$ and produce an additional 0.0100 mol of $CH_3COOH$. To evaluate the impact of these changes, we need to calculate the initial numbers of moles of $CH_3COOH$ and $CH_3COO^-$ ions in both buffers and calculate the changes in the concentrations of the reactants and products. We can predict from Figure 16.3 that the more concentrated buffer will experience a smaller change in pH upon the addition of acid.

**Solve** The initial quantities of $CH_3COO^-$ and $CH_3COOH$ in Buffer A are both

$$0.100\ \cancel{L} \times \frac{1.00\ \text{mol}}{\cancel{L}} = 0.100\ \text{mol}$$

The initial quantities of $CH_3COO^-$ and $CH_3COOH$ in Buffer B are both

$$0.100\ \cancel{L} \times \frac{0.150\ \text{mol}}{1\ \cancel{L}} = 0.0150\ \text{mol}$$

We can use a table to see how the amounts of $CH_3COO^-$ and $CH_3COOH$ in Buffer A change upon the addition of $H_3O^+$ ions. As in Sample Exercise 16.6, we calculate the changes in moles:

| Reaction | $CH_3COO^-(aq)$ + | $H_3O^+(aq)$ → | $CH_3COOH(aq) + H_2O(\ell)$ |
|---|---|---|---|
| | $CH_3COO^-$ (mol) | $H_3O^+$ (mol) | $CH_3COOH$ (mol) |
| Initial | 0.100 | 0.0100 | 0.100 |
| Change | −0.0100 | −0.0100 | +0.0100 |
| After | 0.0900 | 0 | 0.110 |

Likewise, we can track the changes in the quantities in Buffer B:

| Reaction | $CH_3COO^-(aq)$ | + | $H_3O^+(aq)$ | → | $CH_3COOH(aq) + H_2O(\ell)$ |
|---|---|---|---|---|---|
| | $CH_3COO^-$ (mol) | | $H_3O^+$ (mol) | | $CH_3COOH$ (mol) |
| Initial | 0.0150 | | 0.0100 | | 0.0150 |
| Change | −0.0100 | | −0.0100 | | +0.0100 |
| After | 0.0050 | | 0 | | 0.0250 |

We can substitute the mole ratio of $CH_3COO^-$ to $CH_3COOH$ in each buffer for the $[CH_3COO^-]/[CH_3COOH]$ ratio in the Henderson–Hasselbalch equation because both components are dissolved in the same volume of buffer. Using these substitutions, we can calculate the pH:

$$\text{Buffer A: pH} = \text{p}K_a + \log\frac{[CH_3COO^-]}{[CH_3COOH]} = 4.75 + \log\frac{0.090\ \cancel{\text{mol}}}{0.110\ \cancel{\text{mol}}} = 4.66$$

$$\text{Buffer B: pH} = \text{p}K_a + \log\frac{[CH_3COO^-]}{[CH_3COOH]} = 4.75 + \log\frac{0.0050\ \cancel{\text{mol}}}{0.0250\ \cancel{\text{mol}}} = 4.05$$

**Think About It** Adding the same quantity of acid produced a pH change of 0.09 units in Buffer A and a pH change of 0.70 units in an equal volume of Buffer B. As predicted, the buffer with a higher concentration—Buffer A—can better resist a change in pH because the relative changes in the concentrations of its components are smaller.

**Practice Exercise** Calculate the change in pH when 0.24 mol of $OH^-$ is added to 1.00 L of (a) a 1.16 *M* solution of $NaH_2PO_4$ containing 1.16 *M* $Na_2HPO_4$ and (b) a 0.58 *M* solution of $NaH_2PO_4$ containing 0.58 *M* $Na_2HPO_4$.

Does the mole ratio of acid to conjugate base make a difference in how well a buffer controls pH? Suppose that you wish to prepare an acidic buffer and you know that contamination by basic substances is more likely than contamination by acids. You might use a 1:1 mixture if you could find a weak acid with a p$K_a$ that matched the target pH—but what if you found another conjugate acid–base pair that required the acid component to be three times as concentrated as the base to achieve the desired pH? Would the second buffer, with the greater proportion of acid in it, better control pH against additions of base than the 1:1 buffer? Sample Exercise 16.8 helps answer this question.

### SAMPLE EXERCISE 16.8 Effect of Base:Acid Ratio on Buffer Capacity LO3

Two pH 3.75 buffers are prepared. Buffer A contains equimolar concentrations of formic acid (p$K_a$ = 3.75) and sodium formate. Buffer B is prepared using a different weak acid (p$K_a$ = 4.23) and conjugate base pair, but with the same overall concentration of conjugate pair components as in Buffer A. To achieve the target pH of 3.75, the [base]/[acid] mole ratio for Buffer B is 1:3. What is the change in pH of Buffer A when enough NaOH is added to neutralize ¼, ½, and ¾ of the formic acid? What is the change in pH when the same three quantities of NaOH are added to an equal volume of Buffer B? Assume the NaOH additions do not significantly increase the volumes of the buffers.

**Collect, Organize, and Analyze** The composition of the buffer changes due to the following reaction with the added hydroxide ions:

$$HCOOH(aq) + OH^-(aq) \rightarrow HCOO^-(aq) + H_2O(\ell)$$

We are not given concentrations but rather mole ratios of the weak acid and its conjugate base in each buffer. If we let *x* be the initial concentrations of formic acid and sodium formate in Buffer A, then the initial concentrations of the acid and base components in Buffer B must be 1.5*x* and 0.5*x*, respectively, to achieve the 1:3 ratio of [base]/[acid]. The three additions of NaOH will reduce the concentrations of the acid components in both buffers by −0.25*x*, −0.50*x*, and −0.75*x* and will increase the concentrations of the basic components by +0.25*x*, +0.50*x*, and +0.75*x*.

**Solve** Setting up Henderson–Hasselbalch equations with the initial [acid] and [base] values of the two buffers and the three changes in [acid] and [base] due to the three additions of NaOH:

Buffer A, 25% addition: $pH = pK_a + \log\frac{[base]}{[acid]} = 3.75 + \log\frac{\cancel{x}(1.00 + 0.25)}{\cancel{x}(1.00 - 0.25)} = 3.97$

50% addition: $pH = pK_a + \log\frac{[base]}{[acid]} = 3.75 + \log\frac{\cancel{x}(1.00 + 0.50)}{\cancel{x}(1.00 - 0.50)} = 4.23$

75% addition: $pH = pK_a + \log\frac{[base]}{[acid]} = 3.75 + \log\frac{\cancel{x}(1.00 + 0.75)}{\cancel{x}(1.00 - 0.75)} = 4.60$

Buffer B, 25% addition: $pH = pK_a + \log\frac{[base]}{[acid]} = 4.23 + \log\frac{\cancel{x}(0.50 + 0.25)}{\cancel{x}(1.50 - 0.25)} = 4.01$

50% addition: $pH = pK_a + \log\frac{[base]}{[acid]} = 4.23 + \log\frac{\cancel{x}(0.50 + 0.50)}{\cancel{x}(1.50 - 0.50)} = 4.23$

75% addition: $pH = pK_a + \log\frac{[base]}{[acid]} = 4.23 + \log\frac{\cancel{x}(0.50 + 0.75)}{\cancel{x}(1.50 - 0.75)} = 4.45$

The changes in pH compared to the target pH of 3.75 are depicted in **Figure 16.4**.

**Think About It** Buffer B's greater capacity to neutralize additions of base is evident when the 0.75*x* addition of NaOH reacts with 75% of the acid in Buffer A but only 50% of the acid in Buffer B. Although Buffer B has more capacity to neutralize very large additions of base, Figure 16.4 shows that Buffer A controls pH better when small quantities of base are added.

**Practice Exercise** Small volumes of strong acid, each containing 0.015 mol of $H_3O^+(aq)$ ions, are added to 1.00-liter samples of two basic buffers. Buffer A contains 0.150 mol of diethylamine and 0.150 mol of its conjugate acid. Buffer B contains 0.200 mol of methylamine and 0.100 mol of its conjugate acid. Before performing the calculation, predict which one will experience the greater change in pH. Calculate the changes in pH in the two buffers due to the addition of the strong acid.

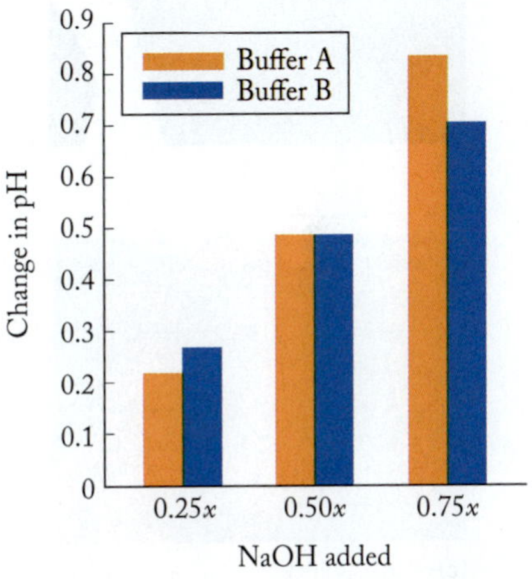

**FIGURE 16.4** Changes in pH produced by adding NaOH to two buffers. Buffer A contains a 1:1 mole ratio of weak acid and its conjugate base. Buffer B contains the same total concentration of acid and conjugate base, but the mole ratio of base:acid is 1:3.

## CONCEPT TEST

Which representation in **Figure 16.5** depicts the aqueous solution of formic acid and sodium formate that would undergo the smallest pH change if a small quantity of strong acid or strong base were added to the solution?

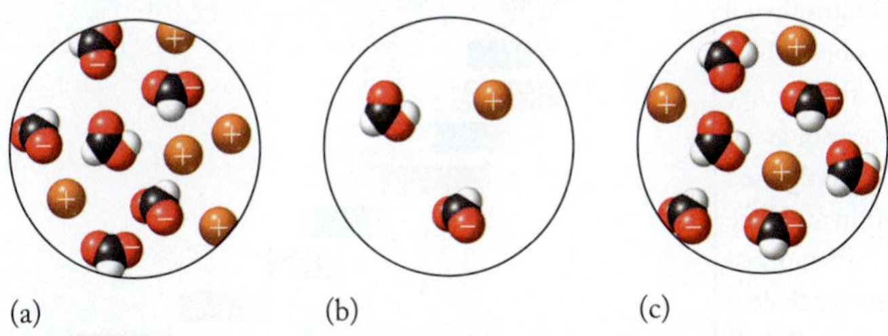


**FIGURE 16.5** Three aqueous solutions of formic acid and sodium formate. The solvent has been omitted for clarity.

*(Answers to Concept Tests are in the back of the book.)*

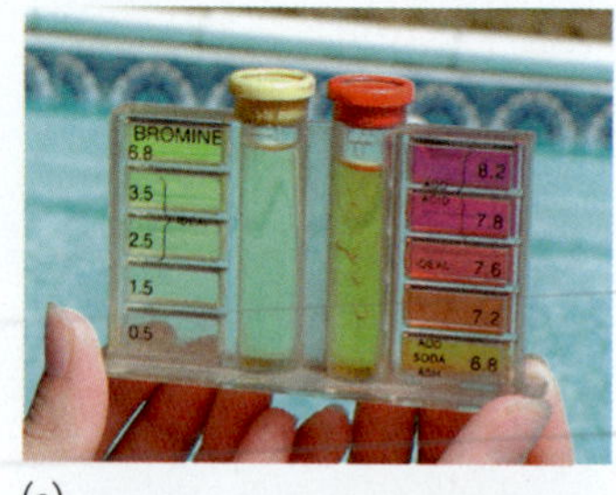


(a)

(b)

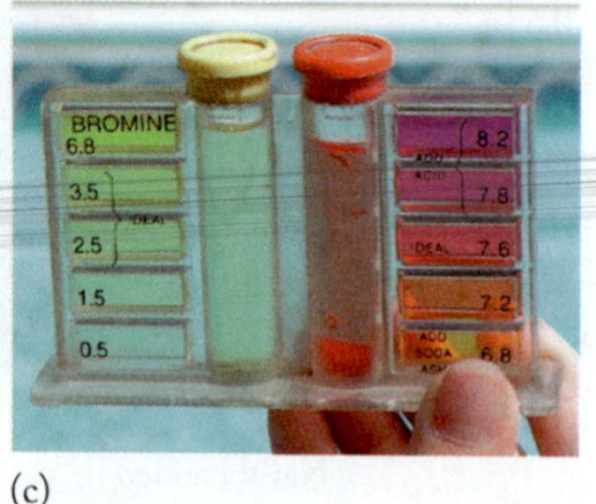


(c)

**FIGURE 16.6** Many pool test kits include the pH indicator phenol red. A few drops are added to a sample of pool water collected in the tube with the red cap. (a) After a rainstorm, the pH of the pool water is 6.8 (or less), as indicated by the yellow color of the sample. (b) Sodium carbonate is added to the pool to raise its pH. (c) A follow-up test produces a red-orange color, indicating that the pH of the pool has been properly adjusted.

## 16.4 Indicators and Acid–Base Titrations

Outdoor swimming pool operators routinely check the pH of pool water to make sure it is close to 7.3, the average pH of our eyes. They often add sodium carbonate to raise the water's pH after acidic summertime precipitation sends it below 7.3. To determine how much sodium carbonate to add, they test the pH of the water by using a kit that includes a **pH indicator** (**Figure 16.6**): a substance that changes color as its pH changes.

One such substance is phenol red, which is yellow as an un-ionized acid and violet when completely ionized to form its conjugate base. Phenol red is well-suited to testing pool pH because its $pK_a = 7.6$, which is close to the desired pool pH of 7.3. This means that one drop of the indicator in a pool sample with a pH of 8.6 (one pH unit above the indicator's $pK_a$) turns violet because most of the phenol red has ionized, as predicted by the Henderson–Hasselbalch equation:

$$8.6 = 7.6 + \log\frac{[\text{base}]}{[\text{acid}]} \quad \text{or} \quad \log\frac{[\text{base}]}{[\text{acid}]} = 1.0 \quad \text{and} \quad \frac{[\text{base}]}{[\text{acid}]} = 10^{1.0} = 10$$

At a sample pH of 6.6, where the [base]/[acid] ratio is only 1:10, phenol red is largely un-ionized, and the sample is yellow. In the pH range between 6.6 and 8.6, the color changes from yellow to orange to red to violet as the pH increases. These color changes allow pool water pH to be determined (and adjusted) to within about 0.1 unit. Most pH indicators can be used to determine pH values in ranges that span about one pH unit above and below their $pK_a$ values, as shown in **Figure 16.7**.

Such indicators are also useful for detecting the even larger changes in pH that occur at the equivalence points in acid–base titrations. We introduced acid–base titration methods in Section 8.5. Here we summarize the principal steps involved:

1. Assemble a titration apparatus (such as the one shown in **Figure 16.8**).
2. Accurately transfer a known volume of sample to the flask.
3. Either add a few drops of an indicator solution to the sample or insert the probe of a pH meter.
4. Fill the burette with a solution (the *titrant*) of known concentration of a substance that reacts with a solute (the *analyte*) in the sample.
5. Slowly add titrant to the sample and monitor the change in pH. When the volume of titrant needed to completely react the analyte has been added, the equivalence point has been reached, as indicated by either a change in

**STEPWISE ANIMATION**

Indicators

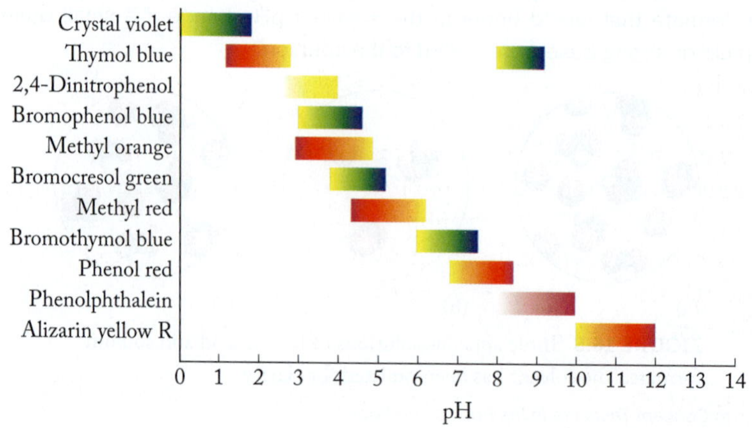


**FIGURE 16.7** A pH indicator is useful within a range of one pH unit above and below the $pK_a$ value of the indicator. This array of indicators could be used to determine pH values from 0 to 12.

pH indicator color or a large change in pH as measured by a pH meter. The volume of titrant needed to reach the equivalence point depends on the concentration of analyte in the sample.

Choosing an appropriate pH indicator and using good lab technique help ensure that no significant difference exists between the theoretical equivalence point and the observed end point.

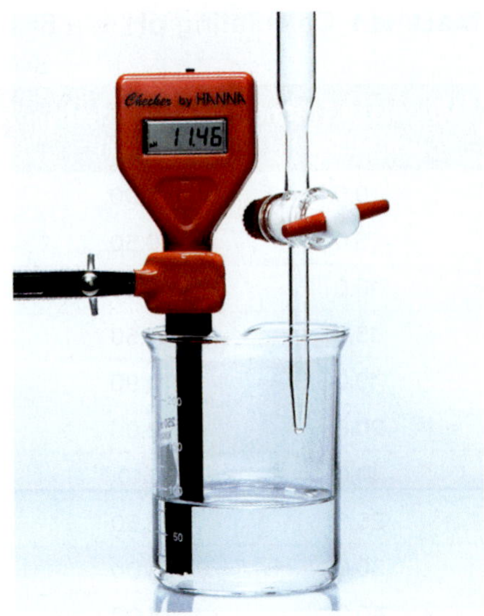


**FIGURE 16.8** A digital pH meter is used to measure pH during a titration.

## Titrating a Strong Acid with a Strong Base

To better understand why a large change in pH happens at the equivalence point, let's begin by calculating the changes in pH that occur during the titration of a strong acid with a strong base. (Later in this section, we will turn to titrations involving weak acids or weak bases.) Suppose we have a 20.0 mL sample of HCl with an unknown concentration that we titrate using 0.100 *M* NaOH. The neutralization reaction is

$$HCl(aq) + NaOH(aq) \rightarrow NaCl(aq) + H_2O(\ell)$$

If 20.0 mL of the 0.100 M NaOH are required to reach the equivalence point, then the quantity of NaOH added is

$$20.0\ \text{m}\cancel{\text{L}} \times \frac{0.100\ \text{mol NaOH}}{\cancel{\text{L}}} = 2.00\ \text{mmol NaOH}$$

Note that we expressed this value in millimoles because the "milli" portion of the milliliter volume did not cancel out. Millimoles are a more convenient unit than moles when working with milliliter volumes of reactant and product solutions because they minimize the use of exponents and simplify conversions between quantities and concentrations.

The 1:1 stoichiometry of the neutralization reaction means that 2.00 mmol of HCl were originally in the sample. At this point, the titration mixture consists of 2.00 mmol of $Na^+$ ions and 2.00 mmol of $Cl^-$ ions dissolved in a total solution volume of 40.0 mL (that is, 20.0 mL of sample + 20.0 mL of titrant). Sodium chloride is a neutral salt, so pH at the equivalence point should be 7.00. To view the pH changes that occur before and after the equivalence point, let's calculate the initial pH of the HCl sample and then its pH at nine additional points over the course of the titration with 0.100 *M* NaOH (**Table 16.1**).

**CONNECTION** In Section 8.5, we learned that the *equivalence point* in an acid–base titration is reached when just enough titrant has been added to react with all the analyte in a sample and that the *end point* of the titration is reached when the indicator used to detect the equivalence point changes color.

**CHEMTOUR**
Acid–Base Titrations

To determine the initial pH of the sample, we know that 2.00 mmol of HCl were present in a volume of 20.0 mL, which means that the concentration was

$$\frac{2.00\ \cancel{\text{m}}\text{mol HCl}}{20.0\ \cancel{\text{m}}\text{L}} = 0.100\ M\ \text{HCl}$$

and the initial pH $= -\log[H_3O^+] = -\log(0.100) = -(-1) = 1$.

Now let's consider the titration as NaOH is added, but before reaching the equivalence point. If 20.0 mL of titrant is needed to completely neutralize the HCl in the sample, then adding less than 20.0 mL of titrant means that NaOH is the limiting reactant and the sample still contains some unreacted HCl. The pH can be calculated by determining the unreacted [HCl]. Because HCl is a strong monoprotic acid, the concentration of HCl equals the concentration of $H_3O^+$ in the sample, and pH $= -\log[H_3O^+]$. For example, adding 5.0 mL of 0.100 *M* NaOH titrant adds

$$5.0\ \text{m}\cancel{\text{L}} \times \frac{0.100\ \text{mol NaOH}}{\cancel{\text{L}}} = 0.50\ \text{mmol NaOH} = 0.50\ \text{mmol OH}^-\ \text{ions}$$

**pH indicator** a water-soluble weak organic acid that changes color as it ionizes.

**TABLE 16.1 Calculating pH in a Strong Acid–Strong Base Titration**

Sample = 20.0 mL of 0.100 *M* HCl; titrant = 0.100 *M* NaOH

| Volume of Titrant Added (mL) | $OH^-$ Added (mmol) | $H_3O^+$ Unreacted (mmol) | $OH^-$ in Excess (mmol) | Total Volume (mL) | $[H_3O^+]$ (*M*) | $[OH^-]$ (*M*) | pH |
|---|---|---|---|---|---|---|---|
| 0.0 | 0.00 | 2.00 | | 20.0 | 0.100 | | 1.00 |
| 5.0 | 0.50 | 1.50 | | 25.0 | 0.0600 | | 1.22 |
| 10.0 | 1.00 | 1.00 | | 30.0 | 0.0333 | | 1.48 |
| 15.0 | 1.50 | 0.50 | | 35.0 | 0.0143 | | 1.85 |
| 19.0 | 1.90 | 0.10 | | 39.0 | 0.00256 | | 2.59 |
| 20.0 | 2.00 | | 0.00 | 40.0 | | | 7.00 |
| 21.0 | 2.10 | | 0.10 | 41.0 | | 0.0024 | 11.38 |
| 25.0 | 2.50 | | 0.50 | 45.0 | | 0.011 | 12.04 |
| 30.0 | 3.00 | | 1.00 | 50.0 | | 0.020 | 12.30 |
| 40.0 | 4.00 | | 2.00 | 60.0 | | 0.033 | 12.52 |

These hydroxide ions neutralize 0.50 mmol $H_3O^+$ ions according to the net ionic equation for the neutralization reaction:

$$H_3O^+(aq) + OH^-(aq) \rightarrow 2\ H_2O(\ell)$$

which means that (2.00 − 0.50) = 1.50 mmol of $H_3O^+$ ions remain unreacted in solution. Meanwhile, the sample now contains 5.0 mL of titrant in addition to the original 20.0 mL of sample. Therefore

$$[H_3O^+] = \frac{1.50\ \text{mmol}\ H_3O^+}{(20.0 + 5.0)\ \text{mL}} = 0.0600\ M$$

and the pH of the reaction mixture is −log(0.0600) = 1.22. These calculations can be repeated to generate the data and the pH values (highlighted in red) in Table 16.1 for the addition of 10.0 mL, 15.0 mL, and 19.0 mL of titrant.

At the equivalence point, 20.0 mL of NaOH have been added, and the pH is 7.00, highlighted in green in Table 16.1. After the equivalence point, adding 21.0 mL, 25.0 mL, 30.0 mL, and 40.0 mL of titrant delivers more $OH^-$ ions than are needed to neutralize the 2.00 mmol of HCl in the sample. The excess $[OH^-]$ increases the pH. For example, when 21.0 mL of NaOH has been added—1.0 mL more than needed to reach the equivalence point, for a total volume of 41.0 mL of titrant plus sample—the additional 1.0 mL of titrant adds

$$1.0\ \text{mL} \times \frac{0.100\ \text{mol NaOH}}{\text{L}} = 0.10\ \text{mmol NaOH} = 0.10\ \text{mmol}\ OH^-\ \text{ions}$$

and produces an $[OH^-]$ in the sample of

$$\frac{0.10\ \text{mmol}\ OH^-}{41.0\ \text{mL}} = 0.0024\ M$$

Calculating pOH: pOH = $-\log[OH^-]$ = 2.62. Calculating pH: pH = 14.00 − 2.62 = 11.38. Similar calculations for titrant volumes of 25.0 mL, 30.0 mL, and 40.0 mL of 0.100 *M* NaOH generate the remaining rows of data and the pH values highlighted in blue in Table 16.1.

A plot of the pH data in Table 16.1 versus the volume of titrant produces the titration curve shown in red (**Figure 16.9**). Adding titrant at the beginning and

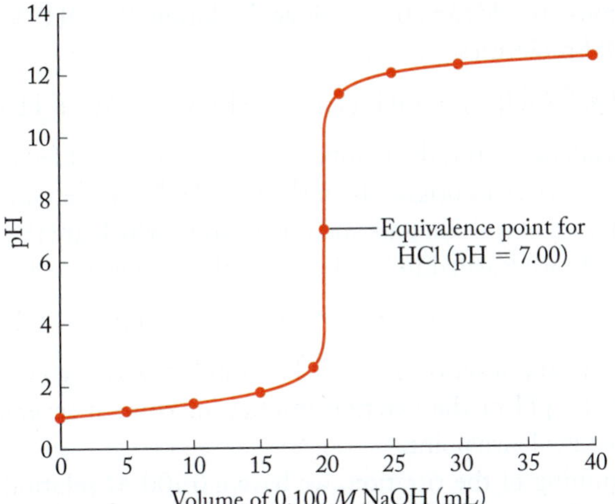


**FIGURE 16.9** The titration curve of 20.0 mL of 0.100 *M* HCl with 0.100 *M* NaOH as the titrant.

end of the titration produces very small changes in pH, but we see a remarkably steep rate of change at the equivalence point. This rapid change makes sense because the pH of the reaction mixture before the equivalence point depends on decreasing concentrations of a strong acid, but the pH after the equivalence point depends on increasing concentrations of a strong base. *A single drop* (0.05 mL) of 0.100 *M* NaOH titrant added to a 40.0 mL reaction mixture could produce most of this change in pH. This sensitivity, along with good lab technique, means that a pH titration is a highly precise, analytical method for determining the concentration of an acidic or basic substance in solution.

## CONCEPT **TEST**

The images shown in **Figure 16.10** depict the solute particles present during the titration of a sample of HCl(*aq*) with LiOH(*aq*). Put the images in order from the beginning to the end of the titration. Which image depicts the equivalence point?

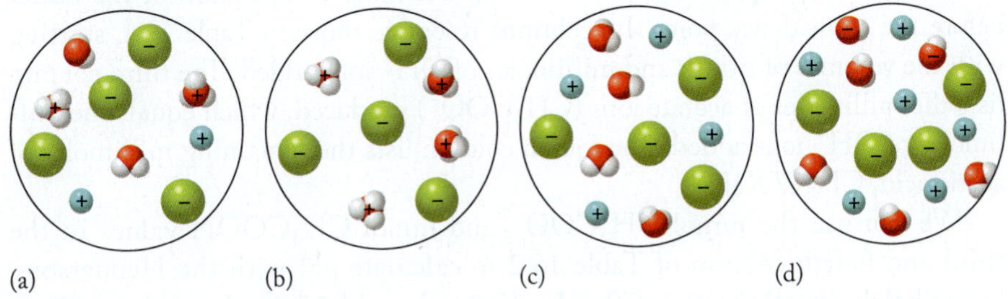


**FIGURE 16.10** Four views of the particles present during a titration of HCl(*aq*) with LiOH(*aq*). The solvent has been omitted for clarity.

## Titrating a Weak Acid with a Strong Base

Now let's examine how the titration of a weak acid with a strong base differs from the titration of a strong acid with a strong base. How does the pH change during the titration of 20.0 mL of 0.100 *M* acetic acid with 0.100 *M* NaOH? Remember that acetic acid is a weak acid and only partially ionizes in a 0.100 *M* solution:

$$CH_3COOH(aq) + H_2O(\ell) \rightleftharpoons CH_3COO^-(aq) + H_3O^+(aq)$$

As hydroxide ions are added to the acetic acid solution, the $OH^-$ ions deprotonate the $CH_3COOH$ molecules:

$$CH_3COOH(aq) + OH^-(aq) \rightarrow CH_3COO^-(aq) + H_2O(\ell)$$

At the equivalence point, the number of moles of $OH^-$ added equals the number of moles of acetic acid originally in the sample. Now the reaction mixture is an aqueous solution of sodium ions and acetate ions, which are the conjugate base of a weak acid. As we learned in Section 15.8, these aqueous solutions are basic:

$$CH_3COO^-(aq) + H_2O(\ell) \rightleftharpoons CH_3COO^-(aq) + OH^-(aq)$$

**CHEMTOUR**
Titrations of Weak Acids

which means that the reaction mixture has a pH > 7 at the equivalence point. Let's calculate the pH of the titration reaction mixture at several points before and after this equivalence point.

At the beginning of the titration, we have a 0.100 *M* solution of acetic acid, which has a $K_a$ value of $1.76 \times 10^{-5}$. Setting up a RICE table for the acid ionization reaction:

| Reaction (R) | $CH_3COOH(aq)$ + $H_2O(\ell)$ ⇌ | $CH_3COO^-(aq)$ + | $H_3O^+(aq)$ |
|---|---|---|---|
| | $[CH_3COOH]$ (*M*) | $[CH_3COO^-]$ (*M*) | $[H_3O^+]$ (*M*) |
| Initial (I) | 0.100 | 0 | 0 |
| Change (C) | $-x$ | $+x$ | $+x$ |
| Equilibrium (E) | $0.100 - x$ | $x$ | $x$ |

and inserting the equilibrium terms into the $K_a$ expression for the reaction:

$$K_a = 1.76 \times 10^{-5} = \frac{[CH_3COO^-][H_3O^+]}{[CH_3COOH]} = \frac{x \cdot x}{0.100 - x} \approx \frac{x^2}{0.100}$$

$$x = [H_3O^+] = 1.33 \times 10^{-3}\ M \text{ and } pH = -\log(1.33 \times 10^{-3}) = 2.88$$

Additions of titrant before the equivalence point convert some, but not all, of the acetic acid in the sample into acetate ions. The presence of a weak acid and its conjugate base means that the reaction mixture can function as a pH buffer. The data in **Table 16.2** provide a record of the changing composition of the buffer before the equivalence point. Its columns resemble those in Table 16.1, starting with the volumes of titrant and millimoles of $OH^-$ ions added. The third column lists the millimoles of acetate ions ($CH_3COO^-$) produced, which equals the millimoles of $OH^-$ ions added. The fourth column lists the remaining millimoles of acetic acid, $CH_3COOH$.

We can use the mmol $CH_3COO^-$ and mmol $CH_3COOH$ values in the third and fourth column of Table 16.2 to calculate pH with the Henderson–Hasselbalch equation after 5.0 mL, 10.0 mL, and 15.0 mL of titrant have been added because comparable concentrations of acetic acid and acetate ions are present. For example, after 5.0 mL of titrant have been added, the quantities of $OH^-$ ions added, acetic acid reacted, and acetate ions produced are all $5.0\ \text{mL} \times \frac{0.100\ \text{mol}}{\text{L}} = 0.50$ mmol. The acetic acid sample initially contained $20.0\ \text{mL} \times \frac{0.100\ \text{mol}\ CH_3COOH}{\text{L}} = 2.00$ mmol $CH_3COOH$. After reacting with 5.0 mL of titrant, the quantity of $CH_3COOH$ remaining is (2.00 − 0.50) = 1.50 mmol. Using the calculated quantities of acetate ions and acetic acid in the Henderson–Hasselbalch equation to calculate pH:

$$pH = pK_a + \log\frac{[\text{base}]}{[\text{acid}]} = 4.75 + \log\frac{\dfrac{0.50\ \text{mmol}\ CH_3COO^-}{25.0\ \text{mL}}}{\dfrac{1.50\ \text{mmol}\ CH_3COOH}{25.0\ \text{mL}}} = 4.75 + \log(0.333) = 4.27$$

**TABLE 16.2** Calculating pH in a Weak Acid–Strong Base Titration

Sample = 0.100 *M* $CH_3COOH$; titrant = 0.100 *M* NaOH

| Volume of Titrant Added (mL) | $OH^-$ Added (mmol) | $CH_3COO^-$ Produced (mmol) | $CH_3COOH$ Remaining (mmol) | Total Volume (mL) | $OH^-$ in Excess (mmol) | pH |
|---|---|---|---|---|---|---|
| 0.0 | 0.00 | 0.00 | 2.00 | 20.0 | | 2.88* |
| 1.0 | 0.10 | 0.10 | 1.90 | 21.0 | | 3.51* |
| 2.0 | 0.20 | 0.20 | 1.80 | 22.0 | | 3.81* |
| 5.0 | 0.50 | 0.50 | 1.50 | 25.0 | | 4.27** |
| 10.0 | 1.00 | 1.00 | 1.00 | 30.0 | | 4.75** |
| 15.0 | 1.50 | 1.50 | 0.50 | 35.0 | | 5.23** |
| 19.0 | 1.90 | 1.90 | 0.10 | 39.0 | | 6.03* |
| 20.0 | 2.00 | 2.00 | 0.00 | 40.0 | 0.00 | 8.73* |
| 21.0 | 2.10 | | | 41.0 | 0.10 | 11.39*** |
| 25.0 | 2.50 | | | 45.0 | 0.50 | 12.05*** |
| 30.0 | 3.00 | | | 50.0 | 1.00 | 12.30*** |
| 40.0 | 4.00 | | | 60.0 | 2.00 | 12.52*** |

*RICE tables and $K_a$ used to calculate equilibrium pH values.

**pH calculated using the Henderson–Hasselbalch equation.

***pH calculated from excess $OH^-$ added after the equivalence point.

Note that the total volume values in the concentration terms above—(20.0 mL of original sample + 5.0 mL of titrant) = 25.0 mL—apply to both terms and cancel out. Thus, the concentration ratio in the Henderson–Hasselbalch equation can also be expressed as a ratio of moles, or here, millimoles. This equivalence was used to calculate the pH values for additions of 10.0 mL and 15.0 mL of NaOH titrant in Table 16.2.

*It is important to note that* when calculating the pH values after additions of 1.0 mL, 2.0 mL, and 19.0 mL of NaOH titrant, we should not use the Henderson–Hasselbalch equation because the concentrations of the two buffer components in these samples differ by a factor of 10 or more. This difference increases the likelihood that the equilibrium concentrations of the buffer components, particularly the one present in the smaller concentration, will differ significantly from the initial values used in the Henderson–Hasselbalch equation.

For example, to calculate the pH after the addition of 1.0 mL of NaOH titrant, we set up a RICE table based on the acid ionization reaction and a total sample volume of 21.0 mL so that $[CH_3COOH_{initial}]$ = 1.90 mmol/21.0 mL = 0.0905 *M* and $[CH_3COO^-{}_{initial}]$ = 0.10 mmol/21.0 mL = 0.0048 *M*:

| Reaction (R) | $CH_3COOH(aq) + H_2O(\ell) \rightleftharpoons$ | $CH_3COO^-(aq) +$ | $H_3O^+(aq)$ |
|---|---|---|---|
| | $[CH_3COOH]$ (*M*) | $[CH_3COO^-]$ (*M*) | $[H_3O^+]$ (*M*) |
| Initial (I) | 0.0905 | 0.0048 | 0 |
| Change (C) | $-x$ | $+x$ | $+x$ |
| Equilibrium (E) | $0.0905 - x$ | $0.0048 + x$ | $x$ |

Inserting the equilibrium terms into the $K_a$ expression and solving for $x$:

$$K_a = 1.76 \times 10^{-5} = \frac{[CH_3COO^-][H_3O^+]}{[CH_3COOH]} = \frac{(0.0048 + x) \cdot x}{0.0905 - x}$$

$$x = [H_3O^+] = 3.11 \times 10^{-4}\ M \text{ and pH} = -\log(3.11 \times 10^{-4}) = 3.51$$

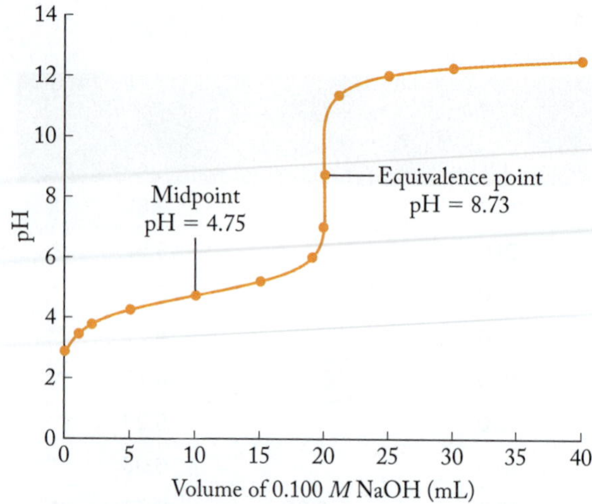


**FIGURE 16.11** The titration curve of 20.0 mL of 0.100 $M$ acetic acid with 0.100 $M$ NaOH as the titrant.

Note that the value of $x$ is (0.000311/0.0048) = 0.065, or 6.5%, of the initial concentration of acetate ions. Although relatively small, this 6.5% difference between initial and equilibrium concentrations is significant—a conclusion supported by the results of using the Henderson–Hasselbalch equation to calculate this pH:

$$\text{pH} = \text{p}K_a + \log\frac{[\text{base}]}{[\text{acid}]} = 4.75 + \log\frac{0.10\ \text{mmol}\ CH_3COO^-}{1.90\ \text{mmol}\ CH_3COOH}$$

$$= 4.75 + \log(1/19) = 3.47$$

Thus, the results obtained by the two methods differ by 3.51 − 3.47 = 0.04 pH units, which may seem like a small difference, but remember: *pH is a log scale.* A decrease of 0.04 pH units corresponds to an increase in $[H_3O^+]$ of $10^{0.04}$, or about 10%. To avoid similar errors, we use the quantities of $CH_3COOH$ and $CH_3COO^-$ in Table 16.2 when 2.0 mL and 19.0 mL of NaOH titrant have been added to calculate initial concentrations of these species and then solve for pH as we just did for 1.0 mL.

An important point in the titration curve of acetic acid (**Figure 16.11**), or any weak acid, lies halfway to the equivalence point. At this *midpoint* in the titration, which was reached after 10.0 mL of titrant was added in the acetic acid titration, half of the acid initially in the sample has been converted into its conjugate base. Therefore, the concentrations of the acid and its conjugate base are equal. If we insert this equality into the log term of the Henderson–Hasselbalch equation, the value of the term is zero. Therefore, the midpoint pH is equal to the p$K_a$ of the acid. This equality holds for any monoprotic weak acid, no matter its initial concentration, and it can be a useful way to identify the acid in an unknown sample.

At the equivalence point, 2.0 mmol of acetic acid have reacted with NaOH to produce 2.0 mmol of acetate ions. The concentration of this sodium acetate solution is $[CH_3COO^-]$ = 2.0 mmol/40.0 mL = 0.050 $M$. We can use this concentration to set up a RICE table:

| Reaction (R) | $CH_3COO^-(aq) + H_2O(\ell) \rightleftharpoons CH_3COOH(aq) + OH^-(aq)$ | | |
|---|---|---|---|
| | $[CH_3COO^-]$ ($M$) | $[CH_3COOH]$ ($M$) | $[H_3O^+]$ ($M$) |
| Initial (I) | 0.050 | 0 | 0 |
| Change (C) | $-x$ | $+x$ | $+x$ |
| Equilibrium (E) | $0.050 - x$ | $x$ | $x$ |

The $K_b$ value for acetate can be determined from the $K_a$ value for acetic acid:

$$K_b = \frac{K_w}{K_a} = \frac{1.00 \times 10^{-14}}{1.76 \times 10^{-5}} = 5.68 \times 10^{-10}$$

The corresponding equilibrium constant expression can be used to solve for $[OH^-]$ at equilibrium and then to calculate pOH followed by pH:

$$K_b = 5.68 \times 10^{-10} = \frac{[CH_3COOH][OH^-]}{[CH_3COO^-]} = \frac{x^2}{0.050 - x}$$

If we assume that $x << 0.050$, solving for $x = 5.33 \times 10^{-6} = [OH^-]$ gives pOH = 5.27, and pH = 8.73.

After the equivalence point, the reaction mixture consists of weakly basic acetate ions and increasing concentrations of a strong base, hydroxide ions. The strong base controls pH, and the titration curve (Figure 16.11) after the equivalence point resembles the curve in Figure 16.9 because pH in this region depends only on the quantity of $OH^-$ ions added by the same titrant, 0.100 *M* NaOH, to the same total sample volume.

### CONCEPT TEST

The images in **Figure 16.12** depict the particles present during a titration of HClO(*aq*) with KOH(*aq*). Which depicts the weak acid before the titration? Which represents the midpoint of the titration? Which represents the equivalence point?

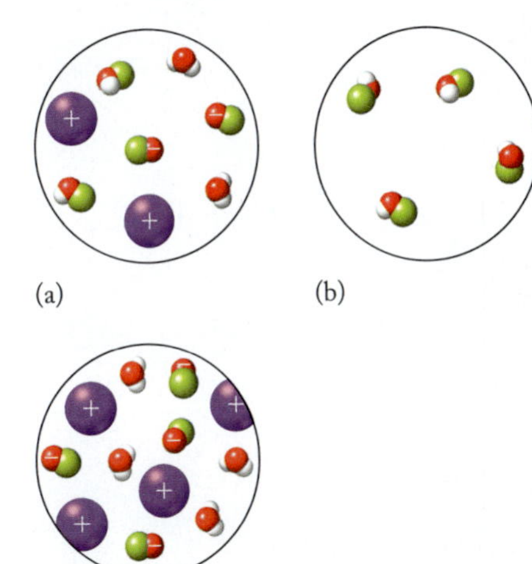


**FIGURE 16.12** Three views of the particles present during a titration of HClO(*aq*) with KOH(*aq*). The solvent has been omitted for clarity.

### SAMPLE EXERCISE 16.9 Relating the Strengths and Concentrations of Acids to Their Titration Curves — LO4

The titration curves in **Figure 16.13** show the changes in pH that occur when 20.0 mL samples of the following acids are each titrated with 0.200 *M* KOH: (a) 0.200 *M* HCl; (b) 0.250 *M* HBr; (c) 0.200 *M* HCOOH; (d) 0.200 *M* $ClCH_2COOH$. Which sample produced which titration curve?

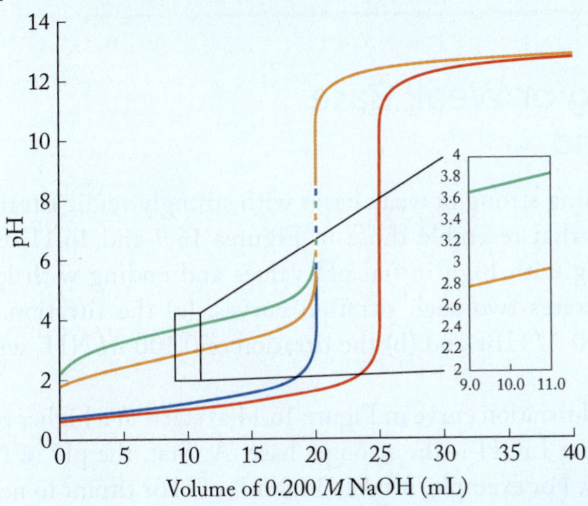


**FIGURE 16.13** Titration curves of four different acids with 0.200 *M* KOH as the titrant.

**Collect, Organize, and Analyze** We are asked to match four titration curves to the four acidic solutions that produced them when each solution was titrated using the same strong base. Solutions (a) and (b) both contain strong acids but at different concentrations. Solutions (c) and (d) contain weak acids at the same concentration. The midpoint pH in the titration of a weak acid is the same as its $pK_a$ value. The following information is available in Table A5.1 of Appendix 5:

| Formula | Name | $K_a$ | $pK_a$ |
|---|---|---|---|
| HCOOH | Formic acid | $1.77 \times 10^{-4}$ | 3.75 |
| $ClCH_2COOH$ | Chloroacetic acid | $1.40 \times 10^{-3}$ | 2.85 |

**Solve** The two strong acids should have the two lowest initial pH values, which are depicted in the red and blue titration curves. The solution that produced the red curve required 25.0 mL/20.0 mL, or 1.25 times as much titrant to reach its equivalence point, which means it contained 1.25 times as much acid. The concentration of the HBr sample is 0.250/0.200 = 1.25 times the concentration of the HCl sample. The volumes of the samples are the same, so the red curve must be HBr (b) and the blue curve was produced by HCl (a).

The midpoints of the orange and green curves were reached when 10.0 mL of titrant had been added. The pH values of these points on the orange and green curves (see the region near the midpoints of the inset graph) are just above pH = 2.8 and below pH = 3.8, respectively, which closely match the $pK_a$ values of the weak acids (2.85 and 3.75). Therefore, the orange curve is chloroacetic acid (d) and the green curve is formic acid (c).

**Think About It** pH titrations are used most often to determine the concentration of an acidic or basic substance in solution, but this exercise demonstrates the qualitative analytical capability of these titrations. They can help identify which weak acid is present in a sample, or at least determine how weak (or strong) it is.

**Practice Exercise** (a) Where would the initial pH be in Figure 16.13, relative to the red and blue curves, for a 0.100 *M* HCl solution? (b) Use Table A5.1 to identify a weak acid for which a 0.200 *M* solution titrated with 0.200 *M* NaOH would reach its midpoint at a higher pH than the green curve in Figure 16.13.

CONCEPT **TEST**

Why is titrating a weak acid with a strong base better than titrating a weak acid with a weak base?

## Titrating a Strong or Weak Base with a Strong Acid

Titrating samples containing strong or weak bases with strongly acidic titrants produces titration curves that resemble those in Figures 16.9 and 16.11, but they are inverted: starting with high initial pH values and ending with low ones. **Figure 16.14** illustrates two such titration curves: (a) the titration of 0.100 *M* LiOH with 0.100 *M* HBr and (b) the titration of 0.100 *M* $NH_3$ with 0.100 *M* HBr.

As expected, the LiOH titration curve in Figure 16.14(a) starts at a higher pH than the $NH_3$ curve because LiOH is the stronger base. At first, the pH of the sample decreases gradually. For example, it takes 10.0 mL of HBr titrant to neutralize half of the $OH^-$ ions in the original sample, but reducing $[OH^-]$ by half increases pOH by $-\log(½) = 0.30$ and decreases pH by only 0.30 units. However, the pH changes more rapidly near the equivalence point as each drop of titrant reacts a larger percentage of the dwindling quantity of hydroxide ions remaining in the sample, similar to the rate of *increase* in pH in the strong acid–strong base titration that we saw in Figure 16.9. At the equivalence point, the pH of the reaction mixture is 7.00 because all the $OH^-$ ions in the original sample have been neutralized, and the mixture is an aqueous solution of LiBr, which is a neutral salt. Beyond the equivalence point, pH is controlled by increasing concentration of $H_3O^+$ ions from additional volumes of HBr titrant. Figure 16.14(a) also includes particulate views of the titration reaction mixture at the beginning of the titration and after 10.0 mL, 20.0 mL, and 30.0 mL of 0.1 *M* HBr have been added. Solvent molecules have been omitted for clarity.

By contrast, the pH of the $NH_3$ sample in Figure 16.14(b) decreases relatively steeply with the first additions of titrant, which produce relatively large increases in the concentration of ammonia's conjugate acid, $NH_4^+$, but the pH changes then become smaller until the titration approaches its equivalence point. In the region of gradual change, $NH_3$ and $NH_4^+$ form a pH buffer that resists pH change just as the mixture of acetic acid and acetate ions do in the titration in Figure 16.11. As in the acetic acid titration, the midpoint in the $NH_3$ titration curve is significant because it is the point at which half of the $NH_3$ initially in the

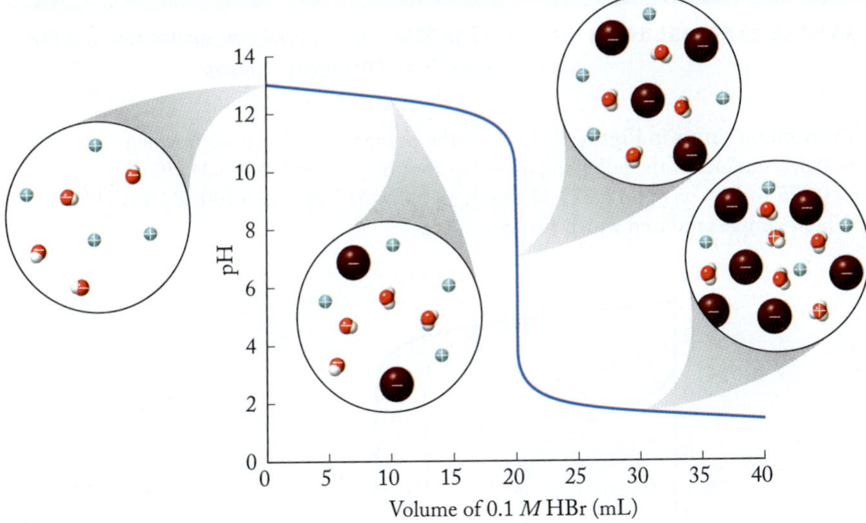


(a)

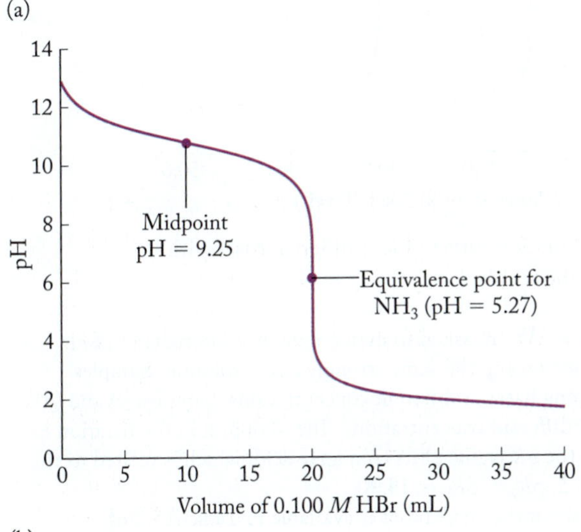


(b)

**FIGURE 16.14** (a) The titration curve of 20.0 mL of 0.100 $M$ LiOH with 0.100 $M$ HBr as the titrant. The four particulate views show how the number of $OH^-$ ions decreases as titrant is added before the equivalence point and the number of $H_3O^+$ ions increases as titrant is added after the equivalence point. The increasing number of $Br^-$ ions is proportional to the volume of titrant added, and the number of $Li^+$ ions is unchanged because neither $Li^+$ nor $Br^-$ is involved in the titration reaction; they are spectator ions. Note that solvent has been omitted for clarity in (a). (b) The titration curve of 20.0 mL of 0.100 $M$ $NH_3$ with 0.100 $M$ HBr as the titrant. The midpoint pH is equal to the p$K_a$ of ammonia's conjugate acid, $NH_4^+$. The pH at the equivalence point is slightly acidic because the sample is a solution of weakly acidic $NH_4^+$ ions and $Br^-$ ions that do not react with water.

sample has been converted into $NH_4^+$ ions. Therefore, $[NH_3] = [NH_4^+]$, and according to the Henderson–Hasselbalch equation,

$$\text{pH} = \text{p}K_a + \log\frac{[NH_3]}{[NH_4^+]} = \text{p}K_a + \log(1) = \text{p}K_a$$

That is, the pH at the midpoint of the titration is equal to the p$K_a$ of the ammonium ion. This p$K_a$ value can be calculated from the p$K_b$ value of ammonia, which is 4.75 (see Table A5.3), using Equation 16.5: p$K_a$ = 14.00 − 4.75 = 9.25. Thus, the pH at the midpoint of the titration is 9.25.

The pH value at the equivalence point is 5.27. This slightly acidic pH makes sense because the product of the titration reaction is an acidic salt, $NH_4Br$, which reacts somewhat with water, forming $H_3O^+$ ions:

$$NH_4^+(aq) + H_2O(\ell) \rightleftharpoons NH_3(aq) + H_3O^+(aq)$$

Beyond their equivalence points, the NaOH and $NH_3$ curves in Figure 16.14 are identical because all the base has been reacted, and the pH is now determined only by the moles of HBr added after the equivalence point and the total volume of the titration reaction mixtures.

## SAMPLE EXERCISE 16.10 Relating the Strengths and Concentrations of Bases to Their Titration Curves LO4

The titration curves in **Figure 16.15** show the changes in pH that occur when 20.0 mL samples of the following bases are each titrated with 0.100 *M* HCl: (a) 0.075 *M* $Ba(OH)_2$, (b) 0.125 *M* KOH, (c) 0.075 *M* $NH_3$, (d) 0.100 *M* $CH_3NH_2$. Which sample produced which titration curve?

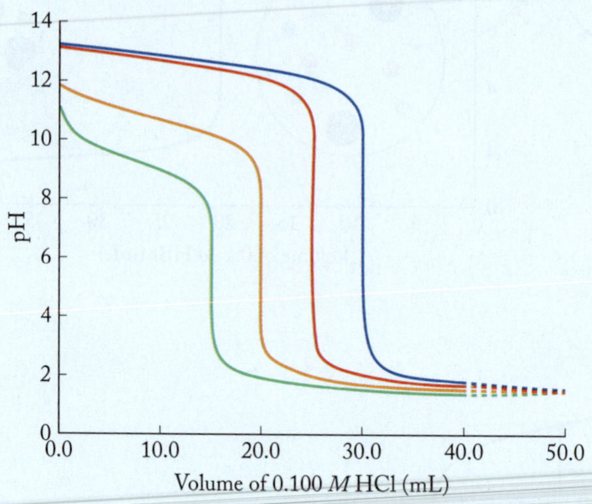


**FIGURE 16.15** Titration curves of four different bases with 0.100 *M* HCl as the titrant.

**Collect, Organize, and Analyze** We are asked to match four titration curves to four basic samples that are each titrated using the same strongly acidic solution. Samples (a) and (b) contain different strong bases at different concentrations. Samples (c) and (d) contain different weak bases at different concentrations. The midpoint in the titration of a weak base is the pH equal to the $pK_a$ value of its conjugate acid, which is related to the $pK_b$ of the base by Equation 16.5: $pK_a + pK_b = 14.00$.

The following information about the weak bases is available in Table A5.3 of Appendix 5:

| Formula | Name | $K_b$ | $pK_b$ |
|---|---|---|---|
| $NH_3$ | Ammonia | $1.76 \times 10^{-5}$ | 4.75 |
| $CH_3NH_2$ | Methylamine | $4.4 \times 10^{-4}$ | 3.36 |

**Solve** The red and blue curves start at the highest pH values, so they must be the curves for the strong bases KOH and $Ba(OH)_2$. The blue curve required more titrant to reach its equivalence point, which means that it contains the higher initial concentration of $OH^-$ ions. The KOH sample appears to have the higher concentration, but KOH produces one mole of hydroxide ions, whereas $Ba(OH)_2$ produces two moles of hydroxide ions. Therefore, a 0.075 *M* $Ba(OH)_2$ solution is 0.150 *M* in $OH^-$ ions. Therefore, the blue curve is (a), $Ba(OH)_2$, and the red curve is (b), KOH.

The orange curve begins at a higher pH and requires more titrant to reach the equivalence point than the green curve. Methylamine has a higher $K_b$ value than that of ammonia, which means it is a stronger base, and the methylamine sample is also more concentrated than the ammonia sample. Therefore, the orange curve is (d), methylamine, and the green curve is (c), ammonia.

**Think About It** The assignment of the two weak bases to the orange and green curves is supported by the curves' midpoint pH values: near 10.5 for the orange curve and a little more than 9.0 for the green one. These pH values should be close to the p$K_a$ values of the conjugate acids of the bases, and they are: 14.00 − 3.36 = 10.64 for methylamine, and 14.00 − 4.75 = 9.25 for ammonia.

**Practice Exercise** (a) What would the concentration of a KOH solution need to be to have an initial pH higher than that of the blue curve in Figure 16.15? (b) Use Table A5.3 to identify a weak base for which a 0.075 *M* solution would reach its equivalence point with a smaller volume of 0.100 *M* HCl titrant than the green curve in Figure 16.15.

## CONCEPT TEST

The images shown in **Figure 16.16** depict the particles present during a titration of $NH_3(aq)$ with $HCl(aq)$. Which represents the start of the titration? Which represents the region where the solution best functions as a buffer? Which represents the equivalence point?

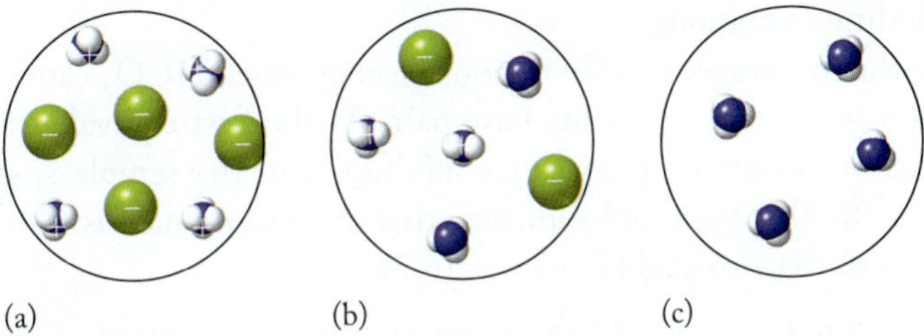


**FIGURE 16.16** Three views of the particles present during a titration of $NH_3$ with HCl. The solvent has been omitted for clarity.

## Titrations with Multiple Equivalence Points

So far, the titration curves in this section have each had only one equivalence point. An alkalinity titration, by contrast (**Figure 16.17**), has two. In environmental science, the alkalinity of a water sample is a measure of its capacity to neutralize additions of acid. Titrating the sample with strong acid provides a way to determine that capacity.

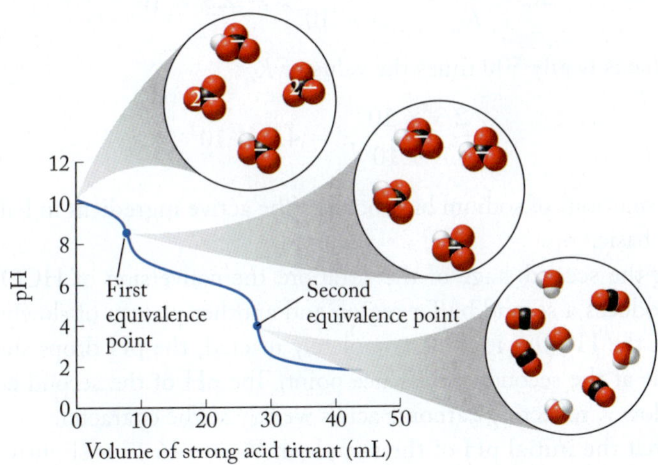


**FIGURE 16.17** An alkalinity titration curve can have two equivalence points. The first marks the complete conversion of any carbonate in the sample into bicarbonate, whereas the second marks the conversion of bicarbonate into carbonic acid, which decomposes to $CO_2$ and $H_2O$.

If carbonate ions are present in the sample, the first additions of titrant convert them into bicarbonate ions,

$$CO_3^{2-}(aq) + H_3O^+(aq) \rightarrow HCO_3^-(aq) + H_2O(\ell)$$

until all the carbonate has been converted into bicarbonate. The titration has then reached its first equivalence point, which is marked by a sharp drop in pH, though not as large a change as observed in the titrations in Figure 16.14. In the second stage of the titration, the bicarbonate ions formed in the first stage plus any bicarbonate ions present in the original sample react with additional acidic titrant, forming carbonic acid, $H_2CO_3(aq)$:

$$HCO_3^-(aq) + H_3O^+(aq) \rightarrow H_2CO_3(aq) + H_2O(\ell)$$

As we have seen, carbonic acid produced in the second stage decomposes to carbon dioxide and water:

$$H_2CO_3(aq) \rightarrow CO_2(g) + H_2O(\ell)$$

In fact, bubbles of $CO_2$ gas may be seen forming in samples during the second stage of alkalinity titrations.

During the first stage, a buffer region exists in which $HCO_3^-$ and $CO_3^{2-}$ ions function as a weak acid–conjugate base pair. At the first equivalence point, the dominant carbon-containing species is $HCO_3^-$ and the sample is still slightly basic (pH > 8). This basic pH indicates that the bicarbonate is a stronger base (Equation 16.6) than an acid (Equation 16.7):

$$HCO_3^-(aq) + H_2O(\ell) \rightleftharpoons H_2CO_3(aq) + OH^-(aq) \qquad (16.6)$$

$$HCO_3^-(aq) + H_2O(\ell) \rightleftharpoons CO_3^{2-}(aq) + H_3O^+(aq) \qquad (16.7)$$

We can confirm the dominance of the reaction in Equation 16.6 by considering the two-step acid ionization equilibria of carbonic acid. Equation 16.7 represents the second ionization step, and its equilibrium constant is $K_{a_2} = 4.7 \times 10^{-11}$. The $K_b$ of the base ionization described in Equation 16.6 is linked to the first ionization of carbonic acid:

$$H_2CO_3(aq) + H_2O(\ell) \rightleftharpoons HCO_3^-(aq) + H_3O^+(aq) \qquad K_{a_1} = 4.3 \times 10^{-7}$$

Converting the $K_{a_1}$ of $H_2CO_3$ into the $K_b$ of $HCO_3^-$ by using Equation 15.19:

$$K_a \times K_b = K_w$$

$$K_b = \frac{K_w}{K_{a_1}} = \frac{1.0 \times 10^{-14}}{4.3 \times 10^{-7}} = 2.3 \times 10^{-8}$$

This $K_b$ value is nearly 500 times the value of $K_{a_2}$:

$$\frac{2.3 \times 10^{-8}}{4.7 \times 10^{-11}} = 4.9 \times 10^2$$

Therefore, solutions of sodium bicarbonate (the active ingredient in baking soda) are weakly basic.

During the second stage of the titration, the conversion of $HCO_3^-$ ions to $H_2CO_3$ produces a second buffer region and another plateau of slowly changing pH. When the $HCO_3^-$ ions are completely reacted, the pH drops sharply for a second time at the second equivalence point. The pH of the second equivalence point is below 7, reflecting carbonic acid's weakly acidic character.

Note that the initial pH of the sample in Figure 16.17 is slightly above 10, which is quite basic and above the pH range tolerated by many species of aquatic life. Such highly basic water may be found in arid regions such as the U.S. Southwest, where rocks containing $CaCO_3$ and other basic compounds are in contact with water. Seawater and most freshwater samples are not that basic, which means that the dominant carbonate species in them is bicarbonate. The alkalinity titration curves of these waters may have only one equivalence point, coinciding with the pH of the second equivalence point in Figure 16.17.

**SAMPLE EXERCISE 16.11** Interpreting the Results of a Titration with Multiple Equivalence Points I **LO4**

If the volume of titrant needed to reach the first equivalence point in the titration shown in Figure 16.17 is 9.00 mL, and a total of 27.00 mL is required to reach the second equivalence point, what was the ratio of carbonate to bicarbonate in the original sample?

**Collect, Organize, and Analyze** Carbonate ions ($CO_3^{2-}$) in the sample react with $H_3O^+$ ions from the titrant and are converted to bicarbonate ions ($HCO_3^-$) in the first stage of the titration. Bicarbonate ions—including any in the original sample plus all those produced in the first-stage reaction—are converted to carbonic acid, $H_2CO_3$, in the second stage.

**Solve** The volume of titrant needed to titrate the $CO_3^{2-}$ in the sample (9.00 mL) is exactly half the additional volume (27.00 − 9.00 = 18.00 mL) needed to titrate the $HCO_3^-$ in the reaction mixture. However, 9.00 mL of the titrant reacted in the second stage was needed just to titrate the $HCO_3^-$ produced in the first stage. This means that the volume of titrant needed to titrate the $HCO_3^-$ in the original sample was only (18.00 − 9.00) = 9.00 mL. Therefore, the original sample contained equal concentrations of $CO_3^{2-}$ and $HCO_3^-$ ions.

**Think About It** It would be tempting to interpret the volumes of titrant reacted in the two stages to mean that the original sample had twice as much bicarbonate as carbonate. Remember that the $HCO_3^-$ ions titrated in the second stage come from the original sample and from $HCO_3^-$ ions produced from $CO_3^{2-}$ ions during the first stage of the titration.

**Practice Exercise** Is the midpoint pH in the first stage of an alkalinity titration always the same as the $pK_{a_2}$ of carbonic acid? Explain why or why not.

We can detect both equivalence points in an alkalinity titration by using a pH electrode, or we can use appropriate color indicators. To detect the first equivalence point, we need an indicator with a $pK_a$ near the pH of the solution at the first equivalence point, which is about 8.5. One candidate in Figure 16.7 is phenolphthalein ($pK_a = 9.7$), which is pink in its basic form and colorless at low pH. Phenol red would not be a good choice for the titration in Figure 16.17 because it changes color between pH 6.8 and pH 8.4. This range is just below the pH of the first equivalence point and well above the pH of the second equivalence point.

To detect the second equivalence point, we could add bromocresol green ($pK_a = 4.6$) after the first equivalence point has been reached, but we should not add it earlier because its blue-green color in basic solutions would obscure the pink-to-colorless transition of phenolphthalein. We need not be concerned about the phenolphthalein obscuring the bromocresol green end point because phenolphthalein is colorless in acidic solutions.

**FIGURE 16.18** The water in Sapphire Pool in Yellowstone National Park is slightly alkaline, due mostly to carbonate and bicarbonate ions. It is also crystal clear and very hot.

### SAMPLE EXERCISE 16.12 Interpreting the Results of a Titration with Multiple Equivalence Points II — LO4

A 100.0 mL sample of water from the Sapphire Pool in Yellowstone National Park (**Figure 16.18**) is titrated with 0.0300 *M* HCl. A few drops of phenolphthalein are added at the beginning of the titration, and the solution turns pink. Reaching the pink-to-clear equivalence point takes 5.91 mL of titrant. Then a few drops of bromocresol green are added, and an additional 26.02 mL of titrant is required before the blue-green color changes to yellow. What were the initial concentrations of carbonate and bicarbonate in the sample?

**Collect and Organize** We are asked to determine the concentrations of $CO_3^{2-}$ and $HCO_3^-$ ions in a water sample from the results of a titration. These determinations are based on the volumes of titrant needed to reach two equivalence points. In the first stage, $CO_3^{2-}$ ions in the sample are converted to $HCO_3^-$ ions:

$$(1)\qquad H_3O^+(aq) + CO_3^{2-}(aq) \rightarrow HCO_3^-(aq) + H_2O(\ell)$$

In the second stage, $HCO_3^-$ ions are converted to $H_2CO_3$:

$$(2)\qquad H_3O^+(aq) + HCO_3^-(aq) \rightarrow H_2CO_3(aq) + H_2O(\ell)$$

**Analyze** According to the stoichiometries of the reactions, it takes one mole of HCl to titrate one mole of carbonate to bicarbonate in the first stage, and it takes one mole of HCl to titrate one mole of bicarbonate to carbonic acid in the second stage. The $HCO_3^-$ ions titrated in stage 2 include any in the original sample plus all the $HCO_3^-$ ions produced in stage 1. The difference between the titrant volumes needed to reach the two equivalence points, (26.02 − 5.91) = 20.11 mL, is the volume of acid required to react with the $HCO_3^-$ present in the original sample. This difference is between three and four times the volume of titrant reacted in stage 1, so we can predict that three to four times as much $HCO_3^-$ was in the original sample as $CO_3^{2-}$.

**Solve** Calculating the concentrations of $CO_3^{2-}$ and $HCO_3^-$ in the original sample from the volumes and molarity of the HCl titrant reacted in stages 1 and 2:

$$[CO_3^{2-}] = \frac{\text{mol } CO_3^{2-}}{\text{L solution}}$$

$$= \frac{5.91 \text{ mL titrant} \times \dfrac{0.0300 \text{ mmol HCl}}{\text{mL titrant}} \times \dfrac{1 \text{ mol HCl}}{1000 \text{ mmol HCl}} \times \dfrac{1 \text{ mmol } CO_3^{2-}}{1 \text{ mmol HCl}}}{100.0 \text{ mL solution} \times \dfrac{1 \text{ L}}{1000 \text{ mL}}}$$

$$= 1.77 \times 10^{-3}\ M$$

$$[HCO_3^-] = \frac{\text{mol } HCO_3^-}{\text{L solution}}$$

$$= \frac{20.11 \text{ mL titrant} \times \dfrac{0.0300 \text{ mmol HCl}}{\text{mL titrant}} \times \dfrac{1 \text{ mol HCl}}{1000 \text{ mmol HCl}} \times \dfrac{1 \text{ mmol } HCO_3^-}{1 \text{ mmol HCl}}}{100.0 \text{ mL solution} \times \dfrac{1 \text{ L}}{1000 \text{ mL}}}$$

$$= 6.03 \times 10^{-3}\ M$$

**Think About It** The titration results confirm that the bicarbonate concentration in the original sample was just over three times the carbonate concentration.

**Practice Exercise** Suppose you titrate a 100.0 mL sample of water from another pool in Yellowstone National Park with a ratio of carbonate to bicarbonate of 3:1. If the total volume of titrant needed to reach the second equivalence point is 21.00 mL, what volume of titrant was used to reach the first equivalence point?

**CONCEPT TEST**

In a titration that initially contains both $CO_3^{2-}$ and $HCO_3^-$, the volume of titrant required to reach the first equivalence point is less than that required to titrate from the first equivalence point to the second. Why?

## 16.5 Lewis Acids and Bases

Until now, we have used the Brønsted–Lowry definition of acids ($H^+$ ion donors) and bases ($H^+$ ion acceptors). The time has come to expand our concept of acids and bases to include acid–base interactions that may or may not involve the transfer of $H^+$ ions. Let's begin by revisiting what happens when ammonia gas dissolves in water:

$$NH_3(g) + H_2O(\ell) \rightleftharpoons NH_4^+(aq) + OH^-(aq)$$

**Figure 16.19(a)** shows a Brønsted–Lowry interpretation of this reaction: in donating $H^+$ ions to ammonia, $H_2O$ acts as a Brønsted–Lowry acid. In accepting $H^+$ ions, $NH_3$ acts as a Brønsted–Lowry base.

Another way to view this reaction is illustrated in **Figure 16.19(b)**. Instead of focusing on the transfer of a hydrogen ion, we consider the reaction as one in which one reactant donates and the other reactant accepts a *pair of electrons*. In this view, the N atom in $NH_3$ donates its lone pair of electrons to one of the H atoms in $H_2O$. In the process, one of the H—O bonds in $H_2O$ is broken in such a way that the bonding pair of electrons remains with the O atom. The donated lone pair from the N atom forms a fourth N—H covalent bond. The result is the same as in the Brønsted–Lowry model of acid–base behavior: a molecule of $NH_3$ bonds to a $H^+$ ion, forming an $NH_4^+$ ion, and a molecule of $H_2O$ loses a $H^+$ ion, becoming a $OH^-$ ion.

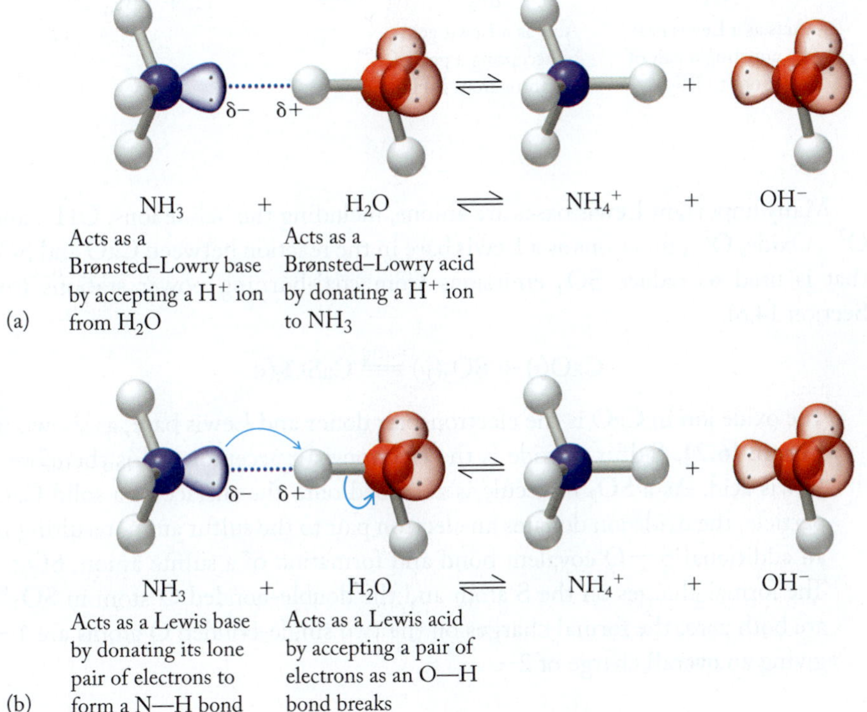


**FIGURE 16.19** (a) The Brønsted–Lowry view of the reaction between $H_2O$ (proton donor) and $NH_3$ (proton acceptor). (b) In the Lewis view of the reaction, $H_2O$ acts as a Lewis acid (an electron-pair acceptor) and $NH_3$ acts as a Lewis base (an electron-pair donor).

**Lewis base** a substance that donates a lone pair of electrons in a chemical reaction.

**Lewis acid** a substance that accepts a lone pair of electrons in a chemical reaction.

**coordinate bond** a covalent bond between two atoms in which one of the two provides both electrons in the bonding pair.

When viewed from the perspective of the electron pair:

- A **Lewis base** is a substance that donates a lone pair of electrons in a chemical reaction.
- A **Lewis acid** is a substance that accepts a lone pair of electrons in a chemical reaction.

These definitions are named after their developer, Gilbert N. Lewis, who also pioneered research into the nature of chemical bonds. The Lewis definition of a base is consistent with the Brønsted–Lowry model we have used because a substance must be able to donate a pair of electrons if it is to bond with a $H^+$ ion. However, the same parallelism does not hold for acids. The Brønsted–Lowry model defines an acid as a hydrogen-ion donor, but the Lewis definition includes species that have no hydrogen ions to donate but that can still accept electrons.

**CONNECTION** Lewis's theories on the nature of covalent bonding were described in Section 4.4.

One such compound is boron trifluoride, $BF_3$. With only six valence electrons, the boron atom in $BF_3$ can accept another pair to complete its octet. $NH_3$ is a suitable electron-pair donor, as shown in **Figure 16.20**. No $H^+$ ions are transferred in this reaction, so it is not an acid–base reaction according to the Brønsted–Lowry model. However, $NH_3$ donates a lone pair of electrons and $BF_3$ accepts it, so it is an acid–base reaction according to the more general Lewis model.

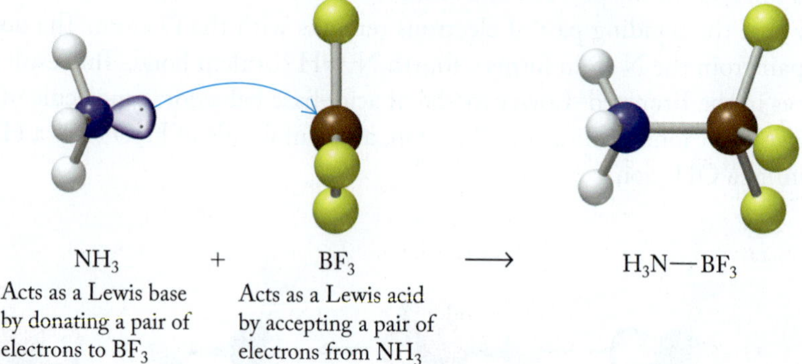


**FIGURE 16.20** The reaction between $NH_3$ and $BF_3$ is an acid–base reaction in the Lewis sense because $NH_3$ donates an electron pair (making it the Lewis base) and $BF_3$ accepts the electron pair (making it the Lewis acid). This is not an acid–base reaction in the Brønsted–Lowry sense, however, because no proton is transferred.

**CONNECTION** The concept of formal charge and how to calculate it were described in Section 4.7.

Many important Lewis bases are anions, including the halide ions, $OH^-$, and $O^{2-}$. Oxide, $O^{2-}$, functions as a Lewis base in the reaction between CaO and $SO_2$ that is used to reduce $SO_2$ emissions from coal-burning power stations (see Section 14.6):

$$CaO(s) + SO_2(g) \rightleftharpoons CaSO_3(s)$$

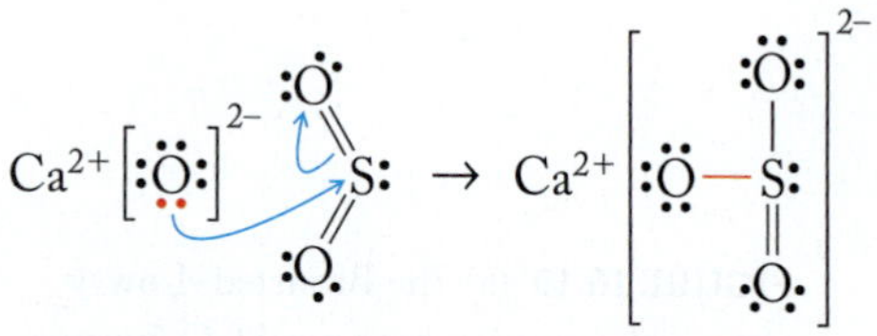

**FIGURE 16.21** In the reaction between CaO and $SO_2$, the oxide ion ($O^{2-}$) is a Lewis base, donating an electron pair to $SO_2$, the Lewis acid.

The oxide ion in CaO is the electron-pair donor and Lewis base, as shown in **Figure 16.21**. Sulfur dioxide is the electron-pair acceptor and is therefore a Lewis acid. As a $SO_2$ molecule is absorbed onto the surface of a solid CaO particle, the oxide ion donates an electron pair to the sulfur atom, resulting in an additional S—O covalent bond and formation of a sulfite anion, $SO_3^{2-}$. The formal charges on the S atom and the double-bonded O atom in $SO_3^{2-}$ are both zero; the formal charges on the two single-bonded O atoms are 1−, giving an overall charge of 2−.

**SAMPLE EXERCISE 16.13** Identifying Lewis Acids and Bases **LO5**

In the following reaction, which species is a Lewis acid and which is a Lewis base?

$$AlCl_3 + Cl^- \rightarrow AlCl_4^-$$

**Collect and Organize** We are given a chemical reaction and asked to identify the Lewis acid (the reactant that accepts a pair of electrons) and the Lewis base (the reactant that donates that pair of electrons).

**Analyze** A $Cl^-$ ion has four lone pairs of electrons in its valence shell, so it can *donate* one of them to form a covalent bond to aluminum. To determine whether $AlCl_3$ can act as a Lewis base, we need to examine its Lewis structure (**Figure 16.22**). This structure accounts for all 24 valence electrons with three Al—Cl single bonds and *no lone pairs.* This means Al has only six valence electrons and the capacity to accept one more pair—that is, to act as a Lewis acid.

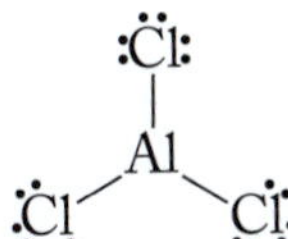

**FIGURE 16.22** Lewis structure of $AlCl_3$.

**Solve** In this reaction, $AlCl_3$ is a Lewis acid and the $Cl^-$ ion is a Lewis base (**Figure 16.23**).

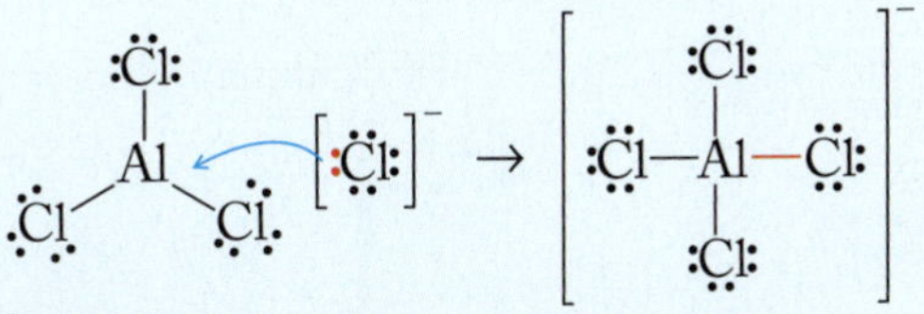

**FIGURE 16.23** The $Cl^-$ ion donates a pair of electrons to $AlCl_3$, forming $AlCl_4^-$.

**Think About It** Drawing the Lewis structure of $AlCl_3$ and determining that the central Al atom has an incomplete octet is the key to identifying its capacity to act as a Lewis acid. Note that these Lewis structures assume that $AlCl_3$ is a molecular compound and not, as with most other metal chlorides, an ionic compound. This assumption is supported by the physical properties of $AlCl_3$ and particularly by the fact that it sublimes at 178°C and 1 atm. Most metal halides do not melt until heated to temperatures many hundreds of degrees higher than that.

**Practice Exercise** In the following reaction, which reactant is the Lewis acid, and which is the Lewis base?

$$CO_2(g) + CaO(s) \rightarrow CaCO_3(s)$$

## 16.6 Formation of Complex Ions

In Chapter 6, we described how ions dissolved in water are *hydrated*—that is, they are surrounded by water molecules oriented with the positive ends of their dipoles directed toward anions and their negative ends directed toward cations (**Figure 16.24**). In some hydrated cations, ion–dipole interactions lead to the sharing of lone-pair electrons on the oxygen atoms of $H_2O$ with empty valence-shell orbitals on the cations. This interaction between a lone pair and an empty orbital is another example of Lewis acid–base behavior. These shared electron pairs meet our definition of covalent bonds, but these particular bonds are called *coordinate* covalent bonds, or simply **coordinate bonds.**

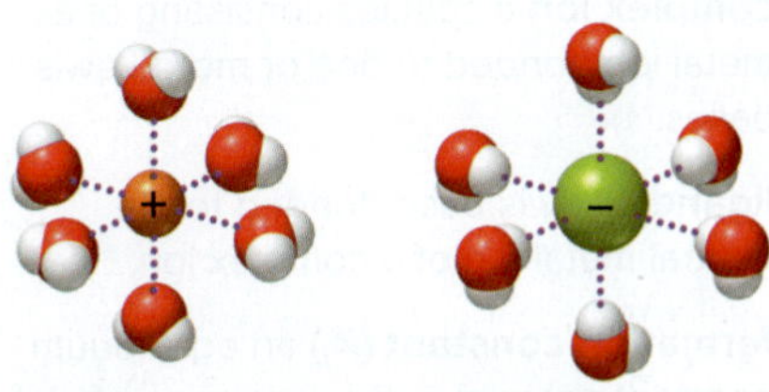

**FIGURE 16.24** Ion–dipole interactions in a hydrated cation and a hydrated anion.

**CONNECTION** Spheres of hydration around cations and anions were introduced in Section 6.3.

Coordinate bonds form when either a molecule or an anion donates a lone pair of electrons to an empty valence-shell orbital of an atom, cation, or molecule. Once formed, a coordinate bond is indistinguishable from any other kind of covalent bond. When a cation forms coordinate bonds with a cluster of molecular or ionic electron-pair donors, the resulting structure is called a **complex ion**. The electron-pair donors in complex ions are called **ligands**. For example, when six molecules of water form coordinate bonds to a $Ni^{2+}$ cation in an aqueous solution, the resulting complex ion is written as $Ni(H_2O)_6^{2+}$ to show that it has six ligands (six water molecules bonded to the metal).

We can characterize the formation of complex ions by using the same mathematical tools we have used to understand acid–base equilibria. These tools are appropriate because complex formation processes are reversible, and many of them reach chemical equilibrium rapidly. We begin this investigation with two aqueous solutions, one containing copper(II) sulfate ($CuSO_4$) and the other containing $NH_3$ (**Figure 16.25**). The $CuSO_4$ solution is robin's-egg blue, the color characteristic of $Cu^{2+}(aq)$ ions dissolved in water, whereas the

**FIGURE 16.25** The beaker on the right contains a solution of $Cu^{2+}(aq)$ which is a characteristic robin's-egg blue. The beaker on the left contains a colorless solution of ammonia. When the two solutions are mixed in the test tube in the middle, the color changes to the dark blue color of the complex ion $Cu(NH_3)_4^{2+}$. The solvent has been omitted from the particulate views for clarity.

**complex ion** a species consisting of a metal ion bonded to one or more Lewis bases.

**ligand** a Lewis base bonded to the central metal ion of a complex ion.

**formation constant ($K_f$)** an equilibrium constant describing the formation of a metal complex from a free metal ion and its ligands.

ammonia solution is colorless. When the solutions are mixed, the robin's-egg blue turns a dark navy blue, as shown in the test tube in Figure 16.25. This is the color of $Cu(NH_3)_4^{2+}$ complex ions. The change in color provides visual evidence that the following equilibrium lies far to the right, favoring complex ion formation:

$$Cu^{2+}(aq) + 4\ NH_3(aq) \rightleftharpoons Cu(NH_3)_4^{2+}(aq)$$

This conclusion is supported by the large equilibrium constant for the reaction:

$$K_f = \frac{[Cu(NH_3)_4^{2+}]}{[Cu^{2+}][NH_3]^4} = 5.0 \times 10^{13}$$

The equilibrium constant $K_f$ is called a **formation constant** because it describes the formation of a complex ion. The $K_f$ values for $Cu(NH_3)_4^{2+}$ and other complex ions are listed in Table A5.5 in Appendix 5.

For the general case in which one mole of metal ions ($M^{m+}$) combines with $n$ moles of ligand ($X^{x-}$) to form the complex ion $MX_n^{(m-nx)+}$, the formation constant expression is

$$K_f = \frac{[MX_n^{(m-nx)+}]}{[M^{m+}][X^{x-}]^n}$$

Formation constants can be used to calculate the concentration of complex ions in solution or to calculate the concentration of free, uncomplexed metal ions, $M^{m+}(aq)$, in equilibrium with a given (usually larger) concentration of a ligand. Because $K_f$ values are usually very large, equilibrium concentrations of uncomplexed metal ions are usually very small. One approach to calculating the concentration of an uncomplexed metal ion is to consider the reverse of the formation reaction and calculate how much $Cu(NH_3)_4^{2+}(aq)$ dissociates. The equilibrium constant for the dissociation reaction ($K_d$) is the reciprocal of the original equilibrium constant.

$$Cu(NH_3)_4^{2+}(aq) \rightleftharpoons Cu^{2+}(aq) + 4\ NH_3(aq) \quad K_d = \frac{1}{K_f} = \frac{[Cu^{2+}][NH_3]^4}{[Cu(NH_3)_4^{2+}]}$$

Using $K_d$ to calculate the concentration of uncomplexed copper ion, $[Cu^{2+}]$, is illustrated in Sample Exercise 16.14.

**SAMPLE EXERCISE 16.14** Calculating the Concentration of Free Metal in Equilibrium with a Complex Ion **LO6**

Ammonia gas is dissolved in a $1.00 \times 10^{-4}$ *M* solution of $CuSO_4$ to give an equilibrium concentration of $[NH_3] = 1.60 \times 10^{-3}$ *M*. Calculate the concentration of $Cu^{2+}(aq)$ ions in the solution.

**Collect and Organize** The concentration of $CuSO_4$ means that $[Cu^{2+}]$ before complex formation is $1.00 \times 10^{-4}$ *M*. The equilibrium concentration of the ligand ($NH_3$) is $1.60 \times 10^{-3}$ *M*. The values of $[NH_3]$, $[Cu^{2+}]$, and $[Cu(NH_3)_4^{2+}]$ at equilibrium are related by the formation constant expression:

$$K_f = \frac{[Cu(NH_3)_4^{2+}]}{[Cu^{2+}][NH_3]^4} = 5.0 \times 10^{13}$$

**Analyze** Because $K_f$ is large, we can assume that essentially all the $Cu^{2+}$ ions are converted to complex ions and that $[Cu(NH_3)_4^{2+}] = 1.00 \times 10^{-4}$ *M*. Only a tiny concentration of free $Cu^{2+}$ ions, *x*, remains at equilibrium. We can calculate $[Cu^{2+}]$ by determining how much $Cu(NH_3)_4^{2+}$ dissociates:

$$Cu(NH_3)_4^{2+}(aq) \rightleftharpoons Cu^{2+}(aq) + 4\,NH_3(aq)$$

The equilibrium constant, $K_d$, for this reaction is the reciprocal of $K_f$:

$$K_d = \frac{1}{K_f} = \frac{[Cu^{2+}][NH_3]^4}{[Cu(NH_3)_4^{2+}]} = \frac{1}{5.0 \times 10^{13}} = 2.0 \times 10^{-14}$$

We can construct a RICE table incorporating the concentrations and concentration changes of the products and reactants. Given the small value of $K_d$, we should obtain a $[Cu^{2+}]$ value at equilibrium that is much less than $1.00 \times 10^{-4}$ *M*.

**Solve** First, we complete the row of equilibrium concentrations of $Cu^{2+}$, $NH_3$, and $Cu(NH_3)_4^{2+}$ in the RICE table:

| Reaction (R) | $Cu(NH_3)_4^{2+}(aq)$ | $\rightleftharpoons$ | $Cu^{2+}(aq)$ | + | $4\,NH_3(aq)$ |
|---|---|---|---|---|---|
| | $[Cu(NH_3)_4^{2+}]$ (*M*) | | $[Cu^{2+}]$ (*M*) | | $[NH_3]$ (*M*) |
| Initial (I) | $1.00 \times 10^{-4}$ | | 0 | | $1.60 \times 10^{-3}$ |
| Change (C) | $-x$ | | $+x$ | | 0 |
| Equilibrium (E) | $1.00 \times 10^{-4} - x$ | | $x$ | | $1.60 \times 10^{-3}$ |

Next, we make the simplifying assumption that *x* is much smaller than $1.00 \times 10^{-4}$ *M*. Therefore, we can ignore the *x* term in the equilibrium value of $[Cu(NH_3)_4^{2+}]$:

$$K_d = \frac{[Cu^{2+}][NH_3]^4}{[Cu(NH_3)_4^{2+}]}$$

$$= \frac{(x)(1.60 \times 10^{-3})^4}{1.0 \times 10^{-4} - x} \approx \frac{(x)(1.60 \times 10^{-3})^4}{1.00 \times 10^{-4}} = 2.0 \times 10^{-14}$$

$$x = 3.1 \times 10^{-7} = [Cu^{2+}]$$

**Think About It** This result validates our simplifying assumption and confirms our prediction that $[Cu^{2+}]$ at equilibrium is much less than $[Cu^{2+}]$ initially. In fact, more than 99% of the copper(II) in the solution is present as $Cu(NH_3)_4^{2+}$.

**Practice Exercise** Calculate the equilibrium concentration of $Ag^+(aq)$ in a solution that is initially 0.100 *M* $AgNO_3$ and 0.800 *M* $NH_3$ after the following reaction takes place:

$$Ag^+(aq) + 2\,NH_3(aq) \rightleftharpoons Ag(NH_3)_2^+(aq) \qquad K_f = 1.7 \times 10^7$$

## 16.7 Hydrated Metal Ions as Acids

**CONNECTION** We discussed periodic trends in electronegativity in Section 4.2.

In Chapter 15, we saw that the strength of an oxoacid depends on the electronegativity of its central atom. For example, the relative strengths of the three hypohalous acids (HOCl > HOBr > HOI) correspond to the relative electronegativities of their halogen atoms: Cl > Br > I. The more electronegative the halogen, the more it draws electron density away from the oxygen atom in the polar O—H bond (see Table 15.5). These shifts make the negative ion formed by deprotonation better able to bear a negative charge because of increased delocalization.

A similar shift in electron density occurs in hydrated metal ions having the generic formula $M(H_2O)_6{}^{n+}$ when $n \geq 2$. The electrons in the O—H bonds of the water molecules surrounding the metal ions are attracted to the positively charged ions. The resulting distortion in electron density increases the likelihood of one of these O—H bonds ionizing and donating a $H^+$ ion to a neighboring molecule of water:

$$M(H_2O)_6{}^{n+}(aq) + H_2O(\ell) \rightleftharpoons M(H_2O)_5(OH)^{(n-1)+}(aq) + H_3O^+(aq)$$

**Figure 16.26** provides a molecular view of this reaction, using $Fe^{3+}(aq)$ as the central ion. Similar reactions allow other hydrated metal ions, particularly those with charges of 3+, to function as Brønsted–Lowry acids. The $K_a$ values of several metal ions are listed in **Table 16.3**. Note how much stronger the 3+ ions are than the 2+ ions because of the greater electron-withdrawing power of the more highly charged central ions.

Figure 16.26 shows how one of the six water molecules of hydration surrounding a $Fe^{3+}$ ion is converted into a hydroxide ion as a result of the acid ionization reaction. This lowers the charge of the complex ion (but does *not* reduce the iron ion) from 3+ to 2+. If the pH of a solution of $Fe(H_2O)_5(OH)^{2+}$ ions is raised by adding a small quantity of a strong base such as NaOH, the ions undergo an additional acid ionization reaction, forming $Fe(H_2O)_4(OH)_2{}^+$ ions:

$$Fe(H_2O)_5(OH)^{2+}(aq) + OH^-(aq) \rightleftharpoons Fe(H_2O)_4(OH)_2{}^+(aq) + H_2O(\ell)$$

At still higher pH, solid iron(III) hydroxide forms:

$$Fe(H_2O)_4(OH)_2{}^+(aq) + OH^-(aq) \rightleftharpoons Fe(H_2O)_3(OH)_3(s) + H_2O(\ell)$$

For simplicity, we usually write the formula of iron(III) hydroxide as $Fe(OH)_3(s)$, even though each formula unit also contains three water molecules of hydration. We will learn much more about the coordination chemistry of transition metals in Chapter 23.

**STEPWISE ANIMATION**
Hydrated Metal Ions

**TABLE 16.3** $K_a$ Values of Hydrated Metal Ions

| Ion | $K_a$ |
|---|---|
| $Fe^{3+}(aq)$ | $3 \times 10^{-3}$ |
| $Cr^{3+}(aq)$ | $1 \times 10^{-4}$ |
| $Al^{3+}(aq)$ | $1 \times 10^{-5}$ |
| $Cu^{2+}(aq)$ | $3 \times 10^{-8}$ |
| $Pb^{2+}(aq)$ | $3 \times 10^{-8}$ |
| $Zn^{2+}(aq)$ | $1 \times 10^{-9}$ |
| $Co^{2+}(aq)$ | $2 \times 10^{-10}$ |
| $Ni^{2+}(aq)$ | $1 \times 10^{-10}$ |

Acid strength (increasing upward)

## CONCEPT TEST

Is $Fe^{2+}(aq)$ a stronger or weaker Lewis acid than $Fe^{3+}(aq)$? Where would the $K_a$ of $Fe^{2+}(aq)$ be listed in Table 16.3 relative to that of $Fe^{3+}(aq)$?

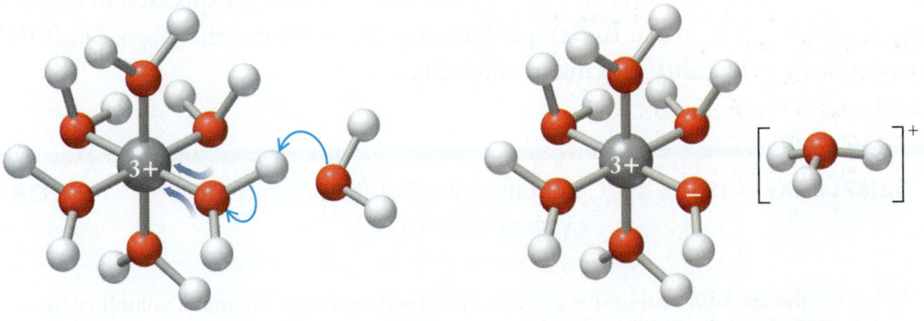

$$Fe(H_2O)_6{}^{3+}(aq) + H_2O(\ell) \rightleftharpoons Fe(H_2O)_5(OH)^{2+}(aq) + H_3O^+(aq)$$

**FIGURE 16.26** A hydrated $Fe^{3+}$ cation draws electron density away from the water molecules of its inner coordination sphere, which makes it possible for one or more of these molecules to donate a $H^+$ ion to a water molecule outside the sphere. The hydrated $Fe^{3+}$ ion is left with one fewer $H_2O$ ligand and a bond between the complex ion and the $OH^-$ ion.

$Cr^{3+}$ and $Al^{3+}$ display similar behavior, but unlike $Fe^{3+}$ they are more soluble in strongly basic solutions than in weakly basic solutions. Why? Because solid $Cr(OH)_3$ and $Al(OH)_3$ may accept additional $OH^-$ ions at high pH, forming soluble anionic complex ions:

$$Cr(OH)_3(s) + OH^-(aq) \rightleftharpoons Cr(OH)_4{}^-(aq) = Cr(H_2O)_2(OH)_4{}^-(aq)$$

$$Al(OH)_3(s) + OH^-(aq) \rightleftharpoons Al(OH)_4{}^-(aq) = Al(H_2O)_2(OH)_4{}^-(aq)$$

Zinc hydroxide, $Zn(OH)_2$, is the only other transition metal hydroxide soluble at high pH because it forms $Zn(OH)_4{}^{2-}$ ions.

Nearly all transition metals exist as $M^{n+}(aq)$ ions only in strongly acidic solutions. They exist as complex ions, such as $M(H_2O)_5(OH)^{(n-1)+}$, in aqueous solutions that range from slightly acidic to slightly basic ($3 < pH < 9$). This pH range includes most environmental waters and biological fluids.

## 16.8 Solubility Equilibria

**FIGURE 16.27** Milk of magnesia is milky because it is a *suspension* (not a solution) of $Mg(OH)_2$.

Recall from Section 15.3 that the common strong bases include the hydroxides of alkaline earth elements (group 2). One exception is $Mg(OH)_2$, which is a weak base because it has limited solubility in water. Magnesium hydroxide is the active ingredient in the antacid called *milk of magnesia*. This liquid appears "milky" (**Figure 16.27**) because it is an aqueous *suspension* (not solution) of solid white $Mg(OH)_2$. We can express the limited solubility of solid $Mg(OH)_2$ with the following equation:

$$Mg(OH)_2(s) \rightleftharpoons Mg^{2+}(aq) + 2\,OH^-(aq)$$

Because $Mg(OH)_2$ is a solid, its effective concentration does not change as long as some of it is present in the system. Therefore, the equilibrium constant for the dissolution of $Mg(OH)_2$ is

$$K_{sp} = [Mg^{2+}][OH^-]^2$$

where $K_{sp}$ represents an equilibrium constant called the **solubility-product constant**, or simply the **solubility product**.

The $K_{sp}$ values of $Mg(OH)_2$ and other slightly soluble compounds are listed in Table A5.4 of Appendix 5. We can use these values to calculate the concentrations of these compounds in aqueous solutions. Two terms are widely used to describe how much of a solid dissolves in a solvent: *solubility*, often expressed in grams of solute per 100 mL of solution, and *molar solubility*, expressed in moles of solute per liter of solution. In Sample Exercise 16.15, we use the $K_{sp}$ of $Mg(OH)_2$ to calculate its solubility and molar solubility.

**solubility-product constant** (also called **solubility product, $K_{sp}$**) an equilibrium constant that describes the formation of a saturated solution of a slightly soluble salt.

**SAMPLE EXERCISE 16.15** Calculating the Solubility of an Ionic Compound from $K_{sp}$ **LO8**

What are the solubility (in grams per 100 mL of solution) and the molar solubility (in moles of solute per liter of solution) of $Mg(OH)_2$ at 25°C?

**Collect, Organize, and Analyze** The $K_{sp}$ of $Mg(OH)_2$ is $5.6 \times 10^{-12}$ (Table A5.4), and the $K_{sp}$ expression is

$$K_{sp} = [Mg^{2+}][OH^-]^2 = 5.6 \times 10^{-12}$$

For every mole of $Mg(OH)_2$ that dissolves, one mole of $Mg^{2+}$ ions and two moles of $OH^-$ ions go into solution. Converting molar solubility—mol of $Mg(OH)_2$/L of solution—to g/100 mL is a matter of multiplying molar solubility by 58.32 g/mol, the molar mass of $Mg(OH)_2$, and accounting for the smaller volume. We can use a RICE table to determine the equilibrium concentrations of dissolved $Mg^{2+}$ ions and $OH^-$ ions.

**Solve** First, we construct a RICE table where $x$ is the molar solubility of the magnesium ions at 25°C, so $2x$ is the molar solubility of the hydroxide ions:

| Reaction (R) | $Mg(OH)_2(s)$ | $\rightleftharpoons$ | $Mg^{2+}(aq)$ | + | $2\ OH^-(aq)$ |
|---|---|---|---|---|---|
| | | | $[Mg^{2+}]$ (*M*) | | $[OH^-]$ (*M*) |
| Initial (I) | — | | 0 | | 0 |
| Change (C) | — | | $+x$ | | $+2x$ |
| Equilibrium (E) | — | | $x$ | | $2x$ |

The solid $Mg(OH)_2$ has limited solubility, so we do not include a concentration value for it in the RICE table. Remember from Chapter 14 that solids are not included in the equilibrium expression; we can insert the terms in the bottom row of the RICE table into the $K_{sp}$ expression and solve for $x$:

$$K_{sp} = [Mg^{2+}][OH^-]^2 = (x)(2x)^2 = (x)(4x^2) = 4x^3 = 5.6 \times 10^{-12}$$

$$x = 1.1 \times 10^{-4}\ M$$

Converting this molar solubility to g/100 mL:

$$\frac{1.1 \times 10^{-4}\ \cancel{\text{mol}}}{\cancel{\text{L}}} \times \frac{58.32\ \text{g}}{1\ \cancel{\text{mol}}} \times \frac{1\ \cancel{\text{L}}}{1000\ \cancel{\text{mL}}} \times 100\ \cancel{\text{mL}} = 6.4 \times 10^{-4}\ \text{g}$$

**Think About It** Note that the entire algebraic expression for $[OH^-]$, $2x$, is squared in this calculation: $(2x)^2 = 4x^2$. Forgetting to square the coefficient is a common mistake. Also note that the molar solubility of $Mg(OH)_2$ is a much larger value than its solubility product. This difference is true for all sparingly soluble ionic compounds because $K_{sp}$ values are the products of small concentration values multiplied together, producing even smaller $K_{sp}$ values.

**Practice Exercise** What is the solubility of $Ca_3(PO_4)_2$ at 25°C expressed in mol/L and in g/100 mL?

## CONCEPT TEST

In Sample Exercise 16.15, we calculated molar solubility from $K_{sp}$, but the reverse calculation is also possible. If the molar solubility of $BaF_2$ is $1.10 \times 10^{-3}\ M$, what is the $K_{sp}$ for barium fluoride?

## SAMPLE EXERCISE 16.16 Evaluating the Common-Ion Effect on Solubility LO8

The mineral barite (**Figure 16.28**) is mostly barium sulfate ($BaSO_4$) and is widely used in industry and in medical imaging. Calculate the molar solubility at 25°C of $BaSO_4$ in (a) pure water and (b) seawater in which the concentration of sulfate ions is 2.8 g/L.

**FIGURE 16.28** Crystalline barite (barium sulfate).

**Collect and Organize** We want to calculate the molar solubility of $BaSO_4$ in both pure water and in seawater that already contains sulfate ions. The $K_{sp}$ value of $BaSO_4$ given in Table A5.4 is $1.08 \times 10^{-10}$.

**Analyze** According to the dissolution reaction for barium sulfate,

$$BaSO_4(s) \rightleftharpoons Ba^{2+}(aq) + SO_4^{2-}(aq)$$

One mole of $Ba^{2+}$ ions and one mole of $SO_4^{2-}$ ions form from each mole of $BaSO_4$ that dissolves. If $x$ mol/L of $BaSO_4$ dissolves in pure water, then $[Ba^{2+}] = [SO_4^{2-}] = x$. However, seawater has a background concentration of $SO_4^{2-}$, and we can use a RICE table to calculate its effect on solubility. According to Le Châtelier's principle and the common-ion effect, the sulfate ion already in seawater should shift the dissolution equilibrium to the left, which means that less $BaSO_4$ should dissolve in seawater than in pure water.

**Solve**

a. In pure water,

$$K_{sp} = [Ba^{2+}][SO_4^{2-}] = (x)(x) = 1.08 \times 10^{-10}$$

$$x = 1.04 \times 10^{-5}\ M$$

b. In seawater, we first need to calculate the value of $[SO_4^{2-}]$ before any $BaSO_4$ dissolves:

$$[SO_4^{2-}]_{\text{initial}} = \frac{2.8\ \text{g}}{\text{L}} \times \frac{1\ \text{mol}}{96.06\ \text{g}} = \frac{0.0291\ \text{mol}}{\text{L}}$$

Now we can use this value as the initial amount of $[SO_4^{2-}]$ present in a RICE table:

| Reaction (R) | $BaSO_4(s)$ | $\rightleftharpoons$ | $Ba^{2+}(aq)$ | + | $SO_4^{2-}(aq)$ |
|---|---|---|---|---|---|
| | | | $[Ba^{2+}]$ ($M$) | | $[SO_4^{2-}]$ ($M$) |
| Initial (I) | — | | 0 | | 0.291 |
| Change (C) | — | | $+x$ | | $+x$ |
| Equilibrium (E) | — | | $x$ | | $0.0291 + x$ |

Incorporating these values into the $K_{sp}$ expression gives

$$K_{sp} = [Ba^{2+}][SO_4^{2-}] = (x)(0.0291 + x) = 1.08 \times 10^{-10}$$

Solving for $x$ is simplified if we assume that the $K_{sp}$ of $BaSO_4$ is so small that we can ignore its contribution to the total $SO_4^{2-}$ concentration. Therefore,

$$(x)(0.0291 + x) \approx (x)(0.0291) = 1.08 \times 10^{-10}$$

$$x = 3.7 \times 10^{-9}\ M$$

**Think About It** The calculated molar solubility of $BaSO_4$ in seawater is much less than the initial $[SO_4^{2-}]$ value, so our simplifying assumption was justified. The lower solubility of $BaSO_4$ in seawater is another illustration of the common-ion effect: the dissolution of $BaSO_4$ is suppressed by the $SO_4^{2-}$ ions already present in seawater.

**Practice Exercise** What is the molar solubility of $MgCO_3$ in alkaline spring water at 25°C in which $[CO_3^{2-}] = 0.0075\ M$?

The results of Sample Exercise 16.16 illustrate how the common-ion effect can suppress the solubility of an ionic compound. Other perturbations to solubility equilibria can actually promote solubility, such as when the anion of the compound is the conjugate base of a weak acid. The molar solubilities of such compounds increase in acidic solutions, as we see in Sample Exercise 16.17.

## SAMPLE EXERCISE 16.17 Calculating the Effect of pH on Solubility LO8

Fluorite is a fluorescent mineral (**Figure 16.29**) composed mostly of calcium fluoride and is the principal source fluoride ions for use in industry and medicine. What is the molar solubility of $CaF_2$ at 25°C in (a) pure water and (b) an acidic buffer in which $[H_3O^+]$ is a constant 0.050 *M*?

**Collect and Organize** We are asked to calculate the solubility of $CaF_2$ in both pure water and in an acidic buffer. The dissolution process is described by the following equilibrium:

$$(1) \quad CaF_2(s) \rightleftharpoons Ca^{2+}(aq) + 2\,F^-(aq) \qquad K_{sp} = 5.3 \times 10^{-9}$$

**Analyze** To account for the effect of acid on the solubility of a fluoride salt in a solution in which $[H_3O^+] = 0.050\ M$, we should expect the fluoride ion to react with the $H_3O^+$ ions in solution:

$$(2) \quad H_3O^+(aq) + F^-(aq) \rightleftharpoons HF(aq) + H_2O(\ell)$$

Note that reaction (2) is the inverse of the acid dissociation equilibrium for HF, shown in reaction (3):

$$(3) \quad HF(aq) + H_2O(\ell) \rightleftharpoons H_3O^+(aq) + F^-(aq) \qquad K_a = 6.8 \times 10^{-4}$$

which means that the equilibrium constant for reaction (2) equals the inverse of the $K_a$ value for HF:

$$K_2 = \frac{[HF]}{[H_3O^+][F^-]} = \frac{1}{6.8 \times 10^{-4}} = 1.47 \times 10^3$$

As reaction (2) proceeds and $F^-$ ions react, the equilibrium in reaction (1) will shift to the right, increasing $CaF_2$ solubility. Therefore, we can anticipate an increase in the solubility of $CaF_2$ when acid is present.

**Solve**

a. Let $x$ be the molar solubility of $CaF_2$ in pure water. According to the stoichiometry of reaction (1), $[Ca^{2+}] = x$ mol/L and $[F^-] = 2x$ mol/L . Inserting these variables in the $K_{sp}$ expression:

$$K_{sp} = [Ca^{2+}][F^-]^2 = (x)(2x)^2 = 5.3 \times 10^{-9}$$
$$4x^3 = 5.3 \times 10^{-9}$$
$$x = 1.1 \times 10^{-3}\ M$$

b. In the acidic buffer, $[H_3O^+] = 0.050\ M$, and the $F^-$ and $H_3O^+$ ions combine as shown in reaction (2). If we assume that $[H_3O^+]$ is a constant 0.050 *M*, then

$$K = \frac{[HF]}{[0.050][F^-]} = 1.47 \times 10^3$$

$$\frac{[HF]}{[F^-]} = 73.5$$

$$(4) \quad [HF] = 73.5[F^-]$$

According to this calculation, most of the $F^-$ ions produced when $CaF_2$ dissolves are converted into HF. The proportion that remain in solution as $F^-(aq)$ ions is defined by the following ratio:

$$(5) \quad \frac{[F^-]}{[F^-] + [HF]}$$

Here the numerator is the concentration of $F^-(aq)$ ions at equilibrium and the denominator is the concentration of all the $F^-$ ions produced when $CaF_2$ dissolved.

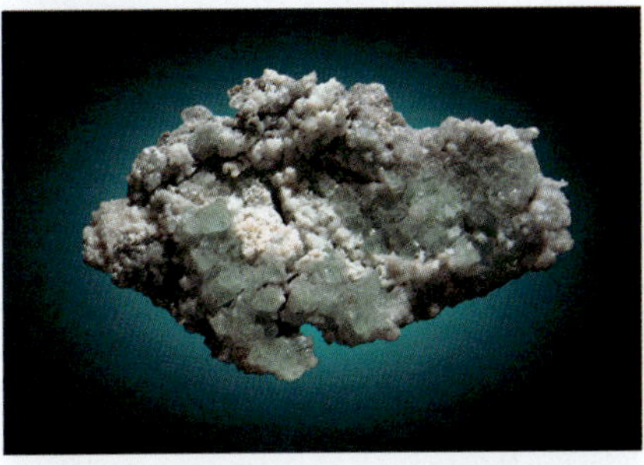

(a)

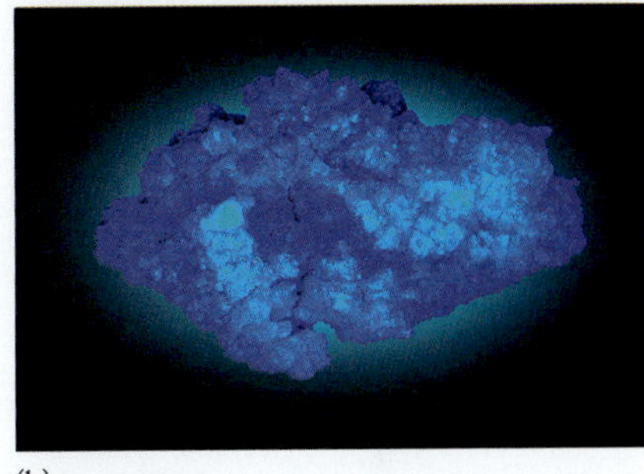

(b)

**FIGURE 16.29** A sample of fluorescent fluorite (calcium fluoride) under (a) normal illumination and (b) under black light.

Combining Equations (4) and (5) to calculate the fraction of the fluoride species that are $F^-(aq)$ ions and not bonded in molecules of HF:

$$\frac{[F^-]}{[F^-] + [HF]} = \frac{[F^-]}{[F^-] + 73.5[F^-]} = \frac{[F^-]}{74.5[F^-]} = 0.0134$$

If $x$ is the molar solubility of $CaF_2$ in the acid, then $x$ mol/L of $Ca^{2+}$ and $2x$ mol/L of $F^-$ are produced. However, most of the fluoride ions are converted into HF. The $F^-(aq)$ ion concentration is only $(0.0134)(2x) = 0.0268x$. Inserting this value in the $K_{sp}$ expression:

$$K_{sp} = [Ca^{2+}][F^-]^2 = (x)(0.0268x)^2 = (7.18 \times 10^{-4})x^3$$

$$(7.18 \times 10^{-4})x^3 = 5.3 \times 10^{-9}$$

$$x = 1.9 \times 10^{-2}$$

**Think About It** Comparing the results from parts a and b shows that the molar solubility of $CaF_2$ is about 4 times higher in the acidic buffer, as we predicted, because most of the $F^-$ ions produced when $CaF_2$ dissolves in the buffer are converted into molecules of HF. This conversion removes a product from the $K_{sp}$ equilibrium mixture. According to Le Châtelier's principle, the result will be a shift in the position of the equilibrium to the right, in favor of forming product and increasing solubility.

**Practice Exercise** What is the molar solubility of $Fe(OH)_2$ at 25°C in a buffer with pH 6.00?

## CONCEPT TEST

Given aqueous solutions of sodium fluoride, hydrofluoric acid, and nitric acid, which would increase the solubility of $PbF_2$? Which would decrease the solubility of $PbF_2$? Which would have little effect?

Table A5.4 shows that the $K_{sp}$ values of the hydroxides of many metals, including all transition metals, are very small. However, these tiny $K_{sp}$ values apply to equilibrium concentrations of free metal ions. As we have seen in this chapter, many metals form stable complex ions in aqueous solution, reducing their free metal ion concentration. Let's consider, for example, the solubility of $Al^{3+}$ ions in pH 7.00 water. The $K_{sp}$ of $Al(OH)_3$ is

$$K_{sp} = [Al^{3+}][OH^-]^3 = 1.3 \times 10^{-33}$$

Solving the $K_{sp}$ expression for $[Al^{3+}]$ and inserting $[OH^-] = 1.0 \times 10^{-7}$, we get

$$[Al^{3+}] = \frac{K_{sp}}{[OH^-]^3} = \frac{1.3 \times 10^{-33}}{(1.0 \times 10^{-7})^3} = 1.3 \times 10^{-12}\,M \quad (16.8)$$

This very small value seems to imply that no aluminum salt is soluble in water because the $Al^{3+}(aq)$ ions that it releases as it dissolves would immediately precipitate as $Al(OH)_3$. However, that conclusion is incorrect. $Al(NO_3)_3$, like all nitrate salts, is quite soluble in water. Its solubility can be explained by the acidic properties of hydrated $Al^{3+}$ ions, $Al(H_2O)_6{}^{3+}$, which make $Al(NO_3)_3$ an acidic salt. The pH of 0.1 $M$ $Al(NO_3)_3$ is about 3.0. At this pH, pH, $[OH^-] = 1 \times 10^{-11}$. Inserting this value into Equation 16.8 and solving for $[Al^{3+}]$, we get

$$[Al^{3+}] = \frac{K_{sp}}{[OH^-]^3} = \frac{1.3 \times 10^{-33}}{(1.0 \times 10^{-11})^3} = 1.3\,M$$

This maximum value is more than 10 times the concentration of $Al^{3+}$ ions in a 0.1 $M$ solution of $Al(NO_3)_3$.

## $K_{sp}$ and $Q$

We can also use $K_{sp}$ values to predict whether a particular concentration of an ionic compound is possible or whether a precipitate will form when the solutions of two salts are mixed. In making these predictions, it is convenient to use the concept of the reaction quotient $Q$ that we developed in Chapter 14. When applied to the equilibrium governing a slightly soluble salt, $Q$ is sometimes called the *ion product* because it is the product of the concentrations of the ions in solution after each is raised to a power equal to its subscript in the formula of the compound. If the calculated $Q$ value is greater than the $K_{sp}$ of the compound ($Q > K_{sp}$), the reaction will favor reactant formation, and the compound will precipitate (or never dissolve in the first place). If $Q < K_{sp}$, the reaction will favor product formation, and the compound will be soluble and will not precipitate.

**SAMPLE EXERCISE 16.18** Predicting Whether a Precipitate Will Form When Two Solutions Are Mixed **LO8**

Lead(II) chloride is a white pigment used in 15th-century European sculpture. Will $PbCl_2$ precipitate when 275 mL of a 0.134 $M$ solution of $Pb(NO_3)_2$ is added to 125 mL of a 0.0339 $M$ solution of NaCl?

**Collect and Organize** We are asked whether $PbCl_2$ will precipitate when two solutions containing $Pb^{2+}$ ions and $Cl^-$ ions are mixed. The solubility product of $PbCl_2$ is

$$K_{sp} = [Pb^{2+}][Cl^-]^2 = 1.70 \times 10^{-5}$$

**Analyze** To determine whether a precipitate forms, we need to calculate the ion product $Q$ and compare its value to $K_{sp}$. If $Q > K_{sp}$, $PbCl_2$ will precipitate; if $Q < K_{sp}$, it will not precipitate.

**Solve** First, we calculate the concentrations of the lead ions and chloride ions in the two solutions immediately after they are mixed. Mixing the solutions dilutes both, so the volumes of the solutions (275 mL and 125 mL) must be added to get the final solution volume (400 mL = 0.400 L).

$$Pb^{2+}(aq): \quad 0.134\,\frac{\text{mol}}{\cancel{\text{L}}} \times 0.275\,\cancel{\text{L}} = 0.0369\text{ mol}$$

$$[Pb^{2+}] = \frac{0.0369\text{ mol}}{0.400\text{ L}} = 0.0921\,M$$

$$Cl^-(aq): \quad 0.0339\,\frac{\text{mol}}{\cancel{\text{L}}} \times 0.125\,\cancel{\text{L}} = 0.00424\text{ mol}$$

$$[Cl^-] = \frac{0.00424\text{ mol}}{0.400\text{ L}} = 0.0106\,M$$

The value of $Q$ is

$$Q = [Pb^{2+}][Cl^-]^2 = (0.0921)(0.0106)^2 = 1.03 \times 10^{-5}$$

$Q < K_{sp}$, so *no* precipitate forms.

**Think About It** Lead(II) chloride was categorized as an *insoluble* compound in Table 8.4. In this scenario, $PbCl_2$ does not precipitate because the solutions of $Pb^{2+}$ and $Cl^-$ ions are too dilute to supply the concentrations required for the precipitate to form.

**Practice Exercise** Will calcium fluoride precipitate when 175 mL of a $4.78 \times 10^{-3}$ $M$ solution of $Ca(NO_3)_2$ is added to 135 mL of a $7.35 \times 10^{-3}$ $M$ solution of KF?

We can use differences in the solubilities of ionic compounds to selectively separate ions, particularly cations, in solution. For example, suppose an aqueous solution contains 0.10 $M$ $Ca^{2+}$ ion and 0.020 $M$ $Mg^{2+}$ ion. Is it possible to selectively remove the $Mg^{2+}$ ions from solution by precipitating them as $Mg(OH)_2$ while leaving the $Ca^{2+}$ ions in solution? This approach might work because $Mg(OH)_2$ is much less soluble than $Ca(OH)_2$, as indicated by their $K_{sp}$ values:

$$K_{sp} = [Mg^{2+}][OH^-]^2 = 5.6 \times 10^{-12}$$

$$K_{sp} = [Ca^{2+}][OH^-]^2 = 5.5 \times 10^{-6}$$

One way to address the question of ion separation involves calculating the maximum concentration of $OH^-$ ions that will *not* cause the 0.10 $M$ $Ca^{2+}$ ion to precipitate and then determining whether that concentration is high enough to precipitate all of the $Mg^{2+}$ ions. We can calculate the target $[OH^-]$ value from the $K_{sp}$ of calcium hydroxide:

$$K_{sp} = 5.5 \times 10^{-6} = [Ca^{2+}][OH^-]^2 = (0.10)(x)^2$$

$$x = \sqrt{\frac{5.5 \times 10^{-6}}{0.10}} = 7.4 \times 10^{-3}\ M$$

Now we need to determine whether all the $Mg^{2+}$ ions in solution would precipitate as $Mg(OH)_2$ if $[OH^-]$ were $6.9 \times 10^{-3}$ $M$. To do this, we calculate the $[Mg^{2+}]$ that would be in equilibrium with $6.9 \times 10^{-3}$ $M$ $OH^-$ ions:

$$K_{sp} = 5.6 \times 10^{-12} = [Mg^{2+}][OH^-]^2 = (x)(6.9 \times 10^{-3})^2$$

$$x = \frac{5.6 \times 10^{-12}}{(6.9 \times 10^{-3})^2} = 1.2 \times 10^{-7}\ M$$

The original solution was 0.020 $M$ $Mg^{2+}$ ions, and $1.2 \times 10^{-7}$ $M$ represents $[(1.2 \times 10^{-7})/0.020](100\%)$, or only 0.00060%, of the original quantity of $Mg^{2+}$ ions in the sample remaining in solution. This tiny percentage means that the removal of $Mg^{2+}$ is complete. (In general, reducing a solute's concentration to 0.1% of its original value or less is considered complete removal.) This separation works as long as we keep $[OH^-]$ below about $10^{-3}$ $M$—that is, as long as we don't let the pH go much above 11. This approach of selectively precipitating $Mg(OH)_2$ has been used to separate these ions in seawater, where $[Ca^{2+}] = 0.0106$ $M$ and $[Mg^{2+}] = 0.054$ $M$.

**SAMPLE EXERCISE 16.19** Separating Anions in Solution **LO9**

Both lead(II) chloride and lead(II) fluoride are slightly soluble salts. A solution of lead(II) nitrate is added to a solution that is 0.275 $M$ in both $Cl^-(aq)$ and $F^-(aq)$. Can we use this method to separate the two halide ions? If complete precipitation is defined as less than 0.1% of a particular ion being left in solution, is the precipitation of the first salt complete before the second salt begins to precipitate?

**Collect and Organize** We are given a solution that contains two ions that form slightly soluble lead(II) salts, and we are asked whether one ion can be completely removed before the second one starts to precipitate when lead(II) ion is added to the solution. We have the $K_{sp}$ values for both salts and the initial concentrations of both ions.

**Analyze** The equilibrium constant expressions for both ions are

$$K_{sp} = [Pb^{2+}][Cl^-]^2 = 1.7 \times 10^{-5}$$
$$K_{sp} = [Pb^{2+}][F^-]^2 = 3.3 \times 10^{-8}$$

The $K_{sp}$ of $PbCl_2$ is $(1.7 \times 10^{-5}/3.3 \times 10^{-8}) = 520$ times the $K_{sp}$ of $PbF_2$, so $PbF_2$ should precipitate first when $Pb^{2+}$ ions are added to a solution containing the same concentrations of $F^-$ and $Cl^-$ ions. We need to determine the maximum $[Pb^{2+}]$ that could be added to precipitate $PbF_2$ but not cause $PbCl_2$ to precipitate. When we determine that value, we can calculate $[F^-]$ remaining in solution to determine whether the precipitation of $F^-$ as $PbF_2$ was complete.

**STEPWISE ANIMATION**
Selective Precipitation

**Solve** The maximum concentration of $Pb^{2+}$ in the solution that will not cause the chloride ion to precipitate is

$$K_{sp} = 1.7 \times 10^{-5} = [Pb^{2+}][Cl^-]^2 = (x)(0.275)^2$$

$$x = \frac{1.7 \times 10^{-5}}{(0.275)^2} = 2.2 \times 10^{-4}\ M$$

The concentration of $F^-(aq)$ in the solution at this concentration of lead(II) ion is

$$K_{sp} = 3.3 \times 10^{-8} = (2.2 \times 10^{-4})(x)^2$$

$$x = \sqrt{\frac{3.3 \times 10^{-8}}{2.2 \times 10^{-4}}} = 0.012\ M$$

The original solution was 0.275 $M$ in $F^-$ ions. A residual concentration of 0.012 $M$ $F^-$ represents $(0.012/0.275) \times 100\% = 4.4\%$ of the original $[F^-]$, which is greater than our 0.1% residual value that represents complete removal. Therefore, precipitation of $PbF_2(s)$ is incomplete and we cannot use this method to separate the two ions.

**Think About It** This attempt at selective precipitation did not work because the ratio of the $K_{sp}$ values was only 520 *and* because the halide concentration terms in the $K_{sp}$ expressions were both squared. If you look closely at the calculations, you can see that these squared concentration terms produce a $F^-/Cl^-$ ion ratio at equilibrium that was only the square root of their $K_{sp}$ ratio: $(1/500)^{0.5} = 0.045$, or 4.5%.

**Practice Exercise** A water sample contains barium ions (0.0375 $M$) and calcium ions (0.0667 $M$). Can they be completely separated by selective precipitation of $CaF_2$?

We end this chapter with a Sample Exercise that integrates acid–base and solubility equilibria by revisiting a topic introduced in Section 16.1: ocean acidification. In this exercise, we use the ionization equilibria of carbonic acid, as we have done in several other exercises, but this time we use different $K_{a_1}$ and $K_{a_2}$ values. Why? Because the $K$ values we have used until now have all been *theoretical* values, which means they apply to ideal solutions in which each solute ion behaves as freely and independently as though it were the only ion present. Theoretical $K$ values work best for very dilute solutions—they don't perform as well for solutions as concentrated (salty) as seawater. They don't account, for example, for ion pair formation. Therefore, we use *apparent* $K'_{a_1}$ and $K'_{a_2}$ values at 25°C in Sample Exercise 16.20 that apply to chemical equilibria in typical seawater, which contains 35 grams of dissolved sea salts per kilogram of seawater. Their symbols contain a prime (′) after the $K$ to indicate that their values apply only to that particular sample.

**CONNECTION** Ion pair formation in 1.0 $M$ solutions of ionic compounds significantly reduces the number of free ions in these solutions, as described in Chapter 11.

**SAMPLE EXERCISE 16.20** Integrating Concepts: Evaluating the Impact of Ocean Acidification

As we discussed at the beginning of this chapter, increasing concentrations of atmospheric $CO_2$ have increased the concentration of $CO_2$ dissolved in the sea. As we saw in Equations 16.1–16.3, an increase in $[CO_2]$ shifts the following equilibria to the right:

(1) $H_2CO_3(aq) + H_2O(\ell) \rightleftharpoons HCO_3^-(aq) + H_3O^+(aq) \quad pK'_{a_1} = 5.85$

(2) $HCO_3^-(aq) + H_2O(\ell) \rightleftharpoons CO_3^{2-}(aq) + H_3O^+(aq) \quad pK'_{a_2} = 9.00$

These shifts raise oceanic $[H_3O^+]$ and lower pH.

a. The average pH of the Pacific Ocean near Hawaii dropped from 8.12 to 8.07 between 1989 and 2014. By how much did the average acidity ($[H_3O^+]$) of the ocean increase? Express your answer as a percentage of the 1989 acidity.
b. Ocean pH may drop by another 0.30 unit by the end of the 21st century. If we assume that today's concentrations of most other dissolved ions (for example, $[Ca^{2+}] = 0.0106$ *M* and $[HCO_3^-] + [CO_3^{2-}] = 0.00211$ *M*)) remain the same, will the skeletal structures of marine organisms that are composed of $CaCO_3$ be soluble or insoluble in seawater in 2100? The $K'_{sp}$ value for aragonite (the crystalline form of $CaCO_3$ in coral and seashells) is $6.46 \times 10^{-7}$.

**Collect and Organize** We are asked to convert a decrease in pH into an increase in acidity and to evaluate how that increase affects the solubility of $CaCO_3$. We know the concentration of $Ca^{2+}$ ions and the total ($[CO_3^{2-}] + [HCO_3^{2-}]$) value as well as the appropriate $K'_{sp}$ value for $CaCO_3$ in seawater. $CO_3^{2-}$ and $HCO_3^{2-}$ are a conjugate acid–base pair whose concentrations are linked by reaction (2) and the Henderson–Hasselbalch equation based on it:

$$pH = pK_a + \log\frac{[base]}{[acid]} = 9.00 + \log\frac{[CO_3^{2-}]}{[HCO_3^-]}$$

**Analyze** The ratio of the acidity of the seawater near Hawaii in 2014 compared to 1989 can be calculated using the equation $[H_3O^+] = 10^{-pH}$. To determine whether $CaCO_3$ will dissolve in seawater in 2100, we need to calculate the $[CO_3^{2-}]/[HCO_3^-]$ ratio in equilibrium with seawater at pH $8.07 - 0.30 = 7.77$ and then use that ratio and the total carbonate and bicarbonate value to calculate $[CO_3^{2-}]$. The product of that value times $[Ca^{2+}]$ will give us a $Q$ value that can be compared with the $K'_{sp}$ of $CaCO_3$ to determine whether seawater will be saturated with $CaCO_3$.

**Solve**

a. Converting the decrease in pH values to an increase in $[H_3O^+]$, expressed as a ratio of the 2014 value to the 1989 value:

$$\frac{[H_3O^+]_{2014}}{[H_3O^+]_{1989}} = \frac{10^{-8.07}\ M}{10^{-8.12}\ M} = 10^{0.05} = 1.12$$

Therefore, the 2014 sample contains 1.12 times as much acidity as the 1989 sample, which means acidity increased 12% during the 25 years before 2014.

b. Calculating the oceanic $[CO_3^{2-}]/[HCO_3^{2-}]$ ratio in 2100:

$$7.77 = 9.00 + \log\frac{[CO_3^{2-}]}{[HCO_3^-]}$$

$$\frac{[CO_3^{2-}]}{[HCO_3^-]} = 0.0589$$

If we let $x$ be $[CO_3^{2-}]$, then

$$\frac{x}{(0.00211 - x)} = 0.0589$$

$$x = 1.17 \times 10^{-4}\ M$$

Calculating $Q$ for $CaCO_3$:

$$Q = [Ca^{2+}][CO_3^{2-}] = (0.0106)(1.17 \times 10^{-4}) = 1.24 \times 10^{-6}$$

This value of $Q$ is $(1.24 \times 10^{-6}/6.46 \times 10^{-7})$, or 1.92 times the value of $K'_{sp}$, which means seawater will still be supersaturated with respect to $CaCO_3$.

**Think About It** The results of the calculations suggest that corals and seashells should still be able to form their skeletal structures in pH 7.77 seawater because it will still be supersaturated with respect to $CaCO_3$. However, the predicted degree of supersaturation (92%) in 2100 will have decreased from about 300% in 1989. (You are invited to repeat the part b calculation by using pH 8.12 to confirm this value.) Moreover, some climate change models predict even greater decreases in oceanic pH by 2100. There is cause for concern.

## SUMMARY

**LO1** As predicted by Le Châtelier's principle, adding conjugate base to a solution of a weak acid inhibits ionization of the acid, causing the pH of the solution to rise. Adding conjugate acid to a solution of a weak base lowers the pH of the solution. These shifts are examples of the **common-ion effect.** (Section 16.2)

**LO2** A **pH buffer** is a solution that resists changes in pH because it consists of either a weak acid and a salt of its conjugate base or a weak base and a salt of its conjugate acid. The acidic component should have a $pK_a$ close to the desired pH of the buffer. (Section 16.3)

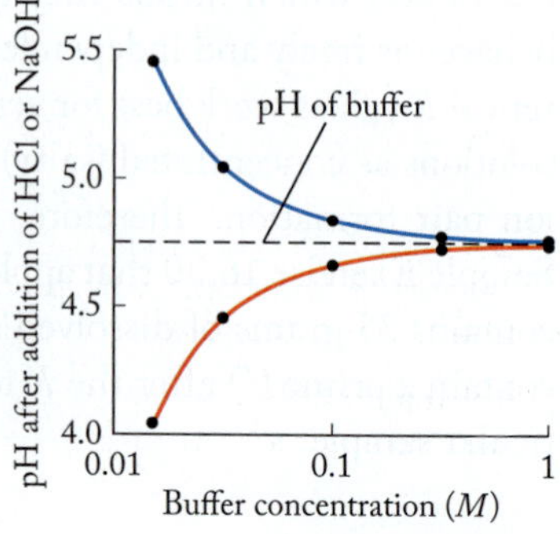

**LO3** A pH buffer has the capacity to resist pH change because its acid component neutralizes additions of bases and its base component neutralizes additions of acids. (Section 16.3)

**LO4** Color **pH indicators** or pH electrodes are used to detect the equivalence points in pH titrations, which are used to determine the concentrations of acids or bases in aqueous samples. (Section 16.4)

**LO5** A **Lewis base** is a substance that donates pairs of electrons to a **Lewis acid**, defined as an electron-pair acceptor. The donated electron pair forms a **coordinate bond**. In some Lewis acid–Lewis base reactions, other bonds will break as the new one forms. (Sections 16.5 and 16.6)

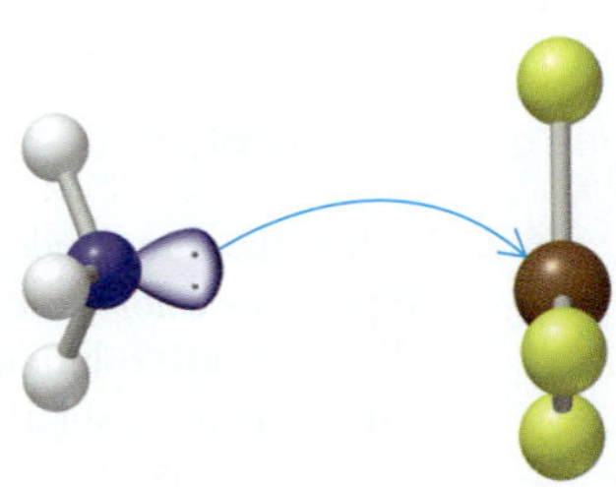

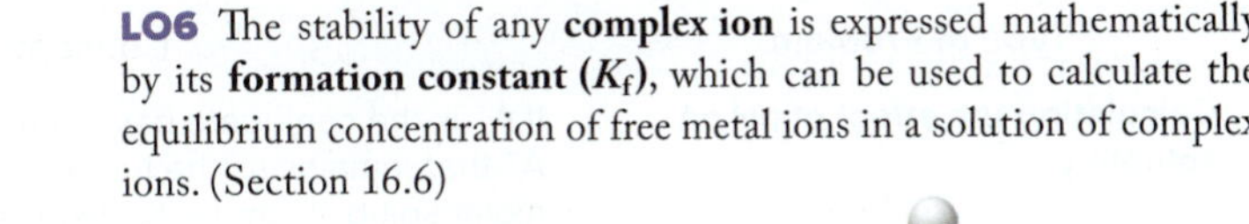

**LO6** The stability of any **complex ion** is expressed mathematically by its **formation constant ($K_f$)**, which can be used to calculate the equilibrium concentration of free metal ions in a solution of complex ions. (Section 16.6)

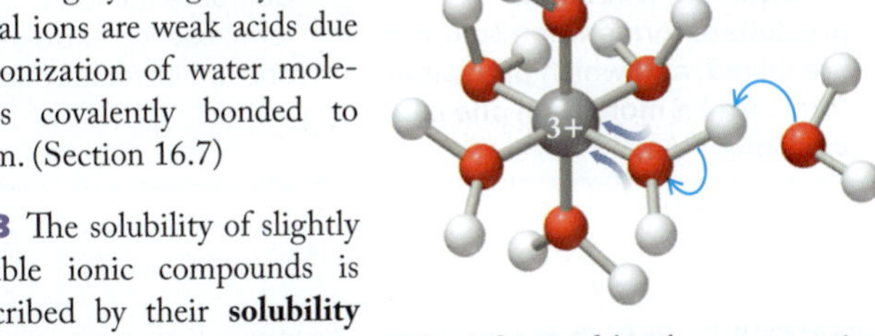


**LO7** Highly charged hydrated metal ions are weak acids due to ionization of water molecules covalently bonded to them. (Section 16.7)

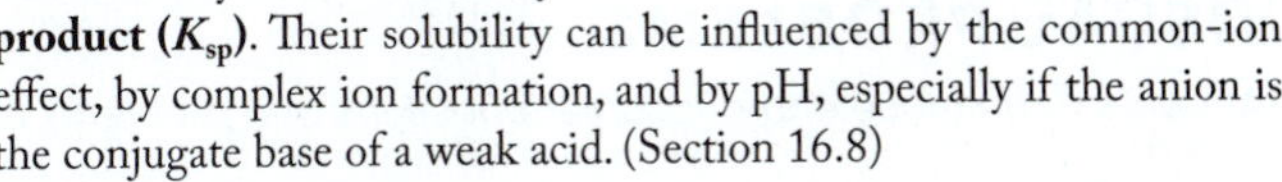

**LO8** The solubility of slightly soluble ionic compounds is described by their **solubility product ($K_{sp}$)**. Their solubility can be influenced by the common-ion effect, by complex ion formation, and by pH, especially if the anion is the conjugate base of a weak acid. (Section 16.8)

**LO9** The relative solubilities of two slightly soluble ionic compounds can be used to selectively precipitate one from solution while the other remains soluble. (Section 16.8)

## PARTICULATE **PREVIEW WRAP-UP**

$NH_4^+$ ions can donate protons (Brønsted–Lowry acid), whereas $NH_3$ molecules and $OH^-$ ions can accept protons (Brønsted–Lowry bases). A molecule of $H_2O$ can both donate and accept a proton. $H_2O$, $NH_3$, and $OH^-$ each have at least one lone pair on the central atom that they can donate (Lewis bases).

## PROBLEM-SOLVING SUMMARY

| Type of Problem | Concepts and Equations | Sample Exercises |
|---|---|---|
| **Calculating pH of a solution of a weak base and its conjugate acid (or a weak acid and its conjugate base)** | Insert the concentrations of the base and acid components and the p$K_a$ of the acid in the Henderson–Hasselbalch equation: $$\text{pH} = \text{p}K_a + \log\frac{[\text{base}]}{[\text{acid}]} \quad (16.4)$$ | **16.1, 16.2** |
| **Preparing a buffer of given pH** | Select a weak acid with a p$K_a$ within one pH unit of the target pH. Add a $Na^+$ salt of its conjugate base in a proportion calculated using the Henderson–Hasselbalch equation. | **16.3–16.5** |
| **Evaluating the effect of concentration and [base]:[acid] ratio on buffer capacity** | Assume that additions of strong acid or strong base react completely with the base or acid components of the buffer, respectively. Calculate pH from the concentrations of the components that remain using the Henderson–Hasselbalch equation. | **16.6–16.8** |
| **Interpreting results of acid–base titrations** | Use the volume and molarity of the titrant needed to reach the equivalence point to calculate the number of moles that react with the analyte and, from the stoichiometry of the titration reaction, the number of moles of analyte in the sample. | **16.9–16.12** |
| **Identifying Lewis acids and bases** | Lewis bases donate pairs of electrons, whereas Lewis acids accept these pairs of electrons. | **16.13** |
| **Using formation constants to calculate the concentration of free or complexed ion** | Set up a RICE table based on the complex formation reaction. Let $x$ be the concentration of free (not complexed) metal ion. Solve for $x$, which is usually much smaller than the concentration of the complex. | **16.14** |
| **Calculating the solubility of an ionic compound from its $K_{sp}$** | Express the concentrations of the cation and anion in the $K_{sp}$ expression in terms of $x$ moles of the compound that dissolve in one liter of solution. | **16.15, 16.16** |

| Type of Problem | Concepts and Equations | Sample Exercises |
|---|---|---|
| **Calculating the effect of pH on solubility** | If $A^-$ is the conjugate base of a weak acid HA, calculate the fraction of $A^-$ that remains as the free ion. Use this fraction as the coefficient for molar solubility in the $K_{sp}$ expression. | **16.17** |
| **Determining whether a precipitate forms when solutions are mixed, and which precipitate forms first if more than one is possible** | Compare the ion product $Q$ to $K_{sp}$ to determine if a precipitate will form; use $K_{sp}$ expressions to calculate maximum concentrations of one ion in solution that will not cause another ion to precipitate. | **16.18, 16.19** |

## VISUAL PROBLEMS

*(Answers to boldface end-of-chapter questions and problems are in the back of the book.)*

**16.1.** The graph in Figure P16.1 shows the titration curves of a 1 *M* solution of a weak acid with a strong base and a 1 *M* solution of a strong acid with the same base. Which curve is which?

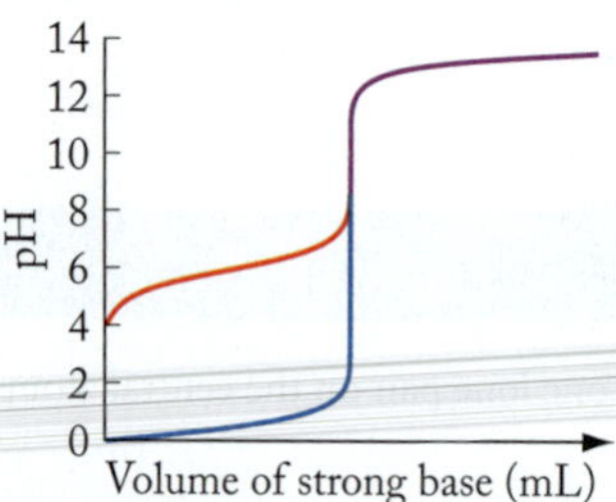


**FIGURE P16.1**

16.2. Estimate the $pK_a$ of the weak acid in Problem 16.1.

**16.3.** Suppose you have four color indicators to choose from to detect the equivalence point of the titration reaction represented by the red curve in Figure P16.1. The $pK_a$ values of the four indicators are 3.3, 5.0, 7.0, and 9.0. Which indicator would be the best one to choose?

16.4. Explain why the slope of the red titration curve in Figure P16.1 is nearly flat in the region extending from about ¼ to ¾ of the way from the start of the titration to its equivalence point.

**16.5.** One of the titration curves in Figure P16.5 represents the titration of an aqueous sample of $Na_2CO_3$ with strong acid; the other represents the titration of an aqueous sample of $NaHCO_3$ with the same acid. Which curve is which?

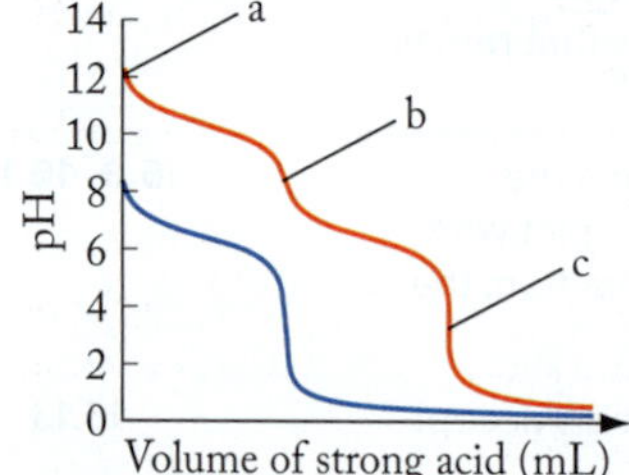


**FIGURE P16.5**

16.6. Identify the principal carbon-containing species in solution at points a, b, and c on the red titration curve in Figure P16.5.

**16.7.** Each of the three beakers in Figure P16.7 contains a few drops of the color indicator bromothymol blue, which is yellow in acidic solutions and blue in basic solutions. One beaker contains a solution of ammonium chloride, one contains ammonium acetate, and the third contains sodium acetate. Which beaker contains which salt?

**FIGURE P16.7**

*16.8. The images in Figure P16.8 depict solute particles present during the titration of an aqueous sample of sodium hydroxide with hydrochloric acid. Arrange the images in order, beginning with the original sample and ending with a particulate view of the titration mixture well after the equivalence point. Which image depicts the equivalence point? Solvent molecules have been omitted for clarity.

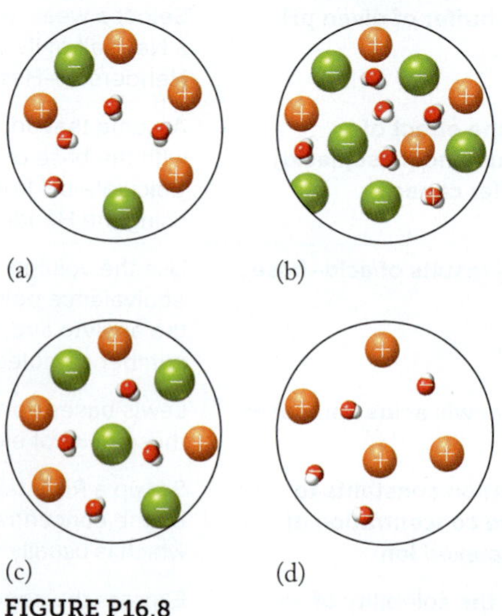


**FIGURE P16.8**

**16.9.** a. Is the aqueous solution depicted in Figure P16.9 a weak acid, a weak base, or a buffer? Solvent molecules have been omitted for clarity.

b. What particles depicted in Figure P16.9 will change, and how will they change, upon adding a strong base such as NaOH?

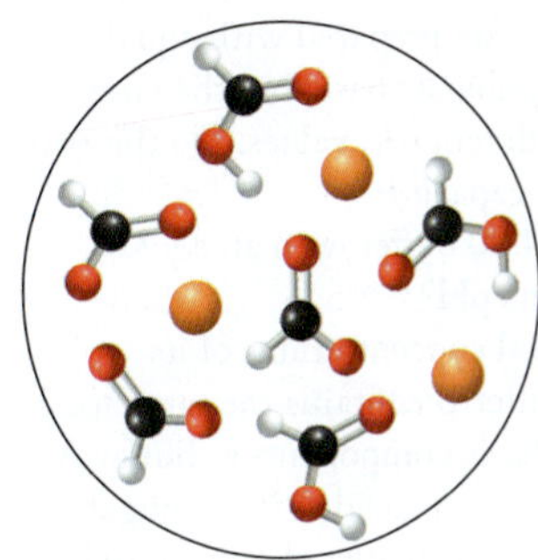

**FIGURE P16.9**

16.10. Use representations [A] through [I] in Figure P16.10 to answer questions a–f. (Solvent molecules have been omitted for clarity of the representations.)

a. Which solution depicts a strong acid? Which depicts a strong acid after titration to its equivalence point with NaOH?

b. Which represents a buffer with equal concentrations of a weak acid and its conjugate base?

c. Which represents the buffer from part b in this question after the addition of strong acid?

d. Which represents the buffer from part b in this question after the addition of strong base?

e. Which structure shows the protonated form of the indicator methyl red? Which shows the deprotonated form of methyl red?

f. If methyl red is red at lower pH and yellow at higher pH, match the flasks in [C] and [G] to the structures in [A] and [I].

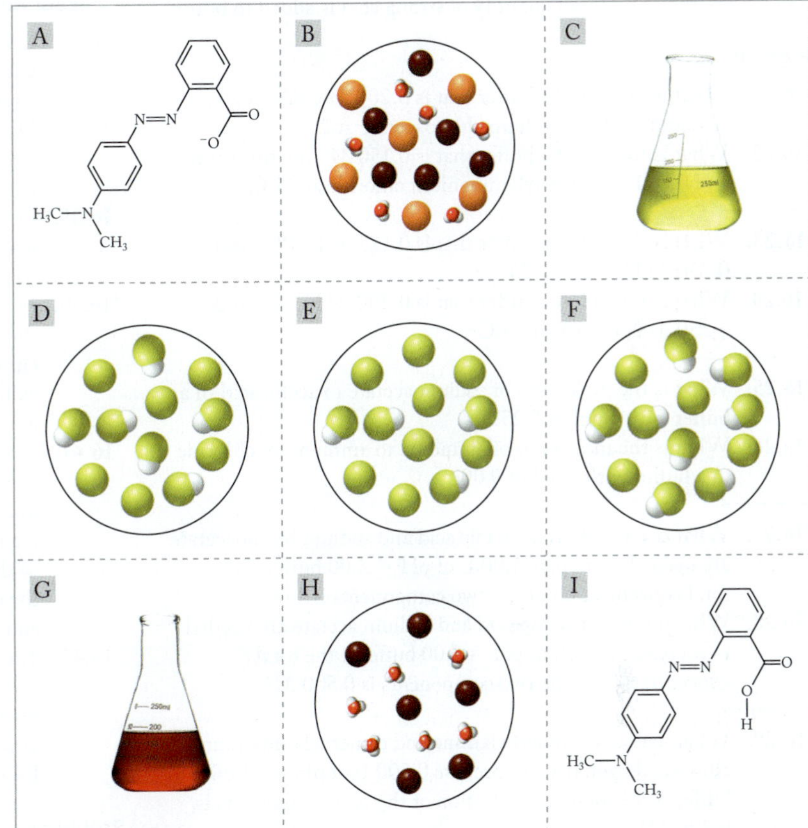


**FIGURE P16.10**

## QUESTIONS AND PROBLEMS

*Note*: Tables A5.1, A5.3, A5.4, and A5.5 in Appendix 5 contain $K_a$, $K_b$, $pK_a$, $pK_b$, $K_{sp}$, and $K_f$ values that may be useful in answering the questions and solving the problems in these sections.

### *The Common-Ion Effect and pH Buffers*

**Concept Review**

**16.11.** Why does a solution of a weak acid and its conjugate base control pH better than a solution of the weak acid alone?

16.12. Why does a solution of a weak base and its conjugate acid control pH better than a solution of the weak base alone?

**16.13.** Identify a suitable buffer system to maintain a pH of 3.0 in an aqueous solution.

16.14. Identify a suitable buffer system to maintain a pH of 9.0 in an aqueous solution.

**16.15.** What does *buffer capacity* mean?

16.16. What effect does adding more NaF have on the pH and buffer capacity of an aqueous solution that is initially 1.0 *M* HF and 0.50 *M* NaF?

**16.17.** Three buffers are separately prepared using equal concentrations of formic acid and sodium formate, hydrofluoric acid and sodium fluoride, and acetic acid and sodium acetate. Rank the three buffers from highest to lowest pH.

16.18. Equal volumes of two buffers are prepared with equal concentrations of acid and conjugate base, but they use different weak acids with different p$K_a$ values. Do the two buffers have the same buffer capacity?

**16.19.** How does diluting a pH = 4.00 buffer with an equal volume of pure water affect its pH?

*16.20. Buffer A contains nearly equal concentrations of its conjugate acid–base pair. Buffer B contains the same total concentration of acidic and basic components as buffer A, but B has twice as much of its weak acid as its conjugate base. Which buffer experiences a smaller change in pH when

a. the same small quantity of strong base is added to both?
b. the same small quantity of strong acid is added to both?

**Problems**

**16.21.** What is the pH of a buffer that is 0.200 *M* chloroacetic acid and 0.100 *M* sodium chloroacetate at 25°C?

16.22. What is the pH of a buffer that is 0.150 *M* dimethylamine and 0.100 *M* dimethylammonium chloride at 25°C?

---

**16.23.** What is the pH of a buffer that is 0.125 *M* $H_3PO_4$ and 0.250 *M* $H_2PO_4^-$ at 25°C?

16.24. What is the pH of a buffer that is 0.200 *M* $H_2SO_3$ and 0.250 *M* $NaHSO_3$ at 25°C?

---

**16.25.** What is the mole ratio of sodium acetate to acetic acid in a buffer with a pH of 5.25?

16.26. What is the mole ratio of ammonia to ammonium chloride in a buffer with a pH of 9.00?

---

**16.27.** What masses of bromoacetic acid and sodium bromoacetate are needed to prepare 1.00 L of pH = 3.00 buffer if the total concentration of the two components is 1.00 *M*?

16.28. What masses of acetic acid and sodium acetate are needed to prepare 125 mL of pH = 5.00 buffer if the total concentration of the two components is 0.500 *M*?

---

**16.29.** What masses of dimethylamine and dimethylammonium chloride do you need to prepare 0.500 L of pH = 11.00 buffer if the total concentration of the two components is 0.300 *M*?

16.30. What masses of ethylamine and ethylammonium chloride do you need to prepare 1.000 L of pH = 11.00 buffer if the total concentration of the two components is 0.500 *M*?

---

**16.31.** What is the pH at 25°C of a solution that results from mixing equal volumes of 0.05 *M* solution of ammonia and a 0.025 *M* solution of hydrochloric acid?

16.32. What is the pH at 25°C of a solution that results from mixing equal volumes of a 0.05 *M* solution of acetic acid and a 0.025 *M* solution of sodium hydroxide?

---

***16.33.** What volume of 0.500 *M* NaOH must be added to 0.500 L of 0.300 *M* acetic acid to raise its pH to 4.00 at 25°C?

*16.34. What volume of 2.00 *M* NaOH must be added to 1.00 L of 1.00 *M* $H_3PO_4$ to prepare a buffer with a pH of 7.19 at 25°C?

---

**16.35.** A buffer consists of 0.120 *M* $HNO_2$ and 0.150 *M* $NaNO_2$ at 25°C.

a. What is the pH of the buffer?
b. What is the pH after adding 1.00 mL of 11.6 *M* HCl to 1.00 L of the buffer solution?

*16.36. A buffer is prepared by mixing 50.0 mL of 0.200 *M* NaOH with 100.0 mL of 0.175 *M* acetic acid.

a. What is the pH of the buffer?
b. What is the pH of the buffer after 1.00 g NaOH is dissolved in it?

## Indicators and Acid–Base Titrations

**Concept Review**

**16.37.** Do all titrations of samples of strong monoprotic acids with solutions of strong bases have the same pH at their equivalence points? Explain why or why not.

16.38. Do all titrations of samples of weak monoprotic acids with solutions of strong bases have the same pH at their equivalence points? Explain why or why not.

**16.39.** Describe two properties of phenolphthalein that make it a good choice as an indicator for detecting the first equivalence point in an alkalinity titration.

*16.40. Phenolphthalein can be used as a color indicator to detect the equivalence points of titrations of samples containing either weak acids or strong acids, even though the pH values of the equivalence points vary depending on the identity of the acid. Explain how this is possible.

**16.41.** In the titration of a solution of a weak monoprotic acid with a standard solution of NaOH, the pH halfway to the equivalence point was 4.44. In the titration of a second solution of the same acid, exactly twice as much of the standard solution of NaOH was needed to reach the equivalence point. What was the pH halfway to the equivalence point in this titration?

16.42. The pH of a solution of a strong monoprotic acid is lower than the pH of an equal concentration of a weak monoprotic acid, yet equal volumes of both require the same volume of basic titrant to reach the equivalence point. Explain why.

**Problems**

**16.43.** A 25.0 mL sample of 0.100 *M* acetic acid is titrated with 0.125 *M* NaOH at 25°C. What is the pH of the solution after 10.0 mL, 20.0 mL, and 30.0 mL of the base have been added?

16.44. A 25.0 mL sample of a 0.100 *M* solution of aqueous trimethylamine is titrated with a 0.100 *M* solution of HCl at 25°C. What is the pH of the solution after 10.0 mL, 25.0 mL, and 40.0 mL of acid have been added?

---

**16.45.** Sketch a titration curve for the titration of 50.0 mL of 0.200 *M* $HNO_2$ with 1.00 *M* NaOH. What is the pH of the sample after 2.50 mL, 5.00 mL, 7.50 mL, and 10.00 mL of titrant have been added?

*16.46. Sketch a titration curve for the titration of 40.0 mL of a 0.100 *M* solution of phosphoric acid with a 0.100 *M* solution of NaOH. What is the pH of the titration reaction mixture after 20.0 mL, 40.0 mL, and 60.0 mL of titrant have been added?

---

**16.47.** What volume of 0.100 $M$ HCl is required to titrate 250 mL of 0.0100 $M$ $Na_2CO_3$ to the first equivalence point?

16.48. What volume of 0.0100 $M$ HCl is required to titrate 250 mL of 0.0100 $M$ $Na_2CO_3$ and 250 mL of 0.0100 $M$ $HCO_3^-$?

---

***16.49.** **Window Cleaner** (a) What is the concentration of ammonia in a popular window cleaner if 25.34 mL of 1.162 $M$ HCl is needed to titrate a 10.00 mL sample of the cleaner? (b) Suppose that the sample was diluted to about 50 mL with deionized water before the titration to make a pH electrode easier to mount in it. How did this dilution affect the volume of titrant needed?

16.50. **Hot Springs** In an alkalinity titration of a 100.0 mL sample of water from a hot spring, 1.32 mL of a 0.100 $M$ solution of HCl is needed to reach the first equivalence point and a total of 6.62 mL is needed to reach the second equivalence point. If the alkalinity of the spring water is due only to the presence of carbonate and bicarbonate, what are the concentrations of each?

---

**16.51.** For each titration, predict whether the pH of the equivalence point is less than, equal to, or greater than 7.
a. Quinine titrated with nitric acid
b. Pyruvic acid titrated with calcium hydroxide
c. Hydrobromic acid titrated with strontium hydroxide

16.52. For each titration, predict whether the pH of the equivalence point is less than, equal to, or greater than 7.
a. HCN titrated with $Ca(OH)_2$
b. LiOH titrated with HI
c. $C_5H_5N$ titrated with $HNO_3$

---

**16.53.** When 100 mL of 0.0125 $M$ ascorbic acid is titrated with 0.010 $M$ NaOH, how many equivalence points will the titration curve have, and what pH indicator(s) could be used? Refer to Figure 16.7 for the colors of indicators.

16.54. **Cabbage Juice** Red cabbage juice is a sensitive acid–base indicator; its colors range from red at acidic pH to yellow in alkaline solutions. What color would red cabbage juice have at the equivalence point when 25 mL of a 0.10 $M$ solution of acetic acid is titrated with 0.10 $M$ NaOH?

## *Lewis Acids and Bases*

### Concept Review

**16.55.** Are all Lewis bases also Brønsted–Lowry bases? Explain why or why not.

16.56. Are all Brønsted–Lowry bases also Lewis bases? Explain why or why not.

**16.57.** Are all Brønsted–Lowry acids also Lewis acids? Explain why or why not.

16.58. Why is $BF_3$ a Lewis acid but not a Brønsted–Lowry acid?

### Problems

**16.59.** Draw Lewis structures that show how electron pairs move and bonds form and break during the autoionization of water. Label the appropriate $H_2O$ molecules as the Lewis acid and Lewis base.

16.60. Draw Lewis structures that show how electron pairs move and bonds form and break in the following reaction, and identify the Lewis acid and Lewis base.

$$MgO(s) + CO_2(g) \rightarrow MgCO_3(s)$$

---

**16.61.** Draw Lewis structures that show how electron pairs move and bonds form and break in the following reaction, and identify the Lewis acid and Lewis base.

$$SO_2(g) + H_2O(\ell) \rightarrow H_2SO_3(aq)$$

16.62. Draw Lewis structures that show how electron pairs move and bonds form and break in the following reaction, and identify the Lewis acid and Lewis base.

$$SeO_3(s) + H_2O(\ell) \rightarrow H_2SeO_4(aq)$$

---

**16.63.** Draw Lewis structures that show how electron pairs move and bonds form and break in the following reaction, and identify the Lewis acid and Lewis base.

$$B(OH)_3(aq) + 2\,H_2O(\ell) \rightleftharpoons B(OH)_4^-(aq) + H_3O^+(aq)$$

*16.64. Draw Lewis structures that show how electron pairs move and bonds form and break in the following reaction, and identify the Lewis acid and Lewis base.

$$SbF_5(\ell) + HF(g) \rightarrow HSbF_6(s)$$

(*Note*: $HSbF_6$ is an ionic compound and one of the strongest Brønsted–Lowry acids known.)

## *Formation of Complex Ions*

### Concept Review

**16.65.** When $CaCl_2$ dissolves in water, which molecules or ions occupy the inner coordination sphere around the $Ca^{2+}$ ions?

16.66. When $AgNO_3$ dissolves in water, which molecules or ions occupy the inner coordination sphere around the $Ag^+$ ions?

***16.67.** **Washing Glassware** A lab technician cleaning glassware that contains residues of AgCl washes the glassware with an aqueous solution of ammonia. The AgCl, which is insoluble in water, rapidly dissolves in the ammonia solution. Why?

*16.68. The procedure used in Problem 16.67 dissolves AgCl but not AgI. Why?

### Problems

**16.69.** A solution is prepared in which 0.0100 mol of $Ni(NO_3)_2$ and 1.00 mol of $NH_3$ are dissolved in a total volume of 1.00 L. What is the concentration of $Ni(H_2O)_6^{2+}$ ions in the solution at equilibrium?

16.70. A 1.00 L solution contains $1.00 \times 10^{-4}$ $M$ $Cu(NO_3)_2$ and $1.00 \times 10^{-3}$ $M$ ethylenediamine. What is the concentration of $Cu(H_2O)_6^{2+}$ ions in the solution at equilibrium?

---

***16.71.** Suppose a solution contains 0.0100 mol of $Co(NO_3)_2$, 0.200 mol of $NH_3$, and 0.200 mol of ethylenediamine in a total volume of 200.0 mL. What is the concentration of $Co(H_2O)_6^{2+}$ ions in the solution?

*16.72. If 1.00 mL of 0.0100 $M$ $AgNO_3$, 1.00 mL of 0.100 $M$ NaBr, and 1.00 mL of 0.100 $M$ NaCN are diluted to 250 mL with deionized water in a volumetric flask and shaken vigorously, will the contents of the flask be cloudy or clear? Support your answer with the appropriate calculations.

## Hydrated Metal Ions as Acids

### Concept Review

**16.73.** Which, if any, aqueous solutions of the following chloride compounds are acidic? (a) $CaCl_2$; (b) $CrCl_3$; (c) NaCl; (d) $FeCl_3$

16.74. If 0.100 $M$ aqueous solutions of each of these compounds were prepared, which one would have the lowest pH? (a) $BaCl_2$; (b) LiCl; (c) KCl; (d) $TiCl_4$

**16.75.** A student prepares 0.100 $M$ solutions of $FeCl_2$ and $FeCl_3$. Which solution has the higher pH?

16.76. As an aqueous solution of KOH is slowly added to a stirred solution of $AlCl_3$, the mixture becomes cloudy but then clears when more KOH is added.

a. Explain the chemical changes responsible for the changes in the appearance of the mixture.

b. Would you expect to observe the same changes if KOH were added to a solution of $FeCl_3$? Explain why or why not.

**16.77.** Chromium(III) hydroxide is amphiprotic. Write chemical equations showing how an aqueous suspension of this compound reacts to the addition of a strong acid and the addition of a strong base.

16.78. Zinc hydroxide is amphiprotic. Write chemical equations showing how an aqueous suspension of this compound reacts to the addition of a strong acid and the addition of a strong base.

**16.79.** **Refining Aluminum** To remove impurities such as calcium and magnesium carbonates and iron(III) oxides from aluminum ore (which is mostly $Al_2O_3$), the ore is treated with a strongly basic solution. In this treatment, $Al^{3+}$ dissolves but the other metal ions do not. Why?

*16.80. Exactly 1.00 g of $FeCl_3$ is dissolved in each of four 0.500 L samples: 1 $M$ $HNO_3$, 1 $M$ $HNO_2$, 1 $M$ $CH_3COOH$, and pure water. Is the concentration of $Fe(H_2O)_6{}^{3+}$ ions the same in all four solutions? Explain why or why not.

### Problems

**16.81.** What is the pH of 0.25 $M$ $Al(NO_3)_3$?

16.82. What is the pH of 0.50 $M$ $CrCl_3$?

---

**16.83.** What is the pH of 0.100 $M$ $Fe(NO_3)_3$?

16.84. What is the pH of 1.00 $M$ $Cu(NO_3)_2$?

---

**16.85.** Sketch the titration curve (pH vs. volume of 0.50 $M$ NaOH added) for a 25 mL sample of 0.25 $M$ $FeCl_3$.

16.86. Sketch the titration curve that results from adding 0.50 $M$ NaOH to a sample containing 0.25 $M$ $KFe(SO_4)_2$.

## Solubility Equilibria

### Concept Review

**16.87.** What is the difference between *molar solubility* and *solubility product*?

16.88. Give an example of how the common-ion effect limits the dissolution of a sparingly soluble ionic compound.

**16.89.** Which cation will precipitate first as a carbonate mineral from an equimolar solution of $Mg^{2+}$, $Ca^{2+}$, and $Sr^{2+}$?

16.90. If the solubility of a compound increases with increasing temperature, does $K_{sp}$ increase or decrease?

**16.91.** The $K_{sp}$ of strontium sulfate increases from $3.4 \times 10^{-7}$ at 37°C to $3.8 \times 10^{-7}$ at 77°C. Is the dissolution of strontium sulfate endothermic or exothermic?

16.92. Identify any of the following solids that are more soluble in acidic solution than in neutral water: $CaCl_2$, $Ba(HCO_3)_2$, $PbSO_4$, $Cu(OH)_2$. Explain your choices.

**16.93.** **Chemistry of Tooth Decay** Tooth enamel is composed of a mineral known as hydroxyapatite, which has the formula $Ca_5(PO_4)_3(OH)$. Explain why tooth enamel can be eroded by acidic substances released by bacteria growing in the mouth.

16.94. **Fluoride and Dental Hygiene** Fluoride ions in drinking water and toothpaste convert hydroxyapatite in tooth enamel into fluorapatite:

$$Ca_5(PO_4)_3(OH)(s) + F^-(aq) \rightleftharpoons Ca_5(PO_4)_3F(s) + OH^-(aq)$$

Why is fluorapatite less susceptible than hydroxyapatite to erosion by acids?

### Problems

**16.95.** If the $[Ba^{2+}]$ in a saturated solution of barium sulfate is $1.04 \times 10^{-5}$ $M$, what is the $K_{sp}$ value of barium sulfate at this temperature?

16.96. If only 0.160 g of $Ca(OH)_2$ dissolves in 0.100 L of water, what is the $K_{sp}$ value for calcium hydroxide at this temperature?

---

**16.97.** What are the equilibrium concentrations of $Cu^+$ and $Cl^-$ in a saturated solution of copper(I) chloride at 25°C?

16.98. What are the equilibrium concentrations of $Pb^{2+}$ and $F^-$ in a saturated solution of lead(II) fluoride at 25°C?

---

**16.99.** What is the solubility of calcite ($CaCO_3$) in grams per milliliter at a temperature at which its $K_{sp} = 9.9 \times 10^{-9}$?

16.100. What is the solubility of silver iodide in grams per milliliter at 25°C?

---

**16.101.** What is the pH at 25°C of a saturated solution of silver hydroxide?

16.102. **pH of Milk of Magnesia** What is the pH at 25°C of a saturated solution of magnesium hydroxide (the active ingredient in the antacid milk of magnesia)?

**16.103.** Suppose you have 100 mL of each of the following solutions. In which one will the most $CaCO_3$ dissolve? (a) 0.1 *M* NaCl; (b) 0.1 *M* $Na_2CO_3$; (c) 0.1 *M* NaOH; (d) 0.1 *M* HCl

16.104. In which of the following solutions will $CaF_2$ be most soluble? (a) 0.010 *M* $Ca(NO_3)_2$; (b) 0.01 *M* NaF; (c) 0.001 *M* NaF; (d) 0.10 *M* $Ca(NO_3)_2$

**16.105.** **Composition of Seawater** The average concentration of sulfate in surface seawater is about 0.028 *M*. The average concentration of $Sr^{2+}$ is $9 \times 10^{-5}$ *M*. Is the concentration of strontium in the sea significantly controlled by the insolubility of its sulfate salt?

16.106. **Fertilizing the Sea to Combat Climate Change** Some scientists have proposed adding iron(III) compounds to large expanses of the open ocean to promote the growth of phytoplankton that would in turn remove $CO_2$ from the atmosphere through photosynthesis. If we assume that the average pH of open ocean water is 8.07, what is the maximum value of $[Fe^{3+}]$ in seawater if the $K_{sp}$ value of $Fe(OH)_3$ is $1.1 \times 10^{-36}$?

**16.107.** Will a precipitate form when 25.0 mL of 0.500 *M* $Ca(NO_3)_2$ is added to 25.0 mL of 0.250 *M* NaF at 25°C?

16.108. Will a precipitate form if 125 mL of 0.100 *M* sodium chloride is added to 50.0 mL of 0.100 *M* lead(II) nitrate at 25°C?

**16.109.** A solution is 0.010 *M* in both $Br^-$ and $SO_4^{2-}$. A 0.250 *M* solution of lead(II) nitrate is slowly added to it by using a burette.

a. Which anion will precipitate first?
b. What is the concentration of the first anion when the second one starts to precipitate at 25°C?

*16.110. Solution A is 0.0200 *M* in $Ag^+$ ions and $Pb^{2+}$ ions. You have access to two other solutions: (B) 0.250 *M* NaCl and (C) 0.250 *M* NaBr.

a. Which solution, B or C, would be the better one to add to solution A to separate $Ag^+$ ions from $Pb^{2+}$ by selective precipitation?
b. Using the solution you selected in part a, is the separation of the two ions complete?

## Additional Problems

***16.111.** **pH of Baking Soda** A cook dissolves a teaspoon of baking soda ($NaHCO_3$) in a cup of water and then discovers that the recipe calls for a tablespoon, not a teaspoon. If the cook adds two more teaspoons of baking soda to make up the difference, does the additional baking soda change the pH of the solution? Explain why or why not.

16.112. **Antacid Tablets** Antacids contain a variety of bases such as $NaHCO_3$, $MgCO_3$, $CaCO_3$, and $Mg(OH)_2$. Only $NaHCO_3$ has appreciable solubility in water.

a. Write a net ionic equation for the reaction of each base with stomach acid (aqueous HCl).
b. Explain how substances sparingly soluble in water can act as effective antacids.

**16.113.** When silver oxide dissolves in water, the following reaction occurs:

$$Ag_2O(s) + H_2O(\ell) \rightarrow 2\,Ag^+(aq) + 2\,OH^-(aq)$$

If a saturated aqueous solution of silver oxide is $1.6 \times 10^{-4}$ *M* in hydroxide ion, what is the $K_{sp}$ of silver oxide?

*16.114. Why does adding $CaCl_2$ to a $HPO_4^{2-}/PO_4^{3-}$ buffer increase the ratio of $HPO_4^{2-}$ ions to $PO_4^{3-}$ ions?

***16.115.** **Greenhouse Gases and Ocean pH** Some climate models predict that the pH of the oceans will decrease by as much as 0.77 pH unit due to increases in atmospheric carbon dioxide.

a. Use the appropriate chemical reactions and equilibria to explain how increasing atmospheric $CO_2$ produces a decrease in oceanic pH.
b. How much more acidic (in terms of $[H_3O^+]$) would the oceans be if their pH dropped this much?
c. Oceanographers are concerned about the impact of a drop in oceanic pH on the survival of oysters. Why?

*16.116. A 125.0 mg sample of an unknown monoprotic acid was dissolved in 100.0 mL of distilled water and titrated with a 0.050 *M* solution of NaOH. The pH of the solution was monitored throughout the titration, and the following data were collected.

| Volume of $OH^-$ Added (mL) | pH | Volume of $OH^-$ Added (mL) | pH |
|---|---|---|---|
| 0 | 3.09 | 22 | 5.93 |
| 5 | 3.65 | 22.2 | 6.24 |
| 10 | 4.10 | 22.6 | 9.91 |
| 15 | 4.50 | 22.8 | 10.2 |
| 17 | 4.55 | 23 | 10.4 |
| 18 | 4.71 | 24 | 10.8 |
| 19 | 4.94 | 25 | 11.0 |
| 20 | 5.11 | 30 | 11.5 |
| 21 | 5.37 | 40 | 11.8 |

a. What is the $K_a$ value for the acid?
b. What is the molar mass of the acid?

**16.117.** On the basis of the location of the equivalence points in the titration curve in Figure P16.117, which of the following statements is true about the alkalinity of the sample?

a. It is due only to the presence of carbonate ions.
b. It is due mostly to the presence of carbonate ions.
c. It is due only to the presence of bicarbonate ions.
d. It is due mostly to the presence of bicarbonate ions.
e. The ratio of bicarbonate to carbonate ions in the sample is exactly 50:50 (on a mole basis).

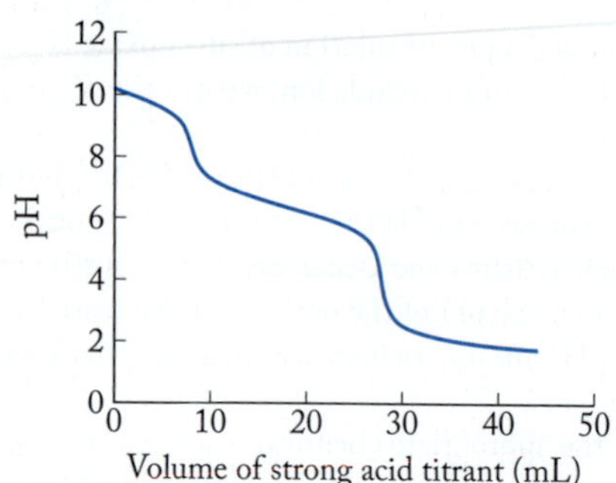

**FIGURE P16.117**

16.118. A pH 3.00 buffer is prepared by mixing solutions of nitrous acid and sodium nitrite. The total concentration of nitrous acid and sodium nitrate in the buffer is 0.100 *M*. Suppose 1.00 milliliter of 1.00 *M* HCl is added to a 100 mL sample of the buffer and 1.00 milliliter of 1.00 *M* NaOH is added to another 100 mL sample. In which sample would adding strong acid or base produce the greater change in pH? Explain your selection.

**16.119.** For each set of three acids, which one should you use to make a buffer with the given pH?

a. Select from acetic acid ($pK_a = 4.75$), fluoroacetic acid ($pK_a = 2.59$), and hypochlorous acid ($pK_a = 7.54$) to make a buffer with a pH of 5.2.
b. Select from formic acid ($pK_a = 3.75$), hypobromous acid ($pK_a = 7.54$), and boric acid ($pK_a = 9.27$) to make a buffer with a pH of 8.0.

16.120. Your task is to prepare a buffer that has a pH of exactly 8.00. The substances you have available are $H_3PO_4$ ($\mathcal{M}$ = 98.0 g/mol), $NaH_2PO_4$ ($\mathcal{M}$ = 120.0 g/mol), $Na_2HPO_4$ ($\mathcal{M}$ = 142.0 g/mol), and $Na_3PO_4$ ($\mathcal{M}$ = 163.9 g/mol). Which two of these substances, and how many grams of each, would you use to prepare exactly one liter of a buffer in which the total concentration of acid and basic components is 0.100 *M*?

**16.121.** Estimate the $K_a$ values of the following indicators:

a. Bromophenol blue, whose transition color occurs at a pH of about 3.8.
b. Bromocresol green, whose transition color occurs at a pH of about 4.3.
c. Alizarin yellow R, whose transition color occurs at a pH of about 10.9.

16.122. Figure 16.7 depicts an array of indicators. Use this figure to:

a. Select a suitable indicator to use in detecting the second equivalence point in the titration of an aqueous solution of oxalic acid with 0.200 *M* NaOH.
b. Select a suitable indicator to use in detecting the first equivalence point in the titration of a sample of phosphoric acid

**16.123.** Calculate the pH at the equivalence point for titrating 10.0 mL of 0.100 *M* formic acid ($K_a = 1.77 \times 10^{-4}$) with 0.100 *M* NaOH and for titrating 10.0 mL of boric acid ($K_{a_1} = 5.4 \times 10^{-10}$) with the same solution. Should the same indicator be used for both titrations?

16.124. The titration curve in Figure P16.124 shows the results of the titration of a 0.100 *M* aqueous solution of propionic acid ($CH_3CH_2COOH$ = HPr) with 0.100 *M* NaOH.

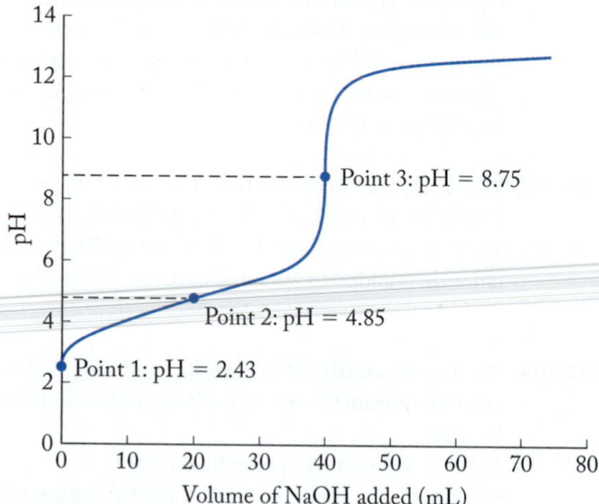

**FIGURE P16.124**

a. At the points indicated on the titration curve, identify the relative concentrations of
Point 1: $[Pr^-]$ and $[H_3O^+]$
Point 2: [HPr] and $[Pr^-]$
Point 3: [HPr] initially present at point 1 and $[OH^-]$ added.
b. Estimate the $pK_a$ of propionic acid.
c. By referring to the amount of sodium hydroxide solution added (*x*-axis), define the buffer region of the titration curve.
d. Refer to Figure 16.7 and select an appropriate indicator to determine the equivalence point of the titration.

**16.125.** Suppose that reaching the first equivalence point in titrating a sample of sulfurous acid requires exactly 15.00 mL of NaOH titrant.

a. How much more titrant (in milliliters) is required to reach the second equivalence point?
b. What is the pH of the titration mixture after 7.50 mL of titrant have been added?
c. What is the pH of the titration mixture after a total of 22.50 mL of titrant have been added?
d. Identify the two most abundant ions in the titration reaction mixture at the first equivalence point.
e. Identify the two most abundant ions in the titration reaction mixture at the second equivalence point.

16.126. Consider the titration of 21.5 mL of 0.120 *M* phenol with 0.250 *M* NaOH.

a. Calculate the pH at the equivalence point.
b. Which indicator from Figure 16.7 would be the best choice for this titration?
c. Calculate the pH for the titration in question after 0 mL, 2.5 mL, 5.0 mL, 7.5 mL, 9.8 mL, 10 mL, 10.2 mL, 10.6 mL, 10.8 mL, 11 mL, 12.5 mL, 15 mL, 17.5 mL, and 20 mL of base have been added.
d. Sketch the titration curve (pH vs. mL of base added) for this titration and indicate which major species are present in solution when pH = $pK_a$, at the equivalence point, and when the titration is complete.

**16.127.** Consider the titration of 15.8 mL of 367 m*M* pyridine with 0.500 *M* HCl.

a. Calculate the pH at the equivalence point.
b. Which indicator from Figure 16.7 would be the best choice for this titration?
c. Calculate the pH for the titration in question after 0.0 mL, 1.0 mL, 3.0 mL, 7.0 mL, 9.0 mL, 10.0 mL, 10.5 mL, 11.0 mL, 11.5 mL, 11.8 mL, 12.5 mL, 15 mL, and 20 mL of base have been added.
d. Sketch the titration curve (pH vs. mL of base added) for this titration and indicate which major species are present in solution when pH = $pK_a$, at the equivalence point, and when the titration is complete.

16.128. In each of the following pairs of compounds, which is less soluble in water?

a. AgI ($K_{sp} = 8.5 \times 10^{-17}$) or AgCl ($K_{sp} = 1.8 \times 10^{-10}$)
b. $SrF_2$ ($K_{sp} = 4.3 \times 10^{-9}$) or $MgF_2$ ($K_{sp} = 5.2 \times 10^{-9}$)
c. $MnCO_3$ ($K_{sp} = 2.3 \times 10^{-1}$) or $PbCO_3$ ($K_{sp} = 7.4 \times 10^{-14}$)

**16.129.** Cobalt(II) and zinc(II) carbonate have nearly identical $K_{sp}$ values. Do they have the same molar solubility? Do they have the same solubility in g/L?

16.130. The $K_{sp}$ values for silver iodide and silver phosphate differ by less than 10%. Do they have the same molar solubility?

**16.131.** Which has a higher pH, a saturated solution of magnesium hydroxide ($K_{sp} = 5.6 \times 10^{-12}$) or a saturated solution of calcium hydroxide ($K_{sp} = 4.7 \times 10^{-6}$)?

16.132. Many monoprotic organic acids (HAs) have $pK_a$ values between 4.0 and 5.0. This means that these acids exist in human serum (pH = 7.30) as:

a. Only molecules of HA
b. About 100 times as many HA molecules as $A^-$ ions
c. About 10 times as many HA molecules as $A^-$ ions
d. About a 50:50 mix of HA molecules and $A^-$ ions
e. Nearly all $A^-$ ions

**16.133. Digestive Enzymes** The activity of many enzymes is pH dependent. For example, the optimum activity of two common digestive enzymes, trypsin and pepsin, occurs at pH 6.5 and pH 1.5, respectively. Using the data in Appendix 5 as a guide, which buffer would you choose for each enzyme to maximize its activity?

16.134. **Diabetes** Diabetics can have elevated levels of two acids in their blood: β-hydroxybutyric acid ($pK_a = 4.72$) and acetoacetic acid ($pK_a = 3.58$). Their structures are shown in Figure P16.134. The presence of these acids lowers the pH of blood and can serve as a way to diagnose diabetes.

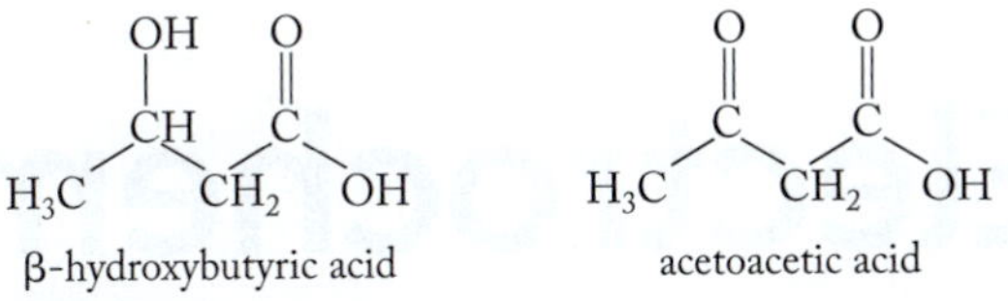

β-hydroxybutyric acid    acetoacetic acid

**FIGURE P16.134**

a. Which acid yields a solution with the lower pH if 0.100 *M* solutions are prepared?
b. Which acid will dissociate to a greater degree?
c. What is the pH of a solution that contains 15.8 m*M* of acetoacetic acid and 10.8 m*M* sodium acetoacetate?
d. What is the pH of a solution of 90 mg/L of β-hydroxybutyric acid and 90 mg/L of β-hydroxybutyrate anion after 100 μL of 0.100 *M* HCl has been added?

**16.135. Hot Chicken** Chickens do not have sweat glands, so when they become overheated, they cool themselves by panting (taking short, quick breaths). This lowers the concentration of $CO_2$ in their blood. If chickens pant too much, this causes the shells on chickens' eggs to become so thin that they break easily. Eggshells are about 95% calcium carbonate. Explain how panting affects shell thickness by considering the equilibrium between $CO_2(aq)$ and $CO_3^{2-}(aq)$ in blood.

16.136. **Canine Dentistry** Teeth in both humans and many animals are made of calcium hydroxyphosphate, $Ca_5(PO_4)_3OH$. Veterinarians have noticed that dogs' saliva has a pH that is more basic than the saliva of humans. This fact is suggested as one reason dogs are less subject to tooth decay than people. Explain this suggestion.

# 17

# Electrochemistry

## The Quest for Clean Energy

ZERO-EMISSION VEHICLES
Propulsion for most all-electric vehicles such as this one comes from electric motors powered by lithium ion batteries.

## PARTICULATE REVIEW

### *Redox: Metal versus Nonmetal*

In Chapter 17, we investigate how chemical energy is transformed through redox reactions into electrical energy. Rust forms on abandoned cars, such as the one in this photo, through a series of redox reactions between iron metal in the car and oxygen gas in the atmosphere.

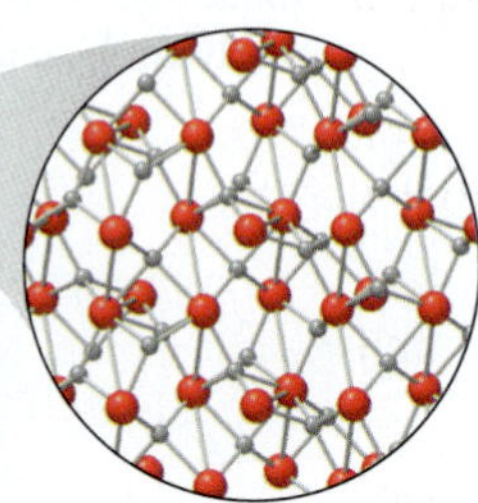

- Write a balanced chemical equation describing the formation of $Fe_2O_3$—a principal component of rust—from iron metal and oxygen gas.
- Which element is oxidized in this reaction? Which element is reduced?
- Describe the direction of electron transfer in the formation of rust.

(Review Section 8.7 if you need help.)

*(Answers to Particulate Review questions are in the back of the book.)*

PARTICULATE **PREVIEW**

### *Redox: Electricity and Clean Fuel*

Not all redox reactions involve the oxidation and reduction of metals. The redox chemistry in most fuel cells involves hydrogen gas and oxygen gas as pictured here. As you read Chapter 17, look for ideas that will help you answer these questions:

- Write chemical equations describing the oxidation and reduction half-reactions pictured here and combine the half-reactions to write a chemical equation describing the overall fuel-cell reaction.
- Describe the flow of ions within a fuel cell and how it produces a flow of electrons that can propel a vehicle.
- Why are vehicles powered by fuel cells and electric motors called "zero emission" vehicles?

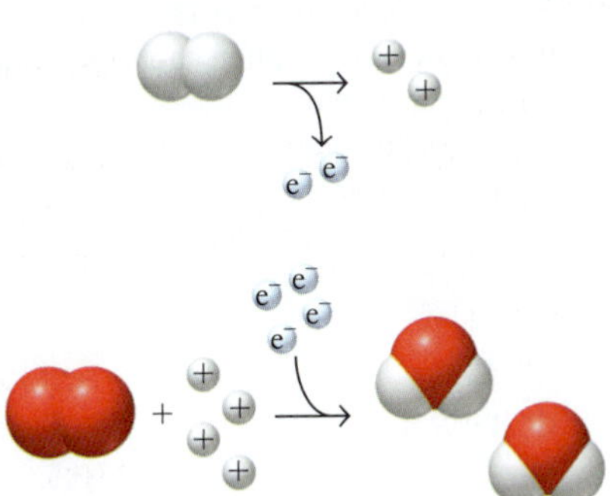

## Learning Outcomes

**LO1** Use cell diagrams to describe the components of electrochemical cells and their roles in interconverting chemical and electrical energy
**Sample Exercises 17.1, 17.2**

**LO2** Use the standard reduction potentials of two half-reactions to decide which occurs at the cathode and which at the anode and to calculate the standard potential of an electrochemical cell based on those half-reactions
**Sample Exercise 17.3**

**LO3** Interconvert an electrochemical cell's standard potential with the change in standard free energy of the cell reaction
**Sample Exercise 17.4**

**LO4** Use the Nernst equation to calculate nonstandard cell potentials
**Sample Exercise 17.5**

**LO5** Interconvert a cell's standard potential with the value of the equilibrium constant of the cell reaction under standard conditions
**Sample Exercise 17.6**

**LO6** Interconvert masses of reactants in cell reactions with quantities of electrical charge
**Sample Exercises 17.7, 17.8**

## 17.1 Running on Electrons: Redox Chemistry Revisited

The 21st century has seen wide swings in the price of crude oil and products derived from it, such as gasoline. These fluctuations, coupled with growing concerns over climate change and other pollution problems, have invigorated the development of electric propulsion systems for cars and trucks. Some vehicles, called hybrids, are propelled by combinations of electric motors powered by rechargeable batteries and small gasoline engines. Others, including the one in this chapter's opening photo, are powered only by electric motors and banks of high-performance batteries or by another source of mobile electrical power: fuel cells.

Batteries and fuel cells convert chemical energy into electrical energy; motors then convert that electrical energy into mechanical energy. These two processes together are more efficient than converting chemical energy directly to mechanical energy in vehicles powered by gasoline and diesel engines. Moreover, electric propulsion earns these vehicles the designation "zero emission" vehicles. This description applies because the chemical energy in their batteries or fuel cells is converted into mechanical work with essentially no release of atmospheric pollutants, including the nitrogen oxides that contribute to photochemical smog and the greenhouse gases, such as $CO_2$, that contribute to climate change. However, these vehicles may still depend on fossil fuels to generate the electricity needed to charge their batteries or to synthesize the hydrogen gas consumed in their fuel cells.

**CONNECTION** In Chapters 13–15, we examined the chemistry of photochemical smog and acid rain formation. These atmospheric pollution problems and the escalating threat of climate change have been linked to energy production based on the combustion of fossil fuels.

In this chapter, we examine the chemical reactions that energize modern batteries and fuel cells. These devices are based on **electrochemistry**, the branch of chemistry that links chemical reactions to the production or consumption of electrical energy. At the heart of electrochemistry are chemical reactions in which electrons are transferred between substances at electrode surfaces. In other words, electrochemistry is based on *red*uction and *ox*idation, or *redox*, chemistry.

**electrochemistry** the branch of chemistry that examines the transformations between chemical and electrical energy.

The principles of redox reactions, which were introduced in Section 8.7, can be summarized as follows:

- A redox reaction consists of two complementary processes: the *reduction* of a substance that gains electrons and the *oxidation* of a substance that loses electrons.
- Reduction and oxidation happen simultaneously so that the number of electrons gained during reduction matches the number lost during oxidation.
- A substance that is easily oxidized is one that readily gives up electrons. This electron-donating power makes the substance an effective reducing agent. In any redox reaction, the *reducing agent is always oxidized.*
- A substance that readily accepts electrons and is thereby reduced is an effective oxidizing agent. In any redox reaction, the *oxidizing agent is always reduced.*

CONNECTION Redox reactions, including balancing redox equations using half-reactions, were discussed in detail in Section 8.7.

We begin our exploration of electrochemistry by reviewing the oxidation and reduction half-reactions involved in a redox reaction between copper and zinc. When a strip of Zn metal is placed in a solution of $CuSO_4$, as shown in **Figure 17.1**, electrons spontaneously transfer from Zn atoms to $Cu^{2+}$ ions, forming $Zn^{2+}$ ions and Cu atoms. The shiny zinc surface turns dark brown as a textured layer of copper metal accumulates on it, and the distinctive blue color of $Cu^{2+}(aq)$ ions fades as these ions gain electrons and become atoms of copper metal.

The electron transfer that occurs in this spontaneous reaction can be represented through two simultaneous half-reactions—an oxidation half-reaction that consumes atoms of zinc and a reduction half-reaction that consumes copper(II) ions:

$$Zn(s) \rightarrow Zn^{2+}(aq) + 2\,e^-$$

$$Cu^{2+}(aq) + 2\,e^- \rightarrow Cu(s)$$

As the reaction proceeds, each mole of zinc atoms loses 2 moles of electrons and each mole of copper(II) ions gains 2 moles of electrons. The number of moles of

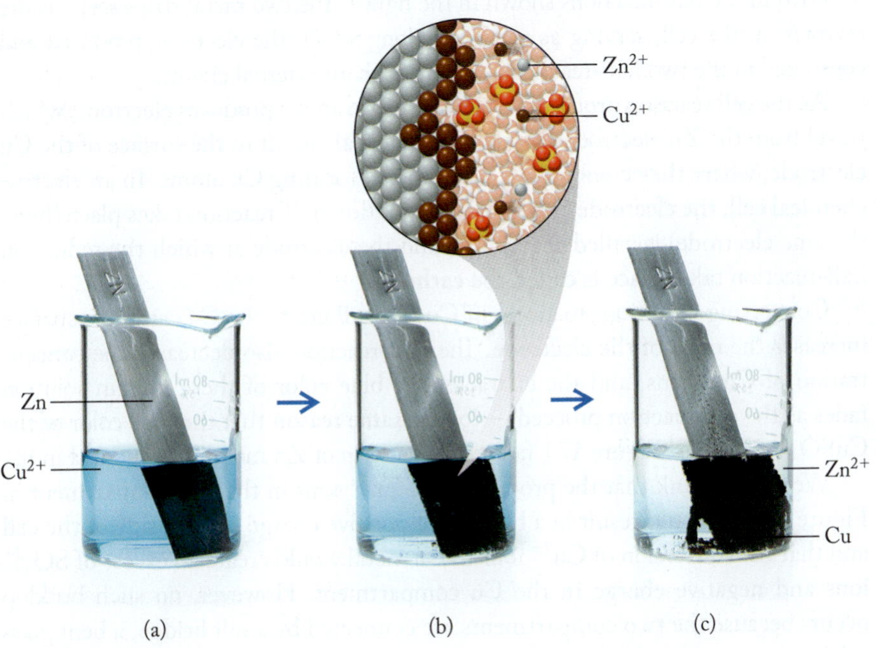


**FIGURE 17.1** A strip of zinc is immersed in an aqueous solution of blue copper(II) sulfate. It becomes encrusted with a dark layer of copper as $Cu^{2+}$ ions are reduced to Cu atoms and Zn atoms are oxidized to colorless $Zn^{2+}$ ions. The solution becomes a paler blue as $Cu^{2+}$ ions are reduced.

electrons lost and gained are the same in the two half-reactions. Therefore, writing a net ionic equation to describe the overall redox reaction is simply a matter of adding the two half-reactions together:

$$\begin{array}{rl} & Zn(s) \rightarrow Zn^{2+}(aq) + 2\,e^- \\ + & Cu^{2+}(aq) + 2\,e^- \rightarrow Cu(s) \\ \hline & Zn(s) + Cu^{2+}(aq) + \cancel{2\,e^-} \rightarrow Cu(s) + Zn^{2+}(aq) + \cancel{2\,e^-} \end{array}$$

Canceling out the equal numbers of electrons gained and lost, we obtain the following equation, which we saw in Chapter 8:

$$Zn(s) + Cu^{2+}(aq) \rightarrow Cu(s) + Zn^{2+}(aq) \qquad (8.26)$$

Combining half-reactions is a convenient way to write net ionic equations for redox reactions. In this chapter, we use a valuable resource in this equation-writing process: the table of common half-reactions in Appendix 6. Note that all the half-reactions are written as *reduction* half-reactions, that is, half-reactions in which electrons are gained. A separate table of *oxidation* half-reactions is not needed because any reduction half-reaction can always be reversed to obtain the corresponding oxidation half-reaction in which electrons are lost.

## 17.2 Electrochemical Cells

**electrochemical cell** an apparatus that converts chemical energy into electrical work or electrical work into chemical energy.

**anode** an electrode at which an oxidation half-reaction (loss of electrons) takes place.

**cathode** an electrode at which a reduction half-reaction (gain of electrons) takes place.

Having seen how the redox reaction between Zn metal and $Cu^{2+}$ ions is the net result of two distinct half-reactions, one involving the oxidation of Zn metal and the other the reduction of $Cu^{2+}$ ions, let's now physically separate the two half-reactions by using a device called an **electrochemical cell**. Doing so forces the electrons lost by the Zn atoms to flow through an external circuit before they can be acquired by $Cu^{2+}$ ions. **Figure 17.2(a)** depicts an electrochemical cell based on the $Zn/Cu^{2+}$ reaction. One compartment in the $Zn/Cu^{2+}$ cell contains a strip of zinc metal immersed in 1.00 *M* $ZnSO_4$; the other contains a strip of copper metal immersed in 1.00 *M* $CuSO_4$. Sulfate ions are spectator ions in this reaction and therefore are not included in the half-reactions shown in the figure. The two metal strips serve as the *electrodes* in the cell, serving as pathways along which the electrons produced and consumed in the two half-reactions flow through an external circuit.

As the cell reaction proceeds, oxidation of Zn atoms produces electrons, which travel from the Zn electrode through the external circuit to the surface of the Cu electrode, where they combine with $Cu^{2+}$ ions, forming Cu atoms. In an electrochemical cell, the electrode at which the oxidation half-reaction takes place (here, the zinc electrode) is called the **anode**, and the electrode at which the reduction half-reaction takes place is called the **cathode**.

**CONNECTION** We introduced Beer's law and the relationship between the color of a solution and the concentration of the solute that imparts that color in Section 8.2.

Converting $Cu^{2+}$ ions to atoms of Cu that adhere to the Cu cathode's surface increases the mass of the electrode. The half-reaction also decreases the concentration of $Cu^{2+}$ ions, and the characteristic blue color of these ions in solution fades as the cell reaction proceeds—for the same reason that the blue color of the $CuSO_4$ solution in Figure 17.1 fades when a strip of Zn metal is immersed in it.

We might think that the production of $Zn^{2+}$ ions in the left compartment in Figure 17.2(a) would result in a buildup of positive charge on that side of the cell and that the conversion of $Cu^{2+}$ ions to Cu metal would create an excess of $SO_4^{2-}$ ions and negative charge in the Cu compartment. However, no such buildup occurs because the two compartments are connected by a salt bridge, a bent glass tube containing a solution of an electrolyte made of ions not involved in either

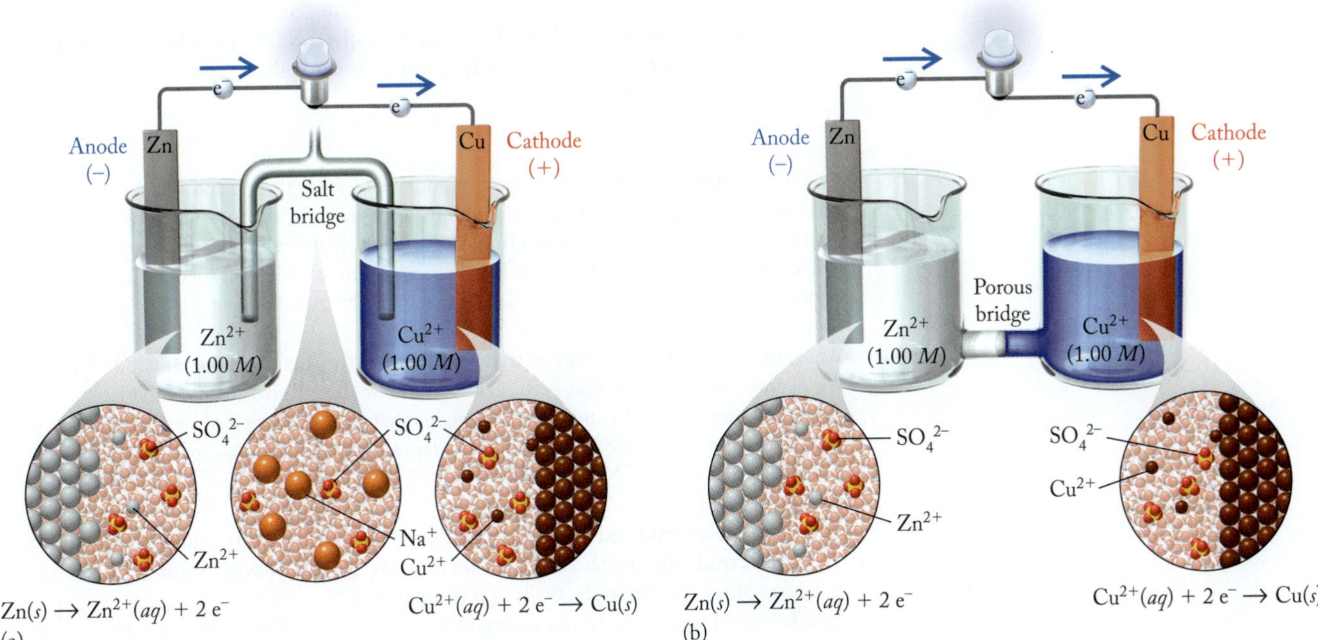


**FIGURE 17.2** (a) This electrochemical cell consists of two compartments: the one on the left contains a zinc metal anode immersed in a 1.00 *M* solution of $ZnSO_4$, and the one on the right contains a copper metal cathode immersed in a 1.00 *M* solution of $CuSO_4$. A salt bridge made of either glass or plastic provides an electrical connection through which the ions of an electrolyte, such as $Na_2SO_4$, can migrate into either compartment to balance electrical charges. (b) In this apparatus, an ion-permeable bridge made of porous glass or ceramic replaces the salt bridge, allowing ions from the background electrolyte ($Na_2SO_4$) to migrate between the two compartments. Background electrolyte ions in (b) are not included for clarity.

**CHEMTOUR**
Zinc–Copper Cell

half-reaction, such as $Na_2SO_4$. As the cell reaction proceeds, sodium ions migrate through the salt bridge toward the copper compartment and sulfate ions migrate toward the zinc compartment. The charges on the migrating $Na^+$ and $SO_4^{2-}$ ions balance the losses and gains of ions in the two compartments and offset the charges on the electrons that flow through the external circuit from the anode to the cathode. Therefore, no charge accumulates in either compartment.

To avoid the need for a salt bridge, we can add $Na_2SO_4$ as a *background electrolyte* directly to each compartment and then connect the compartments with a bridge made of porous glass or a polymeric material as shown in **Figure 17.2(b)**. This procedure separates the contents of the cells but allows $Na^+$ and $SO_4^{2-}$ ions to migrate between the compartments.

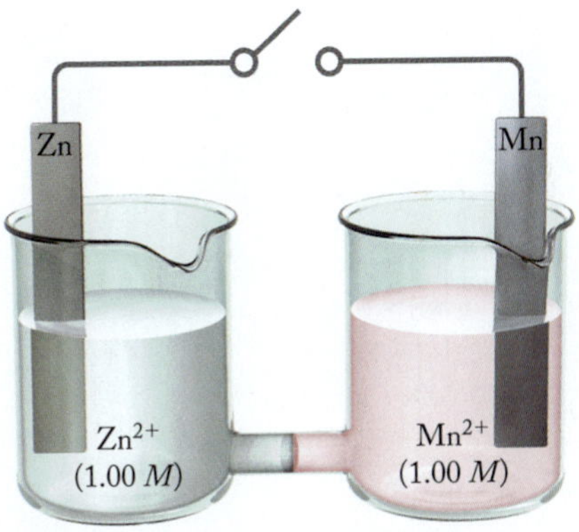

**FIGURE 17.3** An electrochemical cell consisting of a zinc electrode in 1.00 *M* $Zn^{2+}(aq)$, a Mn electrode in 1.00 *M* $Mn^{2+}(aq)$, and a porous bridge between the two solutions.

### SAMPLE EXERCISE 17.1 Describing an Electrochemical Cell LO1

When the switch in the external circuit in **Figure 17.3** is closed, connecting the Zn and Mn electrodes, the following cell reaction occurs spontaneously:

$$Mn(s) + Zn^{2+}(aq) \rightarrow Mn^{2+}(aq) + Zn(s)$$

a. In Figure 17.3, which electrode is the anode and which is the cathode?
b. In which direction do the electrons flow?

c. If the solutions in both beakers also contain dissolved $Na_2SO_4$ as a background electrolyte, in which directions do these ions flow through the porous bridge between the beakers?
d. Which electrode gains mass?

**Collect and Organize** We know the composition of a voltaic cell and the overall cell reaction, and we are asked to identify which electrode is the cathode and which is the anode, in which directions the electrons and ions flow, and which electrode gains mass. The cathode is the electrode at which reduction (a gain of electrons) occurs, and the anode is the electrode at which oxidation (a loss of electrons) occurs.

**Analyze** The cell reaction tells us that Mn atoms at the surface of the electrode form $Mn^{2+}$ ions, which means that each reacting atom loses two electrons. At the same time, $Zn^{2+}$ ions form Zn atoms, which means that each reacting $Zn^{2+}$ ion acquires two electrons from the electrode surface and is deposited as Zn atom on the surface.

**Solve**

a. Mn atoms are oxidized (lose electrons) as they form $Mn^{2+}$ ions, which makes the Mn electrode the anode. $Zn^{2+}$ ions are reduced (gain electrons) as they form Zn atoms at the Zn electrode, making it the cathode.
b. Electrons liberated by the oxidation of Mn atoms at the Mn anode don't accumulate on the electrode surface but instead flow through the external circuit to the Zn cathode, where they are consumed in the reduction of $Zn^{2+}$ ions to Zn atoms.
c. The external flow of electrons and their negative electrical charges from the Mn electrode to the Zn electrode is offset by flows of dissolved $Na^+$ ions through the porous bridge toward the Zn compartment and dissolved $SO_4^{2-}$ ions through the bridge toward the Mn compartment.
d. Reduction of $Zn^{2+}$ ions results in the production of more Zn on the surface of the Zn electrode, so its mass increases.

**Think About It** Zinc atoms are oxidized to $Zn^{2+}$ ions in the Cu/Zn cell shown in Figure 17.2, but $Zn^{2+}$ ions are reduced to Zn atoms in the Mn/Zn cell in Figure 17.3. This reversal of the Zn half-reaction makes sense because Mn is a more reactive metal than Zn (see the activity series of metals in Table 8.8), but Cu is a less reactive metal than Zn.

**Practice Exercise** Identify the anode, cathode, and the direction of electron and ion flow in an electrochemical cell containing a silver electrode immersed in 1.0 m*M* $AgNO_3$ in one compartment and a gold electrode immersed in 1.0 m*M* $Au(NO_3)_3$ in the other compartment. A salt bridge containing a solution of $NaNO_3$ connects the two compartments, and the following net ionic equation describes the cell reaction.

$$3\,Ag(s) + Au^{3+}(aq) \rightarrow 3\,Ag^+(aq) + Au(s)$$

*(Answers to Practice Exercises are in the back of the book.)*

## Cell Diagrams

Figure 17.2 shows the physical reality of a $Zn/Cu^{2+}$ electrochemical cell, but we could use a more compact way to represent the components of the cell. A **cell diagram** uses a string of chemical formulas and symbols to show how the components of the cell are physically and chemically connected. A cell diagram does not convey stoichiometry, so any coefficients in the balanced equation for the cell reaction do not appear in the cell diagram.

**cell diagram** symbols that show how the components of an electrochemical cell are connected.

We follow these steps when writing a cell diagram:

1. Write the chemical symbol of the anode at the far left of the diagram and the symbol of the cathode at the far right. Double vertical lines in the middle

represent the connecting bridge between the anode and cathode compartments. For the $Zn/Cu^{2+}$ cell:

$$Zn(s) \ldots\ldots\ldots \| \ldots\ldots\ldots Cu(s)$$

2. Working inward from the electrodes toward the connecting bridge, use vertical lines to indicate phase changes (for example, between solid electrodes and aqueous solutions). Represent the electrolytes surrounding each electrode by using the symbols of the ions or compounds that are changed by the cell reaction. Use commas to separate species in the same phase:

$$Zn(s) \mid Zn^{2+}(aq) \,\|\, Cu^{2+}(aq) \mid Cu(s)$$

3. If known, use the concentrations of the dissolved species in place of their (*aq*) phase symbols, and add the partial pressures of any gases within their (*g*) phase symbols:

$$Zn(s) \mid Zn^{2+}(1.00\ M) \,\|\, Cu^{2+}(1.00\ M) \mid Cu(s)$$

### SAMPLE EXERCISE 17.2 Diagramming an Electrochemical Cell — LO1

The photograph in **Figure 17.4(a)** shows what happens when an object made of copper (the Statue of Liberty) is exposed to air and ocean spray for over a century. The distinctive pale green color of the statue is due to the products of copper corrosion, including $Cu(OH)_2$, that form when copper is oxidized by atmospheric oxygen. The process can be modeled in the laboratory by using an electrochemical cell (**Figure 17.4b**) in which a copper electrode is immersed in a solution containing $OH^-$ ions. The second compartment contains an electrode made of chemically inert platinum, whose only role in the cell is to transport electrons away from the cell. (We revisit this use of Pt electrodes in Section 17.5.) The Pt electrode is immersed in a solution that contains $OH^-$ ions and a stream of air bubbles around the electrode, as shown in Figure 17.4(b). Write a balanced chemical equation and a cell diagram for this cell reaction.

**Collect and Organize** We have a cell reaction in which electrons spontaneously flow from a copper electrode in contact with a solution containing 1.00 $M$ $OH^-$ ions, through an external circuit, to a platinum electrode in contact with a solution of 0.010 $M$ $OH^-$ ions and a stream of air at 1.00 atm that is 21.0% $O_2$ by volume (see Table 9.1). The half-reactions provided for this cell in Figure 17.4(b) are:

$$O_2(g) + 2\,H_2O(\ell) + 4\,e^- \rightarrow 4\,OH^-(aq)$$

$$Cu(s) + 2\,OH^-(aq) \rightarrow Cu(OH)_2(s) + 2\,e^-$$

In a cell diagram, the anode and the species involved in the oxidation half-reaction are on the left, and the cathode and the species involved in the reduction half-reaction are on the right. We use single lines to separate phases and a double line to represent the porous bridge separating the two compartments of the cell.

**Analyze** Copper is the anode and platinum is the cathode in Figure 17.4(b). Two moles of electrons are produced in the anode half-reaction, but four moles of electrons are consumed in the cathode half-reaction at which $O_2$ is reduced at the Pt electrode. Therefore, we need to multiply the copper half-reaction by 2 before combining the two equations.

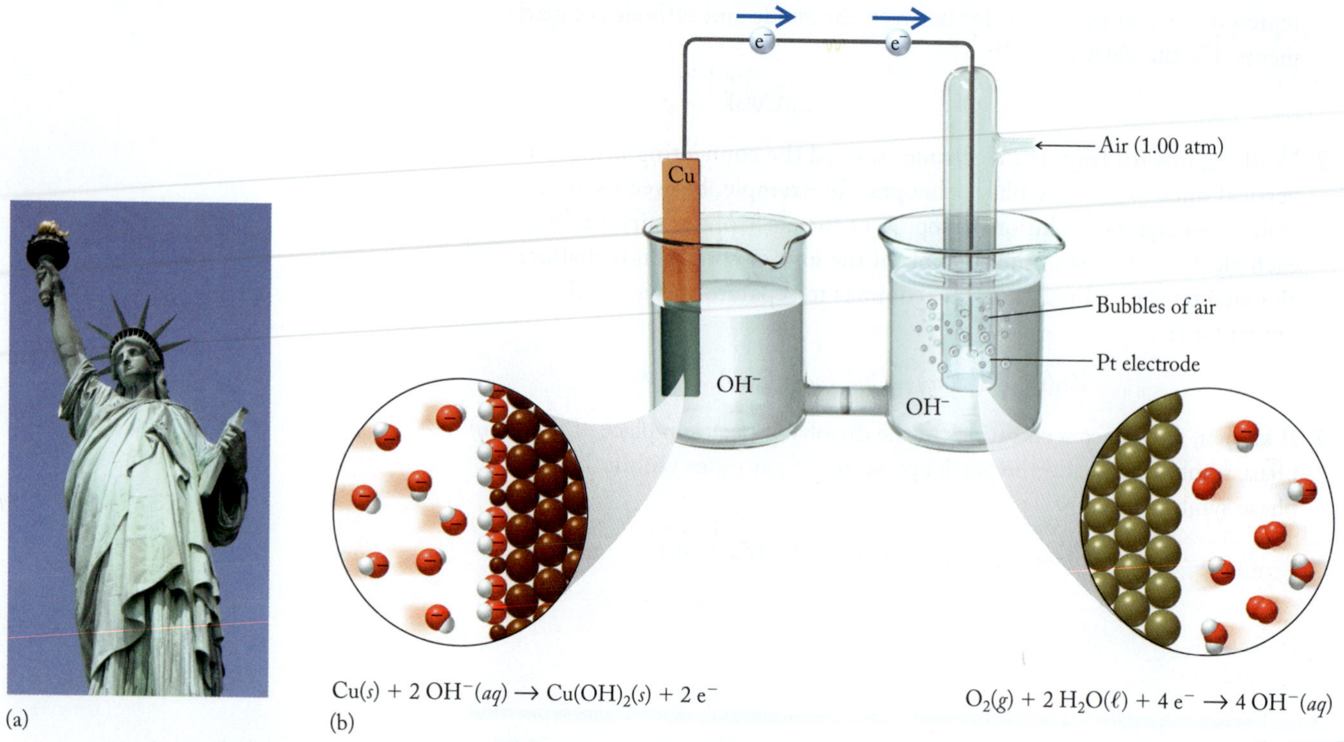


(a)

$Cu(s) + 2\ OH^-(aq) \rightarrow Cu(OH)_2(s) + 2\ e^-$
(b)

$O_2(g) + 2\ H_2O(\ell) + 4\ e^- \rightarrow 4\ OH^-(aq)$

**FIGURE 17.4** (a) The light blue-green patina of the Statue of Liberty is caused by the accumulation of copper(II) compounds on the surface of the copper sheets that make up its exterior. (b) A $Cu/O_2$ electrochemical cell. Solvent water molecules are omitted from the particulate views for clarity.

**Solve** Multiplying the $Cu^{2+}$ half-reaction by 2 and adding it to the $O_2$ half-reaction, we have

$$O_2(g) + 2\ H_2O(\ell) + 4\ e^- \rightarrow 4\ OH^-(aq)$$
$$2\ Cu(s) + 4\ OH^-(aq) \rightarrow 2\ Cu(OH)_2(s) + 4\ e^-$$
$$O_2(g) + 2\ H_2O(\ell) + 2\ Cu(s) + \cancel{4\ OH^-(aq)} + \cancel{4\ e^-} \rightarrow \cancel{4\ OH^-(aq)} + 2\ Cu(OH)_2(s) + \cancel{4\ e^-}$$

or

$$O_2(g) + 2\ H_2O(\ell) + 2\ Cu(s) \rightarrow 2\ Cu(OH)_2(s)$$

The equation is balanced, and we are finished with the first part of the exercise.

Applying the rules for writing a cell diagram (anode on the left, cathode on the right, bridge in the middle), we see that Cu is oxidized at the anode and $O_2$ is reduced at the Pt cathode:

$$Cu(s)\ .\ .\ .\ .\ .\ .\ .\ \|\ .\ .\ .\ .\ .\ .\ .\ Pt(s)$$

Adding electrode boundaries and the formulas of the substances produced and consumed in the cell reaction:

$$Cu(s)\ |\ Cu(OH)_2(s)\ \|\ O_2(g)\ |\ Pt(s)$$

Including the hydroxide ions in the solutions surrounding the electrodes and substituting known concentrations (1.00 *M*, 0.010 *M*) and partial pressure (1.00 atm × 0.210), we obtain a complete cell diagram:

$$Cu(s)\ |\ Cu(OH)_2(s)\ |\ OH^-(1.00\ M)\ \|\ O_2(g, 0.210\ atm)\ |\ OH^-(0.010\ M)\ |\ Pt(s)$$

**Think About It** The sequence of the phases in the cell diagram is linked to the progress of the reaction. Translating the cell diagram into words: Cu atoms at a copper anode are oxidized to copper(II) ions that form solid $Cu(OH)_2$ in an aqueous solution containing $OH^-$ ions. Meanwhile, oxygen gas is reduced to $OH^-$ ions at a platinum cathode.

**Practice Exercise** Write a balanced chemical equation and the cell diagram for an electrochemical cell that has a copper cathode immersed in a solution of $Cu^{2+}$ ions and an aluminum anode immersed in a solution of $Al^{3+}$ ions.

**CONCEPT TEST**

Why doesn't the mass of the platinum electrode change during the reaction in Figure 17.4(b)?

*(Answers to Concept Tests are in the back of the book.)*

## 17.3 Standard Potentials

**Table 17.1** lists selected half-reactions in order of their **standard reduction potentials ($E°$)**, expressed in volts (V), at 25°C. The table includes all the reduction potentials used in the examples and exercises in this chapter. A more complete list is found in Table A6.1 of Appendix 6. The superscript (°) has its usual thermodynamic meaning—all reactants and products are in their standard states; that is, the concentrations of all dissolved substances are 1 *M* and the partial pressures of all gases are 1 bar (≈1 atm).

The more positive the value of $E°$, the greater the probability that the reduction half-reaction will couple with an oxidation half-reaction to produce a spontaneous redox reaction. The most positive $E°$ value in Table 17.1 (and Appendix 6) is for the reduction of fluorine:

$$F_2(g) + 2\,e^- \rightarrow 2\,F^-(aq) \qquad E° = +2.866\ V$$

meaning that fluorine is the most easily reduced substance in Table 17.1. It also means that $F_2$ is the strongest oxidizing agent in the table. It can oxidize any of the substances on the product side of the half-reactions lower in the table. Likewise, $F^-$ is the weakest reducing agent.

**CONNECTION** Oxidizing and reducing agents were introduced in Section 8.7.

The half-reaction at the bottom of the table with the most negative $E°$ value

$$Li^+(aq) + e^- \rightarrow Li(s) \qquad E° = -3.05\ V$$

is least likely to proceed as written because $Li^+$ is the weakest oxidizing agent in the table. However, that also means that Li is the strongest reducing agent in this table, so the reverse reaction

$$Li(s) \rightarrow Li^+(aq) + e^-$$

occurs more readily than the reverse of any other half-reaction in Table 17.1 (and Appendix 6). Lithium metal is a powerful reducing agent that can reduce any of the substances on the reactant side of the half-reactions in Table 17.1. Substances with very negative $E°$ values at the bottom of the table include several of the major cations in biological systems and environmental waters, such as $Ca^{2+}$, $Na^+$, and $Mg^{2+}$. Their negative $E°$ values tell us that these ions are not easily reduced to their free metals in aqueous solutions. They are chemically very stable, which explains the presence of these cations in nature—in fact, none of these elements occurs in nature in the atomic state. Instead, they exist as the cations listed. Their position at the bottom of Table 17.1 is also consistent with their position *at the top* of the activity series of metals in Table 8.8. Recall that in Table 8.8, metals are ranked according to their strengths as reducing agents—that is, how readily they are oxidized.

**standard reduction potential ($E°$)** the potential of a reduction half-reaction in which all reactants and products are in their standard states at 25°C.

**TABLE 17.1** Selected Standard Reduction Potentials at 25°C

| | Half-Reaction | $n$ | $E°$ (V) | |
|---|---|---|---|---|
| Strongest oxidizing agent | $F_2(g) + 2\ e^- \rightarrow 2\ F^-(aq)$ | 2 | 2.866 | Weakest reducing agent |
| | $PbO_2(s) + HSO_4^-(aq) + 3\ H^+(aq) + 2\ e^- \rightarrow PbSO_4(s) + 2\ H_2O(\ell)$ | 2 | 1.685 | |
| | $MnO_4^-(aq) + 8\ H^+(aq) + 5\ e^- \rightarrow Mn^{2+}(aq) + 4\ H_2O(\ell)$ | 5 | 1.507 | |
| | $Cl_2(g) + 2\ e^- \rightarrow 2\ Cl^-(aq)$ | 2 | 1.358 | |
| | $O_2(g) + 4\ H^+(aq) + 4\ e^- \rightarrow 2\ H_2O(\ell)$ | 4 | 1.229 | |
| | $Hg^{2+}(aq) + 2\ e^- \rightarrow Hg(\ell)$ | 2 | 0.851 | |
| | $Ag^+(aq) + e^- \rightarrow Ag(s)$ | 1 | 0.800 | |
| | $Fe^{3+}(aq) + e^- \rightarrow Fe^{2+}(aq)$ | 1 | 0.770 | |
| | $NiO(OH)(s) + H_2O(\ell) + e^- \rightarrow Ni(OH)_2(s) + OH^-(aq)$ | 1 | 0.52 | |
| | $O_2(g) + 2\ H_2O(\ell) + 4\ e^- \rightarrow 4\ OH^-(aq)$ | 4 | 0.401 | |
| | $Ag_2O(s) + H_2O(\ell) + 2\ e^- \rightarrow 2\ Ag(s) + 2\ OH^-(aq)$ | 2 | 0.342 | |
| | $Cu^{2+}(aq) + 2\ e^- \rightarrow Cu(s)$ | 2 | 0.342 | |
| | $Sn^{4+}(aq) + 2\ e^- \rightarrow Sn^{2+}(aq)$ | 2 | 0.154 | |
| | $2\ MnO_2(s) + H_2O(\ell) + 2\ e^- \rightarrow Mn_2O_3(s) + 2\ OH^-(aq)$ | 2 | 0.15 | |
| | **$2\ H^+(aq) + 2\ e^- \rightarrow H_2(g)$** | **2** | **0.000** | |
| | $Cu(OH)_2(s) + 2\ e^- \rightarrow Cu(s) + 2\ OH^-(aq)$ | 2 | −0.222 | |
| | $Ni^{2+}(aq) + 2\ e^- \rightarrow Ni(s)$ | 2 | −0.257 | |
| | $PbSO_4(s) + H^+(aq) + 2\ e^- \rightarrow Pb(s) + HSO_4^-(aq)$ | 2 | −0.356 | |
| | $Cd^{2+}(aq) + 2\ e^- \rightarrow Cd(s)$ | 2 | −0.403 | |
| | $Zn^{2+}(aq) + 2\ e^- \rightarrow Zn(s)$ | 2 | −0.762 | |
| | $Cd(OH)_2(s) + 2\ e^- \rightarrow Cd(s) + 2\ OH^-(aq)$ | 2 | −0.81 | |
| | $2\ H_2O(\ell) + 2\ e^- \rightarrow H_2(g) + 2\ OH^-(aq)$ | 2 | −0.828 | |
| | $FeO(OH)(s) + H_2O(\ell) + 3\ e^- \rightarrow Fe(s) + 3\ OH^-(aq)$ | 3 | −0.87 | |
| | $Mn^{2+}(aq) + 2\ e^- \rightarrow Mn(s)$ | 2 | −1.185 | |
| | $ZnO(s) + H_2O(\ell) + 2\ e^- \rightarrow Zn(s) + 2\ OH^-(aq)$ | 2 | −1.249 | |
| | $Al(OH)_3(s) + 3\ e^- \rightarrow Al(s) + 3\ OH^-(aq)$ | 3 | −2.31 | |
| | $Mg^{2+}(aq) + 2\ e^- \rightarrow Mg(s)$ | 2 | −2.37 | |
| | $Mg(OH)_2(s) + 2\ e^- \rightarrow Mg(s) + 2\ OH^-(aq)$ | 2 | −2.690 | |
| | $Na^+(aq) + e^- \rightarrow Na(s)$ | 1 | −2.71 | |
| | $Ca^{2+}(aq) + 2\ e^- \rightarrow Ca(s)$ | 2 | −2.868 | |
| Weakest oxidizing agent | $Li^+(aq) + e^- \rightarrow Li(s)$ | 1 | −3.05 | Strongest reducing agent |

## CONCEPT TEST

Given the positions of the reactants and products in Table 17.1, predict which of the following reactions are spontaneous under standard conditions:

a. $Cu(s) + Cl_2(g) \rightarrow Cu^{2+}(aq) + 2\ Cl^-(aq)$

b. $2\ Ag^+(aq) + Zn(s) \rightarrow 2\ Ag(s) + Zn^{2+}(aq)$

c. $ZnO(s) + H_2(g) \rightarrow Zn(s) + H_2O(\ell)$

We can use the standard reduction potentials to calculate the **standard cell potentials, $E^\circ_{cell}$**, of electrochemical cells at 25°C. Standard cell potentials are measures of the **electromotive force (emf)** generated by cell reactions, which reflects how forcefully cells pump electrons from their anodes through external circuits and into their cathodes. By pumping electrons, these electrochemical cells convert the chemical energy of a spontaneous cell reaction into electrical work. They are called **voltaic cells** in honor of Italian physicist Alessandro Volta (1745–1827; **Figure 17.5**) who may have built the first operational battery. They can also be called **galvanic cells** after the physicist Luigi Galvani (1737–1798), a contemporary of Volta.

Consider the hypothetical case in which two half-reactions both have the same standard reduction potential. That is, no difference would exist between their relative ability to function as oxidizing agents, and no electrons would spontaneously flow. For two half-reactions with different standard reduction potentials, the larger the difference between those potentials, the larger the $E^\circ_{cell}$ of the electrochemical cell that could be built from them. Therefore, $E^\circ_{cell}$ can be determined by calculating the difference between the standard reduction potentials of a voltaic cell's cathode and anode:

$$E^\circ_{cell} = E^\circ_{cathode} - E^\circ_{anode} \tag{17.1}$$

As in our definition of $E^\circ$, the superscript (°) in Equation 17.1 indicates that all reactants and products are in their standard states. If we want to denote a generalized cell potential in which reactants and products are not necessarily in their standard states, we simply refer to the **cell potential ($E_{cell}$)**, without the superscript.

Let's use Equation 17.1 to calculate $E^\circ_{cell}$ for the $Zn/Cu^{2+}$ voltaic cell in Figure 17.2 at 25°C. The standard reduction potential of the cathode half-reaction is

$$Cu^{2+}(aq) + 2\,e^- \rightarrow Cu(s) \qquad E^\circ = 0.342\ V$$

To obtain the standard potential for the oxidation half-reaction at the zinc anode, we start with the standard reduction potential of $Zn^{2+}$ ions in Table 17.1:

$$Zn^{2+}(aq) + 2\,e^- \rightarrow Zn(s) \qquad E^\circ = -0.762\ V$$

Now we use Equation 17.1 to calculate $E^\circ_{cell}$:

$$\begin{aligned} E^\circ_{cell} &= E^\circ_{cathode} - E^\circ_{anode} \\ &= 0.342\ V - (-0.762\ V) = 1.104\ V \end{aligned}$$

This is the cell potential we measure if we connect a device called a voltmeter across the two electrodes, as shown in **Figure 17.6**. The overall positive value of $E^\circ_{cell}$ correlates with a spontaneous reaction, whereas a negative value for $E^\circ_{cell}$ indicates a nonspontaneous reaction. We will explore the connection between $E^\circ_{cell}$ and $\Delta G$ in the next section.

To use Equation 17.1, we need to know which half-reaction occurs at the cathode (reduction) and which occurs at the anode (oxidation). This decision is based on the data in Table 17.1. In the $Zn/Cu^{2+}$ cell, the value of $E^\circ$ for the reduction of $Cu^{2+}$ ions to Cu metal is 0.342 V, which is greater than the value of $E^\circ$ for reducing $Zn^{2+}$ ions to Zn metal (−0.762 V). Therefore, in the $Zn/Cu^{2+}$ voltaic cell, $Cu^{2+}$ ions are reduced and Zn metal is oxidized. We can generalize this observation to the cell reaction of any voltaic cell: the half-reaction with the more

**STEPWISE ANIMATION**
Electricity and Water Analogy

**FIGURE 17.5** Alessandro Volta is credited with building the first battery in 1798. It consisted of a stack of alternating layers of zinc, blotter paper soaked in salt water, and silver.

**CHEMTOUR**
Cell Potential

**standard cell potential ($E^\circ_{cell}$)** a measure of how forcefully an electrochemical cell, in which all reactants and products are in their standard states, can pump electrons through an external circuit.

**electromotive force (emf)** also called voltage, the force pushing electrons through an electrical circuit.

**voltaic cell** or **galvanic cell** an electrochemical cell in which chemical energy is transformed into electrical work by a spontaneous redox reaction.

**cell potential ($E_{cell}$)** the electromotive force with which an electrochemical cell can pump electrons through an external circuit.

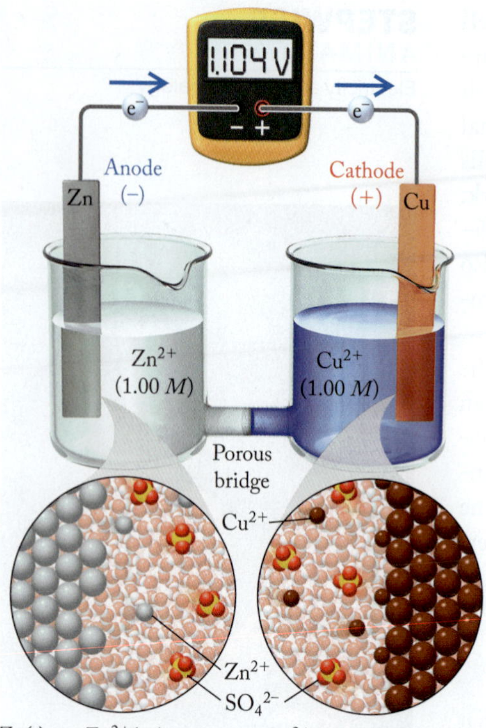


**FIGURE 17.6** A voltmeter displays a cell potential of 1.104 V between a Zn electrode immersed in a 1.00 *M* solution of $Zn^{2+}$ ions and a Cu electrode immersed in a 1.00 *M* solution of $Cu^{2+}$ ions.

positive value of $E°$ runs as a reduction, and the other one runs in reverse as an oxidation.

Let's now examine what happens when an electrochemical cell is built using two half-reactions in which different numbers of electrons are gained and lost. This occurs in a type of battery that has a virtually limitless supply of one of its reactants. It is called the zinc–air battery (**Figure 17.7**), and it powers devices in which small battery size and mass are high priorities, such as hearing aids. Most of the internal volume of one of these batteries is occupied by an anode consisting of a paste of zinc particles packed in an aqueous solution of KOH. As in alkaline batteries (Sample Exercise 17.3), the anode half-reaction is

$$Zn(s) + 2\,OH^-(aq) \rightarrow ZnO(s) + H_2O(\ell) + 2\,e^-$$

which is the reverse of the standard reduction reaction:

$$ZnO(s) + H_2O(\ell) + 2\,e^- \rightarrow Zn(s) + 2\,OH^-(aq) \qquad E° = -1.249\text{ V}$$

The cathode consists of porous carbon supported by a metal screen. Air diffuses through small holes in the battery and across a layer of plastic film that lets gases pass through but keeps electrolyte from leaking out. As air passes through the cathode, oxygen is reduced to hydroxide ions:

$$O_2(g) + 2\,H_2O(\ell) + 4\,e^- \rightarrow 4\,OH^-(aq) \qquad E° = 0.401\text{ V}$$

To write the net ionic equation for the overall cell reaction, we need to multiply the oxidation (anode) half-reaction by 2 before combining it with the reduction (cathode) half-reaction:

$$2[Zn(s) + 2\,OH^-(aq) \rightarrow ZnO(s) + H_2O(\ell) + 2\,e^-]$$
$$O_2(g) + 2\,H_2O(\ell) + 4\,e^- \rightarrow 4\,OH^-(aq)$$
$$2\,Zn(s) + \cancel{4\,OH^-(aq)} + O_2(g) + \cancel{2\,H_2O(\ell)} + \cancel{4\,e^-} \rightarrow 2\,ZnO(s) + \cancel{2\,H_2O(\ell)} + \cancel{4\,OH^-(aq)} + \cancel{4\,e^-}$$

or

$$2\,Zn(s) + O_2(g) \rightarrow 2\,ZnO(s)$$
$$E°_{cell} = E°_{cathode} - E°_{anode} = 0.401\text{ V} - (-1.249\text{ V}) = 1.650\text{ V}$$

Note that when we multiplied the anode half-reaction by 2 and added it to the cathode half-reaction, we did not multiply the $E°$ of the anode half-reaction by 2. $E°$ is an *intensive* property of a half-reaction or a complete cell reaction, so it does not change when the quantities of reactants and products change. Thus, a zinc–air battery the size of a pea has the same $E°_{cell}$ as one the size of a book (such as those being developed for electric vehicles). However, the amount of

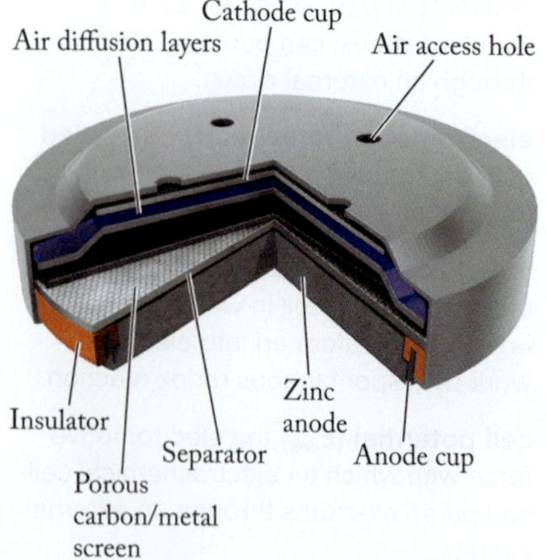


**FIGURE 17.7** Most of the internal volume of a zinc–air battery is occupied by the anode: a paste of Zn particles in an aqueous solution of KOH surrounded by a metal cup that serves as the negative terminal of the battery. Oxygen from the air is the reactant at the cathode. Air enters through holes in an inverted metal cup that serves as the positive terminal of the battery. Once inside the battery, air diffuses through layers of gas-permeable plastic film that let air in but keep electrolyte from leaking out. Oxygen in the air is reduced at the porous carbon/metal cathode to $OH^-$ ions that migrate toward the anode, where they are consumed in the Zn oxidation half-reaction.

electrical work that a zinc–air battery can do does depend on how much zinc is inside it because, as we are about to see, the electrical work that a voltaic cell can do depends on both cell potential and the quantity of charge it can deliver at that potential.

## CONCEPT TEST

The balanced chemical equation for an aluminum–air battery is written as

$$4\ Al(s) + 3\ O_2(g) \rightarrow 2\ Al_2O_3(s)$$

The reaction between aluminum metal and oxygen gas that yields $Al_2O_3$ represents the formation reaction for $Al_2O_3$ (see Section 10.7) when written as

$$2\ Al(s) + \tfrac{3}{2}\ O_2(g) \rightarrow Al_2O_3(s)$$

Do these two reactions have the same $E^\circ_{cell}$ value?

**CHEMTOUR**
Alkaline Battery

### SAMPLE EXERCISE 17.3 Identifying Anode and Cathode Half-Reactions and Calculating the Value of $E^\circ_{cell}$ — LO2

The standard reduction potentials of the half-reactions in single-use alkaline batteries (**Figure 17.8**) are

$$ZnO(s) + H_2O(\ell) + 2\ e^- \rightarrow Zn(s) + 2\ OH^-(aq) \qquad E^\circ = -1.249\ V$$

$$2\ MnO_2(s) + H_2O(\ell) + 2\ e^- \rightarrow Mn_2O_3(s) + 2\ OH^-(aq) \qquad E^\circ = 0.15\ V$$

What is the net ionic equation for the cell reaction and the value of $E^\circ_{cell}$ at 25°C ?

**Collect, Organize, and Analyze** To calculate $E^\circ_{cell}$ by using Equation 17.1, we first need to decide which half-reaction occurs at the cathode and which one occurs at the anode. The $MnO_2$ half-reaction has the more positive $E^\circ$, making it our reduction half-reaction. We must reverse the ZnO half-reaction, turning it into an oxidation half-reaction. Both half-reactions involve the transfer of two electrons, so they may be combined by simply adding them together.

**Solve** The oxidation half-reaction at the anode is

$$Zn(s) + 2\ OH^-(aq) \rightarrow ZnO(s) + H_2O(\ell) + 2\ e^-$$

and the reduction half-reaction at the cathode is

$$2\ MnO_2(s) + H_2O(\ell) + 2\ e^- \rightarrow Mn_2O_3(s) + 2\ OH^-(aq)$$

Combining these half-reactions to obtain the overall cell reaction, we get

$$2\ MnO_2(s) + \cancel{H_2O(\ell)} + Zn(s) + \cancel{2\ OH^-(aq)} + \cancel{2\ e^-} \rightarrow Mn_2O_3(s) + \cancel{2\ OH^-(aq)} + ZnO(s) + \cancel{H_2O(\ell)} + \cancel{2\ e^-}$$

Simplifying gives us the net ionic equation for the cell reaction:

$$2\ MnO_2(s) + Zn(s) \rightarrow Mn_2O_3(s) + ZnO(s)$$

The overall $E^\circ_{cell}$ for this reaction is obtained using Equation 17.1:

$$E^\circ_{cell} = E^\circ_{cathode} - E^\circ_{anode} = 0.15\ V - (-1.249\ V) = 1.40\ V$$

**Think About It** The $E^\circ_{cell}$ value is reasonable because the potential of most alkaline batteries is nominally 1.5 V. In this cell reaction, the net ionic equation is also the complete molecular equation.

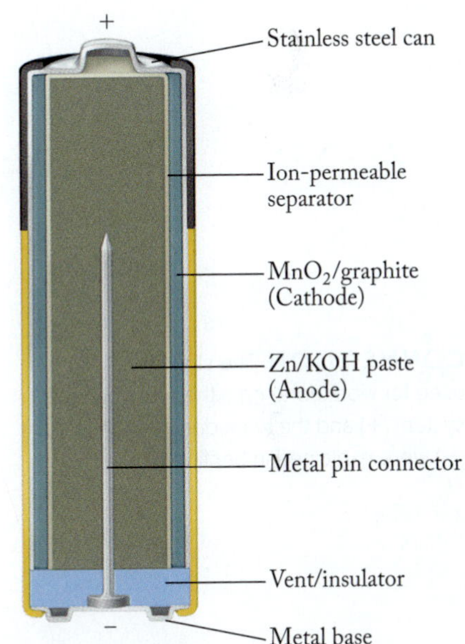


**FIGURE 17.8** This cross-section view shows the principal components of a single-use alkaline battery. A paste of Zn particles and KOH electrolyte is connected to the metal base and serves as the anode. The cathode consists of $MnO_2$ particles mixed with graphite for better conductivity to the stainless steel can and top terminal. A separator between the electrodes allows $OH^-$ ions produced in the cathode half-reaction to migrate to the anode, where they combine with the $Zn^{2+}$ ions produced by the anode half-reaction, forming ZnO and $H_2O$.

**Practice Exercise** The half-reactions in NiCad (nickel–cadmium) batteries (see Table 17.1) are

$$Cd(OH)_2(s) + 2\ e^- \rightarrow Cd(s) + 2\ OH^-(aq) \qquad E° = -0.81\ V$$

$$NiO(OH)(s) + H_2O(\ell) + e^- \rightarrow Ni(OH)_2(s) + OH^-(aq) \qquad E° = 0.52\ V$$

Write the net ionic equation for the cell reaction and calculate the value of $E°_{cell}$ at 25°C.

## 17.4 Chemical Energy and Electrical Work

When the Zn and Cu electrodes in Figure 17.6 are connected to a digital voltmeter—the Zn electrode to the negative terminal of the meter and the Cu electrode to the positive terminal—the meter reads 1.104 V at 25°C . These connections tell us that the battery is pumping electrons from the Zn electrode through the external circuit to the Cu electrode, and the reading tells us how much electromotive force (emf) is pushing the electrons through the circuit. Under standard conditions, this emf is the same as the standard cell potential of the cell ($E°_{cell}$); under any other conditions, it is simply $E_{cell}$.

When a voltaic cell pumps electrons through an external circuit, those moving electrons do electrical work, such as lighting a lightbulb or turning an electric motor. The change in chemical free energy ($\Delta G_{cell}$) that accompanies a spontaneous cell reaction is a measure of the electrical work ($w_{elec}$) that may result:

$$\Delta G_{cell} = w_{elec} \tag{17.2}$$

The sign of $\Delta G_{cell}$ is negative because the reaction in a voltaic cell *must be spontaneous*, which means that the free energy of the system (cell reaction mixture) decreases as reactants become products. The sign of $w_{elec}$ also is negative because it represents work done by the system on its surroundings, which has a negative value according to the sign convention summarized in Table 10.1 and Figure 10.4.

**CONNECTION** The sign conventions used for work done *on* a thermodynamic system (+) and the work done *by* the system (−) were explained in Section 10.1.

The work done by a voltaic cell on its surroundings is the product of the quantity of electric charge ($Q$) that the cell pushes through an external circuit times the force (emf) pushing that charge. That force is the same as the cell potential, so the connection between $E_{cell}$ and $w_{elec}$ is

$$w_{elec} = -QE_{cell} \tag{17.3}$$

where the negative sign reflects the fact that work done by a voltaic cell on its surroundings (the external circuit) corresponds to energy lost by the cell.

The quantity of charge is proportional to the number of electrons flowing through the circuit. As noted in Chapter 2, the magnitude of the charge on a single electron is $1.602 \times 10^{-19}$ coulomb (C). The magnitude of electric charge on one mole of electrons is

$$\frac{1.602 \times 10^{-19}\ C}{\cancel{e^-}} \times \frac{6.022 \times 10^{23}\ \cancel{e^-}}{1\ mol\ e^-} = \frac{9.65 \times 10^4\ C}{mol\ e^-}$$

This quantity of charge, $9.65 \times 10^4$ C/mol, is called the **Faraday constant (*F*)**, after Michael Faraday (1791–1867), the English chemist and physicist who discovered that redox reactions take place when electrons are transferred from one species to another. The quantity of charge $Q$ flowing through an electrical circuit is the product of the number of moles of electrons, $n$, times the Faraday constant:

**Faraday constant ($F$)** the magnitude of electric charge in one mole of electrons; its value to three significant figures is $9.65 \times 10^4$ C/mol.

$$Q = nF \tag{17.4}$$

Combining Equations 17.3 and 17.4 gives us an equation relating $w_{\text{elec}}$ and $E_{\text{cell}}$:

$$w_{\text{elec}} = -nFE_{\text{cell}} \tag{17.5}$$

Equations 17.2 and 17.5 relate the quantity of electrical work that a voltaic cell does on its surroundings to the change in free energy in the cell, which under standard conditions is

$$\Delta G^\circ_{\text{cell}} = -nFE^\circ_{\text{cell}} \tag{17.6}$$

The product on the right side of Equation 17.6 is the equivalent of energy because the units are

$$\cancel{\text{moles}} \times \frac{\text{coulombs}}{\cancel{\text{mole}}} \times \text{volts} = \text{coulombs-volts}$$

A coulomb-volt is the same quantity of energy as a joule:

$$1 \text{ coulomb-volt} = 1 \text{ joule}$$

$$1 \text{ C} \cdot \text{V} = 1 \text{ J}$$

The chemical reaction inside the cell will be spontaneous if the sign of $\Delta G_{\text{cell}}$ is negative, so the negative sign on the right side of Equation 17.6 indicates that $E_{\text{cell}}$ of any voltaic cell must have a positive value. If $E_{\text{cell}}$ is negative, then $\Delta G_{\text{cell}}$ will be positive and the cell reaction will be spontaneous in the opposite direction.

Let's use Equation 17.6 to calculate the change in standard free energy of the $Zn/Cu^{2+}$ cell reaction. We begin with the standard cell potential calculated in Section 17.3:

$$E^\circ_{\text{cell (Zn/Cu}^{2+})} = 1.104 \text{ V}$$

We can convert this standard cell potential into a change in standard free energy ($\Delta G^\circ_{\text{cell}}$ by using Equation 17.6 under standard conditions so that $\Delta G = \Delta G^\circ$ and $E = E^\circ$. We need to use $n = 2$ because two moles of electrons are transferred in the half-reactions for this cell:

$$\begin{aligned}\Delta G^\circ_{\text{cell}} &= -nFE^\circ_{\text{cell}} \\ &= -\left(2 \cancel{\text{ mol}} \times \frac{9.65 \times 10^4 \text{ C}}{\cancel{\text{mol}}} \times 1.104 \text{ V}\right) \\ &= -2.13 \times 10^5 \text{ C} \cdot \text{V} = -2.13 \times 10^5 \text{ J} = -213 \text{ kJ}\end{aligned}$$

To put this value in perspective, the $Zn/Cu^{2+}$ reaction produces, on a mole-for-mole basis, nearly as much useful energy as the combustion of hydrogen gas:

$$H_2(g) + \tfrac{1}{2}O_2(g) \rightarrow H_2O(g) \qquad \Delta G^\circ = -228.6 \text{ kJ}$$

**CONCEPT TEST**

When a rechargeable battery, such as the one used to start a car's engine, is recharged, an external source of electrical power forces the voltaic cell reaction to run in reverse. What are the signs of $\Delta G_{\text{cell}}$ and $E_{\text{cell}}$ during the recharging process?

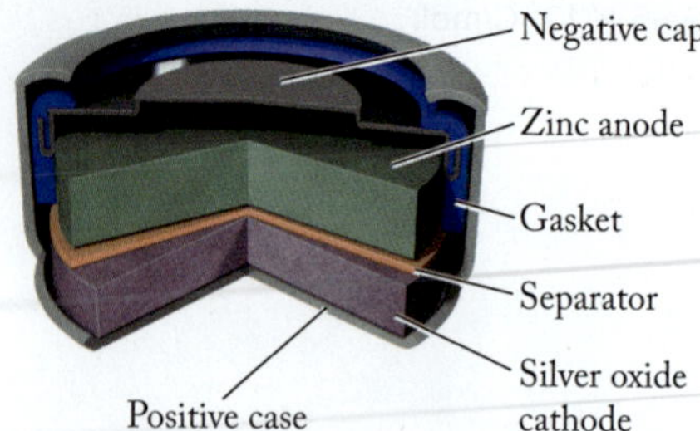


**FIGURE 17.9** Many of the button batteries that power small electronic devices incorporate a Zn anode and a $Ag_2O$ cathode separated by a membrane containing KOH electrolyte.

### SAMPLE EXERCISE 17.4 Relating $\Delta G^\circ_{cell}$ and $E^\circ_{cell}$ LO3

Many of the "button" batteries used in electric watches consist of a Zn anode and a $Ag_2O$ cathode separated by a membrane soaked in a concentrated solution of KOH (**Figure 17.9**). $Ag_2O$ is reduced to Ag metal at the cathode, and Zn is oxidized to solid ZnO at the anode. Write the net ionic equation for the reaction and, using the appropriate standard reduction potentials from Table 17.1, calculate the values of $E^\circ_{cell}$ and $\Delta G^\circ_{cell}$ at 25°C.

**Collect and Organize** We know the reactants and products of the anode and cathode reactions and that the reactions occur in a basic solution. Equations 17.1 and 17.6 should be useful in calculating $E^\circ_{cell}$ and $\Delta G^\circ_{cell}$ from the appropriate standard potentials.

**Analyze** The half-reaction at the cathode is based on the reduction of $Ag_2O$ to Ag. The appropriate half-reaction in Table 17.1 is

$$(1)\quad Ag_2O(s) + H_2O(\ell) + 2\,e^- \rightarrow 2\,Ag(s) + 2\,OH^-(aq) \qquad E^\circ_{cathode} = 0.342\ V$$

We must reverse the anode's oxidation half-reaction to find an entry in Table 17.1 in which ZnO is the reactant and Zn is the product:

$$(2)\quad ZnO + H_2O(\ell) + 2\,e^- \rightarrow Zn(s) + 2\,OH^-(aq) \qquad E^\circ_{anode} = -1.249\ V$$

However, to write the net ionic equation, we need to use the actual oxidation half-reaction at the anode—the reverse of this half-reaction—and combine it with the cathode's reduction half-reaction. The two half-reactions involve transferring the same number of electrons, so combining Equation (1) and the reverse of Equation (2) simply means adding them together. The value of $E^\circ_{cell}$ will be about [0.34 V − (−1.25 V)], or 1.6 V. This value is half again as large as the $E^\circ_{cell}$ of the $Zn/Cu^{2+}$ cell. Therefore, the magnitude of its $\Delta G^\circ_{cell}$ value should be 50% larger than −213 kJ/mol, or about −300 kJ/mol.

**Solve** Reversing the ZnO half-reaction and adding it to the $Ag_2O$ half-reaction, we get

$$Zn(s) + 2\,OH^-(aq) \rightarrow ZnO(s) + H_2O(\ell) + 2\,e^-$$
$$Ag_2O(s) + H_2O(\ell) + 2\,e^- \rightarrow 2\,Ag(s) + 2\,OH^-(aq)$$

$$Ag_2O(s) + \cancel{H_2O(\ell)} + Zn(s) + \cancel{2\,OH^-(aq)} + \cancel{2e^-} \rightarrow 2\,Ag(s) + \cancel{2\,OH^-(aq)} + ZnO(s) + \cancel{H_2O(\ell)} + \cancel{2\,e^-}$$

or

$$Ag_2O(s) + Zn(s) \rightarrow 2\,Ag(s) + ZnO(s)$$

Calculating $E^\circ_{cell}$ by using Equation 17.1:

$$E^\circ_{cell} = E^\circ_{cathode} - E^\circ_{anode} = 0.342\ V - (-1.249\ V) = 1.591\ V$$

Calculating $\Delta G^\circ_{cell}$ by using Equation 17.6:

$$\Delta G^\circ_{cell} = -nFE^\circ_{cell}$$
$$= -\left(2\ \cancel{mol} \times \frac{9.65 \times 10^4\ C}{\cancel{mol}} \times 1.591\ V\right)$$
$$= -3.07 \times 10^5\ C \cdot V = -3.07 \times 10^5\ J = -307\ kJ$$

**Think About It** The positive value of $E^\circ_{cell}$ and the negative value of $\Delta G^\circ_{cell}$ are expected because voltaic cell reactions are spontaneous. The calculated values are close to those we estimated.

**Practice Exercise** The alkaline batteries used in LED flashlights (Sample Exercise 17.3) produce a cell potential of 1.50 V. What is the value of $\Delta G_{cell}$?

Some final thoughts about the $\Delta G^\circ_{cell}$ value calculated in Sample Exercise 17.4 are in order. First, the value is based on the reaction of one mole of $Ag_2O$ and one mole of Zn, which corresponds to 232 g of $Ag_2O$ and 65 g of Zn. The energy stored in a button battery (Figure 17.9), which has a mass of only 1 g or 2 g, would be a tiny fraction of our calculated value. Also, note that no ions appear in the net ionic equation. Because all the reactants and products in the silver oxide battery reaction are solids, its net ionic equation and its molecular equation are identical.

**standard hydrogen electrode (SHE)** a reference electrode based on the half-reaction $2\ H^+(aq) + 2\ e^- \rightarrow H_2(g)$, which produces a standard electrode potential defined to be 0.000 V.

## 17.5 A Reference Point: The Standard Hydrogen Electrode

**STEPWISE ANIMATION**
Standard Hydrogen Electrode (SHE)

We can measure the value of $E_{cell}$ by using a voltmeter, but how do we measure the individual electrode potentials of the cathode and anode? The answer is that we arbitrarily assign a value of zero volts to the standard potential for the reduction of hydrogen ions to hydrogen gas:

$$2\ H^+(aq) + 2\ e^- \rightarrow H_2(g) \qquad E^\circ = 0.000\ V \qquad (17.7)$$

An electrode that generates this reference potential, called the **standard hydrogen electrode (SHE)**, consists of a platinum electrode in contact with a solution of a strong acid ($[H^+] = 1.00\ M$) and hydrogen gas at a pressure of 1.00 bar (**Figure 17.10**). The platinum is unchanged—neither oxidized nor reduced—by the electrode reaction. Rather, it serves as a chemically inert conveyor of electrons. Electrons may be consumed at the electrode surface if $H^+$ ions are reduced to hydrogen gas, or they may flow away from the surface into the Pt electrode if hydrogen gas is oxidized to $H^+$ ions. The potential of the SHE, which is the same for both half-reactions, is defined to be 0.000 V.

To write the cell diagram for a cell in which the SHE serves as the anode, we represent the SHE half of the cell as follows:

$$Pt(s) \mid H_2(g, 1.00\ bar) \mid H^+(1.00\ M) \parallel$$

This diagram indicates that the anode half-reaction involves the oxidation of $H_2$ gas to $H^+$ ions. If the SHE is the cathode, then its half of the cell is diagrammed as follows:

$$\parallel H^+(1.00\ M) \mid H_2(g, 1.00\ bar) \mid Pt(s)$$

This diagram indicates that the cathode half-reaction involves the reduction of $H^+$ ions to $H_2$ gas.

The standard reduction potential of the SHE is 0.000 V, so the measured $E_{cell}$ of any voltaic cell in which a SHE is one of the two electrodes—either cathode or anode—is the potential produced by the other electrode. This means that if we attach a voltmeter to the cell, the meter reading is the electrode potential of the other electrode. Suppose that a voltaic cell consists of a strip of zinc metal immersed in a 1.00 $M$ solution of $Zn^{2+}$ ions in one compartment and a SHE in the other (**Figure 17.11a**). Also suppose that a voltmeter is connected to the cell so that it measures the potential at which the cell pumps electrons from the zinc electrode to the SHE. This direction of electron flow means that the zinc electrode is the cell's anode and the SHE is the cathode of the cell. At 25°C, the meter reads 0.762 V. We know that the value of $E^\circ_{cathode}$

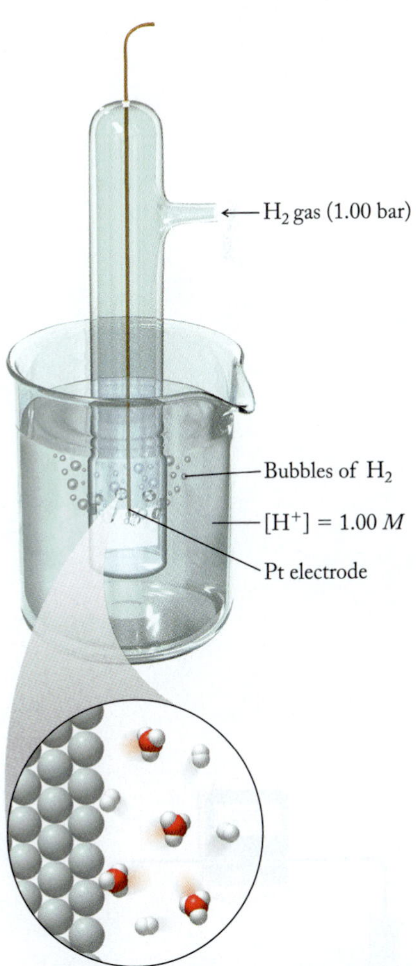


**FIGURE 17.10** The standard hydrogen electrode consists of a platinum electrode immersed in a 1.00 $M$ solution of $H^+(aq)$ and bathed in a stream of pure $H_2$ gas at a pressure of 1.00 bar. Its potential is the same (0.000 V) whether $H^+(aq)$ ions are reduced or $H_2$ gas is oxidized. Solvent water molecules omitted for clarity.

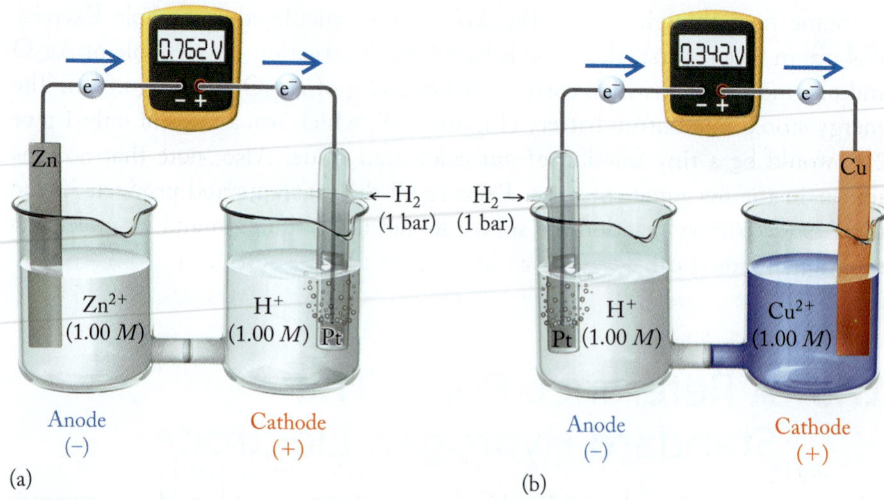


**FIGURE 17.11** The standard hydrogen electrode (SHE) allows us to determine the standard potential of any half-reaction. (a) When coupled to a Zn electrode under standard conditions, the SHE is the cathode and the Zn electrode is the anode. When the SHE is connected to the positive terminal of a voltmeter and the Zn electrode to the negative terminal, the meter measures a cell potential of 0.762 V. (b) When coupled to a Cu electrode under standard conditions, the SHE is the anode, the Cu electrode is the cathode, and the meter measures a cell potential of 0.342 V.

is that of the SHE (0.000 V) and that $E^\circ_{\text{anode}}$ is $E^\circ_{\text{Zn}}$. Inserting these values and symbols into Equation 17.1:

$$\begin{aligned} E^\circ_{\text{cell}} &= E^\circ_{\text{cathode}} - E^\circ_{\text{anode}} \\ E^\circ_{\text{cell}} &= E^\circ_{\text{SHE}} - E^\circ_{\text{Zn}} \\ 0.762\text{ V} &= 0.000\text{ V} - E^\circ_{\text{Zn}} \\ E^\circ_{\text{Zn}} &= -0.762\text{ V} \end{aligned}$$

This value is equal to the standard reduction potential of $Zn^{2+}$ in Table 17.1:

$$Zn^{2+}(aq) + 2\,e^- \rightarrow Zn(s) \qquad E^\circ = -0.762\text{ V}$$

In **Figure 17.11(b)**, the SHE is coupled to a copper electrode immersed in a 1.00 *M* solution of $Cu^{2+}$ ions. In this cell, electrons flow from the SHE through an external circuit to the copper electrode at a cell potential of 0.342 V at 25°C. The direction of current flow means that the electrons are consumed at the copper electrode, making it the cathode. The value of $E^\circ$ for the copper half-reaction is calculated as follows:

$$\begin{aligned} E^\circ_{\text{cell}} &= E^\circ_{\text{cathode}} - E^\circ_{\text{anode}} \\ &= E^\circ_{\text{Cu}} - E^\circ_{\text{SHE}} \\ 0.342\text{ V} &= E^\circ_{\text{Cu}} - 0.000\text{ V} \\ E^\circ_{\text{Cu}} &= 0.342\text{ V} \end{aligned}$$

This half-reaction potential matches the value of $E^\circ$ for the reduction of $Cu^{2+}$ to Cu metal.

**FIGURE 17.12** A Ni/$Ni^{2+}$ half-cell connected by a voltmeter to a standard hydrogen electrode.

## CONCEPT TEST

A cell consists of a SHE in one compartment and a Ni electrode immersed in a 1.00 *M* solution of $Ni^{2+}$ ions in the other. If a voltmeter is connected to the electrode, as shown in **Figure 17.12**, what will be the value in the voltmeter's display?

# 17.6 The Effect of Concentration on $E_{cell}$

Reactions stop when one of the reactants is completely consumed. This concept was the basis for our discussion of limiting reactants in Chapter 7. However, a commercial battery usually stops operating at its rated cell potential—1.5 V for a flashlight battery—before its reactants are completely consumed. This happens because the cell potential of a voltaic cell is determined by the concentrations of the reactants and products.

## The Nernst Equation

In 1889, German chemist Walther Nernst (1864–1941) derived an expression, now called the **Nernst equation**, which describes the dependence of cell potentials on reactant and product concentrations. We can reconstruct his derivation starting with Equation 14.19, which relates the change in free energy, $\Delta G$, of any reaction to its change in free energy under standard conditions, $\Delta G°$:

$$\Delta G = \Delta G° + RT \ln Q \quad (14.19)$$

Note that $Q$ in this equation is the *reaction quotient*, the mass action expression for a reaction that is not necessarily at equilibrium. (It is not related to the $Q$ we use to represent electric charge.) Recall from Section 14.9 that increasing concentrations of products and decreasing concentrations of reactants as a reversible spontaneous reaction proceeds cause the value of $Q$ to increase until the positive value of $RT \ln Q$ offsets the negative value of $\Delta G°$. At that point, $\Delta G = 0$ and the reaction has reached chemical equilibrium.

**CONNECTION** In Chapter 14, we discussed the relationship between free-energy changes and the reaction quotient, Q.

We can write an expression analogous to Equation 14.19 that relates $E_{cell}$ to $E°_{cell}$ by substituting $-nFE_{cell}$ for $\Delta G_{cell}$ and $-nFE°_{cell}$ for $\Delta G°_{cell}$:

$$-nFE_{cell} = -nFE°_{cell} + RT \ln Q$$

Dividing all terms by $-nF$ gives

$$E_{cell} = E°_{cell} - \frac{RT \ln Q}{nF} \quad (17.8)$$

This is the equation Walther Nernst developed in 1889. We can obtain a useful form of Equation 17.8 if we insert values for $R$ [8.314 J/(mol · K)] and $F$ ($9.65 \times 10^4$ C/mol), make $T = 298$ K, and convert the natural logarithm to a base-10 logarithm: $\ln Q = 2.303 \log Q$. With these changes, the Nernst equation becomes

$$E_{cell} = E°_{cell} - \frac{0.0592 \text{ V}}{n} \log Q \quad (17.9)$$

Equation 17.9 enables us to predict how the potential of a voltaic cell changes as the cell reaction proceeds and the concentrations of products inside the cell increase and the concentrations of reactants decrease. As they do, $Q$ increases, so the value of $0.0592/n \times \log Q$ increases, too. The negative sign in front of this term in Equation 17.9 means that the value of $E_{cell}$ decreases as reactants are converted into products. Eventually, $E_{cell}$ approaches zero; the cell reaction has then achieved chemical equilibrium, and the cell can no longer pump electrons through an external circuit. In other words, the cell is dead.

**Nernst equation** an equation relating the potential of a cell (or half-cell) reaction to its standard potential ($E°$) and to the concentrations of its reactants and products.

Let's apply the Nernst equation to the zinc/copper cell we first saw in Figure 17.2. We know that electrons flow from the Zn anode to the Cu cathode

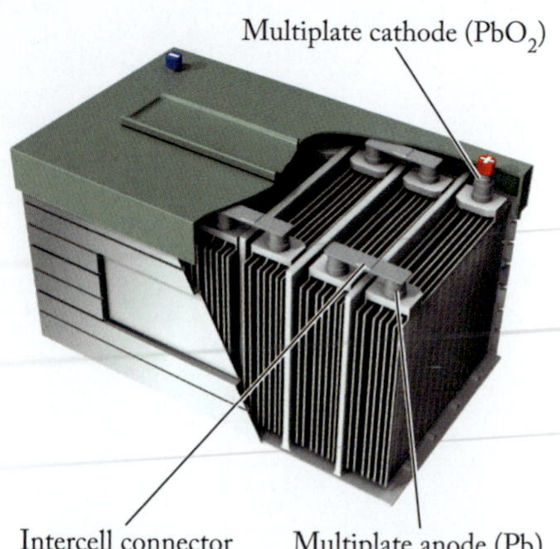


**FIGURE 17.13** The lead–acid battery that supplies power to start most motor vehicles contains six cells. Each has an anode made of lead and a cathode made of $PbO_2$ immersed in a background electrolyte of 4.5 $M$ $H_2SO_4$. The electrodes are formed into plates and held in place by grids made of a lead alloy. The grids connect the cells in series so that the operating potential of the battery (12.0 V) is the sum of the six cell potentials (each 2.0 V).

as Zn atoms are oxidized to $Zn^{2+}$ ions and $Cu^{2+}$ ions are reduced to Cu atoms. Therefore, the initial concentration of 1.00 $M$ $Zn^{2+}$ ions increases as the same initial concentration of $Cu^{2+}$ ions decreases, which means the values of $Q$

$$Q = \frac{[\text{products}]}{[\text{reactants}]} = \frac{[Zn^{2+}(aq)]}{[Cu^{2+}(aq)]}$$

and log $Q$ increase (the $Q$ expression has no terms for solid copper and zinc). Subtracting an increasing value of $0.0592/n \times \log Q$ from a positive value of $E^\circ_{cell}$ produces a decreasing value of $E_{cell}$. Eventually, nearly all the $Cu^{2+}$ ions are consumed and an equivalent number of $Zn^{2+}$ ions are produced, which drives $[Cu^{2+}]$ toward 0.00 $M$ and $[Zn^{2+}]$ toward 2.00 $M$. When the value of $0.0592/n \times \log Q$ reaches $E^\circ_{cell}$ (1.104 V), the value of $E_{cell}$ reaches 0.000 V and no further changes in $[Cu^{2+}]$ and $[Zn^{2+}]$ occur.

The lead–acid batteries used to start most car engines (**Figure 17.13**) are voltaic cells, so let's calculate how much their cell potentials drop with use. Typical commercial batteries each contain six electrochemical cells. Their anodes are made of lead (Pb) and their cathodes are made of lead(IV) oxide ($PbO_2$). Both electrodes are immersed in 4.5 $M$ $H_2SO_4$. The value of a fully charged cell is about 2.0 V. The six cells are connected in series so that the operating potential of the battery is the sum of the six cell potentials, or 12.0 V.

As the battery discharges, $PbO_2(s)$ is reduced to $PbSO_4(s)$ at the cathodes,

$$PbO_2(s) + 3\,H^+(aq) + HSO_4^-(aq) + 2\,e^- \rightarrow PbSO_4(s) + 2\,H_2O(\ell)$$

and Pb($s$) is oxidized to $PbSO_4(s)$ at the anodes,

$$Pb(s) + HSO_4^-(aq) \rightarrow PbSO_4(s) + H^+(aq) + 2\,e^-$$

The reduction half-reaction consumes two moles of electrons, and the oxidation half-reaction involves the loss of two moles of electrons for each mole of Pb.

The net ionic equation for the overall cell reaction is the sum of the two half-reactions:

$$PbO_2(s) + Pb(s) + 2\,H^+(aq) + 2\,HSO_4^-(aq) \rightarrow 2\,PbSO_4(s) + 2\,H_2O(\ell)$$

**STEPWISE ANIMATION**

Lead–Acid Battery

Using the values of $E^\circ_{cathode}$ and $E^\circ_{anode}$ from Table 17.1, we see that the value of $E^\circ_{cell}$ is

$$E^\circ_{cell} = E^\circ_{cathode} - E^\circ_{anode} = 1.685\text{ V} - (-0.356\text{ V}) = 2.041\text{ V}$$

The $Q$ expression contains only terms for $H^+$ and $HSO_4^-$ because the other reactants and the products are either pure solids or $H_2O$. As the battery discharges, $[H^+]$ and $[HSO_4^-]$ decrease, and so does the value of $E_{cell}$ calculated from the Nernst equation:

$$E_{cell} = 2.041\text{ V} - \frac{0.0592\text{ V}}{2}\log\frac{1}{[H^+]^2[HSO_4^-]^2}$$

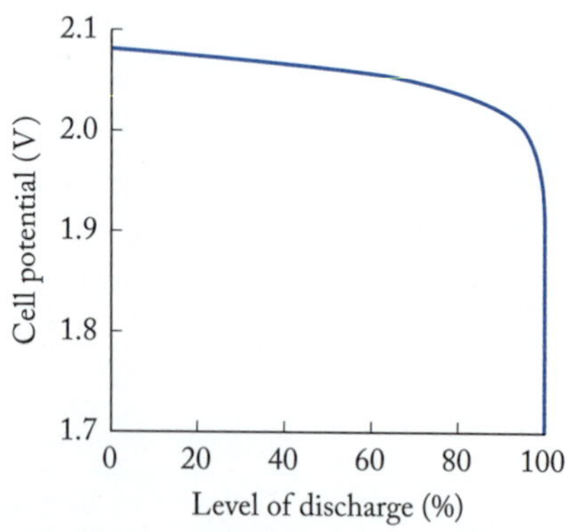


**FIGURE 17.14** The potential of a cell in a lead–acid battery decreases as reactants are converted into products, but the change in potential is small until the battery is nearly completely discharged. Although the value of the standard cell potential is 2.041 V, a fully charged commercial battery has a slightly higher potential (shown here as 2.08 V) due to the use of more concentrated sulfuric acid.

However, the decrease in $E_{cell}$ is gradual, not falling below 2.0 V until the battery is about 97% discharged, as shown in **Figure 17.14**. The gradual decrease is consistent with the logarithmic relationship between $Q$ and $E_{cell}$. If $[H_2SO_4]$ decreased by an order of magnitude—for example, from 1.00 $M$ to 0.100 $M$—the value of $E_{cell}$ would decrease by less than 6%: from 2.041 V to

$$E_{cell} = 2.041\text{ V} - \frac{0.0592\text{ V}}{2}\log\frac{1}{(0.100)^4} = 1.923\text{ V}$$

The logarithmic relationship between $Q$ and $E_{cell}$ means that most batteries can deliver current at a cell potential close to their fully charged potential until they are nearly completely discharged.

**SAMPLE EXERCISE 17.5** Calculating $E_{cell}$ from $E^\circ_{cell}$ and the Concentrations of Reactants and Products **LO4**

The standard potential ($E^\circ_{cell}$) of a voltaic cell based on the $Zn/Cu^{2+}$ ion reaction

$$Zn(s) + Cu^{2+}(aq) \rightarrow Zn^{2+}(aq) + Cu(s)$$

is 1.104 V. What is the value of $E_{cell}$ at 25°C when $[Cu^{2+}] = 0.100\ M$ and $[Zn^{2+}] = 1.90\ M$?

**Collect, Organize, and Analyze** We are given the standard cell potential and are asked to determine the value of $E_{cell}$ when $[Cu^{2+}] = 0.100\ M$ and $[Zn^{2+}] = 1.90\ M$. The Nernst equation (Equation 17.9) enables us to calculate $E_{cell}$ values for different concentrations of reactants and products. The only term in the numerator of the $Q$ expression for this cell reaction is $[Zn^{2+}]$ and the only term in the denominator is $[Cu^{2+}]$ because no terms for pure solids, such as $Zn(s)$ and $Cu(s)$, appear in reaction quotients. Each $Cu^{2+}$ ion acquires two electrons, and each Zn atom donates two, so $n = 2$ in the Nernst equation. The value of $[Zn^{2+}]$ is greater than $[Cu^{2+}]$, which makes $Q > 1$. The negative sign in front of the $0.0592/n \times \log Q$ term in Equation 17.9 means that the calculated value of $E_{cell}$ should be less than the value of $E^\circ_{cell}$.

**Solve** Substituting the values of $[Zn^{2+}]$ and $[Cu^{2+}]$ in the Nernst equation gives

$$E_{cell} = E^\circ_{cell} - \frac{0.0592\ \text{V}}{n}\log Q = E^\circ_{cell} - \frac{0.0592\ \text{V}}{n}\log\frac{[Zn^{2+}]}{[Cu^{2+}]} = 1.104\ \text{V} - \frac{0.0592\ \text{V}}{2}\log\frac{1.90}{0.100}$$

$$E_{cell} = 1.104\ \text{V} - \frac{0.0592\ \text{V}}{2}(1.28) = 1.066\ \text{V}$$

**Think About It** The calculated $E_{cell}$ value is only 0.038 V less than $E^\circ_{cell}$ because the logarithmic dependence of cell potential on reactant and product concentrations minimizes the impact of changing concentrations.

**Practice Exercise** The standard cell potential of the zinc–air battery (Figure 17.7) is 1.65 V. If at 25°C the partial pressure of oxygen in the air diffusing through its cathode is 0.21 atm, what is the cell potential at 25°C ? Assume that the cell reaction is

$$2\ Zn(s) + O_2(g) \rightarrow 2\ ZnO(s)$$

**STEPWISE ANIMATION**
Concentration Cell

One implication of the Nernst equation is that we can construct a voltaic cell whose potential is derived only from different concentrations of the same reactant. **Figure 17.15** shows such a *concentration cell*. This one is composed of two compartments that each contain a copper electrode and a solution of $CuSO_4$ and are joined by a porous bridge. In this cell, the concentration of the $CuSO_4$ solution in compartment 1 is higher than that in compartment 2, as indicated by the darker blue color of the solution in compartment 1.

How can this cell function as a voltaic cell (and have a positive $E_{cell}$ value) when the standard reduction potential in both compartments is the same? If $E_{cell}$ is to be greater than zero when $E^\circ_{cell} = 0$, then the value of the log $Q$ term in Equation 17.9 must be less than zero. The $Q$ value for the cell in Figure 17.15 is the ratio of the $[Cu^{2+}]$ values in the two compartments. If $Q < 0$, then the smaller $[Cu^{2+}]$ value (in compartment 2) must be in the numerator of $Q$ and the larger $[Cu^{2+}]$ value (in compartment 1) must be in its denominator. For example, if the concentration of $Cu^{2+}$ ions in compartment 1 is exactly 5 times the concentration in compartment 2, then:

$$E_{cell} = 0 - \frac{0.0592\ \text{V}}{2}\log\frac{[Cu^{2+}]_{\text{compartment 2}}}{[Cu^{2+}]_{\text{compartment 1}}} = -\frac{0.0592\ \text{V}}{2}\log\frac{x}{5x} = 0.0207\ \text{V}$$

**FIGURE 17.15** An electrochemical cell based on different concentrations of $Cu^{2+}$ ions.

Keep in mind that $Q$ is the ratio of [products]/[reactants]. Therefore, as the spontaneous cell reaction proceeds, $Cu^{2+}$ ions are produced in compartment 2 (which has the lower $[Cu^{2+}]$ value) and consumed in compartment 1. Over time, the difference in $[Cu^{2+}]$ values between the two compartments decreases until finally the two become equal. At that point, $Q = 1$ and $E_{cell} = E^\circ_{cell} = 0$. According to Equation 17.6, if $E_{cell} = 0$ then $\Delta G_{cell} = 0$, which means that the cell reaction has achieved chemical equilibrium.

**CONCEPT TEST**

Does the mass of either electrode change as the cell reaction proceeds in the concentration cell in Figure 17.15? Explain your answer.

## $E°$ and $K$

When the cell reaction of a voltaic cell reaches chemical equilibrium, $\Delta G_{cell} = E_{cell} = 0$ and $Q = K$. Equation 17.9 then becomes

$$0 = E^\circ_{cell} - \frac{0.0592\ \text{V}}{n} \log K$$

which we can rearrange to

$$\log K = \frac{nE^\circ_{cell}}{0.0592\ \text{V}} \quad (17.10)$$

We can use Equation 17.10 to calculate the equilibrium constant for any redox reaction at 25°C, not just those in electrochemical cells. For the more general case, we substitute $E^\circ_{rxn}$ for $E^\circ_{cell}$:

**CHEMTOUR**
Cell Potential, Equilibrium, and Free Energy

$$\log K = \frac{nE^\circ_{rxn}}{0.0592\ \text{V}} \quad (17.11)$$

**SAMPLE EXERCISE 17.6** Calculating the Values of $K$ and $\Delta G°$ for a Redox Reaction from the Standard Potentials of Its Half-Reactions **LO5**

Many procedures for determining mercury concentrations in environmental samples begin by oxidizing all the chemical forms of mercury in the samples to $Hg^{2+}$ and then reducing $Hg^{2+}$ to elemental Hg with $Sn^{2+}$ ions. Use the appropriate standard reduction potentials from Table 17.1 to calculate the values of $K$ and $\Delta G^\circ_{rxn}$ at 25°C for the following reaction:

$$Sn^{2+}(aq) + Hg^{2+}(aq) \rightarrow Sn^{4+}(aq) + Hg(\ell)$$

**Collect and Organize** We are asked to calculate the values of $K$ and $\Delta G^\circ_{rxn}$ for a redox reaction from the standard potentials of its half-reactions. Three equations will help us: Equation 17.11 relates the equilibrium constant for any redox reaction to its standard potential, $E^\circ_{rxn}$, which has the $E^\circ_{cell}$ value of the voltaic cell based on the redox reaction. Equation 17.6 relates the values of standard change in free energy of an electrochemical cell reaction to the standard potential of the cell. Equation 17.1 relates the value of $E^\circ_{cell}$ to the standard reduction potentials of the half-reactions occurring at the cathode and anode of an electrochemical cell. Finally, Table 17.1 lists two reduction half-reactions involving our reactants and products:

$$Hg^{2+}(aq) + 2\,e^- \rightarrow Hg(\ell) \qquad E^\circ = 0.851\ \text{V}$$
$$Sn^{4+}(aq) + 2\,e^- \rightarrow Sn^{2+}(aq) \qquad E^\circ = 0.154\ \text{V}$$

**Analyze** Using Equations 17.1 and 17.6 to solve this exercise requires transforming them from equations based on $E^\circ_{cell}$ values into corresponding equations for any redox reaction using $E^\circ_{rxn}$, as we did above in developing Equation 17.11. Oxidation occurs at anodes and reduction occurs at cathodes, so we obtain the following versions of Equations 17.1 and 17.6: $E^\circ_{rxn} = E^\circ_{reduction} - E^\circ_{oxidation}$ and $\Delta G^\circ_{rxn} = nFE^\circ_{rxn}$. In the redox reaction in this exercise, $Hg^{2+}$ ions are reduced by $Sn^{2+}$ ions. Therefore, the Hg half-reaction proceeds as written above, but the Sn half-reaction must run in reverse as an oxidation, and its standard reduction potential must be subtracted from that of the Hg half-reaction to calculate $E^\circ_{rxn}$. The difference between the two half-reaction potentials is about +0.7 V. Therefore, the right side of Equation 17.11 should be about $(2 \times 0.7)/0.06 \approx 23$, and the value of $K$ should be about $10^{23}$. The value of $\Delta G^\circ_{rxn}$ for the reaction should be about $(2 \times 10^5 \times 0.7) = 1.4 \times 10^5$ J.

**Solve** Calculating $E^\circ_{rxn}$ by using the modified version of Equation 17.1:

$$E^\circ_{rxn} = E^\circ_{reduction} - E^\circ_{oxication} = E^\circ_{Hg} - E^\circ_{Sn} = 0.851 \text{ V} - 0.154 \text{ V} = 0.697 \text{ V}$$

Using this $E^\circ_{rxn}$ value and $n = 2$ in Equation 17.11 to calculate $K$:

$$\log K = \frac{nE^\circ_{rxn}}{0.0592 \text{ V}} = \frac{2(0.697 \cancel{\text{V}})}{0.0592 \cancel{\text{V}}} = 23.55$$

$$K = 10^{23.55} = 4 \times 10^{23}$$

Using the calculated $E^\circ_{rxn}$ value in the modified version of Equation 17.6 to calculate $\Delta G^\circ_{rxn}$:

$$\Delta G^\circ_{rxn} = -nFE^\circ_{rxn} = -(2 \cancel{\text{mol e}^-})\left(\frac{9.65 \times 10^4 \text{ C}}{1 \cancel{\text{mol e}^-}}\right)(0.697 \text{ V})$$

$$= -1.35 \times 10^5 \text{ J} = -135 \text{ kJ}$$

**Think About It** The calculated values are close to those we estimated and are reasonable. Note how a relatively small positive $E^\circ_{rxn}$ value (<1 V) corresponds to a huge equilibrium constant, indicating that the reaction essentially goes to completion.

**Practice Exercise** Use the appropriate standard reduction potentials from Table 17.1 to calculate the values of $K$ and $\Delta G^\circ_{rxn}$ at 25°C for the following reaction:

$$5\ Fe^{2+}(aq) + MnO_4^-(aq) + 8\ H^+(aq) \rightarrow 5\ Fe^{3+}(aq) + Mn^{2+}(aq) + 4\ H_2O(\ell)$$

Before we end our discussion of how to derive equilibrium constant values at 25°C from $E^\circ_{cell}$ values, note that measuring the potential of an electrochemical reaction allows us to calculate equilibrium constant values that may be too large or too small to determine from the values of equilibrium concentrations of reactants and products. A value of $K$ as large as that calculated in Sample Exercise 17.6 could not be obtained by analyzing the composition of an equilibrium reaction mixture because the concentrations of the reactants would be too small to be determined accurately. Similarly, a cell potential of about −1 V would correspond to a tiny $K$ value and concentrations of products too small to be determined quantitatively.

**Table 17.2** summarizes how the values of $K$ and $E^\circ_{cell}$ are related to each other and to the change in free energy ($\Delta G^\circ_{cell}$) of a cell reaction under standard conditions. Note that spontaneous electrochemical reactions have $E^\circ_{cell} > 0$ and $K > 1$. The connection between positive cell potential and reaction spontaneity applies even under nonstandard conditions. Also keep in mind that small positive values of $E^\circ_{cell}$ (only a fraction of a volt, for example) correspond to very large $K$ values and to cell reactions that go nearly to completion.

**TABLE 17.2** Relationships between $K$, $E^\circ_{cell}$, and $\Delta G^\circ_{cell}$ Values of Electrochemical Reactions

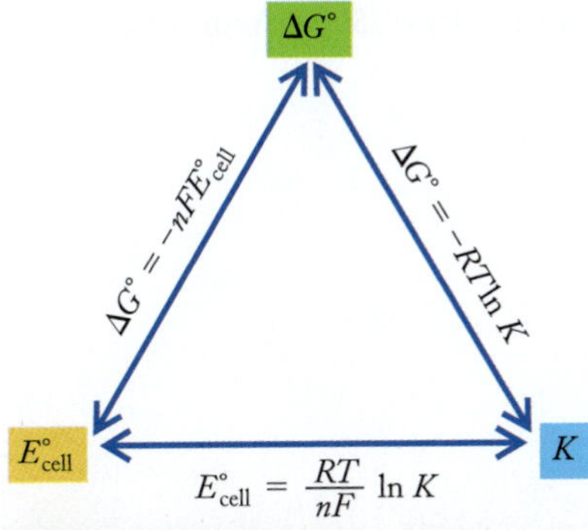


| $K$ | $E^\circ_{cell}$ | $\Delta G^\circ_{cell}$ | Favors Formation of |
|---|---|---|---|
| <1 | <0 | >0 | Reactants |
| >1 | >0 | <0 | Products |
| 1 | 0 | 0 | Neither |

## 17.7 Relating Battery Capacity to Quantities of Reactants

An important performance characteristic of a battery is its capacity to do electrical work—that is, to deliver electric charge at the labeled cell potential. This capacity—the amount of electrical work done—is defined by Equation 17.5,

$$w_{\text{elec}} = -nFE_{\text{cell}}$$

where $nF$ is the quantity of electric charge delivered, in coulombs (C).

Another important unit in electricity is the *ampere* (A), which is the SI base unit of electric current. An ampere is defined as a current of 1 coulomb per second:

$$1 \text{ ampere} = 1 \text{ coulomb/second}$$

which we can rearrange to

$$1 \text{ coulomb} = 1 \text{ ampere} \cdot \text{second}$$

Multiplying both sides of the latter equation by volts, and recalling that 1 joule of electrical energy is equivalent to 1 coulomb-volt of electrical work, we get

$$1 \text{ (coulomb)(volt)} = 1 \text{ joule} = 1 \text{ (ampere} \cdot \text{second)(volt)} \quad (17.12)$$

**FIGURE 17.16** The electrical energy rating of these rechargeable nickel–metal hydride AA batteries is 2500 mA · h at 1.2 V.

Joules are small energy units, so battery capacities are usually expressed in energy units with time intervals longer than seconds. For example, the *energy ratings* of rechargeable AA batteries (**Figure 17.16**) are often expressed in ampere-hours or milliampere-hours at the rated cell potential.

We need even bigger units to express the capacities of the large battery packs used in hybrid vehicles. They are the *watt* (W), the SI unit of power, and the *kilowatt-hour*, a unit of energy equal to more than 3 million joules, as shown in the following unit conversions:

$$1 \text{ watt} = 1 \text{ joule/second} = 1 \text{ ampere} \cdot \text{volt}$$

$$1 \text{ kilowatt} = 1000 \text{ W} = 1000 \text{ J/s}$$

$$\begin{aligned} 1 \text{ kilowatt} \cdot \text{hour} &= (1000 \text{ W})(1 \text{ h}) \\ &= (1000 \text{ J/}\cancel{\text{s}})(60 \cancel{\text{s}}/\cancel{\text{min}})(60 \cancel{\text{min}}/\cancel{\text{h}})(1 \cancel{\text{h}}) \\ &= 3.6 \times 10^6 \text{ J} \end{aligned}$$

**CONCEPT TEST**

Which energy unit—watt-second (joule), watt-minute, watt-hour, or kilowatt-hour—would be most appropriate for expressing (a) the energy capacity of a cell phone battery, (b) the energy capacity of a laptop battery, and (c) the monthly electricity bill for a student apartment?

### Nickel–Metal Hydride Batteries

Hybrid vehicles such as the Toyota Prius (**Figure 17.17**) are powered by combinations of small gasoline engines and electric motors. The motors are often powered by battery packs made of dozens of nickel–metal hydride (NiMH) cells. At the cathodes in these cells, $NiO(OH)$ is reduced to $Ni(OH)_2$; at the anodes, which are made of one or more transition metals, hydrogen atoms are oxidized to $H^+$ ions,

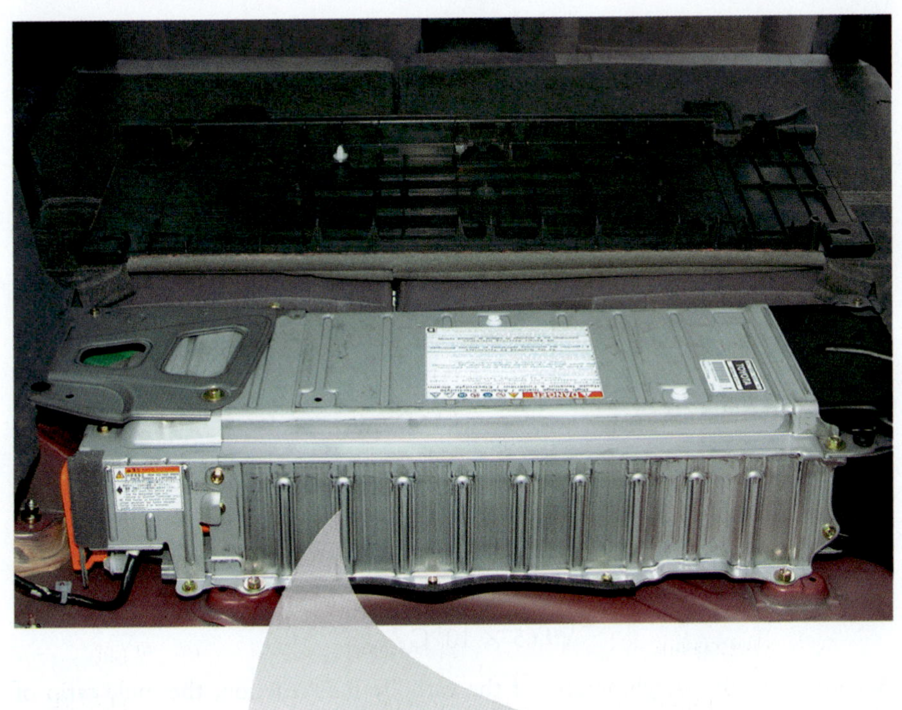

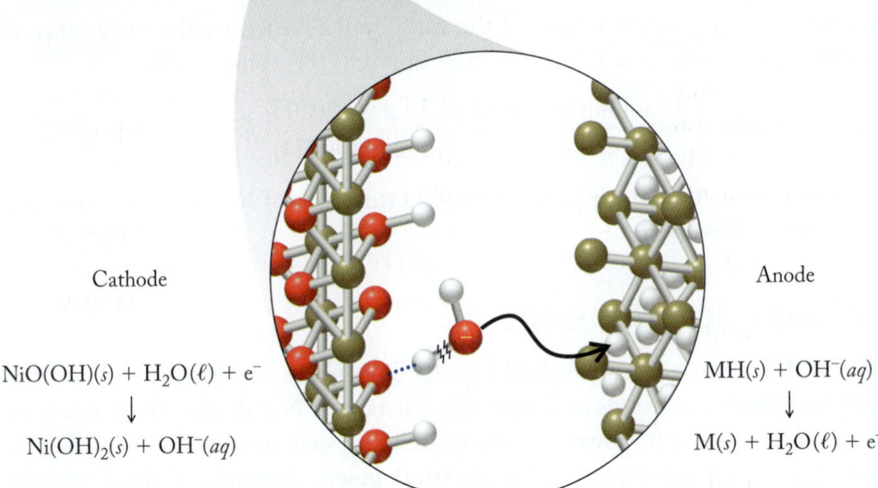


**FIGURE 17.17** The Toyota Prius and other hybrid vehicles are powered by combinations of gasoline engines and electric motors. (a) Electricity for the motor of this Prius is stored in a nickel–metal hydride battery pack under the back seat. (b) The cell of a nickel–metal hydride battery contains a cathode made of NiO(OH), which is reduced to $Ni(OH)_2$ as electrons are consumed and $OH^-$ ions are released that migrate to the anode through a porous membrane soaked with an aqueous solution of KOH. At the anode, $OH^-$ ions react with H atoms held in the crystal structure of compounds composed of two or more metals, such as $LaNi_5$. During this reaction, the H atoms are oxidized to $H^+$ ions that combine with the incoming $OH^-$ ions, forming molecules of $H_2O$. The anode's structural compounds are called metal hydrides.

which react with $OH^-$ to form water. The electrodes are separated by a porous membrane soaked in aqueous KOH.

The cathode half-reaction is

$$NiO(OH)(s) + H_2O(\ell) + e^- \rightarrow Ni(OH)_2(s) + OH^-(aq) \qquad E^\circ = 0.52\ V$$

At the anode, hydrogen is present as a metal hydride. To write the anode half-reaction, we use the generic formula MH, where M stands for one or more metals that form hydrides. In a basic background electrolyte, the anode oxidation half-reaction is

$$MH(s) + OH^-(aq) \rightarrow M(s) + H_2O(\ell) + e^-$$

The standard potential of this half-reaction depends on the chemical properties of MH. The overall cell reaction from these two half-reactions is

$$MH(s) + NiO(OH)(s) \rightarrow M(s) + Ni(OH)_2(s)$$

The value of $E^\circ_{cell}$ for the NiMH battery cannot be calculated precisely because we have only an approximate value of $E^\circ_{anode}$. Most NiMH cells are rated at about 1.2 V.

Now let's relate the electrical energy stored in a battery (in other words, its energy rating) to the quantities of reactants needed to produce that energy. Consider the rechargeable AA NiMH batteries in Figure 17.16, which are rated to deliver 2500 milliampere · hours, or 2.5 ampere · hours, of electrical charge at 1.2 V. How much NiO(OH) must be converted to $Ni(OH)_2$ to deliver this much charge? To answer the question, we need to convert the quantity of charge into a number of moles of electrons and then convert that into an equivalent number of moles of reactant and finally into a mass of reactant. Recall that an ampere (A) is defined as a coulomb (C) per second, which means that 1 C = 1 A · s. Therefore, the quantity of electrical charge delivered by the NiMH battery is

$$2.5\ \cancel{\text{A}\cdot\text{h}} \times \frac{1\ \text{C}}{1\ \cancel{\text{A}}\cdot\text{s}} \times \frac{60\ \cancel{\text{min}}}{1\ \cancel{\text{h}}} \times \frac{60\ \text{s}}{1\ \cancel{\text{min}}} = 9.0 \times 10^3\ \text{C}$$

The Faraday constant tells us that one mole of charge is equivalent to $9.65 \times 10^4$ C, so the number of moles of charge, which is equal to the number of moles of electrons that flow from the battery, is

$$9.0 \times 10^3\ \cancel{\text{C}}\left(\frac{1\ \text{mol e}^-}{9.65 \times 10^4\ \cancel{\text{C}}}\right) = 0.0933\ \text{mol e}^-$$

According to the stoichiometry of the cathode half-reaction, the mole ratio of NiO(OH) to electrons is 1:1, so the mass of NiO(OH) consumed is

$$0.0933\ \cancel{\text{mol e}^-}\left(\frac{1\ \cancel{\text{mol NiO(OH)}}}{1\ \cancel{\text{mol e}^-}}\right)\left(\frac{91.70\ \text{g NiO(OH)}}{1\ \cancel{\text{mol NiO(OH)}}}\right) = 8.6\ \text{g NiO(OH)}$$

The mass of an AA battery is about 30 g, so this mass of NiO(OH) accounts for about 30% of the total.

## Lithium–Ion Batteries

The NiMH batteries used in hybrid vehicles do not have the capacity to power the vehicles at highway speeds or for extended distances. Nor do these batteries have the energy capacity to power all-electric vehicles such as the Chevrolet Bolt in this chapter's opening photo. The electrical power demands of these vehicles require batteries with much greater ratios of energy capacity to battery size. The technology of choice in these applications is the lithium–ion battery (**Figure 17.18**), the same kind of battery that powers laptop computers, cell phones, and digital cameras.

In a lithium–ion battery, $Li^+$ ions are typically stored in a graphite anode, though other anode materials such as nanowires made of silicon have been recently developed. During discharge, these ions migrate through a nonaqueous electrolyte to a porous cathode. These cathodes are made of transition metal oxides or phosphates that can form stable complexes with $Li^+$ ions. One popular cathode material is cobalt(IV) oxide. Lithium–ion batteries with these cathodes and graphite anodes have cell potentials of about 3.6 V (three times that of a NiMH battery). The cell reaction is

$$Li_{1-x}CoO_2(s) + Li_xC_6(s) \rightarrow 6\ C(s) + LiCoO_2(s) \qquad (17.13)$$

In a fully charged cell, $x = 1$, which makes the cathode lithium-free $CoO_2$. As the cell discharges and $Li^+$ ions migrate from the carbon anode to the cobalt oxide cathode, the value of $x$ falls toward zero. To balance this flow of positive charges inside the cell, electrons flow from the anode to the cathode through an external circuit. When fully discharged, the cathode is $LiCoO_2$, and the

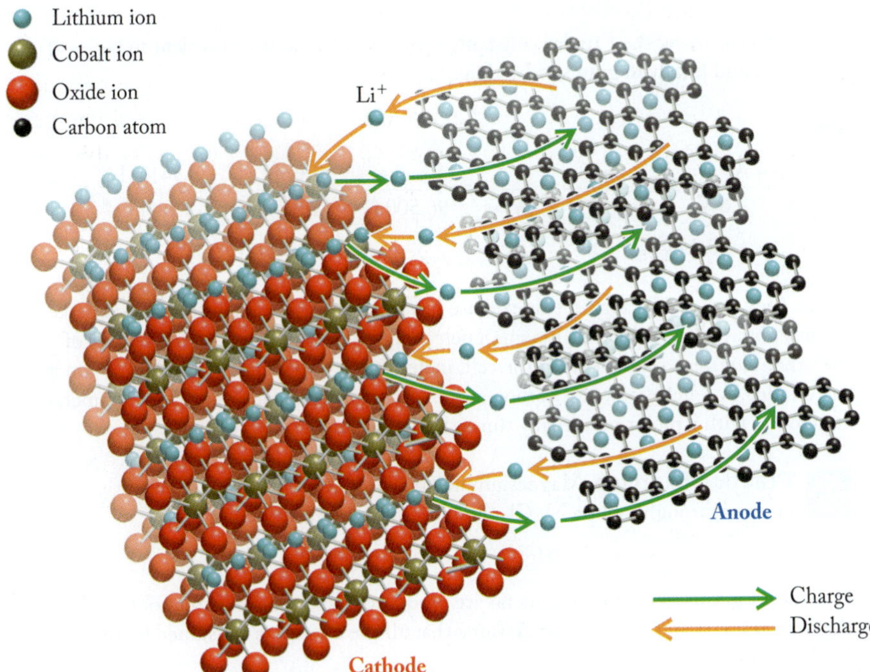


**FIGURE 17.18** Most lithium-ion batteries contain $Li^+$ ions stored in graphite layers of the anode. During discharge, these ions travel to the cathode, which contains a transition metal compound such as $CoO_2$. During recharging, the direction of ion migration reverses.

oxidation number of Co is reduced to +3. The electrodes in a lithium–ion battery may react with oxygen and water, so the electrolytes (for example, $LiPF_6$) are dissolved in polar organic solvents, such as tetrahydrofuran, ethylene carbonate, or propylene carbonate (**Figure 17.19**).

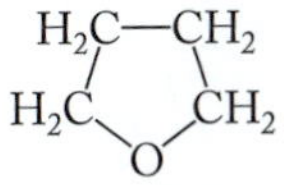

$H_2C$—$CH_2$, $H_2C$, $CH_2$, O

Tetrahydrofuran

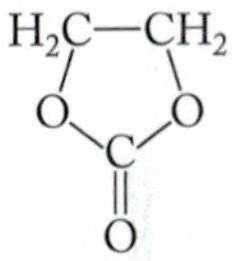

$H_2C$—$CH_2$, O, C, O, O

Ethylene carbonate

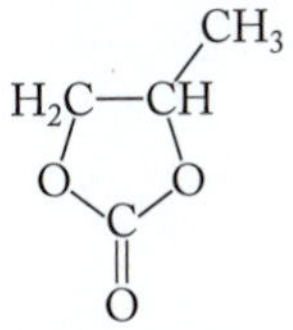

$CH_3$, $H_2C$—CH, O, C, O, O

Propylene carbonate

**FIGURE 17.19** Polar organic compounds such as these molecules are the solvents for the electrolytes in a lithium–ion battery.

## SAMPLE EXERCISE 17.7 Relating the Mass of a Reactant in a Cell Reaction to a Quantity of Electrical Charge — LO6

The capacity of the lithium–ion battery in a digital camera is 3.4 W · h at 3.6 V. How many grams of $Li^+$ ions must migrate from anode to cathode to produce this much electrical energy?

**Collect and Organize** We are asked to relate the electrical energy generated by an electrochemical cell to the mass of the ions involved in generating that energy. We know the cell potential and its capacity in the energy unit of watt-hours. Given these starting points and our need to calculate moles and then grams of $Li^+$ ions, the following equivalencies may be useful:

$$1 \text{ watt (W)} = 1 \text{ ampere-volt (A} \cdot \text{V)}$$

$$1 \text{ coulomb (C)} = 1 \text{ ampere-second (A} \cdot \text{s)}$$

We may also need to use the Faraday constant, $9.65 \times 10^4$ C/mol.

**Analyze** The energy capacity of the battery is the product of the charge (electrons) it can deliver times the cell potential pumping that charge. This exercise focuses on the quantity of charge, so we need to separate the contribution that cell potential makes to the energy rating. To do that, we need to divide the energy rating (3.4 W · h = 3.4 A · V · h) by the battery's cell potential (3.6 V), which translates into slightly less than 1 A · h. Converting this value into a mass of $Li^+$ ions involves unit conversions that take us from ampere-hours to ampere-seconds (equivalent to coulombs), to moles of electrons by using the Faraday constant, to moles of $Li^+$ ions (equivalent to moles of electrons), and finally to grams of $Li^+$ ions.

**Solve**

$$\frac{3.4\ \text{A}\cdot\text{V}\cdot\text{h}}{3.6\ \text{V}} \times \frac{3600\ \text{s}}{\text{h}} \times \frac{1\ \text{C}}{\text{A}\cdot\text{s}} \times \frac{1\ \text{mol e}^-}{96{,}500\ \text{C}} \times \frac{1\ \text{mol Li}^+}{1\ \text{mol e}^-} \times \frac{6.941\ \text{g Li}^+}{1\ \text{mol Li}^+}$$

$$= 0.24\ \text{g Li}^+\ \text{ions}$$

**Think About It** The battery in this exercise has a mass of about 22 g, so $Li^+$ ions make up only about 1% of its mass. This small percentage is reasonable given the masses of the other required components of the cell, including an anode in which each $Li^+$ ion is surrounded by a hexagon of six carbon atoms and a cathode made of a transition metal compound with a molar mass many times that of Li metal.

**Practice Exercise** Magnesium metal is produced by passing an electric current through molten $MgCl_2$. The reaction at the cathode is

$$\text{Mg}^{2+}(\ell) + 2\ \text{e}^- \rightarrow \text{Mg}(\ell)$$

How many grams of magnesium metal are produced if an average current of 63.7 A flows for 4 hours and 45 minutes? Assume that all the current is consumed by the half-reaction shown.

## 17.8 Corrosion: Unwanted Electrochemical Reactions

We began this chapter with a Particulate Review question based on the corrosion of the metal surface of a car. In this section, we examine this process and some of the half-reactions that contribute to it in more detail. We have seen how half-reactions are physically separated in electrochemical cells. It turns out that they are often separated in corrosion reactions as well; in this respect, the chemistry of corrosion is much like the electrochemical reactions in voltaic cells. **Corrosion** is the deterioration of metals due to spontaneous redox reactions. This definition is reflected in several of the factors that promote corrosion:

1. *The presence of water.* Metals left out in the rain and snow rust or corrode more rapidly than those under cover. For example, objects made from iron spontaneously rust in moist air, as iron reacts with $O_2$ from the air in a series of reactions that includes

$$4\ \text{Fe}(s) + 3\ \text{O}_2(g) + 2\ \text{H}_2\text{O}(\ell) \rightarrow 4\ \text{FeO(OH)}(s)$$

2. *The presence of electrolytes.* Just as electrolytes carry electrical current between anodes and cathodes and facilitate cell reactions, corrosion is much more rapid in, for example, seawater than in freshwater.
3. *Contact between dissimilar metals.* Metals corrode more rapidly when in contact with other metals that are less likely to be oxidized—that is, other metals that have higher reduction potentials.

**corrosion** a process in which a metal deteriorates through spontaneous electrochemical reactions.

Let's explore the impact of these factors by using the Statue of Liberty as a model. As the statue was built, its exterior copper sheets were attached to and supported by an interior network of iron beams (**Figure 17.20**). The French

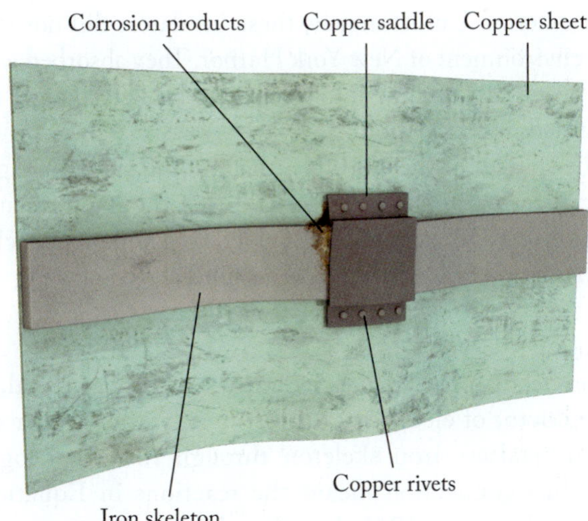


**FIGURE 17.20** When the Statue of Liberty was built, the sheets of copper covering the statue were supported by an iron skeleton that corroded near the points of contact with the copper saddles that held the sheets and skeleton together.

designers of the statue knew that these two metals in contact with each other might someday pose a corrosion problem because the two have very different electrochemical properties. The difference is reflected in the standard reduction potentials of these elements when they oxidize under neutral to slightly basic conditions, as occur in marine environments:

$$Cu(OH)_2(s) + 2\,e^- \rightarrow Cu(s) + 2\,OH^-(aq) \qquad E^\circ = -0.222\text{ V}$$

$$FeO(OH)(s) + H_2O(\ell) + 3\,e^- \rightarrow Fe(s) + 3\,OH^-(aq) \qquad E^\circ = -0.87\text{ V}$$

As we discussed in Section 17.3, the greater $E^\circ$ of $Cu(OH)_2$ means that it is more easily reduced under standard conditions than FeO(OH), and Fe is more easily oxidized than Cu. We can confirm this by reversing the iron half-reaction and combining it with the copper half-reaction. We also need to multiply the iron half-reaction by 2 and the copper half-reaction by 3 to balance the loss and gain of electrons:

$$2[Fe(s) + 3\,OH^-(aq) \rightarrow FeO(OH)(s) + H_2O(\ell) + 3\,e^-]$$

$$3[Cu(OH)_2(s) + 2\,e^- \rightarrow Cu(s) + 2\,OH^-(aq)]$$

$$2\,Fe(s) + \cancel{6\,OH^-}(aq) + 3\,Cu(OH)_2(s) + \cancel{6\,e^-} \rightarrow 2\,FeO(OH)(s) + 2\,H_2O(\ell) + 3\,Cu(s) + \cancel{6\,OH^-}(aq) + \cancel{6\,e^-}$$

or

$$2\,Fe(s) + 3\,Cu(OH)_2(s) \rightarrow 2\,FeO(OH)(s) + 3\,Cu(s) + 2\,H_2O(\ell) \quad (17.14)$$

The resulting equation could be that of an electrochemical cell that has a copper cathode coated in $Cu(OH)_2$ and an iron anode coated in FeO(OH). We obtain the standard potential for the reaction by substituting the reduction potentials for each half-reaction (Table 17.1) into Equation 17.1:

$$E^\circ_{cell} = E^\circ_{cathode} - E^\circ_{anode} = -0.222\text{ V} - (-0.87\text{ V}) = 0.65\text{ V}$$

The reaction has a positive standard cell potential, which means that under standard conditions, iron in contact with copper will spontaneously oxidize to FeO(OH) as $Cu(OH)_2$ is reduced.

To suppress the reaction in Equation 17.14, insulators made of asbestos mats soaked in shellac (Figure 17.20) were used to separate the statue's iron skeleton from its copper exterior. Unfortunately, these insulators did not stand up to the humid, marine environment of New York Harbor. They absorbed water vapor and seawater spray, eventually turning into electrolyte-soaked sponges that promoted rather than retarded iron oxidation.

What are the sources of the $Cu(OH)_2$ that are the oxidizing agents in Equation 17.14? Recall from Sample Exercise 17.2 that the light green patina of the Statue of Liberty is due to the presence of copper(II) compounds such as $Cu(OH)_2$, formed by the oxidation of Cu metal by atmospheric oxygen:

$$O_2(g) + 2\,H_2O(\ell) + 2\,Cu(s) \rightarrow 2\,Cu(OH)_2(s) \qquad (17.15)$$

This reaction occurs on an expansive surface of copper metal, which is also an excellent conductor of electricity. Unfortunately, this copper surface was in contact with the statue's iron skeleton through the water-logged, ion-rich asbestos mats. This connection meant the reactions in Equations 17.14 and 17.15 were linked. Equation 17.16 describes the resulting interior and exterior reactions:

$$2[2\,Fe(s) + 3\,Cu(OH)_2(s) \rightarrow 2\,FeO(OH)(s) + 3\,Cu(s) + 2\,H_2O(\ell)] \qquad (17.14)$$

$$3[O_2(g) + 2\,H_2O(\ell) + 2\,Cu(s) \rightarrow 2\,Cu(OH)_2(s)] \qquad (17.15)$$

$$4\,Fe(s) + \cancel{6\,Cu(OH)_2(s)} + 3\,O_2(g) + 2\,\cancel{6}\,H_2O(\ell) + \cancel{6\,Cu(s)} \rightarrow$$
$$4\,FeO(OH)(s) + \cancel{6\,Cu(s)} + \cancel{4\,H_2O(\ell)} + \cancel{6\,Cu(OH)_2(s)}$$

or

$$4\,Fe(s) + 3\,O_2(g) + 2\,H_2O(\ell) \rightarrow 4\,FeO(OH)(s) \qquad (17.16)$$

Note how the copper half-reaction has disappeared from the overall reaction. Thus, the statue's copper exterior functions as a giant electron delivery system, allowing electrons to flow from iron atoms—as they oxidize to FeO(OH) inside the statue—to molecules of atmospheric $O_2$ on the exterior surface. The overall effect was a dramatic acceleration of the rate of the air oxidation of the original iron skeleton. The reaction in Equation 17.16 and others like it that formed other components of rust resulted in severe deterioration of the skeletal network that held up the Statue of Liberty, so much so that in the 1980s the iron skeleton had to be replaced with one made of corrosion-resistant stainless steel.

Deterioration of the Statue of Liberty was not an isolated incident. According to one industrial estimate, the direct and indirect cost of corrosion and corrosion control to the U.S. economy exceeds $1 trillion annually. Worldwide, the cost was more than $2 trillion. These figures include the cost to repair or replace corroded equipment and structures and to protect them against corrosion. The latter category includes money spent on protective coatings, including paint and chemical or electrochemical modification of metal surfaces to make them less reactive.

Another widely used method for inhibiting corrosion involves chemically bonding metal oxide coatings to metal surfaces. For example, the zinc coating on galvanized nails oxidizes, forming a layer of solid ZnO that protects the underlying metal from further oxidation. A method called bluing is widely used to protect steel tools, gun barrels, wood stoves, and other steel materials. The name comes from the distinctive dark blue color of a surface layer of magnetite ($Fe_3O_4$), which can be reaction-bonded to steel. Formation of a protective oxide layer is also the mechanism that makes stainless steel resistant to corrosion. Though the main component in all forms of stainless steel is iron, it also contains chromium. Oxidation of surface Cr atoms forms a durable protective layer of $Cr_2O_3$. Similarly, objects made of aluminum are protected by a surface layer of $Al_2O_3$ that strongly adheres to the underlying metal and inhibits further oxidation.

Another way to protect metal structures, especially those that contact seawater (**Figure 17.21**), involves attaching to them objects made of even more reactive metals. These objects are called *sacrificial anodes*. As their name implies, their role is to form a voltaic cell with the protected structure in which the object is the anode and the structure is the cathode. That way, the sacrificial anodes oxidize and the structure does not. This preservation technique is called *cathodic protection*. Many sacrificial anodes used in the marine industry are made of zinc or aluminum–magnesium alloys.

$$Zn(s) + H_2O(\ell) \rightarrow ZnO(s) + 2\,H^+(aq) + 2\,e^-$$

$$Mg(s) + H_2O(\ell) \rightarrow MgO(s) + 2\,H^+(aq) + 2\,e^-$$

$$Al(s) + 3\,H_2O(\ell) \rightarrow Al(OH)_3(s) + 3\,H^+(aq) + 3\,e^-$$

These metals have the benefit of forming oxide or hydroxide coatings as they oxidize in weakly basic seawater, which slows the rate at which they oxidize further, thereby requiring replacement less often.

(a)

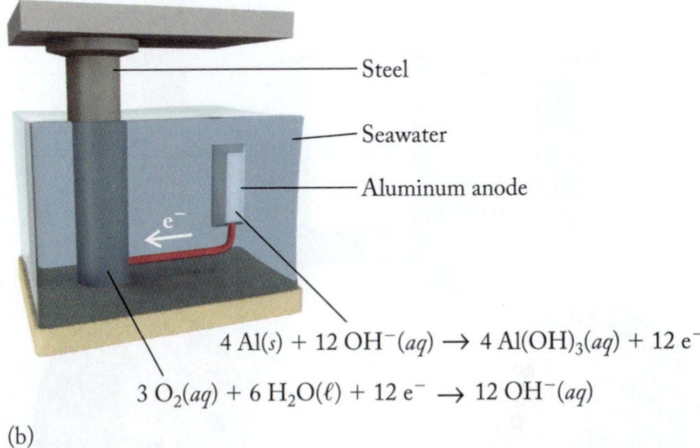


(b)

**FIGURE 17.21** (a) Most metal structures immersed in seawater have serious corrosion problems unless (b) they have cathodic protection.

## 17.9 Electrolytic Cells and Rechargeable Batteries

Lead–acid, NiMH, and lithium–ion batteries are rechargeable, which means that their spontaneous ($\Delta G < 0$) cell reactions, which convert chemical energy into electrical energy, can be forced to run in reverse. During recharging, an external electrical power supply is used to pump electrons into what were once anodes, turning them into cathodes, and pull electrons away from what were once cathodes, turning them into anodes. This expenditure of electrical energy forces the electrode's spontaneous half-reactions to run in reverse, regenerating the substances that were spontaneously oxidized at the anodes and spontaneously reduced at the cathodes as the batteries discharged. Formation of the battery's high-energy reactants from the low-energy products they formed as the battery discharged means that the reverse reaction increases the chemical energy of the cell ($\Delta G > 0$). In other words, electrical energy forces a nonspontaneous cell reaction to occur.

**STEPWISE ANIMATION**
Voltaic vs. Electrolytic Cells

Recharging a battery is an example of **electrolysis**, which is defined as any chemical reaction driven by electricity. During recharging, a battery is transformed from a voltaic cell into an **electrolytic cell** in which electrical energy is transformed into chemical energy ($\Delta G > 0$), as shown in **Figure 17.22**. To explore this transformation, let's revisit the lead–acid battery used to start car engines (Figure 17.13). When this battery discharges, electrons generated by an oxidation half-reaction flow out from the negative terminal through an external circuit and into the positive terminal, where they are consumed by a reduction half-reaction. To recharge the battery, electrons must flow in the opposite direction. Thus, the

**electrolysis** a process in which electrical energy is used to drive a nonspontaneous chemical reaction.

**electrolytic cell** a device in which an external source of electrical energy does work on a chemical system, turning a reactant(s) into a higher-energy product(s).

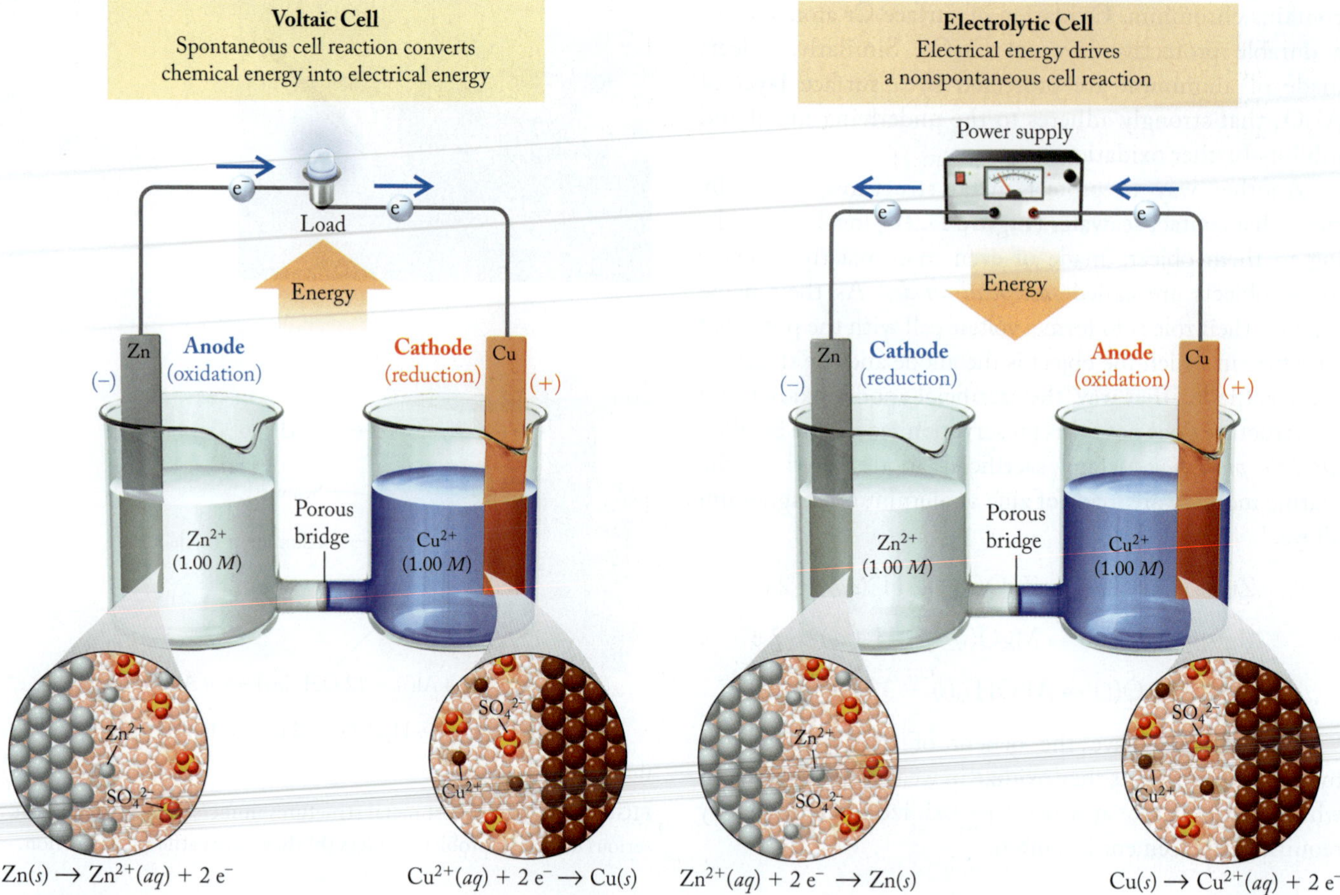


**FIGURE 17.22** Voltaic versus electrolytic cells. (a) In a voltaic cell, a spontaneous reaction produces electrical energy and does electrical work on its surroundings, such as lighting an LED flashlight bulb. (b) In an electrolytic cell, an external supply of electrical energy does work on the chemical system in the cell, driving a nonspontaneous reaction.

Pb electrodes connected to the negative battery terminal, which serve as anodes during discharge, become cathodes during recharge. Any $PbSO_4$ that forms on these Pb electrodes during discharge is reduced back to Pb metal during recharge:

$$PbSO_4(s) + H^+(aq) + 2\,e^- \rightarrow Pb(s) + HSO_4^-(aq)$$

Similarly, the $PbO_2$ electrodes connected to the positive terminal that serve as cathodes during discharge become anodes during recharge, as any $PbSO_4$ that formed on them during discharge is oxidized back to $PbO_2$ during recharge:

$$PbSO_4(s) + 2\,H_2O(\ell) \rightarrow PbO_2(s) + 3\,H^+(aq) + HSO_4^-(aq) + 2\,e^-$$

**SAMPLE EXERCISE 17.8** Calculating the Time to Recharge a Battery **LO6**

Recharging a AA NiMH battery involves converting $Ni(OH)_2$ to $NiO(OH)$. If a charger supplies a current of 1.00 A, how many minutes does oxidizing 0.649 g of $Ni(OH)_2$ to $NiO(OH)$ take?

**Collect and Organize** We are asked to calculate the time required for a charging current to oxidize a given mass of $Ni(OH)_2$ to $NiO(OH)$. During the discharge of a NiMH battery, the spontaneous cathode half-reaction is

$$NiO(OH)(s) + H_2O(\ell) + e^- \rightarrow Ni(OH)_2(s) + OH^-(aq) \qquad E^\circ = 0.52\text{ V}$$

**Analyze** During recharging, the nickel half-reaction runs in reverse:

$$Ni(OH)_2(s) + OH^-(aq) \rightarrow NiO(OH)(s) + H_2O(\ell) + e^-$$

One mole of electrons is produced for each mole of $Ni(OH)_2$ consumed. Our first steps are to convert 0.649 g of $Ni(OH)_2$ into moles of $Ni(OH)_2$ and then into moles of electrons. The Faraday constant can be used to convert moles of electrons to coulombs of charge. A coulomb is the same as an ampere-second, so dividing by the charging current will yield seconds of charging current.

**Solve**

$$0.649\ \text{g Ni(OH)}_2 \times \frac{1\ \text{mol Ni(OH)}_2}{92.71\ \text{g Ni(OH)}_2} \times \frac{1\ \text{mol e}^-}{1\ \text{mol Ni(OH)}_2} \times \frac{9.65 \times 10^4\ \text{C}}{1\ \text{mol e}^-}$$

$$\times \frac{1\ \text{A}\cdot\text{s}}{1\ \text{C}} \times \frac{1}{1.00\ \text{A}} = 676\ \text{s}$$

Thus, the charger must deliver 1.00 A of current for 676 s, or

$$676\ \text{s} \times \frac{1\ \text{min}}{60\ \text{s}} = 11.3\ \text{min}$$

**Think About It** A charging time of 11.3 min may seem short, but the quantity of the $Ni(OH)_2$ to be oxidized (0.649 g) is much less than the quantity of NiO(OH) in a fully charged AA NiMH battery (8.6 g, as calculated in Section 17.7), so this battery was only slightly discharged.

**Practice Exercise** Suppose that a car's starter motor draws 225 amperes of current for 3.5 s to start the car's engine. What mass of Pb is oxidized in the battery to supply this much electricity?

Electrolysis is used in many processes other than recharging batteries. Electrolytic cells are used to electroplate thin layers of silver, gold, and other metals onto objects, giving these objects the appearance, resistance to corrosion, and other properties of the electroplated metal at a fraction of the cost of fabricating the entire object out of that metal (**Figure 17.23**).

To appreciate the utility of electroplating, consider the need to coat a medallion made of a less expensive metal with a layer of gold. The medallion is immersed in an aqueous solution containing $Au(CN)_4^-$ ions and then electroplated using a current of 4.50 A for 33.0 minutes, as $Au(CN)_4^-$ ions are reduced to Au metal on the medallion's surface:

$$\text{Au(CN)}_4^-(aq) + 3\ \text{e}^- \rightarrow \text{Au}(s) + 4\ \text{CN}^-(aq)$$

How much gold is electroplated on the medallion and, assuming uniform coverage, how thick a layer of gold is deposited if the surface area of the medallion is 62.2 cm$^2$ (the same as an Olympic gold medal)? First, we can calculate the mass of the gold by (1) combining the current and time values to obtain a quantity of electrical charge, (2) converting electrical charge into moles of electrons by using

**FIGURE 17.23** The Oscar statuettes given out at the annual Academy Awards are made of cast bronze and immersed in a plating bath containing $Au(CN)_4^-$ ions. When connected to the negative terminal of an electric power supply, the statuette becomes the cathode in an electrolytic cell as the half-reaction $Au(CN)_4^-(aq)$ + 3 e$^-$ → Au($s$) + 4 CN$^-$ ($aq$) produces a layer of pure gold on the statuette.

the Faraday constant, (3) converting moles of electrons into moles of gold by using the stoichiometry of the reduction half-reaction, and (4) converting mol Au into g Au by multiplying by $\mathcal{M}_{Au}$:

$$m_{Au} = 4.50\ \cancel{A} \times \frac{1\ \cancel{C}}{1\ \cancel{A \cdot s}} \times 33.0\ \cancel{min} \times \frac{60\ s}{1\ \cancel{min}} \times \frac{1\ \cancel{mol\ e^-}}{9.65 \times 10^4\ \cancel{C}}$$

$$\times \frac{1\ \cancel{mol\ Au}}{3\ \cancel{mol\ e^-}} \times \frac{196.97\ g\ Au}{1\ \cancel{mol\ Au}} = 6.06\ g\ Au$$

We can then find the surface coverage by dividing this mass by the density of Au (19.3 g/cm$^3$) and then dividing again by the given surface area of the medallion:

$$6.06\ \cancel{g}\ Au \times \frac{1.00\ cm^3}{19.3\ \cancel{g}} \times \frac{1}{62.2\ \cancel{cm^2}} = 0.00505\ cm,\ or\ 50.5\ \mu m\ Au$$

To put this mass of electroplated gold in perspective, if Olympic gold medals were solid gold (which they're not), they would contain 164 g of Au, or 164 g/6.06 g = 27 times as much as on the electroplated gold medallion.

In the chemical industry, the electrolysis of molten salts is used to produce highly reactive substances such as sodium, chlorine, and fluorine, as well as alkali and alkaline earth metals and aluminum. When NaCl, for instance, is heated to just above its melting point (that is, above 800°C), it becomes an ionic liquid that can conduct electricity. If a large enough potential is applied to carbon electrodes immersed in the molten NaCl, the sodium ions are attracted to the negative electrode and are reduced to sodium metal, whereas the chloride ions are attracted to the positive electrode and oxidized to $Cl_2$ gas:

$$2\ Na^+(\ell) + 2\ Cl^-(\ell) \rightarrow 2\ Na(\ell) + Cl_2(g)$$

A final note about anode and cathode polarity: The reactions in voltaic cells are spontaneous ($\Delta G < 0$). These cells pump electric current through external circuits and electrical devices with a force equal to their cell potentials. The anode in a voltaic cell is the negative electrode because an oxidation half-reaction supplies negatively charged electrons to the external circuit. Electrons flow through the circuit and into the positive battery terminal, which is connected to the cathode, where these electrons are consumed in a reduction half-reaction.

The nonspontaneous ($\Delta G > 0$) reactions in electrolytic cells require electrical energy from an external power supply. The negative terminal of such a power supply is connected to the cathode of the cell, pumping electrons into the cathode, where they are consumed in a reduction half-reaction. Electrons generated at the cell's anode in an oxidation half-reaction flow from the cell toward the positive terminal of the power supply. As shown in Figure 17.22, the cathode of an electrolytic cell is the negative electrode, but the cathode of a voltaic cell is the positive electrode. Similarly, the anode of an electrolytic cell is the positive electrode, but the anode of a voltaic cell is the negative electrode. These "pole reversals" make sense if we keep in mind these two fundamental definitions:

- Anodes are electrodes where oxidation takes place.
- Cathodes are electrodes where reduction takes place.

## CONCEPT **TEST**

The electrolysis of molten NaCl produces liquid Na metal at the cathode and $Cl_2$ gas at the anode. However, the electrolysis of an aqueous solution of NaCl produces gases at both cathode and anode. Identify the gas produced at the cathode.

## 17.10 Fuel Cells

**Fuel cells** are promising energy sources for many applications, such as powering office buildings, cruise ships, and electric vehicles. Fuel cells are voltaic cells, but they are different from batteries in that their supplies of reactants are constantly renewed. Therefore, they do not "discharge"; they never run down, and they don't die unless their fuel supply is cut off. From a thermodynamic perspective, batteries are closed systems and fuel cells are open systems.

**fuel cell** a voltaic cell based on the oxidation of a continuously supplied fuel; the reaction is the equivalent of combustion, but chemical energy is converted directly into electrical energy.

**CHEMTOUR**
Fuel Cell

**CONNECTION** In Chapter 10, *open* thermodynamic systems are defined as those that exchange matter and energy with their surroundings; *closed* systems exchange energy but not matter.

In a typical fuel cell, electrons are supplied to an external circuit by the oxidation of $H_2$ at the anode, and electrons are consumed by the reduction of $O_2$ at the cathode. The fuel cell reaction, then, is

$$2\,H_2(g) + O_2(g) \rightarrow 2\,H_2O(\ell)$$

The fuel cells used to power electric vehicles consist of metallic or graphite electrodes separated by a hydrated polymeric material called a *proton-exchange membrane* (PEM). The PEM serves as both an electrolyte and a barrier that prevents crossover and mixing of the fuel and oxidant (**Figure 17.24**). The surfaces of the electrodes are coated with transition metal catalysts to speed up the electrode half-reactions. Platinum catalysts promote H—H bond breaking during the oxidation of $H_2$ gas to $H^+$ ions at the anode,

$$H_2(g) \rightarrow 2\,H^+(aq) + 2\,e^- \qquad E^\circ = 0.000\text{ V}$$

and a platinum–nickel alloy with the formula $Pt_3Ni$ is particularly effective in catalyzing the formation of free O atoms from $O_2$ molecules, which is part of the reduction half-reaction at the cathode:

$$O_2(g) + 4\,H^+(aq) + 4\,e^- \rightarrow 2\,H_2O(\ell) \qquad E^\circ = 1.229\text{ V}$$

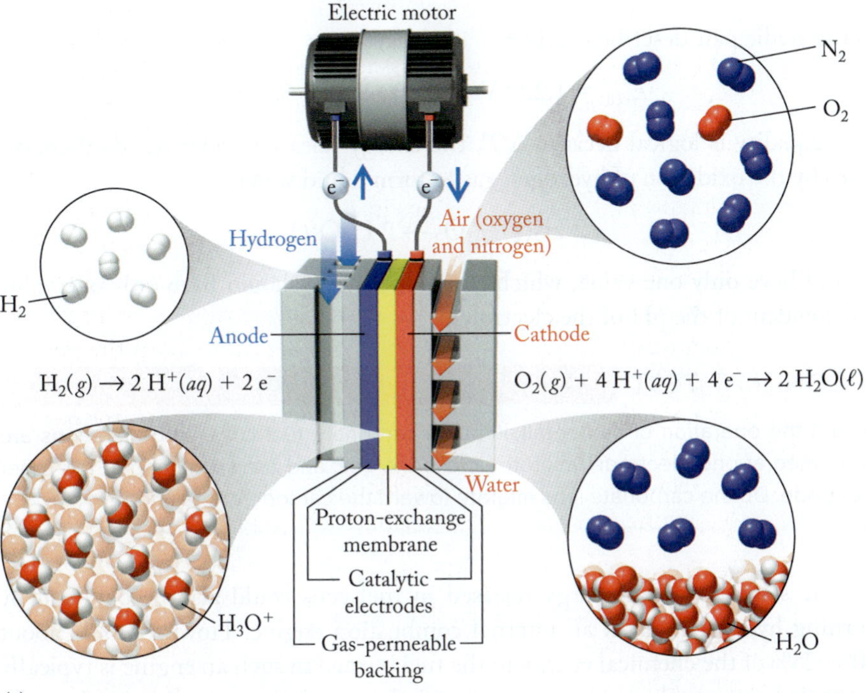

**FIGURE 17.24** (a) Most fuel cells used in vehicles have a proton-exchange membrane between the halves of the cell. Hydrogen gas diffuses to the anode, and oxygen gas diffuses to the cathode. These electrodes are made of a porous material, such as graphite, that has a relatively high surface area for a given mass of material. Catalysts on the electrode surfaces also increase the rates of the two half-reactions. (b) The fuel tanks in a 2019 Honda Clarity fuel cell vehicle can hold 5.5 kg of $H_2$ gas, which gives the car a driving range of about 500 km (300 miles).

Hydrogen ions that form at the anode migrate through the PEM to the cathode, where they are consumed at the cathode. This migration of positive charges inside the fuel cell drives the flow of electrons in the electrical device attached to it.

A single PEM fuel cell typically has a cell potential of about 1.0 V. When hundreds of these cells are assembled into fuel cell *stacks*, they can produce more than 100 kW of electrical power. That is enough to give a midsize car such as the one in Figure 17.24 a top speed of more than 160 km/h (100 mph).

PEM fuel cells are well suited for use in vehicles because they are compact, lightweight, and operate at temperatures of 60°C to 80°C. The performance of nearly all fuel cells is better at above-ambient temperatures because the higher rates of their half-reactions at these temperatures mean that they generate more electrical power.

Other fuel cells use basic electrolytes such as concentrated KOH. Pure $O_2$ is supplied to a cathode made of porous graphite containing a nickel catalyst, and $H_2$ gas is supplied to a graphite anode containing nickel(II) oxide. Hydroxide ions formed during $O_2$ reduction at the cathode,

$$O_2(g) + 2\,H_2O(\ell) + 4\,e^- \rightarrow 4\,OH^-(aq) \qquad E^\circ = 0.401\text{ V}$$

migrate through the cell to the anode, where they react with $H_2$ as it is oxidized to $H^+$ and then react with hydroxide ion to form water:

$$H_2(g) + 2\,OH^-(aq) \rightarrow 2\,H_2O(\ell) + 2\,e^-$$

This oxidation half-reaction is the reverse of the following reduction half-reaction from Table 17.1:

$$2\,H_2O(\ell) + 2\,e^- \rightarrow H_2(g) + 2\,OH^-(aq) \qquad E^\circ = -0.828\text{ V}$$

Note that this basic pair of standard electrode potentials yields the same $E^\circ_{\text{cell}}$ value,

$$E^\circ_{\text{cell}} = 0.401\text{ V} - (-0.828\text{ V}) = 1.229\text{ V}$$

as the acidic pair described earlier:

$$E^\circ_{\text{cell}} = 1.229\text{ V} - 0.000\text{ V} = 1.229\text{ V}$$

This equality is logical because $\Delta G^\circ$, the energy released under standard conditions by the oxidation of hydrogen gas to form liquid water,

$$2\,H_2(g) + O_2 \rightarrow 2\,H_2O(\ell)$$

should have only one value, which means that $E^\circ_{\text{cell}}$ should have only one value independent of the pH of the electrolyte.

## CONCEPT **TEST**

During the operation of molten alkali metal carbonate fuel cells, carbonate ions are generated at one electrode, migrate across the cell, and are consumed at the other electrode. Do the carbonate ions migrate toward the cathode or the anode?

The same chemical energy released in fuel cells could also be obtained by burning hydrogen gas in an internal combustion engine. However, only about 20%–25% of the chemical energy in the fuel burned in such an engine is typically converted into mechanical energy; most is lost to the surroundings as heat. In contrast, fuel cell technologies can convert up to about 80% of the energy released in a fuel cell's redox reaction into electrical energy. The electric motors they power are also about 80% efficient at converting electrical energy into mechanical energy.

Thus, the overall conversion efficiency of a fuel cell–powered car is theoretically as high as (80% × 80%), or 64%. The measured efficiency of the propulsion system of the car in Figure 17.24 is about 60%, which is still more than twice that of an internal combustion engine. In addition, $H_2$-fueled vehicles emit only water vapor; they produce no oxides of nitrogen, no carbon monoxide, and no $CO_2$, hence the name, "zero-emission" vehicles.

As fuel cells become even more efficient and less expensive, the principal limit on their use in passenger cars will be the availability and cost of hydrogen fuel. Although some people worry about the safety of storing hydrogen in a high-pressure tank in a car, $H_2$ has a higher ignition temperature than gasoline and spreads through the air more quickly, reducing the risk of fire. Still, hydrogen in air burns over a much wider range of concentrations than gasoline, and its flame is almost invisible.

**CONNECTION** Processes for producing hydrogen gas by reacting methane and other fuels with steam were discussed in Chapter 14.

**SAMPLE EXERCISE 17.9** Integrating Concepts: The Electrolysis of Salt

Electrolysis of sodium chloride is used industrially to produce sodium metal and chlorine gas. One process uses a Downs cell (**Figure 17.25**), which has a carbon anode and an iron cathode and uses molten NaCl as the electrolyte and source of the reactants in the cell reaction. Calcium chloride is added to the sodium chloride (melting point, 801°C; density, 2.16 g/cm$^3$) to lower its melting point so that the process can be carried out at temperatures around 600°C. The products of the electrolysis are liquid sodium (melting point, 97.8°C; density, 0.93 g/cm$^3$) and chlorine gas. An iron screen in the cell separates the electrodes so that the two products do not contact each other.

a. Write chemical equations for the half-reactions at the anode and cathode and for the overall cell reaction.
b. Use the appropriate standard reduction potentials from Table 17.1 to estimate the minimum voltage that must be applied to the cell to electrolyze molten NaCl.
c. Migration of which ions carries electric charge between the anodes and cathodes in a Downs cell?

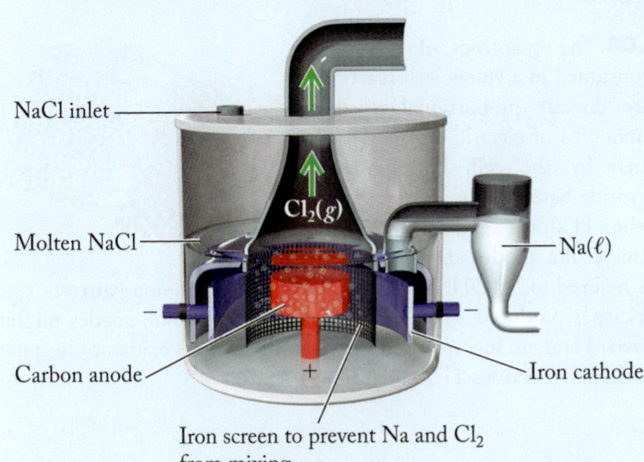


**FIGURE 17.25** The Downs cell is the primary unit used to produce sodium commercially. It is also a minor source of industrially produced chlorine. An iron screen prevents contact between the two products.

d. Calcium chloride is in the reaction mixture, but calcium metal is not produced in the cell. Why?
e. Why is keeping the two products separated important?
f. In a Downs cell, the electrodes are usually positioned at the bottom of the molten NaCl bath, but the products are collected at the top. Why?

**Collect and Organize** We are given a description of a Downs cell, the conditions under which it is run, and the physical properties of the reactants and products of the electrolytic process taking place in it. We are asked to write the cell half-reactions and overall reaction and to estimate the voltage needed to electrolyze molten NaCl. We are also asked how electricity flows through the cell, why $Ca^{2+}$ ions are not also reduced at the cathode, why elemental sodium and chlorine need to be kept separated, and why they are harvested at the top of the cell even though the electrodes at which they are generated are near the bottom.

**Analyze** Sodium ions are reduced at the cathode to elemental sodium, and chloride ions are oxidized at the anode to chlorine gas. Forming these reactive elements from a stable compound such as NaCl requires significant electrical energy and, very likely, a cell potential of several volts. Contact between the products would allow them to react and re-form NaCl. The spontaneity of this reaction is indicated by the $\Delta G^\circ_f$ of NaCl in Appendix 4 (Table A4.3): −384.2 kJ/mol. Sodium ions will be preferentially reduced over $Ca^{2+}$ ions if the standard reduction potential of $Na^+$ ions is less negative than that of $Ca^{2+}$ ions. The manner of how the products are recovered from the electrolysis cell depends on their physical properties, especially their densities relative to the reactant, molten NaCl.

**Solve**

a. The anode half-reaction is

$$2\,Cl^-(\ell) \rightarrow Cl_2(g) + 2\,e^-$$

The cathode half-reaction is

$$Na^+(\ell) + e^- \rightarrow Na(\ell)$$

b. Calculating $E^\circ_{cell}$ requires the appropriate $E^\circ_{cathode}$ and $E^\circ_{anode}$ values. Table 17.1 lists the following half-reactions and $E^\circ$ values:

$$Cl_2(g) + 2\,e^- \rightarrow 2\,Cl^-(aq) \qquad E^\circ = 1.358\ V$$

$$Na^+(aq) + e^- \rightarrow Na(s) \qquad E^\circ = -2.71\ V$$

Calculating the difference between the standard reduction potentials,

$$E^\circ_{cell} = E^\circ_{cathode} - E^\circ_{anode} = -2.71\ V - 1.358\ V = -4.07\ V$$

The negative $E^\circ_{cell}$ value is a measure of how much voltage must be applied to drive the nonspontaneous cell reaction—that is, 4.07 V. However, the actual minimum voltage will not be exactly 4.07 V. For one thing, the sodium and chloride ions are in a nonaqueous environment of molten NaCl, and their concentrations are probably not 1 *M*. However, we have seen that cell potentials change relatively little as the concentrations of reactants and products change, so 4 volts is a reasonable estimate of the voltage needed to drive the reaction in a Downs cell.

c. Migration of $Cl^-$ ions toward the positively charged anode, where they are oxidized to $Cl_2$ gas, carries part of the electrolysis current through the cell. The rest is carried by the migration of $Na^+$ ions toward the negatively charged cathode, where they are reduced to Na metal.

d. The standard reduction potential of $Ca^{2+}$ ions (Table 17.1) is −2.868 V, which is more negative than the standard reduction potential of $Na^+$ ions (−2.71 V). Therefore, $Na^+$ ions are preferentially reduced in the presence of $Ca^{2+}$ ions.

e. The $\Delta G^\circ_f$ of solid NaCl is −384.2 kJ/mol, which indicates that elemental sodium and chlorine spontaneously react to form NaCl. Therefore, these two electrolysis products must be kept away from each other.

f. Liquid sodium is less dense than molten sodium chloride and, as it forms in the cell, it floats on top of the molten salt. Chlorine is a gas that bubbles out of molten NaCl and is collected in hoods above the liquid.

**Think About It** The Downs cell is an electrolytic cell that requires a large input of electrical energy to drive the nonspontaneous cell reaction. Sodium and chlorine are both so reactive that special care and extreme caution are required to collect and transport them.

## SUMMARY

**LO1** In an **electrochemical cell**, an oxidation half reaction at the **anode** releases electrons, which flow through an external circuit to the **cathode** where they are consumed in a reduction half reaction. Migration of the ions in the cell's electrolyte between the cathode and anode compartments completes the electrical circuit. A **cell diagram** shows how the components of the cathode and anode compartments of the cell are connected. (Section 17.2)

**LO2** The difference between the **standard reduction potentials ($E^\circ$)** of a cell's cathode and anode half-reactions is equal to the **standard cell potential ($E^\circ_{cell}$)**. The value of $E^\circ_{cell}$ is a measure of the **electromotive force (emf)** with which a **voltaic cell** can pump electrons through an external circuit under standard conditions. If reactants and products are not in their standard states, this value is simply called the **cell potential ($E_{cell}$)**. All standard cell potentials are referenced to the cell potential of the **standard hydrogen electrode (SHE)**: $E^\circ_{SHE} = 0.000$ V. (Sections 17.3 and 17.5)

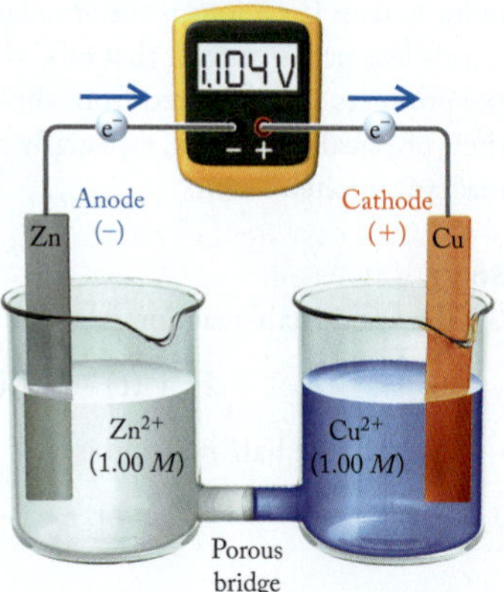


**LO3** A voltaic cell has a positive cell potential ($E_{cell} > 0$) and its cell reaction has a negative change in free energy ($\Delta G^\circ_{cell} < 0$). This decrease in free energy in a voltaic cell is available to do work in an external electrical circuit. The **Faraday constant (*F*)** relates the quantity of electric charge to the number of moles of electrons and indirectly to the number of moles of reactants. (Section 17.4)

**LO4** The potential of a voltaic cell decreases as reactants turn into products. The **Nernst equation** describes how cell potential changes with concentration changes. (Section 17.6)

**LO5** The potential of a voltaic cell approaches zero as the cell reaction approaches chemical equilibrium, at which point $E_{cell} = 0$ and $Q = K$. (Section 17.6)

**LO6** The quantities of reactants consumed in a voltaic cell reaction are directly proportional to the coulombs of electric charge delivered by the cell. Nickel–metal hydride batteries supply electricity when H atoms are oxidized to $H^+$ ions at the anodes and NiO(OH) is reduced to $Ni(OH)_2$ at the cathodes. In lithium–ion batteries, electricity is produced when $Li^+$ ions stored in graphite anodes migrate toward and are incorporated into transition metal oxide or phosphate cathodes. (Sections 17.7 and 17.9)

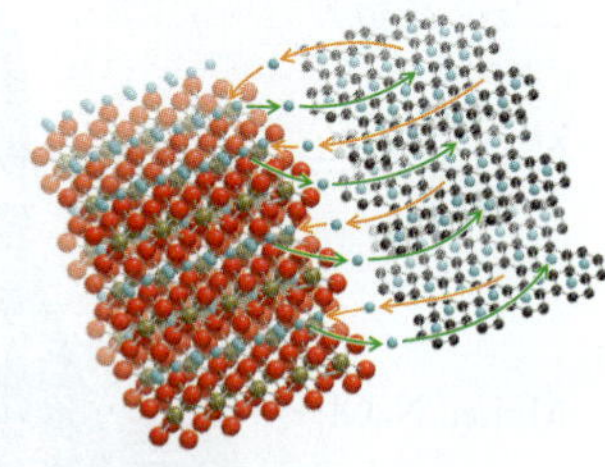

## PARTICULATE PREVIEW WRAP-UP

- The two half-reactions describing the oxidation of hydrogen and reduction of oxygen are

$$H_2(g) \rightarrow 2\,H^+(aq) + 2\,e^-$$
$$O_2(g) + 4\,H^+(aq) + 4\,e^- \rightarrow 2\,H_2O(\ell)$$

  which sum to this balanced chemical equation:

$$2\,H_2(g) + O_2(g) \rightarrow 2\,H_2O(\ell)$$

- In an operating fuel cell, $H^+$ ions produced by oxidation of $H_2$ gas at the anode migrate through a membrane to the cathode, where they are consumed during reduction of $O_2$ gas. The electromotive force generated by this production and consumption of $H^+$ ions pumps electrons from the anode to the cathode through an external circuit that includes the vehicle's electric motor.
- Hydrogen fuel cells produce only water, and batteries and electric motors do not emit significant quantities of any atmospheric pollutants during normal operation.

## PROBLEM-SOLVING SUMMARY

| Type of Problem | Concepts and Equations | Sample Exercises |
|---|---|---|
| **Describing an electrochemical cell** | Oxidation occurs at the anode and reduction occurs at the cathode. Electrons flow through an external circuit from anode to cathode as positive and negative ions flow between the cells compartments, balancing the charge imbalance created by the external flow of electrons. | **17.1** |
| **Diagramming an electrochemical cell** | Start with the symbol for the anode on the left; use double vertical lines to separate it from the symbol for the cathode. Using single lines to separate phases, list species in anode compartments and then those in the cathode compartment; separate species in the same phase with commas. Insert concentrations and/or partial pressures if known. | **17.2** |
| **Identifying anode and cathode half-reactions and calculating the value of $E^\circ_{cell}$** | The half-reaction with the more positive standard reduction potential is the cathode half-reaction. $E^\circ_{cell} = E^\circ_{cathode} - E^\circ_{anode}$ (17.1) | **17.3** |
| **Relating $\Delta G^\circ_{cell}$ and $E^\circ_{cell}$** | $\Delta G^\circ_{cell} = -nFE^\circ_{cell}$ (17.6) where $n$ is the number of moles of electrons transferred in the cell reaction and $F$ is the Faraday constant, $9.65 \times 10^4$ C/mol. | **17.4** |
| **Calculating $E_{cell}$ from $E^\circ_{cell}$ and the concentrations of reactants and products** | $E_{cell}$ at 25°C is related to $E^\circ_{cell}$ and the cell reaction quotient by the Nernst equation: $E_{cell} = E^\circ_{cell} - \frac{0.0592\text{ V}}{n}\log Q$ (17.9) | **17.5** |
| **Calculating $K$ and $\Delta G^\circ$ for a redox reaction from the standard potentials of its half-reactions** | Equation 17.1 can be modified to apply to any redox reaction: $E^\circ_{rxn} = E^\circ_{reduction} - E^\circ_{oxidation}$ $E^\circ_{rxn}$ is related to $K$ at 25°C by $\log K = \frac{nE^\circ_{rxn}}{0.0592\text{ V}}$ (17.11) A modified Equation 17.6 relates $\Delta G^\circ_{rxn}$ and $E^\circ_{rxn}$: $\Delta G^\circ_{rxn} = -nFE^\circ_{rxn}$ | **17.6** |
| **Relating the mass of a reactant in a cell reaction to a quantity of electrical charge** | Determine the ratio of moles of reactants to moles of electrons transferred; use the Faraday constant to relate coulombs of charge to moles of electrons. | **17.7** |
| **Calculating the time to oxidize a quantity of reactant** | Use the Faraday constant and the relation between moles of electrons and moles of reactants to describe an electrolytic process. | **17.8** |

# VISUAL PROBLEMS

*(Answers to boldface end-of-chapter questions and problems are in the back of the book.)*

**17.1.** In the voltaic cell shown in Figure P17.1, the greater density of a concentrated solution of $CuSO_4$ allows a less concentrated solution of $ZnSO_4$ solution to be (carefully) layered on top of it. Why is a porous separator not needed in this cell?

**FIGURE P17.1**

17.2. In the voltaic cell shown in Figure P17.2, the concentrations of $Cu^{2+}$ and $Cd^{2+}$ are 1.00 *M*. Use the standard potentials in Appendix 6 to answer the following questions.

a. Which electrode is the anode, and which is the cathode?
b. What are the directions of electron flow and ion flow?
c. Which electrode gains mass?

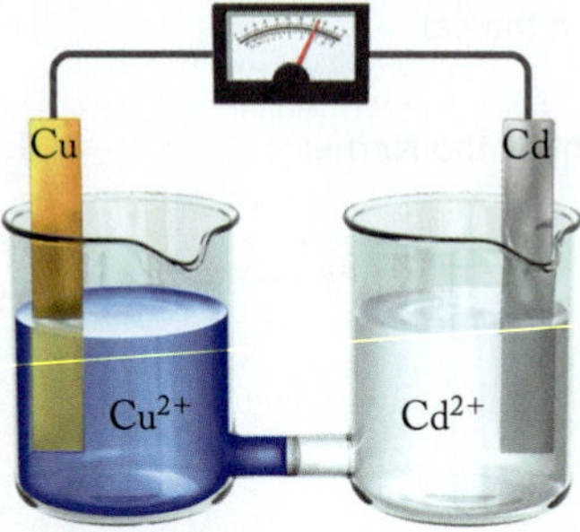


**FIGURE P17.2**

**17.3.** In the voltaic cell shown in Figure P17.3, $[Ag^+] = [H^+] = 1.00$ *M*. Use the standard potentials in Appendix 6 to answer the following questions.

a. Which electrode is the anode, and which is the cathode?
b. What are the directions of electron flow and ion flow?
c. Which electrode gains mass?

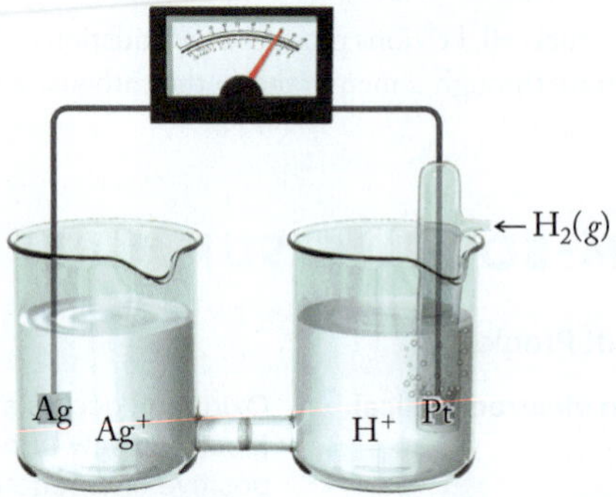


**FIGURE P17.3**

*17.4. In many electrochemical cells, the electrodes are metals that carry electrons to and from the cell but are not chemically changed by the cell reaction. Each of the highlighted clusters in the periodic table in Figure P17.4 consists of three metals. Which of the highlighted clusters is best suited to form inert electrodes?

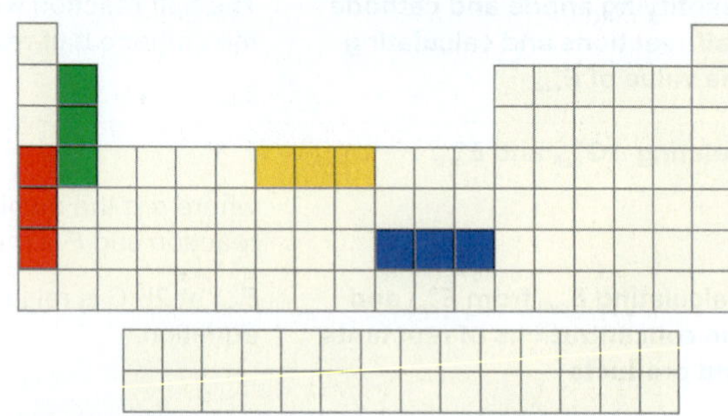

**FIGURE P17.4**

**17.5.** Which of the four curves in Figure P17.5 best represents the dependence of the potential of a lead–acid battery on the concentration of sulfuric acid? Note that the scale of the *x*-axis is logarithmic.

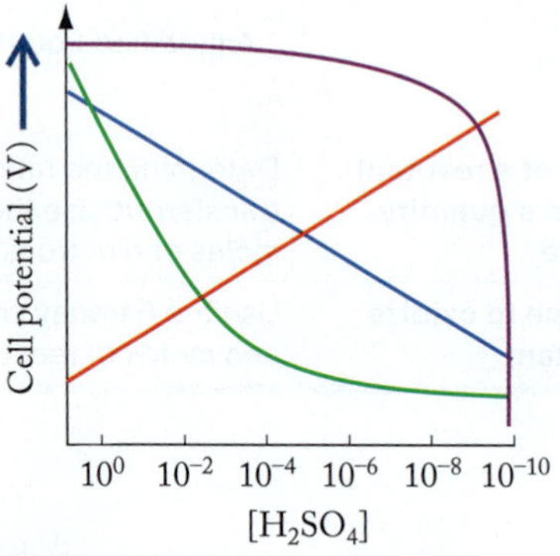


**FIGURE P17.5**

17.6. From front to back, the sizes of the batteries in Figure P17.6 are AAA, AA, C, and D, respectively. The performance of batteries like these is often expressed in units such as (a) volts, (b) watt-hours, or (c) milliampere-hours. Which of the values differ significantly between the four batteries?

**FIGURE P17.6**

**17.7.** The apparatus in Figure P17.7 is used for the electrolysis of water. Hydrogen and oxygen gas are collected in the two inverted burets. An inert electrode at the bottom of the left buret is connected to the negative terminal of a 6-volt battery; the electrode in the buret on the right is connected to the positive terminal. A small quantity of sulfuric acid is added to speed up the electrolytic reaction.

a. What are the half-reactions at the left and right electrodes and their standard potentials?

b. Why does sulfuric acid make the electrolysis reaction go more rapidly?

Overall cell reaction

$H_2O(\ell) \rightarrow H_2(g) + \frac{1}{2} O_2(g)$

**FIGURE P17.7**

17.8. An electrolytic apparatus identical to the one shown in Figure P17.7 is used to electrolyze water, but the reaction is sped up by the addition of sodium carbonate instead of sulfuric acid.

a. What are the half-reactions and the standard potentials for the electrodes on the left and right?

b. Why does sodium carbonate make the electrolysis reaction go more rapidly?

**17.9. Redox Flow Batteries** Redox flow batteries are like fuel cells in that the reactants are continuously delivered to an electrochemical cell from large storage tanks. In the cell shown in Figure P17.9, an ion-exchange membrane separates the reactants from each other; the arrows indicate the direction of flow for the reactants. The reactions in one type of redox flow battery can be diagrammed as

$$C(s) \mid V^{2+}(aq), V^{3+}(aq) \parallel VO_2^{+}(aq), VO^{2+}(aq) \mid C(s)$$

a. Write balanced half-reactions for the processes that occur at the anode and cathode in acid solution.

b. Write a balanced chemical equation for the overall reaction and determine the number of electrons transferred.

c. Label the diagram in Figure P17.9 by putting the appropriate reaction in each compartment and showing the direction of electron flow.

d. How does the value of $E_{cell}$ (1.25 V) change with pH?

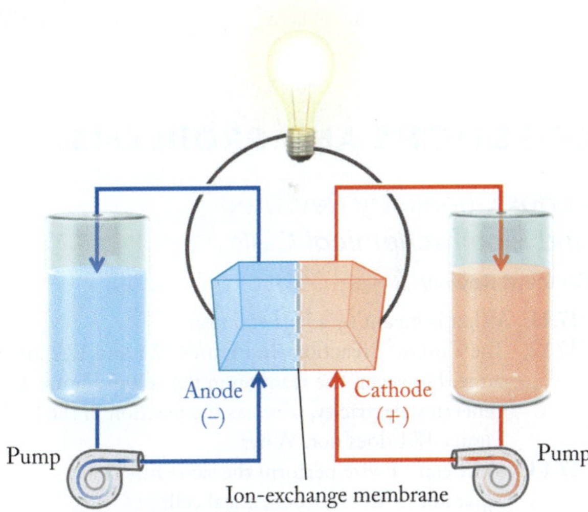


**FIGURE P17.9**

17.10. Use representations [A] through [I] in Figure P17.10 to answer questions a–f. [E] shows silver deposited onto copper. Solvent water molecules and counter ions are omitted from images [B] and [H] for clarity. If the materials in [A], [C], [G], and [I] were connected by a salt bridge and external circuit to generate an electrochemical cell that produced [E],

a. Which metal would be the cathode? Which solution would surround it?
b. Which metal would be the anode? Which solution would surround it?
c. When the reaction in [E] is finished, what will happen to the light blue color of the solution?
d. How could [E] be produced without a salt bridge or external circuit?
e. Of the four particulate images [B], [D], [F], and [H] in Figure P17.10, which two correspond to the solutions [G] and [I]?
f. Which particulate image represents the copper wire with silver deposited onto it? What does image [H] depict?

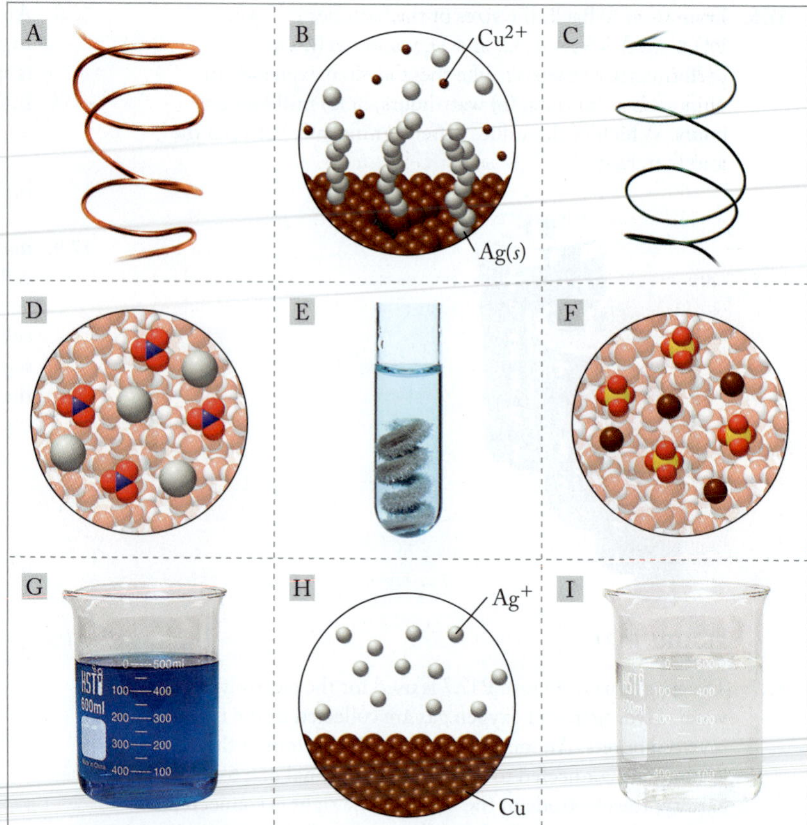


FIGURE P17.10

## QUESTIONS AND PROBLEMS

### Redox Chemistry Revisited and Electrochemical Cells

#### Concept Review

**17.11.** What is meant by a half-reaction?

17.12. The $Zn/Cu^{2+}$ reactions in Figures 17.1 and 17.2 are the same. However, the reaction in the cell in Figure 17.2 generates electricity, whereas the reaction in the beaker in Figure 17.1 does not. Why?

**17.13.** Why can't a wire perform the same function as a porous separator in an electrochemical cell?

17.14. In a voltaic cell, why is the cathode labeled the positive terminal and the anode the negative terminal?

#### Problems

**17.15.** Balance the following half-reactions by adding the appropriate number of electrons. Identify the oxidation half-reactions and the reduction half-reactions.

a. $Br_2(\ell) \rightarrow 2\ Br^-(aq)$
b. $Pb(s) + 2\ Cl^-(aq) \rightarrow PbCl_2(s)$
c. $O_3(g) + 2\ H^+(aq) \rightarrow O_2(g) + H_2O(\ell)$
d. $H_2S(g) \rightarrow S(s) + 2\ H^+(aq)$

17.16. Balance the following half-reactions by adding the appropriate number of electrons. Which are oxidation half-reactions, and which are reduction half-reactions?

a. $TiO^{2+}(aq) + 2\ H^+(aq) \rightarrow Ti^{3+}(aq) + H_2O(\ell)$
b. $S_4^{2+}(aq) \rightarrow 2\ S_2^{2-}(aq)$
c. $VO_2^{\ +}(aq) + 2\ H^+(aq) \rightarrow VO^{2+}(aq) + H_2O(\ell)$
d. $Fe(CN)_6^{\ 3-}(aq) \rightarrow Fe(CN)_6^{\ 4-}(aq)$

---

***17.17.** **Groundwater Chemistry** Write a half-reaction for the oxidation of magnetite ($Fe_3O_4$) to hematite ($Fe_2O_3$) in acidic groundwater.

*17.18. Write a half-reaction for the oxidation of the manganese in $MnCO_3$ to $MnO_2$ in neutral groundwater, where the principal carbonate species is $HCO_3^{\ -}$.

---

**17.19.** **Lithium–Iron Sulfide Batteries** Voltaic cells with nonaqueous electrolytes and based on the oxidation of Li(s) to $Li_2S(s)$ and the reduction of $FeS_2(s)$ to Fe(s) and $S^{2-}$ ions provide the same voltage as traditional alkaline batteries, but the voltaic cells offer more storage capacity.

a. Write half-reactions for the cell's anode and cathode.
b. Write a balanced cell reaction.
c. Diagram the cell.

17.20. **Zinc–Nickel Batteries** Replacing cadmium with zinc in a NiCad battery avoids the use of the toxic element cadmium. A voltaic cell based on the reaction between $NiO(OH)(s)$ and $Zn(s)$ in alkaline electrolyte ($OH^-$) produces $Ni(OH)_2(s)$ and $ZnO(s)$.
a. Write half-reactions for the cell's anode and cathode.
b. Write a balanced cell reaction.
c. How many electrons are transferred in the cell reaction?

**17.21.** **Lithium–Air Batteries** In recent years, engineers have been working on a lithium–air battery (Li–$O_2$) as an alternative energy source for electric vehicles.
a. Write the two half-reactions for the battery and diagram the cell in acidic solution.
b. Write the two half-reactions for the battery and diagram the cell in basic solution.

17.22. **Sodium–Sulfur Batteries** Less expensive materials make sodium–sulfur batteries attractive for electric vehicles. The balanced chemical equation for this battery is:

$$2\ Na(\ell) + 3\ S(\ell) \rightarrow Na_2S_3(s)$$

a. Determine the number of electrons transferred in the cell reaction.
b. Draw the cell diagram.
*c. Draw a Lewis structure for the product, $Na_2S_3$, which contains an $S_3^{2-}$ anion.

## *Standard Potentials*

### Concept Review

**17.23.** Is oxygen or chlorine a stronger oxidizing agent under standard conditions? Use the standard reduction potentials in Table A6.1 to support your answer.

*17.24. Why is $O_2$ a stronger oxidizing agent in acid than in base? Use standard reduction potentials from Appendix 6 to support your answer.

**17.25.** Two half-reactions both have reduction potentials $E^\circ < 0$. Can these half-reactions be combined to make a voltaic cell? What if both half-reactions have $E^\circ > 0$?

17.26. A student is designing a new battery based on the reduction potentials in Appendix 6. If choosing one half-reaction with $E^\circ < 0$ and one half-reaction with $E^\circ > 0$, will constructing a voltaic cell with $E^\circ_{cell} > 0$ *always* be possible?

### Problems

**17.27.** Using standard reduction potentials from Appendix 6 and the following equations for half-reactions:

$$Cu^{2+}(aq) + 2\ e^- \rightarrow Cu(s)$$

$$Co^{2+}(aq) + 2\ e^- \rightarrow Co(s)$$

$$Hg^{2+}(aq) + 2\ e^- \rightarrow Hg(s)$$

a. Identify the combination of half-reactions that would lead to the largest value of $E^\circ_{cell}$.
b. Identify the combination of half-reactions that would lead to the smallest positive value of $E^\circ_{cell}$.

17.28. If a piece of silver and a piece of copper are placed in 1.00 L of a solution in which $[Ag^+] = [Cu^{2+}] = 1.00\ M$,
a. Will the following reaction proceed spontaneously?

$$2\ Ag(s) + Cu^{2+}(aq) \rightarrow 2\ Ag^+(aq) + Cu(s)$$

b. Will the reverse of the above reaction proceed spontaneously?

**17.29.** In a voltaic cell such as the $Zn/Cu^{2+}$ cell in Figure 17.2, the copper half of the cell is replaced with a Ni electrode immersed in a 1.00 $M$ solution of $NiSO_4$.
a. Is the standard potential of this Ni/Zn cell greater than, the same as, or less than 1.10 V?
b. Which electrode in the Ni/Zn cell is the anode?

17.30. In a voltaic cell such as the $Zn/Cu^{2+}$ cell in Figure 17.2, the copper half of the cell is replaced with a silver wire in contact with 1.00 $M$ $Ag^+(aq)$.
a. What is the value of $E^\circ_{cell}$?
b. Which electrode is the anode?

## *Chemical Energy and Electrical Work*

### Concept Review

**17.31.** The negative sign in Equation 17.3 ($w_{elec} = -QE_{cell}$) seems to indicate that a voltaic cell with a *positive* cell potential does *negative* electrical work. How is this possible?

*17.32. Mechanical work ($w$) is done by exerting a force ($F$) to move an object through a distance ($d$) according to the equation $w = F \times d$. Explain how this definition of work relates to electrical work.

### Problems

**17.33.** Starting with the appropriate standard free energies of formation from Appendix 4, calculate the values of $\Delta G^\circ$ and $E^\circ_{cell}$ of the following reactions:
a. $2\ Cu^+(aq) \rightarrow Cu^{2+}(aq) + Cu(s)$
b. $Ag(s) + Fe^{3+}(aq) \rightarrow Ag^+(aq) + Fe^{2+}(aq)$

17.34. Starting with the appropriate standard free energies of formation from Appendix 4, calculate the values of $\Delta G^\circ$ and $E^\circ_{cell}$ of the following reactions:
a. $FeO(s) + H_2(g) \rightarrow Fe(s) + 2\ H_2O(\ell)$
b. $2\ Pb(s) + O_2(g) + 2\ H_2SO_4(aq) \rightarrow 2\ PbSO_4(s) + 2\ H_2O(\ell)$

**17.35.** **Flashlights** For many years, the 1.50 V batteries used to power flashlights were based on the following cell reaction:

$$Zn(s) + 2\ NH_4Cl(s) + 2\ MnO_2(s) \rightarrow Zn(NH_3)_2Cl_2(s) + Mn_2O_3(s) + H_2O(\ell)$$

What is the value of $\Delta G_{cell}$?

17.36. **Laptops** The first generation of laptop computers was powered by nickel–cadmium (NiCad) batteries, which generated 1.20 V on the basis of the following cell reaction:

$$Cd(s) + 2\ NiO(OH)(s) + 2\ H_2O(\ell) \rightarrow Cd(OH)_2(s) + 2\ Ni(OH)_2(s)$$

What is the value of $\Delta G_{cell}$?

**17.37.** **Nickel–Sodium Batteries** Researchers in England are developing a battery for electric vehicles on the basis of the reaction between $NiCl_2(s)$ and $Na(s)$:

$$2\ Na(s) + NiCl_2(s) \rightarrow Ni(s) + 2\ NaCl(s)$$

The cells in the battery produce 2.58 V.

a. Assign oxidation numbers to each element in the nickel and sodium compounds.
b. How many electrons are transferred in the overall reaction?
c. What is the value of $\Delta G_{cell}$?

17.38. Thomas Edison, the inventor of the incandescent lightbulb, developed a voltaic cell that delivers 1.4 V of cell potential in an alkaline electrolyte on the basis of the following cell reaction:

$$3\ Fe(s) + 8\ NiO(OH)(s) + 4\ H_2O(\ell) \rightarrow 8\ Ni(OH)_2(s) + Fe_3O_4(s)$$

a. Assign oxidation numbers to each element in each of the nickel and iron compounds.
b. How many electrons are transferred in the overall reaction?
c. What is the value of $\Delta G_{cell}$?

## A Reference Point: The Standard Hydrogen Electrode

### Concept Review

**17.39.** What is the function of platinum in the standard hydrogen electrode?

*17.40. How would replacing $H_2$ with $O_2$ in the standard hydrogen electrode (making it a standard *oxygen* electrode) affect the standard reduction potential values in Appendix 6?

### Problems

***17.41.** For each pair of half-reactions, write a balanced equation for a spontaneous cell reaction, noting which half-reaction takes place as written at each cathode and which proceed in reverse at each anode. Calculate the values of $E^\circ_{cell}$ and $\Delta G^\circ$. How would $E^\circ_{cell}$ values change if the standard potentials of their anode and cathode half-reactions were referenced to an electrode with an $E^\circ$ of 0.222 V rather than a SHE?

a. $Cl_2(g) + 2\ e^- \rightarrow 2\ Cl^-(aq)$
$Br_2(\ell) + 2\ e^- \rightarrow 2\ Br^-(aq)$
b. $Zn^{2+}(aq) + 2\ e^- \rightarrow Zn(s)$
$Ni^{2+}(aq) + 2\ e^- \rightarrow Ni(s)$
c. $I_2(s) + 2\ e^- \rightarrow 2\ I^-(aq)$
$NO_3^-(aq) + H_2O(\ell) + 2\ e^- \rightarrow NO_2^-(aq) + 2\ OH^-(aq)$

*17.42. Voltaic cells based on the following pairs of half-reactions are constructed. For each pair, write a balanced equation for the cell reaction, and identify which half-reaction takes place at each anode and cathode. Calculate the values of $E^\circ_{cell}$ and $\Delta G^\circ$. How would $E^\circ_{cell}$ values change if the standard potentials of their anode and cathode half-reactions were referenced to an electrode with an $E^\circ$ of 0.245 V rather than a SHE?

a. $Cd^{2+}(aq) + 2\ e^- \rightarrow Cd(s)$
$Ag^+(aq) + e^- \rightarrow Ag(s)$
b. $AgBr(s) + e^- \rightarrow Ag(s) + Br^-(aq)$
$MnO_2(s) + 4\ H^+(aq) + 2\ e^- \rightarrow Mn^{2+}(aq) + 2\ H_2O(\ell)$
c. $PtCl_4^{2-}(aq) + 2\ e^- \rightarrow Pt(s) + 4\ Cl^-(aq)$
$AgCl(s) + e^- \rightarrow Ag(s) + Cl^-(aq)$

## The Effect of Concentration on $E_{cell}$

### Concept Review

**17.43.** Why does the operating cell potential of most batteries change little until the battery is nearly discharged?

17.44. Would the reduction potential of the standard hydrogen electrode increase, decrease, or stay the same if the concentration of hydronium ions in the solution inside the electrode was lowered from 1.00 *M* to 0.100 *M*?

### Problems

**17.45.** Calculate the $E_{cell}$ value at 25°C for the cell based on the reaction

$$Fe^{3+}(aq) + Cr^{2+}(aq) \rightarrow Fe^{2+}(aq) + Cr^{3+}(aq)$$

when $[Fe^{3+}] = [Cr^{2+}] = 1.50 \times 10^{-3}\ M$ and $[Fe^{2+}] = [Cr^{3+}] = 2.5 \times 10^{-4}\ M$.

17.46. Calculate the $E_{cell}$ value at 25°C for the cell based on the reaction

$$Cu(s) + 2\ Ag^+(aq) \rightarrow Cu^{2+}(aq) + 2\ Ag(s)$$

when $[Ag^+] = 2.56 \times 10^{-3}\ M$ and $[Cu^{2+}] = 8.25 \times 10^{-4}\ M$.

---

**17.47.** Using the appropriate standard potentials from Appendix 6, determine the equilibrium constant for the following reaction at 25°C:

$$Fe^{3+}(aq) + Cr^{2+}(aq) \rightarrow Fe^{2+}(aq) + Cr^{3+}(aq)$$

17.48. Using the appropriate standard potentials from Appendix 6, determine the equilibrium constant at 25°C for the following reaction between $MnO_2$ and $Fe^{2+}$ in acidic solution:

$$4\ H^+(aq) + MnO_2(s) + 2\ Fe^{2+}(aq) \rightarrow Mn^{2+}(aq) + 2\ Fe^{3+}(aq) + 2\ H_2O(\ell)$$

---

**17.49. California Condors** The continued recovery of endangered California condors is threatened by lead poisoning because hunters still use ammunition containing lead, and some of their kills are eaten by the condors. Consumed lead may dissolve in a reaction with a condor's stomach acid:

$$Pb(s) + 2\ H^+(aq) \rightarrow Pb^{2+}(aq) + H_2(g)$$

a. Use the appropriate standard reduction potentials in Appendix 6 to calculate $E^\circ_{rxn}$ for this reaction.
b. What is the value of $E_{rxn}$ in stomach acid with a pH of 0.80 if the value of $[Pb^{2+}]$ is $2.00 \times 10^{-6}\ M$?
c. What are the values of $\Delta G^\circ$ and *K* for the reaction at 25°C?

17.50. **Glucose Metabolism** The standard potentials for the reduction of nicotinamide adenine dinucleotide ($NAD^+$) and oxaloacetate (reactants in the multistep metabolism of glucose) are as follows:

$$NAD^+(aq) + 2\,H^+(aq) + 2\,e^- \rightarrow NADH(aq) + H^+(aq) \qquad E° = -0.320\text{ V}$$

$$\text{Oxaloacetate}(aq) + 2\,H^+(aq) + 2\,e^- \rightarrow \text{malate}(aq) \qquad E° = -0.166\text{ V}$$

a. Calculate the standard potential for the following reaction:

$$\text{Oxaloacetate}(aq) + NADH(aq) + H^+(aq) \rightarrow \text{malate}(aq) + NAD^+(aq)$$

b. Calculate the equilibrium constant for the reaction at 25°C.

**17.51.** Permanganate ion can oxidize sulfite to sulfate in basic solution as follows:

$$2\,MnO_4^-(aq) + 3\,SO_3^{2-}(aq) + H_2O(\ell) \rightarrow 2\,MnO_2(s) + 3\,SO_4^{2-}(aq) + 2\,OH^-(aq)$$

Determine the potential for the reaction ($E_{rxn}$) at 25°C when the concentrations of the reactants and products are $[MnO_4^-] = 0.150\ M$, $[SO_3^{2-}] = 0.256\ M$, $[SO_4^{2-}] = 0.178\ M$, and $[OH^-] = 0.0100\ M$. Will the value of $E_{rxn}$ increase or decrease as the reaction proceeds?

*17.52. Manganese dioxide is reduced by iodide ion in acid solution as follows:

$$MnO_2(s) + 2\,I^-(aq) + 4\,H^+(aq) \rightarrow Mn^{2+}(aq) + I_2(aq) + 2\,H_2O(\ell)$$

Determine the $E_{rxn}$ value of the reaction at 25°C when the initial concentrations of the components are $[I^-] = 0.225\ M$, $[H^+] = 0.900\ M$, $[Mn^{2+}] = 0.100\ M$, and $[I_2] = 0.00114\ M$. If the solubility of iodine in water is approximately 0.114 *M*, will the value of $E_{rxn}$ increase or decrease as the reaction proceeds?

**17.53.** A copper penny dropped into a solution of nitric acid produces a mixture of nitrogen oxides. The following reaction describes the formation of NO, one of the products:

$$3\,Cu(s) + 8\,H^+(aq) + 2\,NO_3^-(aq) \rightarrow 2\,NO(g) + 3\,Cu^{2+}(aq) + 4\,H_2O(\ell)$$

a. Starting with the appropriate standard potentials from Appendix 6, calculate $E°_{rxn}$ for this reaction.
b. Calculate $E_{rxn}$ at 25°C when $[H^+] = 0.100\ M$, $[NO_3^-] = 0.0250\ M$, $[Cu^{2+}] = 0.0375\ M$, and the partial pressure of NO = 0.00150 atm.

17.54. Chlorine dioxide ($ClO_2$) is produced by the following reaction of chlorate ($ClO_3^-$) with $Cl^-$ in acid solution:

$$2\,ClO_3^-(aq) + 2\,Cl^-(aq) + 4\,H^+(aq) \rightarrow 2\,ClO_2(g) + Cl_2(g) + 2\,H_2O(\ell)$$

a. Determine $E°$ for the reaction.
b. The reaction at 25°C produces a mixture of gases in the reaction vessel in which $P_{ClO_2} = 2.0$ atm and $P_{Cl_2} = 1.00$ atm. Calculate $[ClO_3^-]$ if, at equilibrium, $[H^+] = [Cl^-] = 10.0\ M$.

**17.55.** What is the $E_{cell}$ value at 25°C of a voltaic cell such as the one in Figure 17.15 in which one compartment contains a Zn electrode immersed in 0.200 *M* $Zn(NO_3)_2$, and the other compartment contains a Zn electrode immersed in $5.00 \times 10^{-4}\ M\ Zn(NO_3)_2$. Which electrode is the anode? The solutions in both compartments are also 1.00 *M* in $NaNO_3$.

17.56. What is the $E_{cell}$ value at 25°C of a voltaic cell such as the one in Figure 17.15 in which one compartment contains a Ag/AgCl electrode immersed in a saturated solution of KCl, and the other contains a Ag/AgCl electrode immersed in 1.00 *M* of KCl? Which electrode is the cathode? The molar solubility of KCl at 25°C is 4.81 *M*.

## Relating Battery Capacity to Quantities of Reactants

### Concept Review

**17.57.** One 12-volt lead–acid battery has a higher ampere-hour rating than another. Which of the following parameters are likely to be different for the two batteries?
a. Individual cell potentials
b. Anode half-reactions
c. Total masses of electrode materials
d. Number of cells
e. Electrolyte composition
f. Combined surface areas of their electrodes

17.58. In a voltaic cell based on the $Zn/Cu^{2+}$ cell reaction,

$$Zn(s) + Cu^{2+}(aq) \rightarrow Cu(s) + Zn^{2+}(aq)$$

the reaction has exactly one mole of each reactant and product. A second cell based on the following cell reaction:

$$Cd(s) + Cu^{2+}(aq) \rightarrow Cu(s) + Cd^{2+}(aq)$$

also has exactly one mole of each reactant and product. Which of the following statements about these two cells is true?
a. Their cell potentials are the same.
b. The masses of their electrodes are the same.
c. The quantities of electric charge that they can produce are the same.
d. The quantities of electric energy that they can produce are the same.

### Problems

***17.59.** Which of the following voltaic cells, **A** or **B**, will produce the greater quantity of electric charge per gram of anode material?

Cell A: $Cd(s) + 2\,NiO(OH)(s) + 2\,H_2O(\ell) \rightarrow 2\,Ni(OH)_2(s) + Cd(OH)_2(s)$

Cell B: $4\,Al(s) + 3\,O_2(g) + 6\,H_2O(\ell) + 4\,OH^-(aq) \rightarrow 4\,Al(OH)_4^-(aq)$

*17.60. Which of the following voltaic cells, **C** or **D**, will produce the greater quantity of electric charge per gram of anode material?

Cell C: $Zn(s) + MnO_2(s) + 4\,H^+(aq) \rightarrow Zn^{2+}(aq) + Mn^{2+}(aq)$

Cell D: $Li(s) + MnO_2(s) \rightarrow LiMnO_2(s) \qquad E°_{cell} = 3.15\text{ V}$

*17.61. Which of the following voltaic cell reactions, **E** or **F**, delivers more electrical energy per gram of anode material at 25°C ?

Reaction E: $Zn(s) + 2\ NiO(OH)(s) + H_2O(\ell) \rightarrow 2\ Ni(OH)_2(s) + ZnO(s)$

Reaction F: $Li(s) + MnO_2(s) \rightarrow LiMnO_2(s)$ $E^\circ_{cell} = 3.15$ V

*17.62. Which of the following voltaic cell reactions, **G** or **H**, delivers more electrical energy per gram of anode material at 25°C ?

Reaction G: $Zn(s) + Ni(OH)_2(s) \rightarrow ZnO(s) + H_2O(\ell) + Ni(s)$

Reaction H: $2\ Zn(s) + O_2(g) \rightarrow 2\ ZnO(s)$

## Corrosion: Unwanted Electrochemical Reactions

### Concept Review

**17.63.** When the iron skeleton of the Statue of Liberty was replaced with stainless steel, the asbestos mats that had separated the skeleton from the copper exterior were replaced with Teflon spacers. Why was Teflon a good choice?

17.64. What does a sacrificial anode do to protect a metal structure, and why is the process called *cathodic* protection?

**17.65.** **Aquarium Windows** The windows of the giant ocean tank at the New England Aquarium (Figure P17.65) are held in place with aluminum frames. What would be a good material to use to make sacrificial anodes for the frames?

**FIGURE P17.65**

17.66. Why is corrosion a bigger problem for metal structures in contact with seawater than in contact with fresh water?

## Electrolytic Cells and Rechargeable Batteries

### Concept Review

**17.67.** The positive terminal of a voltaic cell is the cathode. However, the cathode of an electrolytic cell is connected to the negative terminal of a power supply. Explain this difference in polarity.

17.68. The anode in an electrochemical cell is defined as the electrode where oxidation takes place. Why is the anode in an electrolytic cell connected to the positive (+) terminal of an external supply, whereas the anode in a voltaic cell battery is connected to the negative (−) terminal?

**17.69.** The salts obtained from the evaporation of seawater can act as a source of halogens, principally $Cl_2$ and $Br_2$, through the electrolysis of the molten alkali metal halides. As the potential of the anode in an electrolytic cell is increased, which of these two halogens forms first?

17.70. In the electrolysis described in Problem 17.69, why is it necessary to use molten salts rather than seawater itself?

**17.71.** **Quantitative Analysis** Electrolysis can be used to determine the concentration of $Cu^{2+}$ in a given volume of solution by electrolyzing the solution in a cell equipped with a platinum cathode. If all the $Cu^{2+}$ is reduced to Cu metal at the cathode, the increase in mass of the electrode provides a measure of the concentration of $Cu^{2+}$ in the original solution. To ensure the complete (99.99%) removal of the $Cu^{2+}$ from a solution in which $[Cu^{2+}]$ is initially about 1.0 *M*, must the potential of the cathode (versus SHE) be more negative or less negative than 0.34 V (the standard potential for $Cu^{2+} + 2\ e^- \rightarrow Cu$)?

17.72. A high school chemistry student wishes to show how water can be separated into hydrogen and oxygen by electrolysis. She knows that the reaction will proceed faster if an electrolyte is added to the water. She has access to 2.00 *M* solutions of the following compounds: $H_2SO_4$, HBr, NaI, NaCl, $Na_2SO_4$, and $Na_2CO_3$. Which one(s) should she use? Explain your selection(s).

### Problems

**17.73.** Suppose the current from a battery is used to electroplate an object with silver. Calculate the mass of silver that would be deposited by a battery that delivers 1.7 A · h of charge.

17.74. A battery charger used to recharge the NiMH batteries in a digital camera can deliver as much as 0.50 A of current to each battery. If recharging one battery takes 100 min, how much $Ni(OH)_2$ (in grams) is oxidized to NiO(OH)?

**17.75.** A quantity of electric charge deposits 0.732 g of Ag(*s*) from an aqueous solution of silver nitrate. When that same quantity of charge is passed through a solution of a gold salt, 0.446 g of Au(*s*) is formed. What is the oxidation state of the gold ion in the salt?

17.76. What quantity of electric charge is required to deposit 0.750 g of Pt(*s*) from a solution containing the $[PtCl_4]^{2-}$ ion within a time of 2.5 hours?

**17.77.** A NiMH battery containing 4.10 g of NiO(OH) is 50% discharged when it is connected to a charger with an output of 2.00 A at 1.3 V. How long does recharging the battery take?

*17.78. How much time is needed to deposit a coating of gold 1.00 μm thick on a disk-shaped medallion 4.0 cm in diameter and 2.0 mm thick at a constant current of 85 A? The density for the electroplating process is 19.3 g/cm³. The electroplating solution contains gold(III).

*17.79. **Oxygen Supply in Submarines** Nuclear submarines can stay under water nearly indefinitely because they can produce their own oxygen by the electrolysis of water.

a. How many liters of $O_2$ at 25°C and 1.00 bar are produced in 1 h in an electrolytic cell operating at a current of 25 A?

b. Could seawater be used as the source of oxygen in this electrolysis? Explain why or why not.

17.80. In the electrolysis of water, how much time is needed to produce 125 L of $H_2$ at 20°C and a pressure of 750 torr by using an electrolytic cell through which the current is 52 mA?

## Fuel Cells

### Concept Review

**17.81.** Describe two advantages of hybrid (gasoline engine–electric motor) power systems over all-electric systems based on fuel cells. Describe two disadvantages.

17.82. Describe three factors limiting the more widespread use of cars and other vehicles powered by fuel cells.

**17.83.** Methane can serve as the fuel for electric cars powered by fuel cells. Carbon dioxide is a product of the fuel cell reaction. All cars powered by internal combustion engines burning natural gas (mostly methane) produce $CO_2$. Why are electric vehicles powered by fuel cells likely to produce less $CO_2$ per kilometer driven?

17.84. To make fuel cells easier to refuel, several manufacturers are developing converters that turn readily available fuels—such as natural gas, propane, and methanol—into $H_2$ and $CO_2$. Although vehicles with such power systems are not truly "zero emission," they still offer significant environmental benefits over vehicles powered by internal combustion engines. Describe a few of these benefits.

### Problems

**17.85.** Fuel cells with molten alkali metal carbonates as electrolytes can use methane as a fuel. The methane is first converted into hydrogen in a two-step process:

$$CH_4(g) + H_2O(g) \rightarrow CO(g) + 3\,H_2(g)$$
$$CO(g) + H_2O(g) \rightarrow H_2(g) + CO_2(g)$$

a. Assign oxidation numbers to carbon and hydrogen in the reactants and products.

b. Using the standard free energy of formation values from Appendix 4, calculate the standard free-energy changes in the two reactions and the overall $\Delta G°$ for the formation of $H_2 + CO_2$ from methane and steam.

*17.86. Molten carbonate fuel cells fueled with $H_2$ convert as much as 60% of the free energy released by the formation of water from $H_2$ and $O_2$ into electrical energy. Determine the quantity of electrical energy obtained from converting one mole of $H_2$ into $H_2O(\ell)$ in such a fuel cell.

## Additional Problems

**17.87.** Suppose that a scale existed for expressing electrode potentials in which the standard potential for the reduction of water in base

$$2\,H_2O(\ell) + 2\,e^- \rightarrow H_2(g) + 2\,OH^-(aq)$$

was assigned an $E°$ value of 0.000 V. How would standard potential values on this new scale differ from those in Appendix 6?

17.88. Sacrificial anodes are attached to steel structures in contact with seawater to inhibit their corrosion. Use the contents of Table 8.8 to explain why zinc sacrificial anodes are so effective.

**17.89.** **Lithium–Sulfur Dioxide Batteries** The U.S. military uses batteries based on the reduction of liquid sulfur dioxide at a carbon cathode in certain communications equipment. Lithium metal is used as the anode. The overall cell reaction is

$$2\,Li(s) + 2\,SO_2(\ell) \rightarrow Li_2S_2O_4(s)$$

a. Write half-reactions for the anode and cathode reactions.

b. How many electrons are transferred in the cell reaction?

c. Draw a Lewis structure for the $S_2O_4^{2-}$ anion.

*17.90. **Rusting Shipwrecks** The *Titanic* sank in the waters of the North Atlantic on its maiden voyage in 1912. When its wreck was finally located in 1985, much of the vessel was covered by *rusticles* (Figure P17.90) composed of the minerals lepidocrocite and goethite, both having the formula FeO(OH). The reactions responsible for the formation of FeO(OH) include:

(i) $2\,Fe(s) + O_2(g) + 2\,H_2O(\ell) \rightarrow 2\,Fe(OH)_2(s)$

(ii) $4\,Fe(OH)_2(s) + O_2(g) + 2\,H_2O(\ell) \rightarrow 4\,Fe(OH)_3(s)$

(iii) $Fe(OH)_3(s) \rightarrow FeO(OH)(s) + H_2O(\ell)$

a. Which of these reactions does not involve redox?

b. Use the following standard reduction potential:

$$Fe(OH)_3(s) + e^- \rightarrow Fe(OH)_2(s) + OH^-(aq) \qquad E° = -0.56\text{V}$$

to calculate $E°_{rxn}$ for reaction (ii).

**FIGURE P17.90**

*17.91. **Lithium–Ion Batteries** Scientists at the University of Texas at Austin and at MIT developed a cathode material for lithium–ion batteries based on $LiFePO_4$, which is the composition of the cathode when the battery is fully discharged. Batteries with this cathode are more powerful than those of the same mass with $LiCoO_2$ cathodes. They are also more stable at high temperatures.

a. What is the formula of the $LiFePO_4$ cathode when the battery is fully charged?
b. Is Fe oxidized or reduced as the battery discharges?
c. Is the cell potential of a lithium–ion battery with an iron phosphate cathode likely to differ from one with a cobalt oxide cathode? Explain your answer.

*17.92. What is the minimum (least negative) cathode potential (versus SHE) needed to electroplate silver onto cutlery in a solution of $Ag^+$ and $NH_3$ in which most of the silver ions are present as the complex $Ag(NH_3)_2{}^+$ and the concentration of $Ag^+(aq)$ is only $3.50 \times 10^{-5}$ *M*?

*17.93. In the cell reaction in a NiMH battery: $MH(s) + NiO(OH)(s) \rightarrow M(s) + Ni(OH)_2(s)$, what is the change in oxidation state of (a) Ni, (b) H, and (c) M?

17.94. **Clinical Chemistry** The concentrations of $Na^+$ ions in red blood cells (11 m*M*) are lower than those in the surrounding plasma (140 m*M*). Calculate the electromotive force (emf) across the cell membrane at 37°C due to this concentration gradient.

17.95. Elemental uranium may be produced from uranium dioxide by the following two-step process:

$$UO_2(s) + 4\,HF(g) \rightarrow UF_4(s) + 2\,H_2O(\ell)$$
$$UF_4(s) + 2\,Mg(s) \rightarrow U(s) + 2\,MgF_2(s)$$

a. Identify the reducing agent.
b. Identify the element that is reduced.
c. Using data from the table of standard reduction potentials in Appendix 6, find the maximum $E°$ value for the reduction of $UF_4$ for the second reaction.
d. Will 1.00 g of Mg be enough to produce 1.00 g of U?

*17.96. **Electrolysis of Seawater** Magnesium metal is obtained by the electrolysis of molten $Mg^{2+}$ salts obtained from evaporated seawater.

a. Does elemental Mg form at the cathode or anode?
b. Do you think the principal ingredient in sea salt (NaCl) needs to be separated from the $Mg^{2+}$ salts before electrolysis? Explain your answer.
c. Would electrolysis of an aqueous solution of $MgCl_2$ also produce elemental Mg?
d. If your answer to part (c) was no, what would the products of electrolysis be?

*17.97. **Silverware Tarnish** Low concentrations of hydrogen sulfide in air react with silver to form $Ag_2S$, more familiar to us as tarnish. Silver polish contains aluminum metal powder in a basic suspension.

a. Write a balanced net ionic equation for the redox reaction between $Ag_2S$ and Al metal that produces Ag metal and $Al(OH)_3$.
b. Calculate $E°$ for the reaction. *Hint*: Derive $E°$ values for the half-reactions in which $Ag_2S$ is reduced to Ag metal and $Al(OH)_3$ is reduced to Al metal. Then replace the $[Ag^+]$ and $[Al^{3+}]$ terms in the Nernst equations for these two half-reactions with terms based on the $K_{sp}$ values of $Ag_2S$ and $Al(OH)_3$ and the concentrations of sulfide and hydroxide ions (both of which are 1 *M* under standard conditions).

*17.98. **Flint, Michigan, Water Supply** In 2015, the city of Flint's municipal water supply was found to contain dangerous levels of dissolved $Pb^{2+}$ ions. The problem arose when the city began using water from the Flint River as its water source and made changes in the water purification process. The source of lead was corrosion of the city's water pipes.

a. Write a balanced redox equation describing the reaction between dissolved oxygen and Pb metal that produces dissolved $Pb^{2+}$ ions in an acidic solution.
b. Calculate $E°_{rxn}$ for the reaction in part a on the basis of the appropriate half-reactions in Appendix 6, and determine whether the reaction is spontaneous under standard conditions at 25°C.
c. How does increasing pH affect the value of $E_{rxn}$ for the reaction in part a?
d. The pH of the water in Flint decreased from 8.0 in December 2014 to 7.3 in August 2015. By how much did this increase in pH change $E_{rxn}$ for the reaction in part a if $[Pb^{2+}] = 1.00 \times 10^{-6}$ *M* and $P_{O_2} = 0.20$ atm at 25°C? (*Hint*: Use Henry's Law to calculate $[O_2]$.)

***17.99.** **Olympic Games** The diving competition at the 2016 Olympic Games in Rio de Janeiro was plagued by green, turbid, and odorous water in the diving pool (Figure P17.99). Chlorine was used as a disinfectant in the pool water. One hypothesis for the green color is that 160 liters of a 30% hydrogen peroxide solution (9.8 $M$ $H_2O_2$) was added to the pool by mistake, deactivating the chlorine disinfectant and allowing green algae to grow.

a. When chlorine dissolves in slightly alkaline pool water, it hydrolyzes, forming hypochlorite and chloride ions:

$$Cl_2(g) + 2\,OH^-(aq) \rightarrow ClO^-(aq) + Cl^-(aq) + H_2O(\ell)$$

Use the following half-reactions in to calculate $E^\circ_{rxn}$ for the reaction between $ClO^-$ ions and $H_2O_2$.

$$O_2(g) + 2\,H_2O(\ell) + 2\,e^- \rightarrow H_2O_2(aq) + 2\,OH^-(aq) \qquad E^\circ = -0.15\text{ V}$$

$$ClO^-(aq) + H_2O(\ell) + 2\,e^- \rightarrow ClO^-(aq) + 2\,OH^-(aq) \qquad E^\circ = 0.89\text{ V}$$

b. Calculate the value of $K$ for the reaction between $ClO^-$ ions and $H_2O_2$.

c. If the swimming pool had dimensions of 25.0 m × 25.0 m × 10.0 m and contained 2.0 mg/L of $ClO^-$ ions before $H_2O_2$ was added, what was the concentration of $ClO^-$ ions after the reaction with $H_2O_2$?

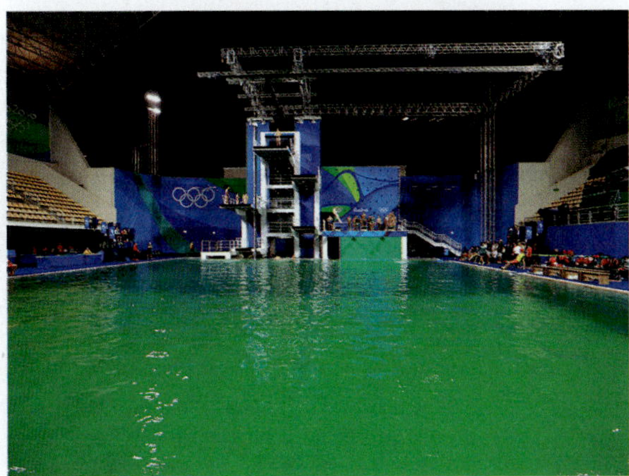

**FIGURE P17.99**

*17.100. The Particulate Review at the beginning of this chapter includes a photograph of a rusting automobile. The following reaction plays an important role in rust formation:

$$2\,Fe(s) + O_2(g) + 2\,H_2O(\ell) \rightarrow 2\,Fe(OH)_2(s)$$

a. Separate this reaction into two half-reactions describing the oxidation of Fe and reduction of $O_2$ in a slightly basic solution.

b. What is the $E^\circ$ value for the Fe half-reaction you wrote in part a? (*Hint*: Use the Nernst equation to combine the half-reaction describing the reduction of $Fe^{2+}$ ions to Fe metal and its $E^\circ$ value in Appendix 6 with the $K_{sp}$ expression for $Fe(OH)_2$ and its value in Table A5.4 of Appendix 5.)

c. What are the values of $E^\circ_{rxn}$ and $\Delta G^\circ_{rxn}$ for the overall reaction above?

# 18

# The Solid State

## A Particulate View

**STRONG ENOUGH TO REACH THE SKY** One World Trade Center, the tallest building in the Western Hemisphere, is clad in stainless steel, an alloy of iron in which about 1/4 of the atoms are chromium and nickel.

PARTICULATE **REVIEW**

### *Ionic Compounds versus Metals*

In Chapter 18, we explore the structure and properties of solids. Most metals and ionic compounds are solids at room temperature, including platinum metal and potassium iodide, whose structures are shown here. Answer the following questions about these substances:

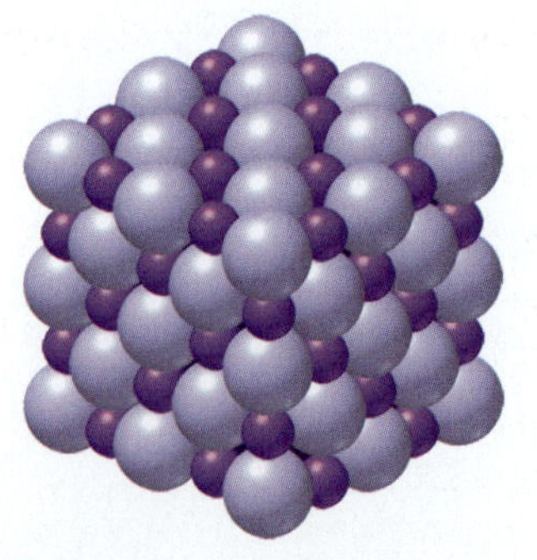

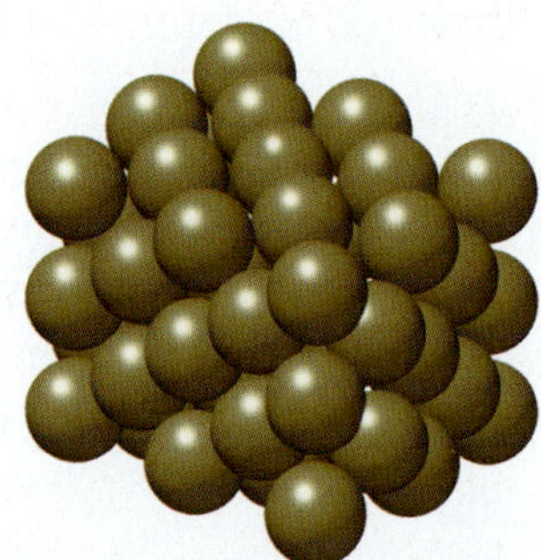

- Describe the nature of the bonds that hold the particles together in each solid.
- Which types of particles experience electrostatic attractions in potassium iodide? Which types of particles experience electrostatic repulsions?
- Which types of particles experience electrostatic interactions in platinum?

(Review Section 4.1 if you need help.)

*(Answers to Particulate Review questions are in the back of the book.)*

PARTICULATE **PREVIEW**

### *Three-Dimensional Networks: Structure and Properties*

Two allotropes of carbon, A and B, are depicted here. As you read Chapter 18, look for ideas that will help you answer these questions:

- Which allotrope has a planar repeating pattern? Which allotrope has a nonplanar repeating pattern?
- Which allotrope consists of carbon atoms covalently bonded to each other? Which allotrope has *both* covalent bonds *and* weak interactions between layers?
- One allotrope is extremely hard and is used as an abrasive; the other is soft and is used as a lubricant. Which is which?

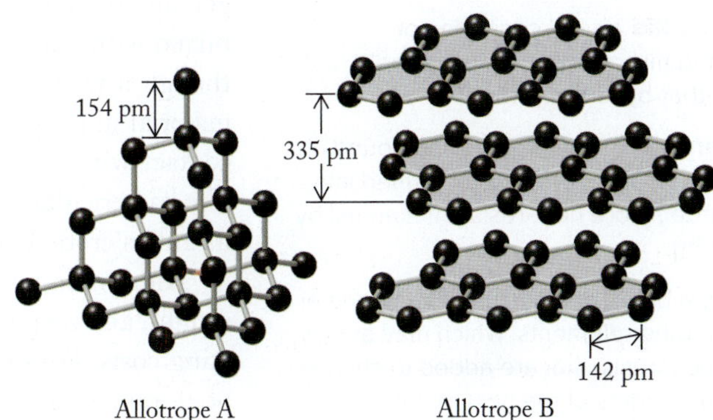

## Learning Outcomes

**LO1** Relate unit cell dimensions to the radii of the particles that make up the cell
**Sample Exercises 18.1, 18.5**

**LO2** Relate the densities of solids to the masses of the particles in their unit cells and the dimensions of the cells
**Sample Exercises 18.2, 18.3, 18.6**

**LO3** Distinguish between substitutional and interstitial alloys
**Sample Exercise 18.4**

**LO4** Use band theory to explain the conductivity of metals and semiconductors

**LO5** Distinguish between ionic solids, molecular solids, ceramics, and covalent network solids

**LO6** Use the Bragg equation to calculate the interlayer spacing in crystalline solids
**Sample Exercise 18.7**

## 18.1 Stronger, Tougher, Harder

People have been using metals for tools, weapons, currency, and jewelry for thousands of years. For most of that time, only three elements—copper, silver, and gold—were used because these are the only ones found as pure metals in Earth's crust. Even so, most silver and copper occur not as pure metals but in ores such as argentite ($Ag_2S$), chalcopyrite ($CuFeS_2$), and chalcocite ($Cu_2S$). **Ores** are naturally occurring compounds or mixtures of compounds from which elements can be extracted. Most ores are composed of **ionic solids** consisting of monatomic or polyatomic ions held together by ionic bonds. Earth's crust also yields materials used to make **ceramics**, solid inorganic compounds or mixtures of compounds that have been heated in such a way as to transform them into a harder and more heat-resistant material.

**STEPWISE ANIMATION**
Alloys

Gold rings and other gold jewelry may look like pure gold, but most are not. Gold is one of the softer metals, so it is very easy to bend jewelry made of pure gold. In contrast, jewelry made of gold blended with other metals such as palladium, silver, or copper is more resistant to physical damage and can last a lifetime. Any mixture of a host metal with one or more other elements is called an **alloy**, and there are many thousands of them. Some are solid solutions, which means they are homogeneous at the atomic level just as liquid solutions are, and others are heterogeneous mixtures. Varying the proportions of their constituent metals modifies the properties of alloys.

One of the first alloys was bronze, a mixture of copper and tin that was used by ancient Greeks and others to make tools and weapons that were harder, stronger, and more durable than those made of copper alone. Brass, an alloy of copper mixed with zinc instead of tin, has long been used in doorknobs and hinges even though it tarnishes and loses its shiny luster over time. Brass remains a popular material in hospitals and in doctors' offices because it kills bacteria that come in contact with it, whereas stainless steel, which resists corrosion better, does not. The bactericidal action of brass is due to copper, which has long been known for its antimicrobial properties.

Aluminum alloys are ideal for making aircraft because they are strong, lightweight, and corrosion resistant. Lighter aircraft consume less fuel, reducing operating costs. For example, the wing of a Boeing 747-400 airplane is made mostly of aluminum alloyed with copper and magnesium, whereas the Boeing 777 uses aluminum alloys containing approximately 8% zinc in addition to copper and magnesium. Even sports equipment has benefited from the development of new alloys: a patented five-metal alloy of nickel, zirconium, titanium, copper, and

**ore** a naturally occurring compound or mixture of compounds from which elements can be extracted.

**ionic solid** a solid consisting of monatomic or polyatomic ions held together by ionic bonds.

**ceramic** a solid inorganic compound or mixture that has been transformed into a harder, more heat-resistant material by heating.

**alloy** a blend of a host metal and one or more other elements, which may or may not be metals, that are added to change the properties of the host metal.

**crystalline solid** a solid made of an ordered array of atoms, ions, or molecules.

beryllium is advertised as producing a stronger, lighter, more resilient golf club that enables a golfer to transfer more energy from the swing into the ball.

In this chapter, we explore the links between the physical properties of solids at the macroscopic level and the structures of these solids at the atomic level. We start with metals and their alloys and conclude with ceramics, addressing questions such as the following: Why do metals bend? Why are they such good conductors of heat and electricity? Why are alloys so much stronger, tougher, and harder than the pure metals from which they are made?

## 18.2 Structures of Metals

Most elements are metals, which means they are typically hard, shiny, malleable (easily shaped), ductile (easily drawn out), and able to conduct electricity. In this section, we take a detailed look at how the atoms in metals are arranged and explore how their atomic structure accounts for their physical properties.

### Stacking Patterns

When a metallic element is heated above its melting point and then allowed to slowly cool, it usually solidifies into a **crystalline solid**—that is, a solid in which atoms are arranged in an ordered three-dimensional array called a *crystal lattice*. Think of a crystal lattice as stacked layers of atoms (designated *a*, *b*, *c*, . . .) that are packed tightly together. In the most tightly packed arrangements, each atom touches six others in its layer (denoted layer *a*) as shown in **Figure 18.1(a).** The blue and red dots between the atoms (seen as yellow spheres) represent the points above which the atoms in additional layers will be centered. The atoms represented by purple spheres in the second (*b*) layer (**Figure 18.1b**) nestle into the spaces created by the first layer that were marked by blue dots in Figure 18.1(a), in much the same way that oranges in a fruit-stand display or cannonballs at a 16th-century fort (**Figure 18.2**) might be stacked together.

The atoms in a third layer nestle among those in the second in one of two alignments. They may sit directly above the atoms in the *a* layer, as shown in **Figure 18.1(c).** This arrangement produces an *ababab* ... stacking pattern throughout the crystal. However, the atoms in the third layer may nestle between the atoms in the second layer in such a way that they are not directly above the atoms in the *a* layer, but rather above the red dots between them, creating a third (*c*) layer aligned with neither the *a* nor the *b* layer (**Figure 18.1d**). When this happens and a fourth layer of atoms is directly above those in the *a* layer, we have an *abcabc* ... stacking pattern. Other patterns are possible, but we will focus on these two.

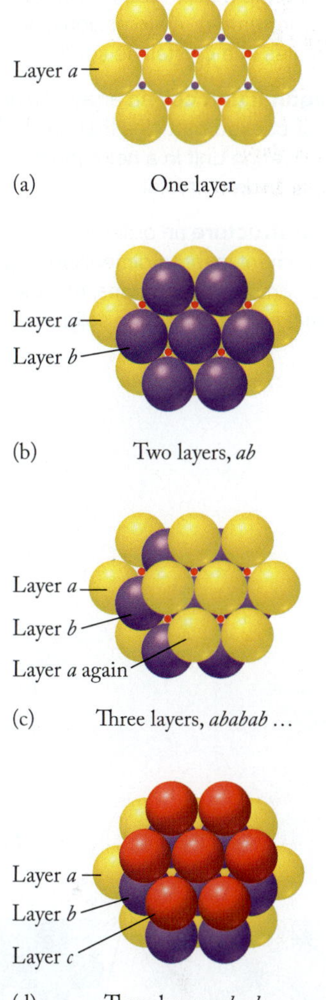


**FIGURE 18.1** The *ababab* ... and *abcabc* ... stacking patterns represent two equally efficient ways to stack layers of atoms (or any particles of equal size).

(a)

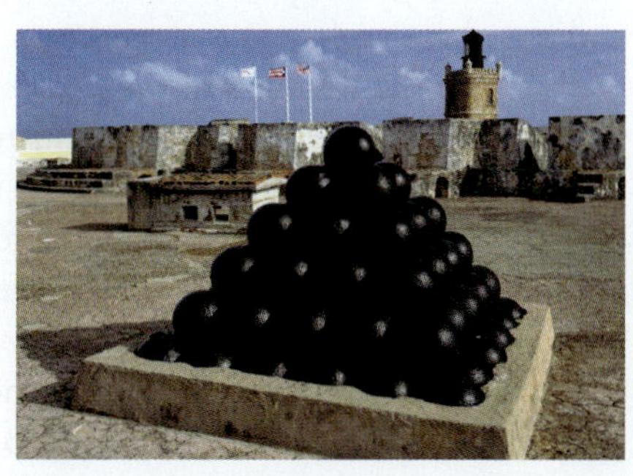

(b)

**FIGURE 18.2** (a) Stacks of oranges in a grocery store and (b) cannonballs at El Morro in San Juan, Puerto Rico, illustrate closest-packed arrays of spherical objects.

**hexagonal closest-packed (hcp)** a crystal lattice in which the layers of atoms or ions have an *ababab* ... stacking pattern.

**unit cell** the basic repeating unit of the arrangement of atoms, ions, or molecules in a crystalline solid.

**hexagonal unit cell** an array of nine closest-packed particles that are the repeating unit in a hexagonal closest-packed crystal.

**crystal structure** an ordered arrangement in three-dimensional space of the particles (atoms, ions, or molecules) that make up a crystalline solid.

## Stacking Patterns and Unit Cells

Stacking patterns determine the shapes of the crystals that metals form. To explore why, let's take a closer look at a cluster of atoms in the *ababab* ... stacking pattern (**Figure 18.3**). This cluster forms a *hexagonal* (six-sided) prism of closely packed atoms. In fact, they are as tightly packed as they can be, so the crystal structure is called **hexagonal closest-packed (hcp)**. In these lattices, the cluster highlighted in Figure 18.3(c) serves as an atomic-scale building block—a pattern of atoms repeated over and over again in all three dimensions in the elements.

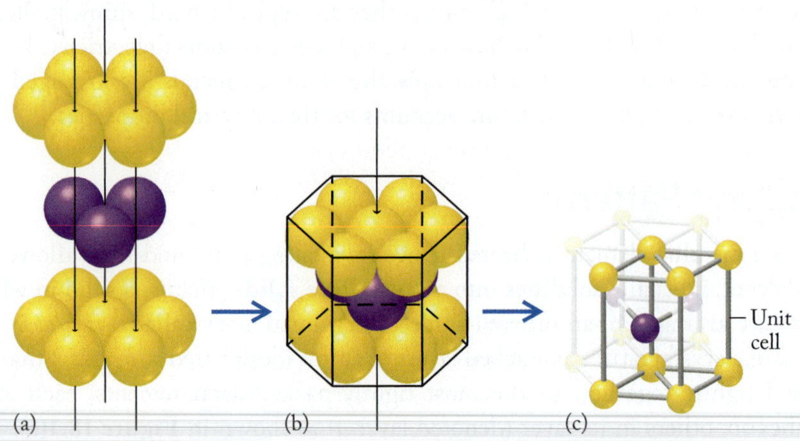


**FIGURE 18.3** (a, b) A hexagonal closest-packed (hcp) crystal structure and (c) its hexagonal unit cell represent a highly efficient way to pack atoms in a crystalline solid. Sixteen metals, including all those in groups 3 and 4, have the hcp crystal structure.

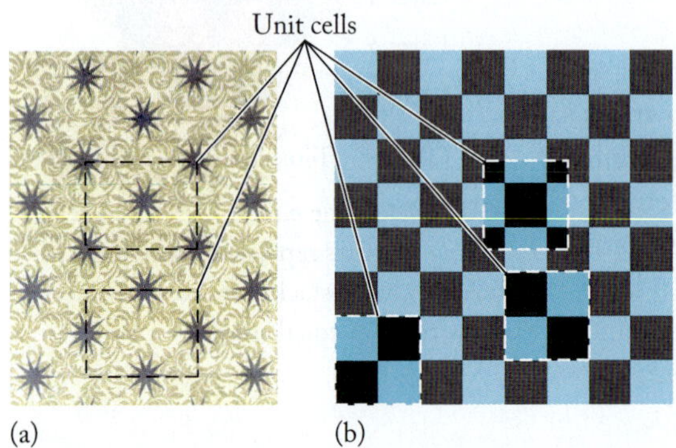


**FIGURE 18.4** Two-dimensional repeating patterns. (a) The repeating patterns outlined with the dashed lines represent "unit cells" for this wrapping-paper design. (b) The portions of the squares inside the white-dashed lines show three ways to represent the repeating pattern in this checkerboard. Can you sketch an alternative way?

Unit Cell

We call each of these building blocks a **unit cell**, and the structure in Figure 18.3 has a **hexagonal unit cell**. A unit cell represents the minimum repeating pattern that describes the array of atoms forming the crystal lattice of any crystalline solid, including metals. Think of unit cells as three-dimensional microscopic analogs of the two-dimensional repeating pattern in fabrics, wrapping paper, or even a checkerboard. Look closely at **Figure 18.4** to confirm that the outlined portions represent the minimum repeating patterns in the paper and in the checkerboard. A unit cell has the same role in the **crystal structure** of a solid—that is, the arrangement in three-dimensional space of the atoms, ions, or molecules of which the solid is composed.

What is the unit cell in the *abcabc* ... stacking pattern? Look at what happens when we take the 14-atom cluster in **Figure 18.5(a)**, rotate it, and tip it 45° to obtain the orientation shown in **Figure 18.5(b)**. The black outline in Figure 18.5(b) shows that the atoms form a cube: one atom at each of the eight corners of the cube and one at the center of each of the six faces. Because the atoms are stacked together as closely as possible, this crystal lattice is called **cubic closest-packed (ccp)**, and the corresponding unit cell is called a **face-centered cubic (fcc) unit cell**. Because this unit cell is a cube, the edges are all of equal length, and the angle between any two edges is 90°. Note in

Figure 18.5(b) how the atom in the center of each face touches each of the atoms at the four corners of that face. However, none of the corner atoms touch any other corner atoms.

So far we have introduced two *closest-packed* crystal lattices—hexagonal and cubic—and their associated unit cells (hexagonal and face-centered cubic). The hcp and ccp crystal lattices represent the most efficient ways of arranging solid spheres of equal radius. We can express the **packing efficiency** as the percentage of the total volume of the unit cell occupied by the spheres:

$$\text{Packing efficiency (\%)} = \frac{\text{volume occupied by spheres}}{\text{volume of unit cell}} \times 100\% \qquad (18.1)$$

$$= \frac{V_{\text{atoms}}}{V_{\text{unit cell}}} \times 100\%$$

For both hcp and ccp crystal lattices, the packing efficiency is approximately 74%. We will come back to this calculation at the end of the next subsection, but first let's look at some other packing arrangements.

Stacking patterns also exist in which the atoms are packed less efficiently than in the hcp and ccp lattices. For instance, we can arrange the atoms in an *a* layer so that each atom touches four adjacent atoms, an arrangement called *square packing* (**Figure 18.6a**). If we add a second layer of spheres directly above the first, we create the *aaa* ... stacking pattern shown in **Figure 18.6(b)** that is called *cubic packing*. The three-dimensional repeating pattern of this arrangement is called a **simple cubic (sc) unit cell** (**Figure 18.6c**). It is the least efficiently packed of the cubic unit cells and is rare among metals: only radioactive polonium (Po) forms a simple cubic unit cell.

If each atom in a second layer is nestled in the space created by four atoms in a square-packed *a* layer (**Figure 18.7a**), we have two layers in an *ab* stacking pattern. If the atoms in the third layer are directly above those in the first, then we have an *ababab* ... stacking pattern based on layers of square-packed atoms. The simplest three-dimensional repeating unit of this pattern is called a **body-centered cubic (bcc) unit cell**. It consists of portions of nine atoms, one at each of the eight corners of a cube and one in the middle of the cube (**Figure 18.7b**). All the group 1 metals and many transition metals have bcc unit cells. **Table 18.1** summarizes the different stacking patterns, packing efficiencies, and unit cells described in this section.

**cubic closest-packed (ccp)** a crystal structure composed of face-centered cubic unit cells and layers of particles having an *abcabc* ... stacking pattern.

**face-centered cubic (fcc) unit cell** an array of closest-packed particles that has eight of the particles at the corners of a cube and six of them at the centers of each face of the cube.

**packing efficiency** the percentage of the total volume of a unit cell occupied by the spheres.

**simple cubic (sc) unit cell** a cell with atoms only at the eight corners of a cube.

**body-centered cubic (bcc) unit cell** a cell with atoms at the eight corners of a cube and at the center of the cell.

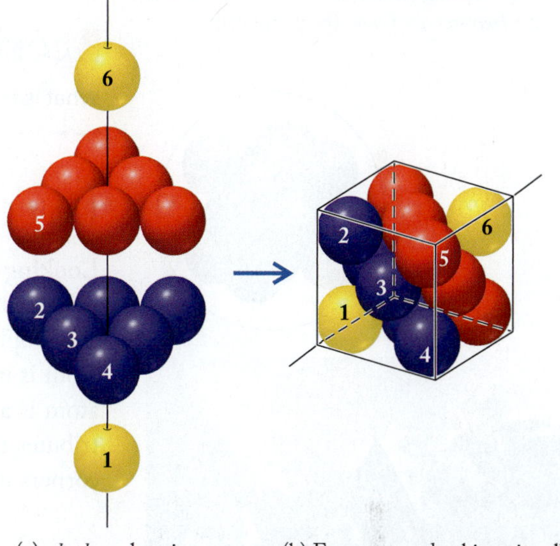


(a) *abcabc* ... layering (b) Face-centered cubic unit cell

**FIGURE 18.5** (a) The stacking pattern *abcabc* ... has a face-centered cubic (fcc) unit cell. (b) The shape of the unit cell is more easily seen when the layers are tipped 45° and rotated. Note that atoms at adjacent corners do not touch each other, but the three atoms along the diagonal of any face of the cube—atoms 2, 3, and 4 here—do touch each other.

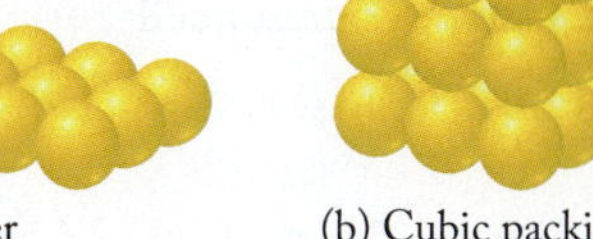

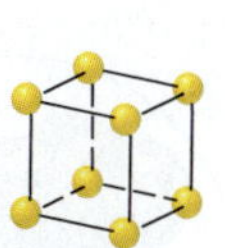

(a) Square-packed *a* layer (b) Cubic packing (c) Simple cubic unit cell

**FIGURE 18.6** (a) In a square-packed *a* layer, each atom (like the one in the middle) touches four others. (b) If the atoms in all other layers are directly above those in the *a* layer, the stacking pattern is called *cubic packing*. (c) The repeating unit of this pattern is called a *simple cubic* (sc) unit cell with eight atoms at the eight corners of a cube.

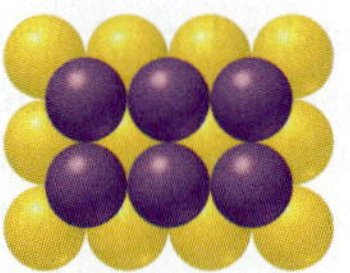

(a) Two square-packed layers in an *ab* pattern

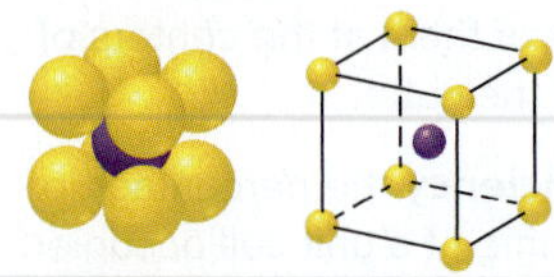

(b) Body-centered cubic unit cell

**FIGURE 18.7** (a) The atoms represented by purple spheres in the *b* layer nestle into the spaces between the square-packed atoms (yellow spheres) in the *a* layer. (b) Atoms in the third layer are directly above those in the first, producing an *ababab* ... stacking pattern and a *body-centered cubic* (bcc) unit cell.

**TABLE 18.1** Summary of Unit Cell Types, Stacking Patterns, and Packing Efficiencies

| Unit Cell | Stacking Pattern | Number of Nearest Neighbors | Packing Efficiency (%) |
|---|---|---|---|
| Hexagonal | *ababab...* | 12 | 74 |
| Face-centered cubic | *abcabc...* | 12 | 74 |
| Body-centered cubic | *ababab...* | 8 | 68 |
| Simple cubic | *aaaa...* | 6 | 52 |

Solids with cubic unit cells form crystals that often contain cubes or four-sided prisms (**Figure 18.8**), whereas crystalline solids with hexagonal unit cells tend to form hexagonal crystals (**Figure 18.9**). As **Figure 18.10** shows, the periodic table contains elements that exist in all the crystal structures we have discussed in this section, as well as structures we will discuss in later sections.

CONCEPT **TEST**

What is the difference between a crystal lattice and a unit cell?

**FIGURE 18.8** Palladium (and other metals) with fcc unit cells may form crystals that contain four-sided prisms such as those in this photomicrograph.

## Unit Cell Dimensions

Looking at both whole-atom and cutaway views of the cubic unit cells allows us to determine how many equivalent atoms are in each type of cell. Let's start with the simple cubic unit cell (**Figure 18.11a**). Note how only a fraction of each corner atom is inside the unit cell boundary. In a crystal lattice with this unit cell, each atom is a corner atom in eight unit cells (**Figure 18.12a**). Thus, each atom contributes the equivalent of one-eighth of an atom to the unit cell. There are eight corners in a cube, so there is a total of

$$\frac{\frac{1}{8}\text{ corner atom}}{\cancel{\text{corner}}} \times \frac{8\ \cancel{\text{corners}}}{\text{unit cell}} = 1\ \frac{\text{corner atom}}{\text{unit cell}}$$

This calculation applies to the corner atoms in any type of cubic unit cell. Note that the two corner atoms along each edge in Figure 18.11(a) touch each other. Therefore, the edge length, $\ell$, in the simple cubic unit cell is equal to twice the atomic radius: $\ell = 2r$.

In an fcc unit cell (**Figure 18.11b**), there are eight corner atoms and one atom in the center of each of the six faces. Each face atom is shared by the two unit cells that abut each other at that face (**Figure 18.12b**). Therefore, each unit cell "owns" half of each face atom, making a total of

$$\frac{\frac{1}{2}\text{ face atom}}{\cancel{\text{face}}} \times \frac{6\ \cancel{\text{faces}}}{\text{unit cell}} = 3\ \frac{\text{face atoms}}{\text{unit cell}}$$

**FIGURE 18.9** Materials with hexagonal unit cells tend to form hexagonal crystals. Two such materials are in this photo: quartz crystals (see also Figure 18.39) coated with a thin layer of titanium.

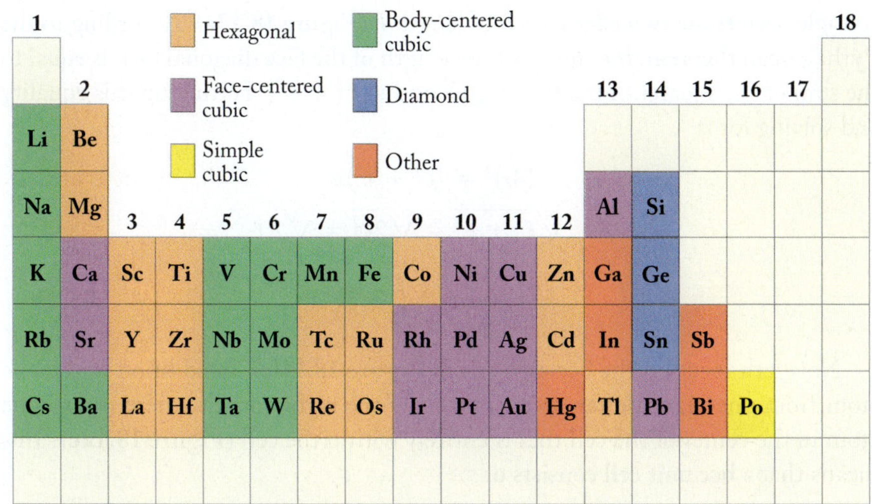


**FIGURE 18.10** Crystal structures of metals and metalloids. The unit cells for the five elements designated "Other" are more complicated and beyond the scope of this book.

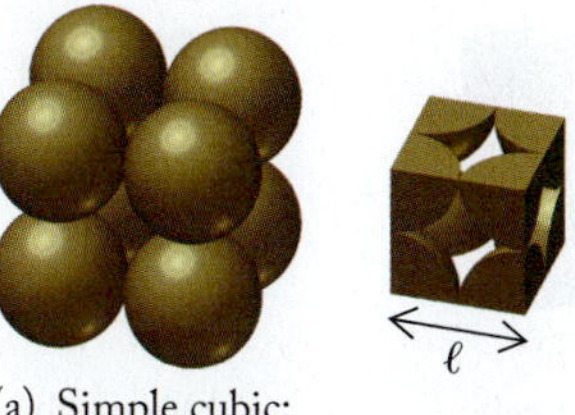


(a) Simple cubic: Atoms touch along edge

(b) Face-centered cubic: Atoms touch along face diagonal

(c) Body-centered cubic: Atoms touch along body diagonal

**FIGURE 18.11** Whole-atom and cutaway views of cubic unit cells. (a) In a simple cubic unit cell, each corner atom of the unit cell is part of eight unit cells. Atoms along each edge touch. (b) In a face-centered cubic unit cell, the face atoms are part of two unit cells. Atoms along the face diagonal touch. (c) In a body-centered cubic unit cell, one atom in the center lies entirely in one unit cell. The atoms along the body diagonal touch.

Moreover, every cubic unit cell owns the equivalent of one corner atom, so an fcc unit cell consists of

$$1 \text{ corner atom} + 3 \text{ face atoms} = 4 \text{ atoms per fcc unit cell}$$

To relate the dimensions of the fcc unit cell to the size of the atoms in it, note in the cutaway view of Figure 18.11(b) that although the corner atoms do not touch each other, adjacent atoms along the face diagonal do touch each other. A face diagonal spans the radius, $r$, of two corner atoms and the diameter (2 radii $= 2r$) of a face atom. Therefore, the length of a face diagonal is $1 + 2 + 1 = 4$ atomic radii $= 4r$. A face diagonal connects the ends of two edges and forms a right

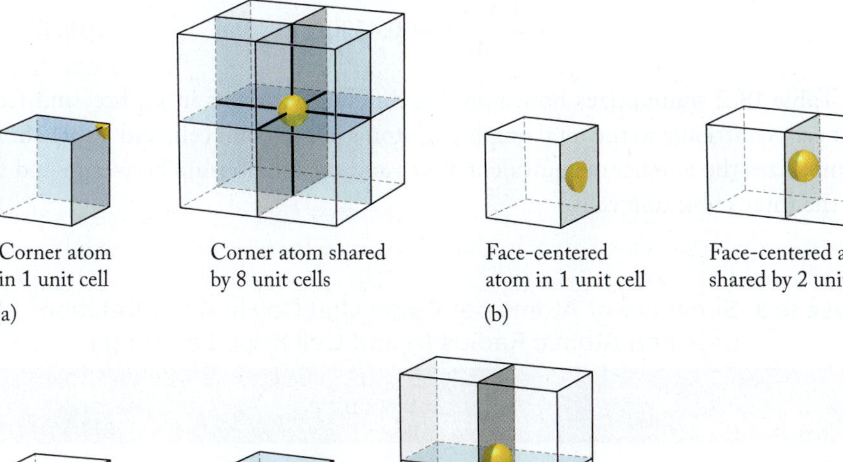


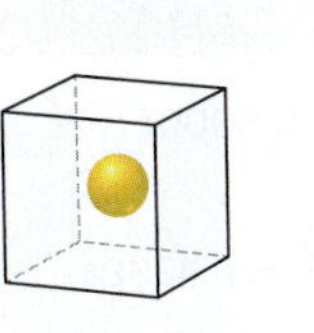


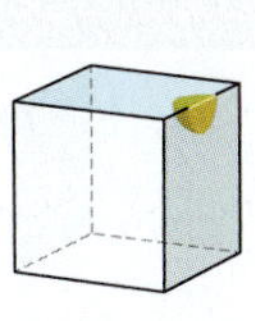


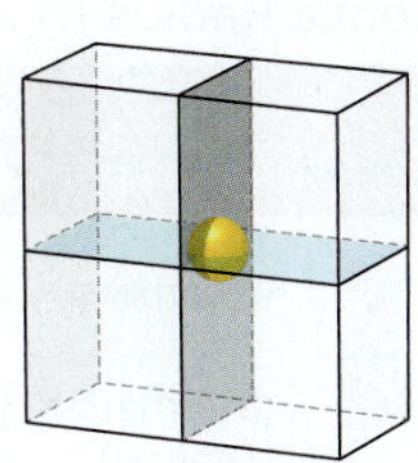


**FIGURE 18.12** Crystal lattices illustrating (a) corner atoms shared by eight unit cells, (b) face atoms shared by two unit cells, (c) center atoms entirely in one unit cell, and (d) edge atoms shared by four unit cells.

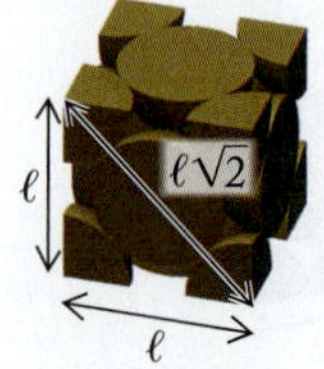

(a) fcc unit cell dimensions

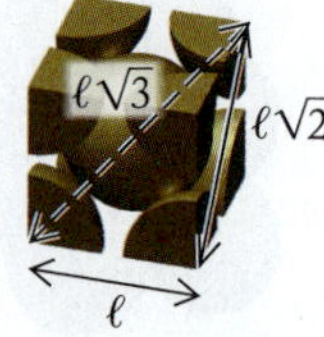

(b) bcc unit cell dimensions

**FIGURE 18.13** (a) In a face-centered cubic unit cell, the face diagonal forms a right triangle with two adjoining edges each of length $\ell$. Applying the Pythagorean theorem to this triangle yields $r = 0.3536\ell$. (b) In a body-centered cubic unit cell, the atoms touch along a body diagonal. Applying the Pythagorean theorem to the right triangle formed by an edge, a face diagonal, and a body diagonal yields $r = 0.4330\ell$.

triangle with those two edges, each of length $\ell$ (**Figure 18.13a**). According to the Pythagorean theorem, the square of the length of the face diagonal $(4r)^2$ is equal to the sum of the squares of the two edge lengths $(\ell^2 + \ell^2)$. Setting up this equality and solving for $r$:

$$(4r)^2 = (\ell^2 + \ell^2)$$

$$4r = \sqrt{\ell^2 + \ell^2} = \sqrt{2\ell^2} = \sqrt{2}\,\ell$$

$$r = \frac{\sqrt{2}}{4}\,\ell = 0.3536\ell \qquad (18.2)$$

Now let's focus on the bcc unit cell in **Figure 18.11(c)**. In addition to the one atom from one-eighth of an atom at each of the eight corners, there is also one atom in the center of the cell that is entirely within the cell (**Figure 18.12c**). This means that a bcc unit cell consists of

$$1 \text{ corner atom} + 1 \text{ center atom} = 2 \text{ atoms per bcc unit cell}$$

Adjacent atoms in a bcc cell do not touch along the edges or along any face diagonal. However, each corner atom does touch the atom in the center of the cell, which means that the atoms touch along a *body diagonal* that runs between opposite corners through the center of the cube. In the cutaway view in **Figure 18.13(b)**, the diagonal runs from the bottom-left corner of the front face to the top-right corner of the rear face. It spans (1) the radius of the bottom-left corner atom, (2) the diameter (2 radii) of the central atom, and (3) the radius of the top-right atom, making the length of the body diagonal equivalent to $4r$.

We can again apply the Pythagorean theorem, this time to a right triangle whose legs are an edge ($\ell$) and a face diagonal ($\sqrt{2}\,\ell$) and whose hypotenuse is the body diagonal ($4r$). Squaring the hypotenuse and setting it equal to the sum of the squares of the other two sides:

$$(4r)^2 = \ell^2 + (\sqrt{2}\,\ell)^2$$

$$4r = \sqrt{\ell^2 + 2\ell^2} = \sqrt{3\ell^2} = \sqrt{3}\,\ell$$

$$r = \frac{\sqrt{3}}{4}\,\ell = 0.4330\ell \qquad (18.3)$$

**Table 18.2** summarizes how atoms in different locations in sc, bcc, and fcc unit cells contribute to the total number of atoms in each unit cell, and **Table 18.3** summarizes the number of equivalent atoms and the relationship between $r$ and $\ell$ for the three cubic unit cells.

**TABLE 18.2 Contributions of Atoms to Unit Cells**

| Atom Position | Contribution to Unit Cell |
|---|---|
| Center | 1 atom |
| Face | $\frac{1}{2}$ atom |
| Edge | $\frac{1}{4}$ atom |
| Corner | $\frac{1}{8}$ atom |

**TABLE 18.3 Summary of Atoms per Cubic Unit Cell and the Relation between Atomic Radius ($r$) and Cell Edge Length ($\ell$)**

| Unit Cell Type | Atoms per Cell | Relationship between $r$ and $\ell$ |
|---|---|---|
| Simple cubic | (8 corners × $\frac{1}{8}$) = 1 | $r = \frac{\ell}{2} = 0.5000\ell$ |
| Body-centered cubic | (8 corners × $\frac{1}{8}$) + (1 center) = 2 | $r = \frac{\ell\sqrt{3}}{4} = 0.4330\ell$ |
| Face-centered cubic | (8 corners × $\frac{1}{8}$) + (6 faces × $\frac{1}{2}$) = 4 | $r = \frac{\ell\sqrt{2}}{4} = 0.3536\ell$ |

**SAMPLE EXERCISE 18.1** Calculating Atomic Radii from Unit Cell Dimensions **LO1**

The structure of the most stable form of iron at room temperature, called *ferrite*, is shown in **Figure 18.14**. The unit cell in the figure has an edge length of 287 pm. What is the radius in picometers of the iron atoms in the cell? Check your answer against the data in Appendix 3.

**Collect, Organize, and Analyze** The unit cell in Figure 18.14 is body-centered cubic, which means that the radii ($r$) of the atoms in the cell are related to the cell's edge length ($\ell$ = 287 pm) by Equation 18.3.

**Solve**

$$r = 0.4330\ell$$
$$= 0.4330 \times 287 \text{ pm} = 124 \text{ pm} \quad (18.3)$$

**Think About It** The calculated value is close to the average atomic radius of iron atoms (126 pm) in Table A3.1 in Appendix 3.

**Practice Exercise** At 1070°C, the most stable form of iron is *austenite* (**Figure 18.15**). The edge length of its unit cell is 361 pm. What is the atomic radius of iron in austenite?

*(Answers to Practice Exercises are in the back of the book.)*

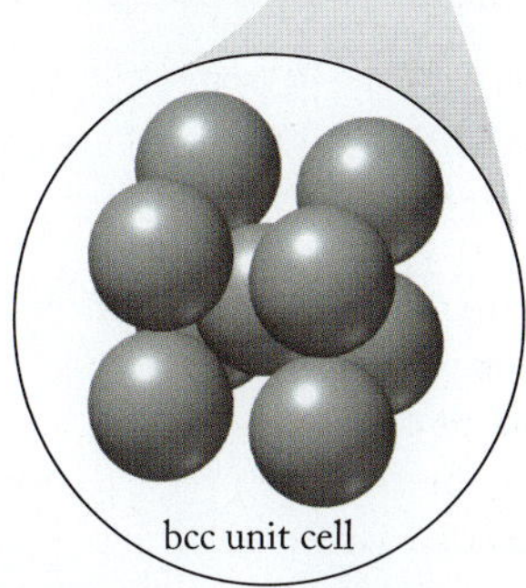


**FIGURE 18.14** The pattern in this newly cut surface of the meteorite that caused the Odessa crater in Texas about 50,000 years ago is due to crystals of ferrite, which grew when the meteor formed from molten iron 4.5 billion years ago. The microview shows the bcc unit cell of ferrite.

**SAMPLE EXERCISE 18.2** Using Unit Cell Dimensions to Calculate Density **LO2**

Calculate the density of iron (ferrite) in grams per cubic centimeter, given that its bcc unit cell has an edge length of 287 pm.

**Collect and Organize** We are asked to calculate the mass-to-volume ratio of ferrite iron, which has a bcc unit cell. There are two iron atoms per unit cell and $6.022 \times 10^{23}$ atoms/mol. The molar mass of Fe is 55.845 g/mol. The volume ($V$) of a cube of edge length $\ell$ is $\ell^3$.

**Analyze** We have to assume that the density of the unit cell matches the density of solid ferrite. To calculate the density of the unit cell, we need to calculate the mass of two Fe atoms, starting with the molar mass of Fe and dividing by the Avogadro constant to calculate the mass of each Fe atom in grams. The volume of the cell is the cube of its edge length (287 pm)$^3$. An object made of iron sinks when immersed in water, so the density of iron must be greater than 1 g/cm$^3$.

**Solve** Calculating the mass, $m$, of two Fe atoms:

$$m = \frac{55.845 \text{ g Fe}}{1 \cancel{\text{mol Fe}}} \times \frac{1 \cancel{\text{mol Fe}}}{6.022 \times 10^{23} \cancel{\text{atoms Fe}}} \times 2 \cancel{\text{atoms Fe}} = 1.855 \times 10^{-22} \text{ g Fe}$$

The volume, $V$, of the cell in cubic centimeters is

$$V = \ell^3 = (287 \cancel{\text{pm}})^3 \times \frac{(10^{-10} \text{ cm})^3}{(1 \cancel{\text{pm}})^3} = 2.364 \times 10^{-23} \text{ cm}^3$$

The density of the cell is

$$d = \frac{m}{V} = \frac{1.855 \times 10^{-22} \text{ g}}{2.364 \times 10^{-23} \text{ cm}^3} = 7.85 \text{ g/cm}^3$$

**FIGURE 18.15** The fcc unit cell of austenite.

**Think About It** The calculated density is larger than 1 g/cm$^3$, as expected, and close to the value (7.874 g/cm$^3$) in Appendix 3. Our assumption that density of the unit cell must equal the density of a bulk sample was reasonable because the latter is composed of a crystal lattice composed of bcc unit cells.

**Practice Exercise** Silver and gold both crystallize in face-centered cubic unit cells with edge lengths of 407.7 pm and 407.0 pm, respectively. Calculate the density of each metal and compare your answers with the densities listed in Appendix 3.

Earlier in this section, we noted that ccp and hcp are the most efficient packing schemes for spheres. Now that we know more about the fcc unit cell, let's take another look at the calculation of packing efficiency by using Equation 18.1. The unit cell for a ccp arrangement of solid spheres of radius $r$ is an fcc unit cell that contains four equivalent spheres. The volume occupied by each of the four spheres is

$$V_{\text{sphere}} = \frac{4}{3}\pi r^3$$

From Equation 18.2, $r = \sqrt{2}\,\ell/4$, and the unit cell edge $\ell$ is

$$\ell = 4r/\sqrt{2} = 2.828r$$

Expressing the volume of the unit cell in terms of $r$:

$$V_{\text{unit cell}} = \ell^3 = (2.828r)^3 = 22.62r^3$$

Substituting $V_{\text{sphere}} \times 4$ spheres and $V_{\text{unit cell}}$ into Equation 18.1:

$$\text{Packing efficiency (\%)} = \frac{V_{\text{spheres}}}{V_{\text{unit cell}}} \times 100\% = \frac{\left(\frac{4}{3}\pi \cancel{r^3}\right) \times 4}{22.62\cancel{r^3}} \times 100\% = 74.07\%$$

This means that the most efficient packing of spheres results in about (100 − 74) = 26% of the total volume being empty.

## SAMPLE EXERCISE 18.3 Using Atomic Radii and Unit Cell Structure to Calculate Density and Packing Efficiency LO2

Polonium is the only metal that crystallizes in a simple cubic structure. If the atomic radius of polonium is 167 pm, (a) what is the density of polonium in grams per cubic centimeter and (b) what is the packing efficiency of Po atoms?

**Collect and Organize** We are asked to calculate the density and packing efficiency of Po from its unit cell geometry and atomic radius. Po crystallizes in a simple cubic structure, so polonium atoms touch along the edge. A simple cubic unit cell contains the equivalent of a single atom. The molar mass of the most stable isotope of Po is 209 g/mol. The volume of a cube of edge length $\ell$ is $V = \ell^3$. The volume of a spherical atom of radius $r$ is $\frac{4}{3}\pi r^3$.

**Analyze** As in Sample Exercise 18.2, we assume that the density of the unit cell is the same as the density of solid Po. The density of the Po unit cell is the mass of one Po atom divided by the cell volume, which is the edge length (2 atomic radii) cubed. We need to divide the molar mass of polonium by the Avogadro constant to obtain the mass of a single Po atom in grams. The density of Po should be similar to that of Fe (~8 g/cm$^3$) because Po atoms are more massive than Fe atoms but are not packed as efficiently. Calculating the packing efficiency with Equation 18.1 involves calculating the volume of a Po atom from its atomic radius and dividing by the volume of the unit cell.

**Solve**

a. The mass, $m$, of a Po atom is

$$m = \frac{209 \text{ g Po}}{1 \text{ mol Po}} \times \frac{1 \text{ mol Po}}{6.022 \times 10^{23} \text{ atoms Po}} \times 1 \text{ atom Po} = 3.47 \times 10^{-22} \text{ g Po}$$

and the volume of the unit cell in cubic centimeters is

$$\ell = 2r = 2(167 \text{ pm}) \times \frac{10^{-10} \text{ cm}}{1 \text{ pm}} = 3.34 \times 10^{-8} \text{ cm}$$

$$V_{\text{unit cell}} = \ell^3 = (3.34 \times 10^{-8} \text{ cm})^3 = 3.73 \times 10^{-23} \text{ cm}^3$$

The density of the unit cell (and bulk Po) is

$$d = \frac{m}{V_{\text{unit cell}}} = \frac{3.47 \times 10^{-22} \text{ g}}{3.73 \times 10^{-23} \text{ cm}^3} = 9.30 \text{ g/cm}^3$$

b. The volume occupied by a Po atom is

$$V_{\text{atoms}} = \frac{4}{3}\pi r^3 = \frac{4}{3}\pi\left(167 \text{ pm} \times \frac{10^{-10} \text{ cm}}{1 \text{ pm}}\right)^3 = 1.95 \times 10^{-23} \text{ cm}^3$$

and the packing efficiency is

$$\text{Packing efficiency (\%)} = \frac{V_{\text{atoms}}}{V_{\text{unit cell}}} \times 100\% = \frac{1.95 \times 10^{-23} \text{ cm}^3}{3.73 \times 10^{-23} \text{ cm}^3} \times 100 = 52.3\%$$

**Think About It** As predicted, the calculated density of polonium is similar to that of iron and close to the Po value in Table A3.2 of Appendix 3: 9.32 g/cm$^3$. The calculated packing efficiency matches the reference value for simple cubic structures in Table 18.1.

**Practice Exercise** Tungsten filaments were widely used in incandescent lightbulbs. Tungsten has an atomic radius of 139 pm and a bcc unit cell. Calculate the density and packing efficiency of the atoms in tungsten. Compare your answer with the density listed in Appendix 3.

## 18.3 Alloys

Around 6000 years ago, metal technology took a giant leap forward when artisans in the Middle East and perhaps other parts of the world discovered how to convert copper ore, principally $CuFeS_2$, into copper metal. The process involved pulverizing the ore and then baking it in ovens. Baking initiated a chemical reaction with $O_2$ (from air), which converted the Cu in $CuFeS_2$ into CuO. In the second step in the process, CuO reacted with carbon monoxide, produced by burning wood or charcoal (mostly carbon) in a furnace with an insufficient supply of air:

$$CuO(s) + CO(g) \rightarrow Cu(s) + CO_2(g) \quad (18.4)$$

One disadvantage of primitive copper tools and weapons is that the metal is malleable, which means that copper objects are easily bent and damaged. The malleability of Cu (and other metals) is due to the relatively weak metallic bonds between their atoms, which make it possible for the atoms in one layer, under stress, to slip past atoms in an adjacent layer (**Figure 18.16**). When the stress is relieved and the atoms stop slipping, many have different atoms as their nearest neighbors, but the overall crystal structure is still the same. The ease with which copper atoms slip past each other made it easy for prehistoric metalworkers to hammer copper metal into spear points and shields, but it also meant that those objects could easily be damaged in battle.

**FIGURE 18.16** Copper and other metals are malleable because their atoms are stacked in layers that can slip past each other under stress. Slippage is possible because of the diffuse nature of metallic bonds and the relatively weak interactions between pairs of atoms in adjoining layers.

## Substitutional Alloys

About 5500 years ago, people living around the Aegean Sea discovered that mixing molten tin and copper produced bronze, an alloy much stronger than either tin or copper alone. Its discovery ushered in the Bronze Age.

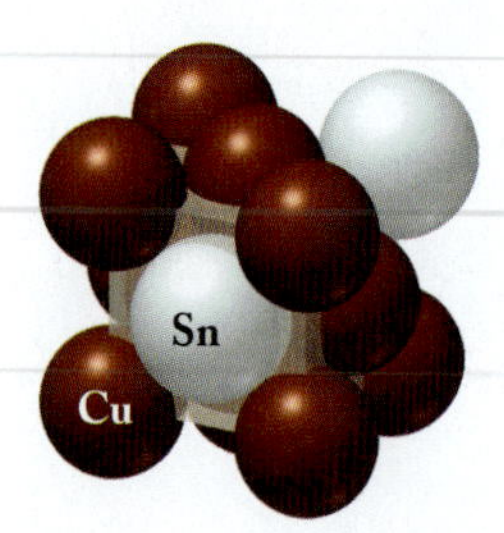


(a) (b)

**FIGURE 18.17** Two atomic-scale views of one type of bronze, a substitutional alloy. (a) A layer of closest-packed copper (Cu) atoms interspersed with a few atoms of tin (Sn). (b) One possible unit cell for bronze. Here, tin atoms have replaced one corner Cu atom and one face Cu atom.

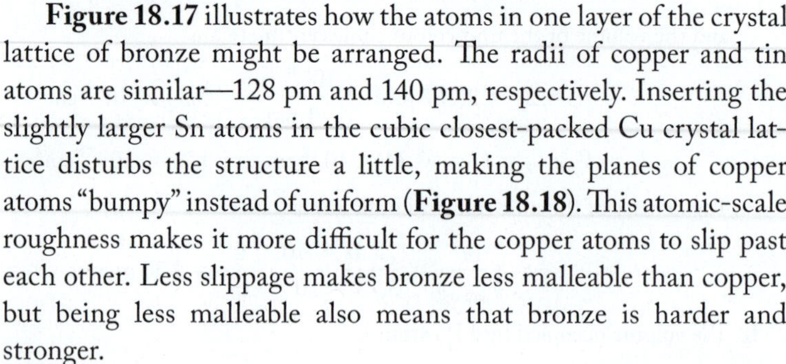

**Figure 18.17** illustrates how the atoms in one layer of the crystal lattice of bronze might be arranged. The radii of copper and tin atoms are similar—128 pm and 140 pm, respectively. Inserting the slightly larger Sn atoms in the cubic closest-packed Cu crystal lattice disturbs the structure a little, making the planes of copper atoms "bumpy" instead of uniform (**Figure 18.18**). This atomic-scale roughness makes it more difficult for the copper atoms to slip past each other. Less slippage makes bronze less malleable than copper, but being less malleable also means that bronze is harder and stronger.

Like the mixtures discussed in Chapter 1, alloys can be classified according to their composition as homogeneous or heterogeneous mixtures. Bronze is a *homogeneous alloy*, a solid solution in which the atoms of the added element(s) (here, tin) are randomly but uniformly distributed among the atoms of the host metal (copper). *Heterogeneous alloys* consist of matrices of atoms of host metals interspersed with small "islands" made up of individual atoms of other elements. Alloy compositions may vary over limited ranges. In contrast, *intermetallic compounds* have a reproducible stoichiometry and constant composition (just like chemical compounds) but are still commonly referred to as alloys and are considered a subgroup within homogeneous alloys. An example of an intermetallic compound is $Ag_3Sn$, which is used in dental fillings. It is a homogeneous mixture of silver and tin atoms in exactly a 3:1 ratio.

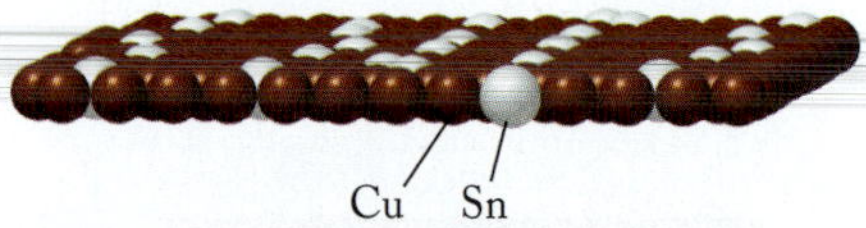


**FIGURE 18.18** The larger Sn atoms in bronze disturb the Cu crystal lattice, producing atomic-scale bumps in the slip plane between layers of Cu atoms. These bumps make it more difficult for the layers of atoms to slide by each other when an external force is applied.

Alloys are also classified by how the minor elements fit into the crystal structure of the host metals. A **substitutional alloy** is one in which atoms of the minor element(s) replace host atoms in the crystal lattice. Bronze is a *homogeneous, substitutional alloy* in which the tin concentration can be as high as 30% by mass. Substitutional alloys may form between metals that have atomic radii within about 15% of each other. It is helpful if both metals have the same crystal lattice as well, but many alloys, such as bronze, form despite having different crystal lattices for the pure metals. Other substitutional alloys include brass (zinc alloyed with copper) and pewter (tin alloyed with copper and antimony). These alloys are stronger than pure copper, and they are more resistant to corrosion.

**CONNECTION** The classification of homogeneous and heterogeneous mixtures and the difference between compounds and mixtures were described in Chapter 1.

Most modern substitutional alloys are ferrous alloys, so called because the host metal is iron. The rust-resistant stainless steels, which contain about 10% nickel and up to 20% chromium, are an important class of ferrous alloys. When atoms of Cr on the surface of a piece of stainless steel combine with oxygen, they form a layer of $Cr_2O_3$ that bonds tightly to the surface and protects the metallic material beneath from further oxidation. This resistance to surface discoloration due to corrosion means that these alloys "stain less" than pure iron.

## Interstitial Alloys

**substitutional alloy** an alloy in which atoms of the nonhost element replace host atoms in the crystal lattice.

**interstitial alloy** an alloy in which the nonhost atoms occupy spaces between atoms of the host.

The Bronze Age began to wane about 3000 years ago when metalworkers were able to design furnaces hot enough to produce molten iron from the reduction of iron oxides. As in the conversion of CuO to Cu (Equation 18.4), these furnaces used carbon monoxide as the reducing agent:

$$Fe_2O_3(s) + 3\,CO(g) \rightarrow 2\,Fe(s) + 3\,CO_2(g) \qquad (18.5)$$

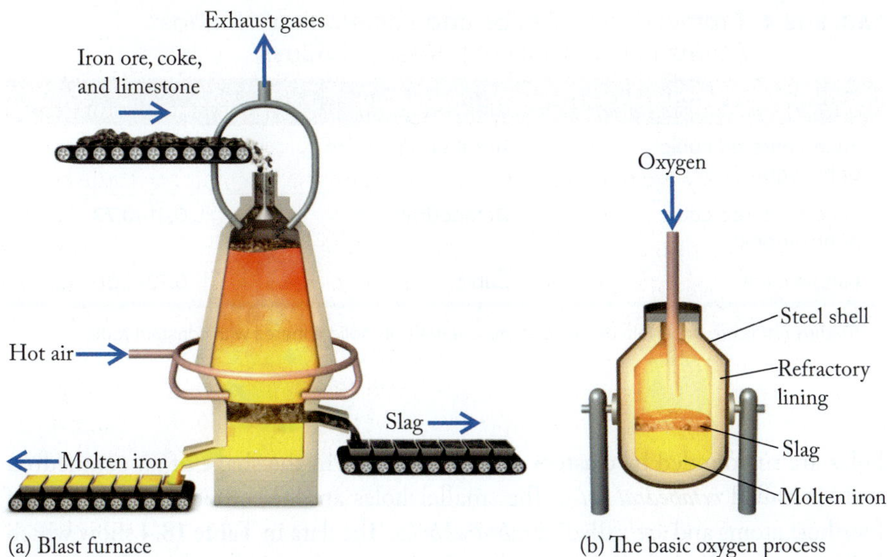


**FIGURE 18.19** (a) Blast furnaces operate continuously at temperatures near 1600°C to convert iron ore into iron. Blasts of hot air inject $O_2(g)$ into the furnace, which converts C to CO. Limestone is added to react with Si and P impurities. The products of these reactions become part of the slag layer. (b) Molten iron from a blast furnace is further purified in a second furnace, where pure $O_2(g)$ is injected instead of air.

Iron replaced bronze as the metallic material of choice for fabricating tools and weapons because iron ore is much more abundant in Earth's crust and because tools and weapons made of iron and ferrous alloys are much stronger than those made of bronze.

Today, iron ore is reduced in enormous blast furnaces (**Figure 18.19**) that operate at about 1600°C. Iron ore, hot carbon (*coke*), and limestone are added to the top of the furnace. Solid impurities, called *slag*, float on top of the molten iron, which is harvested from the bottom. Blast furnaces get their name from blasts of hot air injected through nozzles near the bottom of the furnace and that suspend the reactants until iron reduction is complete. It may take as long as 8 hours for a batch of reactants to fall to the bottom of a blast furnace. On their way down, $O_2$ in the hot-air blasts partially oxidizes the coke to carbon monoxide, and the CO reduces the iron in iron ore, as described in Equation 18.5.

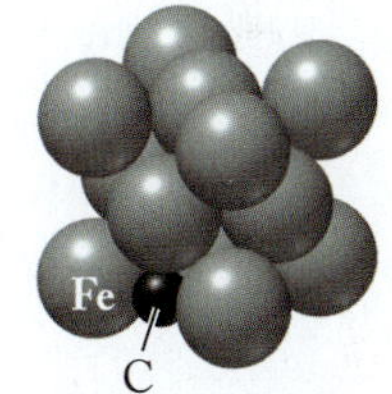


**FIGURE 18.20** Carbon steel is an interstitial alloy of carbon in iron.

Limestone ($CaCO_3$) in the reaction mixture decomposes to calcium oxide:

$$CaCO_3(s) \rightarrow CaO(s) + CO_2(g) \tag{18.6}$$

which reacts with silica impurities in the ore, forming calcium silicate:

$$CaO(s) + SiO_2(s) \rightarrow CaSiO_3(s) \tag{18.7}$$

Calcium silicate becomes part of the slag that floats on the denser molten iron at the bottom of the furnace.

When the molten iron cools to its melting point of 1538°C, it crystallizes in a body-centered cubic structure before undergoing a phase transition at around 1390°C to austenite, a form of solid iron made up of face-centered cubic unit cells. The spaces, or *holes*, between iron atoms in austenite can accommodate carbon atoms, forming an **interstitial alloy**, so named because the carbon atoms occupy spaces, or *interstices*, between the iron atoms (**Figure 18.20**).

All interstices in a crystal lattice are not equivalent. Indeed, holes of two sizes occur between the atoms in any closest-packed crystal lattice (**Figure 18.21**). The larger

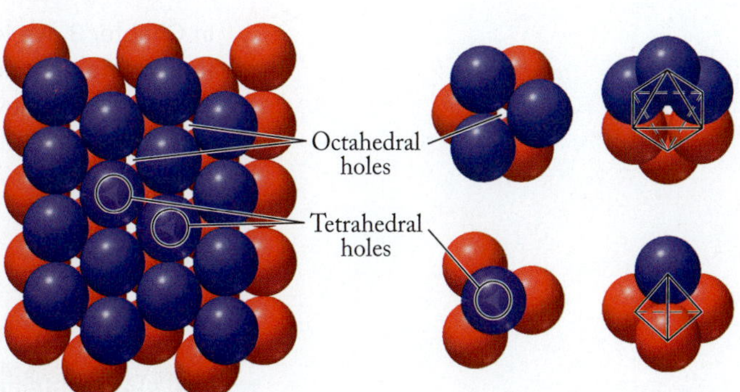


**FIGURE 18.21** Closest-packed atoms in adjacent layers of a crystal lattice produce octahedral holes surrounded by six host atoms and tetrahedral holes surrounded by four host atoms. Octahedral holes are larger than tetrahedral holes and can accommodate larger nonhost atoms in interstitial alloys. Note that all the atoms are identical here; the colors are only to distinguish one atom from another.

**TABLE 18.4** Atomic Radius Ratios and Locations of Nonhost Atoms in Unit Cells of Interstitial Alloys

| Unit Cell Type | Hole Type | $r_{nonhost}/r_{host}$ [a] |
|---|---|---|
| Face-centered cubic or hexagonal | Tetrahedral | 0.22–0.41 |
| Face-centered cubic or hexagonal | Octahedral | 0.41–0.73 |
| Simple cubic | Cubic | 0.73–1.00 |

[a]Radius ratios are uncertain because atoms are not truly solid spheres with constant radii.

holes are surrounded by clusters of six host atoms in the shape of an octahedron and are called *octahedral holes*. The smaller holes are located between clusters of four host atoms and are called *tetrahedral holes*. The data in **Table 18.4** show which holes are more likely to be occupied on the basis of the relative sizes of minor and host atoms. According to Appendix 3, the atomic radii of C and Fe are 77 pm and 126 pm, respectively. According to Table 18.4, the ratio 77/126 = 0.61 means that C atoms should fit in the octahedral holes of austenite, as shown in Figure 18.20, but not in the tetrahedral holes.

As austenite with its fcc unit cell cools to room temperature, it converts into the crystalline solid form of iron called ferrite, which is body-centered cubic. The holes in ferrite are smaller than those in austenite, so many fewer carbon atoms can be accommodated in a ferrite structure than in an austenite structure. The carbon that cannot be accommodated precipitates as clusters of carbon atoms, or it reacts with iron to form iron carbide, $Fe_3C$. The clusters of carbon and $Fe_3C$ disrupt ferrite's crystalline lattice and inhibit the host iron atoms from slipping past each other when a stress is applied. This resistance to slippage, which is much like that experienced by the copper atoms in bronze (Figure 18.18), makes iron–carbon alloys, known as *carbon steel*, much harder and stronger than pure iron. In general, higher carbon concentrations correlate with stronger steel. However, there is a trade-off in this relationship. As **Table 18.5** notes, increased strength and hardness come at the cost of increased brittleness.

**CONNECTION** In Chapter 5, we described how atoms surrounded by four or six bonding pairs of electrons and no lone pairs have tetrahedral or octahedral molecular geometries, respectively.

## CONCEPT TEST

In Chapter 3, we learned that atomic radius increases down a group and decreases from left to right across each row of the periodic table of the elements. Which would you predict to have the larger radius ratio, a lithium–aluminum alloy or a lithium–magnesium alloy?

**TABLE 18.5** Effect of Carbon Content on the Properties of Steel

| Carbon Content (%) | Designation | Properties | Uses |
|---|---|---|---|
| 0.05–0.19 | Low carbon | Malleable, ductile | Nails, cables |
| 0.20–0.49 | Medium carbon | High strength | Construction girders |
| 0.20–3.00 | High carbon | Hard but brittle | Cutting tools |

**SAMPLE EXERCISE 18.4** Predicting the Crystal Structure of a Two-Element Alloy **LO3**

Sterling silver, which is 93% Ag and 7% Cu by mass, is widely used in jewelry. The presence of Cu inhibits tarnishing and strengthens the alloy. Is this copper–silver alloy a substitutional or an interstitial alloy? Silver and copper metal form face-centered cubic unit cells.

**Collect, Organize, and Analyze** Atoms of two metals may form a substitutional alloy if their atoms are of similar size (within 15%) and if they form crystal structures similar to those of pure elements. A host metal with an fcc unit cell may form an interstitial alloy with another element if its atomic radius is 73% or less than that of the host element. According to the data in Appendix 3, the atomic radii of Ag and Cu are 144 pm and 128 pm, respectively.

**Solve** The ratio of the atomic radii of Cu to Ag is 128 pm/144 pm = 0.89 = 89%, which tells us that Cu atoms are too big to fit into either the tetrahedral or octahedral holes in the crystal lattice of metallic silver. Thus, an interstitial alloy is impossible, and sterling silver must be a substitutional alloy.

**Think About It** The atomic radii of Ag and Cu differ by only (144 − 128)/144 = 0.11, or 11%, and the unit cells of both metals are face-centered cubic. Therefore, we expect copper and silver to form a substitutional alloy.

**Practice Exercise** Magnalium is an aluminum (atomic radius = 143 pm) alloy that contains 5% magnesium (atomic radius = 160 pm). It is stronger and less dense than pure aluminum and resists corrosion better. Is magnalium an interstitial or substitutional alloy?

Table 18.4 lists a third type of hole, a *cubic hole*, which accommodates atoms as big as host atoms. What is a cubic hole? It is the space between the eight corner spheres of a simple cubic unit cell. The only metal in Figure 18.10 with a simple cubic unit cell is the radioactive metal polonium, and little is known of its tendencies to form alloys.

Let's take another look at chromium-containing stainless steel alloys. The carbon atoms in steel fit in octahedral holes, but Cr atoms ($r$ = 128 pm) substitute for Fe ($r$ = 126 pm) in the bcc (ferrite) structure of iron. The fact that chromium also crystallizes with a bcc unit cell favors a substitutional alloy. Although metallurgists do not think of chromium steel in this way, another way to view the crystal structures of these alloys is that they consist of simple cubic unit cells of Fe atoms with either an Fe or a Cr atom in the cubic hole at the center of each cell.

## Biomedical Alloys

Every year, thousands of patients suffering from clogged arteries undergo a procedure called balloon angioplasty. It involves inserting a balloon and a small metal support, called a *stent*, into a patient's artery to increase its diameter and to keep it that way after the balloon is removed. Increasingly, stents are manufactured from an alloy—actually an intermetallic compound with the formula NiTi—called nitinol. Nitinol is particularly useful because objects made of NiTi change their crystal structure and macroscopic shape with changing temperature. As shown in **Figure 18.22**, a tube made of woven nitinol wire can be stretched, thereby shrinking the tube's diameter. If the temperature of the tube is less than 20°C, it retains

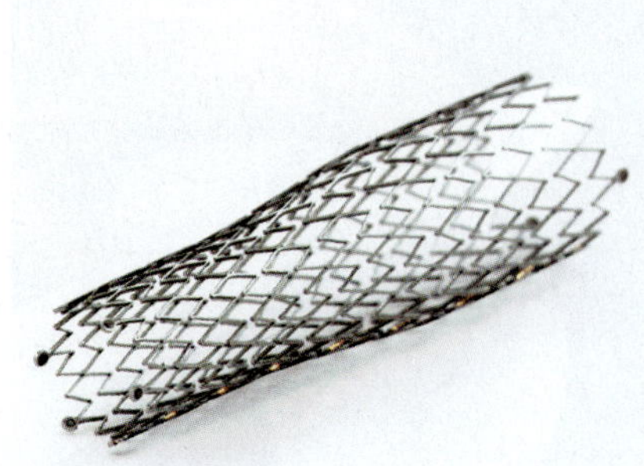

**FIGURE 18.22** Shape memory alloys are used in stents for heart patients. A tube of woven NiTi wire can be stretched out at low temperatures so that it can be inserted into an artery. At body temperature, the stent assumes its original, larger-diameter shape and keeps the artery open.

this shape. However, if the tube is warmed to 37°C, or body temperature, the alloy undergoes a transition in its crystal structure that causes it to assume its original shape. Thus, a small-diameter, stretched-out stent can be inserted into an artery where the temperature of the body (aided by an inflating balloon) allows the stent to reacquire and then maintain its original larger diameter.

Alloys with good strength and corrosion resistance are essential for the manufacture of surgical tools and biomedical implants. Steel containing 13% to 16.5% chromium and 0.4% to 0.6% carbon by mass is appropriate for applications requiring "surgical steel." Dentists use chromium- and nickel-containing alloys in orthodontia.

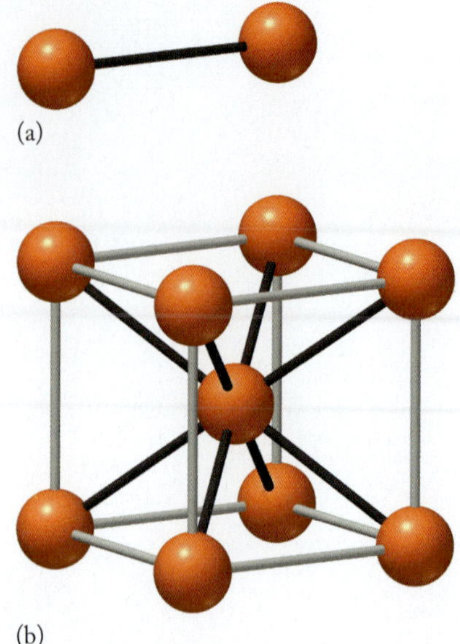


**FIGURE 18.23** Covalent bonds differ from metallic bonds. (a) This ball-and-stick model of the molecule $Na_2$ is based on the two atoms sharing their 3*s* electrons and forming a covalent bond. (b) The atoms in solid sodium metal are actually arranged in a body-centered cubic structure in which each atom is bonded to eight others. As a result, the bonds in sodium and other metals are much more diffuse than the covalent bonds in small molecules.

**CONNECTION** In Chapter 5, we discussed how valence bond theory and molecular orbital theory explain the formation of chemical bonds.

## 18.4 Metallic Bonds and Conduction Bands

In the previous section, we explored the malleability (the ability to be shaped) of metals and how this property can be modified by the formation of alloys. One reason that metals have both malleability and ductility (the ability to be drawn out) is that the bonds between atoms in a solid are weak. Most metals also can conduct electricity. In this section, we explore why they have these properties by examining metals at the atomic level and by exploring models describing the bonds that hold metal atoms together.

According to valence bond theory, a covalent bond forms between two atoms when partially filled atomic orbitals—one from each atom—overlap. Our focus in Chapter 5 was on covalent bonding in gas-phase molecules. In this section, we explore the bonds that form between the densely packed atoms in metallic solids. Dense packing means that the valence orbitals of atoms overlap with orbitals of many nearby atoms. This large number of interactions makes metals strong. At the same time, sharing a limited number of valence electrons with many bonding partners makes the bond linking any two metal atoms relatively weak.

To understand this point, consider the bond that would form if two sodium atoms bonded in a molecule of $Na_2$. As we learned in Chapter 3, sodium has the electron configuration [Ne]$3s^1$. When the partially filled 3*s* orbitals of the two Na atoms overlap, the result should be a diatomic molecule held together by a single covalent bond (**Figure 18.23a**). However, the Na atoms in solid sodium adopt a body-centered cubic structure (**Figure 18.23b**), where each Na atom is surrounded by and bonded to eight other Na atoms, each in turn bonded to eight of its neighbors. This means that each Na atom shares its 3*s* electron with eight other atoms, not just one. Inevitably, this dispersion of bonding electrons weakens the Na—Na bond between each pair of atoms. The weakness of these bonds contributes to the unusual softness of sodium metal: you can literally cut it with a knife (**Figure 18.24**).

**FIGURE 18.24** Solid sodium is a soft metal that can be cut with a knife because the metallic bonds between sodium atoms are weak.

In Chapter 4, we described metal atoms "floating" in seas of mobile bonding electrons, where the electrons are shared by all the nuclei in the sample. The diffuse nature of metallic bonding described in the preceding paragraph certainly fits the sea-of-electrons model, but a more sophisticated approach, called **band theory**, better explains the bonding in metals and other solids. Let's apply band theory, which is an extension of molecular orbital theory, to explain the bonding between atoms in sodium metal.

When the 3*s* atomic orbitals on two Na(*g*) atoms overlap to form $Na_2$, the atomic orbitals combine to form two molecular orbitals with different energies above and below the initial value (**Figure 18.25**). This is analogous to the formation of lower-energy bonding and higher-energy antibonding molecular orbitals (Section 5.7). If another two Na atoms join the first two to form a molecule of $Na_4$, the 3*s* atomic orbitals of four Na atoms combine to form four molecular orbitals. In both of these molecules, the lower-energy orbitals are filled with the available 3*s* electrons and the upper orbitals are empty, as shown in Figure 18.25. If we apply this model to the enormous number of atoms in a piece of solid Na, an equally enormous number of molecular orbitals are created. The lower-energy half of them is occupied by electrons, whereas the higher-energy half is empty. There are so many of these orbitals that they form a continuous *band* of energies with no gap between the occupied lower half and the empty upper half. Because this band of molecular orbitals was formed by combining valence-shell orbitals, it is called a **valence band**. This proximity between the energy of the occupied lower portion of the valence band and the empty upper portion means that valence electrons can move easily from the filled lower portion to the empty upper portion, where they can move from one empty orbital to the next and flow throughout the solid.

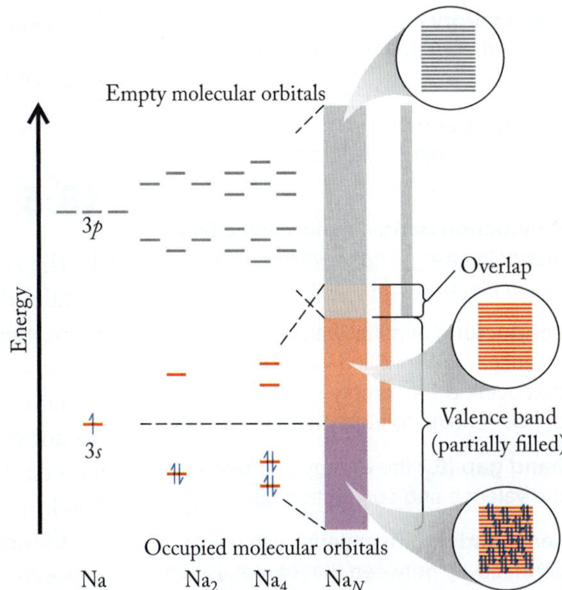


**FIGURE 18.25** The half-filled 3*s* atomic orbitals of increasing numbers of sodium atoms combine to form molecular orbitals. As more molecular orbitals form, their energies get closer together until a continuous energy band is formed—a valence band that is half-filled with electrons. The electrons can move from the filled half (purple) to the slightly higher energy upper half (orange), where they can migrate throughout the solid. The valence band also overlaps with the empty conduction band composed of the 3*p* atomic orbitals.

As the band derived from the 3*s* orbitals of Na broadens with the addition of more atoms, it overlaps with an empty band created by the 3*p* orbitals. This overlap does not change our explanation of the electrical conductivity of sodium metal, but the overlap of bands derived from atomic orbitals with different quantum number ℓ becomes important in explaining the electrical conductivity of some metals. Consider, for example, the excellent conductivity of zinc. Its electron configuration, $[Ar]3d^{10}4s^2$, tells us that all its valence-shell electrons reside in filled 4*s* orbitals, which means that the valence band in solid zinc should be filled (**Figure 18.26**). With no empty space in the valence band to accommodate additional electrons, it might seem that the valence electrons in Zn would be immobile. They are actually mobile, however, and band theory explains why. The energy band produced by combining empty 4*p* orbitals, called a **conduction band**, is also empty and is broad enough to overlap the valence band. This overlap means that electrons from the valence band can move to the conduction band, where they can migrate from atom to atom in solid zinc, thereby conducting electricity.

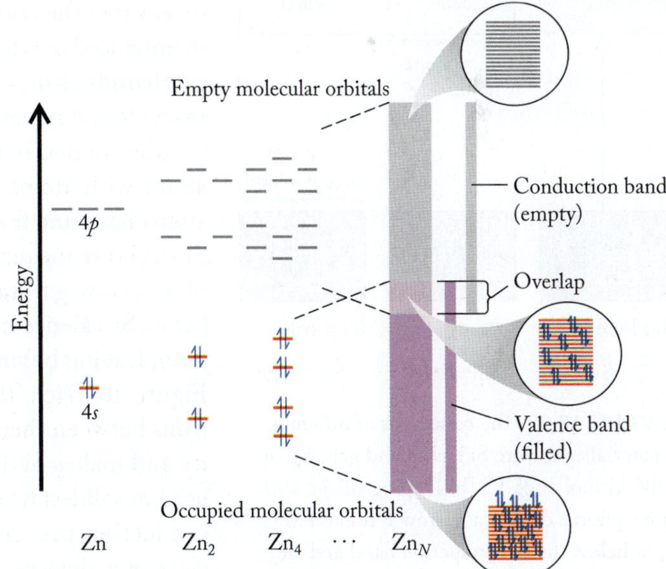


**FIGURE 18.26** As the filled 4*s* atomic orbitals of an increasing number of Zn atoms overlap, they form a filled valence band (purple). An empty conduction band (gray) is produced by combining the empty 4*p* orbitals. The valence and conduction bands overlap, and electrons move easily from the filled valence band to the empty conduction band.

Before we end this discussion, consider that the two views of valence bands provided in Figures 18.25 and 18.26 are not mutually exclusive. The overlap of filled valence and empty conduction bands in Zn shown in Figure 18.26 can also be viewed as a partially filled valence band (as in Figure 18.25) if we assume that molecular orbitals can form from the mixing of the filled 4*s* and empty 4*p* orbitals of many Zn atoms. Either model supports the same conclusion: the valence electrons in zinc metal are mobile, making it a good conductor. We can also define any material with a partially filled valence band or a filled valence band that overlaps with an empty conduction band as an electrical **conductor**.

**band theory** an extension of molecular orbital theory that describes bonding in solids.

**valence band** a band of orbitals that are filled or partially filled by valence electrons.

**conduction band** an unoccupied band higher in energy than a valence band in which electrons can migrate.

**conductor** a material with partially filled valence bands or filled valence bands that overlap with empty conduction bands, leading to highly mobile electrons.

**band gap ($E_g$)** the energy gap between the valence and conduction bands.

**semiconductor** a material with electrical conductivity between that of metals and insulators that can be chemically altered to increase its electrical conductivity.

**n-type semiconductor** a semiconductor containing an electron-rich dopant.

**CONNECTION** We introduced metalloids in Section 2.3, when we described the structure of the periodic table.

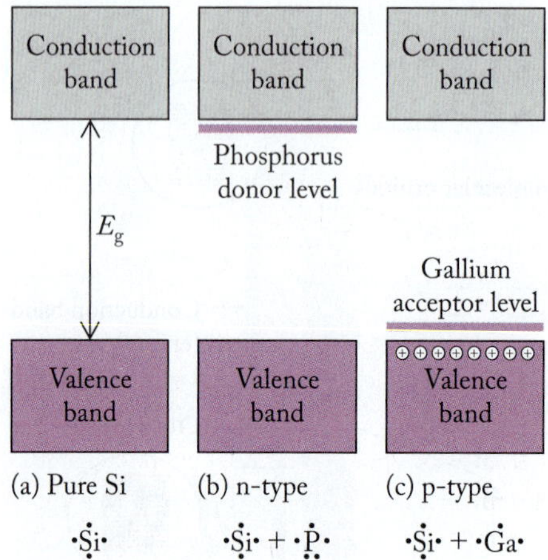


**FIGURE 18.27** The band gaps of different materials. (a) Pure Si has a band gap, $E_g$, of 106 kJ/mol at 25°C. (b) Doping silicon with phosphorus creates a narrow P donor level just below the Si conduction band and an n-type semiconductor. (c) Doping silicon with gallium creates a narrow Ga acceptor level just above the Si valence electrons and a p-type semiconductor. Valence electrons move into the acceptor level, leaving behind positively charged holes (⊕) in the valence band that enhance conductivity.

**CONCEPT TEST**

Use band theory to explain the electrical conductivity of magnesium metal.

## 18.5 Semiconductors

To the right of the metals in the periodic table is a "staircase" of elements—the metalloids—that tend to have the physical properties of metals and the chemical properties of nonmetals. The metalloids are not as good at conducting electricity as metals, but they are better at it than nonmetals. We can use band theory to explain this intermediate behavior. In metalloids, conduction and valence bands do not overlap but instead are separated by an energy gap. In silicon, the most abundant metalloid, band theory predicts an energy gap, or **band gap ($E_g$)**, of 106 kJ/mol at 25°C (**Figure 18.27a**).

Generally, only a few valence-band electrons in Si have enough energy to move to the conduction band, which limits silicon's ability to conduct electricity and makes it a **semiconductor**. However, we can enhance the conductivity of solid Si, or of any other elemental metalloid, by replacing some of the Si atoms with atoms of an element having a similar atomic radius but a different number of valence electrons. The replacement process is called *doping*, and the added element is called a *dopant*. In effect, doped semiconductors represent substitutional alloys.

Suppose the dopant is a group 15 element such as phosphorus. Each P atom has one more valence electron than the atom of silicon it replaces (the Lewis symbols of Si and P are compared in Figure 18.27). The energy of these additional electrons is different from the energy of the silicon electrons. They populate a narrow band labeled the phosphorus *donor level* (**Figure 18.27b**) that is only about 4 kJ/mol below the Si conduction band. This small energy difference means that the donor electrons can easily reach the Si conduction band, resulting in enhanced conductivity. Phosphorus-doped silicon is an example of an **n-type semiconductor**, so called because the dopant donates **n**egative charges (electrons) to the structure of the host element.

The conductivity of solid silicon can also be enhanced by replacing some Si atoms with atoms of a group 13 element, such as gallium (**Figure 18.27c**). Ga atoms have one fewer valence electron than Si atoms, so substituting Ga in the Si structure means fewer valence electrons in the solid. The result is the creation of a narrow gallium *acceptor level* located about 7 kJ/mol above the Si valence band. Si valence electrons can move from the valence band into the acceptor level, leaving behind positively charged "holes" represented by the ⊕ symbols in Figure 18.27(c). The presence of the **p**ositive holes and the migration of electrons between them, filling one hole and creating another, enhances conductivity and makes Si doped with Ga a **p-type semiconductor**. The semiconductors used in solid-state electronics are combinations of n-type and p-type. The semiconducting properties of silicon and doped silicon make possible the myriad electronic devices in our lives, including the cell phones and computers we rely on so heavily.

To put these band gap values into perspective: an electrical insulator such as NaCl has an enormous band gap ($6.8 \times 10^5$ kJ/mol) that prevents significant numbers of electrons from moving from its valence band to the conduction band. The observation of low electrical conductivity allows us to define any material with a wide energy gap between the valence and conduction bands as an **insulator**. By contrast, a metallic conductor such as Zn, with overlapping valence and conduction bands, has essentially no band gap and high electrical conductivity.

Doping is not the only way to change the conductivity of metalloids. Alloys prepared from combinations of group 13 and group 15 elements, such as gallium arsenide, also behave as semiconductors. Most of these alloys adopt a structure in which the group 13 element occupies half of the tetrahedral holes in the cubic closest-packed arrangement of the group 15 element. Semiconductors made of GaAs have the same average number of valence electrons per atom as silicon [(3 + 5)/2 = 4], but their band gaps are larger. Larger-band-gap semiconductors such as GaAs are useful in the power amplifiers that allow voice and data transmission in satellite communications and cell phones. Gallium arsenide also can emit infrared radiation ($\lambda$ = 874 nm) when connected to an electrical circuit. This emission is used in many devices, including remote control units for televisions and DVD players. When electrical energy is applied to the material, electrons are raised to the conduction band. When they fall back to the valence band, they emit radiation.

If aluminum is substituted for gallium in GaAs, the band gap increases, and the wavelength of emitted light decreases. For example, a material with the empirical formula $AlGaAs_2$ emits orange-red light ($\lambda$ = 620 nm). Many of the multicolored indicator lights in electronic devices, called light-emitting diodes, or LEDs (**Figure 18.28a**), use gallium arsenide, indium phosphide, and other semiconducting materials composed of group 13 and group 15 elements. Historically, the most challenging color to produce from LEDs was blue. This was finally achieved by using both indium-substituted gallium nitride ($In_xGa_{1-x}N$) and gallium nitride itself (GaN). One way of generating familiar white light from monochromatic LEDs is to mix the colors emitted by red, green, and blue LEDs. The availability of a broad spectrum of colors in the visible region of the electromagnetic spectrum allowed the development of LED lightbulbs (**Figure 18.28b**) based on gallium nitride. These bulbs offer higher efficiency and longer lifetimes than incandescent and fluorescent lightbulbs.

Transition metal–containing alloys such as cadmium sulfide or cadmium selenide also behave as semiconductors. In Chapter 3, we described the use of CdS and CdSe quantum dots to produce the colors of ultra-high-definition television sets. In these materials, cadmium $Cd^{2+}$ ions occupy tetrahedral holes in hexagonal closest-packed crystal structures of $S^{2-}$ or $Se^{2-}$ ions.

**p-type semiconductor** a semiconductor containing an electron-poor dopant.

**insulator** a material with a large energy gap between its valence and conduction bands.

**CONNECTION** Emission spectra obtained from gas-discharge tubes were discussed in Chapter 3.

(a)

(b)

**FIGURE 18.28** Light from light-emitting diodes (LEDs) used in (a) indicator lights and (b) lightbulbs is generated by semiconductors made from group 13 and group 15 semiconducting materials.

**CONCEPT TEST**

Gallium arsenide (GaAs) can be made an n-type or a p-type semiconductor by replacing some of the As atoms with another element. Which element, Se or Sn, would form an n-type semiconductor with GaAs?

## 18.6 Structures of Some Crystalline Nonmetals

In Figure 18.10, the color key identifies the unit cells of the group 14 elements silicon, germanium, and tin as "diamond." Carbon is a nonmetal (also from group 14), but the arrangement of carbon atoms in diamond (**Figure 18.29a**) is also found in some metallic materials. In addition, some diamonds may include metal impurities in the crystal lattice that give rise to distinctive colors.

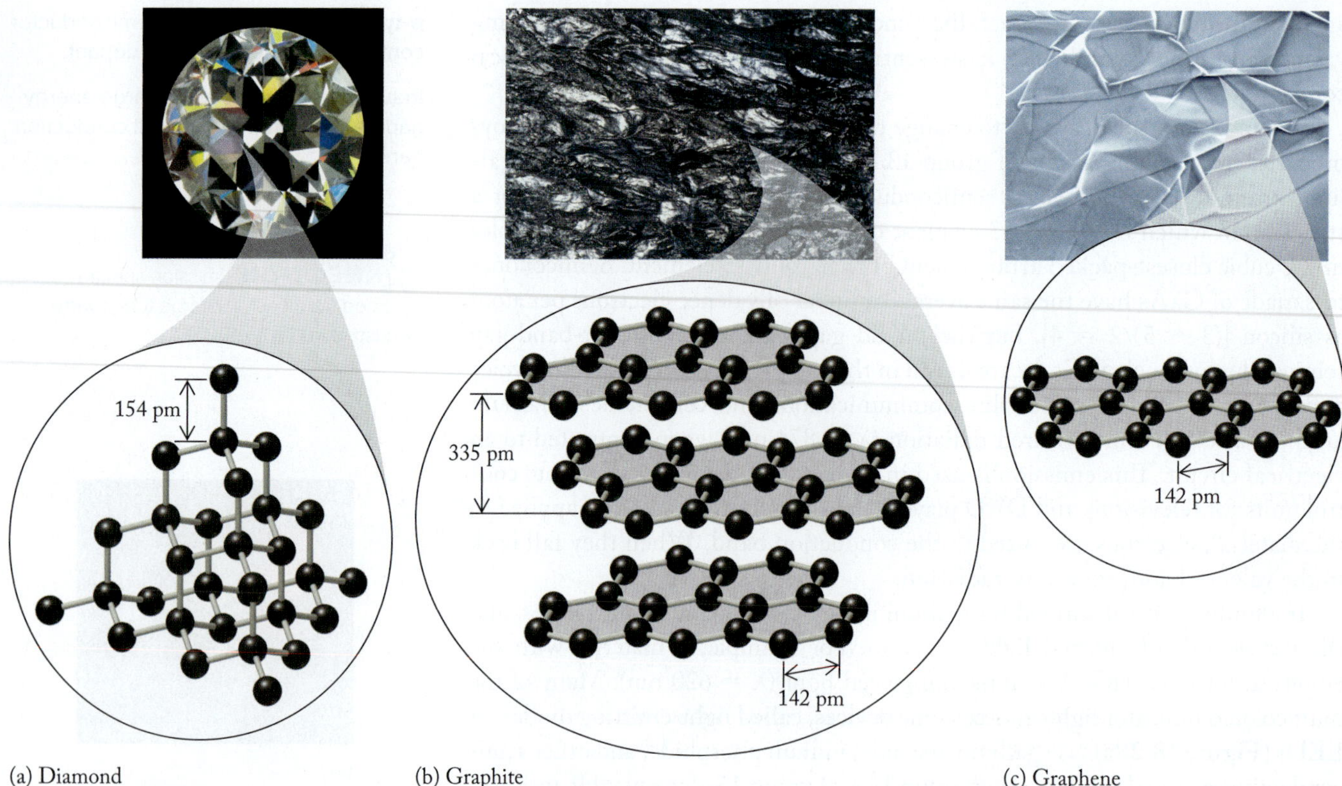


**FIGURE 18.29** Two allotropes of carbon. (a) Diamond is a three-dimensional covalent network solid in which each carbon atom is connected by σ bonds to four adjacent carbon atoms. (b) Graphite is a collection of layers of carbon atoms connected by σ bonds and delocalized π bonds. (c) Graphene consists of single layers of graphitic carbon atoms.

Diamond is one of three allotropes of carbon, the other two being graphite (Figure 18.29a) and fullerenes (**Figure 18.30a**). Diamond is classified as a crystalline **covalent network solid** because it consists of atoms held together in an extended three-dimensional network of covalent bonds. Each carbon atom in diamond forms bonds by overlapping one of its $sp^3$ orbitals with an $sp^3$ orbital in each of four neighboring carbon atoms, creating a network of carbon tetrahedra. The atoms in these tetrahedra are connected by localized σ bonds, making diamond a poor electrical conductor. The sigma-bond network is extremely rigid, however, making diamond the hardest natural material known. The atoms of other group 14 elements also form covalent network solids based on the diamond crystal lattice.

**CONNECTION** Allotropes are structurally different forms of the same physical state of an element, as explained in Section 4.5.

**CHEMTOUR**
Allotropes of Carbon

Natural diamond forms from graphite under intense heat (>1700 K) and pressure (>50,000 atm) deep in Earth. Industrial diamonds synthesized at high temperatures and pressures from graphite or any other source rich in carbon are used as abrasives and for coating the tips and edges of cutting tools. These tools resist overheating due to friction because vibrations of the carbon atoms in diamond at high temperature provide a pathway to efficiently dissipate heat, making diamond a better thermal conductor than any metal. Optimizing these conditions allows the production of gem-quality diamonds up to a record 10-carat stone. Synthetic diamonds of various colors are now available commercially for engagement rings and other jewelry. Choosing a synthetic diamond is seen as an ethical choice that avoids the purchase of "conflict diamonds."

**covalent network solid** a solid consisting of atoms held together by extended arrays of covalent bonds.

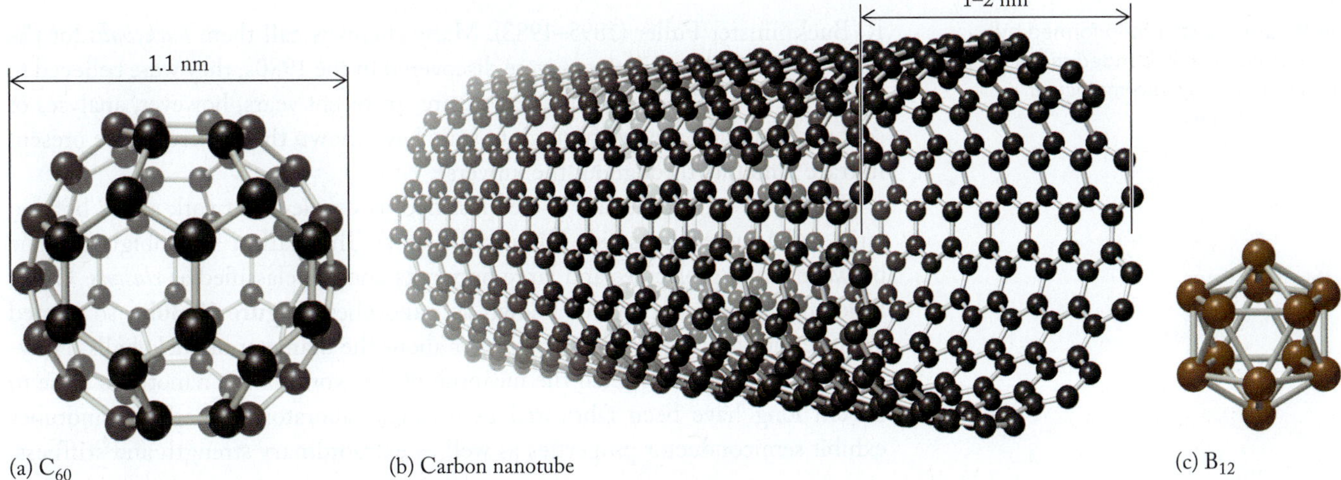


**FIGURE 18.30** The atoms in some solids form clusters, a class of structures between covalent network solids and molecular solids. In this category are fullerenes, an allotrope of carbon that includes structures known as (a) buckyballs and (b) nanotubes. (c) A crystalline form of boron consists of clusters of 12 boron atoms.

By far, the most abundant allotrope of carbon is graphite, another covalent network solid. Graphite is a principal ingredient in soot and smoke and is used to make pencils, lubricants, and gunpowder. Graphite contains sheets of carbon atoms connected by overlapping $sp^2$ orbitals to three neighboring carbon atoms in a two-dimensional covalent network of six-membered rings (**Figure 18.29b**). Each carbon–carbon $\sigma$ bond is 142 pm, which is shorter than the C—C $\sigma$ bond in diamond (154 pm). Overlapping unhybridized $p$ orbitals on the carbon atoms form a network of $\pi$ bonds that are delocalized across the plane defined by the rings. The mobility of these delocalized electrons makes graphite a good conductor of electricity in this plane. However, graphite is a poor conductor in the direction perpendicular to this plane because the layers of fused rings are 335 pm apart (Figure 18.29b). This distance is much too long to be a covalent bonding distance, which means the sheets are held together only by London dispersion forces and not by shared electrons. However, the relatively weak interactions between adjacent sheets allow them to slide past each other, making graphite soft, flexible, and a good lubricant.

**CONNECTION** We introduced London dispersion forces between molecules in Chapter 6.

Another form of carbon is both the simplest and perhaps the most challenging to study: a single sheet of hexagonally bonded carbon atoms (**Figure 18.29c**). Material with this structure is called *graphene*. Isolating a single layer of graphitic carbon atoms is not difficult, but preserving it as a purely two-dimensional material is. It tends to curl and buckle, forming three-dimensional structures with less surface area. One of the first successful approaches to isolating a single sheet of graphene is elegant in its simplicity: single layers of C atoms can be pulled away from graphite by using adhesive tape. Graphene is both the thinnest material known and the strongest—hundreds of times stronger than steel. It conducts heat better than any other known material and conducts electricity better than any metal. It is also nearly transparent, even though its carbon atoms and bonding electrons are so densely packed that atoms of helium gas cannot pass through it.

Yet another allotrope of carbon consists of networks of five- and six-atom carbon rings that form molecules of 60, 70, or more $sp^2$-hybridized carbon atoms. They look like miniature soccer balls (Figure 18.30a) and are called *fullerenes* because their shape resembles the geodesic domes designed by American architect

**molecular solid** a solid formed by neutral, covalently bonded molecules held together by intermolecular attractive forces.

R. Buckminster Fuller (1895–1983). Many chemists call them *buckyballs* for the same reason. When fullerenes were discovered in the 1980s, they were believed to be a form of carbon rarely found in nature. In recent years, however, analyses of soot and emission spectra from giant stars have shown that fullerenes are present in trace amounts throughout the universe.

Fullerenes are too small to be classified as covalent network solids but too large to be molecular solids (discussed below). They fall in an ambiguous zone between small molecules and large networks and are classified as *clusters*. Other fullerenes include structures known as nanotubes (**Figure 18.30b**), so named because they are 1–2 nm in diameter, or about the diameter of buckyballs. However, some fullerenes stretch the meaning of the word *cluster*: nanotubes close to 20 cm long have been fabricated in research laboratories. Carbon nanotubes exhibit semiconductor properties as well as extraordinary strength and stiffness.

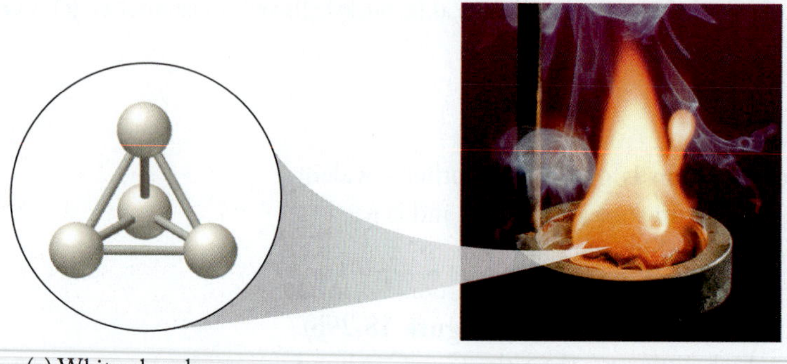
(a) White phosphorus

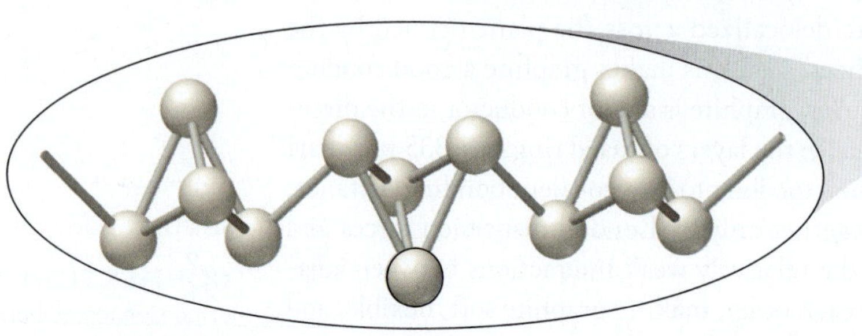
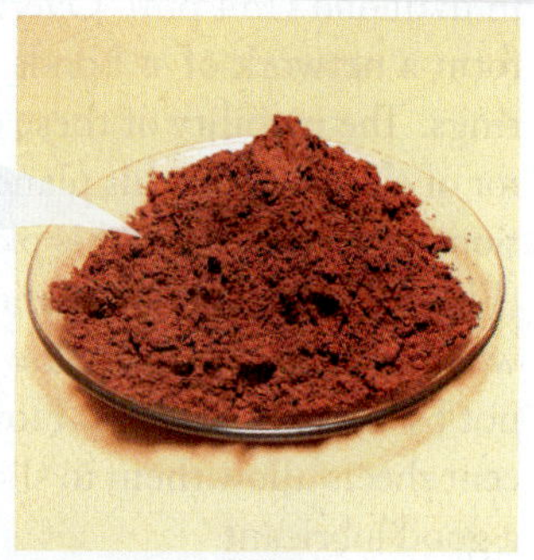
(b) Red phosphorus

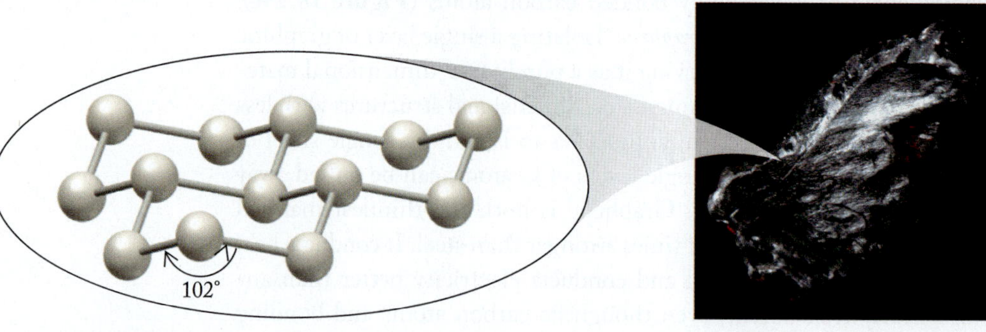

(c) Black phosphorus

**FIGURE 18.31** The two most common allotropes of phosphorus are based on $P_4$ tetrahedra. (a) White phosphorus consists of discrete $P_4$ molecules. (b) Red phosphorus consists of chains of $P_4$ tetrahedra. When heated at high pressure, white phosphorus is transformed into (c) black phosphorus.

Some nonmetals, such as boron, also form clusters. One form of boron contains closest-packed arrays of 12-vertex, 20-sided *icosahedra* (singular, *icosahedron*; **Figure 18.30c**) composed of 12 boron atoms.

Crystalline **molecular solids** consist of molecules held together by intermolecular forces. Ice, $CO_2$, glucose, and most organic molecules crystallize as molecular solids. One of the two most common allotropes of phosphorus, white phosphorus, is a molecular solid consisting of $P_4$ tetrahedra arranged in a cubic array (**Figure 18.31a**). White phosphorus is a waxy, soft material that can be cut with a knife. Because it burns in air, it is commonly stored in water. It gives off a yellow-green light in a phenomenon called *phosphorescence*.

The other common phosphorus allotrope, red phosphorus, is not a molecular solid but rather a covalent network solid made of chains of $P_4$ tetrahedra connected by covalent P—P bonds (**Figure 18.31b**). Both red phosphorus and white phosphorus melt to give the same liquid consisting of symmetrical $P_4$ tetrahedral molecules. Heating white phosphorus at 200°C and high pressure yields black phosphorus. One form of black phosphorus contains layers of phosphorus atoms with a bond angle of 102° (**Figure 18.31c**). Like graphene, materials composed of a few layers of black phosphorus may find applications in molecular electronics and even in medicine in the next decade.

Sulfur has more allotropic forms than any other element. Most are molecular solids. This variety of forms exists because sulfur atoms form cyclic (ring) molecules of different sizes, which means that different crystalline arrangements of the molecules are possible. The most common allotropes of sulfur consist of puckered rings (**Figure 18.32**) containing eight covalently bonded sulfur atoms. London dispersion forces hold one ring to another in solid sulfur in staggered stacking patterns. The weakness of these interactions is the reason elemental sulfur is soft and melts at only 115.21°C.

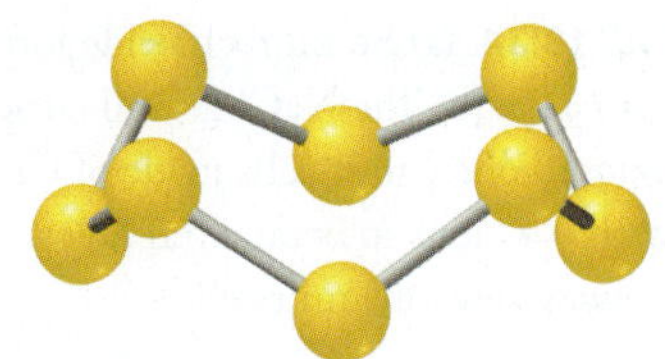

**FIGURE 18.32** One form of sulfur is a molecular solid based on puckered $S_8$ rings.

**CONCEPT TEST**

A crystal structure of solid fluorine is shown in **Figure 18.33**. Is this form of $F_2$ better described as a molecular solid or a network solid?

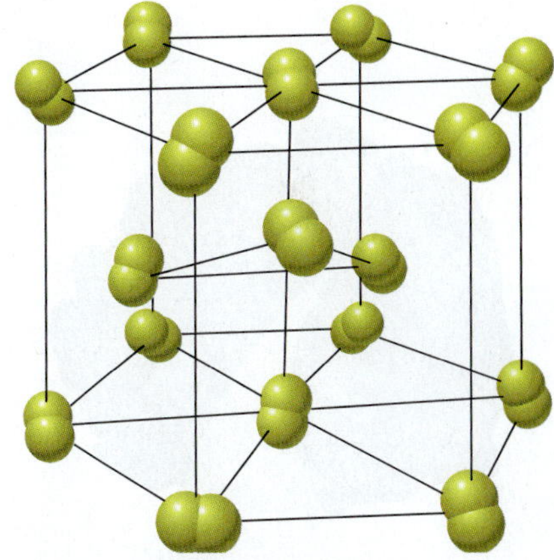

**FIGURE 18.33** An expanded view of the crystal structure of solid $F_2$ at 54 K.

## 18.7 Salt Crystals: Ionic Solids

Most of Earth's crust is composed of ionic solids consisting of monatomic or polyatomic ions held together by ionic bonds. Most of these solids are crystalline. The simplest crystal structures are those of binary salts, such as NaCl (**Figure 18.34a**). The cubic shape of large NaCl crystals is a reflection of the cubic shape of the NaCl unit cell.

The unit cell of NaCl (**Figure 18.34b**) is a face-centered cubic arrangement of $Cl^-$ ions at the corners and in the center of each face, with the smaller $Na^+$ ions occupying the 12 octahedral holes along the edges of the unit cell and the single octahedral hole in the middle of the cell. The $Na^+$ ions fit into the octahedral holes because the radius ratio of $Na^+$ to $Cl^-$ is

$$\frac{r_+}{r_-} = \frac{102 \text{ pm}}{181 \text{ pm}} = 0.564$$

This ratio is too large for $Na^+$ to occupy a tetrahedral hole (see Table 18.4) but well within the range for occupying an octahedral hole.

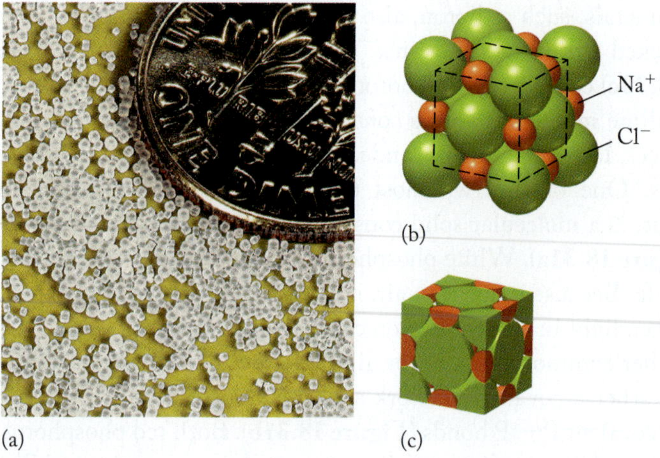


**FIGURE 18.34** (a) Sodium chloride forms cubic crystals. (b) The NaCl crystal lattice is composed of fcc unit cells made of $Cl^-$ ions with $Na^+$ ions in octahedral holes. (c) Cutaway view of a unit cell.

Let's take an inventory of the ions in an isolated unit cell of NaCl (**Figure 18.34c**). Like the fcc unit cell in Figure 18.11(b), this unit cell contains portions of 14 $Cl^-$ ions: one at each corner and one in each of the six faces of the cube. As in the analysis of Figure 18.11, accounting for partial contributions gives us a total of four $Cl^-$ ions in the unit cell. To count the $Na^+$ ions, note that one $Na^+$ ion fits into each of the 12 octahedral holes along the edges of the cell. Because each $Na^+$ ion along an edge is shared by four unit cells, only one-fourth of each $Na^+$ ion on an edge is in each cell (see Figure 18.12d), for a total of three $Na^+$ ions. Where is the fourth $Na^+$ ion needed to match the empirical formula of NaCl? There is also an octahedral hole in the middle of the unit cell that is occupied by a $Na^+$ ion. The $Na^+$ ion in the center belongs completely to the unit cell. Therefore, the total number of $Na^+$ ions in the unit cell is

$$\left(12 \times \frac{1}{4}\right) + 1 = 4\ Na^+ \text{ ions}$$

The ratio of $Na^+$ to $Cl^-$ ions in the unit cell is therefore 4:4, or 1:1, consistent with the chemical formula NaCl. Because the four $Na^+$ ions occupy all the octahedral holes in the unit cell, this calculation also tells us that each fcc unit cell contains the equivalent of four octahedral holes.

Note in Figure 18.34(c) that adjacent $Cl^-$ ions along any face diagonal do not touch each other the way adjacent face-diagonal metal atoms do in Figure 18.11(b) because the $Cl^-$ ions have to spread out a little to accommodate the $Na^+$ ions in the octahedral holes. Sodium and chloride ions touch along each edge of the unit cell, however, which means that each $Na^+$ ion touches six $Cl^-$ ions and that each $Cl^-$ ion touches six $Na^+$ ions. This arrangement of positive and negative ions is common enough among binary ionic compounds to be assigned its own name: the *rock salt structure*.

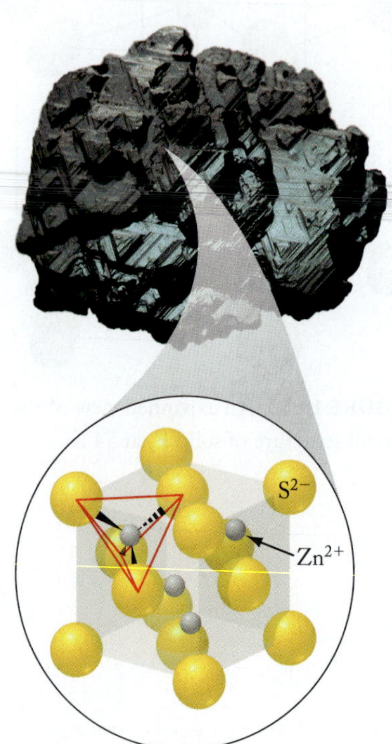


**FIGURE 18.35** Many crystals of the mineral sphalerite (ZnS), like the largest one in this photograph, have a tetrahedral shape. The crystal lattice of sphalerite is based on an fcc unit cell of $S^{2-}$ ions with $Zn^{2+}$ ions in four of the eight tetrahedral holes. In the expanded view of the sphalerite unit cell, each $Zn^{2+}$ ion is in a tetrahedral hole, such as the one outlined in red, which is formed by one corner $S^{2-}$ ion and three face-centered $S^{2-}$ ions.

In other binary ionic solids, the smaller ion is small enough to fit into the tetrahedral holes formed by the larger ions. For example, in the unit cell of the mineral sphalerite (zinc sulfide), the $S^{2-}$ anions (ionic radius, 184 pm) are arranged in an fcc unit cell (**Figure 18.35**), and half of the eight tetrahedral holes inside the cell are occupied by $Zn^{2+}$ cations (74 pm). Therefore, the unit cell contains four $Zn^{2+}$ ions that balance the charges on the four $S^{2-}$ ions. This pattern of half-filled tetrahedral holes in an fcc unit cell is called the *sphalerite structure*. This is also the unit cell for most of the group 13–15 semiconductors described in Section 18.5.

**FIGURE 18.36** The mineral fluorite ($CaF_2$) forms cubic crystals. The crystal lattice of $CaF_2$ is based on an fcc array of $Ca^{2+}$ ions, with $F^-$ ions occupying all eight tetrahedral holes. Because they are bigger than $Ca^{2+}$ ions, the $F^-$ ions do not fit in the tetrahedral holes of a cubic closest-packed array of $Ca^{2+}$ ions. Instead, the $Ca^{2+}$ ions, while maintaining an fcc unit cell arrangement, spread out to accommodate the larger $F^-$ ions. Note how adjacent $Ca^{2+}$ ions along any face diagonal do not touch each other the way they do in the ideal fcc unit cell in Figure 18.11(b).

The crystal structure of the mineral fluorite ($CaF_2$) is based on an fcc unit cell of smaller $Ca^{2+}$ ions at the cube's eight corners and six face centers, with all eight tetrahedral holes filled by larger $F^-$ ions. Because there is a total of four $Ca^{2+}$ ions in the unit cell and the eight $F^-$ ions are all completely inside the cell, this arrangement satisfies the 1:2 mole ratio of $Ca^{2+}$ ions to $F^-$ ions. This structure is so common that it too has its own name: the *fluorite structure* (**Figure 18.36**). Other compounds having this structure are $SrF_2$, $BaCl_2$, and $PbF_2$.

Some compounds in which the cation-to-anion mole ratio is 2:1 have an *anti-fluorite structure.* In the crystal lattices of these compounds, which include $Li_2O$ and $K_2S$, the smaller cations occupy the tetrahedral holes in an fcc unit cell formed by cubic closest-packing of the larger anions.

**CONNECTION** Periodic trends in atomic radii were discussed in Chapter 3.

### SAMPLE EXERCISE 18.5 Calculating Ionic Radii from Unit Cell Dimensions — LO1

The unit cell of lithium chloride (LiCl) contains an fcc arrangement of $Cl^-$ ions (**Figure 18.37**). In LiCl, the $Li^+$ cations (radius, 76 pm) are small enough to allow adjacent $Cl^-$ ions to touch along any face diagonal.

a. If the edge length of the LiCl fcc cell is 513 pm, what is the radius of the $Cl^-$ ion?
b. On the basis of the value calculated for the radius of the $Cl^-$ ion, what type of hole does the $Li^+$ ion occupy?

**Collect and Organize** We are given the edge length of a LiCl unit cell, the radius of the $Li^+$ cation, and a picture of the unit cell. The unit cell is an fcc array of $Cl^-$ ions that touch along the face diagonal.

**Analyze** The face diagonal of an fcc unit cell is equal to four times the radius of the spheres (chloride ions) that make up the cell, and it's also equal to $\sqrt{2}$ times the cell

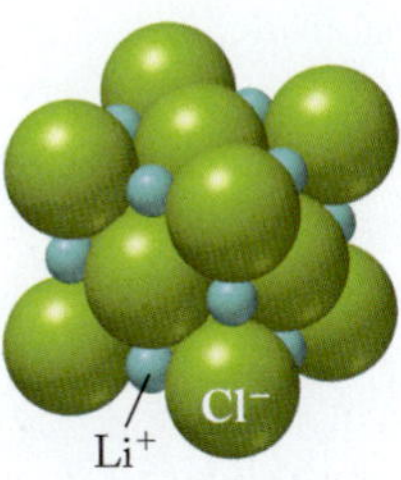


**FIGURE 18.37** The fcc unit cell of LiCl.

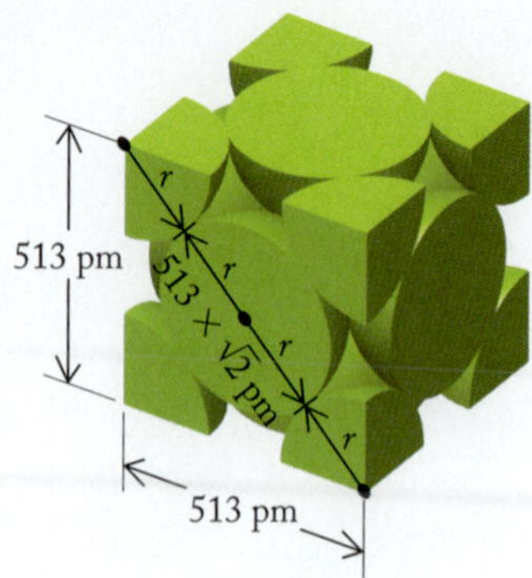


**FIGURE 18.38** The $Cl^-$ ions in the LiCl unit cell allow us to calculate its dimensions.

length of 513 pm (**Figure 18.38**). The radius ratio of $Li^+$ to $Cl^-$ ions and Table 18.4 allow us to predict which type of hole $Li^+$ occupies. We expect $Li^+$ ions to be smaller than $Cl^-$ ions, but they may not be small enough to fit in tetrahedral holes.

**Solve**

a. By substituting the edge length into Equation 18.2, we find that the radius of $Cl^-$ is

$$r = (513 \times \sqrt{2})/4 \text{ pm} = 181 \text{ pm}$$

b. The ratio of the radius of a $Li^+$ ion to the radius of a $Cl^-$ ion is

$$76 \text{ pm}/181 \text{ pm} = 0.42$$

According to Table 18.4, the $Li^+$ cations occupy octahedral holes in the lattice formed by the larger $Cl^-$ anions.

**Think About It** The average ionic radius value in Figure 3.36 for $Cl^-$ ions is 181 pm, so our calculation is correct. Figure 18.37 shows the structure of LiCl to be similar to the rock salt structure of NaCl in Figure 18.34(b).

**Practice Exercise** If we assume that the $Cl^-$ radius in NaCl is also 181 pm, what is the radius of the $Na^+$ ion in NaCl if the edge length of the NaCl unit cell is 564 pm but the anions do *not* touch along any face diagonal?

## SAMPLE EXERCISE 18.6 Calculating the Density of a Salt from Its Unit Cell Dimensions — LO2

What is the density of LiCl if the edge length of its fcc unit cell is 513 pm?

**Collect and Organize** We are given the edge length of an fcc unit cell of LiCl and are asked to calculate its density. Density is the ratio of mass to volume, and the volume of a cubic cell is the cube of its edge length: $V = \ell^3$.

**Analyze** As in Sample Exercise 18.2, we assume that the density of the unit cell is the same as the density of the crystalline solid. The fact that LiCl has an fcc unit cell (like NaCl) means that there are four $Cl^-$ ions and four $Li^+$ ions in the cell. The density of the unit cell is the sum of the masses of these eight ions divided by the volume of the cell, $(513 \text{ pm})^3$. As in Sample Exercise 18.2, we need to divide the molar masses of Li and Cl by the Avogadro constant to obtain the masses of individual atoms (or monatomic ions) of Li and Cl in grams.

**Solve** Calculating the mass of four $Cl^-$ ions,

$$m = \frac{35.45 \text{ g } Cl^-}{1 \text{ mol } Cl^-} \times \frac{1 \text{ mol } Cl^-}{6.022 \times 10^{23} \text{ ions } Cl^-} \times 4 \text{ ions } Cl^- = 2.355 \times 10^{-22} \text{ g } Cl^-$$

and the mass of four $Li^+$ ions,

$$m = \frac{6.941 \text{ g } Li^+}{1 \text{ mol } Li^+} \times \frac{1 \text{ mol } Li^+}{6.022 \times 10^{23} \text{ ions } Li^+} \times 4 \text{ ions } Li^+ = 0.4610 \times 10^{-22} \text{ g } Li^+$$

Combining the masses of the two kinds of ions in the unit cell,

$$2.355 \times 10^{-22} \text{ g} + 0.4610 \times 10^{-22} \text{ g} = 2.816 \times 10^{-22} \text{ g}$$

The volume of the cell in cubic centimeters is

$$V = \ell^3 = (513 \text{ pm})^3 \times \frac{(10^{-10} \text{ cm})^3}{(1 \text{ pm})^3} = 1.350 \times 10^{-22} \text{ cm}^3$$

Taking the ratio of mass to volume, we have

$$d = \frac{m}{V} = \frac{2.816 \times 10^{-22}\ \text{g}}{1.350 \times 10^{-22}\ \text{cm}^3} = 2.09\ \text{g/cm}^3$$

**Think About It** The result is reasonable because most minerals have densities greater than the density of water (1 g/cm$^3$) but are less dense than common metals (>4 g/cm$^3$).

**Practice Exercise** What is the density of NaCl if the edge length of its fcc unit cell is 564 pm?

Ionic compounds such as NaCl are insulators, not conductors, even though aqueous solutions of NaCl conduct electricity nicely. Why is NaCl an insulator? The $Cl^-$ ions have filled 3*s* and 3*p* atomic orbitals, but the 3*s* orbital of the $Na^+$ ions is empty. Therefore, the orbitals of all $Cl^-$ ions form filled 3*s* and 3*p* valence bands while the $Na^+$ orbitals form an empty 3*s* band. The electronegativity difference between Cl (3.0) and Na (0.9) means that the filled bands from the chloride ion lie at a much lower energy ($6.8 \times 10^5$ kJ/mol lower) than the empty $Na^+$ band. This enormous band gap explains why NaCl, like most ionic solids, is an insulator.

## 18.8 Ceramics: Useful, Ancient Materials

The use of *ceramic* materials preceded metal technology by many thousands of years. The first ceramics were probably made of *clay*, the fine-grained soil produced by the physical and chemical weathering of igneous rocks (rocks of volcanic origin). Moist clay is easily molded into a desired shape and then hardened over fires or in wood-burning kilns, an ancient process still in use today.

In this section, we examine some of the physical properties of both primitive earthenware (ceramics fired at low temperatures) and modern ceramic materials (typically fired at high temperatures) and relate those properties to the chemical composition of these materials and to the chemical changes that occur when they are heated. Most ceramics behave as electrical insulators.

### Polymorphs of Silica

One of the most abundant families of minerals found in igneous rocks has the chemical composition $SiO_2$. The correct chemical name is silicon dioxide, but the more common name is *silica*. Silica is a covalent network solid in which each silicon atom is covalently bonded to four oxygen atoms, forming a tetrahedron with an oxygen atom at each corner and the silicon atom at the center (**Figure 18.39**). Each corner oxygen atom is covalently bonded to two silicon atoms, thereby linking the tetrahedra into an extended three-dimensional network. Because each oxygen atom is bonded to two silicon atoms, each silicon atom gets only half "ownership" of the four oxygen atoms to which it is bonded—hence the formula $SiO_2$.

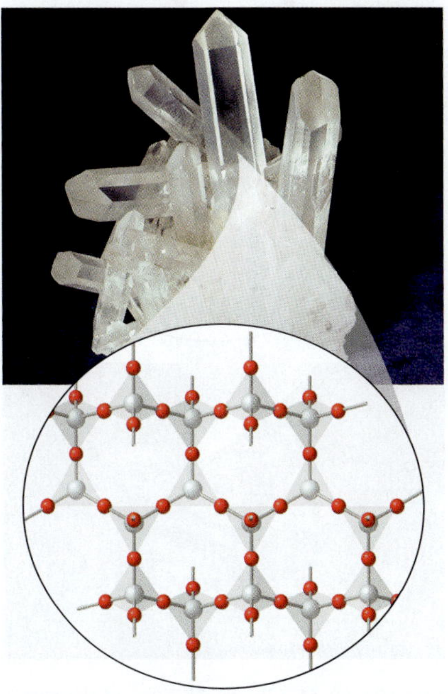

**FIGURE 18.39** Quartz crystals are hexagonal. Their crystal structure features hexagonal arrays of silicon–oxygen tetrahedra that form an extended three-dimensional network. (To make the structure easier to see, all atoms are drawn smaller than the volume they actually occupy.)

At least eight minerals have the empirical formula $SiO_2$. The members of a family of substances with the same empirical formula but different crystal

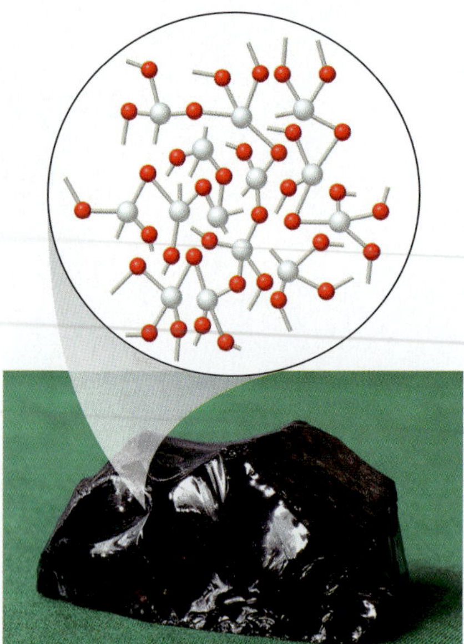

**FIGURE 18.40** Obsidian (volcanic glass) is an unusual form of silica in that it is not crystalline. Obsidian contains mostly amorphous silica with random arrangements of silicon and oxygen atoms.

structures and properties are called *polymorphs*. The most abundant silica polymorph is quartz, a type of $SiO_2$ that can form impressively large, nearly transparent crystals (Figure 18.39). Note how the hexagonal ordering of the $SiO_2$ tetrahedra translates into hexagonal crystals.

Most, but not all, silica polymorphs are crystalline. When lava containing molten $SiO_2$ flows from a volcano into a sea or lake, it cools so quickly that the Si and O atoms may not have enough time to form an ordered crystal lattice as the lava solidifies. The solid formed in this way is an *amorphous* (disordered, noncrystalline) polymorph of silica known as either volcanic glass or obsidian (**Figure 18.40**). *Glass* is a term scientists and engineers use to describe any solid that has either no crystalline structure or only very tiny crystals surrounded by disordered arrays of atoms. This definition also applies to laboratory glassware and the drinking glasses we use at home.

## Ionic Silicates

In addition to covalent silica, igneous rocks also contain ionic minerals made of silicon and oxygen. These minerals have some of the tetrahedral crystal structure of silica, but not all the oxygen corner atoms are bonded to two Si atoms. Instead, some of the O atoms have an extra electron. The result is a *silicate* anion. One of the most common ionic silicates is chrysotile, a type of asbestos that consists of sheets of linked silicon–oxygen tetrahedra that form hexagonal clusters of six tetrahedra each (**Figure 18.41**). Each tetrahedron has three O atoms that it shares with other tetrahedra and one O atom—the one with the extra electron—that it does not share. Thus, the basic tetrahedral unit consists of one Si atom and $[3(\frac{1}{2}) + 1] = 2.5$ oxygen atoms as well as a negative charge. This gives the sheet the empirical formula $SiO_{2.5}^-$. We generally use whole-number subscripts in chemical formulas. Here, multiplying $SiO_{2.5}^-$ by 2 gives us the empirical formula $Si_2O_5^{2-}$. The subscript *n* in the formula in Figure 18.41 indicates that there are many empirical formula units in a single crystal.

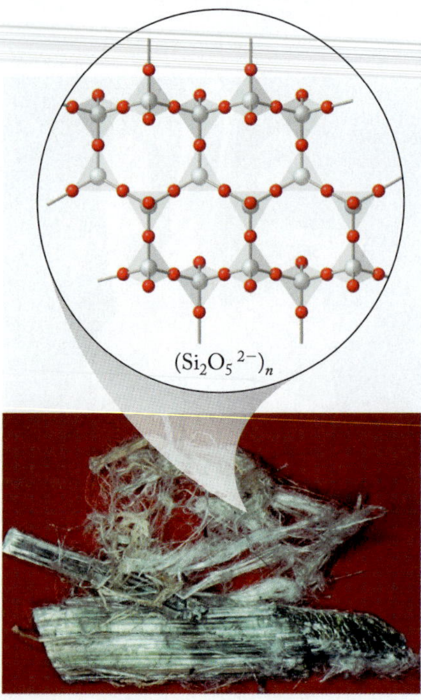


**FIGURE 18.41** Chrysotile, one of the two principal forms of asbestos formerly used in building construction as thermal insulation, is an ionic silicate compound. The ease with which thin fibers of chrysotile can flake off is related to its layered crystal lattice and the relatively weak intermolecular interactions between layers. The O atoms bonded to only one Si atom have an extra electron and a negative charge.

Silicate minerals are neutral materials, so they must contain cations to balance the negative charges on the $Si_2O_5^{2-}$ layers. When this cation is $Al^{3+}$, the minerals are called *aluminosilicates*. One of the most common aluminosilicates is the clay mineral kaolinite (**Figure 18.42**). At least a little kaolinite is found in practically every soil, but rich deposits of nearly pure, brilliantly white kaolinite are found in highly weathered soils. For centuries, these deposits have been mined for the clay used to make fine china and white porcelain. Today, the greatest demand for kaolinite is in the production of the glossy white paper used in magazines and books (including this one).

The metal ions found in most igneous rocks—$Na^+$, $K^+$, $Ca^{2+}$, $Mg^{2+}$, and $Fe^{3+}$—are largely absent in kaolinite. Their absence indicates that kaolinite deposits form under acidic weathering conditions, during which $H^+$ ions displace other cations from ion-exchange sites. For example, $-O^-Na^+$ sites exchange $H^+$ for $Na^+$, leaving behind $-OH$ groups such as those shown in Figure 18.42.

The strong ionic interactions and hydrogen bonds in kaolinite make it hard to separate its layers, and water molecules cannot penetrate between them. For thousands of years, this property has made kaolinite pots handy vessels for carrying water. For the same reason, kaolinite does not expand when water is added, nor does it shrink as much as most other clays when dehydrated at high temperatures, which makes it a desirable starting material for ceramics. Finally, moist kaolinite is *plastic*, meaning that it can be molded into a shape, and it keeps that shape during heating and cooling.

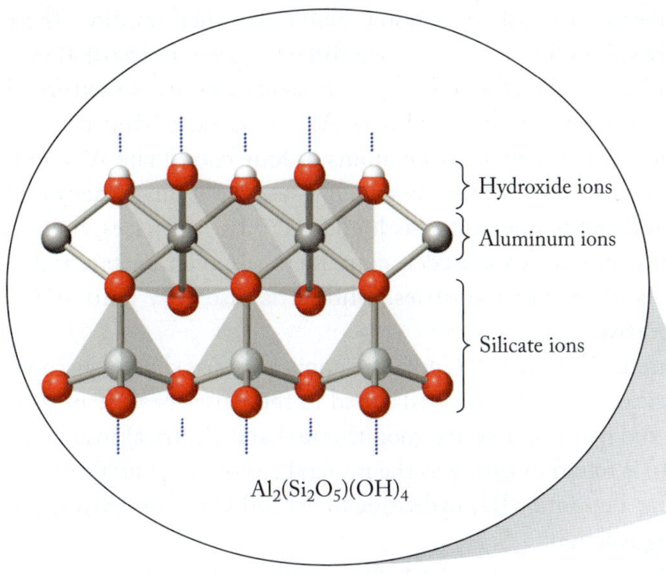

**FIGURE 18.42** An edge-on view of the structure of kaolinite shows a top layer of $OH^-$ ions bonded to a middle layer of $Al^{3+}$ ions followed by a bottom layer of silicate ions ($Si_2O_5^{2-}$). The structure is repeated in subsequent layers, and the empirical formula of this crystal lattice is $Al_2(Si_2O_5)(OH)_4$. This enormous kaolinite mine in Bulgaria is one of the sources of kaolinite.

CONCEPT **TEST**

Magnesium ion, $Mg^{2+}$, can substitute for $Al^{3+}$ in kaolinite to form $Mg_xAl_y(Si_2O_5)(OH)_4$. What are the values of $x$ and $y$ if $x = y$?

**CONNECTION** Ion-exchange reactions were discussed in Section 11.2.

## From Clay to Ceramic

Creating ceramic objects from kaolinite and other clays takes several steps. First, moist clay is formed into pots, bricks, and other objects on a potter's wheel or in molds or presses. Drying at just above 100°C removes much of the water that made the clay plastic. Further heating to about 450°C removes water that was adsorbed onto the surfaces of the clay particles or between the layers of nonkaolinite clays. Above 450°C, kaolinite begins to decompose, releasing molecules of $H_2O$ as some of the aluminum ions that were bonded to octahedra of six oxide ions in kaolinite (Figure 18.42) end up bonded to tetrahedra of only four oxide ions in a substance called metakaolin ($Al_2Si_2O_7$), which has an amorphous structure. The overall decomposition reaction, which is complete at temperatures near 900°C, is described by the equation

$$Al_2Si_2O_5(OH)_4(s) \xrightarrow{450^\circ C-900^\circ C} Al_2Si_2O_7(s) + 2\ H_2O(g)$$

The next compositional change occurs just below 1000°C when $Al_2Si_2O_7$ decomposes into a mixture of yet another disordered aluminosilicate, $Al_4Si_3O_{12}$, and $SiO_2(s)$:

$$2\ Al_2Si_2O_7(s) \xrightarrow{\sim 950^\circ C} Al_4Si_3O_{12}(s) + SiO_2(s)$$

At even higher temperatures, $Al_4Si_3O_{12}$ continues forming mixtures of compounds with increasing aluminum content and $SiO_2$:

$$Al_4Si_3O_{12}(s) \rightarrow Al_4Si_2O_{10}(s) + SiO_2(s)$$

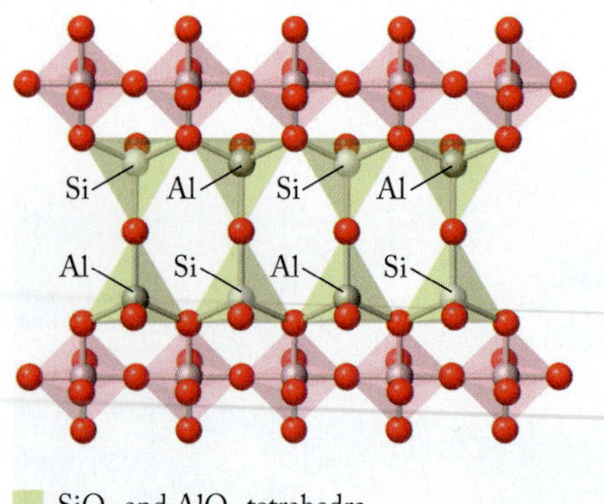


**FIGURE 18.43** One of many possible crystal structures of a family of ceramic materials called mullite.

$Al_4Si_2O_{10}$ is a member of a family of ceramic materials called mullite. These materials are actually solid solutions containing different proportions of aluminum and silicon oxides. At the atomic level, their structures are a mixture of layers that contain octahedra of six-coordinate $Al^{3+}$ ions each bonded to six oxide ions, or tetrahedra with either silicon atoms or four-coordinate $Al^{3+}$ ions at their centers (**Figure 18.43**). Mullite is widely used in the manufacture of products that must tolerate temperatures as high as 1700°C: furnaces, boilers, ladles, and kilns. These products are used as containers of molten metals and in the glass, chemical, and cement industries. Mullite is also very hard and is widely used as an abrasive.

Other ceramics are used in high-temperature applications as diverse as cookware and fireplace bricks. Ceramics are well suited to these uses because of their high melting points and because they are good thermal and electrical insulators. Pure alumina ($Al_2O_3$) is found in nature as the mineral corundum, but it can also be prepared by heating aluminum(III) hydroxide to ~1200°C to form $Al_2O_3$ and water in the following reaction:

$$2\ Al(OH)_3(s) \rightarrow Al_2O_3(s) + 3\ H_2O(g)$$

The crystal structure of alumina consists of a hexagonal closest-packed arrangement of oxide ions ($O^{2-}$) with $Al^{3+}$ in octahedral holes. To account for the observed stoichiometry in $Al_2O_3$, two-thirds of the octahedral holes are occupied. Alumina has myriad uses, including in orthodontia, where colorless alumina brackets offer a more attractive option to traditional metal braces. Avoiding chromium- and nickel-containing dental alloys is becoming increasingly important as some patients develop allergic reactions to these metals. Two familiar gemstones, ruby and topaz, are made of alumina in which some $Al^{3+}$ ions are replaced by $Cr^{3+}$ and $Fe^{3+}$ ions, respectively. In sapphires, trace quantities of $Fe^{2+}$ and $Ti^{4+}$ ions substitute for pairs of $Al^{3+}$ ions in the alumina.

## 18.9 X-ray Diffraction: How We Know Crystal Structures

X-ray Diffraction

Unit cell dimensions are determined using **X-ray diffraction (XRD)**. X-rays are well suited to the task of crystal structure determination because the wavelengths of X-rays ($10^2$ pm to $10^3$ pm) are similar to the distances between the centers of neighboring atoms and ions in crystals. To see how XRD works, we will consider the atoms in a crystalline metal, but keep in mind that our description of XRD also applies to the particles in any crystalline solid.

Suppose that a beam of monochromatic (same wavelength) X-rays is directed at the surface of a single metallic crystal (**Figure 18.44a**). Some of the X-rays will collide with the atoms in the crystal's surface layer and be scattered away from the atoms by these collisions, while other X-rays will pass through the surface layer but then collide with and be scattered by atoms in the second layer (**Figure 18.44b**). Now suppose that a detector is mounted in such a way that it can detect those X-rays that have an angle of scatter equal to the beam's angle of incidence, $\theta$. These X-rays will have undergone a total change in direction, called their *angle of diffraction*, that is the sum of their angles of incidence and scatter, or $2\theta$.

**X-ray diffraction (XRD)** a technique for determining the arrangement of atoms or ions in a crystal by analyzing the pattern that results when X-rays are scattered after bombarding the crystal.

Such X-rays can be detected only if they and many others like them undergo *constructive interference*. This term applies to electromagnetic waves of the same wavelength that are in phase with each other; that is, the crests of the waves are

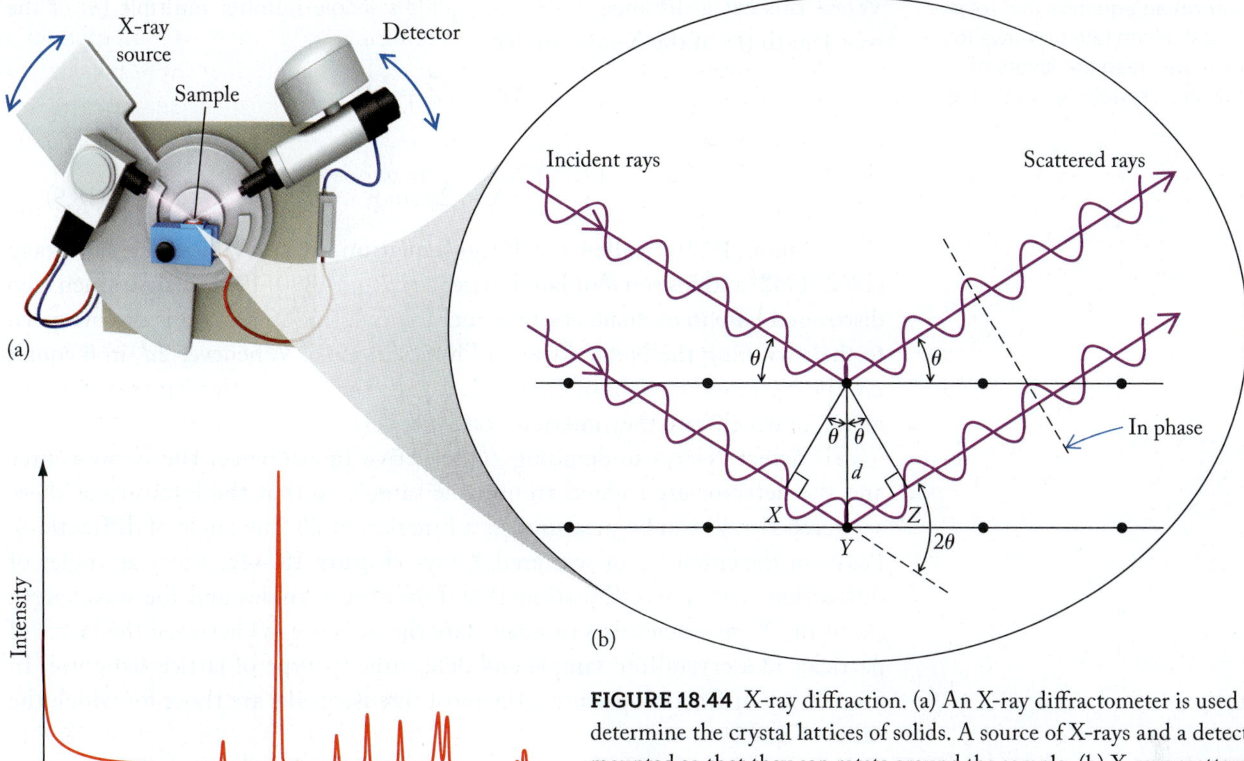


**FIGURE 18.44** X-ray diffraction. (a) An X-ray diffractometer is used to determine the crystal lattices of solids. A source of X-rays and a detector are mounted so that they can rotate around the sample. (b) X-rays scattered from the surface layer of atoms and from the next layer interfere constructively when they are in phase, as they are at the distance, $d$, shown here. The angle between the incident and scattered X-rays is 2θ. (c) Moving the source and detector around the sample produces a scan such as this one for quartz. The peaks at different values of 2θ represent constructive interference among X-rays scattered from different layers of atoms in the solid.

exactly aligned, as are the troughs (bottoms) of the waves. Scattered X-rays that are out of phase with each other undergo *destructive interference* and are not detected. The result is a pattern of white spots where constructive interference occurred (**Figure 18.45**).

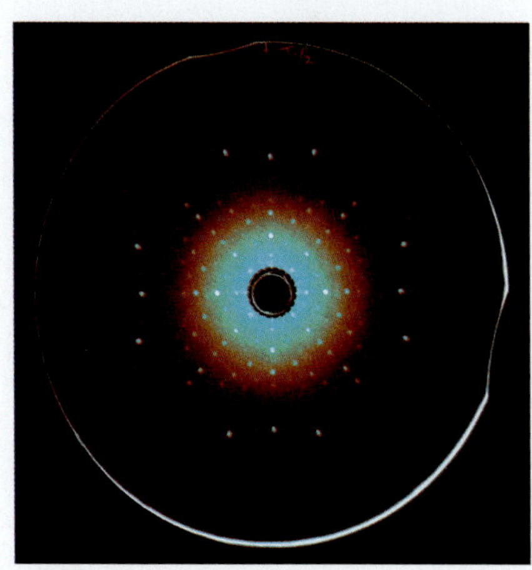

**FIGURE 18.45** The constructive and destructive interference of diffracted X-rays results in a distinctive pattern of spots that can be translated into a three-dimensional structure of the compound. This image was produced by X-ray diffraction of a crystal of an enzyme, bovine superoxide dismutase.

To be in phase with an X-ray scattering off a surface atom, an X-ray scattering off a second-layer atom must travel an extra distance that is some whole-number multiple of the wavelength of the X-rays. This extra distance is the sum of the lengths of line segments $\overline{XY}$ and $\overline{YZ}$ in Figure 18.44(b). The two right triangles incorporating these line segments share a hypotenuse, $d$. Geometry tells us that the angles opposite $\overline{XY}$ and $\overline{YZ}$ are both equal to θ. According to trigonometry, the ratio of either $\overline{XY}$ or $\overline{YZ}$ to $d$ is

$$\frac{\overline{XY}}{d} = \sin\theta \qquad \frac{\overline{YZ}}{d} = \sin\theta$$

which means

$$d\sin\theta = \overline{XY} \qquad d\sin\theta = \overline{YZ} \tag{18.8}$$

We are interested in $\overline{XY} + \overline{YZ}$, the extra distance traveled by the second ray. We therefore add Equations 18.8 to get

$$\overline{XY} + \overline{YZ} = d\sin\theta + d\sin\theta = 2d\sin\theta$$

**Bragg equation** an equation that relates the angle of diffraction ($2\theta$) of X-rays to the spacing ($d$) between the layers of ions or atoms in a crystal: $n\lambda = 2d \sin \theta$.

When this extra distance $\overline{XY} + \overline{YZ}$ equals a whole-number multiple ($n$) of the wavelength ($\lambda$) of the X-rays, we have

$$\overline{XY} + \overline{YZ} = n\lambda$$

or

$$n\lambda = 2d \sin \theta \qquad (18.9)$$

Equation 18.9 is called the **Bragg equation**, after William Henry Bragg (1862–1942) and his son William Lawrence Bragg (1890–1971), Englishmen who discovered how to evaluate crystal structures by using X-rays. This discovery led to their winning the Nobel Prize in Physics in 1915. Whenever $2d \sin \theta$ equals $n\lambda$, the crests and troughs of the two X-rays are in phase as they emerge from the metal, in which case they interfere constructively.

To detect X-rays undergoing constructive interference, the X-ray source and the detector are rotated around the sample so that the intensity of these scattered X-rays can be measured as a function of $2\theta$ (the angle of diffraction). Peaks in the intensity of scattered X-rays (**Figure 18.44c**) occur at angles of diffraction that satisfy Equation 18.9. From these angles and the wavelength ($\lambda$) of the X-rays, scientists can calculate the distance ($d$) between the layers of particles in a crystalline sample and determine its type of lattice structure. In most X-ray diffraction patterns, the most intense peaks are those for which the value of $n$ is 1.

### SAMPLE EXERCISE 18.7 Determining Interlayer Distances by X-ray Diffraction — LO6

An XRD analysis of a sample of copper has a major peak at $2\theta = 24.64°$ and much smaller peaks at $2\theta = 50.54°$ and $79.62°$. What distance $d$ between layers of Cu atoms produces this diffraction pattern if the wavelength of the X-rays is 154 pm?

**Collect and Organize** We are given a series of diffraction angles $2\theta$ and are asked to find the distance between layers of atoms in a sample of copper metal. Equation 18.9 enables us to calculate this distance when we know the angles of diffraction associated with constructive interference in a beam of reflected X-rays.

**Analyze** We need to rearrange Equation 18.9 to solve for $d$, the parameter we are after:

$$d = \frac{n\lambda}{2 \sin \theta}$$

We are given the value of $\lambda$ and can get values of $\theta$ from the given values of $2\theta$, but we need to determine the value(s) of $n$.

**Solve** The key to determining $n$ is to look for a pattern in the $\theta$ values. Notice that the higher values of $\theta$ are approximately two and three times the lowest value:

$$24.64°/2 = 12.32 \qquad 50.54°/2 = 25.27 \qquad 79.62°/2 = 39.81$$

$$\frac{25.27°}{12.32°} = 2.051 \qquad \frac{39.81°}{12.32°} = 3.231$$

This pattern and the intensities of the peaks suggest that the values of $n$ for this set of data are 1, 2, and 3, so let's use these combinations of $n$ and $\theta$ to see whether they all give the same value of $d$:

$$d = \frac{(1)(154 \text{ pm})}{2 \sin 12.32^\circ} = \frac{154 \text{ pm}}{(2)(0.2134)} = 361 \text{ pm}$$

$$d = \frac{(2)(154 \text{ pm})}{2 \sin 25.27^\circ} = \frac{308 \text{ pm}}{(2)(0.4269)} = 361 \text{ pm}$$

$$d = \frac{(3)(154 \text{ pm})}{2 \sin 39.81^\circ} = \frac{462 \text{ pm}}{(2)(0.6402)} = 361 \text{ pm}$$

We do indeed get the same value of $d$, so $d = 361$ pm is the distance between Cu atoms that produced the three peaks.

**Think About It** These consistent results mean that our assumption about the values of $n$ for the three values of $\theta$ was correct. Moreover, the value of $d$ is in the range of the edge lengths of unit cells we have seen in several Sample Exercises in this chapter and is therefore also reasonable.

**Practice Exercise** For crystalline CsCl, an X-ray diffraction analysis using X-rays of wavelength 142.4 pm has a prominent peak at $2\theta = 19.9^\circ$. What is the spacing between the ion layers that produced the peak?

XRD is a powerful tool for analyzing the atomic-level structure of a wide range of materials, including metals, minerals, polymers, plastics, ceramics, pharmaceuticals, and semiconductors. The technique is indispensable for scientific research and industrial production. The link between particulate structure and macroscopic properties is robust. Understanding the ordered arrangements of atoms and ions that characterize pure metals, alloys, semiconductors, ionic compounds, and ceramics enables us to design stronger, tougher, harder, and more heat-resistant solid materials.

## SAMPLE EXERCISE 18.8 Integrating Concepts: Imperfect Crystals and Film Photography

Most of the crystals we have discussed and illustrated in this chapter have been perfect, in the sense that all sites in the crystal structure that should be occupied by an atom or ion *are* occupied. However, many crystals we find in the real world are imperfect, and several types of defects have been recognized as important. Scientists Walter Schottky (1886–1976) and Yakov Frenkel (1894–1952) studied defects known as *point defects* because they occur at specific points (locations) within a crystal structure. A Schottky defect forms when positively and negatively charged ions both leave their normal sites in the structure, creating vacancies at those sites. A Frenkel defect forms when an atom or ion leaves its location in the structure and occupies a site not usually occupied by a particle. Crystals of silver bromide (AgBr), having the rock salt structure like NaCl, may display both kinds of defects.

Silver bromide is used in black-and-white film photography. When light falls on a grain of AgBr embedded in film, a bromide ion loses an electron, which combines with a silver ion in a Frenkel defect and produces a silver atom. Later the film is bathed in a developer that reduces more silver ions to intensify the image. After the image is developed, the film is *fixed*, which involves dissolving any remaining silver bromide in an aqueous solution (about 0.2 $M$) of sodium thiosulfate, $Na_2S_2O_3$.

a. If a Schottky defect occurs in a unit cell of silver bromide (**Figure 18.46a**), does the overall stoichiometry of the unit cell change? If the defect occurs consistently throughout the structure, does the density of the silver bromide change?
b. If a $Ag^+$ ion that normally occupies an octahedral hole in the cubic closest-packed array of $Br^-$ ions forms a Frenkel defect by moving to a tetrahedral hole, does the overall composition of the unit cell change? If the defect occurs consistently throughout a sample of AgBr, do its formula and its density change?

c. Why is AgBr, which is insoluble in water, soluble in a solution of sodium thiosulfate?

d. Silver ions are recovered from photographic processing solutions by passing them through columns containing steel wool (Fe). The reaction between silver(I) ions and Fe produces silver metal. Write a balanced net ionic equation describing this reaction and suggest why it is spontaneous.

**Collect and Organize** We are asked whether Schottky and Frenkel defects in AgBr cause changes in the stoichiometry of the unit cell and the density of the bulk material. We are also asked to explain the solubility of AgBr in a solution that contains thiosulfate ions and the chemistry behind the recovery of silver as the metal by reacting silver(I) compounds with iron metal.

**Analyze** To determine the impact of Schottky and Frenkel defects on AgBr, we start with a unit cell of AgBr (Figure 18.46a) and then incorporate the changes described. By determining the ratio of the ions in defective unit cells and the number of ions per unit cell, we can determine whether there will be changes in the overall formula and density of a sample. To explain the solubility of AgBr in thiosulfate solutions, we need to consider how $Ag^+$ and $Br^-$ ions interact with thiosulfate ($S_2O_3^{2-}$) ions. According to Table A5.5 in Appendix 5, $Ag^+$ ions combine with $S_2O_3^{2-}$ ions to produce a complex ion, $Ag(S_2O_3)_2^{3-}$, which has a formation constant of $4.7 \times 10^{13}$. The $K_{sp}$ of AgBr (Table A5.4) is $5.4 \times 10^{-13}$. Our solution to this problem will need to link these two equilibria. Finally, conversion of silver(I) to Ag metal with iron must be a redox process. Checking the standard reduction potentials in Table A6.1 in Appendix 6, we find the following relevant half-reactions:

$$Ag^+(aq) + e^- \rightarrow Ag(s) \qquad E^\circ = 0.7996\ V$$
$$Fe^{2+}(aq) + 2\ e^- \rightarrow Fe(s) \qquad E^\circ = -0.447\ V$$

The positive $E^\circ$ value of the silver half-reaction and the negative $E^\circ$ value of the iron half-reaction indicate that Fe may be a reducing agent strong enough to convert the silver(I) to silver metal.

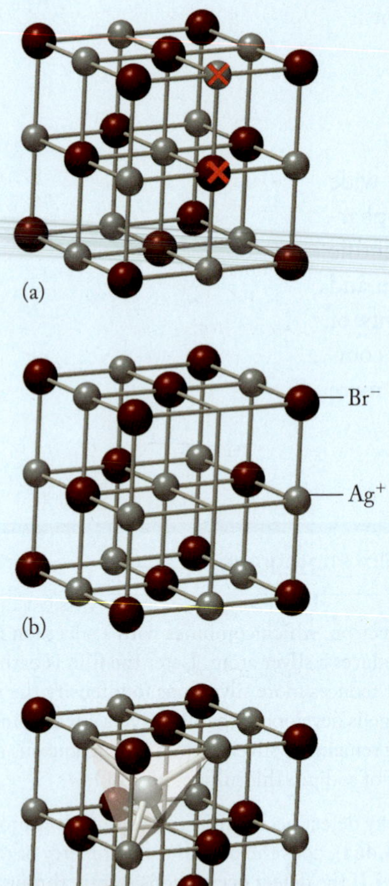


**FIGURE 18.46** (a) A perfect crystal structure of AgBr, without defects. (b) One $Ag^+$ ion and one $Br^-$ ion are missing from the front face, resulting in a Schottky defect. (c) One $Ag^+$ ion has moved from its normal location at the center to the highlighted tetrahedral hole, resulting in a Frenkel defect.

**Solve**

a. A Schottky defect in AgBr occurs when a pair of $Ag^+$ and $Br^-$ ions (marked with red *X*'s in Figure 18.46a) is lost, producing a unit cell like the one shown in **Figure 18.46(b)**. Though the composition of this particular unit cell has changed, because the missing ions occupied different positions that contributed differently to the total ion count in the cell, the overall ratio of silver ions to bromide ions in the solid should still be 1:1. However, if enough ion pairs are lost without changing the cell dimensions, the density of the solid will decrease.

b. If a silver ion moves from the center (octahedral) site to an unoccupied tetrahedral site (**Figure 18.46c**), the composition of the cell does not change, so neither the formula of the solid nor its density changes in the presence of Frenkel defects.

c. The $K_{sp}$ of AgBr can be used to calculate the concentration of $Ag^+$ ions in equilibrium with solid AgBr as a result of the reaction $AgBr(s) \rightleftharpoons Ag^+(aq) + Br^-(aq)$:

$$K_{sp} = 5.4 \times 10^{-13} = [Ag^+][Br^-] = (x)(x) = x^2$$
$$x = [Ag^+] = 7.3 \times 10^{-7}\ M$$

Let's use this concentration to determine whether adding 0.2 *M* thiosulfate ions will result in the formation of $Ag(S_2O_3)_2^{3-}$ complex ions. Inserting the known concentrations into the formation constant expression and solving for $[Ag(S_2O_3)_2^{3-}]$,

$$K_f = \frac{[Ag(S_2O_3)_2^{3-}]}{[Ag^+][S_2O_3^{2-}]^2} = \frac{x}{(7.3 \times 10^{-7})(0.2)^2} = 4.7 \times 10^{13}$$
$$x = [Ag(S_2O_3)_2^{3-}] = 1.4 \times 10^6\ M$$

This very large concentration, as compared to $[Ag^+]$, means that any $Ag^+$ ions produced when AgBr dissolves will form complex ions, effectively removing $Ag^+$ ions from solution. According to Le Châtelier's principle, removing one of the product ions shifts the $K_{sp}$ equilibrium described above toward making more dissolved ions, which keeps on happening until all the AgBr has dissolved (if there is an adequate supply of thiosulfate ions).

d. The two half-reactions in the Analyze section can be combined to write a net ionic equation describing the reaction between silver(I) and iron if we (1) reverse the iron half-reaction and (2) multiply the silver half-reaction by 2 to balance the loss and gain of electrons:

$$\begin{array}{lr} 2\,Ag^+(aq) + \cancel{2e^-} \rightarrow 2\,Ag(s) & E^\circ = 0.7996\text{ V} \\ Fe(s) \rightarrow Fe^{2+}(aq) + \cancel{2e^-} & E^\circ = 0.447\text{ V} \\ \hline 2\,Ag^+(aq) + Fe(s) \rightarrow 2\,Ag(s) + Fe^{2+}(aq) & E^\circ_{rxn} = 1.247\text{ V} \end{array}$$

The positive reaction potential indicates that the reaction is spontaneous *under standard conditions*—that is, when $[Ag^+] = [Fe^{2+}] = 1\ M$. As noted above, the concentration of $Ag^+$ ions is much less than 1 *M*. However, we learned in Chapter 17 that the potentials of half-reactions change relatively little even as the concentrations of reactants and products change a lot. Therefore, we can confidently predict that the reaction will have a positive reaction potential and be spontaneous.

**Think About It** Physical and chemical properties of crystalline materials are closely related to their structure, and structural defects contribute to these properties as well. Other ionic substances with large size differences between the anion and the cation may also demonstrate Frenkel defects. AgCl, with a smaller anion (181 pm for $Cl^-$ vs. 195 pm for $Br^-$), is also photoreactive, but not to the same extent as AgBr because the $Ag^+$ ion (126 pm) is too big to fit into the tetrahedral holes in an array of $Cl^-$ ions.

## SUMMARY

**LO1** Many metallic crystals are based on crystal lattices of the **cubic closest-packed (ccp)** and **hexagonal closest-packed (hcp)** types, which are the two most efficient ways of packing atoms in a solid. **Crystalline solids** contain repeating **unit cells**, which include **simple cubic (sc)**, **body-centered cubic (bcc)**, or **face-centered cubic (fcc)**. The dimensions of the unit cell in a crystalline solid can be used to determine the radius of the atoms or ions. (Sections 18.2 and 18.6)

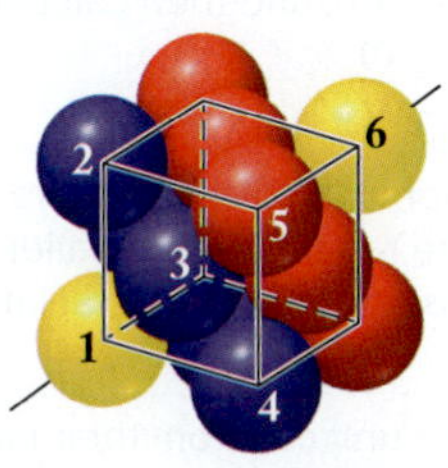


**LO2** The dimensions of a crystalline solid's unit cell can be used to calculate the density of the solid. (Sections 18.2 and 18.6)

**LO3** **Alloys** are blends of a host metal and one or more other elements (which may or may not be metals) added to enhance the properties of the host, including strength, hardness, and corrosion resistance. In **substitutional alloys**, atoms of the added elements replace atoms of the host metal in its crystal lattice. In **interstitial alloys**, atoms of added elements are located in the tetrahedral and/or octahedral holes between atoms of the host metal. Aluminum and aluminum alloys are highly desirable for applications requiring corrosion resistance and low mass. (Section 18.3)

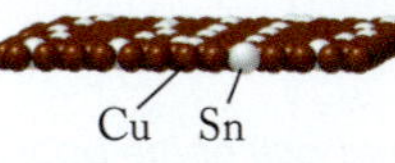


**LO4** Most metals are malleable and ductile. The electrical conductivity of metals can be explained by **band theory** as the ease with which valence electrons can gain mobility by moving into empty energy levels in multiatom structures. Metalloids are **semiconductors**, intermediate in electrical conducting ability between **conductors** and **insulators**. In semiconductors, the filled valence band and empty conduction band are separated by a **band gap ($E_g$)**. Substituting electron-rich atoms into a semiconductor results in **n-type semiconductors**. Substituting electron-poor atoms results in **p-type semiconductors**. Both types of substitution increase the conductivity of the semiconductor by decreasing its band gap. (Sections 18.4 and 18.5)

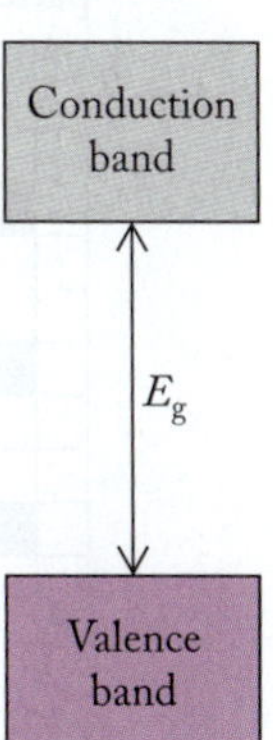


**LO5** Many **ionic solids** consist of crystals in which some number of cations or anions form the unit cell and the opposite ion occupies octahedral or tetrahedral holes. Two allotropes of carbon are the **covalent network solids** graphite and diamond. Many nonmetals form **molecular solids**, including sulfur, which forms puckered rings of eight sulfur atoms, and phosphorus, which forms $P_4$ tetrahedra. Heating selected solid inorganic compounds (such as clays) to high temperature creates **ceramics**, heat- and chemical-resistant materials that are electrical insulators due to the large energy gap between their filled valence and empty conduction bands. The polymorphs of silica consist of tetrahedra made up of four O atoms at the corners surrounding a central Si atom. (Sections 18.6–18.8)

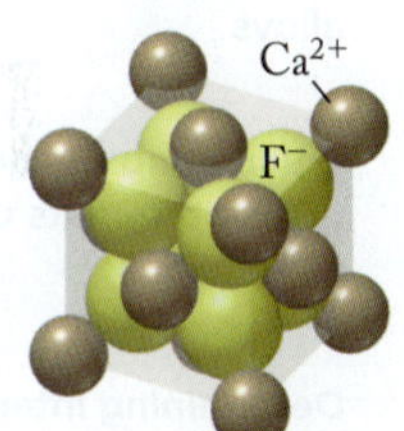


**LO6** **X-ray diffraction (XRD)** is an analytical method that records constructive interference of X-rays reflecting off different layers of atoms or ions in a crystalline solid. The distances between layers are calculated using the **Bragg equation**. X-ray diffraction makes it possible to determine the crystal structure of crystalline solids. (Section 18.9)

## PARTICULATE PREVIEW WRAP-UP

Allotrope B is planar, as each carbon atom is covalently bonded to three other carbon atoms in a trigonal planar geometry and is attracted to the planar layers above and below through London dispersion forces.

Allotrope A is nonplanar, as each carbon atom is covalently bonded to four other carbon atoms in a tetrahedral geometry. Allotrope A is hard and abrasive because it has a rigid three-dimensional atomic structure. Allotrope B is soft and slippery because the layers of atoms are not covalently bonded and can slip past each other.

## PROBLEM-SOLVING SUMMARY

| Type of Problem | Concepts and Equations | Sample Exercises |
|---|---|---|
| **Calculating atomic or ionic radii from unit cell dimensions** | Determine the length, $\ell$, of an edge, face diagonal, or body diagonal along which adjacent atoms touch in a unit cell. Use the relationship between $\ell$ and the atomic/ionic radius, $r$, to calculate the value of $r$: $r = \ell/2$ for sc unit cells, $r = 0.3536\ell$ for fcc unit cells, and $r = 0.4330\ell$ for bcc unit cells. | **18.1, 18.5** |
| **Using unit cells to calculate the density of a solid** | Determine the mass of the atoms in the unit cell from their molar mass and the volume of the unit cell from the edge length, and then calculate the density: $$d = \frac{m}{V}$$ | **18.2, 18.3** |
| **Predicting the crystal structure of two-element alloys** | Compare the radii of the alloying elements. Similarly sized radii (within 15%) suggest a substitutional alloy. When the radius of the smaller atom is <73% of the radius of the larger atom in any closest-packed crystal lattice, an interstitial alloy may form. | **18.4** |
| **Calculating the density of a salt from its unit cell dimensions** | Determine the masses of the ions in the unit cell from their molar masses and the volume of the unit cell from the edge length. Then apply $$d = \frac{m}{V}$$ | **18.6** |
| **Determining interlayer distances by X-ray diffraction** | Apply the Bragg equation, $$n\lambda = 2d \sin \theta \quad (18.9)$$ | **18.7** |

## VISUAL PROBLEMS

*(Answers to boldface end-of-chapter questions and problems are in the back of the book.)*

**18.1.** In Figure P18.1, which drawings are analogous to a crystalline solid, and which are analogous to an amorphous solid?

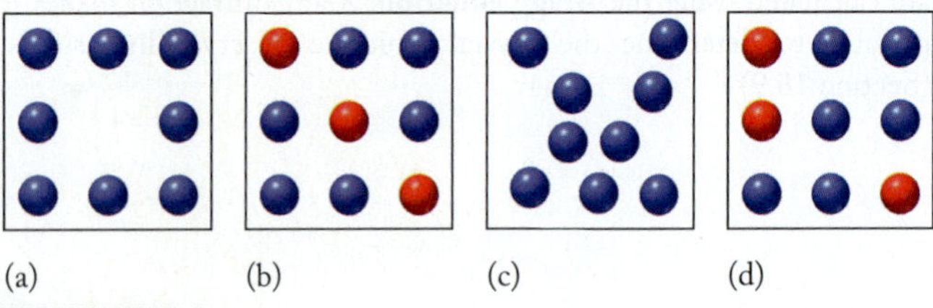

(a) (b) (c) (d)

**FIGURE P18.1**

18.2. The unit cells in Figure P18.2 continue infinitely in two dimensions. Draw a square around a unit cell in each pattern. How many light squares and how many dark squares are in each unit cell?

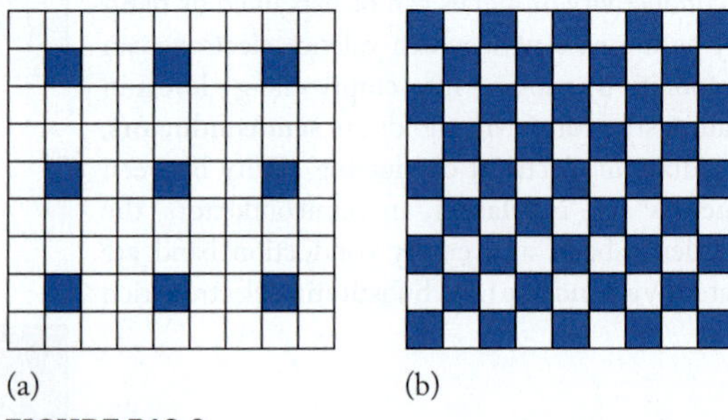

(a) (b)

**FIGURE P18.2**

**18.3.** The unit cells in Figure P18.3 continue infinitely in two dimensions. Outline a unit cell in each pattern. How many white hexagons and how many blue triangles are in each unit cell?

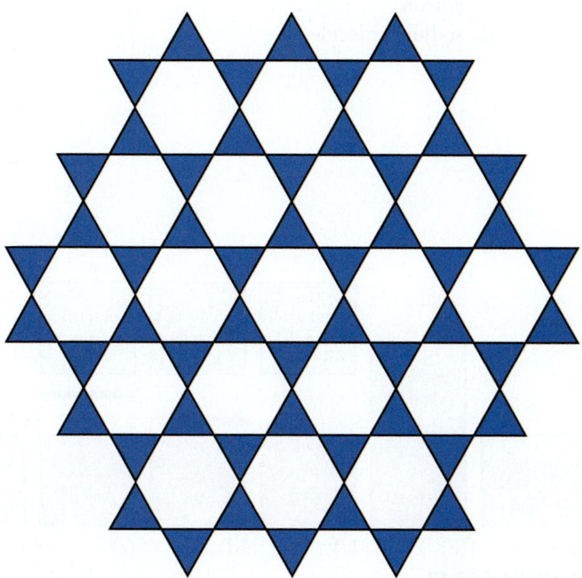

**FIGURE P18.3**

18.4. The pattern in Figure P18.4 continues indefinitely in three dimensions. Draw a box around the unit cell. If the red circles represent element A and the blue circles represent element B, what is the chemical formula of the compound?

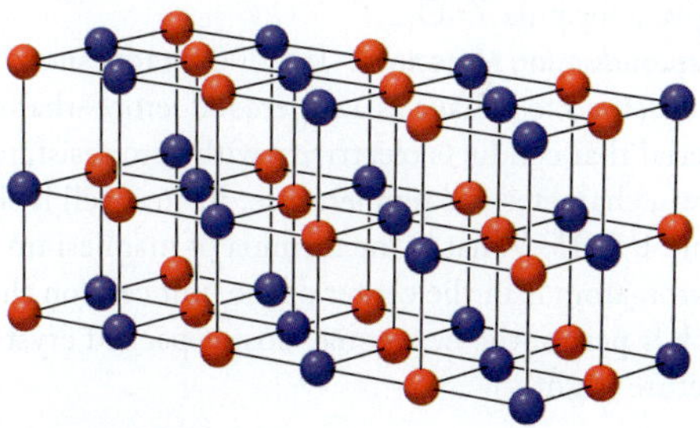

**FIGURE P18.4**

**18.5.** How many cations (A) and anions (B) are in the unit cell in Figure P18.5?

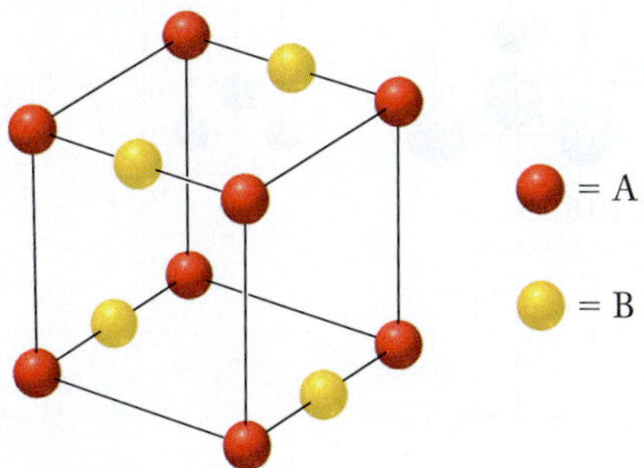


**FIGURE P18.5**

18.6. How many atoms of elements A and B are in the unit cell in Figure P18.6? The blue spheres are in cell faces.

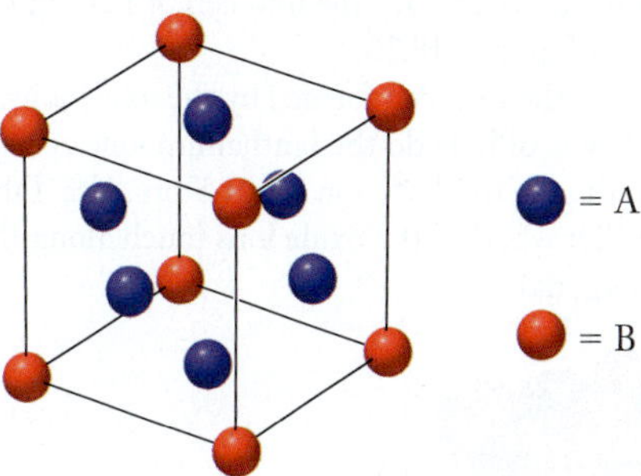


**FIGURE P18.6**

**18.7.** Calculate the empirical formula and density of nitinol, the shape memory alloy composed of nickel and titanium having the unit cell shown in Figure P18.7. The atomic radii of Ni and Ti are 124 pm and 147 pm, respectively.

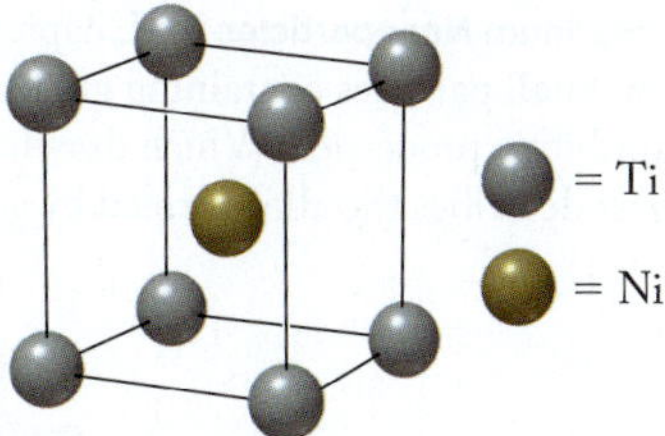


**FIGURE P18.7**

18.8. Calculate the empirical formula and density of a rose gold alloy composed of 25% Cu and 75% Au having the unit cell in Figure P18.8. The atomic radii of Cu and Au are 128 pm and 144 pm, respectively.

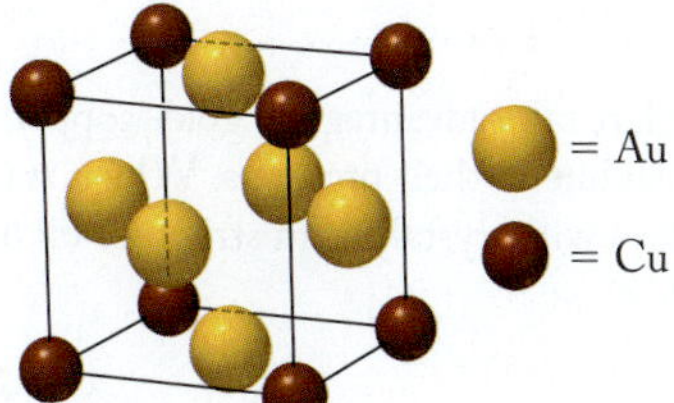


**FIGURE P18.8**

**18.9. Cheaper Thermal Conductors** Boron arsenide (Figure P18.9) has 65% of the thermal conductivity of diamond and 3 times that of copper metal. It is also much cheaper than diamond, making it attractive for use in similar applications.

a. Describe the unit cell formed by the boron atoms.

b. What type of hole do the arsenic atoms occupy?

*c. Use Table 18.4 to predict whether the boron atoms touch along the unit cell diagonal.

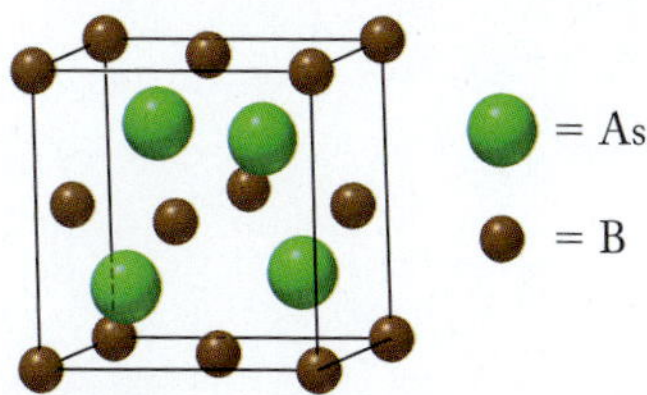


**FIGURE P18.9**

*18.10. **New Superconductors** In 2018, researchers produced thin films of lanthanum oxide, LaO, which showed superconductivity at 5 K. The unit cell of LaO in the films is shown in Figure P18.10.

a. Describe the unit cell formed by the oxide ions.
b. What type of hole do the lanthanum ions occupy?
*c. The radius of the $La^{2+}$ ion is 121.5 pm. Use Table 18.4 to predict whether the oxide ions touch along the unit cell diagonal.

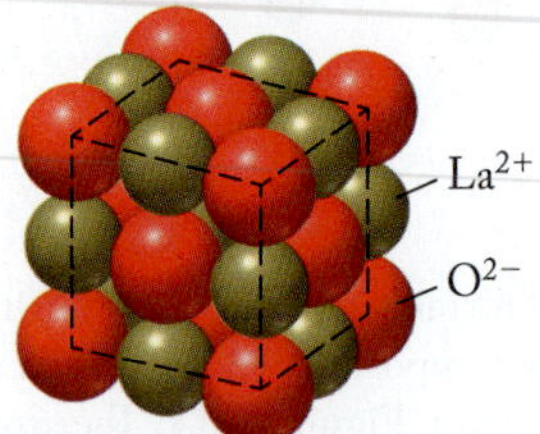


**FIGURE P18.10**

**18.11. Antibiotic Gold–Platinum Nanoparticles** In Chapter 2, we described how small particles containing gold and platinum have antibiotic properties. Which drawing in Figure P18.11 best describes the alloy formed by gold and platinum?

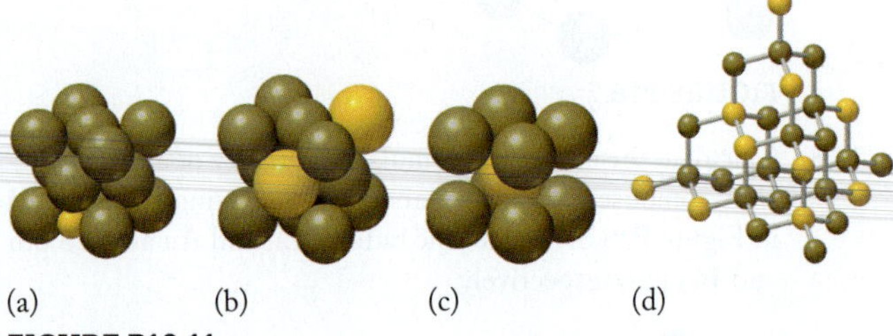


**FIGURE P18.11**

18.12. **Rose Gold** Jewelers take advantage of gold–copper alloys to bring a reddish hue to their products. Why is it unlikely that Au–Cu alloys will crystallize in structure (a) or (d) in Figure P18.12?

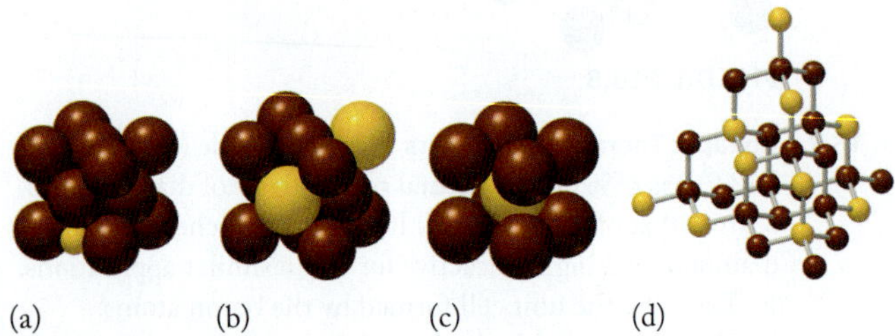


**FIGURE P18.12**

**18.13.** Which drawing in Figure P18.13 best explains the electrical conductivity of the following substances?

a. copper metal
b. magnesium metal
c. silicon
d. sodium chloride

Conduction band
Conduction band
Conduction band
Conduction band
Conduction band
Valence band
Valence band
Valence band
Valence band
Valence band
Valence band
(a) (b) (c) (d) (e) (f)

**FIGURE P18.13**

18.14. Which drawing in Figure P18.13 best explains the electrical conductivity of the following substances?

a. GaAs
b. Silicon doped with phosphorus
c. Silicon doped with gallium
*d. Zinc oxide, where some of the oxide ions are missing to give a formula $ZnO_{1-x}$

***18.15. Superconducting Materials I** In 2000, magnesium boride was observed to behave as a superconductor—that is, a material that conducts electricity with zero resistance below a characteristic temperature. Its unit cell is shown in Figure P18.15. What is the formula of magnesium boride? A boron atom is in the center of the unit cell (on the left), which is part of the hexagonal closest-packed crystal structure (right).

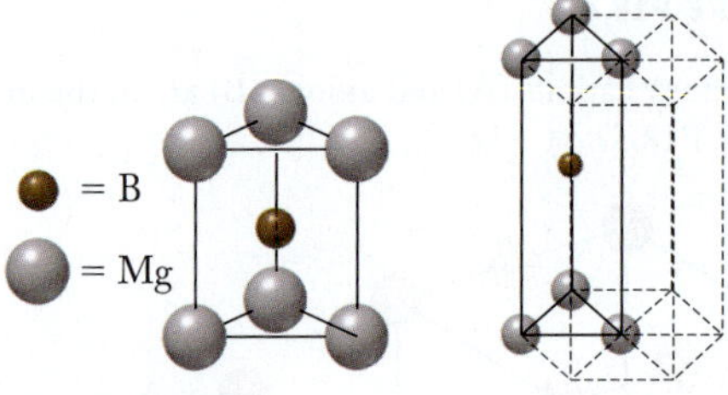


**FIGURE P18.15**

*18.16. **Superconducting Materials II** The 1987 Nobel Prize in Physics was awarded to J. G. Bednorz and K. A. Müller for their discovery of superconducting ceramic materials such as $YBa_2Cu_3O_7$. Figure P18.16 shows the unit cell of another yttrium–barium–copper oxide.

a. What is the chemical formula of this compound?

b. Removing four oxygen atoms from the unit cell in the figure produces the unit cell of $YBa_2Cu_3O_7$. Does it make a difference which oxygen atoms are removed?

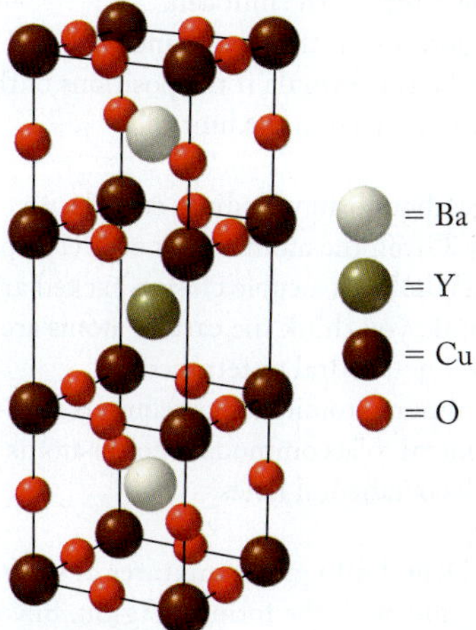


**FIGURE P18.16**

**18.17.** In the unit cell in Figure P18.17, the small spheres are in tetrahedral holes formed by the large spheres.

a. How many of the large spheres and how many of the small ones are assignable to the unit cell?

b. If the radius of the large spheres is 140 pm, what (theoretically) is the maximum size of the small spheres?

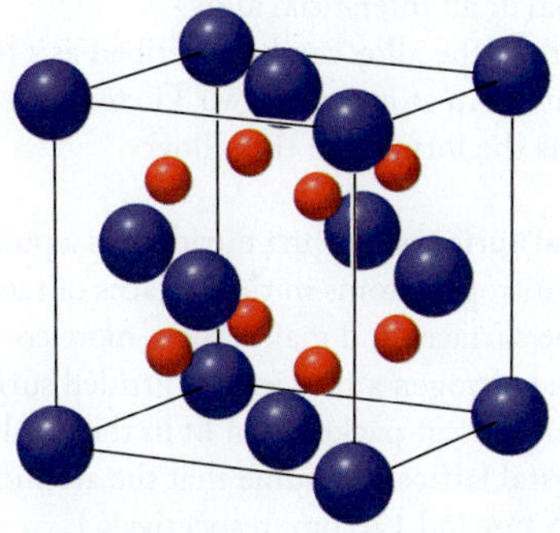

**FIGURE P18.17**

18.18. Use representations [A] through [I] in Figure P18.18 to answer questions a–f.

a. Two forms of iron are ferrite and austenite. One has a body-centered cubic structure, and the other takes a face-centered cubic structure. Which representations depict these forms of iron?

b. White phosphorus is a molecular solid, whereas red phosphorus is a covalent network solid. Which representations depict white and red phosphorus?

c. Which representations depict allotropes of carbon?

d. Which representation depicts the unit cell of LiCl? Of $CaF_2$?

e. Which representation depicts the rock salt structure? The fluorite structure?

f. Which process is depicted in representation [F]?

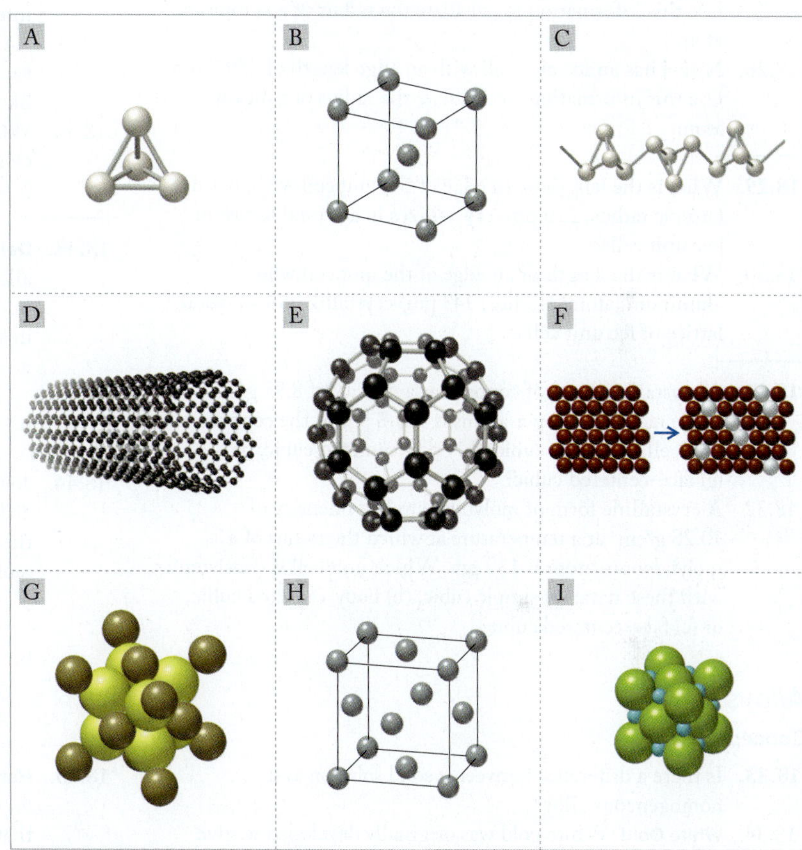


**FIGURE P18.18**

## QUESTIONS AND PROBLEMS

### *Structures of Metals*

#### Concept Review

**18.19.** Explain the difference between cubic closest-packed and hexagonal closest-packed arrangements of identical spheres.

*18.20. Is it possible to have a closest-packed crystal lattice with four different repeating layers, *abcdabcd* ...?

**18.21.** Which unit cell has the greater packing efficiency, simple cubic or body-centered cubic?

18.22. Consult Figure 18.11 to predict which unit cell has the greater packing efficiency, body-centered cubic or face-centered cubic.

**18.23.** The unit cell in iron metal is either fcc or bcc, depending on temperature (see Sample Exercise 18.1). Are the fcc and the bcc forms of iron allotropes? Explain your answer.

*18.24. At low temperatures, the unit cell of calcium metal is found to be fcc. At higher temperatures, the unit cell of calcium metal is bcc. What might be a reason for this temperature dependence?

**Problems**

**18.25.** Derive the edge length in bcc and fcc unit cells in terms of the radius ($r$) of the atoms in the cells in Figure 18.11.

*18.26. Derive the length of the body diagonal in simple cubic and fcc unit cells in terms of the radius ($r$) of the atoms in the unit cells in Figure 18.11.

---

**18.27. Fluorescent Lighting** Europium is a lanthanide element that is used in the phosphors in fluorescent lamps. Metallic Eu forms bcc unit cells with an edge length of 240.6 pm. Use this information to calculate the radius of a europium atom.

18.28. Nickel has an fcc unit cell with an edge length of 350.7 pm. Use this information to calculate the radius of a nickel atom.

---

**18.29.** What is the length of an edge of the unit cell when barium (atomic radius, 222 pm) crystallizes in a crystal lattice of bcc unit cells?

18.30. What is the length of an edge of the unit cell when aluminum (atomic radius, 143 pm) crystallizes in a crystal lattice of fcc unit cells?

---

**18.31.** A crystalline form of copper has a density of 8.95 g/cm$^3$. If the radius of copper atoms is 127.8 pm, is the copper unit cell (a) simple cubic, (b) body-centered cubic, or (c) face-centered cubic?

18.32. A crystalline form of molybdenum has a density of 10.28 g/cm$^3$ at a temperature at which the radius of a molybdenum atom is 139 pm. Which unit cell is consistent with these data: (a) simple cubic, (b) body-centered cubic, or (c) face-centered cubic?

## Alloys

**Concept Review**

**18.33.** Is there a difference between a solid solution and a homogeneous alloy?

18.34. **White Gold** White gold was originally developed to give the appearance of platinum. One formulation of white gold contains 25% nickel and 75% gold. Which is more malleable, white gold or pure gold?

***18.35.** Explain why an alloy that is 28% Cu and 72% Ag melts at a lower temperature than the melting points of either Cu or Ag.

18.36. Can an alloy be both substitutional and interstitial?

**18.37.** A substitutional alloy is prepared using two metals: Mo and W, with the same unit cell (bcc) and atomic radius ($r$ = 139 pm). Will the density of the alloy be less than, greater than, or the same as the density of pure molybdenum?

18.38. The nibs on some fountain pens contained a material called Osmiroid, an alloy composed of osmium and iridium. Using only the periodic table as a guide, do you expect osmium and iridium to form a substitutional or an interstitial alloy?

**Problems**

**18.39.** The unit cell of an intermetallic compound consists of a face-centered cube that has an atom of element X at each corner and an atom of element Y at the center of each face.
a. What is the formula of the compound?
b. What would be the formula if the positions of the two elements were reversed in the unit cell?

18.40. The bcc unit cell of an intermetallic compound has atoms of element A at the corners of the unit cell and an atom of element B at the center of the unit cell.
a. What is the formula of the compound?
b. What would be the formula if the positions of the two elements were reversed in the unit cell?

---

**18.41.** Vanadium and carbon form vanadium carbide, an interstitial alloy. Given the atomic radii of V (135 pm) and C (77 pm), which holes in a cubic closest-packed array of vanadium atoms do you think the carbon atoms are more likely to occupy—octahedral or tetrahedral?

18.42. What is the minimum atomic radius required for a cubic closest-packed metal to accommodate boron atoms ($r$ = 88 pm) in its octahedral holes?

---

**18.43. Dental Fillings** Dental fillings are mixtures of several alloys, including one with the formula $Ag_3Sn$. Silver (radius, 144 pm) and tin (radius, 140 pm) both crystallize in an fcc unit cell.
a. Is this alloy likely to be a substitutional alloy or an interstitial alloy?
*b. Draw an fcc unit cell for $Ag_3Sn$ that corresponds to an *intermetallic* alloy of silver and tin.

18.44. **Joint Replacement** A titanium–gold alloy discovered in 2016 is four times harder than pure titanium, making the new material attractive for use in knee and hip replacements.
a. Is this titanium–gold alloy more likely to be a substitutional or an interstitial alloy?
b. The structure of the alloy can be described as a bcc arrangement of Au atoms with two Ti atoms on each face. What is the formula of this alloy?

---

***18.45. Hardening Metal Surfaces** Plasma nitriding is a process for embedding nitrogen atoms in the surfaces of metals that hardens the surfaces and makes them more corrosion resistant. Do the nitrogen atoms in the nitrided surface of a sample of cubic closest-packed iron fit in the octahedral holes of the crystal lattice? (Assume that the atomic radii of N and Fe are 75 pm and 126 pm, respectively.)

*18.46. **Hydrogen Storage** Several crystalline transition metals (including titanium, zirconium, and hafnium) can store hydrogen as metal hydrides for use as fuel in a hydrogen-powered vehicle. Are the H atoms (radius, 37 pm) more likely to be in the tetrahedral or octahedral holes of these three metals whose atomic radii are 147 pm, 160 pm, and 159 pm, respectively?

**18.47.** A metallurgist allows a molten mixture of 96 mg of molybdenum and 184 mg of tungsten to cool to 25°C, forming a crystalline alloy.
a. The unit cell of the alloy contains eight Mo atoms on the corners and one W atom in the center. What is the empirical formula of the alloy?
b. Is the unit cell consistent with the masses of the metals used in the reaction?
c. Calculate the density of the alloy, given that the atomic radius of both metals is 139 pm.

18.48. Building on the success in Problem 18.47, the metallurgist halved the amount of molybdenum and multiplied the amount of tungsten in the mixture 1.5 times to obtain a different alloy.
a. The unit cell of this alloy contains four W atoms on alternate corners and one additional W atom in the center. Mo atoms occupy the remaining four corners in the unit cell. What is the empirical formula of the alloy?
b. Calculate the density of the alloy, given that the atomic radius of both metals is 139 pm.
*c. Would the density of the alloy change if the tungsten and molybdenum atoms at the corners of the unit cell were randomly distributed among all corners of the unit cell?

**18.49.** An interstitial alloy with an fcc unit cell contains one atom of B for every five atoms of host element A. What fraction of the octahedral holes is occupied in this alloy?

18.50. If the B atoms in the alloy described in Problem 18.49 occupy tetrahedral holes in A, what percentage of the holes would they occupy?

**18.51.** If the unit cell of a substitutional alloy of copper and tin has the same edge length as the unit cell of copper, will the alloy have a greater density than that of copper?

18.52. If the unit cell of an interstitial alloy of vanadium and carbon has the same edge length as the unit cell of vanadium, will the alloy have a greater density than that of vanadium?

## Metallic Bonds and Conduction Bands

### Concept Review

**18.53.** How does the sea-of-electrons model (Section 4.1) explain the high electrical conductivity of gold?

*18.54. How does band theory explain the high electrical conductivity of mercury?

**18.55.** The melting and boiling points of sodium metal are much lower than those of sodium chloride. What does this difference reveal about the relative strengths of metallic bonds and ionic bonds of the alkali metals?

*18.56. Which metal do you expect to have the higher melting point, Al or Na? Explain your answer.

**18.57.** Some scientists believe that the solid hydrogen that forms at very low temperatures and high pressures may conduct electricity. Is this hypothesis supported by band theory?

18.58. Would you expect solid helium to conduct electricity?

## Semiconductors

### Concept Review

**18.59.** Which groups in the periodic table contain metals with filled valence bands?

18.60. Rank the following in order of increasing band gap: semiconductor, insulator, and conductor.

**18.61.** Why is it important to keep phosphorus out of silicon chips during their manufacture?

18.62. How might doping silicon with germanium affect the conductivity of silicon?

***18.63.** Antimony (Sb) combines with sulfur to form the semiconductor compound $Sb_2S_3$. In which group of the periodic table might you find elements for doping $Sb_2S_3$ to form a p-type semiconductor?

*18.64. In which group of the periodic table might you find elements for doping $Sb_2S_3$ to form an n-type semiconductor?

### Problems

**18.65.** Thin films of doped diamond hold promise as semiconductor materials. Trace amounts of nitrogen impart a yellow color to otherwise colorless pure diamonds.
a. Are nitrogen-doped diamonds examples of semiconductors that are p-type or n-type?
b. Draw a picture of the band structure of diamond to indicate the difference between pure diamond and N-doped (nitrogen-doped) diamond.
*c. N-doped diamonds absorb violet light at about 425 nm. What is the magnitude of $E_g$ that corresponds to this wavelength?

18.66. **Hope Diamond** Trace amounts of boron give diamonds (including the Smithsonian's Hope Diamond) a blue color (Figure P18.66).
a. Are boron-doped diamonds examples of semiconductors that are p-type or n-type?
b. Draw a picture of the band structure of diamond to indicate the difference between pure diamond and B-doped diamond.
*c. What is the band gap in energy if blue diamonds absorb red-orange light with a wavelength of 675 nm?

**FIGURE P18.66**

*18.67. The nitride ceramics AlN, GaN, and InN are all semiconductors used in the microelectronics industry. Their band gaps are 580.6 kJ/mol, 322.1 kJ/mol, and 192.9 kJ/mol, respectively. Which, if any, of these energies correspond to radiation in the visible region of the spectrum?

*18.68. Calculate the wavelengths of light emitted by the semiconducting phosphides AlP, GaP, and InP, which have band gaps of 241.1 kJ/mol, 216.0 kJ/mol, and 122.5 kJ/mol, respectively, and are used in the type of light source shown in Figure P18.68.

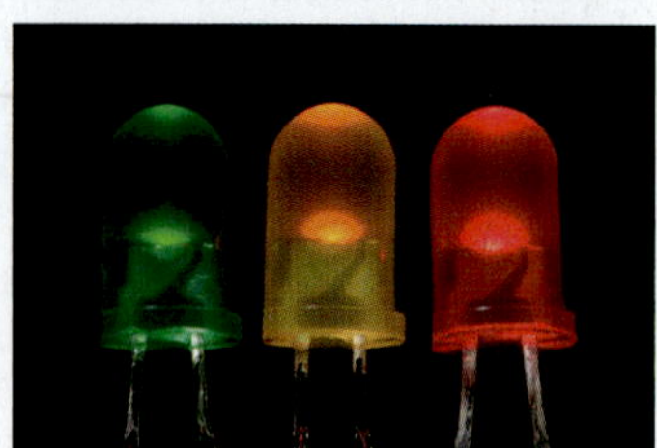

**FIGURE P18.68**

## Structures of Some Crystalline Nonmetals

### Concept Review

**18.69.** Molecules of $S_8$ are not flat octagons. Why?

*18.70. Selenium exists either as $Se_8$ rings or in a structure with helical chains of Se atoms. Are these two structures of selenium allotropes? Explain your answer.

***18.71.** If the carbon atoms in graphite are replaced by alternating B and N atoms, does the resulting structure contain puckered rings, as in black phosphorus, or flat rings, as in graphene (Figure P18.71)?

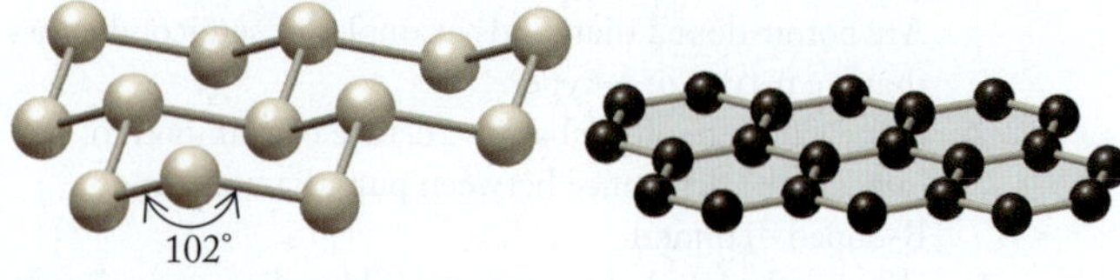


(a) Black phosphorus (b) Graphene

**FIGURE P18.71**

*18.72. Cyclic allotropes of sulfur containing up to 20 sulfur atoms have been isolated and characterized. Propose a reason why the bond angles in $S_n$ (where $n = 10, 12, 18,$ and 20) are all close to 106°.

### Problems

**18.73. Semiconducting Phosphorus** A cubic form of phosphorus is drawing renewed interest as a semiconductor. The distance between atoms in this cubic form of phosphorus is 238 pm (Figure P18.73).

a. Calculate the density of this form of phosphorus.

b. Is the cubic hole in cubic P large enough to hold either a silicon or a boron atom?

*c. Would the material(s) from part b behave as n-type or p-type semiconductors?

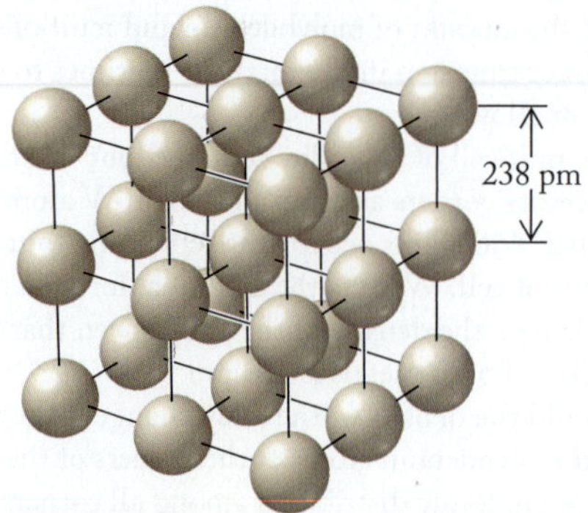


**FIGURE P18.73**

18.74. Which drawing in Figure P18.74 best describes the result of doping cubic phosphorus (see Problem 18.73) with another element?

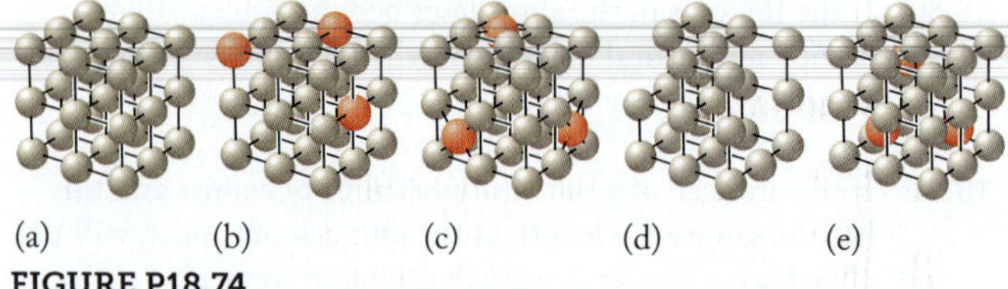


**FIGURE P18.74**

**18.75.** Solid methane crystallizes in a face-centered cubic unit cell where we treat each $CH_4$ molecule as a sphere (Figure P18.75).

a. Calculate the density of solid methane assuming the radii of the methane spheres are 199.4 pm.

b. How many equivalent carbon and hydrogen atoms are in the unit cell?

**FIGURE P18.75**

18.76. **Ice under Pressure** Kurt Vonnegut's novel *Cat's Cradle* describes an imaginary, high-pressure form of ice called "ice-nine." If ice-nine has a cubic closest-packed arrangement of oxygen atoms with hydrogen atoms in the appropriate holes, what type of hole will accommodate the H atoms?

## Salt Crystals: Ionic Solids

### Concept Review

**18.77.** Crystals of both LiCl and KCl have the rock salt structure. In the unit cell of LiCl, adjacent $Cl^-$ ions touch each other. In KCl, they don't. Why?

18.78. Can $CaCl_2$ have the rock salt structure?

***18.79.** In some books, the unit cell of CsCl is described as body-centered cubic (Figure P18.79); in others, as simple cubic (see Figure 18.11a). Explain how CsCl crystals might be described by either unit cell type.

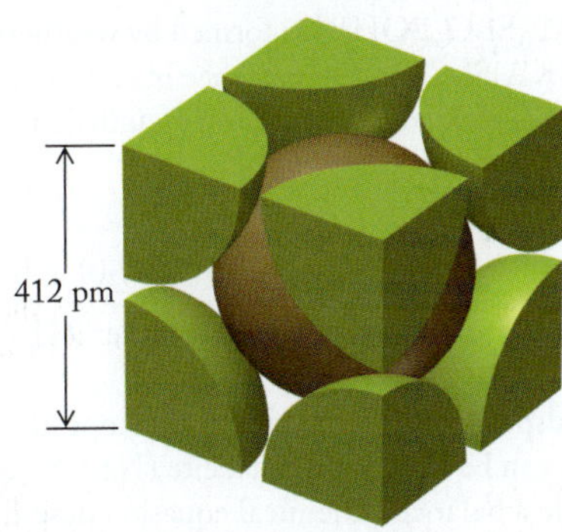


**FIGURE P18.79**

18.80. In the crystals of ionic compounds, how do the relative sizes of the ions influence the location of the smaller ions?

***18.81.** Instead of describing the unit cell of NaCl as an fcc array of $Cl^-$ ions with $Na^+$ ions in octahedral holes, might we describe it as an fcc array of $Na^+$ ions with $Cl^-$ ions in octahedral holes? Explain why or why not.

18.82. Why isn't crystalline sodium chloride considered a network solid? Why isn't sodium chloride considered an alloy of sodium and chlorine?

***18.83.** As the cation–anion radius ratio increases for an ionic compound with the rock salt crystal structure, is the calculated density more likely to be greater than or less than the measured value?

18.84. As the cation–anion radius ratio increases for an ionic compound with the rock salt crystal structure, is the length of the edge of the unit cell calculated from ionic radii likely to be greater than or less than the observed edge length?

### Problems

**18.85.** What is the formula of the oxide that crystallizes with $Fe^{3+}$ ions in one-fourth of the octahedral holes, $Fe^{3+}$ ions in one-eighth of the tetrahedral holes, and $Mg^{2+}$ in one-fourth of the octahedral holes of a cubic closest-packed arrangement of oxide ions ($O^{2-}$)?

18.86. What is the chemical formula of the compound that crystallizes in a simple cubic arrangement of fluoride ions with $Ba^{2+}$ ions occupying half of the cubic holes?

---

**18.87.** A compound of uranium and oxygen consists of a cubic closest-packed arrangement of uranium ions with oxide ions in all the tetrahedral holes. What is the formula of this compound?

18.88. A mixture of gallium and arsenic is a widely used semiconductor. The arsenide ions are in a cubic closest-packed arrangement, and half the tetrahedral holes are occupied by the gallium ions. What is the formula of this compound?

---

**18.89. The Vinland Map** Yale University has a map, believed to date from the 1400s, of a landmass labeled "Vinland" (Figure P18.89). The map is thought to be evidence of early Viking exploration of North America. Debate over the map's authenticity centers on yellow stains on the map paralleling the black ink lines. One analysis suggests that the yellow color is from the mineral anatase, a form of $TiO_2$ that was not used in 15th-century inks.

a. The crystal structure of anatase is approximated by a ccp arrangement of oxide ions with titanium(IV) ions in holes. Which type of hole are $Ti^{4+}$ ions likely to occupy? (The radius of $Ti^{4+}$ is 60.5 pm.)

b. What fraction of these holes is likely to be occupied?

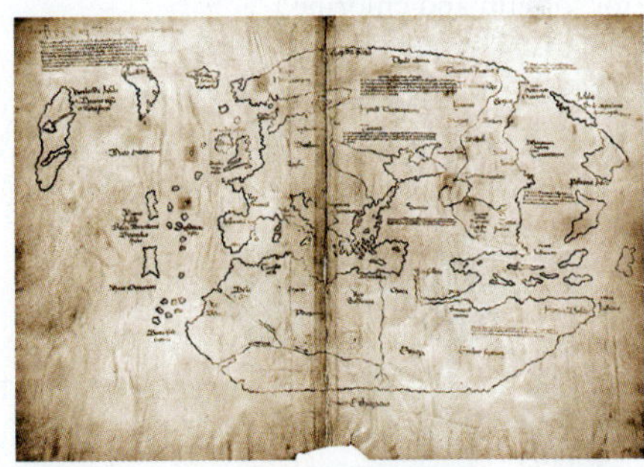

**FIGURE P18.89**

*18.90. The crystal structure of olivine—$M_2SiO_4$ (M = Mg, Fe)—can be viewed as a ccp arrangement of oxide ions with silicon(IV) ions in tetrahedral holes and metal ions in octahedral holes.

a. What fraction of each type of hole is occupied?

b. The unit cells of $Mg_2SiO_4$ and $Fe_2SiO_4$ have volumes of $2.91 \times 10^{-26}$ $cm^3$ and $3.08 \times 10^{-26}$ $cm^3$, respectively. Why is the volume of $Fe_2SiO_4$ larger?

---

***18.91.** The cadmium(II) sulfide (CdS) mineral hawleyite has a sphalerite structure, and its density at 25°C is 4.83 g/$cm^3$. A hypothetical form of CdS with the rock salt structure would have a density of 5.72 g/$cm^3$. Why should the rock salt structure of CdS be denser? The ionic radii of $Cd^{2+}$ and $S^{2-}$ are 95 pm and 184 pm, respectively.

18.92. There are two crystalline forms of manganese(II) sulfide (MnS): the α form has a rock salt structure, whereas the β form has a sphalerite structure.
a. Describe the differences between the two structures of MnS.
b. The ionic radii of $Mn^{2+}$ and $S^{2-}$ are 67 pm and 184 pm, respectively. Which type of hole in a ccp lattice of sulfide ions could theoretically accommodate a $Mn^{2+}$ ion?

**18.93.** The unit cell of rhenium trioxide ($ReO_3$) consists of a cube with a rhenium atom at each of the 8 corners and an oxygen atom on each of the 12 edges. The atoms touch along the edge of the unit cell. The radii of Re and O atoms in $ReO_3$ are 137 pm and 73 pm, respectively.
a. Draw the unit cell of $ReO_3$ and show that it is consistent with the empirical formula.
b. Calculate the density of $ReO_3$.

18.94. **New Forms of Salt** Sodium chloride is the prototypical example of the rock salt structure, but in 2013 researchers discovered a new compound containing sodium and chlorine. Under high pressure, NaCl reacts with excess sodium metal to produce the compound shown in Figure P18.94.
a. What is the empirical formula of this compound?
b. Calculate the density of this compound, assuming that the atoms touch along the edges, using the atomic radii for sodium and chlorine.

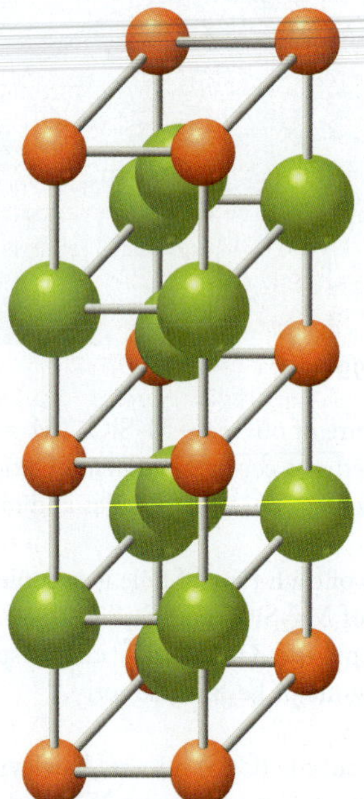

**FIGURE P18.94**

**18.95.** Magnesium oxide crystallizes in the rock salt structure. Its density is 3.60 g/cm$^3$. What is the edge length of the fcc unit cell of MgO?

18.96. Crystalline potassium bromide (KBr) has a rock salt structure and a density of 2.75 g/cm$^3$. Calculate the edge length of its unit cell.

## Ceramics: Useful, Ancient Materials

### Concept Review

**18.97.** Which of the following properties are associated with ceramics and which are associated with metals? (a) ductile; (b) thermal insulator; (c) electrically conductive; (d) malleable

18.98. Many ceramics such as $TiO_2$ are electrical insulators. What differences are there in the band structure of $TiO_2$ compared with Ti metal that account for the different electrical properties?

**18.99.** Replacing $Al^{3+}$ ions in kaolinite [$Al_2(Si_2O_5)(OH)_4$] with $Mg^{2+}$ ions yields the mineral antigorite. What is its formula?

18.100. What is the formula of the silicate mineral talc, obtained by replacing $Al^{3+}$ ions in pyrophyllite [$Al_4Si_8O_{20}(OH)_4$] with $Mg^{2+}$ ions?

### Problems

**18.101.** Kaolinite [$Al_2(Si_2O_5)(OH)_4$] is formed by weathering of the mineral $KAlSi_3O_8$ in the presence of carbon dioxide and water, as described by the following unbalanced reaction:

$$KAlSi_3O_8(s) + H_2O(\ell) + CO_2(g) \rightarrow Al_2(Si_2O_5)(OH)_4(s) + SiO_2(s) + K_2CO_3(aq)$$

Balance the reaction and determine whether it is a redox reaction.

18.102. Albite, a feldspar mineral with an ideal composition of $NaAlSi_3O_8$, can be converted to jadeite ($NaAlSi_2O_6$) and quartz. Write a balanced chemical equation describing this transformation.

**18.103.** Under the high pressures in Earth's crust, the mineral anorthite ($CaAl_2Si_2O_8$) is converted to a mixture of three minerals: grossular [$Ca_3Al_2(SiO_4)_3$], kyanite ($Al_2SiO_5$), and quartz ($SiO_2$). (a) Write a balanced chemical equation describing this transformation. (b) Determine the charges and formulas of the silicate anions in anorthite, grossular, and kyanite.

*18.104. The calcium silicate mineral grossular is also formed under pressure in a reaction between anorthite ($CaAl_2Si_2O_8$), gehlenite ($Ca_2Al_2SiO_7$), and wollastonite ($CaSiO_3$):

$$\underset{\text{Anorthite}}{CaAl_2Si_2O_8} + \underset{\text{Gehlenite}}{Ca_2Al_2SiO_7} + \underset{\text{Wollastonite}}{CaSiO_3} \rightarrow \underset{\text{Grossular}}{Ca_3Al_2(SiO_4)_3}$$

a. Balance this chemical reaction.
b. Express the composition of gehlenite the way mineralogists often do: as the percentage of the metal and metalloid oxides in it—that is, % CaO, % $Al_2O_3$, and % $SiO_2$.

**18.105.** The ceramic material barium titanate ($BaTiO_3$) is used in devices that measure pressure. The radii of $Ba^{2+}$, $Ti^{4+}$, and $O^{2-}$ are 135 pm, 60.5 pm, and 140 pm, respectively. If the $O^{2-}$ ions are in a closest-packed structure, which hole(s) can accommodate the metal cations?

18.106. The mixed metal oxide $LiMnTiO_4$ has a structure with cubic closest-packed oxide ions and metal ions in both octahedral and tetrahedral holes. Which metal ion is most likely to be found in the tetrahedral holes? The ionic radii of $Li^+$, $Mn^{3+}$, $Ti^{4+}$, and $O^{2-}$ are 76 pm, 67 pm, 60.5 pm, and 140 pm, respectively.

## X-ray Diffraction: How We Know Crystal Structures

### Concept Review

**18.107.** Why does an amorphous solid not produce an XRD scan with sharp peaks?

18.108. Why can we not use X-ray diffraction to determine the structures of compounds in solution?

**18.109.** Why are X-rays rather than microwaves chosen for diffraction studies of crystalline solids?

18.110. The radiation sources used in X-ray diffraction can be changed. Figure P18.110 shows a diffraction pattern made by a short-wavelength source. How would changing to a longer-wavelength source affect the pattern?

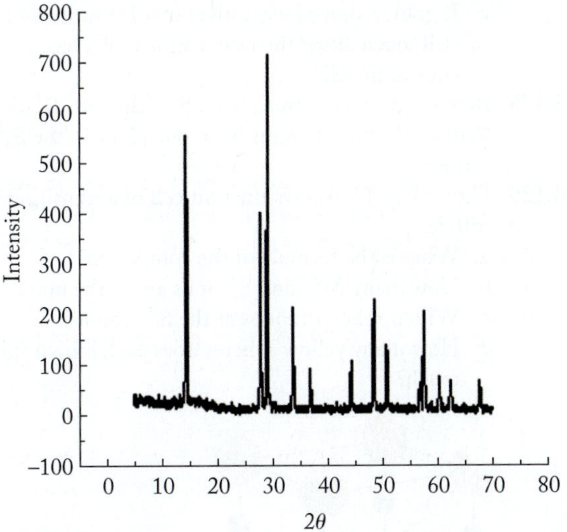


**FIGURE P18.110**

***18.111.** Why might a crystallographer (a scientist who studies crystal structures) use different X-ray wavelengths to determine a crystal structure? *Hint*: Consider the mechanical limits inherent in the design of the instrument depicted in Figure 18.44 and how those limits impact the $2\theta$ scanning range.

*18.112. Where in earlier chapters did we use the interaction between matter and electromagnetic radiation to acquire insights into atomic or molecular structure?

### Problems

**18.113.** The spacing between the layers of ions in sylvite (the mineral form of KCl) is larger than in halite (NaCl). Which crystal will diffract X-rays of a given wavelength through larger $2\theta$ values?

18.114. Silver halides are used in black-and-white photography. In which compound would you expect to see a larger distance between ion layers, AgCl or AgBr? Which compound would you expect to diffract X-rays through larger values of $2\theta$ if the same wavelength of X-ray were used?

---

**18.115.** Galena, Illinois, is named for the rich deposits of lead(II) sulfide (PbS) found nearby. When a crystal of PbS is exposed to X-rays with $\lambda = 71.2$ pm, the resulting XRD pattern contains strong peaks at $2\theta = 13.98°$ and 21.25°. Determine the values of $n$ to which these peaks correspond, and calculate the spacing between the crystal layers.

18.116. **Pigments in Ceramics** Cobalt(II) oxide is used as a pigment in ceramics. It has the same type of crystal structure as NaCl. When cobalt(II) oxide is exposed to X-rays with $\lambda = 154$ pm, the XRD pattern contains strong peaks at $2\theta = 42.38°$, 65.68°, and 92.60°. Determine the values of $n$ to which these peaks correspond, and calculate the spacing between the crystal layers.

---

**18.117.** Pyrophyllite $[Al_2Si_4O_{10}(OH)_2]$ is a silicate mineral with a layered structure. The distance between the layers is 1855 pm. What is the smallest angle of diffraction of X-rays with $\lambda = 154$ pm from this solid?

18.118. Minnesotaite $[Fe_3Si_4O_{10}(OH)_2]$ is a silicate mineral with a layered structure similar to that of kaolinite. The distance between the layers in minnesotaite is 1940 ± 10 pm. What is the smallest angle of diffraction of X-rays with $\lambda = 154$ pm from this solid?

## Additional Problems

**18.119.** A unit cell consists of a cube that has an ion of element X at each corner, an ion of element Y at the center of the cube, and an ion of element Z at the center of each face. What is the formula of the compound?

18.120. An element crystallizes in the cubic closest-packed structure. The length of an edge of the unit cell is 408 pm. The density of the element is 19.27 g/cm$^3$. Identify the element.

***18.121.** What is the packing efficiency of the Si atoms in pure Si if the radius of one Si atom is 117 pm? The density of pure silicon is 2.33 g/cm$^3$.

*18.122. **The Composition of Light-Emitting Diodes** The colored lights on many electronic devices are light-emitting diodes (LEDs). One of the compounds used to make them is aluminum phosphide (AlP), which crystallizes in a sphalerite crystal structure.

a. If AlP were an ionic compound, would the ionic radii of $Al^{3+}$ and $P^{3-}$ be consistent with the size requirements of the ions in a sphalerite crystal structure?

b. If AlP were a covalent compound, would the atomic radii of Al and P be consistent with the size requirements of atoms in a sphalerite crystal structure?

**18.123.** Under the appropriate reaction conditions, small cubes of molybdenum, 4.8 nm on a side, can be deposited on carbon surfaces. These "nanocubes" are made of bcc arrays of Mo atoms.

a. If the edge of each nanocube corresponds to 15 unit cell lengths, what is the effective radius of a molybdenum atom in these structures?

b. What is the density of each molybdenum nanocube?

c. How many Mo atoms are in each nanocube?

18.124. In the fullerene known as buckminsterfullerene, molecules of $C_{60}$ form a cubic closest-packed array of spheres with a unit cell edge length of 1410 pm.

a. What is the density of crystalline $C_{60}$?

b. If we treat each $C_{60}$ molecule as a sphere of 60 carbon atoms, what is the radius of the $C_{60}$ molecule?

c. $C_{60}$ reacts with alkali metals to form $M_3C_{60}$ (where M = Na or K). The crystal structure of $M_3C_{60}$ contains cubic closest-packed spheres of $C_{60}$ with metal ions in holes. If the radius of a $K^+$ ion is 138 pm, which type of hole is a $K^+$ ion likely to occupy? What fraction of the holes will be occupied?

d. Under certain conditions, a different substance, $K_6C_{60}$, can be formed in which the $C_{60}$ molecules have a bcc unit cell. Calculate the density of a crystal of $K_6C_{60}$.

**18.125.** The structure of LiCl is shown in Figure P18.125.

a. Use the ionic radii in Figure 3.36 and the data in Table 18.3 to calculate the density of LiCl.

b. How would the structure change if we treated the particles as neutral atoms? Assume an fcc lattice composed of one type of atom with the other type of atom in holes.

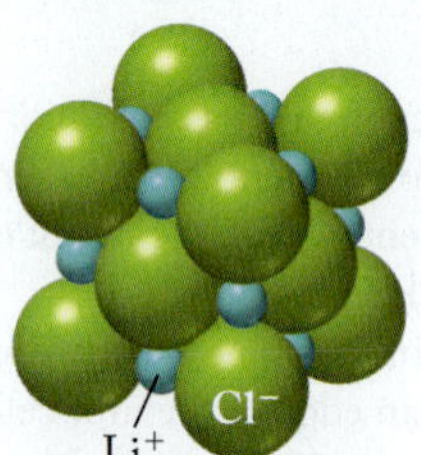


**FIGURE P18.125**

18.126. The unit cell of an alloy with a 1:1 ratio of magnesium and strontium is identical to the unit cell of CsCl. The unit cell edge length of MgSr is 390 pm.

a. What is the density of MgSr?

b. Find the atomic radii of Mg and Sr in Appendix 3. Do atoms touch along the body diagonal of the unit cell?

*c. Why doesn't the formula of the alloy allow us to distinguish between a Mg atom in the cubic hole of simple cubic unit cell of Sr atoms and a Sr atom in the cubic hole of a simple cubic arrangement of Mg atoms?

*d. MgSr is a good electrical conductor. Do you expect a MgSr alloy to have a partially filled valence band or overlapping conduction and valence bands?

**18.127.** Substitutional alloys may form when the difference in atomic radii between the alloying elements is less than 15%.

a. Predict which of the following alloys has the greatest mismatch in atomic radii: AuZn, AgZn, or CuZn.

b. Find the atomic radii of Cu, Ag, Au, and Zn in Appendix 3, and calculate the percent difference in their atomic radii. Are all three alloys expected to form substitutional alloys?

c. If gold is alloyed with silver in a 1:1 ratio, do the atoms still touch along the face diagonal of a face-centered cubic unit cell?

*18.128. Removing two electrons from $S_8$ yields the dication $S_8^{2+}$. Will all the sulfur atoms be in one plane in the $S_8^{2+}$ cation?

**18.129.** Figure P18.129 shows the unit cell of a transition metal sulfide.

a. What is the formula of the compound?

b. How many $M^{2+}$ and $S^{2-}$ ions are in the unit cell?

c. Which spheres represent the $S^{2-}$ anions?

d. How many yellow spheres does each brown sphere touch?

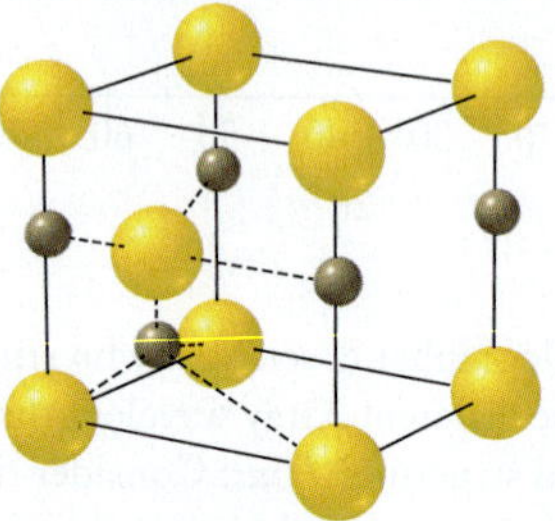

**FIGURE P18.129**

18.130. A transition metal sulfide crystallizes in the unit cell shown in Figure P18.130, where the metal ($M^{n+}$) ions occupy half of the tetrahedral holes in a cubic closest-packed arrangement of sulfide ions.
a. What is the formula of the compound?
b. How many equivalent $M^{n+}$ and $S^{2-}$ ions are in the unit cell?
c. How does the unit cell in Figure P18.130 differ from the unit cell in Figure P18.129?
d. What percentage of the unit cell is empty space?

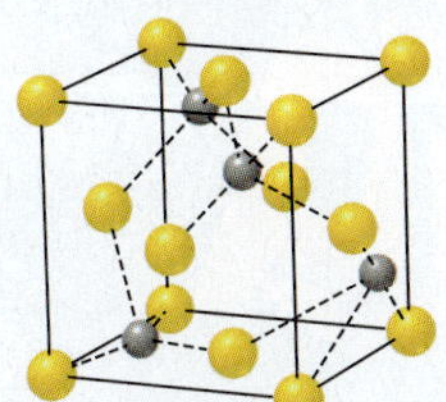
**FIGURE P18.130**

***18.131.** The alloy $Cu_3Al$ crystallizes in a bcc unit cell. Propose a way that the Cu and Al atoms could be allocated between bcc unit cells that is consistent with the formula of the alloy.

18.132. **Forms of Phosphorus** White phosphorus is highly reactive material consisting of independent $P_4$ tetrahedra. Red phosphorus is much less reactive, with a structure consisting of linked $P_4$ tetrahedra (see Figure 18.31b). What is the size of the largest particle that can fit inside a $P_4$ tetrahedron? Does any element fit this criterion?

# 19

# Organic Chemistry

## Fuels, Pharmaceuticals, and Modern Materials

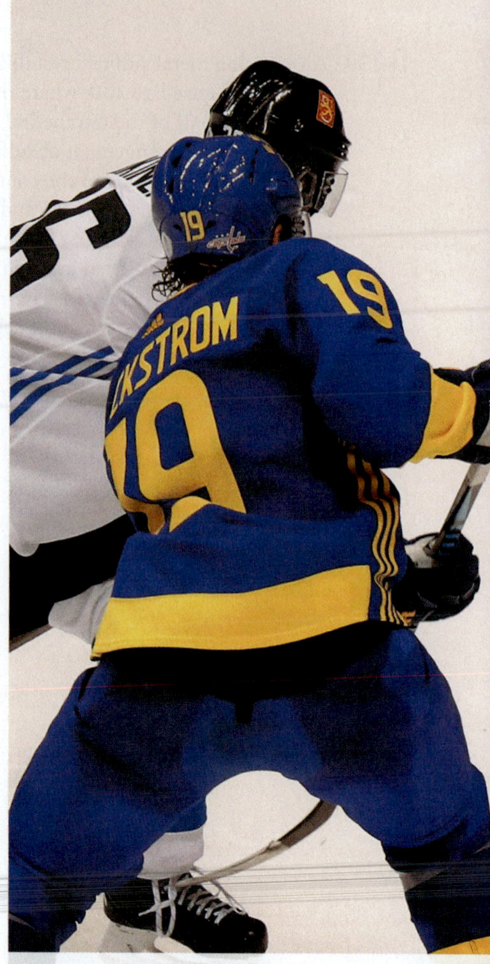


**TOUGH MATERIALS** Physical properties of polymers, such as the impact resistance of a hockey goalie's mask, are linked directly to the composition and structure of the molecules from which they are made.

PARTICULATE **REVIEW**

### *Carbon–Carbon Bonds*

In Chapter 19, we explore the many ways carbon atoms bond to each other and to atoms of other elements in organic compounds, including the three hydrocarbons shown here.

- Draw the Lewis structure of each compound.
- Which hydrocarbon has the strongest carbon–carbon bond? Which has the longest carbon–carbon bond?
- Which compound contains only sigma bonds?

(Review Section 6.2 if you need help.)

*(Answers to Particulate Review questions are in the back of the book.)*

PARTICULATE **PREVIEW**

### *Same Composition, Different Structures*

Compound A and compound B have the same composition but different structures. As you read Chapter 19, look for ideas that will help you answer these questions:

- What word describes two compounds that have the same composition but different structures?
- What functional group is present in each molecule?
- The functional group in compound A can be converted into the functional group in compound B by reacting with a molecule that contains yet a third functional group. What is that functional group?

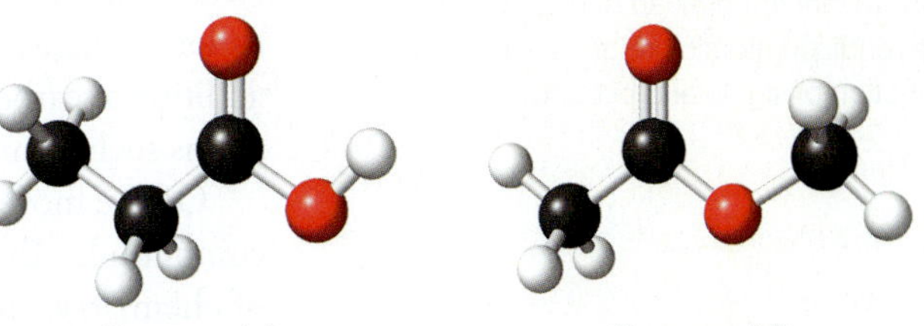

Compound A    Compound B

## Learning Outcomes

**LO1** Relate the names and molecular structures of alkanes
**Sample Exercises 19.1, 19.2**

**LO2** Recognize and name constitutional isomers of alkanes and draw their molecular structures
**Sample Exercises 19.3, 19.4**

**LO3** Explain the differences in the chemical reactivities of alkanes, alkenes, and alkynes on the basis of the bonding between carbon atoms in their molecules
**Sample Exercise 19.5**

**LO4** Recognize and name constitutional isomers and stereoisomers of alkenes and draw their molecular structures
**Sample Exercise 19.6**

**LO5** Identify the monomers that react and the polymers that are produced in addition and condensation polymerization reactions
**Sample Exercises 19.7, 19.9, 19.10**

**LO6** Relate the properties of polymers to their molecular structures
**Sample Exercise 19.8**

**LO7** Explain the properties of organic compounds on the basis of the size of their molecules and the reactivity of their functional groups

# 19.1 Carbon: The Stuff of Daily Life

Fans at a hockey game sit on molded plastic seats and munch on hot dogs wrapped in plastic wrap while watching players who wear helmets made of impact-resistant polymeric materials. Plastics, cling wrap, and hockey helmets are made from compounds derived from crude oil or natural gas—or, due to the growing green chemistry movement, from ethanol derived from renewable, sustainable resources like sugarcane. Either way, these materials are carbon based; that is, they are organic compounds. The plastic seat is strong, lightweight, and capable of bearing considerable mass without breaking. The sandwich wrap must be flexible yet prevent oxygen from reaching the hot dog and bun before you open it, and the players' helmets must be lightweight but still able to withstand the impact of a high-speed hockey puck. All these physical properties can be designed into the materials at the molecular level because they are directly linked to the structure, composition, and size of the molecules.

**CONNECTION** In Section 6.3, we introduced *organic compounds* as molecules that always contain carbon atoms and almost always contain hydrogen atoms.

**CONNECTION** When we explored valence bond theory in Chapter 5, we learned that the overlap of hybrid and unhybridized atomic orbitals leads to the formation of sigma and pi bonds.

Carbon atoms can form up to four covalent bonds, including bonds to other carbon atoms, by using combinations of single, double, and triple bonds. Much of the variety in the chemistry of carbon is derived from this property, and the resulting compounds may contain a few carbon atoms or many thousands. Additional structural variety comes from the many patterns that carbon atoms can adopt, such as long chains, heavily branched structures, and planar and puckered rings. In addition to carbon and hydrogen, organic compounds can also contain nonmetallic atoms such as nitrogen, oxygen, sulfur, phosphorus, or the halogens.

Of the more than 150 million compounds known, the majority are organic compounds. They are so varied in size, shape, and properties that an entire field of chemistry, called **organic chemistry**, is devoted to their study. This chapter is a brief introduction to organic chemistry. The designation *organic* for carbon-containing compounds was once limited to substances produced by living organisms, but that definition has been broadened for two reasons. First, since 1828, when Friedrich Wöhler (1800–1882) discovered how to prepare urea (**Figure 19.1**) in the laboratory "without the intervention of a kidney," scientists have learned how to synthesize many materials previously thought to be the products only of living systems. Second, chemists have learned how to synthesize many carbon-based materials that have never been produced by living systems. Chemical Abstracts Service (CAS), an organization that tracks and collects all chemical information published worldwide, adds thousands of new compounds, most of them organic, to its registry of unique chemical substances each day.

With so many substances to deal with, some organizing principles are necessary. We use two of the most common ones in this chapter, and we have already introduced both of them elsewhere in this text. The first is based on *functional groups*: subunits of only a few atoms that confer particular chemical and physical properties to organic compounds. The second is based on molecular size. Molecules having masses less than 1000 u are typically considered small molecules, whereas much larger molecules with masses up to and exceeding 1,000,000 u are called *macromolecules*.

The sizes and shapes of molecules and the functional groups in them combine to determine the physical and chemical properties of materials at the microscopic and bulk levels. In addition, the shape of many macromolecules in solution depends on their environment and their association with other molecules. The role of intermolecular interactions on properties of bulk materials is also an important theme here in Chapter 19 and in Chapter 20 on biochemistry.

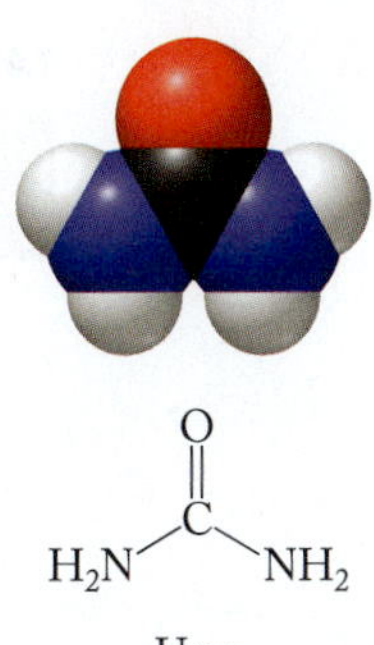


**FIGURE 19.1** Urea was the first naturally occurring organic compound synthesized in the laboratory. It was prepared in 1828 by Friedrich Wöhler.

**CONNECTION** We first used the term *functional group* in Section 6.3 to describe the groups of atoms in molecules that are most responsible for the substance's physical and chemical properties.

## Families Based on Functional Groups

We have discussed several functional groups in prior chapters, including the –COOH group in carboxylic acids and the hydroxyl (–OH) group in alcohols. In this chapter, we examine these and the other functional groups summarized in **Table 19.1**. When one is discussing functional groups, the convention is to use **R** to represent the entire molecule except for the functional group. (Do not confuse it with the italic *R* representing the universal gas constant.)

Throughout this book, we have seen many examples of organic compounds containing these functional groups. We focus on the functional groups in this chapter because much of organic chemistry deals with chemical reactions of organic substances and how those reactions can be used to modify substances, to find synthetic routes to known materials produced by nature, and to find ways to make new materials that have desirable properties. Understanding the characteristic reactivity of specific functional groups enables chemists to carry out these tasks in a logical fashion. Introducing the unique reactivities of the functional groups listed in Table 19.1 is beyond the scope of this text, but we will illustrate a small set of reactions of molecules bearing some of these functional groups that are used in the industrial production of polymers.

## Monomers and Polymers

Many macromolecules consist of long chains made up of chemically bonded subunits called **monomers**; the overall macromolecule is called a **polymer**. The mass boundaries that distinguish small molecules from large ones are arbitrary, and medium-sized organic molecules called *oligomers* inhabit the realm between small molecules and polymers. Organizing molecules by their size works because of the predictable ways that properties vary as composition remains constant but size increases.

Organic compounds composed of small molecules have constant composition and well-defined properties such as melting points and boiling points. In contrast, many synthetic polymers do not have constant composition in terms of the

**organic chemistry** the study of compounds containing C—C and/or C—H bonds.

**R** a symbol in a general formula standing for an organic group that has one available bond; it is used to indicate the part of a molecule that is not the functional group of interest.

**monomer** a small molecule that bonds with other monomers to form polymers.

**polymer** a very large molecule with high molar mass; the root word *meros* is Greek for "part" or "unit," so *polymer* literally means "many units."

**TABLE 19.1** Functional Groups of Organic Compounds

| Name | Structural Formula of Group[a] | Example and Name | |
|---|---|---|---|
| Alkane | R—H | $CH_3CH_2CH_3$ | Propane |
| Alkene | >C=C< | $H_2C=CH_2$ | Ethylene (ethene) |
| Alkyne | —C≡C— | H—C≡C—H | Acetylene (ethyne) |
| Aromatic | e.g., $C_6$ ring with alternating double bonds | $C_6H_6$ | Benzene |
| Amine | R—$NH_2$<br>R—NHR′<br>R—NRR′ | $H_3C$—$NH_2$ | Methylamine |
| Alcohol | R—OH | $CH_3CH_2$—OH | Ethanol |
| Ether | R—O—R′ | $H_3C$—O—$CH_3$ | Dimethyl ether |
| Aldehyde | R—C(=O)—H | $H_3C$—C(=O)—H | Acetaldehyde |
| Ketone | R—C(=O)—R′ | $H_3C$—C(=O)—$CH_3$ | Acetone |
| Carboxylic acid | R—C(=O)—OH | $H_3C$—C(=O)—OH | Acetic acid |
| Ester | R—C(=O)—OR′ | $H_3C$—C(=O)—$OCH_3$ | Methyl acetate |
| Amide | R—C(=O)—$NH_2$ [b] | $H_3C$—C(=O)—$NH_2$ [b] | Acetamide |

[a]In compounds with two R groups, the two may be the same (R = R′) or they may be different (R ≠ R′).

[b]One or both H atoms may be replaced by R– groups: –NHR or NRR′.

*Note*: Functional groups are highlighted in yellow.

number of bonded monomers that make them up. They may not have well-defined physical properties because they are mixtures of molecules that are similar but not identical: they may differ in the number and arrangement of monomers in their molecular structures. In this chapter, we explore the structures of organic polymers and the intermolecular forces between them that define the properties of these polymers and the many familiar materials made from them.

**saturated hydrocarbon** an alkane that has the maximum ratio of hydrogen atoms to carbon atoms.

**unsaturated hydrocarbon** a hydrocarbon that does not have the maximum ratio of hydrogen atoms to carbon atoms.

## 19.2 Alkanes

Recall from Chapter 6 that *hydrocarbons* are compounds whose molecules are composed only of carbon and hydrogen atoms, and *alkanes* are hydrocarbons whose molecules contain only single bonds. Alkanes such as pentane ($C_5H_{12}$), as shown in the left column of **Figure 19.2**, are referred to as *acyclic* because they lack rings. (We will look at alkanes that contain rings later in this section.) Acyclic alkanes are also called **saturated hydrocarbons** because they contain the maximum ratio of hydrogen atoms to carbon atoms. Hydrocarbons containing lower hydrogen-to-carbon ratios are **unsaturated hydrocarbons.**

### Drawing Organic Molecules

Figure 19.2 shows six ways we represent the molecular structure of organic compounds, using the examples of the five-carbon alkane pentane and the organic solvent acetone. Lewis structures show all the bonds in the molecule including

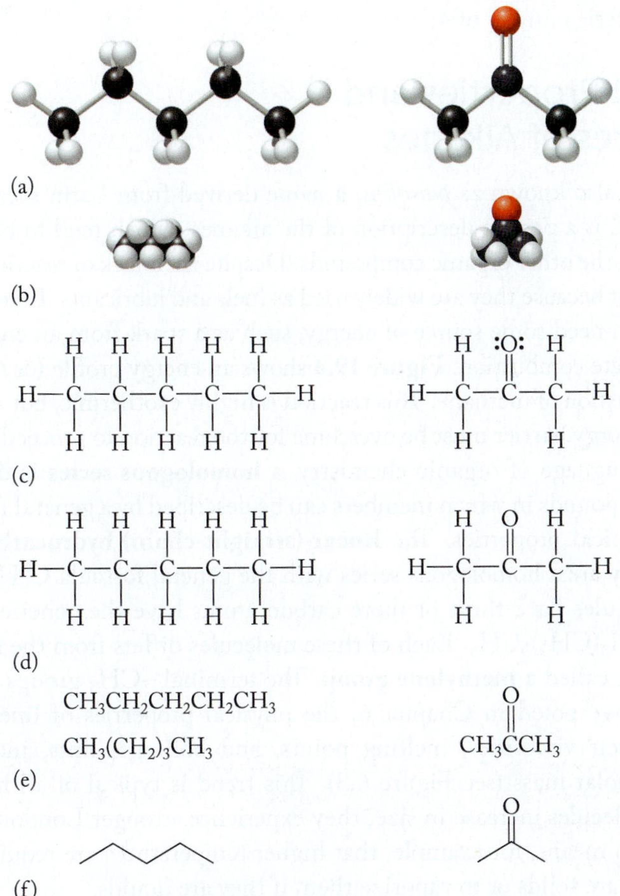


**FIGURE 19.2** Different representations of the molecular structures of pentane and acetone: (a) ball-and-stick models; (b) space-filling models; (c) Lewis structures; (d) Kekulé structures; (e) condensed structures; (f) carbon-skeleton structures.

**CONNECTION** We introduced Lewis structures in Section 4.4. Kekulé structures are the same as the structural formulas we have been drawing throughout the text since Chapter 1.

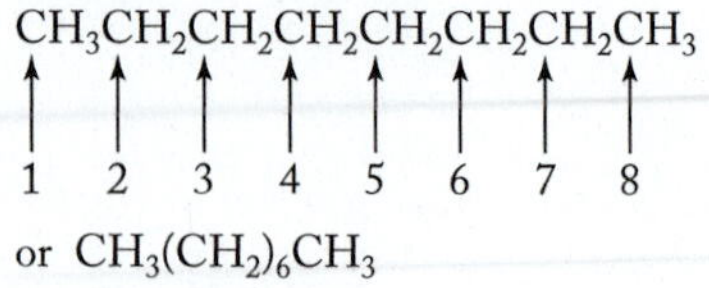


(a) Condensed structure

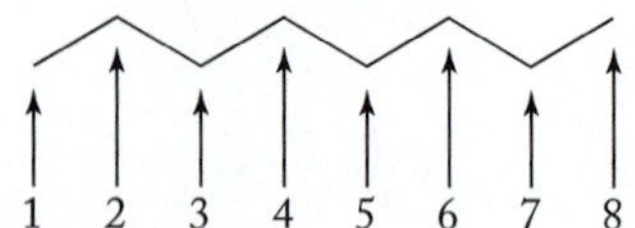


(b) Carbon-skeleton structure

**FIGURE 19.3** When we are converting from (a) a condensed structure to (b) a carbon-skeleton structure, the carbon atoms are represented by junctions between lines, and each junction is assumed to have enough H atoms to give that carbon atom four bonds.

any lone pairs on the atoms (Figure 19.2c) but do not depict the three-dimensional shape of molecules. Structural formulas of organic molecules that show all the bonds with lines but omit any lone pairs are called **Kekulé structures** (Figure 19.2d), after August Kekulé (1829–1896), the German chemist who first used this method to illustrate molecules. Alkanes do not have any lone pairs, so there is no difference between the Lewis structure for pentane and its Kekulé structure, but there is a difference for acetone because the lone pairs on the oxygen atom are omitted in its Kekulé structure.

Writing Lewis or Kekulé structures for large organic molecules can be tedious, so chemists use various shorter notations to convey structures in organic chemistry, including the *condensed structures* shown in Figure 19.2(e). These structures do not show the individual bonds between atoms, as in Lewis structures. Sometimes subscripts are used to indicate the number of times a particular subgroup, written in parentheses, is repeated. For example, the condensed structure of pentane can be written as $CH_3(CH_2)_3CH_3$. The numerical subscript "3" after the parenthetical $-CH_2-$ group means that three of these groups connect the two terminal $-CH_3$ groups in this compound.

The most minimal notation, shown in Figure 19.2(f), is the *carbon-skeleton structure*, in which no letters are used for carbon and hydrogen atoms, but atoms other than C and H are shown. **Figure 19.3** shows how to convert from a condensed structure to a carbon-skeleton structure. Short line segments are drawn at angles to each other, and each line segment represents one carbon–carbon bond in the molecule. The angles represent the bond angle between the two carbon atoms (109.5° for $sp^3$-hybridized carbon atoms in alkanes). Each end of the zigzag line is a $-CH_3$ group, and every intersection of two line segments is a $-CH_2-$ group. It is assumed that each carbon atom is bonded to enough hydrogen atoms to give it a steric number of 4.

**CONNECTION** We learned in Chapter 6 that the larger clouds of electrons in larger molecules are more polarizable, which means that they experience stronger London dispersion forces.

## Physical Properties and Structures of Alkanes

Alkanes are also known as *paraffins*, a name derived from Latin meaning "little affinity." This is a perfect description of the alkanes, which tend to be much less reactive than the other organic compounds. Despite their lack of reactivity, alkanes are important because they are widely used as fuels and lubricants. Like most fuels, alkanes often need some source of energy, such as a spark from an engine's spark plug, to initiate combustion. **Figure 19.4** shows an energy profile for the progress of the combustion of methane. This reaction is highly exothermic, but a significant activation energy barrier must be overcome for the reaction to proceed.

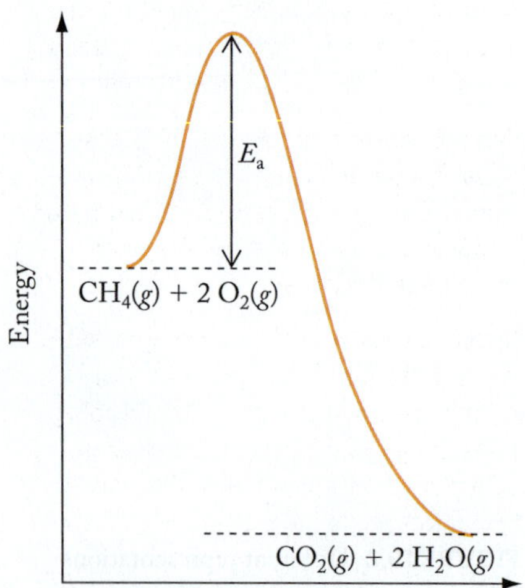


**FIGURE 19.4** The energy profile for the combustion of methane illustrates that a large activation energy barrier, $E_a$, must be overcome before the reaction can proceed.

In the language of organic chemistry, a **homologous series** is defined as a series of compounds in which members can be described by a general formula and similar chemical properties. The **linear (straight-chain) hydrocarbons** in the alkane family are a homologous series with the general formula $C_nH_{2n+2}$. Those whose molecules have three or more carbon atoms have the generic condensed structure $CH_3(CH_2)_nCH_3$. Each of these molecules differs from the next by one $-CH_2-$ unit, called a **methylene group**. The terminal $-CH_3$ groups are **methyl groups**. As we noted in Chapter 6, the physical properties of linear alkanes, including their viscosities, melting points, and boiling points, increase with increasing molar mass (see Figure 6.3). This trend is typical of all homologous series: as molecules increase in size, they experience stronger London dispersion forces, which means, for example, that higher temperatures are required to melt them if they are solids or to vaporize them if they are liquids.

**SAMPLE EXERCISE 19.1** Drawing Structures of Alkanes **LO1**

Write the condensed structure and draw the carbon-skeleton structure for the following linear alkanes: (a) the 3-carbon propane and (b) the 12-carbon dodecane.

**Collect and Organize** Figure 19.2 summarizes the differences between condensed and carbon-skeleton structures, which we need to apply to the hydrocarbons with 3 and 12 carbon atoms. Condensed structures show the symbol for each element in a molecule and subscripts indicating the numbers of atoms of each element, but they do not show C—H or single C—C bonds. Carbon-skeleton structures use short zigzag line segments to represent the carbon–carbon bonds in molecules. The ends of these segments represent carbon atoms.

**Analyze** Each propane and dodecane chain has methyl groups ($-CH_3$) at both ends and one or more methylene groups ($-CH_2-$) in between. We can group bonded methylene groups in a condensed structure into one term by using parentheses followed by a subscript showing the number of $-CH_2-$ groups.

**Solve**

a. Propane has 3 carbon atoms, so the condensed structure is $CH_3CH_2CH_3$. The carbon-skeleton structure consists of two segments representing the two carbon–carbon bonds:

b. Dodecane is a continuous chain of 12 carbon atoms that has two methyl groups at the ends and 10 methylene groups in between. The condensed structure is therefore $CH_3CH_2CH_2CH_2CH_2CH_2CH_2CH_2CH_2CH_2CH_2CH_3$ or, gathering together the 10 $-CH_2-$ groups, $CH_3(CH_2)_{10}CH_3$. The carbon-skeleton structure must contain as many ends and intersections (12) as there are carbon atoms in the molecule, or 11 segments in all:

**Think About It** Condensed and carbon-skeleton structures must both contain the same number of atoms and the same number of bonds, even though the bonds are not shown in the former and not all the atoms or bonds are shown in the latter.

**Practice Exercise** Draw the carbon-skeleton structure of hexane, $CH_3(CH_2)_4CH_3$, and a condensed structure of heptane:

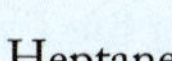

Heptane

*(Answers to Practice Exercises are in the back of the book.)*

## Structural Isomers Revisited

The alkane family would be huge even if it consisted only of straight-chain molecules, but another structural possibility makes the family even larger. We first saw this possibility in Chapter 6 when we compared the boiling points of alkanes having the same molecular formula ($C_5H_{12}$) but different shapes (see Figure 6.4). Similarly, the straight-chain molecule shown in the first column of **Table 19.2** is not the only shape possible for a four-carbon alkane because with four or more carbon atoms per molecule, *branched* structures also exist, as shown in the second column of the table. A **branch** is a side chain attached to the main (longest) chain.

**Kekulé structure** a structure that uses lines to show all the bonds in a covalently bonded molecule but does not show lone pairs on the atoms.

**homologous series** a set of related organic compounds that differ from one another by the number of common subgroups, such as $-CH_2-$, in their molecular structures.

**linear (straight-chain) hydrocarbon** a hydrocarbon in which the carbon atoms are bonded in one continuous line. Linear alkane chains have a methyl group at each end with methylene groups connecting them.

**methylene group** ($-CH_2-$) a structural unit that can make two additional bonds.

**methyl group** ($-CH_3$) a structural unit that can make only one additional bond.

**branch** a side chain bonded to the main (longest) chain in an organic molecule.

**TABLE 19.2** Comparing Constitutional Isomers

| | Butane | 2-Methylpropane |
|---|---|---|
| Condensed Structure | $CH_3CH_2CH_2CH_3$ | $CH_3CH(CH_3)CH_3$ or $CH_3CHCH_3$ with $CH_3$ bonded to the central carbon |
| Carbon-Skeleton Structure | | |
| Melting Point (K) | 135 | 114 |
| Boiling Point (K) at $P$ = 1 atm | 273 | 261 |

The molecular formula of both structures in Table 19.2 is $C_4H_{10}$, but they represent compounds with different properties and names (butane and 2-methylpropane). At this point, it is not important to be able to name them, but you do need to be able to recognize that they are *constitutional isomers* (also called *structural isomers*), compounds that have the same molecular formula but have their atoms connected in different bonding patterns. This difference makes constitutional isomers chemically distinct.

**CONNECTION** Constitutional isomers were introduced in Chapter 6.

2-Methylpropane is an example of a branched hydrocarbon, which means that its molecular structure contains a main chain (the longest carbon chain in the molecule) and at least one side chain or branch. We can illustrate constitutional isomerism with either condensed or carbon-skeleton structures, as shown in Table 19.2. There are two ways to show the location of the methyl group. One way is to use condensed structures and to put the methyl group in parentheses next to the carbon to which it is bonded: $CH_3CH(CH_3)CH_3$. We can also show the carbon–carbon bond between $CH_3$ and the center carbon of the three-carbon main chain. In this instance, it may actually be easier to recognize the constitutional isomers of $C_4H_{10}$ by using the carbon-skeleton structures, which show that butane has no carbon atoms bonded to three other carbon atoms, but 2-methylpropane does.

**SAMPLE EXERCISE 19.2** Identifying the Longest Chain in Organic Molecules **LO1**

Determine the number of carbon atoms in the longest chain in the following three branched alkanes:

a. $CH_3CH_2CH_2CH_2CHCH_3$ with $CH_2CH_3$ bonded to the fifth carbon

b. $CH_3CHCHCH_2CH_2CH_3$ with $CH_2CH_3$ bonded to the second carbon and $CH_3$ bonded to the third carbon

c. $(CH_3)_2CHCH(CH_3)C(CH_3)_2CH_2CH_3$

**Collect and Organize** We are asked to identify the longest carbon chain, or *main chain*, in three compounds that are represented by condensed structures.

**Analyze** We assume that there is a bond between each adjacent pair of carbon atoms in the condensed structures. Carbon bonds to side chains are shown using single lines in structures a and b. In structure c, groups in parentheses represent side-chain groups shown next to the carbon atoms to which they are bonded. We need to determine which carbon atoms belong to the main chain and which belong to the side chains.

**Solve** For a and b, we start at one end of any branch and assign numbers to each carbon atom. If the chain branches, we must choose one branch to follow. Later, we may wish to return to the compound and number the carbon atoms starting from a different end of the molecule or taking a different branch. For c, we first expand this form of condensed structure to more clearly show the branches.

a. Starting with the carbon atom farthest to the left and numbering consecutively from left to right, we reach a branch at carbon atom 5. If we continue straight across the page, we find that the chain contains 6 carbon atoms. If we take the branch at carbon atom 5, however, we find that the chain contains 7 carbon atoms. This makes the longest chain in the compound 7 carbon atoms long.

```
 1    2    3    4    5  6                1    2    3    4    5
CH3CH2CH2CH2CHCH3                  CH3CH2CH2CH2CHCH3
                |                                  |
              CH2CH3                             CH2CH3
                                                  6    7
```

b. There are four places to start counting carbon atoms in structure b. Some possible ways to number the molecule are as follows:

```
    CH3                        CH3                         5
    |                          |                          CH3
 1  2 |3  4   5   6         5  |4  3   2   1               |4  3   2   1
CH3CHCHCH2CH2CH3         CH3CHCHCH2CH2CH3         CH3CHCHCH2CH2CH3
  |                          |                           |
CH2CH3                     CH2CH3                      CH2CH3
                            6    7
```

The longest chain in this molecule (the main chain) also has 7 carbon atoms.

c. We draw the condensed structure and then number the carbon atoms. Here, there are two ways to number the molecule that result in the longest chain having 6 carbon atoms.

```
                                               1
      CH3  CH3                                CH3  CH3
 1    |2  3  4|  5   6                         |2  3  4|  5   6
CH3CHCHCCH2CH3                          CH3CHCHCCH2CH3
        |   |                                    |   |
      H3C  CH3                                 H3C  CH3
```

**Think About It** The main chain in an organic compound may not always be the one running horizontally across the page.

**Practice Exercise** Determine the number of carbon atoms in the longest chain in all three of the following branched alkanes:

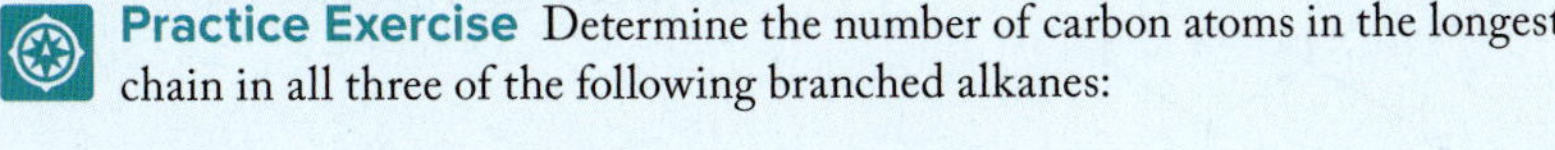


c. $CH_3CH_2C(CH_3)_2CH(CH_2CH_3)CH_2CH(CH_2CH_3)_2$

How do we determine whether two condensed structures or carbon-skeleton structures represent two compounds, a single compound, or two constitutional isomers? Our first step is to translate each structure into a molecular formula. If the molecular formulas are different, the structures represent two compounds. If the molecular formulas are the same, we must compare the way the atoms are connected in the two structures to determine whether they are the same compound or constitutional isomers. Draw the structures with the longest chain horizontal, and then check to see whether the same side chains are attached at the same positions along the longest chain. In doing so, we may have to reverse or rotate one of the structures. For example, structures 1 and 2 in the figure below may seem different, but rotating structure 2 by 180° shows that it is the same as structure 1. The two represent the same compound: a four-carbon main chain with a one-carbon side chain connected to the carbon atom next to a terminal carbon.

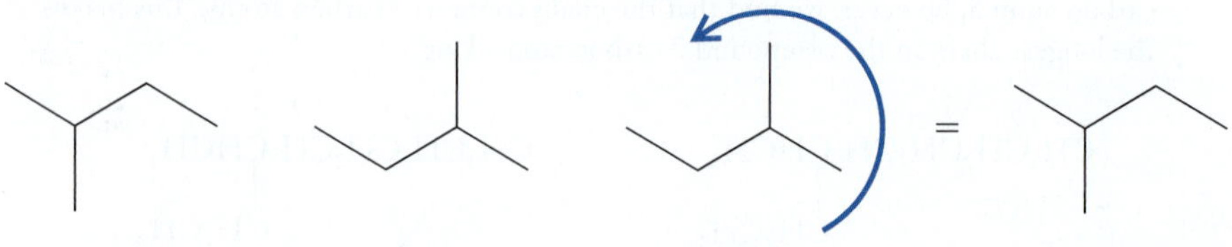

(1) (2) 180° rotation = After rotation

## SAMPLE EXERCISE 19.3 Recognizing Constitutional Isomers LO2

Do the two structures in each set describe the same compound, a pair of constitutional isomers, or two compounds with different molecular formulas?

a. $(CH_3)_2CHCH_2CH(CH_3)_2$

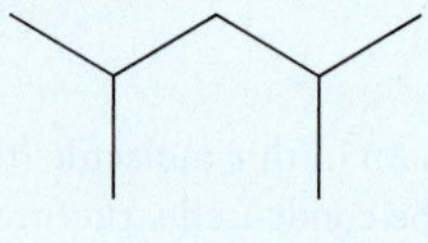

b.

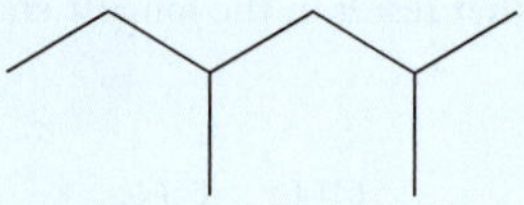

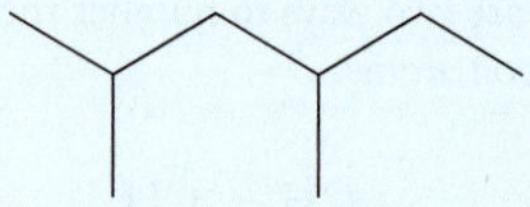

c.

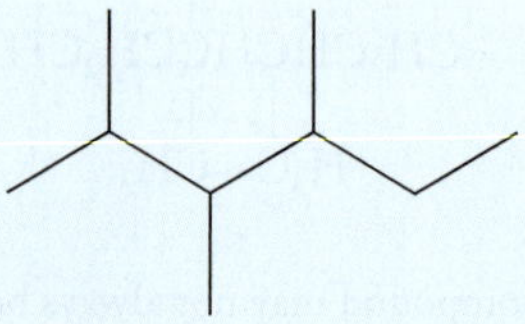

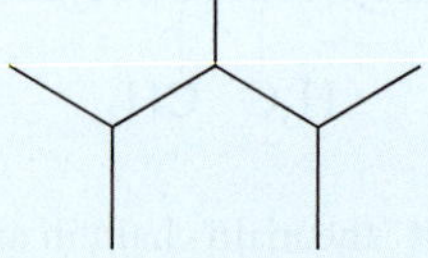

d. 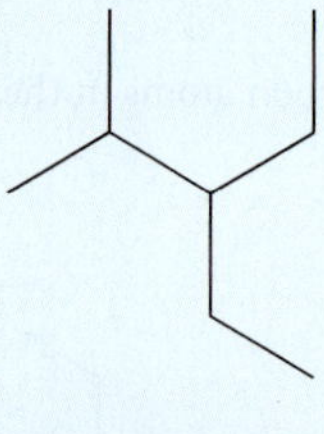 $CH_3CH_2C(CH_3)_2CH(CH_3)_2$

**Collect and Organize** Identical compounds have the same molecular formula and the same connectivity and spatial arrangement of the atoms. Constitutional isomers have the same molecular formula but different connectivity. Compounds with different molecular formulas are different compounds.

**Analyze** In each pair, we check first to see whether the molecular formulas are the same. That means we count the number of carbon and hydrogen atoms. If the molecular formulas are the same, we may have one compound drawn two ways or a pair of constitutional isomers. If the carbon skeletons are the same in any pair of drawings, the drawings represent the same hydrocarbon.

**Solve**

a. The molecular formulas are the same: $C_7H_{16}$. Converting the condensed structure to a carbon-skeleton structure gives us

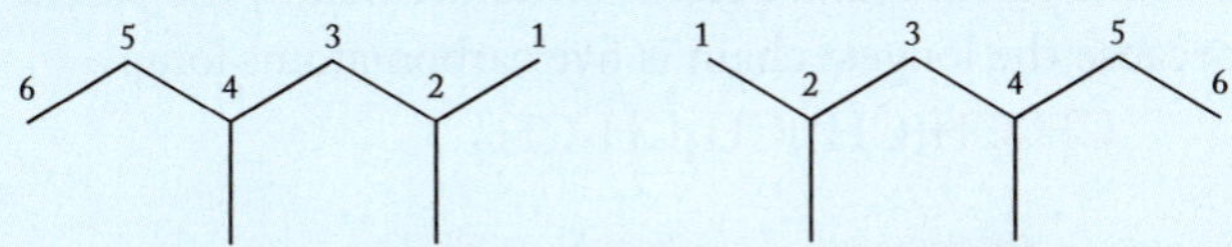


This structure is identical to the carbon-skeleton structure in this set. Therefore, the condensed structure and carbon-skeleton structure represent the same molecule.

b. Both molecules in this set have the molecular formula $C_8H_{18}$. Both have six carbon atoms in the longest chain. Both have a methyl group attached to the second carbon atom from one end of the chain and another methyl group attached to the third carbon atom from the other end. Therefore, the two structures represent the same compound:

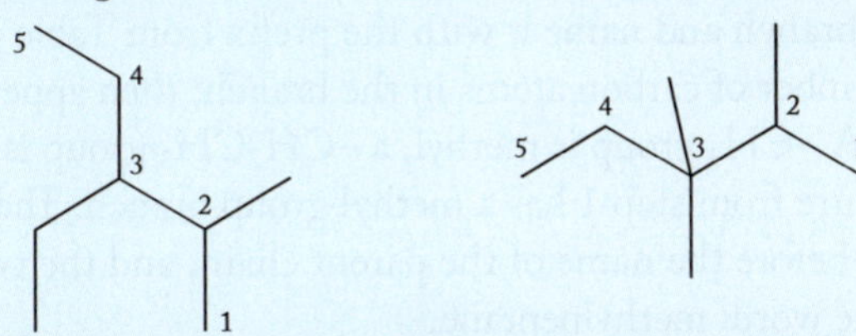

c. The left structure contains nine carbon atoms, but the right structure contains only eight. Therefore, the two structures represent two compounds.

d. The molecular formulas are the same: $C_8H_{18}$. Rotating the carbon-skeleton structure 180° and converting the condensed structure to a carbon-skeleton structure gives us

Both structures have a longest chain of five carbon atoms and a methyl group attached to C2 (the number 2 carbon atom). The structure on the left, however, has an ethyl group ($-CH_3CH_2$) on C3, whereas the structure on the right has two methyl groups ($-CH_3$). Thus, the two structures represent a pair of constitutional isomers.

**Think About It** Even if two compounds have the same molecular formula, they are not necessarily identical. We must determine whether they are constitutional isomers by comparing the connections between atoms.

**Practice Exercise** Do the two structures in each set describe the same compound, two compounds, or a pair of constitutional isomers?

a.

b.

c.

## Naming Alkanes

Now that we know how to draw alkanes, let's look at a few simple rules for naming them. The nomenclature system follows the same pattern for all families of organic compounds, so we begin with rules for naming alkanes and will add a few more for other hydrocarbons. These rules are called IUPAC rules after the organization (the International Union of Pure and Applied Chemistry) that defined them. Appendix 7 contains a more extensive treatment of nomenclature for organic compounds.

The names of the $C_1$ through $C_{10}$ alkanes, in which all carbon atoms are bonded to no more than two other carbon atoms, start with the prefixes listed in **Table 19.3** followed by *-ane*. However, we have seen that straight-chain alkanes have constitutional isomers in which at least one carbon atom is bonded to more than two others. We use the following five steps to name these branched alkanes:

**TABLE 19.3 Prefixes for Naming Alkanes and Other Hydrocarbons**

| Prefix | Condensed Structure of Alkane | Name of Alkane |
|---|---|---|
| Meth- | $CH_4$ | Methane |
| Eth- | $CH_3CH_3$ | Ethane |
| Prop- | $CH_3CH_2CH_3$ | Propane |
| But- | $CH_3(CH_2)_2CH_3$ | Butane |
| Pent- | $CH_3(CH_2)_3CH_3$ | Pentane |
| Hex- | $CH_3(CH_2)_4CH_3$ | Hexane |
| Hept- | $CH_3(CH_2)_5CH_3$ | Heptane |
| Oct- | $CH_3(CH_2)_6CH_3$ | Octane |
| Non- | $CH_3(CH_2)_7CH_3$ | Nonane |
| Dec- | $CH_3(CH_2)_8CH_3$ | Decane |

1. Select the longest chain of carbon atoms and use the prefix in Table 19.3 to name this as the *parent chain*. For the structure below, the parent chain is pentane because the longest chain is five carbon atoms long:

   $CH_3CH(CH_3)CH_2CH_2CH_3$

   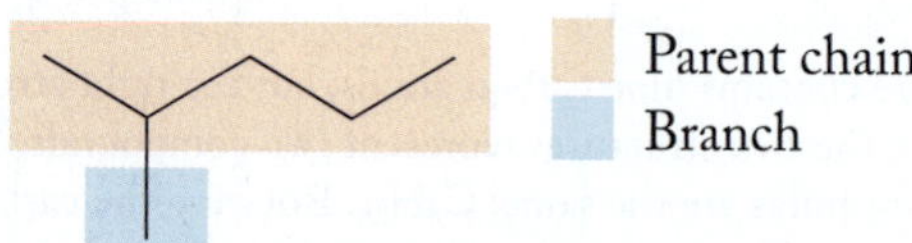
   

2. Identify each branch and name it with the prefix from Table 19.3 that defines the number of carbon atoms in the branch; then append the suffix *-yl* to the prefix. A $-CH_3$ group is methyl, a $-CH_2CH_3$ group is ethyl, and so on. The structure from step 1 has a methyl group branch. The name of the branch comes before the name of the parent chain, and the two are written together as one word: methylpentane.
3. To indicate where along the parent chain the branch is attached, we number the carbon atoms in the parent chain so that the branch (or branches, if there are more than one) has the lowest possible number. In the molecule from step 1, we start numbering from the left so that the methyl group is on carbon atom 2 in the parent chain (instead of on C4):

   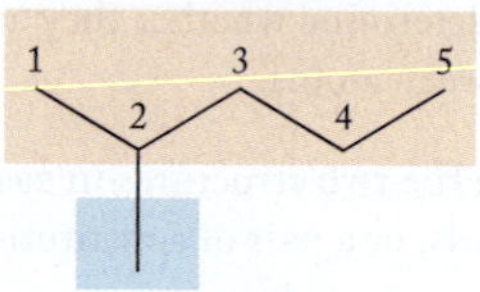
   

   We indicate the position of the branch by a 2 followed by a hyphen in front of the name: 2-methylpentane.
4. If the same group is attached more than once to the parent chain, we use the prefixes *di-*, *tri-*, and *tetra-* to indicate the number of groups present: 2, 3, and 4, respectively. (In this text, we will not need to name any complexes having more than four of the same group in a molecule.) The position of each group is indicated by the appropriate number before the group name, with the numbers separated by commas. Thus, the name of the following structure is 2,4-dimethylpentane:

   $\underset{1}{CH_3}\underset{2}{CH}(CH_3)\underset{3}{CH_2}\underset{4}{CH}(CH_3)\underset{5}{CH_3}$

   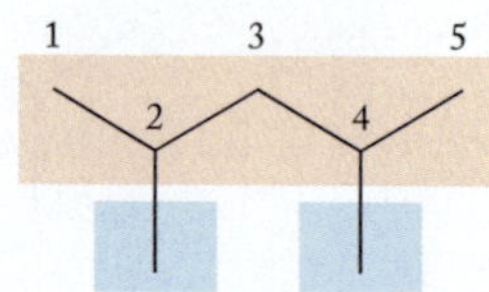
   

5. If different groups are attached to a parent chain, they are named in alphabetical order (for example, *e*thyl before *m*ethyl), as in 4-ethyl-3-methylheptane:

$CH_3CH_2CH(CH_3)CH(CH_2CH_3)CH_2CH_2CH_3$

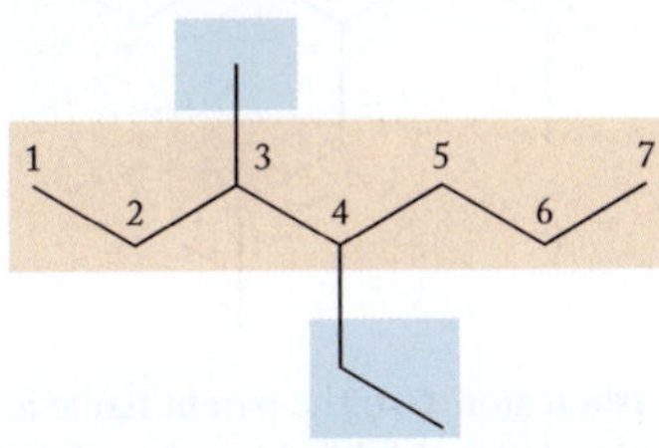


Numerical prefixes such as *di* and *tri* are not considered when determining alphabetical order. Thus, "ethyl" would precede "dimethyl" in the name of an alkane with one ethyl and two methyl groups. Note that we begin numbering the parent chain in 4-ethyl-3-methylheptane at the left end, as described in step 3, so that the first branch location has the lowest possible number. For example, the following compound is 5-ethyl-2-methyloctane, not 4-ethyl-7-methyloctane:

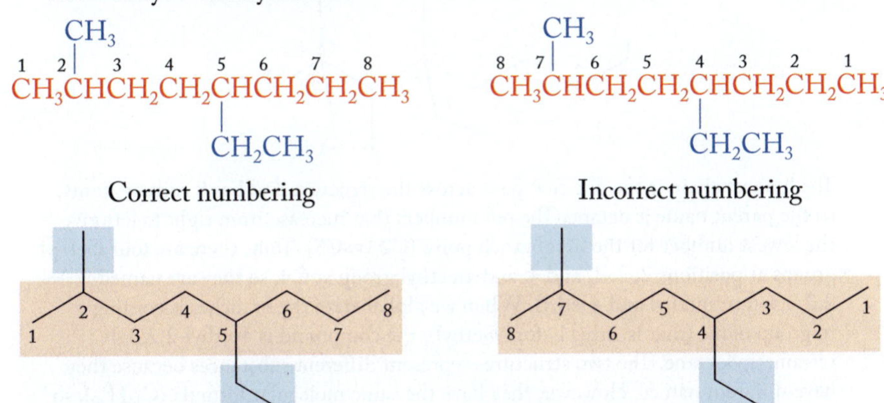


**STEPWISE ANIMATION**
Naming Branched Alkanes

In practice, two or more paths of equal length may compete to be selected as the parent chain. We select the path in step 1 that (a) has the greatest number of side chains and (b) results in the lowest possible numbers for any branches as described in step 3. Therefore, we may need to do these two steps together and not in the order in which they are presented here.

**SAMPLE EXERCISE 19.4** Naming Branched-Chain Alkanes **LO2**

Name the following branched-chain alkanes:

a. $CH_3CH_2C(CH_3)_2CH_2CH(CH_2CH_2CH_3)CH_2CH_2CH_2CH_3$
b. $CH_3CH_2CH_2CH_2C(CH_3)(CH_2CH_3)CH(CH_3)C(CH_3)_3$

Are they identical compounds, constitutional isomers, or compounds with different molecular formulas?

**Collect, Organize, and Analyze** We are given two condensed structures. Identical compounds have identical formulas and identical names, constitutional isomers have identical formulas but different names, and different compounds have different molecular formulas and different names. To simplify naming the compounds, we should redraw them as carbon-skeleton structures and then apply the rules for naming.

**Solve**

a. The carbon-skeleton structure for the molecule is

The longest chain has 9 carbon atoms, so the parent name is nonane. If we begin numbering from the left, the chain with black numbers has three branches located at carbon atoms numbered 3, 3, and 5, whereas numbering from the right with red numbers results in branches at carbon atoms 5, 8, and 8. Because the lowest black number is lower than the lowest red number, we use the black numbering scheme. The branches are two methyl groups on C3 and one propyl group on C5, so they are named 3,3-dimethyl and 5-propyl. When we alphabetize the branches according to group name (that is, *m*ethyl before *p*ropyl), the compound is 3,3-dimethyl-5-propylnonane.

b. The carbon-skeleton structure for this molecule is

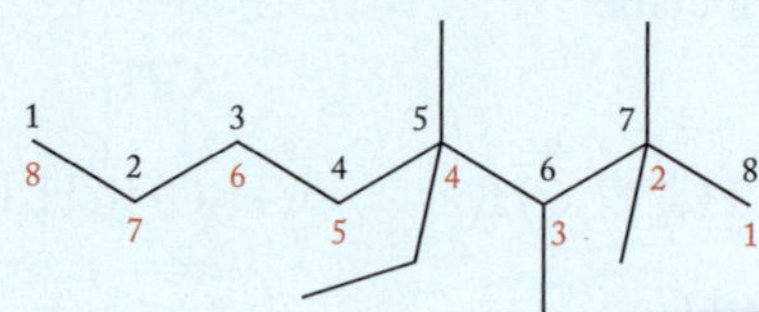


The longest chain is the one that goes across the structure and has 8 carbon atoms, so the parent name is octane. The red numbers that increase from right to left give the lowest number for the first branch point (C2 vs. C5). Thus, there are four methyl groups at positions 2, 2, 3, and 4, and an ethyl group at C4, so they are named 2,2,3,4-tetramethyl and 4-ethyl. When we alphabetize the branches according to group name (that is, *e*thyl before *m*ethyl), the compound is 4-ethyl-2,2,3,4-tetramethyloctane. The two structures represent different substances because they have different names. However, they have the same molecular formula ($C_{14}H_{30}$), so they are constitutional isomers.

**Think About It** Although being able to name molecules correctly is an important skill for chemists, it is most useful for us in this course to simply understand what the names convey in terms of identifying and distinguishing specific substances.

**Practice Exercise** Name both compounds and determine whether they are unrelated compounds or constitutional isomers.

a. $CH_3CH_2CH_2CH(CH_2CH_3)CH(CH_2CH_3)CH(CH_3)CH_2CH_3$
b. $(CH_3)_2CHCH_2CH(CH_3)CH(CH_3)CH_2CH(CH_3)_2$

## Cycloalkanes

Alkanes can form ring structures (**Figure 19.5**). These **cycloalkanes** have the general formula $C_nH_{2n}$, which is different from the general formula for straight-chain alkanes ($C_nH_{2n+2}$) because a cycloalkane has one more carbon–carbon bond and two fewer hydrogen atoms per molecule than a linear or branched alkane with the same number of carbon atoms. One mole of a cycloalkane can, at least in theory, react with one mole of hydrogen to make one mole of a straight-chain alkane. In practice, however, only cycloalkanes with three-carbon and four-carbon rings react with hydrogen.

**cycloalkane** a ring-containing alkane with the general formula $C_nH_{2n}$.

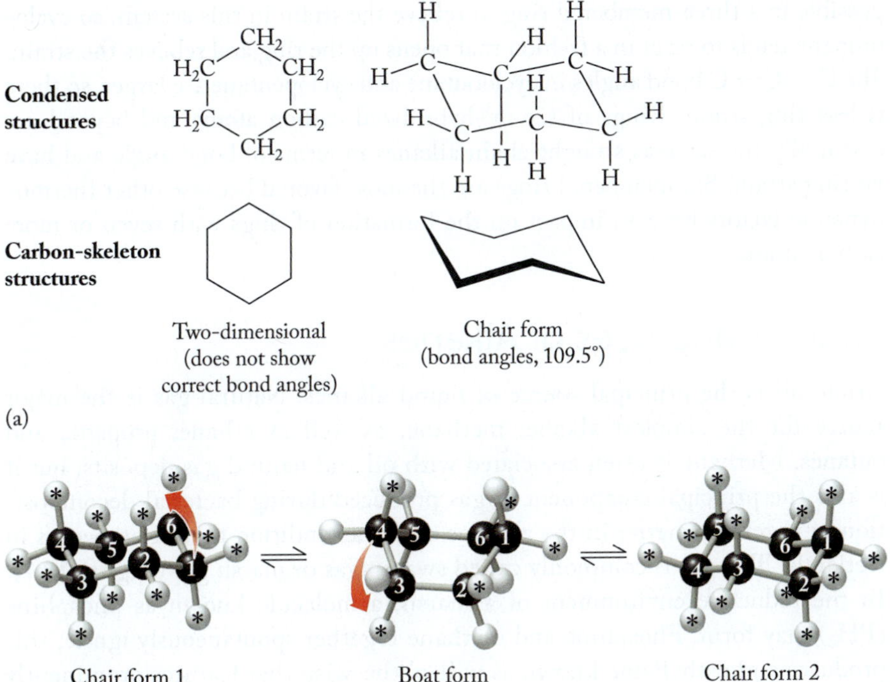


**FIGURE 19.5** (a) The six-membered ring of cyclohexane may be drawn either flat or in styles that show its actual three-dimensional puckered conformations. (b) The most stable are two chair-shaped conformations shown in the left and right structures. A molecule can flip back and forth between them, but it has to temporarily adopt less stable shapes, such as the boat conformation shown in the middle structure, to do so.

**CHEMTOUR**
Structure of Cyclohexane

The left column of Figure 19.5(a) shows the condensed structure and carbon-skeleton structure of cyclohexane drawn in two dimensions, where the C—C—C bond angles appear to be 120°. Because the carbon atoms have $sp^3$ hybrid orbitals, we expect the angles to be 109.5°, and indeed they are in this molecule. A more accurate representation of the ring is shown in the adjacent structures, in which the ring is puckered instead of flat and the bond angles are 109.5°.

There are two possible conformations of cyclohexane that have the greatest structural stability. They are called *chair* forms, and a molecule of cyclohexane spends more than 99% of its time in one chair form or the other. It can also flip back and forth between the two chair conformations, as shown in Figure 19.5(b). However, to get from one chair conformation to the other, the molecule must pass through higher-energy transition-state conformations. One of these high-energy states is called the *boat* conformation, in which repulsion between the two hydrogen atoms across the ring from each other adds to the internal energy and reduces stability. Note how the H atoms in the left chair structure that are marked with an asterisk are above and below the ring (these are called *axial* positions), but these same H atoms are in positions more parallel to the ring (called *equatorial* positions) when the molecule flips to the chair structure on the right.

Cycloalkanes with fewer than six carbon atoms are possible, but they are not as stable as cyclohexane. One of the important features that affects reactivity is the C—C—C bond angle. For example, cyclopropane (a $C_3$ ring) exists but is a highly reactive species because the interior ring angles are 60°, far from the ideal bond angle of 109.5° for an $sp^3$-hybridized carbon atom (**Figure 19.6**). No puckering is

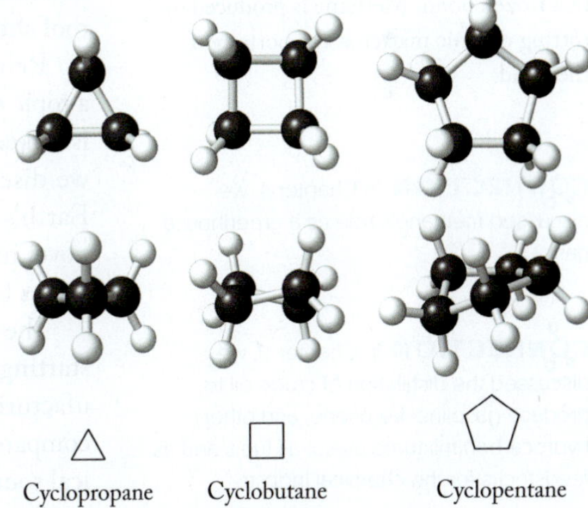


**FIGURE 19.6** Cyclic alkanes have the general formula $C_nH_{2n}$ and are considered unsaturated hydrocarbons. Shown here are cyclopropane ($n$ = 3), cyclobutane ($n$ = 4), and cyclopentane ($n$ = 5). The side views of the ball-and-stick models show that the carbon atoms in a molecule of cyclopropane are in the same plane, but carbon atoms in molecules of cyclobutane and cyclopentane are not.

possible in a three-membered ring to relieve the strain in this system, so cyclopropane tends to react in a fashion that opens up the ring and relieves the strain. The C—C—C bond angles in cyclobutane and cyclopentane are larger, so there is less ring strain. Rings of six $sp^3$-hybridized carbon atoms and beyond are essentially the same as straight-chain alkanes in terms of bond angle and have no ring strain. Six-membered rings are the most favored because other thermodynamic factors have an impact on the formation of rings with seven or more carbon atoms.

## Sources and Uses of Alkanes

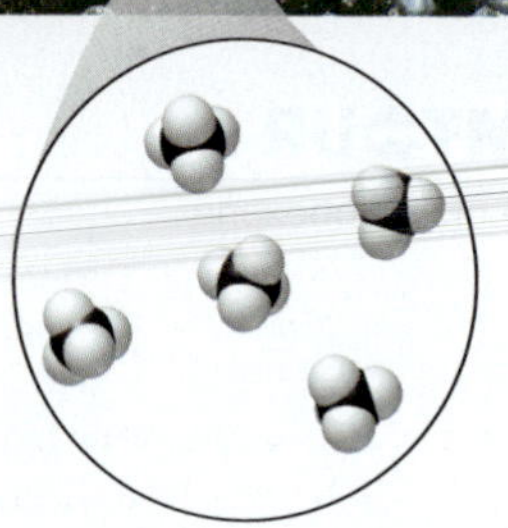

**FIGURE 19.7** Methane bubbles trapped in a frozen pond. Methane is produced by rotting organic matter at the bottom of the pond.

**CONNECTION** In Chapter 4, we discussed methane's role as a greenhouse gas.

**CONNECTION** In Chapter 11, we discussed the distillation of crude oil to produce gasoline, kerosene, and other hydrocarbon mixtures useful as fuels and as feedstocks for the chemical industry.

**fracking** the process of injecting liquids at high pressures into rock formations to force open existing fissures in the formations for the purpose of extracting oil or gas.

Crude oil is the principal source of liquid alkanes. Natural gas is the major source for the simplest alkane, methane, as well as ethane, propane, and butanes. Methane is often associated with oil and natural gas deposits, but it is also the principal component of gas produced during bacterial decomposition of vegetable matter in the absence of air, a condition that often occurs in wetlands. This gas is commonly called swamp gas or marsh gas (**Figure 19.7**). In the reductive environment of a marsh, a molecule known as phosphine ($PH_3$) may form. Phosphine and methane together spontaneously ignite; this produces a ghostly flame known as will-o'-the-wisp that features prominently in some legends and gothic mysteries. Methane can also be produced in coal mines, where it can be especially dangerous because it creates explosive mixtures with humid air, giving rise to another common name for the gas: firedamp.

By far, the most common use of alkanes is as fuels. Combustion reactions between alkanes and oxygen provide heat to power vehicles, generate electricity, warm our homes, and prepare our meals. Gasoline, kerosene, and diesel fuels are mostly mixtures of alkanes. Alkanes with higher boiling points are viscous liquids and used as lubricating oils. Low-melting-point solid alkanes ($C_{20}$–$C_{40}$) are used in candles and in manufacturing matches. Very heavy hydrocarbon gums and solid residues ($C_{36}$ and up) are used as asphalt for paving roads and making roof shingles.

Reliance on fossil fuels to meet the energy needs of industrialized countries is a topic of considerable debate. Release of fossil carbon as $CO_2$, a greenhouse gas, is linked to a recent increase in the concentration of the gas in the atmosphere, as we discussed in Chapter 7 (see Figure 7.5), and to climate change. Moreover, Earth's supplies of fossil fuels are inevitably limited, although widespread use of novel recovery methods such as hydraulic fracturing, or **fracking,** has increased access to them.

The alkanes found in natural gas and distilled from crude oil also serve as the starting material for synthesizing more complex molecules in the chemical manufacturing industry. However, this reliance on oil and gas is changing as more companies develop green chemistry processes that begin with renewable, biological sources of starting materials.

Some alkanes have medicinal use. Mineral oil is a mixture of $C_{15}$–$C_{24}$ alkanes used as a skin ointment ("baby oil") to treat diaper rash and to alleviate some forms of eczema. The oil is used in many cosmetics, creams, and ointments. Taken orally, mineral oil acts as a laxative. As we continue our study of organic compounds, we will return to the themes of fuels, pharmaceuticals, and materials at the bulk level and how composition, structure, and size at the microscopic level determine physical and chemical properties.

## 19.3 Alkenes and Alkynes

In Section 19.2, we focused on alkanes, whose molecules contain only single bonds. In this section, we consider the structures and properties of the two other major categories of hydrocarbons: **alkenes,** whose molecules have one or more carbon–carbon double bonds, and **alkynes,** which have one or more carbon–carbon triple bonds (**Figure 19.8**).

Alkenes and alkynes are unsaturated hydrocarbons that can combine with $H_2$ to form alkanes in a process called **hydrogenation.** A molecule with one C=C bond is described as having one *degree of unsaturation*. It combines with one molecule of $H_2$ to form one molecule of alkane (**Figure 19.9**), whereas a molecule with one C≡C bond has *two* degrees of unsaturation and combines with two molecules of $H_2$. A molecule with one double bond and one triple bond has three degrees of unsaturation, so it needs three molecules of $H_2$ to form one molecule of alkane.

Alkene | Alkyne

(a)

(b)

$H_2C{=}CHCH_2CH_2CH_3$ | $HC{\equiv}CCH_2CH_2CH_3$

$H_2C{=}CH(CH_2)_2CH_3$ | $HC{\equiv}C(CH_2)_2CH_3$

(c)

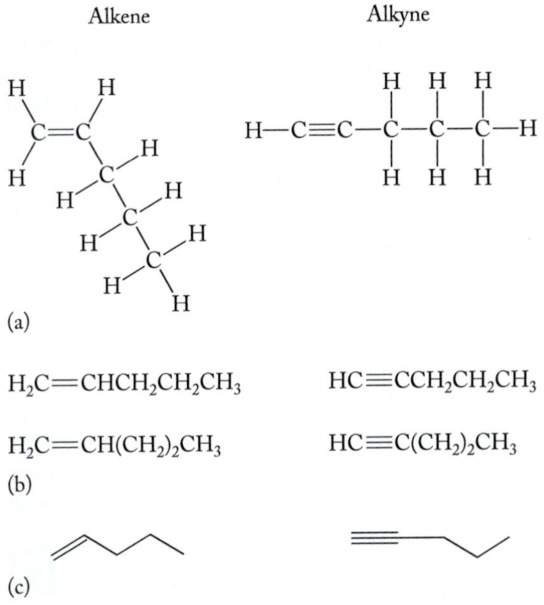

**FIGURE 19.8** (a) Lewis (or Kekulé) structures, (b) condensed structures, and (c) carbon-skeleton structures of a $C_5$ alkene and a $C_5$ alkyne.

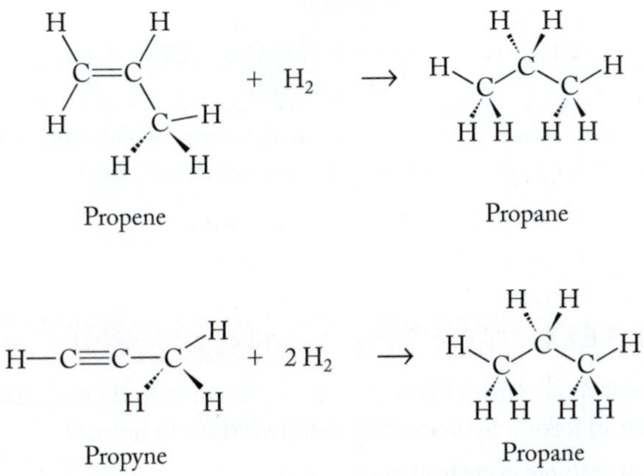

Propene + $H_2$ → Propane

Propyne + $2\,H_2$ → Propane

**FIGURE 19.9** The hydrogenation of propene requires one mole of $H_2$ per mole of propene, whereas the hydrogenation of propyne requires two moles of $H_2$ per mole of propyne.

**CONNECTION** In Chapter 5, we used valence bond theory and hybrid orbitals to model a double bond as one sigma bond and one pi bond, whereas a triple bond forms due to one sigma bond and two pi bonds.

### SAMPLE EXERCISE 19.5 Distinguishing among Alkanes, Alkenes, and Alkynes — LO3

Suppose three gas tanks are filled to the same pressure with propane, propene, and propyne and stored in a basement stockroom. Torrential rain from a hurricane floods the stockroom, and the labels on the tanks wash off. Design an experiment based on pressure measurements and the hydrocarbons' reactivities with $H_2$ gas to determine which tank contains propane, which contains propene, and which contains propyne.

**Collect, Organize, and Analyze** We need to evaluate the capacity of three hydrocarbons to combine with hydrogen to distinguish propane, propene, and propyne from one another. Each propene molecule contains a C=C double bond that should react with one molecule of $H_2$, and each propyne molecule contains a C≡C triple bond that should react with two molecules of $H_2$. Propane is saturated and should not react with hydrogen.

**alkene** a hydrocarbon containing one or more carbon–carbon double bonds.

**alkyne** a hydrocarbon containing one or more carbon–carbon triple bonds.

**hydrogenation** a chemical reaction between molecular hydrogen ($H_2$) and another substance.

**Solve** A gas-tight syringe is used to transfer 1.0 millimole of each hydrocarbon (25 mL at $T = 25°C$ and $P = 1.0$ bar) into a 50 mL rigid reaction vessel already filled 2.0 millimoles of $H_2$ gas. The pressure inside each reaction vessel is used to monitor the progress of each reaction. When all the vessel pressures no longer change, the following results will identify which tank contains which hydrocarbon:

1.0 mmol propane + 2.0 mmol $H_2 \xrightarrow{\text{no reaction}}$
(no change in pressure)

1.0 mmol propene + 2 mmol $H_2 \rightarrow$ 1.0 mmol propane + 1.0 mmol $H_2$ left over
(pressure decreases by 1/3)

1.0 mmol propyne + 2.0 mmol $H_2 \rightarrow$ 1.0 mmol propane
(pressure decreases by 2/3)

**Think About It** Determining the quantity of hydrogen that reacts with a hydrocarbon is a way to distinguish saturated from unsaturated hydrocarbons and to determine their degree of unsaturation.

**Practice Exercise** The labels have fallen off two containers in the lab. One label has structure A printed on it, and the other has structure B printed on it, as shown below:

$$CH_3{-}CH_2{-}CH{=}CH{-}CH{=}CH{-}CH_3$$
Compound A

$$CH_3{-}CH{=}CH{-}CH{=}CH{-}CH{=}CH_2$$
Compound B

Describe how to use hydrogenation reactions to determine which label belongs on which container. What is the structure of the product in each reaction?

## CONCEPT TEST

Could we use information from hydrogenation reactions to distinguish a hydrocarbon with one triple bond from a hydrocarbon with two double bonds?

*(Answers to Concept Tests are in the back of the book.)*

Simple alkenes occur at trace concentrations in crude oil, but alkenes with more complex molecular structures are abundant in plant products, including pine oil, oil of celery, and oil of ginger (**Figure 19.10**). Alkynes are also rare

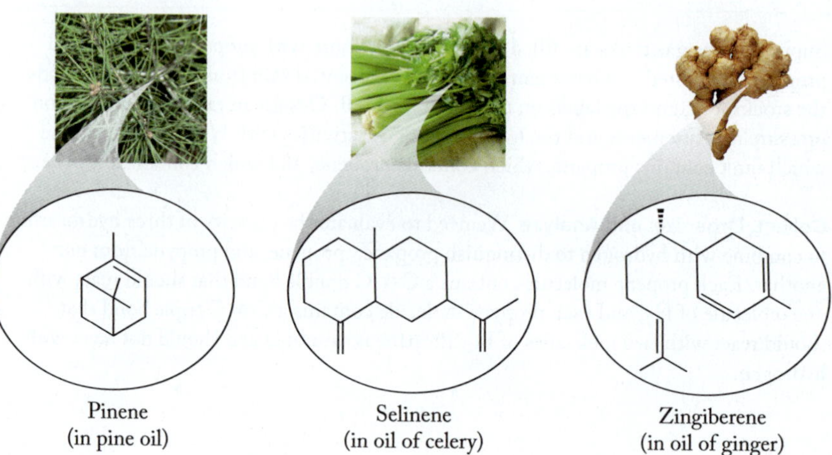


**FIGURE 19.10** Some naturally occurring alkenes.

ingredients in crude oil, but some with value as pharmaceuticals or natural pesticides are produced by plants (**Figure 19.11**).

The simplest alkyne is ethyne, $C_2H_2$, which is better known by its common name, acetylene. It is the fuel used in the high-temperature torches used to cut through steel and other metals. Like other alkynes, acetylene may be synthesized by the partial oxidation of an alkane. For example, controlled oxidation of methane produces two fuels: acetylene and hydrogen gas.

$$6\ CH_4(g) + O_2(g) \rightarrow 2\ HC{\equiv}CH(g) + 2\ CO(g) + 10\ H_2(g) \quad (19.1)$$

This reaction illustrates the reduced state of the carbon atoms in methane and all alkanes. For another example, consider the progression of oxidation states of carbon from ethane to ethene to ethyne. The oxidation number of the carbon atoms in $C_2H_6$ (where each H is assigned an O.N. of +1) is −3. In $C_2H_4$, the O.N. of carbon is −2, and in $C_2H_2$ it's −1. Their capacity to be reduced makes the $sp^2$ carbon atoms in C=C double bonds and the $sp$ carbon atoms in C≡C triple bonds more reactive than the $sp^3$ carbon atoms in alkanes. Actually, the carbon–carbon double bonds in alkenes are among the most versatile functional groups in all of organic chemistry.

Alkenes and alkynes, like alkanes, can be arranged in homologous series with predictable patterns in their physical properties. For example, the melting and boiling points of these compounds (**Table 19.4**) increase with increasing molar mass, just as with the alkanes.

Alkene and alkyne molecules may contain more than one C=C or C≡C bond, as we saw in Figures 19.10 and 19.11. In particular, many molecules produced by living systems have several double

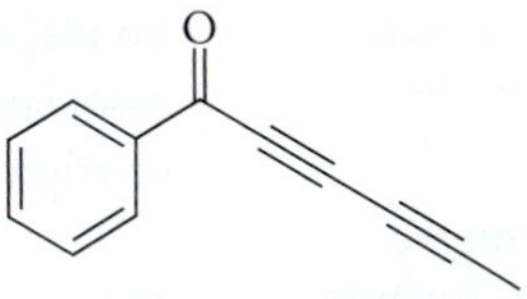

(a) Capillin

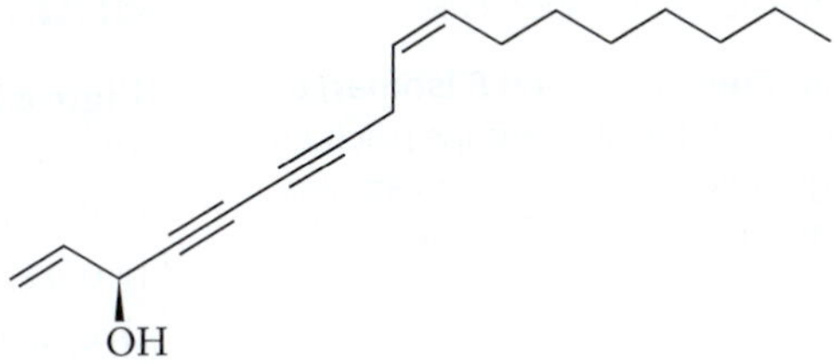

(b) Falcarinol

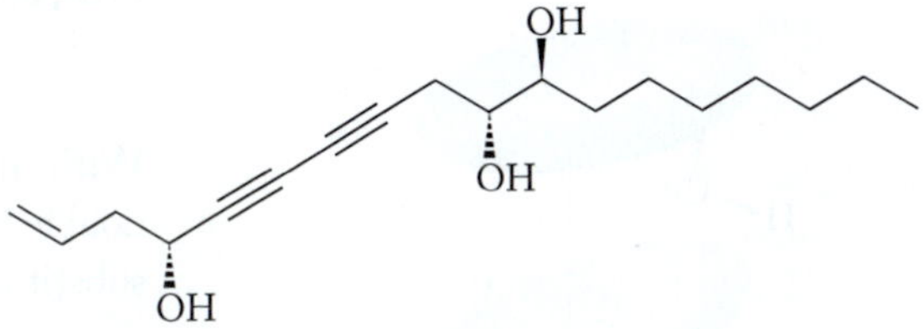

(c) Panaxytriol

**FIGURE 19.11** (a) Capillin is an antifungal produced by Oriental wormwood plants. (b) Falcarinol from carrots and English ivy has antifungal and antitumor activity. (c) Panaxatriol is a potent antitumor drug isolated from ginseng.

**TABLE 19.4 Melting Points and Normal Boiling Points of Homologous Series of Alkenes and Alkynes**

| Condensed Structure | Melting Point (K)[a] | Normal Boiling Point (K) |
|---|---|---|
| $H_2C{=}CHCH_3$ | 88 | 226 |
| $H_2C{=}CHCH_2CH_3$ | 88 | 267 |
| $H_2C{=}CH(CH_2)_2CH_3$ | 135 | 303 |
| $H_2C{=}CH(CH_2)_3CH_3$ | 133 | 336 |
| $H_2C{=}CH(CH_2)_4CH_3$ | 154 | 367 |
| $H_2C{=}CH(CH_2)_5CH_3$ | 169 | 396 |
| $H_2C{=}CH(CH_2)_6CH_3$ | 192 | 419 |
| $H_2C{=}CH(CH_2)_7CH_3$ | 207 | 444 |
| $HC{\equiv}CCH_3$ | 171 | 250 |
| $HC{\equiv}CCH_2CH_3$ | 147 | 281 |
| $HC{\equiv}C(CH_2)_2CH_3$ | 168 | 313 |
| $HC{\equiv}C(CH_2)_3CH_3$ | 141 | 344 |
| $HC{\equiv}C(CH_2)_4CH_3$ | 192 | 373 |

[a]Melting points increase with molar mass but also depend on how molecules fit into crystal lattices. Melting points of alkenes with even numbers of carbon atoms form one series that follows this trend; alkenes with odd numbers of carbon atoms form another series.

**addition reaction** a reaction in which two molecules couple to form one product.

**cis isomer** (also called ***Z* isomer**) a molecule with two like groups (such as two R groups or two hydrogen atoms) on the same side of the molecule.

**trans isomer** (also called ***E* isomer**) a molecule with two like groups (such as two R groups or two hydrogen atoms) on opposite sides of the molecule.

bonds, as we will see in Chapter 20. Here, we will concentrate on the properties associated with small molecules containing only one or a small number of double or triple bonds.

## Chemical Reactivities of Alkenes and Alkynes

**Figure 19.12** shows the electron distributions in the π bonds in an alkene and an alkyne. Electron density in these bonds is greatest around the bonding axis, which makes π electrons more accessible to reactants than the electrons in the σ bonds between carbon atoms. For this reason, unsaturated hydrocarbons are more reactive than saturated ones.

The different reactivities of alkanes and alkenes with hydrogen halides (HCl, HBr, and HI) illustrate this difference. With an alkane, there is no reaction:

$$HX(g) + H_3C{-}CH_3(g) \rightarrow \text{no reaction} \qquad (19.2)$$

With an alkene, however, one molecule of hydrogen halide reacts with each double bond to produce an alkyl halide (an alkane in which a halogen has been substituted for one of the hydrogen atoms):

$$\underset{\text{Hydrogen halide}}{HX(g)} + \underset{\text{Alkene}}{H_2C{=}CH_2(g)} \rightarrow \underset{\text{Alkyl halide}}{CH_3CH_2X} \qquad (19.3)$$

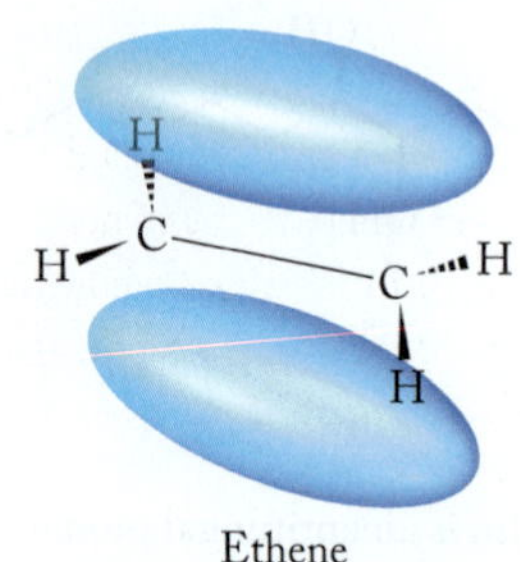


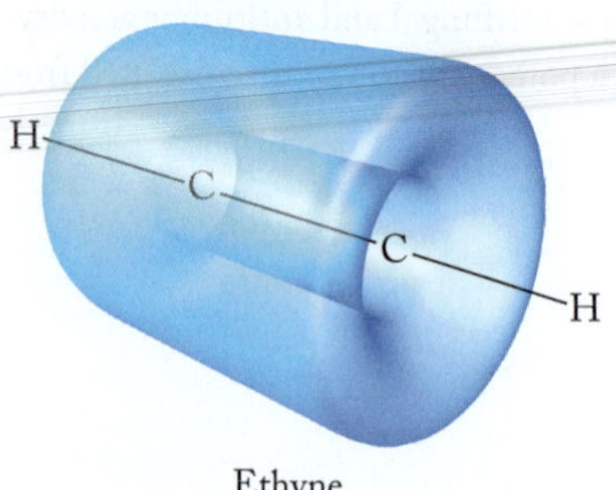


**FIGURE 19.12** Shapes of the π bonds in ethene and ethyne.

(At room temperature and 1 atm of pressure, $CH_3CH_2X$ is a liquid when X = Br or I, and it is a gas when X = F or Cl.) This reaction is called an **addition reaction** because two reactants combine (that is, they add together) to form one product.

Alkynes react with *two* molecules of a hydrogen halide to produce an alkane bearing two halogen atoms:

$$2\,HBr(g) + HC{\equiv}CH(g) \rightarrow CH_3CHBr_2(\ell) \qquad (19.4)$$

Note how the reactivity patterns and the mole ratios of these addition reactions match the patterns and stoichiometry of the hydrogenation reactions we discussed at the beginning of this section. Recall that alkanes don't react with $H_2$, but each C=C double bond in an alkene can combine with one molecule of $H_2$, and each C≡C triple bond in an alkyne can combine with two molecules of $H_2$. In other words, the degree of unsaturation of an alkene or alkyne is directly linked to its capacity to engage in other addition reactions. Because both double and triple bonds react with many reagents in addition to hydrogen halides, alkenes and alkynes are useful substances in the industrial production of other compounds.

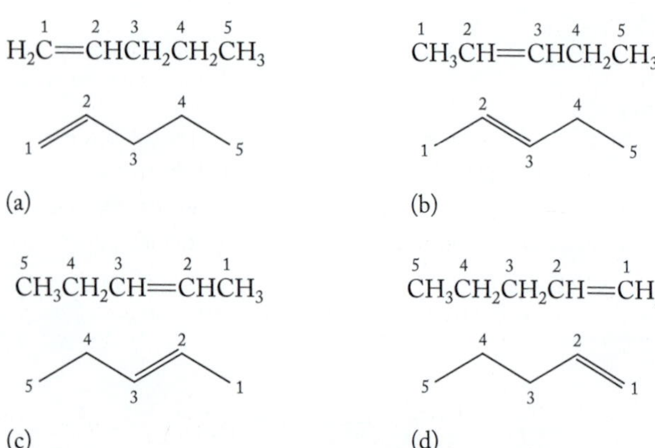


**FIGURE 19.13** Four possible constitutional isomers of pentene, $C_5H_{10}$. Notice that structures (a) and (d) are identical, as are (b) and (c).

## Isomers of Alkenes and Alkynes

Molecules that contain alkene and alkyne functional groups can have straight or branched chains with the same types of constitutional isomers we saw with alkanes. In addition, the location of the double or triple bond in a molecule can distinguish one constitutional isomer from another. For example, **Figure 19.13** shows four possible structures for straight-chain isomers of pentene, the alkene that contains five carbon atoms and one double bond. If we apply the test used in Section 19.2 for alkanes, we find that structures (a) and (d) in Figure 19.13

both represent the same molecule. This is easier to see if we look at the carbon-skeleton structures in the figure. In both cases, the double bond is between C1 and C2. Recall with branched-chain alkanes that we number the carbon atoms from the end that assigns the lowest number to the carbon attached to the first branch. The same rule applies to functional groups like the double bond shown here. Thus, the carbon atoms in structure (d) must be numbered from right to left to give the double bond the lowest possible numbers.

By similar reasoning, structures (b) and (c) in Figure 19.13 both represent the same molecule, which is a constitutional isomer of the molecule represented by (a) and (d): the two compounds have the same chemical formula, but their double bonds are in a different location. Drawing the carbon-skeleton structure of (b) or (c), however, presents us with a new challenge. After we draw the first three atoms of structure (b), as in **Figure 19.14(a)**, and draw a straight dashed line through the double bond, we see that we have two options for how to orient the rest of the molecule with respect to the double bond. We can place the bond between C3 and C4 on the same side as the methyl group at C1 (**Figure 19.14b**) or on the opposite side (**Figure 19.14c**).

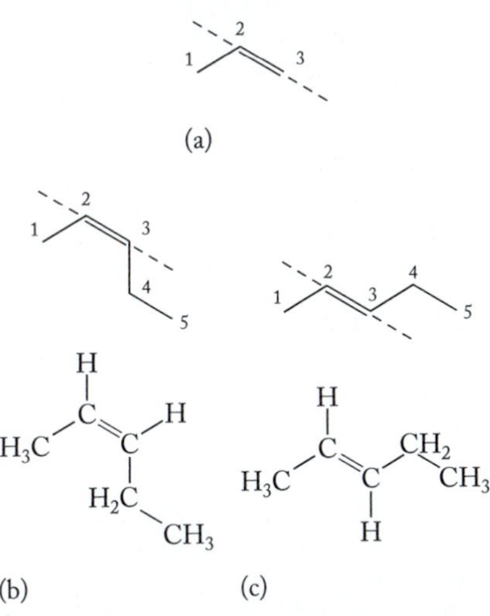


**FIGURE 19.14** (a) The first three atoms of a carbon chain with a double bond between C2 and C3. (b) The chain continues on the same side of the double bond as C1 in the cis isomer. (c) The chain continues on the opposite side of the double bond as C1 in the trans isomer.

These two molecules are isomers of each other because they have the same molecular formula but different structures and therefore different properties. The isomer in Figure 19.14(b) is called either the ***Z* isomer** (*Z* stands for the German word *zusammen*, or "together") or the **cis isomer** (*cis* is Latin for "on this side"), which here translates to "the methyl group and the chain after the double bond are both *together* or *on the same side* of the structure." The isomer in Figure 19.14(c) is called either the ***E* isomer** (*E* for *entgegen*, or "opposite") or **trans isomer** (*trans* is Latin for "across"). Because these isomers are characterized by differences in the spatial arrangement of their atoms, and not by how those atoms are connected, they are *stereoisomers*. The *cis* and *trans* prefixes are widely used in the names of alkenes with simple structures; the *E*/*Z* system is needed for more complex molecules in which more than two different substituents are attached to a double bond.

**CONNECTION** Stereoisomers were introduced in Chapter 5.

Cis/trans isomers exist because there is no free rotation about the double bond. Recall from Chapter 5 that a double bond is formed from the overlap of two unhybridized *p* orbitals on adjacent carbon atoms. As **Figure 19.15** shows, a carbon atom joined in a double bond cannot rotate freely about the bond axis without eliminating orbital overlap and breaking the bond. Breaking a π bond in one mole of an alkene costs about 290 kJ of energy, and that much energy is unavailable to molecules at room temperature. This situation gives rise to restricted rotation about a carbon–carbon double bond and to the existence of stereoisomers.

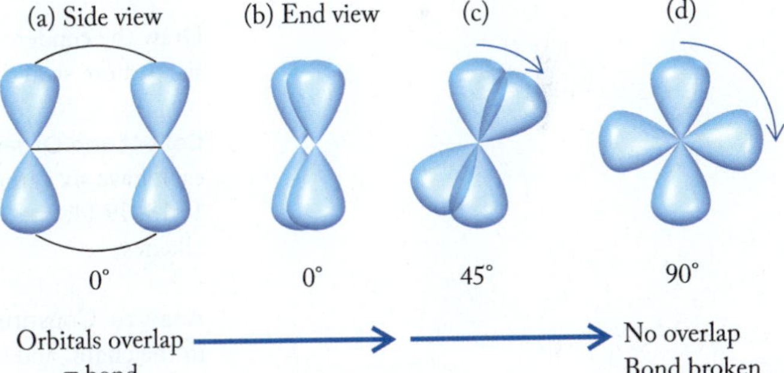


**FIGURE 19.15** (a) To form a π bond, $p_z$ orbitals overlap to establish a region of shared electron density above and below the plane of the carbon–carbon bond. (b) If you look down the bond axis, the orbitals line up. (c) If you rotate one carbon while keeping the other fixed, the orbitals are no longer parallel and do not overlap. (d) If you rotate one of the two bonded C atoms far enough, the π bond breaks.

Now let's examine the stereoisomers of structure (c) from Figure 19.14. In **Figure 19.16(a)**, the two hydrogen atoms are on the same side of the double bond; this is the cis isomer. In **Figure 19.16(b)**, the two hydrogen atoms are on opposite sides of the double bond; this is the trans isomer. Comparing the stereoisomers in Figures 19.14 and 19.16 shows that the two cis structures are identical and the two trans structures are identical. Therefore, the straight-chain alkenes with five carbon atoms exist as three isomers: structure (a) from Figure 19.13 plus the cis and trans isomers of structure (b). All three isomers are chemically distinct molecules.

**CONNECTION** In Chapter 10, we introduced rotations about single bonds as one of the types of motion molecules experience as part of their overall kinetic energy.

## CONCEPT TEST

Which of the following alkenes has cis and trans isomers?

a. $CH_2{=}CHCH_2CH_3$ b. 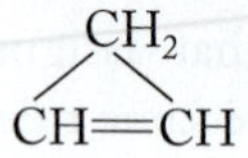 cyclopropene ($CH_2$ bonded to $CH{=}CH$) c. $(CH_3)_2C{=}C(CH_3)_2$

d. $(CH_3)_2C{=}CH_2$ e. $CH_3CH{=}CHCH_3$

## CONCEPT TEST

Why do alkynes not have cis and trans isomers, whereas alkenes do?

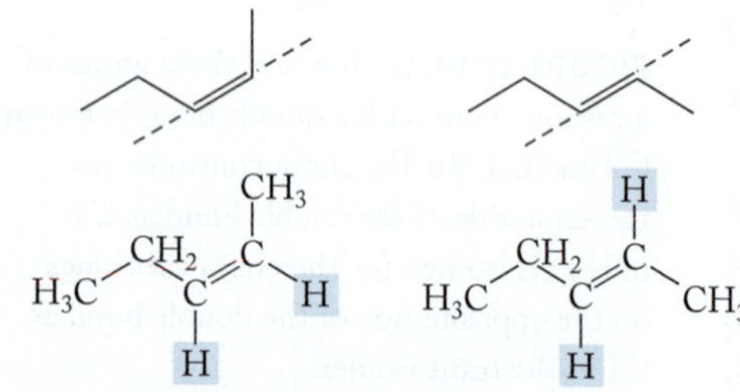

(a) The hydrogen atoms on the double bond are cis.

(b) The hydrogen atoms on the double bond are trans.

**FIGURE 19.16** The positions of the H atoms in alkenes can be used to distinguish between (a) cis and (b) trans isomers.

## Naming Alkenes and Alkynes

To name straight-chain alkenes and alkynes, the prefixes in Table 19.3 are used to identify the length of the chain. The suffix *-ene* is appended if the compound is an alkene, and the suffix *-yne* is used if it is an alkyne. The carbon atoms in the chain are numbered so that the first carbon atom in the double or triple bond has the lowest number possible, and that number precedes the name, followed by a hyphen. Stereoisomers are identified by writing *cis-* or *trans-* before the number. Thus, the compounds in Figure 19.16(a) and (b) are *cis*-2-pentene and *trans*-2-pentene, respectively.

### SAMPLE EXERCISE 19.6 Drawing the Molecular Structures and Naming the Stereoisomers and Constitutional Isomers of an Alkene — LO4

Draw the condensed structures and carbon-skeleton structures of the five isomers of the six-carbon, straight-chain alkene containing one double bond. Name each isomer.

**Collect and Organize** We are asked to draw and name five isomers of molecules that each have six carbon atoms in a single chain and one C=C double bond. Figures 19.8, 19.13, 19.14, and 19.16 show examples of condensed and carbon-skeleton structures of alkenes.

**Analyze** Constitutional isomers, which depend on where the double bond is located in the chain, and stereoisomers, which depend on the orientation of the groups about the double bond (cis/trans isomers), are both possible. A straight-chain, six-carbon alkene has a maximum of five places where a C=C bond can be placed; however, some locations may result in identical molecules. Not all alkenes have cis and trans isomers.

**Solve** Let's start with the constitutional isomers, which have the double bond at different locations, and then draw the stereoisomers (cis/trans isomers) where possible. If all six carbon atoms are in one straight chain, then a double bond can be located between C1 and C2, C2 and C3, or C3 and C4. These isomers are named 1-hexene, 2-hexene, and 3-hexene, respectively, where the number indicates the location of the double bond along the chain. Chains with a double bond between C4 and C5 or C5 and C6 are identical to the isomers with a double bond between C2 and C3 or C1 and C2, respectively.

1-Hexene does not have stereoisomers because the carbon atoms that form the double bond have three H atoms and only one nonhydrogen atom attached. Only 2-hexene and 3-hexene can have stereoisomers. Both *cis*-2-hexene and *cis*-3-hexene have two R groups on the same side of the double bond, whereas *trans*-2-hexene and *trans*-3-hexene have R groups on opposite sides of the double bond.

1-Hexene: $H_2C=CHCH_2CH_2CH_2CH_3$

*trans*-2-Hexene: H and $CH_2CH_2CH_3$ on one C; $H_3C$ and H on the other (trans)

*trans*-3-Hexene: $CH_3CH_2$ and H on one C; H and $CH_2CH_3$ on the other (trans)

*cis*-2-Hexene: H and H on the double-bond carbons; $H_3C$ and $CH_2CH_2CH_3$ (cis)

*cis*-3-Hexene: $CH_3CH_2$ and $CH_2CH_3$ on the same side; H and H (cis)

**Think About It** The number of constitutional isomers of a straight-chain alkene depends on the number of independent locations for its double bond. An alkene with only a terminal double bond does not have stereoisomers.

**Practice Exercise** Draw and name the five isomers of the molecule with the following carbon skeleton and one carbon–carbon double bond:

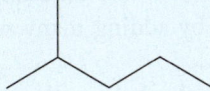

## Polymers of Alkenes

**CHEMTOUR** Polymers

Polyethylene (PE) is the most widely used plastic in the world, with a global production of 100 million metric tons in 2018. PE is a polymer formed from ethene, commonly called ethylene, in a polymerization reaction that can be written this way:

$$n\,H_2C{=}CH_2 \rightarrow \left[CH_2{-}CH_2\right]_n \qquad (19.5)$$

The physical properties of ethylene and three of the common forms of PE made from it are listed in **Table 19.5**. Note that ethylene is a gas at room

**TABLE 19.5** Some Physical Properties of the Monomer Ethylene and Three Polymers Made from It

| | Ethene (Ethylene) | Polyethylene (PE) Polymers | | |
|---|---|---|---|---|
| | | LDPE (low density) | HDPE (high density) | UHMWPE (ultrahigh molecular weight) |
| Molecular Mass (u) | 28 | 100,000–500,000 | 100,000–500,000 | 3,000,000–6,000,000 |
| Physical State at 25°C, 1 bar | Gas | Flexible solid | Rigid solid | Very tough, abrasion-resistant solid |
| Density ($g/cm^3$) at 25°C, 1 atm | 0.0012 | 0.910–0.940 | ≥0.941 | 0.930–0.935 |
| Structure | $H_2C{=}CH_2$ | Branched chains | Long straight chains | Very long straight chains |
| Melting Point (°C) | −196 | Softens from 85°C to 110°C | Softens from 120°C to 180°C | Softens from 130°C to 136°C |
| Boiling Point (°C) | −104 | Decomposes at 430°C–470°C | Decomposes at 430°C–470°C | Decomposes at 440°C–500°C |
| Common Use | Synthesis of PE; accelerates ripening of fruit | Grocery bags | Milk jugs and toys | Artificial joints and bulletproof vests |

temperature, but the three polymers made from it are solids. All three have the same chemical composition as the monomer, but they have different physical properties because of the different sizes and shapes of their enormous molecules.

**CONCEPT TEST**

Which type of intermolecular force between PE molecules is likely to be the dominant one?

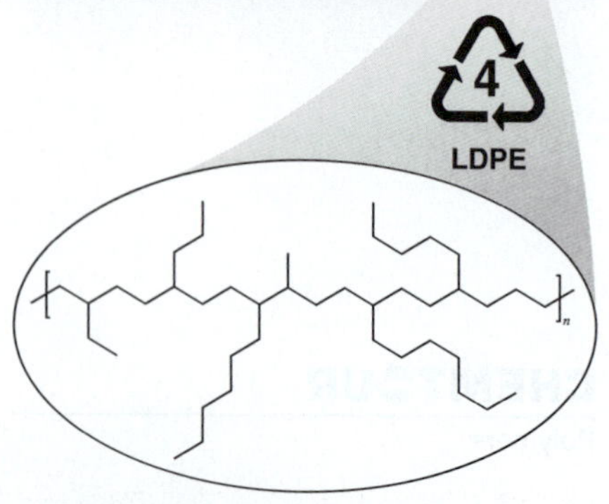


**FIGURE 19.17** Plastic shopping bags such as these are made of low-density polyethylene (LDPE), which consists of branched-chain polymers.

PE is a **homopolymer**, which means that it is composed of only one type of monomer. The condensed structure could be written $CH_3(CH_2)_nCH_3$, but there are so many more methylene groups than methyl groups that the repeating unit is often written as shown in Equation 19.5 to highlight the structure and composition of the monomer. Polyethylene is also an example of an **addition polymer**—that is, a polymer constructed by adding many molecules together to form the polymer chain.

In most products made of polyethylene, the $n$ in Equation 19.5 ranges from 1000 to almost 1 million. Polyethylene has a wide range of properties that depend on the value of $n$ and on whether the polymer chains are straight or branched. In the particular type of low-density PE (LDPE) used in grocery bags, $n = 3500$ to 15,000 and the chains are branched (**Figure 19.17**). When the bagger at the grocery store asks, "Paper or plastic?" the plastic in question is LDPE.

The molar mass of both the flexible LDPE polymer and the rigid, translucent high-density polyethylene (HDPE) used in milk bottles and toys ranges from 100,000 g/mol to 500,000 g/mol. The difference in properties between the two materials arises in large part from the structure of the individual chains. In HDPE the chains are straight with few branches (**Figure 19.18**).

What role does the presence or absence of branches play in determining properties? Think of the branched polymer as a tree branch with lots of smaller branches attached to it. The polymer can have branches that come out of the plane of the paper, so it has three dimensions. In contrast, the straight-chain polymer is like a long, straight pole. Suppose you had a pile of 100 tree branches and a pile of 100 poles, and your task was to stack each pile into the smallest possible volume to fit into a truck. You can certainly pile the branches on top of one another, but they will not fit together neatly, and probably the best you can do is to make the pile a bit more compact. In contrast, you can stack the poles into a very compact pile.

The same situation arises with the branched and linear molecules of polyethylene. The density of LDPE is lower than that of HDPE, primarily because the branched molecules of LDPE stack less efficiently. Lower density, coupled with weaker intermolecular forces between polymer strands that are farther apart, makes LDPE more deformable and softer. HDPE is more rigid and even has some regions that are crystalline because the packing is so uniform and the polymers are more rigidly held in place. These different structures at the molecular level mean that objects made of HDPE and LDPE must be separated before they can be processed for recycling, as indicated by their different recycling symbols.

Ultrahigh-molecular-weight PE (UHMWPE; $n > 100{,}000$) is an even tougher material because its molecules are both straight and significantly larger than the molecules in HDPE. UHMWPE is used as a coating on some artificial ball-and-socket joints for hip replacements and to separate the prosthetic ends of bones in knee replacement surgery (**Figure 19.19a**). It is extremely resistant to abrasion and makes the joints last longer. When woven into threads, it is used in bulletproof vests and body armor; when produced in sheets, it is used as synthetic ice for skating rinks in climates not conducive to maintaining natural ice (**Figure 19.19b**). The different forms of polyethylene illustrate how the size and shape of its molecules affect the physical properties of a material.

**FIGURE 19.18** Many translucent plastic bottles are made of high-density polyethylene (HDPE), which consists of straight polymers that stack together efficiently, making the bottles rigid and tough.

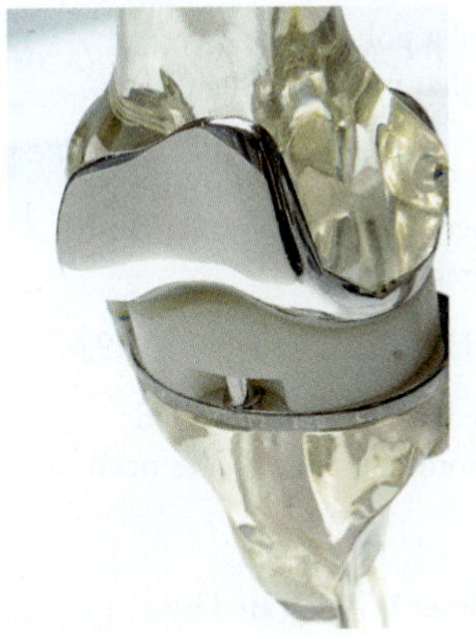

(a)

(b)

**FIGURE 19.19** (a) Devices used in knee-replacement surgery often incorporate a pad made of durable ultrahigh-molecular-weight polyethylene (UHMWPE) to cushion the joint. (b) Warm places such as Hobart, Australia, use the polymer for ice rinks.

**homopolymer** a polymer composed of only one kind of monomer unit.

**addition polymer** a macromolecule prepared by adding monomers to a growing polymer chain.

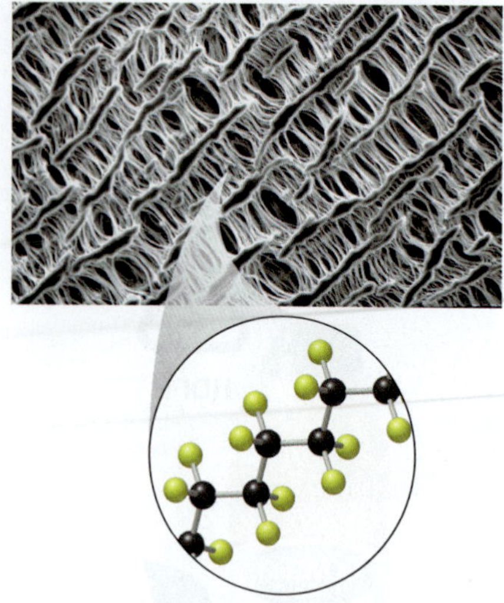

FIGURE 19.20 Fabrics made of woven strands of poly(tetrafluoroethylene) (Teflon) are used during surgical procedures to promote tissue growth. The strong, chemically inert mesh supports tissue that grows into its pores.

Chemical composition also plays a role in determining properties. If the hydrogen atoms in PE are all replaced with fluorine atoms, the resultant polymer is chemically unreactive, can withstand high temperatures, and has a low coefficient of friction, which prevents other things from sticking to it (**Figure 19.20**). This polymer is Teflon, $(\text{–CF}_2\text{–CF}_2\text{–})_n$, a linear polymer most familiar for its use as a nonstick surface in cookware. Teflon tubing is also used in the grafts inserted into small-diameter blood vessels during vascular surgery on limbs. The analogous material cannot be formed with chlorine, but a polymer does exist in which every other $\text{–CH}_2\text{–}$ in the polymer chain is $\text{–CCl}_2\text{–}$. Its repeating monomer unit is $(\text{–CH}_2\text{–CCl}_2\text{–})_n$. The polymer was originally developed as a coating for fighter planes to protect them from the elements and as mildew-resistant insoles for combat boots used in jungles. Later it was developed commercially as the thin, flexible plastic known as Saran wrap, but due to environmental concerns involving industrial processes using chlorine, as well as consumer concerns about the presence of halogenated materials in food wrapping, it has now been replaced in that product with a form of LDPE.

Hydrocarbons containing two or more double bonds are often used to manufacture polymers. For example, polymerization of butadiene yields a stretchy, synthetic rubber useful in rubber bands (**Figure 19.21a**). Polyisoprene, prepared from 2-methyl-1,3-butadiene, is used in surgical gloves (**Figure 19.21b**).

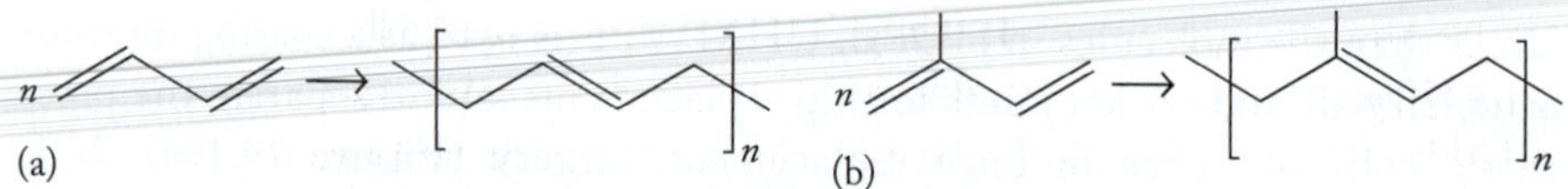


FIGURE 19.21 Monomer units and polymer structures for (a) butadiene and (b) a methylated butadiene known as isoprene.

FIGURE 19.22 Polypropylene is a common material for furniture, containers, clothing, lighting fixtures, and even objects of art. In addition to being moldable into many shapes, polypropylene is a good thermal insulator and does not absorb water easily.

**SAMPLE EXERCISE 19.7** Identifying Monomers **LO5**

Polypropylene, $\text{+CH}_2\text{CH(CH}_3)\text{+}_n$, is an addition polymer used in the manufacture of fabrics, ropes, and other materials (**Figure 19.22**). Draw condensed and carbon-skeleton structures of the monomer used to prepare polypropylene, and name the monomer.

**Collect and Organize** We are given a condensed structure of a polymer, and we are asked to identify the monomer used to prepare it. We know that polypropylene is an addition polymer, so the monomer must be an alkene.

**Analyze** To understand the relationship between the polymer and the monomer from which it is made, let's look at Equation 19.5 in the reverse direction:

$$\text{+CH}_2\text{—CH}_2\text{)+}_n \rightarrow n\,\text{CH}_2\text{=CH}_2 \tag{19.6}$$

Breaking the blue bonds in Equation 19.6 and making the red bond a double bond illustrates the relationship between polyethylene and its monomer, ethylene. We need to apply a similar analysis to polypropylene.

**Solve** The relationship between polypropylene and its monomer is illustrated by a corresponding equation:

$$\text{+CH}_2\text{—CH(CH}_3)\text{+}_n \rightarrow n\,\text{CH}_2\text{=CH(CH}_3) \tag{19.7}$$

Breaking the two blue bonds in Equation 19.7 and making the red C–C bond a double bond yields the alkene shown in **Figure 19.23**. The official name of this three-carbon

$CH_3CH{=}CH_2$

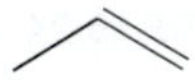

**FIGURE 19.23** The monomer propene is polymerized to make polypropylene.

alkene is propene (though it is also known as propylene). There is no need to precede the name of the monomer with a number to indicate the position of the C=C bond because it has to be between C1 and C2.

**Think About It** The constitutional difference between propylene and ethylene is the presence of a $-CH_3$ group bonded to one of the two $sp^2$ carbon atoms in propylene instead of the H atom found in ethylene.

**Practice Exercise** Draw the condensed and Kekulé structures of the monomer used to make poly(methyl methacrylate), PMMA, a polymer used in shatterproof transparent plastic that can take the place of glass:

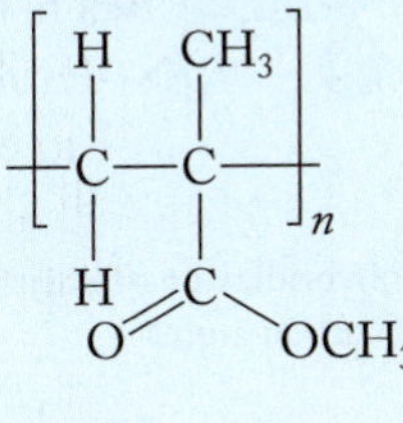

$$\left[ -\overset{\displaystyle H}{\underset{\displaystyle H}{C}} - \overset{\displaystyle CH_3}{\underset{\displaystyle C(=O)OCH_3}{C}} - \right]_n$$

All addition polymers based on addition reactions of monosubstituted ethylene are called **vinyl polymers** because the $CH_2{=}CH-$ subunit is called the **vinyl group**, a name derived from *vinum* (Latin for "wine"). The name was given to the group by 18th-century chemists who prepared ethylene ($CH_2{=}CH_2$) from ethanol ($CH_3CH_2OH$), the alcohol in wine and other beverages. Polyvinyl chloride (PVC) is widely used in commercial articles, such as computer cases and plastic pipes for plumbing. Classic vinyl phonograph records are also made from PVC. The condensed structures for the monomer and the polymer are

$$n\,CH_2{=}CHCl \rightarrow {+}CH_2CHCl{+}_n \quad (19.8)$$

Vinyl polymers are the world's second–largest selling plastics, and polymers in this category are extraordinarily versatile. You probably encounter 5 to 10 vinyl polymers before you leave your apartment or dorm room in the morning: vinyl shower curtains, vinyl drain pipes, vinyl flooring, and vinyl insulation around electrical conduits, among many others. As we explore more organic functional groups, the basic concepts developed for the vinyl polymers will apply to polymers in other categories: the features of functional group, size, and shape determine the chemical and physical properties of these extraordinarily useful materials.

## 19.4 Aromatic Compounds

Among the components of gasoline that play an important role in increasing its octane rating (a measure of the ignition temperature of the fuel and its ability to resist engine "knock") is the class of compounds called *aromatic* hydrocarbons. Aromatic compounds are used as solvents, dyes, and starting materials for the production of pharmaceuticals and polymers. They are manufactured from petroleum, but renewable sources such as biomass are being developed. Many bacteria produce aromatic compounds, and work is ongoing to enhance their production of compounds needed by industry. Technologies are also being developed to produce aromatic compounds directly from wood, agricultural waste, and cellulose.

As their class name implies, aromatic hydrocarbons have distinctive odors such as the smell of mothballs and recently cleaned public restrooms. To a

**vinyl polymer** one of the family of polymers formed from monomers containing the subgroup $CH_2{=}CH-$.

**vinyl group** the subgroup $CH_2{=}CH-$.

**FIGURE 19.24** Different views of the bonding in benzene. (a) Carbon-skeleton structures showing resonance and double-bond delocalization. (b) Hexagonal array of σ bonds. (c) The unhybridized $p_z$ orbitals on the $sp^2$-hybridized carbon atoms. (d) Delocalized π electrons above and below the ring.

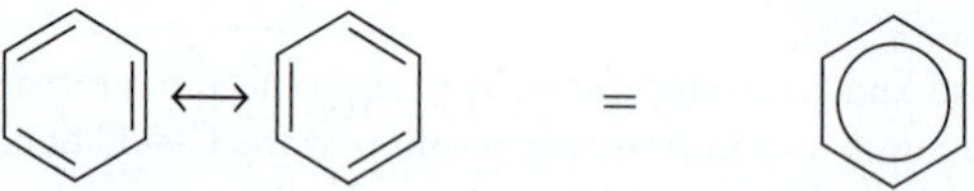


(a) Carbon-skeleton structures showing resonance forms of benzene

Skeletal symbol of benzene ring

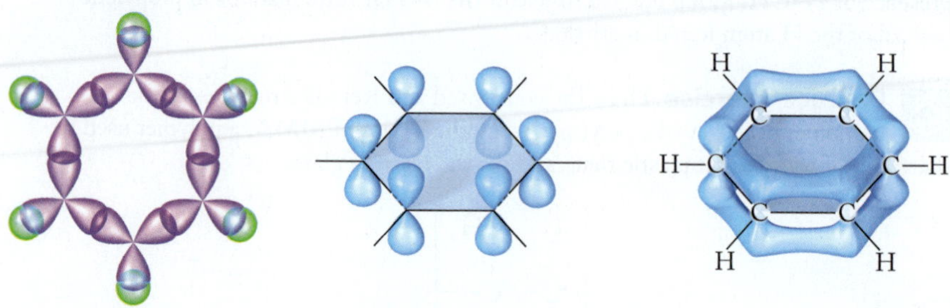


(b) Sigma bonds in benzene

(c) Unhybridized $p$ orbitals of carbon atoms

(d) Delocalized π cloud of electrons above and below plane of ring

**CONNECTION** In Chapters 4 and 5, we described bonding in the benzene molecule by using Lewis theory and valence bond theory. We defined aromatic compounds in Section 5.5.

chemist, *aromaticity* means more than odor; it means cyclic, planar molecules with $sp^2$-hybridized carbon atoms joined by a combination of σ and π bonds (**Figure 19.24**). Aromatic compounds may be considered relatives of alkenes because we often draw them using resonance structures having alternating single and double bonds. However, because their chemical and physical properties are unique and distinct from those of alkenes because of electron delocalization, they merit designation as a separate family.

The most common aromatic compound is benzene, $C_6H_6$. The different ways we view the bonding in benzene by using Lewis theory and valence bond theory are summarized in Figure 19.24. As noted in Section 4.5, the three pairs of delocalized π electrons in benzene's planar ring lead to considerable resonance stability in this molecule, and this is true of all other aromatic molecules as well. Some aromatic molecules have other elements besides carbon in their rings, and some rings have more than, or fewer than, six atoms. However, to be aromatic, the number of π electrons in the ring must be either six or another number that satisfies the expression $4n + 2$, where $n$ is 0 or any positive integer.

The stability of aromatic systems has an impact on their chemical reactivity. For example, in Section 19.3, we saw that alkenes react rapidly with HBr, forming brominated alkanes. Benzene, however, does not react at all with HBr.

## Constitutional Isomers of Aromatic Compounds

Toluene

1, 2, 3, 4, 5, 6

1,2-Dimethylbenzene

1, 2, 3, 4, 5, 6

1,3-Dimethylbenzene

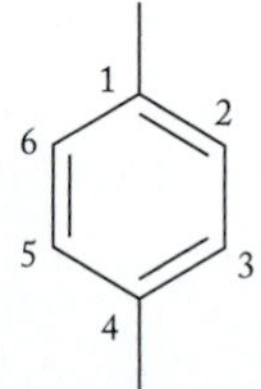


1,4-Dimethylbenzene

**FIGURE 19.25** Toluene (or methylbenzene) and the constitutional isomers of dimethylbenzene.

Many compounds can be formed by replacing the hydrogen atoms in an aromatic ring with other substituents. For example, when one methyl group replaces a hydrogen atom in benzene, we get methylbenzene. All the positions around the benzene ring are equivalent, so we do not have to include a position number in the name of methylbenzene, which is also known by the common name toluene. However, there are three different ways of attaching two methyl groups to a benzene ring, so there are three constitutional isomers of dimethylbenzene, also known as xylene. We distinguish between these isomers by numbering the carbon atoms to give the substituents the lowest possible numbers (**Figure 19.25**). Toluene and xylenes are used in inks, glues, and disinfectants.

The molecules of some aromatic compounds contain benzene rings that share one or more of their hexagonal sides. Three of these polycyclic aromatic hydrocarbons (PAHs) are naphthalene, anthracene, and phenanthrene (**Figure 19.26**). Extensive delocalization of the $\pi$ electrons over all the rings makes PAHs particularly stable. In addition to being found in fossil fuels, PAHs may be formed during the incomplete combustion of hydrocarbons and are present in particularly high concentrations in the soot from incinerators and diesel engines. PAHs have also been identified in interstellar dust clouds and in blackened portions of grilled meat. When introduced into the environment, these compounds persist and are among the longest-lived of the hydrocarbons.

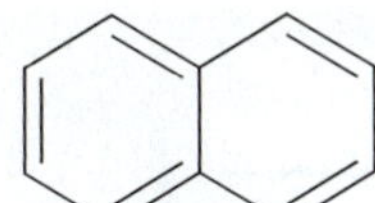

Naphthalene

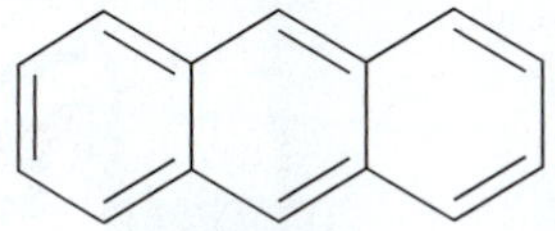

Anthracene

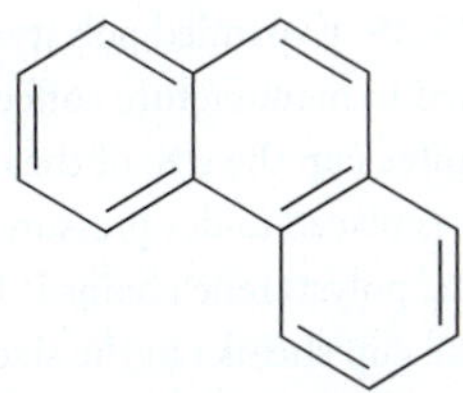

Phenanthrene

**FIGURE 19.26** Three polycyclic aromatic hydrocarbons (PAHs): naphthalene, anthracene, and phenanthrene.

**CONCEPT TEST**

Why isn't 1,3,5-hexatriene considered an aromatic compound?

1,3,5-Hexatriene

## Polymers Containing Aromatic Rings

Individual aromatic rings, as well as fused rings, are flat molecules. Intermolecular interactions between the rings causes them to stack (**Figure 19.27**), which gives rise to useful properties in materials that incorporate aromatic systems. Replacing one hydrogen atom in benzene with a vinyl group gives the monomer styrene, and the polymer made from this monomer is polystyrene (PS):

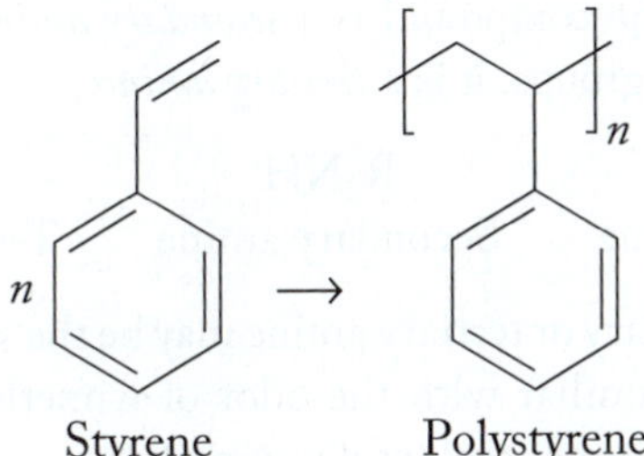


Solid PS is a transparent, colorless, hard, inflexible plastic. In this form, it is used for compact disc cases and plastic cutlery. A more common form of PS, however, is the *expanded solid* made by blowing $CO_2$ or pentane gas into molten polystyrene, which then expands and retains voids in its structure when it solidifies. One form of this expanded PS is the material used in coffee cups and take-out food containers (**Figure 19.28**).

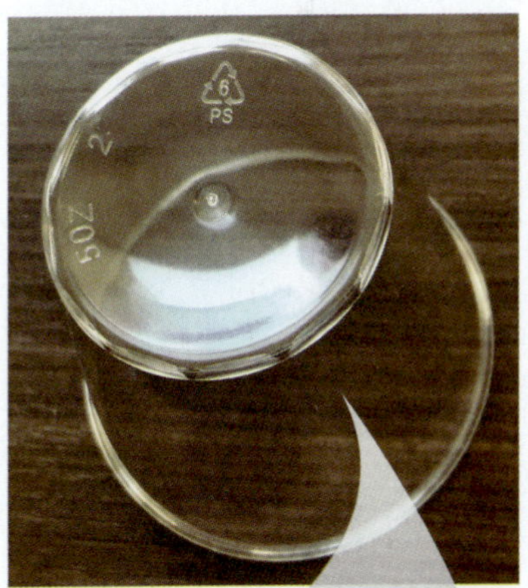


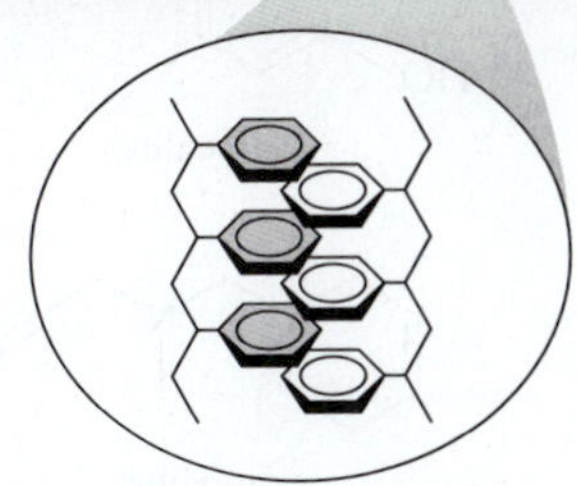

**FIGURE 19.27** The aromatic rings on neighboring chains in polystyrene stack together, strengthening the material.

**FIGURE 19.28** In its nonexpanded form, polystyrene is rigid and strong, suitable for making plastic utensils. In its expanded form, polystyrene is used in carry-out food containers, packing materials, and thermal insulation in buildings.

**FIGURE 19.29** Expanded polystyrene is often used to manufacture coffee cups. When a coffee cup the size of the one on the left is placed under pressure, air between the polystyrene chains is forced out, and the cup shrinks to the size of the one on the right.

The difference in properties between transparent, colorless, inflexible nonexpanded PS and opaque, white, pliable expanded PS can be explained by considering the role of the aromatic ring in aligning the polymer chains. Branches in the chains have the same effect that we saw with polyethylene, but the aromatic rings and their tendency to stack provide additional interactions that make chain alignment more favorable energetically. The aromatic rings along two neighboring chains can stack, and this stacking makes nonexpanded PS rigid. When the chains are blown apart by a gas, the stacking is disrupted and the chains open to make cavities that fill with air, making expanded PS a good thermal insulator and packing material. The presence of air in expanded PS is illustrated in **Figure 19.29**.

## 19.5 Amines

Nitrogen atoms are the defining components of functional groups in another important family of organic molecules called **amines**. In organic compounds, nitrogen atoms—and any atoms other than carbon, hydrogen, or metals—are called **heteroatoms** and are shown in Kekulé structures (without showing the lone pairs), condensed structures, and carbon-skeleton structures. For example, notice the heteroatom N at the junction of three lines in the carbon-skeleton structures of diphenhydramine (trade name, Benadryl) and pargyline in **Figure 19.30**.

Amines containing an alkyl or aromatic group are thought of as being derived from ammonia, $NH_3$. If one hydrogen atom in ammonia is replaced by an R group, the compound is called a *primary amine.* If two hydrogen atoms are replaced by R groups, the compound is a *secondary amine*; if all three hydrogen atoms are replaced by R groups, it is a *tertiary amine*:

| $RNH_2$ | $R_2NH$ | $R_3N$ |
|---|---|---|
| Primary amine | Secondary amine | Tertiary amine |

The R groups in a secondary or tertiary amine may be the same or different organic subunits. You may be familiar with the odor of trimethylamine, $(CH_3)_3N$, the compound responsible for the smell of decaying fish.

The amine functional group is found in many natural products and drugs. Some relatively simple examples include amphetamine (Figure 19.30), a stimulant that also contains an aromatic group, and Benadryl, an antihistamine used to treat the symptoms associated with allergies. Another amine, adrenaline, is produced by our bodies in glands near the kidneys and plays an important role in our nervous system.

Amines are organic bases, and their basicity is their defining chemical characteristic. They all react with water to some extent to produce hydroxide ions and protonated cations, just like ammonia:

$$NH_3(aq) + H_2O(\ell) \rightleftharpoons NH_4^+(aq) + OH^-(aq) \quad (19.9)$$

$$RNH_2(aq) + H_2O(\ell) \rightleftharpoons RNH_3^+(aq) + OH^-(aq) \quad (19.10)$$

Amines also react readily with acids such as hydrochloric acid to form water-soluble salts:

$$RNH_2(aq) + HCl(aq) \rightarrow RNH_3^+(aq) + Cl^-(aq) \quad (19.11)$$

Many pharmaceuticals (such as Benadryl) that contain the amine functional group are sold as hydrochloride salts to improve their solubility in water.

Amine groups are polar and can form hydrogen bonds with water, which also promotes their solubility in water. By contrast, the portions of amine molecules

Amphetamine

Diphenhydramine (Benadryl)

Adrenaline

Pargyline

**FIGURE 19.30** Amphetamine, a drug known to produce increased wakefulness and focus; diphenhydramine (brand name, Benadryl), an antihistamine; adrenaline, a hormone produced in our bodies and involved in the fight or flight response; and pargyline, a drug used to treat hypertension. Each of these physiologically active compounds contains the amine functional group.

composed of carbon and hydrogen are nonpolar and diminish aqueous solubility because the larger the nonpolar portion, the stronger the dispersion forces between amine (solute) molecules that inhibit their solubility in water.

Microorganisms of the genus *Methanosarcina* convert primary, secondary, and tertiary methylamines to methane, carbon dioxide, and ammonia:

$$4\,CH_3NH_2(aq) + 2\,H_2O(\ell) \rightarrow 3\,CH_4(g) + CO_2(g) + 4\,NH_3(aq) \quad (19.12)$$

$$2\,(CH_3)_2NH(aq) + 2\,H_2O(\ell) \rightarrow 3\,CH_4(g) + CO_2(g) + 2\,NH_3(aq) \quad (19.13)$$

$$4\,(CH_3)_3N(aq) + 6\,H_2O(\ell) \rightarrow 9\,CH_4(g) + 3\,CO_2(g) + 4\,NH_3(aq) \quad (19.14)$$

These reactions describe pathways by which methane can be produced from the decay of biomass, serving as potential sources of fuel for heating and transportation. Amines represent only a small fraction of the material in plants, however, and the industrial development of fuel production from amines has been slow, mainly because fossil fuels are still plentiful enough and cheap enough to make processing amines cost-ineffective. This economic imbalance may change as fossil fuels are depleted and become more expensive.

**amine** an organic compound that contains a group with the general formula $RNH_2$, $R_2NH$, or $R_3N$, where R is any organic subgroup.

**heteroatom** any atom other than carbon or hydrogen in an organic compound.

**CONNECTION** We saw in Section 15.3 that amines may be strong or weak bases.

CONCEPT **TEST**

Identify each of the substances in Figure 19.30 as a primary, secondary, or tertiary amine.

## 19.6 Alcohols, Ethers, and Reformulated Gasoline

In 1995, air-quality regulations went into effect in many U.S. cities mandating reductions in atmospheric pollutants from gasoline-fueled engines. The regulations led to the widespread use of gasoline reformulated to contain additives to promote complete combustion and boost octane ratings. These additives are often organic compounds that contain oxygen in addition to hydrogen and carbon. Oxygen atoms in an organic compound are heteroatoms and are components of functional groups in two important families of organic molecules: alcohols and ethers.

**CONNECTION** We considered how hydrogen bonding and dispersion forces affect solubility in Figures 6.18 and 6.19.

### Alcohols: Methanol and Ethanol

*Alcohols* have the general formula R—OH, where R is any alkyl group. The R group can be a straight chain, branched chain, or ring. Like the N atoms in amines, the O atoms in alcohols are shown in carbon-skeleton structures. The chemical and physical properties of alcohols can be understood if we recognize that the structure of an alcohol has features similar to those of both an alkane and water:

| R—H | H—OH | R—OH |
|---|---|---|
| Alkane | Water | Alcohol |

Recall from Section 6.4 that if the R group in the molecule is small, the alcohol behaves like water; as the R group gets larger, however, the alcohol behaves more like a hydrocarbon (**Table 19.6**). The polar −OH group makes one end of these compounds "waterlike." As the number of carbon atoms increases, however, a greater proportion of the alcohol molecule is "oil-like." Therefore, the solubilities of these alcohols in water decrease until about $C_8$, beyond which their solubilities are comparable to those of the corresponding hydrocarbons.

**TABLE 19.6** Solubilities of a Homologous Series of Alcohols in Water at 20°C

| Condensed Structure | Water Solubility (g/100 mL) |
|---|---|
| $CH_3OH$ | Miscible |
| $CH_3CH_2OH$ | Miscible |
| $CH_3(CH_2)_2OH$ | Miscible |
| $CH_3(CH_2)_3OH$ | 7.9 |
| $CH_3(CH_2)_4OH$ | 2.3 |
| $CH_3(CH_2)_5OH$ | 0.6 |
| $CH_3(CH_2)_6OH$ | 0.2 |
| $CH_3(CH_2)_7OH$ | 0.05 |

As the names of the two simplest alcohols—methanol and ethanol—indicate, the chemical names of alcohols end in *-ol*; this suffix identifies the compound as an alcoh*ol*. Methanol ($CH_3OH$) is also known as methyl alcohol or wood alcohol. The latter name comes from one former source of this alcohol: it was made by collecting the vapors given off when wood was heated to the point of decomposition in the absence of oxygen. Methanol is a widely used industrial organic chemical. It is the starting material in the preparation of several organic compounds used to make polymers.

As noted earlier, ethanol ($CH_3CH_2OH$), also known as ethyl alcohol, is the alcohol in alcoholic beverages. It is formed by the fermentation of sugar from an amazing variety of vegetable sources. Indeed, any plant matter containing enough sugar may be used to produce ethanol. Grains are commonly used, from which ethanol derives its trivial name *grain alcohol*. Ethanol may be the oldest organic chemical used by humans, and it is still one of the most important. For industrial purposes, ethanol is prepared by the reaction of ethylene and steam:

$$H_2C{=}CH_2(g) + H_2O(g) \rightarrow CH_3CH_2OH(g)$$

Most of the gasoline sold in the United States and Canada contains ethanol to promote complete combustion and reduce air pollution. Most of this ethanol is produced by the fermentation of sugar derived from corn, with annual production of ethanol for fuel use alone exceeding 108 billion liters in 2018. There are, however, several challenges limiting the greater use of ethanol or other oxygenated compounds as automobile fuels. Their enthalpies of combustion are significantly less negative than those of the hydrocarbons in gasoline. The change in free energy that accompanies the combustion of ethanol is also less negative than the combustion of the same volume of the hydrocarbons. Therefore, there is less energy available to do useful work, such as propelling cars and trucks down the road. On the plus side, burning ethanol reduces $CO_2$ emissions by an average of 34% compared to gasoline and reduces CO emissions by as much as 30%.

To produce ethanol, considerable energy, fertilizer, irrigation water, and valuable farmland are needed to grow and harvest corn, to convert cornstarch into sugar and sugar into ethanol, and to separate the ethanol from the fermentation mixture. It is estimated that more than two-thirds of the energy released in the combustion of ethanol derived from corn is consumed in its production. Ethanol produced in this way is more expensive than gasoline. However, advances that consume less water and require less energy are being made in producing ethanol from grass, wood, and crop residues.

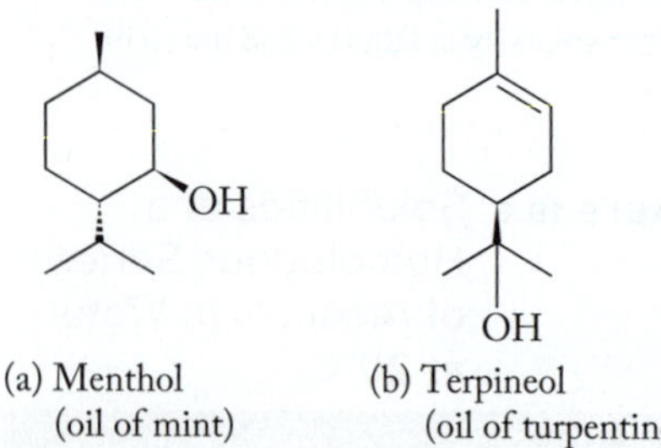


**FIGURE 19.31** (a) Menthol and (b) terpineol are two naturally occurring alcohols present in mint leaves and pine needles, respectively.

Alcohols are also components of many natural products and consumer goods derived from them. The distinctive odor of mint leaves comes from menthol, which is an alcohol, as is terpineol (oil of turpentine), an oil distilled from the resin of pine trees (**Figure 19.31**). Notice the similarity in the carbon-skeleton structures of these two compounds: both contain a six-carbon ring and an –OH group. Only the location of the OH group and the presence of a C=C double bond distinguish these two compounds.

## Ethers: Diethyl Ether

Ethers have the general formula R—O—R, where R is any alkyl group or an aromatic ring. Just as with alcohols, we can think of ethers as water molecules in which the two H atoms have been replaced by two organic groups (R and R′, which may be the same or different):

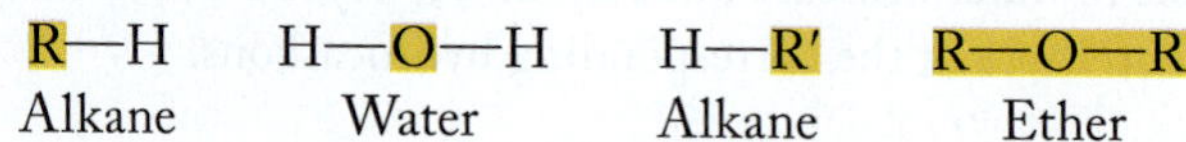

**TABLE 19.7** Functional Groups Affect Physical Properties

| | Molar Mass (g/mol) | Normal Boiling Point (K) | Solubility in Water (g/100 mL at 20°C) |
|---|---|---|---|
| $CH_3CH_2—O—CH_2CH_3$ 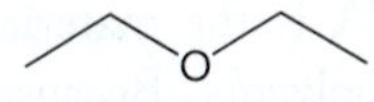 Diethyl ether | 74 | 308 | 6.9 |
| $CH_3CH_2CH_2CH_2CH_3$  Pentane | 72 | 309 | 0.0038 |
| $CH_3CH_2CH_2CH_2OH$  Butanol | 74 | 390 | 7.9 |

Because the C—O—C bond angle is close to the tetrahedral bond angle of 109.5°, the bond dipoles of the two C—O bonds in an ether do not cancel, which means that ethers are polar molecules. This structural feature gives rise to the properties of typical ethers: their water solubilities are comparable to those of alcohols of similar molar mass, but their boiling points are lower: about the same as alkanes of comparable molar mass (**Table 19.7**) because molecules of ethers can't hydrogen bond with each other.

The most important ether industrially is diethyl ether, $CH_3CH_2OCH_2CH_3$. You may have heard of this as the substance simply called "ether" that has had wide use in medicine as an anesthetic since 1842. Although exactly how an anesthetic dulls nerves and puts patients to sleep is still unclear, certain properties of diethyl ether play a role in determining its behavior as a medicinal agent. Diethyl ether has a high vapor pressure at room temperature, so a patient can inhale it. Because diethyl ether has a significant solubility in water, it is soluble in blood, which means that once inhaled, it can be easily transported throughout the body. Its low polarity and short saturated hydrocarbon chains combine to make it soluble in cell membranes, where it blocks stimuli coming into nerves. Ether has the unfortunate side effect of inducing nausea and headaches, and it has been replaced by newer anesthetics in modern hospitals. For many years, however, ether was the anesthetic of choice for surgical procedures.

Diethyl ether is extremely flammable, which was a liability in an operating room. This property becomes an advantage, however, when ether is sprayed into diesel engines to start them when it is too cold for diesel fuel to ignite. During the 1990s, another ether, methyl tert-butyl ether—best known by the abbreviation MTBE (**Figure 19.32**)—was added to gasoline to promote complete combustion. Unlike the nonpolar hydrocarbons in gasoline, MTBE is soluble in water. Consequently, gasoline spills, leaks from storage tanks, and releases from watercraft produced extensive MTBE contamination of groundwater and drinking water. After toxicity tests showed MTBE to be a possible carcinogen (cancer-causing agent), several states—including California, where more than 25% of the world's production of MTBE was used in gasoline—banned the use of MTBE as a gasoline additive. Most oil companies stopped adding MTBE to their gasolines in 2006. These changes raised the question of which oxygenated additive would replace MTBE. The leading candidate to date has been ethanol.

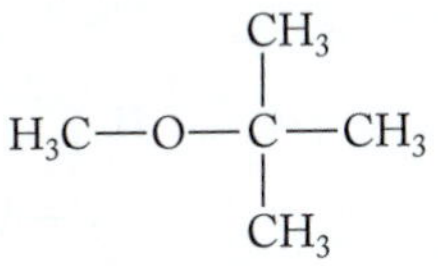

$$H_3C—O—C(CH_3)_2—CH_3$$

MTBE

**FIGURE 19.32** Methyl *tert*-butyl ether (*tert*- is an abbreviation for *tertiary*, referring to a carbon atom bonded to three other carbon atoms), or MTBE, was used in the 1990s as an oxygenated fuel additive.

## Polymers of Alcohols and Ethers

More than 1 million metric tons of the addition polymer poly(vinyl alcohol) (PVA) are used worldwide each year to make adhesives, emulsions, and materials known as sizing, which change the surface properties of textiles and paper to make them less porous, less absorbent, and smoother. PVA is the material of choice for laboratory gloves that are resistant to organic solvents. Because its chains are studded with –OH groups (**Figure 19.33a**), its surface is very polar and very waterlike, and hydrocarbon solvents that are not soluble in water do not penetrate PVA barriers.

PVA is also impermeable to carbon dioxide, and this property has led to its use in soda bottles, in which it is blended with the polymer poly(ethylene terephthalate) (PETE; **Figure 19.33b**). The two polymers do not mix but separate into layers (**Figure 19.33c**). The PETE makes the bottle strong enough to bear pressure changes due to temperature changes and survive the impact of falling off tables. $CO_2$, the dissolved gas that makes soda fizz, passes readily through PETE but not through the PVA layers, so the soda does not go flat. Polymers with different properties are often combined to create new materials with desired properties.

The synthesis of poly(vinyl alcohol) starts with polymerizing the monomer vinyl acetate to make poly(vinyl acetate) (PVAC), which is then reacted with water to replace the acetate groups with –OH groups. This replacement reaction turns PVAC into PVA, as shown in **Figure 19.33(d)**.

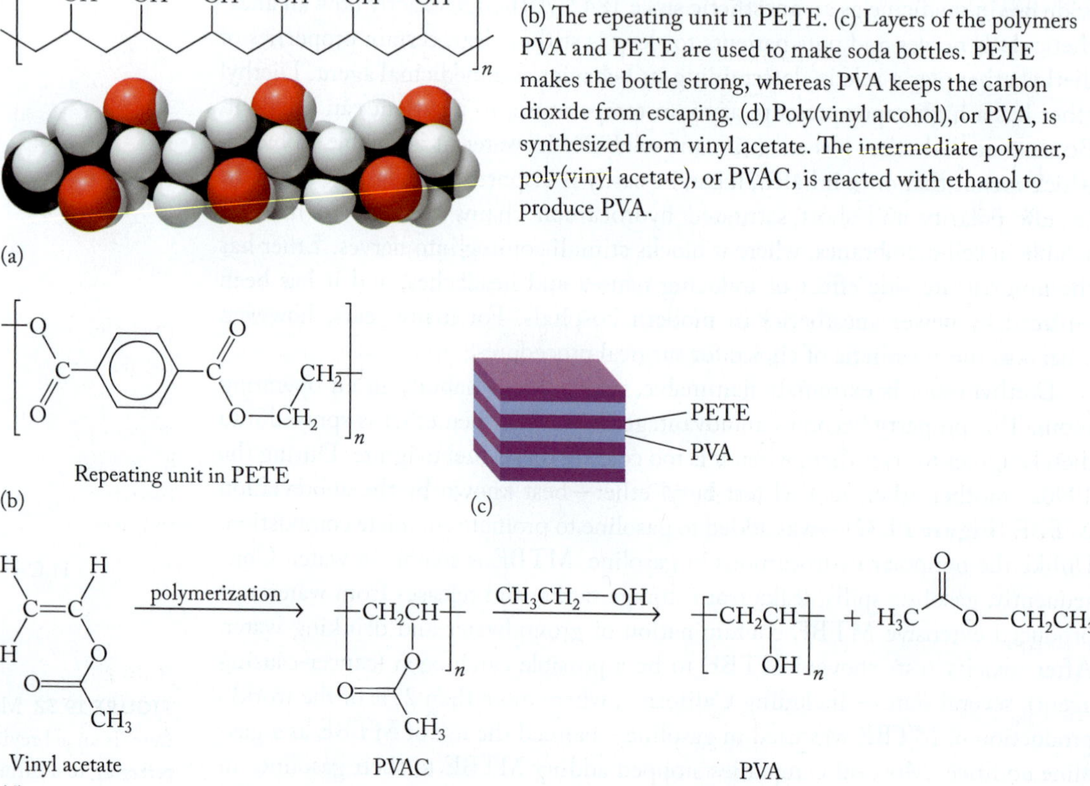


**FIGURE 19.33** (a) Representative section of a chain of PVA. (b) The repeating unit in PETE. (c) Layers of the polymers PVA and PETE are used to make soda bottles. PETE makes the bottle strong, whereas PVA keeps the carbon dioxide from escaping. (d) Poly(vinyl alcohol), or PVA, is synthesized from vinyl acetate. The intermediate polymer, poly(vinyl acetate), or PVAC, is reacted with ethanol to produce PVA.

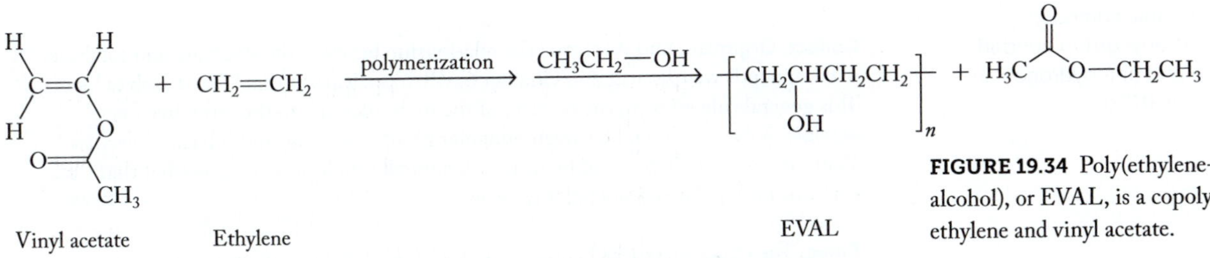

Vinyl acetate + Ethylene $\xrightarrow{\text{polymerization}}$ $\xrightarrow{CH_3CH_2—OH}$ $\left[CH_2CH(OH)CH_2CH_2\right]_n$ (EVAL) + $H_3C—C(=O)—O—CH_2CH_3$

FIGURE 19.34 Poly(ethylene-co-vinyl alcohol), or EVAL, is a copolymer of ethylene and vinyl acetate.

The blend of PVA and PETE in early plastic soda bottles was a physical mixture of the two polymers. New materials can also be made by combining different monomer units in one polymer molecule. This type of molecule is called a **copolymer** when two different monomers are combined and a **heteropolymer** when three or more different monomers are combined. One example of an addition copolymer is a material called EVAL, made from ethylene and vinyl acetate (**Figure 19.34**). Food usually deteriorates in the presence of oxygen, but packages made of EVAL provide an excellent barrier to oxygen while retaining the flavor and fragrance of the packaged food.

Monomers forming hetero- or copolymers can combine in different ways. If we represent the monomer units making up a copolymer with the letters A and B, one possible way they can combine is an arrangement called an *alternating copolymer*:

$$\left[A—B—A—B—A—B—A—B—A—B\right]$$

Another possibility is called a *block copolymer*:

$$\left[A—A—A—A—B—B—B—B—A—A—A—A—B—B—B—B\right]$$

Finally, a *random copolymer* is also possible:

$$\left[A—A—B—A—B—B—A—B—A—A—A—A—B—B—A—B—B—B\right]$$

EVAL is a random copolymer of the monomers ethylene and vinyl acetate.

Commercially important polymers made from ethers include poly(ethylene glycol) (PEG) and poly(ethylene oxide) (PEO), which are made of the same subunit (**Figure 19.35**). PEG is a low-molar-mass liquid oligomer made from ethylene glycol, and PEO is a higher-molar-mass solid made from ethylene oxide. As a polyether, PEG has properties closely related to those of diethyl ether—namely, it is soluble in both polar and nonpolar liquids. It is a common component in toothpaste because it interacts with both water and the water-insoluble materials in the paste and keeps the toothpaste a uniform consistency both in the tube and during use. PEGs of many lengths are finding increasing use as attachments to pharmaceutical agents to improve their solubility and biodistribution in the body.

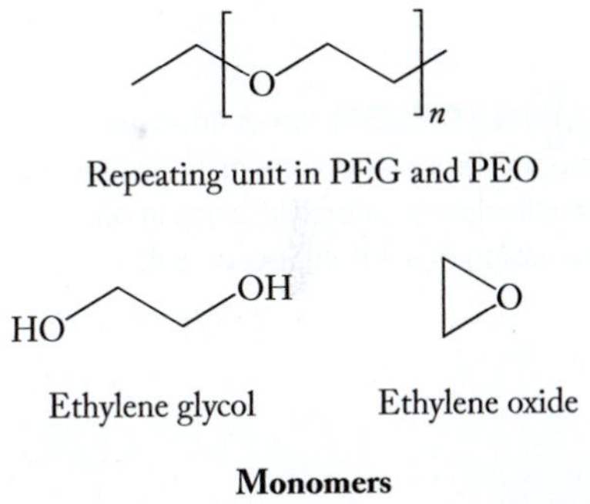

Repeating unit in PEG and PEO

HO ⁄⁄ OH — Ethylene glycol; Ethylene oxide

**Monomers**

FIGURE 19.35 Poly(ethylene glycol) (PEG) and poly(ethylene oxide) (PEO) have the same repeating unit. The two polymers differ only in their molar masses. PEG is typically made from ethylene glycol, whereas ethylene oxide is the monomer of choice for making PEO.

**SAMPLE EXERCISE 19.8** Assessing Properties of Polymers **LO6**

The polymer PEG (Figure 19.35) is used to blend materials that are not soluble in each other. It is soluble both in water and in benzene, a nonpolar solvent. Describe the structural features of PEG that make it soluble in these two liquids of very different polarities.

**copolymer** a macromolecule formed from the chemical combination of two different monomers.

**heteropolymer** a polymer made of three or more different monomer units.

**aldehyde** an organic compound containing a carbonyl group bonded to one R group and one hydrogen; its general formula is RCHO.

**Collect, Organize, and Analyze** The relationship between the structure and solubility of compounds was discussed in Chapter 6, where we learned that "like dissolves like." This general rule refers to the polarity of the molecules and to the attractive forces between polar groups and between nonpolar groups on solute and solvent molecules. Water is a polar molecule and benzene is a nonpolar molecule, so we predict that PEG contains both polar and nonpolar regions.

**Solve** The structure of PEG consists of $-CH_2CH_2-$ groups connected by oxygen atoms. The oxygen atoms are capable of hydrogen bonding with water molecules, so the attractive force between PEG and water is due to hydrogen bonding. Benzene is nonpolar and is attracted to the $-CH_2CH_2-$ groups in the polymer. Nonpolar groups interact via dispersion forces, so those forces must be responsible for the solubility of PEG in nonpolar benzene.

**Think About It** As predicted, PEG contains both polar and nonpolar regions, allowing it to be solvated by both polar solvents such as water and nonpolar solvents such as benzene.

**Practice Exercise** When PEG is added to soft drinks, it keeps $CO_2$ in solution longer after the soda is poured. What intermolecular attractive forces between PEG and $CO_2$ might make this use possible?

## 19.7 Aldehydes, Ketones, Carboxylic Acids, Esters, and Amides

**CONNECTION** We defined the carbonyl group and ketones in Section 6.3. We introduced carboxylic acids in our discussion of acids in Section 8.4.

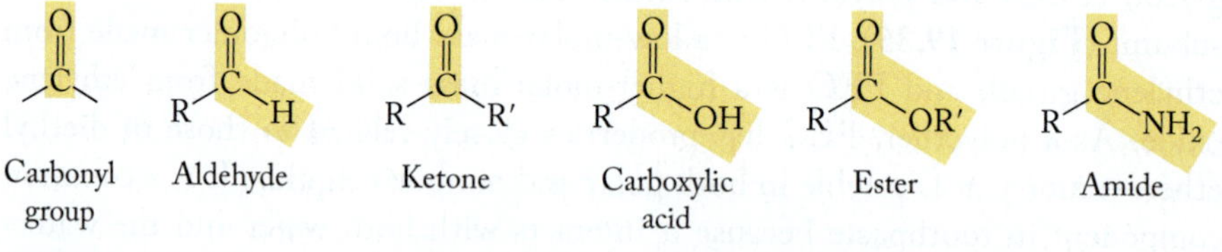

Five functional groups—aldehydes, ketones, carboxylic acids, esters, and amides—all contain a carbon atom double-bonded to an oxygen atom (**Figure 19.36**). This configuration of carbon and oxygen atoms is known as a carbonyl group.

The carbonyl groups in aldehydes and ketones are bonded only to R groups or H atoms. Carboxylic acids, as their name implies, are acidic, and their chemistry is determined by the $-COOH$ subunit, referred to as a *carboxylic acid group*. Esters and amides can be made from carboxylic acids by reacting them with alcohols and amines.

Carbonyl group, Aldehyde, Ketone, Carboxylic acid, Ester, Amide

**FIGURE 19.36** The carbonyl group is found in five important functional groups: aldehydes, ketones, carboxylic acids, esters, and amides. The R and R′ groups may be any organic group.

### Aldehydes and Ketones

Aldehydes and ketones resemble each other in both chemical and physical properties. An **aldehyde** contains a carbonyl group bound to one R group and one hydrogen atom; its general formula is RCHO or R(C=O)H. A *ketone* contains a carbonyl group bound to two R groups; its general formula is RCOR or R(C=O)R. The R groups may be the same, as in acetone, $CH_3C(O)CH_3$, or different, as in 2-heptanone (**Figure 19.37**). 2-Heptanone is found in cloves, blue cheese, and many fruits and dairy products.

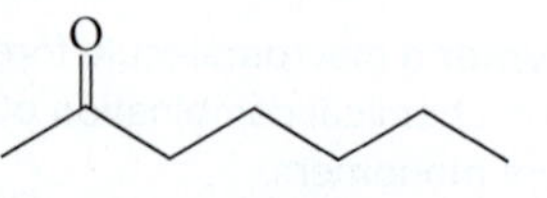


2-Heptanone

**FIGURE 19.37** The ketone 2-heptanone is found in many plants and dairy products.

The C═O double bond in the carbonyl group accounts for the reactivity of aldehydes and ketones. Unlike the C═C bond in an alkene, the C═O bond is polar: its more electronegative oxygen atom pulls electron density toward itself (**Figure 19.38**). Other polar species tend to react with the carbonyl group when electron-rich (δ−) regions of their molecules approach its electron-poor (δ+) carbon atom.

**CONNECTION** The formation of polar bonds between atoms of elements with different electronegativities is described in Chapter 4.

**FIGURE 19.38** The electron distribution in a carbonyl group is skewed toward the oxygen end of the bond because oxygen is more electronegative than carbon.

Because aldehydes and ketones are polar, they tend to parallel the ethers with respect to water solubility. They cannot form hydrogen bonds with other aldehyde and ketone molecules because they contain only carbon-bonded hydrogen atoms, so they have lower boiling points than alcohols of comparable molar mass.

Because of the C═O bond, the carbon atom in a carboxyl group is in a higher oxidation state than the C—OH bond's carbon atom in alcohols, and indeed many of the smaller aldehydes and ketones are made by oxidizing alcohols of the same carbon number. Aldehydes and ketones do not polymerize through their carbonyl groups. Many polymers have carbonyl functional groups as part of their structure, but these groups themselves do not react to form long chains.

We have already seen several examples of aldehydes and ketones in earlier chapters. In Chapter 5, we were introduced to formaldehyde and acrolein, two aldehydes shown in **Figure 19.39**. Acetone, $CH_3C(O)CH_3$, is a widely used solvent found in nail polish remover. The flavors and aromas of ginger (zingerone), spearmint (carvone), and cinnamon (cinnamaldehyde) all come from compounds that contain a carbonyl group.

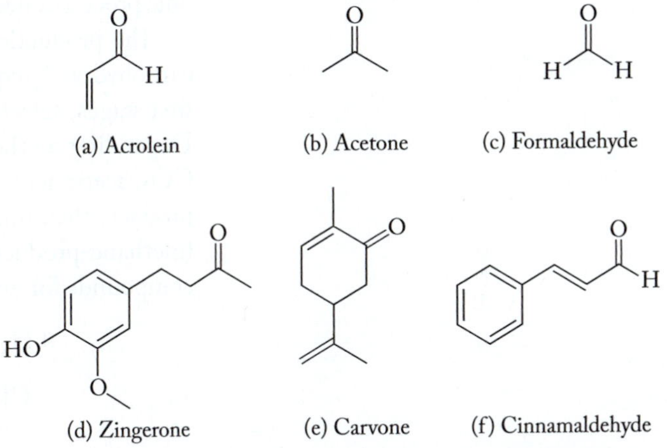

(a) Acrolein (b) Acetone (c) Formaldehyde

(d) Zingerone (e) Carvone (f) Cinnamaldehyde

**FIGURE 19.39** The ketone and aldehyde functional groups are common among organic compounds: (a) acrolein is found in barbeque smoke, (b) acetone is used in nail polish remover, (c) aqueous solutions of formaldehyde are used to preserve biological specimens, (d) zingerone is found in the spice ginger, (e) carvone is found in the leaves of spearmint, and (f) cinnamon owes its flavor and odor to cinnamaldehyde.

**CONCEPT TEST**

Besides the carbonyl group, what other functional groups are present in zingerone, carvone, and cinnamaldehyde (Figure 19.39)?

## Carboxylic Acids

Carboxylic acids are organic compounds that are proton donors, which means that they are Brønsted–Lowry acids. The R group in RCOOH may be any organic subunit. There is extensive hydrogen bonding between molecules of carboxylic acids. The hydrogen atom on the –COOH group of one molecule can form a hydrogen bond with either O atom on a nearby carboxylic acid group (**Figure 19.40**). This interaction results in high boiling points relative to those of other organic compounds of comparable molar mass.

**CONNECTION** We discussed quantitative aspects of equilibria involving weak acids and weak bases in Chapter 15.

Donating a proton leaves the carboxylic acid with a negatively charged oxygen atom whose electron density is delocalized over the entire carboxylate ($-COO^-$) group. This delocalization contributes to the stability of the carboxylate anion. The common carboxylic acids are weak acids, which means that they are present in aqueous solutions as mostly neutral molecules, a small fraction of which are ionized, donating $H^+$ ions to molecules of water (**Figure 19.41**).

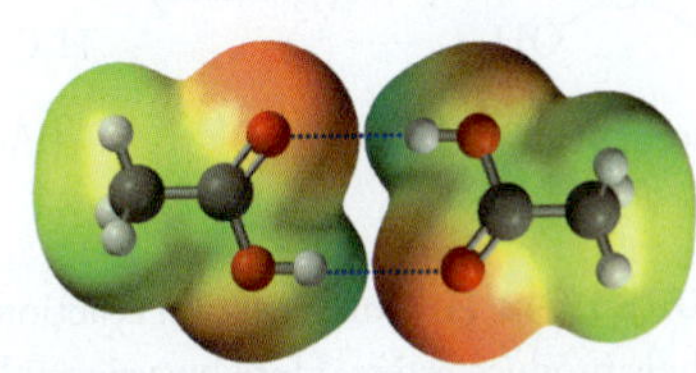

**FIGURE 19.40** The high boiling points of carboxylic acids are the result of strong hydrogen bonds between neighboring molecules.

Vinegar is a dilute aqueous solution of the carboxylic acid acetic acid. Large quantities of vinegar are produced commercially by the air oxidation of ethanol in the presence of enzymes from *Acetobacter* bacteria. Bacteria can also convert acetic acid and other constituents in biomass to methane. For example, the digestive

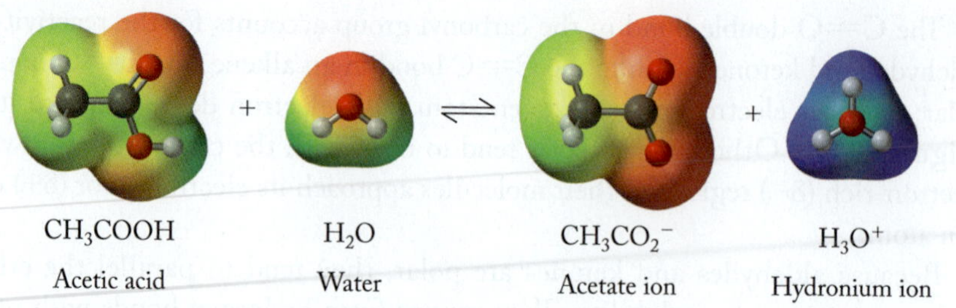

**FIGURE 19.41** Carboxylic acids such as acetic acid (found in vinegar) are weak acids in water.

**CONNECTION** We described the timescale for the formation of fossil fuels in Section 7.1.

systems of cows introduce about 100–200 liters of methane to the atmosphere per day per animal. Translating this process to an industrial scale is an attractive future source of hydrocarbons, provided that the complexities of bacterial action can be adapted for large-scale production. If so, then converting organic matter into hydrocarbon fuel would be possible without waiting millennia for the anaerobic processes deep within Earth to do so.

The production of methane from plant residues that are mostly cellulose (a carbohydrate) requires the sequential action of several types of bacteria. In the first stages, selected bacteria break up cellulose into mixtures of small molecules. Depending on the bacterial strain, these small-molecule products include $H_2$ and $CO_2$, acetic acid, formic acid, or methanol or some other small alcohol. All these products then undergo reactions promoted by the metabolism of **methanogenic** (methane-producing) **bacteria**, which consume hydrogen and simple organic compounds for energy and produce methane gas in the process:

$$4\,H_2(g) + CO_2(g) \rightarrow CH_4(g) + 2\,H_2O(\ell)$$

$$\underset{\text{Acetic acid}}{CH_3COOH(aq)} \rightarrow CH_4(g) + CO_2(g)$$

$$4\,\underset{\text{Formic acid}}{HCOOH(aq)} \rightarrow CH_4(g) + 3\,CO_2(g) + 2\,H_2O(\ell)$$

$$4\,\underset{\text{Methanol}}{CH_3OH(aq)} \rightarrow 3\,CH_4(g) + CO_2(g) + 2\,H_2O(\ell)$$

Methanogenic bacteria have a measurable effect on Earth's atmosphere and climate because methane is a potent greenhouse gas, trapping about 20 times more heat per molecule than carbon dioxide.

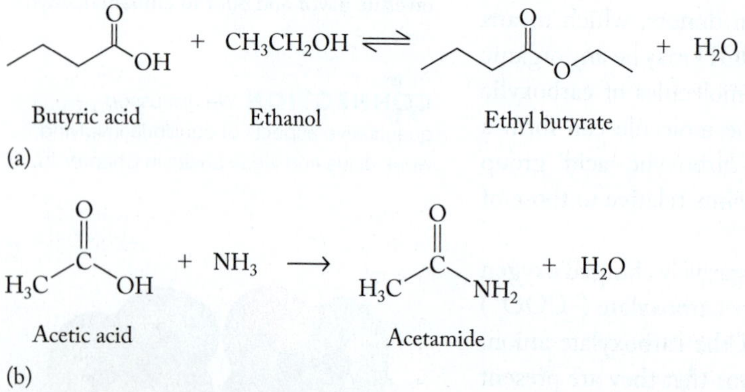

**FIGURE 19.42** (a) Condensation reactions between carboxylic acids and alcohols produce esters. Here butyric acid reacts with ethanol, forming ethyl butyrate. (b) Condensation reactions between carboxylic acids and ammonia (or amines) produce amides. Here acetic acid reacts with ammonia, forming acetamide.

## Esters and Amides

Several chemical families are closely related to the carboxylic acids. We consider only two of them here: esters and amides. In **esters**, the –COOH group of a carboxylic acid becomes a –COOR group, where R can be any organic group. The presence of these groups makes esters polar, and their boiling points are comparable to those of aldehydes and ketones of similar size. An ester is prepared by the *esterification* of an acid with an alcohol (**Figure 19.42a**). Esterification is a **condensation reaction**: two molecules combine ("condense") to create a larger molecule while a small molecule (typically water) is also formed.

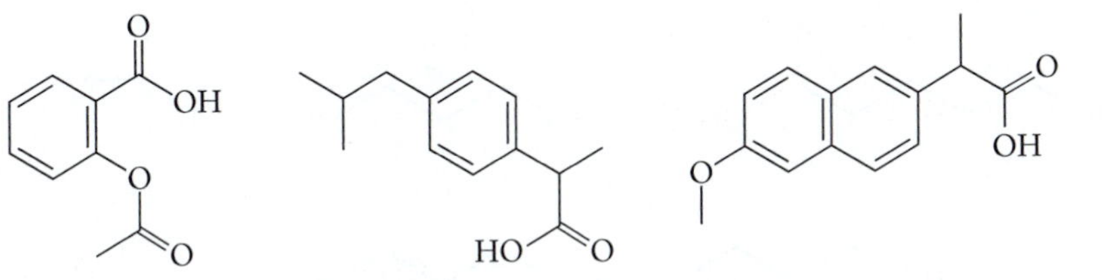

(a) Aspirin (b) Ibuprofen (c) Naproxen

**FIGURE 19.43** Three pain medications, all of which contain carboxylic acid functional groups. (a) Aspirin was the first medication available in tablet form. (b) Ibuprofen has fewer side effects than aspirin. (c) Naproxen is used to manage mild to moderate pain. All three belong to a class of compounds called nonsteroidal anti-inflammatory drugs (NSAIDs).

Esters often have pleasant fragrances that are much different from those of the acids from which they are derived. For example, the carboxylic acid butyric acid, with a straight chain of four carbon atoms, is responsible for the odor of rancid butter. The ethyl ester of butyric acid (ethyl butyrate) is responsible for the aroma of ripe pineapple. Esters are widely used to provide pleasant scents for personal products such as shampoos and soaps.

Many medications consist of molecules that contain carboxylic acid and ester groups. **Figure 19.43** illustrates the carbon-skeleton structures of three common pain relievers—aspirin, ibuprofen, and naproxen—each of which contains a carboxylic acid group. Aspirin also contains an ester group and naproxen, an ether.

**Amides** are made in condensation reactions between carboxylic acids and either ammonia or a primary or secondary amine (**Figure 19.42b**). Amides are polar, and hydrogen atoms bonded to nitrogen can form hydrogen bonds with the oxygen atom of an adjacent amide group. These hydrogen bonds cause the boiling points of amides to be higher than those of esters of comparable molar mass.

## Polyesters and Polyamides

Most of the compounds we have examined have been monofunctional, which means that they have only one functional group that identifies their family. With the polymers of carboxylic acids and their derivatives, we enter the world of *difunctional* molecules, which are molecules with two functional groups. The key point to remember is that the functional groups for the most part still retain their individual chemical reactivity, even if they are in a molecule with another functional group. Also remember that the same features we listed for all other polymers still apply here: for polymers, function is determined by composition, structure, and size.

Look again at the esterification reaction in Figure 19.42(a). The –COOH group of the acid reacts with the –OH group of the alcohol to form a carbon–oxygen single bond and release a molecule of water. What would happen at the molecular level if we had a single compound that contained a carboxylic acid functional group at one end and an alcohol functional group at the other (**Figure 19.44**)? The carboxylic acid group of one molecule could react with the alcohol group of another molecule in a condensation reaction to generate a molecule that has a carboxylic acid group at one end, an alcohol at the other end, and an ester linkage in between. If this reaction happens repeatedly, a monomer containing one carboxylic acid and one hydroxyl group (a hydroxy acid) polymerizes, as shown in Figure 19.44, to form a polyester, a **condensation polymer**. PETE (Figure 19.33b) is a condensation polymer, too. In general, condensation polymers are formed by the reaction of monomers that produce a polymer and water or another small molecule. In addition to its use in plastic soda bottles, PETE is used extensively in medicine to make artificial heart valves and grafts for arteries.

**methanogenic bacteria** bacteria using simple organic compounds and hydrogen for energy; their respiration produces methane, carbon dioxide, and water, depending on the compounds they consume.

**ester** an organic compound in which the –OH of a carboxylic acid group is replaced by –OR, where R can be any organic group.

**condensation reaction** two molecules combining to form a larger molecule and a small molecule (typically water).

**amide** an organic compound in which the –OH of a carboxylic acid group is replaced by $-NH_2$, –NHR, or $-NR_2$, where R can be any organic group.

**condensation polymer** a macromolecule formed by a reaction of monomers, producing a polymer and water or another small molecule.

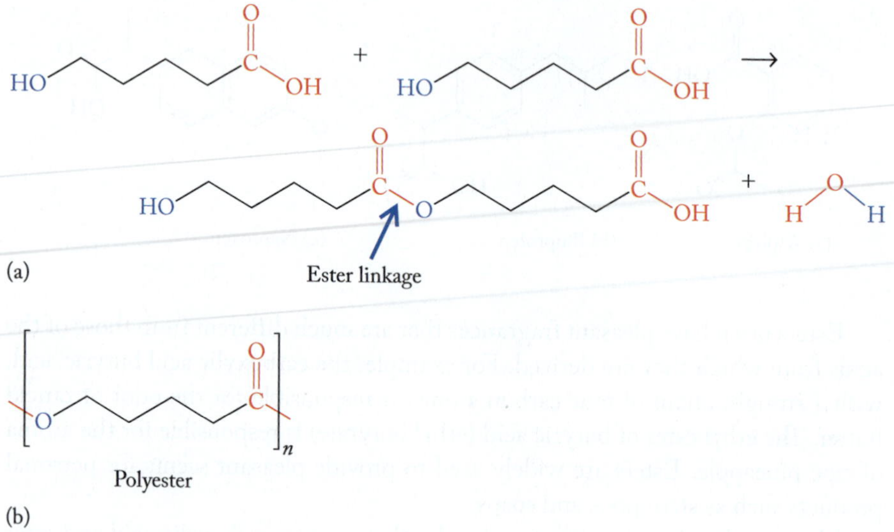

**FIGURE 19.44** (a) Synthesis of an ester from a condensation reaction between two identical difunctional molecules, each one containing an alcohol group and a carboxylic acid group. The reaction repeats over and over, forming (b) the polyester made up of the repeating monomer unit shown.

(a) Ester linkage

(b) Polyester

A copolymer formed by the reaction of glycolic acid and lactic acid (**Figure 19.45**) is used to support the growth of skin cells used in grafts for burn victims. This condensation polymer is also used to make dissolving sutures. Esterification reactions used to make polyesters can be reversed by the addition of water, breaking the ester linkage and forming alcohol and acid functional groups. We examine this process in greater detail in Chapter 20.

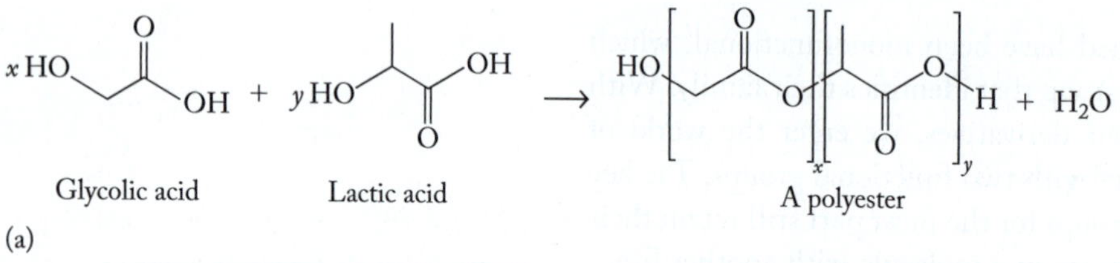

Glycolic acid Lactic acid A polyester

(a)

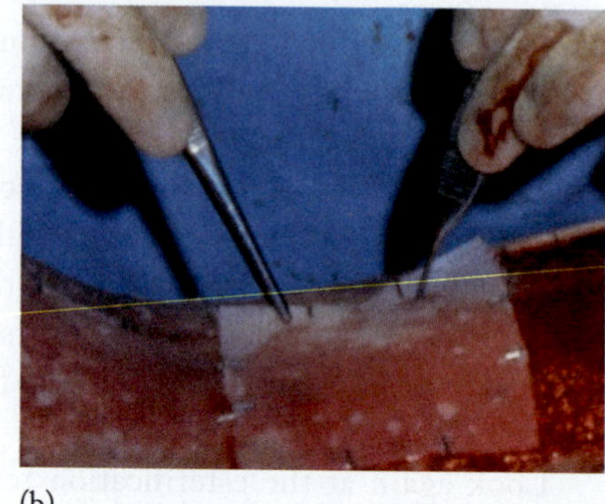

(b)

**FIGURE 19.45** (a) The condensation polymer prepared from glycolic acid and lactic acid is used to make sutures that dissolve and artificial skins that protect against infection while promoting regrowth of skin cells. (b) Synthetic skin being applied to a burn patient.

**SAMPLE EXERCISE 19.9** Making a Polyester **LO5**

Show how a polyester can be synthesized from the difunctional alcohol $HO(CH_2)_3OH$ and the difunctional carboxylic acid $HOOC(CH_2)_3COOH$.

**Collect, Organize, and Analyze** An ester is the product of a reaction between a carboxylic acid and an alcohol. A polyester is a polymer with a repeating unit containing an ester functional group. We are given an alcohol and a carboxylic acid to react to make the ester monomer. Because the starting materials are difunctional, the alcohol can react with two molecules of carboxylic acid, and the acid can react with two molecules of alcohol.

**Solve** The reaction is

$$\mathrm{HOCH_2CH_2CH_2OH} + \mathrm{HO(O{=})CCH_2CH_2CH_2C({=}O)OH} \rightarrow$$

$$\mathrm{HOCH_2CH_2CH_2O(O{=})CCH_2CH_2CH_2C({=}O)OH} + \mathrm{H_2O}$$

The two OH groups shown in blue react to form an ester at one end of the carboxylic acid. The product molecule has an alcohol group on one end (shown in red) that can react with another molecule of carboxylic acid and a carboxylic acid group (green) on the other end that can react with another molecule of alcohol. Continuing these condensation reactions results in the formation of a polymer whose repeating unit is

$$\left[\mathrm{-CH_2CH_2CH_2O(O{=})CCH_2CH_2CH_2C({=}O)O-}\right]_n$$

**Think About It** The repeating unit in the polyester contains one section that came from the alcohol and a second section that came from the carboxylic acid, which makes sense because esters are formed from alcohols and acids.

**Practice Exercise** A difunctional molecule may contain two functional groups, such as this one with an alcohol group and a carboxylic acid group:

$$\mathrm{HO{-}CH_2CH_2CH_2C({=}O)OH}$$

Draw the repeating unit of the polyester made from this molecule.

## CONCEPT TEST

Clothes made from the polyester fabric known as Dacron may be less comfortable in hot weather than clothes made of cotton (also a polymer) because Dacron does not absorb perspiration as effectively as cotton. On the basis of the repeating units of these two polymers (**Figure 19.46**), suggest a structural reason why cotton absorbs perspiration (water) better than Dacron. Which other polymer in this chapter has the same repeating unit as Dacron?

Combining difunctional molecules in a condensation reaction can be used to make *polyamides*, another class of useful synthetic polymers. The functional groups are a carboxylic acid and an amine. The difunctional monomer units can be identical (**Figure 19.47a**), each containing one carboxylic acid group and one amine group, or they can be different (**Figure 19.47b**), with one monomer containing two carboxylic acid groups (a dicarboxylic acid) and the other containing two amine groups (a diamine).

Nylon-6,6 is a polyamide made from the monomers shown in Figure 19.47(b). Each monomer contains six carbon atoms, which is what the digits in the name represent.

$$\left[\mathrm{-(O{=})C{-}C_6H_4{-}C({=}O){-}OCH_2CH_2O-}\right]_n$$

Repeating unit in Dacron

(a)

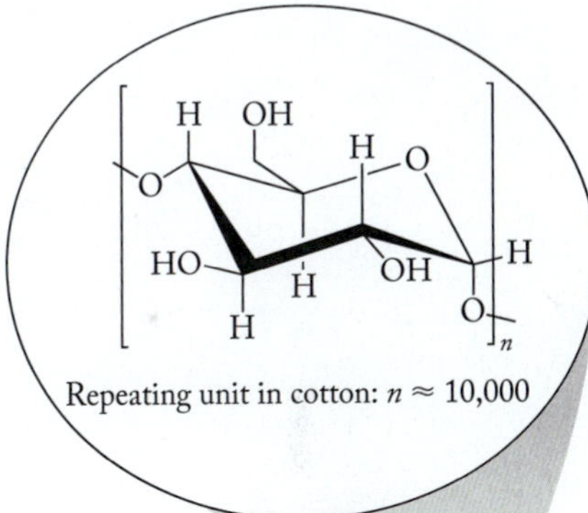


Repeating unit in cotton: $n \approx 10{,}000$

(b)

**FIGURE 19.46** According to the molecular structures of (a) Dacron and (b) cotton, why is cotton the better material for making T-shirts worn during strenuous exercise?

**FIGURE 19.47** (a) Synthesis of a polyamide from two identical monomers, each containing a carboxylic acid functional group and an amine functional group. (b) Synthesis of the polyamide nylon-6,6 from nonidentical monomers: adipic acid (a dicarboxylic acid) and hexamethylenediamine (a diamine).

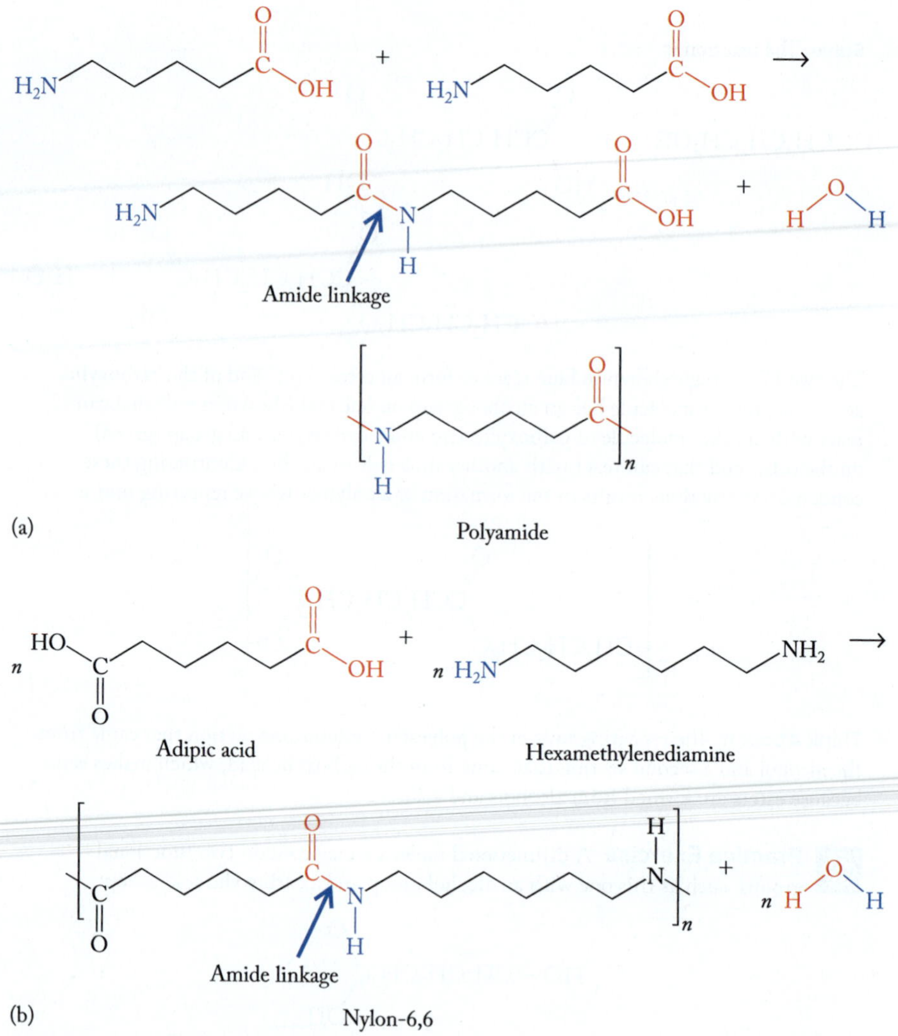

Amide linkage

(a) Polyamide

Adipic acid

Hexamethylenediamine

Amide linkage

(b) Nylon-6,6

**SAMPLE EXERCISE 19.10** Identifying Monomers **LO5**

Another form of nylon is nylon-6, with the single 6 indicating that the polymer is made from the reaction of a series of identical six-carbon monomers. Draw the condensed structure of a six-carbon molecule that could polymerize to make nylon-6.

**Collect, Organize, and Analyze** Amides form from carboxylic acids and amines. By analogy to Sample Exercise 19.9 and its accompanying Practice Exercise, we can suggest a difunctional molecule that has a carboxylic acid on one end, an amine on the other, and enough carbon atoms in between to form a polyamide with a repeating unit six carbon atoms long.

**Solve** We can build the required monomer by starting with one of the functional groups. It doesn't matter which one, so let's begin with the amine:

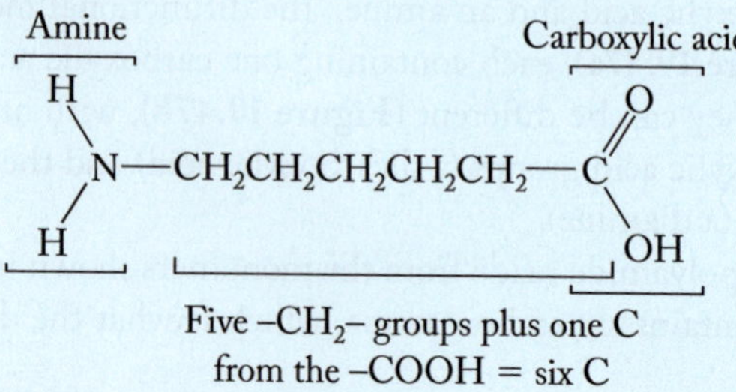

Amine — Carboxylic acid

$$H_2N{-}CH_2CH_2CH_2CH_2CH_2{-}C({=}O)OH$$

Five $-CH_2-$ groups plus one C from the $-COOH$ = six C

We then add a chain of five $-CH_2-$ units because the name nylon-6 indicates that six carbon atoms separate the ends of the repeating unit. Finally, we add the carboxylic acid functional group as the second functional group and the sixth carbon atom in the chain.

**Think About It** Many nylons with different properties can be made by varying the length of the carbon chain in a monomer such as the one in this exercise or by varying the lengths of the chains in both the difunctional acid and difunctional amine in Figure 19.47(b).

**Practice Exercise** (a) Draw the carbon-skeleton structures of two monomers that could react with each other to make nylon-5,4. (b) Draw the carbon-skeleton structure of the repeating unit in the polymer. (*Note*: The first number refers to the carboxylic acid monomer; the second refers to the amine monomer.)

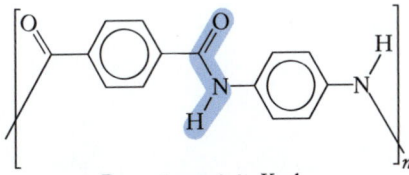

Benzene-1,4-dicarboxylic acid
(terephthalic acid)

1,4-Diaminobenzene

Repeating unit in Kevlar

**FIGURE 19.48** The monomers used to make Kevlar are a dicarboxylic acid and a diamine. The amide bond in the repeating unit is highlighted.

Polymers of nylon make long, straight fibers that are strong and excellent for weaving into fabrics. Nylon is flexible and stretchable because the hydrocarbon chains can stretch and bend like a spring or a Slinky toy. To produce an even stronger nylon, researchers had to find some way to reduce the ability of the chains to form coils. This was achieved by using monomer units containing functional groups that made it difficult for the chains to bend. One product of this work was a polyamide called Kevlar, invented by Stephanie Kwolek of DuPont in 1965. Kevlar is formed from a dicarboxylic acid of benzene and a diamine of benzene (**Figure 19.48**). When these two monomers polymerize, the flat, rigid aromatic rings keep the chains straight.

Two additional intermolecular interactions orient the chains and hold them tightly together (**Figure 19.49**). First, the –NH hydrogen atoms form hydrogen bonds with the oxygen atoms of carbonyl groups on adjacent chains. Second, the rings stack on top of one another (as in polystyrene) and provide additional interactions that hold the chains together in parallel arrays. The result is a fiber that is strong but still flexible. Fabrics and helmets made of Kevlar resist puncture, even by bullets and hockey pucks fired at them, and they are resistant to flames and reactive chemicals.

We have seen how the macroscopic properties of Kevlar and other polymers are influenced by their microscopic structure. **Table 19.8** reinforces this message with a list of the functional groups in several addition and condensation polymers and examples of familiar materials made from them.

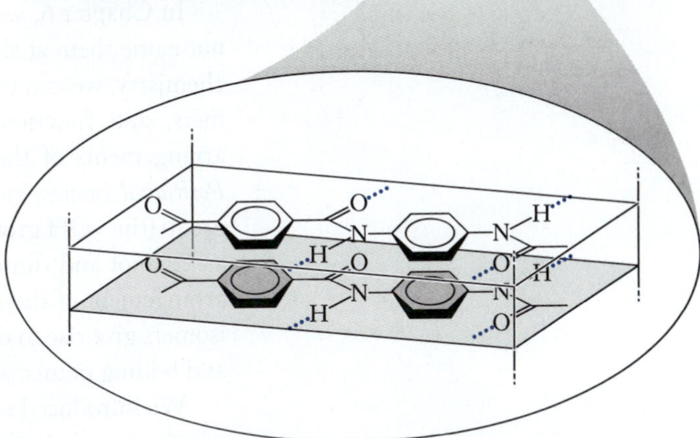

**FIGURE 19.49** A bullet fired point-blank does not puncture fabric made of Kevlar. The strength is due in part to the strong interactions between the functional groups in the molecular structure of Kevlar.

## 19.8 A Brief Survey of Isomers

The existence of isomers is partly responsible for the enormous number and wide variety of organic compounds in the world. We have already discussed isomers in Chapters 5 and 6 and in several sections of this

**TABLE 19.8** Common Polymers and Their Uses

| Name | Abbreviation | Functional Group | Use |
|---|---|---|---|
| **Addition Polymers** | | | |
| Polyethylene | PE | Alkane | Plastic bags and films |
| Poly(tetrafluoroethylene) | Teflon | Fluoroalkane | Nonstick coatings |
| Polyvinyl chloride | PVC | Chloroalkane | Drain pipes |
| Poly(methyl methacrylate) | PMMA | Alkane and ester | Shatter-resistant glass, such as Plexiglas, Lucite |
| Polystyrene | PS | Aromatic hydrocarbon | Dishes, insulation |
| Poly(vinyl alcohol) | PVA | Alcohol | Gloves, bottles |
| **Condensation Polymers** | | | |
| Poly(ethylene glycol) | PEG | Ether | Pharmaceuticals, consumer products |
| Poly(ethylene terephthalate) | PETE | Ester | Plastic bottles |
| Nylon | | Amide | Clothing |
| Kevlar | | Amide | Protective equipment |
| Dacron | | Ester | Clothing |

chapter. This brief survey collects all the information we have presented so far, and **Figure 19.50** summarizes the relationships between the types of isomers we have mentioned. Remember that isomers are molecules that have the same number and same kinds of atoms but differ in how those atoms are arranged. Because the molecules' structures are not the same, their physical and chemical properties differ as well.

**CONNECTION** We introduced stereoisomers in Section 5.6 and constitutional isomers in Section 6.2.

As Figure 19.50 shows, isomers fall into two general categories: constitutional isomers and stereoisomers. Constitutional isomers, which we introduced in Chapter 6 and revisited with alkanes in Section 19.2, differ in how the atoms that make them up are connected. Stereoisomers have their atoms connected in the same way, but the three-dimensional arrangement of their atoms differs. Let's review constitutional isomers first.

In Chapter 6, we saw three kinds of constitutional isomers, although we did not name them at the time. Now that we have a better understanding of organic chemistry, we can categorize them more clearly as chain isomers, positional isomers, and functional isomers. *Chain isomers* are molecules having different arrangements of their carbon skeletons, such as butane and 2-methylpropane. *Positional isomers* such as propanol and isopropanol have the same functional group (the –OH group) bonded to different carbon atoms. *Functional isomers* such as ethanol and dimethyl ether have different functional groups because of the arrangement of their atoms. The different shapes and polarities of constitutional isomers give rise to differences in their physical properties, such as melting points and boiling points, as we saw in Figure 6.4.

We introduced stereoisomers in Chapter 5 in our discussion of chirality. We saw them again in Section 19.3 when discussing alkenes. Stereoisomers are different from constitutional isomers because their atoms have the same connectivity: the same atoms are connected to each other in the same way. What makes

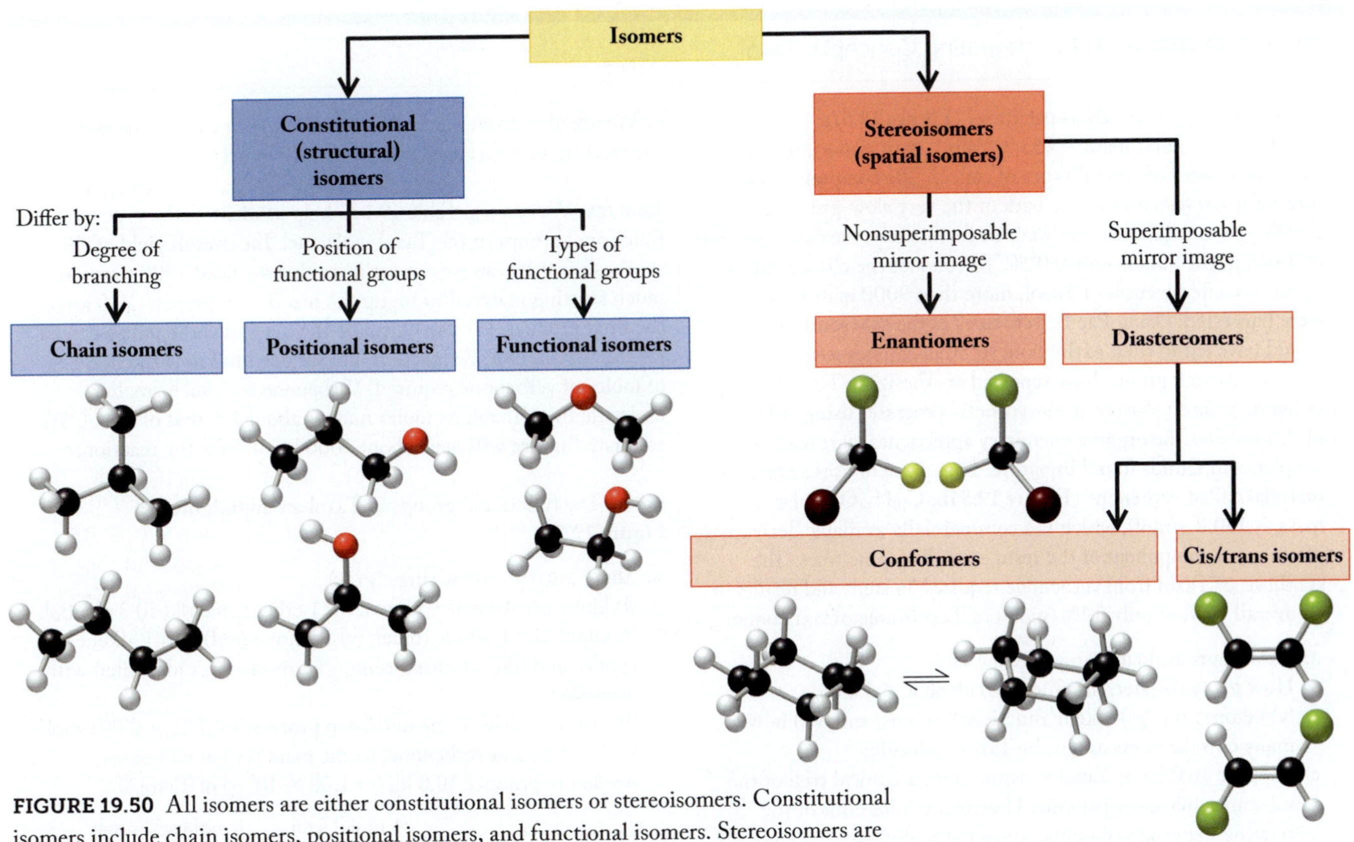


**FIGURE 19.50** All isomers are either constitutional isomers or stereoisomers. Constitutional isomers include chain isomers, positional isomers, and functional isomers. Stereoisomers are further categorized as enantiomers (optically active compounds) or diastereomers.

stereoisomers special is that the arrangement of the atoms in three-dimensional space differs. That's why the prefix *stereo-* is used in the name; it means that we need to consider what the molecule looks like in three dimensions.

Stereoisomers can be *enantiomers* or *diastereomers*. Enantiomers are mirror-image isomers, whereas diastereomers are stereoisomers that are *not* enantiomers. Let's first review enantiomers.

Several structural features can give rise to enantiomers, but the most important one for our purposes is the presence of a carbon atom that has four different atoms or groups of atoms attached to it. This arrangement about a carbon atom makes a molecule nonsuperimposable on its mirror image, much like your right and left hands. Such a molecule is chiral. This molecule and its mirror image rotate beams of plane-polarized light in opposite directions, as we saw in Figure 5.39. This behavior gives rise to another name for enantiomers: *optically active compounds*, or *optical isomers*.

Diastereomers include the cis and trans isomers that can occur because of the C=C double bonds in alkenes. *Conformers*, such as the chair and boat conformations of cyclohexane, are also diastereomers. Other structural features may give rise to enantiomers and diastereomers as well, but they are beyond the scope of this book.

## SAMPLE EXERCISE 19.11 Integrating Concepts: Taxol

Taxol, known generically as paclitaxel (**Figure 19.51a**; $C_{47}H_{51}NO_{14}$, molar mass = 853.9 g/mol), is an important drug in the treatment of several types of cancer. The compound was originally harvested from the bark of the very slow-growing Pacific yew tree (*Taxus brevifolia*). Unfortunately, the concentration of Taxol in the bark is only 0.01%. To conduct the clinical trials to test the effectiveness of Taxol, more than 9000 mature trees were harvested. Using Pacific yew trees as the sole source of Taxol would have led to their extinction, so other sources were needed. Several research groups have reported synthesizing Taxol with either organic chemistry or biosynthetic processes using cultures of *Taxus* cells. The organic chemistry approaches start with other, simpler compounds found in nature. One of them uses a starting material called verbenone (**Figure 19.51b**; $C_{10}H_{14}O$, molar mass = 150.2 g/mol), which is a commercially available derivative of pinene, a component of the resin exuded by pine trees. The synthesis of Taxol from verbenone requires 36 steps and results in an overall yield of only 0.1% (moles of Taxol/mole of verbenone).

a. How many aromatic rings are in Taxol?
b. How many (i) ester, (ii) ether, (iii) alcohol, (iv) amide, (v) ketone, and (vi) alkene functional groups, and (vii) how many chiral centers are in the Taxol molecule?
c. Suppose 10.0 kg of Taxol is required for a clinical trial of the substance in cancer patients. How many kilograms of the starting material verbenone is needed to prepare that amount?

**Collect and Organize** We have the structural formula and molar mass of Taxol, which consists of a variety of functional groups and chiral carbon centers. We know the amount of Taxol we need for a study and the overall yield of the 36-step synthesis used to prepare it. We are also given the structure and molar mass of verbenone, the starting material used in step 1 of the process.

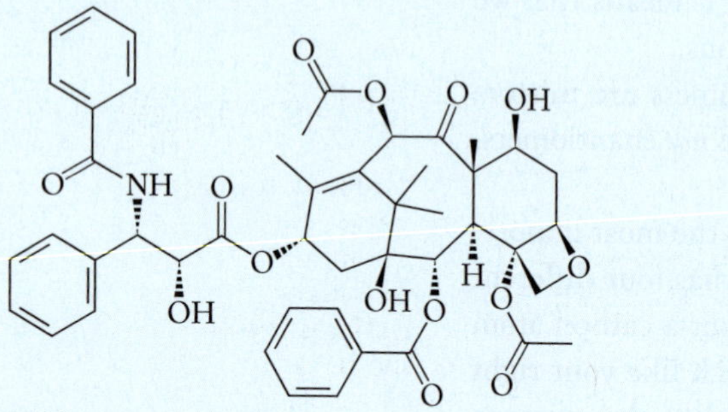

**Analyze** We can use Table 19.1 to help us identify the types of functional groups in the Taxol molecule. The overall yield of the synthesis is 0.1%, so we can estimate that we need 1000 times as much starting material to set up the reaction to prepare 10.0 kg of the final product. We can calculate the mass of verbenone we need for the first step of the reaction, but we need to know the number of moles of verbenone required. Verbenone is a much smaller molecule than Taxol; its molar mass is about 1/5 that of Taxol. We estimate that we will need about 2000 kg of it for the reaction.

**Solve** The functional groups in Taxol are highlighted in **Figure 19.52**.

a. There are 3 aromatic rings (gray).
b. (i) There are 4 esters (yellow); (ii) 1 ether (green); (iii) 3 alcohols (orange); (iv) 1 amide (blue); (v) 1 ketone (red); (vi) 1 alkene (pink), and (vii) 11 chiral centers (carbon atoms identified with asterisks).
c. The overall yield of the multistep process is 0.1%, or 0.001 mol of Taxol/mol of verbenone, so the mass (kg) of verbenone needed to produce 10.0 kg (or $1.00 \times 10^4$ g) of Taxol is

$$1.00 \times 10^4 \text{ g Taxol} \times \frac{1 \text{ mol Taxol}}{853.9 \text{ g Taxol}} \times \frac{1 \text{ mol verbenone}}{0.001 \text{ mol Taxol}}$$

$$\times \frac{150.24 \text{ g verbenone}}{1 \text{ mol verbenone}} \times \frac{1 \text{ kg}}{10^3 \text{ g}} = 2 \times 10^3 \text{ kg verbenone}$$

Thus, we need 2000 kilograms of verbenone to produce just 10 kilograms of Taxol.

**Think About It** The answer matches our initial estimate (to one significant figure, which is all we are entitled to, given the 0.1% yield of the process). Like many natural substances with complex structures, Taxol is a challenging substance to prepare synthetically. The overall yield of a series of reactions is equal to the product of the yields of 36 individual reactions. If the average yield of the reactions is $n$, then $n^{36} = 0.001$. Solving for $n$, we find that the average yield of each reaction is about 83%, which is good for most synthetic work.

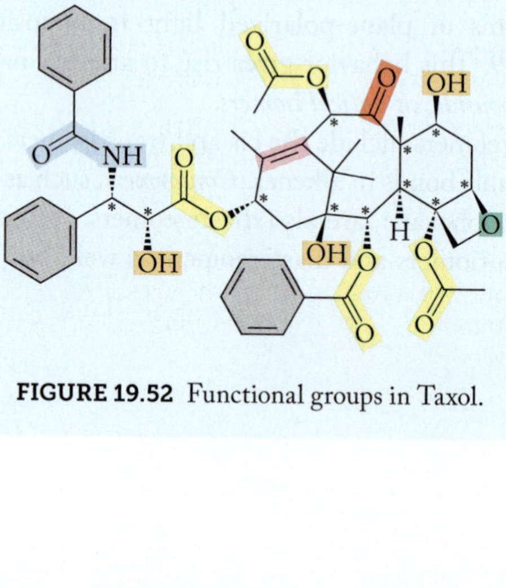

(a) Taxol

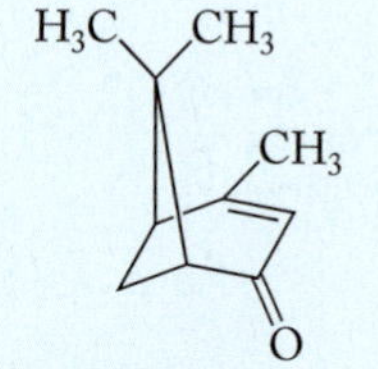

(b) Verbenone, the starting material for 36-step synthesis of taxol

**FIGURE 19.51** One of the laboratory syntheses of (a) Taxol begins with (b) the starting material verbenone.

**FIGURE 19.52** Functional groups in Taxol.

## SUMMARY

**LO1** Alkanes are hydrocarbons that contain only C—C single bonds. **Cycloalkanes** are alkanes containing rings of carbon atoms. Acyclic alkanes are **saturated hydrocarbons** because their molecules contain the maximum number of hydrogen atoms per carbon atom. IUPAC naming rules relate the names of alkanes to their molecular structures. (Section 19.2)

**LO2** Alkanes with the same formula but different bonding patterns are constitutional isomers. IUPAC naming rules relate the names and molecular structures of these isomers. (Section 19.2)

**LO3** The –C═C– double bonds in **alkenes** and the –C≡C– triple bonds in **alkynes** mean that they are **unsaturated hydrocarbons** and that they can engage in **hydrogenation** and **addition reactions** that alkanes cannot. (Section 19.3)

**LO4** Alkenes may have both constitutional isomers and **cis** or **trans** stereoisomers, depending on the arrangement of the groups around the double bond. (Section 19.3)

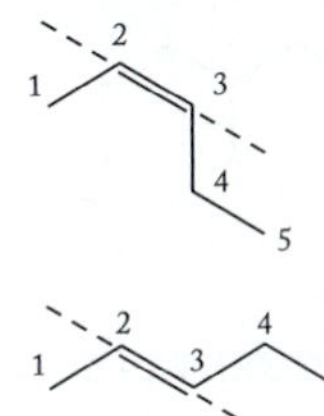


**LO5** Alkenes can be polymerized to **homopolymers**, most of which are **addition polymers**. Alcohols and carboxylic acids react in **condensation reactions** to form **heteropolymers** called polyesters; alcohols and amines react similarly to produce polyamides. The properties of the polymers depend on the identity and arrangement of the monomer units in the polymers. (Sections 19.3, 19.4, 19.6, 19.7)

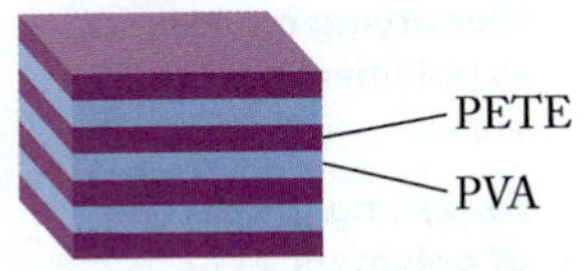


**LO6** Polymers made from ethers may be soluble in both polar and nonpolar liquids and are increasingly used to improve the biodistribution of pharmaceuticals in the body. The identity of monomer units in both natural and synthetic polymers determines their behavior as bulk materials. (Section 19.6)

**LO7** The presence of **heteroatoms** in the functional groups of many organic compounds strongly influences their chemical properties and allows them to interact with polar inorganic compounds, such as water, in ways that nonpolar hydrocarbons cannot. (Sections 19.5 and 19.7)

## PARTICULATE PREVIEW WRAP-UP

Compounds A and B are isomers. Compound A contains a carboxylic acid functional group; compound B contains an ester functional group.

A carboxylic acid can be converted into an ester by reacting with an alcohol.

## PROBLEM-SOLVING SUMMARY

| Type of Problem | Concepts and Equations | Sample Exercises |
|---|---|---|
| **Naming alkanes and drawing their molecular structures** | Start at one end of any branch and assign numbers to each carbon atom. If the chain branches, choose one branch to follow. Repeat the process to explore other side chains and identify the longest chain. The ends of alkane molecules are methyl groups; between those ends, gather all methylene groups inside parentheses to create the final condensed structure. To create a carbon-skeleton structure, use lines to depict single covalent bonds between carbon atoms. The presence of enough H atoms to complete the valence of the carbon atoms is assumed. | **19.1, 19.2** |
| **Recognizing constitutional isomers** | Establish that the compounds have the same molecular formula, and if they do, look for different arrangements of C—C bonds. | **19.3, 19.4** |
| **Distinguishing among alkanes, alkenes, and alkynes** | Acyclic alkanes are saturated hydrocarbons and do not react with $H_2$. Alkenes contain at least one C═C double bond that can combine with a molecule of $H_2$. Alkynes contain at least one C≡C triple bond that can combine with two molecules of $H_2$. | **19.5** |
| **Identifying and naming stereoisomers and constitutional isomers of alkenes** | If molecules have the same formula, look for different arrangements of their bonds. Molecules with two groups on the same side of a C═C bond are cis isomers; those with groups on opposite sides are trans isomers. [cis: R, R above; H, H below — trans: R, H above; H, R below] | **19.6** |

| Type of Problem | Concepts and Equations | Sample Exercises |
|---|---|---|
| **Identifying monomers in polymers and vice versa** | Find the smallest portion of the polymer that is repeated. For condensation polymers, combine monomers with alcohol or amine functional groups and carboxylic acid functional groups to form polymers with ester or amide linkages. | **19.7, 19.9, 19.10** |
| **Assessing properties of polymers** | Evaluate the polarity of the functional groups in the polymer and assess the relative importance of all types of intermolecular forces possible in the molecules. | **19.8** |

## VISUAL PROBLEMS

*(Answers to boldface end-of-chapter questions and problems are in the back of the book.)*

**19.1.** How many degrees of unsaturation does each of the hydrocarbons shown in Figure P19.1 have?

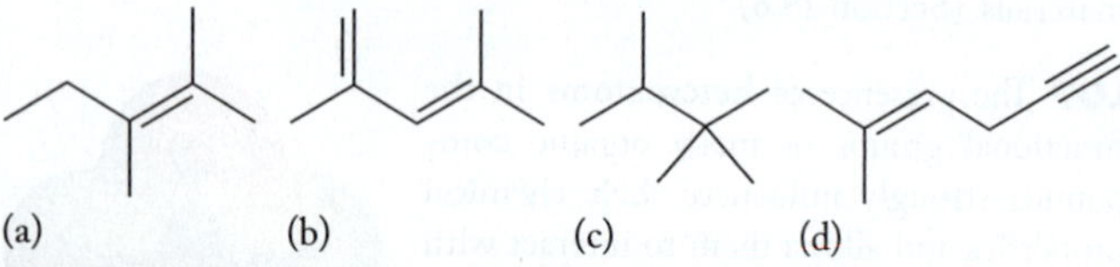

**FIGURE P19.1**

19.2. Which of the hydrocarbons in Figure P19.2 are constitutional isomers of each other?

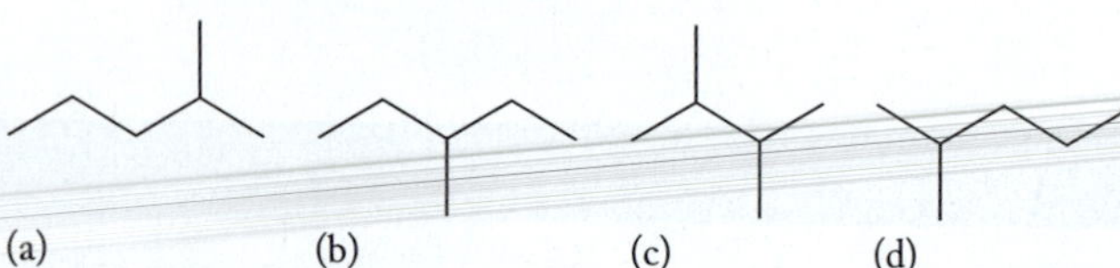

**FIGURE P19.2**

**19.3.** Figure P19.3 shows the carbon-skeleton structures of four organic compounds found in nature as fragrant oils. Which are alkenes?

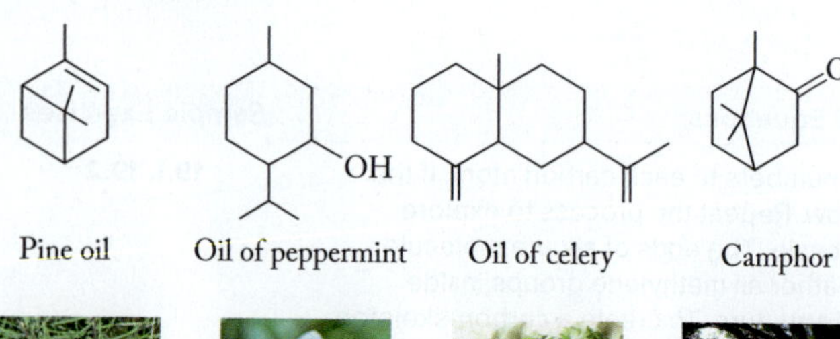

**FIGURE P19.3**

19.4. Which of the three molecules shown in Figure P19.4—acrylonitrile (found in barbeque smoke), capillin (an antifungal drug), and pargyline (an antihypertensive drug)—does *not* contain the alkyne functional group?

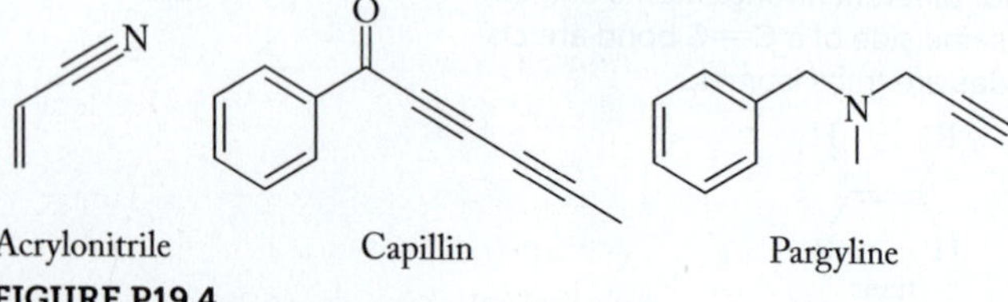


**FIGURE P19.4**

**19.5.** Which molecules in Figure P19.5 are considered aromatic compounds?

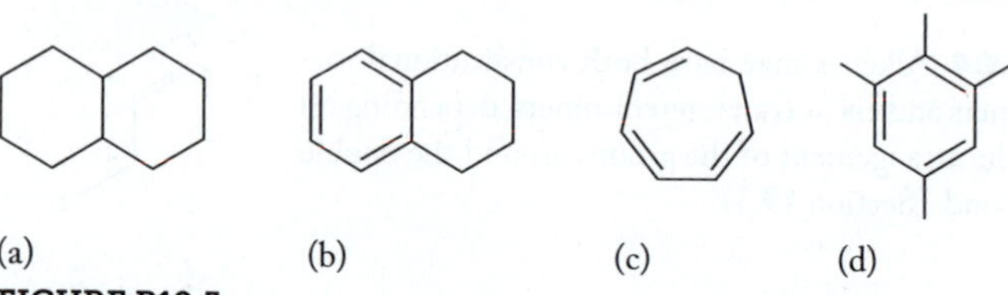

**FIGURE P19.5**

19.6. Benzyl acetate, carvone, and cinnamaldehyde are all naturally occurring oils. Their carbon-skeleton structures are shown in Figure P19.6. Which ones contain an aromatic ring?

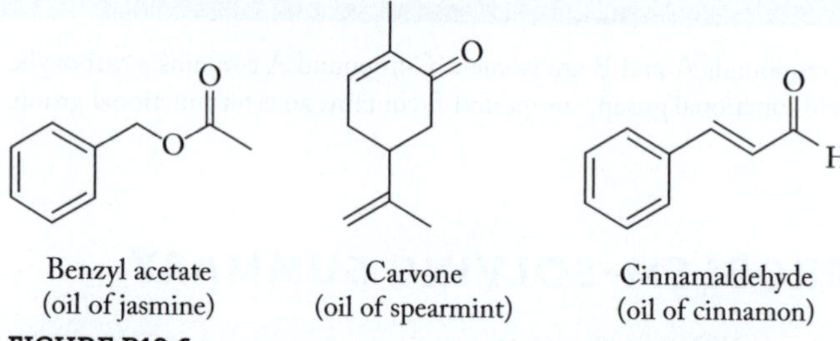


**FIGURE P19.6**

**19.7.** In polypropylene, $\mathrm{-\!\!\!+\!CH_2CH(CH_3)\!+\!\!\!-_n}$, the methyl groups can be in one of two positions with respect to each other (Figure P19.7): (a) on the same side or (b) on opposite sides of the carbon–carbon backbone. Many possible variations exist for (b). One form of the polymer in bulk is rigid and resists deformation; the other is soft and rubbery. Suggest which structure gives rise to which set of properties and explain your choice.

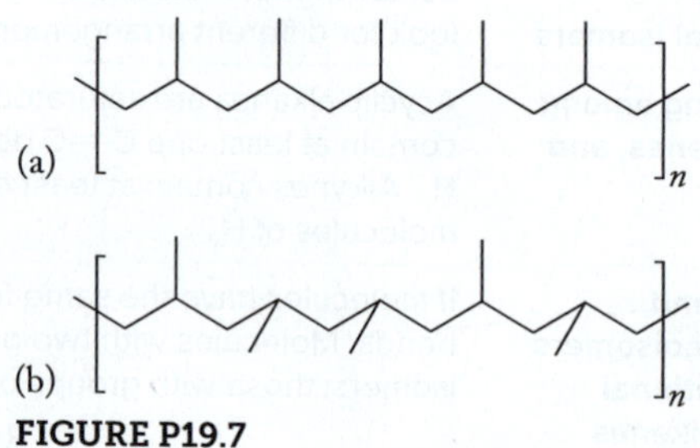


**FIGURE P19.7**

*19.8. The three polymers shown in Figure P19.8 are widely used in the plastics industry. In which of them are the intermolecular forces per mole of monomer the strongest?

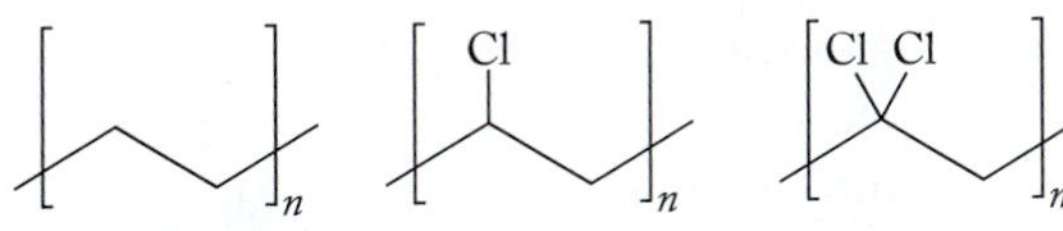


(a) Polyethylene (b) Poly(vinyl chloride) (c) Poly(1,1-dichloroethylene)

**FIGURE P19.8**

***19.9. Silly Putty** Silly Putty is a condensation polymer of dihydroxydimethylsilane (Figure P19.9). Draw the condensed structure of the repeating monomer unit in Silly Putty.

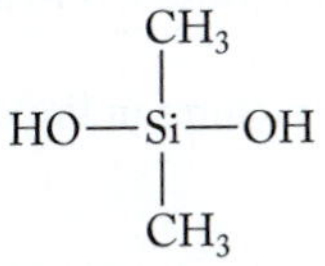

$$\begin{array}{c} CH_3 \\ | \\ HO—Si—OH \\ | \\ CH_3 \end{array}$$

Dihydroxydimethylsilane

**FIGURE P19.9**

19.10. **Orlon and Acrilon Fibers** Figure P19.10 shows the carbon-skeleton structure of polyacrylonitrile, which is marketed as Orlon and Acrilon. Draw the Kekulé structure of the monomeric reactant that produces this polymer.

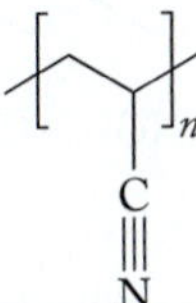


Polyacrylonitrile

**FIGURE P19.10**

**19.11.** Rubber is a polymer of isoprene. It is sometimes called polyisoprene. Polyisoprene comes in two forms (Figure P19.11): *cis*-polyisoprene is the soft, flexible material we associate with the term *rubber*; gutta-percha, or *trans*-polyisoprene, is a much harder material.

a. Draw the monomeric units of *cis*- and *trans*-polyisoprene.

b. Consider the structures of the polymers and comment on why one form is rubbery and the other hard.

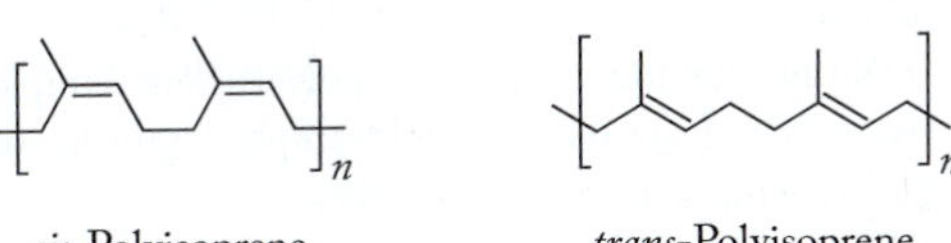


*cis*-Polyisoprene *trans*-Polyisoprene

**FIGURE P19.11**

19.12. Use representations [A] through [I] in Figure P19.12 to answer questions a–f.

a. Identify each functional group in [B], [D], [E], [F] and [H].

b. Which polymers were formed by addition?

c. Which polymers were formed by condensation?

d. Which functional groups reacted to form the condensation polymers?

e. Which polymer structures are likely to result in rigid, hard materials?

f. Which polymer structures are likely to result in flexible, stretchable, soft materials?

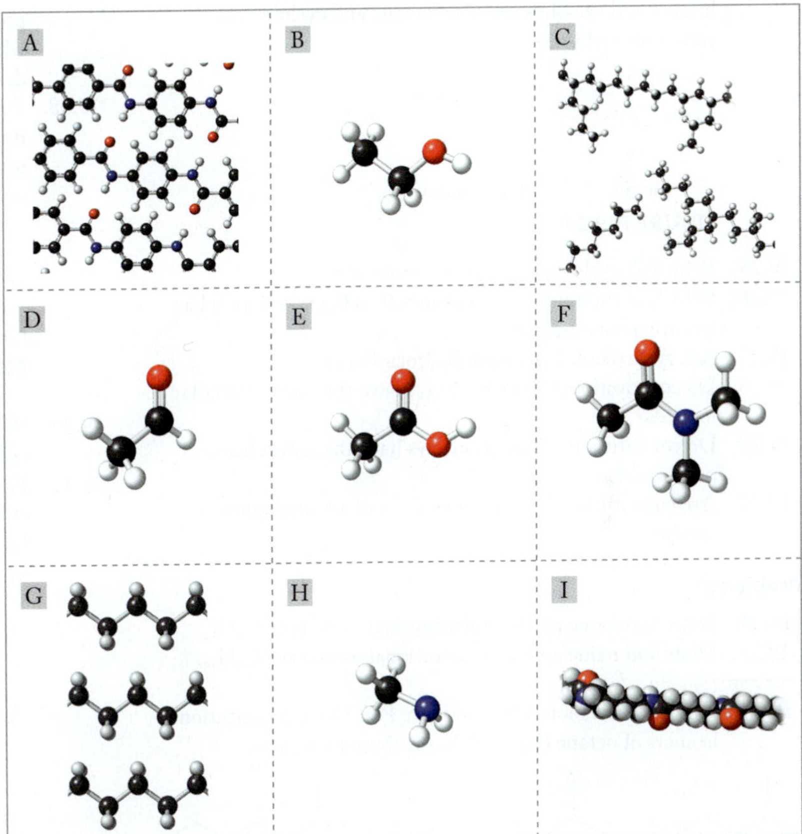


**FIGURE P19.12**

# QUESTIONS AND PROBLEMS

## Carbon: The Stuff of Daily Life

### Concept Review

**19.13.** List the different sets of hybrid orbitals (valence bond theory) used to describe bonding in organic compounds. What combination of single and multiple bonds is possible with each hybridized set?

19.14. What is the principal difference between an oligomer and a polymer formed from the same monomer?

***19.15.** Is the interstitial alloy tungsten carbide (WC) considered an organic compound?

*19.16. Calcium carbide, $CaC_2$, was once used in miner's lamps. Reaction of $CaC_2$ with water yields acetylene that, when ignited, gives light. Is calcium carbide considered an organic compound?

**19.17.** Which of the functional groups in Table 19.1 are polar?

19.18. Which of the compounds in Table 19.1 should be water soluble?

**19.19.** If the average molar mass of a polyethylene sample A is twice that of sample B, which sample begins to soften at a higher temperature? Explain why.

19.20. Which of the following properties of polyethylene increases as the number of monomer units per molecule of the polymer increases? (a) melting point; (b) viscosity; (c) density; (d) C:H ratio; (e) fuel value

## Alkanes

### Concept Review

**19.21.** Do linear and branched alkanes with the same number of carbon atoms all have the same empirical formula?

19.22. If an alkane and a cycloalkane have equal numbers of carbon atoms per molecule, do they have the same number of hydrogen atoms?

**19.23.** What is the hybridization of carbon in alkanes?

19.24. Figure P19.24 shows the carbon-skeleton structures of hexane and cyclohexane. Are hexane and cyclohexane constitutional isomers?

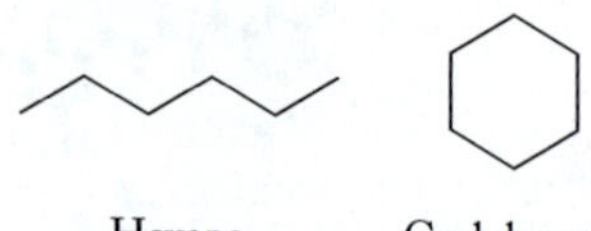

Hexane Cyclohexane

**FIGURE P19.24**

**19.25.** Why isn't cyclohexane a planar molecule?

19.26. Which of the simple cycloalkanes ($C_nH_{2n}$; $n = 3$ to 8) has a nearly planar geometry?

**19.27.** Are cycloalkanes saturated hydrocarbons?

19.28. Do constitutional isomers always have the same molecular formula?

**19.29.** Do constitutional isomers always have the same chemical properties?

19.30. Are constitutional isomers members of a homologous series?

### Problems

**19.31.** Draw and name all the constitutional isomers of $C_5H_{12}$.

19.32. Draw and name all the constitutional isomers of $C_6H_{14}$.

**19.33.** Which of the molecules in Figure P19.33 are constitutional isomers of octane ($C_8H_{18}$)? Name these molecules.

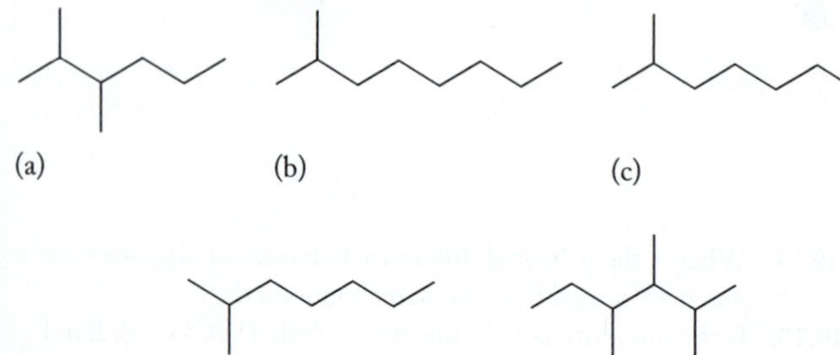


**FIGURE P19.33**

19.34. Which of the molecules in Figure P19.34 are constitutional isomers of heptane ($C_7H_{16}$)? Name these molecules.

(a) (b) (c) (d)

(e) (f)

**FIGURE P19.34**

**19.35.** Convert the carbon-skeleton structures in Problem 19.33 to molecular formulas.

19.36. Convert the carbon-skeleton structures in Problem 19.34 to molecular formulas.

**19.37. Crude Oil** Place the following molecules, all of which are products of the distillation of crude oil, in the order in which they would appear in the distillate during simple distillation: $C_6H_{14}$, $C_{18}H_{38}$, $C_{12}H_{26}$, $C_9H_{20}$.

*19.38. Figure P19.38 shows the carbon-skeleton structures of three hydrocarbons. Rank them in order of decreasing intermolecular forces between molecules of (a), between molecules of (b), and between molecules of (c).

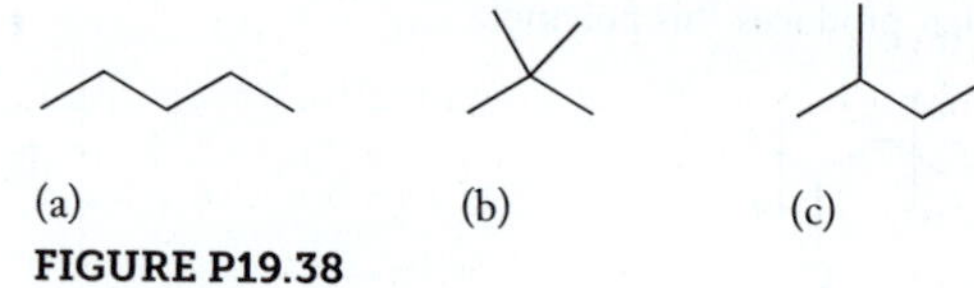


**FIGURE P19.38**

**19.39.** Which has a higher hydrogen-to-carbon ratio: hexane or cyclohexane?

*19.40. For linear alkanes, each methylene group ($-CH_2-$) in the structure contributes 658 kJ/mol to the heat of combustion. Heats of combustion for some cycloalkanes are listed below as a function of ring size. Explain the trend in the data.

| Ring size | Heat of combustion per $-CH_2-$ group (kJ/mol) |
|---|---|
| 3 | 696 |
| 4 | 686 |
| 5 | 663 |
| 6 | 658 |

## Alkenes and Alkynes

### Concept Review

**19.41.** Can combustion analysis distinguish between an alkene and a cycloalkane containing the same number of carbon atoms?

19.42. Can their reaction with hydrogen distinguish between an alkene and a cycloalkane containing the same number of carbon atoms?

**19.43.** Why don't the alkenes in Figure P19.43 have cis and trans isomers?

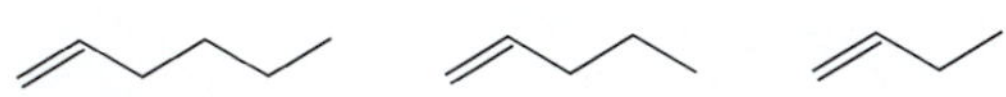

**FIGURE P19.43**

19.44. Why don't alkynes have cis and trans isomers?

***19.45.** Figure P19.45 shows the carbon-skeleton structure of carvone, which is found in oil of spearmint. Why doesn't the molecule carvone have cis and trans isomers?

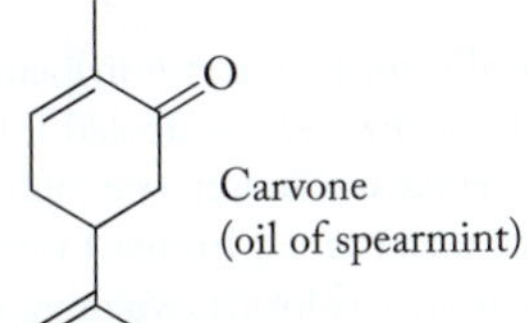


Carvone
(oil of spearmint)

**FIGURE P19.45**

*19.46. Figure P19.46 shows the carbon-skeleton structure of the antifungal compound capillin. Are the $\pi$ electrons in capillin delocalized?

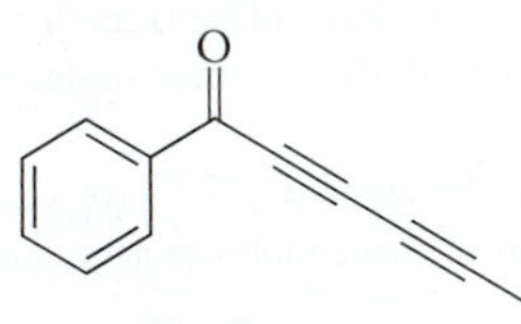


Capillin

**FIGURE P19.46**

**19.47.** Ethylene reacts quickly with HBr at room temperature, but polyethylene is chemically unreactive toward HBr. Explain why these related substances have such different properties.

*19.48. Polymerization of butadiene ($CH_2{=}CHCH{=}CH_2$) does not yield the same polymer as polymerization of ethylene ($CH_2{=}CH_2$).

a. How could we convert poly(butadiene) into poly(ethylene)?

b. Predict the reactivity of poly(butadiene) with HBr.

## Problems

**19.49.** Using the average bond strengths given in Appendix 4, estimate the molar heat of hydrogenation, $\Delta H_{\text{hydrogenation}}$, for the conversion of $C_2H_4$ to $C_2H_6$.

$$H_2C{=}CH_2(g) + H_2(g) \rightarrow CH_3CH_3(g)$$

19.50. Using the average bond strengths given in Appendix 4, estimate the molar heat of hydrogenation, $\Delta H_{\text{hydrogenation}}$, for the conversion of $C_2H_2$ to $C_2H_6$.

$$CH{\equiv}CH(g) + 2\,H_2(g) \rightarrow CH_3CH_3(g)$$

**19.51. Cinnamon** Label the isomers of cinnamaldehyde (oil of cinnamon) in Figure P19.51 as cis or trans and as *E* or *Z*.

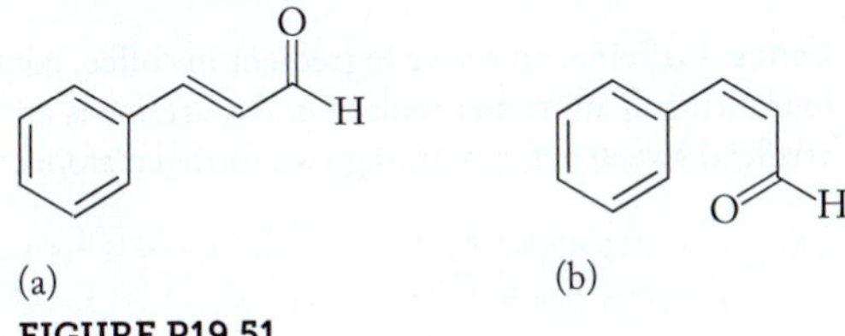


**FIGURE P19.51**

19.52. Prostaglandins, naturally occurring compounds in our bodies that cause inflammation and other physiological responses, are formed from arachidonic acid, an unsaturated hydrocarbon containing four C=C bonds and a carboxylic acid functional group. The stereoisomer containing all cis double bonds is shown in Figure P19.52. How many stereoisomers other than this one are possible? Draw the isomer containing all trans double bonds.

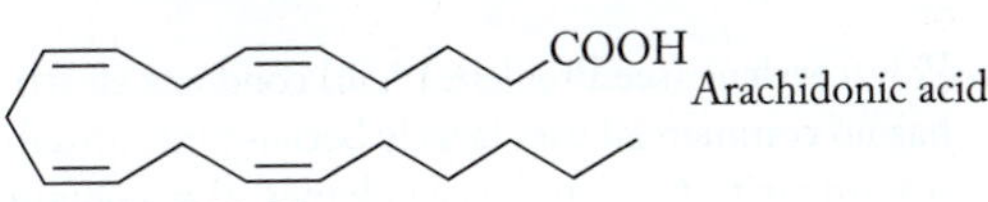


**FIGURE P19.52**

***19.53.** Given the following two reactions and thermodynamic data from Appendix 4, estimate $\Delta H_{\text{rxn}}$ for the hydrogenation of acetylene ($C_2H_2$) with one mole of hydrogen gas to make ethylene ($C_2H_4$).

$$\underset{\text{Acetylene}}{HC{\equiv}CH(g)} + 2\,H_2(g) \rightarrow \underset{\text{Ethane}}{CH_3CH_3(g)}$$

$$\underset{\text{Ethylene}}{H_2C{=}CH_2(g)} + H_2(g) \rightarrow \underset{\text{Ethane}}{CH_3CH_3(g)}$$

*19.54. The heat of hydrogenation of *cis*-2-butene is −119.7 kJ/mol, whereas that of the trans isomer is −115.5 kJ/mol. Draw both isomers and the product of the hydrogenation reaction, and locate them on the graph of relative energy given in Figure P19.54. The condensed structural formula of 2-butene is $CH_3CH{=}CHCH_3$.

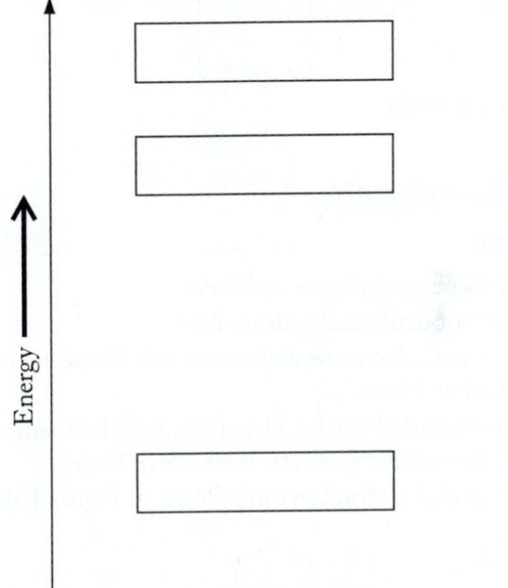


**FIGURE P19.54**

**19.55. Making Glue** Wood glue, or "carpenter's glue," is made of poly(vinyl acetate). Draw the carbon-skeleton structure of this polymer. The monomer is shown in Figure P19.55.

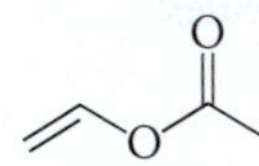


Vinyl acetate

**FIGURE P19.55**

*19.56. The 2000 Nobel Prize in Chemistry was awarded for research on the electrically conductive polymer polyacetylene.

a. Draw the carbon-skeleton structure of three monomeric units of the addition polymer that results from the polymerization of acetylene, HC≡CH.

b. There are two possible stereoisomers of polyacetylene. Describe the two isomeric forms.

---

***19.57.** Polyacetylene (see Problem 19.56) conducts electricity but has no commercial uses largely because it is unstable in air, reacting with $O_2$ to produce polymers that contain carbonyl groups and ether linkages (Figure P19.57). Suggest why these changes in composition and structure decrease the conductivity of the polymer.

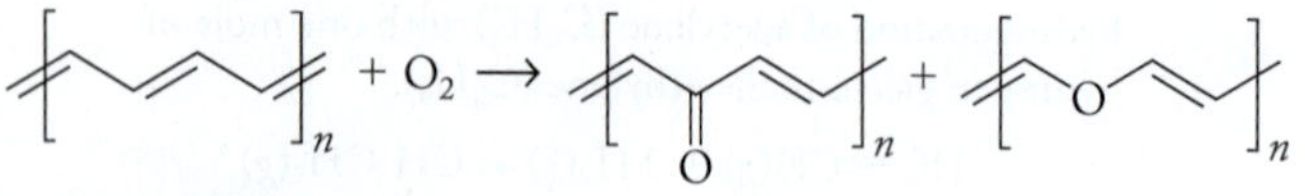

**FIGURE P19.57**

19.58. Substituted polyacetylene can be made by replacing hydrogen atoms with functional groups on the backbone of the polymer (Figure P19.58). The backbone may bend when this is done, decreasing the conductivity of the polymer. Why does bending the backbone of the chain decrease conductivity?

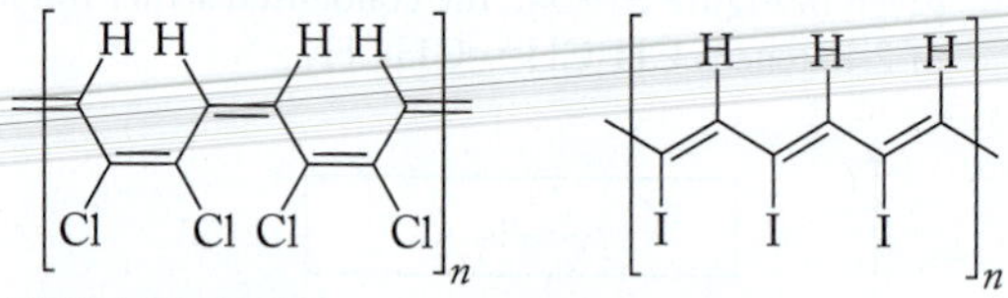

**FIGURE P19.58**

## Aromatic Compounds

### Concept Review

**19.59.** Why is benzene a planar molecule?

19.60. Why are aromatic molecules stable?

**19.61.** Do tetramethylbenzene and pentamethylbenzene have constitutional isomers?

19.62. Why aren't butadiene ($C_4H_6$) and 1,3-cyclohexadiene ($C_6H_8$) considered aromatic molecules? Their carbon-skeleton structures are shown in Figure P19.62.

Butadiene 1,3-Cyclohexadiene

**FIGURE P19.62**

***19.63.** Pyridine (Figure P19.63) has the molecular formula $C_6H_5N$. Is pyridine an aromatic molecule?

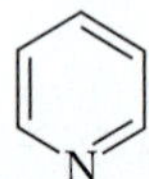


Pyridine

**FIGURE P19.63**

*19.64. Is cyclooctatetraene (Figure P19.64) an aromatic compound?

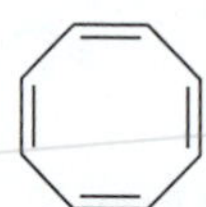

Cycloocatatetraene

**FIGURE P19.64**

### Problems

**19.65.** Draw all the constitutional isomers of trimethylbenzene.

*19.66. The three isomeric trichlorobenzenes are labeled I, II, and III. As a result of a chemical reaction, one methyl group ($-CH_3$) is added to each isomer. Isomer I forms one monomethyl compound, isomer II forms two monomethyl compounds, and isomer III forms three monomethyl compounds. Draw the structures of the original trichlorobenzene isomers.

---

**19.67.** Calculate the fuel values of gaseous benzene ($C_6H_6$) and ethylene gas ($C_2H_4$). Does one mole of benzene have a higher or lower fuel value than that of three moles of ethylene?

19.68. Does one mole of gaseous benzene ($C_6H_6$) have a higher or lower fuel value than that of three moles of acetylene gas ($C_2H_2$)?

## Amines

### Concept Review

**19.69.** Explain why methylamine ($CH_3NH_2$) is more soluble in water than butylamine [$CH_3(CH_2)_3NH_2$].

19.70. Combustion of hydrocarbons in air yields carbon dioxide and water. What other product is expected in the complete combustion of amines?

### Problems

**19.71. Are You Hungry?** Serotonin and amphetamine both contain the amine functional group (Figure P19.71). Serotonin is responsible, in part, for signaling that we have had enough to eat. Amphetamine, an addictive drug, can be used as an appetite suppressant. Identify the primary and secondary amine functional groups in these molecules.

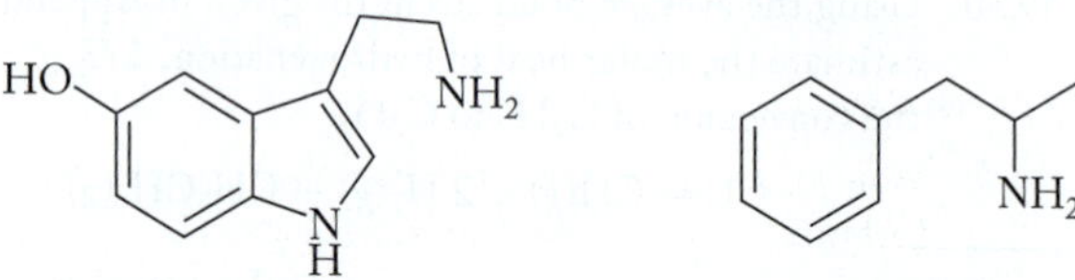


**FIGURE P19.71**

19.72. **Coffee** Caffeine, an active ingredient in coffee, contains four nitrogen atoms per molecule. Aspartame is an artificial sweetener containing two nitrogen atoms.

Structures of caffeine and aspartame are shown in Figure P19.72. Which nitrogen atoms represent secondary amines and which ones represent tertiary amines?

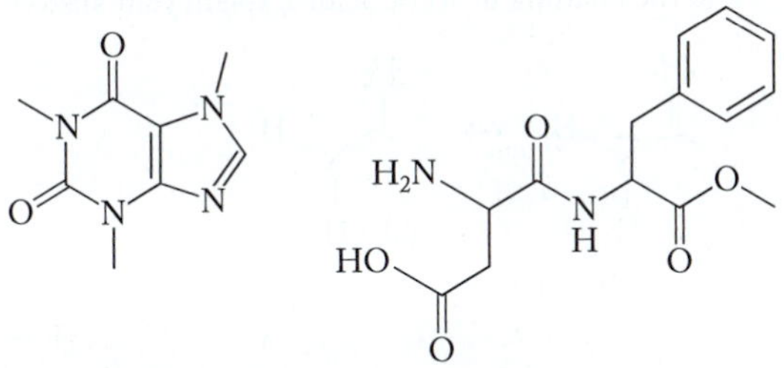

Caffeine Aspartame

**FIGURE P19.72**

**19.73.** **Renewable Energy** Microorganisms of the genus *Methanosarcina* convert amines to methane. Their action helps make methane a renewable energy source. Determine the standard enthalpy of the following reaction from the appropriate standard enthalpies of formation ($\Delta H^\circ_{f,CH_3NH_2} = -23.0$ kJ/mol):

$$4\,CH_3NH_2(g) + 2\,H_2O(\ell) \rightarrow 3\,CH_4(g) + CO_2(g) + 4\,NH_3(g)$$

19.74. Determine the $\Delta H^\circ_{rxn}$ values of the following combustion reactions of methylamine ($\Delta H^\circ_{f,CH_3NH_2} = -23.0$ kJ/mol):

$$4\,CH_3NH_2(g) + 13\,O_2(g) \rightarrow 4\,CO_2(g) + 4\,NO_2(g) + 10\,H_2O(\ell)$$

$$4\,CH_3NH_2(g) + 6\,O_2(g) \rightarrow 4\,CO_2(g) + 4\,NH_3(g) + 4\,H_2O(\ell)$$

***19.75.** Methylamine is a weak base.

a. Use information in Appendix 5 to sketch the titration curve for the titration of 125 mL of a 0.015 *M* solution of methylamine with 0.100 *M* HCl.

b. Label the curve with the pH of the analyte solution, the p$K_a$ of the analyte, and the pH and titrant volumes halfway to the equivalence point and at the equivalence point.

c. Draw the structures of the species present in the solution at the equivalence point.

*19.76. The p$K_b$ values in Appendix 5 tell us that methylamine is a stronger base than ammonia and that dimethylamine is even stronger. Use the differences in their molecular structures to explain this trend in the strengths of these three bases.

## Alcohols, Ethers, and Reformulated Gasoline

### Concept Review

**19.77.** Why are the fuel values of ethanol and dimethyl ether (Figure P19.77) lower than that of ethane?

Dimethyl ether Ethanol

**FIGURE P19.77**

19.78. Would you expect the fuel value of alcohols to increase or decrease as the number of carbon atoms in the alcohol increases?

**19.79.** Why do ethers typically boil at lower temperatures than alcohols with the same molecular formula?

19.80. Which do you expect to be more soluble in water, MTBE or 2,2-dimethylbutane (Figure P19.80)? Explain your answer.

MTBE 2,2-Dimethylbutane

**FIGURE P19.80**

***19.81.** **Clean Skin** Disposable wipes used to clean the skin before getting an immunization shot contain ethanol. After your arm is wiped, your skin feels cold. Why?

*19.82. **Dry Gas** In the winter months in cold climates, water condensing in a vehicle's gas tank reduces engine performance. An auto mechanic recommends adding "dry gas" to the tank during your next fill-up. Dry gas is typically an alcohol that dissolves in gasoline and absorbs water. On the basis of the structures shown in Figure P19.82, which product would you predict would do a better job—methanol or 2-propanol?

$CH_3OH$ Methanol 2-Propanol

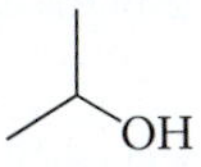

**FIGURE P19.82**

### Problems

**19.83.** Which of the compounds in Figure P19.83 are alcohols and which ones are ethers? Place them in order of increasing boiling point.

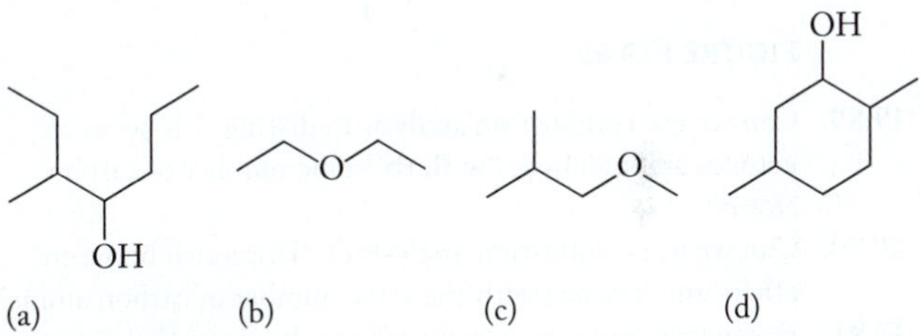


(a) (b) (c) (d)

**FIGURE P19.83**

19.84. Which of the compounds in Figure P19.84 are alcohols and which ones are ethers? Place them in order of increasing vapor pressure at 25°C.

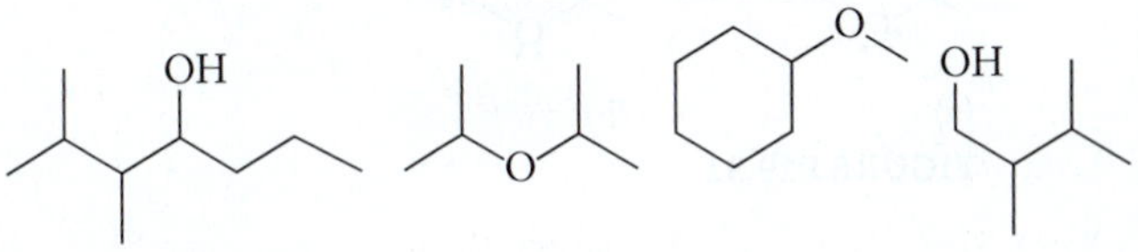

(a) (b) (c) (d)

**FIGURE P19.84**

## *Aldehydes, Ketones, Carboxylic Acids, Esters, and Amides*

### Concept Review

**19.85.** Explain why carboxylic acids tend to be more soluble in water than aldehydes with the same number of carbon atoms.

19.86. In reference books, diethyl ether (Figure P19.86) is usually listed as "slightly soluble" in water, but 2-butanone is listed as "very soluble." Why is 2-butanone more soluble?

Diethyl ether 2-Butanone

**FIGURE P19.86**

**19.87.** Are butanal and 2-butanone (Figure P19.87) constitutional isomers?

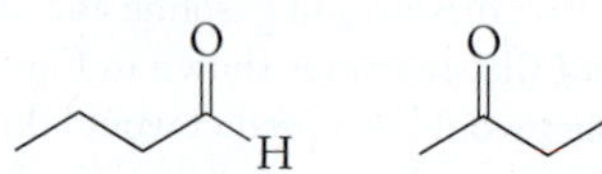

Butanal 2-Butanone

**FIGURE P19.87**

19.88. **Apples** Both of the esters shown in Figure P19.88 are found in apples and contribute to the flavor and aroma of the fruit. Are the two compounds identical, constitutional isomers, or stereoisomers, or do they have different molecular formulas?

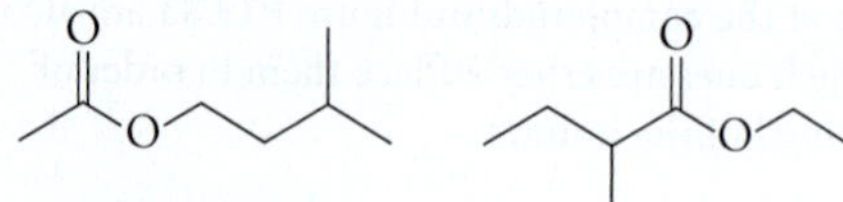

**FIGURE P19.88**

**19.89.** Can we use combustion analysis to distinguish between ketones and aldehydes with the same number of carbon atoms?

19.90. Can we use combustion analysis to distinguish between ethers and ketones with the same number of carbon atoms?

**19.91.** Resonance forms for acetic acid are shown in Figure P19.91. Which one contributes more to bonding? Explain your choice.

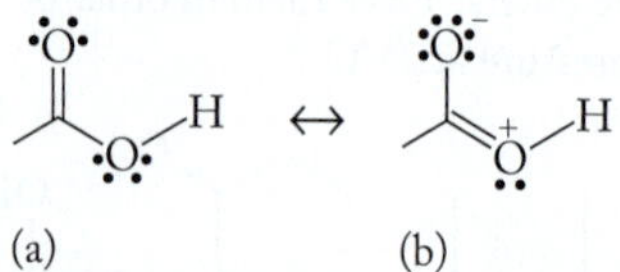

(a) (b)

**FIGURE P19.91**

19.92. Figure P19.92 shows resonance forms for acetamide and acetic acid. Does resonance form (a) contribute more to the bonding in acetamide than resonance form (b) contributes to the bonding in acetic acid? Explain your answer.

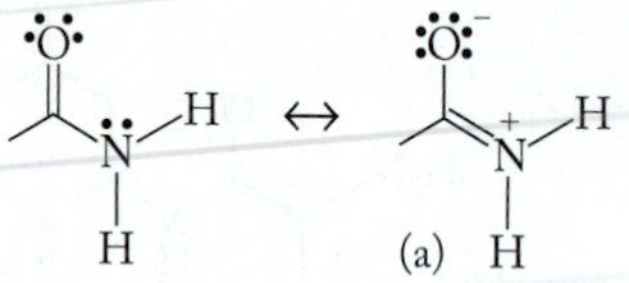

Acetamide

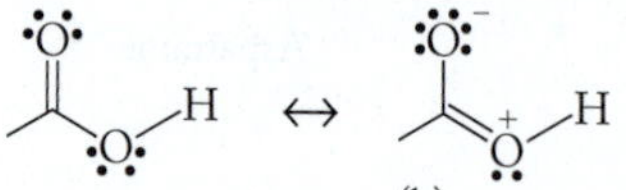

Acetic acid

**FIGURE P19.92**

**19.93.** What distinguishes an amine from an amide?

19.94. Explain why nylon reacts with acid, whereas polyethylene is inert to acid.

### Problems

**19.95.** Which of the compounds in Figure P19.95 are constitutional isomers of the aldehyde $C_5H_{10}O$?

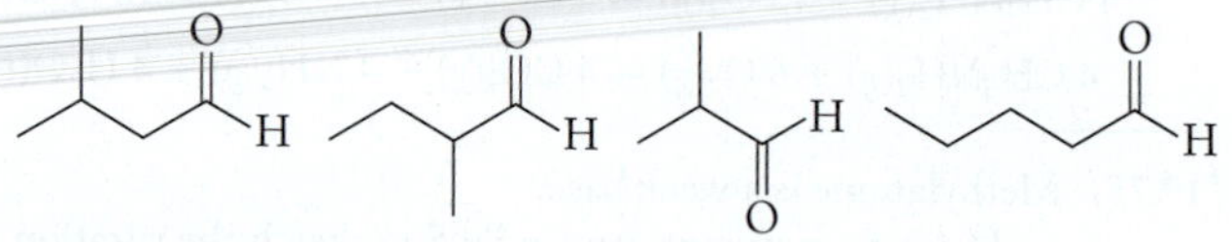

(a) (b) (c) (d)

**FIGURE P19.95**

19.96. Each of the natural products in Figure P19.96 contains more than one functional group. Which of the compounds is an aldehyde?

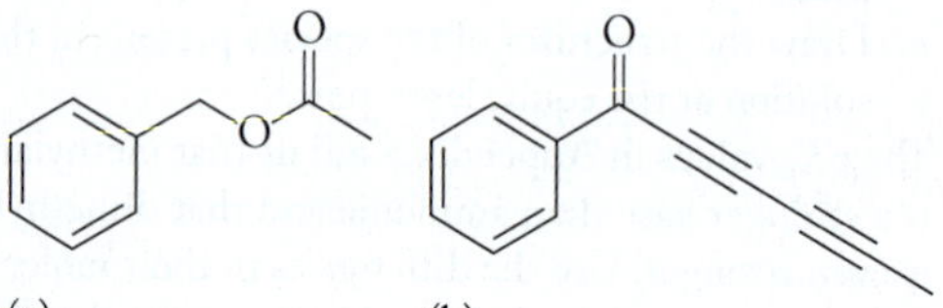

(a) (b) 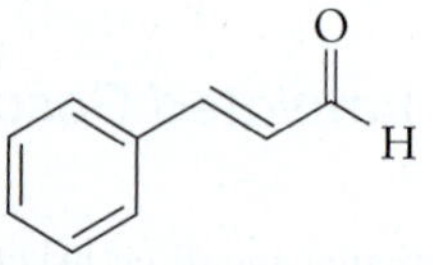 (c)

**FIGURE P19.96**

**19.97.** Which of the compounds in Figure P19.97 is a ketone?

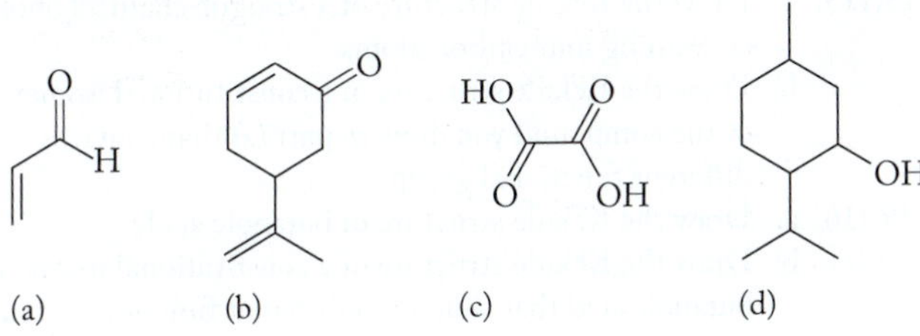


**FIGURE P19.97**

19.98. Propanal and acetone (2-propanone) have the same molecular formula, $C_3H_6O$, but different structures (Figure P19.98). Which compound is a ketone?

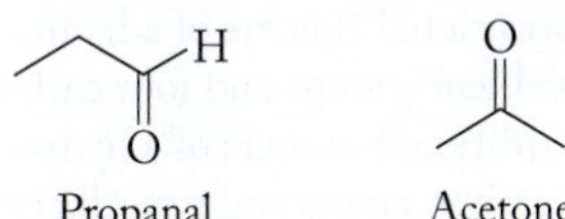


**FIGURE P19.98**

**19.99.** Plot the carbon-to-hydrogen ratio in aldehydes with one to six carbon atoms as a function of the number of carbon atoms. Does this graph correlate better with the plot of C:H ratios for alkanes or for alkenes?

19.100. Plot the carbon-to-hydrogen ratio in carboxylic acids with one to six carbon atoms as a function of the number of carbon atoms. Does this graph correlate better with the plot of C:H ratios for alkanes or for alkenes?

**19.101.** Esters are responsible for the odors of fruits, including pineapples, bananas, and apples. Figure P19.101 shows three esters from these fruits. Identify the alcohol and carboxylic acid that react to form these compounds.

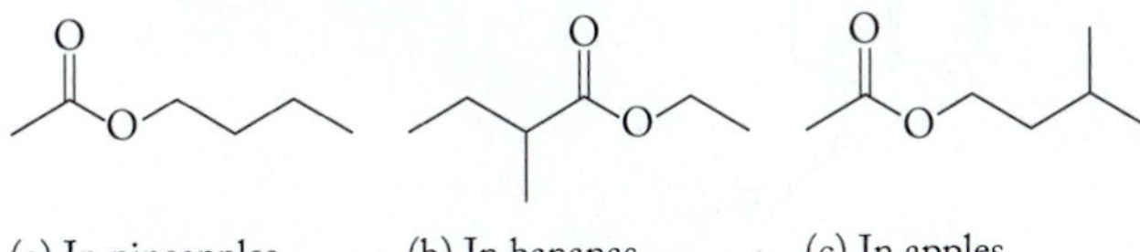


**FIGURE P19.101**

19.102. **Beeswax** The waxy material found in beehives (Figure P19.102) is an ester formed from an alcohol and a carboxylic acid with a long hydrocarbon chain. Identify the alcohol and acid in beeswax.

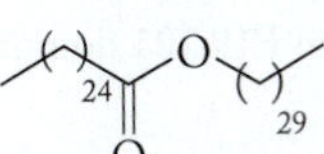


**FIGURE P19.102**

**19.103. Stimulants and Tranquilizers** Nicotine is a stimulant found in tobacco, whereas diazepam (trade name, Valium) is a tranquilizer. Both molecules contain two nitrogen atoms in addition to other functional groups. Do the nitrogen atoms shown in blue in Figure P19.103 belong to an amine or an amide?

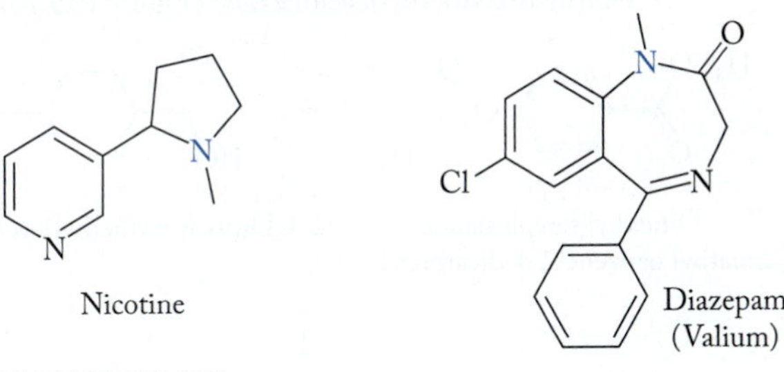


**FIGURE P19.103**

19.104. **Hot Peppers** Piperine and capsaicin are ingredients of peppers that give "heat" to spicy foods. Do the nitrogen atoms shown in blue on the cover of this book belong to an amine or an amide?

**19.105.** Reactions between 1,6-diaminohexane, $H_2N(CH_2)_6NH_2$, and different dicarboxylic acids, $HOOC(CH_2)_nCOOH$, are used to prepare polymers that have a structure similar to that of nylon. How many carbon atoms were in the dicarboxylic acids used to prepare the polymers with the repeating units shown in Figure P19.105?

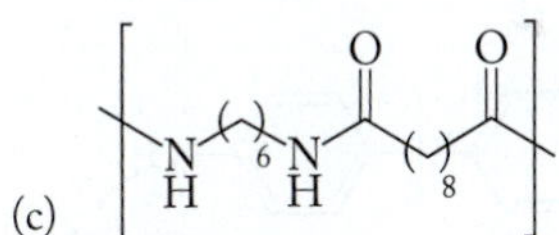


**FIGURE P19.105**

19.106. The two polymers in Figure P19.106 have the same empirical formula.

a. What pairs of monomers could be used to make each of them?
b. How might the physical properties of these two polymers differ?

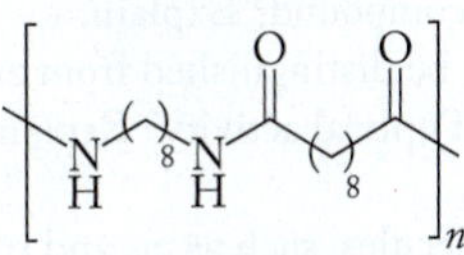


Polymer I

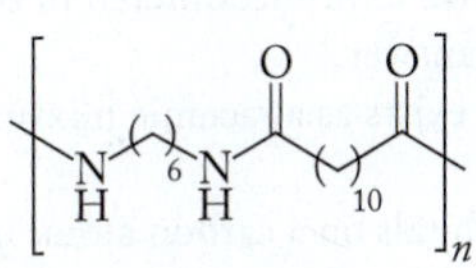


Polymer II

**FIGURE P19.106**

**19.107.** The polyester called Kodel is made with polymeric strands prepared by the reaction of dimethyl terephthalate with 1,4-di(hydroxymethyl)cyclohexane (Figure P19.107).

Dimethyl terephthalate (dimethyl benzene-1,4-dicarboxylate) + 1,4-Di(hydroxymethyl)cyclohexane

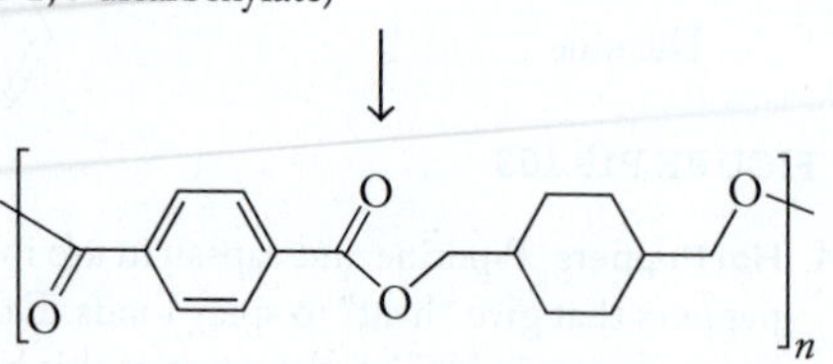

Kodel

**FIGURE P19.107**

a. Is Kodel a condensation polymer or an addition polymer? What is the other product of the reaction?

*b. Dacron (Figure 19.46) is made from dimethyl terephthalate and ethylene glycol. What properties of Kodel fibers might make them better than Dacron as a clothing material?

19.108. Lexan is a polymer belonging to the class of materials called polycarbonates. Figure P19.108 shows the polymerization reaction for Lexan.

a. What other compound is formed in the polymerization reaction?

*b. Why is Lexan called a polycarbonate?

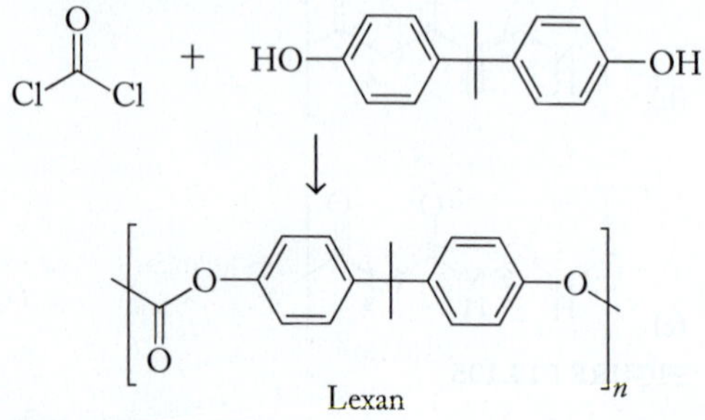

Lexan

**FIGURE P19.108**

## A Brief Survey of Isomers

### Concept Review

**19.109.** Can the terms *enantiomer*, *achiral*, and *optically active* all be used to describe a single compound? Explain.

*19.110. Could a racemic mixture be distinguished from an achiral compound on the basis of optical activity? Explain your answer.

**19.111.** Can stereoisomers of molecules, such as cis and trans RCH═CHR, also have optical isomers? (R may be any of the functional groups we have encountered in this textbook.) Explain your answer.

19.112. The compound $C_3H_6Br_2$ exists as a racemic mixture. Draw its structure.

**19.113.** Which type of hybrid orbitals on a carbon atom, *sp*, $sp^2$, or $sp^3$, can give rise to enantiomers?

*19.114. Could an oxygen atom in an alcohol, ketone, or ether be a chiral center in a molecule?

### Problems

**19.115.** a. Draw the Kekulé structure of a straight-chain alcohol containing four carbon atoms.

b. Draw the Kekulé structure of a constitutional isomer of the compound you drew in part (a) that contains a different functional group.

19.116. a. Draw the Kekulé structure of butanoic acid.

b. Draw the Kekulé structure of a constitutional isomer of butanoic acid that contains a different functional group.

**19.117.** a. Draw all the constitutional isomers of a saturated hydrocarbon with four carbon atoms.

b. Draw all the constitutional isomers of a hydrocarbon that contains one alkene group and four carbon atoms.

c. Draw all the constitutional isomers of a hydrocarbon that contains two alkene groups and four carbon atoms.

19.118. a. Draw all the constitutional isomers of a hydrocarbon that contains five carbon atoms and two alkene groups.

b. Draw all the constitutional isomers of a hydrocarbon that contains five carbon atoms, one alkene group, and one alkyne group.

**19.119. Nutmeg** The compound myristicin (Figure P19.119) is in part responsible for the flavor of nutmeg. Limited animal studies on myristicin have shown that it may have a small influence on sexual behavior, but large doses cause unpleasant side effects.

a. The structure shown for myristicin has four double bonds. Are cis and trans isomers of myristicin possible?

*b. How many structural isomers are possible as determined solely from the arrangement of the groups on the aromatic ring in myristicin?

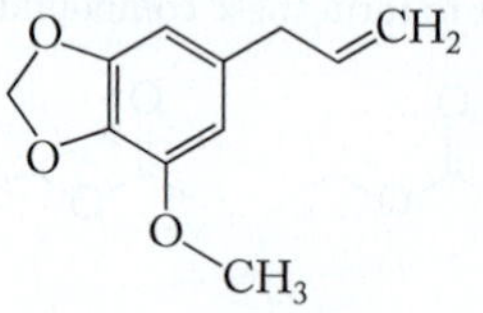

Myristicin

**FIGURE P19.119**

*19.120. The condensed structure of a molecule is $CH_3CH{=}C(CH_3)CH(OH)CH_2CH_3$. Draw all the possible stereoisomers of this compound.

**19.121.** Which of the molecules in Figure P19.121 do not have any possible constitutional isomers?

(a) (b) (c)

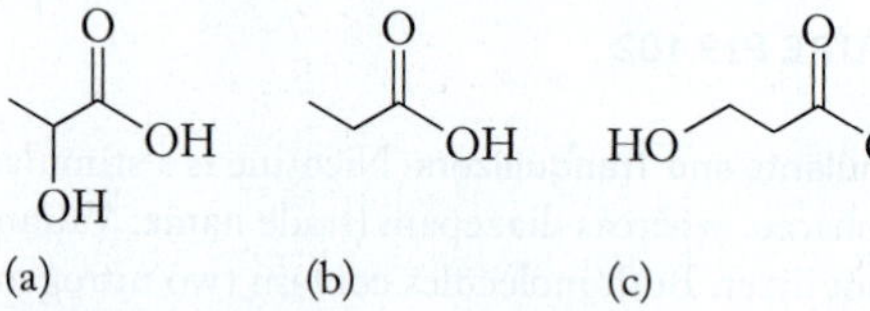

**FIGURE P19.121**

19.122. Which of the molecules in Figure P19.122 do not have any possible stereoisomers?

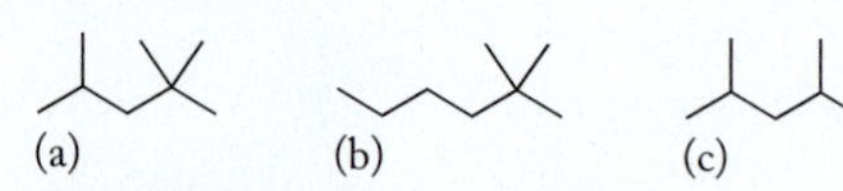

**FIGURE P19.122**

## Additional Problems

**19.123. Lou Gehrig's Disease** Edaravone (Figure P19.123) is a drug that was approved in 2018 in Canada to treat amyotrophic lateral sclerosis (ALS), more commonly known as Lou Gehrig's disease.

a. How many carbon atoms are in edaravone?
b. Identify the functional groups present in edaravone.

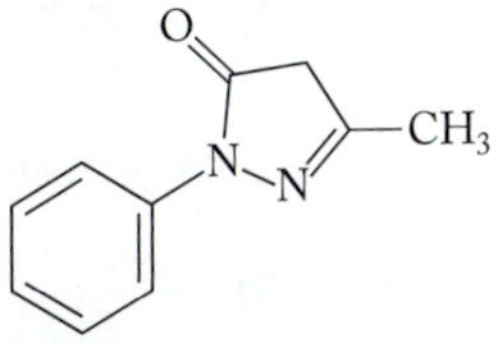

**FIGURE P19.123**

19.124. **Salsa** Salsa has antibacterial properties because it contains dodecanal (Figure P19.124), a compound found in the cilantro used to make salsa.

a. How many carbon atoms are in dodecanal?
b. What functional groups are present in dodecanal?
c. What types of isomerism are possible in dodecanal?

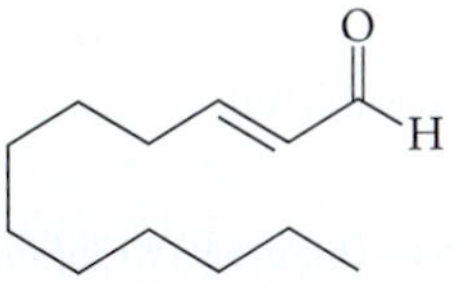

**FIGURE P19.124**

***19.125. Treating Muscular Dystrophy** In 2017, the Food and Drug Administration approved the use of deflazacort to slow the progress of Duchenne muscular dystrophy, a fatal disease that affects boys. The structure of deflazacort (Figure P19.125) contains six oxygen atoms. Identify the functional group for each of the six oxygen atoms.

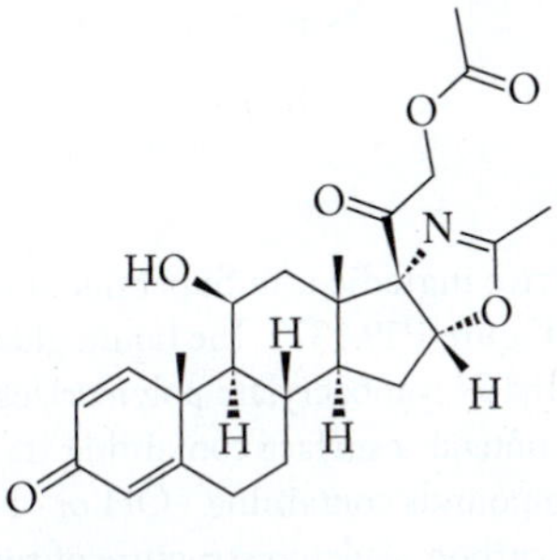

**FIGURE P19.125**

19.126. **Glaucoma Medication** A new drug to reduce eye pressure for patients suffering from glaucoma was approved for use in the United States in 2017. Figure P19.126 shows the structure of this medication, latanoprostene (trade name, Vyzulta). Identify each of the functional groups present in this medication, and determine whether the stereochemistry at the carbon–carbon double bond depicts the cis or the trans isomer.

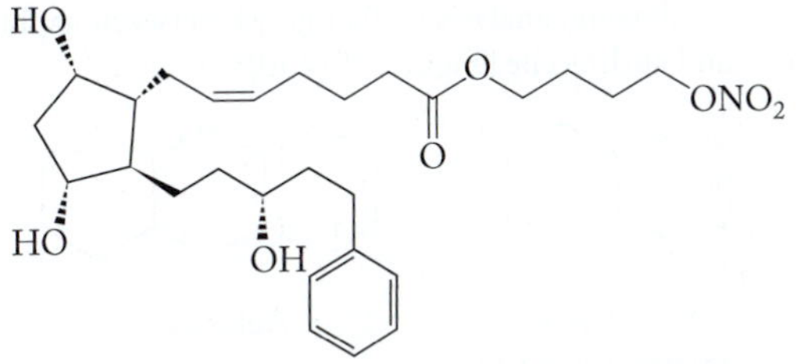

**FIGURE P19.126**

**19.127.** Benzene rings with two methyl groups are called xylenes. 1,4-Dimethylbenzene (*para*-xylene, or *p*-xylene) and 1,3-dimethylbenzene (*ortho*-xylene, or *o*-xylene) have similar boiling points (138°C and 139°C, respectively) but different melting points (13°C and −48°C, respectively). Explain this observation.

19.128. **Perfume from Whales** Ambergris comes from the intestines of sperm whales and was once used to make perfumes. Ambrein (Figure P19.128) is a constituent of ambergris that, when injected in rats, causes an increase in their sexual activity.

a. What functional groups are present in ambrein?
b. How many chiral centers does ambrein have?
c. Are any cis/trans isomers possible for ambrein?

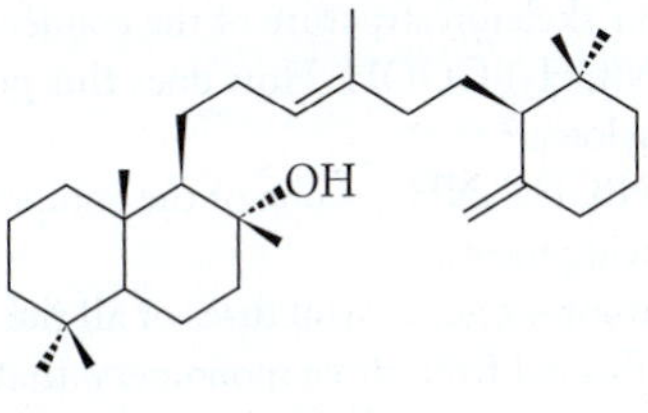

**FIGURE P19.128**

**19.129. Turmeric** Turmeric is commonly used as a spice in Indian and Southeast Asian dishes. Turmeric contains a high concentration of curcumin (Figure P19.129), a potential anticancer drug and a possible treatment for cystic fibrosis.

a. Are the substituents on the C═C bonds (nonaromatic) in cis or trans configurations?
b. Draw two other stereoisomers of this compound.
c. List all the types of hybridization of the carbon atoms in curcumin.

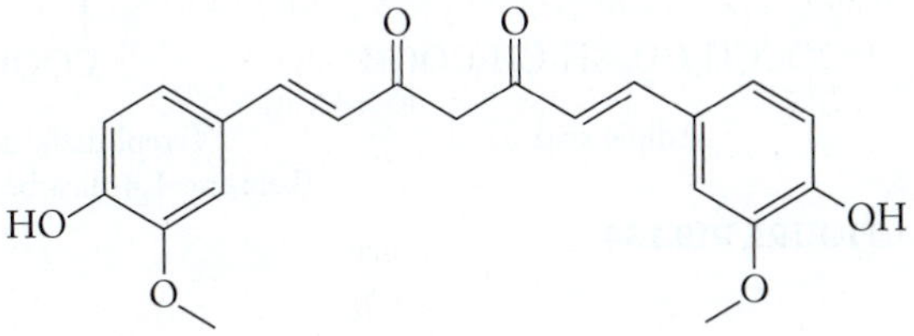

**FIGURE P19.129**

19.130. Polycyclic aromatic hydrocarbons are potent carcinogens. They are produced during the combustion of fossil fuels and have been found in meteorites. Can we use combustion analysis to distinguish between naphthalene and anthracene (Figure P19.130)?

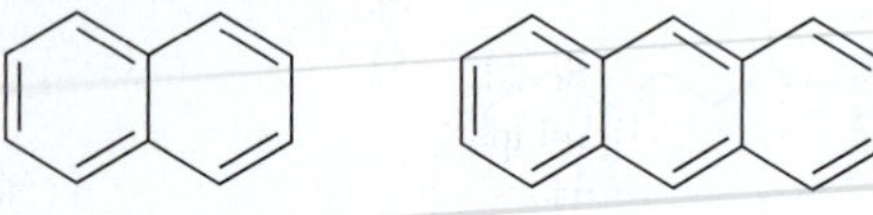

Naphthalene Anthracene

**FIGURE P19.130**

**19.131.** Identify the reactants in the polymerization reactions that produce the polymers shown in Figure P19.131.

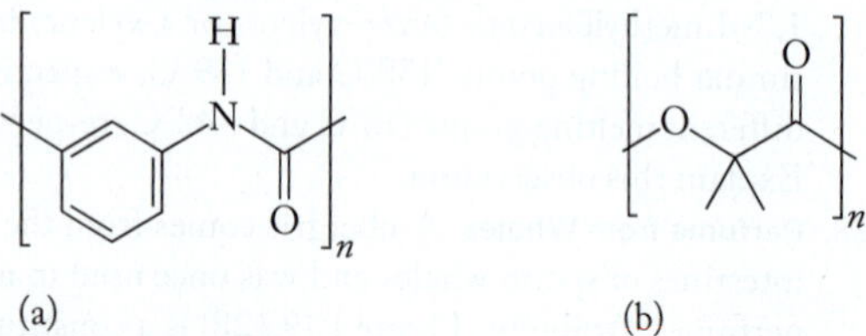


**FIGURE P19.131**

*19.132. **Raincoats** "Waterproof" nylon garments have a coating to prevent water from penetrating the hydrophilic fibers. Which functional groups in the nylon molecule make it hydrophilic?

**19.133.** Draw the carbon-skeleton structure of the condensation polymer of $H_2N(CH_2)_6COOH$. How does this polymer compare with nylon-6?

*19.134. Putrescine, $H_2N(CH_2)_4NH_2$, is one of the compounds that form in rotting meat.

a. Draw the carbon-skeleton structures of all the trimers (a molecule formed from three monomers) that can be formed from putrescine, adipic acid, and terephthalic acid (Figure P19.134). The three monomers forming the trimer do not have to be different.

b. A chemist wishes to make a polymer containing putrescine and a 1:1 ratio of adipic acid to terephthalic acid. What should be the mole ratio of the three reactants?

$HOOCCH_2CH_2CH_2CH_2COOH$

Adipic acid

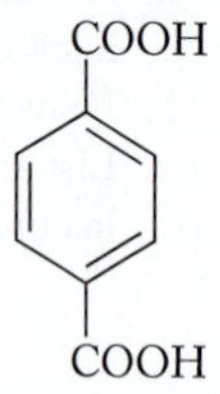

COOH

COOH

Terephthalic acid (benzene-1,4-dicarboxylic acid)

**FIGURE P19.134**

***19.135.** Polymer chemists can modify the physical properties of polystyrene by copolymerizing divinylbenzene with styrene (Figure P19.135). The resulting polymer has strands of polystyrene cross-linked with divinylbenzene. Predict how the physical properties of the copolymer might differ from those of 100% polystyrene.

Divinylbenzene (DVB) Styrene (S)

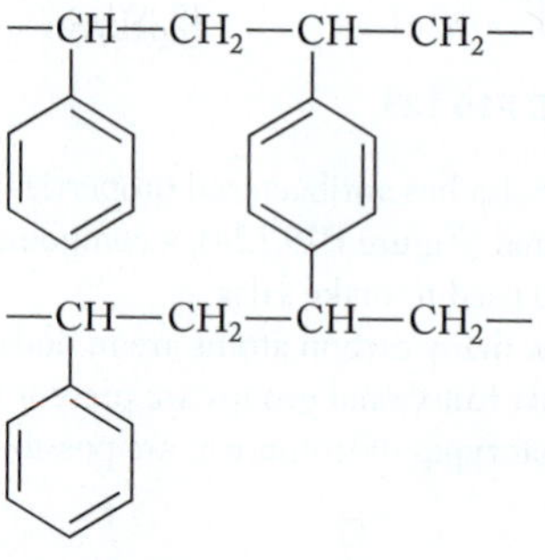

$-CH-CH_2-CH-CH_2-$

$-CH-CH_2-CH-CH_2-$

S cross-linked with DVB

**FIGURE P19.135**

19.136. Maleic anhydride and styrene (Figure P19.136) form a polymer with alternating units of each monomer.

a. Draw two repeating monomer units of the polymer.

b. From the structure of the copolymer, predict how its physical properties might differ from those of polystyrene.

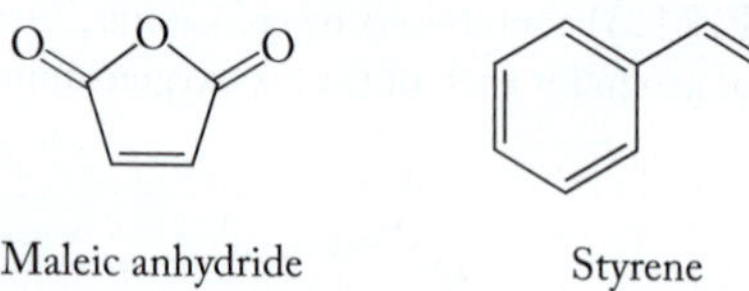


Maleic anhydride Styrene

**FIGURE P19.136**

**19.137. Superglue** The active ingredient in Superglue is methyl 2-cyanoacrylate (Figure P19.137). The liquid glue hardens rapidly when methyl 2-cyanoacrylate polymerizes. This happens when it contacts a surface containing traces of water or other compounds containing –OH or –NH– groups. Draw the carbon-skeleton structure of two repeating monomer units of poly(methyl 2-cyanoacrylate).

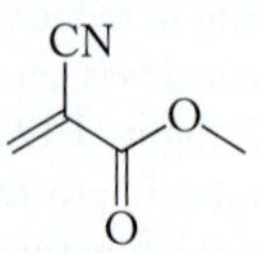

CN

O

O

Methyl 2-cyanoacrylate

**FIGURE P19.137**

*19.138. Silicones are polymeric materials with the formula $[R_2SiO]_n$ (Figure P19.138). They are prepared by the reaction of $R_2SiCl_2$ with water to yield the polymer and aqueous HCl. Imagine this reaction as taking place in two steps: (1) water reacts with one mole of $R_2SiCl_2$ to produce a new monomer and one mole of HCl(*aq*); (2) one new monomer molecule reacts with another new monomer molecule to eliminate one molecule of HCl and make a dimer with a Si—O—Si bond.

a. Suggest two balanced equations describing these reactions that occur over and over again to produce a silicone polymer.
b. Why are silicones water repellent?

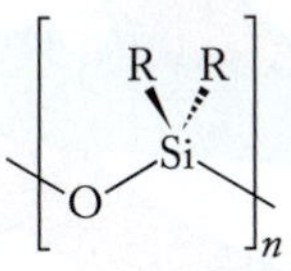

Silicone

**FIGURE P19.138**

**19.139.** Molecules of piperine and capsaicin (see cover) contain amide functional groups.

a. Draw the amine and the carboxylic acid that could react to form these two compounds.
b. Are the nonaromatic double bonds in these molecules cis or trans?
c. Name the functional groups that contain the oxygen atoms in these compounds.

19.140. The heats of combustion of two constitutional isomers are the same when estimated from average bond energies, but they are different when determined experimentally. Why?

# 20

# Biochemistry

## The Compounds of Life

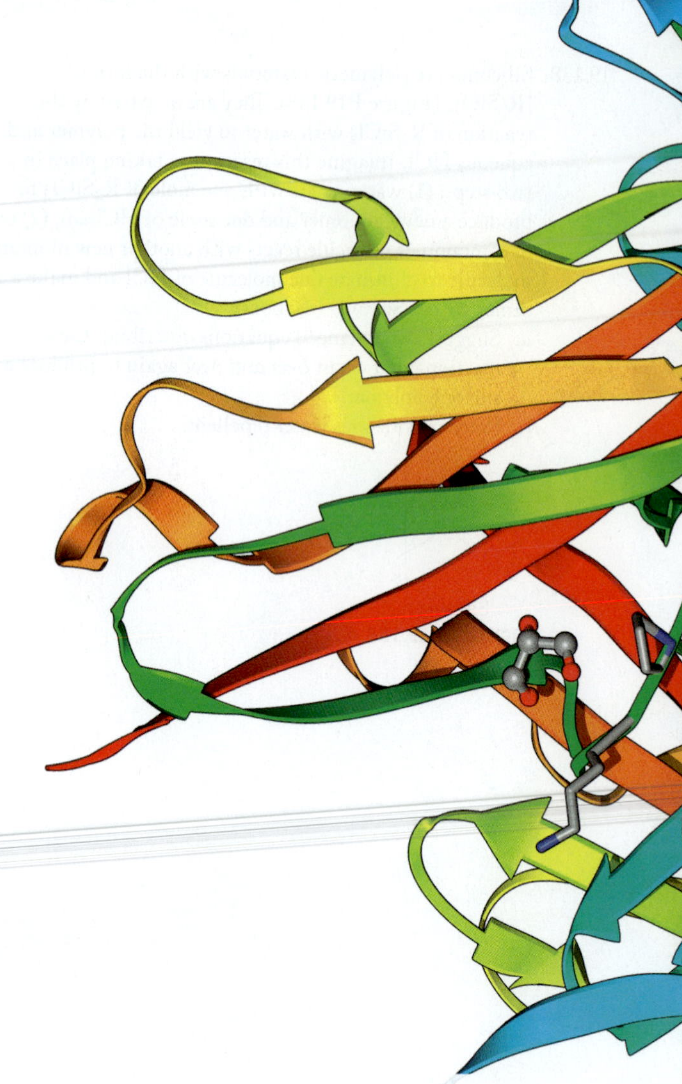

**BIOLOGICAL MACROMOLECULES** Titin is the largest known protein, consisting of over 34,000 amino acids. Its elasticity contributes to the function of skeletal muscles.

PARTICULATE **REVIEW**

### *Protons versus Hydrogen Bonds*

In Chapter 20, we investigate the properties of proteins and the role of hydrogen bonding in defining protein structure and function.

- Hydrogen bonding is an important intermolecular force among which of the three molecules shown here?
- Identify any acidic hydrogen atoms in the three molecules.
- Which molecule(s) could function as a Lewis base?

(Review Sections 6.3, 8.4, and 16.5 if you need help.)

*(Answers to Particulate Review questions are in the back of the book.)*

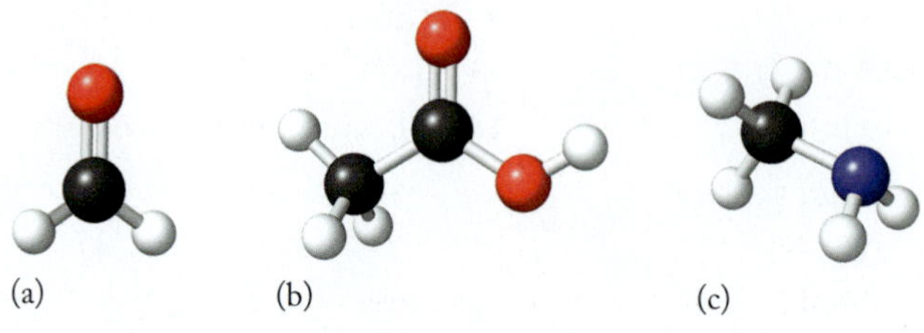

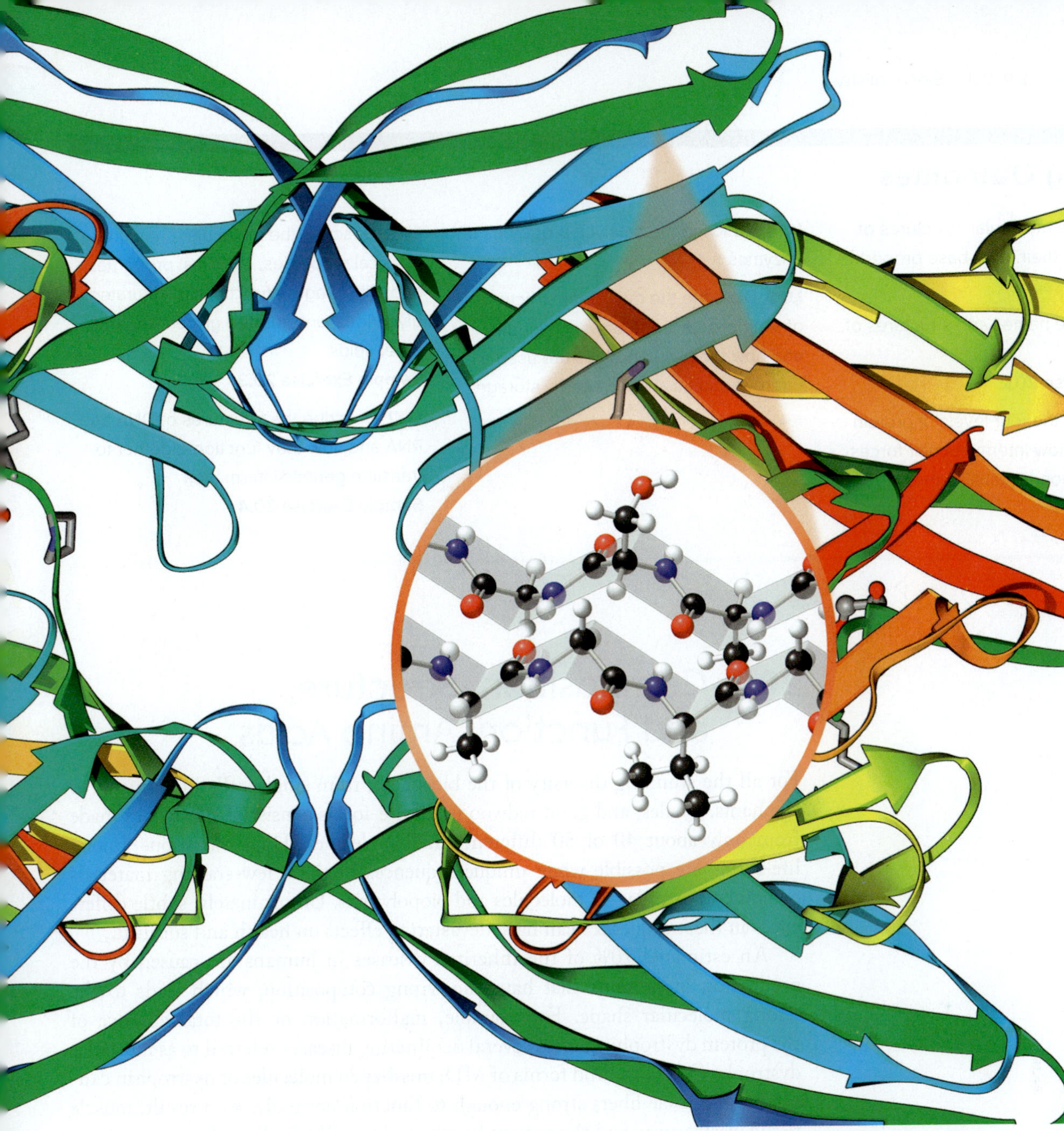

PARTICULATE **PREVIEW**

### *Functional Groups and pH*

The molecule shown here contains two different functional groups. As you read Chapter 20, look for ideas that will help you answer these questions:

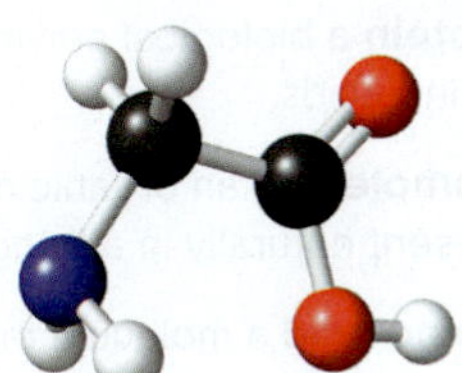

- What are the two functional groups in this molecule?
- What happens to each of these functional groups in moderately acidic (pH = 3) aqueous solutions?
- What happens to each of these functional groups in moderately basic (pH = 11) aqueous solutions?

## Learning Outcomes

**LO1** Relate the molecular structures of amino acids to their acid–base properties
**Sample Exercise 20.1**

**LO2** Relate the names and structures of small peptides
**Sample Exercise 20.2**

**LO3** Describe the four levels of protein structure and how intermolecular forces and covalent bonds stabilize these structures

**LO4** Describe the catalytic properties of enzymes

**LO5** Describe the molecular structures of simple sugars and polysaccharides and how these compounds are used as energy sources and for energy storage

**LO6** Describe the molecular structure, physical properties, chemical properties, and physiological functions of saturated glycerides, unsaturated glycerides, and other lipids
**Sample Exercise 20.3**

**LO7** Describe the structures of DNA and RNA and how they function together to translate genetic information
**Sample Exercise 20.4**

# 20.1 Composition, Structure, and Function: Amino Acids

For all the stunning diversity of the biosphere, from single-celled organisms to elephants, whales, and giant redwoods, all life-forms consist of substances made from only about 40 or 50 different small molecules. Huge variations among life-forms are possible when unique sequences of these few starting materials combine to form larger molecules and biopolymers. Unfortunately, subtle differences in these sequences can have devastating effects on health and survival.

An estimated 70% of the inherited diseases in humans are caused by the production of proteins that have the wrong composition, which leads to the wrong molecular shape. For example, malformation or the total absence of the protein dystrophin causes several debilitating diseases referred to as muscular dystrophy (MD). In mild forms of MD, misshapen molecules of dystrophin cannot build muscle fibers strong enough to function normally. As a result, muscle tissue wastes away and the patient becomes physically disabled. In a severe form of the disease, called Duchenne MD (DMD), functional dystrophin is absent, and the disability often results in early death. Understanding the mechanisms that produce dystrophin has led to promising therapies, some of which involve grafting healthy muscle cells that can synthesize normal dystrophin into the tissues of MD patients. For DMD patients, treatments involve injecting stem cells that fuse with and genetically complement dystrophic muscle. Similar approaches have yielded new drugs and cell therapies for other inherited and contagious diseases.

**protein** a biological polymer made of amino acids.

**biomolecule** an organic molecule present naturally in a living system.

**amino acid** a molecule that contains at least one amine group and one carboxylic acid group.

**essential amino acid** any of the 10 amino acids that make up peptides and proteins but are not synthesized in the human body and must be obtained through the food we eat.

In this chapter, we explore the composition, structure, and function of proteins and three other major classes of biomolecular compounds: carbohydrates, lipids, and nucleic acids. The functions of these compounds are similar in all life-forms in which they occur. Many of these molecules are large and complex, but the knowledge we have developed about the chemical behavior of small molecules will serve as a useful framework on which to build an understanding of the behavior of the larger assemblies of molecules that form cells. The first small molecules we explore are the amino acids, the monomer units that combine to form proteins.

## Amino Acids: The Building Blocks of Proteins

**Proteins** are the most abundant class of **biomolecules** in all animals, including humans. Proteins account for about half of the mass of the human body that is not water. They are the major component in skin, muscles, cartilage, hair, and nails. Most of the enzymes that catalyze biochemical reactions are proteins, as are the molecules that transport oxygen to our cells and many of the hormones that regulate cell function and growth. Most proteins are biopolymers with molar masses of $10^5$ g/mol or more.

The molecular structure and biological function of large biomolecules depend on the identities of their small-molecule building blocks and the sequence in which those small molecules occur. The molecular building blocks of proteins are **amino acids**, so named because each of them contains at least one amine ($-NH_2$) group and at least one carboxylic acid (–COOH) group. The amino acids in proteins are called α-amino acids because each of their molecules contains an α-carbon atom that is directly attached to both the carbonyl group of the carboxylic acid group and to the nitrogen atom of the amine group (**Figure 20.1**). The α-carbon is the carbon atom in a structure that is directly attached to a functional group.

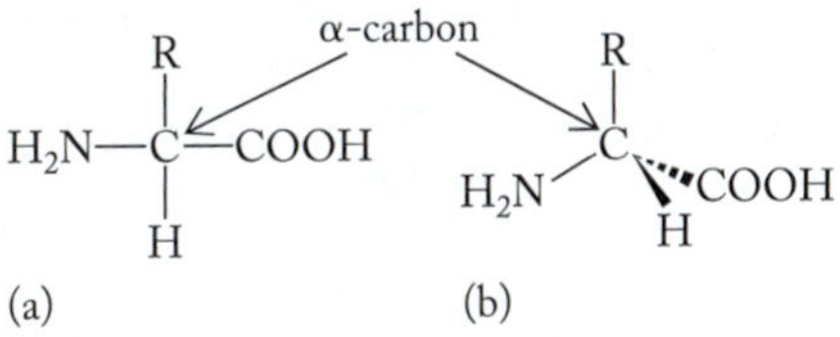

α-carbon

(a) R, $H_2N$—C—COOH, H

(b) R, $H_2N$, C, COOH, H

**FIGURE 20.1** (a) General structure of an α-amino acid; (b) three-dimensional structure.

**CONNECTION** The weakly basic properties of amines were described in Section 15.3.

**CONCEPT TEST**

According to valence bond theory, what is the hybridization of the α-carbon atom in the amino acid in Figure 20.1, and what are the approximate angles between the four bonds surrounding it?

*(Answers to Concept Tests are in the back of the book.)*

**CONNECTION** As in Chapter 19, we use the generic symbol R to represent any group of covalently bonded atoms that is connected to the main structure of a molecule.

In addition to its single bonds to the $-NH_2$ and –COOH groups, the α-carbon atom in each of the amino acids that make up human proteins is bonded to a hydrogen atom and one of 20 common *R groups*. The structures of these R groups, often called *side-chain* groups, are highlighted in pink in the amino acid structures in **Table 20.1**. The amino acids are arranged in four categories according to the polarities or acid–base properties of their R groups. In the first category, which contains nine amino acids, the R groups contain mostly carbon and hydrogen atoms and are nonpolar. The R groups of the remaining 11 amino acids contain at least one heteroatom (O, N, or S) and are polar. Of these amino acids, two (aspartic acid and glutamic acid) have R groups that contain carboxylic acid functional groups, and three (histidine, lysine, and arginine) contain groups with nitrogen atoms that are weakly basic.

Our bodies can synthesize 10 of the 20 amino acids in Table 20.1, but the other 10 must be present in the food we eat. These 10 are marked with a superscript *b* and are referred to as the **essential amino acids**. Most proteins from animal sources, including those in meats, eggs, and dairy products, contain all the essential amino acids needed by the human body and in close to the needed proportions. These foods are sometimes referred to as sources of *complete proteins*. In contrast, most plant proteins from foods such as legumes and vegetables are incomplete proteins, so vegetarians must be careful to eat a combination of foods that provides all the essential amino acids. A good example is beans and rice (**Figure 20.2**), a traditional dish in Louisiana Creole and Latin American cuisine that provides a balance of these amino acids: rice has all of them but lysine, and beans lack only methionine.

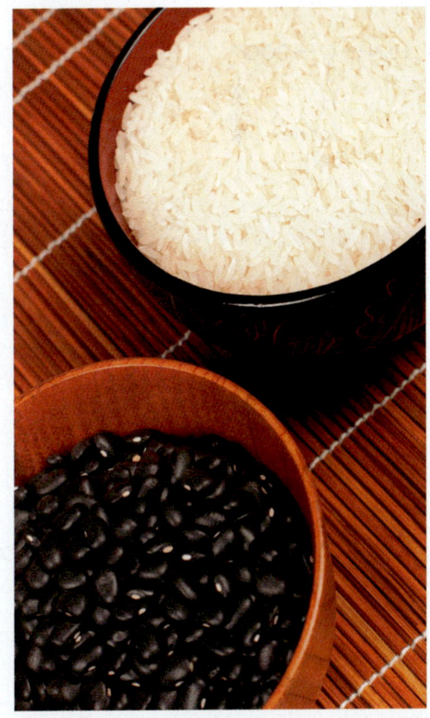

**FIGURE 20.2** A meal of beans and rice provides all the essential amino acids.

**TABLE 20.1 Structures and Abbreviations of the 20 Common Amino Acids**[a]

### Nonpolar R Groups

| | | | | |
|---|---|---|---|---|
| Glycine (Gly) | Alanine (Ala) | Valine[b] (Val) | Leucine[b] (Leu) | Isoleucine[b] (Ile) |
| Proline (Pro) | Phenylalanine[b] (Phe) | Tryptophan[b] (Trp) | Methionine[b] (Met) | |

### Polar R Groups

| | | | | | |
|---|---|---|---|---|---|
| Serine (Ser) | Threonine[b] (Thr) | Cysteine (Cys) | Tyrosine (Tyr) | Asparagine (Asn) | Glutamine (Gln) |

### Acidic R Groups

| | |
|---|---|
| Aspartic acid (Asp) | Glutamic acid (Glu) |

### Basic R Groups

| | | |
|---|---|---|
| Histidine[b] (His) | Lysine[b] (Lys) | Arginine[b] (Arg) |

[a]R groups in pink, ionized forms at pH near 7 shown.

[b]One of the 10 essential amino acids for adults; Roger L. Miesfeld and Megan M. McEvoy, *Biochemistry* (Norton 2017), p. 874.

## Chirality

The carbon atom in the generic structure of an amino acid (Figure 20.1) is bonded to four different groups. As a result, these α-carbon atoms are *chiral centers* (see Section 5.6). It also means that all the amino acids in Table 20.1, except glycine (R = H), are chiral compounds: their molecular structures are not superimposable on their mirror images, as illustrated for alanine in **Figure 20.3**. Instead, each amino acid and its mirror image constitute a pair of enantiomers.

For historical reasons, amino acid enantiomers are designated by the prefixes D- (for *dextro-*, right) and L- (*levo-*, left). These labels refer to how the four groups bonded to each chiral carbon atom are oriented in three-dimensional space. However, they do not refer to the rotation of plane-polarized light (see Figure 5.39) as it passes through a solution of the amino acid. In other words, there is no connection between the D- and L- prefixes in the names of amino acids and the *dextrorotary* and *levorotary* enantiomers that are designated with (+) and (−) signs according to their optical properties. All the chiral amino acids in the proteins in our bodies are L-enantiomers, even though nine of the 19 are actually dextrorotary.

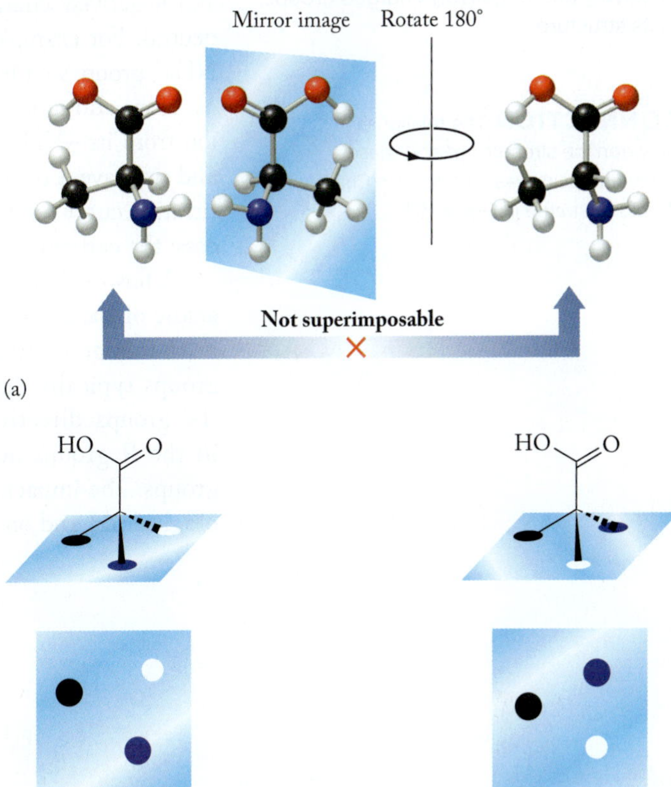


**FIGURE 20.3** Like most α-amino acids, alanine is chiral, which means that its molecular structure is not superimposable on its mirror image. (a) Rotating the mirror image 180° so that the $-CH_3$ and $-COOH$ groups overlap does not produce a structure in which $-NH_2$ and $-H$ overlap. (b) To visualize alanine's chirality another way, keep the carboxylic acid groups on the two molecules aligned and make "footprints" of the other three groups on two pieces of paper. No matter how you turn them, you cannot superimpose the two patterns.

## Zwitterions

If we dissolve an amino acid in a solution that already contains a strong acid (and therefore has a low pH), the carboxylic acid group on each molecule does not ionize. Each amine group, however, being a weak base, accepts a $H^+$ ion to form a protonated $-NH_3^+$ group. The result is a molecular ion with an overall positive charge, as illustrated for alanine in **Figure 20.4(a)**.

Now suppose we add a strong base to this solution to neutralize the strong acid and to raise the pH to about 7.4, close to the pH of human blood and of the fluids in most of our tissues. At this *physiological pH*, the $-COOH$ groups on the α-carbon atoms in amino acids are mostly ionized because they all have $pK_a$ values in the range from 1.8 to 2.8. These values mean that the $-COOH$ groups are essentially completely ionized at pH 7.4. Meanwhile, most of the basic amine groups bonded to the α-carbon atoms of amino acid accept $H^+$ ions and form $-NH_3^+$ groups at pH 7.4. This combination of negatively charged $-COO^-$ groups and positively charged $-NH_3^+$ groups means that these amino acid molecules are

**CONNECTION** Enantiomers and chirality were discussed in Section 5.6.

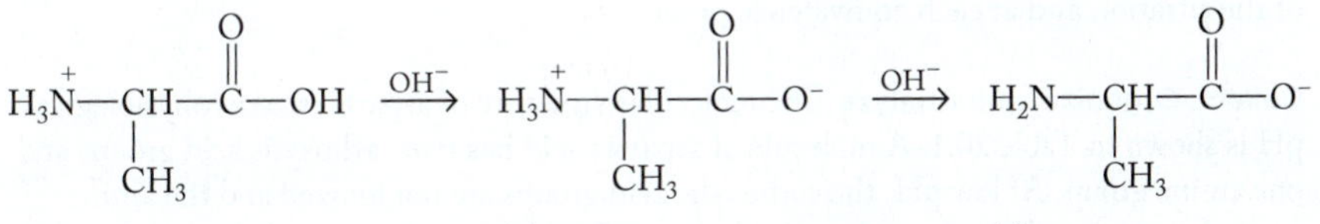

$$\underset{\text{(a)}}{H_3\overset{+}{N}-\underset{\underset{CH_3}{|}}{CH}-\overset{\overset{O}{\|}}{C}-OH} \xrightarrow{OH^-} \underset{\text{(b)}}{H_3\overset{+}{N}-\underset{\underset{CH_3}{|}}{CH}-\overset{\overset{O}{\|}}{C}-O^-} \xrightarrow{OH^-} \underset{\text{(c)}}{H_2N-\underset{\underset{CH_3}{|}}{CH}-\overset{\overset{O}{\|}}{C}-O^-}$$

**FIGURE 20.4** (a) At low pH, alanine (like many other amino acids) exists as a 1+ ion. (b) At physiological pH (7.4), the $-COOH$ group is ionized and the molecule becomes a zwitterion with an overall charge of zero. (c) In basic solutions, the charge decreases to 1− as the $-NH_3^+$ group deprotonates.

**zwitterion** a molecule that has both positively and negatively charged groups in its structure.

**CONNECTION** The relationship between the strengths of acid–base conjugate pairs was discussed in Chapter 15 and illustrated in Figure 15.8.

**zwitterions.** This term is used to describe molecules that contain both positively and negatively charged functional groups even when the molecules are, overall, neutral. For example, at pH 7.4, alanine contains one $-COO^-$ group and one $-NH_3^+$ group, so it has no net electrical charge (**Figure 20.4b**).

If we add more base to raise the pH well above 7.4, alanine loses a hydrogen ion from its $-NH_3^+$ group (we say that it *deprotonates*) to form an $-NH_2$ group, and we have a molecular ion with an overall charge of 1− (**Figure 20.4c**). The titration curve for alanine in **Figure 20.5** shows this two-step neutralization process: the carboxylic acid is neutralized first, the protonated amine second.

A few of the common amino acids in Table 20.1 have R groups that contain acidic or basic units. These groups have p$K_a$ values around 4 for the acids and in the range of 10.5 to 12.5 for the protonated amines, so the acids in the R groups typically deprotonate at pH values closer to physiological pH than do the groups directly attached to the α-carbon atom. The protonated amines in the R groups deprotonate at pH values lower than those of the α-amino groups. The impact of this behavior on the charges of the amino acids at physiological pH and on titration curves is the subject of Sample Exercise 20.1.

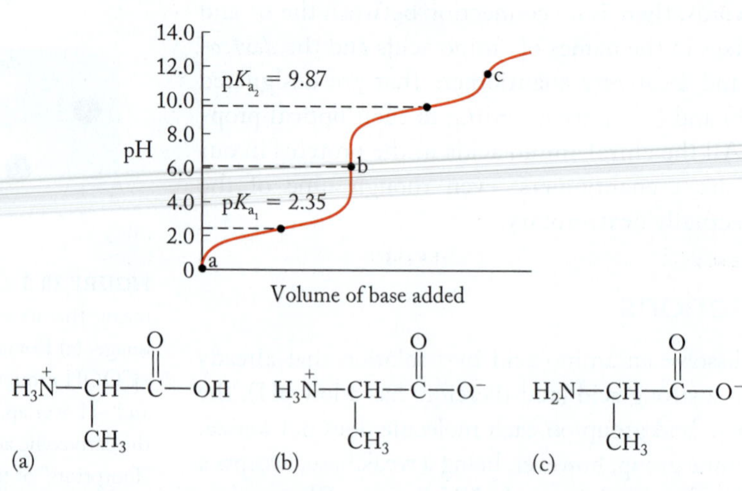


**FIGURE 20.5** The titration curve of alanine resembles that of a weak diprotic acid. The carboxylic acid group (p$K_a$ = 2.35) is neutralized first. The protonated amine group (p$K_a$ = 9.87) is neutralized in the second step of the titration. The dominant forms of alanine present in solution are shown at (a), the starting point, and at (b) and (c), the two equivalence points.

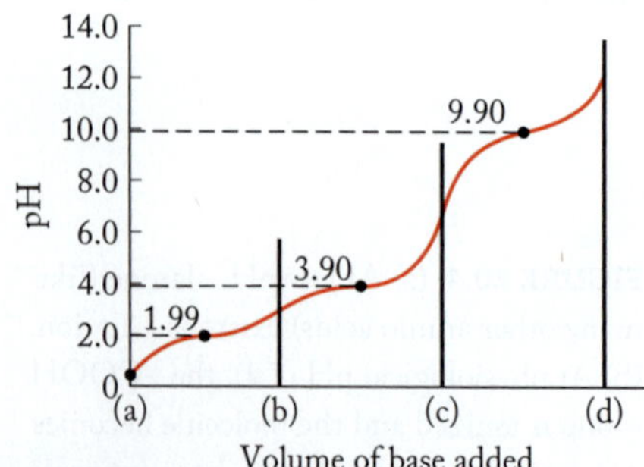


**FIGURE 20.6** Titration curve of aspartic acid. The vertical lines identify equivalence points in the titration.

**SAMPLE EXERCISE 20.1** Interpreting Acid–Base Titration Curves of Amino Acids **LO1**

The $-NH_2$ groups in α-amino acids are weak bases, and their conjugate acids ($-NH_3^+$) are even weaker acids. **Figure 20.6** shows that the titration curve for aspartic acid has three equivalence points at $pK_{a_1} = 1.99$, $pK_{a_2} = 3.90$, and $pK_{a_3} = 9.90$. Draw the molecular structures of the principal form of aspartic acid that is in solution at the start of the titration and at each equivalence point.

**Collect, Organize, and Analyze** The molecular structure of aspartic acid at physiological pH is shown in Table 20.1. A molecule of aspartic acid has two carboxylic acid groups and one amine group. At low pH, the carboxylic acid groups are not ionized and the amine group is protonated. As the pH is raised, the –COOH groups ionize and then, under basic conditions, the protonated amine group releases its proton. The stronger of the two acid groups will ionize first. The –COOH group bonded to the α-carbon atom is only one carbon atom away from a protonated $NH_3^+$ group, which pulls electron density away from the –COOH group and stabilizes the resulting $COO^-$ group. The side-chain –COOH group is two carbon atoms away, so the stability of this carboxylate ion is less affected by the $NH_3^+$ group. The –COOH group bonded to the α-carbon atom should ionize first.

**Solve** At the beginning of the titration (a), the fully protonated 1+ ion is present:

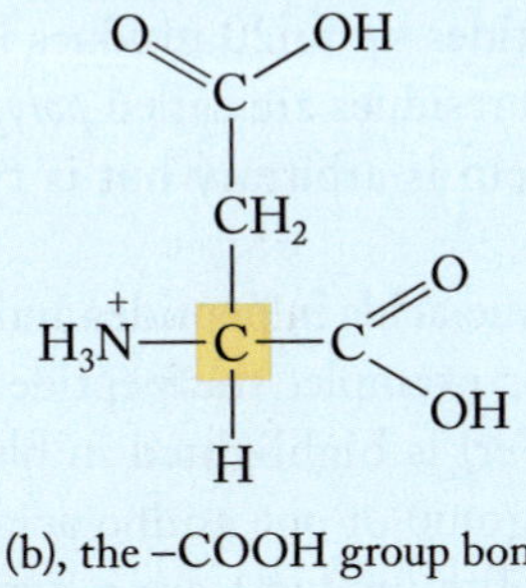

At the first equivalence point (b), the –COOH group bonded to the α-carbon atom (highlighted in yellow) is ionized:

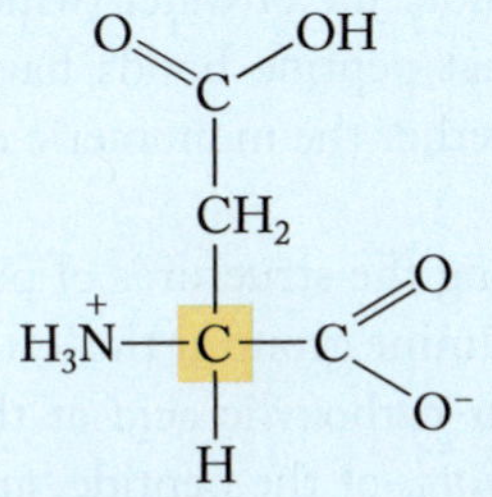

At the second equivalence point (c), the side-chain –COOH group is ionized:

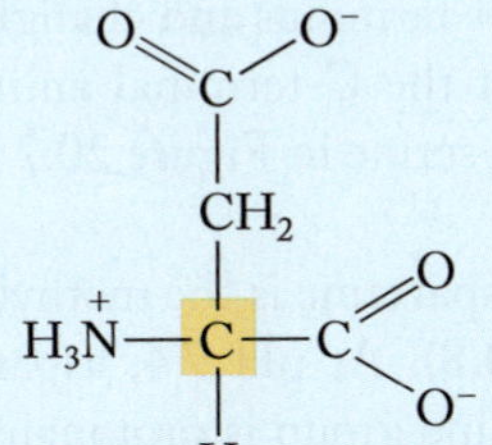

And at the third equivalence point (d) in the titration, the amine group is deprotonated:

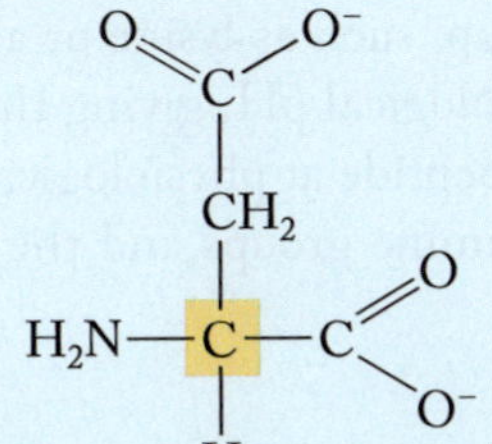

**Think About It** The $pK_a$ values of the two carboxylic acid groups in the first two steps of the titration are 1.99 and 3.90. They differ by nearly 2 pH units, which means that the α-COOH group is almost $10^2$ times stronger an acid than the side-chain –COOH group, due to the presence of the α-$NH_3^+$ group.

**Practice Exercise** Sketch the titration curve for lysine starting at pH 1.0. Draw the molecular structures of the principal form of lysine that is in solution at pH 1.0 at each equivalence point.

*(Answers to Practice Exercises are in the back of the book.)*

**peptide** a compound of two or more amino acids joined by peptide bonds. Small peptides containing up to 20 amino acids are *oligopeptides*; the term *polypeptide* is used for chains longer than 20 amino acids but shorter than proteins.

**peptide bond** the result of a condensation reaction between the carboxylic acid group of one amino acid and the amine group of another.

**CHEMTOUR**
Condensation of Biological Polymers

**CONNECTION** The formation of amide linkages in reactions between carboxylic acids and amines was described in Section 19.7 and illustrated in Figure 19.47.

## Peptides

Amino acids react to form chainlike molecules with a wide range of sizes. Amino acid *residues* (the name we give to amino acids that are part of a larger molecule) make up each link in the chain. The longest chains, some with molar masses as high as 3 million g/mol, are called proteins, but the shortest chains, called **peptides**, are only a few amino acid residues long. The smallest peptides contain only two or three amino acid residues and are called *dipeptides* and *tripeptides*, respectively. Peptides up to 20 residues long are called *oligopeptides*. Those made of more than 20 residues are called *polypeptides*. The size at which a polypeptide becomes a protein is arbitrary but is typically set at about 50–75 amino acid residues.

The bond linking the amino acids in peptides and proteins is called a **peptide bond**, or *peptide linkage*. For example, the peptide bond that links molecules of valine (Val) and serine (Ser) is highlighted in blue in **Figure 20.7**. It forms when the α-carboxylic acid group of one amino acid (here, Val) reacts with the α-amine group of another. The products are a peptide bond linking the two amino acids together and a molecule of water (which makes the reaction a condensation reaction). Note that peptide bonds have the same structure as the amide linkages that hold together the monomeric units of synthetic polyamides such as nylon.

The convention for drawing the structures of peptides begins by placing the amino acid that has a free α-amine group at the left end of the peptide chain and the amino acid with a free α-carboxylic acid at the right end. The left end is called the *amine* (or *N-*) *terminus* of the peptide, and the right end is called the *carboxylic acid* (or *C-*) *terminus*. A peptide's name is formed from the names of its amino acids, starting at the N-terminus and changing the ending of each amino acid's name to *-yl* for all but the C-terminal amino acid. Thus, the dipeptide formed by linking valine and serine in Figure 20.7 is called valylserine (abbreviated ValSer).

The artificial sweetener aspartame is the methyl ester of the dipeptide aspartylphenylalanine (**Figure 20.8**). At pH 7.4, aspartame exists as a zwitterion because the aspartic acid amine group is protonated and the carboxylic acid in its R group is ionized. In contrast, its parent dipeptide has a net charge of 1− because both –COOH groups are ionized at that pH. In an amino acid that has an amine group in its R group, such as lysine or arginine, that amine group is probably protonated at physiological pH, giving the amino acid a net charge of 1+. The overall charge on a peptide at physiological pH is the sum of the positive charges on protonated amine groups and the negative charges of ionized carboxylic acid groups.

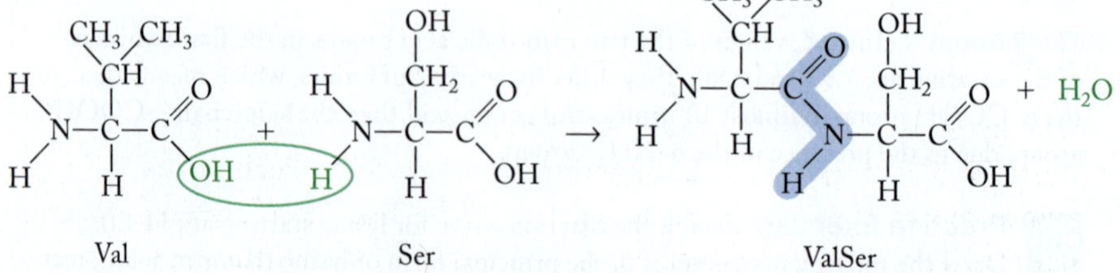

Val + Ser → ValSer + $H_2O$

**FIGURE 20.7** When the carboxylic acid group of valine forms a peptide bond with the amine group of serine, the products are the dipeptide valylserine and water.

**SAMPLE EXERCISE 20.2** Drawing and Naming Peptides **LO2**

a. Name the four dipeptides that can be prepared from alanine and glycine.
b. Draw their molecular structures in solution at pH 7.4.

**Collect, Organize, and Analyze** Amino acid structures are in Table 20.1. Four different peptides can be made from alanine and glycine by changing their sequence from the N-terminus to the C-terminus: AlaAla, GlyGly, AlaGly, and GlyAla. At pH 7.4, the carboxylic acid terminus –COOH groups should be ionized and the amine terminus $-NH_2$ groups should be protonated.

**Solve**

a. The names of the four peptides with the amino acid sequences AlaAla, GlyGly, GlyAla, and AlaGly are alanylalanine, glycylglycine, glycylalanine, and alanylglycine, respectively.
b. At pH 7.4, the $-NH_2$ groups of amino acids are protonated and the –COOH groups are ionized, so the principal forms of these dipeptides are

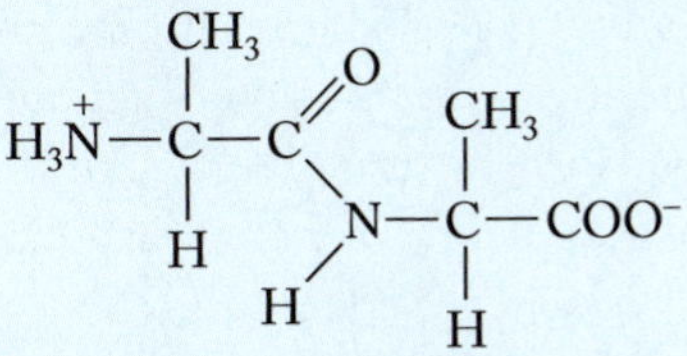

Alanylalanine (AlaAla)

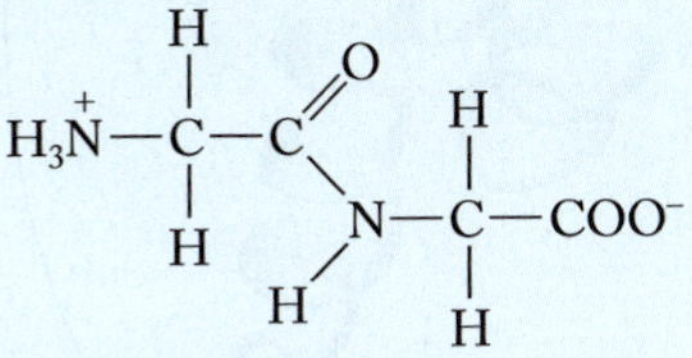

Glycylglycine (GlyGly)

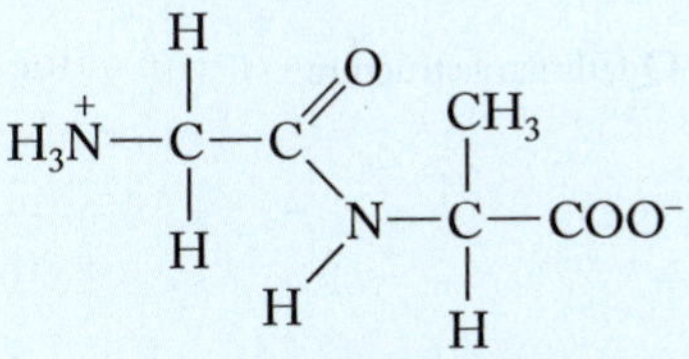

Glycylalanine (GlyAla)

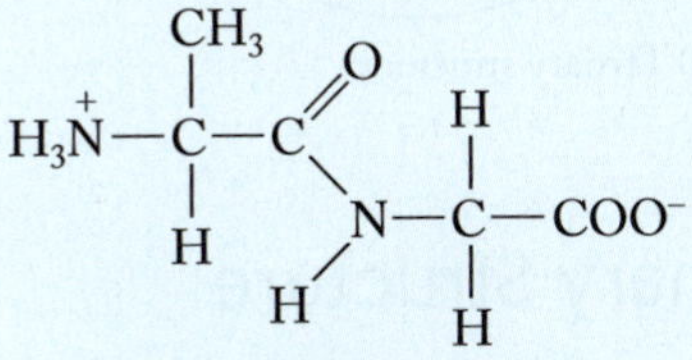

Alanylglycine (AlaGly)

**Think About It** Only the N-terminal $-NH_2$ and C-terminal –COOH groups can be protonated or ionized in these dipeptides. Therefore, all four of the dipeptides are zwitterionic with a net charge of (1+) + (1–) = 0 at pH 7.4.

**Practice Exercise** How many different tripeptides can be synthesized from one molecule each of Gly, Ala, and Val?

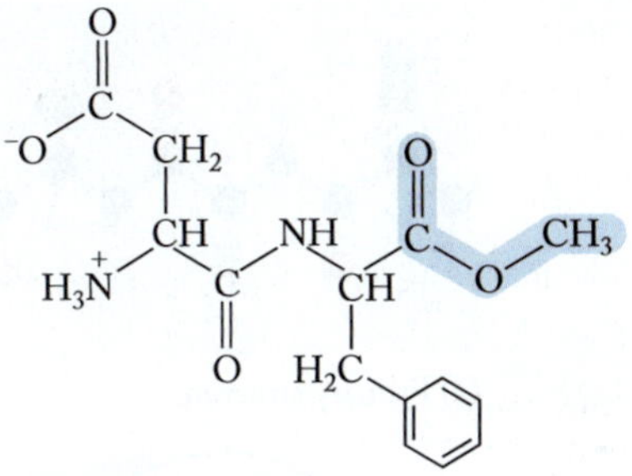

Aspartame

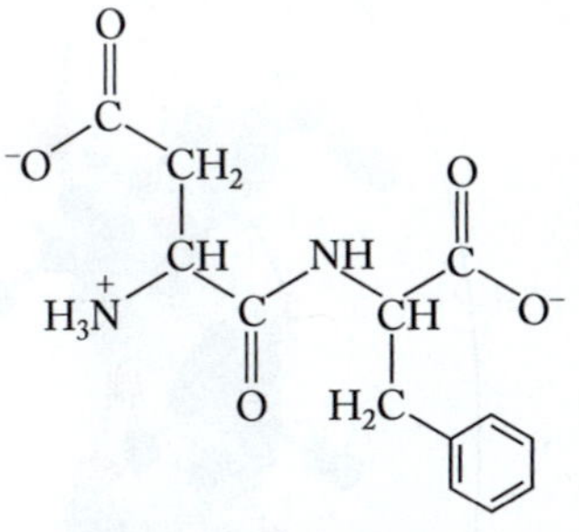

Aspartylphenylalanine

**FIGURE 20.8** Aspartame is the methyl ester (highlighted in blue) of the dipeptide aspartylphenylalanine and has one fewer ionized –COOH group than its parent dipeptide. Therefore, the overall charge of a molecule of aspartame at pH 7.4 is zero, whereas the charge on aspartylphenylalanine is 1–.

## 20.2 Protein Structure and Function

At the beginning of this chapter, we mentioned the catastrophic impact of a malformed muscle protein. The functions of all proteins, such as the ability of enzymes to catalyze biochemical reactions, are also closely linked to their structures. These large biomolecules must assume particular three-dimensional conformations to interact with other molecules and to fulfill their biological mission.

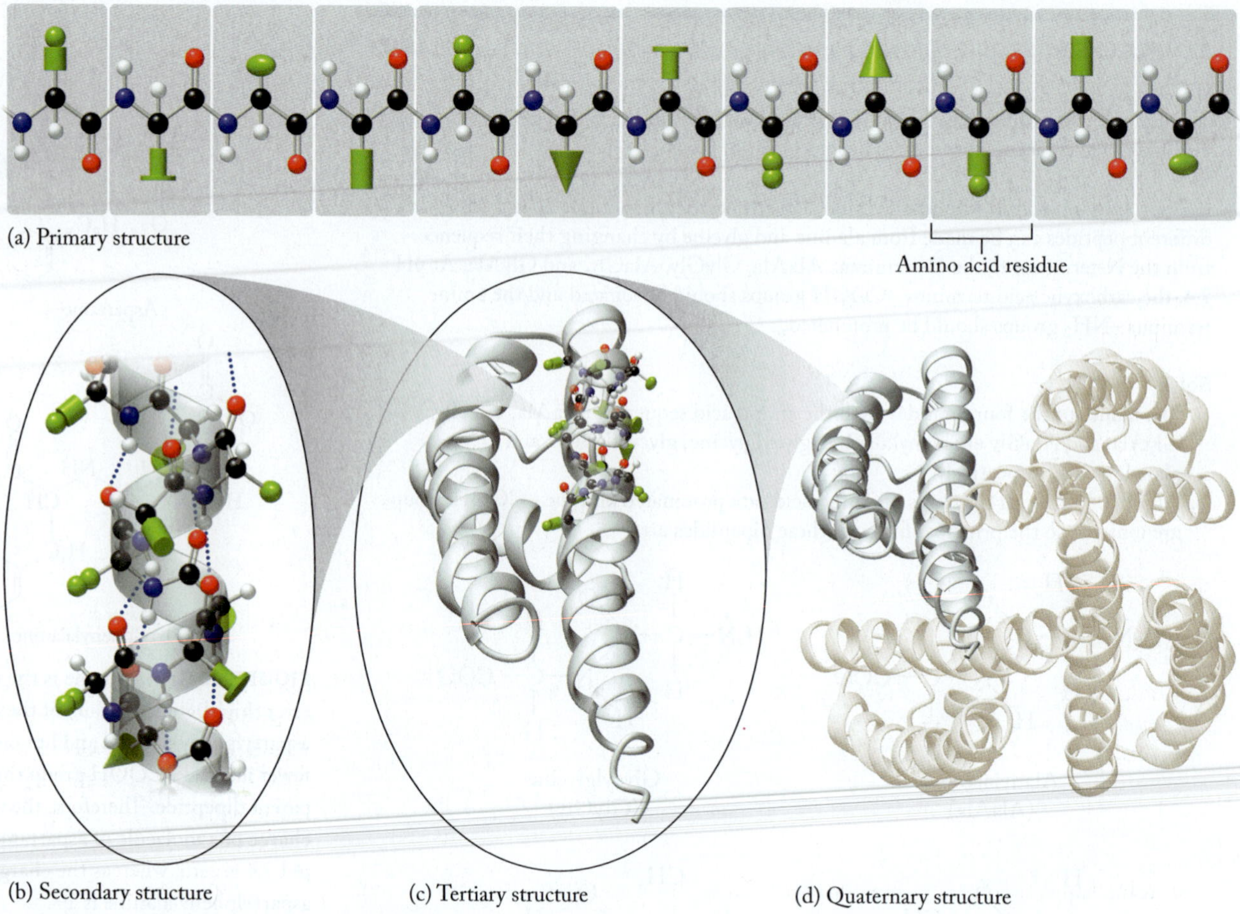

(a) Primary structure

Amino acid residue

(b) Secondary structure

(c) Tertiary structure

(d) Quaternary structure

**FIGURE 20.9** The four levels of protein structure. (a) A protein's primary structure is its amino acid sequence. The green shapes represent different R groups. (b) Secondary structure (here, an α helix) describes the three-dimensional pattern adopted by segments of the protein strand. (c) Tertiary structure is the overall shape of the molecule as segments of it bend and fold. (d) Quaternary structure refers to the overall shape adopted by multiple protein strands that assemble into a single unit.

## Primary Structure

The **primary (1°) structure** of a protein is the sequence of the amino acids in it, starting with the N-terminus (**Figure 20.9a**). If two proteins are made up of the same number and type of amino acids but have different amino acid sequences, they are different proteins.

Changing only one amino acid can dramatically alter a protein's function. In hemoglobin, for example, the sixth amino acid from the N-terminus of a protein strand that is 146 amino acids long is normally glutamic acid. In some people, valine substitutes for glutamic acid at this position. This one substitution alters the solubility of the protein and causes the red blood cell to take on a sickle shape instead of the normal, plump disc shape (**Figure 20.10**). These sickled cells do not

**CONNECTION** We discussed in Chapter 6 how ion–dipole and dipole–dipole interactions are the key to understanding the solubility of solutes in water.

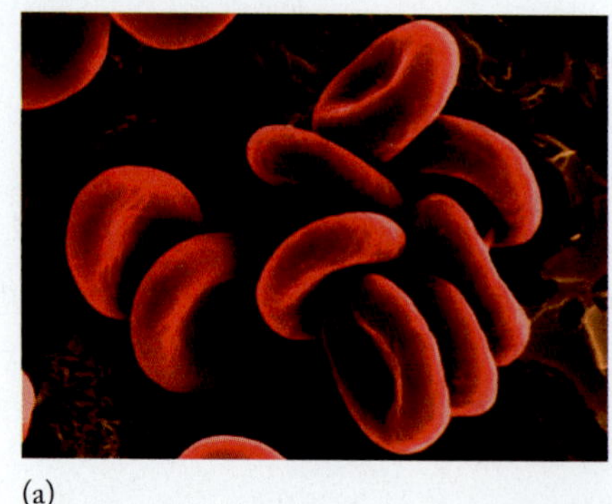

(a)

(b)

**FIGURE 20.10** (a) Normal red blood cells are plump discs, whereas (b) those in patients suffering from sickle-cell anemia are distorted.

pass through capillaries easily and may impede blood circulation. They do not last as long as normal blood cells, either. These factors diminish the blood's capacity to carry oxygen—a characteristic of the disease called sickle-cell anemia.

Why does switching valine for glutamic acid affect the solubility of hemoglobin? The answer lies in the R groups of these two amino acids (**Figure 20.11**). The side-chain –COOH group of glutamic acid is ionized at physiological pH, and strong ion–dipole interactions with water molecules enhance the protein's solubility. However, if valine, with its nonpolar isopropyl R group, replaces glutamic acid, that strong intermolecular interaction with water is lost. The valine creates a hydrophobic region on the surface of the protein, which causes deoxygenated hemoglobins to stick to each other, deforming the cells.

Although sickle-cell anemia is a debilitating disease, it provides a survival advantage in regions where malaria is endemic. Having sickle-cell anemia does not protect people from contracting malaria or make them invulnerable to the parasite that causes it. However, children infected with malaria are more likely to survive the illness if they have sickle-cell anemia. Why sickle-cell anemia has this effect is not completely understood.

Primary structure

Normal protein:

Val - His - Leu - Thr - Pro - Glu - Lys - . . .

Abnormal protein:

Val - His - Leu - Thr - Pro - Val - Lys - . . .

(a)

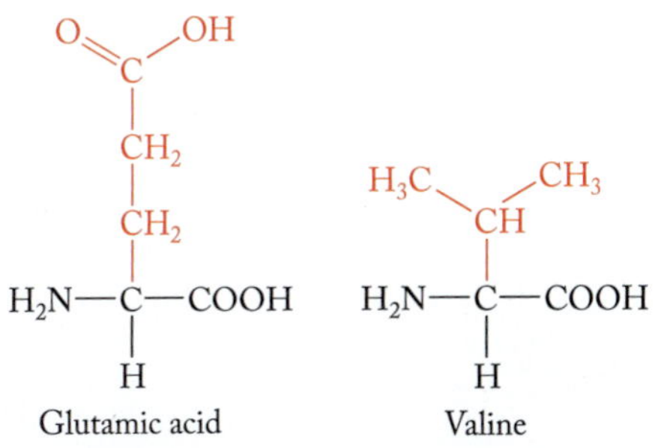

Glutamic acid: HOOC–$CH_2$–$CH_2$–CH($NH_2$)–COOH

Valine: $(H_3C)_2CH$–CH($NH_2$)–COOH

(b)

**FIGURE 20.11** (a) The primary structure of the amine end of a protein in normal hemoglobin and in the abnormal hemoglobin responsible for sickle-cell anemia. (b) The replacement of glutamic acid (which has a hydrophilic R group) by valine (which has a hydrophobic R group) is responsible for the disease.

CONCEPT **TEST**

Which other amino acids besides valine might diminish hemoglobin solubility when substituted for glutamic acid?

## Secondary Structure

For proteins to do what they are supposed to do, they must fold into a particular three-dimensional conformation. The first stage in the folding process is called a protein's **secondary (2°) structure**. The secondary structure of a protein reflects the hydrophobic interactions and intermolecular forces dominated by hydrogen bonds between peptide linkages, dipole–dipole interactions, and hydrogen bonds between polar side groups. Among the most common geometric patterns that segments of amino acid chains make (**Figure 20.9b**) is the **α helix**, a coiled arrangement with the R groups pointing outward. The helical structure is maintained by hydrogen bonds between –NH groups on one part of the chain and C=O groups on nearby amino acids. The α helix resembles a spring. Muscle tissue that stretches and contracts is made largely of α-helical proteins, as are keratins, the proteins in hair and fingernails.

Another common 2° structure is the **β-pleated sheet**. These sheets are assemblies of multiple amino acid chains aligned side by side. The pleats are caused by the tetrahedral molecular geometries of the atoms along the chains (**Figure 20.12**). Adjacent chains are linked by hydrogen bonds, and the collection of side-by-side zigzag chains forms a continuous sheet. R groups extend above and below the pleats. β-Pleated sheets may stack on top of one another like two pieces of corrugated roofing. Stacked sheets are held together by the same interactions that hold all proteins together, including ion–ion interactions, hydrogen bonding, and London dispersion forces, depending on the pairs of R groups involved.

The proteins that make up strands of silk form thin, planar crystals of β-pleated sheets that are only a few nanometers on a side. Enormous numbers of these crystals form long arrays of sheets stacked together like nanoscale pancakes. Hydrogen bonds hold the stacks together and reinforce adjacent sheets. As a result of these interactions, strands of silk are stronger than strands of steel with the same mass; indeed, silk is one of the toughest known materials—natural or synthetic.

**CHEMTOUR**
Fiber Strength and Elasticity

**primary (1°) structure** the sequence in which amino acid monomers occur in a protein chain.

**secondary (2°) structure** the pattern of arrangement of segments of a protein chain.

**α helix** a coil in a protein chain's secondary structure.

**β-pleated sheet** a puckered two-dimensional array of protein strands held together by hydrogen bonds.

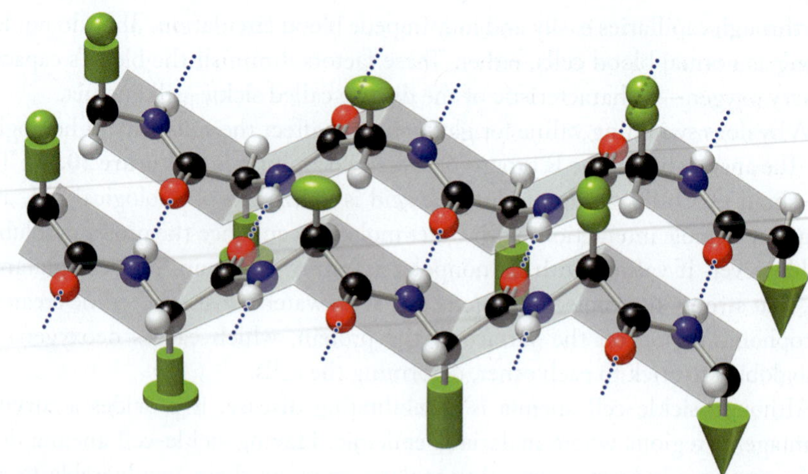

**FIGURE 20.12** Each amino acid chain in a β-pleated sheet is folded in a zigzag pattern. Adjacent chains in a sheet are held together by hydrogen bonds (blue dotted lines). The R groups (green shapes) extend above and below the sheet, linking it to adjacent sheets via intermolecular forces.

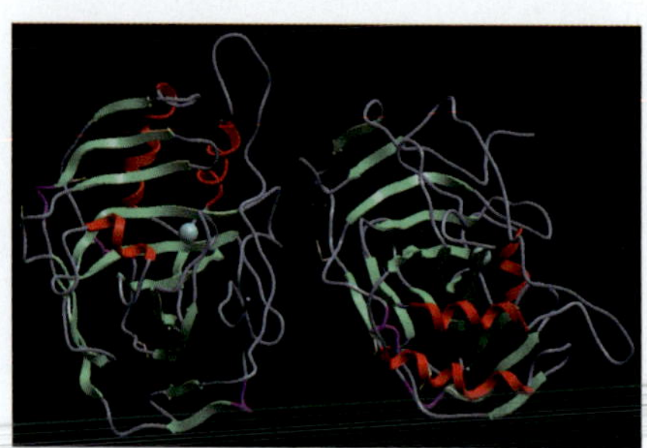

**FIGURE 20.13** Two views of carbonic anhydrase. The structure of the protein has α-helical regions (red), β-pleated sheet regions (green), and other structures (blue). The light blue sphere in the center is a zinc ion.

Some single-strand proteins exist as α helices on their own but form β-pleated sheets when they clump together in multistrand aggregates. One consequence of this clumping is the formation of insoluble protein deposits called plaque. Abnormal accumulation of plaque can be a serious health risk: plaque formed by a protein called amyloid β has been linked to the onset of Alzheimer's disease.

Large protein molecules may contain two or more types of 2° structure. In describing a protein, scientists may indicate the percentage of amino acids involved in each type—for example, 50% α helix, 30% β-pleated sheet, and 20% other structures. **Figure 20.13** shows a model of a protein called carbonic anhydrase, which illustrates all three types.

**CONCEPT TEST**

Why do proteins denature (lose their secondary structure) when they are heated?

## Tertiary and Quaternary Structure

Large proteins have structure beyond the 1° and 2° levels, folding back on themselves as a result of interactions between R groups on amino acids that can be far apart along the protein chain. These interactions may be ion–ion, dipole–dipole (hydrogen bonding), or London dispersion forces (**Figure 20.14**). They may even involve the formation of covalent bonds—specifically, disulfide bonds. The –SH groups on two cysteine residues may undergo an oxidation half-reaction that results in the formation of a disulfide linkage that holds two parts of the protein strand together by means of an intrastrand –S—S– covalent bond (**Figure 20.15**). Interactions and reactions such as these determine **tertiary (3°) structure**, the overall three-dimensional shape of the protein that is key to its biological activity (**Figure 20.9c**). Hydrophobic R groups tend to reside in the interiors of proteins in the aqueous media in living cells, whereas hydrophilic groups (as we saw in normal hemoglobin) are oriented toward the outside, where they interact with nearby molecules of water. Hydrophobic interactions are the

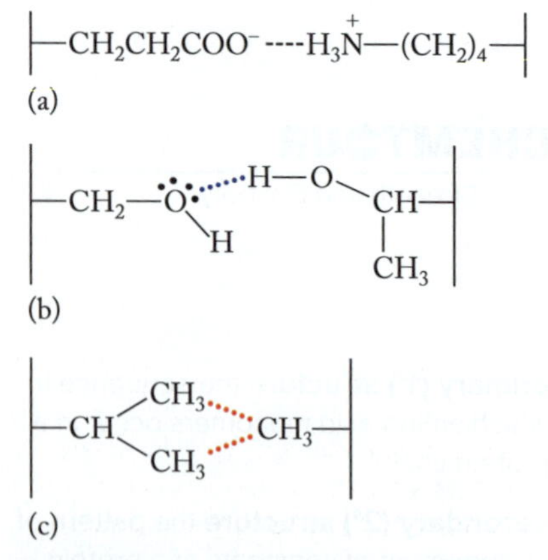

$—CH_2CH_2COO^- \cdots H_3\overset{+}{N}—(CH_2)_4—$

(a)

$—CH_2—\ddot{O}(H) \cdots H—O—CH(CH_3)—$

(b)

$—CH(CH_3)(CH_3) \cdots CH_3—$

(c)

**FIGURE 20.14** Interactions that influence the tertiary and quaternary structures of proteins include (a) ion–ion interactions between acidic and basic R groups, (b) hydrogen bonding (dipole–dipole interactions), and (c) London dispersion forces between nonpolar side chains.

primary cause of protein folding and compaction, but all the other modes of interaction help a large protein form its unique 3° structure.

Hemoglobin and some other proteins exhibit an even higher order of structure. One hemoglobin unit (**Figure 20.16a**) contains four protein strands, each of which enfolds a porphyrin ring containing one $Fe^{2+}$ ion. The combination of four protein strands to make one hemoglobin assembly is an example of **quaternary (4°) structure** (**Figure 20.9d**). The strands in these structures are held together by many of the same intermolecular forces and covalent bonds that determine tertiary structures, such as disulfide linkages that help protein strands with α-helical 2° structures coil around each other to make even larger coils (**Figure 20.16b**). When protein strands are held together mostly by intermolecular forces, as they are in the keratin in skin tissue, the structures are flexible and elastic. If they are restrained by many covalent bonds, as in Figure 20.16(b), they produce tissues that are hard and less flexible, such as fingernails and the beaks of birds.

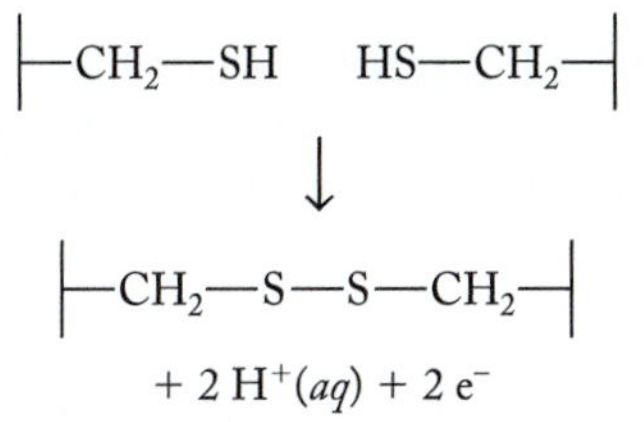

$$\text{—CH}_2\text{—SH} \quad \text{HS—CH}_2\text{—}$$
$$\downarrow$$
$$\text{—CH}_2\text{—S—S—CH}_2\text{—}$$
$$+ 2\,H^+(aq) + 2\,e^-$$

**FIGURE 20.15** The tertiary structure of some proteins is stabilized by intrastrand –S—S– bonds that form between the –SH groups on the side chains of cysteine residues.

**CONNECTION** All the types of intermolecular forces described in Chapter 6 are involved in the intramolecular interactions that give proteins their unique 2° and 3° structures.

## Enzymes: Proteins as Catalysts

The chemical reactions involved with metabolism—both *catabolism* (the breakdown of molecules) and *anabolism* (the synthesis of complex materials from simple feedstocks)—are mediated largely by proteins called **enzymes.** Enzymes are biological catalysts. Both catabolism and anabolism are organized in sequences of reactions called *metabolic pathways,* and each step in a metabolic pathway is catalyzed by a specific enzyme. For example, carbonic anhydrase (Figure 20.13) is an enzyme that speeds up the reaction shown in Equation 20.1:

$$CO_2(aq) + H_2O(\ell) \rightleftharpoons HCO_3^-(aq) + H^+(aq) \qquad (20.1)$$

Note that this reaction is reversible, and remember that a catalyst speeds up a reversible reaction in both directions. In the presence of carbonic anhydrase, this reaction proceeds about 10 million times faster than in its absence. Without carbonic anhydrase, we would not be able to expel carbon dioxide fast enough to survive (as the reaction in Equation 20.1 runs in reverse). The interconversion of

**CONNECTION** Recall from Section 14.7 that catalysts lower the activation energy barrier for a reaction and thereby increase the rate of both the forward and reverse reactions in an equilibrium.

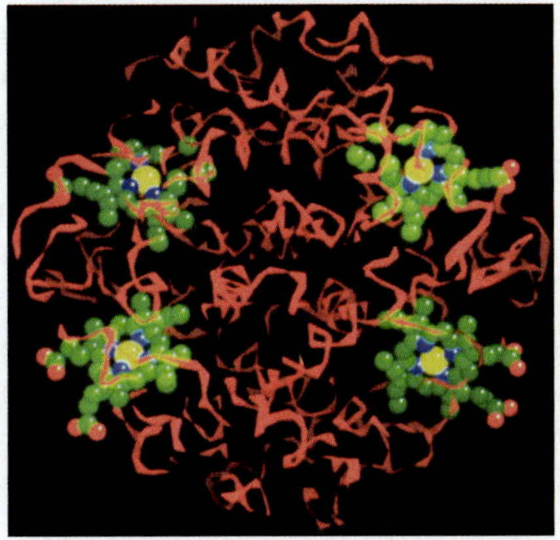

(a) Hemoglobin

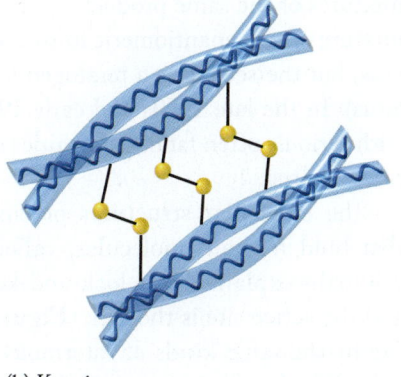

(b) Keratin

**FIGURE 20.16** Quaternary structure of proteins. (a) Four protein chains form a single unit in the quaternary structure of hemoglobin. The iron-containing porphyrins are bright green. (b) Pairs of α-helical chains (blue) wound together and linked by interstrand –S—S– bonds (yellow) stabilize the quaternary structure of the keratin in hair and fingernails.

**tertiary (3°) structure** the three-dimensional, biologically active structure of the protein that arises because of interactions between R groups on amino acids.

**quaternary (4°) structure** the larger structure functioning as a single unit that results when two or more proteins associate.

**enzyme** a protein that catalyzes a reaction.

**CONNECTION** In Chapter 15, we examined the bicarbonate buffering system and the role of breathing in maintaining physiological pH.

**CONNECTION** In Section 13.6, we discussed CFCs, inorganic homogeneous catalysts that contribute to the destruction of ozone. Enzymes are bio-organic homogeneous catalysts that selectively speed up biochemical reactions.

**CONNECTION** A racemic mixture contains equal proportions of the two enantiomers of a chiral compound, as we discussed in Section 5.6.

$CO_2$ and $HCO_3^-$ is a reaction of extreme importance in the physiological system because it is responsible for maintaining a relatively constant pH.

One molecule of carbonic anhydrase can catalyze the hydrolysis and ionization of more than a million molecules of $CO_2$ in 1 second. This value is called the *turnover number* for the enzyme; in general, the higher the turnover number, the faster the enzyme-catalyzed reaction proceeds. Turnover numbers for enzymes typically range from $10^3$ to $10^7$. The lower the activation energy of the catalyzed reaction, the higher the turnover number. As we learned in Chapter 13, molecules must collide with the proper orientation for reactions to proceed. Carbonic anhydrase is essentially a perfect enzyme because it catalyzes the hydrolysis reaction nearly every time it collides with $CO_2$. Your saliva contains carbonic anhydrase, and this is why carbonated beverages fizz and release $CO_2$ quickly in your mouth, whereas it takes a much longer time for them to release dissolved $CO_2$ when they are simply open to the air.

Enzymes are highly selective: each catalyzes a particular reaction involving a particular reactant. For example, an enzyme called lactase catalyzes only the reaction by which lactose, the sugar in milk, is broken down during digestion. People who lack this enzyme cannot metabolize the sugar; they are said to be *lactose intolerant*. If they consume dairy products, unmetabolized lactose passes into their large intestines where bacteria ferment it, causing unpleasant and painful abdominal disturbances.

Enzyme selectivity sometimes extends to enantiomeric pairs, which is the main reason why natural products tend to be optically pure materials, whereas the same products synthesized in the laboratory tend to be racemic mixtures. In the pharmaceutical industry, research is currently focused on using enzymes outside living systems to produce enantiomerically pure materials needed as drugs. This research area, called **biocatalysis**, uses enzymes to catalyze chemical reactions run in industrial-sized reactors. Enzyme-catalyzed reactions are frequently run at lower temperatures (between 4°C and 40°C) and pressures and use less energy than noncatalyzed processes. Therefore, their development and use are being encouraged in many industrial processes as part of the green chemistry initiative.

Synthetic reaction pathways catalyzed by enzymes can significantly reduce the production cost of, for example, biological pharmaceuticals. However, biocatalytic reactions often run best in very dilute solutions, which limits production. A key issue driving interest in such processes is that an enantiomerically pure pharmaceutical is likely to be more potent and produce fewer side effects than racemic mixtures of the same product. As an example, the drug thalidomide is a racemic mixture of two enantiomeric forms: one is very effective in treating morning sickness, but the second is a teratogen (an agent that disturbs the development of a fetus). In the late 1950s and early 1960s, more than 10,000 children were born with serious, often fatal, deformities as a result of their mothers having taken the racemic drug.

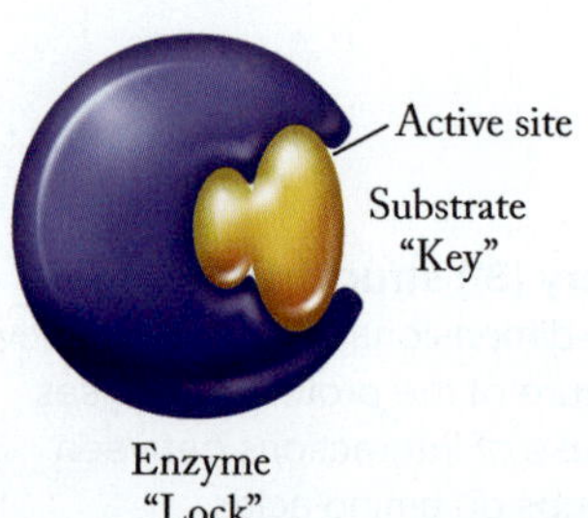


**FIGURE 20.17** In the lock-and-key model of enzyme activity, the substrate (yellow) fits exactly into the active site (purple) of the enzyme that catalyzes a chemical reaction involving the substrate.

The molecular structures of enzymes contain regions called **active sites** that bind reactant molecules, called **substrates**. The action of enzymes was originally explained by a lock-and-key analogy in which the substrate is the key and the active site is the lock (**Figure 20.17**). The substrate is held in the active site by the same kinds of intermolecular interactions that hold other biomolecules together. Some enzymes become covalently bound to intermediates in the catalytic process. Once in the active site, the substrate is converted into product through a reaction having a lower energy transition state than it would without the enzyme. The reaction of a substrate S with an enzyme E produces an *enzyme–substrate* (E–S) *complex* that decomposes, forming products, P, and regenerating the enzyme:

$$E + S \rightleftharpoons E\text{–}S \rightarrow E + P$$

The lock-and-key analogy, however, does not fully account for all enzyme behavior. A more accurate view is provided by the *induced-fit model*, which assumes that as an E–S complex forms, the active site undergoes subtle changes in its shape to more precisely fit the three-dimensional structure of the transition state. A simplified illustration of such an interaction is shown in **Figure 20.18(a)**.

The induced-fit model helps explain the behavior of compounds called **inhibitors** that can diminish or destroy the effectiveness of enzymes. An inhibitor may bind to an active site and block it from interacting with the substrate (**Figure 20.18b**), or it may disable the enzyme by preventing it from assuming its active shape (**Figure 20.18c**).

Inhibitors play important roles in regulating the rates of reactions that are catalyzed by enzymes. For example, in a multistep reaction pathway, the product of a later step may inhibit an enzyme that catalyzes an earlier reaction. This kind of negative feedback keeps the sequence of reactions from running too quickly and perhaps jeopardizing the health of the organism due to the accumulation of undesirable products or intermediates. Enzyme inhibitors also may be used to fight disease. Powerful drugs have been developed that inhibit the enzymes that allow viruses to break down proteins. Several such drugs have been particularly effective in treating HIV.

**CONNECTION** The phenomenon of molecular recognition in biological systems based on complementary molecular shapes was introduced in Chapter 5.

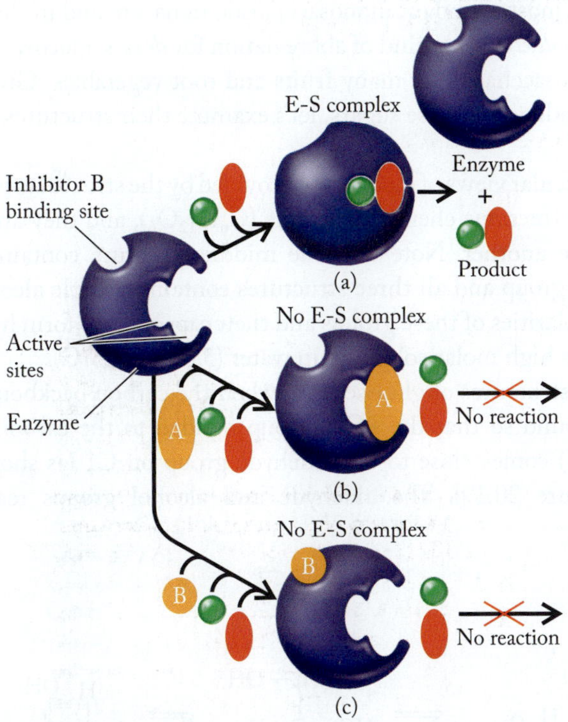

**FIGURE 20.18** (a) The induced-fit model assumes that the shape of the enzyme (purple) changes to allow formation of the enzyme–substrate (E–S) complex. (b) Inhibitor A blocks the enzyme's active sites, preventing the E–S complex from forming. (c) Inhibitor B causes a change elsewhere in the enzyme's structure that prevents the active sites from attaining the shape they need to form the E–S complex.

**biocatalysis** the strategy of using enzymes to catalyze reactions on a large scale.

**active site** the location on an enzyme where a reactive substance binds.

**substrate** the reactant that binds to the active site in an enzyme-catalyzed reaction.

**inhibitor** a compound that diminishes or destroys enzyme's ability to catalyze a reaction.

**carbohydrate** an organic molecule with the generic formula $C_x(H_2O)_y$.

**monosaccharide** a single-sugar unit and the simplest carbohydrate.

**polysaccharide** a polymer of monosaccharides.

**glycosidic bond** a C—O—C bond between sugar molecules.

# 20.3 Carbohydrates

**Carbohydrates** are a class of biological compounds with the generic formula $C_x(H_2O)_y$. Their name and formula suggest that these compounds are *hydrates of carbon*, but we will see that this label does not fit their molecular structures. The smallest carbohydrates are **monosaccharides** (the name means "one sugar"), which react to form more complex carbohydrates called **polysaccharides.** Many organisms use monosaccharides as their main energy source but convert them to polysaccharides for energy storage. Starch is the most abundant energy-storage polysaccharide in plants. Polysaccharides in the form of cellulose also provide structural support in plants. Plants produce more than 100 billion tons of cellulose each year—the woody parts of trees are more than 50% cellulose, and cotton is 99% cellulose.

The principal building block of both starch and cellulose is the monosaccharide glucose (**Figure 20.19**). The different properties and functions of these polysaccharides come from the subtle differences in the molecular geometries of the glucose monomers in their structures and how the monomers are bonded. Molecules of most monosaccharides, including glucose, contain several chiral centers, so multiple isomers exist. This complexity gives rise to the third major function of carbohydrates: molecular recognition. For example, combinations of carbohydrates and proteins, called *glycoproteins*, on the surfaces of blood cells determine the blood type of an individual.

## Molecular Structures of Glucose and Fructose

Glucose is the most abundant monosaccharide in nature and in the human body. It is also called *dextrose*—a kind of abbreviation for *dextro*-glucose. Fructose is the principal monosaccharide in many fruits and root vegetables. Given the importance and abundance of these sugars, let's examine their structures and properties more closely.

**CONNECTION** We learned in Chapter 19 that molecules of aldehydes and ketones contain functional groups with C═O double bonds.

Three molecular views of glucose are provided by the structures in Figure 20.19. All three have the same chemical formula ($C_6H_{12}O_6$), and they are all structural isomers of one another. Note that the middle structure contains an aldehyde (R—C(O)H) group and all three structures contain multiple alcohol (R—OH) groups. The polarities of these groups and their capacities to form hydrogen bonds give glucose its high molar solubility in water (5.0 *M* at 25°C).

The cyclic structures of glucose form when the carbon backbone of the linear form curls around so that the –OH group bonded to the carbon atom labeled number 5 (C5) comes close to the aldehyde group on C1 (as shown by the red arrow in Figure 20.19). The aldehyde and alcohol groups react to form a

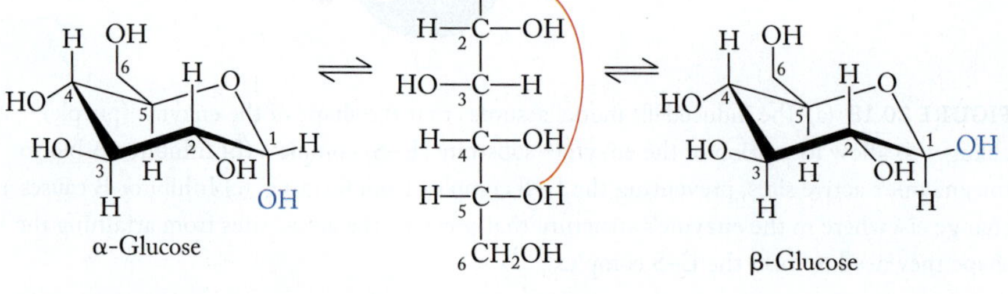


**FIGURE 20.19** An equilibrium exists between the linear structure of glucose and the two cyclic forms, α-glucose and β-glucose. The difference between the two cyclic structures is the orientation of the –OH group on C1 (highlighted in blue).

six-membered ring made up of five carbon atoms (C1 to C5) and the oxygen atom of the C5 alcohol. A small but significant difference exists between the two cyclic structures. In the molecule on the left, called α-glucose, the –OH group on C1 points down. In the molecule on the right, β-glucose, the C1 –OH group points up. These two orientations are possible because the carbon chain in the middle structure can form a cyclic structure in two ways: the –OH group on C5 can approach C1 from either of the two sides of the plane defined by the C1 carbon atom and the O and H atoms bonded to it. When the C5 –OH group approaches C1, as shown in **Figure 20.20**, the α isomer is produced. Approaching C1 from the other side yields the β isomer. C1 is an aldehyde and is not a chiral center in the linear form, but the cyclization creates a new chiral center in the cyclic molecule.

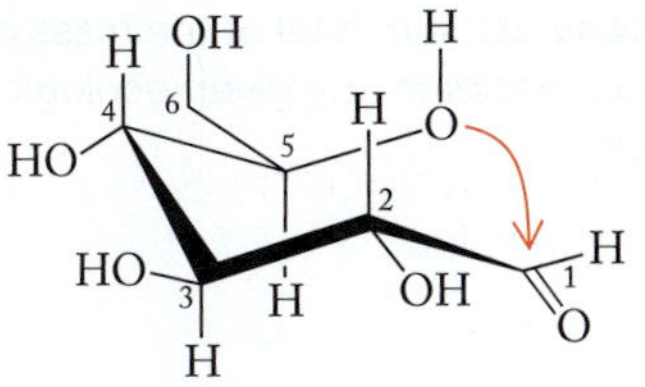

The β form is slightly more stable than the α form and accounts for 64% of glucose molecules in aqueous solution; the α form accounts for the remaining 36%. Both cyclic forms are more stable than the straight-chain form, which exists only as an intermediate between the two cyclic forms. The energy barrier to flipping conformations is relatively small, so glucose molecules in solution are constantly opening and closing in a dynamic structural equilibrium.

**Figure 20.21** shows the structures of the linear and cyclic forms of fructose, which also has the formula $C_6H_{12}O_6$. The C═O group in the linear form of fructose is not on the terminal carbon atom, so fructose is a ketone rather than an aldehyde. It forms a five-membered ring, not a six-membered ring, when the –OH group at C5 reacts with the carbonyl carbon atom at the C2 position. The product is a ring with two –$CH_2OH$ groups that are either on the same side of the ring (α-fructose) or on opposite sides (β-fructose).

**FIGURE 20.20** The –OH group on C5 may approach the aldehyde on C1 from either side of the plane defined by the C, H, and O atoms in the aldehyde group. For the approach shown here, the product will be the α isomer.

## Disaccharides and Polysaccharides

Sucrose, or ordinary table sugar, is a *disaccharide* ("two sugars") that consists of one molecule of α-glucose bonded to one molecule of β-fructose (**Figure 20.22**). The bond between them forms when the –OH group on the C1 carbon atom of glucose reacts with the C2 carbon atom of fructose, producing a C—O—C **glycosidic bond**, or *glycosidic linkage*. Water also is produced, making this reaction another example of a condensation reaction. The glycosidic bond in sucrose is called an α,β-1,2 linkage because of the orientations (α and β) of the two –OH groups involved and their positions (C1 and C2) in the cyclic structures of the two monosaccharides.

**CHEMTOUR**
Formation of Sucrose

When glucose molecules form glycosidic linkages with the –OH groups on C1 and C4, they can form long-chain polysaccharides. If the starting monomer is α-glucose, the bonds between them are α-1,4 glycosidic linkages, and the product

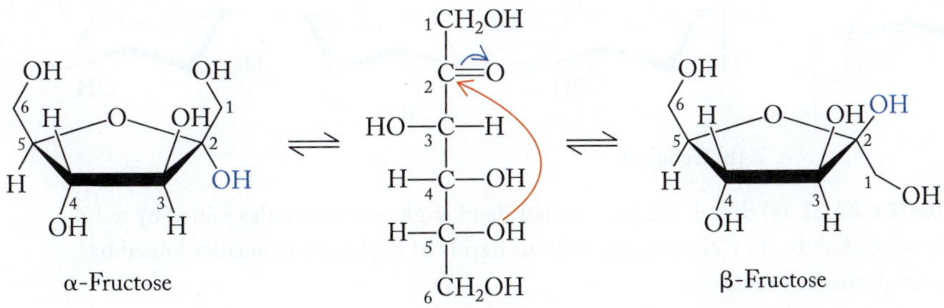

α-Fructose

β-Fructose

**FIGURE 20.21** The cyclization of fructose involves the –OH group on C5 and the ketone on C2.

**biomass** the sum total of the mass of organic matter in any given ecological system.

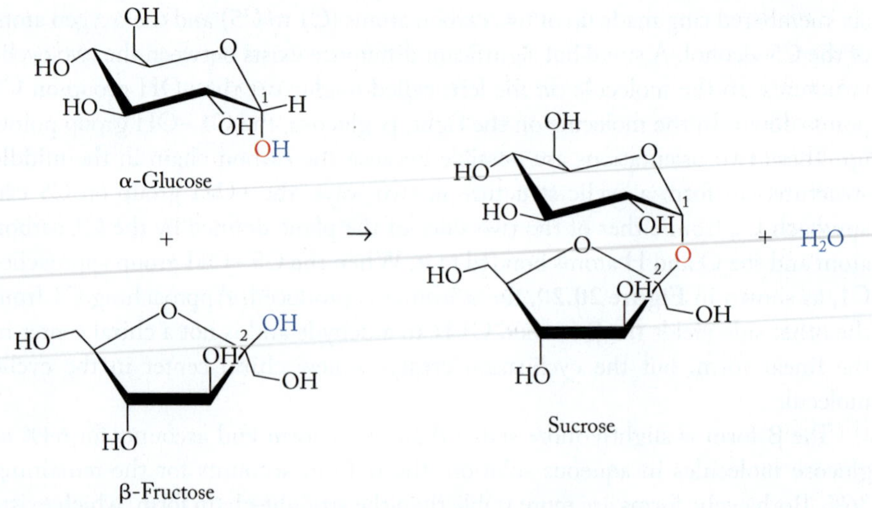

**FIGURE 20.22** α-Glucose and β-fructose (both $C_6H_{12}O_6$) react to form a molecule of sucrose ($C_{12}H_{22}O_{11}$)—ordinary table sugar—plus a molecule of water.

of their formation is starch (**Figure 20.23a**). The conversion of α-glucose into starch is an effective way for plants to store energy because formation of the α-1,4 linkage is reversible. With the aid of digestive enzymes, α-1,4 bonds can be hydrolyzed and starch can be converted back into glucose by plants that make the starch or by animals that eat the plants.

The cellulose that plants synthesize to build stems and other organs has a structure (**Figure 20.23b**) slightly different from that of starch because the building blocks of cellulose are β-glucose instead of α-glucose, so the monomers are linked by β-1,4 glycosidic bonds. This structural difference between α- and β-glycosidic bonds is important because it enables starch to coil and make granules for efficient storage, whereas cellulose forms structural fibers.

## CONCEPT TEST

Cellobiose is a disaccharide made from the degradation of cellulose. We cannot digest cellobiose. Which of the two structures in **Figure 20.24** represents a molecule of cellobiose?

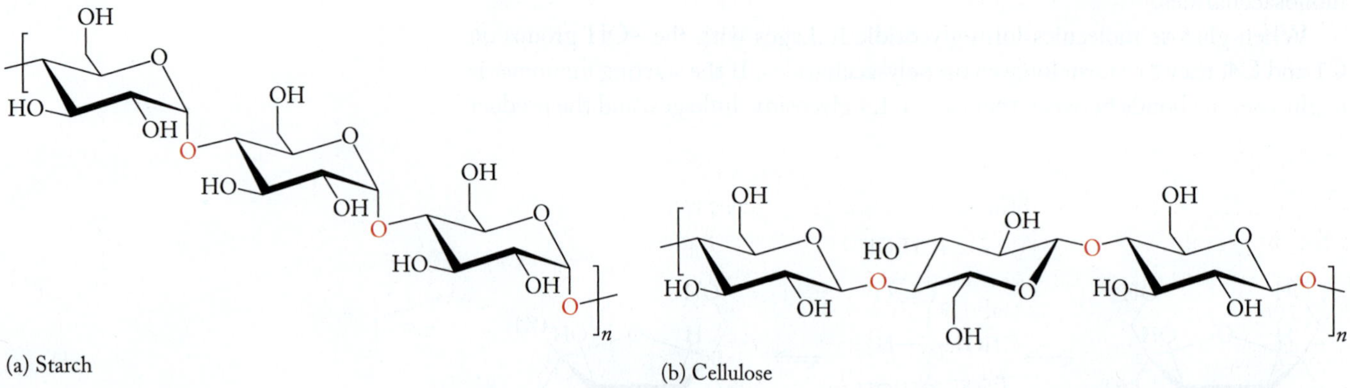

**FIGURE 20.23** (a) Starch is a polysaccharide of α-glucose molecules joined by α-1,4 glycosidic bonds. (b) Cellulose is a polysaccharide of β-glucose molecules joined by β-1,4 glycosidic bonds.

Carbohydrates in plants are a major part of the total organic matter in any given ecological system—that is, they make up much of the system's **biomass**. People have used various forms of biomass, including wood and animal dung, as fuel for thousands of years. More recently, we have begun to convert biomass into a liquid *biofuel*, ethanol, for use as a gasoline additive or substitute. Ethanol can be produced from several renewable sources such as sugarcane, sugar beets, corn, and wheat. Ethanol contains oxygen, which improves the combustion of a hydrocarbon fuel such as gasoline and reduces carbon monoxide emissions. An important industrial application of starch hydrolysis is the conversion of cornstarch into glucose and then, through fermentation, into ethanol:

$$C_6H_{12}O_6(aq) \rightarrow 2\ CH_3CH_2OH(aq) + 2\ CO_2(g) \qquad (20.2)$$

This exothermic reaction provides energy to the yeast cells, whose biological processes drive fermentation.

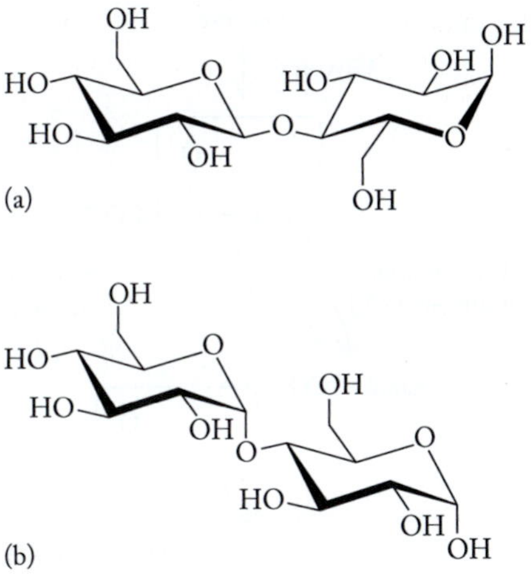

Unlike grazing animals, humans cannot digest cellulose because our digestive tracts do not contain microorganisms that have enzymes called cellulases, which catalyze hydrolysis of β-glycosidic bonds. The challenge of reproducing what cellulose-eating bacteria do in a laboratory or on an industrial scale is the focus of an enormous research effort as scientists try to develop efficient procedures for converting cellulose to glucose and then to ethanol. This research has focused on more efficient, less energy-intensive ways to break apart cellulose fibers. In addition, scientists are genetically engineering microorganisms like those in cattle stomachs to increase the supply of cellulases.

If this research succeeds, it will address several major problems associated with ethanol as a gasoline additive or alternative fuel. First, it will lower the cost of production. Today, producing ethanol from cornstarch costs more than producing gasoline from crude oil. One reason for this is that most of the mass of a corn plant, or any plant, is cellulose, not starch. Ethanol production from cornstarch is also energy intensive. More than 70% of the energy contained in ethanol is expended in producing it; therefore, the net energy value of ethanol from corn is less than 30%. If ethanol could be efficiently produced from cellulose instead of starch, its net energy value could be as high as 80%. Finally, the use of edible cornstarch in fuel production has driven up the price of foods derived from corn. The impacts of this inflation have been felt worldwide and have been particularly painful in developing countries. This competition with food has led to shifting research from the use of agricultural crops toward nonfood sources such as cassava and sweet sorghum. However, the use of agricultural land and the consumption of increasingly scarce water resources for ethanol production raise further concerns.

(a) (b)

**FIGURE 20.24** Two disaccharide structures.

## Glycolysis Revisited

Most cells, including those in human bodies, use glucose as a fuel in glycolysis (**Figure 20.25**), a series of reactions that oxidizes glucose to pyruvate ion, the conjugate base of pyruvic acid. Pyruvate sits at a metabolic crossroads and can be converted into different products depending on the type of cell in which it is generated, the enzymes present, and the availability of oxygen (**Figure 20.26**). In yeast cells growing under low-oxygen conditions, pyruvate is converted into ethanol and $CO_2$. Another series of reactions, called the

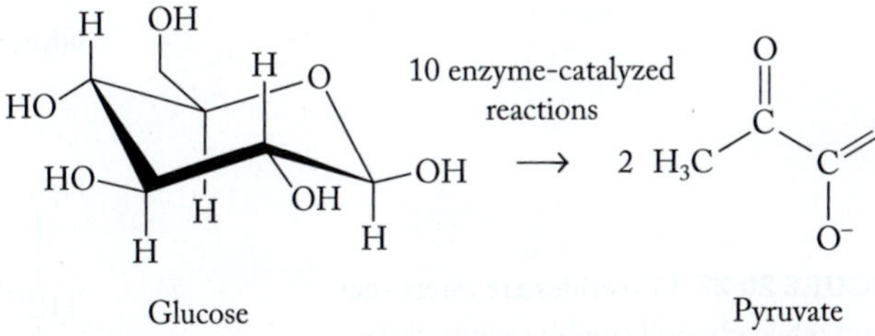

10 enzyme-catalyzed reactions

Glucose → 2 Pyruvate

**FIGURE 20.25** During glycolysis, molecules of glucose are broken down to pyruvate ions.

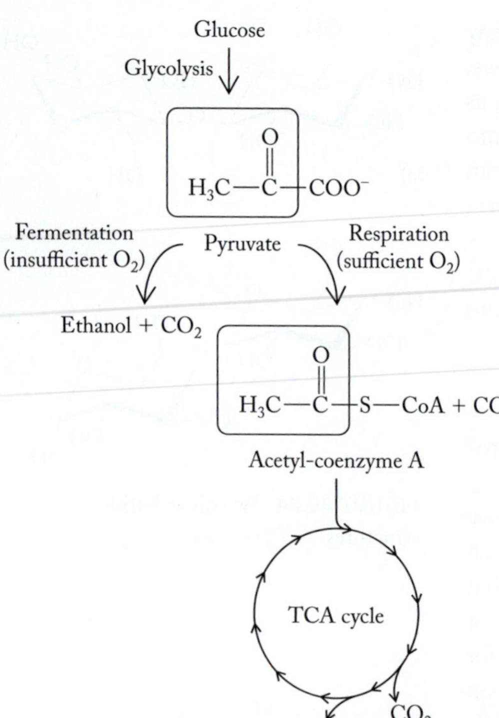


**FIGURE 20.26** The fate of the pyruvate ions formed during glycolysis depends on the partial pressure of $O_2$ in the system. In the presence of sufficient $O_2$, the oxidation of pyruvate proceeds via the TCA cycle. When there is insufficient dissolved $O_2$ available, as in fermentation, the pyruvate may be converted to ethanol.

**CONNECTION** In Section 12.8, we introduced glycolysis in a discussion of the thermodynamics of coupled reactions.

**tricarboxylic acid (TCA) cycle**, occurs when enough dissolved $O_2$ is present; the TCA cycle is fundamental to converting glucose to energy in humans and other animals. A key step before the TCA cycle occurs when pyruvate loses $CO_2$ and forms an acetyl group, which then combines with coenzyme A. The resulting product, acetyl-coenzyme A, is a reactant in many biosynthetic pathways, including the production and breakdown of cholesterol.

**CONCEPT TEST**

A newspaper article contained the wording, "Primarily made by the liver, cholesterol begins with tiny pieces of sugar . . ." What does this statement mean at the molecular level?

## 20.4 Lipids

Unlike carbohydrates and proteins, **lipids** are not biopolymers. Instead, lipids are best identified by a common physical property—they are all hydrophobic—rather than a common structural subunit. As a result, lipids do not dissolve in water but are soluble in nonpolar solvents, and they are oily to the touch. Because they are insoluble in water, they are ideal components of cell membranes, which separate the aqueous solutions within cells from the aqueous environments outside them. This group of compounds includes fats, oils, waxes, cholesterol-like compounds called sterols, and some other related compounds. An important class of lipids called **glycerides** are esters formed between glycerol and long-chain fatty acids (**Figure 20.27**). The three –OH groups on glycerol allow for mono-, di-, and triglycerides, with triglycerides the most abundant. Glycerides account for more than 98% of the lipids in the fatty tissues of mammals.

**Table 20.2** lists some common fatty acids. The most abundant ones have an even number of carbon atoms because the fatty acids are built from two-carbon units. Their biosynthesis begins, as does the biosynthesis of cholesterol, with the conversion of pyruvate to acetyl-coenzyme A. Most fatty acids contain between 14 and 22 carbon atoms. Some have no carbon–carbon double or triple bonds and are called *saturated* because they are bonded to the maximum possible number of hydrogen atoms. Other fatty acids have carbon–carbon double

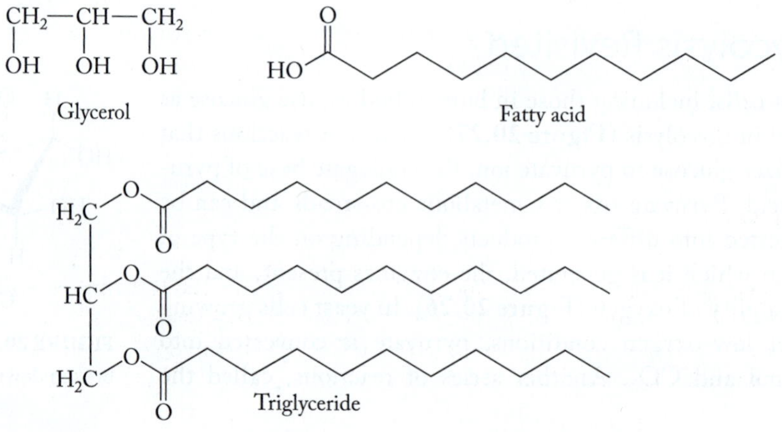

Glycerol

Fatty acid

Triglyceride

**FIGURE 20.27** Glycerides are esters that form when glycerol combines with fatty acids. When all three –OH groups on glycerol react to form ester bonds, the product is a *tri*glyceride.

**TABLE 20.2** Names, Formulas, and Structures of Common Fatty Acids

| Common Name (chemical name) (source) | Formula |
|---|---|
| **SATURATED FATTY ACIDS** | |
| Lauric acid (dodecanoic acid) (coconut oil) | $CH_3(CH_2)_{10}COOH$ |
| Myristic acid (tetradecanoic acid) (nutmeg butter) | $CH_3(CH_2)_{12}COOH$ |
| Palmitic acid (hexadecanoic acid) (animal and vegetable fats) | $CH_3(CH_2)_{14}COOH$ |
| Stearic acid (octadecanoic acid) (animal and vegetable fats) | $CH_3(CH_2)_{16}COOH$ |
| **UNSATURATED FATTY ACIDS** | |
| Oleic acid (*cis*-9-octadecenoic acid) (animal and vegetable fats) | $CH_3(CH_2)_7CH{=}CH(CH_2)_7COOH$ |
| Linoleic acid (*cis,cis*-9,12-octadecadienoic acid) (linseed oil, cottonseed oil) | $CH_3(CH_2)_4CH{=}CHCH_2CH{=}CH(CH_2)_7COOH$ |
| Linolenic acid (*cis,cis,cis*-9,12,15-octadecatrienoic acid) (linseed oil) | $CH_3CH_2CH{=}CHCH_2CH{=}CHCH_2CH{=}CH(CH_2)_7COOH$ |

and/or triple bonds and are called *unsaturated* because they are not bonded to the maximum possible number of hydrogen atoms. The double and triple bonds in unsaturated fatty acids can react with hydrogen molecules to form additional bonds to hydrogen atoms.

Triglycerides are metabolized by enzymes that hydrolyze the esters and release glycerol and the fatty acids bonded to it. Glycerol enters the metabolic pathway for glucose. The fatty acids are oxidized in a series of reactions known as β-oxidation, a process that removes two carbon atoms at a time. For example, stearic acid (the saturated $C_{18}$ fatty acid) is transformed into a $C_2$ fragment and the $C_{16}$ acid, palmitic acid. Palmitic acid yields another $C_2$ fragment and myristic acid, and so forth, until the fatty acid is completely degraded. Electrons released from this oxidative process are eventually donated to $O_2$. The energy released by this process powers metabolism.

**CONNECTION** The structures of the hydrocarbon chains in saturated and unsaturated fatty acids are consistent with our definitions of saturated and unsaturated hydrocarbons from Chapter 19.

**SAMPLE EXERCISE 20.3** Identifying Triglycerides **LO6**

How many different triglycerides (including structural isomers and stereoisomers) can be made from glycerol combining with two different fatty acids (symbolized by the letters X and Y to simplify their structures) if each molecule of triglyceride contains at least one molecule of each fatty acid?

**tricarboxylic acid (TCA) cycle** a series of reactions that continues the oxidation of the pyruvate formed in glycolysis.

**lipid** a class of hydrophobic, water-insoluble, oily organic compounds that are common structural materials in cells.

**glyceride** lipid consisting of esters formed between fatty acids and the alcohol glycerol.

**Collect and Organize** We are asked to determine how many different triglycerides can be made from glycerol and two different fatty acids. A triglyceride contains three fatty acid units.

**Analyze** Each fatty acid may bond to one of three –OH groups in glycerol. Each triglyceride has at least one X and one Y residue, which makes two combinations of X and Y possible: $X_2Y$ and $Y_2X$. Each of these formulas has two structural isomers, depending on whether the single fatty acid in the formula is bonded to the middle carbon or to one of the end carbon atoms. Finally, if a structure has an X on one end carbon atom and a Y on the other, then the middle carbon atom is a chiral center, which means that there are two enantiomeric forms of that compound.

**Solve** We can generate four different molecular structures by attaching X and Y in four different sequences to the glycerol –OH groups:

| (1) | (2) | (3) | (4) |
|---|---|---|---|
| $H_2C{-}O{-}X$ | $H_2C{-}O{-}X$ | $H_2C{-}O{-}Y$ | $H_2C{-}O{-}Y$ |
| $HC{-}O{-}X$ | $HC{-}O{-}Y$ | $HC{-}O{-}Y$ | $HC{-}O{-}X$ |
| $H_2C{-}O{-}Y$ | $H_2C{-}O{-}X$ | $H_2C{-}O{-}X$ | $H_2C{-}O{-}Y$ |

Structures 1 and 3 have chiral centers (highlighted in yellow), so each of these isomers has two enantiomeric forms. Therefore, a total of six different triglycerides are possible.

**Think About It** The central carbon atoms in structures 1 and 3 are chiral because they are each bonded to a H atom, an O atom, and to two C atoms that are themselves bonded to different fatty acids: X and Y.

**Practice Exercise** How many different triglycerides can be made from glycerol and one molecule each of three different fatty acids symbolized by the letters A, B, and C?

## Fats, Oils, and Lipoproteins

Lipids are an important energy source in our diets, providing more energy per gram than carbohydrates or proteins. **Fats** are glycerides composed primarily of saturated fatty acids and come primarily from animal sources. They are solids at room temperature. **Oils** are glycerides composed predominantly of unsaturated fatty acids, come primarily from plants, and are liquids at room temperature. Consuming too much saturated fat was once thought to be associated with coronary heart disease, but many recent studies indicate that no significant association exists between saturated fat intake and cardiovascular risk. Saturated fats also were thought to raise cholesterol levels in blood, but that link also has been questioned.

**fat** a solid triglyceride containing primarily saturated fatty acids.

**oil** a liquid triglyceride containing primarily unsaturated fatty acids.

**lipoprotein** a soluble protein that combines with and transports fat or other lipids in the blood plasma.

**phospholipid** a molecule of glycerol with two fatty acid chains and one polar group containing a phosphate; phospholipids are major constituents of cell membranes.

**lipid bilayer** a double layer of molecules whose polar head groups interact with water molecules and whose nonpolar tails interact with each other.

Olive oil, one of the main ingredients in what is known as the Mediterranean diet, is a liquid composed of glycerides containing more than 80% oleic acid, an unsaturated fatty acid. Oils can be converted into solid, saturated glycerides by hydrogenation. For example, in the hydrogenation of corn oil—a mixture of mostly two unsaturated fatty acids (oleic and linoleic acids)—hydrogen is added to convert some or all of the $-CH{=}CH-$ subunits into $-CH_2{-}CH_2-$ subunits. The resulting solid can be whipped with skim milk, coloring agents, and vitamins to produce the food spread we know as margarine.

A problem with hydrogenating vegetable oil arises when the oils are only partially hydrogenated. Partial hydrogenation alters the molecular structure around the remaining C═C bonds, changing them from their natural cis isomers into trans isomers (**Figure 20.28**). Trans unsaturated fatty acids (or trans fats) such as elaidic acid tend to be solids at room temperature because their molecules pack together more uniformly than do molecules of cis unsaturated fatty acids (cis fats). Consumption specifically of trans fats is associated with increased levels of cholesterol in the blood as well as other health risks.

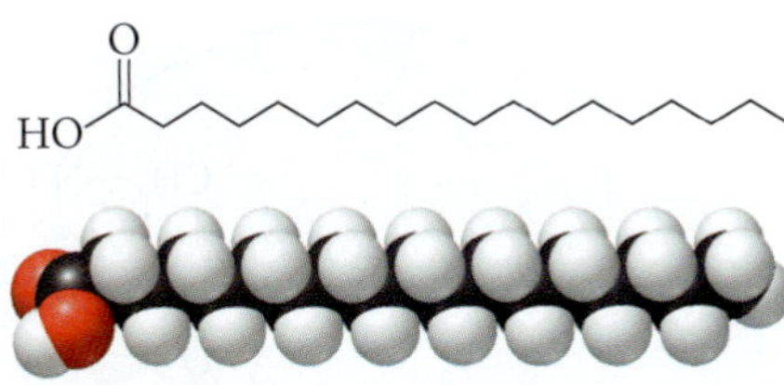


(a) Stearic acid: a saturated fatty acid

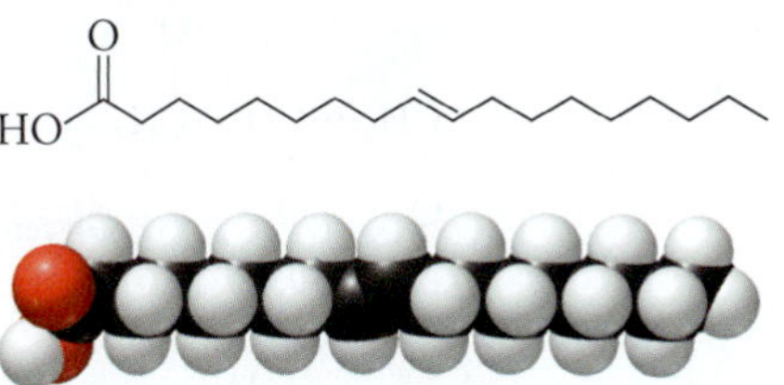


(b) Elaidic acid: a trans unsaturated fatty acid

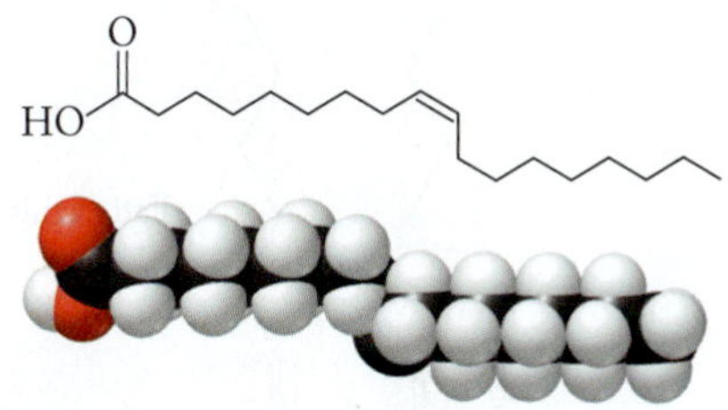


(c) Oleic acid: a cis unsaturated fatty acid

**FIGURE 20.28** Types of fatty acids. (a) Stearic acid, a saturated $C_{18}$ fatty acid. (b) Elaidic acid, an unsaturated $C_{18}$ fatty acid (trans isomer). (c) Oleic acid, an unsaturated $C_{18}$ fatty acid (cis isomer).

The molecule-level relationships between proteins and cholesterol and fats are beyond the scope of this text, but you should be aware of the basic issues. Cholesterol is not soluble in blood, so it must be transported through the bloodstream by soluble carriers called **lipoproteins**. Two types of lipoproteins carry cholesterol to and from cells: low-density lipoprotein (LDL) and high-density lipoprotein (HDL). The main structural difference between LDL and HDL is their composition. Approximately 50% by weight of LDL particles is cholesterol and only 25% is protein. HDL particles consist of 20% cholesterol by weight and 50% protein. The remaining mass of both carriers is made up of other fats and water-insoluble molecules.

Low- and high-density lipoproteins deliver cholesterol to different parts of the body. Low-density lipoproteins—the primary carriers of cholesterol—transport cholesterol to cells throughout the body, where it is used to produce cell membranes and hormones. LDL particles can bind to artery walls. This sets off a cascade of events that may form atherosclerotic plaques (**Figure 20.29**). LDL is absolutely essential to health, but it is referred to as "bad" cholesterol because elevated levels of LDL are associated with a higher rate of atherosclerosis and increased risk of cardiovascular disease.

High-density lipoproteins are also absolutely essential to maintaining human health because they carry cholesterol in the direction opposite to LDL—that is, away from the heart and other organs—and deliver it back to the liver, where it is broken down and excreted. HDL cholesterol is considered "good" cholesterol because it helps remove LDL cholesterol from the arteries. Both HDL and LDL are essential to human health—the balance between the two is what matters in determining "good" and "bad."

Cis and trans fats differ in their influence on the amounts of LDL and HDL in the blood. Cis fats promote HDL cholesterol, whereas trans fats increase LDL and decrease HDL. Consequently, trans fats are considered harmful to cardiovascular health. Trans fats that come from unnatural sources such as hydrogenated oils in processed foods seem especially detrimental in terms of changing the HDL:LDL ratio. As a result, in 2013, the U.S. Food and Drug Administration (FDA) announced that it was requiring the food industry to phase out artificial trans fats.

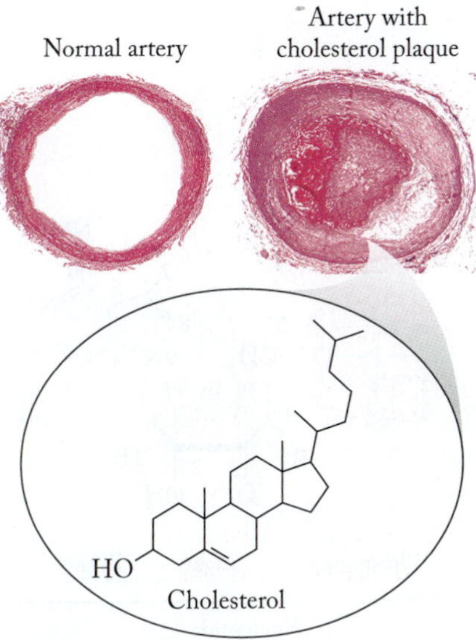


**FIGURE 20.29** Cholesterol deposits are responsible for restricted blood flow, which results in a variety of sometimes catastrophic cardiovascular problems.

## Phospholipids

Cells contain other types of lipids in addition to triglycerides. One type, **phospholipids** (**Figure 20.30a**), plays a key role in cell structure. A phospholipid molecule consists of a glycerol molecule bonded to two fatty acid chains and to one phosphate group that is also bonded to a polar substituent. The presence of nonpolar fatty acid chains and a polar region in the same molecule makes phospholipids ideal for forming cell membranes. In an aqueous medium, phospholipids form a **lipid bilayer**, a double layer enclosing each cell and isolating its interior from the outside environment. The phospholipid molecules of the bilayer align so that the nonpolar groups interact with each other within the membrane while the polar groups interact with water molecules outside the membrane (**Figure 20.30b**). Membranes exist both to isolate the contents of cells and to serve as the locus of communication between intracellular and extracellular processes.

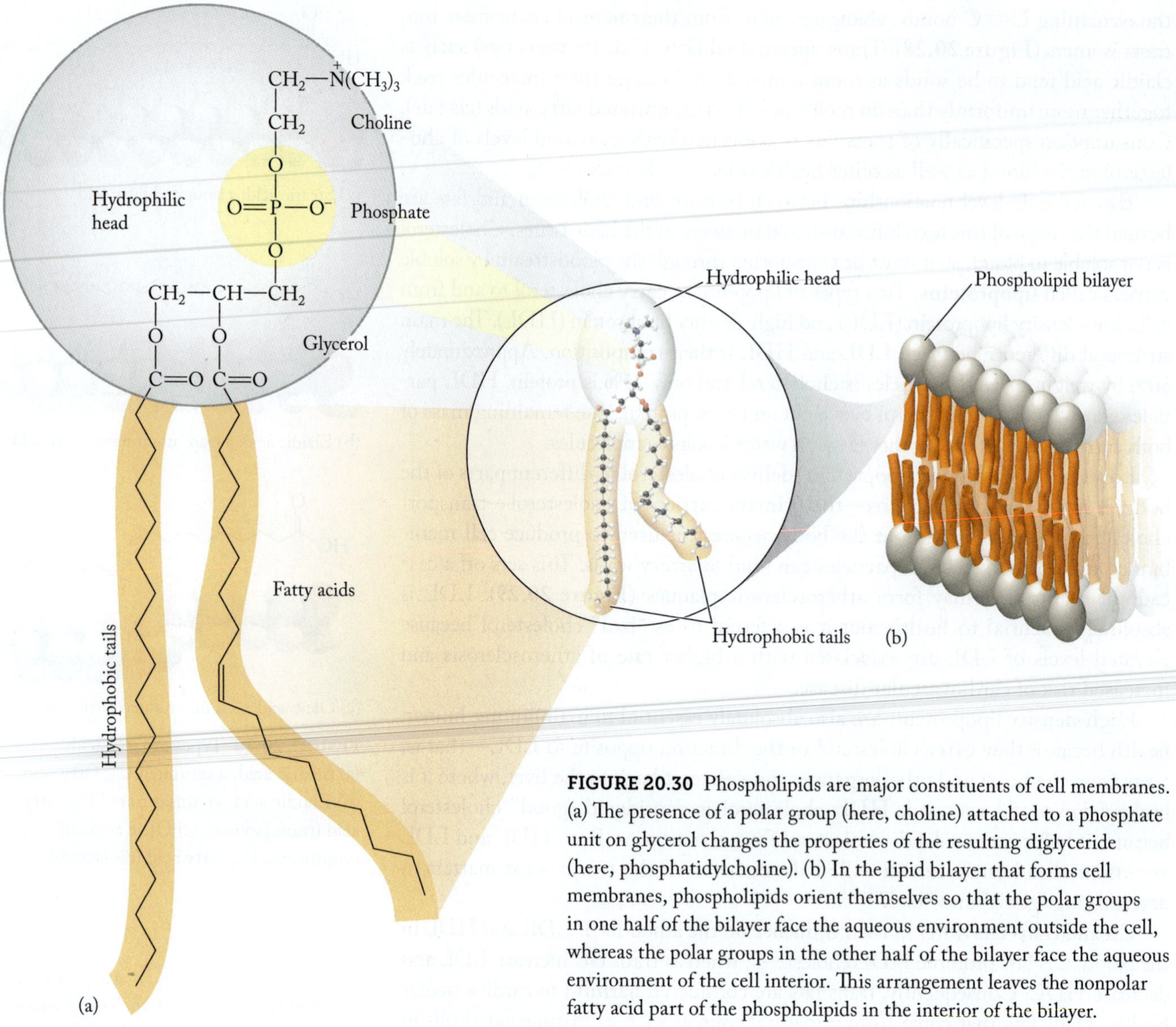

FIGURE 20.30 Phospholipids are major constituents of cell membranes. (a) The presence of a polar group (here, choline) attached to a phosphate unit on glycerol changes the properties of the resulting diglyceride (here, phosphatidylcholine). (b) In the lipid bilayer that forms cell membranes, phospholipids orient themselves so that the polar groups in one half of the bilayer face the aqueous environment outside the cell, whereas the polar groups in the other half of the bilayer face the aqueous environment of the cell interior. This arrangement leaves the nonpolar fatty acid part of the phospholipids in the interior of the bilayer.

## 20.5 Nucleotides and Nucleic Acids

**Nucleic acids** are our fourth class of biomolecules and third class of biopolymers. We focus on two types: deoxyribonucleic acid (DNA) and ribonucleic acid (RNA). Even though nucleic acids make up only about 1% of an organism's mass, they code for the enzymes that control the metabolic activity of all its cells. DNA carries the genetic blueprint of an organism, and a variety of RNAs use that DNA blueprint to guide the production of proteins.

A nucleic acid is a polymer of monomeric units called **nucleotides**. Each nucleotide unit is in turn composed of three subunits: a five-carbon sugar, a phosphate group, and a nitrogen-containing base (**Figure 20.31**). The phosphate group in each nucleotide is attached to the carbon in the side chain of the sugar called the 5′ carbon atom. (The numbers 1′ through 5′ refer to the position of carbon atoms in the sugar molecule.) The nitrogen-containing base is attached to the sugar's 1′ carbon atom. The sugar in Figure 20.31 is called *ribose*, which makes this a nucleotide in a strand of *ribo*nucleic acid, or RNA. If the sugar were

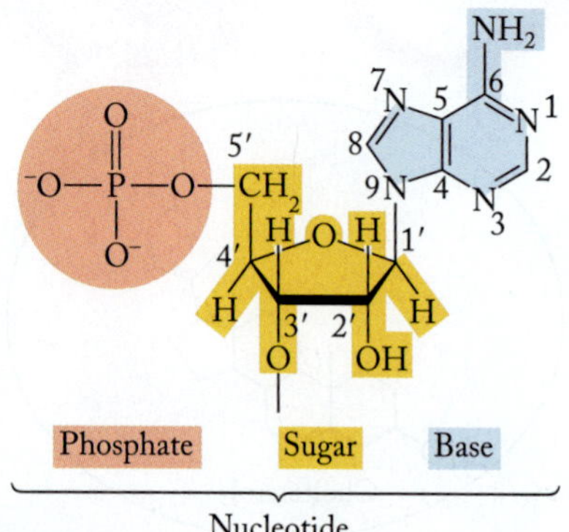


FIGURE 20.31 A nucleotide consists of a phosphate group and a nitrogen-containing base that are both bonded to a five-carbon sugar.

*deoxyribose* instead, a H atom would be there in place of the –OH group on the 2′ carbon atom, and the nucleotide would be a building block of *deoxyribo*nucleic acid, or DNA.

The nitrogen-containing base in Figure 20.31 is called adenine (A). Structural formulas of adenine and the other four bases in nucleic acids—cytosine (C), guanine (G), thymine (T), and uracil (U)—are shown in **Figure 20.32**. The point of attachment of the sugar–phosphate groups on each base is indicated by –R. In addition to the difference in their sugars, RNA and DNA also differ in one of the bases in their nucleotides: RNA contains A, C, G, and *U*, whereas DNA contains A, C, G, and *T*.

In a polymeric strand of nucleic acid, each phosphate is also linked to the 3′ carbon atom in the sugar of the monomer that precedes it in the chain, as shown for a strand of DNA in **Figure 20.33**. Both DNA and RNA strands are synthesized in the cell from the 5′ to the 3′ direction (downward in Figure 20.33). The structures of DNA and RNA are often written using only the single-letter labels of their bases, beginning with the free phosphate group on the 5′ end of the chain and reading toward the free 3′ hydroxyl group at the other terminus.

When scientists first isolated DNA and analyzed its composition, they made a pivotal observation about the abundance of the nitrogen-containing bases. A typical molecule of DNA consists of thousands of nucleotides, and the percentages of the four bases in different samples of DNA can vary widely. However, the percentage of A in a sample always matches the percentage of T. Likewise, the percentage of C always matches that of G. This result makes sense if the bases are paired because a molecule of A can form two hydrogen bonds to a molecule of T, whereas a molecule of C can form three hydrogen bonds with a molecule of G. Therefore, A–T and G–C pairings maximize the number of hydrogen bonds possible (**Figure 20.34a**). More important, the base pairs assembled this way all have the same width and fit together in a regular structure.

The normal structure of DNA features two strands of nucleotides wrapped around each other in a form called a *double helix* (**Figure 20.34b**). The nucleotide backbone is on the outside of the spiraling strands, whereas hydrogen bonds between the complementary bases on the inside of the double helix keep the two strands together. The base pairs are parallel to each other and perpendicular to

**nucleic acid** one of a family of large molecules, which includes deoxyribonucleic acid (DNA) and ribonucleic acid (RNA), that contains the genetic blueprint of an organism and controls the production of proteins.

**nucleotide** a monomer unit from which nucleic acids are made.

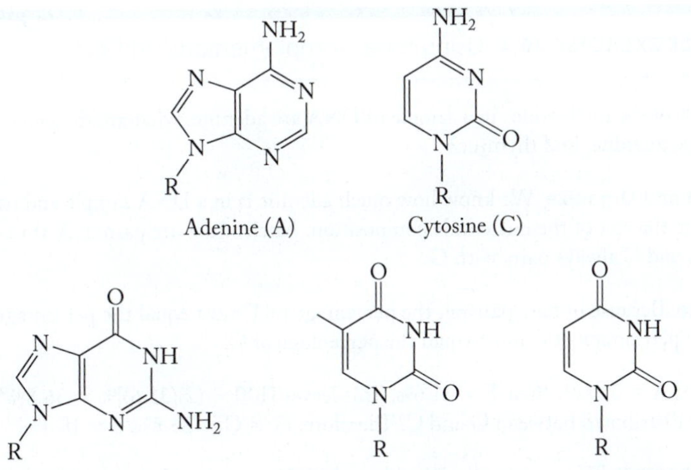

Adenine (A) Cytosine (C)

Guanine (G) Thymine (T) Uracil (U)

**FIGURE 20.32** Structural formulas of the five nitrogen-containing bases in nucleotides. The nucleotides in DNA contain A, C, G, and T; those in RNA contain A, C, G, and U. The R group identifies the point of attachment of the sugar residue.

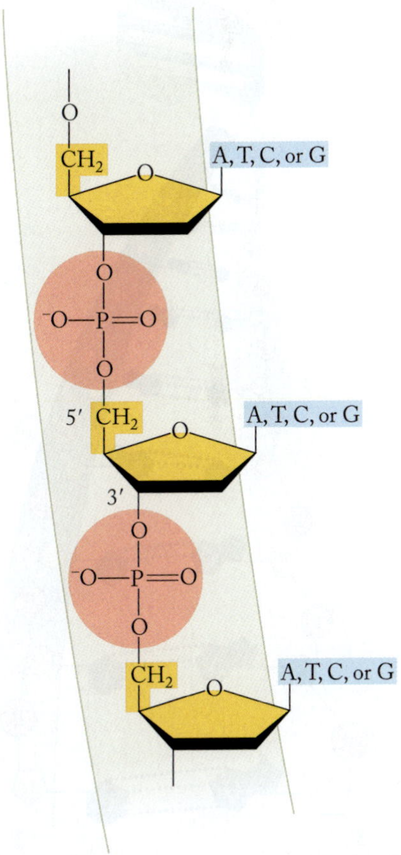


**FIGURE 20.33** The backbone of the polymer chain in DNA consists of alternating sugar units (yellow) and phosphate units (pink). The bases (blue) are attached to the backbone through the 1′ carbon atom of the sugar unit.

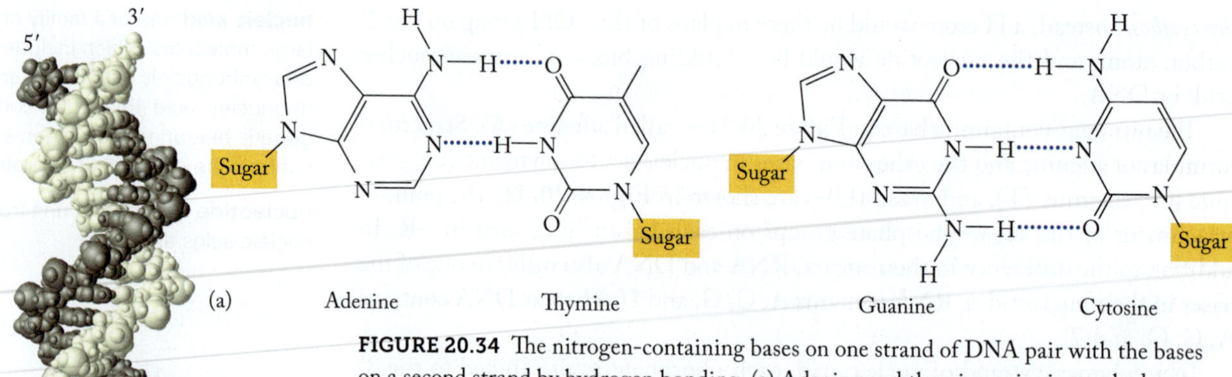


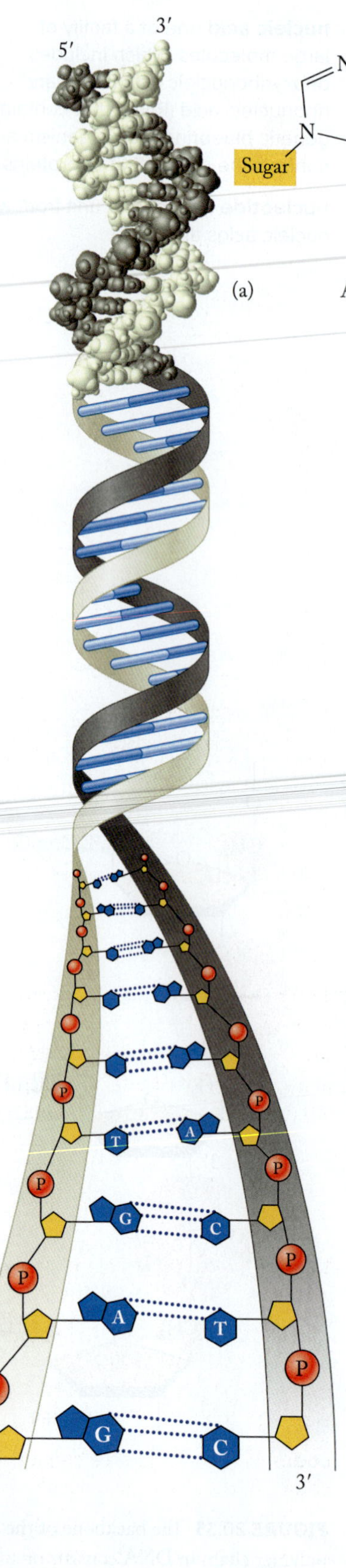


**FIGURE 20.34** The nitrogen-containing bases on one strand of DNA pair with the bases on a second strand by hydrogen bonding. (a) Adenine and thymine pair via two hydrogen bonds; guanine and cytosine pair via three hydrogen bonds. (b) DNA as a double helix with the sugar–phosphate backbone on the outside and the base pairs on the inside.

the helical axis. The fidelity of this base pairing—A always with T, C always with G—gives DNA a mechanism to copy itself. If a pair of complementary strands is unzipped into two single strands, each strand provides a template on which a new complementary strand can be synthesized via the process called **replication** (**Figure 20.35**).

During replication, the two strands are separated, forming a structure called the *replication fork*. The fork advances through the DNA as the process proceeds. One of the two resulting strands, the leading strand, is unzipped in the 3′ to 5′ direction, which allows the new complementary strand to be continuously synthesized in the 5′ to 3′ direction. The replication of the second strand, called the lagging strand, is more complicated. It proceeds in short fragments that are later assembled into a continuous strand.

DNA's double-stranded structure is also the key to its ability to preserve genetic information. The two strands carry the same information, much like an old-fashioned photograph and its negative. Genetic information is duplicated every time a DNA molecule is replicated, a process that is essential whenever a cell divides into two new cells.

**replication** the process by which one double-stranded DNA forms two new DNA molecules, each one containing one strand from the original molecule and one new strand.

### SAMPLE EXERCISE 20.4 Using Base Complementarity in DNA — LO7

If 31.6% of the nucleotides in a sample of DNA are adenine, what are the percentages of cytosine, guanine, and thymine?

**Collect and Organize** We know how much adenine is in a DNA sample and need to calculate the rest of the nucleotide composition. Nucleotides are paired: A always pairs with T, and C always pairs with G.

**Analyze** Because of base pairing, the percentage of T must equal the percentage of A, and the percentage of C must equal the percentage of G.

**Solve** If A = 31.6%, then T = 31.6%. This leaves [100 − (2)(31.6)]% = 36.8% to be equally distributed between G and C. Therefore, G = C = 36.8%/2 = 18.4%.

**Think About It** The percentages should total 100%, and they do.

**Practice Exercise** Determine the sequence of the complementary strand on the double helix formed by each of the following sequences of nucleotides: (a) CGGTATCCGAT; (b) TTAAGCCGCTAG.

## From DNA to New Proteins

Proteins are formed from amino acids in accordance with the *genetic code* contained in the base sequences of DNA strands. The bases A, T, G, and C are the alphabet in this code, and the "words" in the code are three-letter combinations of these four letters, with each word representing a particular amino acid or a signal to begin or end protein synthesis. Using four letters to write three-letter words means that $4^3 = 64$ combinations are possible—more than enough to encode the 20 amino acids found in cells. The genetic code specifies the protein's primary structure—the sequence of amino acids in proteins.

Protein synthesis begins with a process called **transcription** (**Figure 20.36a**), in which double-stranded DNA unwinds and its genetic information guides the synthesis of a single strand of a molecule called **messenger RNA (mRNA).** This strand of mRNA has the complementary base sequence of the original DNA, although it does so using the bases A, C, G, and U, not A, C, G, and T:

DNA: . . . ACGTTGAGC . . .
mRNA: . . . UGCAACUCG . . .

The mRNA carries the three-letter words of the DNA in the form of three-base sequences called **codons** (**Table 20.3**), from the nucleus of the cell into the cytoplasm, where the mRNA binds with a cellular structure called a ribosome.

At the ribosome, the genetic information in the mRNA directs the synthesis of particular proteins in a process called **translation.** Another type of RNA, called **transfer RNA (tRNA),** plays a key role in translation. The cell has 20 forms of tRNA, one for each amino acid. **Figure 20.36(b)** shows how tRNA works. The first codon in this piece of an mRNA strand is AUG, which codes for the amino acid methionine (see Table 20.3). In the cytoplasm surrounding the ribosome, molecules of tRNA are reversibly bonded to molecules of

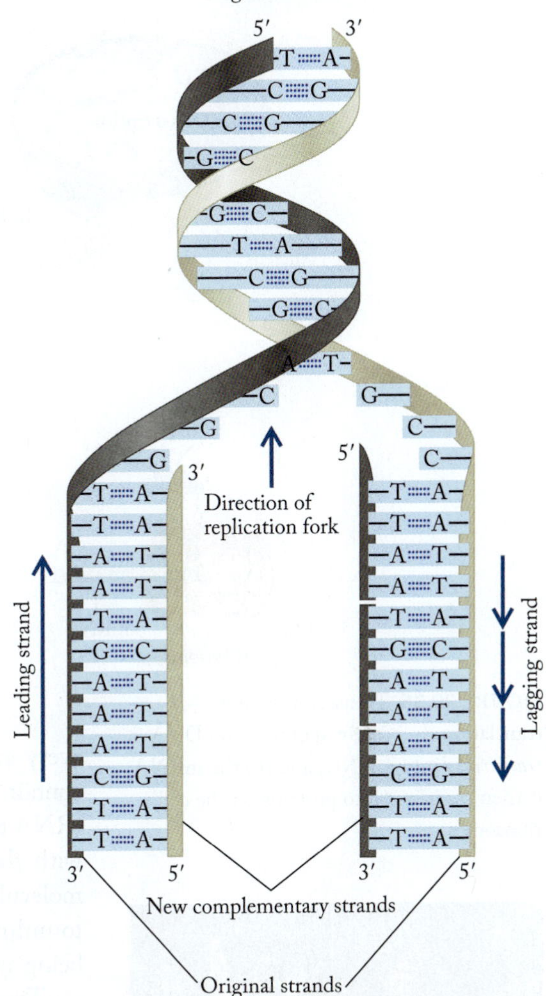


**FIGURE 20.35** When DNA replicates, the two strands of a short portion of the double helix are unzipped, and each strand serves as a template for a new complementary strand, thus producing two new DNA molecules. Each new DNA molecule contains one original strand and one new strand.

**TABLE 20.3** mRNA Codons

| Amino Acid | Codon | Amino Acid | Codon |
|---|---|---|---|
| **Ala** | GCU, GCC, GCA, GCG | **Leu** | UUA, UUG, CUU, CUC, CUA, CUG |
| **Arg** | CGU, CGC, CGA, CGG, AGA, AGG | **Lys** | AAA, AAG |
| **Asn** | AAU, AAC | **Met** | AUG |
| **Asp** | GAU, GAC | **Phe** | UUU, UUC |
| **Cys** | UGU, UGC | **Pro** | CCU, CCC, CCA, CCG |
| **Gln** | CAA, CAG | **Ser** | UCU, UCC, UCA, UCG, AGU, AGC |
| **Glu** | GAA, GAG | **Thr** | ACU, ACC, ACA, ACG |
| **Gly** | GGU, GGC, GGA, GGG | **Trp** | UGG |
| **His** | CAU, CAC | **Tyr** | UAU, UAC |
| **Ile** | AUU, AUC, AUA | **Val** | GUU, GUC, GUA, GUG |
| **Start** | AUG | **Stop** | UAG, UGA, UAA |

**transcription** the process of copying the information in DNA to mRNA.

**messenger RNA (mRNA)** the form of RNA that carries the code for synthesizing proteins from DNA to the site of protein synthesis in a cell.

**codon** a three-nucleotide sequence that codes for a specific amino acid.

**translation** the process of assembling proteins from the information encoded in mRNA.

**transfer RNA (tRNA)** the form of RNA that delivers amino acids, one at a time, to polypeptide chains being assembled by the ribosome–mRNA complex.

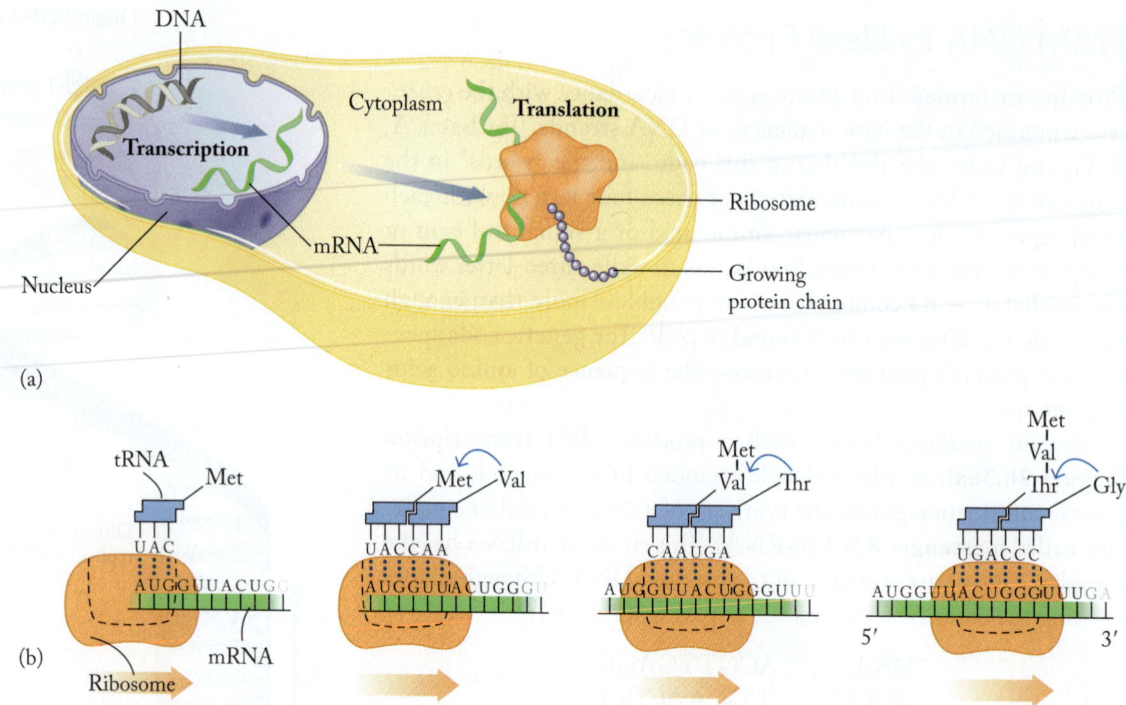


**FIGURE 20.36** Transcription and translation. In protein synthesis, (a) DNA is *transcribed* into mRNA, and (b) the mRNA is then *translated* into proteins on the cell's ribosomes.

every amino acid. The particular tRNA molecules bonded to methionine also contain the *anticodon* UAC, the complement of AUG, at a site that allows the tRNA to interact with mRNA. As Figure 20.36(b) shows, the segment of mRNA with the AUG codon links with the complementary anticodon on the tRNA molecule bonded to methionine. In doing so, the methionine is put into a position to unlink from the tRNA and to be the first amino acid residue in the protein being synthesized.

The sequence of events described in the preceding paragraph is repeated many times. In this illustration, the next codon, GUU, links up with a molecule of tRNA that has a CAA binding site and a molecule of valine in tow. In this way, valine moves into position to become the next amino acid residue in the protein and to form a peptide bond with the N-terminal methionine. Valine is followed by threonine, which is followed by glycine, and so on, until a "stop" codon signals the end of the translation process.

**CONCEPT TEST**

If a GUU codon attracts valine to the translation site, does a UUG codon do the same thing? Explain your answer.

## 20.6 From Biomolecules to Living Cells

**FIGURE 20.37** The apparatus used by Miller and Urey to simulate the synthesis of amino acids in the atmosphere of early (prebiotic) Earth. Discharges between the tungsten electrodes were meant to provide the sort of energy that might have come from lightning.

We end this chapter by addressing two fundamental questions about the major classes of biomolecules and their roles in sustaining life: how were they first formed on prebiotic Earth, and how did they assemble into living cells? Experiments conducted in the 1950s at the University of Chicago by chemistry professor Harold Urey (1893–1981) and his student Stanley Miller (1930–2007) showed that amino acids could form from $H_2O$, $CH_4$, $NH_3$, and $H_2$ (**Figure 20.37**).

Although the reactants the two scientists chose are now thought to be different from those present on the early Earth, the result still stands: inorganic molecules can react to produce the organic molecules found in living systems.

Evidence also indicates that some biomolecules may have reached Earth from extraterrestrial origins. In 2006, the NASA spacecraft *Stardust* returned to Earth with samples collected from the tail of a comet believed to have formed at about the same time as the solar system (**Figure 20.38**). Analyses disclosed the presence of glycine in the comet dust. This was not the first experiment to detect amino acids in space. A class of meteorites called carbonaceous chondrites contains isovaline and other amino acids. Interestingly, some of these amino acids are not racemic mixtures but instead are more than 50% L-amino acids, the enantiomeric form that dominates our biosphere. These observations have led some researchers to suggest that L-amino acids are somehow "favored" and that life on Earth was "seeded" from elsewhere.

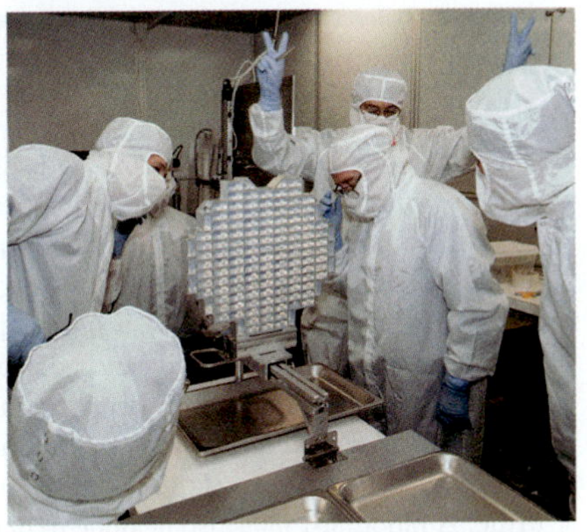

**FIGURE 20.38** NASA scientists who analyzed the samples collected by the *Stardust* spacecraft were careful to avoid contaminating it with biological material from Earth. This meant isolating themselves from the sample as they prepared it for analysis.

As we have seen in this chapter, RNA is needed to guide the assembly of amino acids into proteins. Since the 1990s, groups of scientists have explored the possibility that strands of RNA may have formed spontaneously from solutions of nucleotides in contact with mineral surfaces that guided the self-assembly of the nucleotides into long strands of RNA. Oligonucleotides made in this way might have been able to catalyze their own replication. This ability to self-replicate is crucial to life, and the observation that molecules can speed up their own replication on a clay surface suggests that processes essential to the formation of living cells could have happened spontaneously. Much controversy still exists about these ideas, but the capacity of RNA to both store information like DNA *and* act as a catalyst like an enzyme has motivated the *RNA world hypothesis*. This hypothesis proposes that a world filled with cellular or precellular life based on RNA alone predates the current world of life based on DNA and proteins.

Researchers also are testing the theory that life on Earth may have evolved near deep-ocean hydrothermal vents. Entire ecosystems have been discovered at these locations since they were first explored in the 1970s. They are sustained by geothermal and chemical energy rather than energy from the sun. Life may have actually begun in such environments, with hydrothermal energy driving reactions in which inorganic compounds such as carbon dioxide and hydrogen sulfide formed small organic compounds. As with the reactions on the surfaces of clay minerals, the synthesis reactions at hydrothermal vents may have been catalyzed and guided by reactants adsorbed on solid compounds such as FeS and $MnO_2$, which pour into the sea in dense black clouds near some vents (**Figure 20.39**). Among the known products of these reactions are acetate ions ($CH_3COO^-$). Acetate is a key intermediate in many biosynthetic pathways in living organisms. In modern bacteria, the systems that make acetate depend on a catalyst made of iron, nickel, and sulfur that has a structure much like that of particles produced by "black smokers" on the ocean floor.

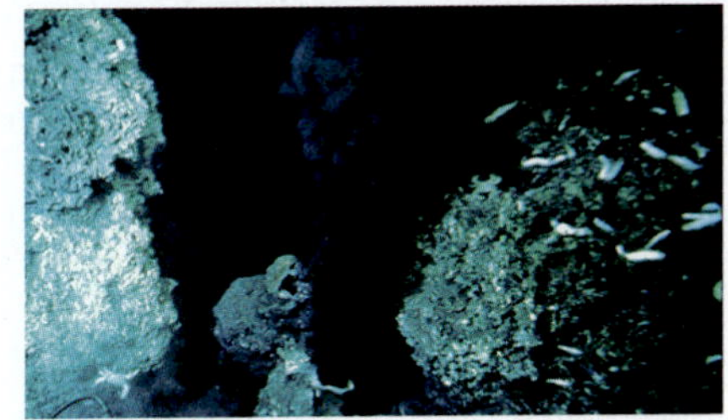

**FIGURE 20.39** Black clouds of transition metal oxides and sulfides flow into the sea through chimneys such as this one at a deep-ocean hydrothermal vent. Some scientists believe that these particles may have guided and catalyzed the formation of the first self-replicating molecules on Earth.

To take the next step toward forming living cells, large biomolecules must have assembled themselves into even larger structures, such as membranes, that allow cells and structures within them to collect materials and retain them at concentrations different from those in the surrounding medium. Molecules in these assemblies are not necessarily connected by covalent bonds but rather are held together by the intermolecular interactions we have discussed in this chapter.

## SAMPLE EXERCISE 20.5 Integrating Concepts: PKU Screening in Infants

Phenylketonuria (PKU) is a rare genetic disorder in which a baby is born without the ability to properly metabolize the amino acid phenylalanine (Phe). The gene for the liver enzyme phenylalanine hydroxylase, which is necessary to convert Phe to the amino acid tyrosine (Tyr) (**Figure 20.40a**), is mutated, and phenylalanine hydroxylase is absent or significantly decreased in activity. As a result, Phe accumulates, hindering brain development and causing mental retardation. Eventually Phe is converted into phenylpyruvate (**Figure 20.40b**), which may be detected in the urine. A blood test may also be done to determine the amount of Phe present; normal levels are less than 2.0 mg/dL of blood. A different test using mass spectrometry (**Figure 20.40c**) can determine the ratio of Phe to Tyr; in a healthy infant, the ratio should be less than 2.5 ([Phe]/[Tyr] $< 2.5$). In previous work, researchers have shown that IQ is impacted when Phe levels exceed 800 μmol/L.

a. Describe the molecular changes in Phe in terms of functional groups and charge when it is converted into Tyr and phenylpyruvate.
b. What should be the concentration of Tyr (in mg/dL) in a healthy infant if the concentration of Phe reaches 2 mg/dL?
c. Express the concentrations of Tyr and Phe from part b in mol/L.
d. Assume that the level of Tyr does not change from the value you calculated in part c. What is the [Phe]/[Tyr] ratio when [Phe] = 800 μmol/L?

**Collect and Organize** We are given the structural formulas for Phe, Tyr, and phenylpyruvate. We also know normal Phe levels in mg/dL, the normal [Phe]/[Tyr] ratio, and the level at which the amount of Phe damages mental abilities in μmol/L. We need to convert the numbers to one set of units to make the comparisons requested.

**Analyze** We can determine the molecular changes caused by metabolism from the structural formulas in Figure 20.40. We need to calculate the molecular masses of Phe and Tyr to convert between mg/dL and mol/L. The value of [Tyr] in part b will be $\frac{1}{25} \times 2.0$ mg/dL, or a little less than 1 mg/dL. From their molecular structures, we can estimate that the molar masses of Tyr and Phe will be between 100 g/mol and 200 g/mol. Therefore, the molar concentrations in part c that are equivalent to 1 or 2 mg/dL (or 10–20 mg/L = 0.01–0.02 g/L) will be about (0.01 g/L)/(100 g/mol), or $\sim 10^{-4}$ mol/L.

**Solve**

a. Phenylalanine is a neutral amino acid with a nonpolar R group. It contains an aromatic ring in addition to its amine and carboxylic acid functional groups. Phenylalanine hydroxylase converts Phe into Tyr, which has an –OH group on the aromatic ring and hence a polar R group. Tyr is also neutral. In the absence of phenylalanine hydroxylase, Phe is converted into phenylpyruvate, which is not an amino acid. Phenylpyruvate contains a ketone group and a carboxylic acid group that is ionized at physiological pH, so it is negatively charged.
b. If the concentration of Phe reaches 2 mg/dL, and if the Phe/Tyr ratio is to be <2.5, then the *minimum* concentration of Tyr ($x$) in a healthy infant would be

$$\frac{2.0 \text{ mg/dL}}{x} = 2.5, \quad x = 0.80 \text{ mg/dL}$$

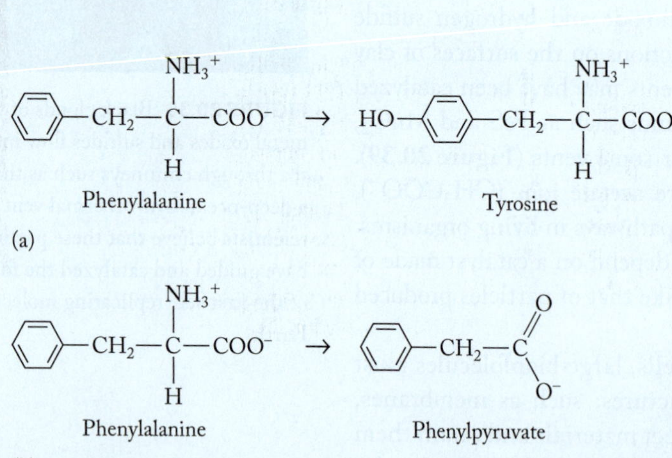


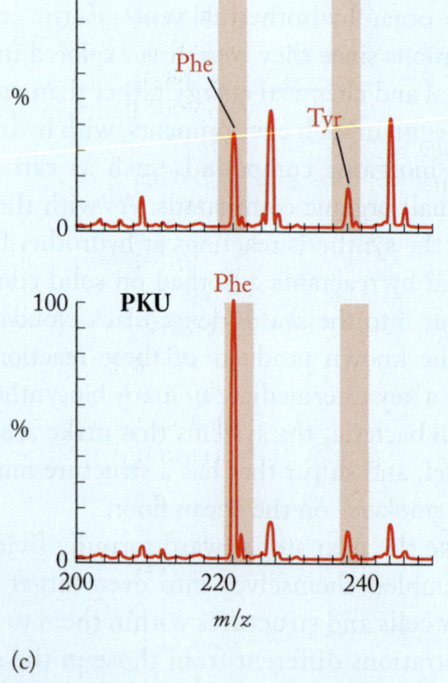


**FIGURE 20.40** (a) The amino acid phenylalanine is normally converted into the amino acid tyrosine. (b) In PKU, phenylalanine is instead converted into phenylpyruvate. (c) Amino acid profiles from a mass spectrometer of normal blood (top) and blood from a newborn infant with PKU (bottom). Adapted from S. A. Banta and R. D. Steiner in *Journal of Perinatal and Neonatal Nursing*, 18(1), 41 (2004).

c. Using the structural formulas given, Phe is $C_9H_{11}NO_2$, with a molecular mass of 165.19 g/mol, and Tyr is $C_9H_{11}NO_3$, with a molecular mass of 181.19 g/mol. In units of mol/L, the concentrations of Phe and Tyr from part b are

$$[\text{Phe}] = \frac{2.0\ \cancel{\text{mg}}}{\cancel{\text{dL}}} \times \frac{1\ \cancel{\text{g}}}{1000\ \cancel{\text{mg}}} \times \frac{1\ \text{mol}}{165.19\ \cancel{\text{g}}} \times \frac{10\ \cancel{\text{dL}}}{1\ \text{L}}$$

$$= 1.2 \times 10^{-4}\ M$$

$$[\text{Tyr}] = \frac{0.80\ \cancel{\text{mg}}}{\cancel{\text{dL}}} \times \frac{1\ \cancel{\text{g}}}{1000\ \cancel{\text{mg}}} \times \frac{1\ \text{mol}}{181.19\ \cancel{\text{g}}} \times \frac{10\ \cancel{\text{dL}}}{1\ \text{L}}$$

$$= 4.4 \times 10^{-5}\ M$$

d. Converting [Phe] = 800 μmol/L into molarity:

$$800\ \frac{\cancel{\mu\text{mol}}}{\text{L}} \times \frac{1\ \text{mol}}{10^6\ \cancel{\mu\text{mol}}} = 8.0 \times 10^{-4}\ M$$

If we assume [Tyr] = $4.4 \times 10^{-5}$, then the [Phe]/[Tyr] ratio is

$$\frac{8.0 \times 10^{-4}\ \cancel{M}}{4.4 \times 10^{-5}\ \cancel{M}} = 18$$

**Think About It** Mandatory screening programs identify infants with PKU at birth. Early diagnosis is essential to managing this disorder. Infants diagnosed with PKU require a lifelong diet extremely low in Phe. This diet is challenging to maintain because Phe occurs in significant amounts in many common foods, including milk and eggs. Many food products that contain phenylalanine have warning labels.

## SUMMARY

**LO1** In 19 of the 20 **amino** acids in human **proteins,** the α-carbon atom is a chiral center. The acid–base properties of amino acids are also important for structural and functional reasons. At physiological pH (7.4), most amino acids exist as **zwitterions.** (Section 20.1)

**LO2** Amino acids linked by **peptide bonds** form **peptides** and proteins. (Sections 20.1 and 20.2)

**LO3** Protein molecules have four levels of structure. **Primary (1°) structure** is the amino acid sequence. **Secondary (2°) structure** is the shape the amino acid chain takes (**α helix, β-pleated sheet,** or **random coil**). **Tertiary (3°) structure** is the three-dimensional shape that results from attractive forces between amino acids located in various parts of the chain. **Quaternary (4°) structure** results when two or more protein strands associate with each other to make larger functional units. (Section 20.2)

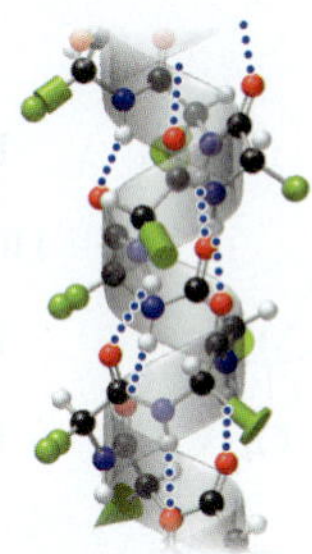

**LO4** Proteins called **enzymes** mediate the chemical reactions involved in metabolism. The structure of an enzyme contains an **active site** that binds to a **substrate** and catalyzes a reaction involving it. The *induced-fit model* suggests that the binding of a substrate to an enzyme changes the shape of the enzyme so that the reaction can take place. **Biocatalysis** seeks to replicate the selectivity of enzymes in industrial settings and is important in green chemistry processes. (Section 20.2)

**LO5 Carbohydrates** have the generic formula $C_x(H_2O)_y$. **Monosaccharides** are joined into **polysaccharides** through **glycosidic bonds.** The isomers α-glucose and β-glucose form the polysaccharides starch and cellulose, respectively. Organisms derive energy from glycolysis and the **tricarboxylic acid (TCA) cycle.** (Section 20.3)

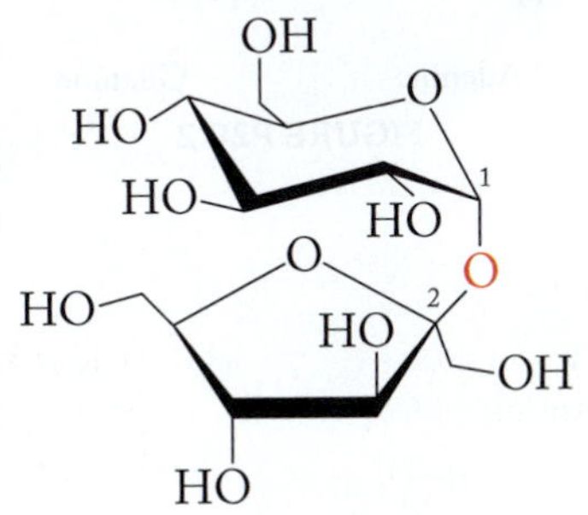


**LO6** All **lipids** are hydrophobic compounds. **Glycerides** are esters of the alcohol glycerol and up to three long-chain fatty acids. **Fats** are solid glycerides composed primarily of saturated fatty acids; **oils** are mostly unsaturated fatty acids and are liquids at room temperature. **Lipoproteins** are combinations of water-soluble proteins and lipids that transport fats and other lipids in the bloodstream. **Phospholipids** form **lipid bilayer** cell membranes. (Section 20.4)

**LO7** The **nucleic acids** DNA and RNA are made of chains of **nucleotides.** DNA consists of two nucleotide chains coiled into a double helix and connected with hydrogen bonds between complementary bases—A with T, C with G. RNA, in which the base pairings are A with U and C with G, is involved in protein synthesis through **messenger RNA (mRNA)** and **transfer RNA (tRNA).** DNA and RNA contain an organism's genetic information and control protein synthesis through **transcription** and **translation.** Genetic information is transmitted to new cells during cell division in a process called **replication.** (Section 20.5)

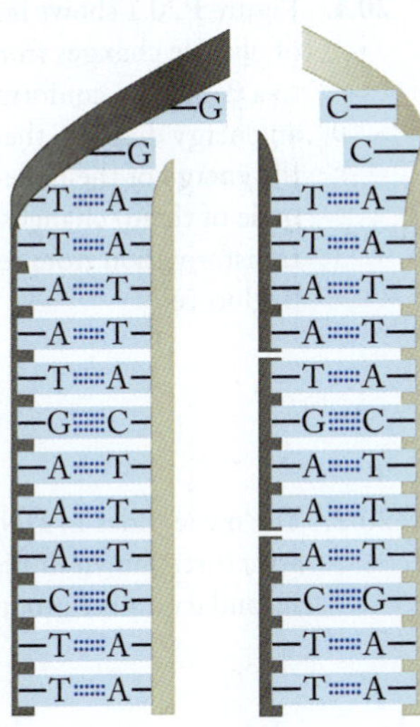

## PARTICULATE PREVIEW WRAP-UP

The two functional groups are a carboxylic acid and a primary amine. At pH $< 3$, both will be protonated: $-COOH$ and $-NH_3^+$. At pH $> 11$ the carboxylic acid will be ionized: $-COO^-$, and the amine group will not be protonated: $-NH_2$.

## PROBLEM-SOLVING SUMMARY

| Type of Problem | Concepts and Equations | Sample Exercises |
|---|---|---|
| **Interpreting acid–base titration curves of amino acids** | At low pH, all amino acids have at least two ionizable H atoms, one each from $-COOH$ and $-NH_3^+$. Side-chain carboxylic acid and amine groups may also impart acidic and basic strength to amino acids. | **20.1** |
| **Drawing and naming peptides** | Connect the α-amine of one amino acid to the α-carboxylic acid of another with a peptide bond. Starting with the free amine (N-terminus) on the left, name each amino acid residue by changing the ending of the name of the parent amino acid to *-yl* in all but the last (C-terminal) amino acid. | **20.2** |
| **Identifying triglycerides** | The $-COOH$ groups of fatty acids react with the $-OH$ groups in glycerol to form triglycerides and water. | **20.3** |
| **Using base complementarity in DNA** | Identify the base pairs: A pairs with T; G pairs with C. The percentage of T should equal the percentage of A, and the percentage of C should equal the percentage of G. | **20.4** |

## VISUAL PROBLEMS

*(Answers to boldface end-of-chapter questions and problems are in the back of the book.)*

**20.1.** Figure P20.1 shows how a molecule of glucose changes from an α-glucose to a β-glucose conformation. Sketch an energy diagram that shows how the energy of the molecule (or a mole of them) changes during the transformation from α-glucose to β-glucose.

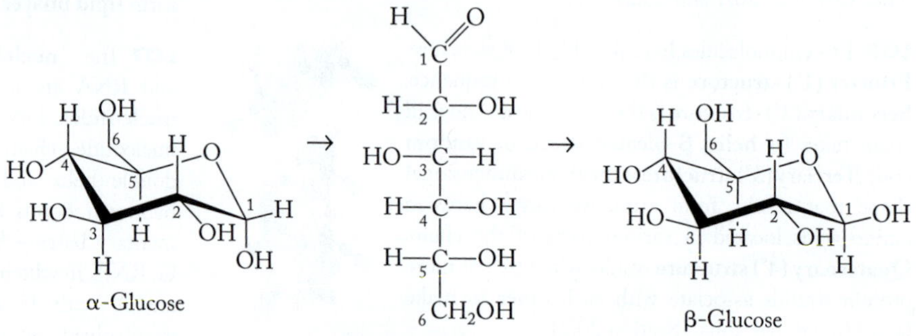

α-Glucose → open chain → β-Glucose

**FIGURE P20.1**

20.2. The nucleotides in DNA contain the bases with the structures shown in Figure P20.2. How many primary and secondary amine groups are in each structure?

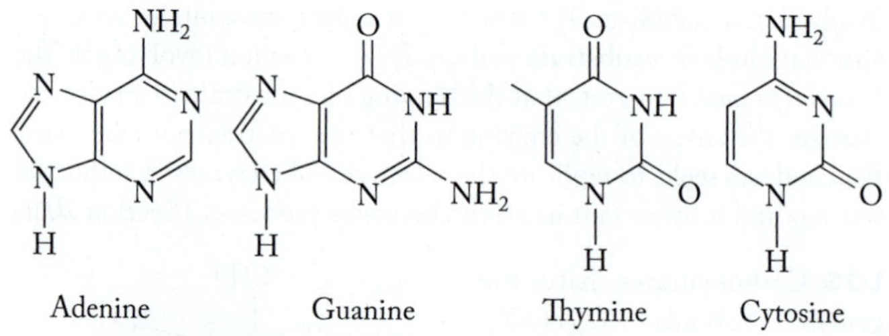

Adenine Guanine Thymine Cytosine

**FIGURE P20.2**

**20.3. Olive Oil** Olive oil contains triglycerides such as those shown in Figure P20.3.

a. Which of the fatty acids in these triglycerides is/are saturated?

b. Are the unsaturated fatty acids likely to be cis or trans isomers?

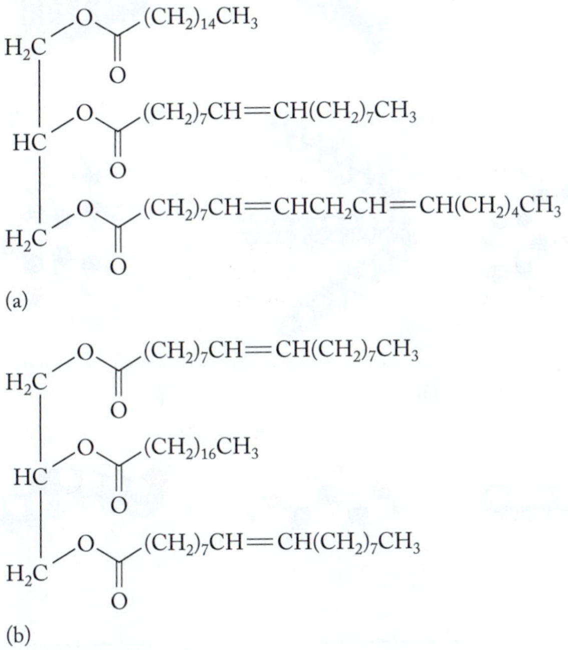

(a)
$H_2C{-}O{-}C(=O){-}(CH_2)_{14}CH_3$
$HC{-}O{-}C(=O){-}(CH_2)_7CH{=}CH(CH_2)_7CH_3$
$H_2C{-}O{-}C(=O){-}(CH_2)_7CH{=}CHCH_2CH{=}CH(CH_2)_4CH_3$

(b)
$H_2C{-}O{-}C(=O){-}(CH_2)_7CH{=}CH(CH_2)_7CH_3$
$HC{-}O{-}C(=O){-}(CH_2)_{16}CH_3$
$H_2C{-}O{-}C(=O){-}(CH_2)_7CH{=}CH(CH_2)_7CH_3$

**FIGURE P20.3**

20.4. The two compounds shown in Figure P20.4 are both amino acids. Which of them is an α-amino acid? Explain your choice.

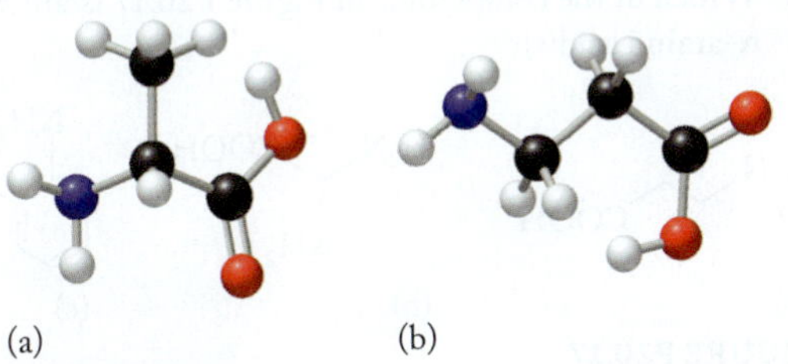


**FIGURE P20.4**

**20.5. Natural Painkillers** The human brain produces polypeptides called *endorphins* that help in controlling pain. The endorphin in Figure P20.5 is called enkephalin. Identify the five amino acids that make up enkephalin.

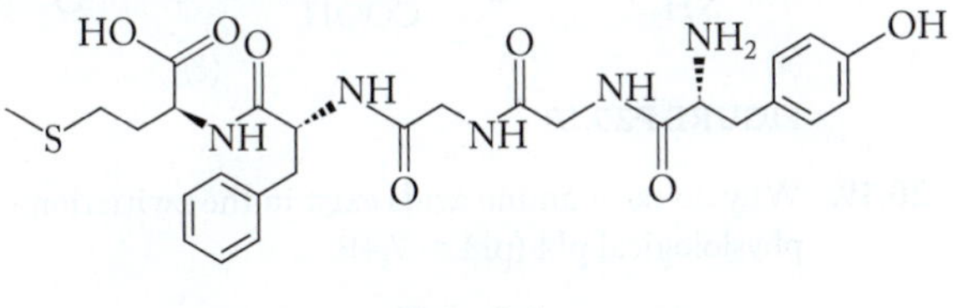

Enkephalin

**FIGURE P20.5**

20.6. **Regulating Blood Pressure** Angiotensin II is a polypeptide that regulates blood pressure. Which amino acids make up the structure of angiotensin II shown in Figure P20.6?

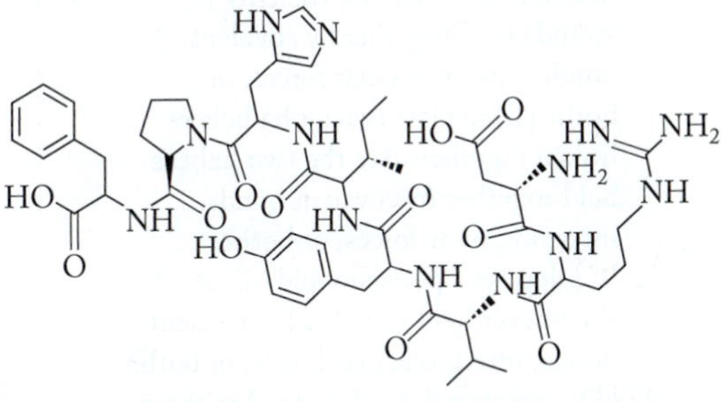

Angiotensin II

**FIGURE P20.6**

**20.7. Trans Fats** The role of trans fats in human health has been extensively debated both in the scientific community and in the popular press. Which of the molecules in Figure P20.7 are trans fats?

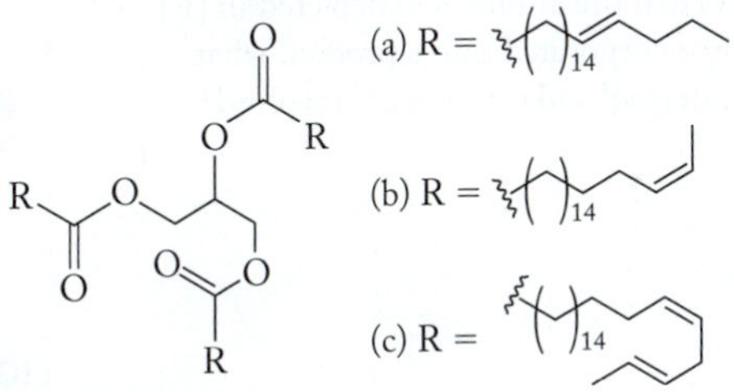

**FIGURE P20.7**

20.8. **Cocoa Butter** Cocoa butter (Figure P20.8) is a key ingredient in chocolate. Cocoa butter is a triglyceride that results from the esterification of glycerol with three different fatty acids. Identify the fatty acids produced by the hydrolysis of cocoa butter.

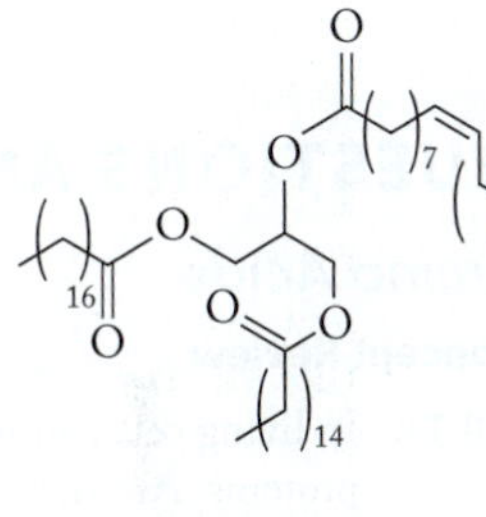

Cocoa butter

**FIGURE P20.8**

**20.9. Sucralose** The molecular structure of the artificial sweetener sucralose (trade name, Splenda) is shown in Figure P20.9. Advertising for this product claims that it is made from sugar, implying that it is a natural product. What sugar might it be made from? Comment on the implication that it is a "natural" product.

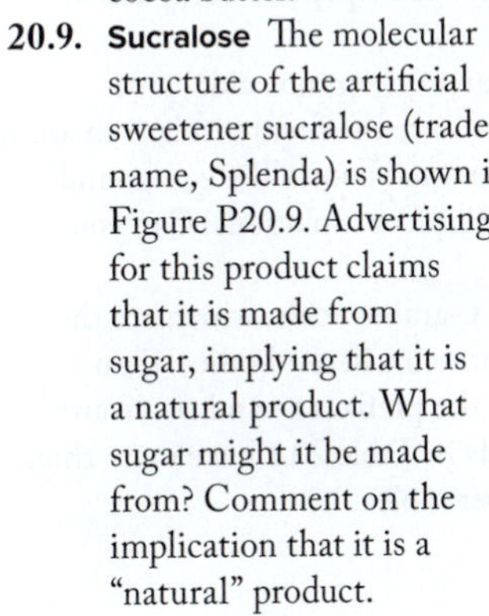

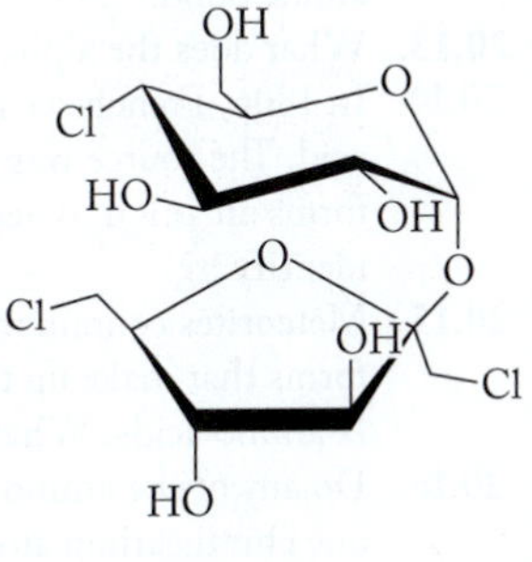

Sucralose

**FIGURE P20.9**

20.10. Use representations [A] through [I] in Figure P20.10 to answer questions a–f.

a. [A] depicts two strands of DNA in a double helix. Are the two strands held together by covalent bonds, intermolecular forces, or both? [B] depicts two alpha helices linked together. Are the two helices held together by covalent bonds, intermolecular forces, or both?

b. [C] depicts a phospholipid bilayer. Is the bilayer held together by covalent bonds, intermolecular forces, or both?

c. [D] depicts a disaccharide. Are the two sugars held together by covalent bonds, intermolecular forces, or both?

d. Which of the single molecules depicted in the matrix are an important structural component of [C]?

e. Which of the single molecules depicted in the matrix react to form [E]?

f. When the amino acid depicted in [F] is incorporated into a protein, what intermolecular forces are involved?

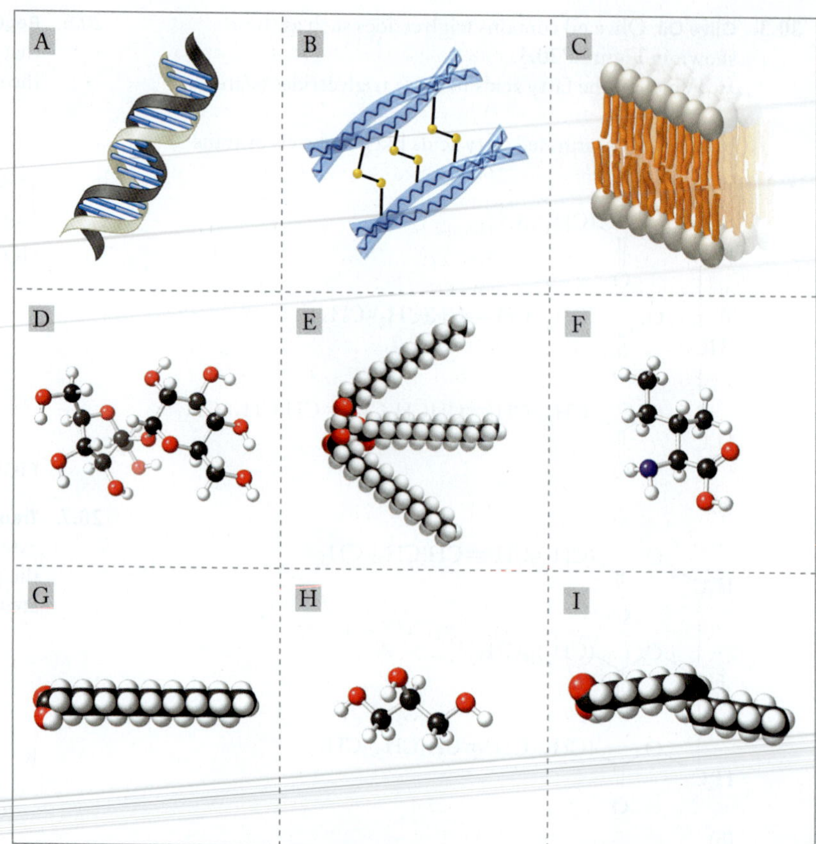


**FIGURE P20.10**

## QUESTIONS AND PROBLEMS

### *Amino Acids*

**Concept Review**

**20.11.** In living cells, amino acids combine to make peptides and proteins. Are these processes accompanied by increases or decreases in the entropy of the reaction system?

20.12. What is the difference between a peptide bond and an amide bond?

**20.13.** What does the alpha mean in α-amino acid?

20.14. In 1806, French scientists were the first to isolate an amino acid. The source was asparagus shoots. The compound forms an anion in neutral aqueous solutions. Can you identify it?

**20.15.** Meteorites contain more L-amino acids, which are the forms that make up the proteins in our bodies, than D-amino acids. What do the prefixes L- and D- mean?

20.16. Do any of the amino acids in Table 20.1 have more than one chiral carbon atom per molecule?

**20.17.** Which of the compounds in Figure P20.17 is/are *not* an α-amino acid(s)?

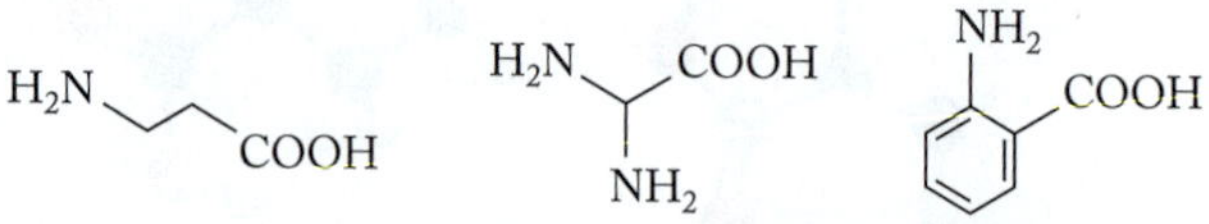

(a) $H_2N$, COOH (b) $H_2N$, COOH, $NH_2$ (c) $NH_2$, COOH

**FIGURE P20.17**

20.18. Which of the compounds in Figure P20.18 are α-amino acids?

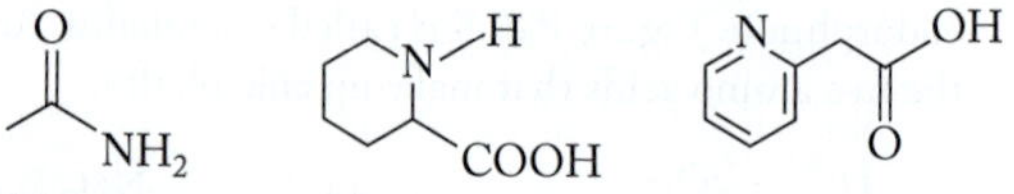

(a) O, $NH_2$ (b) N–H, COOH (c) N, OH, O

**FIGURE P20.18**

**20.19.** Why do most amino acids exist in the zwitterionic form at physiological pH (pH ≈ 7.4)?

**20.20.** A simple organic acid with no functional groups other than an alkyl chain and a carboxylic acid typically has a p$K_a$ between 4 and 5. The p$K_a$ values of amino acids are all between 1.7 and 2.4. Why are amino acids more acidic than simple carboxylic acids?

**20.21.** ***Into the Wild*** The author of the book *Into the Wild* postulated that the main character in the book died in the Alaskan wilderness because he accidentally ate seeds containing a toxin called β-ODAP (Figure P20.21). This substance is a neurotoxin; its mechanism of action is not fully understood, but it is referred to as a "glutamic acid mimic." Compare the structure of β-ODAP to that of glutamic acid and describe the similarities between the two molecules.

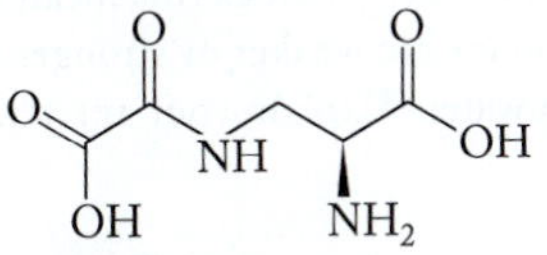

**FIGURE P20.21**

**20.22.** **Metabolic Disease in Dogs** Cystinuria is a metabolic disease that occurs in some breeds of dogs. It is characterized by the presence of kidney stones made of *cystine*, a dimer formed when two *cysteine* residues covalently link through a –S—S– bond (see Figure 20.15).

a. Draw the structure of the dipeptide Cys-Cys.
b. Is cystine a dipeptide? Why or why not?

## Problems

**20.23.** Draw all possible structures of the peptides produced from condensation reactions of the following L-amino acids:

a. Alanine + serine
b. Alanine + phenylalanine
c. Alanine + valine

**20.24.** Draw all possible structures of the peptides produced from condensation reactions of the following L-amino acids:

a. Methionine + alanine + glycine
b. Methionine + valine + alanine
c. Serine + glycine + tyrosine

---

**20.25.** Identify the amino acids in the dipeptides shown in Figure P20.25.

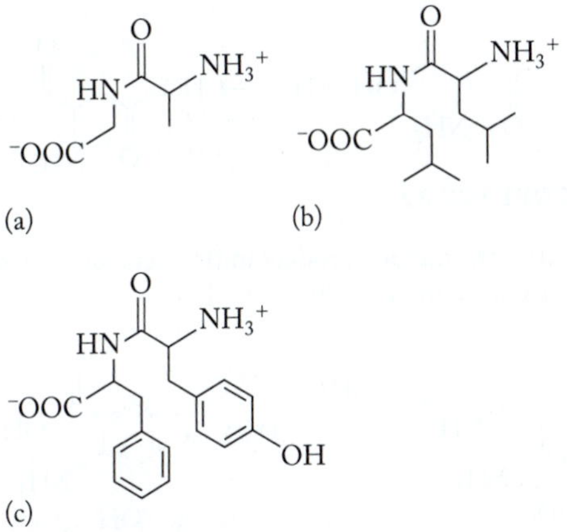

(a) (b)

(c)

**FIGURE P20.25**

**20.26.** Identify the amino acids in the tripeptides shown in Figure P20.26.

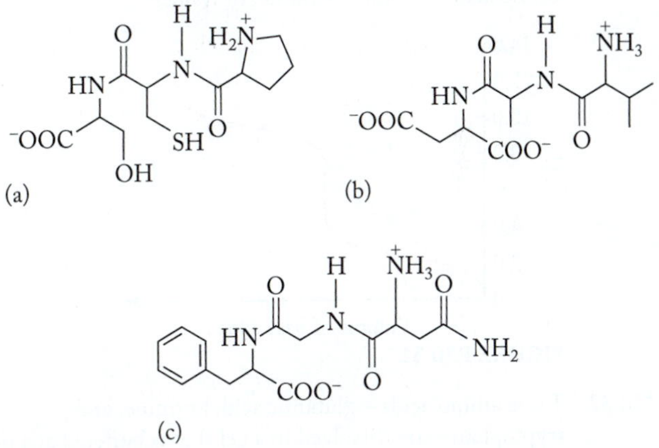

(a) (b)

(c)

**FIGURE P20.26**

---

**20.27.** Ion–ion interactions are particularly effective at stabilizing tertiary structures of proteins. Suggest a pair of amino acid residues that would be attracted to each other via ion–ion interactions at pH 7.4.

**20.28.** Use Table 20.1 to identify two amino acids that are not neutral at pH 7.4, despite the presence of $COO^-$ and $NH_3^+$ groups.

**20.29.** Identify the missing product in the metabolic reaction shown in Figure P20.29.

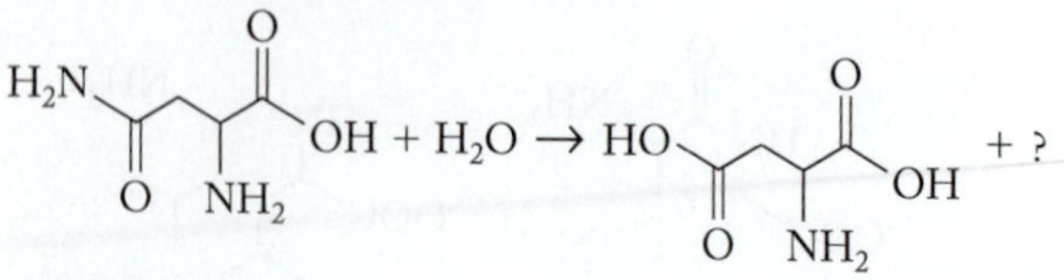

H$_2$N, O, O, NH$_2$, OH + H$_2$O → HO, O, O, NH$_2$, OH + ?

**FIGURE P20.29**

20.30. Identify the missing product in the metabolic reaction shown in Figure P20.30.

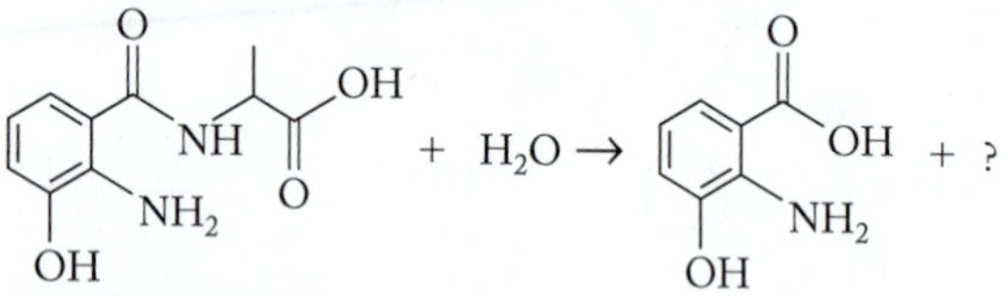

O, NH, OH, O, NH$_2$, OH + H$_2$O → O, OH, NH$_2$, OH + ?

**FIGURE P20.30**

---

**20.31.** Figure P20.31 shows the titration curve of which of these amino acids: leucine, histidine, or lysine?

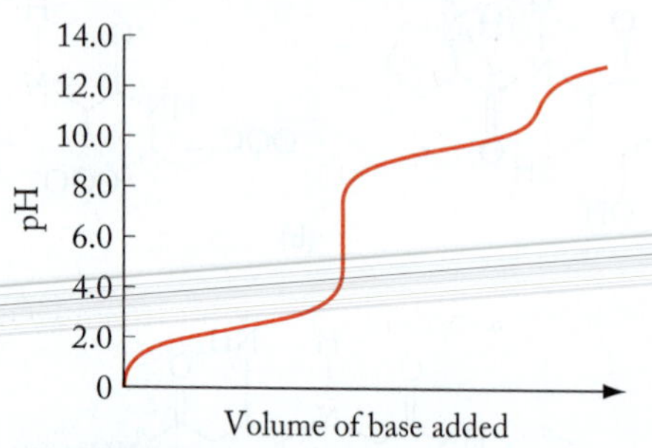


**FIGURE P20.31**

*20.32. Three amino acids—glutamic acid, arginine, and tryptophan—are dissolved in a gel that is buffered at a pH of 5.9. Two electrodes are placed in the gel and an electric current is applied.

a. Toward which electrode does each amino acid migrate?
b. Draw the forms of each amino acid present in the gel at a pH of 5.9.

## Protein Structure and Function

### Concept Review

**20.33.** Which of the four levels of protein structure is most closely associated with the sequence of amino acids in a protein?

20.34. Which type of intermolecular interaction plays the dominant role in holding strands of proteins together in β-pleated sheets and stabilizing α helices?

**20.35.** Which level of protein structure is associated with ion–ion interactions and disulfide bond formation?

20.36. **Hard-Boiled Eggs** The protein in egg whites is ovalbumin. When an egg is hard-boiled, which is *least* affected in ovalbumin: its primary, secondary, tertiary, or quaternary structure? Explain your answer.

**20.37.** Describe the induced-fit theory of enzyme activity.

*20.38. What will happen if an enzyme is added to a solution in which the substrate and product are in equilibrium?

**20.39.** Describe the role that molecular structure plays in the specificity of enzyme activity.

20.40. The rates of most reactions increase with increasing temperature, but above a critical temperature the rate of an enzyme-mediated reaction decreases with increasing temperature. Explain why.

**20.41.** When protein strands fold back onto themselves in forming stable tertiary structures, lysine residues are often paired up with glutamic acid residues. Why?

20.42. When the α-helical region in a protein unfolds, breaking the hydrogen bonds between the amino acids requires approximately 2 kJ/mol of energy. Does this mean that the hydrogen bonds in proteins are weaker or stronger than the hydrogen bonds in water? Explain your answer.

## Carbohydrates

### Concept Review

**20.43.** What are the structural differences between starch and cellulose?

20.44. Do cellulose fibers resemble proteins in α-helical, β-pleated sheet, or globular structure?

**20.45.** Is the fuel value of glucose in the linear form the same as that in the cyclic form?

*20.46. Without doing the actual calculation, estimate the fuel values of glucose and starch by considering average bond energies. Do you predict the fuel values of the two substances to be the same or different?

**20.47.** Describe in your own words the function of carbohydrates in the diet.

20.48. Do polysaccharides have a quaternary structure? Explain your answer.

### Problems

**20.49.** Draw a diagram similar to Figure 20.19 that shows how the linear molecule in Figure P20.49 forms a six-atom ring.

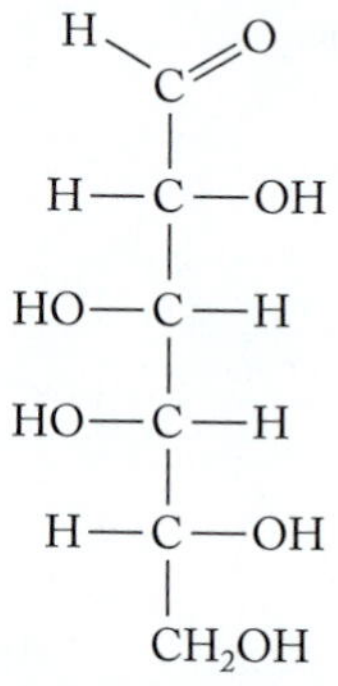

H–C=O
H—C—OH
HO—C—H
HO—C—H
H—C—OH
CH$_2$OH

Galactose

**FIGURE P20.49**

20.50. Draw a diagram similar to Figure 20.21 that shows how the linear molecule in Figure P20.50 forms a five-atom ring.

Ribose

**FIGURE P20.50**

**20.51.** Which of the structures in Figure P20.51 are α isomers of a monosaccharide and which ones are β isomers?

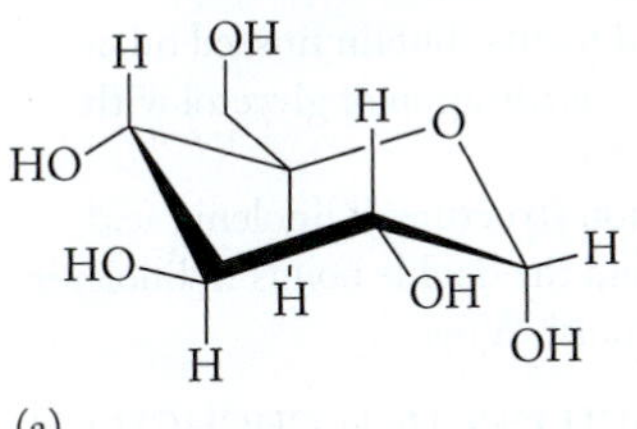

(a)

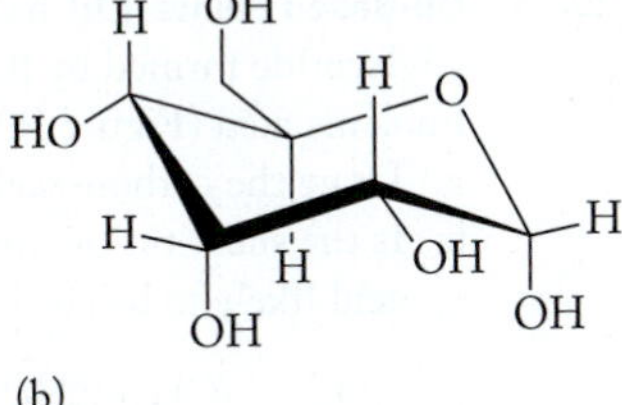

(b)

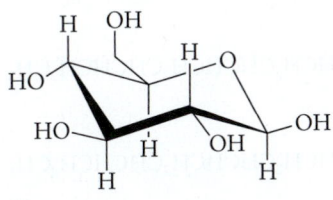

(c)

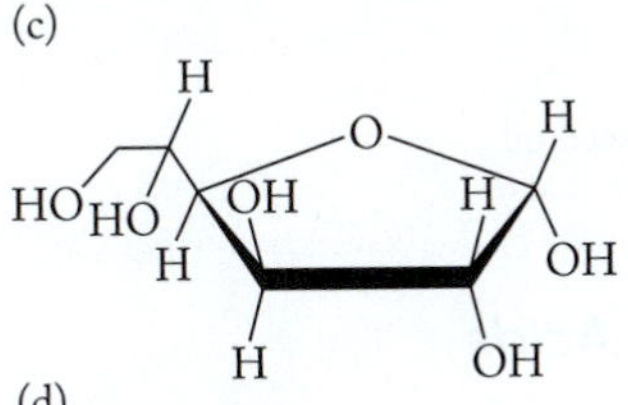

(d)

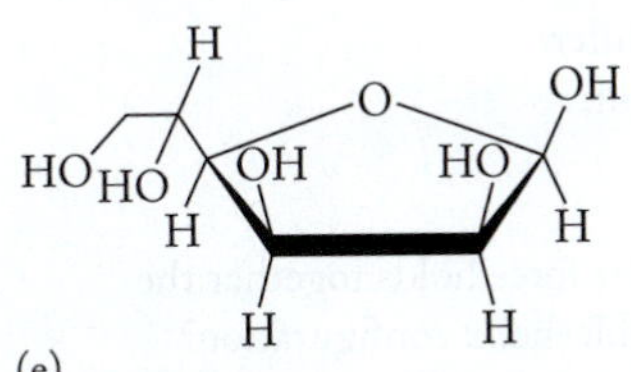

(e)

(f)

**FIGURE P20.51**

20.52. Which of the structures in Figure P20.52 are α isomers of a monosaccharide and which ones are β isomers?

(a)

(b)

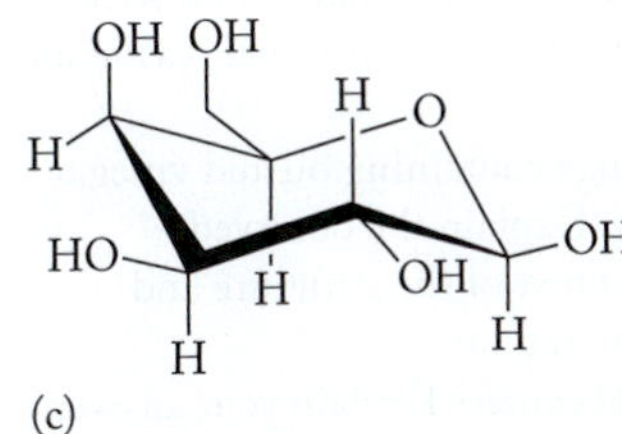

(c)

(d)

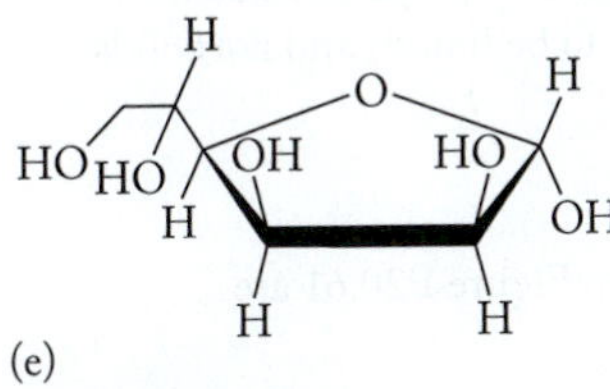

(e)

(f)

**FIGURE P20.52**

**20.53.** Which of the saccharides in Figure P20.53 is digestible by humans?

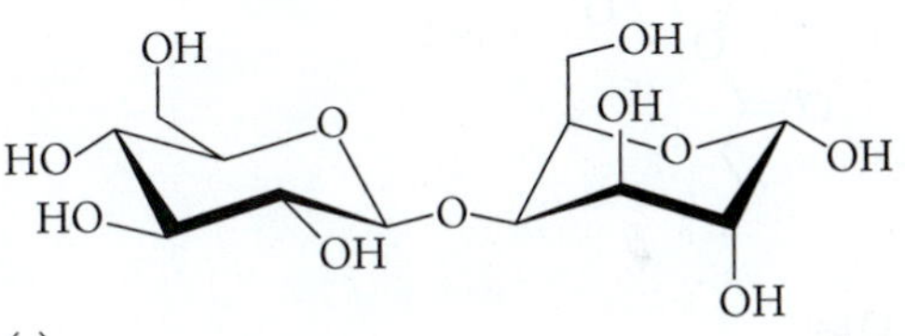

(a)

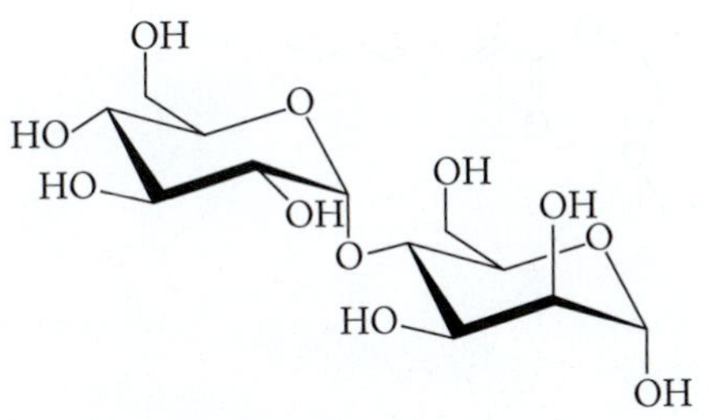

(b)

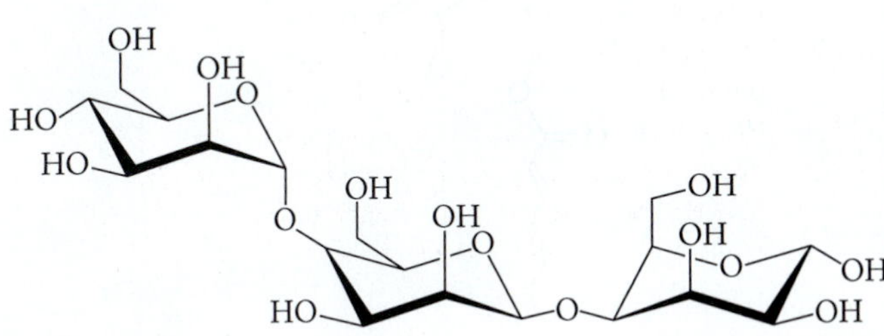

(c)

**FIGURE P20.53**

*20.54. For any of the disaccharides in Problem 20.53 that are *not* digestible by humans, draw an isomer that would be.

## Lipids

### Concept Review

**20.55.** What is the difference between a saturated and an unsaturated fatty acid?

20.56. Why are the average fuel values of fats higher than those of carbohydrates and proteins?

**20.57.** **Polar Exploration** Some Arctic explorers have eaten sticks of butter on their explorations. Give a nutritional reason for this unusual cuisine.

20.58. **Salad Dressing** Salad dressings containing oil and vinegar quickly separate on standing. Explain the observed separation of layers in the context of the structure and properties of aqueous vinegar and oil.

**20.59.** Do triglycerides have a chiral center? Explain your answer.

*20.60. Using your knowledge of molecular geometry and intermolecular forces, explain why polyunsaturated triglycerides are more likely to be liquids and not solids than saturated triglycerides.

### Problems

**20.61.** Which of the triglycerides in Figure P20.61 are unsaturated fats?

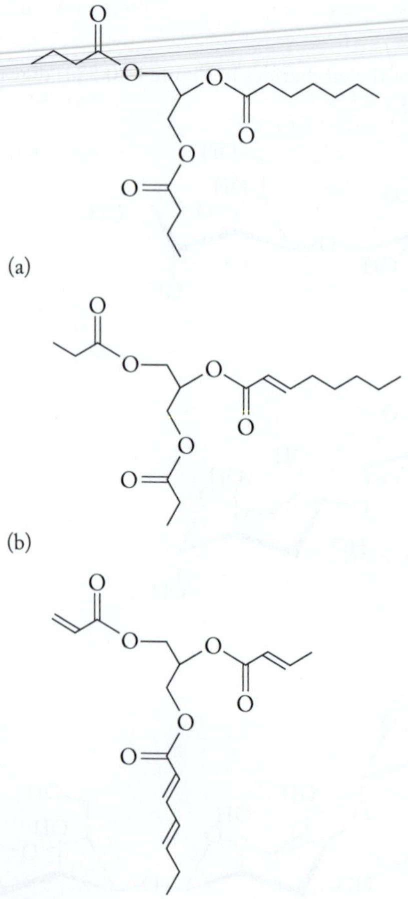

(a)

(b)

(c)

**FIGURE P20.61**

20.62. For each of the pairs of fatty acids in Figure P20.62, indicate whether they are structural isomers, stereoisomers, or unrelated compounds.

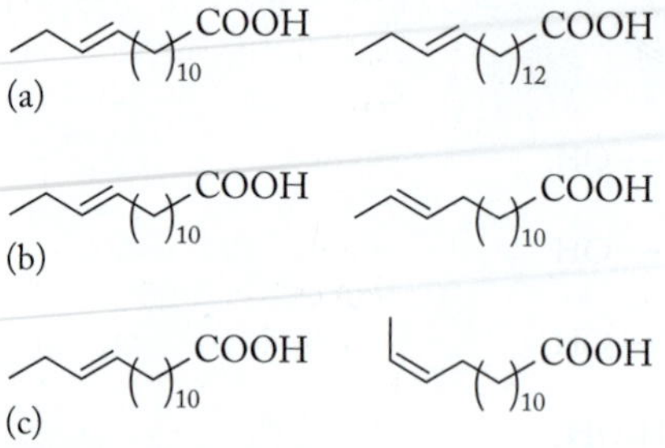

(a) COOH, 10; COOH, 12

(b) COOH, 10; COOH, 10

(c) COOH, 10; COOH, 10

**FIGURE P20.62**

**20.63.** Draw the carbon-skeleton structures of the three fats formed by the reaction of glycerol with (a) octanoic acid ($C_7H_{15}COOH$), (b) decanoic acid ($C_9H_{19}COOH$), and (c) dodecanoic acid ($C_{11}H_{23}COOH$).

20.64. **Oil-Based Paints** Oil-based paints contain linseed oil, a triglyceride formed by the esterification of glycerol with linolenic acid (Figure P20.64).

a. Draw the carbon-skeleton structure of linolenic acid.

*b. Is the substitution around the double bonds in linolenic acid likely to be cis or trans? Why?

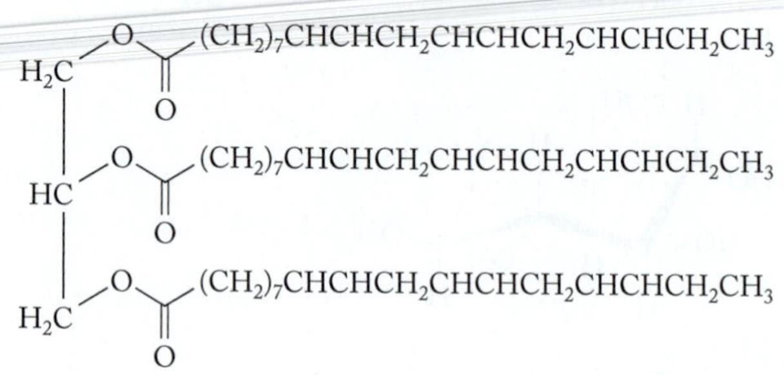

$H_2C$ O (CH$_2$)$_7$CHCHCH$_2$CHCHCH$_2$CHCHCH$_2$CH$_3$ O

HC O (CH$_2$)$_7$CHCHCH$_2$CHCHCH$_2$CHCHCH$_2$CH$_3$ O

$H_2C$ O (CH$_2$)$_7$CHCHCH$_2$CHCHCH$_2$CHCHCH$_2$CH$_3$ O

Linseed oil

**FIGURE P20.64**

## Nucleotides and Nucleic Acids

### Concept Review

**20.65.** How do DNA and RNA differ:

a. in molecular composition?

b. in structure?

c. in function?

20.66. What kind of intermolecular force holds together the strands of DNA in the double-helix configuration?

**20.67.** DNA is a highly charged polyanion. If a solution of DNA is heated, the DNA will separate into individual strands, a process called denaturation. If the salt concentration of the solution is increased, the temperature at which denaturation occurs increases. Suggest a reason why.

*20.68. Because of base pairing of the two strands in DNA, the percentage of T must equal the percentage of A, and the percentage of C must equal the percentage of G. In contrast, in RNA no relationship exists between the quantities of the four bases. What does this fact suggest about the structure of RNA?

**Problems**

**20.69.** Draw the structure of adenosine 5′-monophosphate, the "A" ribonucleotide in a strand of RNA.

20.70. Draw the structure of deoxythymidine 5′-monophosphate, the "T" nucleotide in a strand of DNA.

**20.71.** During the replication of DNA, a segment of an original strand has the sequence T-C-G-G-T-A. What is the sequence of the double-stranded helix formed in replication?

20.72. During transcription, a segment of the strand of DNA that is transcribed has the sequence T-C-G-G-T-A. What is the corresponding sequence of nucleotides on the messenger RNA that is produced in transcription?

## Additional Problems

**20.73. Treating Infections** The major component of the antibiotic ointment bacitracin is a cyclic polypeptide called bacitracin A. It was first isolated in 1943 from a knee scrape from a girl named Margaret Tracy, after whom it is named. Bacitracin is effective topically and is used to treat skin, eye, and wound infections. Figure P20.73 shows the structure of bacitracin A.

a. How many amino acids are found in the structure of bacitracin A?

b. Which, if any, of the constituent amino acids are among the 20 α-amino acids found in human proteins?

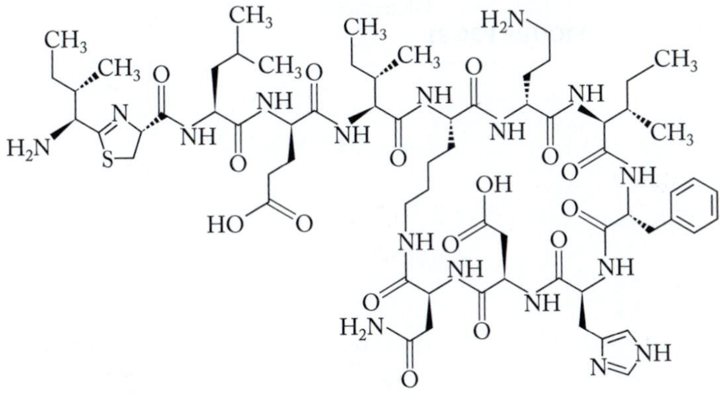

**FIGURE P20.73**

20.74. **Amino Acid Supplements** Supplements containing BCAAs (branched-chain amino acids) have been studied for their effects on athletic performance (in terms of the perception of exertion) and mental fatigue. Current research does not support that they have either of these effects, but they are used medically to slow muscle wasting in bedridden patients. BCAAs are amino acids having hydrocarbon side chains with a branch, a carbon atom bound to more than two other carbon atoms. Consult Table 20.1 and draw the structures of the BCAAs among the common amino acids.

**20.75.** Homocysteine (Figure P20.75) is formed during the metabolism of amino acids. A mutation in some people's genes leads to high concentrations of homocysteine in the blood and a consequent increase in their risk of heart disease and their incidence of bone fractures in old age.

a. What is the structural difference between homocysteine and cysteine?

b. Cysteine is a chiral compound. Is homocysteine chiral?

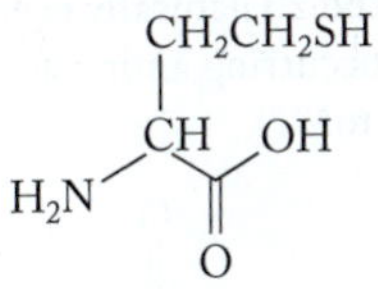

Homocysteine

**FIGURE P20.75**

20.76. **Molecules in Meteorites** Some scientists believe life on Earth can be traced to amino acids and other molecules brought to Earth by comets and meteorites. In 2004, a new class of amino acids called diamino acids (Figure P20.76) was found in the Murchison meteorite.

a. Which of these diamino acids is not an α-amino acid?

b. Which of these amino acids is chiral?

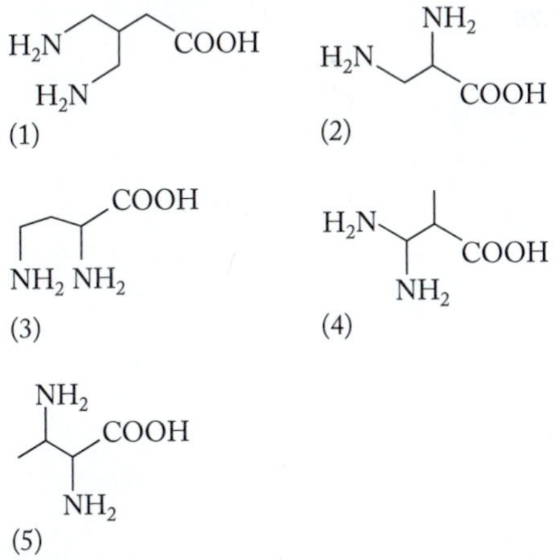

(1) (2) (3) (4) (5)

**FIGURE P20.76**

**20.77. Ackee** Ackee, the national fruit of Jamaica, is a staple in many Jamaican diets. Unfortunately, a potentially fatal sickness known as Jamaican vomiting disease is caused by consuming unripe ackee fruit, which contains the amino acid hypoglycin (Figure P20.77). Is hypoglycin an α-amino acid?

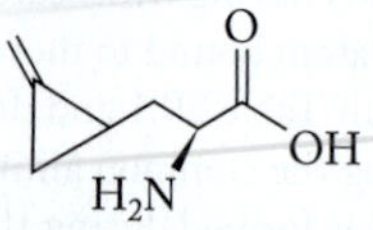

Hypoglycin

**FIGURE P20.77**

20.78. ***Escherichia coli*** In late 2003, researchers at The Scripps Research Institute reported the development of genetically modified *E. coli* that could incorporate five new amino acids into proteins. These five amino acids, shown in Figure P20.78, are *not* among the 20 naturally occurring amino acids. Which naturally occurring amino acids are these compounds most similar to?

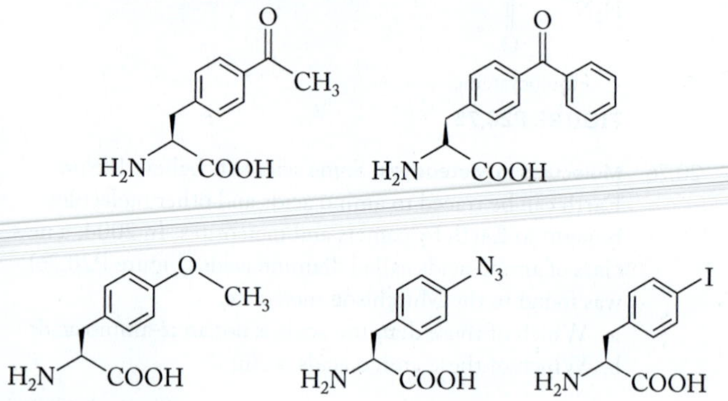

**FIGURE P20.78**

**20.79. Creatine** Creatine (Figure P20.79) is an amino acid produced by the human body. Bodybuilders sometimes take creatine supplements to help gain muscle strength. A 2003 study reported that creatine may boost memory and cognitive thinking.

a. Is creatine an α-amino acid?

b. Draw the two dipeptides that can be formed from glycine and creatine.

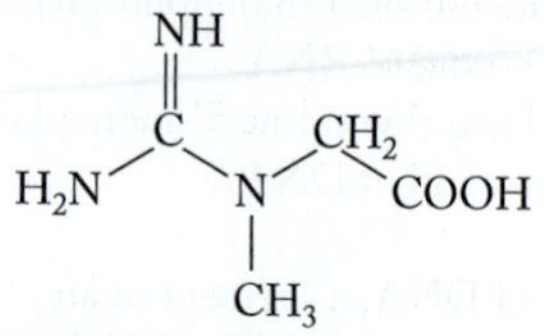

Creatine

**FIGURE P20.79**

20.80. Cytochrome c is an enzyme involved in oxidation–reduction reactions and is an intermediate in apoptosis, a controlled form of killing cells in response to infection or DNA damage. An elemental analysis of cytochrome c determined that it is 0.43% iron and 1.48% sulfur by mass. What is the minimum molecular weight of cytochrome c, and what is the minimum number of sulfur atoms per molecule?

**20.81.** Glutathione (Figure P20.81) is an essential molecule in the human body. It acts as an activator for enzymes and protects lipids from oxidation. Which three amino acids combine to make glutathione?

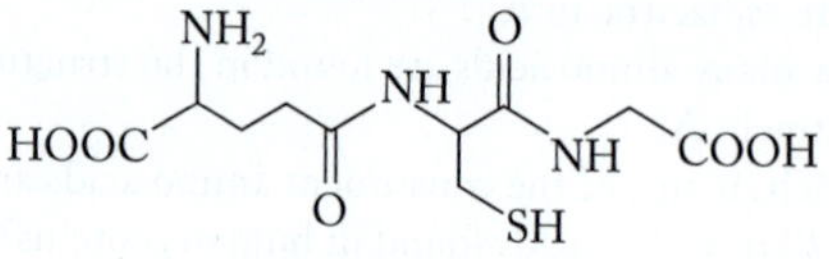

Glutathione

**FIGURE P20.81**

20.82. Adding ethanol to an aqueous solution of a globular protein causes the protein to denature. Ethanol also disrupts the structure of cell membranes. What interactions could be responsible for both of these effects?

***20.83.** **Opioids** The structures of morphine and meperidine (trade name, Demerol), two powerful pain medicines, are shown in Figure P20.83. They interact similarly with receptors because they share similar structural features, including one part of the molecule that is flat or planar and another part that can bind to the receptor site. Identify these two structural features.

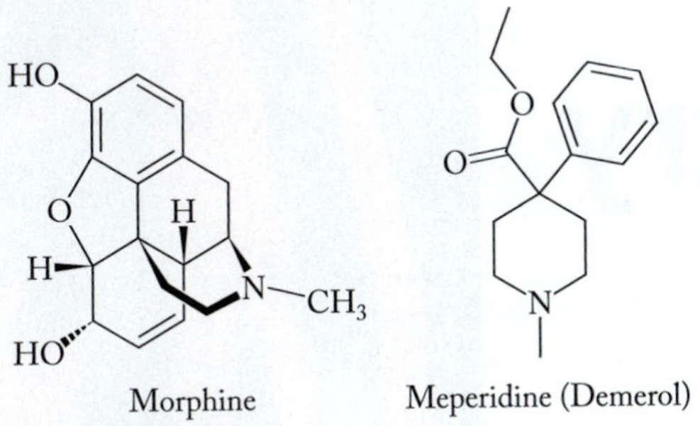

Morphine Meperidine (Demerol)

**FIGURE P20.83**

*20.84. **Antihistamines** Figure P20.84 shows the structure of the histamine molecule that causes sneezing and itching in allergy sufferers. Antihistamines are a class of molecules that reduce allergy symptoms by preferentially binding to the same receptor sites as histamines. What structural features would permit these molecules to bind to the same receptor?

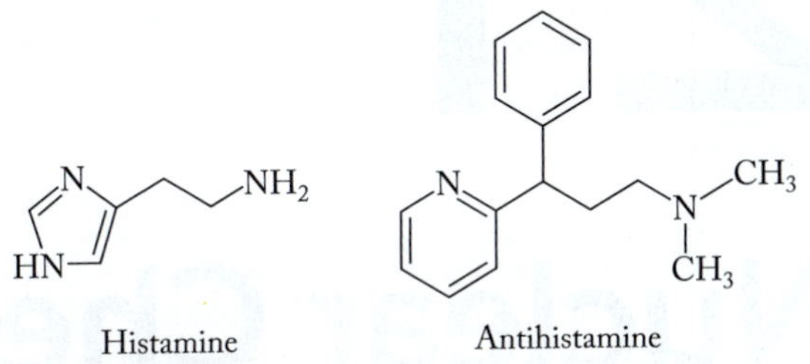

Histamine Antihistamine

**FIGURE P20.84**

# 21

# Nuclear Chemistry

## The Risks and Benefits

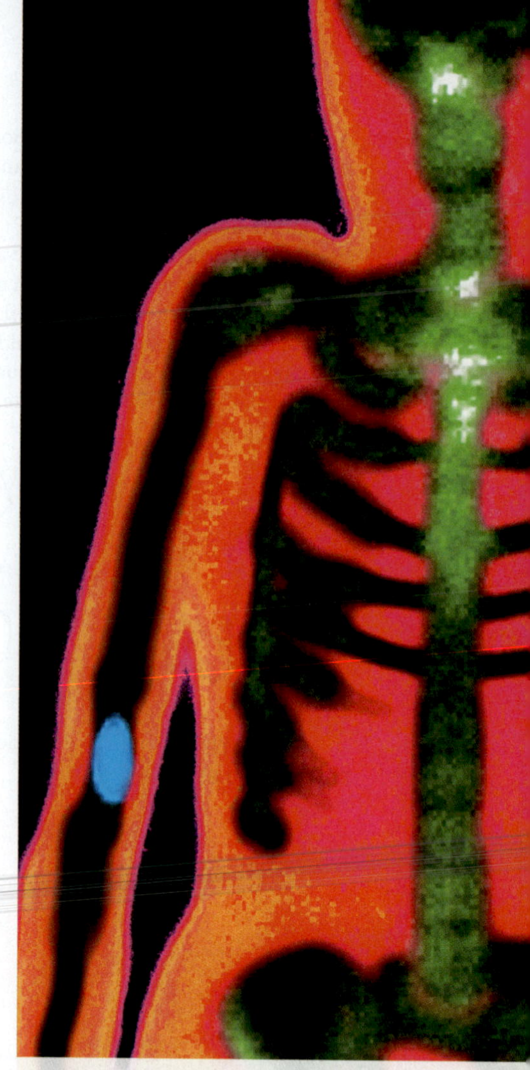

**GAMMA SCANS OF THE HUMAN SKELETON** Front and back views of a patient who was injected with a radionuclide that accumulated in bone tissues and emitted gamma rays.

PARTICULATE **REVIEW**

### *Isotopes Revisited*

In Chapter 21, we investigate the stability and properties of radioactive nuclei. Three nuclides are depicted here.

- How many protons and how many neutrons does each nuclide contain?
- What is the mass of each nuclide in u?
- Which two nuclides are isotopes of one another? (Review Section 2.2 if you need help.)

*(Answers to Particulate Review questions are in the back of the book.)*

1

2

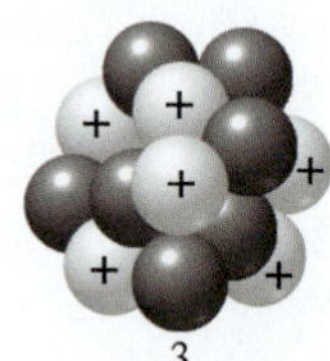

3

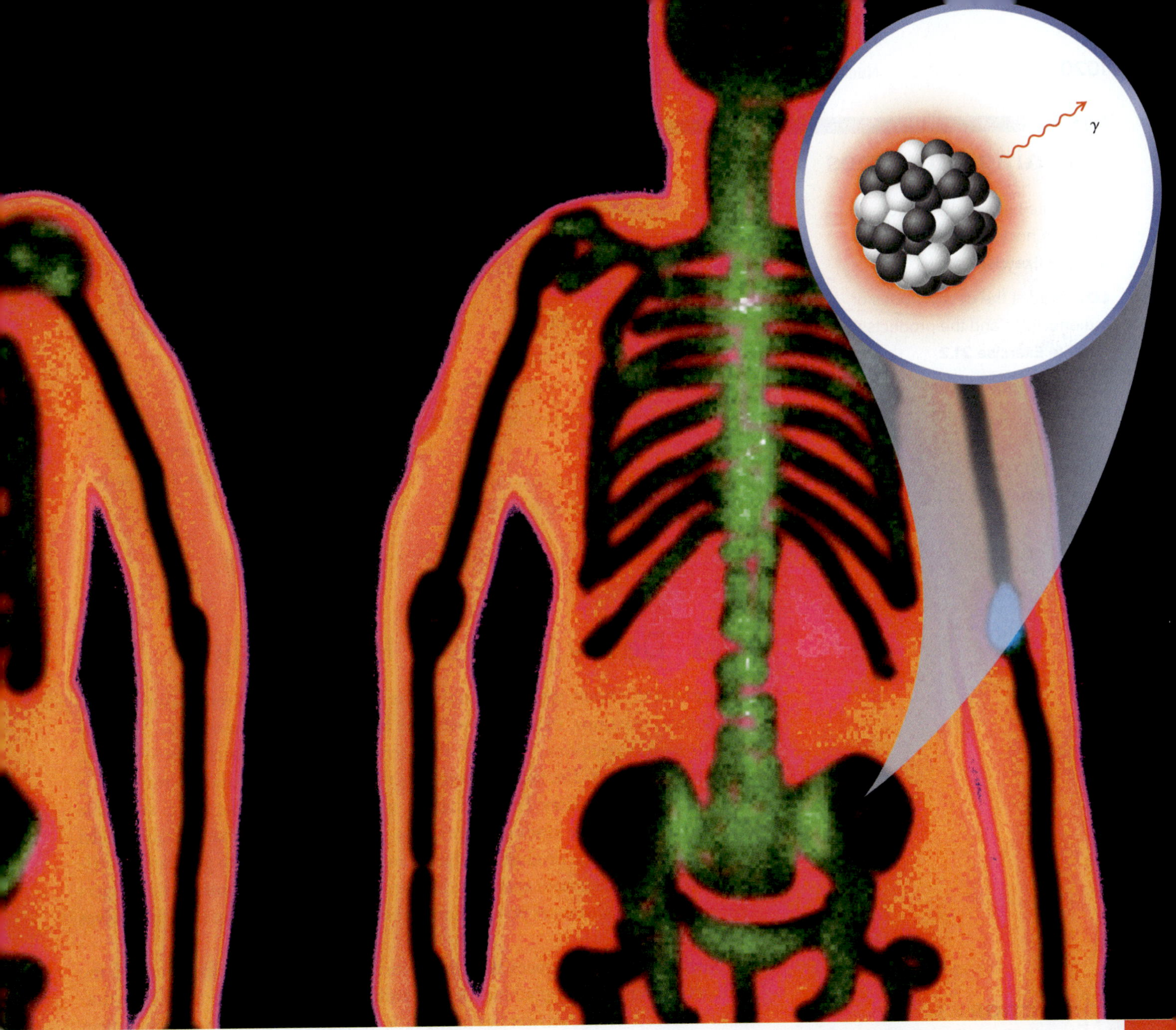

PARTICULATE **PREVIEW**

## *Unstable Nuclides versus Stable Nuclides*

Radiocarbon dating uses the amount of carbon-14 in an artifact to determine its age. As you read Chapter 21, look for ideas that will help you answer these questions:

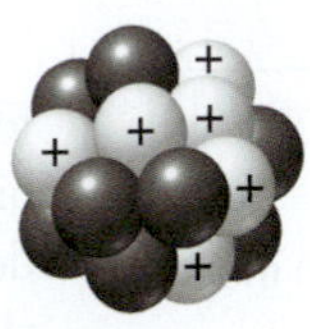

- Does the carbon-14 nucleus depicted here have more protons or more neutrons?
- How might the ratio of neutrons to protons affect the decay of an unstable nuclide?
- What nuclide is produced when carbon-14 undergoes decay? What is the neutron-to-proton ratio of this new nuclide?

## Learning Outcomes

**LO1** Write balanced equations to describe nuclear reactions
**Sample Exercise 21.1**

**LO2** Predict the decay modes of radionuclides and the products they form
**Sample Exercise 21.2**

**LO3** Relate the quantity of a radionuclide remaining in a sample to the quantity initially in the sample, the age of the sample, and the half-life of the nuclide
**Sample Exercises 21.3, 21.4**

**LO4** Calculate the binding energy of a nucleus and the energy released in a nuclear reaction from the masses of the products and reactants
**Sample Exercises 21.5, 21.6**

**LO5** Describe how elements are synthesized in the cores of giant stars

**LO6** Compare nuclear fusion and nuclear fission

**LO7** Relate the level of radioactivity in a sample of a radionuclide to the nuclide's quantity and half-life, and describe how radioactivity levels are measured
**Sample Exercise 21.7**

**LO8** Describe the dangers of exposure to nuclear radiation and calculate effective radiation doses
**Sample Exercise 21.8**

**LO9** Describe some medical applications of radionuclides and the properties of the nuclides that make them suitable for these applications

## 21.1 Decay Modes of Radionuclides

Throughout this book, we have seen that the identities of atoms remain unchanged in chemical reactions, in keeping with the law of conservation of mass. We now turn to *nuclear* reactions, in which the identities of atoms do change—because their nuclei change. The field of chemistry that studies these kinds of reactions is called **nuclear chemistry**. In this chapter, we will see why some nuclei are stable and others are not, the kinds of spontaneous nuclear reactions that unstable nuclei undergo, and how the products of these reactions can be used to generate electrical power and to diagnose and treat disease. We also address some of the dangers that radioactive substances pose to human health and how we can shield ourselves from them.

Nuclear chemistry traces back to Henri Becquerel's 1896 discovery of the radioactivity emitted by pitchblende, a uranium-containing mineral. Why is uranium radioactive? The answer lies in the ratio of neutrons to protons in the nucleus of the uranium atom. The values of the atomic masses and mass numbers of the elements in the periodic table tell us about the ratios of neutrons to protons in the nuclei of their stable isotopes. The lighter elements have atomic masses that are about twice their atomic numbers and have neutron-to-proton ratios close to one. For example, $^{12}C$ has six neutrons and six protons, and most oxygen atoms have eight neutrons and eight protons.

**CONNECTION** In Section 2.1, we introduced radioactivity and defined alpha (α) and beta (β) particles, and we described how Rutherford's study of particles emitted by radioactive uranium led to his model of the structure of the atom.

**STEPWISE ANIMATION**
Belt of Stability

As $Z$ (atomic number) increases, however, so do the ratios of neutrons to protons. This trend is illustrated in **Figure 21.1**, where the green dots represent combinations of neutrons and protons that form stable nuclides. The band of green dots runs diagonally through the graph, defining the **belt of stability**. Note how the belt curves upward away from the purple straight line, which represents a neutron-to-proton ratio of 1:1. This curvature shows that the neutron-to-proton ratio increases from about 1:1 for the lightest stable nuclides to about 1.5:1 for the most massive ones.

**nuclear chemistry** the study of reactions that involve changes in the nuclei of atoms.

**belt of stability** the region on a graph of number of neutrons versus number of protons that includes all stable nuclei.

**radioactive decay** the spontaneous disintegration of unstable particles accompanied by the release of radiation.

The nuclides represented by orange dots in Figure 21.1 are *radionuclides*. They can be classified as either neutron rich or neutron poor depending on their neutron-to-proton ratio relative to the belt of stability. Radionuclides are unstable and undergo **radioactive decay**, the spontaneous disintegration of radioactive nuclei accompanied by the release of nuclear radiation. Their mode of radioactive decay depends on whether they are above or below the belt of stability. Four principal pathways—beta (β) decay, alpha (α) decay, positron emission, and electron capture—represent the most common decay mechanisms for unstable nuclei.

**CHEMTOUR**
Modes of Radioactive Decay

## Beta (β) Decay

Radionuclides above the belt of stability are neutron rich and tend to undergo decay reactions that reduce their neutron-to-proton ratio. For example, when $^{14}C$ undergoes radioactive decay, a neutron in its nucleus spontaneously disintegrates, producing a proton that remains in the nucleus and a high-speed, high-energy

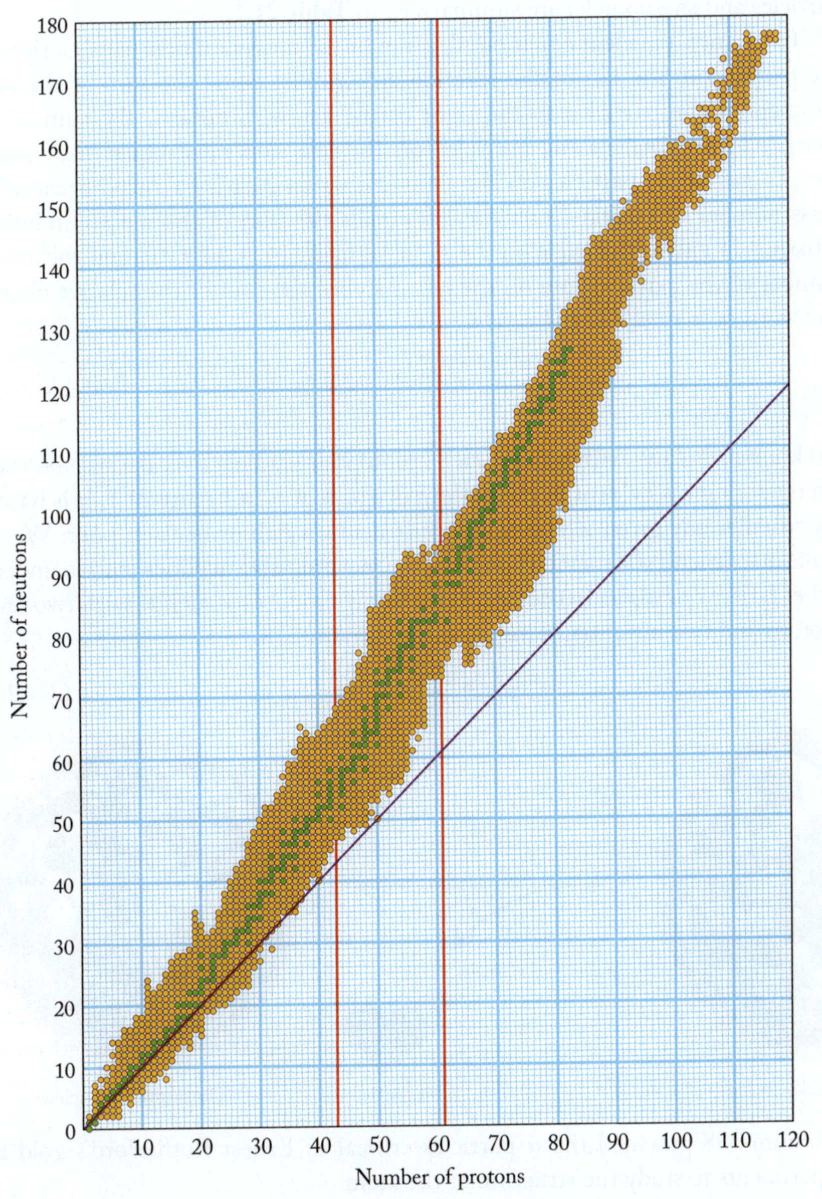


**FIGURE 21.1** The belt of stability. Green dots represent stable combinations of protons and neutrons. Orange dots represent known radioactive (unstable) nuclides. Nuclides that fall along the purple line have equal numbers of neutrons and protons. Note that no stable nuclides (no green dots) exist for $Z = 43$ (technetium) and $Z = 61$ (promethium), as indicated by the two vertical red lines.

**beta (β) decay** the process by which a neutron in a neutron-rich nucleus decays into a proton and a high-energy electron (β particle).

**alpha (α) decay** a nuclear reaction in which an unstable nuclide spontaneously emits an α particle.

electron called a β particle, which is ejected from the nucleus in the process known as **beta (β) decay**:

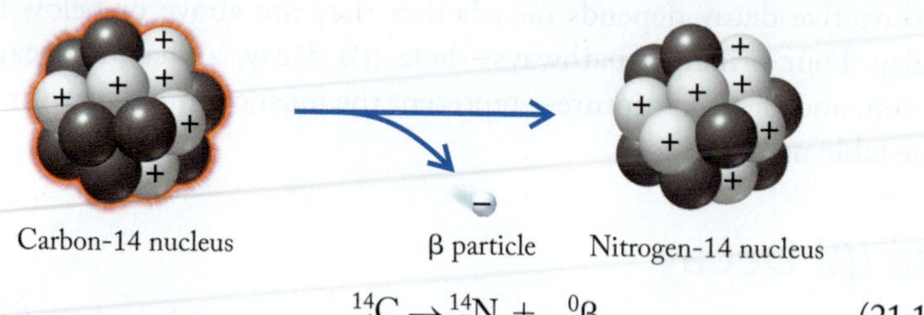


$$^{14}_{6}C \rightarrow {}^{14}_{7}N + {}^{0}_{-1}\beta \qquad (21.1)$$

The shimmering red shadow around the carbon-14 nucleus above indicates that it is unstable and releases energy as it decays. In writing Equation 21.1, we follow the rules described in Section 2.2 for writing nuclide symbols—namely, superscripts for mass numbers and subscripts for the atomic numbers of nuclei (or the relative charges of subatomic particles). The symbols and masses of subatomic particles and small nuclei are summarized in **Table 21.1**.

**TABLE 21.1** Symbols and Masses of Subatomic Particles and Small Nuclei

| Particle | Symbol | Mass (kg) |
|---|---|---|
| Neutron | $^{1}_{0}n$ | $1.67493 \times 10^{-27}$ |
| Proton | $^{1}_{1}p$ or $^{1}_{1}H$ | $1.67262 \times 10^{-27}$ |
| Electron (β particle) | $^{0}_{-1}\beta$ or $^{0}_{-1}e$ | $9.10939 \times 10^{-31}$ |
| Deuteron | $^{2}_{1}D$ or $^{2}_{1}H$ | $3.34370 \times 10^{-27}$ |
| α Particle | $^{4}_{2}\alpha$ or $^{4}_{2}He$ | $6.64465 \times 10^{-27}$ |
| Positron | $^{0}_{1}\beta$ | $9.10939 \times 10^{-31}$ |

To balance a nuclear equation, the sum of the mass numbers (superscripts) of the particles to the left of the reaction arrow must equal the sum of the mass numbers of the particles to the right of the arrow. Similarly, the sum of the charges (subscripts) of the particles on the left side must equal the sum of the charges of the particles on the right. Equation 21.1 is balanced because the mass number of carbon-14 on the left equals the sum of the mass numbers of nitrogen-14 and a β particle on the right (namely, $14 = 14 + 0$), and the atomic number of the carbon atom on the left matches the sum of the relative charges on the right (namely, $6 = 7 - 1$).

**CHEMTOUR**
Balancing Nuclear Equations

## Alpha (α) Decay

All known nuclides with more than 83 protons are radioactive. Because no stable reference point is evident in the pattern of green dots in Figure 21.1, it is hard to say whether any given $Z > 83$ nuclide is neutron rich or neutron poor. We can make one general statement, though: these most massive nuclides tend to undergo either β decay or **alpha (α) decay**. Alpha decay produces a nuclide with two fewer protons and two fewer neutrons, as with uranium-238:

$$^{238}_{92}U \rightarrow {}^{234}_{90}Th + {}^{4}_{2}\alpha \qquad (21.2)$$

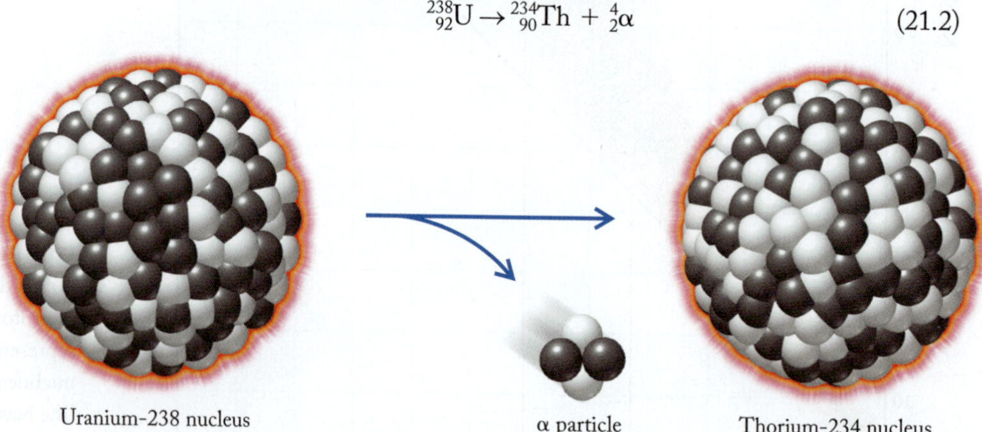


Uranium-238 provided the α particles critical to Ernest Rutherford's gold foil experiments to study the structure of the atom.

Among the most massive radioactive isotopes, one radioactive decay process often leads to another in what is referred to as a *radioactive decay series*. Consider the decay series that begins with the α decay of $^{238}U$ to $^{234}Th$ (**Figure 21.2**). Thorium has no stable isotopes and undergoes two β decay steps to produce $^{234}U$. In a series of later α decay steps, $^{234}U$ turns into thorium-230, radium-226, radon-222, polonium-218, and finally lead-214. Although some isotopes of lead ($Z = 82$) are stable, $^{214}Pb$ is not one of them. Therefore, the radioactive decay series continues as shown at the bottom left of Figure 21.2 and does not end until the stable nuclide $^{206}Pb$ is produced. Rutherford and other scientists studied the uranium decay series for many years in the early 20th century, identifying new isotopes of existing elements and discovering some new elements as well.

## CONCEPT TEST

In the $^{234}U$ radioactive decay series, five α decay steps in a row transform $^{234}U$ into $^{214}Pb$. Given the shape of the belt of stability, why does it make sense that the product of these α decay steps would be a neutron-rich nuclide that undergoes β decay?

*(Answers to Concept Tests are in the back of the book.)*

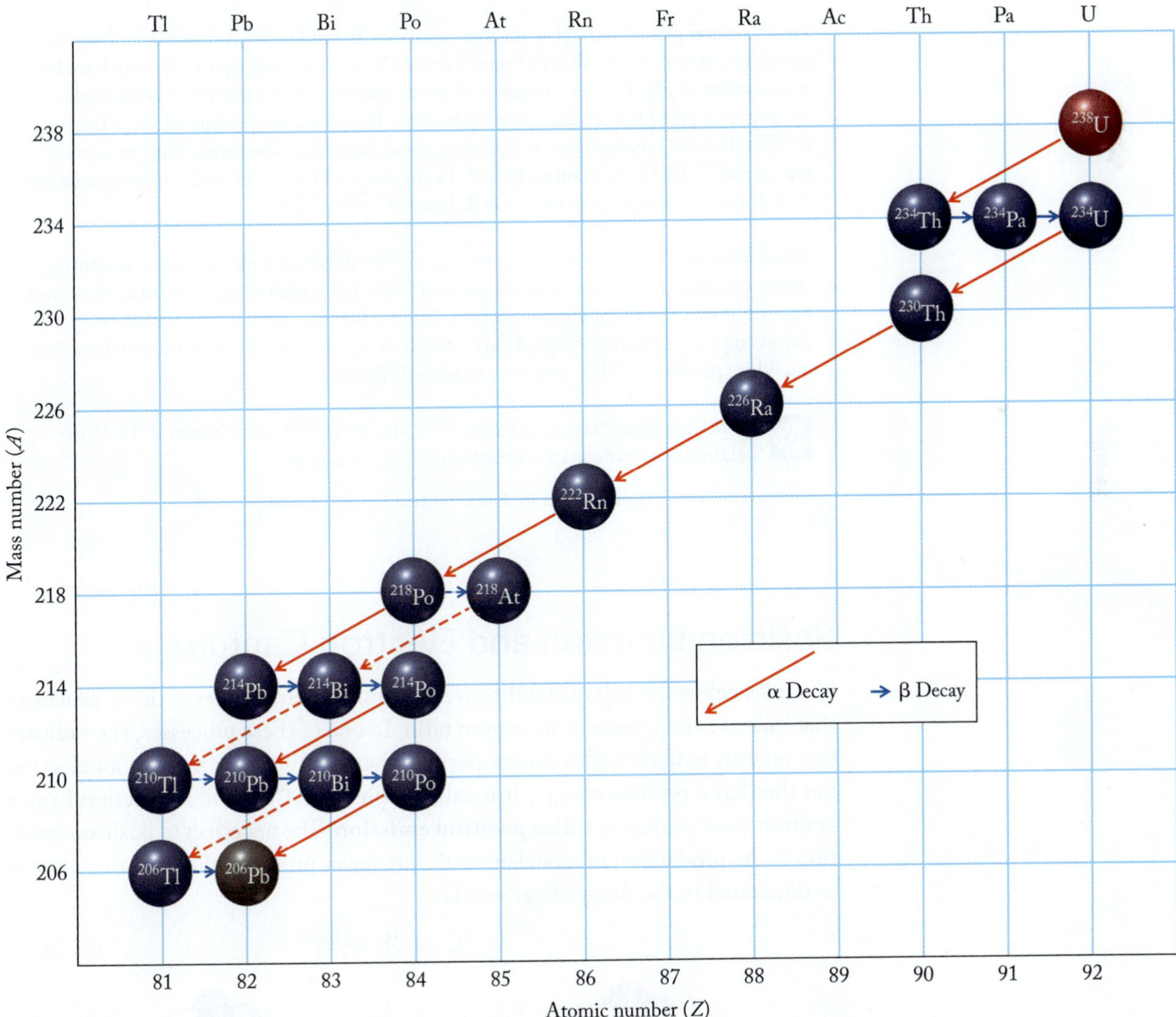


**FIGURE 21.2** The uranium-238 radioactive decay series, which ends with stable lead-206. The dashed arrows are alternative pathways representing less than 1% of the decay events in this series. (Note that the vertical axis in this figure is mass number, not the number of neutrons, as in Figure 21.1.)

**SAMPLE EXERCISE 21.1** Completing and Balancing Nuclear Equations **LO1**

Starting with 7 tons of pitchblende, Pierre and Marie Curie isolated a few milligrams of a new element, which they named polonium. Polonium-218 undergoes both α and β decay. Write balanced nuclear equations describing these nuclear reactions. Use symbols of the form ${}^{A}_{Z}X$ to represent the nuclides and any subatomic particles that form.

**Collect, Organize, and Analyze** Polonium (Po) has atomic number 84, α particles are helium-4 nuclei, and β particles are high-speed electrons. We can combine the element symbols and atomic numbers with the given mass numbers to write symbols of the nuclides involved in the reaction. When we are writing a balanced nuclear equation, the sum of the subscripts (the atomic numbers) of the nuclides on the left side of the reaction arrow must equal the sum of the subscripts on the right. The sums of the superscripts (the mass numbers) must also match.

**Solve** The symbols of the nuclides involved in the reactions are

$$\alpha \text{ decay:} \quad {}^{218}_{84}\text{Po} \rightarrow {}^{4}_{2}\text{He} + ?$$

$$\beta \text{ decay:} \quad {}^{218}_{84}\text{Po} \rightarrow {}^{0}_{-1}\beta + ?$$

Let's first complete the α decay nuclear equation. The unknown product must have an atomic number of 82 (so that the subscripts of the products add up to 84), which makes it an isotope of Pb. The mass number of the unknown product must be 214 (so that the superscripts of the products add up to 218). Therefore, the product of the α decay is ${}^{214}_{82}\text{Pb}$. For the β decay nuclear equation, the unknown product must have an atomic number of 85 (so that the subscripts of the products add up to 84) and a mass number of 218. Therefore, the product from the β decay of ${}^{218}\text{Po}$ is ${}^{218}_{85}\text{At}$.

**Think About It** The multiple decay modes of ${}^{218}\text{Po}$ illustrate a pattern of uncertainty that is common among the most massive nuclides. Even after α decay occurs, we cannot be certain that the slightly more neutron-rich product nuclide will undergo β decay. For example, ${}^{234}\text{U}$ may undergo 5 successive α decays (Figure 21.2) before a radioactive nuclide is produced (${}^{214}\text{Pb}$) that undergoes only β decay.

**Practice Exercise** Lead-214, ${}^{214}\text{Pb}$, decays to ${}^{214}\text{Bi}$ and then to ${}^{210}\text{Tl}$. Write balanced nuclear equations describing these reactions.

*(Answers to Practice Exercises are in the back of the book.)*

## Positron Emission and Electron Capture

Nuclides below the belt of stability are neutron poor and undergo decay processes that increase their neutron-to-proton ratio. In one of these processes, the radioactive nucleus emits a high-velocity particle that has the same mass as an electron but that has a positive charge. It is called a **positron** (${}^{0}_{1}\beta$), and its ejection from a neutron-poor nucleus is called **positron emission**. The net effect of positron emission is the production of a nucleus with one fewer proton and one more neutron, as illustrated in the decay of carbon-11:

$${}^{11}_{6}\text{C} \rightarrow {}^{11}_{5}\text{B} + {}^{0}_{1}\beta \qquad (21.3)$$

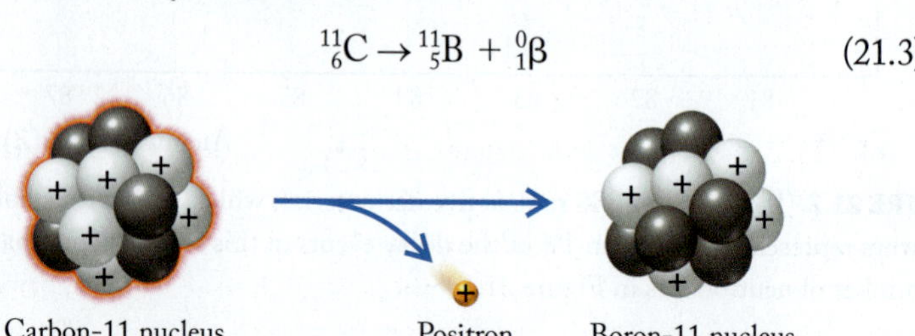

The boron-11 produced in this reaction is a stable isotope. In fact, 80.2% of all boron atoms in nature are boron-11.

Positrons belong to a group of subatomic particles that have the opposite charge but the same mass as particles typically found in atoms. In addition to these electrons with positive charges, protons with negative charges also exist, called *antiprotons*, ${}_{-1}^{1}p$. These charge opposites are particles of **antimatter**.

**positron** a particle with the mass of an electron but with a positive charge.

**positron emission** the spontaneous emission of a positron from a neutron-poor nucleus.

**antimatter** particles that are the charge opposites of normal subatomic particles.

**electron capture** a nuclear reaction in which a neutron-poor nucleus draws in one of its surrounding electrons, which transforms a proton in the nucleus into a neutron.

Particles of matter and their antimatter opposites are like mortal enemies. If they collide, they instantly annihilate each other. In their mutual destruction, they cease to exist as matter, and all their mass is converted to energy in the form of two or more gamma (γ) rays:

$$ {}_{1}^{0}\beta + {}_{-1}^{0}\beta \rightarrow 2\,\gamma \qquad (21.4)$$

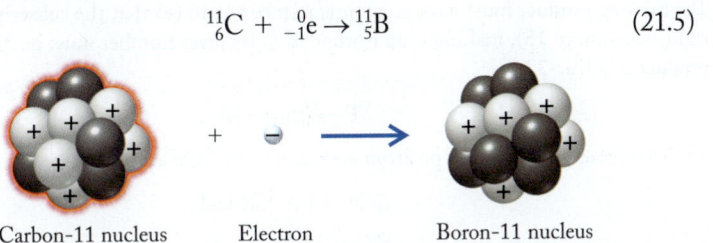

The yellow "sunburst" here symbolizes the energy released in the process depicted.

All nuclear reactions are accompanied by gamma ray emission. Gamma rays represent high-energy electromagnetic radiation that has essentially no mass. Gamma rays are also generated in the nuclear furnaces of stars and permeate outer space. Those that reach Earth are absorbed by the gases in our atmosphere. In the process, molecular gases are broken up into their component atoms, and atomic nuclei may be broken up into their subatomic particles. Oncologists also use radioactive sources that emit gamma rays to disrupt the molecules in cancer cells and kill them.

**CONNECTION** Gamma rays are the highest-energy form of electromagnetic radiation (see Figure 3.1).

The neutron-to-proton ratio of a neutron-poor nucleus can be increased another way: it can capture one of the inner-shell electrons of its atom. When it does, the negatively charged electron combines with a positively charged proton. The product of this combination reaction is a neutron. The effect of this **electron capture** process on the nucleus is the same as positron emission: the number of protons decreases by one and the number of neutrons increases by one. So, when carbon-11 undergoes electron capture, the product is identical to the product formed by positron emission, namely, boron-11:

$$ {}_{6}^{11}C + {}_{-1}^{0}e \rightarrow {}_{5}^{11}B \qquad (21.5)$$

Carbon-11 nucleus + Electron → Boron-11 nucleus

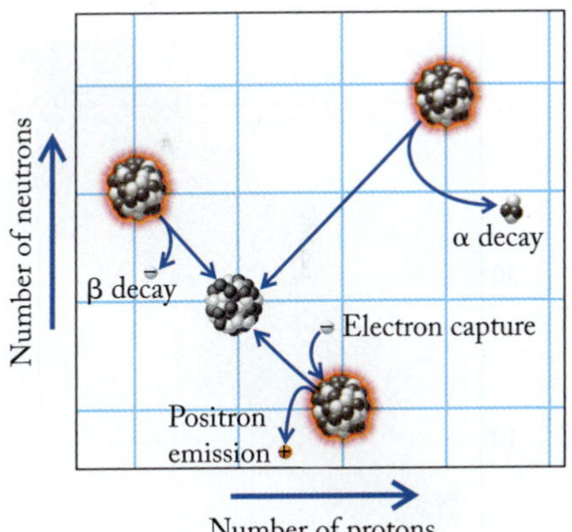


**FIGURE 21.3** Radioactive decay results in predictable changes in the number of protons and neutrons in a nucleus. In α decay, the nucleus loses 2 neutrons and 2 protons, resulting in a decrease of 2 in atomic number and 4 in mass number. Beta decay leads to an increase of 1 proton at the expense of 1 neutron, so the atomic number increases by 1 but the mass number is unchanged. In positron emission and electron capture, the number of protons decreases by 1 and the number of neutrons increases by 1, so the atomic number decreases by 1 but the mass number remains the same.

**Table 21.2** illustrates the impact of being neutron rich, neutron poor, or neither on the stability of isotopes of carbon. Two are stable: $^{12}C$ and $^{13}C$. The isotopes with mass numbers greater than 13 are neutron rich and undergo β decay, whereas those with mass numbers less than 12 are neutron poor and undergo either positron emission or electron capture. **Figure 21.3** summarizes

**TABLE 21.2** Isotopes of Carbon and Their Radioactive Decay Products

| Name | Symbol | Mode(s) of Decay | Half-Life | Natural Abundance (%)[a] |
|---|---|---|---|---|
| Carbon-10 | ${}^{10}_{6}C$ | Positron emission | 19.45 s | — |
| Carbon-11 | ${}^{11}_{6}C$ | Positron emission, electron capture | 20.3 min | — |
| Carbon-12 | ${}^{12}_{6}C$ | (Stable) | (Stable) | 98.89 |
| Carbon-13 | ${}^{13}_{6}C$ | (Stable) | (Stable) | 1.11 |
| Carbon-14 | ${}^{14}_{6}C$ | β decay | 5730 yr | — |
| Carbon-15 | ${}^{15}_{6}C$ | β decay | 2.4 s | — |
| Carbon-16 | ${}^{16}_{6}C$ | β decay | 0.74 s | — |

[a]"—" means none or very close to zero.

the changes in atomic number and mass number that result from the various modes of decay. Taken together, Figures 21.1 and 21.3 allow us to predict the stability of a nucleus with respect to radioactive decay and to identify likely decay pathways.

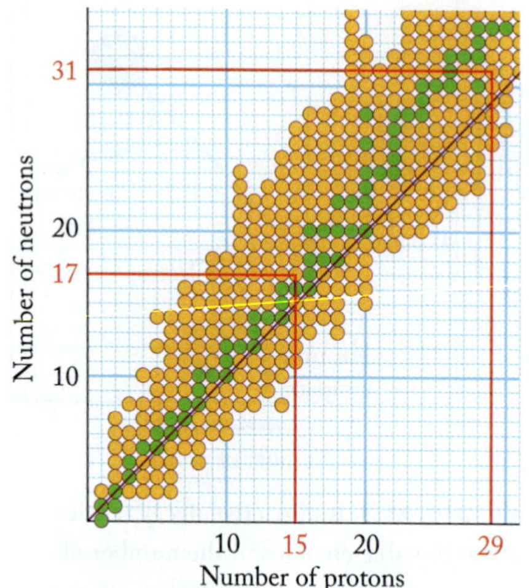


**FIGURE 21.4** Region in the belt of stability for $Z \leq 30$.

**SAMPLE EXERCISE 21.2** Predicting the Modes and Products of Radioactive Decay **LO2**

Predict the mode of radioactive decay of $^{32}P$ and $^{60}Cu$, two radionuclides used in biomedical research and treatment. Identify the nuclide produced in each decay process.

**Collect, Organize, and Analyze** The mode of decay of each radionuclide will depend on whether it is neutron rich or neutron poor. In **Figure 21.4**, $^{32}P$ (17 neutrons + 15 protons) is represented by the orange dot directly above the green dot for $^{31}P$ (the one and only stable phosphorus nuclide), which means that $^{32}P$ is radioactive and neutron rich. Neutron-rich radioisotopes of lighter elements undergo β decay. The orange dot for $^{60}Cu$ lies *below* the green dots for the two stable copper isotopes ($^{63}Cu$ and $^{65}Cu$), which means that $^{60}Cu$ is a neutron-poor isotope and suggests that either positron emission or electron capture is a likely decay pathway. In the balanced nuclear equations, regardless of the nuclide or mode of decay, the sums of the superscripts on the left and right sides must be equal, as must the sums of the subscripts.

**Solve** A β particle must be one product of the decay reaction for $^{32}P$, giving the incomplete nuclear equation:

$${}^{32}_{15}P \rightarrow ? + {}^{0}_{-1}\beta$$

The missing product must have an atomic number of 16 (so that the subscripts on the right side sum to 15), making it an isotope of S. Its mass number must be 32, so the product is sulfur-32:

$${}^{32}_{15}P \rightarrow {}^{32}_{16}S + {}^{0}_{-1}\beta$$

The analogous equations for positron emission from $^{60}Cu$ are

$${}^{60}_{29}Cu \rightarrow ? + {}^{0}_{+1}\beta \text{ and}$$

$${}^{60}_{29}Cu \rightarrow {}^{60}_{28}Ni + {}^{0}_{+1}\beta$$

The atomic number of copper-60 must equal the sum of the subscripts (29 = 28 + 1) on the right side of the equation, and the mass number of copper-60 must equal the sum of the superscripts (60 = 60 + 0).

Neutron-poor isotopes such as copper-60 can also decay by electron capture. Using Equation 21.5 as a model, we can write another nuclear equation for the decay of $^{60}Cu$:

$$^{60}_{29}Cu + {}^{0}_{-1}e \rightarrow {}^{60}_{28}Ni$$

Note that the *total* number of nucleons (neutrons and protons) is equal for copper-60 and nickel-60 because they have the same mass number, consistent with converting one proton and one electron in a Cu atom into one neutron.

**Think About It** By emitting a β particle, the $^{32}P$ nucleus has one additional proton and one fewer neutron, thereby reducing its neutron "richness" and becoming a stable isotope of sulfur. Because both positron emission and electron capture produce the same product in the decay of $^{60}Cu$, predicting which pathway will dominate is difficult. Because the number of protons increases by either pathway, the neutron-to-proton ratio increases and the product isotope, nickel-60, is stable, as indicated by a green dot in Figure 21.4.

**Practice Exercise** Predict the modes of radioactive decay of $^{28}Al$ and $^{18}F$. Identify the nuclide produced in each decay process.

## 21.2 Rates of Radioactive Decay

In Section 21.1, we examined how radionuclides undergo radioactive decay; in this section, we focus on how rapidly they decay. We have defined radioactive decay as the spontaneous disintegration of unstable nuclei, but, as with chemical reactions, spontaneous does not necessarily mean rapid.

### First-Order Radioactive Decay

In Chapter 13, we used the integrated rate law for first-order reactions to relate the concentration of reactant X, [X], at any time $t$ during a first-order reaction to its initial concentration, $[X]_0$, and the rate constant ($k$) of the reaction:

$$\ln \frac{[X]_t}{[X]_0} = -kt \qquad (13.16)$$

To apply this relationship to radioactive decay reactions, we replace the concentration terms with the numbers of radioactive atoms at time 0, $N_0$, and at any later time $t$, $N_t$:

$$\ln \frac{N_t}{N_0} = -kt \qquad (21.6)$$

**CHEMTOUR** Half-Life

The kinetics of radioactive decay processes are often expressed using their half-lives (**Figure 21.5**) rather than rate constants, so we use Equation 13.19: $t_{1/2} = 0.693/k$, or $k = 0.693/t_{1/2}$, to substitute for $k$ in Equation 21.6:

$$\ln \frac{N_t}{N_0} = -0.693 \frac{t}{t_{1/2}} \qquad (21.7)$$

Note that the fraction $t/t_{1/2}$ represents the number of half-lives (including fractions of half-lives) that have passed since the sample contained $N_0$ atoms of the radionuclide. The following exercise illustrates how useful Equation 21.7 can be.

**CONNECTION** We introduced the concept of *half-life*, $t_{1/2}$, and how its value is inversely proportional to the rate constant, $k$, of a first-order reaction in Chapter 13.

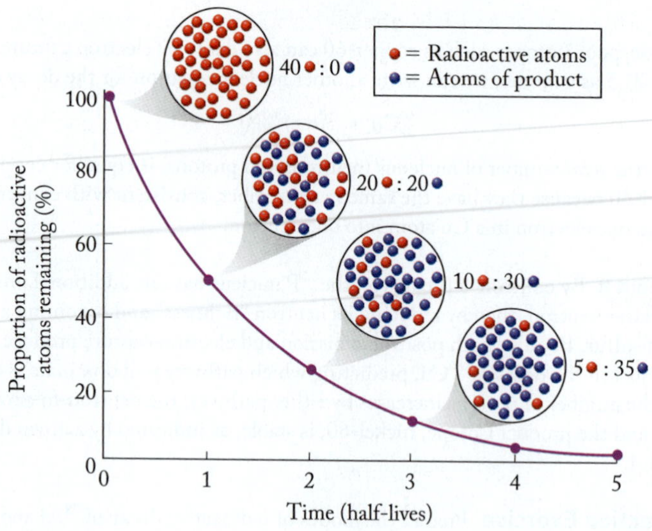


**FIGURE 21.5** Radioactive decay follows first-order kinetics. For example, if a sample initially contains 40 radioactive atoms, it will contain only one-half that number after a time interval equal to one half-life. Half of the remaining half, or 10 radioactive atoms, remains after two half-lives, and so on.

**SAMPLE EXERCISE 21.3** Calculating the Quantity of a Radionuclide Remaining in a Sample **LO3**

Free neutrons are radioactive, undergoing β decay with a half-life of 10.25 minutes. Starting with a population of $7.25 \times 10^5$ free neutrons, how many are left after 6.67 minutes?

**Collect, Organize, and Analyze** Equation 21.7 relates quantities of radioactive particles to decay times. In this problem, the initial number of neutrons ($N_0$) is $7.25 \times 10^5$, $t = 6.67$ min, $t_{1/2} = 10.25$ min, and we need to solve for $N_t$. The value of $t$ is less than half of $t_{1/2}$; so fewer than half of the initial number of neutrons will have decayed after 6.67 min. In other words, most of the neutrons will still be present.

**Solve** Substituting the data into Equation 21.7 and solving for $N_t$:

$$\ln \frac{N_t}{N_0} = -0.693 \frac{t}{t_{1/2}}$$

$$\ln N_t = \ln N_0 - 0.693 \frac{t}{t_{1/2}} = \ln (7.25 \times 10^5) - 0.693 \left(\frac{6.67 \text{ min}}{10.25 \text{ min}}\right) = 13.043$$

$$N_t = e^{13.043} = 4.62 \times 10^5 \text{ neutrons}$$

**Think About It** The value of $N_t$ is reasonable because, as predicted, less than half of the initial quantity of free neutrons decayed in 6.67 minutes.

**Practice Exercise** Cesium-131 is a short-lived radionuclide ($t_{1/2} = 9.7$ d) used to treat prostate cancer. If the therapeutic strength of the radionuclide is directly proportional to the number of nuclei present, how much therapeutic strength does a cesium-131 source lose over exactly 60 days? Express your answer as a percentage of the strength the source had at the beginning of the first day.

## Radiometric Dating

The rate of radioactive decay has a practical application in **radiometric dating**, a term used to describe methods for determining the age of objects from the tiny concentrations of radionuclides that occur naturally in them and the rates of radioactive decay of these nuclides. The concept was invented in the early 1900s by Ernest Rutherford, who had already recognized that radioactive decay processes have characteristic half-lives. Rutherford proposed to use this concept to determine the age of rocks and even the age of Earth itself. The basis for his initial attempt was the emission of α particles from uranium ore. He correctly suspected that α particles were part of helium atoms, and so he proposed to determine the age of uranium ore samples by determining the concentration of helium gas trapped inside them.

Rutherford's helium method did not yield very accurate results, but it did inspire a young American chemist, Bertram Boltwood (1870–1927), who had determined that the decay of radioactive uranium involves a series of decay events ending with the formation of stable lead (illustrated for $^{238}U$ in Figure 21.2). In 1907, Boltwood published the results of dating 43 samples of uranium-containing minerals on the basis of the ratio of lead to uranium in them. The ages he reported spanned hundreds of millions to more than a billion years and probably represent the first successful attempt at radiometric dating.

Much more recently, the development of the mass spectrometer for accurately determining the abundances of individual isotopes of elements, coupled with more accurate half-life values for decay events such as those in Figure 21.2, has allowed scientists to use the ratio of $^{206}Pb$ to $^{238}U$ in geological samples to determine their ages with a precision of about ±1%. Other methods, including one based on the decay of $^{235}U$ to $^{207}Pb$ ($t_{1/2} = 7.0 \times 10^6$ yr), may be used to analyze the same samples, providing multiple independent determinations that mutually assure more accurate results. These analyses have shown that the oldest rocks on Earth are more than 4.0 billion years old and that meteorites that formed as the solar system formed are 4.5 billion years old.

**CONNECTION** In Section 2.6, we saw that mass spectrometry can be used to determine the abundances of the isotopes of elements in a sample.

The radiometric methods described above yield reliable results only when the sample is a closed system, which means that the only loss of the radionuclide is through radioactive decay and that all the nuclides produced by the decay processes remain in the sample. In addition, those decay processes must be the only source of the product nuclides. For these reasons, scientists must exercise care in selecting the types of samples they subject to radiometric dating analysis. For example, the presence of the mineral zircon ($ZrSiO_4$) in a geological sample is good news for scientists interested in using radiometric dating because $U^{4+}$ ions readily substitute for $Zr^{4+}$ ions as crystals of $ZrSiO_4$ solidify from the molten state, but $Pb^{2+}$ ions do not. Therefore, the only source of $^{206}Pb$ and $^{207}Pb$ in a zircon sample should be the decay of $^{238}U$ and $^{235}U$, respectively.

**CONNECTION** We examined the crystal structures of silicate minerals in Section 18.8.

In 1947, American chemist Willard Libby (1908–1980) developed **radiocarbon dating**, a radiometric dating technique for determining the age of artifacts from prehistory and early civilizations. The method is based on determining the carbon-14 content of samples derived from plants or the animals that consumed them. Carbon-14 originates in the upper atmosphere, where cosmic rays break apart the nuclei of atoms, forming free protons and neutrons. When one of these neutrons collides with a nitrogen-14 atom, they form an atom of radioactive carbon-14 and a proton:

$$^{14}_{7}N + ^{1}_{0}n \rightarrow ^{14}_{6}C + ^{1}_{1}p$$

Atmospheric carbon-14 reacts with oxygen, forming $^{14}CO_2$. The atmospheric concentration of $^{14}CO_2$ amounts to only about $10^{-12}$ of all the molecules of $CO_2$ in the air. These traces of radioactive $CO_2$, along with the stable forms $^{12}CO_2$ and $^{13}CO_2$, are incorporated into the structures of green plants during photosynthesis. The tiny fraction of the plant's mass that is $^{14}C$ gets even tinier after a plant dies,

**radiometric dating** a method for determining the age of an object on the basis of the quantity of a radioactive nuclide and/or the products of its decay that the object contains.

**radiocarbon dating** a method for establishing the age of a carbon-containing object by measuring the amount of radioactive carbon-14 remaining in the object.

or after a part of it stops growing and photosynthesizing, because $^{14}C$ undergoes β decay (with a half-life of 5730 years), as we described in Section 21.1:

$$^{14}_{6}C \rightarrow ^{14}_{7}N + ^{0}_{-1}\beta$$

If we can determine the $^{14}C$ content ($N_t$) of an object of historical interest, such as a piece of wood from an ancient building, charcoal from a prehistoric campfire, or papyrus from an early Egyptian scroll, and if we know (or can predict) its $^{14}C$ content when the material in it was alive ($N_0$), then we can apply Equation 21.7 to determine its age if we rearrange the terms to isolate $t$ on the left side:

$$t = -\frac{t_{1/2}}{0.693} \ln \frac{N_t}{N_0} \quad (21.8)$$

Substitution for $N_0$, $N_t$, and $t_{1/2}$ in Equation 21.8 yields the time elapsed since the material was alive, as shown in the following Sample Exercise.

**SAMPLE EXERCISE 21.4** Radiocarbon Dating **LO3**

The $^{14}C$ content of a wooden harpoon handle found in the remains of an Inuit encampment in western Alaska is 61.9% of the $^{14}C$ content of the same type of wood from a recently cut tree. How old is the harpoon?

**Collect and Organize** We are asked to calculate the age of a sample that contains 61.9% of the $^{14}C$ in a modern sample of the same material. The half-life of carbon-14 is 5730 years. Equation 21.8 provides the age, $t$, of the artifact if we know the ratio of the $^{14}C$ in it today to its initial $^{14}C$ content.

**Analyze** The $^{14}C$ content of the modern sample can be used as a surrogate for the initial $^{14}C$ content of the artifact. Therefore, 61.9% (or 0.619) represents the ratio $N_t/N_0$. This value is greater than 0.5, which means that the age of the sample is less than one half-life (5730 yr).

**Solve**

$$t = -\frac{t_{1/2}}{0.693} \ln \frac{N_t}{N_0}$$

$$= -\frac{5730 \text{ yr}}{0.693} \ln(0.619)$$

$$= 3966 \text{ yr} = 3.97 \times 10^3 \text{ yr}$$

**Think About It** The resulting age is less than one half-life, which is reasonable because the sample contained more than half the original carbon-14 content. The result is expressed with three significant figures to match that of the starting composition (61.9%).

**Practice Exercise** The Old Testament describes the construction of the Siloam Tunnel, used to carry water into Jerusalem under the reign of King Hezekiah (727–698 BCE). An inscription on the tunnel has been interpreted as evidence that the tunnel was not built until 200–100 BCE. $^{14}C$ dating (in 2003) indicated a date close to 700 BCE. What is the ratio of $^{14}C$ in a wooden object made in 100 BCE to one made from the same kind of wood in 700 BCE?

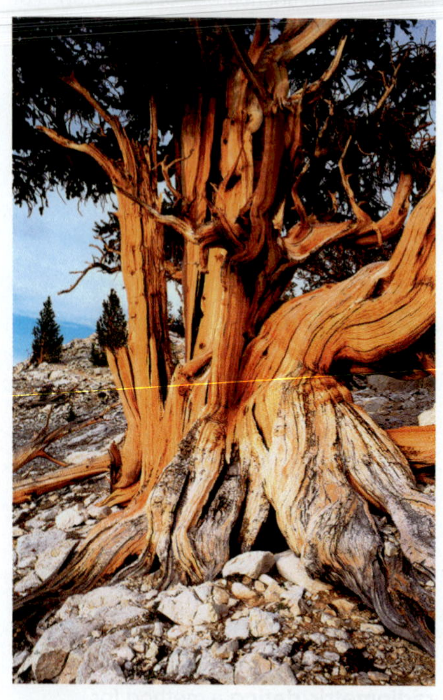

**FIGURE 21.6** Radiocarbon dating relies on knowing the atmospheric concentration of carbon-14 over time. Ancient living trees, such as bristlecone pines in the American Southwest, provide a record of atmospheric carbon-14 levels over thousands of years. The ages of the rings can be determined by counting them, and their carbon-14 content can be determined by mass spectrometry.

The accuracy of radiocarbon dating can be checked by determining the $^{14}C$ content of the annual growth rings of very old trees, such as the bristlecone pines that grow in the American Southwest (**Figure 21.6**). When scientists plot

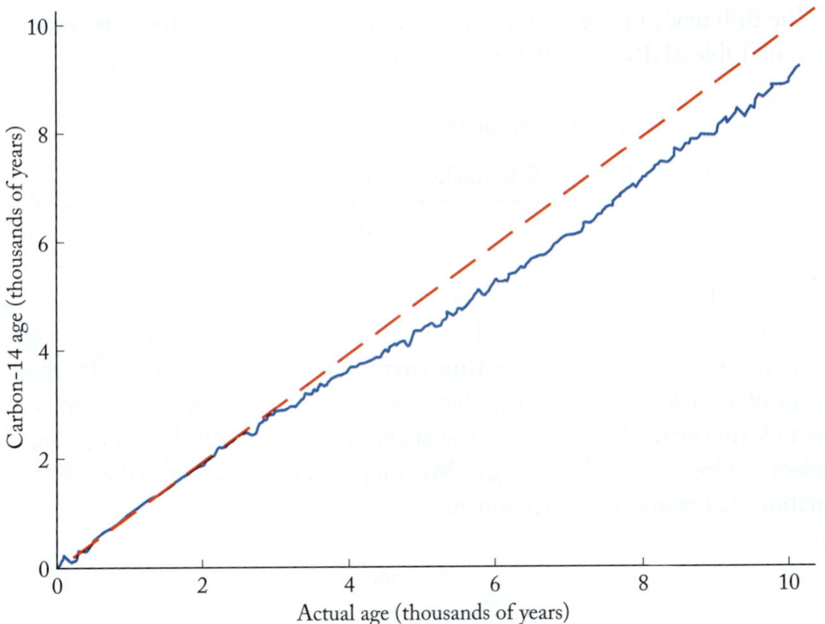


**FIGURE 21.7** Calibration curves for radiocarbon dating allow scientists to accurately calculate the ages of archaeological objects. If the rate of $^{14}C$ production in the upper atmosphere were constant, then the age of objects on the basis of their $^{14}C$ content would match their actual age—a condition represented by the red dashed line. However, analyses of tree rings, corals, and lake sediments indicate that the rate of $^{14}C$ production in the upper atmosphere varies, so a correct plot of $^{14}C$ age versus actual age produces the jagged blue line. This plot allows scientists to convert $^{14}C$ ages into actual ages.

the radiocarbon ages of these rings against their actual ages obtained by counting rings starting from the outer growth layer of the tree (representing $t = 0$), they find that the two sets of ages do not agree exactly, as shown in **Figure 21.7**. Several reasons account for this lack of agreement, including variability in the rates of $^{14}C$ production due to changing intensity of the cosmic rays striking Earth's upper atmosphere. To ensure accurate $^{14}C$ results, scientists must correct for these and other variations, and they must be careful to avoid contaminating ancient samples with modern carbonaceous material. With proper analytical technique, radiocarbon dating results are generally accurate to within 40 yr for samples that are 500–50,000 years old.

**CONCEPT TEST**

How might the increased consumption of fossil fuels over the last century affect the $^{14}C$ content of growing plant tissues?

## 21.3 Energy Changes in Radioactive Decay

To understand why some nuclei are more stable than others, we need to explore the nature of the energy that keeps the nucleons in atomic nuclei together. By the 1930s, scientists had discovered that the mass of a stable nucleus is always less than the sum of the separate masses (see Table 21.1) of its nucleons. For example, a helium-4 nucleus consists of two neutrons and two protons corresponding to a mass:

$$\begin{aligned} \text{Mass of 2 neutrons} &= 2(1.67493 \times 10^{-27}\text{ kg}) \\ +\ \text{Mass of 2 protons} &= 2(1.67262 \times 10^{-27}\text{ kg}) \\ \hline \text{Total mass of nucleons} &= 6.69510 \times 10^{-27}\text{ kg} \end{aligned}$$

**mass defect ($\Delta m$)** the difference between the mass of a stable nucleus and the masses of its individual nucleons.

**binding energy (BE)** the energy that holds the nucleons together in a nucleus.

**strong nuclear force** the fundamental force of nature that keeps quarks together in subatomic particles and nucleons together in atomic nuclei.

The difference between this value and the actual mass of the $^4$He nucleus, as listed in Table 21.1 at $6.64465 \times 10^{-27}$, is

$$\begin{aligned} \text{Total mass of nucleons} &= 6.69510 \times 10^{-27}\ \text{kg} \\ -\ \text{Mass of } ^4_2\text{He nucleus} &= 6.64465 \times 10^{-27}\ \text{kg} \\ \hline &= 0.05045 \times 10^{-27}\ \text{kg} \end{aligned}$$

or $5.045 \times 10^{-29}$ kg. This tiny difference in mass is called the **mass defect ($\Delta m$)** of the $^4$He nucleus. It represents the energy that holds the four nucleons together in the nucleus—that is, the **binding energy (BE)** of the nucleus. The binding energy of a nucleus is the energy that would be released if free nucleons were to fuse to form the nucleus. It is also the energy needed to split the nucleus into free nucleons. How big is this energy? We can calculate it using Albert Einstein's equation that relates mass ($m$) and energy ($E$):

$$E = mc^2 \tag{21.9}$$

where $c$ is the speed of light in a vacuum, $2.998 \times 10^8$ m/s. Let's use Equation 21.9 to calculate the binding energy of an α particle ($^4$He nucleus) on the basis of its mass defect of $5.045 \times 10^{-29}$ kg.

$$\begin{aligned} \text{BE} &= (\Delta m)c^2 \\ &= 5.045 \times 10^{-29}\ \text{kg} \times (2.998 \times 10^8\ \text{m/s})^2 \\ &= 4.534 \times 10^{-12}\ \text{kg} \cdot (\text{m/s})^2 = 4.534 \times 10^{-12}\ \text{J} \end{aligned}$$

This amount of energy may seem very small, but it is the binding energy of just a single helium-4 nucleus. Its value per mole of $^4$He nuclei is equivalent to billions of kilojoules:

$$\frac{4.534 \times 10^{-12}\ \text{J}}{1\ \cancel{\text{atom}}} \times \frac{6.022 \times 10^{23}\ \cancel{\text{atoms}}}{1\ \text{mol}} \times \frac{1\ \text{kJ}}{1000\ \cancel{\text{J}}} = 2.730 \times 10^9\ \text{kJ/mol}$$

By comparison, consider that the bond dissociation energy of $H_2$, 436 kJ/mol, is less than one six-millionth the binding energy of a mole of He atoms. This enormous amount of energy means that $^4$He is a very stable nuclide.

Most of the stable nuclei with larger numbers of nucleons than $^4$He have even larger binding energies. This makes sense because a nucleus with many protons near each other must be held together by an enormous energy to overcome the electrostatic repulsion between all the positively charged particles. In such proximity, nucleons come under the influence of a fundamental force of nature known as the **strong nuclear force**. It operates only over very small distances, such as the diameters of atomic nuclei, but it is 100 times stronger than the repulsions the protons experience. The strong nuclear force binds nucleons together and stabilizes atomic nuclei.

To make comparisons of nuclear binding energies more meaningful, they are usually divided by the number of nucleons in each nucleus. Expressing binding energy on a per-nucleon basis allows us to compare the relative stabilities of different nuclide of similar atomic number. When we plot binding energy per nucleon values against mass number, we get the curve shown in **Figure 21.8**.

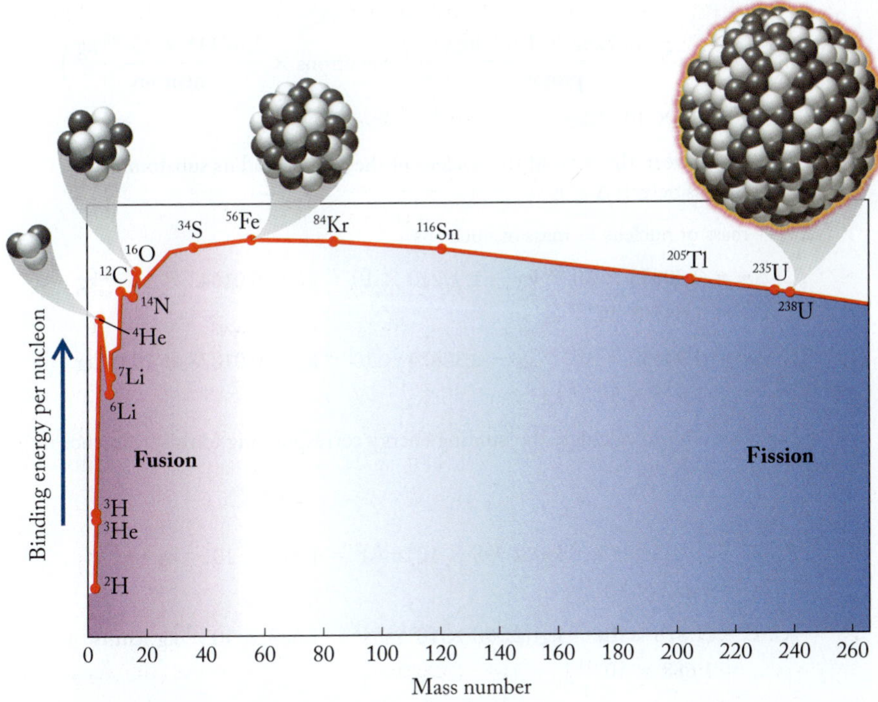


**FIGURE 21.8** The stability of a nucleus is directly proportional to its binding energy per nucleon, which reaches a maximum at $^{56}$Fe.

**SAMPLE EXERCISE 21.5** Calculating the Binding Energy of an Isotope **LO4**

Recall from Table 21.2 that $^{12}$C is a stable isotope of carbon but $^{14}$C is radioactive. Calculate the binding energy per nucleon of these two nuclei, given their atomic masses are 12.00000 u and 14.00324 u, respectively.

**Collect and Organize** We are asked to calculate the binding energy of two nuclei: $^{12}$C and $^{14}$C. We are given their atomic masses, but we will also need the masses of a single neutron ($1.67493 \times 10^{-27}$ kg) and a single proton ($1.67262 \times 10^{-27}$ kg) from Table 21.1. We will also need a conversion factor to change the masses of $^{12}$C and $^{14}$C from u to kg: $1.66054 \times 10^{-27}$ kg = 1 u.

**Analyze** We must calculate the mass defect between the masses of the nuclei of the two isotopes and the masses of their subatomic particles. Using Equation 21.9, we convert the mass defect to energy. This energy represents the binding energy of each nucleus. We predict that the binding energy of the stable isotope of carbon, $^{12}$C, will be greater than that of the radioactive (unstable) isotope, $^{14}$C.

**Solve** First we convert the atomic mass of one atom of each isotope from u to kg:

$$^{12}_{6}\text{C: } 12.00000 \text{ u} \times \frac{1.66054 \times 10^{-27} \text{ kg}}{1 \text{ u}} = 1.99265 \times 10^{-26} \text{ kg}$$

$$^{14}_{6}\text{C: } 14.00324 \text{ u} \times \frac{1.66054 \times 10^{-27} \text{ kg}}{1 \text{ u}} = 2.32529 \times 10^{-26} \text{ kg}$$

Because we will be comparing the mass of the $^{12}$C and $^{14}$C nuclei, we need to subtract the mass of 6 electrons from the atomic masses of the $^{12}$C and $^{14}$C atoms:

$$^{12}_{6}\text{C: } 1.99265 \times 10^{-26} \text{ kg} - 6(9.10939 \times 10^{-31} \text{ kg}) = 1.99210 \times 10^{-26} \text{ kg}$$
$$^{14}_{6}\text{C: } 2.32529 \times 10^{-26} \text{ kg} - 6(9.10939 \times 10^{-31} \text{ kg}) = 2.32474 \times 10^{-26} \text{ kg}$$

Next, we calculate the mass of the protons and neutrons in the nucleus of one atom of each isotope:

$$^{12}_{6}\text{C: } \left(6 \text{ protons} \times \frac{1.67262 \times 10^{-27} \text{ kg}}{\text{proton}}\right) + \left(6 \text{ neutrons} \times \frac{1.67493 \times 10^{-27} \text{ kg}}{\text{neutron}}\right)$$
$$= 2.00853 \times 10^{-26} \text{ kg}$$

$$^{14}_{6}\text{C: } \left(6 \text{ protons} \times \frac{1.67262 \times 10^{-27}\text{ kg}}{\text{proton}}\right) + \left(8 \text{ neutrons} \times \frac{1.67493 \times 10^{-27}\text{ kg}}{\text{neutron}}\right)$$
$$= 2.34352 \times 10^{-26}\text{ kg}$$

The difference between the mass of the nucleus of the isotope and its subatomic particles (the mass defect, $\Delta m$) is

$$\Delta m = \text{mass of nucleus} - \text{mass of nucleons}$$

$$^{12}_{6}\text{C: } \Delta m = 2.00853 \times 10^{-26}\text{ kg} - 1.99210 \times 10^{-26}\text{ kg} = 0.01643 \times 10^{-26}\text{ kg}$$
$$= 1.643 \times 10^{-28}\text{ kg}$$

$$^{14}_{6}\text{C: } \Delta m = 2.34352 \times 10^{-26}\text{ kg} - 2.32474 \times 10^{-26}\text{ kg} = 0.01878 \times 10^{-26}\text{ kg}$$
$$= 1.878 \times 10^{-28}\text{ kg}$$

Using Equation 21.9 to calculate the binding energy corresponding to this difference in mass:

$$E = mc^2$$

$$^{12}_{6}\text{C: } E = (1.643 \times 10^{-28}\text{ kg})(2.998 \times 10^8\text{ m/s})^2 = 1.477 \times 10^{-11}\text{ kg} \cdot \text{m}^2/\text{s}^2$$
$$= 1.477 \times 10^{-11}\text{ J}$$

$$^{14}_{6}\text{C: } E = (1.878 \times 10^{-28}\text{ kg})(2.998 \times 10^8\text{ m/s})^2 = 1.688 \times 10^{-11}\text{ kg} \cdot \text{m}^2/\text{s}^2$$
$$= 1.688 \times 10^{-11}\text{ J}$$

To calculate the binding energy per nucleon, we divide each value of $E$ by the number of nucleons:

$$^{12}_{6}\text{C: } E = \frac{1.477 \times 10^{-11}\text{ J}}{12 \text{ nucleons}} = 1.231 \times 10^{-12}\text{ J/nucleon}$$

$$^{14}_{6}\text{C: } E = \frac{1.688 \times 10^{-11}\text{ J}}{14 \text{ nucleons}} = 1.206 \times 10^{-12}\text{ J/nucleon}$$

**Think About It** The difference in mass between the mass of a nucleus and its subatomic particles reflects the binding energy. Stable nuclei should have larger binding energies per nucleon than those of unstable nuclei. The binding energy per nucleon of radioactive $^{14}$C ($1.206 \times 10^{-12}$ J/nucleon) is indeed less than the binding energy per nucleon of stable $^{12}$C ($1.231 \times 10^{-12}$ J/nucleon), as predicted.

**Practice Exercise** A "free" neutron undergoes β decay, forming a proton and a high-energy electron (a β particle). How much energy is released in this nuclear reaction? Express your answer in kJ/mol.

## 21.4 Making New Elements

Rutherford's experiments with gold foil led him to bombard other elements with α particles. In a 1919 experiment, Rutherford reported that passing α particles through nitrogen yielded two products: oxygen-17 and free protons. Bombardment of nuclei by α particles became a popular method for transmuting elements in the 1920s and 1930s. In 1933, French chemists Irène (1897–1956) and Frédéric (1900–1958) Joliot-Curie synthesized the first radionuclide not found in nature, phosphorus-30, by bombarding aluminum-27 with α particles:

$$^{27}_{13}\text{Al} + ^{4}_{2}\text{He} \rightarrow ^{30}_{15}\text{P} + ^{1}_{0}\text{n}$$

In 1940 elements 93 (neptunium) and 94 (plutonium) were produced at the University of California, Berkeley, by bombarding uranium-238 with neutrons. Neither of these elements occur naturally, and their synthesis opened the door for extending the periodic table. Transmutation by neutron bombardment is easier than by α bombardment because both α particles and atomic nuclei have positive charges and so repel each other. The α particles must be traveling at high speeds to overcome this repulsion in a successful bombardment. Because neutrons have no charge, however, they are more readily captured by nuclei. For example, when nuclei of uranium-238 are bombarded with neutrons, a nuclear reaction may occur in which two β particles are ejected, leading to the formation of plutonium-239:

$$^{238}_{92}\mathrm{U} + {}^{1}_{0}\mathrm{n} \rightarrow {}^{239}_{94}\mathrm{Pu} + 2\,{}^{0}_{-1}\beta$$

Between 1944 and 1961, the Berkeley research team synthesized elements through $Z = 103$, lawrencium, by bombarding actinide nuclei with different combinations of neutrons and α particles. The research team was led by American chemist Glenn T. Seaborg (1912–1999; **Figure 21.9**), who won the Nobel Prize in Chemistry in 1951 for his team's ability to first synthesize and then characterize the chemical properties of these *transuranic* elements.

**FIGURE 21.9** Glenn T. Seaborg became the first living scientist to have an element named after him in 1994, when element 106 was named seaborgium.

**STEPWISE ANIMATION**
Transmutation

The methods used by Seaborg's team did not work for synthesizing elements with $Z > 103$: nuclides more massive than californium-249 are highly unstable and rapidly lose α or β particles, so they are not useful target materials for making even more massive nuclides. However, by bombarding californium-249 targets with carbon, nitrogen, and oxygen nuclei instead of α particles, scientists synthesized isotopes of rutherfordium ($Z = 104$), dubnium ($Z = 105$), and seaborgium ($Z = 106$):

$$^{249}_{98}\mathrm{Cf} + {}^{12}_{6}\mathrm{C} \rightarrow {}^{257}_{104}\mathrm{Rf} + 4\,{}^{1}_{0}\mathrm{n}$$

$$^{249}_{98}\mathrm{Cf} + {}^{15}_{7}\mathrm{N} \rightarrow {}^{260}_{105}\mathrm{Dd} + 4\,{}^{1}_{0}\mathrm{n}$$

$$^{249}_{98}\mathrm{Cf} + {}^{18}_{8}\mathrm{O} \rightarrow {}^{263}_{106}\mathrm{Sg} + 4\,{}^{1}_{0}\mathrm{n}$$

In recent years, scientists have reported synthesizing nuclides that have as many as 118 protons by bombarding targets as massive as californium-249 with medium-mass nuclei, such as calcium-48. In January 1999, for example, an atom with 114 protons and 175 neutrons was synthesized and lasted for 30 s (most supermassive nuclides have half-lives that are fractions of a second) before undergoing a series of α decay events that yielded isotopes of elements with atomic numbers 112, 110, and 108. Some of these "supermassive" elements, and the bombarding ions and target nuclides used to make them, are listed in **Table 21.3**.

**TABLE 21.3** Synthesis of Supermassive Nuclides

| Bombarding Ion | Target | Nuclide Synthesized | Year First Synthesized |
|---|---|---|---|
| $^{62}$Ni | $^{208}$Pb | $^{269}_{110}$Ds | 1994 |
| $^{64}$Ni | $^{209}$Bi | $^{272}_{111}$Rg | 1994 |
| $^{69}$Zn | $^{208}$Pb | $^{277}_{112}$Cn | 1996 |
| $^{70}$Zn | $^{209}$Bi | $^{278}_{113}$Nh | 2003 |
| $^{48}$Ca | $^{244}$Pu | $^{289}_{114}$Fl | 1999 |
| $^{48}$Ca | $^{243}$Am | $^{288}_{115}$Mc | 2003 |
| $^{48}$Ca | $^{248}$Cm | $^{293}_{116}$Lv | 2000 |
| $^{48}$Ca | $^{249}$Bk | $^{293}_{117}$Ts | 2009 |
| $^{48}$Ca | $^{249}$Cf | $^{294}_{118}$Og | 2002 |

Confirmation of new unstable elements can take decades. The syntheses of elements 113 (nihonium), 115 (moscovium), 117 (tennessine), and 118 (oganesson) were recognized in January 2016.

Why bother to make such short-lived elements? Their mere existence, no matter how brief, can be a source of insight into the nature of nuclear structure. Their behavior illustrates the competition between the strong nuclear force, which holds nucleons together, and the electrostatic repulsion between protons. Supermassive elements are pieces of a puzzle that someday may tell us whether the size of atoms has an upper limit. Our ability to create new elements in the laboratory raises the following question: where did the stable elements found on Earth come from?

## 21.5 Nuclear Fusion and the Origin of the Elements

Theoretical physicists believe that our universe began about 13.8 billion years ago with an enormous release of energy that has come to be known as the Big Bang. Current theories suggest that within a few microseconds of the Big Bang, much of its energy had transformed into matter made of elementary particles such as electrons and **quarks** (**Figure 21.10**). Less than a millisecond later, the universe had expanded and "cooled" to a temperature of $10^{12}$ K, which allowed quarks to combine with one another to form neutrons and protons. Thus, in less than a second, the universe contained the three types of subatomic particles that would eventually make up atoms.

**CHEMTOUR**
Fusion of Hydrogen

### Primordial Nucleosynthesis

By about 4 minutes after the Big Bang, the universe had expanded and cooled to about $10^9$ K. In this hot, dense subatomic "soup," colliding neutrons and protons began to fuse in a process called primordial **nucleosynthesis**. When a proton and a neutron collide and fuse, they form a *deuteron* (${}^{2}_{1}\text{D}$), which is the nucleus of the deuterium isotope of hydrogen:

**CHEMTOUR**
The Synthesis of Elements

$${}^{1}_{1}\text{H} + {}^{1}_{0}\text{n} \rightarrow {}^{2}_{1}\text{D}$$

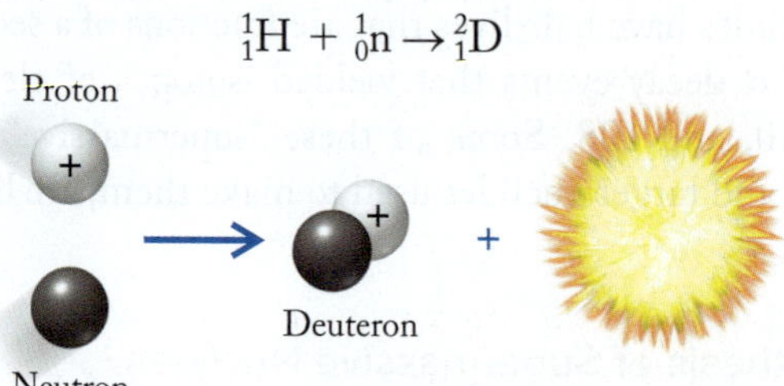


Deuteron formation proceeded rapidly in the minutes after the Big Bang, consuming most of the free neutrons in the universe. However, no sooner did deuterons form than they, too, were consumed in collisions with one another, fusing to make ${}^{4}_{2}\text{He}$ nuclei (α particles) in a process called **nuclear fusion**:

$$2\,{}^{2}_{1}\text{D} \rightarrow {}^{4}_{2}\text{He}$$

Deuteron
Helium-4 nucleus

**quarks** elementary particles that combine to form neutrons and protons.

**nucleosynthesis** the natural formation of nuclei because of fusion and other nuclear processes.

**nuclear fusion** a nuclear reaction in which subatomic particles or atomic nuclei collide at very high speeds and fuse, forming more massive nuclei and releasing energy.

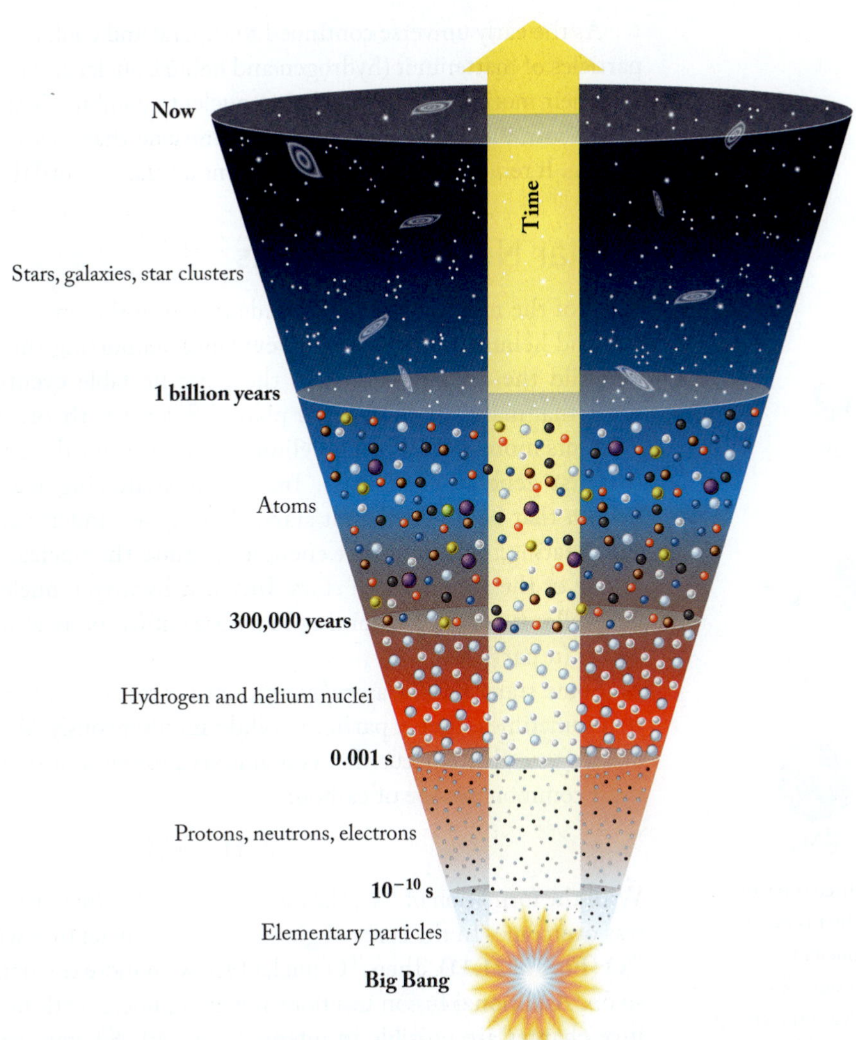


**FIGURE 21.10** Time line for energy and matter transformations believed to have occurred since the universe began. In this model, protons, neutrons, and electrons were formed from quarks in the first millisecond after the Big Bang, followed by hydrogen and helium nuclei. Atoms of $^1$H and $^4$He did not form until after 300,000 years of expansion and cooling, and other elements did not form until the first galaxies appeared around 1 billion years after the Big Bang. According to this model, our solar system, our planet, and all life-forms on it are composed of elements synthesized in stars that were born, burned brightly, and then disappeared millions to billions of years after the Big Bang.

By about 5 minutes after the Big Bang, the matter of the universe had become 75% (by mass) protons and 25% α particles. Nucleosynthesis then stopped. Why? Why didn't α particles and protons, for example, fuse to make $^5$Li, as shown in the equation below?

$$^{4}_{2}\text{He} + ^{1}_{1}\text{H} \xrightarrow{?} ^{5}_{3}\text{Li}$$

Or why didn't two α particles fuse to make a nucleus of $^8$Be?

$$2\ ^{4}_{2}\text{He} \xrightarrow{?} ^{8}_{4}\text{Be}$$

Neither process took place because neither $^5$Li nor $^8$Be is a stable nuclide. In fact, no stable nuclides with mass numbers of 5 or 8 appear in Figure 21.1. These nuclides do not exist because they have no more binding energy per nucleon than $^4$He. In other words, no energy would be released if two $^4$He nuclei were to fuse to form $^8$Be. Similarly, $^8$Be nuclei would require no input of energy to spontaneously decompose into $^4$He nuclei. Actually, $^8$Be nuclei never form in the first place.

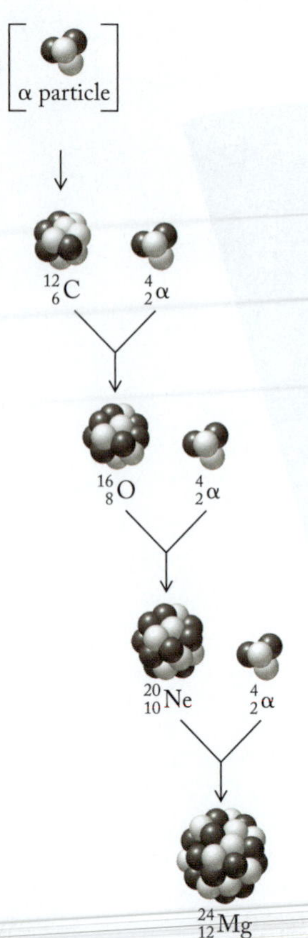


**FIGURE 21.11** Fusion of three α particles forms carbon-12, followed by fusion of successively more massive nuclei to form oxygen-16, neon-20, magnesium-24, and so on. The fusion processes release the energy that fuels the nuclear furnaces of stars today.

As the early universe continued to expand and cool, the kinetic energies of the particles of matter in it (hydrogen and helium nuclei and free electrons) decreased and their motions slowed, allowing nuclei to combine with free electrons to produce neutral atoms. The result was a universe that was 75% hydrogen and 25% helium. It remained that way for millions of years—until the first galaxies formed.

## Stellar Nucleosynthesis

Most of the matter detected and identified in the universe today is still hydrogen and helium, which is strong evidence supporting the Big Bang theory. But how did the other elements in the periodic table eventually form, including those that make up most of our planet? Scientists theorize that the synthesis of elements more massive than helium had to wait until nuclear fusion resumed in the first generation of stars. Inside the coalescing masses of hydrogen and helium that became the first stars, these gases underwent enormous compression heating, becoming hot enough to ignite the nuclear furnaces that are the source of the energy in all stars. Initially, hydrogen nuclei serve as the fuel for the stellar furnaces, combining in a series of fusion reactions that forms helium nuclei (α particles).

Some stars, known as *red giants*, have cores so extraordinarily hot and dense that sometimes three α particles collide simultaneously. When they do, they fuse, forming a stable nucleus that contains six protons and six neutrons, which is the most common isotope of carbon:

$$3\,^{4}_{2}\text{He} \rightarrow {}^{12}_{6}\text{C}$$

With the formation of $^{12}$C, the barrier that had halted primordial nucleosynthesis was overcome. In the cores of giant stars, $^{12}$C nuclei fuse with α particles to form $^{16}$O (**Figure 21.11**). Then $^{16}$O nuclei fuse with more α particles to form $^{20}$Ne, and so on. Additional fusion reactions involving nuclei with increasingly greater positive charges are possible in intensely hot ($10^9$ K) stars because the nuclei have enough kinetic energy, and are moving fast enough, to overcome the electrostatic repulsion experienced by particles with large positive charges.

For billions of years, fusion reactions of this sort have simultaneously fueled the nuclear furnaces of stars and produced isotopes as heavy as $^{56}$Fe (**Figure 21.12**). However, once the core of a star turns into iron, the star is in trouble because fusion reactions involving iron nuclei do not release energy; instead, the reactions consume it. This happens because the binding energy per nucleon (and therefore the nuclear stability) reaches a maximum with $^{56}$Fe (see Figure 21.8). Thus, a star with an iron core has essentially run out of fuel. Its nuclear furnace goes out, and the star begins to cool and collapse into itself.

As the star collapses, compression reheats its core to above $10^9$ K. At such temperatures, nuclei begin to disintegrate into free protons and neutrons. Free neutrons may collide and fuse with atomic nuclei in a process called **neutron capture**. If a stable nucleus captures enough neutrons, it becomes unstable. For example, if a nucleus of $^{56}$Fe captures three neutrons, it forms the unstable, neutron-rich nuclide $^{59}$Fe, which undergoes β decay to cobalt-59:

$$^{56}_{26}\text{Fe} + 3\,^{1}_{0}\text{n} \rightarrow {}^{59}_{26}\text{Fe} \rightarrow {}^{59}_{27}\text{Co} + {}^{0}_{-1}\beta$$

The combination of repeated neutron capture and β decay events in the cores of collapsing stars produces the most massive stable nuclides in the periodic table.

**neutron capture** the absorption of a neutron by a nucleus.

Eventually, the enormous heating that occurs when a massive star collapses produces a gigantic explosion—what cosmologists call a *supernova*. In addition to

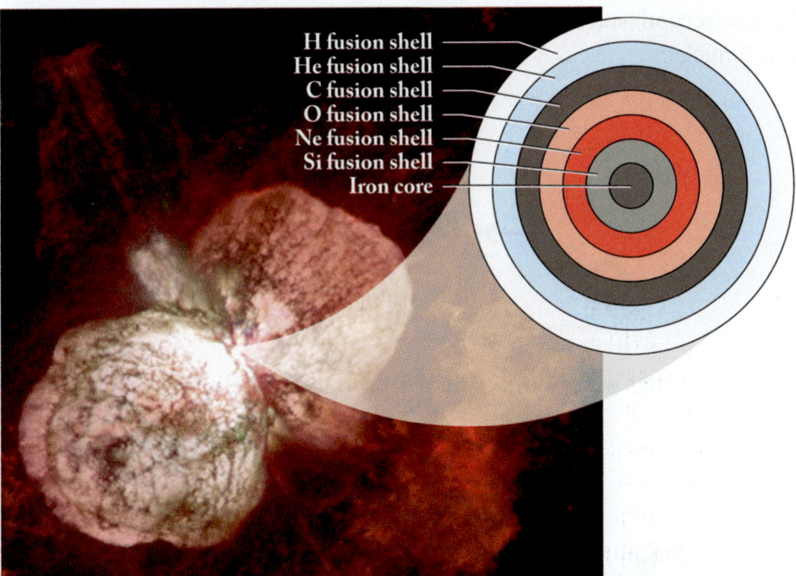


**FIGURE 21.12** The star Eta Carinae is believed to be evolving toward an explosion. The outer regions are still fueled by energy released as hydrogen isotopes fuse, but the star is increasingly hotter and denser closer to the center. This central heating allows the fusion of larger nuclei and produces $^{56}Fe$ in the core.

finishing the job of synthesizing the elemental building blocks found in the universe—up to and including the isotopes of uranium—a supernova serves as its own element-distribution system, blasting its inventory of elements throughout its galaxy (**Figure 21.13**). The legacies of supernovas are found in the elemental composition of later-generation stars such as our sun and in the planets that orbit the stars. Indeed, all the matter in our solar system—including us—is essentially debris from the demolition of ancient exploding stars.

**FIGURE 21.13** The Crab Nebula in the constellation Taurus is actually the debris field of a supernova that was observed on Earth in the year 1054.

## Nucleosynthesis in Our Sun

The energy of our sun is derived from a nuclear fusion process that involves more steps than the process that took place during primordial nucleosynthesis, when protons (hydrogen nuclei) and neutrons fused to form deuterons. This is because free neutron concentrations in the sun are far lower there than they were in the primordial universe. Instead, colliding protons may fuse to form a deuteron and a positron:

$$2\,{}^{1}_{1}\text{H} \rightarrow {}^{2}_{1}\text{D} + {}^{0}_{1}\beta \qquad (21.10)$$

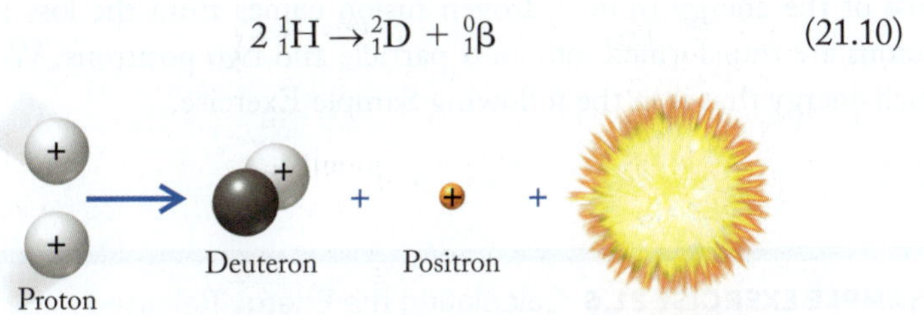


**CONCEPT TEST**

What particle is formed when a proton fuses with an electron?

In the second stage of solar fusion, protons fuse with deuterons to form helium-3 nuclei:

$$ {}_{1}^{1}\mathrm{H} + {}_{1}^{2}\mathrm{D} \rightarrow {}_{2}^{3}\mathrm{He} \qquad (21.11)$$

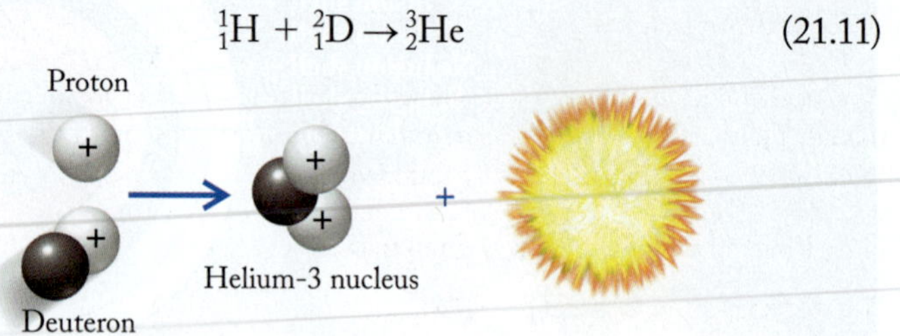


Finally, fusion of two helium-3 nuclei produces a helium-4 nucleus and two protons:

$$2\,{}_{2}^{3}\mathrm{He} \rightarrow {}_{2}^{4}\mathrm{He} + 2\,{}_{1}^{1}\mathrm{H} \qquad (21.12)$$

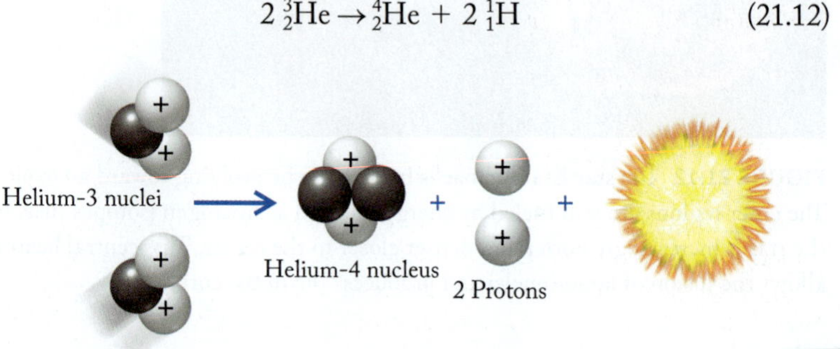


Deuterium and $^{3}$He nuclei are intermediates in the hydrogen-fusion process because they are made in one step but then consumed in another. To write an overall equation for solar fusion, we combine Equations 21.10–21.12, multiplying Equations 21.10 and 21.11 by 2 to balance the production and consumption of the intermediate particles:

$$\begin{array}{l} 2\,[2\,{}_{1}^{1}\mathrm{H} \rightarrow {}_{1}^{2}\mathrm{D} + {}_{1}^{0}\beta] \\ +\ 2\,[{}_{1}^{1}\mathrm{H} + {}_{1}^{2}\mathrm{D} \rightarrow {}_{2}^{3}\mathrm{He}] \\ +\ \quad 2\,{}_{2}^{3}\mathrm{He} \rightarrow {}_{2}^{4}\mathrm{He} + 2\,{}_{1}^{1}\mathrm{H} \\ \hline 4\,\cancel{6}\,{}_{1}^{1}\mathrm{H} + \cancel{2\,{}_{1}^{2}\mathrm{D}} + \cancel{2\,{}_{2}^{3}\mathrm{He}} \rightarrow \cancel{2\,{}_{1}^{2}\mathrm{D}} + \cancel{2\,{}_{2}^{3}\mathrm{He}} + {}_{2}^{4}\mathrm{He} + \cancel{2\,{}_{1}^{1}\mathrm{H}} + 2\,{}_{1}^{0}\beta \end{array}$$

This reduces to

$$4\,{}_{1}^{1}\mathrm{H} \rightarrow {}_{2}^{4}\mathrm{He} + 2\,{}_{1}^{0}\beta \qquad (21.13)$$

Annihilation reactions between the positrons produced in Equation 21.13 and electrons in the matter surrounding the reactants release considerable energy, but most of the energy from hydrogen fusion comes from the loss in mass as four protons are transformed into an α particle and two positrons. We calculate how much energy that is in the following Sample Exercise.

**SAMPLE EXERCISE 21.6** Calculating the Energy Released in a Nuclear Reaction **LO4**

How much energy in joules is released by the overall fusion process in which four protons undergo nuclear fusion, producing an α particle and two positrons?

**Collect and Organize** We are asked to calculate the energy released in Equation 21.13 above. The energies associated with nuclear reactions are related to differences in the masses of the reactant and product particles by Equation 21.9, $E = mc^2$.

**Analyze** Given the values of the masses in Table 21.1, the difference in mass will probably be less than $10^{-27}$ kg. When multiplied by the square of the speed of light $(2.998 \times 10^8 \text{ m/s})^2 \approx 10^{17}$, the calculated value of $E$ should be less than $10^{-10}$ J.

**Solve** First we calculate the change in mass:

$$\begin{aligned}\Delta m &= (m_{\alpha \text{ particle}} + 2\, m_{\text{positron}}) - 4\, m_{\text{proton}} \\ &= [6.64465 \times 10^{-27} + (2 \times 9.10939 \times 10^{-31})]\text{ kg} - (4 \times 1.67262 \times 10^{-27})\text{ kg} \\ &= -4.40081 \times 10^{-29}\text{ kg}\end{aligned}$$

The energy corresponding to this loss in mass is calculated using Equation 21.9, where $m = -4.40081 \times 10^{-29}$ kg:

$$\begin{aligned}E &= mc^2 \\ &= -4.40081 \times 10^{-29}\text{ kg} \times (2.998 \times 10^8\text{ m/s})^2 \\ &= -3.955 \times 10^{-12}\text{ kg} \cdot (\text{m/s})^2 = -3.955 \times 10^{-12}\text{ J}\end{aligned}$$

**Think About It** The decrease in mass translates into energy lost by the reaction system to its surroundings. As predicted, the absolute value of this energy is much less than $10^{-10}$ J, which seems like a very small value compared to the world's energy needs. However, this value applies to the formation of a single α particle. If we multiply it by the Avogadro constant and convert to kilojoules per mole, a unit we typically use in thermochemistry, we get

$$\frac{-3.955 \times 10^{-12}\text{ J}}{1\ \alpha\text{ particle}} \times \frac{6.022 \times 10^{23}\ \alpha\text{ particles}}{1\text{ mol}} \times \frac{1\text{ kJ}}{1000\text{ J}} = -2.382 \times 10^9\text{ kJ/mol}$$

To put this value in perspective, it is about $10^7$ times the change in free energy from the combustion of one mole of hydrogen gas.

**Practice Exercise** For decades, scientists and engineers have sought to harness the enormous energy released during hydrogen fusion for peaceful purposes by using high-speed collisions between deuterium and tritium ($^3$H) nuclei to produce helium-4 nuclei through Equation 21.14 below. How much energy is released in the nuclear reaction described by Equation 21.14? Express your answer in kJ/mol. The mass of a tritium nucleus is $5.00736 \times 10^{-27}$ kg.

$$^{2}_{1}\text{H} + {}^{3}_{1}\text{H} \rightarrow {}^{4}_{2}\text{He} + {}^{1}_{0}\text{n} \qquad (21.14)$$

## 21.6 Nuclear Fission

When an atom of uranium-235 captures a neutron, it does not undergo nuclear fusion. Instead, the nucleus of the unstable product, uranium-236, splits into two lighter nuclei in a process called **nuclear fission**. Several uranium-235 fission reactions can occur, including these three:

$$^{235}_{92}\text{U} + {}^{1}_{0}\text{n} \rightarrow {}^{141}_{56}\text{Ba} + {}^{92}_{36}\text{Kr} + 3\,{}^{1}_{0}\text{n}$$

$$^{235}_{92}\text{U} + {}^{1}_{0}\text{n} \rightarrow {}^{137}_{52}\text{Te} + {}^{97}_{40}\text{Zr} + 2\,{}^{1}_{0}\text{n}$$

$$^{235}_{92}\text{U} + {}^{1}_{0}\text{n} \rightarrow {}^{138}_{55}\text{Cs} + {}^{96}_{37}\text{Rb} + 2\,{}^{1}_{0}\text{n}$$

In all these reactions, the sums of the masses of the products are slightly less than the sums of the masses of the reactants. As with hydrogen fusion, this difference in mass is converted to energy.

**nuclear fission** a nuclear reaction in which the nucleus of an element splits into two lighter nuclei, usually accompanied by the release of one or more neutrons and energy.

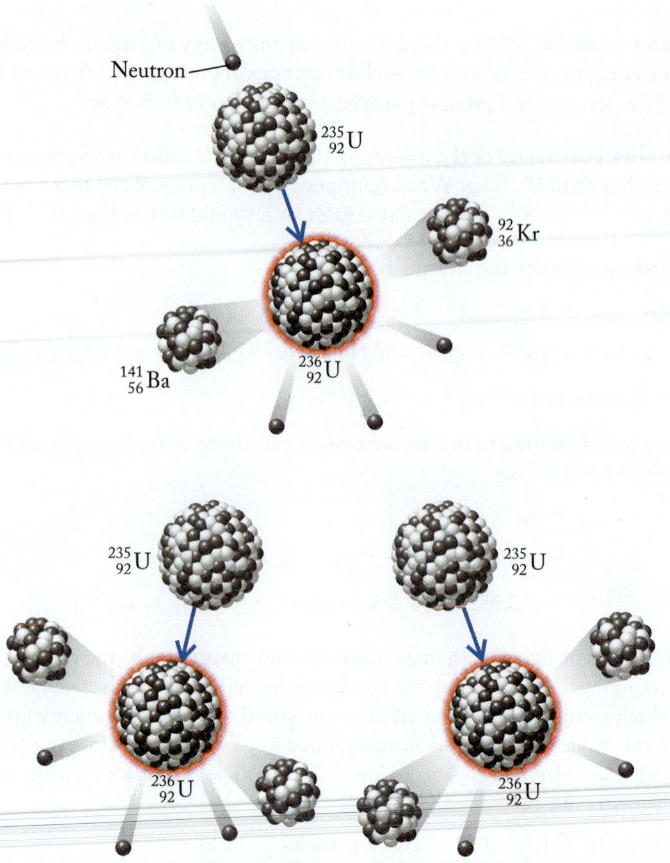


**FIGURE 21.14** Each fission event in the chain reaction of a uranium-235 nucleus begins when the nucleus captures a neutron, forming an unstable uranium-236 nucleus that then splits apart (fissions) in one of several ways. In the first process shown here, the uranium-236 nucleus splits into krypton-92, barium-141, and three neutrons. If, on average, at least one of the three neutrons from each fission event causes the fission of another uranium-235 nucleus, then the process is sustained in a chain reaction.

**STEPWISE ANIMATION**

Induced Fission and Chain Reactions

Fission reactions may also produce additional neutrons that smash into other uranium-235 nuclei and initiate more fission events in a **chain reaction** (**Figure 21.14**). The reaction proceeds if enough uranium-235 nuclei are present to absorb the neutrons being produced. On average, at least one neutron from each fission event must cause another nucleus to split apart for the chain reaction to be self-sustaining. The quantity of fissionable material needed to ensure that every fission event produces another is called the **critical mass**. For uranium-235, the critical mass is about 1 kg of the pure isotope.

Uranium-235 is the most abundant fissionable isotope, but it makes up only 0.72% of the uranium in the principal uranium ore, pitchblende (**Figure 21.15a**). The uranium in nuclear reactors must be at least 3% to 4% uranium-235, and enrichment to about 85% is needed for nuclear weapons. The most common method for enriching uranium ore involves extracting the uranium in a process that yields a material called yellowcake, which is mostly $U_3O_8$ (**Figure 21.15b**). This oxide is then converted to $UF_6$, which despite having a molar mass more than 300 g/mol is a volatile solid that sublimes at 56°C. The volatility of this nonpolar molecular compound can be explained by the relatively weak London dispersion forces experienced by its compact, symmetrical molecules (**Figure 21.16**). Fissionable $^{235}UF_6$ is separated from $^{238}UF_6$ on the basis of their slightly different densities. Elaborate centrifuge systems are used to exploit this difference and speed up the separation (**Figure 21.15c**).

Harnessing the energy released by nuclear fission to generate electricity began in the mid-20th century. In a typical nuclear power plant (**Figure 21.17**), fuel

**chain reaction** a self-sustaining series of fission reactions in which the neutrons released when nuclei split apart initiate additional fission events and sustain the reaction.

**critical mass** the minimum quantity of fissionable material needed to sustain a chain reaction.

**breeder reactor** a nuclear reactor in which fissionable material is produced during normal reactor operation.

(a)

(b)

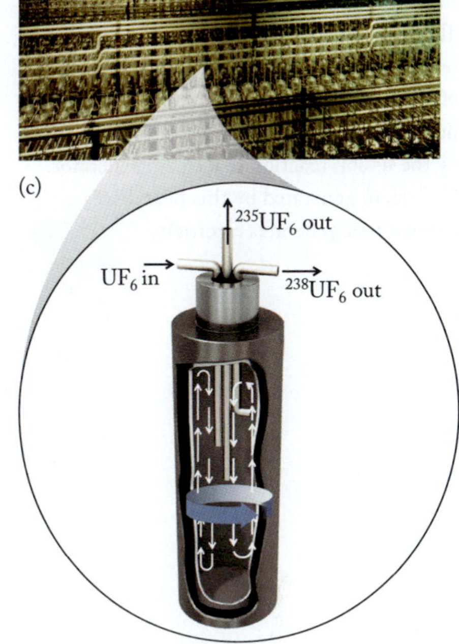


(c)

**FIGURE 21.15** Preparing uranium fuel. (a) A piece of pitchblende, source of the uranium fuel for nuclear reactors. (b) Pitchblende ore is ground up and extracted with strong acid. The uranium compounds (mostly $U_3O_8$) obtained from the extract are called yellowcake. (c) Uranium oxides are converted to volatile $UF_6$, which is centrifuged at very high speed to separate $^{235}UF_6$ from $^{238}UF_6$. The less dense and less abundant $^{235}UF_6$ is enriched near the center of the centrifuge cylinder and separated from the heavier $^{238}UF_6$.

rods containing 3% to 4% uranium-235 are interspersed with rods of boron or cadmium that control the rate of the fission by absorbing some of the neutrons produced during fission. Pressurized water flows around the fuel and control rods, removing the heat created during fission and transferring it to a steam generator. The water also acts as a moderator, slowing down the neutrons and thereby allowing for their more efficient capture by $^{235}U$ atoms.

In 1952, the first **breeder reactor** was built, so named because in addition to producing energy to make electricity, the reactor makes ("breeds") its own fuel. The reactor starts out with a mixture of plutonium-239 and uranium-238. As the plutonium fissions and the energy from those reactions is collected to produce electricity, some of the neutrons produced sustain the fission chain reaction just as in the reactor depicted in Figure 21.17, whereas others convert the uranium into more plutonium fuel:

$$^{238}_{92}U + {}^{1}_{0}n \rightarrow {}^{239}_{92}U + \gamma \rightarrow {}^{239}_{94}Pu + 2\ {}^{0}_{-1}\beta$$

In less than 10 years of operation, a breeder reactor can make enough plutonium-239 to refuel itself and another reactor. Unfortunately, only about half a kilogram of plutonium-239 is needed to make an atomic bomb, and it has a long half-life: $2.4 \times 10^4$ years. Understandably, extreme caution and tight security surround the handling of plutonium fuel and the transportation and storage of nuclear wastes containing even small amounts of plutonium. Health and safety concerns related to reactor operation and spent-fuel disposal are the principal reasons why the United States has no breeder power stations, although they have been built in at least seven other countries.

**FIGURE 21.16** Uranium hexafluoride is a volatile solid that sublimes at a relatively low temperature of 56°C because it is composed of compact, symmetrical molecules that experience relatively weak London dispersion forces despite their considerable mass.

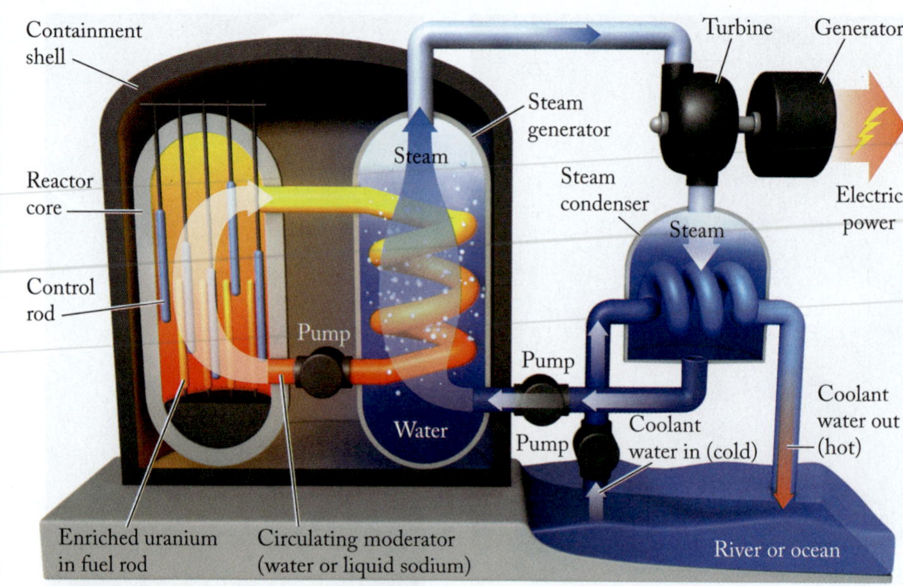


**FIGURE 21.17** A pressurized, water-cooled nuclear power plant uses fuel rods containing uranium enriched to about 4% uranium-235. The fission chain reaction is regulated with control rods and a moderator that is either water or liquid sodium. The moderator slows down the neutrons released by fission so that they can be more efficiently captured by other uranium-235 nuclei. It also transfers the heat produced by the fission reaction to a steam generator. The steam generated by this heat drives a turbine that generates electricity.

**CONCEPT TEST**

Nuclear reactors powered by the energy released by the fission of uranium-235 have been operating since the 1950s, but a reactor powered by the energy released by the fusion of hydrogen has yet to be built. Why is building a fusion reactor taking so long?

## 21.7 Measuring Radioactivity

French scientist Henri Becquerel discovered radioactivity in 1896 when he observed that uranium and other substances produce radiation that fogs photographic film. Photographic film is still used to detect radioactivity in the film dosimeter badges worn by people working with radioactive materials to record their exposure to radiation. Detectors called **scintillation counters** use materials called *phosphors* to absorb energy released during radioactive decay. The phosphors then release the absorbed energy as visible light, the intensity of which is a measure of the amount of radiation in the sample.

**STEPWISE ANIMATION**
Geiger Counter

Radioactivity also can be measured with **Geiger counters**, which detect the common products of radioactivity—$\alpha$ particles, $\beta$ particles, and $\gamma$ rays—on the basis of their abilities to ionize atoms (**Figure 21.18**). A sealed metal cylinder, filled with gas (usually argon) and a positively charged electrode, has a window that allows nuclear radiation to enter. Once inside the cylinder, the particles and $\gamma$ rays ionize argon atoms into $Ar^+$ ions and free electrons. If an electrical potential difference is applied between the cylinder shell and the central electrode, free electrons migrate toward the positive electrode, whereas argon ions migrate toward the negatively charged shell. This ion migration produces a pulse of electrical current whenever radiation enters the cylinder. The current is amplified and read out to a meter and a speaker that makes a clicking sound.

**scintillation counter** an instrument that determines the level of radioactivity in samples by measuring the intensity of light emitted by phosphors in contact with the samples.

**Geiger counter** a portable device for determining nuclear radiation levels by measuring how much the radiation ionizes the gas in a sealed detector.

**becquerel (Bq)** the SI unit of radioactivity; 1 Bq = 1 decay event per second.

**curie (Ci)** non-SI unit of radioactivity; 1 Ci = $3.70 \times 10^{10}$ decay events per second.

One measure of radioactivity in a sample is the number of decay events per unit time. This parameter is called the radioactivity ($A$) of the sample. The SI unit of radioactivity is the **becquerel (Bq)**, named in honor of Henri Becquerel and

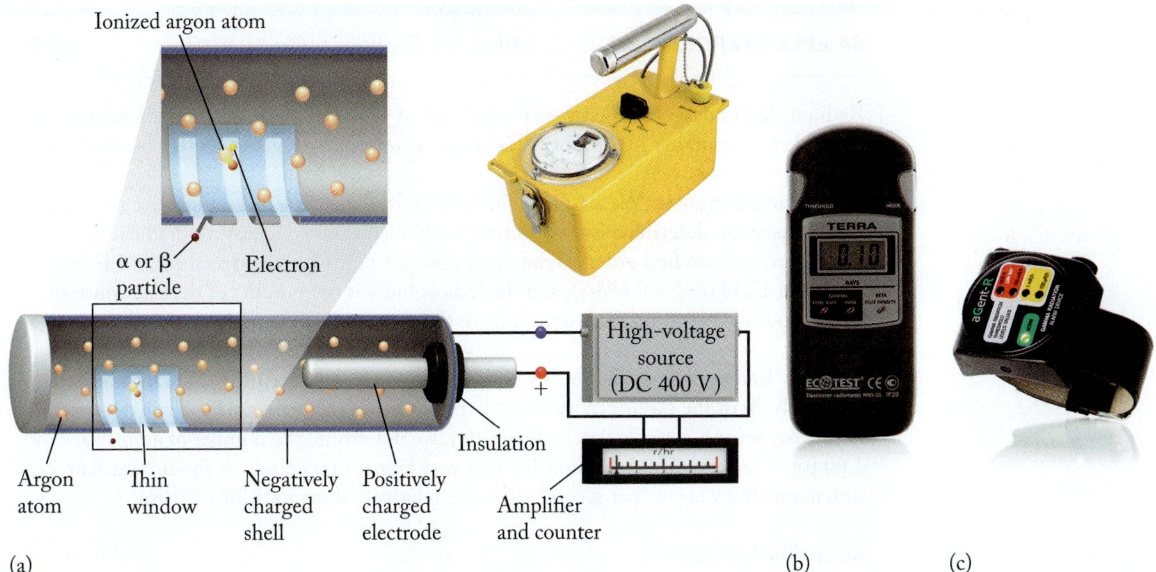


**FIGURE 21.18** (a) In a Geiger counter, a particle produced by radioactive decay passes through a thin window usually made of beryllium or a plastic film. Inside the tube, the particle collides with atoms of argon gas and ionizes them. The resulting argon cations migrate toward the negatively charged tube housing, and the electrons migrate toward a positive electrode, creating a pulse of current through the tube. The current pulses are amplified and recorded through a meter and a speaker that produces an audible "click" for each pulse. (b) When radiation hits the silicon dioxide detector in this dosimeter, electrons are ejected, creating holes. The change in conductivity of the $SiO_2$ correlates with the amount of radiation absorbed. (c) This device looks like a wristwatch but is actually a γ-ray detector, illustrating advances in the miniaturization of radiation detection equipment.

equal to one decay event per second. An older radioactivity unit is the **curie (Ci)**, named in honor of Marie and Pierre Curie, where

$$1\ \text{Ci} = 3.70 \times 10^{10}\ \text{Bq} = 3.70 \times 10^{10}\ \text{decay events/s}$$

Both the becquerel and the curie quantify the rate at which a radioactive substance decays, which provides a measure of how much of it is in a sample.

As we noted in Section 21.2, all decay processes follow first-order kinetics. Recall from Section 13.3 that the rate constant $k$ of a first-order reaction is related to the concentration of a reactant R according to the equation

$$\text{Rate} = k[\text{R}]$$

**STEPWISE ANIMATION**
Activity Example

The same mathematical relationships apply to radioactive decay processes, except that we refer to radioactivity ($A$) instead of rate of decay and to the number of atoms ($N$) of a radionuclide in a sample instead of its concentration:

$$A = kN \tag{21.15}$$

Because radioactivity is the number of decay events per second, the units of the decay rate constant $k$ are decay events per atom per second. Scientists usually express the quantity of a radionuclide in a sample in terms of its radioactivity because it defines how the substance may be used and how it should be handled. This allows us to substitute radioactivity, $A$, for $N$ in Equation 21.8:

$$t = -\frac{t_{1/2}}{0.693} \ln \frac{A_t}{A_0} \tag{21.16}$$

The radiocarbon dating methods described in Section 21.2 rely on Equation 21.16 and the measured radioactivity of $^{14}C$ in samples to determine their ages.

**SAMPLE EXERCISE 21.7** Calculating the Radioactivity of a Sample **LO7**

Radium-223 undergoes β decay with a half-life of 11.4 days. What is the radioactivity of a sample that contains 1.00 μg of $^{223}$Ra? Express your answer in becquerels and in curies.

**Collect and Organize** We are given the half-life and quantity of a radioactive substance and are asked to determine its radioactivity—that is, its rate of decay. All radioactive decay processes are first order, so the decay rate constant $k$ is related to the half-life by Equation 13.19 ($t_{1/2} = 0.693/k$), and the radioactivity is the product of the rate constant and the number of atoms of radionuclide in the sample (Equation 21.15; $A = kN$).

**Analyze** Before using Equation 13.19, we must convert the half-life units into seconds because both of the radioactivity units that we need to calculate are based on decay events per second. To calculate radioactivity, we determine the number of atoms in 1.00 μg of radium. This will probably be a very large number, which should translate into many decay events per second given the relatively short half-life of $^{223}$Ra.

**Solve** The half-life is

$$11.4\ \text{d} \times \frac{24\ \text{h}}{1\ \text{d}} \times \frac{60\ \text{min}}{1\ \text{h}} \times \frac{60\ \text{s}}{1\ \text{min}} = 9.85 \times 10^5\ \text{s}$$

Using this value in Equation 13.19 and solving for $k$,

$$k = \frac{0.693}{9.85 \times 10^5\ \text{s}}$$

$$= 7.04 \times 10^{-7}\ \text{s}^{-1} = 7.04 \times 10^{-7}\ \text{decay events/(atom} \cdot \text{s)}$$

The number of atoms ($N$) of $^{223}$Ra is

$$N = 1.00\ \mu\text{g} \times \frac{1\ \text{g}}{10^6\ \mu\text{g}} \times \frac{1\ \text{mol Ra}}{223\ \text{g Ra}} \times \frac{6.022 \times 10^{23}\ \text{atoms Ra}}{1\ \text{mol Ra}} = 2.70 \times 10^{15}\ \text{atoms Ra}$$

Inserting these values of $k$ and $N$ into Equation 21.15,

$$A = kN = \frac{7.04 \times 10^{-7}\ \text{decay events}}{\text{atom} \cdot \text{s}} \times 2.70 \times 10^{15}\ \text{atoms Ra} = 1.90 \times 10^9\ \text{decay events/s}$$

Because 1 Bq = 1 decay event/s, the radioactivity of the sample is $1.90 \times 10^9$ Bq. Expressing radioactivity in curies,

$$\frac{1.90 \times 10^9\ \text{decay events}}{\text{s}} \times \frac{1\ \text{Ci}}{3.70 \times 10^{10}\ \text{decay events/s}} = 0.0514\ \text{Ci}$$

**Think About It** We expected many decay events per second because the sample contained many radioactive atoms, which resulted in a high level of radioactivity.

**Practice Exercise** In March 2011, an earthquake and tsunami off the coast of northeastern Japan crippled nuclear reactors at a power station in Fukushima. The resulting explosions and fires released $^{133}$Xe into the atmosphere. Determine the radioactivity in 1.00 μg of this radionuclide ($t_{1/2} = 5.25$ d) in becquerels and in millicuries.

## 21.8 Biological Effects of Radioactivity

The γ rays and many of the α and β particles produced by nuclear reactions have more than enough energy to tear chemical bonds apart, producing odd-electron radicals or free electrons and cations. Consequently, these rays and particles are classified as **ionizing radiation**. Other examples are X-rays and short-wavelength ultraviolet rays. The ionization of atoms and molecules in living tissue can lead to radiation sickness, cancer, birth defects, and death. Not aware of these hazards, some of the scientists who first worked with radioactive materials suffered for it. Marie Curie died of aplastic anemia caused by her many years of radiation exposure, and leukemia claimed her daughter Irène Joliot-Curie, who continued the research program her parents started.

In medicine, the term *ionizing radiation* is limited to photons and particles that have enough energy to remove an electron from water, forming a radical cation:

$$H_2O(\ell) \xrightarrow{1216\ \text{kJ/mol}} H_2O^+(aq) + e^-$$

The logic behind this definition is that the human body is composed largely of water. Therefore, water molecules are the most abundant ionizable targets when we are exposed to nuclear radiation. The radical cation reacts with another water molecule in the body to form a hydronium ion and a hydroxyl free radical:

$$H_2O^+(aq) + H_2O(\ell) \rightarrow H_3O^+(aq) + OH(aq)$$

The rapid reactions of free radicals with biomolecules can threaten the integrity of cells and the health of the entire organism.

Radiation-induced alterations to the biochemical machinery that controls cell growth are most likely to occur in tissues in which cells grow and divide rapidly. One such tissue is bone marrow, where billions of white blood cells are produced each day to fortify the body's immune system. Molecular damage to bone marrow can lead to leukemia, an uncontrolled production of nonfunctioning white blood cells that spread throughout the body, crowding out healthy cells. Ionizing radiation can also cause molecular alterations in the genes and chromosomes of sperm and egg cells, increasing the chances of birth defects in offspring.

**ionizing radiation** high-energy products of radioactive decay that can ionize molecules.

**gray (Gy)** the SI unit of absorbed radiation; 1 Gy = 1 J/kg of tissue.

**relative biological effectiveness (RBE)** a factor that accounts for the differences in physical damage caused by different types of radiation.

**sievert (Sv)** SI unit used to express the amount of biological damage caused by ionizing radiation.

## Radiation Dosage

The biological impact of ionizing radiation depends on how much of it an organism absorbs. If the radiation is coming from one radioactive source, then the amount absorbed depends on the radioactivity of the source and the energy of the radiation that is produced per decay event. Tables of radioactive isotopes often include information about their modes of decay and the energies of the particles and gamma rays they emit.

*Absorbed dose* is the quantity of ionizing radiation absorbed by a unit mass of living tissue. The SI unit of absorbed dose is the **gray (Gy)**. One gray is equal to the absorption of 1 J of radiation energy per kilogram of body mass:

$$1\ \text{Gy} = 1\ \text{J/kg}$$

Grays express dosage, but they do not indicate the amount of *tissue damage* that dosage causes. Different products of nuclear reactions affect living tissue differently. Exposure to 1 Gy of $\gamma$ rays produces about the same amount of tissue damage as exposure to 1 Gy of $\beta$ particles. However, 1 Gy of $\alpha$ particles, which move about 10 times more slowly than $\beta$ particles but have nearly $10^4$ times the mass, causes 20 times as much damage as 1 Gy of $\gamma$ rays. Neutrons cause 3–5 times as much damage as $\gamma$ rays. To account for these differences, values of **relative biological effectiveness (RBE)** have been established for the various forms of ionizing radiation (**Table 21.4**). When the absorbed dose in grays is multiplied by an RBE factor, the product is called the *effective* dose, a measure of tissue damage. The SI unit of effective dose is the **sievert (Sv)**.

**TABLE 21.4** RBE[a] Values of Nuclear Radiation

| Radiation | RBE |
|---|---|
| $\gamma$ Rays | 1.0 |
| $\beta$ Particles | 1.0–1.5 |
| Neutrons | 3–5 |
| Protons | 10 |
| $\alpha$ Particles | 20 |

[a]RBE, relative biological effectiveness.

**TABLE 21.5** Units for Expressing Quantities of Ionizing Radiation

| Parameter | SI Unit | Description | Alternative Common Unit | Description |
|---|---|---|---|---|
| Radioactivity | Becquerel (Bq) | 1 decay event/s | Curie (Ci) | $3.70 \times 10^{10}$ decay events/s |
| Ionizing energy absorbed | Gray (Gy) | 1 J/kg of tissue | Rad | 0.01 J/kg of tissue |
| Amount of tissue damage | Sievert (Sv) | 1 Gy × RBE[a] | Rem | 1 rad × RBE |

[a]RBE, relative biological effectiveness.

**Table 21.5** summarizes the various units used to express quantities of radiation and their biological impact. Two non-SI units are listed that predate their SI counterparts but are still often used. They are *radiation absorbed dose*, or *rad*, which is equivalent to 0.01 Gy, and the *rem* for tissue damage, which is an acronym for *roentgen equivalent man*. One rem is the product of one rad of ionization times the appropriate RBE factor. One hundred rems are in 1 Sv.

**STEPWISE ANIMATION**
Radiation Penetration

The RBE of 20 for α particles may lead you to believe that these particles pose the greatest health threat from radioactivity. Not exactly. Alpha particles are so big that they have little penetrating power; they are stopped by a sheet of paper, clothing, or even a layer of dead skin (**Figure 21.19**). However, if you ingest or inhale an α emitter, tissue damage can be severe because the relatively massive α particles do not have to penetrate far to cause cell damage. Gamma rays are considered the most dangerous form of radiation emanating from a source outside the body because they have the greatest penetrating power.

The effects of exposure to different single effective doses of radiation are summarized in **Table 21.6**. To put these data in perspective, the effective dose from a typical dental X-ray is about 25 μSv, or about 1/2000 of the lowest exposure level cited in the table.

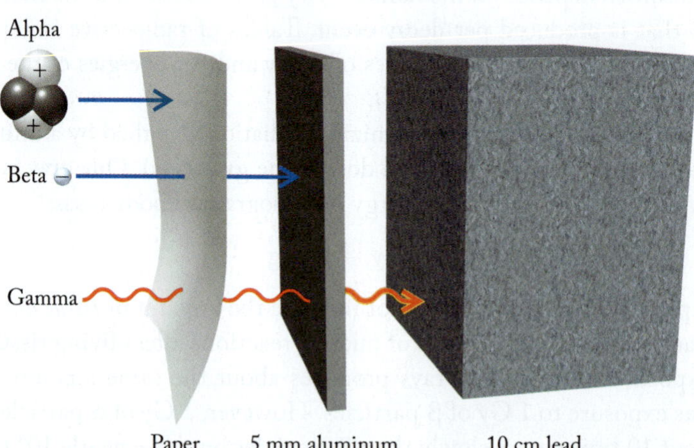


**FIGURE 21.19** The tissue damage caused by α particles, β particles, and γ rays depends on their ability to penetrate materials shielding you from their source. Alpha particles are stopped by paper or clothing but are extremely dangerous if formed inside the body because they do not have to travel far to cause cell damage. Stopping gamma rays requires a thick layer of lead or several meters of concrete or soil.

**TABLE 21.6** Acute Effects of Single Whole-Body Effective Doses of Ionizing Radiation

| Effective Dose (Sv) | Toxic Effect |
|---|---|
| 0.05–0.25 | No acute effect, possible carcinogenic or mutagenic damage to DNA |
| 0.25–1.0 | Temporary reduction in white blood cell count |
| 1.0–2.0 | Radiation sickness: fatigue, vomiting, diarrhea, impaired immune system |
| 2.0–4.0 | Severe radiation sickness: intestinal bleeding, bone marrow destruction |
| 4.0–10.0 | Death, usually through infection, within weeks |
| >10.0 | Death within hours |

**FIGURE 21.20** The ruins of the nuclear reactor at Chernobyl, Ukraine, which exploded in 1986.

Widespread exposure to very high levels of radiation occurred after the 1986 explosion at the Chernobyl nuclear reactor in what is now Ukraine (**Figure 21.20**). Many plant workers and first responders were exposed to more than 1.0 Sv of radiation. At least 30 of them died in the weeks after the accident. Many of the more than 300,000 workers who cleaned up the area around the reactor exhibited symptoms of radiation sickness, and at least 5 million people in Ukraine, Belarus, and Russia were exposed to fallout in the days after the accident. Studies conducted in the early 1990s uncovered high incidences of thyroid cancer in children in southern Belarus due to $^{131}I$ released in the Chernobyl accident, and children born in the region nearly a decade after the accident had unusually high rates of mutations in their DNA because of their parents' exposure to ionizing radiation. Genetic damage was also widespread among plants and animals living in the region.

Radiation exposure was not confined to Ukraine and Belarus. After the accident, a cloud of radioactive material spread rapidly across northern Europe, and within 2 weeks increased levels of radioactivity were detected throughout the Northern Hemisphere (**Figure 21.21**). The accident produced a global increase in human exposure to ionizing radiation estimated to be equivalent to 0.05 mSv per year.

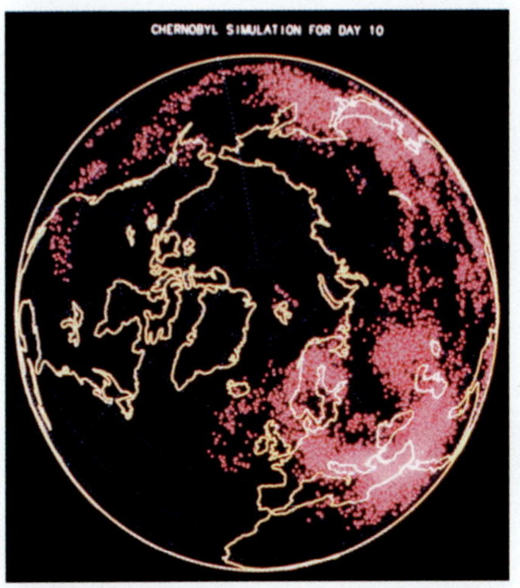


**FIGURE 21.21** Radioactive fallout (shown in pink) from the Chernobyl accident in 1986 was detected throughout the Northern Hemisphere.

An earthquake and tsunami in March 2011 crippled nuclear reactors at a power station in Fukushima, Japan, which released about 1/3 of the radiation released by the Chernobyl disaster (**Figure 21.22**). No fatalities due to acute exposure occurred, but as many as 500 deaths may someday be traceable to diseases caused by exposure to radiation released from the Fukushima reactors.

**FIGURE 21.22** The cleanup after the 2011 accident at the Fukushima nuclear power plant included washing away radioactive debris and dust from streets and sidewalks. The worker in the background is monitoring the level of the radioactivity.

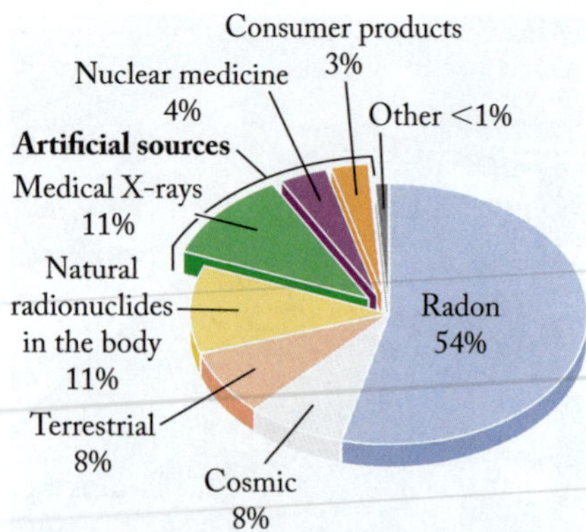


**FIGURE 21.23** Sources of radiation exposure of the U.S. population. On average, a person living in the United States is exposed to 0.0036 Sv of radiation each year. More than 80% of this exposure comes from natural sources—mainly radon in the air and water. Artificial sources account for about 18% of the total exposure.

## Evaluating the Risks of Radiation

To put global radiation exposure from Chernobyl and Fukushima in perspective, we need to consider typical annual exposure levels. For many people, the principal source of radiation is radon gas in indoor air and in well water (**Figure 21.23**). Like all noble gases, radon is chemically inert. Unlike the others, all its isotopes are radioactive. The most common isotope, radon-222, is produced when uranium-238 in rocks and soil decays to lead-206 (see Figure 21.2). The radon gas formed in this decay series percolates upward and can enter a building through cracks and pores in its foundation.

If you breathe radon-contaminated air and then exhale before it decays, no harm is done. If radon-222 decays inside the lungs, however, it emits an α particle that can damage lung tissue. The nuclide produced by the α decay of $^{222}Rn$ is radioactive polonium-218, which may become attached to tissue in the respiratory system and undergo a second α decay, forming lead-214:

$$^{222}_{86}Rn \rightarrow {}^{218}_{84}Po + {}^{4}_{2}\alpha \qquad t_{1/2} = 3.8 \text{ d}$$

$$^{218}_{84}Po \rightarrow {}^{214}_{82}Pb + {}^{4}_{2}\alpha \qquad t_{1/2} = 3.1 \text{ min}$$

As we have seen, α particles are the most damaging product of nuclear decay when formed inside the body. How big a threat does radon pose to human health? Concentrations of indoor radon depend on local geology (**Figure 21.24**) and on how gastight building foundations are. Many buildings contain concentrations of radon in the range of 1 pCi per liter of air. How hazardous are such tiny concentrations? No simple answer is evident. The U.S. Environmental Protection Agency has established 4 pCi/L as an "action level," meaning that people occupying houses with higher concentrations should take measures to minimize their exposure.

This action level is based on studies of the incidence of lung cancer in workers in uranium mines. These workers are exposed to radon concentrations (and concentrations of other radionuclides) that are much higher than the concentrations in homes and other buildings. However, many scientists believe that people exposed to very low levels of radon for many years are as much at risk as miners exposed to high levels of radiation for shorter periods. Some researchers use a model that assumes a linear relation between radon exposure

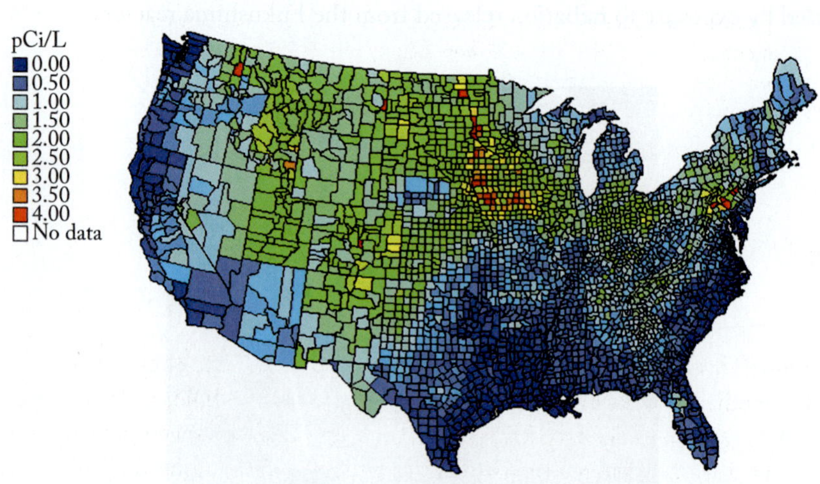


**FIGURE 21.24** Levels of radon gas in soils and rocks across the United States.

and the risk of lung cancer. This model is represented by the red line in **Figure 21.25**. According to this dose–response model, an estimated 15,000 Americans die of lung cancer each year because of exposure to indoor radon. This number comprises 10% of all lung-cancer fatalities and 30% of those among nonsmokers.

Is this linear model valid? Perhaps—but some scientists believe that a threshold exposure may exist, below which radon poses no significant threat to public health. They advocate an S-shaped dose–response curve, shown by the blue line in Figure 21.25. Notice that the risk of death from cancer in the S-shaped curve is much lower than in the linear response model at low radiation exposure but rises rapidly above a critical value.

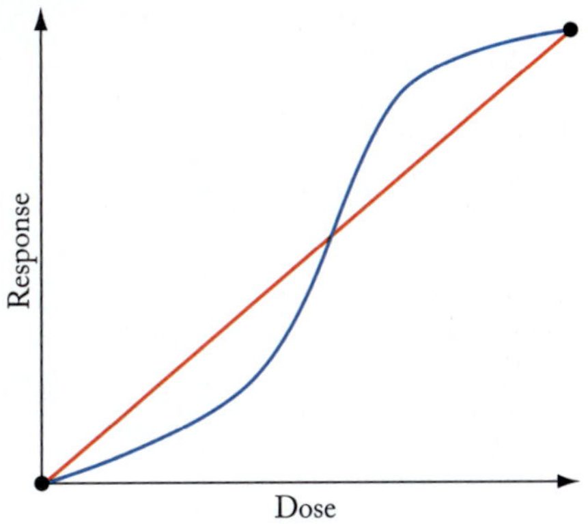


**FIGURE 21.25** The risk of death from radiation-induced cancer may follow one of two models. In the linear response model (red line), risk is directly proportional to the radiation exposure. In the S-shaped model (blue line), risk remains low below a critical threshold and then increases rapidly as the exposure increases. In the S-shaped model, the risk is less than that for the linear model at low doses but is higher at higher doses.

**SAMPLE EXERCISE 21.8** Calculating Effective Dose **LO8**

A person living in a home where the air radon concentration is 4.0 pCi/L receives an annual absorbed dose of ionizing radiation equivalent to an estimated 0.40 mGy. What is the person's annual effective dose, in millisieverts, from this radon? Use information from Figure 21.23 to compare this annual effective dose of radon with the average annual effective dose of radon estimated for persons living in the United States.

**Collect, Organize, and Analyze** We are given an absorbed radiation dose of 0.40 mGy. Radon isotopes emit $\alpha$ particles, which we know from Table 21.4 have a relative biological effectiveness of 20. The effective dose caused by an absorbed dose of ionizing radiation is the absorbed dose multiplied by the RBE of the radiation.

**Solve**

$$0.40 \text{ mGy} \times 20 = 8.0 \text{ mSv}$$

The caption for Figure 21.23 tells us that the average American is exposed to 3.6 mSv of radiation per year, with 54% of that amount, or 1.9 mSv, from radon. The person living in the home described in the problem receives an effective dose from radon that is slightly more than four times the average value.

**Think About It** The calculated value is more than twice the average annual effective dose of 3.6 mSv from all sources of radiation. The U.S. National Research Council has estimated that a nonsmoker living in air contaminated with 4.0 pCi/L of radon has a 1% chance of dying from lung cancer due to this exposure. The cancer risk for a smoker is close to 5%.

**Practice Exercise** A dental X-ray for imaging impacted wisdom teeth produces an effective dose of 15 $\mu$Sv. If a dental X-ray machine emits X-rays with an energy of $6.0 \times 10^{-17}$ J each, how many of these X-rays must be absorbed per kilogram of tissue to produce an effective dose of 15 $\mu$Sv? Assume that the RBE of these X-rays is 1.2.

## 21.9 Medical Applications of Radionuclides

Radionuclides are used to both detect and treat diseases, and they are key agents in the medical fields of diagnostic and therapeutic radiology. In diagnostic radiology, radionuclides are used alongside magnetic resonance imaging (MRI) and other imaging systems that involve only nonionizing radiation. Therapeutic radiology, however, is based almost entirely on the ionizing radiation that comes from nuclear processes.

## Therapeutic Radiology

Because ionizing radiation causes the most damage to cells that grow and divide rapidly, it is a powerful tool in the fight against cancer. In the early days of radiochemistry, radium was prescribed in cancer treatment and for ailments that included skin lesions, lethargy, and arthritis. Modern radiation therapy consists of exposing cancerous tissue to $\gamma$ radiation. Often the radiation source is external to the patient, but sometimes it is encased in a platinum capsule and surgically implanted in a cancerous tumor. The platinum provides a chemically inert outer layer and acts as a filter, absorbing $\alpha$ and $\beta$ particles emitted by the radionuclide but allowing $\gamma$ rays to pass into the tumor.

A nuclide's chemical properties can be exploited to direct it to a tumor site. For example, most iodine in the body is concentrated in the thyroid gland, so an effective therapy against thyroid cancer starts with the ingestion of potassium iodide containing radioactive iodine-131. Some radionuclides used in cancer therapy are listed in **Table 21.7**.

Surgically inaccessible tumors can be treated with beams of $\gamma$ rays from a radiation source outside the body. Unfortunately, $\gamma$ radiation destroys both cancer cells and healthy ones. Thus, patients receiving radiation therapy often suffer symptoms of radiation sickness, including nausea and vomiting (the tissues that make up intestinal walls are especially susceptible to radiation-induced damage), fatigue, weakened immune response, and hair loss. To reduce the severity of these side effects, radiologists must carefully control the dosage a patient receives.

## Diagnostic Radiology

The transport of radionuclides in the body and their accumulation in certain organs serve as ways to assess organ function. A tiny quantity of a radioactive isotope is used, together with a much larger amount of a stable isotope of the same element. The radioactive isotope is called a *tracer*, and the stable isotope is the *carrier*. For example, the circulatory system can be imaged by injecting into the blood a solution of sodium chloride containing a trace amount of $^{24}NaCl$. Circulation is monitored by measuring the $\gamma$ rays emitted by $^{24}Na$ as it decays.

The ideal isotope for medical imaging is one that has a half-life about equal to the length of time required to perform the imaging measurements. It should emit moderate-energy $\gamma$ rays but no $\alpha$ particles or $\beta$ particles that might cause tissue damage. Sodium-24 (a $\gamma$ emitter with a half-life of 15 hours) meets both these criteria. **Table 21.8** lists several other radionuclides used in medical imaging.

**TABLE 21.7** Some Radionuclides Used in Radiation Therapy

| Nuclide | Radiation | Half-Life | Treatment |
|---|---|---|---|
| $^{32}P$ | $\beta$ | 14.3 d | Leukemia therapy |
| $^{60}Co$ | $\beta, \gamma$ | 5.3 yr | Cancer therapy |
| $^{131}I$ | $\beta$ | 8.1 d | Thyroid therapy |
| $^{131}Cs$ | $\gamma$ | 9.7 d | Prostate cancer therapy |
| $^{192}Ir$ | $\beta, \gamma$ | 74 d | Coronary disease |

**TABLE 21.8** Selected γ-Emitting Radionuclides Used for Medical Imaging

| Nuclide | Half-Life (h) | Use |
|---|---|---|
| $^{67}Ga$ | 78 | Tumors in the brain and other organs |
| $^{99}Tc$ | 6.0 | Bones, circulatory system, various organs |
| $^{123}I$ | 13.3 | Thyroxine production in thyroid gland |
| $^{201}Tl$ | 73 | Coronary arteries, heart muscle |

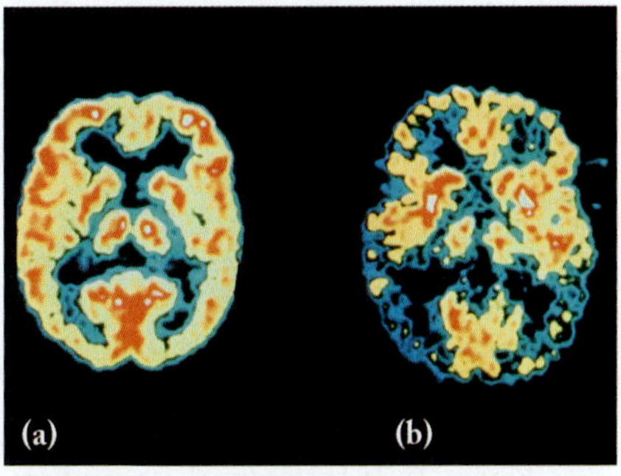


**FIGURE 21.26** Positron emission tomography (PET) is used to monitor cell activity in organs such as the brain. (a) Brain function in a healthy person. The red and yellow regions indicate high brain activity; blue and black indicate low activity. (b) Brain function in a patient suffering from Alzheimer's disease.

*Positron emission tomography* (PET) is a powerful tool for diagnosing organ and cell function. PET uses short-lived, neutron-poor, positron-emitting radionuclides. For example, a patient might be administered a solution of a glucose derivative in which some of the sugar molecules' atoms have been replaced with atoms of $^{11}C$, $^{15}O$, or $^{18}F$. The rate at which glucose is metabolized in various regions of the brain is monitored by detecting the γ rays produced by positron–electron annihilations (Equation 21.4). Unusual patterns in PET images of brains (**Figure 21.26**) can indicate damage from strokes, schizophrenia, depression, Alzheimer's disease, and even nicotine addiction in tobacco smokers.

## SAMPLE EXERCISE 21.9 Integrating Concepts: Radium Girls and Safety in the Workplace

Radium was discovered by Pierre and Marie Curie in 1898. By 1902, the new element had its first practical use. Radium compounds were mixed with zinc sulfide (ZnS) to make paint that glowed in the dark as α particles emitted by the decay of $^{226}Ra$ ($t_{1/2} = 1.60 \times 10^3$ years) caused ZnS crystals to emit a greenish fluorescence (**Figure 21.27**). The paint was used to make dials for watches, clocks, and instruments used on naval vessels and, a few years later, in military and civilian airplanes.

By 1914, U.S. companies were making radium-painted dials and employing young women in their late teens and early 20s as dial painters. Soon after their employment, many of the women became very sick, suffering from anemia and other symptoms we now associate with overexposure to nuclear radiation. Some developed bone cancer and more than 100 of them, who became known around the world as the Radium Girls, died. Their deaths were linked to the practice of "pointing" the fine paint brushes they used, which meant using their lips to make fine points on the brushes to help them paint the tiny numerals and hands on watch faces (**Figure 21.28**). In doing so, they ingested some of

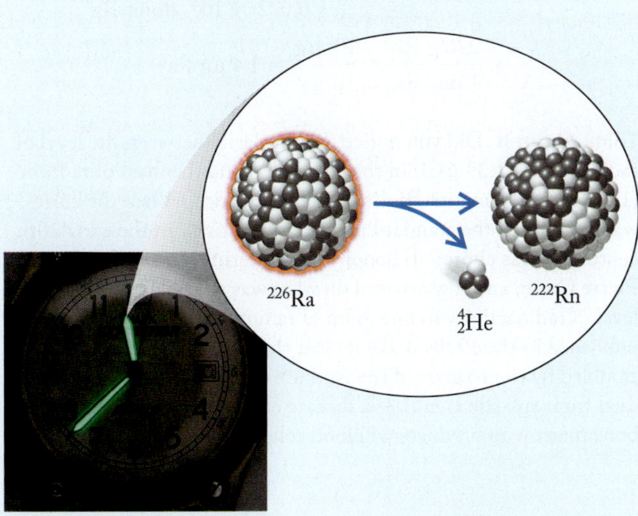


**FIGURE 21.27** During the 20th century, many millions of watches and clocks had dials that glowed in the dark as high-energy α particles emitted by $^{226}Ra$ caused crystals of ZnS to fluoresce.

**FIGURE 21.28** This editorial cartoon appeared in Sunday newspapers on February 28, 1926. It portrayed the deadly consequences of young women "pointing" their brushes with their lips as they painted the dials of watches with paint that contained radioactive radium.

the radioactive paint. Tests later determined that about 20% of the radium ingested was incorporated into their bones, where it attacked the bone marrow and caused malignancies known as osteosarcomas.

a. Suggest a reason why radium was concentrated in the victims' bones.
b. Studies of radiation levels and the incidence of cancer in more than 1000 female dial painters yielded the results in the following table. Which of the two dose–response curves in Figure 21.25 better fits these results?

| Radium Exposure (μg ingested) | Occurrence of Malignancy (% of workers exposed) |
|---|---|
| 1 | 0 |
| 3 | 0 |
| 10 | 0 |
| 30 | 0 |
| 100 | 5 |
| 300 | 53 |
| 1000 | 85 |

c. The green luminescence of radium watch dials began to fade after a few years. Was this loss in luminosity due to decreased radioactivity in the paint? Explain why or why not.
d. In one study, the levels of radioactivity in pocket watches with radium-painted dials were found to be between 0.6 μCi and 1.39 μCi per watch. How many micrograms of $^{226}$Ra produce 1.39 μCi of radioactivity?

**Collect and Organize** We are asked (a) why ingested $^{226}$Ra concentrates in bones, (b) whether malignancy in dial painters was proportional to their exposure to $^{226}$Ra radiation or followed an S-shaped dose–response curve, (c) why the luminosity of radium-activated paint fades after a few years, and (d) how many micrograms of $^{226}$Ra are needed to produce 1.39 μCi of radiation. One curie (Ci) is equal to $3.70 \times 10^{10}$ decay events/s. The half-life ($t_{1/2}$) of $^{226}$Ra is 1600 years and is related to the first-order rate constant ($k$) of the decay reaction by the equation $t_{1/2} = 0.693/k$. The level of radioactivity ($A$) in a sample of radium is equal to the product of the rate constant and the number ($N$) of $^{226}$Ra atoms: $A = kN$.

**Analyze** Radium is a group 2 element and should have chemical and biochemical properties like those of the other elements in that group, including its association with biological tissues. High concentrations of another group 2 element, calcium, occur in teeth and bones. Worker exposure levels in the preceding table cover a wide range, but exposure of up to nearly 100 μg of $^{226}$Ra caused few malignancies, whereas concentrations above 100 μg caused many. The half-life of $^{226}$Ra is so long that the radioactivity of a sample decreases little over a few years or even over many decades. Relating a half-life expressed in years to a level of radioactivity expressed in a multiple of decay events per second will require converting units of time and then quantities of radioactive atoms to moles and then micrograms.

**Solve**

a. Radium probably accumulates in bones because its chemistry is like that of calcium, which means that $^{226}Ra^{2+}$ ions are likely to take the place of $Ca^{2+}$ ions in bone tissue.
b. Malignancies did not occur among the dial painters who ingested less than 100 μg of $^{226}$Ra; however, the percentage of women who suffered from them increased sharply with exposure between 100 μg and 1000 μg. This pattern is described by the S-shaped (blue) curve in Figure 21.25.
c. Given the 1600-year half-life of $^{226}$Ra, the loss of watch dial luminescence was not the result of depleted radioactivity. Rather, it must have been due to less efficient conversion of the energy of radioactive decay into visible light by ZnS crystals.
d. Let's first convert the half-life of $^{226}$Ra into a decay rate constant in units of $s^{-1}$:

$$k = \frac{0.693}{t_{1/2}} = \frac{0.693}{1.60 \times 10^3\ \cancel{\text{yr}}} \times \frac{1\ \cancel{\text{yr}}}{365.25\ \cancel{\text{d}}} \times \frac{1\ \cancel{\text{d}}}{24\ \cancel{\text{h}}} \times \frac{1\ \cancel{\text{h}}}{3600\ \text{s}}$$
$$= 1.372 \times 10^{-11}\ \text{s}^{-1}$$

Next, we solve the equation $A = kN$ for $N$, and we use the above rate constant and the radioactivity of the watch to calculate the number of $^{226}$Ra atoms in the dial:

$$N = \frac{A}{k} = \frac{1.39\ \cancel{\mu\text{Ci}}}{1.372 \times 10^{-11}\ \text{s}^{-1}} \times \frac{1\ \cancel{\text{Ci}}}{10^6\ \cancel{\mu\text{Ci}}}$$
$$\times \frac{3.70 \times 10^{10}\ \text{atoms Ra s}^{-1}}{\cancel{\text{Ci}}} = 3.75 \times 10^{15}\ \text{atoms of Ra}$$

The corresponding mass in micrograms is

$$3.75 \times 10^{15}\ \cancel{\text{atoms Ra}} \times \frac{1\ \cancel{\text{mol Ra}}}{6.022 \times 10^{23}\ \cancel{\text{atoms Ra}}}$$
$$\times \frac{226\ \cancel{\text{g}}\ \text{Ra}}{1\ \cancel{\text{mol Ra}}} \times \frac{10^6\ \mu\text{g}}{1\ \cancel{\text{g}}} = 1.4\ \mu\text{g Ra}$$

**Think About It** Did you notice the similarity between the level of radioactivity (1.39 μCi) in the watch dial and the mass of radium (1.4 μg) producing it? This is not a coincidence. When the curie was adopted as the standard unit of radioactivity in the early 20th century, it was chosen to honor the pioneering work of Marie and Pierre Curie, and it was based on what was then believed to be the level of radioactivity in one gram of radium. Newspaper articles published in the 1920s make it clear that Marie Curie was deeply troubled by the tragedy of the Radium Girls. Sadly, in 1934 she died from aplastic anemia—a disease caused by the inability of bone marrow to produce red blood cells.

## SUMMARY

**LO1** **Nuclear chemistry** is the study and application of reactions that involve changes in atomic nuclei. The sum of the mass numbers of the reactants in a nuclear equation must be equal to the sum of the mass numbers of the products. The sum of the atomic numbers or charges (subscripts) of the reactants must also equal the sum of the atomic numbers or charges of the products. (Section 21.1)

**LO2** Stable nuclei have neutron-to-proton ratios that fall within a range of values called the **belt of stability**. Unstable nuclides undergo **radioactive decay**. Neutron-rich nuclides (mass number greater than the average atomic mass) undergo **β decay**; neutron-poor nuclides undergo **positron emission** or **electron capture.** When a particle of matter encounters a particle of **antimatter**, both are converted into energy (they annihilate each other), yielding γ rays. Very large nuclides ($Z > 83$) may undergo β decay or **α decay**. (Section 21.1)

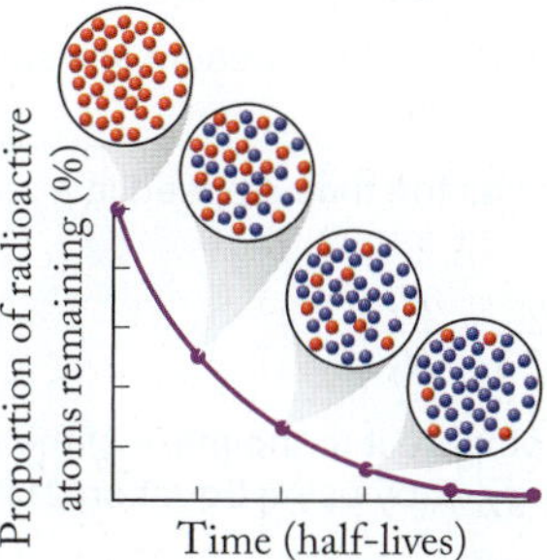


**LO3** Radioactive decay follows first-order kinetics, so the rate of decay of a radionuclide is directly proportional to the rate constant of the decay process and the quantity of radionuclide in a sample, and inversely proportional to the half-life ($t_{1/2}$) of the nuclide. **Radiometric dating** is used to determine the age of an object on the basis of its content of a radionuclide and/or its decay product. **Radiocarbon dating** involves determining the amount of radioactive carbon-14 that remains in an object derived from plant or animal tissue to calculate the age of the object. To improve the accuracy of the technique, scientists calibrate the results of radiometric analysis of samples of known age, such as growth rings in ancient trees. (Section 21.2)

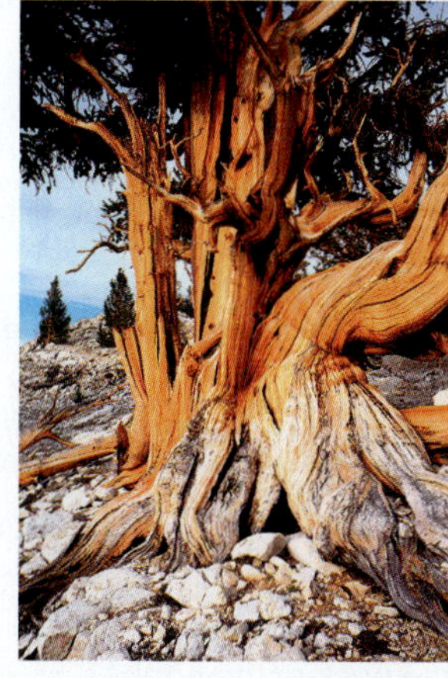

**LO4** The **mass defect ($\Delta m$)** of a nucleus is the difference between its mass and the sum of the masses of its nucleons. **Binding energy (BE)** is the energy released when the nucleons combine to form a nucleus. It is also the energy needed to split the nucleus into its nucleons. Binding energy per nucleon is a measure of the relative stability of a nucleus. The energy released in nuclear reactions is calculated using Einstein's equation: $E = mc^2$. (Section 21.3)

**LO5** Neutrons, protons, and electrons formed within seconds of the Big Bang. During primordial **nucleosynthesis**, protons and neutrons fused to produce nuclei of helium. After galaxies formed, the nuclei of atoms with $Z \leq 26$ formed when the nuclei of lighter elements fused in the cores of giant stars (stellar nucleosynthesis). The nuclei of elements with $Z > 26$ formed by a combination of **neutron capture**, β decay, and other nuclear reactions that occurred during supernovas (explosions of giant stars), which distributed the elements throughout galaxies for possible inclusion in later-generation stars and in planets such as our own. Stellar nucleosynthesis continues today. (Sections 21.4 and 21.5)

**LO6** **Nuclear fusion** occurs when subatomic particles or atomic nuclei collide and fuse. Neutron absorption by uranium-235 and a few other massive isotopes may lead to **nuclear fission** into lighter nuclei accompanied by the release of energy, which can be harnessed to generate electricity. A **chain reaction** happens when the neutrons released during fission collide with other fissionable nuclei. They require a **critical mass** of a fissionable isotope. A **breeder reactor** is used to make plutonium-239 from uranium-238 while also producing energy to make electricity. (Sections 21.5 and 21.6)

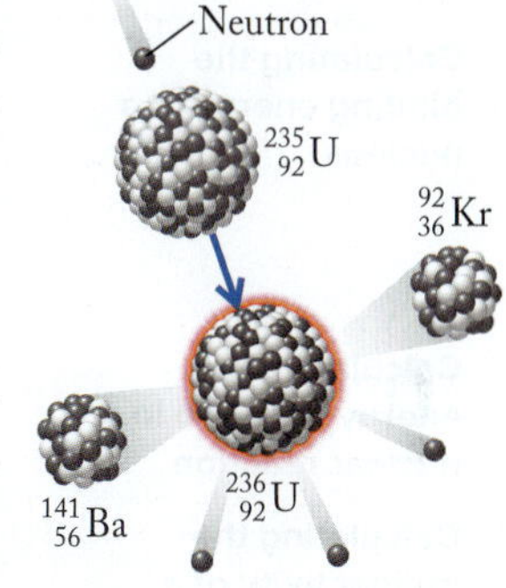


**LO7** **Scintillation counters** and **Geiger counters** are used to measure levels of nuclear radiation. Radioactivity is the number of decay events per unit time. Common units are the **becquerel (Bq)** (1 decay event/s) and the **curie (Ci)** (1 Ci = $3.70 \times 10^{10}$ Bq). (Section 21.7)

**LO8** Alpha particles, β particles, and γ rays have enough energy to break up molecules into electrons and cations and are examples of **ionizing radiation** that can damage body tissue and DNA. The quantity of ionizing radiation energy absorbed per kilogram of body mass is called the *absorbed dose* and is expressed in **grays (Gy)**: 1 Gy = 1.00 J/kg. The effective dose of any type of ionizing radiation is the product of the absorbed dose in grays and the **relative biological effectiveness (RBE)** of the radiation; the unit of effective dose is the **sievert (Sv)**. Alpha particles have a larger RBE than those of β particles and γ rays but have the least penetrating power. (Section 21.8)

**LO9** Short-lived radioactive isotopes that have suitable chemical properties and emit relatively low-energy nuclear radiation are used as tracers in the human body to map biological activity and diagnose diseases. Other radioactive isotopes that produce higher-energy radiation are used to destroy cancerous tissue. (Section 21.9)

## PARTICULATE PREVIEW WRAP-UP

Carbon-14 has more neutrons than protons: six protons and eight neutrons. With a neutron-to-proton ratio greater than 1, it will decay to reduce that ratio. When $^{14}C$ radioactively decays by emitting a β particle, it produces a nitrogen nuclide with a 1:1 neutron:proton ratio (seven protons and seven neutrons).

## PROBLEM-SOLVING SUMMARY

| Type of Problem | Concepts and Equations | Sample Exercises |
|---|---|---|
| **Completing and balancing nuclear equations** | Add nuclides or subatomic particles as needed to equalize the sum of the mass numbers (superscripts) and the sum of the atomic numbers or particle charges (subscripts) on the left and right sides of the equation. | **21.1** |
| **Predicting the modes and products of radioactive decay** | Neutron-rich nuclides tend to undergo β decay; neutron-poor nuclides undergo positron emission or electron capture. | **21.2** |
| **Calculations involving half-lives** | $\ln \frac{N_t}{N_0} = -0.693 \frac{t}{t_{1/2}}$ (21.7)<br>where $N_t/N_0$ is the ratio of the quantity of radionuclide present in a sample at time $t$ ($N_t$) to the quantity at $t = 0$ ($N_0$). | **21.3** |
| **Radiometric dating** | Rearranging Equation 21.7 to solve for $t$ yields the time elapsed in a radioactive decay process:<br>$t = -\frac{t_{1/2}}{0.693} \ln \frac{N_t}{N_0}$ (21.8) | **21.4** |
| **Calculating the binding energy of a nucleus** | Calculate the mass of the nucleons and compare it to the mass of the nucleus. Convert the difference in mass, $m$, to energy, $E$, by using Equation 21.9:<br>$E = mc^2$ (21.9)<br>where $c$ is the speed of light in a vacuum, $2.998 \times 10^8$ m/s. When the units of $m$ are kg and the units of $c$ are m/s, $E$ is in joules because $1\text{ J} = 1\text{ kg} \cdot (\text{m/s})^2$. | **21.5** |
| **Calculating the energy released in a nuclear reaction** | Substitute into a modified form of Equation 21.9, $E = \Delta mc^2$, where $\Delta m$ is the loss in mass as reactants form products. | **21.6** |
| **Calculating the radioactivity of a sample** | Insert the values of $k$ and $N$ into Equation 21.15:<br>$A = kN$ (21.15)<br>where $k = 0.693/t_{1/2}$. | **21.7** |
| **Calculating effective dose** | Substitute values for absorbed dose and relative biological effectiveness (RBE) in the equation:<br>Effective dose = absorbed dose × RBE | **21.8** |

## VISUAL PROBLEMS

*(Answers to boldface end-of-chapter questions and problems are in the back of the book.)*

**21.1.** Which of the following nuclear processes do Figures P21.1(a) and P21.1(b) represent?
a. primordial nucleosynthesis
b. synthesis of a supermassive nuclide
c. solar fusion
d. fission
e. β decay

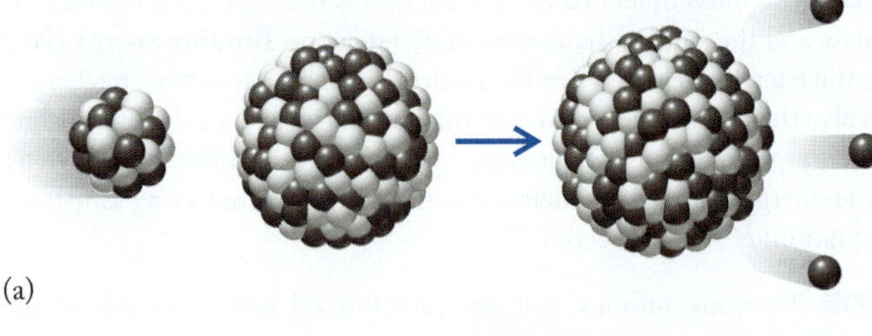

(a)

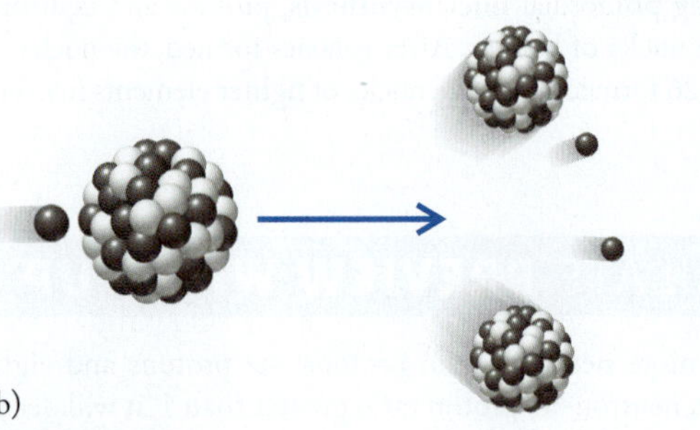

(b)

**FIGURE P21.1**

21.2. Which of the graphs in Figure P21.2 illustrates the decay of radon-222?

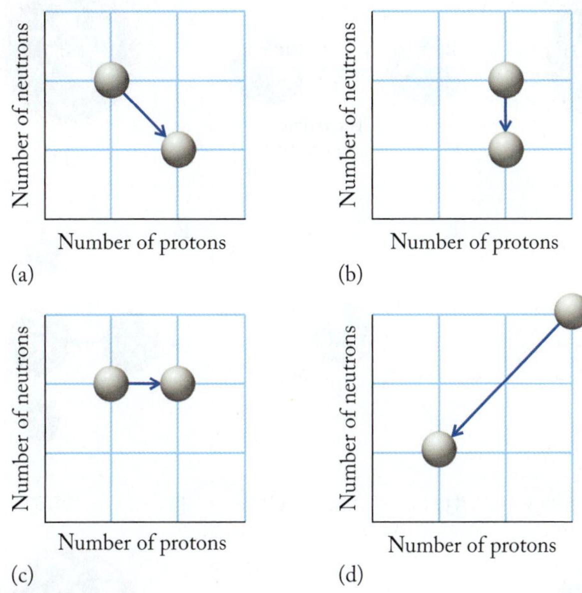


**FIGURE P21.2**

**21.3.** Which of the graphs in Figure P21.2 illustrates the decay of carbon-14?

21.4. Which of the graphs in Figure P21.2 illustrates the overall effect of neutron capture followed by β decay?

**21.5.** In 1932, James Chadwick discovered neutrons by bombarding $^{9}Be$ nuclei with α particles, which produced an unstable fusion product that decayed, forming $^{12}C$ and emitting a neutron. Which of the graphs in Figure P21.2 describes this decay process?

21.6. Which of the curves in Figure P21.6 represents the decay of an isotope that has a half-life of 2.0 days?

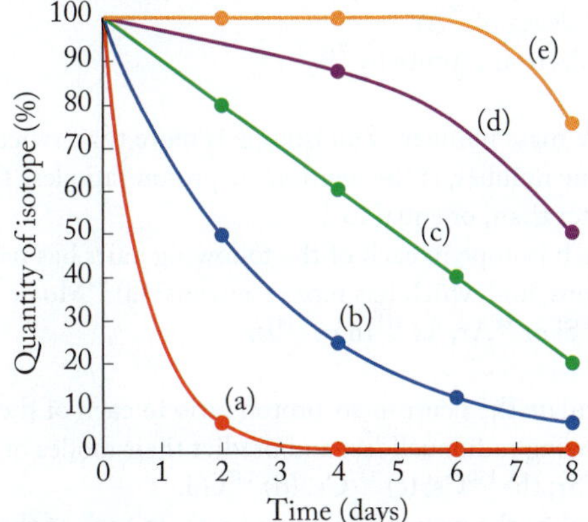


**FIGURE P21.6**

**21.7.** Which of the curves in Figure P21.6 do not represent radioactive decay?

21.8. Isotopes in a nuclear decay series emit particles with a positive charge and particles with a negative charge. The two kinds of particles penetrate a column of water, as shown in Figure P21.8. Is the "X" particle the positive or the negative one?

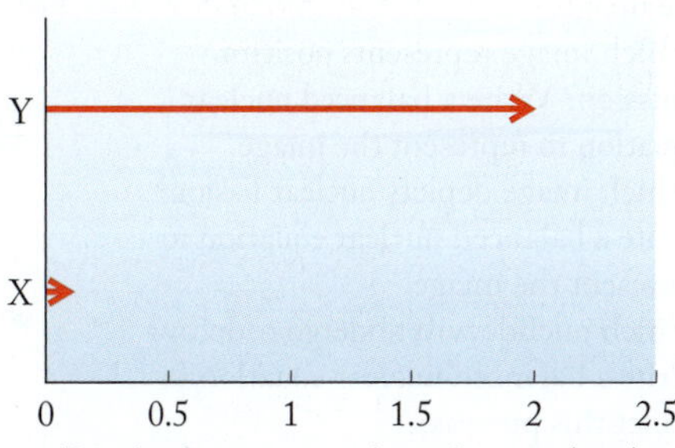


**FIGURE P21.8**

**21.9.** Radioactive polonium-218 emits three forms of radiation that follow the paths illustrated by the colored arrows in Figure P21.9 when passing through an electrical field. Which form of radiation follows each of the three paths?

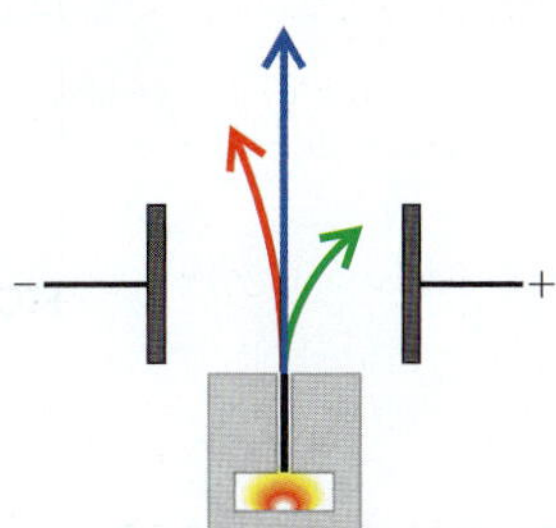


**FIGURE P21.9**

21.10. Use representations [A] through [I] in Figure P21.10 to answer questions a–f.
  a. Which image depicts β decay? Write a balanced nuclear equation to represent the image.
  b. Which image depicts fusion? Write a balanced nuclear equation to represent the image.
  c. Which image represents positron emission? Write a balanced nuclear equation to represent the image.
  d. Which image depicts nuclear fission? Write a balanced nuclear equation to represent the image.
  e. Which nuclide will undergo α decay? Write a balanced nuclear equation to depict this process.
  f. Which nuclide will undergo β decay? Write a balanced nuclear equation to depict this process.

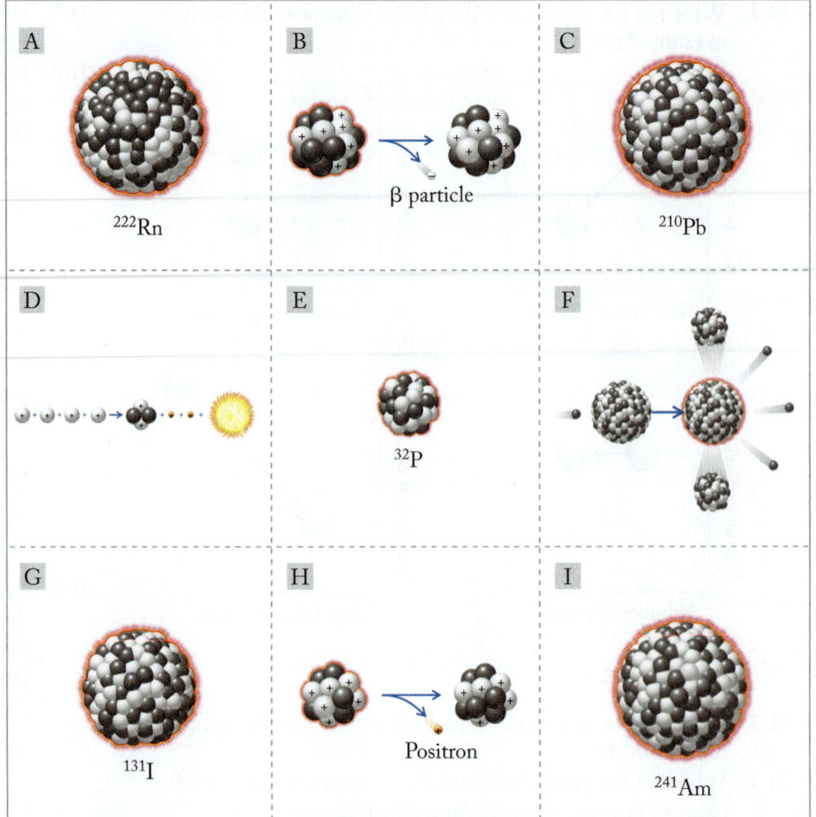


**FIGURE P21.10**

# QUESTIONS AND PROBLEMS

## *Decay Modes of Radionuclides*

### Concept Review

**21.11.** What is the effect of β decay on the ratio of neutrons to protons in a nucleus?

21.12. Explain how the product of β decay has a higher atomic number than that of the radionuclide from which the product forms.

**21.13.** How can the belt of stability be used to predict the probable decay mode of an unstable nuclide?

21.14. Compare positron-emission and electron-capture processes.

**21.15.** The ratio of neutrons to protons in stable nuclei increases with increasing atomic number. Use this trend to explain why multiple α decay steps in the $^{238}$U decay series are often followed by β decay.

*21.16. Chlorine-36 nuclei may undergo β decay *or* positron emission. How is it possible that one isotope can decay by pathways that we associate with neutron-rich nuclei on the one hand *and* neutron-poor nuclei on the other?

### Problems

**21.17.** Compounds containing $^{64}$Cu are useful for measuring blood flow in the brain. Copper-64 decays to $^{64}$Zn and to $^{64}$Ni. What are the modes of decay in these two reactions?

21.18. Write a balanced nuclear equation for
  a. α decay of $^{232}$Th
  b. β decay of $^{42}$K
  c. β decay of $^{35}$Ar
  d. Electron capture by $^{7}$B

**21.19.** If the mass number of an isotope is more than twice the atomic number, is the neutron-to-proton ratio less than, greater than, or equal to 1?

21.20. Which isotope in each of the following pairs has more protons, and which has more neutrons? (a) $^{92}$Mo or $^{92}$Zr; (b) $^{28}$Si or $^{28}$Ar; (c) $^{111}$In or $^{114}$In

**21.21.** Calculate the neutron-to-proton ratio in each of the following radionuclides, and predict their modes of decay: (a) $^{90}$Sr; (b) $^{130}$Cs; (c) $^{137}$Cs, (d) $^{145}$Gd.

21.22. Calculate the neutron-to-proton ratio in each of the following radionuclides, and predict their modes of decay: (a) $^{238}$U; (b) $^{186}$Re; (c) $^{86}$Y, (d) $^{220}$Rn.

**21.23.** Iridium-192 can be inserted into tumors, where it undergoes both β decay and electron capture. Write balanced nuclear equations describing the decay of $^{192}$Ir.

21.24. Radioactive arsenic-74 may undergo β decay or positron emission. Which nuclides are produced by each of these decay pathways?

**21.25.** **Elements in a Supernova** The isotopes $^{56}Co$ and $^{44}Ti$ were detected in supernova SN 1987A. Predict the decay pathways for these radioactive isotopes.

21.26. Nine isotopes of sulfur have mass numbers ranging from 30 to 38. Five of the nine are radioactive: $^{30}S$, $^{31}S$, $^{35}S$, $^{37}S$, and $^{38}S$. Which of these isotopes do you expect to decay by β decay?

## Rates of Radioactive Decay

### Concept Review

**21.27.** Explain why radiocarbon dating is not reliable for artifacts and fossils older than about 50,000 years.

21.28. Which of the following statements about $^{14}C$ dating are true?
a. The amount of $^{14}C$ in all objects is the same.
b. Carbon-14 is unstable and is readily lost from the atmosphere.
c. The ratio of $^{14}C$ to $^{12}C$ in the atmosphere is a constant.
d. Living tissue will absorb $^{12}C$ but not $^{14}C$.

**21.29.** Why is $^{40}K$ dating ($t_{1/2} = 1.28 \times 10^9$ years) useful only for rocks older than 300,000 years?

21.30. Where does the $^{14}C$ found in plants come from?

### Problems

**21.31.** What percentage of a sample's original radioactivity remains after three half-lives?

21.32. What percentage of a sample's original radioactivity remains after six half-lives?

---

**21.33.** What is the half-life of $^{11}C$ if 87.5% of a sample of it decays in 60.9 minutes?

21.34. A sample of $^{89}Zr$ is delivered to a hospital on Monday at 9 a.m. for use in labeling antibodies. By Wednesday at 5 p.m., 74.4% of it remains. What is the half-life of $^{89}Zr$?

---

**21.35.** **Fukushima Disaster** Explosions at a disabled nuclear power station in Fukushima, Japan, in 2011 may have released more cesium-137 ($t_{1/2} = 30.2$ years) into the ocean than any other single event. How long will the radioactivity of this radionuclide take to decay to 5.0% of the level released in 2011?

21.36. Spent fuel removed from nuclear power stations contains plutonium-239 ($t_{1/2} = 2.41 \times 10^4$ years). How long will a sample of this radionuclide take to reach a level of radioactivity that is 1.5% of the level it had when it was removed from a reactor?

---

**21.37.** **First Humans in South America** Archaeologists continue to debate the origins and dates of arrival of the first humans in the Western Hemisphere. Radiocarbon dating of charcoal from a cave in Chile was used to establish the earliest date of human habitation in South America as 8700 years ago. What fraction of the $^{14}C$ initially present remained in the charcoal after 8700 years?

21.38. **Early Financial Records** Native Americans living along the north coast of Peru used knotted cotton strands called *quipu* (Figure P21.38) to record financial transactions and governmental actions. Compared with the fibers of cotton plants growing today, what is the ratio of carbon-14 to carbon-12 in an Incan quipu from 1425 CE?

**FIGURE P21.38**

---

***21.39.** **The Ages of Sequoia Trees** Figure P21.39 shows a slice of the trunk of a giant sequoia tree cut down in 1891 in what is now Kings Canyon National Park. It contained 1342 annual growth rings. If samples of the tree were removed for radiocarbon dating today, what would be the difference in the $^{14}C/^{12}C$ ratio in the innermost (oldest) ring compared with that ratio in the youngest ring?

**FIGURE P21.39**

*21.40. **Dating Volcanic Eruptions** Geologists who study volcanoes can develop historical profiles of previous eruptions by determining the $^{14}C/^{12}C$ ratios of charred plant remains trapped in old magma and ash flows. If the uncertainty in determining these ratios is 0.1%, could radiocarbon dating distinguish between debris from the eruptions of Mt. Vesuvius that occurred in the years 472 and 512? (*Hint*: Calculate the $^{14}C/^{12}C$ ratios for samples from the two dates.)

---

**21.41.** **Age of Mammoths** Figure P21.41 shows a carved mammoth tusk uncovered at an ancient camp site in the Ural Mountains in 2001. The $^{14}C/^{12}C$ ratio in the tusk was only 1.19% of that in modern elephant tusks. How old is the mammoth tusk?

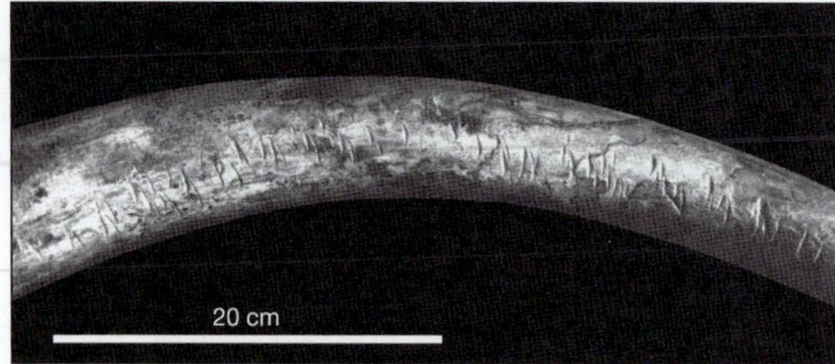


**FIGURE P21.41**

21.42. **The Destruction of Jericho** The Bible describes the Exodus as a period of 40 years that began with plagues in Egypt and ended with the destruction of Jericho. Archeologists seeking to establish the exact dates of these events have proposed that the plagues coincided with a huge eruption of the volcano Thera in the Aegean Sea.

a. Radiocarbon dating suggests that the eruption occurred around 1360 BCE, though other records place the eruption of Thera in the year 1628 BCE. What is the percent difference in the $^{14}C$ decay rate in biological samples from these two dates?

b. Radiocarbon dating of blackened grains from the site of ancient Jericho provides a date of 1315 BCE ± 13 years for the fall of the city. What is the $^{14}C/^{12}C$ ratio in the blackened grains compared with that of grain harvested last year?

## Energy Changes in Radioactive Decay

### Concept Review

**21.43.** Why do all nuclear reactions release energy?

21.44. Which isotope of phosphorus should have the larger binding energy: $^{31}P$, which is stable, or $^{32}P$, which is radioactive with a half-life of 14.3 days?

### Problems

**21.45.** Substitution of carbon-11 for some of the carbon-12 atoms in glucose yields a useful compound for imaging brain function.

a. Write a balanced nuclear equation for the decay of $^{11}C$.

b. Calculate the binding energy of $^{11}C$. The atomic mass of $^{11}C$ is $1.82850 \times 10^{-26}$ kg.

21.46. Modified glucose in which an –OH group in its molecular structure is replaced with an atom of radioactive $^{18}F$ (Figure P12.46) is a widely used radiochemical imaging agent.

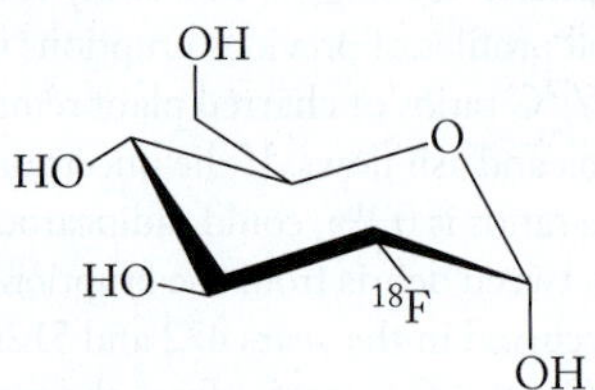


**FIGURE P21.46**

a. Write a balanced nuclear equation for the decay of $^{18}F$.

b. Calculate the binding energy of $^{18}F$. The atomic mass of $^{18}F$ is $2.98915 \times 10^{-26}$ kg.

---

**21.47.** Calculate the binding energy in kJ/mol of the two naturally occurring isotopes of chlorine: $^{35}Cl$ (34.9689 u) and $^{37}Cl$ (36.9659 u).

21.48. Which isotope has a larger binding energy: $^{10}B$ (10.0129 u) or $^{11}B$ (11.0093 u)?

## Making New Elements

### Problems

**21.49.** Write a balanced nuclear equation describing how bombarding a $^{209}Bi$ target with α particles could produce $^{211}At$.

21.50. Bombarding a $^{239}Pu$ target with α particles produces $^{242}Cm$ and another particle.

a. Use a balanced nuclear equation to identify the other particle.

b. The synthesis of which other nuclide described in this chapter involves the same subatomic particles?

## Nuclear Fusion and the Origin of the Elements

### Concept Review

**21.51.** A key stage in primordial nucleosynthesis involved the fusion of protons and neutrons, forming deuterons: ${}^{1}_{1}H + {}^{1}_{0}n \rightarrow {}^{2}_{1}D$. This was followed by the fusion of two deuterons to form ${}^{4}_{2}He$ nuclei (α particles): $2\,{}^{2}_{1}D \rightarrow {}^{4}_{2}He$. Fusion reactions in our sun and other stars also convert ${}^{1}_{1}H$ nuclei into ${}^{4}_{2}He$ nuclei, but not by the above process. Why not?

21.52. Why is energy released in a nuclear fusion process when the product is an element preceding iron in the periodic table?

**21.53.** **Components of Solar Wind** Most ions that flow out from the sun in the solar wind are hydrogen ions. The ions of which element should be next most abundant?

*21.54. **Nucleosynthesis in Giant Stars** A star needs a core temperature of about $10^7$ K for hydrogen fusion to occur. Core temperatures above $10^8$ K are needed for helium fusion. Why does helium fusion require much higher temperatures?

**21.55.** **Origins of the Elements** Our sun contains carbon even though its core is not hot or dense enough to sustain carbon synthesis through the triple-alpha process. Where could the carbon have come from?

21.56. Early nucleosynthesis produced a universe that was more than 99% hydrogen and helium, with less than 1% lithium. Why were the other elements not formed?

### Problems

**21.57.** Calculate the energy and wavelength of the two gamma rays released by the annihilation of a proton and an antiproton.

21.58. Calculate the energy released and the wavelength of the two photons emitted in the annihilation of an electron and a positron.

**21.59.** Calculate the energy released in each of the following reactions from the masses of the isotopes: $^{2}$H (2.0146 u), $^{4}$He (4.00260 u), $^{10}$B (10.0129 u), $^{12}$C (12.000 u), $^{14}$N (14.00307 u), $^{16}$O (15.99491 u), $^{24}$Mg (23.98504 u), $^{28}$Si (27.97693 u).

a. $^{14}N + {}^{14}N \rightarrow {}^{28}Si$

b. $^{10}B + {}^{16}O + {}^{2}H \rightarrow {}^{28}Si$

c. $^{16}O + {}^{12}C \rightarrow {}^{28}Si$

d. $^{24}Mg + {}^{4}He \rightarrow {}^{28}Si$

21.60. All the following fusion reactions produce $^{32}$S. Calculate the energy released in each reaction from the masses of the isotopes: $^{4}$He (4.00260 u), $^{6}$Li (6.01512 u), $^{12}$C (12.000 u), $^{14}$N (14.00307 u), $^{16}$O (15.99491 u), $^{24}$Mg (23.98504 u), $^{28}$Si (27.97693 u), $^{32}$S (31.97207 u).

a. $^{16}O + {}^{16}O \rightarrow {}^{32}S$

b. $^{28}Si + {}^{4}He \rightarrow {}^{32}S$

c. $^{14}N + {}^{12}C + {}^{6}Li \rightarrow {}^{32}S$

d. $^{24}Mg + 2\,{}^{4}He \rightarrow {}^{32}S$

**21.61.** **Hydrogen Bombs** In January 2016, North Korea claimed to have tested a small hydrogen bomb based on the fusion of deuterium and tritium (see Equation 21.14). The large amount of tritium required to sustain that reaction in a hydrogen bomb is produced by the reaction between a neutron and a lithium-6 nucleus that yields helium and tritium. Calculate the energy change for the reaction shown in Figure P21.61. *Note*: The masses of tritium and lithium-6 nuclei are $5.00827 \times 10^{-27}$ kg and $9.98841 \times 10^{-27}$ kg, respectively.

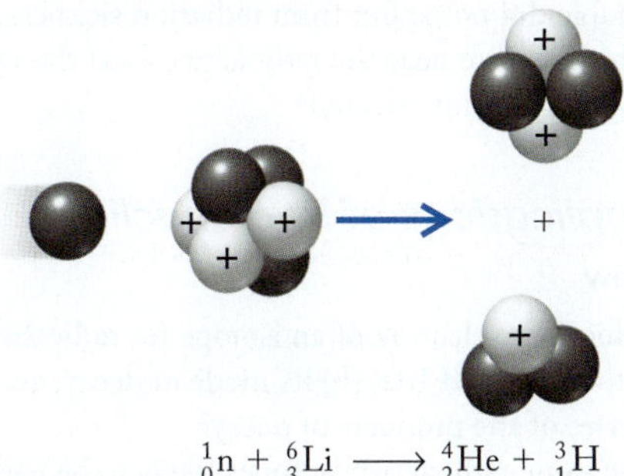

$^{1}_{0}n + {}^{6}_{3}Li \longrightarrow {}^{4}_{2}He + {}^{3}_{1}H$

**FIGURE P21.61**

21.62. **Tokamak Radiochemistry** In 2009, construction began on ITER (originally an acronym for International Thermonuclear Experimental Reactor), a project to build the world's largest nuclear fusion reactor. When it is operational, ITER will use the fusion reactions described in Problem 21.61. Because the natural abundance of lithium-6 is only about 6.0%, ITER planned to use the more abundant $^{7}$Li isotope. How much energy is released per nucleus of tritium produced during the reaction shown in Figure P21.62? *Note*: The masses of tritium and lithium-7 nuclei are $5.00827 \times 10^{-27}$ kg and $1.16504 \times 10^{-26}$ kg, respectively.

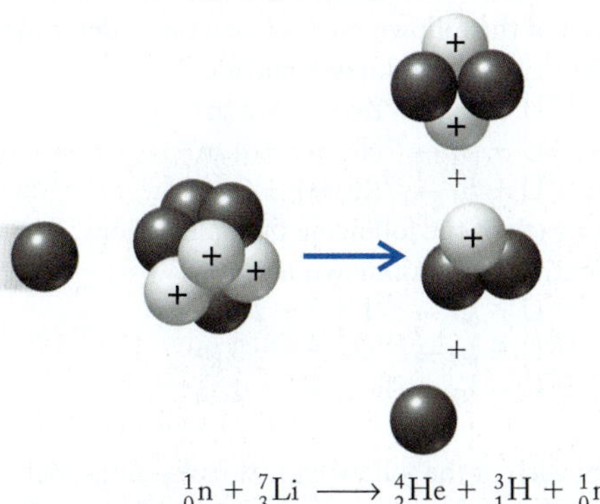

$^{1}_{0}n + {}^{7}_{3}Li \longrightarrow {}^{4}_{2}He + {}^{3}_{1}H + {}^{1}_{0}n$

**FIGURE P21.62**

**21.63.** What nuclide is produced in the core of a giant star by each of the following fusion reactions? Assume that each reaction has only one product.

a. $^{12}C + {}^{4}He \rightarrow$

b. $^{20}Ne + {}^{4}He \rightarrow$

c. $^{32}S + {}^{4}He \rightarrow$

21.64. What nuclide is produced in the core of a giant star by each of the following fusion reactions? Assume that each reaction has only one product.

a. $^{28}Si + {}^{4}He \rightarrow$

b. $^{40}Ca + {}^{4}He \rightarrow$

c. $^{24}Mg + {}^{4}He \rightarrow$

**21.65.** What nuclide is produced in the core of a collapsing giant star by each of the following reactions?

a. $^{96}_{42}Mo + 3\,{}^{1}_{0}n \rightarrow ? + {}^{0}_{-1}\beta$

b. $^{118}_{50}Sn + 3\,{}^{1}_{0}n \rightarrow ? + {}^{0}_{-1}\beta$

c. $^{108}_{47}Ag + {}^{1}_{0}n \rightarrow ? + {}^{0}_{-1}\beta$

21.66. What nuclide is produced in the core of a collapsing giant star by each of the following reactions?

a. $^{65}_{29}Cu + 3\,{}^{1}_{0}n \rightarrow ? + {}^{0}_{-1}\beta$

b. $^{68}_{30}Zn + 2\,{}^{1}_{0}n \rightarrow ? + {}^{0}_{-1}\beta$

c. $^{88}_{38}Sr + {}^{1}_{0}n \rightarrow ? + {}^{0}_{-1}\beta$

## Nuclear Fission

### Concept Review

**21.67.** How is the rate of energy release controlled in a nuclear reactor?

21.68. How does a breeder reactor create fuel and energy at the same time?

***21.69.** Why are neutrons always by-products of the fission of most massive nuclides? (*Hint*: Look closely at the neutron-to-proton ratios shown in Figure 21.1.)

21.70. Seaborgium (Sg; element 106) is prepared by bombarding curium-248 with neon-22, which produces two isotopes, $^{265}$Sg and $^{266}$Sg. Write balanced nuclear reactions for the formation of both isotopes. Are these reactions better described as fusion or fission processes?

### Problems

**21.71.** The fission of uranium produces dozens of isotopes. For each of the following fission reactions, determine the identity of the unknown nuclide:
a. $^{235}U + ^{1}_{0}n \rightarrow ^{96}Zr + ? + 2\,^{1}_{0}n$
b. $^{235}U + ^{1}_{0}n \rightarrow ^{99}Nb + ? + 4\,^{1}_{0}n$
c. $^{235}U + ^{1}_{0}n \rightarrow ^{90}Rb + ? + 3\,^{1}_{0}n$

21.72. For each of the following fission reactions, determine the identity of the unknown nuclide:
a. $^{235}U + ^{1}_{0}n \rightarrow ^{137}I + ? + 2\,^{1}_{0}n$
b. $^{235}U + ^{1}_{0}n \rightarrow ^{137}Cs + ? + 3\,^{1}_{0}n$
c. $^{235}U + ^{1}_{0}n \rightarrow ^{141}Ce + ? + 2\,^{1}_{0}n$

**21.73.** For each of the following fission reactions, determine the identity of the unknown nuclide:
a. $^{235}U + ^{1}_{0}n \rightarrow ^{131}I + ? + 2\,^{1}_{0}n$
b. $^{235}U + ^{1}_{0}n \rightarrow ^{103}Ru + ? + 3\,^{1}_{0}n$
c. $^{235}U + ^{1}_{0}n \rightarrow ^{95}Zr + ? + 3\,^{1}_{0}n$

21.74. For each of the following fission reactions, determine the identity of the unknown nuclide:
a. $^{235}U + ^{1}_{0}n \rightarrow ^{147}Pm + ? + 2\,^{1}_{0}n$
b. $^{235}U + ^{1}_{0}n \rightarrow ^{94}Kr + ? + 2\,^{1}_{0}n$
c. $^{235}U + ^{1}_{0}n \rightarrow ^{95}Sr + ? + 3\,^{1}_{0}n$

## *Measuring Radioactivity; Biological Effects of Radioactivity*

### Concept Review

**21.75.** What is the difference between a *level* of radioactivity and a *dose* of radioactivity?

21.76. What are some of the molecular effects of exposure to radioactivity?

**21.77.** Describe the dangers of exposure to radon-222.

21.78. **Food Safety** Periodic outbreaks of food poisoning from meat contaminated with *Escherichia coli* bacteria have renewed the debate about irradiation as an effective treatment of food. In one newspaper article on the subject, the following statement appeared: "Irradiating food destroys bacteria by breaking apart their molecular structure." How would you improve or expand on this explanation?

### Problems

**21.79.** **Radiation Exposure from Dental X-rays** Dental X-rays expose patients to about 5 μSv of radiation. Given an RBE of 1 for X-rays, how many grays of radiation does 5 μSv represent? For a 50 kg person, how much energy does 5 μSv correspond to?

*21.80. **Radiation Exposure at Chernobyl** Some workers responding to the explosion at the Chernobyl nuclear power plant were exposed to 5 Sv of radiation, killing many of them. If the exposure was primarily in the form of γ rays with an energy of $3.3 \times 10^{-14}$ J and an RBE of 1, how many γ rays did an 80 kg person absorb?

***21.81.** **Strontium-90 in Milk** In the years immediately after the explosion at the Chernobyl nuclear power plant, the concentration of $^{90}$Sr in cow's milk in southern Europe was slightly elevated. Some samples contained as much as 1.25 Bq/L of $^{90}$Sr radioactivity. The half-life of strontium-90 is 28.8 years.
a. Write a balanced nuclear equation describing the decay of $^{90}$Sr.
b. How many atoms of $^{90}$Sr are in a 200 mL glass of milk with 1.25 Bq/L of $^{90}$Sr radioactivity?
c. Why would strontium-90 be more concentrated in milk than other foods, such as grains, fruits, or vegetables?

*21.82. **Radium Watch Dials** If exactly 1.00 μg of $^{226}$Ra was applied to the glow-in-the-dark dial of a wristwatch made in 1914, how radioactive is the watch today? Express your answer in microcuries and becquerels. The half-life of $^{226}$Ra is $1.60 \times 10^3$ years.

**21.83.** In 1999, the U.S. Environmental Protection Agency set a maximum radon level for drinking water at 4.0 pCi per milliliter.
a. How many decay events occur per second in a milliliter of water for this level of radon radioactivity?
b. If the above radioactivity were due to the decay of $^{222}$Rn ($t_{1/2}$ = 3.8 d), how many $^{222}$Rn atoms would be in 1.0 mL of water?

21.84. **Death of a Spy** A former Russian spy died from radiation sickness in 2006 after dining at a London restaurant where he apparently ingested polonium-210. The other people at his table did not suffer from radiation sickness, even though they were near the radioactive food the victim ate. Why were they not affected?

## *Medical Applications of Radionuclides*

### Concept Review

**21.85.** How does the selection of an isotope for radiotherapy relate to (a) its half-life, (b) its mode of decay, and (c) the properties of the products of decay?

21.86. Are the same radioactive isotopes likely to be used for both imaging and cancer treatment? Why or why not?

### Problems

**21.87.** Predict the most likely mode of decay for the following isotopes used as imaging agents in nuclear medicine: (a) $^{197}$Hg (kidney); (b) $^{75}$Se (parathyroid gland); (c) $^{18}$F (bone).

21.88. Predict the most likely mode of decay for the following isotopes used as imaging agents in nuclear medicine: (a) $^{133}$Xe (cerebral blood flow); (b) $^{57}$Co (tumor detection); (c) $^{51}$Cr (red blood cell mass); (d) $^{67}$Ga (tumor detection).

**21.89.** A 1.00 mg sample of $^{192}$Ir was inserted into the artery of a heart patient. After 30 days, 0.756 mg remained. What is the half-life of $^{192}$Ir?

21.90. A sample of dysprosium-165 with a radioactivity of 1100 counts per second was injected into the knee of a patient suffering from rheumatoid arthritis. After 24 hr, the radioactivity had dropped to 1.14 counts per second. Calculate the half-life of $^{165}$Dy.

**21.91.** **Treatment of Tourette's Syndrome** Tourette's syndrome is a condition whose symptoms include sudden movements and vocalizations. Iodine isotopes are used in brain imaging of people suffering from Tourette's syndrome. To study the uptake and distribution of iodine in cells, researchers treated mammalian brain cells in culture with a solution containing $^{131}I$ with an initial radioactivity of 108 counts per minute. The cells were removed after 30 days, and the remaining solution had a radioactivity of 4.1 counts per minute. Did the brain cells absorb any $^{131}I$ ($t_{1/2}$ = 8.1 d)?

21.92. **Mercury Test of Kidney Function** A patient is administered mercury-197 to evaluate kidney function. Mercury-197 has a half-life of 65 hours. What fraction of an initial dose of mercury-197 remains after 6 days?

---

**21.93.** Rhodium-105 is an isotope under investigation in diagnostic applications. The half-life of $^{105}Rh$ is 35.4 hr, which is long enough for transport from the supplier to a hospital. A supplier ships 250 mg of $^{105}RhCl_3$ overnight (12 hr).

a. What percentage of the $^{105}Rh$ remains upon arrival?
b. How long will 95% of the $^{105}Rh$ take to decay?

21.94. **Treating Prostate Cancer** Palladium-103 is used to treat prostate cancer by inserting a small (1 mm × 5 mm) cylindrical piece of $^{103}Pd$ directly into the tumor. How long will 85% of the $^{103}Pd$ take to decay, given that $t_{1/2}$ = 17 d?

---

***21.95.** **Boron Neutron-Capture Therapy** In boron neutron-capture therapy (BNCT), a patient is given a compound (Figure P21.95) containing $^{10}B$ that accumulates inside cancer tumors. Then the tumors are irradiated with neutrons, which are absorbed by $^{10}B$ nuclei. The product of neutron capture is an unstable form of $^{11}B$ that undergoes α decay to $^{7}Li$.

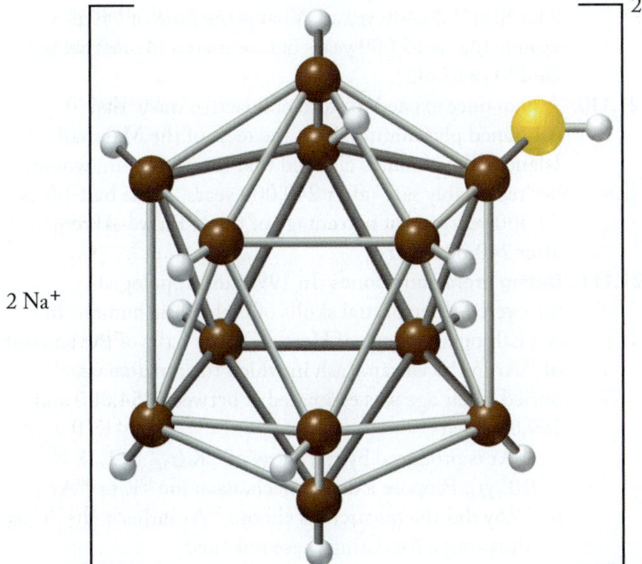


**FIGURE P21.95**

a. Write a balanced nuclear equation for the neutron absorption and α decay process.
b. Calculate the energy released by each nucleus of boron-10 that captures a neutron and undergoes α decay, given the following masses of the particles in the process: $^{10}B$ (10.0129 u), $^{7}Li$ (7.01600 u), $^{4}He$ (4.00260 u), and $^{1}n$ (1.00866 u).
c. Why is the formation of a nuclide that undergoes α decay a particularly effective cancer therapy?

21.96. **Balloon Angioplasty and Arteriosclerosis** Balloon angioplasty is a common procedure for unclogging arteries in patients suffering from arteriosclerosis. Iridium-192 therapy is being tested as a treatment to prevent reclogging of the arteries. In the procedure, a thin ribbon containing pellets of $^{192}Ir$ is threaded into the artery. The half-life of $^{192}Ir$ is 74 days. How long will 99% of the radioactivity from 1.00 mg of $^{192}Ir$ take to disappear?

## Additional Problems

**21.97.** Thirty years before the creation of antihydrogen, TV producer Gene Roddenberry (1921–1991) proposed to use this form of antimatter to fuel the powerful "warp" engines of the fictional starship *Enterprise.*

a. Why would antihydrogen have been a particularly suitable fuel?
b. Describe the challenges of storing such a fuel on a starship.

21.98. Tiny concentrations of radioactive tritium ($^{3}_{1}H$) occur naturally in rain and groundwater. The half-life of $^{3}_{1}H$ is 12 years. If tiny concentrations of tritium could be determined accurately, could the isotope be used to determine whether a bottle of wine with the year 1969 on its label contained wine made from grapes that were grown in 1969? Explain your answer.

**21.99.** In Section 21.5, we state that "no energy would be released if two $^{4}He$ nuclei were to fuse to form $^{8}Be$. Similarly, $^{8}Be$ nuclei would require no input of energy to spontaneously decompose into $^{4}He$ nuclei." Verify this statement by calculating the binding energy of $^{8}Be$ and comparing it to that of $^{4}He$.

21.100. How much energy is required to remove a neutron from the nucleus of an atom of carbon-13 (mass = 13.00335 u)? (*Hint*: The mass of an atom of carbon-12 is exactly 12.00000 u.)

**21.101.** **Smoke Detectors** Americium-241 ($t_{1/2}$ = 433 yr) is used in smoke detectors. The α particles from this isotope ionize nitrogen and oxygen in the air, creating an electric current. When smoke is present, the current decreases, setting off the alarm.

a. Is a smoke detector more like a Geiger counter or a scintillation counter?
b. How long will the radioactivity of a sample of $^{241}Am$ take to drop to 1% of its original radioactivity?
c. Why are smoke detectors containing $^{241}Am$ safe to handle without protective equipment?

*21.102. **Potassium–Argon Dating** Potassium-40 is a radioactive isotope of potassium, a common element in terrestrial rocks, which decays to $^{40}Ar$ with a half-life of $1.28 \times 10^9$ years. By measuring the ratio of $^{40}Ar$ to $^{40}K$, geologists can determine the age of ancient rocks.

a. Balance the nuclear equations for the decay of $^{40}K$ by identifying the missing isotope or particle.

$$^{40}K \rightarrow {}^{40}Ar + ?$$

$$^{40}K \rightarrow ? + {}^{0}_{-1}\beta$$

b. Why might $^{40}K$ decay by two pathways?

c. Only about 11% of the $^{40}K$ decays to $^{40}Ar$. If the ratio of $^{40}Ar$ to $^{40}K$ in a rock is found to be 0.435, how old is the rock?

d. Why don't geologists measure the $^{40}Ca$:$^{40}K$ ratio instead?

**21.103. Synthesizing a New Element** In 2006, an international team of scientists confirmed the synthesis of a total of three atoms of Og (oganesson) in experiments run in 2002 and 2005. They bombarded a $^{249}Cf$ target with $^{48}Ca$ nuclei.

a. Write a balanced nuclear equation describing the synthesis of $^{294}_{118}Og$.

b. The synthesized isotope of Og undergoes α decay ($t_{1/2}$ = 0.9 ms). What nuclide is produced by the decay process?

c. The nuclide produced in part b also undergoes α decay ($t_{1/2}$ = 10 ms). What nuclide is produced by this decay process?

d. The nuclide produced in part c also undergoes α decay ($t_{1/2}$ = 0.16 s). What nuclide is produced by this decay process?

e. If you had to select an element that occurs in nature and that has physical and chemical properties similar to those of Og, which element would it be?

21.104. Which element in the following series will be present in the greatest amount after 1 year?

$$^{214}_{83}Bi \xrightarrow{\alpha} {}^{210}_{81}Tl \xrightarrow{\beta} {}^{210}_{82}Pb \xrightarrow{\beta} Bi \longrightarrow$$

$t_{1/2}$ = 20 min   1.3 min   20 yr   5 d

**21.105.** Radiation exposure ionizes water to $H_2O^+$, which reacts to form $H_3O^+$ and OH. Draw Lewis structures for these three molecules or molecular ions.

*21.106. **Dating Cave Paintings** Cave paintings in Gua Saleh Cave in Borneo have been dated by measuring the amount of $^{14}C$ in calcium carbonate deposits that formed over the pigments used in the paint. The source of the carbonate ion was atmospheric $CO_2$.

a. What is the ratio of the $^{14}C$ radioactivity in calcium carbonate that formed 9900 years ago to that in calcium carbonate formed today?

b. The archaeologists also used a second method, uranium–thorium dating, to confirm the age of the paintings by measuring trace quantities of these elements present as contaminants in the calcium carbonate. Shown below are two candidates for the U–Th dating method. Which isotope of uranium do you suppose was chosen? Explain your answer.

| $^{235}_{92}U$ | $^{234}_{92}U$ |
|---|---|
| ↓ $t_{1/2} = 7.04 \times 10^8$ yr | ↓ $t_{1/2} = 2.24 \times 10^5$ yr |
| $^{231}_{90}Th$ | $^{230}_{90}Th$ |
| ↓ $t_{1/2}$ = 25.6 h | ↓ $t_{1/2} = 7.7 \times 10^4$ yr |
| $^{231}_{91}Pa$ | $^{226}_{91}Pa$ |
| ↓ $t_{1/2} = 3.25 \times 10^4$ yr | ↓ $t_{1/2}$ = 1600 yr |

**21.107.** The synthesis of new elements and specific isotopes of known elements in linear accelerators involves fusing smaller nuclei.

a. An isotope of platinum can be prepared from nickel-64 and tin-124. Write a balanced equation for this nuclear reaction. (You may assume that no neutrons are ejected in the fusion reaction.)

b. Substituting tin-132 for tin-124 increases the rate of the fusion reaction 10 times. Which isotope of Pt is formed in this reaction?

21.108. **Radon in Drinking Water** A sample of drinking water collected from a suburban Boston municipal water system in 2002 contained 0.5 pCi/L of radon. Assume that this level of radioactivity was due to the decay of $^{222}Rn$ ($t_{1/2}$ = 3.8 d).

a. What was the level of radioactivity (Bq/L) of this nuclide in the sample?

b. How many decay events per hour would occur in 2.5 L of the water?

**21.109. Stone Age Skeletons** The first attempt at radiocarbon dating six skeletons discovered in an Italian cave at the beginning of the 20th century indicated an age of 15,000 years. Redetermination of the age in 2004 indicated an older age for two bones of between 23,300 and 26,400 years. What is the ratio of $^{14}C$ in a sample that is 15,000 years old compared to one that is 25,000 years old?

21.110. A plan once existed to store radioactive waste that contained plutonium-239 in the reefs of the Marshall Islands. The planners claimed that the plutonium would be "reasonably safe" after 240,000 years. If the half-life is 24,400 years, what percentage of the $^{239}Pu$ would remain after 240,000 years?

**21.111. Dating Prehistoric Bones** In 1997, anthropologists uncovered three partial skulls of prehistoric humans in the Ethiopian village of Herto. On the basis of the amount of $^{40}Ar$ in the volcanic ash in which the remains were buried, their age was estimated at between 154,000 and 160,000 years.

a. $^{40}Ar$ is produced by the decay of $^{40}K$ ($t_{1/2} = 1.28 \times 10^9$ yr). Propose a decay mechanism for $^{40}K$ to $^{40}Ar$.

b. Why did the researchers choose $^{40}Ar$ rather than $^{14}C$ as the isotope for dating these remains?

*21.112. The carbon-14 radioactivity in papyrus growing along the Nile River today is 231 Bq per kilogram of carbon. If a papyrus scroll found near the Great Pyramid at Cairo has a carbon-14 radioactivity of 127 Bq per kilogram of carbon, how old is the scroll?

**21.113.** Thorium-232 slowly decays to bismuth-212 ($t_{1/2} = 1.4 \times 10^{10}$ yr). Bismuth-212 decays to lead-208 by two pathways: first β decay and then α decay, or α decay and then β decay. The intermediate nuclide in the second pathway is thallium-208. The thallium-208 can be separated from a sample of thorium nitrate by passing a solution of the sample through a filter pad containing ammonium phosphomolybdate. The radioactivity of $^{208}$Tl trapped on the filter is measured as a function of time. In one such experiment, the following data were collected:

| Time (s) | Counts/min |
|---|---|
| 60 | 62 |
| 120 | 40 |
| 180 | 35 |
| 240 | 22 |
| 300 | 16 |
| 360 | 10 |

Use the data in the table to determine the half-life of $^{208}$Tl.

21.114. The following nuclear equations are based on successful attempts to synthesize supermassive elements. Complete each equation by writing the symbol of the supermassive nuclide that was synthesized.

a. $^{58}_{26}\text{Fe} + ^{209}_{83}\text{Bi} \rightarrow ? + ^{1}_{0}\text{n}$
b. $^{64}_{28}\text{Ni} + ^{209}_{83}\text{Bi} \rightarrow ? + ^{1}_{0}\text{n}$
c. $^{62}_{28}\text{Ni} + ^{208}_{82}\text{Pb} \rightarrow ? + ^{1}_{0}\text{n}$
d. $^{22}_{10}\text{Ne} + ^{249}_{97}\text{Bk} \rightarrow ? + 4\,^{1}_{0}\text{n}$
e. $^{58}_{26}\text{Fe} + ^{208}_{82}\text{Pb} \rightarrow ? + ^{1}_{0}\text{n}$

**21.115.** The absorption of a neutron by $^{11}$B produces a radioactive nuclide that decays by either α decay or β decay. Write balanced nuclear equations describing the decay reactions.

21.116. An atom of darmstadtium-269 was synthesized in 2003 by bombarding a $^{208}$Pb target with $^{62}$Ni nuclei. Write a balanced nuclear equation describing the synthesis of $^{269}$Ds.

**21.117.** The origins of the two naturally occurring isotopes of boron, $^{11}$B and $^{10}$B, are unknown. Both isotopes may be formed from collisions between protons and carbon, oxygen, or nitrogen in the aftermath of supernova explosions. Write balanced nuclear equations describing how $^{10}$B might be formed from such collisions with $^{12}$C and $^{14}$N.

21.118. **Isotopes in Geochemistry** The relative abundances of the stable isotopes of the elements are not entirely constant. For example, in some geological samples (soils and rocks), the ratio of $^{87}$Sr to $^{86}$Sr is affected by the presence of a radioactive isotope of another element, which slowly undergoes β decay to produce more $^{87}$Sr. What is this other isotope?

**21.119.** Rhodium-105 is made by neutron bombardment of $^{104}$Ru, which decays to $^{105}$Rh with a half-life of 4.4 hr.

a. Write a balanced nuclear equation for the formation of $^{105}$Rh.
b. Calculate how long 99% of the $^{105}$Rh will take to decay.

21.120. **Biblical Archaeology** In 2012, Biblical scholar Karen King at Harvard University suggested that a piece of ancient papyrus believed to date from 400 CE (Figure P21.120) portrays Jesus as having a wife. Other researchers have argued that the document is a forgery. Their argument relies on both linguistic interpretation and results of radiocarbon dating. Carbon-14 testing of the papyrus revealed a date in the 8th century CE.

a. By what percentage is the ratio of $^{14}$C to $^{12}$C in a sample from 750 CE smaller than the ratio in modern papyrus?
b. How much smaller is this ratio in papyrus from 400 CE than in papyrus from 750 CE?

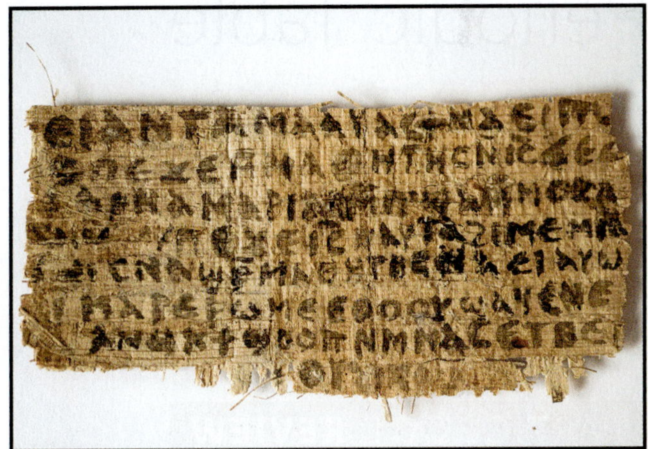

**FIGURE P21.120**

# 22

# The Main Group Elements

## Life and the Periodic Table

**DIETARY SUPPLEMENTS** Calcium is included in many vitamins and dietary supplements to promote healthy bones and teeth.

PARTICULATE **REVIEW**

### *Nitrogen, Sulfur, and Phosphorus: One, Two, or All Three?*

In Chapter 22, we survey the roles of main group elements in living organisms. In addition to carbon, hydrogen, and oxygen, atoms of nitrogen, sulfur, and phosphorus are fundamental building blocks for important biological molecules in plants and animals.

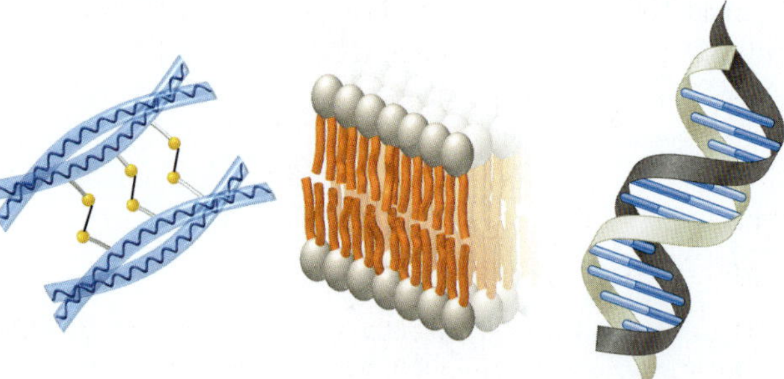

- Which of these structures contains sulfur atoms? Where are the sulfur atoms?
- Which two structures contain phosphorus atoms? Where are the phosphorus atoms?
- All three structures contain nitrogen atoms. List the functional group(s) that contains nitrogen atoms in each structure.

(Review Chapter 20 if you need help.)

*(Answers to Particulate Review questions are in the back of the book.)*

PARTICULATE **PREVIEW**

### *Charge versus Size: Selective Transport*

Shown here are representations of four essential cations involved in selective ion transport through channels in cell membranes: $Ca^{2+}$, $H_3O^+$, $K^+$, and $Na^+$. The size of each cation is listed. As you read Chapter 22, look for ideas that will help you answer these questions:

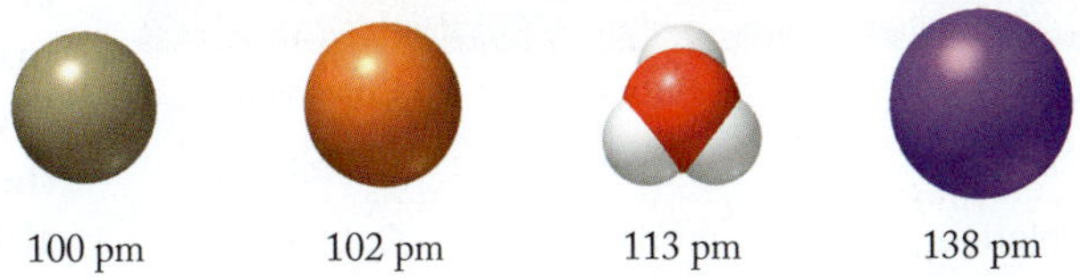

100 pm

102 pm

113 pm

138 pm

- Identify each of the four ions pictured.
- The $Na^+$ channel is selective, allowing only $Na^+$ cations and one of the other three cations to pass through it. Which other ion—$Ca^{2+}$, $H_3O^+$, or $K^+$—is most likely to undergo selective transport through the sodium channel?
- Which of the other three ions, if any, is likely to pass through the $K^+$ channel?

## Learning Outcomes

**LO1** Distinguish between essential and nonessential elements and between major, trace, and ultratrace elements

**LO2** Describe the pathways and calculate the free energy and electrochemical potential driving ion transport across cell membranes
**Sample Exercise 22.1**

**LO3** Balance equations, draw structures, and carry out calculations relevant to the behavior of major, trace, and ultratrace elements and their compounds in vivo
**Sample Exercises 22.2–22.4**

**LO4** Describe the function of the essential and nonessential group 14–16 elements in the human body

**LO5** Describe the health hazards posed by some radioactive isotopes of main group elements and the use of others in medical diagnosis and therapy
**Sample Exercise 22.5**

**LO6** Describe how compounds containing main group elements are used in therapy and in other biomedical applications

## 22.1 Main Group Elements and Human Health

How many of the elements in the periodic table are in the human body? Moreover, which of them are important to the health of humans?

Roughly one-third of the 90 naturally occurring elements have an identifiable role in human health and in organisms in general. **Essential elements** are defined as those that have a beneficial physiological effect, including those whose absence impairs functioning of the organism (**Table 22.1**). Some—such as carbon, hydrogen, oxygen, nitrogen, sulfur, and phosphorus—are the principal constituents of all plants and animals. The alkali metal cations $Na^+$ and $K^+$ act as charge carriers, maintain osmotic pressure, and transmit nerve impulses. The alkaline earth cation $Mg^{2+}$ is important in photosynthesis, and its family member $Ca^{2+}$ ions are components of structural tissues such as bones and teeth. Chloride ions balance the charge of $Na^+$ and $K^+$ ions to maintain electrical neutrality in living cells. Other main group elements, such as iodine and selenium, are required in tiny amounts in our bodies to regulate metabolism and as components of the enzymes that catalyze biochemical reactions.

The essential elements are further classified as **major**, **trace**, or **ultratrace essential elements**. Major essential elements are present at concentrations greater than 1 mg of element per gram of body mass. Almost all foods are rich in compounds containing carbon, hydrogen, oxygen, nitrogen, sulfur, and phosphorus. Salt is perhaps the most familiar dietary source of sodium and chloride ions, although both are ubiquitous in food. Vegetables such as broccoli and brussels sprouts and fruits such as bananas are rich in potassium. Calcium is found in dairy products and is often added to orange juice as a dietary supplement.

Many **nonessential elements** are also present in the body but have no known function (**Table 22.2**). Some are useful in medicine as either diagnostic tools or therapeutic agents. Some radioactive isotopes may be used as imaging agents for organs and tumors, for example, whereas others are used to treat disease. Drugs containing lithium ions are used to treat depression. Compounds of bismuth act as mild antibacterial agents for treatment of diarrhea, and

**TABLE 22.1** Essential Elements Found in the Human Body

| Major (>1 mg/g of body mass) | Trace (1–1000 μg/g of body mass) | Ultratrace (<1 μg/g of body mass) |
|---|---|---|
| Calcium | Fluorine | Chromium |
| Carbon | Iodine | Cobalt |
| Chlorine | Iron | Copper |
| Hydrogen | Silicon | Manganese |
| Magnesium | Zinc | Molybdenum |
| Nitrogen | | Nickel |
| Oxygen | | Selenium |
| Phosphorus | | Vanadium |
| Potassium | | |
| Sodium | | |
| Sulfur | | |

antimony compounds represent one of the few options for patients suffering from the tropical disease leishmaniasis, caused by protozoan parasites.

The presence of a nonessential element sometimes has a **stimulatory effect**, which means that consuming small amounts of the element causes increased activity or growth in an organism. The effect may be beneficial or not, and often the mechanism of the effect is not understood. For example, small amounts of the nonessential element antimony promote growth in some mammals when added to their diets. Nonessential elements, and even toxic elements, are often incorporated into our bodies because their chemical properties are similar to those of an essential element. For example, $Rb^+$ ions are retained by the human body because they are similar to $K^+$ ions in size, charge, and chemistry, making rubidium the most abundant nonessential element in humans.

Oxygen, in the form of $O_2$ gas, occurs in the body in elemental form; oxygen is also incorporated into many compounds, such as $H_2O$, and many ions, such as $HCO_3^-$. When we speak of any element in the body other than oxygen, however, we usually refer to an ion or compound containing that element rather than the pure element. For example, when we describe calcium as an essential element, we are referring to calcium ions, $Ca^{2+}$, not calcium metal.

**Table 22.3** compares the elemental compositions of the human body, the universe, Earth's crust, and seawater. The composition of our bodies most closely resembles the composition of seawater. The match would be even closer if not for the biological processes in the sea that remove essential elements such as nitrogen

**TABLE 22.3** Percent Composition[a] of the Universe, Earth's Crust, Seawater, and the Human Body

| Element | Universe | Crust | Seawater | Human Body |
|---|---|---|---|---|
| Hydrogen | 91 | 0.22 | 66 | 63 |
| Oxygen | 0.57 | 47 | 33 | 25.5 |
| Carbon | 0.021 | 0.019 | 0.0014 | 9.5 |
| Nitrogen | 0.042 | | | 1.4 |
| Calcium | | 3.5 | 0.006 | 0.31 |
| Phosphorus | | | | 0.22 |
| Chlorine | | | 0.33 | 0.03 |
| Potassium | | 2.5 | 0.006 | 0.06 |
| Sulfur | 0.001 | 0.034 | 0.017 | 0.05 |
| Sodium | | 2.5 | 0.28 | 0.01 |
| Magnesium | 0.002 | 2.2 | 0.033 | 0.01 |
| Helium | 9.1 | | | |
| Silicon | 0.003 | 28 | | |
| Aluminum | | 7.9 | | |
| Neon | 0.003 | | | |
| Iron | 0.002 | 6.2 | | |
| Bromine | | | 0.0005 | |
| Titanium | | 0.46 | | |
| All other elements | < 0.1 | < 0.1 | < 0.1 | < 0.1 |

[a]Compositions are expressed as the percentage of the total number of atoms. Because of rounding, the totals do not sum to 100%.

**TABLE 22.2** Nonessential Elements Found in the Human Body

| Stimulatory | Unknown Role | No Role |
|---|---|---|
| Boron | Antimony | Barium |
| Titanium | Arsenic | Bromine |
| | | Cesium |
| | | Germanium |
| | | Rubidium |
| | | Strontium |

**essential element** an element present in tissue, blood, or other body fluids that has a physiological function.

**nonessential element** an element present in humans that has no known function.

**stimulatory effect** increased activity, growth, or other biological response to the presence of a nonessential element.

**major essential element** an essential element present in the body in average concentrations greater than 1 mg of element per gram of body mass.

**trace essential element** an essential element present in the body in average concentrations between 1 μg and 1000 μg of element per gram of body mass.

**ultratrace essential element** an essential element present in the body in average concentrations less than 1 μg of element per gram of body mass.

**TABLE 22.4 Dietary Reference Intakes (DRIs) and Recommended Dietary Allowances (RDAs) for Selected Essential Elements**[a]

| Element | DRI | RDA |
|---|---|---|
| Calcium | 1000 mg | 1200 mg |
| Chlorine | 2300 mg | 2300 mg |
| Chromium | 25–35 μg | 35 μg |
| Copper | 900 μg | 900 μg |
| Fluorine | 3–4 mg | 4 mg |
| Iodine | 150 μg | 150 μg |
| Iron | 8–18 mg | 18 mg |
| Magnesium | 420 mg | 320–400 mg |
| Manganese | 1.8–2.3 mg | 2–5 mg |
| Molybdenum | 45 μg | 45 μg |
| Phosphorus | 700 mg | 700 mg |
| Potassium | 4700 mg | 4700 mg |
| Selenium | 55 μg | 55 μg |
| Sodium | 1500 mg | 1500 mg |
| Zinc | 8–11 mg | 11 mg |

[a]DRI and RDA values in milligrams or micrograms per day from the U.S. Department of Agriculture (2009) and from the Council on Responsible Nutrition (CRN) for 19- to 30-year-olds.

and phosphorus and that store others in solid structures such as the $CaCO_3$ that makes up corals and mollusk shells.

Our diet should supply us with enough of all essential elements. In the United States and Canada, these quantities are called *dietary reference intake* (DRI) values. They are based on the recommendations of the Food and Nutrition Board of the National Academy of Sciences and are often updated in response to research. For many essential elements, DRI values have replaced the *recommended dietary allowance* (RDA) values you may be familiar with from labels on food and vitamin packages (**Figure 22.1**). **Table 22.4** compares the DRI and RDA values for several major, trace, and ultratrace essential elements.

Some elements, such as radon, beryllium, and lead, are toxic. As described in Section 21.8, inhaled radon gas poses serious health hazards when it undergoes α decay inside the body. Beryllium toxicity is most often encountered in industrial settings where beryllium-contaminated dust is inhaled; the $Be^{2+}$ ion replaces $Mg^{2+}$ in the body, where it inhibits $Mg^{2+}$-catalyzed RNA and DNA synthesis in cells. Lead ions, $Pb^{2+}$, are incorporated into teeth and bones because they are similar to $Ca^{2+}$ ions in size and charge. Lead also interferes with the functioning of enzymes that require calcium ions and thereby causes chronic neurological problems and blood-based disorders, especially in children. This issue became a major news story early in 2016 when it was discovered that a budget-cutting decision to switch drinking-water sources in Flint, Michigan, had exposed as many as 8000 children under age 6 to unsafe levels of lead. In June 2014, the city's source of water was changed from Lake Huron to the Flint River. The river water proved more corrosive than the lake water and leached lead from old water pipes, causing lead levels to rise to more than 100 ppb; the maximum acceptable concentrations of lead in drinking water in the U.S. and Canada are 15 and 10 ppb, respectively. Unsafe levels of lead have been identified in other cities with aging water systems.

Here in Chapter 22, we review the periodic properties of the main group elements and survey the roles of selected elements in the human body, as well as their importance to good health. At the same time, we call on the knowledge and skills you have acquired in your study of general chemistry to solve problems that link concepts from prior chapters to the central question of this chapter: What are the

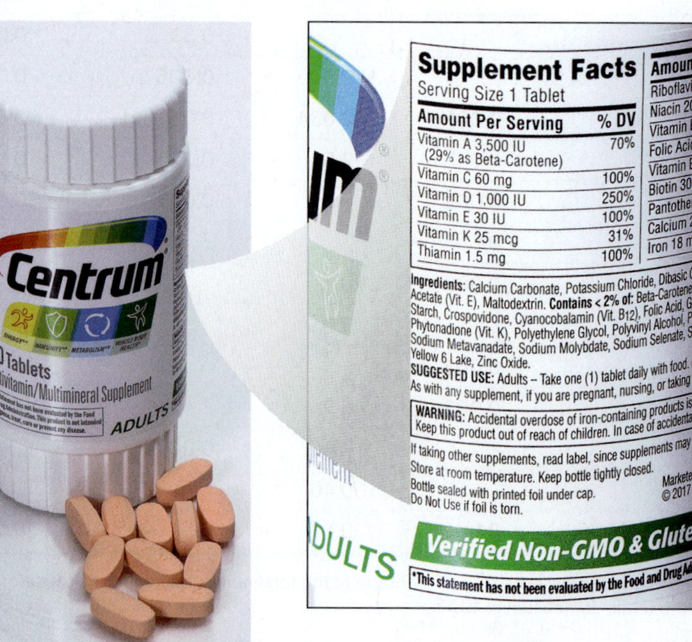


**FIGURE 22.1** The labels on multivitamin supplements may not list DRI or RDA values but rather *% daily values* (DVs). Daily values are based on RDA or DRI values, but inconsistencies can occur, particularly among the ultratrace essential elements.

roles of the main group elements in the chemistry of life? The roles of transition metal ions such as $Fe^{2+}$, $Fe^{3+}$, and $Zn^{2+}$ in biology are addressed in Chapter 23.

## 22.2 Periodic Properties of Main Group Elements

The main group or representative elements are found in groups 1, 2, and 13–18 in the periodic table. These eight groups, comprising 44 naturally occurring elements, include five elements for which no stable isotopes are known: Fr, Ra, Po, At, and Rn. Relatively little is known about the chemistry of francium or astatine, both of which are exceedingly rare. Estimates indicate that the outermost kilometer of Earth's crust contains at most 44 mg of At and only 15 g of Fr. They have no proven uses, so we will largely ignore them in our discussions. Before we turn to the role of the main group elements in life, it is useful to review what we have already learned about the physical and chemical properties of these elements and to look for periodic trends in these properties.

**CONNECTION** We explored the origin of some periodic trends in Sections 3.10–3.12.

The atomic radii of the main group elements increase with increasing atomic number within each group and decrease left to right across each period. In other words, the largest elements are found to the left and toward the bottom of the periodic table (**Figure 22.2a**). First ionization energies and electronegativities decrease down a group and increase left to right across a period, so that the main group elements with the greatest electronegativities are located in the upper right corner of the periodic table (**Figure 22.2b**). An overall trend toward more negative electron affinities is seen left to right across a period, although several anomalies are also observed. Each of these periodic trends reflects the changes in the effective nuclear charge and shielding of the valence electrons by core electrons as atomic numbers increase and inner shells fill with electrons.

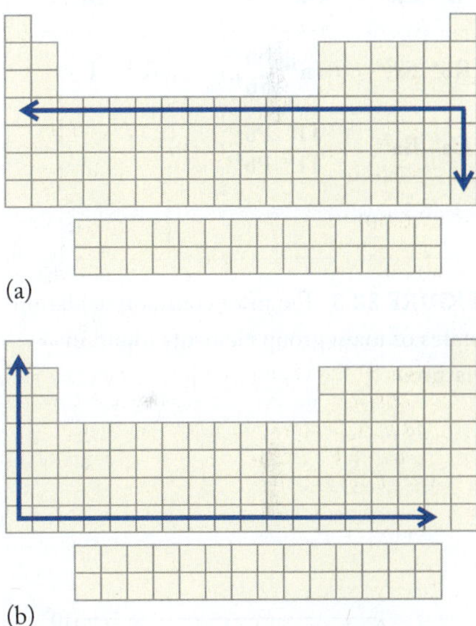


**FIGURE 22.2** The symbols of the main group elements with (a) the largest atomic radii are located to the left and toward the bottom of the periodic table. (b) Those with the greatest electronegativities and first ionization energies are located to the right and toward the top.

A survey of the main group elements reveals a variety of physical properties. The alkali metal and alkaline earth elements are all solids, whereas all the group 18 elements (the noble gases) are gases at standard temperature and pressure. All group 17 elements (the halogens) exist as diatomic molecules. Group 17 is also the only group that contains elements in all three phases of matter at room temperature: bromine is one of two liquid elements in the periodic table, iodine is a volatile solid, and the remaining elements are gases. Groups 13–16 include seven elements classified as metalloids: B, Si, Ge, As, Sb, Te, and At, as well as elements with metallic properties: Al, Ga, In, Tl, Sn, Pb, Bi, and Po. The remaining elements in these groups behave as nonmetals with the properties described in Chapter 2: they are gases ($N_2$, $O_2$) or brittle solids that are poor conductors of electricity (C, P, S, and Se). For groups 13–16, metallic properties increase down a group.

The melting and boiling points of the main group elements decrease down the metals in groups 1 and 2, but this trend is reversed for the nonmetals in groups 17 and 18 (**Table 22.5**). These trends can be understood in terms of the

**TABLE 22.5** Periodic Trends for the Main Group Elements

| | TREND FOR GROUP NUMBER: | | | | | | | |
|---|---|---|---|---|---|---|---|---|
| Property | 1 | 2 | 13 | 14 | 15 | 16 | 17 | 18 |
| Melting point | Decreases going down | Decreases going down | No single trend | No single trend | No single trend | No single trend | Increases going down | Increases going down |
| Boiling point | Decreases going down | Decreases going down | No single trend | Decreases going down | No single trend | No single trend | Increases going down | Increases going down |

**CONNECTION** Ionic compounds were defined in Section 1.6, and organic compounds were introduced in Section 6.3. Covalent bonding was discussed in detail in Chapters 4 and 5.

**CONNECTION** The organization of the periodic table was introduced in Chapter 2. Periodic properties were discussed in Chapter 3, and the stability of nuclei was addressed in Chapter 21.

types of forces holding the atoms together. The increase in the size of the metallic elements when descending groups 1 and 2 leads to weaker metallic bonds and lower boiling points. In groups 17 and 18, however, increasing size leads to greater London dispersion forces as the polarizability of the atoms increases. The trends in melting points for groups 13–16 do not fit a clear pattern, in part because the properties of the elements change from nonmetallic to metallic down these groups.

Of the main group elements, only hydrogen, carbon, nitrogen, oxygen, sulfur, and the six noble gases exist in nature in elemental form. The remaining elements occur exclusively in ionic and covalent compounds. The group 1 and 2 elements readily lose their valence electrons, forming 1+ and 2+ cations, respectively, and familiar ionic compounds with halide anions such as NaCl and KI (**Figure 22.3**). Halogen compounds of groups 13–17 generally contain covalent bonds; however, the heavier group 13 and 14 elements form insoluble ionic salts such as $PbCl_2$ and TlCl. Main group metallic elements form widely used oxides, including CaO (quicklime, used in manufacturing steel), and $Al_2O_3$ (alumina, used in orthodontics and abrasives).

| 1 | 2 | 13 | 14 | 15 | 16 | 17 | 18 |
|---|---|---|---|---|---|---|---|
| $H^+$ | | | | | | | |
| $Li^+$ | | | | $N^{3-}$ | $O^{2-}$ | $F^-$ | |
| $Na^+$ | $Mg^{2+}$ | $Al^{3+}$ | | $P^{3-}$ | $S^{2-}$ | $Cl^-$ | |
| $K^+$ | $Ca^{2+}$ | $Ga^{3+}$ | | | $Se^{2-}$ | $Br^-$ | |
| $Rb^+$ | $Sr^{2+}$ | $In^{3+}$ | $Sn^{2+}$ $Sn^{4+}$ | | $Te^{2-}$ | $I^-$ | |
| $Cs^+$ | $Ba^{2+}$ | $Tl^+$ $Tl^{3+}$ | $Pb^{2+}$ $Pb^{4+}$ | | | | |
| | | | | | | | |

**FIGURE 22.3** The most common oxidation states of main group elements found in nature.

The lighter elements of groups 13–17 tend to form covalent bonds rather than ionic bonds. The result is a library of more than 150 million organic compounds, covalently bonded substances composed primarily of carbon, nitrogen, hydrogen, and oxygen. More than 95% of the new compounds registered in a given year are classified as organic compounds.

Hydrogen, the smallest and lightest element, is difficult to classify. Most periodic tables include hydrogen in group 1 on the basis of its electron configuration of a half-filled *s* orbital and on the dissociation of acids to protons and anions, analogous to the dissolution of alkali metal salts to 1+ cations and anions. Hydrogen also has a complete valence shell after gaining an electron to form a hydride ion, $H^-$, which is isoelectronic with He, a noble gas. Compounds called metal hydrides form between hydrogen and group 1, 2, or 13 metals, and they behave as salts containing a metal cation and a hydride anion. However, hydrogen is most commonly covalently bonded in compounds to oxygen (as water, $H_2O$) or to another group 14–16 element.

## 22.3 Major Essential Elements

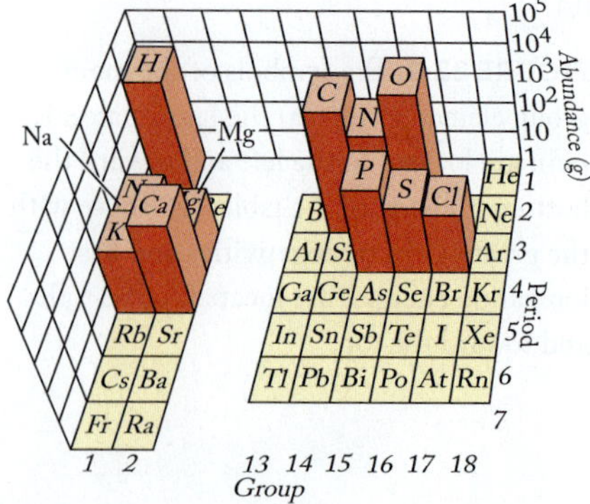


**FIGURE 22.4** The 11 elements shown in red are the major essential elements. Their abundances in a 70-kg adult human range from 35 g of magnesium to 46 kg of oxygen.

The 11 elements shown in red in **Figure 22.4** and listed in the first column of Table 22.1 are the major essential elements. Together they account for more than 99% of the mass of the human body. Oxygen is the most abundant element by mass, followed by carbon and hydrogen. Although life depends on the presence of elemental oxygen in the form of $O_2$ gas, much of the oxygen in our bodies is combined with hydrogen in water molecules.

The most abundant elements in the human body include seven nonmetals: C, H, O, Cl, S, P, and N. They are the building blocks for most of the body's molecular compounds and its principal polyatomic ions, $HCO_3^-$, $SO_4^{2-}$, and $H_2PO_4^-$, which are dissolved in body fluids. The average concentrations of the four major metals in the human body—$Ca^{2+}$, $K^+$, $Na^+$, and $Mg^{2+}$—are listed in **Table 22.6**. Here in Section 22.3 we explore some of the roles that sodium, potassium, magnesium, calcium, chlorine, nitrogen, phosphorus, and sulfur play in the biochemistry of the human body. As we do, we revisit several chemical principles discussed in earlier chapters.

## Sodium and Potassium

Regulated concentrations of sodium and potassium ions are crucial to cell function. For some people, consuming too much $Na^+$ has been linked to hypertension (high blood pressure). For a constant concentration of these two alkali metal ions to be maintained in body fluids, the ions must be able to move into and out of cells. As noted in Section 20.4, the membrane surrounding a typical cell is a lipid bilayer, with polar groups containing phosphate groups on the two surfaces of the cell membrane and nonpolar fatty acids oriented toward the interior of the membrane. Direct diffusion of $Na^+$ and $K^+$ through the lipid bilayer is difficult because these polar cations do not dissolve in the nonpolar interior.

As **Figure 22.5** shows, the cell membrane is pierced by **ion channels**, which are groups of protein complexes that allow selective transport of ions. The ion channels control which ions pass through the membrane, on the basis of the size and charge of the ion as well as the shape of the protein. For example, the protein complexes of the potassium ion channel have amino acids oriented so that favorable ion–dipole interactions occur only for ions with the radius of a $K^+$ ion but not for $Na^+$ or any other cation. The sodium ion channel is also selective, excluding $K^+$ and $Ca^{2+}$ even though the radii of $Na^+$ and $Ca^{2+}$ differ by only 2 pm. Another difference between the $Na^+$ and $K^+$ channels is the ability of $H_3O^+$ to pass through $Na^+$ channels but not $K^+$ channels.

Living organisms also contain oxygen-rich molecules such as the Lewis base nonactin (**Figure 22.6**) that bond to $Ca^{2+}$, $K^+$, $Na^+$, and $Mg^{2+}$ ions (Lewis acids) through strong ion–dipole forces. The resulting complex ions consist of a polar, charged alkali metal ion encapsulated in a nonpolar exterior. The complex ion has both a polar portion and a nonpolar portion; therefore it does not require a channel for passage through a cell membrane. Instead, the complex ion carries its alkali metal cation through both the polar and nonpolar regions of the bilayer, providing an alternative to ion channels for the transport of these ions. We revisit the bonding between Lewis acids and bases in more detail in Chapter 23.

In addition to ion channels and diffusion of complex ions, alkali metal cations can be transported by a third mechanism, one involving $Na^+$–$K^+$ ion pumps. An **ion pump** is a system of membrane proteins that exchange ions inside the cell

**ion channel** a group of helical proteins that penetrate cell membranes and allow selective transport of ions.

**ion pump** a system of membrane proteins that exchange ions inside the cell with those in the intercellular fluid.

**TABLE 22.6** Average Concentration of Four Metallic Elements in the Human Body

| Element | Concentration (mg/g of Body Mass) |
|---|---|
| Calcium | 15.0 |
| Potassium | 2.0 |
| Sodium | 1.5 |
| Magnesium | 0.5 |

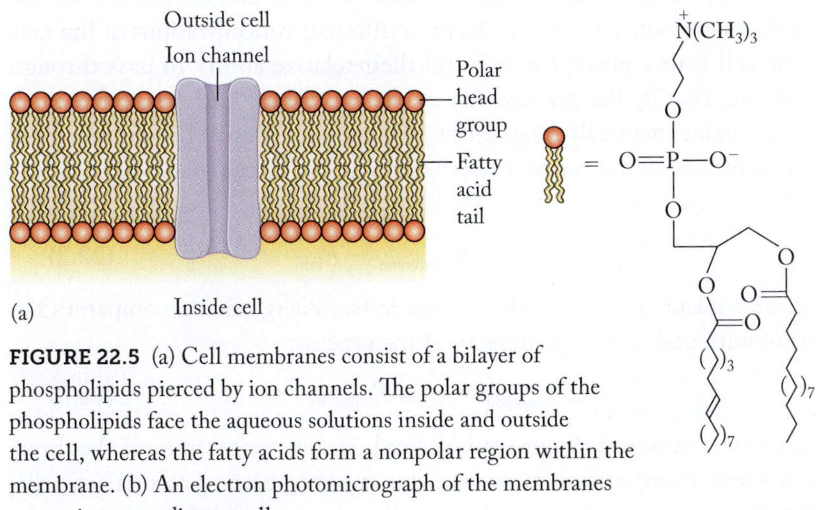


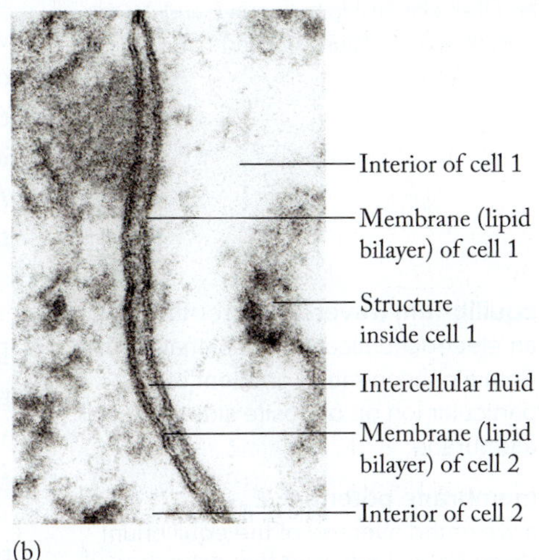


**FIGURE 22.5** (a) Cell membranes consist of a bilayer of phospholipids pierced by ion channels. The polar groups of the phospholipids face the aqueous solutions inside and outside the cell, whereas the fatty acids form a nonpolar region within the membrane. (b) An electron photomicrograph of the membranes separating two adjacent cells.

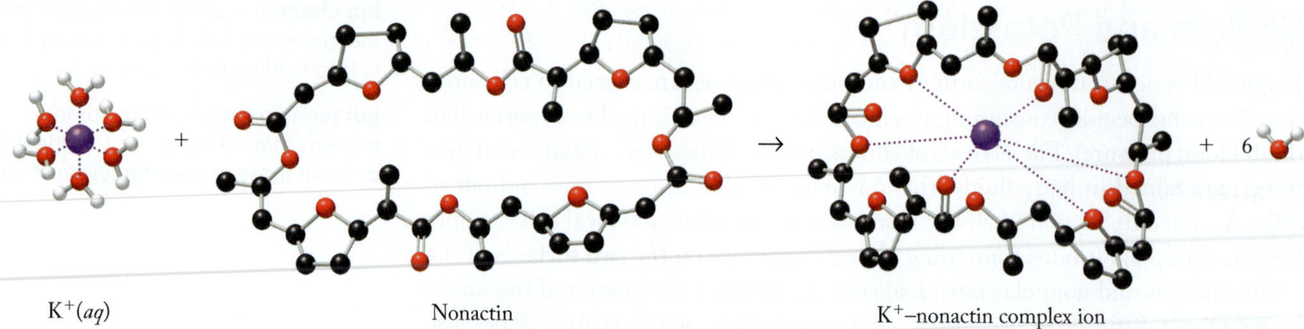


**FIGURE 22.6** In living organisms, ligands such as nonactin can form complex ions with any one of the four alkali or alkaline earth major essential ions and carry the ions directly through cell membranes. An ion channel is *not* required in this transport pathway.

(such as $Na^+$) with those in the intercellular fluid (such as $K^+$). Unlike diffusion or transport through ion channels, transport via the $Na^+$–$K^+$ pump requires energy, which is provided by the hydrolysis of adenosine triphosphate (ATP) to adenosine diphosphate (ADP). An example of how the $Na^+$–$K^+$ ion pump works is the response of a nerve cell to touch. Stimulating the nerve cell causes $Na^+$ to flow into the cell and $K^+$ to flow out via ion channels; this two-way flow of ions produces the nerve impulse. The ion pump then "recharges" the system by pumping $Na^+$ out of the cell and $K^+$ into the cell so that another impulse can immediately be transmitted along the nerve.

**CONNECTION** In Section 6.3, we described how alkali metal cations dissolved in water are surrounded by six water molecules (a sphere of hydration). Each water molecule is oriented so that the oxygen atoms point toward the cation (see Figure 6.17). In Sections 16.5 and 16.6, we described this interaction as an example of a Lewis acid (cation, electron-pair acceptor) interacting with a Lewis base or ligand (water, electron-pair donor).

Unequal concentrations of $Na^+$, $K^+$, and other ions on opposite sides of a cell membrane produce an electrochemical **equilibrium potential** or **reversal potential ($E_{ion}$)** for each ion that can be calculated using the following variation of the Nernst equation (introduced in Section 17.6) in which $n$ now represents the *charge* on one of the ions and the $Q$ term has been replaced with *the ratio of the concentrations of the ion outside the cell versus inside the cell*:

$$E_{ion} = \frac{0.0592\ \text{V}}{n} \log \frac{[M^{x+}]_{outside}}{[M^{x+}]_{inside}} \tag{22.1}$$

**CONNECTION** The Nernst equation and the relationship between $E$ and $\Delta G$ are described in Sections 17.6 and 17.4, respectively.

A related overall **membrane potential ($E_{membrane}$)** represents contributions from all the major ions, both cations and anions, inside and outside a cell. The values of the membrane potential reflect the different concentrations of the ions outside the cell versus inside the cell and their relative ability to pass through the membrane, that is, the *permeability* of the membrane with respect to each ion. $E_{membrane}$ values typically range from −50 mV to −70 mV. The driving force $E_{transport}$ that pushes an ion across a cell membrane can be calculated from Equation 22.2:

$$E_{transport} = E_{membrane} - E_{ion} \tag{22.2}$$

Equation 22.3 relates $E_{transport}$ to the change in free energy that accompanies the transport of ions, and to the spontaneity of the process:

$$\Delta G_{transport} = -nFE_{transport} \tag{22.3}$$

**equilibrium (reversal) potential, $E_{ion}$** an electrochemical potential that results from a concentration gradient of a particular ion on opposite sides of a cell membrane.

**membrane potential, $E_{membrane}$** a weighted average of the equilibrium (reversal) potentials of the major ions based on concentration and permeability of the individual ions.

Ion channels, transporters, and pumps, and the ion concentrations they control, are a finely tuned system that regulates the membrane potential of cells. Rapidly multiplying normal cells such as fertilized eggs and differentiating stem cells have membrane potentials much less negative than the typical values and may be as low as zero to −10 mV. It has recently been suggested that membrane potential plays a role in wound healing, and some cancer cells have membrane potentials that are also in the 0 mV to −10 mV range.

### SAMPLE EXERCISE 22.1 Calculating $E_{ion}$ for Ion Transport **LO2**

Nerve cells contain structures called axons whose function relies on concentration differences on opposite sides of a membrane. The concentration of $Na^+$ inside a squid axon is 0.050 *M*, in comparison with $[Na^+] = 0.440\ M$ in the fluid surrounding the cell.

a. Calculate the equilibrium potential $E_{ion}$ for the $Na^+$ ion at 25°C.
b. If the membrane potential $E_{membrane} = -0.050$ V, is the transport of $Na^+$ from inside to outside the cell membrane spontaneous?

**Collect, Organize, and Analyze** We are given the concentrations of $Na^+$ on opposite sides of a cell membrane, and we are asked to calculate $E_{ion}$ for $Na^+$ and to determine whether the transport across the membrane is spontaneous. Equation 22.1 allows us to calculate the equilibrium potential that arises from an unequal concentration of an ion on either side of a permeable membrane. If $[Na^+_{outside}] > [Na^+_{inside}]$, then the log term in Equation 22.3 will be greater than 1 and $E_{ion}$ will be positive. Using Equations 22.2 and 22.3, we can calculate the change in free energy during transport. If $\Delta G$ is negative, the process is spontaneous; if positive, it is nonspontaneous.

**Solve**

a. Substitution into Equation 22.3 gives:

$$E_{ion} = \frac{0.0592\ V}{n} \log \frac{[Na^+]_{outside}}{[Na^+]_{inside}} = \frac{0.0592\ V}{1+} \log \frac{0.440\ mol/L}{0.050\ mol/L} = 0.056\ V$$

b. Using Equation 22.4, we can calculate $E_{transport}$, which is the driving force available to push the ions across the membrane:

$$E_{transport} = E_{membrane} - E_{ion} = -0.050\ V - (0.056\ V) = -0.106\ V$$

Substituting $E_{transport}$ − 0.106 V into Equation 22.3:

$$\Delta G_{transport} = -nFE_{transport} = (1\ mol)(96{,}500\ \cancel{C}/mol)(-0.106\ \cancel{V})(1\ J/\cancel{C}-\cancel{V})$$
$$= 10{,}200\ J = 10.2\ kJ$$

$\Delta G_{transport} > 0$, therefore, the transport process is nonspontaneous.

**Think About It** The equilibrium potential for $Na^+$ under these conditions is negative, consistent with our prediction. The transport of $Na^+$ from the inside to the outside of the cell membrane is spontaneous and goes against the concentration gradient under these conditions. The membrane potential (−0.050 V) would not have to vary by much to change the transport from spontaneous to nonspontaneous.

**Practice Exercise** The concentration of $K^+$ inside a frog muscle cell is 124 m*M* and 2.3 m*M* in the fluid surrounding the cell. Calculate the equilibrium potential for the $K^+$ ion at 298 K. If the membrane potential $E_{membrane}$ is −73 mV, will $K^+$ ions flow spontaneously through open $K^+$ ion channels into the cell from the fluid surrounding it?

*(Answers to Practice Exercises are in the back of the book.)*

### CONCEPT TEST

Is ion transport through an ion pump a spontaneous or nonspontaneous process?

*(Answers to Concept Tests are in the back of the book.)*

## Magnesium and Calcium

The biological roles of $Mg^{2+}$ and $Ca^{2+}$ are more varied than those of $Na^+$ and $K^+$. Calcium is a major component of teeth and bones. A prolonged deficiency of calcium can lead to osteoporosis (a disease characterized by low bone density), whereas high concentrations of calcium in muscle cells contribute to cramps. Most kidney stones are made of calcium oxalate or calcium phosphate. Magnesium deficiencies can reduce physical and mental capacity because of the role of $Mg^{2+}$ in transferring phosphate groups to and from ATP ($Mg^{2+}$ binds to ATP); slowing this transfer diminishes the amount of energy available to cells. Ion pumps maintain the cellular concentrations of $Mg^{2+}$ and $Ca^{2+}$.

Magnesium is a component of chlorophyll (**Figure 22.7**), which is one of several molecules used by plants to collect and capture light energy across the visible portion (400 nm to 700 nm) of the electromagnetic spectrum. Chlorophylls from different plants vary slightly in composition, but all contain magnesium (a Lewis acid) bonded to four nitrogen atoms (Lewis bases) by coordinate bonds. The presence of magnesium in chlorophyll does not account for the green color of the molecule, nor does it play a direct role in absorption of sunlight. The $Mg^{2+}$ ion orients the molecules in positions that allow energy to be transferred to the reaction centers where $H_2O$ is consumed and $O_2$ is produced during photosynthesis. $Mg^{2+}$ ions play important roles in ATP hydrolysis and ADP phosphorylation. The many $Mg^{2+}$-mediated ATP $\rightarrow$ ADP processes include transferring phosphate to glucose in the conversion of glucose to pyruvate and driving $Na^+$–$K^+$ ion pumps.

**CONNECTION** Lewis acids, Lewis bases and coordinate bonds were described in Sections 16.5 and 16.6.

To some extent, calcium ions also can mediate ATP hydrolysis, but these ions play other roles in the cell. They are necessary to trigger muscle contractions, for example—the calcium ions used for this purpose are stored in proteins. Recall that the action of $Na^+$–$K^+$ pumps is to generate nerve impulses. One effect of nerve impulses is to trigger the release of $Ca^{2+}$ ions from their storage proteins into the intracellular fluid. In a multistep process, muscle cells contract and relax as $Ca^{2+}$ ions are released. Once the muscle action is complete, the ions are returned to their storage proteins in a process coupled to $Mg^{2+}$-mediated ATP hydrolysis.

**CONNECTION** The role of ATP and ADP in metabolism was described in Section 12.8.

Of Na, K, Mg, and Ca—the alkali metal and alkaline earth major essential elements (see Figure 22.4)—only calcium plays a major role in forming teeth and bones. Mammalian bones are a *composite material*—a material containing a mixture of substances. About 30% of dry bone mass is elastic protein fibers. The rest of the mass consists of calcium compounds, including the mineral hydroxyapatite, $Ca_5(PO_4)_3(OH)$, which is also a principal component of teeth. Hydroxyapatite crystals are bound to the protein fibers in bone through phosphate groups.

**CONNECTION** The catabolism of glucose was described in Sections 12.8 and 20.3.

The shells of marine organisms are mostly calcium carbonate ($CaCO_3$) in a matrix of proteins and polysaccharides. Some magnesium is incorporated into the calcium carbonate outer shell of marine organisms that can carry out photosynthesis, such as algae and phytoplankton.

## Chlorine

Of all the halogens, only chlorine (as chloride ions) is present in concentrations large enough to be considered a major essential element in humans. Chloride ions are the most abundant anions in the human body and are involved in many

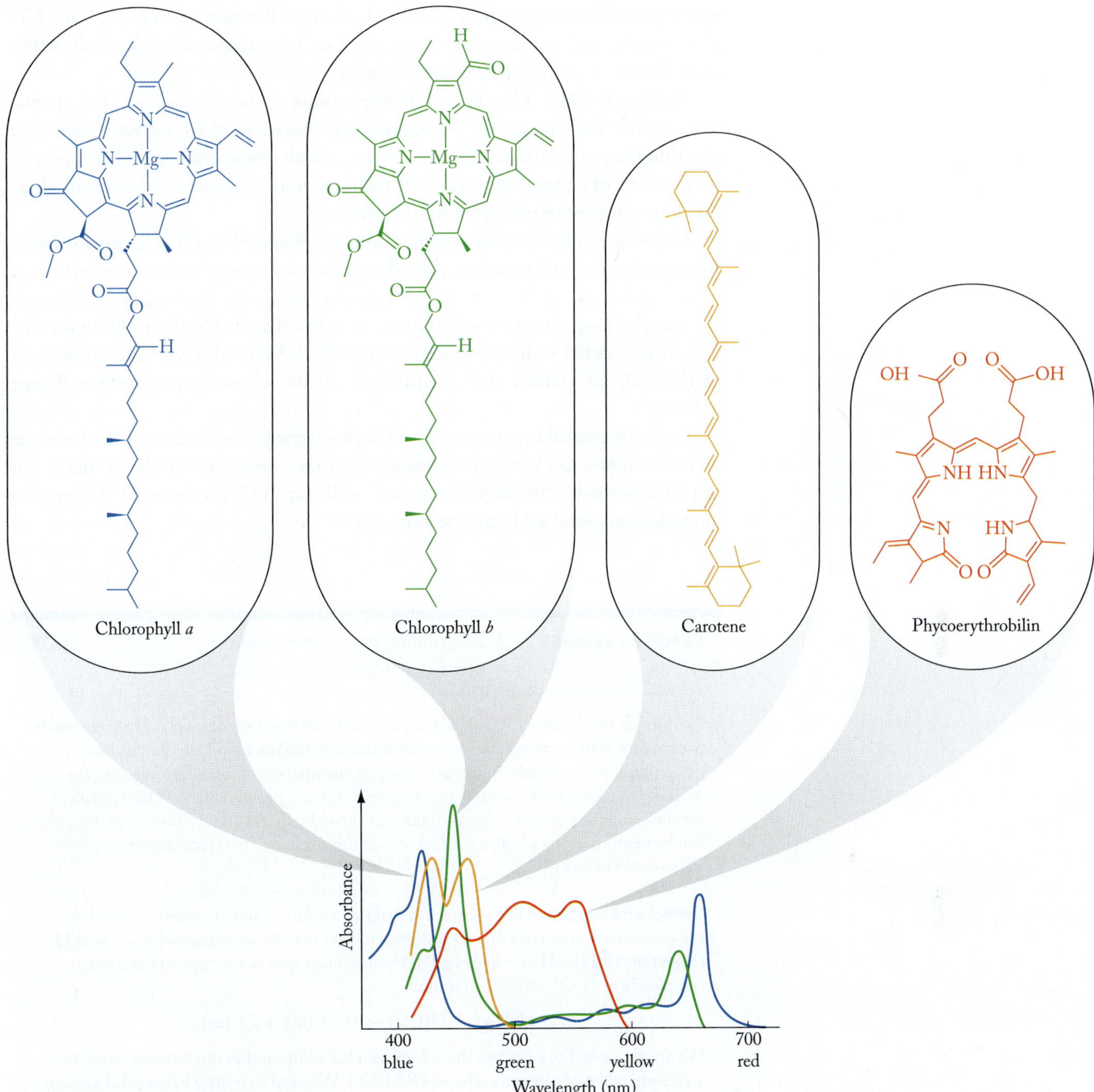


**FIGURE 22.7** Photosynthetic bacteria, green plants, and algae use a variety of molecules to absorb all the visible wavelengths in sunlight. Among them, only chlorophylls *a* and *b* contain magnesium and absorb blue-green and red-orange light. Carotene also absorbs in the blue-green region, whereas phycoerythrobilin absorbs a broad range of wavelengths from 400 nm to 600 nm.

processes. The concentration of chloride ions in the human body (1.5 mg per gram of body mass) is slightly less than one-tenth of the concentration of $Cl^-$ in seawater (19 mg per gram of water) but about 12 times greater than in Earth's crust (0.13 mg per gram of crust). Like the major essential cations, chloride ions are transported into and out of cells primarily via ion channels and ion pumps. To maintain electrical neutrality in a cell, the transport of alkali metal cations is accompanied by the transport of chloride anions. The *cotransport* of $Na^+$ and $Cl^-$ is essential in kidney function, where the ions are reabsorbed by the body rather than eliminated with liquid waste products.

Malfunctioning chloride ion channels cause cystic fibrosis, a lethal genetic disease that causes patients to accumulate mucus in their airways such that breathing becomes difficult. The discovery of high concentrations of $Na^+$ and $Cl^-$ in the sweat of cystic fibrosis patients led to an understanding of the role of chloride ion transport in patients with this disease.

Chloride ions also play a major role in eliminating $CO_2$ from the body. Because it is nonpolar, carbon dioxide produced during glucose catabolism can pass from muscle cells (for example) into red blood cells, moving easily through the largely nonpolar cell membranes of these cells. Inside the red blood cells, $CO_2$ is converted to bicarbonate ion ($HCO_3^-$). When $HCO_3^-$ is pumped out of the cell, $Cl^-$ enters the cell through an ion channel to maintain charge balance.

Chloride ion concentrations are high in gastric juices because of the presence of hydrochloric acid, which catalyzes digestive processes in the stomach. In response to food in the digestive system, cells tap ATP for the needed energy to pump hydrochloric acid into the stomach.

### SAMPLE EXERCISE 22.2 Calculating the Concentration of HCl in Stomach Acid — LO3

Acid reflux (sometimes called heartburn, though the heart is not involved) affects many people. It results from acid in the stomach leaking into the esophagus and causing discomfort. Stomach acid is primarily an aqueous solution of HCl. (a) Calculate the molarity of hydrochloric acid in gastric juice that has a pH of 0.80. (b) One treatment for the symptoms of acid reflux is to take an antacid tablet. What volume of gastric juice can be neutralized by a 750-mg tablet of calcium carbonate (a typical size for an over-the-counter antacid)?

**Collect and Organize** We are given the pH of a solution and are asked to calculate the concentration of HCl that corresponds to that pH. As we explained in Section 15.5 (Equation 15.12), $pH = -\log[H_3O^+]$. Hydrochloric acid is a strong acid and ionizes completely to $H_3O^+$ and $Cl^-$ in water:

$$HCl(aq) + H_2O(\ell) \rightarrow H_3O^+(aq) + Cl^-(aq)$$

We are also asked to calculate the volume of HCl solution that can be neutralized by a 750-mg tablet of calcium carbonate ($CaCO_3$). We need to write a balanced chemical equation for the neutralization reaction.

**Analyze** The equation describing the ionization of hydrochloric acid indicates that 1 mole of $H_3O^+$ ions is formed for every mole of HCl present. The pH of gastric juice falls between 1 and 0, so $[H_3O^+]$ will be between $10^{-1}$ (= 0.1) $M$ and $10^0$ (= 1) $M$.

The net ionic equation for the neutralization reaction is

$$CaCO_3(s) + 2\ H_3O^+(aq) \rightarrow Ca^{2+}(aq) + CO_2(g) + 3\ H_2O(\ell)$$

This equation indicates that 2 moles of $H_3O^+$ are consumed for every mole of $CaCO_3$. We are told that the tablet size is typical of an antacid tablet, so common sense suggests that the volume of acid this tablet can neutralize will not be excessively large (greater than 1 L) or small (less than 10 mL): too large a tablet would be a waste of antacid, and too small a tablet would not relieve the symptoms.

**Solve**

a. Substitution into Equation 15.12 gives

$$pH = -\log[H_3O^+] = 0.80$$

We take the antilog of both sides to solve for $[H_3O^+]$:

$$[H_3O^+] = 10^{-0.80} = 0.16\ M\ H_3O^+$$

Therefore, the concentration of HCl is

$$0.16\ M\ \cancel{H_3O^+} \times \frac{1\ \text{mol HCl}}{1\ \cancel{\text{mol}\ H_3O^+}} = 0.16\ M\ \text{HCl}$$

b. First, we calculate the number of moles of $CaCO_3$ present in 750 mg:

$$0.750\ \cancel{g\ CaCO_3} \times \frac{1\ \text{mol}\ CaCO_3}{100.09\ \cancel{g\ CaCO_3}} = 7.49 \times 10^{-3}\ \text{mol}\ CaCO_3$$

Next, we use the stoichiometry of the neutralization reaction to calculate the volume of 0.16 *M* HCl that this quantity of $CaCO_3$ can neutralize:

$$7.49 \times 10^{-3}\ \cancel{\text{mol}\ CaCO_3} \times \frac{2\ \cancel{\text{mol}\ H_3O^+}}{1\ \cancel{\text{mol}\ CaCO_3}} \times \frac{1\ L}{0.16\ \cancel{\text{mol}\ H_3O^+}}$$

$$= 9.36 \times 10^{-2}\ L = 94\ \text{mL of}\ 0.16\ M\ \text{HCl}$$

**Think About It** A concentration of 0.16 *M* seems reasonable because it is indeed within the range of values predicted for a solution with $pH < 1$. The volume of 0.16 *M* acid that a 750-mg tablet of $CaCO_3$ can neutralize is also reasonable; 94 mL represents about 3 ounces of gastric juice.

**Practice Exercise** Calculate the pH of a solution prepared by mixing 10.0 mL of 0.160 *M* HCl with 15.0 mL of water. How much antacid containing $4.00 \times 10^2$ mg of $Mg(OH)_2$ in 5.00 mL of water is needed to neutralize this volume of acid?

**CONCEPT TEST**

Taking an antacid tablet is often enough to treat an occasional case of mild acid reflux. Another remedy is a drug such as omeprazole (one brand of which is Prilosec), which inhibits a cell's proton pumps by binding to the site of the pump and disabling it for more than 24 hours. Is the equilibrium constant for the binding of a proton pump inhibitor likely to be less than or greater than 1?

## Nitrogen

Nitrogen is a major essential element found primarily in proteins but also in DNA and RNA. Nitrogen is available in the atmosphere as $N_2$, and soil and water contain nitrate ions, but neither of these forms of nitrogen can be directly incorporated into amino acids, the building blocks of proteins. The biosynthesis of amino acids requires ammonia or ammonium ions. For example, glycine ($NH_2CH_2COOH$), the simplest amino acid, is formed by reaction of $CO_2$ and $NH_3$ in the presence of the appropriate enzyme. Interconversion of nitrogen-containing compounds in the environment is described by the nitrogen cycle (**Figure 22.8**). Certain bacteria use enzymes called *nitrogenases* to convert $N_2$ to $NH_3$. Plants convert $NO_3^-$ ions to $NO_2^-$ and then to $NH_3$ by using enzymes called *reductases*. Ultimately, the chemical reactions in these organisms begin a food chain that supplies the essential amino acids for human diets.

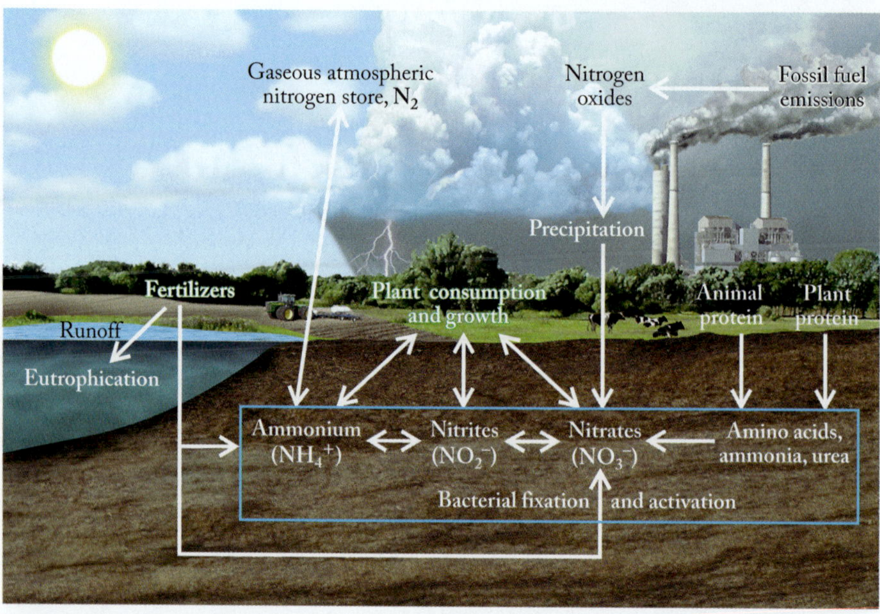


**FIGURE 22.8** The nitrogen cycle: Enzymes interconvert the nitrogen-containing molecules and ions found in nature. Bacteria convert atmospheric nitrogen to ammonium ion ($NH_4^+$), which is oxidized to nitrite ($NO_2^-$) and nitrate ($NO_3^-$) ions before being reduced back to $N_2$.

The reactions in **Figure 22.9** are all redox reactions that illustrate the range of oxidation numbers found among nitrogen compounds in the nitrogen cycle. By definition, each atom in nitrogen gas ($N_2$) is assigned an oxidation number of zero. When $N_2$ is converted to $NH_3$, the oxidation number of N decreases to −3, a reduction. The oxidation numbers of N in nitrite ($NO_2^-$) and nitrate ($NO_3^-$) are +3 and +5, respectively, as a result of oxidation.

Modern agriculture depends heavily on the use of nitrogen-based fertilizers, which include ammonia and salts containing ammonium salts and nitrate ions. Most fertilizer used in the 21st century is derived from the high-temperature, high-pressure Haber–Bosch process. Considerable research is in progress to emulate the nitrogen cycle in Figure 22.8 and produce fertilizer more efficiently. In Chapter 13, we encountered a different cycle for nitrogen in the environment: converting $N_2$ and $O_2$ to NO and $NO_2$ in the engines of automobiles. Dinitrogen monoxide ($N_2O$) is a greenhouse gas. The nitrogen atoms in these volatile nitrogen oxides are assigned oxidation numbers of +1, +2, and +4 for $N_2O$, NO, and $NO_2$, respectively.

**CONNECTION** The Haber–Bosch process for making ammonia was discussed in Chapter 14.

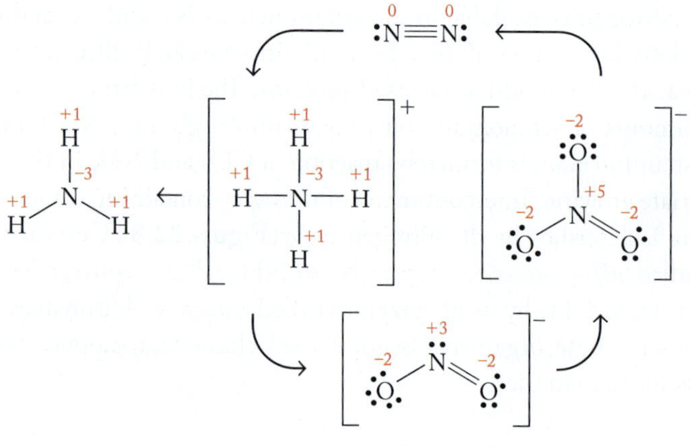

**FIGURE 22.9** Nitrogen-containing molecules and ions in the nitrogen cycle exhibit oxidation numbers ranging from −3 to +5.

**SAMPLE EXERCISE 22.3** Writing a Balanced Chemical Equation Describing the Reaction of Nitrate Reductases **LO3**

Nitrate ion can be reduced to ammonia by enzymes called nitrate reductases. The first step is to convert nitrate ion to nitrite ion. Assign oxidation numbers to the elements in these ions, and write a balanced net ionic equation for the conversion of nitrate ions into nitrite ions in a basic solution:

$$NO_3^-(aq) \rightarrow NO_2^-(aq)$$

**Collect, Organize, and Analyze** We need to assign oxidation numbers according to the guidelines in Section 8.7. We know that the oxidation numbers of nitrogen and oxygen in each polyatomic ion must add up to the charge on the ion. Oxygen in compounds usually has an oxidation number of −2.

**Solve** The oxidation number of nitrogen is unknown, so we call it *x*. The oxidation number of nitrogen in $NO_3^-$ is

$$x + 3(-2) = -1$$
$$x = +5$$

The oxidation number of nitrogen in $NO_2^-$ is

$$x + 2(-2) = -1$$
$$x = +3$$

In the preliminary expression,

$$NO_3^-(aq) \rightarrow NO_2^-(aq)$$

the number of nitrogen atoms is balanced; we balance oxygen by adding water:

$$NO_3^-(aq) \rightarrow NO_2^-(aq) + H_2O(\ell)$$

Then we balance hydrogen by adding hydrogen ions:

$$2\,H^+(aq) + NO_3^-(aq) \rightarrow NO_2^-(aq) + H_2O(\ell)$$

and balance charge by adding electrons:

$$2\,e^- + 2\,H^+(aq) + NO_3^-(aq) \rightarrow NO_2^-(aq) + H_2O(\ell)$$

We switch to a basic solution by adding just enough $OH^-$ ions to neutralize any hydrogen ions; we must add the same number of $OH^-$ ions to both sides of the equation:

$$2\,OH^-(aq) + 2\,e^- + 2\,H^+(aq) + NO_3^-(aq) \rightarrow NO_2^-(aq) + H_2O(\ell) + 2\,OH^-(aq)$$

The hydrogen ions combine with the hydroxide ions to form water, and we cancel the species that are the same on both sides:

$$2\,e^- + 2\,H_2O(\ell) + NO_3^-(aq) \rightarrow NO_2^-(aq) + \cancel{H_2O(\ell)} + 2\,OH^-(aq)$$

This gives us a final net ionic equation for the reduction half-reaction:

$$2\,e^- + H_2O(\ell) + NO_3^-(aq) \rightarrow NO_2^-(aq) + 2\,OH^-(aq)$$

**Think About It** Assigning oxidation numbers is a convenient way of identifying which element is reduced or oxidized in a half-reaction and of determining how many electrons are gained or lost. The balanced half-reaction confirms that electrons are added to nitrate to reduce it to nitrite ion.

**Practice Exercise** The reduction half-reaction catalyzed by one type of nitrogenase produces 1 mole of $H_2(g)$ for every 2 moles of $NH_4^+(aq)$ from molecular nitrogen under acidic conditions. Write a balanced equation for this half-reaction.

In humans and other mammals, excess nitrogen is converted to urea in the liver and excreted via the kidneys. Plants use urea as a source of ammonia by the action of *ureases* through this reaction:

$$H_2N{-}C({=}O){-}NH_2 + H_2O \rightarrow 2NH_3 + CO_2$$

Unlike reactions catalyzed by nitrogenases and nitrate reductases, the conversion of urea to ammonia and carbon dioxide is *not* a redox reaction. It is a hydrolysis reaction, similar to the reaction of nonmetal oxides with water. Acidic solutions are observed when $NH_4^+$, $NO_2$, and $N_2O_5$ dissolve in water:

$$NH_4^+(aq) + H_2O(\ell) \rightleftharpoons NH_3(aq) + H_3O^+(aq)$$

$$2\,NO_2(g) + H_2O(\ell) \rightarrow HNO_2(aq) + HNO_3(aq)$$

$$N_2O_5(g) + H_2O(\ell) \rightarrow 2\,HNO_3(aq)$$

Hydrolysis of nitrite ion, however, leads to weakly basic solutions:

$$NO_2^-(aq) + H_2O(\ell) \rightleftharpoons HNO_2(aq) + OH^-(aq)$$

## Phosphorus and Sulfur

Phosphorus and sulfur are major essential elements found primarily in proteins and DNA, but they are also present in polyatomic anions prevalent in the environment. In comparison with the nitrogen cycle, the biological phosphorus cycle contains only a single major species: the phosphate ion ($PO_4^{3-}$) and its conjugate acids $HPO_4^{2-}$ and $H_2PO_4^-$, along with phosphate esters. Reduction of phosphorus(V) in $PO_4^{3-}$ to phosphine ($PH_3$) does occur in swamps. An oxygen-free environment is needed because $PH_3$ spontaneously ignites in humid air, yielding phosphoric acid:

$$PH_3(g) + 2\,O_2(g) \rightarrow H_3PO_4(\ell)$$

Gases analogous to those in the nitrogen cycle, such as NO and $NO_2$, are absent from the phosphorus cycle. Slow weathering of insoluble phosphate minerals by weak acids in soil introduces phosphate ion into the environment, where plants eventually take it up. Some of this phosphate is incorporated into biominerals such as hydroxyapatite [$Ca_5(PO_4)_3(OH)$], as seen previously in our discussion of calcium.

**CONCEPT TEST**

Why are phosphates more likely to precipitate with cations from aqueous solution than nitrates?

Phosphate ion is in equilibrium with its conjugate acid, $HPO_4^{2-}$, which hydrolyzes, in turn, to $H_2PO_4^-$:

$$PO_4^{3-}(aq) + H_2O(\ell) \rightleftharpoons HPO_4^{2-}(aq) + OH^-(aq)$$

$$HPO_4^{2-}(aq) + H_2O(\ell) \rightleftharpoons H_2PO_4^-(aq) + OH^-(aq)$$

Aqueous phosphate ion can be transported into cells, where it is incorporated into organic molecules such as ATP, ADP, glucose-6-phosphate, and nucleic acids

(**Figure 22.10**). Notice the carbon–oxygen bond between the monosaccharide and the phosphate group in all four molecules in Figure 22.10. In glucose-6-phosphate and nucleic acids, this new bond forms through a condensation reaction of an –OH group on the sugar molecule with an $HPO_4^{2-}$ ion, producing water. ADP is converted to ATP through a condensation reaction between $HPO_4^{2-}$ and a phosphate group on ADP.

The processes in Figure 22.10 are all reversible; the P–O bond can be hydrolyzed, releasing $HPO_4^{2-}$. For ATP, this reaction is exothermic and provides the energy for many cellular processes. Given the importance of phosphorus to life, significant amounts of phosphates are used as fertilizer in agriculture. Agricultural runoff may stimulate rapid growth of algae in freshwater ponds and lakes. The explosive growth of algae can use up all the dissolved oxygen, killing other higher aquatic organisms.

The sulfur cycle in **Figure 22.11** illustrates the array of sulfur compounds found in the environment. We have already encountered volatile sulfur oxides, $SO_2$ and $SO_3$, in several contexts, such as the combustion of coal in Section 7.2. Like the nonmetal oxides of groups 14 and 15, $SO_2$ and $SO_3$ dissolve in water to produce the weak acid $H_2SO_3$ and the strong acid $H_2SO_4$, respectively. The sulfate ions produced by the dissociation of $H_2SO_4$, the sulfate ions derived from minerals in soil, and the sulfate ions produced by the oxidation of $H_2S$ are all absorbed by plants. Sulfuric acid is one of the most important industrial chemicals produced in the world.

**CONNECTION** The role of nitrogen and sulfur oxides in acid rain is discussed in Chapters 7, 8, and 15.

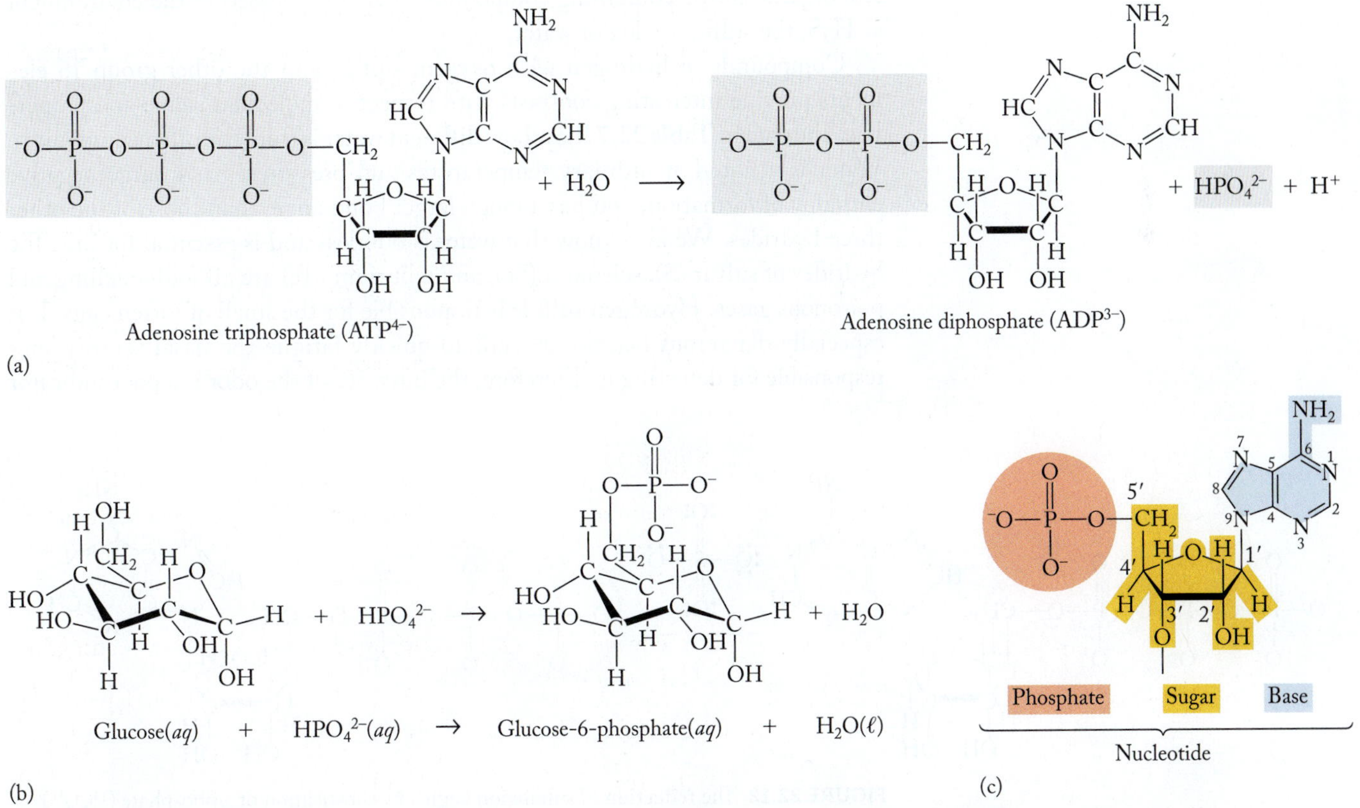

**FIGURE 22.10** Condensation reactions between $HPO_4^{2-}(aq)$ and –OH groups yield phosphate esters such as (a) ATP and ADP, (b) glucose-6-phosphate, and (c) nucleic acids.

**FIGURE 22.11** The key components in the sulfur cycle are sulfide ion ($S^{2-}$) in sediments and metal sulfides, sulfur oxides ($SO_2$ and $SO_3$), hydrogen sulfide ($H_2S$), and sulfate ($SO_4^{2-}$).

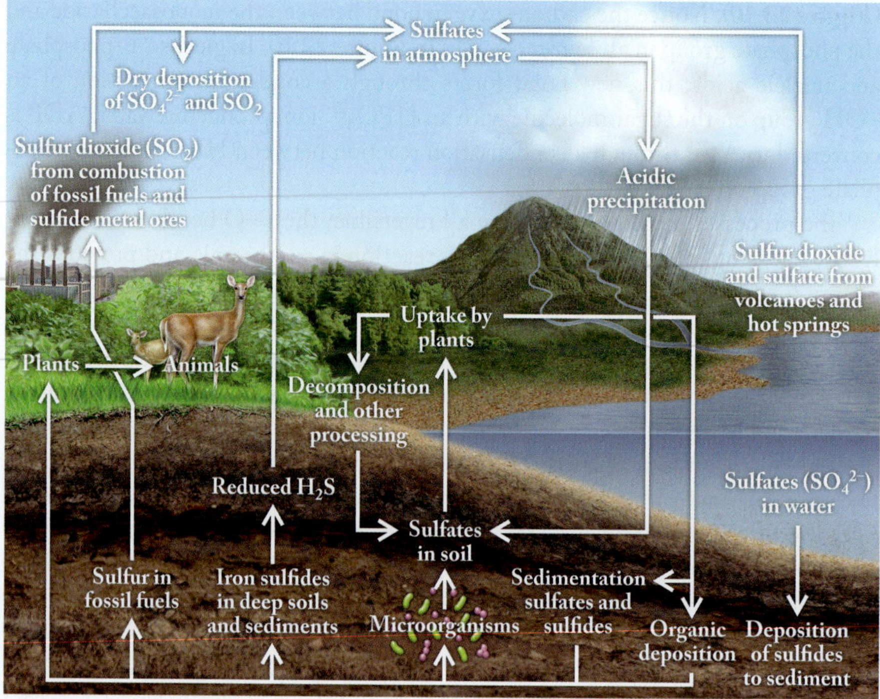


The enzyme ATP sulfurylase promotes the introduction of $SO_4^{2-}$ into ATP as adenosine-5′-phosphosulfate (APS; **Figure 22.12**). The sulfur in APS eventually finds its way into sulfur-containing amino acids (cysteine and methionine) and organic sulfur-containing compounds. It is also released to the environment as $H_2S$, the sulfur analog of water.

Compounds of hydrogen with oxygen, sulfur, and the other group 16 elements provide interesting contrasts with respect to molecular shape and properties. The data in **Table 22.7** show how different water is from the other compounds. Water is a liquid at ordinary temperatures and pressures, has a large negative enthalpy of formation, and has a much larger bond angle than those of the other three hydrides. We also know that water is odorless and is essential for life. The hydrides of sulfur (S), selenium (Se), and tellurium (Te) are all foul-smelling and poisonous gases. Hydrogen sulfide is responsible for the smell of rotten eggs. It is especially dangerous because it tends to quickly fatigue the nasal sensory sites responsible for detecting it. Therefore, the intensity of the odor is a poor indicator

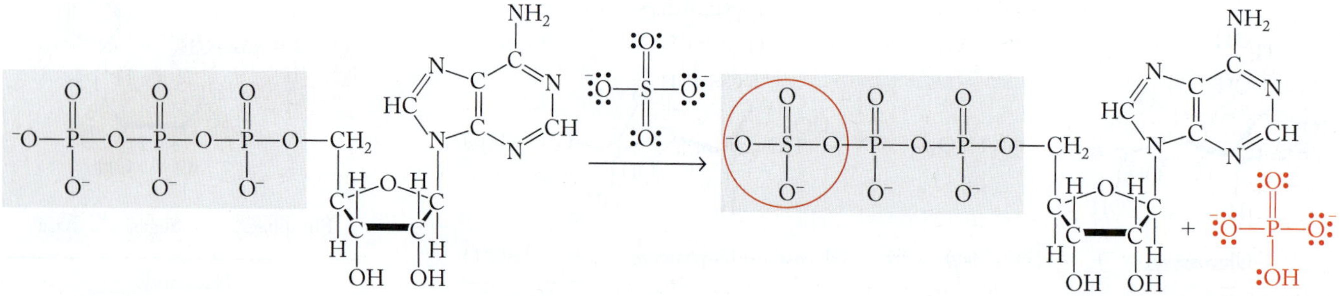

**FIGURE 22.12** The reduction of sulfate ion begins by substitution of a phosphate ($PO_4^{3-}$) group on ATP by sulfate.

**TABLE 22.7** Properties of Group 16 Dihydrides

| Hydride[a] | Melting Point (°C) | Boiling Point (°C) | Bond Length (pm) | Bond Angle (degrees) | Enthalpy of Formation (kJ/mol) |
|---|---|---|---|---|---|
| $H_2O$ | 0 | 100 | 96 | 104.5 | −285.8 |
| $H_2S$ | −86 | −60 | 134 | 92 | −20.17 |
| $H_2Se$ | −66 | −41 | 146 | 91 | 73.0 |
| $H_2Te$ | −51 | −4 | 169 | 90 | 99.6 |

[a]$H_2Po$ is excluded; too little is known of its chemistry. Polonium has no stable isotopes and is present on Earth only in very small quantities.

of the concentration of $H_2S$ in the air. Headache and nausea begin at air concentrations of $H_2S$ as low as 5 ppm, and exposure to 100 ppm leads to paralysis and may result in death.

Similarly, organic compounds containing sulfur have properties that differ from those of their oxygen-containing counterparts. Many such compounds also have characteristic odors. Methanol is an alcohol with the formula $CH_3OH$. It is a liquid at room temperature and has an odor usually described as slightly alcoholic. Methanethiol ($CH_3—SH$) is a gas at room temperature and has the pungent odor of rotten cabbage. It is produced in the intestinal tract of animals by the action of bacteria on proteins and is one of the sulfur compounds responsible for the characteristic aroma of a feedlot or a barnyard.

Let's compare compounds with two carbon atoms: ethanol and ethanethiol. Ethanol ($CH_3CH_2OH$) is a liquid at room temperature. The corresponding sulfur compound ($CH_3CH_2SH$; ethanethiol) is a more volatile liquid that has a penetrating and unpleasant odor, like that of powerful green onions. The human nose can detect the presence of ethanethiol at levels as low as 1 ppb (part per billion) in the air. This gives rise to its use as an odorant in natural gas. Natural gas has no odor, and natural gas leaks are such enormous fire hazards that ethanethiol is added to natural gas streams to make even small leaks detectable.

If we rearrange the atoms in ethanol and ethanethiol, we produce two new compounds. For ethanol we get dimethyl ether ($CH_3—O—CH_3$), a colorless gas used in refrigeration systems. Its counterpart, dimethyl sulfide ($CH_3—S—CH_3$), is one of the compounds responsible for the "low tide" smell of ocean shorelines.

Three of the sulfur compounds described—hydrogen sulfide, methanethiol, and dimethyl sulfide—are referred to as volatile sulfur compounds (VSCs) by dentists. They are produced by bacteria in the mouth and are the principal compounds responsible for bad breath. One of the reasons the odors of these compounds differ from those of their oxygen counterparts is that their molecular sizes and shapes are slightly different. Also, their polarities differ because of the electronegativity difference between oxygen ($\chi = 3.5$) and sulfur ($\chi = 2.5$). Recall that we discussed the importance of molecular shape in determining the extent of interaction of a compound with receptors in nasal membranes in Section 5.5. In part, the vast differences in odor and sensory detectability of these compounds are due to their shapes and electron distributions.

Not all (but many) sulfur compounds have an odor. The characteristic and unpleasant smell of urine produced by some people after eating asparagus results from the inability of their bodies to convert sulfur compounds (**Figure 22.13**) into odor-free sulfate ions. Not all people can convert the sulfur compounds in

**FIGURE 22.13** Structures of some of the volatile sulfur compounds responsible for the smell of "asparagus" urine. Compounds shown here have strong and unpleasant odors.

$H_3C{-}SH$ Methanethiol

$H_3C{-}S{-}S{-}CH_3$ Dimethyl disulfide

$H_3C{-}S{-}CH_2{-}S{-}CH_3$ Bis(methylthio)methane

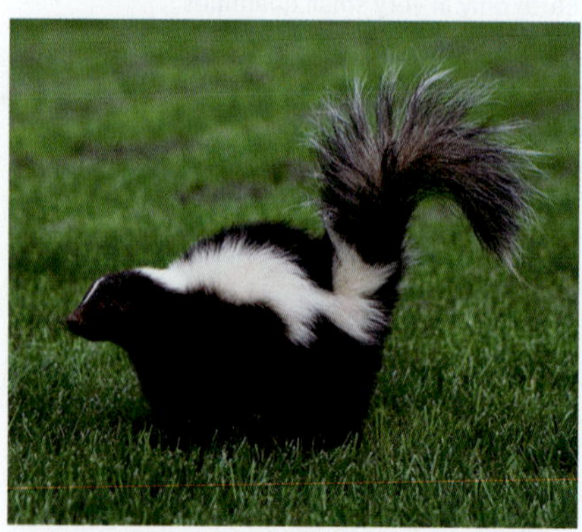

**FIGURE 22.14** The pungent smell of skunk spray is due to butanethiol, $CH_3(CH_2)_3SH$.

asparagus to sulfate, and not all people can smell the odiferous sulfur compounds. Apparently, genetic differences determine how we metabolize these compounds and how well we can sense their odors.

The odor of skunk is due mostly to butanethiol (**Figure 22.14**), and the odor of well-used athletic shoes is due primarily to the presence of sulfur compounds produced by bacteria. Not all sulfur compounds have aromas as unpleasant as these, however. A compound with the formula $C_{10}H_{18}S$, commonly known as the grapefruit mercaptan, is responsible for the aroma of grapefruit (**Figure 22.15a**). If the orientation of the substituents at one of the carbon atoms in the molecule is switched, the resulting molecule has the same Lewis structure but no aroma at all (**Figure 22.15b**). The carbon atom circled in red in Figure 22.15 is a chiral site, and the two compounds represent an enantiomeric pair (see Section 5.6).

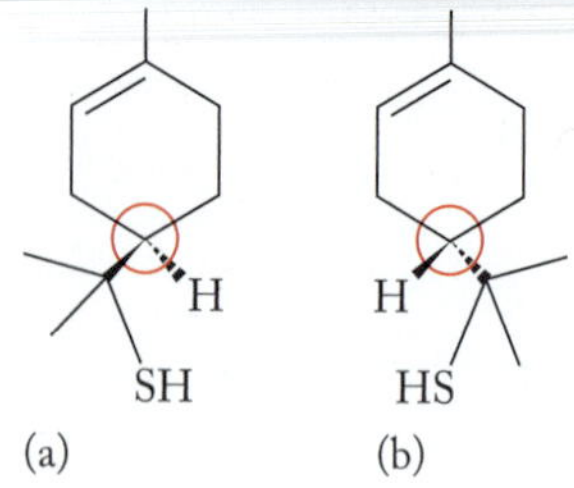


**FIGURE 22.15** (a) The compound known by the common name grapefruit mercaptan is responsible for the aroma of grapefruit. (b) If the orientation of the substituents at one of the carbon atoms in the molecule is switched, the resulting molecule has the same Lewis structure but no aroma at all. The carbon atoms circled in red identify the chiral carbon in each molecule.

## SAMPLE EXERCISE 22.4 Drawing the Lewis Structures of Molecules in the Sulfur Cycle — LO3

Dimethyl sulfide ($CH_3SCH_3$) is one of the products of the sulfur cycle in Figure 22.9. In marine environments, bacteria can oxidize dimethyl sulfide to dimethyl sulfoxide [$(CH_3)_2SO$].

a. Draw Lewis structures for $CH_3SCH_3$ and $(CH_3)_2SO$, and determine the molecular geometry about the S atom in each.
b. Methanethiol ($CH_3SH$) is the simplest thiol and has a boiling point of 6°C. Dimethyl sulfide has a boiling point of 38°C. Explain the difference in boiling point between $CH_3SH$ and $CH_3SCH_3$.
c. Which hybrid orbitals does sulfur use in bonding to carbon in $CH_3SCH_3$ and $CH_3SH$?

**Collect and Organize** We need to draw Lewis structures for two sulfur compounds, determine their geometry, and compare their boiling points. Guidelines for drawing Lewis structures were introduced in Section 4.4. We are also asked to describe the hybrid orbitals of S in these compounds. Hybrid orbitals were described in Section 5.4. The effect of structure on boiling points was discussed in Section 6.2.

**Analyze** The Lewis structure for a molecule depends on the total number of valence electrons available, distributed over the atoms so that each atom (except H) has a complete octet. Atoms with $Z > 12$ may have more than an octet of valence electrons if such an arrangement leads to lower formal charges on the atoms. The arrangement of the bonding and lone pairs on the central atom allows us to predict intermolecular forces such as dipole–dipole interactions, hydrogen bonds, and London dispersion forces, all of which can affect the boiling point of a substance. We can account for observed molecular geometries by combining *s* and *p* orbitals to form hybrid atomic orbitals such as *sp*, $sp^2$, and $sp^3$.

**Solve** (a) $CH_3SCH_3$ has 20 valence electrons: 6 from the S atom, 6 from the H atoms, and 8 from the C atoms. These electrons can be distributed in six C—H single bonds and two S—C single bonds, with four electrons remaining as two nonbonding pairs on S. $(CH_3)_2SO$ has an oxygen atom bonded to sulfur, which brings an additional six valence electrons to the molecule, giving $(CH_3)_2SO$ 26 valence electrons. Sharing one of the S lone pairs with O will complete the octets of both S and O but will leave O with a formal charge of −1 and sulfur with a formal charge of +1. Forming a S=O double bond makes the formal charges on both S and O equal to zero but requires that sulfur have more than an octet. The two structures for $(CH_3)_2SO$ represent resonance forms:

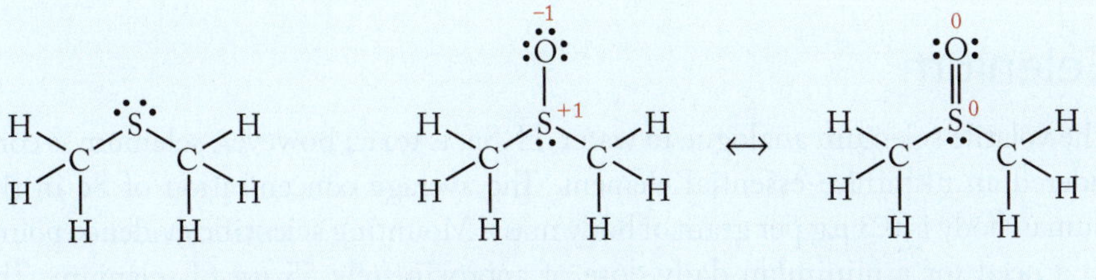

(b) $CH_3SH$ has 14 valence electrons: 6 from the S atom, 4 from the H atoms, and 4 from the C atom. The electrons are distributed in three C—H single bonds, one C—S single bond, and one S—H single bond. As in $CH_3SCH_3$, two nonbonding pairs are left on S. Both $CH_3SCH_3$ and $CH_3SH$ contain a sulfur atom surrounded by two bonding pairs and two nonbonding pairs of electrons, for a total of four electron pairs. The electron-pair geometry is tetrahedral, and the molecular geometry is bent:

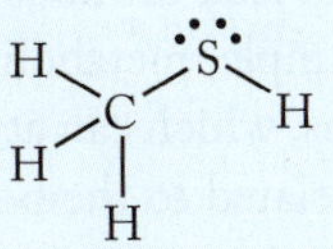

Both $CH_3SH$ and $CH_3SCH_3$ are polar as a result of their molecular geometry and experience dipole–dipole interactions. Both molecules interact through London dispersion forces as well. One might expect that the stronger dipole–dipole forces in $CH_3SH$ would lead to a higher boiling point than that for $CH_3SCH_3$, but we observe the opposite. We conclude that the London dispersion forces in $CH_3SCH_3$ have a greater effect than the dipole–dipole interactions in $CH_3SH$, leading to $CH_3SCH_3$ boiling at a temperature about 32°C higher than $CH_3SH$.

(c) The similar geometry for $CH_3SCH_3$ and $CH_3SH$, with each molecule's central sulfur atom having two bonded atoms and two lone pairs, is consistent with $sp^3$ hybrid orbitals on the sulfur atom in both cases.

**Think About It** Methanethiol and dimethyl sulfide differ only in that $CH_3SCH_3$ has a $CH_3$ group in place of a hydrogen atom. (This relationship between thiols and sulfides corresponds to the relationship between alcohols and ethers.) Sulfur does not have more than an octet of valence electrons in these compounds because using the lone pairs on S to form multiple bonds is not needed. The boiling points of $CH_3SCH_3$ and $CH_3SH$ reveal an important observation: many weaker bonds (London dispersion forces) in $CH_3SCH_3$ can outweigh a few stronger forces (dipole–dipole forces) in $CH_3SH$.

**Practice Exercise** The nitrogen cycle in Figure 22.6 involves both neutral compounds such as $NO_2$ and polyatomic ions such as $NO_2^-$. Draw Lewis structures for both species and determine whether they have the same molecular geometry about nitrogen. Identify which hybrid orbitals contain nitrogen lone pairs in $NO_2$ and $NO_2^-$.

## 22.4 Trace and Ultratrace Essential Elements

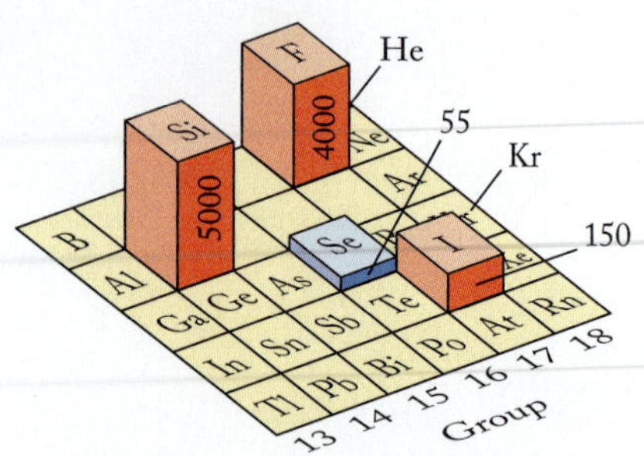


**FIGURE 22.16** Silicon, fluorine, and iodine are trace essential elements, and selenium is an ultratrace essential element. The remaining labeled elements are nonessential. The vertical bars show DRI values for these elements in micrograms per day.

**Figure 22.16** shows the DRI values of four main group essential elements: silicon, selenium, fluorine, and iodine. Silicon, fluorine, and iodine are present in the body in average concentrations between 1 μg and 1000 μg of element per gram of body mass and are considered trace elements. Selenium is considered ultratrace, meaning that it is present in an average concentration of less than 1 μg of element per gram of body mass.

### Selenium

The volatile selenium analogue to water, $H_2Se$, is toxic; however, selenium is considered an ultratrace essential element. The average concentration of Se in the human body is 0.3 μg per gram of body mass. Mounting scientific evidence points to a need for a minimum daily dose of approximately 55 μg of selenium. The effects of selenium toxicity, however, are apparent in people who ingest more than 500 μg per day. Most of the selenium we need is obtained from selenium-rich produce, such as garlic, mushrooms, and asparagus, or from fish. Selenium occurs in the body as the amino acid selenocysteine (**Figure 22.17**) and is incorporated into enzymes.

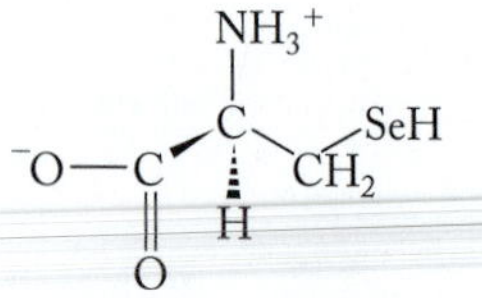

**FIGURE 22.17** Selenocysteine is the selenium-containing analogue of the amino acid cysteine. Much of the selenium in the human body is found in proteins containing selenocysteine.

Selenocysteine is an antioxidant. Our bodies need oxygen to survive yet living in an oxygen-rich atmosphere can lead to the formation of potentially dangerous oxidizing agents in cells. For example, metabolism of fatty acids forms oxidizing agents called alkyl hydroperoxides, which can attack the lipid bilayer of cell membranes. Aging is believed to be related to the body's inability to inhibit oxidative degradation of tissue. Selenocysteine participates in a series of reactions that decompose these alkyl hydroperoxides.

### Fluorine and Iodine

Fluorine is a trace essential element, and fluoride ions have significant benefits for dental health. Tooth enamel is composed of the mineral hydroxyapatite, $Ca_5(PO_4)_3(OH)$, which is essentially insoluble in water:

$$Ca_3(PO_4)_3(OH)(s) \rightleftharpoons Ca_5(PO_4)_3^{+}(aq) + OH^-(aq) \qquad K_{sp} = 2.4 \times 10^{-59}$$

When hydroxyapatite comes into contact with weak acids in your mouth, this equilibrium shifts to the right as the acid reacts with the hydroxide ions. This shift effectively increases the solubility of hydroxyapatite, so that your tooth enamel becomes pitted, and dental caries, or cavities, form. Fluoride ions reduce the likelihood of caries by displacing the $OH^-$ ions in hydroxyapatite to form fluorapatite:

$$Ca_3(PO_4)_3(OH)(s) + F^-(aq) \rightleftharpoons Ca_5(PO_4)_3F(s) + OH^-(aq) \qquad K = 8.48$$

The solubility of fluorapatite depends less on pH than the solubility of hydroxyapatite does, so changing tooth enamel to fluorapatite makes your teeth more resistant to decay. This is why toothpaste contains fluoride compounds and why fluoride is added to drinking water in many communities in North America and Europe.

**CONNECTION** The solubility product ($K_{sp}$) was introduced in Section 16.8. Le Châtelier's principle was discussed in Section 14.7.

Of all the trace essential elements, iodine may have the best-defined role in human health. The body concentrates iodide ions in the thyroid gland, where they

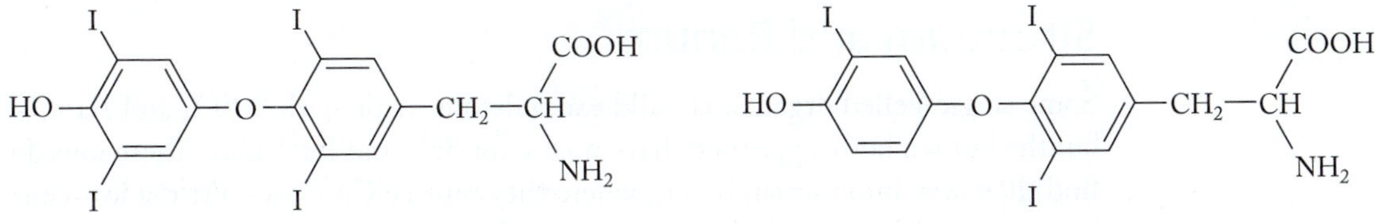

Thyroxine

3,5,3′-Triiodothyronine

**FIGURE 22.18** Thyroxine and 3,5,3′-triiodothyronine, two iodine-containing hormones found in the thyroid gland, regulate metabolism.

are incorporated into two hormones—thyroxine and 3,5,3′-triiodothyronine (**Figure 22.18**)—whose role is to regulate energy production and use. The conversion of thyroxine to 3,5,3′-triiodothyronine is catalyzed by selenocysteine-containing proteins. A deficiency of iodine or of either hormone can cause fatigue or feeling cold and can ultimately lead to an enlarged thyroid gland, a condition known as goiter. To help prevent iodine deficiency, table salt sold in the United States and many other countries is "iodized" with a small amount of sodium iodide. An excess of either hormone can cause a person to feel hot and is linked to Graves' disease, an autoimmune disease. The immune system in a patient with Graves' disease attacks the thyroid gland and causes it to overproduce the two hormones.

### Silicon

The role of silicon in biological systems is less clear than for selenium and the halides. In mammals, a lack of the trace essential element silicon stunts growth. The presence of silicon as silicic acid, $Si(OH)_4$, is believed to reduce the toxicity of $Al^{3+}$ ions in organisms by precipitating the aluminum as aluminosilicate minerals. Amorphous silica, $SiO_2$, is found in the exoskeletons of diatoms and in the cell membranes of some plants, such as the tips of stinging nettles.

## 22.5 Nonessential Elements

Ten elements listed in Table 22.2 found in the human body are classified as nonessential. Here in Section 22.5, we discuss how some of these elements may end up in our bodies, working our way from left to right across the periodic table.

### Rubidium and Cesium

Rubidium is generally regarded as nonessential in humans, yet it is the 15th-most-abundant element in the body. The body is believed to retain $Rb^+$ because its size and chemistry are similar to that of $K^+$. Like the other cations of group 1, cesium ions ($Cs^+$) are also readily absorbed by the body. Cesium cations have no known function, although they can substitute for $K^+$ and interfere with potassium-dependent functions. Usually, the concentration of cesium in the environment is low, so exposure to $Cs^+$ is not a health concern. The nuclear accident at Chernobyl in 1986, however, released significant quantities of radioactive $^{137}Cs$ into the environment. The ability of $Cs^+$ to substitute for $K^+$ led to the incorporation of $^{137}Cs^+$ into plants, which rendered crops grown in the immediate area unfit for human consumption because of the radiation hazard posed by this long-lived ($t_{1/2} \approx 30$ yr) β emitter.

## Strontium and Barium

Some single-celled organisms build exoskeletons made with $SrSO_4$ and $BaSO_4$, but the human body appears to have no use for $Sr^{2+}$ and $Ba^{2+}$ ions. These ions do find their way into human bones, where they replace $Ca^{2+}$ ions. At the low concentrations of $Sr^{2+}$ and $Ba^{2+}$ that are typically present in the human body, these elements appear to be benign. As with radioactive $^{137}Cs$, however, incorporation of $^{90}Sr$ ($t_{1/2}$ = 29 yr) in bones can lead to leukemia. Atmospheric testing of nuclear weapons over the Pacific Ocean and in sparsely populated regions of the American West in the 1950s released $^{90}Sr$ into the environment. The full extent of the toxic effects of the fallout from these tests did not become apparent for several decades.

**CONNECTION** The biological effects of types of nuclear radiation were described in Section 21.8.

## Germanium

Germanium is a nonessential element and is barely detectable in the human body. Bis(carboxyethyl)germanium sesquioxide (**Figure 22.19**) has been touted as a nutritional supplement, but its efficacy remains in doubt.

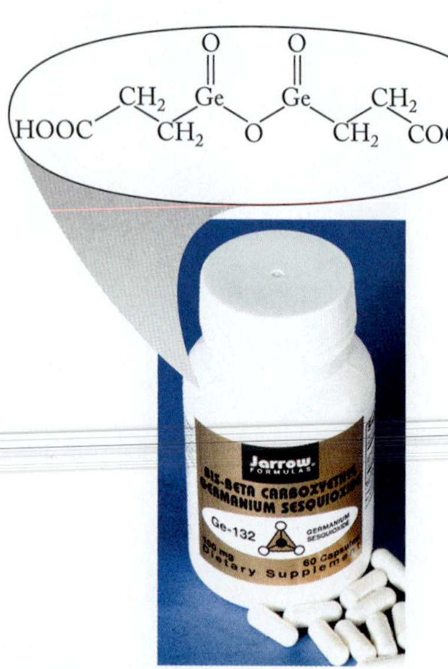


**FIGURE 22.19** Bis(carboxyethyl) germanium sesquioxide has been sold as a nutrition supplement, but its benefits are not well established. The claims made for many dietary supplements remain controversial, and scientists have not fully explored them.

## Antimony

The role of antimony is also poorly understood. Most antimony compounds are toxic because they cause liver damage. However, ultratrace amounts of antimony may have a stimulatory effect, and selected antimony compounds have been used medically as antiparasitic agents, as discussed in Section 22.6.

## Bromine

Bromine has no known function in the human body but is consumed in foods such as grains, nuts, and fish in amounts ranging from 2 mg to 8 mg per day, leading to average concentrations of $Br^-$ in blood of about 6 mg/L. Bromide ions have sedative and anticonvulsive properties but becomes toxic at concentrations around 100 mg/L, limiting its use to veterinary medicine. Bromide ion concentrations in seawater typically range from 65 mg/L to 80 mg/L. A select group of aquatic species can metabolize $Br^-$ into bromomethane ($CH_3Br$) and other brominated organic compounds.

**CONCEPT TEST**

Looking at groups 1, 2, 14, 15, and 17, what periodic trend do you see in the location of the nonessential elements compared with the essential elements in the same group?

# 22.6 Elements for Diagnosis and Therapy

So far, we have talked about the biological roles of several essential and nonessential main group elements found in our bodies. Some of these elements are also useful in diagnosing or treating diseases, as are some of the other elements in groups 1, 2, and 13–18 that we have not mentioned (**Figure 22.20**). Here in Section 22.6, we describe some applications of radioactive isotopes in diagnosing

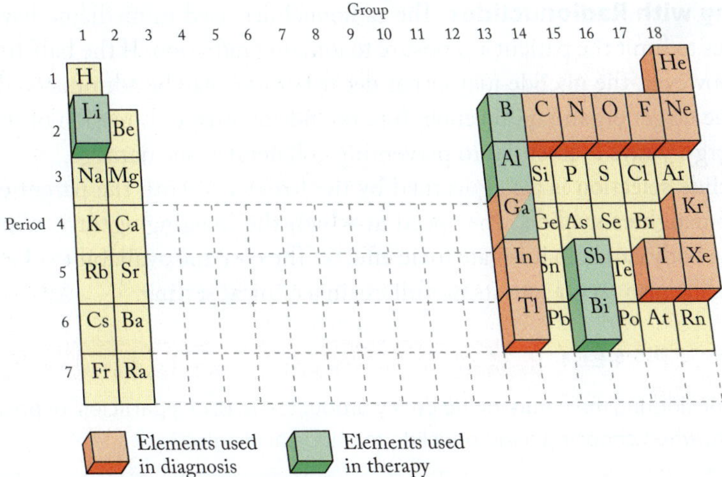


**FIGURE 22.20** The elements shown in red are used in diagnostic imaging, and those shown in green are used in therapy. Gallium is used in both diagnostic imaging and therapy.

diseases. We also explore how compounds of essential and nonessential elements have found application in treating various illnesses.

Any diagnostic or therapeutic compound that is injected intravenously must be soluble enough in blood to be delivered to the target. While in transit, the compound must be stable enough not to undergo chemical reactions that result in its precipitation or rapid elimination from the body. A medicinal chemist can also take advantage of substances that occur naturally in the body, such as antibodies, to carry a diagnostic or therapeutic metal ion to its target. We focus here on the main group elements and their compounds (groups 1, 2, and 13–18). Medical applications of compounds containing the transition elements (groups 3–12) are described in Chapter 23.

## Diagnostic Applications

Physicians in the 21st century have an array of imaging agents to help in diagnosing disease. Some methods use radionuclides with short half-lives that emit easily detectable gamma rays. Examples include using iodine-131 to image the thyroid gland (**Figure 22.21**) and using neutron-poor isotopes such as carbon-11 and fluorine-18 for positron emission tomography (PET). Not all imaging depends on radionuclides, however. In magnetic resonance imaging (MRI), for instance, which can diagnose soft-tissue injuries, stable isotopes of gadolinium (a lanthanide) are heavily used as contrast agents to enhance images despite their toxicity to kidneys.

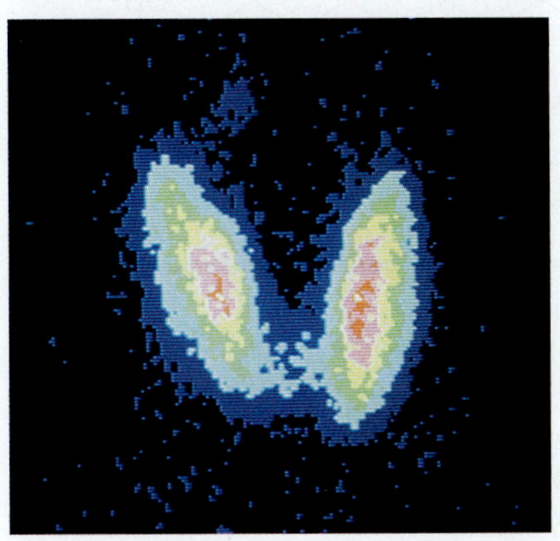

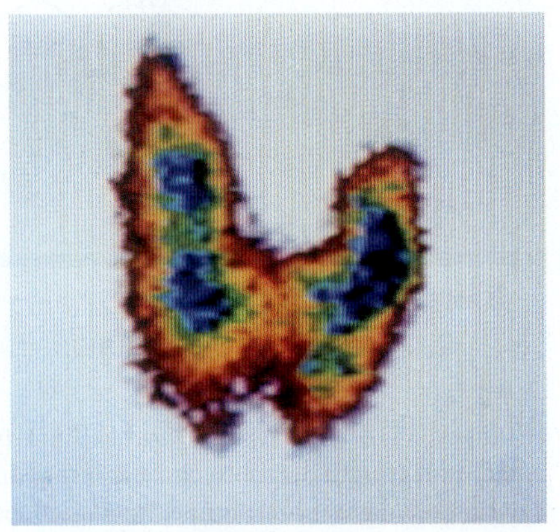

**FIGURE 22.21** Gamma radiation that accompanies the decay of iodine-131 can be used to image the two butterfly-shaped lobes of the thyroid gland.

**Imaging with Radionuclides** The radionuclides used in medicine have short half-lives to limit the patient's exposure to ionizing radiation. If the half-life is too short, however, the nuclide may either decay before it can be administered or not reach the target organ rapidly enough to provide an image. Emission of relatively low-energy $\gamma$ rays is essential to preventing collateral tissue damage.

**CONNECTION** Chapter 21 gives a more detailed discussion of nuclear chemistry and nuclear medicine, including an assessment of the effects of different types of radiation on living tissue.

Nuclide selection is also governed by the toxicity of both the parent element and the daughter nuclide. The speed at which the imaging agent is eliminated from the body can help mitigate toxic effects. The cost and availability of a particular nuclide also factor into its usefulness in a clinical setting.

### CONCEPT TEST

Why is considering the nature of the decay products—$\alpha$, $\beta$, or $\gamma$ particles, or positrons—important when choosing a radionuclide for medical imaging?

**Gallium, Indium, and Thallium** Gallium-66, gallium-67, gallium-68, and indium-111 are used as imaging agents for tumors and leukemia. All four nuclides decay by electron capture, and the $\gamma$ radiation emitted in this nuclear reaction produces the images. All three gallium isotopes also decay by positron emission, which makes compounds containing these isotopes attractive for PET scans. Their half-lives range from just over 1 h for gallium-68 to 78 h for gallium-67. The indium-111–containing compound ibritumomab tiuxetan (brand name, Zevalin) is currently used to treat some forms of non-Hodgkin's lymphoma. It is formed by reacting indium(III) chloride with the molecule in **Figure 22.22**. The $In^{3+}$ ion acts as a Lewis acid and forms coordinate bonds of the type described in Section 16.6. The ibritumomab in Figure 22.22 is an antibody used to transport the radioactive indium to its target.

The use of the gamma emitter thallium-201 ($t_{1/2}$ = 73 h) in diagnosing heart disease presents an interesting case for balancing the risks and benefits of using a particular isotope in medicine. Although thallium compounds are among the most toxic metal-containing compounds known, the nanogram quantities required for diagnosis pose few, if any, health hazards, meaning that the benefits outweigh the risks.

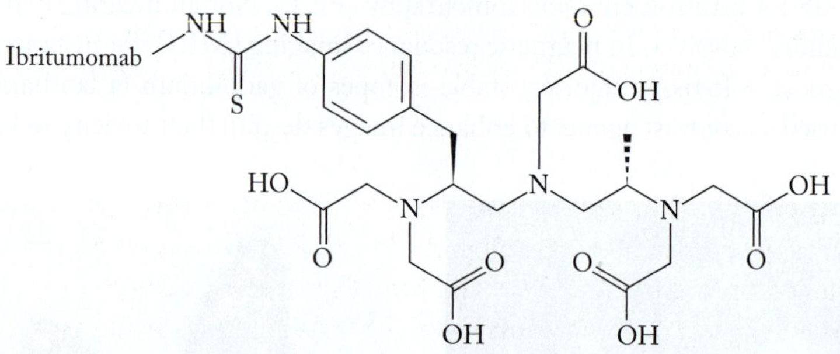

**FIGURE 22.22** The drug ibritumomab tiuxetan (brand name, Zevalin) contains the large organic molecule tiuxetan covalently bonded to an antibody called ibritumomab. Adding indium-111 results in a useful imaging agent.

### SAMPLE EXERCISE 22.5 Calculating Quantities of Radioactive Isotopes LO5

Indium-111 ($t_{1/2}$ = 2.805 d) and gallium-67 ($t_{1/2}$ = 3.26 d) are both used in radioimaging to diagnose chronic infections. Which isotope decays faster? If we start with 10.0 mg of each isotope, how much of each remains after 24 h (1 day)?

**Collect and Organize** We are given the half-lives of two radionuclides and are asked to predict which one will decay faster and to calculate how much of each isotope remains after 24 h of decay.

**Analyze** An isotope with a shorter half-life decays faster. Radioactive decay follows first-order kinetics. Quantitatively, the relationship between half-life and the amount of material remaining is described by Equation 21.7:

$$\ln \frac{N_t}{N_0} = -0.693 \frac{t}{t_{1/2}}$$

Here $N_0$ and $N_t$ refer to the amount of material present initially and at time $t$, respectively. If our prediction for the relative decay rates of the two isotopes is correct, then more of the isotope with the longer half-life should remain after 24 h. We need to complete two calculations to determine the amount of each sample present after 24 h.

**Solve** Indium-111 has the shorter half-life, so it should decay faster.

For the amount of indium-111 remaining after 24 h, we have

$$\ln \frac{N_t}{10.0 \text{ mg}} = \frac{(-0.693)(1 \cancel{\text{d}})}{(2.805 \cancel{\text{d}})} = -0.247$$

Taking the antilog of both sides, we get

$$\frac{N_t}{10.0 \text{ mg}} = 0.781$$

$$N_t = (0.781)(10.0 \text{ mg}) = 7.81 \text{ mg}$$

For gallium-67:

$$\ln \frac{N_t}{10.0 \text{ mg}} = \frac{(-0.693)(1 \cancel{\text{d}})}{(3.26 \cancel{\text{d}})} = -0.213$$

$$\frac{N_t}{10.0 \text{ mg}} = 0.808$$

$$N_t = (0.808)(10.0 \text{ mg}) = 8.08 \text{ mg}$$

**Think About It** We predicted that indium-111 would decay faster, which means that after 24 h the quantity of this isotope should be less than the quantity of gallium-67, and it is.

**Practice Exercise** Two radioactive isotopes of bismuth are used to treat cancer. The half-lives are 61 min for bismuth-212 and 46 min for bismuth-213. If we start with 25.0 mg of each isotope, how much of each sample remains after 24 h?

**Imaging with Noble Gases and MRI** None of the noble gas elements are essential to the human body, although the World Anti-Doping Agency (WADA) has included both Xe and Ar on the list of banned substances for athletes at Olympic and other sporting events since 2014. Apparently, in addition to behaving as an anesthetic, xenon increases blood's oxygen-carrying capacity, providing an advantage in aerobically demanding sports. Similar effects are believed to occur with argon.

The lack of chemical reactivity of the group 18 elements and their ease of introduction into the body by inhalation, however, make selected isotopes of the noble gases—including helium-3, krypton-83, and xenon-129—attractive as agents for enhancing MRI images, particularly of the lungs. Krypton-83 affords greater sensitivity than xenon-129, allowing for better resolution in the images and, in principle, requiring less gas.

Helium-3 has no known side effects and is preferable to xenon-129 for MRI, but it is present in only trace natural abundance. This isotope is obtained from β decay of tritium ($^{3}$H):

$$^{3}_{1}\text{H} \rightarrow ^{3}_{2}\text{He} + ^{\ 0}_{-1}\beta$$

Neon-19 has been used in PET despite its short half-life (17.5 s). A patient positioned in a PET scanner breathes air containing a small amount of this isotope. Positron emission from the neon is recorded, and an image is created.

## Therapeutic Applications

We next examine therapeutic agents that contain metallic and heavier main group elements in addition to carbon, hydrogen, nitrogen, oxygen, and sulfur.

**Lithium, Boron, Aluminum, and Gallium** Because of the similar size of $Li^{+}$ (76 pm) and $Mg^{2+}$ (72 pm), lithium ions can compete with magnesium ions in biological systems. The substitution of lithium for magnesium may account for its toxicity at high concentrations. Nevertheless, lithium carbonate is used to treat bipolar disorder, and other lithium compounds have been used to treat hyperactivity. However, the use of lithium-containing drugs must always be carefully monitored.

Of the elements in group 13, only boron and aluminum have been detected in humans. The role of boron in our bodies is not fully understood, but it appears to play a role in nucleic acid synthesis and carbohydrate metabolism. Selected boron compounds appear to concentrate in human brain tumors. This property has opened the door to a treatment known as boron neutron-capture therapy (BNCT). Once a suitable boron compound has been injected and has made its way to a tumor, irradiating the tumor with low-energy neutrons leads to the nuclear reaction

$$^{10}_{5}\text{B} + ^{1}_{0}\text{n} \rightarrow ^{7}_{3}\text{Li} + ^{4}_{2}\text{He}$$

**CONNECTION** The RBE of radioactive particles is a factor that accounts for the differences in physical damage caused by different types of radiation. It was introduced in Section 21.8.

The α particles generated in the reaction have a short penetration depth but high relative biological effectiveness (RBE), so they can kill the tumor cells without harming surrounding tissue. Identifying compounds suitable for BNCT remains an area of active research.

Aluminum is found in some antacids as aluminum hydroxide, $Al(OH)_3$, or aluminum carbonate, $Al_2(CO_3)_3$. Some brands of baking powder contain sodium aluminum sulfate, $NaAl(SO_4)_2 \cdot 12\ H_2O$. Most aluminum in the human body can be traced to these sources. Aluminum is not considered essential to humans, but low-aluminum diets have been observed to harm goats and chickens. High concentrations of aluminum are clearly toxic; the effects are most noticeable in patients with impaired kidney function. The role of aluminum in Alzheimer's disease has been extensively debated but remains unresolved.

Simple gallium compounds such as gallium(III) nitrate and gallium(III) chloride, either alone or in combination with other drugs, have shown activity on bladder and ovarian cancers. The similar ionic radii of $Ga^{3+}$ (62 pm) and $Fe^{3+}$ (64.5 pm) allow gallium to block DNA synthesis by replacing iron in a protein called transferrin and in other enzymes. Because gallium compounds accumulate in tumors faster than in healthy tissue, the disruption of DNA synthesis in the tumor cells inhibits tumor growth.

**Antimony and Bismuth** Antimony compounds are generally considered toxic. It has been reported, for instance, that exposing infants to antimony compounds used as fire retardants in mattresses may contribute to sudden infant death syndrome (SIDS). However, this element does appear to have a medical use. Leishmaniasis, an insect-borne disease characterized by the formation of boils or skin lesions, is resistant to most treatments, but some patients have been successfully treated with sodium stibogluconate, one of the few applications of antimony compounds in human health.

Popular over-the-counter remedies for indigestion, diarrhea, and other gastrointestinal disorders contain bismuth subsalicylate (**Figure 22.23**). The bismuth in these compounds acts as a mild antibacterial agent that reduces the number of diarrhea-causing bacteria.

The human body requires about 30 elements to function properly. To manage the problems of disease and injury, scientists, physicians, engineers, and scores of other people have turned to the properties of these and many other elements in the periodic table to develop treatments and to enhance quality of life. Here in Chapter 22 we have introduced the roles of the main group elements in establishing and maintaining living systems.

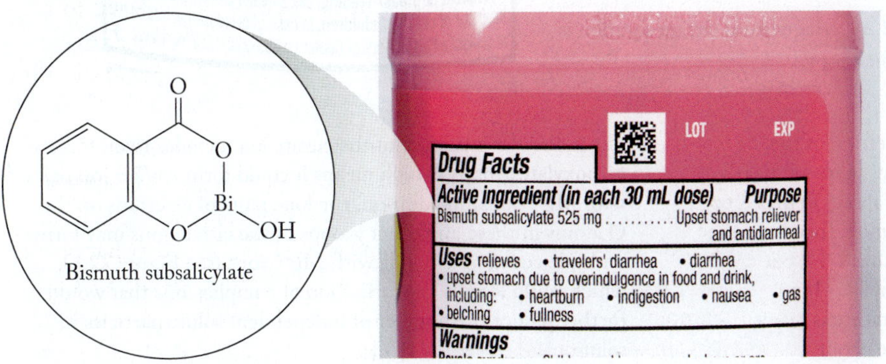


**FIGURE 22.23** Bismuth subsalicylate is found in some antacids.

## SAMPLE EXERCISE 22.6 Integrating Concepts: Colonoscopy

Many doctors recommend that patients over 50 years old have a procedure called a colonoscopy to screen for colorectal cancer. To prepare for the procedure, many patients drink about 230 mL of a magnesium citrate solution ($MgC_6H_6O_7$; $\mathcal{M}$ = 214.42 g/mol), which serves as a laxative that clears the large intestine. Magnesium citrate is an electrolyte composed of magnesium(II) ions and citrate ions, whose carbon-skeleton molecular structure is shown in **Figure 22.24**.

As indicated on the label in Figure 22.24, the solution contains 1.745 g magnesium citrate per fluid ounce (fl. oz.).

a. What is the molarity of this solution? There are 29.57 mL/fl. oz.
b. What is the osmotic pressure ($\Pi$) of the solution at 37°C? Its van 't Hoff factor ($i$) is 1.09.
c. Most electrolytes composed of 2+ cations and 2− anions have van 't Hoff factors closer to 2 than to 1 at the concentration calculated in part (a). Propose a reason why the $i$ value for sodium citrate is so low. In developing your answer, consider the structure of citrate ions and how they might interact with $Mg^{2+}$ ions in aqueous solutions.
d. The wall of the small intestines act as a semipermeable membrane, which allows water to pass from the tissues surrounding the intestine into the fluid passing through it, or water can flow from inside the intestine into the surrounding cells. If the osmotic pressure in these cells is about 3.5 atm, in which direction will osmosis occur if the fluid passing through the intestine has an ionic strength near that of the solution in the bottle?

**Collect and Organize** We are asked to calculate the molar concentration and osmotic pressure of a solution of magnesium citrate. These two properties of the solution are related by Equation 11.13: $\Pi = iMRT$. We are also asked if water will flow into, or out from, the small intestine if the fluid in it has nearly the same osmotic pressure as the $MgC_6H_6O_7$ solution. We know the osmotic pressure of the fluid in the surrounding cells is about 3.5 atm.

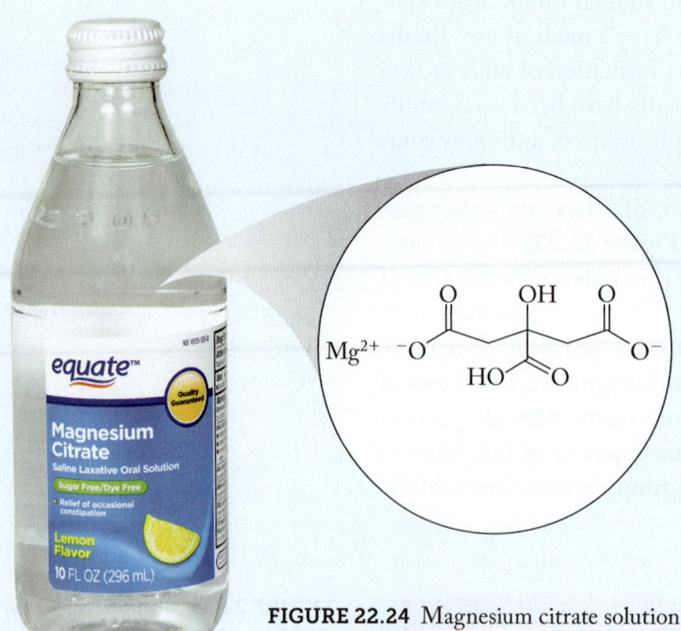


**Drug Facts**

**Active ingredient (in each fl oz)** — **Purpose**

Magnesium citrate 1.745g.......................................Saline laxative

**Uses** ■ relieves occasional constipation (irregularity) ■ generally produces bowel movement in 1/2 to 6 hours

**Warnings**

**Do not use** laxative products for longer than 1 week unless directed by a doctor.

**Ask a doctor before use if you have**

■ kidney disease ■ a magnesium restricted diet ■ a sodium restricted diet ■ stomach pain, nausea, or vomiting ■ noticed a sudden change in bowel habits that lasts more than 1 week

**Ask a doctor or pharmacist before use if you are** presently taking a prescription drug.

**When using this product,** do not exceed the maximum recommended daily dosage in a 24 hour period.

**Stop use and ask a doctor if** you have rectal bleeding or no bowel movement after use. These could be signs of a serious condition.

**If pregnant or breast-feeding,** ask a health professional before use.

**Keep out of reach of children.** In case of overdose, get medical help or contact a Poison Control Center (1-800-222-1222) right away.

**FIGURE 22.24** Magnesium citrate solution is used in colonoscopy.

**Analyze** We first need to convert the concentration of the $MgC_6H_6O_7$ solution from g/fl oz into molarity. There are nearly 2 g $MgC_6H_6O_7$/fl oz in the solution; there are about 30 mL/fl oz, and $MgC_6H_6O_7$ has a molar mass near 200 g/mol. Therefore, the molarity of the solution should be about 2 g/fl oz × 1 fl oz/30 mL × 1000 mL/L × 1 mol/200 g = 0.3 mol/L. The van 't Hoff factor (*i*) is about 1, and the temperature is 37 + 273 = 310 K, so the value of $\Pi$ should be about 0.3 mol/L × 0.08 L · atm/(mol · K) × 300 K = 7 atm. This value is about twice that of the fluids surrounding the small intestine.

**Solve**

a. Calculating the molarity of the $MgC_6H_6O_7$ solution:

$$\frac{1.745\text{ g}}{1\text{ fl oz}} \times \frac{1\text{ fl oz}}{29.57\text{ mL}} \times \frac{1000\text{ mL}}{1\text{ L}} \times \frac{1\text{ mol}}{214.42\text{ g}} = 0.275\text{ mol/L}$$

b. Calculating the osmotic pressure of this solution:

$$\Pi = iMRT = 1.09 \times 0.275\,\frac{\text{mol}}{\text{L}} \times 0.08206\,\frac{\text{L}\cdot\text{atm}}{\text{mol}\cdot\text{K}} \times 310\text{ K} = 7.63\text{ atm}$$

c. The molecular structure of the citrate ion includes two carboxylate groups, which means it could form 2+/2− ion pairs with $Mg^{2+}$ ions. In addition, the lone pairs of electrons on the O atoms in these and other groups in the citrate ions may form multiple coordinate bonds with $Mg^{2+}$ ions (see Figure 22.6), which could lead to the formation of complex ions that would further reduce the number of independent solute particles in solution.

d. The osmotic pressure of the $MgC_6H_6O_7$ solution is more than twice that in the cells surrounding the walls of the small intestine. This means water will spontaneously flow from those cells into the intestine if the fluid inside the intestine has nearly the same osmotic pressure as the $MgC_6H_6O_7$ solution.

**Think About It** Osmotic laxatives such as magnesium citrate work by drawing water into the intestines. This softens the fecal matter produced during digestion, which makes bowel movements easier, and it also stimulates the muscles in the large intestine, which helps speed up the clearance process.

## SUMMARY

**LO1** **Essential elements** have a physiological function in the body. **Nonessential elements** are present in the body but have no known functions. Some may have **stimulatory effects**. Essential elements are categorized as **major, trace**, or **ultratrace essential elements** depending on their concentrations in the body. (Sections 22.1 and 22.2)

**LO2** Transport of $Na^+$ and $K^+$ across cell membranes involves **ion pumps** or selective transport through **ion channels**. Chloride ion is the most abundant anion in the human body, facilitating transport of alkali metal cations and elimination of $CO_2$. Differences in concentrations of ions inside and outside cell membranes give rise to a **membrane potential**, which determines the permeability of ions through the membrane. (Section 22.3)

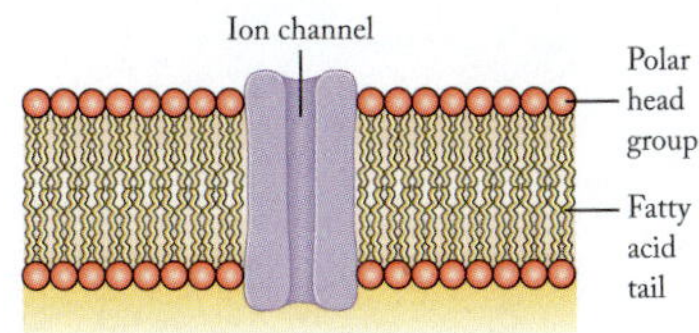


**LO3** Acid–base chemistry and redox reactions have great significance in living systems. The principles of structure and electron distribution in molecules apply to interactions *in vivo* just as they do in the laboratory. (Sections 22.3–22.6)

**LO4** The most abundant elements in the human body include seven nonmetals: C, H, O, Cl, S, P, and N; the four major metals in the human body are $Ca^{2+}$ $K^+$, $Na^+$, and $Mg^{2+}$. Nonessential elements from groups 1 and 2 may be absorbed into cells or substituted into bone or other tissues because they are similar in size and charge to essential elements. Some nonessential elements are used in nutritional supplements, whereas others have therapeutic properties in low concentrations but are toxic at higher levels. (Sections 22.3–22.5)

**LO5** Radionuclides with short half-lives that emit low-energy γ rays are used to assess function and diagnose disease. The selection of a radionuclide for medical use is governed by the toxicities of the element and its daughter nuclides, its radioactive half-life, and how quickly it is eliminated from the body. (Section 22.6)

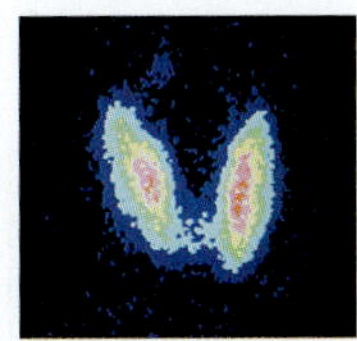

**LO6** Main group elements beyond the essential elements are useful in a variety of compounds with therapeutic value. Lithium salts are used in treating depression, whereas aluminum compounds find use as antacids. Fluoride in toothpaste helps prevent cavities. (Section 22.6)

## PARTICULATE **PREVIEW WRAP-UP**

From left to right, the ions are $Ca^{2+}$, $Na^+$, $H_3O^+$, and $K^+$. The $H_3O^+$ ion is most likely to undergo selective transport through the sodium ion channel as a result of its similar size and charge. None of the other three ions is likely to pass through the $K^+$ ion channel because they are different in size from the $K^+$ ion.

## PROBLEM-SOLVING SUMMARY

| Type of Problem | Concepts and Equations | Sample Exercises |
|---|---|---|
| **Calculating equilibrium potential ($E_{ion}$) for ions and $\Delta G$ for ion transport** | Calculate $E_{ion}$ by using a modified form of the Nernst equation: $$E_{ion} = \frac{0.0592\ V}{n} \log \frac{[M^{x+}]_{outside}}{[M^{x+}]_{inside}} \quad (22.1)$$ Calculate $\Delta G$ by using the relationships between $\Delta G$ and $E$: $$E_{transport} = E_{membrane} - E_{ion} \quad (22.2)$$ $$\Delta G_{transport} = -nFE_{transport} \quad (22.3)$$ | 22.1 |
| **Calculating an acid concentration from its pH** | Relate the pH of a solution to the $[H_3O^+]$ by the equation $$pH = -\log[H_3O^+]$$ | 22.2 |
| **Assigning oxidation numbers and writing half-reactions** | Use the guidelines in Section 8.7. | 22.3 |
| **Drawing Lewis structures for molecules** | Use the guidelines in Section 4.4. | 22.4 |
| **Calculating quantities of radioactive isotopes** | Use the equation $$\ln \frac{N_t}{N_0} = -0.693 \frac{t}{t_{1/2}}$$ where $N_0$ and $N_t$ are the amounts of material present initially and at time $t$, respectively. | 22.5 |

## VISUAL PROBLEMS

*(Answers to boldface end-of-chapter questions and problems are in the back of the book.)*

**22.1.** Which part of Figure P22.1 best describes the periodic trend in the size of monatomic cation radii? (Arrows point in the direction of *increasing* radii.)

22.2. Which part of Figure P22.1 best describes the periodic trend in the size of monatomic anion radii? (Arrows point in the direction of *increasing* radii.)

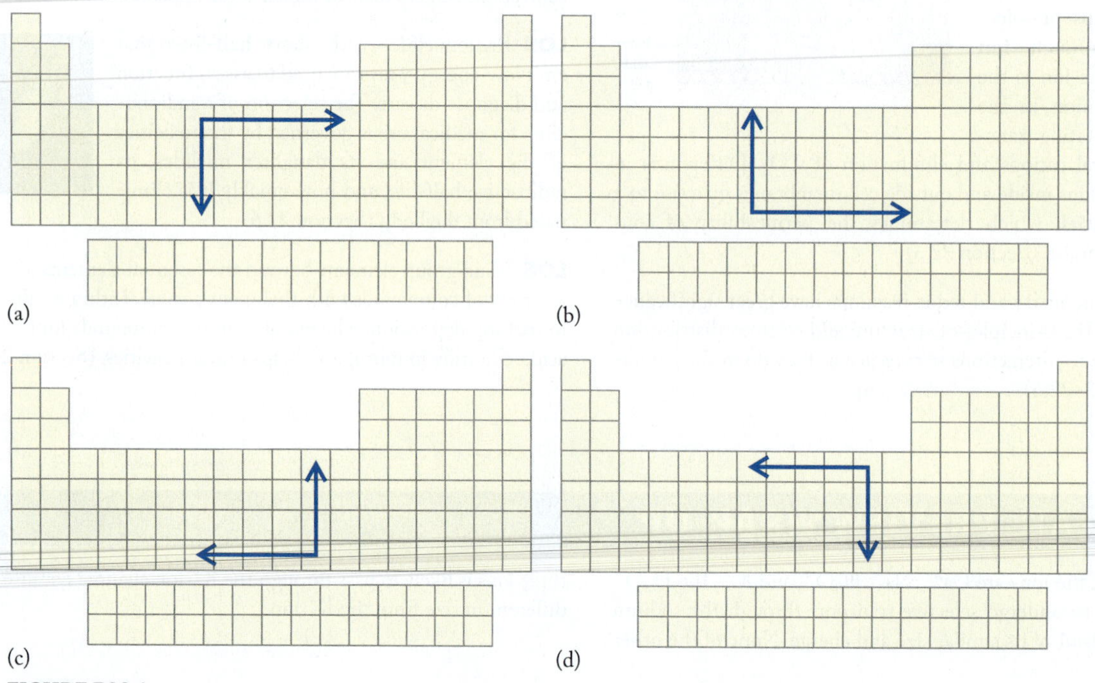


**FIGURE P22.1**

**22.3.** Which of the two groups highlighted in the periodic table in Figure P22.3 typically forms ions that have *larger* radii than those of the corresponding neutral atoms?

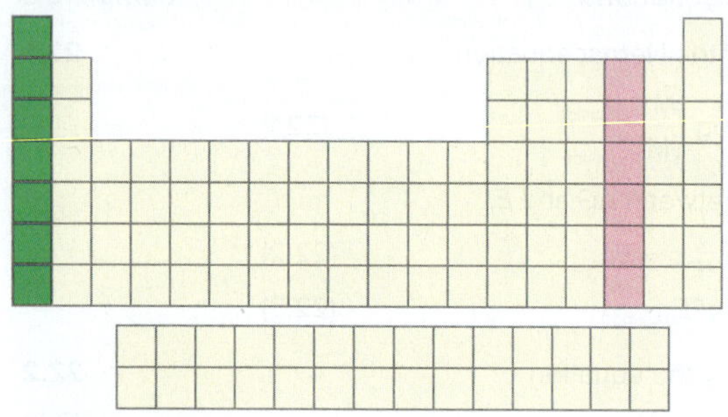

**FIGURE P22.3**

22.4. Which of the two groups highlighted in the periodic table in Figure P22.4 typically forms ions that have *smaller* radii than those of the corresponding neutral atoms?

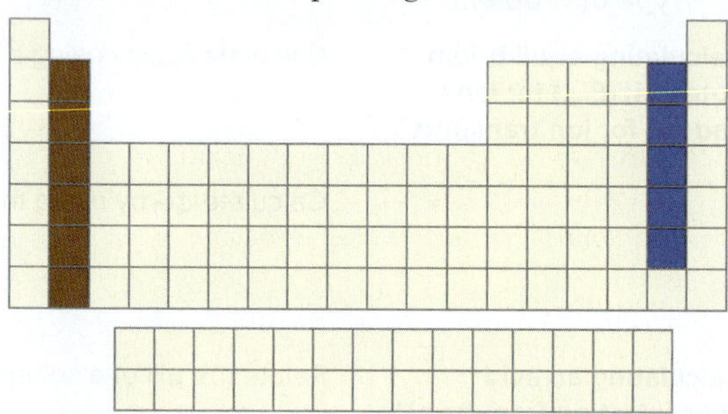

**FIGURE P22.4**

**22.5.** As we saw in Chapter 17, the free energy ($\Delta G$) of an electrochemical cell reaction is related to cell potential by the equation $\Delta G = -nFE$. In Figure P22.5, two solutions of $Na^+$ of different concentrations are separated by a semipermeable membrane. Calculate $\Delta G$ for the transport of $Na^+$ from the side with higher concentration to the side with lower concentration.

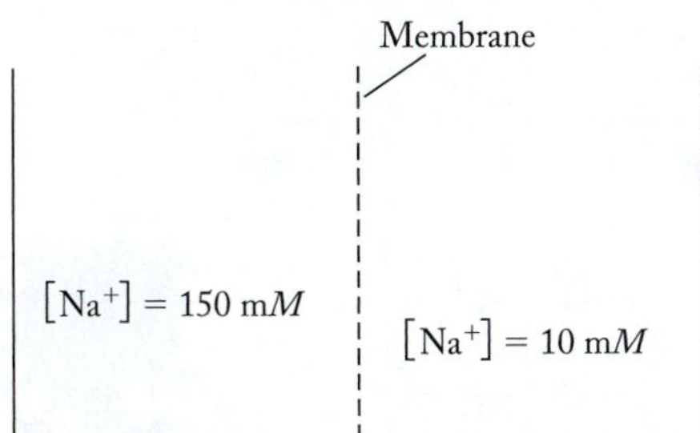


**FIGURE P22.5**

22.6. Two solutions of $K^+$ are separated by a semipermeable membrane in Figure P22.6. Calculate $\Delta G$ for the transport of $K^+$ from the side with lower concentration to the side with higher concentration.

Membrane

$[K^+] = 100\ mM$ | $[K^+] = 12\ mM$

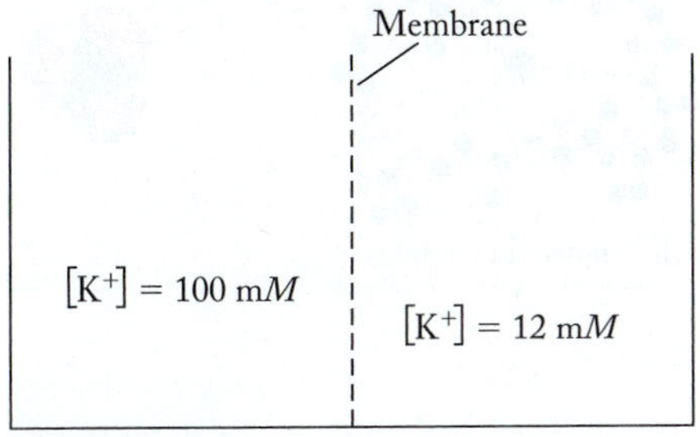

**FIGURE P22.6**

**22.7.** Describe the molecular geometry around each germanium atom in the compound shown in Figure P22.7.

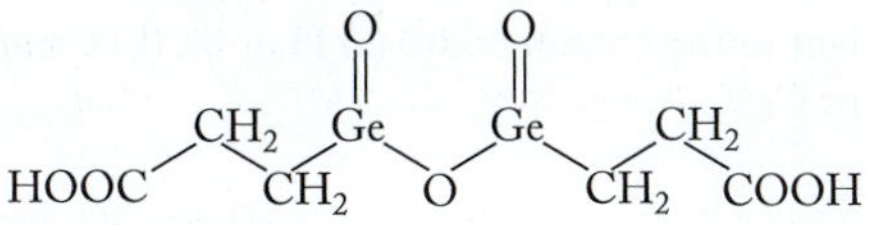

HOOC–$CH_2$–$CH_2$–Ge(=O)–O–Ge(=O)–$CH_2$–$CH_2$–COOH

**FIGURE P22.7**

22.8. Selenocysteine can exist as two enantiomers (stereoisomers). Identify the atom in Figure P22.8 responsible for the two enantiomers.

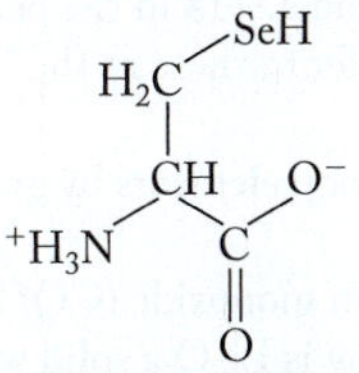

SeH–$H_2C$–CH($^+H_3N$)–C(=O)–$O^-$

**FIGURE P22.8**

***22.9.** The relative sizes of the main group atoms and ions are shown in Figure P22.9. Using this figure as a guide, which of the following polyatomic ions is likely to be the largest: sulfate, phosphate, or perchlorate?

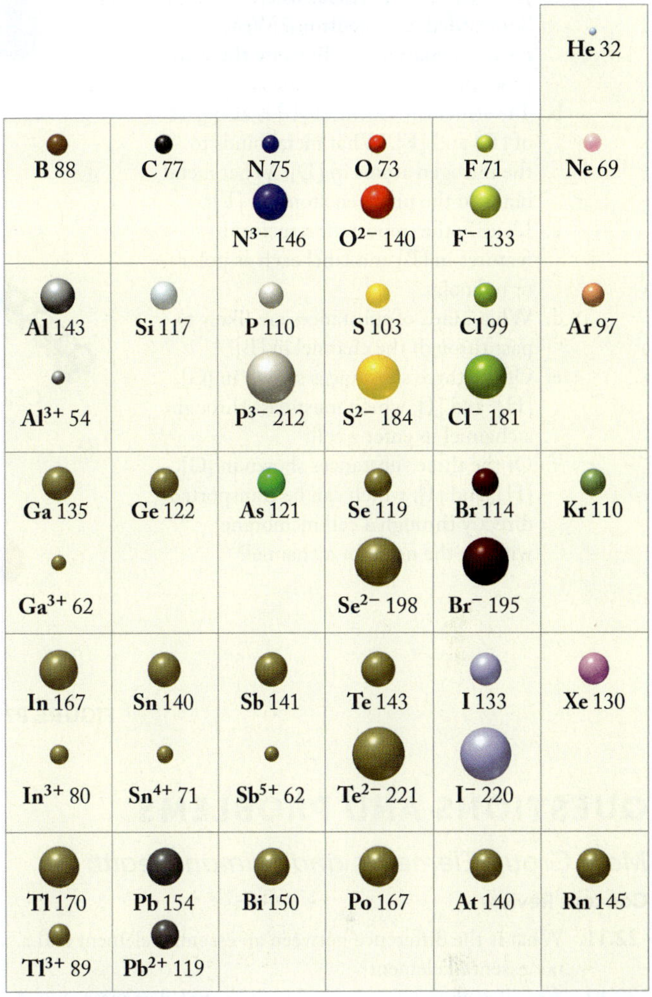

| | | | | | |
|---|---|---|---|---|---|
| | | | | | He 32 |
| B 88 | C 77 | N 75<br>$N^{3-}$ 146 | O 73<br>$O^{2-}$ 140 | F 71<br>$F^-$ 133 | Ne 69 |
| Al 143<br>$Al^{3+}$ 54 | Si 117 | P 110<br>$P^{3-}$ 212 | S 103<br>$S^{2-}$ 184 | Cl 99<br>$Cl^-$ 181 | Ar 97 |
| Ga 135<br>$Ga^{3+}$ 62 | Ge 122 | As 121 | Se 119<br>$Se^{2-}$ 198 | Br 114<br>$Br^-$ 195 | Kr 110 |
| In 167<br>$In^{3+}$ 80 | Sn 140<br>$Sn^{4+}$ 71 | Sb 141<br>$Sb^{5+}$ 62 | Te 143<br>$Te^{2-}$ 221 | I 133<br>$I^-$ 220 | Xe 130 |
| Tl 170<br>$Tl^{3+}$ 89 | Pb 154<br>$Pb^{2+}$ 119 | Bi 150 | Po 167 | At 140 | Rn 145 |

**FIGURE P22.9**

22.10. Use representations [A] through [I] in Figure P22.10 to answer questions a–f.
a. Which nuclide is the product of β decay from tritium? Which will produce two α particles when bombarded with neutrons? Write nuclear equations to illustrate these two processes.
b. The structure shown in [E] is also part of [D] and [F]. What metal binds to the nitrogen atoms in [D]? What metal binds to the nitrogen atoms in [F]?
c. Identify the important structural features in [B] and label each as polar or nonpolar.
d. What kinds of substances are likely to pass through the channel in [B]?
e. Of the three substances shown in [G], [H], and [I], which must pass through a channel to enter a cell?
f. Of the three substances shown in [G], [H], and [I], which can be transported directly through a cell membrane, without the need for a channel?

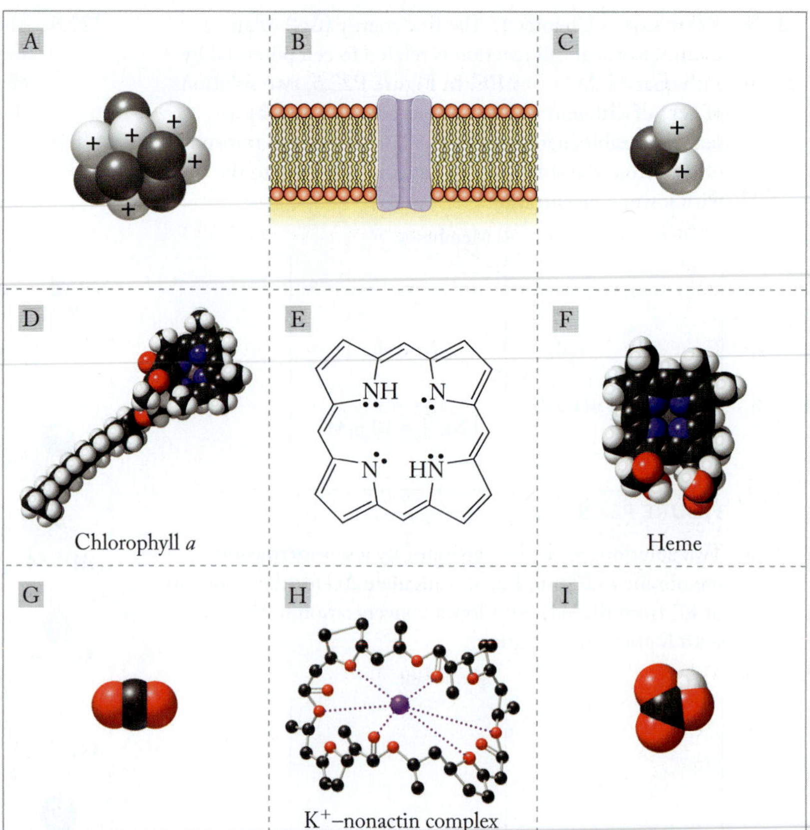


**FIGURE P22.10**

# QUESTIONS AND PROBLEMS

## *Main Group Elements and Human Health*

### Concept Review

**22.11.** What is the difference between an essential element and a nonessential element?

22.12. Are all essential elements major essential elements?

**22.13.** What is the main criterion that distinguishes major, trace, and ultratrace essential elements from one another?

22.14. Should trace essential elements also be considered stimulatory?

### Problems

**22.15.** The concentrations of very dilute solutions are sometimes expressed as parts per million. Express the concentration of each of the following trace and ultratrace essential elements in parts per million:
a. Fluorine, 110 mg in 70 kg
b. Silicon, 525 mg/kg
c. Iodine, 0.043 g in 100 kg

22.16. In the human body, the concentrations of ultratrace essential elements are even lower than those of trace essential elements and therefore are sometimes expressed in parts per billion. Express the concentrations of each of the following elements in parts per billion:
a. Bromine, 6 mg/L
b. Boron, 0.014 g/100 kg
c. Selenium, 5.0 mg/70 kg

**22.17.** Which element in each of the following pairs is more abundant in the human body? (a) silicon or oxygen; (b) iron or oxygen; (c) carbon or aluminum

22.18. Which element in each of the following pairs is more abundant in the human body? (a) H or Si; (b) Ca or Fe; (c) N or Cr

## *Periodic and Chemical Properties of Main Group Elements*

### Concept Review

**22.19.** In Chapter 2, we defined main group elements as those elements found in groups 1, 2, and 13–18 in the periodic table. Why do some chemists refer to these as the "*s*-block" and "*p*-block" elements?

22.20. Why do we classify the main group elements by group rather than by period?

**22.21.** Lithium oxide ($Li_2O$) and carbon monoxide (CO) have nearly the same molar mass. Why is $Li_2O$ a solid with a high melting point, whereas CO is a gas?

22.22. The nonradioactive group 17 elements are found as diatomic molecules, $X_2$ (X = F, Cl, Br, or I). Why is $Br_2$ a liquid at room temperature, whereas $Cl_2$ is a gas?

**22.23.** Which of the following properties can be used to distinguish a metal from a metalloid: atomic radius, electrical conductivity, and/or molar mass?

22.24. Which of the following cannot be measured: ionization energy, electron affinity, ionic radius, atomic radius, or electronegativity?

**22.25.** Why is $Be^{2+}$ more likely than $Ca^{2+}$ to displace $Mg^{2+}$ in biomolecules?

22.26. PbS, $PbCO_3$, and PbCl(OH) have limited solubility in water. Which of them is/are more likely to dissolve in acidic solutions?

### Problems

**22.27.** Which ion channel—$K^+$ or $Na^+$—must accommodate the larger cation?

22.28. Which ion is larger: $Cl^-$ or $I^-$?

---

**22.29.** Place the following ions in order of increasing ionic radius: $Mg^{2+}$, $Li^+$, $Al^{3+}$, and $Cl^-$.

22.30. Place the following ions in order of increasing ionic radius: $Br^-$, $O^{2-}$, $K^+$, and $Ca^{2+}$.

---

**22.31.** Place the following elements in order of increasing electronegativity: K, S, F, and Mg.

22.32. When we compare any two main group elements, is the element with the smaller atomic radius always more electronegative?

---

**22.33.** Why can we estimate the electron affinity of Cl atoms by measuring the ionization energy of a $Cl^-$ anion?

22.34. Place the following ions in order of increasing ionization energy: $Na^+$, $S^{2-}$, $F^-$, and $Mg^+$.

## *Major Essential Elements*

### Concept Review

**22.35.** **Ion Transport in Cells** Describe three ways in which ions of major essential elements (such as $Na^+$ and $K^+$) enter and exit cells.

22.36. Which transport mechanism for ions requires ATP: diffusion, ion channels, or ion pumps?

**22.37.** Why is diffusion across cell membranes difficult for ions?

22.38. Why does $Sr^{2+}$ substitute for $Ca^{2+}$ in bones?

**22.39.** Which alkali metal ion is $Rb^+$ most likely to substitute for?

22.40. Why don't alkaline earth metal cations substitute for alkali metal cations in cases where the ionic radii are similar?

***22.41.** Why might nature have selected calcium carbonate over calcium sulfate as the major exoskeleton material in shells?

22.42. Bromide ion and fluoride ion are nonessential elements in the body. Do you expect their concentrations to be more similar to the concentrations of major essential elements or to the concentrations of ultratrace essential elements?

### Problems

**22.43.** **Osmotic Pressure of Red Blood Cells** One of the functions of the alkali metal cations $Na^+$ and $K^+$ in cells is to maintain the cells' osmotic pressure. The concentration of NaCl in red blood cells is approximately 11 m*M*. Calculate the osmotic pressure of this solution at body temperature (37°C).

22.44. Calculate the osmotic pressure exerted by a 92 m*M* solution of KCl in a red blood cell at body temperature (37°C).

---

***22.45.** **Electrochemical Potentials across Cell Membranes** Very different concentrations of $Na^+$ ions exist in red blood cells (11 m*M*) and the blood plasma (160 m*M*) surrounding those cells. Solutions with two concentrations separated by a membrane constitute a concentration cell. Calculate the electrochemical potential created by the unequal concentrations of $Na^+$ at 298 K.

22.46. The concentration of $K^+$ in red blood cells is 92 m*M*, and the concentration of $K^+$ in plasma is 10 m*M*. Calculate the electrochemical potential created by the two concentrations of $K^+$ at 298 K.

---

***22.47.** **Calcium Ion Pumps** Transport of $Ca^{2+}$ across cell membranes is critical for cardiac muscle cells. The $[Ca^{2+}]$ in these cells increases by $\approx 10^2$ when the heart contracts.

a. What electrochemical potential ($E^\circ_{\text{ion}}$) does this difference in concentration correspond to at 298 K?

b. How much ATP must be hydrolyzed to achieve this change in concentration given that the hydrolysis of ATP is described by the equation:

$$ATP^{4-}(aq) + H_2O(\ell) \rightarrow ADP^{3-}(aq) + HPO_4^{2-}(aq) + H^+(aq) \quad \Delta G^\circ = -34.5 \text{ kJ}$$

*22.48. Removing excess $Na^+$ from a cell by an ion pump requires energy. How many moles of ATP must be hydrolyzed to overcome a cell potential of $-0.07$ V? The hydrolysis of 1 mole of ATP provides 34.5 kJ of energy.

---

**22.49.** **Genetic Engineering** Because plants require phosphate ($PO_4^{3-}$) to grow, applying fertilizers that contain phosphate promotes the growth of the desired crops as well as weeds, requiring herbicides to kill the latter. In 2018, geneticists announced the development of a strain of cotton that could grow on *phosphite* ($HPO_3^{2-}$) rather than phosphate, whereas common weeds could not.

a. Draw Lewis structures for both anions and assign oxidation states to phosphorus in each. (The phosphite ion contains a P–H bond).

*b. Why might phosphite-based fertilizers make it possible to use less herbicide?

22.50. **Naming Molecular Ions** The name "phosphite" for $HPO_3^{2-}$ is misleading. By analogy to the nomenclature for oxoanions in Chapter 4, what molecular formula and charge does the term "phosphite" suggest? Draw a Lewis structure for this molecular ion.

## *Trace and Ultratrace Essential Elements; Nonessential Elements*

### Concept Review

**22.51.** What danger to human health is posed by $^{137}Cs$ ($t_{1/2} \approx 30$ yr)?

22.52. Why is $^{137}Cs$ ($t_{1/2} \approx 30$ yr) considered dangerous to human health, whereas naturally occurring $^{40}K$ ($t_{1/2} = 1.28 \times 10^6$ yr) is benign?

**22.53.** What are the likely signs of $\Delta S$ and $\Delta G$ for the dissolution of tooth enamel?

22.54. Why does fluorapatite resist acid better than hydroxyapatite if both are insoluble in water?

***22.55.** Why do superoxide ions ($O_2^-$) act as strong oxidizing agents?

*22.56. Why are thyroxine and 3,5,3′-triiodothyronine (Figure 22.18) considered amino acids? Why aren't they *essential* amino acids?

## Problems

**22.57. The Aroma of Coffee** More than 1000 compounds have been identified in roasted coffee beans, many of which contribute to the aroma of brewed coffee when the beans are ground and extracted with water. Four sulfur compounds are included among the known aroma compounds in coffee (Figure P22.57)

a. Which compound do you expect to have the highest vapor pressure?
b. If we replaced the S with O, would you expect the vapor pressure of the resulting compounds to increase or decrease? Explain.

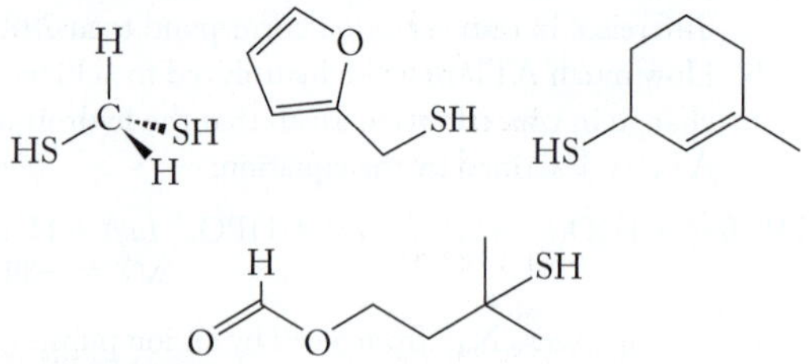

**FIGURE P22.57**

22.58. Identify the functional group that contains the sulfur atom in Figure P22.57 and determine whether any of these compounds are chiral. Are these compounds likely to be polar or nonpolar?

**22.59.** Calculate the pH of a $1.00 \times 10^{-3}$ *M* solution of selenocysteine ($pK_{a_1} = 2.21$; $pK_{a_2} = 5.43$).

22.60. Calculate the pH of a $1.00 \times 10^{-3}$ *M* solution of cysteine ($pK_{a_1} = 1.7$; $pK_{a_2} = 8.3$). Is selenocysteine a stronger acid than cysteine?

**22.61. Composition of Tooth Enamel** Tooth enamel contains the mineral hydroxyapatite. Hydroxyapatite reacts with fluoride ion in toothpaste to form fluorapatite. The equilibrium constant for the reaction between hydroxyapatite and fluoride ion is $K = 8.48$. Write the equilibrium constant expression for the following reaction. In which direction does the equilibrium lie?

$$Ca_5(PO_4)_3(OH)(s) + F^-(aq) \rightleftharpoons Ca_5(PO_4)_3(F)(s) + OH^-(aq)$$

*22.62. **Effects of Excess Fluoridation on Teeth** Too much fluoride might cause calcium fluoride to form, according to the reaction

$$Ca_5(PO_4)_3(OH)(s) + 10\,F^-(aq) \rightleftharpoons 5\,CaF_2(s) + 3\,PO_4^{3-}(aq) + OH^-(aq)$$

Write the equilibrium constant expression for the reaction. Given the $K_{sp}$ values for the following two reactions, calculate $K$ for the reaction between $Ca_5(PO_4)_3(OH)$ and fluoride ion that forms $CaF_2$.

$$Ca_5(PO_4)_3(OH)(s) \rightleftharpoons 5\,Ca^{2+}(aq) + 3\,PO_4^{3-}(aq) + OH^-(aq) \quad K_{sp} = 2.3 \times 10^{-59}$$

$$CaF_2(s) \rightleftharpoons Ca^{2+}(aq) + 2\,F^-(aq) \quad K_{sp} = 3.9 \times 10^{-11}$$

**22.63.** Tooth enamel is actually a composite material containing both hydroxyapatite and a calcium phosphate, $Ca_8(HPO_4)_2(PO_4)_4 \cdot 6\,H_2O$ ($K_{sp} = 1.1 \times 10^{-47}$).

a. Is this calcium mineral more or less soluble than hydroxyapatite ($K_{sp} = 2.3 \times 10^{-59}$)?
b. Calculate the solubility in moles per liter of hydroxyapatite [$Ca_5(PO_4)_3(OH)$], where $K_{sp} = 2.3 \times 10^{-59}$ in water at 25°C and pH = 7.00.
c. What is the solubility of hydroxyapatite at pH = 5.00?

22.64. The $K_{sp}$ of actual tooth enamel is reported to be $1 \times 10^{-58}$.

a. Does this mean that tooth enamel is more soluble than pure hydroxyapatite ($K_{sp} = 2.3 \times 10^{-59}$)?
b. Does the measured value of $K_{sp}$ for tooth enamel support the idea that tooth enamel is a mixture of hydroxyapatite [$Ca_5(PO_4)_3(OH)$] and a calcium phosphate, $Ca_8(HPO_4)_2(PO_4)_4 \cdot 6\,H_2O$ ($K_{sp} = 1.1 \times 10^{-47}$)?
c. Calculate the solubility in moles per liter of $Ca_8(HPO_4)_2(PO_4)_4 \cdot 6\,H_2O$ ($K_{sp} = 1.1 \times 10^{-47}$) in water at 25°C and pH = 7.00.

**22.65. Lead in Drinking Water** The municipal water systems of many older cities, including Flint, Michigan, still rely on water pipes made of lead or lead alloys. In principle, this is not as dangerous as it seems because lead corrodes, forming a protective coating of lead(II) carbonate ($PbCO_3$; Figure P22.65).

a. Calculate the solubility of $PbCO_3$ at neutral pH.
b. The EPA limit for dissolved $Pb^{2+}$ is 15 ppb. Does your answer to part (a) meet this standard?
c. How does the solubility of $PbCO_3$ depend on pH?

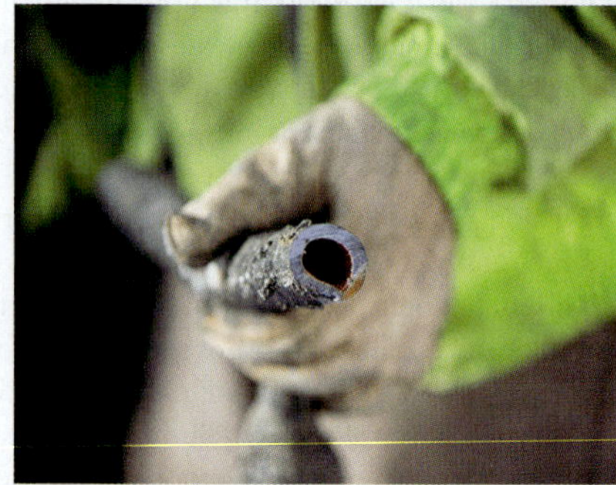

**FIGURE P22.65**

22.66. **Removing Lead from Water** To reduce the concentration of $Pb^{2+}$ in drinking water, many treatment plants add phosphate ion ($PO_4^{3-}$) to the water to precipitate lead(II) phosphate.

a. Is $Pb_3(PO_4)_2$ ($K_{sp} = 3.0 \times 10^{-44}$) more or less soluble than $PbCO_3$?
b. Why does adding phosphate also regulate the pH of the solution?
c. Calculate the pH of a 0.125 *M* solution of $Na_3PO_4$.

**22.67.** All the group 16 elements form compounds with the generic formula $H_2E$ (E = O, S, Se, or Te). Which compound is the most polar? Which compound is the least polar?

*22.68. All the group 15 elements form compounds with the generic formula $H_3E$ (E = N, P, As, Sb, or Bi). Which compound is the most polar? Which compound do you predict to have the smallest H–E–H bond angle?

## Elements for Diagnosis and Therapy

### Concept Review

**22.69.** When choosing an isotope for imaging, why is considering both the isotope's decay mode and half-life important?

22.70. Why might an α emitter be a good choice for radiation therapy?

***22.71.** What advantage does a β emitter have over an α emitter for imaging?

22.72. Why do we sometimes use radioisotopes of toxic elements, such as thallium, for imaging?

**22.73.** Why does $^{213}Bi$ undergo β decay, whereas $^{111}In$ decays by electron capture?

22.74. Several isotopes of arsenic are used in medical imaging. Which isotope, $^{72}As$ or $^{77}As$, is more likely to be useful for PET imaging?

**22.75.** The World Anti-Doping Agency (WADA) added xenon and argon to the list of banned substances in 2014. Which intermolecular forces account for the solubility of Xe and Ar in blood?

22.76. Helium is used in scuba gear to prevent nitrogen narcosis. Do you expect the solubility of He in blood to be greater than or less than the solubility of Xe and Ar in blood?

### Problems

**22.77.** **PET Imaging with Gallium** A patient is injected with a 5 μ*M* solution of gallium citrate containing $^{68}Ga$ ($t_{1/2}$ = 9.4 h) for a PET study. How long is it before the activity of the $^{68}Ga$ drops to 5% of its initial value?

22.78. Iodine-123 ($t_{1/2}$ = 13.3 h) has replaced iodine-131 ($t_{1/2}$ = 8.1 d) in diagnosing thyroid conditions. How long is it before the activity of $^{123}I$ drops to 5% of its initial value?

---

**22.79.** Bismuth ions ($Bi^{3+}$) form acidic solutions with a p$K_a$ of 1.5.
a. Write a balanced net ionic equation for the reaction of $Bi^{3+}$ with water to form a dication, $Bi(H_2O)_5OH^+$.
b. What is the pH of a 0.00364 *M* solution of $Bi^{3+}$?
c. Is this solution more or less acidic than an equimolar solution of vinegar? (*Note*: No calculation is necessary to answer this question.)

22.80. Some medicines used in treating depression contain lithium carbonate.
a. Draw the Lewis structure for $Li_2CO_3$.
b. Explain whether $Li_2CO_3$ hydrolyzes in water.
c. Calculate the solubility of $Li_2CO_3$ in water in moles/L.

---

**22.81.** Aluminum hydroxide is used in some antacids. Write a balanced net ionic equation for the reaction of aluminum hydroxide with HCl.

22.82. Aluminum carbonate is used in some antacids. Write a balanced net ionic equation for the reaction of aluminum carbonate with hydrochloric acid.

---

**22.83.** **Maalox** The antacid known as Maalox contains a mixture of magnesium and aluminum hydroxides.
a. Which substance will neutralize more acid on a per-mole basis?
b. Does the same substance also neutralize more acid on a per-gram basis?
c. How many grams of each are required to neutralize 115 mL of 0.75 *M* stomach acid?

22.84. **Tums** Sodium bicarbonate and calcium carbonate both act as antacids and are found in common stomach remedies such as Tums.
a. Which substance will neutralize more acid on a per-mole basis?
b. Does the same substance also neutralize more acid on a per-gram basis?
c. How many grams of each are required to neutralize 115 mL of 0.75 *M* stomach acid?

---

**22.85.** ***Il Campanello*** The 1836 opera *Il Campanello* (*The Night Bell*) by Gaetano Donizetti features the marital troubles of a wealthy pharmacist and his much younger wife. The former turns to a mixture of antimony(III) chloride and mercury(II) sulfide to solve his problems.
a. In which group of the periodic table is antimony found?
b. Draw a Lewis structure for antimony(III) chloride.
c. Is antimony(III) chloride more likely to behave as a Lewis acid or a Lewis base?

22.86. The chemistry of antimony(III) chloride is similar to that of the other elements in its group. Antimony(III) chloride can be converted to $Sb_2O_3$, which has been used as a flame retardant.
a. Write a balanced chemical equation for the reaction of antimony(III) oxide with water to form $H_3SbO_3$.
b. Write a balanced chemical equation for the reaction of antimony(III) oxide with sodium hydroxide to form $NaSbO_2$ and water.
c. Identify the role of $Sb_2O_3$ in parts (a) and (b) as either an acid or a base.

# 23

# Transition Metals

## Biological and Medical Applications

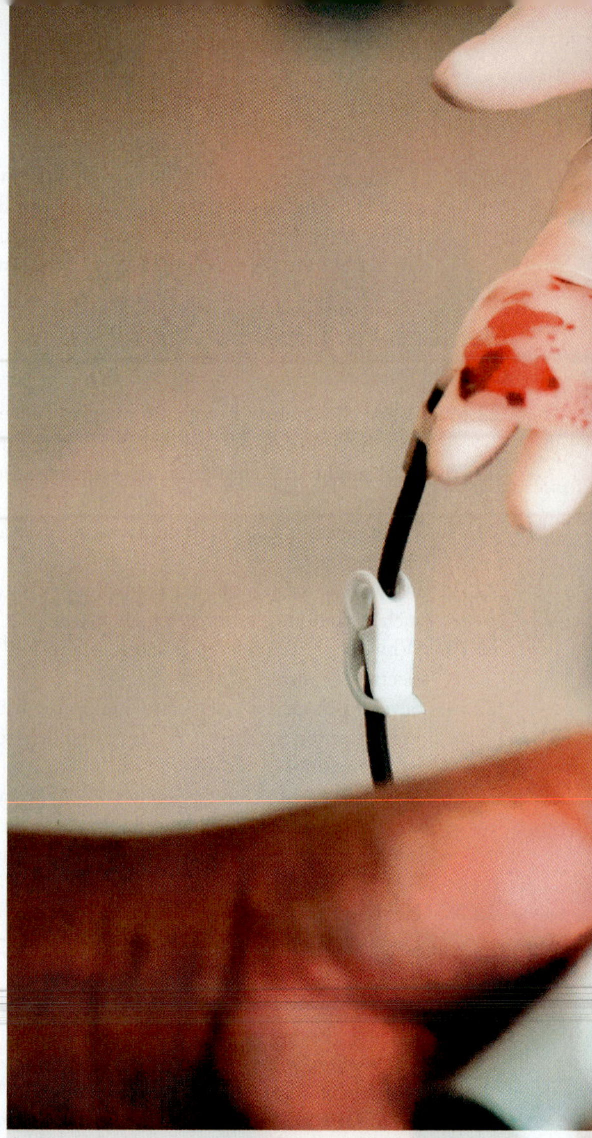

RED COLOR OF BLOOD The red color of blood comes from the heme group, a molecule containing a ring with four nitrogen atoms that bind to a central $Fe^{2+}$ ion.

## PARTICULATE **REVIEW**

### *Lewis Acid or Lewis Base?*

In Chapter 23, we discuss the bonding and structure of compounds and complex ions formed by transition metals and learn about their presence in biological systems and their applications to medicine. The compounds shown here are ammonia, borane, and water.

- Draw the Lewis structure for each compound.
- Which compound(s) include at least one lone pair of electrons on the central atom?
- Which compound(s) can function as a Lewis acid? Which can function as a Lewis base?

(Review Chapters 4, 8, and 16 if you need help.)

*(Answers to Particulate Review questions are in the back of the book.)*

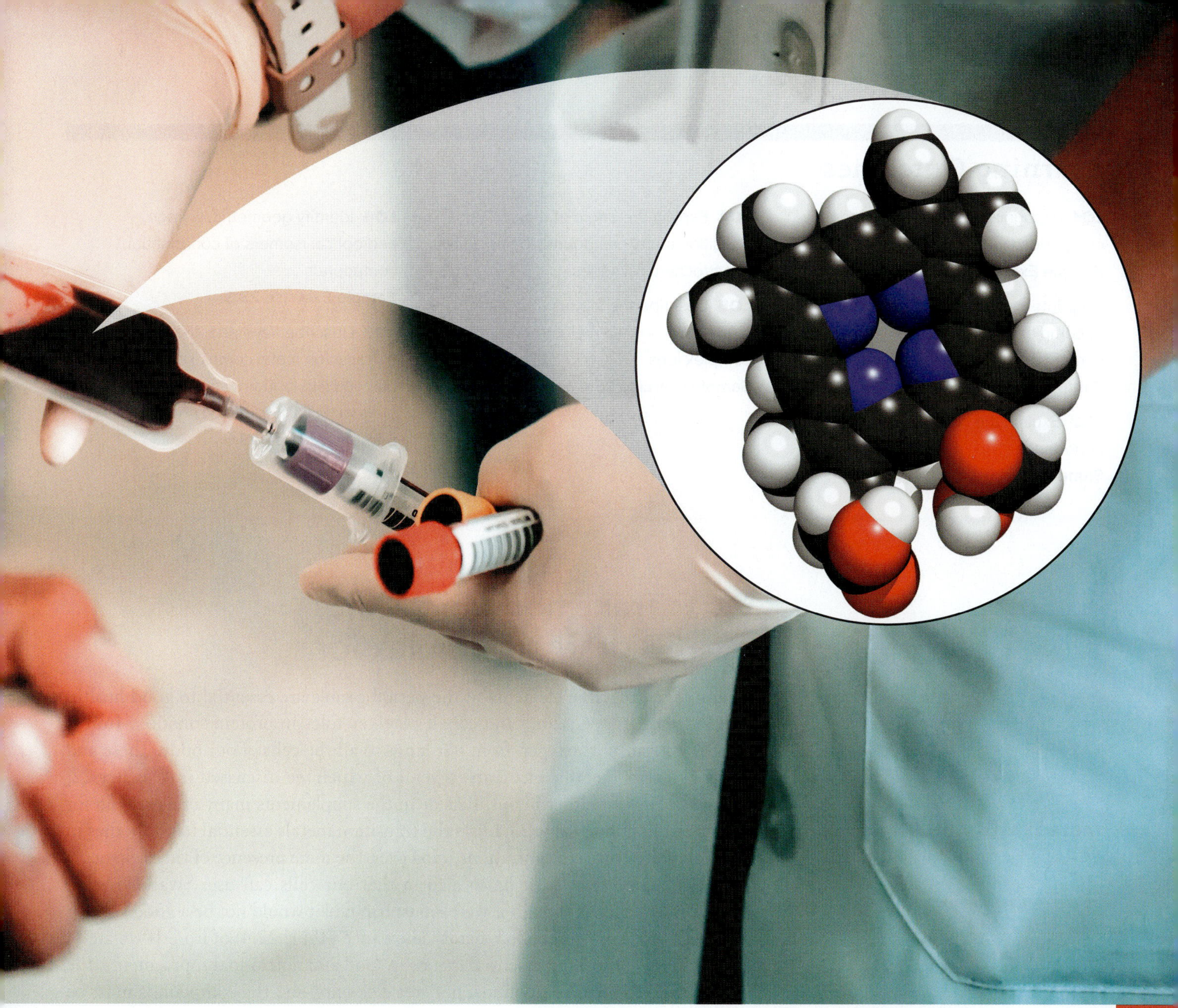

PARTICULATE **PREVIEW**

### *One Molecule, One Bond versus One Molecule, Two Bonds*

Here are two complex ions that contain metal cations bonded to different molecules. As you read Chapter 23, look for ideas that will help you answer these questions:

- Atoms of what element are bonded to the $Cu^{2+}$ ion and the $Ni^{2+}$ ion in the two complexes to the right?
- What molecule is bonded to the $Cu^{2+}$ ion? How many of these molecules form coordinate bonds with the ion?
- What molecule is bonded to the $Ni^{2+}$ ion? How many bonds does each molecule form?

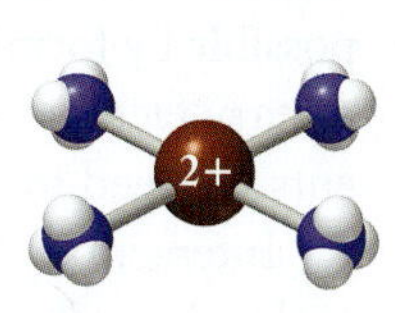

$Cu^{2+}$ complex

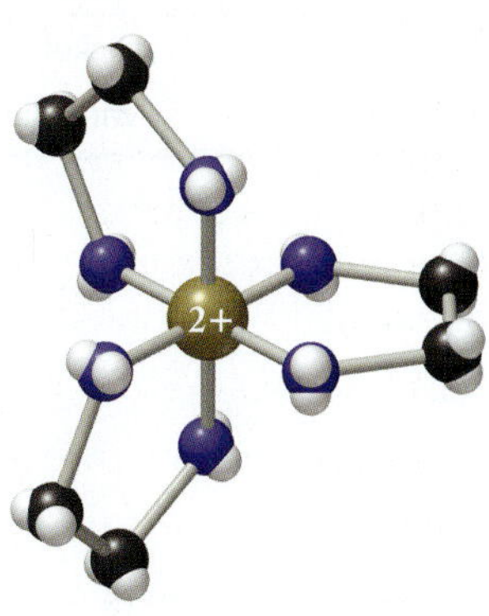

$Ni^{2+}$ complex

## Learning Outcomes

**LO1** Recognize complex ions and their counterions in chemical formulas
**Sample Exercise 23.1**

**LO2** Interconvert the names and formulas of complex ions and coordination compounds
**Sample Exercise 23.2**

**LO3** Explain the chelate effect and its importance
**Sample Exercise 23.3**

**LO4** Explain the origin of the colors of transition metal compounds by using the spectrochemical series

**LO5** Describe the factors that lead to high-spin or low-spin electronic states of complex ions
**Sample Exercise 23.4**

**LO6** Identify geometric, linkage, and optical isomers of coordination compounds
**Sample Exercise 23.5**

**LO7** Describe several of the roles of transition metal complexes in biochemistry and how they are used as diagnostic or therapeutic compounds
**Sample Exercise 23.6**

## 23.1 Transition Metals in Biology: Complex Ions

Many of the metallic elements in the periodic table are essential to good health. For example, copper, zinc, and cobalt play key roles in protein function. Iron is needed to transport oxygen from our lungs to all the cells of our body. These and other essential metallic elements, several of which we discussed in Chapter 22, should be present either in our diets or in the supplements many of us rely on for balanced nutrition. **Table 23.1** lists the transition metals essential to our bodies in trace and ultratrace concentrations. However, the mere presence of these elements is not enough—they must be in a form that our cells can use. Swallowing an 18-mg steel pellet as though it were an aspirin tablet would not be a good way for you to get your recommended dietary allowance (Table 22.4) of iron. If we are to benefit from consuming essential metals in food and nutritional supplements, the metals need to be in compounds, not free elements, and the compounds must be bioavailable to the body.

**TABLE 23.1 Essential Transition Elements Found in the Human Body**

| Trace (1–1000 μg/g of Body Mass) | Ultratrace (<1 μg/g of Body Mass) |
|---|---|
| Iron | Chromium |
| Zinc | Cobalt |
| | Copper |
| | Manganese |
| | Molybdenum |
| | Nickel |
| | Vanadium |

All the metallic transition elements essential to human health occur in nature in ionic compounds, but not all ionic forms are absorbed equally well. For example, most of the iron in fish, poultry, and red meat is readily absorbed because it is present in a form called *heme iron*. The iron in plants, however, is mostly nonheme and is not as readily absorbed. Eating a meal that includes both meat and vegetables improves the absorption of the nonheme iron in the vegetables. Fruits, vegetables, and other foods high in vitamin C increase the availability of iron, possibly by forming complex ions of the type discussed in Section 16.6. All these dietary factors work together at the molecular level to provide us with the nutrients we need to survive.

Interactions between transition metals and accompanying nonmetal ions and molecules influence the solubility of the metals, which is a key factor for making them chemically reactive and biologically available to plants and animals. These interactions also influence other properties, including the wavelengths of visible light the metals absorb and therefore the colors of their compounds and solutions. Here in Chapter 23, we explore how the chemical environment of transition metal ions in solids and in solutions affects their physical, chemical, and biological properties. We answer questions such as why many, but not all, metal compounds have distinctive colors, and how, by forming complex ions with biomolecules, transition metals play key roles in many biological processes.

We begin by examining the interactions between transition metal ions and the other ions and molecules that surround them in solids and solutions. To understand these interactions, we need to review the definitions of Lewis acids and Lewis bases we first used in Section 16.5:

- A *Lewis base* is a substance that *donates* a lone pair of electrons in a chemical reaction.
- A *Lewis acid* is a substance that *accepts* a lone pair of electrons in a chemical reaction.

**CONNECTION** Lewis's pioneering theories of the nature of covalent bonding were described in Chapter 4, and the formation of complex ions between metal ions (Lewis acids) and ligands (Lewis bases) was discussed in Section 16.6.

In Section 6.3, we described how ions dissolved in water are *hydrated*, that is, surrounded by water molecules oriented with their positive dipoles directed toward anions and their negative dipoles directed toward cations (see Figure 6.17). When these ion–dipole interactions lead to the sharing of lone-pair electrons with empty valence-shell orbitals on the cations, they meet our definition of covalent bonds and then are called *coordinate covalent bonds*, or simply *coordinate bonds*. Much of the chemistry of the transition metals is associated with their ability to form coordinate bonds with molecules or anions.

**CONNECTION** We discussed in Chapter 8 how the electrical conductivity of aqueous solutions depends on the concentrations of dissolved ions in the solutions.

As we saw in Section 16.6, molecules or anions that function as Lewis bases and form coordinate bonds with metal cations are called *ligands*. The resulting species, which are composed of central metal ions and the surrounding ligands, are called *complex ions* or simply *complexes*. Direct bonding to a central cation means that the ligands in a complex occupy the **inner coordination sphere** of the cation. Take another look at Sample Exercise 14.11, in which we discussed the equilibrium between two forms of cobalt(II), one pink and one blue (see Figure 14.11), in an aqueous solution of HCl (**Figure 23.1**). Both the pink and blue forms are complexes; the pink cobalt(II) species has six water ligands in its inner coordination sphere, whereas the blue species has four chloride ions. The charge on each complex ion is the sum of the charges of the metal ion and the ligands: 2+ for $Co(H_2O)_6^{2+}$ because the charge of the cobalt ion is 2+ and water molecules are neutral, and 2− for $CoCl_4^{2-}$ because the sum of the 2+ charge on the cobalt ion and four 1− charges on the chloride ions is 2−.

The foundation of our understanding of the bonding and structure of complex ions comes from the pioneering research of Swiss chemist Alfred Werner (1866–1919), for which he was awarded the Nobel Prize in Chemistry in 1913. Some of Werner's research addressed the unusual behavior of different compounds formed by dissolving cobalt(II) chloride in aqueous ammonia and oxidizing it to

**inner coordination sphere** the ligands that are bound directly to a metal via coordinate bonds.

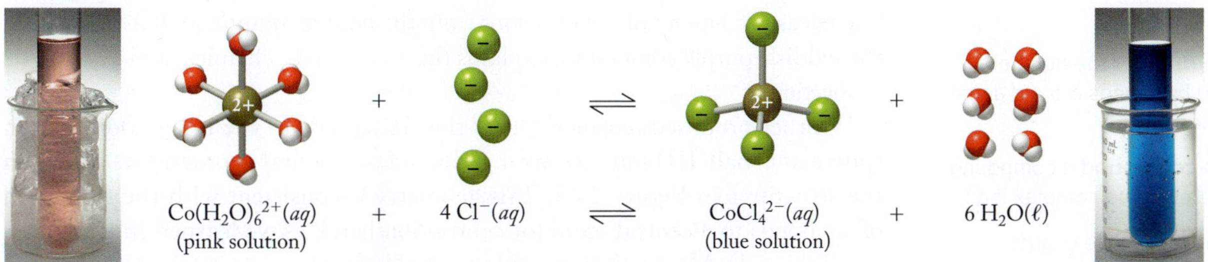


**FIGURE 23.1** Equilibrium between two forms of cobalt in aqueous HCl solution.

(a) $[Co(NH_3)_6]Cl_3$

(b) $[Co(NH_3)_5Cl]Cl_2$

**FIGURE 23.2** Two compounds of cobalt(III) chloride and ammonia.

cobalt(III) by bubbling air through the solution. One of the redox reactions produces an orange compound (**Figure 23.2a**) that contains 3 moles of $Cl^-$ ions for every 1 mole of $Co^{3+}$ ions and 6 moles of ammonia. Another reaction produces a reddish-purple compound (**Figure 23.2b**) that has the same proportions of $Cl^-$ and $Co^{3+}$ ions, but with only 5 moles of ammonia per mole of cobalt(III). Both compounds are water-soluble solids that react with aqueous solutions of $AgNO_3$, forming solid AgCl. However, 1 mole of the orange compound produces 3 moles of solid AgCl, whereas 1 mole of the reddish-purple one produces only 2 moles of AgCl. Results like these inspired Werner to study the electrical conductivity of aqueous solutions of the two compounds. He found that the orange compound was the better conductor, indicating that it produced more ions in solution than the reddish-purple one.

Werner concluded that the differences in the two compounds' composition, in their capacities to react with $Ag^+$ ions, and in their electrolytic properties are all caused by the presence of two types of Co–Cl bonds inside them. He proposed that these bonds were all ionic in the orange compound but that only two-thirds of them were ionic in the reddish-purple compound. The remaining one-third were coordinate covalent bonds.

## CONCEPT TEST

How would the electrical conductivity and freezing points of 0.010 *m* aqueous solutions of the two compounds in Figure 23.2 differ?

*(Answers to Concept Tests are in the back of the book.)*

This capacity to form two kinds of bonds means that $Co^{3+}$ and other transition metal ions have two kinds of bonding capacity, or *valence*. The first kind involves ionic bonds and is based on the number of electrons a metal atom loses when it forms an ion. This valence is equivalent to its oxidation number, which is +3 for cobalt in both the orange and reddish-purple compounds. The second kind of valence is based on the capacity of metal ions to form coordinate bonds. This property corresponds to an ion's **coordination number**, the number of sites around a central metal ion where bonds to ligands form.

When the two compounds in Figure 23.2a and b dissolve in water, they form an orange solution and a reddish-purple solution. The respective solutions contain the ions shown in **Figure 23.3**. The orange solution contains $[Co(NH_3)_6]^{3+}$ and three chloride ions, whereas the reddish-purple solution contains $[Co(NH_3)_5Cl]^{2+}$ and two chloride ions. Notice how brackets in the formulas of these **coordination compounds** set off the complex ions from the ionically bonded chloride **counterions**, which balance the charges on the complex ions: 3+ for the orange one (Figure 23.3a), but only 2+ for the reddish-purple one because it contains a negatively charged $Cl^-$ ion within its inner coordination sphere (Figure 23.3b). The release of 3 moles of chloride ions from the orange compound, but only 2 from the reddish-purple compound, explains the compounds' chemical and electrolytic properties.

**coordination number** the number of sites occupied by ligands around a metal ion in a complex.

**coordination compound** a compound made up of at least one complex ion.

**counterion** an ion whose charge balances the charge of a complex ion in a coordination compound.

Werner proposed, correctly, that the six ligands in the inner coordination sphere of cobalt(III) ions are arranged in an octahedral geometry, as shown in the structures in Figure 23.3. This geometry is consistent with the formation of six bonds to a central atom (or central ion here), as we learned in Chapter 5 (see Table 5.1). Many other transition metal ions form six coordinate bonds and octahedral complex ions. Some are listed in **Table 23.2**, as are complex ions

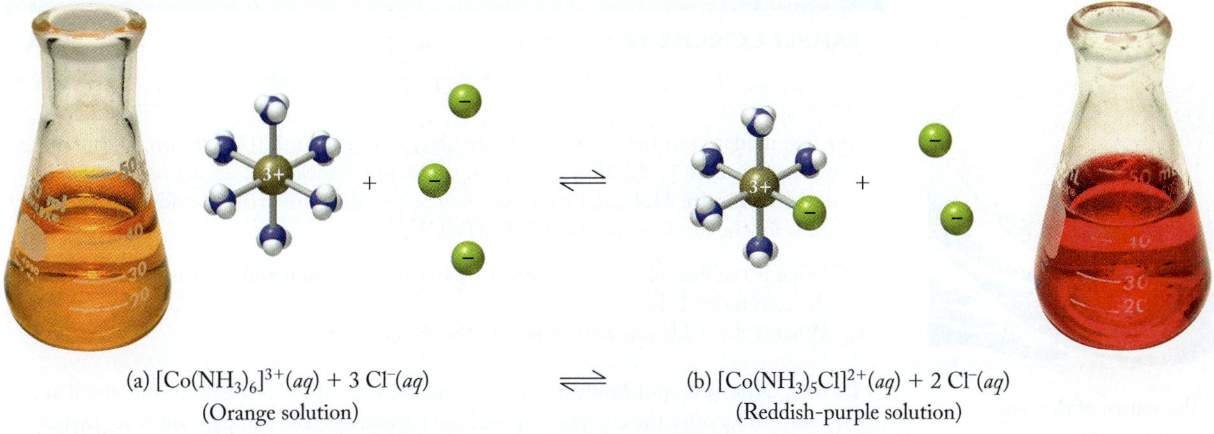


**FIGURE 23.3** Structures of cobalt(III) complex ions and counterions in aqueous solution.

that have only two or four ligands. Those with two have linear structures, whereas those with four may be tetrahedral or square planar. Both geometries occur in cobalt(II) ions: $CoCl_4^{2-}$ is tetrahedral, whereas $Co(CN)_4^{2-}$ is square planar.

## CONCEPT TEST

In the coordination compound $Na_3[Fe(CN)_6]$, which ions occupy the inner coordination sphere of the $Fe^{3+}$ ion, and which ions are counterions? Of the four cobalt complexes described in this section, which would have conductivity in aqueous solution similar to that of $Na_3[Fe(CN)_6]$?

**TABLE 23.2** Common Coordination Numbers and Shapes of Complex Ions

| Coordination Number | Steric Number | Shape | Structure | Examples |
|---|---|---|---|---|
| 6 | 6 | Octahedral | | $Fe(H_2O)_6^{3+}$<br>$Ni(H_2O)_6^{2+}$<br>$Co(H_2O)_6^{3+}$ |
| 4 | 6 | Square planar | | $Pt(NH_3)_4^{2+}$ |
| 4 | 4 | Tetrahedral | | $Zn(H_2O)_4^{2+}$ |
| 2 | 2 | Linear | | $Ag(NH_3)_2^{+}$ |

**FIGURE 23.4** The source of the blue color in *The Great Wave off Kanagawa*, a woodblock print by the Japanese artist Hokusai, is the pigment Prussian blue.

**SAMPLE EXERCISE 23.1** Assigning Oxidation States in Coordination Compounds **LO1**

The first modern synthetic color, Prussian blue, was made in 1704 in Berlin, Germany. It was heavily used in the 19th-century Japanese woodblock print *The Great Wave off Kanagawa* (**Figure 23.4**), and it continues to be popular with artists worldwide. The formula for the insoluble pigment is $Fe_4[Fe(CN)_6]_3$.

a. What is the formula of the complex ion in this compound and what is its charge if the counterion is $Fe^{3+}$?
b. What is the oxidation state of iron in the complex ion?

**Collect, Organize, and Analyze** We have the formula of a coordination compound and are asked to identify the complex ion and the counterion. The complex ion is contained in brackets, the four $Fe^{3+}$ counterions are outside the brackets, and the compound must be electrically neutral.

**Solve**

a. The combined charge of four $Fe^{3+}$ is 12+. To achieve electrical neutrality, the combined charge of the three complex anions must be 12−. Therefore, the charge of each anion is (12−)/3 = 4−, and the formula for the anion is $[Fe(CN)_6]^{4-}$.

b. Six cyanide ligands, $CN^-$, occupy the inner coordination sphere for a total charge of 6−. Therefore, the oxidation state of Fe must be 2+ for the charge on the complex ion to equal 4−.

**Think About It** Complex ions may have positive or negative charges. Correspondingly, counterions will have charges opposite to those of the complex ions to ensure electrical neutrality.

**Practice Exercise** A coordination compound of ruthenium, $[Ru(NH_3)_4Cl_2]Cl$, has shown some activity against leukemia in animal studies. Identify the complex ion, determine the oxidation number of the metal, and identify the counterion.

*(Answers to Practice Exercises are in the back of the book.)*

## 23.2 Naming Complex Ions and Coordination Compounds

The names of complex ions and coordination compounds tell us the identity and oxidation state of the central ion, the names and numbers of ligands, the charge in the case of complex ions, and the identity of counterions. To convey all this information, we need to follow some naming rules.

### Complex Ions with a Positive Charge

1. Start with the identities of the ligand(s). Names of common ligands appear in **Table 23.3**. If more than one kind of ligand is present, list the names alphabetically.
2. Use the usual prefix(es) in front of the name(s) written in step 1 to indicate the number of each type of ligand (**Table 23.4**).
3. Write the name of the metal ion with a Roman numeral indicating its oxidation state.

**TABLE 23.3** Names and Structures of Common Ligands

| Ligand | Name within Complex Ion | Structure | Charge | Number of Donor Groups |
|---|---|---|---|---|
| Iodide | Iodo | $I^-$ | 1– | 1 |
| Bromide | Bromo | $Br^-$ | 1– | 1 |
| Chloride | Chloro | $Cl^-$ | 1– | 1 |
| Fluoride | Fluoro | $F^-$ | 1– | 1 |
| Nitrite | Nitro | 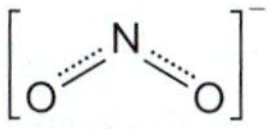 $[O{\cdots}N{\cdots}O]^-$ | 1– | 1 |
| Hydroxide | Hydroxo | $[O{-}H]^-$ | 1– | 1 |
| Water | Aqua | 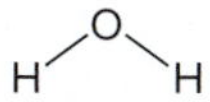 H–O–H | 0 | 1 |
| Pyridine (py) | Pyridyl | 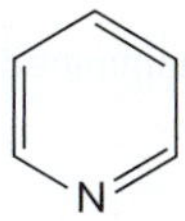 N | 0 | 1 |
| Ammonia | Ammine | $NH_3$ | 0 | 1 |
| Ethylenediamine (en) | (same)[a] | 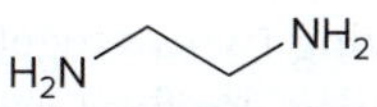 $H_2N$–CH₂CH₂–$NH_2$ | 0 | 2 |
| 2,2'-Bipyridine (bipy) | Bipyridyl | 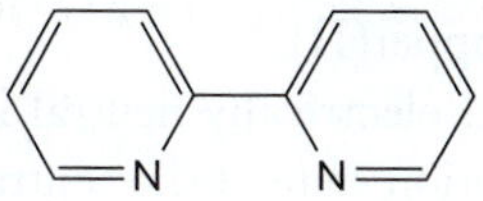 N N | 0 | 2 |
| 1,10-Phenanthroline (phen) | (same)[a] | 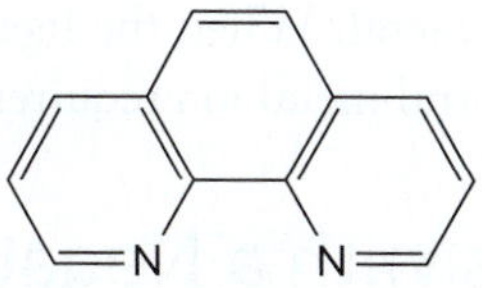 N N | 0 | 2 |
| Cyanide[b] | Cyano | $[C{\equiv}N]^-$ | 1– | 1 |
| Carbon monoxide[b] | Carbonyl | $C{\equiv}O$ | 0 | 1 |

[a]The names of some electrically neutral ligands in compexes are the same as the names of the molecules.

[b]Carbon atoms are the lone pair donors in these ligands.

**TABLE 23.4** Common Prefixes Used in the Names of Complex Ions

| Number of Ligands | Prefix |
|---|---|
| 2 | di- |
| 3 | tri- |
| 4 | tetra- |
| 5 | penta- |
| 6 | hexa- |

**STEPWISE ANIMATION**

Naming Coordination Compounds

Examples:

| Formula | Name | Structure |
|---|---|---|
| $Ni(H_2O)_6^{2+}$ | Hexaaquanickel(II) | 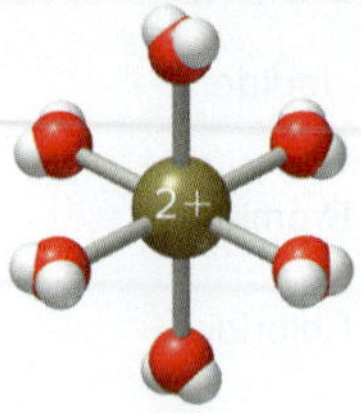 |
| $Co(NH_3)_6^{3+}$ | Hexaamminecobalt(III) | 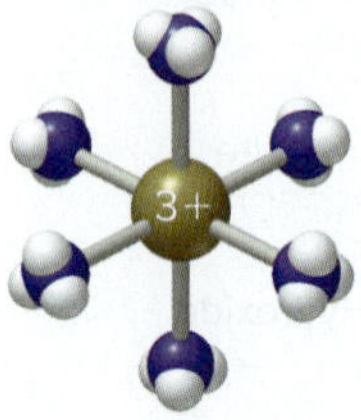 |
| $Cu(NH_3)_4(H_2O)_2^{2+}$ | Tetraamminediaquacopper(II) | 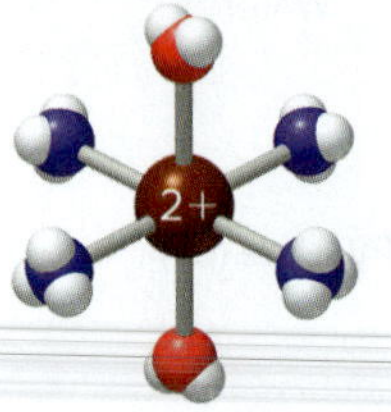 |

It may seem strange having two *a*'s together in these names, but it is consistent with current naming rules. Prefixes are ignored in determining alphabetical order, which is why *ammine* comes before *aqua* rather than *di* before *tetra* in tetraamminediaquacopper(II).

The ligands are all electrically neutral in these three examples. This makes determining the oxidation state of the central metal ion a simple task because the charge on the complex ion is the same as the charge on the metal ion, which is the oxidation state of the metal. When the ligands are anions, determining the oxidation state of the central metal ion requires us to account for these charges.

## Complex Ions with a Negative Charge

1. Follow the steps for naming positively charged complexes.
2. Add *-ate* to the name of the central metal ion to indicate that the complex ion carries a negative charge (just as we use *-ate* to end the names of oxoanions). For some metals, the base name changes, too. The two most common examples are iron, which becomes *ferrate*, and copper, which becomes *cuprate*.

Examples:

| Formula | Name | Structure |
|---|---|---|
| $Fe(CN)_6^{3-}$ | Hexacyanoferrate(III) | |
| $[Fe(H_2O)(CN)_5]^{3-}$ | Aquapentacyanoferrate(II) | |
| $[VCl_4(NH_3)_2]^-$ | Diamminetetrachlorovanadate(III) | |

In the first two examples, we must determine the oxidation state of Fe. We start with the charge on the complex ion and then take into account the charges on the ligand anions to calculate the charge on the metal ion. For example, the overall charge of the aquapentacyanoferrate(II) ion is 3−. It contains five $CN^-$ ions. To reduce the combined charge of 5− from these cyanide ions to an overall charge of 3−, the charge on Fe must be 2+. The most common ionic charges for some transition metals are shown in **Figure 23.5**.

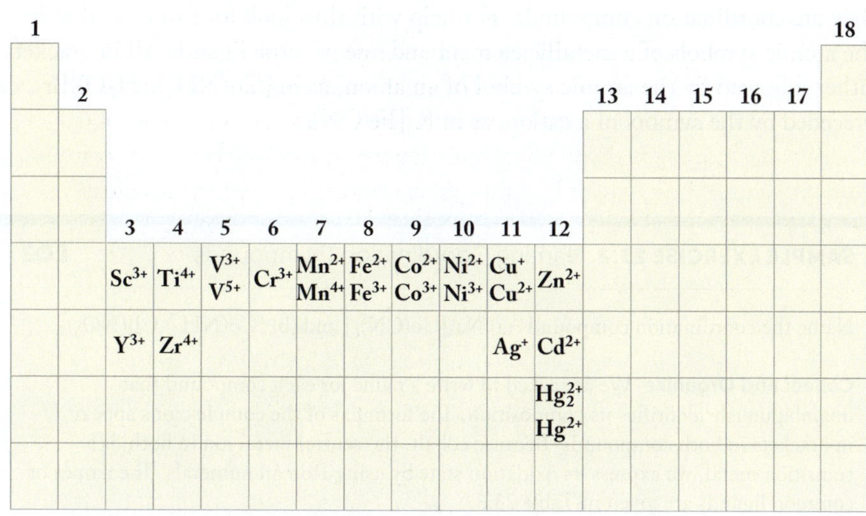


**FIGURE 23.5** Common ionic charges of some transition metals.

CONCEPT **TEST**

What is the name of the complex anion with the formula $PtCl_4^{2-}$?

## Coordination Compounds

1. If the counterion of the complex ion is a cation, the cation's name goes first, followed by the name of the anionic complex ion.
2. If the counterion of the complex ion is an anion, the name of the cationic complex ion goes first, followed by the name of the anion.

Examples:

| Formula | Name | Structure |
|---|---|---|
| $[Ni(NH_3)_6]Cl_2$ | Hexaamminenickel(II) chloride | $[Ni(NH_3)_6]Cl_2$ |
| $K_3[Fe(CN)_6]$ | Potassium hexacyanoferrate(III) | $K_3[Fe(CN)_6]$ |
| $[Co(NH_3)_5(H_2O)]Br_2$ | Pentaammineaquacobalt(II) bromide | $[Co(NH_3)_5(H_2O)]Br_2$ |

A key to naming coordination compounds is to recognize from their formulas that they are coordination compounds. For help with this, look for formulas that have the atomic symbols of a metallic element and one or more ligands, all in brackets, either followed by the atomic symbol of an anion, as in $[Co(NH_3)_5(H_2O)]Br_2$, or preceded by the symbol of a cation, as in $K_3[Fe(CN)_6]$.

**SAMPLE EXERCISE 23.2** Naming Coordination Compounds **LO2**

Name the coordination compounds (a) $Na_4[Co(CN)_6]$ and (b) $[Co(NH_3)_5Cl](NO_3)_2$.

**Collect and Organize** We are asked to write a name for each compound that unambiguously identifies its composition. The formulas of the complex ions appear in brackets in both compounds. Because cobalt, the central metal ion in both, is a transition metal, we express its oxidation state by using Roman numerals. The names of common ligands are given in Table 23.3.

**Analyze** Taking an inventory of the ligands and counterions is useful:

| | | LIGAND | | | |
|---|---|---|---|---|---|
| Compound | Counterion | Formula | Name | Number | Prefix |
| $Na_4[Co(CN)_6]$ | $Na^+$ | $CN^-$ | Cyano | 6 | Hexa- |
| $[Co(NH_3)_5Cl](NO_3)_2$ | $NO_3^-$ | $NH_3$ | Ammine | 5 | Penta- |
| | | $Cl^-$ | Chloro | 1 | — |

The oxidation state of each cobalt ion can be calculated by setting the sum of the charges on all the ions in both compounds equal to zero:

a. Ions: (4 $Na^+$ ions) + (1 $Co^x$ ion) + (6 $CN^-$ ions)

Charges: 4+ + $x$ + 6− = 0

$x$ = 2+

b. Ions: (1 $Co^x$ ion) + (1 $Cl^-$ ion) + (2 $NO_3^-$ ions)

Charges: $x$ + 1− + 2− = 0

$x$ = 3+

**Solve**

a. Because the counterion, sodium, is a cation, its name comes first. The complex ion is an anion. To name it, we begin with the ligand cyano, to which we add the prefix *hexa-* and write *hexacyano*. This is followed by the name of the transition metal ion: hexacyano*cobalt*. We add *-ate* to the ending of the name of the complex ion because it is an anion: hexacyanocobalt*ate*. We then add a Roman numeral to indicate the oxidation state of the cobalt: hexacyanocobaltate(*II*). Putting it all together, we get sodium hexacyanocobaltate(II).

b. The complex ion is the cation in this compound, and we begin by naming the ligands directly attached to the metal ion in alphabetical order: ammine and chloro. We indicate the number (5) of $NH_3$ ligands with the appropriate prefix: *penta*amminechloro. We name the metal next and indicate its oxidation state with a Roman numeral: pentaamminechloro*cobalt(III)*. Finally, we name the anionic counterion: *nitrate*. Putting it all together, we obtain the name: pentaamminechlorocobalt(III) nitrate.

The structures of the complex ions in the named coordination compounds are shown in **Figure 23.6**.

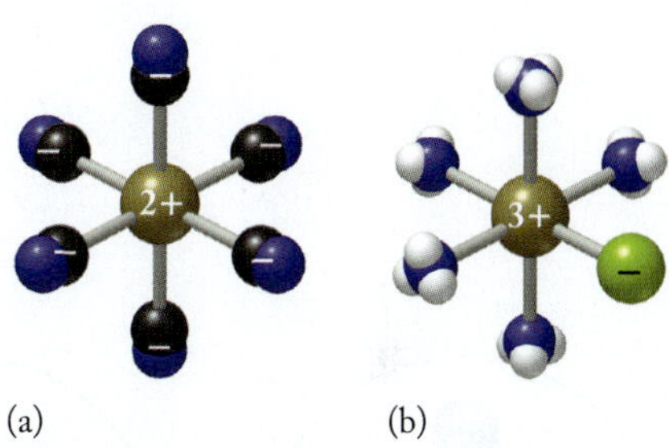


**FIGURE 23.6** The structures of the complex ions in (a) sodium hexacyanocobaltate(II) and (b) pentaamminechlorocobalt(III) nitrate.

**Think About It** Naming coordination compounds requires us to write a name for each compound that unambiguously identifies its composition. The names sodium hexacyanocobaltate(II) and pentaamminechlorocobalt(III) nitrate are unique to the compounds $Na_4[Co(CN)_6]$ and $[Co(NH_3)_5Cl](NO_3)_2$, respectively.

**Practice Exercise** Identify the ligands and counterions in (a) $[Zn(NH_3)_4]Cl_2$ and (b) $[Co(NH_3)_4(H_2O)_2](NO_2)_2$, and name each compound.

# 23.3 Polydentate Ligands and Chelation

Ligands are electron-pair donors—that is, Lewis bases. Let's explore the strengths of several ligands as Lewis bases by considering their affinity for $Ni^{2+}(aq)$ ions. Suppose we dissolve crystals of nickel(II) chloride hexahydrate, $NiCl_2 \cdot 6\,H_2O$, in water. The dot connecting the two halves of the formula and the prefix *hexa* indicate that each $Ni^{2+}$ ion in crystals of nickel(II) chloride is surrounded by six water molecules. Four of the six water molecules are bonded directly to the $Ni^{2+}$ ion, as

**monodentate ligand** a species that forms only a single coordinate bond to a metal ion in a complex.

**polydentate ligand** a species that can form more than one coordinate bond per molecule.

**chelation** the interaction of a metal with a polydentate ligand (chelating agent); pairs of electrons on one molecule of the ligand occupy two or more coordination sites on the central metal.

shown in **Figure 23.7.** The other two are near the ion within the crystal structure, but not bonded to the ion. These other two are called *water of crystallization*. Lime green crystals of $NiCl_2 \cdot 6\ H_2O$ form green aqueous solutions in which six water molecules are bonded to each $Ni^{2+}$ ion, as shown in Figure 23.7. The formula of these fully hydrated $Ni(H_2O)_6{}^{2+}$ ions is often abbreviated $Ni^{2+}(aq)$.

Now let's bubble colorless ammonia gas through a green solution of $Ni(H_2O)_6{}^{2+}$ ions. As shown in the middle beaker in Figure 23.7(b), the green solution turns blue. The color change means that different ligands are bonded to the $Ni^{2+}$ ions. We may conclude that $NH_3$ molecules have displaced at least some $H_2O$ molecules around the $Ni^{2+}$ ions. If all the molecules of $H_2O$ are displaced, the complex $Ni(NH_3)_6{}^{2+}$ is formed. The following chemical equation describes this change:

$$Ni(H_2O)_6{}^{2+}(aq) + 6\ NH_3(g) \rightleftharpoons Ni(NH_3)_6{}^{2+}(aq) + 6\ H_2O(\ell) \qquad K_f = 5 \times 10^8$$

This *ligand displacement* reaction illustrates that $Ni^{2+}$ ions have a greater affinity for molecules of $NH_3$ than for molecules of $H_2O$, as do many other transition metal ions. We may conclude that ammonia is inherently a better electron-pair donor and hence a stronger Lewis base than water. This conclusion is reasonable because we saw in Chapter 15 that ammonia was also a stronger Brønsted–Lowry base than $H_2O$.

Next we add the compound ethylenediamine (see Table 23.3) to the blue solution of $Ni(NH_3)_6{}^{2+}$ ions. The solution changes color again, from blue to purple (Figure 23.7b), indicating yet another change in the ligands surrounding the $Ni^{2+}$ ions. Molecules of ethylenediamine displace ammonia molecules from the inner coordination sphere of $Ni^{2+}$ ions. This affinity of $Ni^{2+}$ ions for ethylenediamine molecules is reflected in the large formation constant for $Ni(en)_3{}^{2+}$ (where "en" represents ethylenediamine):

$$Ni(H_2O)_6{}^{2+}(aq) + 3\ en(aq) \rightleftharpoons Ni(en)_3{}^{2+}(aq) + 6\ H_2O(\ell) \qquad K_f = 1.1 \times 10^{18}$$

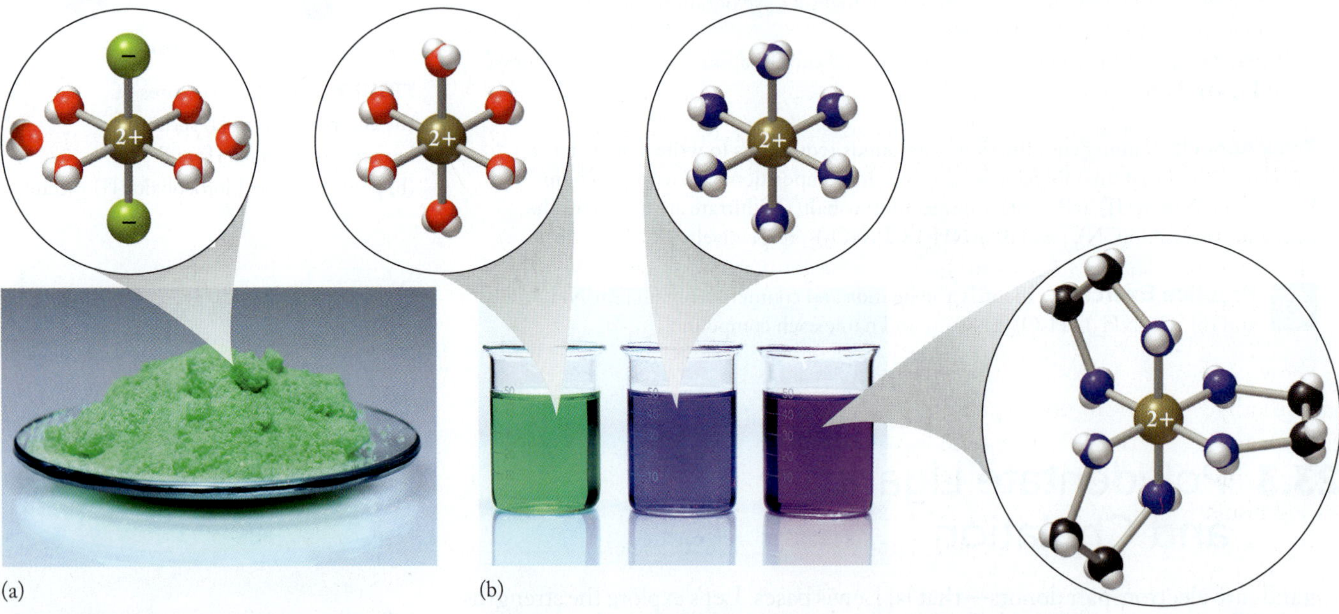


**FIGURE 23.7** Structures of the complex ions in solid and dissolved nickel(II) chloride. (a) Solid nickel(II) chloride hexahydrate. (b) When it dissolves in water, the resulting solution has the same green color, indicating that each $Ni^{2+}$ ion (gold sphere) must be surrounded by $H_2O$ molecules both in the solid and in the solution. When ammonia gas is bubbled through a solution of $Ni(H_2O)_6{}^{2+}$, the color changes to blue as $NH_3$ replaces $H_2O$ in the $Ni^{2+}$ ion's inner coordination sphere. When ethylenediamine (en) is added to a solution of $Ni(NH_3)_6{}^{2+}$, the color turns from blue to purple as the ethylenediamine displaces the ammonia ligands and the $Ni(en)_3{}^{2+}$ complex forms.

This value is more than $10^9$ times the $K_f$ value for $Ni(NH_3)_6^{2+}$. Why should the affinity of $Ni^{2+}$ ions for ethylenediamine be so much greater than their affinity for ammonia? After all, in both ligands the coordinate bonds are formed by lone pairs of electrons on N atoms. To answer this question, we need to look at the differences between ligands such as ammonia, which occupy only one site on a metal ion, and ligands such as ethylenediamine, which occupy two or more sites.

Many ligands in Table 23.3 can donate only one pair of electrons to a single metal ion. Even atoms with more than one lone pair usually donate only one pair at a time to a given metal ion because the other lone pair or pairs are oriented away from the metal ion. Because these ligands have effectively only one donor group, they are called **monodentate ligands**, which literally means "single-toothed."

Certain molecules larger than ammonia and water may be able to donate more than one lone pair of electrons and therefore form more than one coordinate bond to a central metal ion. Ligands in this category are called **polydentate ligands**, or more specifically *bidentate*, *tridentate*, and so on. One group of polydentate ligands is the polyamines, which include ethylenediamine, a bidentate ligand that has the structure:

$$H_2\ddot{N}\text{—}CH_2\text{—}CH_2\text{—}\ddot{N}H_2$$

The lone pairs on the two $-NH_2$ groups are separated from each other by two $-CH_2-$ groups. This combination means that a molecule of ethylenediamine can partially encircle a metal ion so that both lone pairs can bond to the same metal ion.

The structure of an ethylenediamine complex of $Ni^{2+}(aq)$ is shown in **Figure 23.8(a)**. The two orbitals that share the lone pairs of electrons from a molecule of ethylenediamine must be on the same side of the $Ni^{2+}$ ion. However, two more ethylenediamine molecules can bond to other pairs of bonding sites, displacing additional pairs of water molecules and forming a complex in which the $Ni^{2+}$ ion is surrounded by three bidentate ethylenediamine molecules, as shown in **Figure 23.8(b)**.

Each ethylenediamine molecule forms a five-atom ring with the metal ion. If the ring were a regular pentagon (meaning that all bond lengths and bond angles were the same), each of its bond angles would be 108°. These pentagons are not perfect, but each ring's preferred octahedral bond angles of 90° for the N–Ni–N bond, and of 107° to 109° for all the other bonds, are accommodated with only a little strain on the ideal bond angles. The compound in Figure 23.8(b) is also chiral, a topic we address in Section 23.6.

An even larger ligand, diethylenetriamine ($H_2NCH_2CH_2NHCH_2CH_2NH_2$), is shown in **Figure 23.9(a)**. The lone pairs of electrons on its three nitrogen atoms give diethylenetriamine the capacity to form three coordinate bonds to a metal ion, meaning that this is a tridentate ligand (**Figure 23.9b**).

The interaction of a metal ion with a ligand having multiple donor atoms is called **chelation** (pronounced *key-LAY-shun*). The word comes from the Greek *chele*, meaning "claw." The polydentate ligands that take part in these interactions are called *chelating agents*.

When we added ethylenediamine to the blue solution of $Ni(NH_3)_6^{2+}$ ions in Figure 23.7(b), molecules of ethylenediamine (en) displaced ammonia molecules from the inner coordination sphere of $Ni^{2+}$ ions, as described in the following chemical equation:

$$Ni(NH_3)_6^{2+}(aq) + 3\ en(aq) \rightleftharpoons Ni(en)_3^{2+}(aq) + 6\ NH_3(aq) \tag{23.1}$$

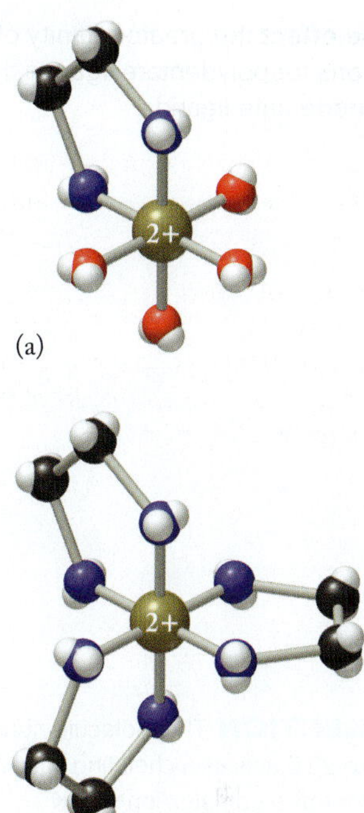


**FIGURE 23.8** (a) The bidentate ligand ethylenediamine has two N atoms that can each donate a pair of electrons to empty orbitals of adjacent octahedral bonding sites on the same $Ni^{2+}(aq)$ ion (gold sphere), displacing two molecules of water. (b) Three ethylenediamine molecules occupy all six octahedral coordination sites of a $Ni^{2+}$ ion.

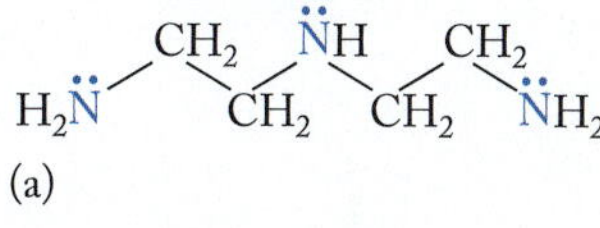


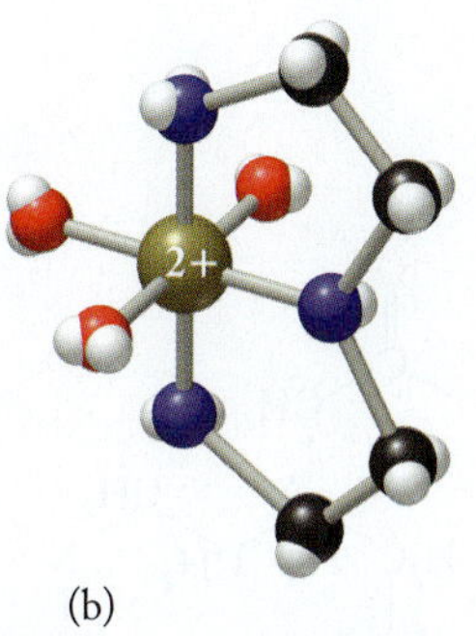


**FIGURE 23.9** Tridentate chelation. (a) The three amine groups in the tridentate ligand diethylenetriamine are all potential electron-pair donor groups. (b) When these groups donate their lone pairs of electrons to a $Ni^{2+}(aq)$ ion (gold sphere), they occupy three of the six coordination sites on the ion.

**chelate effect** the greater affinity of metal ions for polydentate ligands than for monodentate ligands.

The color change tells us that this reaction as written is spontaneous. As we discussed in Chapter 12, spontaneous reactions are those in which free energy decreases (that is, $\Delta G < 0$). Furthermore, under standard conditions, the change in free energy $\Delta G^\circ$ is related to the changes in enthalpy and entropy that accompany the reaction:

$$\Delta G^\circ = \Delta H^\circ - T\Delta S^\circ$$

The displacement of $NH_3$ by ethylenediamine is exothermic, but only slightly ($\Delta H^\circ = -12$ kJ/mol). More important, $\Delta S^\circ = +185$ J/mol · K. At 25°C, then,

$$T\Delta S^\circ = 298 \cancel{\text{K}} \times \frac{185 \cancel{\text{J}}}{\text{mol} \cdot \cancel{\text{K}}} \times \frac{1 \text{ kJ}}{1000 \cancel{\text{J}}} = 55.1 \text{ kJ/mol}$$

Such a large increase in entropy occurs because 4 moles of reactants are present but 7 moles of products in Equation 23.1. Nearly doubling the number of moles of aqueous products over reactants translates into a large increase in entropy. It is this positive $\Delta S^\circ$, more than the negative $\Delta H^\circ$ value, that drives the reaction and makes it spontaneous. Entropy changes drive many complexation reactions that involve polydentate ligands. The entropy-driven affinity of metal ions for polydentate ligands is called the **chelate effect.**

**CONNECTION** The molecule nonactin in Figure 22.6 acts as a chelating ligand when it forms a complex ion with $K^+$.

Many chelating agents have more than one kind of electron pair–donating group. *Aminocarboxylic acids* represent one family of such compounds. The most important of them is ethylenediaminetetraacetic acid (EDTA), which is shown in **Figure 23.10(a)**. One molecule of EDTA contains two amine (nitrogen-containing) groups and four carboxylic acid (–COOH) groups. When the acid groups release their $H^+$ ions, they form four carboxylate anions, $-COO^-$, in which either of the O atoms can donate a pair of electrons to a central metal ion. When O atoms on all four groups do so and the two amine groups do as well, six octahedral bonding sites around the metal ion can be occupied, as shown in Figure **23.10(b)**.

EDTA forms very stable complex ions and is used as a metal ion *sequestering agent*—that is, as a chelating agent that binds metal ions so tightly that they are "sequestered" and prevented from reacting with other substances. For example, EDTA is used as a preservative in many beverages and prepared foods because it sequesters iron, copper, zinc, manganese, and other transition metal ions often present in these foods that can catalyze the degradation of ingredients in the foods. Many foods are fortified with ascorbic acid (vitamin C), which is particularly vulnerable to metal-catalyzed degradation because it is also a polydentate

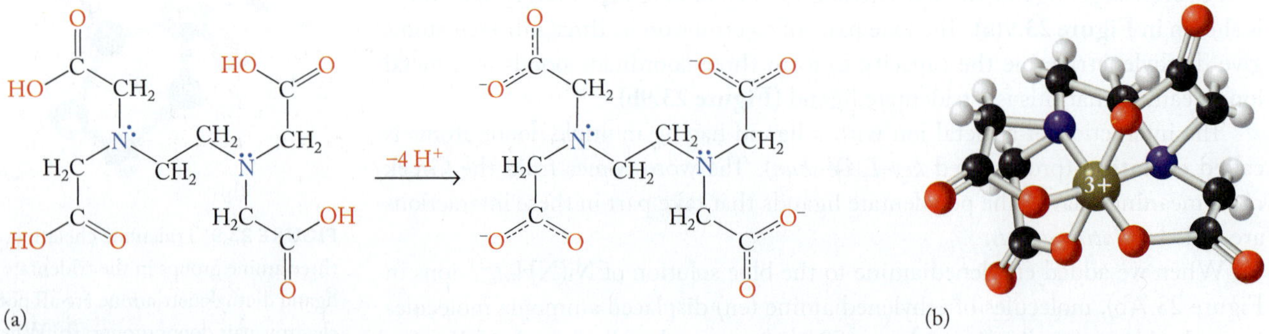

**FIGURE 23.10** (a) In the hexadentate ligand EDTA, the six donor groups are the two amine groups and the four carboxylic acid groups. The acid groups ionize to form carboxylate anions. (b) All six Lewis base groups in ionized EDTA can form a coordinate bond with the same metal ion, such as $Co^{3+}$ (the gold sphere) shown here. In the process, they form four 5-membered rings.

ligand and is more likely to be oxidized when chelated to one of the above metal ions. EDTA effectively shields vitamin C from these ions. EDTA is also a common additive to shampoos because it can remove metal cations from hard water.

**CONNECTION** Hard water is discussed in Section 11.2.

**SAMPLE EXERCISE 23.3** Identifying the Potential Electron Pair Donor Groups in a Molecule **LO3**

How many donor groups does nitrilotriacetic acid (NTA) have?

```
          O
          ‖
          C—OH
         /
       CH₂
         \
HO        N—CH₂
  \      /     \
   C—CH₂        C=O
   ‖           /
   O         HO
```

**Collect, Organize, and Analyze** We need to examine the molecular structure of this polydentate ligand to find electron pairs that can be donated. The three N–C single bonds use three $sp^3$ orbitals from the N atom, with the fourth $sp^3$ orbital containing a lone pair of electrons. Ionization of the three carboxylic acid groups results in three carboxylate groups in the molecule, each carboxylate group capable of donating one nonbonding pair of electrons from one of its oxygen atoms to a metal atom.

**Solve** The central N atom and an O atom from each of the three carboxylate groups form a total of four coordinate bonds. Therefore, NTA is potentially a tetradentate ligand with four donor groups. If we also consider the lone pairs on the oxygen atoms of the C=O group, NTA could coordinate a total of six oxygens to a transition metal cation, making NTA a hexadentate ligand. If we coordinate the nitrogen in NTA through its lone pair, one can imagine a seven-coordinate, heptadentate NTA ligand. The three carboxylate groups, though, are all bonded to a single N through a $CH_2$ group, so the bond distances and bond angles in NTA may prevent coordination numbers greater than four.

**Think About It** The tetradentate capacity of NTA is reasonable because, like EDTA, it is an aminocarboxylic acid. It has one fewer amino group and one fewer carboxylic acid group than the hexadentate EDTA.

**Practice Exercise** How many potential donor groups are in citric acid, a component of citrus fruits and a widely used preservative in the food industry?

```
            O   OH
             \\ /
              C      O
   O          |      ‖
   ‖          C      C
   C         /|\    / \
HO/ \CH₂/    | \CH₂/   OH
             OH
```

**CONNECTION** In Section 3.3, we first used the equation $E = h\nu = hc/\lambda$ when discussing the energy of light in the electromagnetic spectrum (Figure 3.1).

## 23.4 Crystal Field Theory

Why does the formation of complex ions change the color of solutions of transition metals? The colors of transition metal compounds and ions in solution are due to transitions of $d$-orbital electrons. Let's explore these transitions by using $Cr^{3+}$ as our model transition metal ion (**Figure 23.11**).

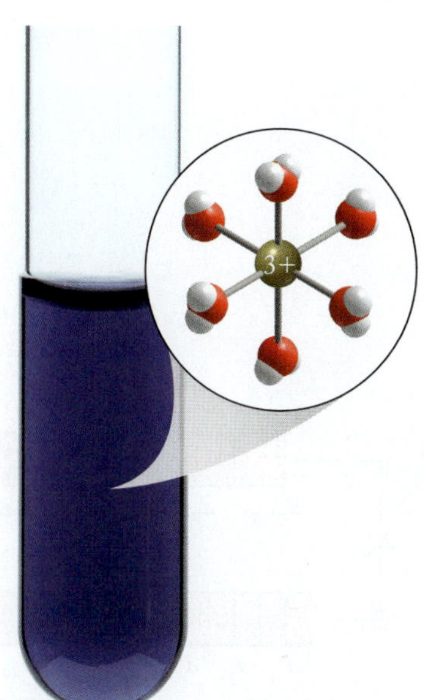


**FIGURE 23.11** When chromium(III) nitrate dissolves in water, the resulting solution has a distinctive violet color due to the presence of $Cr(H_2O)_6^{3+}$ ions.

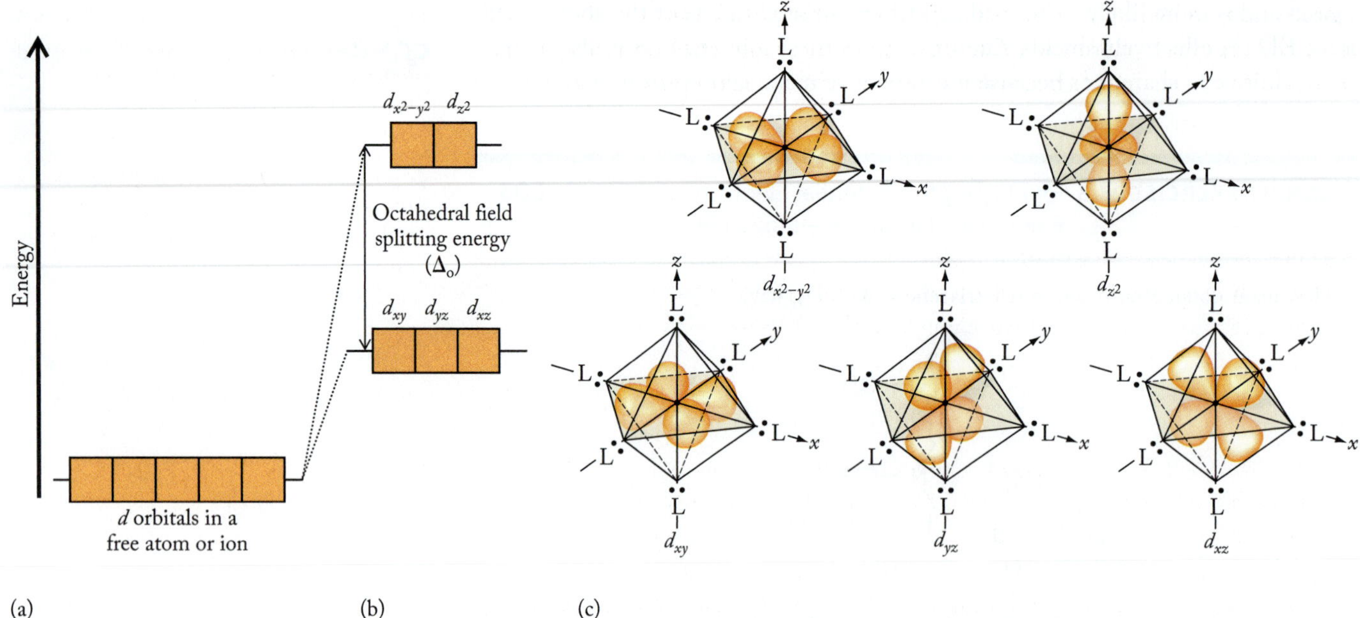


**FIGURE 23.12** Octahedral crystal field splitting. (a) In an atom or ion in the gas phase, all orbitals in a subshell are degenerate, as shown here for the five 3*d* orbitals. (b) When an ion is part of a complex ion in a compound or solution, repulsions between electrons in the ion's *d* orbitals and ligand electrons raise the energy of the orbitals, as shown here for an octahedral field. (c) The greatest repulsion is experienced by electrons in the $d_{x^2-y^2}$ and $d_{z^2}$ orbitals because the lobes of these orbitals are directed toward the corners of the octahedron and so are closest to the lone pairs on the ligands (L). The lobes of the lower-energy $d_{xy}$, $d_{yz}$, and $d_{xz}$ orbitals are directed toward points that lie between the corners of the octahedron, so electrons in them experience less repulsion.

A $Cr^{3+}$ ion has the electron configuration [Ar]$3d^3$ in the gas phase. When a $Cr^{3+}$ ion (or any atom or ion) is in the gas phase, all the orbitals in a given subshell have the same energy (**Figure 23.12a**). However, when a $Cr^{3+}$ ion is in an aqueous solution and surrounded by an octahedral array of water molecules in $Cr(H_2O)_6{}^{3+}$, the energies of its 3*d* orbitals are no longer equivalent. The $3d_{xy}$, $3d_{yz}$, and $3d_{xz}$ orbitals experience some increase in energy, but the energies of the $3d_{x^2-y^2}$ and $3d_{z^2}$ orbitals increase even more (**Figure 23.12b**) because the lobes of the $3d_{x^2-y^2}$ and $3d_{z^2}$ orbitals point directly toward the $H_2O$ molecules' oxygen atoms at the corners of the octahedron formed by the ligands and are repelled by the electrons on those O atoms (**Figure 23.12c**). The energies of the $3d_{xy}$, $3d_{yz}$, and $3d_{xz}$ orbitals are not raised as much because the lobes of these three orbitals do not point directly toward the corners of the octahedron, so they experience weaker electron repulsion.

This process of changing degenerate (equal-energy) *d*-orbitals into orbitals with different energies is known as **crystal field splitting**, and the difference in energy created by crystal field splitting is called **crystal field splitting energy (Δ)**. The name was originally used to describe splitting of *d*-orbital energies in minerals, but the theory is also routinely applied to species in aqueous solutions.

In a $Cr(H_2O)_6{}^{3+}$ ion, three electrons are distributed among five 3*d* orbitals. According to Hund's rule, each of the three electrons should occupy one of the three lower-energy orbitals, leaving the two higher-energy orbitals unoccupied, as shown in **Figure 23.13(a)**. The energy difference between the two subsets of orbitals is symbolized by $\Delta_o$, where the subscript "o" indicates that the energy split was caused by an *o*ctahedral array of electron repulsions.

What if an aqueous $Cr^{3+}$ ion absorbs a photon whose energy is equal to $\Delta_o$? As the photon is absorbed, a 3*d* electron moves from a

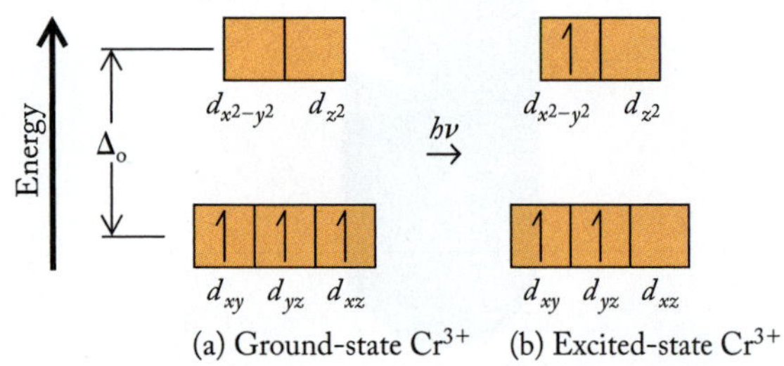


**FIGURE 23.13** A $Cr^{3+}$ ion, [Ar]$3d^3$, in an octahedral field can absorb a photon of light that has energy ($h\nu$) equal to $\Delta_o$. This energy raises a 3*d* electron from (a) one of the lower-energy *d* orbitals to (b) one of the higher-energy *d* orbitals.

lower-energy orbital to a higher-energy orbital (**Figure 23.13b**). The wavelength $\lambda$ of the absorbed photon is related to the energy difference between the two groups of orbitals—in other words, to the crystal field splitting energy—as follows:

$$E = \frac{hc}{\lambda} = \Delta_o \qquad (23.2)$$

The energy and wavelength of a photon are inversely proportional to each other (see Equation 3.4 in Section 3.3). Therefore, the larger the crystal field splitting in a complex ion, the shorter the wavelength of the photons the ion absorbs.

## Crystal Field Splitting and Color

The size of the energy gap between split *d* orbitals often corresponds to radiation in the visible region of the electromagnetic spectrum. This means that the colors of solutions of metal complexes depend on the strengths of metal–ligand interactions that affect $\Delta_o$. When white light (which contains all colors of visible light) passes through a solution containing complex ions, the ions may absorb energy corresponding to one or more colors. So, the light leaving the solution and reaching our eyes is missing.

The color we perceive for any transparent object is not the color it absorbs but rather the color(s) that it transmits. To relate the color of a solution to the wavelengths of light it absorbs, we need to consider complementary colors as defined by a simple color wheel (**Figure 23.14**). For example, red and green are complementary colors, so a solution that absorbs green light appears red to us.

Aqueous solutions of $Cu^{2+}$ (see Figure 17.1) are blue because $Cu(H_2O)_6{}^{2+}$ absorbs orange light, the color that is complementary to blue. A solution of $Cu(NH_3)_4{}^{2+}$ ions also has a distinctive deep blue color (**Figure 23.15**). The spectrum of a solution containing $Cu(NH_3)_4{}^{2+}$ features a broad absorption band between 580 nm and 590 nm that spans yellow, orange, and red wavelengths. Once again, the complementary color is in the blue-violet region. If more than one color is absorbed, then our brain processes the bands of transmitted colors and signals to us as the average of these colors. The violet color of the solution containing $Cr^{3+}(aq)$ in Figure 23.11 is the result of color averaging, as are the colors of the aqueous $Ni^{2+}$, $Ni(NH_3)_6{}^{2+}$, and $Ni(en)_3{}^{2+}$ solutions in Figure 23.7.

The nickel(II) and chromium(III) complexes we have examined so far were octahedral, and their central ions had a coordination number of 6. In the solution of $Cu(NH_3)_4{}^{2+}$, however, the copper(II) ion has a coordination number of 4. This means that the deep blue color of this complex ion is caused by a different crystal field.

**crystal field splitting** the separation of a set of *d* orbitals into subsets with different energies as a result of interactions between electrons in those orbitals and lone pairs of electrons in ligands.

**crystal field splitting energy ($\Delta$)** the difference in energy between subsets of *d* orbitals split by interactions in a crystal field.

**CONNECTION** Crystal field theory is an example of molecular orbital theory, which was introduced in Section 5.7.

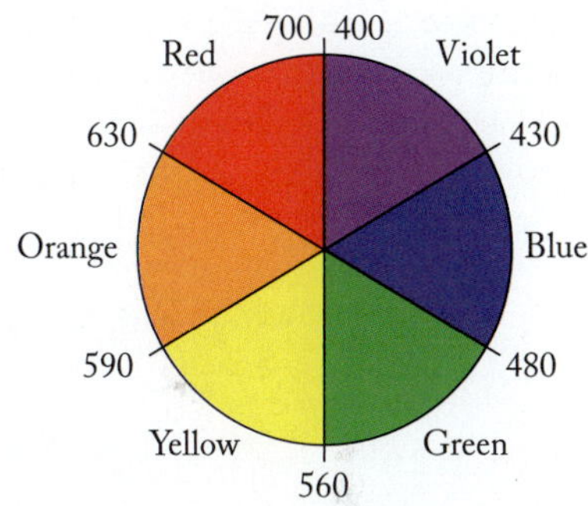


**FIGURE 23.14** A color wheel. Colors on opposite sides of the wheel are complementary to each other. When we look at a solution or object that absorbs light corresponding to a given color, we see the complementary color. Wavelengths are in nanometers.

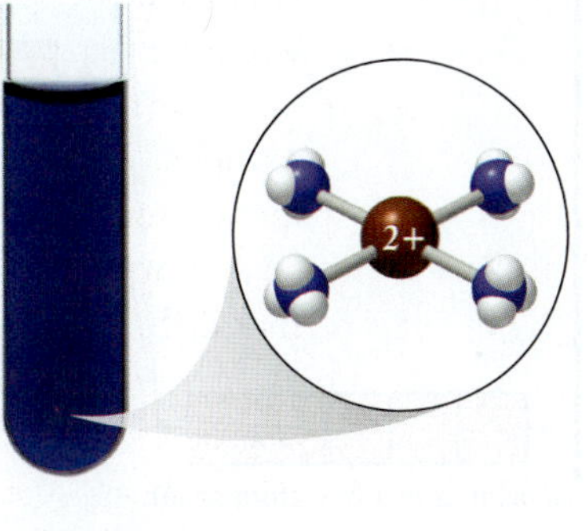


**FIGURE 23.15** The visible light transmitted by a solution of $Cu(NH_3)_4{}^{2+}$ ions is missing much of the yellow, orange, and red portions of the visible spectrum because of a broad absorption band centered at 590 nm. Our eyes and brain perceive the transmitted colors as navy blue.

**spectrochemical series** a list of ligands rank-ordered by their ability to split the energies of the *d* orbitals of transition metal ions.

Four ligands around a central metal have either a tetrahedral arrangement or a square planar arrangement (Table 23.2). Square planar geometries tend to be limited to the transition metal ions with nearly filled valence-shell *d* orbitals, particularly those with $d^8$ or $d^9$ electron configurations. $Cu^{2+}$ has the electron configuration [Ar]$3d^9$, and the $Cu(NH_3)_4{}^{2+}$ complex is square planar—which means that the strongest interactions occur between the 3*d* orbitals on the central ion and the nitrogen atom lone pairs at the four corners of the equatorial plane of the octahedron, as shown in **Figure 23.16**. The $d_{x^2-y^2}$ orbital has the strongest interactions and the highest energy because its lobes are oriented directly at the four corners of the plane. The $d_{xy}$ orbital has slightly less energy because its lobes, although in the *xy* plane, are directed 45° away from the corners. Electrons in the three *d* orbitals with most of their electron density out of the *xy* plane (that is, $d_{z^2}$, $d_{xz}$, and $d_{yz}$) interact even less with the lone pairs of the ligand and thus have even lower energies.

**CHEMTOUR**
Crystal Field Splitting

Finally, let's consider the *d* orbital crystal field splitting that occurs in a tetrahedral complex (**Figure 23.17a**). In this geometry, the greatest electron–electron repulsions are experienced in the $d_{xy}$, $d_{yz}$, and $d_{xz}$ orbitals because the lobes of these orbitals are oriented most directly to the corners of the tetrahedron, which are occupied by ligand electron pairs (**Figure 23.17b**). The two other *d* orbitals ($d_{x^2-y^2}$ and $d_{z^2}$) are less affected because their lobes do not point toward the corners. The difference in energy between the two subsets of *d* orbitals in a tetrahedral geometry is labeled $\Delta_t$.

Before ending this discussion on the colors of transition metal ions, let's revisit the color changes we saw in Figure 23.7, when first ammonia and then ethylenediamine were added to a solution of $Ni^{2+}$ ions. Let's think in terms of the colors these solutions *absorb*. In the visible region of the electromagnetic spectrum, a green solution of $Ni^{2+}$ ions absorbs colors at the red end of the spectrum. Blue solutions of $Ni(NH_3)_6{}^{2+}$ and violet solutions of $Ni(en)_3{}^{2+}$ absorb at shorter wavelengths, as shown in **Table 23.5**. The sequence runs from longest wavelength

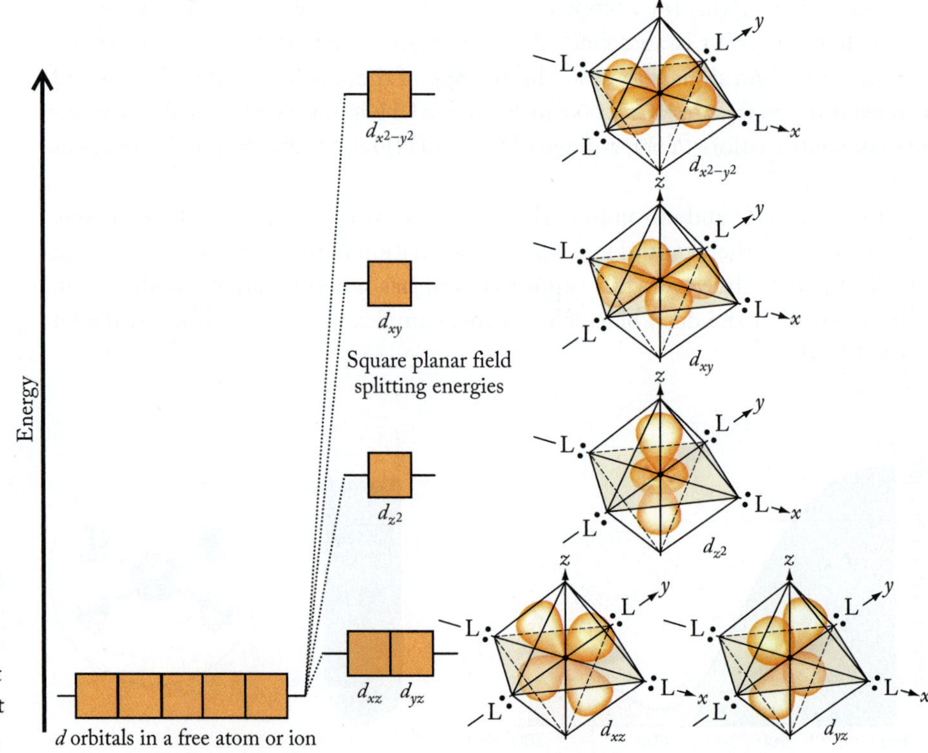


**FIGURE 23.16** Square planar crystal field splitting. The *d* orbitals of a transition metal ion in a square planar field are split into several energy levels depending on the relative orientations of the metal orbitals and ligand electrons at the four corners of the square. The $d_{x^2-y^2}$ orbital has the highest energy because its lobes are directed right at the four corners of the square plane.

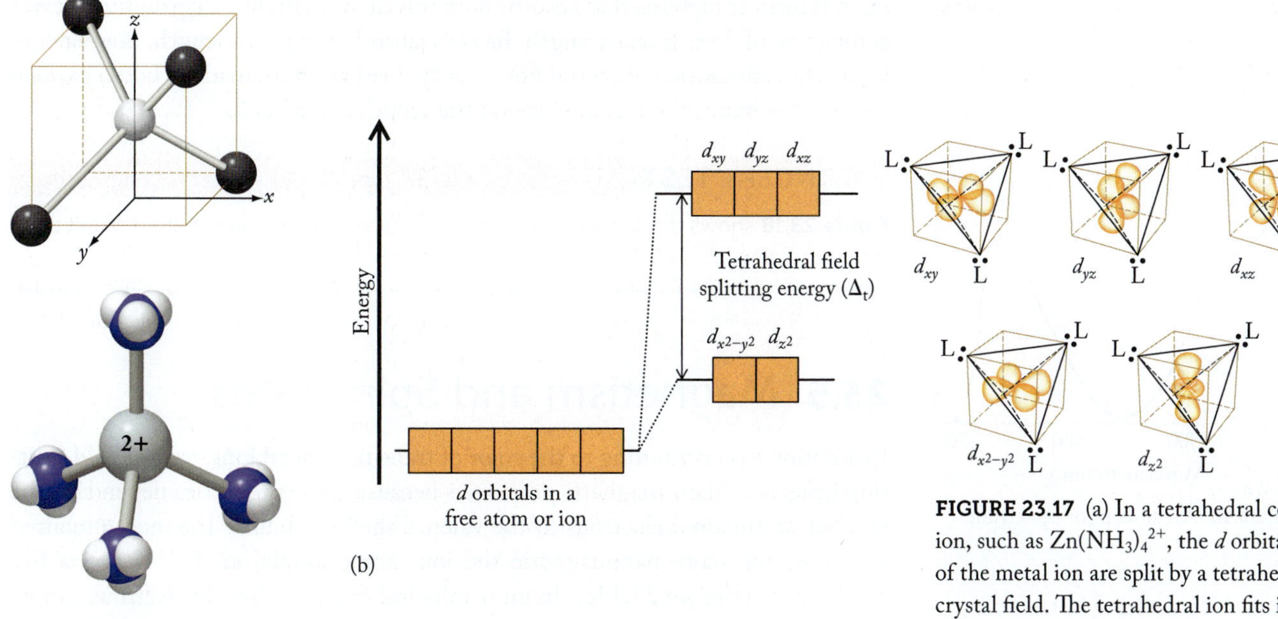


**FIGURE 23.17** (a) In a tetrahedral complex ion, such as $Zn(NH_3)_4^{2+}$, the *d* orbitals of the metal ion are split by a tetrahedral crystal field. The tetrahedral ion fits inside a cube so that $NH_3$ molecules occupy opposite corners on the top and bottom face of the cube. The $NH_3$ ligands lie between the *x*, *y*, and *z* axes. (b) The lobes of the higher-energy orbitals—$d_{xy}$, $d_{yz}$, and $d_{xz}$—are closer to the ligands at the four corners of the tetrahedron than the lobes of the lower-energy orbitals—$d_{x^2-y^2}$ and $d_{z^2}$—are. (One of the four corners of the tetrahedron is hidden in these drawings.)

(about 725 nm for red) to shortest (about 545 nm for green). Radiant energy is inversely proportional to wavelength (Equation 23.2); therefore, the crystal field splitting of the *d* orbitals of $Ni^{2+}$ ions is en > $NH_3$ > $H_2O$. Like the solution of $Cr^{3+}$ ions in Figure 23.11, solutions of $Ni^{2+}$, $Ni(NH_3)_6^{2+}$, and $Ni(en)_3^{2+}$ have additional absorbances at wavelengths slightly longer and slightly shorter than those shown in Table 23.5. Because these absorbances are broad, the color we observe for solutions of $Ni^{2+}$ coordination complexes is the result of color averaging, not just the complementary color we might predict from the color wheel in Figure 23.14.

Chemists use the parameter *field strength* to describe the relative magnitude of the split in the energies of the *d* orbitals in metal ions, ranking ligands in what is called a **spectrochemical series. Table 23.6** contains one such series. As the field strength of the ligand increases from the bottom to the top of Table 23.6, the crystal field splitting energy (Δ) increases. Consequently, high–field strength

**TABLE 23.6** Spectrochemical Series of Some Common Ligands

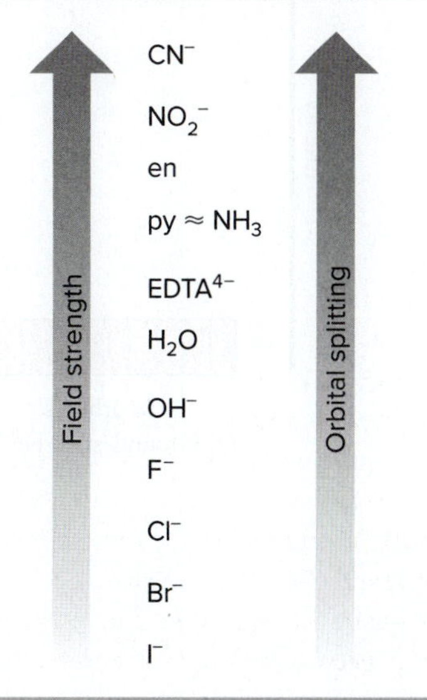


**TABLE 23.5** Light Transmitted and Absorbed by Three $Ni^{2+}$ Complexes

| Complex | $Ni(H_2O)_6^{2+}$ | $\xrightarrow{NH_3}$ | $Ni(NH_3)_6^{2+}$ | $\xrightarrow{en}$ | $Ni(en)_3^{2+}$ |
|---|---|---|---|---|---|
| Appearance | Green | | Blue | | Violet |
| Absorbs | Red | | Yellow | | Green |
| Absorbed λ (nm) | 725 | > | 570 | > | 545 |
| $E = hc/\lambda$ (J × $10^{19}$) | 2.7 | < | 3.5 | < | 3.6 |

ligands form complexes that absorb short-wavelength, high-energy light, whereas complexes of low–field strength ligands absorb long-wavelength, low-energy light. The refinements of crystal field theory that have been made to better explain the spectrochemical series are beyond the scope of this book.

**CONCEPT TEST**

**Figure 23.18** shows the absorption spectrum of a complex found in nature. What color is this complex ion?

Absorption

400 500 600 700

Wavelength (nm)

**FIGURE 23.18** An absorption spectrum.

## 23.5 Magnetism and Spin States

**CONNECTION** We introduced the magnetic behavior of matter in Section 5.7 in our discussion of molecular orbital theory.

In addition to contributing to the color of transition metal ions, crystal field splitting influences their magnetic properties because these properties depend on the number of unpaired electrons in the valence shell *d* orbitals. The more unpaired electrons, the more paramagnetic the ion. For example, an $Fe^{3+}$ ion has five 3*d* electrons (**Figure 23.19a**). In an octahedral field, the five 3*d* electrons can be distributed among these orbitals in two ways. One arrangement conforms to Hund's rule and has a single electron in each orbital, leaving them all unpaired (**Figure 23.19b**) and giving rise to the green color in aquamarine (**Figure 23.19d**). However, when $\Delta_o$ is large, as shown in **Figure 23.19(c)**, all five electrons occupy the three lower-energy orbitals. This pattern of electron distribution occurs when the energy of repulsion between two electrons in the same orbital is less than the energy needed to promote an electron to a higher-energy orbital. In this configuration, only one electron is unpaired.

The configuration with all five electrons unpaired is called the *high-spin state* because the spin on all five electrons is in the same direction, resulting in the maximum magnetic field produced by the spins. The configuration with only one electron unpaired is called the *low-spin state*. Both configurations are paramagnetic because both have at least one unpaired electron, but material containing

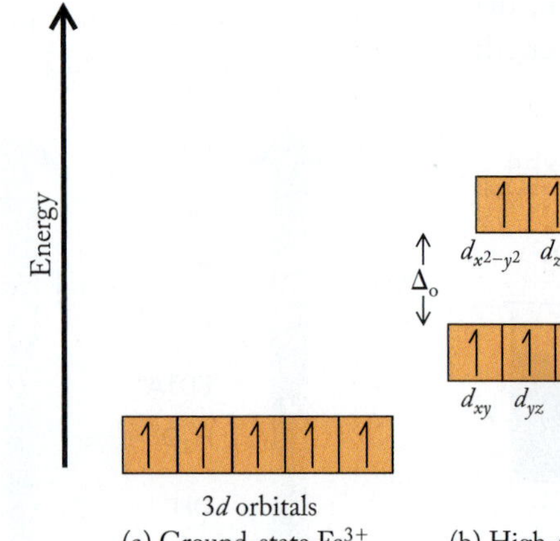


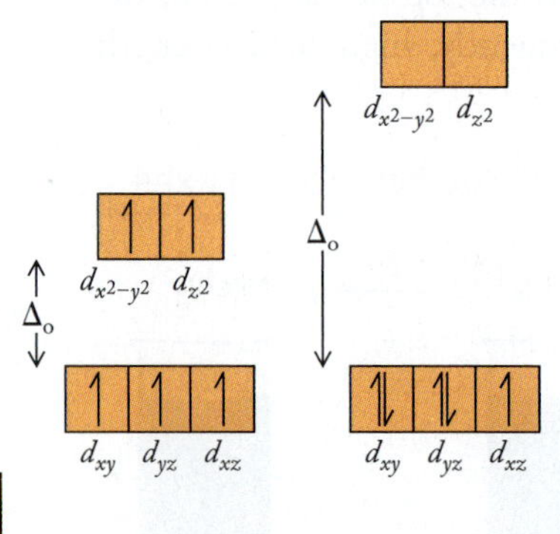

**FIGURE 23.19** Low-spin and high-spin complexes. (a) The ground state of a free $Fe^{3+}$ ion has a degenerate, half-filled set of 3*d* orbitals. (b) A weak octahedral field ($\Delta_o <$ electron-pairing energy) produces the high-spin state: five unpaired electrons, each in its own orbital. (c) In a strong octahedral field ($\Delta_o >$ electron-pairing energy), the energies of the 3*d* orbitals are split enough to produce the low-spin state: two sets of paired electrons, one unpaired electron, and two empty, higher energy orbitals. (d) The $Fe^{3+}$ ions in crystals of aquamarine are high spin.

high-spin iron(III) ions would be more strongly attracted to an external magnet than a material containing low-spin iron(III) ions.

Not all transition metal ions can have both high-spin and low-spin states. A $Cr^{3+}$ ion in an octahedral field, for instance, has only three $3d$ electrons (Figure 23.13), so each electron is unpaired whether the orbital energies are split a lot or only a little. Therefore, $Cr^{3+}$ ions and any ion with fewer than four $d$ electrons will have only one spin state. Metal ions with eight or more electrons in $d$ orbitals also have only one spin state because no more than one or two orbitals can contain unpaired electrons no matter how the electrons are distributed.

### SAMPLE EXERCISE 23.4 Predicting Spin States LO5

Determine which of the following ions can have high-spin and low-spin configurations when surrounded by six Lewis bases in an octahedral complex: (a) $Mn^{4+}$; (b) $Mn^{2+}$; (c) $Cu^{2+}$.

**Collect and Organize** Mn and Cu are in groups 7 and 11 of the periodic table, respectively, so their atoms have 7 and 11 valence electrons. In an octahedral field, a set of five $d$ orbitals splits into a low-energy subset of three orbitals and a high-energy subset of two orbitals.

**Analyze** To determine whether high-spin and low-spin states are possible, we need to determine the number of $d$ electrons in each ion. Then we need to distribute them among sets of $d$ orbitals split by an octahedral field to see whether the ions can have different spin states. The electron configurations for manganese and copper atoms are $[Ar]3d^5 4s^2$ and $[Ar]3d^{10} 4s^1$, respectively. When the transition metals Mn and Cu form cations, their atoms lose their $4s$ electrons first and then their $3d$ electrons. Therefore, the numbers of $d$ electrons in the ions are 3 in $Mn^{4+}$, 5 in $Mn^{2+}$, and 9 in $Cu^{2+}$. It is likely that the ion with the fewest $d$ electrons ($Mn^{4+}$) and the one with nearly the most $d$ electrons possible ($Cu^{2+}$) will each have only one spin state.

**Solve**

a. $Mn^{4+}$: Following Hund's rule and putting three electrons into the lowest-energy $3d$ orbitals available and keeping them as unpaired as possible gives this orbital distribution of electrons (**Figure 23.20a**). $Mn^{4+}$ has only one spin state.

b. $Mn^{2+}$: Two options are available for distributing five electrons among the five $3d$ orbitals (**Figure 23.20b**). Thus, $Mn^{2+}$ can have a high-spin (on the left) or a low-spin (on the right) configuration in an octahedral field.

c. $Cu^{2+}$: The nine $3d$ electrons fill the lower-energy orbitals and nearly fill the higher-energy ones. Only one arrangement is possible, so $Cu^{2+}$ has only one spin state (**Figure 23.20c**).

**Think About It** In an octahedral field, metal ions with 4, 5, 6, or 7 $d$ electrons can exist in high-spin and low-spin states. Those ions with 3 or fewer $d$ electrons have only one spin state, in which all the electrons are unpaired and in the lower-energy set of orbitals. Ions with 8 or more $d$ electrons have only one spin state because their lower-energy set of orbitals is filled. The magnitude of the crystal field splitting energy, $\Delta_o$, determines which spin state that an ion with 4, 5, 6, or 7 $d$ electrons occupies.

**Practice Exercise** Which of the following ions can have high-spin and low-spin configurations when part of an octahedral complex: (a) $V^{4+}$; (b) $Cr^{3+}$; (c) $Ni^{3+}$? Are any of the possible spin configurations diamagnetic?

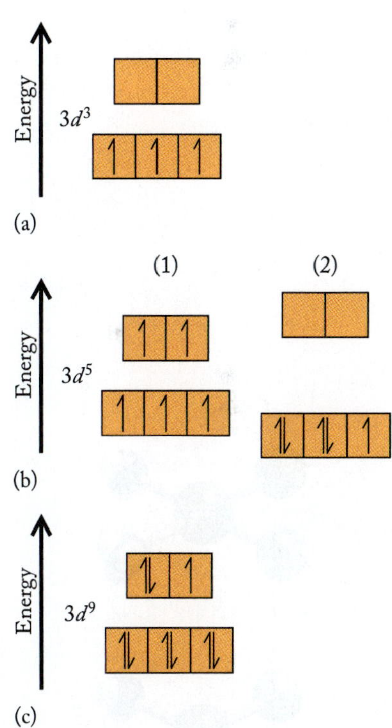


**FIGURE 23.20** Options for electron distribution.

Whether a transition metal ion is in a high-spin state or a low-spin state depends on whether less energy is needed to promote an electron to a higher-energy orbital or to overcome the repulsion experienced by two electrons sharing the same lower-energy orbital. Several factors affect the size of $\Delta_o$. We have already discussed a major one in the context of the spectrochemical series shown in Table 23.6: the different field strengths of different ligands. Because molecules containing nitrogen atoms with lone pairs of electrons are stronger field-splitting ligands than $H_2O$ molecules, hydrated metal ions (small $\Delta_o$) are more likely to be in high-spin states, and metals surrounded by nitrogen-containing ligands (large $\Delta_o$) are more likely to be in low-spin states. Another factor affecting spin state is the oxidation state of the metal ion. The higher the oxidation number (and ionic charge), the stronger the attraction of the electron pairs on the ligands for the ion. Greater attraction leads to more ligand–*d* orbital interaction and therefore to a larger $\Delta_o$.

Complexes of transition metals in the fifth and sixth rows tend to be low spin because their 4*d* and 5*d* orbitals extend farther from the nucleus than do 3*d* orbitals. These larger *d* orbitals overlap more and interact more strongly with the lone pairs of electrons on the ligands, leading to greater crystal field splitting.

Our discussion of high-spin and low-spin states has focused entirely on *d* orbitals split by octahedral fields. What about spin states in tetrahedral fields? Almost all tetrahedral complexes are high spin because tetrahedral fields are weaker than octahedral fields. Weaker field strength means less *d*-orbital splitting—not enough to offset the energies associated with pairing two electrons in the same orbitals. Therefore, Hund's rule is obeyed.

**CONCEPT TEST**

Explain the following:

a. $Mn(py)_6^{2+}$ is a high-spin complex ion, but $Mn(CN)_6^{4-}$ is low spin.

b. $Fe(NH_3)_6^{2+}$ is high spin, but $Ru(NH_3)_6^{2+}$ is low spin.

## 23.6 Isomerism in Coordination Compounds

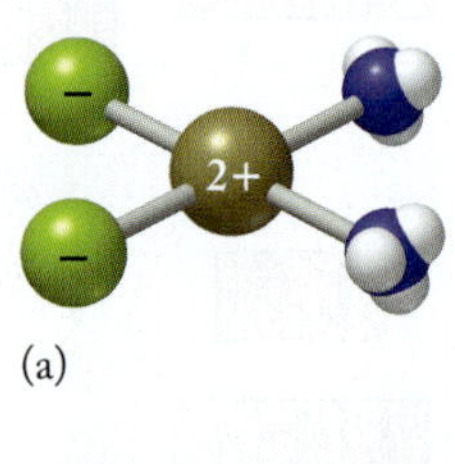


(a)

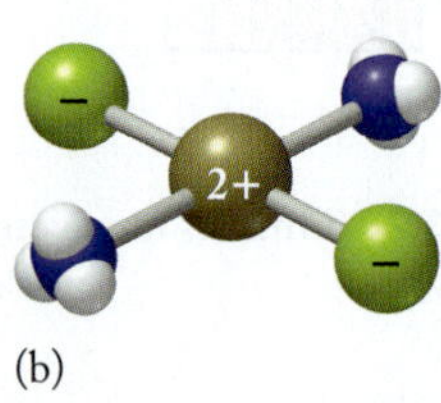


(b)

**FIGURE 23.21** Two ways to orient the $Cl^-$ ions and $NH_3$ molecules around a $Pt^{2+}$ ion (gold sphere) in the square planar coordination compound $Pt(NH_3)_2Cl_2$. (a) The two members of each pair of ligands are on the same side of the square in *cis*-diamminedichloroplatinum(II). (b) The two members of each pair are at opposite corners in *trans*-diamminedichloroplatinum(II).

The coordination compound $Pt(NH_3)_2Cl_2$ was first described in 1849 and is now widely used as a chemotherapeutic agent to treat several cancers. Both molecules of ammonia and the two chloride ions are coordinately bonded to the $Pt^{2+}$ ion, so the name of the compound is diamminedichloroplatinum(II). No counterions are present because the sum of the charges on the ligands and the platinum is zero and the complex is neutral. The compound is square planar, and when we draw it, the ligands about the central metal ion can be arranged in two ways: the two chloride ions and the two ammonia ligands can be at adjacent corners of the square (**Figure 23.21a**) or at opposite corners (**Figure 23.21b**). Coordination compounds such as these two that have the same composition and the same connections between parts but differ in the three-dimensional arrangement of those parts are stereoisomers.

The two molecules in Figure 23.21 have different physical and chemical properties, and we must find a way to distinguish between them when we name them. Isomer (a), with two members of each pair of ligands at adjacent corners, is called *cis*-diamminedichloroplatinum(II). Isomer (b), with pairs of ligands at opposite corners, is *trans*-diamminedichloroplatinum(II). By analogy to the stereoisomers of alkenes we saw in Chapter 19, these types of stereoisomers are called cis/trans isomers.

**FIGURE 23.22** The reaction of *cis*-diamminedichloroplatinum(II) with DNA. (a) After the drug is administered, one of its chloride ions is displaced by a molecule of water, turning each molecule into a 1+ ion. (b) The ion is attracted to a nitrogen-containing base on a strand of DNA, which displaces the water molecule and forms a coordinate bond to Pt. (c) A nearby base forms a bond to Pt by displacing the chloride ion. Forming bonds to the two DNA bases distorts the DNA molecule so much that it cannot function properly. (d) A magnified view of the bonds that form between platinum(II) and nitrogen atoms in DNA. We explored the structure and function of DNA in Section 20.5.

To illustrate the importance of stereoisomerism in coordination compounds, consider this: *cis*-diamminedichloroplatinum(II) is a widely used anticancer drug with the common name *cisplatin*, but the trans isomer is much less effective in fighting cancer. The therapeutic power of cisplatin comes from its structurally specific reactions with DNA. During these reactions, the two $Cl^-$ ions of $Pt(NH_3)_2Cl_2$ are replaced by two nitrogen-containing bases on a strand of DNA in the nucleus of a cell, as shown in **Figure 23.22**. The ability of cisplatin to cross-link these bases distorts the molecular shape of the DNA (Figure 23.22d) and disrupts its normal function. Most important, this kind of cell DNA damage inhibits the ability of cancerous cells to grow and replicate. The trans isomer forms complexes with other intracellular compounds more readily than with DNA, so it is much less effective as an anticancer agent.

**CONNECTION** We introduced the concept of isomers in Section 5.6 when we discussed chirality. Enantiomers, also called optical isomers, are one type of stereoisomer.

Cis/trans isomers are also possible in octahedral complexes containing more than one type of ligand. For example, two stereoisomers are possible for $[Co(NH_3)_4Cl_2]Cl$ (**Figure 23.23**). The two chloro ligands in $[Co(NH_3)_4Cl_2]^+$ are either on the same side of the complex with a 90° Cl–Co–Cl bond angle (the cis isomer) or across from each other so that the Cl–Co–Cl angle is 180° (the trans isomer). *cis*-Tetraamminedichlorocobalt(III)

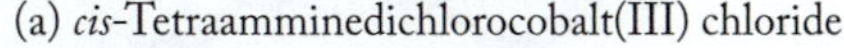

(a) *cis*-Tetraamminedichlorocobalt(III) chloride

(b) *trans*-Tetraamminedichlorocobalt(III) chloride

**FIGURE 23.23** Structures of the complex ions in the two stereoisomers of the coordination compound with the formula $[Co(NH_3)_4Cl_2]Cl$: (a) *cis*-tetraamminedichlorocobalt(III) chloride and (b) *trans*-tetraamminedichlorocobalt(III) chloride.

chloride is violet, whereas *trans*-tetraamminedichlorocobalt(III) chloride is green.

## Enantiomers and Linkage Isomers

In the octahedral cobalt(III) complex ion containing two ethylenediamine molecules and two chloride ions, two ways are available to arrange the chloride ions: on adjacent bonding sites (the cis isomer) or on opposite sides of the octahedron (the trans isomer). Although these two molecules are cis/trans isomers, the cis isomer (**Figure 23.24**) also illustrates another kind of stereoisomerism that is possible in complex ions and coordination compounds.

Figure 23.24 shows that *cis*-$Co(en)_2Cl_2^+$ is chiral: it has a mirror image that is not identical to the original. The difference is demonstrated by the fact that no way exists to rotate the mirror image so that its atoms align exactly with those in the original. In other words, the two structures are not superimposable. We encountered this phenomenon in Chapters 5 and 19 and noted that such nonsuperimposable stereoisomers are called *enantiomers*. If you look closely at Figure 23.8b, you will recognize a similar situation. The arrangement of the three ethylenediamine ligands around $Ni^{2+}$ makes the compound chiral, too.

Naming *cis*-$Co(en)_2Cl_2^+$ also requires an additional rule. The complex ion is named *cis*-dichlorobis(ethylenediamine)cobalt(III). Note the prefix *bis-* just before (ethylenediamine). In naming complex ions containing a polydentate ligand, we use *bis-* instead of *di-* to indicate that two molecules of the ligand are present and to avoid the use of two *di-* prefixes in the same ligand name. Other prefixes used in this fashion are *tris-* for three polydentate ligands and *tetrakis-* for four.

### CONCEPT TEST

Would four different ligands arranged in a square planar geometry produce a chiral complex ion? What about four different ligands in a tetrahedral geometry?

A third kind of isomerism called *linkage isomerism* occurs in coordination compounds when a ligand can bind to a central metal ion by using either of two possible electron pair–donating atoms. The two complexes of $Co^{3+}$ shown in

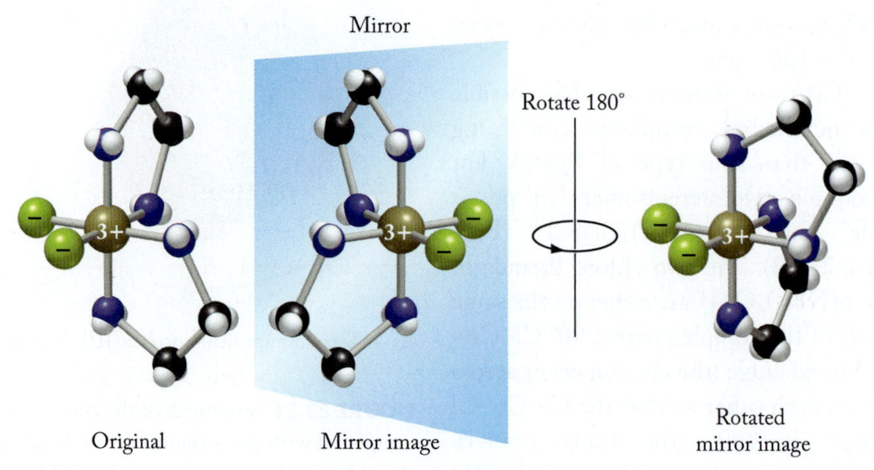


**FIGURE 23.24** The complex ion *cis*-dichlorobis(ethylenediamine)cobalt(III) is chiral, which means that its mirror image is *not* superimposable on the original complex. To illustrate this point, we rotate the mirror image 180° about its vertical axis so that it looks as much like the original as possible. However, note that the top ethylenediamine ligand is located behind the plane of the page in the original but in front of the plane of the page in the rotated mirror image. The mirror images, therefore, are *not* superimposable.

**Figure 23.25** are a pair of linkage isomers. One complex contains the thiocyanate ($SCN^-$) ion as a ligand, where the S atom covalently bonds to the metal (Figure 23.25a). The other contains the isothiocyanate ($NCS^-$) ion as a ligand, in which the N atom forms the bond (Figure 23.25b). Another ligand that forms linkage isomers with metals is $NO_2^-$, called *nitro* when the N atom donates an electron pair and *nitrito* when an O atom is the donor. Because these pairs of molecules have different connectivities, linkage isomers are *not* stereoisomers. Instead, linkage isomers are a type of constitutional isomer.

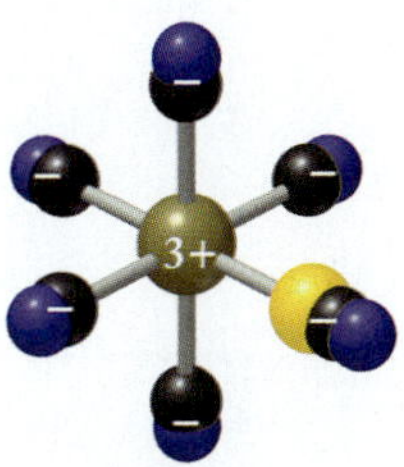


(a) $[Co(CN)_5SCN]^{3-}$

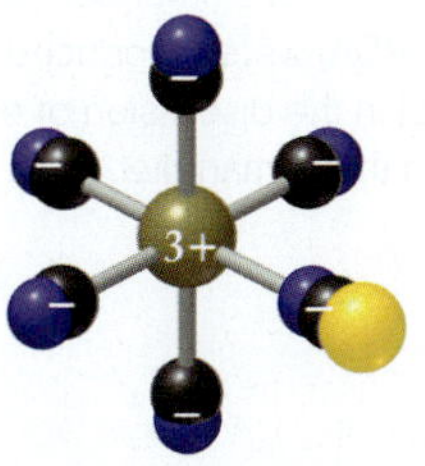


(b) $[Co(CN)_5NCS]^{3-}$

**FIGURE 23.25** The $SCN^-$ ligand is bound to the $Co^{3+}$ ion through the S atom in (a) the pentacyanothiocyanatocobaltate(III) ion and through the N atom in (b) the pentacyanoisothiocyanatocobaltate(III) ion.

**CONNECTION** We discussed constitutional isomers, which have the same molecular formula but different connections between their atoms, in Sections 6.2 and 19.2.

## SAMPLE EXERCISE 23.5 Identifying Isomers of Coordination Compounds LO6

Sketch the structures and name the isomers of $Ni(NH_3)_4(NO_2)_2$.

**Collect and Organize** We are given the formula and asked to name and draw the structures of the isomers of a coordination compound. The $Ni^{2+}$ complexes we have seen so far in the chapter have all been octahedral. Ammonia molecules and $NO_2^-$ ions are both monodentate ligands, but $NO_2^-$ can bond to Lewis bases through either the lone pair on N or through one of the lone pairs on O.

**Analyze** The formula contains no brackets, so the $NO_2^-$ ions are not counterions; they must be covalently bonded to the Ni ion. Two $NO_2^-$ ions are present, so the charge on Ni must be $2^+$. The total number of ligands is 6, which confirms that the compound is octahedral.

**Solve** The nitrite ions can be oriented two ways relative to the four ammonia ligands: opposite each other, with a $O_2N–Ni–NO_2$ bond angle of 180°, or on the same side of the octahedron, with a $O_2N–Ni–NO_2$ bond angle of 90°:

| Top | Upper left | Upper right | Center | Lower left | Lower right | Bottom |
|---|---|---|---|---|---|---|
| $NO_2$ | $H_3N$ | $NH_3$ | Ni | $H_3N$ | $NH_3$ | $NO_2$ |
| $NH_3$ | $H_3N$ | $NO_2$ | Ni | $H_3N$ | $NO_2$ | $NH_3$ |

The first case represents a *trans* arrangement of the $NO_2^-$ ligands, whereas the second is a *cis* arrangement. The first isomer is a *trans*-tetraamminedinitronickel(II); the second is *cis*-tetraamminedinitronickel(II).

Both names include the word *nitro* to signify coordination of the $NO_2^-$ group through N, but the nitrite ligand can also bond through the O, which changes the ligand name to *nitrito* and generates two additional linkage isomers for both the trans and cis isomers.

| Top | Upper left | Upper right | Center | Lower left | Lower right | Bottom |
|---|---|---|---|---|---|---|
| ONO | $H_3N$ | $NH_3$ | Ni | $H_3N$ | $NH_3$ | $NO_2$ |
| ONO | $H_3N$ | $NH_3$ | Ni | $H_3N$ | $NH_3$ | ONO |
| $NH_3$ | $H_3N$ | ONO | Ni | $H_3N$ | $NO_2$ | $NH_3$ |
| $NH_3$ | $H_3N$ | ONO | Ni | $H_3N$ | ONO | $NH_3$ |

The names for these four isomers are *trans*-tetraammine(nitrito)(nitro)nickel(II), *trans*-tetraamminedinitritonickel(II); *cis*-tetraammine(nitrito)(nitro)nickel(II), and *cis*-tetraamminedinitritonickel(II), respectively.

**Think About It** We can draw other tetraamminedichloronickel(II) structures that may not look exactly like these two structures. If we flip or rotate them, however, they will match one of the two structures shown.

**Practice Exercise** Sketch the stereoisomers of $[CoBr_2(en)(NH_3)_2]^+$ and name them.

# 23.7 Coordination Compounds in Biochemistry

**CONNECTION** Dietary reference intake (DRI) and recommended dietary allowance (RDA) were introduced in Section 22.1 in the discussion of essential elements in the human diet.

In Section 23.1, we noted that metals essential to human health must be present in foods in forms the body can absorb. Here in Section 23.7, we explore some biological polydentate ligands that help metal ions participate in processes essential to nutrition and good health. DRI/RDA values for trace essential elements, including iron and zinc, are in the range of 10–20 mg/d. DRI/RDA values have been determined for some, but not all, of the ultratrace essential elements, ranging from 45 μg/d for molybdenum to up to 5 mg/d for manganese. Compounds of iron and zinc are present in the body in average concentrations between 1 μg and 1000 μg of element per gram of body mass and are considered trace elements. Other transition metals (chromium, cobalt, copper, manganese, molybdenum, nickel, and vanadium) are considered ultratrace, meaning essential elements present in the body in average concentrations less than 1 μg of element per gram of body mass. Here we survey the roles of some of the other ultratrace transition metals in biology.

## Manganese and Photosynthesis

Let's begin with photosynthesis, a chemical process at the foundation of our food chain. Green plants can harness solar energy because they contain large biomolecules we collectively call *chlorophyll.* All molecules of chlorophyll contain ring-shaped tetradentate ligands called *chlorins* (**Figure 23.26a**). The structures of chlorins are similar to those of **porphyrins**, another class of tetradentate ligands found in biological systems (**Figure 23.26b**). We saw porphyrins in Section 20.2 as the $Fe^{2+}$-binding molecules in hemoglobin (Figure 20.16a). Chlorins and porphyrins are members of a larger category of polydentate compounds known as **macrocyclic ligands**. (*Macrocycle* means, literally, "big ring.")

Two of the four nitrogen atoms in porphyrins and chlorins are $sp^3$ hybridized and bound to hydrogen atoms, whereas the other two are $sp^2$ hybridized with no hydrogen atoms. When either compound forms coordinate bonds with a metal ion $M^{n+}$ (**Figure 23.26c**), the two hydrogen atoms ionize, giving the ring a charge of 2− and the complex ion an overall charge of $(n - 2)$. The lone pairs of electrons on the N atoms in the ionized structure are oriented toward the ring center. These lone pairs can occupy either the four equatorial coordination sites in an octahedral complex ion or all four coordination sites in a square planar complex ion. In octahedral complex ions, each central metal ion still has two axial sites available for bonding to other ligands. Depending on the charge of the central ion, the coordination compound may be either ionic (a complex ion) or electrically neutral.

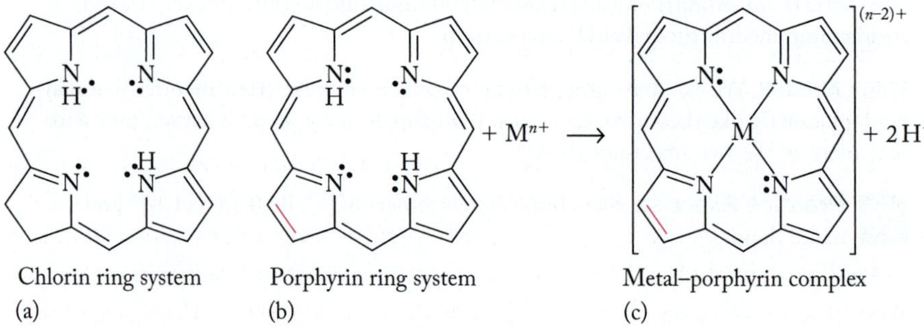

**FIGURE 23.26** Skeletal structures of (a) chlorin and (b) porphyrin. The principal difference between the structures is a C=C double bond in the porphyrin structure (shown in red) that is a single bond in chlorin rings. The innermost atoms in each ring are four nitrogen atoms with lone pairs of electrons. All four N atoms form coordinate bonds with a metal ion, as shown with the porphyrin ring in (c).

Chlorin ring system (a)

Porphyrin ring system (b) $+ M^{n+} \longrightarrow$

Metal–porphyrin complex (c) $^{(n-2)+}$ $+ 2H^+$

Porphyrin and chlorin rings are widespread in nature and play many biochemical roles. Their chemical and physical properties depend on

1. the identity of the central metal ion,
2. the species that occupy the axial coordination sites of octahedral complexes, and
3. the number and identity of organic groups attached to the outside of the ring.

**porphyrin** a type of tetradentate macrocyclic ligand.

**macrocyclic ligand** a ring containing multiple electron-pair donors that bind to a metal ion.

The role of the major essential element $Mg^{2+}$ in photosynthesis was described in Section 22.3. Another metal, the ultratrace essential element manganese, also plays a role in the production of oxygen during photosynthesis. To examine that role, let's write an equation for photosynthesis that is slightly different from the equation we are used to seeing. Rather than write the formula $C_6H_{12}O_6$ for glucose, we use the generic carbohydrate formula $(CH_2O)_n$ so that the coefficient is 1 for all other species in the reaction (rather than 6):

$$H_2O(\ell) + CO_2(g) \rightarrow 1/n\,(CH_2)_n(aq) + O_2(g)$$

Writing the equation in this form makes it easier to see how the overall reaction between water and carbon dioxide involves electron transfer: photosynthesis is a redox reaction where oxygen is oxidized and carbon is reduced.

This redox process requires manganese-containing enzymes in which Mn ions in the +3 and +4 oxidation states mediate the transfer of electrons from water in photosynthesis. Although the exact structure of these manganese compounds remains undetermined, two manganese(III) ions and two manganese(IV) ions are believed to be present at the site of $O_2$ production. Copper-containing enzymes are also involved in the series of reactions that make up photosynthesis.

CONCEPT **TEST**

Manganese ions involved in photosynthesis are often surrounded by six ligands in an octahedral geometry. Is reduction of a manganese(IV) ion to manganese(III) in an octahedral complex (see Section 23.5) accompanied by a change in spin state of the Mn ion?

## Transition Metals in Enzymes

Transition metal cations can form complexes by bonding to the nitrogen atoms of the amino acids in proteins and other Lewis bases in biological systems. Many of the enzymes in our bodies contain zinc or iron ions; these enzymes, along with others containing transition metal ions, are called *metalloenzymes.*

**Zinc** The enzyme carbonic anhydrase catalyzes the reaction between water and carbon dioxide to form bicarbonate ions:

$$H_2O(\ell) + CO_2(g) \rightleftharpoons HCO_3^-(aq) + H^+(aq)$$

The α-form of carbonic anhydrase contains 260 amino acid residues and a zinc ion at the active site. Notice in **Figure 23.27** that the zinc ion is coordinately bonded to three nitrogen atoms on histidine side chains and to one molecule of water. The presence of these ligands and a fourth histidine nearby facilitates ionization of the water molecule. Ionization leaves a $OH^-$ ion attached to the $Zn^{2+}$ ion and a $H^+$ ion bonded to the side-chain nitrogen atom of the fourth histidine. In addition, a space just the right size and shape to accommodate a $CO_2$ molecule is next to the active site. When a $CO_2$ molecule in this space bonds to a hydroxide

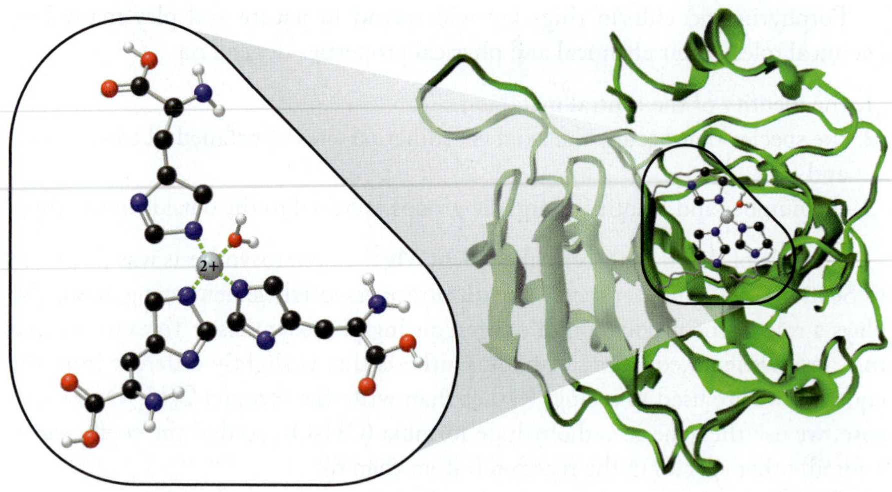


**FIGURE 23.27** The active site of one form of carbonic anhydrase consists of a zinc ion (gray sphere) bonded to three histidine molecules and one $H_2O$ molecule. The $OH^-$ ion produced when this $H_2O$ ionizes reacts with a molecule of $CO_2$, forming an $HCO_3^-$ ion.

ion, a $HCO_3^+$ ion forms. As the bicarbonate ion pulls away, another water molecule occupies the fourth coordination site on the $Zn^{2+}$ ion, another $CO_2$ molecule enters, the histidine is protonated, and the catalytic cycle repeats. This reaction is important because it helps eliminate $CO_2$ from cells during respiration and mediates the uptake of $CO_2$ during photosynthesis in some plants.

**Iron** Iron-containing peroxidases and catalases are integral to transferring oxygen to biomolecules. For example, plants use a fatty acid peroxidase to catalyze the stepwise degradation of fatty acids, where R represents a long hydrocarbon chain. One $-CH_2-$ group at a time is removed from the fatty acid by using hydrogen peroxide:

$$\underset{\text{Fatty acid}}{R{-}CH_2{-}COOH(aq)} + \underset{\substack{\text{Hydrogen}\\ \text{peroxide}}}{2\,H_2O_2(aq)} \xrightarrow[\text{peroxidase}]{\text{fatty acid}} 3\,H_2O(\ell) + \underset{\text{Aldehyde}}{R{-}CHO(aq)} + CO_2(aq)$$

Subsequent oxidation of the aldehyde back to a carboxylic acid yields a new fatty acid with one fewer $-CH_2-$ groups than the original fatty acid:

$$\underset{\text{Aldehyde}}{2\,R{-}CHO(aq)} + O_2(aq) \rightarrow \underset{\text{Fatty acid}}{2\,R{-}COOH(aq)}$$

Iron-containing enzymes also catalyze the reduction of nitrite ($NO_2^-$) and sulfite ($SO_3^{2-}$) ions:

$$NO_2^-(aq) + 6\,e^- + 8\,H^+(aq) \xrightarrow{\text{nitrite reductase}} NH_4^+(aq) + 2\,H_2O(\ell)$$

$$SO_3^{2-}(aq) + 6\,e^- + 7\,H^+(aq) \xrightarrow{\text{sulfite reductase}} HS^-(aq) + 3\,H_2O(\ell)$$

**CONNECTION** Fatty acids were described in Section 20.4.

In each case, the iron in the enzyme is oxidized, supplying the electrons needed for reduction.

Proteins called *cytochromes* also contain one or more heme groups (**Figure 23.28a**). Cytochromes mediate oxidation and reduction processes connected with energy production in cells. The heme group conveys electrons as the half-reaction

$$Fe^{3+} + e^- \rightleftharpoons Fe^{2+}$$

rapidly and reversibly consumes or releases electrons needed in the biochemical reactions that sustain life. Cytochromes catalyze electron transport in photosynthesis and the metabolism of glucose to $CO_2$ and water. Their role in electron

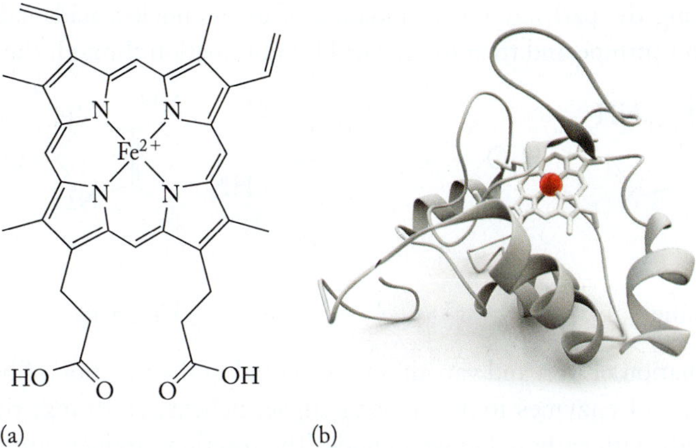


**FIGURE 23.28** (a) Heme is the specific porphyrin that binds iron in the oxygen-transport protein hemoglobin and in many enzymes of great significance in all life on Earth. (b) The structure of cytochrome proteins, such as cytochrome *c* shown here, includes one or more heme complexes (shown in gray) that mediate energy production and redox reactions in living cells. Different cytochromes have different axial ligands occupying the fifth and sixth octahedral coordination sites, and different groups in the protein may be attached to the porphyrin ring.

transport makes them key participants in the in vivo reactions that produce ATP, which is required for intracellular energy production and transfer in living things. As catalysts for oxidation and reduction, cytochromes are essential for the removal of toxic substances and for the process of programmed cell death (also known as apoptosis), both of which are required for the health of multicellular organisms. Cytochrome *c* (**Figure 23.28b**) is a component of the electron transport chain in mitochondria that is ubiquitous in life forms ranging in complexity from microbes to human beings. Its amino acid sequence is very similar or the same across a spectrum of plants, animals, and many unicellular organisms. Consequently, studies of the cytochrome *c* molecule are seminal in evolutionary biology.

**CONNECTION** The role of adenosine triphosphate (ATP) as an energy source in biology is discussed in Section 12.8.

Many other kinds of cytochrome proteins have different substituents on the porphyrin rings and different axial ligands, each of which influences the function of the complex. This last point has been repeated several times in this chapter: the chemical properties and biological functions of transition metals essential to living organisms are linked to their molecular environments and to the formation of stable complex ions with ligands that are strong electron donors—that is, strong Lewis bases.

### CONCEPT TEST

Which of the following small molecules could *not* function as a Lewis base and hence would not be expected to bond to iron in a heme protein? CO; $H_2$; $NO_2$; $H_2S$

**Molybdenum and Vanadium** Many metalloenzymes are involved in transformations of nitrogen. Molybdenum-containing reductases are responsible for converting $NO_3^-$ ions to $NO_2^-$ ions and then to $NH_3$ in the nitrogen cycle (see Figure 22.8). The active site of sulfite oxidase, which converts $SO_3^{2-}$ ions to $SO_4^{2-}$ ions in the sulfur cycle (see Figure 22.11), also contains molybdenum. Sometimes more than one type of transition metal is found in a metalloenzyme. For example, both molybdenum and iron are required by xanthine oxidase, an important enzyme along the pathway for degradation of excess nucleic acids (adenine and guanine) to xanthine and then to uric acid for elimination through the kidneys.

$$\text{Xanthine} + H_2O + O_2 \xrightarrow{\text{xanthine oxidase}} \text{Uric acid} + H_2O_2$$

Xanthine    Uric acid

The combination of iron and vanadium is essential to the function of haloperoxidases, a class of enzymes found in some algae, lichens, and fungi that replace C—H bonds with carbon–halogen bonds. The reaction products are thought to function in the defense systems of these organisms.

**Copper** Copper-containing proteins perform several functions in both plants and animals, including oxygen transport in mollusks such as clams and oysters. Copper is also an essential element in the enzymes azurin and plastocyanin, which mediate electron transfer during photosynthesis.

Reactions catalyzed by xanthine oxidase may produce other reactive oxygen species besides $H_2O_2$. One of them is the superoxide ion, $O_2^-$. Superoxide ion is a strong oxidizing agent and must be eliminated to prevent cell damage. The removal of superoxide begins with the action of superoxide dismutases, which convert superoxide to hydrogen peroxide:

$$2\,O_2^-(aq) + 2\,H^+(aq) \xrightarrow{\text{superoxide dismutase}} H_2O_2(aq) + O_2(aq)$$

Researchers have isolated superoxide dismutases containing a variety of transition metals, including a copper–zinc enzyme. The hydrogen peroxide produced in this reaction is decomposed to water and oxygen by iron-containing catalase.

**Nickel** Ureases, the enzymes responsible for converting urea to ammonia in plants, contain nickel(II). Nickel is also found with iron in enzymes called *hydrogenases*. Hydrogenases oxidize hydrogen gas to protons:

$$H_2(g) \xrightarrow{\text{hydrogenase}} 2\,H^+(aq) + 2\,e^-$$

Clusters containing iron and nickel ions play key roles in the function of CO dehydrogenase, an enzyme that catalyzes the formation of acetyl-CoA, a key component in the tricarboxylic acid cycle discussed in Section 20.3. Finally, nickel-containing enzymes are among the enzymes responsible for methane generation by bacteria.

**FIGURE 23.29** Coenzyme $B_{12}$ contains cobalt(III) (gold sphere), an ultratrace essential metal.

**Cobalt and Coenzymes** Many enzymes require the presence of a **coenzyme**, an organic compound that cocatalyzes a biochemical reaction. The coenzyme $B_{12}$ (**Figure 23.29**) contains the ultratrace essential element cobalt(III) and is a derivative of vitamin $B_{12}$. The cobalt(III) in coenzyme $B_{12}$ is easily reduced to cobalt(II) and even cobalt(I) in enzyme-catalyzed redox reactions. The change in oxidation state of Co allows for facile transfer of methyl groups, as in the conversion of methionine to homocysteine:

$$\text{Methionine} \longrightarrow \text{Homocysteine}$$

Methionine    Homocysteine

**coenzyme** an organic molecule that, like an enzyme, accelerates the rate of biochemical reactions.

Coenzyme $B_{12}$ is also critical to the function of *mutases*, enzymes that catalyze the rearrangement of the skeleton of a molecule, as in the interconversion of glutamate and methylaspartate:

$$\text{Glutamate } ({}^{-}OOC\text{–}CH(\overset{+}{N}H_3)\text{–}CH_2\text{–}CH_2\text{–}COO^{-}) \underset{\text{coenzyme } B_{12}}{\overset{\text{glutamate mutase}}{\rightleftharpoons}} \text{Methylaspartate } ({}^{-}OOC\text{–}CH(\overset{+}{N}H_3)\text{–}CH(CH_3)\text{–}COO^{-})$$

**Chromium** Chromium in the +3 oxidation state is an ultratrace essential element in our diets. It is involved in regulating glucose levels in the blood through a molecule called chromodulin. Chromodulin is a polypeptide containing only 4 of the 20 naturally occurring amino acids: glycine, cysteine, glutamic acid, and aspartic acid. Four $Cr^{3+}$ ions are bound to the peptide chain. Cereals and grains contain enough chromium for our daily needs, but certain plants (such as shepherd's purse) concentrate chromium and have been used as herbal remedies to treat diabetes.

Chromium in the +6 oxidation state, as found in chromate ions ($CrO_4^{2-}$), is acutely toxic and also carcinogenic. Chromate ion enters cells through ion channels that transport $SO_4^{2-}$ ions. Once inside, $CrO_4^{2-}$ is reduced to $Cr^{3+}$, which binds to the phosphate backbone of DNA.

## 23.8 Coordination Compounds in Medicine

Transition metal coordination compounds are becoming important in both the diagnosis and treatment of diseases (**Figure 23.30**). Any diagnostic or therapeutic compound injected intravenously must be soluble enough in blood to be delivered to the target. While in transit, the compound must be stable enough not to undergo chemical reactions that result in its precipitation or rapid elimination from the body. Occasionally, the compound can be in the form of a simple salt, but more often a metal ion is introduced as a coordination complex or coordination compound. Ligands used in forming biologically active coordination complexes include amino acids and simple anions such as the citrate ion. Chelating ligands such as diethylenetriaminepentaacetate ($DTPA^{5-}$; **Figure 23.31**) are often used in biological applications. A medicinal chemist can also take advantage of substances that occur naturally in the body, such as antibodies, to transport a diagnostic or therapeutic metal ion to its target.

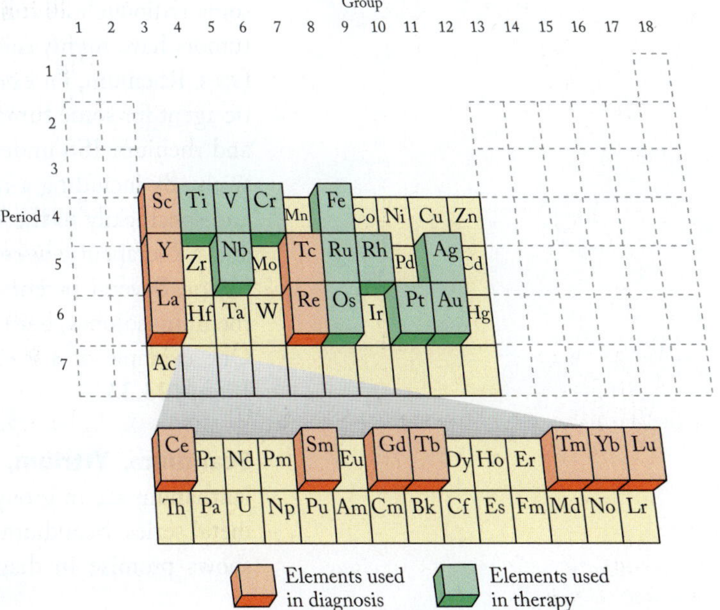


**FIGURE 23.30** Radioisotopes of the elements shown in red are used in diagnostic imaging, and those shown in green are used in therapy.

### Transition Metals in Diagnosis

As discussed in Section 21.9, radionuclides of main group elements with short half-lives that emit easily detectable gamma rays are useful in positron emission tomography (PET). An equally wide array of transition metal isotopes is available for imaging. Magnetic resonance imaging (MRI), for instance,

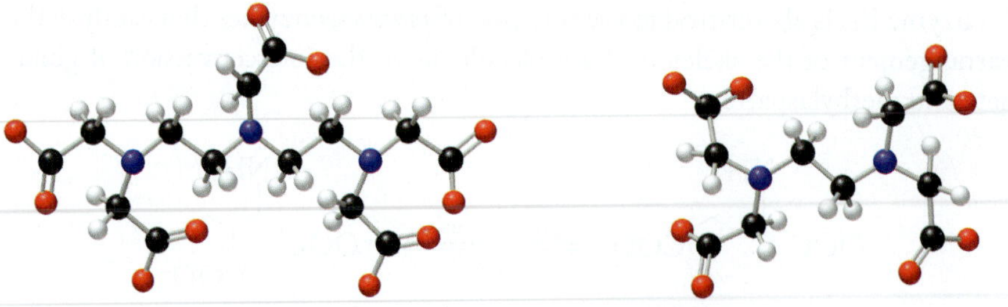

**FIGURE 23.31** Diethylenetriaminepentaacetate ($DTPA^{5-}$) and ethylenediaminetetraacetate ($EDTA^{4-}$) are often used as chelating ligands for diagnostic and therapeutic agents based on transition metals. These ions form stable complex ions with 2+ and 3+ metal cations. The solubilities of the complex ions are typically much greater than those of the hydrated ions at physiological pH (7.4).

Diethylenetriaminepentaacetate ($DTPA^{5-}$) Ethylenediaminetetraacetate ($EDTA^{4-}$)

can be used to diagnose soft-tissue injuries and uses stable isotopes of gadolinium to enhance images.

**CONNECTION** Chapter 21 gave a more detailed discussion of nuclear chemistry and nuclear medicine, including an assessment of how different types of radiation affect living tissue.

**Technetium and Rhenium** Technetium, just below manganese in group 7 of the periodic table, is the most widely used radioactive isotope in imaging. Technetium is unusual in that it does not exist naturally on Earth in easily measurable amounts because it has no stable isotopes; in other words, all technetium isotopes are radioactive. These isotopes can be produced in nuclear reactors for use in medicine. The $^{99m}Tc$ used in hospitals for imaging is prepared in technetium generators in which a stable isotope of molybdenum, $^{98}Mo$ (23.78% natural abundance), is bombarded with neutrons. Technetium-99 has a half-life greater than 20,000 years; when it is produced in a nuclear reactor, however, its nucleus is in an excited state, called a *metastable* nucleus. The metastable state, designated by adding the letter "m" to the mass number, as in technetium-99m or $^{99m}Tc$, has a half-life of 6 hours and decays to the more stable technetium-99 nucleus.

Technetium-99m has been widely used as an imaging agent because it has a short half-life and emits low-energy $\gamma$ rays. Patients can be injected with a variety of technetium compounds, depending on the target organ. For imaging the heart, the coordination compounds shown in **Figure 23.32** are used.

The ability to target the delivery of radionuclides to a particular organ opens the possibility for selective irradiation of a tumor located in that organ. Therefore, some radionuclides can be used to not only image but also treat cancers. Certain tumors have highly selective receptor sites for particular molecules on their surfaces. Rhenium, for example, is being studied as both an imaging and a therapeutic agent for some tumors, including breast, liver, and skin cancer. Rhenium-186 and rhenium-188 undergo $\beta$ decay with half-lives of 3.72 d and 17.0 h, respectively. By including a radioactive rhenium ion in a molecule that binds strongly and specifically to these receptors, physicians can deliver both an imaging agent and a therapeutic agent to the tumor. In principle, the $\beta$ particles destroy the tumor. Several patents have been issued for the use of compounds containing rhenium isotopes, but therapies based on these compounds remain experimental. One example of a rhenium compound used in these applications is shown in **Figure 23.33**.

**Scandium, Yttrium, and Lanthanide Elements** Scandium, yttrium, and lanthanum are in group 3 in the periodic table, the first group in the transition metal series. Scandium-46 has been used to image the spleen, and scandium-47 shows promise in diagnosing breast cancer. Among the available radioactive

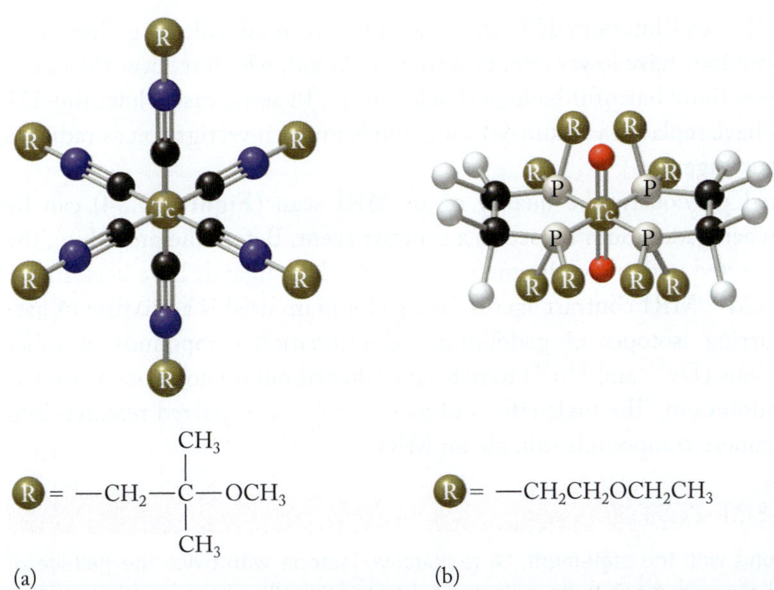


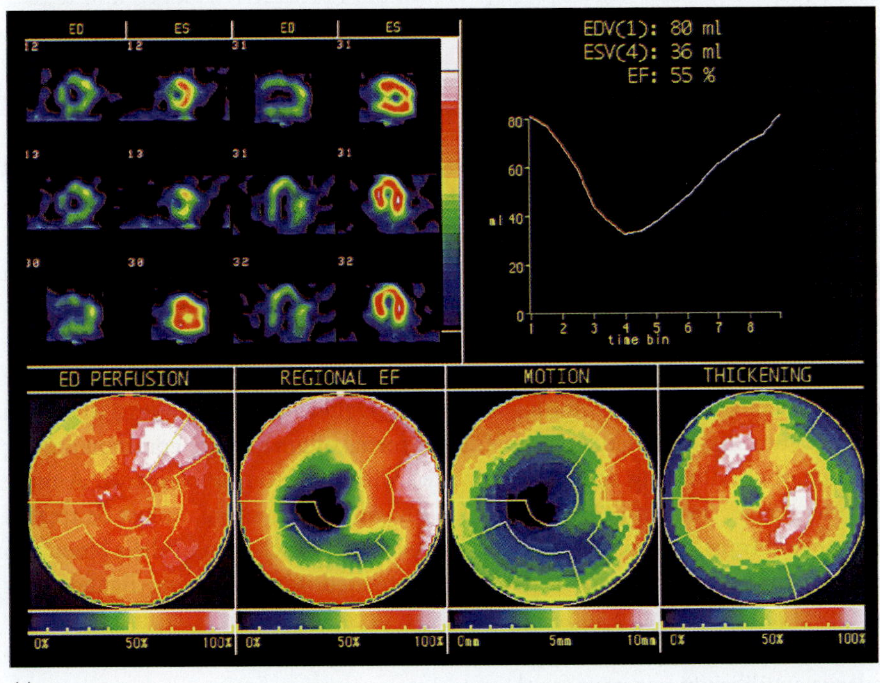


**FIGURE 23.32** (a) Cardiolite $[Tc(CNR)_6]$ and (b) Myoview $[TcO_2(RPCH_2CH_2PR)_2]$ are used for imaging the heart. (c) Images of a patient's heart after intravenous injection with a technetium-containing drug. The images along the bottom show four measurements of heart function. Radioactive technetium compounds emit gamma rays that allow the blood to be tracked as it is pumped through the heart.

isotopes of yttrium, yttrium-90 has been applied to imaging a variety of tumors, including intestinal, breast, and thyroid cancers. Chelating ligands are used to complex the $^{90}Y^{3+}$ cation and deliver it to the tumor, where it binds strongly to receptors on the surface of the tumor. Yttrium-90 agents for treatment of non-Hodgkin's lymphoma are in clinical trials.

Radioactive isotopes of the lanthanides have seen extensive use in scintigraphic imaging, a procedure that uses a scintillation counter (see Section 21.7) to measure the light emitted when radiation strikes a phosphor-coated screen in the instrument. The intensity of the emitted light is translated into a three-dimensional image of the organ of interest, a method similar to the PET technique described in Section 21.9. Among the nuclides used for this purpose are cerium-141, samarium-153, gadolinium-153, terbium-160, thulium-170,

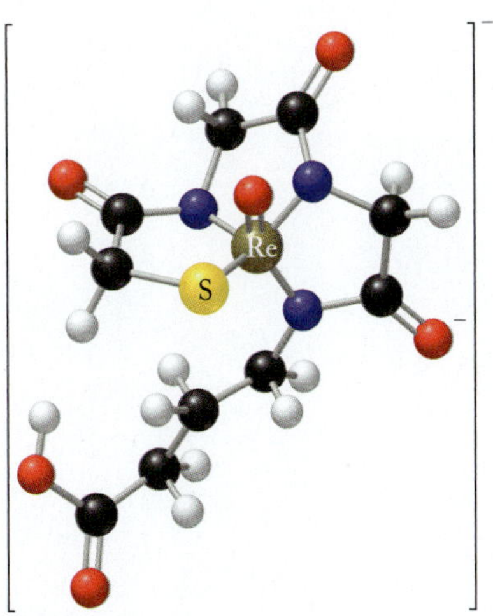


**FIGURE 23.33** This rhenium–mercaptoacetylglycylglycyl-γ-amino acid complex is used to attach $^{186}Re$ or $^{188}Re$ to monoclonal antibodies or peptides.

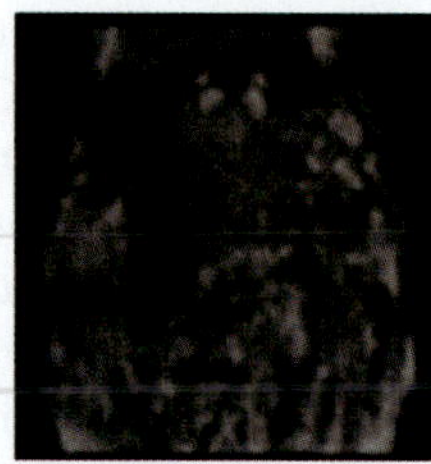

(a) Unenhanced

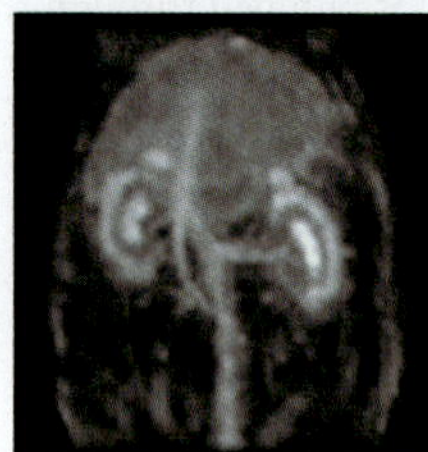

(b) Enhanced

**FIGURE 23.34** MRI scans made (a) without a contrast agent and (b) enhanced by a gadolinium contrast agent.

ytterbium-169, and lutetium-177. These nuclides are preferable to gallium isotopes because they have lower retention times in blood, which reduces the possibility of potentially harmful biological side effects. In some cases, lutetium-177 derivatives have replaced yttrium-90 compounds under investigation as radioactive antitumor agents.

As noted previously, the quality of an MRI scan (**Figure 23.34**) can be improved when gadolinium is used as a contrast agent. Before the procedure, the patient is injected with a gadolinium compound. Many ligands have been investigated for $Gd^{3+}$ MRI contrast agents. The gadolinium used is a mixture of naturally occurring isotopes of gadolinium. Coordination compounds of other lanthanide ions ($Dy^{3+}$ and $Ho^{3+}$) have been evaluated but do not work as well as those of gadolinium. The toxic effects of gadolinium have spurred research into using manganese compounds suitable for MRI.

**CONCEPT TEST**

What is wrong with the statement, "A radioactive isotope with twice the half-life of another will allow an image to be collected in half the time"?

## Transition Metals in Therapy

Like main group metal compounds described in Chapter 22, coordination compounds of the transition metals are finding their way into our expanding arsenal of useful therapeutic agents. We have already described the use of platinum compounds to treat cancer and the potential for chromium compounds to help mitigate diabetes. We now examine additional therapeutic agents that contain transition metal elements.

**Iron** The body must regulate the amount of iron in cells in order to produce enough hemoglobin to maintain good health. Deficiencies in iron lead to several diseases broadly classified as anemia. Mild forms of anemia are common among women of childbearing age and are treated with oral iron supplements. A more serious genetic form of anemia, however, known as thalassemia, must be treated with blood transfusions that leave patients with too much iron in their blood. To remove the excess iron, these patients are treated with chelating ligands that complex some of the iron and transport it out of the red blood cells and eventually out of the body in what is known as chelation therapy. Chelation therapy relies on a large equilibrium constant for the reaction

$$\mathrm{FeL}(aq) + \mathrm{L}'(aq) \rightleftharpoons \mathrm{FeL}'(aq) + \mathrm{L}(aq)$$

where iron coordinated to ligand L (such as hemoglobin) in the blood is treated with ligand L′, which also binds iron. If the formation constant for the complex between iron and L′ is greater than that for iron and L, then the equilibrium constant for the equation is greater than 1, and formation of the iron–L′ complex is favored. The ligand L′ is chosen so that the complex formed between iron and L′ is eliminated from the body, thereby reducing the concentration of iron and relieving the symptoms caused by thalassemia treatments. Often the chelating ligand leads to chiral metal complexes of the kind described in Section 23.6. One enantiomer of an iron complex often works better than the other in the treatment of diseases such as thalassemia. Chelation therapy is also the treatment of choice for heavy metal poisoning from toxic metals, including lead and mercury.

**SAMPLE EXERCISE 23.6** Calculating the Equilibrium Constant of a Ligand Exchange Reaction **LO7**

The protein transferrin is involved in transporting iron into cells. Iron accumulation in the human body has been implicated in diseases such as Parkinson's, Alzheimer's, and thalassemia. The chelating ligand deferoxamine (DFO) is used to treat thalassemia; the complex between iron and DFO is eliminated from the body. Given the formation constants for the complexation of iron(III) by transferrin (Equation 1) and the reaction of DFO with iron (Equation 2), calculate the equilibrium constant for the ligand exchange reaction between DFO and transferrin (Equation 3).

$$Fe^{3+}(aq) + \text{transferrin}(aq) \rightleftharpoons Fe(\text{transferrin})^{3+}(aq) \qquad K_{f,1} = 4.7 \times 10^{20} \quad (1)$$

$$Fe^{3+}(aq) + DFO(aq) \rightleftharpoons Fe(DFO)^{3+}(aq) \qquad K_{f,2} = 4.0 \times 10^{30} \quad (2)$$

$$Fe(\text{transferrin})^{3+}(aq) + DFO(aq) \rightleftharpoons Fe(DFO)^{3+}(aq) + \text{transferrin}(aq) \qquad K_3 = ? \quad (3)$$

**Collect and Organize** We are given the formation constants of two reactions involving iron and different ligands and are asked to calculate the equilibrium constant for the ligand exchange reaction. You may want to review Section 16.6 for calculations involving formation constants of complexes, as well as Chapters 14 and 15, where we manipulated equilibrium constants.

**Analyze** Transferrin is a reactant in Equation 1 and a product in the exchange reaction (3). Therefore, we need to reverse Equation 1 before adding it to Equation 2 to obtain Equation 3:

$$Fe(\text{transferrin})^{3+}(aq) \rightleftharpoons \cancel{Fe^{3+}(aq)} + \text{transferrin}(aq)$$
$$\cancel{Fe^{3+}(aq)} + DFO(aq) \rightleftharpoons Fe(DFO)^{3+}(aq)$$
$$\overline{Fe(\text{transferrin})^{3+}(aq) + DFO(aq) \rightleftharpoons Fe(DFO)^{3+}(aq) + \text{transferrin}(aq)}$$

Reversing a reaction means taking the reciprocal of its equilibrium constant. Combining the reverse of Equation 1 with Equation 2 to obtain Equation 3 means multiplying $(1/K_{f,1})$ and $K_{f,2}$ together to obtain $K_3$. The reciprocal of $K_{f,1}$ has a value of about $10^{-20}$. Therefore, the value of $K_3$ should be about $10^{-20} \times 10^{30}$, or about $10^{10}$.

**Solve** Multiplying $(1/K_{f,1})$ by $K_{f,2}$ to obtain $K_3$:

$$K_3 = \frac{1}{4.7 \times 10^{20}} \times (4.0 \times 10^{30}) = 8.5 \times 10^{9}$$

**Think About It** The equilibrium constant for the ligand exchange between the $Fe(\text{transferrin})^{3+}$ complex and DFO is indeed $\sim 10^{10}$ and illustrates why DFO is an effective treatment for thalassemia—because of the large value for the equilibrium constant, the equilibrium in Equation 3 lies toward the products (to the right), removing iron from the blood.

**Practice Exercise** The equilibrium constants for the reactions of penicillamine and methionine with methyl mercury are

$$CH_3Hg^+ + \text{penicillamine} \rightleftharpoons CH_3Hg(\text{penicillamine})^+ \qquad K_f = 6.3 \times 10^{13}$$

$$CH_3Hg^+ + \text{methionine} \rightleftharpoons CH_3Hg(\text{methionine})^+ \qquad K_f = 2.5 \times 10^{7}$$

Calculate the equilibrium constant for the exchange reaction:

$$CH_3Hg(\text{methionine})^+ + \text{penicillamine} \rightleftharpoons CH_3Hg(\text{penicillamine})^+ + \text{methionine}$$

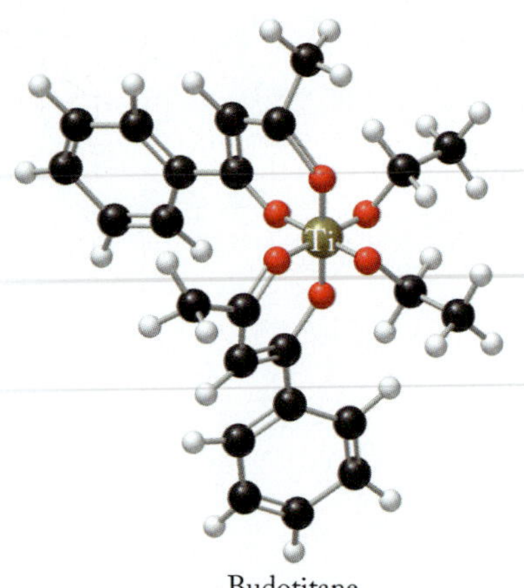


**FIGURE 23.35** Budotitane is a titanium(IV) complex that shows activity against colon cancer.

**Titanium, Vanadium, and Niobium Cancer Drugs** Compounds of a surprisingly large number of transition metals demonstrate antitumor activity. Titanium(IV) complexes such as budotitane (**Figure 23.35**) show promise against colon cancer. Encouraging results against breast, lung, and colon cancers were observed with a series of compounds with formulas $(C_5H_5)_2TiCl_2$, $(C_5H_5)_2VCl$, and $(C_5H_5)_2NbCl_2$. The delivery of titanium and many other transition metal complexes to cells is believed to be mediated by the protein transferrin, described in Sample Exercise 23.6. The precise way in which these compounds inhibit the growth of tumor cells is not yet understood. One working hypothesis proposes a mechanism similar to one for platinum compounds that we will discuss shortly. Unfortunately, at therapeutically useful doses the potential for liver damage outweighs the benefits of these compounds. These compounds contain carbon–metal bonds and belong to a class of substances called **organometallic compounds**.

**Chromium, Vanadium, and Controlling Blood Sugar** Insulin is a hormone needed for proper glucose metabolism and protein synthesis. People with diabetes either cannot produce insulin or can produce the hormone but cannot use it effectively. The suggestion that vanadium plays a role in insulin production has prompted investigation of vanadium compounds as oral diabetes drugs. Redox activity and selective inhibition of enzymes at the insulin receptor site are possible modes of action for vanadium coordination compounds. Encouraging results from animal studies using bis(allixinato)oxovanadium(IV) (**Figure 23.36a**) have been reported, but the compound is not currently approved for use in humans. Chromium(III) compounds have been shown to lower fasting blood sugar levels and reduce the amount of insulin required by some diabetics (**Figure 23.36b**).

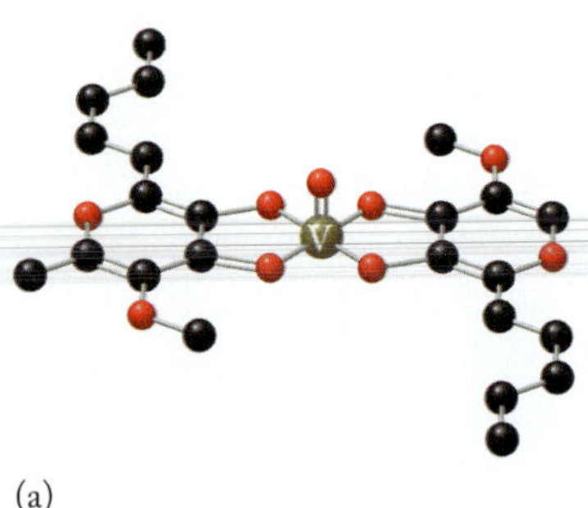


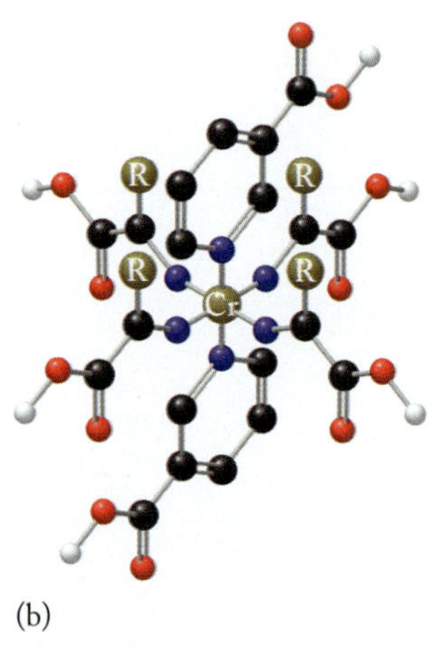


**FIGURE 23.36** (a) Vanadium(IV) compounds such as bis(allixinato) oxovanadium(IV) can act as insulin mimics. (b) The chromium(III) complex in a glucose tolerance factor contributes to our bodies' ability to regulate insulin levels. The light-brown spheres are the side-chain R groups of the amino acids in the structure. (To simplify the structures, the hydrogen atoms bonded to carbon atoms are not shown.)

**Platinum Group and Coinage Metals** The period 5 and period 6 elements in groups 8, 9, and 10 (Ru, Rh, Pd, Os, Ir, and Pt) are often referred to as the *platinum group metals*. The group 11 elements of these two periods (Ag and Au) along with Cu are called the *coinage metals*. We now explore examples of soluble compounds of these metals used in medications for arthritis, cancer, and other diseases.

The serendipitous discovery in the 1960s that *cis*-diamminedichloroplatinum(II) (cisplatin in **Figure 23.37**) is effective in treating testicular, ovarian, and other cancers spurred the development of a host of cancer drugs based on both platinum group metals and coinage metals. The results of this research include a compound known as carboplatin that shows the same activity as cisplatin, whose mechanism of activity is shown in Figure 23.22, but has fewer side effects. In addition to platinum compounds, the antitumor activities of complexes of gold, rhodium, ruthenium, and silver have been explored. The effectiveness of all these drugs lies in their ability to bind to the nitrogen-containing bases in DNA and inhibit cell replication in tumors. If their cells cannot divide, tumors cannot grow. The greater toxicity of rhodium compounds than those of platinum and ruthenium has limited their clinical application.

Selected osmium compounds reduce inflammation in joints resulting from arthritis, although the use of such compounds has diminished with the development of other anti-inflammatory agents. The therapeutic effects of aqueous solutions of osmium tetroxide ($OsO_4$) were first investigated in the 1950s. The use of osmium tetroxide was superseded by the use of glucose polymers

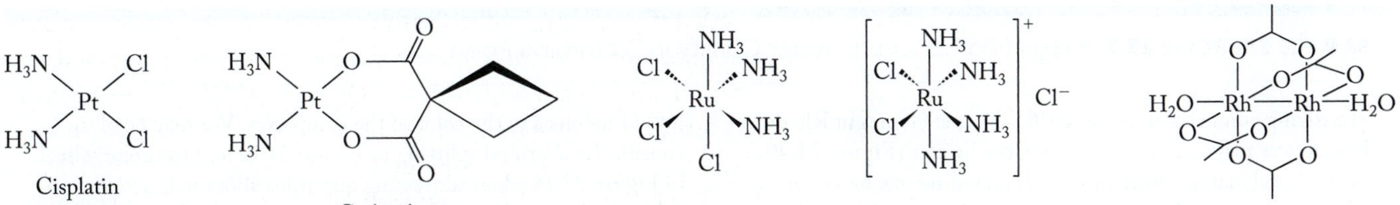

Cisplatin (Platinol)

Carboplatin

**FIGURE 23.37** Cisplatin and carboplatin are effective antitumor agents. The ruthenium and rhodium compounds show activity against leukemia but have not yet seen widespread use.

containing osmium, known as osmarins. The use of osmarins reduces the toxic effects of osmium and illustrates how even toxic metals can be adapted to therapy.

Although the historic use of gold for medicinal purposes dates back millennia, the effective use of gold-containing pharmaceuticals originated with the discovery in the 1920s and 1930s that a gold thiosulfate compound, sanochrysin (**Figure 23.38a**), alleviates the symptoms of rheumatoid arthritis. The most commonly used gold drugs for the treatment of arthritis today are myocrisin (trade name, Myochrysine) and auranofin (trade name, Ridaura) (**Figure 23.38b, c**). Myochrysine is injected; auranofin can be taken orally.

[R = —$CH_2(CH_2COONa)COONa$] [L = $P(CH_2CH_3)_3$]

(a) Sanochrysin (b) Myochrysine (c) Auranofin

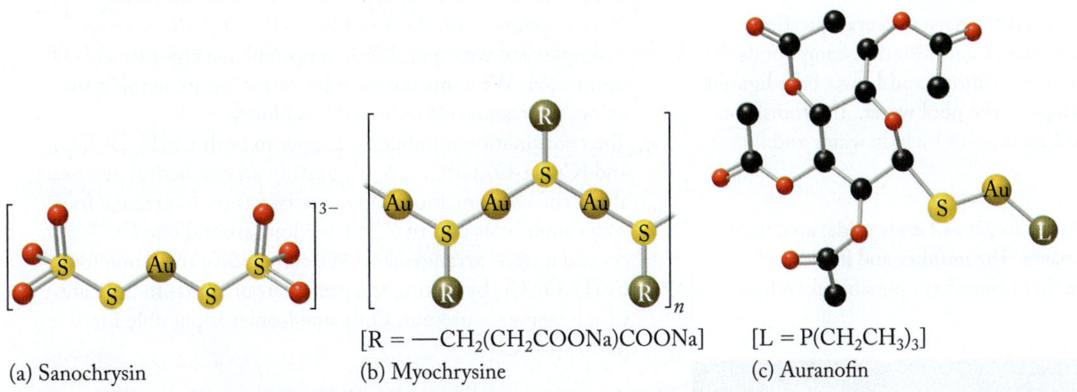

**FIGURE 23.38** (a) Sanochrysin, (b) Myochrysine, and (c) auranofin are three gold-containing drugs used to treat arthritis.

Eye drops containing silver salts are used to treat eye infections. Silver sulfadiazine (**Figure 23.39**) is a broad-spectrum "sulfa" drug used to prevent and treat bacterial and fungal infections. It is also an active ingredient in creams used to treat thermal and chemical burns.

Although present in smaller amounts in the human body than the main group elements, the transition metals play crucial roles in our physiological system and in those of many organisms. Their rich and diverse chemistry has led to many applications of both essential and nonessential transition metal compounds to maintaining health and diagnosing and treating disease. Without a doubt, the smorgasbord of the periodic table will continue to provide us with useful substances and materials.

Silver sulfadiazine

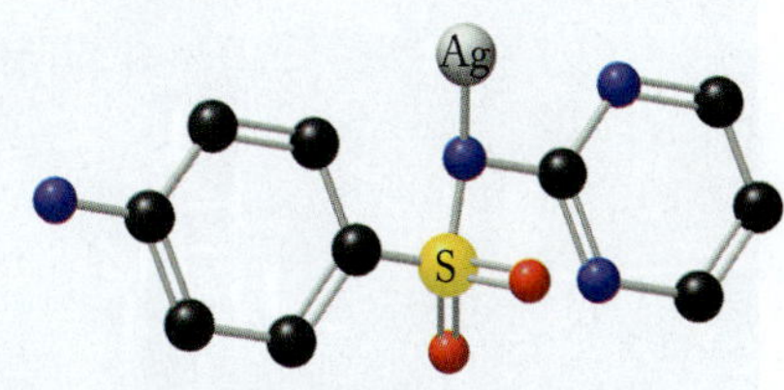

**FIGURE 23.39** Silver sulfadiazine is an effective antibiotic when applied to burns. (H atoms are not shown.)

**organometallic compound** a molecule containing direct carbon–metal covalent bonds.

## SAMPLE EXERCISE 23.7 Integrating Concepts: Water Quality in Swimming Pools

The diving competition at the 2016 Olympic Games in Rio de Janeiro was plagued by smelly green pool water (**Figure 23.40**). Several explanations were proposed, including one based on the addition of copper(II) sulfate to kill the algae growing under the tropical conditions. The color in the water was attributed to the formation of green copper(II) chloride complexes and the odor from the conversion of sulfate to hydrogen sulfide. (The concentration of $Cl^-$ ions in the pool water was derived, in part, from chlorine-based disinfectants added to the water.)

a. Aqueous solutions of copper(II) sulfate contain blue $[Cu(H_2O)_6]^{2+}$ ions. How do the neutral water molecules bond to the $Cu^{2+}$ ion?
b. Two possible copper(II) chloride complexes are $Cu(H_2O)_4Cl_2$ and $[Cu(H_2O)_5Cl]Cl$. How would we name these two coordination complexes?
c. How many isomers can be drawn for the copper(II) complexes in part (b)?
d. Are any of these complexes paramagnetic?
e. What is the origin of the blue color in solutions of $[Cu(H_2O)_6]^{2+}$?
f. Why does the color of the soluble copper complexes change from blue to green in the presence of chloride ion?
g. Does the conversion of $SO_4^{2-}$ to $H_2S$ require a reducing agent or an oxidizing agent?

**Collect and Organize** We need to answer several questions about the structure and bonding in coordination compounds formed between transition metal cations and Lewis base ligands that may account for the color of the pool water. The transition metal ion is $Cu^{2+}$, and the Lewis bases include water and $Cl^-$ ions.

**Analyze** Transition metal cations act as Lewis acids, accepting electron pairs from Lewis bases. The number and identity of the ligands determines whether isomers are possible as well as providing clues to the color of the complexes. We may need to consult the *d*-orbital splitting in Figure 23.12 and the color wheel in Figure 23.14 when addressing questions about magnetism and color in copper complexes.

**FIGURE 23.40** The aquatic events at the 2016 Olympic Games were plagued by green, odorous water.

**Solve**

a. The Lewis structure of water tells us that it has two lone pairs of electrons, either one of which can form a coordinate covalent bond with a $Cu^{2+}$ ion.

$H–\ddot{O}–H$ $[Cu(H_2O)_6]^{2+}$ (octahedral: $OH_2$ ligands around Cu)

b. The two complexes $Cu(H_2O)_4Cl_2$ and $[Cu(H_2O)_5Cl]Cl$ have different numbers of water and chloride ligands in their inner coordination sphere, which leads to their having different names. When naming coordination complexes containing two or more different ligands, we list the names of the ligands in alphabetical order, we indicate how many of each ligand is present with the prefixes found in Table 23.4, and we include the oxidation state of the metal using a Roman numeral. The correct name for $Cu(H_2O)_4Cl_2$ is tetraaquadichlorocopper(II). The compound with the formula $[Cu(H_2O)_5Cl]Cl$ contains a complex ion with a positive charge and one chloride as a counterion. We name the complex cation before naming the anion: pentaaquachlorocopper(II) chloride.
c. The coordination number for copper in both $Cu(H_2O)_4Cl_2$ and $[Cu(H_2O)_5Cl]Cl$ is 6, suggesting an octahedral geometry about the copper cation. There are two ways to arrange four water molecules and two chloride ions around one $Cu^{2+}$: a cis and a trans arrangement. We can modify the name for $Cu(H_2O)_4Cl_2$ by adding the prefix *cis*- or *trans*- to indicate which isomer is present. Only one isomer is possible for $[Cu(H_2O)_5Cl]^+$:

Pentaaquachloro-copper(II) | *cis*-Tetraaquadichloro-copper(II) | *trans*-Tetraaquadichloro-copper(II)

d. The magnetic properties of all three coordination compounds are determined by the nine valence electrons in the five 3*d* orbitals of each of their $Cu^{2+}$ ions. The odd number of these electrons means that one of the *d* orbitals in each ion has an unpaired electron, which makes each compound paramagnetic.
e. The colors of these compounds are linked to crystal field splitting of their $Cu^{2+}$ ions' *d* orbital energies. As seen in

Figure 23.12, the energy of $d_{z^2}$ or $d_{x^2-y^2}$ orbitals is higher than that of the $d_{xy}$, $d_{xz}$, and $d_{yz}$ orbitals. Absorption of visible light of the proper wavelength by $[Cu(H_2O)_6]^+$ leads to promoting an electron from one of the fully occupied lower energy orbitals to the partially filled higher energy orbital. The color we see is the complementary color to the wavelengths absorbed by $[Cu(H_2O)_6]^+$ (see Figure 23.14). Here, absorption of orange light accounts for the blue color.

f. The magnitude of $\Delta_o$ depends on the nature of the ligands that surround $Cu^{2+}$, as shown in the spectrochemical series (Table 23.6). Chloride is below water in Table 23.6, which means that $\Delta_o$ for $Cu(H_2O)_4Cl_2$ and $[Cu(H_2O)_5Cl]^+$ is less than that for $[Cu(H_2O)_6]^+$. The result is that the energy of light required to promote an electron from the $d_{xy}$, $d_{xz}$, and $d_{yz}$ set of orbitals to the $d_{z^2}$ or $d_{x^2-y^2}$ orbitals will be *less* for copper chloride complexes than that for $[Cu(H_2O)_6]^+$. Lower-energy light corresponds to a *longer* wavelength, so copper chloride complexes will absorb the red region of the electromagnetic spectrum shown in Figure 23.14. The complementary color of red is green.

g. The oxidation number of sulfur decreases from +6 in $SO_4^{2-}$ to −2 in $H_2S$, which means sulfur is reduced, which means chemical or biochemical reducing agents must have been present in the pool or in the water treatment system through which pool water was circulated.

**Think About It** If copper sulfate was added to the diving pool water at the 2016 Olympic Games, the observed green color is consistent with the formation of copper(II) chloride complexes. This observation does not prove, however, that these complexes were the origin of the green color.

## SUMMARY

**LO1** Molecules or anions that function as Lewis bases and form coordinate bonds with metal cations are called *ligands*. The resulting species, which are composed of central metal ions and the surrounding ligands, are called *complex ions* or simply complexes. Direct bonding to a central cation means that the ligands in a complex occupy the **inner coordination sphere** of the cation. (Section 23.1)

**LO2** The names of complex ions and coordination compounds indicate the identities and numbers of ligands, the identity and oxidation state of the central metal ion, and the identity and number of **counter-ions**. (Section 23.2)

**LO3** A **monodentate ligand** donates one pair of electrons in a complex ion; a **polydentate ligand** donates more than one pair in a process called **chelation**. Polydentate ligands are particularly effective at forming complex ions. This phenomenon is called the **chelate effect**. EDTA is a particularly effective sequestering agent, which is a chelating agent that prevents metal ions in solution from reacting with other substances. (Section 23.3)

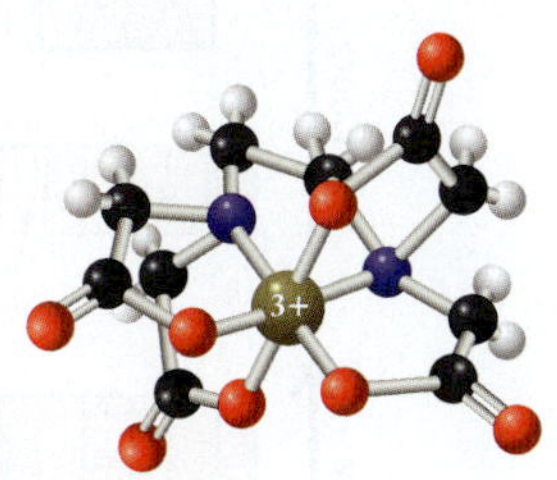


**LO4** The colors of transition metal compounds can be explained by the interactions between electrons in different *d* orbitals and the lone pairs of electrons on surrounding ligands. These interactions create **crystal field splitting** of the energies of the *d* orbitals. A **spectrochemical series** ranks ligands on the basis of their field strength and the wavelengths of electromagnetic radiation absorbed by their complex ions. (Section 23.4)

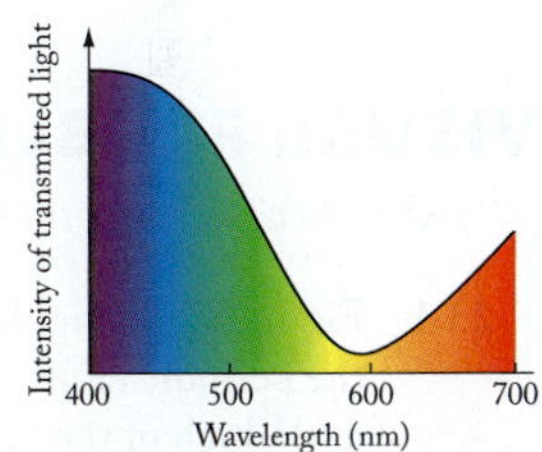


**LO5** Strong repulsions between electrons and large values of **crystal field splitting energy (Δ)** can lead to electron pairing in lower-energy orbitals and an electron configuration called a low-spin state. Metals and their ions with their *d* electrons evenly distributed across all the *d* orbitals in the valence shell represent a high-spin state. (Section 23.5)

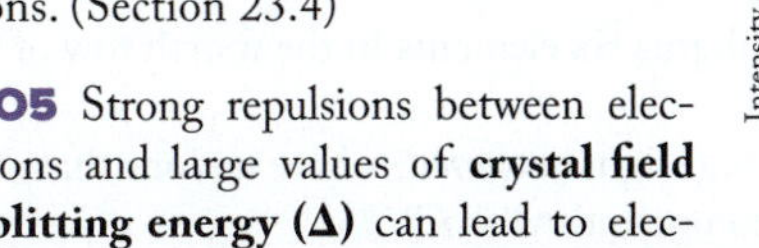

**LO6** Complex metal ions containing more than one type of ligand may form different kinds of isomers. When one type occupies two adjacent corners of a square planar complex, the complex is a *cis* isomer; when the same ligand occupies opposite corners, it is a *trans* isomer. When ligands contain two different Lewis base atoms, linkage isomers are possible. Octahedral complexes may form as a pair of enantiomers. (Sections 23.6 and 23.7)

**LO7** Many enzymes contain transition metal ions. Soluble **coordination compounds** and **organometallic compounds** of the transition metals appear in enzymes. Such compounds are used for imaging and in medications for arthritis, cancer, and other diseases. (Section 23.8)

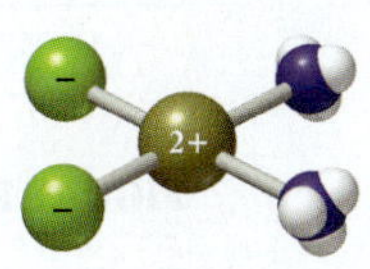


## PARTICULATE **PREVIEW WRAP-UP**

Nitrogen atoms bond to both metal ions. Four ammonia molecules are bonded to the $Cu^{2+}$ ion. Three ethylenediamine molecules are bonded to the $Ni^{2+}$ ion, with each ethylenediamine molecule forming two bonds.

## PROBLEM-SOLVING SUMMARY

| Type of Problem | Concepts and Equations | Sample Exercises |
|---|---|---|
| **Interpreting and writing formulas of coordination compounds** | Identify complex ions, counterions, and their charges. Determine the oxidation states of metals in complex ions. Use square brackets around each complex ion. | **23.1** |
| **Naming coordination compounds** | Follow the naming rules in Section 23.2. | **23.2** |
| **Identifying the potential electron pair–donor groups in a molecule** | Examine the molecular structure of a compound and find lone pairs that can be donated to a metal. | **23.3** |
| **Predicting spin states** | Sketch a *d*-orbital diagram based on crystal field splitting. Fill the lowest energy orbitals with one valence electron each. If there are two ways of adding the remaining valence electrons to the diagram, then multiple spin states are possible. | **23.4** |
| **Identifying stereoisomers of coordination compounds** | If ligands of one type are all on the same side of the complex ion, it is a cis isomer. If ligands of one type are on opposite sides, it is a trans isomer. | **23.5** |
| **Calculating the equilibrium constant for a ligand exchange reaction** | Use the rules for manipulating equilibrium constant expressions to calculate the equilibrium constant for a ligand exchange reaction. | **23.6** |

## VISUAL PROBLEMS

*(Answers to boldface end-of-chapter questions and problems are in the back of the book.)*

**23.1.** Figure P23.1 highlights six elements in the fourth row of the periodic table.

a. Which of the highlighted elements have cations that form colored compounds with $Cl^-$?
b. Which of the highlighted transition metals form $M^{2+}$ cations that cannot have high-spin and low-spin states?
c. Which of the highlighted transition metals have $M^{2+}$ cations that form colorless tetrahedral complex ions?

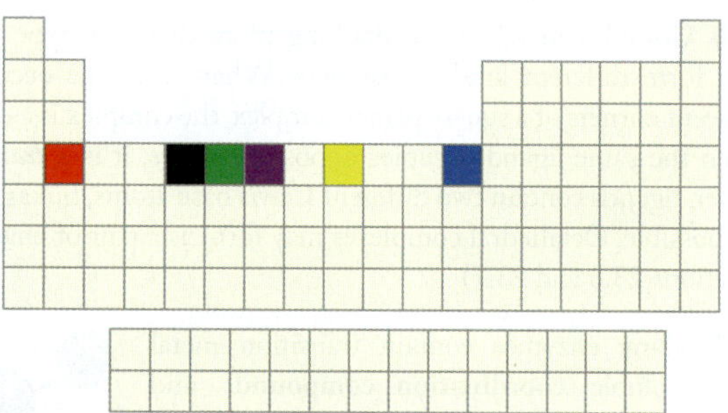

**FIGURE P23.1**

23.2. Predict the number of unpaired electrons for each of the transition metal complexes in Figure P23.2.

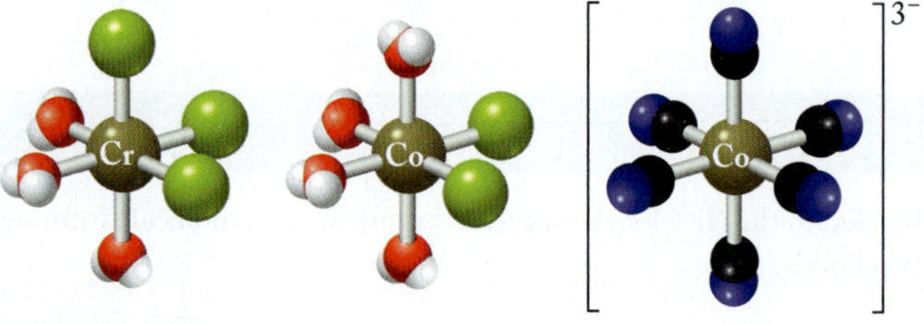


**FIGURE P23.2**

**23.3.** Smoky quartz has distinctive lavender and purple colors due to the presence of manganese impurities in crystals of silicon dioxide. Which of the orbital diagrams in Figure P23.3 best describes the $Mn^{2+}$ ion in a tetrahedral field?

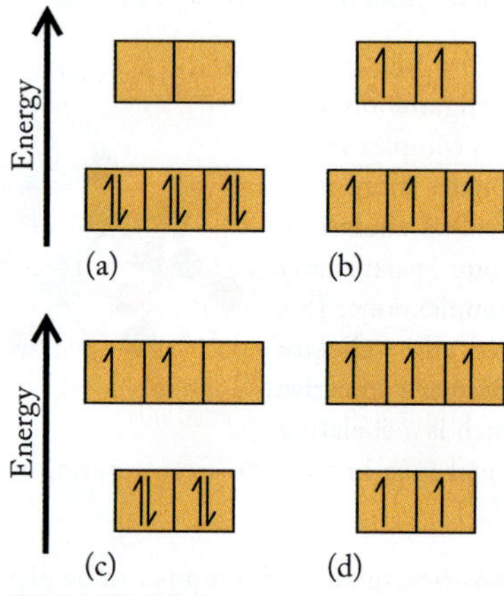


**FIGURE P23.3**

23.4. The magnetic properties of the nickel(II) complex $L_2NiX_2$, where L is a neutral Lewis base, depend on the identity of the X group. When X = Cl, the complex has no unpaired electrons, but it has two when X = Br. Which of the orbital diagrams in Figure P23.4 best explains the difference between the two complexes?

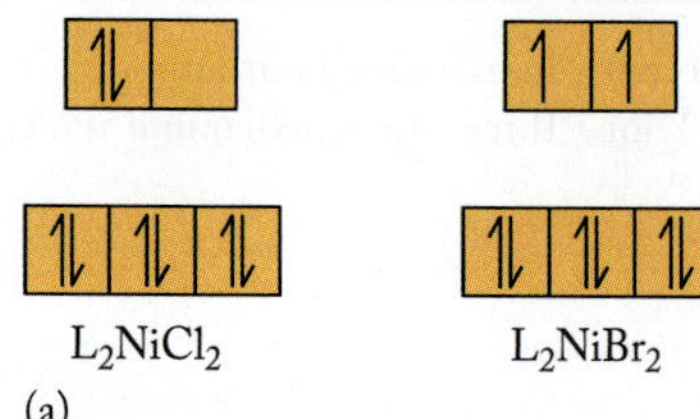

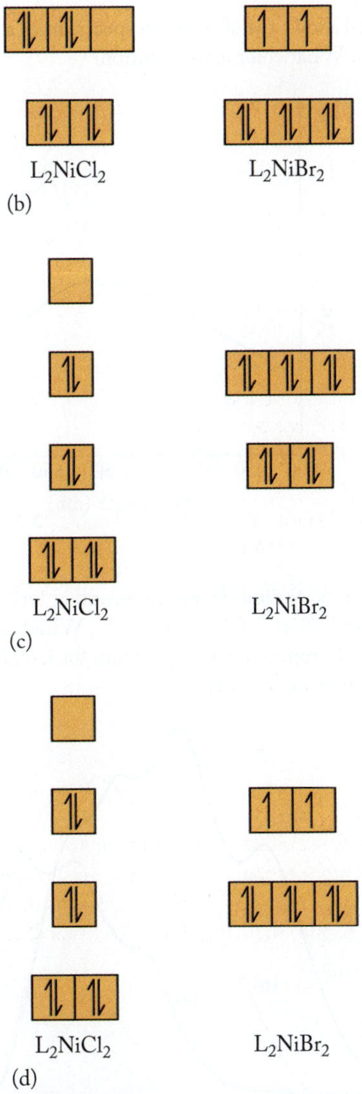

$L_2NiCl_2$ $L_2NiBr_2$

(b)

$L_2NiCl_2$ $L_2NiBr_2$

(c)

$L_2NiCl_2$ $L_2NiBr_2$

(d)

**FIGURE P23.4**

23.5. **Platinum Drugs** Researchers have great interest in developing new cancer drugs containing platinum(IV). Which of the compounds in Figure P23.5 contain ligands that are in a trans arrangement? In a cis arrangement? Both?

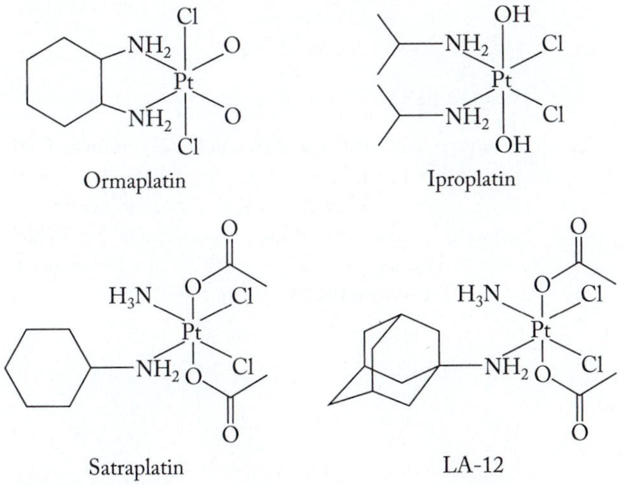

Ormaplatin

Iproplatin

Satraplatin

LA-12

**FIGURE P23.5**

23.6. For each pair of complexes in Figure P23.6, indicate whether the complexes are (i) identical, (ii) isomers, or (iii) neither.

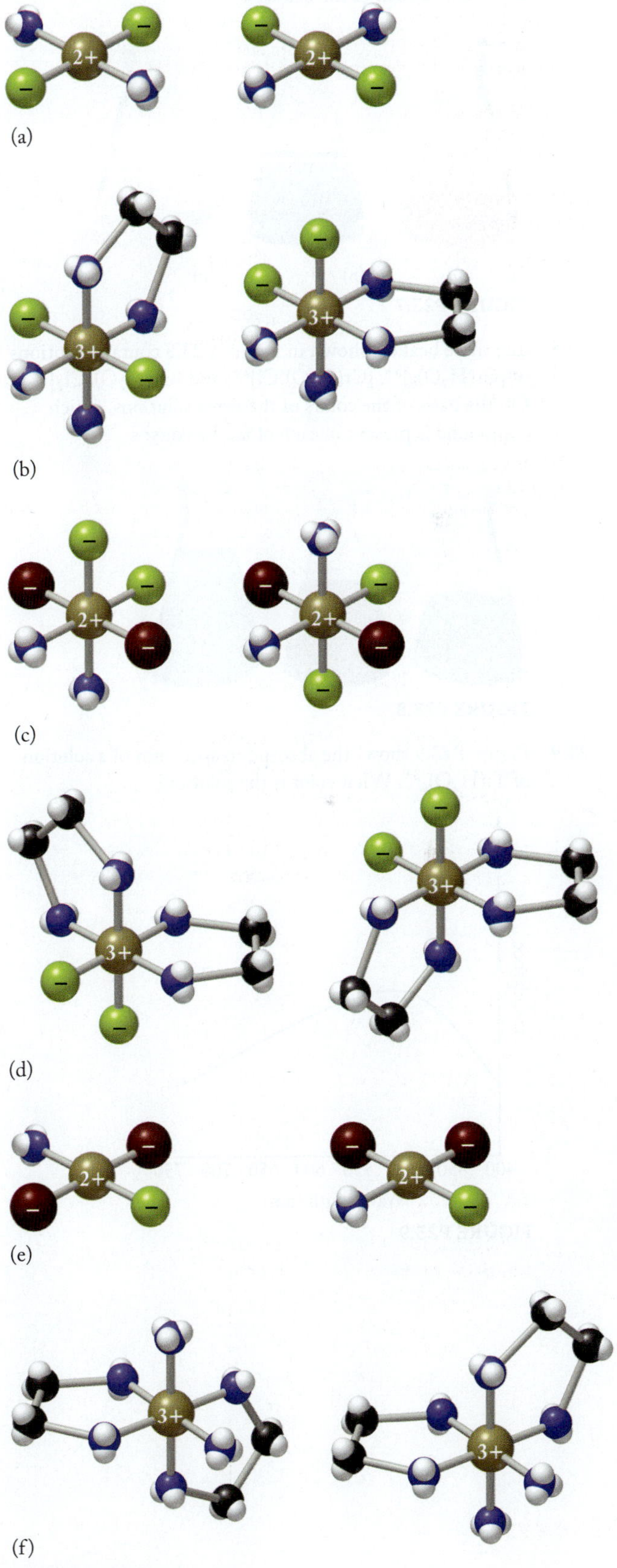

(a)

(b)

(c)

(d)

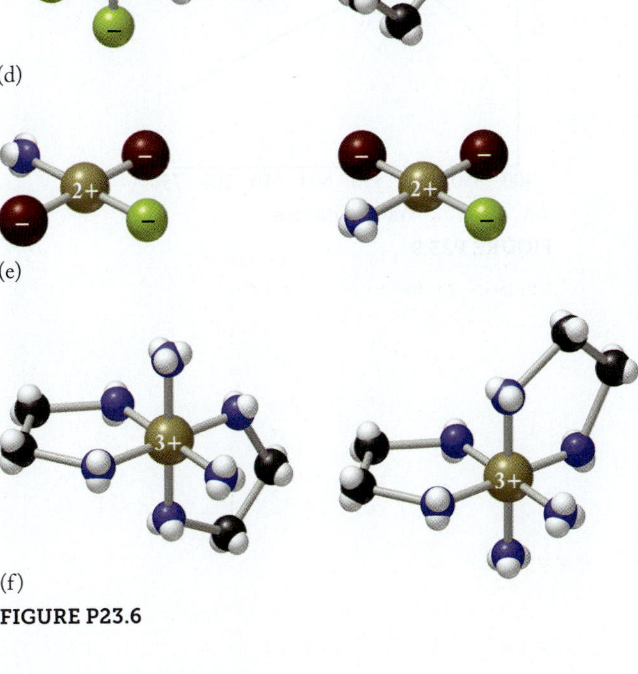

(e)

(f)

**FIGURE P23.6**

**23.7.** The three beakers shown in Figure P23.7 contain solutions of $[Co(H_2O)_6]^{3+}$, $[Co(NH_3)_6]^{3+}$, and $[Co(CN)_6]^{3-}$. On the basis of the colors of the three solutions, which compound is present in each of the beakers?

**FIGURE P23.7**

23.8. The three beakers shown in Figure P23.8 contain solutions of $[Cr(H_2O)_6]^{3+}$, $[Cr(H_2O)_5Cl]^{2+}$, and $[Cr(H_2O)_4Cl_2]^{+}$. On the basis of the colors of the three solutions, which compound is present in each of the beakers?

**FIGURE P23.8**

**23.9.** Figure P23.9 shows the absorption spectrum of a solution of $Ti(H_2O)_6^{3+}$. What color is the solution?

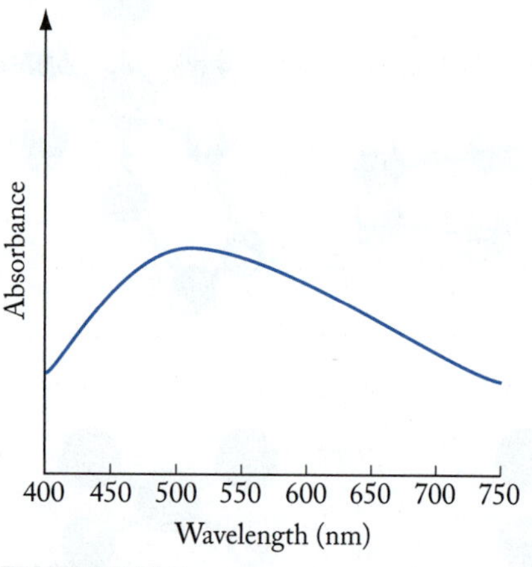


**FIGURE P23.9**

23.10. Figure P23.10 shows the absorption spectrum of a solution of $TiCl_3(py)_3$. What color is the solution?

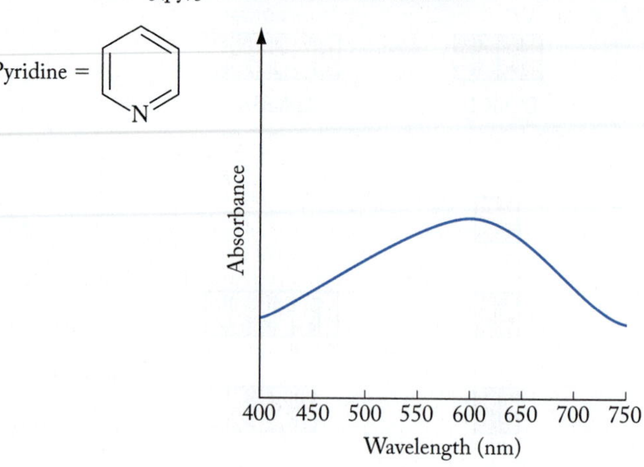


**FIGURE P23.10**

***23.11.** The periodic trend for iodide, bromide, and chloride in the spectrochemical series is $Cl^- > Br^- > I^-$. Which curve in Figure P23.11 represents the spectrum for $CoCl_4^{2-}$? Which curve represents $CoI_4^{2-}$?

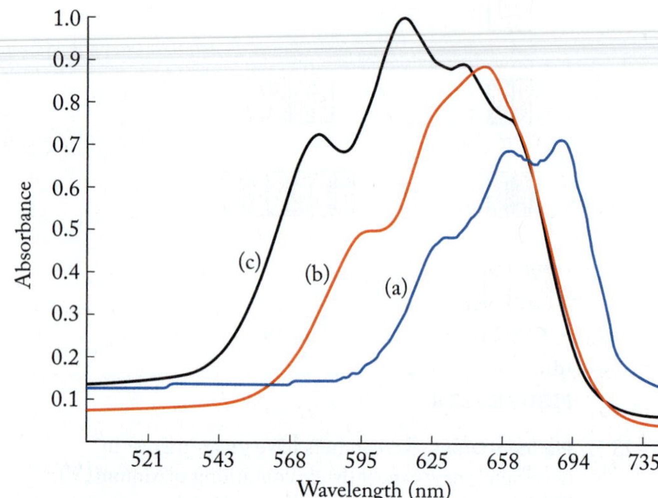


**FIGURE P23.11**

23.12. Use representations [A] through [I] in Figure P23.12 to answer questions a–f.

a. What are the donor atoms in each ligand depicted?
b. Which ligands are monodentate? Which ligands are bidentate?
c. What is the coordination number for nickel in $Ni(CN)_4^{2-}$? In $Ni(dmg)_2$?
d. If the field strength of cyanide is greater than that of dmg, which complex forms the yellow solution in [E]: $Ni(CN)_4^{2-}$ or $Ni(dmg)_2$? Which forms the red solution?
e. Which spectrum corresponds to the red solution?
f. Which orbital diagram depicts the *d*-orbital energies for $Ni(CN)_4^{2-}$? For $Ni(dmg)_2$? (Both are diamagnetic.)

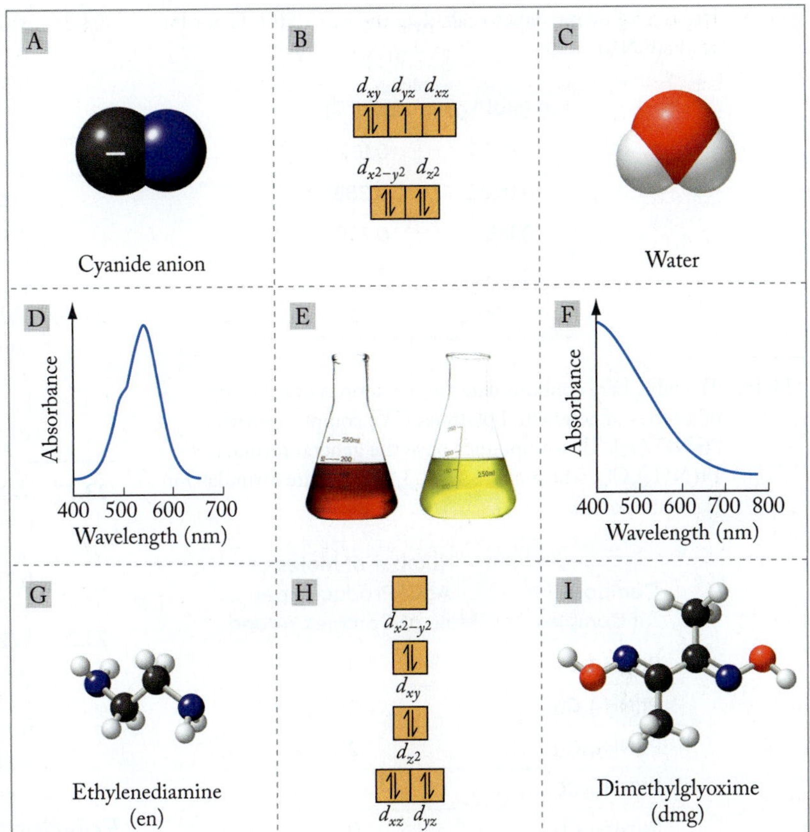


**FIGURE P23.12**

# QUESTIONS AND PROBLEMS

## *Complex Ions*

### Concept Review

**23.13.** When NaCl dissolves in water, which molecules or ions occupy the inner coordination sphere around the $Na^+$ ions?

23.14. When $Cr(NO_3)_3$ dissolves in water, which of the following species are nearest the $Cr^{3+}$ ions?

a. Other $Cr^{3+}$ ions
b. $NO_3^-$ ions
c. $H_2O$ molecules with the O atoms closest to the $Cr^{3+}$
d. $H_2O$ molecules with the H atoms closest to the $Cr^{3+}$

**23.15.** The coordination compound $Fe_4[Fe(CN)_6]_3$ contains a $Fe^{3+}$ cation and a complex anion. What is the likely oxidation state for Fe in the anion?

23.16. The coordination compound $Co_3[Cr(CN)_6]_2$ contains a $Co^{2+}$ cation and a complex anion. What is the likely oxidation state for Cr in the anion?

### Problems

**23.17.** Use the following data to calculate the van 't Hoff factor for $K_3[Fe(CN)_6]$.

| *m* (mol/kg) | Δ*T* (°C) |
|---|---|
| 0.0301 | 0.180 |
| 0.0780 | 0.334 |
| 0.160 | 0.807 |
| 0.247 | 1.19 |
| 0.338 | 1.58 |
| 0.414 | 1.90 |
| 0.535 | 2.41 |
| 0.664 | 3.00 |
| 0.754 | 3.43 |

23.18. Use the following data to calculate the van 't Hoff factor for $K_4[Fe(CN)_6]$.

| *m* (mol/kg) | Δ*T* (°C) |
|---|---|
| 0.0272 | 0.162 |
| 0.0702 | 0.385 |
| 0.145 | 0.709 |
| 0.224 | 1.02 |
| 0.309 | 1.32 |

**23.19.** The table here contains data from reactions of solutions of a series of octahedral platinum(IV) complexes with $AgNO_3(aq)$. The compounds have the general formula $Pt(NH_3)_xCl_4$, where $x = 6, 5, 4, 3$, or 2. Write formulas for each compound.

| Composition of Complex | Number of Moles of AgCl Produced per Mole of Complex Added |
|---|---|
| $Pt(NH_3)_6Cl_4$ | 4 |
| $Pt(NH_3)_5Cl_4$ | 3 |
| $Pt(NH_3)_4Cl_4$ | 2 |
| $Pt(NH_3)_3Cl_4$ | 1 |
| $Pt(NH_3)_2Cl_4$ | 0 |

23.20. The compositions of three compounds of chromium(III) and chloride ion are known: $Cr(H_2O)_6Cl_3$, $Cr(H_2O)_5Cl_3$, and $Cr(H_2O)_4Cl_3$. The following table summarizes their properties.

| Compound | Color of Aqueous Solution | Number of Moles of AgCl Produced per Mole of Complex Added |
|---|---|---|
| A | Dark green | 1 |
| B | Gray-blue | 3 |
| C | Light green | 2 |

Write the correct formulas for compounds A, B, and C.

## Naming Complex Ions and Coordination Compounds

### Problems

**23.21.** Name the complex ions of platinum(IV) in Problem 23.19.

23.22. Name the complex ions of chromium(III) in Problem 23.20.

**23.23.** What are the names of the following complex ions?
a. $Cr(NH_3)_6^{3+}$
b. $Co(H_2O)_6^{3+}$
c. $Ni(CN)_5^{3-}$

23.24. What are the names of the following complex ions?
a. $Cu(NH_3)_2^{+}$
b. $Ti(H_2O)_4(OH)_2^{2+}$
c. $[Cr(en)(OH)_4]^-$

**23.25.** What are the names of the following coordination compounds?
a. $K_2CoBr_4$
b. $NH_4[Zn(H_2O)(OH)_3]$
c. $[Fe(NH_3)_5Cl]Cl_2$

23.26. What are the names of the following coordination compounds?
a. $Na_2CoI_4$
b. $BaCuCl_4$
c. $[Ni(NH_3)_4(H_2O)_2]SO_4$

**23.27.** What are the names of the following coordination compounds?
a. $[Zn(en)_2]SO_4$
b. $[Ni(NH_3)_5(H_2O)]Cl_2$
c. $K_4[Fe(CN)_6]$

23.28. What are the names of the following coordination compounds?
a. $(NH_4)_3[Co(CN)_6]$
b. $[Co(en)_2Cl](NO_3)_2$
c. $[Fe(H_2O)_4(OH)_2]Cl$

## Polydentate Ligands and Chelation

### Concept Review

**23.29.** What is meant by the term *sequestering agent*? What properties make a substance an effective sequestering agent?

23.30. The structures of two compounds that each contain two $-NH_2$ groups are shown in Figure P23.30. The one on the left is ethylenediamine, a bidentate ligand. Does the molecule on the right have the same ability to donate two pairs of electrons to a metal ion? Explain why you think it does or does not.

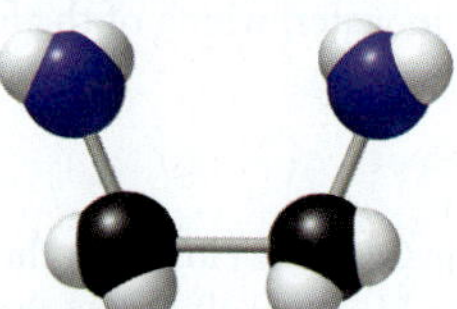
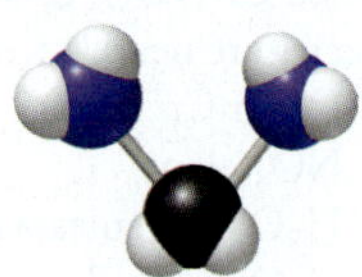

**FIGURE P23.30**

**23.31.** How does the chelating ability of an aminocarboxylic acid vary with changing pH?

*23.32. **Food Preservative** The EDTA that is widely used as a food preservative is added to food—not as the undissociated acid but rather as the calcium disodium salt: $Na_2[CaEDTA]$. This salt is actually a coordination compound with a $Ca^{2+}$ ion at the center of a complex ion. Draw the structure of this compound.

## Crystal Field Theory

### Concept Review

**23.33.** Explain why the compounds of most of the first-row transition metals are colored.

23.34. Unlike the compounds of most transition metal ions, those of $Ti^{4+}$ are colorless. Why?

**23.35.** Why is the $d_{xy}$ orbital higher in energy than the $d_{xz}$ and $d_{yz}$ orbitals in a square planar crystal field?

23.36. On average, the *d* orbitals of a transition metal ion in an octahedral field are higher in energy than they are when the ion is in the gas phase. Why?

### Problems

**23.37.** Aqueous solutions of one of the following complex ions of chromium(III) are violet; solutions of the other are yellow. Which is which? (a) $Cr(H_2O)_6^{3+}$; (b) $Cr(NH_3)_6^{3+}$

23.38. Which of the following complex ions should absorb the shortest wavelengths of electromagnetic radiation? (a) $CuCl_4^{2-}$; (b) $CuF_4^{2-}$; (c) $CuI_4^{2-}$; (d) $CuBr_4^{2-}$

---

**23.39.** The octahedral crystal field splitting energy $\Delta_o$ of $Co(phen)_3^{3+}$ is $5.21 \times 10^{-19}$ J/ion. What is the color of a solution of this complex ion?

23.40. The octahedral crystal field splitting energy $\Delta_o$ of $Co(CN)_6^{3-}$ is $6.74 \times 10^{-19}$ J/ion. What is the color of a solution of this complex ion?

---

**23.41.** Solutions of $NiCl_4^{2-}$ and $NiBr_4^{2-}$ absorb light at 702 nm and 756 nm, respectively. In which ion is the split of *d*-orbital energies greater?

23.42. Chromium(III) chloride forms six-coordinate complexes with bipyridine, including *cis*-$[Cr(bipy)_2Cl_2]^+$, which reacts slowly with water to produce *cis*-$[Cr(bipy)_2(H_2O)Cl]^{2+}$ and *cis*-$[Cr(bipy)_2(H_2O)_2]^{3+}$. In which of these complexes should $\Delta_o$ be the largest?

## Magnetism and Spin States

### Concept Review

**23.43.** What determines whether a transition metal ion is in a *high-spin* configuration or a *low-spin* configuration?

23.44. Would you expect a solution of a high-spin complex of a transition metal ion to be the same color as a solution of a low-spin complex? Why?

### Problems

**23.45.** How many unpaired electrons are in the following transition metal ions in an octahedral field? High-spin $Fe^{2+}$, $Cr^{2+}$, $Co^{2+}$, and $Mn^{3+}$

23.46. Which of the following cations can have either a high-spin or a low-spin electron configuration in an octahedral field? $Fe^{2+}$, $Co^{3+}$, $Mn^{2+}$, and $Cr^{3+}$

---

***23.47.** A solid compound containing iron(II) in an octahedral crystal field has four unpaired electrons at 298 K. When the compound is cooled to 80 K, the same sample appears to have no unpaired electrons. How do you explain this change in the compound's properties?

23.48. The iron(II) compound $Fe(bipy)_2(SCN)_2$ is paramagnetic, but the corresponding cyanide compound $Fe(bipy)_2(CN)_2$ is diamagnetic. Why do these two compounds have different magnetic properties?

---

**23.49.** The manganese minerals pyrolusite ($MnO_2$) and hausmannite ($Mn_3O_4$) contain manganese ions surrounded by oxide ions.

a. What are the charges of the Mn ions in each mineral?
b. In which of these compounds could high-spin and low-spin Mn ions be present?

23.50. **Dietary Supplement** Chromium picolinate is an over-the-counter diet aid sold in many pharmacies. The $Cr^{3+}$ ions in this coordination compound are in an octahedral field. Is the compound paramagnetic or diamagnetic?

---

***23.51.** **Refining Cobalt** One method for refining cobalt involves the formation of the complex ion $CoCl_4^{2-}$. This anion is tetrahedral. Is this complex paramagnetic or diamagnetic?

*23.52. Why is $Ni(CN)_4^{2-}$ diamagnetic, but $NiCl_4^{2-}$ is paramagnetic?

## Isomerism in Coordination Compounds

### Concept Review

**23.53.** What do the prefixes *cis*- and *trans*- mean in the context of an octahedral complex ion?

23.54. What do the prefixes *cis*- and *trans*- mean in the context of a square planar complex?

**23.55.** What minimum number of donor groups are required in order to have stereoisomers of a square planar complex?

23.56. With respect to your answer to Problem 23.55, do all square planar complexes with this many types of donor groups have stereoisomers?

### Problems

**23.57.** Does the complex $Co(en)(H_2O)_2Cl_2$ have stereoisomers?

23.58. Does the complex ion $Fe(en)_3^{3+}$ have stereoisomers?

---

***23.59.** Sketch the stereoisomers of the square planar complex ion $CuCl_2Br_2^{2-}$. Are any of these isomers chiral?

*23.60. Sketch the stereoisomers of the complex ion $Ni(en)Cl_2(CN)_2^{2-}$. Are any of these isomers chiral?

## Coordination Compounds in Biochemistry

### Concept Review

**23.61.** Enzymes are large proteins.

a. What is the function of enzymes?
b. Are all proteins enzymes?

*23.62. Why is $Cd^{2+}$ more likely than $Cr^{2+}$ to replace $Zn^{2+}$ in an enzyme such as carbonic anhydrase?

***23.63.** How does an enzyme affect the activation energy of a biochemical reaction?

*23.64. Why might reductases also be described as reducing agents?

***23.65.** When a transition metal ion such as $Cu^{2+}$ is incorporated into a metalloenzyme, is the formation constant likely to be much greater than 1 ($K >> 1$) or much less than 1 ($K << 1$)?

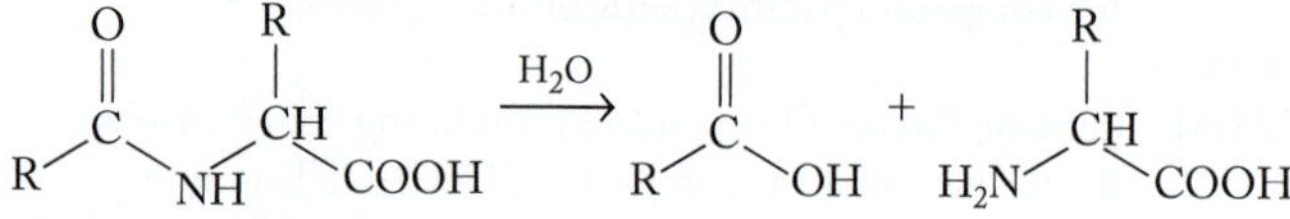

$$Cu^{2+} + \text{protein} \rightleftharpoons \text{metalloenzyme} \qquad K = \frac{[\text{metalloenzyme}]}{[Cu^{2+}][\text{protein}]}$$

*23.66. Carbon monoxide poisoning derives from competitive binding of CO versus $O_2$ to the $Fe^{2+}$ in the coordination sphere of hemoglobin. If CO can be displaced by breathing pure $O_2$, which of the following is likely to be correct under these conditions: $K_{O_2}/K_{CO} < 1$ or $K_{O_2}/K_{CO} > 1$?

**23.67.** What is the likely sign of $\Delta S$ for the reaction in Figure P23.67, which is catalyzed by carboxypeptidase?

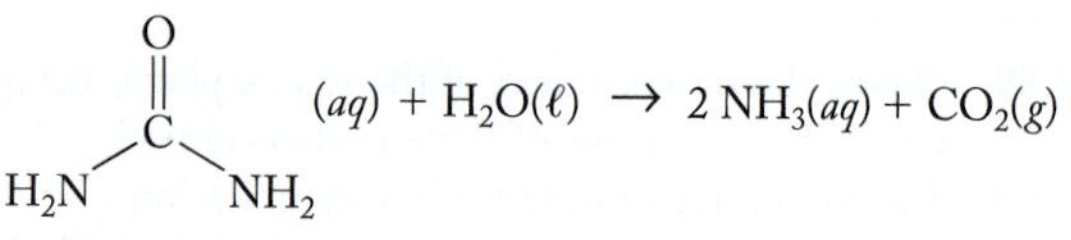

**FIGURE P23.67**

23.68. What is the likely sign of $\Delta G$ for the reaction in Problem 23.67?

### Problems

***23.69.** The activation energy for the uncatalyzed decomposition of hydrogen peroxide at 20°C is 75.3 kJ/mol. In the presence of the enzyme catalase, the activation energy is reduced to 29.3 kJ/mol. By using the following form of the Arrhenius equation, $RT\ln(k_1/k_2) = E_{a_2} - E_{a_1}$, how much faster is the catalyzed reaction?

*23.70. **Enzymatic Activity of Urease** Urease catalyzes the decomposition of urea to ammonia and carbon dioxide (Figure P23.70). The rate constant for the uncatalyzed reaction at 20°C and pH = 8 is $k = 3 \times 10^{-10}\ s^{-1}$. A urease isolated from the jack bean increases the rate constant to $k = 3 \times 10^4\ s^{-1}$. By using the $RT\ln(k_1/k_2) = E_{a_1} - E_{a_2}$ form of the Arrhenius equation, calculate the difference between the activation energies.

$$H_2NC(O)NH_2(aq) + H_2O(\ell) \rightarrow 2\ NH_3(aq) + CO_2(g)$$

**FIGURE P23.70**

***23.71.** The initial rate of an enzyme-catalyzed reaction depends on the concentration of substrate, as shown in Figure P23.71. What is the apparent reaction order at the far-right side of the graph?

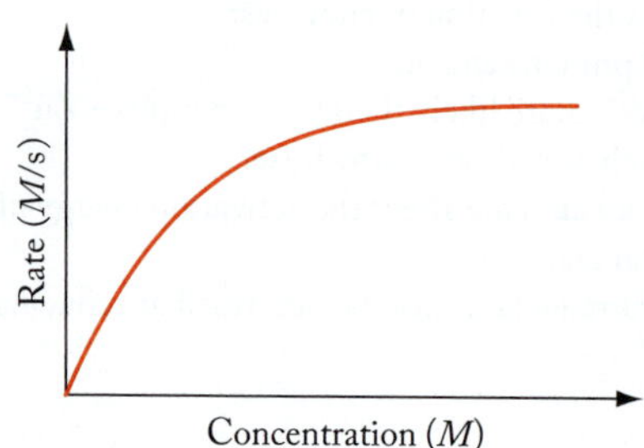


**FIGURE P23.71**

23.72. The rate of an enzyme-catalyzed reaction depends on the concentration of substrate. Which line in Figure P23.72 has the highest reaction rate?

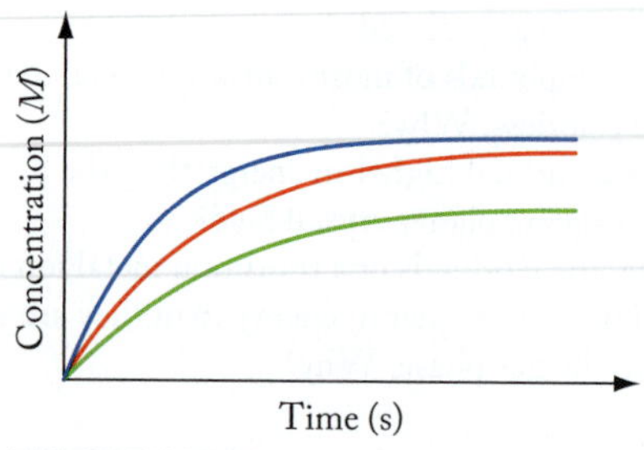


**FIGURE P23.72**

## Coordination Compounds in Medicine

### Concept Review

**23.73.** Under what circumstances might a coordination complex of a toxic metal be useful in medicine?

23.74. **Cancer Treatment** Brachytherapy is a cancer treatment that involves surgically implanting a small capsule of $^{192}$Ir into a tumor. Iridium-192 lies between two stable isotopes of iridium, $^{193}$Ir and $^{191}$Ir. How does this account for the fact that $^{192}$Ir decays by β decay, electron capture, and positron emission?

**23.75.** Gadolinium-153 decays by electron capture. What type of radiation does $^{153}$Gd produce that makes it useful for imaging?

23.76. Gadolinium-153 and samarium-153 both have the same mass number. Why might $^{153}$Gd decay by electron capture, whereas $^{153}$Sm decays by emitting β particles?

**23.77.** How do platinum- and ruthenium-containing drugs fight cancer?

*23.78. Many transition metal complexes are brightly colored. Why might the titanium(IV) compound budotitane be colorless?

***23.79.** Is the glucose tolerance factor that contains chromium(III) paramagnetic or diamagnetic?

23.80. Coordination complexes containing paramagnetic transition metal ions make the best MRI contrast agents.

a. Which of the first-row transition metal cations with a 2+ charge have the most unpaired valence electrons in the gas phase?

b. Which of the first-row transition metal cations with a 2+ charge have the most unpaired valence electrons in octahedral coordination complexes?

### Problems

**23.81.** The complexation of mercury(II) ion with methionine,

$$Hg^{2+} + \text{methionine} \rightleftharpoons \text{Hg(methionine)}^{2+}$$

has a formation constant of $\log K_f = 14.2$, whereas the formation constant for the $Hg^{2+}$ complex with penicillamine,

$$Hg^{2+} + \text{penicillamine} \rightleftharpoons \text{Hg(penicillamine)}^{2+}$$

is $\log K_f = 16.2$. Calculate the equilibrium constant for the reaction

$$\text{Hg(methionine)}^{2+} + \text{penicillamine} \rightleftharpoons \text{Hg(penicillamine)}^{2+} + \text{methionine}$$

23.82. The complexation of mercury(II) ion with cysteine in aqueous solution,

$$Hg^{2+} + \text{cysteine} \rightleftharpoons Hg(\text{cysteine})^{2+}$$

has a formation constant of log $K_f$ =14.2, whereas the formation constant for the $Hg^{2+}$ complex with glycine,

$$Hg^{2+} + \text{glycine} \rightleftharpoons Hg(\text{glycine})^{2+}$$

is log $K_f$ = 10.3. Calculate the equilibrium constant for the reaction

$$Hg(\text{cysteine})^{2+} + \text{glycine} \rightleftharpoons Hg(\text{glycine})^{2+} + \text{cysteine}$$

**23.83.** The equilibrium constant of the reaction

$$CH_3Hg(\text{penicillamine})^+(aq) + \text{cysteine}(aq) \rightleftharpoons CH_3Hg(\text{cysteine})^+(aq) + \text{penicillamine}(aq)$$

is $K = 0.633$. Calculate the equilibrium concentrations of cysteine and penicillamine if we start with a 1.00 *M* solution of cysteine and a 1.00 m*M* solution of $CH_3Hg(\text{penicillamine})^+$.

23.84. The equilibrium constant of the reaction

$$CH_3Hg(\text{glutathione})^+(aq) + \text{cysteine}(aq) \rightleftharpoons CH_3Hg(\text{cysteine})^+(aq) + \text{glutathione}(aq)$$

is $K = 5.0$. Calculate the equilibrium concentrations of cysteine and glutathione if we start with a 1.20 m*M* solution of cysteine and a 1.20 m*M* solution of $CH_3Hg(\text{glutathione})^+$.

**23.85.** The two platinum compounds shown in Figure P23.85 have been studied in cancer therapy.

a. What is the difference between the two compounds?
b. Name the two complex ions.
c. Do both compounds have the same orbital diagram?
d. Sketch the orbital diagram for the platinum compound(s), including the appropriate number of electrons.

OH; $H_3N$, $N_3$; Pt; $H_3N$, $N_3$; OH

OH; $N_3$, $NH_3$; Pt; $H_3N$, $N_3$; OH

**FIGURE P23.85**

23.86. Anticancer treatments based on coordination complexes of ruthenium may have less severe side effects than those based on cisplatin and other platinum-based medications. The complex shown in Figure P23.86 was tested against a colon cancer cell line with promising results.

a. Draw the other stereoisomer of $[Ru(py)_2Cl_4]^-$.
b. Name both isomers.
c. Sketch the orbital diagram for the ruthenium compound, including the appropriate number of electrons.

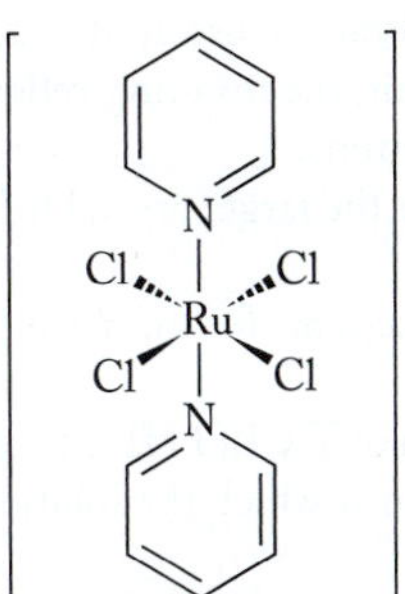


**FIGURE P23.86**

***23.87.** Two square planar gold and platinum complexes with medicinal properties are shown in Figure P23.87.

a. Do these two complexes have the same number of *d* electrons in their orbital diagram?
b. Are these compounds diamagnetic or paramagnetic?

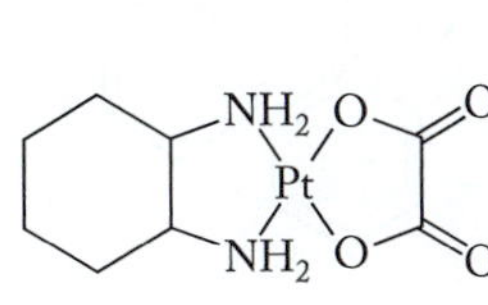


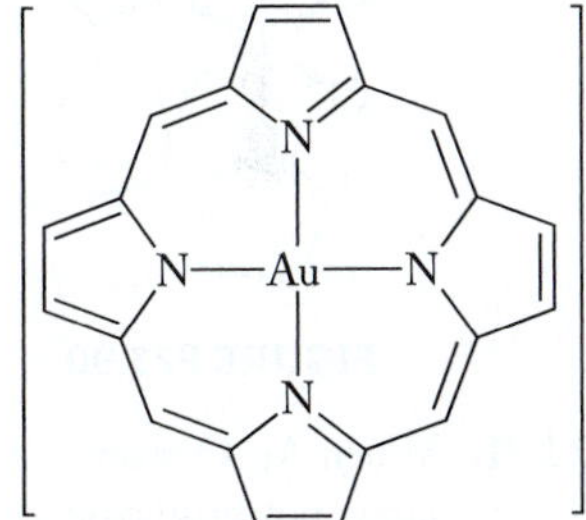


**FIGURE P23.87**

*23.88. In Section 5.5, we saw that polycyclic aromatic compounds can intercalate into DNA. Compounds such as the rhodium and ruthenium complexes shown in Figure P23.88 behave similarly toward DNA.

a. What is the hybridization at the nitrogen and the carbon atoms in these ligands?
b. Why are these ligands planar?
c. Are these compounds diamagnetic or paramagnetic?

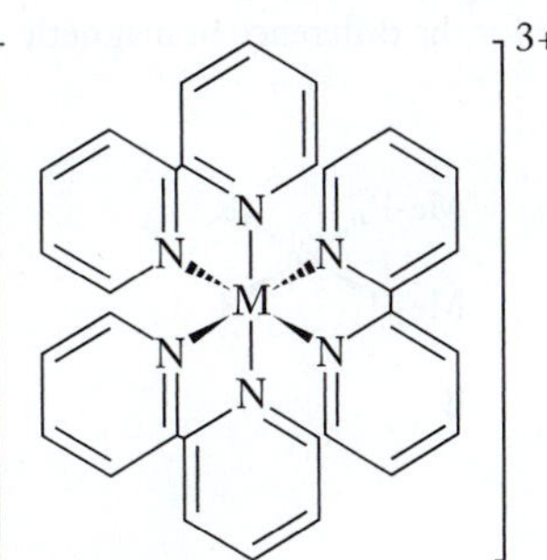


M = Ru, Rh

**FIGURE P23.88**

## Additional Problems

**23.89.** Dissolving cobalt(II) nitrate in water gives a beautiful purple solution. This cobalt(II) complex has three unpaired electrons. When cobalt(II) nitrate is dissolved in aqueous ammonia and oxidized with air, the resulting yellow complex has no unpaired electrons.

a. Which cobalt complex has the larger crystal field splitting energy $\Delta_o$?

b. Draw molecular orbital diagrams for the *d* orbitals of both complexes.

23.90. Adding pyridine to a solution of $TiCl_3(THF)_3$ (Figure P23.90) causes the wavelength at which the solution absorbs light to decrease.

a. Is this consistent with the formation of $TiCl_3(py)_3$?

b. Figure P23.90 shows one possible arrangement of three chloride ligands and three THF or three pyridine ligands. Draw another geometric isomer of these compounds.

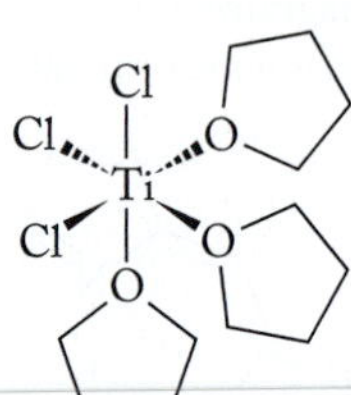


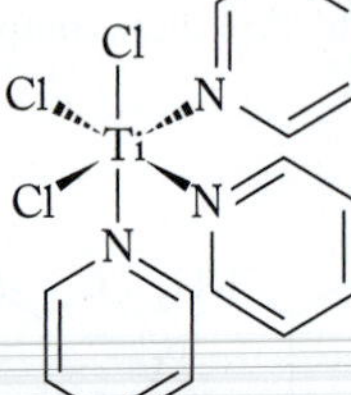


**FIGURE P23.90**

**23.91.** When $Ag_2O$ reacts with peroxodisulfate ($S_2O_8^{2-}$) ion (a powerful oxidizing agent), AgO is produced. Crystallographic and magnetic analyses of AgO suggest that it is not simply silver(II) oxide but rather a blend of silver(I) and silver(III) in a square planar environment. The $Ag^{2+}$ ion is paramagnetic but, like AgO, $Ag^+$ and $Ag^{3+}$ are diamagnetic. Explain why.

*23.92. Two iron compounds **A** and **B** in Figure P23.92 both contain $Fe^{2+}$ (R is a monoanion and the phosphorus ligands are neutral). Compound **A** has a tetrahedral geometry at iron and four unpaired electrons. Compound **B** has a square planar geometry at iron and two unpaired electrons. Account for the difference in magnetic properties.

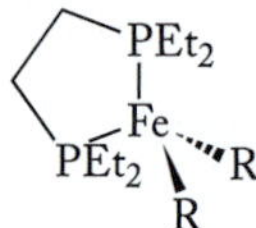


Me3P, Me3P, Fe, R, R

**A** **B**

**FIGURE P23.92**

**23.93.** **Ruthenium Drugs** One of the earliest ruthenium compounds to be investigated for medicinal use was *trans*-$RuCl_2(DMSO)_4$, where DMSO represents dimethyl sulfoxide, $(CH_3)_2S{=}O$. When dissolved in water, some of the ligands are replaced by water to form $[RuCl(DMSO)_2(H_2O)_3]^+$:

$$trans\text{-}RuCl_2(DMSO)_4 + H_2O \rightarrow [RuCl(DMSO)_2(H_2O)_3]Cl + 2\ DMSO$$

a. The product has an octahedral geometry (coordination number = 6) and can exist as two geometric isomers. Draw these isomers.

*b. Why might the starting material and product have linkage isomers?

23.94. When the compound in Figure P23.94 is exposed to light at 400 nm, it undergoes linkage isomerism.

a. Draw the linkage isomer.

b. Are either or both of the isomers chiral?

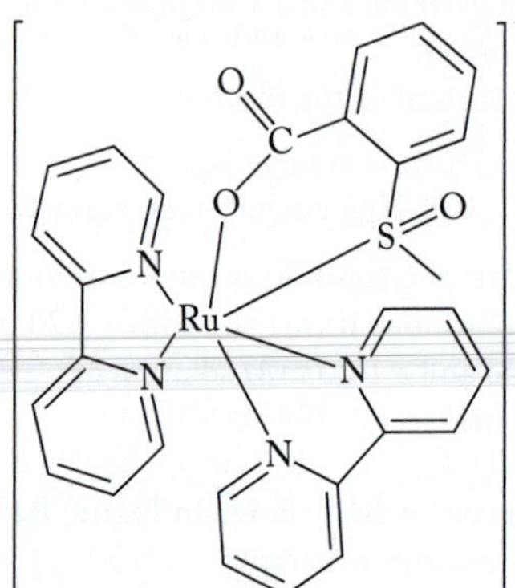


**FIGURE P23.94**

**23.95.** Manganese(II) chloride was one of the first compounds to be investigated as an MRI contrast agent. How many unpaired electrons does the complex have when $MnCl_2$ dissolves in water to form the coordination compound $MnCl_2(H_2O)_4$?

23.96. Gadolinium(III) ions are used in contrast agents for MRI because they have unpaired electrons. What is the electron configuration for $Gd^{3+}$, and how many unpaired electrons does it have?

# Appendix 1

## Mathematical Procedures

### Working with Scientific Notation

Quantities that scientists work with are sometimes very large, such as Earth's mass, and other times very small, such as the mass of an electron. It is easier to work with these values when they are expressed in scientific notation.

The general form of standard scientific notation is a value between 1 and 10 multipled by 10 raised to an integral power. According to this definition, $598 \times 10^{22}$ kg (Earth's mass) is not in standard scientific notation, but $5.98 \times 10^{24}$ kg is. It is good practice to use and report values in standard scientific notation.

1. **To convert an "ordinary" number to standard scientific notation,** move the decimal point to the left for a large number, or to the right for a small one, so that the decimal point is located after the first nonzero digit.
   A. For example, to express Earth's average density (5517 kg/m$^3$) in scientific notation requires moving the decimal point three places to the left. Doing so is the same as dividing the number by 1000, or $10^3$. To keep the value the same we multiply it by $10^3$. So, Earth's density in standard scientific notation is $5.517 \times 10^3$ kg/m$^3$.
   B. If we move the decimal point of a value less than 1 to the right to express it in scientific notation, then the exponent is a negative integer equal to the number of places we moved the decimal point to the right. For example, the value of $R$ used in solving ideal gas law problems is 0.08206 L · atm/(mol · K). Moving the decimal point two places to the right converts the value of $R$ to scientific notation: $8.206 \times 10^{-2}$ L · atm/(mol · K).
   C. Another value of $R$, 8.314 J/(mol · K), does not need an exponent, though it could be written $8.314 \times 10^0$ J/(mol · K) because any value raised to the zero power is equal to 1.

2. **For calculations with numbers in scientific notation,** most calculators have a function key for entering the exponents of values expressed in scientific notation. In many calculators it is labeled "E" or "EE" or "Exp." To enter, for example, the speed of light in meters per second, $2.998 \times 10^8$, we enter the value before the exponent, 2.998, followed by the exponent key and then the value of the exponent (8). To enter a value with a negative exponent, use the sign-change key. Sometimes it is labeled "(−)" or "+/−" or "±."

### Working with Logarithms

A logarithm to the base 10 has the following form:

$$\log_{10} x = \log x = a, \text{ where } x = 10^a$$

We usually abbreviate the logarithm function "log" if the logarithm is to the base 10, which means the scale in which log 10 = 1.

A logarithm to the base $e$, called a *natural logarithm*, has the following form:

$$\log_e x = \ln x = b, \text{ where } x = e^b$$

Scientific calculators have (LOG) and (LN) keys, so it is easy to convert a number into its log or ln form. The directions below apply to most calculators.

**Sample Exercise 1** Find the logarithm to the base 10 of 2.247 (log 2.247).

**Solution** In some calculators you enter 2.247 first and then press the (LOG) key. In others, such as the TI 84/89 series, you press the (LOG) key first followed by 2.247 and then the (ENTER) key. Either way, the answer should be 0.3516 (to four significant figures).

**Sample Exercise 2** Find the natural logarithm of 2.247 (ln 2.247).

**Solution** Follow the same procedure as in Sample Exercise 1 except use the (LN) key. The answer should be 0.8096. This answer is (0.8096/0.3516) = 2.303 times larger than log value. That is,

$$\ln x = 2.303 \log x$$

**Sample Exercise 3** Find the log of $6.0221 \times 10^{23}$.

**Solution** Following the procedures described above for entering a value with an exponent into your calculator and then taking its log to the base 10, you should obtain the value 23.77974796. We know the original value to five significant figures; the log value should have the same precision, but how do we express it? You might think the log value should be 23.780; however, the 23 to the left of the decimal point reflects the value of the exponent in the original value, and exponents don't count in determining the number of significant figures in a value. Therefore, the value of the logarithm to five significant figures is 23.77975. To understand better why this is so, calculate the log of 6.0221. You should get 0.77975 to five significant figures. Note how the difference between the two log values (23.77975 and 0.77975) is simply the value of the exponent in the first value.

**Combining Logs** The following equations summarize how logarithms of the products or quotients of two or more values are related to the individual logs of those values:

$$\text{logarithm } (ab) = \text{logarithm } a + \text{logarithm } b$$

and

$$\text{logarithm } (a/b) = \text{logarithm } a - \text{logarithm } b$$

## Converting Logarithms into Numbers

If we know the value of log $x$, what is the value of $x$? This question is frequently asked when working with pH, which is the negative log of the concentration of hydrogen ions, $[H_3O^+]$, in solution:

$$\text{pH} = -\log[H_3O^+]$$

Suppose the pH of a solution of a weak acid is 2.50. The concentration of $H_3O^+$ is related to this pH value as follows:

$$2.50 = -\log[H_3O^+]$$

or

$$-2.50 = \log[H_3O^+]$$

To find the value of $[H_3O^+]$, we enter 2.5 into the calculator and press the appropriate key to change its sign to −2.5. The next step depends on the type of calculator. If yours has a ($10^x$) key (often accessible using a second function key), use it to find the value of $10^{-2.5}$, which is the value we are looking for. The corresponding keystrokes with many graphing calculators are ($10^x$), (−), 2.5, (ENTER). Some calculators, including the virtual one in many Windows operating systems, have an ($x^y$) key. We can use it for this problem by entering 10 and pushing the ($x^y$) key, then entering 2.5 and pushing the (+/−) key followed by the (=) key. All of these approaches do the same calculation, taking 10 to the −2.50 power, and give the same answer, $[H_3O^+] = 3.2 \times 10^{-3}$ to two significant figures. (Remember, pH is a log value, so the digit before the decimal point is not significant.)

## Solving Quadratic Equations

If the terms in an equation can be rearranged so that they take the form

$$ax^2 + bx + c = 0$$

they have the form of a quadratic equation. The value(s) of $x$ can be determined from the values of the coefficients $a$, $b$, and $c$ by using the equation

$$x = \frac{-b \pm \sqrt{b^2 - 4ac}}{2a}$$

For example, if the solution to a problem yields the following expression where $x$ is the concentration of a solute:

$$x^2 + 0.112x - 1.2 \times 10^{-3} = 0$$

Then the value of $x$ can be determined as follows:

$$x = \frac{-b \pm \sqrt{b^2 - 4ac}}{2a}$$

$$= \frac{-0.112 \pm \sqrt{(0.112)^2 - 4(1)(-1.2 \times 10^{-3})}}{2(1)}$$

$$= \frac{-0.112 \pm \sqrt{0.01254 + 0.0048}}{2}$$

$$= \frac{-0.112 \pm 0.132}{2} = +0.010 \text{ or } -0.122$$

In this example, the negative value for $x$ satisfies the equation, but it has no meaning because we cannot have negative concentration values; therefore we use only the +0.010 value.

## Expressing Data in Graphical Form

Fitting curves to plots of experimental data is a powerful tool in determining the relationships between variables. Many natural phenomena obey exponential functions. For example, the rate constant ($k$) of a chemical reaction increases exponentially with increasing absolute temperature ($T$). This relationship is described by the Arrhenius equation:

$$k = A\,e^{-E_a/RT}$$

where $A$ is a constant for a particular reaction (called the frequency factor), $E_a$ is the activation energy of the reaction, and $R$ is the ideal gas constant. Taking the natural logarithms of both sides of the Arrhenius equation gives

$$\ln k = \ln A - \left(\frac{E_a}{RT}\right)$$

This equation fits the general equation of a straight line ($y = mx + b$) if ($\ln k$) is the $y$-variable and ($1/T$) is the $x$-variable. Plotting ($\ln k$) versus ($1/T$) should give a straight line with a slope equal to $-E_a/R$. The slopes of these plots are negative because the activation energies, $E_a$, of chemical reactions are positive. The data for a reaction given in columns 2 and 4 of Table A1.1 are plotted in Figure A1.1. The program that generated the graph also gives us the equation of the straight line that best fits the data. The slope (−1281 K) of this line is used to calculate the value of $E_a$:

$$-1281 \text{ K} = -\frac{E_a}{R}$$

$$E_a = -(-1281 \text{ K})[8.314 \text{ J/(mol} \cdot \text{K)}]$$

$$= 10{,}650 \text{ J/mol} = 10.65 \text{ kJ/mol}$$

## Expressing Precision and Accuracy

The precision in the results of a set of replicate measurements or analyses is determined by calculating the arithmetic mean ($\bar{x}$) of the data set and its standard deviation ($s$), which is a measure of the variation between the mean and the results of each measurement or analysis ($x_i$). The equation for calculating $s$ is

$$s = \sqrt{\frac{\sum_i (x_i - \bar{x})^2}{n - 1}}$$

**TABLE A1.1** Rate Constant $k$ as a Function of Temperature $T$

| Temperature $T$ (K) | $1/T$ (K$^{-1}$) | Rate Constant $k$ | ln $k$ |
|---|---|---|---|
| 500 | 0.0020 | 0.030 | −3.51 |
| 550 | 0.0018 | 0.38 | −0.97 |
| 600 | 0.0017 | 2.9 | 1.06 |
| 650 | 0.0015 | 17 | 2.83 |
| 700 | 0.0014 | 75 | 4.32 |

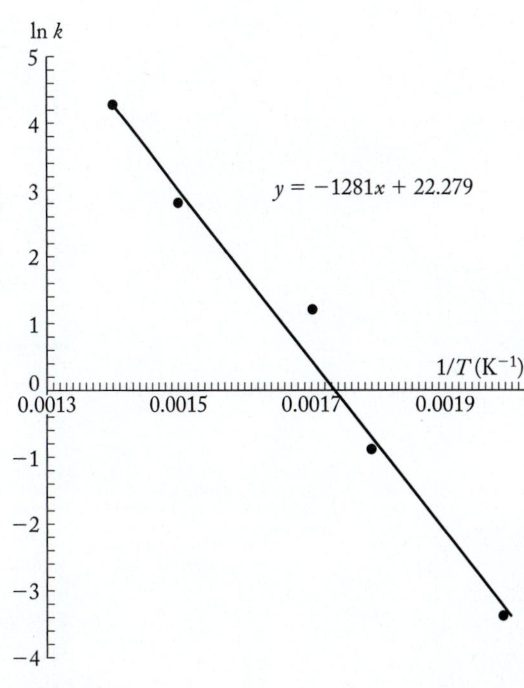


**FIGURE A1.1**

where $n$ is the number of data points. To calculate the confidence interval, which is the range that is predicted to contain the true mean value ($\mu$) of the data set, we use this equation:

$$\mu = \bar{x} \pm \frac{ts}{\sqrt{n}}$$

where $t$ is a statistic that depends on the value of $n$ and the degree of certainty, or *confidence level*, in the prediction. Table A1.2 contains values of $t$ for confidence levels of 90.0, 95.0, 99.0, and 99.9%.

**TABLE A1.2** Values of *t*

| | CONFIDENCE LEVEL | | | |
|---|---|---|---|---|
| $n - 1$ | 90.0% | 95.0% | 99.0% | 99.9% |
| 1 | 6.314 | 12.71 | 63.66 | 636.62 |
| 2 | 2.920 | 4.303 | 9.925 | 31.599 |
| 3 | 2.353 | 3.182 | 5.841 | 12.924 |
| 4 | 2.132 | 2.776 | 4.604 | 8.610 |
| 5 | 2.015 | 2.571 | 4.032 | 6.869 |
| 6 | 1.943 | 2.447 | 3.707 | 5.959 |
| 7 | 1.895 | 2.365 | 3.499 | 5.408 |
| 8 | 1.860 | 2.306 | 3.355 | 5.041 |
| 9 | 1.833 | 2.262 | 3.250 | 4.781 |
| 10 | 1.812 | 2.228 | 3.169 | 4.587 |
| 11 | 1.796 | 2.201 | 3.106 | 4.437 |
| 12 | 1.782 | 2.179 | 3.055 | 4.318 |
| 13 | 1.771 | 2.160 | 3.012 | 4.221 |
| 14 | 1.761 | 2.145 | 2.977 | 4.140 |
| 15 | 1.753 | 2.131 | 2.947 | 4.073 |
| 16 | 1.746 | 2.120 | 2.921 | 4.015 |
| 17 | 1.740 | 2.110 | 2.898 | 3.965 |
| 18 | 1.734 | 2.101 | 2.878 | 3.922 |
| 19 | 1.729 | 2.093 | 2.861 | 3.883 |
| 20 | 1.725 | 2.086 | 2.845 | 3.850 |
| 25 | 1.708 | 2.060 | 2.787 | 3.725 |
| 30 | 1.697 | 2.042 | 2.750 | 3.646 |
| 40 | 1.684 | 2.021 | 2.704 | 3.551 |
| 60 | 1.671 | 2.000 | 2.660 | 3.460 |
| 80 | 1.664 | 1.990 | 2.639 | 3.416 |
| 100 | 1.660 | 1.984 | 2.626 | 3.390 |
| $\infty$ | 1.645 | 1.960 | 2.576 | 3.291 |

# Combining Vector Quantities

Unequal sharing of the electrons in covalent bonds between atoms of different elements produces partial negative and positive charges on the ends of those bonds. This separation of partial charge creates a *bond dipole*. The quantity called a *dipole moment* assigns a value to the magnitude of the negative and positive charges that are separated and to the distance between the charges. Polar bonds have dipole moments and so, too, do polar molecules.

Dipole moments are *vector* quantities. A vector has two components: *magnitude*, which is the product of the partial charges times the distance between them, and *direction*, which is usually indicated by an arrow, ⟷, pointing toward the negative end of a polar bond, or toward the negative side of a polar molecule. Adding the dipole moments of all the polar bonds in a molecule allows us to predict whether the molecule itself is polar—that is, whether it has a permanent dipole.

For example, a molecule of $CO_2$ has two polar C═O bonds with dipole moments directed toward the oxygen atoms at each end of the molecule. The two dipole moments are equivalent because they involve the same kinds of atoms separated by the same kind of bond, as shown in the Lewis structure below. However, the linear shape of the molecule means that the direction of one C═O dipole is opposite the direction of the other, so the vectors corresponding to the two dipole moments offset each other, as shown in the graph in the middle image. As a result, a molecule of $CO_2$ has no permanent dipole, which means that carbon dioxide is a nonpolar compound.

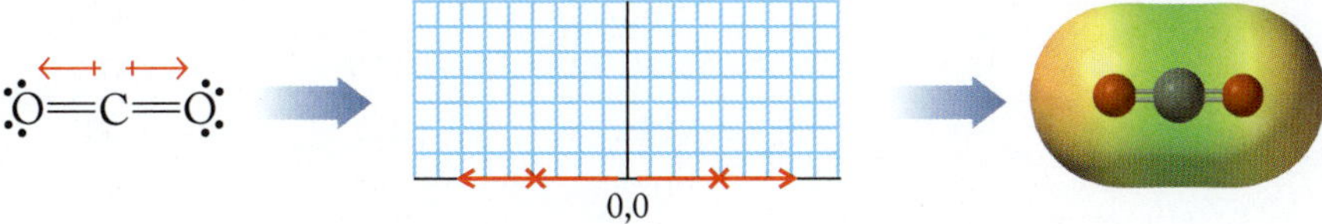


On the other hand, a molecule of $H_2O$ has a bent molecular geometry, as shown in the Lewis structure below. Therefore, the dipole moments of the two identical O—H bonds *do not completely* offset each other: their *x* components (shown in blue in the graph below) point directly toward each other, and they do offset each other. However, their *y* components (also shown in blue) are pointed in the same direction, and they add together as shown, giving the molecule a permanent dipole with a positive end centered between the H atoms and a negative pole in the O atom:

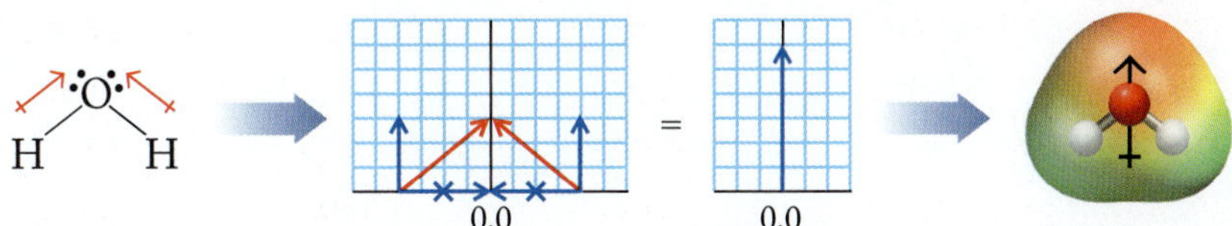


The above method of adding the *x* and *y* components of the vectors corresponding to dipole moments can be extended to three-dimensional molecular structures such as the trigonal pyramidal structure of $NH_3$. In the figure below, three: N—H bond dipoles are shown for a molecule of $NH_3$ drawn with its N atom at the top. Its three H atoms form a trigonal base in a two-dimensional plane as shown in the vector diagram. One component of each bond dipole vector is directed from the H atom toward the center of the base, directly below the N atom. All these vectors are spaced equally about the center and have the same magnitude. Therefore, they offset each other. However, the vertical components of all three dipole moments point in the same direction (upward), which means they are additive, giving $NH_3$ a permanent dipole centered between the three H atoms and pointing toward the N atom.

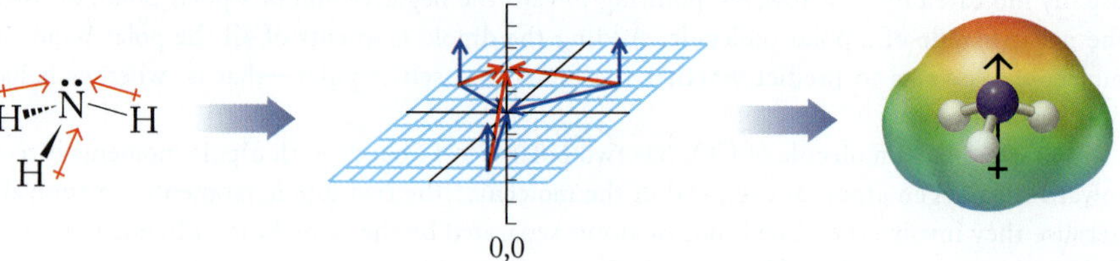

# Appendix 2

## SI Units and Conversion Factors

**TABLE A2.1** Six SI Base Units

| SI Base Quantity | Unit | Symbol |
|---|---|---|
| length | meter | m |
| mass | kilogram | kg |
| time | second | s |
| amount of substance | mole | mol |
| temperature | kelvin | K |
| electric current | ampere | A |

**TABLE A2.2** Some SI-Derived Units

| SI-Derived Quantity | Unit | Symbol | Dimensions |
|---|---|---|---|
| electric charge | coulomb | C | A · s |
| electric potential | volt | V | J/C |
| force | newton | N | kg · m/s$^2$ |
| frequency | hertz | Hz | s$^{-1}$ |
| momentum | newton-second | N · s | kg · m/s |
| power | watt | W | J/s |
| pressure | pascal | Pa | N/m$^2$ |
| radioactivity | becquerel | Bq | s$^{-1}$ |
| speed or velocity | meter per second | m/s | m/s |
| energy | joule (newton-meter) | J (N · m) | kg · m$^2$/s$^2$ |

**TABLE A2.3** SI Prefixes

| Prefix | Symbol | Multiplier | Prefix | Symbol | Multiplier |
|---|---|---|---|---|---|
| deci | d | $10^{-1}$ | deka | da | $10^{1}$ |
| centi | c | $10^{-2}$ | hecto | h | $10^{2}$ |
| milli | m | $10^{-3}$ | kilo | k | $10^{3}$ |
| micro | μ | $10^{-6}$ | mega | M | $10^{6}$ |
| nano | n | $10^{-9}$ | giga | G | $10^{9}$ |
| pico | p | $10^{-12}$ | tera | T | $10^{12}$ |
| femto | f | $10^{-15}$ | peta | P | $10^{15}$ |
| atto | a | $10^{-18}$ | exa | E | $10^{18}$ |
| zepto | z | $10^{-21}$ | zetta | Z | $10^{21}$ |

**TABLE A2.4** Special Units and Conversion Factors

| Quantity | Unit | Symbol | Conversion[a] |
|---|---|---|---|
| energy[a] | electron-volt | eV | 1 eV = $1.6022 \times 10^{-19}$ J |
| energy[a] | kilowatt-hour | kWh | 1 kWh = 3600 kJ |
| energy | calorie | cal | 1 cal = 4.184 J |
| mass | pound | lb | 1 lb = 453.59 g |
| mass[a] | unified atomic mass unit | u | 1 u = $1.66054 \times 10^{-27}$ kg |
| length | angstrom | Å | 1 Å = $10^{-8}$ cm = $10^{-10}$ m |
| length | inch | in | 1 in = 2.54 cm |
| length | mile | mi | 1 mi = 5280 ft = 1.6093 km |
| pressure | atmosphere | atm | 1 atm = $1.01325 \times 10^{5}$ Pa |
| pressure | torr | torr | 1 torr = 1/760 atm |
| temperature | Celsius scale | °C | $T$(°C) = $T$(K) − 273.15 |
| temperature | Fahrenheit scale | °F | $T$(°F) = $\frac{9}{5}$ $T$(°C) + 32 |
| volume | liter | L | 1 L = 1 $dm^3$ = $10^{-3}$ $m^3$ |
| volume | cubic centimeter | $cm^3$, cc | 1 $cm^3$ = 1 mL = $10^{-3}$ L |
| volume | cubic foot | $ft^3$ | 1 $ft^3$ = 7.4805 gal |
| volume | gallon (U.S.) | gal | 1 gal = 3.785 L |

[a]From http://physics.nist.gov/cuu/constants/.

**TABLE A2.5** Physical Constants[a]

| Quantity | Symbol | Value |
|---|---|---|
| acceleration due to gravity (Earth) | $g$ | 9.807 m/$s^2$ |
| Avogadro constant | $N_A$ | $6.0221 \times 10^{23}$ $mol^{-1}$ |
| Bohr radius | $a_0$ | $5.29 \times 10^{-11}$ m |
| Boltzmann constant | $k_B$ | $1.3806 \times 10^{-23}$ J/K |
| electron charge-to-mass ratio | $-e/m_e$ | $1.7588 \times 10^{11}$ C/kg |
| elementary charge | $e$ | $1.602 \times 10^{-19}$ C |
| Faraday constant | $F$ | $9.65 \times 10^{4}$ C/mol |
| mass of an electron | $m_e$ | $9.10938 \times 10^{-31}$ kg |
| mass of a neutron | $m_n$ | $1.67493 \times 10^{-27}$ kg |
| mass of a proton | $m_p$ | $1.67262 \times 10^{-27}$ kg |
| molar volume of ideal gas at 0°C and 1 atm | $V_m$ | 22.4 L/mol |
| Planck constant | $h$ | $6.626 \times 10^{-34}$ J · s |
| speed of light in vacuum | $c$ | $2.998 \times 10^{8}$ m/s |
| universal gas constant | $R$ | 8.314 J/(mol · K)<br>0.08206 L · atm/(mol · K) |

[a]From http://physics.nist.gov/cuu/constants/.

# Appendix 3

## The Elements and Their Properties

**TABLE A3.1** Ground-State Electron Configurations, Atomic Radii, and First Ionization Energies of the Elements

| Element | Symbol | Atomic Number $Z$ | Ground-State Configuration | Atomic Radius (pm) | First Ionization Energy (kJ/mol) |
|---|---|---|---|---|---|
| hydrogen | H | 1 | $1s^1$ | 37 | 1312.0 |
| helium | He | 2 | $1s^2$ | 32 | 2372.3 |
| lithium | Li | 3 | $[He]2s^1$ | 152 | 520.2 |
| beryllium | Be | 4 | $[He]2s^2$ | 112 | 899.5 |
| boron | B | 5 | $[He]2s^22p^1$ | 88 | 800.6 |
| carbon | C | 6 | $[He]2s^22p^2$ | 77 | 1086.5 |
| nitrogen | N | 7 | $[He]2s^22p^3$ | 75 | 1402.3 |
| oxygen | O | 8 | $[He]2s^22p^4$ | 73 | 1313.9 |
| fluorine | F | 9 | $[He]2s^22p^5$ | 71 | 1681.0 |
| neon | Ne | 10 | $[He]2s^22p^6$ | 69 | 2080.7 |
| sodium | Na | 11 | $[Ne]3s^1$ | 186 | 495.3 |
| magnesium | Mg | 12 | $[Ne]3s^2$ | 160 | 737.7 |
| aluminum | Al | 13 | $[Ne]3s^23p^1$ | 143 | 577.5 |
| silicon | Si | 14 | $[Ne]3s^23p^2$ | 117 | 786.5 |
| phosphorus | P | 15 | $[Ne]3s^23p^3$ | 110 | 1011.8 |
| sulfur | S | 16 | $[Ne]3s^23p^4$ | 103 | 999.6 |
| chlorine | Cl | 17 | $[Ne]3s^23p^5$ | 99 | 1251.2 |
| argon | Ar | 18 | $[Ne]3s^23p^6$ | 97 | 1520.6 |
| potassium | K | 19 | $[Ar]4s^1$ | 227 | 418.8 |
| calcium | Ca | 20 | $[Ar]4s^2$ | 197 | 589.8 |
| scandium | Sc | 21 | $[Ar]3d^14s^2$ | 162 | 633.1 |
| titanium | Ti | 22 | $[Ar]3d^24s^2$ | 147 | 658.8 |
| vanadium | V | 23 | $[Ar]3d^34s^2$ | 135 | 650.9 |
| chromium | Cr | 24 | $[Ar]3d^54s^1$ | 128 | 652.9 |
| manganese | Mn | 25 | $[Ar]3d^54s^2$ | 127 | 717.3 |
| iron | Fe | 26 | $[Ar]3d^64s^2$ | 126 | 762.5 |
| cobalt | Co | 27 | $[Ar]3d^74s^2$ | 125 | 760.4 |
| nickel | Ni | 28 | $[Ar]3d^84s^2$ | 124 | 737.1 |
| copper | Cu | 29 | $[Ar]3d^{10}4s^1$ | 128 | 745.5 |
| zinc | Zn | 30 | $[Ar]3d^{10}4s^2$ | 134 | 906.4 |
| gallium | Ga | 31 | $[Ar]3d^{10}4s^24p^1$ | 135 | 578.8 |
| germanium | Ge | 32 | $[Ar]3d^{10}4s^24p^2$ | 122 | 762.2 |
| arsenic | As | 33 | $[Ar]3d^{10}4s^24p^3$ | 121 | 947.0 |
| selenium | Se | 34 | $[Ar]3d^{10}4s^24p^4$ | 119 | 941.0 |

*Continued on next page*

**TABLE A3.1 Ground-State Electron Configurations, Atomic Radii, and First Ionization Energies of the Elements *(Continued)***

| Element | Symbol | Atomic Number $Z$ | Ground-State Configuration | Atomic Radius (pm) | First Ionization Energy (kJ/mol) |
|---|---|---|---|---|---|
| bromine | Br | 35 | $[Ar]3d^{10}4s^2 4p^5$ | 114 | 1139.9 |
| krypton | Kr | 36 | $[Ar]3d^{10}4s^2 4p^6$ | 110 | 1350.8 |
| rubidium | Rb | 37 | $[Kr]5s^1$ | 247 | 403.0 |
| strontium | Sr | 38 | $[Kr]5s^2$ | 215 | 549.5 |
| yttrium | Y | 39 | $[Kr]4d^1 5s^2$ | 180 | 599.8 |
| zirconium | Zr | 40 | $[Kr]4d^2 5s^2$ | 160 | 640.1 |
| niobium | Nb | 41 | $[Kr]4d^4 5s^1$ | 146 | 652.1 |
| molybdenum | Mo | 42 | $[Kr]4d^5 5s^1$ | 139 | 684.3 |
| technetium | Tc | 43 | $[Kr]4d^5 5s^2$ | 136 | 702.4 |
| ruthenium | Ru | 44 | $[Kr]4d^7 5s^1$ | 134 | 710.2 |
| rhodium | Rh | 45 | $[Kr]4d^8 5s^1$ | 134 | 719.7 |
| palladium | Pd | 46 | $[Kr]4d^{10}$ | 137 | 804.4 |
| silver | Ag | 47 | $[Kr]4d^{10}5s^1$ | 144 | 731.0 |
| cadmium | Cd | 48 | $[Kr]4d^{10}5s^2$ | 151 | 867.8 |
| indium | In | 49 | $[Kr]4d^{10}5s^2 5p^1$ | 167 | 558.3 |
| tin | Sn | 50 | $[Kr]4d^{10}5s^2 5p^2$ | 140 | 708.6 |
| antimony | Sb | 51 | $[Kr]4d^{10}5s^2 5p^3$ | 141 | 833.6 |
| tellurium | Te | 52 | $[Kr]4d^{10}5s^2 5p^4$ | 143 | 869.3 |
| iodine | I | 53 | $[Kr]4d^{10}5s^2 5p^5$ | 133 | 1008.4 |
| xenon | Xe | 54 | $[Kr]4d^{10}5s^2 5p^6$ | 130 | 1170.4 |
| cesium | Cs | 55 | $[Xe]6s^1$ | 265 | 375.7 |
| barium | Ba | 56 | $[Xe]6s^2$ | 222 | 502.9 |
| lanthanum | La | 57 | $[Xe]5d^1 6s^2$ | 187 | 538.1 |
| cerium | Ce | 58 | $[Xe]4f^1 5d^1 6s^2$ | 182 | 534.4 |
| praseodymium | Pr | 59 | $[Xe]4f^3 6s^2$ | 182 | 527.2 |
| neodymium | Nd | 60 | $[Xe]4f^4 6s^2$ | 181 | 533.1 |
| promethium | Pm | 61 | $[Xe]4f^5 6s^2$ | 183 | 535.5 |
| samarium | Sm | 62 | $[Xe]4f^6 6s^2$ | 180 | 544.5 |
| europium | Eu | 63 | $[Xe]4f^7 6s^2$ | 208 | 547.1 |
| gadolinium | Gd | 64 | $[Xe]4f^7 5d^1 6s^2$ | 180 | 593.4 |
| terbium | Tb | 65 | $[Xe]4f^9 6s^2$ | 177 | 565.8 |
| dysprosium | Dy | 66 | $[Xe]4f^{10}6s^2$ | 178 | 573.0 |
| holmium | Ho | 67 | $[Xe]4f^{11}6s^2$ | 176 | 581.0 |
| erbium | Er | 68 | $[Xe]4f^{12}6s^2$ | 176 | 589.3 |
| thulium | Tm | 69 | $[Xe]4f^{13}6s^2$ | 176 | 596.7 |
| ytterbium | Yb | 70 | $[Xe]4f^{14}6s^2$ | 193 | 603.4 |
| lutetium | Lu | 71 | $[Xe]4f^{14}5d^1 6s^2$ | 174 | 523.5 |
| hafnium | Hf | 72 | $[Xe]4f^{14}5d^2 6s^2$ | 159 | 658.5 |
| tantalum | Ta | 73 | $[Xe]4f^{14}5d^3 6s^2$ | 146 | 761.3 |
| tungsten | W | 74 | $[Xe]4f^{14}5d^4 6s^2$ | 139 | 770.0 |
| rhenium | Re | 75 | $[Xe]4f^{14}5d^5 6s^2$ | 137 | 760.3 |
| osmium | Os | 76 | $[Xe]4f^{14}5d^6 6s^2$ | 135 | 839.4 |

**TABLE A3.1** Ground-State Electron Configurations, Atomic Radii, and First Ionization Energies of the Elements *(Continued)*

| Element | Symbol | Atomic Number $Z$ | Ground-State Configuration | Atomic Radius (pm) | First Ionization Energy (kJ/mol) |
|---|---|---|---|---|---|
| iridium | Ir | 77 | $[Xe]4f^{14}5d^{7}6s^{2}$ | 136 | 878.0 |
| platinum | Pt | 78 | $[Xe]4f^{14}5d^{9}6s^{1}$ | 139 | 868.4 |
| gold | Au | 79 | $[Xe]4f^{14}5d^{10}6s^{1}$ | 144 | 890.1 |
| mercury | Hg | 80 | $[Xe]4f^{14}5d^{10}6s^{2}$ | 151 | 1007.1 |
| thallium | Tl | 81 | $[Xe]4f^{14}5d^{10}6s^{2}6p^{1}$ | 170 | 589.4 |
| lead | Pb | 82 | $[Xe]4f^{14}5d^{10}6s^{2}6p^{2}$ | 154 | 715.6 |
| bismuth | Bi | 83 | $[Xe]4f^{14}5d^{10}6s^{2}6p^{3}$ | 150 | 703.3 |
| polonium | Po | 84 | $[Xe]4f^{14}5d^{10}6s^{2}6p^{4}$ | 167 | 812.1 |
| astatine | At | 85 | $[Xe]4f^{14}5d^{10}6s^{2}6p^{5}$ | 140 | 924.6 |
| radon | Rn | 86 | $[Xe]4f^{14}5d^{10}6s^{2}6p^{6}$ | 145 | 1037.1 |
| francium | Fr | 87 | $[Rn]7s^{1}$ | 242 | 380 |
| radium | Ra | 88 | $[Rn]7s^{2}$ | 211 | 509.3 |
| actinium | Ac | 89 | $[Rn]6d^{1}7s^{2}$ | 188 | 499 |
| thorium | Th | 90 | $[Rn]6d^{2}7s^{2}$ | 179 | 587 |
| protactinium | Pa | 91 | $[Rn]5f^{2}6d^{1}7s^{2}$ | 163 | 568 |
| uranium | U | 92 | $[Rn]5f^{3}6d^{1}7s^{2}$ | 156 | 587 |
| neptunium | Np | 93 | $[Rn]5f^{4}6d^{1}7s^{2}$ | 155 | 597 |
| plutonium | Pu | 94 | $[Rn]5f^{6}7s^{2}$ | 159 | 585 |
| americium | Am | 95 | $[Rn]5f^{7}7s^{2}$ | 173 | 578 |
| curium | Cm | 96 | $[Rn]5f^{7}6d^{1}7s^{2}$ | 174 | 581 |
| berkelium | Bk | 97 | $[Rn]5f^{9}7s^{2}$ | 170 | 601 |
| californium | Cf | 98 | $[Rn]5f^{10}7s^{2}$ | 186 | 608 |
| einsteinium | Es | 99 | $[Rn]5f^{11}7s^{2}$ | 186 | 619 |
| fermium | Fm | 100 | $[Rn]5f^{12}7s^{2}$ | 167 | 627 |
| mendelevium | Md | 101 | $[Rn]5f^{13}7s^{2}$ | 173 | 635 |
| nobelium | No | 102 | $[Rn]5f^{14}7s^{2}$ | 176 | 642 |
| lawrencium | Lr | 103 | $[Rn]5f^{14}6s^{1}7s^{2}$ | 161 | — |
| rutherfordium | Rf | 104 | $[Rn]5f^{14}6d^{2}7s^{2}$ | 157 | — |
| dubnium | Db | 105 | $[Rn]5f^{14}6d^{3}7s^{2}$ | 149 | — |
| seaborgium | Sg | 106 | $[Rn]5f^{14}6d^{4}7s^{2}$ | 143 | — |
| bohrium | Bh | 107 | $[Rn]5f^{14}6d^{5}7s^{2}$ | 141 | — |
| hassium | Hs | 108 | $[Rn]5f^{14}6d^{6}7s^{2}$ | 134 | — |
| meitnerium | Mt | 109 | $[Rn]5f^{14}6d^{7}7s^{2}$ | 129 | — |
| darmstadtium | Ds | 110 | $[Rn]5f^{14}6d^{8}7s^{2}$ | 128 | — |
| roentgenium | Rg | 111 | $[Rn]5f^{14}6d^{9}7s^{2}$ | 121 | — |
| copernicium | Cn | 112 | $[Rn]5f^{14}6d^{10}7s^{2}$ | 122 | — |
| nihonium | Nh | 113 | $[Rn]5f^{14}6d^{10}7s^{2}7p^{1}$ | 136 | — |
| flerovium | Fl | 114 | $[Rn]5f^{14}6d^{10}7s^{2}7p^{2}$ | 143 | — |
| moscovium | Mc | 115 | $[Rn]5f^{14}6d^{10}7s^{2}7p^{3}$ | 162 | — |
| livermorium | Lv | 116 | $[Rn]5f^{14}6d^{10}7s^{2}7p^{4}$ | 175 | — |
| tennessine | Ts | 117 | $[Rn]5f^{14}6d^{10}7s^{2}7p^{5}$ | 165 | — |
| oganesson | Og | 118 | $[Rn]5f^{14}6d^{10}7s^{2}7p^{6}$ | 157 | — |

**TABLE A3.2** Miscellaneous Physical Properties of the Naturally Occurring Elements[a]

| Element | Symbol | Atomic Number | Physical State[b,c] | Density[d] (g/cm$^3$) | Melting Point (°C) | Boiling Point (°C) |
|---|---|---|---|---|---|---|
| hydrogen | H | 1 | gas | 0.000090 | −259.14 | −252.87 |
| helium | He | 2 | gas | 0.000179 | <−272.2 | −268.93 |
| lithium | Li | 3 | solid | 0.534 | 180.5 | 1347 |
| beryllium | Be | 4 | solid | 1.848 | 1283 | 2484 |
| boron | B | 5 | solid | 2.34 | 2300 | 3650 |
| carbon | C | 6 | solid (gr) | 1.9–2.3 | ~3350 | sublimes |
| nitrogen | N | 7 | gas | 0.00125 | −210.00 | −195.8 |
| oxygen | O | 8 | gas | 0.00143 | −218.8 | −182.95 |
| fluorine | F | 9 | gas | 0.00170 | −219.62 | −188.12 |
| neon | Ne | 10 | gas | 0.00090 | −248.59 | −246.08 |
| sodium | Na | 11 | solid | 0.971 | 97.72 | 883 |
| magnesium | Mg | 12 | solid | 1.738 | 650 | 1090 |
| aluminum | Al | 13 | solid | 2.6989 | 660.32 | 2467 |
| silicon | Si | 14 | solid | 2.33 | 1414 | 2355 |
| phosphorus | P | 15 | solid (wh) | 1.82 | 44.15 | 280 |
| sulfur | S | 16 | solid | 2.07 | 115.21 | 444.60 |
| chlorine | Cl | 17 | gas | 0.00321 | −101.5 | −34.04 |
| argon | Ar | 18 | gas | 0.00178 | −189.3 | −185.9 |
| potassium | K | 19 | solid | 0.862 | 63.28 | 759 |
| calcium | Ca | 20 | solid | 1.55 | 842 | 1484 |
| scandium | Sc | 21 | solid | 2.989 | 1541 | 2380 |
| titanium | Ti | 22 | solid | 4.54 | 1668 | 3287 |
| vanadium | V | 23 | solid | 6.11 | 1910 | 3407 |
| chromium | Cr | 24 | solid | 7.19 | 1857 | 2671 |
| manganese | Mn | 25 | solid | 7.3 | 1246 | 1962 |
| iron | Fe | 26 | solid | 7.874 | 1538 | 2750 |
| cobalt | Co | 27 | solid | 8.9 | 1495 | 2870 |
| nickel | Ni | 28 | solid | 8.902 | 1455 | 2730 |
| copper | Cu | 29 | solid | 8.96 | 1084.6 | 2562 |
| zinc | Zn | 30 | solid | 7.133 | 419.53 | 907 |
| gallium | Ga | 31 | solid | 5.904 | 29.76 | 2403 |
| germanium | Ge | 32 | solid | 5.323 | 938.25 | 2833 |
| arsenic | As | 33 | solid (gy) | 5.727 | 614 | sublimes |
| selenium | Se | 34 | solid (gy) | 4.79 | 221 | 685 |
| bromine | Br | 35 | liquid | 3.12 | −7.2 | 58.78 |
| krypton | Kr | 36 | gas | 0.00373 | −157.36 | −153.22 |
| rubidium | Rb | 37 | solid | 1.532 | 39.31 | 688 |
| strontium | Sr | 38 | solid | 2.64 | 777 | 1382 |
| yttrium | Y | 39 | solid | 4.469 | 1526 | 3336 |
| zirconium | Zr | 40 | solid | 6.506 | 1855 | 4409 |
| niobium | Nb | 41 | solid | 8.57 | 2477 | 4744 |
| molybdenum | Mo | 42 | solid | 10.22 | 2623 | 4639 |

**TABLE A3.2** Miscellaneous Physical Properties of the Naturally Occurring Elements[a] *(Continued)*

| Element | Symbol | Atomic Number | Physical State[b,c] | Density[d] (g/cm$^3$) | Melting Point (°C) | Boiling Point (°C) |
|---|---|---|---|---|---|---|
| technetium | Tc | 43 | solid | 11.50 | 2157 | 4538 |
| ruthenium | Ru | 44 | solid | 12.41 | 2334 | 3900 |
| rhodium | Rh | 45 | solid | 12.41 | 1964 | 3695 |
| palladium | Pd | 46 | solid | 12.02 | 1555 | 2963 |
| silver | Ag | 47 | solid | 10.50 | 961.78 | 2212 |
| cadmium | Cd | 48 | solid | 8.65 | 321.07 | 767 |
| indium | In | 49 | solid | 7.31 | 156.60 | 2072 |
| tin | Sn | 50 | solid (wh) | 7.31 | 231.9 | 2270 |
| antimony | Sb | 51 | solid | 6.691 | 630.63 | 1750 |
| tellurium | Te | 52 | solid | 6.24 | 449.5 | 998 |
| iodine | I | 53 | solid | 4.93 | 113.7 | 184.4 |
| xenon | Xe | 54 | gas | 0.00589 | −111.75 | −108.0 |
| cesium | Cs | 55 | solid | 1.873 | 28.44 | 671 |
| barium | Ba | 56 | solid | 3.5 | 727 | 1640 |
| lanthanum | La | 57 | solid | 6.145 | 920 | 3455 |
| cerium | Ce | 58 | solid | 6.770 | 799 | 3424 |
| praseodymium | Pr | 59 | solid | 6.773 | 931 | 3510 |
| neodymium | Nd | 60 | solid | 7.008 | 1016 | 3066 |
| promethium | Pm | 61 | solid | 7.264 | 1042 | ~3000 |
| samarium | Sm | 62 | solid | 7.520 | 1072 | 1790 |
| europium | Eu | 63 | solid | 5.244 | 822 | 1596 |
| gadolinium | Gd | 64 | solid | 7.901 | 1314 | 3264 |
| terbium | Tb | 65 | solid | 8.230 | 1359 | 3221 |
| dysprosium | Dy | 66 | solid | 8.551 | 1411 | 2561 |
| holmium | Ho | 67 | solid | 8.795 | 1472 | 2694 |
| erbium | Er | 68 | solid | 9.066 | 1529 | 2862 |
| thulium | Tm | 69 | solid | 9.321 | 1545 | 1946 |
| ytterbium | Yb | 70 | solid | 6.966 | 824 | 1194 |
| lutetium | Lu | 71 | solid | 9.841 | 1663 | 3393 |
| hafnium | Hf | 72 | solid | 13.31 | 2233 | 4603 |
| tantalum | Ta | 73 | solid | 16.654 | 3017 | 5458 |
| tungsten | W | 74 | solid | 19.3 | 3422 | 5660 |
| rhenium | Re | 75 | solid | 21.02 | 3186 | 5596 |
| osmium | Os | 76 | solid | 22.57 | 3033 | 5012 |
| iridium | Ir | 77 | solid | 22.42 | 2446 | 4130 |
| platinum | Pt | 78 | solid | 21.45 | 1768.4 | 3825 |
| gold | Au | 79 | solid | 19.3 | 1064.18 | 2856 |
| mercury | Hg | 80 | liquid | 13.546 | −38.83 | 356.73 |
| thallium | Tl | 81 | solid | 11.85 | 304 | 1473 |
| lead | Pb | 82 | solid | 11.35 | 327.46 | 1749 |
| bismuth | Bi | 83 | solid | 9.747 | 271.4 | 1564 |

*Continued on next page*

**TABLE A3.2** Miscellaneous Physical Properties of the Naturally Occurring Elements[a] *(Continued)*

| Element | Symbol | Atomic Number | Physical State[b,c] | Density[d] ($g/cm^3$) | Melting Point (°C) | Boiling Point (°C) |
|---|---|---|---|---|---|---|
| polonium | Po | 84 | solid | 9.32 | 254 | 962 |
| astatine | At | 85 | solid | unknown | 302 | 337 |
| radon | Rn | 86 | gas | 0.00973 | −71 | −61.7 |
| francium | Fr | 87 | solid | unknown | 27 | 677 |
| radium | Ra | 88 | solid | 5 | 700 | 1737 |
| actinium | Ac | 89 | solid | 10.07 | 1051 | ~3200 |
| thorium | Th | 90 | solid | 11.72 | 1750 | 4788 |
| protactinium | Pa | 91 | solid | 15.37 | 1572 | unknown |
| uranium | U | 92 | solid | 19.05 | 1132 | 3818 |

[a]For relative atomic masses and alphabetical listing of the elements, see the flyleaf at the front of this volume.
[b]Normal state at 25°C and 1 atm.
[c]Allotropes: gr = graphite, gy = gray, wh = white.
[d]Liquids and solids at 25°C and 1 atm; gases at 0°C and 1 atm (STP).

**TABLE A3.3** A Selection of Stable Isotopes[a]

| Isotope $^AX$ | Natural Abundance (%) | Atomic Number $Z$ | Neutron Number $N$ | Mass Number $A$ | Atomic Mass (u) | Binding Energy per Nucleon (MeV)[b] |
|---|---|---|---|---|---|---|
| $^{1}$H | 99.985 | 1 | 0 | 1 | 1.007825 | — |
| $^{2}$H | 0.015 | 1 | 1 | 2 | 2.014000 | 1.160 |
| $^{3}$He | 0.000137 | 2 | 1 | 3 | 3.016030 | 2.572 |
| $^{4}$He | 99.999863 | 2 | 2 | 4 | 4.002603 | 7.075 |
| $^{6}$Li | 7.5 | 3 | 3 | 6 | 6.015121 | 5.333 |
| $^{7}$Li | 92.5 | 3 | 4 | 7 | 7.016003 | 5.606 |
| $^{9}$Be | 100.0 | 4 | 5 | 9 | 9.012182 | 6.463 |
| $^{10}$B | 19.9 | 5 | 5 | 10 | 10.012937 | 6.475 |
| $^{11}$B | 80.1 | 5 | 6 | 11 | 11.009305 | 6.928 |
| $^{12}$C | 98.90 | 6 | 6 | 12 | 12.000000 | 7.680 |
| $^{13}$C | 1.10 | 6 | 7 | 13 | 13.003355 | 7.470 |
| $^{14}$N | 99.634 | 7 | 7 | 14 | 14.003074 | 7.476 |
| $^{15}$N | 0.366 | 7 | 8 | 15 | 15.000108 | 7.699 |
| $^{16}$O | 99.762 | 8 | 8 | 16 | 15.994915 | 7.976 |
| $^{17}$O | 0.038 | 8 | 9 | 17 | 16.999131 | 7.751 |
| $^{18}$O | 0.200 | 8 | 10 | 18 | 17.999160 | 7.767 |
| $^{19}$F | 100.0 | 9 | 10 | 19 | 18.998403 | 7.779 |
| $^{20}$Ne | 90.48 | 10 | 10 | 20 | 19.992435 | 8.032 |
| $^{21}$Ne | 0.27 | 10 | 11 | 21 | 20.993843 | 7.972 |
| $^{22}$Ne | 9.25 | 10 | 12 | 22 | 21.991383 | 8.081 |
| $^{23}$Na | 100.0 | 11 | 12 | 23 | 22.989770 | 8.112 |
| $^{24}$Mg | 78.99 | 12 | 12 | 24 | 23.985042 | 8.261 |
| $^{25}$Mg | 10.00 | 12 | 13 | 25 | 24.985837 | 8.223 |
| $^{26}$Mg | 11.01 | 12 | 14 | 26 | 25.982593 | 8.334 |

**TABLE A3.3** A Selection of Stable Isotopes[a] *(Continued)*

| Isotope $^{A}X$ | Natural Abundance (%) | Atomic Number $Z$ | Neutron Number $N$ | Mass Number $A$ | Atomic Mass (u) | Binding Energy per Nucleon (MeV)[b] |
|---|---|---|---|---|---|---|
| $^{27}$Al | 100.0 | 13 | 14 | 27 | 26.981538 | 8.331 |
| $^{28}$Si | 92.23 | 14 | 14 | 28 | 27.976927 | 8.448 |
| $^{29}$Si | 4.67 | 14 | 15 | 29 | 28.976495 | 8.449 |
| $^{30}$Si | 3.10 | 14 | 16 | 30 | 29.973770 | 8.521 |
| $^{31}$P | 100.0 | 15 | 16 | 31 | 30.973761 | 8.481 |
| $^{32}$S | 95.02 | 16 | 16 | 32 | 31.972070 | 8.493 |
| $^{33}$S | 0.75 | 16 | 17 | 33 | 32.971456 | 8.498 |
| $^{34}$S | 4.21 | 16 | 18 | 34 | 33.967866 | 8.584 |
| $^{36}$S | 0.02 | 16 | 20 | 36 | 35.967080 | 8.575 |
| $^{35}$Cl | 75.77 | 17 | 18 | 35 | 34.968852 | 8.520 |
| $^{37}$Cl | 24.23 | 17 | 20 | 37 | 36.965903 | 8.570 |
| $^{36}$Ar | 0.337 | 18 | 18 | 36 | 35.967545 | 8.520 |
| $^{38}$Ar | 0.063 | 18 | 20 | 38 | 37.962732 | 8.614 |
| $^{40}$Ar | 99.600 | 18 | 22 | 40 | 39.962384 | 8.595 |
| $^{39}$K | 93.258 | 19 | 20 | 39 | 38.963707 | 8.557 |
| $^{41}$K | 6.730 | 19 | 22 | 41 | 40.961825 | 8.576 |
| $^{40}$Ca | 96.941 | 20 | 20 | 40 | 39.962591 | 8.551 |
| $^{42}$Ca | 0.647 | 20 | 22 | 42 | 41.958618 | 8.617 |
| $^{43}$Ca | 0.135 | 20 | 23 | 43 | 42.958766 | 8.601 |
| $^{44}$Ca | 2.086 | 20 | 24 | 44 | 43.955480 | 8.658 |
| $^{46}$Ca | 0.004 | 20 | 26 | 46 | 45.953689 | 8.669 |
| $^{48}$Ca | 0.187 | 20 | 28 | 48 | 47.952533 | 8.666 |
| $^{45}$Sc | 100.0 | 21 | 24 | 45 | 44.955910 | 8.619 |
| $^{46}$Ti | 8.0 | 22 | 24 | 46 | 45.952629 | 8.656 |
| $^{47}$Ti | 7.3 | 22 | 25 | 47 | 46.951764 | 8.661 |
| $^{48}$Ti | 73.8 | 22 | 26 | 48 | 47.947947 | 8.723 |
| $^{49}$Ti | 5.5 | 22 | 27 | 49 | 48.947871 | 8.711 |
| $^{50}$Ti | 5.4 | 22 | 28 | 50 | 49.944792 | 8.756 |
| $^{51}$V | 99.750 | 23 | 28 | 51 | 50.943962 | 8.742 |
| $^{50}$Cr | 4.345 | 24 | 26 | 50 | 49.946046 | 8.701 |
| $^{52}$Cr | 83.789 | 24 | 28 | 52 | 51.940509 | 8.776 |
| $^{53}$Cr | 9.501 | 24 | 29 | 53 | 52.940651 | 8.760 |
| $^{54}$Cr | 2.365 | 24 | 30 | 54 | 53.938882 | 8.778 |
| $^{55}$Mn | 100.0 | 25 | 30 | 55 | 54.938049 | 8.765 |
| $^{54}$Fe | 5.9 | 26 | 28 | 54 | 53.939612 | 8.736 |
| $^{56}$Fe | 91.72 | 26 | 30 | 56 | 55.934939 | 8.790 |
| $^{57}$Fe | 2.1 | 26 | 31 | 57 | 56.935396 | 8.770 |
| $^{58}$Fe | 0.28 | 26 | 32 | 58 | 57.933277 | 8.792 |
| $^{59}$Co | 100.0 | 27 | 32 | 59 | 58.933200 | 8.768 |
| $^{204}$Pb | 1.4 | 82 | 122 | 204 | 203.973020 | 7.880 |

*Continued on next page*

**TABLE A3.3** A Selection of Stable Isotopes[a] *(Continued)*

| Isotope $^AX$ | Natural Abundance (%) | Atomic Number $Z$ | Neutron Number $N$ | Mass Number $A$ | Atomic Mass (u) | Binding Energy per Nucleon (MeV)[b] |
|---|---|---|---|---|---|---|
| $^{206}Pb$ | 24.1 | 82 | 124 | 206 | 205.974440 | 7.875 |
| $^{207}Pb$ | 22.1 | 82 | 125 | 207 | 206.975872 | 7.870 |
| $^{208}Pb$ | 52.4 | 82 | 126 | 208 | 207.976627 | 7.868 |
| $^{209}Bi$ | 100.0 | 83 | 126 | 209 | 208.980380 | 7.848 |

[a]Selection is complete through cobalt-59. Where natural abundances do not add to 100%, the differences are made up by radioactive isotopes with exceedingly long half-lives: potassium-40 (0.0117%, $t_{1/2} = 1.3 \times 10^9$ yr); vanadium-50 (0.250%, $t_{1/2} > 1.4 \times 10^{17}$ yr).
[b]1 MeV (mega electron-volt) = $1.6022 \times 10^{-13}$ J.

**TABLE A3.4** A Selection of Radioactive Isotopes

| Isotope $^AX$ | Decay Mode[a] | Half-Life $t_{1/2}$ | Atomic Number $Z$ | Neutron Number $N$ | Mass Number $A$ | Atomic Mass (u) | Binding Energy per Nucleon (MeV)[b] |
|---|---|---|---|---|---|---|---|
| $^{3}H$ | $\beta^-$ | 12.3 yr | 1 | 2 | 3 | 3.01605 | 2.827 |
| $^{8}Be$ | $\alpha$ | $\sim 7 \times 10^{-17}$ s | 4 | 4 | 8 | 8.005305 | 7.062 |
| $^{14}C$ | $\beta^-$ | $5.7 \times 10^3$ yr | 6 | 8 | 14 | 14.003241 | 7.520 |
| $^{22}Na$ | $\beta^+$ | 2.6 yr | 11 | 11 | 22 | 21.994434 | 7.916 |
| $^{24}Na$ | $\beta^-$ | 15.0 hr | 11 | 13 | 24 | 23.990961 | 8.064 |
| $^{32}P$ | $\beta^-$ | 14.3 d | 15 | 17 | 32 | 31.973907 | 8.464 |
| $^{35}S$ | $\beta^-$ | 87.2 d | 16 | 19 | 35 | 34.969031 | 8.538 |
| $^{59}Fe$ | $\beta^-$ | 44.5 d | 26 | 33 | 59 | 58.934877 | 8.755 |
| $^{60}Co$ | $\beta^-$ | 5.3 yr | 27 | 33 | 60 | 59.933819 | 8.747 |
| $^{90}Sr$ | $\beta^-$ | 29.1 yr | 38 | 52 | 90 | 89.907738 | 8.696 |
| $^{99}Tc$ | $\beta^-$ | $2.1 \times 10^5$ yr | 43 | 56 | 99 | 98.906524 | 8.611 |
| $^{109}Cd$ | EC | 462 d | 48 | 61 | 109 | 108.904953 | 8.539 |
| $^{125}I$ | EC | 59.4 d | 53 | 72 | 125 | 124.904620 | 8.450 |
| $^{131}I$ | $\beta^-$ | 8.04 d | 53 | 78 | 131 | 130.906114 | 8.422 |
| $^{137}Cs$ | $\beta^-$ | 30.3 yr | 55 | 82 | 137 | 136.907073 | 8.389 |
| $^{222}Rn$ | $\alpha$ | 3.82 d | 86 | 136 | 222 | 222.017570 | 7.695 |
| $^{226}Ra$ | $\alpha$ | 1600 yr | 88 | 138 | 226 | 226.025402 | 7.662 |
| $^{232}Th$ | $\alpha$ | $1.4 \times 10^{10}$ yr | 90 | 142 | 232 | 232.038054 | 7.615 |
| $^{235}U$ | $\alpha$ | $7.0 \times 10^8$ yr | 92 | 143 | 235 | 235.043924 | 7.591 |
| $^{238}U$ | $\alpha$ | $4.5 \times 10^9$ yr | 92 | 146 | 238 | 238.050784 | 7.570 |
| $^{239}Pu$ | $\alpha$ | $2.4 \times 10^4$ yr | 94 | 145 | 239 | 239.052157 | 7.560 |

[a]Modes of decay include alpha emission ($\alpha$), beta emission ($\beta^-$), positron emission ($\beta^+$), and electron capture (EC).
[b]1 MeV (mega electron-volt) = $1.6022 \times 10^{-13}$ J.

# Appendix 4

## Chemical Bonds and Thermodynamic Data

**TABLE A4.1** Average Lengths and Energies of Covalent Bonds

| Atom | Bond | Bond Length (pm) | Bond Energy (kJ/mol) |
|---|---|---|---|
| H | H—H | 75 | 436 |
| | H—F | 92 | 567 |
| | H—Cl | 127 | 431 |
| | H—Br | 141 | 366 |
| | H—I | 161 | 299 |
| C | C—C | 154 | 348 |
| | C═C | 134 | 614 |
| | C≡C | 120 | 839 |
| | C—H | 110 | 413 |
| | C—N | 147 | 293 |
| | C═N | 127 | 615 |
| | C≡N | 116 | 891 |
| | C—O | 143 | 358 |
| | C═O | 123 | 743[a] |
| | C≡O | 113 | 1072 |
| | C—F | 133 | 485 |
| | C—Cl | 177 | 328 |
| | C—Br | 179 | 276 |
| | C—I | 215 | 238 |
| N | N—N | 147 | 163 |
| | N═N | 124 | 418 |
| | N≡N | 110 | 945 |
| | N—H | 104 | 391 |
| | N—O | 136 | 201 |
| | N═O | 122 | 607 |
| | N≡O | 106 | 678 |
| O | O—O | 148 | 146 |
| | O═O | 121 | 498 |
| | O—H | 96 | 463 |
| S | S—O | 151 | 265 |
| | S═O | 143 | 523 |
| | S—S | 204 | 266 |
| | S—H | 134 | 347 |
| F | F—F | 143 | 155 |
| Cl | Cl—Cl | 200 | 243 |
| Br | Br—Br | 228 | 193 |
| I | I—I | 266 | 151 |

[a]The bond energy of C═O in $CO_2$ is 799 kJ/mol.

**TABLE A4.2** Critical Temperatures ($T_c$) and van der Waals Parameters ($a$, $b$) of Real Gases

| Gas[a] | Molar Mass (g/mol) | $T_c$ (K) | $a$ ($L^2 \cdot atm/mol^2$) | $b$ (L/mol) |
|---|---|---|---|---|
| $H_2O$ | 18.015 | 647.14 | 5.46 | 0.0305 |
| $Br_2$ | 159.808 | 588 | 9.75 | 0.0591 |
| $CCl_3F$ | 137.367 | 471.2 | 14.68 | 0.1111 |
| $Cl_2$ | 70.906 | 416.9 | 6.343 | 0.0542 |
| $CO_2$ | 44.010 | 304.14 | 3.59 | 0.0427 |
| Kr | 83.798 | 209.41 | 2.325 | 0.0396 |
| $CH_4$ | 16.043 | 190.53 | 2.25 | 0.0428 |
| $O_2$ | 31.999 | 154.59 | 1.36 | 0.0318 |
| Ar | 39.948 | 150.87 | 1.34 | 0.0322 |
| $F_2$ | 37.997 | 144.13 | 1.171 | 0.0290 |
| CO | 28.010 | 132.91 | 1.45 | 0.0395 |
| $N_2$ | 28.013 | 126.21 | 1.39 | 0.0391 |
| $H_2$ | 2.016 | 32.97 | 0.244 | 0.0266 |
| He | 4.003 | 5.19 | 0.0341 | 0.0237 |

[a]Listed in descending order of critical temperature.

**TABLE A4.3** Thermodynamic Properties at 25°C

| Substance[a,b] | Molar Mass (g/mol) | $\Delta H_f^\circ$ (kJ/mol) | $S^\circ$ [J/(mol · K)] | $\Delta G_f^\circ$ (kJ/mol) |
|---|---|---|---|---|
| **ELEMENTS AND MONATOMIC IONS** | | | | |
| $Ag^+(aq)$ | 107.87 | 105.6 | 72.7 | 77.1 |
| Ag(*g*) | 107.87 | 284.9 | 173.0 | 246.0 |
| Ag(*s*) | 107.87 | 0.0 | 42.6 | 0.0 |
| $Al^{3+}(aq)$ | 26.982 | −531 | −321.7 | −485 |
| Al(*g*) | 26.982 | 330.0 | 164.6 | 289.4 |
| Al(*s*) | 26.982 | 0.0 | 28.3 | 0.0 |
| Al(ℓ) | 26.982 | 10.6 | 39.6 | −1.2 |
| Ar(*g*) | 39.948 | 0.0 | 154.8 | 0.0 |
| Au(*g*) | 196.97 | 366.1 | 180.5 | 326.3 |
| Au(*s*) | 196.97 | 0.0 | 47.4 | 0.0 |
| B(*g*) | 10.811 | 565.0 | 153.4 | 521.0 |
| B(*s*) | 10.811 | 0.0 | 5.9 | 0.0 |
| $Ba^{2+}(aq)$ | 137.33 | −537.6 | 9.6 | −560.8 |
| Ba(*g*) | 137.33 | 180.0 | 170.2 | 146.0 |
| Ba(*s*) | 137.33 | 0.0 | 62.8 | 0.0 |
| Be(*g*) | 9.0122 | 324.0 | 136.3 | 286.6 |
| Be(*s*) | 9.0122 | 0.0 | 9.5 | 0.0 |
| $Br^-(aq)$ | 79.904 | −121.6 | 82.4 | −104.0 |
| Br(*g*) | 79.904 | 111.9 | 175.0 | 82.4 |
| $Br_2(g)$ | 159.808 | 30.9 | 245.5 | 3.1 |
| $Br_2(\ell)$ | 159.808 | 0.0 | 152.2 | 0.0 |
| C(*g*) | 12.011 | 716.7 | 158.1 | 671.3 |
| C(*s*, diamond) | 12.011 | 1.9 | 2.4 | 2.9 |

**TABLE A4.3** Thermodynamic Properties at 25°C *(Continued)*

| Substance[a,b] | Molar Mass (g/mol) | $\Delta H_f^\circ$ (kJ/mol) | $S^\circ$ [J/(mol · K)] | $\Delta G_f^\circ$ (kJ/mol) |
|---|---|---|---|---|
| C(*s*, graphite) | 12.011 | 0.0 | 5.7 | 0.0 |
| $Ca^{2+}$(*aq*) | 40.078 | −542.8 | −55.3 | −553.6 |
| Ca(*g*) | 40.078 | 177.8 | 154.9 | 144.0 |
| Ca(*s*) | 40.078 | 0.0 | 41.6 | 0.0 |
| $Cl^-$(*aq*) | 35.453 | −167.2 | 56.5 | −131.2 |
| Cl(*g*) | 35.453 | 121.3 | 165.2 | 105.3 |
| $Cl_2$(*g*) | 70.906 | 0.0 | 223.0 | 0.0 |
| $Co^{2+}$(*aq*) | 58.933 | −58.2 | −113 | −54.4 |
| $Co^{3+}$(*aq*) | 58.933 | 92 | −305 | 134 |
| Co(*g*) | 58.933 | 424.7 | 179.5 | 380.3 |
| Co(*s*) | 58.933 | 0.0 | 30.0 | 0.0 |
| Cr(*g*) | 51.996 | 396.6 | 174.5 | 351.8 |
| Cr(*s*) | 51.996 | 0.0 | 23.8 | 0.0 |
| $Cs^+$(*aq*) | 132.91 | −258.3 | 133.1 | −292.0 |
| Cs(*g*) | 132.91 | 76.5 | 175.6 | 49.6 |
| Cs(*s*) | 132.91 | 0.0 | 85.2 | 0.0 |
| $Cu^+$(*aq*) | 63.546 | 71.7 | 40.6 | 50.0 |
| $Cu^{2+}$(*aq*) | 63.546 | 64.8 | −99.6 | 65.5 |
| Cu(*g*) | 63.546 | 337.4 | 166.4 | 297.7 |
| Cu(*s*) | 63.546 | 0.0 | 33.2 | 0.0 |
| $F^-$(*aq*) | 18.998 | −332.6 | −13.8 | −278.8 |
| F(*g*) | 18.998 | 79.4 | 158.8 | 62.3 |
| $F_2$(*g*) | 37.997 | 0.0 | 202.8 | 0.0 |
| $Fe^{2+}$(*aq*) | 55.845 | −89.1 | −137.7 | −78.9 |
| $Fe^{3+}$(*aq*) | 55.845 | −48.5 | −315.9 | −4.7 |
| Fe(*g*) | 55.845 | 416.3 | 180.5 | 370.7 |
| Fe(*s*) | 55.845 | 0.0 | 27.3 | 0.0 |
| $H^+$(*aq*) | 1.0079 | 0.0 | 0.0 | 0.0 |
| H(*g*) | 1.0079 | 218.0 | 114.7 | 203.3 |
| $H_2$(*g*) | 2.0158 | 0.0 | 130.6 | 0.0 |
| He(*g*) | 4.0026 | 0.0 | 126.2 | 0.0 |
| $Hg_2{}^{2+}$(*aq*) | 401.18 | 172.4 | 84.5 | 153.5 |
| $Hg^{2+}$(*aq*) | 200.59 | 171.1 | −32.2 | 164.4 |
| Hg(*g*) | 200.59 | 61.4 | 175.0 | 31.8 |
| Hg(*ℓ*) | 200.59 | 0.0 | 75.9 | 0.0 |
| $I^-$(*aq*) | 126.90 | −55.2 | 111.3 | −51.6 |
| I(*g*) | 126.90 | 106.8 | 180.8 | 70.2 |
| $I_2$(*g*) | 253.81 | 62.4 | 260.7 | 19.3 |
| $I_2$(*s*) | 253.81 | 0.0 | 116.1 | 0.0 |
| $K^+$(*aq*) | 39.098 | −252.4 | 102.5 | −283.3 |
| K(*g*) | 39.098 | 89.0 | 160.3 | 60.5 |
| K(*s*) | 39.098 | 0.0 | 64.7 | 0.0 |
| $Li^+$(*aq*) | 6.941 | −278.5 | 13.4 | −293.3 |

*Continued on next page*

**TABLE A4.3** Thermodynamic Properties at 25°C *(Continued)*

| Substance[a,b] | Molar Mass (g/mol) | $\Delta H_f^\circ$ (kJ/mol) | $S^\circ$ [J/(mol · K)] | $\Delta G_f^\circ$ (kJ/mol) |
|---|---|---|---|---|
| Li(*g*) | 6.941 | 159.3 | 138.8 | 126.6 |
| $Li^+$(*g*) | 6.941 | 685.7 | 133.0 | 648.5 |
| Li(*s*) | 6.941 | 0.0 | 29.1 | 0.0 |
| $Mg^{2+}$(*aq*) | 24.305 | −466.9 | −138.1 | −454.8 |
| Mg(*g*) | 24.305 | 147.1 | 148.6 | 112.5 |
| Mg(*s*) | 24.305 | 0.0 | 32.7 | 0.0 |
| $Mn^{2+}$(*aq*) | 54.938 | −220.8 | −73.6 | −228.1 |
| Mn(*g*) | 54.938 | 280.7 | 173.7 | 238.5 |
| Mn(*s*) | 54.938 | 0.0 | 32.0 | 0.0 |
| N(*g*) | 14.007 | 472.7 | 153.3 | 455.5 |
| $N_2$(*g*) | 28.013 | 0.0 | 191.5 | 0.0 |
| $Na^+$(*aq*) | 22.990 | −240.1 | 59.0 | −261.9 |
| Na(*g*) | 22.990 | 107.5 | 153.7 | 77.0 |
| $Na^+$(*g*) | 22.990 | 609.3 | 148.0 | 574.3 |
| Na(*s*) | 22.990 | 0.0 | 51.3 | 0.0 |
| Ne(*g*) | 20.180 | 0.0 | 146.3 | 0.0 |
| $Ni^{2+}$(*aq*) | 58.693 | −54.0 | −128.9 | −45.6 |
| Ni(*g*) | 58.693 | 429.7 | 182.2 | 384.5 |
| Ni(*s*) | 58.693 | 0.0 | 29.9 | 0.0 |
| O(*g*) | 15.999 | 249.2 | 161.1 | 231.7 |
| $O_2$(*g*) | 31.999 | 0.0 | 205.0 | 0.0 |
| $O_3$(*g*) | 47.998 | 142.7 | 238.8 | 163.2 |
| P(*g*) | 30.974 | 314.6 | 163.1 | 278.3 |
| $P_4$(*s*, red) | 123.895 | −17.6 | 22.8 | −12.1 |
| $P_4$(*s*, white) | 123.895 | 0.0 | 41.1 | 0.0 |
| $Pb^{2+}$(*aq*) | 207.2 | −1.7 | 10.5 | −24.4 |
| Pb(*g*) | 207.2 | 195.2 | 162.2 | 175.4 |
| Pb(*s*) | 207.2 | 0.0 | 64.8 | 0.0 |
| $Rb^+$(*aq*) | 85.468 | −251.2 | 121.5 | −284.0 |
| Rb(*g*) | 85.468 | 80.9 | 170.1 | 53.1 |
| Rb(*s*) | 85.468 | 0.0 | 76.8 | 0.0 |
| S(*g*) | 32.065 | 277.2 | 167.8 | 236.7 |
| $S_8$(*g*) | 256.520 | 102.3 | 430.2 | 49.1 |
| $S_8$(*s*) | 256.520 | 0.0 | 32.1 | 0.0 |
| Sc(*g*) | 44.956 | 377.8 | 174.8 | 336.0 |
| Sc(*s*) | 44.956 | 0.0 | 34.6 | 0.0 |
| Si(*g*) | 28.086 | 450.0 | 168.0 | 405.5 |
| Si(*s*) | 28.086 | 0.0 | 18.8 | 0.0 |
| Sn(*g*) | 118.71 | 301.2 | 168.5 | 266.2 |
| Sn(*s*, gray) | 118.71 | −2.1 | 44.1 | 0.1 |
| Sn(*s*, white) | 118.71 | 0.0 | 51.2 | 0.0 |
| $Sr^{2+}$(*aq*) | 87.62 | −545.8 | −32.6 | −559.5 |
| Sr(*g*) | 87.62 | 164.4 | 164.6 | 130.9 |
| Sr(*s*) | 87.62 | 0.0 | 52.3 | 0.0 |

**TABLE A4.3** Thermodynamic Properties at 25°C *(Continued)*

| Substance[a,b] | Molar Mass (g/mol) | $\Delta H_f^\circ$ (kJ/mol) | $S^\circ$ [J/(mol · K)] | $\Delta G_f^\circ$ (kJ/mol) |
|---|---|---|---|---|
| Ti(*g*) | 47.867 | 473.0 | 180.3 | 428.4 |
| Ti(*s*) | 47.867 | 0.0 | 30.7 | 0.0 |
| V(*g*) | 50.942 | 514.2 | 182.2 | 468.5 |
| V(*s*) | 50.942 | 0.0 | 28.9 | 0.0 |
| W(*s*) | 183.84 | 0.0 | 32.6 | 0.0 |
| $Zn^{2+}$(*aq*) | 65.38 | −153.9 | −112.1 | −147.1 |
| Zn(*g*) | 65.38 | 130.4 | 161.0 | 94.8 |
| Zn(*s*) | 65.38 | 0.0 | 41.6 | 0.0 |
| **POLYATOMIC IONS** | | | | |
| $CH_3COO^-$(*aq*) | 59.045 | −486.0 | 86.6 | −369.3 |
| $CO_3^{2-}$(*aq*) | 60.009 | −677.1 | −56.9 | −527.8 |
| $C_2O_4^{2-}$(*aq*) | 88.020 | −825.1 | 45.6 | −673.9 |
| $CrO_4^{2-}$(*aq*) | 115.994 | −881.2 | 50.2 | −727.8 |
| $Cr_2O_7^{2-}$(*aq*) | 215.988 | −1490.3 | 261.9 | −1301.1 |
| $HCOO^-$(*aq*) | 45.018 | −425.6 | 92 | −351.0 |
| $HCO_3^-$(*aq*) | 61.017 | −692.0 | 91.2 | −586.8 |
| $HSO_4^-$(*aq*) | 97.072 | −887.3 | 131.8 | −755.9 |
| $MnO_4^-$(*aq*) | 118.936 | −541.4 | 191.2 | −447.2 |
| $NH_4^+$(*aq*) | 18.038 | −132.5 | 113.4 | −79.3 |
| $NO_3^-$(*aq*) | 62.005 | −205.0 | 146.4 | −108.7 |
| $OH^-$(*aq*) | 17.007 | −230.0 | −10.8 | −157.2 |
| $PO_4^{3-}$(*aq*) | 94.971 | −1277.4 | −222 | −1018.7 |
| $SO_4^{2-}$(*aq*) | 96.064 | −909.3 | 20.1 | −744.5 |
| **INORGANIC COMPOUNDS** | | | | |
| AgCl(*s*) | 143.32 | −127.1 | 96.2 | −109.8 |
| AgI(*s*) | 234.77 | −61.8 | 115.5 | −66.2 |
| $AgNO_3$(*s*) | 169.87 | −124.4 | 140.9 | −33.4 |
| $Al_2O_3$(*s*) | 101.961 | −1675.7 | 50.9 | −1582.3 |
| $B_2H_6$(*g*) | 27.669 | 35.0 | 232.0 | 86.6 |
| $B_2O_3$(*s*) | 69.622 | −1263.6 | 54.0 | −1184.1 |
| $BaCO_3$(*s*) | 197.34 | −1216.3 | 112.1 | −1137.6 |
| $BaSO_4$(*s*) | 233.39 | −1473.2 | 132.2 | −1362.2 |
| $CaCO_3$(*s*) | 100.087 | −1206.9 | 92.9 | −1128.8 |
| $CaCl_2$(*s*) | 110.984 | −795.4 | 108.4 | −748.8 |
| $CaF_2$(*s*) | 78.075 | −1228.0 | 68.5 | −1175.6 |
| CaO(*s*) | 56.077 | −634.9 | 38.1 | −603.3 |
| $Ca(OH)_2$(*s*) | 74.093 | −985.2 | 83.4 | −897.5 |
| $CaSO_4$(*s*) | 136.142 | −1434.5 | 106.5 | −1322.0 |
| CO(*g*) | 28.010 | −110.5 | 197.7 | −137.2 |
| $CO_2$(*g*) | 44.010 | −393.5 | 213.8 | −394.4 |
| $CO_2$(*aq*) | 44.010 | −412.9 | 121.3 | −386.2 |
| $CS_2$(*g*) | 76.143 | 115.3 | 237.8 | 65.1 |
| $CS_2$(ℓ) | 76.143 | 87.9 | 151.0 | 63.6 |

*Continued on next page*

**TABLE A4.3** Thermodynamic Properties at 25°C *(Continued)*

| Substance[a,b] | Molar Mass (g/mol) | $\Delta H_f^\circ$ (kJ/mol) | $S^\circ$ [J/(mol · K)] | $\Delta G_f^\circ$ (kJ/mol) |
|---|---|---|---|---|
| $CsCl(s)$ | 168.358 | −443.0 | 101.2 | −414.6 |
| $CuSO_4(s)$ | 159.610 | −771.4 | 109.2 | −662.2 |
| $Cu_2S(s)$ | 159.16 | −79.5 | 120.9 | −86.2 |
| $FeCl_2(s)$ | 126.750 | −341.8 | 118.0 | −302.3 |
| $FeCl_3(s)$ | 162.203 | −399.5 | 142.3 | −334.0 |
| $FeO(s)$ | 71.844 | −271.9 | 60.8 | −255.2 |
| $Fe_2O_3(s)$ | 159.688 | −824.2 | 87.4 | −742.2 |
| $HBr(g)$ | 80.912 | −36.3 | 198.7 | −53.4 |
| $HCl(g)$ | 36.461 | −92.3 | 186.9 | −95.3 |
| $HCN(g)$ | 27.02 | 135.1 | 201.81 | 124.7 |
| $HF(g)$ | 20.006 | −273.3 | 173.8 | −275.4 |
| $HI(g)$ | 127.912 | 26.5 | 206.6 | 1.7 |
| $HNO_2(g)$ | 47.014 | −79.5 | 254.1 | −46.0 |
| $HNO_3(g)$ | 63.013 | −135.1 | 266.4 | −74.7 |
| $HNO_3(\ell)$ | 63.013 | −174.1 | 155.6 | −80.7 |
| $HNO_3(aq)$ | 63.013 | −206.6 | 146.0 | −110.5 |
| $HgCl_2(s)$ | 271.50 | −224.3 | 146.0 | −178.6 |
| $Hg_2Cl_2(s)$ | 472.09 | −265.4 | 191.6 | −210.7 |
| $H_2O(g)$ | 18.015 | −241.8 | 188.8 | −228.6 |
| $H_2O(\ell)$ | 18.015 | −285.8 | 69.9 | −237.2 |
| $H_2S(g)$ | 34.082 | −20.17 | 205.6 | −33.01 |
| $H_2O_2(g)$ | 34.015 | −136.3 | 232.7 | −105.6 |
| $H_2O_2(\ell)$ | 34.015 | −187.8 | 109.6 | −120.4 |
| $H_2SO_4(\ell)$ | 98.079 | −814.0 | 156.9 | −690.0 |
| $H_2SO_4(aq)$ | 98.079 | −909.2 | 20.1 | −744.5 |
| $KBr(s)$ | 119.002 | −393.8 | 95.9 | −380.7 |
| $KCl(s)$ | 74.551 | −436.5 | 82.6 | −408.5 |
| $KHCO_3(s)$ | 100.115 | −963.2 | 115.5 | −863.6 |
| $K_2CO_3(s)$ | 138.205 | −1151.0 | 155.5 | −1063.5 |
| $LiBr(s)$ | 86.845 | −351.2 | 74.3 | −342.0 |
| $LiCl(s)$ | 42.394 | −408.6 | 59.3 | −384.4 |
| $Li_2CO_3(s)$ | 73.891 | −1215.9 | 90.4 | −1132.1 |
| $MgCl_2(s)$ | 95.211 | −641.3 | 89.6 | −591.8 |
| $Mg(OH)_2(s)$ | 58.320 | −924.5 | 63.2 | −833.5 |
| $MgO$ | 40.30 | −601.1 | 27.0 | −630.9 |
| $MgSO_4(s)$ | 120.369 | −1284.9 | 91.6 | −1170.6 |
| $MnO_2(s)$ | 86.937 | −520.0 | 53.1 | −465.1 |
| $CH_3COONa(s)$ | 82.034 | −708.8 | 123.0 | −607.2 |
| $NaBr(s)$ | 102.894 | −361.1 | 86.82 | −349.0 |
| $NaCl(s)$ | 58.443 | −411.2 | 72.1 | −384.2 |
| $NaCl(g)$ | 58.443 | −181.4 | 229.8 | −201.3 |
| $Na_2CO_3(s)$ | 105.989 | −1130.7 | 135.0 | −1044.4 |
| $NaHCO_3(s)$ | 84.007 | −950.8 | 101.7 | −851.0 |
| $NaNO_3(s)$ | 84.995 | −467.9 | 116.5 | −367.0 |

**TABLE A4.3** Thermodynamic Properties at 25°C *(Continued)*

| Substance[a,b] | Molar Mass (g/mol) | $\Delta H_f^\circ$ (kJ/mol) | $S^\circ$ [J/(mol · K)] | $\Delta G_f^\circ$ (kJ/mol) |
|---|---|---|---|---|
| $NaOH(s)$ | 39.997 | −425.6 | 64.5 | −379.5 |
| $Na_2SO_4(s)$ | 142.043 | −1387.1 | 149.6 | −1270.2 |
| $NF_3(g)$ | 71.002 | −132.1 | 260.8 | −90.6 |
| $NH_3(aq)$ | 17.031 | −80.3 | 111.3 | −26.50 |
| $NH_3(g)$ | 17.031 | −46.1 | 192.5 | −16.5 |
| $NH_4Cl(s)$ | 53.491 | −314.4 | 94.6 | −203.0 |
| $NH_4NO_3(s)$ | 80.043 | −365.6 | 151.1 | −183.9 |
| $N_2H_4(g)$ | 32.045 | 95.35 | 238.5 | 159.4 |
| $N_2H_4(\ell)$ | 32.045 | 50.63 | 121.52 | 149.3 |
| $NiCl_2(s)$ | 129.60 | −305.3 | 97.7 | −259.0 |
| $NiO(s)$ | 74.60 | −239.7 | 38.0 | −211.7 |
| $NO(g)$ | 30.006 | 90.3 | 210.7 | 86.6 |
| $NO_2(g)$ | 46.006 | 33.2 | 240.0 | 51.3 |
| $N_2O(g)$ | 44.013 | 82.1 | 219.9 | 104.2 |
| $N_2O_3(g)$ | 76.01 | 86.6 | 314.7 | 142.4 |
| $N_2O_4(g)$ | 92.011 | 11.1 | 304.2 | 99.8 |
| $NOCl(g)$ | 65.459 | 51.7 | 261.7 | 66.1 |
| $PCl_3(g)$ | 137.33 | −288.07 | 311.7 | −269.6 |
| $PCl_3(\ell)$ | 137.33 | −319.6 | 217 | −272.4 |
| $PF_5(g)$ | 125.96 | −1594.4 | 300.8 | −1520.7 |
| $PH_3(g)$ | 33.998 | 5.4 | 210.2 | 13.4 |
| $PbCl_2(s)$ | 278.1 | −359.4 | 136.0 | −314.1 |
| $PbSO_4(s)$ | 303.3 | −920.0 | 148.5 | −813.0 |
| $SO_2(g)$ | 64.065 | −296.8 | 248.2 | −300.1 |
| $SO_3(g)$ | 80.064 | −395.7 | 256.8 | −371.1 |
| $ZnCl_2(s)$ | 136.30 | −415.1 | 111.5 | −369.4 |
| $ZnO(s)$ | 81.37 | −348.0 | 43.9 | −318.2 |
| $ZnSO_4(s)$ | 161.45 | −982.8 | 110.5 | −871.5 |
| **ORGANIC COMPOUNDS** | | | | |
| $CCl_4(g)$ | 153.823 | −102.9 | 309.7 | −60.6 |
| $CCl_4(\ell)$ | 153.823 | −135.4 | 216.4 | −65.3 |
| $CH_4(g)$ | 16.043 | −74.8 | 186.2 | −50.8 |
| $CH_3COOH(g)$ | 60.053 | −432.8 | 282.5 | −374.5 |
| $CH_3COOH(\ell)$ | 60.053 | −484.5 | 159.8 | −389.9 |
| $CH_3OH(g)$ | 32.042 | −200.7 | 239.9 | −162.0 |
| $CH_3OH(\ell)$ | 32.042 | −238.7 | 126.8 | −166.4 |
| $C_2H_2(g)$ | 26.038 | 226.7 | 200.8 | 209.2 |
| $C_2H_4(g)$ | 28.054 | 52.4 | 219.5 | 68.1 |
| $C_2H_6(g)$ | 30.070 | −84.67 | 229.5 | −32.9 |
| $CH_3CH_2OH(g)$ | 46.069 | −235.1 | 282.6 | −168.6 |
| $CH_3CH_2OH(\ell)$ | 46.069 | −277.7 | 160.7 | −174.9 |
| $CH_3CHO(g)$ | 44.053 | −166 | 266 | −133.7 |
| $C_3H_8(g)$ | 44.097 | −103.8 | 269.9 | −23.5 |

*Continued on next page*

**TABLE A4.3** Thermodynamic Properties at 25°C *(Continued)*

| Substance[a,b] | Molar Mass (g/mol) | $\Delta H_f^\circ$ (kJ/mol) | $S^\circ$ [J/(mol · K)] | $\Delta G_f^\circ$ (kJ/mol) |
|---|---|---|---|---|
| $CH_3(CH_2)_2CH_3(g)$ | 58.123 | −125.6 | 310.0 | −15.7 |
| $CH_3(CH_2)_2CH_3(\ell)$ | 58.123 | −147.6 | 231.0 | −15.0 |
| $CH_3COCH_3(\ell)$ | 58.079 | −248.4 | 199.8 | −155.6 |
| $CH_3COCH_3(g)$ | 58.079 | −217.1 | 295.3 | −152.7 |
| $CH_3(CH_2)_2CH_2OH(\ell)$ | 74.122 | −327.3 | 225.8 | |
| $(CH_3CH_2)_2O(\ell)$ | 74.122 | −279.6 | 172.4 | |
| $(CH_3CH_2)_2O(g)$ | 74.122 | −252.1 | 342.7 | |
| $(CH_3)_2C{=}C(CH_3)_2(\ell)$ | 84.161 | 66.6 | 362.6 | −69.2 |
| $(CH_3)_2NH(\ell)$ | 45.084 | −43.9 | 182.3 | |
| $(CH_3)_2NH(g)$ | 45.084 | −18.5 | 273.1 | |
| $(CH_3CH_2)_2NH(\ell)$ | 73.138 | −103.3 | | |
| $(CH_3CH_2)_2NH(g)$ | 73.138 | −71.4 | | |
| $(CH_3)_3N(\ell)$ | 59.111 | −46.0 | 208.5 | |
| $(CH_3)_3N(g)$ | 59.111 | −23.6 | 287.1 | |
| $(CH_3CH_2)_3N(\ell)$ | 101.191 | −134.3 | | |
| $(CH_3CH_2)_3N(g)$ | 101.191 | −95.8 | | |
| $C_6H_6(g)$ | 78.114 | 82.9 | 269.2 | 129.7 |
| $C_6H_6(\ell)$ | 78.114 | 49.0 | 172.9 | 124.5 |
| $C_6H_{12}O_6(s)$ | 180.158 | −1274.4 | 212.1 | −910.1 |
| $CH_3(CH_2)_6CH_3(\ell)$ | 114.231 | −249.9 | 361.1 | 6.4 |
| $CH_3(CH_2)_6CH_3(g)$ | 114.231 | −208.6 | 466.7 | 16.4 |
| $C_{12}H_{22}O_{11}(s)$ | 342.300 | −2221.7 | 360.2 | −1543.8 |
| $HCOOH(\ell)$ | 46.026 | −424.7 | 129.0 | −361.4 |

[a]Substances are arranged alphabetically by chemical formula within each class: (1) elements and monatomic ions; (2) polyatomic ions; (3) inorganic compounds (including CO and $CO_2$); (4) organic compounds (hydrocarbon-based).

[b]Symbols denote standard enthalpy of formation ($\Delta H_f^\circ$), standard third-law entropy ($S^\circ$), and standard Gibbs free energy of formation ($\Delta G_f^\circ$). Entropies in aqueous solution are referred to $S^\circ[H^+(aq)] = 0$, not to absolute zero.

**TABLE A4.4** Vapor Pressure of Water as a Function of Temperature

| $T$ (°C) | $P$ (torr) |
|---|---|
| 0.0 | 4.579 |
| 10.0 | 9.209 |
| 20.0 | 17.535 |
| 25.0 | 23.756 |
| 30.0 | 31.824 |
| 40.0 | 55.324 |
| 60.0 | 149.4 |
| 70.0 | 233.7 |
| 90.0 | 525.8 |
| 100 | 760.0 |
| 105 | 906.0 |

# Appendix 5

## Equilibrium Constants

**TABLE A5.1** Ionization Constants of Selected Acids at 25°C

| Acid | Step | Aqueous Equilibrium[a] | $K_a$ | p$K_a$ |
|---|---|---|---|---|
| acetic | 1 | $CH_3COOH(aq) + H_2O(\ell) \rightleftharpoons H_3O^+(aq) + CH_3COO^-(aq)$ | $1.76 \times 10^{-5}$ | 4.75 |
| arsenic | 1 | $H_3AsO_4(aq) + H_2O(\ell) \rightleftharpoons H_3O^+(aq) + H_2AsO_4^-(aq)$ | $5.5 \times 10^{-3}$ | 2.26 |
| | 2 | $H_2AsO_4^-(aq) + H_2O(\ell) \rightleftharpoons H_3O^+(aq) + HAsO_4^{2-}(aq)$ | $1.7 \times 10^{-7}$ | 6.77 |
| | 3 | $HAsO_4^{2-}(aq) + H_2O(\ell) \rightleftharpoons H_3O^+(aq) + AsO_4^{3-}(aq)$ | $5.1 \times 10^{-12}$ | 11.29 |
| ascorbic | 1 | $H_2C_6H_6O_6(aq) + H_2O(\ell) \rightleftharpoons H_3O^+(aq) + HC_6H_6O_6^-(aq)$ | $9.1 \times 10^{-5}$ | 4.04 |
| | 2 | $HC_6H_6O_6^-(aq) + H_2O(\ell) \rightleftharpoons H_3O^+(aq) + C_6H_6O_6^{2-}(aq)$ | $5 \times 10^{-12}$ | 11.3 |
| benzoic | 1 | $C_6H_5COOH(aq) + H_2O(\ell) \rightleftharpoons H_3O^+(aq) + C_6H_5COO^-(aq)$ | $6.25 \times 10^{-5}$ | 4.20 |
| boric | 1 | $H_3BO_3(aq) + H_2O(\ell) \rightleftharpoons H_3O^+(aq) + H_2BO_3^-(aq)$ | $5.4 \times 10^{-10}$ | 9.27 |
| | 2 | $H_2BO_3^-(aq) + H_2O(\ell) \rightleftharpoons H_3O^+(aq) + HBO_3^{2-}(aq)$ | $<10^{-14}$ | >14 |
| bromoacetic | 1 | $CH_2BrCOOH(aq) + H_2O(\ell) \rightleftharpoons H_3O^+(aq) + CH_2BrCOO^-(aq)$ | $2.0 \times 10^{-3}$ | 2.70 |
| butanoic | 1 | $CH_3CH_2CH_2COOH(aq) + H_2O(\ell) \rightleftharpoons H_3O^+(aq) + CH_3CH_2CH_2COO^-(aq)$ | $1.5 \times 10^{-5}$ | 4.82 |
| carbonic | 1 | $H_2CO_3(aq) + H_2O(\ell) \rightleftharpoons H_3O^+(aq) + HCO_3^-(aq)$ | $4.3 \times 10^{-7}$ | 6.37 |
| | 2 | $HCO_3^-(aq) + H_2O(\ell) \rightleftharpoons H_3O^+(aq) + CO_3^{2-}(aq)$ | $4.7 \times 10^{-11}$ | 10.33 |
| chloric | 1 | $HClO_3(aq) + H_2O(\ell) \rightleftharpoons H_3O^+(aq) + ClO_3^-(aq)$ | ~1 | ~0 |
| chloroacetic | 1 | $CH_2ClCOOH(aq) + H_2O(\ell) \rightleftharpoons H_3O^+(aq) + CH_2ClCOO^-(aq)$ | $1.4 \times 10^{-3}$ | 2.85 |
| chlorous | 1 | $HClO_2(aq) + H_2O(\ell) \rightleftharpoons H_3O^+(aq) + ClO_2^-(aq)$ | $1.1 \times 10^{-2}$ | 1.96 |
| citric | 1 | $CH_2(COOH)C(OH)(COOH)CH_2COOH(aq) + H_2O(\ell) \rightleftharpoons H_3O^+(aq) + CH_2(COOH)C(OH)(COO^-)CH_2COOH(aq)$ | $7.4 \times 10^{-4}$ | 3.13 |
| | 2 | $CH_2(COOH)C(OH)(COO^-)CH_2COOH(aq) + H_2O(\ell) \rightleftharpoons H_3O^+(aq) + CH_2(COO^-)C(OH)(COO^-)CH_2COOH(aq)$ | $1.7 \times 10^{-5}$ | 4.77 |
| | 3 | $CH_2(COO^-)C(OH)(COO^-)CH_2COOH(aq) + H_2O(\ell) \rightleftharpoons H_3O^+(aq) + CH_2(COO^-)C(OH)(COO^-)CH_2COO^-(aq)$ | $4.0 \times 10^{-7}$ | 6.40 |
| dichloroacetic | 1 | $CHCl_2COOH(aq) + H_2O(\ell) \rightleftharpoons H_3O^+(aq) + CHCl_2COO^-(aq)$ | $5.5 \times 10^{-2}$ | 1.26 |
| ethanol | 1 | $CH_3CH_2OH(aq) + H_2O(\ell) \rightleftharpoons H_3O^+(aq) + CH_3CH_2O^-(aq)$ | $1.3 \times 10^{-16}$ | 15.9 |
| fluoroacetic | 1 | $CH_2FCOOH(aq) + H_2O(\ell) \rightleftharpoons H_3O^+(aq) + CH_2FCOO^-(aq)$ | $2.6 \times 10^{-3}$ | 2.59 |
| formic | 1 | $HCOOH(aq) + H_2O(\ell) \rightleftharpoons H_3O^+(aq) + HCOO^-(aq)$ | $1.77 \times 10^{-4}$ | 3.75 |
| germanic | 1 | $H_2GeO_3(aq) + H_2O(\ell) \rightleftharpoons H_3O^+(aq) + HGeO_3^-(aq)$ | $9.8 \times 10^{-10}$ | 9.01 |
| | 2 | $HGeO_3^-(aq) + H_2O(\ell) \rightleftharpoons H_3O^+(aq) + GeO_3^{2-}(aq)$ | $5 \times 10^{-13}$ | 12.3 |
| hydr(o)azoic | 1 | $HN_3(aq) + H_2O(\ell) \rightleftharpoons H_3O^+(aq) + N_3^-(aq)$ | $1.9 \times 10^{-5}$ | 4.72 |
| hydrobromic | 1 | $HBr(aq) + H_2O(\ell) \rightarrow H_3O^+(aq) + Br^-(aq)$ | $\gg$1 (strong) | <0 |
| hydrochloric | 1 | $HCl(aq) + H_2O(\ell) \rightarrow H_3O^+(aq) + Cl^-(aq)$ | $\gg$1 (strong) | <0 |
| hydrocyanic | 1 | $HCN(aq) + H_2O(\ell) \rightleftharpoons H_3O^+(aq) + CN^-(aq)$ | $6.2 \times 10^{-10}$ | 9.21 |
| hydrofluoric | 1 | $HF(aq) + H_2O(\ell) \rightleftharpoons H_3O^+(aq) + F^-(aq)$ | $6.8 \times 10^{-4}$ | 3.17 |
| hydr(o)iodic | 1 | $HI(aq) + H_2O(\ell) \rightarrow H_3O^+(aq) + I^-(aq)$ | $\gg$1 (strong) | <0 |

*Continued on next page*

**TABLE A5.1** Ionization Constants of Selected Acids at 25°C *(Continued)*

| Acid | Step | Aqueous Equilibrium[a] | $K_a$ | $pK_a$ |
|---|---|---|---|---|
| hydrosulfuric | 1 | $H_2S(aq) + H_2O(\ell) \rightleftharpoons H_3O^+(aq) + HS^-(aq)$ | $8.9 \times 10^{-8}$ | 7.05 |
| | 2 | $HS^-(aq) + H_2O(\ell) \rightleftharpoons H_3O^+(aq) + S^{2-}(aq)$ | $\sim 10^{-19}$ | $\sim 19$ |
| hypobromous | 1 | $HBrO(aq) + H_2O(\ell) \rightleftharpoons H_3O^+(aq) + BrO^-(aq)$ | $2.3 \times 10^{-9}$ | 8.64 |
| hypochlorous | 1 | $HClO(aq) + H_2O(\ell) \rightleftharpoons H_3O^+(aq) + ClO^-(aq)$ | $2.9 \times 10^{-8}$ | 7.54 |
| hypoiodous | 1 | $HIO(aq) + H_2O(\ell) \rightleftharpoons H_3O^+(aq) + IO^-(aq)$ | $2.3 \times 10^{-11}$ | 10.64 |
| iodic | 1 | $HIO_3(aq) + H_2O(\ell) \rightleftharpoons H_3O^+(aq) + IO_3^-(aq)$ | $1.7 \times 10^{-1}$ | 0.77 |
| iodoacetic | 1 | $CH_2ICOOH(aq) + H_2O(\ell) \rightleftharpoons H_3O^+(aq) + CH_2ICOO^-(aq)$ | $7.6 \times 10^{-4}$ | 3.12 |
| lactic | 1 | $CH_3CHOHCOOH(aq) + H_2O(\ell) \rightleftharpoons H_3O^+(aq) + CH_3CHOHCOO^-(aq)$ | $1.4 \times 10^{-4}$ | 3.85 |
| maleic (*cis*-butenedioic) | 1 | $HOOCCH{=}CHCOOH(aq) + H_2O(\ell) \rightleftharpoons H_3O^+(aq) + HOOCCH{=}CHCOO^-(aq)$ | $1.2 \times 10^{-2}$ | 1.92 |
| | 2 | $HOOCCH{=}CHCOO^-(aq) + H_2O(\ell) \rightleftharpoons H_3O^+(aq) + {}^-OOCCH{=}CHCOO^-(aq)$ | $4.7 \times 10^{-7}$ | 6.33 |
| malonic | 1 | $HOOCCH_2COOH(aq) + H_2O(\ell) \rightleftharpoons H_3O^+(aq) + HOOCCH_2COO^-(aq)$ | $1.5 \times 10^{-3}$ | 2.82 |
| | 2 | $HOOCCH_2COO^-(aq) + H_2O(\ell) \rightleftharpoons H_3O^+(aq) + {}^-OOCCH_2COO^-(aq)$ | $2.0 \times 10^{-6}$ | 5.70 |
| nitric | 1 | $HNO_3(aq) + H_2O(\ell) \rightarrow H_3O^+(aq) + NO_3^-(aq)$ | $\gg 1$ (strong) | $<0$ |
| nitrous | 1 | $HNO_2(aq) + H_2O(\ell) \rightleftharpoons H_3O^+(aq) + NO_2^-(aq)$ | $4.0 \times 10^{-4}$ | 3.40 |
| oxalic | 1 | $HOOCCOOH(aq) + H_2O(\ell) \rightleftharpoons H_3O^+(aq) + HOOCCOO^-(aq)$ | $5.9 \times 10^{-2}$ | 1.23 |
| | 2 | $HOOCCOO^-(aq) + H_2O(\ell) \rightleftharpoons H_3O^+(aq) + {}^-OOCCOO^-(aq)$ | $6.4 \times 10^{-5}$ | 4.19 |
| perchloric | 1 | $HClO_4(aq) + H_2O(\ell) \rightarrow H_3O^+(aq) + ClO_4^-(aq)$ | $\gg 1$ (strong) | $<0$ |
| periodic | 1 | $HIO_4(aq) + H_2O(\ell) \rightleftharpoons H_3O^+(aq) + IO_4^-(aq)$ | $2.3 \times 10^{-2}$ | 1.64 |
| phenol | 1 | $C_6H_5OH(aq) + H_2O(\ell) \rightleftharpoons H_3O^+(aq) + C_6H_5O^-(aq)$ | $1.3 \times 10^{-10}$ | 9.89 |
| phosphoric | 1 | $H_3PO_4(aq) + H_2O(\ell) \rightleftharpoons H_3O^+(aq) + H_2PO_4^-(aq)$ | $6.9 \times 10^{-3}$ | 2.16 |
| | 2 | $H_2PO_4^-(aq) + H_2O(\ell) \rightleftharpoons H_3O^+(aq) + HPO_4^{2-}(aq)$ | $6.4 \times 10^{-8}$ | 7.19 |
| | 3 | $HPO_4^{2-}(aq) + H_2O(\ell) \rightleftharpoons H_3O^+(aq) + PO_4^{3-}(aq)$ | $4.8 \times 10^{-13}$ | 12.32 |
| propanoic | 1 | $CH_3CH_2COOH(aq) + H_2O(\ell) \rightleftharpoons H_3O^+(aq) + CH_3CH_2COO^-(aq)$ | $1.4 \times 10^{-5}$ | 4.85 |
| pyruvic | 1 | $CH_3C(O)COOH(aq) + H_2O(\ell) \rightleftharpoons H_3O^+(aq) + CH_3C(O)COO^-(aq)$ | $2.8 \times 10^{-3}$ | 2.55 |
| sulfuric | 1 | $H_2SO_4(aq) + H_2O(\ell) \rightarrow H_3O^+(aq) + HSO_4^-(aq)$ | $\gg 1$ (strong) | $<0$ |
| | 2 | $HSO_4^-(aq) + H_2O(\ell) \rightleftharpoons H_3O^+(aq) + SO_4^{2-}(aq)$ | $1.2 \times 10^{-2}$ | 1.92 |
| sulfurous | 1 | $H_2SO_3(aq) + H_2O(\ell) \rightleftharpoons H_3O^+(aq) + HSO_3^-(aq)$ | $1.7 \times 10^{-2}$ | 1.77 |
| | 2 | $HSO_3^-(aq) + H_2O(\ell) \rightarrow H_3O^+(aq) + SO_3^{2-}(aq)$ | $6.2 \times 10^{-8}$ | 7.21 |
| thiocyanic | 1 | $HSCN(aq) + H_2O(\ell) \rightleftharpoons H_3O^+(aq) + SCN^-(aq)$ | $\gg 1$ (strong) | $<0$ |
| trichloroacetic | 1 | $CCl_3COOH(aq) + H_2O(\ell) \rightleftharpoons H_3O^+(aq) + CCl_3COO^-(aq)$ | $2.3 \times 10^{-1}$ | 0.64 |
| trifluoroacetic | 1 | $CF_3COOH(aq) + H_2O(\ell) \rightleftharpoons H_3O^+(aq) + CF_3COO^-(aq)$ | $5.9 \times 10^{-1}$ | 0.23 |
| water | 1 | $H_2O(aq) + H_2O(\ell) \rightleftharpoons H_3O^+(aq) + OH^-(aq)$ | $1.0 \times 10^{-14}$ | 14.00 |

[a]The formulas of most carboxylic acids are written in an RCOOH format to highlight their molecular structures.

**TABLE A5.2** Acid Ionization Constants of Hydrated Metal Ions at 25°C

| Free Ion | Hydrated Ion | $K_a$ |
|---|---|---|
| $Fe^{3+}$ | $Fe(H_2O)_6{}^{3+}$ | $3 \times 10^{-3}$ |
| $Sn^{2+}$ | $Sn(H_2O)_6{}^{2+}$ | $4 \times 10^{-4}$ |
| $Cr^{3+}$ | $Cr(H_2O)_6{}^{3+}$ | $1 \times 10^{-4}$ |
| $Al^{3+}$ | $Al(H_2O)_6{}^{3+}$ | $1 \times 10^{-5}$ |
| $Cu^{2+}$ | $Cu(H_2O)_6{}^{2+}$ | $3 \times 10^{-8}$ |
| $Pb^{2+}$ | $Pb(H_2O)_6{}^{2+}$ | $3 \times 10^{-8}$ |
| $Zn^{2+}$ | $Zn(H_2O)_6{}^{2+}$ | $1 \times 10^{-9}$ |
| $Co^{2+}$ | $Co(H_2O)_6{}^{2+}$ | $2 \times 10^{-10}$ |
| $Ni^{2+}$ | $Ni(H_2O)_6{}^{2+}$ | $1 \times 10^{-10}$ |

**TABLE A5.3** Ionization Constants of Selected Bases at 25°C

| Base | Aqueous Equilibrium | $K_b$ | $pK_b$ |
|---|---|---|---|
| ammonia | $NH_3(aq) + H_2O(\ell) \rightleftharpoons NH_4{}^+(aq) + OH^-(aq)$ | $1.76 \times 10^{-5}$ | 4.75 |
| aniline | $C_6H_5NH_2(aq) + H_2O(\ell) \rightleftharpoons C_6H_5NH_3{}^+(aq) + OH^-(aq)$ | $4.0 \times 10^{-10}$ | 9.40 |
| diethylamine | $(CH_3CH_2)_2NH(aq) + H_2O(\ell) \rightleftharpoons (CH_3CH_2)_2NH_2{}^+(aq) + OH^-(aq)$ | $8.6 \times 10^{-4}$ | 3.07 |
| dimethylamine | $(CH_3)_2NH(aq) + H_2O(\ell) \rightleftharpoons (CH_3)_2NH_2{}^+(aq) + OH^-(aq)$ | $5.9 \times 10^{-4}$ | 3.23 |
| ethylamine | $CH_3CH_2NH_2(aq) + H_2O(\ell) \rightleftharpoons CH_3CH_2NH_3{}^+(aq) + OH^-(aq)$ | $5.6 \times 10^{-4}$ | 3.25 |
| methylamine | $CH_3NH_2(aq) + H_2O(\ell) \rightleftharpoons CH_3NH_3{}^+(aq) + OH^-(aq)$ | $4.4 \times 10^{-4}$ | 3.36 |
| nicotine (1) | $+ H_2O \rightleftharpoons$ $+ OH^-$ | $1.0 \times 10^{-6}$ | 6.0 |
| (2) | $+ H_2O \rightleftharpoons$ $+ OH^-$ | $1.3 \times 10^{-11}$ | 10.9 |
| pyridine | $C_5H_5N(aq) + H_2O(\ell) \rightleftharpoons C_5H_5NH^+(aq) + OH^-(aq)$ | $1.7 \times 10^{-9}$ | 8.77 |
| quinine (1) | $+ H_2O \rightleftharpoons$ $+ OH^-$ | $3.3 \times 10^{-6}$ | 5.48 |
| (2) | $+ H_2O \rightleftharpoons$ $+ OH^-$ | $1.4 \times 10^{-10}$ | 9.9 |
| trimethylamine | $(CH_3)_3N(aq) + H_2O(\ell) \rightleftharpoons (CH_3)_3NH^+(aq) + OH^-(aq)$ | $6.46 \times 10^{-5}$ | 4.19 |
| urea | $H_2NCONH_2(aq) + H_2O(\ell) \rightleftharpoons H_2NCONH_3{}^+(aq) + OH^-(aq)$ | $1.3 \times 10^{-14}$ | 13.9 |

**TABLE A5.4** Solubility-Product Constants at 25°C

| Cation | Anion | Heterogeneous Equilibrium[a] | $K_{sp}$[b] |
|---|---|---|---|
| aluminum | hydroxide | $Al(OH)_3(s) \rightleftharpoons Al^{3+}(aq) + 3\,OH^-(aq)$ | $1.3 \times 10^{-33}$ |
| | phosphate | $AlPO_4(s) \rightleftharpoons Al^{3+}(aq) + PO_4^{3-}(aq)$ | $9.84 \times 10^{-21}$ |
| barium | carbonate | $BaCO_3(s) \rightleftharpoons Ba^{2+}(aq) + CO_3^{2-}(aq)$ | $2.58 \times 10^{-9}$ |
| | fluoride | $BaF_2(s) \rightleftharpoons Ba^{2+}(aq) + 2\,F^-(aq)$ | $1.84 \times 10^{-7}$ |
| | sulfate | $BaSO_4(s) \rightleftharpoons Ba^{2+}(aq) + SO_4^{2-}(aq)$ | $1.08 \times 10^{-10}$ |
| calcium | carbonate | $CaCO_3(s) \rightleftharpoons Ca^{2+}(aq) + CO_3^{2-}(aq)$ | $2.8 \times 10^{-9}$ |
| | fluoride | $CaF_2(s) \rightleftharpoons Ca^{2+}(aq) + 2\,F^-(aq)$ | $5.3 \times 10^{-9}$ |
| | hydroxide | $Ca(OH)_2(s) \rightleftharpoons Ca^{2+}(aq) + 2\,OH^-(aq)$ | $5.5 \times 10^{-6}$ |
| | phosphate | $Ca_3(PO_4)_2(s) \rightleftharpoons 3\,Ca^{2+}(aq) + 2\,PO_4^{3-}(aq)$ | $2.07 \times 10^{-29}$ |
| | sulfate | $CaSO_4(s) \rightleftharpoons Ca^{2+}(aq) + SO_4^{2-}(aq)$ | $4.93 \times 10^{-5}$ |
| cobalt(II) | carbonate | $CoCO_3(s) \rightleftharpoons Co^{2+}(aq) + CO_3^{2-}(aq)$ | $1.4 \times 10^{-3}$ |
| | phosphate | $Co_3(PO_4)_2(s) \rightleftharpoons 3\,Co^{2+}(aq) + 2\,PO_4^{3-}(aq)$ | $2.05 \times 10^{-7}$ |
| | sulfide | $CoS(s) \rightleftharpoons Co^{2+}(aq) + S^{2-}(aq)$ | $2.0 \times 10^{-25}$ |
| copper(I) | bromide | $CuBr(s) \rightleftharpoons Cu^+(aq) + Br^-(aq)$ | $6.27 \times 10^{-9}$ |
| | chloride | $CuCl(s) \rightleftharpoons Cu^+(aq) + Cl^-(aq)$ | $1.72 \times 10^{-7}$ |
| | iodide | $CuI(s) \rightleftharpoons Cu^+(aq) + I^-(aq)$ | $1.27 \times 10^{-12}$ |
| copper(II) | phosphate | $Cu_3(PO_4)_2(s) \rightleftharpoons 3\,Cu^{2+}(aq) + 2\,PO_4^{3-}(aq)$ | $1.4 \times 10^{-37}$ |
| | hydroxide | $Cu(OH)_2(s) \rightleftharpoons Cu^{2+}(aq) + 2\,OH^-(aq)$ | $2.2 \times 10^{-20}$ |
| iron(II) | carbonate | $FeCO_3(s) \rightleftharpoons Fe^{2+}(aq) + CO_3^{2-}(aq)$ | $3.13 \times 10^{-11}$ |
| | fluoride | $FeF_2(s) \rightleftharpoons Fe^{2+}(aq) + 2\,F^-(aq)$ | $2.36 \times 10^{-6}$ |
| | hydroxide | $Fe(OH)_2(s) \rightleftharpoons Fe^{2+}(aq) + 2\,OH^-(aq)$ | $4.87 \times 10^{-17}$ |
| | sulfide | $FeS(s) \rightleftharpoons Fe^{2+}(aq) + S^{2-}(aq)$ | $6.3 \times 10^{-18}$ |
| lead | bromide | $PbBr_2(s) \rightleftharpoons Pb^{2+}(aq) + 2\,Br^-(aq)$ | $6.60 \times 10^{-6}$ |
| | carbonate | $PbCO_3(s) \rightleftharpoons Pb^{2+}(aq) + CO_3^{2-}(aq)$ | $7.4 \times 10^{-14}$ |
| | chloride | $PbCl_2(s) \rightleftharpoons Pb^{2+}(aq) + 2\,Cl^-(aq)$ | $1.7 \times 10^{-5}$ |
| | fluoride | $PbF_2(s) \rightleftharpoons Pb^{2+}(aq) + 2\,F^-(aq)$ | $3.3 \times 10^{-8}$ |
| | iodide | $PbI_2(s) \rightleftharpoons Pb^{2+}(aq) + 2\,I^-(aq)$ | $9.8 \times 10^{-9}$ |
| | sulfate | $PbSO_4(s) \rightleftharpoons Pb^{2+}(aq) + SO_4^{2-}(aq)$ | $2.53 \times 10^{-8}$ |
| lithium | carbonate | $Li_2CO_3(s) \rightleftharpoons 2\,Li^+(aq) + CO_3^{2-}(aq)$ | $2.5 \times 10^{-2}$ |
| magnesium | carbonate | $MgCO_3(s) \rightleftharpoons Mg^{2+}(aq) + CO_3^{2-}(aq)$ | $6.82 \times 10^{-6}$ |
| | fluoride | $MgF_2(s) \rightleftharpoons Mg^{2+}(aq) + 2\,F^-(aq)$ | $5.16 \times 10^{-11}$ |
| | hydroxide | $Mg(OH)_2(s) \rightleftharpoons Mg^{2+}(aq) + 2\,OH^-(aq)$ | $5.61 \times 10^{-12}$ |
| manganese(II) | carbonate | $MnCO_3(s) \rightleftharpoons Mn^{2+}(aq) + CO_3^{2-}(aq)$ | $2.34 \times 10^{-11}$ |
| | hydroxide | $Mn(OH)_2(s) \rightleftharpoons Mn^{2+}(aq) + 2\,OH^-(aq)$ | $1.9 \times 10^{-13}$ |
| mercury(I) | bromide | $Hg_2Br_2(s) \rightleftharpoons Hg_2^{2+}(aq) + 2\,Br^-(aq)$ | $6.40 \times 10^{-23}$ |
| | carbonate | $Hg_2CO_3(s) \rightleftharpoons Hg_2^{2+}(aq) + CO_3^{2-}(aq)$ | $3.6 \times 10^{-17}$ |
| | chloride | $Hg_2Cl_2(s) \rightleftharpoons Hg_2^{2+}(aq) + 2\,Cl^-(aq)$ | $1.43 \times 10^{-18}$ |
| | iodide | $Hg_2I_2(s) \rightleftharpoons Hg_2^{2+}(aq) + 2\,I^-(aq)$ | $5.2 \times 10^{-29}$ |
| | sulfate | $Hg_2SO_4(s) \rightleftharpoons Hg_2^{2+}(aq) + SO_4^{2-}(aq)$ | $6.5 \times 10^{-7}$ |
| mercury(II) | hydroxide | $Hg(OH)_2(s) \rightleftharpoons Hg^{2+}(aq) + 2\,OH^-(aq)$ | $3.2 \times 10^{-26}$ |
| | iodide | $HgI_2(s) \rightleftharpoons Hg^{2+}(aq) + 2\,I^-(aq)$ | $2.9 \times 10^{-29}$ |
| nickel(II) | carbonate | $NiCO_3(s) \rightleftharpoons Ni^{2+}(aq) + CO_3^{2-}(aq)$ | $1.42 \times 10^{-7}$ |
| | phosphate | $Ni_3(PO_4)_2(s) \rightleftharpoons 3\,Ni^{2+}(aq) + 2\,PO_4^{3-}(aq)$ | $4.74 \times 10^{-32}$ |
| | sulfide | $NiS(s) \rightleftharpoons Ni^{2+}(aq) + S^{2-}(aq)$ | $1 \times 10^{-24}$ |
| silver | bromide | $AgBr(s) \rightleftharpoons Ag^+(aq) + Br^-(aq)$ | $5.35 \times 10^{-13}$ |
| | carbonate | $Ag_2CO_3(s) \rightleftharpoons 2\,Ag^+(aq) + CO_3^{2-}(aq)$ | $8.46 \times 10^{-12}$ |
| | chloride | $AgCl(s) \rightleftharpoons Ag^+(aq) + Cl^-(aq)$ | $1.77 \times 10^{-10}$ |
| | chromate | $Ag_2CrO_4(s) \rightleftharpoons 2\,Ag^+(aq) + CrO_4^{2-}(aq)$ | $1.12 \times 10^{-12}$ |
| | hydroxide | $AgOH(s) \rightleftharpoons Ag^+(aq) + OH^-(aq)$ | $2.0 \times 10^{-8}$ |
| | iodide | $AgI(s) \rightleftharpoons Ag^+(aq) + I^-(aq)$ | $8.52 \times 10^{-17}$ |
| | phosphate | $Ag_3PO_4(s) \rightleftharpoons 3\,Ag^+(aq) + PO_4^{3-}(aq)$ | $8.89 \times 10^{-17}$ |
| | sulfate | $Ag_2SO_4(s) \rightleftharpoons 2\,Ag^+(aq) + SO_4^{2-}(aq)$ | $1.20 \times 10^{-5}$ |
| | sulfide | $Ag_2S(s) \rightleftharpoons 2\,Ag^+(aq) + S^{2-}(aq)$ | $6.3 \times 10^{-50}$ |
| strontium | carbonate | $SrCO_3(s) \rightleftharpoons Sr^{2+}(aq) + CO_3^{2-}(aq)$ | $5.60 \times 10^{-10}$ |
| | fluoride | $SrF_2(s) \rightleftharpoons Sr^{2+}(aq) + 2\,F^-(aq)$ | $4.33 \times 10^{-9}$ |
| | sulfate | $SrSO_4(s) \rightleftharpoons Sr^{2+}(aq) + SO_4^{2-}(aq)$ | $3.44 \times 10^{-7}$ |
| zinc | carbonate | $ZnCO_3(s) \rightleftharpoons Zn^{2+}(aq) + CO_3^{2-}(aq)$ | $1.46 \times 10^{-10}$ |
| | hydroxide | $Zn(OH)_2(s) \rightleftharpoons Zn^{2+}(aq) + 2\,OH^-(aq)$ | $3.0 \times 10^{-17}$ |

[a]Equilibrium is between solid phase and aqueous solution.
[b]From Dean, J. *Lange's Handbook of Chemistry* (The McGraw-Hill Companies, 1998).

**TABLE A5.5** Formation Constants of Complex Ions at 25°C

| Complex Ion | Aqueous Equilibrium | $K_f$ |
|---|---|---|
| $[Ag(NH_3)_2]^+$ | $Ag^+(aq) + 2\ NH_3(aq) \rightleftharpoons Ag(NH_3)_2^+(aq)$ | $1.7 \times 10^7$ |
| $[AgCl_2]^-$ | $Ag^+(aq) + 2\ Cl^-(aq) \rightleftharpoons AgCl_2^-(aq)$ | $2.5 \times 10^5$ |
| $[Ag(CN)_2]^-$ | $Ag^+(aq) + 2\ CN^-(aq) \rightleftharpoons Ag(CN)_2^-(aq)$ | $1.0 \times 10^{21}$ |
| $[Ag(S_2O_3)_2]^{3-}$ | $Ag^+(aq) + 2\ S_2O_3^{2-}(aq) \rightleftharpoons Ag(S_2O_3)_2^{3-}(aq)$ | $4.7 \times 10^{13}$ |
| $[AlF_6]^{3-}$ | $Al^{3+}(aq) + 6\ F^-(aq) \rightleftharpoons AlF_6^{3-}(aq)$ | $4.0 \times 10^{19}$ |
| $[Al(OH)_4]^-$ | $Al^{3+}(aq) + 4\ OH^-(aq) \rightleftharpoons Al(OH)_4^-(aq)$ | $7.7 \times 10^{33}$ |
| $[Au(CN)_2]^-$ | $Au^+(aq) + 2\ CN^-(aq) \rightleftharpoons Au(CN)_2^-(aq)$ | $2.0 \times 10^{38}$ |
| $[Co(NH_3)_6]^{2+}$ | $Co^{2+}(aq) + 6\ NH_3(aq) \rightleftharpoons Co(NH_3)_6^{2+}(aq)$ | $7.7 \times 10^4$ |
| $[Co(NH_3)_6]^{3+}$ | $Co^{3+}(aq) + 6\ NH_3(aq) \rightleftharpoons Co(NH_3)_6^{3+}(aq)$ | $5.0 \times 10^{31}$ |
| $[Co(en)_3]^{2+}$ | $Co^{2+}(aq) + 3\ en(aq) \rightleftharpoons Co(en)_3^{2+}(aq)$ | $8.7 \times 10^{13}$ |
| $[Co(C_2O_4)_3]^{4-}$ | $Co^{2+}(aq) + 3\ C_2O_4^{2-}(aq) \rightleftharpoons Co(C_2O_4)_3^{4-}(aq)$ | $4.5 \times 10^6$ |
| $[Cu(NH_3)_4]^{2+}$ | $Cu^{2+}(aq) + 4\ NH_3(aq) \rightleftharpoons Cu(NH_3)_4^{2+}(aq)$ | $5.0 \times 10^{13}$ |
| $[Cu(en)_2]^{2+}$ | $Cu^{2+}(aq) + 2\ en(aq) \rightleftharpoons Cu(en)_2^{2+}(aq)$ | $3.2 \times 10^{19}$ |
| $[Cu(CN)_4]^{2-}$ | $Cu^{2+}(aq) + 4\ CN^-(aq) \rightleftharpoons Cu(CN)_4^{2-}(aq)$ | $1.0 \times 10^{25}$ |
| $[Cu(C_2O_4)_2]^{2-}$ | $Cu^{2+}(aq) + 2\ C_2O_4^{2-}(aq) \rightleftharpoons Cu(C_2O_4)_2^{2-}(aq)$ | $1.7 \times 10^{10}$ |
| $[Fe(C_2O_4)_3]^{4-}$ | $Fe^{2+}(aq) + 3\ C_2O_4^{2-}(aq) \rightleftharpoons Fe(C_2O_4)_3^{4-}(aq)$ | $6 \times 10^6$ |
| $[Fe(C_2O_4)_3]^{3-}$ | $Fe^{3+}(aq) + 3\ C_2O_4^{2-}(aq) \rightleftharpoons Fe(C_2O_4)_3^{3-}(aq)$ | $3.3 \times 10^{20}$ |
| $[HgCl_4]^{2-}$ | $Hg^{2+}(aq) + 4\ Cl^-(aq) \rightleftharpoons HgCl_4^{2-}(aq)$ | $1.2 \times 10^{15}$ |
| $[Ni(NH_3)_6]^{2+}$ | $Ni^{2+}(aq) + 6\ NH_3(aq) \rightleftharpoons Ni(NH_3)_6^{2+}(aq)$ | $5.5 \times 10^8$ |
| $[PbCl_4]^{2-}$ | $Pb^{2+}(aq) + 4\ Cl^-(aq) \rightleftharpoons PbCl_4^{2-}(aq)$ | $2.5 \times 10^1$ |
| $[Zn(NH_3)_4]^{2+}$ | $Zn^{2+}(aq) + 4\ NH_3(aq) \rightleftharpoons Zn(NH_3)_4^{2+}(aq)$ | $2.9 \times 10^9$ |
| $[Zn(OH)_4]^{2-}$ | $Zn^{2+}(aq) + 4\ OH^-(aq) \rightleftharpoons Zn(OH)_4^{2-}(aq)$ | $2.8 \times 10^{15}$ |

# Appendix 6

## Standard Reduction Potentials

**TABLE A6.1** Standard Reduction Potentials at 25°C

| Half-Reaction | $n$ | $E°$ (V) |
|---|---|---|
| $F_2(g) + 2\ e^- \rightarrow 2\ F^-(aq)$ | 2 | 2.866 |
| $H_2N_2O_2(s) + 2\ H^+(aq) + 2\ e^- \rightarrow N_2(g) + 2\ H_2O(\ell)$ | 2 | 2.65 |
| $O(g) + 2\ H^+(aq) + 2\ e^- \rightarrow H_2O(\ell)$ | 2 | 2.421 |
| $Cu^{3+}(aq) + e^- \rightarrow Cu^{2+}(aq)$ | 1 | 2.4 |
| $XeO_3(s) + 6\ H^+(aq) + 6\ e^- \rightarrow Xe(g) + 3\ H_2O(\ell)$ | 6 | 2.10 |
| $O_3(g) + 2\ H^+(aq) + 2\ e^- \rightarrow O_2(g) + H_2O(\ell)$ | 2 | 2.076 |
| $OH(g) + e^- \rightarrow OH^-(aq)$ | 1 | 2.02 |
| $Co^{3+}(aq) + e^- \rightarrow Co^{2+}(aq)$ | 1 | 1.92 |
| $H_2O_2(\ell) + 2\ H^+(aq) + 2\ e^- \rightarrow 2\ H_2O(\ell)$ | 2 | 1.776 |
| $N_2O(g) + 2\ H^+(aq) + 2\ e^- \rightarrow N_2(g) + H_2O(\ell)$ | 2 | 1.766 |
| $Ce(OH)^{3+}(aq) + H^+(aq) + e^- \rightarrow Ce^{3+}(aq) + H_2O(\ell)$ | 1 | 1.70 |
| $Au^+(aq) + e^- \rightarrow Au(s)$ | 1 | 1.692 |
| $PbO_2(s) + SO_4{}^{2-}(aq) + 4\ H^+(aq) + 2\ e^- \rightarrow PbSO_4(s) + 2\ H_2O(\ell)$ | 2 | 1.691 |
| $PbO_2(s) + HSO_4{}^-(aq) + 3\ H^+(aq) + 2\ e^- \rightarrow PbSO_4(s) + 2\ H_2O(\ell)$ | 2 | 1.685 |
| $MnO_4{}^-(aq) + 4\ H^+(aq) + 3\ e^- \rightarrow MnO_2(s) + 2\ H_2O(\ell)$ | 3 | 1.673 |
| $NiO_2(s) + 4\ H^+(aq) + 2\ e^- \rightarrow Ni^{2+}(aq) + 2\ H_2O(\ell)$ | 2 | 1.678 |
| $HClO(\ell) + H^+(aq) + e^- \rightarrow \frac{1}{2}\ Cl_2(g) + H_2O(aq)$ | 1 | 1.63 |
| $Ce^{4+}(aq) + e^- \rightarrow Ce^{3+}(aq)$ | 1 | 1.61 |
| $Mn^{3+}(aq) + e^- \rightarrow Mn^{2+}(aq)$ | 1 | 1.542 |
| $MnO_4{}^-(aq) + 8\ H^+(aq) + 5\ e^- \rightarrow Mn^{2+}(aq) + 4\ H_2O(\ell)$ | 5 | 1.507 |
| $BrO_3{}^-(aq) + 6\ H^+(aq) + 5\ e^- \rightarrow \frac{1}{2}\ Br_2(\ell) + 3\ H_2O(\ell)$ | 5 | 1.52 |
| $ClO_3{}^-(aq) + 6\ H^+(aq) + 5\ e^- \rightarrow \frac{1}{2}\ Cl_2(g) + 3\ H_2O(\ell)$ | 5 | 1.47 |
| $PbO_2(s) + 4\ H^+(aq) + 2\ e^- \rightarrow Pb^{2+}(aq) + 2\ H_2O(\ell)$ | 2 | 1.455 |
| $Au^{3+}(aq) + 3\ e^- \rightarrow Au(s)$ | 3 | 1.40 |
| $Cl_2(g) + 2\ e^- \rightarrow 2\ Cl^-(aq)$ | 2 | 1.358 |
| $Cr_2O_7{}^{2-}(aq) + 14\ H^+(aq) + 6\ e^- \rightarrow 2\ Cr^{3+}(aq) + 7\ H_2O(\ell)$ | 6 | 1.33 |
| $MnO_2(s) + 4\ H^+(aq) + 2\ e^- \rightarrow Mn^{2+}(aq) + 2\ H_2O(\ell)$ | 2 | 1.23 |
| $O_2(g) + 4\ H^+(aq) + 4\ e^- \rightarrow 2\ H_2O(\ell)$ | 4 | 1.229 |
| $IO_3{}^-(aq) + 6\ H^+(aq) + 5\ e^- \rightarrow \frac{1}{2}\ I_2(s) + 3\ H_2O(\ell)$ | 5 | 1.195 |
| $IO_3{}^-(aq) + 6\ H^+(aq) + 6\ e^- \rightarrow I^-(aq) + 3\ H_2O(\ell)$ | 6 | 1.085 |
| $Br_2(\ell) + 2\ e^- \rightarrow 2\ Br^-(aq)$ | 2 | 1.066 |
| $HNO_2(\ell) + H^+(aq) + e^- \rightarrow NO(g) + H_2O(\ell)$ | 1 | 1.00 |
| $VO_2{}^+(aq) + 2\ H^+(aq) + e^- \rightarrow VO^{2+}(aq) + H_2O(\ell)$ | 1 | 0.991 |
| $NO_3{}^-(aq) + 4\ H^+(aq) + 3\ e^- \rightarrow NO(g) + 2\ H_2O(\ell)$ | 3 | 0.96 |
| $2\ Hg^{2+}(aq) + 2\ e^- \rightarrow Hg_2{}^{2+}(aq)$ | 2 | 0.92 |

**TABLE A6.1** Standard Reduction Potentials at 25°C *(Continued)*

| Half-Reaction | *n* | *E*° (V) |
|---|---|---|
| $ClO^-(aq) + H_2O(\ell) + 2\,e^- \rightarrow Cl^-(aq) + 2\,OH^-(aq)$ | 2 | 0.89 |
| $HO_2^-(aq) + H_2O(\ell) + 2\,e^- \rightarrow 3\,OH^-(aq)$ | 2 | 0.88 |
| $Hg^{2+}(aq) + 2\,e^- \rightarrow Hg(\ell)$ | 2 | 0.851 |
| $Ag^+(aq) + e^- \rightarrow Ag(s)$ | 1 | 0.800 |
| $Hg_2^{2+}(aq) + 2\,e^- \rightarrow 2\,Hg(\ell)$ | 2 | 0.797 |
| $Fe^{3+}(aq) + e^- \rightarrow Fe^{2+}(aq)$ | 1 | 0.770 |
| $PtCl_4^{2-}(aq) + 2\,e^- \rightarrow Pt(s) + 4\,Cl^-(aq)$ | 2 | 0.73 |
| $O_2(g) + 2\,H^+(aq) + 2\,e^- \rightarrow H_2O_2(\ell)$ | 2 | 0.68 |
| $MnO_4^-(aq) + 2\,H_2O(\ell) + 3\,e^- \rightarrow MnO_2(s) + 4\,OH^-(aq)$ | 3 | 0.59 |
| $H_3AsO_4(s) + 2\,H^+(aq) + 2\,e^- \rightarrow H_3AsO_3(aq) + H_2O(\ell)$ | 2 | 0.559 |
| $I_2(s) + 2\,e^- \rightarrow 2\,I^-(aq)$ | 2 | 0.536 |
| $Cu^+(aq) + e^- \rightarrow Cu(s)$ | 1 | 0.521 |
| $2\,NiO(OH)(s) + 2\,H_2O(\ell) + 2\,e^- \rightarrow 2\,Ni(OH)_2(s) + 2\,OH^-(aq)$ | 2 | 0.52 |
| $H_2SO_3(\ell) + 4\,H^+(aq) + 4\,e^- \rightarrow S(s) + 3\,H_2O(\ell)$ | 4 | 0.449 |
| $Ag_2CrO_4(s) + 2\,e^- \rightarrow 2\,Ag(s) + CrO_4^{2-}(aq)$ | 2 | 0.447 |
| $O_2(g) + 2\,H_2O(\ell) + 4\,e^- \rightarrow 4\,OH^-(aq)$ | 4 | 0.401 |
| $Fe(CN)_6^{3-}(aq) + e^- \rightarrow Fe(CN)_6^{4-}(aq)$ | 1 | 0.36 |
| $Ag_2O(s) + H_2O(\ell) + 2\,e^- \rightarrow 2\,Ag(s) + 2\,OH^-(aq)$ | 2 | 0.342 |
| $Cu^{2+}(aq) + 2\,e^- \rightarrow Cu(s)$ | 2 | 0.342 |
| $BiO^+(aq) + 2\,H^+(aq) + 3\,e^- \rightarrow Bi(s) + H_2O(\ell)$ | 3 | 0.32 |
| $AgCl(s) + e^- \rightarrow Ag(s) + Cl^-(aq)$ | 1 | 0.222 |
| $HSO_4^-(aq) + 3\,H^+(aq) + 2\,e^- \rightarrow H_2SO_3(\ell) + H_2O(\ell)$ | 2 | 0.17 |
| $Sn^{4+}(aq) + 2\,e^- \rightarrow Sn^{2+}(aq)$ | 2 | 0.154 |
| $Cu^{2+}(aq) + e^- \rightarrow Cu^+(aq)$ | 1 | 0.153 |
| $2\,MnO_2(s) + H_2O(\ell) + 2\,e^- \rightarrow Mn_2O_3(s) + 2\,OH^-(aq)$ | 2 | 0.15 |
| $S(s) + 2\,H^+(aq) + 2\,e^- \rightarrow H_2S(g)$ | 2 | 0.141 |
| $HgO(s) + H_2O(\ell) + 2\,e^- \rightarrow Hg(\ell) + 2\,OH^-(aq)$ | 2 | 0.0977 |
| $AgBr(s) + e^- \rightarrow Ag(s) + Br^-(aq)$ | 1 | 0.095 |
| $Ag(S_2O_3)_2^{3-}(aq) + e^- \rightarrow Ag(s) + 2\,S_2O_3^{2-}(aq)$ | 1 | 0.01 |
| $NO_3^-(aq) + H_2O(\ell) + 2\,e^- \rightarrow NO_2^-(aq) + 2\,OH^-(aq)$ | 2 | 0.01 |
| $2\,H^+(aq) + 2\,e^- \rightarrow H_2(g)$ | 2 | 0.000 |
| $Pb^{2+}(aq) + 2\,e^- \rightarrow Pb(s)$ | 2 | −0.126 |
| $CrO_4^{2-}(aq) + 4\,H_2O(\ell) + 3\,e^- \rightarrow Cr(OH)_3(s) + 5\,OH^-(aq)$ | 3 | −0.13 |
| $Sn^{2+}(aq) + 2\,e^- \rightarrow Sn(s)$ | 2 | −0.136 |
| $AgI(s) + e^- \rightarrow Ag(s) + I^-(aq)$ | 1 | −0.152 |
| $CuI(s) + e^- \rightarrow Cu(s) + I^-(aq)$ | 1 | −0.185 |
| $N_2(g) + 5\,H^+(aq) + 4\,e^- \rightarrow N_2H_5^+(aq)$ | 4 | −0.23 |
| $Ni^{2+}(aq) + 2\,e^- \rightarrow Ni(s)$ | 2 | −0.257 |
| $PbSO_4(s) + H^+(aq) + 2\,e^- \rightarrow Pb(s) + HSO_4^-(aq)$ | 2 | −0.356 |
| $Co^{2+}(aq) + 2\,e^- \rightarrow Co(s)$ | 2 | −0.277 |
| $Ag(CN)_2^-(aq) + e^- \rightarrow Ag(s) + 2\,CN^-(aq)$ | 1 | −0.31 |
| $Cd^{2+}(aq) + 2\,e^- \rightarrow Cd(s)$ | 2 | −0.403 |
| $Cr^{3+}(aq) + e^- \rightarrow Cr^{2+}(aq)$ | 1 | −0.41 |

*Continued on next page*

**TABLE A6.1** Standard Reduction Potentials at 25°C *(Continued)*

| Half-Reaction | *n* | *E*° (V) |
|---|---|---|
| $Fe^{2+}(aq) + 2\ e^- \rightarrow Fe(s)$ | 2 | −0.447 |
| $2\ CO_2(g) + 2\ H^+(aq) + 2\ e^- \rightarrow H_2C_2O_4(s)$ | 2 | −0.49 |
| $Ni(OH)_2(s) + 2\ e^- \rightarrow Ni(s) + 2\ OH^-(aq)$ | 2 | −0.72 |
| $Cr^{3+}(aq) + 3\ e^- \rightarrow Cr(s)$ | 3 | −0.74 |
| $Zn^{2+}(aq) + 2\ e^- \rightarrow Zn(s)$ | 2 | −0.762 |
| $Cd(OH)_2(s) + 2\ e^- \rightarrow Cd(s) + 2\ OH^-(aq)$ | 2 | −0.81 |
| $2\ H_2O(\ell) + 2\ e^- \rightarrow H_2(g) + 2\ OH^-(aq)$ | 2 | −0.828 |
| $SO_4^{2-}(aq) + H_2O(\ell) + 2\ e^- \rightarrow SO_3^{2-}(aq) + 2\ OH^-(aq)$ | 2 | −0.92 |
| $N_2(g) + 4\ H_2O(\ell) + 4\ e^- \rightarrow 4\ OH^-(aq) + N_2H_4(\ell)$ | 4 | −1.16 |
| $Mn^{2+}(aq) + 2\ e^- \rightarrow Mn(s)$ | 2 | −1.185 |
| $Zn(OH)_2(s) + 2\ e^- \rightarrow Zn(s) + 2\ OH^-(aq)$ | 2 | −1.249 |
| $Al^{3+}(aq) + 3\ e^- \rightarrow Al(s)$ | 3 | −1.662 |
| $Mg^{2+}(aq) + 2\ e^- \rightarrow Mg(s)$ | 2 | −2.37 |
| $Na^+(aq) + e^- \rightarrow Na(s)$ | 1 | −2.71 |
| $Ca^{2+}(aq) + 2\ e^- \rightarrow Ca(s)$ | 2 | −2.868 |
| $Ba^{2+}(aq) + 2\ e^- \rightarrow Ba(s)$ | 2 | −2.912 |
| $K^+(aq) + e^- \rightarrow K(s)$ | 1 | −2.95 |
| $Li^+(aq) + e^- \rightarrow Li(s)$ | 1 | −3.05 |

# Appendix 7

## Naming Organic Compounds

While organic chemistry was becoming established as a discipline within chemistry, many compounds were given trivial names that are still commonly used and recognized. We refer to many of these compounds by their nonsystematic names throughout this book, and their names and structures are listed in Table A7.1.

**TABLE A7.1** Organic Compounds and Their Commonly Used Nonsystematic Names

| Name | Formula | Structure |
|---|---|---|
| ethylene | $C_2H_4$ | $H_2C{=}CH_2$ |
| acetylene | $C_2H_2$ | HC≡CH |
| benzene | $C_6H_6$ | (benzene ring) |
| toluene | $C_6H_5CH_3$ | (benzene ring with $CH_3$) |
| ethyl alcohol | $CH_3CH_2OH$ | $H_3C{-}CH_2{-}OH$ |
| acetone | $CH_3COCH_3$ | $H_3C{-}C({=}O){-}CH_3$ |
| acetic acid | $CH_3COOH$ | $H_3C{-}C({=}O){-}OH$ |
| formaldehyde | $CH_2O$ | $H{-}C({=}O){-}H$ |

The International Union of Pure and Applied Chemistry (IUPAC) has proposed a set of rules for the systematic naming of organic compounds. When naming compounds or drawing structures based on names, we need to keep in mind that the IUPAC system of nomenclature is based on two fundamental ideas: (1) the name of a compound must indicate how the carbon atoms in the skeleton are bonded together, and (2) the name must identify the location of any functional groups in the molecule.

## Alkanes

Table A7.2 contains the prefixes used for carbon chains ranging in size from $C_1$ to $C_{20}$ and gives the names for compounds consisting of unbranched chains. The name of a compound consists of a prefix identifying the number of carbons in the chain and a suffix defining the type of hydrocarbon. The suffix *-ane* indicates that the compounds are alkanes and that all carbon–carbon bonds are single bonds.

**TABLE A7.2** Prefixes for Naming Carbon Chains

| Prefix | Example | Name | Prefix | Example | Name |
|---|---|---|---|---|---|
| meth | $CH_4$ | methane | undec | $C_{11}H_{24}$ | undecane |
| eth | $C_2H_6$ | ethane | dodec | $C_{12}H_{26}$ | dodecane |
| pro | $C_3H_8$ | propane | tridec | $C_{13}H_{28}$ | tridecane |
| but | $C_4H_{10}$ | butane | tetradec | $C_{14}H_{30}$ | tetradecane |
| pent | $C_5H_{12}$ | pentane | pentadec | $C_{15}H_{32}$ | pentadecane |
| hex | $C_6H_{14}$ | hexane | hexadec | $C_{16}H_{34}$ | hexadecane |
| hept | $C_7H_{16}$ | heptane | heptadec | $C_{17}H_{36}$ | heptadecane |
| oct | $C_8H_{18}$ | octane | octadec | $C_{18}H_{38}$ | octadecane |
| non | $C_9H_{20}$ | nonane | nonadec | $C_{19}H_{40}$ | nonadecane |
| dec | $C_{10}H_{22}$ | decane | eicos | $C_{20}H_{42}$ | eicosane |

## Branched-Chain Alkanes

The alkane drawn here is used to illustrate each step in the naming rules:

$CH_3$
$CH_3CH_2CHCH_2CHCHCH_2CH_2CH_3$
$CH_3$ $CH_2CH_3$

1. **Identify and name the longest continuous carbon chain.**

$CH_3$
$CH_3CH_2CHCH_2CHCHCH_2CH_2CH_3$ Nonane
$CH_3$ $CH_2CH_3$

2. **Identify the groups attached to this chain and name them.** Names of substituent groups consist of the prefix from Table A7.2 that identifies the length of the group and the suffix *-yl* that identifies it as an alkyl group.

methyl-
$CH_3$
$CH_3CH_2CHCH_2CHCHCH_2CH_2CH_3$
$CH_3$ $CH_2CH_3$
methyl- ethyl-

3. **Number the carbon atoms in the longest chain,** starting at the end nearest a substituent group. Doing this identifies the points of attachment of the alkyl groups with the lowest possible numbers.

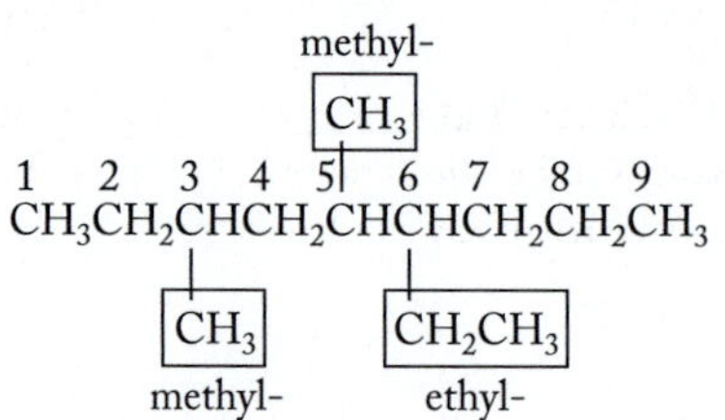

4. **Designate the location and identity of each substituent group with a number,** followed by a hyphen, and its name.

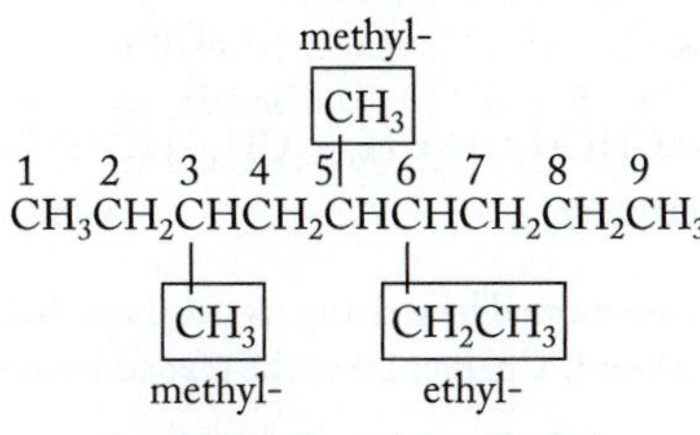


3-methyl-, 5-methyl-, 6-ethyl-

5. **Put together the complete name by listing the substituent groups in alphabetical order.** If more than one of a given type of substituent group is present, prefixes *di-*, *tri-*, *tetra-*, and so forth, are appended to the names, but these numerical prefixes are not considered when determining the alphabetical order. The name of the last substituent group is written together with the name identifying the longest carbon chain.

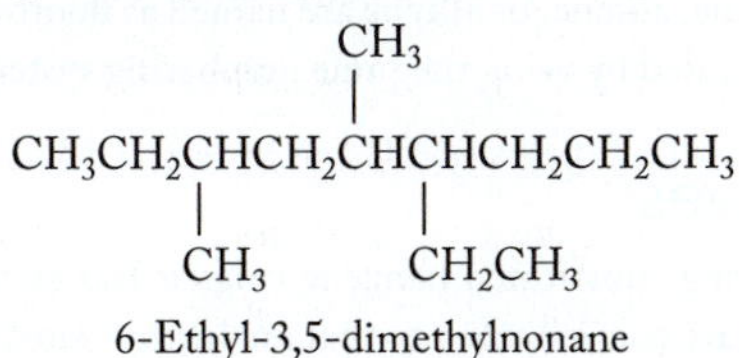


6-Ethyl-3,5-dimethylnonane

## Cycloalkanes

The simplest examples of this class of compounds consist of one unsubstituted ring of carbon atoms. The IUPAC names of these compounds consist of the prefix *cyclo-* followed by the parent name from Table A7.2 to indicate the number of carbon atoms in the ring. As an illustration, the names, formulas, and line structures of the first three cycloalkanes in the homologous series are

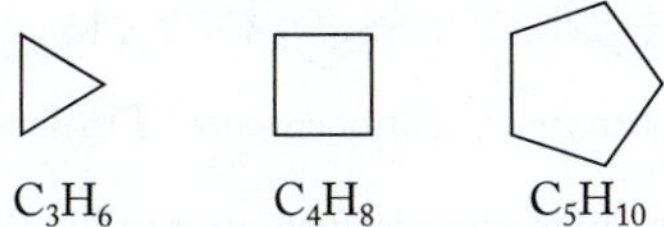

$C_3H_6$ $C_4H_8$ $C_5H_{10}$

Cyclopropane Cyclobutane Cyclopentane

## Alkenes and Alkynes

Alkenes have carbon–carbon double bonds and alkynes have carbon–carbon triple bonds as functional groups. The names of these types of compounds consist of (1) a parent name that identifies the longest carbon chain that includes the double or triple bond, (2) a suffix that identifies the class of compound, and (3) names of any substituent groups attached to the longest carbon chain. The suffix *-ene* identifies an alkene; *-yne* identifies an alkyne.

The alkene and alkyne drawn here are used to illustrate each step in the naming rules:

$$\begin{array}{l} CH_3 \\ | \\ CH_3CHCH{=}CHCH_2CH_2CH_3 \end{array} \qquad \begin{array}{l} CH_3 \\ | \\ CH_3CHC{\equiv}CCH_2CH_2CH_3 \end{array}$$

1. **To determine the parent name,** identify the longest chain that contains the unsaturation. Name the parent compound with the prefix that defines the number of carbons in that chain and the suffix that identifies the class of compound.

$$\begin{array}{l} CH_3 \\ | \\ \boxed{CH_3CHCH{=}CHCH_2CH_2CH_3} \\ \text{Heptene} \end{array} \qquad \begin{array}{l} CH_3 \\ | \\ \boxed{CH_3CHC{\equiv}CCH_2CH_2CH_3} \\ \text{Heptyne} \end{array}$$

2. **Number the parent chain from the end nearest the unsaturation so that the first carbon in the double or triple bond has the lowest number possible.** (If the unsaturation is in the middle of a chain, the location of any substituent group is used to determine where the

numbering starts.) The smaller of the two numbers identifying the carbon atoms involved in the unsaturation is used as the locator of the multiple bond.

$$\overset{1}{C}H_3\overset{2}{C}H(CH_3)\overset{3}{C}H{=}\overset{4}{C}H\overset{5}{C}H_2\overset{6}{C}H_2\overset{7}{C}H_3 \qquad \overset{1}{C}H_3\overset{2}{C}H(CH_3)\overset{3}{C}{\equiv}\overset{4}{C}\overset{5}{C}H_2\overset{6}{C}H_2\overset{7}{C}H_3$$

3-Heptene 3-Heptyne

3. **Stereoisomers of alkenes are named by writing *cis*- or *trans*- before the number identifying the location of the double bond.** Chapter 20 in the text addresses naming stereoisomers.

4. **The rules for naming substituted alkanes are followed to name and locate any other groups on the chain.**

$$CH_3CH(CH_3)CH{=}CHCH_2CH_2CH_3 \qquad CH_3CH(CH_3)C{\equiv}CCH_2CH_2CH_3$$

2-Methyl-3-heptene 2-Methyl-3-heptyne

Halogens attached to an alkane, alkene, or alkyne are named as fluoro- (F–), chloro- (Cl–), bromo- (Br–), or iodo- (I–) and are located by using the same numbering system described for alkyl groups.

## Benzene Derivatives

Naming compounds containing substituted benzene rings is less systematic than naming hydrocarbons. Many compounds have common names that are incorporated into accepted names, but for simple substituted benzene rings, the following rules may be applied.

1. **For monosubstituted benzene rings,** a prefix identifying the group is appended to the parent name benzene:

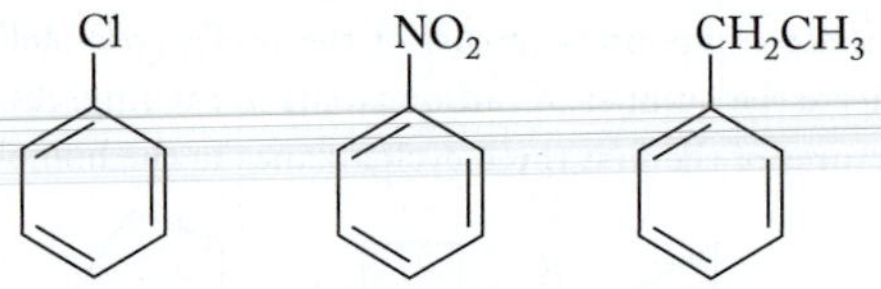


Chlorobenzene Nitrobenzene Ethylbenzene

2. **For disubstituted benzene rings,** three isomers are possible. The relative position of the substituent groups is indicated by numbers in IUPAC nomenclature, but the set of prefixes shown are very commonly used as well:

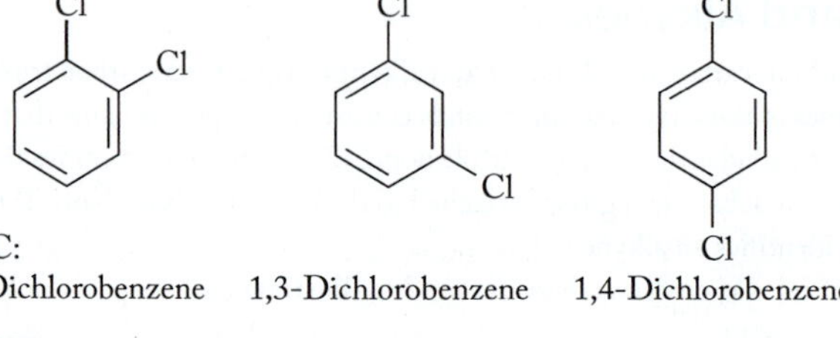


IUPAC:
1,2-Dichlorobenzene 1,3-Dichlorobenzene 1,4-Dichlorobenzene

Common:
*ortho*-Dichlorobenzene *meta*-Dichlorobenzene *para*-Dichlorobenzene
*o*-Dichlorobenzene *m*-Dichlorobenzene *p*-Dichlorobenzene

3. **When three or more groups are attached to a benzene ring,** the lowest possible numbers are assigned to locate the groups with respect to each other.

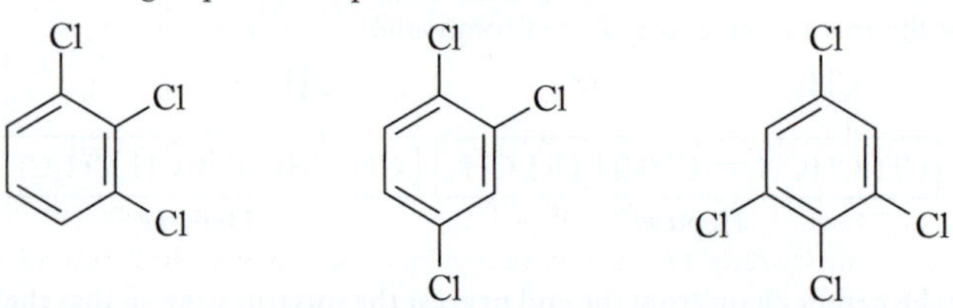


1,2,3-Trichlorobenzene 1,2,4-Trichlorobenzene 1,2,3,5-Tetrachlorobenzene

(*Note*: Not 1,3,4-trichlorobenzene; and not 1,3,4,5-tetrachlorobenzene.)

## Hydrocarbons Containing Other Functional Groups

The same basic principles developed for naming alkanes apply to naming hydrocarbons with functional groups other than alkyl groups. The name must identify the carbon skeleton, locate the functional group, and contain a suffix that defines the class of compound. The following examples give the suffixes for some common functional groups; when suffixes are used, they replace the final *-e* in the name of the parent alkane. Other functional groups may be identified by including the name of the class of compounds in the name of the molecule.

**Alcohols: Suffix *-ol***

$CH_3CH_2CH_2OH$ — 1-Propanol

$CH_3CH(OH)CH_3$ — 2-Propanol

$CH_3CH_2CH_2CH(OH)CH_3$ — 2-Pentanol

**Aldehydes: Suffix *-al***

| | HCHO | $CH_3CHO$ |
|---|---|---|
| IUPAC: | Methanal | Ethanal |
| Common: | Formaldehyde | Acetaldehyde |

Because the aldehyde group can only be on a terminal carbon, no number is necessary to locate it on the carbon chain.

**Ketones: Suffix *-one*** The location of the carbonyl is given by a number, and the chain is numbered so that the carbonyl carbon has the lowest possible value. Many ketones also have common names generated by identifying the hydrocarbon groups on both sides of the carbonyl group.

| | $H_3C-C(=O)-CH_3$ | $H_3C-C(=O)-CH_2CH_3$ |
|---|---|---|
| IUPAC: | Propan-2-one | Butan-2-one |
| Common: | Acetone | Methyl ethyl ketone |

**Carboxylic Acids: Suffix *-oic acid*** The carboxylic acid group is by definition carbon 1, so no number identifying its location is included in the name.

| | $H_3C-C(=O)-OH$ | $H_3C-CH=CH-C(=O)-OH$ |
|---|---|---|
| IUPAC: | Ethanoic acid | *trans*-2-Butenoic acid |
| Common: | Acetic acid | |

**Salts of Carboxylic Acids** Salts are named with the cation first, followed by the anion name of the acid from which *-ic acid* is dropped and the suffix *-ate* is added. The sodium salt of acetic acid is sodium acetate.

| $H_3C-C(=O)-OH$ | $H_3C-C(=O)-O^-$ | $Na^+\ {}^-O-C(=O)-CH_3$ |
|---|---|---|
| Acetic acid | Acetate ion | Sodium acetate |

**Esters** Esters are viewed as derivatives of carboxylic acids. They are named in a manner analogous to that of salts. The alkyl group comes first, followed by the name of the carboxylate anion.

$R-O-C(=O)-R$ (R: Alkyl; O–C(=O)–R: Carboxylate)

$CH_3CH_2-O-C(=O)-CH_3$ — Ethyl acetate

**Amides** Amides are also derivatives of carboxylic acids. They are named by replacing *-ic acid* (of the common names) or *-oic acid* (of the IUPAC names) with *-amide.*

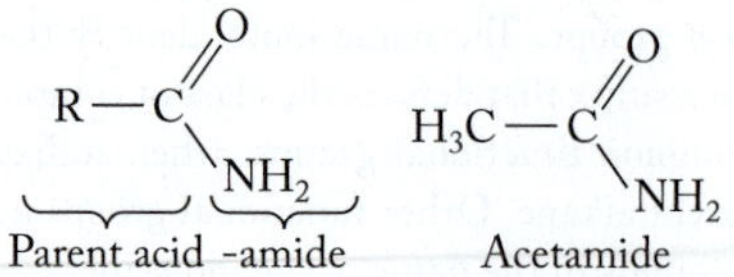

$$\underbrace{\mathrm{R{-}C}}_{\text{Parent acid}}\!\underbrace{(=\mathrm{O})\mathrm{NH_2}}_{\text{-amide}} \qquad \mathrm{H_3C{-}C(=O)NH_2}$$

Parent acid -amide

Acetamide (ethanamide)

**Ethers** Ethers are frequently named by naming the two groups attached to the oxygen and following those names by the word *ether.*

$CH_3OCH_3$ — Dimethyl ether

$CH_3CH_2OCH_2CH_3$ — Diethyl ether

**Amines** Aliphatic amines are usually named by listing the group or groups attached to the nitrogen and then appending *-amine* as a suffix. They may also be named by prefixing *amino-* to the name of the parent chain.

$H_3C{-}NH_2$ — Methylamine

$H_3C{-}NH(CH_2CH_3)$ — Ethylmethylamine

$H_2NCH_2CH_2OH$ — 2-Aminoethanol

This brief summary will enable you to understand the names of organic compounds used in this book. IUPAC rules are much more extensive than this and can be applied to all varieties of carbon compounds, including those with multiple functional groups. It is important to recognize that the rules of systematic nomenclature do not necessarily lead to a unique name for each compound, but they do always lead to an unambiguous one. Furthermore, common names are still used frequently in organic chemistry because the systematic alternatives do not improve communication. Remember that the main purpose of chemical nomenclature is to identify a chemical species by means of written or spoken words. Anyone who reads or hears the name should be able to deduce the structure and thereby the identity of the compound.

# Glossary

## A

**absolute zero (0 K)** The zero point on the Kelvin temperature scale; theoretically the lowest temperature possible. (Ch. 1)

**absorbance ($A$)** A measure of the quantity of light that a sample absorbs. (Ch. 8)

**accessible microstate** A unique arrangement of the positions and momenta of the particles in a thermodynamic system. (Ch. 12)

**accuracy** Agreement between an experimentally measured value and the true value. (Ch. 1)

**acid (Brønsted–Lowry acid)** A proton donor. (Ch. 8)

**activated complex** A short-lived species formed in a chemical reaction in which molecules have both the proper orientation and enough energy to react with each other. (Ch. 13)

**activation energy ($E_a$)** The minimum energy that molecules need to react when they collide. (Ch. 13)

**active site** The location on an enzyme where a reactive substance binds. (Ch. 20)

**activity series** A high-to-low ranking of metals on the basis of their strengths as reducing agents. (Ch. 8)

**addition polymer** A macromolecule prepared by adding monomers to a growing polymer chain. (Ch. 19)

**addition reaction** A reaction in which two molecules couple to form one product. (Ch. 19)

**alcohol** An organic compound whose molecular structure includes a hydroxyl group bonded to a carbon atom that is not bonded to any other functional group(s). (Ch. 6)

**aldehyde** An organic compound containing a carbonyl group bonded to one R group and one hydrogen; its general formula is RCHO. (Ch. 19)

**alkali metal** An element in group 1 of the periodic table. (Ch. 2)

**alkaline earth metal** An element in group 2 of the periodic table. (Ch. 2)

**alkane** A hydrocarbon in which each carbon atom is bonded to four other carbon or hydrogen atoms. (Ch. 6)

**alkene** A hydrocarbon containing one or more carbon–carbon double bonds. (Ch. 19)

**alkyne** A hydrocarbon containing one or more carbon–carbon triple bonds. (Ch. 19)

**allotropes** Different molecular forms of the same element, such as oxygen ($O_2$) and ozone ($O_3$). (Ch. 4)

**alloy** A blend of a host metal and one or more other elements, which may or may not be metals, that are added to change the properties of the host metal. (Ch. 18)

**alpha (α) decay** A nuclear reaction in which an unstable nuclide spontaneously emits an α particle. (Ch. 21)

**alpha (α) particle** A particle that is emitted during radioactive decay and is equivalent in mass and charge to a $^4$He nucleus. (Ch. 2)

**α helix** A coil in a protein chain's secondary structure. (Ch. 20)

**amide** An organic compound in which the –OH of a carboxylic acid group is replaced by $-NH_2$, –NHR, or $-NR_2$, where R can be any organic group. (Ch. 19)

**amine** An organic compound that contains a group with the general formula $RNH_2$, $R_2NH$, or $R_3N$, where R is any organic subgroup. (Ch. 19)

**amino acid** A molecule that contains at least one amine group and one carboxylic acid group. (Ch. 20)

**Amontons's law** The principle that the pressure of a fixed quantity of gas is proportional to its absolute temperature if its volume does not change. (Ch. 9)

**amphiprotic** Describes a substance that can behave as either a proton acceptor or a proton donor. (Ch. 8)

**analyte** The substance whose concentration is to be determined in a chemical analysis. (Ch. 8)

**angular** (also called *bent*) Molecular geometry about a central atom with a steric number of 3 and one lone pair or a steric number of 4 and two lone pairs. (Ch. 5)

**angular momentum quantum number ($\ell$)** An integer having any value from 0 to $(n - 1)$ that defines the shape of an orbital. (Ch. 3)

**anion** A negatively charged ion. (Ch. 2)

**anode** An electrode at which an oxidation half-reaction (loss of electrons) takes place. (Ch. 17)

**antibonding orbital** A term in MO theory describing regions of electron density in a molecule that destabilize the molecule because they do not increase the electron density between nuclear centers. (Ch. 5)

**antimatter** Particles that are the charge opposites of normal subatomic particles. (Ch. 21)

**aromatic compound** A cyclic, planar compound with delocalized π electrons above and below the plane of the molecule. (Ch. 5)

**Arrhenius acid** A compound that produces $H_3O^+$ ions in aqueous solution. (Ch. 15)

**Arrhenius base** A compound that produces $OH^-$ ions in aqueous solution. (Ch. 15)

**Arrhenius equation** An equation relating the rate constant of a reaction to the absolute temperature ($T$), the activation energy of the reaction ($E_a$), and the frequency factor ($A$). (Ch. 13)

**atom** The smallest particle of an element that cannot be chemically or mechanically divided into smaller particles. (Ch. 1)

**atomic absorption spectra** Characteristic patterns of dark lines produced when an external source of radiation passes through free gaseous atoms. (Ch. 3)

**atomic emission spectra** Characteristic patterns of bright lines produced when atoms are vaporized in high-temperature flames or electrical discharges. (Ch. 3)

**atomic number ($Z$)** The number of protons in the nucleus of an atom. (Ch. 2)

**aufbau principle** The concept of building up ground-state atoms so that their electrons occupy the lowest-energy orbitals available. (Ch. 3)

**autoionization** The process that produces equal and very small concentrations of $H_3O^+$ and $OH^-$ ions in pure water. (Ch. 15)

**average atomic mass** The weighted average of the masses of all isotopes of an element, calculated by multiplying the natural abundance of each isotope by its mass in unified atomic mass units and then summing the products. (Ch. 2)

**Avogadro constant ($N_A$)** The number of elementary entities—atoms, ions, or molecules—in one mole of a substance. The exact value of the constant is $6.02214076 \times 10^{23}$ mol$^{-1}$. (Ch. 2)

**Avogadro's law** The principle that the volume of a gas at constant temperature and pressure is proportional to the quantity (number of moles) of the gas. (Ch. 9)

# B

**band gap ($E_g$)** The energy gap between the valence and conduction bands. (Ch. 18)

**band theory** An extension of molecular orbital theory that describes bonding in solids. (Ch. 18)

**barometer** An instrument that measures atmospheric pressure. (Ch. 9)

**base (Brønsted–Lowry base)** A proton acceptor. (Ch. 8)

**becquerel (Bq)** The SI unit of radioactivity; 1 Bq = 1 decay event per second. (Ch. 21)

**Beer's law** The relation of the absorbance of a solution ($A$) to concentration ($c$), the light's path length ($b$), and the solute's molar absorptivity ($\varepsilon$) by the equation $A = \varepsilon bc$. (Ch. 8)

**belt of stability** The region on a graph of number of neutrons versus number of protons that includes all stable nuclei. (Ch. 21)

**bent** (also called *angular*) Molecular geometry about a central atom with a steric number of 3 and one lone pair or a steric number of 4 and two lone pairs. (Ch. 5)

**beta (β) decay** The process by which a neutron in a neutron-rich nucleus decays into a proton and a high-energy electron (β particle). (Ch. 21)

**beta (β) particle** A particle that is emitted during radioactive decay and is equivalent in mass and charge to a high-energy electron. (Ch. 2)

**β-pleated sheet** A puckered two-dimensional array of protein strands held together by hydrogen bonds. (Ch. 20)

**bimolecular step** A step in a reaction mechanism involving a collision between two molecules (or atoms or ions). (Ch. 13)

**binding energy (BE)** The energy that holds the nucleons together in a nucleus. (Ch. 21)

**biocatalysis** The strategy of using enzymes to catalyze reactions on a large scale. (Ch. 20)

**biomass** The sum total of the mass of organic matter in any given ecological system. (Ch. 20)

**biomolecule** An organic molecule present naturally in a living system. (Ch. 20)

**body-centered cubic (bcc) unit cell** A cell with atoms at the eight corners of a cube and at the center of the cell. (Ch. 18)

**boiling point elevation** An increase in the normal boiling point of solutions relative to that of their pure solvents. (Ch. 11)

**bomb calorimeter** A constant-volume device used to measure the energy released during a combustion reaction. (Ch. 10)

**bond angle** The angle (in degrees) defined by lines joining the centers of two atoms to a third atom to which they are chemically bonded. (Ch. 5)

**bond dipole** Separation of electrical charge created when atoms with different electronegativities form a covalent bond. (Ch. 5)

**bond energy** The energy needed to break one mole of a specific covalent bond in the gas phase; also called *bond strength.* (Ch. 4)

**bond length** The distance between the nuclei of two atoms joined in a bond. (Ch. 4)

**bond order** The number of bonds between atoms: 1 for a single bond, 2 for a double bond, and 3 for a triple bond. (Ch. 4)

**bonding capacity** The number of covalent bonds an atom forms to have an octet of electrons in its valence shell. (Ch. 4)

**bonding orbital** A term in MO theory describing regions of increased electron density between nuclear centers that serve to hold atoms together in molecules. (Ch. 5)

**bonding pair** A pair of electrons that two atoms share. (Ch. 4)

**Born–Haber cycle** A series of steps with corresponding enthalpy changes that describes the formation of an ionic solid from its constituent elements. (Ch. 10)

**Boyle's law** The principle that the volume of a fixed quantity of gas at constant temperature is inversely proportional to its pressure. (Ch. 9)

**Bragg equation** An equation that relates the angle of diffraction ($2\theta$) of X-rays to the spacing ($d$) between the layers of ions or atoms in a crystal: $n\lambda = 2d \sin \theta$. (Ch. 18)

**branch** A side chain bonded to the main (longest) chain in an organic molecule. (Ch. 19)

**breeder reactor** A nuclear reactor in which fissionable material is produced during normal reactor operation. (Ch. 21)

**buffer capacity** The quantity of acid or base that a pH buffer can neutralize while keeping its pH within a desired range. (Ch. 16)

# C

**calibration curve** A graph showing how a measurable property, such as absorbance, varies for a set of standard samples of known concentration that can later be used to identify an unknown concentration from a measured absorbance. (Ch. 8)

**calorimeter** A device used to measure the absorption or release of energy by a phase change or chemical process. (Ch. 10)

**calorimeter constant ($C_{\text{calorimeter}}$)** The heat capacity of a calorimeter. (Ch. 10)

**calorimetry** The experimental determination of the quantity of energy transferred during a phase change or chemical process. (Ch. 10)

**capillary action** The rise of a liquid in a narrow tube as a result of adhesive forces between the liquid and the tube and cohesive forces within the liquid. (Ch. 6)

**carbohydrate** An organic molecule with the generic formula $C_x(H_2O)_y$. (Ch. 20)

**carbonyl group** A functional group that consists of a carbon atom with a double bond to an oxygen atom. (Ch. 6)

**carboxylic acid** A compound containing the $-COOH$ functional group. (Ch. 8)

**catalyst** A substance added to a reaction that increases the rate of the reaction but is not consumed in the process. (Ch. 13)

**cathode** An electrode at which a reduction half-reaction (gain of electrons) takes place. (Ch. 17)

**cathode rays** Streams of electrons emitted by the cathode in a partially evacuated tube. (Ch. 2)

**cation** A positively charged ion. (Ch. 2)

**cell diagram** Symbols that show how the components of an electrochemical cell are connected. (Ch. 17)

**cell potential ($E_{\text{cell}}$)** The electromotive force with which an electrochemical cell can pump electrons through an external circuit. (Ch. 17)

**ceramic** A solid inorganic compound or mixture that has been transformed into a harder, more heat-resistant material by heating. (Ch. 18)

**chain reaction** A self-sustaining series of fission reactions in which the neutrons released when nuclei split apart initiate additional fission events and sustain the reaction. (Ch. 21)

**chalcogen** An element in group 16 of the periodic table. (Ch. 2)

**Charles's law** The principle that the volume of a fixed quantity of gas at constant pressure is directly proportional to its absolute temperature. (Ch. 9)

**chelate effect** The greater affinity of metal ions for polydentate ligands than for monodentate ligands. (Ch. 23)

**chelation** The interaction of a metal with a polydentate ligand (chelating agent); pairs of electrons on one molecule of the ligand occupy two or more coordination sites on the central metal. (Ch. 23)

**chemical bond** A force that holds two atoms or ions in a molecule or a compound together. (Ch. 1)

**chemical energy** Potential energy stored in chemical bonds. (Ch. 10)

**chemical equation** A description of the identities and proportions of the reactants and the products in a chemical reaction. (Ch. 7)

**chemical equilibrium** A dynamic process in which the concentrations of reactants and products remain constant over time and the rate of a reaction in the forward direction matches its rate in the reverse direction. (Ch. 14)

**chemical formula** A notation that uses the symbols of the elements to represent the elemental composition of a pure substance; subscripts indicate the relative number of atoms of each element in the substance. (Ch. 1)

**chemical kinetics** The study of the rates of chemical reactions and the factors that influence reaction rates. (Ch. 13)

**chemical property** A property of a substance that can be observed only by reacting the substance with something else to form another substance. (Ch. 1)

**chemical reaction** The conversion of one or more substances into one or more different substances; also called *chemical change*. (Ch. 1)

**chemistry** The study of the composition, structure, and properties of matter and of the energy consumed or given off when matter undergoes a change. (Ch. 1)

**chiral** Describes a molecule that is not superimposable on its mirror image. (Ch. 5)

**chromatography** A process involving a stationary and a mobile phase for separating a mixture of substances according to their different affinities for the two phases. (Ch. 1)

**cis isomer** (also called *Z isomer*) A molecule with two like groups (such as two R groups or two hydrogen atoms) on the same side of the molecule. (Ch. 19)

**Clausius–Clapeyron equation** A relationship between the vapor pressure of a substance at different temperatures and its enthalpy of vaporization. (Ch. 11)

**codon** A three-nucleotide sequence that codes for a specific amino acid. (Ch. 20)

**coenzyme** An organic molecule that, like an enzyme, accelerates the rate of biochemical reactions. (Ch. 23)

**colligative properties** Characteristics of solutions that depend on the concentration, not the identity, of particles dissolved in the solvent. (Ch. 11)

**collision theory** A model that describes how collisions between gas-phase atoms or molecules cause chemical reactions to occur. (Ch. 13)

**combination reaction** A reaction in which two or more substances form a single product. (Ch. 7)

**combined gas law** The principle that the ratio $PV/T$ for a given quantity of gas is a constant. (Ch. 9)

**combustion** A rapid reaction between fuel and oxygen that produces energy. (Ch. 7)

**combustion analysis** A laboratory procedure in which a substance is burned completely in oxygen to produce known compounds whose masses are used to determine the composition of the original material. (Ch. 7)

**common-ion effect** The shift in the position of an equilibrium caused by the presence or addition of an ion taking part in the reaction. (Ch. 16)

**complex ion** A species consisting of a metal ion bonded to one or more Lewis bases. (Ch. 16)

**compound** A pure substance composed of two or more elements that are chemically bonded in fixed proportion. (Ch. 1)

**condensation polymer** A macromolecule formed by a reaction of monomers, producing a polymer and water or another small molecule. (Ch. 19)

**condensation reaction** Two molecules combining to form a larger molecule and a small molecule (typically water). (Ch. 19)

**conduction band** An unoccupied band higher in energy than a valence band in which electrons can migrate. (Ch. 18)

**conductor** A material with partially filled valence bands or filled valence bands that overlap with empty conduction bands, leading to highly mobile electrons. (Ch. 18)

**confidence interval** A range of values that has a specified probability of containing the true value of a measurement. (Ch. 1)

**conjugate acid** The acid formed when a Brønsted–Lowry base accepts a $H^+$ ion. (Ch. 15)

**conjugate acid–base pair** A Brønsted–Lowry acid and base that differ from each other only by a $H^+$ ion: acid $\rightleftharpoons$ conjugate base + $H^+$. (Ch. 15)

**conjugate base** The base formed when a Brønsted–Lowry acid donates a $H^+$ ion. (Ch. 15)

**constitutional isomer** (also called *structural isomer*) One of a set of compounds with the same molecular formula but different connections between the atoms in their molecules. (Ch. 6)

**conversion factor** Fraction in which the numerator is equivalent to the denominator but is expressed in different units, making the fraction equivalent to 1. (Ch. 1)

**coordinate bond** A covalent bond between two atoms in which one of the two provides both electrons in the bonding pair. (Ch. 16)

**coordination compound** A compound made up of at least one complex ion. (Ch. 23)

**coordination number** The number of sites occupied by ligands around a metal ion in a complex. (Ch. 23)

**copolymer** A macromolecule formed from the chemical combination of two different monomers. (Ch. 19)

**core electrons** The electrons in the filled, inner shells in an atom or ion; they are not involved in chemical reactions. (Ch. 3)

**corrosion** A process in which a metal deteriorates through spontaneous electrochemical reactions. (Ch. 17)

**coulombic attraction** (also called *electrostatic potential energy* [$E_{el}$]) The energy a charged particle has because of its position relative to another charged particle; $E_{el}$ is directly proportional to the product of the charges of the particles and inversely proportional to the distance between them. (Ch. 4)

**counterion** An ion whose charge balances the charge of a complex ion in a coordination compound. (Ch. 23)

**covalent bond** A bond created by two atoms sharing one or more pairs of electrons. (Ch. 4)

**covalent network solid** A solid consisting of atoms held together by extended arrays of covalent bonds. (Ch. 18)

**critical mass** The minimum quantity of fissionable material needed to sustain a chain reaction. (Ch. 21)

**critical point** A specific temperature and pressure at which the liquid and gas phases of a substance have the same density and are indistinguishable from each other. (Ch. 6)

**crystal field splitting energy (Δ)** The difference in energy between subsets of *d* orbitals split by interactions in a crystal field. (Ch. 23)

**crystal field splitting** The separation of a set of *d* orbitals into subsets with different energies as a result of interactions between electrons in those orbitals and lone pairs of electrons in ligands. (Ch. 23)

**crystal lattice** An ordered three-dimensional array of particles. (Ch. 4)

**crystal structure** An ordered arrangement in three-dimensional space of the particles (atoms, ions, or molecules) that make up a crystalline solid. (Ch. 18)

**crystalline solid** A solid made of an ordered array of atoms, ions, or molecules. (Ch. 18)

**cubic closest-packed (ccp)** A crystal structure composed of face-centered cubic unit cells and layers of particles having an *abcabc* . . . stacking pattern. (Ch. 18)

**curie (Ci)** Non-SI unit of radioactivity; 1 Ci = $3.70 \times 10^{10}$ decay events per second. (Ch. 21)

**cycloalkane** A ring-containing alkane with the general formula $C_nH_{2n}$. (Ch. 19)

# D

**dalton (Da)** A unit of mass equal to 1 unified atomic mass unit. (Ch. 2)

**Dalton's law of partial pressures** The principle that the total pressure of a mixture of gases is the sum of the partial pressures of all the gases in the mixture. (Ch. 9)

**degree of ionization** The ratio of the quantity of a substance that is ionized to the concentration of the substance before ionization; when expressed as a percentage, called *percent ionization*. (Ch. 15)

**density (*d*)** The ratio of the mass ($m$) of an object to its volume ($V$). (Ch. 1)

**deposition** The transformation of a gas (vapor) directly into a solid. (Ch. 1)

**diamagnetic** Describes a substance with no unpaired electrons that is weakly repelled by a magnetic field. (Ch. 5)

**diffusion** The spread of one substance (usually a gas or liquid) through another. (Ch. 9)

**dilution** The process of lowering the concentration of a solution by adding more solvent. (Ch. 8)

**dipole** A pair of oppositely charged poles separated by a distance. (Ch. 4)

**dipole moment (μ)** A measure of the degree to which a molecule aligns itself in an applied electric field; a quantitative expression of the polarity of a molecule. (Ch. 5)

**dipole–dipole interaction** An attraction between regions of polar molecules that have partial charges of opposite sign. (Ch. 6)

**dipole–induced dipole interaction** An attraction between a polar molecule and the oppositely charged pole it temporarily induces in another molecule. (Ch. 6)

**distillation** A process using evaporation and condensation to separate a mixture of substances with different volatilities. (Ch. 1)

**double bond** A bond formed when two atoms share two pairs of electrons. (Ch. 4)

# E

**E isomer** (also called *trans isomer*) A molecule with two like groups (such as two R groups or two hydrogen atoms) on opposite sides of the molecule. (Ch. 19)

**effective nuclear charge ($Z_{eff}$)** The attraction toward the nucleus experienced by an electron in an atom; the positive charge on the nucleus is reduced by the extent to which other electrons in the atom shield the electron from the nucleus. (Ch. 3)

**effusion** The process by which a gas escapes from its container through a tiny hole into a region of lower pressure. (Ch. 9)

**electrochemical cell** An apparatus that converts chemical energy into electrical work or electrical work into chemical energy. (Ch. 17)

**electrochemistry** The branch of chemistry that examines the transformations between chemical and electrical energy. (Ch. 17)

**electrode** A solid electrical conductor used to make contact with a solution or other nonmetallic component of an electrical circuit. (Ch. 8)

**electrolysis** A process in which electrical energy is used to drive a nonspontaneous chemical reaction. (Ch. 17)

**electrolyte** A solute that produces ions in solution, enabling its solutions to conduct electricity. (Ch. 8)

**electrolytic cell** A device in which an external source of electrical energy does work on a chemical system, turning a reactant(s) into a higher-energy product(s). (Ch. 17)

**electromagnetic radiation** Any form of radiant energy in the electromagnetic spectrum. (Ch. 3)

**electromagnetic spectrum** A continuous range of radiant energy that includes gamma rays, X-rays, ultraviolet radiation, visible light, infrared radiation, microwaves, and radio waves. (Ch. 3)

**electromotive force (emf)** Also called *voltage*, the force pushing electrons through an electrical circuit. (Ch. 17)

**electron** A subatomic particle that has a negative charge and negligible mass. (Ch. 2)

**electron affinity (EA)** The energy change that occurs when one mole of electrons combines with one mole of atoms or ions in the gas phase. (Ch. 3)

**electron capture** A nuclear reaction in which a neutron-poor nucleus draws in one of its surrounding electrons, which transforms a proton in the nucleus into a neutron. (Ch. 21)

**electron configuration** The distribution of electrons among the orbitals of an atom or ion. (Ch. 3)

**electron transition** Movement of an electron between energy levels. (Ch. 3)

**electron-pair delocalization** The spreading out of electron density over several atoms. (Ch. 4)

**electron-pair geometry** The three-dimensional arrangement of bonding pairs and lone pairs of electrons about a central atom. (Ch. 5)

**electronegativity** A relative measure of an atom's ability to attract electrons to itself within a bond. (Ch. 4)

**electrostatic potential energy ($E_{el}$)** (also called *coulombic attraction*) The energy a charged particle has because of its position relative to another charged particle; $E_{el}$ is directly proportional to the product of the charges of the particles and inversely proportional to the distance between them. (Ch. 4)

**element** A pure substance that cannot be separated into simpler substances. (Ch. 1)

**elementary step** A molecular-level view of a single process taking place in a chemical reaction. (Ch. 13)

**empirical formula** A chemical formula in which the subscripts represent the simplest whole-number ratio of the atoms or ions in a compound. (Ch. 1)

**enantiomer** One of a pair of optical isomers of a compound. (Ch. 5)

**end point** The point in a titration at which a color change or other signal indicates that enough titrant has been added to react with all the analyte in the sample. (Ch. 8)

**endothermic process** A process in which energy in the form of heat flows from the surroundings into the system. (Ch. 10)

**energy profile** A graph showing the changes in chemical energy during a reaction. (Ch. 13)

**energy** The capacity to do work ($w$). (Ch. 1)

**enthalpy ($H$)** A measure of the total energy of a system; the sum of the internal energy and the pressure–volume product of a system: $H = E + PV$. (Ch. 10)

**enthalpy change ($\Delta H$)** The energy absorbed by an endothermic process or given off by an exothermic process that takes place at constant pressure. (Ch. 10)

**enthalpy of fusion ($\Delta H_{fus}$)** The energy required to convert one mole of a solid substance at its melting point into the liquid state; also called the *heat of fusion*. (Ch. 10)

**enthalpy of reaction ($\Delta H_{rxn}$)** The enthalpy change that accompanies a chemical reaction; also called the *heat of reaction*. (Ch. 10)

**enthalpy of solution ($\Delta H_{solution}$)** The overall change in enthalpy that occurs when a solute is dissolved in a solvent; also called the *heat of solution*. (Ch. 10)

**enthalpy of vaporization ($\Delta H_{vap}$)** The energy required to convert one mole of a liquid substance at its boiling point into the vapor state; also called the *heat of vaporization*. (Ch. 10)

**entropy ($S$)** A measure of how dispersed the energy in a system is at a specific temperature.

**enzyme** A protein that catalyzes a reaction. (Ch. 20)

**equilibrium (reversal) potential, $E_{ion}$** An electrochemical potential that results from a concentration gradient of a particular ion on opposite sides of a cell membrane. (Ch. 22)

**equilibrium constant ($K$)** The value of the ratio of concentration (or partial pressure) terms in the equilibrium constant expression at a specific temperature. (Ch. 14)

**equilibrium constant expression** The ratio of the equilibrium concentrations or partial pressures of products to reactants, with each term raised to a power equal to the coefficient of that substance in the balanced chemical equation for the reaction. (Ch. 14)

**equivalence point** The point in a titration at which just enough titrant has been added to react with all the analyte in the sample. (Ch. 8)

**essential amino acid** Any of the 10 amino acids that make up peptides and proteins but are not synthesized in the human body and must be obtained through the food we eat. (Ch. 20)

**essential element** An element present in tissue, blood, or other body fluids that has a physiological function. (Ch. 22)

**ester** An organic compound in which the –OH of a carboxylic acid group is replaced by –OR, where R can be any organic group. (Ch. 19)

**ether** An organic compound that contains an oxygen atom with single bonds to two carbon atoms. (Ch. 6)

**excited state** Any energy state above the ground state. (Ch. 3)

**exothermic process** A process in which energy in the form of heat flows from a system into its surroundings. (Ch. 10)

**extensive property** A property that varies with the amount of substance. (Ch. 1)

## F

**face-centered cubic (fcc) unit cell** An array of closest-packed particles that has eight of the particles at the corners of a cube and six of them at the centers of each face of the cube. (Ch. 18)

**family** (also called *group*) (of elements) All the elements in a column of the periodic table. (Ch. 2)

**Faraday constant ($F$)** The magnitude of electric charge in one mole of electrons; its value to three significant figures is $9.65 \times 10^4$ C/mol. (Ch. 17)

**fat** A solid triglyceride containing primarily saturated fatty acids. (Ch. 20)

**filtration** A process for separating solid particles from a liquid or gaseous sample by passing the sample through a porous material that retains the solid particles. (Ch. 1)

**first law of thermodynamics** The principle that the energy gained or lost by a system must equal the energy lost or gained by the surroundings. Energy cannot be created or destroyed. (Ch. 10)

**formal charge (FC)** The value calculated for an atom in a molecule or polyatomic ion by determining the difference between the number of valence electrons in the free atom and the sum of the lone-pair electrons plus half the electrons in the atom's bonding pairs. (Ch. 4)

**formation constant ($K_f$)** An equilibrium constant describing the formation of a metal complex from a free metal ion and its ligands. (Ch. 16)

**formation reaction** A reaction in which one mole of a substance is formed from its component elements in their standard states. (Ch. 10)

**formula mass** The mass in u of one formula unit of an ionic compound. (Ch. 2)

**formula unit** The smallest electrically neutral unit of an ionic compound. (Ch. 2)

**fracking** The process of injecting liquids at high pressures into rock formations to force open existing fissures in the formations for the purpose of extracting oil or gas. (Ch. 19)

**fractional distillation** A method of separating a mixture of compounds on the basis of their different boiling points. (Ch. 11)

**Fraunhofer lines** A set of dark lines in the otherwise continuous solar spectrum. (Ch. 3)

**free energy** A measure of the maximum amount of work that a thermodynamic system can perform. (Ch. 12)

**free radical** An atom, ion, or molecule with unpaired electrons. (Ch. 4)

**freezing point depression** A decrease in the normal freezing point of solutions relative to that of their pure solvents. (Ch. 11)

**frequency ($\nu$)** The number of crests of a wave that pass a stationary point of reference per second. (Ch. 3)

**frequency factor ($A$)** The product of the frequency of molecular collisions and a factor that expresses the probability that the relative orientation of the molecules is appropriate for a reaction to occur. (Ch. 13)

**fuel cell** A voltaic cell based on the oxidation of a continuously supplied fuel; the reaction is the equivalent of combustion, but chemical energy is converted directly into electrical energy. (Ch. 17)

**fuel density** The quantity of energy released during the complete combustion of a particular volume of a liquid fuel. (Ch. 10)

**fuel value** The quantity of energy released during the complete combustion of 1 g of a substance. (Ch. 10)

**functional group** A group of atoms in an organic compound's molecular structure that imparts characteristic physical and chemical properties. (Ch. 6)

## G

**galvanic cell** (also called *voltaic cell*) An electrochemical cell in which chemical energy is transformed into electrical work by a spontaneous redox reaction. (Ch. 17)

**gas** A phase of matter that has neither definite volume nor definite shape and that expands to fill its container; also called *vapor*. (Ch. 1)

**Geiger counter** A portable device for determining nuclear radiation levels by measuring how much the radiation ionizes the gas in a sealed detector. (Ch. 21)

**Gibbs free energy ($G$)** The maximum amount of energy released by a process occurring at constant temperature and pressure that is available to do useful work. (Ch. 12)

**glyceride** Lipid consisting of esters formed between fatty acids and the alcohol glycerol. (Ch. 20)

**glycolysis** A series of reactions that converts glucose into pyruvate; a major anaerobic (no oxygen required) pathway for the metabolism of glucose in the cells of almost all living organisms. (Ch. 12)

**glycosidic bond** A C—O—C bond between sugar molecules. (Ch. 20)

**Graham's law of effusion** The principle that the rate of effusion of a gas is inversely proportional to the square root of its molar mass. (Ch. 9)

**gray (Gy)** The SI unit of absorbed radiation; 1 Gy = 1 J/kg of tissue. (Ch. 21)

**green chemistry** Laboratory practices that reduce or eliminate the use or generation of hazardous substances. (Ch. 6)

**ground state** The most stable, lowest-energy state of a particle. (Ch. 3)

**group** (also called *family*) (of elements) All the elements in a column of the periodic table. (Ch. 2)

**Grubbs' test** A statistical test used to detect an outlier in a set of data. (Ch. 1)

## H

**half-life ($t_{1/2}$)** The reaction time interval during which the concentration of a reactant decreases by half. (Ch. 13)

**half-reaction** One of the halves of an oxidation–reduction reaction; one half-reaction is the oxidation component, and the other is the reduction component. (Ch. 8)

**halogen** An element in group 17 of the periodic table. (Ch. 2)

**heat** The transfer of energy between objects that occurs because of differences in their temperatures. (Ch. 1)

**heat capacity ($C_P$)** The energy required to raise the temperature of an object by 1°C at constant pressure. (Ch. 10)

**Heisenberg uncertainty principle** The principle that one cannot simultaneously know the exact position and the exact momentum of an electron. (Ch. 3)

**Henderson–Hasselbalch equation** An equation used to calculate the pH of a solution in which the concentrations of acid and conjugate base are known. (Ch. 16)

**Henry's law** The principle that the concentration of a sparingly soluble gas in a liquid is proportional to the partial pressure of the gas. (Ch. 11)

**hertz (Hz)** The SI unit of frequency with units of reciprocal seconds: $1\text{ Hz} = 1\text{ s}^{-1}$ = 1 cycle per second (cps). (Ch. 3)

**Hess's law** The principle that the enthalpy of reaction ($\Delta H_{rxn}$) for a process that is the sum of two or more reactions is equal to the sum of the $\Delta H_{rxn}$ values of the constituent reactions; also called *Hess's law of constant heat of summation*. (Ch. 10)

**heteroatom** Any atom other than carbon or hydrogen in an organic compound. (Ch. 19)

**heterogeneous catalyst** A catalyst in a different phase from that of the reactants. (Ch. 13)

**heterogeneous equilibrium** An equilibrium in which reactants and products are in more than one phase. (Ch. 14)

**heterogeneous mixture** A mixture in which the components are not distributed uniformly, so that the mixture contains regions of different compositions. (Ch. 1)

**heteropolymer** A polymer made of three or more different monomer units. (Ch. 19)

**hexagonal closest-packed (hcp)** A crystal lattice in which the layers of atoms or ions have an *ababab* . . . stacking pattern. (Ch. 18)

**hexagonal unit cell** An array of nine closest-packed particles that are the repeating unit in a hexagonal closest-packed crystal. (Ch. 18)

**homogeneous catalyst** A catalyst in the same phase as the reactants. (Ch. 13)

**homogeneous equilibrium** An equilibrium in which reactants and products are in the same phase. (Ch. 14)

**homogeneous mixture** A mixture in which the components are distributed uniformly and the composition and appearance are uniform. (Ch. 1)

**homologous series** A set of related organic compounds that differ from one another by the number of common subgroups, such as $-CH_2-$, in their molecular structures. (Ch. 19)

**homopolymer** A polymer composed of only one kind of monomer unit. (Ch. 19)

**Hund's rule** The principle that the lowest-energy electron configuration of an atom has the maximum number of unpaired electrons, all of which have the same spin, in degenerate orbitals. (Ch. 3)

**hybrid atomic orbital** In valence bond theory, one of a set of equivalent orbitals about an atom created when specific atomic orbitals are mixed. (Ch. 5)

**hybridization** In valence bond theory, the mixing of atomic orbitals to generate new sets of orbitals that may form sigma bonds with other atoms. (Ch. 5)

**hydrocarbon** An organic compound whose molecules contain only carbon and hydrogen atoms. (Ch. 6)

**hydrogen bond** A strong dipole–dipole interaction, which occurs between a hydrogen atom bonded to a N, O, or F atom and another N, O, or F atom. (Ch. 6)

**hydrogenation** A chemical reaction between molecular hydrogen ($H_2$) and another substance. (Ch. 19)

**hydronium ion ($H_3O^+$)** An $H^+$ ion plus a water molecule, $H_2O$; the form in which the hydrogen ion is found in an aqueous solution. (Ch. 8)

**hydrophilic** Describes a "water-loving," or attractive, interaction between a solute and water that increases water solubility. (Ch. 6)

**hydrophobic** Describes a "water-fearing," or repulsive, interaction between a solute and water that decreases water solubility. (Ch. 6)

**hydroxyl group** A functional group that consists of an oxygen atom with a single bond to a hydrogen atom. (Ch. 6)

**hypothesis** A tentative and testable explanation for an observation or a series of observations. (Ch. 1)

## I

**ideal gas** A gas whose behavior is predicted by the linear relations defined by the combined gas law. (Ch. 9)

**ideal gas equation** The principle relating the pressure, volume, number of moles, and temperature of an ideal gas, expressed by the equation $PV = nRT$, where $R$ is the universal gas constant; also called the *ideal gas law*. (Ch. 9)

**ideal solution** A solution that obeys Raoult's law. (Ch. 11)

**immiscible liquids** Combinations of liquids that do not mix with, or dissolve in, each other.

**induced dipole** (also called *temporary dipole*) The separation of charge produced in an atom or molecule by a momentary uneven distribution of electrons. (Ch. 6)

**inhibitor** A compound that diminishes or destroys enzyme's ability to catalyze a reaction. (Ch. 20)

**inner coordination sphere** The ligands that are bound directly to a metal via coordinate bonds. (Ch. 23)

**insulator** A material with a large energy gap between its valence and conduction bands. (Ch. 18)

**integrated rate law** A mathematical expression that describes the change in concentration of a reactant in a chemical reaction with time. (Ch. 13)

**intensive property** A property that is independent of the amount of substance. (Ch. 1)

**intermediate** A species produced in one step of a reaction and consumed in a later step. (Ch. 13)

**internal energy (*E*)** The sum of all the kinetic and potential energies of all the components of a system. (Ch. 10)

**interstitial alloy** An alloy in which the nonhost atoms occupy spaces between atoms of the host. (Ch. 18)

**ion** A particle consisting of one or more atoms that has a net positive or negative charge. (Ch. 1)

**ion channel** A group of helical proteins that penetrate cell membranes and allow selective transport of ions. (Ch. 22)

**ion exchange** A process by which one ion is displaced by another. (Ch. 11)

**ion pair** A cluster formed when a cation and an anion associate with each other in solution. (Ch. 11)

**ion pump** A system of membrane proteins that exchange ions inside the cell with those in the intercellular fluid. (Ch. 22)

**ion–dipole interaction** An attractive force between an ion and a molecule that has a permanent dipole. (Ch. 6)

**ionic bond** A bond resulting from the electrostatic attraction of a cation to an anion. (Ch. 4)

**ionic compound** A compound that consists of a characteristic ratio of positive ions and negative ions. (Ch. 1)

**ionic solid** A solid consisting of monatomic or polyatomic ions held together by ionic bonds. (Ch. 18)

**ionization energy (IE)** The amount of energy needed to remove one mole of electrons from one mole of ground-state atoms or ions in the gas phase. (Ch. 3)

**ionizing radiation** High-energy products of radioactive decay that can ionize molecules. (Ch. 21)

**isoelectronic** Describes atoms or ions that have identical electron configurations. (Ch. 3)

**isomer** One of a group of compounds having the same chemical formula but different molecular structures. (Ch. 5)

**isotopes** Atoms of an element containing the same number of protons but different numbers of neutrons. (Ch. 2)

## J

**joule (J)** The SI unit of energy, equivalent to 1 kg • $(m/s)^2$. (Ch. 1)

## K

**Kekulé structure** Aa structure that uses lines to show all the bonds in a covalently bonded molecule but does not show lone pairs on the atoms. (Ch. 19)

**ketone** An organic compound that contains a carbonyl group bonded to two other carbon atoms. (Ch. 6)

**kinetic energy (KE)** The energy of an object in motion because of its mass ($m$) and its speed ($u$). (Ch. 1)

**kinetic molecular theory (KMT)** A model that explains the behavior of gases on the basis of the motion of the particles that make them up. (Ch. 9)

## L

**lattice energy (*U*)** The energy released when one mole of an ionic compound forms from its free ions in the gas phase. (Ch. 4)

**law of conservation of energy** The principle that energy cannot be created or destroyed but can change from one form to another. (Ch. 1)

**law of conservation of mass** The principle that the sum of the masses of the reactants in a chemical reaction is equal to the sum of the masses of the products. (Ch. 7)

**law of constant composition** The principle that all samples of a given compound have the same elemental composition. (Ch. 1)

**law of definite proportions** The principle that a compound always contains the same proportion of its component elements. (Ch. 1)

**law of mass action** The principle relating the balanced chemical equation of a reversible reaction to its mass action expression (or equilibrium constant expression). (Ch. 14)

**law of multiple proportions** The principle that, when two masses of one element react with a given mass of another element to form two compounds, the two masses of the first element have a ratio of two small whole numbers. (Ch. 1)

**Le Châtelier's principle** The principle that a system at equilibrium responds to a stress in such a way that it relieves that stress. (Ch. 14)

**leveling effect** The observation that all strong acids completely ionize in aqueous solutions, forming $H_3O^+$ ions; strong bases are likewise leveled in water and are completely converted into their conjugate acids and $OH^-$ ions. (Ch. 15)

**Lewis acid** A substance that accepts a lone pair of electrons in a chemical reaction. (Ch. 16)

**Lewis base** A substance that donates a lone pair of electrons in a chemical reaction. (Ch. 16)

**Lewis structure** A two-dimensional representation of the bonds and lone pairs of valence electrons in an ionic or molecular compound. (Ch. 4)

**Lewis symbol** The chemical symbol for an element surrounded by one or more dots representing valence electrons; also called a *Lewis dot symbol*. (Ch. 4)

**ligand** A Lewis base bonded to the central metal ion of a complex ion. (Ch. 16)

**limiting reactant** A reactant that is consumed completely in a chemical reaction. The amount of product formed depends on the amount of the limiting reactant available. (Ch. 7)

**linear** Molecular geometry about a central atom with a steric number of 2 and no lone pairs of electrons. (Ch. 5)

**linear (straight-chain) hydrocarbon** A hydrocarbon in which the carbon atoms are bonded in one continuous line. Linear alkane chains have a methyl group at each end with methylene groups connecting them. (Ch. 19)

**lipid** A class of hydrophobic, water-insoluble, oily organic compounds that are common structural materials in cells. (Ch. 20)

**lipid bilayer** A double layer of molecules whose polar head groups interact with water molecules and whose nonpolar tails interact with each other. (Ch. 20)

**lipoprotein** A soluble protein that combines with and transports fat or other lipids in the blood plasma. (Ch. 20)

**liquid** A phase of matter that occupies a definite volume but flows to assume the shape of its container. (Ch. 1)

**London dispersion force** An intermolecular force between atoms or molecules caused by the presence of temporary dipoles in the molecules. (Ch. 6)

**lone pair** A pair of electrons not shared. (Ch. 4)

## M

**macrocyclic ligand** A ring containing multiple electron-pair donors that bind to a metal ion. (Ch. 23)

**magnetic quantum number ($m_\ell$)** An integer defining the orientation of an orbital in space; it may have any value from $-\ell$ to $+\ell$, where $\ell$ is the angular momentum quantum number. (Ch. 3)

**main group elements** (also called *representative elements*) The elements in groups 1, 2, and 13–18 of the periodic table. (Ch. 2)

**major essential element** An essential element present in the body in average concentrations greater than 1 mg of element per gram of body mass. (Ch. 22)

**manometer** An instrument for measuring the pressure exerted by a gas. (Ch. 9)

**mass (*m*)** The property that defines the quantity of matter in an object. (Ch. 1)

**mass action expression** An expression equivalent in form to the equilibrium constant expression, but applied to reaction mixtures that may or may not be at equilibrium. (Ch. 14)

**mass defect ($\Delta m$)** The difference between the mass of a stable nucleus and the masses of its individual nucleons. (Ch. 21)

**mass number (*A*)** The number of nucleons in an atom. (Ch. 2)

**mass spectrometer** A device that separates, weighs, and counts ions according to their ratio of mass (*m*) to charge (*z*), *m*/*z*. (Ch. 2)

**mass spectrum** A graph of data from a mass spectrometer featuring peaks whose heights are measures of the numbers of ions of different masses produced from a sample in the spectrometer's ion source. (Ch. 2)

**matter** Anything that has mass and occupies space. (Ch. 1)

**matter wave** The wave associated with any moving particle. (Ch. 3)

**mean (arithmetic mean)** An average calculated by summing all the values in a series and then dividing the sum by the number of values. (Ch. 1)

**mean free path** The average distance that a particle can travel through air or any gas before colliding with another particle. (Ch. 9)

**membrane potential, $E_{\text{membrane}}$** A weighted average of the equilibrium (reversal) potentials of the major ions based on concentration and permeability of the individual ions. (Ch. 22)

**meniscus** The concave or convex surface of a liquid. (Ch. 6)

**messenger RNA (mRNA)** The form of RNA that carries the code for synthesizing proteins from DNA to the site of protein synthesis in a cell. (Ch. 20)

**metallic bond** A bond consisting of the nuclei of metal atoms surrounded by a "sea" of shared electrons. (Ch. 4)

**metalloids** Elements that tend to have some properties of metals and some properties of nonmetals. (Ch. 2)

**metals** Elements that are typically shiny, malleable, ductile solids that conduct heat and electricity well and tend to form positive ions. (Ch. 2)

**meter (m)** The standard unit of length, equivalent to 39.37 inches. (Ch. 1)

**methanogenic bacteria** Bacteria using simple organic compounds and hydrogen for energy; their respiration produces methane, carbon dioxide, and water, depending on the compounds they consume. (Ch. 19)

**methyl group** ($-CH_3$) A structural unit that can make only one additional bond. (Ch. 19)

**methylene group** ($-CH_2-$) A structural unit that can make two additional bonds. (Ch. 19)

**miscible** Capable of being mixed in any proportion (without reacting chemically). (Ch. 6)

**mixture** A combination of pure substances in various proportions in which the individual substances retain their chemical identities and can be separated from one another by a physical process. (Ch. 1)

**molality (*m*)** Concentration expressed as the number of moles of solute per kilogram of solvent. (Ch. 11)

**molar absorptivity ($\varepsilon$)** A measure of how well a compound or ion absorbs light. (Ch. 8)

**molar heat capacity ($c_P$)** The energy required to raise the temperature of one mole of a substance by 1°C at constant pressure. (Ch. 10)

**molar mass ($\mathcal{M}$)** The mass of one mole of a substance. (Ch. 2)

**molarity (*M*)** The number of moles of solute per liter of solution: $M = n/V$; also called *molar concentration*. (Ch. 8)

**mole (mol)** An amount of a substance that contains a number of particles (atoms, ions, molecules, or formula units) equal to the Avogadro constant. (Ch. 2)

**mole fraction ($x_i$)** The ratio of the number of moles of a particular component *i* in a mixture to the total number of moles in the mixture. (Ch. 9)

**molecular equation** A balanced equation describing a reaction in solution in which the reactants and products are written as neutral compounds. (Ch. 8)

**molecular formula** A chemical formula that indicates how many atoms of each element are in one molecule of a pure substance. (Ch. 1)

**molecular geometry** The three-dimensional arrangement of the atoms in a molecule. (Ch. 5)

**molecular ion ($M^+$)** An ion formed in a mass spectrometer when a molecule loses an electron after being bombarded with high-energy electrons. The molecular ion has a charge of 1+ and has essentially the same molecular mass as the molecule from which it came. (Ch. 2)

**molecular mass** The mass in u of one molecule of a molecular compound. (Ch. 2)

**molecular orbital (MO) theory** A bonding theory based on the mixing of atomic orbitals of similar shapes and energies to form molecular orbitals that belong to the molecule as a whole. (Ch. 5)

**molecular orbital** A region of characteristic shape and energy where electrons in a molecule are located. (Ch. 5)

**molecular orbital diagram** In MO theory, an energy-level diagram showing the relative energies and electron occupancy of the molecular orbitals for a molecule. (Ch. 5)

**molecular recognition** The process by which molecules interact with other molecules to produce a biological effect. (Ch. 5)

**molecular solid** A solid formed by neutral, covalently bonded molecules held together by intermolecular attractive forces. (Ch. 18)

**molecularity** The number of molecules (or atoms or ions) involved in an elementary step in a reaction. (Ch. 13)

**molecule** A collection of chemically bonded atoms. (Ch. 1)

**monodentate ligand** A species that forms only a single coordinate bond to a metal ion in a complex. (Ch. 23)

**monomer** A small molecule that bonds with other monomers to form polymers. (Ch. 19)

**monoprotic acid** An acid that has one ionizable hydrogen atom per molecule. (Ch. 15)

**monosaccharide** A single-sugar unit and the simplest carbohydrate. (Ch. 20)

## N

**n-type semiconductor** A semiconductor containing an electron-rich dopant. (Ch. 18)

**nanoparticle** Approximately spherical sample of matter with dimensions less than 100 nm ($1 \times 10^{-7}$ m). (Ch. 2)

**natural abundance** The proportion of a particular isotope, usually expressed as a percentage, relative to all the isotopes of that element in a natural sample. (Ch. 2)

**Nernst equation** An equation relating the potential of a cell (or half-cell) reaction to its standard potential ($E°$) and to the concentrations of its reactants and products. (Ch. 17)

**net ionic equation** A balanced equation that describes the actual reaction taking place in aqueous solution; it is obtained by eliminating the spectator ions from the overall ionic equation. (Ch. 8)

**neutralization reaction** A reaction that takes place when an acid reacts with a base and produces a solution of a salt in water. (Ch. 8)

**neutron** An electrically neutral (uncharged) subatomic particle with a mass number of 1. (Ch. 2)

**neutron capture** The absorption of a neutron by a nucleus. (Ch. 21)

**noble gases** The elements in group 18 of the periodic table. (Ch. 2)

**node** A location in a standing wave that experiences no displacement. In the context of orbitals, nodes are locations at which electron density goes to zero. (Ch. 3)

**nonelectrolyte** A molecular substance that does not ionize when it dissolves in water. (Ch. 8)

**nonessential element** An element present in humans that has no known function. (Ch. 22)

**nonmetals** Elements with properties opposite those of metals, including poor conductivity of heat and electricity. (Ch. 2)

**nonpolar covalent bond** A bond characterized by an even distribution of charge; the two atoms equally share the electrons in the bond. (Ch. 4)

**nonspontaneous process** A process that requires continuous outside intervention to proceed. (Ch. 12)

**nuclear chemistry** the study of reactions that involve changes in the nuclei of atoms. (Ch. 21)

**nuclear fission** A nuclear reaction in which the nucleus of an element splits into two lighter nuclei, usually accompanied by the release of one or more neutrons and energy. (Ch. 21)

**nuclear fusion** A nuclear reaction in which subatomic particles or atomic nuclei collide at very high speeds and fuse, forming more massive nuclei and releasing energy. (Ch. 21)

**nucleic acid** One of a family of large molecules, which includes deoxyribonucleic acid (DNA) and ribonucleic acid (RNA), that contains the genetic blueprint of an organism and controls the production of proteins. (Ch. 20)

**nucleon** A proton or neutron in a nucleus. (Ch. 2)

**nucleosynthesis** The natural formation of nuclei because of fusion and other nuclear processes. (Ch. 21)

**nucleotide** A monomer unit from which nucleic acids are made. (Ch. 20)

**nucleus** (of an atom) The positively charged center of an atom that contains nearly all the atom's mass. (Ch. 2)

**nuclide** A specific isotope of an element. (Ch. 2)

## O

**octahedral** Molecular geometry about a central atom with a steric number of 6 and no lone pairs of electrons, in which all six sites are equivalent. (Ch. 5)

**octet rule** The tendency of atoms of main group elements to make bonds by gaining, losing, or sharing electrons to achieve a valence shell containing eight electrons, or four electron pairs. (Ch. 4)

**oil** A liquid triglyceride containing primarily unsaturated fatty acids. (Ch. 20)

**orbital** A region around the nucleus of an atom where the probability of finding an electron is high; each orbital is defined by the square of the wave function ($\Psi^2$). (Ch. 3)

**orbital diagram** A depiction of the arrangement of electrons in an atom or ion, using boxes to represent orbitals. (Ch. 3)

**ore** A naturally occurring compound or mixture of compounds from which elements can be extracted. (Ch. 18)

**organic chemistry** The study of compounds containing C—C and/or C—H bonds. (Ch. 19)

**organic compound** A molecule containing carbon atoms whose structure typically consists of carbon–carbon bonds and carbon–hydrogen bonds and that may include one or more heteroatoms such as oxygen, nitrogen, sulfur, phosphorus, or the halogens. (Ch. 6)

**organometallic compound** A molecule containing direct carbon–metal covalent bonds. (Ch. 23)

**osmosis** The spontaneous flow of solvent molecules through a semipermeable membrane from a more dilute solution into a more concentrated one. (Ch. 11)

**osmotic pressure ($\Pi$)** The pressure applied across a semipermeable membrane to stop the flow of water from the compartment containing pure solvent or a less concentrated solution to the compartment containing a more concentrated solution. (Ch. 11)

**outlier** A data point that is distant from the other observations. (Ch. 1)

**overall ionic equation** A balanced equation that shows all the species, both ionic and molecular, present in a reaction occurring in aqueous solution. (Ch. 8)

**overall reaction order** The sum of the exponents of the concentration terms in the rate law. (Ch. 13)

**overlap** A term in valence bond theory describing bonds arising from two orbitals on different atoms that occupy the same region of space. (Ch. 5)

**oxidation** A chemical change in which an element loses electrons; the oxidation number of the element increases. (Ch. 8)

**oxidation number (O.N.)** A numerical value (which can be positive, negative, or zero) based on the number of electrons that an atom gains or loses when it forms an ion or that it shares when it forms a covalent bond with an atom of another element; also called *oxidation state*. (Ch. 8)

**oxidizing agent** A reactant that accepts electrons from another in a redox reaction, thereby oxidizing the other reactant; the oxidizing agent is reduced in the reaction. (Ch. 8)

**oxoacid** A compound composed of oxoanions bonded to $H^+$ ions. (Ch. 4)

**oxoanion** A polyatomic anion that contains at least one nonoxygen central atom bonded to one or more oxygen atoms. (Ch. 4)

## P

**p-type semiconductor** A semiconductor containing an electron-poor dopant. (Ch. 18)

**packing efficiency** The percentage of the total volume of a unit cell occupied by the spheres. (Ch. 18)

**paramagnetic** Describes a substance with unpaired electrons that is attracted to a magnetic field. (Ch. 5)

**partial pressure** The contribution to the total pressure made by a component in a mixture of gases. (Ch. 9)

**Pauli exclusion principle** The principle that no two electrons in an atom can have the same set of four quantum numbers. (Ch. 3)

**peptide** A compound of two or more amino acids joined by peptide bonds. Small peptides containing up to 20 amino acids are *oligopeptides*; the term *polypeptide* is used for chains longer than 20 amino acids but shorter than proteins. (Ch. 20)

**peptide bond** The result of a condensation reaction between the carboxylic acid group of one amino acid and the amine group of another. (Ch. 20)

**percent composition** The composition of a compound expressed in terms of the percentage by mass of each element in the compound. (Ch. 7)

**percent ionization** The ratio of the quantity of a substance that is ionized to the concentration of the substance before ionization expressed as a percentage. (Ch. 15)

**percent yield** The ratio, expressed as a percentage, of the actual yield of a chemical reaction to the theoretical yield. (Ch. 7)

**period** (of elements) All the elements in a row of the periodic table. (Ch. 2)

**periodic table of the elements** A chart of the elements in order of their atomic numbers and in a pattern based on their physical and chemical properties. (Ch. 2)

**permanent dipole** Permanent separation of electrical charge in a molecule resulting from unequal distributions of bonding and/or lone pairs of electrons. (Ch. 5)

**pH** The negative logarithm of the hydronium ion concentration in an aqueous solution. (Ch. 15)

**pH buffer** A solution that resists changes in pH when acids or bases are added to it; typically, a solution of a weak acid and its conjugate base. (Ch. 16)

**pH indicator** A water-soluble weak organic acid that changes color as it ionizes. (Ch. 16)

**phase diagram** A graphical representation of the dependence of the stabilities of the physical states of a substance on temperature and pressure. (Ch. 6)

**phospholipid** A molecule of glycerol with two fatty acid chains and one polar group containing a phosphate; phospholipids are major constituents of cell membranes. (Ch. 20)

**phosphorylation** A reaction resulting in the addition of a phosphate group to an organic molecule. (Ch. 12)

**photochemical smog** A mixture of gases formed in the lower atmosphere when sunlight interacts with compounds produced in internal combustion engines and with other pollutants. (Ch. 13)

**photoelectric effect** The release of electrons from a material as a result of electromagnetic radiation striking it. (Ch. 3)

**photoelectron spectroscopy (PES)** A technique for determining the binding energies of electrons with an atom. (Ch. 3)

**photoelectron spectrum** A graph plotting the number of electrons ejected by a metal as a function of binding energy. (Ch. 3)

**photon** A quantum of electromagnetic radiation. (Ch. 3)

**photosynthesis** A set of chemical reactions driven by the energy of sunlight that convert carbon dioxide and water into oxygen and a variety of molecules containing C, H, and O. (Ch. 7)

**physical process** A transformation of a sample of matter, such as a change in its physical state, that does not alter the chemical identity of any substance in the sample; also called *physical change.* (Ch. 1)

**physical property** A property of a substance that can be observed without changing the substance into another substance. (Ch. 1)

**pi ($\pi$) bond** A covalent bond in which electron density is greatest above and below the internuclear axis. (Ch. 5)

**pi ($\pi$) molecular orbital** In MO theory, an orbital formed by the mixing of atomic orbitals oriented above and below, or in front of and behind, the internuclear axis. (Ch. 5)

**Planck constant ($h$)** The proportionality constant between the energy and frequency of electromagnetic radiation expressed in $E = h\nu$; $h = 6.626 \times 10^{-34}$ J · s. (Ch. 3)

**pOH** The negative logarithm of the hydroxide ion concentration in an aqueous solution. (Ch. 15)

**polar covalent bond** A bond resulting from unequal sharing of pairs of electrons between atoms. (Ch. 4)

**polarizability** The relative ease with which the electron cloud in a molecule, ion, or atom can be distorted, inducing a temporary dipole. (Ch. 6)

**polyatomic ion** A charged group of atoms joined by covalent bonds. (Ch. 4)

**polydentate ligand** A species that can form more than one coordinate bond per molecule. (Ch. 23)

**polymer** A very large molecule with high molar mass; the root word *meros* is Greek for "part" or "unit," so *polymer* literally means "many units." (Ch. 19)

**polyprotic acid** An acid that has two or more ionizable hydrogen atoms per molecule. (Ch. 15)

**polysaccharide** A polymer of monosaccharides. (Ch. 20)

**porphyrin** A type of tetradentate macrocyclic ligand. (Ch. 23)

**positron** A particle with the mass of an electron but with a positive charge. (Ch. 21)

**positron emission** The spontaneous emission of a positron from a neutron-poor nucleus. (Ch. 21)

**potential energy (PE)** The energy stored in an object because of its position or composition. (Ch. 1)

**precipitate** A solid product formed from a reaction in solution. (Ch. 8)

**precipitation reaction** A reaction in which soluble reactants form a product that has limited solubility. (Ch. 8)

**precision** The extent to which repeated measurements of the same variable agree. (Ch. 1)

**pressure ($P$)** The ratio of a force to the surface area over which the force is applied. (Ch. 6)

**pressure–volume ($P$–$V$) work** The work associated with the expansion or compression of a gas. (Ch. 10)

**primary (1°) structure** The sequence in which amino acid monomers occur in a protein chain. (Ch. 20)

**principal quantum number ($n$)** A positive integer describing the relative size and energy of an atomic orbital or group of orbitals in an atom. (Ch. 3)

**products** Substance(s) formed as a result of a chemical reaction. (Ch. 7)

**protein** A biological polymer made of amino acids. (Ch. 20)

**proton** A subatomic particle in the nuclei of atoms that has a positive charge and a mass number of 1. (Ch. 2)

**protonated** Describes the molecular structure of a base that has accepted a proton ($H^+$). (Ch. 15)

**pseudo–first order** Describes a reaction in which all the reactants but one are present at such high concentrations that they do not decrease significantly during the reaction, so that the reaction rate is controlled by the concentration of the limiting reactant. (Ch. 13)

**pure substance** Matter that has a constant composition and cannot be broken down to simpler matter by any physical process. (Ch. 1)

## Q

**quantized** Having values restricted to whole-number multiples of a specific base value. (Ch. 3)

**quantum** (plural, *quanta*) The smallest discrete quantity of a particular form of energy. (Ch. 3)

**quantum mechanics** (also called *wave mechanics*) A mathematical description of the wavelike behavior of particles on the atomic level. (Ch. 3)

**quantum number** One of four related numbers that specify the energy, shape, and orientation of orbitals in an atom and the spin orientation of electrons in the orbitals. (Ch. 3)

**quantum theory** A model of matter and energy based on the principle that energy is absorbed or emitted in discrete packets, or quanta. (Ch. 3)

**quarks** Elementary particles that combine to form neutrons and protons. (Ch. 21)

**quaternary (4°) structure** The larger structure functioning as a single unit that results when two or more proteins associate. (Ch. 20)

## R

**R** A symbol in a general formula standing for an organic group that has one available bond; it is used to indicate the part of a molecule that is not the functional group of interest. (Ch. 19)

**racemic mixture** A sample containing equal amounts of both enantiomers of a compound. (Ch. 5)

**radioactive decay** The spontaneous disintegration of unstable particles accompanied by the release of radiation. (Ch. 21)

**radioactivity** The spontaneous emission of high-energy radiation and particles by materials. (Ch. 2)

**radiocarbon dating** A method for establishing the age of a carbon-containing object by measuring the amount of radioactive carbon-14 remaining in the object. (Ch. 21)

**radiometric dating** A method for determining the age of an object on the basis of the quantity of a radioactive nuclide and/or the products of its decay that the object contains. (Ch. 21)

**radionuclide** A radioactive (unstable) nuclide. (Ch. 2)

**Raoult's law** The principle that the vapor pressure of a solution is the sum of the vapor pressures of the volatile components of the solution, which are each the product of the vapor pressure of the pure component and its mole fraction in the solution. (Ch. 11)

**rate constant** The proportionality constant that relates the rate of a reaction to the concentrations of reactants. (Ch. 13)

**rate law** An equation that defines the experimentally determined relation between the concentration of reactants in a chemical reaction and the rate of that reaction. (Ch. 13)

**rate-determining step** The slowest step in a multistep chemical reaction. (Ch. 13)

**reactants** Substance(s) consumed during a chemical reaction. (Ch. 7)

**reaction mechanism** A set of steps that describes how a reaction occurs at the molecular level; the mechanism must be consistent with the rate law for the reaction. (Ch. 13)

**reaction order** An experimentally determined number defining the dependence of the reaction rate on the concentration of a reactant. (Ch. 13)

**reaction quotient ($Q$)** The numerical value of the mass action expression for any values of the concentrations (or partial pressures) of reactants and products; at equilibrium, $Q = K$. (Ch. 14)

**reaction rate** A measure of how rapidly a reaction occurs; it is related to rates of change in the concentrations of reactants and products over time. (Ch. 13)

**reducing agent** A reactant that donates electrons to another in a redox reaction, thereby reducing the other reactant; the reducing agent is oxidized in the reaction. (Ch. 8)

**reduction** A chemical change in which an element gains electrons; the oxidation number of the element decreases. (Ch. 8)

**relative biological effectiveness (RBE)** A factor that accounts for the differences in physical damage caused by different types of radiation. (Ch. 21)

**replication** The process by which one double-stranded DNA forms two new DNA molecules, each one containing one strand from the original molecule and one new strand. (Ch. 20)

**representative elements** (also called *main group elements*) The elements in groups 1, 2, and 13–18 of the periodic table. (Ch. 2)

**resonance** A characteristic of electron distributions in which two or more equivalent Lewis structures can be drawn for one compound. (Ch. 4)

**resonance stabilization** The stability of a molecular structure resulting from the delocalization of its electrons. (Ch. 4)

**resonance structure** One of two or more Lewis structures with the same arrangement of atoms but different arrangements of bonding pairs of electrons. (Ch. 4)

**reverse osmosis (RO)** A purification process in which solvent is forced through semipermeable membranes, leaving dissolved impurities behind. (Ch. 11)

**reversible process** A process that happens so slowly that an incremental change can be reversed by another tiny change, restoring the original state of the system with no net flow of energy between the system and its surroundings. (Ch. 12)

**root-mean-square speed ($u_{rms}$)** The square root of the average of the squared speeds of all the particles in a population of gas particles. (Ch. 9)

## S

**salt** The product of a neutralization reaction, made up of the cation of the base in the reaction plus the anion of the acid. (Ch. 8)

**saturated hydrocarbon** An alkane that has the maximum ratio of hydrogen atoms to carbon atoms. (Ch. 19)

**saturated solution** A solution that contains the maximum concentration of a solute possible at a given temperature. (Ch. 8)

**Schrödinger wave equation** A description of how the electron matter wave varies with location and time around the nucleus of a hydrogen atom. (Ch. 3)

**scientific law** A concise and generally applicable statement of a fundamental scientific principle. (Ch. 1)

**scientific methods** Approaches to acquiring knowledge by observing phenomena, developing a testable hypothesis, and carrying out additional experiments that test the validity of the hypothesis. (Ch. 1)

**scientific theory** A concise, extensively tested explanation of widely observed natural phenomena. (Ch. 1)

**scintillation counter** An instrument that determines the level of radioactivity in samples by measuring the intensity of light emitted by phosphors in contact with the samples. (Ch. 21)

**second law of thermodynamics** The principle stating that the total entropy of the universe increases in any spontaneous process. (Ch. 12)

**secondary (2°) structure** The pattern of arrangement of segments of a protein chain. (Ch. 20)

**seesaw** Molecular geometry about a central atom with a steric number of 5 and one lone pair of electrons in an equatorial position; the atoms occupy two axial sites and two equatorial sites. (Ch. 5)

**semiconductor** A material with electrical conductivity between that of metals and insulators that can be chemically altered to increase its electrical conductivity. (Ch. 18)

**SI units** A set of base and derived units used worldwide to express distances and quantities of matter and energy. (Ch. 1)

**sievert (Sv)** SI unit used to express the amount of biological damage caused by ionizing radiation. (Ch. 21)

**sigma ($\sigma$) bond** A covalent bond in which the highest electron density lies between the two atoms along the internuclear axis. (Ch. 5)

**sigma ($\sigma$) molecular orbital** In MO theory, the orbital that results in the highest electron density between the two bonded atoms. (Ch. 5)

**significant figures** All the certain digits in a measured value plus one estimated digit. The greater the number of significant figures, the greater the certainty with which the value is known. (Ch. 1)

**simple cubic (sc) unit cell** A cell with atoms only at the eight corners of a cube. (Ch. 18)

**single bond** A bond that results when two atoms share one pair of electrons. (Ch. 4)
**solid** A phase of matter that has a definite shape and volume. (Ch. 1)
**solubility** The maximum quantity of a substance that can dissolve in a given volume of solution. (Ch. 6)
**solubility-product constant** An equilibrium constant that describes the formation of a saturated solution of a slightly soluble salt; also called *solubility product*, $K_{sp}$. (Ch. 16)
**solute** Any component in a solution other than the solvent. A solution may contain one or more solutes. (Ch. 6)
**solution** Another name for a *homogeneous mixture*. Solutions are often liquids, but they may also be solids or gases. (Ch. 1)
**solvent** The component of a solution for which the largest number of moles is present. (Ch. 6)
***sp* hybrid orbitals** Two hybrid orbitals on opposite sides of the hybridized atom formed by mixing one *s* and one *p* orbital. (Ch. 5)
**$sp^2$ hybrid orbitals** Three hybrid orbitals in a trigonal planar orientation formed by mixing one *s* and two *p* orbitals. (Ch. 5)
**$sp^3$ hybrid orbitals** Four hybrid orbitals with a tetrahedral orientation produced by mixing one *s* and three *p* orbitals. (Ch. 5)
**specific heat ($c_s$)** The energy required to raise the temperature of 1 g of a substance by 1°C at constant pressure. (Ch. 10)
**spectator ion** An ion present in a reaction vessel when a chemical reaction takes place but is unchanged by the reaction; spectator ions appear in an overall ionic equation but not in a net ionic equation. (Ch. 8)
**spectrochemical series** A list of ligands rank-ordered by their ability to split the energies of the *d* orbitals of transition metal ions. (Ch. 23)
**sphere of hydration** The cluster of water molecules surrounding an ion in an aqueous solution. (Ch. 6)
**spin quantum number ($m_s$)** Either $+\frac{1}{2}$ or $-\frac{1}{2}$, indicating the spin orientation of an electron. (Ch. 3)
**spontaneous process** A process that proceeds without continuous outside intervention. (Ch. 12)
**square planar** Molecular geometry about a central atom with a steric number of 6 and two lone pairs of electrons that occupy axial sites; the atoms occupy four equatorial positions. (Ch. 5)
**square pyramidal** Molecular geometry about a central atom with a steric number of 6 and one lone pair of electrons; as typically drawn, the atoms occupy four equatorial sites and one axial site. (Ch. 5)
**standard atmosphere (atm)** The average pressure at sea level on Earth. (Ch. 6)
**standard cell potential ($E^\circ_{cell}$)** A measure of how forcefully an electrochemical cell, in which all reactants and products are in their standard states, can pump electrons through an external circuit.
**standard conditions** In thermodynamics: a pressure of 1 bar (~1 atm) and some specified temperature, assumed to be 25°C unless otherwise stated; for solutions, a concentration of 1 *M* is specified. (Ch. 10)
**standard deviation (*s*)** A measure of the amount of variation, or dispersion, in a set of related values. (Ch. 1)
**standard enthalpy of formation ($\Delta H^\circ_f$)** The enthalpy change of a formation reaction; also called the *standard heat of formation*. (Ch. 10)
**standard enthalpy of reaction ($\Delta H^\circ_{rxn}$)** The enthalpy change associated with a reaction that takes place under standard conditions; also called the *standard heat of reaction*. (Ch. 10)
**standard free energy of formation ($\Delta G^\circ_f$)** The change in free energy associated with the formation of one mole of a compound in its standard state from its elements in their standard states. (Ch. 12)
**standard hydrogen electrode (SHE)** A reference electrode based on the half-reaction $2\,H^+(aq) + 2\,e^- \rightarrow H_2(g)$, which produces a standard electrode potential defined to be 0.000 V. (Ch. 17)
**standard molar entropy ($S^\circ$)** The absolute entropy of one mole of a substance in its standard state. (Ch. 12)
**standard reduction potential ($E^\circ$)** The potential of a reduction half-reaction in which all reactants and products are in their standard states at 25°C. (Ch. 17)
**standard solution** A solution of accurately known concentration that is used in chemical analysis. (Ch. 8)
**standard state** The most stable form of a substance under a pressure of 1 bar (~1 atm) and some specified temperature (usually 25°C). (Ch. 10)
**standard temperature and pressure (STP)** 0°C and 1 bar as defined by IUPAC; 0°C and 1 atm are commonly used in the United States. (Ch. 9)
**standing wave** A wave confined to a given space, with a wavelength λ related to the length *L* of the space by $L = n(\lambda/2)$, where *n* is a whole number. (Ch. 3)
**state function** A property of an entity based solely on its chemical or physical state or both but not on how it achieved that state. (Ch. 10)
**stereoisomers** Molecules with the same formula and bonding order, but with different spatial arrangements of their atoms. (Ch. 5)
**steric number (SN)** The sum of the number of atoms bonded to a central atom plus the number of lone pairs of electrons on the central atom. (Ch. 5)
**stimulatory effect** Increased activity, growth, or other biological response to the presence of a nonessential element. (Ch. 22)
**stoichiometric yield** (also called *theoretical yield*) The maximum amount of product possible in a chemical reaction for the given quantities of reactants. (Ch. 7)
**stoichiometry** The mole ratios among the reactants and products in a chemical reaction. (Ch. 7)
**strong acid** An acid that completely ionizes in aqueous solution. (Ch. 8)
**strong base** A base that completely dissociates into ions in aqueous solution. (Ch. 8)
**strong electrolyte** A substance that dissociates completely when it dissolves in water. (Ch. 8)
**strong nuclear force** The fundamental force of nature that keeps quarks together in subatomic particles and nucleons together in atomic nuclei. (Ch. 21)
**structural formula** A representation of a molecule that uses short lines between the symbols of elements to show chemical bonds between atoms. (Ch. 1)
**structural isomer** (also called *constitutional isomer*) One of a set of compounds with the same molecular formula but different connections between the atoms in their molecules. (Ch. 6)
**subatomic particles** The neutrons, protons, and electrons in an atom. (Ch. 2)
**sublimation** The transformation of a solid directly into a gas (vapor). (Ch. 1)
**substitutional alloy** An alloy in which atoms of the nonhost element replace host atoms in the crystal lattice. (Ch. 18)
**substrate** The reactant that binds to the active site in an enzyme-catalyzed reaction. (Ch. 20)
**supercritical fluid** A substance in a state that is above the temperature and pressure at the critical point, at which the liquid and vapor phases are indistinguishable. (Ch. 6)
**supersaturated solution** A solution that contains more than the maximum quantity of solute predicted to be soluble in a given volume of solution at a given temperature. (Ch. 8)
**surface tension** The ability of the surface of a liquid to resist an external force. (Ch. 6)

**surroundings** Everything in a thermochemical study that is not part of the system. (Ch. 10)

**system** The part of the universe that is the focus of a thermochemical study. (Ch. 10)

## T

**T-shaped** Molecular geometry about a central atom with a steric number of 5 and two lone pairs of electrons that occupy equatorial positions; the atoms occupy two axial sites and one equatorial site. (Ch. 5)

**temporary dipole** (also called *induced dipole*) The separation of charge produced in an atom or molecule by a momentary uneven distribution of electrons. (Ch. 6)

**termolecular step** A step in a reaction mechanism involving a collision among three molecules (or atoms or ions). (Ch. 13)

**tertiary (3°) structure** The three-dimensional, biologically active structure of the protein that arises because of interactions between R groups on amino acids. (Ch. 20)

**tetrahedral** Molecular geometry about a central atom with a steric number of 4 and no lone pairs of electrons. (Ch. 5)

**theoretical yield** (also called *stoichiometric yield*) The maximum amount of product possible in a chemical reaction for the given quantities of reactants. (Ch. 7)

**thermal energy** The portion of the total internal energy of a system that is proportional to its absolute temperature. (Ch. 10)

**thermochemical equation** The chemical equation of a reaction that includes the change in enthalpy that accompanies the reaction. (Ch. 10)

**thermochemistry** The study of the changes in energy that accompany chemical reactions. (Ch. 10)

**thermodynamics** The study of energy and its transformations. (Ch. 10)

**third law of thermodynamics** The principle stating that the entropy of a perfect crystal is zero at absolute zero. (Ch. 12)

**threshold frequency ($\nu_0$)** The minimum frequency of light required to produce the photoelectric effect. (Ch. 3)

**titrant** The standard solution added to the sample in a titration. (Ch. 8)

**titration** An analytical method of determining the concentration of a solute in a sample by reacting the solute with a solution of known concentration. (Ch. 8)

**trace essential element** An essential element present in the body in average concentrations between 1 μg and 1000 μg of element per gram of body mass. (Ch. 22)

**trans isomer** (also called *E isomer*) A molecule with two like groups (such as two R groups or two hydrogen atoms) on opposite sides of the molecule. (Ch. 19)

**transcription** The process of copying the information in DNA to mRNA. (Ch. 20)

**transfer RNA (tRNA)** The form of RNA that delivers amino acids, one at a time, to polypeptide chains being assembled by the ribosome–mRNA complex. (Ch. 20)

**transition metals** The elements in groups 3–12 of the periodic table. (Ch. 2)

**transition state** A high-energy state between reactants and products in a chemical reaction. (Ch. 13)

**translation** The process of assembling proteins from the information encoded in mRNA. (Ch. 20)

**tricarboxylic acid (TCA) cycle** A series of reactions that continues the oxidation of the pyruvate formed in glycolysis. (Ch. 20)

**trigonal bipyramidal** Molecular geometry about a central atom with a steric number of 5 and no lone pairs of electrons, in which three atoms occupy equatorial sites and two other atoms occupy axial sites above and below the equatorial plane. (Ch. 5)

**trigonal planar** Molecular geometry about a central atom with a steric number of 3 and no lone pairs of electrons. (Ch. 5)

**trigonal pyramidal** Molecular geometry about a central atom with a steric number of 4 and one lone pair of electrons. (Ch. 5)

**triple bond** A bond formed when two atoms share three pairs of electrons. (Ch. 4)

**triple point** The temperature and pressure at which all three phases of a substance coexist. Freezing and melting, boiling and liquefaction, and sublimation and deposition all proceed at the same rate, so no net change takes place in the system. (Ch. 6)

## U

**ultratrace essential element** An essential element present in the body in average concentrations less than 1 μg of element per gram of body mass. (Ch. 22)

**unified atomic mass unit (u)** The unit used to express the relative masses of atoms and subatomic particles that is exactly 1/12 the mass of 1 atom of carbon with 6 protons and 6 neutrons in its nucleus. (Ch. 2)

**unimolecular step** A step in a reaction mechanism involving only one molecule (or atom or ion) on the reactant side. (Ch. 13)

**unit cell** The basic repeating unit of the arrangement of atoms, ions, or molecules in a crystalline solid. (Ch. 18)

**universal gas constant** The constant $R$ in the ideal gas equation; its value and units depend on the units used for the variables in the equation. (Ch. 9)

**unsaturated hydrocarbon** A hydrocarbon that does not have the maximum ratio of hydrogen atoms to carbon atoms. (Ch. 19)

**unsaturated solution** A solution that contains less than the maximum quantity of solute predicted to be soluble in a given volume of solution at a given temperature. (Ch. 8)

## V

**valence** The capacity of the atoms of an element to form chemical bonds. (Ch. 4)

**valence band** A band of orbitals that are filled or partially filled by valence electrons. (Ch. 18)

**valence bond theory** A quantum mechanics–based theory of bonding that assumes covalent bonds form when half-filled orbitals on different atoms overlap or occupy the same region in space. (Ch. 5)

**valence electrons** The electrons in the outermost occupied shell of an atom; they have the most influence on the atom's chemical behavior. (Ch. 3)

**valence shell** The outermost occupied shell of an atom. (Ch. 3)

**valence-shell electron-pair repulsion theory (VSEPR)** A model predicting the arrangement of valence electron pairs around a central atom that minimizes their mutual repulsion to produce the lowest energy orientations. (Ch. 5)

**van 't Hoff factor ($i$)** The ratio of the concentration of solute particles in a solution to the concentration of particles that would be there if the solute did not dissociate. (Ch. 11)

**van der Waals equation** An equation describing how the pressure, volume, and temperature of a quantity of a real gas are related; it includes terms that account for the incompressibility of gas particles and interactions between them. (Ch. 9)

**van der Waals force** Any interaction between neutral atoms and molecules, including hydrogen bonds, other dipole–dipole interactions, and London dispersion forces; the term does not apply to interactions involving ions. (Ch. 9)

**vapor pressure ($P_{vap}$)** The pressure exerted by a gas in equilibrium with its liquid phase at a given temperature. (Ch. 11)

**vinyl group** The subgroup $CH_2{=}CH-$. (Ch. 19)

**vinyl polymer** One of the family of polymers formed from monomers containing the subgroup $CH_2{=}CH-$. (Ch. 19)

**viscosity** A measure of a fluid's resistance to flow. (Ch. 6)

**volatile** Having a significant vapor pressure at a given temperature. (Ch. 11)

**volatility** A measure of how readily a substance vaporizes. (Ch. 1)

**voltaic cell** (also called *galvanic cell*) An electrochemical cell in which chemical energy is transformed into electrical work by a spontaneous redox reaction. (Ch. 17)

**wave function ($\Psi$)** A solution to the Schrödinger wave equation. (Ch. 3)

**wave mechanics** (also called *quantum mechanics*) A mathematical description of the wavelike behavior of particles on the atomic level. (Ch. 3)

**wavelength ($\lambda$)** The distance from crest to crest or trough to trough on a wave. (Ch. 3)

**weak acid** A weak electrolyte that only partially ionizes in aqueous solution. (Ch. 8)

**weak base** A base that is a weak electrolyte and so has a limited capacity to accept protons. (Ch. 8)

**weak electrolyte** A substance that only partly ionizes when it dissolves in water. (Ch. 8)

**work ($w$)** The exertion of a force ($F$) through a distance ($d$): $w = F \times d$. (Ch. 1)

**work function ($\phi$)** The amount of energy needed to dislodge an electron from the surface of a material. (Ch. 3)

**X-ray diffraction (XRD)** A technique for determining the arrangement of atoms or ions in a crystal by analyzing the pattern that results when X-rays are scattered after bombarding the crystal. (Ch. 18)

***Z* isomer** (also called *cis isomer*) A molecule with two like groups (such as two R groups or two hydrogen atoms) on the same side of the molecule. (Ch. 19)

**zwitterion** A molecule that has both positively and negatively charged groups in its structure. (Ch. 20)

# Answers
## to Particulate Review, Concept Tests, and Practice Exercises

# Chapter 1

## Particulate Review

Image (b) depicts liquid mercury; image (a) depicts solid mercury; image (c) depicts mercury vapor.

## Concept Tests

p. 9 Mass and volume are extensive; density and hardness are intensive.
p. 13 (a) Filtration; (b) chromatography; (c) distillation
p. 15 (a) Sublimation; (b) deposition
p. 17 1.49 or 49%
p. 18 $CH_4O$ or $CH_3OH$
p. 23 3, 3, 4
p. 26 (a) 77.74 km/day, or 48.30 miles/day; (b) The trip took 65,259 minutes, which is known to five significant figures: one more than the distance hiked (3523 km). Therefore, distance is the weak link.

## Practice Exercises

1.1. (a) and (c) are physical; (b) and (d) are chemical.
1.2. (a) Only oxygen gas is a pure substance.
1.3. (a) Gas in the upper part of the figure turns into a solid in the lower part as a result of deposition; (b) sublimation.
1.4. Statistics (a) and (e) are exact numbers; statistics (b), (c), and (d) have inherent uncertainty.
1.5. (a) 1.130 g/cm$^3$; (b) more dense
1.6. 8.91 g/cm$^3$
1.7. Store A
1.8. 93 cm$^3$/s and 5.7 in$^3$/s
1.9. $T(\text{K})_{\text{low}} = 40$ K and $T(\text{K})_{\text{high}} = 396$ K; $T(°\text{F})_{\text{low}} = -387°\text{F}$ and $T(°\text{F})_{\text{high}} = 253°\text{F}$
1.10. 161 J
1.11. Mean = 36.38; standard deviation = 0.38; the 95% confidence interval = 36.38 ± 0.44, which becomes 36.4 ± 0.4.
1.12. The mean and standard deviation values are 193.7 ± 18.7 mg/dL. Testing 215 mg/dL as an outlier gives a calculated $Z$ value of 1.15, which is slightly less than the reference $Z$ values (1.155) for $n = 3$ from Table 1.7 at both the 95% and 99% confidence levels. Therefore, the value 215 should *not* be considered an outlier.

# Chapter 2

## Particulate Review

- Four atoms are in one molecule of hydrogen peroxide ($H_2O_2$).
- The right panel depicts two oxygen ($O_2$) molecules.
- The ratio of H:O is 1:1 in $H_2O_2$.
- The ratio of H:O is 2:1 in $H_2O$.

## Concept Tests

p. 50 top plate
p. 52 An α particle would be attracted to, and combine with, an electron.
p. 57 The concept of atomic number was unknown in the mid-19th century.
p. 65 One box of tissues provides a convenient number (500) of them, just as one mole of a substance is a convenient number of particles because their mass is equal to the molar mass of the substance.
p. 69 One gram of silver (Ag) contains more atoms than 1 gram of gold because Ag atoms have less mass, so there are more of them per gram.
p. 72 (a) The peak at $m/z = 38$ can be assigned to molecules made from a hydrogen atom with a mass of 1 u and a chlorine atom with a mass of 37 u. (b) The relative abundance of chlorine isotopes indicates that approximately three of every four chlorine atoms will have a mass of 35 u, whereas the fourth chlorine atom will have a mass of 37 u. The relative abundance of the isotopes is reflected in the relative intensity of the peaks.

## Practice Exercises

2.1. (a) $^{56}_{26}Fe$; (b) $^{15}_{7}N$; (c) $^{37}_{17}Cl$; (d) $^{39}_{19}K$
2.2.

| | Symbol | Protons | Neutrons | Electrons |
|---|---|---|---|---|
| (a) | $^{40}Ca^{2+}$ | 20 | 20 | 18 |
| (b) | $^{79}Br^{-}$ | 35 | 44 | 36 |
| (c) | $^{35}Cl^{-}$ | 17 | 18 | 18 |
| (d) | $^{23}Na^{+}$ | 11 | 12 | 10 |

2.3. (a) As, arsenic; (b) Mg, magnesium; (c) Hg, mercury
2.4. 83.80 u
2.5. 37.0% $^{191}Ir$; 63.0% $^{193}Ir$
2.6. $1.5 \times 10^{10}$ atoms
2.7. 180.15 g/mol
2.8. 5.25 μm
2.9. 0.0541 mol of C, or $3.26 \times 10^{22}$ atoms of C
2.10. $1.81 \times 10^{-3}$ mol, or $1.09 \times 10^{21}$ molecules
2.11. A molecular-ion peak is present at $m/z = 288$ u, and the molecular mass of testosterone is 288.42 u. Those values match.

# Chapter 3

## Particulate Review

- 79 protons, 118 neutrons, and 79 electrons
- Image (a)
- Image (b) includes a nucleus that could deflect an alpha particle.

## Concept Tests

p. 88 UV rays have higher frequencies than IR radiation.
p. 90 (c)
p. 92 b and d
p. 99 c > a > b > d
p. 115 $ns^2np^5$
p. 115 (a) Ultraviolet radiation
p. 121 There is less electron-electron repulsion in a $4f^75d^1$ configuration because all these orbitals are only half-filled.
p. 126 Because the valence electrons are farther from the nucleus and shielded from it by more inner-shell electrons
p. 129 The Al spectrum's largest binding energy is much greater than that in the Li spectrum because the $1s$ electrons in Al experience a larger effective nuclear charge due to a nucleus of 13 protons in Al versus a nucleus of 3 protons in Li. The smallest binding energies are similar because the larger nuclear charge of an Al atom is offset by its larger size and the greater number of inner shell electrons that shield the atom's $3p$ electron from its nuclear charge.
p. 130 The magnitude of IE and EA both increase with increasing $Z$ across a row (except for group 18). EA values do not display clear trends within groups, whereas IE values decrease with increasing $Z$.

## Practice Exercises

3.1. $\lambda = 3.30$ m
3.2. $\lambda = 397$ nm $= 3.97 \times 10^{-7}$ m
3.3. $\lambda = 262$ nm
3.4. 374.98 nm; no, that wavelength is in the ultraviolet region.
3.5. Prediction: Less energy is required to remove an electron from the hydrogen atom in the $n = 3$ state versus that for a hydrogen atom in the $n = 1$ state. Calculated value: $2.420 \times 10^{-19}$ J
3.6. $\lambda = 3.3 \times 10^{-10}$ m
3.7. $\Delta x \geq 7 \times 10^{-11}$ m
3.8. For $n = 5$ there are $n^2 = 25$ orbitals. The subshells are $5s$, $5p$, $5d$, $5f$, and $5g$.
3.9.

| $n$ | $\ell$ | $m_\ell$ | $m_s$ |
|---|---|---|---|
| 3 | 1 | −1 | $+\frac{1}{2}$ |
| 3 | 1 | 0 | $+\frac{1}{2}$ |
| 3 | 1 | 1 | $+\frac{1}{2}$ |

3.10. Co = $[Ar]3d^74s^2$
3.11. $K^+ = [Ar]$; $I^- = [Kr]4d^{10}5s^25p^6 = [Xe]$; $Ba^{2+} = [Xe]$; $S^{2-} = [Ne]3s^23p^6 = [Ar]$; $Al^{3+} = [Ne]$; $K^+$ and $S^{2-}$ are isoelectronic with Ar.
3.12. $Mn^{3+} = [Ar]3d^4$; $Mn^{4+} = [Ar]3d^3$
3.13. (a) $Na^+ < F^- < Cl^-$; (b) $Al^{3+} < Mg^{2+} < P^{3-}$
3.14. Ne > Ca > Cs
3.15. (a) The peaks at 1.09 MJ/mol and 0.58 MJ/mol would not be present because these peaks belong to the $3s$ and $3p$ electrons, respectively. The electron configuration for $Al^{3+}$ cations is $1s^22s^22p^6$. (b) The binding energies of the electrons in $Al^{3+}$ cations would be larger than those of the electrons in the Al atoms in Figure 3.38b because they experience a larger effective nuclear charge.

# Chapter 4

## Particulate Review

- a. $N_2$; b. $O_2$; c. $H_2O$
- a. 7 + 7 = 14, b. 8 + 8 = 16, c. 8 + (2 × 1) = 10
- a. 5 + 5 = 10, b. 6 + 6 = 12, c. 6 + (2 × 1) = 8

## Concept Tests

p. 149 a. RbCl < KCl < NaCl
b. KF < CaO < MgO
p. 153 $\longleftrightarrow$ (crossed arrow)
C—O
p. 160 Hydrofluoric acid
p. 161 Bonding capacity = 18 − group number
p. 177 −1
p. 184 Neither vibration is infrared active because the bonds are not polar.

## Practice Exercises

4.1. Should be more negative than KCl; $-3.25 \times 10^{-18}$ J
4.2. Be and Cl; No, $\Delta\chi = 1.5$, so the bond is polar covalent.
4.3. (a) Tetraphosphorus decoxide; (b) carbon monoxide; (c) nitrogen trichloride; (d) $SF_6$; (e) ICl; (f) $Br_2O$
4.4. (a) $SrCl_2$; (b) MgO; (c) NaF; (d) $CaBr_2$
4.5. $MnCl_2$; $MnO_2$; chromium(III) oxide; lead(II) sulfide
4.6. (a) $Sr(NO_3)_2$; (b) $KHCO_3$; (c) $BaCrO_4$
4.7. (a) Calcium phosphate; (b) magnesium sulfite; (c) lithium nitrite; (d) sodium hypochlorite; (e) potassium permanganate
4.8. (a) Hypobromous acid; (b) bromous acid; (c) carbonic acid
4.9. $Na^+$ [:Ö:]$^{2-}$ $Na^+$
4.10.
```
    H
    |
H — C — H
    |
    H
```
4.11.
```
:Cl̈ — P̈ — C̈l:
       |
      :C̈l:
```
4.12.
```
[     H     ]+
[     |     ]
[ H — N — H ]
[     |     ]
[     H     ]
```
4.13. :Ö=C=Ö:
4.14.
[:N̈=N=N̈:]$^-$ ⟷ [:N≡N—N̈:]$^-$ ⟷ [:N̈—N≡N:]$^-$

[Ö=N=Ö]$^+$ ⟷ [:O≡N—Ö:]$^+$ ⟷ [:Ö—N≡O:]$^+$

4.15.
```
        :Ö:                    :Ö:
        ||                      |
H — Ö — N — Ö:   ⟷   H — Ö — N = Ö:
```
No, the N—O bond on the left is a single bond and should be longer than the other two, which have a length between that of N—O single and double bonds.

4.16. 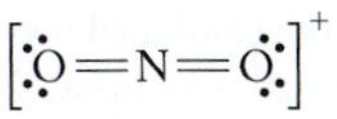

4.17. 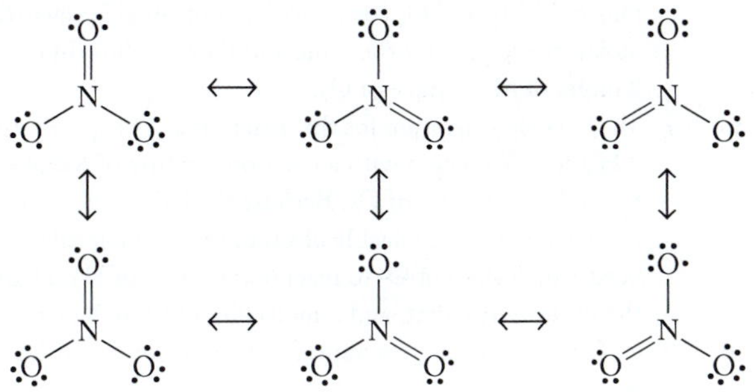

4.18. 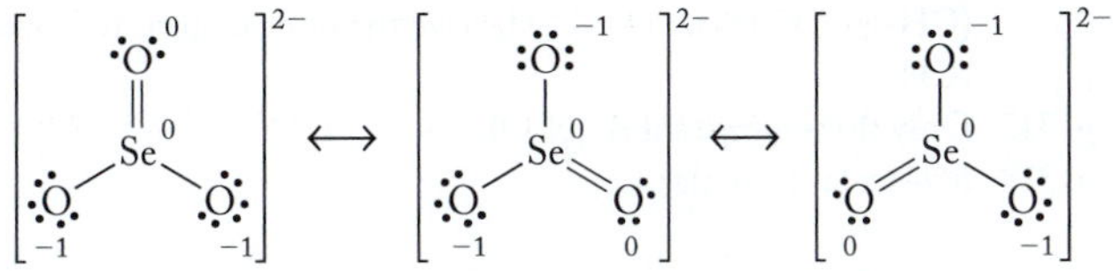

# Chapter 5

## Particulate Review

- 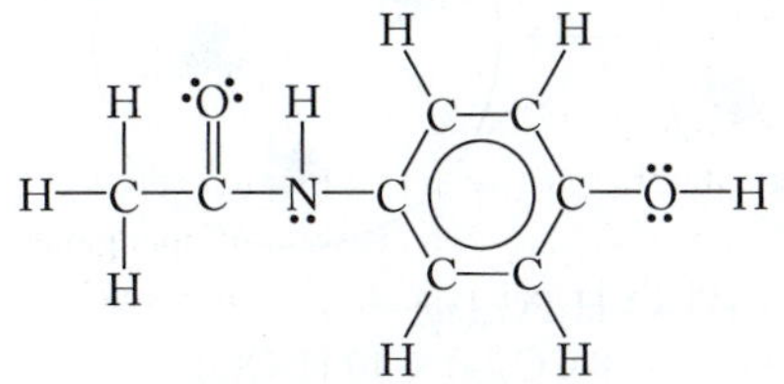
- Three pairs of delocalized $\pi$ electrons.
- The O—H, N—H, and C=O bonds.

## Concept Tests

p. 201 SN = 2 or 3 for less than an octet; SN = 4 for exactly an octet; SN = 5 or 6 for an expanded octet

p. 208 Slightly smaller bond angles in the seesaw molecular geometry

p. 212 Sulfur is less electronegative than oxygen.

p. 219 It has no unhybridized *p* orbitals with which to form $\pi$ bonds.

p. 220 In structures (a) and (c) because the double bonds are conjugated

p. 227 If the muscarine came from mushrooms, it would be one enantiomer and its solution would be optically active. Because the solution did not rotate the plane of polarized light, the muscarine present was a racemic mixture, which had to be the product of a laboratory synthesis. The coroner concluded that the victim was poisoned by someone who had access to synthetic muscarine.

p. 233 (a) $Be_2$ and $Ne_2$; (b) $B_2$

p. 236 No, bond order is lower because one more electron is in an antibonding orbital and one fewer is in a bonding orbital.

## Practice Exercises

5.1. Tetrahedral; 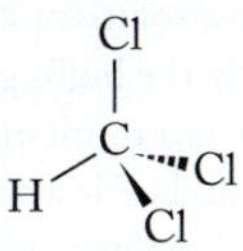

5.2. $N_2O$ has the largest bond angle; $NO_2^-$ has the smallest.

5.3. (a) Tetrahedral molecular geometry with average bond angles ~109.5°C; (b) T-shaped molecular geometry with ~90° and 180° bond angles; (c) square planar molecular geometry with 90° bond angles

5.4. (a) No, because of its linear molecular geometry. (b) Yes, it does. Even though the bonds between the oxygen atoms are nonpolar, the shape of the molecule is bent, with a lone pair of electrons on the central atom.

5.5. $CCl_4$ and $PH_3$

5.6. (a) 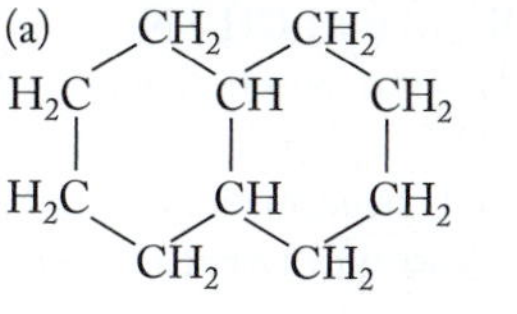

Achiral

(b) 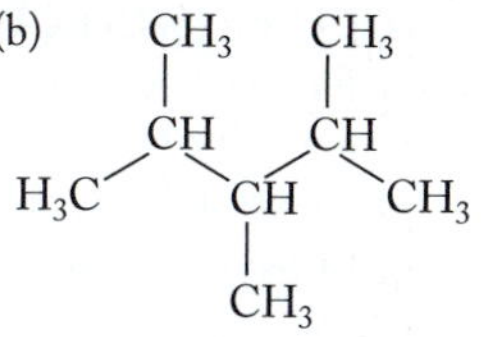

Achiral

(c) 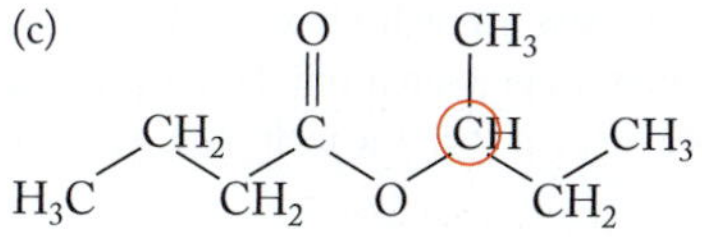

Chiral

(d) 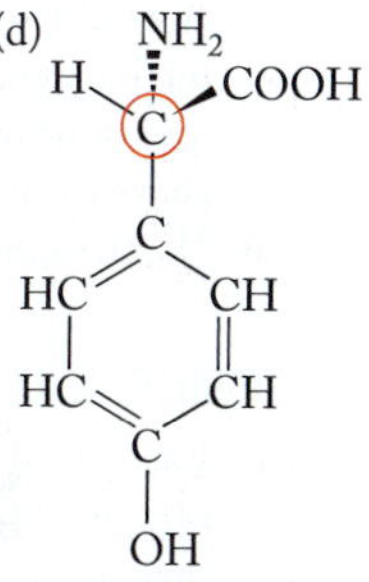

Chiral

(e) 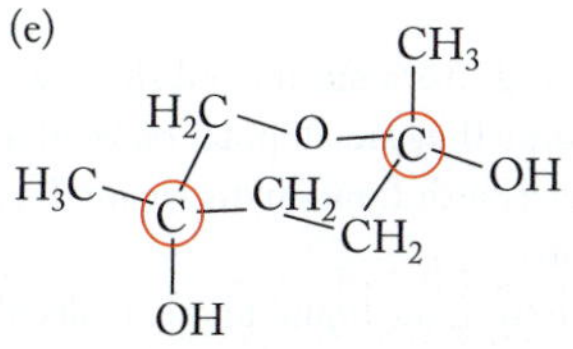

Chiral

5.7. Bond order is 0.5.

5.8. Molecules experiencing an increase in bond order are $B_2$, $C_2$, and, if they existed, $Be_2$ and $Ne_2$.

5.9. 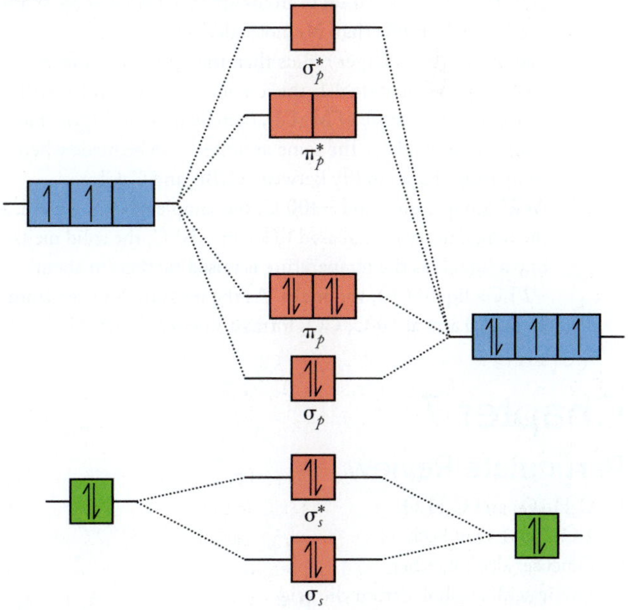


C Bond order = 3 O

# Chapter 6

## Particulate Review

- $H_2O$
- $CO_2$
- $O_3$

## Concept Tests

p. 257 $CCl_4$ is larger and experiences larger London dispersion forces.
p. 258 (a) $CH_3Br$; (b) $CH_3CH_2OH$; (c) $(CH_3CH_2)_3N$
p. 261 To form hydrogen bonds, H atoms must be bonded to N, O, or F atoms. The H atoms in $CH_3F$ are bonded to C atoms.
p. 262 Molecules of $N_2H_4$ can form H bonds with each other; $C_2H_6$ molecules cannot and experience only London dispersion forces.
p. 269 The solubilities of these alcohols decrease with increasing number of C atoms per molecule.
p. 273 A weighted wire can pass through a block of ice without melting it because downward pressure from the weights melts the ice just below the wire, but then the melt water refreezes just above the wire.
p. 276 No, the adhesive forces between mercury and glass are too weak.

## Practice Exercises

6.1. Each of the four compounds has a tetrahedral shape with carbon as the central atom. They are all polar molecules, and therefore they will all have both London dispersion forces and dipole–dipole interactions.
6.2. To enter the vapor phase from the liquid phase, molecules of ethylene glycol must break more hydrogen bonds than molecules of isopropanol. Therefore, ethylene glycol has a higher boiling point.
6.3. The dipole–induced dipole interactions between molecules of water and He atoms are weaker than those between molecules of water and $N_2$ because He atoms are much smaller ($m$ = 4 u) and less polarizable than $N_2$ molecules ($m$ = 28 u).
6.4. The $Br^-$ ion has a larger radius than that of the $Cl^-$ ion, so according to Equation 4.1, the ion–ion forces of $MgBr_2$ will be weaker than those of $MgCl_2$, thereby making $MgBr_2$ more soluble than $MgCl_2$. The same argument can be made when comparing the solubility between $CaBr_2$ and $CaCl_2$.
6.5. At 25 atm pressure and −100°C, the sample of $CO_2$ is solid. As the temperature is increased to about −50°C, the solid melts into a liquid; as the temperature is raised further (to about −20°C), liquid $CO_2$ vaporizes. As the pressure is raised from 25 to 100 atm at 50°C, $CO_2$ forms a supercritical fluid.

# Chapter 7

## Particulate Review

- $C_6H_{12}O_6$ and $C_3H_6O_3$
- 1 C:2 H:1 O in both
- Glucose: alcohol, ether
  Lactic acid: alcohol, carboxylic acid

## Concept Tests

p. 297 When writing balanced equations, we cannot change the subscripts because doing so changes the identity of the substance.
p. 305 Yes, the total mass of product equals the combined masses of the reactants consumed (law of conservation of mass). No, the masses of Mg and $O_2$ consumed are not equal because their molar masses are not the same and they combine in the ratio 2 moles of Mg:1 mole of $O_2$.
p. 307 The balanced equation for this reaction is $2\ H_2(g) + O_2(g) \rightarrow 2\ H_2O(g)$. We begin with a reaction mixture of 6 molecules of $H_2$ and 6 molecules of $O_2$. Because the balanced equation tells us that we need 1 molecule of $O_2$ to react 2 molecules of $H_2$, we need 3 molecules of $O_2$ to react 6 molecules of $H_2$. Hydrogen is the limiting reactant, and 3 molecules of $O_2$ will be in excess.
p. 309 (a) C is the limiting reactant. (b) 0.5 mole of A and 1.5 mole of B are left over.
p. 314 (a) $C_2H_4$ and (b) $C_{20}H_{40}$ have the same empirical formula ($CH_2$); (c) $C_2H_2$ and (d) $C_6H_6$ have the same empirical formula (CH).
p. 317 Only (b) isopropanol ($C_3H_8O$)
p. 318 Molecular formula

## Practice Exercises

7.1.

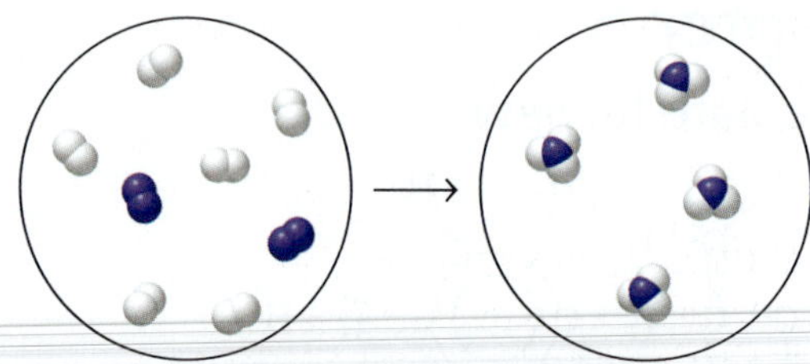

7.2. $P_4O_{10}(s) + 6\ H_2O(\ell) \rightarrow 4\ H_3PO_4(aq)$
7.3. $2\ C_4H_{10}(g) + 13\ O_2(g) \rightarrow 8\ CO_2(g) + 10\ H_2O(g)$
7.4. 3.58 g of $O_2$
7.5. 321 g of $Cu_2S$; 129 g of $SO_2$
7.6. The fuel–oxygen mixture is fuel-lean.
7.7. 22.1 g of $NaBH_4$
7.8. 75%
7.9. Percent composition of tetracycline, $C_{22}H_{24}N_2O_8$: C = 59.45%; H = 5.443%; N = 6.305%; O = 28.80%
7.10. $Cu_2S$
7.11. $FeCr_2O_4$
7.12. Empirical formula: $P_2O_5$; molecular formula: $P_4O_{10}$
7.13. $C_8H_8O_3$

# Chapter 8

## Particulate Review

- Solid iron(II) nitrate consists of $Fe^{2+}$ cations (represented by particle a) and $NO_3^-$ anions (represented by particle c) in a ratio of one cation for every two anions.

## Concept Tests

p. 338 b > a > c > d (b is the most concentrated solution)
p. 343 a < d < c < e < b (a is the most dilute solution)
p. 347 The concentrations of the three solutions are equal; therefore, any differences in how brightly the bulbs glow can be attributed to the number of ions present as a result of differences in dissociation, not concentration. LiF is a strong electrolyte and completely dissociates into $Li^+$ cations and $F^-$ anions. HF is a weak acid and partially dissociates into a few $H^+$ and $F^-$ ions

but mostly remains as HF molecules. $CH_3OH$ is a nonelectrolyte and does not dissociate at all, remaining as intact $CH_3OH$ molecules. Therefore, the number of ions decreases from LiF > HF > $CH_3OH$.

p. 352 (a) $H_2E^{2-}$; (b) $H_4E$

p. 354 (a) weak acid; (b) neither acid nor base; (c) strong base; (d) strong acid

p. 360 Soluble: a. $CaCl_2$, b. $K_3PO_4$, d. $Na_2S$. Slightly soluble: c. $Fe(OH)_3$ and e. ZnO.

p. 363 React aqueous lead(II) nitrate with the stoichiometric amount of aqueous potassium dichromate. Separate the solid lead chromate from the aqueous reaction mixture by filtration. Wash filtered solid to remove residual potassium nitrate and allow the solid to dry.

p. 366 Iron is oxidized.

p. 370 Reactions a and b are redox reactions. Reaction c is not. In reaction a, $Br_2$ is the oxidizing agent and $Sn^{2+}$ is the reducing agent. In reaction b, $F_2$ is the oxidizing agent and $H_2O$ is the reducing agent.

## Practice Exercises

8.1. 1.88 $M$ $MgCl_2$

8.2. 0.109 $M$ $K^+$

8.3. 4.48 g of $NaCH_3CH(OH)CO_2$

8.4. $V_{initial} = 2.50$ mL

8.5. $1.16 \times 10^{-4}$ $M$

8.6. a. $CH_3COOH(aq) + NaOH(aq) \rightarrow CH_3COONa(aq) + H_2O(\ell)$
b. $CH_3COOH(aq) + Na^+(aq) + OH^-(aq) \rightarrow Na^+(aq) + CH_3COO^-(aq) + H_2O(\ell)$
c. $CH_3COOH(aq) + OH^-(aq) \rightarrow CH_3COO^-(aq) + H_2O(\ell)$

8.7. a. $H_3PO_4(aq) + 3\ NaOH(aq) \rightarrow 3\ H_2O(\ell) + Na_3PO_4(aq)$
b. $H_3PO_4(aq) + 3\ Na^+(aq) + 3\ OH^-(aq) \rightarrow 3\ H_2O(\ell) + 3\ Na^+(aq) + PO_4^{3-}(aq)$
c. $H_3PO_4(aq) + 3\ OH^-(aq) \rightarrow 3\ H_2O(\ell) + PO_4^{3-}(aq)$

8.8. (a) strong electrolyte, neither; (b) strong acid, strong electrolyte; (c) weak base, weak electrolyte; (d) nonelectrolyte, neither

8.9. The lemon juice is 0.292 $M$ $C_6H_8O_7$; 100 mL of juice contains 5.62 g of $C_6H_8O_7$.

8.10. 0.0987 $M$

8.11. (a) No precipitate forms. (b) $Hg_2Cl_2$ precipitates from the mixture. The net ionic equation is $Hg_2^{2+}(aq) + 2\ Cl^-(aq) \rightarrow Hg_2Cl_2(s)$.

8.12. 26.7 mmol/L of $SO_4^{2-}$

8.13. 0.174 g of HgS

8.14. (a) +4; (b) +1; (c) +5

8.15. The element oxygen is reduced as its oxidation number changes from 0 to −2. The element sulfur is oxidized as its oxidation number changes from +4 to +6. $O_2$ is the oxidizing agent, and $SO_2$ is the reducing agent.

8.16. (a) Yes. The reaction is a redox reaction because the oxidation numbers for iron and palladium change from reactants to products. (b) $2\ Fe(s) + 3\ Pd^{2+}(aq) \rightarrow 2\ Fe^{3+}(aq) + 3\ Pd(s)$.

8.17. $Li(s)$ is most easily oxidized. $Au^{3+}(aq)$ is most easily reduced.

8.18. Aluminum nitrate and metallic silver
Molecular equation: $Al(s) + 3\ AgNO_3(aq) \rightarrow Al(NO_3)_3(aq) + 3\ Ag(s)$
Net ionic equation: $Al(s) + 3\ Ag^+(aq) \rightarrow Al^{3+}(aq) + 3\ Ag(s)$
Oxidation: $Al(s) \rightarrow Al^{3+}(aq) + 3\ e^-$
Reduction: $Ag^+(aq) + e^- \rightarrow Ag(s)$

8.19. $4\ Fe(OH)_2(s) + O_2(g) + 2\ H_2O(\ell) \rightarrow 4\ Fe(OH)_3(s)$

# Chapter 9

## Particulate Review

- $CH_4(g) + 2\ O_2(g) \rightarrow CO_2(g) + 2\ H_2O(g)$
- $CH_4$ is the limiting reactant.
- Seven moles, assuming that 4 moles of water vapor are produced. If the water condenses, 3 moles of gas are present: 1 mol of $O_2$ and 2 mol of $CO_2$.

## Concept Tests

p. 398 (c)

p. 403 (c)

p. 409 (c)

p. 413 (c)

p. 420 (a) i; (b) iv

p. 425 Yes

p. 427 HBr molecules experience dipole–dipole interactions, so they would deviate more from ideal behavior than $H_2$ molecules, which experience only London dispersion forces.

p. 428 The van der Waals pressure correction factor $a$ of $SO_2$ is nearly 2 times that for $CO_2(g)$ because bent, polar $SO_2$ experiences collectively greater van der Waals forces, including dipole–dipole interactions, than linear, nonpolar $CO_2$.

## Practice Exercises

9.1. Ar

9.2. $u_{rms,He} = 1.36 \times 10^3$ m/s, or 2.65 times faster than $N_2$

9.3. $\Delta h = 127$ mmHg

9.4. 7.0 L

9.5. $V_2/V_1 = 1.61$

9.6. 246 kPa (36 psi)

9.7. $V_2 = 2.5 \times 10^3$ L

9.8. (a) $9.41 \times 10^5$ g He; (b) $6.77 \times 10^6$ g air $-$ $9.4 \times 10^5$ g He $= 5.8 \times 10^6$ g buoyancy

9.9. The $O_2$ balloon in air will sink to the floor.

9.10. $\mathcal{M} = 44.0$ g/mol; $CO_2$

9.11. $1.7 \times 10^2$ g $NaN_3$

9.12. $P_{O_2} = 5.24 \times 10^{-2}$ atm; outside air must be compressed 3.82 times

9.13. 2.1 mg $H_2$

9.14. $P_{ideal} = 252$ atm for tank of $CH_4$ ($d = 275$ g/L) at 25°C; $P_{real} = 212$ atm

# Chapter 10

## Particulate Review

- Solid = (c); liquid = (a); gas = (b)
- The phase change from (a) to (b) is evaporation, so energy must be added to overcome the hydrogen bonding between water molecules.
- The phase change from (a) to (c) is freezing, so energy would be released; the phase change from (c) to (a) is melting, so energy must be added.

## Concept Tests

p. 450 a. A cooling pot of water with a tight lid is a closed system; a cooling pot of water without a lid is an open system.
b. The pot of water without the lid cools faster because it can exchange matter with the surroundings and also exchanges energy faster.

p. 457 Water has the highest enthalpy of vaporization in Table 10.2 because $H_2O$ experiences the strongest intermolecular forces of attraction—namely, multiple hydrogen bonds with neighboring molecules.

p. 466 a. Less energy is transferred to the water, so the calculated $c_s$ will be too low.
b. Drops of hot water add more energy to the cool water, so the calculated $c_s$ will be too high.
c. Heating other materials means less energy is transferred to the water, so the calculated $c_s$ will be too low.
d. Energy transfer from the system would lower the final temperature, leading to a calculated $c_s$ that is too low.

p. 467 Combustion of 2 mol of $H_2$ gives $\Delta H_{rxn} = -572$ kJ.

p. 477 The combustion reaction for C(*s*, graphite) is C(*s*, graphite) + $O_2(g) \rightarrow CO_2(g)$. The $\Delta H_f^\circ$ values for C(*s*, graphite) and $O_2(g)$ are zero because they are elements in their standard states. Therefore, when Equation 10.17 is used to calculate the enthalpy change of this combustion reaction, $\Delta H^\circ_{rxn}$ equals the $\Delta H_f^\circ$ of $CO_2(g)$.

p. 487 Fuel densities (kJ/mL) of $C_3$ to $C_{10}$ alkanes increase with molar mass even though fuel values (kJ/g) decrease because the densities of the alkanes increase with molar mass.

p. 490 Combustion of 1 g of glucose releases less energy than combustion of 1 g of sucrose because of the greater carbon content of sucrose (144.12/342.30, or 42% C; glucose is 72.06/180.16, or 40% C).

## Practice Exercises

10.1. a. The match is the system, $q < 0$, and the process is exothermic.
b. The wax is the system, $q < 0$, and the process is exothermic.
c. The dry ice (solid $CO_2$) is the system, $q > 0$, and the process (sublimation) is endothermic.

10.2. $w = 1.56 \times 10^7\ \text{L} \cdot \text{atm} = 1.58 \times 10^9\ \text{J} = 1.58 \times 10^6$ kJ

10.3. $\Delta E = 32$ J

10.4. $1.13 \times 10^3$ g or 1.13 kg

10.5. 16.1°C

10.6. $\Delta H_{rxn} = -1.36$ kJ

10.7. 17.7 kJ

10.8.
$$2\ CH_4(g) + 3\ O_2(g) \rightarrow 2\ CO(g) + 4\ H_2O(g) \qquad H_{comb} = -1038\ \text{kJ}$$
$$2\ CO(g) + O_2(g) \rightarrow 2\ CO_2(g) \qquad \Delta H_{comb} = -566\ \text{kJ}$$
$$2\ CH_4(g) + 4\ O_2(g) \rightarrow 2\ CO_2(g) + 4\ H_2O(g) \qquad 2 \times \Delta H_{comb} = -1604\ \text{kJ}$$
$$\Delta H_{comb} = -802\ \text{kJ}$$

10.9. a. $Ca(s) + C(s, \text{graphite}) + \frac{3}{2} O_2(g) \rightarrow CaCO_3(s)$
b. $2\ C(s, \text{graphite}) + 2\ H_2(g) + O_2(g) \rightarrow CH_3COOH(\ell)$
c. $K(s) + Mn(s) + 2\ O_2(g) \rightarrow KMnO_4(s)$

10.10. $\Delta H^\circ_{rxn} = -41.2$ kJ

10.11. $\Delta H_{rxn} = -36$ kJ

10.12. $U = -3792$ kJ

10.13. $1.1 \times 10^3$ L of kerosene per second

10.14. $C_{calorimeter} = 11.2$ kJ/°C

# Chapter 11

## Particulate Review

Ethanol molecules interact via London dispersion forces, hydrogen bonds, and dipole–dipole interactions. Dimethyl ether molecules interact via London dispersion forces and dipole–dipole interactions. The hydrogen bonds between molecules of ethanol and molecules of water make ethanol more soluble in water.

## Concept Tests

p. 513 Toward the KCl

p. 519 c < b < a

p. 520 Gasoline

p. 523 Dimethyl ether, because its molecules should experience weaker dipole–dipole interactions in the liquid phase than acetone molecules do. Therefore, its vapor pressure should be higher than that of acetone in a 50:50 mixture.

p. 525 (a) Acetone and ethanol, because acetone molecules cannot form hydrogen bonds with each other but can with ethanol. Therefore, a mixture of the two substances should have lower vapor pressure than predicted by Raoult's law.

p. 527 The vapor pressure of a substance depends only on its identity, not on how much of it there is in the liquid phase, so vapor pressure is an intensive property.

p. 530 The solution is mostly water, which has a density of 1 kg/L.

## Practice Exercises

11.1. 27.6 atm

11.2. 27 atm

11.3. $6.40 \times 10^4$ g/mol

11.4. $\Delta H_{vap} = 28.4$ kJ/mol

11.5. 1.6

11.6. $P_{solution} = 70$ torr

11.7. 0.840 *m*

11.8. 2.4 *m*

11.9. −6.9°C

11.10. $i = 1.10$

11.11. 180 g/mol

11.12. 4.6 g of $CO_2$

# Chapter 12

## Particulate Review

The two phase changes are (a) freezing and (b) melting. Process (a) is spontaneous at −5°C. Process (a) is exothermic; process (b) is endothermic.

## Concept Tests

p. 552 a, c, and d

p. 558 (b) $10^3$; $W = 4.47 \times 10^3$

p. 565 $\Delta S_{univ} > 0$

p. 570 The reaction is very slow at room temperature and pressure.

p. 570 No, the $\Delta G^\circ_{rxn}$ values for combustion of the three $C_8$ alkanes are not the same because each compound has a different $\Delta G_f^\circ$ value; the reaction stoichiometry is the same for all three, so the other values in Equation 12.12 for calculating $\Delta G^\circ_{rxn}$ do not change.

p. 571 a. Yes, it is spontaneous ($\Delta G < 0$).
b. No, spontaneity tells you nothing about the rate of reaction.

p. 573

$T\Delta S$
not spontaneous
Energy
$\Delta H$
spontaneous
Temperature (K)

## Practice Exercises

12.1. (a) Increase; (b) decrease; (c) increase
12.2. a > b > d > c (benzene has the lowest $S°$)
12.3. Prediction: $S_{sys}$ decreases; $\Delta S°_{rxn} = -242.6$ J/K
12.4. $\Delta G°_{rxn} = -1484.7$ kJ
12.5. a. $\Delta S°_{rxn}$ is expected to be negative.
b. $\Delta S°_{rxn} = -326.4$ J/K
c. $\Delta H°_{rxn} = -571.6$ kJ
d. The reaction is spontaneous at 298 K and 1 bar.
12.6. $\Delta G°_{rxn} = 142.2$ kJ
12.7. The reaction is spontaneous at $T > 465$ K.
12.8. $\Delta G°_{rxn} = -196$ kJ

# Chapter 13

## Particulate Review

$N_2(g) + O_2(g) \rightarrow 2\,NO(g)$
Mixture (a) is at the highest temperature, has the highest average velocities for $N_2$ and $O_2$, and has the most frequent collisions.

## Concept Tests

p. 593 (a) Positive ($\Delta S°_{rxn} > 0$); (b) There are the same number of moles of gaseous reactants and products.
p. 601 (d)
p. 604 44 *M*/s
p. 618 Plot the four calculated $k'$ values ($y$-axis) versus $[O_3]$ ($x$-axis); the slope of the curve $= k$.
p. 625 None of them are true.
p. 627 There are *n* activation energies (and *n* transition states).
p. 632 Similar rate laws suggest similar reaction mechanisms.

## Practice Exercises

13.1. (a) [CO] decreases twice as rapidly as $[O_2]$.
(b) $\dfrac{\Delta[CO_2]}{\Delta t} = -2\dfrac{\Delta[O_2]}{\Delta t}$
13.2. $\dfrac{\Delta[A]}{\Delta t} = -4.0\ M\,s^{-1}$; $\dfrac{\Delta[B]}{\Delta t} = -2.5\ M\,s^{-1}$; $\dfrac{\Delta[D]}{\Delta t} = 3.0\ M\,s^{-1}$
13.3. $1.2 \times 10^{-6}\ M\,s^{-1}$
13.4. Rate $= k[NO][NO_3]$; $k = 1.57 \times 10^{10}\ M^{-1}\,s^{-1}$
13.5. The decomposition of $H_2O_2$ is first order in $[H_2O_2]$; $k = 8.30 \times 10^{-4}\ s^{-1}$
13.6. (a) $k = 4.33 \times 10^{-4}\ yr^{-1}$ (b) 96% (c) $1.06 \times 10^4$ yr
13.7. The reaction is second order in $[NO_2]$; $k = 0.751\ M^{-1}\,s^{-1}$
13.8. $t_{1/2} = 8.83 \times 10^{-3}$ s
13.9. The pseudo–first-order rate constant is $k' = 6.11 \times 10^{-4}\ \mu s^{-1}$. The second-order rate constant is $k = 7.2 \times 10^6\ M^{-1}\,\mu s^{-1}$ or $7.2 \times 10^{12}\ M^{-1}\,s^{-1}$.
13.10. $E_a = 6.9$ kJ/mol; $A = 1.17 \times 10^{10}\ M^{-1}\,s^{-1}$
13.11. 21 times larger
13.12. Rate $= k_{overall}[A]^2[B]^2$
13.13. No, none of the rate laws for the elementary steps matches the overall rate law.
13.14. Yes, $NO_2$ acts as a catalyst: it speeds up the reaction, but it's not, overall, consumed by it.

# Chapter 14

## Particulate Review

The reaction taking place in the system is $H_2O(g) + CO(g) \rightleftharpoons H_2(g) + CO_2(g)$. Adding $CO(g)$ to the system increases collisions between $H_2O$ molecules and CO molecules, increasing the rate of the forward reaction and forming additional products. Removing $CO_2(g)$ from the system decreases collisions between $H_2$ molecules and $CO_2$ molecules, decreasing the rate of the reverse reaction, and forming additional products.

## Concept Tests

p. 661 3.0
p. 663 b. $[CO_2] = [H_2] > [CO] = [H_2O]$
p. 675 Zone a in Figure 14.4, where $Q < K$.
p. 677 No term for liquid water appears because its "concentration" is considered constant.
p. 685 Removing ammonia, adding nitrogen, adding hydrogen, decreasing the volume of the system, lowering the temperature
p. 692 e. More B than A is present.

## Practice Exercises

14.1. $K_c = \dfrac{[CO][H_2]^3}{[CH_4][H_2O]}$ $\quad K_p = \dfrac{(P_{CO})(P_{H_2})^3}{(P_{CH_4})(P_{H_2O})}$
14.2. $K_c = [CH_3OH]/[CO][H_2]^2 = 2.9 \times 10^2$
14.3. $K_p = 2.7 \times 10^4$
14.4. $K_p = 1.4 \times 10^{-5}$
14.5. $K_c = 0.32$
14.6. $K_{c,overall} = 1.7 \times 10^2$
14.7. This reaction is not at equilibrium and proceeds to the right.
14.8. $K_p = P_{CO_2} \cdot P_{H_2O}$; $K_c = [CO_2][H_2O]$; $K_c = 1.4 \times 10^{-9}$.
14.9. a. When the reaction is cooled and water vapor condenses, one product is removed from the reaction mixture and the equilibrium shifts to the right, forming more $SO_2$.
b. When $SO_2$ gas dissolves in liquid water as it condenses, products are removed and the equilibrium shifts to the right, forming more products.
c. When $O_2$ is added, the concentration of one reactant increases and the equilibrium shifts to the right, forming more products.
14.10. Increasing the pressure shifts the equilibrium in the reaction to the products (that is, $NH_3$), the side of the reaction equation that has fewer moles of gas.
14.11. The value of *K* for the endothermic reaction increases with increasing reaction temperature.
14.12. $P_{HI} = 0.156$ atm

14.13. $[NO_2] = 0.058\ M$ and $[N_2O_4] = 0.016\ M$ at equilibrium.
14.14. $K_p = 7.8 \times 10^2$
14.15. $K_{p,298\,K} = 2.96 \times 10^{-37}$ and $K_{p,2000\,K} = 9.2 \times 10^{-13}$

# Chapter 15

## Particulate Review

The covalent bond between the H atom and the Cl atom breaks in HCl(*g*). The covalent bond that breaks in $CH_3COOH(\ell)$ is between the O atom and the H atom. In both cases, the H atoms form covalent bonds with water to form hydronium ions, $H_3O^+(aq)$. The intermolecular forces that form in both solutions are ion–dipole forces. HCl(*g*) dissociates completely: the HCl(*aq*) solution consists primarily of hydronium ions, solvated chloride ions, and water molecules. $CH_3COOH(\ell)$ dissociates to a small extent: the acetic acid solution consists primarily of $CH_3COOH$ molecules and water molecules because $CH_3COOH$ is a weak electrolyte.

## Concept Tests

p. 715 Reduced concentrations of dissolved $CO_2$ in the blood shift both equilibria to the left, which lowers the concentration of $H_3O^+$ ions and makes the blood more basic.

p. 719 The polarity of the H—X bond, where X is a nonmetal in row 3, increases as the electronegativity of the nonmetal increases across the row from left to right, that is, from Si to Cl. The more polar the H—X bond, the more readily that bond ionizes to produce hydronium ions.

p. 725 Chloramine is a weaker base than ammonia because the Cl atom draws electron density away from the $-NH_2$ group and the lone pair of electrons that accepts $H^+$ ions.

p. 729 a. 13.77 = strongly basic; b. 10.03 = weakly basic; c. 7.00 = neutral; d. 4.37 = weakly acidic; e. 0.22 = strongly acidic

p. 731 The solution of the weak acid has a higher pH.

p. 734 An acid with a larger $K_a$ value is more ionized than an acid with a smaller $K_a$ value at the same concentration, so its percent ionization is larger.

p. 736 a. C; b. A

p. 743 Phosphoric acid is a weak acid, so the most abundant particles are molecules of phosphoric acid, $H_3PO_4$. The next-most-abundant particles are equal concentrations of $H_2PO_4^-$ and $H_3O^+$ ions. The products of the second and third ionization steps are much less abundant.

p. 746 The values of $K_{a_2}$ and $K_{a_3}$ for phosphoric acid are much smaller than its $K_{a_1}$ value, so the second and third ionization steps should not significantly affect the pH of 0.100 $M$ $H_3PO_4$. For citric acid, because the value of $K_{a_2}$ is close to that of $K_{a_1}$, an effect is likely to occur.

p. 747 (a) $NH_4F$ or $NH_4NO_3$; (b) KF or $KCH_3COO$ or $CaF_2$ or $CaCH_3COO$; (c) $KNO_3$ or $NH_4CH_3COO$ or $Ca(NO_3)_2$

## Practice Exercises

15.1. $HClO_2 > CH_2ClCOOH > C_6H_5COOH > HN_3 > HCN$

15.2. Trifluoroacetic > difluoroacetic > fluoroacetic > chloroacetic > bromoacetic acid. The first three acids are ranked by the number of electronegative F atoms per molecule that withdraw electron density away from the carboxylic acid group and the resultant carboxylate anion; the last three are ranked by electronegativity of the halogen in the molecule from most electronegative (fluorine) to least (bromine).

15.3. $H_2SO_4$ is the strongest (more electronegative central atom and the most O atoms per molecule); $H_2SeO_3$ is the weakest.

15.4. Acetic acid and the acetate ion are one conjugate acid–base pair; $H_2O$ and $H_3O^+$ are a conjugate base–acid pair.

$$\underset{\text{acid}}{CH_3COOH(aq)} + \underset{\text{base}}{H_2O(\ell)} \rightleftharpoons \underset{\text{conjugate base}}{CH_3COO^-(aq)} + \underset{\text{conjugate acid}}{H_3O^+(aq)}$$

15.5. $S^{2-} > ClO^- > CH_3COO^- > HSO_3^- > Br^-$

15.6. Greatest percent increase: b. pH 7 → pH 4; greatest percent decrease: d. pH 5 → pH 9.

15.7. The pH at the start of the industrial revolution was 8.18, so the value has dropped by 0.11 pH units. The ocean has a higher $[H_3O^+]$ today (1.29 times more acidic) and a lower pH.

15.8. $[H_3O^+] = 2.0 \times 10^{-12}\ M$; $[OH^-] = 5.0 \times 10^{-3}\ M$.

15.9. pH = 2.23; percent ionization = 12%; $K_a = 7.9 \times 10^{-4}$.

15.10. Percent ionization = 7.2%; $K_b = 5.6 \times 10^{-4}$.

15.11. pH = 12.65

15.12. pH = 3.11

15.13. pH = 11.96

15.14. pH = 7.52

15.15. pH = 0.68; percent ionization = 5.4%, which is smaller than in a solution with half the concentration of acid (9.8%). This makes sense because the degree of ionization of a weak acid, such as $HSO_4^-$, tends to increase as acid concentration decreases.

15.16. pH = 5.60

15.17. $SO_4^{2-}(aq) + H_2O(\ell) \rightleftharpoons HSO_4^-(aq) + OH^-(aq)$

15.18. pH = 11.65

15.19. 0.77 pH units less acidic

# Chapter 16

## Particulate Review

- $CaCO_3$, $Li_2O$, KCl
- (a) lithium oxide; (b) potassium chloride; (c) calcium carbonate
- $Li^+$ and $OH^-$ ions would be present in the highest concentration (2 $M$) because $O^{2-}$ ions react with water, forming $OH^-$ ions.

## Concept Tests

p. 777 Image (c)

p. 781 The sample of HCl(*aq*) is (b). As LiOH(*aq*) is added in the titration, the solution becomes (a), then (c), then (d). The equivalence point is (c).

p. 785 (b) is the weak acid at the start of the titration, (a) is the midpoint, and (c) is the equivalence point.

p. 786 A smaller change in pH would occur at the equivalence point with a weak base—so small that the volume corresponding to it would be difficult to detect precisely.

p. 789 At the beginning = image (c), best buffer = image (b), equivalence point = image (a).

p. 793 Reaching the first equivalence point takes less titrant than the second: Neutralization of the $CO_3^{2-}$ during the first step produces more $HCO_3^-$, which adds to the $HCO_3^-$ present initially in the sample that reacts in the second step, making the second plateau wider than the first.

p. 799 $Fe^{2+}(aq)$ is a weaker acid than $Fe^{3+}(aq)$ because ions with greater charges are stronger acids. The $K_a$ of $Fe^{2+}(aq)$ should be smaller, so $Fe^{2+}(aq)$ would be below $Fe^{3+}(aq)$ in Table 16.3.

p. 801 $K_{sp} = 1.8 \times 10^{-7}$

p. 804 Nitric acid would increase solubility, sodium fluoride would decrease solubility, and hydrofluoric acid would have little effect.

## Practice Exercises

16.1. pH = 3.24
16.2. pH = 10.16
16.3. (a) and (b)
16.4. Dimethylamine ($pK_b = 3.23$); $\frac{[CH_3NH_2]}{[CH_3NH_3^+]} = 0.95$
16.5. 0.61 kg of sodium ascorbate; 0.34 kg of ascorbic acid
16.6. Initial pH = 4.97; after addition of NaOH, pH = 5.07
16.7. After addition of $OH^-$, (a) the pH of the 1.16 $M$ buffer increased by 0.18 pH unit; (b) the pH of the 0.58 $M$ buffer increased by 0.38 pH unit.
16.8 Buffer B should experience the greater change because 50:50 acid–conjugate base buffers (such as buffer A) are generally the best at controlling pH. The pH of buffer A decreases by 0.087 pH units; the pH of buffer B decreases by 0.21 pH units.
16.9. (a) The red and blue curves show the initial pH of a 0.200 $M$ strong acid solution. For a 0.100 $M$ strong acid, the lower concentration would be represented by an initial pH greater than that of the red and blue curves. (b) The midpoint is reached at the $pK_a$ of the weak acid; therefore, any weak acid in Table A5.1 that has a $K_a$ smaller than that of formic acid—for example, butanoic acid—would have a midpoint higher than that of the green curve.
16.10. (a) The KOH must have a concentration greater than 0.150 $M$. (b) Any weak base with a $K_b$ smaller than that of $NH_3$—for example, pyridine.
16.11. It would be the same only if no bicarbonate were initially in the sample. In that case, $[CO_3^{2-}] = [HCO_3^-]$ at the midpoint because half of the carbonate originally in the sample has combined with hydrogen ions to form bicarbonate ions, and $pH = pK_{a_2} + \log 1$.
16.12. 9.00 mL
16.13. The oxide ion from CaO is the Lewis base and $CO_2$ is the Lewis acid.
16.14. $[Ag^+] = 1.6 \times 10^{-8}\ M$
16.15. molar solubility = $7.2 \times 10^{-7}\ M$; solubility = $2.2 \times 10^{-5}$ g/100 mL
16.16. $9.1 \times 10^{-4}\ M$
16.17. 0.49 $M$
16.18. Yes
16.19. No, $Ba^{2+}$ and $Ca^{2+}$ ions in solution cannot be completely separated by selective precipitation with $F^-$.

# Chapter 17

## Particulate Review

- $4\ Fe(s) + 3\ O_2(g) \rightarrow 2\ Fe_2O_3(s)$
- Iron is oxidized, and oxygen is reduced.
- Electrons transfer from iron to oxygen, forming $Fe^{3+}$ ions and $O^{2-}$ ions.

## Concept Tests

p. 827 Platinum carries electrons as the cell reaction proceeds, but it is neither oxidized nor reduced in the process.
p. 828 Reactions a and b are spontaneous.
p. 831 The two $E^\circ_{cell}$ values are the same because cell potential is an intensive property of the cell reaction.
p. 833 $\Delta G_{cell} > 0$; $E_{cell} < 0$
p. 836 0.257 V
p. 840 The mass of the anode decreases as Cu metal on the surface of the electrode is oxidized to $Cu^{2+}$ ions, and the mass of the cathode increases by an equal amount as $Cu^{2+}$ ions are reduced to atoms of Cu metal that accumulate on the electrode's surface.
p. 842 (a) watt-hour; (b) kilowatt-hour; (c) kilowatt-hour
p. 852 $H_2$
p. 854 Negatively charged $CO_3^{2-}$ ions would migrate toward the anode.

## Practice Exercises

17.1. Silver is the anode and gold is the cathode. Electrons flow from the Ag anode through an external circuit to the Au cathode. Nitrate ions flow toward the Ag compartment, and $Na^+$ ions flow toward the Au compartment.
17.2. The net ionic equation describing the redox reaction is
$3\ Cu^{2+}(aq) + 2\ Al(s) \rightarrow 3\ Cu(s) + 2\ Al^{3+}(aq)$
The cell diagram is
$Al(s) \mid Al^{3+}(aq) \parallel Cu^{2+}(aq) \mid Cu(s)$.
17.3. The net ionic equation is
$$Cd(s) + 2\ NiO(OH)(s) + 2\ H_2O(\ell) \rightarrow Cd(OH)_2(s) + 2\ Ni(OH)_2(s)$$
$$E^\circ_{cell} = 1.33\ V$$
17.4. $\Delta G_{cell} = -290$ kJ
17.5. $E_{cell} = 1.64$ V
17.6. $K = 2 \times 10^{62}$; $\Delta G^\circ_{rxn} = -356$ kJ
17.7. 137 g of Mg
17.8. 0.85 g of Pb

# Chapter 18

## Particulate Review

- The potassium and iodide ions in potassium iodide are held together by ionic bonds. The platinum atoms are held together by metallic bonds.
- The $K^+$ and $I^-$ ions experience electrostatic attractions in KI. $K^+$ ions are repelled by other $K^+$ ions, and $I^-$ ions are repelled by other $I^-$ ions.
- Neighboring platinum atoms experience electrostatic interactions between positively charged nuclei and negatively charged valence electrons on neighboring atoms.

## Concept Tests

p. 874 A crystal lattice extends indefinitely in three dimensions. A unit cell is the smallest repeating pattern in the arrangement of the particles in the lattice.
p. 882 Lithium–aluminum
p. 886 The filled valence band overlaps with the empty conduction band, allowing valence electrons to migrate into the conduction band, where they are free to move when an electrical field is applied.
p. 887 Selenium
p. 891 Fluorine molecules are arranged at the lattice points of a hcp structure, so $F_2$ is a molecular solid.
p. 897 $x = y = 1.2$, or $Mg_{1.2}Al_{1.2}(Si_2O_5)(OH)_4$

## Practice Exercises

18.1. 128 pm

18.2. The density of silver is 10.57 g/cm$^3$, which is close to the 10.50 g/cm$^3$ value in Appendix 3; the density of gold is 19.41 g/cm$^3$, which is close to the 19.3 g/cm$^3$ value in Appendix 3.

18.3. $d$ = 18.5 g/cm$^3$, which differs from the value in Appendix 3 (19.3 g/cm$^3$) by about 4%. The packing efficiency = 68.0%.

18.4. It's a substitutional alloy.

18.5. 101 pm

18.6. 2.16 g/cm$^3$

18.7. $d$ = 412 pm

# Chapter 19

## Particulate Review

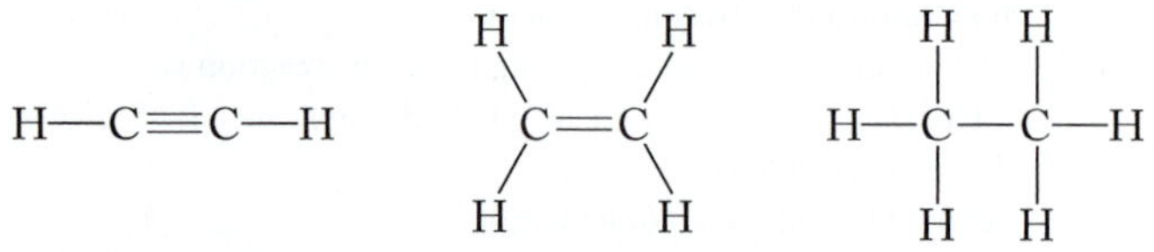

H—C≡C—H

$C_2H_2$ has the strongest carbon–carbon bond. $C_2H_6$ has the longest carbon–carbon bond and contains only sigma bonds.

## Concept Tests

p. 934 No. Both would consume 3 mol of $H_2$ per mol of compound.

p. 938 (e)

p. 938 Alkynes have a single substituent on each carbon atom of the triple bond and have a linear geometry.

p. 940 London dispersion forces

p. 945 The $\pi$ electrons are localized in the double bonds.

p. 947 amphetamine = primary amine; Benadryl = tertiary amine; adrenaline = secondary amine; pargyline = tertiary amine.

p. 953 Zingerone: alcohol, ether, aromatic ring; carvone: carbon–carbon double bonds; cinnamaldehyde: aromatic ring, carbon–carbon double bonds.

p. 957 Cotton has more –OH groups in its structure, which interact with water helping wick perspiration away from the skin and promoting evaporative cooling. Dacron and PETE have the same repeating units.

## Practice Exercises

19.1.  $CH_3(CH_2)_5CH_3$

19.2. (a) 7; (b) 9; (c) 8

19.3. (a) constitutional isomers; (b) different compounds; (c) constitutional isomers

19.4. (a) 4,5-diethyl-3-methyloctane; (b) 2,4,5,7-tetramethyloctane. The two compounds are unrelated.

19.5. Compound A reacts with 2 mol of $H_2$ per mol; compound B reacts with 3 mol of $H_2$ per mol. The product of each reaction is heptane: $CH_3CH_2CH_2CH_2CH_2CH_2CH_3$.

19.6.

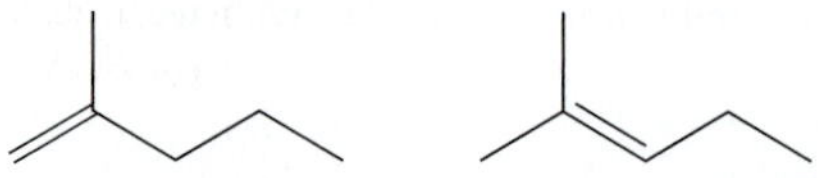

2-Methyl-1-pentene 2-Methyl-2-pentene

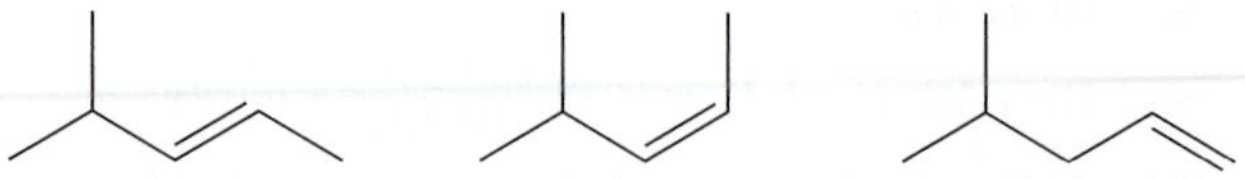

*trans*-4-Methyl-2-pentene *cis*-4-Methyl-2-pentene 4-Methyl-1-pentene

19.7. 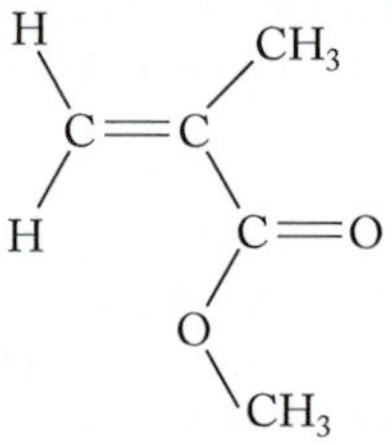 $H_2C{=}C(CH_3)C(O)OCH_3$

19.8. Dipole–induced dipole interactions and London dispersion forces.

19.9.

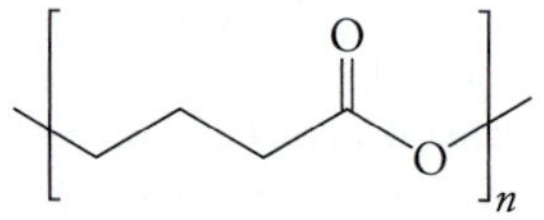

19.10. The two monomers are

(a)

And the repeat unit of the polymer is

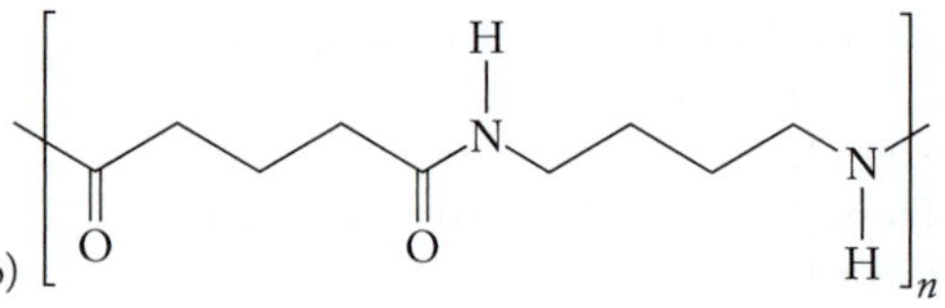

(b)

# Chapter 20

## Particulate Review

Hydrogen bonds form between molecules of (b) and between molecules of (c). Molecule (b) contains an acidic hydrogen atom, which is highlighted in red in the formula of the compound: $CH_3COOH$. Molecule (c) has a primary amine (–$NH_2$) group, which can act as a Lewis base by donating the lone pair of electrons on its nitrogen atom to an electron-pair acceptor. Molecule (a) has two lone pairs of electrons on the carbonyl oxygen, so it can also act as a weak Lewis base.

## Concept Tests

p. 979 $sp^3$ hybrid; ideal bond angles, 109.5°

p. 987 Any with nonpolar R groups, for example, leucine, isoleucine.

p. 988 Increased internal energy at high temperature overcomes the strength of the intermolecular forces responsible for a protein's secondary structure. This allows a more random secondary structure.

p. 994 (a)

p. 996 The breakdown of glucose by glycolysis produces acetyl-coenzyme A, which is a small molecule used in the biosynthesis of cholesterol.

p. 1004 No. The direction in which the code is read matters.

## Practice Exercises

20.1.

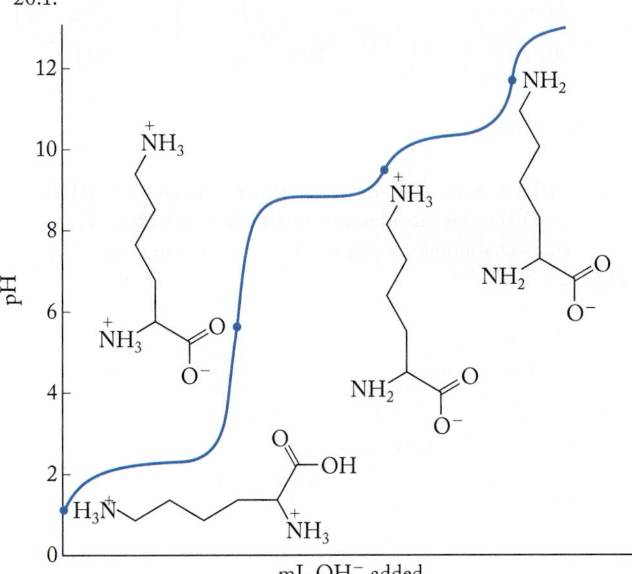


20.2. 6

20.3. 6

20.4. (a) GCCATAGGCTA (b) AATTCGGCGATC

# Chapter 21

## Particulate Review

From left to right, nuclide 1 has a mass of 11 u (5 protons + 6 neutrons), as does nuclide 2 (6 protons + 5 neutrons). Nuclide 3 has a mass of 13 u (6 protons + 7 neutrons). Nuclides 2 and 3 are isotopes.

## Concept Tests

p. 1023 The n–p ratio for stable isotopes increases with atomic number. For heavy isotopes it is about 1.5:1. Losing α particles increases this ratio, creating neutron-rich nuclides that undergo β decay.

p. 1031 Increasing (C-14–depleted) $CO_2$ in the air by burning fossil fuels will reduce the $^{14}C/^{12}C$ ratio in the air and in plants.

p. 1039 Neutron

p. 1044 To fuse nuclei, their coulombic repulsion must be overcome. Doing so means that nuclei must collide at very high velocities, which requires very high temperatures. Artificial fusion has not been carried out except in a hydrogen bomb.

## Practice Exercises

21.1. $^{214}_{82}Pb \rightarrow {}^{214}_{83}Bi + {}^{0}_{-1}\beta$

$^{214}_{83}Bi \rightarrow {}^{210}_{81}Tl + {}^{4}_{2}He$

21.2. $^{28}_{13}Al \rightarrow {}^{28}_{14}Si + {}^{0}_{-1}\beta$ (β decay)

$^{18}_{9}F \rightarrow {}^{18}_{8}O + {}^{0}_{1}\beta$ (positron emission)

$^{18}_{9}F + {}^{0}_{-1}e \rightarrow {}^{18}_{8}O$ (electron capture)

21.3. 98.6%

21.4. 1.07:1

21.5. $7.572 \times 10^7$ kJ/mol

21.6. $1.704 \times 10^9$ kJ/mol

21.7. $A = 6.92 \times 10^9$ Bq; $A = 187$ mCi

21.8. $2.1 \times 10^{11}$ X-rays

# Chapter 22

## Particulate Review

Sulfur atoms are in keratin: disulfide bonds are located between the α-helices.

Phosphorus atoms are in the phospholipid bilayer in the phosphate units in the hydrophilic head-groups. Phosphorus atoms are also in DNA in the phosphate units connecting the ribose units in the backbone.

Keratin has nitrogen atoms in the amide groups of amino acids; phospholipids contain nitrogen atoms in the quaternary amine groups in the hydrophilic head-group; DNA contains nitrogen atoms in the nucleotide bases.

## Concept Tests

p. 1075 Nonspontaneous—they require energy to pump ions.

p. 1079 $K > 1$ for the drug to be effective.

p. 1082 Solubility rules and $K_{sp}$ values indicate that all nitrates are soluble, but many phosphates are not. This finding is borne out by the presence of phosphate in biominerals such as bone and teeth.

p. 1090 Essential elements tend to be the first or second elements in a group. Nonessential elements are generally located toward the bottom of the periodic table. They have larger atomic numbers and are less abundant than the essential elements in the same group.

p. 1092 To avoid tissue damage, the nuclide should not emit high-energy α or β particles, which have greater RBE.

## Practice Exercises

22.1. $E_{K^+} = -103$ mV. No, $K^+$ ions will not flow spontaneously into the cell because $E_{transport}$ out of the cell $= -73$ mV $-$ $(-103$ mV$) = +0.030$ mV. This positive potential corresponds to $\Delta G < 0$, making outward flow spontaneous.

22.2. pH = 1.194; the volume of $Mg(OH)_2$ solution required to neutralize the acid solution is 0.584 mL.

22.3. $N_2(g) + 10\,H^+(aq) + 8\,e^- \rightarrow 2\,NH_4^+(aq) + H_2(g)$

22.4.

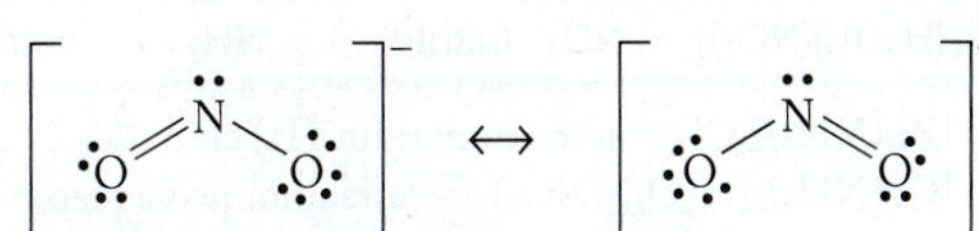


Both species have bent molecular geometry about the nitrogen atoms; the nitrogen atoms are $sp^2$ hybridized, and lone pairs are in $sp^2$ hybrid orbitals.

22.5. Bismuth-212: $2.0 \times 10^{-6}$ mg; bismuth-213: $9.5 \times 10^{-9}$ mg

# Chapter 23

## Particulate Review

The Lewis structures for ammonia ($NH_3$), borane ($BH_3$), and water ($H_2O$) are:

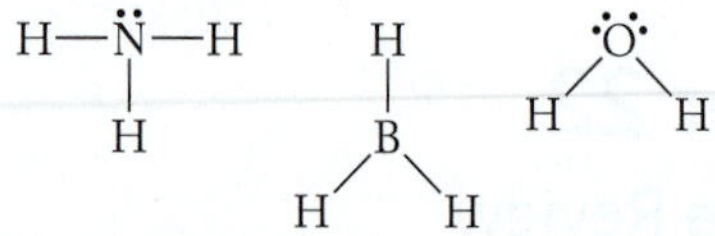

Ammonia has one lone pair on its central atom, and water has two lone pairs on its central atom; therefore, both can function as Lewis bases. Borane has no lone pairs on the central boron atom and an incomplete octet; it can function as a Lewis acid.

## Concept Tests

p. 1108 Solutions of the orange compound will have greater electrical conductivity and a lower freezing point than solutions of the reddish-purple compound because they form four and three ions, respectively, per formula unit in aqueous solutions.

p. 1109 $CN^-$ ions occupy the inner coordination of $Fe^{3+}$; $Na^+$ is the counterion. $Na_3[Fe(CN)_6]$ would have the same conductivity as $[Co(NH_3)_6]Cl_3$.

p. 1114 Tetrachloroplatinate(II)

p. 1124 Green

p. 1126 (a) $CN^-$ is a stronger field ligand than pyridine. (b) $Ru^{2+}$ ions are larger than $Fe^{2+}$ and have a larger $\Delta_o$; their 4*d* electrons interact more with ligand lone pairs than do the 3*d* electrons of $Fe^{2+}$.

p. 1128 No for square planar; yes for tetrahedral.

p. 1131 Yes. Manganese(IV) has three unpaired *d* electrons with only one possible distribution; no other option is possible. Manganese(III) has four *d* electrons; depending on the strength of the ligands, it can be either high spin (4 unpaired electrons) or low spin (2 unpaired electrons).

p. 1133 $H_2$

p. 1138 A longer half-life means slower decay; collecting a sufficient signal to get an image will take longer.

## Practice Exercises

23.1. $[Ru(NH_3)_4Cl_2]^+$ is the complex ion; Ru is present as $Ru^{3+}$; chloride is the counterion.

23.2.

| | | LIGAND | | | | | | | | |
|---|---|---|---|---|---|---|---|---|---|---|
| Compound | Counterion | Formula | Name | Number | Prefix | Formula | Name | Number | Prefix | $M^{n+}$ |
| $[Zn(NH_3)_4]Cl_2$ | $Cl^-$ (chloride) | $NH_3$ | ammine | 4 | tetra- | | | | | $2^+$ |
| $[Co(NH_3)_4(H_2O)_2](NO_2)_2$ | $NO_2^-$ (nitrite) | $NH_3$ | ammine | 4 | tetra- | $H_2O$ | aqua | 2 | di- | $2^+$ |

a. $[Zn(NH_3)_4]Cl_2$ = tetraamminezinc(II) chloride

b. $[Co(NH_3)_4(H_2O)_2](NO_2)_2$ = tetraamminediaquacobalt(II) nitrite

23.3. Six if we count the six oxygens in the $-CO_2^-$ groups, or three if we consider each $-CO_2^-$ as one group.

23.4. $Ni^{3+}$ can have either a high-spin or a low-spin configuration; none of the ions is diamagnetic.

23.5.

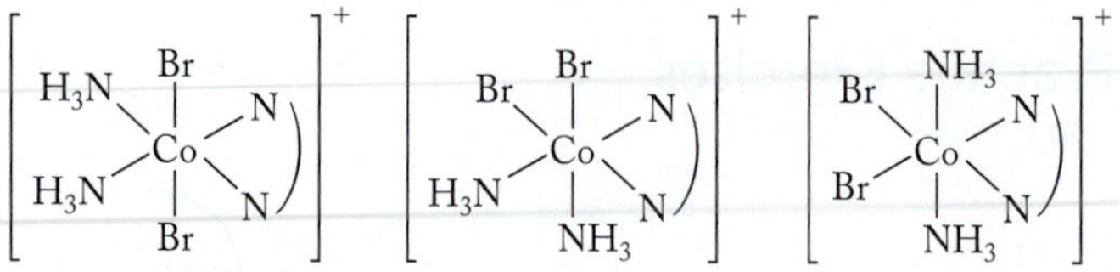

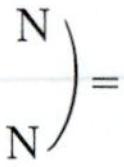

N–N = $H_2NCH_2CH_2NH_2$

*cis*-Diammine-*trans*-dibromo(ethylenediamine)cobalt(III)
*cis*-Diammine-*cis*-dibromo(ethylenediamine)cobalt(III)
*trans*-Diammine-*cis*-dibromo(ethylenediamine)cobalt(III)

23.6. $2.5 \times 10^6$

# Answers

## to Selected End-of-Chapter Questions and Problems

## Chapter 1

1.1. (a) A pure compound in the gas phase; (b) A mixture of blue element atoms and red element atoms: blue atoms are in the gas phase; red atoms are in the solid phase.

1.3. (b)

1.5. $CO_2$ molecules would be at the bottom of the box in a tightly packed regular pattern showing that it is in the solid phase.

1.7. (a) $CH_2O$; (b) $C_2H_6O_2$; (c) $CCl_4$

1.9. Sample A is accurate but not precise; Sample B is precise but not accurate.

1.11. A hypothesis is a tentative explanation of an observation or set of observations; a scientific theory describes a natural phenomenon with a concise explanation that has been extensively tested and explains why certain phenomena are always observed.

1.13. The law of multiple proportions is supported by the atomic theory by connecting the ideas of mass ratios of the elements in a compound to indivisibility of atoms. If atoms of different atoms have different masses, then the compounds that are made up of atoms should have proportional mass ratios. Where different mass ratios occur, we have a different compound made up of a different whole-number ratio of the atoms.

1.15. Proust's law needed to have corroborating evidence to fully support it. At the time, experiments to prepare a compound of tin with oxygen gave compositions that varied. The compounds they prepared later turned out to be mixtures of two compounds of tin oxide.

1.17. *Theory* in normal conversation is someone's idea or opinion or speculation that can be changed.

1.19. Yes, disproving a hypothesis would take only one counterexample.

1.21. A Snickers bar (b) and an uncooked hamburger (d)

1.23. Orange juice (with pulp) (d) and tomato juice (e)

1.25. Soil particles are not volatile, but water is; we can boil water, but not the soil. Therefore, yes, distillation can be used to remove soil particles from water. This, however, would require a lot of heat energy to accomplish.

1.27. One chemical property of gold is its resistance to corrosion (oxidation). Gold's physical properties include its density, color, melting temperature, and electrical and thermal conductivity.

1.29. We can distinguish between table sugar, water, and oxygen by examining their physical states (sugar is a solid, water is a liquid, and oxygen is a gas) and by their densities, melting points, and boiling points.

1.31. Density, melting point, thermal and electrical conductivity, and softness (a–d) are all physical properties, whereas tarnishing and reaction with water (e and f) are both chemical properties.

1.33. Extensive properties will change with the size of the sample and therefore cannot be used to identify a substance.

1.35. Carbon dioxide is a nonflammable gas (a chemical property) and is denser than air (a physical property). Therefore, $CO_2$ fire extinguishing properties are due to both its physical and its chemical properties.

1.37. In both liquid and ice, the water molecules are touching each other. In ice, however, the water molecules arrange themselves in a rigid hexagonal arrangement; in the liquid, the molecules can move around each other and their arrangement has no long-range structure.

1.39. Because of their freedom of movement, gases have the greatest particle motion; because of the restriction of their solid lattice, solids have the least particle motion.

1.41. The snow, instead of melting, sublimes directly to form water vapor.

1.43. Energy is the capacity to work; work is force times distance; energy makes work possible.

1.45. (a); (b); (c)

1.47. 13 times more energy

1.49. SI units can be easily converted into a larger or smaller unit by multiplying or dividing by multiples of 10. English units are based on other number multiples and thus are more complicated to manipulate.

1.51. Because the Celsius scale is based on the freezing and boiling points of a common liquid—water.

1.53. The absolute temperature scale (Kelvin scale) has no negative temperatures, and its zero value is placed at the lowest possible temperature.

1.55. (a) 1000 mL/1 L; (b) 1000 m/1 km; (c) 1 kg/1 $\times$ $10^5$ cg; (d) 1 $\times$ $10^6$ $cm^3$/1 $m^3$

1.57. (a) 2; (b) 4.0 km, 4.0 $\times$ $10^5$ cm, 1.6 $\times$ $10^5$ in, 1.3 $\times$ $10^4$ ft

1.59. (a) 4; (b) 3; (c) 16.8 m/s; (d) 37.7 mph

1.61. 17 hours

1.63. 19.0 mL

1.65. 58.0 $cm^3$

1.67. 5.1 $g/cm^3$

1.69. Yes

1.71. 109 g; 3.85 oz

1.73. (a) $9.2 \times 10^2$; (b) $1.293 \times 10^{-16}$; (c) $1.53 \times 10^{-23}$; (d) $3.73 \times 10^{-6}$

1.75. 54.1°F; 285.4 K

1.77. −89.2°C; 183.9 K

1.79. The $T_c$ for $YBa_2Cu_3O_7$ is already expressed in kelvins; $T_c$ = 93.0 K. The $T_c$ of $Nb_3Ge$ converted to K is 23.2 K. The $T_c$ of $HgBa_2CaCu_2O_6$ converted to K is 127.0 K. The superconductor with the highest $T_c$ is $HgBa_2CaCu_2O_6$.

1.81. Because we determine through Grubbs' test whether a particular data point is an outlier, we test only one data point at a time.

1.83. The standard deviation has a greater range.

1.85. (a) The mean and standard deviations, respectively, for each manufacturer are as follows: Manufacturer 1, 0.511, 0.00469; Manufacturer 2, 0.513, 0.00100; Manufacturer 3, 0.501, 0.000866. (b) The 95% confidence interval for each manufacturer is as follows: Manufacturer 1, 0.511 ± 0.00582; Manufacturer 2, 0.513 ± 0.001241; Manufacturer 3, 0.501 ± 0.0001075. (c) The manufacturer that is both precise and accurate is Manufacturer 3; the manufacturer that is precise but not accurate is Manufacturer 2.
1.87. This point is an outlier.
1.89. 0.031 mg/L
1.91. Both mixtures a and b react so that neither sodium nor chlorine is left over.
1.93. Day 11
1.95. 16 times more administered than prescribed
1.97. (a) None; (b) Thermometer B
1.99. 32,460 miles/h

# Chapter 2

2.1. Purple (H, hydrogen)
2.3. Purple (H, hydrogen)
2.5. (a) Green (Au, gold); (b) Blue (Na, sodium); (c) Lilac (Cl, chlorine); (d) Red (Ne, neon)
2.7. Red arrow, alpha; green arrow, beta
2.9. Dichloromethane; the mass ion peaks at *m/z* 84, 86, and 88 show the presence of chlorine isotopes ($^{35}$Cl and $^{37}$Cl), as do the peaks at *m/z* 50 and 52, which show the loss of one Cl from $CH_2Cl_2$ (84 − 35 = 50 and 86 − 53 = 52). Cyclohexane, having no chlorine atoms, would not show this isotopic pattern.
2.11. (a) Blue (K) and green (Ag); (b) Gray (Mg); (c) Yellow (Sc); (d) Purple (I); (e) Red (O)
2.13. Rutherford concluded that the positive charge in the atom could not be spread out (the pudding) in the atom but must result from a concentration of charge in the center of the atom (the nucleus). Most of the particles were deflected only slightly or passed directly through the gold foil, so he reasoned that the nucleus must be small compared to the size of the entire atom. The negatively charged electrons do not deflect the particles, and Rutherford reasoned that the electrons took up the rest of the space of the atom outside the nucleus.
2.15. The fact that cathode rays were deflected by a magnetic field indicated that the rays were streams of charged particles.
2.17. Helium got in the ore through the alpha decay of the radioactive uranium ore and its products.
2.19. Absorption of β particles by gold atoms would have transmuted the gold atoms into lighter elements when the β particles interacted with protons in the nucleus to form a neutron.
2.21. Greater than 1
2.23. Hydrogen ($^1$H)
2.25.

| | Atom | Mass Number | Atomic Number = Number of Protons | Number of Neutrons = Mass Number − Atomic Number | Number of Electrons = Number of Protons |
|---|---|---|---|---|---|
| (a) | $^{14}$C | 14 | 6 | 8 | 6 |
| (b) | $^{59}$Fe | 59 | 26 | 33 | 26 |
| (c) | $^{90}$Sr | 90 | 38 | 52 | 38 |
| (d) | $^{210}$Pb | 210 | 82 | 128 | 82 |

2.27. The neutron-to-proton ratio for each of these elements is (a) $^4$He, 2/2 = 1.00; (b) $^{23}$Na, 12/11 = 1.09; (c) $^{59}$Co, 32/27 = 1.19; (d) $^{197}$Au, 118/79 = 1.49. As the atomic number (*Z*) increases, the neutron-to-proton ratio increases.
2.29.

| Symbol | $^{23}$Na | $^{89}$Y | $^{118}$Sn | $^{197}$Au |
|---|---|---|---|---|
| Number of protons | 11 | 39 | 50 | 79 |
| Number of neutrons | 12 | 50 | 68 | 118 |
| Number of electrons | 11 | 39 | 50 | 79 |
| Mass number | 23 | 89 | 118 | 197 |

2.31.

| Symbol | $^{37}Cl^-$ | $^{23}Na^+$ | $^{81}Br^-$ | $^{226}Ra^{2+}$ |
|---|---|---|---|---|
| Number of protons | 17 | 11 | 35 | 88 |
| Number of neutrons | 20 | 12 | 46 | 138 |
| Number of electrons | 18 | 10 | 36 | 86 |
| Mass number | 37 | 23 | 81 | 226 |

2.33. Group 2, RO; group 3, $R_2O_3$; group 4, $RO_2$
2.35. Mendeleev based his groups on chemical reactivity. No compounds of the noble gases existed to indicate their presence as a group.
2.37. C, N, and O
2.39. (a) palladium (Pd); (b) rhodium (Rh); (c) platinum (Pt)
2.41. Three (Na, Mg, and Al)
2.43. A *weighted average* takes into account the proportion of each value in the group of values to be averaged.
2.45. The mass of the electron is small in comparison to the mass of the atoms, so the mass of the ions (whether anion or cation) is exceedingly close to the mass of the neutral atoms.
2.47. $^{40}$Ar
2.49. Because we are given that the $^{195}$Pt isotope is not 100% abundant, we must conclude that the other isotopes with masses greater than 195 have natural abundances in equal proportion to those isotopes with masses lower than 195 in order that the weighted average atomic mass calculates to 195.08 u.
2.51. 1.0078506 u; this is, to 5 significant digits, exactly the value on the periodic table in the text (1.0079 u).
2.53. Yes
2.55. 47.95 u
2.57. (a) NaI, 149.89 u; (b) CaS, 72.143 u; (c) $Al_2O_3$, 101.961 u; (d) $NH_4Cl$, 53.492 u
2.59. (a) 1; (b) 2; (c) 2; (d) 2
2.61. (c) $NO_2$ < (d) HClO < (e) $C_4H_{10}$ < (a) $CS_2$ < (b) HI.
2.63. A dozen is too small a unit to express the very large number of atoms, ions, or molecules present in laboratory quantities such as a mole.
2.65. (a) $7.3 \times 10^{-10}$ mol of Ne; (b) $7.0 \times 10^{-11}$ mol of $CH_4$; (c) $4.2 \times 10^{-12}$ mol of $O_3$; (d) $8.1 \times 10^{-15}$ mol of $NO_2$
2.67. (a) 1 mol; (b) 1 mol; (c) 2 mol; (d) 2 mol
2.69. (a) $Fe_2O_3$; (b) $N_2O_5$; (c) $SO_2$
2.71. (a) 64.063 g/mol; (b) 47.997 g/mol; (c) 44.009 g/mol; (d) 108.009 g/mol
2.73. (a) 152.148 g/mol; (b) 164.203 g/mol; (c) 148.204 g/mol; (d) 132.161 g/mol
2.75. 0.1244 mol
2.77. 0.25 mol; 10 g
2.79. (a) NO; (b) $CO_2$; (c) $O_2$

2.81. (c) < (e) < (b) < (d) < (a)

2.83. In mass spectrometry, a gaseous substance is ionized with high-energy electrons to form +1 cations. When one electron has been displaced, the ion that forms is the *molecular* ion, $M^+$. This and other cations are separated by means of an external magnetic field based on their mass ($m$)-to-charge ($z$) ratio by the detector. Where the $M^+$ molecule appears on the $m/z$ axis is directly related to the molecular mass. Often, but not always, this the peak that appears at the highest $m/z$ ratio.

2.85. Yes

2.87. (a) 222 u; (b) 296 u; (c) 316 u; (d) 450 u

2.89. The very low intensity peaks at 35, 36, and 37 u are due to the low abundances of $^{33}S$, $^{34}S$, and $^{35}S$ as are assigned to $H_2{}^{33}S$, $H_2{}^{34}S$, and $H_2{}^{35}S$, respectively. In addition, the peak at 36 u could be partially due to the loss of one H atoms from $H_2{}^{35}S$; likewise, the peak at 35 u could be partially due to the loss of one H atom from $H_2{}^{34}S$. The largest peak at 34 u is primarily due to $H_2{}^{32}S$ because of the high natural abundance of $^{32}S$, but a very small amount might also be due to the loss of one H atom from $H_2{}^{33}S$. The peak at 33 is mainly due to the loss of one H atom from $H_2{}^{32}S$, but a tiny portion might be due to the loss of two H atoms from $H_2{}^{35}S$. Finally the peak at 32 is due to the loss of two H atoms from $H_2{}^{32}S$.

2.91. (a) Electrons; (b) The negatively charged electrons were attracted to the positively charged plate as the electrons passed through the electric field; (c) If the polarities of the plates were switched, the electron would still be deflected toward the positively charged plate, which would now be at the opposite side of the screen.

2.93. Blue, 9; yellow, 16; green, 25; red, 33; lilac, 73

2.95. (a) CdS, 144.48 g/mol; CdSe, 191.37 g/mol
(b) $2.7 \times 10^7$ atoms of Se
(c) $3.3 \times 10^{-13}$ g of Cd; $9.5 \times 10^{-16}$ g of S

2.97. 303 g/mol

2.99. (a) 0.7580 mol of C; (b) $4.565 \times 10^{23}$ atoms of C

# Chapter 3

3.1. (a) True purple (Na), red (Cr), and orange (Au)
(b) Dark purple (Ne)
(c) Orange (Au)
(d) Red (Cr)
(e) Dark purple (Ne) and green (Cl)

3.3. (a) Green (Cl); (b) Purple (Na), red (Cr), and yellow (Au)

3.5. Blue (Rb), green (Sr), and orange (Y)

3.7. (c)

3.9. (a)

3.11. (a) Purple
(b) Red
(c) Red

3.13. All these forms of light have perpendicular, oscillating electric and magnetic fields that travel together through space.

3.15. The lead shield protects the parts of our bodies that might be exposed to X-rays but are not being imaged. Lead is a very high-density metal with many electrons, which interact with X-rays and absorb nearly all the X-rays before they can reach our bodies.

3.17. No, it still emits infrared radiation.

3.19. $6.06 \times 10^{14}$/s

3.21. (a) 0.13 m; (b) 0.354 m

3.23. The X-rays emitted from copper have a higher frequency than that of X-rays emitted from iron.

3.25. 8.317 min

3.27. The absorption spectrum consists of dark lines at wavelengths specific to that element. The emission spectrum has bright lines on a dark background with the lines appearing at the same wavelengths as the dark lines in the absorption spectrum.

3.29. Because each element shows distinctive and unique absorption and emission lines, the bright emission lines observed for the pure elements could be matched to the many dark absorption lines in the spectrum of sunlight. This approach can be used to deduce the sun's elemental composition.

3.31. The quantum is the smallest indivisible amount of radiant energy that an atom can absorb or emit.

3.33. No

3.35. $6.62 \times 10^{-19}$ J

3.37. (b) Elevator doors open only at discrete elevations and not at heights in between those elevations.

3.39. $7.71 \times 10^{-19}$ J

3.41. No

3.43. Potassium; $8.04 \times 10^5$ m/s

3.45. $3.17 \times 10^{18}$ photons/s

3.47. The $n = 1$ to $n = \infty$ arrow in Figure 3.18 represents the energy required to ionize a ground state H atom. It is the longest arrow in the figure. A photon with this energy would have a shorter wavelength than any of those emitted by any of the other transition arrows.

3.49. The difference between $n$ levels determines emission energy.

3.51. (d)

3.53. 1875 nm; infrared

3.55. No, because for hydrogen the transition is in the ultraviolet region, and as $Z$ increases the wavelength will shorten further.

3.57. (a) No; (b) Yes

3.59. For a baseball, which is a massive particle, the wavelength of it traveling at 44 m/s is very, very small, so its behavior in being thrown is not abnormal.

3.61. (b) and (c)

3.63. (a) 10.8 nm
(b) 0.180 nm
(c) $8.2 \times 10^{-28}$ nm
(d) $3.7 \times 10^{-54}$ nm

3.65. $\Delta x \geq 1.3 \times 10^{-13}$ m

3.67. The Bohr model orbit showed the quantized nature of the electron in the atom as a particle moving around the nucleus in concentric orbits. In quantum theory, an orbital is a region of space where the probability of finding the electron is high. The electron is not viewed as a particle, but as a wave, and it is not confined to a clearly defined orbit; rather, we refer to the probability of the electron's being at various locations around the nucleus.

3.69. As $n$ increases, the electron occupies an $s$ orbital farther from the nucleus.

3.71. (a) 1; (b) 4; (c) 9; (d) 16; (e) 25

3.73. 3, 2, 1, 0

3.75. (a) $6s$
(b) $3d$
(c) $2p$
(d) $5g$

3.77. (a) 2
(b) 2
(c) 10
(d) 2

3.79. (b)

3.81. Degenerate orbitals have the same energy.

3.83. As we start from an argon core of electrons, we move to potassium and calcium, which are located in the *s* block on the periodic table. Not until Sc, Ti, V, and so on, do we begin to fill electrons into the 3*d* shell.

3.85. (c) $3s <$ (a) $3d <$ (d) $4p <$ (b) $7f$

3.87. $Li^+$: $1s^2$ or [He]
Ca: $[Ar]4s^2$
$F^-$: $[He]2s^22p^6$ or [Ne]
$Mg^{2+}$: $[He]2s^22p^6$ or [Ne]
$Al^{3+}$: $[He]2s^22p^6$ or [Ne]

3.89. (b) for Mn; (d) for $Mn^{2+}$

3.91. (a) K: $[Ar]4s^1$; 1 unpaired $e^-$
(b) $K^+$: [Ar]; 0 unpaired $e^-$
(c) $S^{2-}$: $[Ne]3s^23p^6$ or [Ar]; 0 unpaired $e^-$
(d) N: $[He]2s^22p^3$; 3 unpaired $e^-$
(e) Ba: $[Xe]6s^2$; 0 unpaired $e^-$
(f) $Ti^{4+}$: [Ar] or $[Ne]3s^23p^6$; 0 unpaired $e^-$
(g) Al: $[Ne]3s^23p^1$; 1 unpaired $e^-$

3.93. (a) Ti; 2 unpaired $e^-$
(b) Cr; 6 unpaired $e^-$
(c) Cu; 1 unpaired $e^-$
(d) Pd; 0 unpaired $e^-$

3.95. $Cl^-$, 0 unpaired $e^-$

3.97. Carbon

3.99. (a) and (d)

3.101. 5*p*; yes

3.103. (c)

3.105. (a) The first ionization energy is the energy to remove a 2*s* electron, not a 1*s* electron.
(b) No, the second ionization of Be is removing the second 2*s* electron, not a 1*s* electron.

3.107. A sodium atom loses its entire valence shell when it forms a $Na^+$ ion. However, a Cl atom gains a valence electron when it forms a $Cl^-$ ion, which increases $e^-$–$e^-$ repulsion, creating a larger valence shell.

3.109. (a) Al > P > Cl > Ar
(b) Sn > Ge > Si > C
(c) K > Na > Li > $Li^+$
(d) $Cl^-$ > Cl > F > Ne

3.111. (a) As the atomic number increases down a group, electrons are added to higher *n* levels, leading to a decrease in ionization energy.
(b) As the atomic number increases across a period, the effective nuclear charge increases. This means that the ionization energy increases across a period of elements.

3.113. Because less energy would be required to ionize a larger atom, the wavelength of light required would be longer for atoms with higher atomic numbers than for atoms with lower atomic numbers in the same group on the periodic table.

3.115. (a) I < Br < Cl < F
(b) Na < Li < Mg < Be
(c) N < O < F < Ne

3.117.

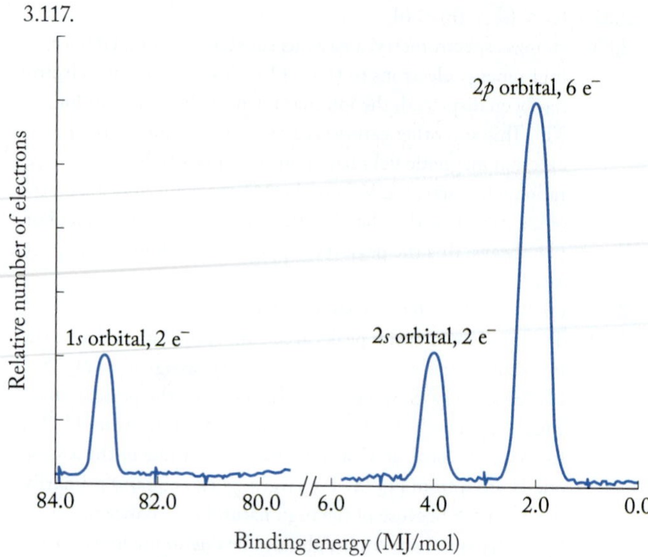


Neon has the electron configuration $1s^22s^22p^6$.

3.119. No, sodium has a negative (favorable) electron affinity but is never found as an anion in nature. It is always a cation in salts such as NaCl and $Na_2CO_3$.

3.121. Electrons are added farther from the nucleus as halogen atoms get larger. This means that the electrons are not as strongly attracted to the nucleus, which results in less negative electron affinities.

3.123. (a) 0.242 m; $8.22 \times 10^{-25}$ J/photon
(b) Paschen lines

3.125. (a) Sirius XM radio
(b) Sirius XM radio
(c) Sirius XM radio

3.127. (a) As we move across the fourth period (K–Kr), the energy separation between the 2*s* and 2*p* orbitals increases. Because energy is inversely proportional to wavelength, the wavelength for the $2p \rightarrow 2s$ transition would decrease.
(b) As we move down a given column in the periodic table, the energy separation between the 2*s* and 2*p* orbitals would increase. Because energy is inversely proportional to wavelength, the wavelength for the $2p \rightarrow 2s$ transition would decrease.

3.129. (a) A silver atom forms a 1+ ion through the loss of its single 5*s* electron. Palladium and cadmium both lose two 5*s* electrons to form 2+ cations.
(b) The heavier group 13 elements may form 1+ cations through the loss of one *np* electron and form 3+ cations through the loss of one *np* electron and two *ns* electrons.
(c) The heavier group 14 elements may form 2+ cations through the loss of two *np* electrons and form 4+ cations through the loss of two *np* electron and two *ns* electrons. The group 4 elements may lose the two *ns* electrons to form 2+ cations and may lose both the *ns* and the two $(n-1)d$ electrons to form 4+ cations.

3.131. (a) $Sn^{2+}$: $[Kr]4d^{10}5s^2$
$Sn^{4+}$: $[Kr]4d^{10}$
$Mg^{2+}$: $[He]2s^22p^6$ or [Ne]
(b) Cadmium has the same electron configuration as $Sn^{2+}$, and neon has the same electron configuration as $Mg^{2+}$.
(c) $Cd^{2+}$

3.133. Yes

# Chapter 4

4.1. (e) $Al^{3+}$

4.3. In the three structures shown, the arrangement of atoms is different; therefore, these are not resonance structures. The structure that is one of the possible resonance forms for the thiocyanate ion, $SCN^-$, is the middle structure.

4.5. (d)

4.7. (a)

4.9. All three modes are infrared active.

4.11. The blue curve represents the interaction between potassium and chloride ions. The red curve represents the interaction between potassium and fluoride ions.

4.13. Ionic compounds have attractive forces between oppositely charged ions. In ionic bonding, the electrons of one atom are not shared. In covalent compounds, bonds are formed through electron sharing.

4.15. Protons and electrons attract each other; electrons repel each other, as do protons.

4.17. The cation and anion attract each other in an ionic compound. The strength of that attraction is affected by the magnitude of the charge on each ion and by the distance between centers of the cation and anion (that is, their ionic radii).

4.19. $-6.94 \times 10^{-20}$ J

4.21. $TiO_2$

4.23. $CsBr < KBr < SrBr_2$

4.25. If the electronegativity difference is 2.0 or greater, the bond between the atoms is ionic; if the electronegativity difference is below 0.4, the bond is covalent; if the electronegativity difference is between 0.4 and 2.0, the bond is polar covalent.

4.27. Among the nonmetals and metalloids, those with the smallest atoms tend to be the most electronegative. This pattern makes sense because atomic sizes decrease and electronegativities increase with greater attraction between atoms' valence electrons and their nuclei.

4.29. A polar covalent bond is one in which the electrons are shared, but not equally, by the atoms. It is different from an ionic bond in that ionic bonds do not share electrons.

4.31. $Cs < K < Br < F$

4.33. The polar bonds (with $\Delta\chi > 0.4$) and the atoms with the greater electronegativity (underlined) are $C—\underline{O}$ and $\underline{N}—H$.

4.35. 

$$\overset{\longleftarrow\!\!+}{\underset{\delta+\quad\delta-}{N—H}} \qquad \overset{+\!\!\longrightarrow}{\underset{\delta+\quad\delta-}{C—O}}$$

4.37. Binary compounds of (b) Al and Cl and (c) C and O have polar covalent bonds. The binary compound of (d) Ca and O has ionic bonds.

4.39. Roman numerals indicate the charge on the transition metal cation.

4.41. $XO_2^{2-}$

4.43. (a) $NO_3$, nitrogen trioxide
(b) $N_2O_5$, dinitrogen pentoxide
(c) $N_2O_4$, dinitrogen tetroxide
(d) $NO_2$, nitrogen dioxide
(e) $N_2O_3$, dinitrogen trioxide
(f) NO, nitrogen monoxide
(g) $N_2O$, dinitrogen monoxide
(h) $N_4O$, tetranitrogen monoxide

4.45. (a) $Na_2S$, sodium sulfide
(b) $SrCl_2$, strontium chloride
(c) $Al_2O_3$, aluminum oxide
(d) LiH, lithium hydride

4.47. (a) cobalt(II) oxide
(b) cobalt(III) oxide
(c) cobalt(IV) oxide

4.49. (a) $BrO^-$
(b) $SO_4^{2-}$
(c) $IO_3^-$
(d) $NO_2^-$

4.51. (a) nickel(II) carbonate
(b) sodium cyanide
(c) lithium hydrogen carbonate
(d) calcium hypochlorite

4.53. (a) hydrofluoric acid
(b) sulfurous acid
(c) $H_3PO_4$
(d) $HNO_2$

4.55. (a) $K_2S$
(b) $K_2Se$
(c) $Rb_2SO_4$
(d) $RbNO_2$
(e) $MgSO_4$

4.57. (a) manganese(II) sulfide
(b) vanadium(II) nitride
(c) chromium(III) sulfate
(d) cobalt(II) nitrate
(e) iron(III) oxide

4.59. (b) $Na_2SO_3$

4.61. (a) sulfuric acid
(b) sodium sulfate
(c) copper(I) sulfate
(d) sulfur trioxide
(e) $H_2SO_3$
(f) $SO_2$
(g) $SO_3^{2-}$

4.63. The number of valence electrons for groups 1 and 2 are equal to the group number; the number of valence electrons for groups 13–18 are equal to the group number −10.

4.65. An H atom can form only one bond.

4.67. K· ·Mg· $\cdot\dot{\underset{\cdot\cdot}{P}}\cdot$

4.69. $K^+$ $Al^{3+}$ $[:\ddot{\underset{\cdot\cdot}{N}}:]^{3-}$ $[:\ddot{\underset{\cdot\cdot}{I}}:]^-$

4.71. (a) 10
(b) 8
(c) 8
(d) 10

4.73. (a) $:C{\equiv}O:$
(b) $\ddot{\underset{\cdot\cdot}{O}}{=}\ddot{\underset{\cdot\cdot}{O}}$
(c) $[:\ddot{\underset{\cdot\cdot}{Cl}}—\ddot{\underset{\cdot\cdot}{O}}:]^-$
(d) $[:C{\equiv}N:]^-$

4.75. (a) $CCl_4$: C bonded to four :Cl: atoms

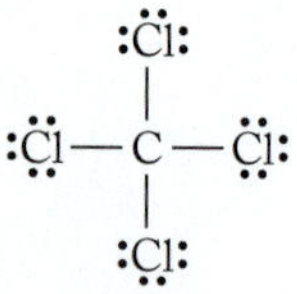

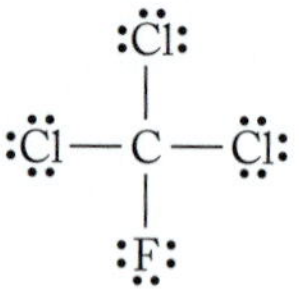

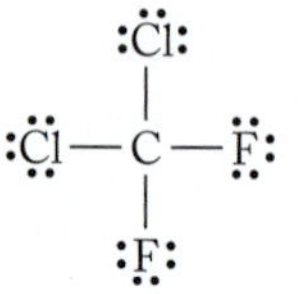

(b) H—B—H with H bonded to B

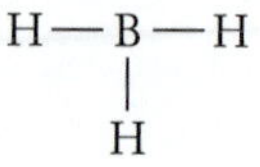

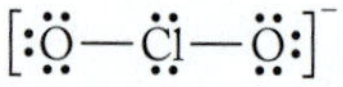

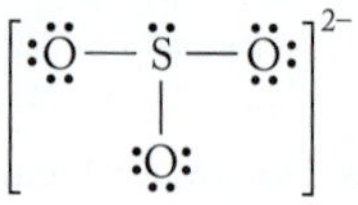

(c) $SiF_4$: Si bonded to four :F: atoms

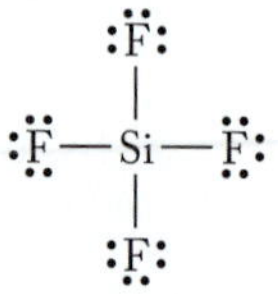

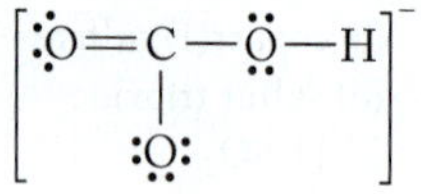

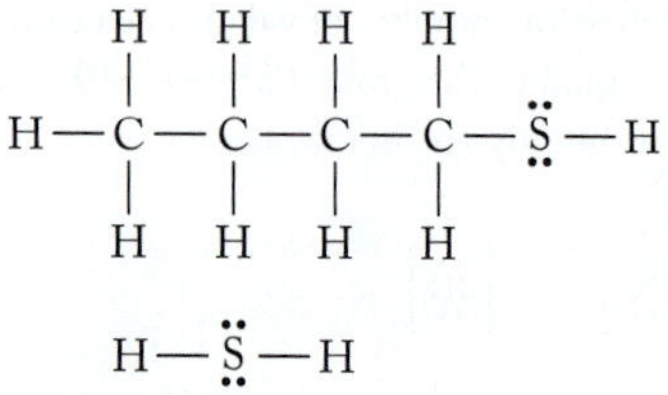

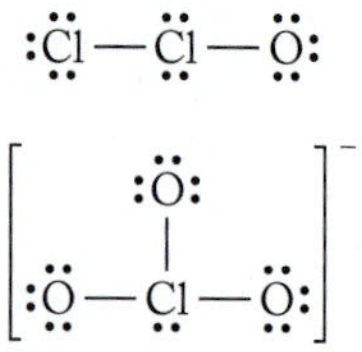

(d) $[BH_4]^-$: [H—B—H with H above and below B]$^-$

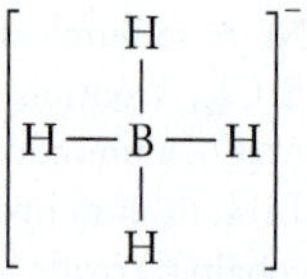

(e) $[PH_4]^+$: [H—P—H with H above and below P]$^+$

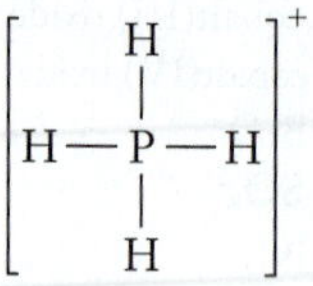

4.77. (a) C bonded to three Cl and one F

(b) C bonded to two Cl and two F

(c) C bonded to one Cl and three F

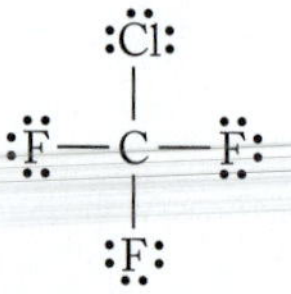

(d) Cl—C(Cl)(F)—C(Cl)(F)—F

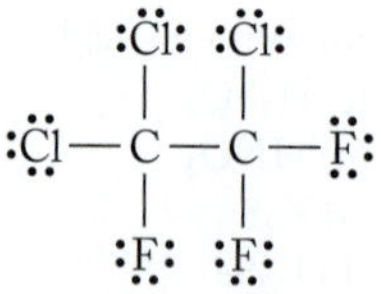

(e) F—C(Cl)(F)—C(Cl)(F)—F

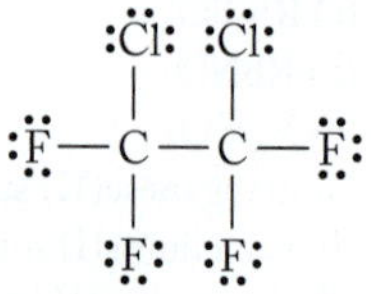

4.79. (a) $[:\ddot{O}—\ddot{Cl}—\ddot{O}:]^-$ (c) $[:O=C(—\ddot{O}—H)(—:O:)]^-$

(b) $[:\ddot{O}—\ddot{S}(—:O:)—\ddot{O}:]^{2-}$

4.81. H—C(H)(H)—C(H)(H)—C(H)(H)—C(H)(H)—$\ddot{S}$—H

H—$\ddot{S}$—H

4.83. $:\ddot{Cl}—\ddot{Cl}—\ddot{O}:$

$[:\ddot{O}—\ddot{Cl}(—:\ddot{O}:)—\ddot{O}:]^-$

4.85. H—C(H)(H)—$\ddot{O}$—H

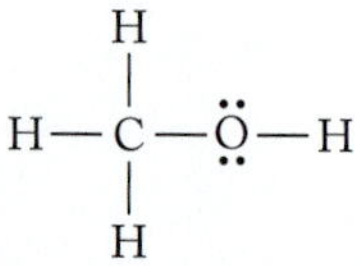

4.87. Resonance stabilizes a molecule or polyatomic ion.

4.89. The resonance Lewis structures of $NO_2$ both contain a single and double N—O bond. From a formal charge perspective, the structures are equivalent.

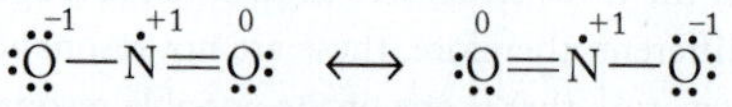

$:\ddot{O}^{-1}—\ddot{N}^{+1}=O^{0}: \leftrightarrow :O^{0}=\ddot{N}^{+1}—\ddot{O}^{-1}:$

The resonance forms of $CO_2$ show that one is dominant (the one in which all formal charges are zero), and the other forms contribute little to the bonding in $CO_2$.

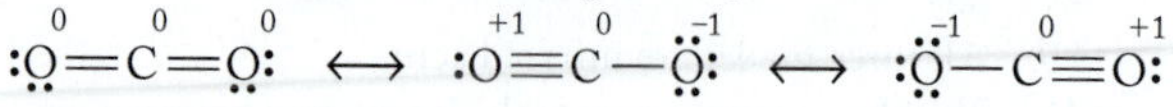

$:O^{0}=C^{0}=O^{0}: \leftrightarrow :O^{+1}\equiv C^{0}—\ddot{O}^{-1}: \leftrightarrow :\ddot{O}^{-1}—C^{0}\equiv O^{+1}:$

4.91. Cyclobutadiene ($C_4H_4$): H–C=C–H / H–C=C–H ring $\leftrightarrow$ ring with double bonds shifted

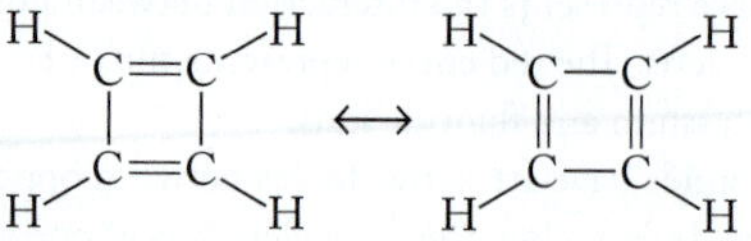

493.

$N_2O_2$:

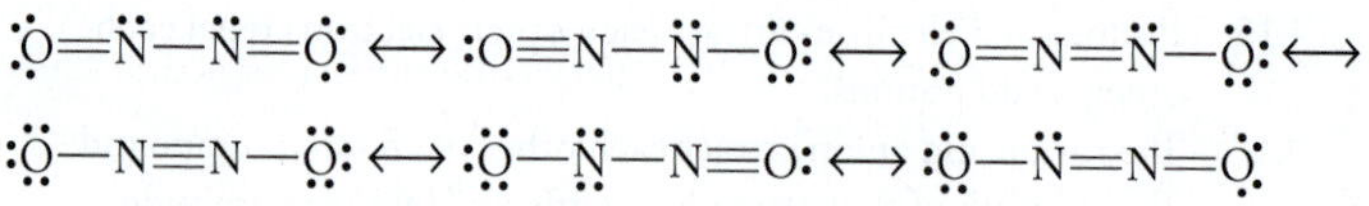

$:O=\ddot{N}—\ddot{N}=O: \leftrightarrow :O\equiv N—\ddot{N}—\ddot{O}: \leftrightarrow :O=N=\ddot{N}—\ddot{O}: \leftrightarrow$

$:\ddot{O}—N\equiv N—\ddot{O}: \leftrightarrow :\ddot{O}—\ddot{N}—N\equiv O: \leftrightarrow :\ddot{O}—\ddot{N}=N=O:$

$N_2O_3$:

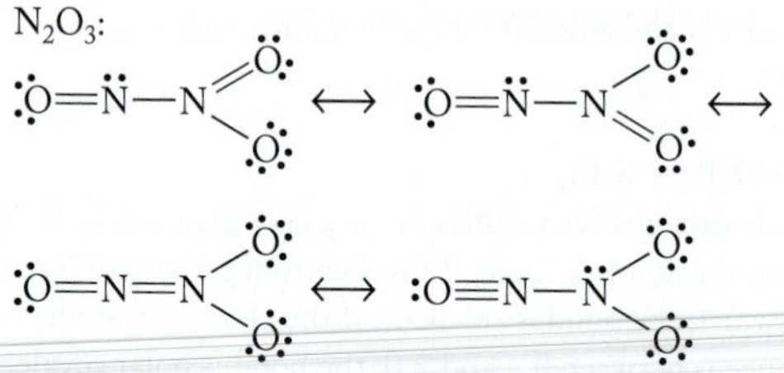

$O=\ddot{N}—N(=O)(—O) \leftrightarrow O=\ddot{N}—N(—O)(=O) \leftrightarrow$

$O=N=N(—O)(—O) \leftrightarrow :O\equiv N—\ddot{N}(—O)(—O)$

4.95.

$H—C\equiv N—\ddot{O}: \leftrightarrow H—\ddot{C}—N\equiv O: \leftrightarrow H—\ddot{C}=\ddot{N}=O:$

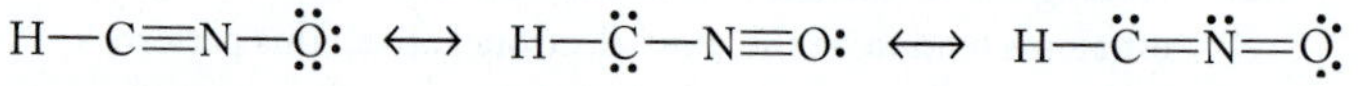

4.97. Resonance structures of $N_2O_5$: O₂N—O—NO₂ with the various N=O / N—O and central N=O arrangements (structures shown as Lewis diagrams):

$:\ddot{O}=N(-\ddot{O}:)-\ddot{O}-N(=\ddot{O})-\ddot{O}: \leftrightarrow \cdots$; $O_2N{=}O{-}NO_2 \leftrightarrow \cdots$; $O_2N{-}O{=}NO_2 \leftrightarrow \cdots$; $O_2N{=}O{=}NO_2$

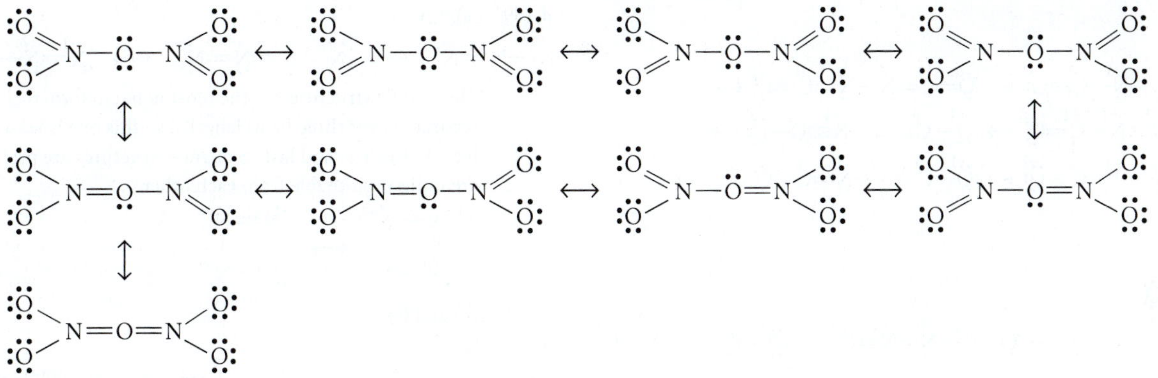

4.99. The bond in $O_2$ is a double bond (bond order of 2), whereas the bond in $O_3$, because of resonance, has a bond order of 1.5.

4.101. $NO^+ < NO_2^- < NO_3^-$

4.103. $NO_3^- < NO_2^- < NO^+$

4.105. C—C bonds in ethane are longer than those in acetylene.

4.107. The best possible structure for a molecule judging by formal charges is the structure in which the formal charges are minimized and the negative formal charges are on the most electronegative atoms in the structure.

4.109. No

4.111. $\overset{0}{H}-\overset{+1}{N}\equiv\overset{-1}{C}:$ $\quad \overset{0}{H}-\overset{0}{C}\equiv\overset{0}{N}:$

The formal charges are zero for all the atoms in HCN, whereas in HNC the carbon atom, with a lower electronegativity than that of N, has a −1 formal charge.

4.113. $(\overset{0}{H})_2\overset{+1}{N}=\overset{0}{C}=\overset{-1}{\ddot{N}}: \leftrightarrow (\overset{0}{H})_2\overset{0}{\ddot{N}}-\overset{0}{C}\equiv\overset{0}{N}:$

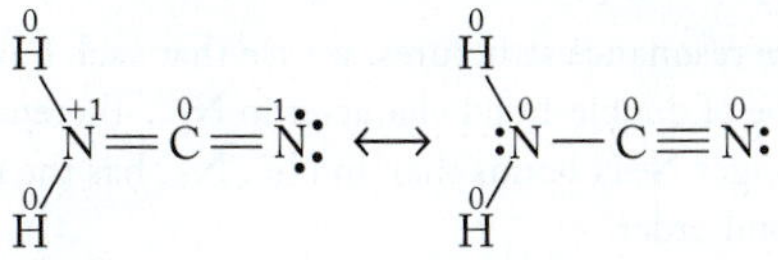

The preferred structure is the one with the C triple-bonded to N.

4.115. $:\overset{-2}{\ddot{N}}-\overset{+2}{O}\equiv\overset{0}{N}: \leftrightarrow :\overset{-1}{\ddot{N}}=\overset{+2}{O}=\overset{-1}{\ddot{N}}: \leftrightarrow :\overset{0}{N}\equiv\overset{+2}{O}-\overset{-2}{\ddot{N}}:$

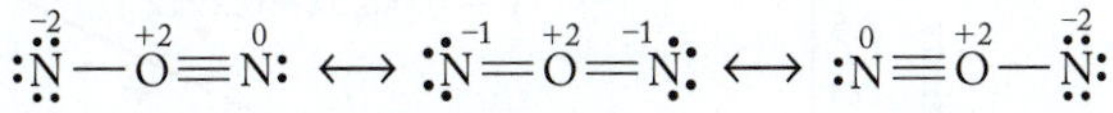

Because oxygen is more electronegative than nitrogen, none of these structures is likely to be stable because the formal charge on O is positive.

4.117. (a) $H_3\overset{0}{C}-\overset{+1}{N}(-\overset{-1}{\ddot{O}}:)(=\overset{0}{\ddot{O}}) \leftrightarrow H_3\overset{0}{C}-\overset{+1}{N}(=\overset{0}{\ddot{O}})(-\overset{-1}{\ddot{O}}:)$ (all H formal charges 0)

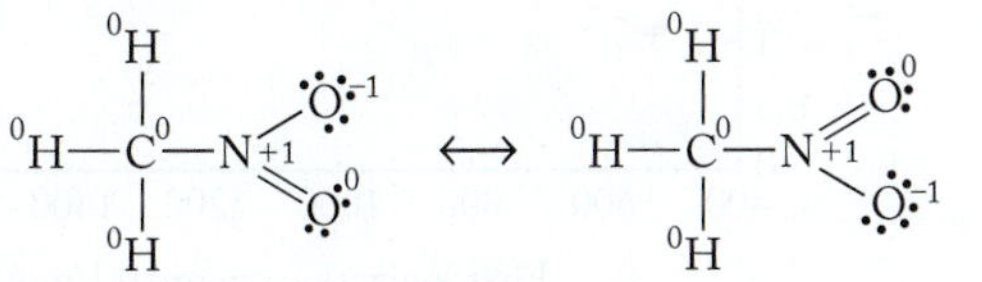

(b) $:\overset{-1}{C}\equiv\overset{+1}{N}-\overset{+1}{N}(=\overset{0}{O})(-\overset{-1}{O}) \leftrightarrow :\overset{-1}{C}\equiv\overset{+1}{N}-\overset{+1}{N}(-\overset{-1}{O})(=\overset{0}{O})$

$:\overset{0}{N}\equiv\overset{0}{C}-\overset{+1}{N}(=\overset{0}{O})(-\overset{-1}{O}) \leftrightarrow :\overset{0}{N}\equiv\overset{0}{C}-\overset{+1}{N}(-\overset{-1}{O})(=\overset{0}{O})$

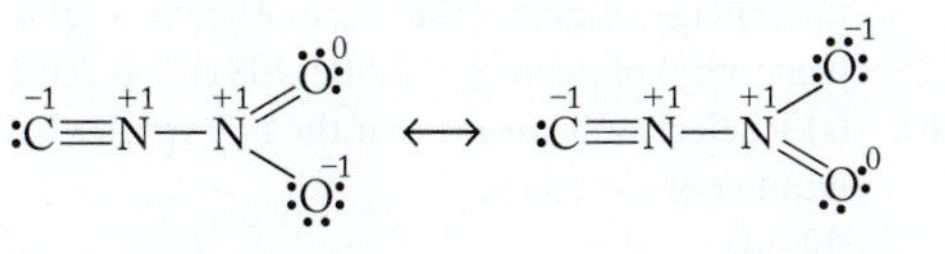

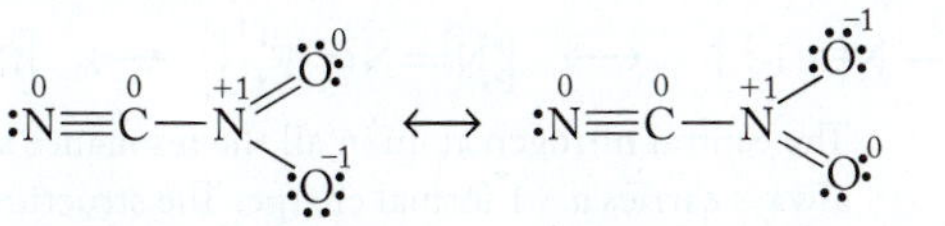

Formal charges are minimized in structures with a terminal N atom, so they are preferred.

(c) No

4.119. Yes

4.121. (a) $SF_6$, (b) $SF_5$, and (c) $SF_4$

4.123. (a) 12; (b) 10; (c) 8; (d) 12

4.125. $SeF_4$ (Se with four F and one lone pair) and $[SeF_5]^-$ (Se with five F and one lone pair)

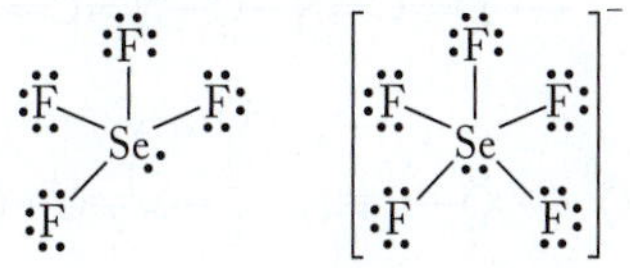

In both structures, Se has more than 8 valence electrons.

4.127. $\overset{0}{O}=\overset{0}{Cl}(=\overset{0}{O})-\overset{0}{\ddot{Cl}}:$

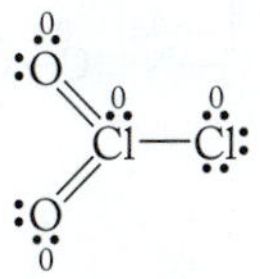

The central chlorine atom has an expanded octet.

4.129. (c) $ClO_4$, (d) $ClO_3$, and (e) $ClO_2$

4.131. (d)

4.133. Like the panes of glass in a greenhouse, the greenhouse gases in the atmosphere are transparent to visible light. Once the visible light warms the surface of Earth and is reemitted as infrared (lower energy) light, the greenhouse gases absorb the infrared light, in the same way that the panes of glass do not allow the heat to escape from the greenhouse.

4.135. The N—O bond would be expected to absorb IR radiation.

4.137. Yes, CO can absorb IR radiation because stretching its bond gives rise to a fluctuating electric field.

4.139. Infrared radiation—with its longer wavelengths and lower energy than UV radiation—causes chemical bonds only to stretch and bend, but not to break.

4.141. The stretching frequency of the CO triple bond is higher than that of the CO double bond due to the greater strength of the triple bond.

4.143. (a) Group 14; (b) Group 15; (c) Group 16; (d) Group 17

4.145. $\ddot{O}=C(-\ddot{Cl}:)_2$

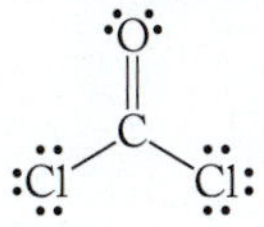

4.147.

(a)

$\overset{+1}{:O}\equiv\overset{0}{C}-\overset{-1}{N}-\overset{-1}{N}-\overset{0}{C}\equiv\overset{+1}{O:} \leftrightarrow \overset{0}{O}=\overset{0}{C}=\overset{0}{N}-\overset{0}{N}=\overset{0}{C}=\overset{0}{O} \leftrightarrow$

$\overset{0}{O}=\overset{-1}{C}-\overset{+1}{N}\equiv\overset{+1}{N}-\overset{-1}{C}=\overset{0}{O} \leftrightarrow :\overset{-1}{O}-\overset{0}{C}\equiv\overset{+1}{N}-\overset{+1}{N}\equiv\overset{0}{C}-\overset{-1}{O}: \leftrightarrow$

$\overset{0}{O}=\overset{0}{C}=\overset{0}{N}-\overset{+1}{N}\equiv\overset{0}{C}-\overset{-1}{O}: \leftrightarrow :\overset{-1}{O}-\overset{0}{C}\equiv\overset{+1}{N}-\overset{0}{N}=\overset{0}{C}=\overset{0}{O}$

(b)

$:Br-N=O$

$\overset{+1}{:O}\equiv\overset{0}{C}-\overset{0}{N}=\overset{0}{N}-\overset{-1}{O}: \leftrightarrow \overset{0}{O}=\overset{-1}{C}-\overset{0}{N}=\overset{+1}{N}=\overset{0}{O} \leftrightarrow :\overset{-1}{O}-\overset{-1}{C}=\overset{0}{N}-\overset{+1}{N}\equiv\overset{+1}{O}:$

(c)

$\overset{+1}{:O}\equiv\overset{0}{C}-\overset{-1}{N}-\overset{0}{C}(=\overset{0}{O})-\overset{-1}{N}-\overset{0}{C}\equiv\overset{+1}{O:} \leftrightarrow \overset{0}{O}=\overset{0}{C}=\overset{0}{N}-\overset{0}{C}(=\overset{0}{O})-\overset{0}{N}=\overset{0}{C}=\overset{0}{O} \leftrightarrow$

$\overset{0}{O}=\overset{0}{C}=\overset{0}{N}-\overset{0}{C}(-\overset{-1}{O}:)=\overset{+1}{N}=\overset{0}{C}=\overset{0}{O} \leftrightarrow :\overset{-1}{O}-\overset{0}{C}\equiv\overset{+1}{N}-\overset{0}{C}(=O)-\overset{+1}{N}\equiv\overset{0}{C}-\overset{-1}{O}: \leftrightarrow$

$:\overset{-1}{O}-\overset{0}{C}\equiv\overset{+1}{N}-\overset{0}{C}(=\overset{0}{O})-\overset{0}{N}=\overset{0}{C}=\overset{0}{O} \leftrightarrow \overset{0}{O}=\overset{0}{C}=\overset{0}{N}-\overset{0}{C}(=\overset{0}{O})-\overset{+1}{N}\equiv\overset{0}{C}-\overset{-1}{O}: \leftrightarrow$

$\overset{0}{O}=\overset{0}{C}=\overset{+1}{N}=\overset{0}{C}(-\overset{-1}{O}:)-\overset{0}{N}=\overset{0}{C}=\overset{0}{O}$

4.149. $O=Cl(=O)(=O)-Cl(=O)(=O)=O$ $\quad$ $O=Cl(=O)(=O)-O-Cl(=O)(=O):$

4.151. (a) $\cdot C\equiv N:$; the more likely structure for cyanogen is the one that contains the C—C bond,

$:\overset{0}{N}\equiv\overset{0}{C}-\overset{0}{C}\equiv\overset{0}{N}: \quad :\overset{-1}{C}\equiv\overset{+1}{N}-\overset{+1}{N}\equiv\overset{-1}{C}:$.

(b) Oxalic acid would be expected to retain the C—C bond from the cyanogen from which it is formed in the reaction of cyanogen with water. This is consistent with the structure for cyanogen predicted through formal charge analysis.

4.153. $\overset{0}{N}\equiv\overset{0}{C}-\overset{0}{S}(-\overset{0}{F}:)(-\overset{0}{F}:)-\overset{0}{F}:$

4.155. $[TeF_5O]^{2-}$: $\overset{-1}{Te}$ bonded to five :F: and one $\overset{-1}{O}$

4.157. $Ca^{2+}$ $[O=C(-O:)(-O:)]^{2-}$ $\quad Mg^{2+}$ $2[:O-H]^-$

4.159. (a), (b)

$:\overset{0}{N}\equiv\overset{+1}{N}-\overset{0}{N}=\overset{-1}{N}: \leftrightarrow :\overset{-1}{N}=\overset{+1}{N}=\overset{+1}{N}=\overset{-1}{N}: \leftrightarrow :\overset{-1}{N}=\overset{0}{N}-\overset{+1}{N}\equiv\overset{0}{N}:$

The middle structure has the most nonzero formal charges separated over three bond lengths, so this one is least preferred. The first and last resonance structures are preferred and are indistinguishable from each other.

(c) cyclic $N_4$: $\overset{0}{N}-\overset{0}{N}$ / $\overset{0}{N}-\overset{0}{N}$ with N=N sides $\leftrightarrow$ $\overset{0}{N}=\overset{0}{N}$ / $\overset{0}{N}=\overset{0}{N}$ with N–N sides

4.161. (b) and (c)

4.163. (a), (b)

$[:\overset{-2}{N}-\overset{0}{N}=\overset{0}{N}-\overset{+1}{N}\equiv\overset{0}{N}:]^- \leftrightarrow [:\overset{-2}{N}-\overset{0}{N}=\overset{+1}{N}=\overset{+1}{N}=\overset{-1}{N}:]^- \leftrightarrow$

$[:\overset{-2}{N}-\overset{+1}{N}\equiv\overset{+1}{N}-\overset{0}{N}=\overset{-1}{N}:]^- \leftrightarrow [:\overset{-1}{N}=\overset{0}{N}-\overset{+1}{N}=\overset{+1}{N}-\overset{-2}{N}:]^- \leftrightarrow$

$[:\overset{0}{N}\equiv\overset{+1}{N}-\overset{0}{N}=\overset{0}{N}-\overset{-2}{N}:]^- \leftrightarrow [:\overset{-1}{N}=\overset{-1}{N}-\overset{0}{N}=\overset{+1}{N}=\overset{-1}{N}:]^- \leftrightarrow$

$[:\overset{-1}{N}=\overset{+1}{N}=\overset{0}{N}-\overset{0}{N}=\overset{-1}{N}:]^- \leftrightarrow [:\overset{0}{N}\equiv\overset{+1}{N}-\overset{-1}{N}-\overset{0}{N}=\overset{-1}{N}:]^- \leftrightarrow$

$[:\overset{-1}{N}=\overset{0}{N}-\overset{-1}{N}-\overset{+1}{N}\equiv\overset{0}{N}:]^-$

The structures that contribute most have the lowest formal charges (the last four structures shown).

(c) $N_3^-$ has the Lewis structures:

$[:\overset{-2}{N}-\overset{+1}{N}\equiv\overset{0}{N}:]^- \leftrightarrow [:\overset{-1}{N}=\overset{+1}{N}=\overset{-1}{N}:]^- \leftrightarrow [:\overset{0}{N}\equiv\overset{+1}{N}-\overset{-2}{N}:]^-$

From these resonance structures, we see that each bond is predicted to be of double-bond character in $N_3^-$. Therefore, $N_5^-$ has two longer N–N bonds than in $N_3^-$. $N_3^-$ has the higher average bond order.

4.165.

y = 0.0025x – 0.2912

Electronegativity

0 0.5 1 1.5 2 2.5 3 3.5 4 4.5

400 600 800 1000 1200 1400 1600 1800

First ionization energy (kJ/mol)

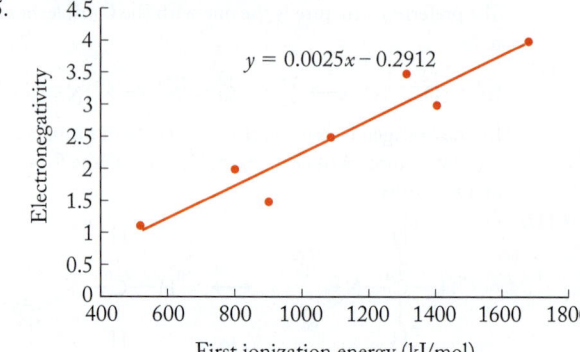

Using the equation for the best-fit line where $x$ = the ionization energy of neon (2081 kJ/mol) gives a value of $y$ (electronegativity) of neon: $y = 0.002(2081) - 0.2912 = 4.9$

4.167. (a) Isoelectronic means that the two species have the same number of electrons.

(b)–(d)

$[:\overset{-2}{N}-\overset{+1}{N}\equiv\overset{+2}{F}:]^+ \leftrightarrow [:\overset{-1}{N}=\overset{+1}{N}=\overset{+1}{F}:]^+ \leftrightarrow [:\overset{0}{N}\equiv\overset{+1}{N}-\overset{0}{F}:]^+$

The central nitrogen atom in all the resonance structures always carries a +1 formal charge. The structure on the right should contribute the most to bonding in the molecule.

(e) Yes, the fluorine could be the central atom in the molecule, but this would place significant positive formal charge on the fluorine atom (the most electronegative element). These structures are unlikely:

$[:\overset{-2}{N}-\overset{+3}{F}\equiv\overset{0}{N}:]^+ \leftrightarrow [:\overset{-1}{N}=\overset{+3}{F}=\overset{-1}{N}:]^+ \leftrightarrow [:\overset{0}{N}\equiv\overset{+3}{F}-\overset{-2}{N}:]^+$.

4.169.
```
    H       H
    |       |
H—C—Ö—C—H
    |       |
    H       H
```

4.171.
```
    H   H   H   H
    |   |   |   |
H—C—C—C—C—H
    |   |   |   |
    H   H   H   H
```

# Chapter 5

5.1. (a)

5.3. $N_2F_2$ and NCCN are planar; no delocalized $\pi$ electrons are present in any of these molecules.

5.5. More

5.7. Yes, a carbon in the ring structure is connected to four different groups, so the compound is chiral.

5.9. The axial F–Re–axial F bond angle is 180°. The axial F–Re–equatorial F angle is 90°. The equatorial F–Re–equatorial F bonds are all 72°.

5.11. Because the electrons take up most of the space in the atom and because the nucleus is located in the center of the electron cloud, the electron clouds repel each other before the nuclei get close enough to interact.

5.13. Because they both have the same steric number around the central nitrogen atom.

5.15. From the Lewis structures, we see that the central atoms in these two molecules have a different number of electron pairs. Boron in $BH_3$ has three bond pairs, making its molecular geometry trigonal planar, whereas nitrogen has three bond pairs plus one lone pair, making its geometry trigonal pyramidal.

5.17. The steric number of the carbon atom in $CH_4$ is 4, which means a tetrahedral molecular geometry and bond angles of 109.5°. The steric number of the carbon atom in $CH_2O$ is 3, which means trigonal planar molecular geometry and bond angles near 120°.

5.19. (a) Tetrahedral
(b) Trigonal pyramidal
(c) Bent
(d) Tetrahedral

5.21. (a) and (c)

5.23. (a) Tetrahedral
(b) Trigonal planar
(c) Bent
(d) Square pyramidal

5.25. (a) Tetrahedral
(b) Tetrahedral
(c) Trigonal planar
(d) Linear

5.27. $O_3$ and $SO_2$

5.29. $SCN^-$ and $CNO^-$

5.31. (c) $H_2S$ < (a) $NH_2Cl$ < (b) $CCl_4$

5.33. S̈=S=Ö: Bent

:Ö=S̈=S̈=Ö: Bent at each S atom

or

:Ö=S(=O)=S̈: Trigonal planar

5.35. $[XeF_5]^-$ Pentagonal bipyramidal

5.37. Lewis structure of sarin: $(CH_3)_2CH$–O–P(=O)($CH_3$)(F), formal charges 0 on O, P, O

The geometry around the P atom in sarin is tetrahedral.

5.39. A polar bond is only between two atoms in a molecule. Molecular polarity takes into account all the individual bond polarities and the geometry of the molecule. A polar molecule has a permanent, measurable dipole moment.

5.41. Yes

5.43. C—Cl

5.45. C=O

5.47. Polar molecules are (b) $CHCl_3$, (d) $H_2S$, and (e) $SCl_2$. Molecules with no dipole moment are (a) $CCl_4$ and (c) $CS_2$.

5.49. All the molecules (a–c) are polar.

5.51. (a) $CBrF_3$
(b) $CHF_2Cl$

5.53. (a) $H_2O$, $H_2S$, and $SO_2$ are polar; $CO_2$ is nonpolar.
(b) Because $H_2O$, $H_2S$, and $SO_2$ have bent molecular geometries.

5.55. To form hybrid orbitals, atomic orbitals must have similar size (energy) and the same phase.

5.57. (a) $sp^3$
(b) $sp^2$
(c) $sp^3$
(d) $sp$
(e) $sp$

5.59. Both Lewis structures of $N_2F_2$ have $sp^2$-hybridized orbitals on N. Each F atom is $sp^3$ hybridized. In acetylene, $C_2H_2$, the carbon atoms are $sp$ hybridized.

5.61.

| $CO_2$ | $NO_2$ | $O_3$ |
|---|---|---|
| $sp$ | $sp^2$ | $sp^2$ |

5.63. $[:\ddot{Cl}—\ddot{Cl}—\ddot{Cl}:]^+$

Bent molecular geometry; $sp^3$ hybridization

5.65. No

5.67. Yes

5.69. Yes, in resonance structures the electron distribution is blurred across all the resonance forms, which in essence defines delocalized electrons.

5.71. One N atom has trigonal pyramidal geometry. The other N atom has trigonal planar geometry. No, the hybridization of both N atoms is not the same.

5.73.

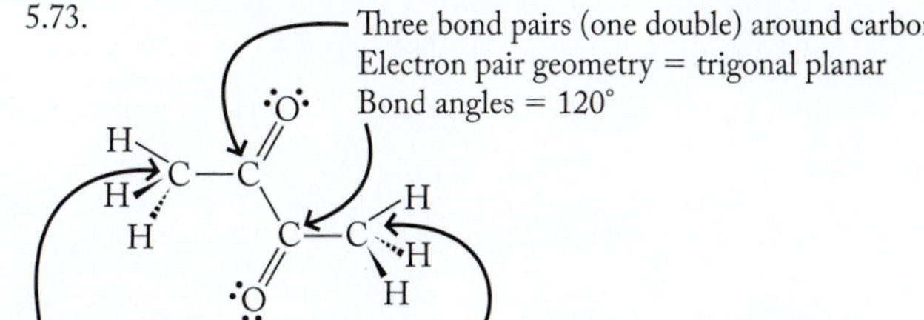

5.75.

5.77.

conjugated double bonds, $sp^2$

5.79. (a), (c), and (d)

5.81. No, $sp$-hybridized carbon is linear with only two bonds to each carbon, and to be chiral, the carbon must be bonded to four different groups.

5.83. Homogeneous

5.85. (a) and (c)

5.87. (a)

5.89.

Saccharin

Sodium cyclamate

Aspartame

5.91.

5.93. Molecular orbital theory

5.95. No

5.97.

5.97.

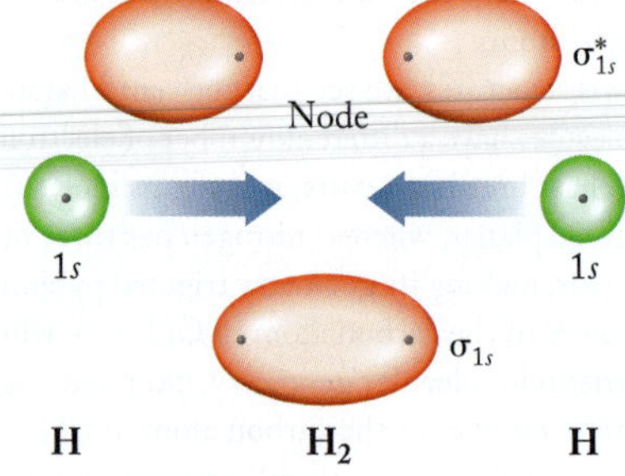


5.99. $N_2^+$ bond order (BO) = 2.5
$O_2^+$ BO = 2.5
$C_2^+$ BO = 1.5
$Br_2^{2-}$ BO = 0
All species with nonzero BO ($N_2^+$, $O_2^+$, and $C_2^+$) are expected to exist.

5.101. (a) $N_2^+$, (b) $O_2^+$, and (c) $C_2^+$

5.103. (b) $N_2^{2-}$, (c) $O_2^{2-}$, and (d) $Br_2^{2-}$

5.105. (a) $B_2$ and (b) $C_2$

5.107. No

5.109.

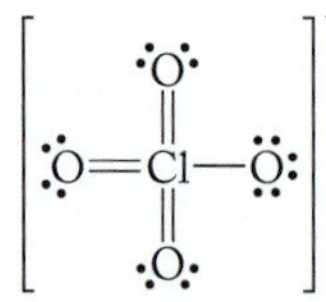

Tetrahedral; $sd^3$ hybridization ($p$ orbitals on Cl are needed to form the pi bonds to O)

5.111. Both the S and N atoms have SN = 4 for an electron-pair geometry of tetrahedral. The presence of a lone pair on N gives this atom trigonal pyramidal geometry and the nitrogen atom is $sp^3$ hybridized. The steric number for S is also 4, which means that the hybridization would be assigned as $sp^3$.

5.113. (Structure: urea, O=C(NH–H)(NH–H), with lone pairs on O and N)

Electron pair geometry = trigonal planar
Planar

Electron pair geometry = tetrahedral
Not planar

5.115. $[NH_4]^+$ (H–N–H with H above and below) Tetrahedral

$[ClO_4]^-$ (O=Cl(=O)(=O)–O, formal charges 0 on Cl, the three double-bonded O's and the Cl; −1 on the single-bonded O) Tetrahedral

5.117.

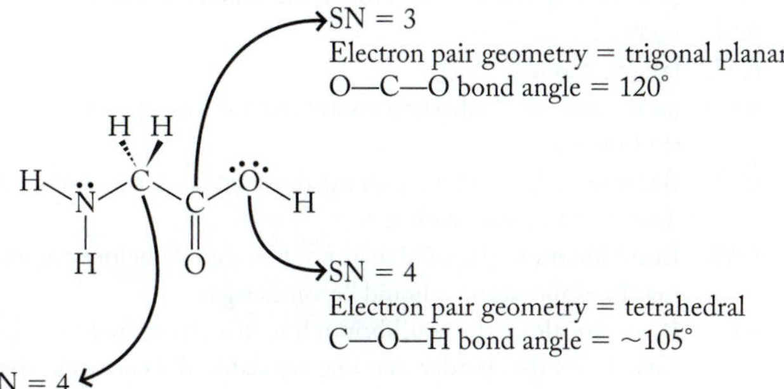


SN = 4
Electron pair geometry = tetrahedral
N—C—C bond angle = 109.5°

5.119. (a) No, neither of the two molecules is linear.
(b) Cl-O-Cl-O would have a permanent dipole.

5.121. (a) $[:\!\ddot{Cl}=\ddot{O}\!:]^+$

(b) BO = 2

5.123. (Structure: $H_3PO_4$, O=P(–OH)$_3$ with lone pairs on the O atoms)

Phosphorus has $sp^2d$ hybridization (one $p$ orbital on P must be used to π-bond to an oxygen); doubly bonded oxygen is $sp^2$ hybridized; O–H oxygen atoms are $sp^3$ hybridized; the molecular geometry around P is tetrahedral.

5.125. (a) (Structure: $H_2\ddot{N}–\ddot{N}H_2$)

(b) $sp^3$

(c) Hydrazine, $NH_2NH_2$ is polar because the presence of a lone pair of electrons and two N—H bonds on each N atom ensures that the molecule is asymmetrical.

5.127. (a), (b)

$\ddot{B}^{-2}—N^{+1}\equiv O^{+1} \longleftrightarrow :B^{-1}=N^{+1}=O^{0} \longleftrightarrow B^{0}\equiv N^{+1}—\ddot{O}^{-1}$

The structure in which boron is triply bonded to nitrogen is the best description; the −1 formal charge is on the most electronegative atom. All these structures have boron with an incomplete octet.

(c) Because SN = 2 around the central nitrogen atom, the molecular geometry is linear.

5.129. (a) and (b)

$H_3C—\ddot{N}=C=\ddot{S}$ (all formal charges 0) $\longleftrightarrow H_3C—\ddot{N}^{-1}—C\equiv S^{+1} \longleftrightarrow H_3C—N^{+1}\equiv C—\ddot{S}^{-1}$ (formal charges 0 on all H and on the C atoms)

Structure 1 is likely to contribute the most to bonding. The methyl ($CH_3$) carbon is tetrahedral.

(c) The isothiocyanate (NCS) carbon is linear; the methyl ($CH_3$) carbon is tetrahedral.

5.131. BO = 0

5.133. $N_2O_5$ and $N_2O_3$; $N_2O_2$ (depending on its actual structure)

5.135.

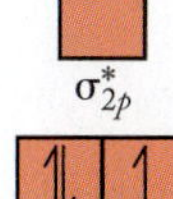

$\sigma^*_{2p}$

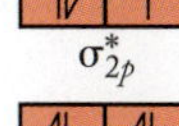

$\pi^*_{2p}$

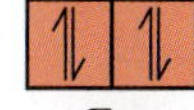

$\pi_{2p}$

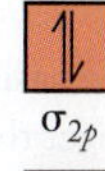

$\sigma_{2p}$

$\sigma^*_{2s}$

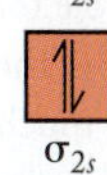

$\sigma_{2s}$

Bond order in $F_2^+$ = 1.5

5.137. (Structures: $H_3C—\ddot{N}(CH_3)_2$ and $H_3Si—N(=SiH_3)(SiH_3)$)

With a lone pair on N in trimethylamine, the molecular geometry is trigonal pyramidal. If the lone pair on nitrogen double-bonds with the Si atom, then the SN = 3 and the molecular geometry of trisilylamine is trigonal planar. Silicon can form this double bond by expanding its octet because it has available 3$d$ orbitals to do so.

5.139.

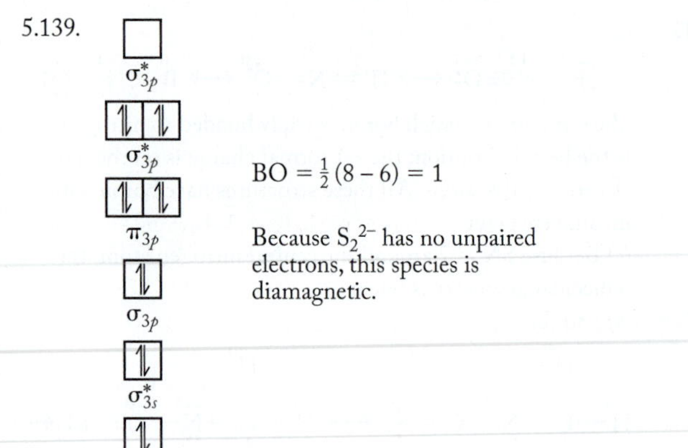


## Chapter 6

6.1. (a)

6.3. Ammonia boils higher than $PH_3$ because it is more polar and is hydrogen-bonded.

6.5. (a) Ketone
(b) Alcohol
(c) Ether

6.7. Solid

6.9. Solid to liquid

6.11. No

6.13. Liquid to solid

6.15. Solid; dispersion forces

6.17. Because a straight-chain alkane has greater dispersion forces due to its greater surface area

6.19. (a) $C_2Cl_6$
(b) $C_3H_8$
(c) $CS_2$

6.21. Fuel oil

6.23. Both $CH_2Cl_2$ and $CH_2F_2$ are polar, and on the basis of dipole–dipole interactions alone we would expect that the more polar molecule, $CH_2F_2$, would have the higher boiling point. However, because Cl is a much larger atom than F, the dispersion forces adding to the dipole–dipole forces in $CH_2Cl_2$ are much greater than those in $CH_2F_2$. The lower dipole moment in $CH_2Cl_2$ is overtaken by the much greater dispersion forces. Therefore, the total intermolecular forces for $CH_2Cl_2$ are greater than the stronger dipole moment but weaker dispersion forces in $CH_2F_2$.

6.25. (a) $CCl_4$
(b) $C_3H_8$

6.27. Molecule (a) has an asymmetrical, trigonal pyramidal structure, which means that it has a permanent dipole and experiences dipole–dipole forces that nonpolar molecule (b) does not. However, molecule (b) experiences stronger dispersion forces than molecule (a) because (b) has a larger molecular mass and more electrons.

6.29. The water molecule is oriented around an anion so as to point to the partially positive hydrogen atoms toward the anion.

6.31. Because of the full positive or negative charge on the ion, the ion–dipole interaction is generally stronger than the dipole–dipole interaction.

6.33. The charge buildup on H (partially positive) and the electronegative element (partially negative) means that the X—H bond is polar. It is still a dipole–dipole interaction except that it is noticeably stronger than other dipole–dipole interactions.

6.35. Because S—H bonds are not as polar and are not hydrogen-bonded as O—H bonds are

6.37. The H in methane has just a single bond to the relatively low-electronegativity C atom, and therefore the carbon–hydrogen bond is not polar enough to exhibit hydrogen bonding. In methanol, however, one of the H atoms is bonded to oxygen, which is second to fluorine in electronegativity. That is the H that shows hydrogen bonding in methanol.

6.39. (b) $CF_2Cl_2$ and (d) $CFCl_3$

6.41. (a) Methanol and (d) acetic acid

6.43. $Cl^-$

6.45. *Miscible* and *soluble* are nearly the same in that both describe one substance dissolving into another. *Miscible*, however, refers to two liquids mixing, whereas *soluble* refers generally to a solid dissolving in a liquid. In addition, the solubility of a substance in a solvent may be limited and so may form a saturated solution or precipitate from the solution if the concentration is too high, but two miscible liquids are soluble in each other in all proportions.

6.47. Hydrophilic substances dissolve in water. Hydrophobic substances do not dissolve, or are immiscible, in water.

6.49. (a) is more polar and therefore more soluble in water.

6.51. (a) NaCl

6.53. (a), (c), and (d)

6.55. (a) Because the hydrocarbons are not too hydrophobic
(b) Butanol

6.57. Because each structure is charged, water can form strong ion–dipole interactions with them.

6.59. In sublimation, the solid does not first liquefy before evaporating. In evaporation, a liquid becomes a gas.

6.61. If you are along the equilibrium line in a phase diagram, the two phases that border that line are stable and coexist at that pressure–temperature combination.

6.63. (a) Solid phase
(b) Gas phase

6.65. Yes

6.67. About 310°C

6.69. The water vaporizes from liquid to gas.

6.71. −57°C at 5.1 atm or −78°C at 1 atm

6.73. Liquid; it will vaporize into a gas if the pressure is lowered to 0.50 atm.

6.75. (a) Yes
(b) 100 atm

6.77. A needle floats on water but not on methanol because of water's high surface tension. This is because water can hydrogen-bond through two O—H bonds with other water molecules, whereas methanol has only one O—H bond through which to form strong hydrogen bonds.

6.79. In the pipes, water that expands on freezing may create enough pressure on the wall of the pipes to cause them to burst.

6.81. The cohesive forces in mercury are stronger than the adhesive forces of the mercury to the glass.

6.83. Molecules in the bulk liquid are "pulled" by all the other liquid molecules surrounding them, and they are therefore "suspended" in the bulk liquid. Molecules on the surface of a liquid, however, are pulled only by molecules under and beside them, creating a tight film of molecules on the surface.

6.85. Water

6.87. London dispersion forces

6.89. Although the dispersion forces between methanol molecules are greater than those between water molecules because methanol has more electrons and greater molar mass, water can form two hydrogen bonds compared with methanol's one hydrogen bond. This greater number of stronger interactions between water molecules raises the boiling point of water above that of methanol.

6.91. (a) $CH_4$

6.93. Increase

6.95. Contract

6.97. (b) and (d)

6.99. (a) $CH_3CH_2OCH_2CH_3$
(b) $CH_3CH_2SCH_2CH_3$
(c) $CH_3CH_2CH_2CH_2OH$

# Chapter 7

7.1. Empirical, $NO_2$; molecular, $N_2O_4$

7.3. (b) and (d)

7.5. (a) Y
(b) $Y_2$
(c) $Y_2$
(d) $XY_2$

7.7. 5.00 g

7.9. Reactant B

7.11. 2*x*

7.13. 1.5 times more

7.15. Yes

7.17. 2 mol

7.19. (a) $N_2(g) + O_2(g) \rightarrow 2\,NO(g)$
(b) $2\,N_2(g) + O_2(g) \rightarrow 2\,N_2O(g)$
(c) $NO(g) + NO_3(g) \rightarrow 2\,NO_2(g)$
(d) $4\,NO(g) + O_2(g) + 2\,H_2O(\ell) \rightarrow 4\,HNO_2(\ell)$

7.21. (a) $N_2O_5(g) + Na(s) \rightarrow NaNO_3(s) + NO_2(g)$
(b) $N_2O_4(g) + H_2O(\ell) \rightarrow HNO_3(aq) + HNO_2(aq)$
(c) $3\,NO(g) \rightarrow N_2O(g) + NO_2(g)$

7.23. (a) $C_3H_8(g) + 5\,O_2(g) \rightarrow 3\,CO_2(g) + 4\,H_2O(g)$
(b) $2\,C_4H_{10}(g) + 13\,O_2(g) \rightarrow 8\,CO_2(g) + 10\,H_2O(g)$
(c) $2\,C_6H_6(\ell) + 15\,O_2(g) \rightarrow 12\,CO_2(g) + 6\,H_2O(g)$
(d) $2\,C_8H_{18}(\ell) + 25\,O_2(g) \rightarrow 16\,CO_2(g) + 18\,H_2O(g)$

7.25. (a) $C_2H_4(g) + 3\,O_2(g) \rightarrow 2\,CO_2(g) + 2\,H_2O(g)$
(b) $2\,C_3H_6(g) + 9\,O_2(g) \rightarrow 6\,CO_2(g) + 6\,H_2O(g)$
(c) $2\,C_4H_{10}(g) + 13\,O_2(g) \rightarrow 8\,CO_2(g) + 10\,H_2O(g)$
(d) $C_4H_8(g) + 6\,O_2(g) \rightarrow 4\,CO_2(g) + 4\,H_2O(g)$

7.27. (a) $2\,SO_2(g) + O_2(g) \rightarrow 2\,SO_3(g)$
(b) $2\,H_2S(g) + 3\,O_2(g) \rightarrow 2\,SO_2(g) + 2\,H_2O(g)$
(c) $16\,H_2S(g) + 8\,SO_2(g) \rightarrow 3\,S_8(s) + 16\,H_2O(g)$

7.29. No, both equations give the same numerical answer for the amount of $CO_2$ produced in the reaction for a given amount of $C_2H_6$ because in both the molar ratio of $CO_2$ to $C_2H_6$ is 2:1.

7.31.

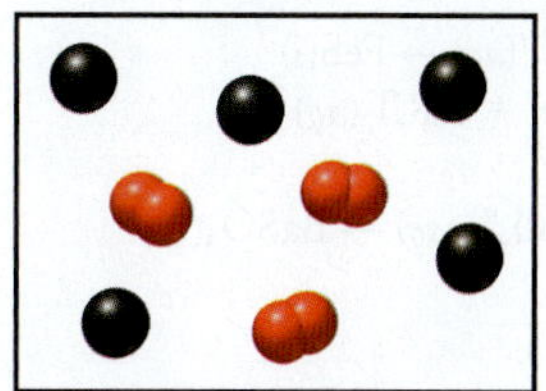

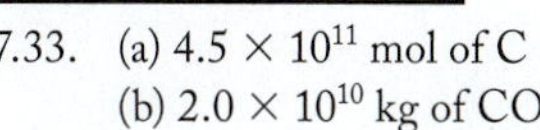

7.33. (a) $4.5 \times 10^{11}$ mol of C
(b) $2.0 \times 10^{10}$ kg of $CO_2$

7.35. (a) $2\,NaHCO_3(s) \rightarrow CO_2(g) + H_2O(g) + Na_2CO_3(s)$
(b) 6.55 g of $CO_2$

7.37. 29 g of $O_2$

7.39. (a) 1.48 kg
(b) 1.108 kg

7.41. 346 g

7.43. The ratio of the reactants in the balanced chemical equation also is important. If A and B were necessary in equal molar ratios, then yes, with fewer moles of A, A would be the limiting reactant. However, if more than 1 mol of B were required for every mole of A and fewer moles of A than B were present, we could not unequivocally say that A was the limiting reactant.

7.45. Theoretical yield is the greatest amount of a product possible from a reaction and assumes that the reaction goes to 100% completion. The percent yield is the observed experimental yield divided by the theoretical yield and multiplied by 100.

7.47. The actual and the percent yields will be lower.

7.49. 40

7.51. (a) 5.0 g
(b) 34.5 g
(c) 8.0 g

7.53. 87% efficient; 0.60 metric tons escaped

7.55. (a) $CHCl_3$
(b) 308 g
(c) 529 g

7.57. (a) $C_6H_{12}O_6(aq) \rightarrow 2\,C_2H_5OH(aq) + 2\,CO_2(g)$
(b) 77.1%

7.59. An empirical formula shows the lowest whole-number ratio of atoms in a substance. A molecular formula shows the actual numbers of each kind of atom that compose one molecule of the substance.

7.61. No, lighter elements may be present in quantities large enough to be a greater percentage of the mass in a molecular formula than a heavier element.

7.63. (a) 74.19% Na, 25.81% O
(b) 57.48% Na, 40.00% O, 2.52% H
(c) 27.37% Na, 1.20% H, 14.30% C, 57.13% O
(d) 43.38% Na, 11.33% C, 45.28% O

7.65. Pyrene, $C_{16}H_{10}$

7.67. No

7.69. $Ti_6Al_4V$

7.71. $Mg_3Si_2H_4O_9$

7.73. $CuCl_2O_8$

7.75. Tartaric acid

7.77. Empirical formula, CHN; molecular formula, $C_5H_5N_5$

7.79. The excess of oxygen is required in combustion analysis to ensure the complete reaction of the hydrogen and carbon to form water and carbon dioxide.

7.81. Yes

7.83. $C_3H_6O_2$

7.85. Empirical, $C_2H_3$; molecular, $C_{20}H_{30}$

7.87. $C_9H_8O_4$

7.89. (a) Calcium triphosphate hydroxide
(b) 39.89%
(c) Decreases slightly

7.91. (a) No
(b) $C_5H_{10}O_5(s) + 5\,O_2(g) \rightarrow 5\,CO_2(g) + 5\,H_2O(\ell)$
$2\,C_7H_{12}O_7(s) + 13\,O_2(g) \rightarrow 14\,CO_2(g) + 12\,H_2O(\ell)$

7.93. (a) 45 g
(b) 15 g
(c) 4.66 cm$^3$

7.95. (a) $a = 1, b = 3$; charge on U is 6+
(b) $c = 3, d = 8$; charge on U is 5.33+
(c) $x = 2, y = 2, z = 6$.

7.97. (a) FeS, $Fe^{2+}$ and $S^{2-}$; $FeS_2$, $Fe^{4+}$ and $S^{2-}$
(b) FeS is iron(II) sulfide; $FeS_2$ is iron(II) persulfide, a compound composed of $Fe^{2+}$ with $S_2^{2-}$.
(c) 0.26 g

7.99. (a) $3\ FeO(s) + H_2O(\ell) \rightarrow Fe_3O_4(s) + H_2(g)$
(b) $12\ FeO(s) + 2\ H_2O(\ell) + CO_2(aq) \rightarrow 4\ Fe_3O_4(s) + CH_4(g)$

7.101. (a) 55 mol of ethanol
(b) 110 mol of carbon dioxide

7.103. Re

7.105. 82.4%

7.107. (a) $2\ SO_2(g) + 2\ CaO(s) + O_2(g) \rightarrow 2\ CaSO_4(s)$
(b) 2.13 metric tons

7.109. $Mg_2SiO_4$

# Chapter 8

8.1. Yellow

8.3. Green, $SO_4^{2-}$; blue, $Na^+$

8.5. From 2+ to 4+

8.7. (a) $N_2O_5$ > (b) $N_2O_4$ = (e) $NO_2$ > (d) $N_2O_3$ > (c) NO

8.9. (a) $2\ H^+(aq) + SO_4^{2-}(aq) + Ba^{2+}(aq) + 2\ OH^-(aq) \rightarrow BaSO_4\ (s) + 2\ H_2O(\ell)$
(b) b

8.11. (a) 5.6 *M* $BaCl_2$
(b) 1.00 *M* $Na_2CO_3$
(c) 1.30 *M* $C_6H_{12}O_6$
(d) 5.92 *M* $KNO_3$

8.13. (a) 0.29 *M* $NaNO_3$
(b) 0.23 *M* NaCl
(c) 1.8 *M* $Na_2SO_4$
(d) 0.084 *M* $Na_3PO_4$

8.15. (a) 11.7 g NaCl
(b) 4.99 g $CuSO_4$
(c) 6.41 g $CH_3OH$

8.17. 2.72 g

8.19. (a) $1.9 \times 10^{-2}$ mol
(b) $4.07 \times 10^{-4}$ mol
(c) $2.72 \times 10^{-4}$ mol
(d) $4.00 \times 10^{-1}$ mol

8.21. Orchard sample: $3.4 \times 10^{-4}$ mmol/L
Residential area sample: $5.6 \times 10^{-5}$ mmol/L
Residential after-storm sample: $3.2 \times 10^{-2}$ mmol/L

8.23. $1.2 \times 10^{-3}$ *M*

8.25. (b) $AgNO_3$ and (c) $Fe(NO_3)_2 \cdot 6\ H_2O$

8.27. $4.57 \times 10^{-2}$ *M* $Mg^{2+}$

8.29. (a) $1.81 \times 10^{-2}$ *M* $Na^+$
(b) $2.7 \times 10^{-1}$ m*M* LiCl
(c) $1.28 \times 10^{-2}$ m*M* $Zn^{2+}$

8.31. 1.95 *M*

8.33. The concentration of the adult-strength medication is 1.8 mg/mL; 23 mL is needed to prepare the child-strength cough syrup.

8.35. 12.3 mL

8.37. 0.81

8.39. Table salt produces $Na^+$ and $Cl^-$ ions in solution when it dissolves. Sugar does not dissociate into ions because it is not a salt. Ions are required to conduct electricity.

8.41. In order of decreasing conductivity, 1.0 *M* $Na_2SO_4$ (c) > 1.2 *M* KCl (b) > 1.0 *M* NaCl (a) > 0.75 *M* LiCl (d).

8.43. (a) 0.025 *M*
(b) 0.050 *M*
(c) 0.075 *M*

8.45. A strong acid ionizes completely, whereas a weak acid only partially ionizes.

8.47. A base accepts $H^+$ from another species (an acid) in solution.

8.49. Strong bases include NaOH, KOH, CsOH, LiOH, RbOH, $Ba(OH)_2$, $Sr(OH)_2$, and $Ca(OH)_2$; weak bases include $NH_3$, $CH_3NH_2$, and $C_5H_5N$.

8.51. (a) The acid is $H_2SO_4$; the base is $Ca(OH)_2$.
Net ionic equation: $H^+(aq) + HSO_4^-(aq) + Ca(OH)^2(s) \rightarrow Ca^{2+}(aq) + SO_4^{2-}(aq) + 2\ H_2O(\ell)$
(b) $PbCO_3$ is the base; $H_2SO_4$ is the acid.
Net ionic equation: $PbCO_3(s) + H^+(aq) + HSO_4^-(aq) \rightarrow PbSO_4(s) + CO_2(g) + 2\ H_2O(\ell)$
(c) $Ca(OH)_2$ is the base; $CH_3COOH$ is the acid.
Net ionic equation: $Ca(OH)_2(s) + 2\ CH_3COOH(aq) \rightarrow Ca^{2+}(aq) + 2\ CH_3COO^-(aq) + 2\ H_2O(\ell)$

8.53. (a) Molecular equation: $Mg(OH)_2(s) + H_2SO_4(aq) \rightarrow MgSO_4(aq) + 2\ H_2O(\ell)$
Net ionic equation: $Mg(OH)_2(s) + H^+(aq) + HSO_4^-(aq) \rightarrow Mg^{2+}(aq) + SO_4^{2-}(aq) + 2\ H_2O(\ell)$
(b) Molecular equation: $MgCO_3(s) + 2\ HCl(aq) \rightarrow MgCl_2(aq) + H_2O(\ell) + CO_2(g)$
Net ionic equation: $MgCO_3(s) + 2\ H^+(aq) \rightarrow Mg^{2+}(aq) + H_2O(\ell) + CO_2(g)$
(c) Molecular equation: $NH_3(g) + HCl(g) \rightarrow NH_4Cl(s)$. This is also the net ionic equation.

8.55. $PbCO_3(s) + 2\ H^+(aq) \rightarrow Pb^{2+}(aq) + CO_2(g) + H_2O(\ell)$; $Pb(OH)_2(s) + 2\ H^+(aq) \rightarrow Pb^{2+}(aq) + 2\ H_2O(\ell)$

8.57. (a) 5.00 mL
(b) 31.5 mL
(c) 215 mL

8.59. 500 mL

8.61. 556.0 m*M*; 19.23 g/kg

8.63. A saturated solution contains the maximum concentration of a solute. A supersaturated solution *temporarily* contains *more* than the maximum concentration of a solute at a given temperature.

8.65. A saturated solution may not be a concentrated solution if the solute is only sparingly or slightly soluble in the solution. In that case, the saturated solution is a dilute solution.

8.67. (a) Barium sulfate, (e) lead hydroxide, and (f) calcium phosphate

8.69. (a) Balanced reaction: $Pb(NO_3)_2(aq) + Na_2SO_4(aq) \rightarrow PbSO_4(s) + 2\ NaNO_3(aq)$
Net ionic equation: $Pb^{2+}(aq) + SO_4^{2-}(aq) \rightarrow PbSO_4(s)$
(b) No precipitation reaction occurs.
(c) Balanced reaction: $FeCl_2(aq) + Na_2S(aq) \rightarrow FeS(s) + 2\ NaCl\ (aq)$
Net ionic equation: $Fe^{2+}(aq) + S^{2-}(aq) \rightarrow FeS(s)$
(d) Balanced reaction: $MgSO_4(aq) + BaCl_2(aq) \rightarrow MgCl_2(aq) + BaSO_4(s)$
Net ionic equation: $Ba^{2+}(aq) + SO_4^{2-}(aq) \rightarrow BaSO_4(s)$

8.71. $CaCO_3$

8.73. $2.11 \times 10^{-2}$ g

8.75. $5.4 \times 10^{-2}$ g

8.77. The number of electrons gained or lost is directly related to the change in oxidation number of a species. Gaining electrons makes the oxidation number more negative; losing electrons makes the oxidation number more positive.

8.79. (a) −1
(b) +1
(c) −2
(d) −3

8.81. Silver

8.83. A half-reaction is either the reduction component or the oxidation component of a redox reaction.

8.85. (a) +1
(b) +5
(c) +7

8.87. No change

8.89. (a) $Br_2(\ell) + 2\,e^- \rightarrow 2\,Br^-(aq)$; reduction half-reaction
(b) $Pb(s) + 2\,Cl^- \rightarrow PbCl_2(s) + 2\,e^-$; oxidation half-reaction
(c) $O_3(g) + 2\,H^+(aq) + 2\,e^- \rightarrow O_2(g) + H_2O(\ell)$; reduction half-reaction
(d) $2\,H_2SO_3(aq) + H^+(aq) + 2\,e^- \rightarrow HS_2O_4^-(aq) + 2\,H_2O(\ell)$; reduction half-reaction

8.91. (a) $MnO_2(s) + 2\,H^+(aq) + 2\,HCl(aq) \rightarrow Mn^{2+}(aq) + 2\,H_2O(\ell) + Cl_2(g)$; oxidizing agent, $MnO_2$; reducing agent, HCl; oxidized element, Cl; reduced element, Mn
(b) $I_2(s) + 2\,S_2O_3^{2-}(aq) \rightarrow 2\,I^-(aq) + S_4O_6^{2-}(aq)$; oxidizing agent, $I_2$; reducing agent, $S_2O_3^{2-}$; oxidized element, S; reduced element, I
(c) $MnO_4^-(s) + 8\,H^+(aq) + 5\,Fe^{2+}(aq) \rightarrow Mn^{2+}(aq) + 5\,Fe^{3+}(aq) + 4\,H_2O(\ell)$; oxidizing agent, $MnO_4^-$; reducing agent, $Fe^{2+}$; oxidized element, Fe; reduced element, Mn

8.93. 0.25 mol

8.95.

| (a) Reactants | Products |
|---|---|
| $SiO_2$: Si = +4, O = −2 | $Fe_2SiO_4$: Fe = +2, Si = +4, O = −2 |
| $Fe_3O_4$: Fe = +8/3, O = −2 | $O_2$: O = 0 |

Oxygen is oxidized ($O^{2-}$ to $O_2$) and iron is reduced ($Fe^{3+}$ to $Fe^{2+}$).

| (b) Reactants | Products |
|---|---|
| $SiO_2$: Si = +4, O = −2 | $Fe_2SiO_4$: Fe = +2, Si = +4, O = −2 |
| Fe: Fe = 0 | |
| $O_2$: O = 0 | |

Iron is oxidized ($Fe^0$ to $Fe^{2+}$) and oxygen is reduced ($O_2$ to $O^{2-}$).

| (c) Reactants | Products |
|---|---|
| FeO: Fe = +2, O = −2 | $Fe(OH)_3$: Fe = +3, O = −2, H = +1 |
| $O_2$: O = 0 | |
| $H_2O$: H = +1, O = −2 | |

Iron is oxidized ($Fe^{2+}$ to $Fe^{3+}$) and oxygen is reduced ($O_2$ to $O^{2-}$).

8.97. (a) 0.25 mol
(b) 0.17 mol

8.99. $NH_4^+(aq) + 2\,O_2(g) \rightarrow NO_3^+(aq) + 2\,H^+(aq) + H_2O(\ell)$

8.101. $2\,Fe(OH)_2^+(aq) + Mn^{2+}(aq) \rightarrow 2\,Fe^{2+}(aq) + 2\,H_2O(\ell) + MnO_2(s)$

8.103. $2\,H_2O(\ell) + 4\,Ag(s) + 8\,CN^-(aq) + O_2(g) \rightarrow 4\,Ag(CN)_2^-(aq) + 4\,OH^-(aq)$

8.105. (a) $2\,ClO_3^-(aq) + SO_2(g) \rightarrow 2\,ClO_2(g) + SO_4^{2-}(aq)$
(b) $4\,H^+(aq) + 2\,ClO_3^-(aq) + 2\,Cl^-(aq) \rightarrow 2\,ClO_2(g) + 2\,H_2O(\ell) + Cl_2(g)$
(c) $12\,H^+(aq) + 12\,ClO_3^-(aq) + Cl_2(g) \rightarrow 14\,ClO_2(g) + 6\,H_2O(\ell) + O_2(g)$

8.107. Zinc and aluminum

8.109. Scandium would be placed above aluminum and vanadium would be placed below aluminum. (Only if Mg reduces $Sc^{3+}$ ions. What if it doesn't?)

8.111. (a) 41.6 g
(b) 2.93 L

8.113. (a) 11.7 *M*
(b) 42.7 mL
(c) 1.72 kg

8.115. (a) $2\,OH^-(aq) + 2\,H_2O(\ell) + 3\,S_2O_4^{2-}(aq) + 2\,CrO_4^{2-}(aq) \rightarrow 6\,SO_3^{2-}(aq) + 2\,Cr(OH)_3(s)$
(b) Sulfur is oxidized; chromium is reduced.
(c) Oxidizing agent = $CrO_4^{2-}$; reducing agent = $S_2O_4^{2-}$
(d) 38.7 g

8.117. (a) Balanced equation: $4\,Ag(s) + 2\,H_2S(g) + O_2(g) \rightarrow 2\,Ag_2S(s) + 2\,H_2O(g)$
(b) $3\,Ag_2S(s) + 3\,H_2O(\ell) + 3\,OH^- + 2\,Al(s) \rightarrow 6\,Ag(s) + 3HS^-(g) + 2\,Al(OH)_3(s)$

8.119. (a) $H_3PO_4$
(b) $H_2SeO_3$
(c) $H_3BO_3$

8.121. $2\,H^+(aq) + ClO^-(aq) + 2\,I^-(aq) \rightarrow Cl^-(aq) + H_2O(\ell) + I_2(aq)$
$I_2(aq) + 2\,S_2O_3^{2-}(aq) \rightarrow 2\,I^-(aq) + S_4O_6^{2-}(aq)$

8.123. (a) $NaClO_4$; $NH_4ClO_4$
(b) 29.75 kg
(c) $7.40 \times 10^9$ L
(d) The Massachusetts lab

8.125. (a) $3\,CH_2O \rightarrow CO_2 + CH_3CH_2OH$
(b) $CH_3CH_2OH + O_2 \rightarrow CH_3COOH + H_2O$
(c) $CH_2O$: C = 0
$CO_2$: C = +4
$C_2H_5OH$: C = −4 over two carbon atoms, so oxidation number on each carbon = −2
$HC_2H_3O_2$: C = 0
(d) 1.11 mol, or 66.7 g of acetic acid

8.127. (a) The first reaction is a redox reaction.
(b) $2\,H^+(aq) + SO_4^{2-}(aq) + CaCO_3(s) \rightarrow CaSO_4(s) + H_2O(\ell) + CO_2(g)$
(c) $SO_4^{2-}(aq) + CaCO_3(s) \rightarrow CaSO_4(s) + CO_3^{2-}(aq)$

8.129. (c) and (d)

# Chapter 9

9.1. (a) ii
(b) ii

9.3. The motion has no effect.

9.5. (a) The pressure doubles.
(b) The frequency of collisions doubles.
(c) The most probable speed does not change.

9.7. (a) a
(b) b
(c) b

9.9. $SO_2$, curve 1; curve 2, propane

9.11. (a) The red line
(b) Slope = 0.0366 $cm^3$/K; pressure = 2240 atm

9.13. Line 2

9.15. Line 2

9.17. (a) Mount Everest; (b) San Diego; (c) Tau Tona gold mine

9.19. The left test tube

9.21. The lower blue curve

9.23. The root-mean-square speed ($u_{rms}$) is the square root of the arithmetic mean of the squares of the speeds of all the particles in a gas sample. It is the speed of a particle that has the average kinetic energy of all the particles in the sample.

9.25. No, only temperature affects the root-mean-square speed.

9.27. The rank order in terms of increasing root-mean-square speed is $N_2O_5 < N_2O_4 < NO_2 < NO$.

9.29. 32 g/mol

9.31. (a) $CH_4$, 681 m/s; $C_2H_6$, 497 m/s; $C_3H_8$, 411 m/s
(b) $C_4H_{10}$, 359 m/s; $C_5H_{12}$, 298 m/s
(c) $CH_4$

9.33. 509 m/s

9.35. Br

9.37. Force is a push or pull on an object that can cause it to accelerate. The pressure on an object is the ratio of the force applied to it divided by the surface area of the object over which the force is applied.

9.39. The ethanol barometer

9.41. For a sharpened blade, the force is distributed over a smaller area than for a dull blade.

9.43. (a) 10 atm
(b) 100 mmHg
(c) 1.0 atm
(d) 562 mmHg

9.45. (a) 814.6 mmHg
(b) 1.072 atm
(c) 1086 mbar

9.47. As temperature increases, the speed of gas particles increases and they will more often collide with the container walls and with more force, thereby increasing pressure.

9.49. The balloonist should decrease the temperature.

9.51. 1.45 atm

9.53. 6.6 atm

9.55. 25%

9.57. (a) 3.06 atm
(b) 21 m

9.59. Raising the temperature in (b) produces a 12.8% increase, which is greater than the 7.9% increase due to lowering the pressure in (a).

9.61. (a) No change
(b) Decrease of ¼ original volume
(c) Increase in volume by 17%

9.63. (a) 120 L
(b) No
(c) 1220 L
(d) 1270 L

9.65. 1120 mmHg

9.67. STP is defined as 1 atm and 0°C (273 K); $V = 22.4$ L.

9.69. 0.67 mol

9.71. 2.36 atm

9.73. $4.20 \times 10^4$ g

9.75. 1%

9.77. (a) 1.0 mol
(b) Helium

9.79. No, because the density of a gas is directly proportional to its molar mass.

9.81. Density (a) increases with increasing pressure and (b) increases with decreasing temperature.

9.83. (a) 9.08 g/L
(b) In the basement

9.85. (a) 2.6 g/L
(b) $SO_2$

9.87. (a) 28.0 g/mol
(b) CO

9.89. 394 g

9.91. $6.2 \times 10^2$ g

9.93. 3.8 L

9.95. The partial pressure of a gas is the pressure that a particular gas in a mixture of gasses contributes to the total pressure.

9.97. (a) 0.20
(b) $N_2$, 16.3 atm; $H_2$, 4.6 atm; $CH_4$, 2.3 atm

9.99. Hydrogen

9.101. 0.0190 mol

9.103. (a) Greater than
(b) Less than
(c) Greater than

9.105. 67.1% more

9.107. 13%

9.109. At low temperatures, the gas particles move more slowly and their collisions become inelastic; they stick together due to the weak attractive forces between them. The particles, therefore, do not act separately to contribute to the pressure in the container, and the pressure is lower than would be expected by the ideal gas law. Also, the gas particles take up real volume in the container, and as the pressure increases the volume of the particles takes up a greater volume of the free space in the container. That raises the pressure–volume product above what we would expect from the ideal gas law (in a plot of $PV/RT$ vs. $P$).

9.111. Because $b$ is a measure of the volume that the gas particles occupy, $b$ increases as the atomic numbers of the noble gases increase.

9.113. Because Ar has more electrons and is more polarizable than He, it has stronger London forces between its atoms than He, so $a$ for Ar is larger.

9.115. $H_2$

9.117. (a)

```
     H              H              H
     |              |              |
 H—Si—H         H—C—H          H—C—Cl:
     |              |              |
     H              H              H
```

All of these molecules are tetrahedral.
(b) $SiH_4$, 4.320 $L^2 \cdot atm/mol^2$; $CH_4$, 2.253 $L^2 \cdot atm/mol^2$; $CH_3Cl$, 7.471 $L^2 \cdot atm/mol^2$.

9.119. (a) $P = 1590$ atm
(b) $P = 595$ atm
(c) At high pressure, the volume correction term ($b$) in the van der Waals equation increases more for hydrogen than the correction term ($a$) for attractive forces.

9.121. (a) 70.3 mL
(b) Yes, although gases heat up when compressed, the compression in an engine happens many times per second and a steady-state temperature will be achieved in the cylinder.

9.123. 0.25 m/s

9.125. 18 kg

9.127. (a) $CH_3COOH$ and $(CH_3)_3N$
(b) HCl and $(CH_3)_3N$
(c) No

9.129. 2.7 atm

9.131. (a) 38.4 L of $N_2$
(b) 192 L of $CO_2$
(c) 1.74 g/L

9.133. 70.0 g; 2.40 g additional needed at 10°C

9.135. 0.567 atm

# Chapter 10

10.1. Internal energy increases.

10.3. Highest fuel value, (c) $C_5H_{12}$; lowest fuel value, (b) $C_6H_{12}$

10.5. $q$ is negative, $w$ is positive, $\Delta E$ is negative; 67% yield.

10.7. (a) $CH_4(g) + H_2O_2(g) \rightarrow CO(g) + 3\ H_2(g)$
(b) −206.1 kJ/mol
(c) Energy flows out from the reaction mixture to the surroundings.

10.9. (a) 69 kJ
(b) 343°C
(c) 0.60 kJ/K
(d) 0.66 kJ/K

10.11. (a) 243 kJ
(b) −1537 kJ
(c) −347 kJ

10.13. Energy makes work possible.

10.15. The value of a state function is independent of the path; only the initial and final values are important.

10.17. Because the potential energy increases as those particles are moved farther apart, those particles attract each other.

10.19. The internal energy of the gas sample can be increased by raising the temperature or by increasing the pressure through compression.

10.21. No

10.23. (a) Exothermic
(b) Endothermic
(c) Exothermic

10.25. Energy is absorbed from the surroundings. Thus, $q$ increases and therefore $\Delta E > 0$.

10.27. $9.25 \times 10^5$ L · atm, or $9.37 \times 10^7$ J

10.29. (a) 150.0 J
(b) 6.3 kJ
(c) $-1.23 \times 10^4$ kJ

10.31. −275.5 kJ

10.33. (b); $w$ is negative.

10.35. A change in enthalpy is the sum of the change of internal energy and the product of the system's pressure and change in volume.

10.37. If the system transfers energy to the surroundings, the energy of the system will be less after the process than at the start of the process.

10.39. Negative

10.41. (a) Negative
(b) Positive

10.43. Specific heat is specified for a gram of the substance. Heat capacity does not take into account how much of a substance is present; the value is defined for a given object.

10.45. Because to vaporize water you have to completely break the strong intermolecular hydrogen bonds between water molecules, not just loosen them, which occurs during melting.

10.47. Chloroform < ethanol < water

10.49. The amount of energy required to be removed to decrease the temperature of that phase by 1°C

10.51. Water's heat capacity is higher than that of air, which means that water carries away more energy from the engine for every Celsius degree rise in temperature. So water is a good choice to cool automobile engines.

10.53. 29.3 kJ

10.55.

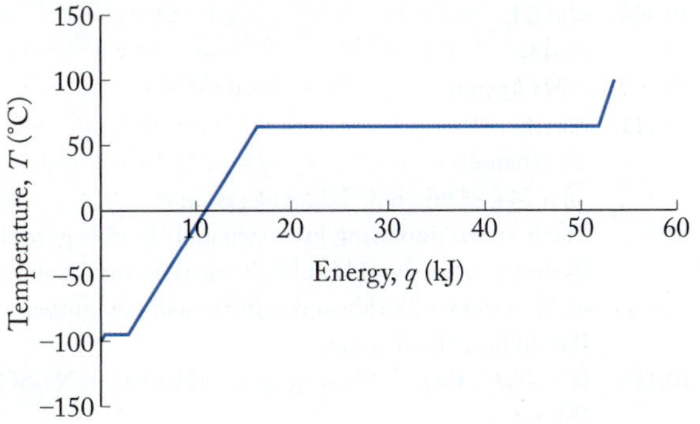

10.57. 103 g

10.59. −47.5°C

10.61. To know how much energy (generated or absorbed by the system) is required to change the temperature of the surroundings (the calorimeter) in order to calculate the heat capacity or final temperature of the system in an experiment

10.63. Yes

10.65. 8.044 kJ/°C

10.67. 4875 kJ/mol is produced

10.69. 3.07°C

10.71. −464 kJ

10.73. When we apply Hess's law, all the energy is accounted for in the reaction; energy is neither created nor destroyed when using Hess's law.

10.75. −297 kJ

10.77. 28.0 kJ

10.79. (a) Add the reverse of reaction iii to that of reaction ii
(b) No

10.81. Because $CO_2$ formation is energetically more favored.

10.83. No; the standard enthalpy of formation for the most stable physical state of the element is zero. Other allotropes of the element will have enthalpies different from that of the most stable allotrope.

10.85. We must account for all the bonds that break and all the bonds that form in the reaction. To do so, we must have a balanced chemical reaction.

10.87. If the compounds are in the solid or liquid phase, interactions between molecules may slightly change the bond energy for a given bond.

10.89. (a) and (c)

10.91. (a) −1167 kJ
(b) No

10.93. $NH_4NO_3(g) \rightarrow N_2O(g) + 2\ H_2O(g)$; −35.9 kJ

10.95. (a) −29.71 kJ
(b) −1094 kJ/mol
(c) $C_2H_6O(\ell) + 3\ O_2(g) \rightarrow 2\ CO_2(g) + 3\ H_2O(\ell)$
(d) −550.4 kJ

10.97. (a) −93 kJ
(b) 90 kJ
(c) 96 kJ

10.99. 554 kJ

10.101. −610 kJ

10.103. When the ion–dipole forces between the ions and the solvent molecules are stronger than the ion–ion forces between the cations and anions in the salt, the enthalpy of solution will be exothermic. If, however, they are weaker than the ion–ion forces, the enthalpy of solution will be endothermic.

10.105. The energy per gram that a fuel releases on burning

10.107. (a) $CH_4$
(b) $H_2$
10.109. −717 kJ/mol
10.111. (a) ethanol
(b) ethanol
(c) 8.74 g of ethanol, 5.22 g of gasoline
10.113. The first reaction using hydrogen and chlorine gases is exothermic, so we should cool that reaction; the second reaction using sodium chloride and sulfuric acid is endothermic, so we should heat that reaction.
10.115. (a) $2\ NaOH(aq) + H_2SO_4(aq) \rightarrow 2\ H_2O(\ell) + Na_2SO_4(aq)$
(b) No
(c) −57.1 kJ/mol $H_2SO_4$
10.117. 26.0°C
10.119. (a) −1255.5 kJ/mol
(b) 48.22 kJ/g
10.121. Exothermic
10.123. (a) $2\ CH_3OH(g) + 2\ N_2(g) \rightarrow 2\ HCN(g) + 2\ NH_3(g) + O_2(g)$
(b) 307 kJ
10.125. −1400 kJ
10.127. 119.3 kJ
10.129. (a) $N_2H_2(\ell) \rightarrow N_2(g) + 2\ H_2(g)$; $\Delta H^\circ_{rxn} = -\Delta H^\circ_f$.
(b) −622.2 kJ/mol
(c) −19.41 kJ/g
(d) 19,820 kJ/L
(e) Greater than

# Chapter 11

11.1. (a)
11.3. (a) NaCl
(b) $MgCl_2$
(c) $C_6H_{12}O_6$
(d) $K_3PO_4$
11.5. $bp_X \approx 20°C$ and $bp_Y \approx 40°C$; Y has stronger intermolecular forces.
11.7. Line A (red)
11.9. (a) For (i) the water flows from the pure water side to the proteins side; for (ii) water flows from the KCl solution side to the proteins side.
(b) For (i) sodium ions flow from the proteins side to the pure water side; for (ii) the sodium ions flow from the proteins side to the KCl solution side.
(c) For (i) potassium ions flow from the proteins side to the pure water side; for (ii) the potassium ions flow from the KCl solution side to the proteins side.
11.11. (a)
11.13. A semipermeable membrane is a boundary between two solutions through which some molecules may pass but others cannot. Usually, small molecules may pass through, but large molecules are excluded.
11.15. There is a net flow of solvent across a semipermeable membrane from the more dilute solution side to the more concentrated solution side to balance the concentration of solutes on both sides of the membrane.
11.17. Reverse osmosis results in a net transfer of solvent across a semipermeable membrane from a region of higher solute concentration to a region of lower solute concentration. Because reverse osmosis goes against the natural flow of solvent across the membrane, the key component needed is a pump to apply pressure to the more concentrated side of the membrane. Other components needed include a containment system, piping to introduce and remove the solutions, and a tough semipermeable membrane that can withstand the high pressures needed.
11.19. Acetic acid is a weak electrolyte; it dissociates only partially, giving $i > 1$, but not completely so that $i < 2$.
11.21. The larger the van't Hoff factor, the greater the electrical conductivity.
11.23. In using a cation exchanger, 2+ cations in water (such as $Mg^{2+}$ and $Ca^{2+}$) are exchanged for $H^+$ cations on the resin; in using an anion exchanger, the anions in the water are exchanged for $OH^-$ ions on the resin. Because of the release of $H^+$ and $OH^-$ into the water, the ions neutralize each other and produce deionized water.
11.25. (a) From solution A to solution B
(b) From solution B to solution A
(c) From solution A to solution B
11.27. (a) 57.5 atm
(b) 0.682 atm
(c) 52.9 atm
(d) 46.5 atm
11.29. (a) $2.75 \times 10^{-2}\ M$
(b) $1.11 \times 10^{-3}\ M$
(c) $1.00 \times 10^{-2}\ M$
11.31. 2.86 atm
11.33. 94.1 g/mol
11.35. Protein A is 2680 g/mol; protein B is 3290 g/mol.
11.37. When the average kinetic energy of the liquid molecules increases, more of the molecules can escape the liquid phase and enter the gas phase. The presence of more molecules in the gas phase increases the vapor pressure. Vapor pressure is an intensive property.
11.39. (c) < (b) < (a)
11.41. 41.0 kJ/mol
11.43. For isooctane, 0.105 atm, or 80.0 torr
For tetramethylbutane, 0.0483 atm, or 36.7 torr
11.45. The components of crude oil can be separated by fractional distillation, which uses differences in boiling points of the compounds.
11.47. $CH_2Cl_2$
11.49. 18.5 torr
11.51. The vapor pressure of water in the compartment of pure water is greater than that of the seawater because the seawater contains dissolved solids that lower its vapor pressure. Therefore, water evaporates faster from the pure water than from the seawater. The water in the gas phase condenses at the same rate into both containers, but once in the seawater the water evaporates more slowly. Over time, this leads to the transfer of water from the pure water compartment to the seawater compartment because the seawater will always contain dissolved salts even after dilution and, therefore, evaporate more slowly than pure water.
11.53. Molarity is the moles of the solute in 1 L of solution. Molality is the moles of solute in 1 kg of solvent.
11.55. The greater the concentration of dissolved solutes in water (as present in seawater), the lower the freezing point of the water. The presence of nonvolatile solutes shifts the solid–liquid line on the phase diagram to lower temperature.
11.57. $\chi_{water} = 0.70$; $P_{soln} = 17$ torr.
11.59. (a) 0.58 $m$
(b) 0.18 $m$
(c) 1.12 $m$
11.61. (a) 1.34 $m$
(b) 2.61 $m$

(c) 17.4 $m$
(d) 2.05 $m$

11.63. (a) 307 g
(b) 86.8 g
(c) 28.8 g

11.65. $6.5 \times 10^{-5}$ $m$ $NH_3$; $8.7 \times 10^{-6}$ $m$ $NO_2^-$; $2.195 \times 10^{-2}$ $m$ $NO_3^-$

11.67. 3.1°C

11.69. $2.52 \times 10^{-2}$ $m$

11.71. −1.89°C

11.73. (a) 0.5 $m$ $CaCl_2$
(b) 0.0450 $m$ NaI

11.75. (a) 0.5 $m$ $CaCl_2$
(b) 0.400 $m$ $NH_4NO_3$

11.77. Molar mass = 194 g/mol; molecular formula = $C_8H_{10}N_4O_2$

11.79. The reaction of $CO_2$ with water changes dissolved $CO_2$ into carbonic acid. Once this occurs, more $CO_2$ can dissolve in water.

11.81. The bubbles must be dissolved gases ($N_2$, $O_2$, $CO_2$) released from the water on heating.

11.83. $3.7 \times 10^{-2}$ mol/(L · atm)

11.85. (a) $2.7 \times 10^{-3}$ $M$
(b) $2.3 \times 10^{-2}$ $M$

11.87. Yes

11.89. For 0.0935 $m$ $NH_4Cl$, $i = 1.85$.
For 0.0378 $m$ $(NH_4)_2SO_4$, $i = 2.46$.

11.91. (a) 228 g/mol
(b) $C_{15}H_{16}O_2$
(c) $C_{15}H_{16}O_2$

11.93. 4270 g/mol

11.95. (a)

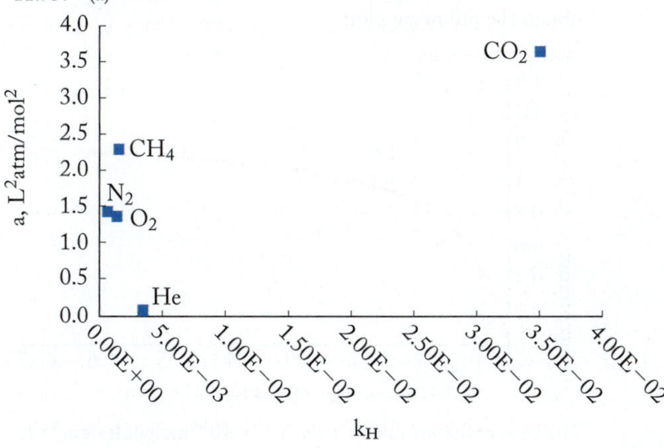


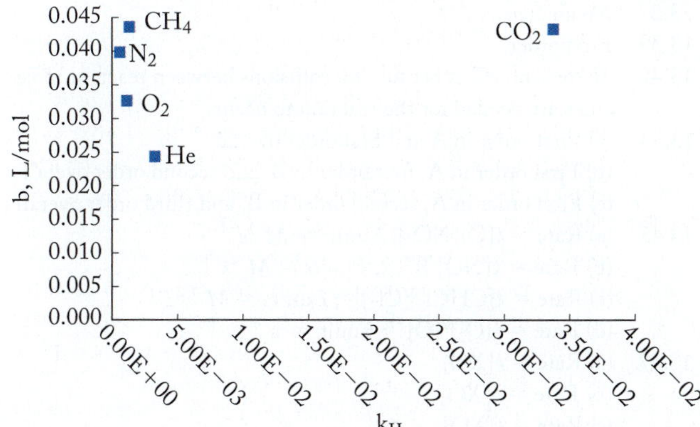


$CO_2$ is the outlier.
(b) When $CO_2$ dissolves in water, it reacts with water to form carbonic acid. This reaction reduced the amount of $CO_2$ dissolved, so more of that gas then dissolves in the water.

## Chapter 12

12.1. (a) 3
(b) 6

12.3. Tire on the right has greater pressure; tire on the right also has greater entropy.

12.5. No, the distribution of gases is not affected by gravity because the particles have enough energy to move throughout the container. This is not an isolated system because we can add heat to the tank to increase the temperature of the gases inside.

12.7. Spontaneous at low temperature

12.9. Condensation, freezing, and deposition

12.11. The sign is reversed.

12.13. Eight microstates; the most likely microstates have sums of +1 and −1.

12.15. $4.47 \times 10^3$

12.17. (b) < (a) < (c)

12.19. (d) $Cr(NO_3)_3$

12.21. (a) $S_8(g)$
(b) $S_2(g)$
(c) $O_3(g)$

12.23. Fullerenes

12.25. (a) $CH_4(g) < CF_4(g) < CCl_4(g)$
(b) $CH_2O(g) < CH_3CHO(g) < CH_3CH_2CHO(g)$
(c) $HF(g) < H_2O(g) < NH_3(g)$

12.27. $CO_2(s) \rightarrow CO_2(g)$; the entropy change is greater than zero.

12.29. $\Delta S_{sys}$ is positive; $\Delta S_{surr}$ is negative.

12.31. (a) and (b)

12.33. Decreases

12.35. $\Delta S_{surr}$ must be more positive than 66.0 J/K.

12.37. $\Delta S^\circ_{rxn}$ will be positive.

12.39. $\Delta S^\circ_{rxn} < 0$ because in precipitation reactions solid products will have a lower molar entropy than the dispersed ions or molecules in the solution from which the precipitate forms.

12.41. (a) 24.9 J/K
(b) −146.4 J/K
(c) −73.2 J/K
(d) −175.8 J/K

12.43. 218.9 J/(mol · K)

12.45. When $\Delta G$ is positive, the reaction is nonspontaneous; when $\Delta G$ is negative, the reaction is spontaneous.

12.47. The magnitude of $\Delta H$ is generally greater than the magnitude of $T\Delta S$; therefore, an exothermic reaction with negative $\Delta H$ is often dominant in determining spontaneity.

12.49. $\Delta S$ is positive; $\Delta H$ is positive; $\Delta G$ is negative

12.51. (a) and (d)

12.53. For NaBr, −18 kJ/mol; for NaI, −30 kJ/mol

12.55. $\Delta G^\circ_{rxn} = 91.4$ kJ

12.57. −35.3 kJ

12.59. −90.3 kJ

12.61. No, if $\Delta S_{rxn}$ were positive, the exothermic reaction would be spontaneous at all temperatures.

12.63. 981.3 K, or 708.2°C

12.65. $\Delta H^\circ_{rxn} = 51.5$ kJ; $\Delta S^\circ_{rxn} = 123.1$ J/K; 418 K, or 145°C

12.67. (a) Spontaneous only at low temperatures
(b) Spontaneous only at low temperatures
(c) Spontaneous at all temperatures

12.69. (a) −29.2 kJ
(b) 29.5 kJ; not spontaneous

12.71. The spontaneous reaction must have a more negative or less positive free energy than the nonspontaneous reaction.

12.73. ATP has the important function of energy storage and transfer. The cell must make this important molecule, and so the overall process must be spontaneous. Even if some of the glycolysis steps are nonspontaneous, at least some glycolysis steps must be spontaneous to produce ATP.

12.75. (a) 142.2 kJ
(b) −28.6 kJ
(c) $CH_4(g) + 2\ H_2O(g) \rightarrow CO_2(g) + 4\ H_2(g)$
(d) 113.6 kJ; nonspontaneous

12.77. −51.5 kJ

12.79. 83.5 J/mol · K

12.81. $T = 618$ K, or 344.8°C

12.83. −630.4 kJ; spontaneous

12.85. 56.5 kJ

12.87. (a) and (c) are spontaneous because the free energy of the products is lower than the free energy of the reactants.

12.89. 9.58 J/(mol · K)

12.91. (a) $\Delta H$ is negative; $\Delta S$ is positive
(b) The reverse reaction would have $+\Delta H$ and $-\Delta S$ and therefore would never be spontaneous.

# Chapter 13

13.1. $[N_2O]$ is represented by the green curve and $[O_2]$ is represented by the red curve.

13.3. (b)

13.5. (a) 1
(b) 5
(c) 2
(d) 4
(e) 3

13.7. (b)

13.9. (c)

13.11. (b)

13.13. (b) Greater than

13.15. The rate of a reaction is increased with increasing temperature because (1) the particles collide more often and have increased opportunities to react and (2) the particles collide with more kinetic energy and so are more likely to exceed the activation energy for the reaction.

13.17. CO has the highest root-mean-square speed because it has the lowest molar mass; all of these have the same average kinetic energy because kinetic energy is directly proportional to temperature and does not depend on molar mass.

13.19. NO is a component of car and truck emissions that peak during the morning commute. NO is oxidized by atmospheric $O_2$, forming $NO_2$, whose concentration peaks later in the morning. Photodecomposition of $NO_2$ produces O atoms that combine with $O_2$ molecules, forming $O_3$, whose concentration peaks in the afternoon.

13.21. In the evening the sunlight (and UV radiation) is less intense, so the photochemical breakdown of $NO_2$ does not occur to as great an extent as after the morning rush hour.

13.23. The average rate is the rate averaged over a defined time interval, whereas the instantaneous rate is the rate at a specific moment.

13.25. (a) The rates are the same.
(b) The rate of formation of $HNO_2$ is two-thirds the rate of consumption of $O_2$.
(c) The rate of consumption of $NH_3$ is two-thirds the rate of consumption of $O_2$.

13.27. (a) $2.9 \times 10^{-6}\ M/s$
(b) $-5.7 \times 10^{-6}\ M/s$

13.29. (a) Rate $= \frac{\Delta[CO_2]}{\Delta t} = -\frac{2}{3}\frac{\Delta[CO]}{\Delta t}$
(b) Rate $= \frac{\Delta[COS]}{\Delta t} = -\frac{\Delta[SO_2]}{\Delta t}$
(c) Rate $= \frac{\Delta[CO]}{\Delta t} = 3\frac{\Delta[SO_2]}{\Delta t}$

13.31. (a) $1.2 \times 10^{7}\ M/s$
(b) $2.9 \times 10^{4}\ M/s$

13.33. Between 0 and 100 μs: $1.4 \times 10^{-5}\ M/\mu s$
Between 200 and 300 μs: $5.5 \times 10^{-6}\ M/\mu s$

13.35. We obtain the following plot for the change in concentration of ClO versus time:

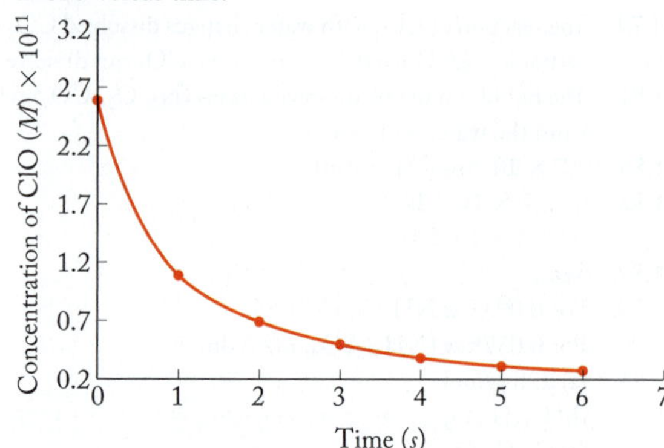


The instantaneous rate at 1 s is $-6.3 \times 10^{10}$ molecules/cm$^3$ · s. For the change in concentration of $Cl_2O_2$ versus time, we obtain the following plot:

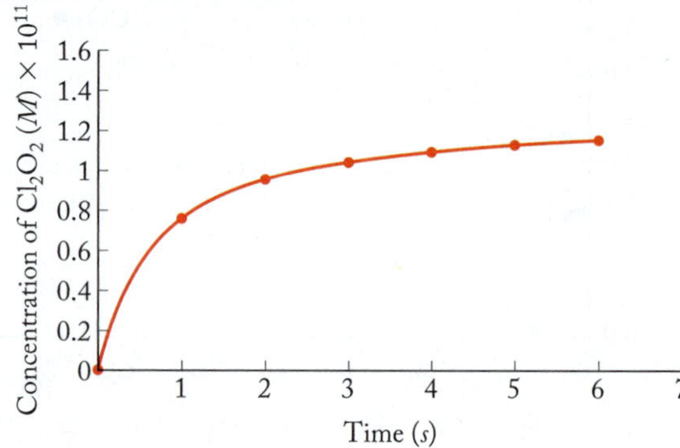


The instantaneous rate at 1 s is $3.1 \times 10^{10}$ molecules/cm$^3$ · s.

13.37. No change

13.39. First order

13.41. There is no effect because no collisions between reactant molecules are needed for the reaction to occur.

13.43. (a) First order in A and first order overall
(b) First order in A, first order in B, and second order overall
(c) First order in A, second order in B, and third order overall

13.45. (a) Rate $= k[O][NO_2]$; $k$ units $= M^{-1}s^{-1}$
(b) Rate $= k[NO]^2[Cl_2]$; $k$ units $= M^{-2}s^{-1}$
(c) Rate $= k[CHCl_3][Cl_2]^{1/2}$; $k$ units $= M^{-1/2}s^{-1}$
(d) Rate $= k[O_3]^2[O]^{-1}$; $k$ units $= s^{-1}$

13.47. (a) Rate $= k[XO]$
(b) Rate $= k[XO]^2$
(c) Rate $= k[XO]$
(d) Rate $= k[XO]^0 = k$

13.49. We need to determine the change in the rate when only [NO] or [ClO] is changed.
13.51. (a) Rate = $k[NO_2][O_3]$
(b) $4.9 \times 10^{-11}$ *M*/s
(c) $4.9 \times 10^{-11}$ *M*/s
(d) The rate doubles.
13.53. (c)
13.55. Rate = $k[NO][NO_2]$
13.57. Rate = $k[ClO_2][OH^-]$; $k = 14$ *M*/s
13.59. Rate = $k[NO]^2[H_2]$; $k = 6.32$ *M*/s
13.61. 0.051 *M* $N_2O_5$ remains, 90% has reacted
13.63. (a) Rate = $k[N_2O]$
(b) 4 half-lives
13.65. 0.32/μ*M* · min
13.67. Rate = $k[NH_3]$; $k = 0.0030$/s
13.69. (a) Rate = $k[^{32}P]$
(b) 0.0485/day
(c) 14.3 days
(d) 95.0 days
13.71. (a) $k = 5.40 \times 10^{-12}$ cm³/molecules · s
(b) $t_{1/2} = 0.712$ s
13.73. Rate = $k[C_{12}H_{22}O_{11}][H_2O] = k'[C_{12}H_{22}O_{11}]$;
$k' = 5.19 \times 10^{-5}$/s
13.75. The larger the activation energy, the slower the rate of the reaction.
13.77. When the energy of the products is lower than the energy of the reactants
13.79. No.
13.81. (a) $E_a = 314$ kJ/mol
(b) $A = 5.03 \times 10^{10}$
(c) $k = 1.06 \times 10^{-44}/M^{1/2} \cdot s$
13.83. (a) $E_a = 39.1$ kJ/mol; $A = 1.27 \times 10^{12}/M \cdot s$
(b) $k = 1.54 \times 10^4/M \cdot s$
13.85. No.
13.87. Pseudo–first-order kinetics occurs when one of the reactants is in concentration high enough that its concentration does not change appreciably during the reaction.

13.89.

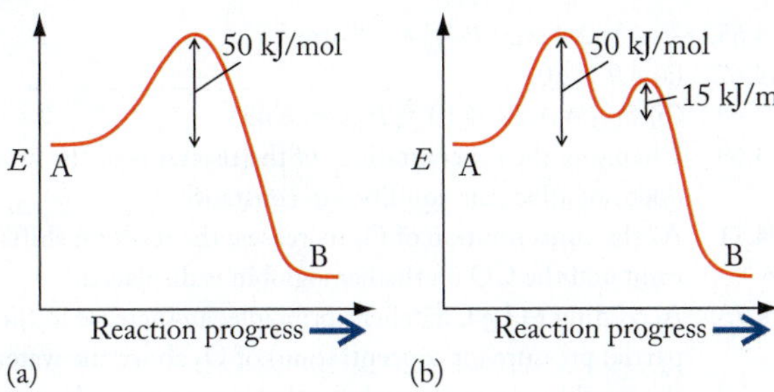

(a) (b)

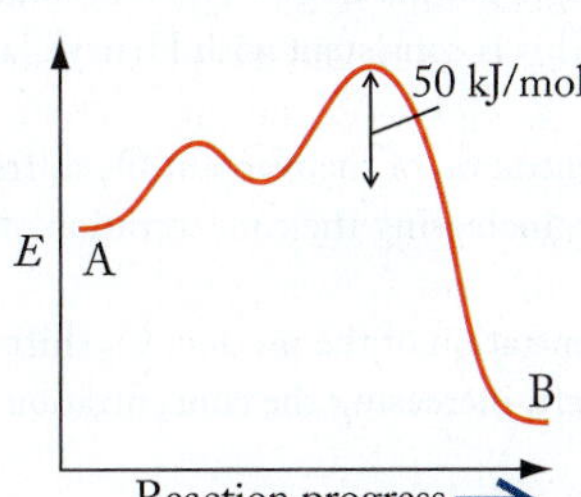

(c)

13.91. (a) Rate = $k[SO_2Cl_2]$; unimolecular
(b) Rate = $k[NO_2][CO]$; bimolecular
(c) Rate = $k[NO_2]^2$; bimolecular
13.93. (a) $2\ H_2O_2(aq) \rightarrow 2\ H_2O(aq) + O_2(g)$
(b) Rate = $k[H_2O_2][I^-]$
(c) $I^-$
(d) $IO^-$
13.95. Yes; the second step is rate-determining.
13.97. The first step
13.99. Photochemical decomposition: (a)
Thermal decomposition: (b) or (c)
13.101. Yes.
13.103. Yes.
13.105. The concentration of a homogeneous catalyst may not appear in the rate law if the catalyst is not involved in the rate-limiting step. However, if the catalyst is involved in the slowest step of the mechanism, it can appear in the rate law.
13.107. NO is the catalyst.
13.109. The reaction of $O_3$ with Cl has the larger rate constant.
13.111. When the concentration of a reactant ($O_2$ for the combustion reaction) increases, the rate of combustion increases.
13.113. The bodily reactions that use $O_2$ are slower at colder temperatures.
13.115. Yes, we could use other times, not just $t = 0$, as long as the rate of the reverse reaction is still much slower than that of the forward reaction.
13.117. In this plot $1/[X] - 1/[X]_0$ divided by $t - t_0$ is the slope of the line, which corresponds to $k$, the reaction rate constant.
13.119. The rate of consumption of $O_3$ is the same as the rate of formation of $N_2O_5$ and $O_2$ and one-half the rate of consumption of $NO_2$.
13.121. (a) Yes.
(b) $E_a = 62.5$ kJ/mol
(c) Rate = $1.2 \times 10^{-12}$ *M*/s
(d) At 10°C (283 K), $k = 21/M \cdot s$; at 35°C (308 K), $k = 1.8 \times 10^2/M \cdot s$
13.123. (a) Rate = $k[Na(H_2O)_6{}^+]$
(b)

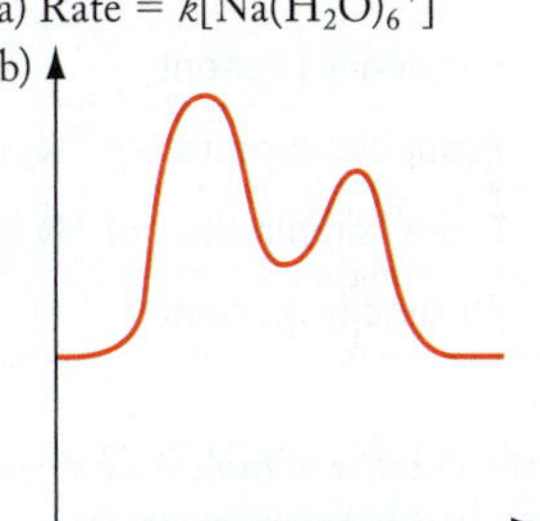

13.125. (a) Rate = $k[NO][ONOO^-]$; $k = 1.30 \times 10^{-3}/M \cdot s$
(b)
$$[\overset{0}{\ddot{O}}=\overset{0}{\ddot{N}}-\overset{0}{\ddot{O}}-\overset{-1}{\ddot{O}}:]^- \leftrightarrow [:\overset{-1}{\ddot{O}}-\overset{0}{N}=\overset{+1}{\ddot{O}}-\overset{-1}{\ddot{O}}:]^- \leftrightarrow [:\overset{-1}{\ddot{O}}-\overset{-1}{\ddot{N}}-\overset{+1}{\ddot{O}}=\overset{0}{\ddot{O}}]^-$$
The first resonance form is preferred.
(c) −55 kJ
13.127. (a) Second order
(b) No.

# Chapter 14

14.1. Reaction C $\rightleftharpoons$ D has the larger $k_f$, the smaller $k_r$, and the larger $K_c$.
14.3. $K_c = K_p = 2$
14.5. As temperature increases for the reaction, fewer products form, which reduces $K$, and the reaction is exothermic.

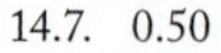
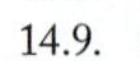

14.7. 0.50

14.9. (a)

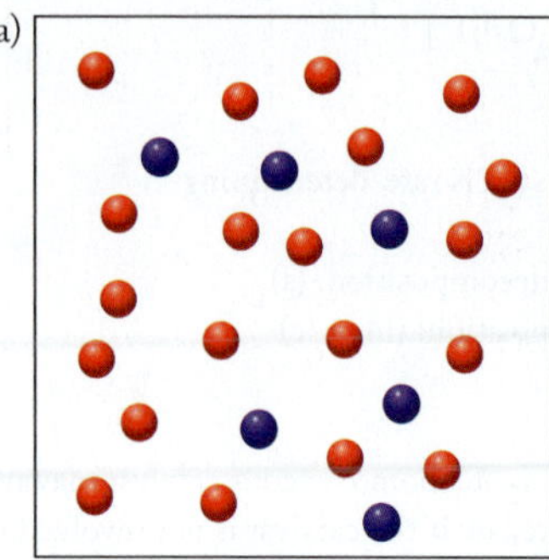

(b)

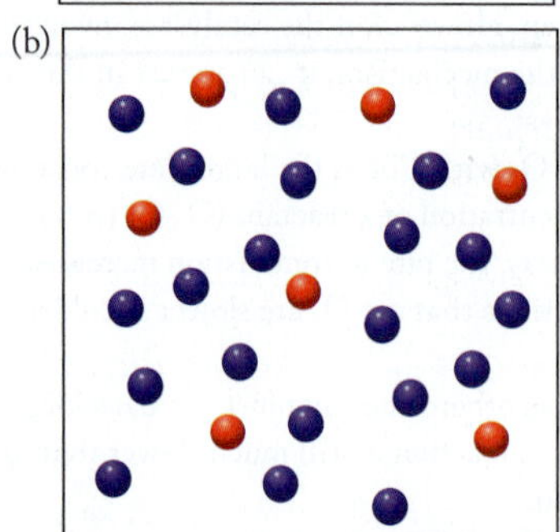

14.11. Yes, a reaction that is not reversible has a very, very large equilibrium constant.

14.13. Greater than 1

14.15.

| Molar Mass | Compound | How Present |
|---|---|---|
| 28 | $^{14}N_2$ | Originally present |
| 29 | $^{15}N^{14}N$ | From decomposition of $^{15}N^{14}NO$ |
| 30 | $^{15}N_2$ | From decomposition of $^{15}N_2O$ |
| 32 | $O_2$ | Originally present |
| 44 | $^{14}N_2O$ | From combination of $^{14}N_2$ and $O_2$ |
| 45 | $^{15}N^{14}NO$ | From combination of $^{15}N^{14}N$ and $O_2$ |
| 46 | $^{15}N_2O$ | Originally present |

14.17. 0.333

14.19. When $\Delta n = 0$; when the number of moles of gaseous products equals the number of moles of gaseous reactants.

14.21. (a) $K_c = \dfrac{[N_2O_4]}{[N_2][O_2]^2}$ and $K_p = \dfrac{(P_{N_2O_4})}{(P_{N_2})(P_{O_2})^2}$

(b) $K_c = \dfrac{[NO_2][N_2O]}{[NO]^3}$ and $K_p = \dfrac{(P_{NO_2}P_{N_2O})}{(P_{NO})^3}$

(c) $K_c = \dfrac{[N_2]^2[O_2]}{[N_2O]^2}$ and $K_p = \dfrac{(P_{N_2})^2(P_{O_2})}{(P_{N_2O})^2}$

14.23. (a) $K_c = \dfrac{[SO_3]^2}{[SO_2]^2[O_2]}$; $K_p = \dfrac{(P_{SO_3})^2}{(P_{SO_2})^2(P_{O_2})}$

(b) $K_c = \dfrac{[CO]}{[CO_2]}$; $K_p = \dfrac{(P_{CO})}{(P_{CO_2})}$

(c) $K_c = \dfrac{[CO_2][H_2]}{[H_2O][CO]}$; $K_p = \dfrac{(P_{CO_2})(P_{H_2})}{(P_{H_2O})(P_{CO})}$

14.25. 0.068

14.27. 0.50

14.29. 1.5

14.31. 73.5

14.33. 822

14.35. 0.10

14.37. When the coefficients of a reaction are scaled up or down, the new value of the equilibrium constant is the first $K$ raised to the power of the scaling constant.

14.39. 11.0

14.41. $K_{c,\text{forward}} = \dfrac{[NO_2]^2}{[NO][NO_3]}$; $K_{c,\text{reverse}} = \dfrac{[NO][NO_3]}{[NO_2]^2}$;

$K_{c,\text{reverse}} = \dfrac{1}{K_{c,\text{forward}}}$

14.43. $K_{c(\text{rxn2})} = [K_{p(\text{rxn1})}/(RT)^{\Delta n}]^2$

14.45. (a) 0.0104

(b) 9.34

(c) 96.6

14.47. 7.4

14.49. The reaction quotient is the ratio of the concentrations of the products raised to their stoichiometric coefficients to the concentrations of reactants raised to their stoichiometric coefficients. The reaction quotient has the same form as the equilibrium constant expression, but the reaction concentrations (or partial pressures) are not necessarily at their equilibrium values.

14.51. The system is at equilibrium.

14.53. No, $Q < K$, so the reaction proceeds to the right to reach equilibrium.

14.55. Mixtures b and c are at equilibrium.

14.57. $Q > K$, so the reaction will proceed to the left.

14.59. (a)

14.61. The concentrations of pure solids ($CaCO_3$ and CaO) do not change during the reaction and so they do not appear in the equilibrium constant expression.

14.63. (a) $K_c = [Cu^{2+}][S^{2-}]$

(b) $K_c = \dfrac{[H_3O^+][Cl^-]}{[HCl]}$

(c) $K_c = [Al^{3+}]^2[OH^-]^6$

(d) $K_c = \dfrac{[NH_4^+][OH^-]}{[NH_3]}$

14.65. $P_{CO} = 2.4$ atm, $P_{CO_2} = 3.8$ atm

14.67. (a) $4.9 \times 10^2$

(b) $P_{CO} = 1.32 \times 10^{-2}$; $P_{CO_2} = 6.49$

14.69. Changing the concentrations of the reactants or the products does not affect the equilibrium constant.

14.71. As the concentration of $O_2$ increases, the reaction shifts to the right and the CO on the hemoglobin is displaced.

14.73. According to Le Châtelier's principle, an increase in the partial pressure (or concentration) of $O_2$ above the water shifts the equilibrium to the right so that more oxygen becomes dissolved in the water. This is consistent with Henry's law.

14.75. (b) and (d)

14.77. (a) Increasing the concentration of the reactant $O_3$ shifts the equilibrium to the right, increasing the concentration of the product $O_2$.

(b) Increasing the concentration of the product $O_2$ shifts the equilibrium to the left, increasing the concentration of the reactant $O_3$.

(c) Decreasing the volume of the reaction to 1/10 its original volume shifts the equilibrium to the left, increasing the concentration of the reactant $O_3$.

14.79. The equilibrium shifts to the left.

14.81. (b)

14.83. (a) $P_{PCl_5} = 0.024$ atm; $P_{PCl_3} = 1.036$ atm; $P_{Cl_2} = 0.536$ atm
(b) The partial pressure of $PCl_5$ increases and the partial pressure of $PCl_3$ decreases.

14.85. $[H_2O] = [Cl_2O] = 3.76 \times 10^{-3}\ M$; $[HOCl] = 1.13 \times 10^{-3}\ M$

14.87. $9 \times 10^5$

14.89. (a) $P_{NO} = 0.272$ atm; $P_{NO_2} = 7.98 \times 10^{-3}$ atm
(b) $P_T = 0.416$ atm

14.91. $P_{N_2} = 0.75$ atm; $P_{O_2} = 0.17$ atm; $P_{NO} = 0.080$ atm

14.93. 5.75 $M$

14.95. $P_{CO} = P_{Cl_2} = 0.258$ atm, $P_{COCl_2} = 0.00680$ atm

14.97. $[CO] = [H_2O] = 0.031\ M$, $[CO_2] = [H_2] = 0.069\ M$

14.99. $[NO] = 0.027\ M$, $[NO_2] = 0.277\ M$; $[N_2O_3] = 0.098\ M$

14.101. Yes.

14.103. To the right.

14.105. (c)

14.107. $7.24 \times 10^{-19}$

14.109. (a) $-197.8$ kJ
(b) $7.4 \times 10^{24}$
(c) $-142$ kJ/mol from $K_P$; $-142.0$ kJ/mol from Appendix 4

14.111. Exothermic

14.113. Exothermic

14.115. $1.3 \times 10^{-31}$

14.117. $-115$ kJ/mol

14.119. The reaction is endothermic:
At 1500 K, $K_p = 5.5 \times 10^{-11}$.
At 2500 K, $K_p = 4.0 \times 10^{-3}$.
At 3000 K, $K_p = 0.40$.
This reaction does not favor products even at very high temperature, so this is not a viable source of CO and is not a remedy to decrease $CO_2$ as a contributor to global warming. Also, the process produces poisonous CO gas.

14.121. $9 \times 10^{-22}\ M$

14.123. $P_{SO_2} = 9.2 \times 10^{-74}$ atm

# Chapter 15

15.1. Red line

15.3. The bar with the lowest percent ionization

15.5. Piperidine is the stronger base because the presence of oxygen (high electronegativity) on the morpholine withdraws electrons, making the N atom less basic.

15.7. a. Basic
b. The amine functional group

15.9. The ionizable protons are those bonded to the oxygen atoms.

15.11. HF acts as the acid; $H_2O$ acts as the base.

15.13. $H_2O$ acts as the acid; $NH_3$ acts as the base.

15.15. (a) HCl is the acid; NaOH is the base.
(b) HCl is the acid; $MgCO_3$ is the base.
(c) $H_2SO_4$ is the acid; $NH_3$ is the base.

15.17. $NO_2^-$; $OCl^-$; $H_2PO_4^-$; $NH_3$

15.19. $H_2SO_4$ is the conjugate acid; $SO_4^{2-}$ is the conjugate base.

15.21. Sulfur is more electronegative than selenium. The higher electronegativity of the sulfur atom stabilizes the anion $HSO_4^-$ more than the anion $HSeO_4^-$.

15.23. (a) $H_2SO_3$
(b) $H_2SeO_4$

15.25. 0.65 $M$

15.27. 0.0410 $M$

15.29. Because the pH function is a $-\log$ function, as $[H^+]$ increases, the value of $-\log[H^+]$ decreases.

15.31. When $[H_3O^+]$ is greater than 1.0 $M$

15.33.

```
[   H  H      ]-     [   H  H  H     ]+
[   |  |      ]      [   |  |  |     ]
[ H—C—C—Ö:    ]      [ H—C—C—O—H     ]
[   |  |  ..  ]      [   |  |  ..    ]
[   H  H      ]      [   H  H        ]
```

15.35. (a) pH = 2.28; pOH = 11.72; acidic
(b) pH = 8.42; pOH = 5.58; basic
(c) pH = 5.14; pOH = 8.86; acidic
(d) pH = 0.00; pOH = 14.00; basic

15.37. (a) $1.2 \times 10^{-11}\ M$
(b) $7.6 \times 10^{-11}\ M$
(c) $2.2 \times 10^{-12}\ M$
(d) $3.4 \times 10^{-10}\ M$

15.39. (a) pH = 0.810; pOH = 13.190
(b) pH = 2.301; pOH = 11.699
(c) pH = 2.380; pOH = 11.620
(d) pH = 2.204; pOH = 11.796

15.41. 7.003

15.43. (a) In order of decreasing $[H_3O]^+$: HCl > $HNO_2$ > $CH_3COOH$ > HClO
(b) In order of increasing acid strength: HClO < $CH_3COOH$ < $HNO_2$ < HCl

15.45. $HNO_3$ dissociates completely into $H^+$ and $NO_3^-$ ions in water, each in 1.0 $M$ concentration for a total ion concentration of 2.0 $M$. $HNO_2$, however, only partially dissociates in water and so produces fewer ions in solution. $HNO_3$, therefore, with more dissolved ions in solution, is a better conductor of electricity.

15.47. $K_a = \dfrac{[H_3O^+][F^-]}{[HF]}$

15.49. (a) Water
(b) Water

15.51 $H_2O$ is the Brønsted–Lowry acid, $CH_3NH_2$ is the Brønsted–Lowry base.

15.53. $8.91 \times 10^{-4}$

15.55. 3.26%; $K_a = 1.37 \times 10^{-4}$

15.57. 0.0126%

15.59. 2.3 times

15.61. (a) Weaker
(b) 10.86
(c) $1.00 \times 10^{-4}\ M$

15.63. (a) 9.73
(b) 9.30

15.65. With each successive ionization, it becomes more difficult to remove $H^+$ from a species that is more negatively charged.

15.67. Ge is less electronegative than C.

15.69. 0.12

15.71. 2.80

15.73. 10.27

15.75. Because $F^-$ reacts with water to produce the weak acid HF and $OH^-$, making the solution basic, but $Cl^-$ does not react with water and the solution is neutral.

15.77. Ammonium nitrate

15.79. The citric acid in the lemon juice neutralizes the volatile trimethylamine to make a nonvolatile dissolved salt.

15.81. 3.32

15.83. 6.02

15.85. 10.58

15.87. (a) $CH_3COOH$ and $HClO_4$
(b) $Ca(OH)_2$
(c) $CH_3COOH$ and $HClO_4$
(d) $Ca(OH)_2$ and $CH_3NH_2$

15.89. Yes, all Arrhenius bases are Brønsted–Lowry bases because Arrhenius bases are all proton acceptors. No, not all Brønsted–Lowry bases are Arrhenius bases because not all (such as $CH_3NH_2$) release $OH^-$ into the aqueous solution.

15.91. Reaction of sulfur with oxygen generates $SO_x$: $4\ S(s) + 5\ O_2(g) \rightarrow 2\ SO_2(g) + 2\ SO_3(g)$
Reaction of $SO_x$ with water produces a weak acid ($H_2SO_3$) and a strong acid ($H_2SO_4$): $SO_2(g) + H_2O(\ell) \rightarrow H_2SO_3(aq)$; $SO_3(g) + H_2O(\ell) \rightarrow H_2SO_4(aq)$
Reaction of $H_2SO_4$ with $CaCO_3$ forms $H_2O$, $CO_2$, and soluble $CaSO_4$: $H_2SO_4(aq) + CaCO_3(s) \rightarrow H_2O(\ell) + CO_2(g) + CaSO_4(aq)$

15.93. $1.4 \times 10^9$ L

15.95. (a) Basic because Prozac has an amine group, which will pick up a proton from water to form ammonium cations and $OH^-$
(b) Amine group
(c) Because it is charged and water molecules form stronger ion–dipole forces around the molecule than the dipole–induced dipole forces between the neutral molecule and water

15.97. (a)
F
C
F C C F
C C
F F
(b) Because the presence of five very electronegative F atoms on the carbon ring stabilizes the anion formed when the proton is lost

15.99. (a) No
(b) Acid = $H_2SO_4$; base = $HNO_3$; conjugate acid = $H_3O^+$; conjugate base = $HSO_4^-$

15.101. Most acidic = 1.0 *M* $H_2SO_3$; most basic = 0.30 *M* $Na_2SO_3$

15.103. (a) The N in the 5-member ring of nicotine is a stronger base because it is $sp^3$-hybridized and therefore has less "*s* character" than the *sp*-hybridized N in the 6-member ring. More *s* character means that the electrons of the basic atom are held more tightly and therefore are less able to be donated in an acid–base reaction.
(b) 4.50

15.105. 8.45

15.107. (a) $ClCH_2COO^-$; $K_b = 7.14 \times 10^{-12}$
(b) $NH_3$; $K_b = 1.76 \times 10^{-5}$
(c) $CN^-$; $K_b = 1.61 \times 10^{-5}$
(d) $CH_3CH_2O^-$; $K_b = 76.9$
$CH_3CH_2O^-$ is the strongest conjugate base

15.109. (a) $CO_3^{2-}(aq) + H^+(aq) + HSO_4^{2-}(aq) \rightarrow CO_2(g) + H_2O(\ell) + SO_4^{2-}(aq)$; acid = $H_2SO_4$; base = $Na_2SO_3$
(b) $CH_3COOH(aq) + OH^-(aq) \rightarrow CH_3COO^-(aq) + H_2O(\ell)$; acid = $CH_3COOH$, base = $Mg(OH)_2$
(c) $CaO(s) + H_2O(\ell) \rightarrow Ca^{2+}(aq) + 2\ OH^-(aq)$; acid = $H_2O$; base = CaO
(d) $SH^-(aq) + H^+(aq) \rightarrow H_2S(g)$; acid = $HNO_3$; base = NaSH

## Chapter 16

16.1. The blue titration curve represents the titration of a 1 *M* solution of strong acid. The red titration curve represents the titration of a 1 *M* solution of weak acid.

16.3. The indicator with a $pK_a$ of 9.0

16.5. The red titration curve represents the titration of $Na_2CO_3$; the blue titration curve represents the titration of $NaHCO_3$.

16.7. Ammonium chloride is dissolved in the yellow solution, sodium acetate is dissolved in the blue solution, and ammonium acetate is dissolved in the middle solution.

16.9. (a) Buffer
(b) The weak acid molecules of HCOOH will react with NaOH to produce more $HCOO^-$ and $H_2O$.

16.11. The presence of both the base and the conjugate acid in a buffer means that the system can neutralize small additions of either acid or base with only slight changes in pH. The weak acid alone has little capacity to control pH against additions of acid.

16.13. Citric acid/citrate ions, hydrofluoric acid/fluoride ions, or iodoacetic acid/iodoacetate ions.

16.15. Buffer capacity is the quantity of acid or base that a buffer can neutralize while keeping its pH within a desired range.

16.17. Acetic acid/sodium acetate > formic acid/sodium formate > hydrofluoric acid/sodium fluoride.

16.19. No significant change in pH should occur.

16.21. 2.55

16.23. 2.46

16.25. 1:0.32

16.27. 92.6 g of bromoacetic acid; 54.5 g of sodium bromoacetate

16.29. 7.69 g of dimethylammonium chloride; 2.51 g of dimethylamine

16.31. 9.25

16.33. 45.3 mL

16.35. (a) 3.50
(b) 3.42

16.37. Yes, because at the equivalence point only ions that do not react with water are formed; the pH will always be neutral.

16.39. (1) It changes color near the pH of the first equivalence point.
(2) It is colorless at low pH, which means that it will not obscure the color change of a second indicator added after the first equivalence point to detect the second equivalence point.

16.41. 4.44

16.43. After 10.0 mL of $OH^-$ has been added, pH = 4.754; after 20.0 mL of $OH^-$ has been added, pH = 8.750; after 30.0 mL of $OH^-$ has been added, pH = 12.356.

16.45.

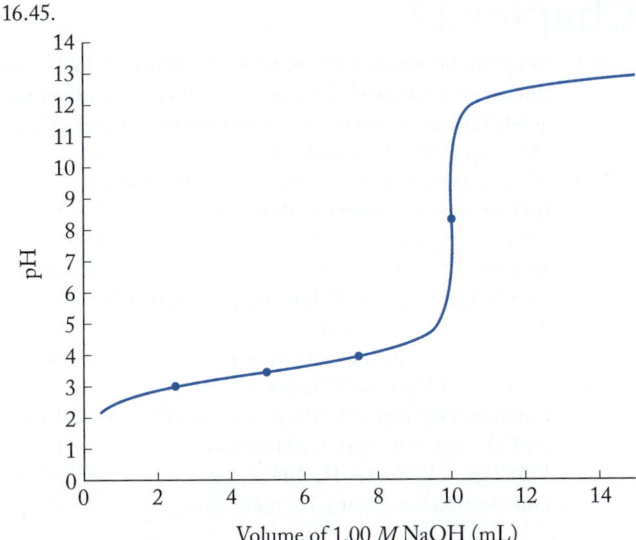


After 2.50 mL, pH = 2.92; after 5.00 mL, pH = 3.40; after 7.50 mL, pH = 3.87; after 10.00 mL, pH = 8.31.

16.47. 25.0 mL

16.49. (a) 2.949 $M$
(b) No effect

16.51. (a) pH < 7
(b) pH > 7
(c) pH = 7

16.53. Two equivalence points; phenol red or phenolphthalein for first and alizarin yellow R for second.

16.55. No, because a substance can be a Lewis base if it can donate an electron pair but does not accept a proton; $Cl^-$ is an example.

16.57. Yes, because Brønsted–Lowry acids must be able to accept pairs of electrons so that the protons they donate can form bonds to Brønsted–Lowry bases.

16.59.

H—Ö: (H) + H—Ö: (H) → [H—Ö—H (H)]$^+$ + [:Ö: (H)]$^-$

Lewis base   Lewis acid

16.61.

O=S(=O)—O—H + :Ö(H)—H → HO—S(=O)—Ö—H

Lewis acid   Lewis base

16.63.

H—Ö—B(OH)(OH) + :Ö(H)—H → [H—Ö—B(OH)(OH)—Ö—H]$^-$ + $H^+$

Lewis acid   Lewis base

16.65. Water molecules

16.67. Because $Ag^+$ ions form a complex with ammonia, $Ag(NH_3)_2^+$, which is soluble.

16.69. $2.64 \times 10^{-9}$ $M$

16.71. $9.35 \times 10^{-15}$ $M$

16.73. b and d

16.75. $FeCl_2$

16.77. In basic solution: $Cr(OH)_3(s) + OH^-(aq) \rightleftharpoons Cr(OH)_4^-(aq)$
In acidic solution: $Cr(OH)_3(s) + 3\,H^+(aq) \rightleftharpoons Cr^{3+}(aq) + 3\,H_2O(\ell)$

16.79. $Al(OH)_3$ reacts with $OH^-$ in solution to form soluble $Al(OH)_4^-$. The other ions do not form this type of soluble complex ion.

16.81. 2.80

16.83. 1.80

16.85.

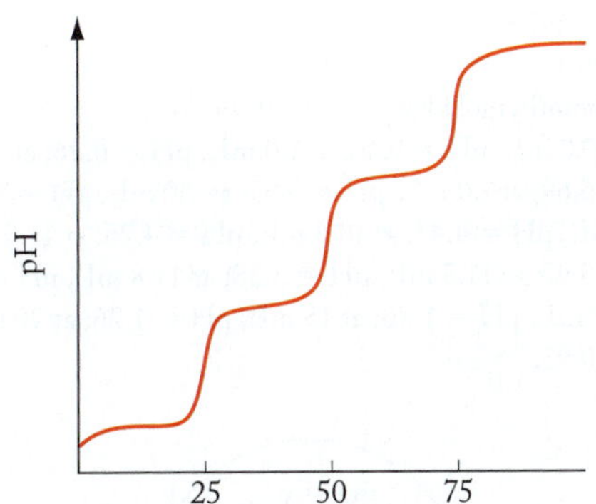


16.87. Molar solubility is the quantity (moles) of substance that dissolves in a liter of solution. The solubility product is the equilibrium constant for the dissolution of a substance.

16.89. $Sr^{2+}$

16.91. Endothermic

16.93. Acidic substances react with the $OH^-$ released on dissolution of hydroxyapatite. The equilibrium is shifted to the right, dissolving more hydroxyapatite.

16.95. $1.08 \times 10^{-10}$

16.97. $[Cu^+] = [Cl^-] = 1.01 \times 10^{-3}$ $M$

16.99. $9.96 \times 10^{-6}$ g/mL

16.101. 10.091

16.103. (d)

16.105. Probably, because the product of the $[Sr^{2+}]$ and $[SO_4^{2-}]$ values exceeds the $K_{sp}$ of $SrSO_4$.

16.107. Yes

16.109. (a) $SO_4^{2-}$
(b) $2.7 \times 10^{-7}$ $M$

16.111. Later additions of $HCO_3^-$ react with water to form bicarbonate's conjugate acid ($H_2CO_3$) and its conjugate base ($CO_3^{2-}$) in the same proportions as the first addition, so pH does not change.

16.113. $6.6 \times 10^{-16}$

16.115. (a) The acidity of the oceans increases (lower pH) because of increased ionization of the weak acid carbonic acid.

$$CO_2(g) \rightleftharpoons CO_2(aq)$$
$$CO_2(aq) + H_2O(\ell) \rightleftharpoons H_2CO_3(aq)$$
$$H_2CO_3(aq) \rightleftharpoons H^+(aq) + HCO_3^-(aq)$$

(b) The acidity of the ocean will be about six times greater.
(c) Because oyster shells are composed of $CaCO_3$, they are likely to dissolve to a greater extent as the pH decreases.

16.117. (e)

16.119. (a) Acetic acid
(b) Hypobromous acid

16.121. (a) $1.6 \times 10^{-5}$
(b) $5.0 \times 10^{-5}$
(c) $1.3 \times 10^{-11}$

16.123. Formic acid, pH = 8.22; boric acid, pH = 10.98. These are different enough that different indicators should be used.

16.125. (a) 15.00 mL more
(b) 1.77
(c) 7.21
(d) $Na^+$ and $HSO_3^-$
(e) $Na^+$ and $SO_3^{2-}$

16.127. (a) 2.95
(b) Bromothymol blue
(c) At 0.0 mL, pH = 9.40; at 1.0 mL, pH = 6.26; at 3.0 mL, pH = 5.69; at 7.0 mL, pH = 5.05; at 9.0 mL, pH = 5.05; at 10.0 mL, pH = 4.43; at 10.5 mL, pH = 4.25; at 11.0 mL, pH = 3.97; at 11.5 mL, pH = 3.28; at 11.8 mL, pH = 2.40; at 12.5 mL, pH = 1.80; at 15 mL, pH = 1.26; at 20 mL, pH = 0.93.
(d)

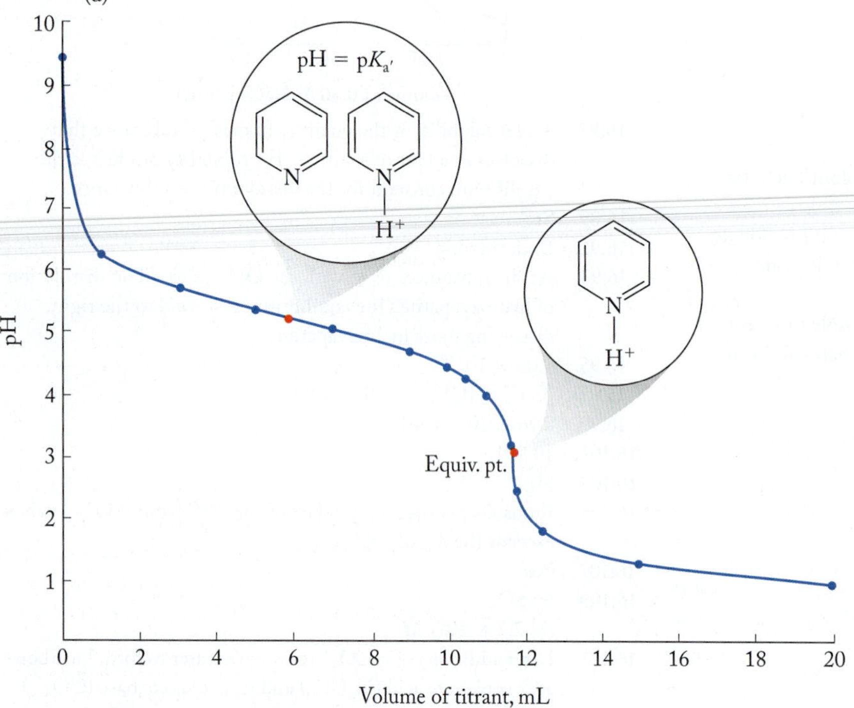


16.129. They have the same molar solubility but not the same solubility in g/L.

16.131. Calcium hydroxide

16.133. For trypsin, carbonic acid/sodium bicarbonate; for pepsin, oxalic acid/sodium oxalate.

16.135. Lowering $CO_2$ will lower the amount of $CO_3^{2-}$ in the blood, which will lower the $CaCO_3$ in shells.

# Chapter 17

17.1. With careful layering, the metal electrode in each half-cell is immersed in a solution of the metal's cations. Ions can migrate across the interface between the two solutions, which allows the cell reaction to proceed without a porous bridge.

17.3. (a) Ag is the cathode; Pt in the SHE is the anode.
(b) Electrons flow from the SHE to Ag.
(c) Ag

17.5. Blue line

17.7. (a) $2\,H^+(aq) + 2\,e^- \rightarrow H_2(g) \quad E^\circ_{cathode} = 0.000\ V$
$2\,H_2O(\ell) \rightarrow O_2(g) + 4\,H^+(aq) + 4\,e^- \quad E^\circ_{cathode} = 1.229\ V$
(b) To increase the conductivity of the solution

17.9. (a) Anode: $V^{2+}(aq) \rightarrow V^{3+}(aq) + e^-$
Cathode: $VO_2^+(aq) + 2\,H^+(aq) + e^- \rightarrow VO^{2+}(aq) + H_2O(\ell)$
(b) $VO_2^+(aq) + V^{2+}(aq) + 2\,H^+(aq) \rightarrow VO^{2+}(aq) + V^{3+}(aq) + H_2O(\ell)$
One electron is transferred in the reaction.

(c)

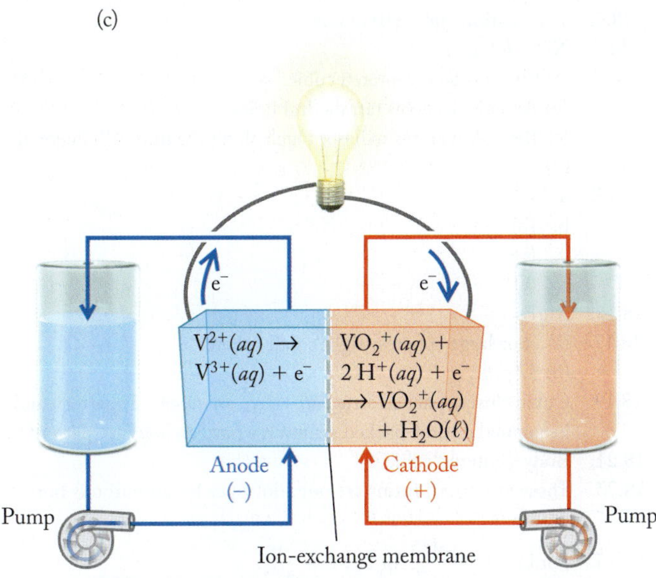


(d) As pH increases, $E_{cell}$ decreases.

17.11. A half-reaction is the oxidation or the reduction portion of a redox reaction.

17.13. Ions migrate between the compartments in an electrochemical cell, not free electrons such as those that conduct electricity in wires.

17.15. (a) $Br_2(\ell) + 2e^- \rightarrow 2\ Br^-(aq)$; reduction
(b) $Pb(s) + 2\ Cl^-(aq) \rightarrow PbCl_2(s) + 2\ e^-$; oxidation
(c) $O_3(g) + 2\ H^+(aq) + 2\ e^- \rightarrow O_2(g) + H_2O(\ell)$; reduction
(d) $H_2S(g) \rightarrow S(s) + 2\ H^+(aq) + 2\ e^-$; oxidation

17.17. $2\ Fe_3O_4(s) + H_2O(\ell) \rightarrow 3\ Fe_2O_3(s) + 2\ H^+(aq) + 2\ e^-$

17.19. (a) Anode: $S^{2-} + 2\ Li \rightarrow Li_2S + 2\ e^-$
Cathode: $4\ e^- + FeS_2 \rightarrow Fe + 2\ S^{2-}$
(b) $4\ Li(s) + FeS_2(s) \rightarrow 2\ Li_2S(s) + Fe(s)$
(c) $Li(s) \mid LiS_2(s) \parallel FeS_2(s) \mid Fe(s)$

17.21. (a) $Li(s) \rightarrow Li^+(aq) + e^-$
$O_2(g) + 4\ H^+(aq) + 4\ e^- \rightarrow 2\ H_2O(\ell)$
$Li(s) \mid Li^+(aq) \parallel H_2O(\ell) \mid O_2(g)$
(b) $Li(s) \rightarrow Li^+(aq) + e^-$
$O_2(g) + 2\ H_2O(\ell) + 4\ e^- \rightarrow 4\ OH^-(aq)$
$Li(s) \mid Li^+(aq) \parallel H_2O(\ell), O_2(g) \mid OH^-(aq)$

17.23. Chlorine is the stronger oxidizing agent because its standard reduction potential ($E° = 1.358$ V) is greater than the standard reduction potential of $O_2$ in 1 $M$ acid ($E° = 1.229$ V).

17.25. Yes in both cases, as long as one of the reduction potentials is more positive (or less negative) than the other.

17.27. (a) $Hg^{2+}(aq) + 2\ e^- \rightarrow Hg(\ell)$ with $Co(s) \rightarrow Co^{2+}(aq) + 2\ e^-$; $E°_{cell} = 1.128$ V
(b) $Hg^{2+}(aq) + 2\ e^- \rightarrow Hg(\ell)$ with $Cu(s) \rightarrow Cu^{2+}(aq) + 2\ e^-$; $E°_{cell} = 0.509$ V

17.29. (a) Less than 1.10 V
(b) The Zn electrode is the anode.

17.31. Because the voltaic cell does work on the surroundings, which is defined as negative work.

17.33. (a) $\Delta G° = -34.5$ kJ; $\Delta E°_{cell} = 0.358$ V
(b) $\Delta G° = 2.9$ kJ; $\Delta E°_{cell} = -0.030$ V

17.35. −290 kJ

17.37. (a) $NiCl_2$: $Ni^{2+}$, $Cl^-$
Ni: $Ni^0$
Na: $Na^0$
NaCl: $Na^+$, $Cl^-$
(b) 2
(c) −498 kJ

17.39. It is an inert electrode that serves as an electron carrier.

17.41. (a) $Cl_2(g) + 2\ Br^-(aq) \rightarrow 2\ Cl^-(aq) + Br_2(g)$; $\Delta E°_{cell} = 0.292$ V; $\Delta G° = -56.4$ kJ/mol; at the cathode, $Cl_2$ is reduced and at the anode $Br^-$ is oxidized; no change in cell voltage.
(b) $Ni^{2+}(aq) + Zn(s) \rightarrow Ni(s) + Zn^{2+}(aq)$; $\Delta E°_{cell} = 0.505$ V; $\Delta G° = -97.5$ kJ/mol; at the cathode, $Ni^{2+}$ is reduced and at the anode, Zn is oxidized; no change in cell voltage.
(c) $I_2(s) + NO_2^-(aq) + 2\ OH^-(aq) \rightarrow 2\ I^-(aq) + NO_3^-(aq) + H_2O(\ell)$; $\Delta E°_{cell} = 0.526$ V; $\Delta G° = -101.5$ kJ/mol; at the cathode, $I_2$ is reduced and at the anode, $NO_2^-$ is oxidized; no change in cell voltage.

17.43. Voltage of a battery (a voltaic cell) is governed by the Nernst equation.

$$E_{cell} = E°_{cell} - \frac{RT}{nF} \ln Q$$

As a battery discharges, the value of $Q$ increases and the value of $E°_{cell}$ decreases, but not by very much because of the log relationship between $q$ and $E°_{cell}$. Not until nearly all the reactants have been turned into products does the drop in $E_{cell}$ become significant.

17.45. 1.27 V

17.47. $8.56 \times 10^{19}$

17.49. (a) 0.126 V
(b) 0.271 V
(c) −52.30 kJ/mol; $K = 1.47 \times 10^9$

17.51. $E_{rxn} = 1.54$ V; $E_{rxn}$ will decrease.

17.53. (a) 0.62 V
(b) 0.61 V

17.55. 0.118 V; the cell with the lower concentration of $Zn(NO_3)_2$ is the cathode.

17.57. (c) and (f)

17.59. Cell B

17.61. Cell F

17.63. Teflon is a good electrical insulator and, when used as a spacer, prevents corrosion by preventing redox reactions between the copper exterior and the stainless steel skeleton.

17.65. A metal, such as magnesium, that has a more negative reduction potential than aluminum.

17.67. In a voltaic cell, the electrons are produced at the anode, so a negative (−) charge builds up there; in an electrolytic cell, electrons are being forced onto the cathode so that it builds up negative (−) charge.

17.69. $Br_2$

17.71. More negative

17.73. 6.8 g

17.75. +3

17.77. 18.0 minutes

17.79. (a) 58 L
(b) No, because $Cl_2$ and $Br_2$ would be produced.

17.81. A hybrid vehicle uses a relatively inexpensive fuel (gasoline) in the internal combustion engine and has good fuel economy but still gives off emissions. A fuel-cell vehicle does not give off emissions (the reaction produces $H_2O$) but requires a more expensive and explosive fuel (hydrogen); moreover, current battery technologies incorporate materials that are still very expensive and bulky.

17.83. Electric engines are more efficient by converting more of the energy into motion instead of losing it as heat.

17.85. (a) $\overset{-4+1}{CH_4}(g) + \overset{+1}{H_2}O(g) \rightarrow \overset{+2}{C}O(g) + 3\ \overset{0}{H_2}(g)$

$\overset{+2}{C}O(g) + \overset{+1}{H_2}O(g) \rightarrow \overset{0}{H_2}(g) + \overset{+4}{C}O_2(g)$

(b) For the reaction of $CH_4$ with $H_2O$, $\Delta G^\circ_{rxn} = 142.2$ kJ. For the reaction of CO with $H_2O$, $\Delta G^\circ_{rxn} = -28.6$ kJ. For the overall reaction, $\Delta G^\circ_{overall} = \Delta G^\circ_{rxn_1} + \Delta G^\circ_{rxn_2} = 113.6$ kJ.

17.87. All the values in Appendix 6 would increase by 0.8277 V.

17.89. (a) $Li \rightarrow Li^+ + e^-$

$2\ e^- + 2\ SO_2 \rightarrow S_2O_4^{2-}$

(b) 2

(c)

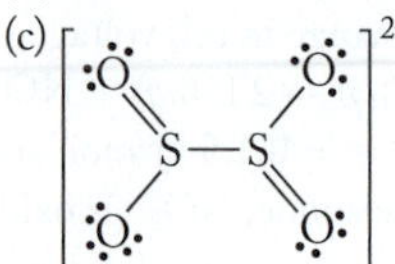

17.91. (a) $FePO_4$

(b) Reduced

(c) Yes, because the reduction of $Co^{4+}$ to $Co^{3+}$ is likely to have a different potential as $CoO_2$ is reduced to $LiCoO_2$.

17.93. (a) +3 to +2

(b) −1 to +1

(c) +1 to 0

17.95. (a) Mg

(b) U

(c) −2.37 V

(d) Yes

17.97. (a) $3\ Ag_2S(s) + 2\ Al(s) + 6\ OH^-(aq) \rightarrow 6\ Ag(s) + 3\ S^{2-}(aq) + 2\ Al(OH)_3(s)$

(b) 1.66 V

17.99. (a) 0.74 V

(b) $1 \times 10^{35}$

(c) Because the value of $K$ is very large, virtually no hypochlorite remained in the pool.

# Chapter 18

18.1. (a), (b), and (d) are crystalline; (c) is amorphous.

18.3.

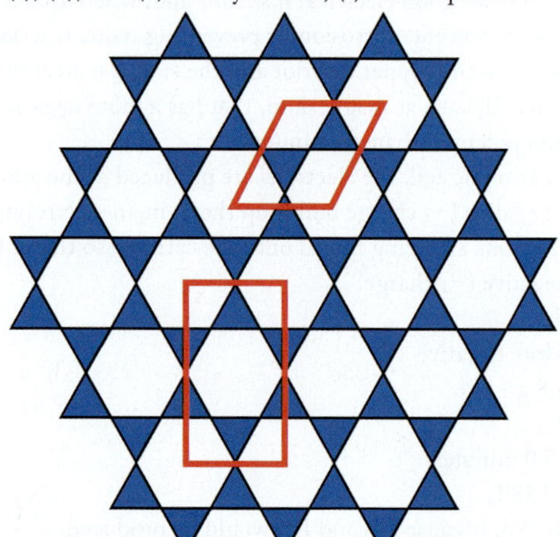

In the rectangular unit cell are two light hexagons and four dark triangles; in the other unit cell there are one light hexagon and two dark triangles.

18.5. 1 (A) cation and 1 (B) anion

18.7. NiTi; 6.5 g/cm³

18.9. (a) Boron is face-centered cubic.

(b) Arsenic occupies tetrahedral holes.

(c) The boron atoms will not touch along the unit cell diagonal.

18.11. (b)

18.13. a. (a)

b. (b)

c. (c)

d. (f)

18.15. $MgB_2$

18.17. (a) Four large spheres, eight small spheres

(b) 57 pm

18.19. Cubic closest-packed structures have an *abcabc*. . . pattern and hexagonal closest-packed structures have an *abab*. . . pattern.

18.21. Body-centered cubic

18.23. These structural forms are not allotropes because iron is not molecular.

18.25. For bcc, $\ell = \dfrac{4r}{\sqrt{3}}$; for fcc, $\ell = \dfrac{4r}{\sqrt{2}}$.

18.27. 104.2 pm

18.29. 513 pm

18.31. (c)

18.33. No

18.35. Because the metallic bonding between Cu and Ag in the alloy is weaker due to a mismatch of their atomic sizes

18.37. Greater than

18.39. (a) $XY_3$

(b) $YX_3$

18.41. Octahedral holes

18.43. (a) Substitutional alloy

(b)

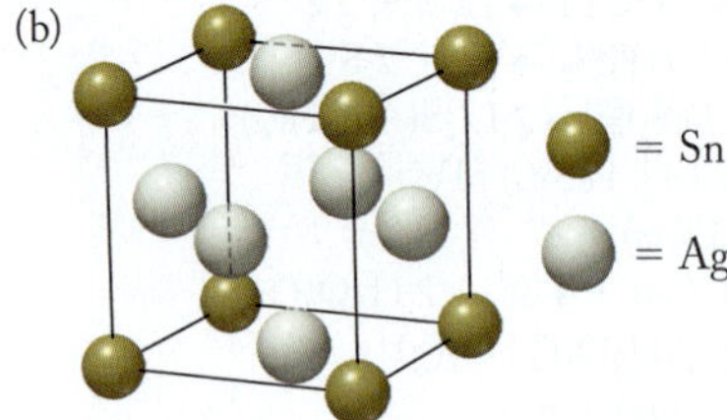


18.45. Yes

18.47. (a) MoW

(b) Yes

(c) 14.1 g/cm³

18.49. One-fifth

18.51. Yes, the alloy is more dense.

18.53. Applying an electrical potential across a metal causes its mobile valence electrons to move toward the positive potential.

18.55. Metallic bonds are weaker than ionic bonds.

18.57. Yes

18.59. Groups 2 and 12

18.61. Phosphorus, with one more valence electron than silicon, would give pure silicon a higher conductivity and so would change the electrical nature of the silicon chips.

18.63. Group 14

18.65. (a) n-type
(b)

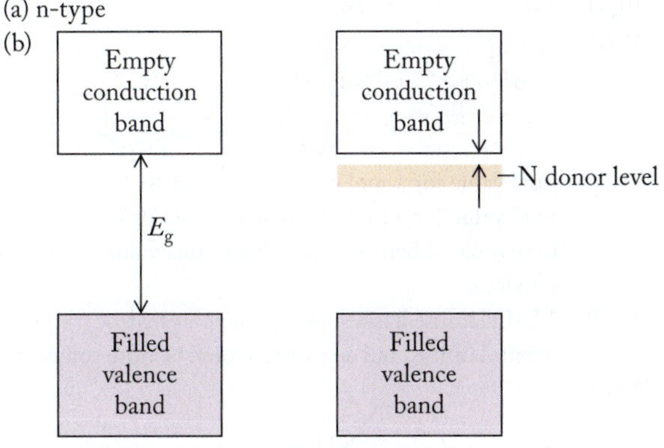


(c) $4.68 \times 10^{-19}$ J

18.67. InN, 192.9 kJ/mol

18.69. Each S atom has a bent geometry due to $sp^3$ hybridization, and therefore the ring is not flat.

18.71. The ring of BN atoms is flat due to $sp^2$ hybridization of the B and N atoms.

18.73. (a) 3.81 g/cm$^3$
(b) Boron will fit; silicon is too large.
(c) Both B and Si as dopants in P would behave as p-dopants to give p-type semiconductor behavior.

18.75. (a) 0.60 g/cm$^3$
(b) 4 carbons + 16 hydrogens

18.77. $K^+$ ions are larger than $Li^+$ ions and do not fit well into the octahedral holes of the fcc lattice.

18.79. As drawn in Figure P18.79, $Cs^+$ ions appear to occupy the centers of bcc unit cells of $Cl^-$ ions. However, the crystal can also be viewed as interpenetrating simple cubic cells composed entirely of $Cs^+$ ions or $Cl^-$ ions.

18.81. We might, but $Cl^-$ ions are much larger than $Na^+$ ions, so they don't fit into the holes of an fcc unit cell of $Na^+$ ions.

18.83. Less than

18.85. $MgFe_2O_4$

18.87. $UO_2$

18.89. (a) Octahedral
(b) Half

18.91. This rock salt arrangement is denser than the sphalerite arrangement because in sphalerite, the lattice of $S^{2-}$ ions must expand to accommodate the $Cd^{2+}$ ions in smaller tetrahedral holes.

18.93. (a)

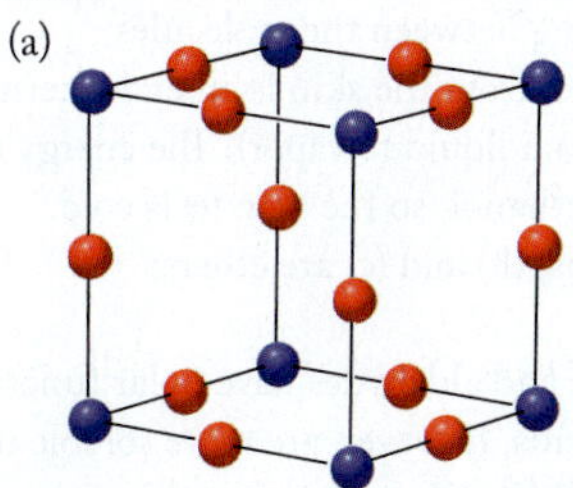

(b) 5.25 g/cm$^3$

18.95. 421 pm

18.97. Ceramics are (b) thermal insulators. (a) Ductility, (c) electrical conductivity, and (d) malleability describe metals.

18.99. $Mg_3(Si_2O_5)(OH)_4$

18.101. $2\ KAlSi_3O_8(s) + 2\ H_2O(\ell) + CO_2(g) \rightarrow Al_2(Si_2O_5)(OH)_4(s) + 4\ SiO_2(s) + K_2CO_3(aq)$; this is not a redox reaction.

18.103. (a) $3\ CaAl_2Si_2O_8(s) \rightarrow Ca_3Al_2(SiO_4)_3(s) + 2\ Al_2SiO_5(s) + SiO_2(s)$
(b) Anorthite contains $Si_2O_8^{8-}$ ions.
Grossular contains $SiO_4^{4-}$ ions.
Kyanite appears to contain $SiO_5^{6-}$ ions, but it actually contains an array of $SiO_4^{4-}$ and $O_2^-$ ions.

18.105. Cubic holes can accommodate $Ba^{2+}$. Octahedral holes can accommodate $Ti^{4+}$.

18.107. An amorphous solid has no regular, repeating lattice to diffract X-rays.

18.109. X-rays have wavelengths of the order of the separation of atoms in crystals. Microwaves have wavelengths too long to be diffracted by crystal lattices.

18.111. If a crystallographer uses a shorter λ wavelength, the data set can be collected over a smaller scanning range.

18.113. Halite

18.115. The values of $n$ are 2 ($\theta = 6.99°$) and 3 ($\theta = 10.62°$). The average lattice spacing is $d = 582$ pm and 580 pm, respectively.

18.117. $2\theta = 4.76°$

18.119. $XYZ_3$

18.121. 33.5%

18.123. (a) 139 pm
(b) 9.96 g/cm$^3$
(c) 6750 Mo atoms

18.125. (a) 2.10 g/cm$^3$
(b) The structure might have an fcc array of Li atoms with smaller Cl atoms in octahedral holes.

18.127. (a) AuZn
(b) CuZn, 4.58%; AgZn, 7.19%; AuZn, 7.19%; yes, all will be expected to form substitutional alloys.
(c) Yes

18.129. (a) MS
(b) 2 $M^{2+}$ and 2 $S^{2-}$
(c) Large yellow spheres
(d) 4

18.131. Every other bcc unit cell of Cu atoms could have an Al atom in its center. That way, half the bcc unit cells would have 2 Cu atoms each and half would have 1 Cu and 1 Al atom, for a total of 3 Cu atoms and 1 Al atom.

# Chapter 19

19.1. (a) One
(b) Two
(c) None
(d) Three

19.3. Pine oil and oil of celery

19.5. (b) and (d)

19.7. Form (a) is the rigid polymer because its regular structure allows the chain to pack in a regular way; form (b) is the rubbery polymer because its random arrangement of side chains make it more difficult to pack in a uniform manner.

19.9.

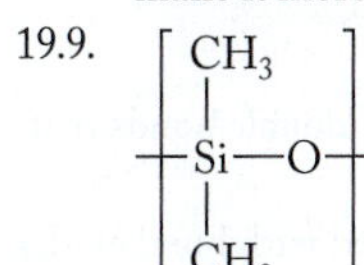

19.11. For *cis*-polyisoprene:

For *trans*-polyisoprene:

19.13. *sp* (triple bond and single bond); $sp^2$ (double bond and two single bonds); $sp^3$ (four single bonds)

19.15. No

19.17. Amine, alcohol, ether, aldehyde, ketone, carboxylic acid, ester, and amide

19.19. Sample A because of its higher average molar mass and stronger dispersion forces

19.21. Yes

19.23. $sp^3$

19.25. The structure of cyclohexane shows that C atoms are $sp^3$ hybridized with bond angles of 109.5°. It cannot be a planar molecule.

19.27. No

19.29. No

19.31.

Pentane

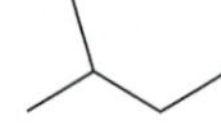

2-Methylbutane

2,2-Dimethylpropane

19.33. (a) 2,3-dimethylhexane; (c and d) 2-methylheptane

19.35. (a) $C_8H_{18}$
(b) $C_9H_{20}$
(c) $C_8H_{18}$
(d) $C_8H_{18}$
(e) $C_9H_{20}$

19.37. Earliest to latest: $C_6H_{14} < C_9H_{20} < C_{12}H_{26} < C_{18}H_{38}$

19.39. Hexane

19.41. No

19.43. When the double bond is "terminal" (occurs at the end or beginning of the carbon chain), three like groups (H) are present, so no cis and trans isomers are possible.

19.45. The C=C double bond outside the ring does not show cis–trans isomerism because the terminal carbon atom does not have two dissimilar groups. The C=C double bond in the ring of carbon atoms is cis in the structure of carvone. This bond cannot be trans or the ring of 6 carbon atoms would not be possible.

19.47. Ethylene has a C=C bond with which HBr is reactive, but polyethylene has only saturated C—C bonds that do not react with HBr.

19.49. −124 kJ

19.51. (a) is trans, *E*; (b) is cis, *Z*.

19.53. −174.30 kJ

19.55.

19.57. Because they lack the conjugated C=C double bonds that give polyacetylene its conductivity.

19.59. In benzene, each C atom is $sp^2$ hybridized with bond angles of 120°. This geometry at each of the carbon atoms in the ring makes benzene a planar molecule.

19.61. Tetramethylbenzene has three constitutional isomers; pentamethylbenzene has no constitutional isomers.

19.63. Yes

19.65.

19.67. Fuel value for 1 mol of benzene = 41.83 kJ/g.
Fuel value for 3 mol of ethylene = 50.30 kJ/g.
One mole of benzene has a lower fuel value than 3 mol of ethylene.

19.69. Methylamine has a smaller nonpolar hydrocarbon chain than *n*-butylamine, and so methylamine is more soluble in water.

19.71.

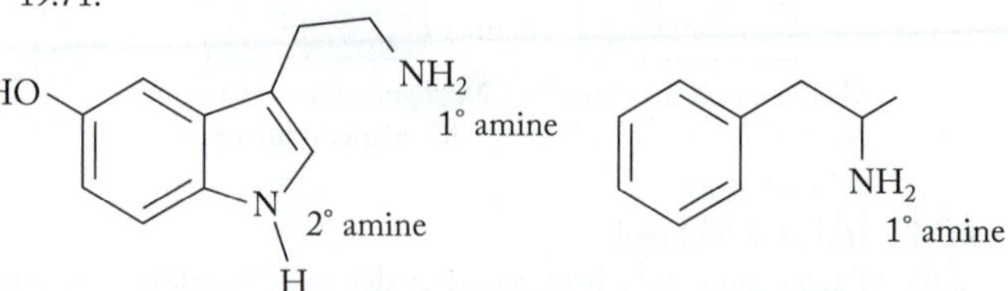


19.73. −138.7 kJ

19.75.

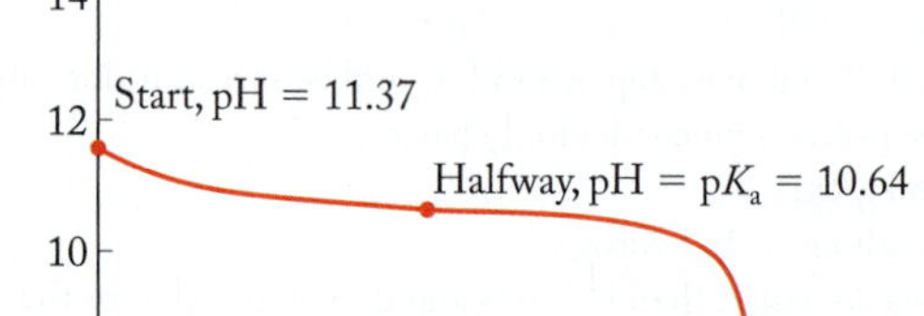

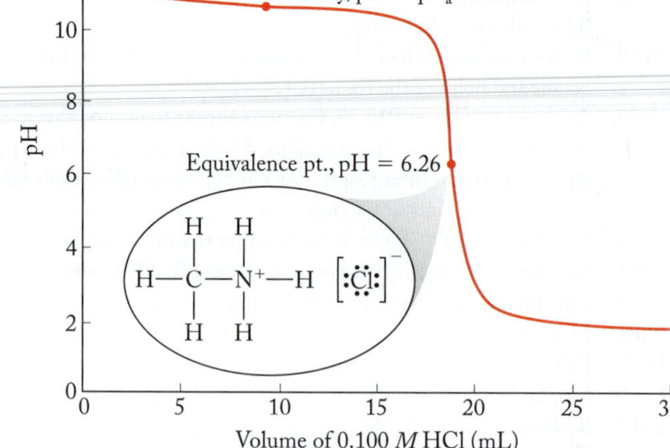


19.77. The more oxygenated the fuel, the lower the fuel value.

19.79. Ethers have lower boiling points than alcohols because they have weaker dipole–dipole forces than the alcohols, which have hydrogen bonding between the molecules.

19.81. Evaporation of ethanol from the skin is an endothermic process (phase change from liquid to vapor). The energy transfers from the skin to the ethanol, so the skin feels cold.

19.83. (a) and (d) are alcohols, (b) and (c) are ethers;
(b) < (c) < (d) < (a)

19.85. Both carboxylic acids and aldehydes have polar functional groups. Carboxylic acids, however, are more soluble in water because they form hydrogen bonds with water and partially ionize when they dissolve in water.

19.87. Yes

19.89. No

19.91. Structure a because all the formal charges are zero.

19.93. An amide includes a carbonyl (C=O) as part of its functional group in addition to the $-NH_2$ group.

19.95. (a), (b), and (d)

19.97. (b)

19.99.

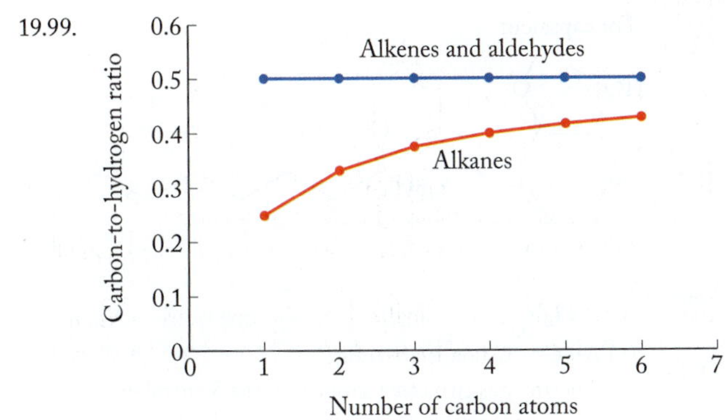


The plot of C:H ratio versus number of C atoms for aldehydes correlates exactly to that of alkenes and poorly to that of alkanes.

19.101. (a) Pineapples

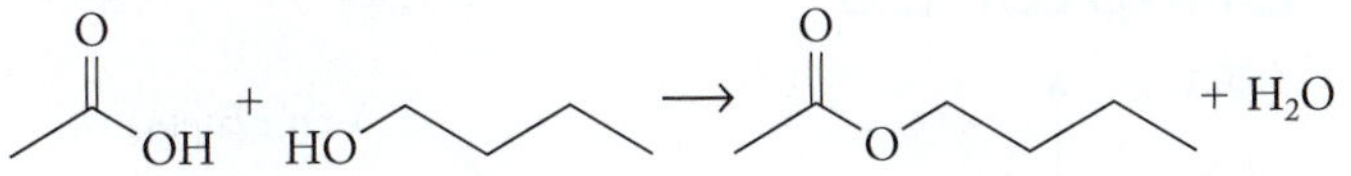

Acetic acid + *n*-Butanol → (ester) + $H_2O$

(b) Bananas

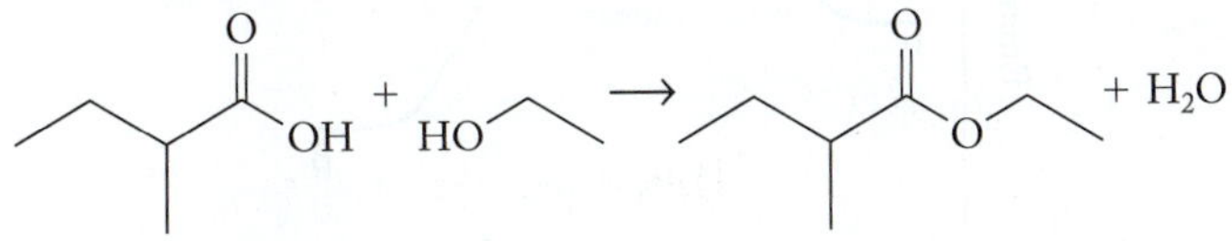

2-Methylbutanoic acid + Ethanol → (ester) + $H_2O$

(c) Apples

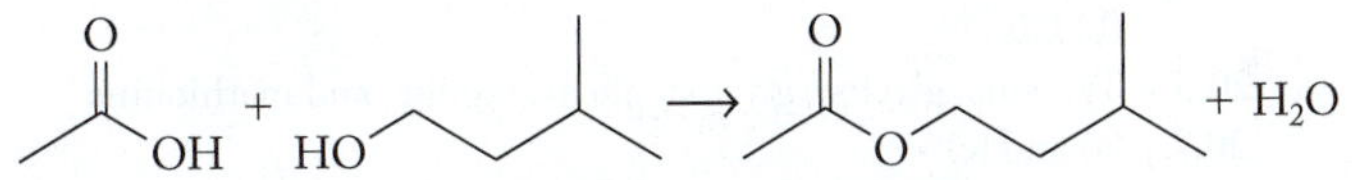

Acetic acid + 3-Methylbutanol → (ester) + $H_2O$

19.103. Nicotine's highlighted N atom is in a tertiary amine group; valium's highlighted N atom is in an amide group.

19.105. (a) 6
(b) 8
(c) 10

19.107. (a) Condensation; methanol
(b) Because of the presence of the 6-membered ring, Kodel might be better able to accept nonpolar organic dyes.

19.109. No; *enantiomer* and *optically active* can describe the same chiral molecule, but *achiral* cannot.

19.111. Yes. If R contains a chiral center, the cis or trans isomer of RCH=CHR would have optical isomers.

19.113. $sp^3$

19.115. (a)

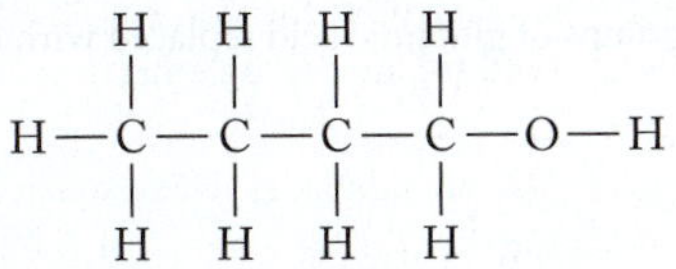

```
    H  H  H  H
    |  |  |  |
H — C — C — C — C — O — H
    |  |  |  |
    H  H  H  H
```

(b)

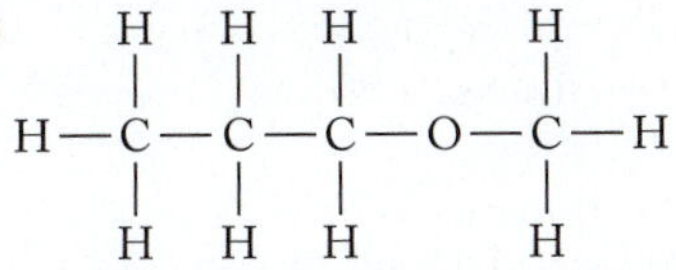

```
    H  H  H       H
    |  |  |       |
H — C — C — C — O — C — H
    |  |  |       |
    H  H  H       H
```

19.117. (a)

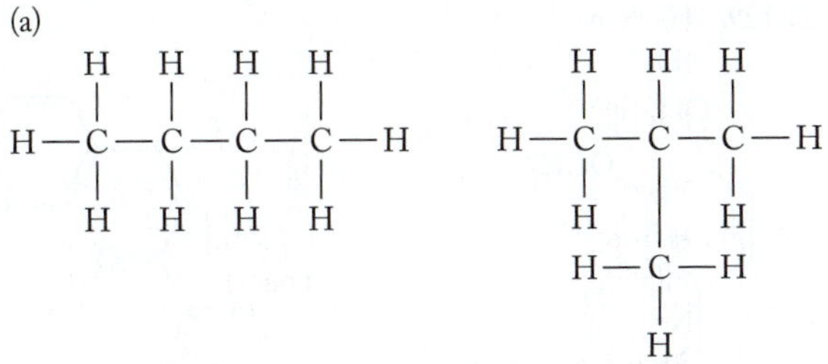

```
    H   H   H   H               H   H   H
    |   |   |   |               |   |   |
H — C — C — C — C — H       H — C — C — C — H
    |   |   |   |               |   |   |
    H   H   H   H               H   |   H
                                H — C — H
                                    |
                                    H
```

(b)

```
    H   H   H   H               H   H   H   H
    |   |   |   |               |   |   |   |
H — C = C — C — C — H       H — C — C = C — C — H
            |   |               |           |
            H   H               H           H

                                H       H
                                |       |
                            H — C = C — C — H
                                    |   |
                                    |   H
                                H — C — H
                                    |
                                    H
```

(c)

```
    H   H   H   H               H       H   H
    |   |   |   |               |       |   |
H — C = C — C = C — H       H — C = C = C — C — H
                                            |
                                            H
```

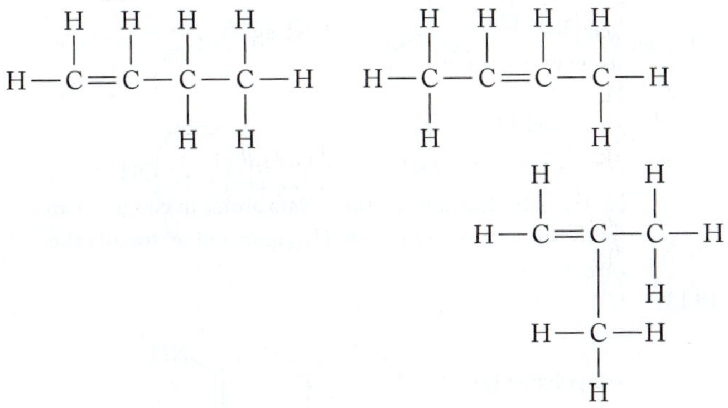

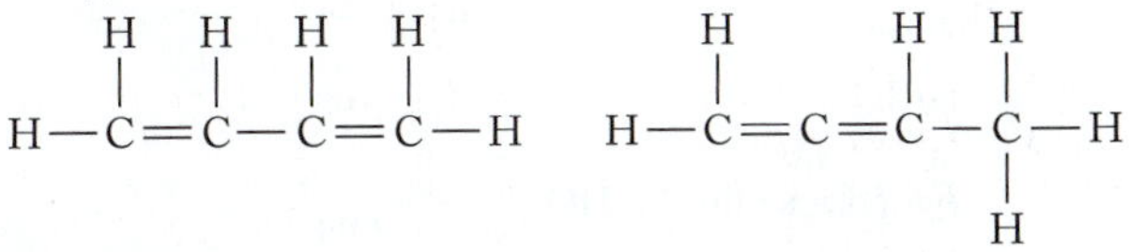

19.119. (a) No
(b) 6

19.121. None

19.123. (a) 10
(b) Aromatic, alkane, amide

19.125.

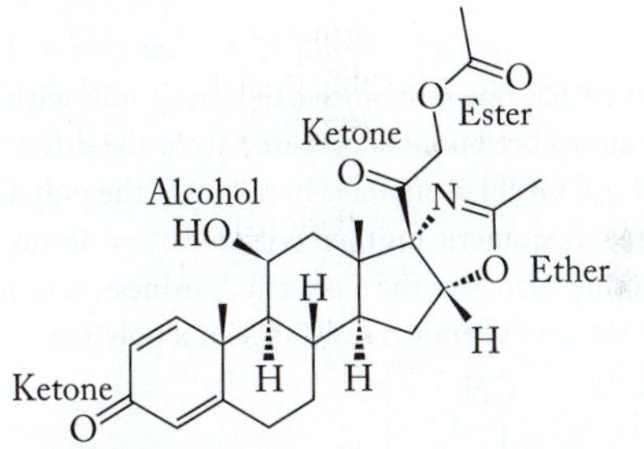

Ester
Ketone
Alcohol
HO
O
N
O Ether
H
H
H
H
Ketone
O

19.127. 1,4-Dimethylbenzene has a symmetrical structure, and so it packs efficiently in the solid state and therefore has a much higher melting point.

19.129. (a) Trans

(b)

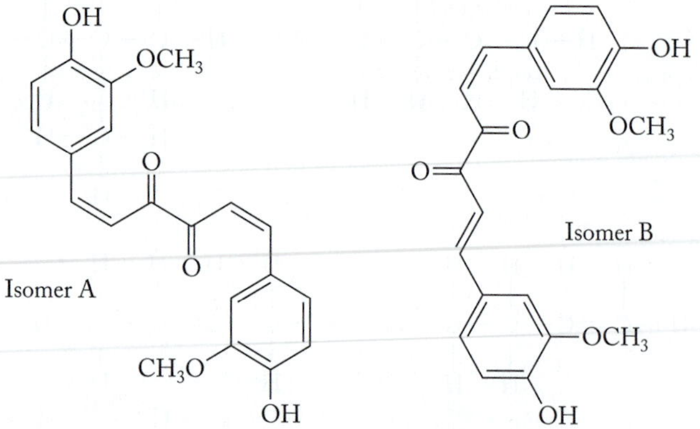

Isomer A

Isomer B

(c) The hybridizations on the carbon atoms in curcumin are $sp^3$ for the carbon of the $-OCH_3$ group and $sp^2$ for all other carbon atoms.

19.131.

For polymer (a)

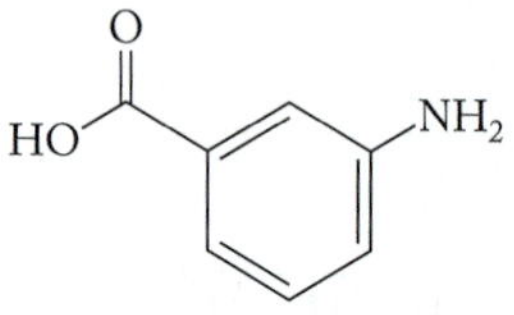

For polymer (b)

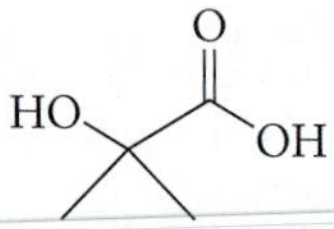

19.133.

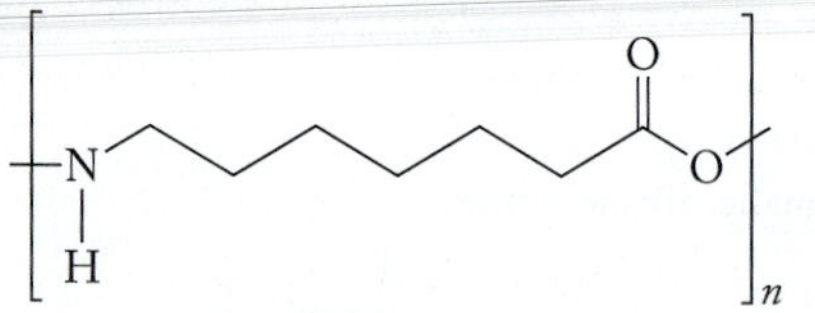

This polymer has one monomeric repeating unit with 7 carbon atoms because it is prepared from the difunctional $H_2N(CH_2)_6COOH$ monomer. In nylon-6, the polymer also has a single monomeric unit but with 6 carbon atoms.

19.135. Cross-linking increases the strength, hardness, melting (softening) point, and chemical resistance of a polymer.

19.137.

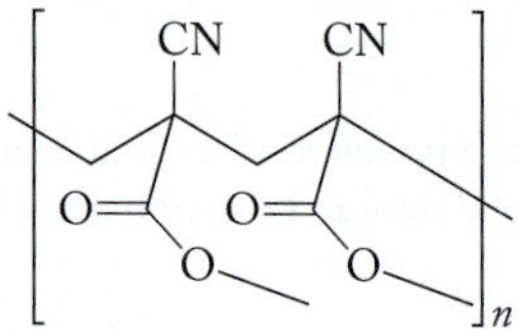

19.139. (a)

For piperine

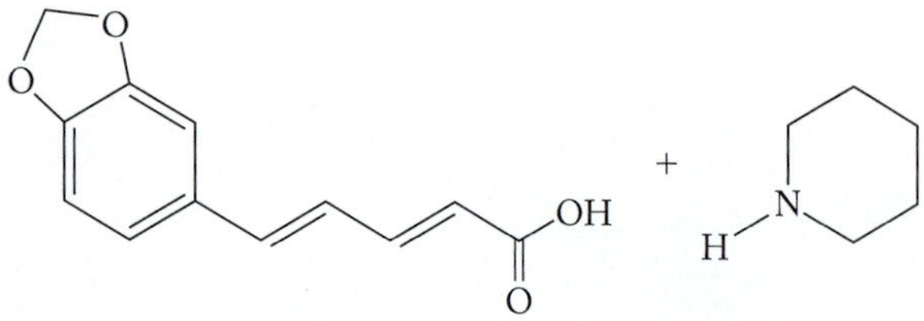

For capsaicin

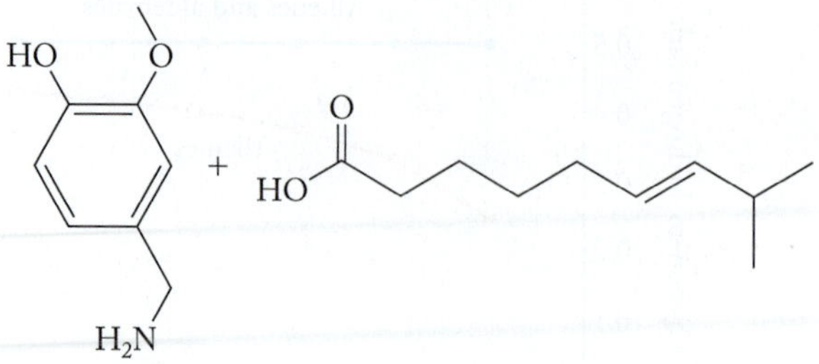

(b) All C=C bonds are trans.

(c) Piperine has two ether groups (in the 5-membered ring) and the amide group. Capsaicin has an ether group, an alcohol group, and the amide group.

# Chapter 20

20.1.

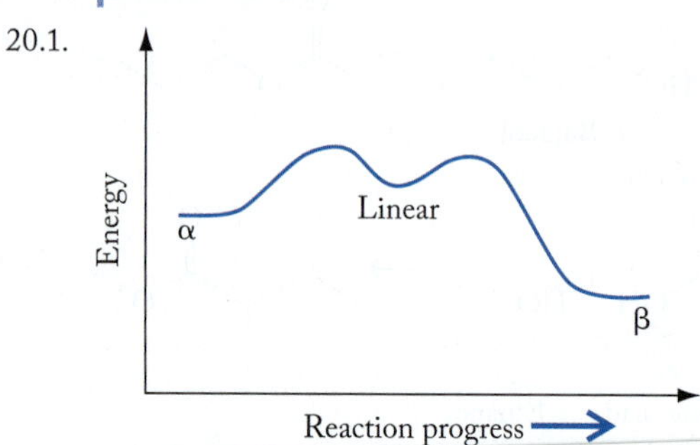


20.3. (a) Palmitic acid in (a); stearic acid in (b)
(b) Cis

20.5. Tyrosine, glycine, glycine, phenylalanine, and methionine

20.7. (a) and (c)

20.9. Sucrose. The difference in the structures is that in sucralose, three –OH groups on sucrose have been replaced by Cl atoms. Being derived from sucrose implies that the sugar is natural, but the presence of Cl atoms on sugars is not natural.

20.11. Decreases

20.13. The α refers to the single carbon atom in amino acids to which both $-NH_2$ and $-COOH$ groups are bonded.

20.15. D- and L- refer to how the four groups on a chiral carbon are oriented.

20.17. (a) and (c)

20.19. Most amino acids are zwitterions at pH ≈ 7.4 because the amino group will be protonated and the carboxylic acid group will be deprotonated, giving $RC\overset{+}{N}H_3COO^-$.

20.21. Both structures have carboxylic acids and an amide group. β-ODAP, however, has an additional ketone group with one of the $CH_2$ groups of glutamic acid replaced with an amine group.

20.23.

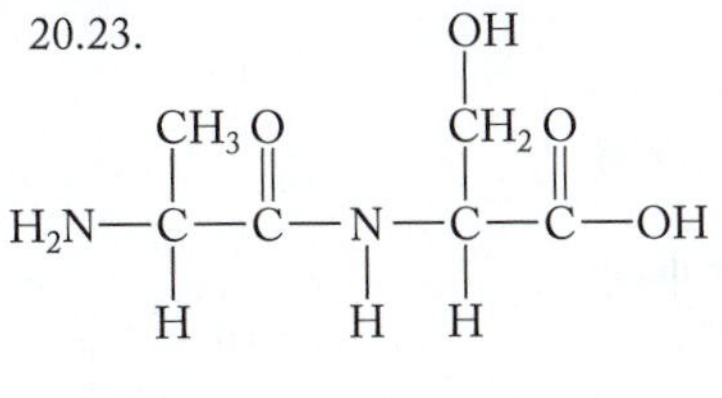

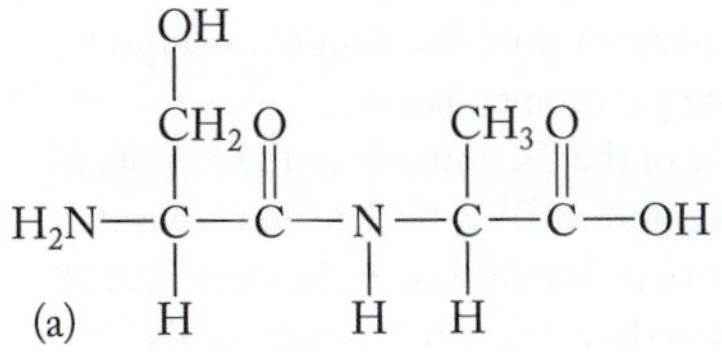

```
              OH
              |
    CH3 O     CH2 O
    |   ||    |   ||
H2N—C—C—N—C—C—OH
    |       |   |
    H       H   H

    OH
    |
    CH2 O     CH3 O
    |   ||    |   ||
H2N—C—C—N—C—C—OH
    |       |   |
(a) H       H   H
```

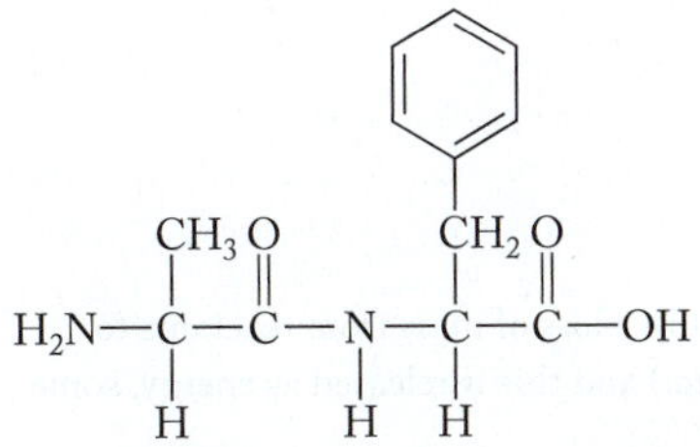

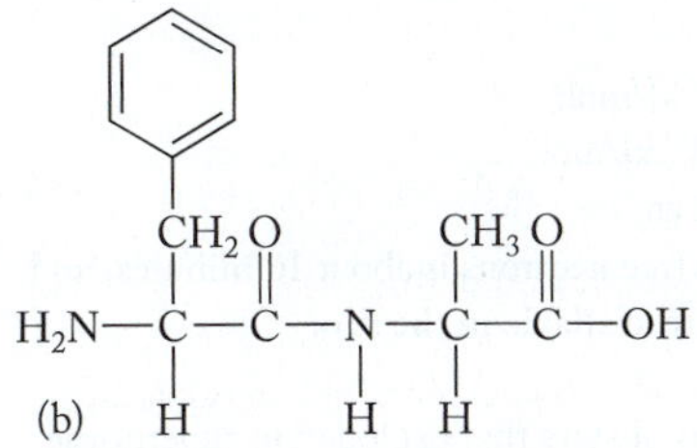

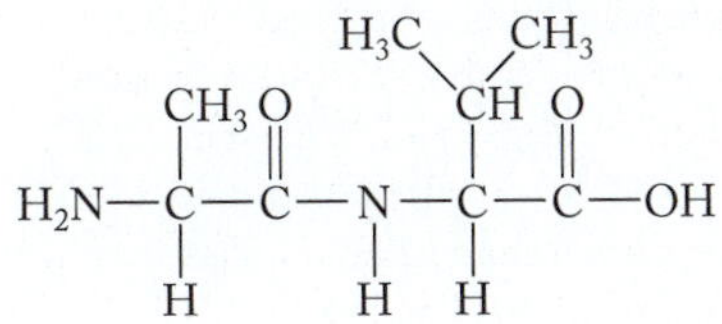

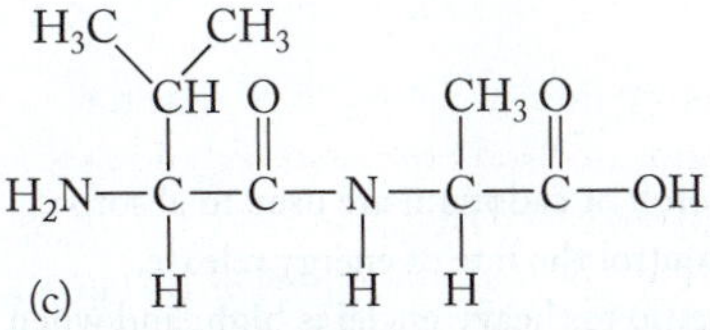

β-Galactose

```
    CH3 O     CH2(C6H5) O
    |   ||    |   ||
H2N—C—C—N—C—C—OH
    |       |   |
    H       H   H

    CH2(C6H5) O    CH3 O
    |   ||    |   ||
H2N—C—C—N—C—C—OH
    |       |   |
(b) H       H   H

              H3C  CH3
                 CH  O
    CH3 O        |   ||
    |   ||    |   ||
H2N—C—C—N—C—C—OH
    |       |   |
    H       H   H

H3C  CH3
   CH  O      CH3 O
    |   ||    |   ||
H2N—C—C—N—C—C—OH
    |       |   |
(c) H       H   H
```

20.25. (a) Alanine + glycine
(b) Leucine + leucine
(c) Tyrosine + phenylalanine

20.27. Glutamic acid and lysine

20.29. $NH_3$

20.31. Leucine

20.33. Primary structure

20.35. Tertiary structure

20.37. In this model, the binding site changes conformation slighty as it binds to the substrate.

20.39. The substrate and the active site must be complementary to one another with regard to both sterics and charge.

20.41. Lysine contains two amino groups, one of which is on a long carbon tail. This can react with the carboxylic acid on the carbon tail of glutamic acid to form a bridge.

20.43. Starch has α-glycosidic bonds, whereas cellulose has β-glycosidic bonds. Starch coils into granules, and cellulose forms linear molecules.

20.45. No

20.47. Carbohydrates are fuels supplying the energy necessary for the operation of organs and tissues, physical activity, and brain function.

20.49.

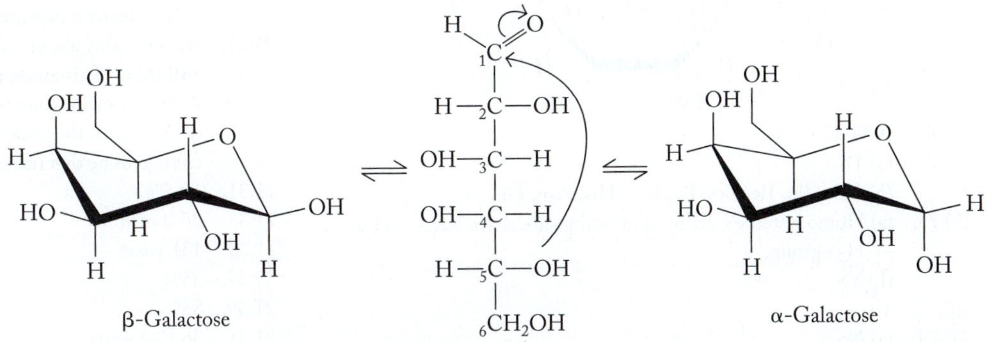

```
      H   O
       \ //
        1C
        |
    H—2C—OH
        |
   OH—3C—H
        |
   OH—4C—H
        |
    H—5C—OH
        |
       6CH2OH
```

α-Galactose

20.51. The α isomers are (a), (b), and (d). The β isomers are (c), (e), and (f).

20.53. (a)

20.55. Saturated fatty acids have all C—C single bonds in their structure; unsaturated fatty acids have at least one C=C double bond.

20.57. Fatty acids have a high fuel value, and eating sticks of butter affords Arctic explorers with more energy per gram of food than carbohydrates or proteins.

20.59. If the two fatty acids linked to the glycerol at C-1 and C-3 are different, then, yes, the triglyceride has a chiral center.

20.61. (b) and (c)

20.63.

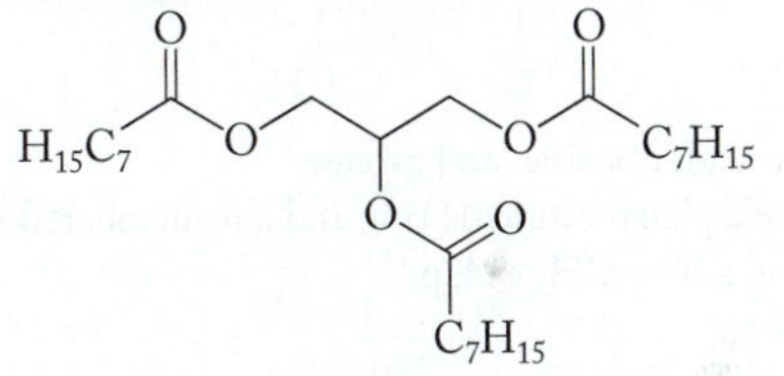

$H_{15}C_7$, $C_7H_{15}$, $C_7H_{15}$

(a) Glycerol with octanoic acid

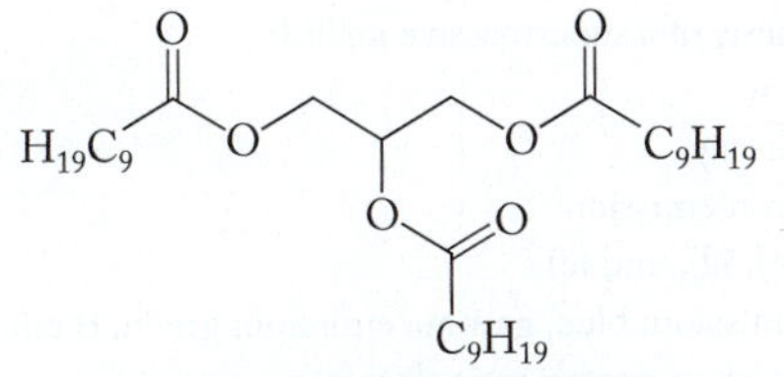

$H_{19}C_9$, $C_9H_{19}$, $C_9H_{19}$

(b) Glycerol with decanoic acid

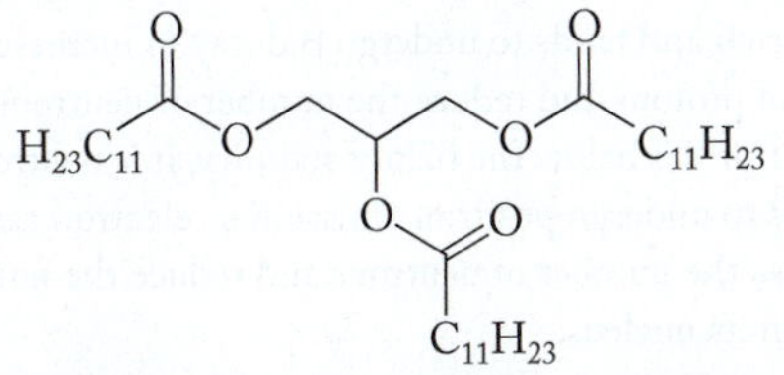

$H_{23}C_{11}$, $C_{11}H_{23}$, $C_{11}H_{23}$

(c) Glycerol with dodecanoic acid

20.65. (a) DNA is composed of A, C, G, and T; RNA is composed of A, C, G, and U.
(b) DNA is double-stranded; RNA is single-stranded.
(c) DNA is used for information storage; RNA makes that information usable to make proteins or transcribe DNA.

20.67. The increased salt concentration enables DNA to keep its negative charges on the outside of the molecule and the hydrogen-bonded base pairs on the inside.

20.69.

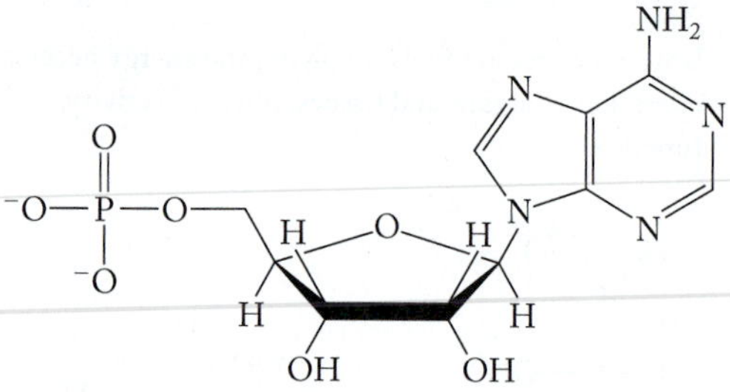

$NH_2$ N N N N; $^-O-P(=O)(-O^-)-O-$ H O H H H OH OH

20.71. A-G-C-C-A-T

20.73. (a) 11
(b) Val, Glu, Ile, Lys, Ile, Phe, His, Asp, Gln

20.75. (a) Homocysteine's sulfur-containing side chain has an extra $-CH_2-$ group.
(b) Yes

20.77. Yes

20.79. (a) No
(b) When the $-NH_2$ group of glycine reacts with the $-COOH$ group of creatine:

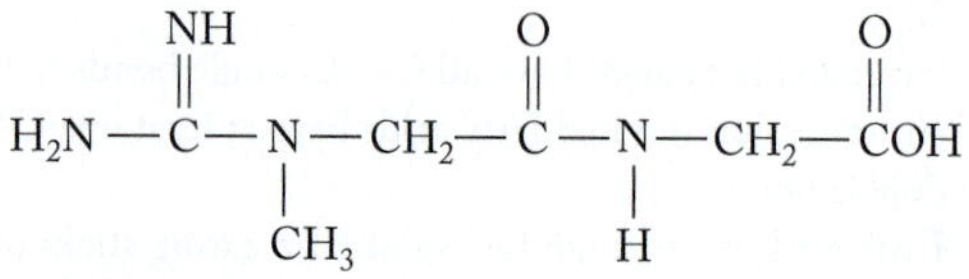

$$H_2N-\overset{NH}{\overset{\|}{C}}-\underset{CH_3}{\underset{|}{N}}-CH_2-\overset{O}{\overset{\|}{C}}-\underset{H}{\underset{|}{N}}-CH_2-\overset{O}{\overset{\|}{C}}OH$$

When the $-NH_2$ group of creatine reacts with the $-COOH$ group of glycine:

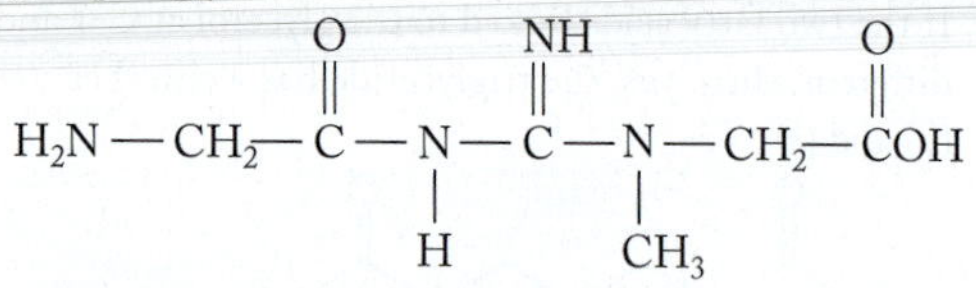

$$H_2N-CH_2-\overset{O}{\overset{\|}{C}}-\underset{H}{\underset{|}{N}}-\overset{NH}{\overset{\|}{C}}-\underset{CH_3}{\underset{|}{N}}-CH_2-\overset{O}{\overset{\|}{C}}OH$$

20.81. Glutamic acid, cysteine, and glycine

20.83. Both have a planar aromatic ring and a 6-membered ring containing a $N-CH_3$ group.

# Chapter 21

21.1. (a) Synthesis of a supermassive nuclide
(b) Fission

21.3. (a) Beta decay

21.5. (b) Neutron emission

21.7. Curves (c), (d), and (e)

21.9. Red, α emission; blue, gamma emission; green, β emission

21.11. The neutron-to-proton ratio decreases.

21.13. If the nuclide lies in the belt of stability, it is not radioactive and is stable. If it lies above the belt of stability, it is neutron-rich and tends to undergo β decay to increase the number of protons and reduce the number of neutrons in its nucleus. If it lies below the belt of stability, it is neutron-poor and tends to undergo positron emission or electron capture to increase the number of neutrons and reduce the number of protons in its nucleus.

21.15. Alpha decay increases the neutron-to-proton ratio to produce less-stable isotopes, which can then be made more stable through β emission to decrease the neutron-to-proton ratio.

21.17. $^{64}Cu$ to $^{64}Zn$ is β decay; $^{64}Zn$ to $^{64}Ni$ is either positron emission or electron capture.

21.19. Greater than 1

21.21. (a) n/p = 1.37; β decay
(b) n/p = 1.36; positron decay
(c) n/p = 1.49; β decay
(d) n/p = 1.27; positron decay

21.23. (a) $^{192}Ir \rightarrow {}^{0}_{-1}\beta + {}^{192}Pt$
(b) $^{192}Ir + {}^{0}_{-1}e \rightarrow {}^{192}Os$

21.25. $^{56}Co$ and $^{44}Ti$ are both neutron-poor and should undergo either electron capture or positron emission.

21.27. Because only about 0.2% of the $^{14}C$ initially in the sample is still there. This amount is too small to be detected accurately.

21.29. If the rocks are younger than 300,000 years, less than 0.02% of the $^{40}K$ in the rocks has decayed, which is too small a change to be determined accurately.

21.31. 12.5%

21.33. 20.2 min

21.35. 131 years

21.37. 35%

21.39. 85%

21.41. 36,640 years

21.43. Nuclear reactions result in a loss of mass from reactants to products (the mass defect) and this is released as energy, some in the form of heat.

21.45. (a) $^{11}C \rightarrow {}^{11}B + {}^{0}_{1}\beta$
(b) $1.1760 \times 10^{-11}$ J

21.47. For $^{35}Cl$, $2.8768 \times 10^{10}$ kJ/mol; for $^{37}Cl$, $3.110375 \times 10^{10}$ kJ/mol.

21.49. $^{209}_{83}Bi + {}^{4}_{2}\alpha \rightarrow {}^{211}_{85}At + 2\,{}^{1}_{0}n$

21.51. Because the half-life of free neutrons is about 10 minutes, and therefore no neutrons are available in the sun.

21.53. Helium

21.55. From earlier generations of stars that exploded in supernovae

21.57. $\Delta E = 3.01 \times 10^{-10}$ J; $\lambda = 1.32 \times 10^{-15}$ m, or $1.32 \times 10^{-6}$ nm

21.59. (a) $4.37 \times 10^{-12}$ J
(b) $6.80 \times 10^{-12}$ J
(c) $2.69 \times 10^{-12}$ J
(d) $1.60 \times 10^{-12}$ J

21.61. $9.36 \times 10^{-13}$ J

21.63. (a) $^{16}O$
(b) $^{24}Mg$
(c) $^{36}Ar$

21.65. (a) $^{99}Tc$
(b) $^{121}Sb$
(c) $^{109}Cd$

21.67. Control rods made of boron or cadmium are used to absorb the excess neutrons to control the rate of energy release.

21.69. The neutron-to-proton ratio for heavy nuclei is high, and when the nuclide undergoes fission to form smaller nuclides, it must emit neutrons because the fission products require a lower neutron-to-proton ratio for stability.

21.71. (a) $^{138}_{52}Te$
(b) $^{133}_{51}Sb$
(c) $^{143}_{55}Cs$

21.73. (a) $^{103}_{39}Y$
(b) $^{130}_{48}Cd$
(c) $^{138}_{52}Te$

21.75. The level of radioactivity is the number of decay events per unit of time. The dose is a measure of exposure to ionizing radiation over time.

21.77. Radon-222 is a gas. If inhaled, it may undergo α decay to solid polonium-218 while in the lungs. When radon-222 decays to polonium-218 while in the lungs, it may continue to emit α radiation. Alpha radiation is one of the most damaging kinds of radiation when it comes in contact with biological issues. Exposure increases the risk of lung cancer.

21.79. 5 μSv = 5 μGy; 250 μJ

21.81. (a) $^{90}_{38}Sr \rightarrow ^{0}_{-1}\beta + ^{90}_{39}Y$
(b) $3.28 \times 10^8$
(c) Strontium-90 is found in milk and not other foods because it is chemically similar to calcium, and milk is rich in calcium.

21.83. (a) 0.15 decays/s
(b) $7.0 \times 10^4$

21.85. (a) The half-life should be long enough to effect treatment of the cancerous cells but not so long as to damage healthy tissues.
(b) Because α radiation does not penetrate far beyond a tumor, the α decay mode is best.
(c) Products should be nonradioactive, if possible, or have short half-lives and be flushable from the body by normal cellular and biological processes.

21.87. (a) Positron emission or electron capture
(b) Positron emission or electron capture
(c) Positron emission or electron capture

21.89. 74.3 days

21.91. Yes

21.93. (a) 79.1%
(b) 153 h

21.95. (a) $^{10}_{5}B + ^{1}_{0}n \rightarrow ^{7}_{3}Li + ^{4}_{2}\alpha$
(b) $4.43 \times 10^{-13}$ J
(c) Alpha particles have a high RBE and do not penetrate into healthy tissue if the radionuclide is placed inside a tumor.

21.97. (a) Besides releasing a large amount of energy to power the starship *Enterprise* when combined with antihydrogen, hydrogen is an abundant fuel in the universe and therefore could easily react with any antihydrogen produced.
(b) Antihydrogen would react with all matter, so it would have to be contained—which would be difficult except maybe with a magnetic or energy field.

21.99. Binding energy for $^{8}Be = 9.0522 \times 10^{-12}$ J; binding energy for 2 atoms of $^{4}He = 9.06701 \times 10^{-12}$ J.

21.101. (a) Geiger counter
(b) 2877 years
(c) Smoke detectors are safe to handle because the $^{241}Am$ is an α emitter, and α particles do not travel more than a few inches in air and cannot penetrate the first layer of skin.

21.103. (a) $^{249}_{98}Cf + ^{48}_{20}Ca \rightarrow ^{249}_{118}Og + 3\,^{1}_{0}n$
(b) $^{290}_{116}Lv$
(c) $^{286}_{114}Fl$
(d) $^{282}_{112}Cn$
(e) Because $^{294}Og$ is a member of the noble gas family, it has chemical and physical properties similar to those of naturally occurring radon.

21.105. $[\mathrm{H{-}\ddot{O}\cdot}$ (with H bonded to O)$]^+$ $[\mathrm{H{-}\ddot{O}{-}H}$ (with H bonded to O)$]^+$ $\mathrm{H{-}\ddot{O}\cdot}$

21.107. (a) $^{64}_{28}Ni + ^{124}_{50}Sn \rightarrow ^{188}_{78}Pt$
(b) $^{196}_{78}Pt$

21.109. 3.35

21.111. (a) $^{40}_{19}K \rightarrow ^{40}_{18}Ar + ^{0}_{1}\beta$
(b) Because the half-life of $^{40}K$ is so much longer than that of $^{14}C$

21.113. 118 s

21.115. $^{12}_{5}B \rightarrow ^{8}_{3}Li + ^{4}_{2}\alpha$
$^{12}_{5}B \rightarrow ^{12}_{6}C + ^{0}_{-1}\beta$

21.117. $^{12}_{6}C + ^{1}_{1}p \rightarrow ^{10}_{5}B + ^{3}_{2}He$
$^{14}_{7}N + ^{1}_{1}p \rightarrow ^{10}_{5}B + ^{5}_{3}Li$

21.119. (a) $^{104}_{44}Ru + ^{1}_{0}n \rightarrow ^{105}_{45}Rh + ^{0}_{-1}\beta$
(b) 29 h

# Chapter 22

22.1. (d)

22.3. Pink, group 16

22.5. −6.7 kJ

22.7. Both are trigonal planar.

22.9. Phosphate

22.11. Essential elements have beneficial physiological effects, whereas nonessential elements have no known function.

22.13. The amount present in the body

22.15. (a) 1.6 ppm
(b) 525 ppm
(c) 0.43 ppm

22.17. (a) Oxygen
(b) Oxygen
(c) Carbon

22.19. Because groups 1 and 2 have valence electrons in the *s* orbitals and groups 13–18 have valence electrons in the *p* orbitals.

22.21. $Li_2O$ is an ionic solid with strong ionic forces between ions, whereas CO is a covalent molecule with much weaker dipole–dipole forces between molecules.

22.23. Electrical conductivity

22.25. $Be^{2+}$ is smaller and would interact more strongly with binding sites with full negative charges. $Ca^{2+}$ ions might not even fit into the sites.

22.27. Potassium

22.29. $Li^+ < Al^{3+} < Mg^{2+} < Cl^-$

22.31. K < Mg < S < F

22.33. The ionization of a $Cl^-$ ion is the reverse process of a Cl atom acquiring an electron.

22.35. Diffusion, ion channels, and ion pumps

22.37. The inner portion of the phospholipid bilayer is nonpolar.

22.39. $K^+$

22.41. $CaCO_3$ is less soluble than $CaSO_4$.

22.43. 0.56 atm

22.45. 0.072 V

22.47. (a) 0.062 V
(b) 0.345 mol

22.49. (a) $[\mathrm{H{-}P({=}O)(O^-){-}O^-}]^{2-}$ (H bonded to P, with :O: double-bonded above, :Ö: below, and :Ö: to the right) $[\mathrm{O{=}P(O^-)(O^-){-}O^-}]^{3-}$ (P with four O atoms)
(b) This might be due to the P–H bond's being more reactive.

22.51. $Cs^+$ can interfere with $K^+$-dependent functions. It is also a β emitter and could cause radiation sickness and cancer.

22.53. $\Delta S$, positive; $\Delta G$, negative

22.55. Superoxide ions ($O^{2-}$) are easily reduced (with a reduction potential of about 1.5 V) by a two-electron process to become $O_2^{2-}$.

22.57. (a) $CH_4S_2$
(b) The vapor pressure would decrease because O–H groups will hydrogen-bond.

22.59. 3.05

22.61. $K = \frac{[OH^-]}{[F^-]}$; the equilibrium lies to the right.

22.63. (a) More soluble
(b) $8.7 \times 10^{-8}\ M$
(c) $6.24 \times 10^{-7}\ M$

22.65. (a) $2.72 \times 10^{-8}\ M$
(b) No
(c) As the pH decreases (the solution becomes more acidic), the solubility of $PbCO_3$ will increase.

22.67. $H_2O$ is the most polar; $H_2Te$ is the least polar.

22.69. Different decay products have different energies and different relative biological effectiveness values.

22.71. Beta particles are more penetrating and cause less damage to tissues.

22.73. $^{213}Bi$ has a high neutron-to-proton ratio; $^{111}In$ has a low neutron-to-proton ratio.

22.75. Dipole–induced dipole

22.77. 41 h

22.79. (a) $Bi^{3+}(aq) + 6\ H_2O(\ell) \rightarrow Bi(H_2O)_5(OH)^{2+}(aq) + H^+(aq)$
(b) 2.04
(c) More acidic

22.81. $Al(OH)_3(s) + 3\ H^+(aq) \rightarrow Al^{3+}(aq) + 3\ H_2O(\ell)$

22.83. (a) Aluminum hydroxide
(b) Yes
(c) 2.24 g of $Al(OH)_3$; 2.52 g of $Mg(OH)_2$

22.85. (a) Group 15
(b)

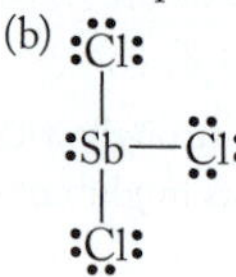


(c) Lewis acid

# Chapter 23

23.1. (a) Vanadium (black), chromium (green), and cobalt (yellow)
(b) $V^{2+}$ (black) and $Zn^{2+}$
(c) Zinc (blue)

23.3. (d)

23.5. All the compounds have ligands in both the cis and trans arrangement.

23.7. (a) $[Co(NH_3)_6]^{3+}$; (b) $[Co(H_2O)_6]^{3+}$; (c) $[Co(CN)_6]^{3-}$

23.9. Red

23.11. (c) is $CoCl_4^{2-}$ and (a) is $CoI_4^{2-}$

23.13. Water

23.15. +2

23.17. 1.50

23.19. $[Pt(NH_3)_6]Cl_4$, $[Pt(NH_3)_5Cl]Cl_3$, $[Pt(NH_3)_4Cl_2]Cl_2$, $[Pt(NH_3)_3Cl_3]Cl$, $[Pt(NH_3)_2Cl_4]$

23.21. Hexaammineplatinum(IV) chloride; pentaamminechloroplatinum(IV) chloride; tetraamminedichloroplatinum(IV) chloride; triamminetrichloroplatinum(IV) chloride; diamminetetrachloroplatinum(IV)

23.23. (a) Hexaamminechromium(III)
(b) Hexaaquacobalt(III)
(c) Pentacyanonickelate(II)

23.25. (a) Potassium tetrabromocolbaltate(II)
(b) Ammonium aquatrihydroxoozincate(II)
(c) Pentaamminechloroiron(III) chloride

23.27. (a) Bis(ethylenediamine)zinc(II) sulfate
(b) Pentaammineaquanickel(II) chloride
(c) Potassium hexacyanoferrate(II)

23.29. A sequestering agent is a multidentate ligand that separates metal ions from other substances so that they can no longer react. Properties that make a sequestering agent effective include strong bonds formed between the metal and the ligand and large formation constants.

23.31. As pH increases, the chelating ability increases because carboxylic acid groups ionize, and amino groups deprotonate, providing additional coordination sites.

23.33. When the transition metals bond to ligands, their *d*-orbital energies split. If *d* to *d* transitions are possible, the compound is likely to be colored.

23.35. The lobes of the $d_{xy}$ orbital lie in the plane of the ligands' lone pair of electrons. Repulsion between the lone pairs and electrons in the $d_{xy}$ orbitals raises the electrons' energy.

23.37. The yellow solution contains (b) $Cr(NH_3)_6^{3+}$. The violet solution contains (a) $Cr(H_2O)_6^{3+}$.

23.39. Yellow, because the absorption band extends into the violet end of the visible spectrum.

23.41. $NiCl_4^{2-}$

23.43. The magnitude of the crystal field splitting energy compared with the pairing energy of the electrons in a lower energy *d* orbital

23.45. $Fe^{2+}$ has 4 unpaired electrons; $Cr^{2+}$ has 4 unpaired electrons; $Co^{2+}$ has 3 unpaired electrons; $Mn^{3+}$ has 4 unpaired electrons.

23.47. Cooling the complex to 80K must induce greater *d*-orbital splitting so that all six electrons are paired up in the $d_{xy}$, $d_{xz}$, and $d_{yz}$ orbitals in the octahedral geometry.

23.49. (a) $Mn^{4+}$ in $MnO_2$; 2 $Mn^{3+}$ and 1 $Mn^{2+}$ in $Mn_3O_4$
(b) Both low-spin and high-spin configurations are possible in $Mn_3O_4$ ($d^4$ and $d^5$) but not in $MnO_2$ ($d^3$).

23.51. Paramagnetic

23.53. For an octahedral geometry, *cis-* means that two ligands are side by side and have a 90° bond angle between them. Ligands that are *trans-* with respect to each other have a 180° bond angle between them.

23.55. Two of the same ligand

23.57. Yes

23.59.

| Cis: $[CuBr_2Cl_2]^{2-}$ (Br, Cl top; Br, Cl bottom) | Trans: $[CuBr_2Cl_2]^{2-}$ (Cl, Br top; Br, Cl bottom) |
|---|---|

No, neither isomer is chiral.

23.61. (a) To catalyze reactions by lowering the activation energy
(b) No

23.63. Lowers the activation energy

23.65. Much greater than 1

23.67. Positive

23.69. $1.57 \times 10^8$ times

23.71. Zero order

23.73. To kill targeted disease cells such as cancer

23.75. Gamma radiation

23.77. They bind to DNA to prevent replication.

23.79. Paramagnetic

23.81. 126

23.83. [penicillamine] = $9.99 \times 10^{-4}$ *M*; [cysteine] = 1.00 *M*

23.85. (a) One has the $N_3$ ligands *cis* and the other has the $N_3$ ligands *trans.*

(b) *cis*-diaaminediazido *trans*-hydroxoplatinum(IV); *trans*-diaaminediazido *trans*-hydroxoplatinum(IV)

(c) Yes

(d)

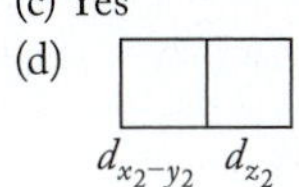


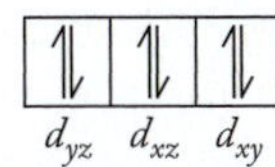


23.87. (a) Yes, $Pt^{2+}$ and $Au^{3+}$ both have a $d^8$ electron configuration.

(b) Diamagnetic

23.89. (a) $[Co(NH_3)_6]^{3+}$

(b)

↑ ↑          — —

↑↓ ↑↓ ↑          ↑↓ ↑↓ ↑↓

$[Co(H_2O)_6]^{2+}$          $[Co(H_2O)_6]^{3+}$

23.91. $Ag^{2+}$ has 9 *d* electrons, leaving an unpaired electron in the $d_{x^2-y^2}$ orbital to make it paramagnetic. $Ag^{3+}$ has 8 *d* electrons and $Ag^+$ has 10 *d* electrons. Both have all electrons paired, so those silver ions are diamagnetic.

23.93. (a)

$OH_2$, $H_2O$, $OS(CH_3)_2$, Ru, $(H_3C)_2SO$, Cl, $OH_2$

$OH_2$, $(H_3C)_2SO$, $OS(CH_3)_2$, Ru, $H_2O$, Cl, $OH_2$

(b) In DMSO, the sulfur atom has a lone pair of electrons that could also be a donor as well as the oxygen atom.

# Credits

## Front Matter

Page v: (top) dpa picture alliance archive/Alamy Stock Photo; (bottom) Sonja Pacho/Corbis/VCG/Getty Images; p. vi: (top) Hemis/Alamy Stock Photo; (bottom) Laurel/Alamy Stock Photo; p. vii: (top) Phil Dotson/Science Source; (center) Adam Pretty/Getty Images; (bottom) Emmanuel LATTES/Alamy Stock Photo; p. viii: (top) NASA/NOAA/GSFC/Suomi NPP/VIIRS/Norman Kuring; (bottom) Khoroshunova Olga/Shutterstock; p. ix: (top) Susan Schmitz/Shutterstock; (center) JEAN-MARIE LIOT/AFP/Getty Images; (bottom) Reinhard, H./picture alliance/Arco Images G/Newscom; p. x: (top) Kyodo/Newscom; (center) Design Pics Inc/Alamy Stock Photo; (bottom) azndc/Getty Images; p. xi: (top) WaterFrame/Alamy Stock Photo; (bottom) Jim Motavalli; p. xii: (top) Rudi Van Starrex/Getty Images; (bottom) Gregory Shamus/Getty Images; p. xiii: (top) Dennis Kunkel Microscopy/Science Source; (bottom) LADA/Science Source; p. xiv: (top) Michael Neelon (misc)/Alamy Stock Photo; (bottom) addkm/Shutterstock; p. xviii: Gilbert photo courtesy of Christina Gilbert; Kirss photo courtesy of Paige Scannell; Foster photo courtesy of Kathryn Dyer; Bretz photo courtesy of Youngstown State University.

## Chapter 1

Pages 2–3: James D. Morgan/Getty Images; p. 4: Geoff Tompkinson/Getty Images; p. 5: (a) Anthony Bradshaw/Getty Images; (b) Abramova Elena/Shutterstock; (c) GIPhotoStock/Science Source; (d) dlerick/Getty Images; p. 9: (top) npix/Getty Images; (bottom) Andy Clarke/Science Source; p. 11: Aquacone courtesy of Solar Solutions, Inc.; p. 12: (top left) SeaWiFS Project, NASA/Goddard Space Flight Center, and ORBIMAGE; (top right) The Natural History Museum, London/Science Source; (bottom) Turtle Rock Scientific/Science Source; p. 13: (both) Turtle Rock Scientific/Science Source; p. 14: (all) David Wrobel/Visuals Unlimited; p. 15: (top left) Dorling Kindersley/Getty Images; (middle left) Arenacreative/Dreamstime; p. (bottom left) Podisu/Dreamstime; (top right) Lisovskaya Natalia/Shutterstock; p. 15: (middle right) Katherine Fawssett/Getty Images; (bottom right) Erik Page Photography/Getty Images; p. 16: dpa picture alliance archive/Alamy Stock Photo; p. 18: Dirk Wiersma/Science photo Library/Science Source; p. 21: (both) Shutterstock; p. 22: (both) Courtesy of Ohaus; p. 27: Canadian Press Images; p. 28: Gladskikh Tatiana/Shutterstock; p. 29: Filip Tot/Alamy Stock Photo; p. 32: Jarun Ontakrai/Shutterstock; p. 37: Aquacone courtesy of Solar Solutions, Inc.; p. 42: AP Photo/Neal Hamberg; p. 44: Brian Hartshorn/Alamy; p. 45: (left) Shutterstock; (right) NASA/JHUAPL/SwRI.

## Chapter 2

Pages 46–47: Sonja Pacho/Corbis/VCG/Getty Images; p. 48: The Royal Institution, London/Bridgeman Images; p. 50: DeAgostini/Getty Images; p. 63: USGS/ZUMA Press/Newscom; p. 65: Science Source; p. 66: (top) KarSol/Shutterstock; (bottom) RuslanK/Shutterstock; p. 69: FRANCOIS LENOIR/REUTERS/Newscom; p. 70: (top) AP Photo/Erik S. Lesser; (bottom) Adrien Deneu/Morpho/Safran; p. 74: Science Source; p. 80: Björn Wylezich/Alamy Stock Photo; p. 82: Courtesy of NanoTrade s.r.o.; p. 83: Smithsonian Institution, Washington DC, US/Bridgeman Images.

## Chapter 3

Pages 84–85: Hemis/Alamy Stock Photo; p. 87: (a) Digital Vision/Getty Images; (b) Tom Grundy/Alamy Stock Photo; p. 87: (c) Andrew Holt/Getty Images; (d) U.S. Army NVESD; (e) Tim Graham/Getty Images; p. 89: (3.05a) INTERFOTO/Alamy Stock Photo; (3.05b) Department of Physics, Imperial College London/Science Source; (3.06a) Richard Treptow/Science Source; (3.06aa) Ted Kinsman/Science Source; (3.06bb) Richard Treptow/Science Source; (3.06b) Ted Kinsman/Science Source; (3.06cc) Richard Treptow/Science Source; (3.06c) Ted Kinsman/Science Source; p. 91: (top, all) Ted Kinsman/Science Source; (bottom) Science Source; p. 92: (top) James Leynse/Getty Images; (bottom) ROBYN BECK/AFP/Getty Images; p. 94: (both) U.S. Army NVESD; p. 102: Westend61 GmbH/Alamy Stock Photo; p. 115: (top): David Taylor/Science Source; (bottom): Dorling Kindersley/Getty Images; p. 132: Tetra Images/Alamy Stock Photo; p. 133: (left) Richard Treptow/Science Source; (right) Ted Kinsman/Science Source; p. 136: SPL/Science Source; p. 137: (left) medeni/Alamy Stock Photo; (right) Jacobs Stock Photography Ltd/Getty Images; (bottom) Arterra Picture Library/Alamy Stock Photo.

## Chapter 4

Pages 144–145: Laurel/Alamy Stock Photo; p. 150: (bottom left) GIPhotoStock/Science Source; (bottom right) Charles D. Winters/Science Source; (bottom center) Charles D. Winters/Science Source; p. 152: Roman Samokhin/Dreamstime.com; p. 157: (both) Martyn F. Chillmaid/Science Source; p. 169: W. Perry Conway/Getty Images; p. 191: (left) GIPhotoStock/Science Source; (middle) Ed Endicott/Alamy Stock Photo; (right) GIPhotoStock/Science Source; p. 195: B.A.E. Inc./Alamy Stock Photo.

## Chapter 5

Pages 196–197: Image Source/Getty Images; p. 198: (top): Schöning/ullstein bild via Getty Images; (bottom): KonstantinGushcha/ Getty Images; p. 201: (all) Editorial Image, LLC/Alamy Stock Photo; p. 219: Courtesy of Thomas Gilbert; p. 222: Photo Researchers/Science Source; p. 226: Science Stock Photography/Science Source; p. 227: (top) Phil Dotson/Science Source; (bottom) Incredible Arctic/Shutterstock; p. 233: Diomedia/Yoav Levy; p. 241: Diomedia/Yoav Levy; p. 245: Science Photo Library – MARK GARLICK/Getty Images; p. 246: U.S. Geological Survey via AP; p. 247: Kellis/Shutterstock; p. 249: TunedIn by Westend61/Shutterstock; p. 250: EBlokhina/Getty Images.

## Chapter 6

Pages 252–253: Adam Pretty/Getty Images; p. 258: Science Source; p. 269: Photo Researchers/Science Source; p. 273: Charles D. Winters/Science Source; p. 275: (top) Martin Shields/Science Source; (bottom) Jeff Daly/Visuals Unlimited; p. 276: Sinclair Stammers/Science Source; p. 277: (top left) Peter Gardner/Dorling Kindersley/Science Source; (top right) Daniel/phcimages/Cover Images/Newscom; (bottom) pacaypalla/Getty Images; p. 280: Jeff Daly/Visuals Unlimited.

## Chapter 7

Pages 288–289: Emmanuel LATTES/Alamy Stock Photo; p. 297: (both) TopFoto/The Image Works; p. 301: vovan/Shutterstock; p. 302: robert8/Shutterstock; p. 306: stargatechris/Alamy Stock Photo; p. 309: DOD Collection/Alamy Stock Photo; p. 313: Mark A. Schneider/Science Source; p. 315: (top) Millard H. Sharp/Science Source; (bottom) John Cancalosi/Getty Images; p. 323: Millard H. Sharp/Science Source; p. 328: kali9/Getty Images; p. 329: (top) Comstock Images /Getty Images; (bottom) Olga Miltsova/Alamy Stock Photo; p. 330: Martin Shields/Science Source; p. 331: Keith Homan/Alamy Stock Photo.

## Chapter 8

Pages 334–335: NASA/NOAA/GSFC/Suomi NPP/VIIRS/Norman Kuring; p. 339: LeighSmithImages/Alamy Stock Photo; p. 341: Photo Researchers/Science Source; p. 342: Science Source; p. 343: (top) Courtesy of Ohaus; (bottom) Turtle Rock Scientific/Science Source; p. 344: (top) Courtesy of Thermoscientific; (bottom) Turtle Rock Scientific/Science Source; p. 345: (all) Turtle Rock Scientific/Science Source; p. 347: SPL/Science Source; p. 352: Science Source; p. 356: (all) Turtle Rock Scientific/Science Source; p. 358: Dirk Ercken/Shutterstock; p. 361: (all) Turtle Rock Scientific/Science Source; p. 365: (8.17-all) Charles D. Winters/Science Source; (top right): Javier Trueba/MSF/Science Source; (bottom right): Richard Thom/Visuals Unlimited; (inset) molekuul.be/Alamy Stock Photo; p. 366: HFeather/Getty Images; p. 369: (all) GIPhotoStock/Science Source; p. 371: NASA; p. 372: (all) Charles D. Winters/Science Source; p. 374: (all) Turtle Rock Scientific/Science Source; p. 377: (top) Bill Ross/Getty Images; (bottom) USDA Natural Resources Conservation Service; p. 378: Turtle Rock Scientific/Science Source; p. 380: (all) Turtle Rock Scientific/Science Source; p. 381: (left) Science Source; (right) TUMS and the shape of the TUMS bottle are trademarks of GlaxoSmithKline. This image was provided courtesy of GlaxoSmithKline.; p. 382: (left) Turtle Rock Scientific/Science Source; (center) Charles D. Winters/Science Source; (right) Charles D. Winters/Science Source; p. 388: (both) Courtesy Thomas Gilbert; p. 389: P. Rona/NOAA; p. 393: Joel Arem/Science Source.

## Chapter 9

Pages 394–395: Khoroshunova Olga/Shutterstock; p. 395: Science Source; p. 397: (left) Bill Stormont/Getty Images; (right) British Antarctic Survey/Science Source; (bottom) EXTREME-PHOTOGRAPHER/Getty Images; p. 398: (both) Science Source; p. 399: iofoto/Getty Images; p. 405: (left) Maciej Bledowski/Alamy Stock Photo; (middle) Hubert Stadler/Getty Images; (right) Bill Ross/Getty Images; p. 406: Sam Ogden/Science Source; p. 407: Stan Pritchard/Alamy Stock Photo; p. 408: (top) LOC/Getty Images; (bottom) Matt Anderson Photography/Getty Images; p. 417: Spencer Sutton/Science Source; p. 418: (top) nycshooter/Getty Images; (center) George W. Kling; (bottom) Thierry Orban/Getty Images; p. 419: (all) Martyn F. Chillmaid/Science Source; p. 423: Art Directors & TRIP/Alamy; p. 425: Turtle Rock Scientific/Science Source; p. 428: Fuse/Getty Images; p. 429: Stephen B. Goodwin/Shutterstock; p. 430: Nicholas Burningham/Dreamstime.com; p. 431: (left) Art Directors & TRIP/Alamy; (right) Turtle Rock Scientific/Science Source; p. 434: (both) Lester V. Bergman/Getty Images; p. 436: Arnold Media/Getty Images; p. 437: Jonathan Blair/Getty Images; p. 438: AFP/Getty Images; p. 439: F. Jack Jackson/Alamy; p. 441: DHE/Hugh Peterswald/Icon Sportswire/Newscom; p. 443: National Museums Scotland.

## Chapter 10

Pages 444–445: Susan Schmitz/Shutterstock; p. 446: Richard and Ellen Thane/Getty Images; p. 447: NASA; p. 451: Phoenixns/shutterstock; p. 454: TREVOR COLLENS/UPI Photo Service/Newscom; p. 455: Alex Menendez/AP Photo; p. 458: Johner Images/Getty Images; p. 461: Sprokop/Dreamstime.com; p. 480: DigitalVues/Alamy; p. 481: (all) Courtesy of Thomas Gilbert; p. 483: (a & b) Andrew Lambert Photography/Science Source; (c) Charles D. Winters/Science Photo Library/Science Source; p. 488: John Raoux/AP Images; p. 490: (a) Granger; (b & c) Courtesy of Alcoa; p. 498: (top b and h): Charles D. Winters/Science Source; (top e) Science Source; (bottom) Stephen Coburn/Shutterstock.

## Chapter 11

Pages 506–507: JEAN-MARIE LIOT/AFP/Getty Images; p. 509: (left) David M. Phillips/Science Source; (center) David M. Phillips/Science Source; (right) SPL/Science Source; p. 514: Courtesy of Katadyn; p. 518: W.W. Norton; p. 520: Stephen VanHorn/Shutterstock; p. 521: Martyn F. Chillmaid/Science Source; p. 527: Hemis/Alamy Stock Photo; p. 536: Stock Connection Blue/Alamy Stock Photo; p. 543: Steve Hawkins Photography/Alamy Stock Photo; p. 543: PixMarket/Shutterstock; p. 544: Susan McAnnally/Alamy Stock Photo.

## Chapter 12

Pages 548–549: Stratos Giannikos/Shutterstock; p. 551: Turtle Rock Scientific/Science Source; p. 552: Science Source; p. 560: (left) Jon Stokes /Science Source; (right) Ken Lucas/Visuals Unlimited; p. 565: B. & C. Alexander/Science Source; p. 575: (left) Reinhard, H./picture alliance/Arco Images G/Newscom; (right) CIMAT/Science Photo Library/Science Source; p. 582: (both) cherezoff/Shutterstock; p. 583: (b): Portland Press Herald/Getty Images; (d and f) Turtle Rock Scientific/Science Source; (h) Colin Anderson Productions pty ltd/Getty Images; p. 585: Reika/Shutterstock.

## Chapter 13

Pages 590–591: The Yomiuri Shimbun via AP Images; p. 591: Kyodo/Newscom; p. 595: (a) Kateleigh/Dreamstime.com; (b) SPL/Science Source; (c) Charles D. Winters/Science Source; p. 634: Image courtesy the TOMS science team & and the Scientific Visualization Studio, NASA GSFC; p. 637: Dorling Kindersley ltd/Alamy Stock Photo; p. 643: (all) Image courtesy the TOMS science team & and the Scientific Visualization Studio, NASA GSFC; p. 647: Lawrence Migdale/Science Source.

## Chapter 14

Pages 656–657: Design Pics Inc/Alamy Stock Photo; p. 661: Xiao Lu Chu/Getty Images; p. 676: ITAR-TASS News Agency/Alamy Stock Photo; p. 677: Courtesy of Thermo Fisher Scientific; p. 678: Turtle Rock Scientific/Science Source; p. 679: Charles D. Winters/Science Source; p. 681: (all) Turtle Rock Scientific/Science Source; p. 684: (both) Charles D. Winters/Science Source; p. 700: (both) Charles D. Winters/Science Source; p. 703: (b, d, f) Turtle Rock Scientific/Science Source; (h) Photo Researchers/Science Source.

## Chapter 15

Pages 712–713: azndc/Getty Images; p. 738: wwing/Getty Images; p. 738: www.pqpictures.co.uk/Alamy Stock Photo; p. 748: Ed Endicott/Alamy Stock Photo.

## Chapter 16

Pages 762–763: WaterFrame/Alamy Stock Photo; p. 778: (all) Courtesy of Thomas Gilbert; p. 779: Turtle Rock Scientific/Science Source; p. 792: Wizard8492/Shutterstock; p. 796: Martyn F. Chillmaid/Science Source; p. 800: David R. Frazier Photolibrary, Inc./Alamy Stock Photo; p. 801: Albert Russ/Shutterstock; p. 803: (both) Charles D. Winters/Science Source; p. 810: Turtle Rock Scientific/Science Source; p. 811: (both) SPL/Science Source.

## Chapter 17

Pages 818–819: Jim Motavalli; p. 818: Jeff Greenberg/UIG via Getty Images; p. 821: (all) GIPhotoStock/Science Source; p. 826: yenwen/iStock; p. 829: Mary Evans Picture Library/Alamy; p. 842: Paul Mogford/Alamy Stock Photo; p. 843: MAILLAC/REA/Redux; p. 849: Richard Goldberg/Dreamstime.com; p. 851: JOHN GRESS/REUTERS/Newscom; p. 853: Stephen Barnes/Energy/Alamy Stock Photo; p. 856: GIPhotoStock/Science Source; p. 859: (top) Gaspar R Avila/Alamy Stock Photo; (bottom) sciencephotos/Alamy Stock Photo; p. 860: (a, c, g, i) Science Source; (e) Charles D. Winters/Science Source; p. 864: age fotostock/Alamy Stock Photo; p. 865: Emory Kristof, National Geographic Image Collection; p. 867: Mike Egerton/ZUMA Press/Newscom.

## Chapter 18

Pages 868–869: Rudi Van Starrex/Getty Images; p. 871: Dorling Kindersley/Getty Images; p. 871: image-broker/Alamy; p. 874: (top): Manfred Kage/Science Source; (bottom) Sebastian Janicki/Shutterstock; p. 877: David Parker/Science Source; p. 879: Kris Mercer/Alamy; p. 883: Alexey Kamenskiy/Getty Images; p. 884: Charles D. Winters/Science Source; p. 887: (left) David J. Green – technology/Alamy Stock Photo; (right) doomu/Shutterstock; p. 888: (left) Jon Stokes/Science Source; (middle): Ken Lucas/Visuals Unlimited; (right) Andre Geim & Kostya Novoselov/Science Source; p. 890: (top) Charles D. Winters/Science Source; (bottom) Theodore W. Gray; p. 892: (top) Science Source; (bottom) Raul Gonzalez Perez/Science Source; p. 893: Harry Taylor/Dorling Kindersley/Science Source; p. 895: De Agostini Picture Library/Getty Images; p. 896: (top) WILDLIFE GmbH/Alamy Stock Photo; (bottom): Arthur Hill/Visuals Unlimited, Inc.; p. 897: Ashley Cooper/Alamy; p. 899: Omikron/Science Source; p. 909: Smithsonian Institution, Washington DC, US/Bridgeman Images; p. 910: SphinxHK/Shutterstock; p. 911: VCG Wilson/Corbis via Getty Images.

## Chapter 19

Pages 916–917: Gregory Shamus/Getty Images; p. 932: Dr. Keith Wheeler/Science Source; p. 934: (left) Eric and David Hosking/Getty Images; (center) Image Professionals GmbH/Alamy Stock Photo; (right) Digital Vision/Getty Images; p. 940: Lynnette Peizer/Alamy; p. 941: (top) CANADA/Newscom; (bottom left): Bsip/Photoshot/ZUMA Press/Newscom; (bottom right) Manfred Gottschalk/Getty Images; p. 942: (top) Courtesy of Dr. Ann Schmierer; (bag) Emily Spence/Lexington Herald-Leader/MCT/Newscom; (carpet) Fotosearch; (rope) studiomode/Alamy; (chairs) Elizabeth Whiting & Associates/Alamy Stock Photo; p. 945: (left) Stephen Stickler/Getty Images; (right) W.W. Norton; p. 946: Courtesy of Elizabeth Hinchey; p. 956: Southern Illinois University/Science Source; p. 957: Creatas/Getty Images; p. 959: Courtesy of DuPont; p. 964: (l-r) Eric and David Hosking/Getty Images; John Madere/Getty Images; Image Professionals GmbH/Alamy Stock Photo; Douglas Peebles/Getty Images.

## Chapter 20

Pages 976–977: LAGUNA DESIGN/Getty Images; p. 979: Olaf Speier/Shutterstock; p. 986: (left) Dennis Kunkel Microscopy/Science Source; (right) Omikron/Science Source; p. 988: Courtesy of N.I.S.T.; p. 989: Chemical Design/Science Source; p. 999: Biophoto Associates/Science Source; p. 1004: Roger Ressmeyer/Corbis/VCG/Getty Images; p. 1005: (top) NASA; (bottom) W.R. Normak, courtesy USGS.

## Chapter 21

Pages 1018–1019: LADA/Science Source; p. 1030: Rob Blakers/Getty Images; p. 1035: Corbis via Getty Images; p. 1039: (top) NASA/HST/J. Morse/K. Davidson; (bottom) X-ray: NASA/CXC/ASU/J.Hester et al.; Optical: NASA/ESA/ASU/J.Hester & A.Loll; Infrared: NASA/JPL-Caltech/Univ. Minn./R.Gehrz; p. 1043: (a) Astrid & Hanns-Frieder Michler/Science Source; (b) Tom Tracey Photography/Alamy; (c) MEIGNEUX/SIPA/Newscom; (bottom): Argonne National Laboratory, US DOE Office of Environmental Management; p. 1045: (a) Iridiumphotographics/Getty Images; (b & c) Courtesy of Ecotest; p. 1049: (top) REUTERS/Newscom; (center) Science Source; (bottom) Xinhua/eyevine/Redux; p. 1053: (top) Dr. Robert Friedland/Science Source; (bottom left) Jeff J Daly/Alamy Stock Photo; (bottom right) The American Weekly, 1926; p. 1055: (left) Rob Blakers/Getty Images; (right) X-ray: NASA/CXC/ASU/J. Hester et al.; Optical: NASA/ESA/ASU/J.Hester & A.Loll; Infrared: NASA/JPL-Caltech/Univ. Minn./R.Gehrz; p. 1059: (top) Science and Society/Superstock; (bottom) The Natural History Museum, London/Science Source; p. 1060: Pavel, et al; "Human Presence in the European Arctic Nearly 40,000 Years ago," Nature, 413:64,2001. © 2001, Rights Managed by Nature Publishing Group.; p. 1065: SIPA USA/Karen L. King/Harvard via Sipa/Newscom.

## Chapter 22

Pages 1066–1077: Michael Neelon (misc)/Alamy Stock Photo; p. 1070: (both) Photo Researchers/Science Source; p. 1073: DENNIS KUNKEL MICROSCOPY/Science Source; p. 1086: (right) Darren Muir/Stocksy; (left) Tierfotoagentur/Alamy Stock Photo; p. 1090: Photo Researchers/Science Source; p. 1091: (left) Simon Fraser/Medical Physics, RVI, Newcastle upon Tyne/Science Source; (right) Southern Illinois University/Science Source; p. 1095: Photo Researchers/Science Source; p. 1096: (both) Photo Researchers/Science Source; p. 1097: Simon Fraser/Medical Physics, RVI, Newcastle upon Tyne/Science Source; p. 1102: LAURA MCDERMOTT/The New York Times/Redux.

## Chapter 23

Pages 1104–1105: addkm/Shutterstock; p. 1107: (both) Charles D. Winters/Science Source; p. 1108: (both) Courtesy of Paige Scannell; p. 1109: (both) Courtesy of Paige Scannell; p. 1110: The Metropolitan Museum of Art/Art Resource, NY; p. 1116: (both) Turtle Rock Scientific/Science Source; p. 1119: Turtle Rock Scientific/Science Source; p. 1121: GIPhotoStock/Science Source; p. 1123: Turtle Rock Scientific/Science Source; p. 1124: Vaughan Fleming/Science Source; p. 1127: (both) Courtesy of Paige Scannell; p. 1133: Phantatomix/Science Source; p. 1137: Zephry/Science Photo Library/Science Source; p. 1138: (a &b) From "Small molecular gadolinium (III) complexes as MRI contrast agents for diagnostic imaging," Chan Kannie Wai-Yan and Wong wing-Tak, Coordination Chemistry Reviews, Sept. 2007, Elsevier B.V., Copyright Clearance Center; p. 1142: Mike Egerton/ZUMA Press/Newscom; p. 1146: (both) Courtesy of Paige Scannell; p. 1147: (e, both) SPL/Science Source.

# Index

*Note*: Material in figures or tables is indicated by *italic* page numbers. Footnotes are indicated by n after the page number.

## A

## B

# D

## E

# F

## G

## H

## J



## K



## L

## M

## N

# P

# S

# T

## W



## X

## Y



## Z